◆ and the Learning Assessments

SECTION 3.1 Learning Assessment

■ Mathematical Reasoning and Proof

42. **a.** Determine how many symbols are needed to write each number as an Egyptian numeral.

    i. 27   **ii.** 149

    iii. 2,540   **iv.** 34,508

    b. Suggest a rule that can be used to determine how many Egyptian symbols are needed to represent any number as an Egyptian numeral.

43. **a.** If you were given a numeration system you had never seen before, how would you know whether it was positional?

    b. Suppose the system was positional. How might you determine its base?

> Reasoning and Proof questions assess the students' ability to implement mathematical reasoning with regard to the concepts of the section.

■ Mathematical Communication

46. Look up the definition of three in a dictionary. Do you think the definition adequately describes the quantity of three? Why or why not?

47. Explain to another person how to select the symbols needed to write 345 as a(n):

    a. Egyptian numeral.

    b. Roman numeral.

    c. Mayan numeral.

> Mathematical Communication questions ask students to communicate mathematical content to others through writing or group discussion.

In addition to the five processes, the text concentrates on developing and understanding ideas associated with elementary mathematics. It also builds a connection between these ideas and the elementary classroom. For instance, tables located at the beginning of each chapter give specific content objectives adapted from the *Common Core State Standards for Mathematics*.

Table 3.1 Classroom learning objectives	K	1	2	3	4	5	6	7	8
Write numerals for the numbers 0 to 20.	X								
Read and write multidigit whole numbers using base-ten numerals, number names, and expanded form.		X	X						
Compose and decompose numbers from 11 to 19 into 10 ones and some further ones.	X								
Compare numbers between 1 and 10 by using matching and counting strategies.	X								
Compare multidigit numbers based on meanings of the digits in each place.		X	X		X				
Understand that a two-digit number represents an amount of tens and ones and a three-digit number represents an amount of hundreds, tens, and ones.		X	X						
Recognize that a digit in one place represents 10 times what it represents in the place to its right and 1/10 of what it represents in the place to its left.					X	X			
Represent whole numbers on a number line.			X						

Source: Adapted from the *Common Core State Standards for Mathematics* (Common Core State Standards Initiative, 2010).

Mathematics for Elementary School Teachers

A PROCESS APPROACH

Mark A. Freitag
Augusta State University

BROOKS/COLE
CENGAGE Learning

Australia • Brazil • Japan • Korea • Mexico • Singapore • Spain • United Kingdom • United States

BROOKS/COLE
CENGAGE Learning™

Mathematics for Elementary School Teachers: A Process Approach
Mark A. Freitag, Augusta State University

Executive Editor: Charlie Van Wagner

Senior Acquisitions Editor: Marc Bove

Developmental Editors: Stefanie Beeck, Donald Gecewicz

Assistant Editor: Lauren Crosby

Editorial Assistant: Ryan Furtkamp

Media Editor: Bryon Spencer

Senior Brand Manager: Gordon Lee

Senior Market Development Manager: Danae April

Content Project Manager: Tanya Nigh

Art Director: Vernon Bocs

Manufacturing Planner: Rebecca Cross

Rights Acquisitions Specialist: Tom McDonough

Production and Composition: MPS Limited

Photo Researcher: Scott Rosen, Q2A/Bill Smith

Text Researcher: Pablo D'Stair

Copy Editor: Fred Dahl

Illustrator: Graphic World

Text and Cover Designer: Diane Beasley

Cover Image: Image Source/Corbis

For product information and technology assistance, contact us at
Cengage Learning Customer & Sales Support, 1-800-354-9706.
For permission to use material from this text or product,
submit all requests online at **www.cengage.com/permissions.**
Further permissions questions can be e-mailed to
permissionrequest@cengage.com.

Library of Congress Control Number: 2011942673

ISBN-13: 978-0-618-61008-2

ISBN-10: 0-618-61008-1

Brooks/Cole
20 Davis Drive
Belmont, CA 94002-3098
USA

Cengage Learning is a leading provider of customized learning solutions with office locations around the globe, including Singapore, the United Kingdom, Australia, Mexico, Brazil, and Japan. Locate your local office at **www.cengage.com/global.**

Cengage Learning products are represented in Canada by Nelson Education, Ltd.

To learn more about Brooks/Cole, visit **www.cengage.com/brooks/cole**

Purchase any of our products at your local college store or at our preferred online store **www.CengageBrain.com.**

Printed in the United States of America
1 2 3 4 5 6 7 16 15 14 13 12

TO MY GOD,

IT'S BEEN A LONG, SOMETIMES BUMPY ROAD.

THANKS FOR GIVING ME THE PERSEVERANCE

TO SEE IT THROUGH TO THE END.

ABOUT THE AUTHOR

Mark A. Freitag is currently an associate professor at Augusta State University. He has a PhD in mathematics education from the University of Georgia and an MS in mathematics from Memphis State University. He has taught mathematics content courses for preservice elementary and middle grades teachers for over 15 years in a variety of settings from small private colleges to large public universities. Dr. Freitag's scholarly interests revolve around creating strategies to help preservice teachers learn to effectively communicate mathematics through writing and dialogue. He is also interested in developing ways to help students become more independent in their learning by improving their reading comprehension.

Contents

Preface

Welcome to the text *Mathematics for Elementary School Teachers: A Process Approach.* The textbook and its accompanying supplements were written to provide materials that serve the diverse needs of mathematical content courses for preservice K–8 teachers. In this textbook, you will find a number of pedagogical and content features that are specifically designed to help preservice teachers understand the mathematics they will teach.

Traditionally, textbooks for preservice elementary teachers have focused on problem solving. However, problem solving is not the only process through which mathematics is learned. It is also learned through mathematical reasoning, communication, representation, and connections. Recent trends in mathematics education now advocate implementing all five processes as a vital part of learning and doing mathematics. Consequently, preservice teachers need to have concrete experiences with these processes that they will be required to teach.

The goal of this textbook is to treat each of the processes equitably by using an approach in which the five processes serve as the central pedagogical theme. Most of the examples, exercises, and activities are designed to either model the processes or to have students directly engaged in working with them. By doing so, preservice teachers will not only come to understand the different processes, but also appreciate them as integral to learning and doing mathematics. This broader view can enable preservice teachers to give their students a more well-rounded and holistic view of mathematics once they enter the classroom.

Pedagogical Features

This textbook contains a number of pedagogical features that are designed to promote learning. To begin, the text contains detailed explanations of the content written in a voice that makes the ideas accessible to students, while remaining mathematically correct. The content is arranged in a logical order, and the formatting highlights the main points, making them easy to find. Specifically, definitions and theorems are clearly identified and separated from other content by gold and blue text boxes. Examples are numbered, named, and separated by red text boxes.

Definition of Integer Addition

Let a and b be any two integers.

1. If a and b have the same sign, then $a + b$ is the sum of the absolute values of a and b and takes the sign of a and b.
2. If a and b have different signs, then $a + b$ is the difference of the smaller absolute value from the larger absolute value and takes the sign of the number with the larger absolute value.

The definition is well-defined because it describes how to compute the sum of any two integers. Unfortunately, it is unavoidably complicated. However, we can summarize it as follows: To find the sum of two integers, find the sum or difference of two whole numbers and then determine the sign.

Example 6.5 Using the Definition of Integer Addition

Use the definition of integer addition to compute each sum.

a. $(-18) + (-23)$ **b.** $(-567) + 983$

Solution

a. $(-18) + (-23) = -(|-18| + |-23|) = -(18 + 23) = -41.$
b. $(-567) + 983 = |983| - |-567| = 983 - 567 = 416.$

Money is another effective model for integer operations. Deposits represent positive integers, and withdrawals represent negative integers.

Example 6.6 Representing Integers with Money

Application

Billy overdrew his bank account by $115 but soon after deposited $250. He then wrote a check for $95. Write an expression using integer addition to find his present balance.

Solution We write the sequence of transactions as $(-115) + 250 + (-95)$. Adding the numbers shows that Billy's account is presently at $(-\$115) + (\$250) + (-\$95) = \$40.$

Like whole-number addition, integer addition satisfies several properties.

Theorem 6.1 Properties of Integer Addition

If a, b, and c are integers, then integer addition satisfies the following properties:

Closure property: $a + b$ is a unique integer.
Commutative property: $a + b = b + a.$
Associative property: $a + (b + c) = (a + b) + c.$
Identity property: $a + 0 = a = 0 + a.$

Integer addition also satisfies another property related to adding opposites. Specifically, when two opposites are added, the result is always zero. For instance, if we add $5 + (-5)$ or $(-5) + 5$ on the number line, the net result in either case is zero (Figure 6.5). This is true for any pair of opposites, which leads us to the **additive inverse property.**

Making the Five Mathematical Processes Accessible and Visible

The five mathematical processes identified in the *NCTM Principles and Standards* serve as a framework through which the content is stated, explained, demonstrated, and engaged. One way this process approach can be seen is in the examples because many of them fall into one of five following categories:

Problem Solving

- **Problem-Solving** examples demonstrate how problems can be solved using the concepts and procedures identified in the section. The problems often integrate multiple topics or extend mathematical ideas in some way. The examples use Polya's four-step method, and when possible, identify and implement new problem-solving strategies.

Reasoning & Proof

- **Reasoning and Proof** examples demonstrate various aspects of mathematical reasoning. They include, but are not limited to, investigating and creating conjectures, recognizing and extending patterns, and proving mathematical assertions.

Communication

- **Communication** examples illustrate how to interpret formal mathematical language, use informal language to describe a concept or procedure, or analyze the mathematical thinking of others.

Representation

- **Representation** examples show how to use different representations and make translations between them. They can also demonstrate the strengths and weaknesses of different representations and how different representations reveal different aspects of the mathematics.

Connection

- **Connection** examples link the content of the section to other mathematical ideas, both within and outside the section or chapter. They also connect the content to the K–8 classroom.

Applications

- **Application** examples make additional connections by linking the mathematics to other disciplines and to the uses of mathematics in everyday life.

Engaging and Assessing Students' Mastery of the Content

The **Learning Assessments** at the end of each section are key opportunities for students to engage the mathematics that they are learning. They are specifically designed to help students' use the five mathematical processes and deepen their understanding of the content. They also provide an excellent opportunity for instructors to assess their students' mastery of the content. Each Learning Assessment asks between 60 and 80 questions that are subdivided into the following categories. Answers to highlighted questions are located in the back of the textbook.

SECTION 3.2   Learning Assessment

■ **Understanding the Concepts**

1. How is the base of the decimal system connected to the size of the groups?
2. What is place value, and why is it a key feature of the decimal system?
3. a. Why must the decimal system, as a grouping system, also be additive?
   b. Why must the decimal system, as a positional system, also be multiplicative?

■ **Representing the Mathematics**

10. Write the decimal numeral that represents each of the following.
    a. Six hundreds, eight tens, and two ones.
    b. Six thousands, eight hundreds, five tens, and seven ones.
    c. Two hundred thousands, three ten thousands, seven thousands, and three tens.
    d. Five millions, four thousands, and seven ones.

- **Understanding the Concepts** questions assess students' understanding of the concepts explored in the section. These items may ask students to:
  - Describe and explain concepts using informal language.
  - Compare and contrast key ideas of the section.
  - Assess key aspects and necessary conditions associated with definitions, properties, and theorems.

- **Representing the Mathematics** questions ask students to show their understanding of the content by using the mathematical representations addressed in the section. These items may ask students to:
  - Demonstrate effective use of a representation.
  - Analyze strengths and weaknesses of different representations.
  - Make translations between representations.
  - Demonstrate how different representations can reveal important aspects of the mathematics.

- **Exercises** give students the opportunity to practice and develop the basic procedural and computational skills associated with the content. They also connect these skills to simple applications of the content. Many of the exercises are given in pairs, one of which has the answer in the back of the text.

- **Problems and Applications** are more involved than the exercises in that they require creative thinking and the use of different problem-solving strategies. The problems and applications assess students' knowledge by asking students to:
  - Integrate multiple topics.
  - Extend the content of the section.
  - Make use of the mathematical content in a new way.
  - Rely on a specific strategy to solve the problem.
  - Use different strategies to solve the same problem and then compare the solutions.
  - Create and assess different representations.
  - Solve problems in nonmathematical or real-world contexts.

- **Mathematical Reasoning and Proof** questions assess the students' ability to implement mathematical reasoning with regard to the concepts of the section. These items may ask students to use reasoning skills to:
  - Discern and extend patterns in mathematical examples.
  - Make, investigate, and test mathematical conjectures.
  - Evaluate mathematical arguments and proofs.
  - Verify properties using proof.

- **Mathematical Communication** questions ask students to communicate mathematical content to others through writing or group discussion. These items may ask students to:
  - Organize and consolidate their thinking about a particular concept.
  - Discuss misconceptions that may have occurred with the content.
  - Analyze and evaluate the mathematical strategies and thinking of others.
  - Complete a writing assignment about a particular topic.
  - Discuss a particular topic with a group of peers.
  - Create mathematical tasks to be given to another to complete.

- **Building Connections** questions help students to build connections within the mathematics as well as between mathematics and the K–8 classroom. These items may ask students to:
  - Explain how mathematical ideas are connected to each other to produce a coherent whole.
  - Connect mathematical ideas both within and outside the section.
  - Show how conceptual ideas are linked to procedures.
  - Connect the mathematics of the section to occurrences in everyday life.
  - Connect the mathematics of the section to the K–8 classroom.
  - Link mathematics to other disciplines.

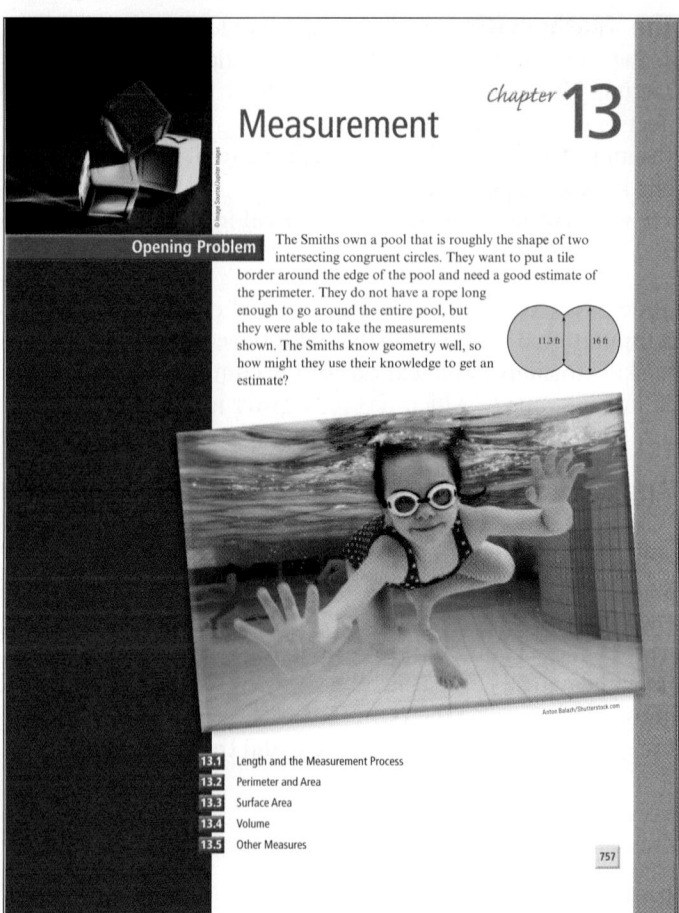

Special Features Helping Students Engage the Content

In addition to the Learning Assessments, this textbook contains a number of other pedagogical features that are designed to test students' acquisition of the concepts in ways that appeal to different learning styles:

◀ **Opening Problems** provide an application of the content in the chapter on its first page. Answers are given directly after the chapter review exercises.

Check Your Understanding in-text quizzes provide an opportunity for students to monitor their understanding of the content through a small set of quick skill-check exercises. These concept checks also help moderate the flow of the material by further subdividing each section into manageable subsections of common content. Answers to all items are given after the chapter review exercises. ▼

Talk About It questions give students the opportunity to discuss the content they have just learned. Instructors can use these questions as part of a lecture or they can be given to small groups of students to discuss. ▶

Activities are designed to give students an opportunity to directly engage the material. They often make use of hands-on manipulatives to solve problems, reason mathematically, or build connections. They are designed to take a minimal amount classroom time, usually taking only 5 to 10 minutes. Answers to all activities are given after the chapter review exercises.

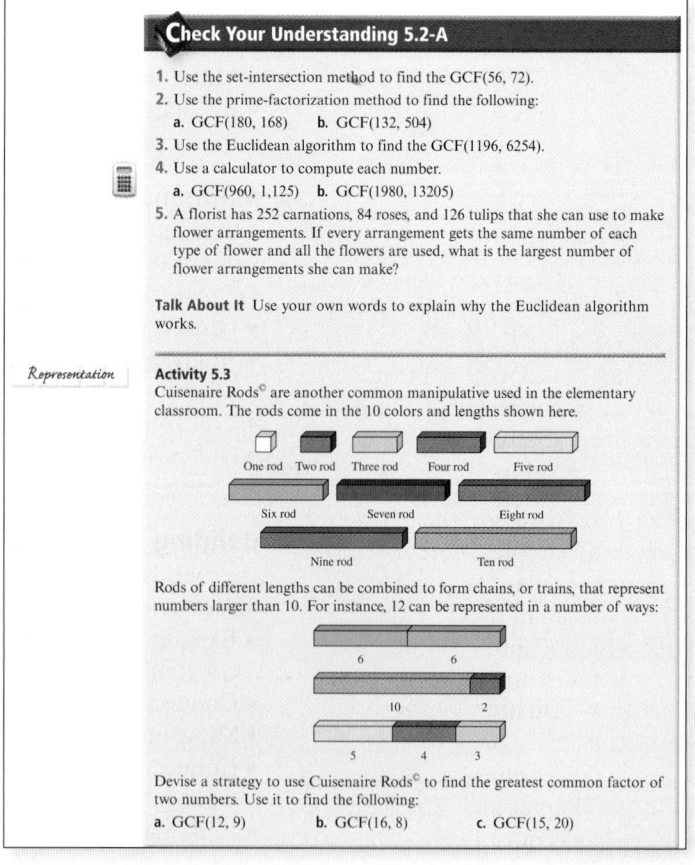

Reinforcing the Concepts Through Chapter Reviews

Students are given the opportunity to review the content in three ways:

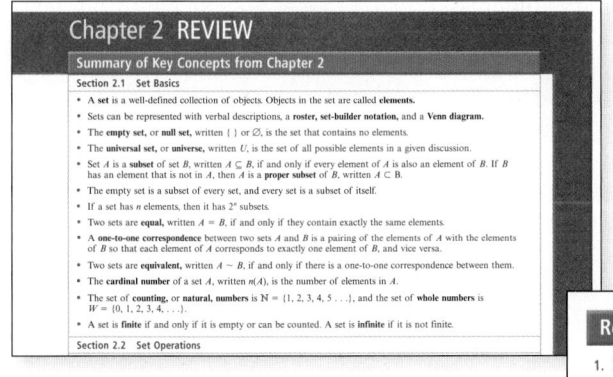

A **Summary of Key Concepts** occurs the end of each chapter and provides a brief overview of the most important ideas found in the chapter.

Review Exercises provide students another opportunity to work the different exercises and problems encountered throughout the chapter. Answers to the highlighted questions can be found in the back of the textbook.

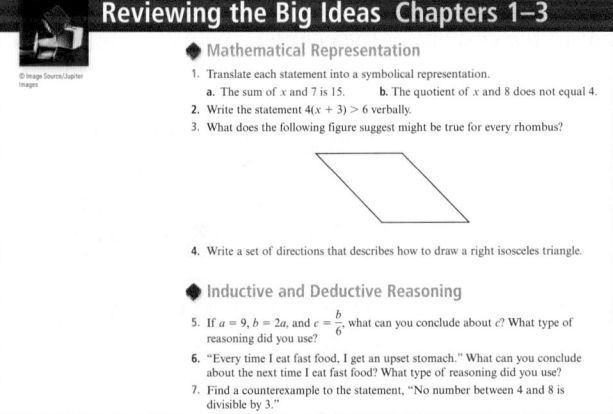

Reviewing the Big Ideas are cross-chapter reviews that provide students with another opportunity to work with the content from previous chapters. Each review clearly identifies the major themes from three chapters. This not only integrates the big ideas of the content, but also shows students which content is essential to master and take away from the course. Six to ten assessment items are given for each major theme, allowing students to test their mastery of the content by revisiting the key ideas in each area. Answers to highlighted questions can be found in the back of the textbook.

Placing Concepts in Context for Preservice Teachers

Placing the concepts in context, particularly a classroom context, is an important part of this text. Beyond the examples and exercises that do so, several features in the text are specifically designed to accomplish this task:

Getting Started sections introduce each chapter and are designed to help preservice teachers build a connection between the content they are learning and the content they will be teaching in the K–8 classroom. Relevant *NCTM Standards* are identified, and a progression of topics is given. Each Getting Started section also contains a table of **Classroom learning objectives** that are adapted from the *Common Core State Standards for Mathematics*. The tables list the learning objectives that are relevant to the content of the chapter and then indicate the grade level at which the objective is most likely to be taught.

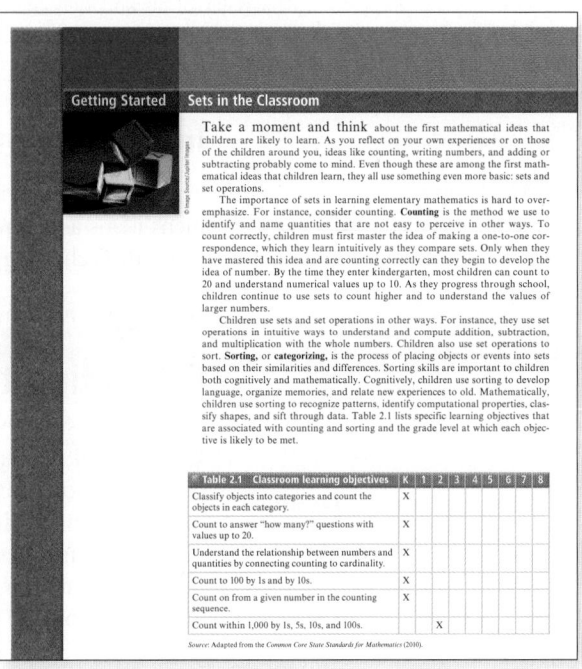

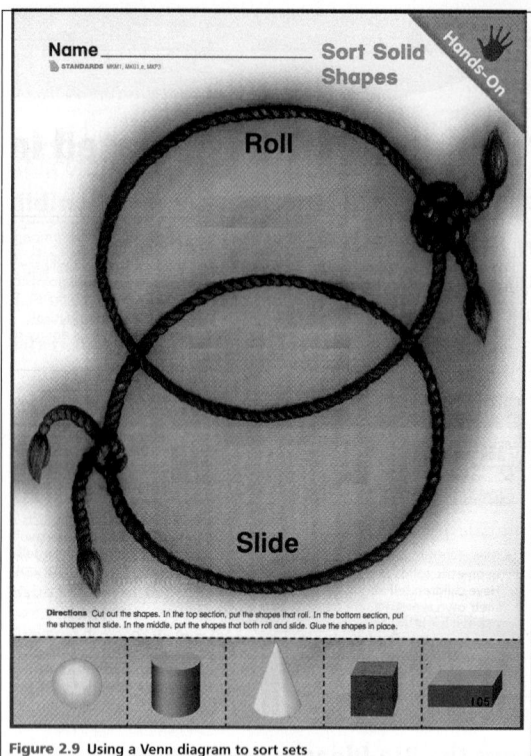

Figure 2.9 Using a Venn diagram to sort sets
Source: From *Mathematics Grade K: Houghton Mifflin Mathematics Georgia* (p. 105). Copyright © by Houghton Mifflin Company, Inc. Reproduced by permission of the publisher, Houghton Mifflin

◀ **Student Pages** from elementary textbooks provide further connections to the elementary classroom by demonstrating how concepts are often presented to elementary students.

Historical Notes provide a context for the content by identifying historical figures or events relevant to the content of the chapter. ▼

Historical Note

One of the first mathematicians to make major contributions in the area of set theory was Georg Cantor (1845–1918). Born in St. Petersburg, Russia, he was educated in Germany during what many consider to be the golden age of mathematical scholarship. Cantor not only defined many of the notions associated with sets, he also demonstrated how they could be used to solve problems in other mathematical areas. Much of his research revolved around the measurement of infinite sets. When Cantor showed that a cardinal number could be assigned to an infinite set, he caused a great debate among mathematicians that eventually ostracized him from the mathematical community. Not until late in his life was the genius of his work fully appreciated.

Georg Cantor

Integration of Technology

The text also integrates relevant technology into the discussions, examples, and assessment items as appropriate. Specific icons indicate when technology is discussed or is to be used. Technology used in this text includes, but is not limited to, calculators of various types and computer software such as spreadsheets and geometry software.

Expanding the Course Through Explorations

Explorations Manual

To assist instructors who use a more hands-on approach to teaching, an extensive Exploration Activities manual accompanies the text. The manual contains a variety of explorations for each section, which are referenced in the text by an icon in the margin. Some explorations deal directly with the content of the chapter, often making use of relevant manipulatives or other hands-on activities. Other explorations extend the content of the section either mathematically or by building a connection to the K–8 classroom. Most of the explorations can be done individually or with groups and should take 30–45 minutes to complete.

Content Features

The content of the book was developed using the philosophy that elementary teachers are teachers of mathematics who play a crucial role in their students' future studies in mathematics. Not only must elementary teachers be trained in appropriate teaching methodologies, but they must also have a deep understanding of the mathematics that they will teach. For this reason, the content of the book is directly related to the mathematics taught in grades K–8. The purpose is not to reteach elementary mathematics. Rather, the intent is to look at the content from a theoretical or generalized point of view, so that elementary and middle-grades teachers can better understand the concepts and processes behind the mathematics they will teach. In short, the book focuses on the "why" behind the mathematics in addition to the "how."

Two primary sources were used to shape the content of the text. The first was the *NCTM Principles and Standards for School Mathematics*. The *Principles and Standards* has been widely regarded as one of the most influential documents in mathematics education in recent years. Because it has served as the foundation for state education standards and nationally used curriculum materials, the five Content Standards were used for the general framework of the text. Specifically:

- The Number and Operations Standard is addressed in Chapters 2–8.
- The Algebra Standard is addressed in Chapter 9.
- The Geometry Standard is addressed in Chapters 10–12.
- The Measurement Standard is addressed in Chapter 13.
- The Data Analysis and Probability Standard is addressed in Chapters 14 and 15.

The second source was the *Common Core State Standards for Mathematics*. This national set of standards provides a clear and detailed list of mathematical expectations that every child should achieve in grades K–8. Because it has been widely adopted across the United States, it was used to identify specific content objectives to be included in the text. By using the *Principles and Standards* and the *Common Core*, not only are the content objectives in the text current, but they also reflect the content K–8 teachers are most likely to teach.

The following list describes how the *NCTM Principles and Standards*, the *Common Core*, and the structure of the mathematics itself were used to shape the content of individual chapters.

Chapter 1 Mathematical Processes

Chapter 1 introduces the five processes that play an important role in learning mathematics: communication, representation, reasoning and proof, problem solving, and connections. Four of the processes involve important mathematical ideas in their own right, so they are discussed in detail throughout the chapter. Specifically, Chapter 1 begins with a discussion of mathematical communication and the role that mathematical representations play in communicating mathematics efficiently and effectively. It then provides an overview of two key kinds of reasoning: (1) inductive reasoning and its connection to patterns and (2) deductive reasoning and its connection to proof. The chapter concludes with a discussion of Polya's problem-solving method and the strategies commonly used to solve problems in the elementary and middle grades. The fifth process, mathematical connections, is content specific and is addressed throughout other chapters. Once introduced, the five processes are used as the pedagogical framework for the rest of the text.

Chapter 2 Sets

Chapter 2 begins the content on number and numeration by introducing sets. Although not formally taught in the K–8 curriculum, sets and set operations form the foundation for much of elementary mathematics. Without an intuitive understanding of these ideas, students may find it difficult to make sense of number, numeration, and numerical operations. The chapter begins with the basic notions and relationships of sets. It then turns to set operations and how to apply those operations in sorting.

Chapter 3 Numbers and Numeration

Chapter 3 begins with a brief history of numeration. The historical discussion introduces basic concepts of numeration such as place value and bases. It also provides a context for understanding the power of the decimal system. The characteristics of the decimal system are then developed and generalized to numeration systems with other bases.

Chapter 4 Whole-Number Computation

Chapter 4 develops the whole-number operations of addition, subtraction, multiplication, exponents, and division. Different models are used to represent and develop the intuitive notions behind the operations and their properties. The chapter then considers procedures for computing with large numbers, including written algorithms, mental strategies, and estimation. Chapter 4 concludes with whole-number operations in other base numeration systems.

Chapter 5 Basic Number Theory

Chapter 5 introduces many notions from basic number theory, including divisibility, primes, composites, greatest common factors, and least common multiples. These ideas are not only important for learning number sense, but are also useful in subsequent work with the rational numbers. The chapter concludes with a discussion of modular and clock arithmetic.

Chapter 6 The Integers

Chapter 6 expands the set of whole numbers to the set of integers. It begins with the basic notions of the integers such as opposites and absolute value, and then moves to integer operations. Several different models are used to develop an intuitive notion of integer operations and their properties.

Chapter 7 Fractions and the Rational Numbers

Chapter 7 expands the set of integers to the set of rational numbers. The chapter begins with a detailed discussion of the rational numbers in fractional form and the properties that separate them from other sets of numbers. From these basics, the chapter then develops rational-number operations and their properties by extending ideas used with whole-number operations.

Chapter 8 Decimals, Real Numbers, and Proportional Reasoning

Chapter 8 concludes the content on number sets by developing the decimals and the set of the real numbers. It connects the rational numbers to the decimals and the decimal operations to whole-number algorithms for large numbers. The chapter then applies real-number computation to proportional reasoning. Specifically, Chapter 8 introduces several methods for using proportional reasoning to solve proportion and percent problems.

Chapter 9 Algebraic Thinking

Chapter 9 discusses three key components of algebraic thinking relevant to the K–8 curriculum. It begins with numerical patterns, which are used to introduce functions and their representations. The chapter then discusses linear functions and constant change, which is often the focus of functional thinking in the primary grades. Chapter 9 goes on to discuss linear equations and inequalities. The chapter concludes with a section on mathematical modeling and problem solving with functions and equations.

Chapter 10 Geometrical Shapes

Chapter 10 begins a discussion of geometrical shapes and their properties. It starts by describing and stating the properties of the basic shapes of geometry: points, lines, planes, and angles. It then moves on to define and identify the properties of triangles, quadrilaterals, other polygons, and circles. Chapter 10 concludes by describing and stating the properties of geometric surfaces and solids.

Chapter 11 Congruence, Similarity, and Constructions

Chapter 11 continues the discussion of geometry by introducing congruence and similarity. These relationships are used to analyze triangles and other shapes and to solve problems. The chapter goes on to introduce and verify basic geometrical constructions, which are then used to construct other shapes and solve problems.

Chapter 12 Coordinate and Transformation Geometry

Chapter 12 concludes the content on geometry by introducing two topics related to spatial reasoning: coordinate and transformation geometry. It begins by using the Cartesian coordinate system to represent and study geometrical figures. It then uses coordinate geometry to study basic transformations, including translations, rotations, reflections, and glide reflections. The last two sections apply transformations in a study of congruence, similarity, symmetry, and geometrical patterns.

Chapter 13 Measurement

Chapter 13 discusses the basic concepts of measurement. It begins with an overview of the measurement process using the attribute of length. It then goes on to discuss specific measurable attributes that are commonly taught in grades K–8, including perimeter, area, surface area, and volume. In doing so, the English and metric systems are used, and conversions between the systems are made.

Chapter 14 Statistical Thinking

Chapter 14 provides an overview of statistical thinking by covering four basic steps included in most statistical studies. It begins with a discussion of how to formulate questions and collect data. It then moves into representing and analyzing data using statistical graphs and descriptive statistics. Chapter 14 concludes with a section devoted to the misuses of statistical graphs and descriptive statistics.

Chapter 15 Probability

Chapter 15 begins with making predictions from data by using experimental probability and simulations. It then goes on to discuss theoretical probability, odds, and mathematical expectation. The chapter concludes with a section on counting techniques such as the Fundamental Counting Principle, permutations, and combinations.

SUPPLEMENTS

FOR THE STUDENT	FOR THE INSTRUCTOR
	Instructor Edition (ISBN: 978-1-133-10399-8) The *Instructor Edition* features an appendix containing the answers to all problems in the book.
Student Solutions Manual (ISBN: 978-1-285-42023-3) Author: Mark A. Freitag Go beyond the answers. See what it takes to get there and improve your grade! This manual provides step-by-step solutions to selected problems in the text, giving you the information you need to truly understand how these problems are solved.	**Instructor's Resource Manual** (ISBN: 978-1-285-42027-1) Author: Mark A. Freitag The *Instructor's Resource Manual* provides detailed solutions to all problems in the text.

Exploration Activities
(ISBN: 978-1-133-96315-8)
Author: Mark A. Freitag
      Linda B. Crawford
Exploration Activities contains a variety of explorations for each section, which are referenced in the text by an icon in the margin [Explorations Manual]. Some explorations deal directly with the content of the chapter, often making use of relevant manipulatives or other hands-on activities. Other explorations extend the content of the section either mathematically or by building a connection to the K–8 classrooms. Most of the explorations can be done individually or with groups and should take about 30–45 minutes to complete.
	Exploration Activities
	(ISBN: 978-1-133-96315-8)
	Author: Mark A. Freitag
	      Linda B. Crawford
	To facilitate those instructors that use a more hands-on approach to teaching, an extensive Exploration Activities manual accompanies the text. The manual contains a variety of explorations for each section, which are referenced in the text by an icon in the margin [Explorations Manual]. Some explorations deal directly with the content of the chapter, often making use of relevant manipulatives or other hands-on activities. Other explorations extend the content of the section either mathematically or by building a connection to the K–8 classroom. Most of the explorations can be done individually or with groups and should take about 30–45 minutes to complete.

	PowerLecture with ExamView (ISBN: 978-1-285-42085-1) This DVD provides you with dynamic media tools for teaching. Create, deliver, and customize tests (both print and online) in minutes with ExamView Computerized Testing Featuring Algorithmic Equations. Easily build solution sets for homework or exams using the Solution Builder's online solutions manual. Microsoft PowerPoint lecture slides and figures from the book are included on this DVD.
	Solution Builder This online instructor database offers complete worked solutions to all exercises in the text, allowing you to create customized, secure solutions printouts (in PDF format) matched exactly to the problems you assign in class. For more information, visit www.cengage.com/solutionbuilder.

Acknowledgments

I would like to offer my thanks to the many people who took the time to review the manuscript and offer a number of valuable suggestions.

Jim Brandt, Southern Utah University
Carol Castellon, University of Illinois-Urbana-Champaign
Linda Cooke, University of South Carolina Upstate
Monette Elizalde, Palo Alto College
Melinda Gann, Mississippi College
Beth Greene Costner, Winthrop University
Rita Eisele, Eastern Washington University
Sue Glascoe, Mesa Community College
Leigh Hollyer, University of Houston Central Campus
Kelly Jackson, Camden County College
Pete Johnson, Eastern Connecticut State University
Honey Kirk, Palo Alto College
Sue McMillen, Buffalo State College
Elsa Medina, California Polytechnic State University
Perla Myers, University of San Diego
Cheryl Chute Miller, State University of New York at Potsdam
Nancy Ressler, Oakton Community College
Connie Schrock, Emporia State University
Rosa Seyfried, Harrisburg Area Community College
Agnes Tuska, California State University, Fresno

Special thanks go to Linda Crawford for testing the text in her courses, for being my coauthor on the explorations manual, and for always providing me with a different perspective to the content and to teaching. Thanks also go to Carol Rychly for reviewing the entire text on her own time and always lending me her ear, and to Sam Robinson for encouragement and patience while I completed the project.

I also would like to express my gratitude to the many people who made this project a reality. Thanks to Acquisitions Editor Marc Bove for overseeing the development process and getting us successfully through many ups and downs. Thanks to Developmental Editor Don Gecewicz for his comments and suggestions that helped in shaping and improving the overall integrity of the text. Thanks to everyone at Cengage Learning involved in the production process, who made the text come out so well. Special thanks goes to Tanya Nigh for overseeing the production process, to Stefanie Beeck for double- and triple-checking everything, and to Leslie Lahr for doing a great job with the art manuscript.

I would also like to thank everyone who worked on the text at Macmillan Publishing Solutions. Special thanks goes to Project Manager Lindsay Schmonsees, who did a wonderful job overseeing and coordinating the entire production process. Thanks to Chris Ufer and all the artists at Graphic World, who took my pitiful drawings and turned them into something special. Thanks to all those involved in copyediting and fact checking. Your comments and efforts were greatly appreciated.

Finally, and most importantly, I would like to thank my wonderful wife Amy, and my children Ashley, Caleb, and Maggie. The road has not always been easy, but thanks for your love, your support, and your patience in going through it with me.

Mathematical Processes

Opening Problem

Can you use two lines to divide the face of a clock into three parts so that the sum of the numbers in a part is equal to the sum in any other part?

It is not hard to see that our society depends increasingly on science, technology, and information. Almost every day brings new reports about the latest advances in science and medicine. Computers, cellular phones, and other technology now play an important role in our workplaces, homes, schools, and cars, giving us continuous access to a wealth of information. As our society becomes more centered on technology and information, we need more than ever to understand mathematics and to learn how to use it in powerful ways. Unfortunately, many people view mathematics only as a set of rules that tell us how to compute and not as the diverse subject that it really is. To change this limited view, we must show students the true nature of mathematics from the start of their school experience.

The role of elementary teachers in mathematics education is particularly important because they help students develop a solid foundation for future studies in mathematics. In addition to having a dramatic impact on what students learn, what elementary teachers do in the classroom affects how their students succeed in mathematics for years to come. For this reason, these teachers must not only be trained in appropriate teaching techniques, but they must also have a deep understanding of the mathematical ideas and processes they will teach.

For most of this book, we will concentrate on developing an understanding of ideas associated with elementary mathematics. However, in this chapter, we focus on mathematical processes. A **mathematical process** is any activity that students learn by doing mathematics or through which they learn and use mathematics. In the past, we have focused primarily on problem solving. Now, however, the National Council of Teachers of Mathematics (NCTM) recommends teaching four additional processes: mathematical communication, representation, reasoning and proof, and connections. Each process is important enough to merit a standard of its own in the *NCTM Principles and Standards for School Mathematics*. Each standard describes a specific process and suggests ways to implement it in the classroom. Despite their separate treatments, the five mathematical processes are closely connected, so we cannot engage in one without engaging in the others. The specific goals associated with each process standard are given in Table 1.1.

Table 1.1   NCTM Process Standards

NCTM Process Standard	Instructional programs from prekindergarten through grade 12 should enable all students to:
Communication Standard	■ Organize and consolidate their mathematical thinking through communication. ■ Communicate their mathematical thinking coherently and clearly to peers, teachers, and others. ■ Analyze and evaluate the mathematical thinking and strategies of others. ■ Use the language of mathematics to express mathematical ideas precisely.
Representation Standard	■ Create and use representations to organize, record, and communicate mathematical ideas. ■ Select, apply, and translate among mathematical representations to solve problems. ■ Use representations to model and interpret physical, social, and mathematical phenomena.
Reasoning and Proof Standard	■ Recognize reasoning and proof as fundamental aspects of mathematics. ■ Make and investigate mathematical conjectures. ■ Develop and evaluate mathematical arguments and proofs. ■ Select and use various types of reasoning and methods of proof.
Problem-Solving Standard	■ Build new mathematical knowledge through problem solving. ■ Solve problems that arise in mathematics and in other contexts. ■ Apply and adapt a variety of appropriate strategies to solve problems. ■ Monitor and reflect on the process of mathematical problem solving.
Connections Standard	■ Recognize and use connections among mathematical ideas. ■ Understand how mathematical ideas interconnect and build on one another to produce a coherent whole. ■ Recognize and apply mathematics in contexts outside of mathematics.

For the most part, four of these processes—mathematical communication, representation, reasoning and proof, and problem solving—are independent of specific mathematical topics, so they can be incorporated at any grade level and learned through any mathematical content. This being said, these four processes involve important mathematical ideas in their own right. Because they warrant additional attention, in Chapter 1 we:

- Consider how different representations are used to communicate mathematics.
- Look at different types of mathematical reasoning and how reasoning is used to make and verify mathematical conjectures.
- Identify the steps in the problem-solving process and use different strategies to solve a variety of mathematical problems.

The fifth process—developing mathematical connections—is difficult to discuss without relating it to specific mathematical content. In the upcoming chapters, we will take time to show how mathematical ideas are connected among themselves, to other disciplines, and to real-world situations. By building these connections, we will gain a deeper understanding of mathematics and a better appreciation of its power.

SECTION 1.1 Communication Through Mathematical Representation

Thinking of mathematics as an important part of communication may seem strange. Yet, to realize its importance, all we have to do is examine the communication around us. As we talk with others, listen to the radio, watch television, or read magazines, we are likely to come across many phrases like the following:

- I'll be there in half an hour.
- I have an 85 average in English.
- The fish I caught was about a foot long.
- Put your chairs in a circle.

As these examples illustrate, mathematics is crucial for conveying ideas involving quantities, measures, shapes, patterns, and relationships.

Just as mathematics is essential to communication, communication is essential to doing and learning mathematics. It is through communication that we

- Describe, organize, and add to what we know and understand about mathematical concepts.
- Develop and refine our ability to solve problems and reason mathematically.
- Share our ideas publicly and receive feedback on our thinking.
- Learn to analyze and evaluate the mathematical thinking of others.

So, if we are encouraged to communicate about mathematics, we will learn to communicate mathematically and come to understand and own the mathematics we are learning.

If communication is an important part of mathematics, then it makes sense to look at what it means to communicate mathematically. Mathematical communication, like other communication, can be done in a number of ways and using a variety of tools. However, mathematical communication differs from other communication in that it relies on a wide variety of representations. We cannot separate mathematical communication from mathematical representations because the representations are the tools we use to communicate mathematically.

Simply put, **representations** are ways of portraying mathematical objects, ideas, and relationships that allow us to analyze mathematics from different points of view. This characteristic makes them a fundamental part of how we understand and think about mathematics. If we can learn to use a variety of representations and understand

Explorations
Manual
1.1

how they complement one another, then we have a valuable set of tools for improving our ability to think mathematically. Although mathematical representations can vary in complexity and form, there are basically five types: verbal, numerical, symbolic, visual, and tactile. Even though we address each type of representation individually, it is common and often necessary to use several representations together to clarify ideas.

Verbal Representations

Explorations Manual 1.2

As the name implies, **verbal representations** use words to describe mathematics. Words are important in mathematics because their flexibility allows us to describe a variety of complex ideas completely and precisely. We use them to define concepts, state mathematical facts, justify reasoning, and even describe computational procedures. Whenever we use verbal representations, we need to be careful with our wording because, without such care, our communication can be unclear and lead to unintended results.

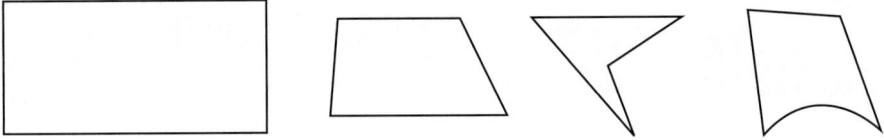

Example 1.1   Communicating Unclearly

Communication

Miss Aldrich, a first-year kindergarten teacher, introduced her class to squares. As part of her lesson, she had students draw a picture of a square, which she described as a four-sided figure. When she collected the drawings, she was surprised to see many figures like the following. What went wrong?

Solution Based on Miss Aldrich's description of a square, each figure has four sides and is correct. Yet none of them is a square because Miss Aldrich was not careful in her wording. Specifically, she needed to state that the sides must be straight, the same length, and perpendicular to one another. Given these criteria, the only shape her students could have drawn is a square.

Although precise language is important, it can make mathematical communication difficult for at least two reasons. First, we may have to use a number of technical terms that we do not normally use in our everyday language. Learning mathematics is much like learning a foreign language, in that we must learn new terms and use them correctly. Second, precise language can make our verbal representations long and awkward. For this reason, we often use other representations, like numbers and symbols, to make our communications efficient and clear.

Numerical Representations

Numerical representations use numbers to communicate mathematics. We generally use them in three ways.

1. **Identification numbers** are used to label or identify objects. Examples are social security numbers, driver's license numbers, and bank account numbers.
2. **Ordinal numbers** are used to assign an order to a group of objects. Examples are the placement of runners in a race or the class rank of an individual.
3. **Cardinal numbers** are used to express quantities. Examples are *3* pencils, *35* students, or *24* hours.

Example 1.2   Identifying Different Kinds of Numbers

Representation   State whether the numbers in each sentence are identification, ordinal, or cardinal.

a. Tammy's phone number is 342-6211.

b. On the fifth day of Christmas, my true love gave to me five golden rings.

Solution

a. The number "342-6211" is an identification number because it identifies a particular phone.

b. "Fifth" is an ordinal number because it describes the order of the days. The "five" in "five golden rings" is a cardinal number because it describes a quantity.

∎

Example 1.2 shows that numbers take on more meaning when placed in a context. In fact, we can get a surprising amount of information from a set of numbers if we simply know the context in which they occur.

Example 1.3   Analyzing Numbers in a Context

A 50-point test is given to a class of 15 students with the following results:

$$24 \ \ 31 \ \ 32 \ \ 33 \ \ 33 \ \ 41 \ \ 42 \ \ 44 \ \ 44 \ \ 44 \ \ 46 \ \ 46 \ \ 47 \ \ 48 \ \ 49$$

If a student must have a 35 or better to pass, what information can we get from the list of numbers without making any additional computations?

Solution We can draw several conclusions from the numbers. Because a score of 35 or better is passing, 10 students passed and 5 failed. The lowest score was a 24, the highest was a 49, and the most frequently occurring score was a 44. Based on this information alone, we can conclude that the class as a whole did well, even though some students need to improve.

∎

Although simple numerical representations can hold a tremendous amount of information, they can present at least two problems. First, it can be difficult to express general facts, properties, and relationships using only numbers. Second, as the set of numbers grows, using them in computations and analyzing them for trends or relationships becomes more difficult.

One way to handle large sets of numbers is to organize them in a table, which we can make by hand or with a spreadsheet. A **spreadsheet** is essentially a large electronic table that has the ability to store, organize, and make computations on large sets of numbers. For instance, Table 1.2 shows how a monthly household budget can be organized with a spreadsheet. Each row is labeled with a number and each column with a letter, making it possible to identify any particular **cell** in the table. For example, cell 4C is the cell where the fourth row meets the third column. In Table 1.2, cell 4C is highlighted and has a value of $325.

In addition to numbers, cells in a spreadsheet can also contain words and formulas. This capability makes a spreadsheet a particularly powerful computational tool. When we enter a formula, we can change values, and the computer automatically makes the related changes in the spreadsheet. Later in this book, we will use spreadsheets in powerful ways.

Table 1.2   An example spreadsheet

	A	B	C	D	E	F
1	**Monthly Expenses**					
2		**January**	**February**	**March**	**April**	**May**
3	Mortgage	$950	$950	$950	$950	$950
4	Utilities	$345	$325	$305	$280	$255
5	Food	$625	$615	$630	$585	$610
6	Transportation	$360	$325	$360	$340	$345
7	Clothing	$50	$75	$30	$25	$45
8	Entertainment	$150	$135	$120	$175	$95
9	Miscellaneous	$225	$235	$195	$140	$375
10						
11	**Total**	**$2,695**	**$2,660**	**$2,590**	**$2,495**	**$2,675**

Check Your Understanding 1.1-A

1. Give a verbal description of the following figure. What mathematical terms do need to you use?

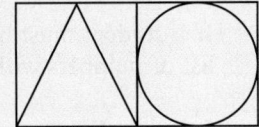

2. State whether the numbers in each sentence are identification, ordinal, or cardinal numbers.

   a. Joann and her partner placed third in the golf tournament.

   b. Lottery ticket number 38-45611 was worth $200,000.

   c. I have three pens, all from First Montgomery Bank.

3. Write three different pairs of numbers, each of which satisfies the following conditions.

   a. The second is double the first.

   b. The first is half the second plus two.

Talk About It   Why is communication an important part of learning mathematics? What does it mean to communicate mathematically?

Communication

Activity 1.1
Write a set of directions that explains how to draw a rectangle. Do not use the words "rectangle," "quadrilateral," or "right angle." Then, without revealing their purpose, give the directions to another person to follow. Were the directions clear enough for the person to draw a rectangle? What does this activity indicate about the need for mathematical terms to communicate ideas?

Symbolic Representations

Symbolic representations use symbols to communicate mathematics. They can give us the efficiency that we lose with words and the generality that we lose with numbers. We use them to represent:

- *Mathematical concepts.* Symbols are used to represent both abstract and concrete concepts, as well as general and specific ideas. Examples are:

    ∞ to represent infinity.
    $\overline{x}$ to represent the arithmetic mean of a set of numbers.
    $\triangle ABC$ to represent a triangle with vertices A, B and C.

- *Mathematical operations.* Almost every mathematical operation has a special symbol associated with it. Examples are:

    $+$ to represent addition.
    $-$ to represent subtraction.
    $\sqrt{}$ to represent the square root.

- *Mathematical relationships.* Symbols are also used to represent relationships between mathematical expressions and ideas. Examples are:

    $=$ to represent that two expressions have the same value.
    $<$ to represent that one quantity is smaller than another.
    $\cong$ to represent that two shapes are congruent.

- *Commonly used words or phrases.* Many words and phrases commonly used in mathematics are given their own symbols. Examples are:

    $\therefore$ to represent "therefore."
    iff   to represent "if and only if."
    : or | to represent "such that."

Algebraic expressions are symbolic representations that provide a compact way to represent relationships, patterns, and problems. Many students have trouble with algebraic expressions because they view them as abstract ideas that are governed by their own set of rules. In truth, algebraic expressions are an extension of numerical expressions and follow the rules of basic arithmetic. The only real difference between numerical and algebraic expressions is the use of variables.

A **variable** is a symbol representing a quantity that can take on different values. We use them in various ways in algebraic expressions. One use is to represent a specific but **unknown quantity** in problem-solving situations. In this case, we use an algebraic expression as a tool to find the missing value. For instance, in the equation $8x - 2 = 18$, the variable x represents an unknown value that makes the statement true. Using variables in this way is a key component of algebraic thinking, which is a topic we examine in Chapter 9.

A variable can also serve as a **general representative** from a set of numbers. In this case, the variable can be replaced with any number from a given set, and the value of the variable can change based on the needs and the context of the expression. We use variables in this sense in at least three ways.

1. *To represent generalized numerical properties.* The variables represent elements from the set for which a property holds true. For example, the commutative property of addition states that, for any two real numbers a and b, $a + b = b + a$.

2. *To express relationships.* The variables express a relationship between two or more quantities. For example, in $y = 2x - 5$, the value of y depends on x. Whatever value x takes on, the value of y is twice x minus 5.

3. *To represent values in standardized formulas.* Variables are often used in formulas to represent well-established relationships between two or more quantities. In such cases, letters are typically chosen to match the quantities they represent. For instance, in the distance formula, $D = rt$, D represents distance, r represents rate, and t represents time.

If we use a variable to express a relationship, then we can find numbers that satisfy the relationship by evaluating the expression. To **evaluate** an expression, we replace the variables with numbers and perform any computations.

Example 1.4   Evaluating Expressions with Variables

Evaluate each expression at the given value or values.

a. $y = 2x - 5$ at $x = 3$.
b. $D = rt$ at $r = 50$ and $t = 5$.
c. $y = 2n + 1$ at $n = 0, 1, 2,$ and 3.

Solution

a. To evaluate $y = 2x - 5$ at $x = 3$, replace x with 3, and compute:
$y = 2(3) - 5 = 6 - 5 = 1$.

b. In this expression, replace r with 50 and t with 5, and compute:
$D = rt = (50)(5) = 250$.

c. We are asked to evaluate one equation at four numbers. To do so, we make four separate calculations.

$$\text{For } n = 0, y = 2n + 1 = 2(0) + 1 = 0 + 1 = 1$$
$$\text{For } n = 1, y = 2n + 1 = 2(1) + 1 = 2 + 1 = 3$$
$$\text{For } n = 2, y = 2n + 1 = 2(2) + 1 = 4 + 1 = 5$$
$$\text{For } n = 3, y = 2n + 1 = 2(3) + 1 = 6 + 1 = 7$$

Even though symbols make mathematical writing more efficient, they can make mathematics more difficult to read. Consequently, the ability to translate between symbols and words is an essential skill for learning mathematics. In most cases, the information needed to make the translation is given in the definition of the concept, operation, or relationship. This means that it is important that we become familiar with mathematical definitions and their related symbols. Even though a symbol may have a formal translation in its definition, it may also be associated with other words and phrases. For example, consider the basic operations and relationships given in Table 1.3.

Table 1.3   Symbols and words for common mathematical ideas

Mathematical Idea	Symbol	Associated Word/Phrases
Addition	+	Add, sum, more than, plus
Subtraction	−	Subtract, minus, less than, from
Multiplication	× or ·	Times, multiply, of, product
Division	÷ or /	Divides, quotient, ratio, into
Equality	=	Is, equals, the same as
Greater than	>	Bigger than, more than

| Example 1.5 | Translating Verbal Expressions into Symbols |

Representation

Translate each verbal statement into numbers and symbols.

a. The sum of three and y is greater than ten.

b. The product of four and x plus two is twenty-two.

c. The ratio of four and x is the same as the ratio of y and three.

Solution

a. Proceeding one word at a time, we represent the expression as $3 + y > 10$.

b. The language in this statement is vague, and it leads to two possible interpretations. Mathematically, we could use the fact that multiplication is computed before addition and translate the expression as $4x + 2 = 22$. Grammatically, we could use "and" as a key word that separates the factors and translate the expression as $4(x + 2) = 22$. Generally, we prefer the mathematical interpretation, so we translate the expression as $4x + 2 = 22$. To avoid ambiguity, we often use the phrase "the quantity of" to indicate a group of two or more symbols.

c. Because "ratio" means to divide, $\dfrac{4}{x} = \dfrac{y}{3}$.

Visual Representations

Visual representations use pictures to communicate mathematics. They include simple diagrams, algebraic graphs, geometrical figures, and statistical graphs. They are particularly powerful in that they can convey large amounts of information and allow us to see concepts and relationships in ways that other representations do not. We use visual representations to illustrate concepts, identify properties, devise problem-solving strategies, justify reasoning, and describe data. In some cases, we can even identify mathematical facts just by looking at pictures.

| Example 1.6 | Using Pictures to Identify the Properties of Shapes |

Representation

The following shapes are all examples of parallelograms. Based on what you see, what might be true for all parallelograms?

Solution Just by looking at the pictures, we can identify several possible facts about parallelograms. For instance, in each shape, the opposite sides are parallel and the same length, and the opposite angles have the same measure.

Pictures become even more powerful when we use technology to make them dynamic. Geometry software now enables us to label, measure, and manipulate diagrams in different ways. This capability increases our ability to explore geometrical situations and verify mathematical conjectures by allowing us to view and analyze many examples in a short period of time.

For instance, in Example 1.6 we used only four examples to surmise that the opposite angles in a parallelogram have the same measure. We can explore this conjecture by using geometry software, first by constructing a parallelogram and then by allowing the program to measure the angles. As we drag one of the vertices, we create many different parallelograms, and the software automatically recalculates the angle measures for each one. Figure 1.1 gives three possible examples, all of which show that opposite angles in a parallelogram have the same measure.

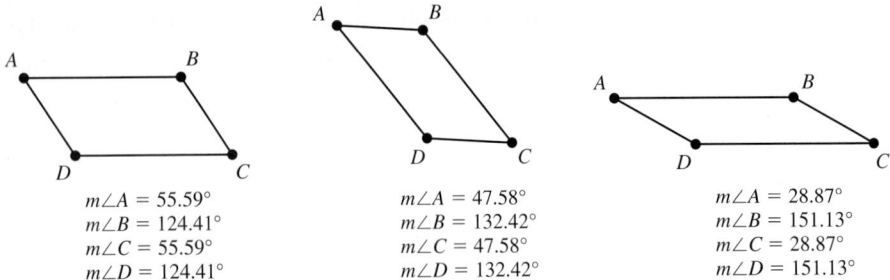

$m\angle A = 55.59°$
$m\angle B = 124.41°$
$m\angle C = 55.59°$
$m\angle D = 124.41°$

$m\angle A = 47.58°$
$m\angle B = 132.42°$
$m\angle C = 47.58°$
$m\angle D = 132.42°$

$m\angle A = 28.87°$
$m\angle B = 151.13°$
$m\angle C = 28.87°$
$m\angle D = 151.13°$

Figure 1.1 Angle measures of three parallelograms

Although the software cannot guarantee that our conjecture is true, its dynamic capability supplies many examples as evidence that it is.

Another way to convey more information through visual representations is to combine them with verbal and numerical representations. Consider the next example.

Example 1.7   Analyzing a Statistical Graph

Application

Use the statistical graph to answer each question.

a. What gender age group had the most number of fatalities in 2010?

b. How many females between the ages of 20 and 24 died in car accidents for every 100,000 people?

c. How many people between the ages of 25 and 29 died in car accidents for every 100,000 people?

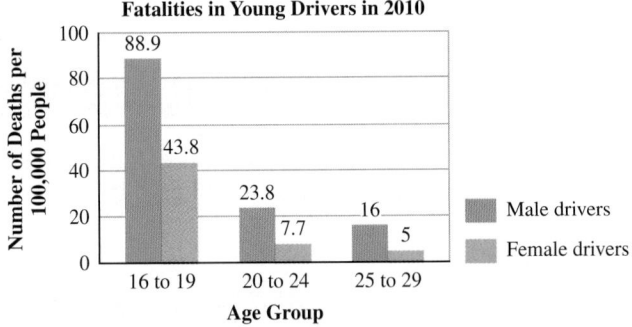

Solution This statistical graph is a visual representation, but it has verbal and numerical components. We must rely on all of them to answer the questions.

a. The gender age group with the most fatalities is males from age 16 to 19.

b. For every 100,000 people, an average of 7.7 females from age 20 to 24 died in car accidents.

c. For every 100,000 people, an average of $16 + 5 = 21$ people from age 25 to 29 died in car accidents.

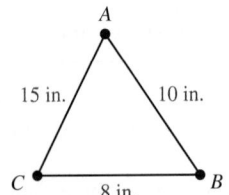

Figure 1.2 A misleading picture of a triangle

Even though pictures can be powerful, they can also be misleading, particularly if they are not drawn carefully or offer only a limited view. For instance, consider the triangle in Figure 1.2. Visually, sides $\overline{AB}$ and $\overline{AC}$ seem to be close to the same length. Side $\overline{AB}$ may even be slightly longer than side $\overline{AC}$. However, the numbers in the figure tell us that side $\overline{AC}$ is actually 5 in. longer than side $\overline{AB}$.

Tactile Representations

Tactile representations, or manipulatives, communicate mathematics through physical models that can be touched, moved, and arranged in different ways. Some are quite simple; yet, when we place them in a context, they become powerful tools for learning. For instance, when children first learn to add, they do so by combining and counting two sets of objects (Figure 1.3). Here, the manipulatives can be any convenient objects, such as coins, blocks, paperclips, or other counters.

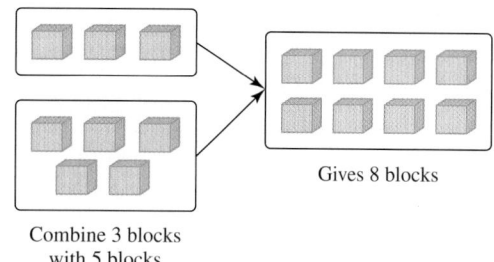

Combine 3 blocks
with 5 blocks

Gives 8 blocks

Figure 1.3 Manipulatives used to add 3 + 5 = 8

Other manipulatives are more complex or serve a specific purpose. We will encounter many such manipulatives and learn how they are used to represent elementary mathematics. Some common classroom manipulatives are shown in Figure 1.4.

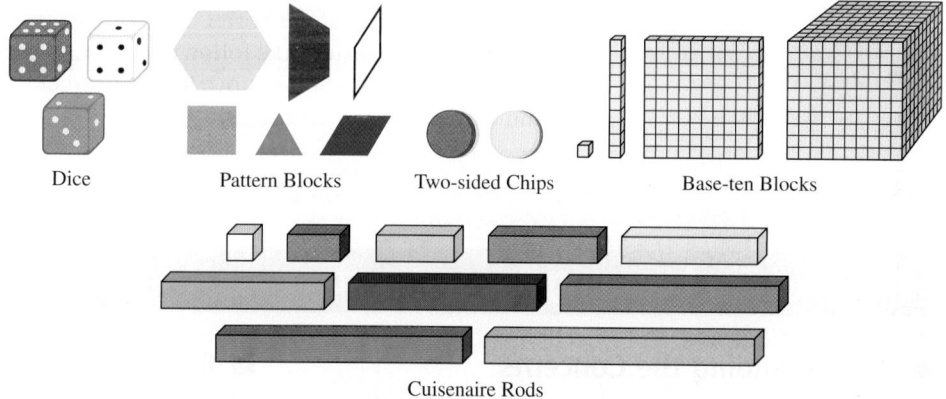

Dice   Pattern Blocks   Two-sided Chips   Base-ten Blocks

Cuisenaire Rods

Figure 1.4 Common classroom manipulatives

Although tactile representations are powerful learning tools, they can have limitations. In some cases, they are awkward to use, or they are limited in their ability to represent every possible situation. For example, using individual blocks to add 59 to 68 is not helpful simply because the number of blocks is too large.

Check Your Understanding 1.1-B

1. Evaluate each algebraic expression at the given values.

   a. $y = 3x + 5$ at $x = 4$ **b.** $m = 60h$ at $h = 5$ **c.** $A = lw$ at $l = 25$ and $w = 9$

2. Translate each verbal statement into numbers and symbols.

   a. Twelve subtracted from y is greater than or equal to seven.

   b. Four times the quantity of x plus nine is twenty-one.

   c. A number divided by five equals thirty.

3. Use the pie chart to answer each question.

   a. What continent has the largest percentage of the world's total land area?

   b. What continent has the smallest percentage of the world's total land area?

c. What two continents together make up about 40% of the world's total land area?

d. Order the continents by area from largest to smallest.

Percent of Total Land Area

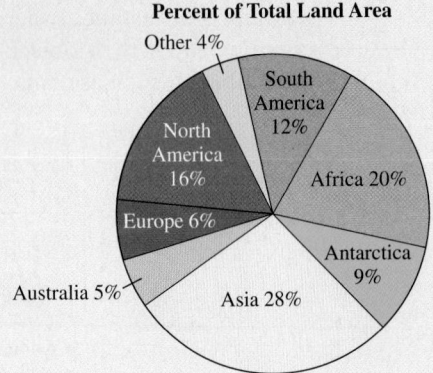

Talk About It What different roles can a variable have in an algebraic expression? What advantages does this give symbolic representations over other representations?

Problem Solving

Activity 1.2
Simple manipulatives are often used in problems. For instance, consider the following arrangement of toothpicks. Can you remove four toothpicks and leave only four triangles?

SECTION 1.1　Learning Assessment

■ Understanding the Concepts

1. Why is mathematics an important part of communication?

2. What are mathematical representations, and why are they important to learning mathematics?

3. Make a table that compares the five kinds of mathematical representations. Include a description of each representation along with its advantages and disadvantages.

4. Explain the difference between an identification number, an ordinal number, and a cardinal number.

5. Describe four different uses of symbolic representations.

6. What are variables, and what purposes do they serve in algebraic expressions?

7. What does it mean to evaluate an algebraic expression at a given value?

■ Representing the Mathematics

8. In each case, make a table of four pairs of numbers that satisfy the given relationship. Then represent the relationship with an algebraic expression using the variables x and y.

a. The first plus the second equals zero.

b. The second is three times the first.

c. The second is three more than the first.

d. The first is the square of the second.

9. Each table demonstrates a numerical relationship between x and y. Represent the relationship verbally and symbolically.

a.

x	y
1	1
2	4
3	9
4	16
5	25

b.

x	y
5	4
4	3
3	2
2	1
1	0

c.

x	y
1	2
2	4
3	6
4	8
5	10

d.

x	y
1	1
2	1/2
3	1/3
4	1/4
5	1/5

10. Give a verbal description of each relationship. Then make a table that shows four pairs of numbers that satisfy the given condition.

a. $x + y = 6$ **b.** $3x - 4 = y$

c. $x \cdot y = 24$ **d.** $\dfrac{x}{y} = 1$

11. Give a verbal description of each symbolic statement.

a. $x + 3 = 6$ **b.** $6 - y = 2x$

c. $4x + 1 < 10$ **d.** $18 \div x = 6$

12. Give a verbal description of each symbolic statement.

a. $11 - x > 4$ **b.** $3(x + 4) = 7$

c. $14 - 3x \le y$ **d.** $\dfrac{(x - 1)}{y} \le 3$

13. Translate each verbal statement into numbers and symbols.

a. The sum of five and x is less than three.

b. The difference of twice p and five is four.

c. The product of 6 and the quantity x plus two is less than nine.

d. The ratio of six and x is the same as y plus one.

14. Translate each verbal statement into numbers and symbols.

a. The product of three and four is twelve.

b. The difference of the quantity of x plus seven and four is the same as nine.

c. The quotient of x and y is less than four.

d. Four times the quantity of x minus five is the same as ten.

15. Give a verbal description of each figure.

a. 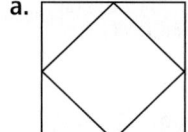   **b.**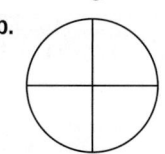

c. 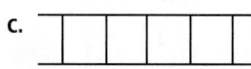   **d.**

16. Draw a picture that satisfies the given description.

a. A square inside a circle so that each corner of the square is on the circle

b. A triangle on top of a rectangle

c. Two circles beside each other touching at one point

d. Which of the preceding descriptions are precise enough to allow for only one drawing? For those that are not, draw two different pictures that satisfy the description. What does this imply about the need for precise language in mathematical descriptions?

17. Draw a visual representation that is misleading, and briefly explain why it is so.

■ Exercises

18. Write a set of directions that describes how to add 58 to 67.

19. Without using the word "triangle," write a precise description of the following figure.

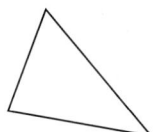

20. Consider the following rectangle. How would you tell someone to take $\dfrac{1}{6}$ of it without using the word "one-sixth"?

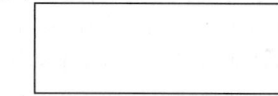

21. State whether the numbers in each sentence are identification, ordinal, or cardinal.

a. I found 10 pollywogs swimming in the pond.

b. Carol just cashed check number 315.

c. We charged $42.50 to credit card number 0123456789.

22. State whether the numbers in each sentence are identification, ordinal, or cardinal.

a. The first place car finished the race at a speed of 204 mph.

b. There are 360 students enrolled in Biology 101.

c. My address is 735 North 7th Street.

23. On her last 7 rounds of golf, Linda made the following scores: 97, 98, 95, 94, 97, 98, 93. Linda plans to play today. If lower numbers represent better scores, what score do you think Linda would need to consider it a good round?

24. The following set of numbers represents the scores of 10 people on an IQ test: 95, 100, 110, 115, 120, 120, 120, 130, 135, 150. If the average IQ score is 115, what do these numbers reflect about the IQs of these 10 people?

25. The following table represents the approximate trout population in a lake. What does the table show about trends in the trout population over seven years?

Year	2005	2006	2007	2008	2009	2010	2011
Trout population	59,000	55,000	47,000	48,000	51,000	58,000	63,000

26. The following table shows the number of speeding tickets given in a town over the course of one year. Use the information to answer the following questions.

   a. In what month were the fewest tickets given?

   b. In what month were the most tickets given?

   c. In how many months were over 100 tickets given?

   d. Is there a relationship between the month and the number of tickets given? If so, describe it.

Month	Number of Tickets	Month	Number of Tickets
January	78	July	189
February	81	August	167
March	85	September	131
April	103	October	95
May	120	November	88
June	151	December	72

27. The following portion of a spreadsheet shows the test grades of five students. Use the information in the spreadsheet to answer each question.

   a. State the number that occurs in each cell.

      i. 3B   **ii.** 5E   **iii.** 4D   **iv.** 8F

   b. What does column C represent?

   c. What does row 4 represent?

   d. Which student had the highest test score? The lowest?

   e. Which student is doing the best?

	A	B	C	D	E	F
1	Student	Test 1	Test 2	Test 3	Test 4	Total
2	Alicia	73	65	67	74	279
3	Belinda	52	71	69	82	274
4	Cole	95	73	76	84	328
5	Douglas	76	41	39	52	208
6	Evan	85	71	70	74	300
7						
8	Points Possible	100	80	80	90	350

28. Find y by evaluating the expression $y = x + 4$ at each value of x.

   a. $x = 3$ **b.** $x = 0$ **c.** $x = 10$ **d.** $x = 9$

29. Find u by evaluating the expression $u = 7t - 6$ at each value of t.

   a. $t = 0$ **b.** $t = 2$ **c.** $t = 5$ **d.** $t = 9$

30. Find V by evaluating the expression $V = 3 \cdot l \cdot w$ for each pair of l and w.

   a. $l = 2$, $w = 5$ **b.** $l = 4$, $w = 3$

   c. $l = 6$, $w = 6$ **d.** $l = 0$, $w = 7$

31. Evaluate $y = 3x^2 - 4$ at $x = 0, 1, 2,$ and 3.

32. What mathematical truths about equilateral triangles might be evident just by looking at the following one?

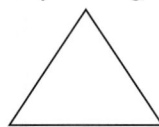

33. Which is the longest side of the figure? Explain how you know.

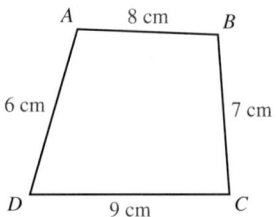

34. In the following triangle, which do you think is the largest angle, $\angle A$, $\angle B$, or $\angle C$? Explain how you made your decision.

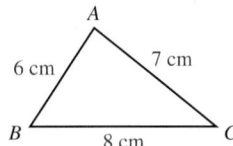

35. The following chart gives the average family size in the United States for every decade from 1940 to 1990. Answer the questions based on the information in the chart.

   a. In what decade was the average family size the largest? The smallest?

   b. Between what two consecutive decades was the decrease in family size the largest?

   c. What trend does the chart suggest about the average size of the family?

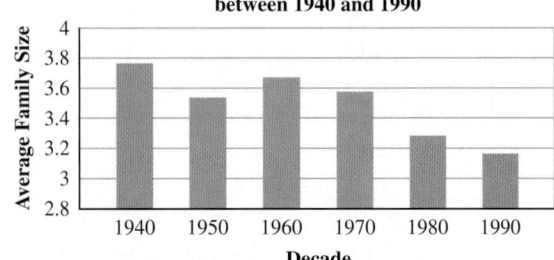

Source: U.S. Bureau of Census

36. Consider the following arrangement of 4 nickels and 3 pennies.

If only 2 adjacent coins can be switched, what is the fewest number of switches needed to place the coins into the following arrangements?

a.

b.

c.

37. A quarter, a nickel, and a penny are stacked on one another as shown. Moving only one coin at a time so that no bigger coin is stacked on top of a smaller coin, move the stack from box A to box C. What is the lowest number of moves it takes?

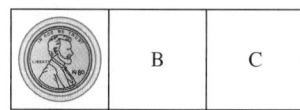

38. The following diagram shows an arrangement of six dice.

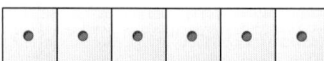

If a change of one die to an adjacent face is considered a flip, how many flips must be made to show the following arrangements?

a.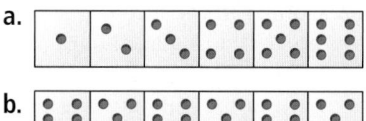

b.

■ Mathematical Communication

39. Represent 25 in as many ways as possible. You can use models, words, pictures, and so on. Compare your list with those of others in your class. How many did you come up with? What does this suggest about representations and their ability to show mathematical concepts?

40. Write two or three sentences describing situations that include identification, ordinal, and cardinal numbers. Give your sentences to a partner, and ask your partner to classify each type of number used in your sentences.

41. Write down instructions for drawing a square without using the words "square" or "rectangle." Ask a friend to follow the directions without telling what they are for. Did your friend have difficulty

following them? What does this imply about communicating mathematically through writing?

42. Tell someone how to draw the following figure without watching what the person is drawing. Once the figure is complete, compare the drawing to the original. If the figure is drawn correctly, discuss what made your communication effective. If the figure is drawn incorrectly, discuss what could have made your communication better.

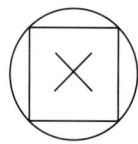

43. Get with two or three of your peers. Have each person write a set of directions that describes how to get to your classroom from the entrance of the building. When each person is done, take turns sharing your directions with the group.

a. How are the sets of directions the same? How are they different?

b. Take a few moments and combine the separate directions into one. Give them to someone who does not know where your classroom is and see whether he or she can follow your directions correctly.

44. Variables often prove difficult for children to learn. Do an Internet search to find different reasons as to why this is so.

■ Building Connections

45. Each of the following words is used in mathematics and in common language. Write two sentences for each word, one expressing the word in a mathematical situation and the other in a nonmathematical one. How is the mathematical usage of the word similar to the everyday usage of the word? How is it different?

a. Triangle   **b.** Mean   **c.** Parallel   **d.** Multiply

46. One mathematical symbol can take on different meanings depending on context. Can you think of at least two different meanings for each of the following symbols?

a. +   **b.** −   **c.** ( )   **d.** /

47. Many people view mathematics as a language with a vocabulary, syntax, and symbolism all its own. Take a few moments to search the Internet to find out more about this view. To what degree do you agree with what you find?

48. Read the Communication Standards for grades Pre-K–2 and grades 3–5.

a. Do the Standards make sense or have meaning to you?

b. To what extent do you agree with them?

c. Write a short report on what the Standards suggest mathematical communication should look like in the classroom in each of these grade bands. Do you think this is attainable? Why or why not?

49. The NCTM Standards state that a mathematical representation "refers both to process and to product—in other words, to the act of capturing a mathematical concept or relationship in some form and to the form itself" (NCTM, 2000, p. 67). Explain the meaning of this statement in your own words. To what extent do you agree or disagree with this portrayal of a mathematical representation?

50. Think of a time when you have struggled to explain an idea to someone else. Did your struggle help you understand the idea better? If so, how? What implications might this have for teaching mathematics?

51. Check an elementary-level mathematics textbook for evidence of promoting communication. Describe the ways in which the textbook encourages students to communicate about the mathematics they are learning.

SECTION 1.2 Inductive Reasoning and Patterns

Much of mathematics revolves around making and verifying **conjectures,** which are educated guesses about what we believe to be true. In most sciences, conjectures are made from observations and verified through experiments. As a consequence, the validity of a conjecture depends necessarily on the quality of the observation and the experiment. If either is flawed, the conjecture remains in doubt. In mathematics, however, we make and verify conjectures through **mathematical reasoning,** which is a logical and systematic way of thinking that uses a specific set of rules and assumptions.

When you read the introduction to the chapter, you may have found it surprising that mathematical reasoning is advocated at all grade levels. However, mathematical reasoning is a defining characteristic of mathematics, and at its core lies the assumption that mathematical ideas should make sense, be supported by logical reasons, and lead to a better understanding of mathematics. So, if we want children to understand and do mathematics, then we must teach them how to reason mathematically. Because this is not always easy, including reasoning at the earliest grades gives children more opportunities to learn and to use reasoning skills. In general, elementary children learn two kinds of reasoning skills: inductive and deductive reasoning. We discuss inductive reasoning in this section and deductive reasoning in the next.

Inductive Reasoning

In the classroom, children first learn to reason mathematically by making conjectures based on the patterns they see in situations involving numbers, shapes, and colors. To make these conjectures, they use **inductive reasoning.**

Definition of Inductive Reasoning

Inductive reasoning is the process of drawing a general conclusion from observations and patterns seen in specific examples.

Inductive reasoning is a two-step process. The first step is to look for a pattern among a set of related examples. The **pattern** can be anything that has a perceptual structure that we identify by analyzing the examples for similarities and differences. Once we have found a pattern and have made sure that it holds for every example, the second step is to make a generalization. A **generalization** is a statement indicating that the pattern will likely hold true for all other similar situations.

For example, consider the following facts about the family members of a girl named Jill.

Jill's father has black hair.
Jill's brother has black hair.

Jill's mother has black hair.
Jill's sister has black hair.

Jill's family members are specific examples that we can examine for a pattern. Because every member of her family has black hair, it makes sense to extend the pattern to Jill. As a result, we make the generalization that Jill also has black hair.

Pattern recognition is a key component of inductive reasoning. So given the wide variety of patterns in the world, we use inductive reasoning in many different situations. To better understand how inductive reasoning is used in the elementary classroom, we can look at the patterns children are likely to encounter in school: repetitive patterns, patterns of common traits, and growing patterns.

Repetitive Patterns

Repetitive patterns are patterns with a core component that continuously repeats. They are generally easy to recognize, and we often classify them according to their underlying structures.

Example 1.8   Naming Repetitive Patterns

Connections

Use the naming scheme on the student page in Figure 1.5 to classify the eight given patterns.

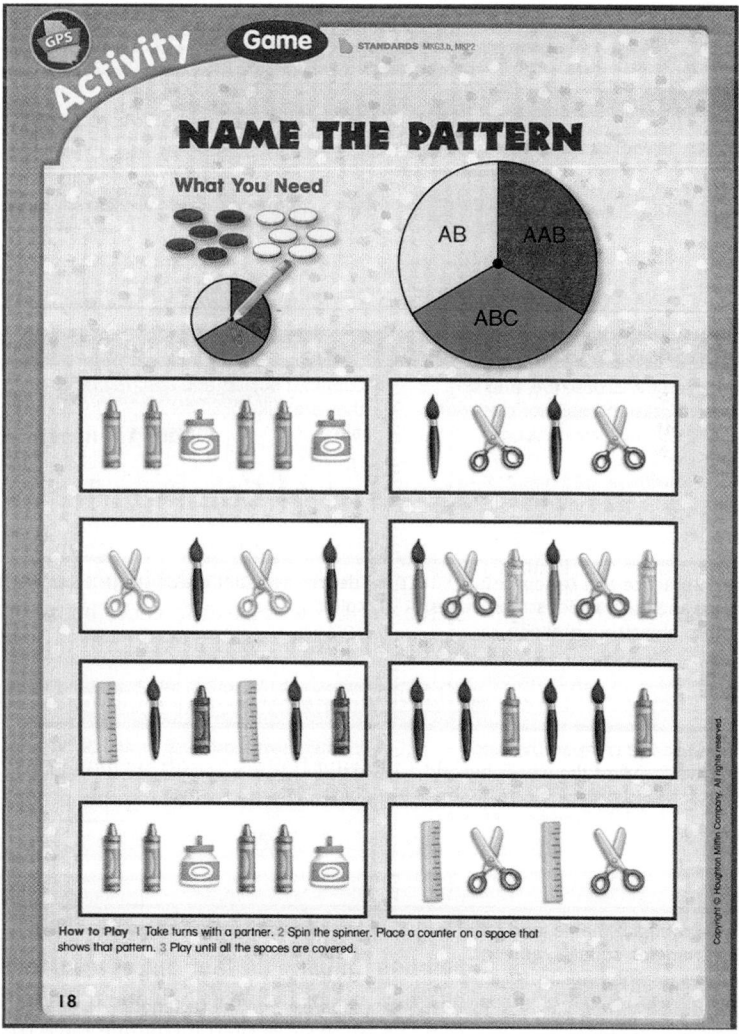

Figure 1.5 Repetitive patterns
Source: From *Mathematics Grade K: Houghton Mifflin Mathematics Georgia* (p. 18). Copyright © by Houghton Mifflin Company, Inc. Reproduced by permission of the publisher, Houghton Mifflin Harcourt Publishing Company.

Solution A common way to classify repeating patterns is to use capital letters to name specific pattern structures. In this case, we have three structures: an *AB* pattern, an *AAB* pattern, and an *ABC* pattern. *AB* patterns have two objects that repeat in succession. The three *AB* patterns are:

AAB patterns use two objects repeating in a two-to-one ratio. The three *AAB* patterns are:

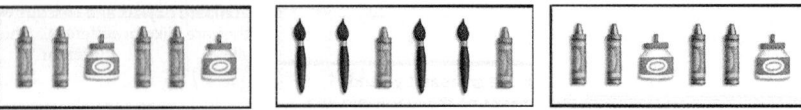

ABC patterns have three shapes that repeat in succession. The two *ABC* patterns are:

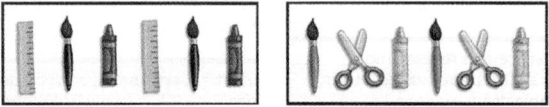

We can also make repetitive patterns from geometric shapes. Some, like border patterns, are linear in nature, and others, like tessellations, cover the entire plane (Figure 1.6). Describing these repetitive patterns relies heavily on symmetry and transformations. These kinds of patterns are discussed fully in Chapter 12.

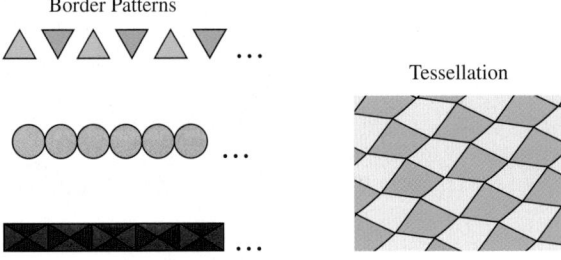

Figure 1.6 Repetitive geometrical patterns

Patterns of Common Traits

Explorations
Manual
1.3

Another kind of pattern often seen in mathematics is a **pattern of common traits.** The examples in the pattern exhibit some common characteristic, which we find by looking at the similarities and differences between them. Patterns of common traits are often more subtle than repetitive patterns and therefore can be more difficult to identify.

Example 1.9   **Identifying a Pattern of Common Traits in Shapes**

Reasoning

The following figures are all regular polygons. Based on what you see in these shapes, what might hold true for all regular polygons?

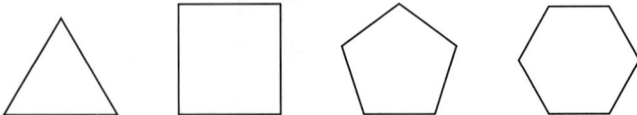

Solution Initially, the four shapes seem to have nothing in common. Each has a different number of sides, different side lengths, and different angle measures. However, notice that the three sides of the triangle have the same length, as do the sides of the square, pentagon, and hexagon. From this pattern, we might conclude that the sides of any regular polygon are the same length. Also notice that the three angles in the triangle have the same measure, as do the angles in the square, pentagon, and hexagon. From this pattern, we might conclude that the angles in any regular polygon have the same measure.

If the pattern involves numbers, we may be able to write our generalization in the form of an algebraic expression. Doing so gives us an efficient way of representing patterns that hold for an infinite number of possibilities.

Example 1.10	**Writing a Generalization for a Pattern**

Reasoning

What do the following sums suggest might hold true for addition? Use an algebraic expression to write your generalization.

$2 + 4 = 6$	$1 + 7 = 8$	$4 + 2 = 6$	$7 + 1 = 8$	$10 + 5 = 15$
$3 + 4 = 7$	$5 + 10 = 15$	$4 + 3 = 7$	$5 + 6 = 11$	$6 + 5 = 11$

Solution Initially, the sums may seem to have no pattern. However, upon a closer look, we notice that the sums occur in pairs.

$3 + 4 = 7$	$5 + 6 = 11$	$4 + 2 = 6$	$7 + 1 = 8$	$10 + 5 = 15$
$4 + 3 = 7$	$6 + 5 = 11$	$2 + 4 = 6$	$1 + 7 = 8$	$5 + 10 = 15$

In each case, we can see that, although the order of the numbers has changed, the sum remains the same. Because the sums involve randomly chosen counting numbers, it seems reasonable the pattern will hold true for all such sums. To write the generalization, we pick two variables, say a and b, to represent the numbers in the sum. We then write $a + b = b + a$.

■

Check Your Understanding 1.2-A

1. Use the naming scheme in Example 1.8 to characterize each repetitive pattern.
   a. 1, 2, 3, 1, 2, 3, 1, 2, 3, . . .   b. 1, 2, 1, 2, 1, 2, . . .
   c. 4, 5, 5, 4, 5, 5, 4, 5, 5, . . .   d. 9, 8, 7, 9, 8, 7, 9, 8, 7, . . .

2. The last three times you and your roommate ordered pizza, she wanted only pepperoni. You plan to order pizza tonight. What do you think will be your roommate's request?

3. What conclusion can you draw about multiplication based on the pattern you observe in the following five facts?

   $2 \times 4 = 8$ $5 \times 2 = 10$ $3 \times 2 = 6$ $2 \times 11 = 22$ $9 \times 2 = 18$

Talk About It What role do mathematical representations play in establishing patterns?

Reasoning

Activity 1.3
Pattern blocks are manipulatives that can be used to make geometrical patterns. They come in the following six basic shapes and colors. Use a set of pattern blocks to make a repetitive pattern that is:
   a. One-dimensional.   b. Two-dimensional.

Pattern Blocks

Growing Patterns

Inductive reasoning also helps us identify growing patterns. A **growing pattern** is one that continuously increases or decreases in a predictable fashion.

| **Example 1.11** | **Extending Growing Patterns** |

Write the next three numbers in each pattern.

a. 2, 4, 6, 8, . . .   **b.** $1, \frac{1}{2}, \frac{1}{3}, \frac{1}{4}, \frac{1}{5}, \ldots$ **c.** 5, 25, 125, . . .

Solution In each case, we identify the pattern by analyzing the changes between consecutive numbers. Once we know the pattern, we can predict the next three numbers.

a. This pattern is consecutive even numbers. The next numbers are 10, 12, and 14.

b. In this pattern, the numerators of the fractions are always 1, and the denominators are consecutive counting numbers. The next three numbers are $\frac{1}{6}, \frac{1}{7}$, and $\frac{1}{8}$.

c. This pattern is made from the consecutive powers of 5. The next three numbers are $5^4 = 625$, $5^5 = 3{,}125$, and $5^6 = 15{,}625$.

Explorations
Manual
1.4

As the last example shows, growing patterns can grow larger or smaller and do so at different rates. The patterns in the last example are called **sequences,** which are simply ordered lists of numbers. Sequences are closely connected to algebraic thinking, so we discuss them in greater detail in Chapter 9. The next example demonstrates how inductive reasoning and growing patterns can be used to make a computation.

| **Example 1.12** | **Using a Pattern to Compute a Sum** |

Reasoning

Find the sum of the first 100 odd numbers.

Solution We can approach the task in at least two ways. One is to add the numbers directly. But even with a calculator, this takes time, and the possibility is good that we would eventually press a wrong key. Another approach is to compute several smaller sums and then examine them for a pattern that we can extend to find larger sums. The sums involving the first five odd numbers are as follows:

$$1 = 1$$
$$1 + 3 = 4$$
$$1 + 3 + 5 = 9$$
$$1 + 3 + 5 + 7 = 16$$
$$1 + 3 + 5 + 7 + 9 = 25$$

Each sum is a perfect square, so we surmise that the sum of the first n odd numbers is equal to n^2. Consequently, if $n = 100$, the sum of the first 100 odd numbers is $n^2 = 100^2 = 10{,}000$.

A particularly interesting growing pattern is **Pascal's triangle,** which is an infinite triangular array of numbers generated in the following way. Three 1s are arranged in a triangle. Then rows are added to the bottom of the triangle using the following rules:

1. Every row begins and ends with a 1.
2. Entries in between the 1s are computed by adding the two entries that are diagonally above in the preceding row.

Figure 1.7a shows how we use the rules to generate the first four rows of Pascal's triangle. The first eight rows of Pascal's triangle are then shown in Figure 1.7b.

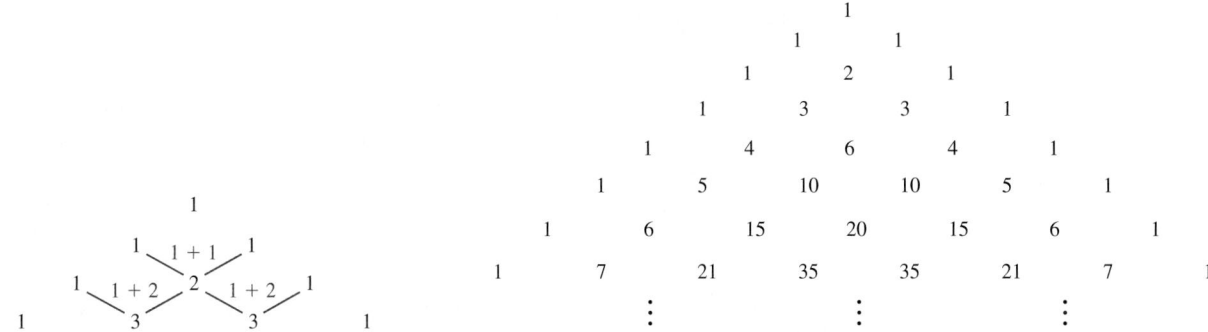

Figure 1.7a Generating Pascal's triangle   **Figure 1.7b** The first eight rows of Pascal's triangle

Pascal's triangle is interesting because it contains a surprising number of patterns. For instance, the outer diagonals of the triangle are all 1s, and the second diagonals give the counting numbers. We can also see that every diagonal that goes down from left to right has an equivalent counterpart that goes down from right to left.

| Example 1.13 | Finding and Generalizing a Pattern in Pascal's Triangle |

Reasoning

Make a generalization about the sum of the numbers in each row of Pascal's triangle.

Solution The following diagram shows the sum of the numbers in each row of Pascal's triangle. From the sums, we get the sequence 1, 2, 4, 8, . . . , which are exactly the powers of 2. Consequently, we surmise that the sum of the numbers in each row is equal to 2^{n-1}, where n is the number of the row.

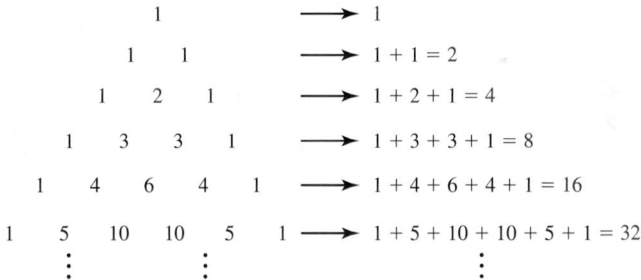

Other patterns can be seen by considering a slightly different version of Pascal's triangle. If we place every number in a hexagon and then shade different numbers, we get some interesting results. For instance, Figure 1.8 shows the geometric pattern when the multiples of 3 are shaded.

Multiples of 3

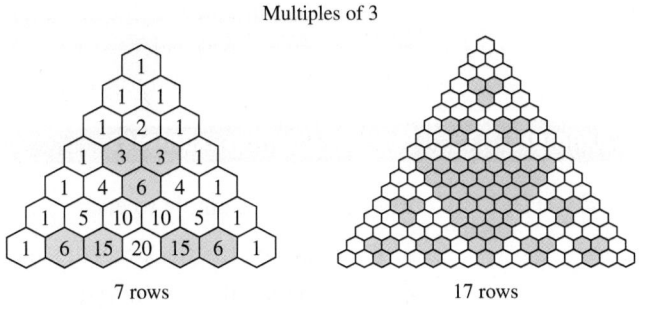

7 rows   17 rows

Figure 1.8 The multiples of 3 in Pascal's triangle

Counterexamples

We have used inductive reasoning to analyze patterns and to make generalizations in a variety of mathematical situations. Although inductive reasoning can lead to new discoveries, two problems can occur when we base our generalizations on a limited number of examples. First, we may be able to make several different generalizations from the same set of examples. For instance, consider the sequence 31, 28, 31, 30, ... In this pattern, every other number is 31, and the other numbers seem to increase by two. If we continue this pattern, the next four numbers would be 31, 32, 31, 34. However, if we think outside the box, the sequence could also represent the number of days in January, February, March, and April. In this case, the next four numbers would represent the number of days in May, June, July, and August, or 31, 30, 31, 31. In such situations, we can always look at more examples to help us better identify the pattern in question.

Second, looking at a limited number of examples may lead to incorrect generalizations. Consider the next example.

Example 1.14   Making an Incorrect Generalization

Communication

Melinda is a fourth-grade student who is learning about divisibility for the first time. She examines several odd numbers and notices the following facts:

3 is divisible by only two numbers: 1 and 3.

5 is divisible by only two numbers: 1 and 5.

7 is divisible by only two numbers: 1 and 7.

11 is divisible by only two numbers: 1 and 11.

From these statements, she concludes that every odd number is divisible by only two numbers, one and the number itself. Is she correct?

Solution Melinda has drawn a reasonable conclusion from the four examples she has observed. Unfortunately, her generalization is incorrect because many odd numbers are divisible by more than two numbers. For instance, 9 is odd and is divisible by three numbers: 1, 3, and 9. Likewise, 15 is odd and is divisible by four numbers: 1, 3, 5, and 15.

As the last example shows, we should always test our generalizations with other examples. If just one example does not work, then the generalization must be false. Any such example is called a **counterexample.**

Definition of Counterexample

A **counterexample** is a specific example that shows a generalization is false.

Example 1.15   Finding Counterexamples

Reasoning

Find a counterexample to each statement.

a. Every day of the week has the letter "s" in its spelling.

b. The product of an even and an odd number is odd.

c. If a number is divisible by 3, then the number is odd.

Solution

a. Friday is a day of the week that does not have an "s" in its spelling.

b. There are many counterexamples. For instance, if we multiply 2, an even number, by 3, an odd number, we get $2 \cdot 3 = 6$, an even number.

c. Twelve is an even number that is divisible by 3.

Example 1.16	Extending a Pattern

Reasoning

Consider the following list of equations. Use inductive reasoning to predict the next three equations. Does the pattern always hold?

$$1^2 = 1$$
$$11^2 = 121$$
$$111^2 = 12{,}321$$
$$1{,}111^2 = 1{,}234{,}321$$

Solution Based on the three examples, the next three equations will be:

$$11{,}111^2 = 123{,}454{,}321$$
$$111{,}111^2 = 12{,}345{,}654{,}321$$
$$1{,}111{,}111^2 = 1{,}234{,}567{,}654{,}321$$

This fact is easily verified using a calculator. If we try more products, we find that the tenth product does not follow the pattern because $1{,}111{,}111{,}111^2 = 1{,}234{,}567{,}900{,}987{,}654{,}321$. This means the pattern holds only for the first nine equations.

Finding counterexamples is important in mathematical reasoning because they prove that a generalization is false. If a counterexample is hard to find, we may be more confident in our generalization and even consider it to be **inductively strong.** For example, suppose Geoffrey has gone golfing every day for 30 years and has never hit a hole in one. Based on his past record, we are confident that he will not hit a hole in one the next time he goes golfing. Even though our conclusion is inductively strong because it is based on many examples, it may still be incorrect if he hits a hole in one on his next outing. Consequently, no matter how much inductive evidence we have for our generalization, we must remember that *inductive reasoning never guarantees that a generalization is true in every instance.* To make such a guarantee, we need a different kind of reasoning called deductive reasoning.

Check Your Understanding 1.2-B

1. Consider the following arrangements of dots. How many dots are needed to make the seventh arrangement in the sequence?

2. What is the fourth number in the 12th row of Pascal's triangle?

3. Find a counterexample to each statement.

  a. All boys' names begin with the letter B.

  b. All geometric shapes have three sides.

  c. The product of two numbers is always greater than the two numbers.

Talk About It If a generalization is considered inductively strong, is the generalization always true? Explain your answer.

Reasoning

Activity 1.4
Each column in the following table follows the same specific pattern. Find the pattern, and use it to fill in the missing numbers. How does inductive reasoning play a role in completing this task?

1	3	6		2		2
2	2				7	
3	5		7	7	8	
6	10	40	28			63

SECTION 1.2 Learning Assessment

■ Understanding the Concepts

1. What is mathematical reasoning, and why is it important to learning mathematics?

2. What two steps are involved in inductive reasoning?

3. Describe the characteristics of each type of pattern.

   a. Repetitive   **b.** Common traits   **c.** Growing

4. Why is an algebraic expression a powerful way to represent a generalization?

5. Describe how to generate Pascal's triangle.

6. What two problems can arise from inductive reasoning?

7. What are counterexamples, and why are they important in mathematical reasoning?

8. What does it mean for a conjecture to be inductively strong? Does it mean that the conjecture is always true?

■ Representing the Mathematics

9. Use the letters *A*, *B*, and *C* to name each repetitive pattern.

   a.

   b.

10. Use the letters *A*, *B*, and *C* to name each repetitive pattern.

    a. 3, 1, 3, 1, 3, 1, . . .

    b. 3, 1, 2, 3, 1, 2, . . .

    c. 1, 2, 2, 1, 1, 2, 2, 1, . . .

    d. 1, 2, 3, 2, 1, 2, 3, 2, . . .

c.

d.

11. What do these three patterns have in common?

    1, 2, 1, 2, 2, 1, 2, 2, 2, 1, 2, 2, 2, 2, 1, . . .

12. Use the shapes ♣, ♦, ♥, ♠ to make two different:

    a. *AB* patterns.

    b. *AAB* patterns.

    c. *ABC* patterns.

13. Use the shapes ● and ■ to make a growing pattern.

14. Consider the following sequences. Draw a sequence of dots to represent each list so that each arrangement in the sequence forms the same basic geometrical shape.

    a. 1, 4, 9, 16, . . .

    b. 1, 3, 6, 10, . . .

15. Draw the figure that would come next in this sequence:

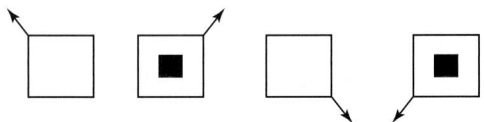

16. Draw the figure that would come next in this sequence:

■ Exercises

17. Every time I vacuum, I get warm. What can I conclude about the next time I vacuum?

18. If it has been sunny the last three days, what might I conclude about tomorrow?

19. Steve notices that his father gets home late from work every Monday night. What can he conclude about next Monday night?

20. Use the following information to determine what Tom might eat on Thursday night.

    On Monday, Tom ate steak, green beans, and mashed potatoes.

    On Tuesday, Tom ate chicken, carrots, and French fries.

    On Wednesday, Tom ate ham, peas, and a baked potato.

21. Give the next three numbers in each sequence.

    a. 3, 6, 9, 12, . . .

    b. 1, 10, 100, 1,000, . . .

    c. $\dfrac{1}{2}, \dfrac{2}{3}, \dfrac{3}{4}, \dfrac{4}{5}, \ldots$

22. Give the next three letters in each sequence.

    a. B, C, D, F, G, . . .

    b. J, F, M, A, . . .

    c. E, L, E, M, . . .

23. Give the first four numbers of a sequence that is growing larger by a constant value of 3.

24. Give the first four numbers of a sequence that is growing smaller by a constant factor of $\dfrac{1}{2}$.

25. What two equations come next in each list?

    a.
    $$4^2 = 16$$
    $$34^2 = 1,156$$
    $$334^2 = 1,11,556$$
    $$3,334^2 = 11,115,556$$

    b.
    $$7 \times 9 = 63$$
    $$77 \times 99 = 7,623$$
    $$777 \times 999 = 776,223$$
    $$7,777 \times 9,999 = 77,762,223$$

    c.
    $$2,222 \times 9 = 19,998$$
    $$3,333 \times 9 = 29,997$$
    $$4,444 \times 9 = 39,996$$
    $$5,555 \times 9 = 49,995$$

26. Generate the eighth, ninth, and tenth rows of Pascal's triangle.

27. In what row of Pascal's triangle is:

    a. 15 the third number?

    b. 21 the sixth number?

    c. 16 the second number?

    d. 84 the seventh number?

28. Consider Pascal's triangle.

    a. What is the 5th number in the 6th row?

    b. What is the 2nd number in the 10th row?

    c. What is the 3rd number in the 11th row?

29. What is the largest number in each of the following rows of Pascal's triangle?

    a. 5th   b. 8th   c. 10th

30. Another pattern in Pascal's triangle is often referred to as the "hockey stick" pattern. Start at any 1 and proceed down the diagonal ending at any number. The sum of the numbers inside the stick will equal the number that is below the last number and not in the same diagonal. Three sticks are shown here. Find a hockey stick that has a sum of:

    a. 55.   b. 126.   c. 4.   d. 220.

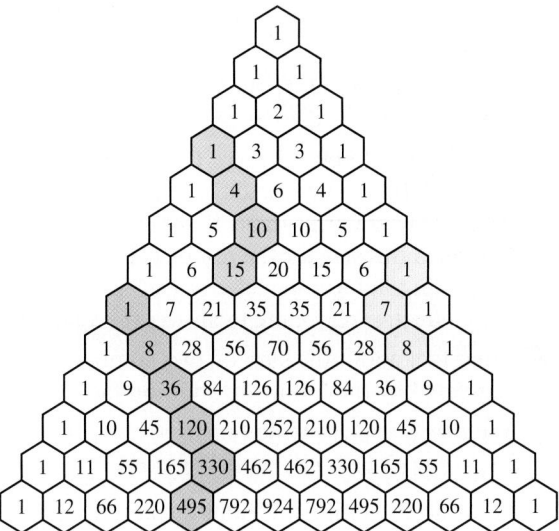

31. Find a counterexample to each generalization.

    a. All books are fiction.

    b. Every month of the year has 31 days in it.

c. Every English word contains an a, e, i, o, or u.

d. All U.S. states that begin with C border an ocean.

32. Find a counterexample to each generalization.

a. All numbers are less than 10.

b. Every number divisible by 2 is also divisible by 4.

c. The square of every number is not equal to the number itself.

d. The sum of any two numbers is odd.

33. Consider the following list of equations. Use inductive reasoning to predict the next three equations. Will the pattern always hold? If not, give the first equation for which the pattern changes.

$$3{,}367 \times 3 = 10{,}101$$
$$3{,}367 \times 6 = 20{,}202$$
$$3{,}367 \times 9 = 30{,}303$$
$$3{,}367 \times 12 = 40{,}404$$

34. Consider the following list of equations. Use inductive reasoning to predict the next three equations. Will the pattern always hold? If not, give the first equation for which the pattern changes.

$$(1 \times 9) + 2 = 11$$
$$(12 \times 9) + 3 = 111$$
$$(123 \times 9) + 4 = 1{,}111$$
$$(1{,}234 \times 9) + 5 = 11{,}111$$

■ Mathematical Reasoning and Proof

35. Amy has three children who are girls. When she has her next baby, is it guaranteed to be a girl? Explain.

36. Use A, B, and C to describe all possible repeating patterns with no more than three objects in the repeating part of the pattern.

37. Maggie loves to pick flowers. On Monday she picks one flower, on Tuesday she picks two, and on Wednesday she picks three. If she continues to pick flowers in this way for the rest of the week, how many flowers in total will she have picked? Explain how you used inductive reasoning to answer this question.

38. What fact about the sum of two numbers is suggested by the following addition facts?

$$3 + 5 = 8 \qquad 7 + 9 = 16 \qquad 1 + 5 = 6$$
$$11 + 3 = 14 \qquad 7 + 7 = 14$$

39. What do the following sums suggest about addition? Use an algebraic expression to write your generalization.

$$3 + 0 = 3 \qquad 0 + 4 = 4 \qquad 11 + 0 = 11$$
$$0 + 5 = 5 \qquad 0 + 0 = 0$$

40. What do the following products suggest about multiplication? Use an algebraic expression to write your generalization.

$$3 \cdot 1 = 3 \qquad 1 \cdot 6 = 6 \qquad 14 \cdot 1 = 14$$
$$1 \cdot 8 = 8 \qquad 1 \cdot 1 = 1$$

41. What do the following sums suggest about addition? Use an algebraic expression to write your generalization.

$$2 + (3 + 5) = 10 = (2 + 3) + 5$$
$$4 + (7 + 1) = 12 = (4 + 7) + 1$$
$$3 + (5 + 9) = 17 = (3 + 5) + 9$$
$$2 + (2 + 5) = 9 = (2 + 2) + 5$$

42. a. Use the following examples to make a conjecture about the product of two odd numbers.

$$1 \cdot 3 = 3 \quad 5 \cdot 7 = 35 \quad 3 \cdot 7 = 21 \quad 7 \cdot 7 = 49$$
$$3 \cdot 3 = 9 \quad 3 \cdot 11 = 33 \quad 9 \cdot 5 = 45 \quad 11 \cdot 11 = 121$$

b. Test your conjecture to see whether it holds true with other products of odd numbers.

c. Do you think your conjecture will always hold true? Why?

43. Consider the following quotients.

$$4 \div 2 = 2 \qquad 8 \div 2 = 4 \qquad 20 \div 2 = 10$$
$$28 \div 2 = 14 \qquad 36 \div 2 = 18$$

a. Based on what you see, write a conjecture about dividing a natural number by 2.

b. Is your conjecture true for all natural numbers? If not, provide a counterexample.

44. Consider the conjecture, "The sum of any 3 consecutive natural numbers can be divided by 3."

a. Create several examples that test the truthfulness of the conjecture.

b. Based on your examples, do you think the conjecture is true or false? Explain.

45. a. How many blocks are needed to build the following pyramid?

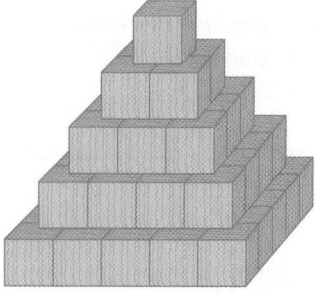

b. Explain how you used inductive reasoning to obtain your answer.

46. Consider the following successive arrangements of toothpicks.

    a. How many toothpicks are needed to make the 10th arrangement?

    b. Explain how you used inductive reasoning to obtain your answer.

47. A rubber ball is dropped from a building, and the heights of its first three bounces are shown in the figure.

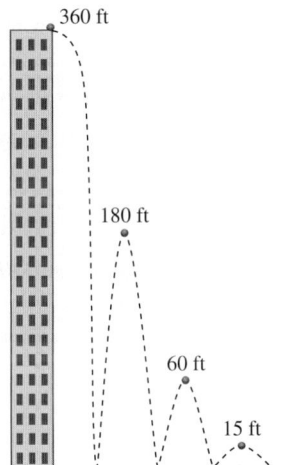

How high will the ball go after the 4th and 5th bounces?

48. **a.** How many numbers are in each of the following rows of Pascal's triangle?

    i. third   **ii.** fourth   **iii.** fifth   **iv.** sixth

    b. Make a generalization about the numbers in each row of Pascal's triangle based on the pattern you see in part (a). Use it to predict how many numbers will be in each of the following rows of Pascal's triangle.

    i. 35th   **ii.** 75th   **iii.** 153rd   **iv.** 1,000th

49. Find the rows in Pascal's triangle in which the second numbers are 3, 5, and 7, respectively. What do you notice about the other numbers in these rows? Does this hold true for the rows in which the second number is 4, 6, or 9? What pattern do you think this might establish in Pascal's triangle?

■ Mathematical Communication

50. Use what you have learned to write a set of directions that instructs someone to reason inductively. Compare your steps with one of your

peer's. Discuss how they are similar or different. Finally, use your discussion to formulate a common procedure.

51. Write a story problem that requires a student to use a growing number pattern to make a count.

52. Summer is given the following two sequences. She classifies one as an *AB* pattern and the other as an *AC* pattern. Is she correct? Why or why not?

Pattern 1 ...

Pattern 2  ...

53. Mario has computed the following problems and noticed that all the answers are odd. He concludes that dividing by 3 must always lead to an odd answer. Is his conclusion correct? If not, how might you correct his reasoning?

$$3 \div 3 = 1 \qquad 9 \div 3 = 3 \qquad 15 \div 3 = 5$$

54. Amelia is asked to make a growing pattern that grows smaller. She responds with the following drawing. Did she correctly answer the question? Why or why not?

■ Building Connections

55. Read the NCTM Reasoning and Proof Standards for grades K–2 and 3–5. Based on what you read, how might inductive reasoning be used in the elementary grades?

56. Inductive reasoning and pattern recognition play an important role in the cognitive development of children. Research the topic, and write a short report on your findings.

57. Intelligence and aptitude tests often include a number of inductive reasoning questions. Why do you think this is an important part of assessing someone's intelligence or ability to learn?

58. In our everyday lives, we settle for generalizations that are true most of the time. For instance, if the weather is cold, we expect the heat in our homes to come on automatically. Make a list of five other such generalizations or common expectations. Explain how we use inductive reasoning to make them.

SECTION 1.3   Deductive Reasoning and Proof

In Section 1.2, we used inductive reasoning to make generalizations based on patterns that occurred in sets of specific examples. Given the generalization, the next step is to determine whether it is true for all cases. A counterexample renders the generalization false. However, if a counterexample is hard to find, then we may be able to make an argument that the generalization is true.

In mathematics, an **argument** is a sequence of statements that explains how to arrive at a **conclusion** from a set of facts called the **premises** or **hypotheses.** A good argument depends on two things. First, the premises, which are the foundation of the argument, must be true in and of themselves. If they are not, then we cannot guarantee the truthfulness of the conclusion. Second, the premises must be arranged in such a way that they lead logically to the conclusion. The process by which we do this is called **deductive reasoning.**

Definition of Deductive Reasoning

Deductive reasoning is the process of reasoning from a set of known facts or general principles to a specific conclusion.

Deductive arguments are particularly powerful because if we use true statements to create a logically correct argument, then the argument *guarantees* that the conclusion is true. However, arguments must be done with care and precision. Consequently, to make correct deductive arguments, we need to understand the different kinds of statements used in arguments and how to put them together so that they logically make sense.

Statements and Negations

The foundation of any deductive argument is made with **statements,** which are sentences that are either true or false. Not every sentence is a statement. For instance, questions, exclamations, and sentences with relative terms are not statements because we cannot determine whether they are true or false.

Example 1.17   Identifying Statements

Which of the following sentences are statements?

a. You look great!   **b.** Tasha is 5 ft 6 in.   **c.** $x - y = 2.$

Solution

a. The sentence contains the relative term "great." Because your looks are relative to whoever is looking at you, we cannot know whether this statement is true or false. It is not a statement.

b. This sentence is a statement because we can determine whether it is true or false by measuring Tasha.

c. Because we do not know the exact values of x and y, we cannot determine whether the sentence is true or false. It is not a statement.

Because a statement is either true or false, we can write another statement, called the **negation,** which takes on the opposite **truth value.** If a statement is true, its negation is false. If a statement is false, its negation is true. In most cases, we can write a negation by adding or removing the word "not" to the original statement.

Example 1.18	**Writing Negations**

Communication

Write the negation of each statement.

a. The tree is over 30 ft tall. **b.** Tim does not swim. **c.** $2 + 3 \neq 5$.

Solution

a. The negation is, "The tree is not over 30 ft tall."

b. Negate the statement by removing the word "not" or by saying "Tim swims."

c. Negate the statement by changing the "$\neq$" to "$=$": $2 + 3 = 5$.

Table 1.4 Truth values for the negation of *p*

p	$\sim p$
T	F
F	T

The truth values of a statement and its negation can be summarized in a **truth table** like Table 1.4. Here, p represents a simple statement, and the symbol "$\sim$" represents the word "not." Hence, the notation $\sim p$ is read "not p."

Compound Statements

A **compound statement** combines two simple statements by using a connecting word like "and" or "or." Examples are:

Jose got an A in math *and* a B in English.

Kiesha plays field hockey, *or* she swims.

Table 1.5 Truth values for the conjunction

p	q	$p \wedge q$
T	T	T
T	F	F
F	T	F
F	F	F

Many students have trouble using compound statements correctly because they do not understand the subtle difference between "and" and "or." In a **conjunction,** which is a compound statement using "and," both simple statements must be true for the entire statement to be true. Otherwise, it is false. To understand why, consider the conjunction, "Jose got an A in math and a B in English." The statement is true only if Jose gets an A in math *and* he gets a B in English. If he does not get either grade, then the conjunction must be false. Table 1.5 contains the truth values of the conjunction. Because each simple statement, p and q, can take on one of two truth values, we must list four different combinations of truth values. The symbol "$\wedge$" represents the word "and."

It can be useful to write compound statements in symbols to determine their truth values. To do so, we can use any lowercase letter that best represents each simple statement.

Example 1.19	**Writing Conjunctions Symbolically**

Representation

Let e represent "2 is even" and o represent "3 is odd." Rewrite each statement symbolically, and then find its truth value.

a. 2 is even, and 3 is odd.

b. 2 is not even, and 3 is odd.

c. It is not the case that 2 is even and 3 is odd.

Solution

a. $e \wedge o$. Because e and o are true, $e \wedge o$ is true.

b. $\sim e \wedge o$. Because $\sim e$ is false and o is true, $\sim e \wedge o$ is false.

c. The phrase "It is not the case" indicates that the entire statement is to be negated, so we represent it as $\sim(e \wedge o)$. Because $e \wedge o$ is true, $\sim(e \wedge o)$ is false.

Students can also have trouble with compound statements that use "or," perhaps because we use "or" in two different ways in our everyday language. In some cases, we use "or" in an exclusive sense. For instance, we may ask a child whether he wants a peanut butter and jelly or a turkey sandwich for lunch. Here, the "or" indicates a choice between one or the other, but not both. We can also use "or" in an inclusive sense. For instance, if the child chooses turkey, then we may ask whether he wants lettuce or tomato. In this case, the choice is between one or the other or possibly both. In mathematics, "or" is always inclusive, and we call such statements **disjunctions.**

Logically, it is not necessary for both simple statements to be true for the disjunction to be true. To understand why, consider the statement, "I have lettuce or I have tomato on my sandwich." The disjunction is true if both simple statements are true: I have both lettuce and tomato on my sandwich. However, the statement is also true if I have lettuce and no tomato or I have tomato and no lettuce. Consequently, a disjunction is true if either one or both of the simple statements is true. It is false only if both simple statements are false. Table 1.6 contains the truth values of a disjunction. We use the symbol "$\lor$" to represent the word "or."

■ **Table 1.6 Truth values for the disjunction**

p	q	$p \lor q$
T	T	T
T	F	T
F	T	T
F	F	F

Example 1.20 Writing Disjunctions Symbolically

Representation

Let g represent "Georgia borders Florida" and s represent "Florida is a southern state." Rewrite each statement symbolically, and then find its truth value.

a. Georgia borders Florida, or Florida is a southern state.

b. Georgia does not border Florida, or Florida is not a southern state.

c. Georgia does not border Florida, or Florida is a southern state.

Solution

a. $g \lor s$. Because g and s are true, $g \lor s$ is true.

b. $\sim g \lor \sim s$. Because $\sim g$ and $\sim s$ are both false, $\sim g \lor \sim s$ is false.

c. $\sim g \lor s$. Because $\sim g$ is false, but s is true, $\sim g \lor s$ is true.

Conditional Statements

Conditional statements, or **implications,** generally take the form, "If p, then q." The "if" part of the statement is called the hypothesis, and the "then" part is called the conclusion. Other ways of stating conditional statements are given in Table 1.7.

■ **Table 1.7 Different ways to state a conditional**

Equivalent Conditional Statements	Statement Form
If I do the dishes, then my wife is happy.	If p, then q.
If I do the dishes, my wife is happy.	If p, q.
My wife is happy if I do the dishes.	q, if p.
My doing the dishes implies that my wife is happy.	p implies q.
When I do the dishes, my wife is happy.	When p, q.

Conditional statements earn their name because the hypothesis gives a condition that is necessary for the conclusion to occur. In many senses, a conditional statement entails a cause-and-effect relationship, where p is the cause of the effect q. We can use this fact to determine the truth values of conditional statements.

Example 1.21	Finding the Truth Values of a Conditional Statement

Reasoning

Consider the statement, "If you work hard, then I will give you a raise." Under what conditions is the statement true?

Solution The statement makes a promise that if you work hard, then I will give you a raise. If you work hard and I do give you a raise, then the promise has been kept and the conditional statement is true. However, if you work hard and I do not give you a raise, the promise has been broken and the conditional statement is false. The promise can be upheld only if you work hard. If not, then the promise goes unbroken, whether or not you are given a raise. So, if the hypothesis is false, then the conditional is true regardless of whether the conclusion is true or false.

Table 1.8 Truth values for the conditional

p	q	$p \rightarrow q$
T	T	T
T	F	F
F	T	T
F	F	T

Example 1.21 shows that the conditional is false only when the hypothesis is true and the conclusion is false. Otherwise, it is true. The truth values are summarized in Table 1.8. We use the symbol "$\rightarrow$" to represent the conditional.

Three important statements are associated with the conditional, which we form by switching or negating the hypothesis and conclusion. For instance, if the conditional is "If p, then q," we form the **converse** by switching the hypothesis and conclusion: "If q, then p." For the **inverse,** the hypothesis and conclusion remain in their original position, but we negate them: "If not p, then not q." In the **contrapositive,** we do both. We switch the hypothesis and conclusion, and negate them: "If not q, then not p." These four forms of the conditional are summarized in Table 1.9.

Table 1.9 Four forms of the conditional

	Example	Statement Form	Symbolic Form
Conditional	If it is raining, then it is cloudy.	If p, then q.	$p \rightarrow q$
Converse	If it is cloudy, then it is raining.	If q, then p.	$q \rightarrow p$
Inverse	If it is not raining, then it is not cloudy.	If not p, then not q.	$\sim p \rightarrow \sim q$
Contrapositive	If it is not cloudy, then it is not raining.	If not q, then not p.	$\sim q \rightarrow \sim p$

Example 1.22	Writing Different Forms of a Conditional

Communication

Consider the conditional statement, "If a shape is a rectangle, then it has four sides." Write the converse, inverse, and contrapositive of the statement both symbolically and verbally.

Solution Let r represent "The shape is a rectangle" and f represent "The shape has four sides." The three statements are:

Converse: $\quad f \rightarrow r \quad$ If a shape has four sides, then it is a rectangle.

Inverse: $\quad \sim r \rightarrow \sim f \quad$ If a shape is not a rectangle, then it does not have four sides.

Contrapositive: $\quad \sim f \rightarrow \sim r \quad$ If a shape does not have four sides, then it is not a rectangle.

Note in Example 1.22 that the four forms of the conditional do not take on the same truth values. For instance, the converse of the statement is false because a rhombus is a figure with four sides but is not a rectangle. The inverse is also false because a rhombus is not a rectangle, but it does have four sides. The contrapositive, however, happens to be true. Specifically, if a shape does not have four sides, then it cannot be a rectangle. In the next example, we summarize the truth values of each form of the conditional.

Example 1.23   **Finding the Truth Values of Different Forms of the Conditional**

Find the truth values for the converse, inverse, and contrapositive of the conditional $p \rightarrow q$.

Solution Refer to Table 1.8. For the converse, we use the truth values for the general conditional, but look from q to p. For the inverse, we first negate the truth values of p and q, and then use the general conditional. For the contrapositive, we negate the truth values of p and q, and then use the general conditional in reverse. The following table summarizes the results:

p	q	Conditional $(p \rightarrow q)$	Converse $(q \rightarrow p)$	Inverse $(\sim p \rightarrow \sim q)$	Contrapositive $(\sim q \rightarrow \sim p)$
T	T	T	T	T	T
T	F	F	T	T	F
F	T	T	F	F	T
F	F	T	T	T	T

The table in the last example shows that the conditional and the contrapositive have exactly the same truth values. Likewise, the converse and inverse have exactly the same truth values. In general, when two statements, p and q, have the same truth values, we say that they are **logically equivalent** and write $p \equiv q$. If two statements are logically equivalent, then one statement may be exchanged for the other. This fact is useful when making deductive arguments because arguing from the contrapositive is sometimes easier than arguing from the conditional.

Even though the conditional and the converse do not have the same truth values, both can be true in some situations. For instance, consider the following situation in which the conditional and its converse are true:

Conditional: If it is my birthday, then it is June 10.

Converse: If it is June 10, then it is my birthday.

In situations like this, we can combine the statements by writing them as a **biconditional.** To do so, we connect the simple statements in the conditional using the phrase "if and only if":

Biconditional: It is my birthday if and only if it is June 10.

Likewise, a biconditional statement can be written as the conjunction of two conditional statements. For instance, we can rewrite the statement, "A number is even if and only if it is divisible by 2," as "If a number is even, then it is divisible by 2, and if a number is divisible by 2, then it is even."

Symbolically, we use the symbol "↔" to represent a biconditional. Because it is logically equivalent to the conjunction of a conditional and its converse, the biconditional can also be written as $p \leftrightarrow q \equiv (p \rightarrow q) \wedge (q \rightarrow p)$.

Example 1.24   Finding the Truth Values of a Biconditional

Find the truth values of the biconditional $p \leftrightarrow q$.

Solution Proceed by using the statement $(p \to q) \wedge (q \to p)$ and truth tables. First, use the general conditional and Table 1.8 to find the truth values of $p \to q$ and $q \to p$. Next, combine these truth values using the truth values for the conjunction. The following table summarizes the results.

	q	$p \to q$	$q \to p$	$(p \to q) \wedge (q \to p)$
T	T	T	T	T
T	F	F	T	F
F	T	T	F	F
F	F	T	T	T

The last example shows that a biconditional statement is true when both simple statements have the same truth values. Otherwise, it is false. Mathematically, biconditionals are important because many definitions and theorems are stated this way.

Check Your Understanding 1.3-A

1. Let s represent "You like soda" and c represent "You like chips." Represent each statement symbolically, and then determine its truth value if both s and c are true.
   a. You like soda, and you do not like chips.
   b. You do not like soda, or you do not like chips.
   c. It is not the case that you like soda or you like chips.
   d. If you do not like chips, then you like soda.
2. Write the converse, inverse, and contrapositive of each conditional statement.
   a. If you read your e-mail, then you got my message.
   b. If $x = 5$, then $2x = 10$.
3. Write the statement "You are a dog if and only if you have fur" as the conjunction of two conditionals.

Talk About It As a teacher, how would you explain the difference between the word "and" and the word "or"?

Communication

Activity 1.5
Conditional statements can be rewritten using the word "all" or "every." For instance, the statement, "If the animal is a gorilla, then it is hairy," can be rewritten, as "All gorillas are hairy."
a. Rewrite the following statements using "all."
   i. If you are a human, then you are a mammal.
   ii. If you teach in elementary school, then you teach mathematics.
   iii. If a geometric figure is a triangle, then it has three sides.
b. Rewrite the following statements using "if, then."
   i. All dogs have fur.   ii. Every square is a rectangle.
   iii. All numbers divisible by 2 are even.
c. Is there an advantage to writing the statement one way or the other?

We can now use the different kinds of statements to form deductive arguments. Although elementary students do not learn formal reasoning methods, they can and should be encouraged to make simple deductive arguments. For example, we might ask a first-grade student how she knows that 3 + 4 = 7. She might explain how she counted out two groups, one with three counters and another with four, combined them, and then counted a total of seven objects (Figure 1.9). In giving this explanation, she has made a grade-appropriate deductive argument. In general, children can make simple deductive arguments using both direct and indirect reasoning.

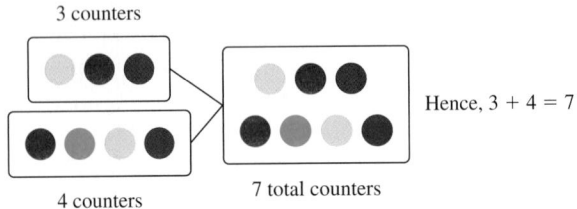

Figure 1.9 A grade appropriate argument for 3 + 4 = 7

Direct Reasoning

Explorations
Manual
1.5

For the most part, we reason directly by using the cause-and-effect relationships expressed in conditional statements. In the simplest case, we argue directly from a single conditional. To do so, we assume that the cause-and-effect relationship of the conditional is true; that is, the cause always leads to the effect. Next, we assume that the cause happens, so it logically follows that the effect must also happen. In other words, if we assume that the conditional "If p, then q" is true and that the hypothesis p happens, then the conclusion q must also happen. Formally, we call this kind of argument **affirming the hypothesis,** or **modus ponens.** Table 1.10 gives an example both verbally and symbolically. Here, we use the symbol "$\therefore$" to represent the word "therefore."

Table 1.10 Affirming the hypothesis		
Verbal Example	**Argument Form**	**Symbolic Representation**
If Camille is 18, then she is old enough to vote.	If p, then q.	$p \rightarrow q$
Camille is 18.	p.	p
Therefore, she is old enough to vote.	Therefore, q.	$\therefore q$

Example 1.25   Completing Direct Arguments

Reasoning

Fill in the missing statement in each argument.

a. If you watch 4 hours of television a day, then you are lazy.

   You watch 4 hours of television a day.

   Therefore, _____.

b. Living in a college dorm implies you are a freshman.

   _____.

   Therefore, you are a freshman.

Solution

a. The missing statement is the conclusion: "you are lazy."

b. The missing statement is the hypothesis of the conditional: "You are living in a college dorm."

When reasoning directly, we can make a more complex argument by stringing several conditionals together. When doing so, the conclusion of one conditional must be the hypothesis of the next. Then, if the conditional statements are true and the first hypothesis happens, we get a chain reaction through which we can conclude that the last consequence also happens. This kind of argument is much like a series of dominoes. If the dominoes are set up correctly, then once the first domino falls, all the dominoes must fall—including the last, as shown in Figure 1.10.

Explorations Manual 1.6

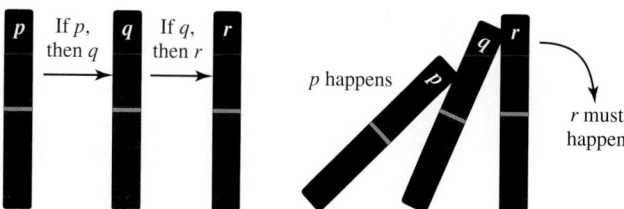

Figure 1.10 Arguing with several conditionals

We call this kind of argument a **hypothetical syllogism.** Table 1.11 gives an example, both verbally and symbolically.

Table 1.11 Hypothetical syllogism		
Verbal Example	**Argument Form**	**Symbolic Representation**
If you turn on the light, you will see the bed.	If p, then q.	$p \rightarrow q$
If you see the bed, you will not stub your toe.	If q, then r.	$q \rightarrow r$
Therefore, if you turn on the light, you will not stub your toe.	Therefore, if p, then r.	$\therefore p \rightarrow r$

Explorations Manual 1.7

The idea of a hypothetical syllogism extends to any number of conditional statements. When using this type of argument, we must be sure that the conditional statements are arranged in the appropriate order so that the chain reaction can occur.

Example 1.26   Forming a Direct Argument

Reasoning

Rearrange the following statements to form a direct argument.

If Stephanie reads well, she will pass her reading test.

If she has extra reading practice, she will learn to read well.

Therefore, if Stephanie gets to school early, she will pass her reading test.

If Stephanie gets to school early, she will get extra reading practice.

Solution We first identify the conclusion of the argument, which gives us the first hypothesis and the last consequence. We then match the consequence of one statement to the hypothesis of the next to get the following argument.

If Stephanie gets to school early, she will get extra reading practice.

If she has extra reading practice, she will learn to read well.

If Stephanie reads well, she will pass her reading test.

Therefore, if Stephanie gets to school early, she will pass her reading test.

| Example 1.27 | **Using Direct Reasoning to Find a Number** |

Reasoning

What is the value of x, if $u = 4$, $v = 2u$, $w = 3v$, and $x = 4w$?

Solution From the given information, we can set up a string of conditional statements from which we can deduce the value of x. Specifically, we have

$$\text{If } u = 4, \text{ then } v = 2u = 2(4) = 8.$$
$$\text{If } v = 8, \text{ then } w = 3v = 3(8) = 24.$$
$$\text{If } w = 24, \text{ then } x = 4w = 4(24) = 96.$$

Explorations
Manual
1.8

When arguing from a conditional, we must be careful not to make a **converse error**. In this logical mistake, we assume that the conditional "If p, then q" is true. However, instead of reasoning from p to q, we incorrectly claim that if q happens, then p must follow. For example, the logic of the argument in Table 1.12 does not seem quite right. For instance, you may be an excellent softball player, which makes you an athlete. However, playing softball does not guarantee that you can swim.

| Table 1.12   Converse error |

Verbal Example	Error Form	Symbolic Representation
If you swim well, then you are an athlete.	If p, then q.	$p \rightarrow q$
You are an athlete.	q.	q
Therefore, you swim well.	Therefore, p.	$\therefore p$

Indirect Reasoning

We can also reason indirectly, which means we reach the conclusion in a roundabout way. One way to do so is to again use the cause-and-effect relationship of the conditional. However, rather than arguing directly from the conditional, we use the contrapositive. If we assume that the effect did not happen, then it logically follows that the cause did not happen either. In other words, if we assume that the conditional "If p, then q" is true and that conclusion q did not happen, then the hypothesis p did not happen either. This kind of argument is called **denying the conclusion,** or **modus tollens**. Table 1.13 gives an example both verbally and symbolically.

| Table 1.13   Denying the conclusion |

Verbal Example	Argument Form	Symbolic Representation
If you ate an apple, then you ate a serving of fruit.	If p, then q.	$p \rightarrow q$
You did not eat a serving of fruit.	Not q.	$\sim q$
Therefore, you did not eat an apple.	Therefore, not p.	$\therefore \sim p$

Example 1.28	Completing Indirect Arguments

Reasoning   Fill in the missing statement in each argument.

a. If you are young, then you drive fast.
You do not drive fast.
Therefore, _____.

b. _____.
You do not like animals.
Therefore, you do not own a dog.

Solution

a. The missing statement is the conclusion: "You are not young."

b. The second statement is the negation of the original consequence, and the conclusion is the negation of the original hypothesis. Consequently, the missing statement must be "If you own a dog, then you like animals."

When arguing from a contrapositive, we must be careful not to make an **inverse error.** In this logical mistake, we assume that the conditional "If p, then q" is true. However, instead of reasoning from $\sim q$ to $\sim p$, we incorrectly claim that if $\sim p$ happens, then $\sim q$ must follow. For example, the logic of the argument in Table 1.14 does not seem quite right. For instance, you may be over 40, which means you are not young. However, being over 40 does not guarantee that you drive slowly.

Table 1.14   Inverse error

Verbal Example	Error Form	Symbolic Representation
If you are young, then you drive fast.	If p, then q.	$p \rightarrow q$
You are not young.	Not p.	$\sim p$
Therefore, you do not drive fast.	Therefore, not q.	$\therefore \sim q$

Another way to reason indirectly is to use the **process of elimination.** Here, the reasoning is based on a true disjunction where one or the other condition must be true. Eliminating one of the possibilities as false means that the other must be true. Specifically, if we assume that p or q is true and p is false, then q must be true. Table 1.15 gives an example of this argument form.

Table 1.15   Process of elimination

Verbal Example	Argument Form	Symbolic Representation
You drink either tea or soda.	p or q.	$p \vee q$
You do not drink tea.	Not p.	$\sim p$
Therefore, you drink soda.	Therefore, q.	$\therefore q$

We can extend the process of elimination to any number of choices. We then eliminate choices that are impossible, illogical, or known to be incorrect until only one choice remains.

Example 1.29 **Using Indirect Reasoning to Solve a Problem**

Reasoning

Fill in the following grid so that the numbers 1 through 4 appear in each row, column, and 2 × 2 square only once.

1			4
3			2
	1	2	3
2			1

Solution Begin with the 2 × 2 square in the lower right-hand corner. Because it already has a 1, 2, and 3, a 4 must go into the open box. Next, consider the 2 × 2 square in the lower left-hand corner. The box in the upper left cannot be a 3 because there is already a 3 in that row and column. This means it must be a 4, and a 3 goes into the lower right-hand box.

1			4
3			2
4	1	2	3
2	3	4	1

Now consider the 2 × 2 square in the upper left-hand corner. The lower right box cannot be 2 because a 2 is in that row. So it must be 4, making the box in the upper right a 2. Likewise, the 2 × 2 square in the upper right-hand corner must have a 3 in the upper left box and a 1 in the lower left box.

1	2	3	4
3	4	1	2
4	1	2	3
2	3	4	1

Check Your Understanding 1.3-B

1. Fill in the missing statement in each argument.

a. If it is dark, then you turn a light on.

_____.

Therefore, it is not dark.

b. If a figure is a square, then it has 4 sides.

This figure is a square.

Therefore, _____.

c. _____.

If you teach elementary school, then you are smart.

You like children.

Therefore, you are smart.

2. What is the value of y if $u = 5$, $v = 4u$, $w = 6v$, $x = \dfrac{w}{8}$, and $y = 5x$?

3. Write each argument symbolically, and then determine whether it uses correct logic. If it is correct, state the type of argument. If not, state the type of error.

   a. If you watch 4 hours of television a day, then you are lazy.

   You are lazy.

   Therefore, you watch 4 hours of television a day.

   b. Living in the dorm implies you are a freshman.

   You live in the dorm.

   Therefore, you are a freshman.

Talk About It In mathematics, we often want to prove conjectures. What does it mean to prove something? What role does deductive reasoning play in mathematical proof?

Problem Solving

Activity 1.6
Sudoku is a popular puzzle game that uses deductive reasoning extensively. The goal is to fill in the grid so that the numbers 1 through 9 appear in each row, column, and 3 × 3 square only once. Complete the following Sudoku puzzle.

8		1	7	4	9			
3					1	7	8	
	7			6	3			
4			1			2	7	6
2								1
1	5	9		2				4
			5	9			4	
	9	2	4					5
			1	7	6	9		8

SECTION **1.3** Learning Assessment

■ Understanding the Concepts

1. What two things are necessary for an argument to guarantee its conclusion?

2. What is the difference between:

   a. A conjunction and a disjunction?

   b. An exclusive or and an inclusive or?

3. Summarize the truth values of each kind of statement.

   a. Negation   **b.** Conjunction

   c. Disjunction   **d.** Conditional

4. What is the intuitive relationship between the parts of a conditional statement?

5. If we are given a conditional, how do we form its:

   a. Converse?   **b.** Inverse?   **c.** Contrapositive?

6. a. What does it mean for two statements to be logically equivalent?

   b. What is a biconditional statement logically equivalent to?

7. What does it mean to reason directly? Indirectly?

8. Describe how to make an argument:

   a. Directly from a conditional.

   b. Indirectly from a conditional.

   c. Indirectly from a disjunction.

9. Use your own words to describe a hypothetical syllogism.

10. What two common errors can occur when arguing from a conditional?

■ Representing the Mathematics

11. Write the negation of each statement.

    a. John is a good swimmer.

    b. Susan went to work this morning.

    c. The weather is warm today.

    d. I like cookies and ice cream.

12. Let s represent "Samantha is 5 ft 6 in." and a represent "Samantha is athletic." Represent each statement symbolically.

    a. Samantha is not 5 ft 6 in.

    b. Samantha is 5 ft 6 in., and Samantha is athletic.

    c. If Samantha is athletic, then Samantha is 5 ft 6 in.

    d. Samantha is not athletic, or Samantha is not 5 ft 6 in.

13. Let l represent "I love children" and t represent "I teach elementary school." Represent each statement symbolically.

    a. I do not teach elementary school.

    b. I love children, or I teach elementary school.

    c. If you do not teach elementary school, then you love children.

    d. You teach elementary school if and only if you love children.

14. Let t represent "Thanksgiving is my favorite holiday" and f represent "I like the Fourth of July." Translate the following symbolic statements to verbal ones.

    a. $\sim t$ b. $t \wedge f$ c. $t \vee f$

    d. $\sim(t \vee f)$ e. $\sim f \to t$

15. Let e represent "Lunch is at 11:30" and d represent "I have lunch duty." Translate the following symbolic statements to verbal ones.

    a. $\sim d$ b. $e \vee d$ c. $\sim e \wedge \sim d$

    d. $\sim(e \to d)$ e. $e \leftrightarrow d$

16. Identify the hypothesis and conclusion in each conditional. Then write the statement in "if, then" form.

    a. You can sleep in a five-star hotel if you have money.

    b. The fact that you lift weights implies you are strong.

    c. When two even numbers are added, the result is even.

    d. $x + y = 8$, if $x = 6$, and $y = 2$.

17. Write the converse, inverse, and contrapositive of each statement verbally and symbolically.

    a. If it is the Fourth of July, then you go on a picnic.

    b. You like young children if you teach in elementary school.

18. Write the converse, inverse, and contrapositive of each statement verbally and symbolically.

    a. If $2 \cdot x = 6$, then $x = 3$.

    b. A shape is a pentagon if it has five sides.

19. In each case, the contrapositive of a conditional is given. Write the original conditional statement.

    a. If I do not take a nap tomorrow, then I slept last night.

    b. If two angles are not supplementary, then the sum of their angles is not 180°.

    c. If $x - 6 \neq 4$, then $x \neq 10$.

20. Write each biconditional statement as two "if, then" statements.

    a. A number is divisible by 5 if and only if it ends in a 0 or 5.

    b. A rectangle is a square if and only if the sides are all the same length.

21. Write each pair of "if, then" statements as a single biconditional statement.

    a. If a triangle has three sides the same length, then it is equilateral.

    If a triangle is equilateral, then it has three sides the same length.

    b. If $x + 4 = 9$, then $x = 5$.

    If $x = 5$, then $x + 4 = 9$.

22. Give a verbal interpretation of each symbolic statement.

    a. $p \to q \equiv \sim q \to \sim p$

    b. $q \to p \equiv \sim p \to \sim q$

    c. $p \leftrightarrow q \equiv (p \to q) \wedge (q \to p)$

■ Exercises

23. Which of the following sentences are statements? For those that are not, explain why.

    a. Susan has a dog.

    b. $3 + 4 = 8$.

    c. Tomorrow is Christmas.

    d. $x - 3 = 9$.

    e. Yeah!!

    f. Which team won?

24. Which of the following sentences are statements? For those that are not, explain why.

    a. Fourteen is divisible by 2.

    b. Who are you?

    c. It rains every Monday.

    d. I hope Sam gets here!

    e. $x + 4 < 7$.

    f. This is not a true statement.

25. If $x = 3$ and $y = 7$, determine whether each statement is true or false.

    a. $x < y$.   **b.** $x + y = 9$.

    c. $y - x = 4$.   **d.** $y - x > x$.

26. If $x = 6$ and $y = 4$, determine whether each statement is true or false.

    a. $x \cdot y$ is positive.   **b.** $x + y = y + x$.

    c. $x + 0 = y$.   **d.** $x \ne y$.

27. If p is true and q is false, find the truth value of each statement.

    a. $\sim q$ **b.** $p \wedge q$ **c.** $p \vee q$

    d. $p \to q$ **e.** $\sim(p \wedge q)$

28. If p is false and q is true, find the truth value of each statement.

    a. $\sim(\sim q)$ **b.** $\sim(p \vee \sim q)$ **c.** $q \to p$

    d. $\sim p \to \sim q$ **e.** $q \to (p \vee q)$

29. If p is true and q is false, find the truth value of each statement.

    a. $p \vee \sim q$ **b.** $\sim p \to q$

    c. $(p \to q) \vee \sim p$ **d.** $(p \wedge q) \to \sim q$

30. Make a truth table that shows all of the truth values for the given statements.

    a. $p \wedge \sim q$ **b.** $\sim p \wedge \sim q$ **c.** $p \to \sim q$

31. Make a truth table that shows all of the truth values for the given statements.

    a. $\sim(\sim p \wedge q)$ **b.** $(p \wedge q) \to p$

    c. $(p \vee q) \to (p \wedge q)$

In Exercises 32–39, write each argument symbolically, and then determine whether it uses correct logic. If it is correct, state the type of argument. If not, state the type of error.

32. If you hit your thumb with a hammer, then it will hurt.

Your thumb hurts.

Therefore, you hit your thumb with a hammer.

33. If you mop the floor, it will be wet.

If the floor is wet, then you will slip.

Therefore, if you mop the floor, you will slip.

34. If you like Mozart, then you like classical music.

You do not like classical music.

Therefore, you do not like Mozart.

35. If my sister and I share a room, then we get along.

My sister and I do not share a room.

Therefore, we do not get along.

36. Tamika will watch cartoons or sports.

She does not watch sports.

Therefore, Tamika will watch cartoons.

37. If you make your bed, then you are neat.

You are neat.

Therefore, you make your bed.

38. If a shape is a square, then it is a rectangle.

A shape is a quadrilateral if it is a rectangle.

Therefore, when a shape is a square, it is a quadrilateral.

39. If this number is divisible by 6, then it is divisible by 3.

This number is not divisible by 6.

Therefore, this number is not divisible by 3.

In Exercises 40–45, fill in the missing statement in each argument.

40. If Tom is tall, then he plays basketball.

If Tom plays basketball, then he is popular.

Tom is tall.

Therefore, _____

41. _____

This animal is not a mouse.

Therefore, it must be a hamster.

42. _____

If Juanita can get an A in this course, then she will have a 4.0 this semester.

Juanita studies hard.

Therefore, she will have a 4.0 this semester.

43. If you are an honest person, then you do not cheat on tests.

You cheat on tests.

Therefore, _____

44. If an animal is a duck, then it likes to swim on the water.

_____

This animal is a duck.

Therefore, this animal likes to each fish.

45. _____

You are a child.

Therefore, you like cookies.

■ Mathematical Reasoning and Proof

46. Two commonly used laws of logic are called DeMorgan's laws. Symbolically, they are given as $\sim(p \wedge q) \equiv \sim p \vee \sim q$ and $\sim(p \vee q) \equiv \sim p \wedge \sim q$.

    a. State in your own words the meaning of each of DeMorgan's laws.

    b. Use DeMorgan's laws to negate the following statements.

        i. The test will be Monday, or the test will be Tuesday.

    ii. It is cold, and it is wet.

    iii. I was born in a small town, and I moved to the big city.

    iv. I like corn, or I do not like peas.

47. Rearrange the statements so that they form a direct argument.

Therefore, if Kayeaka gets a job, she will have a bad spring break.

If she breaks her leg, Kayeaka will have a bad spring break.

Kayeaka will make enough money if she gets a job.

If Kayeaka makes enough money, then she is going skiing on spring break.

When she goes skiing, she will break her leg.

48. Rearrange the statements so that they form a direct argument.

When he takes a job in the city, he can live in his dream house.

Mike will graduate if he passes his finals.

Therefore, if Mike takes a nap, then he'll live in his dream house.

If Mike studies tonight, then he'll pass his finals.

If Mike takes a nap, he can study tonight.

He can take a job in the city if he graduates.

49. Sherlock Holmes is a fictional character that uses deductive reasoning to solve crimes. Suppose that Holmes is told that four crooks—Adams, Brooks, Cole, and Dent—each stole one of four jewels: a diamond, a ruby, a sapphire, and an emerald. Help Holmes determine which crook stole which jewel by using the following information:

Adams had access to the diamond and sapphire, but not to the ruby or emerald.

Brooks had access to only the diamond.

Cole had access to every jewel except the sapphire.

Dent had access to every jewel except the sapphire and ruby.

50. Jimmy picked a new pet from a snake, a turtle, a rabbit, and a dog. If his pet is not furry and has legs, which animal did he pick?

51. What even number is less than 25 and is named when counting by sevens?

52. Fill in the following grid so that the shapes $\bigcirc, \square, \bigstar$, and $\triangle$ appear in each row, column, and 2 × 2 square only once.

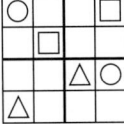

53. Many mathematical conjectures make claims of existence; that is, they claim that at least one number, concept, or relationship satisfies the conditions of the conjecture. Such statements are called **existential statements** and are relatively simple to prove because they can be justified by just one example that works. Provide an example that proves each statement to be true.

    a. There exists a number x that, when multiplied by 3, equals 18.

    b. There is a number x such that x times any other number is x.

    c. There exists a number x such that $x^2 = x$.

    d. There exists an even number that is divisible by only two numbers.

54. Provide an example that proves each statement to be true.

    a. There exists a number x such that $x + 7 = 9$.

    b. There exists a number x such that $3x < 13$ and $4x > 15$.

    c. There exists numbers x and y such that $x + y = 15$ and $x - y = 5$.

    d. There exists a number x such that $x^2 < x$.

55. Prove that there exists two counting numbers, a and b, such that $\dfrac{1}{a} + \dfrac{1}{b} = 1$.

56. Prove that there exists counting numbers a, b, and c such that $a^2 + b^2 = c^2$.

57. In most instances, we cannot prove a conjecture to be true simply by showing a single true example. However, we can prove the conjecture by showing that all possibilities for the conjecture hold true. This technique is called the **method of exhaustion** because it literally exhausts all possibilities under consideration. Use this method to prove that each of the following conjectures is true.

    a. $x^2 \geq x$, whenever x is 1, 2, 3, 4, or 5.

    b. $x^3 \geq x$, whenever x is 1, 2, 3, 4, or 5.

    c. Each of the numbers 2 through 10 can be written as the sum of exactly two or three odd numbers.

58. Use the method of exhaustion to prove that every even number from 2 to 26 can be written as the sum of three or fewer perfect squares. (For example, $2 = 1^2 + 1^2$, $4 = 2^2$, and $6 = 2^2 + 1^2 + 1^2$.)

59. Lewis Carroll was the pen name of English mathematician Charles Dodgson. Dodgson is probably most famous for writing *Alice in Wonderland* but is also known for writing children's logic puzzles. Several of his puzzles follow. Determine the correct conclusion for each argument.

a. No ducks waltz.

No officer ever declines a waltz.

All my poultry are ducks.

Therefore, _____

b. No experienced person is incompetent.

Jenkins is always blundering.

No competent person is always blundering.

Therefore, _____

c. All puddings are nice.

This dish is a pudding.

No nice things are wholesome.

Therefore, _____

d. Babies are illogical.

Nobody is despised who can manage a crocodile.

Illogical persons are despised.

Therefore, _____

Mathematical Communication

60. Write an example argument that reasons directly by:

a. Affirming the hypothesis.

b. Hypothetical syllogism.

61. Write an example argument that reasons indirectly by:

a. Denying the conclusion.

b. Process of elimination.

62. Deductive arguments and proof can be considered an important part of mathematical communication. Why do you think this is so?

63. Consider the sentences, "My car is red" and "My car is blue." Are these statements negations of one another? Why or why not?

64. Wes and Anna are having dessert. Anna says she does not like brownies and ice cream. When Wes brings her ice cream, she is happy with the choice. How is this possible?

65. When children learn grammar, they are taught not to use two negatives in the same sentence. Why is it illogical to use two negatives in a sentence?

Building Connections

66. Read the NCTM Reasoning and Proof Standards for grades K–2 and 3–5. Based on what you read, how and why should deductive reasoning be used in the elementary grades?

67. How do inductive and deductive reasoning complement each other when we are reasoning mathematically?

68. Conjectures are an important part of many scientific disciplines. If a conjecture is made from observations of a repeated scientific experiment, has inductive or deductive reasoning been used? Is it possible to verify such a conjecture to be absolutely true? Why or why not?

69. An automobile dealer claims that every car that leaves his lot will be trouble free for its next 5,000 mi.

a. On what kind of reasoning is such a claim made?

b. Would you be inclined to except the claim as true? What information would you want to know to help you decide?

c. Look through your local newspaper or a magazine to see whether you can find three other examples of advertising using this type of reasoning.

SECTION 1.4 Mathematical Problem Solving

In today's fast-moving and technological society, the ability to solve problems is now an essential job skill. Many workplace problems are quantitative in nature, so they require mathematical ideas and procedures to solve them. As future teachers, we must help our students learn to be not only good problem solvers, but also good *mathematical* problem solvers.

In mathematics, there is a difference between solving exercises and solving problems. With **exercises,** we apply a known procedure or fact in a systematic way to arrive at the answer. Although exercises are useful for learning ideas and procedures, they seldom require creative thinking. **Problems,** on the other hand, are more challenging because we do not know how to solve them in advance. We can use no one set of skills or ideas for every problem, so we may need to think creatively or use our mathematical knowledge in new and unfamiliar ways. So solving problems is not always easy. Fortunately, we can do some things to become better problem solvers.

First, we can develop good habits and a good mental attitude toward problem solving. Good problem solvers:

- Take time to understand the problem.
- Can change directions as needed.
- Learn to use a variety of tools.
- Reflect on their progress.
- Are patient and take time to explore.
- Keep track of their work.
- Adjust their planning as necessary.
- Do not give up easily!

Second, we can view problem solving as a process with a sequence of steps that we can follow to find a solution. In the elementary grades, the process of problem solving is often taught using the method developed by George Polya.

Polya's Problem-Solving Method

George Polya first established his problem-solving method in his classic book *How to Solve It*. His method is popular in school mathematics because it is specifically designed to help students learn how to solve problems by asking appropriate questions. Polya divides problem solving into four steps.

Step 1: Understanding the Problem

The first step is to make sense of the problem by identifying the principal parts. When doing so, we might ask the following questions:

- What is the goal of the problem?
- What are the unknowns?
- What information are we given?
- What information do we need but are not given?
- Do we understand the vocabulary?
- Is the problem similar to another we have solved? If so, how?

In many problems, the answers to these questions are easy to find, giving us a good start toward a solution. However, in other problems, we may have to read the problem several times before finding all of the relevant information.

Step 2: Devising a Plan

The next step is to devise a plan for solving the problem. The plan should come not only from our understanding of the problem, but also from our common sense, mathematical knowledge, and previous experience. Devising a plan can be the most difficult part of the process. For this reason, it helps to be familiar with the strategies that we often use in mathematics. Some common strategies are:

Strategy 1: Draw a picture.
Strategy 2: Guess and check.
Strategy 3: Make a list or table.
Strategy 4: Work backward.

Strategy 5: Look for a pattern.
Strategy 6: Use deductive reasoning.
Strategy 7: Use a formula.
Strategy 8: Write an equation.

In the classroom, many of these strategies receive specific instructional attention so students understand how and when to implement them. Notice how many of the strategies rely on mathematical representations and reasoning. This is another indication of the strong connection among the five mathematical processes.

Step 3: Carrying Out the Plan

We carry out the plan by using the appropriate representations and making the necessary computations. We continue until we either reach an answer or find that we need to change our plan. Because our original plan may not lead to a solution, we should always keep an accurate record of our work so that we can monitor progress. This step requires patience and persistence. If a problem proves to be particularly difficult, we should put it aside and try again later. Often, a short break leads to fresh insights.

Step 4: Looking Back

Looking back at a solution is an important step that even the best problem solvers often skip. This last step gives us an opportunity to check the solution and allows us to review and consolidate what we have learned. In checking our solution, we may ask ourselves the following questions:

- Does the solution answer the question?
- Is the solution correct or even reasonable?
- Did we include the details of the solution, such as appropriate labels?
- Is there another, more efficient way to solve the problem?
- Can we use the result or method in another problem?

Even though we presented these four steps sequentially, we often move back and forth among them as we solve a problem. For instance, as we devise a plan, we often get a better understanding of the problem. When we carry out the plan, we often find that it will not work. So we may need to devise a new plan. Consequently, we seldom solve problems linearly, but rather skip or revisit steps, as shown in Figure 1.11.

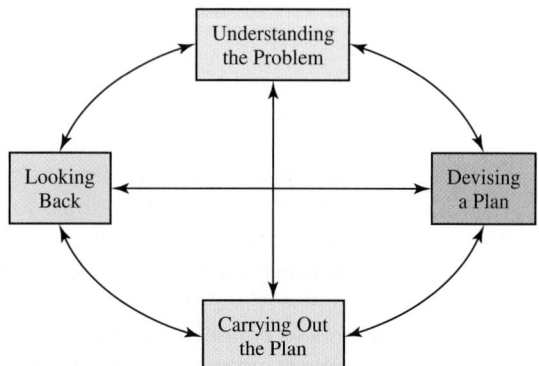

Figure 1.11 The problem-solving process

In the following sections, we use Polya's method and different strategies to solve a variety of problems, giving you the opportunity to see how other people solve problems. Nevertheless, the best way to learn how to solve problems is to practice, so we encourage you to try the problems in the Learning Assessment.

Strategy 1: Draw a Picture

Explorations Manual 1.9

Drawing a picture is a useful strategy when the problem involves geometry, measurement, or a physical situation. In some situations, the picture itself is the solution. In others, pictures help us gain insight into the problem and how we might use other strategies in the solution.

Example 1.30	Wiring a Small College Campus

Problem Solving   A small college campus plans to set up a new telephone network in which each building is connected to every other building by a separate cable. If there are six buildings, how many different cables must be run?

Solution

Understanding the Problem. Our goal is to find out how many telephone cables must be run if each distinct pair of buildings is connected with a single line. We are told that there are six buildings on campus.

Devising a Plan. Running wires between buildings is a physical situation that we can represent by *drawing a picture*. We use a square for each building and then draw lines between them to represent the cables. After drawing the picture, we simply count the lines for the answer.

Carrying Out the Plan. We draw six squares, connect each distinct pair of squares with a line, and then carefully count the lines. Because there are 15 lines, 15 cables must be run to connect the campus.

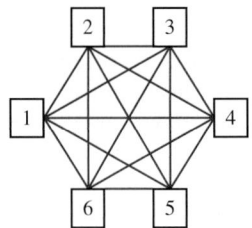

Looking Back. We can check our answer in two ways. One is to recount the lines using a different method. The picture shows that each building is connected to five wires. For six buildings, there must be $6 \cdot 5 = 30$ connections. However, each wire makes two connections, so there must be $30 \div 2 = 15$ wires. Another method is to count wires one building at a time. To connect building 1 to the five other buildings, we need five wires. To connect building 2, we need only four wires because it is already connected to building 1. Continuing in this way, we need three wires to connect building 3, two wires to connect building 4, and only one wire to connect buildings 5 and 6. So, it takes $5 + 4 + 3 + 2 + 1 = 15$ wires to connect the campus.

Strategy 2: Guess and Check

Whenever there seems to be no clear-cut way to solve a problem, we can always make a guess and check to see if it works. Although this is a legitimate strategy, it is not always efficient. It works well only when the number of possibilities is limited or when we can test answers in a systematic way. We can also make it more efficient by analyzing our previous guesses so that our next guesses are more effective. By using a series of selective trials, we can close in on the final answer.

| **Example 1.31** | **Finding the Least Positive Difference** |

Problem Solving

Use each of the digits 2, 3, 5, 6, 8, and 9 only once to find the least positive difference between two three-digit numbers.

Solution

Understanding the Problem. Our goal is to find two three-digit numbers so that the difference between them is positive and as small as possible. In doing so, we can use each digit (2, 3, 5, 6, 8, and 9) only once.

Devising a Plan. We proceed by *guessing and checking,* analyzing each solution to see whether a better guess can be made.

Carrying Out the Plan. Because the difference must be positive, the top number must be greater than the bottom. This leads us to make guesses with the larger digits on top. Possible trials include:

$$\begin{array}{cccc} 869 & 698 & 896 & 986 \\ -325 & -235 & -352 & -532 \\ \hline 544 & 463 & 544 & 454 \end{array}$$

The last guess is the smallest of the four. When we examine why it is the smallest, we see that the numbers in this difference are the closest together. So we look to find even smaller differences by trying number combinations that are closer together.

$$\begin{array}{ccc} 952 & 683 & 356 \\ -863 & -592 & -298 \\ \hline 89 & 91 & 58 \end{array}$$

Clearly, the new guesses are better, but we are still not sure whether 58 is the smallest difference. When we examine the differences, we see that the digits in the hundreds place are consecutive. We also see that the smallest difference occurs when the last two digits in the top number are small and the last two digits in the bottom number are big. Combining these ideas leads us to guess:

$$\begin{array}{c} 623 \\ -598 \\ \hline 25 \end{array}$$

This is the least possible difference, so our numbers are 623 and 598.

Looking Back. We can see how important it is to analyze our guesses. Not only did analysis get us to the solution quicker, but it also kept us from repeating guesses. We could have reduced the number of guesses even further by first thinking about the nature of the problem. Because we wanted the difference to be small, it makes sense to have the numbers close together. With this in mind, we may have even tried 623 and 598 on our first guess.

■

Strategy 3: Make a List or Table

Making a table or a list is useful with problems that ask how many outcomes satisfy a given condition. We carefully make the list or table and then count the outcomes. The key to this strategy is to list all the outcomes and not to duplicate any. So we should try to create the list or table systematically or sequentially. A table or list can also help organize information, often enabling us to find patterns or other useful information that we can use to solve the problem. Figure 1.12 shows a student page that illustrates how a list is used in a problem-solving situation in the elementary grades.

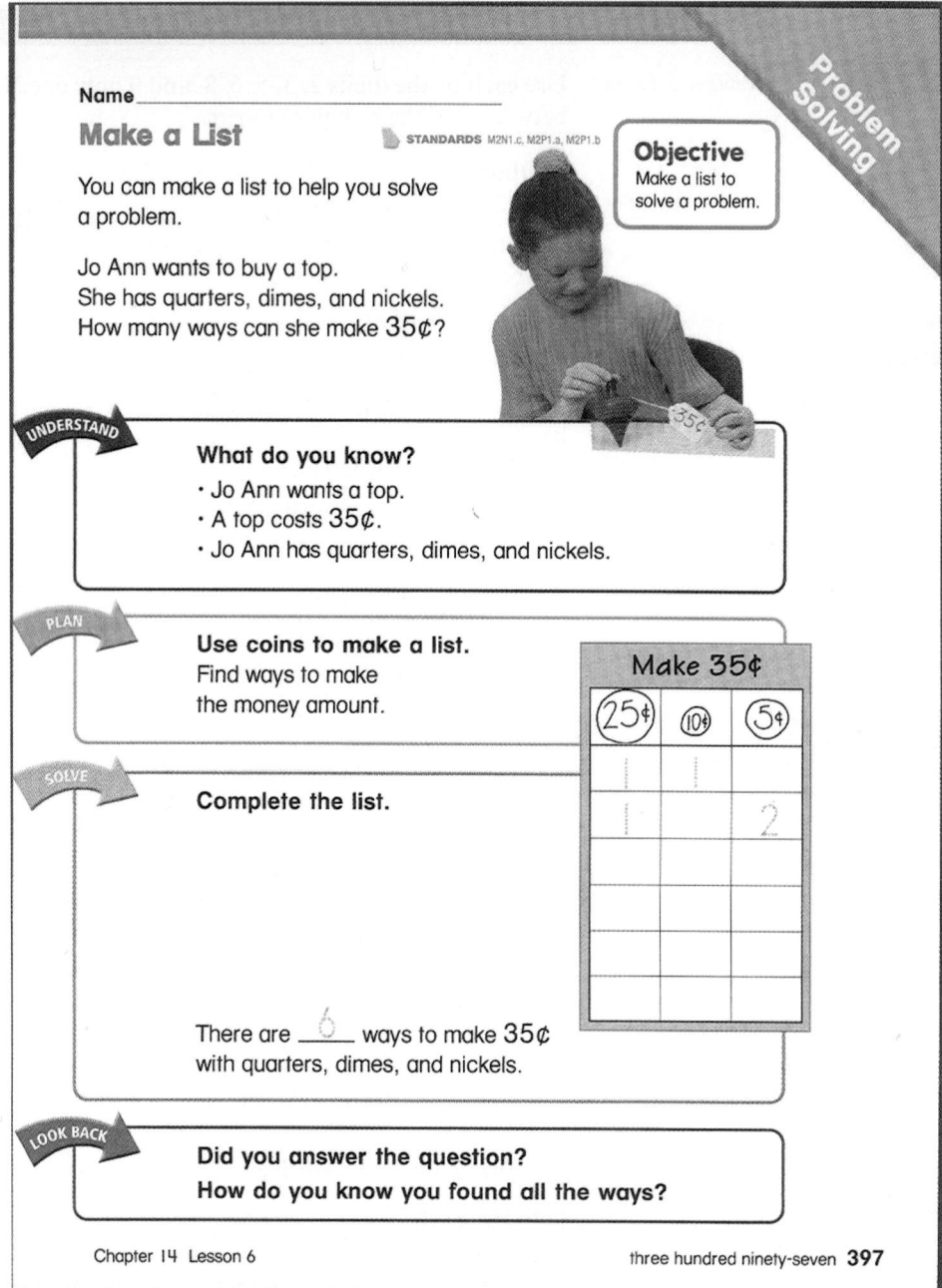

Figure 1.12 Using a list to solve a problem
Source: From *Mathematics Grade 2: Houghton Mifflin Mathematics Georgia* (p. 397). Copyright © by Houghton Mifflin Company, Inc. Reproduced by permission of the publisher, Houghton Mifflin Harcourt Publishing Company.

Strategy 4: Work Backward

In some problems, we know the end result, and our task is to find the starting point or the method by which the result was obtained. In these situations, working backward is often the best strategy. We start with the answer and then use the conditions of the problem to figure out the steps taken to achieve it.

Example 1.32	Counting Bacteria in a Culture

Problem Solving   A culture of bacteria is created one morning at 6 a.m. A lab technician begins to monitor the culture at 9 a.m. and notices that the population triples every hour. At 12 p.m., she applies an antibiotic that kills half the existing population every hour.

Finally, at 5 p.m., she counts the bacteria and finds 36,450 living bacteria. How many bacteria were in the culture at 6 a.m.?

Solution

Understanding the Problem. We need to find how many bacteria were in the culture at 6 a.m. The bacteria tripled each hour until an antibiotic was applied. After that, the number of bacteria was halved each hour.

Devising a Plan. *Working backward* is our strategy of choice. We start with the final count and find the number of bacteria at each hour from 5 p.m. back to 6 a.m.

Carrying Out the Plan. From 12 p.m. to 5 p.m., the number of bacteria is halved each hour, which we undo by multiplying by 2. Similarly, from 6 a.m. to 12 p.m., the number of bacteria tripled each hour, which we undo by dividing by 3. We use a table to track our progress and find that the initial culture had 1,600 bacteria.

Hour	Number of Bacteria	Hour	Number of Bacteria
5 p.m.	36,450	11 a.m.	$1,166,400 \div 3 = 388,800$
4 p.m.	$36,450 \cdot 2 = 72,900$	10 a.m.	$388,800 \div 3 = 129,600$
3 p.m.	$72,900 \cdot 2 = 145,800$	9 a.m.	$129,600 \div 3 = 43,200$
2 p.m.	$145,800 \cdot 2 = 291,600$	8 a.m.	$43,200 \div 3 = 14,400$
1 p.m.	$291,600 \cdot 2 = 583,200$	7 a.m.	$14,400 \div 3 = 4,800$
12 p.m.	$583,200 \cdot 2 = 1,166,400$	6 a.m.	$4,800 \div 3 = 1,600$

Looking Back. We can check our work by reversing each step in the table. First, we undo the six divisions with six multiplications. The computation is simpler if we realize that $3 \cdot 3 \cdot 3 \cdot 3 \cdot 3 \cdot 3 = 3^6 = 729$. So, there are $1,600 \cdot 729 = 1,166,400$ bacteria at 12 p.m. Similarly, we can undo the five multiplications by dividing the number of bacteria at 12 p.m. by $2^5 = 32$. As a result, there are $1,166,400 \div 32 = 36,450$ bacteria at 5 p.m.

Check Your Understanding 1.4-A

1. A farmer plans to put a fence around a rectangular field that is 50 yards wide. If the length of the field is three times the width, how many yards of fencing does he need?

2. How many other four-digit numbers have the same digits as 2,562?

3. Consider the following diagram. What number belongs in the oval?

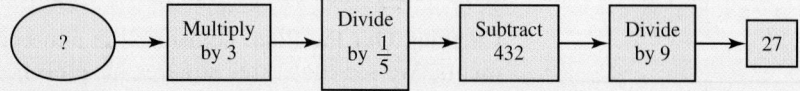

Talk About It Review Polya's problem-solving method, and then think about how you have solved mathematical problems in the past. How does your past approach to problem solving compare to Polya's? If it is different, how so?

Problem Solving

Activity 1.7

Many times something that we might consider to be a "party game" is actually mathematical. For instance, consider the following problem, which is related to a branch of mathematics called knot theory. First, cut two pieces of string, each about 4 ft long. Next, have a partner tie the ends of one string around both of your wrists. Loop the other piece of string over the string tied around your wrists, and then tie its ends around your partner's wrists. You should be connected as shown:

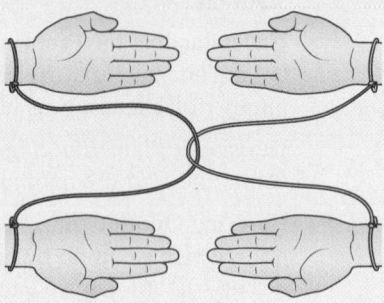

Your task is to separate yourselves from one another without untying, cutting, or removing the strings from your wrists in any way.

Strategy 5: Look for a Pattern

As explained in Section 1.2, patterns play an important role in mathematical reasoning. The same is true in problem solving because we often use inductive reasoning to search for patterns in problem situations. As the following example shows, the strategy of looking for a pattern is particularly useful in making a prediction or a generalization from a list of numbers, data, or shapes.

Example 1.33   Counting Squares

Problem Solving   Consider the following three successive arrangements of squares. How many squares are needed to make the tenth arrangement?

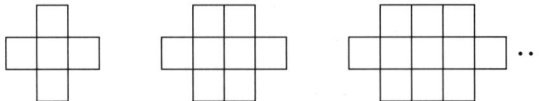

Solution

Understanding the Problem. The goal is to find the number of squares needed to make the tenth arrangement. We are given the first three arrangements. By counting, we know that 5 squares are needed for the first, 8 for the second, and 11 for the third.

Devising a Plan. One way to approach the problem is to try to *find a pattern* in the number of squares added to each successive arrangement. If we can find such a pattern, we can then extend it to find the number of squares in the tenth arrangement.

Carrying Out the Plan. Consider the number of squares in the three given arrangements. We see that three squares have been added from the first to the second and three more have been added from the second to the third. Continuing this pattern, we add three to the number of squares until we reach the tenth arrangement. The following table helps us keep track of our progress. As the last entry shows, we need 32 squares to make the tenth arrangement.

Arrangement	Number of Squares	Arrangement	Number of Squares
1	5	6	17 + 3 = 20
2	5 + 3 = 8	7	20 + 3 = 23
3	8 + 3 = 11	8	23 + 3 = 26
4	11 + 3 = 14	9	26 + 3 = 29
5	14 + 3 = 17	10	29 + 3 = 32

Looking Back. One way to check the solution is to draw the tenth arrangement by adding three squares to the end of each successive arrangement. As shown in the figure, the tenth arrangement consists of 32 squares.

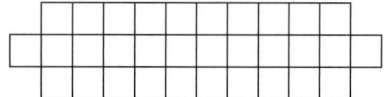

Strategy 6: Use Deductive Reasoning

Explorations Manual 1.10

Deductive reasoning is another powerful tool for solving problems. Although often used in conjunction with other problem-solving strategies, deductive reasoning can be sufficient in and of itself. We can use it when we need to explain a process or make a statement based on a collection of specified conditions.

Example 1.34	Finding a Counterfeit Coin

Problem Solving   Twelve coins look identical. If one is a heavy counterfeit, how can you find it in only three weighings on a balance scale?

Solution

Understanding the Problem. There are two parts to this problem: (1) Determine which coin is the heavy counterfeit and (2) describe how to isolate it from the others. We are permitted only three weighings on a balance scale. On a balance scale, weights are placed on both sides of the balance. If the weights are equal, the sides of the balance remain at the same height. If not, the heavier side is lower than the lighter side.

Devising a Plan. We can use *deductive reasoning* to isolate the counterfeit coin. However, because we have only a limited amount of information, we must think creatively to get to a solution.

Carrying Out the Plan. We begin by placing six coins on each side of the balance. The side with the heavy counterfeit must go down. This first weighing confines the counterfeit to one of six coins. Next, we place three of the six coins containing the counterfeit on either side of the balance. Again, the side with the heavy counterfeit goes down, confining the counterfeit to one of three coins in the second weighing.

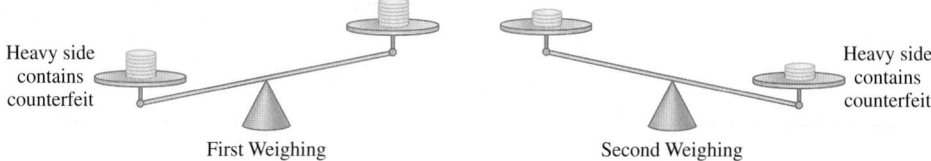

Although we cannot weigh each pair of the remaining coins directly, we can pick two coins, weigh them, and know which is the counterfeit. If the heavy counterfeit is on the balance, its side of the scale will go down. If not, the scale will balance, and the heavy counterfeit must be the coin not on the scale.

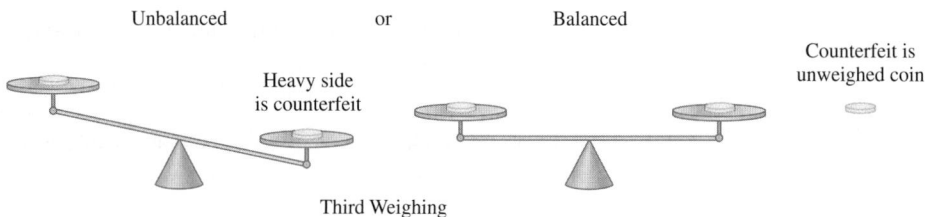

Looking Back. We have found the counterfeit coin and outlined the method for doing so. Crucial to the process is recognizing that not every set of coins can be weighed. We could have used this insight at any of the three weighings. For instance, we could have weighed two sets of four coins in the first weighing. If either set contained the counterfeit, that side of the balance would have gone down. If the two sets balanced, then the counterfeit must be in the unweighed set. Either way, only four coins remain to be weighed. Once the heavy set of four is identified, the second and third weighings can be done directly. ■

Strategy 7: Use a Formula

Many academic disciplines, such as chemistry, physics, and finance, use formulas to solve problems. A **formula** is an algebraic expression that represents a general fact, principle, or relationship. Once we know that a specific fact or principle is relevant to the problem, we can apply the related formula to arrive at the answer.

Example 1.35	Measuring Sod

Problem Solving

Avril built a circular patio that fits perfectly inside a square courtyard measuring 16 ft on a side. If he wants to plant sod in the four corners of the square, about how many square feet of sod does he need to buy?

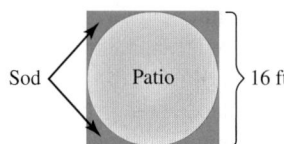

Sod — Patio — 16 ft

Solution

Understanding the Problem. We are asked to find the amount of sod needed to plant the four corners of the patio. This amount includes the area inside the square but outside the circle. The square has sides 16 ft long. The radius of the circle is not given, but we do know the circle fits perfectly inside the square.

Devising a Plan. Both the area of a square and a circle can be measured by *using a formula*. The area of a square is the length of the side squared: $A_{square} = s^2$. The area of a circle is π times the radius squared: $A_{circle} = \pi r^2$. If we find the two areas, we can compute the amount of sod by subtracting the area of the circle from that of the square.

Carrying Out the Plan. Because the side of the square is $s = 16$ ft, its area is

$$A_{square} = s^2 = 16^2 = 256 \text{ ft}^2$$

To compute the area of the circle, we must know its radius. Because the circle fits perfectly inside the square, the radius of the circle is half the side of the square, or $r = 8$ ft. Hence,

$$A_{circle} = \pi r^2 = \pi(8)^2 = 64\pi \approx 64(3.14) \approx 201 \text{ ft}^2$$

By subtracting, we determine that Avril must buy

$$A_{square} - A_{circle} = 256 - 201 = 55 \text{ ft}^2 \text{ of sod}$$

Looking Back. One way to check our work is to use estimation. If we divide the diagram into 16 congruent squares, we see that the shaded region has an area that is slightly less than four small squares.

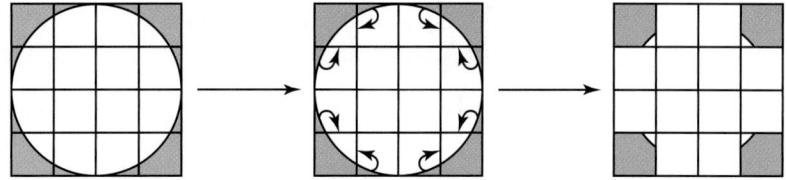

So, the shaded region has an area slightly less than $\frac{4}{16} = \frac{1}{4}$ the area of the square. Because $\frac{1}{4}$ of 256 ft² is 64 ft², our answer of 55 ft² seems reasonable.

Strategy 8: Write an Equation

Writing an equation is a useful strategy when an unknown quantity has to be found. The unknown is represented by a variable, and the information in the problem guides us in writing the equation. This strategy may seem similar to using a formula. However, it differs in that the algebraic equation is specific to the problem and cannot be applied in other situations.

Example 1.36	Finding the Missing Page Numbers in a Book

Problem Solving   Ondra opens a library book and notices that two consecutive sheets are missing. What two page numbers are showing, if the sum of the numbers is 489?

Solution

Understanding the Problem. We are asked to find two page numbers by using the facts that their sum is 489 and that two consecutive sheets are missing. The word "consecutive" means, "One right after another," so four numbered pages are missing.

Devising a Plan. Because we have two unknown quantities, our strategy is to *write an equation*, relying on key words in the problem to guide us. Once we have the equation, we can use basic algebra to find the unknown numbers.

Carrying Out the Plan. We do not need to use different variables to represent the unknown page numbers because we have information connecting the two. Specifically, two consecutive sheets are missing, so the page numbers must be five numbers apart. If x represents the smaller page number, then $x + 5$ represents the larger, and $x + (x + 5) = 489$. Now solve for x.

$$x + (x + 5) = 489 \qquad \text{Combine like terms}$$
$$2x + 5 = 489$$
$$2x + 5 - 5 = 489 - 5 \qquad \text{Subtract 5 from both sides}$$
$$2x = 484$$
$$\frac{2x}{2} = \frac{484}{2} \qquad \text{Divide both sides by 2}$$
$$x = 242$$

If the smaller page is 242, the larger page must be $242 + 5 = 247$.

Looking Back. Checking the solution is relatively easy. First, notice that 242 and 247 are five numbers apart. Second, when we add them, we get $242 + 247 = 489$, our original sum.

Check Your Understanding 1.4-B

1. A woman is traveling to market with a wolf, a goose, and a bag of seed. She must cross a river in a boat that will hold only her and one other object. If left alone, the wolf will eat the goose, and the goose will eat the seed. How many crossings must she make to get them all across safely?

2. The formula for the area (A) of a rectangle is $A = lw$, where l is the length of the rectangle and w is the width. Use this information to find the area of the figure.

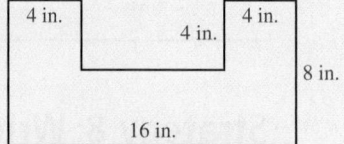

4 in.   4 in.   4 in.   8 in.   16 in.

3. The difference between two positive whole numbers is 12. If the larger is divided by the smaller, the quotient is 3. What are the two numbers?

Talk About It How many of the eight strategies in this section have you used before? Have you used others that were not included? Get together with a group of your peers and discuss the different strategies that you have used.

Connections

Activity 1.8

Carl Friedrich Gauss (1777–1855) is considered to be one of the greatest mathematicians of all time. Legend has it that, in one of his first arithmetic classes, Gauss was asked to add the numbers from 1 to 100, which he did almost immediately. Determine a process that Gauss may have used to solve the problem and use it to find the sum. (*Hint:* Use the strategy of solving a problem that is similar to the original, yet simpler to solve. For instance, consider the sum of the numbers from 1 to 10. By solving the simpler problem, you may be able to adapt your strategy for use with the larger problem.)

Carl Friedrich Gauss

Hulton Archive/Getty Images

SECTION 1.4   Learning Assessment

■ Understanding the Concepts

1. What is problem solving, and what does it mean to be a good problem solver?

2. Explain the difference between an exercise and a problem.

3. Understanding the problem involves identifying its principal parts. What is meant by "principal parts?"

4. A plan should arise out of what four things?

5. Why is looking back an important part of problem solving? Why do you think it is so commonly overlooked?

6. Why is drawing a picture a valuable strategy for solving problems in mathematics?

7. What is the key to successfully using a table or a list in problem-solving situations?

8. When is it reasonable to use the following strategies to solve a problem?

   a. Work backward   **b.** Guess and check

9. How are inductive and deductive reasoning useful as problem-solving strategies?

10. What is the difference between using a formula and writing an equation?

■ Exercises

11. Draw a diagram to solve the following problem: A square table can seat 4 people. When two such tables are set end to end, the new table can seat 6 people. If 15 tables are set end to end, how many seats are there?

12. Is it possible to cut a round cake into exactly eight pieces using only three cuts? Assume no pieces can be moved.

13. Given the following arrangement of toothpicks, is it possible to make five triangles by moving just three toothpicks?

14. Use a table or a list to determine the number of ways to make 37¢ using dimes, nickels, and pennies. Why is making a table or list an effective strategy for solving this problem?

15. Place the digits 1, 2, 3, 4, 5, 6 in the circles so that the sum of the numbers on each side is equal. Use guess and check to find your solution.

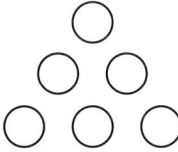

16. Use guess and check to determine whether six Xs can be placed on the grid without having three in a row in any direction.

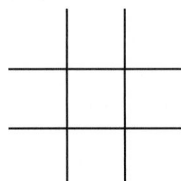

17. a. Work backward to solve the following problem. Derrick gave half his pieces of candy to Cindy. Cindy gave a third of these pieces to Kasey, and Kasey gave a fifth of the candy she received to Bill. If Bill received 5 pieces of candy, how many did Derrick start with?

b. Why is working backward an effective strategy for solving this problem?

c. How might you check your answer to ensure that it is correct?

18. Work backward to find the number that belongs in the oval of the diagram.

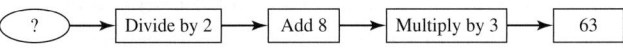

19. What symbol might come next in each of the following patterns?

a. ♠, ♣, ♠, ♠, ♣, ♣, ♠, ♠, ♠, ♣, ♣, ♣, . . .

b. ♦, ♥, ♦, ♥, ♥, ♦, ♥, ♥, ♥, ♦, . . .

c. ♣, ♦, ♠, ♠, ♥, ♥, ♣, ♣, ♣, . . .

d. ←, ↑, →, ↓, ←, . . .

e. +, ★, ✳, . . .

20. How many cuts does it take to cut the length of a log into:

a. 3 equal lengths?

b. 4 equal lengths?

c. 5 equal lengths?

d. What pattern is established between the number of cuts and the number of lengths? Use the pattern to write an expression for the number of cuts needed to cut the log into n equal lengths.

21. Julie has four other members in her family: James, Jonathan, Jenny, and Jill. James is older than Julie, and Jill is not Julie's sister. If Julie is the oldest sibling, who is Julie's mother, father, sister, and brother?

22. The sum of two numbers is 59. If one is two times the other minus 10, what are the two numbers? Why is writing an equation an effective strategy for solving this problem?

23. When Kevin opened his math book, he found a page missing. If the sum of the two showing pages is 171, what are the missing page numbers? Write an equation to find your solution.

24. Use formulas from geometry to find the area of the shaded regions in each diagram.

a.

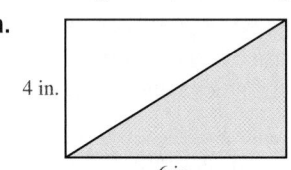

b.

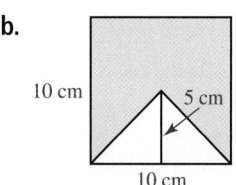

c.

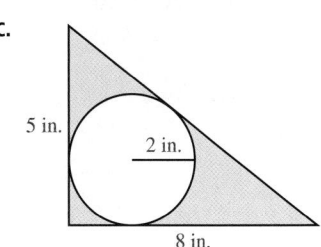

■ Problems and Applications

25. A gardener plants 15 rows of corn. Each row of corn has 30 plants in it, and each plant produces 4 ears of corn. How many ears of corn will the garden produce?

26. Consider the following 6 × 6 grid. How many Xs can be placed in the grid so that there are no more than two Xs in any one row, column, or main diagonal? Two have already been placed.

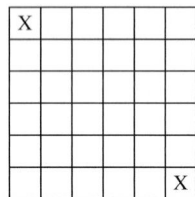

27. A farmer wants to fence in a rectangular field that is 50 ft long and 30 ft wide. He will use boards that are 5 ft long. If each end of a board is to be nailed to a post, how many posts are needed to fence in the field?

28. Hannah, Sarah, and Debra went to the Bahamas on a vacation. The plane trip there took 215 min. They spent a total of 6,327 min on the island, and the plane trip back took 264 min. To the nearest hour, how many days and hours were they gone?

29. How many cubes are needed to build this figure?

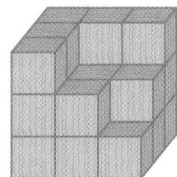

30. The price of a certain stock in 2010 was $15.00 a share. A stock market analyst predicts that the value of the stock will increase by $0.85 every year. In what year will the value of the stock will be over $25.00?

31. Move only three triangles to make the figure on the left look like the figure on the right.

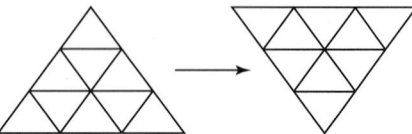

32. Can this figure be drawn without retracing any lines or without lifting your pencil from the paper? Describe the strategy you used and why it led to your conclusion.

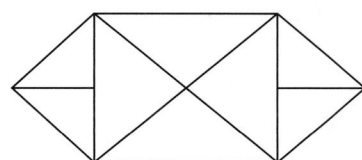

33. Place the numbers 1 through 8 in the boxes so that no two consecutive numbers are next to each other horizontally, vertically, or diagonally.

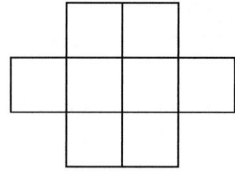

34. Consider the following list of letters. What letter does not belong in the sequence?

A, B, E, C, I, G, O, F, U

35. What letter comes next in each sequence?
   a. 1, B, 3, D, 5, . . .   **b.** Z, A, Y, B, . . .
   c. M, T, W, T, . . .   **d.** J, F, M, A, . . .

36. Marc wrote the numbers from 1 to 100. How many times did he write a 3 or a 7?

37. The two volumes of a book sit side by side on a shelf. A bookworm eats from the inside of the back cover of volume I through to the inside of the front cover of volume II. If each book cover is an $\frac{1}{8}$ in. thick, and if each book has 2 in. worth of pages, how far did the worm burrow?

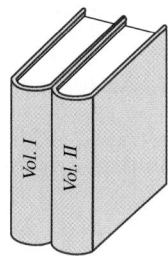

38. Without lifting your pencil from the paper, connect all nine dots using only four lines.

● ● ●
● ● ●
● ● ●

39. How many triangles are in this figure?

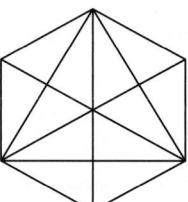

40. Draw six squares in a row on a sheet of paper, and place two nickels at one end and two pennies at the other. If a coin can either move forward one space or jump over a single coin, what is the smallest number of moves needed to switch the pennies and nickels?

41. Three boxes contain fruit. One contains oranges, one contains bananas, and one contains oranges and bananas. If all three boxes have been incorrectly labeled, how can you correct the labeling by drawing a piece of fruit from just one box?

  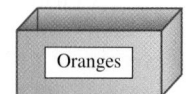

42. Can this figure be divided into four parts, each with the same size and shape?

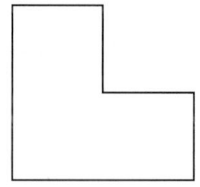

43. The number of hours left in the day is one-third of the number of hours already passed. How many hours have passed?

44. Move one toothpick in the following arrangement to make the computation with Roman numerals correct.

45. Find the area of the shaded region in the following diagram. Assume figure *ABC* is a quarter circle with a 5-in. radius.

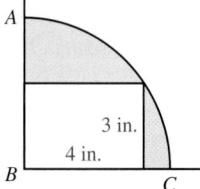

46. Eight gold coins look exactly alike, but one is a heavy counterfeit. How can the counterfeit be found in two weighings on a balance scale?

47. How can four pennies be placed into three cups so that each cup has a different number of pennies and no cup is empty?

48. Two days ago, I was 25 years old. Later next year, I will be 28. How is this possible?

49. Three men—Jack, John, and Jim—are accused of committing a crime. The guilty man always lies, whereas the two innocent men always tell the truth.

When asked who committed the crime, they replied as follows:

> Jack: I am not guilty.
>
> John: Jim is the guilty one.
>
> Jim: Jack committed the crime.

Which man is the liar who committed the crime?

■ Mathematical Communication

50. Choose a problem in this Learning Assessment that you have solved. Write up your solution to the problem using Polya's four-step method.

51. Create a set of three problems that use different strategies from those in this section. Give them to a partner to solve. In solving the problem, did your partner use the strategy that you had in mind? If not, what does this imply about problem solving?

52. Polya's general problem-solving method is not the only one. Explore several Internet sites on problem solving, and keep a record of the methods you find. Select one that is different from Polya's, and then write a one- or two-paragraph summary comparing and contrasting your selected method to Polya's.

53. Charlie is given the following problem: A dartboard has three rings. The bull's-eye is worth 20 points, the middle ring is worth 15 points, and the outer ring is worth 5 points. If three darts are thrown and all hit the dartboard, what are the possible resulting scores? He is at a loss as to where to start. He asks you, his teacher, for guidance. What advice might you give him in solving this problem?

■ Building Connections

54. Read the NCTM Problem Solving Standards for grades K–2 and grades 3–5. What reasons are given for learning mathematics through problem solving? Do some research on your own to find other reasons for learning mathematics through problem solving.

55. Which of the problem-solving strategies presented in this section do you believe are appropriate for grades K–2? For grades 3–5? Compare your thoughts to a set of elementary curriculum materials. Are you surprised with what you find?

56. Consider Polya's ten commandments for teaching, which are given in the historical note at the end of the chapter.

   a. Which of the ten commandments do you think are reasonable to follow in the classroom?

   b. Which of the ten commandments do you follow while learning mathematics? Why do you think doing so is important?

57. How are the other mathematical processes important to problem solving?

In the last section, our main goal was to introduce Polya's method and to illustrate how different problem-solving strategies work. In doing so, we demonstrated only a single strategy with each problem. However, in some cases, we can choose from a variety of strategies to solve a problem. Still in others, we need to combine strategies to arrive at a solution. We now consider these aspects of problem solving.

Solving Problems with Different Strategies

Explorations Manual 1.11

Whenever we set out to solve a problem, we often limit our thinking to just one strategy. However, there can be many ways to solve a problem, and we should use our knowledge and previous experience to look for different approaches. If we do, the problem will be more manageable, and we learn how to be better problem solvers. The next two examples show how different strategies can be used to solve the same problem. Keep in mind that the solutions offered may not be the only ones and that you are encouraged to look for others.

Example 1.37   Counting Outfits

Problem Solving

Maria has just returned from clothes shopping. She bought four blouses that are red, blue, white, and green. She also bought jeans, khakis, a pair of sneakers, and a pair of pumps. If all of her new clothes can be mixed and matched, how many different new outfits does she have?

Solution

Understanding the Problem. We are asked to count Maria's new outfits by finding how many different blouse-slacks-shoe combinations can be made from four blouses, two pairs of pants, and two pairs of shoes.

Devising and Carrying out a Plan. Because our goal is to count the number of outfits, we need a strategy that will allow us to generate the combinations without missing or duplicating any. We can do so in at least three ways.

Strategy 1: Make a Diagram. One way to generate the possible combinations is to make a *tree diagram.* Begin at a point and draw a "branch" for each blouse, and label them r, b, w, and g. Off each blouse branch, draw two branches for each pair of pants, label them j and k. Finally, off each pair of pants, draw two more branches for each pair of shoes, and label them s and p. Then count the number of branches at the last level to find the total number of outfits. Because there are 16 branches, Maria has 16 outfits.

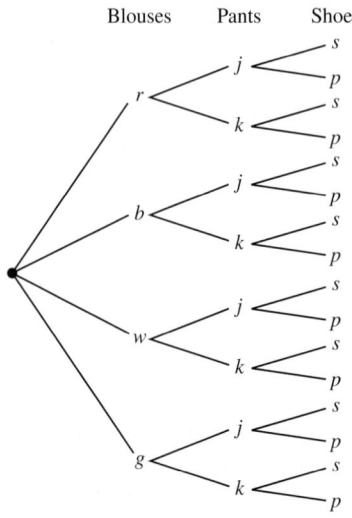

Blouses   Pants   Shoes

Strategy 2: Make a List. Another way to generate the outfits is to make a *list*. To avoid missing or duplicating combinations, work with one blouse at a time and match it to all pants-shoe combinations. This gives us the following list of outfits:

red, jeans, sneakers	blue, jeans, sneakers
red, jeans, pumps	blue, jeans, pumps
red, khakis, sneakers	blue, khakis, sneakers
red, khakis, pumps	blue, khakis, pumps
white, jeans, sneakers	green, jeans, sneakers
white, jeans, pumps	green, jeans, pumps
white, khakis, sneakers	green, khakis, sneakers
white, khakis, pumps	green, khakis, pumps

Again, Maria has 16 new outfits.

Strategy 3: Use Reasoning. A third way to solve the problem is to *use reasoning*. We know that Maria can mix and match any of her new clothes; any of her four new blouses can be worn with any of her two new pairs of pants. In other words, there are

$$\underset{\text{blouses}}{4} \cdot \underset{\text{pants}}{2} = 8 \text{ blouse-pants possibilities}$$

We also know that any of the eight blouse-pants combinations can be worn with either the sneakers or the pumps. As a consequence, Maria has

$$\underset{\text{blouse–pants}}{8} \cdot \underset{\text{shoes}}{2} = 16 \text{ different outfits}$$

Looking Back. Having counted 16 outfits in three different ways, we should be confident that the answer is correct. Also notice that the three strategies are interconnected. For instance, if we match the branches of the tree diagram to the list, we get the same outfits. We can also use the diagram or list to justify why multiplying the number of blouses, pants, and shoes gives the total number of outfits. The last strategy used the **Fundamental Counting Principle.** Informally, it states that if one event can happen m ways and a second event can happen n ways, then the total number of ways the two events can happen is $m \cdot n$. This principle offers a useful strategy for problems that involve counting, and we discuss it in greater detail in Chapter 15.

Example 1.38   Water in a Watercooler

Problem Solving   An office watercooler is filled on a Monday morning. That day, the workers drink a fifth of the water. On Tuesday, they drink half the remaining water, and on Wednesday they drink half the water leftover from Tuesday. If 16 cups remain, how much water did the cooler originally hold?

Solution

Understanding the Problem. Our goal is to find the original amount of water in the cooler. We know that a fifth of the original amount was removed on Monday, half of the remaining amount on Tuesday, and half the remaining amount again on Wednesday. The final amount in the cooler is 16 cups.

Devising and Carrying Out a Plan. Because water is removed from the cooler over three days, our strategy must allow us to proceed in a sequence of steps. Again, we can do so using at least three different strategies.

Strategy 1: Draw a Picture. Removing water from a cooler is a physical situation that we can represent by *drawing a sequence of pictures.* Even though we do not know the original amount, we do know that a fifth of the water is removed on Monday. We can represent this by using a cylinder as the tank, dividing it into fifths, and showing that one-fifth has been removed. On Tuesday, half the remaining amount is removed, which we show by removing two more fifths on our diagram. Similarly, on Wednesday, half the remaining amount from Tuesday is removed, which we represent by removing one more fifth on our diagram. This leaves one-fifth of the original amount, which equals 16 cups. So the tank must have originally held $5 \cdot 16 = 80$ cups.

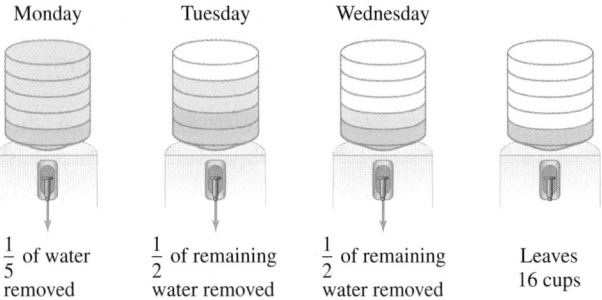

| Monday | Tuesday | Wednesday | |
| $\frac{1}{5}$ of water removed | $\frac{1}{2}$ of remaining water removed | $\frac{1}{2}$ of remaining water removed | Leaves 16 cups |

Strategy 2: Guess and Check. Because the amount of water on any given day is part of the original amount, our problem involves computations with fractions. So another way to solve the problem is to make a *guess and check* it with the appropriate computations. Because of how the water is removed, we know that the original amount must be divisible by 5 once and by 2 twice. In other words, we can limit our guesses to multiples of $5 \cdot 2 \cdot 2 = 20$. We check a guess by multiplying the number of cups by $\frac{4}{5}$ to get the number of cups after Monday. We then multiply by $\frac{1}{2}$ to get the number of cups after Tuesday and by $\frac{1}{2}$ again to get the number of cups after Wednesday. A table helps us keep track of our progress. As the table shows, our fourth guess works, indicating that the watercooler must have originally held 80 cups.

Guess (cups)	Monday (cups)	Tuesday (cups)	Wednesday (cups)
20	$\frac{4}{5}(20) = 16$	$\frac{1}{2}(16) = 8$	$\frac{1}{2}(8) = 4$
40	$\frac{4}{5}(40) = 32$	$\frac{1}{2}(32) = 16$	$\frac{1}{2}(16) = 8$
60	$\frac{4}{5}(60) = 48$	$\frac{1}{2}(48) = 24$	$\frac{1}{2}(24) = 12$
80	$\frac{4}{5}(80) = 64$	$\frac{1}{2}(64) = 32$	$\frac{1}{2}(32) = 16$

Strategy 3: Work Backward. The problem presents a sequence of steps, which can be reversed, so *working backward* is another strategy. On Wednesday, only 16 cups of water were left. Because this is half the water from Tuesday, there were $16 \cdot 2 = 32$ cups on Tuesday. In turn, 32 cups constitute half the water from Monday, so $32 \cdot 2 = 64$ cups remained on Monday. Because 64 cups is $\frac{4}{5}$ of the original amount, the cooler must have started with $\frac{5}{4} \cdot 64 = 80$ cups of water.

Looking Back. All three strategies lead to 80 cups, so we are confident that our answer is correct. In this problem, each strategy relied on making critical connections among pieces of information. To draw a picture, we had to realize that the original tank must be divided into fifths and that each fifth is equivalent to 16 cups. In the guess and check strategy, we had to realize that the number of cups was divisible by 20 to limit our guesses. Finally, when working backward, we had to realize that the 64 cups left after Monday was $\frac{4}{5}$ of the original amount, not $\frac{1}{5}$.

Check Your Understanding 1.5-A

1. Sally, Sue, and Sarah each have a certain number of pens. One has 8, one has 12, and one has 14. If Sally gives 6 pens to Sue, she will have half her original amount, and Sue will have 20. How many pens does each girl have?

2. Eight Little League teams play in Columbia County. If each team plays every other team exactly twice, how many games will be played during the season?

Talk About It   What other strategies can be used to solve the problems given in the last two examples?

Problem Solving

Activity 1.9

In educational settings, we are usually given all the information we need to solve a problem. However, in real life, this is seldom the case. For the following problem, identify the goal and the missing information that you need to solve it. Do an Internet search to find the missing information, and then solve the problem.

A company of soldiers is going on a one-week training exercise in a desert. The soldiers will receive no supplies on the exercise, so they must bring everything they need. How many gallons of water are needed for the entire company to remain properly hydrated throughout the week?

Using Multiple Strategies to Solve Problems

Explorations Manual 1.12

To this point, we have solved problems using a single strategy. However, many problems require several strategies. The strategies explained so far are not independent of one another and can be combined to solve complex problems. Learning how the different strategies complement one another significantly increases our ability to solve problems. Consider the next two examples.

Example 1.39   Solving a Magic Square

Problem Solving

In a magic square, the sum of every row, column, and main diagonal is the same. Complete the following 3 × 3 magic square using all the numbers 1 through 9 only once.

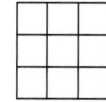

Solution

Understanding the Problem. We want to place the numbers 1 though 9 into the squares so that every row, column, and diagonal has the same sum.

Devising a Plan. Our first inclination may be to use the *guess and check* strategy simply by placing numbers in the squares to see what works. Unfortunately, this may require us to make a large number of guesses before we find the solution. However, if we analyze the problem, we may be able to *find a pattern* or use *deductive reasoning* to limit the number of trials.

Carrying Out the Plan. Because the sum of every row, column, and diagonal must be the same, it makes sense to add large numbers to small numbers to balance out the sums. Consequently, we begin by adding numbers from opposite ends of the following list.

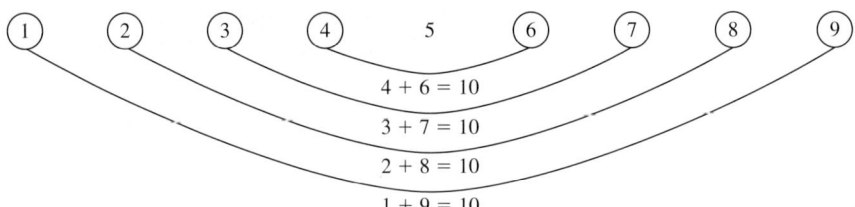

$$4 + 6 = 10$$
$$3 + 7 = 10$$
$$2 + 8 = 10$$
$$1 + 9 = 10$$

Each sum is equal to 10, and 5 is not included in any sum. So 5 may be a good choice to place in the middle square, which is included in most of the sums. If we place 5 in the middle, then the sum we need is likely to be $10 + 5 = 15$. Using this information, we proceed to make our first guess by working with pairs of numbers. Starting with 1 and 9, there are two possible placements: on a diagonal or on the middle row or column. We make our first trial by placing them on a diagonal.

9		
	5	
		1

Because we want to add big numbers to small numbers, the 8 can be placed only directly below or to the right of the 5. We place it below the 5, calling for the 2 to be placed directly to the right of the 9.

9	2	
	5	
	8	1

Next, we place the 3, 4, 6, and 7, each of which can go in only one square given the constraint of adding big numbers to small numbers. After the squares are all filled, we check our guess and see that not all of the rows and columns have the same sum.

9	2	4	= 15
3	5	7	= 15
6	8	1	= 15

‖ ‖ ‖
18 15 12

Because the guess did not work, one of the original placements must be wrong: Either the 5 does not go in the middle, or the 1 and 9 do not go on a diagonal. Before changing the location of the 5, we first try placing the 1 and 9 in the middle column. We then place numbers as before, making sure that small numbers are added to large numbers. When we check this trial, we see that every row, column, and diagonal has a sum of 15, and we have a solution.

2	9	4	= 15
7	5	3	= 15
6	1	8	= 15

$$15 \quad \underset{15}{\parallel} \quad \underset{15}{\parallel} \quad \underset{15}{\parallel} \quad 15$$

Looking Back. We can see how important our analysis was for arriving at a solution in only two trials. Actually, several different arrangements will work. In each case, the even numbers are always on a diagonal, the odd numbers are always in the middle row and column, and 5 is always in the middle square.

Example 1.40   Height of a Bouncing Ball

Problem Solving   A rubber ball is known to bounce to exactly half the height from which it fell. If the ball is dropped from a skyscraper that is 1,024 ft tall, how high will it go after the:

a. Second bounce?   **b.** Sixth bounce?   **c.** *n*th bounce?

Solution

Understanding the Problem. We are asked not only to find the height of the ball after a certain number of bounces, but also to determine how high it goes after any bounce (*n*). The ball falls from an initial height of 1,024 ft and always bounces to exactly half the height from which it fell.

Devising a Plan. This is a physical situation, which we can represent by *drawing a picture* that shows the height to which the ball goes after the first several bounces. A drawing works well for the first two questions, but it becomes more difficult to read as we add bounces. We can then *use a table* to collect information about the bounces and *look for a pattern* that we can extend. The last part of the problem is really asking for a way to relate the number of the bounce to its height. Because we are looking for the height of any bounce, we can represent the number of the bounce with a variable and *write an equation* to represent the needed relationship.

Carrying Out the Plan. We draw a building and label its height as the distance the ball must initially travel. We then draw several bounces, computing the height of the ball after each one. From the following diagram, we can see that the ball reaches a height of 256 ft after the second bounce, answering the first part of the question.

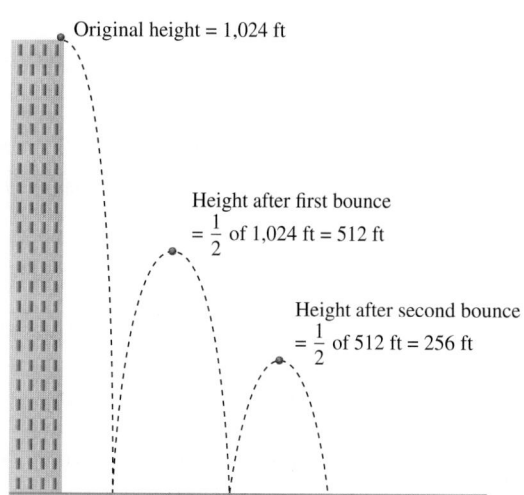

Original height = 1,024 ft

Height after first bounce
$= \frac{1}{2}$ of 1,024 ft = 512 ft

Height after second bounce
$= \frac{1}{2}$ of 512 ft = 256 ft

At this point, the diagram is becoming more complicated and less clear. Although it is useful for understanding the problem and answering the first question, we do not need it to make our computations. From now on, we organize the computations in a table to make them more meaningful.

Bounce Number	Height After Bounce (ft)
Bounce 1	$\frac{1}{2}$(original height) $= \frac{1}{2}(1{,}024) = 512$
Bounce 2	$\frac{1}{2}$(bounce 1) $= \frac{1}{2}(512) = 256$
Bounce 3	$\frac{1}{2}$(bounce 2) $= \frac{1}{2}(256) = 128$
Bounce 4	$\frac{1}{2}$(bounce 3) $= \frac{1}{2}(128) = 64$
Bounce 5	$\frac{1}{2}$(bounce 4) $= \frac{1}{2}(64) = 32$
Bounce 6	$\frac{1}{2}$(bounce 5) $= \frac{1}{2}(32) = 16$

From the table, we can see that the ball rises to a height of 16 ft after the sixth bounce, answering the second part of our problem.

Analyzing the values in the table, we see a relationship between the number of the bounce and the number of times we multiply by $\frac{1}{2}$.

Bounce Number	Height After Bounce (ft)
Bounce 1	$\frac{1}{2}$(original height) $= \frac{1}{2}(1{,}024)$
Bounce 2	$\frac{1}{2}$(bounce 1) $= \frac{1}{2}\left[\frac{1}{2}(1{,}024)\right] = \left(\frac{1}{2}\right)^2(1{,}024)$
Bounce 3	$\frac{1}{2}$(bounce 2) $= \frac{1}{2}\left[\left(\frac{1}{2}\right)^2(1{,}024)\right] = \left(\frac{1}{2}\right)^3(1{,}024)$
Bounce 4	$\frac{1}{2}$(bounce 3) $= \frac{1}{2}\left[\left(\frac{1}{2}\right)^3(1{,}024)\right] = \left(\frac{1}{2}\right)^4(1{,}024)$
Bounce 5	$\frac{1}{2}$(bounce 4) $= \frac{1}{2}\left[\left(\frac{1}{2}\right)^4(1{,}024)\right] = \left(\frac{1}{2}\right)^5(1{,}024)$
Bounce 6	$\frac{1}{2}$(bounce 5) $= \frac{1}{2}\left[\left(\frac{1}{2}\right)^5(1{,}024)\right] = \left(\frac{1}{2}\right)^6(1{,}024)$
Bounce n	$\left(\frac{1}{2}\right)^n(1{,}024)$

The height of the ball after bounce n is given by the formula:

$$\text{Height after bounce } n = \left(\frac{1}{2}\right)^n(1{,}024)$$

This answers the third part of the problem.

Looking Back. We can use our formula to check our first two answers:

$$\text{Height after 2nd bounce} = \left(\frac{1}{2}\right)^2 (1{,}024) = 256 \text{ ft}$$

$$\text{Height after 6th bounce} = \left(\frac{1}{2}\right)^6 (1{,}024) = 16 \text{ ft}$$

In this problem, we used multiple representations to understand and solve the problem and to make our solution more efficient.

Check Your Understanding 1.5-B

1. Is it possible to make a 3×3 magic square using the numbers 2, 4, 6, 8, 10, 12, 14, 16, and 18? If so, what is the sum of each row, column, and diagonal?

2. Phone-on-the-go cellular phone service charges its customers a flat fee of $15 for the first 100 min and then 10¢ per minute for every minute over 100. My Phone cellular phone service charges a flat rate of $25 dollars for the first 100 min and then 5¢ per minute for every minute over 100. How many minutes do you have to use to make My Phone the more cost-effective service?

Talk About It Of the problem-solving strategies in Section 1.4, which strategies do you think complement one another and why?

Problem Solving

Activity 1.10
Here is another problem missing some information. As before, identify the goal of the problem and find the missing information that you need to solve it. Do an Internet search to get the missing information and solve the problem: To the nearest penny, how many pennies would have to be stacked on top of one another to equal the height of the Empire State Building?

SECTION 1.5 Learning Assessment

■ Exercises

1. Use a table or a list to determine how many arrangements can be made with the letters in the word "math." How might multiplication be used to solve the problem more efficiently?

2. A ferret is 15 in. long and has a head and tail that are the same length. If the tail were twice as long as the head, the head and tail together would be as long as the body. How long is each part of the ferret?
   a. Solve the problem by drawing a picture.
   b. Solve the problem by using an equation.

   c. Which strategy did you find easier to use and why?

3. The sum of two consecutive even numbers is 110.
   a. Use guess and check to find the two numbers.
   b. Write an equation to solve the same problem.
   c. What are the advantages and disadvantages of using these two strategies?

4. Is it possible to make a 3×3 magic square using the numbers 1, 2, 3, 4, 6, 7, 8, 9, and 10? If so, what is the sum of each row, column, and diagonal? If not, why not?

5. **a.** Which is a better way to be paid: (1) $1,000,000 for a month's worth of work or (2) 1¢ for the first day, 2¢ for the second, 4¢ for the third, 8¢ for the fourth, and so on but keep only the amount paid on the 31st day?

   b. One way that many people solve the problem is to multiply two by itself enough times to arrive at the answer. Use a pattern to find a more efficient way to solve the problem.

6. Ten insects that are either flies or dragonflies are on a lily pad. Flies have two wings, and dragonflies have four. There are a total of 32 wings on the lily pad.

   a. Use a table to determine how many of each kind of insect is on the lily pad.

   b. What other strategies can be used to solve this problem?

7. Each child in a family has at least three brothers and two sisters. What is the smallest number of children the family might have?

8. The U.S. Open Tennis Tournament has 128 entrants in each of the men's and women's singles competitions. If there are no draws and a loss means elimination from the tournament, how many matches must be played over the course of two weeks to determine the men's and women's champions?

9. Consider the following arrangements of toothpicks. How many toothpicks are needed to make the 15th arrangement?

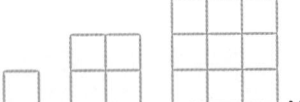

10. Consider the following arrangements of toothpicks. How many toothpicks are needed to make the square that has nine toothpicks per side?

11. Consider the following sequence of squares. How many dots will be on each side of the square that uses 48 dots?

12. Zach is on a 56-mile bike trip. He has completed three-fourths of the last half of the trip when his front tire goes flat. He does not have a repair kit, so he must walk the rest of the way. How far does he have to walk?

13. Juan is playing with a set of blocks and makes the following staircase. How many blocks will he need to make the staircase 8 steps high?

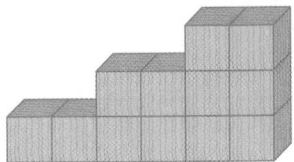

■ Problems and Applications

14. Lydia has the following writing utensils: a blue marker, a red crayon, a green colored pencil, a yellow crayon, a purple marker, a green crayon, a red colored pencil, and a purple colored pencil. If she wants to have one of each given color and utensil type, how many more must she get, and what are they?

15. Jade went grocery shopping and bought two of the items in the following table. She gave the cashier a $10 bill and received 84¢ in change. Which two items did she buy?

Item	Cost
Laundry detergent	$5.27
Case of soda	$2.49
2-lb package of hamburger	$6.39
Gallon of milk	$3.29
Bunch of bananas	$2.77

16. A certain man weighs 300 lb and is told by his doctor that he must lose at least 100 lb. He goes on a strict diet and loses 24 lb the first month, 21 lb the second, and 18 lb the third. If he continues this trend, how long will it take him to lose the needed weight?

17. David has three books from the library. One is three days overdue, one is eight days overdue, and the last is ten days overdue. The library charges 25¢ for books that are overdue by one day, 50¢ for books overdue by two days, 75¢ for books that are overdue three days, and so on. How much is David's fine when he returns the books?

18. A man and his wife are 18 km apart. The man can walk 4 km in an hour, and his wife can walk 3 km. How long will it take them to meet if they begin walking toward each other at the same time?

19. A pen at a local farm contains only sheep and chickens. If 35 animals are in the pen and have a total of 106 legs, how many of each type of animal is in the pen?

20. You have 100 boxes to deliver in a series of stops. At the first stop, you leave one box. At the second, you leave three more than the first. At each stop thereafter, you leave three more boxes than at the previous stop. At what stop will you not have enough boxes to make a delivery?

21. To log onto a computer network, a user must enter a password consisting of three characters. The first must be an a, b, or c, the second must be a number 1 through 5, and the third must be an x, y, or z. How many different passwords are possible?

22. It takes 140 ft^2 tiles to cover the floor of a rectangular room. If the distance around the room is 48 ft, what are the dimensions of the room?

23. Kiesha has several pieces of candy. If she places the candy into piles of 4, 5, or 6, she always has one left over. What is the smallest number of pieces Kiesha can have?

24. In an unfortunate incident, a squirrel fell down a 30-ft well. The squirrel eventually climbed out, but with some difficulty because the sides of the well were slippery. For every minute of effort, the squirrel climbed 4 ft but then had to rest for 1 min, during which time it slipped back 2 ft. How long did it take for the squirrel to climb out?

25. Natasha leaves at 9:00 a.m. to drive 120 mi to a friend's house. She planned on driving 60 mph the entire way but developed car trouble. After developing the trouble, she was only able to drive 15 mph. If she arrived at her friend's house at 1:00 p.m., at what time did she start having car problems?

26. Two train tracks lie adjacent to one another, connecting the cities of New York and Richmond, Virginia. Traveling time between the two cities in either direction is 12 hr. Trains leave from both cities every hour on the hour. If a train leaves New York at 11:00 a.m., how many trains will it pass on its way to Richmond?

27. The numbers 1 through 10 are written on chips and placed in a hat. The chips are then drawn in pairs and not replaced. If the sums of the five pairs are 6, 9, 11, 14, and 15, what are the five pairs?

28. Consider each of the numbers from 100 to 300. How many times will the sum of the first two digits equal the third?

29. A painter is standing on the middle rung of a ladder, painting a house. He notices that he has missed several spots, so he makes the following climbs to cover them: down three rungs, up five rungs, down seven rungs, and up two rungs. After he is finished, he climbs down five rungs to the ground. How many rungs does the ladder have?

30. Kim is assigned to set out the chairs for her high school graduation. The chairs are to be arranged so that there is a center aisle with an equal number of chairs on either side. Each row on a side is to have 12 chairs, and a total of 408 chairs are needed. If Kim has finished the eighth row of the first side, how many more chairs does she need to set out?

31. Mouse plagues often affect the grain-growing regions of southern Australia. In some cases, such as in 1993, there were over 100,000,000 mice in a local area. Suppose that a mouse population quadruples each month when conditions are favorable and becomes a plague when the population reaches 10,000,000 animals. If conditions have been favorable for 10 months and now cause a mouse plague of 10,000,000 animals, about how many mice were in the area at the beginning of the 10 months?

32. A painter has a job in which she paints the numbers on the parking spaces of a large parking lot. If she is paid $1 for each digit she paints and the job pays her $309, how many parking spots are in the lot?

33. Amy loves squash and zucchini, so she plans to plant a large garden containing only these vegetables. She plans to put each plant 4 ft apart in rows that are 4 ft apart. She also wants to leave at least 2 ft between any edge of the garden and the nearest plant. If her garden is 12 ft by 20 ft, how many plants should she put in?

34. Use the numbers 1 through 25 only once to complete the following 5 × 5 magic square.

35. How many different ways are there to get from the corner of 1st Avenue and Elm Street (*A*) to the corner of 5th Avenue and Beech Street (*B*), if a person can only travel east or south?

36. Three sisters—Erica, Eve, and Elizabeth—go on a three-day road trip. Erica paid for all the lodging and spent $525.34. Eve bought all the food and spent $247.89. Elizabeth paid for all the gas and spent $186.45. If the three sisters want to split the cost of the trip evenly, who should give what amount of money to whom?

37. For each sequence of dots, find:

   a. The number of dots needed to make each of the next three figures.

   b. The number of dots needed to make the 15th figure.

   c. A formula that relates the number of dots to the position of the figure in the sequence.

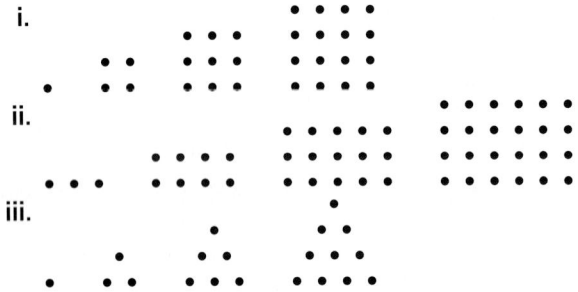

38. An 8-qt container is full of water. Describe how to split the water into two equal portions using only the 8-qt container, a 5-qt container, and a 3-qt container.

39. A cube of wood is 3 in. on each edge. It is to be cut into 1-in. cubes. After each cut, the pieces can be moved or stacked before making the next cut. What is the fewest number of cuts required?

40. Try to use a calculator to find the ones digits of 2^{50} and 2^{100}. Are you able to do it? If not, explain why, and then develop a strategy for solving the problem.

■ Mathematical Communication

41. a. How many rectangles are in the figure?

   b. Describe your method of solving the problem. Did you use one or more of the strategies in this section, or did you come up with your own?

   c. Write a general description of your method so that it can be applied to other problems of this variety.

42. Choose a problem in this Learning Assessment that you solved using two different methods. Write up your solutions using Polya's four-step method.

43. Pick one of the problem-solving strategies covered in Section 1.4. Write a problem that is appropriate for the elementary grades and that you think can be solved using this strategy. If you were to give this problem to a group of students, what other strategies might they use to solve the problem?

■ Building Connections

44. In light of what you have learned in this section, why do you think it is important to introduce elementary students to a variety of problem-solving strategies?

45. Look at a set of elementary mathematics curriculum materials for instances of using multiple strategies in problem-solving situations. Are you surprised with what you found? If so, why?

Historical Note

George Polya is noted for many mathematical discoveries but is best known for his work in problem solving. His work in this area began when, as a personal tutor, he noticed that his pupil did not have the skills to be an effective problem solver. To help his student, Polya spent hours developing a problem-solving method based on the belief that problem solving is a skill that can be taught. Throughout his career, he maintained an interest in mathematics education and used his method in many of his classes. After retiring, he wrote several books pertaining to mathematics education, including *How to Solve It* and *Mathematical Discovery*, Volumes 1 and 2. In his later books, he set forth the following ten commandments for mathematics teachers.

George Polya (1887–1985)

1. Be interested in your subject.
2. Know your subject.
3. Know about the ways of learning: The best way to learn anything is to discover it by yourself.
4. Try to read the faces of your students, try to see their expectations and difficulties, put yourself in their place.
5. Give them not only information, but "know-how," attitudes of mind, the habit of methodical work.
6. Let them learn guessing.
7. Let them learn proving.
8. Look for such features of the problem at hand as may be useful in solving the problems to come—try to disclose the general pattern that lies behind the present concrete situation.
9. Do not give away your whole secret at once—let the students guess before you tell it. Let them find out by themselves as much as is feasible.
10. Suggest it, do not force it down their throats.

Opening Problem Answer

© Istockphoto.com/Zmeel Photography

Yes. We can approach the problem by adding the numbers and dividing by 3: $\dfrac{1 + 2 + \cdots + 12}{3} = \dfrac{78}{3} = 26$. Next, we look for ways to make sums of 26. This gives the following solution.

Chapter 1 REVIEW

Summary of Key Concepts from Chapter 1

Section 1.1   Communication Through Mathematical Representation

- **Mathematical representations** can be verbal, numerical, symbolic, visual, and tactile.
- **Identification numbers** label or identify objects, **ordinal numbers** assign an order to a group of objects, and **cardinal numbers** express quantities.
- **Algebraic expressions** use **variables,** which are symbols that take on different values. Variables are used to represent unknown quantities or as a general representative from a set of numbers.

Section 1.2   Inductive Reasoning and Patterns

- **Inductive reasoning** is the process of drawing a general conclusion from observations and patterns seen in a set of specific examples.
- A **pattern** is anything with a perceptual structure, which is identified by analyzing examples for similarities and differences.

- Different kinds of patterns are **repetitive patterns, patterns of common traits,** and **growing patterns.**
- A **generalization** is a statement indicating that a pattern will likely hold true for all other similar situations.
- A **counterexample** is a specific example showing that a generalization is false.

Section 1.3   Deductive Reasoning and Proof

- An **argument** is a sequence of statements that explains how to arrive at a **conclusion** from a set of facts, called the **premises** or **hypotheses.** A good argument must have true premises and good logic.
- **Deductive reasoning** is the process of reasoning from a set of known facts or general principles to a specific conclusion.
- **Statements** are sentences that are true or false. A **negation** takes on the opposite truth value of a statement.
- **Conjunctions** $(p \wedge q)$ are true only if both simple statements are true. **Disjunctions** $(p \vee q)$ are false only if both simple statements are false.
- A **conditional** statement $(p \rightarrow q)$ is an "if, then" statement that represents a cause-and-effect relationship. It is false only when the hypothesis is true and the conclusion is false.
- Three other forms of the conditional are: the **converse** $(q \rightarrow p)$, the **inverse** $(\sim p \rightarrow \sim q)$, and the **contrapositive** $(\sim q \rightarrow \sim p)$. The conditional and the contrapositive are logically equivalent.
- **Affirming the hypothesis** argues directly from a single conditional. A **hypothetical syllogism** argues directly from several conditionals.
- **Denying the conclusion** argues indirectly from the contrapositive. The **process of elimination** argues indirectly from a disjunction.

Section 1.4   Mathematical Problem Solving

- Polya's problem-solving method involves four steps: understanding the problem, devising a plan, carrying out the plan, and looking back.
- Eight strategies of problem solving are drawing a picture, guessing and checking, making a list or table, working backward, looking for a pattern, using deductive reasoning, using a formula, and writing an equation.

Section 1.5   Problem Solving with Different Strategies

- Many problems can be solved with different strategies. Other problems must be solved using several strategies.

Review Exercises Chapter 1

1. Translate each statement into a symbolic representation.
   a. The difference of 8 from x is 5.
   b. The quotient of the quantity 3 plus x and 7 does not equal 8.
   c. Three-fourths of y is 150.

2. Write each statement verbally.
   a. $2x = 8$.
   b. $3(x - 2) < 2$.
   c. $2(x - 1) = x + 5$.

3. What mathematical truths about isosceles trapezoids might be evident by just looking at the following picture?

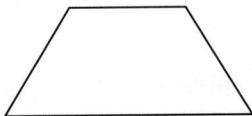

4. The following numbers represent the test scores on a 50-point test. What can be said about the

performance of the class based on a simple observation of their scores?

23 25 27 27 39 42 43 47 48 48

5. Write a set of directions that describes to someone how to subtract 43 from 92.

6. Without using the word "circle" or "round," precisely describe the following figure.

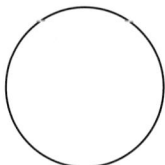

7. Tell someone how to draw the following figure without watching what the person is doing. Once the figure is complete, compare the drawing to the original figure. If the drawing was correct, analyze your communication as to why it was effective. If the drawing was incorrect, analyze your communication to make it better.

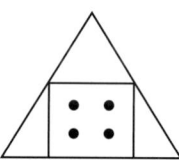

8. The following graph gives the top track speeds on a raceway for the years 1985–2005. What can be said about trends in the top speeds?

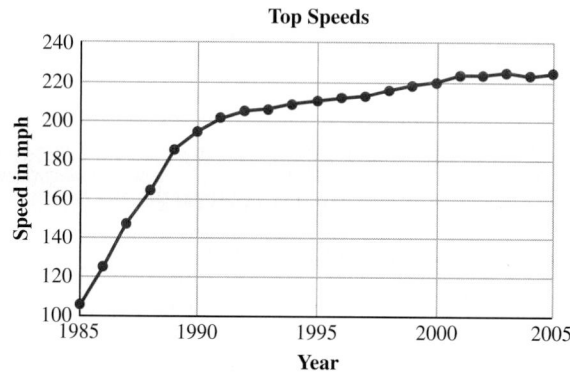

9. I make the statement, "It has rained the entire time on each of our last ten vacations." What can you conclude about our next vacation? What type of reasoning have you used?

10. If $x = 4$, $y = 3x$, and $z = 4y$, what can you conclude about z? What type of reasoning did you use?

11. Two Saturdays ago, I had a pepperoni pizza. Last Saturday, I had a sausage pizza. Tonight, what am I likely to order? What type of reasoning did you use?

12. Find a counterexample for each generalization.
   a. All numbers between 1 and 5 are even.
   b. Every month of the year has an "a" in its spelling.
   c. All numbers between 5 and 10 are divisible by 5.

13. Consider the following list of equations. Use inductive reasoning to predict the next three equations. Will the pattern always hold? If not, give the first equation for which the pattern changes.

$$37,037 \times 3 = 111,111$$
$$37,037 \times 6 = 222,222$$
$$37,037 \times 9 = 333,333$$
$$37,037 \times 12 = 444,444$$

14. Write the negation of each statement.
   a. Carissa likes to wear blue.
   b. I do not own a green truck.

15. Let d represent "I like dogs" and a represent "I like animals." Represent each statement symbolically.
   a. I do not like animals.
   b. I like animals, and I do not like dogs.
   c. If you do not like animals, then you do not like dogs.
   d. You like animals if and only if you like dogs.

16. Let c represent "I have class at 1 p.m." and t represent "I have a test." Translate the following symbolic statements to verbal ones.
   a. $\sim c$ b. $c \wedge t$
   c. $\sim c \vee t$ d. $t \rightarrow c$
   e. $\sim(c \rightarrow t)$

17. Write the biconditional, "$15 \div x = 5$ if and only if $x = 3$," as two "if, then" statements.

18. Write the converse, inverse, and contrapositive of the statement, "If you go to the beach, then you wear sunscreen."

19. If $x = 5$ and $y = 8$, determine whether each statement is true or false.
   a. $x > y$
   b. $x + y = 13$
   c. $y \cdot x = 40$
   d. $y - x > x$

20. If p is false and q is true, find the truth value of each statement.

   a. $\sim q$ b. $p \wedge q$ c. $p \vee q$

   d. $p \rightarrow q$ e. $\sim(p \wedge q)$

21. Make a truth table that shows all the truth values for the statement $\sim p \vee \sim q$.

For Exercises 22–24, fill in the missing statement to make a direct argument.

22. If you are a vegetarian, then you eat broccoli.

   You do not eat broccoli.

   Therefore, _____.

23. If it is ice cream, then it tastes good.

   _____.

   Therefore, if it is ice cream, then it is high in fat.

24. _____.

   It is a square.

   Therefore, it is a rectangle.

25. Use the method of exhaustion to show that $x^2 \le x$, whenever x is 0.25, 0.5, 0.75, or 1.

26. How many different ways are there to write 20 as the sum of 2s, 5s, and 10s?

27. If a boy can eat a hot dog and a half in a minute and a half, how many hot dogs can 6 boys eat in 6 min?

28. A man has $1.19 in coins, but he cannot make change for a dollar, a half dollar, a quarter, a dime, or a nickel. What coins does he have?

29. Jim and Joe each have a certain number of marbles. If Jim gives Joe 5 marbles, Jim and Joe will have exactly the same number of marbles. If Joe gives Jim 5 marbles, then Jim will have twice as many marbles as Joe. How many marbles does each boy have?

30. Consider the following stairway made of cubes. If the stairs have to be 15 steps high, how many cubes are needed to make the entire staircase?

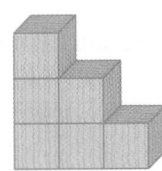

31. How many numbers from 1 to 200 are divisible by 5?

32. The hens in a hen house lay 1,440 eggs a day. How many dozens of eggs are laid in one week?

33. You have 210 boxes to deliver at a series of stops. At the first stop, you leave 1 box. At the second, you leave 2. At the third, you leave 4, and at the fourth stop you leave 8. If you continue to leave packages in this way, at what stop will you not have enough boxes to make a delivery?

34. When David opened his book, he found that two consecutive pages were missing. If the sum of the two page numbers he has opened to is 581, what are the missing page numbers?

35. How many cubes are needed to build this figure?

Answers to Chapter 1   Check Your Understandings and Activities

Check Your Understanding 1.1-A

1. Answers will vary.

2. a. Third is an ordinal number.

   b. 38-45611 is an identification number, and $200,000 is a cardinal number.

   c. Three is a cardinal number, and first is an ordinal number.

3. Possible answers are:

a.

First	Second
1	2
2	4
3	6

b.

First	Second
3	2
4	4
5	6

Check Your Understanding 1.1-B

1. a. $y = 17$.   b. $m = 300$.   c. $A = 225$.

2. a. $y - 12 \ge 7$.   b. $4(x + 9) = 21$.   c. $x \div 5 = 30$.

3. a. Asia.   b. Australia.

   c. Asia and South America.

   d. Asia, Africa, North America, South America, Antarctica, Europe, Australia.

Check Your Understanding 1.2-A

1. a. ABC pattern.   b. AB pattern.

   c. ABB pattern.   d. ABC pattern.

2. She is likely to request only pepperoni on the pizza.

3. Any number multiplied by 2 is an even number.

Check Your Understanding 1.2-B

1. 56 dots.   2. 165.

3. Possible answers are:

   a. Fred is a boy's name that does not begin with B.

   b. A square is a shape that has four sides, not three.

   c. $\frac{1}{2} \times \frac{1}{2} = \frac{1}{4}$, which is smaller than either number.

Check Your Understanding 1.3-A

1. a. $s \wedge \sim c$; False.

   b. $\sim s \vee \sim c$; False.

   c. $\sim (s \vee c)$; False.

   d. $\sim c \rightarrow s$; True.

2. a. *Converse:* If you got my message, then you read your e-mail.

    Inverse: If you did not read your e-mail, then you did not get my message.

    Contrapositive: If you did not get my message, then you did not read your e-mail.

   b. *Converse:* If $2x = 10$, then $x = 5$.

    Inverse: If $x \neq 5$, then $2x \neq 10$.

    Contrapositive: If $2x \neq 10$, then $x \neq 5$.

3. If you are a dog, then you have fur, and if you have fur, then you are a dog.

Check Your Understanding 1.3-B

1. a. You do not turn on a light.   b. It has four sides.

   c. If you like children, then you teach elementary school.

2. $y = 75$.

3. a. Let w represent "You watch 4 hours of television" and l represent "You are lazy." Symbolically, we represent the argument as:

$$w \rightarrow l$$
$$l$$
$$\therefore w$$

   This argument matches the converse error, so we conclude that the logic is incorrect.

   b. Let l represent "Living in the dorm" and f represent "You are a freshman." The argument can then be represented symbolically as:

$$l \rightarrow f$$
$$l$$
$$\therefore f$$

   This argument matches modus ponens, so we conclude that it is logically correct.

Check Your Understanding 1.4-A

1. 400 yd of fencing.

2. 11 other numbers have the same digits as 2,562.

3. 45.

Check Your Understanding 1.4-B

1. It takes 7 crossings.

2. 96 in.^{2}.

3. 6 and 18.

Check Your Understanding 1.5-A

1. Sarah has 8, Sally has 12, and Sue has 14.

2. 56 games.

Check Your Understanding 1.5-B

1. Yes. The sum of each row, column, and diagonal will be 30. One solution is as follows:

4	18	8
14	10	6
12	2	16

2. The user must use 300 min of the My Phone service to make it the better service.

Activity 1.1

Answers will vary.

Activity 1.2

Activity 1.3

a. Answers will vary.

b. Answers will vary.

Activity 1.4

The third number in a column is the sum of the first two, and the fourth number in the column is the product of the second and third. The completed table is as follows:

1	3	6	3	2	1	2
2	2	4	4	5	7	7
3	5	10	7	7	8	9
6	10	40	28	35	56	63

Activity 1.5

a. i. All humans are mammals.

   ii. All elementary school teachers teach mathematics.

   iii. All triangles have three sides.

b. i. If it is a dog, then it has fur.

   ii. If it is a square, then it is a rectangle.

   iii. If a number is divisible by 2, then it is even.

c. Answers will vary.

Activity 1.6

8	2	1	7	4	9	6	5	3
3	4	6	2	5	1	7	8	9
9	7	5	8	6	3	4	1	2
4	8	3	9	1	5	2	7	6
2	6	7	3	8	4	5	9	1
1	5	9	6	2	7	8	3	4
6	1	8	5	9	2	3	4	7
7	9	2	4	3	8	1	6	5
5	3	4	1	7	6	9	2	8

Activity 1.7

Take one person's string and pull it through the loop around one of the other person's wrists by coming from inside and beneath up toward the hand. This creates a loop. Move the loop up over top the hand, and the two people should be free.

Activity 1.8

List the numbers from 1 to 10. Next, observe that adding numbers from opposite ends always results in a sum of 11. Because there are five pairs of numbers, the sum is equal to the product of the number of pairs and 11, or $1 + 2 + 3 + \cdots + 10 = 5 \cdot 11 = 55$. Applying this method to the larger problem, observe that adding numbers from opposite ends of the list is always 101. Because there are 100 numbers, there will be $100 \div 2 = 50$ pairs. Multiplying the number of pairs by 101, the sum of the numbers from 1 to 100 is $1 + 2 + \ldots + 100 = 50 \cdot 101 = 5,050$.

Activity 1.9

The following information is readily available on the Internet. A large company of soldiers has about 180 members, each one of which needs about ten 8-oz glasses of water a day, and there are 128 oz in a gallon. Hence, the company needs about $180 \cdot 8 \cdot 10 = 14,400$ oz of water a day. This is equivalent to $14,400 \div 128 = 112.5$ gal of water. Multiplying by 7, $112.5 \cdot 7 = 787.5$ gal are needed for the week.

Activity 1.10

The following information is readily available on the Internet. One penny is 1.55 mm thick, 1,000 mm = 1 m, and the Empire State Building is 381 m tall. Consequently, the Empire State Building is $381 \times 1,000 = 381,000$ mm tall. This height is equivalent to $381,000 \div 1.55 = 245,806$ pennies.

Chapter 2

Sets

Opening Problem   Madison has 15 songs and wants to create a new playlist on her MP3 player. If she can use any number of songs and the order of the songs does not matter, how many different lists can she create?

2.1   Set Basics

2.2   Set Operations

Sets in the Classroom

© Image Source/Corbis

Take a moment and think about the first mathematical ideas that children are likely to learn. As you reflect on your own experiences or on those of the children around you, ideas like counting, writing numbers, and adding or subtracting probably come to mind. Even though these are among the first mathematical ideas that children learn, they all use something even more basic: sets and set operations.

The importance of sets in learning elementary mathematics is hard to over-emphasize. For instance, consider counting. **Counting** is the method we use to identify and name quantities that are not easy to perceive in other ways. To count correctly, children must first master the idea of making a one-to-one correspondence, which they learn intuitively as they compare sets. Only when they have mastered this idea and are counting correctly can they begin to develop the idea of number. By the time they enter kindergarten, most children can count to 20 and understand numerical values up to 10. As they progress through school, children continue to use sets to count higher and to understand the values of larger numbers.

Children use sets and set operations in other ways. For instance, they use set operations in intuitive ways to understand and compute addition, subtraction, and multiplication with the whole numbers. Children also use set operations to sort. **Sorting**, or **categorizing**, is the process of placing objects or events into sets based on their similarities and differences. Sorting skills are important to children both cognitively and mathematically. Cognitively, children use sorting to develop language, organize memories, and relate new experiences to old. Mathematically, children use sorting to recognize patterns, identify computational properties, classify shapes, and sift through data. Table 2.1 lists specific learning objectives that are associated with counting and sorting and the grade level at which each objective is likely to be met.

Table 2.1 Classroom learning objectives	K	1	2	3	4	5	6	7	8
Classify objects into categories and count the objects in each category.	X								
Count to answer "how many?" questions with values up to 20.	X								
Understand the relationship between numbers and quantities by connecting counting to cardinality.	X								
Count to 100 by 1s and by 10s.	X								
Count on from a given number in the counting sequence.	X								
Count within 1,000 by 1s, 5s, 10s, and 100s.			X						

Source: Adapted from the *Common Core State Standards for Mathematics* (2010).

Because sets provide a basic foundation for many important mathematical ideas, they are a natural place to begin our study of elementary mathematics. Specifically, in Chapter 2, we:

- Describe basic set ideas and their relationship to counting and numbers.
- Present set operations as the fundamental ideas on the basis of which numerical operations and sorting are developed.

SECTION 2.1 Set Basics

Notions from basic set theory are the foundation for many ideas in elementary mathematics, including number and numerical operations. If you were to look up the word "set" in the *Oxford English Dictionary*, you would find over 450 different definitions. In mathematics, we consider only one.

Definition of a Set

A **set** is a well-defined collection of objects. Objects in the set are called **elements.**

The phrase "well-defined" separates sets from other collections. If we take the word "well" to mean good and the word "defined" to mean description, then a set is a collection that has a "good description." In other words, a set is well-defined if we know exactly which objects are or are not in the set. So we must avoid describing sets with relative words that can take on different meanings depending on the situation.

Example 2.1 Identifying Well-Defined Sets

Communication

Determine whether each collection is well-defined. If not, rewrite the description to make it well-defined.

a. The collection of all United States presidents

b. The collection of all good basketball players

Solution

a. This is a well-defined set because the description makes it possible to decide who is or is not in the set. For instance, George Washington is in the set, but Benjamin Franklin is not.

b. The collection of good basketball players is not well-defined because the word "good" is relative. We can make it a well-defined set by changing the description to include only currently playing professional basketball players.

Like other mathematical concepts, sets have their own notation and terminology. In general, we name sets with italicized capital letters and, when possible, list specific elements between braces: { }. For example, $A = \{1, 2, 3\}$ means that set A contains the elements 1, 2, and 3. To describe specific objects in the set, we can use the symbol $\in$ to replace the words "is an element of" and $\notin$ to replace the words "is not an element of." For instance, we can write $1 \in A$ and $4 \notin A$.

Explorations Manual 2.1

Two basic rules govern sets:

1. Elements in a set are generally not listed more than once.
2. The order of the elements in a set is not important.

Explorations Manual 2.2

These two rules imply that sets are concerned only with membership: what is or is not in the set. Once an object is in a set, there is no need to write it more than once or be concerned about the order of the elements. As a practical matter, we generally list elements in the order that is commonly associated with them, such as numerical or alphabetical order.

Braces—{ }—are generally reserved for sets, and they signify that order *is not* important. An error commonly made by students is to use parentheses—( )—instead of braces. This is a mistake in notation because parentheses are used to signify that order *is* important.

Representing Sets

To this point, we have already seen two ways to represent sets: verbal descriptions and rosters. **Verbal descriptions** use words to describe sets. Words can fully describe a set, but the length of the description often makes words awkward to use. A **roster** is simply a list of the elements in the set. When possible, we list all of them. However, if a set is very large, we list enough elements to establish a pattern for identifying other elements in the set. We then replace the unlisted elements with an **ellipsis** (. . .). For instance, in the set $A = \{1, 2, 3, 4, \ldots\}$, the ellipsis shows that A includes every counting number 5 or greater. In the set $B = \{1, 2, 3, \ldots, 99, 100\}$, the ellipsis replaces the numbers 4 through 98. Rosters offer the advantage of being concise and efficient, but they are useful only when listing the elements is possible.

A third way to represent sets is to use **set-builder notation:**

$$\{x \,|\, x \text{ satisfies some condition}\}$$

The vertical bar (|) replaces the words "such that." So we read set-builder notation as "The set of all x such that x satisfies some condition."

Example 2.2   Representing Sets

Representation   Rewrite each set as a roster and in set-builder notation.
   a. The set of primary colors
   b. The set of all lowercase English alphabet letters
   c. The set containing the counting numbers 1 through 5

Solution
   a. {red, blue, yellow} = $\{c \,|\, c \text{ is a primary color}\}$.
   b. $\{a, b, c, \ldots x, y, z\} = \{x \,|\, x \text{ is a lowercase English alphabet letter}\}$.
   c. $\{1, 2, 3, 4, 5\} = \{n \,|\, n = 1, 2, 3, 4, \text{ or } 5\}$.

The use of set-builder notation over a verbal description or a roster may seem strange. However, it can be useful when a verbal description is too lengthy or when a pattern cannot be established for a roster.

Before we discuss a fourth way to represent sets, we need to introduce two important sets. The first is the **empty set.**

Definition of the Empty Set

The **empty set**, or **null set**, written { } or $\varnothing$, is the set that contains no elements.

The empty set is useful because many sets are equivalent to it. For instance, the set of human beings who have traveled to Mars and the set of counting numbers less than zero are both equivalent to the empty set.

We must be careful when using the two symbols for the empty set: { } and $\varnothing$. First, we must not use the symbol $\varnothing$ for zero, because they are two different concepts. One is a set, and the other is a number. Second, we must not represent the empty set as $\{\varnothing\}$, because this set is not empty. It has exactly one element in it: the symbol for the empty set.

If we allow a set with no elements, then it also makes sense to allow a set with every element.

Definition of the Universal Set

The **universal set**, or **universe**, written U, is the set of all possible elements in a given discussion.

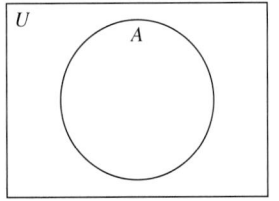

Figure 2.1 Basic Venn diagram

In most situations, the universal set is either given or understood from the context. For example, the implied universe of $A = \{a, b, c, d\}$ is the set of lowercase letters, and the implied universe of $B = \{1, 2, 3, 4, 5\}$ is the set of counting numbers.

Venn diagrams are a fourth way to represent sets that use the notion of universal sets. The standard convention is to represent the universe with a rectangle and sets within the universe as circles. In Figure 2.1, set A (the circle) is in universe U (the rectangle). As we will see, Venn diagrams are a powerful way to represent sets and the relationships among them.

Subsets

Sets can relate to one another in a number of ways. The Venn diagram in Figure 2.1 shows one such relationship: containment. Specifically, any element of A must also be in U, and we say A is a **subset** of U.

Definition of a Subset and a Proper Subset

Set A is a **subset** of set B, written $A \subseteq B$, if and only if, every element of A is also an element of B. If B has an element that is not in A, then A is a **proper subset** of B, written $A \subset B$.

The subtle difference between a subset and a proper subset is analogous to the difference between less than or equal to ($\leq$) and less than ($<$). As less than or equal to allows two numbers to be the same, the subset relationship allows two sets to be the same. The proper subset relationship, like less than, does not.

If one set is not a subset of another, then we use the notation $\nsubseteq$ to say "is not a subset."

Example 2.3 | Identifying Subsets

Consider the sets $A = \{1, 2\}$, $B = \{1, 2, 3\}$, $C = \{2, 3, 1\}$, and $D = \{2, 3, 4\}$. Place a symbol $\subseteq$, $\subset$, or $\nsubseteq$ in the blank to make each statement true.

a. A _____ B **b.** B _____ C **c.** B _____ D

Solution

a. Every element of A is also an element of B, so we can correctly write $A \subseteq B$. However, 3 is an element in B that is not in A, so we can also correctly write $A \subset B$.

b. Every element of B is also in C, and C has no element that is not in B. So $B \subseteq C$.

c. Because 1 is in B but not in D, $B \nsubseteq D$.

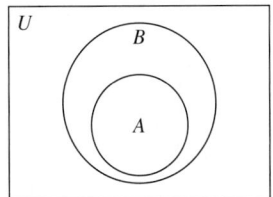

Figure 2.2 Venn diagram of $A \subseteq B$

Example 2.3(a) indicates that if one set is a proper subset of another, then it must also be a subset. Additionally, part b indicates that it is possible for one set to be a subset of another, but not a proper subset.

A Venn diagram can easily show the subset relationship. If $A \subseteq B$, then we place the circle for A inside the circle for B, as shown in Figure 2.2. Specific elements in each set can also be placed in the diagram.

Example 2.4   Drawing a Venn Diagram

Representation

Draw a Venn diagram of the sets $A = \{red\}$, $B = \{c \mid c$ is a primary color$\}$, and $U = \{r \mid r$ is a color of the rainbow$\}$.

Solution Begin with the Venn diagram in Figure 2.2. Red is an element of A, so it must be placed inside the circle for A. This automatically places it inside the circle for B and the rectangle for U. Next, $B = \{c \mid c$ is a primary color$\} = \{red, blue, yellow\}$. The colors blue and yellow are in B but not in A, so they go inside the circle for B but outside the circle for A. Finally, $U = \{r \mid r$ is a color of the rainbow$\} = \{red, orange, yellow, green, blue, indigo, violet\}$. The colors orange, green, indigo, and violet are in U, but are not in A or B. They are placed inside the rectangle for U but outside the circle for B:

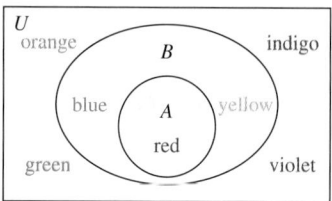

Counting subsets can be another useful problem-solving strategy, as demonstrated in the next example.

Example 2.5   Different Ways to Top a Sundae

Problem Solving

You and a friend visit a local ice cream shop where they have three different sundae toppings; fudge sauce, whipped cream, and cherries. How many different ways can you top a sundae with these toppings?

Solution

Understanding the Problem. Our goal is find out how many ways we can top a sundae using three toppings. It is not necessary to use every topping in every instance.

Devising a Plan. Finding the different ways to top a sundae is equivalent to finding all the subsets of the set of three toppings. Once we have found them, we simply *count the subsets*.

Carrying Out the Plan. Represent the set of toppings as $T = \{f, w, c\}$, where f represents fudge sauce, w represents whipped cream, and c represents cherries. First, a sundae can have no toppings at all. This option corresponds to the empty set: $\emptyset$. Or, a sundae can have just one topping: fudge sauce $\{f\}$, whipped cream $\{w\}$, and cherries $\{c\}$. A sundae can also have exactly two toppings: $\{f, w\}$, $\{f, c\}$, and $\{w, c\}$. Finally, a sundae can have all three toppings: $T = \{f, w, c\}$. After counting the subsets: $\emptyset$, $\{f\}$, $\{w\}$, $\{c\}$, $\{f, w\}$, $\{f, c\}$, $\{w, c\}$, and $\{f, w, c\}$, we find there are eight ways to top the sundae.

Looking Back. The number of subsets demonstrates a pattern. If only one topping is available, say fudge sauce, then there are only two ways to top the sundae: $\emptyset$ and $\{f\}$. With two toppings, say fudge sauce and whipped cream, a sundae can be topped in four ways: $\emptyset$, $\{f\}$, $\{w\}$, and $\{f, w\}$. If we summarize the results for one, two, and three toppings in a table, we have:

Number of toppings	1	2	3
Number of subsets	$2 = 2^1$	$4 = 2^2$	$8 = 2^3$

The numbers of subsets are all powers of 2. If we generalize the pattern, we have that a set of n elements will have 2^n subsets.

SECTION 2.1   Set Basics   **81**

Example 2.5 demonstrates three facts about subsets, which are summarized in the following theorem.

Theorem 2.1   Properties of Subsets

1. The empty set is a subset of every set.
2. Every set is a subset of itself.
3. If set A has n elements, then A has 2^n subsets.

Example 2.6   Proving the Empty Set Is a Subset of Every Set

Reasoning

Explain why the empty set is a subset of every set.

Solution It may seem strange that the empty set is a subset of every set, especially because it has no elements. However, consider the following indirect argument. Suppose $\varnothing \not\subseteq A$. By the definition of a subset, $\varnothing$ must contain an element that is not in A. However, because the empty set contains no elements, there cannot be an element in $\varnothing$ that is not in A. Consequently, $\varnothing \not\subseteq A$ is false, and we conclude that $\varnothing \subseteq A$.

Check Your Understanding 2.1-A

1. Which of the following collections are well-defined? For those that are not, rewrite them to make them well-defined.
   a. The collection of all teenagers weighing 125 lb
   b. The collection of smart students
   c. The collection of all counting numbers

2. Use a roster and set-builder notation to represent each set.
   a. The set of all states in the United States starting with the letter C
   b. The set of all odd numbers less than 10
   c. The set of all multiples of 5

3. Draw a Venn diagram that shows the relationship between $A = \{s, o\}$, $B = \{s, o, m, e\}$, and $U = \{m, o, u, s, e\}$

4. List all the subsets of $B = \{$black, white$\}$. How many of them are proper subsets?

Talk About It What is the difference between $\{1, 2\}$ and $(1, 2)$?

Communication

Activity 2.1
You may have noticed that a set is defined with the word "collection," a synonym for "set." Consequently, we can replace "set" with any number of other words that mean "collection" or "assortment" and still have the same intuitive idea. How many other words can you and a partner think of that are synonymous with "set"?

Equal and Equivalent Sets

Explorations
Manual
2.3

A second set relationship is equality, which indicates that two sets contain exactly the same elements.

Definition of Equal Sets

Two sets are **equal,** written $A = B$, if and only if they contain exactly the same elements.

As an example, consider the sets $A = \{a, b, c\}$ and $B = \{c, a, b\}$. Clearly, they have the same elements, so we write $A = B$. Not only are A and B equal, but each is a subset of the other. So we can think of equal sets in another way: Two sets are equal if and only if they are subsets of each other.

In everyday language, we often use the word "equivalent" to mean equal. However, in set theory, the words have different meanings. To understand equivalent sets, we must understand a **one-to-one correspondence.**

Definition of Equivalent Sets

A **one-to-one correspondence** between two sets A and B is a pairing of the elements of A with the elements of B so that each element of A corresponds to exactly one element of B, and vice versa. Two sets are **equivalent,** written $A \sim B$, if and only if there is a one-to-one correspondence between them.

Example 2.7 | **Identifying a One-to-One Correspondence**

If $A = \{a, b, c\}$, $B = \{1, 2, 3\}$, and $C = \{w, x, y, z\}$, which of the following diagrams demonstrates a one-to-one correspondence?

a. 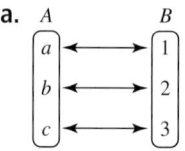 **b.** 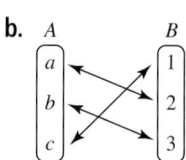 **c.**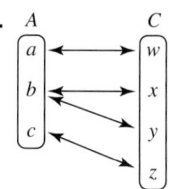

Solution We check for a one-to-one correspondence by examining the arrows between the sets.

a. Because each element of the first set is paired with exactly one element from the second set, and vice versa, this is a one-to-one correspondence. Consequently, $A \sim B$.

b. This diagram shows a different one-to-one correspondence between A and B, again showing $A \sim B$.

c. The diagram does not show a one-to-one correspondence because the letter b is paired with two letters, x and y. There is no way to make a one-to-one correspondence between A and C; either one element from A will be used twice, or one element from C will be unused. As a result, A is not equivalent to C.

Example 2.7 shows the key difference between equal and equivalent sets. Equal sets have exactly the same elements, whereas equivalent sets have the same *number* of elements. As a consequence, all equal sets are equivalent, but not all equivalent sets are equal.

| Example 2.8 | **Counting One-to-One Correspondences** |

Connections

Mrs. Samples wants to assign three groups of students, *A*, *B*, and *C*, to three different centers, Center 1, Center 2, and Center 3. How many different ways can students be assigned to the centers?

Solution When we assign students to centers, we are making a one-to-one correspondence between two sets. If we can find all such one-to-one correspondences, then we will know how many ways the students can be assigned. The possible one-to-one correspondences among the sets are:

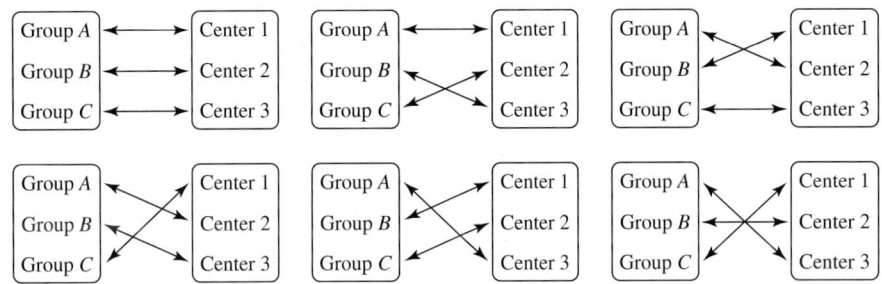

We count six one-to-one correspondences, so the groups of students can be assigned to the centers in six ways.

Another way to approach Example 2.8 is to use the fundamental counting principle. Because we can assign a group to Center 1 in three ways, a group to Center 2 in two ways, and a group to Center 3 in one way, there are:

$$\frac{3}{\text{Center 1}} \times \frac{2}{\text{Center 2}} \times \frac{1}{\text{Center 3}} = 6 \text{ ways to assign students}$$

This not only confirms our original answer, but also gives us a quick way to count one-to-one correspondences among sets with any number of elements.

A one-to-one correspondence is actually a special kind of **function**, which is a rule that tells us how to match the elements in one set with the elements in another. Functions are an important concept in mathematics and a key part of algebraic thinking. We discuss them in detail in Chapter 9.

Counting and the Cardinal Number of a Set

Counting

Explorations Manual 2.4

The idea of a one-to-one correspondence is essential to counting, so it plays a key role in how children develop their concept of number. To count correctly, children must master four skills.

1. *Children must learn the number word sequence.* When first learning to talk, children are often introduced to the word sequence for the counting numbers: "one, two, three," and so on. Through imitation, children not only learn the first few number names, but also their fixed order.

2. *Children must learn to make a one-to-one correspondence between the set of number words and a set of objects to be counted.* Once children know the first few number words, they can begin to count. To do so, they must learn to make a one-to-one correspondence between the set of number words and the set of things to be counted. Children can establish a one-to-one correspondence by moving, touching, or pointing to the objects as they say the number words. For example, to count five blocks, a child must make a one-to-one correspondence between the set of words {"one," "two," "three," "four," "five"} and the set of blocks (Figure 2.3).

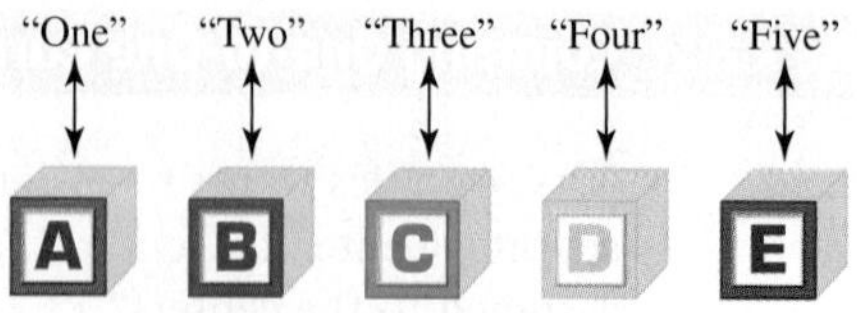

"One" "Two" "Three" "Four" "Five"

Figure 2.3 Counting five blocks

3. *Children must learn to keep track of the objects they have counted.* Making a one-to-one correspondence is not an automatic process. Children can undercount by not including every object or overcount by including some objects more than once.
4. *Children must learn that the last word used in the number sequence is the number of objects in the set.* To bridge the gap between counting and quantity, children must realize that the last word used not only is associated with the last object but also represents the total number of objects in the set.

We call counting with a one-to-one correspondence **rote counting.** Once children can rote-count, they can move on to other counting strategies, such as:

■ *Counting on.* To count on, begin at one place in the number word sequence and rote-count forward from that number. For example, to count on from 11, count "twelve, thirteen, fourteen, ..."
■ *Counting back.* To count back, begin at one place in the number word sequence and count backward. For example, to count back from 15, count, "fourteen, thirteen, twelve, eleven, ..."
■ *Skip counting.* To skip-count, begin at one place in the number word sequence and count on by 2s, 5s, 10s, or other values. For example, to skip-count to 20 by 5s, count, "five, ten, fifteen, twenty."

These strategies help children learn a variety of mathematical ideas and skills, including whole-number operations, counting money, and even telling time.

As children count a widening variety of sets, they eventually realize an abstract idea of quantity is independent of the elements in the set. In other words, children use counting to bridge the gap between sets of objects and the concept of a number.

Cardinal Number

The number of elements in a set is called its **cardinal number.**

Definition of a Cardinal Number

The **cardinal number** of a set A, written $n(A)$, is the number of elements in A.

Example 2.9 Finding Cardinal Numbers

What is the cardinal number of each set?

a. $A = \{a, b, c, \ldots, z\}$. **b.** $B = \{x \mid x$ is a state starting with the letter N$\}$.
c. $C = \{2, 4, 6, \ldots, 100\}$. **d.** $D = \{p \mid p$ is a person over the age of 200$\}$.

Solution

a. Because there are 26 letters in the English alphabet, $n(A) = 26$.
b. The states beginning with the letter N are Nebraska, Nevada, New Hampshire, New Jersey, New Mexico, New York, North Carolina, and North Dakota. As a result, $n(B) = 8$.

SECTION 2.1   Set Basics   85

c. There are 50 even numbers from 2 to 100, so $n(C) = 50$.

d. Set D has no elements, so $n(D) = n(\varnothing) = 0$.

The first set of cardinal numbers children use is $\mathbb{N} = \{1, 2, 3, 4, 5 \ldots\}$, which we call the **counting** or **natural numbers.** If this set is expanded to include zero, the cardinal number of the empty set, then we get the set of all cardinal numbers, or the set of the **whole numbers:** $W = \{0, 1, 2, 3, 4, \ldots\}$.

Both the natural numbers and the whole numbers are infinite sets, which we can intuitively describe as sets that go on without end. However, to define an infinite set, we must first define a finite set.

A set is **finite** if and only if one of two conditions hold.

Definition of a Finite Set

A set is **finite** if and only if:

1. It is empty, or
2. There exists a 1-1 correspondence between the set and a set of the form $\{1, 2, 3 \ldots, n\}$, where n is a natural number.

Example 2.10 | Showing Sets are Finite

Reasoning

Let $A = \{x \mid x < 0, x \in W\}$ and $B = \{a, b, c, d, e\}$. Show that A and B are finite.

Solution Because there are no whole numbers less than 0, $A = \{x \mid x < 0, x \in W\} = \varnothing$. Hence, A is finite by the first part of the definition. To show that set B is finite, we must establish a one-to-one correspondence between B and a set of the form $\{1, 2, 3, \ldots n\}$, where n is a natural number. If $n = 5$, then one such one-to-one correspondence is:

This allows us to conclude that B is finite by the second part of the definition.

An **infinite** set is difficult to describe mathematically. So we define it roundaboutly by using the definition of a finite set.

Definition of an Infinite Set

A set is **infinite** if it is not finite.

By this definition, the sets $O = \{1, 3, 5, 7, \ldots\}$ and $S = \{1, 4, 9, 16, 25, \ldots\}$ must be infinite. Neither set is empty, nor is it possible to make a one-to-one correspondence between them and a set of the form $\{1, 2, 3, 4, \ldots, n\}$, where n is a natural number.

Check Your Understanding 2.1-B

1. What is the cardinal number of each set?

   a. $A = \{1, 3, 5, \ldots, 19\}$.

   b. $B = \{x \mid x$ is a color of the rainbow$\}$.

   c. $C = \{x \mid x$ is a letter in the word "beginning"$\}$.

2. Show three different one-to-one correspondences between the sets $A = \{h, o, m, e\}$ and $B = \{w, o, r, k\}$.

3. Seven students have to be assigned to seven lockers. How many different ways can they be assigned?

Talk About It Surprisingly, we can show that the natural numbers and the whole numbers are the same size. Draw a one-to-one correspondence between the sets, and use what you see to explain why this is true.

Connection

Activity 2.2

Examine a set of curriculum materials from first, second, or third grade. Using what you find, describe how the different counting strategies are used to teach each of the following skills.

   a. Addition   **b.** Subtraction   **c.** Counting money   **d.** Measuring time

---

SECTION 2.1   Learning Assessment

■ Understanding the Concepts

1. How is a set different from other collections of objects?

2. What two "rules" govern sets, and what do they imply about the nature of sets?

3. What is the empty set? The universal set?

4. What is the difference between a subset and a proper subset? Use example sets to explain the difference.

5. **a.** Why is the empty set a subset of every set?

   b. Why is every set a subset of itself?

6. If $A \subseteq B$ and $B \subseteq A$, then what else must be true about A and B? Why?

7. Explain the difference between equal and equivalent sets.

8. **a.** What is a one-to-one correspondence between two sets?

   b. How is the concept of a one-to-one correspondence used in counting?

9. Explain the differences among rote counting, counting on, counting back, and skip counting.

10. What is the difference between the natural numbers and the whole numbers?

11. How do we determine whether a set is finite or infinite?

■ Representing the Mathematics

12. Give a verbal description of each set.

   a. $\{c, u, t, e\}$

   b. {Alabama, Alaska, Arizona, . . . , Wyoming}

   c. $\{1, 3, 5, 7, 9\}$

   d. $\{n \mid 3 \leq n < 7, n \in \mathbb{N}\}$

13. Write each set as a roster and in set-builder notation.

   a. The set of letters used in the word "elementary"

   b. The set of all United States presidents serving from 1990 to 2010

   c. The set of multiples of 10

14. Write each set as a roster and in set-builder notation.

   a. The set of all primary colors

   b. The set of all even numbers less than 20

   c. The set of square numbers

15. Describe the advantages and disadvantages of each way of representing sets.
    a. Verbal descriptions
    b. Rosters
    c. Set-builder notation
    d. Venn diagrams

16. Let $U = \{x \mid x$ is a letter in the word "mathematics"$\}$, $A = \{m, a, t, h\}$, and $B = \{h, a, t\}$. Draw a Venn diagram showing the relationships among the three sets.

17. Let $U = \{1, 2, 3, 4, 5, 6, 7, 8, 9, 10\}$, $A = \{2, 4, 6, 8, 10\}$, and $B = \{2, 6\}$. Draw a Venn diagram showing the relationships among U, A, and B.

18. Draw a Venn diagram showing the relationships among the natural numbers, the whole numbers, and the set of all numbers.

19. Write each statement using correct mathematical notation.
    a. a is an element of the set A.
    b. 1 is not an element of the set containing 2, 3, and 4.
    c. The set containing 1, 2, and 3 is a subset of the set of natural numbers less than or equal to 5.
    d. The set of 1, 2, and 3 is not equal to the set containing a, b, and c.

20. Give a verbal translation of each mathematical statement.
    a. $4 \in D$.
    b. $5 \notin \{1, 2, 3, 4\}$.
    c. $\{purple\} \not\subset \{red, blue, yellow\}$.
    d. $\{1, 2, 3\} \sim \{x, y, z\}$.

21. Draw a diagram that shows that $\{1, 2, 3, 4, 5\}$ is equivalent to $\{a, b, c, d, e\}$.

22. Draw a diagram that shows that $\{1, 2, \ldots, 10\}$ is equivalent to $\{3, 6, \ldots, 30\}$.

23. Why do we not represent the empty set as $\{\varnothing\}$?

■ Exercises

24. State whether each collection is well-defined. If not, explain why.
    a. The collection of all beautiful animals
    b. The collection of all astronauts that have been to the moon
    c. The collection of all whole numbers between 1 and 10
    d. The collection of numbers less than 5
    e. The collection of all natural numbers between 4 and 5

25. Place $\in$ or $\notin$ in the blank to make each statement true.
    a. 3 _____ $\{1, 2, 3\}$.
    b. $\{a\}$ _____ $\{a, b, c\}$.
    c. 25 _____ $\{x \mid x = 5n$, where $n \in \mathbb{N}\}$.
    d. $\{1\}$ _____ $\{\{1\}, \{2\}\}$.

26. Place the best symbol, $\subset$, $\subseteq$, or $\not\subseteq$, in the blank to make each statement true.
    a. $\{a\}$ _____ $\{a, b, c\}$.
    b. $\{3, 4, 5\}$ _____ $\{4, 5, 3\}$.
    c. $\{2, 4, 6, 8 \ldots\}$ _____ $\mathbb{N}$.
    d. $\{x \mid x = 2n - 1$, where $n \in \mathbb{N}\}$ _____ $\{x \mid x = 2n + 1$, where $n \in W\}$.

27. Place one of the symbols $\subseteq$, $=$, or $\sim$, in the blank to make each statement true.
    a. $\{2, 3, 4\}$ _____ $\{1, 2, 3, 4, 5\}$.
    b. $\{1, 2, 3\}$ _____ $\{2, 3, 4\}$.
    c. $\{u, s, e\}$ _____ $\{s, u, e\}$.
    d. $\{x \mid x = 2n - 1$, where $n \in \mathbb{N}\}$ _____ $\{x \mid x = 2n$, where $n \in \mathbb{N}\}$

28. Find all the subsets of $T = \{t, o, e\}$.

29. Find all the proper subsets of $M = \{m, a, t, h\}$.

30. a. If a set has 5 elements, how many subsets does it have?
    b. If a set has 8 elements, how many subsets does it have?
    c. If a set has 10 elements, how many proper subsets does it have?
    d. If a set has 15 elements, how many proper subsets does it have?

31. a. If a set has 4 subsets, how many elements are in the set?
    b. If a set has 16 subsets, how many elements are in the set?
    c. If a set has 31 proper subsets, how many elements are in the set?
    d. If a set has 127 proper subsets, how many elements are in the set?

32. If $A = \{1, 2, 3\}$, $B = \{2, 3\}$, $C = \{3, 2, 1\}$, and $D = \{2, 3, 4\}$, which of the following statements are true and which are false? For those that are false, explain why.
    a. $A \subset B$.   b. $A = C$.
    c. $A \subset C$.   d. $B \subset A$.
    e. $A = D$.   f. $A \subseteq C$.
    g. $B \subseteq D$.   h. $C \subseteq D$.

33. Use the following diagram and determine whether each statement is true or false.
    a. $3 \in A$.   b. $6 \notin B$.

c. $7 \in B$.

d. $A \subseteq U$.

e. $A \subseteq B$.

f. $B \subseteq U$.

g. $U = \{3, 4, 5, 8\}$.

h. $A = \{1, 2, 6, 7\}$.

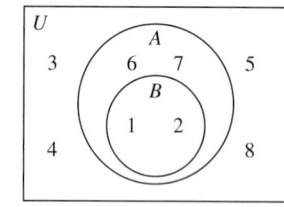

34. Between which of the following sets can a one-to-one correspondence be made?

$A = \{1, 2, 3, 4, 5\}$. $B = \{\diamond, \square, \bigcirc\}$.

$C = \{w, x, y, z\}$. $D = \{ \}$.

$E = \{f, u, n\}$. $F = \{a, b, c, d\}$.

$G = \{\text{red, blue, yellow}\}$.

35. What type of counting has been used to generate each list of numbers?

a. 1, 2, 3, . . . 10   b. 2, 4, 6, 8, . . . 22

c. 8, 9, 10, 11, . . . 20   d. 1, 7, 13, 19, 25

e. 16, 15, 14, 13   f. 18, 15, 12, . . . 3

36. List the set of numbers that will be named in each counting situation.

a. Counting on from 5 to 10

b. Counting back from 17 to 8

c. Skip counting by 3s from 0 to 15

d. Skip counting by 5s from 10 to 50

37. How many numbers will be named in each counting situation?

a. Counting on from 14 to 25

b. Counting back from 39 to 24

c. Skip counting by 5s from 5 to 100

d. Skip counting by 4s from 8 to 56

38. What is the cardinal number of each set?

a. $A = \{1, 2, \ldots, 10\}$.

b. $B = \{1, 3, 5, \ldots, 101\}$.

c. $C = \{5, 10, 15, \ldots, 150\}$.

d. $D = \{1, 4, 9, \ldots, 256\}$.

39. What is the cardinal number of each set?

a. $A = \{x \mid x \in W \text{ and } 4 \le x < 12\}$.

b. $B = \{x \mid x \in \mathbb{N} \text{ and } x < 1\}$.

c. $C = \{x - y \mid x \in \{5, 4, 3\} \text{ and } y \in \{3, 2, 1\}\}$.

d. $D = \{x \cdot y \mid x \in \{1, 2, 3\} \text{ and } y \in \{4, 5\}\}$.

40. Consider the different sets of objects you might find in an elementary classroom. Give two or three examples of such sets that are likely to have the given cardinal number.

a. 1   b. 20   c. 300   d. 0

41. Decide whether each set is finite or infinite.

a. The set of all Canadian provinces beginning with A

b. The set of all children enrolled in kindergarten during 2004

c. The set of people that have been to Mars

d. The set of all minutes in time

42. Decide whether each set is finite or infinite.

a. The set of all natural numbers greater than 1,000

b. The set of all whole numbers between 3 and 4

c. $\{x + y \mid x \in \mathbb{N} \text{ and } y \in W\}$

d. $\{x \cdot y \mid x \in \mathbb{N} \text{ and } y = 0\}$

43. How many ways can 8 runners be assigned to 8 lanes on a track?

44. How many ways can 12 students be assigned to 12 desks?

■ Problems and Applications

45. How many one-to-one correspondences are there between two sets with:

a. 2 elements?   b. 3 elements?

c. 4 elements?   d. n elements?

46. a. Use a calculator to determine how many one-to-one correspondences are between two sets, each with 15 elements.

b. If you could list a one-to-one correspondence every second, how long would it take you to list all the one-to-one correspondences in part a.

47. If we count on from 17 to 65, how many even numbers are named? How many odd numbers are named?

48. How many numbers will be named in skip counting by 7s from 15 to 225?

49. What different numbers can be used to skip count from 0 and arrive at exactly 150? What about 0 to 180?

50. a. If $A \subset B$, and $n(A) = 10$, what is the smallest that $n(B)$ could be? What is the largest it could be?

b. If $A \subseteq B$, and $n(A) = 10$, what is the smallest that $n(B)$ could be? What is the largest it could be?

51. a. If $n(A) = 6$ and $B \subset A$, what are the possible cardinal numbers of B?

b. If $n(A) = 6$ and $B \subseteq A$, what are the possible cardinal numbers of B?

52. Mama Mia's Pizzeria has 5 different toppings that can be put on a pizza: pepperoni, sausage, olives, green peppers, and mushrooms.

a. How many different pizzas can be made using any number of the 5 toppings?

b. How many different pizzas can be made using exactly 3 of the toppings?

53. A committee of 3 is to be formed from 5 senators: Avery, Babbage, Colton, Davis, and Emmett. If Babbage must be one of the senators on the committee, how many different committees can there be?

54. The CEO of a company wants to make a subcommittee from 4 vice presidents: Al, Bob, Carmen, and Debbie. If the subcommittee can be either a group of 2 or a group of 3 people, how many different subcommittees are possible?

■ Mathematical Reasoning and Proof

55. Determine whether each statement is true or false. If false, explain why.
   a. If B is a proper subset of A, then B must be a subset of A.
   b. If B is a subset of A, then B must be a proper subset of A.
   c. If B is equal to A, then B must be equivalent to A.
   d. If B is equivalent to A, then B must be equal to A.

56. Determine whether each statement is true or false. If false, explain why.
   a. If $A \subset B$, then $n(A) < n(B)$.
   b. If $A \subseteq B$, then $n(A) \leq n(B)$.
   c. If $A \sim B$, then $n(A) = n(B)$.
   d. If $A \subset B$, then $A \sim B$.

57. a. If A has two elements, how many proper subsets will it have?
   b. If A has three elements, how many proper subsets will it have?
   c. If A has four elements, how many proper subsets will it have?
   d. If A has n elements, how many proper subsets will it have?

58. a. If A has three elements, how many two-element subsets does A have?
   b. If A has four elements, how many two-element subsets does A have?
   c. If A has five elements, how many two-element subsets does A have?
   d. If A has n elements, find a formula that relates n to the number of two-element subsets of A?

59. Answer each question and explain your thinking.
   a. If $A \subseteq B$, and $B \subseteq C$, then is $A \subseteq C$?
   b. If $a \in B$, and $B \subseteq C$, then is $a \in C$?
   c. If $a \in B$, and $B \not\subseteq C$, then is $a \in C$?

60. Write an argument that explains why every set must be a subset but not a proper subset of itself.

61. Which set is bigger, the set of natural numbers or the set of whole numbers? Explain your thinking.

■ Mathematical Communication

62. Provide two examples of sets that are well-defined and two examples that are not. How did your descriptions of the sets change between those that were well-defined and those that were not?

63. Consider the notion of a proper subset. Do you think the term "proper" is an appropriate adjective for describing this type of subset? Why or why not?

64. In mathematics, we often use the word "is" to represent equality. However, this is not always the case. In each of the following sentences, a verb is underlined. Determine which mathematical symbol ($=$, $\in$, or $\subseteq$) best represents the meaning of the verb.
   a. Sheila is an American.
   b. Motorcycles are motor vehicles.
   c. Dogs are mammals.
   d. Donna is rich.

65. The word "set" has over 450 definitions. How many can you and a partner think of?

66. Four young children try to count a set of 5 colored blocks. Describe the counting mistake made by each child.
   a. Timmy's answer: 4

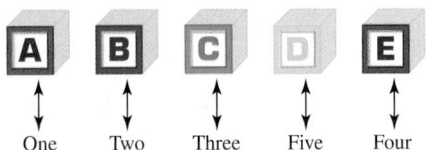

   b. Jodie's answer: 8

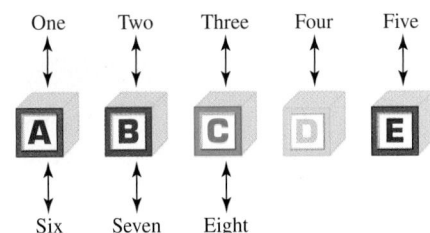

   c. LaToya's answer: 4

   d. Juan's answer: 4

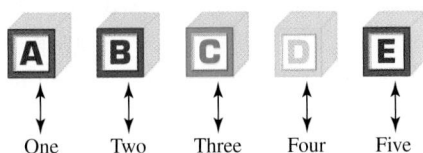

67. Describe how you might teach the idea of a one-to-one correspondence to a young child who is struggling with that aspect of counting.

68. Mathias claims that a set cannot be both a subset and a proper subset of another set. How do you, as his teacher, address his misconception?

69. Some mathematics educators recommend that we teach children to begin counting from 0 rather than 1. What do you think are the advantages and disadvantages of teaching children to count that way?

■ Building Connections

70. Consider yourself as an object to be placed in a set (e.g., you are in the set of students in your class). Make a list of 10 sets that you are a member of.

71. **a.** Does it make sense to draw an analogy between the ideas of a subset and a proper subset and those of less than or equal to and less than? Why or why not?

b. The subset relationship will eventually be connected to the idea of less than in a more formal way. Can you think of how this might be done?

72. Although the ideas of this section are not often taught in the elementary grades, they do serve as the foundation for content that is. Look at several sets of curriculum materials from kindergarten, first grade, and second grade. What ideas from basic set theory do you find?

73. One way to teach children to multiply is with skip counting. What multiplication fact is shown by each list of numbers obtained by skip counting?
    a. 2, 4, 6, 8, 10
    b. 4, 8, 12, 16, 20, 24
    c. 10, 20, 30, 40
    d. 7, 14, . . . , 63
    e. 25, 50, 75
    f. 9, 18, . . . , 81

74. Why do think that it is generally easier for most people to skip count by 2s, 5s, and 10s, rather than by 3s, 4s, or 7s?

SECTION 2.2 Set Operations

Set operations are simply ways of combining or comparing sets to form new sets. They are important to elementary mathematics because they are a foundation for numerical operations. The five basic set operations are union, intersection, set difference, complement, and Cartesian product.

Union and Intersection

If we have two sets, say $A = \{1, 2, 3\}$ and $B = \{3, 4, 5\}$, then one of the easiest ways to make a new set from A and B is to combine them: $\{1, 2, 3, 4, 5\}$. The new set is called the **union** of A and B.

> **Definition of the Union of Two Sets**
>
> The **union** of the two sets A and B, written $A \cup B$, is the set of all elements in A or in B or in both. In set-builder notation, $A \cup B = \{x \mid x \in A \text{ or } x \in B\}$.

"Or" is a key word in the definition because it tells us the elements can come from one set or the other or both. If an element is in both sets, then we list it only once in the union.

Example 2.11 | **Finding Unions**

Let $A = \{1, 2, 3\}$, $B = \{3, 4, 5\}$, and $C = \{5, 6, 7\}$. Find:
a. $A \cup B$.   **b.** $B \cup C$.   **c.** $(A \cup B) \cup C$.

Solution Using the definition, we have:
a. $A \cup B = \{1, 2, 3, 4, 5\}$.   **b.** $B \cup C = \{3, 4, 5, 6, 7\}$.
c. To find the union of three or more sets, first take the union of two sets, and then take the union of the result with the third. Consequently, $(A \cup B) \cup C = \{1, 2, 3, 4, 5\} \cup C = \{1, 2, 3, 4, 5, 6, 7\}$.

Another way to make a new set is to compare sets. One way to do so is to look for the elements that the two sets have in common. For instance, if we compare the sets $A = \{a, b, c, d\}$ and $B = \{b, c, e, f\}$ for common elements, we get the set $\{b, c\}$. This set is called the **intersection** of A and B.

Definition of the Intersection of Two Sets

The **intersection** of two sets A and B, written $A \cap B$, is the set of all elements common to A and B. In set-builder notation, $A \cap B = \{x \mid x \in A \text{ and } x \in B\}$.

"And" is a key word in the definition because it tells us that elements must be in both A and B to be in the intersection.

Example 2.12 **Finding Intersections**

Let $A = \{1, 2, 3\}$, $B = \{2, 3, 4\}$, and $C = \{4, 5, 6\}$. Find:

a. $A \cap B$. **b.** $A \cap C$. **c.** $(A \cap B) \cap C$.

Solution Using the definition, we have:

a. $A \cap B = \{2, 3\}$.

b. $A \cap C = \varnothing$.

c. To find the intersection of three or more sets, first take the intersection of two sets, and then take the intersection of the result with the third: $(A \cap B) \cap C = \{2, 3\} \cap C = \varnothing$.

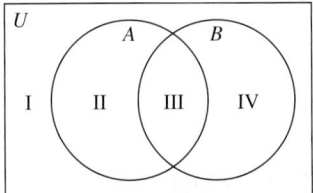

Figure 2.4 Four regions (I–IV) in a two-set Venn diagram

The last example shows us that if two sets have no elements in common, then their intersection is empty. In this case, we call the sets **disjoint** and write $A \cap B = \varnothing$.

We can draw a Venn diagram of the union and intersection of two sets. To do so, draw two overlapping circles for sets A and B (Figure 2.4). This arrangement allows for the fact that some elements may be in A but not in B, some elements may be in B but not in A, and some elements may be common to both. It also allows for the possibility that some elements in the universe may not be in either A or B.

As Figure 2.4 shows, this arrangement divides the diagram into four distinct regions: I, II, III, and IV. To represent a particular set operation, we shade the appropriate regions of the diagram. Because $A \cup B$ is the set of all elements in A or in B, we shade any region in the circle for A or in the circle for B: II, III, and IV (Figure 2.5). Because $A \cap B$ is the set of elements common to A and B, we shade the region common to both circles: III.

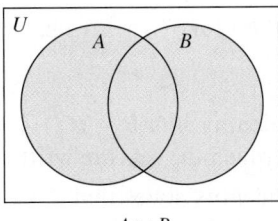

$A \cup B$

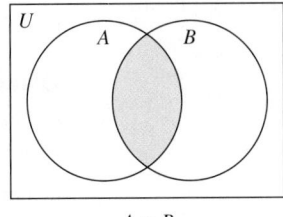

$A \cap B$

Figure 2.5 Venn diagrams of $A \cup B$ and $A \cap B$

If we know the elements in the sets, then we can also place them in the Venn diagram.

Example 2.13 Drawing a Venn Diagram

Representation

Draw a Venn diagram of $U = \{m, a, t, h, e, i, c, s\}$, $A = \{m, a, t, h\}$, and $B = \{m, i, s, t\}$.

Solution We draw overlapping circles and then look to place the elements from each set. The letters m and t are common to A and B, so we place them in the overlapping region of the circles. The elements a and h are in A but not in B, so we place them inside the circle for A but outside the circle for B. Similarly, we place the elements i and s inside the circle for B but outside the circle for A. The remaining letters, e and c, are in the universe but not in A or B. We place them inside the rectangle but outside both circles:

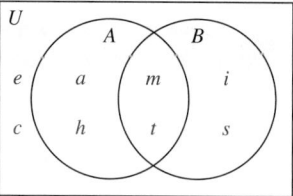

Explorations Manual 2.5

The union and intersection satisfy several **set identities,** which are equations between two sets that are always true. Because set operations are closely linked to numerical operations, many of the identities listed in Theorem 2.2 lead to properties for numerical operations.

Theorem 2.2 Set Identities for the Union and Intersection of Two Sets

For all sets A, B, and C in universe U:

1. *Commutative laws:* $\quad A \cup B = B \cup A \quad$ and $\quad A \cap B = B \cap A$
2. *Associative laws:* $\quad (A \cup B) \cup C = A \cup (B \cup C) \quad$ and
   $\quad (A \cap B) \cap C = A \cap (B \cap C)$
3. *Distributive laws:* $\quad A \cup (B \cap C) = (A \cup B) \cap (A \cup C) \quad$ and
   $\quad A \cap (B \cup C) = (A \cap B) \cup (A \cap C)$
4. *Union and intersection with U:* $\quad A \cup U = U \quad$ and $\quad A \cap U = A$
5. *Union and intersection with $\varnothing$:* $\quad A \cup \varnothing = A \quad$ and $\quad A \cap \varnothing = \varnothing$

Example 2.14 Interpreting Set Identities

Communication

Give a verbal interpretation of each property.

a. $A \cap B = B \cap A$.   **b.** $(A \cup B) \cup C = A \cup (B \cup C)$.

Solution

a. The set identity $A \cap B = B \cap A$ tells us that when we compute an intersection, the order in which the sets are written does not matter; the result is the same.

b. This set identity states that if we combine three sets, it does not matter if we take the union of A and B first or the union of B and C first; the result is the same.

Example 2.15 | **Proving a Set Identity with a Venn Diagram**

Reasoning

Use a Venn diagram to show that $A \cup (B \cap C) = (A \cup B) \cap (A \cup C)$.

Solution In general, two sets are equal if they result in the same Venn diagram. Because our equation involves three sets, we must include three sets in our diagram. The standard practice is to overlap the three circles so that all possible relationships between the sets are represented by the eight numbered regions in the following diagram.

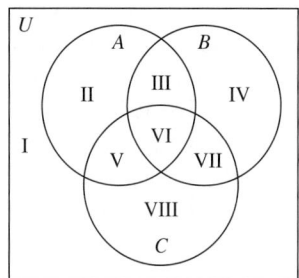

Because A contains the regions {II, III, V, VI}, B contains the regions {III, IV, VI, VII}, and C contains the regions {V, VI, VII, VIII}, we have $A \cup (B \cap C) = A \cup \{VI, VII\} = \{II, III, V, VI, VII\}$. Likewise, $(A \cup B) \cap (A \cup C) = \{II, III, IV, V, VI, VII\} \cap \{II, III, V, VI, VII, VIII\} = \{II, III, V, VI, VII\}$. Because the two sets represent the same regions as shown in the following diagram, we conclude that $A \cup (B \cap C) = (A \cup B) \cap (A \cup C)$.

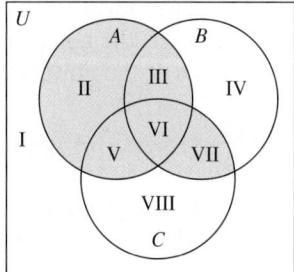

Set Difference and Complement

Another way to compare two sets is not by what they have in common but by what they have that is different. For example, if $A = \{a, b, c, d\}$ and $B = \{b, c, e, f\}$, then A has two elements that are not in B: $\{a, d\}$. This new set is called the **set difference** of B from A.

Definition of the Set Difference

The **set difference** of set B from set A, written $A - B$, is the set of all elements in A that are not in B. In set-builder notation, $A - B = \{x | x \in A$ and $x \notin B\}$.

Example 2.16   Finding Set Differences

Let $A = \{1, 2, 3, 4\}$, $B = \{3, 4, 5\}$, and $C = \{4, 6, 7\}$. Interpret the meaning of each set difference and then find it.

a. $A - B$ **b.** $B - A$ **c.** $B - C$

Solution

a. $A - B$ is the set of all elements in A that are not in B: $\{1, 2\}$.

b. $B - A$ is the set of all elements in B that are not in A: $\{5\}$.

c. $B - C$ is the set of all elements in B that are not in C: $\{3, 5\}$.

In the last example, $A - B$ and $B - A$ are not the same set: $A - B \neq B - A$. This is generally true, which means that unlike the union and the intersection, the order in which the sets are written affects the outcome of the set difference. Figure 2.6 shows the difference between $A - B$ and $B - A$ with Venn diagrams.

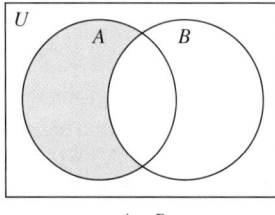

$A - B$ $B - A$

Figure 2.6 Venn diagrams of $A - B$ and $B - A$

The fourth set operation, the **complement** of a set, is similar to the set difference. However, the complement is always relative to the universe rather than to another subset of the universe.

Definition of the Complement of a Set

The **complement** of a set A, written $\overline{A}$, is the set of all elements in the universe U that are *not* in A. In set-builder notation, $\overline{A} = \{x \mid x \in U \text{ and } x \notin A\}$.

Because the set difference is similar to the complement, we often refer to the set difference as the **relative complement.** We can also write the complement as $\overline{A} = U - A$.

Example 2.17   Finding Complements

Let $U = \{1, 2, 3, 4, 5\}$, $A = \{1, 4, 5\}$, $B = \{2, 4\}$, and $C = \{1, 2, 3\}$. Find

a. $\overline{A}$.   **b.** $\overline{B}$.   **c.** $\overline{C}$.

Solution Using the definition, we have:

a. $\overline{A} = \{2, 3\}$.   **b.** $\overline{B} = \{1, 3, 5\}$.   **c.** $\overline{C} = \{4, 5\}$.

In the Venn diagram of the complement, we shade the region inside the rectangle for U but outside the circle for A. The Venn diagrams of $\overline{A}$ and $\overline{A \cup B}$ are shown in Figure 2.7.

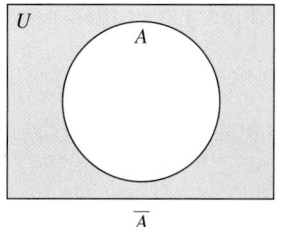 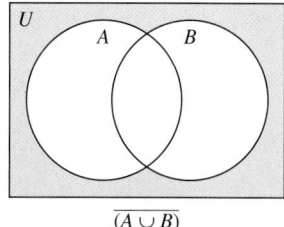

Figure 2.7 Venn diagrams of $\overline{A}$ and $\overline{A \cup B}$

The set difference and complement satisfy several set identities, which are listed in the next theorem.

Theorem 2.3   Set Identities for the Set Difference and Complement

For all sets A and B in universe U:

1. *Double complement law:* $\overline{(\overline{A})} = A$
2. *De Morgan's laws:* $\overline{A \cup B} = \overline{A} \cap \overline{B}$ and $\overline{A \cap B} = \overline{A} \cup \overline{B}$
3. *Alternative representation for a set difference:* $A - B = A \cap \overline{B}$

□

Cartesian Product

The fifth set operation, the **Cartesian product,** is different from the others in that its elements are ordered pairs rather than single objects. An **ordered pair,** (a, b), is a pair of objects in which elements from set A are designated as first and elements from set B are designated as second. If $A = \{1, 2, 3\}$ and $B = \{a, b, c\}$, then examples of ordered pairs are $(1, a)$, $(2, c)$ and $(3, b)$. Because order is important, two ordered pairs are equal if and only if the first components are the same and the second components are the same. For instance, $(1, 2)$ is not equal to $(2, 1)$ because the order within the pairs is different. Order is the key difference between ordered pairs and sets, so a distinction must be made between braces—{ }—which imply that order is not important and parentheses—( )—which imply that it is.

To form the Cartesian product, we make all possible ordered pairs in which the first object is from A and the second is from B.

Definition of the Cartesian Product

The **Cartesian product** of set A with set B, written $A \times B$, is the set of all ordered pairs (a, b) where $a \in A$ and $b \in B$. In set-builder notation, $A \times B = \{(a, b) | a \in A \text{ and } b \in B\}$.

Example 2.18   Finding Cartesian Products

Let $A = \{1, 2\}$, $B = \{a, b\}$, $C = \{\bigcirc, \square, \diamondsuit\}$. Find:

a. $A \times B$.   **b.** $B \times A$.   **c.** $B \times C$.

Solution Using the definition, we have:

a. $A \times B = \{(1, a), (1, b), (2, a), (2, b)\}$.

b. $B \times A = \{(a, 1), (a, 2), (b, 1), (b, 2)\}$.

c. $B \times C = \{(a, \bigcirc), (a, \square), (a, \diamondsuit), (b, \bigcirc), (b, \square), (b, \diamondsuit)\}$.

■

Note several things about the Cartesian product.

- The order of the ordered pairs in the set does not matter as long as the order *within* the ordered pairs is correct.
- The last example shows that the ordered pairs in $A \times B$ are not the same as those in $B \times A$. In general, this means that $A \times B \neq B \times A$ for sets A and B.
- If we try to form a Cartesian product with the empty set, we cannot form ordered pairs because the empty set has no elements to contribute. As a result, $A \times \varnothing = \varnothing$ for any set A.

Unlike the other set operations, we cannot represent the Cartesian product with a Venn diagram. However, we can use it to solve problems.

Example 2.19   Counting Lunches

Application

At lunch, a restaurant serves three types of sandwiches (ham, turkey, and tuna) and four types of beverages (coffee, tea, soda, and milk). If a meal consists of a sandwich and a beverage, what different lunches can be ordered?

Solution Let $S = \{\text{ham, turkey, tuna}\}$ and $B = \{\text{coffee, tea, soda, milk}\}$. We can answer the question by forming the Cartesian product $S \times B$. To ensure that we get all possible ordered pairs, we can use a table like the following one:

	Ham	Turkey	Tuna
Coffee	(ham, coffee)	(turkey, coffee)	(tuna, coffee)
Tea	(ham, tea)	(turkey, tea)	(tuna, tea)
Soda	(ham, soda)	(turkey, soda)	(tuna, soda)
Milk	(ham, milk)	(turkey, milk)	(tuna, milk)

Check Your Understanding 2.2-A

1. Let $U = \{1, 2, 3, 4, 5, 6\}$, $A = \{2, 3, 4, 5\}$ and $B = \{1, 3, 5\}$. Find:
   a. $A \cup B$.   b. $A \cap B$.   c. $A - B$.   d. $B - A$.   e. $\overline{A}$.
2. Let $A = \{\text{red, white, blue}\}$ and $B = \{\text{yellow, green}\}$. Find $A \times B$.
3. Let $U = \{o, u, t, s, i, d, e\}$, $A = \{d, i, e, t\}$, and $B = \{d, o, t, s\}$. Draw a Venn diagram that shows the relationship between U, A, and B.
4. If $U = \{a, b, c, d, e, f, g\}$, $A = \{a, c, e\}$, $B = \{b, c, e, g\}$ and $C = \{a, b, d, g\}$, find:
   a. $A \cup \overline{B}$.   b. $(C - A) \cap B$.   c. $\overline{B} \times \overline{C}$.   d. $\overline{(A - B)} \cup C$.

Talk About It The union, the set difference, and the Cartesian product are closely connected to three numerical operations. What are the operations and how might the set operations be used to compute them?

Reasoning

Activity 2.3
Reconsider Figure 2.4. Use what you see to help you write an equation that uses addition and subtraction to set $n(A \cup B)$ equal to $n(A)$, $n(B)$, and $n(A \cap B)$.

Sorting

In the elementary classroom, set operations are used in a number of intuitive ways, one of which is to sort. **Sorting,** or **categorizing,** is the process of placing objects or events into sets based on the similarities and differences between them. A manipulative that is often used to teach sorting skills is **attribute blocks.** A standard set of attribute blocks contains 60 blocks, each of which has four attributes: shape, color, size, and thickness (Figure 2.8). The notation for the attributes is given.

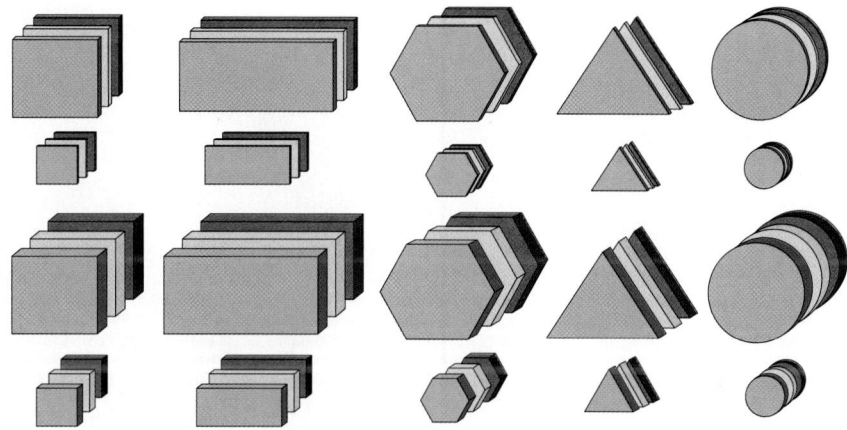

Shape	Color	Size	Thickness
Circle = ○	Red = R	Large = L	Thick = T
Hexagon = ⬡	Blue = B	Small = S	Thin = Th
Rectangle = ▭	Yellow = Y		
Square = □			
Triangle = △			

Figure 2.8 **A set of 60 attribute blocks**

Teachers can use attribute blocks to teach a number of sorting skills. One way to sort is to select objects with a particular characteristic or attribute. For instance, we might ask how many blocks are blue in a standard set of attribute blocks. To answer the question, children have to identify and sort out the 20 different blue blocks.

Example 2.20 **Counting Attribute Blocks**

Using a standard set of 60 attribute blocks, find each cardinal number.

a. $n(\bigcirc)$ **b.** $n(L)$ **c.** $n(Th\,Y)$ **d.** $n(SR\square)$

Solution In each case, we find the blocks with the specified attributes and then count them.

a. $n(\bigcirc) = 12$.

b. $n(L) = 30$.

c. There are 10 thin yellow blocks, so $n(Th\,Y) = 10$.

d. There are 2 small red squares, so $n(SR\square) = 2$.

Explorations Manual 2.6

Another way to sort is to use similarities or differences. When sorting by similarities, we look for objects that two or more sets have in common; that is, we find the intersection of the sets. When sorting by differences, we look for objects that are in one set but not in the other; that is, we find the set difference. As the next example shows, we can use Venn diagrams to help us sort two sets for similarities and differences.

Example 2.21 | **Sorting Shapes**

Connections

The student page in Figure 2.9 shows a sorting activity with shapes. Complete the activity, and describe how set operations were used to do so.

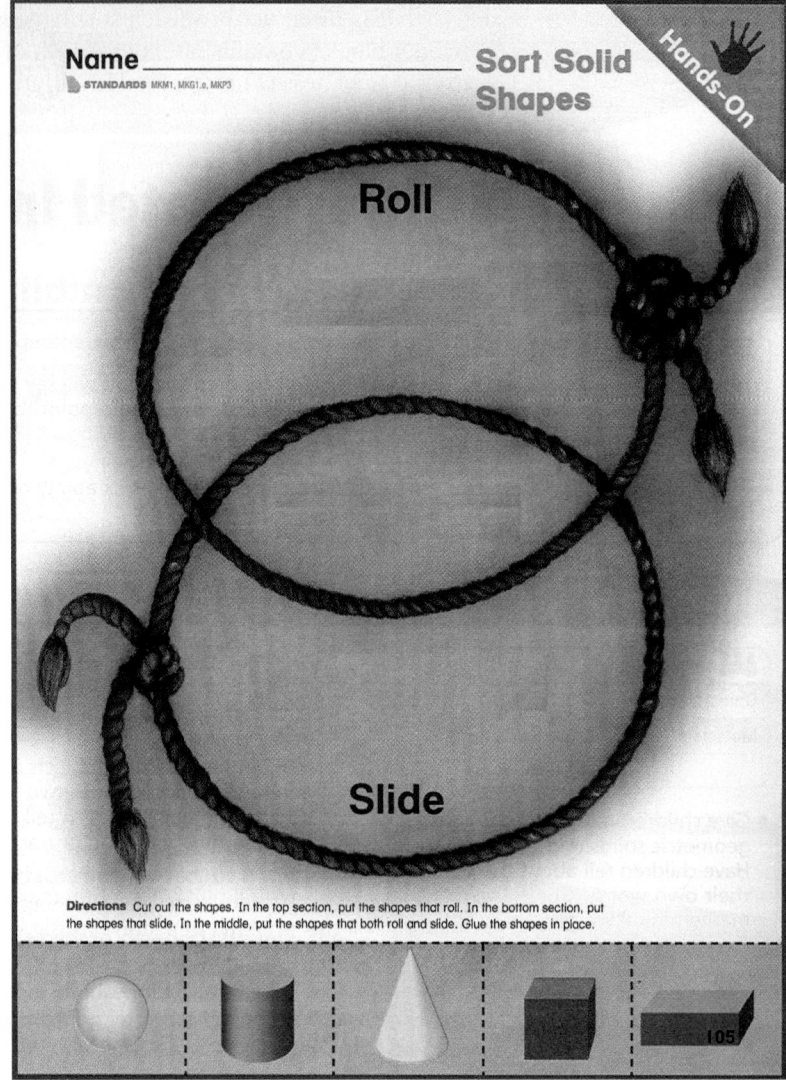

Figure 2.9 Using a Venn diagram to sort sets
Source: From *Mathematics Grade K: Houghton Mifflin Mathematics Georgia* (p. 105). Copyright © by Houghton Mifflin Company, Inc. Reproduced by permission of the publisher, Houghton Mifflin Harcourt Publishing Company.

Solution We intuitively use set operations to sort the objects into the three sets. First, we use the intersection to find the objects that both roll and slide. Next, we use the set difference to find the objects that either roll but do not slide or that slide but do not roll. The completed diagram is shown.

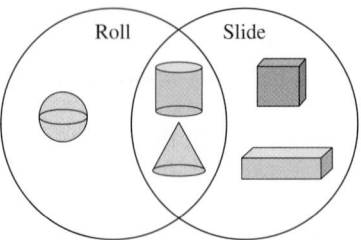

The last example shows how to sort small sets by placing their elements in a Venn diagram. If the sets are larger, we can use counts rather than the objects themselves.

For instance, suppose we wanted to know how many attribute blocks are in each region of the Venn diagram in Figure 2.10(a). A standard set of attribute blocks has 20 red and 30 thick. Of these, 10 are red and thick, so we place a 10 in the intersection of the circles. The red circle consists of two regions that must sum to 20. Consequently, the region for the red but not thick must have $20 - 10 = 10$ elements. Similarly, the region for the thick but not red must have $30 - 10 = 20$ elements. Because we have placed $10 + 10 + 20 = 40$ blocks, the region for the blocks that are neither red nor thick must have $60 - 40 = 20$ elements. Figure 2.10(b) shows the completed diagram.

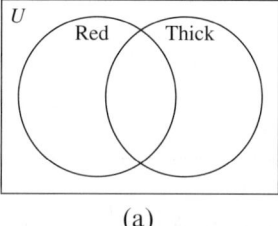 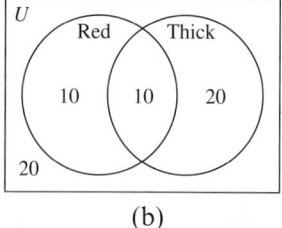

(a)   (b)

Figure 2.10 Sorting the red and thick blocks

Example 2.22   Sorting Three Sets of Attribute Blocks

Use sorting skills and a standard set of attribute blocks to determine the correct number of elements in each region of the Venn diagram.

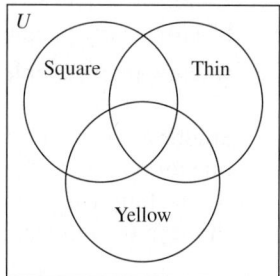

Solution We begin with the intersection of all three sets. Because there are 2 thin, yellow squares, we place a 2 in the intersection of the 3 circles. Next, the intersection of the Square and Thin circles consists of two regions. Because there are 6 thin squares and one region already has 2 elements, the other must have $6 - 2 = 4$ elements. Similarly, there are 8 shapes that are thin and yellow but not squares and 2 yellow squares that are not thin.

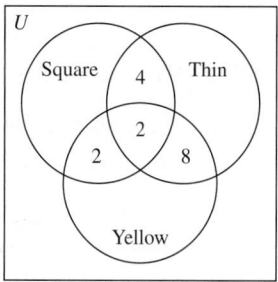

We now fill in the remaining region for squares. The Square circle is divided into 4 regions, of which 3 already have numbers. Because there are 12 squares, the last region must have $12 - 4 - 2 - 2 = 4$ elements. Likewise, there are $30 - 4 - 2 - 8 = 16$ elements in the remaining region of the Thin circle and $20 - 8 - 2 - 2 = 8$ elements in the remaining region of the Yellow circle. Finally, the universe is divided into 8 regions, of which 7 already have numbers. Because there are a total of 60 blocks, the number outside the circles must be $60 - 4 - 4 - 16 - 2 - 2 - 8 - 8 = 16$ elements:

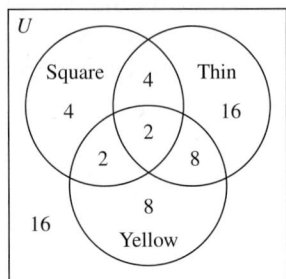

We now apply these sorting skills to solve what is often called a **survey problem.**

Example 2.23 | **Solving a Survey Problem**

Application

In a group of 75 people, 30 are blond, 40 have brown eyes, 25 are tall, 15 are blond and brown-eyed, 10 are tall and brown-eyed, 15 are blond and tall, and 5 are blond, tall, and brown-eyed.

a. How many people have blond hair but not brown eyes?

b. How many people have brown eyes and are tall but do not have blond hair?

c. How many people in the group have none of the characteristics?

Solution There are three sets of people, so we draw a Venn diagram with three circles: one for blond hair, one for brown eyes, and one for tall people. We then place numbers by starting with the intersection of the three groups and working our way backward through the information:

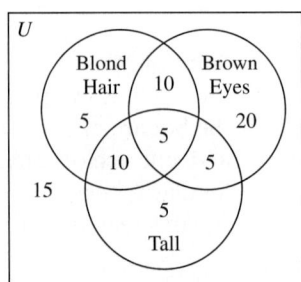

We now answer the questions by finding the appropriate regions in the diagram.

a. Fifteen people have blond hair, but not brown eyes.

b. Five people have brown eyes and are tall but do not have blond hair.

c. Fifteen people have none of the characteristics.

Check Your Understanding 2.2-B

1. Using a standard set of attribute blocks, find each cardinal number.

   a. $n(\bigcirc)$　　　　**b.** $n(S)$　　　　**c.** $n(LT)$　　　　**d.** $n(ThB\square)$

   e. $n(B \cap \triangle)$　　**f.** $n(Th \cap L \cap Y)$　　**g.** $n(Th - \square)$　　**h.** $n(R - S)$

2. A high school homeroom has 35 students. The teacher knows that 20 students are taking mathematics, 15 are taking English, and 10 are taking both. Use a Venn diagram to determine how many students are taking mathematics but not English.

3. Use sorting skills and a standard set of attribute blocks to determine the correct number of elements in each region of the Venn diagram.

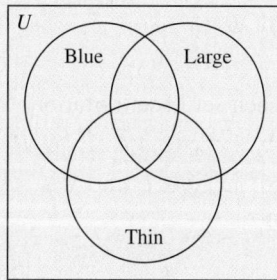

Talk About It   How are sorting skills connected to or used with inductive reasoning and pattern recognition?

Problem Solving

Activity 2.4

Consider the 3 × 3 grid. Use only the 30 small blocks from a standard set of attribute blocks, and fill in the grid so that no two blocks in the same row or column have more than two attributes in common.

SECTION 2.2   Learning Assessment

■ Understanding the Concepts

1. If A and B are sets, explain how to compute each set operation.

   a. $A \cup B$　　　　　**b.** $A \cap B$

   c. $B - A$　　　　　**d.** $\overline{A}$

   e. $A \times B$

2. What does it mean for two sets to be disjoint?

3. How are the set difference and the complement similar? How are they different?

4. What is the difference between a set with two elements and an ordered pair?

5. How are set operations used in sorting?

■ Representing the Mathematics

6. If $C = \{p \,|\, p$ is a person who likes chocolate$\}$ and $D = \{p \,|\, p$ is a person who likes donuts$\}$, give a verbal description of each set.

   a. $C \cap D$　　**b.** $C \cup D$　　**c.** $C - D$　　**d.** $D - C$

7. If A is the set of all people who eat apples and B is the set of all people who eat bananas, write each set in set notation.

   a. The set of all people who eat apples but not bananas

   b. The set of all people who eat apples or bananas

   c. The set of all people who do not eat apples

   d. The set of all people who do not eat apples or bananas

8. Given sets A and B, use the definitions of the set operations to write each set in set-builder notation.

   a. $B \cup A$ **b.** $A \cap B$

   c. $B - A$ **d.** $\overline{B}$

   e. $B \times A$

9. Given sets A and B, write each set in set notation.

   a. $\{x \mid x \in A \text{ or } x \in B\}$ **b.** $\{x \mid x \in A \text{ and } x \in B\}$

   c. $\{x \mid x \in A \text{ and } x \notin B\}$ **d.** $\{x \mid x \in U \text{ and } x \notin A\}$

   e. $\{(a, b) \mid a \in A \text{ and } b \in B\}$

   f. $\{x \mid x \notin A \text{ or } x \notin B\}$

10. Give a verbal interpretation of each set identity.

   a. $A \cup B = B \cup A$.

   b. $(A \cap B) \cap C = A \cap (B \cap C)$.

   c. $A \cup \varnothing = A$.

   d. $A \cup U = U$.

11. Why would each of the following Venn diagrams be a poor choice for representing set operations?

   a.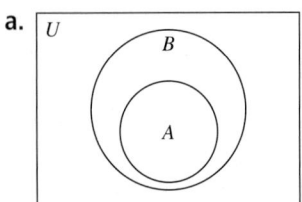

   b.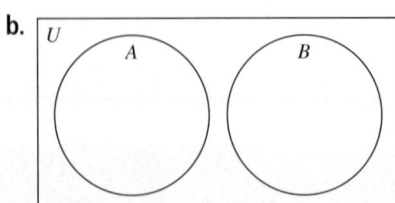

12. Let set $A = \{c, l, a, p\}$ and $B = \{c, l, u, e\}$. If the universal set is the set of letters in the word "applesauce," place the letters from each set in the correct place in the Venn diagram.

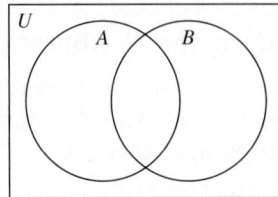

13. Let $B = \{x \mid x$ is a letter in "banana"$\}$, $I = \{x \mid x$ is a letter in "Indiana"$\}$, $P = \{x \mid x$ is a letter in "Panama"$\}$. Place the letters from each set in the correct place in the Venn diagram.

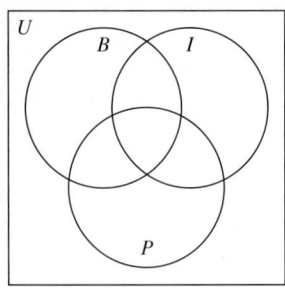

14. Represent each set by shading in the appropriate region(s) in a Venn diagram.

   a. $A \cap \overline{B}$ **b.** $\overline{A \cup B}$ **c.** $\overline{A \cap B}$ **d.** $\overline{A} \cap \overline{B}$

15. Represent each set by shading in the appropriate region(s) in a Venn diagram.

   a. $A \cup (B \cap C)$ **b.** $A \cap B \cap \overline{C}$

   c. $(A - B) \cap \overline{C}$ **d.** $\overline{B - A} \cup C$

16. Use set notation to represent the shaded region in each Venn diagram.

   a.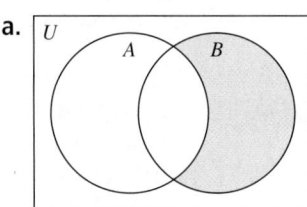

   b.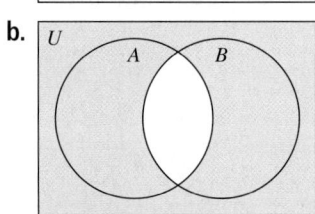

   c.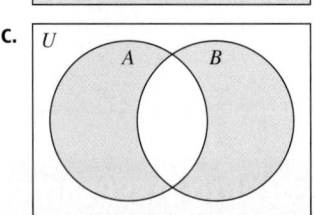

17. Use set notation to represent the shaded region in each Venn diagram.

   a.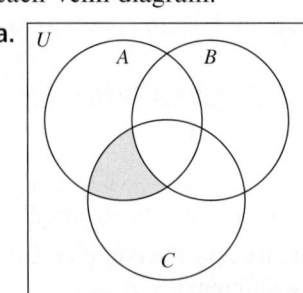

b.

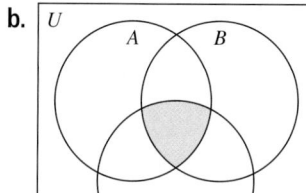

c.

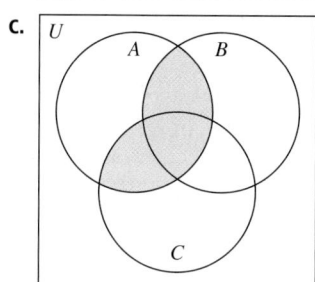

Exercises

18. If $A = \{i, n, t, e, r\}$ and $B = \{s, e, c, t, i, o, n\}$, find:
  a. $A \cup B$.   **b.** $A \cap B$.
  c. $A - B$.   **d.** $B - A$.

19. If $A = \{2, 4, 6, \ldots, 20\}$ and $B = \{3, 6, 9, \ldots, 21\}$, find:
  a. $A \cup B$.   **b.** $A \cap B$.
  c. $A - B$.   **d.** $B - A$.

20. Let $U = \{1, 2, 3, \ldots, 8\}$, $A = \{2, 4, 5, 6\}$, $B = \{3, 5, 7, 8\}$, and $C = \{1, 4, 6, 7, 8\}$. Find:
  a. $\overline{A}$.   **b.** $\overline{B}$.
  c. $\overline{C}$.   **d.** $\overline{A \cup B}$.
  e. $\overline{B \cap C}$.

21. Let $U = \{w, a, s, h, i, n, g, t, o\}$, $A = \{w, a, s, h\}$, $B = \{t, h, i, n, g\}$, and $C = \{s, o, n, g\}$. Find:
  a. $\overline{A}$.   **b.** $\overline{B}$.
  c. $\overline{C}$.   **d.** $\overline{A - B}$.
  e. $\overline{A \cup C}$.

22. If $A = \{1, 2, 3, 4, 5\}$ and $B = \{4, 5, 6, 7\}$, what set operation results in each set?
  a. $\{6, 7\}$ **b.** $\{4, 5\}$
  c. $\{1, 2, 3\}$ **d.** $\{1, 2, \ldots, 7\}$

23. Let $A = \{m, a, t, h\}$, $B = \{i, s\}$, and $C = \{f, u, n\}$. Find:
  a. $A \times B$.   **b.** $A \times C$.
  c. $B \times A$.   **d.** $B \times C$.

24. Let $A = \{k, i, n, d\}$, $B = \{e, r\}$, and $C = \{g, a, r, t, e, n\}$. Find:
  a. $A \times B$.   **b.** $A \times C$.
  c. $B \times A$.   **d.** $B \times C$.

25. Find the sets A and B that were used to make each Cartesian product $A \times B$.
  a. $\{(1, a), (1, b), (1, c), (2, a), (2, b), (2, c), (3, a), (3, b), (3, c)\}$
  b. $\{(1,1), (1,2), (2, 1), (2, 2)\}$
  c. $\{(4, 1), (4, 2), (4, 3), (4, 4)\}$

26. Give an example of two sets A and B that have a Cartesian product with the following number of elements.
  a. 1   **b.** 4   **c.** 10   **d.** 7

27. Find $n(A \times B)$, if
  a. $n(A) = 3$ and $n(B) = 4$.
  b. $n(A) = 2$ and $n(B) = 5$.
  c. $n(A) = 4$ and $n(B) = 7$.
  d. $n(A) = a$ and $n(B) = b$.

28. If $n(A \times B) = 36$, what are possible values for $n(A)$ and $n(B)$?

29. If $A \subseteq B$, determine whether each statement is equal to A, B, or $\varnothing$.
  a. $A \cup B$ **b.** $A \cap B$ **c.** $A - B$

30. If $U = \mathbb{N} = \{1, 2, 3, 4, 5, \ldots\}$, $E = \{x \mid x = 2n,$ where $n \in \mathbb{N}\}$, and $O = \{x \mid x = 2n - 1$, where $n \in \mathbb{N}\}$, determine whether each set is equal to $\mathbb{N}$, E, O, or $\varnothing$.
  a. $E \cup O$ **b.** $E \cap O$
  c. $E \cap \mathbb{N}$ **d.** $O \cup \mathbb{N}$
  e. $\overline{E}$ **f.** $\overline{O}$

31. Using a standard set of 60 attribute blocks, find each cardinal number.
  a. $n(T)$ **b.** $n(\bigcirc)$
  c. $n(R\bigcirc)$ **d.** $n(STh\square)$
  e. $n(LYTh\triangle)$

32. Using a standard set of 60 attribute blocks, find each cardinal number.
  a. $n(R \cup Th)$ **b.** $n(\square \cup T)$
  c. $n(\bigcirc \cap B)$ **d.** $n(Th \cap R)$
  e. $n(Y - L)$ **f.** $n(Th - \triangle)$
  g. $n(\overline{\square})$ **h.** $n(\overline{S})$

33. Using a standard set of 60 attribute blocks, find each cardinal number.
  a. $n(B \cup Th \cup \bigcirc)$ **b.** $n(\square \cup B \cup T \cup L)$
  c. $n(Th \cap R \cap L)$ **d.** $n(\square \cap B \cap L \cap Th)$
  e. $n(\overline{\triangle \cap R})$ **f.** $n(\overline{Th - R})$

34. If $n(U) = 30$, $n(A - B) = 8$, $n(B - A) = 7$, and $n(A \cap B) = 5$, find $n(A)$ and $n(B)$.

35. Using a standard set of attribute blocks, draw a Venn diagram of the set of red blocks and the set of thin blocks. Use the diagram to find $n(R \cap Th)$, $n(R - Th)$, and $n(Th - R)$.

36. Using a standard set of attribute blocks, draw a Venn diagram of the set of large blocks and the set of circular blocks. Use the diagram to find $n(L \cap O)$, $n(L - O)$, and $n(O - L)$.

37. Using a standard set of attribute blocks, draw a Venn diagram of the set of blue blocks, the set of triangular blocks, and the set of thin blocks. Use the diagram to find each number.

a. $n(B \cup \triangle)$

b. $n(B \cap \triangle \cap Th)$

c. $n(B \cap \overline{\triangle} \cap \overline{Th})$

d. $n(\overline{B} \cap \triangle \cup \overline{Th})$

■ Problems and Applications

38. Let $U = \{1, 2, 3, 4, 5, 6, 7, 8\}$, $A = \{1, 2, 3, 7, 8\}$, $B = \{1, 2, 4, 6\}$, and $C = \{3, 4, 5, 7, 8\}$. Find each set.

a. $(A \cup C) \cap B$ **b.** $(A \cap B) \cup (A \cap C)$

c. $\overline{(B - C)}$ **d.** $A \cap B \cap \overline{C}$

e. $\overline{(B - A)} \cup C$ **f.** $(A \cap B) \times C$

39. Use the diagram to find the letters in each set.

a. $(A \cup B) \cap C$ **b.** $(B \cap C) \cup A$

c. $(A \cap B) \cup (A \cap C)$ **d.** $\overline{(A - C)}$

e. $\overline{(B - A)} \cup C$ **f.** $\overline{(A \cup B \cup C)}$

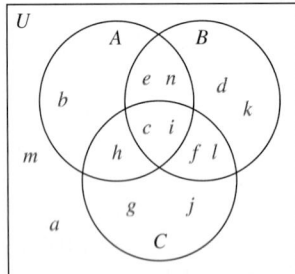

40. If $n(A) = 5$, $n(B) = 6$, and $n(C) = 7$, what is the greatest number of elements that each set could have?

a. $A \cup B$ **b.** $B \cap C$ **c.** $A - C$

41. If $n(A) = 3$, $n(B) = 6$, and $n(C) = 8$, what is the least number of elements that each set could have?

a. $A \cup C$ **b.** $A \cap B$ **c.** $C - B$

42. Write an equation that uses addition and subtraction to set $n(A \cup B \cup C)$ equal to $n(A)$, $n(B)$, $n(C)$, $n(A \cap B)$, $n(A \cap C)$, $n(B \cap C)$, and $n(A \cap B \cap C)$.

43. Draw a Venn diagram of sets A, B, and C so that the given conditions are satisfied.

a. $A \subseteq B$ and $B \cap C = \emptyset$

b. $A \subseteq B$, $C \subseteq B$, and $A \cap C = \emptyset$

c. $A \cap B = \emptyset$, $A \cap C = \emptyset$, and $B \cap C = \emptyset$

44. In this section, we learned how Venn diagrams can be drawn to show all possible relationships between either two or three sets. A Venn diagram can also show all possible relationships among four sets. How can this be done?

45. A local business has 73 employees. Twenty-five of the employees are over the age of 45, 30 are female, and 15 are both.

a. How many employees are male and over 45?

b. How many employees are either over 45 or female but not both?

46. A survey on soda preferences is taken at a local mall. Of the 150 people surveyed, 103 liked cola, 78 liked root beer, and 49 liked both. How many liked neither cola nor root beer?

47. Kara has four tops, which are red, blue, green, and white. She also has a pair of jeans and a skirt.

a. Assuming tops and bottoms can be mixed and matched, use the Cartesian product to make a list of her outfits.

b. Besides forming the Cartesian product, what is another way to find the number of outfits? What principle is this approach based on?

48. The customers at a restaurant were recently surveyed, and 50 ate hamburgers and 65 ate French fries.

a. What is the greatest number of customers that could have bought both?

b. What is the least number of customers that could have bought both?

c. If 120 people were surveyed, what is the fewest number of people that could have ordered neither a hamburger nor French fries? What is the most?

49. In a survey of 100 students, 40 were taking mathematics, 34 were taking economics, 29 were taking chemistry, 17 were taking both mathematics and economics, 14 were taking economics and chemistry, 13 were taking both mathematics and chemistry, and 5 were taking all three.

a. How many were taking only mathematics?

b. How many were taking mathematics and chemistry but not economics?

c. How many were taking none of the subjects?

50. Another soda survey is given at a different mall. This time, 250 people were surveyed: 151 liked cola, 119 liked root beer, 109 liked ginger ale, 75 liked both cola and root beer, 63 liked both cola and ginger ale, 46 liked root beer and ginger ale, and 23 liked all three.

   a. How many liked none of the sodas?

   b. How many liked only ginger ale?

   c. How many liked ginger ale and cola but not root beer?

51. A political poll surveyed people about the president's economic, social, and foreign policies. It was found that 56% of the people agreed with the economic policy, 45% agreed with the social policy, 53% agree with the foreign policy, 25% agreed with both the economic and social policy, 24% agreed with both the social and foreign policy, 21% agreed with both the economic and foreign policy, and 11% agreed with all three.

   a. What percentage of the people did not agree with any of the president's policies?

   b. What percentage of the people agreed with the president on foreign and social policies but not economic policies?

52. Detective Johnson has been assigned to a recent jewelry heist. There are three suspects: Carlson, Friedrich, and Hanson. The crime had to be committed while the guards were switching shifts, which happens at 6:00 a.m. and 6:00 p.m. The thief is believed to have hidden the jewels in one of three places: a river, a vacant lot, or an abandoned building, all of which are close to the crime scene. How many different crime scenarios does detective Johnson have to consider, and what are they?

■ Mathematical Reasoning and Proof

53. Determine whether each statement is true or false for sets A and B. If false, explain why.

   a. $(A \cap B) \subseteq A$.   b. $(A \cup B) \subseteq A$.

   c. $(A \cap B) \subseteq (A \cup B)$.   d. $A \subseteq (A \cup B)$.

   e. $A - B \subseteq B$.   f. $B - A \subseteq (A \cup B)$.

54. If $A \subseteq U$, determine whether each statement is true or false. If false, explain why.

   a. $A \cap \varnothing = A$.   b. $\overline{\varnothing} = U$.

   c. $\overline{U} = \varnothing$.   d. $U - A = \overline{A}$.

   e. $A \cap \overline{A} = A$.   f. $A \cup \overline{A} = U$.

55. Let $A = \{\text{Al, Mike, Dave}\}$ and $B = \{\text{Amy, Sue}\}$. Determine whether each statement is true or false. If false, explain why.

   a. $A \times B$ has 5 elements in it.

   b. Mike $\in A \times B$.

   c. (Al, Amy) $\in A \times B$.

   d. (Dave, Sue) $\subseteq A \times B$.

   e. (Sue, Mike) $\in A \times B$.

   f. $\{(\text{Al, Amy})\} \subseteq A \times B$.

56. a. If $a \in A \cap B$, does it follow that $a \in A$? Explain.

   b. If $a \in A \cup B$, does if follow that $a \in A$? Explain.

   c. If $a \in A \cup B$, does if follow that $a \in A \cap B$? Explain.

57. a. If $n(A \cup B) = n(A)$, what must be true about the relationship between A and B?

   b. If $n(A \cap B) = n(A)$, what must be true about the relationship between A and B?

58. If A and B are finite sets, will $n(A) \leq n(A \cup B)$ always be true? Explain.

59. a. Is it possible to find sets A and B such that $A - B = B - A$? If so, what conditions must the sets satisfy?

   b. Is it possible to find sets A and B such that $A \times B = B \times A$? If so, what conditions must the sets satisfy?

60. Show directly that the set identities $A \cup B = B \cup A$ and $A \cap B = B \cap A$ hold true for the sets $A = \{2, 4, 5, 6\}$ and $B = \{1, 3, 4, 6\}$.

61. Use a Venn diagram and the numbered regions method to show that $\overline{A \cap B} = \overline{A} \cup \overline{B}$.

62. Use a Venn diagram and the numbered regions method to show that $A - B = A \cap \overline{B}$.

63. Use a Venn diagram and the numbered regions method to show that $(A \cap B) \cap C = A \cap (B \cap C)$.

64. Another way to show that two sets are equal using Venn diagrams is to shade the regions represented by the sets. If the two sets result in the same shaded regions, they are equal. Use Venn diagrams with shaded regions to show that each equation is true.

   a. $A \cap (B \cap C) = (A \cap B) \cap C$.

   b. $A \cap (B \cup C) = (A \cap B) \cup (A \cap C)$.

65. Once specific set identities are known, they can be used to show that other sets are equal. The following is a proof that $(A \cup B) - C = (A - C) \cup (B - C)$. State the specific set identity that has been used to make each step in the proof.

$$(A \cup B) - C = (A \cup B) \cap \overline{C}$$
$$= (A \cap \overline{C}) \cup (B \cap \overline{C})$$
$$= (A - C) \cup (B - C)$$

66. Use set identities to show that $A \cup (B - A) = A \cup B$.

67. Use set identities to show that $(A - B) \cup (A \cap B) = A$.

■ Mathematical Communication

68. In mathematics, we often use the word "or" to represent the union of two sets and the word "and" to represent the intersection of two sets. Why is it reasonable to do so?

69. Consider the operations of the set difference and the Cartesian product. Do you think the names "set difference" and "Cartesian product" accurately describe the operations? Why or why not?

70. With several of your classmates, use a standard set of attribute blocks to design an activity that will teach children to sort first by 1 attribute, then by 2, and then by 3.

71. With several of your classmates, design an activity for elementary students that uses a set of attribute blocks and is solved by using the:

   a. Union of two sets.

   b. Intersection of two sets.

   c. Set difference of two sets.

   d. Complement of a set.

72. Do a quick Internet search to find out what problems children are likely to encounter as they learn to sort. Write a paragraph or two to summarize what you find.

73. Write a survey question with three sets like the one in Example 2.23. Give it to a partner to solve.

What information do you need to include for your partner to be able to solve it?

■ Building Connections

74. In what other situations, mathematical or otherwise, do we use the words "union" and "intersection?" Are these situations analogous to the union and intersection as described in the set definitions? If so, how?

75. In Chapter 4, we will see how the union of two sets is used to compute addition. What do you think the identities in Theorem 2.2 might indicate is true about addition?

76. Sorting is an everyday activity that we use in a variety of situations. Describe five activities in which you have recently used sorting. In each case, briefly describe what was sorted, how it was sorted, and why sorting was necessary.

77. Look at several curriculum materials from kindergarten and first grade. Describe the sorting activities you find in the materials.

   a. What set notions are used in the activities?

   b. While set notions are commonly used in sorting activities, they are seldom taught as an important part of sorting. Why do you think this is the case?

Historical Note

One of the first mathematicians to make major contributions in the area of set theory was Georg Cantor (1845–1918). Born in St. Petersburg, Russia, he was educated in Germany during what many consider to be the golden age of mathematical scholarship. Cantor not only defined many of the notions associated with sets, he also demonstrated how they could be used to solve problems in other mathematical areas. Much of his research revolved around the measurement of infinite sets. When Cantor showed that a cardinal number could be assigned to an infinite set, he caused a great debate among mathematicians that eventually ostracized him from the mathematical community. Not until late in his life was the genius of his work fully appreciated.

Georg Cantor

Opening Problem Answer

If we assume that the playlist has at least one song, then the problem is equivalent to counting all the subsets of a set with 15 elements except for the empty set. As a result, Madison can make $2^{15} - 1 = 32{,}768 - 1 = 32{,}767$ different playlists.

Monkey Business Images/Shutterstock.com

Chapter 2 REVIEW

Summary of Key Concepts from Chapter 2

Section 2.1   Set Basics

- A **set** is a well-defined collection of objects. Objects in the set are called **elements.**
- Sets can be represented with verbal descriptions, a **roster, set-builder notation,** and a **Venn diagram.**
- The **empty set,** or **null set,** written { } or $\varnothing$, is the set that contains no elements.
- The **universal set,** or **universe,** written U, is the set of all possible elements in a given discussion.
- Set A is a **subset** of set B, written $A \subseteq B$, if and only if every element of A is also an element of B. If B has an element that is not in A, then A is a **proper subset** of B, written $A \subset B$.
- The empty set is a subset of every set, and every set is a subset of itself.
- If a set has n elements, then it has 2^n subsets.
- Two sets are **equal,** written $A = B$, if and only if they contain exactly the same elements.
- A **one-to-one correspondence** between two sets A and B is a pairing of the elements of A with the elements of B so that each element of A corresponds to exactly one element of B, and vice versa.
- Two sets are **equivalent,** written $A \sim B$, if and only if there is a one-to-one correspondence between them.
- The **cardinal number** of a set A, written $n(A)$, is the number of elements in A.
- The set of **counting,** or **natural, numbers** is $\mathbb{N} = \{1, 2, 3, 4, 5 \ldots\}$, and the set of **whole numbers** is $W = \{0, 1, 2, 3, 4, \ldots\}$.
- A set is **finite** if and only if it is empty or can be counted. A set is **infinite** if it is not finite.

Section 2.2   Set Operations

- The **union** of two sets A and B, written $A \cup B$, is the set of all elements in A or in B or in both.
- The **intersection** of two sets A and B, written $A \cap B$, is the set of all elements common to A and B.
- Two sets, A and B, are **disjoint** if and only if $A \cap B = \varnothing$.
- The union and intersection satisfy several set identities, including the commutative laws, associative laws, distributive laws, and laws involving U and $\varnothing$.
- The **set difference** of set B from set A, written $A - B$, is the set of all elements in A that are not in B.
- The **complement** of a set A, written $\overline{A}$, is the set of all elements in the universe U that are *not* in A.
- An **ordered pair,** (a, b), is a pair of objects in which elements from set A are designated as first and elements from set B are designated as second.
- The **Cartesian product** of set A with set B, written $A \times B$, is the set of all ordered pairs (a, b) where $a \in A$ and $b \in B$.
- **Sorting,** or **categorizing,** is the process of placing objects or events into sets based on characteristics or the similarities and differences between them.

Review Exercises Chapter 2

1. Write each set as a roster and in set-builder notation.
   a. The set of colors in the rainbow
   b. The set of whole numbers less than 5
   c. The set of odd natural numbers between 10 and 20

2. What is the cardinal number of each set?
   a. The set of colors on the United States flag
   b. $\{x | 4 < x < 9, x \in \mathbb{N}\}$
   c. $\{4, 8, 12, \ldots, 32\}$

3. Find all the subsets of $A = \{l, o, w\}$.

4. Demonstrate two one-to-one correspondences between the sets $A = \{1, 2, 3, 4\}$ and $B = \{a, b, c, d\}$.

5. The set of natural numbers is an infinite set. Give an example of one subset of the natural numbers that is finite and another that is infinite.

6. Jim, Janice, and Jane are running for the club offices of president, secretary, and treasurer. How many different ways are there for the election results to occur?

7. Consider a set of 60 attribute blocks and let x represent any block. Determine whether each statement is true or false.

   a. $\{x \,|\, x \text{ is red}\} = \{x \,|\, x \text{ is blue}\}$.

   b. $\{x \,|\, x \text{ is yellow}\} \sim \{x \,|\, x \text{ is red}\}$.

   c. $\{x \,|\, x \text{ is thin}\} \sim \{x \,|\, x \text{ is large}\}$.

   d. $\{x \,|\, x \text{ is thin}\} \subset \{x \,|\, x \text{ is thick}\}$.

   e. $\{x \,|\, x \text{ is thin and red}\} \subseteq \{x \,|\, x \text{ is red}\}$.

   f. $\{x \,|\, x \text{ is thick and a square}\} = \{x \,|\, x \text{ is yellow and thick}\}$.

8. If $A = \{1, 3, 4, 5, 7\}$ and $B = \{2, 4, 5, 6, 7\}$, find:

   a. $A \cup B$.

   b. $A \cap B$.

   c. $A - B$.

   d. $B - A$.

9. Let $U = \{1, 2, 3, \ldots, 8\}$, $A = \{1, 3, 5, 6\}$, $B = \{2, 3, 4, 7\}$, and $C = \{2, 4, 6, 7, 8\}$. Find:

   a. $\overline{A}$.   b. $\overline{B}$.   c. $\overline{C}$.   d. $\overline{A \cup B}$.

10. Let $A = \{\bigcirc, \square\}$ and $B = \{\text{red, white, blue}\}$. Find $A \times B$.

11. Let $U = \{1, 2, 3, 4, 5, 6, 7, 8\}$, $A = \{3, 4, 5, 7, 8\}$, $B = \{1, 2, 3, 5, 6\}$, and $C = \{1, 2, 6, 7, 8\}$. Find:

    a. $(A \cup C) \cap B$.

    b. $(\overline{A - C})$.

    c. $A \cup B \cap \overline{C}$.

    d. $(\overline{B \cap A}) - C$.

12. Draw a Venn diagram of each set.

    a. $\overline{A} \cap B$ b. $\overline{A} - B$

    c. $(A \cup B) \cap (A \cup C)$ d. $A \cap (B \cup \overline{C})$

13. Use set notation to represent each of the following shaded regions.

    a.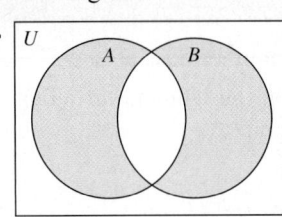

b.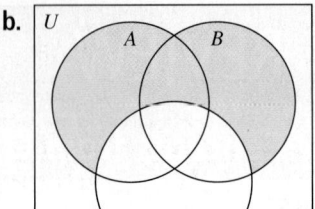

14. If $n(A) = 10$ and $n(B) = 8$, what is $n(A \times B)$?

15. If $n(A) = 3$ and $n(B) = 5$, what are possible numbers for $n(A \cup B)$? For $n(A \cap B)$?

16. Suppose sets A and B are subsets of universe U. If $n(U) = 60$, $n(A) = 23$, $n(B) = 34$, and $n(A \cap B) = 8$, what are $n(A - B)$ and $n(B - A)$?

17. Let $n(U) = 40$, $n(A) = 18$, $n(B) = 15$, and $n(A \cap B) = 7$. Use a Venn diagram to find:

    a. $n(A - B)$.

    b. $n(B - A)$.

    c. $n(\overline{A \cup B})$.

18. Use the following diagram to find:

    a. $n(A \cap B)$.

    b. $n(A \cup B)$.

    c. $n(\overline{A} \cap B)$.

    d. $n(A \cap \overline{B})$.

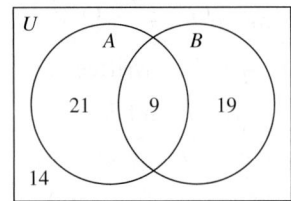

19. Use the following diagram to find:

    a. $n(A \cap B \cap C)$.

    b. $n(B)$.

    c. $n(A \cup C)$.

    d. $n(A \cup \overline{B} \cup \overline{C})$.

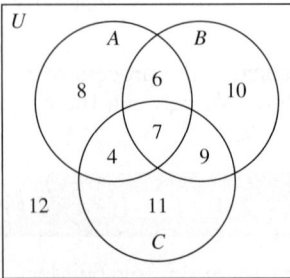

20. In a group of 75 fast-food eaters, 35 liked hamburgers, 25 liked chicken sandwiches, and 10 liked both. How many liked hamburgers but not chicken sandwiches?

21. A cosmetics magazine does a survey that has 145 respondents. Of the respondents, 90 wear lipstick, 80 wear mascara, 75 wear blush, 50 wear lipstick and blush, 60 wear lipstick and mascara, 45 wear mascara and blush, and 35 wear all three.

   a. How many respondents wear only mascara?

   b. How many respondents wear no makeup?

22. An employer examines a pool of 236 applicants and finds that 117 have technical skills, 61 have mathematical skills, 152 have writing skills, 49 have technical and mathematical skills, 31 have mathematical and writing skills, 73 have technical and writing skills, and 22 have all three. If the employer is interested only in applicants with mathematical skills and writing skills but not technical skills, how many applicants must the employer seriously consider?

23. Draw a Venn diagram of sets A, B, and C so that $A \cup B \subseteq C$.

24. **a.** If $A \cup B = A \cup C$, does it follow that $B = C$? Explain.

   b. If $A \cap B = A \cap C$, does it follow that $B = C$? Explain.

25. Show that $\overline{A \cup B} = \overline{A} \cap \overline{B}$ by shading a Venn diagram.

26. Use a Venn diagram and the numbered regions method to show that $A \cap (B \cup C) = (A \cap B) \cup (A \cap C)$.

27. If A, B, and C are sets, use set identities to show $A - (A - B) = A \cap B$.

Answers to Chapter 2   Check Your Understandings and Activities

Check Your Understanding 2.1-A

1. **a.** Well-defined **b.** Not well-defined. One way to rewrite the set is, "The set of all students with a GPA of 3.0." **c.** Well-defined.

2. **a.** {California, Colorado, Connecticut} = $\{x \mid x$ is a state in the United States beginning with C} **b.** {1, 3, 5, 7, 9} = $\{x \mid x$ is a odd number less than 10} **c.** {5, 10, 15, 20, . . .} = $\{x \mid x$ is a multiple of 5}

3.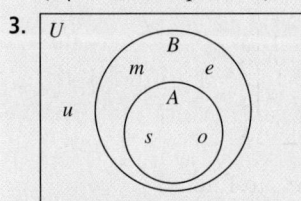

4. Subsets include ∅, {black}, {white}, and {black, white}. Proper subsets include ∅, {black}, and {white}.

Check Your Understanding 2.1-B

1. **a.** 10   **b.** 7   **c.** 5

2. Three possible answers are:

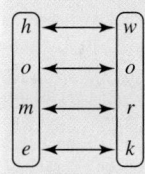 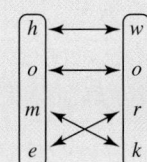 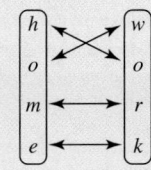

3. 5,040

Check Your Understanding 2.2-A

1. **a.** {1, 2, 3, 4, 5} **b.** {3, 5} **c.** {2, 4} **d.** {1} **e.** {1, 6}

2. $A \times B$ = {(red, yellow), (red, green), (white, yellow), (white, green), (blue, yellow), (blue, green)}.

3.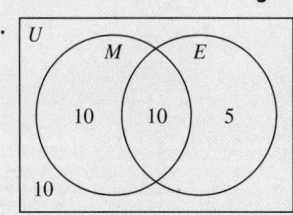

4. **a.** {a, c, d, e, f} **b.** {b, g} **c.** {(a, c), (a, e), (a, f), (d, c), (d, e), (d, f), (f, c), (f, e), (f, f)} **d.** {a, b, c, d, e, f, g} = U

Check Your Understanding 2.2-B

1. **a.** 12   **b.** 30   **c.** 15   **d.** 2
   e. 4   **f.** 5   **g.** 24   **h.** 10

2.

U
M   E
10   10   5
10

Ten are taking mathematics but not English.

3.

U
Blue   Large
5   5
5   10
5
5   10
10
10   Thin

Activity 2.1

Answers will vary.

Activity 2.2

a. Counting on is a common strategy used to teach addition.

b. Counting back is a common strategy used to teach subtraction.

c. Counting on and skip counting are commonly used when counting money.

d. Counting on, counting back, and skip counting are all used when measuring time.

Activity 2.3

$n(A \cup B) = n(A) + n(B) - n(A \cap B)$.

Activity 2.4

Here is one possible solution:

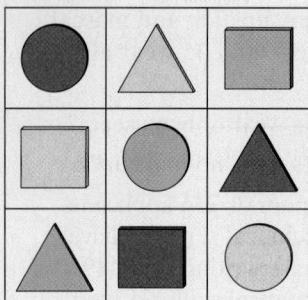

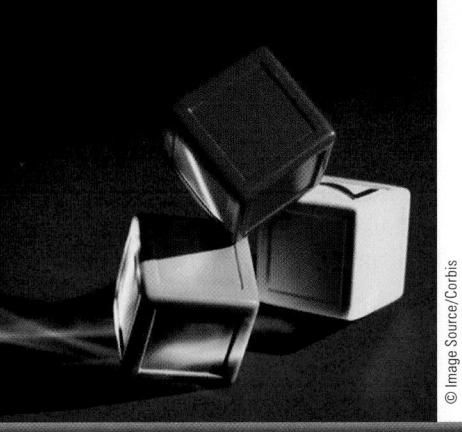

© Image Source/Corbis

Chapter 3

Numbers and Numeration

Opening Problem

All modern computers use a code called ASCII. This binary code represents all letters and symbols with strings of eight digits containing only 0s and 1s. What string of digits represents the word "MATH," if the letters A through Z are represented by the consecutive binary numbers A = 01000001, B = 01000010, C = 01000011, . . . and Z = 01011001?

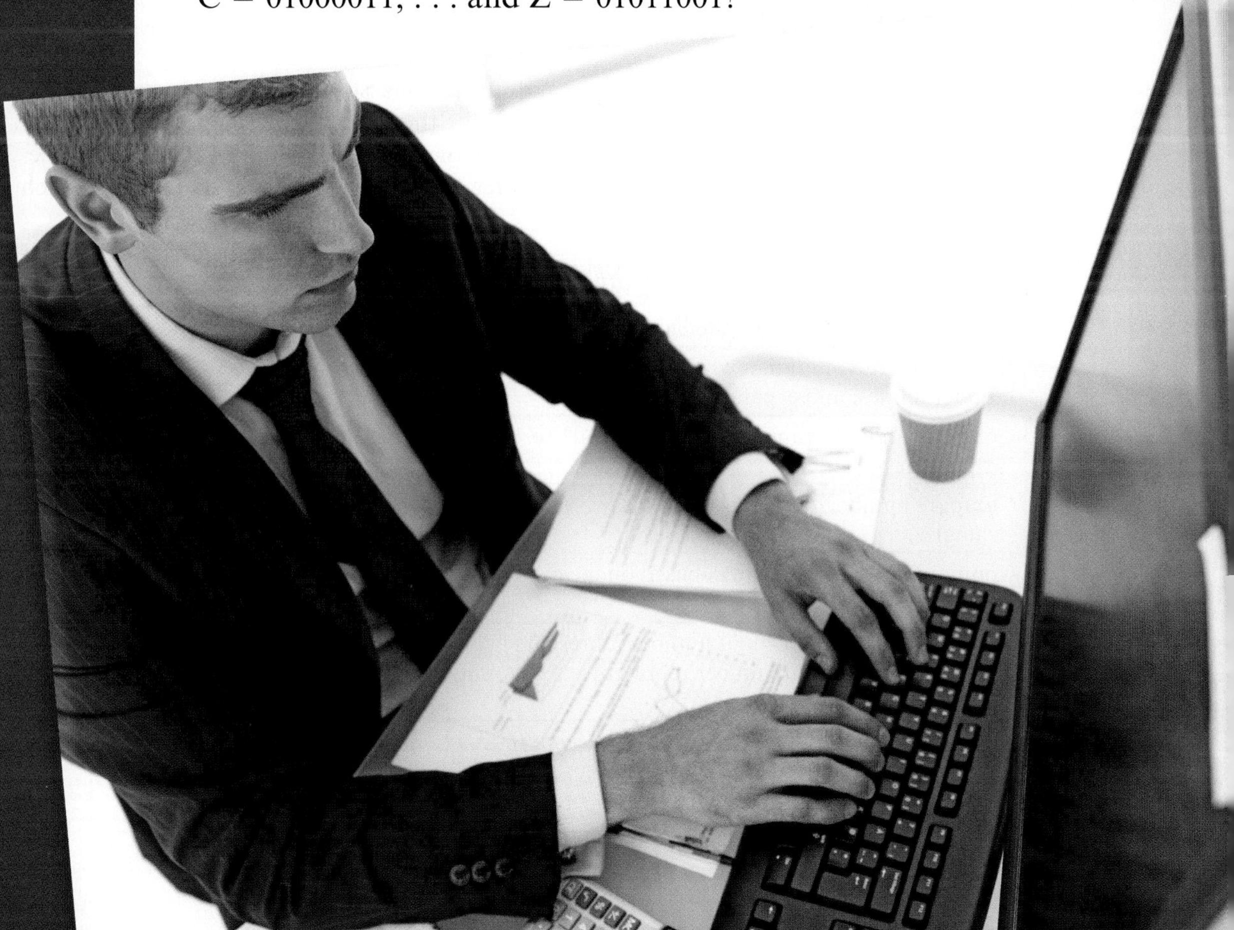

Yuri Arcurs/Shutterstock.com

© Image Source/Corbis

Once children have learned to count, they can use counting to learn about numbers, numeration, and numerical operations. It is essential that children master base-ten numeration and numerical operations because they are the fundamental building blocks for almost all of mathematics. Their importance is reflected in the NCTM Number and Operations Standard.

■ NCTM Number and Operations Standard

Instructional programs from prekindergarten through grade 12 should enable all students to:

- Understand numbers, ways of representing numbers, relationships among numbers, and number systems.
- Understand meanings of operations and how they relate to one another.
- Compute fluently and make reasonable estimates.

Source: NCTM STANDARDS Copyright © 2011 by NATIONAL COUNCIL OF TEACHERS OF MATHEMATICS. Reproduced with permission of NATIONAL COUNCIL OF TEACHERS OF MATHEMATICS.

The Number and Operations Standard covers an extensive part of the elementary curriculum, so we take the next several chapters to discuss it in detail. In this chapter, our focus is on the first goal of the Standard. Table 3.1 gives specific objectives covered under this goal and the grade level at which each objective is likely to be met.

Table 3.1 Classroom learning objectives	K	1	2	3	4	5	6	7	8
Write numerals for the numbers 0 to 20.	X								
Read and write multidigit whole numbers using base-ten numerals, number names, and expanded form.		X	X						
Compose and decompose numbers from 11 to 19 into 10 ones and some further ones.	X								
Compare numbers between 1 and 10 by using matching and counting strategies.	X								
Compare multidigit numbers based on meanings of the digits in each place.		X	X		X				
Understand that a two-digit number represents an amount of tens and ones and a three-digit number represents an amount of hundreds, tens, and ones.		X	X						
Recognize that a digit in one place represents 10 times what it represents in the place to its right and 1/10 of what it represents in the place to its left.					X	X			
Represent whole numbers on a number line.			X						

Source: Adapted from the *Common Core State Standards for Mathematics* (*Common Core State Standards Initiative*, 2010).

The objectives in Table 3.1 encompass two important topics from elementary mathematics: base-ten numeration and number sense. At first, most of what children learn about numbers is learned orally. However, once in school, they must learn to represent numbers with written numerals. In kindergarten, children learn to write the digits of the decimal system and associate them with values. Then, in the first and second grades, children use the digits to write numerals for larger values. In doing so, they learn about place value, groupings, and making exchanges. These ideas are difficult for some students to master, but they must do so because these ideas are not only important in writing numerals but also in computing with large numbers.

As children learn numeration, they also develop number sense. Number sense involves several key ideas, not the least of which is understanding numerical values. Children first develop a concept of small numbers when they count different sets of the same size. As they count, they come to realize that the number of objects in the set is independent of the objects themselves. Once they make this connection, children can apply their concept of number to other sets, as well as learn to order numbers. Initially, they learn to order numbers by counting or by making direct comparisons between sets. However, once children start to work with larger numbers, they turn to using place value.

Another important part of number sense involves representing and using numbers in different ways. If children can be flexible in how they think about numbers, they have a greater ability to apply what they know simply by choosing the representation that is best suited to their needs. In the early grades, this primarily entails making exchanges between base-ten place values.

As adults, most of us take our knowledge of counting and numeration for granted. We forget how long it took and how difficult it was for us to master these ideas. For this reason, we need to take another look at the concepts associated with number and numeration. Consequently, Chapter 3:

- Gives a brief history of numeration to introduce its different features.
- Describes the characteristics and properties of the decimal system.
- Uses other base numeration systems to further illustrate the properties of the decimal system.

SECTION 3.1 A Brief History of Numeration

We begin our study of numeration by making a distinction between numbers and numerals. **Numbers** are abstract ideas of quantity that are difficult to define. We come to an intuitive understanding of their values only by experiencing them in a variety of situations. **Numerals,** on the other hand, are symbolic representations of numbers that take on different forms. For instance, the symbols "3" and "|||" are two different numerals for the number three. Because numerals can take a variety of forms, we must learn how to use different symbols and rules to write them. Any collection of symbols and rules that govern how numerals are written is called a **numeration system.**

Anthropologists believe that most cultures throughout human history have had some concept of number and numeration. Although the abstract notion of numbers has generally remained the same, numeration systems have not. Some have been quite simple. For instance, some Australian aboriginal tribes use specific words for small numbers like one or two but express larger values with a generic term like "many." In the Amazon, some native tribes have words for one and two but then count higher by using phrases like "two-one" and "two-two" for three and four. It is thought that the first numeration systems were verbal like these, which were adequate for the small groups of nomadic people who used them. However, as civilization developed, the need to record large numbers in concise ways caused people to develop written numeration.

The Tally System

In 1937, a record of the earliest known numeral was found in Czechoslovakia on a Paleolithic wolf bone. The bone, shown in Figure 3.1, had 55 notches cut into it. It is believed to have been used to record something like the number of animals in a herd, the number of people in a tribe, or even the cycles of the moon. The markings illustrate a numeral in the tally system. This simple system uses a strict one-to-one correspondence between the objects to be counted and another set of objects that are easy to employ, such as pebbles, shells, or marks on a bone. For example, to represent five objects, we simply make five tallies: ||||| = 5.

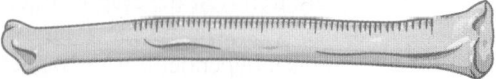

Figure 3.1 Tally marks on a Paleolithic wolf bone

Although simple to use, the tally system works well only with small numbers. Larger numbers require many more tallies, making the numerals difficult to read. For instance, determining how many tallies are shown here is difficult without counting them directly.

|||||||||||||||||||||||

At some point, tally marks became grouped, most commonly in groups of five. With this slight change, the preceding numeral is much easier to recognize as 23.

卌 卌 卌 卌 ||| = 23

Archeological records indicate that the tally system was the only numeration system in use for thousands of years. It was not until the ancient Egyptians that the first improvements in numeration began to appear.

Egyptian Numerals

The Egyptian system dates back to about 3500 B.C.E. The Egyptians improved the tally system by creating a **grouping system.** In this kind of system, once a certain number of objects are counted, they are grouped together and represented with just one symbol. The number of objects needed to make a group is called the **base** of the system. The Egyptians used a **base of ten;** any time a group of ten was counted, the group was represented with a new symbol. Consequently, to write an Egyptian numeral, all that was needed was the appropriate number of symbols for each grouping. The Egyptians used **hieroglyphics,** or picture symbols, to represent their numbers. Table 3.2 shows the specific symbols used for each group.

Table 3.2	Symbols and values of Egyptian numeration		
Egyptian Symbol	**Name**	**Exchange Factor**	**Value**
|	Stroke or staff	Unit	1
∩	Heel bone	= 10 staffs	10
ᕱ	Scroll or coiled rope	= 10 heel bones	100
⌠	Lotus flower	= 10 scrolls	1,000
⌡	Pointing finger	= 10 lotus flowers	10,000
⌢	Tadpole	= 10 pointing fingers	100,000
⍦	Astonished person	= 10 tadpoles	1,000,000
⍓	Rising sun	= 10 astonished persons	10,000,000

To find the value of an Egyptian numeral, we use Table 3.2 to translate each symbol and then add their values. For instance, consider ⌡⌡⌡⌡ᕱᕱᕱ∩∩|||||:

⌡⌡⌡⌡ represents $10{,}000 + 10{,}000 + 10{,}000 + 10{,}000 = 40{,}000$

ᕱᕱᕱ represents $100 + 100 + 100 = 300$

∩∩ represents $10 + 10 = 20$

||||| represents $1 + 1 + 1 + 1 + 1 = 5$

By adding, we find ⌡⌡⌡⌡ᕱᕱᕱ∩∩||||| has a value of $40{,}000 + 300 + 20 + 5 = 40{,}325$.

| Example 3.1 | **Writing Egyptian Numerals** |

Representation

Write each number as an Egyptian numeral.

a. 4,512   **b.** 1,340,256

Solution

a. 4,512 = 4,000 + 500 + 10 + 2

= ℒℒℒℒ𝟿𝟿𝟿𝟿𝟿∩II.

b. 1,340,256 = 1,000,000 + 300,000 + 40,000 + 200 + 50 + 6

= 𝒴ꙍꙍꙍ∩∩∩∩𝟿𝟿∩∩∩∩∩IIIIII.

The last example illustrates several features of the Egyptian system. First, it did not entirely do away with the tally system; tallies were still used to represent the first nine numerals. Second, the Egyptian system was **additive.** In an additive system, the value of a numeral is found by adding the values of each symbol in the numeral. Because the value of an Egyptian numeral depended only on the symbols, the system was **nonpositional.** The symbols could be written in any order, and the value of the numeral would still be clear. For example, the numerals in Figure 3.2 are all legitimate ways to represent 245:

𝟿𝟿∩∩∩∩IIIII   𝟿𝟿∩∩∩∩   IIIII∩∩∩∩𝟿𝟿
                            IIIII

Figure 3.2 Different Egyptian numerals for 245

Although Egyptian numeration was an improvement over the tally system, it had problems. First, certain small numbers still required the use of many symbols. For instance, the Egyptian numeral for 59 required 14 different symbols. Second, the system made certain computations, like multiplication and division, difficult to perform. Despite its shortcomings, the Egyptians used it to construct some of the wonders of the ancient world, including the great pyramids at Giza.

Roman Numerals

Roman numeration is another system worth mentioning because we still use it today in such places as in movie credits and on the faces of old clocks. Like the Egyptian system, this is a grouping system, but it does not use a common base. Instead, all Roman numerals are written using combinations of the seven symbols shown in Table 3.3.

Table 3.3 **Symbols and values of Roman numeration**						
Roman numeral	I	V	X	L	C	D	M
Value	1	5	10	50	100	500	1,000

The Roman system is also additive, but, to condense the length of certain numerals, it is also **subtractive.** If a smaller symbol appears to the left of a larger one, the value of the smaller is subtracted from that of the larger. Subtraction is done only in the following prescribed ways:

I only precedes V or X

X only precedes L or C

C only precedes D or M

As a consequence, no symbol ever appeared in a Roman numeral more than three times in succession. For instance, we would write:

- 4 as IV instead of IIII because 4 = 5 − 1.
- 40 as XL instead of XXXX because 40 = 50 − 10.
- 90 as XC instead of LXXXX, since 90 = 100 − 10.

Using this convention, the first 20 Roman numerals were I, II, III, IV, V, VI, VII, VIII, IX, X, XI, XII, XIII, XIV, XV, XVI, XVII, XVIII, XIX, and XX.

Example 3.2 Interpreting Roman Numerals

Representation

Find the value of each Roman numeral.

a. LVII   **b.** CCCXCIV   **c.** MCMLXXVIII

Solution In each case, work with the symbols in groups from left to right.

a. $\underset{50}{\underline{L}}\ \underset{7}{\underline{VII}} = 50 + 7 = 57.$

b. $\underset{300}{\underline{CCC}}\ \underset{90}{\underline{XC}}\ \underset{4}{\underline{IV}} = 300 + 90 + 4 = 394.$

c. $\underset{1,000}{\underline{M}}\ \underset{900}{\underline{CM}}\ \underset{70}{\underline{LXX}}\ \underset{8}{\underline{VIII}} = 1,000 + 900 + 70 + 8 = 1,978.$

The symbols in Table 3.3 allow us to write values only up to about 3,900. For larger values, the Romans placed a bar over a portion of a numeral to signify that it was to be multiplied by 1,000. For instance, $\overline{IX}CCIII = (9 \times 1,000) + 200 + 3 = 9,203.$

Example 3.3 Writing Roman Numerals

Representation

Write each number in Roman numerals.

a. 109,000   **b.** 1,450,550

Solution

a. Because $109,000 = 109 \times 1,000$ and CIX represents 109, then $\overline{CIX}$ represents 109,000.

b. $1,450,550 = (1,450 \times 1,000) + 550$. Because MCDL represents 1,450, then $\overline{MCDL}$ represents $1,450 \times 1,000$. Next, $550 = 500 + 50$, so we use DL to represent 550. Putting the parts together, we have $1,450,550 = \overline{MCDL}DL.$

Check Your Understanding 3.1-A

1. Find the value of each numeral.

a. ꟷꟷꟷꟷꟷꟷꟷꟷ II

b. ꓕꓕꓕ꘎∩∩∩∩∩IIII

c. ꝺꝺꝺꝺ ꓕꓕℒℒℒℒℒ∩∩IIIIIIIII

d. DCCXLVII

e. $\overline{LXII}CDIV$

2. Write 57 using the tally system.

3. Write each number as an Egyptian and a Roman numeral.

    a. 55   **b.** 6,418   **c.** 1,302,040

Talk About It How are the Egyptian and Roman numeration systems like the numeration system we use? How are they different?

Representation

> ## Activity 3.1
>
> A certain numeration system is a grouping system that is additive and uses a base of ten. For symbols, it uses only lowercase English alphabet letters, where a = 1, b = 10, c = 100, and so on through z, which represents the largest group of ten in the system.
>
> **a.** What is the largest power of ten used in this system?
>
> **b.** What is the numerical value of each of the following words?
>   **i.** cafe   **ii.** feed   **iii.** babble   **iv.** head   **v.** fable
>
> **c.** Find an English word that in this numeration system has a value of:
>   **i.** 110,101   **ii.** 22,000   **iii.** 200,000,010,011

Babylonian Numerals

Another interesting numeration system is the Babylonian system. The Babylonians used a form of writing called **cuneiform,** which was created by making wedge-shaped impressions in clay tablets using a writing utensil called a stylus. A stylus could make only two shapes: a vertical stroke (|) and a wedge (▲). So all writing was done using different combinations of these shapes. (See Figure 3.3.)

Figure 3.3 Cuneiform writing

When writing numerals, the Babylonians used two basic symbols:

- ▼ was used to represent the unit, or 1.

- ◀ was used to represent a group of 10.

Like the Egyptian system, the Babylonian system was additive, so the value of smaller numerals was found by adding the values of each symbol in the numeral. For instance, the first 20 Babylonian numerals are shown in Figure 3.4.

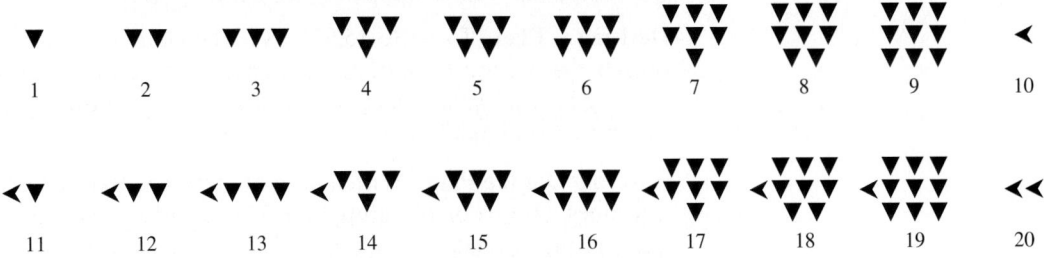

Figure 3.4 First 20 Babylonian numerals

The Babylonians used this convention only for values less than 60. For larger numbers, they chose to create new positions in their numerals rather than new symbols. The Babylonian system was therefore **positional,** or had place value. Having **place value** means that the position of a particular symbol, called a **digit,** determines its value. The numerals for 1 to 59 also served as the digits in the Babylonian system, giving it a base of sixty. The positions were read from right to left and represented increasing powers of 60. A change in position was often indicated by a small space between the digits. The values of the first four positions are given in Table 3.4.

Table 3.4   First four positions of Babylonian numerals

Position	Fourth	Third	Second	First
Value	$60 \times 60 \times 60 = 216{,}000$	$60 \times 60 = 3{,}600$	60	1

The additional positions meant that the system also had to be **multiplicative,** which means the value of any digit is found by multiplying its value by its corresponding place value. By adding all such products, we get the total value for the numeral.

Example 3.4   Interpreting Babylonian Numerals

Representation

Find the value of each Babylonian numeral.

a.   b.

Solution

a. The numeral uses only one place value, so we need to consider only the value of the digit:

$$\text{(numeral)} = (3 \times 10) + (8 \times 1) = 38$$

b. This numeral has two positions. The digit farthest to the right, , is in the ones position, giving it a value of $27 \times 1 = 27$. The digit to the left, , is in the sixties position, giving it a value of $15 \times 60 = 900$. By adding the values, we get

$$\text{(numeral)} = (15 \times 60) + (27 \times 1) = 900 + 27 = 927$$

Example 3.5   Writing a Babylonian Numeral

Problem Solving

Represent 15,971 as a Babylonian numeral.

Solution

Understanding the Problem. Our task is to represent 15,971 as a Babylonian numeral. The Babylonian system was positional, with a base of sixty, and it was both additive and multiplicative.

Devising a Plan. To write 15,971 as a Babylonian numeral, imagine having 15,971 objects that we must regroup in the base of the Babylonian system. We can do so by *using basic arithmetic.* Once we know how many groups are in each place value, we can write the numeral.

Carrying Out the Plan. With a base of sixty, the Babylonian system grouped objects into ones, groups of 60, groups of 3,600, and so on. We need to only consider the ones, groups of 60, and groups of 3,600 because there is no way to make larger groups using only 15,971 objects. To determine how many groups of 3,600 are in

15,971 objects, divide 15,971 by 3,600. Because 15,971 ÷ 3,600 = 4 remainder 1,571, we can make four groups of 3,600 from 15,971 objects with 1,571 objects left over. Next, we determine how many groups of 60 we can make from 1,571 objects. Because 1,571 ÷ 60 = 26 remainder 11, we can make 26 groups of 60 with 11 left over. Using this information, we have

$$\underset{\substack{\text{Position 3}\\(\text{Groups}\\ \text{of 3,600})}}{4} \quad \underset{\substack{\text{Position 2}\\(\text{Groups}\\ \text{of 60})}}{26} \quad \underset{\substack{\text{Position 1}\\(\text{Ones})}}{11} \quad \text{equals} \quad \blacktriangledown\!\blacktriangledown \, \text{<<} \blacktriangledown\!\blacktriangledown\!\blacktriangledown \, \text{<}\blacktriangledown$$

Looking Back. We can check our answer by converting it back: $\blacktriangledown\!\blacktriangledown \, \text{<<} \blacktriangledown\!\blacktriangledown\!\blacktriangledown \, \text{<}\blacktriangledown$ = $(4 \times 3,600) + (26 \times 60) + (11 \times 1) = 14,400 + 1,560 + 11 = 15,971$. Regrouping works well in this problem and may be a strategy to keep in mind as we work with other numeration systems.

∎

Although the Babylonian system was powerful, it had a drawback: It had no real way to represent empty positions. The Babylonians first addressed the problem by leaving a gap for empty positions, but the rule was not always strictly obeyed. Around 300 B.C.E., the symbols ▲ and ◢ were introduced to indicate empty positions in the numeral. Even with this convention, two numbers with the same representation could still be confused. For instance, without a placeholder, only context indicated whether the numeral ◄▼▼◄◄▼▼▼ represented $(12 \times 60) + (24 \times 1) = 744$ or $(12 \times 3,600) + (24 \times 60) = 44,640$.

Mayan Numerals

Explorations
Manual
3.1

Mayan numeration was similar to the Babylonian system in that it was additive, multiplicative, and positional. Small values were written using combinations of the three symbols shown in Table 3.5.

Table 3.5 Symbols and values of Mayan numerals		
Mayan Symbol	⬭	•	———
Value	0	1	5

The symbol, ⬭, was used in two ways. First, it represented **zero,** which is an idea that almost no other ancient numeration system had. Second, it was also used as a **placeholder;** that is, the symbol was put into a numeral to move other nonzero digits out to the appropriate position but did not otherwise add any value. Using these symbols, the first 20 Mayan numerals were written as shown in Figure 3.5.

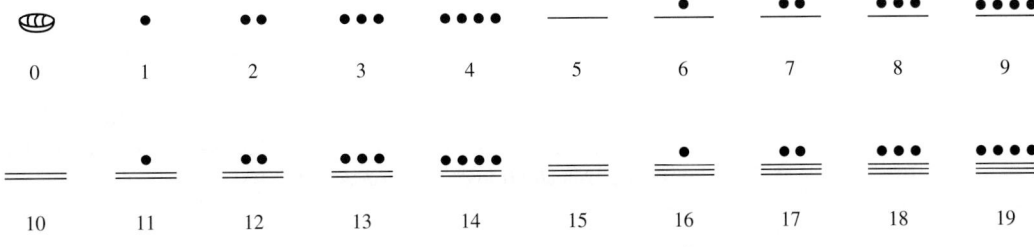

Figure 3.5 First 20 Mayan numerals

For larger numbers, the Mayans extended their numerals to other place values. They primarily used a base of twenty, except for the third position, which had a value of $18 \times 20 = 360$. (It is believed that 360 was an important value for their yearly calendar.) From the fourth position on, the system continued to use a base of twenty. The first four positions and their corresponding values are shown in Table 3.6. They are listed vertically, because the Mayans chose to write their numerals in this way.

Table 3.6 First four positions of Mayan numerals	
Position 4	$360 \times 20 = 7{,}200$
Position 3	$18 \times 20 = 360$
Position 2	20
Position 1	1

Example 3.6 Interpreting Mayan Numerals

Representation

Find the value of each Mayan numeral.

a.

b.

Solution Using Table 3.6, multiply the value of each digit by its place value, and then add the resulting products.

a.

$$\underline{\bullet\bullet} = 7 \times 20 = 140$$

$$\underline{\underline{\bullet\bullet}} = 12 \times 1 = \underline{12}$$
$$152$$

b.

$$\bullet\bullet = 2 \times 360 = 720$$

$$\oplus = 0 \times 20$$

$$\underline{\underline{\bullet\bullet\bullet}} = 18 \times 1 = \underline{18}$$
$$738$$

Example 3.7 Writing a Mayan Numeral

Representation

Represent 36,276 in Mayan numerals.

Solution Using the same process of regrouping as in Example 3.5, we can represent the number as $36{,}276 = (5 \times 7{,}200) + (0 \times 360) + (13 \times 20) + (16 \times 1)$. Translating these values into Mayan digits gives us

$$5 \times 7{,}200 = \overline{}$$

$$0 \times 360 = \oplus$$

$$13 \times 20 = \underline{\underline{\bullet\bullet\bullet}}$$

$$16 \times 1 = \underline{\underline{\bullet}}$$

Explorations
Manual
3.2

These five numeration systems have shown not only the diversity of numerals but also several key features of numeration systems. The next example summarizes and compares the features of these systems.

Example 3.8 Summarizing the Numeration Systems

Connections

Create a table that summarizes the characteristics of the tally, Egyptian, Roman, Babylonian, and Mayan numeration systems.

Solution Numeration systems can use tallies and group objects with different bases. They can be additive, subtractive, and multiplicative. They can also be positional and have a concept of zero. The following table summarizes the characteristics of these numeration systems.

Characteristic	Tally	Egyptian	Roman	Babylonian	Mayan
Uses tallies	Yes	Up to 10	Up to 3	Up to 10	Up to 5
Groups objects using a base	No	Base of ten	Yes, but base is not constant	Base of sixty	Mostly base of twenty
Additive	No	Yes	Yes	Yes	Yes
Subtractive	No	No	Yes	No	No
Multiplicative	No	No	Yes	Yes	Yes
Positional	No	No	No, except with regard to being subtractive	Yes	Yes
Uses zero	No	No	No	No	Yes

The features of the Babylonian and Mayan systems gave these systems more power by enabling users to write large numbers with only a few symbols. Another ancient system, called the Hindu-Arabic system, uses many of the same features as the Babylonian and Mayan systems. However, unlike those two systems, the Hindu-Arabic system is still in use today. Because of its importance both in the United States and around the world, we take a closer look at this system in the next section.

Explorations Manual 3.3

Check Your Understanding 3.1-B

1. Express each Babylonian numeral in decimal notation.

a. ◀◀◀▼▼▼   b. ◀▼▼▼▲◀◀▼▼▼   c. ◀▼▲◀▼▲◀▼

2. Write each Mayan numeral in decimal notation.

a. •••   b. ••••   c.
   ════   ════

3. Write each number as a Babylonian numeral and as a Mayan numeral.

a. 17   b. 75   c. 1,874

Talk About It How are the Babylonian and Mayan numeration systems like the numeration system we use? How are they different?

Problem Solving

Activity 3.2

Consider the following numeration system, which has the properties of the Mayan system: It is a grouping system with one number for its base; it is positional, additive, and multiplicative; and it has a concept of zero. Further, suppose that the capital letters from the English alphabet are the only symbols it uses for digits, where $A = 0$, $B = 1$, $C = 2$, and so on. Answer the following questions:

a. What is the base of the numeration system?

b. What are the first four place values in the system?

c. What is the numerical value of each of these words?

   i. DOG   ii. CAT   iii. HELP   iv. BLUE

SECTION 3.1   Learning Assessment

■ Understanding the Concepts

1. **a.** Explain the difference between a number and a numeral.

   b. Is it possible to write a number? Why or why not?

2. **a.** What does it mean for a numeration system to be a grouping system?

   b. What other characteristics must a grouping numeration system possess?

3. What does it mean for a numeration system to be additive? Subtractive?

4. **a.** What does it mean for a numeration system to be multiplicative?

   b. If a system is additive, must it also be multiplicative?

   c. If a system is multiplicative, must it also be additive?

   d. What other characteristics must a numeration system have to be multiplicative?

5. **a.** What does it mean for a numeration system to be positional?

   b. What role do the digits play in a positional system?

   c. What role does the base play in a positional system?

6. What role does a zero symbol play in numeration?

■ Representing the Mathematics

7. Determine whether each statement relates to a number or a numeral.

   a. Most children find 1 easy to write.

   b. Five is the age when most children begin school.

   c. Children often mistakenly read 0 as "oh."

   d. 4 is bigger than 6.

8. Write five different numerals for the number 5.

9. What advantages do Egyptian numerals offer over tally numerals?

10. What advantages do Babylonian and Mayan numerals offer over Egyptian numerals?

11. What are the disadvantages to using the:

    a. Egyptian system?

    b. Roman system?

    c. Babylonian system?

12. Which of the following are proper representations for the number 34?

    a. ⦀⦀ ⦀⦀ ⦀⦀ ⦀⦀ ⦀⦀ ⦀⦀ ‖‖   **b.** XXLIV

    c. ∩∩∩∩‖‖   **d.** XXXIV

---

e. ⦗⦗⦗▼▼▼   **f.** ‖‖‖∩∩

g. ▼▼▼▲⦗⦗⦗   **h.** XXXIIII

■ Exercises

13. What number is represented by each tally numeral?

    a. 卌 卌 卌 卌 卌 ‖   **b.** 卌 卌 卌 卌 卌 卌 卌 ‖‖‖

14. Write 18 and 29 as tally numerals.

15. What number is represented by each Egyptian or Roman numeral?

    a. 𓏺𓏺∩∩∩∩∩‖‖‖‖   **b.** 𓆼𓏺𓏺𓏺‖‖‖‖

    c. ∩∩∩𓂭𓂭𓂭𓏺∩‖‖‖‖‖‖‖   **d.** XIV

    e. LXXVII   **f.** CCXLIX

16. What number is represented by each Egyptian or Roman numeral?

    a. 𓂭𓂭𓂭𓂭 𓏺𓏺𓏺𓏺𓏺 ∩∩‖‖‖‖‖

    b. 𓆼𓆼𓆼 ∩∩∩∩∩𓂭𓂭 𓏺𓏺𓏺𓏺‖

    c. 𓂭𓃀𓃀 𓂭𓂭𓂭𓂭𓂭 ∩∩∩‖‖‖‖

    d. MMCDLXIII

    e. X̄L̄ĪX̄DCCLXV   **f.** M̄X̄L̄CLXIV

17. Write each number as an Egyptian and as a Roman numeral.

    a. 23   **b.** 57   **c.** 182   **d.** 283

18. Write each number as an Egyptian and as a Roman numeral.

    a. 1,987   **b.** 24,681

    c. 56,703   **d.** 1,040,360

19. Write the Egyptian numerals for each of the numbers 98 through 102.

20. Write the Roman numerals for each of the numbers 148 through 152.

21. What number is represented by each Babylonian or Mayan numeral?

    a. ⦗⦗⦗▼▼▼   **b.** ⦗⦗⦗⦗▼▼▼

    c. ⦗▼⦗⦗▼▼   **d.** ••••

    e. ▬▬ ••   **f.** ▬▬ ••••

22. What number is represented by each Babylonian or Mayan numeral?

    a. ▼⦗⦗▼▼▼ ⦗⦗⦗⦗▼▼▼

    b. ⦗▼▼▼▲⦗▼▼▼ ▼▼▼▲⦗▼▼▼

c. ◄▼▼▲◄◄▼▲◄◄◄▼▼▼

d. [Mayan numeral: two dots over shell]   e. [Mayan numeral: dot / two dots / three dots]   f. [Mayan numeral: four dots over bar over shell]

d. ◄◄▼▼▼ or ◄◄▲▼▼▼

e. [Mayan: two bars over dot, or dot over two bars]

23. Write each number as a Babylonian and as a Mayan numeral.

  a. 34   **b.** 61

  c. 168   **d.** 309

24. Write each number as a Babylonian and as a Mayan numeral.

  a. 2,919   **b.** 9,806

  c. 13,436   **d.** 31,015

25. Write the Babylonian numerals for each of the numbers 658 through 662.

26. Write the Mayan numerals for each of the numbers 118 through 122.

27. How many symbols are needed to write the number 512 as a(n):

  a. Tally numeral?   **b.** Babylonian numeral?

  c. Egyptian numeral?   **d.** Roman numeral?

28. Determine the number of symbols needed to write each number in the given numeration system.

  a. 23,457 in the tally system

  b. 3,560,245 in Egyptian numerals

  c. 451 in Babylonian numerals

  d. 3,982 in Roman numerals

  e. 4,319 in Mayan numerals

  f. 123,456,789 in Egyptian numerals

29. Complete each sequence of consecutive numerals.

  a. ∩IIIIIII, ____, ____, ____

  b. LIX, ____, ____, ____

  c. ◄◄◄▼▼▼ over ▼ , ____, ____, ____

  d. [Mayan: two dots over bar], ____, ____, ____

30. Fill in the blanks in each sequence of consecutive numerals.

  a. ____, ____, ꝑ, ____

  b. ____, CD, ____, ____,

  c. ____, ____, ____, ◄◄▼▼ over bar

  d. ____, ____, [shell] ____,

31. For each pair of numerals, which represents the larger quantity?

  a. ⅢⅢ ⅢⅢ IIII or ⅢⅢ ⅢⅢ III

  b. X̄ or IX

  c. ꝑꝑꝑꝑ∩∩∩ or ∩∩∩∩ꝑꝑꝑ

32. For each pair of numerals, which represents the larger quantity?

  a. [Mayan: two dots] or ⅢⅢ ⅢⅢ IIII

  b. XLII or ◄◄◄▼▼▼

  c. [Mayan: three bars over dot over two bars] or ▼▼▼◄▼▼▼

  d. ꝑꝑꝑꝑ∩∩∩∩∩∩I or CDXLVI

  e. XXXILXVII or ∩∩∩ꝑ∩∩∩IIII over ∩∩∩IIII

  f. ◄◄◄▼▼▼ over ▼ ◄◄▼▼▼ over ▼▼▼ or ꝑꝑꝑ∩∩∩

■ Problems and Applications

33. Complete the chart by converting each number into the other numeration systems.

Egyptian	ꝑꝑ∩II		
Babylonian		◄◄◄◄▼▼▼ over ▼▼	
Roman			XLIV
Mayan	• over ••		

34. Which numeration system, the Egyptian or the Babylonian, would use fewer symbols to represent the number 125? What about the Egyptian or the Roman?

35. **a.** What is the smallest Egyptian numeral that makes use of exactly 22 symbols?

  b. What is the smallest Roman numeral that makes use of exactly 5 symbols?

36. Roman numerals are often used to express copyright dates on movies. In the last five years, what year has required the most symbols when written in Roman numerals?

37. In the Chinese rod or bamboo numeration system, numerals are made using arrangements of small sticks. The system is positional, has a base of 10, and uses blank spaces to represent zero. It has two sets of digits, which are used in alternating positions, as shown in the table.

	1	2	3	4	5	6	7	8	9
Symbols for ones, hundreds, ten thousands . . .	I	II	III	IIII	IIIII	T	TT	TTT	TTTT
Symbols used for tens, thousands, . . .	⊤	=	≡	≣	≣	⊥	⊥	⊥	⊥

1	2	3	4	5	6	7	8	9	10
/	//	///	////	\	/\	/\	//\	///\	\\
11	**12**	**13**	**14**	**15**	**16**	**17**	**18**	**19**	**20**
/\	//\	///\	////\	\\	/\\	//\\	///\\	////\\	\\\

Numerals are written from right to left. For instance 23,491 is written as II≡IIII⊥I. Represent each number using the Chinese rod system.

  a. 341   **b.** 8,401   **c.** 12,749   **d.** 351,781

38. What number is represented by each numeral in the Chinese rod system?

  a. — III⊥T

  b. T⊥T⊥T

  c. ⊥ =
    ≣ ≡IIII

  d. TTT TTT II

39. One ancient Greek numeration system was based on the Greek alphabet, which originally contained 27 symbols. The Greeks assigned different values to the letters, as shown in the table.

1	α	Alpha	10	ι	Iota	100	ρ	Rho
2	β	Beta	20	κ	Kappa	200	σ	Sigma
3	γ	Gamma	30	λ	Lambda	300	τ	Tau
4	δ	Delta	40	μ	Mu	400	υ	Upsilon
5	ε	Epsilon	50	ν	Nu	500	φ	Phi
6	ς	Vau	60	ξ	Xi	600	χ	Chi
7	ζ	Zeta	70	o	Omicron	700	ψ	Psi
8	η	Eta	80	π	Pi	800	ω	Omega
9	θ	Theta	90	$\backslash$	Koppa	900	⅄	Sampi

This enabled them to write numbers up to 1,000. For larger values, the same symbols were used, but a comma was placed in front of the number, signifying multiplication by 1,000. For example, $\chi\nu\varsigma$ had a value of 656, and ,$\zeta\tau\lambda$ had a value of 7,330. What number is represented by each Greek numeral?

  a. $\nu\,\beta$ **b.** $o\,\theta$ **c.** $\varphi\,\nu\,\varepsilon$ **d.** $\omega\,\delta$

  e. ,η⅄$\backslash$ **f.** ,$\pi\,\psi\,\lambda\,\alpha$ **g.** ,$\rho\rho\kappa\varsigma$ **h.** ,$\chi\,\mu\,\delta$

40. Using the information in the last exercise, write the following as Greek numerals.

  a. 54   **b.** 73   **c.** 341   **d.** 2,345   **e.** 10,809

41. The numerals for the Urnfield culture originated in central Europe about 1200 B.C.E. Not everything is known about this system, but the first 20 numerals shown in the following table seem to indicate a grouping system.

  a. What does the base appear to be?

  b. How might the numbers 21 through 25 be written?

■ Mathematical Reasoning and Proof

42. a. Determine how many symbols are needed to write each number as an Egyptian numeral.

  i. 27   **ii.** 149

  iii. 2,540   **iv.** 34,508

  b. Suggest a rule that can be used to determine how many Egyptian symbols are needed to represent any number as an Egyptian numeral.

43. a. If you were given a numeration system you had never seen before, how would you know whether it was positional?

  b. Suppose the system was positional. How might you determine its base?

44. a. Write the first 10 counting numbers in each of the Egyptian, Roman, Babylonian, and Mayan systems. How are the first 10 numerals in each system alike? How are they different?

  b. Based on what you see in the first 10 numerals, which system appears to be most efficient for writing small numbers? Why?

45. The Egyptian and Roman numeration systems require only one symbol to write the numeral for 1,000, whereas the Babylonian requires 11 and the Mayan requires 9. Are the Egyptian and Roman systems always a more efficient way of writing large numbers? If so, explain. If not, provide a counterexample.

■ Mathematical Communication

46. Look up the definition of three in a dictionary. Do you think the definition adequately describes the quantity of three? Why or why not?

47. Explain to another person how to select the symbols needed to write 345 as a(n):

  a. Egyptian numeral.

  b. Roman numeral.

  c. Mayan numeral.

48. Create your own numeration system that uses the properties discussed in this section. Write several numerals from your numeration system on a piece of paper. Ask one of your peers not only to determine the properties of the numeration system but also to translate your numerals into decimal numerals.

49. An interesting ancient numeration system was that of finger numerals. In this system, different gestures had different values, and any number could be communicated by adding the collective values in a sequence of gestures. Here are the gestures and their values:

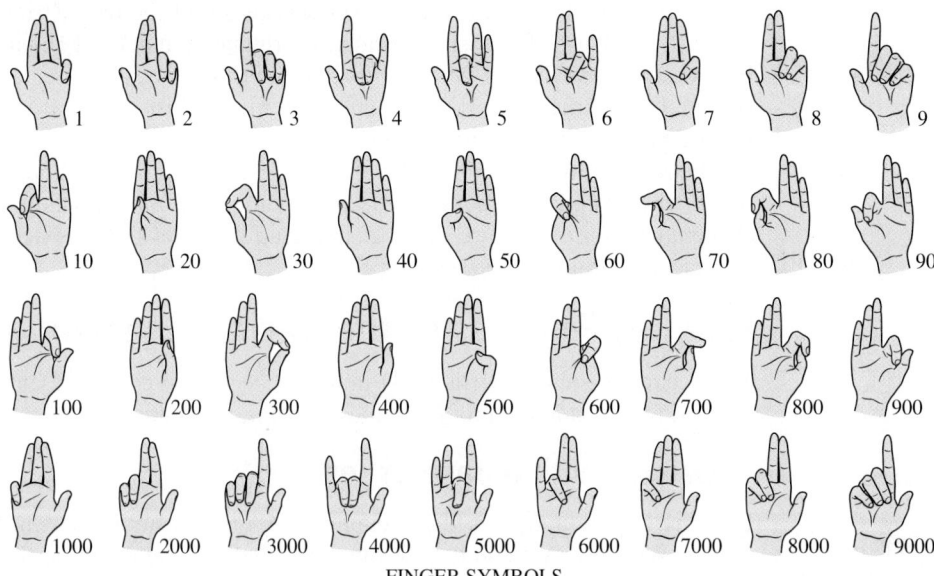

FINGER SYMBOLS

With one of your peers, communicate several different numerals in this way. When you are done, discuss the advantages and disadvantages of the system.

50. When first learning to write numerals, children often make the mistake of writing the digits in the wrong order. For instance, the child might write 53 for 35.

    a. Would this same problem occur in the Egyptian system? The Roman? The Mayan?

    b. What characteristic of numeration systems causes this problem?

51. The Roman numerals for small values are often taught in the elementary and middle grades. Based on your experience in working with the Roman numerals, what problems do you think children will have in learning them? What might you do as a teacher to help your students better understand the additive and subtractive features of this system?

■ **Building Connections**

52. Tally and Roman numerals are still in use today. Describe situations in which you have either used or encountered numerals from these systems.

53. Grouping tallies by 5s made counting objects quicker.

    a. When you count the following tallies, what counting strategies do you use?

    卌 卌 卌 卌 卌 卌 卌 | |

    b. Devise a strategy that goes beyond grouping by 5 and makes counting tallies even easier.

54. Archaeologists believe that written numeration came before and eventually led to written language. Why do you think this might be the case?

55. In this section, we say that a concept of zero was not a part of every numeration system. Think of several situations in which you have used zero. What difficulties might you encounter if zero were not available?

56. The numeration system used by the Incas of Peru is another interesting system. Use the Internet to research this numeration system to find out its properties. Of the systems discussed in this section, which is the Inca system most like?

57. Search through several sets of curriculum materials for grades K–4. Look for uses of the ancient numeration systems discussed in this section. Which ones do you find? What role do you think they serve in the elementary mathematics curriculum?

SECTION 3.2   The Hindu-Arabic or Decimal System

Most of the ancient numeration systems discussed in the last section are no longer in use today. However, one ancient system has stood the test of time. Formally, it is called the **Hindu-Arabic system** because it originated in ancient India and migrated to Europe by way of North Africa. More commonly, it is called the **decimal** or **base-ten system.** One of the greatest achievements in human history, it is currently used in the United States and throughout the world.

Because learning numerical concepts and operations without numerals is difficult, writing and understanding decimal numerals is an important topic in the early elementary grades. To use decimal numeration to its fullest extent, children must master the features of the system.

Five Key Features of the Decimal System

The decimal system is particularly powerful because it can represent very large numbers using only a few symbols. It does so using five features.

1. Groups with a Base of Ten

Explorations
Manual
3.4

Like the Egyptian system, the decimal system is a grouping system that uses a base of ten. Whenever we count 10 groups of anything, we combine them and consider them one. To demonstrate this idea, we can count with a useful representation of the decimal system called **base-ten,** or **Dienes blocks.** Begin by counting **units,** or **ones,** which are represented by small cubes. Once we count 10 units, we exchange them for one group of ten, or one **rod** or **long** (Figure 3.6).

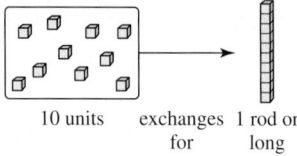

10 units   exchanges   1 rod or
                for   long

Figure 3.6 Exchanging 10 ones for 1 ten

As we count on, we continue to exchange 10 ones for groups of ten until we reach 10 tens. We then exchange the 10 tens for 1 hundred, or one **flat** or **square** (Figure 3.7).

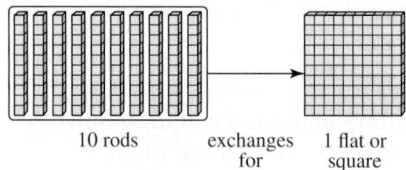

10 rods   exchanges   1 flat or
               for   square

Figure 3.7 Exchanging 10 tens for 1 hundred

In the same way, once we count 10 hundreds, we exchange them for 1 thousand, or one **block** or **cube** (Figure 3.8).

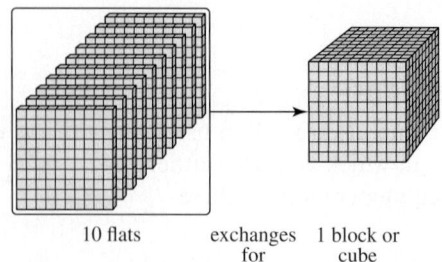

10 flats   exchanges   1 block or
               for   cube

Figure 3.8 Exchanging 10 hundreds for 1 thousand

Because we can continue the process indefinitely, the groups of the decimal system include:

Ones, or units
Tens   = 10 ones
Hundreds   = 10 tens or 100 ones
Thousands   = 10 hundreds or 1,000 ones
Ten thousands = 10 thousands or 10,000 ones . . .

2. Use of Place Value

The decimal system is also positional; that is, it has **place value,** which means the position of a digit in a numeral determines its value. The place values in the decimal system correspond to groups of ten and are listed in ascending order from right to left. Because the number of groups is infinite, so is the number of place values. The first six are shown in Figure 3.9.

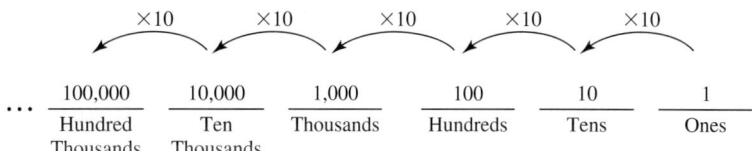

...	100,000	10,000	1,000	100	10	1
	Hundred Thousands	Ten Thousands	Thousands	Hundreds	Tens	Ones

Figure 3.9 The first six place values of the decimal system

Figure 3.9 also illustrates how each place value is 10 times the value of the place to the right. Consequently, we can generate the place values by repeatedly multiplying by 10. We can also express the place values with exponents, which are a notational shortcut to repeated multiplication. Exponents are discussed in greater detail in Chapter 4, but here they are a convenient way to shorten our notation. In general, if n is a natural number, we define the exponent a^n to be $\underbrace{a \times a \times \cdots \times a}_{n \text{ factors}}$. With this notation, 100

becomes $100 = 10 \times 10 = 10^2$, and 1,000 becomes $1,000 = 10 \times 10 \times 10 = 10^3$. We also define $a^0 = 1$ as long as $a \neq 0$, and use 10^0 instead of 1 for the first place value. Table 3.7 summarizes the ways we can represent decimal place values.

Table 3.7 Representations of decimal place values						
Position	Sixth	Fifth	Fourth	Third	Second	First
Name	Hundred thousands	Ten thousands	Thousands	Hundreds	Tens	Ones
Value in Units	100,000	10,000	1,000	100	10	1
Power of 10	10^5	10^4	10^3	10^2	10^1	10^0

3. Use of 10 Digits

All decimal numerals are written using only 10 digits: 0, 1, 2, 3, 4, 5, 6, 7, 8, and 9. The digits have taken on many different forms over time, some of which are shown in Figure 3.10. Not until the invention of the printing press did they take on the standard forms known and used today.

Brahmi, Second Century   Devanagari, Eighth Century

West Arab Gobar, Tenth Century   Spain, 976

Figure 3.10 Ancient versions of the decimal digits

The digits serve two purposes. First, when standing alone, each digit represents a value less than 10. Specifically, 1 has a value of one, 2 has a value of two, and so on. Second, the digits enable us to write the numerals for values larger than 9. There are two reasons why we do not need more than 10 digits to do so. One, because the decimal system has a base of ten, we always make an exchange once we have a group of ten. So there is no need for digits that represent values of 10 or larger. Two, because the decimal system has place value, we can use digits repeatedly. For instance, if we place a 2 in the ones place value, it has a value of $2 \times 1 = 2$. However, if we place a 2 in the hundreds place value, it has a value of $2 \times 100 = 200$ (Figure 3.11).

$$\underline{2^{\nearrow^{2 \times 100 = 200}}} \quad \underline{} \quad \underline{2^{\nearrow^{2 \times 1 = 2}}}$$
$$\text{Hundreds} \quad \text{Tens} \quad \text{Ones}$$

Figure 3.11 Different values of the digit 2

4. Use of Zero

The digit for zero, 0, is particularly special in the decimal system. It is used not only for the cardinal number of the empty set but also as a **placeholder.** When a placeholder is used in a place value, it tells us that no group of that size is included in the value of the number. In other words, we use placeholders to move nonzero digits out to the appropriate place value. Given these four features, we can write decimal numerals.

Example 3.9 | **Writing Decimal Numerals**

Representation

Write the decimal numeral that represents each set of values.

a. Eight tens and five ones   **b.** Seven thousands, six tens, and three ones

Solution Writing the numeral is a matter of placing the correct digit in the appropriate position. Doing so gives us the following numerals.

a. 85   **b.** 7,063

Example 3.9 demonstrates one way to **compose** numerals; that is, we find the value of the numeral by combining the values of the groups within it. To **decompose** a numeral, we break it into smaller groups that have a total value equal to the number. To do that, we need the fifth feature of the decimal system.

5. Additive and Multiplicative

As a natural consequence of its groups and places values, the decimal system must also be **additive** and **multiplicative.** To find the value of any numeral, multiply each digit by its corresponding place value and then add the products. For example, to find the value of 63,284:

$$63,284 = 60,000 + 3,000 + 200 + 80 + 4$$
$$= (6 \times 10,000) + (3 \times 1,000) + (2 \times 100) + (8 \times 10) + (4 \times 1)$$

The last expression is written in **expanded notation,** which can be a powerful learning tool because it explicitly connects the value of a digit to its place value. The place values can also be written in exponential form in expanded notation.

Example 3.10   Writing Numerals in Expanded Notation

Write each number in expanded notation.

a. 458   **b.** 19,014

Solution Each number is written in both forms of expanded notation.

a. $458 = (4 \times 100) + (5 \times 10) + (8 \times 1)$
$= (4 \times 10^2) + (5 \times 10^1) + (8 \times 10^0).$

b. $19,014 = (1 \times 10,000) + (9 \times 1,000) + (0 \times 100) + (1 \times 10) + (4 \times 1)$
$= (1 \times 10^4) + (9 \times 10^3) + (0 \times 10^2) + (1 \times 10^1) + (4 \times 10^0).$

Modeling Decimal Numerals

Several common models can be used to represent decimal numerals. In addition to base-ten blocks, we can also use bundled sticks or an abacus.

Base-Ten Blocks

As the next example demonstrates, we can use base-ten blocks to represent decimal numerals simply by selecting the appropriate number of each kind of shape. For instance, because 27 = 2 tens + 7 ones, we represent it with 2 rods and 7 units. Similarly, because 2,056 = 2 thousands + 0 hundreds + 5 tens + 6 ones, we represent it with 2 blocks, 5 rods, and 6 units (Figure 3.12).

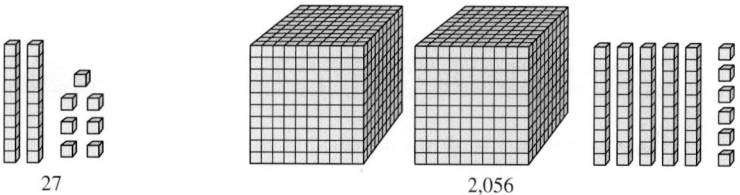

27   2,056

Figure 3.12 Base-ten blocks for 27 and 2,056

In the classroom, children often compose numerals by putting base-ten blocks on place value work mats (Figure 3.13). Doing so not only connects the size of the group to its position in the numeral but also connects the number of each of the groups to a digit.

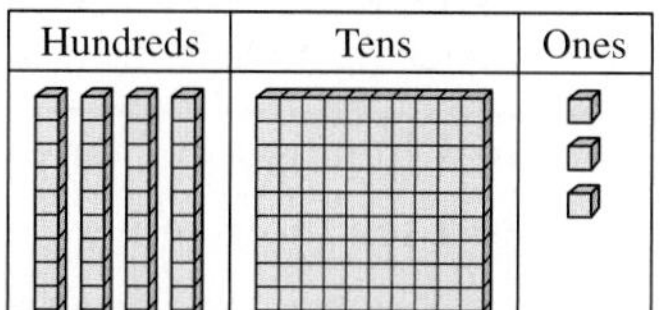

Hundreds	Tens	Ones

Figure 3.13 A place value work mat

Bundled Sticks

Another way to represent decimal numerals is to use **bundled sticks.** A single toothpick or popsicle stick represents the unit. To make a group of ten, we bundle 10 sticks with a rubber band or string. Similarly, to make a hundred, we bundle 10 tens together. The appropriate number of each kind of bundle represents specific numerals. For instance, the bundled stick representation of 153 is shown in Figure 3.14. In the classroom, bundled sticks give children direct experience in making groups of 10.

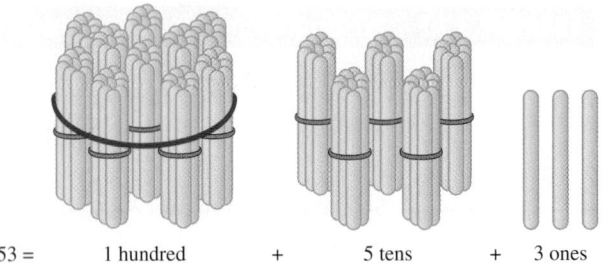

$153 =$ 1 hundred   +   5 tens   +   3 ones

Figure 3.14 Bundled sticks

Abacus

Unlike base-ten blocks or bundled sticks, an **abacus** does not show every unit in the numeral but rather only the number of groups in each place value. On a typical abacus, the place values are represented by a series of wires or posts. The appropriate number of colored beads are placed on the posts to represent each digit in the numeral. In another version of this model, we place colored chips on a place value work mat. In Figure 3.15, both versions are used to represent 2,024.

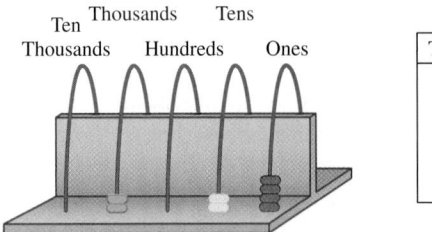

Thousands	Hundreds	Tens	Ones
● ●		○ ○	●● ●●

Figure 3.15 Two versions of an abacus

Example 3.11   Using an Abacus

Representation   What numeral is shown on each abacus?

a.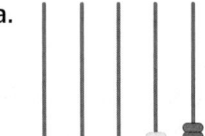

b.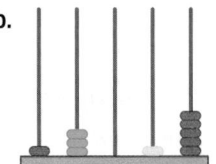

Solution We count the number of chips in each place value, which are in ascending order from right to left. We then form the numeral.

a. 1 hundred + 2 tens + 3 ones = 123.

b. 1 ten thousand + 3 thousands + 0 hundreds + 1 ten + 5 ones = 13,015.

Making Exchanges

In addition to composing and decomposing numerals according to the digits in various place values, numbers can be composed or decomposed in other ways. These methods often require us to be flexible in how we make exchanges between groups. They also constitute an important part of developing number sense and computing with large numbers. Consider the next three examples.

Example 3.12   Reducing Base-Ten Blocks

Reduce the blocks to a minimal set, and then write the numeral.

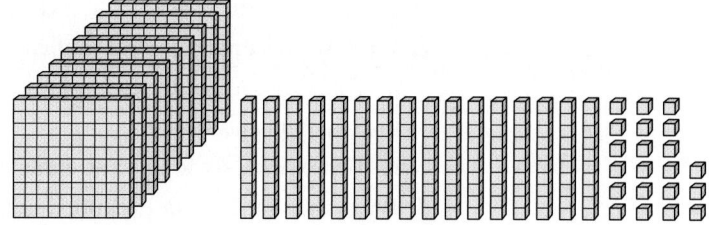

Solution Starting with the units, we exchange 20 units for 2 longs, 10 longs for 1 flat, and finally 10 flats for 1 block. After the exchanges, we have 1 block, 8 longs, and 1 unit, or 1,081.

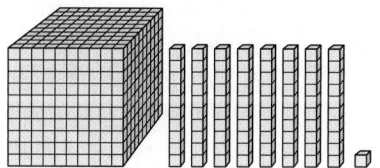

Example 3.13   Decomposing Numbers

a. How many tens are in 250?

b. How many hundreds are in 1,800?

Solution We decompose numbers not by digits and place values, but rather by specific group sizes.

a. 250 = 1 hundred + 1 hundred + 5 tens = 10 tens + 10 tens + 5 tens = 25 tens.

b. 1,800 = 1 thousand + 8 hundred = 10 hundreds + 8 hundreds = 18 hundreds.

Example 3.14   A Student Error in Decomposing

Connections

Caleb, a second-grade student, is learning to regroup. When completing an assignment, he wrote the following expressions:

$$4 \text{ tens} + 15 \text{ ones} = 4 \text{ tens} + 5 \text{ ones}$$
$$2 \text{ hundreds} + 12 \text{ tens} + 6 \text{ ones} = 2 \text{ hundreds} + 2 \text{ tens} + 6 \text{ ones}$$
$$3 \text{ hundreds} + 6 \text{ tens} + 10 \text{ ones} = 3 \text{ hundreds} + 6 \text{ tens} + 0 \text{ ones}$$

What has he done wrong?

Solution Caleb has recognized that he can regroup and has removed the appropriate group of ten in each case. However, he has failed to count the newly made group as part of the numeral. In doing so, he has actually changed its value.

Check Your Understanding 3.2-A

1. Describe the groups represented by each digit in each decimal numeral.
   a. 13   b. 6,329   c. 14,109
2. Write the decimal numeral that each statement represents.
   a. Three flats, eight longs, and four units
   b. Five flats and six longs
   c. Two blocks, six longs, and nine units
3. Write each number in expanded notation.
   a. 410   b. 8,601   c. 12,358   d. 200,601
4. Decompose 1,540 in five different ways.

Talk About It What would be the consequences of changing the base in the decimal system from ten to five?

Reasoning

Activity 3.3
a. What is the smallest number of base-ten blocks needed to represent each numeral?
   i. 17   ii. 327   iii. 2,458   iv. 3,091
b. Based on your answers to the first part of this problem, write a statement that explains how to find the smallest number of base-ten blocks needed to represent any numeral.

Naming Decimal Numerals

Before entering kindergarten, most children know numeral names up to 10. Once in school, they learn to name two-digit numerals in the following way:

- 10, 11, and 12 have unique names called "ten," "eleven," and "twelve."
- 13 through 19 are named in the reverse order of other two-digit numerals and are called "thirteen" (three-ten), "fourteen" (four-ten), "fifteen" (five-ten), and so on.
- 20 through 99 are named by the multiple of ten (i.e., twenty, thirty, forty, . . .), followed by the name of the ones digit. For instance, 27 is named "twenty-seven," and 63 is named "sixty-three."

We name three-digit numerals by the digit in the hundreds place, followed by "hundred," and the name of the two-digit numeral in the tens and ones places. For example, 347 is named "three hundred forty-seven," and 658 is named "six hundred fifty-eight."

If a numeral has four or more digits, we use commas to separate it into groups of three digits called **periods**. Each period is named for the smallest place value in the period. The first six periods are shown in Figure 3.16.

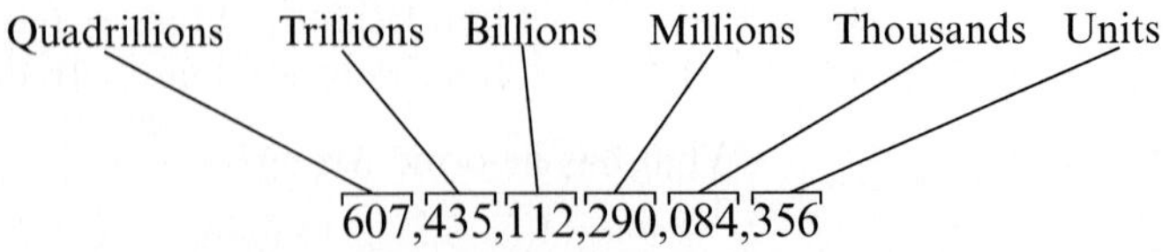

Quadrillions   Trillions   Billions   Millions   Thousands   Units

607,435,112,290,084,356

Figure 3.16 Periods of decimal numerals

To name large numbers, we read the one-, two-, or three-digit numeral in each period and then follow it with the name of the period. A common mistake is to insert the word "and" between periods. However, the only time "and" is ever used in naming a numeral is to separate whole objects from parts of the whole.

Example 3.15	Writing Decimal Numeral Names

Communication   Write the name of each numeral.

   a. 34,567   **b.** 23,609,231,010

Solution

a. 34,567 is named "thirty-four thousand, five hundred sixty-seven."

b. 23,609,231,010 is named "twenty-three billion, six hundred nine million, two hundred thirty-one thousand, ten."

Ordering Numbers

To **order** numbers, we compare their values to determine either that they are the same or that one is smaller than the other. If two numbers, a and b, have the same value, we say they are **equal** and write $a = b$. If not, we write $a \neq b$. If the numbers are not equal, then one must be smaller, or **less than** the other.

Definition of Less Than

Let a and b be whole numbers, where $a = n(A)$ and $b = n(B)$. a is **less than** b, written $a < b$, if and only if A is equivalent to a proper subset of B.

The idea behind the definition is simple and intuitive. It tells us to make a one-to-one correspondence between two sets of the appropriate size. If the numbers are not equal, then one set will have objects left over. So it must represent the larger value.

Example 3.16	Using the Definition of Less Than

Reasoning   Use the sets $A = \{1, 2, 3\}$ and $B = \{a, b, c, d\}$ to demonstrate that 3 is less than 4.

Solution The following diagram shows that, when we try to make a one-to-one correspondence between sets A and B, one object from B remains unpaired. In other words, we are able to make a one-to-one correspondence only between A and a proper subset of B. Because $n(A) = 3$ and $n(B) = 4$, we conclude, by the definition of less than, that $3 < 4$.

If a is less than b, we can also say that b is **greater than** a and write $b > a$. Two other order relationships are **less than or equal to,** written $a \leq b$, and **greater than or equal to,** written $b \geq a$. Less than or equal to is similar to less than, but it allows two numbers to have the same value. Greater than or equal to is similar.

Another way to compare numbers is to use the **whole number line.** This visual representation of the whole numbers is drawn as a one-directional arrow pointing to the right (Figure 3.17). The furthest point to the left is marked as 0, and other whole numbers are marked off in ascending order at equal intervals to the right. The farther we go to the right on a number line, the larger the values become. Consequently, one number is less than another if it occurs to the left on the number line. For instance, $4 < 7$ because 4 is to the left of 7.

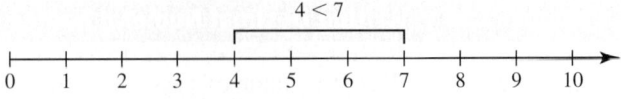

Figure 3.17 Whole number line

Ordering numbers by means of sets or a number line is impractical with large numbers. A more efficient way is to compare digits in each place value.

Example 3.17 Using Place Value to Order Numbers

Which number is larger: 3,452 or 3,442?

Solution When we compare each place value directly, we find the numbers differ in the tens place value. Because 4 is less than 5, we conclude that $3,442 < 3,452$.

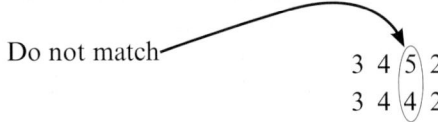

Explorations Manual 3.5

Understanding the relative size of numbers is an important part of number sense. Generally, the more we work with numbers, the better we understand how big or small they are. However, very large numbers are far removed from our common experience, so we are less likely to understand how big they are. In such cases, we can use a common frame of reference to portray how big certain numbers really are.

Example 3.18 Illustrating the Size of Large Numbers

Representation

Use base-ten blocks to illustrate the relative sizes of 1 to 10,000 and of 1 to 100,000.

Solution One way to illustrate the relative size of these numbers is to place a unit next to sets of base-ten blocks that represent 10,000 and 100,000. To make a group of 10,000, stack 10 blocks on top of one another to make a **megalong.** To make a group of 100,000, place 10 mega-longs side by side to make a **megaflat.** Setting a unit next to these shapes illustrates the relative size of 1 to 10,000 and of 1 to 100,000.

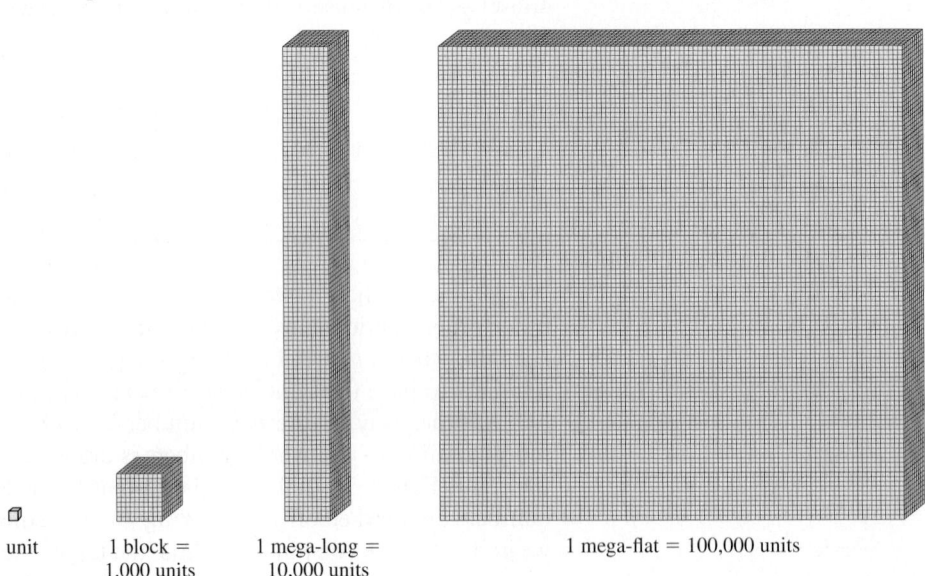

1 unit   1 block = 1,000 units   1 mega-long = 10,000 units   1 mega-flat = 100,000 units

Check Your Understanding 3.2-B

1. Write the name of each numeral.
   a. 438   **b.** 23,006   **c.** 12,071,029
2. Use place value to explain why 347 is less than 357.
3. A stack of 100 index cards is roughly an inch thick. If 1,000,000,000 index cards are stacked on top of one another, how tall is the stack?

Talk About It Young children often develop the misconception that equals means to compute an operation. Why is this a reasonable misconception? What are its causes, and how might you, as a teacher, address the difficulty?

Connections

Activity 3.4

In Example 3.18, we learned how to make groups of 10,000 and 100,000 using base-ten blocks. To make a group of 1,000,000, put 10 megaflats together to make a **megablock.** In most cases, a unit is about the size of a cubic centimeter, making the megablock about 1 m³. If you were to represent 1,000,000,000 with base-ten blocks, what shape and size would it be?

SECTION **3.2** Learning Assessment

■ Understanding the Concepts

1. How is the base of the decimal system connected to the size of the groups?
2. What is place value, and why is it a key feature of the decimal system?
3. **a.** Why must the decimal system, as a grouping system, also be additive?
   b. Why must the decimal system, as a positional system, also be multiplicative?
4. **a.** Why does the decimal system need only 10 digits?
   b. What two purposes do the digits serve in the decimal system?
5. What is the role of a placeholder in writing decimal numerals?
6. What does it mean to compose a numeral? To decompose a numeral?
7. What is a period in a decimal numeral, and how are specific periods named?
8. What role does a one-to-one correspondence play in determining whether one number is less than another?
9. What is the difference between less than and less than or equal to?

■ Representing the Mathematics

10. Write the decimal numeral that represents each of the following.
    a. Six hundreds, eight tens, and two ones.
    b. Six thousands, eight hundreds, five tens, and seven ones.
    c. Two hundred thousands, three ten thousands, seven thousands, and three tens.
    d. Five millions, four thousands, and seven ones.
11. Describe the groups represented by each digit in each decimal numeral.
    a. 23   **b.** 671   **c.** 8,299   **d.** 60,012
12. Draw a diagram of the base-ten blocks needed to represent each decimal numeral.
    a. 45   **b.** 138   **c.** 501
    d. 1,293   **e.** 4,020
13. Write the numeral that corresponds to each set of base-ten blocks?
    a. 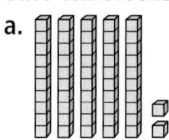   **b.**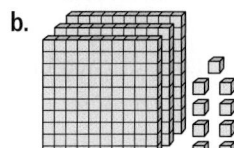

c.

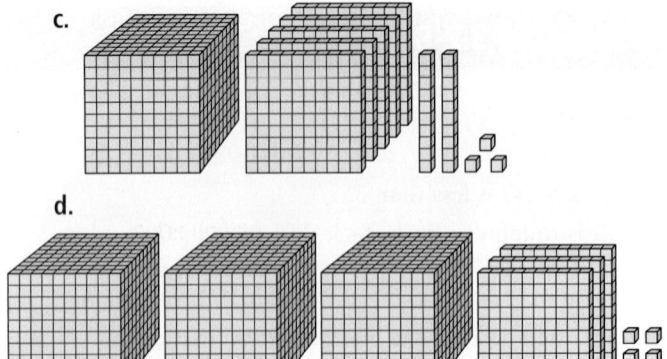

d.

14. Use base-ten blocks to answer each question.
    a. How many units are in a long? In a flat? In a block?
    b. How many longs are in a flat? In a block?
    c. How many flats are in a block?

15. Use base-ten blocks to answer each question.
    a. Six longs are equal to how many units?
    b. Four flats are equal to how many longs? How many units?
    c. One block and four flats are equal to how many longs? How many units?
    d. Fifteen blocks are equal to how many flats? How many longs?

16. Use a set of bundled sticks to represent each numeral.
    a. 9   b. 26   c. 153   d. 204

17. What numeral is represented in each diagram of an abacus?

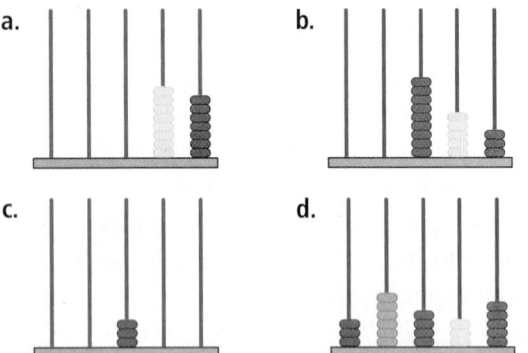

18. Draw a picture of an abacus that represents each numeral.
    a. 19   b. 241   c. 3,452   d. 12,983

19. How is modeling a numeral with base-ten blocks different from modeling a numeral with an abacus? What are the advantages and disadvantages of each model?

20. Write the name of each numeral.
    a. 53   b. 129   c. 4,572
    d. 269,018   e. 231,764,920   f. 123,409,001,230

21. Write the numeral for each number name.
    a. Four hundred ninety-nine
    b. Three thousand, two hundred eighty-seven
    c. Eight hundred one thousand, nine hundred six
    d. Forty-five million, three hundred thirty-seven thousand, one hundred thirteen

22. The number 236 can be represented with base-ten blocks by using 2 flats, 3 longs, and 6 units. List five other ways to represent 236 using base-ten blocks.

23. In each set of base-ten blocks, reduce the number of blocks by making all exchanges possible. Write the numeral represented by the blocks.

a.
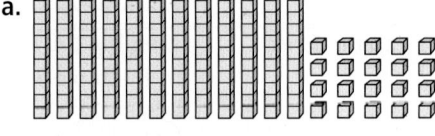

b.
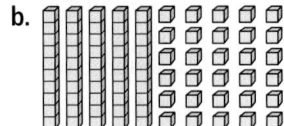

c.
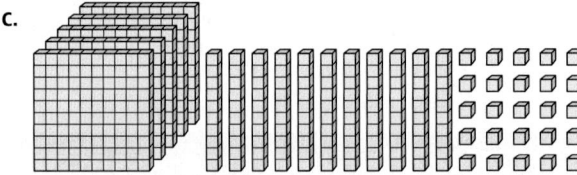

24. Decompose the number 2,100 in five different ways.

■ Exercises

25. Give the place value of each underlined digit in the following numerals.
    a. 24$\underline{7}$1   b. 34,58$\underline{9}$ c. 1,0$\underline{9}$1,257   d. 4,$\underline{2}$13,$\underline{9}$14

26. Write each numeral in expanded notation.
    a. 457   b. 3,581   c. 40,201   d. 200,390

27. Write each numeral in expanded notation using exponents.
    a. 219   b. 4,561   c. 16,004   d. 301,490

28. Condense each number written in expanded notation.
    a. $(3 \times 100) + (4 \times 10) + (8 \times 1)$
    b. $(4 \times 10,000) + (4 \times 100) + (4 \times 1)$
    c. $(1 \times 10^5) + (2 \times 10^4) + (8 \times 10^2)$
    d. $(7 \times 10^5) + (7 \times 10^4) + (2 \times 10^3) + (3 \times 10^1) + (8 \times 10^0)$

29. Condense each number written in expanded notation.
    a. $(9 \times 100,000) + (6 \times 10,000) + (3 \times 1,000) + (1 \times 10)$
    b. $(6 \times 1,000) + (4 \times 1)$

c. $(5 \times 10^2) + (3 \times 10^1) + (7 \times 10^0)$

d. $(2 \times 10^4) + (9 \times 10^3) + (1 \times 10^1) + (1 \times 10^0)$

30. Answer each question.

   a. How many tens are in 160?

   b. How many hundreds are in 2,500?

   c. How many thousands are in 124,000?

   d. How many hundreds are in 316,000?

31. Answer each question.

   a. How many ones are in 316?

   b. How many tens are in 3,150?

   c. How many hundreds are in 34,500?

   d. How many thousands are in 2,301,000?

32. Fill in the blank with the correct number.

   a. 2 tens + 4 ones = 1 ten + _____ ones.

   b. 3 hundreds + _____ tens + 6 ones = 4 hundreds + 1 ten + 6 ones.

   c. 12 tens + 8 ones = _____ hundred + 2 tens + 8 ones.

33. Fill in the blank with the correct number.

   a. 3 tens + 16 ones = 4 tens + _____ ones.

   b. 1 hundred + 25 tens + 4 ones = _____ hundreds + 5 tens + 4 ones.

   c. 135 ones = _____ hundreds + _____ tens + 5 ones.

34. Place <, >, or = in the blank to make each statement true.

   a. 5 _____ 15   b. 9 _____ 9   c. 18 _____ 17

35. Use place value to order the following numbers from smallest to greatest.

   3,456 4,356 2,456 5,436 6,534 5,634

36. Use place value to order the following numbers from smallest to greatest.

   5,788 7,588 8,578 8,857 5,878 7,858

■ Problems and Applications

37. a. What is the largest number that can be made using the digits 3, 4, 6, 8, and 9 if they are used only once? The smallest?

   b. If the digits are used more than once, what is the largest four-digit number that can be made? The smallest?

38. If a unit in base-ten blocks is 1 cm³, give the dimensions of each of the following.

   a. Long   b. Flat

   c. Block   d. A shape representing 100,000 units

39. a. If a thousand is ten hundreds, how big is a thousand hundreds?

   b. If a million is a thousand thousands, how big is a thousand millions?

40. Which is bigger: a million billion or a billion million?

41. A candy company packages its boxes of chocolates in the following way:

   24 candies = 1 box,   12 boxes = 1 flat,
   12 flats = 1 crate

   a. How many boxes are in four crates?

   b. How many candies are in three crates?

   c. How many candies are in two crates, three flats, and one box?

   d. How would 14,832 candies be packaged?

42. A standard brick is about 8 in. long. If a building is 900 bricks long, about how many feet is this?

43. A penny is approximately 1.5 mm thick. If there are 1,000 mm in a meter, how tall in meters is a stack of one million pennies?

44. Kenny can text 100 characters in a minute. If he were to text a message that has 35,300 characters, how long would it take him?

45. A quarter cup holds about 100,000 grains of salt. How many cups are needed to hold one billion grains of salt?

■ Mathematical Reasoning and Proof

46. Make two example sets, and use them to write an argument that 5 < 7.

47. Use a number line to explain why 6 > 3.

48. You may have noticed that, in base-ten blocks, the unit and the block have the same basic shape. The only difference is size.

   a. Based on this observation, what might the base-ten blocks look like for numbers such as 10,000, 100,000, 1,000,000, and so on?

   b. Do you notice any relationship between the shapes and the periods of the decimal system?

49. A game is played with base-ten blocks and a regular six-sided die. When the die is rolled, the person collects the number of units shown on the face-up side of the die.

   a. If you wanted to collect exactly enough units to make a long, what is the smallest number of rolls it would take? The largest number of rolls?

   b. If it took three rolls to get a long, list five combinations of numbers that might have been rolled.

   c. If it took four rolls to get a long, list five combinations of numbers that might have been rolled.

   d. To win the game, a person must make enough longs to make a flat. What is the least number of rolls it would take to win? What is the most?

Mathematical Communication

50. The following statement is taken from a math textbook published in 1877. "As we have no single character to represent ten, we express it by writing the unit, 1, at the left of the cipher." What is a cipher?

51. Later in the same text, this statement is given: "Figures occupying different places in a number, as units, tens, hundreds, etc., are said to express different orders of units." Specific orders are given in the following table.

Order	First Order	Second Order	Third Order	Fourth Order
Names	Units	Tens	Hundreds	Thousands

Using this terminology, write the following numerals.

a. Three units of the third order, two of the second, and seven of the first.

b. Eight units of the sixth order, four of the third, five of the second, and nine of the first.

c. What are the eighth, ninth, and tenth orders?

52. Again, in the same text, the following statement is made: "The different orders of units increase from right to left, and decrease from left to right, in a tenfold ratio." Explain in your own words the meaning of this statement.

53. One-to-one correspondences can be applied directly in the classroom when teaching children to order small numbers. Find a partner, and design an activity for young children that uses sets and one-to-one correspondences to order small numbers.

54. Design an activity for elementary school children in which you use base-ten blocks to demonstrate that one number is greater than another.

Specifically, address how base-ten blocks can be used to compare two numbers with different values.

55. When children learn to count to 100, they often pause after they reach a 9 in the ones place (i.e., 29, 39, 49, . . .). Why do you think they do that?

Building Connections

56. a. How is the Hindu-Arabic system like the Egyptian system? The Babylonian system? The Mayan system?

b. Explain the advantages of the Hindu-Arabic numeration system over the Egyptian, Babylonian, and Mayan numeration systems.

57. When reading large numerals, we often make the mistake of using the word "and" to separate periods. The word "and" is used when reading numerals, but only to represent a very specific situation. What is it?

58. a. What role does composing numerals play when adding or multiplying?

b. What role does decomposing numerals play when subtracting?

59. Consider the following denominations of U.S. currency: pennies, dimes, $1 bills, $10 dollar bills, and $100 bills.

a. What characteristics of the decimal system does this group of currency have?

b. In what ways does it differ from the decimal system?

c. What is the smallest number of these denominations needed to represent $512.36?

d. Describe how you might use U.S. currency to teach decimal numeration.

SECTION 3.3   Other Base Numeration Systems

Reading, writing, and understanding decimal numerals may seem simple to us as adults because we have been working with them for a long time. However, for children who are learning them for the first time, it is not so simple. To better help students, elementary teachers should reconsider the features of decimal numeration. One way to do so is to look at other base numeration systems.

In any base numeration system, the value of the base determines the size of the groups, the place values, and the digits used. Consequently, by changing the base, we get another numeration system that has all the features of the decimal system. We call these systems **base-*b* numeration systems.**

Explorations Manual 3.6

Base-Five Numeration

When humans developed the decimal system, they probably chose a base of ten because we have 10 fingers on our hands. However, if we consider the fingers on only one hand, then a base of five makes sense. This numeration system, called **base-five,** has five key features, just like the decimal system.

1. Groups with a Base of Five

Base-five is a grouping system in which we group by fives instead of by tens. To illustrate the change in the grouping process, consider the 33 blocks in Figure 3.18. To regroup the blocks in base-five, we first put them into groups of 5. Doing so gives us 6 groups of 5 and 3 ones.

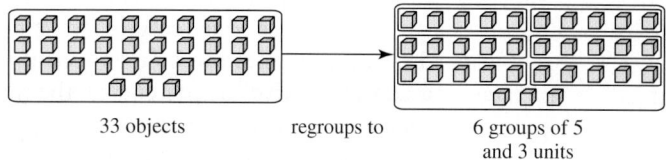

33 objects   regroups to   6 groups of 5 and 3 units

Figure 3.18 Regrouping objects by fives

We can regroup 5 of the 6 groups of 5 into 1 group of 25, giving us 1 group of 25, 1 group of 5, and 3 ones (Figure 3.19).

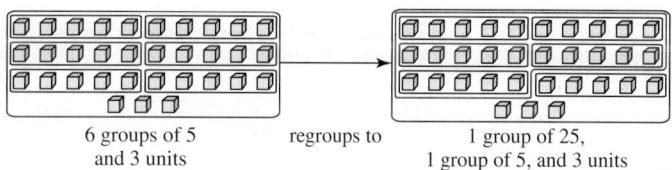

6 groups of 5 and 3 units   regroups to   1 group of 25, 1 group of 5, and 3 units

Figure 3.19 Regrouping objects by fives

Like the decimal system, we can continue to make groups of 5 as the number of objects increases. Consequently, the groups in base-five are:

Ones, or units
Fives   = five 1s
Twenty-fives   = 5 fives or 25 units
One hundred twenty-fives = 5 twenty-fives or 125 units . . .

We can represent certain groups of five with base-five blocks. Although we use the same names, the number of units they represent has changed. Specific base-five blocks are shown in Figure 3.20.

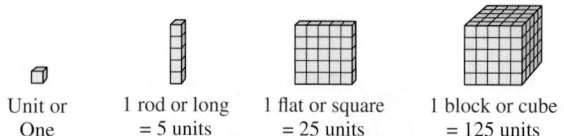

Unit or   1 rod or long   1 flat or square   1 block or cube
One   = 5 units   = 25 units   = 125 units

Figure 3.20 Base-five blocks

2. Use of Place Value

The groups of base-five lead directly to the system's place values, which are again listed in ascending order from right to left. The first seven place values of base-five are given in Table 3.8. Each place value is always five times greater than the place value to the right. Consequently, the place values of base-five are consecutive powers of 5.

Table 3.8 Place values of base-five						
Position	Seventh	Sixth	Fifth	Fourth	Third	Second	First
Value	15,625	3,125	625	125	25	5	1
Power of five	5^6	5^5	5^4	5^3	5^2	5^1	5^0

3. Use of 5 Digits

Base-five is a grouping system, so we always regroup once we reach a group of five of any one size. Because we never have more than four groups, we need only 5 digits, namely 0, 1, 2, 3, and 4.

4. Use of Zero

Base-five also involves the concept of zero, which is used both as a placeholder and to represent the cardinal number of the empty set.

Given these four features, we can write base-five numerals. However, before we do, two things need to be mentioned. First, we identify base-five numerals by writing a subscript "five" at the end of the numeral. If no base is indicated, the numeral is understood to be in base-ten. Second, we cannot use the vocabulary of the decimal system in base-five. Words like "-teen," "hundred," and "thousand," refer to groups of ten, not of five. Consequently, we read each digit individually and then end with the base. For instance, we read the numeral 24_{five} as "two–four, base-five" and the numeral 31_{five} as "three–one, base-five."

Example 3.19 | **Writing Base-Five Numerals**

Write the base-five numeral that represents the given groups of five.

a. Three 5s and four ones

b. Two 25s and three 5s

Solution Using the place values of base-five, we have the following numerals:

a. 34_{five} **b.** 230_{five}

We can also count in base-five. As children learn to count in the decimal system, they realize that each place value cycles through the digits 0–9. Upon reaching 9, the place value to the left increases by 1 while the original place value goes back to 0. We can apply the same idea to base-five, but we cycle only through the digits 0–4. Consequently, the first 15 counting numbers in base-five are

$$1_{\text{five}}, 2_{\text{five}}, 3_{\text{five}}, 4_{\text{five}}, 10_{\text{five}}, 11_{\text{five}}, 12_{\text{five}}, 13_{\text{five}},$$
$$14_{\text{five}}, 20_{\text{five}}, 21_{\text{five}}, 22_{\text{five}}, 23_{\text{five}}, 24_{\text{five}}, \text{and } 30_{\text{five}}$$

Eventually, children can use the idea of cycling digits to start counting at any numeral. We can do the same in base-five.

Example 3.20 | **Counting On in Base-Five**

What are the next five counting numbers after 42_{five}?

Solution By cycling through the digits, we know the next two numbers are 43_{five} and 44_{five}. The next number, however, is neither 45_{five} nor 50_{five} because 5 is not a digit in base-five. To find the next value, we realize that 44_{five} has a value of 4 groups of 5 and 4 ones, or 24 ones. The numeral after it must have a value of 25 ones. Because 25 ones makes 1 group of 25, 0 groups of 5, and 0 ones, the number after 44_{five} must be 100_{five}. As a result, the 5 counting numbers after 42_{five} are 43_{five}, 44_{five}, 100_{five}, 101_{five}, and 102_{five}.

When children first learn larger decimal numbers, they usually have no sense of relative size. The same is likely to be true for us as we work with base-five numerals for the first time. For instance, despite our initial impression, the numeral 4100_{five} does not have a value anywhere near 4,000. To get a better understanding of the values of large base-five numerals, we can decompose them using the fifth feature of base-five.

5. Additive and Multiplicative

Base-five is additive and multiplicative. To find the value of any base-five numeral, multiply each digit by its corresponding place value and then add the products. Unfortunately, this feature is not helpful unless we can translate the value of the base-five numeral into something we know. One way to do so is to use base-five blocks.

Example 3.21 | **Using Base-Five Blocks**

Representation   Use base-five blocks to find the value of each numeral.

a. 413_{five} **b.** 1024_{five}

Solution

a. We represent the numeral 413_{five} with 4 flats, 1 long, and 3 units.

Because each flat contains 25 units and each long contains 5 units, 413_{five} has a value of $(4 \times 25) + (1 \times 5) + (3 \times 1) = 100 + 5 + 3 = 108$.

b. We represent the numeral 1024_{five} with 1 block, 2 longs, and 4 units.

Because each block contains 125 units, 1024_{five} has a value of $(1 \times 125) + (2 \times 5) + (4 \times 1) = 125 + 10 + 4 = 139$.

The last example shows that we can use expanded notation to find the value of a base-five numeral because each of the base-five place values were described using decimal numerals. By multiplying and adding, we not only know the value of the numeral but also convert the base-five numeral into a decimal numeral.

Example 3.22 | **Finding the Value of Base-Five Numerals**

Find the value of each base-five numeral.

a. 231_{five} **b.** 20430_{five}

Solution

a. $231_{\text{five}} = (2 \times 25) + (3 \times 5) + (1 \times 1)$
$= 50 + 15 + 1$
$= 66.$

b. $20430_{\text{five}} = (2 \times 625) + (0 \times 125) + (4 \times 25) + (3 \times 5) + (0 \times 1)$
$= 1{,}250 + 100 + 15$
$= 1{,}365.$

Once we can interpret base-five numerals, we can order them using the same methods we used with the decimal system. With small numbers, we can use a number line. For example, we know that 13_{five} is smaller than 21_{five}, because 13_{five} occurs farther to the left on the base-five number line (Figure 3.21). With large numbers, we can compare place values. For instance, consider 3441_{five} and 3414_{five}. Because the two numbers differ in the fives place value and because 4 is bigger than 1, we know that $3441_{five} > 3414_{five}$.

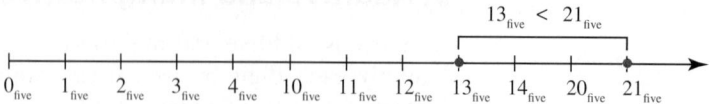

Figure 3.21 Base-five number line

Check Your Understanding 3.3-A

1. What are the next five counting numbers after 443_{five}?

2. Use base-five blocks to find the value of each numeral.

 a. 23_{five} **b.** 231_{five} **c.** 1403_{five}

3. Use expanded notation to find the value of each numeral.

 a. 41_{five} **b.** 440_{five} **c.** 2431_{five}

Talk About It How is counting in base-five like counting in base-ten? How is it different?

Reasoning

Activity 3.5

In the decimal system, we can classify a number as even or odd simply by looking at its last digit. Can we do the same in base-five? If so, why? If not, what other method might we use to determine whether a number is even or odd?

Other Base-*b* Numeration Systems

We can create other base numeration systems by changing the base to any whole number b greater than 1. In doing so:

1. Base-b will be a grouping system with a base of b.
2. Base-b will have place value.
3. Base-b will use b digits.
4. Base-b will have a concept of zero.
5. Base-b will be additive and multiplicative.

When we work with numerals in other bases, we should always look at the subscript first because it tells us the base. Knowing the base enables us to find the groups, place values, and digits. The next example demonstrates how to find the place values.

Example 3.23	Finding Place Values in Base-*b*

What are the first four place values in each base-b numeration system?

a. Base-two   **b.** Base-eight   **c.** Base-eleven   **d.** Base-sixteen

Solution Like the decimal system and base-five, the place values in any base-b numeration system are consecutive powers of the base. Starting with an exponent of zero, we generate the place values of the different systems as shown in the table:

	Fourth	Third	Second	First
a. Base-two	$2^3 = 2 \times 2 \times 2 = 8$	$2^2 = 2 \times 2 = 4$	$2^1 = 2$	$2^0 = 1$
b. Base-eight	$8^3 = 8 \times 8 \times 8 = 512$	$8^2 = 8 \times 8 = 64$	$8^1 = 8$	$8^0 = 1$
c. Base-eleven	$11^3 = 11 \times 11 \times 11 = 1,331$	$11^2 = 11 \times 11 = 121$	$11^1 = 11$	$11^0 = 1$
d. Base-sixteen	$16^3 = 16 \times 16 \times 16 = 4,096$	$16^2 = 16 \times 16 = 256$	$16^1 = 16$	$16^0 = 1$

As the last example shows, we can work with bases larger than ten. However, we need to make an extra consideration with the digits. Consider base-sixteen, which is a base commonly used in computer science. We can have up to 15 ones before we regroup. However, we cannot represent a value of fifteen with the numeral 15_{sixteen} because it represents 1 group of 16 and 5 ones, or 21 ones. The problem arises from the fact that the decimal system digits represent quantities only up to nine. For bases larger than ten, we need digits for the values that are larger than ten but less than the base. A common convention is to use capital letters as digits:

$$A = \text{ten}$$
$$B = \text{eleven}$$
$$C = \text{twelve}$$
$$D = \text{thirteen}$$
$$E = \text{fourteen}$$
$$F = \text{fifteen}$$
$$G = \text{sixteen}$$
$$H = \text{seventeen}, \ldots$$

For example, using this notation, numerals in base systems larger than ten include:

$$3BA_{\text{twelve}} \quad 4A2_{\text{eleven}} \quad BCD_{\text{fourteen}} \quad FD9_{\text{sixteen}}$$

Although the numerals may look strange, they work exactly like any other base numeral.

Example 3.24 Finding the Digits of Base-b

Determine the digits used in each base numeration system.

a. Base-four   **b.** Base-seven   **c.** Base-twelve   **d.** Base-fourteen

Solution In any base-b system, the digits start at 0 and get successively higher until the digit $b - 1$ is reached. For bases larger than ten, we use the digits from the decimal system to represent the quantities 0–9 and then add letters as necessary.

a. Base-four uses the digits 0, 1, 2, and 3.

b. Base-seven uses the digits 0, 1, 2, 3, 4, 5, and 6.

c. In base-twelve, we need to represent the quantities of zero through eleven with single digits. We use 0, 1, 2, . . . , 9 for the quantities zero through nine, A for ten, and B for eleven.

d. In base-fourteen we need single digits for ten, eleven, twelve, and thirteen. So base-fourteen uses 0, 1, 2, . . . , 9, A, B, C, and D.

Once we know the place values and digits, we can use them to count. If the base is small, there are fewer digits and we cycle through them more quickly. If the base is large, there are more digits and we cycle through them more slowly.

Example 3.25 Counting in Base-*b*

List the first 20 counting numbers in each base.

a. Base-seven   **b.** Base-sixteen

Solution

a. Because base-seven uses the digits 0, 1, 2, 3, 4, 5, and 6, the first 20 counting numbers are 1_{seven}, 2_{seven}, 3_{seven}, 4_{seven}, 5_{seven}, 6_{seven}, 10_{seven}, 11_{seven}, 12_{seven}, 13_{seven}, 14_{seven}, 15_{seven}, 16_{seven}, 20_{seven}, 21_{seven}, 22_{seven}, 23_{seven}, 24_{seven}, 25_{seven}, and 26_{seven}.

b. Because base-sixteen uses the digits 0, 1, 2, 3, 4, 5, 6, 7, 8, 9, A, B, C, D, E, and F, the first 20 counting numbers are $1_{sixteen}$, $2_{sixteen}$, $3_{sixteen}$, $4_{sixteen}$, $5_{sixteen}$, $6_{sixteen}$, $7_{sixteen}$, $8_{sixteen}$, $9_{sixteen}$, $A_{sixteen}$, $B_{sixteen}$, $C_{sixteen}$, $D_{sixteen}$, $E_{sixteen}$, $F_{sixteen}$, $10_{sixteen}$, $11_{sixteen}$, $12_{sixteen}$, $13_{sixteen}$, and $14_{sixteen}$.

Example 3.26 Counting On in Base-*b*

Find the next five counting numbers after each numeral.

a. 306_{eight} **b.** $ABA_{thirteen}$

Solution

a. Base-eight uses the digits 0, 1, 2, 3, 4, 5, 6, and 7. Because the digits roll over after we reach 7, the next five counting numbers are 307_{eight}, 310_{eight}, 311_{eight}, 312_{eight}, and 313_{eight}.

b. Base-thirteen uses the digits 0, 1, 2, 3, 4, 5, 6, 7, 8, 9, A, B, and C. Because the digits roll over after C, the next five counting are $ABB_{thirteen}$, $ABC_{thirteen}$, $AC0_{thirteen}$, $AC1_{thirteen}$, and $AC2_{thirteen}$.

We can also use the additive and multiplicative features of base-*b* numeration to help us understand the value of base-*b* numerals. As in base-five, we write the base-*b* numeral in expanded notation, expressing its place values with decimal numerals to help us make sense of its value. If we multiply and then add, we convert the base-*b* numeral to a decimal numeral.

Example 3.27 Converting Base-*b* Numerals to Decimal Numerals

Find the value of each numeral by converting it to a decimal numeral.

a. 541_{six} **b.** $3B2_{twelve}$ **c.** $BAD_{fifteen}$

Solution For each numeral, we first look at the subscript to determine the base and its corresponding place values. Using this information, we write the numeral in expanded notation, multiply, and add.

a. $541_{six} = (5 \times 36) + (4 \times 6) + (1 \times 1) = 180 + 24 + 1 = 205.$

b. $3B2_{twelve} = (3 \times 144) + (11 \times 12) + (2 \times 1) = 432 + 132 + 2 = 566.$

c. $BAD_{fifteen} = (11 \times 225) + (10 \times 15) + (13 \times 1) = 2,475 + 150 + 13 = 2,638.$

In addition to converting a base-*b* numeral to a decimal numeral, we can convert a decimal numeral to another base. One way to do so is to rearrange the corresponding number of blocks into groups consistent with the base. For instance, to convert

26 into a base-four numeral, we arrange 26 blocks into groups of 4, 16, and so on (Figure 3.22). Because 26 blocks can be arranged into 1 group of 16, 2 groups of 4, and 2 ones, we have $26 = 122_{\text{four}}$.

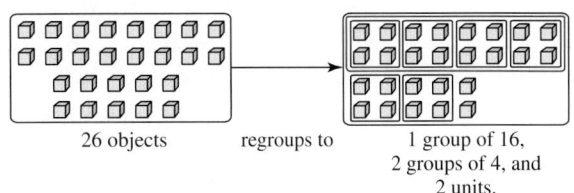

26 objects   regroups to   1 group of 16,
                                    2 groups of 4, and
                                    2 units.

Figure 3.22 Regrouping object by fours

Although the strategy works in any situation, it becomes more difficult to use as the numbers grow larger. However, we can achieve the same results by using division, as demonstrated in the next example.

Example 3.28 | Converting Decimal Numerals to Base-*b*

Make each conversion.

a. 45 to base-five   **b.** 426 to base-twelve

Solution

a. To make the conversion, we must regroup the objects into the groups of base-five, which include ones, groups of 5, groups of 25, groups of 125, and so on. Because there are only 45 objects, we cannot make a group of 125, so we start with groups of 25. To determine how many groups of 25 are in 45 objects, we divide 45 by 25, or

$$45 \div 25 = 1 \text{ with a remainder of } 20$$

This indicates that we can make 1 group of 25 with 20 objects left over. Next, we put the 20 objects into groups of 5 by dividing by 5, or

$$20 \div 5 = 4 \text{ with a remainder } 0$$

This gives 4 groups of 5 with 0 objects left over. By combining the results, we can make 1 group of 25, 4 groups of 5, and 0 ones from 45 objects. Consequently, $45 = 140_{\text{five}}$.

b. We use the same approach for base-twelve, which has ones, groups of 12, and groups of 144. By dividing, we find

$$426 \div 144 = 2 \text{ remainder } 138 \text{ (2 groups of 144 with 138 ones left over)}$$

$$138 \div 12 = 11 \text{ with remainder } 6 \text{ (11 groups of 12 with 6 ones left over)}$$

Because we represent the quantity of 11 with B, $426 = 2B6_{\text{twelve}}$.

Check Your Understanding 3.3-B

1. List the digits used in each numeration system.
   a. Base-three   **b.** Base-eight   **c.** Base-thirteen
2. List the first four place values in each numeration system.
   a. Base-three   **b.** Base-seven   **c.** Base-fourteen

3. Convert each numeral to a decimal numeral.

   a. 10110_{two} **b.** 631_{eight} **c.** AA_{twelve} **d.** $C1B_{fifteen}$

4. Make each conversion.

   a. 89 to base-five   **b.** 134 to base-eight   **c.** 523 to base-sixteen

Talk About It Why does it make sense to use division when regrouping objects from the decimal system into another base numeration system?

Reasoning

Activity 3.6

Determine whether 4451_{six} is less than, greater than, or equal to 2106_{eight}. Describe your method.

SECTION 3.3 Learning Assessment

Understanding the Concepts

1. How is base-five numeration similar to decimal numeration? How is it different?

2. Why does it make little sense to have a base-one numeration system?

3. How are the place values and the digits affected by changing the base of a base-b numeration system?

4. Why are letters used in base numeration systems with a base larger than ten?

5. How is counting in a base-b numeration system similar to counting in the decimal system? How is it different?

6. Why does writing a base-b numeral in expanded notation also provide a way to convert it into a decimal numeral?

7. Explain how to convert a decimal numeral into a base-five numeral.

Representing the Mathematics

8. Write the name of each numeral.

   a. 23_{four} **b.** 451_{six}

   c. 7007_{eight} **d.** $AB3_{twelve}$

9. Give the base-b numeral that represents each of the following.

   a. 4 groups of 25, 3 groups of 5, and 0 ones.

   b. 2 groups of 16, 1 group of 4, and 3 ones.

   c. 5 groups of 343, 3 groups of 49, 4 groups of 7, and 1 one.

   d. 10 groups of 144, 8 groups of 12, and 9 ones.

10. Give the base-b numeral that represents each of the following.

   a. 2 groups of 27, 2 groups of 9, and 1 one.

   b. 5 groups of 216, 4 groups of 6, and 4 ones.

   c. 7 groups of 64 and 5 ones.

   d. 9 groups of 225, 11 groups of 15, and 14 ones.

11. Describe the groups represented by each base-b numeral.

   a. 114_{five} **b.** 2046_{seven}

   c. 8081_{nine} **d.** $ABBA_{twelve}$

12. Describe the groups represented by each base-b numeral.

   a. 1021_{three} **b.** 354_{eight}

   c. $ABC_{thirteen}$ **d.** $CDC_{fourteen}$

13. State the number of units contained in the long, flat, and block of each base.

   a. Base-four   **b.** Base-seven

   c. Base-eleven   **d.** Base-sixteen

14. Draw a diagram of the base-five blocks needed to represent each numeral.

   a. 32_{five} **b.** 104_{five}

   c. 4003_{five} **d.** 2222_{five}

15. Draw a diagram of the base-b blocks needed to represent each numeral.

 a. 1101_{two} **b.** 323_{four}

 c. 561_{seven} **d.** $10A_{\text{twelve}}$

16. Write the numeral represented by each set of base blocks.

 a. 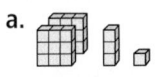 **b.**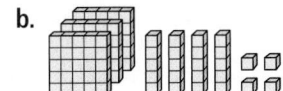

 c.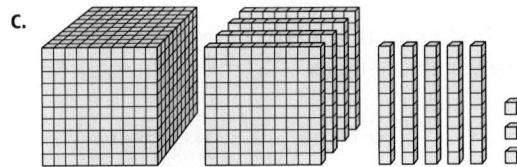

 d.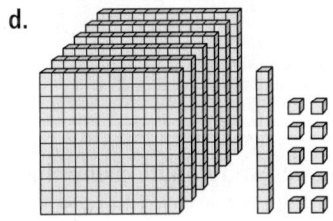

17. What is the smallest number of base-five blocks needed to represent each of the following numbers of units?

 a. 17 **b.** 51

 c. 113 **d.** 392

18. Determine the smallest number of blocks, flats, longs, and units that can be made from the 60 units shown in each of the given bases.

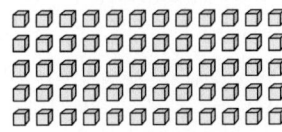

 a. Base-three **b.** Base-five

 c. Base-eight **d.** Base-twelve

■ Exercises

19. Write each numeral in expanded notation using exponents.

 a. 21_{five} **b.** 304_{five} **c.** 331_{five} **d.** 2043_{five}

20. Write each numeral in expanded notation using exponents.

 a. 22_{three} **b.** 125_{six}

 c. 1307_{nine} **d.** $AB3D_{\text{fourteen}}$

21. Condense the expanded notation into a base-b numeral.

 a. $(3 \times 25) + (4 \times 5) + (2 \times 1)$

 b. $(7 \times 64) + (7 \times 8) + (4 \times 1)$

 c. $(1 \times 121) + (10 \times 11) + (10 \times 1)$

 d. $(13 \times 3{,}375) + (14 \times 225) + (0 \times 15) + (9 \times 1)$

22. Condense the expanded notation into a base-b numeral.

 a. $(1 \times 5^2) + (1 \times 5^1) + (4 \times 5^0)$

 b. $(1 \times 3^4) + (2 \times 3^2) + (2 \times 3^0)$

 c. $(5 \times 9^3) + (7 \times 9^2) + (8 \times 9^1)$

 d. $(10 \times 14^3) + (11 \times 14) + (3 \times 14^0)$

23. Give the first five place values of each base.

 a. Base-four **b.** Base-six

 c. Base-twelve **d.** Base-fifteen

24. List the digits used in each base numeration system.

 a. Base-two **b.** Base-nine

 c. Base-eleven **d.** Base-eighteen

25. How many different digits are needed to write numerals in:

 a. Base-twenty?

 b. Base-thirty-five?

 c. Base-fifty?

26. List the first 20 counting numbers in each base.

 a. Base-two **b.** Base-eight **c.** Base-twelve

27. Complete the following table to show the same consecutive counting numbers in each base.

Decimal	23	24	25	26	27	28	29	30	31	32
Base-three										
Base-nine										
Base-twelve										

28. Convert each numeral to a decimal numeral.

 a. 23_{five} **b.** 401_{six}

 c. 2311_{nine} **d.** 4423_{thirteen}

29. Convert each numeral to a decimal numeral.

 a. 1211_{three} **b.** 564_{eight}

 c. $51B_{\text{twelve}}$ **d.** $2AF_{\text{sixteen}}$

30. Write each decimal numeral as a base-five numeral.

 a. 15 **b.** 76 **c.** 128 **d.** 231

31. Make each conversion.

 a. 23 to base-two

 b. 168 to base-eight

 c. 732 to base-twelve

32. Make each conversion.

 a. 47 to base-four

 b. 241 to base-nine

 c. 1,730 to base-sixteen

33. Give the numeral that comes before and after the given numeral.

    a. 122_{three} **b.** 100_{eight} **c.** 88_{nine} **d.** $CD_{fourteen}$

34. Find the missing numerals in each list.

    a. $3_{six}, 4_{six}, 5_{six},$ _____, _____, _____

    b. $175_{eight}, 176_{eight}, 177_{eight},$ _____, _____, _____

    c. _____, _____, _____, $800_{nine}, 801_{nine}, 802_{nine}$

    d. _____, $FF_{sixteen}, 100_{sixteen}, 101_{sixteen},$ _____, _____

35. Find the missing numerals in each list.

    a. $3_{seven}, 5_{seven}, 10_{seven},$ _____, _____, _____

    b. $5_{fifteen}, A_{fifteen}, 10_{fifteen},$ _____, _____, _____

    c. $14_{sixteen}, 16_{sixteen}, 18_{sixteen},$ _____, _____, _____

    d. _____, $10_{eight}, 14_{eight}, 20_{eight},$ _____, _____

■ Problems and Applications

36. What is the largest 3-digit numeral in base-five? 4-digit numeral? 5-digit numeral? n-digit numeral?

37. What is the largest 4-digit numeral in base-four? Base-eight? Base-twelve?

38. Determine the total number of units that are equivalent to the blocks in the given base.

    a. Base-four: 1 flat, 3 longs, 2 units

    b. Base-six: 1 block, 3 flats, 2 longs, five units

    c. Base-seven: 2 blocks, 5 flats, 3 longs, 0 units

    d. Base-sixteen: 10 flats, 9 longs, 15 units

39. The base-ten long is twice as big as the base-five long.

    a. How many times bigger is the base-ten flat than the base-five flat?

    b. How many times bigger is the base-ten block than the base-five block?

40. The base-nine long is three times larger than the base-three long.

    a. How many times bigger is the base-nine flat than the base-three flat?

    b. How many times bigger is the base-nine block than the base-three block?

41. **a.** One rod in base-ten blocks is equivalent to what in base-two blocks?

    b. One flat in base-ten blocks is equivalent to what in base-five blocks?

    c. One block in base-ten blocks is equivalent to what in base-sixteen blocks?

42. How many numerals come before:

    a. 100_{five}? **b.** 100_{eight}? **c.** $100_{sixteen}$?

43. How many two-digit numbers are there in

    a. Base-four? **b.** Base-eight? **c.** Base-twelve?

44. **a.** In base-five, how many numerals have 2 or fewer digits?

    b. How many have 3 or fewer digits?

    c. How many have 4 or fewer digits?

    d. How many have n or fewer digits?

45. Find the missing base that makes each equation true.

    a. $32_{five} = 21$ ____

    b. $2A_{twelve} = 54$ ____

    c. $10110_{two} = 1A$ ____

46. The first four numerals in base-two and base-four are given in the table.

Base-two	Base-four
00	0
01	1
10	2
11	3

    The table shows an interesting relationship that allows us to make quick conversions between the bases by considering sequences of digits. For example, because $3_{four} = 11_{two}$, we can convert 33_{four} to base-two by replacing each 3 with an 11. Hence, $33_{four} = 1111_{two}$.

    a. Use the table to convert the following to base-two.

    i. 12_{four} **ii.** 320_{four}

    iii. 3201_{four} **iv.** 3132_{four}

    b. Use the table to convert the following to base-four.

    i. 1010_{two} **ii.** 111101_{two}

    iii. 10101_{two} **iv.** 1000011_{two}

    c. Why is such a conversion possible between base-two and base-four?

47. A relationship between base-two and base-eight is similar to the one in the last problem.

    a. Make a table of values that can be used for converting between base-two and base-eight.

    b. Use the table from part (a) to convert the following to base-two.

    i. 45_{eight} **ii.** 356_{eight}

    iii. 1072_{eight} **iv.** 30042_{eight}

    c. Use the table from part a to convert the following to base-eight.

    i. 110101_{two} **ii.** 101001_{two}

    iii. 10011101_{two} **iv.** 100001110110_{two}

    d. Name three other bases that share a similar relationship with base-two.

48. The conversions in the last two problems can be done with bases other than two.

    a. What criteria must be true about two bases so that they share this conversion relationship?

    b. List five other pairs of bases that satisfy this conversion relationship.

■ Mathematical Reasoning and Proof

49. Which is bigger: 213_{five} or 223_{five}? Explain how you know.

50. Which is bigger: 1000_{two} or 100_{three}? Explain how you know.

51. Does $100_{five} = 100_{eight}$? Why or why not?

52. Does $21_{five} = B_{sixteen}$? Explain.

53. If the letters w, x, y, and z represent four different digits in a base-b numeration system, write the numeral $wxyz_b$ in expanded notation.

54. **a.** Represent the quantity of six in base-six.

    b. Represent the quantity of eight in base-eight.

    c. Represent the quantity of twelve in base-twelve.

    d. Based on what you observe in parts a, b, and c, make a generalization about the numeral that will always be used to represent the quantity of b in base-b.

55. The first 24 numerals of base-six are listed in the following array. Based on the patterns you see, write the next four rows of the array.

0	1	2	3	4	5
10	11	12	13	14	15
20	21	22	23	24	25
30	31	32	33	34	35

■ Mathematical Communication

56. Why is it inappropriate to use the term "hundred" when reading numerals in other bases?

57. Write down directions for converting a base-b numeral to a decimal numeral. Give the directions along with trial problems to someone who has never done it before. Ask them to make the conversions. Were they successful? Explain why or why not.

■ Building Connections

58. The Babylonian numeration system is close to being a true base-sixty system.

    a. What digits are used to represent the quantities of 15, 25, 42, 59?

    b. What are the first five place values in the Babylonian numeration system? How do these compare to the place values of base-sixty?

    c. What features of base-sixty, if any, are missing from the Babylonian system?

59. Is the Mayan system a true base-twenty numeration system? Why or why not?

60. Consider the definition of less than given in Section 3.2. Does the definition have to be changed for it to be applicable to base-five? If so, what changes must be made? If not, why not?

61. Do other base numeration systems play a role in the elementary or middle grades' mathematics curriculum? Examine a set of curriculum materials to find out.

Historical Note

Near the end of the twelfth century, a man by the name of Leonardo of Pisa, or Fibonacci, came to the forefront of European scholarship. Although from Italy, Fibonacci was educated in North Africa and traveled extensively around the Mediterranean. At some point during his travels, he was exposed to the decimal system, which the Arabs had brought from India. He quickly realized its power and brought it to the attention of Europe in 1202 in his book *Liber Abaci*, or *Book of Counting*. At first, the system was not widely accepted, and it was even outlawed in Florence because the 0 made it too easy to falsify records. The system eventually won out because it facilitated writing and computing with numbers. With the invention of the printing press, the symbols of the decimal system became standardized, and the system became commonplace and widely used throughout Europe. Over the next several centuries, it eventually became the dominant numeration system throughout the world.

Fibonacci

Opening Problem Answer

In ASCII, M = 01001101, A = 01000001, T = 01010100, and H = 01001000. So the word "MATH" is represented by the 32-digit string 01001101010000010101010001001000.

Chapter 3 REVIEW

Summary of Key Concepts from Chapter 3

Section 3.1   A Brief History of Numeration

- **Numbers** are abstract ideas of quantity. **Numerals** are symbolic representations of numbers that take on different forms.

- Any collection of symbols and rules that govern how numerals are written is called a **numeration system.**

- In a **grouping system,** once a certain number of objects are counted, they are grouped and represented with one symbol. The number of objects needed to make a group is called the **base** of the system.

- In an **additive** system, the value of a numeral is found by adding the values of each symbol in the numeral.

- **Place value** means that the position of a **digit** in the numeral determines its value.

- In a **multiplicative** system, the value of a digit is found by multiplying its value by its corresponding place value.

- A **placeholder** is used to move other nonzero digits out to the appropriate position.

Section 3.2   The Hindu-Arabic or Decimal System

- The decimal system has five key features.

   1. It is a grouping system that uses a base of ten.   **2.** It has place value.

   3. It uses 10 digits: 0, 1, 2, 3, 4, 5, 6, 7, 8, and 9.   **4.** It has a concept of zero.

   5. It is additive and multiplicative.

- **Composing** a numeral means finding its value by combining the values of the groups within it. **Decomposing** means breaking a numeral into smaller groups that have a total value equal to the number.

- If two numbers, a and b, have the same value, we say they are **equal** and write $a = b$.

- Let a and b be whole numbers, where $a = n(A)$ and $b = n(B)$. a is **less than** b, written $a < b$, if and only if A is equivalent to a proper subset of B.

Section 3.3   Other Base Numeration Systems

- Other base numeration systems can be made by changing the base to any whole number b greater than 1.

- **Base-b numeration systems** have five key components.

   1. They have a grouping system with a base of b.   **2.** They have place value.

   3. They use b digits 0, 1, 2, 3, . . . , $b - 1$.   **4.** They have a concept of zero.

   5. They are additive and multiplicative.

Review Exercises Chapter 3

1. Determine whether each statement relates to a number or a numeral.

   a. She had a 5 written backward on her paper.

   b. Tomorrow I turn 21.

   c. There are 25 students in my class.

2. Convert each numeral to a Hindu-Arabic numeral.

   a. |||| |||| |||| |||| |||| |||| |||| |||| b. ꙮꙮ𐎛𐎛∩∩∩∩∩|||||||||

   c. ◀◀◀▲◀▼▼▼
      ▲◀▼▼

   d. MCDXXVII

   e. $\overline{\text{XLIIVIII}}$

3. Write each decimal numeral as an Egyptian, Babylonian, Roman, and Mayan numeral.

   a. 56   b. 2,134   c. 41,235

4. Complete each list of consecutive numerals.

   a. ∩∩∩∩∩∩∩∩∩|||||||||, _____, _____,

      _____

   b. _____, _____, _____, ◀▼▼

   c. _____, _____, XCIX, _____, _____

5. How many symbols are needed to write the number 645 as a(n):

   a. Egyptian numeral?

   b. Roman numeral?

   c. Babylonian numeral?

   d. Mayan numeral?

6. What is the smallest Roman numeral that makes use of exactly five symbols?

7. State the place value of the underlined digits in each number.

   a. 3$\underline{4}$,09$\underline{8}$

   b. $\underline{2}$,910,0$\underline{8}$3

8. Write each number in expanded notation.

   a. 1,201

   b. 85,319

   c. 45,981,003

9. Condense each number written in expanded notation to a standard decimal number.

   a. $(4 \times 100) + (6 \times 1)$

   b. $(5 \times 10,000) + (1 \times 1,000) + (8 \times 100) + (2 \times 1)$

   c. $(7 \times 100,000) + (1 \times 10,000) + (4 \times 1,000) + (1 \times 100) + (8 \times 10) + (6 \times 1)$

10. Write the name of each number.

    a. 657

    b. 1,509

    c. 23,981,085

11. Describe the groups represented by each numeral.

    a. 51   b. 861   c. 12,671

12. What is the decimal numeral that represents:

    a. 3 hundreds, 6 tens, and 0 ones?

    b. 4 thousands, 9 hundreds, 9 tens, and 1 one?

    c. 1 hundred thousand, 1 ten thousand, 4 thousands, and 4 tens?

13. List the next three numbers in each sequence.

    a. 5_{nine}, 6_{nine}, 7_{nine}, _____, _____, _____

    b. 331_{four}, 332_{four}, 333_{four}, _____, _____, _____

    c. C_{fifteen}, D_{fifteen}, E_{fifteen}, _____, _____, _____

14. Given the 40 units shown, determine the number of blocks, flats, longs, and units that can be made in each base.

    a. Base-four   b. Base-seven   c. Base-thirteen

15. Determine the total number of units that are equivalent to the following blocks in the given base.

    a. Base-three: 2 flats, 1 long, 0 units

    b. Base-eight: 2 blocks, 4 flats, 0 longs, 4 units

    c. Base-thirteen: 1 block, 4 flats, 11 longs, 6 units

16. Make the following conversions.

    a. 41 to base-three

    b. 125 to base-nine

    c. 471 to base-twelve

17. Write each number in expanded notation, and then convert them to decimal numerals.

    a. 3412_{five} b. 8710_{nine} c. $4AD_{\text{sixteen}}$

18. How many numerals precede 100_{fifteen}?

19. How many different digits are needed to write numerals in:

    a. base-twenty-five

    b. base-thirty

    c. base-one hundred

20. a. In base-three, how many numbers have two or fewer digits?

    b. How many have three or fewer digits?

    c. How many have four or fewer digits?

    d. How many have n or fewer digits?

21. Which is bigger: 41_{five} or 41_{six}?

22. Which is bigger: 135_{six} or 100101_{two}?

Answers to Chapter 3 Check Your Understandings and Activities

Check Your Understanding 3.1-A

1. a. 42 **b.** 30,154 **c.** 424,038 **d.** 747 **e.** 62,404

2. ||||| ||||| ||||| ||||| ||||| ||||| ||||| ||||| ||||| ||||| ||||| ||

3. a. ∩∩∩∩∩IIIII and LV.

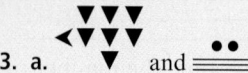

   b.  and $\overline{\text{V}}$ICDXVIII.

   c.  and $\overline{\text{MCCCIIXL}}$.

Check Your Understanding 3.1-B

1. a. 35 **b.** 46,824 **c.** 142,599,611

2. a. 18 **b.** 280 **c.** 1,806

3. a. ▼▼▼ / ◀▼▼▼ / ▼ and ══ (•• ••)

   b. ▼ ◀▼▼▼ ▼▼ and ════ (•••)

   c. ◀◀◀▼◀ ▼▼▼ ▼ and ═══ (•••• / •••)

Check Your Understanding 3.2-A

1. a. 1 ten and 3 ones **b.** 6 thousands, 3 hundreds, 2 tens, and 9 ones **c.** 1 ten thousand, 4 thousands, 1 hundred, 0 tens, and 9 ones

2. a. 384 **b.** 560 **c.** 2,069

3. a. $(4 \times 100) + (1 \times 10) + (0 \times 1)$

   b. $(8 \times 1,000) + (6 \times 100) + (0 \times 10) + (1 \times 1)$

   c. $(1 \times 10,000) + (2 \times 1,000) + (3 \times 100) + (5 \times 10) + (8 \times 1)$

   d. $(2 \times 100,000) + (0 \times 10,000) + (0 \times 1,000) + (6 \times 100) + (0 \times 10) + (1 \times 1)$

4. Answers will vary, but five possible answers are 1,540 ones; 154 tens, 15 hundreds and 40 tens, 1 thousand and 54 tens, or 1 thousand, 5 hundreds, and 4 tens.

Check Your Understanding 3.2-B

1. a. Four hundred thirty-eight

   b. Twenty-three thousand, six

   c. Twelve million, seventy-one thousand, twenty-nine

2. Compare the place values directly, and notice that 347 and 357 differ in the second place value. Because 4 is less than 5, 347 is less than 357.

3. The stack is $1,000,000,000 \div 100 = 10,000,000$ in. tall. This is about 833,333 ft or 158 mi tall.

Check Your Understanding 3.3-A

1. 444_{five}, 1000_{five}, 1001_{five}, 1002_{five}, and 1003_{five}

2. a. 13 **b.** 66

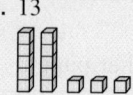 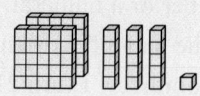

   c. 228

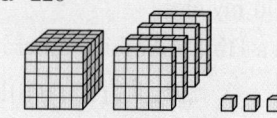

3. a. 21 **b.** 120 **c.** 366

Check Your Understanding 3.3-B

1. a. 0, 1, and 2 **b.** 0, 1, 2, 3, 4, 5, 6, and 7 **c.** 0, 1, 2, 3, 4, 5, 6, 7, 8, 9, A, B, and C

2. a. 1, 3, 9, and 27 **b.** 1, 7, 49, and 343 **c.** 1, 14, 196, and 2,744

3. a. 22 **b.** 409 **c.** 130 **d.** 2,726

4. a. 324_{five} **b.** 206_{eight} **c.** $20B_{\text{sixteen}}$

Activity 3.1

a. 10^{25}

b. i. 110,101 **ii.** 121,000 **iii.** 100,000,010,031 **iv.** 10,011,001 **v.** 100,000,110,011

c. Possible answers are:

   i. face. **ii.** deed. **iii.** label.

Activity 3.2

a. 26

b. Ones, groups of 26, groups of $26 \times 26 = 676$, and groups of $26 \times 26 \times 26 = 17,576$

c. i. 2,398 **ii.** 1,371 **iii.** 126,037 **iv.** 25,536

Activity 3.3

a. i. 8 **ii.** 12 **iii.** 19 **iv.** 13

b. Add the digits to obtain the smallest number of base-ten blocks needed to represent a numeral.

Activity 3.4

A cube that measures 10 m on each side.

Activity 3.5

We cannot look at the last digit in base-five and determine whether the number is even or odd. However, if the sum of the digits in the numeral is even, then the number is even. Likewise, if the sum of the digits is odd, then the number is odd.

Activity 3.6

One way to know is to convert both numbers into base-ten and then compare. Because $4451_{\text{six}} = 1,039$, $2106_{\text{eight}} = 1,094$, and $1,039 < 1,094$ we have $4451_{\text{six}} < 2106_{\text{eight}}$.

◆ Mathematical Representation

1. Translate each statement into a symbolical representation.

   a. The sum of x and 7 is 15. b. The quotient of x and 8 does not equal 4.

2. Write the statement $4(x + 3) > 6$ verbally.

3. What does the following figure suggest might be true for every rhombus?

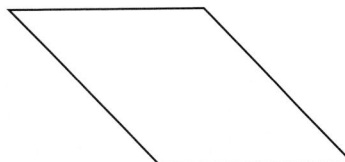

4. Write a set of directions that describes how to draw a right isosceles triangle.

◆ Inductive and Deductive Reasoning

5. If $a = 9$, $b = 2a$, and $c = \dfrac{b}{6}$, what can you conclude about c? What type of reasoning did you use?

6. "Every time I eat fast food, I get an upset stomach." What can you conclude about the next time I eat fast food? What type of reasoning did you use?

7. Find a counterexample to the statement, "No number between 4 and 8 is divisible by 3."

8. Let t represent "I have a toothache" and d represent "I am going to the dentist." Represent each statement symbolically.

   a. I have a toothache, and I am going the dentist.

   b. If I do not have a toothache, then I am not going to the dentist.

9. Write the converse, inverse, and contrapositive of the statement, "If you are a woman, then you like chocolate."

10. If p is false and q is false, find the truth value of each statement.

    a. $\sim p$ b. $p \wedge q$ c. $p \vee q$ d. $p \rightarrow q$ e. $\sim q \rightarrow \sim p$

11. Fill in the missing statement to make a direct argument.

    _____.

    I saw a funny movie.

    Therefore, I am laughing.

◆ Problem Solving

12. How many different ways can 50 be written as the sum of 5s, 10s, and 20s?

13. Consider the following successive arrangements of toothpicks. How many toothpicks are needed to make the fifteenth arrangement?

14. A soda factory produces 100,000 cans of soda in one day. About how many cases of 24 cans are produced in one week?

15. How many triangles are in the following figure?

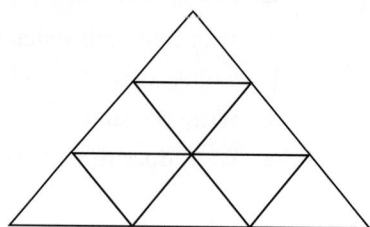

16. A bag of grass seed covers 2,000 ft². Which will require more bags of grass seed to sow: a square yard with sides of length 130 ft or a rectangular yard with a length of 200 ft and a width of 75 ft?

◆ Sets and Set Operations

17. Write the set of whole numbers greater than 5 as a roster and in set-builder notation.

18. Demonstrate two one-to-one correspondences between $A = \{w, x, y, z\}$ and $B = \{2, 4, 6, 8\}$.

19. Assume $A = \{1, 3, 4, 5\}$, $B = \{2, 3, 5, 6\}$, and $U = \{1, 2, \ldots, 6\}$. Find:

  a. $A \cup B$ **b.** $A \cap B$ **c.** $A - B$ **d.** $\overline{B}$

20. Find $A \times B$, if $A = \{s, t, u\}$ and $B = \{\square, \diamond\}$.

21. A sports magazine does a survey that has 230 female respondents. Of the respondents, it was found that 90 swim, 105 run, 80 bike, 35 swim and run, 30 swim and bike, 40 run and bike, and 20 do all three.

  a. How many respondents only run?

  b. How many respondents do none of the activities?

◆ Decimal Numerals and Other Numeration Systems

22. Write each decimal numeral as an Egyptian, Roman, Babylonian, and Mayan numeral.

  a. 47   **b.** 368   **c.** 12,940

23. Write each number in expanded notation

  a. 450   **b.** 13,498   **c.** 120,065

24. What is the decimal numeral that represents:

  a. 5 hundreds, 5 tens, and 3 ones?

  b. 3 thousands, 8 hundreds, 3 tens, and 8 ones?

25. List the next three numbers in each sequence.

  a. 7_{twelve}, 8_{twelve}, 9_{twelve}, _____, _____, _____

  b. 453_{six}, 454_{six}, 455_{six}, _____, _____, _____

26. Convert 94 to base-five.

27. Write each number in expanded notation, and then convert them to a decimal numeral.

  a. 221_{five} **b.** 534_{nine} **c.** $2E0_{\text{sixteen}}$

Whole-Number Computation

Opening Problem

Here is a magic trick to try on your friends. Take your age and multiply it by 2. Add 5, and then multiply the result by 50. Subtract 365, and then add the amount of change in your pocket that is less than one dollar. Finally, add 115. How does the final result relate to your age and the amount of change in your pocket? Can you explain why the trick works?

Jupiterimages

© Image Source/Corbis

In this chapter, we develop whole-number operations using ideas from set theory and numeration. Much of elementary mathematics revolves around the whole-number operations of addition, subtraction, multiplication, and division. These operations are important not only because they are useful in our everyday lives, but also because they serve as the foundation for all later work in mathematics. For these reasons, whole-number computation is included in the second and third goals of the NCTM Number and Operations Standard.

> ### ■ NCTM Number and Operations Standard
>
> Instructional programs from prekindergarten through grade 12 should enable all students to:
>
> - Understand numbers, ways of representing numbers, relationships among numbers, and number systems.
> - Understand meanings of operations and how they relate to one another.
> - Compute fluently and make reasonable estimates.

Source: NCTM STANDARDS Copyright © 2011 by NATIONAL COUNCIL OF TEACHERS OF MATHEMATICS. Reproduced with permission of NATIONAL COUNCIL OF TEACHERS OF MATHEMATICS.

As the second goal indicates, part of computation involves intuitively understanding the meaning of each operation. Children first come to understand the operations by manipulating and counting sets. Then, as they gain experience, they begin to refine their concepts. They come to see the operations in other ways and how they relate to one another. They learn the properties of the operations and how to use them to make computations easier. They also learn how and when to apply the operations in a variety of contexts by solving story problems. Specific objectives for understanding the whole-number operations are listed in Table 4.1.

Table 4.1 Classroom learning objectives	K	1	2	3	4	5	6	7	8
Represent addition and subtraction with objects.	X								
Solve addition and subtraction word problems.	X	X							
Relate counting to addition and subtraction.		X							
Understand subtraction as an unknown-addend problem.		X							
Apply properties of operations as strategies to add and subtract.		X							
Use addition and subtraction to solve one- and two-step word problems.			X						
Interpret products and quotients of whole numbers.				X					
Understand division as an unknown-factor problem.				X					
Use multiplication and division to solve word problems in situations involving equal groups, arrays, and measurement quantities.				X					
Apply properties of operations to multiply and divide.				X					
Solve multistep word problems having whole-number answers using the four operations.					X				
Write and evaluate numerical expressions involving whole-number exponents.						X			

Source: Adapted from the *Common Core State Standards for Mathematics* (*Common Core State Standards Initiative*, 2010).

The third goal of the Number and Operations Standard indicates that children must learn to compute fluently; that is, they need to compute operations efficiently and accurately using a variety of methods. Computational fluency begins with learning the single-digit facts for addition and multiplication, as well as their counterparts for subtraction and division. *Mastery of these facts is crucial because they are the foundation for all other computation.* Without knowing them, students will be unable to compute with large numbers or with the integers, fractions, and decimals. Once students know these basic facts, they continue to work on computational fluency by learning to use a variety of methods for computing with large numbers. In general, such methods include:

- *Manipulatives.* Many manipulatives, like base-ten blocks or bundled sticks, can be used to model and learn the intuitive ideas behind large number computations.
- *Written algorithms.* Although several written algorithms can be used to compute with large numbers, children are usually taught the standard algorithms because they are efficient. Unfortunately, these algorithms seldom reflect the intuitive meaning of the operation, so they are often taught through a progression of steps that are easier for children to understand.
- *Calculators.* Calculators are an important part of the mathematics classroom. Their computational power enables students to focus on understanding the concepts and solving problems, not on procedural skills. Elementary students are likely to use both four-function and fraction calculators.
- *Mental computation.* Mental computation is a way for students to compute quickly without using any computational tools. Children often struggle with mental computation because they try to use the standard algorithms. Although the algorithms work well with a pencil and paper, they are difficult to use mentally. Consequently, children must not only learn new strategies for computing but also develop the number sense necessary to use them effectively.
- *Estimation.* Estimation is closely connected to the use of calculators and mental computation because we use it to judge whether our answers are reasonable and correct. Children need to know how to make good estimations, by learning how to alter numbers to make them more manageable and by using basic computation facts in diverse ways.

Computational fluency with the whole numbers takes several years to develop. Table 4.2 gives specific objectives associated with computational fluency and the grade level at which each objective is likely to be met.

Table 4.2 Classroom learning objectives	K	1	2	3	4	5	6	7	8
Fluently add and subtract within 5.	X								
Given a two-digit number, mentally find 10 more or 10 less than the number.		X							
Use different strategies to add and subtract fluently within 20.		X	X						
Fluently add and subtract multidigit whole numbers using strategies based on place value, properties of operations, and the standard algorithm.		X	X	X	X				
Mentally add or subtract 10 or 100 from a given number 100–900.			X						
Fluently multiply and divide multidigit whole numbers using strategies based on place value, properties of operations, and the standard algorithm.				X	X	X	X		
Use place value to round multidigit whole numbers.				X	X				
Use parentheses, brackets, or braces in numerical expressions and evaluate expressions with these symbols.						X			

Source: Adapted from the *Common Core State Standards for Mathematics* (*Common Core State Standards Initiative*, 2010).

Because elementary teachers spend a good deal of time teaching their students the different facets of whole-number computation, they themselves must understand all that it entails. For this reason, we now discuss the operations of addition, subtraction, multiplication, and division by:

- Defining each operation and discussing its intuitive meaning.
- Stating the properties associated with each operation.
- Demonstrating different methods for computing with large numbers.
- Demonstrating computation in other bases as a way to further develop an understanding of the whole-number operations.

SECTION 4.1 Understanding Whole-Number Addition and Subtraction

Children in the elementary grades will learn whole-number addition, subtraction, multiplication, and division. They typically begin with addition because it is the easiest of the operations to understand and compute.

Understanding Whole-Number Addition

Long before children arrive at school, they are likely to have informal experiences with addition as they manipulate sets during their play. For example, a young boy who is playing with three toy cars and four toy trucks might want to know how many toy vehicles he has in all. To answer the question, he can combine the cars and trucks and count the total (Figure 4.1). In doing so, he has performed addition in an intuitive and meaningful way. He has also used the conceptual ideas that we use to define whole-number addition.

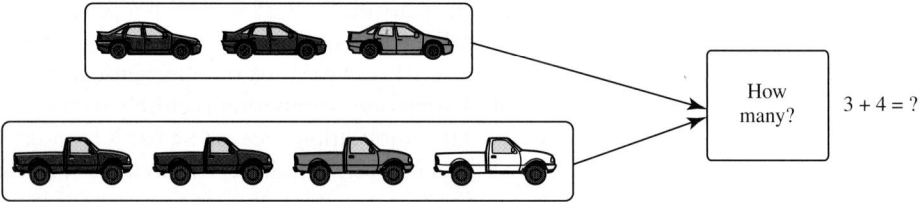

Figure 4.1 Adding by combining sets

Definition of Whole-Number Addition

Suppose a and b are whole numbers where $a = n(A)$ and $b = n(B)$. If A and B are disjoint, finite sets, then $a + b = n(A \cup B)$.

There are several important things to know about the definition. First, we call a and b **addends, terms,** or **summands,** and we read the notation $a + b$ as "a plus b," "a add b," or "the sum of a and b." Second, the definition reflects the intuitive meaning of addition. $n(A)$ and $n(B)$ tell us to count out two sets of the appropriate size, and $n(A \cup B)$ tells us to combine them and count the total. However, for this to work, the sets must be finite and disjoint. If the sets are infinite, then the union is infinite and impossible to count. If the sets are not disjoint, then any elements common to the sets are considered only once in the union. This results in fewer elements than required for the sum. For instance, if we model $4 + 3$ with the disjoint sets $A = \{a, b, c, d\}$ and $B = \{e, f, g\}$, we get the correct number of elements in the union. However, if we use $A = \{a, b, c, d\}$ and $B = \{c, d, e\}$, then c and d are written only once in $A \cup B$, and we encounter our problem.

Disjoint Sets (Correct Sum)	Not Disjoint Sets (Incorrect Sum)
$4 + 3 = n(A \cup B)$	$4 + 3 = n(A \cup B)$
$= n(\{a, b, c, d\} \cup \{e, f, g\})$	$= n(\{a, b, c, d\} \cup \{c, d, e\})$
$= n(\{a, b, c, d, e, f, g\})$	$= n(\{a, b, c, d, e\})$
$= 7$	$= 5$

Modeling Addition

One way children get a better understanding of addition is to compute simple problems using models like the set model and the number line model.

Set Model

We can use almost any set of objects to model the intuitive ideas behind whole-number addition. However, in doing so, we can arrive at the sum by counting in two ways. One way is to count out each set, combine them, and then *count the total.* Another way is to count out the first set and then *count on* the amount of numbers equal to the second addend. Figure 4.2 uses connecting cubes to demonstrate both methods to find the sum 5 + 3 = 8.

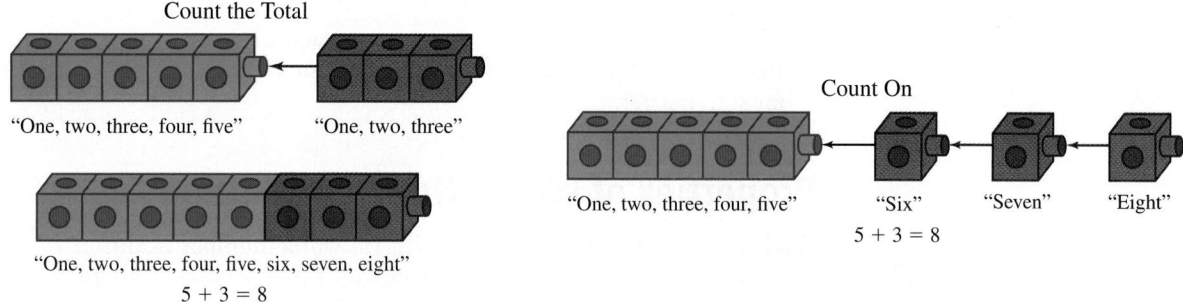

Figure 4.2 **Adding with two versions of the set model**

Number Line or Measurement Model

The set model works well with sets of discrete objects that can be counted. However, we often need to add two continuous measures, such as length or time. In these cases, the continuum of the whole-number line is a better representation of the sum. For example, suppose a woman rides her bike for 6 mi, takes a break, and then rides for 3 more. If we want to know how many total miles she rides, we can represent each addend with a directed arrow on a number line (Figure 4.3). Here, the first arrow begins at 0 and ends at 6. The second arrow begins at 6 and goes three more units. Because our final position is at 9, we conclude that 6 + 3 = 9.

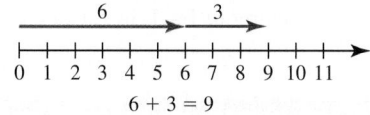

Figure 4.3 **Addition on a number line**

Once children have used models to help them understand addition, they can turn to a **fact table,** like the one shown in Table 4.3, to help them master all 100 single-digit

Table 4.3	Addition fact table								
+	0	1	2	3	4	5	6	7	8	9
0	0	1	2	3	4	5	6	7	8	9
1	1	2	3	4	5	6	7	8	9	10
2	2	3	4	5	6	7	8	9	10	11
3	3	4	5	6	7	8	9	10	11	12
4	4	5	6	7	8	9	10	11	12	13
5	5	6	7	8	9	10	11	12	13	14
6	6	7	8	9	10	11	12	13	14	15
7	7	8	9	10	11	12	13	14	15	16
8	8	9	10	11	12	13	14	15	16	17
9	9	10	11	12	13	14	15	16	17	18

addition facts. Internalizing these facts can be difficult for some children. However, several strategies can make learning them more manageable:

■ *Counting on by 1 and 2.* Facts that involve adding 1 or 2, such as 3 + 1 or 6 + 2, can be computed by counting on one or two numbers.

■ *Adding doubles.* The doubles are facts in which a number is added to itself, like 2 + 2 = 4 and 3 + 3 = 6. They result in the sequence of consecutive even numbers, which many children find easy to master. Once children know the doubles, they can use them to find sums that are one or two more than a double. For instance, we can add 4 + 5 as 4 + 4 + 1 = 8 + 1 = 9.

■ *Sums of 10.* Certain pairs of numbers, such as 3 + 7 and 9 + 1, add up to 10. Because 10 is a benchmark number, these combinations are often easier than most to master.

Properties of Whole-Number Addition

Explorations
Manual
4.1

Students can also use the properties of whole-number addition to master addition facts. We can see several patterns in Table 4.3 that lead to these properties. For instance, notice that all the sums in Table 4.3 are whole numbers. This simple pattern demonstrates the **closure property.**

Theorem 4.1   Closure Property of Whole-Number Addition

The sum of any two whole numbers is a unique whole number.

The closure property guarantees two things. One, the sum of two whole numbers is always another whole number. Two, the sum is unique; we always get exactly one number when we add. We can also apply the closure property to other sets. In general, a set is **closed** with respect to addition if the sum of any two numbers in the set is an element back in the set.

Example 4.1   Identifying Closed Sets

Reasoning   Are the sets E = {2, 4, 6, 8, . . .} and T = {1, 2, 3} closed with respect to addition?

Solution E is the set of all even numbers. Whenever we add two even numbers, the sum is always even. Consequently, E is closed under addition. With the set T, if we add 2 + 2, we get 4, which is not in T. Because the sum of two numbers in T may not be another element of T, we conclude that T is not closed under addition. ■

Another pattern in Table 4.3 is its *symmetry about the main diagonal*; specifically, the values in the upper right triangular part of the table are the same as the values in the lower left. The pattern results from the fact that the value of the sum is not affected by the order in which the numbers are added. For instance, 2 + 4 = 6 and 4 + 2 = 6. We call this property the **commutative property.**

Theorem 4.2   Commutative Property of Whole-Number Addition

If a and b are any whole numbers, then $a + b = b + a$.

The commutative property is true because addition is based on the union of two sets, which is commutative. For instance, if we take the union of a set of two objects with a set of four objects, we get a total of six objects (Figure 4.4). Likewise, if we take the union of a set of four objects with a set of two objects, we again get a set with six objects. As a result, we should be able to use set identities to show that the commutative property is true.

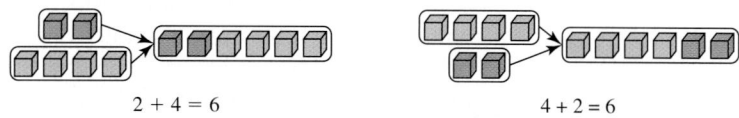

$$2 + 4 = 6 \qquad\qquad 4 + 2 = 6$$

Figure 4.4 Illustrating the commutative property with sets

Example 4.2   Proving the Commutative Property

Reasoning

Use a set identity to prove the commutative property.

Solution Let $a = n(A)$ and $b = n(B)$, where A and B are disjoint, finite sets. Then,

$$
\begin{aligned}
a + b &= n(A \cup B) && \text{Definition of addition} \\
&= n(B \cup A) && \text{Set identity } A \cup B = B \cup A \\
&= b + a && \text{Definition of addition}
\end{aligned}
$$

The commutative property is useful for learning the single-digit addition facts because it cuts the number of facts to be learned almost in half. For instance, once we know $5 + 7 = 12$, we know that $7 + 5 = 12$.

Whole-number addition is a **binary operation** because it allows us to work with only two numbers at a time. However, we often need to add three or more numbers. To do so, we add two numbers and then add the result to the third. In this computation, the order in which the numbers are added does not matter; the result is the same. Consequently, whenever we add, we can change how we group, or associate, numbers without affecting the sum.

Theorem 4.3   Associative Property of Whole-Number Addition

If a, b, and c are any whole numbers, then $a + (b + c) = (a + b) + c$.

Example 4.3   The Associative Property on a Number Line

Representation

Use a number line to show $2 + (5 + 3) = (2 + 5) + 3$.

Solution To show $2 + (5 + 3)$ on a number line, we go right 5 units, then 3, and then 2. Because we finish at 10, we conclude that $2 + (5 + 3) = 2 + 8 = 10$. For $(2 + 5) + 3$, we go right 2 units, then 5, and then 3. Again, we finish at 10, so $(2 + 5) + 3 = 7 + 3 = 10$. Because both sums equal 10, it must be that $2 + (5 + 3) = (2 + 5) + 3$.

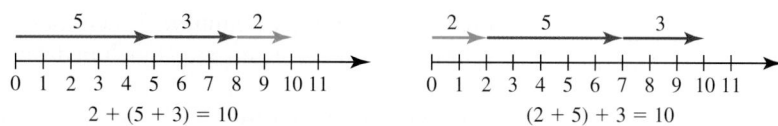

As with the commutative property, we can verify the associative property by using a set identity. To do so, is left to the Learning Assessment at the end of the section.

We can also use the associative property to learn addition facts by using a strategy called *making ten*. Here, we split one addend into two numbers so that a sum of 10 is made. We then add the remaining number to get the result. For instance, to add $8 + 7$, split 7 into $2 + 5$. This gives us $8 + 7 = 8 + (2 + 5) = (8 + 2) + 5 = 10 + 5 = 15$.

We can observe one more property in Table 4.3; specifically, the first row and column in the table are the same as the initial row and column. Because they correspond to adding zero, we conclude that adding zero does not change the value of a number. In other words, if we add zero to a whole number, the whole number maintains its *identity*. This makes the addition facts involving zero generally easy to learn.

Theorem 4.4   Identity Property for Whole-Number Addition

For any whole number a, there exists the unique number 0, called the **additive identity,** such that $a + 0 = a$ and $0 + a = a$.

The property states that zero is the *unique* identity element, so this property holds true for no other number. We can verify the identity property by using sets.

Example 4.4   Proving the Additive Identity Property

Reasoning

Use a set identity to prove the additive identity property.

Solution If we let $a = n(A)$ and $0 = n(\varnothing)$, then

$$a + 0 = n(A \cup \varnothing) \quad \text{Definition of addition}$$
$$= n(A) \quad \text{Set identity } A \cup \varnothing = A$$
$$= a \quad \text{Original assumption}$$

The reverse property, $0 + a = a$ can be shown in a similar fashion.

Less Than Revisited

We now use whole-number addition to take a different approach to less than.

Definition of Less Than Using Whole-Number Addition

For any whole numbers a and b, *a* **is less than** *b,* written *a* < *b,* if and only if there exists a natural number n such that $a + n = b$.

For example, consider the numbers 5 and 8. By the definition, 5 is less than 8 because there exists the natural number 3 such that $5 + 3 = 8$. Limiting n to a natural number is important because, if zero is included, the numbers might be equal. Consequently, if we alter the definition so that n is a whole number, then we have the definition of **less than or equal to ($\leq$).** The definitions for **greater than (>)** and **greater than or equal to ($\geq$)** are similar.

Check Your Understanding 4.1-A

1. Use the set and number line models to represent and find each sum.

    a. 5 + 1   **b.** 9 + 3

2. Is the set $S = \{1, 4, 9, 16, \ldots\}$ closed under addition? Explain.

3. State the property of whole-number addition that is demonstrated in each expression.

    a. 5 + (9 + 3) = (5 + 9) + 3   **b.** $5 + 6 = 11 \in W$

    c. 4 + 0 = 4   **d.** (6 + 2) + 5 = 5 + (6 + 2)

    e. (3 + 1) + 0 = 3 + 1   **f.** 4 + 8 = 8 + 4

4. Verify each inequality by applying the definition of less than.

    a. 3 < 6   **b.** 4 > 2   **c.** 19 < 35   **d.** 2,345 > 692

Talk About It What strategies, other than the ones given in this section, might you as a teacher use to help your students learn the single-digit addition facts?

Representation

Activity 4.1
Consider Table 4.3. Highlight the rows, columns, and diagonals that show the addition facts that can be learned using each strategy.

a. Counting on by 1 or 2

b. Adding doubles

c. Adding doubles plus 1 or 2

d. Sums to 10

e. Using the identity property

f. How many addition facts are encompassed by these strategies?

Understanding Whole-Number Subtraction

Unlike addition, we can approach whole-number subtraction in three ways. Each approach plays an important role in understanding the operation, and the approach used is often dictated by the context of a problem. For this reason, short story problems are used to help students understand the different meanings of subtraction and when to apply them. The most familiar and intuitive approach to subtraction is the take-away approach.

Take-Away Approach

As with addition, children are likely to have their first informal experiences with subtraction as they manipulate sets during their play. For instance, suppose a little girl has 5 cookies and wants to share with a friend. She might ask herself, "If I give my friend 2 cookies, how many will I have left?" (See Figure 4.5.) To answer the question, she can take 2 cookies away and then count what remains. In doing so, she uses the conceptual ideas we use to define the **take-away approach** to whole-number subtraction.

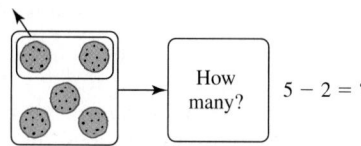

Figure 4.5 Subtracting by using sets

> ### Definition of Whole-Number Subtraction (Take-Away Approach)
>
> Suppose a and b are whole numbers where $a = n(A)$ and $b = n(B)$. If $B \subseteq A$, then $a - b = n(A - B)$.

There are several important things to mention about the definition. First, we call a the **minuend**, b the **subtrahend**, and we read the notation $a - b$ as "a minus b," "a subtract b," or the "difference of b from a." Second, the notation $n(A - B)$ tells us to subtract by removing the elements of B from A and counting what remains. However, this works only if B is a subset of A. If not, then we will not remove the correct number of elements when we compute the set difference. For instance, if we model $5 - 3$ with $A = \{a, b, c, d, e\}$ and $B = \{c, d, e\}$, then $B \subseteq A$, and we get two elements in the set difference. However, if $A = \{a, b, c, d, e\}$ and $B = \{d, e, f\}$, then $B \nsubseteq A$, and we do not remove enough elements when we compute $A - B$.

B a subset of A (Correct Difference)	B not a subset of A (Incorrect Difference)
$5 - 3 = n(A - B)$	$5 - 3 = n(A - B)$
$\quad = n(\{a, b, c, d, e\} - \{c, d, e\})$	$\quad = n(\{a, b, c, d, e\} - \{d, e, f\})$
$\quad = n(\{a, b\})$	$\quad = n(\{a, b, c\})$
$\quad = 2$	$\quad = 3$

Explorations Manual 4.2

Even though the take-away approach to subtraction is defined in terms of sets, we can also use it with measures.

Example 4.5 Writing Subtraction Story Problems

Communication

Write two story problems that use the take-away approach to subtraction: one that uses sets and another that uses measures. Use the appropriate model to represent and solve each problem.

Solution For our problems to reflect the take-away approach, our language must indicate that one amount is removed from another.

For instance, with *sets* we might write, "If Sue has 7 pieces of candy and gives 3 to Jim, how many does she have left?" To find the difference, we can count the sets in two ways. One is to remove the appropriate number of objects and then *count what remains*. The other is to *count back* while removing one object at time. Because we arrive at 4 either way, we conclude $7 - 3 = 4$.

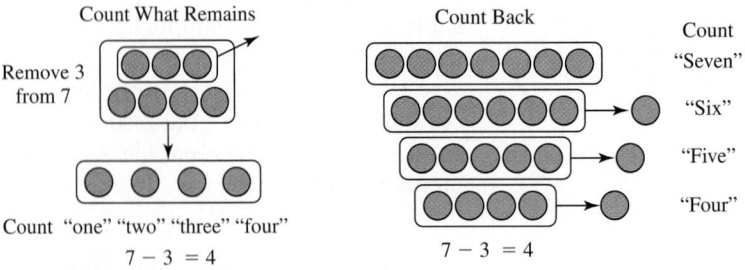

Count What Remains

Remove 3 from 7

Count "one" "two" "three" "four"

$7 - 3 = 4$

Count Back

Count "Seven" "Six" "Five" "Four"

$7 - 3 = 4$

With *measures*, we might write, "If Sarah has a 10-in. string and cuts off 4 in., how long is the remaining piece?" Here, we can represent the difference $10 - 4$ on a number line by drawing the first arrow from 0 to 10 and then a second arrow from 10 backward four spaces. Because the final position is 6, we conclude that $10 - 4 = 6$.

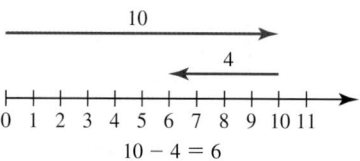

Comparison Approach

In the **comparison approach,** we find the difference by comparing two sets or measures and asking, "How much more does the one have over the other?" With sets, we pair the elements from each set until one has no more left over; then we count any unpaired elements to find the difference. With measures, we ask how much more is the one measure than the other.

Example 4.6   Representing Subtraction Problems

Representation

Use a model to represent and solve each problem.

a. I have 8 apples and 5 oranges. How many more apples do I have than oranges?

b. Julius rode his bike for 9 hr this week and Sean rode his for 5 hr. How many more hours did Julius ride than Sean?

Solution

a. We have two sets of objects, 8 apples and 5 oranges. To find the difference, match one apple to one orange until all oranges have been used. When we count the remaining apples, we find that $8 - 5 = 3$.

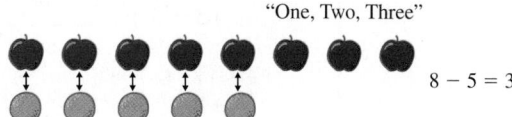

b. We are comparing two measures of time, so we represent the problem on a number line. To do so, we draw two arrows; one from 0 to 9 and another from 0 to 5. Because one arrow goes 4 units beyond the other, $9 - 5 = 4$.

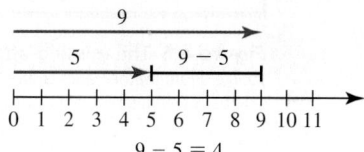

Missing-Addend Approach

The third approach to subtraction makes use of addition.

| Example 4.7 | Connecting Subtraction to Addition |

Connections   Describe how addition is used to compute subtraction on the student page in Figure 4.6.

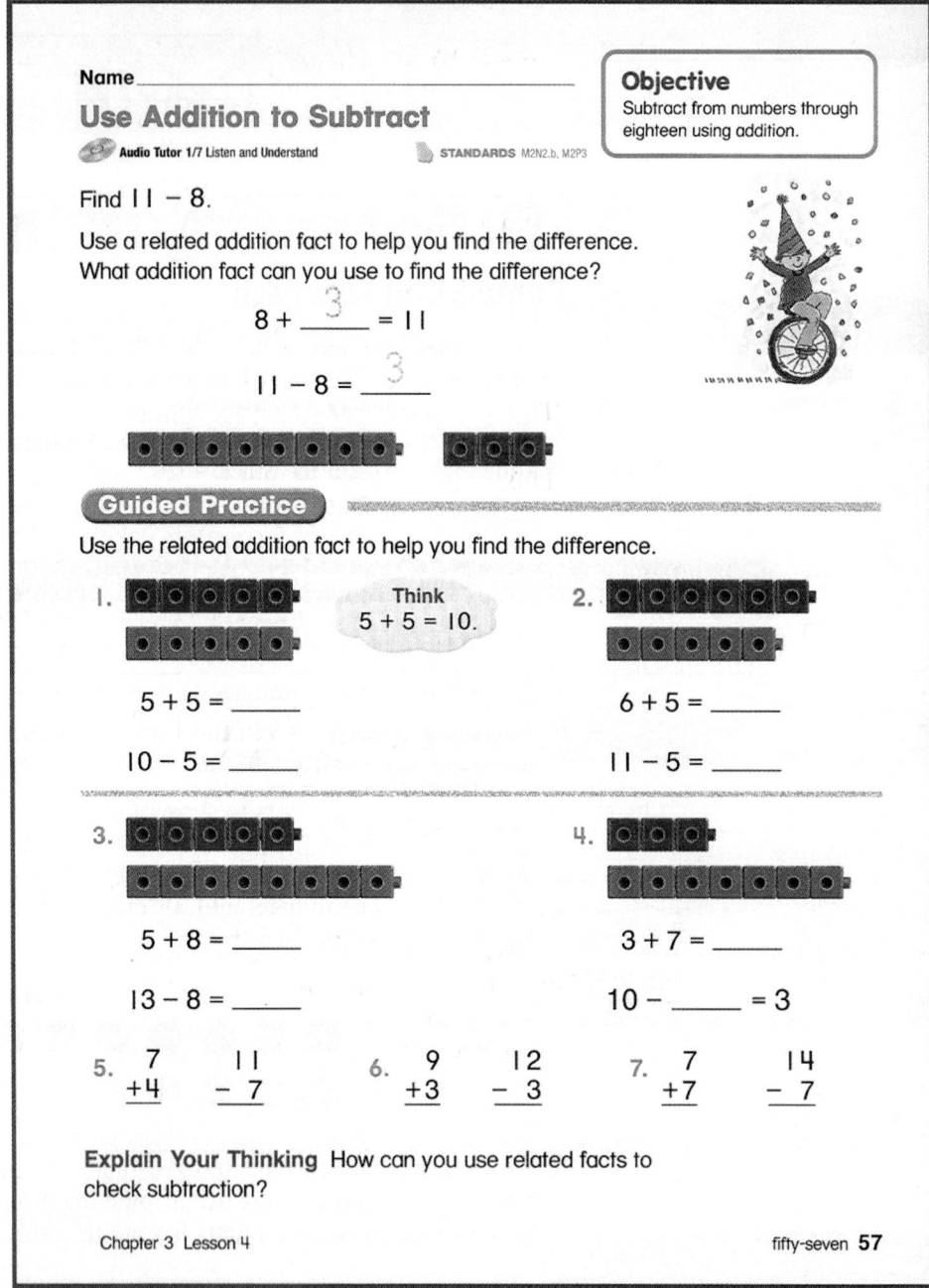

Figure 4.6 The missing-addend approach to subtraction
Source: From *Mathematics Grade 2: Houghton Mifflin Mathematics Georgia* (p. 57). Copyright © by Houghton Mifflin Company, Inc. Reproduced by permission of the publisher, Houghton Mifflin Harcourt Publishing Company.

Solution The student page shows we can compute a difference by using a related addition fact. For instance, the first problem shows that $11 - 8 = 3$ because $8 + 3 = 11$. This connection between addition and subtraction is reinforced by the use of connecting cubes. Combining the red and blue cubes gives us a total number of cubes, so it represents an addition fact. Given the total number of cubes, we can remove either the red or the blue to get a related subtraction fact.

The last example uses the **missing-addend approach** to subtraction. It comes directly from an alternative definition for whole-number subtraction.

Definition of Whole-Number Subtraction (Missing-Addend Approach)

If a and b are any whole numbers, then $a - b = c$ if and only if $a = b + c$ for some whole number c.

To compute a difference using the missing-addend approach, we use a related addition fact in which one addend is missing. When we use our knowledge of addition to find the missing number, we also find the difference. For example, we can think of the problem $8 - 2 = ?$ as $8 = 2 + ?$. Because $2 + 6 = 8$, we conclude $? = 6$ and $8 - 2 = 6$.

Like the other approaches, the missing-addend approach can be used with sets and continuous measures. In either case, we find the difference by asking ourselves, "How much more is needed to give the total value?" Figure 4.7 illustrates how to use the set and number line models to solve two story problems that involve this approach.

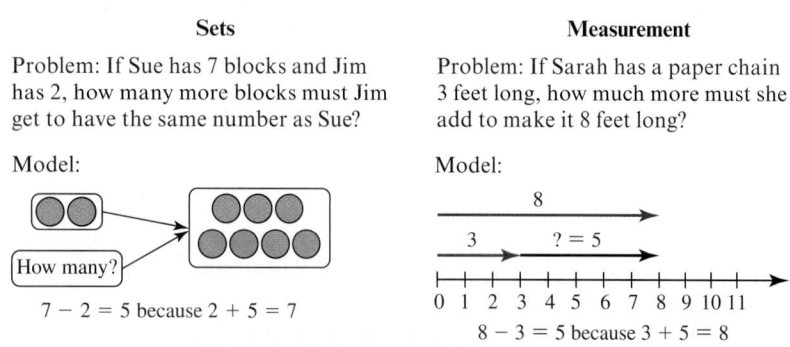

Sets

Problem: If Sue has 7 blocks and Jim has 2, how many more blocks must Jim get to have the same number as Sue?

Model:

$7 - 2 = 5$ because $2 + 5 = 7$

Measurement

Problem: If Sarah has a paper chain 3 feet long, how much more must she add to make it 8 feet long?

Model:

$8 - 3 = 5$ because $3 + 5 = 8$

Figure 4.7 Modeling the missing-addend approach

The missing-addend approach points out a special relationship between addition and subtraction. In addition, we combine two addends to find a sum. In subtraction, we start with a sum and subtract one addend to find the other. In other words, addition undoes subtraction, and subtraction undoes addition. In mathematics, when two operations share this relationship, we call them **inverse operations.** So addition facts can help students master subtraction facts. For instance, once we know that $2 + 5 = 7$, we also know that $7 - 5 = 2$ and $7 - 2 = 5$. These facts, along with the fact that $5 + 2 = 7$, are called a **fact family.** In the classroom, students often use strip diagrams to help them learn fact families. Figure 4.8 shows how each fact in a given family can be practiced by covering up one of the numbers.

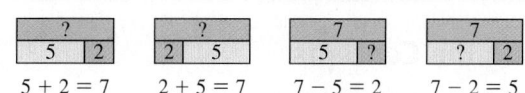

$5 + 2 = 7$ $2 + 5 = 7$ $7 - 5 = 2$ $7 - 2 = 5$

Figure 4.8 Strip diagrams

The missing-addend approach must be used with care because it does not guarantee that the difference of every two whole numbers is another whole number. For instance, consider $2 - 3$. There is no whole number c such that $2 = 3 + c$. In general, if $a < b$, then $a - b$ is not meaningful with the whole numbers. Consequently, the set of whole numbers is not closed under subtraction. The commutative, associative, and identity properties also fail to hold for subtraction.

| Example 4.8 | A Counterexample to the Associative Property |

Reasoning

Find a counterexample showing that the associative property does not hold for whole-number subtraction.

Solution We must find three whole numbers, a, b, and c, such that $a - (b - c) \neq (a - b) - c$. If we let $a = 7$, $b = 3$ and $c = 2$, then $a - (b - c) = 7 - (3 - 2) = 7 - 1 = 6$ and $(a - b) - c = (7 - 3) - 2 = 4 - 2 = 2$. Because $6 \neq 2$, $a - (b - c) \neq (a - b) - c$.

Check Your Understanding 4.1-B

1. Write a story problem that can be solved using sets and the take-away approach to subtraction. Use the set model to represent and solve your problem.
2. Write a story problem that can be solved using a measurement and the comparison approach. Use the number line model to represent and solve your problem.
3. Use the missing-addend approach to find the missing number in each equation.

   a. $7 - 5 = ?$ **b.** $12 - ? = 4$ **c.** $? - 6 = 9$

4. List the other facts in the fact families of the given addition or subtraction fact.

   a. $9 - 3$ **b.** $10 - 5$ **c.** $8 + 3$

5. Find a counterexample showing that the commutative property does not hold for whole-number subtraction.

Talk About It In this section, we looked at some story problems that used sets and others that used measurements. Which kind of problem do you think is more difficult for children to understand? Why do you think this is so?

Reasoning

Activity 4.2
Earlier in this section, we used addition to define the concept of less than. How might we use subtraction to do the same?

SECTION 4.1   Learning Assessment

■ Understanding the Concepts

1. Describe in your own words the conceptual meaning behind whole-number addition.
2. Why must the sets A and B be finite and disjoint in the definition of addition?
3. Describe how each model is used to represent and compute whole-number addition.

   a. Set model
   b. Number line model

4. What does each property imply about whole-number addition?

   a. Closure   **b.** Commutative
   c. Associative   **d.** Identity

5. What does it mean for addition to be a binary operation? What implications does this have for adding more than two numbers?

6. In the definition of less than, why is n a natural number and not a whole number?

7. Explain how to compute subtraction using each of the following approaches.
   a. Take-away
   b. Comparison
   c. Missing-addend

8. Describe how the set and number line models can be used to represent each approach to subtraction.
   a. Take-away
   b. Comparison
   c. Missing-addend

9. Why must $B \subseteq A$ in the definition of the take-away approach to subtraction?

10. What does it mean for addition and subtraction to be inverse operations?

11. What are fact families, and how are they used to learn single-digit addition and subtraction facts?

■ Representing the Mathematics

12. Use the set and number line models to represent and solve each sum.
    a. $1 + 6$
    b. $3 + 8$
    c. $5 + 0$

13. Consider Table 4.3. How is each property shown in the table?
    a. Closure property
    b. Commutative property
    c. Identity property

14. Use the number line model to show that $5 + 6 = 6 + 5$.

15. Use the set model to show that:
    a. $1 + (3 + 6) = (1 + 3) + 6$.
    b. $6 + 0 = 6$.

16. What single-digit addition fact is represented in each of the following?

a.
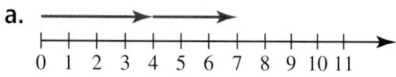

b.
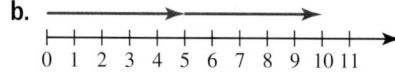

c.
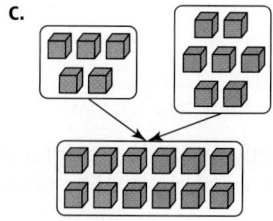

d.
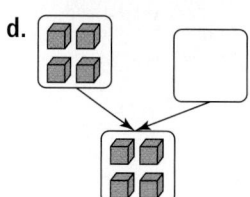

e. I start at "six," and then count "seven, eight, nine."

17. Use the set and number line models to demonstrate the take-away approach with each difference.
    a. $9 - 4$
    b. $13 - 8$
    c. $8 - 8$

18. Use the set and number line models to demonstrate the comparison approach with each difference.
    a. $7 - 5$
    b. $11 - 6$
    c. $4 - 4$

19. Use the set and number line models to demonstrate the missing-addend approach with each difference.
    a. $5 - 1$
    b. $12 - 7$
    c. $9 - 7$

20. What subtraction fact is demonstrated in each of the following?
    a. Start at "eight" and count "seven, six, five, four."
    b. $n(\{a, b, c, d, e\} - \varnothing)$

c.
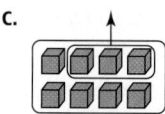

d.
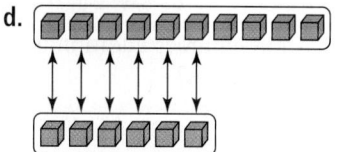

e.
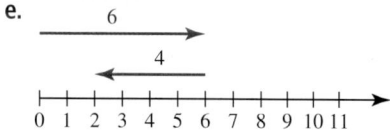

f.
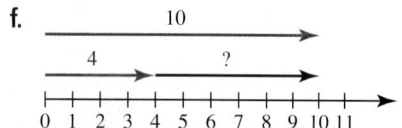

21. Write the facts in the fact family illustrated in each figure.

a. | 11 |
   | 7 | 4 |

b. | 9 |
   | 2 | 7 |

c. | 8 |
   | 4 | 4 |

■ Exercises

22. For each pair of sets A and B, determine whether $n(A) + n(B) = n(A \cup B)$. If so, state the single-digit addition fact it represents.
    a. $A = \{a, b\}$ $B = \{c, d, e\}$.
    b. $A = \{a, b, c\}$ $B = \{c, d, e\}$.
    c. $A = \{a, b, c, d\}$ $B = \varnothing$.
    d. $A = \{a, b, c, d, e\}$ $B = \{c, d, e, f, g\}$.

23. For each pair of sets, determine whether $n(A) - n(B) = n(A - B)$. If so, state the subtraction fact it represents.
    a. $A = \{a, b, c, d, e\}$ $B = \{b, e, d, c\}$.
    b. $A = \{a, b, c, d\}$ $B = \{ \}$.
    c. $A = \{a, b, c, d\}$ $B = \{b, a, c, d\}$.
    d. $A = \{a, b, c, d, e\}$ $B = \{b, c, d, e, f\}$.

24. Use the sets $A = \{1, 2, 3, 4, 5\}$, $B = \{a, b, c, d\}$, and $C = \{x, y, z\}$ and the definition of addition to demonstrate and solve each problem.

    a. $5 + 4$ b. $4 + 3$

    c. $5 + 3$ d. $5 + 4 + 3$

25. Use the sets $A = \{a, b, c, d, e, f, g\}$, $B = \{a, b, c, d\}$, and $C = \{a, b\}$ and the take-away definition of subtraction to demonstrate and solve each problem.

    a. $7 - 4$ b. $7 - 2$ c. $4 - 2$

26. Describe a strategy that can be used to learn each single-digit addition fact.

    a. $4 + 1 = 5$.   b. $7 + 7 = 14$.

    c. $5 + 5 = 10$.   d. $8 + 0 = 8$.

27. Describe a strategy that can be used to learn each single-digit addition fact.

    a. $0 + 5 = 5$.   b. $7 + 8 = 15$.

    c. $4 + 3 = 7$ if $3 + 4 = 7$.   d. $8 + 2 = 10$.

28. Which property of whole-number addition is demonstrated in each equation?

    a. $1 + 0 = 1$.

    b. $4 + 6 = 6 + 4$.

    c. $8 + 7 = 15 \in W$.

    d. $(4 + 3) + 7 = (3 + 4) + 7$.

    e. $(6 + 3) + 9 = 6 + (3 + 9)$.

    f. $(9 + 4) + (7 + 3) = (7 + 3) + (9 + 4)$.

29. If each equation demonstrates a property of whole-number addition, fill in the blank with the appropriate number and then name the property.

    a. $(4 + 7) + 9 = 4 + (\underline{\quad} + 9)$.

    b. $4 + 0 = \underline{\quad}$.

    c. $6 + 3 = 3 + \underline{\quad}$.

    d. $9 + \underline{\quad} = 9$.

    e. $7 + \underline{\quad} = 10 \in W$.

30. Which of the following sets are closed under addition? Explain why they are or are not closed.

    a. $O = \{1, 3, 5, 7, \ldots\}$

    b. $\{0\}$

    c. $T = \{3, 6, 9, 12, \ldots\}$

31. Which of the following sets are closed under addition? Explain why they are or are not closed.

    d. $A = \{4, 8, 12\}$.

    b. $F = \{5, 10, 15, \ldots\}$.

    c. $\{1\}$

32. Verify each inequality by applying the definition of less than.

    a. $5 < 8$.   b. $16 > 11$.

    c. $121 < 228$.   d. $698 > 341$.

33. List the other facts in the fact family of the given addition or subtraction fact.

    a. $7 - 5$ b. $3 + 6$ c. $12 - 6$

34. List the other facts in the fact family of the given addition or subtraction fact.

    a. $4 + 8$ b. $16 - 9$ c. $7 + 7$

35. Use the missing-addend definition to write the following subtraction problems in terms of addition. Solve the problem, and explain your answer in terms of addition.

    a. $13 - 9 = ?$ b. $35 - 23 = ?$ c. $110 - 56 = ?$

36. What approach to subtraction is used in each story problem?

    a. If Kim has 4 pencils and Joyce has 9, how many more pencils does Joyce have?

    b. Julia has studied for 3 hr. How many more hours until she has studied for 6 hr?

    c. If Tim gives Caleb 5 of his 11 baseball cards, how many does he have left?

37. What approach to subtraction is used in each story problem?

    a. Joe is on a 9-mi bike ride. If he has ridden 5 mi, how many miles does he have left to go?

    b. Akeem has solved 9 out 12 addition problems. How many more does he have left to solve?

    c. On a certain test, Carrie got 15 correct and Dwayne got 19 correct. How many more did Dwayne get correct than Carrie?

■ Problems and Applications

38. Using the numbers 1–9 only once, place one number in each circle so that the sum along any line is the same.

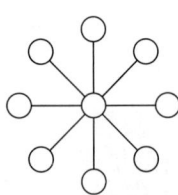

39. Using the numbers 1–9 only once, place one number in each circle so that the sum along any line is the same.

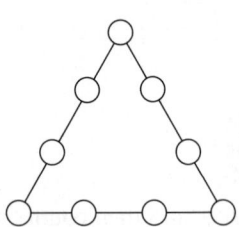

40. Using the numbers 1, 2, 3, 4, 5, 6, 7, 8 and 9 only once, place them in the square so that the sum of every row, column, and diagonal is different.

41. Consider the equation $A + B = CD$, where each of A, B, C, and D represents a different single digit from 0 to 9 and CD represents a two-digit number.
   a. If $A = 6$ and $D = 4$, what are B and C?
   b. If $A = 3$ and $C = 1$, what are possible values for B and D?
   c. Given the constraints of the problem, is it possible for $A = 2$ and $B = 3$? Explain.
   d. What value must C always be? Why?
   e. What is the smallest value that D can be? In this case, what are A and B?
   f. What is the largest value that D can be? In this case, what are A and B?

42. Suppose that A, B, and C are single-digit whole numbers that can only be used once in the equation $A - B = C$.
   a. What is the largest number that A could be?
   b. What is the largest number that B could be?
   c. What is the largest number that C could be?
   d. What is the smallest number that C could be?
   e. Repeat parts a–d, but now assume that the digits can be used more than once.

43. What is the total value of the ninth row in the following pattern?

$$1$$
$$2 + 1$$
$$3 - 2 + 1$$
$$4 + 3 - 2 + 1$$
$$5 - 4 + 3 - 2 + 1$$

44. Place any number of addition or subtraction signs between the digits 9 8 7 6 5 4 3 2 1 so that a sum of 100 is obtained. If two numbers do not have an operation between them, assume they represent a two-digit number.

45. Consider the 100 single-digit addition facts and their corresponding subtraction facts. How many fact families are there?

■ Mathematical Reasoning and Proof

46. Let $n(A) = 4$ and $n(A \cup B) = 7$.
   a. Is $n(B)$ guaranteed to be 3? If not, what other values might it be?
   b. If $n(B) = 3$, what must be true about $n(A \cap B)$?

47. Is it true that every finite subset of the whole numbers is not closed under addition? Why or why not?

48. If a set contains the given number, what other numbers must be in the set for it to be closed under addition?
   a. 2   b. 3   c. 4   d. n

49. A set contains 4, 16, and 32 and is closed under addition.
   a. Must 8 be in the set? Why or why not?
   b. Must 10 be in the set? Why or why not?
   c. List five other numbers that must be in the set.

50. a. Why is the set of whole numbers not closed under subtraction?
   b. Does the set of whole numbers have any subsets that are closed under subtraction? If so, give an example.

51. If we use the missing-addend approach to solve the problem $10 - 6 = x$, which is more correct to write: $10 = 6 + x$, or $10 = x + 6$? Explain.

52. Use the set identity $A \cup (B \cup C) = (A \cup B) \cup C$ to show that $a + (b + c) = (a + b) + c$ for whole numbers a, b, and c.

53. a. Find a counterexample showing that whole-number subtraction does not satisfy the commutative property.
   b. Even though subtraction is not commutative in general, in some instances $x - y = y - x$. What must be true about x and y in such cases?

54. a. Find a counterexample that shows whole-number subtraction does not satisfy the associative property.
   b. Even though subtraction is not associative in general, in some instances $x - (y - z) = [x - y) - z]$. What must be true about x, y, and z in such cases?

55. Even though $a - 0 = a$, the identity property still does not hold for whole number subtraction. Can you explain why?

■ Mathematical Communication

56. Write three short story problems that require the sum of two or more numbers. Give your problems to a partner to solve and critique.

57. Write a short story problem that:
   a. Uses the comparison approach to solve the problem $7 - 5$.
   b. Uses the take-away approach to solve the problem $9 - 4$.

c. Uses the missing-addend approach to solve the problem $11 - 8$.

d. Write a short paragraph explaining how your language was different for each problem.

58. When developing their computational fluency, students commonly have difficulty translating verbal statements into symbolic statements, specifically with recognizing words that represent numerical operations. With one of your peers, make a list of words that indicate the operation of subtraction. How many of the words do you think elementary students are likely to encounter?

59. Write a short paragraph or two explaining how you would use the set model to demonstrate the missing-factor approach to subtraction.

60. Rewrite the definition of less than to change it to a definition for less than or equal to.

61. When Cameron adds $8 + 5$, he does it as $8 + 5 = 8 + (2 + 3) = (8 + 2) + 3 = 10 + 3 = 13$. What properties of addition does he use, and will his method always work?

62. Shalondra is given the following problem: "Tricia has 9 dolls, and Marie has 7. How many more dolls does Tricia have than Marie?" She does not understand why the situation represents subtraction when nothing is being taken away. How do you help her understand?

63. Reconsider the strategies for mastering the 100 single-digit facts. Are there similar strategies for learning the 100 subtraction facts? If so, list them and describe how you would use them in your classroom.

■ Building Connections

64. Write a paragraph or two explaining how sets, numeration, and addition are all connected. Specifically, explain the role of sets and numeration in understanding and computing whole-number addition and subtraction.

65. a. Describe in your own words the meaning of the phrase "computational fluency." What are your impressions on how it might be achieved in the classroom?

b. Read the NCTM Standards vision for how computational fluency is to be achieved in the Numbers and Operations Standards for K–2 and 3–5. Does it match your own perception of computational fluency? If not, what differences did you notice, and did they have an effect on your thoughts?

66. Andrea is playing cards and has three jacks and three red cards. Must she necessarily have six cards in her hand? Why or why not? How does this situation relate to the definition of whole-number addition?

67. Select a set of curriculum materials that span first, second, and third grades. Examine the materials for lessons that involve properties of addition.

a. At what grade level are the properties introduced?

b. Do the curriculum materials give the names of the properties? Does this surprise you?

c. Write a paragraph or two of how the properties of addition are developed and used in the curriculum materials.

68. The missing-factor approach to subtraction is an important connection to algebraic thinking in the elementary grades. Can you explain how?

69. Now that calculators are readily available, do you think it is still important for students to master the basic addition and subtraction facts? What do the NCTM Standards have to say about this question?

SECTION 4.2 Adding and Subtracting Large Numbers

Once children understand addition and subtraction and can compute single-digit facts, they can learn to compute the sums and differences of larger numbers. Methods for doing so include written algorithms, mental computation, estimation, and calculators.

Written Algorithms for Addition

One of the key ways students learn how to add or subtract large numbers is to use written **algorithms,** which are step-by-step procedures that tell us how to complete an operation. Although there are different written algorithms for each of the operations, the goal is to move students toward the **standard algorithms.** These algorithms are highly efficient but seldom correspond to how children think about computation. This can make them difficult for children to understand, so we tend teach them in a series of steps or with concrete models.

Explorations
Manual
4.3

With addition, most written algorithms involve two basic procedures: adding single-digit numbers and **regrouping,** which we also call **trading** or **carrying.** When we regroup, we exchange 10 groups of the same size for one of the next highest place value. Children often begin to explore this idea by adding with base-ten blocks, typically finding the sum by using an extension of the set model.

| **Example 4.9** | **Adding with Base-Ten Blocks** |

Representation

Use base-ten blocks to compute 348 + 275.

Solution We proceed by using the intuitive idea behind the definition of whole-number addition: combining two disjoint sets and counting. Now, however, the objects can take on different values. To begin, we represent each addend with the appropriate base-ten blocks and then combine the sets.

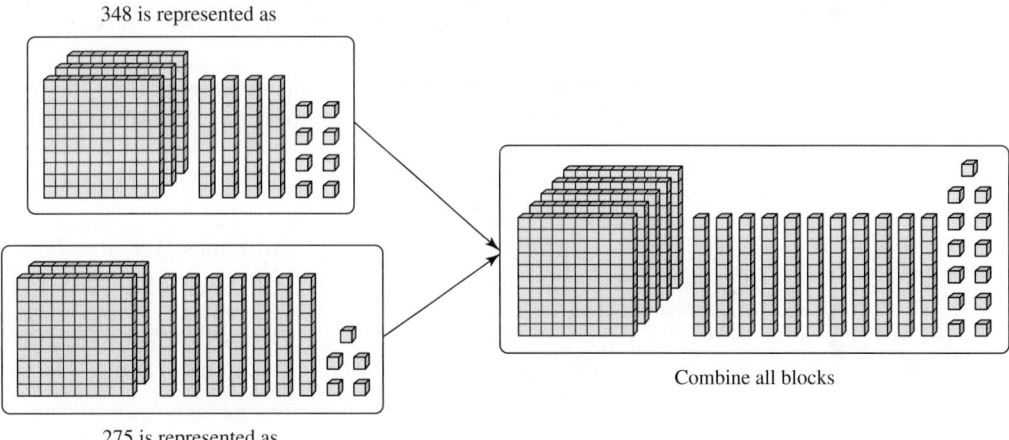

348 is represented as

275 is represented as

Combine all blocks

However, before we count, we regroup the longs and units. Specifically, we trade 10 units for 1 long and 10 longs for 1 flat. This gives us 6 flats, 2 longs, and 3 units, so 348 + 275 = 623.

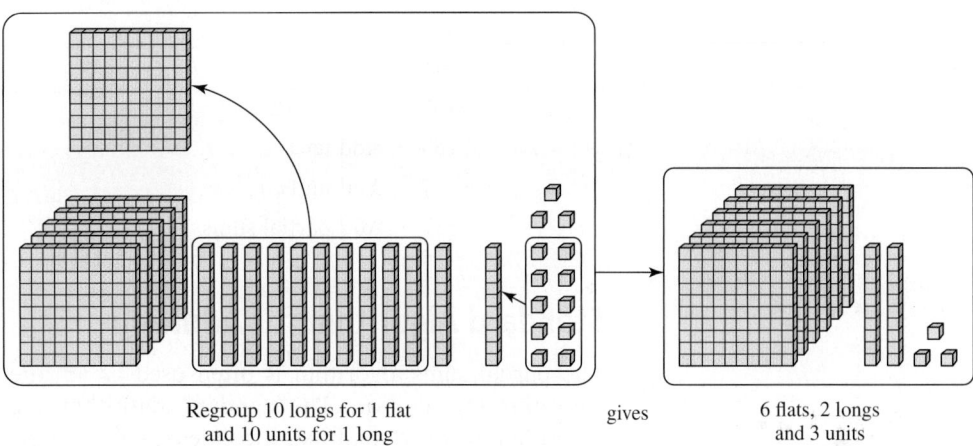

Regroup 10 longs for 1 flat   gives   6 flats, 2 longs
and 10 units for 1 long   and 3 units

In the last example, we computed the sum by adding the digits in the same place value and regrouping as necessary. Numerically, we can express this method by using expanded notation. For instance, to compute $235 + 134$:

$$235 + 134 = (2 \cdot 100 + 3 \cdot 10 + 5) + (1 \cdot 100 + 3 \cdot 10 + 4) \qquad \text{Expanded notation}$$
$$= (2 \cdot 100 + 1 \cdot 100) + (3 \cdot 10 + 3 \cdot 10) + (5 + 4) \qquad \text{Commutative and associative properties}$$
$$= (2 + 1) \cdot 100 + (3 + 3) \cdot 10 + (5 + 4) \qquad \text{Distributive property}$$
$$= 3 \cdot 100 + 6 \cdot 10 + 9 \qquad \text{Single-digit addition}$$
$$= 369 \qquad \text{Simplified form}$$

Partial-Sums Algorithm

If we simplify some of the notation in the previous example, we get what we call the partial sums algorithm. In this algorithm, we add the digits in like place values to obtain **partial sums,** being sure to maintain the value of the digit. Once we have all the partial sums, we add them to get the final sum.

Example 4.10　Adding with Partial Sums

Use the partial-sums algorithm to add $483 + 374$.

Solution We obtain three partial sums by adding the hundreds digits, the tens digits, and the ones digits. Then we add the partial sums for the result.

$$
\begin{array}{r}
483 \\
+374 \\
\hline
7 \\
150 \\
+700 \\
\hline
857
\end{array}
$$

Add ones: $3 + 4 = 7$.
Add tens: $80 + 70 = 150$.
Add hundreds: $300 + 400 = 700$.
Add partial sums.

The last example shows that the value of each partial sum is affected only by the place values of the digits, not by other partial sums. So we can work this algorithm from right to left or from left to right. We can also write the partial sums without placeholders as long as we keep the nonzero digits in the correct place value.

Working left to right

$$
\begin{array}{r}
483 \\
+374 \\
\hline
700 \\
150 \\
+\ \ 7 \\
\hline
857
\end{array}
$$

Add hundreds.
Add tens.
Add units.
Add partial sums.

No placeholders

$$
\begin{array}{r}
483 \\
+374 \\
\hline
7 \\
15 \\
+\ \ 7 \\
\hline
857
\end{array}
$$

Standard Algorithm for Addition

The partial sums algorithm is often used as an intermediate step to the standard algorithm for addition. The standard algorithm is different from the partial sums algorithm in two ways. First, it condenses how the partial sums are written. Second, regrouping is done while adding rather than at the end.

The next example shows the steps in the standard algorithm and how they can be modeled with base-ten blocks.

Example 4.11 Adding with the Standard Algorithm

Representation Use the standard algorithm to compute 367 + 145.

Solution We begin the standard algorithm with the ones digits. Since 7 ones + 5 ones = 12 ones, we can regroup them into 1 ten and 2 ones. We move the ten to the tens place value and write down the 2 ones.

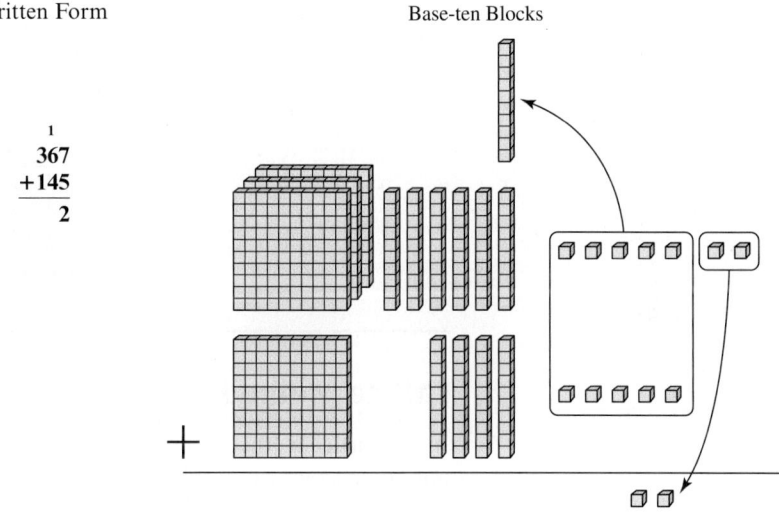

Written Form   Base-ten Blocks

$$\begin{array}{r} {}^{1} \\ 367 \\ +145 \\ \hline 2 \end{array}$$

Next, add the digits in the tens place value. Because 1 ten + 6 tens + 4 tens = 11 tens, we regroup them into 1 hundred and 1 ten. We move the hundred to the hundreds place value and write down the 1 ten.

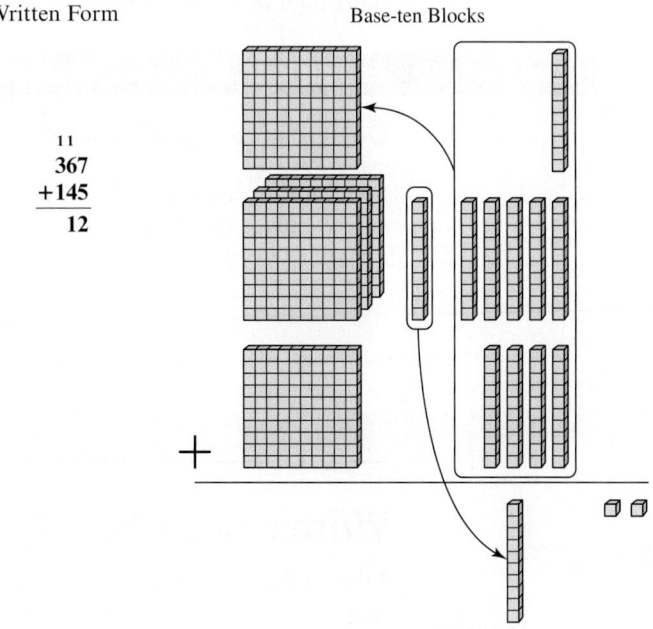

Written Form   Base-ten Blocks

$$\begin{array}{r} {}^{11} \\ 367 \\ +145 \\ \hline 12 \end{array}$$

Finally, add the digits in the hundreds place value. Because 1 hundred + 3 hundreds + 1 hundred = 5 hundreds, we do not have to regroup, and we write down the 5 hundreds. As a result, 367 + 145 = 512.

Written Form   Base-ten Blocks

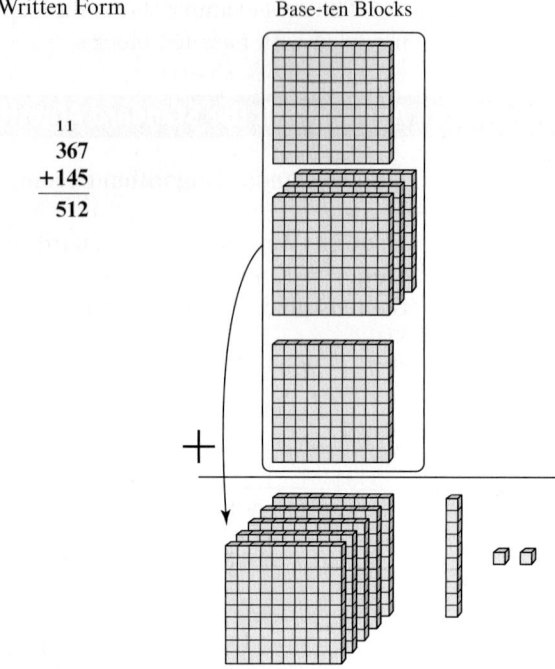

$$\begin{array}{r} ^{1\,1} \\ 367 \\ +145 \\ \hline 512 \end{array}$$

Lattice Method

In the **lattice method,** we write the sums from single-digit facts in a lattice, where the ones digit goes into the bottom corner and any tens digit goes into the top corner.

We find the sum by adding the digits down the diagonals, and regrouping as necessary. The method works because the tens digit in any square takes on the same value as the units digit in the square to the immediate left.

Example 4.12   Adding with the Lattice Method

Use the lattice method to add 658 + 273.

Solution We use single-digit facts to add the numbers in each place value, making sure we place the digits in the appropriate spot in the lattice. We then add down the diagonals. As a result, 658 + 273 = 931.

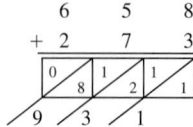

Written Algorithms for Subtraction

Like addition, there are several written algorithms for subtraction. Most of them involve two basic procedures: using basic subtraction facts and exchanging. **Exchanging,** also called **borrowing,** is the opposite of regrouping; one group from a higher place value is exchanged for a group of ten in the next lower place value. Like regrouping, children begin to explore the idea of exchanging by subtracting with base-ten blocks, generally using the take-away approach to subtraction.

Example 4.13 Subtracting with Base-Ten Blocks

Representation

Use base-ten blocks to compute $352 - 198$.

Solution We first represent 352 with 3 flats, 5 longs, and 2 units. From this, remove 198: 1 flat, 9 longs, and 8 units. Initially, we do not have enough longs and units to remove. However, we can exchange 1 flat for 10 longs and 1 long for 10 units.

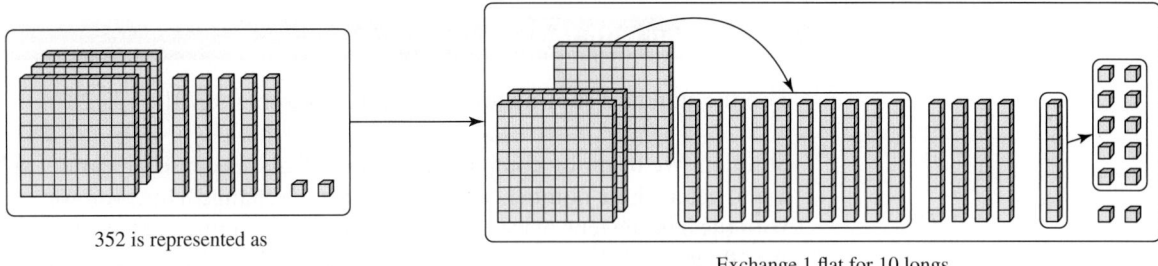

352 is represented as

Exchange 1 flat for 10 longs
and 1 long for 10 units

After the exchange, we take away 1 flat, 9 longs, and 8 units. This leaves us with 1 flat, 5 longs, and 4 units. Consequently, $352 - 198 = 154$.

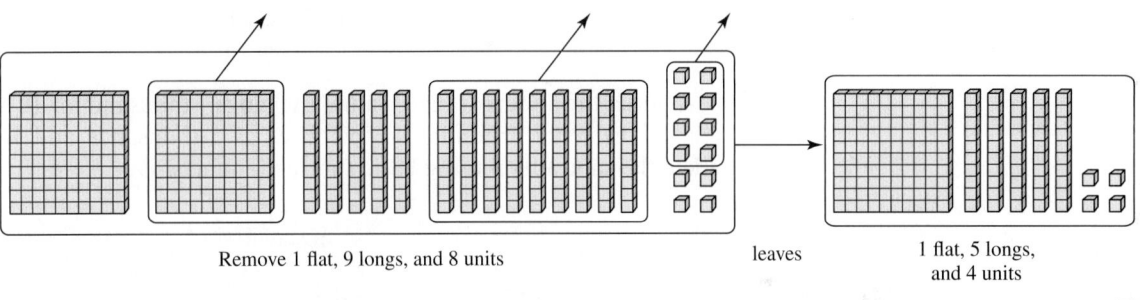

Remove 1 flat, 9 longs, and 8 units   leaves   1 flat, 5 longs,
and 4 units

In the last example, we made all the exchanges first and then subtracted by place value. Numerically, we can express this in two ways. One is to use expanded notation:

$$
\begin{array}{l}
341 \\
-157
\end{array}
\;\rightarrow\;
\begin{array}{l}
3 \cdot 100 + 4 \cdot 10 + 1 \cdot 1 \\
-(1 \cdot 100 + 5 \cdot 10 + 7 \cdot 1)
\end{array}
\qquad \text{Write in expanded notation.}
$$

$$
\;\rightarrow\;
\begin{array}{l}
2 \cdot 100 + 14 \cdot 10 + 1 \cdot 1 \\
-(1 \cdot 100 + 5 \cdot 10 + 7 \cdot 1)
\end{array}
\qquad \text{Exchange 1 hundred for 10 tens.}
$$

$$
\;\rightarrow\;
\begin{array}{l}
2 \cdot 100 + 13 \cdot 10 + 11 \cdot 1 \\
-(1 \cdot 100 + 5 \cdot 10 + 7 \cdot 1)
\end{array}
\qquad \text{Exchange 1 ten for 10 units.}
$$

$$
(2-1) \cdot 100 + (13-5) \cdot 10 + (11-7) \cdot 1 \qquad \text{Subtract place values.}
$$
$$
= 1 \cdot 100 + 8 \cdot 10 + 4 \cdot 1
$$
$$
= 184
$$

The other way is to note the exchanges in a more condensed fashion by placing the exchanged values over specific digits.

$$
\begin{array}{r}
352 \\
-198 \\
\end{array}
\;\rightarrow\;
\begin{array}{r}
{}^{2}\,{}^{14}\,{}^{12} \\
\cancel{3}\,\cancel{5}\,2 \\
-1\ 9\ 8 \\
\hline
1\ 5\ 4
\end{array}
$$

Explorations
Manual
4.5

Explorations
Manual
4.6

Standard Algorithm for Subtraction

The standard algorithm for subtraction uses the same ideas as in the last two written examples, but with two differences. One, the way in which the algorithm is written is condensed. Two, exchanges are made while subtracting specific place values, rather than making all the exchanges at once. As we subtract, we may have to exchange across several place values before we can compute a difference.

Example 4.14 | **Subtracting with the Standard Algorithm**

Use the standard subtraction algorithm to subtract $6{,}701 - 3{,}728$.

Solution We begin by comparing the ones digits. Because we cannot subtract 8 from 1 and get a whole number, we must make an exchange. However, we cannot exchange with the 0 in the tens place, so we must go to the hundreds place. Exchange 1 hundred for 10 tens and then 1 ten for 10 ones. This gives us 6 hundreds, 9 tens, and 11 ones. We now subtract both the units and the tens place values using basic subtraction facts.

$$
\begin{array}{cccc}
 & \overset{9}{\underset{6}{}} & \overset{\cancel{10}}{} & \overset{11}{} \\
6 & \cancel{7} & \cancel{0} & \cancel{1} \\
-3 & 7 & 2 & 8 \\
\hline
 & & 7 & 3
\end{array}
$$

Next, subtract the hundreds place value. Exchange 1 thousand for 10 hundreds, which we add to the other 6 hundreds. This gives us 5 thousands and 16 hundreds. We complete the subtraction using basic subtraction facts.

$$
\begin{array}{cccc}
 & & \overset{9}{} & \\
\overset{5}{} & \overset{16}{} & \overset{\cancel{10}}{} & \overset{11}{} \\
\cancel{6} & \cancel{7} & \cancel{0} & \cancel{1} \\
-3 & 7 & 2 & 8 \\
\hline
2 & 9 & 7 & 3
\end{array}
$$

Equal Additions Algorithm

The **equal additions algorithm** is an interesting algorithm because it uses addition to make the numbers easier to subtract by avoiding situations that make exchanging necessary. It is based on the principle that adding the same amount to the minuend and subtrahend does not affect the difference.

Example 4.15 | **Subtracting by Equal Additions**

Use the equal additions algorithm to subtract $452 - 276$.

Solution Exchanges are needed to subtract the digits in both the ones and the tens place values. To avoid making these exchanges, we add the same number to both 452 and 276. We can use many numbers, but we choose 24 because $276 + 24 = 300$, which is relatively easy to subtract. Completing the computation, we have

$$
\begin{array}{rcl}
452 \ +24 & \rightarrow & 476 \\
-276 \ +24 & & -300 \\
\hline
 & & 176
\end{array}
$$

Check Your Understanding 4.2-A

1. Compute each sum using the partial-sums algorithm.
 a. 421 + 379   **b.** 712 + 318
2. Compute each problem using expanded notation.
 a. 501 + 399   **b.** 172 − 97
3. Compute each problem using the standard algorithm.
 a. 891 + 457   **b.** 2,471 + 458   **c.** 789 − 341   **d.** 2,039 − 1,893
4. Use the lattice method to compute each sum.
 a. 63 + 28   **b.** 815 + 794
5. Use the equal additions algorithm to subtract 287 − 219.

Talk About It Which written algorithm do you think best represents the conceptual ideas behind adding and subtracting large numbers? Which do you think would be the easiest for second- or third-grade students to learn and use?

Representation

Activity 4.3
Use base-ten blocks to model and compute each sum or difference.

a. 121 + 89   **b.** 681 + 385   **c.** 131 − 53   **d.** 205 − 147

Mental Addition and Subtraction

Explorations Manual 4.7

Many modern mathematics curricula now spend increased instructional time on mental computation because research indicates that as much as 80 percent of all computations made by adults are done mentally. We often struggle with mental computation because we tend to use the standard algorithms in our head, which require keeping track of too much information. Different strategies are needed that require us to keep track of less. Common strategies for mental computation are described next.

Counting Strategies

Counting strategies, such as counting on, counting back, and skip counting, are useful when computing sums or differences that involve multiples of numbers like 5, 10, or 100.

| Example 4.16 | Mental Computation with Counting Strategies |

Use counting strategies to compute each problem mentally.

a. 63 + 40   **b.** 671 − 300

Solution

a. Starting at 63, skip-count by 10 four times: 73, 83, 93, and 103. Consequently, 63 + 40 = 103.

b. To find 671 − 300, skip-count backward by 100. Start at 671 and count "571, 471, 371." As a result, 671 − 300 = 371.

Left-to-Right Methods

Another strategy is to work from left to right or from larger place values to smaller ones.

Example 4.17 Buying Jewelry

Application

Kyle is looking at a diamond pendant and earrings for his wife. The pendant costs $679, and the earrings cost $289 for the pair. How can he mentally determine the cost of the jewelry?

Solution Kyle can add the numbers mentally by working from left to right. He first adds the hundreds, or $600 + 200 = 800$. He then adds the tens to this result: $800 + (70 + 80) = 800 + 150 = 950$. Finally, he adds the ones: $950 + (9 + 9) = 950 + 18 = 968$. If he buys both, the jewelry will cost him $968.

Compatible Numbers

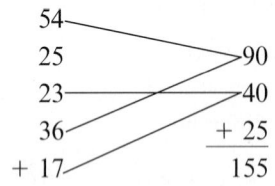

Figure 4.9 Making compatible numbers

In this strategy, we look for two or more **compatible** numbers, that is, numbers that add or subtract to a multiple of 10, 100, or another number that is easy to use in computing. For instance, consider the sum $54 + 25 + 23 + 36 + 17$. Because $54 + 36 = 90$ and $23 + 17 = 40$, we choose to add these numbers first. We then add these sums to any remaining numbers to obtain the final answer (Figure 4.9).

Numbers can be made compatible by means of **compensation.** Here, we split one addend into two numbers so that one of them is compatible with the other addend. We add the compatible numbers, and then add the remaining number.

Example 4.18 Using Compensation

Use compensation to add $157 + 49$.

Solution Split 157 into $156 + 1$ to make a number that is compatible with 49. We then rethink the problem as $(156 + 1) + 49 = 156 + (1 + 49) = 156 + 50 = 206$.

When subtracting, numbers can be made compatible by means of **equal additions.** As in the written algorithm, add the same amount to both numbers, making them easier to subtract by avoiding exchanges.

Example 4.19 Buying Appliances

Application

Merry is out shopping for a new stove and sink. The stove she likes is $649, and the sink is $289. She has saved $850 to buy the two items. If she buys the stove, how much more money does she need for the sink?

Solution Merry can compute the cost by making the two numbers compatible. She adds $1 to $850 and $649 and then finds the difference by subtracting left to right: $(\$850 + \$1) - (\$649 + \$1) = \$851 - \$650 = \$201$. To figure out how much more she needs to purchase the sink, she subtracts $201 from the price of the sink. Again, working from left to right: $\$289 - \$201 = \$88$. She needs another $88 to buy the sink.

Breaking Apart Numbers

Another strategy is to break one number apart by using expanded notation. We then use counting strategies or left-to-right strategies to add or subtract the expanded number one term at a time.

Example 4.20	Mental Computation by Breaking Apart Numbers

Use the breaking-apart strategy to compute each sum or difference.

a. $78 + 35$ **b.** $871 - 634$

Solution

a. Add by thinking $78 + 35 = 78 + (30 + 5) = (78 + 30) + 5 = 108 + 5 = 113$.

b. Subtract by thinking:

$$
\begin{aligned}
871 - 634 &= 871 - (600 + 30 + 4) & \text{Expanded notation} \\
&= (871 - 600) - 30 - 4 & \text{Subtract hundreds} \\
&= (271 - 30) - 4 & \text{Subtract tens} \\
&= 241 - 4 & \text{Subtract units.} \\
&= 237
\end{aligned}
$$

Estimating Addition and Subtraction

In many situations in our lives, we use an estimation to make an informed decision because an exact number is not necessary. For instance, we use estimates to compute how far we can drive on a tank of gas or whether we have enough money to make a purchase. To make such estimates, we change the original numbers to ones that are close in value but that are easier to work with. We do so by either truncating or rounding. With **truncation,** we cut off a number at a certain place value with no regard for the digits that come after. To keep the truncated value close to the original number, placeholders are inserted to move the remaining digits to the proper place values.

Example 4.21	Truncating

Truncate each number to the specified place value.

a. 345 to the hundreds place value

b. 15,921 to the ten thousands place value

Solution

a. 345 truncated to the hundreds place value is 300.

b. 15,921 truncated to the ten thousands place value is 10,000.

Rounding is similar to truncation, but now we consider the digit in the next lowest place value to determine which value the number is closest to. In standard rounding, if the digit is 4 or lower, we round down. If it is 5 or higher, we round up. Again, we use placeholders to move the remaining digits out to the appropriate place value.

Example 4.22 Rounding

Round each number to the specified place value.

a. 561 to the tens place value

b. 4,082 to the hundreds place value

Solution

a. The digit in the ones place value is 1, so we round 561 down to 560.

b. An 8 is the tens place value, so we round 4,082 up to 4,100.

Estimation is not always based on a fixed set of rules; the methods can be changed based on the needs of the situation. For instance, sometimes we may want to overestimate instead of underestimate, or the estimate may need to have a certain degree of accuracy. Other times, it may be useful to ignore standard rounding and round to the nearest even number or to the nearest multiple of 5. Because estimation never leads to an exact answer, the symbol "$\approx$" is used instead of "$=$" to show that the answer is an approximation of the actual value.

Front-End Estimation

Front-end estimates provide the fastest way to estimate a sum or difference. In this method, truncate or round the numbers to a single nonzero digit in the largest place value, and then add or subtract as appropriate.

Example 4.23 Making a Front-End Estimate

Use front-end estimation to approximate $5,492 + 3,487$.

Solution Round 5,492 down to 5,000 and 3,487 down to 3,000. Add the numbers for an estimate: $5,492 + 3,487 \approx 5,000 + 3,000 = 8,000$.

The estimate in the last example is not close to the actual sum: $5,492 + 3,487 = 8,979$. We can improve our estimate by using other digits in the numbers to make an **adjustment.** For instance, 492 and 487 are both about 500. So we can add another $500 + 500 = 1,000$ to our initial sum. Doing so gives us a better estimate of $8,000 + 1,000 = 9,000$.

Example 4.24 Estimating Remaining Tickets

Application

A local arena, which has 7,980 seats, is the venue for a country music concert. If 4,509 tickets have already been purchased, about how many more tickets are available?

Solution We make an initial front-end estimate by subtracting the digits in the thousands place value: $7,000 - 4,000 = 3,000$ tickets. This estimate does not take into account that 7,980 is closer to 8,000 and 4,509 is closer to 4,500. We should adjust our estimate by $1,000 - 500 = 500$ tickets. Consequently, about $3,000 + 500 = 3,500$ tickets are still available.

Compatible Numbers

Another way to make an estimate is to round to compatible numbers. In doing so, we should be flexible in our rounding to make the computation easier or the estimate better. For instance, consider $6,563 - 2,443$. If we round 6,563 down to 6,500 and 2,443 up to 2,500, we get an estimate of $6,500 - 2,500 = 4,000$.

Example 4.25 | **Estimating a Number of Fish**

Application

Nadine is in charge of the fish department at a local pet store. The store needs to have at least 1,500 feeder fish on hand at any given time. The store has three tanks of feeder fish, which currently hold 544, 403, and 364 fish. Are enough feeder fish on hand, or should Nadine order more?

Solution In this situation, Nadine does not need an exact count. She needs to know only whether she has more or less than 1,500 fish. To make an estimate, she can round to compatible numbers, in this case to multiples of 50:

$$
\begin{array}{rll}
544 & \text{is rounded up to} & 550 \\
406 & \text{is rounded down to} & 400 \\
+155 & \text{is rounded down to} & +150 \\
\hline
& & 1,100
\end{array}
$$

Because she only has about 1,100 fish on hand, she needs to order about 400 more. ■

Clustering

Clustering is useful with sums that have addends close in value or when the approximate average of the numbers is easy to compute. We estimate the typical value of the numbers, and then multiply by the amount of numbers.

Example 4.26 | **Estimating Profit**

Application

Fabio owns a small landscaping company. He and his crew completed six jobs this week, which paid the following amounts: $4,890, $5,150, $4,910, $4,850, $5,050, and $5,120. About how much money did he make in total?

Solution Because each job made about $5,000 and he did six jobs, Fabio made about $5,000 \times 6 = \$30,000$. ■

Adding and Subtracting on Calculators

Calculators are important computational tools in and out of the classroom. There are four general types of calculators:

- *Four-function* calculators only add, subtract, multiply, and divide.
- *Scientific* calculators include the four operations and many other features that apply to the sciences and statistics.
- *Fraction* calculators are scientific calculators that have the added ability to compute with fractions and mixed numbers.
- *Graphing* calculators are scientific calculators that have the ability to create and display algebraic and statistical graphs.

Computing large sums and differences becomes easy by using the addition ($+$), and subtraction ($-$) keys on the calculator. If the computation involves several numbers, we can arrive at the answer in a number of ways. One is to store sums and differences in the calculator's memory and recall them as needed. On many calculators, the memory key is labeled STO, MEM, or M+. Once values are stored, the recall function, often labeled RCL or M−, can be used to retrieve the values from the memory, and then they can be added or subtracted as needed.

Example 4.27   Adding on a Calculator

 Write the sequence of keys that are pressed when adding $345 + 671 + 891$ on a calculator:

a. If memory is not used.   **b.** If memory is used.

Solution The sequence of keys on various calculators may differ. We give the sequence of keys for calculators with the STO key. To make reading the key sequences easier, the numbers appear without boxes.

a. To add $345 + 671 + 891$ without using the memory, press: 345 $+$ 671 $+$ 891 ENTER.

b. To add using memory, add $345 + 671$, and then store the sum in the memory. To do that, use the following sequence of keys: 345 $+$ 671 ENTER STO CE/C. To add 891, press 891 $+$ RCL ENTER.

The parentheses keys, $($ and $)$, are grouping symbols, used to tell the calculator which operations to perform first. For instance, in the problem $6{,}698 - (3{,}450 + 2{,}714)$, we can use parentheses to tell the calculator to compute the sum prior to the difference:

$$6698 \;\boxed{-}\; \boxed{(} \; 3450 \; \boxed{+} \; 2714 \; \boxed{)} \; \boxed{ENTER},$$

This gives us an answer of 534.

Check Your Understanding 4.2-B

1. Use mental computation to compute each sum. Explain the method you used.
   a. $57 + 20$ **b.** $123 + 77$ **c.** $571 + 428$

2. Use mental computation to find each difference. Explain the method you used.
   a. $321 - 40$ **b.** $459 - 364$ **c.** $592 - 483$

3. Estimate each sum.
   a. $56 + 61 + 65 + 58$ **b.** $8{,}912 + 7{,}101$ **c.** $10{,}420 + 9{,}013$

4. Estimate each difference.
   a. $4{,}590 - 3{,}478$ **b.** $16{,}875 - 6{,}771$ **c.** $145{,}650 - 129{,}703$

5. List the keys you press on your calculator to add $78 + 53$, store the result in the memory, recall it, and then add 65.

Talk About It Find a partner or form a small group. Make a list of the different words that can be used to indicate that estimation is an appropriate approach to solving a problem (i.e., "about," "nearly," . . .). How many do you come up with?

Application

Activity 4.4

Mike just received his tax refund and wants to buy a new home theater system. He plans to buy a wide-screen television, a DVD player, a surround-sound stereo, and an entertainment center to contain it all. He also plans to get cable television. He shops around and finds several models for each part of the system. Their prices are as follows:

Television	46-in. ($1,562)	50-in. ($2,379)	55-in. ($2,793)
DVD Player	Model A ($165)	Model B ($127)	
Stereo	Model 1 ($452)	Model 2 ($379)	
Cable	Provider 1 (115 channels at $32/month)		
	Provider 2 (165 channels at $53/month)		
Entertainment Center	Oak ($352)	Pine ($267)	Cherry ($547)

a. Provide a high and low estimate for the cost of Mike's system for one year.

b. Calculate the actual highest and lowest costs for Mike's system for one year. How do your estimates compare to the actual values?

SECTION 4.2   Learning Assessment

■ Understanding the Concepts

1. What is regrouping or trading, and why is it an important part of adding large numbers?

2. What is exchanging or borrowing, and why is it an important part of subtracting large numbers?

3. What are the advantages and disadvantages of the standard algorithms over other algorithms for adding and subtracting?

4. **a.** The equal additions algorithm for subtraction is based on what principle?

   b. How does this method make it easier to subtract large numbers?

5. What is the difference between mental computation and computational estimation?

6. **a.** Explain the difference between truncating and rounding.

   b. How are these processes used to make computational estimations?

7. **a.** What are compatible numbers?

   b. What properties of whole-number addition do we use when using compatible numbers to compute mentally?

8. Explain how each strategy is used to mentally compute sums and differences.

   a. Counting strategies

   b. Left-to-right methods

   c. Compatible numbers

   d. Compensation

   e. Breaking apart numbers

9. Explain how each strategy is used to estimate sums or differences.

   a. Front-end estimations

   b. Compensation

   c. Compatible numbers

   d. Clustering

■ Representing the Mathematics

10. State the addition problem represented in each diagram, and then compute it by combing the base-ten blocks.

a.   **b.**

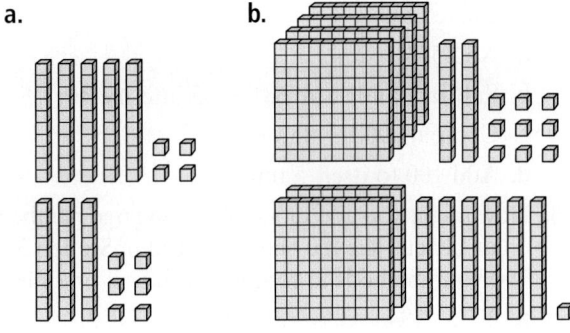

c.

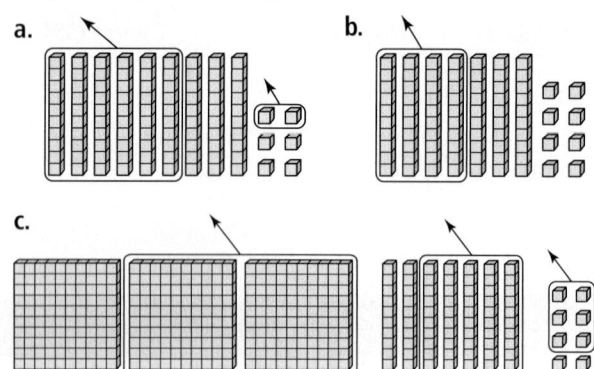

11. State the subtraction problem represented in each diagram, and then compute it.

a.

b.

c.

12. Use base-ten blocks to model and compute each sum or difference.
   a. 47 + 34
   b. 89 + 72
   c. 56 − 24
   d. 93 − 47

13. Use base-ten blocks to model and compute each sum or difference.
   a. 143 + 76
   b. 463 + 281
   c. 184 − 66
   d. 247 − 193

14. Base-ten blocks can also be used with the comparison approach to subtraction. Determine how this is done and then use it to model and subtract each difference.
   a. 47 − 26
   b. 152 − 89
   c. 301 − 223

15. List the keys that you would press on your calculator to:
   a. Add 543 + 254.
   b. Subtract and store 681 − 93 into memory.
   c. Compute 902 − 357 + 78.
   d. Add 100 to itself 5 times.

16. How might you add or subtract two large numbers on an abacus like the ones described in Section 3.2. Use your methods to compute each sum or difference.
   a. 237 + 194
   b. 4,500 + 3,893
   c. 341 − 284
   d. 2,386 − 1,177

■ Exercises

17. Use expanded notation to compute each sum or difference.
   a. 34 + 27
   b. 65 − 32
   c. 155 + 109
   d. 241 − 167

18. Use the partial-sum algorithm to compute each sum.
   a. 65 + 39
   b. 236 + 113
   c. 981 + 676
   d. 1,076 + 547

19. Use the standard algorithm to compute each sum or difference.
   a. 139 + 67
   b. 271 − 59
   c. 278 + 185
   d. 707 − 348

20. Use the standard algorithm to compute each sum or difference.
   a. 6,782 + 987
   b. 3,209 − 903
   c. 8,821 + 7,453
   d. 7,001 − 3,881

21. Use the lattice method to compute each sum.
   a. 56 + 19
   b. 234 + 79
   c. 258 + 493
   d. 593 + 807

22. Use the equal additions method to compute each difference.
   a. 43 − 29
   b. 137 − 98
   c. 763 − 146
   d. 1,204 − 997

23. Use mental computation to compute each sum or difference. State which method you used and why you chose it.
   a. 67 + 40
   b. 57 + 44
   c. 293 − 189
   d. 906 − 450

24. Use mental computation to compute each sum or difference. State which method you used and why you chose it.
   a. 476 + 134
   b. 6,871 + 3,210
   c. 765 − 300
   d. 1,978 − 1,246

25. Truncate each number to the indicated place value.
   a. 419 to the hundreds place value
   b. 3,812 to the tens place value
   c. 67,351 to the hundreds place value
   d. 21,890 to the thousands place value

26. Use standard rounding to round each number to the indicated place value.
   a. 351 to the nearest ten
   b. 45,802 to the nearest ten
   c. 4,672 to the nearest hundred
   d. 30,401 to the nearest thousand

27. Use standard rounding to round each number to the indicated place value.

    a. 13,428 to the nearest hundred

    b. 35,894 to the nearest ten thousand

    c. 144,672 to the nearest thousand

    d. 230,401 to the nearest hundred thousand

28. Approximate each problem with a front-end estimate. Use an adjustment to make the estimate better.

    a. $2,391 + 4,545$ b. $12,010 + 9,874$

    c. $6,709 - 3,451$ d. $13,331 - 9,367$

29. Round to compatible numbers to estimate each problem.

    a. $309 + 211$ b. $378 + 529$

    c. $6,591 - 4,092$ d. $15,097 - 9,970$

30. Use clustering to estimate each sum.

    a. $31 + 29 + 27 + 35 + 32$

    b. $98 + 101 + 104 + 95$

    c. $5,998 + 6,025 + 5,965$

31. Estimate each sum, and then order them from least to greatest.

    $2,408 + 3,499$ $3,101 + 3,110$

    $2,998 + 2,997$ $4,101 + 1,563$

32. Estimate each difference, and then order them from least to greatest.

    $468 - 305$ $972 - 491$

    $823 - 576$ $1,789 - 1,592$

33. Compute each sum or difference on a calculator.

     a. $14,760 + 12,455$ b. $141,768 + 279,051$

    c. $17,910 - 11,989$ d. $478,001 - 391,234$

34. Make each computation on a calculator.

     a. $7,563 + 9,832 - 14,537$

    b. $173 - (23 + 45 + 61)$

    c. $(963 + 412) - (818 + 299)$

    d. $1,789 - (431 + 565) + 213$

■ Problems and Applications

35. Find the missing digits in each sum or difference.

    a.
    ```
       _ 6 3
     +2 _ 8
       7 4 _
    ```

    b.
    ```
       _ _ 3 1
         1 6 4 8
     +4 2 7 _
       8 3 _ 8
    ```

    c.
    ```
       4 _ _ 7
      -2 4 9 1
       1 8 1 _
    ```

    d.
    ```
       6 7 3 4
      -2 _ _ _
       3 8 3 8
    ```

36. Using the digits 1, 2, 5, 6, 8, and 9 only once, place a digit in each blank so that:

    a. The least sum is obtained.

    b. The greatest sum is obtained.

    ```
        _ _ _
     +  _ _ _
    ```

37. Using the digits 2, 3, 5, 6, 8, and 9 only once, place a digit in each blank to so that:

    a. The least difference is obtained.

    b. The greatest difference is obtained.

    ```
        _ _ _
     -  _ _ _ _
    ```

38. a. About what number must be added to 145 to get a sum between 320 and 330?

    b. About what number must be added to 763 to get a sum between 1,850 and 1,870?

39. Make a calculator count backward from 50 to 0 first by 2s and then by 5s. In each case, what is the lowest number of keystrokes needed to complete the task?

40. Another addition algorithm is the *scratch*, or *low-stress*, algorithm. Start with the ones column, adding only one pair of numbers at a time and recording the sum near the second addend. If the sum is 10 or more, the ones digit is placed to the right, and the number is scratched off to represent a group of 10. Take the ones digit from the previous sum, and add the next number. Continue in this way until all the ones digits are added. For instance:

    ```
      2
      4 1
      3 8₉        1 + 8 = 9
      5 6₅        9 + 6 = 15
    + 2 9₄        5 + 9 = 14
          4       Bring down the 4, and take the two
                  scratches to the tens place value.
    ```

    Add the number of scratches, carry them to the tens place, and repeat the process.

    ```
       2
       4₆ 1       2 + 4 = 6
       3₉ 8       6 + 3 = 9
       5₄ 6       9 + 5 = 14
     + 2₆ 9       4 + 2 = 6
       1 6 4      Bring down the 6 and the 1 group
                  of 10 to obtain the result.
    ```

    Use the scratch algorithm to compute each sum.

    a. $54 + 76 + 39$

    b. $135 + 79 + 101$

    c. $239 + 412 + 198 + 65$

41. The *cashier's algorithm* for subtraction is similar to the missing-addend approach. The amount of change the cashier gives back is the difference between the bill and the amount of money given to the cashier. For instance, suppose you hand the cashier $40 for a $27 bill. In making change, the cashier may count out, "28, 29, 30, and 40" while handing back three $1s and a $10.

   a. How does this algorithm represent the missing-addend approach to subtraction?

   b. A customer gives $60 for a $45 bill. What change will the cashier give back, and what might the cashier say while passing back the change?

   c. Repeat part b for a customer who gives $100 for a $63 bill.

42. Another way to compute subtraction is to *add the complement*. The complement is a number that, when added to another number, results in a sum of all 9s. For instance, the complement of 46 is 53 because $46 + 53 = 99$. The algorithm proceeds by adding the complement of the second number to first, discarding the leading one in the sum, and then adding 1. For instance, for the problem $458 - 341$, the method is used as follows:

$$
\begin{array}{rl}
458 & \text{Add the complement of the} \\
+\ 658 & \text{second number to the first.} \\
\hline
\cancel{1}116 & \text{Cross out leading 1.} \\
+\quad 1 & \text{Add 1 to obtain} \\
\hline
117 & \text{the result.}
\end{array}
$$

   Use this algorithm to compute:

   a. $73 - 31$ **b.** $121 - 98$ **c.** $1,487 - 1,298$

43. Arthur is making a turkey sandwich but wants to watch his calories. His sandwich consists of 2 slices of bread (175 cal each), 3 slices of turkey (45 cal each), a slice of tomato (25 cal), lettuce (55 cal), and a teaspoon of mayonnaise (130 cal). How many total calories are in his sandwich?

44. A truck driver must deliver goods to four different towns: Albion, Bartlesville, Caldwell, and Danville. If she starts in Albion, visits each town once in alphabetical order, and then returns to Albion, how many total miles does she drive if the distances between towns are as follows?

Albion to Bartlesville	234 mi
Bartlesville to Caldwell	453 mi
Caldwell to Danville	176 mi
Danville to Albion	341 mi

45. Tammy walks into a store with $325. She makes purchases of $45, $37, and $156. How much money does she have left?

46. The manager of a fast-food restaurant checks the inventory on Friday and then again on the following Monday. She makes the counts as shown.

Food Item	Friday	Monday
Hamburger patties	9,364	2,719
Chicken patties	6,209	3,268
Fish fillets	2,781	912
Hamburger buns	17,437	4,107

If no new stock came in over the weekend, how many of each item was used?

47. Bill is out shopping for clothes and purchases 5 shirts priced $21, $12, $17, $14, and $29 and 2 pairs of pants priced $32 and $41. About how much did his clothes cost?

48. A car dealer goes to a car auction and buys 5 cars at the following prices: $2,550, $3,110, $2,615, $4,795, and $3,785. Estimate the total cost of the cars.

49. About how much time has passed since:

   a. Christopher Columbus landed in North America in 1492?

   b. The signing of the Declaration of Independence in 1776?

   c. Abraham Lincoln was shot in 1865?

   d. Neil Armstrong set foot on the moon in 1969?

50. The chart shows the five states with the largest populations in 2010.

State	Population
California	36,961,664
Texas	24,782,302
New York	19,541,453
Florida	18,537,969
Illinois	12,910,409

Source: U.S. Census Bureau

   a. About how many more people live in California than in New York?

   b. About how many more people live in California than in Illinois?

   c. If the total population of the United States in 2010 was 306,406,893, about how many live outside these five states?

51. Jenny takes one afternoon to balance her checkbook. Over the last two weeks she made the following transactions.

Credits: $45, $327, and $410
Debits: $21, $23, $65, $107, $10, $36, $219, and $89

If her beginning balance was $57, use a calculator to compute her final balance in two different ways.

■ Mathematical Reasoning and Proof

52. Although possible, we generally do not use the missing-addend approach to compute the difference of two large numbers. Why not?

53. **a.** If you truncate the two numbers in a sum, will the result be an overestimate or an underestimate? Explain.

   b. If you truncate the two numbers in a difference, will the result be an overestimate or an underestimate? Explain.

54. How should you round the addends if you want to:

   a. Overestimate a sum?

   b. Underestimate a sum?

55. How should you round the numbers if you want to:

   a. Overestimate a difference?

   b. Underestimate a difference?

56. Sharice subtracts $54 - 38$ as shown.

$$\begin{array}{r} 54 \\ -38 \\ \hline -4 \\ +20 \\ \hline 16 \end{array}$$

Find her method and use it to subtract:

   a. $45 - 26$

   b. $145 - 119$

   c. $901 - 857$

   d. Can you explain why her method works?

■ Mathematical Communication

57. Why do you think the term "regrouping" is now preferred over the term "carrying" to describe the process in addition algorithms? What about the term "exchanging" over "borrowing" in subtraction algorithms?

58. Suppose you were a second-grade teacher. Write a short paragraph or two explaining the steps you would take to teach the standard algorithm for addition.

59. When children first learn to add and subtract with base-ten blocks, they often work from left to right. Why do you think that it might seem natural for them to do so?

60. Three students have incorrectly used the standard algorithm to add the same problem. Identify their mistakes, and then explain how you might correct their errors.

   a. $\begin{array}{r} 47 \\ +69 \\ \hline 1016 \end{array}$ **b.** $\begin{array}{r} 47 \\ +69 \\ \hline 106 \end{array}$ **c.** $\begin{array}{r} {}^{6}\;47 \\ +69 \\ \hline 161 \end{array}$

61. Three students have incorrectly used the standard algorithm to subtract the same problem. Identify their mistakes, and then explain how you might correct their errors.

   a. $\begin{array}{r} 46 \\ -27 \\ \hline 13 \end{array}$ **b.** $\begin{array}{r} 46 \\ -27 \\ \hline 29 \end{array}$ **c.** $\begin{array}{r} 46 \\ -27 \\ \hline 21 \end{array}$

62. Keylon estimates the value of $3,419 + 4,910$ to be 7,000. What method of estimation has he used? How would you, as his teacher, help him to make a better estimate?

63. Marcia adds 19,563 and 15,031 on her calculator and gets a result of 21,094. What went wrong, and how would you tell her to check her computation with estimation?

64. List the sequence of keys you need to press on your calculator to make it skip-count backward from 34 to 4 by 5s. Swap calculators with a partner who has a different kind of calculator. Try the sequence of keys you wrote down for your calculator. Did you get the same answer? If not, determine how to make the other calculator skip-count backward from 34 to 4 by 5s. What does this imply about using calculators in the classroom?

■ Building Connections

65. When you use addition or subtraction in your daily life, which of the methods and tools discussed in this section are you most likely to use? Explain your preference.

66. Devise a method for adding the following two Egyptian numerals without converting them to decimal numerals.

$$\begin{array}{r} \text{999}\cap\cap\cap\cap\cap\text{IIIIIIIII} \\ +\;\text{99}\cap\cap\cap\cap\text{IIIII} \\ \hline \end{array}$$

What written addition algorithm was your method most like?

67. Devise a method for subtracting the following two Egyptian numerals without converting to decimal numerals.

$$\begin{array}{r} \text{ℓℓ999}\cap\cap\text{IIII} \\ \text{999}\cap\cap\cap_{\text{III}} \\ -\text{999}\;\;\cap\cap\;\;{}_{\text{III}} \\ \hline \end{array}$$

   a. What approach to subtraction did your method use?

   b. What written subtraction algorithm was your method most like?

68. With a small group, analyze several third- and fourth-grade textbooks. Write a summary of the addition and subtraction algorithms that you find and how they are presented to students. What do the corresponding instructor textbooks suggest for helping students master the various algorithms?

69. a. Which of the written subtraction algorithms do you find easiest to use?

b. Was it helpful to your understanding of subtraction to compute differences with several different algorithms? How so?

c. How do your answers to these questions apply to your future career as a teacher?

70. There is evidence to suggest that some students first develop mathematics anxiety in second, third, and

fourth grades. One reason for this may be the fact that written algorithms for computing with larger numbers are taught in these grades.

a. Why do you think the algorithms may be the source of anxiety?

b. Do you think the reason has to do with the algorithms themselves or with the way in which the algorithms are taught? Explain.

SECTION 4.3

Understanding Whole-Number Multiplication and Division

After children learn to add and subtract, they begin to work with multiplication and division. As before, they first develop an intuitive understanding of the operations and compute basic facts before they multiply and divide larger numbers.

Understanding Whole-Number Multiplication

In general, there are two approaches to whole-number multiplication: One uses addition, and the other uses the Cartesian product. The most common and intuitive approach to multiplication is the repeated-addition approach.

Repeated-Addition Approach

When working with addition, children are likely to encounter sums in which the addends are all the same number, such as $2 + 2 + 2 + 2$ and $5 + 5 + 5$. They learn to condense these sums by using multiplication as a notational shortcut to repeated addition. For instance, they learn to rewrite $2 + 2 + 2 + 2$ as $4 \cdot 2$ and $5 + 5 + 5$ as $3 \cdot 5$. They do so by using the following definition.

> **Definition of Whole-Number Multiplication (Repeated-Addition Approach)**
>
> If a and b are any whole numbers with $a \neq 0$, then $a \cdot b = \underbrace{b + b + \ldots + b}_{a \text{ addends}}$. If $a = 0$, then $0 \cdot b = 0$ for all b.

The numbers a and b are called **factors** of the product $a \cdot b$, which can also be written as ab or $a \times b$ and read as "a multiplied by b" or "a times b." The last phrase comes from the fact that we add b to itself a "times." Also, the order of the factors changes the interpretation of the product. For instance, we interpret $4 \cdot 3$ as $3 + 3 + 3 + 3$ and $3 \cdot 4$ as $4 + 4 + 4$.

Example 4.28 | **Writing Products as Repeated Sums**

Use repeated addition to compute each product.

a. $5 \cdot 7$ **b.** $1 \cdot 6$ **c.** $8 \cdot 0$

Solution Using the definition:

a. $5 \cdot 7 = 7 + 7 + 7 + 7 + 7 = 35$.   **b.** $1 \cdot 6 = 6$.

c. $8 \cdot 0 = 0 + 0 + 0 + 0 + 0 + 0 + 0 + 0 = 0$.

The repeated-addition approach to multiplication can be modeled in a number of concrete ways, including the set model, the array model, and the number line model.

Set Model. Because multiplication is defined in terms of addition, the set model can be used to represent products. We interpret $a \cdot b$ as combining a sets, each having b elements. After we combine the sets, we count the objects to find the product. For instance, Figure 4.10 shows how to compute $5 \cdot 3$ using 5 sets of 3 marbles each.

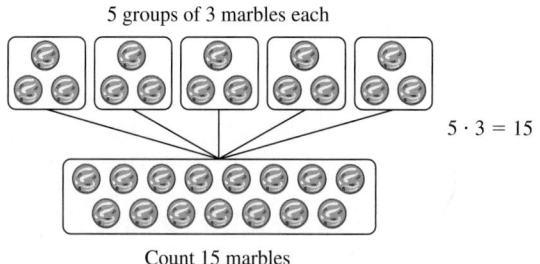

Figure 4.10 **Using sets to compute 5 · 3**

When we use this model, we can also skip-count to find the total. Figure 4.11 illustrates how to skip-count by 5s to compute the product $3 \cdot 5 = 15$.

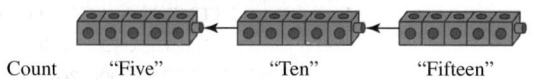

Figure 4.11 **Using skip counting to compute 3 · 5**

Array or Area Model. In the **array model,** the product is represented by a rectangular array of objects, such as coins, blocks, or other counters. The first factor tells us the number of rows in the array, and the second factor tells us the number of columns. Once the array is filled, we count the total number of objects to find the product. Figure 4.12 illustrates how to compute the product of $2 \cdot 5$ using an array of pennies.

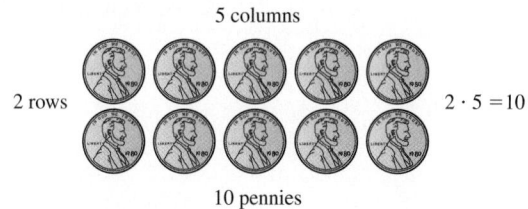

Figure 4.12 **Using an array to compute 2 · 5**

Example 4.29	Counting Tiles

Suzanne wants to tile a rectangular floor space that measures 4 ft by 6 ft. If she uses square foot tiles, how many tiles does she need?

Solution Suzanne can make a rectangular grid with 4 rows of 6 tiles each. When she fills in the grid and counts the tiles, she finds that she needs $4 \cdot 6 = 24$ tiles.

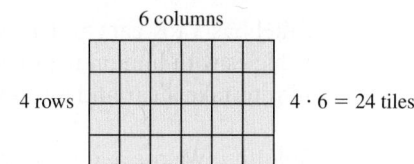

The last example shows that computing a product is equivalent to finding the area of a rectangle with sides of lengths equivalent to the two factors. For this reason, the array model is also called the **area model.**

Number Line Model. In addition to being used with sets of objects, the repeated-addition approach can be used with measures, which we represent by joining arrows of equal length on a number line. The first factor represents the number of arrows to draw, and the second represents the number of units in each arrow. The product is the final position on the number line. For instance, $5 \cdot 2$ is shown in Figure 4.13.

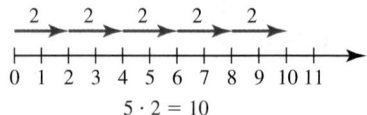

$$5 \cdot 2 = 10$$

Figure 4.13 Using a number line to compute $5 \cdot 2$

Cartesian Product Approach

Another way to approach whole-number multiplication is to use the Cartesian product of two sets.

> **Definition of Whole-Number Multiplication (Cartesian-Product Approach)**
>
> If a and b are any whole numbers where $a = n(A)$ and $b = n(B)$, then $a \cdot b = n(A \cdot B)$.

Example 4.30 Using the Cartesian-Product Approach

Use the definition of the Cartesian-product approach to multiplication to show that $3 \cdot 5 = 15$.

Solution We represent each number with a set, form the Cartesian product, and then count the number of ordered pairs. For instance, if we let $A = \{a, b, c\}$ and $B = \{1, 2, 3, 4, 5\}$, then $n(A) = 3$, $n(B) = 5$, and:

$$3 \cdot 5 = n(A \times B)$$
$$= n(\{a, b, c\} \times \{1, 2, 3, 4, 5\})$$
$$= n(\{(a, 1), (a, 2), (a, 3), (a, 4), (a, 5), (b, 1),$$
$$(b, 2), (b, 3), (b, 4), (b, 5), (c, 1), (c, 2),$$
$$(c, 3), (c, 4), (c, 5)\})$$
$$= 15$$

The Cartesian-product approach to multiplication can be modeled in two ways: with sets and with tree diagrams.

Set Model. In the set model, we make two sets of the appropriate size, form all possible pairs between them, and then count the number of pairs.

Example 4.31 Counting Toy Vehicles

Jaliel has 3 toy cars and 4 toy trucks. His parents tell him that he can take only 2 toy vehicles with him on their upcoming vacation. How many different ways are there for him to take a car and a truck?

Solution We use a table to make the pairings between the set of three cars and the set of four trucks.

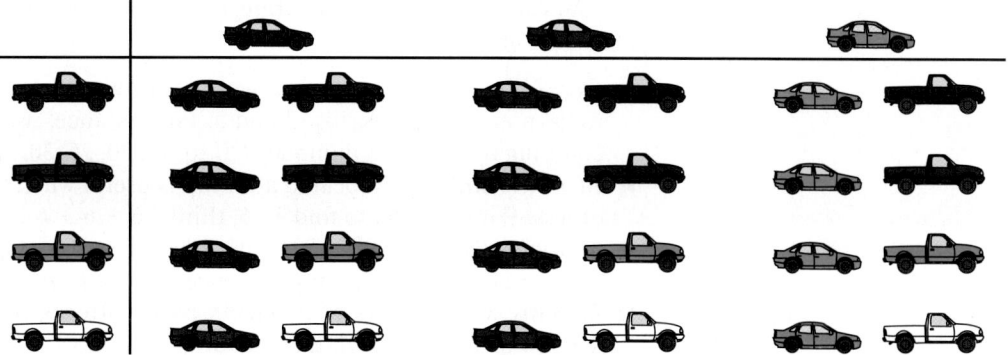

Then we count the pairings to find the product. Because there are 12 pairs, there are 3 · 4 = 12 ways he can take a car and a truck.

Tree Diagrams. Another way to model the Cartesian-product approach is to use a tree diagram. For instance, to model 3 · 2, pick two sets with the appropriate number of elements, say $A = \{1, 2, 3\}$ and $B = \{a, b\}$. We then make a tree diagram like the one in Figure 4.14. Because there are six total branches or pairings, we conclude that 3 · 2 = 6.

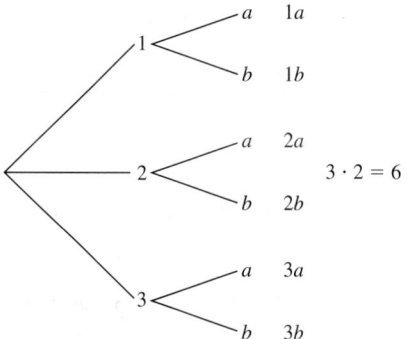

Figure 4.14 Using a tree diagram to compute 3 · 2

. . .

Once children have used the different approaches and models to help them understand multiplication, they can turn to a fact table (Table 4.4) to help them master all 100 single-digit multiplication facts.

Table 4.4	Single-digit multiplication facts								
×	0	1	2	3	4	5	6	7	8	9
0	0	0	0	0	0	0	0	0	0	0
1	0	1	2	3	4	5	6	7	8	9
2	0	2	4	6	8	10	12	14	16	18
3	0	3	6	9	12	15	18	21	24	27
4	0	4	8	12	16	20	24	28	32	36
5	0	5	10	15	20	25	30	35	40	45
6	0	6	12	18	24	30	36	42	48	54
7	0	7	14	21	28	35	42	49	56	63
8	0	8	16	24	32	40	48	56	64	72
9	0	9	18	27	36	45	54	63	72	81

As with addition, several strategies can be used to learn these facts. Some common ones are:

- *Skip Counting.* Skip counting works well with facts that involve numbers with well-known multiples, like 2 and 5. For instance, we can compute $6 \cdot 5$ by skip-counting by 5s for six numbers: "5, 10, 15, 20, 25, 30."
- *Repeated Addition.* Repeated addition is useful when the first factor is three or less. For example, to find $3 \cdot 6$, think "$6 + 6 + 6 = 18$."
- *Multiplication by 2.* Products with 2 are often easy to learn because they are the same as adding doubles. For instance, we can think of $2 \cdot 7$ as $7 + 7 = 14$.
- *Patterns.* Certain sets of products have patterns that can make them easy to learn. For instance, products involving 9 follow a pattern in which the sum of the digits in the product is 9, and the tens digit is always 1 less than the factor multiplying 9. For example, consider $6 \cdot 9$. The tens digit of the product will be 5 because it is one less than 6, and the unknown ones digit must be 4 because $5 + 4 = 9$. As a result, $6 \cdot 9 = 54$.

Properties of Whole-Number Multiplication

Explorations Manual 4.8

Other strategies for learning single-digit multiplication facts are centered on the properties of whole-number multiplication. Because we can define whole-number multiplication using whole-number addition, multiplication will satisfy many of the same properties.

Theorem 4.5 Properties of Whole-Number Multiplication

Closure property.	The product of any two whole numbers is a unique whole number.
Commutative property.	If a and b are any two whole numbers, then $a \cdot b = b \cdot a$.
Associative property.	If a, b, and c are any whole numbers, then $a \cdot (b \cdot c) = (a \cdot b) \cdot c$.
Identity property.	For any whole number a, there exists the unique number 1, called the **multiplicative identity,** such that $a \cdot 1 = a$ and $a = 1 \cdot a$.

The properties for multiplication have the same basic interpretation as those for addition. The closure property guarantees that the product of any two whole numbers is another, unique whole number. This may seem obvious, but not all sets are closed under multiplication. For instance, the set $P = \{2, 3, 5, 7, 11, \ldots\}$ is not closed under multiplication because $2 \cdot 3 = 6$ and 6 is not an element of P.

The commutative property for multiplication states that the order in which the factors are written does not matter; the product will be same. To understand why, consider $4 \cdot 2$ and $2 \cdot 4$ on a number line (Figure 4.15). Four arrows of two units or two arrows of four units lead to the same final position: 8.

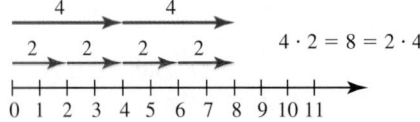

Figure 4.15 **The commutative property on a number line**

The associative property states that the order in which the numbers are multiplied does not matter; the product will be the same. We can illustrate this property using blocks and the array model. For instance, if $a = 2$, $b = 3$, and $c = 5$, we interpret $2 \cdot (3 \cdot 5)$ as 2 layers of blocks arranged in 3×5 arrays. Similarly, we interpret $(2 \cdot 3) \cdot 5$ as 5 layers of blocks arranged in 2×3 arrays. Either way, the array contains 30 blocks (Figure 4.16).

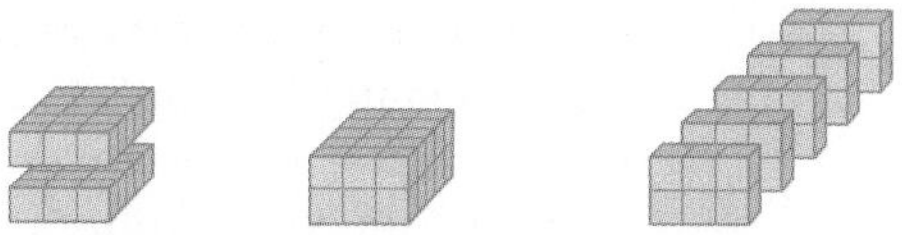

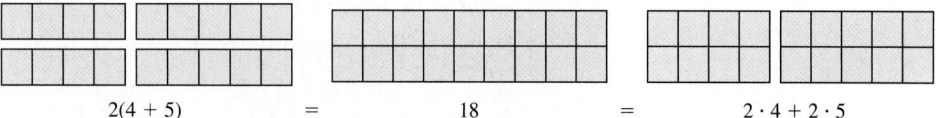

$$2 \cdot (3 \cdot 5) \quad = \quad 30 \quad = \quad (2 \cdot 3) \cdot 5$$

Figure 4.16 The associative property with the array model

The identity property states that multiplying any whole number by 1 will always result in the given whole number, regardless of the order of the factors. To understand why, consider $1 \cdot 4$ and $4 \cdot 1$ on the number line (Figure 4.17). One arrow of 4 units and 4 arrows of 1 unit lead to the same final position: 4. The identity property also states that 1 is the only number for which this is true.

We might ask why the additive and multiplicative identities are not both zero. However, as the next property states, the multiplicative identity cannot be zero because a product involving zero is always equal to zero.

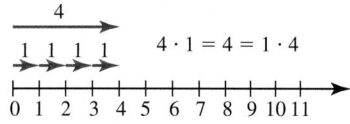

$4 \cdot 1 = 4 = 1 \cdot 4$

Figure 4.17 The identity property on a number line

Theorem 4.6   Zero Multiplication Property

If a is any whole number, then $a \cdot 0 = 0 = 0 \cdot a$.

To understand why the zero multiplication property is true, think of a specific example in terms of repeated addition. For instance, we interpret $3 \cdot 0$ as adding zero to itself 3 times: $3 \cdot 0 = 0 + 0 + 0 = 0$. For the reverse, we interpret $0 \cdot 3$ as adding 3 to itself zero times so not even one 3 is put into the sum. This again results in zero.

One other property applies to both addition and multiplication. It is called the **distributive property of multiplication over addition** because it allows us to "distribute" factors to addends.

Theorem 4.7   Distributive Property of Multiplication over Addition

If a, b, and c are any whole numbers, then $a(b + c) = a \cdot b + a \cdot c$.

One way to illustrate the distributive property is to use the array model. For instance, if $a = 2$, $b = 4$, and $c = 5$, then we represent $2(4 + 5)$ with 2 rows, each with a set of 4 and 5 blocks. We represent $2 \cdot 4 + 2 \cdot 5$ with a 2×4 array and a 2×5 array. In either case, the array requires 18 blocks (Figure 4.18).

$$2(4 + 5) \qquad = \qquad 18 \qquad = \qquad 2 \cdot 4 + 2 \cdot 5$$

Figure 4.18 The distributive property of multiplication over addition with arrays

The next example proves this property more definitively.

Example 4.32   Proof of the Distributive Property

Reasoning   Use repeated addition to prove the distributive property of whole-number multiplication over addition.

Solution If a, b, and c represent whole numbers, then:

$$a \cdot (b + c) = \underbrace{(b + c) + (b + c) + \ldots + (b + c)}_{a \text{ addends}} \qquad \text{Repeated addition}$$

$$= \underbrace{b + b + \ldots + b}_{a \text{ addends}} + \underbrace{c + c + \ldots + c}_{a \text{ addends}} \qquad \text{Commutative and associative properties}$$

$$= a \cdot b + a \cdot c \qquad \text{Repeated addition}$$

Several other comments must be made about the distributive property. First, because multiplication is commutative, the factors can be written in the reverse order: $(b + c) \cdot a = b \cdot a + c \cdot a$. Second, the distributive property of multiplication over addition can be generalized to more than two addends. For instance, $a \cdot (b + c + d) = a \cdot b + a \cdot c + a \cdot d$. Third, as long as $b \geq c$, there is also a distributive property of multiplication over subtraction: $a \cdot (b - c) = a \cdot b - a \cdot c$.

Example 4.33   Using the Distributive Property

Use the distributive property to rewrite each expression.

a. $5 \cdot (3 + 2)$ **b.** $4 \cdot (x + y + 2)$ **c.** $2 \cdot (8 - 3)$

Solution

a. Distribute the 5, or $5 \cdot (3 + 2) = 5 \cdot 3 + 5 \cdot 2 = 15 + 10 = 25$.

b. Distribute the 4, or $4 \cdot (x + y + 2) = 4 \cdot x + 4 \cdot y + 4 \cdot 2 = 4x + 4y + 8$.

c. Distribute the 2, or $2 \cdot (8 - 3) = 2 \cdot 8 - 2 \cdot 3 = 16 - 6 = 10$.

Factoring is the use of the distributive property in reverse. To factor, pull out any common factors in the two addends. For instance, each product in $3 \cdot 5 + 3 \cdot 7$ has a common factor of 3. To factor, pull out the 3, or $3 \cdot 5 + 3 \cdot 7 = 3(5 + 7)$. Writing numbers and expressions in terms of their factors is an important skill in both number theory and algebra.

Check Your Understanding 4.3-A

1. Use repeated addition to compute each product.

   a. $5 \cdot 4$ **b.** $3 \cdot 8$ **c.** $9 \cdot 1$

2. Create sets and use them with the Cartesian-product definition of multiplication to compute $4 \cdot 3$.

3. Determine whether each set is closed under multiplication. Explain your answer.

   a. $T = \{2, 4, 6, 8\}$.   **b.** $O = \{1, 3, 5, 7, \ldots\}$.   **c.** $\{1\}$.

4. State the property of multiplication demonstrated in each statement.

   a. $3 \cdot (4 \cdot 7) = (3 \cdot 4) \cdot 7$.   **b.** $1 \cdot 4 = 4$.   **c.** $8 \cdot 0 = 0$.

   d. $8(3 + 7) = 8 \cdot 3 + 8 \cdot 7$.   **e.** $8 \cdot 9 = 9 \cdot 8$.   **f.** $3 \cdot 6 = 18 \in W$.

Connection

Activity 4.5
Consider Table 4.4. As a teacher, how would you use the multiplication table to demonstrate:

a. Skip-counting by 2s? By 3s?

b. Multiples of 5? Multiples of 7?

c. The multiplicative identity property?

d. The zero multiplication property?

Understanding Whole Numbers and Exponents

Like multiplication, **exponents** are a notational shortcut for repeated multiplication.

Definition of Exponents

If a is any whole number and n is a natural number, then $a^n = \underbrace{a \cdot a \cdot \ldots \cdot a}_{n\ factors}$.

The n is called the **exponent**, or **power**, the a is called the **base**, and we read the expression a^n as, "a raised to nth power." If the exponent is a 2 or a 3, we say the base is squared or cubed, respectively. Specific examples of exponents are:

$$5^3 = 5 \cdot 5 \cdot 5 = 125 \quad 2^6 = 2 \cdot 2 \cdot 2 \cdot 2 \cdot 2 \cdot 2 = 64 \quad y^2 = y \cdot y$$

The definition limits exponents to just the natural numbers. Zero is excluded as a power because multiplying a number by itself zero times has little meaning. However, having a value for a^0 is mathematically useful.

Example 4.34   Finding the Value of 10^0

Problem Solving   Determine a reasonable value for 10^0.

Solution

Understanding the Problem. The goal is to find a reasonable value for 10^0 that is consistent with what we know about exponents. The only information we have is the definition.

Devising a Plan. Even though we have little to go on, we are working with a power of 10. If we consider other powers of ten, we can *look for a pattern* that might help us find an answer.

Carrying Out the Plan. We use the definition to create a list of the powers of 10 from 10^4 down to 10^1.

$$10^4 = 10 \cdot 10 \cdot 10 \cdot 10 = 10,000$$
$$10^3 = 10 \cdot 10 \cdot 10 = 1,000$$
$$10^2 = 10 \cdot 10 = 100$$
$$10^1 = 10$$
$$10^0 = ?$$

A pattern exists between the exponents and the corresponding powers of 10. Specifically, as the exponent decreases by 1, we divide the previous power of 10 by 10. Continuing the pattern seems reasonable, so $10^0 = 10^1 \div 10 = 10 \div 10 = 1$.

Looking Back. The definition of an exponent allows for the base to be any whole number, so we can replace 10 with another nonzero whole number and achieve the same results. Zero itself must be excluded as a base because $0^n = 0$, not 1. As a result, if a is any nonzero whole number, we define $a^0 = 1$.

Definition of a^0

If a is any nonzero whole number, then $a^0 = 1$.

Exponential expressions can be simplified by using the following properties.

Theorem 4.8   Properties of Exponents

Let a, b, m, and n be any whole numbers with $a \neq 0$ and $b \neq 0$. Then:

1. $a^m \cdot a^n = a^{m+n}$ **2.** $a^m \cdot b^m = (a \cdot b)^m$ **3.** $(a^m)^n = a^{m \cdot n}$

Example 4.35 Simplifying Exponents

Use the properties of exponents to simplify each expression.

a. $5^2 \cdot 5^4$ **b.** $4^3 \cdot 2^3$ **c.** $(5^2)^3$ **d.** $(3^4)^4 \cdot 9^3$

Solution

a. $5^2 \cdot 5^4 = 5^{2+4} = 5^6$. **b.** $4^3 \cdot 2^3 = (4 \cdot 2)^3 = 8^3$. **c.** $(5^2)^3 = 5^{2\cdot3} = 5^6$.

d. $(3^4)^4 \cdot 9^3 = (3^4)^4 \cdot (3^2)^3 = 3^{4\cdot4} \cdot 3^{2\cdot3} = 3^{16} \cdot 3^6 = 3^{16+6} = 3^{22}$.

Each property in Theorem 4.8 follows directly from the definition. We prove $a^m \cdot b^m = (a \cdot b)^m$ in the next example and leave the proofs of the others to the Learning Assessment.

Example 4.36 Proof of $a^m \cdot b^m = (a \cdot b)^m$

Reasoning

If a and b are nonzero whole numbers, show that $a^m \cdot b^m = (a \cdot b)^m$.

Solution Assume a and b are nonzero whole numbers. Then:

$$a^m b^m = \underbrace{a \cdot a \cdot \ldots \cdot a}_{m\,factors} \cdot \underbrace{b \cdot b \cdot \ldots \cdot b}_{m\,factors} \qquad \text{Definition of an exponent}$$

$$= \underbrace{(a \cdot b) \cdot (a \cdot b) \cdot \ldots \cdot (a \cdot b)}_{m\,factors} \qquad \text{Commutative and associative properties}$$

$$= (a \cdot b)^m \qquad \text{Definition of an exponent}$$

Understanding Whole-Number Division

The fourth operation commonly taught in the elementary grades is division, which is represented symbolically as $a \div b$, $\frac{a}{b}$, or $b\overline{)a}$. In these representations, a is the **dividend** (the number being divided), and b is the **divisor** (the number doing the dividing). The result of the division is called the **quotient.**

Approaches to Division

In general, we can think of division in three ways: the repeated-subtraction approach, the partitioning or fair-shares approach, and the missing-factor approach.

Repeated-Subtraction Approach. Because we used repeated addition to define multiplication, we can use repeated subtraction—the inverse operation of addition—to define division. Specifically, if a and b are whole numbers, then $a \div b$ is the number of times we can subtract b from a as long as $b \neq 0$.

Example 4.37 Using Repeated Subtraction to Compute a Quotient

Use repeated subtraction to interpret and compute each quotient.

a. $9 \div 3$ **b.** $10 \div 2$

Solution

a. $9 \div 3$ is equal to the number of times 3 can be subtracted from 9. Because this can be done 3 times ($9 - 3 - 3 - 3 = 0$), then $9 \div 3 = 3$.

b. $10 \div 2$ is equal to the number of times 2 can be subtracted from 10. Because this can be done 5 times ($10 - 2 - 2 - 2 - 2 - 2 = 0$), then $10 \div 2 = 5$.

One way to better understand the repeated-subtraction approach to division is to model it with sets. The quotient $a \div b$ represents the number of sets that can be removed from a objects if b objects are in each set and $b \neq 0$. For instance, $15 \div 3$ is the number of sets that can be removed from 15 objects if 3 objects are in each set. Because we can remove 5 sets of 3 objects ($15 - 3 - 3 - 3 - 3 - 3 = 0$), then $15 \div 3 = 5$ (Figure 4.19).

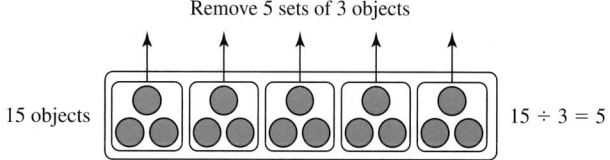

Figure 4.19 Using sets to compute $15 \div 3$ by repeated subtraction

Repeated subtraction can also be represented on a number line. The dividend serves as the starting point, and the divisor indicates how many units to include in each arrow. We then draw backward arrows until we reach 0. The quotient is the number of arrows. For instance, the quotient $8 \div 2 = 4$ is shown in Figure 4.20.

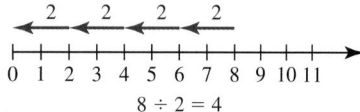

Figure 4.20 Using a number line to compute $8 \div 2$ by repeated subtraction

Partitioning or Fair-Shares Approach. In the **partitioning**, or **fair-shares approach**, $a \div b$ represents the number of objects in each set when a objects are equally distributed among b sets. For example, $20 \div 4$ is equal to the number of objects in each set when 20 objects are equally distributed among 4 sets. Because each of the 4 sets has 5 objects, $20 \div 4 = 5$ (Figure 4.21).

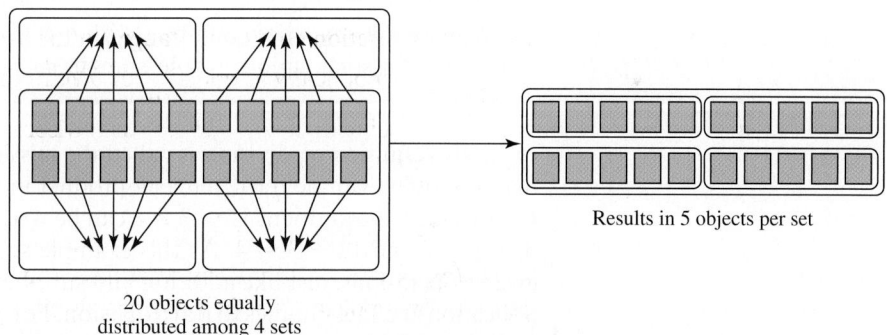

Results in 5 objects per set

20 objects equally
distributed among 4 sets

Figure 4.21 Using sets to compute $20 \div 4$ by partitioning

Example 4.38	Writing Division Story Problems

Communication

Write two story problems for division: one that is solved by using repeated subtraction and another that is solved by partitioning. Use sets to solve each problem.

Explorations
Manual
4.9

Solution To reflect the repeated-subtraction approach, the story problem must indicate that subsets of a certain size are removed until no more are left. For instance: "Fatima has 12 necklaces that she wants to put on hooks. If she hangs 3 necklaces on each hook, how many hooks does she need?" To model the problem, begin with 12 necklaces and remove sets of 3 until you can no longer do so. Because we can make 4 sets, $12 \div 3 = 4$.

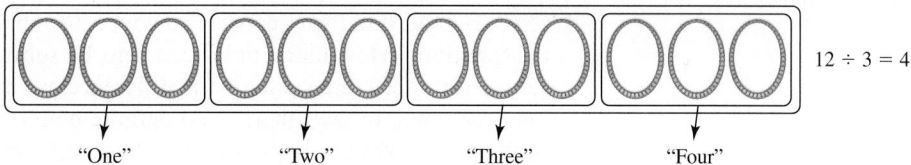

$12 \div 3 = 4$

"One"   "Two"   "Three"   "Four"

In the partitioning approach, the story problem must indicate that objects are removed one at a time until they are equally distributed among a fixed number of sets. For instance, "Fatima has 12 necklaces that she wants to place on 3 hooks. If each hook gets the same number of necklaces, how many are on each hook?" To model the problem, again begin with 12 necklaces. This time, distribute the necklaces to the hooks so that each hook gets the same number. Because 4 necklaces are on each hook, $12 \div 3 = 4$.

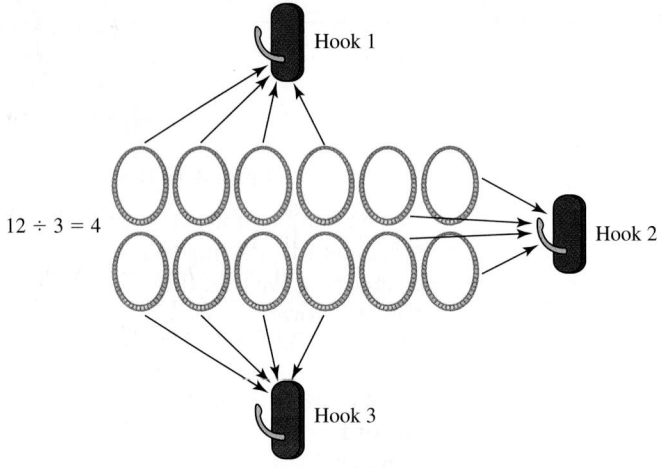

Missing-Factor Approach. Because multiplication can be described as repeated addition and division as repeated subtraction, the two operations have a close connection. In fact, we use can use multiplication to define division.

> ### Definition of Whole-Number Division (Missing-Factor Approach)
>
> If a and b are any whole numbers with $b \neq 0$, then $a \div b = c$ if and only if $a = b \cdot c$ for some unique whole number c.

This definition of division is called the **missing-factor approach** because it shows that a quotient can be rewritten as a product in which one of the factors is missing. For instance, the quotient $12 \div 3 = c$ can be written as $12 = 3 \cdot c$. Because $3 \cdot 4 = 12$, then $c = 4$ and $12 \div 3 = 4$. As this example shows, multiplication and division are inverse operations, just like addition and subtraction. Consequently, we can use multiplication and fact families to learn division. For example, if $4 \cdot 7 = 28$, then $28 \div 7 = 4$ and $28 \div 4 = 7$.

The definition, however, does not guarantee that the quotient of any two whole numbers is another whole number. An example is $7 \div 4 = c$, because there is no whole number c such that $7 = 4 \cdot c$. As a consequence, whole-number division does not satisfy the closure property. Nor does it satisfy the commutative, associative, or identity properties. Counterexamples to these properties are easy to find, so we leave them to the Learning Assessment. A distributive property for division over addition takes the form $(a + b) \div c = (a \div c) + (b \div c)$, provided $c \neq 0$ and the whole-number quotient exists for each division.

Division by Zero is Undefined

The definition of division also states that the divisor can be any whole number except zero. To understand why, consider the repeated-subtraction interpretation of $5 \div 0$ and the question, "How many times can zero be subtracted from 5?" Because subtracting zero never affects the value of the difference, we can subtract it once, twice, or even a hundred times. In fact, there is no definite number of times that we can subtract zero from another number, so we say that division by zero is undefined and do not allow it.

Theorem 4.9   Division by Zero

Division by zero is undefined.

The next example offers a formal proof of Theorem 4.9.

Example 4.39   Division by Zero Is Undefined

Reasoning   Use the definition of division to show that division by zero is undefined.

Solution To show that division by zero is undefined, we use a proof by cases. In other words, we separate the situation into different possibilities, which are then proven separately. If all of the cases hold true, then the theorem holds true.

Case 1: $a \div 0$, with $a \neq 0$. Suppose a is any nonzero whole number. By the definition of division, $a \div 0 = c$ if and only if $a = 0 \cdot c$. However, by the zero multiplication property, there is no whole number that can be multiplied by 0 to get the nonzero number a. As a result, $a \div 0$ is indeterminate or undefined.

Case 2: $0 \div 0$. By the definition of division, $0 \div 0 = c$ if and only if $0 = 0 \cdot c$. According to the zero multiplication property, this equation will be true for any whole number. However, the definition of division says that c must be unique. Because we cannot designate a specific whole number for c, we conclude that $0 \div 0$ is indeterminate or undefined.

Even though division by zero is undefined, we can divide zero by another nonzero whole number. In this case, $0 \div b = c$ if and only if $0 = b \cdot c$. The zero multiplication property tells us that any number multiplied by zero is zero. Because $b \neq 0$, we conclude that c must be, so $0 \div b = 0$. In other words, zero divided by any nonzero number is zero.

Division with Remainders

Even though the definition of division does not allow us to compute every combination of whole numbers, in some situations we do subdivide a set into equivalent subsets and leave some part remaining. For instance, suppose 14 pieces of candy are to be equally distributed among 4 friends. After each person receives 3 pieces, 2 pieces would remain (Figure 4.22).

Explorations Manual 4.10

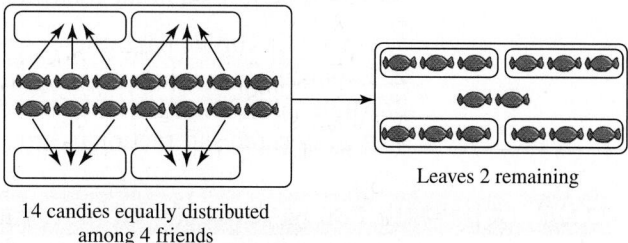

14 candies equally distributed
among 4 friends

Leaves 2 remaining

Figure 4.22 Distributing 14 pieces of candy to four friends

Because these situations are common, we need to extend division beyond its definition to include them. We do so by using the division algorithm.

Theorem 4.10   Division Algorithm for Whole-Number Division

For any whole numbers a and b with $b \neq 0$, there exists unique whole numbers q and r such that $a = bq + r$, where $0 \leq r < b$.

In the formula for the division algorithm, a is the dividend, b is the divisor, q is the quotient, and r is the **remainder.** Formally, the division algorithm states that the quotient times the divisor added to the remainder is equal to the dividend. Informally, it is a way to express any amounts that are "left over" after computing the quotient. Notice that the remainder is always less than the divisor; that is, r is always limited to a value from the set $\{0, 1, 2, \ldots b - 1\}$. If $r = 0$, then we have the earlier definition of division.

Example 4.40   Using the Division Algorithm

Find the quotient and remainder for each problem.

a. $19 \div 5$ **b.** $37 \div 7$

Solution

a. Using repeated subtraction, $19 - 5 - 5 - 5 = 4$. Hence, the quotient is $q = 3$, and the remainder is $r = 4$.

b. Because $7 \cdot 5 = 35$, we have $37 = 35 + 2 = 7 \cdot 5 + 2$. As a result, $q = 5$ and $r = 2$.

Order of Operations

In many situations, we use combinations of addition, subtraction, multiplication, and division. In such cases, the question is in what order should the operations be performed? For instance, given $6 \cdot 4 - 2$, which is computed first: the product or the difference? By trying both, we see that order does matter. If we multiply first, we get $6 \cdot 4 - 2 = 24 - 2 = 22$. If we subtract first, we get $6 \cdot 4 - 2 = 6 \cdot 2 = 12$. To find out which answer is correct, we write the product as repeated addition: $6 \cdot 4 - 2 = 4 + 4 + 4 + 4 + 4 + 4 - 2 = 22$. Because this matches the answer when we computed the product first, it makes sense that we should compute products and quotients before sums and differences.

A similar question arises with exponents. For instance, consider $6^3 \div 3$. Computing the exponent first, we have $6^3 \div 3 = 216 \div 3 = 72$. Computing the quotient first, we get $6^3 \div 3 = 2^3 = 8$. Again, which is correct? Because exponents represent repeated multiplication, the problem can be rewritten as $6^3 \div 3 = 6 \cdot 6 \cdot 6 \div 3 = 72$. Because this matches the answer when the exponent was computed first, it makes sense that we should compute exponents before products and quotients.

Combining these two examples gives us an order for computing operations, which can be suspended by means of parentheses or other grouping symbols. Operations should be performed in the following order:

1. Operations inside parentheses or other grouping symbols
2. Exponents
3. Multiplication and division in order from left to right
4. Addition and subtraction in order from left to right

Example 4.41   Using the Order of Operations

Use the order of operations to compute $(4 + 6)^2 \div 2 \cdot 5 + 7$.

Solution

$$(4 + 6)^2 \div 2 \cdot 5 + 7$$

$= 10^2 \div 2 \cdot 5 + 7$	Add $4 + 6$.
$= 100 \div 2 \cdot 5 + 7$	Square 10.
$= 50 \cdot 5 + 7$	Divide by 2.
$= 250 + 7$	Multiply by 5.
$= 257$	Add 7.

Check Your Understanding 4.3-B

1. Rewrite each expression in exponential form.

    a. $3 \cdot 3 \cdot 3$ **b.** $7 \cdot 7 \cdot 7 \cdot 7 \cdot 7 \cdot 7$ **c.** $2 \cdot a \cdot 2 \cdot a \cdot 2 \cdot a \cdot 2 \cdot a$

2. Use the definitions and properties of exponents to simplify each expression.

    a. $4^3 \cdot 4^3$ **b.** $14 \cdot 3^0$ **c.** $(6^3)^9$ **d.** $8^6 \cdot y^6$

3. Use repeated subtraction to interpret and compute each quotient.

    a. $8 \div 4$ **b.** $14 \div 7$ **c.** $18 \div 3$

4. Use the missing-factor definition of division to write each problem in terms of multiplication and then solve.

    a. $15 \div 3$ **b.** $34 \div 17$ **c.** $78 \div 6$

5. Find the quotient and remainder for each division problem.

    a. $19 \div 6$ **b.** $38 \div 9$ **c.** $67 \div 5$

Talk About It Consider the following story problem: "A ferry can hold 15 cars. How many trips does the ferry need to make to transfer 67 cars to the far side of a lake?" How does the remainder to the division affect the answer to the problem?

Representation

Activity 4.6

Draw a picture or use blocks to explain how the array model for multiplication can be used to represent and compute the following exponents. Based on what you see in the array model, explain why it makes sense to give the exponents of 2 and 3 the special names of "squared" and "cubed."

a. 4^2 **b.** 6^2 **c.** 7^2 **d.** 3^3 **e.** 5^3 **f.** 8^3

SECTION 4.3   Learning Assessment

■ Understanding the Concepts

1. Explain how to compute a product using each approach to multiplication.

    a. Repeated addition   **b.** Cartesian product

2. Why is the word "times" useful in describing multiplication?

3. Describe how each model is used to represent the repeated-addition approach to multiplication.

    a. Set model   **b.** Array model

    c. Number line model

4. What does each property imply about whole-number multiplication?

    a. Closure property

    b. Commutative property

    c. Associative property

    d. Identity property

    e. Distributive property

    f. Zero multiplication property

5. Why is 1 the multiplicative identity and not 0?

6. How are exponents related to multiplication?

7. Explain how to compute a quotient using each approach to division.

    a. Repeated subtraction

    b. Partitioning

    c. Missing factor

8. What does it mean for division by zero to be undefined? Why is this the case?

9. How does the division algorithm expand our ability to compute division?

10. Consider the standard order of operations.

    a. How is it possible to suspend the normal order of operations?

    b. Why are exponents computed before multiplication and division?

    c. Why are multiplication and division computed before addition and subtraction?

Representing the Mathematics

11. Write each product in terms of repeated addition, and then give a verbal interpretation of the repeated sum.

    a. $3 \cdot 11$ **b.** $6 \cdot 3$ **c.** $1 \cdot 7$ **d.** $4 \cdot 0$

12. Write each repeated sum in terms of multiplication.

    a. $6 + 6 + 6 + 6$ **b.** $3 + 3$
    c. 5 **d.** $0 + 0 + 0 + 0 + 0 + 0$

13. Use the set model and the repeated-addition approach to multiplication to represent and solve each product.

    a. $6 \cdot 3$ **b.** $7 \cdot 1$ **c.** $2 \cdot 5$

14. Use the array model to represent and solve each product.

    a. $9 \cdot 3$ **b.** $6 \cdot 5$ **c.** $4 \cdot 8$ **d.** $7 \cdot 2$

15. Use a tree diagram and the sets $A = \{1, 2, 3, 4, 5\}$ and $B = \{a, b, c\}$ to represent and compute the product $5 \cdot 3$.

16. What multiplication fact is demonstrated in each of the following diagrams?

    a.
    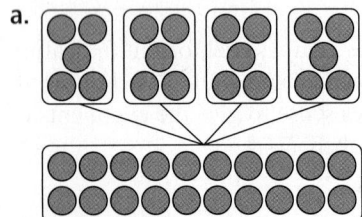

    b. 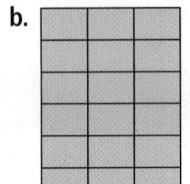   **c.**

    d.
    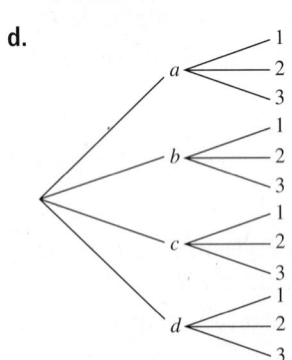

    e.
    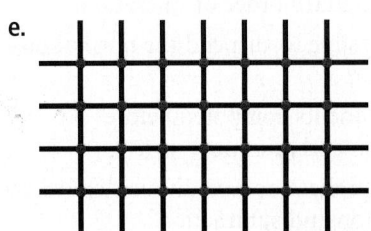

17. How is each property of multiplication shown in Table 4.4?

    a. Commutative
    b. Identity
    c. Zero multiplication

18. Rewrite each exponent in terms of repeated multiplication, and then give a verbal interpretation of the repeated product.

    a. 4^5 **b.** 2^7 **c.** $8^3 x^4$ **d.** $(x - 2)^2$

19. Rewrite each product in exponential form.

    a. $4 \cdot 4 \cdot 4 \cdot 4$
    b. $3 \cdot 3 \cdot 3 \cdot 3 \cdot 3 \cdot 3 \cdot 3 \cdot 3$
    c. $x \cdot y \cdot x \cdot y$
    d. $6 \cdot 6 \cdot 6 \cdot 6 \cdot x \cdot x \cdot x$
    e. $(3 + x)(3 + x)(3 + x)$
    f. $y^2 \cdot y^2 \cdot y^2 \cdot y^2$

20. Give a verbal interpretation of each property for exponents.

    a. $a^m \cdot a^n = a^{m+n}$
    b. $a^m \cdot b^m = (a \cdot b)^m$
    c. $(a^m)^n = a^{m \cdot n}$

21. Rewrite each quotient in terms of repeated subtraction, and then give a verbal interpretation of the repeated difference.

    a. $9 \div 3$ **b.** $28 \div 7$ **c.** $36 \div 4$ **d.** $18 \div 6$

22. Write each division fact in terms of a product with a missing factor. Solve the problem, and explain your answer in terms of multiplication.

    a. $12 \div 4 = ?$ **b.** $49 \div 7 = ?$
    c. $72 \div 9 = ?$ **d.** $39 \div 13 = ?$

23. **a.** Use sets and the repeated-subtraction approach to model $6 \div 2$ and $14 \div 2$.
    b. Use sets and the partitioning approach to model $8 \div 2$ and $15 \div 3$.
    c. Explain how your pictures in part a are different from those in part b.

24. Does the following diagram represent the repeated-subtraction or partitioning approach to division? Explain how you know.

    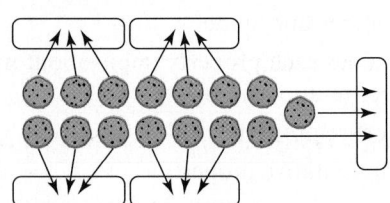

25. Represent each problem using only sums and differences. Use the new representation to compute.

    a. $6 \cdot 4 + 8$ **b.** $4^2 - 3 \cdot 5$

■ Exercises

26. State the property of multiplication that is demonstrated in each statement.

a. $4(2 + 1) = 4 \cdot 2 + 4 \cdot 1$.

b. $7 \cdot 3 = 3 \cdot 7$.

c. $6 \cdot 1 = 6$.

d. $4 \cdot (2 \cdot 7) = (4 \cdot 2) \cdot 7$.

e. $4 \cdot 5 = 20 \in W$.

f. $0 \cdot 9 = 0$.

g. $1(3 + 4) = 3 + 4$.

h. $2(3 + 4 + 5) = 2 \cdot 3 + 2 \cdot 4 + 2 \cdot 5$.

27. Which of the following sets is (are) closed under multiplication? Explain why or why not.

a. $\{2, 4, 6, 8, 10, \ldots\}$
b. $\{1, 3, 5, 7\}$

c. $\{0\}$
d. $\{0, 1\}$

e. $\{2, 3, 6, 9, 12, \ldots\}$
f. $\{1, 3^2, 3^3, 3^4, \ldots\}$

28. Use each set of numbers to verify that $a(b + c) = a \cdot b + a \cdot c$.

a. $a = 4, b = 5, c = 7$
b. $a = 2, b = 8, c = 3$

29. Use the distributive property to rewrite each statement without parentheses. Complete the computation.

a. $5 \cdot (1 + 3)$
b. $(4 + 7) \cdot 9$

c. $2 \cdot (8 - 6)$
d. $3 \cdot (4 + 2 + 7)$

30. Use the distributive property to factor each expression.

a. $4 \cdot 3 + 4 \cdot 2$
b. $3x + 3 \cdot 3$

c. $6x - 6$
d. $3 \cdot (x + 3) + y(x + 3)$

e. $x^2 - 3x$
f. $4x^2 - 2x$

31. Simplify each expression. Leave your answer in exponential form.

a. $6^2 \cdot 6^7$
b. 158^0
c. $(3^3)^4$

d. $2^2 \cdot 5^2$
e. $3^2 \cdot x^2$

32. Simplify each expression. Leave your answer in exponential form.

a. $(9 - 6)^0$
b. $(3^2 \cdot 4^2)^3$

c. $16^2 \cdot 4^3$
d. $25^2 - 5^4$

e. $[(27^3 - 9^2)^0]^3$

33. Find the value of x that makes the given statement true.

a. $4^2 \cdot 4^x = 4^8$
b. $(2^x)^7 = 2^{21}$

c. $6^3 \cdot x^3 = 30^3$
d. $14^x = 1$

34. Determine which number is larger in each pair.

a. 3^2 or 2^3
b. 4^6 or 6^4
c. 8^0 or 0^8

35. Find the quotient and remainder for each division problem.

a. $22 \div 8$
b. $37 \div 7$
c. $49 \div 9$

36. Find the quotient and remainder for each division problem.

a. $83 \div 9$
b. $105 \div 14$
c. $148 \div 12$

37. List the other facts in each fact family.

a. $15 \div 5 = 3$.
b. $8 \cdot 6 = 48$.

c. $75 \div 15 = 5$.
d. $7 \cdot 7 = 49$.

38. Use the order of operations to compute.

a. $8 \cdot 9 + 12 \div 6$
b. $9 + 3^2 \cdot 6$

c. $[(34 - 13) \div 7]^2$

39. Use the order of operations to compute.

a. $6^2 \div 4 - (3 + 5)$
b. $[(3 + 4)^2 - 6]^2$

c. $(9^2 - 63) \cdot (3 + 5) \div 36$

■ Problems and Applications

40. Is 2^x ever equal to x^2? If so, for what number(s)?

41. Which is greater: $3^5 + 3^5$ or 3^{10}?

42. Order the following from greatest to least: 2^4, 15, $3 \cdot 2^2$, $2 \cdot 3^2$, $4^2 + 1$.

43. Find the missing whole number in each expression.

a. $x \div 4 = 9$ remainder 3.

b. $49 \div x = 8$ remainder 1.

c. $56 \div 3 = x$ remainder 2.

44. How many divisions without remainders can be computed using only the numbers from the set $A = \{2, 3, 4, 6, 8, 10\}$?

45. Find values for a, b, and c so that the statement $a \div (b \div c) = (a \div b) \div c$ holds true.

46. a. If $36 \div n = c$ for $n \le 36$, what numbers can n be? What numbers can c be?

b. If $36 \div n = c$ remainder r for $n \le 36$, what numbers can n be? What numbers can c be? What numbers can r be?

47. Place parentheses in each equation to make the given statement true.

a. $36 \div 4 + 5 \cdot 8 = 32$.

b. $13 - 6 \cdot 8 \div 2 = 28$.

c. $6 + 4 + 5 \div 3 = 5$.

48. A rectangular room measures 14 ft by 12 ft. How many tiles are needed to cover the floor if each tile is 1 ft²?

49. An ice cream store has 8 different flavors, 2 different types of cones, and 3 different toppings. How many different ways are there to make an ice cream cone?

50. Ten friends order 4 pizzas. If each pizza has 8 slices and every person gets the same number of slices, how many slices does each person get, and how many are left over?

51. Brooke has 8 different playlists on her mp3 player, each having 15 songs. If she equally divides these songs among 10 playlists, how many songs will be in each?

52. Pick any number. Double it, and then add 14. Divide by 2, and subtract the original number. The result is always 7. Can you explain why?

53. A particular town has 6 subdivisions. Each subdivision has 6 streets, and each street has 6 houses. If 6 people live in each house, how many people live in the town?

■ Mathematical Reasoning and Proof

54. Use the Cartesian product definition of multiplication and the sets $A = \{1, 2, 3, 4, 5\}$ and $B = \{a, b, c\}$ to show that $5 \cdot 3 = 15$.

55. **a.** Is the set of whole numbers without 3 closed under multiplication? Explain.

    b. Is the set of whole numbers without 6 closed under multiplication? Explain.

56. Give an example of a set of numbers that is:

    a. Closed under both addition and multiplication.

    b. Closed under multiplication but not addition.

    c. Not closed under either multiplication or addition.

57. Use properties of the Cartesian product to show that multiplication satisfies the commutative property.

58. Use the Cartesian product definition of multiplication to show that $4 \cdot 0 = 0$.

59. If $a \cdot b = 0$, what must be true about either a or b? Why?

60. If a and b are whole numbers, what must be true about a and b if $a \cdot b = 1$?

61. Let a, b, m, and n be any whole numbers with $a \neq 0$ and $b \neq 0$. Use the definition of an exponent to prove each property.

    a. $a^m \cdot a^n = a^{m+n}$. **b.** $(a^m)^n = a^{m \cdot n}$.

62. The following four expressions represent common mistakes made with multiplication and exponents. Provide a numerical example that shows each is false.

    a. $a \cdot (b \cdot c) = (a \cdot b) \cdot (a \cdot c)$.

    b. $a + (b \cdot c) = (a + b) \cdot (a + c)$.

    c. $(a + b)^n = a^n + b^n$.

    d. $(a - b)^n = a^n - b^n$.

63. For each property, provide a counterexample that shows the property does not hold for whole-number division.

    a. Closure property   **b.** Commutative property

    c. Associative property **d.** Identity property

64. Find values for a, b, and c such that $(a + b) \div c = (a \div c) + (b \div c)$ is true. Is this equation always true? Explain.

65. Let a and n be whole numbers such that $n \leq a$. What are the possible remainders when a is divided by:

    a. $n = 5$? **b.** $n = 6$? **c.** $n = 9$? **d.** Any n?

    e. What fact allows you to guarantee your answer?

■ Mathematical Communication

66. **a.** Write a multiplication story problem that can be solved by using repeated addition.

    b. Write a multiplication story problem that can be solved by using a Cartesian product.

    c. Describe how you had to change your language to indicate each approach.

67. What interpretation of division is represented in each story problem? Explain how you know.

    a. John has 12 oranges. He put them into groups of 3 oranges each. How many groups were there?

    b. Joan paid $1.35 for 3 notepads. What was the cost of one notepad?

    c. Peter walked 9 mi at rate of 3 mi/hr. How many hours did it take Peter to walk the 9 mi?

68. **a.** Write a division story problem that can be solved by using repeated subtraction.

    b. Write a division story problem that can be solved by using partitioning.

    c. Write a division story problem that can be solved by using the missing-factor approach.

    d. Describe how you had to change your language to indicate each approach.

69. A student uses the distributive property to make the following calculation: $4 + (3 \cdot 5) = (4 \cdot 3) + (4 \cdot 5) = 12 + 20 = 32$. How do you respond to what he has written?

70. A student asks you, "Is 15 divided by 3 and 3 divided by 15 the same?" How would you respond?

71. Jocelyn states that $5^3 \cdot 5^2 = 25^5$. Is she correct? Why or why not?

72. Write a story problem that uses the quotient $7 \div 0$. How could you use the context of the story problem to explain that division by zero is undefined?

■ Building Connections

73. Write the first five place values of the decimal system using exponents. What pattern do you notice? Why does it make sense that this particular pattern occurs?

74. Write a paragraph or two explaining how sets, addition, and multiplication are interconnected. Specifically, explain how set operations and their properties are useful in defining and computing multiplication. Similarly, describe how whole-number addition and its properties are useful in defining and computing multiplication.

75. Complete the diagram by drawing arrows to show how the four operations are interconnected. Explain why you chose to draw the arrows as you did.

Addition	Subtraction
Multiplication	Division

76. What kind of numbers must we have in order for division to satisfy the closure property?

77. As children work with the partitioning approach to division, they should begin to see that division is used to represent a part-to-whole relationship. Thinking of division in this way connects division to what kind of numbers?

78. Examine several textbooks from second, third, or fourth grade, specifically looking for story problems involving division. What approach to division do these problems take? Can you find story problems that use the repeated-subtraction approach to division? What about the missing-factor approach and the partitioning approach?

SECTION 4.4 Multiplying and Dividing Large Numbers

We can use a variety of methods to multiply and divide large numbers, including written algorithms, mental computation, estimation, and calculators.

Written Algorithms for Multiplication

Explorations
Manual
4.11

Like addition and subtraction, we can use several written algorithms to multiply large numbers. Although the algorithms take on different forms, they all use the same basic procedures: multiplying single-digit numbers, regrouping, and multidigit addition. We begin by multiplying with base-ten blocks.

Example 4.42 Multiplying with Base-Ten Blocks

Representation

Use base-ten blocks to find the product of $13 \cdot 25$.

Solution We use an extension of the array model in which our objects now take on different values. First, represent 13 with 1 long and 3 units, listed down the left side of the array. Similarly, represent 25 with 2 longs and 5 units, listed across the top. Next, we fill in the array by considering the products of different combinations of base-ten blocks:

- A long by a long = $10 \cdot 10 = 100$ or 1 flat
- A long by a unit = $10 \cdot 1 = 10$ or 1 long
- A unit by a unit = $1 \cdot 1 = 1$ unit

Here is the finished array:

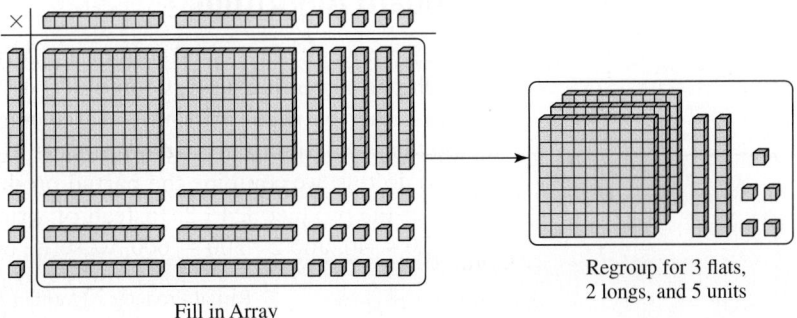

Fill in Array

Regroup for 3 flats,
2 longs, and 5 units

After regrouping, we have 3 flats, 2 longs, and 5 units, so we conclude that $25 \cdot 13 = 325$.

The last example demonstrates two things about multiplication. First, products of large numbers still have repeated-addition interpretations. For instance, $13 \cdot 25$ represents 25 added to itself 13 times. The 13 groups of 25 run down the array. Second, the base-ten blocks show how the place value of two digits can affect the value of their product. Numerically, we can show this effect by writing the product in expanded notation. For example:

$13 \cdot 25 = (10 + 3) \cdot 25$	Expanded notation of 13.
$= 10 \cdot 25 + 3 \cdot 25$	Distributive property.
$= 10 \cdot (20 + 5) + 3 \cdot (20 + 5)$	Expanded notation of 25.
$= 10 \cdot 20 + 10 \cdot 5 + 3 \cdot 20 + 3 \cdot 5$	Distributive property.
$= 200 + 50 + 60 + 15$	Multiplication.
$= 325$	Addition.

Partial Product Algorithm

The array of base-ten blocks and the expanded notation indicate that we compute the product of two large numbers by multiplying each digit in one factor by each digit in the other. If we maintain the place value of each digit as we multiply, then we get **partial products** that we add together to get our final answer. Multiplying in this way is called the **partial products algorithm.** Figure 4.23 shows how to use this algorithm to compute $13 \cdot 25$, as well as the connection between multiplying with base-ten blocks and partial products.

Notice that the partial products are affected only by the place values of the digits and not by other partial products. So the algorithm can be worked from right to left or from left to right. Also, the partial products can be written without placeholders as long as the nonzero digits stay in their correct place values.

Partial Product Algorithm

$$\begin{array}{rl} 25 & \\ \times\ 13 & \\ \hline 15 & = 3 \cdot 5 \\ 60 & = 3 \cdot 20 \\ 50 & = 10 \cdot 5 \\ +\ 200 & = 10 \cdot 20 \\ \hline 325 & \end{array}$$

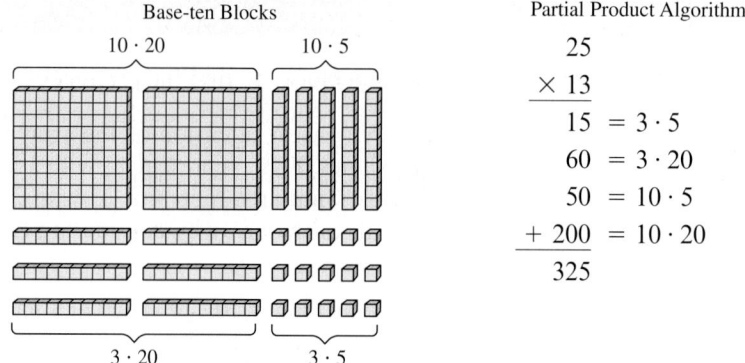

Figure 4.23 Base-ten blocks and partial products

Standard Algorithm

One advantage of the partial products algorithm is that it clearly identifies the value of each partial product as it is computed. However, if each factor has a large number of digits, then we necessarily have a large number of partial products. One way to condense them is to limit the partial products to just one for each digit in the second factor. To do this, we combine the partial products as they are multiplied. For instance, consider the product $324 \cdot 2$. Instead of writing the three partial products of $2 \cdot 4 = 8$, $2 \cdot 20 = 40$, and $2 \cdot 300 = 600$, we write only their sum: 648 (Figure 4.24).

Partial Products Algorithm

$$\begin{array}{rl} 324 & \\ \times\ 2 & \\ \hline 8 & = 2 \times 4 \\ 40 & = 2 \times 20 \\ +\ 600 & = 2 \times 300 \\ \hline 648 & \end{array} \xrightarrow{\text{Condense partial products}} \begin{array}{r} 324 \\ \times 2 \\ \hline 648 \end{array}$$

Figure 4.24 Condensing partial products

This is equivalent to writing each single-digit product in the proper place value. Specifically, the product of $2 \cdot 4 = 8$ is in the ones place, $2 \cdot 2 = 4$ is in the tens place, and $2 \cdot 3 = 6$ is in the hundreds place. As long as we line up the numbers properly, this procedure takes care of the place value, and all we have to do is multiply single-digit numbers. By combining all the condensed partial products from the second factor, we get the **standard algorithm** for multiplication.

Example 4.43 **Multiplying with the Standard Algorithm**

Use the standard algorithm to multiply $371 \cdot 63$.

Solution We multiply 371 by 3, condensing the partial sums as we multiply. Notice when we multiply $3 \cdot 7$, we get a value larger than 10, namely 21. So we must regroup and move the two tens to the next highest place value. Consequently, when we multiply the next pair of digits, an additional 2 has to be added to the product. Here is the completed first step:

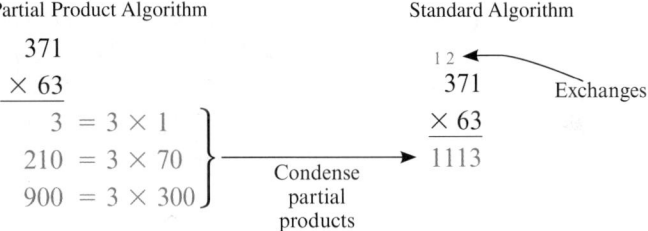

We repeat the process with the 6 in the second factor. However, because it is in the tens place value, it actually has a value of 60. To continue multiplying with a single-digit number, 6, we put a placeholder zero in the ones place value. This moves the product one place value to the left, having the same effect as multiplying it by 10.

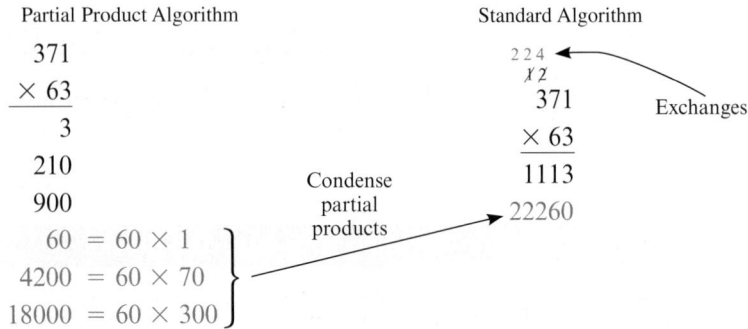

We now add the partial products to obtain the final product.

Partial Product Algorithm

$$
\begin{array}{r}
371 \\
\times\ 63 \\
\hline
3 \\
210 \\
900 \\
60 \\
4200 \\
+\ 18000 \\
\hline
23373
\end{array}
$$

Standard Algorithm

$$
\begin{array}{r}
^{2\,2\,4} \\
^{\not{1}\,\not{2}} \\
371 \\
\times\ 63 \\
\hline
1113 \\
+\ 22260 \\
\hline
23373
\end{array}
$$

Lattice Method

To use the lattice method to multiply large numbers, extend the lattice so that there is one row for each digit in the second factor.

Example 4.44   Multiplying with the Lattice Method

Use the lattice method to multiply 359 · 73.

Solution Because the second factor has two digits, we create two rows in the lattice. List 359 across the top and 73 down the right side, making sure to place 7 on top to get the digits in the correct place values. Next, fill in the lattice by multiplying digit by digit. As in addition, we place the tens digit in the upper triangle and the ones digit in the lower. We then add down each diagonal from right to left, regrouping as necessary.

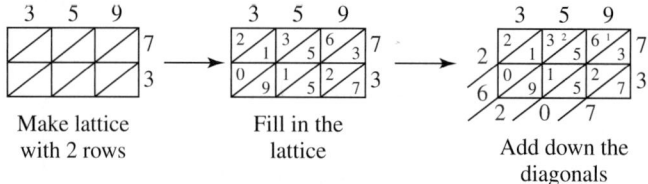

| Make lattice with 2 rows | Fill in the lattice | Add down the diagonals |

The final product is read from left to right around the bottom of the lattice: 359 · 73 = 26,207.

■

Written Algorithms for Division

Explorations Manual 4.12

Many of the written algorithms for whole-number division are based on the repeated-subtraction approach, which can be modeled using base-ten blocks.

Example 4.45   Dividing with Base-Ten Blocks

Representation

Use base-ten blocks and repeated subtraction to divide 288 ÷ 36.

Solution We represent 288 with 2 flats, 8 longs, and 8 units. In order to remove groups of 36 (3 longs and 6 units), we exchange the 2 flats for longs and 4 longs for units.

288 is represented as

Exchange 2 flats for 20 longs
and 4 longs for 40 units

After making the exchanges, we remove groups of 3 longs and 6 units each.

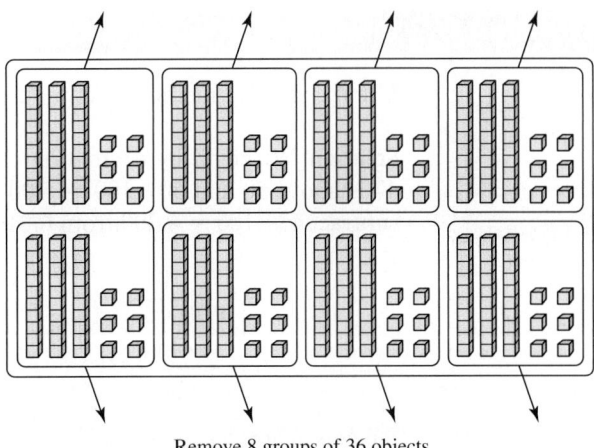

Remove 8 groups of 36 objects

Because we can remove 8 such groups, we conclude that $288 \div 36 = 8$.

Repeated Subtraction Algorithm

We can also use repeated subtraction to compute a quotient numerically.

Example 4.46 Dividing by Repeated Subtraction

Use repeated subtraction to compute $212 \div 33$.

Solution We compute $212 \div 33$ by subtracting 33 from 212 until we reach either zero or a number smaller than 33. Because we can subtract 33 from 212 six times, then $212 \div 33 = 6$ with a remainder of 14.

$$
\begin{array}{r}
33\overline{)212} \\
-33 \\
\hline
179 \\
-33 \\
\hline
146 \\
-33 \\
\hline
113 \\
-33 \\
\hline
80 \\
-33 \\
\hline
47 \\
-33 \\
\hline
14
\end{array}
$$

Scaffold Method

Explorations
Manual
4.13

Explorations
Manual
4.14

Although repeated subtraction works, the large number of differences makes it inefficient. Condensing the number of differences would improve the method, and we do so by using the **scaffold method,** which subtracts multiples of the divisor rather than just the divisor itself. To find the multiples to subtract, we use estimations. Because estimations vary from person to person, the multiples we use will also vary. Even so, we will still arrive at the same quotient in the end.

Example 4.47 | **Dividing with the Scaffold Method**

Use the scaffold method to compute $6{,}016 \div 47$.

Solution We begin by estimating the number of times 47 goes into 6,016. Because 47 is about 50 and $50 \cdot 100 = 5{,}000$, 47 should go into 6,016 at least 100 times. We subtract $47 \cdot 100 = 4{,}700$ from 6,016, and record 100 above the dividend.

$$
\begin{array}{r}
100 \\
\hline
47\overline{)6016} \\
-\ 4700 \\
\hline
1316
\end{array}
$$
Subtract an estimate of $47 \cdot 100 = 4{,}700$.

We continue to estimate and subtract by always asking ourselves how many times 47 goes into the remaining number. We do so until we reach 0 or a number smaller than 47. For each estimate, record the number multiplied by 47 above the dividend. After completing all the differences, add the numbers above the dividend to find the quotient: in this case, $100 + 20 + 7 + 1 = 128$. Consequently, $6{,}016 \div 47 = 128$.

$$
\begin{array}{r}
128 \\
\hline
1 \\
7 \\
20 \\
100 \\
\hline
47\overline{)6016} \\
-\ 4700 \\
\hline
1316 \\
-\ 940 \\
\hline
376 \\
-\ 329 \\
\hline
47 \\
-\ 47 \\
\hline
0
\end{array}
$$
Subtract an estimate of $47 \cdot 100 = 4{,}700$.

Subtract an estimate of $47 \cdot 20 = 940$.

Subtract an estimate of $47 \cdot 7 = 329$.

The scaffold method is popular in the classroom because it maintains place values throughout the algorithm and shows how division relies on both subtraction and multiplication. It also serves as an intermediate algorithm to the standard algorithm.

Standard Algorithm

The scaffold algorithm can be made more efficient in two ways. One is to remove the placeholder zeros. Because placeholders add no value to the computation, they can be removed as long as the nonzero digits are in the correct place. For instance, in the last problem we estimated that 47 went into 6,016 about 100 times. As long as we kept our estimate in the proper place from left to right, we could have also estimated the number of times 47 goes into 601 or the number of times 47 goes into 60 and still arrived at the same answer (Figure 4.25). This means we can condense the algorithm by working with only a few digits at a time.

$$
\begin{array}{ccc}
100 & 10 & 1 \\
47\overline{)6016} & 47\overline{)6016} & 47\overline{)6016} \\
-\ 4700 & -\ 470 & -\ 47
\end{array}
$$

Figure 4.25 Equivalent estimates

Another way to make the scaffold method more efficient is to work with exact values rather than with estimates. Doing so has the advantage of requiring fewer subtractions. We can find the exact number of times one number goes into another by computing products off to the side. If we combine these changes to the scaffold method, we get the **standard algorithm,** which is also known as **long division.**

Example 4.48 Dividing with the Standard Algorithm

Use the standard algorithm to compute $8{,}878 \div 26$.

Solution We first determine how many times 26 will go into 88 by computing several products with 26 to find which is closest to 88.

$$\begin{array}{ccc}
26 & 26 & 26 \\
\underline{\times\, 2} & \underline{\times\, 3} & \underline{\times\, 4} \\
52 & 78 & 104
\end{array}$$

Because 26 goes into 88 three times, we place a 3 above the second 8 and subtract 78 from 88.

$$\begin{array}{r}
3 \\
26\overline{)8878} \\
\underline{-\,78} \\
10
\end{array}$$

We bring down the next digit, 7, and repeat the process. Because 26 goes into 107 four times, we place a 4 above the 7, and subtract $4 \cdot 26 = 104$ from 107.

$$\begin{array}{r}
341 \\
26\overline{)8878} \\
\underline{-\,78} \\
107 \\
\underline{-\,104} \\
3
\end{array}$$

In the final step, we bring down the last 8, and repeat the process. Because 26 goes into 38 only one time, subtract 26 from 38 to complete the division. The quotient is the number above the dividend, and any number left after the final subtraction is the remainder. As a result, $8{,}878 \div 26 = 341$, remainder 12.

$$\begin{array}{r}
341 \\
26\overline{)8878} \\
\underline{-\,78} \\
107 \\
\underline{-\,104} \\
38 \\
\underline{-\,26} \\
12
\end{array}$$

Check Your Understanding 4.4-A

1. Use expanded notation to compute each product.

   a. $28 \cdot 13$ **b.** $57 \cdot 34$

2. Use the lattice method to compute each product.

   a. $98 \cdot 71$ **b.** $471 \cdot 62$

3. Use the standard algorithm to compute each product or quotient.

    a. 231 · 84   **b.** 357 · 93   **c.** 5,271 ÷ 21   **d.** 6,603 ÷ 53

4. Use repeated subtraction to compute each quotient.

    a. 228 ÷ 57   **b.** 312 ÷ 39

5. Use the scaffold method to compute each quotient.

    a. 3,336 ÷ 24   **b.** 8,435 ÷ 35

Talk About It Which of the written algorithms do you think best represents the conceptual ideas behind multiplying large numbers? What about dividing large numbers? Which algorithms do you think are easiest to use?

Representation

Activity 4.7

1. Use base-ten blocks to model and solve each product.

    a. 25 · 6   **b.** 32 · 24

2. Use base-ten blocks and repeated subtraction to model and solve each quotient.

    a. 90 ÷ 18   **b.** 161 ÷ 23

Mental Multiplication and Division

 Explorations Manual 4.15

Many of the strategies for adding and subtracting mentally can also be used with multiplication.

Example 4.49   **Computing Products Mentally**

Compute each product mentally.

a. 36 · 25   **b.** 59 · 8   **c.** 5 · 25 · 9 · 2 · 4

Solution

a. Factor 36 into 9 · 4, and think 36 · 25 = (9 · 4) · 25 = 9 · (4 · 25) = 9 · 100 = 900.

b. The single-digit factor in this problem makes working left to right a reasonable strategy. First multiply 50 · 8 = 400, and then add 9 · 8 = 72. As a result, 59 · 8 = 400 + 72 = 472.

c. Use compatible numbers. By using the commutative and associative properties, we have 5 · 25 · 9 · 2 · 4 = (5 · 2) · (25 · 4) · 9 = 10 · 100 · 9 = 9,000.

Another strategy is to use the distributive property to split one factor into two numbers that are easier to multiply. After we distribute, we compute the two products and then add.

Example 4.50   **Buying Uniforms**

Application

Coach Wilson is buying uniforms for her softball team. She has 17 players, and the uniforms cost $41 each. If she does not have calculator, how can she compute the cost of the uniforms?

Solution Coach Wilson can compute $17 \cdot 41$ by using the distributive property. Specifically, she can think $17 \cdot 41 = (10 + 7) \, 41 = 10 \cdot 41 + 7 \cdot 41 = 410 + 41 \cdot 7$. To compute $41 \cdot 7$, she can work from left to right: $41 \cdot 7 = 280 + 7 = 287$. Finally, she can add the products to find the cost of the uniforms: $41 \cdot 17 = 410 + 287 = 697$.

The strategies for mental division are more limited. The next example shows how two different strategies, *breaking apart the dividend* and *compatible numbers*, can be used when the divisor is a single digit.

Example 4.51   Computing Quotients Mentally

Compute each quotient mentally.

a. $6{,}432 \div 8$ **b.** $258 \div 6$

Solution

a. The number 6,432 can be split into two numbers that are both divisible by 8: 6,400 and 32. Using the distributive property, $6{,}432 \div 8 = (6{,}400 + 32) \div 8 = (6{,}400 \div 8) + (32 \div 8) = 800 + 4 = 804$.

b. In the quotient $258 \div 6$, we can also split the dividend, but this time to make compatible numbers: $258 \div 6 = (240 + 18) \div 6 = (240 \div 6) + (18 \div 6) = 40 + 3 = 43$.

Estimating Multiplication and Division

We can also estimate multiplication and division using the techniques learned in addition and subtraction.

Example 4.52   Using Compatible Numbers

Use compatible numbers to estimate each problem.

a. $974 \cdot 57$ **b.** $6{,}287 \div 83$

Solution

a. Use flexible rounding to round 974 up to 1,000 and 57 down to 50. Then multiply to get the estimate: $974 \cdot 57 \approx 1{,}000 \cdot 50 = 50{,}000$.

b. Round 6,287 up to 6,300 and 83 up to 90 and estimate $6{,}287 \div 83 \approx 6{,}300 \div 90 = 70$.

Example 4.53   Estimating the Number of Microchips

Application

A computer warehouse contains 4,571 crates, each of which holds 3,850 microchips. About how many microchips are in the warehouse?

Solution We first make a front-end estimate: $4{,}571 \cdot 3{,}802 \approx 4{,}000 \cdot 3{,}000 = 12{,}000{,}000$. For a better estimate, we take the second digit of each number into consideration and multiply: $571 \cdot 802 \approx 500 \cdot 800 = 400{,}000$. By adding the two products, we estimate there are about 12,400,000 microchips in the warehouse.

Multiplying and Dividing on Calculators

We can compute products and quotients of large numbers easily and efficiently by using the $\boxed{\times}$ and $\boxed{\div}$ keys on a calculator. Many calculators also have two or more keys dedicated to exponents, including the square key, $\boxed{x^2}$, and a general exponent key, $\boxed{\wedge}$ or $\boxed{y^x}$. Computations involving these single operations are straightforward. However, if a computation contains several operations, we must be careful to consider the order in which the calculator performs them. Most modern calculators now use **algebraic logic;** that is, they are programmed to compute operations according to the standard order. If necessary, we can supersede the order by means of the parentheses keys. Other calculators, like the four-function calculator found on a cellular phone, use **arithmetic logic;** the calculator performs the operations in the order in which they are entered. It is up to the user to enter the operations correctly.

Example 4.54 Using a Calculator

Use a calculator to compute each problem.

a. $1{,}768 \cdot 439$ **b.** $1{,}145{,}628 \div 789$

c. 17^2 **d.** $(16 - 9)^3 \cdot (18 + 21)$

Solution The following sequences of keystrokes will lead to correct answers on most calculators.

a. 1768 $\boxed{\times}$ 439 $\boxed{=}$ displays 776152.

b. 1145628 $\boxed{\div}$ 789 $\boxed{=}$ displays 1452.

c. We can use one of three possible key sequences to obtain the answer: 17 $\boxed{\times}$ 17 $\boxed{=}$, 17 $\boxed{x^2}$ $\boxed{=}$, or 17 $\boxed{\wedge}$ 2 $\boxed{=}$ all display 289.

d. If the calculator has parentheses keys, we can enter this problem as

$$\boxed{(}\ 16\ \boxed{-}\ 9\ \boxed{)}\ \boxed{\wedge}\ 3\ \boxed{\times}\ \boxed{(}\ 18\ \boxed{+}\ 21\ \boxed{)}\ \boxed{=}.$$

Or, if the calculator has memory, we can also enter:

$$16\ \boxed{-}\ 9\ \boxed{=}\ \boxed{\wedge}\ 3\ \boxed{STO}\ \boxed{CE}\ 18\ \boxed{+}\ 21\ \boxed{=}\ \boxed{\times}\ \boxed{RCL}\ \boxed{=}.$$

Either way, the calculator displays 13377.

Computing quotients with remainders is quite common. If we use the division key, $\boxed{\div}$, the answer is usually expressed as a decimal. However, some calculators are equipped with an integer divide key, $\boxed{INT \div}$, which returns the answer as a quotient and a remainder. For instance, if we compute $1{,}480{,}690 \div 367$ using the integer divide key, we press:

$$1480690\ \boxed{2nd}\ \boxed{INT \div}\ 367\ \boxed{=}$$

The calculator displays 4034, remainder 212. If the calculator does not have the $\boxed{INT \div}$ key, we can still use it to find the quotient and remainder.

Example 4.55 Finding a Quotient and Remainder on a Calculator

Use the division key on a calculator to find the quotient and remainder of $98{,}652 \div 45$.

Solution We first divide 98,652 by 45 by pressing 98652 $\boxed{\div}$ 45 $\boxed{=}$, which gives us a quotient of approximately 2192.2667. The whole number part, 2,192, represents the quotient. To find the remainder, take 2,192, multiply it by 45, and subtract the result from 98,652, or 98652 $\boxed{-}$ 2192 $\boxed{\times}$ 45 $\boxed{=}$. The display gives us 12, so $98{,}652 \div 45$ is 2,192, remainder of 12.

Check Your Understanding 4.4-B

1. Compute each product or quotient mentally. Describe the strategy you used.

    a. $32 \cdot 40$ **b.** $45 \cdot 36$ **c.** $3{,}684 \div 6$ **d.** $225 \div 9$

2. Estimate each product or quotient. Describe the strategy you used.

    a. $6{,}782 \cdot 35$ **b.** $34{,}571 \cdot 198$ **c.** $5{,}589 \div 72$ **d.** $809 \div 87$

3. Use a calculator to compute the following.

    a. $6{,}091 \cdot 5{,}363$ **b.** 14^3 **c.** $50{,}700 \div 156$ **d.** $(93 - 47)^2 + 56 \cdot 87$

Talk About It What happens if you enter $8 \div 0$ on a calculator? Why does this occur?

Problem Solving

Activity 4.8
Place parentheses in each equation to make it true.

a. $4 \cdot 3^2 - 8 + 6 = 130$ **b.** $16 \div 4 \div 2 \cdot 8 + 7 = 30$

c. $6 + 3 - 5^2 \cdot 4 = 64$ **d.** $4 + 8 \cdot 15^2 \div 9 = 300$

SECTION 4.4 Learning Assessment

■ Understanding the Concepts

1. a. What basic procedures are used in most written algorithms for multiplication?

    b. What basic procedures are used in most written algorithms for division?

2. How does the standard algorithm for multiplication condense the partial products algorithm for multiplication?

3. What role do placeholder zeros play in the standard algorithm for multiplication?

4. Describe how to use the lattice method to multiply large numbers.

5. a. How does the scaffold algorithm for division condense the repeated subtraction algorithm for division?

    b. How does the standard algorithm for division condense the scaffold algorithm?

6. How are subtraction and multiplication used in the standard algorithm for whole-number division?

7. Describe how the distributive property can be used to compute products mentally?

8. What is the difference between algebraic and arithmetic logic on a calculator?

■ Representing the Mathematics

9. Verbally describe the repeated-addition interpretation of each product.

    a. $14 \cdot 31$ **b.** $37 \cdot 63$ **c.** $135 \cdot 29$

10. Use base-ten blocks and the array model to represent and compute each product. In each case, illustrate how the partial products are shown in your diagram.

    a. $12 \cdot 21$ **b.** $33 \cdot 26$ **c.** $104 \cdot 40$

11. Use base-ten blocks and repeated-addition to model and compute each product.

    a. $2 \cdot 78$ **b.** $3 \cdot 241$ **c.** $2 \cdot 1{,}013$

    d. What are the advantages and disadvantages of using the repeated-addition approach with base-ten blocks?

12. One disadvantage to using base-ten blocks to multiply is that any product over a certain size cannot be represented. For instance, there are no base-ten blocks that represent the product of a long and a block or a flat and a flat.

    a. What is the largest product that can be represented with base-ten blocks? Give a pair of factors that have a product equal to this number.

    b. Give two examples of products that cannot be computed with base-ten blocks.

13. Verbally describe the repeated-subtraction interpretation of each quotient.

    a. $96 \div 12$ **b.** $114 \div 38$ **c.** $1,134 \div 42$

14. Use base-ten blocks and repeated subtraction to compute each quotient.

    a. $23 \overline{)92}$ **b.** $42 \overline{)168}$ **c.** $123 \overline{)615}$

15. Use base-ten blocks and partitioning to compute each quotient.

    a. $3 \overline{)72}$ **b.** $4 \overline{)204}$ **c.** $6 \overline{)192}$

    d. What are the advantages and disadvantages of using the partitioning approach with base-ten blocks?

16. Exponents are commonly represented on a calculator in one of two ways: either $\boxed{\wedge}$ or $\boxed{y^x}$. Why are these reasonable symbols to use for the exponent key?

17. Write a possible sequence of keys you would press on your calculator to compute

    a. 5^3 using the exponent key.

    b. 4^4 without using the exponent key.

18. Write a possible sequence of keys you would press on your calculator to compute:

    a. $25 + 15 \div 5 \cdot 2$ if your calculator has algebraic logic.

    b. $25 + 15 \div 5 \cdot 2$ if your calculator has arithmetic logic.

■ Exercises

19. Use expanded notation to compute each product.

    a. $31 \cdot 12$ **b.** $63 \cdot 42$ **c.** $137 \cdot 68$

20. Use the partial products algorithm to compute each product.

    a. $47 \cdot 18$ **b.** $89 \cdot 27$

    c. $408 \cdot 63$ **d.** $579 \cdot 131$

21. Use the lattice method to compute each product.

    a. $45 \cdot 7$ **b.** $68 \cdot 53$

    c. $362 \cdot 79$ **d.** $432 \cdot 128$

22. Use repeated subtraction to compute each quotient.

    a. $43 \overline{)344}$ **b.** $67 \overline{)268}$

    c. $107 \overline{)545}$ **d.** $253 \overline{)1045}$

23. Use the scaffold method to compute each quotient.

    a. $36 \overline{)3888}$ **b.** $54 \overline{)3402}$

    c. $72 \overline{)4193}$ **d.** $47 \overline{)6504}$

24. Use the standard algorithm to compute each problem.

    a. $114 \cdot 91$ **b.** $476 \cdot 65$

    c. $15 \overline{)4005}$ **d.** $41 \overline{)2304}$

25. Use the standard algorithm to compute each problem.

    a. $1671 \cdot 43$ **b.** $782 \cdot 209$

    c. $37 \overline{)9284}$ **d.** $145 \overline{)3545}$

26. Compute each problem mentally. In each case, describe the method you used.

    a. $24 \cdot 18$ **b.** $43 \cdot 61$

    c. $9,633 \div 3$ **d.** $364 \div 7$

27. Compute each problem mentally. In each case, describe the method you used.

    a. $31 \cdot 27$ **b.** $901 \cdot 12$

    c. $30,450 \div 50$ **d.** $4,860 \div 12$

28. Estimate each product. Describe how you arrived at your estimate.

    a. $415 \cdot 29$ **b.** $18,901 \cdot 499$

    c. $97,031 \cdot 98$ **d.** $53,408 \cdot 987$

29. Estimate each quotient. Describe how you arrived at your estimate.

    a. $79,435 \div 103$ **b.** $5,398 \div 87$

    c. $67,983 \div 398$

30. Use estimation to determine which of the following products is incorrect.

    a. $453 \cdot 26 = 11,778$

    b. $19,362 \cdot 403 = 9,863,886$

    c. $1,036 \cdot 52 = 49,732$

    d. $8,736 \cdot 365 = 3,188,640$

31. Use estimation to order the following products from least to greatest.

    $37 \cdot 51$ $41 \cdot 59$ $29 \cdot 68$ $72 \cdot 41$ $48 \cdot 52$

32. Use a calculator to compute each problem.

    a. $13,410 \cdot 1,289$ **b.** 18^5

    c. $1,298,187 \div 891$ **d.** $12^2 + 18 \cdot 21$

33. Use a calculator to compute the each problem.

    a. $(18 \cdot 40 - 341)^3$ **b.** $13^2 + 63 \cdot 97 \div 9$

    c. $45 \cdot 98^2 - 64 \cdot 119$

34. Use a calculator to find the quotient and remainder in each problem. If your calculator has an integer divide key, do not use it.

    a. $4,956 \div 35$ **b.** $13,490 \div 347$

    c. $145,897 \div 622$

■ Problems and Applications

35. Using each of the digits 9, 8, 6, 5, 3, and 2 only once, place the digits into the empty blanks to obtain the largest possible product: __ __ × __ __ × __ × __ = _____ .

36. Using each of the digits 2, 3, 6, and 8 only once, place them into the blanks to obtain:

    a. The greatest product.   **b.** The least product.

37. Using each of the digits 2, 3, 6, and 9 only once, place them in the blanks to obtain:
   a. The least and greatest quotient.
   b. The least and greatest remainder.

$$\_\overline{)\_\,\_\,\_}$$

38. Find the missing digits.

   a.
   $$\begin{array}{r} 8\_ \\ \times\ 53 \\ \hline 2\_1 \\ 4\_\_0 \\ \hline 4\,6\,1\,1 \end{array}$$

   b.
   $$\begin{array}{r} 163 \\ \times\ \_4 \\ \hline 65\_ \\ 14\_70 \\ \hline 1\_3\_2 \end{array}$$

   c.
   $$\begin{array}{r} 1\_4 \\ 2\,\overline{)3\_2\_} \\ \underline{-6} \\ 6\_ \\ \underline{52} \\ 104 \\ \hline \underline{} \\ 0 \end{array}$$

39. The equation $4{,}781 - (100 \cdot 26) - (80 \cdot 26) - (3 \cdot 26) = 23$ represents a particular quotient. Write the quotient in a standard long division format.

40. Write each of the following division problems as a single equation.

   a.
   $$\begin{array}{r} 10 \\ 27\,\overline{)293} \\ \underline{27} \\ 23 \end{array}$$

   b.
   $$\begin{array}{r} 46 \\ 41\,\overline{)1894} \\ \underline{164} \\ 254 \\ \underline{246} \\ 8 \end{array}$$

41. The Russian peasant algorithm is an interesting algorithm for multiplication. It doubles one factor and halves the other, ignoring remainders until one factor is reduced to 1. In the resulting two columns of numbers, cross out any double that is across from any half that is even. Add the remaining doubles to find the product. For instance, $26 \cdot 45$ is computed as:

Halves	Doubles
~~26~~	~~45~~
13	90
~~6~~	~~180~~
3	360
1	+720
	1170

   a. Use the Russian peasant algorithm to compute
      i. $12 \cdot 31$ ii. $35 \cdot 62$ iii. $82 \cdot 163$
   b. Can you explain why the Russian peasant algorithm works?

42. Find an exponent that makes each equation true.
   a. $18^x = 324$ b. $5^y = 625$
   c. $19^z = 47{,}045{,}881$ d. $4^m = 16{,}777{,}216$

43. Place parentheses in each equation to make it true.
   a. $36 \div 4 + 5 \cdot 8 = 32$ b. $13 - 6 \cdot 8 \div 2 = 28$
   c. $72 \div 4 \times 2 + 12 = 21$ d. $6 + 4 + 5 \div 3 = 5$

44. Consider the six numbers 53, 45, 95, 87, 113, and 38. Use mental math to determine which two have a product closest to 4,000.

45. What is the largest number that can be expressed using three digits and any combination of whole number operations?

46. A beverage delivery truck can hold 450 cases of soda. If each case holds 24 cans, how many cans of soda are in the truck?

47. If there are 50 envelopes per box, 40 boxes per case, and 8 cases per crate, how many envelopes are in 3 crates?

48. A particular restaurant typically serves about 375 eggs a day for breakfast. How many cartons of one dozen eggs must the manager buy to have enough eggs for one day?

49. You have 364 pennies that you want to exchange for quarters. How many quarters do you get, and how many pennies are left over?

50. An estate worth \$566,584 is to be distributed equally to 8 heirs. How much money does each heir receive?

51. Tony's pizza parlor employs 4 cooks, each of whom can make 6 pizzas in half an hour. A large group nearby has ordered 120 pizzas. How long will it take the 4 cooks to make the pizzas?

52. A truck driver can drive from New York to Los Angeles in 3 days, 8 hr, and 25 min. How long will it take him to make 8 such trips? Give your answer in days, hours, and minutes.

53. The first 4 moves in chess can be made in 197,299 different ways. If any one move takes 5 sec, how many days would it take one person to try every possible combination of 4 moves?

54. If March 1 falls on a Monday this year, on what day will it fall three years from now? Assume no leap years.

55. a. Estimate the number of breaths you take in one year.
   b. Estimate the amount of water you drink in a week.
   c. Did you use the same method to make both estimates? If so, describe it.

56. a. Estimate the number of meals you eat in one year.
   b. If you ingest 800 cal in a meal, about how many calories do you eat in a year?

57. a. Randomly select any page in this book, and estimate the number of words on the page. Describe your method of making the estimate.
   b. Use your estimate in part a to estimate the number of words in this book.

■ Mathematical Reasoning and Proof

58. Describe how it is possible to use addition and multiplication to compute the quotient $765 \div 35$.

59. How many different ways can you think of to compute $14 \cdot 6^2$ on a calculator?

60. Use a calculator to find the quotient and remainder in each problem, but do not use the division key to do so. Explain your method for arriving at your answer.

    a. $142 \div 31$ **b.** $1,248 \div 89$ **c.** $45,790 \div 8,965$

61. Use a calculator to find 11^2, 111^2, and $1,111^2$. What pattern do you notice? Use the pattern to find $11,111^2$ and $111,111^2$ without the aid of a calculator.

62. The following quotient has been computed correctly using a somewhat different method of division. Can you determine how the quotient was computed and whether the method will work for all division problems?

$$
\begin{array}{r}
24 \\
32\overline{)768} \\
60 \\
\hline
16 \\
4 \\
\hline
128 \\
120 \\
\hline
8 \\
8 \\
\hline
0
\end{array}
$$

63. Find all numbers x, such that when 98 is divided by x, the remainder is 5.

64. Find all numbers x less than 100, such that when x is divided by 8, the remainder is 3.

65. Provide a reason for each step in the following computation.

$$
\begin{aligned}
53 \cdot 31 &= (50 + 3) \cdot 31 \\
&= 50 \cdot 31 + 3 \cdot 31 \\
&= 50(30 + 1) + 3(30 + 1) \\
&= 50 \cdot 30 + 50 \cdot 1 + 3 \cdot 30 + 3 \cdot 1 \\
&= 1,500 + 50 + 90 + 3 \\
&= 1,643
\end{aligned}
$$

■ Mathematical Communication

66. Find a partner and write a sequence of steps that describe how to multiply two numbers using the standard algorithm. What difficulties do you encounter in describing the algorithm? What might this imply about teaching the algorithm to children? Repeat the process for the standard algorithm for division.

67. **a.** Write a story problem that can be solved using the product of two or more large numbers. What approach to multiplication did you use?

    b. Write a story problem that can be solved using the quotient of two large numbers. What approach to division did you use?

    c. Write a story problem that can be solved using several different operations on large numbers.

68. Luana uses her calculator to calculate $1,824 \div (16 + 8)$ and gets 122. She knows the answer should be smaller. Can you explain her mistake?

69. Describe how to use a calculator to find the remainder of any division problem. Give your description to a friend, and see whether he or she can use it to find the remainder of $9,276 \div 36$.

70. Make up three calculator problems that use several operations. Give them to a partner to solve. Were they solved correctly? Were there differences between the keystrokes you used and the ones your partner used? How so?

71. Madison says that when using the partial products algorithm, she can multiply the digits in any order as long as she keeps track of the place values. Is she correct?

72. Three students compute the same product in the following ways. Determine their errors and explain how you would help correct their mistakes.

    a.
$$
\begin{array}{r}
52 \\
\times 27 \\
\hline
3514 \\
1040 \\
\hline
4554
\end{array}
$$

    b.
$$
\begin{array}{r}
{}^{1} \\
52 \\
\times 27 \\
\hline
364 \\
104 \\
\hline
468
\end{array}
$$

    c.
$$
\begin{array}{r}
{}^{1} \\
52 \\
\times 27 \\
\hline
104 \\
3640 \\
\hline
3744
\end{array}
$$

73. Three students make three different errors in trying to compute the same quotient. Describe the error made by each student.

    a.
$$
\begin{array}{r}
24 \\
9\overline{)1836} \\
18 \\
\hline
36 \\
36 \\
\hline
0
\end{array}
$$

    b.
$$
\begin{array}{r}
402 \\
9\overline{)1836} \\
18 \\
\hline
36 \\
36 \\
\hline
0
\end{array}
$$

    c.
$$
\begin{array}{r}
240 \\
9\overline{)1,836} \\
18 \\
\hline
36 \\
36 \\
\hline
0
\end{array}
$$

■ Building Connections

74. **a.** Do you think calculators should be used to help students learn to multiply and divide? Discuss this question with a group of your peers.

    b. Read through the NCTM Standards Technology Principle. Does what you read change your view? If so, how?

75. Does estimation play a role when multiplying or dividing on a calculator? Why or why not?

76. Think of five different times when you have used mental multiplication or division. What methods did you use to make the computations?

77. When it comes to computing products and quotients in your daily life, which of the methods and tools discussed in this section are you most likely to use? Explain your preference.

78. Select a set of curriculum materials that span second, third, and fourth grades. Examine the

materials for lessons that teach multiplying and dividing large numbers.

a. At what grade level is this topic introduced?

b. Briefly describe the progression of lessons that teach students how two multiply or divide large numbers. Specifically, describe the manipulatives, written algorithms, and technology used in these lessons.

79. The standard algorithm for division is often considered the most difficult algorithm for students to learn. Why do you think this might be so?

SECTION 4.5 Computation in Base-*b*

Many of the ideas related to addition, subtraction, multiplication, and division are difficult for children who are learning them for the first time. We can gain some perspective on how difficult this is for children by adding, subtracting, multiplying, and dividing in other base-*b* numeration systems.

Understanding Computations in Base *b*

We can compute addition, subtraction, multiplication, and division in any base-*b* numeration system. We do so by defining each operation as we did in the decimal system, making only a few minor changes to how we represent the numbers. Specifically, in base-*b* we can define:

- Addition using the union of two disjoint finite sets.
- Subtraction using the set difference or the missing-addend approach.
- Multiplication using the repeated-addend or the Cartesian-product approach.
- Division using the missing-factor approach.

Explorations
Manual
4.16

Although the definitions are only a little different in base-*b*, the resulting sums, differences, products, and quotients look quite different because of the changes in the groups, digits, and place values. So the single-digit facts we learned with decimal operations no longer apply, and we need to relearn how to compute them. Fortunately, the same models we used with the decimal operations can also be used in base-*b*.

Example 4.56 Using Sets to Compute in Base-*b*

Representation

Use the set model to compute each operation.

a. $4_{\text{five}} + 3_{\text{five}}$ **b.** $13_{\text{eight}} - 6_{\text{eight}}$ **c.** $3_{\text{six}} \cdot 5_{\text{six}}$ **d.** $11_{\text{seven}} \div 4_{\text{seven}}$

Solution We first look at the base, which indicates the digits and place values being used. In each problem, we use the set model, as in the decimal system. Now, however, we must regroup the objects to match the base.

a. To compute $4_{\text{five}} + 3_{\text{five}}$, combine a set of 4 units with a set of 3 units. Then regroup by fives and count. In doing so, we have 1 group of five and 2 units, so $4_{\text{five}} + 3_{\text{five}} = 12_{\text{five}}$.

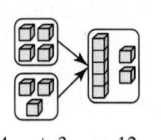

$$4_{\text{five}} + 3_{\text{five}} = 12_{\text{five}}$$

b. Combine a long of 8 units with 3 units to form 13_{eight}. We then take away 6 units and count. Because only 5 remain, $13_{\text{eight}} - 6_{\text{eight}} = 5_{\text{eight}}$.

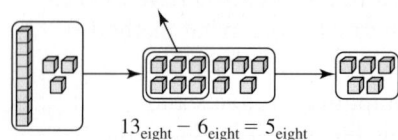

$$13_{\text{eight}} - 6_{\text{eight}} = 5_{\text{eight}}$$

c. To compute $3_{\text{six}} \cdot 5_{\text{six}}$, we use the repeated-addition approach to multiplication. First, form 3 sets of 5 units and combine them. We then regroup by sixes and count. Because we can make 2 groups of six with 3 left over, $3_{\text{six}} \cdot 5_{\text{six}} = 23_{\text{six}}$.

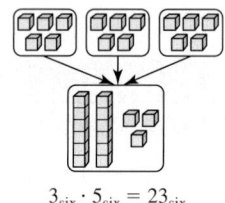

$$3_{\text{six}} \cdot 5_{\text{six}} = 23_{\text{six}}$$

d. To compute $11_{\text{seven}} \div 4_{\text{seven}}$, we use the repeated-subtraction approach. Combine a long of 7 units with 1 unit to make 11_{seven}. We then remove sets of 4 units. Because we can take away 2 sets of 4 units with no units left over, $11_{\text{seven}} \div 4_{\text{seven}} = 2_{\text{seven}}$.

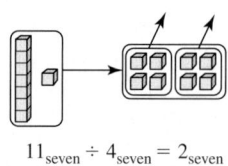

$$11_{\text{seven}} \div 4_{\text{seven}} = 2_{\text{seven}}$$

Example 4.57	**Using a Number Line to Compute in Base-*b***

Representation   Use the number line model to compute each operation.

a. $5_{\text{six}} + 3_{\text{six}}$ **b.** $12_{\text{five}} - 4_{\text{five}}$ **c.** $3_{\text{four}} \cdot 3_{\text{four}}$

Solution We use the number line model as we did for decimal operations. Now, however, we must change the number line to match the given base.

a. Model $5_{\text{six}} + 3_{\text{six}}$ by drawing an arrow 5 units long, followed by an arrow 3 units long. Because we end at 12_{six}, $5_{\text{six}} + 3_{\text{six}} = 12_{\text{six}}$.

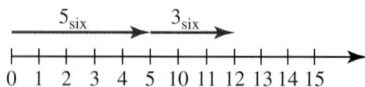

b. Draw an arrow to 12_{five} and then another backward that is 4 units long. Because we end at 3_{five}, $12_{\text{five}} - 4_{\text{five}} = 3_{\text{five}}$.

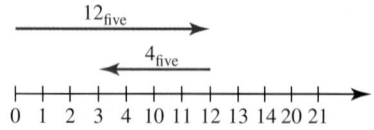

c. Model $3_\text{four} \cdot 3_\text{four}$ by drawing 3 arrows, each 3 units long. Because we end at 21_four, $3_\text{four} \cdot 3_\text{four} = 21_\text{four}$.

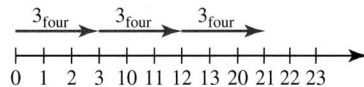

There is another way to perform computations in base-*b*. Because base-*b* numbers can be converted to base-ten and vice versa, we can convert the numbers to base-ten, perform the computation, and then convert the answer back to base-*b*.

Example 4.58　Computing Single-Digit Facts in Base-*b*

Compute each operation by converting to base-ten.

a. $A_\text{sixteen} + C_\text{sixteen}$　**b.** $14_\text{seven} - 6_\text{seven}$　**c.** $7_\text{nine} \cdot 5_\text{nine}$　**d.** $12_\text{four} \div 3_\text{four}$

Solution

a. $A_\text{sixteen} + C_\text{sixteen} = 10 + 12 = 22$. Because there is 1 group of 16 and 6 ones in 22, $A_\text{sixteen} + C_\text{sixteen} = 16_\text{sixteen}$.

b. $14_\text{seven} - 6_\text{seven} = 11 - 6 = 5$. Because $5 = 5_\text{seven}$, $14_\text{seven} - 6_\text{seven} = 5_\text{seven}$.

c. $7_\text{nine} \cdot 5_\text{nine} = 7 \cdot 5 = 35$. Because 35 consists of 3 groups of 9 and 8 ones, $7_\text{nine} \cdot 5_\text{nine} = 38_\text{nine}$.

d. $12_\text{four} \div 3_\text{four} = 6 \div 3 = 2$. Because $2 = 2_\text{four}$, $12_\text{four} \div 3_\text{four} = 2_\text{four}$.

Explorations Manual 4.17

As in the decimal system, we can also learn single-digit facts in base-*b* by generating fact tables for addition and multiplication.

Example 4.59　Fact Tables in Base-*b*

Generate the single-digit fact tables for:

a. Base-eight addition.　　**b.** Base-five multiplication.

Solution

a. Many of the facts in base-eight addition, such as $1_\text{eight} + 1_\text{eight}$ and $3_\text{eight} + 1_\text{eight}$, have the same sum as their base-ten counterparts. This makes them relatively easy to compute. However, for a sum of 8 or greater, we must write the number in base-eight. Table 4.5 shows the completed table. All entries are given in base-eight. Subscripts have been left out for ease in reading.

Table 4.5　Base-eight single-digit addition facts

+	0	1	2	3	4	5	6	7
0	0	1	2	3	4	5	6	7
1	1	2	3	4	5	6	7	10
2	2	3	4	5	6	7	10	11
3	3	4	5	6	7	10	11	12
4	4	5	6	7	10	11	12	13
5	5	6	7	10	11	12	13	14
6	6	7	10	11	12	13	14	15
7	7	10	11	12	13	14	15	16

b. As before, many of the facts in base-five multiplication are the same as their base-ten counterparts. We can compute those that are different by using sets or the number line model. Table 4.6 shows the single-digit multiplication facts for base-five. Again, the subscripts have been left out for ease in reading.

Table 4.6	Base-five single-digit multiplication facts				
$\times$	0	1	2	3	4
0	0	0	0	0	0
1	0	1	2	3	4
2	0	2	4	11	13
3	0	3	11	14	22
4	0	4	13	22	31

Tables like the ones in Example 4.59 show that base-b addition and multiplication satisfy the algebraic properties discussed throughout the chapter. All the entries in Table 4.5 are base-eight numbers, indicating that the base-eight numbers are closed under addition. Also, Table 4.5 is symmetric about the diagonal going down from left to right, indicating that base-eight addition is commutative. The table does not show that base-eight addition is associative. To ascertain that, we would have to test every combination of three numbers. Because of the large number of combinations, we select three numbers at random and show that they satisfy the associative property. For instance, using 2_{eight}, 5_{eight}, and 7_{eight}:

$$(2_{\text{eight}} + 5_{\text{eight}}) + 7_{\text{eight}} = 7_{\text{eight}} + 7_{\text{eight}} = 16_{\text{eight}}$$

and

$$2_{\text{eight}} + (5_{\text{eight}} + 7_{\text{eight}}) = 2_{\text{eight}} + 14_{\text{eight}} = 16_{\text{eight}}$$

Finally, the first row and column in Table 4.5 match the initial row and column, which means that base-eight addition has an identity element: 0. Because we can repeat this process with addition in any base, we conclude that *for any whole number* b $\geq$ *2, base-*b *addition satisfies the closure, commutative, associative, and identity properties.*

Similar arguments can be used with Table 4.6 to show that base-five multiplication is closed, is commutative, is associative, and has an identity element. Also, the zero multiplication property holds because each entry in the first row and column is zero. It is also possible to verify that the distributive property holds. Consequently, *for any whole number* b $\geq$ *2, base-*b *multiplication satisfies the closure, commutative, associative, identity, zero multiplication, and distributive properties.* Subtraction and division in base-b fail to satisfy any of the algebraic properties.

Example 4.60   A Counterexample to the Associative Property

Reasoning Example   Find a counterexample showing that base-nine subtraction is not associative.

Solution We need to find three base-nine numbers, a, b, and c, such that $a - (b - c) \neq (a - b) - c$. If $a = 12_{\text{nine}}$, $b = 6_{\text{nine}}$, and $c = 2_{\text{nine}}$, then:

$$a - (b - c) = 12_{\text{nine}} - (6_{\text{nine}} - 2_{\text{nine}}) = 12_{\text{nine}} - 4_{\text{nine}} = 7_{\text{nine}}$$

and

$$(a - b) - c = (12_{\text{nine}} - 6_{\text{nine}}) - 2_{\text{nine}} = 5_{\text{nine}} - 2_{\text{nine}} = 3_{\text{nine}}$$

Because $7_{\text{nine}} \neq 3_{\text{nine}}$, base-nine subtraction is not associative.

Check Your Understanding 4.5-A

1. Compute each problem by converting to base-ten as necessary. Be sure your answer is expressed in the given base.

    a. $9_{\text{twelve}} + B_{\text{twelve}}$ **b.** $16_{\text{eight}} - 7_{\text{eight}}$ **c.** $4_{\text{seven}} \cdot 6_{\text{seven}}$ **d.** $23_{\text{six}} \div 5_{\text{six}}$

2. Generate tables for the single-digit addition and multiplication facts of base-four.

3. Use the given numbers to demonstrate the given property.

    a. 6_{seven} and 5_{seven} to show base-seven multiplication is commutative.

    b. 2_{five}, 3_{five}, and 4_{five} to show that base-five satisfies the distributive property.

Talk About It Consider the base-eight addition facts given in Table 4.5. What different strategies could you use to learn the 64 facts given in the table?

Representation

Activity 4.9
Part A. Use the set model to solve each problem.

    1. $2_{\text{eight}} + 7_{\text{eight}}$ **2.** $15_{\text{nine}} - 8_{\text{nine}}$ **3.** $4_{\text{five}} \cdot 3_{\text{five}}$

Part B. Use the number line model to solve each problem.

    1. $5_{\text{nine}} + 6_{\text{nine}}$ **2.** $14_{\text{seven}} - 6_{\text{seven}}$ **3.** $5_{\text{six}} \cdot 2_{\text{six}}$

Computing with Large Numbers in Base-*b*

The algorithms presented in this chapter can be used to compute with large numbers in base-*b* in two ways. One is to perform the algorithm using single-digit facts for the given base and operation. This method most closely mimics how children learn whole-number computation. The second way is to convert the numbers to base-ten, perform the calculation, and then convert back. Both methods are shown with the standard algorithms in the next several examples.

Example 4.61 | **Adding Large Numbers in Base-*b***

Use the standard algorithm to compute each sum.

a. $\begin{array}{r} 134_{\text{five}} \\ + \ 423_{\text{five}} \end{array}$ **b.** $\begin{array}{r} AC3_{\text{fourteen}} \\ + \ B1B_{\text{fourteen}} \end{array}$

Solution In each problem, we first look at the base because this indicates the place values and the digits. We then proceed with the standard algorithm.

a. In this problem, we use the facts table for base-five addition:

+	0	1	2	3	4
0	0	1	2	3	4
1	1	2	3	4	10
2	2	3	4	10	11
3	3	4	10	11	12
4	4	10	11	12	13

We first add the digits in the ones place: $4_{five} + 3_{five} = 12_{five}$. We write down the 2, and place the 1 in the next highest place value, signifying we are trading five ones for 1 group of 5.

$$
\begin{array}{r}
{\scriptstyle 1} \\
134_{five} \\
+ \; 423_{five} \\
\hline
2
\end{array}
$$

We repeat the process in the next two place values using facts from the table as a guide. We obtain the following sum:

$$
\begin{array}{r}
{\scriptstyle 1\,1\,1} \\
134_{five} \\
+ \; 423_{five} \\
\hline
1112_{five}
\end{array}
$$

b. This problem is in base-fourteen, so we must remember that A = ten, B = eleven, C = twelve, and D = thirteen. Given the many single-digit addition facts in base-fourteen, we convert the numbers to base-ten rather than using a fact table. We make the conversions as we work with particular place values. Start in the ones place value, and add $3_{fourteen} + B_{fourteen} = 3 + 11 = 14 = 10_{fourteen}$:

$$
\begin{array}{r}
{\scriptstyle 1} \\
AC3_{fourteen} \\
+ \; B1B_{fourteen} \\
\hline
0
\end{array}
$$

Next, add $1_{fourteen} + C_{fourteen} + 1_{fourteen} = 1 + 12 + 1 = 14 = 10_{fourteen}$ and $1_{fourteen} + A_{fourteen} + B_{fourteen} = 1 + 10 + 11 = 22 = 18_{fourteen}$ to complete the problem:

$$
\begin{array}{r}
{\scriptstyle 1\,1\,1} \\
AC3_{fourteen} \\
+ \; B1B_{fourteen} \\
\hline
1800_{fourteen}
\end{array}
$$

Example 4.62 Subtracting Large Numbers in Base-b

Use the standard algorithm to subtract.

a. 4312_{five}
$-\;303_{five}$

b. 3005_{eight}
$-\;2756_{eight}$

Solution We proceed using the standard algorithm for subtraction exactly as we would in base-ten. We compute the first problem by using a fact table and the second by converting to base-ten.

a. This problem is in base-five, so any exchanges we make are in a ratio of 1 to 5. We look at the ones digits and see that 2_{five} is smaller than 3_{five}. So we must exchange 1 group of 5 for 5 ones and add them to the 2 ones already in the place value. This gives us a total of 12_{five} in the ones place value and 0 in the fives place value. We now subtract all place values directly using the base-five addition facts in Example 4.61:

$$
\begin{array}{r}
{\scriptstyle 0\;12} \\
43\cancel{1}2_{five} \\
- \; 303_{five} \\
\hline
4004_{five}
\end{array}
$$

b. This problem is in base-eight, so all exchanges are in a ratio of 1 to 8. To subtract the ones digits, we go to the 3 in the fourth place value and exchange back several places. As we make the exchanges, we write the results in base-ten numerals for easier computation. We then subtract directly:

$$
\begin{array}{r}
{}^{7\,7}\\
2\,\cancel{8}\,\cancel{8}\,13\\
\cancel{3}\,\cancel{0}\,\cancel{0}\,\cancel{3}_{\text{eight}}\\
-\,2\,7\,5\,6_{\text{eight}}\\
\hline
2\,7_{\text{eight}}
\end{array}
$$

Example 4.63   Multiplying Large Numbers in Base-b

Use the standard algorithm to multiply.

a. 231_{four}
   $\times\,21_{\text{four}}$

b. 351_{six}
   $\times\,32_{\text{six}}$

Solution

a. The problem is in base-four, so we use the following table of single-digit multiplication facts.

×	0	1	2	3
0	0	0	0	0
1	0	1	2	3
2	0	2	10	12
3	0	3	12	21

Because the ones digit in the lower factor is 1_{four}, the product of $1_{\text{four}} \cdot 231_{\text{four}} = 231_{\text{four}}$. We put in a placeholder, and multiply $2_{\text{four}} \cdot 231_{\text{four}}$, using the facts from the table and trading as necessary. After multiplying:

$$
\begin{array}{r}
{}^{1}\\
231_{\text{four}}\\
\times\,21_{\text{four}}\\
\hline
231\\
11420
\end{array}
$$

We add in base-four to obtain the product.

$$
\begin{array}{r}
{}^{1}\\
231_{\text{four}}\\
\times\,21_{\text{four}}\\
\hline
231\\
+\,11420\\
\hline
12311_{\text{four}}
\end{array}
$$

b. We proceed through the standard algorithm by multiplying in base-ten and converting to base-six as we go. Begin with the ones digits: $2_{\text{six}} \cdot 1_{\text{six}} = 2_{\text{six}}$. Next: $2_{\text{six}} \cdot 5_{\text{six}} = 2 \cdot 5 = 10 = 14_{\text{six}}$. Write down the 4 and carry the 1. Then compute: $2_{\text{six}} \cdot 3_{\text{six}} + 1_{\text{six}} = 2 \cdot 3 + 1 = 7 = 11_{\text{six}}$:

$$
\begin{array}{r}
{}^{1}\\
351_{\text{six}}\\
\times\,32_{\text{six}}\\
\hline
1142
\end{array}
$$

We put in a placeholder, and repeat the process with the 3 in the bottom factor. After multiplying:

$$
\begin{array}{r}
\overset{2}{\cancel{1}} \\
351_{\text{six}} \\
\times\ 32_{\text{six}} \\
\hline
1142 \\
15330 \\
\end{array}
$$

To complete the operation, we add in base-six:

$$
\begin{array}{r}
\overset{2}{\cancel{1}} \\
351_{\text{six}} \\
\times\ 32_{\text{six}} \\
\hline
1142 \\
+\ 15330 \\
\hline
20512_{\text{six}} \\
\end{array}
$$

Example 4.64   Dividing in Base-*b*

Use the standard algorithm to compute $22_{\text{five}}\overline{)3412_{\text{five}}}$.

Solution One problem with computing division in other bases is that determining how many times one number goes into another can be difficult. To figure this out, we compute all the multiplication facts associated with the problem, and then use them to make comparisons. Because we are in base-five and are dividing by 22_{five}, we have the following five multiplication facts:

$$
\begin{array}{ccccc}
22_{\text{five}} & 22_{\text{five}} & 22_{\text{five}} & 22_{\text{five}} & 22_{\text{five}} \\
\times\ 0_{\text{five}} & \times\ 1_{\text{five}} & \times\ 2_{\text{five}} & \times\ 3_{\text{five}} & \times\ 4_{\text{five}} \\
\hline
0_{\text{five}} & 22_{\text{five}} & 44_{\text{five}} & 121_{\text{five}} & 143_{\text{five}} \\
\end{array}
$$

With these facts, we proceed through the long division algorithm, being sure to subtract in base-five. First, we note that 22_{five} will go into 34_{five} one time, and then we subtract:

$$
\begin{array}{r}
1 \\
22\overline{)3412} \\
-\ 22 \\
\hline
12 \\
\end{array}
\qquad
\begin{array}{l}
=\ 1_{\text{five}} \times 22_{\text{five}} \\
=\ 34_{\text{five}} - 22_{\text{five}} \\
\end{array}
$$

Next, we bring down the 1, and repeat the process. This time, 22_{five} goes into 121_{five} exactly three times. Again we subtract:

$$
\begin{array}{r}
130 \\
22\overline{)3412} \\
-\ 22 \\
\hline
121 \\
-\ 121 \\
\hline
02 \\
\end{array}
\qquad
\begin{array}{l}
\\
=\ 3_{\text{five}} \times 22_{\text{five}} \\
=\ 121_{\text{five}} - 121_{\text{five}} \\
\end{array}
$$

We bring down the remaining 2, but because 2_{five} is less than 22_{five}, the division is complete. Consequently, $3412_{\text{five}} \div 22_{\text{five}} = 130_{\text{five}}$, remainder 2_{five}.

Check Your Understanding 4.5-B

1. Compute each problem using the standard algorithms for addition and subtraction.

    a. $2031_{five} + 414_{five}$ **b.** $571_{eight} + 246_{eight}$

    c. $3013_{four} - 2210_{four}$ **d.** $6742_{nine} - 3811_{nine}$

2. Compute each problem using the standard algorithms for multiplication and division.

    a. $23_{five} \cdot 14_{five}$ **b.** $352_{six} \cdot 34_{six}$

    c. $31_{four}\overline{)3312_{four}}$ **d.** $14_{eight}\overline{)671_{eight}}$

Talk About It How are the standard algorithms for base-*b* addition, subtraction, multiplication, and division similar to those in base-ten? How are they different?

Representation

Activity 4.10

Draw a picture that shows how base-five blocks can be used to model and solve each problem.

    a. $322_{five} + 142_{five}$ **b.** $401_{five} - 322_{five}$ **c.** $23_{five} \cdot 13_{five}$

SECTION 4.5   Learning Assessment

◼ Understanding the Concepts

1. Why do the results of base-*b* operations seem dramatically different from base-ten operations, even though they are computed in the same basic way?

2. How are addition and subtraction in base-*b* similar to addition and subtraction in base-ten? How are they different?

3. Why should it make sense that base-*b* addition satisfies the commutative, associative, and identity properties given that base-*b* addition can be defined using the union of two finite, disjoint sets?

4. How are multiplication and division in base-*b* similar to multiplication and division in base-ten? How are they different?

5. Why should it make sense that base-*b* multiplication satisfies the commutative, associative, and identity properties given that base-*b* multiplication can be defined using repeated addition?

6. Describe two different strategies for computing with large numbers in base-*b*.

7. Is division by zero always undefined in any base-*b*? Explain.

◼ Representing the Mathematics

8. Use the set model to solve each problem. Use the take-away approach for each difference.

    a. $2_{three} + 11_{three}$ **b.** $4_{seven} + 6_{seven}$

    c. $11_{five} - 4_{five}$ **d.** $13_{six} - 5_{six}$

9. Use the set model to solve each problem. Use the comparison approach for each difference.

    a. $4_{six} + 5_{six}$ **b.** $7_{eight} + 7_{eight}$

    c. $12_{three} - 2_{three}$ **d.** $15_{nine} - 7_{nine}$

10. Use the set model to solve each problem. Use repeated addition for each product and repeated subtraction for each quotient.

    a. $2_{four} \cdot 3_{four}$ **b.** $3_{six} \cdot 3_{six}$

    c. $13_{seven} \div 2_{seven}$ **d.** $15_{nine} \div 7_{nine}$

11. Use the set model to solve each problem. Use repeated addition for each product and partitioning for each quotient.

    a. $5_{eight} \cdot 4_{eight}$ **b.** $4_{nine} \cdot 6_{nine}$

    c. $22_{five} \div 4_{five}$ **d.** $40_{six} \div 5_{six}$

12. Draw a picture of the whole-number line in:

   a. Base-two.

   b. Base-six.

   c. Base-nine.

   d. Base-twelve.

   e. How does a base-b number line differ from the base-ten number line?

13. Use the number line model to solve each problem.

   a. $6_{nine} + 8_{nine}$

   b. $14_{six} - 5_{six}$

   c. $4_{five} \cdot 3_{five}$

   d. $13_{four} \div 2_{four}$

14. Use the number line model to solve each problem.

   a. $5_{six} + 5_{six}$

   b. $10_{eight} - 5_{eight}$

   c. $4_{six} \cdot 4_{six}$

   d. $1A_{eleven} \div 7_{eleven}$

15. Determine the base and the computation fact shown in each diagram.

   a.
   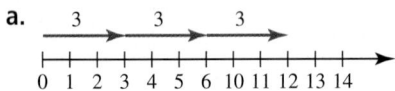

   b.
   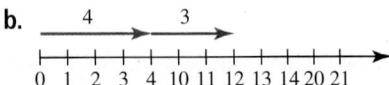

   c.
   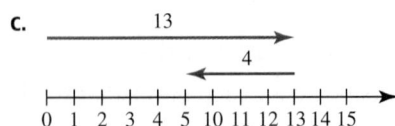

   d.
   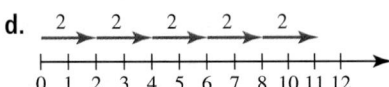

16. This is the facts table for base-six addition. Subscripts have been left out for ease in reading. State how the closure, commutative, and identity properties are shown in the table.

+	0	1	2	3	4	5
0	0	1	2	3	4	5
1	1	2	3	4	5	10
2	2	3	4	5	10	11
3	3	4	5	10	11	12
4	4	5	10	11	12	13
5	5	10	11	12	13	14

17. This is the facts table for base-eight multiplication. Subscripts have been left out for ease in reading. State how the closure, commutative, identity, and zero multiplication properties are shown in the table.

×	0	1	2	3	4	5	6	7
0	0	0	0	0	0	0	0	0
1	0	1	2	3	4	5	6	7
2	0	2	4	6	10	12	14	16
3	0	3	6	11	14	17	22	25
4	0	4	10	14	20	24	30	34
5	0	5	12	17	24	31	36	43
6	0	6	14	22	30	36	44	52
7	0	7	16	25	34	43	52	61

18. Use base-four blocks to solve each problem.

   a. $223_{four} + 132_{four}$ **b.** $201_{four} - 121_{four}$

   c. $33_{four} \cdot 22_{four}$

19. Determine the base and the computation fact shown in each diagram.

   a.
   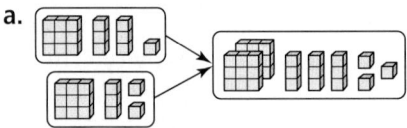

   b.
   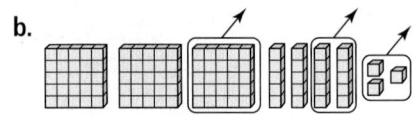

   c.
   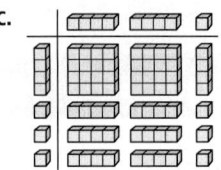

■ Exercises

20. Compute each sum or difference.

   a. $4_{five} + 3_{five}$ **b.** $3_{seven} + 6_{seven}$

   c. $12_{three} - 2_{three}$ **d.** $14_{six} - 5_{six}$

21. Compute each sum or difference.

   a. $5_{nine} + 8_{nine}$ **b.** $6_{fourteen} + B_{fourteen}$

   c. $14_{twelve} - 9_{twelve}$ **d.** $18_{fifteen} - D_{fifteen}$

22. Compute each product or quotient.

   a. $3_{four} \cdot 2_{four}$ **b.** $3_{seven} \cdot 6_{seven}$

   c. $22_{three} \div 2_{three}$ **d.** $41_{six} \div 5_{six}$

23. Compute each product or quotient.

   a. $5_{eleven} \cdot 6_{eleven}$ **b.** $4_{thirteen} \cdot 9_{thirteen}$

   c. $24_{twelve} \div 7_{twelve}$ **d.** $28_{sixteen} \div A_{sixteen}$

24. Construct a table for the single-digit addition facts in each base.

   a. Base-three   **b.** Base-five   **c.** Base-eight

25. Construct a table for the single-digit multiplication facts in each base.

   a. Base-four   **b.** Base-six   **c.** Base-seven

26. List the other facts that are in the fact families of each of the following:
    a. $4_{six} + 3_{six}$
    b. $13_{eight} - 6_{eight}$
    c. $3_{seven} \cdot 5_{seven}$
    d. $20_{six} \div 4_{six}$

27. List the other facts that are in the fact families of each of the following:
    a. $2_{three} + 2_{three}$
    b. $1A_{eleven} - 6_{eleven}$
    c. $5_{nine} \cdot 7_{nine}$
    d. $28_{twelve} \div 4_{twelve}$

28. Use the standard algorithm to compute each sum or difference.
    a. $2101_{three} + 1102_{three}$
    b. $4423_{five} + 232_{five}$
    c. $3102_{four} - 203_{four}$
    d. $5401_{six} - 3520_{six}$

29. Use the standard algorithm to compute each sum or difference.
    a. $568_{nine} + 478_{nine}$
    b. $CEF_{sixteen} + A1F_{sixteen}$
    c. $7008_{nine} - 3413_{nine}$
    d. $A021_{fifteen} - 9BD_{fifteen}$

30. Use the standard algorithm to compute each product or quotient.
    a. $110_{two} \cdot 10_{two}$
    b. $323_{four} \cdot 21_{four}$
    c. $22_{three} \overline{)2102_{three}}$
    d. $43_{five} \overline{)3214_{five}}$

31. Use the standard algorithm to compute each product or quotient.
    a. $512_{six} \cdot 42_{six}$
    b. $631_{seven} \cdot 14_{seven}$
    c. $32_{four} \overline{)3102_{four}}$
    d. $32_{seven} \overline{)3646_{seven}}$

Problems and Applications

32. Use the lattice method to compute each operation.
    a. $1101_{two} + 101_{two}$
    b. $653_{eight} + 477_{eight}$
    c. $212_{three} \cdot 12_{three}$
    d. $341_{five} \cdot 42_{five}$

33. Determine the base that makes each equation true.
    a. $4\_ + 6\_ = 13\_$
    b. $2\_ \cdot 4\_ = 13\_$
    c. $15\_ - 7\_ = 6\_$
    d. $14\_ \div 5\_ = 2\_$

34. Let $x + y = z$, where x, y, and z each represent a different single digit from base-eight.
    a. If $x = 2$ and $y = 4$, what is z?
    b. What is the largest digit that z could be?
    c. What is the smallest digit that z could be?
    d. If z is even, what digits could x be?

35. Find the missing digits in each problem.

    a.
    $$\begin{array}{r} 3\,4\,\_\,6_{seven} \\ +\ 1\,3\,4\,5_{seven} \\ \hline 5\,\_\,0\,\_{seven} \end{array}$$

    b.
    $$\begin{array}{r} 4\,0\,2\,1_{five} \\ -\ 2\,\_\,\_\,\_{five} \\ \hline 1\,1\,0\,3_{five} \end{array}$$

    c.
    $$\begin{array}{r} 2\,3\,\_{four} \\ \times\ 2\,3_{four} \\ \hline 2\,\_\,1\,3 \\ 1\,1\,\_\,\_\,0 \\ \hline 1\,3\,\_\,3\,\_{four} \end{array}$$

36. Use repeated multiplication to compute each exponent in base-five.
    a. $(3_{five})^2$
    b. $(2_{five})^4$
    c. $(2_{five})^7$
    d. $(13_{five})^2$

37. The standard order of operations for base-ten holds in any base. Use the order of operations to compute the following.
    a. $(3_{five} + 4_{five}) \cdot 2_{five}$
    b. $((3_{six})^2 + 4_{six}) - 11_{six}$
    c. $15_{seven} \div 2_{seven} \cdot 4_{seven}$

38. Find all base-eight numbers x, such that $41_{eight} \div x$ has a remainder of 3_{eight}.

39. Solve the following magic square by finding the missing numbers that make every row, column, and diagonal add to the same sum.

20_{five}	3_{five}	
	12_{five}	
11_{five}	21_{five}	

40. Using each of the digits from base-five only once, place the digits into the empty blanks to obtain the largest possible product.
    $$\_\,\_ \times \_ \times \_ \times \_ =$$

Mathematical Reasoning and Proof

41. Find two specific numbers that show base-seven is not closed under subtraction.

42. Find two specific numbers that show base-eight subtraction is not commutative.

43. Find three specific numbers that show base-four subtraction is not associative.

44. Why does multiplying any number in base-*b* by 0_b always result in 0_b?

45. a. Will 0_b always be the additive identity for addition in base-*b*? Why?
    b. Will 1_b always be the multiplicative identity for multiplication in base-*b*? Why?

46. a. Use 4_{five} and 3_{five} to show base-five multiplication is commutative.
    b. Use 4_{six}, 5_{six}, and 2_{six} to show that base-six addition satisfies the associative property.
    c. Use 3_{nine}, 6_{nine}, and 7_{nine} to show that base-nine satisfies the distributive property.

47. Consider standard rounding in base-ten. How might we make changes to use standard rounding in base-six?

Mathematical Communication

48. a. Write the definition of addition in base-five.
    b. Write the missing-addend definition of subtraction in base-six.
    c. Write the repeated-addend definition of multiplication in base-eight.
    d. How are the definitions similar to their base-ten counterparts? How are they different? Specifically, what aspect of the definition, if any, did you need to change to make it consistent with the given base?

49. After performing computation in other bases, do you think differently about computation in base-ten? Describe how your thoughts have changed?

50. a. What aspects of computing do you find most difficult when adding or subtraction in a base other than ten?

    b. What aspects of computing do you find most difficult when multiplying or dividing in a base other than ten?

    c. What might your difficulties imply about teaching children to compute in the decimal system?

■ Building Connections

51. Base-*b* computations have several real-life applications. Search the Internet to see how many different uses of base-*b* computation you can find. Write a paragraph or two describing the uses.

52. Use what you have learned in this section to devise a strategy for adding two Mayan numerals. Write a short paragraph that briefly describes your method, and then use it to add the following two numbers.

Add 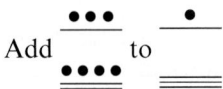 to

53. Use what you have learned in this section to devise a strategy for subtracting two Babylonian numerals. Use your method to subtract the following two numbers.

Subtract

from

54. Use what you have learned in this section to devise a strategy for multiplying two Mayan numerals. Use your method to multiply the following two numbers.

Multiply ⋮ by ⋮

Historical Note

John Von Neumann (1903–1957) is widely recognized as one of the most brilliant mathematicians of the twentieth century. He quickly showed his mathematical aptitude when, by the age of six, he could divide two 8-digit numbers and by the age of eight, he had a solid understanding of calculus. Although he made significant contributions to a wide range of academic disciplines, he is best known for his work on the Manhattan project and the early development of computers. For ten years, he directed the Electronic Computer Project at Princeton's Institute for Advanced Study, where he developed MANIAC (mathematical analyzer, numerical integrator, and computer), which was the fastest computer of its kind at the time. Von Neumann got many of his ideas for computer processing from his own uncanny ability to compute quickly. It is even said that he participated in a computational race against one of his first computers and won.

John Von Neumann

© Everett Collection Inc/Alamy

Opening Problem Answer

Jupiterimages

The first two digits are your age, and the second two digits are the amount of change in your pocket. If we let x represent your age and y the amount of change in your pocket, then we can represent the steps in the trick as follows.

x	Your age.
$2x + 5$	Multiply by 2 and add 5.
$50(2x + 5)$	Multiply by 50.
$100x + 250$	Distribute 50.
$100x - 115$	Subtract 365.
$100x - 115 + y + 115$	Add change and 115.
$100x + y$	Your age plus your change.

Chapter 4 REVIEW

Summary of Key Concepts from Chapter 4

Section 4.1   Understanding Whole-Number Addition and Subtraction

- If A and B are disjoint, finite where $a = n(A)$ and $b = n(B)$, then $a + b = n(A \cup B)$.

- If a, b, and c are whole numbers, then addition satisfies the following properties.

  1. *Closure:* The sum of any two whole numbers is a unique whole number.

  2. *Commutative:* $a + b = b + a$.

  3. *Associative:* $a + (b + c) = (a + b) + c$.

  4. *Identity:* There exists the unique number 0, called the **additive identity,** such that $a + 0 = a$ and $0 + a = a$.

- **a is less than b** if and only if there exists a natural number n such that $a + n = b$.

- *Take-away approach to subtraction:* Suppose $a = n(A)$ and $b = n(B)$. If $B \subseteq A$, then $a - b = n(A - B)$. a is the **minuend,** and b is the **subtrahend.**

- *Comparison approach to subtraction:* The difference is found by comparing two sets or measures to find out how much more one has over the other.

- *Missing-addend approach to subtraction:* $a - b = c$ if and only if $a = b + c$ for some whole number c.

- Addition and subtraction are **inverse operations;** that is, one undoes the other.

Section 4.2   Adding and Subtracting Large Numbers

- Addition algorithms use two basic procedures: adding single-digit numbers and **regrouping,** which is also called **trading** or **carrying.**

- The sum of two large numbers can be computed with base-ten blocks, the partial-sums algorithm, the standard algorithm, and the lattice method.

- Subtraction algorithms involve two basic procedures, using basic subtraction facts and **exchanging** or **borrowing.**

- The difference of two large numbers can be computed with base-ten blocks, the standard algorithm, and the **equal additions** algorithm.

- Strategies for mental addition and subtraction include counting strategies, left-to-right methods, compatible numbers, compensation, equal additions, and breaking apart numbers.

- With **truncation,** a number is "cut off" at a certain place value with no regard for the digits that come after.

- With **rounding,** the digit in the next lowest place value is used to determine which value the number is closer to. If it is 4 or lower, round down. If it is 5 or higher, round up.

- Strategies for making estimates include front-end estimation, compatible numbers, and clustering.

Section 4.3   Understanding Whole-Number Multiplication and Division

- *Repeated-addition approach to multiplication:* If a and b are any whole numbers with $a \neq 0$, then $a \cdot b = \underbrace{b + b + \ldots + b}_{a \text{ addends}}$. If $a = 0$, then $0 \cdot b = 0$ for all b. a and b are **factors.**

- *Cartesian-product approach to multiplication:* If $a = n(A)$ and $b = n(B)$, then $a \cdot b = n(A \times B)$.

- If a, b, and c are whole numbers, then multiplication satisfies the following properties:

  1. *Closure:* The product of any two whole numbers is a unique whole number.

  2. *Commutative:* $a \cdot b = b \cdot a$.

234

3. *Associative:* $a \cdot (b \cdot c) = (a \cdot b) \cdot c$.

4. *Identity:* There exists the unique number 1, called the **multiplicative identity,** such that $a \cdot 1 = a$ and $a = 1 \cdot a$.

5. *Zero multiplication:* $a \cdot 0 = 0 = 0 \cdot a$.

6. *Distributive property of multiplication over addition:* $a \cdot (b + c) = (a \cdot b) + (a \cdot c)$.

- $a^n = \underbrace{a \cdot a \cdot \ldots \cdot a}_{n \text{ factors}}$. n is called the **exponent** or **power,** a is called the **base.**

- $a^0 = 1$ for $a \neq 0$.

- If a, b, m, and n are whole numbers with $a \neq 0$ and $b \neq 0$, then:

  1. $a^m \cdot a^n = a^{m+n}$.

  2. $a^m \cdot b^m = (a \cdot b)^m$.

  3. $(a^m)^n = a^{m \times n}$.

- *Repeated-subtraction approach to division:* If $b \neq 0$, then $a \div b$ is the number of times b can be subtracted from a. a is called the **dividend,** and b is called the **divisor.**

- *Partitioning approach to division:* If $b \neq 0$, then $a \div b$ is the number of objects in each set when a objects are equally distributed among b sets.

- *Missing-factor approach to division:* If $b \neq 0$, then $a \div b = c$ if and only if $a = b \cdot c$ for some unique whole number c.

- Division by zero is undefined.

- *Division algorithm:* For any whole numbers a and b with $b \neq 0$, there exists unique whole numbers q and r such that $a = bq + r$, where $0 \leq r < b$.

- Operations are performed in the following order:

  1. Operations inside parentheses or other grouping symbols

  2. Exponents

  3. Multiplication and division in order from left to right

  4. Addition and subtraction in order from left to right

Section 4.4 Multiplying and Dividing Large Numbers

- Multiplication algorithms involve three basic procedures: multiplying single-digit numbers, regrouping, and multidigit addition.

- The product of two large numbers can be computed with base-ten blocks, the partial product algorithm, the standard algorithm, and the lattice method.

- The quotient of two large numbers can be computed with base-ten blocks, the repeated subtraction algorithm, the scaffold method, and the standard algorithm.

Section 4.5 Computation in Base-*b*

- Base-*b* operations can be defined in the same way as decimal operations.

- For any whole number $b \geq 2$, base-*b* addition satisfies the closure, commutative, associative, and identity properties.

- For any whole number $b \geq 2$, base-*b* multiplication satisfies the closure, commutative, associative, identity, zero multiplication, and distributive properties.

- Algorithms for the decimal operations can be used with base-*b* operations.

Review Exercises Chapter 4

1. Use the set model to represent and solve each problem.

   a. $4 + 5$ **b.** $13 - 8$ **c.** $4 \cdot 5$ **d.** $18 \div 6$

2. Use the number line model to represent and solve each problem.

   a. $3 + 7$ **b.** $11 - 5$ **c.** $2 \cdot 7$ **d.** $12 \div 4$

3. Use base-ten blocks to make each computation.

   a. $345 + 269$ **b.** $510 - 283$ **c.** $24 \cdot 18$

4. List the other facts in each fact family.

   a. $6 + 5$ **b.** $14 - 7$ **c.** $7 \cdot 5$ **d.** $42 \div 7$

5. Which of the following sets is closed under addition? Multiplication? Explain why or why not.

   a. $\{0, 4, 8, 12, \ldots\}$ **b.** $\{4, 5, 6\}$

   c. $\{2, 4, 5, 6, 7, \ldots\}$ **d.** $\{1\}$

6. Identify the property that is demonstrated in each equation.

   a. $5 \cdot 6 = 30 \in W$

   b. $7 \cdot 4 = 4 \cdot 7$

   c. $7(8 + 5) = (7 \cdot 8) + (7 \cdot 5)$

   d. $5 \cdot 1 = 5$

   e. $5 + 0 = 5$ **f.** $(8 + 3) + 6 = (3 + 8) + 6$

   g. $3 \cdot 0 = 0$ **h.** $(4 + 5) + 1 = 4 + (5 + 1)$

7. Use expanded notation to compute each problem.

   a. $571 + 645$ **b.** $718 - 340$ **c.** $56 \cdot 34$

8. Use the lattice method to compute each problem.

   a. $4{,}579 + 2{,}091$ **b.** $451 \cdot 72$

9. Use the standard algorithm to compute each problem.

   a. $560 + 348$ **b.** $3412_{\text{five}} + 3312_{\text{five}}$

   c. $8{,}009 - 5{,}671$ **d.** $6312_{\text{seven}} - 4451_{\text{seven}}$

   e. $781 \cdot 62$ **f.** $231_{\text{four}} \cdot 23_{\text{four}}$

   g. $35\overline{)4560}$ **h.** $32_{\text{four}}\overline{)3312_{\text{four}}}$

10. Compute each problem mentally.

    a. $567 + 345$ **b.** $63 + 45 + 57 + 35$

    c. $472 - 361$ **d.** $1{,}493 - 547$

    e. $620 \cdot 34$ **f.** $215 \cdot 50$

    g. $4{,}563 \div 9$ **h.** $378 \div 6$

11. Round each number to the indicated place value.

    a. 409 to the nearest ten

    b. 19,003 to the nearest ten

    c. 3,871 to the nearest hundred

    d. 46,820 to the nearest thousand

12. Truncate each number to the indicated place value.

    a. 810 to the hundreds place value

    b. 43,927 to the thousands place value

    c. 29,993 to the hundreds place value

13. Use front-end methods with adjustment to estimate each problem.

    a. $5{,}672 + 4{,}623$ **b.** $7{,}904 - 2{,}102$ **c.** $4{,}879 \cdot 52$

14. Use flexible rounding to estimate each problem.

    a. $4{,}987 + 3{,}101$ **b.** $7{,}634 - 5{,}578$ **c.** $689 \cdot 56$

15. Write each expression using one exponent.

    a. $5 \cdot 5 \cdot 5 \cdot 5 \cdot 5 \cdot 5 \cdot 5$ **b.** $x^4 \cdot x^6$

    c. $(4^6)^3$ **d.** $3^3 \cdot 6^3$

    e. $(7^4 \cdot 7^9)^2$ **f.** $(2^5 \cdot 3^5)^7$

16. Use the order of operations to compute each problem.

    a. $(4 + 5)^2 \div 27 \cdot 6$ **b.** $\{[(35 - 21) + 28] \div 7\}^2$

    c. $6 \cdot 4 \div 8 + 16 - 12$

17. Compute each problem using a calculator.

    a. $(5{,}220 \div 36)^2$ **b.** $(546 - 324) \cdot 156$

    c. $25^2 \div 125 \cdot 63$

18. A certain high school has 314 freshman, 365 sophomores, 289 juniors, and 301 seniors. What is the total enrollment at the school?

19. Three brothers go fishing for a long four-day weekend. The three brothers catch a total of 25 fish on the first day, 13 on the second, 29 on the third, and 35 on the fourth. Estimate the total number of fish they caught over the course of the weekend.

20. A total of 3,500 chairs needs to be set up for a high school graduation. If 2,345 chairs have already been put out, how many more chairs are needed?

21. A certain state has a population of 7,569,034 people. One city in the state has a population of 3,782,820. About how many people live in the state but not in the city?

22. Estimate the number of gallons of water you use in a day.

23. How many different pairs of objects can be made between a set of 7 objects and a set of 9 objects?

24. A mail carrier makes 353 stops on a typical day. How many stops does she make in one week? Remember that mail is carried only six days a week.

25. A teacher has 26 students and wants to give each student three crayons. If she has a total of 85 crayons, how many will she have left after giving them to her students?

26. Another teacher has 165 sheets of construction paper to distribute equally to 27 students. How many sheets does each student get, and how many sheets are left over?

Answers to Chapter 4   Check Your Understandings and Activities

Check Your Understanding 4.1-A

1. a.

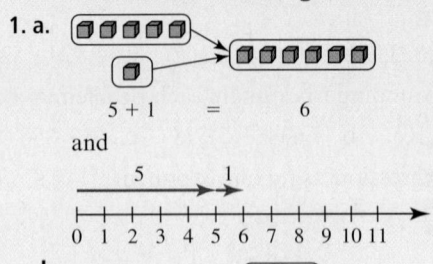

$$5 + 1 = 6$$

and

0 1 2 3 4 5 6 7 8 9 10 11

b.

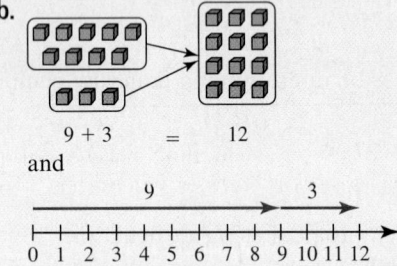

$$9 + 3 = 12$$

and

0 1 2 3 4 5 6 7 8 9 10 11 12

2. No. $1 + 4 = 5$ and $5 \notin S$.

3. a. Associative property.   **b.** Closure property.
   c. Additive identity property.   **d.** Commutative property.
   e. Additive identity property.   **f.** Commutative property.

4. a. $3 < 6$ because there exists the natural number $n = 3$ such that $3 + 3 = 6$.
   b. $4 > 2$ because there exists the natural number $n = 2$ such that $2 + 2 = 4$.
   c. $19 < 35$ because there exists the natural number $n = 16$ such that $19 + 16 = 35$.
   d. $2{,}345 > 692$ because there exists the natural number $n = 1{,}653$ such that $692 + 1{,}653 = 2{,}345$.

Check Your Understanding 4.1-B

1. Answers will vary.

2. Answers will vary.

3. a. If $7 - 5 = ?$, then $5 + ? = 7$. Because $5 + 2 = 7$, then $7 - 5 = 2$.   **b.** If $12 - ? = 4$, then $? + 4 = 12$. Because $8 + 4 = 12$, then $? = 8$.   **c.** If $? - 6 = 9$, then $9 + 6 = ?$ Hence, $? = 15$.

4. a. $9 - 6 = 3$, $3 + 6 = 9$, and $6 + 3 = 9$.   **b.** $5 + 5 = 10$.
   c. $3 + 8 = 11$, $11 - 8 = 3$, and $11 - 3 = 8$.

5. Answers will vary, but one possible answer is to let $a = 6$ and $b = 2$. Then, $a - b = 6 - 2 = 4$ and $b - a = 2 - 6 = -4$. Because $4 \neq -4$, $a - b \neq b - a$.

Check Your Understanding 4.2-A

1. a. $421 + 379 = 700 + 90 + 10 = 800$.   **b.** $712 + 318 = 1{,}000 + 20 + 10 = 1{,}030$.

2. a. $501 + 399 = (5 \cdot 100 + 0 \cdot 10 + 1) + (3 \cdot 100 + 9 \cdot 10 + 9) = (5 + 3) \cdot 100 + (0 + 9) \cdot 10 + (1 + 9) = (8 \cdot 100) + (9 \cdot 10) + 10 = 900$.   **b.** $172 - 97 = (1 \cdot 100 + 7 \cdot 10 + 2) - (9 \cdot 10 + 7) = (1 \cdot 100) + (7 - 9) \cdot 10 + (2 - 7) = (0 \cdot 100) + (16 - 9) \cdot 10 + (12 - 7) = 7 \cdot 10 + 5 = 75$.

3. a. 1,348.   **b.** 2,929.   **c.** 448.   **d.** 146.

4. a.

6	3
+ 2	8

9 1

b.

8	1	5
+ 7	9	4

1 6 0 9

5. $287 - 219 = (287 + 1) - (219 + 1) = 288 - 220 = 68$.

Check Your Understanding 4.2-B

1. a. 77.   **b.** 200.   **c.** 999.

2. a. 281.   **b.** 95.   **c.** 109.

3. a. 240.   **b.** 16,000.   **c.** 19,500.

4. a. 1,100.   **b.** 10,000.   **c.** 15,000.

5. Answers will vary.

Check Your Understanding 4.3-A

1. a. $5 \cdot 4 = 4 + 4 + 4 + 4 + 4 = 20$.
   b. $3 \cdot 8 = 8 + 8 + 8 = 24$.   **c.** $9 \cdot 1 = 1 + 1 + 1 + 1 + 1 + 1 + 1 + 1 + 1 = 9$.

2. Let $A = \{1, 2, 3, 4\}$ and $B = \{a, b, c\}$. Then $4 \cdot 3 = n(A \times B) = n(\{(1, a), (1, b), (1, c), (2, a), (2, b), (2, c), (3, a), (3, b), (3, c), (4, a), (4, b), (4, c)\}) = 12$.

3. a. $T = \{2, 4, 6, 8\}$ is not closed under multiplication because $4 \cdot 6 = 24$ and $24 \notin T$.
   b. $O = \{1, 3, 5, 7, \ldots\}$ is closed under multiplication because an odd number times an odd number is always odd.
   c. $\{1\}$ is closed under multiplication because $1 \cdot 1 = 1$.

4. a. Associative property.   **b.** Multiplicative identity property.   **c.** Zero multiplication property.
   d. Distributive of multiplication over addition property.
   e. Commutative property.
   f. Closure property.

Check Your Understanding 4.3-B

1. a. 3^3.   **b.** 7^6.   **c.** $(2a)^4$.

2. a. 4^6.   **b.** 14.   **c.** 6^{27}.   **d.** $(8y)^6$.

3. a. Because we can subtract 4 from 8 two times, $8 \div 4 = 2$.
   b. Because we can subtract 7 from 14 two times, $14 \div 7 = 2$.
   c. Because we can subtract 3 from 18 six times, $18 \div 3 = 6$.

4. a. If $15 \div 3 = ?$, then $15 = 3 \cdot ?$. Because $3 \cdot 5 = 15$, then $15 \div 3 = 5$.
   b. If $34 \div 17 = ?$, then $34 = 17 \cdot ?$. Because $17 \cdot 2 = 34$, then $34 \div 17 = 2$.
   c. If $78 \div 6 = ?$, then $78 = 6 \cdot ?$. Because $6 \cdot 13 = 78$, then $78 \div 6 = 13$.

5. a. Quotient = 3, remainder = 1.
   b. Quotient = 4, remainder = 2.
   c. Quotient = 13, remainder = 2.

Check Your Understanding 4.4-A

1. a. $28 \cdot 13 = 28(10 + 3) = 28 \cdot 10 + 28 \cdot 3 = (20 + 8)10 + (20 + 8)3 = 20 \cdot 10 + 8 \cdot 10 + 20 \cdot 3 + 8 \cdot 3 = 200 + 80 + 60 + 24 = 364$.
   b. $57 \cdot 34 = 57(30 + 4) = 57 \cdot 30 + 57 \cdot 4 = (50 + 7)30 + (50 + 7)4 = 50 \cdot 30 + 7 \cdot 30 + 50 \cdot 4 + 7 \cdot 4 = 1500 + 210 + 200 + 28 = 1{,}938$.

2. a. $98 \cdot 71 = 6{,}958$. **b.** $471 \cdot 62 = 29{,}202$.

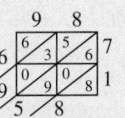

3. a. 19,404. **b.** 33,201. **c.** 251. **d.** 124 remainder 31.

4. a.
$$
\begin{array}{r}
4 \\
57\overline{)228} \\
-57 \\
\hline
171 \\
-57 \\
\hline
114 \\
-57 \\
\hline
57 \\
-57 \\
\hline
0
\end{array}
$$

b.
$$
\begin{array}{r}
8 \\
39\overline{)312} \\
-39 \\
\hline
273 \\
-39 \\
\hline
234 \\
-39 \\
\hline
195 \\
-39 \\
\hline
156 \\
-39 \\
\hline
117 \\
-39 \\
\hline
78 \\
-39 \\
\hline
39 \\
-39 \\
\hline
0
\end{array}
$$

5. a.
$$
\begin{array}{r}
139 \\
9 \\
30 \\
100 \\
24\overline{)3336} \\
-2400 \\
\hline
936 \\
-720 \\
\hline
216 \\
-216 \\
\hline
0
\end{array}
$$

b.
$$
\begin{array}{r}
241 \\
11 \\
30 \\
200 \\
35\overline{)8435} \\
-7000 \\
\hline
1435 \\
-1050 \\
\hline
385 \\
-385 \\
\hline
0
\end{array}
$$

Check Your Understanding 4.4-B

1. a. 1,280. **b.** 1,620. **c.** 614. **d.** 25.

2. a. 245,000. **b.** 6,900,000. **c.** 80. **d.** 9.

3. a. 32,666,033. **b.** 2,744. **c.** 325 **d.** 6,988.

Check Your Understanding 4.5-A

1. a. 18_{twelve}. **b.** 7_{eight}. **c.** 33_{seven}. **d.** 3_{six}.

2. All numbers are in base-four. Subscripts have been left out for ease in reading.

+	0	1	2	3
0	0	1	2	3
1	1	2	3	10
2	2	3	10	11
3	3	10	11	12

×	0	1	2	3
0	0	0	0	0
1	0	1	2	3
2	0	2	10	12
3	0	3	12	21

3. a. $6_{\text{seven}} \cdot 5_{\text{seven}} = 42_{\text{seven}} = 5_{\text{seven}} \cdot 6_{\text{seven}}$.

b. $2_{\text{five}} \cdot (3_{\text{five}} + 4_{\text{five}}) = 2_{\text{five}} \cdot 12_{\text{five}} = 24_{\text{five}}$ and $2_{\text{five}} \cdot 3_{\text{five}} + 2_{\text{five}} \cdot 4_{\text{five}} = 11_{\text{five}} + 13_{\text{five}} = 24_{\text{five}}$.

Check Your Understanding 4.5-B

1. a. 3000_{five}. **b.** 1037_{eight}. **c.** 203_{four}. **d.** 2831_{nine}.

2. a. 432_{five}. **b.** 22132_{six}. **c.** 102_{four} remainder 30_{four}. **d.** 44_{eight} remainder 11_{eight}.

Activity 4.1

a. Facts found by counting on by 1s or 2s are shown in blue.

+	0	1	2	3	4	5	6	7	8	9
0	0	1	2	3	4	5	6	7	8	9
1	1	2	3	4	5	6	7	8	9	10
2	2	3	4	5	6	7	8	9	10	11
3	3	4	5	6	7	8	9	10	11	12
4	4	5	6	7	8	9	10	11	12	13
5	5	6	7	8	9	10	11	12	13	14
6	6	7	8	9	10	11	12	13	14	15
7	7	8	9	10	11	12	13	14	15	16
8	8	9	10	11	12	13	14	15	16	17
9	9	10	11	12	13	14	15	16	17	18

b–c. Adding doubles are in blue, and adding doubles plus 1 or 2 are in red.

+	0	1	2	3	4	5	6	7	8	9
0	0	1	2	3	4	5	6	7	8	9
1	1	2	3	4	5	6	7	8	9	10
2	2	3	4	5	6	7	8	9	10	11
3	3	4	5	6	7	8	9	10	11	12
4	4	5	6	7	8	9	10	11	12	13
5	5	6	7	8	9	10	11	12	13	14
6	6	7	8	9	10	11	12	13	14	15
7	7	8	9	10	11	12	13	14	15	16
8	8	9	10	11	12	13	14	15	16	17
9	9	10	11	12	13	14	15	16	17	18

d–e. Sums to 10 are shown in blue, and using the identity property is in red.

+	0	1	2	3	4	5	6	7	8	9
0	0	1	2	3	4	5	6	7	8	9
1	1	2	3	4	5	6	7	8	9	10
2	2	3	4	5	6	7	8	9	10	11
3	3	4	5	6	7	8	9	10	11	12
4	4	5	6	7	8	9	10	11	12	13
5	5	6	7	8	9	10	11	12	13	14
6	6	7	8	9	10	11	12	13	14	15
7	7	8	9	10	11	12	13	14	15	16
8	8	9	10	11	12	13	14	15	16	17
9	9	10	11	12	13	14	15	16	17	18

f. By counting from the tables, we see these five strategies can help students learn 65 of the 100 facts.

Activity 4.2

Let a and b be whole numbers. $a < b$ if and only if there exists a natural number n, such that $b - a = n$.

Activity 4.3

a. $121 + 89 = 210$.

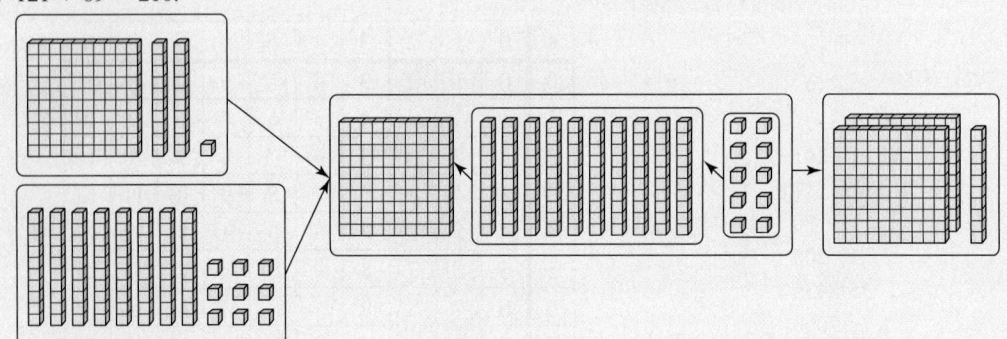

b. $681 + 385 = 1{,}066$.

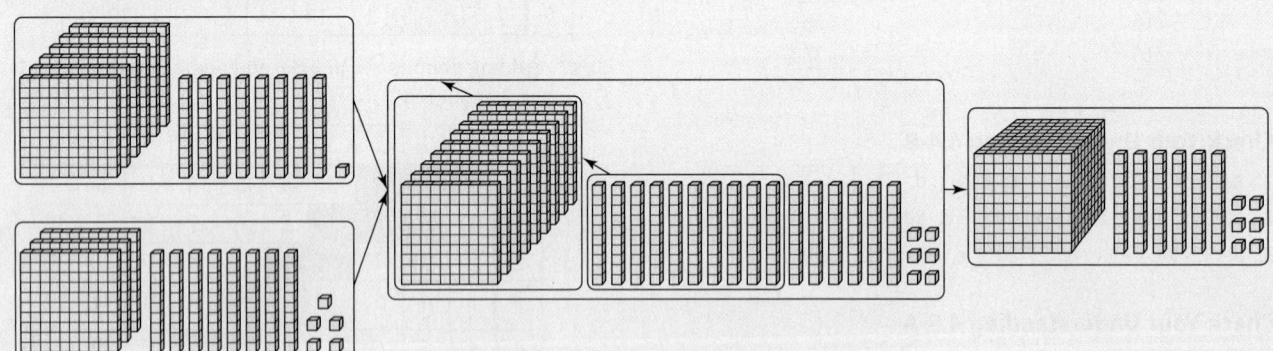

c. $131 - 53 = 78$.

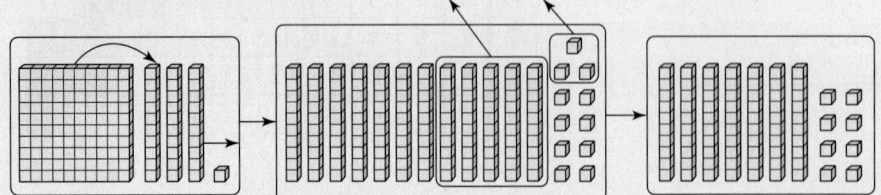

d. $205 - 147 = 58$.

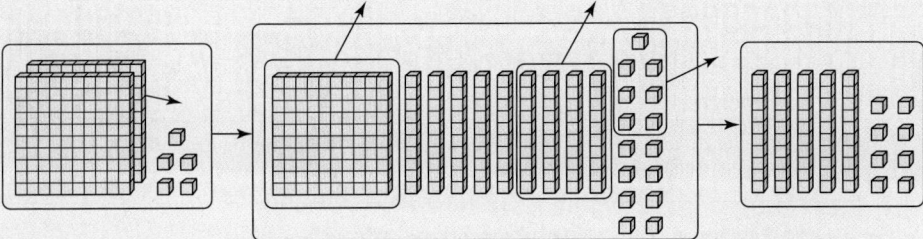

Activity 4.4

a. High estimate = $4,800. Low estimate = $2,780.

b. Actual highest cost is $4,593. This is about $200 below the high estimate. Actual lowest cost is $2,719. This is about $60 dollars below the low estimate.

Activity 4.5

a. Skip counting by 2s is shown by the third row or third column in Table 4.4. Skip counting by 3s is shown by the fourth row or column.

b–c. Multiples of 5 are shown in the sixth row or column, and multiples of 7 are shown in the eighth row or column. The multiplicative identity property is shown in the second row or column of the table.

d. The zero multiplication property is shown in the first row or column of the table.

Activity 4.6

Whenever the power on the base is 2, the rectangular array is always a square. Whenever it is a 3, the array is always a cube. The names of these exponents are "square" and "cube" because they describe the shape.

a. **b.**

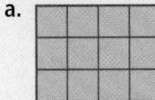 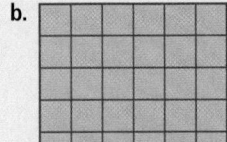

c. **d.**

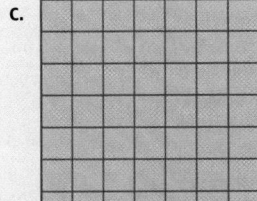 

e. **f.**

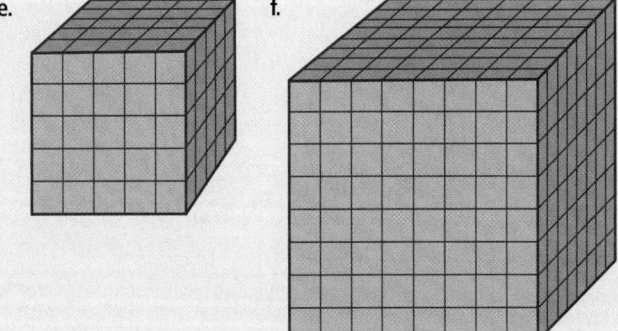

Activity 4.7

1. a. $25 \cdot 6 = 150$.

b. $32 \cdot 24 = 768$.

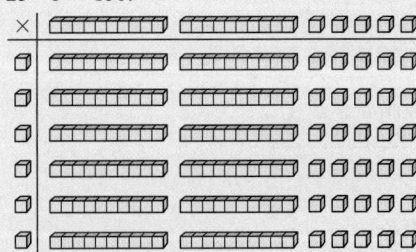

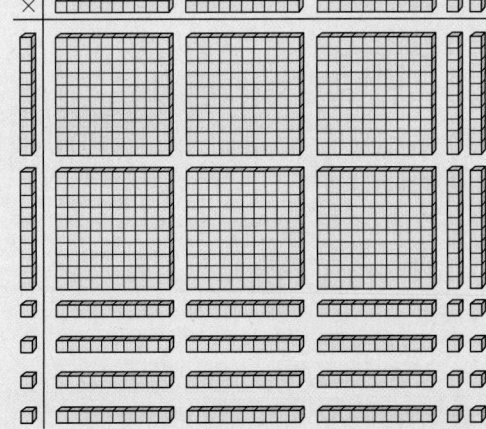

2. a.

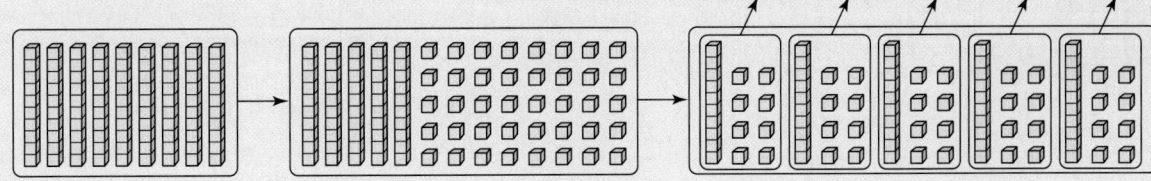

90   Exchange 4 longs for 40 units   Remove 5 groups of 18

b.

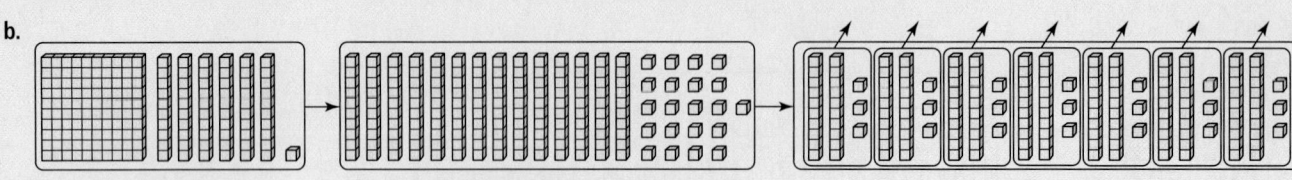

161   Exchange 1 flat for 10 longs and   Remove 7 groups of 23
                       2 longs for 20 units

Activity 4.8

a. $(4 \cdot 3)^2 - (8 + 6) = 130$ **b.** $16 \div 4 \div 2 \cdot (8 + 7) = 30$

c. $(6 + 3 - 5)^2 \cdot 4 = 64$ **d.** $(4 + 8) \cdot 15^2 \div 9 = 300$

Activity 4.9

A. 1. $2_{\text{eight}} + 7_{\text{eight}} = 11_{\text{eight}}$.

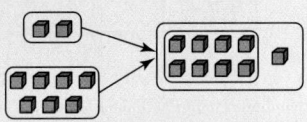

2. $15_{\text{nine}} - 8_{\text{nine}} = 6_{\text{nine}}$.

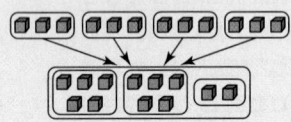

3. $4_{\text{five}} \cdot 3_{\text{five}} = 22_{\text{five}}$.

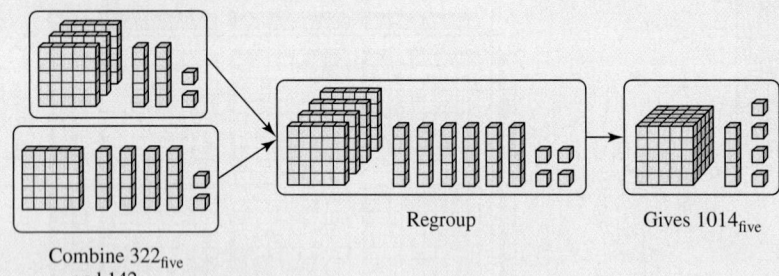

B. 1. $5_{\text{nine}} + 6_{\text{nine}} = 12_{\text{nine}}$.

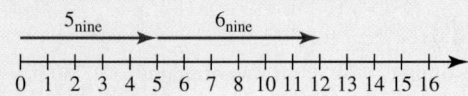

2. $14_{\text{seven}} - 6_{\text{seven}} = 5_{\text{seven}}$.

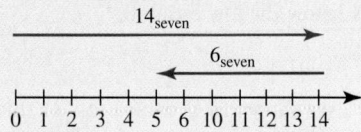

3. $5_{\text{six}} \cdot 2_{\text{six}} = 14_{\text{six}}$.

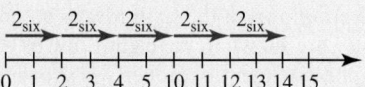

Activity 4.10

a. $322_{\text{five}} + 142_{\text{five}} = 1014_{\text{five}}$.

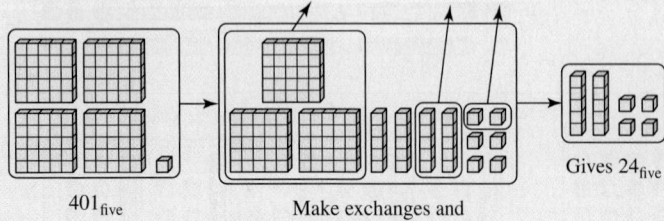

Combine 322_{five} Regroup   Gives 1014_{five}
and 142_{five}

b. $401_{\text{five}} - 322_{\text{five}} = 24_{\text{five}}$.

401_{five} Make exchanges and   Gives 24_{five}
                             remove 322_{five}

c. $23_{\text{five}} \cdot 13_{\text{five}} = 404_{\text{five}}$.

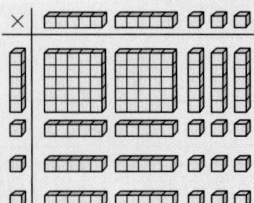

Basic Number Theory

© Image Source/Corbis

Opening Problem Allie is buying snacks for a graduation party. She plans to spend the same amount of money on soda as she does on potato chips. If each bottle of soda costs $0.99 and each bag of chips costs $1.29, what is the least amount of money she can spend on party food? At this cost, how many of each can she get?

© paparazzit/Shutterstock.com

5.1 Divisibility, Primes, and Composites

5.2 Greatest Common Factors and Least Common Multiples

5.3 Modular and Clock Arithmetic

© Image Source/Corbis

In the last three chapters, we have developed many of the basic ideas of number, numeration, and numerical operations. Our focus has been on content related to the goals in the NCTM Number and Operations Standard.

> ### ■ NCTM Number and Operations Standard
>
> Instructional programs from prekindergarten through grade 12 should enable all students to:
>
> - Understand numbers, ways of representing numbers, relationships among numbers, and number systems.
> - Understand meanings of operations and how they relate to one another.
> - Compute fluently and make reasonable estimates.

Source: NCTM STANDARDS Copyright © 2011 by NATIONAL COUNCIL OF TEACHERS OF MATHEMATICS. Reproduced with permission of NATIONAL COUNCIL OF TEACHERS OF MATHEMATICS.

We can now use our understanding of whole-number operations to revisit the first goal. Specifically, we can now use numerical operations to represent numbers in new ways and to study the relationships among them. The study of numbers, their properties, and the relationships among them is called number theory.

Basic number theory plays an important role in the mathematics that is taught in the elementary and middle grades. In the earliest grades, children use number theory to:

- Learn number sense by representing numbers in different ways.
- Classify numbers by simple characteristics, such as evens and odds.
- Learn facts about numerical operations, such as those that involve the sums and products of even and odd numbers.

In the late elementary grades, children begin to learn about factors, multiples, and divisibility as they continue to study multiplication and division. These ideas are important because children can use them to decompose numbers in different ways and to solve problems while working with fractions and algebraic expressions. Table 5.1 lists specific learning objectives that are associated with number theory and that are commonly taught in the elementary and middle grades.

Table 5.1 Classroom learning objectives	K	1	2	3	4	5	6	7	8
Determine whether a group of objects has an odd or even number of members.		X							
Find all factor pairs for a whole number less than 100.				X					
Determine whether a whole number less than 100 is a multiple of a given one-digit number.				X					
Determine whether a whole number less than 100 is prime or composite.				X					
Recognize that a whole number is a multiple of each of its factors.				X					
Find the GCF of two whole numbers less than or equal to 100 and the LCM of two whole numbers less than or equal to 12.							X		

Source: Adapted from the *Common Core State Standards for Mathematics* (*Common Core State Standards Initiative*, 2010).

Number theory topics are often scattered throughout the elementary curriculum because children are taught these ideas as they are needed. As a result, children often do not connect the ideas of number theory, nor do they associate basic number theory with numerical operations. These problems are compounded when the teachers themselves are uncertain about such relationships, causing them to teach number theory without connecting it to other content. For this reason, Chapter 5 provides a holistic view of elementary number theory by:

- Discussing divisibility, factors, multiples, primes, and composites.
- Discussing the greatest common divisor and the least common multiple along with the properties, computations, and applications relevant to them.
- Providing an overview of modular congruence and clock arithmetic.

Number theory is the study of patterns and relationships within the whole numbers. In this section, we decompose whole numbers into their fundamental structures by using ideas associated with multiples, factors, and divisibility. We then use these structures to characterize the whole numbers in new ways.

Multiples, Factors, and Divisibility

Much of number theory revolves around factors and multiples, which are closely connected to multiplication and division. If a and b are whole numbers and $a \cdot b = c$, then a and b are **factors** of c and c is a **multiple** of both a and b. For instance, because $4 \cdot 5 = 20$, 4 and 5 are factors of 20, and 20 is a multiple of both 4 and 5.

Finding multiples of a number is relatively easy because we can generate them by using multiplication. For instance, to find the multiples of 4, we can multiply 4 by every natural number:

$$4 \cdot 1 = 4 \qquad 4 \cdot 2 = 8 \qquad 4 \cdot 3 = 12 \qquad 4 \cdot 4 = 16 \ldots$$

Because the natural numbers are an infinite set, the multiples of four, $\{4, 8, 12, \ldots\}$ must also be an infinite set. Because we can replace 4 with any natural number, every natural number must have an infinite set of multiples.

Finding the factors of a number can be more difficult. If the number is small, we can use rectangular arrays. as we did with multiplication. To compute a product, we formed an array with two factors and then counted the objects in the array. To find factors, we reverse the process. Specifically, if we start with c blocks, and make a rectangular array that has the dimensions of a and b; then a and b must be factors of c, or $c = a \cdot b$.

Example 5.1 Finding Factors with Arrays

Representation

Use rectangular arrays to find the factors of 18.

Solution Using 18 blocks, we form as many rectangular arrays as possible. Because we can form arrays that are $1 \cdot 18$, $2 \cdot 9$, and $3 \cdot 6$, the factors of 18 are 1, 2, 3, 6, 9, and 18.

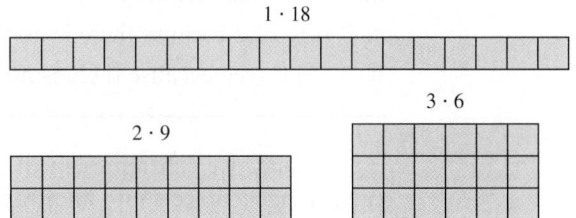

Another way to find factors is to use division. In general, if a is a factor of c and $a \neq 0$, then there exists a whole number b such that $a \cdot b = c$. In this case, b represents an unknown factor, so the missing-factor approach to division tells us that $b = c \div a$. In other words, if $c \div a = b$ with a remainder of zero, then both a and b are factors of c.

Example 5.2 Finding Factors with Division

Use division to find the factors of 36.

Solution To find the factors of 36, we divide it by every number that is less than 36 to find the quotients with a remainder of zero. The first five such quotients are:

$$36 \div 1 = 36 \qquad 36 \div 2 = 18 \qquad 36 \div 3 = 12 \qquad 36 \div 4 = 9 \qquad 36 \div 6 = 6$$

Notice that as the divisor increases, the quotient decreases. Once we reach 6, the divisor and the quotient are the same. We do not need to test more numbers because any factors bigger than 6 must have a quotient that is less than 6, and they already appear in the list. Consequently, the factors of 36 are 1, 2, 3, 4, 6, 9, 12, 18, and 36. ∎

In the last example, the main concern was not in computing a quotient, but rather in knowing whether the division had a remainder of zero. If it does, then we say that one number **divides** the other.

Definition of Divisibility

If a and b are any whole numbers with $b \neq 0$, then **b divides a,** written $b \mid a$, if and only if there exists a whole number c such that $a = b \cdot c$.

Initially, the definition of divisibility may appear to be the same as the missing-factor approach to division, but the two are fundamentally different. Division is an operation that always leads to a number. Divisibility, on the other hand, is a relationship between numbers that can only be true or false. Despite the difference in interpretation, the two are closely connected. Specifically, if the quotient $a \div b$ has a remainder of zero, then we conclude $b \mid a$ is true. If not, we write $b \nmid a$. Other ways to express the relationship of b divides a are:

a is **divisible** by b b is a **divisor** of a

a is a **multiple** of b b is a **factor** of a

Example 5.3   Using the Definition of Divisibility

Use the definition of divisibility to determine whether each statement is true or false.

a. $4 \mid 7$ **b.** $6 \mid 18$ **c.** $0 \mid 4$ **d.** $5 \nmid 9$

Solution

a. $4 \mid 7$ is false because there is no whole number c such that $4 \cdot c = 7$.

b. $6 \mid 18$ is true because there exists the whole number $c = 3$ such that $6 \cdot 3 = 18$.

c. $0 \mid 4$ is false because there is no whole number c such that $0 \cdot c = 4$.

d. $5 \nmid 9$ is true because there is no whole number c such that $5 \cdot c = 9$. ∎

Visually, we can represent divisibility with a rectangular array. If $b \mid a$, then we can arrange a objects into an array with b rows. If $b \nmid a$, we cannot. For instance, the arrays in Figure 5.1 show that 5 divides 30, but not 32.

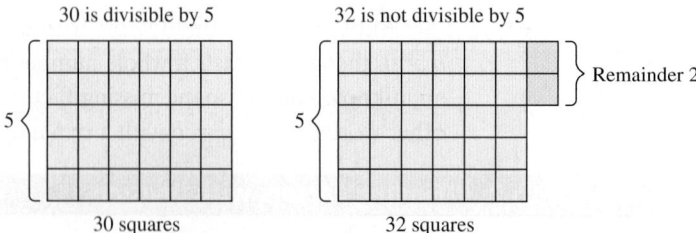

Figure 5.1 Divisibility with rectangular arrays

Divisibility satisfies several properties. For instance, not only does $4 \mid 24$ and $4 \mid 16$, but 4 also divides their sum, $4 \mid 40$, and their difference, $4 \mid 8$. Likewise, because $4 \mid 24$, it also divides any multiple of 24. For example, $4 \mid (2 \cdot 24)$, or $4 \mid 48$, and $4 \mid (3 \cdot 24)$, or $4 \mid 72$.

Although we have shown only one or two numerical examples of each property, they hold true in general. In other words, if a number divides two other numbers, it also divides their sum, their difference, and any multiples of the numbers.

Theorem 5.1 Properties of Divisibility

Let a, b, d, and n be any whole numbers.

1. If $d\,|\,a$ and $d\,|\,b$, then $d\,|\,(a + b)$.
2. If $d\,|\,a$ and $d\,|\,b$, then $d\,|\,(a - b)$ for $a \geq b$.
3. If $d\,|\,a$, then $d\,|\,(n \cdot a)$.

Each property in Theorem 5.1 can be proven more formally by using the definition of divisibility. The proof of part 1 is demonstrated in the next example. The proofs of parts 2 and 3 are similar and are left to the Learning Assessment.

Example 5.4 Proof of Theorem 5.1(1)

Reasoning

Show that if $d\,|\,a$ and $d\,|\,b$, then $d\,|\,(a + b)$.

Solution Suppose $d\,|\,a$ and $d\,|\,b$. By the definition of divisibility, there exists whole numbers j and k, such that $a = d \cdot j$ and $b = d \cdot k$. By adding a and b, we have $a + b = (d \cdot j) + (d \cdot k) = d(j + k)$. If we let $r = j + k$, then r is a whole number by the closure property. As a result, $a + b = d \cdot r$ for some whole number r. Consequently, $d\,|\,(a + b)$ by the definition of divisibility.

Divisibility Tests

One way to find factors quickly and easily is to use **divisibility tests,** which are simple rules that allow us to mentally determine whether one number divides another. Here, we explain some of the more common tests, which we group by method rather than by numerical order.

Divisibility Tests for 2, 5, and 10

- A natural number is divisible by 2 if and only if its ones digit is a 0, 2, 4, 6, or 8.
- A natural number is divisible by 5 if and only if its ones digit is a 0 or 5.
- A natural number is divisible by 10 if and only if its ones digit is a 0.

Example 5.5 Divisibility by 2, 5, and 10

Determine whether each number is divisible by 2, 5, or 10.

a. 4,765　　　**b.** 4,390　　　**c.** 63,153

Solution

a. Because 4,765 ends in a 5, it is divisible by 5, but not by 2 or 10.
b. Because 4,390 ends in a 0, it is divisible by 2, 5, and 10.
c. Because 63,153 ends in a 3, it is not divisible by 2, 5, or 10.

Explorations
Manual
5.1

To understand why these tests work, consider 765 and the test for 5. We can rewrite 765 as $(760 + 5) = (10 \cdot 76) + 5$. Because $5 \mid 10$, then $5 \mid (10 \cdot 76)$ by Theorem 5.1(3). We also know that $5 \mid 5$. Consequently, 5 must divide 765 because it is the sum of two numbers divisible by 5.

The argument can be generalized as follows. Suppose n is a natural number. If we divide n by 10, then the division algorithm tells us that $n = 10q + r$, where q and r are whole numbers and $0 \le r < 10$. Because $5 \mid 10$, then 5 must divide $10q$ because it is a multiple of 10. So the only concern is r, which happens to be the ones digit of n. The only single-digit numbers that are divisible by 5 are 0 and 5. If r is one of these numbers, then n is the sum of two numbers divisible by 5, making it divisible by 5. Otherwise, n is not. The tests for 2 and 10 can be justified in much the same way.

Divisibility Tests for 4 and 8

- A natural number is divisible by 4 if and only if the last two digits represent a number divisible by 4.
- A natural number is divisible by 8 if and only if the last three digits represent a number divisible by 8.

Example 5.6   Divisibility by 4 and 8

Determine whether each number is divisible by 4 or 8.

a. 1,032   **b.** 3,460

Solution

a. Because $4 \mid 32$ and $8 \mid 32$, both 4 and 8 divide 1,032.

b. Because $4 \mid 60$ and $8 \nmid 460$, then 4 divides 3,460, but 8 does not.

Explorations
Manual
5.2

To understand why these tests work, consider a four-digit number n and the divisibility test for 4. We begin by writing n in expanded form: $n = (a \cdot 1{,}000) + (b \cdot 100) + (c \cdot 10) + d$, where a, b, c, and d are single-digit numbers. Because $1{,}000 = 4 \cdot 250$ and $100 = 4 \cdot 25$, then 4 must divide both numbers. It follows from Theorem 5.1 that 4 also divides $(a \cdot 1{,}000)$, $(b \cdot 100)$, and $[(a \cdot 1{,}000) + (b \cdot 100)]$. This leaves the sum $(c \cdot 10) + d$, which is exactly the number given by the last two digits of n. If 4 divides $[(c \cdot 10) + d]$, then 4 must divide n because it is the sum of numbers divisible by 4. If 4 does not divide $[(c \cdot 10) + d]$, then 4 does not divide n.

A similar argument can be applied to the divisibility test for 8. However, because 8 does not divide 10 or 100 but does divide 1,000, we must consider the last three place values of n rather than the last two.

Divisibility for 2, 4, and 8 is closely connected. If a number is not divisible by 2, then it cannot be divisible by 4 or 8 because 4 and 8 are both multiples of 2. Likewise, if a number is not divisible by 4, then it cannot be divisible by 8. On the other hand, if a number is divisible by 8, then it must be divisible by 2 and 4 because 2 and 4 are factors of 8.

Divisibility Tests for 3 and 9

- A natural number is divisible by 3 if and only if the sum of its digits is divisible by 3.
- A natural number is divisible by 9 if and only if the sum of its digits is divisible by 9.

Example 5.7   Divisibility by 3 and 9

Determine whether each statement is true or false.

a. $3 \mid 452$ **b.** $9 \mid 5{,}283$

Solution

a. False. Because $4 + 5 + 2 = 11$ and $3 \nmid 11$, it follows that $3 \nmid 452$.

b. True. Because $5 + 2 + 8 + 3 = 18$ and $9 \mid 18$, it follows that $9 \mid 5{,}283$.

*Explorations
Manual
5.3*

To understand why the tests for 3 and 9 work, consider a four-digit number n and the divisibility test for 3. Again, we write n in expanded form: $n = (a \cdot 1{,}000) + (b \cdot 100) + (c \cdot 10) + (d \cdot 1)$. However, because 3 does not divide any power of 10, 3 will generally not divide any multiple of a power of 10. However, 3 does divide numbers very close to powers of 10. Particularly, 3 divides 9, 99, 999, and so on. Using these numbers, we can rewrite the expanded notation of n:

$$n = (a \cdot 1{,}000) + (b \cdot 100) + (c \cdot 10) + d$$
$$= a(999 + 1) + b(99 + 1) + c(9 + 1) + d$$
$$= a(999) + a(1) + b(99) + b(1) + c(9) + c(1) + d$$
$$= \underbrace{[a(999) + b(99) + c(9)]}_{\text{divisible by 3}} + \underbrace{(a + b + c + d)}_{\text{sum of digits in } n}$$

Because $3 \mid 999$, $3 \mid 99$, and $3 \mid 9$, it follows that $3 \mid [a(999) + b(99) + c(9)]$ by Theorem 5.1. This leaves us with $(a + b + c + d)$, which is exactly the sum of the digits of n. If this sum is divisible by 3, then 3 divides $a(999) + b(99) + c(9) + (a + b + c + d)$, and 3 divides n. If 3 does not divide this sum, it does not divide n. Because 9 also divides 9, 99, 999, and so on, we can justify the test for 9 in the same way. As with 2, 4, and 8, if a number is not divisible by 3, it cannot be divisible by 9, and if a number is divisible by 9, it is divisible by 3.

The next test, divisibility by 11, is somewhat more complicated in its practice.

Divisibility Test for 11

A natural number is divisible by 11 if and only if the sum of the digits in the place values that are even powers of 10 minus the sum of the digits in the place values that are odd powers of 10 is divisible by 11.

To use the divisibility test for 11, we compute the sums of alternating digits, and then subtract the two sums. If the difference is divisible by 11, the original number is divisible by 11. We can justify the test for 11 as we did with the tests for 3 and 9 with only a few minor adjustments.

Example 5.8   Divisibility by 11

Determine whether each number is divisible by 11.

a. 3,197,321   **b.** 405,405,022

Solution

a. The sums of alternating digits are $3 + 9 + 3 + 1 = 16$ and $1 + 7 + 2 = 10$. Because $16 - 10 = 6$ and $11 \nmid 6$, we conclude that $11 \nmid 3{,}197{,}321$.

b. The sums of alternating digits are $4 + 5 + 0 + 0 + 2 = 11$ and $0 + 4 + 5 + 2 = 11$. Because $11 - 11 = 0$ and $11 \mid 0$, we conclude that $11 \mid 405{,}405{,}022$.

The final divisibility test is for 6. If a number is divisible by 2 and 3, it must also be divisible by 6 because $6 = 2 \cdot 3$. If the number is divisible by 2 and not by 3 or by 3 and not by 2, the number is not divisible by 6.

Divisibility Test for 6

A natural number is divisible by 6 if and only if it is divisible by both 2 and 3.

The idea behind the divisibility test for 6 can be used to create other tests. For instance, because $3 \cdot 4 = 12$, a number is divisible by 12 if it is divisible by both 3 and 4. We do have to be careful in using these rules because not every combination of tests will work. For instance, even though $2 \cdot 6 = 12$, divisibility by 2 and 6 does not guarantee divisibility by 12. Specifically, 18 is divisible by 2 and 6, but not by 12. For a number to be divisible by 12, it must have at least one factor of 3 and two factors of 2. We get these factors if we use 3 and 4, but not if we use 2 and 6. The reason is that 2 and 6 have a common factor of 2, but 3 and 4 do not. In general, if two numbers divide another number and have no factors in common, then their product will also divide the number.

Theorem 5.2   Divisibility by Products

If a, b, and n are natural numbers such that $a\,|\,n$, $b\,|\,n$, and the only common factor of a and b is 1, then $(a \cdot b)\,|\,n$.

Example 5.9   Using Other Divisibility Tests

Determine whether each statement is true or false.

a. $15\,|\,45{,}045$ **b.** $45\,|\,24{,}825$

Solution

a. We use the tests for 3 and 5 because $3 \cdot 5 = 15$ and 3 and 5 have only a common factor of 1. Because $3\,|\,45{,}045$ and $5\,|\,45{,}045$, we conclude that $15\,|\,45{,}045$ is true.

b. We use the tests for 5 and 9 because $5 \cdot 9 = 45$ and 5 and 9 have only 1 as a common factor. Because $5\,|\,24{,}825$ but $9\nmid24{,}825$, then $45\,|\,24{,}825$ is false.

Check Your Understanding 5.1-A

1. What are the 10 smallest multiples of 13?
2. List the factors of 56.
3. Determine whether each statement is true or false. Justify your answer.
   a. $15\,|\,45$ **b.** $16\,|\,92$ **c.** $25\nmid225$ **d.** $34\nmid66$
4. Determine whether each statement is true or false. Justify your answer.
   a. If $6\,|\,36$ and $6\,|\,42$, then $6\,|\,(36+42)$.
   b. If $7\,|\,35$ and $7\nmid12$, then $7\nmid(35 \cdot 12)$.
   c. If $3\nmid(x-y)$, then $3\nmid x$ or $3\nmid y$.
5. Determine whether 2, 3, 4, 5, 6, 8, 9, 10, and 11 divide each number.
   a. 2,184   **b.** 2,160   **c.** 13,860   **d.** 360,360

Talk About It Use the number 6,237 to explain why the divisibility test for 9 works.

Problem Solving

Activity 5.1
Make a list of the divisibility tests that can be made by combining two of the divisibility tests for 2 through 11 in a fashion similar to that of the divisibility test for 6.

Even and Odd Numbers

We can now use our basic understanding of divisibility and factors to classify numbers in different ways. One way is to separate the numbers into those that are divisible by a given number and those that are not. For instance, divisibility by 2 can be used to classify the whole numbers as **even** or **odd.**

Definition of Even and Odd

A whole number is **even** if and only if it is divisible by 2.
A whole number is **odd** if and only if it is not divisible by 2.

In set-builder notation, we express the set of even numbers as $E = \{x \mid x = 2n, n \in W\} = \{0, 2, 4, 6, 8, \ldots\}$. Because any odd number is always one more than an even, we express the odd numbers as $O = \{x \mid x = 2n + 1, n \in W\} = \{1, 3, 5, 7, \ldots\}$.

Children work with even and odd numbers as early as kindergarten and first grade. Although they do not know about divisibility in these grades, they can still use rectangular arrays to separate the evens from the odds. See Figure 5.2. An even number can be represented with a rectangular array that has exactly two rows. An odd number cannot.

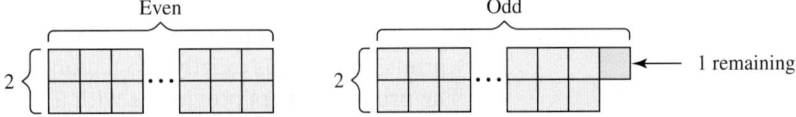

Figure 5.2 Even and odd with rectangular arrays

Once students understand the difference between even and odds, they can explore properties associated with these numbers. For instance, when students learn to add, many of them realize that the sum of two even numbers is even and that the sum of two odd numbers is even, as shown in Figure 5.3.

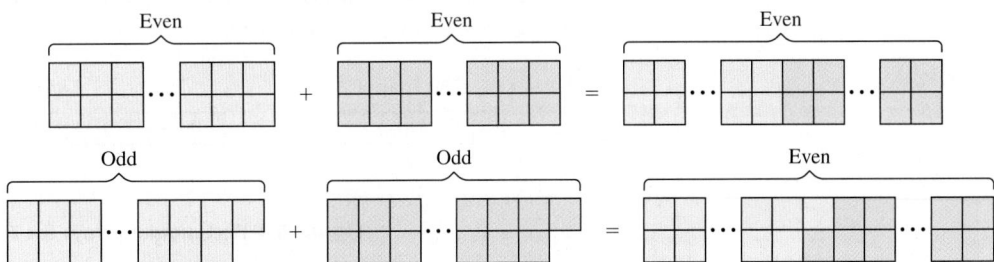

Figure 5.3 Sums of evens and odds

Example 5.10 The Sum of Two Odds

Reasoning

Prove that the sum of two odd numbers is even.

Solution Let m and n be two odd numbers. By the definition of odd, we have $m = 2j + 1$ and $n = 2k + 1$, where j and k are two whole numbers. Then:

$$m + n = (2j + 1) + (2k + 1)$$
$$= 2j + 2k + 2$$
$$= 2(j + k + 1)$$

If we let $r = (j + k + 1)$, then r is a whole number by the closure property and $m + n = 2r$. Consequently, by the definition of an even number, $m + n$ must be even.

Prime and Composite Numbers

Another way to classify the natural numbers is by their number of factors. Table 5.2 shows the first 25 natural numbers classified in this way. The numbers in the second column of the table play a special role in number theory because they have exactly two factors. We commonly refer to them as **primes.**

Table 5.2	Number of factors						
1 Factor	2 Factors	3 Factors	4 Factors	5 Factors	6 Factors	7 Factors	8 Factors
1	2, 3, 5, 7, 11, 13, 17, 19, 23	4, 9, 25	6, 8, 10, 14, 15, 21, 22	16	12, 18, 20		24

Definition of Prime and Composite Numbers

A natural number is **prime** if and only if it has exactly two factors. A natural number is **composite** if and only if it has three or more factors.

Explorations Manual 5.4  The definition and Table 5.2 enable us to learn several things about primes. First, examples of primes include 2, 3, 5, and 7, and examples of composites include 6, 8, 9, and 10. Second, 1 is neither prime nor composite because it has only a single factor: itself. Third, 2 is the only even prime because every other even number is divisible by 2. Fourth, every prime number has exactly two factors, which include 1 and the number itself.

The prime numbers provide us with a powerful way to decompose and characterize the natural numbers. However, before using them in this way, we need to be able to identify which numbers are prime. One way to identify primes is to use rectangular arrays. We have used arrays to find the factors of a number n by making arrays of different sizes using n blocks. If only one such array is possible, then n has only two factors and must be prime. If two or more arrays are possible, then n must be composite. For instance, 7 must be prime because we can make only a 1×7 array from 7 blocks (Figure 5.4). However, with 8 blocks, we can make a 1×8 and a 2×4 array, indicating that 8 must be composite.

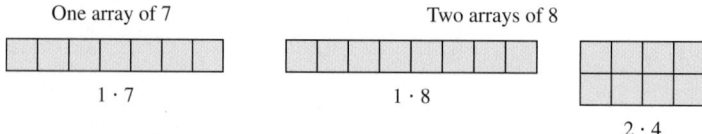

One array of 7

$1 \cdot 7$

Two arrays of 8

$1 \cdot 8$

$2 \cdot 4$

Figure 5.4 Rectangular arrays for 7 and 8

Explorations Manual 5.5 Rectangular arrays work well with small numbers, but become awkward as the numbers get larger. Another way to find slightly larger primes is to use a process called the Sieve of Eratosthenes with a **hundreds chart.** A hundreds chart is a table that lists the numbers 1 through 100 in 10 rows by 10 columns.

Example 5.11 | The Sieve of Eratosthenes

Use the Sieve of Eratosthenes to find all primes less than 100.

Solution The Sieve of Eratosthenes sifts out primes on a hundreds chart by looking at multiples of numbers rather than factors. We begin by crossing out 1 because 1 is not prime. The next number, 2, is circled because it is prime. We then cross out every multiple of 2, or all even numbers because they must be composite. Next, return to the beginning of the list to find the next number that has not been crossed out. Circle it as prime, and then cross out all of its multiples. We repeat the process until all numbers are either circled as prime or crossed out as not prime. The following table shows the results of the process: the prime numbers less than 100.

~~1~~	2	3	~~4~~	5	~~6~~	7	~~8~~	~~9~~	~~10~~
11	~~12~~	13	~~14~~	~~15~~	~~16~~	17	~~18~~	19	~~20~~
~~21~~	~~22~~	23	~~24~~	~~25~~	~~26~~	~~27~~	~~28~~	29	~~30~~
31	~~32~~	~~33~~	~~34~~	~~35~~	~~36~~	37	~~38~~	~~39~~	~~40~~
41	~~42~~	43	~~44~~	~~45~~	~~46~~	47	~~48~~	~~49~~	~~50~~
~~51~~	~~52~~	53	~~54~~	~~55~~	~~56~~	~~57~~	~~58~~	59	~~60~~
61	~~62~~	~~63~~	~~64~~	~~65~~	~~66~~	67	~~68~~	~~69~~	~~70~~
71	~~72~~	73	~~74~~	~~75~~	~~76~~	~~77~~	~~78~~	79	~~80~~
~~81~~	~~82~~	83	~~84~~	~~85~~	~~86~~	~~87~~	~~88~~	89	90
~~91~~	~~92~~	~~93~~	~~94~~	~~95~~	~~96~~	97	~~98~~	~~99~~	~~100~~

The Sieve of Eratosthenes can be used to find the primes in any list of numbers. However, as the numbers grow larger, the process becomes tedious. Another way to determine whether large numbers are prime is to use division. If a number is not divisible by any number smaller than itself, then it must be prime. Again, making all these divisions would be tedious. However, as shown in Example 5.2, small divisors have large quotients, and large divisors have small quotients. Consequently, we do not need to test larger possibilities because, if they were factors, then they would have already appeared when we found smaller factors. Consequently, we do not need to test every number. In fact, we only need to test the primes with squares less than the number.

Theorem 5.3 Prime Divisor Test

Let n be a natural number greater than 1. n is prime if it is not divisible by any prime p where $p^2 \leq n$.

Example 5.12 Using the Prime Divisor Test

Reasoning

Use the prime divisor test to determine whether 127 is prime.

Solution Theorem 5.3 indicates that we need to consider only primes with squares less than 127: 2, 3, 5, 7 and 11. By the divisibility tests for 2, 3, 5, and 11, we know that 127 is not divisible by these numbers. There is no test for 7, so we check it directly and find that $127 \div 7 = 18$, remainder 1. Because 127 is not divisible by 2, 3, 5, 7, or 11, the prime divisor test guarantees that 127 is prime.

To understand why Theorem 5.3 works, suppose n is a composite number such that $n = a \cdot b$, with $1 < a \leq b$. Multiply both sides of the inequality $a \leq b$ by a

to get $a \cdot a = a^2 \le a \cdot b = n$. Take the square root of both sides of the inequality: $a < \sqrt{n}$. This implies that we do not have to test any factor greater than $\sqrt{n}$. However, if p is a prime factor of a, including $p = a$, then p also divides n. Because $p \le a$, then $p^2 \le n$.

The search for large primes has gone on since the beginning of number theory. To date, two of the largest known primes are $2^{6972593} - 1$ and $2^{13466917} - 1$. Primes of this form, $2^n - 1$, are called **Mersenne primes** after the French monk, Marin Mersenne (1588–1648), who discovered the form. Finding large primes is difficult, but with the use of technology and sophisticated computing techniques. even larger primes will undoubtedly be found. We know this because there are infinitely many primes.

Theorem 5.4   Infinitude of the Primes

The set of primes is infinite.

The first known proof of Theorem 5.4 was given by Euclid around 300 B.C.E., and it is considered to be a classic example of proof by contradiction. The proof is based on the premise that there are either a finite or an infinite number of primes. If it can be shown that it is not one possibility, then it must be the other. The proof is given in the next example.

Example 5.13   Infinitely Many Primes

Reasoning   Prove that there are infinitely many primes.

Solution Suppose that there is only a finite number of primes, which include 2, 3, 5, 7, . . . , p, where p is the largest prime. If $N = (2 \cdot 3 \cdot 5 \cdot 7 \cdot \ldots \cdot p) + 1$, then N must be 1, prime, or composite. Because N is the sum of the product of all primes and 1, N is greater than 1, so it cannot be equal to 1. N is also greater than p, which means that N cannot be prime because p is the largest prime. If N is composite, then it must have a prime factor. However, if we divide N by any prime 2, 3, 5, . . . , p, the remainder is always 1. This implies that N cannot be composite. So N cannot be 1, prime, or composite, but this is impossible because N must be one of the three. Consequently, the original assumption that there are only a finite number of primes must be wrong, and we conclude that there are infinitely many primes.

Composite Numbers and Factorizations

In number theory, it is often useful to write numbers as the product of their factors. This is called **factoring,** and the resulting product is called a **factorization** of the number. Every natural number n, including the primes, has at least one factorization: $1 \cdot n = n$. Composite numbers, however, can have many factorizations, which we consider to be different only if they contain different numbers. For instance, 12 has four factorizations:

$$12 = 1 \cdot 12 \qquad 12 = 2 \cdot 6 \qquad 12 = 3 \cdot 4 \qquad 12 = 2 \cdot 2 \cdot 3$$

Of the four factorizations, one consists of all primes, which we call a **prime factorization.** Prime factorizations can be useful, and in many cases, we can use divisibility tests and **factor trees** to find them.

Example 5.14 Using Factor Trees

Use factor trees to find the prime factorizations of 72 and 240.

Solution Factor trees are a visual way to decompose a composite number into its prime factors. Here are possible factor trees for 72 and 240:

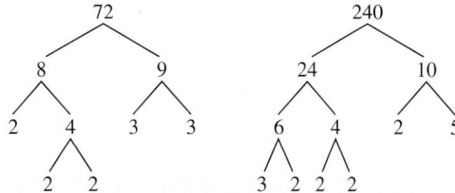

By collecting the primes at the end of each branch, we get the prime factorization of each number: $72 = 2 \cdot 2 \cdot 2 \cdot 3 \cdot 3 = 2^3 \cdot 3^2$ and $240 = 2 \cdot 2 \cdot 2 \cdot 2 \cdot 3 \cdot 5 = 2^4 \cdot 3 \cdot 5$.

As the last example shows, the convention is to write prime factorizations in exponential form with the primes listed in ascending order. Also, composite numbers can have many different factor trees. For instance, two other factor trees for 72 are shown in Figure 5.5. In either case, the prime factorization of 72 is still $2^3 \cdot 3^2$, suggesting that composite numbers have only one prime factorization.

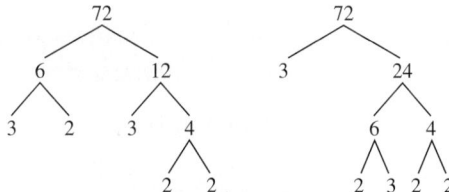

Figure 5.5 Two possible factor trees for 72

Theorem 5.5 Fundamental Theorem of Arithmetic

Each composite number can be expressed as the product of primes in exactly one way, except possibly for the order of the factors.

The Fundamental Theorem of Arithmetic, also called the **Unique Factorization Theorem,** guarantees two things about composite numbers. First, every composite number has exactly one prime factorization. Second, no two composite numbers have the same prime factorization. In other words, the prime factorization completely characterizes a number and its factors. For instance, consider 108. Because $108 = 2^2 \cdot 3^3$, any power of 2 less than or equal to 2^2 divides 108, including $2^0 = 1$, $2^1 = 2$, and $2^2 = 4$. Similarly, $3^0 = 1$, $3^1 = 3$, $3^2 = 9$, and $3^3 = 27$ also divide 108. By considering different combinations of the powers of 2 with the powers of 3, we get all the factors of 108. A table like Table 5.3 helps to organize and compute the possibilities. The table shows that the factors of 108 are 1, 2, 3, 4, 6, 9, 12, 18, 27, 36, 54, and 108.

Table 5.3 Factors of 108

×	3^0	3^1	3^2	3^3
2^0	$2^0 \cdot 3^0 = 1$	$2^0 \cdot 3^1 = 3$	$2^0 \cdot 3^2 = 9$	$2^0 \cdot 3^3 = 27$
2^1	$2^1 \cdot 3^0 = 2$	$2^1 \cdot 3^1 = 6$	$2^1 \cdot 3^2 = 18$	$2^1 \cdot 3^3 = 54$
2^2	$2^2 \cdot 3^0 = 4$	$2^2 \cdot 3^1 = 12$	$2^2 \cdot 3^2 = 36$	$2^2 \cdot 3^3 = 108$

The next theorem tells us how to use the prime factorization to count the factors of a composite number.

Theorem 5.6   Number of Divisors of a Natural Number

If the prime factorization of a number n is $n = p_1^{e_1} \cdot p_2^{e_2} \cdot \ldots \cdot p_m^{e_m}$, then n has $(e_1 + 1)(e_2 + 1) \ldots (e_m + 1)$ factors.

Theorem 5.6 states that to determine how many factors a number has, write its prime factorization in exponential form, add one to each exponent, and then multiply the sums.

Example 5.15   Determining the Number of Divisors

How many divisors does each number have?

a. 360   b. $2^4 \cdot 3^2 \cdot 5 \cdot 7$

Solution

a. Because $360 = 2^3 \cdot 3^2 \cdot 5$, it has $(3 + 1)(2 + 1)(1 + 1) = 4 \cdot 3 \cdot 2 = 24$ factors.

b. $2^4 \cdot 3^2 \cdot 5 \cdot 7$ has $(4 + 1)(2 + 1)(1 + 1)(1 + 1) = 5 \cdot 3 \cdot 2 \cdot 2 = 60$ factors.

Check Your Understanding 5.1-B

1. Use the Sieve of Eratosthenes to classify each number as prime or composite.
   a. 17   b. 27   c. 31   d. 69   e. 91
2. Use Theorem 5.3 to classify each number as prime or composite.
   a. 213   b. 317   c. 241   d. 435   e. 421
3. Use a factor tree to find the prime factorization of each number. Then determine how many factors it has.
   a. 56   b. 125   c. 450   d. 1,680

Talk About It Why do you think it is useful to study factorizations, primes, and composites in the elementary grades?

Problem Solving

Activity 5.2
Part A. Use a table like Table 5.3 to find the factors of 392.
Part B. Devise a strategy for listing all the factors of a number that has three primes in its prime factorization. Use it to find the factors of 600.

SECTION 5.1   Learning Assessment

■ Understanding the Concepts

1. What is the difference between a factor and a multiple?

2. What is the difference between division and divisibility? How are the two ideas connected?

3. Explain how to use the divisibility test for each number.
   a. 2   b. 3   c. 4   d. 6   e. 11

4. Describe several ways that divisibility can be used to classify the whole numbers into different sets?

5. What is the difference between a prime number and a composite number?

6. Why is 1 neither prime nor composite?

7. Describe three methods of finding primes.

8. What does the Fundamental Theorem of Arithmetic imply about composite numbers?

9. How can the prime factorization of a number be used to find all of its factors?

■ Representing the Mathematics

10. Write three statements that are equivalent to "5 is a divisor of 30."

11. Write three statements that are equivalent to "28 is a multiple of 7."

12. Draw a Venn diagram that shows the relationship between the numbers that are divisible by:

   a. 2, 4, and 8. **b.** 2, 5, and 10.

13. Let a, b, c, and n be any whole numbers. Give a verbal interpretation of each statement.

   a. If $c \mid a$ and $c \mid b$, then $c \mid (a + b)$.

   b. If $c \mid a$ and $c \nmid b$, then $c \nmid (a - b)$ for $a \geq b$.

   c. If $c \mid a$, then $c \mid (n \cdot a)$.

14. Use rectangular arrays to find the factors of each number.

   a. 9 **b.** 18 **c.** 24

15. Use rectangular arrays to show that:

   a. $4 \mid 16$, but $4 \nmid 25$.

   b. $7 \mid 21$, but $7 \nmid 32$.

16. Use rectangular arrays to show that the sum of an even and an odd is odd.

17. Use rectangular arrays to explain why 13 is prime and 21 is composite.

18. Make three factor trees of 24.

19. Table 5.2 classifies the first 25 natural numbers by the number of their factors. Make a table that classifies the first 50 natural numbers in this way.

   a. How many numbers have only two factors?

   b. How many numbers have four factors?

   c. Which column(s) in your table has (have) the most numbers in it?

   d. Which column(s) in your table has (have) the least numbers in it?

■ Exercises

20. List the first 10 multiples of 12.

21. List the first 10 multiples of 36.

22. Find all of the factors of each number.

   a. 8 **b.** 40 **c.** 84

23. Find all of the factors of each number.

   a. 35 **b.** 60 **c.** 120

24. Determine whether each statement is true or false.

   a. $8 \mid 48$ **b.** $14 \mid 36$ **c.** $5 \nmid 60$ **d.** $9 \nmid 59$

25. Determine whether each statement is true or false.

   a. If $5 \mid 25$ and $5 \mid 15$, then $5 \mid (25 - 15)$.

   b. If $6 \mid 12$ and $3 \mid 12$, then $(6 + 3) \mid 12$.

   c. If $6 \mid 48$ and $6 \nmid 13$, then $6 \nmid (48 \cdot 13)$.

   d. If $4 \mid 40$ and $4 \nmid 38$, then $4 \nmid (40 + 38)$.

26. Answer each question based on what you know about divisibility.

   a. Can 756 boxes be equally stacked 8 boxes high?

   b. Can 504 eggs be grouped equally into dozens of eggs?

   c. Can 30,576 trees be planted equally into rows of 72 trees?

   d. Can lottery winnings of $2,784,600 be divided equally among 11 people?

27. Determine whether 2, 3, 4, 5, 6, 8, 9, 10, and 11 divide each number.

   a. 135 **b.** 198 **c.** 220 **d.** 248

28. Determine whether 2, 3, 4, 5, 6, 8, 9, 10, and 11 divide each number.

   a. 2,400 **b.** 5,005 **c.** 102,960 **d.** 765,765

29. Place a digit in the blank so that the number 85, 64__ is:

   a. Divisible by 3.

   b. Divisible by 2.

   c. Divisible by 6.

   d. Divisible by 2 and 5.

   e. Divisible by 4, but not 8.

30. Use the Sieve of Eratosthenes to determine whether each number is prime or composite.

   a. 19 **b.** 29 **c.** 67 **d.** 87 **e.** 93

31. Use the Sieve of Eratosthenes to find all prime numbers between 100 and 200.

32. Use the prime factor test to determine whether each number is prime or composite.

   a. 193 **b.** 253 **c.** 473 **d.** 539

33. Use the prime factor test to find the prime numbers between 350 and 360.

34. Fill in the missing numbers in each factor tree.

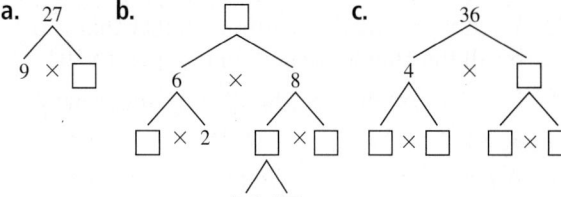

35. Find the prime factorization of each number.
    a. 132   **b.** 245   **c.** 432

36. Find the prime factorization of each number.
    a. 1,008   **b.** 2,016   **c.** 5,940

37. How many factors does each number have?
    a. 45   **b.** 120   **c.** 360

38. How many factors does each number have?
    a. $n = 2^7 \cdot 3^2 \cdot 7^2$
    b. $n = 2^{10} \cdot 3^4 \cdot 5^7$
    c. $n = 3^2 \cdot 5^3 \cdot 7 \cdot 13^4$

39. Which of the following must be a factor of $2^2 \cdot 3^3 \cdot 5$?
    a. 12   **b.** 25   **c.** $2 \cdot 3 \cdot 5^2$ **d.** $1 \cdot 2 \cdot 3^2$

40. If a number is divisible by 24, what other numbers is it divisible by?

41. What is the smallest set of numbers that must be tested to determine whether 413 is prime?

42. What is the smallest set of numbers that must be tested to determine whether 611 is prime?

■ Problems and Applications

43. What is the smallest natural number that has exactly five factors? Six factors? Seven factors?

44. Find a four-digit number that uses the digits 1, 3, 5, and 7 only once such that the number is divisible by 11, the digits in the two highest place values represent a number divisible by 5, and the digits in the two lowest place values represent a prime number.

45. Fill in the blanks so that the given number is divisible by 2, 3, and 4, but not by 5, 8, or 9.

    19, 5 __ __

46. A natural number is divisible by 7 if and only if the number formed by subtracting twice the last digit from the number formed by all the digits but the last is divisible by 7. For instance, consider 2,268. Because $\underbrace{226}_{\substack{\text{All digits} \\ \text{but last}}} - \underbrace{2(8)}_{\substack{\text{2 times} \\ \text{last digit}}} = 210$ and 210 is divisible by 7, then 2,268 is divisible by 7. Use this method to determine whether each number is divisible by 7.
    a. 1,155   **b.** 3,789   **c.** 5,891   **d.** 8,715

47. Find the smallest number that is divisible by three odd primes.

48. What is the smallest natural number that is divisible by all the numbers less than or equal to 10?

49. Find two prime numbers that result in each product.
    a. 22   **b.** 65   **c.** 161   **d.** 899

50. A number has exactly 16 factors and is divisible by 6, 8, and 10. What is the number?

51. The primes 11 and 13, 17 and 19, and 29 and 31 are called twin primes because one is only two more than the other.
    a. What other sets of twin primes exist in the numbers 20 through 100?
    b. Find three pairs of twin primes between 100 and 200.

52. **a.** Two numbers are **amicable** if each number is the sum of the proper factors of the other. Verify that 220 and 284 are amicable.
    b. One number in an amicable pair is 1,184. What is the other number?

53. **a.** A number is **perfect** if it is equal to the sum of all its proper factors. For instance, 6 is perfect because its proper factors are 1, 2, and 3 and $6 = 1 + 2 + 3$. Verify that 8,128 is perfect.
    b. A number is **deficient** if the sum of its proper factors is less than the number, and it is **abundant** if the sum of its proper factors is greater than the number. Determine whether each number is perfect, deficient, or abundant.
    i. 28   **ii.** 35   **iii.** 60   **iv.** 144   **v.** 196

54. Eggs are often sold in containers holding one-and-a-half dozen eggs. Can 1,680 eggs be equally packaged in this way?

55. A social club with 165 members wants to subdivide into committees of equal size. A committee can have no fewer than 6 members and no more than 14 members. If each person can only serve on one committee, how many people serve on each committee?

56. A couple is expecting to have 284 guests at their wedding reception. Each table at the banquet hall seats six people. Will all the tables be full with no extra people?

57. Juan used his debit card to purchase 15 quality pens to give to his friends. When he went to record the transaction in his checkbook, he realized that he had lost the receipt. He recalls that the pens cost $39.65 without tax. Can this number be correct?

58. Can a rectangular pan of brownies of dimensions 9×13 in. be divided into 39 brownies all the same size?

59. Julie works at a bakery and notices that the number of donuts made today, when divided by 6, 7, 8, or 9, always has a remainder of 1. What is the smallest possible number of donuts that were made today?

■ Mathematical Reasoning and Proof

60. The only two consecutive prime numbers are 2 and 3. Can you explain why?

61. Which of the following is (are) true for whole numbers a, b, and n?
    a. $1|n$ for all whole numbers n.
    b. If $a|n$, then $a|n+1$.
    c. If $a|n$, then $a|n^2$.
    d. If $a|b$, then $an|b$.

62. Which of the following is (are) true for whole numbers a and b, where $a > b$?
    a. If a and b are even, then $2|(a-b)$.
    b. If a and b are odd, then $2|(a-b)$.
    c. If a is even and b is odd, then $2|(a-b)$.
    d. If $2|a$ and $3|b$, then $6|(a \cdot b)$.

63. If $d \nmid a$ and $d \nmid b$, can we conclude $d \nmid (a+b)$? Explain.

64. a. Find five numbers that have exactly three factors. Find a pattern, and use it to make a generalization as to which numbers are likely to have exactly three factors.
    b. Find five numbers, each having exactly four factors. Find a pattern, and use it to make a generalization as to which numbers are likely to have exactly four factors.

65. The expression $2^n - 1$, where n is a whole number, often produces a prime number. Compute $2^n - 1$, for $n = 0, 1, 2, 3, 4,$ and 5. Which of the results is (are) prime?

66. Another formula that can yield primes is $n^2 + n + 17$. If n equals $0, 1, 2, 3, 4,$ or 5, verify that $n^2 + n + 17$ is prime.

67. A **Sophie Germain prime** is an odd prime p such that $2p + 1$ is also prime. Check that 3, 5, 11, and 23 are all Sophie Germain primes. Can you find three others?

68. The **Goldbach conjecture,** named after Christian Goldbach (1690–1764), states that every even number greater than 4 can be written as the sum of two odd primes. It is one of the famous unsolved problems in mathematics because no one has been able to prove it or provide a counterexample.
    a. Write the following even numbers as the sum of two odd primes.
       i. 6 ii. 8 iii. 14 iv. 26 v. 50
    b. Is it the case that every even number greater than 4 can be written as *only one* sum of odd prime numbers? If not, provide a counterexample.

69. Use expanded notation and Theorem 5.1 to explain why 340 is divisible by 10.

70. Use expanded notation and Theorem 5.1 to explain why 288 is divisible by 9.

71. Use an argument like the one in Example 5.4 to show that if $d|a$ and $d|b$, then $d|(a-b)$ for $a \geq b$.

72. Use an argument like the one in Example 5.4 to show that if $d|a$, then $d|(n \cdot a)$.

73. Use an argument like the one in Example 5.10 to show that the sum of two even numbers is even.

74. Use an argument like the one in Example 5.10 to show that the product of two odd numbers is odd.

Mathematical Communication

75. Search the Internet to find three other divisibility tests that were not addressed in this section. Choose three of them, and write a short description of how each one works.

76. The word "composite" often means "to be made of separate, yet interconnected parts." Why is this a reasonable word to describe the set of numbers that have three or more factors?

77. Write a story problem that can be solved by using division. Write another that can be solved using divisibility. What changes in language did you have to make to distinguish the approaches?

78. Describe how skip counting can be used to find the multiples of a number.

79. How would you explain to a group of students how to use a calculator to find all the factors of 132?

80. A student claims that because 20 is divisible by 2 and 4, it must be also divisible by 8. If you were the teacher, how would you respond to this student's claim?

81. Thad claims that 12 is even because it is divisible by 2, but it is also odd because it is divisible by 3. How do you respond to Thad's thinking?

Building Connections

82. Divisibility, factors, and multiples play an important part in manipulating fractions. Can you think of how?

83. With a group of your peers, discuss the merits of teaching the divisibility tests to students in the fourth, fifth, and sixth grades. As part of your discussion, talk about what difficulties you think students might encounter in learning this content.

84. The notions of even and odd are often taught before children learn division and divisibility. List some different ways that you could tell a child the difference between an even and an odd number without using division or divisibility?

85. Number theory as a mathematical subject can be traced back to the ancient Greek Pythagoras. Do an Internet search on Pythagoras and his group, the Pythagoreans. What interesting facts about numbers did they discover?

SECTION 5.2   Greatest Common Factors and Least Common Multiples

We now use divisibility and prime numbers to study greatest common factors and least common multiples. Greatest common factors and least common multiples are useful not only in solving problems but also in the upcoming work with fractions.

The Greatest Common Factor

In many situations in our daily lives, we need to consider the factors of two or more numbers. For instance, suppose we are making holiday fruit baskets, and we have 24 apples and 18 oranges. If we want every basket to have the same number of apples and oranges, then the number of baskets we can make must be a factor of 24 and 18. Because the common factors of the two numbers are 1, 2, 3, and 6, we have the following choices:

> 1 basket with 24 apples and 18 oranges
> 2 baskets, each with 12 apples and 9 oranges
> 3 baskets, each with 8 apples and 6 oranges
> 6 baskets, each with 4 apples and 3 oranges

If we want to find the largest number of fruit baskets we can make, we are trying to find not just common factors, but the **greatest common factor** (GCF). The greatest common factor is also called the greatest common divisor (GCD).

Definition of the Greatest Common Factor

If a and b are two whole numbers that are both not equal to zero, then the **greatest common factor** of a and b, written GCF(a, b), is the largest whole number that divides both a and b.

Children often find the GCF confusing because it is defined with two words that seemingly contradict one another. Children often associate the word "factor" with something small, yet the word "greatest" implies something large. In fact, the terms are not contradictory because the GCF is literally the largest factor that two numbers have in common. Note that the GCF is not defined when both whole numbers are zero. In this case, there cannot be a largest common factor because zero is divisible by every natural number, no matter how large. However, if only one of the numbers is 0, then the GCF can be found. Specifically, if a is a nonzero whole number, then the GCF(a, 0) = a because zero is divisible by a and a is its own greatest factor.

Example 5.16   Finding Greatest Common Factors

Find the following:

a. GCF(8, 12)   **b.** GCF(16, 24, 32)   **c.** GCF(5, 11)   **d.** GCF(8, 9)

Solution Because the numbers are small, we use what we know about factors to find each GCF.

a. The common factors of 8 and 12 are 1, 2, and 4. Because 4 is the largest, the GCF(8, 12) = 4.

b. Because 8 is the largest factor of 16, 24, and 32, the GCF(16, 24, 32) = 8.

c. One is the only number that divides 5 and 11, so the GCF(5, 11) = 1.

d. One is the only number that divides 8 and 9, so the GCF(8, 9) = 1.

Example 5.16 demonstrates several facts about the GCF. First, the GCF is always less than or equal to the smaller of the two numbers: $GCF(a, b) \leq a$, if $a \leq b$. Second, if necessary, we can extend the GCF to more than two numbers. Third, the GCF can be equal to 1. This occurs not only when the numbers are prime but also when two composite numbers have no common factor. In the latter case, we consider the numbers to be prime with respect to one another, so we call them **relatively prime.**

Definition of Relatively Prime Numbers

Two nonzero whole numbers, a and b, are **relatively prime,** if the $GCF(a, b) = 1$.

Computing the GCF is relatively easy if the numbers are small or if their factors are well-known. However, if the numbers are large or have many factors, finding the GCF can be more difficult. Fortunately, we can turn to several methods in these circumstances.

Set-Intersection Method

In the set-intersection method, we list the factors of both numbers, and then use the set intersection to find those that are common to both lists. The GCF is then the largest number in the intersection.

Example 5.17 Using the Set Intersection to Find the GCF

Use the set-intersection method to find the GCF(56, 60).

Solution We list the factors of each number:

Factors of $56 = F_{56} = \{1, 2, 4, 7, 8, 14, 28, 56\}$

Factors of $60 = F_{60} = \{1, 2, 3, 4, 5, 6, 10, 12, 15, 20, 30, 60\}$

We then take the intersection of the two sets:

$F_{56} \cap F_{60} = \{1, 2, 4, 7, 8, 14, 28, 56\} \cap \{1, 2, 3, 4, 5, 6, 10, 12, 15, 20, 30, 60\}$
$= \{1, 2, 4\}.$

Because the largest number in $F_{56} \cap F_{60}$ is 4, the GCF(56, 60) = 4.

A Venn diagram is a useful way to visualize this procedure. See Figure 5.6. The GCF will be the largest number in the overlapping region of the circles.

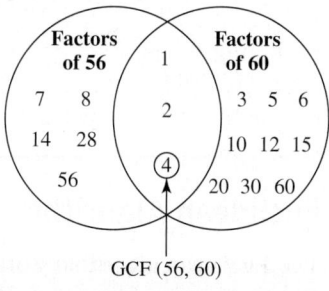

GCF (56, 60)

Figure 5.6 Common factors of 56 and 60

Prime Factorization Method

Explorations
Manual
5.6

The set-intersection method will work with any numbers but becomes awkward when the numbers are large or have many factors. Another, more efficient method is to use prime factorizations. Recall that the prime factorization of a number tells exactly which primes are factors of the number. So, if two numbers are divisible by the same prime, then the GCF must also be divisible by that prime. Consequently, the GCF can be found by taking the product of every prime that appears in both prime factorizations.

Example 5.18 **Using Prime Factorizations to Find the GCF**

Use prime factorizations to find GCF(315, 450).

Solution We first find the prime factorization of each number:
$$315 = 3 \cdot 3 \cdot 5 \cdot 7 \qquad 450 = 2 \cdot 3 \cdot 3 \cdot 5 \cdot 5$$
Both numbers are divisible by two 3s and one 5, and they share no other prime factors. Consequently, the GCF must also be divisible by two 3s and one 5, and we conclude that the GCF(315, 450) = $3 \cdot 3 \cdot 5 = 45$.

We can also work with the prime factorizations in exponential form. In this case, take any prime that is common to both factorizations, raise it to the lowest power that occurs in either factorization, and then multiply to find the GCF. For instance, for $315 = 3^2 \cdot 5 \cdot 7$ and $450 = 2 \cdot 3^2 \cdot 5^2$, the GCF is the product of the common powers of primes: $3^2 \cdot 5 = 45$.

Example 5.19 **Relatively Prime Numbers**

Are 1,568 and 2,025 relatively prime?

Solution One way to make the determination is to look at their prime factorizations:
$$1{,}568 = 2^5 \cdot 7^2 \qquad 2{,}025 = 3^4 \cdot 5^2$$
Because there are no common primes between the factorizations, the GCF(1568, 2025) = 1. As a result, 1,568 and 2,025 are relatively prime.

Example 5.20 **Packaging School Supplies**

Application

A school store plans to make and sell packages that contain pencils, erasers, and notepads. If the store has 600 pencils, 450 erasers, and 300 notepads, what is the largest number of packages that can be made if no items are left over? How many pencils, erasers, and notepads are in each package?

Solution The first question can be answered by finding the GCF(600, 450, 300). The prime factorizations of each number are:
$$600 = 2^3 \cdot 3 \cdot 5^2 \qquad 450 = 2 \cdot 3^2 \cdot 5^2 \qquad 300 = 2^2 \cdot 3 \cdot 5^2$$
By taking the product of the common powers of primes, we find that the store can make GCF(600, 450, 300) = $2 \cdot 3 \cdot 5^2 = 150$ packages. In this case, each package will have $600 \div 150 = 4$ pencils, $450 \div 150 = 3$ erasers, and $300 \div 150 = 2$ notepads.

Euclidean Algorithm

Explorations
Manual
5.7

The **Euclidean algorithm** works well with large numbers and is based on long division and the division algorithm. Recall that the division algorithm states that if we divide two numbers, a and b, we get a quotient q and a remainder r such that $a = bq + r$, where $0 \le r < b$. If d is a factor of both a and b, then d is also a factor of $r = a - bq$ by Theorem 5.1. Consequently, any common factors of a and b are also common factors of b and r. Because this includes the GCF, then GCF(a, b) = GCF(b, r).

Theorem 5.7 **The Division Algorithm and the GCF**

If a and b are any two nonzero whole numbers with $a \ge b$ and $a = bq + r$ for whole numbers q and r, with $r < b$, then GCF(a, b) = GCF(b, r).

Because r is always smaller than a, computing the GCF(b, r) is often easier than computing the GCF(a, b). In the Euclidean algorithm, we use Theorem 5.7 repeatedly to make a sequence of divisions in which the remainder is always divided into the previous divisor. The numbers grow smaller until we get a remainder of 0 or we get two numbers that have a GCF that we can compute directly. In the case of a zero remainder, we use the fact that the GCF(a, 0) = a. Once we have finished the computations, we use Theorem 5.7 and conclude that the GCF of the original numbers is equal to the GCF of the last two numbers.

Example 5.21   Using the Euclidean Algorithm to Find the GCF

Use the Euclidean algorithm to find the GCF(3185, 5005).

Solution We begin the algorithm by dividing 5,005 by 3,185. We then continue to take the remainder from the division and divide it into the previous divisor until we reach the GCF(455, 0). Because GCF(455, 0) = 455, we conclude that GCF(3185, 5005) = GCF(455, 0) = 455.

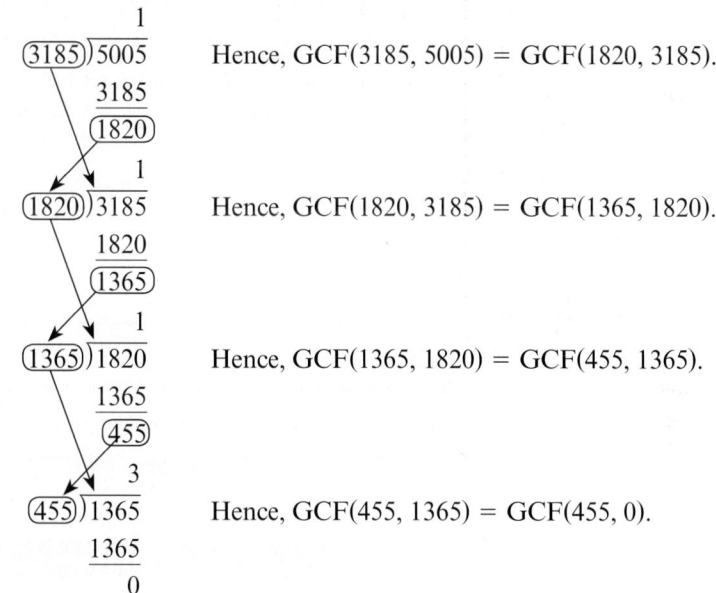

Hence, GCF(3185, 5005) = GCF(1820, 3185).

Hence, GCF(1820, 3185) = GCF(1365, 1820).

Hence, GCF(1365, 1820) = GCF(455, 1365).

Hence, GCF(455, 1365) = GCF(455, 0).

Calculators and the GCF

A calculator can help compute the GCF in at least two ways. For one, a calculator facilitates the divisions in the Euclidean algorithm, particularly if the calculator has an integer divide key, $\boxed{\text{INT} \div}$. Recall that this key returns the quotient and remainder rather than a decimal value. With this key, the following keystrokes can be used to find the GCF(5544, 4914).

$$5544 \; \boxed{\text{INT} \div} \; 4914 \; \boxed{=} \quad \text{displays} \quad \frac{1}{Q} \; \frac{630}{R}$$

$$4914 \; \boxed{\text{INT} \div} \; 630 \; \boxed{=} \quad \text{displays} \quad \frac{7}{Q} \; \frac{504}{R}$$

$$630 \; \boxed{\text{INT} \div} \; 504 \; \boxed{=} \quad \text{displays} \quad \frac{1}{Q} \; \frac{126}{R}$$

$$504 \; \boxed{\text{INT} \div} \; 126 \; \boxed{=} \quad \text{displays} \quad \frac{4}{Q} \; \frac{0}{R}$$

A remainder of 0 means that the last divisor must be the GCF. As a result, GCF(5544, 4914) = 126.

Many scientific and fraction calculators also have a built-in GCF key, or [GCF]. On many calculators, this feature is found under the [MATH] menu. To use it, select the function, key in the two numbers, and then press enter. For instance, we can compute the GCF(27720, 40425) by means of the following keystrokes:

[GCF] 27720 [,] 40425 [)] [ENTER]

The result is GCF(27720, 40425) = 1,155.

Check Your Understanding 5.2-A

1. Use the set-intersection method to find the GCF(56, 72).

2. Use the prime-factorization method to find the following:
   a. GCF(180, 168)   **b.** GCF(132, 504)

3. Use the Euclidean algorithm to find the GCF(1196, 6254).

4. Use a calculator to compute each number.
   a. GCF(960, 1,125)   **b.** GCF(1980, 13205)

5. A florist has 252 carnations, 84 roses, and 126 tulips that she can use to make flower arrangements. If every arrangement gets the same number of each type of flower and all the flowers are used, what is the largest number of flower arrangements she can make?

Talk About It Use your own words to explain why the Euclidean algorithm works.

Representation

Activity 5.3
Cuisenaire Rods© are another common manipulative used in the elementary classroom. The rods come in the 10 colors and lengths shown here.

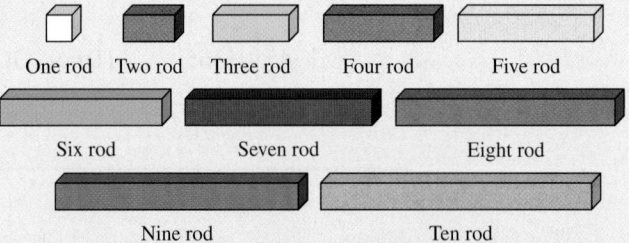

Rods of different lengths can be combined to form chains, or trains, that represent numbers larger than 10. For instance, 12 can be represented in a number of ways:

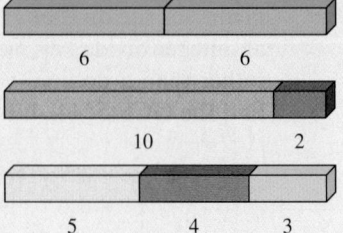

Devise a strategy to use Cuisenaire Rods© to find the greatest common factor of two numbers. Use it to find the following:

a. GCF(12, 9)   **b.** GCF(16, 8)   **c.** GCF(15, 20)

The Least Common Multiple

Like common factors, common multiples are useful in many situations. For instance, suppose we have a science quiz every four days and a math quiz every six days. If we had both today, on what days will we have both again? To answer the question, we list the multiples of each number, and then find those that are common to both.

Multiples of 4: 4, 8, 12, 16, 20, 24, 28, 32, 36, . . .
Multiples of 6: 6, 12, 18, 24, 30, 36, . . .

The common multiples are 12, 24, 36, . . . , which means we will have both quizzes in 12 days, 24 days, and so on. Of course, because we are usually the most concerned about the next time we have both quizzes, the smallest common number to the lists, 12, is of greatest interest. This number is called the **least common multiple (LCM).**

> **Definition of Least Common Multiple**
>
> If a and b are two natural numbers, then the **least common multiple** of a and b, written LCM(a, b), is the smallest natural number divisible by both a and b.

Like the GCF, children can find the LCM confusing because it is defined with two seemingly contradictory words. Children often associate the word "multiple" with something large, yet the word "least" implies something small. In fact, there is no contradiction: The LCM is literally the smallest multiple that two numbers have in common.

Notice in the definition that a and b are limited to the natural numbers; Zero is excluded as a possibility. The reason is that $n \cdot 0 = 0$ for every natural number n, making zero a multiple of every number. So the LCM of zero and any other number will always be zero: LCM(a, 0) = 0 for all a.

Example 5.22 Finding Least Common Multiples

Find the following:

a. LCM(5, 7)   **b.** LCM(4, 5)   **c.** LCM(9, 12)   **d.** LCM(3, 4, 6)

Solution Because the numbers are small, we use what we know about multiples to find each LCM.

a. LCM(5, 7) = 35.   **b.** LCM(4, 5) = 20.
c. LCM(9, 12) = 36.   **d.** LCM(3, 4, 6) = 12.

Example 5.22 demonstrates three facts about least common multiples. First, if a and b are prime or relatively prime, then the LCM is the product of the two numbers: LCM(a, b) = $a \cdot b$. If a and b share a common factor, then the LCM is less than the product of the two numbers: LCM(a, b) < $a \cdot b$. Second, the LCM is always greater than or equal to the largest of a or b. Third, we can find the LCM of more than two numbers.

Set-Intersection Method

The set-intersection method can also be used to find least common multiples. Now, however, instead of factors, we list multiples of both numbers until we find one that is common to both lists. If the numbers are large or relatively prime, then many multiples may have to be listed before we find a common one.

Example 5.23 **Using the Set Intersection to Find the LCM**

Use the set-intersection method to find the LCM(35, 40).

Solution The multiples of 35 and 40 include:

Multiples of 35 = M_{35} = {35, 70, 105, 140, 175, 210, 245, 280, . . . }
Multiples of 40 = M_{40} = {40, 80, 120, 160, 200, 240, 280, . . . }.

Because the first multiple in both lists is 280, then LCM(35, 40) = 280.

The hundreds chart can also be used with the set-intersection method to find LCMs.

Example 5.24 **Using a Hundreds Chart**

Representation   Use the hundreds chart to find the LCM(7, 9).

Solution In the following table, we highlight the multiples of 7 and the multiples of 9 in different colors (blue for the multiples of 7 and red for the multiples of 9). The LCM is the first number at which the two colors coincide. Because this happens at 63, the LCM(7, 9) = 63.

1	2	3	4	5	6	7	8	9	10
11	12	13	14	15	16	17	18	19	20
21	22	23	24	25	26	27	28	29	30
31	32	33	34	35	36	37	38	39	40
41	42	43	44	45	46	47	48	49	50
51	52	53	54	55	56	57	58	59	60
61	62	63	64	65	66	67	68	69	70
71	72	73	74	75	76	77	78	79	80
81	82	83	84	85	86	87	88	89	90
91	92	93	94	95	96	97	98	99	100

Prime Factorization Method

In the prime factorization method, we first find the prime factorization of each number, and write it in exponential form. Then we take any prime that occurs in either factorization, raise it to the highest power to which it occurs, and multiply.

Example 5.25 **Using Prime Factorizations to Find the LCM**

Use the prime factorization method to find the LCM(56, 60).

Solution The prime factorizations of the numbers are $56 = 2^3 \cdot 7$ and $60 = 2^2 \cdot 3 \cdot 5$. The primes that occur in either factorization raised to the highest power include 2^3, 3, 5, and 7. Multiplying these, we get the LCM(56, 60) = $2^3 \cdot 3 \cdot 5 \cdot 7 = 840$.

The prime factorization method works because any multiple of a number must be divisible by any factor of the number. For instance, because 56 is divisible by 2^3 and 7, then any multiple of 56 must also be divisible by 2^3 and 7. Similarly, any multiple of

60 must be divisible by 2^2, 3, and 5. Consequently, any common multiple of 56 and 60 must be divisible by 2^3, 3, 5, and 7. Multiplying these powers of primes gives the LCM. Notice that it is not necessary to include 2^2 in the computation, because 2^3 is a multiple of 2^2. By including a factor of 2^3 in the computation, we automatically include a factor of 2^2.

Example 5.26	Revolving Gears

Application

A machine has three interlocking gears with 76, 24, and 42 teeth, respectively. What is the fewest number of revolutions that the large gear has to make in order for the three gears to return to their original position together?

Explorations Manual 5.8

Solution One revolution of any gear is equivalent to all the teeth on the gear passing by the initial starting point. Any multiple of the number of teeth therefore brings the gear back to its starting position. For all the gears to return to the starting point at the same time, the number of teeth that have passed by must be a common multiple of 76, 24, and 42. Consequently, it is the LCM that requires the fewest number of revolutions. The prime factorization of each number is:

$$76 = 2^2 \cdot 19 \qquad 24 = 2^3 \cdot 3 \qquad 42 = 2 \cdot 3 \cdot 7$$

Because the LCM of the numbers is LCM(76, 24, 42) = $2^3 \cdot 3 \cdot 7 \cdot 19 = 3{,}192$, the large gear must make $3{,}192 \div 76 = 42$ revolutions.

Euclidean Algorithm Method

Another method for computing the LCM uses a particular relationship between the GCF(a, b), the LCM(a, b), and the product $a \cdot b$. We can discover the relationship by looking for a pattern in the specific examples given in Table 5.4. The table indicates that the product $a \cdot b$ is always equal to the product of the GCF(a, b) and LCM(a, b). This insight leads to the next theorem.

Table 5.4	Comparing $a \cdot b$ to the GCF and LCM			
a and b	**GCF(a, b)**	**LCM(a, b)**	**GCF(a, b) $\cdot$ LCM(a, b)**	**$a \cdot b$**
$a = 4, b = 6$	2	12	24	24
$a = 6, b = 9$	3	18	54	54
$a = 5, b = 6$	1	30	30	30
$a = 10, b = 15$	5	30	150	150

Theorem 5.8   The Product of the GCF and LCM

If a and b are two nonzero whole numbers, then GCF(a, b) $\cdot$ LCM(a, b) = $a \cdot b$.

We can provide some justification for Theorem 5.8 by using two specific numbers: $a = 9,450$ and $b = 17,640$. First, we find the prime factorizations for a and b:

$$a = 9,450 = 2 \cdot 3^3 \cdot 5^2 \cdot 7 \qquad b = 17,640 = 2^3 \cdot 3^2 \cdot 5 \cdot 7^2$$

Then, from the prime factorization method:

$$\text{GCF}(a, b) = 2 \cdot 3^2 \cdot 5 \cdot 7 \qquad \text{LCM}(a, b) = 2^3 \cdot 3^3 \cdot 5^2 \cdot 7^2$$

Working with the numbers in exponential form and applying the commutative and associative properties:

$$\begin{aligned}
\text{GCF}(a, b) \cdot \text{LCM}(a, b) &= (2 \cdot 3^2 \cdot 5 \cdot 7) \cdot (2^3 \cdot 3^3 \cdot 5^2 \cdot 7^2) \\
&= (2 \cdot 3^3 \cdot 5^2 \cdot 7) \cdot (2^3 \cdot 3^2 \cdot 5 \cdot 7^2) \\
&= 9,450 \cdot 17,640 \\
&= a \cdot b
\end{aligned}$$

This demonstrates that the GCF(a, b) and the LCM(a, b) provide the powers of primes needed to make the product of a and b. Because this is the case for any nonzero whole numbers, Theorem 5.8 is generally true.

Theorem 5.8 becomes useful in computing the LCM when the numbers involved are particularly large. Because we can use the Euclidean algorithm to find the GCF of large numbers, using Theorem 5.8 to find the LCM is sometimes called the **Euclidean algorithm method.**

Example 5.27 Using the Euclidean Algorithm to Find the LCM

Use the GCF to find the LCM(9438, 10164).

Solution First, we use the Euclidean algorithm to find that the GCF(9438, 10164) = 726. Next, by placing the known values in Theorem 5.8 and rearranging the terms, we have

$$\begin{aligned}
\text{LCM}(9438, 10164) &= \frac{9,438 \cdot 10,164}{\text{GCF}(9438, 10164)} \\
&= \frac{95,927,832}{726} \\
&= 132,132
\end{aligned}$$

Calculators and the LCM

Making the computations in the Euclidean algorithm method is significantly easier with a calculator. Also, many scientific and fraction calculators offer a built-in LCM function. The LCM key is usually found near the GCD key and is used in a similar fashion.

Check Your Understanding 5.2-B

1. Use the set-intersection method to find the LCM(24, 36).
2. Use the prime-factorization method to find each number.
   a. LCM(180, 168)   b. LCM(132, 504)
3. Use Theorem 5.8 to find each number.
   a. LCM(30, 45)   b. LCM(715, 1001)
4. Use a calculator to compute each number.
   a. LCM(845, 1035)   b. LCM(13452, 9876)

Talk About It Why do you think the term "least common multiple" might be confusing for some students learning it for the first time?

Representation

Activity 5.4

Devise a strategy for using trains of Cuisenaire Rods© to find the LCM of two numbers. Use your strategy to find:

a. LCM(4, 6).   **b.** LCM(5, 7).   **c.** LCM(6, 9).

SECTION **5.2**   Learning Assessment

■ Understanding the Concepts

1. **a.** What is the GCF of two numbers _a_ and _b_?

   b. What is the LCM of two numbers _a_ and _b_?

2. If _a_ and _b_ are prime, why is the GCF(_a_, _b_) = 1?

3. **a.** What does it mean for two numbers to be relatively prime?

   b. If two numbers are relatively prime, must they necessarily be prime? Explain.

4. Why is zero a multiple of every number but a factor of no number?

5. How is it possible to always find a common multiple of any two natural numbers _a_ and _b_?

6. How can prime factorizations be used to find the GCF and the LCM of two numbers?

7. What fact allows us to use the Euclidean algorithm to find the GCF of two numbers?

8. **a.** Why is the GCF(_a_, _b_) always less than or equal to the smaller of the two numbers?

   b. Why is the LCM(_a_, _b_) always greater than or equal to the larger of the two numbers?

   c. Why is the GCF(_a_, _b_) always less than or equal to the LCM(_a_, _b_)?

■ Representing the Mathematics

9. Use the factors of 48, 64, and 80 to fill in the Venn diagram.

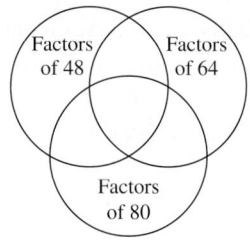

10. Use the multiples of 12, 16, and 24 less than 100 to fill in the Venn diagram.

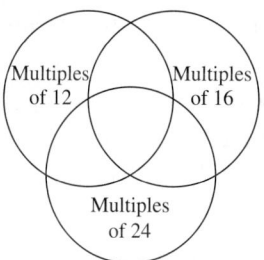

11. Draw a Venn diagram of each pair of sets.

   a. $A = \{x | x$ is a factor of 24$\}$
   $B = \{x | x$ is a factor of 40$\}$

   b. $C = \{x | x$ is a factor of 36$\}$
   $D = \{x | x$ is a factor of 60$\}$

12. Draw a Venn diagram of each pair of sets.

   a. $A = \{x | x$ is a multiple of 4 less than 50$\}$
   $B = \{x | x$ is a multiple of 6 less than 50$\}$

   b. $C = \{x | x$ is a multiple of 15 less than 100$\}$
   $D = \{x | x$ is a multiple of 20 less than 100$\}$

13. Use a hundreds chart to find each number.

   a. LCM(4, 5)   **b.** LCM(6, 8)   **c.** LCM(9, 10)

14. **a.** Use a hundreds chart to find all the common multiples of 5 and 7 less than 100.

   b. Use a hundreds chart to find all the common multiples of 9 and 12 less than 100.

15. Describe the pattern made by the multiples of each number when they are highlighted on a hundreds chart.

   a. 2   **b.** 5   **c.** 3   **d.** 15   **e.** 12

16. What does the following diagram of Cuisenaire© Rods show about 3 and 4?

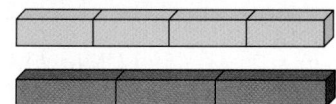

17. Use Cuisenaire© Rods to demonstrate how to find each number.

a. GCF(8, 12)   **b.** GCF(10, 16)

c. LCM(3, 5)   **d.** LCM(2, 7)

■ Exercises

18. Use the set-intersection method to find each number.

a. GCF(24, 36)   **b.** GCF(60, 75)

c. LCM(25, 35)   **d.** LCM(24, 56)

19. Use the set-intersection method to find each number.

a. GCF(32, 40)   **b.** GCF(80, 96)

c. LCM(18, 27)   **d.** LCM(45, 55)

20. Use the prime factorization method to find the GCF and the LCM for each pair of numbers.

a. 135 and 165   **b.** 504 and 220

c. 1,050 and 825

21. Use the prime factorization method to find the GCF and LCM for each pair of numbers.

a. 900 and 975   **b.** 1,452 and 594

c. 2,925 and 2,640

22. Extend the prime factorization method to find each number.

a. GCF(504, 120, 315)   **b.** LCM(504, 120, 315)

23. Find the GCF(a, b) and LCM(a, b) in exponential form if $a = 2^3 \cdot 3^5 \cdot 5^4$ and $b = 2^4 \cdot 3^2 \cdot 5^4$.

24. Find the GCF(a, b) and LCM(a, b) in exponential form if $a = 2^3 \cdot 5^3 \cdot 7$ and $b = 3^6 \cdot 5 \cdot 11^2$.

25. Use the Euclidean algorithm to find each number.

a. GCF(5420, 5332)   **b.** GCF(10352, 1430)

26. Use the Euclidean algorithm to find each number.

a. GCF(15444, 2640)   **b.** GCF(15555, 14205)

27. Which pairs of numbers are relatively prime?

a. 24 and 36   **b.** 19 and 21   **c.** 35 and 49

d. 198 and 423   **e.** 243 and 343

28. Use Theorem 5.8 and the GCF to find the following:

a. LCM(27, 63).   **b.** LCM(345, 275).

c. LCM(261, 1452).

29. The GCF of two numbers is 4, and their product is 96. What are the two numbers?

30. The LCM of two numbers is 18, and their product is 54. What are the two numbers?

31. If GCF(a, b) = $2^3 \cdot 5^2 \cdot 7$, LCM(a, b) = $2^4 \cdot 5^3 \cdot 7^2$, and $a = 2^4 \cdot 5^3 \cdot 7$, what is b?

32. If GCF(a, b) = $2^2 \cdot 3^4 \cdot 5^2$, LCM(a, b) = $2^4 \cdot 3^4 \cdot 5^3$, and $a = 2^2 \cdot 3^4 \cdot 5^3$, what is b?

33. Use a calculator to compute each number.

a. GCF(1850, 1225)   **b.** GCF(1344, 2868)

c. LCM(1234, 2345)   **d.** LCM(2111, 3222)

34. Use a calculator to compute each number.

a. GCF(14590, 13565)   **b.** GCF(43095, 30888)

c. LCM(9870, 3450)   **d.** LCM(17820, 14040)

■ Problems and Applications

35. If $a = 2 \cdot 3 \cdot 4 \cdot 5 \cdot 6 \cdot 7$ and $b = 4 \cdot 5 \cdot 6 \cdot 7 \cdot 8 \cdot 9$, find the GCF($a$, b) and the LCM(a, b).

36. When a number is divided by 4, 5, or 6, the remainder is 3. If the number is also divisible by 7, what is the smallest the number could be?

37. Find three consecutive composite numbers a, b, and c such that LCM(a, b, c) = $a \cdot b \cdot c$.

38. How many numbers less than 30 are relatively prime with 18?

39. Consider the numbers 24, 36, 42, and 54. Which pair of numbers will give:

a. The largest GCF?   **b.** The smallest GCF?

c. The largest LCM?   **d.** The smallest LCM?

40. Two frogs hop along a path one lily pad wide. One frog jumps 2 lily pads at a time, while the other jumps 3 at a time. If the frogs leave shore one after another and there are 24 lily pads, what is the total number of lily pads jumped on by both frogs?

41. Two buses leave the terminal at the same time. One bus completes its route in 45 min, the other in 60 min. How many hours pass before the buses meet back at the terminal?

42. A college professor has two sections of the same course, one with 42 students and the other with 54. He has each class work in small groups, all of the same size. What are the largest groups that can be formed?

43. Two racing cars leave the starting line at the same time. One completes a lap every 100 s, while the other completes a lap every 105 s.

a. How long will it take for the two cars to meet at the starting line?

b. At this point in time, how many laps will each car have completed?

c. Assuming the same rates and a race of 200 laps, how many laps will the slower car have yet to go when the faster car finishes its last lap?

44. Three kitchen timers in a restaurant are all set to start at the same time. The first goes off every 15 min, the second every 20 min, and the third every 25 min. How many minutes will pass before all three timers go off at the same time?

45. Billy has 98 blue blocks and 70 green blocks. He wants to stack them by color so that each stack has the same number of blocks. What is the largest number of blocks that can be placed in a stack?

46. If Jill has every third night off from work, Joan has every fourth night off, and Jackie has every sixth night off, how many nights will the three girls have off together in the next 30 days?

47. A farmer has a large field that he wants to subdivide and fence to make three cow pens. The pens are to be rectangular and measure 52,500 ft², 30,600 ft², and 19,800 ft², respectively. If fencing comes in one-foot lengths, what is the greatest amount of fencing that that three pens can share? How much total fencing is required?

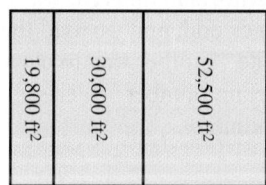

48. A machine has three interlocking gears. The largest has 48 teeth, and each of the other two has 36 teeth. What is the fewest number of revolutions that each gear will make in order for the three gears to return to their original position at the same time?

49. Susan buys tiles that measure 9 × 12 in. What is the smallest square that she can make with these tiles?

■ Mathematical Reasoning and Proof

50. Find all the factors of 32. How can you use the factors of 32 to find all the factors of 64? What about the factors of 96?

51. Determine whether each statement is true or false. If true, explain why. If false, provide a counterexample.

  a. If $GCF(a, b) = 1$, then $LCM(a, b) = a \cdot b$.

  b. If $GCF(a, b) = a$, then $LCM(a, b) = a$.

  c. If $LCM(a, b) = a \cdot b$, then a and b are relatively prime.

  d. If $LCM(a, b) = b$, then $GCF(a, b) = a$.

52. **a.** Find a pair of numbers, a and b, for which the $GCF(a, b) = 1$.

  b. Find a pair of numbers, a and b, for which the $GCF(a, b) = a$.

  c. Find a pair of numbers, a and b, for which the $GCF(a, b) = LCM(a, b)$.

53. Suppose a and b are two nonzero whole numbers. Under what conditions will

  a. $LCM(a, b) < a \cdot b$?

  b. $LCM(a, b) = a \cdot b$?

54. If $c|a$ and $c|b$, why does $c|GCF(a, b)$ and $c|LCM(a, b)$?

55. **a.** If the GCF of two numbers is 2, must the numbers be even?

  b. If two numbers are even, must their GCF be even?

56. If the $GCF(a, b, c) = 1$, must $GCF(a, b) = 1$? Explain.

57. Determine whether each statement is true or false. If true, explain. If false, provide a counterexample.

  a. If two numbers are relatively prime, then both numbers must be odd.

  b. If two numbers are relatively prime, at least one of the numbers must be odd.

  c. If two numbers are relatively prime, at least one of the numbers must be even.

58. Write an argument that explains why 27 and 5,725,720 are relatively prime.

59. **a.** Does every pair of nonzero whole numbers have a least common factor? If so, what is it?

  b. Does every pair of nonzero whole numbers have a greatest common multiple? If so, what is it?

60. Does the $GCF(a, b, c) \cdot LCM(a, b, c) = a \cdot b \cdot c$? If so, explain why. If not, provide a counterexample.

■ Mathematical Communication

61. The greatest common factor is also commonly called the greatest common divisor. Which term do you think better describes the concept and why?

62. Make up several story problems that are solved using the GCF. Describe the language that you used to indicate that the problem should be solved using this concept.

63. Make up several story problems that are solved using the LCM. Describe the language that you used to indicate that the problem should be solved using this concept.

64. Louis has computed the following LCMs:

  $LCM(4, 9) = 36$ $LCM(5, 8) = 40$ $LCM(7, 9) = 63$

  From his examples, he concludes that $LCM(a, b) = a \cdot b$. As his teacher, what would you say about his thinking?

65. Kwan claims that the $LCM(a, 0) = 0$ for all natural numbers a. Is he correct?

66. Michelle is confused as to why the $GCF(a, 0) = a$ and not 0. How would you explain this to her as her teacher?

67. Write a set of directions on how to compute the Euclidean algorithm using a calculator that does not have [INT ÷]. Give your directions to a partner along with a problem to solve. Were they able to complete the problem using your directions? If not, how can you improve your instructions?

Building Connections

68. Examine several sets of curriculum materials from the fourth, fifth, and sixth grades. What methods of computing the GCF and LCM do you find? Which do think would be easiest for students to use?

69. Write a short summary explaining how divisibility and prime numbers play a key role in computing the GCF and LCM of two numbers.

70. Both the GCF and the LCM are useful in working with fractions. How so?

71. **a.** Describe several situations in which you have computed a GCF. How were these situations alike? How were they different?

   b. Describe several situations in which you have computed a LCM. How were these situations alike? How were they different?

SECTION 5.3   Modular and Clock Arithmetic

Earlier in the chapter, we classified the whole numbers in different ways. For instance, we used divisibility by 2 to classify numbers as even or odd, and we used the total number of factors to classify numbers as prime or composite. Another way to classify numbers is to use the remainders after dividing them by a certain number. To classify numbers in this way, we turn to a relation called congruence modulo m.

Congruence Modulo m

Recall that according to the division algorithm, if a and b are whole numbers with $b \neq 0$, then there exists whole numbers q and r such that $a = bq + r$ with $0 \leq r < b$. The last inequality indicates that the remainder r is limited to values that are strictly less than the divisor b. For instance, if we divide by 4, then the remainder must be a 0, 1, 2, or 3. We can quickly verify this by dividing several numbers by 4.

$$8 \div 4 = 2 \text{ remainder } 0 \qquad 12 \div 4 = 3 \text{ remainder } 0$$
$$9 \div 4 = 2 \text{ remainder } 1 \qquad 13 \div 4 = 3 \text{ remainder } 1$$
$$10 \div 4 = 2 \text{ remainder } 2 \qquad 14 \div 4 = 3 \text{ remainder } 2$$
$$11 \div 4 = 2 \text{ remainder } 3 \qquad 15 \div 4 = 3 \text{ remainder } 3 \ldots$$

Because the remainders continue to cycle through 0, 1, 2, and 3 as we divide consecutive numbers by 4, we can use them as another way to classify the whole numbers. Doing so separates the whole numbers into four sets:

$$F_0 = \{0, 4, 8, 12, 16, 20, \ldots\} \qquad \text{(Remainders of 0)}$$
$$F_1 = \{1, 5, 9, 13, 17, 21, \ldots\} \qquad \text{(Remainders of 1)}$$
$$F_2 = \{2, 6, 10, 14, 18, 22, \ldots\} \qquad \text{(Remainders of 2)}$$
$$F_3 = \{3, 7, 11, 15, 19, 23, \ldots\} \qquad \text{(Remainders of 3)}$$

Despite the different numbers, the four sets have one characteristic in common: The difference between any two elements in a given set is always divisible by 4. For instance, consider two elements from F_1, such as 21 and 13. Their difference, $21 - 13 = 8$, is divisible by 4. Likewise, if we consider two elements from F_3, say 47 and 23, their difference, $47 - 23 = 24$, is also divisible by 4. We call this relationship between the numbers in each set **congruence modulo m.**

Definition of Congruence Modulo m

Let a, b, and m be whole numbers with $a \geq b$ and $m \geq 2$. a is **congruent to b modulo m,** written $a \equiv b \bmod m$, if and only if $m \mid (a - b)$.

In the definition, the divisor m is limited to be greater than or equal to 2 because every whole number is divisible by 1, making congruence mod 1 a trivial relationship. Also, congruence mod m is based on divisibility; it is a relationship, not an operation. In other words, it is either true or false and does not necessarily lead to a numerical answer.

Example 5.28   Determining Congruence Mod m

Communication

Use the definition of congruence mod m to interpret each statement, and then determine whether it is true or false.

a. $19 \equiv 7 \bmod 4$ **b.** $27 \equiv 13 \bmod 6$

Solution

a. $19 \equiv 7 \bmod 4$ indicates that, if we subtract 7 from 19, the difference is divisible by 4. Because $19 - 7 = 12$ and $4|12$, then $19 \equiv 7 \bmod 4$ is true.

b. $27 \equiv 13 \bmod 6$ indicates that, if we subtract 13 from 27, the difference is divisible by 6. However, because $27 - 13 = 14$ and $6 \nmid 14$, then $27 \equiv 13 \bmod 6$ is false. In this case, we write $27 \not\equiv 13 \bmod 6$.

As explained earlier, each whole number can be placed into a set based on the remainder when we divide by the number m. These sets are called **equivalence classes** for congruence mod m.

Example 5.29   Finding Equivalence Classes

List the elements in each equivalence class for congruence mod 6.

Solution According to the division algorithm, if we divide by 6, the possible remainders are 0, 1, 2, 3, 4, and 5. Consequently, congruence mod 6 has six equivalence classes, and the smallest element in each set corresponds to one of the remainders. To generate the rest of the elements in each set, repeatedly add 6, the value of m, to the first number. This gives the following six equivalence classes:

$$S_0 = \{0, 6, 12, 18, 24, \ldots\} \qquad S_3 = \{3, 9, 15, 21, 27, \ldots\}$$
$$S_1 = \{1, 7, 13, 19, 25, \ldots\} \qquad S_4 = \{4, 10, 16, 22, 28, \ldots\}$$
$$S_2 = \{2, 8, 14, 20, 26, \ldots\} \qquad S_5 = \{5, 11, 17, 23, 29, \ldots\}.$$

We often use congruence mod m when we use calendars and clocks. In the case of a calendar, it is congruence mod 7.

Example 5.30   Finding Dates in a Month

Application

To prepare for an upcoming concert, Mrs. Kisting wants her fourth- and fifth-grade chorus to meet every Tuesday for the next month. If the first Tuesday of the month occurs on the fifth, what are the dates of the other meetings?

Solution To find the dates of the other meetings, we begin with 5 and add 7 three times. This gives us the 12th, 19th, and 26th, which are also elements from the equivalence class of 5 mod 7.

When we count time in hours, we often use congruence mod 24.

| Example 5.31 | Calculating a Return Time |

Application

Janice lives in New York and is taking a vacation to Florida. It will take her 21 hr to drive from New York to Florida, and then she plans to stay for three and a half days. If she leaves at 11:00 in the morning and stays the full time, at what time can she expect to arrive home?

Solution Janice's trip will require 21 hr to drive to Florida, 84 hr to stay, and 21 hr to drive back to New York. She will be gone for $21 + 84 + 21 = 126$ hr. Because $126 \div 24 = 5$ remainder 6, she will be gone 5 days and 6 hr. By adding 6 hr to her departure time of 11:00 a.m., she can expect to arrive back in New York at 5:00 p.m. five days later. ∎

Clock Arithmetic

Congruence modulo 12 is particularly important in elementary mathematics because it is used to tell time and do arithmetic on a clock. The key difference between clock and whole-number arithmetic is that clock arithmetic uses only the 12 numbers from the set $\{1, 2, 3, \ldots, 12\}$. The numbers are used in a cyclical fashion: Once we reach 12, we start over again at 1. This has a significant impact on the way we add, subtract, multiply, and divide.

Clock Addition

Figure 5.7 Adding $9 \oplus 6$ on a clock

In many senses, **clock addition,** written $\oplus$, is like whole-number addition. For instance, we can compute the sum $9 \oplus 6$ by counting on. We start at 12, count forward 9 hr, and then count on 6 more (Figure 5.7). We necessarily go past 12, so we restart our count at 1 and count on 3 more numbers. We end at 3, so $9 \oplus 6 = 3$.

In whole-number addition, the sum is $9 + 6 = 15$. However, if we divide 15 by 12, we get $15 \div 12 = 1$, remainder 3. The remainder of 3 is equal to the sum in clock addition, implying that we can compute clock addition by using congruence mod 12. Specifically, we compute the sum using whole-number addition, divide by 12, and then take the remainder as the answer.

| Example 5.32 | Computing 12-Hr Clock Addition |

Compute each sum.

a. $3 \oplus 4$ **b.** $8 \oplus 10$ **c.** $5 \oplus 12$

Solution

a. $3 \oplus 4 = 7$.

b. $8 \oplus 10 = 6$ because $8 + 10 = 18$ and $18 \div 12 = 1$, remainder 6.

c. $5 \oplus 12 = 5$ because $5 + 12 = 17$ and $17 \div 12 = 1$, remainder 5. ∎

Because only 12 numbers are used in clock arithmetic, it is possible to generate a table of every single-digit clock addition fact (Table 5.5). This table shows that clock addition satisfies the same properties as whole-number addition. First, because all the numbers in the table are elements of the set $\{1, 2, 3, \ldots, 12\}$, the set is closed under clock addition. Second, Table 5.5 is symmetric about the main diagonal from left to right, so clock addition is commutative. Third, using specific facts from the table, we can demonstrate that clock addition is associative. For instance, $3 \oplus (6 \oplus 7) = 3 \oplus 1 = 4 = 9 \oplus 7 = (3 \oplus 6) \oplus 7$. Fourth, the row and column for 12 match the initial row and column. Consequently, clock addition has an identity element, and it satisfies the additive identity property. In other words, 12 acts as the zero of clock addition.

Table 5.5   12-hr clock addition facts

⊕	1	2	3	4	5	6	7	8	9	10	11	12
1	2	3	4	5	6	7	8	9	10	11	12	1
2	3	4	5	6	7	8	9	10	11	12	1	2
3	4	5	6	7	8	9	10	11	12	1	2	3
4	5	6	7	8	9	10	11	12	1	2	3	4
5	6	7	8	9	10	11	12	1	2	3	4	5
6	7	8	9	10	11	12	1	2	3	4	5	6
7	8	9	10	11	12	1	2	3	4	5	6	7
8	9	10	11	12	1	2	3	4	5	6	7	8
9	10	11	12	1	2	3	4	5	6	7	8	9
10	11	12	1	2	3	4	5	6	7	8	9	10
11	12	1	2	3	4	5	6	7	8	9	10	11
12	1	2	3	4	5	6	7	8	9	10	11	12

In addition to the closure, commutative, associative, and additive identity properties, clock addition satisfies another property. The identity element, 12, appears in every row and column. So, for any number a in the set $\{1, 2, 3, \ldots, 12\}$, there is a unique number b such that $a \oplus b = 12$. This is the **additive inverse property,** and b is the **additive inverse** of a. In general, this property is not true for whole-number addition. If a is a whole number, then there need not exist a whole number b such that $a + b = 0$. For instance, if $a = 3$, then there is no whole number b such that $3 + b = 0$. This is one of the shortcomings of the whole numbers.

Example 5.33   Finding Additive Inverses

Find the additive inverse for each number in clock addition.

a. 3   **b.** 6   **c.** 8

Solution

a. The additive inverse of 3 is 9 because $3 \oplus 9 = 12$.
b. The additive inverse of 6 is 6 because $6 \oplus 6 = 12$.
c. The additive inverse of 8 is 4 because $8 \oplus 4 = 12$.

The ideas and techniques of clock addition can be applied to clocks with any number of hours. For instance, on a 9-hr clock, we cycle through the numbers 1 through 9, rather than 1 through 12. To compute sums, we add the numbers using whole-number addition, divide by 9, and take the remainder as the result. For instance, if we are on a 9-hr clock, then $5 \oplus 8 = 4$ because $5 + 8 = 13$ and $13 \div 9 = 1$, remainder 4 (Figure 5.8).

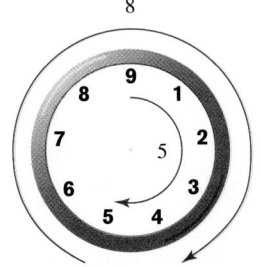

Figure 5.8 Addition on a 9-hr clock

Example 5.34   Computing *m*-Hr Clock Addition

Find each sum for the given *m*-hr clock.

a. $3 \oplus 5$ on a 6-hr clock   **b.** $4 \oplus 9$ on a 10-hr clock

Solution

a. On a 6-hr clock, $3 \oplus 5 = 2$ because $3 + 5 = 8$ and $8 \div 6 = 1$, remainder 2.
b. On a 10-hr clock, $4 \oplus 9 = 3$ because $4 + 9 = 13$ and $13 \div 10 = 1$, remainder 3.

Figure 5.9 Subtraction on a clock

Clock Subtraction

We compute clock subtraction, written $\ominus$, by using either the take-away or the missing-addend approach. With the take-away approach, we count backward, keeping in mind that, upon reaching 1, we continue the count by going back to 12. For instance, to compute $4 \ominus 7$, start at 4 and count backward 7 hr (Figure 5.9). Because we end at 9, we conclude that $4 \ominus 7 = 9$.

In the missing-addend approach, $a \ominus b = c$ if and only if c is the unique number from the set $\{1, 2, 3, \ldots, 12\}$ such that $a = b \oplus c$. For instance, $3 \ominus 10$ is the unique number c in the set $\{1, 2, 3, \ldots, 12\}$ such that $3 = 10 \oplus c$. Based on the clock addition facts in Table 5.5, $c = 5$.

Example 5.35 | Computing 12-Hr Clock Subtraction

Compute each difference using clock subtraction.

a. $9 \ominus 7$ **b.** $6 \ominus 11$ **c.** $8 \ominus 12$

Solution

a. $9 \ominus 7 = 2$.

b. $6 \ominus 11 = c$ if and only if $6 = 11 \oplus c$. Because $11 \oplus 7 = 6$, then $6 \ominus 11 = 7$.

c. Because 12 is the identity in clock arithmetic, subtracting 12 must result in the first number. Hence, $8 \ominus 12 = 8$.

Because clock subtraction always results in a number from the set $\{1, 2, 3, \ldots, 12\}$, the set is closed under clock subtraction. However, like whole-number subtraction, clock subtraction does not satisfy the commutative, associative, or identity properties.

We can use the take-away approach or the missing-addend approach to perform clock subtraction on clocks with any number of hours. The next example demonstrates both methods on m-hr clocks.

Example 5.36 | Computing m-Hr Clock Subtraction

Compute each difference for the given m-hr clock.

a. $2 \ominus 5$ on a 7-hr clock **b.** $3 \ominus 7$ on an 8-hr clock

Solution

a. To use the take-away approach to compute $2 \ominus 5$ on a 7-hr clock, we imagine a 7-hr clock, start at 2, and count back five units: "1, 7, 6, 5, 4." Because we end at 4, $2 \ominus 5 = 4$.

b. To use the missing-addend approach to compute $3 \ominus 7$ on an 8-hr clock, we must find the number that when added to 7 on an 8-hr clock is 3. Because $7 \oplus 4 = 3$, then $3 \ominus 7 = 4$.

Clock Multiplication

We define clock multiplication, written $\otimes$, in terms of repeated clock addition:

$a \otimes b = \underbrace{b \oplus b \oplus \ldots \oplus b}_{a \text{ addends}}$. For instance, $4 \otimes 7 = 7 \oplus 7 \oplus 7 \oplus 7 = [(7 \oplus 7) \oplus 7] \oplus 7 =$

$(2 \oplus 7) \oplus 7 = 9 \oplus 7 = 4$. Because the product uses addition, we can simplify the computation as we did with clock addition. Multiply the numbers using whole-number multiplication, divide by 12, and take the remainder as the answer.

Example 5.37 **Computing 12-Hr Clock Multiplication**

Compute each product using clock multiplication.

a. $3 \otimes 9$ **b.** $5 \otimes 7$ **c.** $6 \otimes 12$

Solution

a. $3 \otimes 9 = 3$ because $3 \cdot 9 = 27$ and $27 \div 12 = 2$, remainder 3.

b. $5 \otimes 7 = 11$ because $5 \cdot 7 = 35$ and $35 \div 12 = 2$, remainder 11.

c. $6 \otimes 12 = 12$ because $6 \cdot 12 = 72$ and $72 \div 12 = 6$, remainder 0. We replace 0 with the zero of clock arithmetic: 12.

As with whole-number multiplication, clock multiplication satisfies the closure, commutative, associative, multiplicative identity, and zero multiplication properties. In the case of the zero multiplication property, 12 acts as zero. In other words, anytime we multiply by 12 in clock multiplication, the answer is always 12.

Like clock addition and subtraction, clock multiplication can be computed on a clock with any number of hours. We compute using whole-number multiplication, divide by the number of hours on the clock, and take the remainder as the answer.

Example 5.38 **Computing *m*-Hr Clock Multiplication**

Compute each product for the given *m*-hr clock.

a. $3 \otimes 4$ on a 10-hr clock   **b.** $3 \otimes 7$ on a 9-hr clock

Solution

a. $3 \otimes 4 = 2$ because $3 \cdot 4 = 12$ and $12 \div 10 = 1$, remainder 2.

b. $3 \otimes 7 = 3$ because $3 \cdot 7 = 21$ and $21 \div 9 = 2$, remainder 3.

Clock Division

We compute clock division, written $\ominus$, by using the missing-factor approach: $a \ominus b = c$ if and only if c is a unique number in the set $\{1, 2, 3, \ldots, 12\}$ such that $a = b \otimes c$. For example, $6 \ominus 5$ is the unique number c from the set $\{1, 2, 3, \ldots, 12\}$ such that $6 = 5 \otimes c$. Using what we know about clock multiplication, we find that $c = 6$. Like whole-number division, not all quotients are possible. For instance, consider $5 \ominus 2$. In this case, there is no number in the set $\{1, 2, 3, \ldots, 12\}$ for which $5 = 2 \otimes c$. Consequently, $5 \ominus 2$ is undefined. We can verify this by looking at every clock product involving 2.

$$
\begin{array}{llll}
2 \otimes 1 = 2 & 2 \otimes 4 = 8 & 2 \otimes 7 = 2 & 2 \otimes 10 = 8 \\
2 \otimes 2 = 4 & 2 \otimes 5 = 10 & 2 \otimes 8 = 4 & 2 \otimes 11 = 10 \\
2 \otimes 3 = 6 & 2 \otimes 6 = 12 & 2 \otimes 9 = 6 & 2 \otimes 12 = 12
\end{array}
$$

Note that several products involving 2 lead to the same number. For instance, both $2 \otimes 5$ and $2 \otimes 11$ are equal to 10. Because $10 \ominus 2$ can lead to two different answers and we do not know which one is correct, $10 \ominus 2$ is another quotient that is undefined. Because clock division is undefined in several situations, we must take care when we compute it.

| Example 5.39 | Computing 12-Hr Clock Division |

Compute each quotient when possible.

a. $8 \oplus 5$ **b.** $5 \oplus 12$ **c.** $9 \ominus 3$

Solution

a. $8 \oplus 5 = 4$ because $5 \otimes 4 = 8$.

b. $5 \oplus 12$ is undefined because there is no number c in the set $\{1, 2, 3, \ldots, 12\}$ such that $5 = 12 \otimes c$. This should make sense because 12 acts as zero in clock arithmetic.

c. $9 \oplus 3$ is undefined because there are three numbers in the set $\{1, 2, 3, \ldots, 12\}$ for which $3 \otimes c = 9$: 3, 7, and 11.

Although not every quotient is possible on a 12-hr clock, all quotients are defined for some m-hr clocks. To understand why, consider Table 5.6, which contains every single-digit multiplication fact for a 7-hr clock. In the table, note that each element from the set $\{1, 2, \ldots, 7\}$ is used exactly once in each row and column, except for 7, which serves as zero in 7-hr clock arithmetic. Consequently, when computing a quotient on a 7-hr clock, we can always obtain an exact answer as long as we are not dividing by 7. For instance,

$$6 \oplus 5 = 4 \text{ because } 5 \otimes 4 = 6$$
$$4 \oplus 6 = 3 \text{ because } 6 \otimes 3 = 4$$
$$3 \oplus 5 = 2 \text{ because } 5 \otimes 2 = 3$$

Table 5.6   7-hr clock multiplication facts

$\otimes$	1	2	3	4	5	6	7
1	1	2	3	4	5	6	7
2	2	4	6	1	3	5	7
3	3	6	2	5	1	4	7
4	4	1	5	2	6	3	7
5	5	3	1	6	4	2	7
6	6	5	4	3	2	1	7
7	7	7	7	7	7	7	7

| Example 5.40 | Computing *m*-Hr Clock Division |

Compute each quotient when possible.

a. $6 \oplus 3$ on a 7-hr clock   **b.** $5 \oplus 4$ on an 6-hr clock
c. $7 \oplus 5$ on an 8-hr clock   **d.** $8 \oplus 6$ on a 9-hr clock

Solution

a. $6 \oplus 3 = 2$ because $3 \otimes 2 = 6$ on a 7-hr clock.

b. $5 \oplus 4$ is undefined because there is no number c such that $4 \otimes c = 5$ on a 6-hr clock.

c. $7 \oplus 5 = 3$ because $5 \otimes 3 = 7$ on an 8-hr clock.

d. $8 \oplus 6$ is undefined because there is no number c such that $6 \otimes c = 8$ on a 9-hr clock.

Check Your Understanding 5.3-A

1. Determine whether each statement is true or false.

    a. $22 \equiv 6 \bmod 5$.   **b.** $48 \equiv 38 \bmod 10$.   **c.** $57 \not\equiv 19 \bmod 6$.

2. Find and list the members in each equivalence class for congruence modulo 5.

3. Use 12-hr clock arithmetic to compute each problem when possible.

    a. $5 \oplus 10$ **b.** $10 \oplus 11$ **c.** $7 \ominus 9$ **d.** $2 \ominus 10$

    e. $3 \otimes 8$ **f.** $5 \otimes 5$ **g.** $1 \oslash 7$ **h.** $2 \oslash 8$

4. Compute each problem for the given *m*-hr clock.

    a. $5 \oplus 7$ on a 9-hr clock   **b.** $6 \oplus 8$ on a 10-hr clock

    c. $6 \ominus 7$ on an 8-hr clock   **d.** $2 \ominus 4$ on a 5-hr clock

5. Compute each problem for the given *m*-hr clock.

    a. $2 \otimes 5$ on a 7-hr clock   **b.** $4 \otimes 4$ on a 6-hr clock

    c. $6 \oslash 5$ on a 9-hr clock   **d.** $3 \oslash 5$ on a 7-hr clock

Talk About It Many children find it easier to tell time on a digital clock than on an analog clock. Why do you think this is true?

Reasoning

Activity 5.5

Construct a fact table for 6-hr clock multiplication. Use it to answer each question.

a. Compute $4 \otimes 5$, $2 \otimes 6$, and $5 \oslash 3$ on a 6-hr clock.

b. Does multiplication on a 6-hr clock satisfy the closure property? Explain.

c. Is multiplication on a 6-hr clock commutative? Explain.

d. Does multiplication on a 6-hr clock satisfy the identity property? If so, what is the identity element, and how do you know?

e. Does division on a 6-hr clock satisfy the closure property? If so, explain. If not, list three undefined quotients.

SECTION 5.3   Learning Assessment

▪ Understanding the Concepts

1. How can remainders be used to classify the whole numbers?

2. Why is $16 \equiv 9 \bmod 7$?

3. How are clock arithmetic and whole-number arithmetic similar? How are they different?

4. Explain how to compute each operation on a 12-hr clock.

    a. Addition   **b.** Subtraction

    c. Multiplication   **d.** Division

5. Explain how to compute each operation on an *m*-hr clock.

    a. Addition   **b.** Subtraction

    c. Multiplication   **d.** Division

6. What does the additive inverse property state is true for clock addition?

7. If we are computing a whole number difference $a - b$, then $a \geq b$. Why is this not the case with clock subtraction?

8. Under what circumstances is a quotient in 12-hr clock division undefined?

▪ Representing the Mathematics

9. Give a verbal interpretation of each statement.

    a. $17 \equiv 7 \bmod 5$.

    b. $21 \equiv 5 \bmod 8$.

    c. $34 \not\equiv 7 \bmod 6$.

10. The following table shows the equivalence classes for congruence mod 6.

1	2	3	4	5	6
7	8	9	10	11	12
13	14	15	16	17	18
19	20	21	22	23	24
25	26	27	28	29	30
...	...	...	...	...	...

   a. How are the equivalence classes represented in the table?

   b. Based on what you see, how might you use a table like this to quickly generate the equivalence classes for congruence mod 5?

11. Draw a diagram that illustrates how to compute each problem on a 12-hr clock.

   a. $7 \oplus 9$ b. $6 \ominus 8$ c. $4 \otimes 3$

12. Draw a diagram that illustrates how to compute each problem on the given clock.

   a. $5 \oplus 5$ on a 6-hr clock   b. $4 \ominus 6$ on a 7-hr clock

   c. $3 \otimes 6$ on a 10-hr clock   d. $5 \otimes 2$ on an 8-hr clock

Exercises

13. Determine whether each statement is true or false.

   a. $31 \equiv 8 \bmod 5$.   b. $43 \equiv 23 \bmod 5$.

   c. $27 \not\equiv 9 \bmod 6$.

14. Determine whether each statement is true or false.

   a. $54 \not\equiv 24 \bmod 6$.   b. $48 \equiv 26 \bmod 11$.

   c. $63 \not\equiv 18 \bmod 11$.

15. Find three numbers that make each statement true.

   a. $x \equiv 2 \bmod 6$.   b. $x \equiv 15 \bmod 9$.

   c. $64 \equiv x \bmod 12$.

16. Find three numbers that make each statement true.

   a. $x \equiv 3 \bmod 4$.   b. $41 \equiv x \bmod 8$.

   c. $27 \equiv 3 \bmod x$.

17. List the members of each equivalence class for congruence mod 3.

18. List the members of each equivalence class for congruence mod 8.

19. List the members of each equivalence class for congruence mod 11.

20. List three whole numbers that are congruent mod 6 to each number.

   a. 0   b. 1   c. 3   d. 4   e. 5

21. List three whole numbers that are congruent mod 14 to each number.

   a. 0   b. 3   c. 7   d. 9   e. 11

22. Compute each sum or difference on a 12-hr clock.

   a. $6 \oplus 7$ b. $12 \oplus 9$ c. $8 \ominus 12$ d. $1 \ominus 9$

23. Compute each sum or difference on a 12-hr clock.

   a. $11 \oplus 8$ b. $6 \ominus 3$ c. $11 \oplus 3 \oplus 8$ d. $1 \ominus 8 \ominus 5$

24. Find the additive inverse of each number in 12-hr clock addition.

   a. 1   b. 5   c. 6   d. 9   e. 11

25. When possible, compute each product or quotient on a 12-hr clock.

   a. $3 \otimes 6$ b. $12 \otimes 7$ c. $11 \oplus 2$ d. $7 \oplus 4$

26. When possible, compute each product or quotient on a 12-hr clock.

   a. $11 \otimes 5$ b. $6 \otimes 3 \otimes 5$ c. $8 \ominus 5$ d. $12 \ominus 7$

27. When possible, compute each problem using 5-hr clock arithmetic.

   a. $3 \oplus 4$ b. $2 \ominus 4$ c. $3 \otimes 3$ d. $5 \ominus 4$

28. When possible, compute each problem using 7-hr clock arithmetic.

   a. $5 \oplus 5$ b. $3 \ominus 5$ c. $2 \otimes 6$ d. $6 \ominus 2$

29. When possible, compute each problem using 8-hr clock arithmetic.

   a. $5 \oplus 7$ b. $3 \ominus 7$ c. $3 \otimes 5$ d. $6 \ominus 4$

30. When possible, compute each problem using 14-hr clock arithmetic.

   a. $10 \oplus 10$ b. $4 \ominus 9$ c. $6 \otimes 9$ d. $3 \ominus 8$

Problems and Applications

31. Find all whole numbers less than 100 that make each statement true.

   a. $x \equiv 3 \bmod 5$.   b. $x \equiv 12 \bmod 7$.

   c. $51 \equiv x \bmod 13$.

32. List the elements in each equivalence class for congruence mod m.

33. Determine the number of equivalence classes for each of the following without actually making a list.

   a. Congruence mod 24

   b. Congruence mod 35

   c. Congruence mod 73

   d. Congruence mod n, $n > 2$

   e. Congruence mod $(2n)$, $n > 2$

   f. Congruence mod n^2, $n > 2$

34. Calculate each problem using 12-hr clock arithmetic.

   a. $(3 \oplus 4) \otimes (6 \ominus 7)$ b. $(8 \ominus 6) \oplus (3 \otimes 5)$

   c. $(9 \ominus 12) \ominus (8 \otimes 7)$

35. We define an exponent in 12-hr clock arithmetic to be $a^n = \underbrace{a \otimes a \otimes \ldots \otimes a}_{n \text{ factors}}$ where a and n are elements

of the set {1, 2, . . ., 12}. Based on this definition, find:

a. 2^3.   **b.** 3^4.   **c.** 12^2.   **d.** 9^4.

36. If August 5 falls on a Wednesday, what are the dates of the Mondays in August?

37. If January 1 falls on a Monday, use congruence modulo 7 to determine what day January 1 will fall on next year if this year is not a leap year.

38. Find out what day your birthday will fall on this year. Use congruence mod 7 to determine the days on which your birthday will fall over the next three years.

39. Many dosages of medicine can be taken only every 8 hr. If Nicole takes a dose at 9:00 a.m., at what times will she be able to take the next two doses?

40. A cargo ship arrives in New York harbor at 10:00 a.m. after leaving Amsterdam, Holland 205 hr ago.

   a. How many days and hours was the ship at sea?

   b. At what time did the ship leave port?

41. Military time uses a 24-hr clock where the numbers 1300, 1400, . . ., 2300 correspond to the 12 p.m. hours. Minutes are expressed in the tens and units place values. For instance 1:30 p.m. is expressed as 1330 in military time.

   a. Convert the following standard times to military time.

      i. 4:00 p.m.   **ii.** 6:30 a.m.

      iii. 2:45 p.m.   **iv.** 10:38 p.m.

   b. Convert the following military times to standard time.

      i. 1500   **ii.** 0845   **iii.** 1630   **iv.** 2015

42. A military cargo jet leaves base at 1300 hours. Its flight to the Middle East will take 19 hr. At what time, in military hours, will it arrive at its destination?

43. Over the course of one weekend, 20 commuter trains are to travel from Boston to Washington D.C., with one train leaving every 3 hr. If the last train arrives in Washington at 7:00 p.m. on a Sunday night and the trip takes 5 hr, at what time did the first train leave Boston?

■ Mathematical Reasoning and Proof

44. Congruence mod 2 divides the set of whole numbers into two equivalence classes. What two well-known sets are they?

45. Explore clock division for several different m-hr clocks where m is less than 12. Based on what you discover, can you determine the values of m for which all divisions are possible?

46. Provide a counterexample showing that each property does not hold for 12-hr clock subtraction.

   a. Commutative property   **b.** Associative property

   c. Identity property

47. Construct a fact table for 8-hr clock addition, and use it to answer each question.

   a. Find $4 \oplus 6$, $7 \oplus 7$, and $3 \ominus 7$.

   b. Does addition on an 8-hr clock satisfy the closure property? Explain.

   c. Does addition on an 8-hr clock satisfy the commutative property? Explain.

   d. Does addition on an 8-hr clock satisfy the additive identity property? If so, what is the additive identity?

48. Construct a fact table for 8-hr clock multiplication, and use it to answer each question.

   a. Find $4 \otimes 5$, $2 \otimes 7$, and $5 \oslash 3$.

   b. Does multiplication on an 8-hr clock satisfy the closure property? Explain.

   c. Is multiplication on an 8-hr clock commutative? Explain.

   d. Does multiplication on an 8-hr clock have an identity element? If so, what is it?

   e. Does multiplication on an 8-hr clock have a zero element? If so, what is it?

49. Operation * is defined on the set {a, b, c, d} by the following table. Use it to answer the questions.

*	a	b	c	d
a	c	a	d	b
b	a	b	c	d
c	d	c	b	a
d	b	d	a	c

   a. Find:

      i. $a * b$.   **ii.** $d * c$.

      iii. $(b * d) * a$.   **iv.** $(c * a) * (d * d)$.

   b. Does * satisfy the closure property on the set {a, b, c, d}? Explain.

   c. Does * satisfy the commutative property? Explain.

   d. Does * have an identity element? If so, what is it?

   e. Does b have an inverse for operation *? How do you know?

50. Operation # is defined on the set {a, b, c, d, e} by the following table. Use it to answer the questions.

#	a	b	c	d	e
a	e	d	b	c	a
b	c	e	d	a	b
c	d	a	e	b	c
d	b	c	a	e	d
e	a	b	c	d	e

a. Find:

    i. *a # e.*   **ii.** *d # b.*

    iii. *(b # c) # e.*   **iv.** *(a # a) # (c # d).*

b. Does # satisfy the closure property on the set {*a, b, c, d, e*}? Explain.

c. Does # satisfy the commutative property? Explain.

d. Does # have an identity element? If so, what is it?

e. Does *b* have an inverse for operation #? How do you know?

■ Mathematical Communication

51. Generate a table of subtraction facts for 12-hr clock subtraction. Use it to explain why the set {1, 2, 3, . . . , 12} is closed under 12-hr clock subtraction.

52. Write down a set of directions explaining how to add on a 12-hr clock without actually making reference to a clock. Give them to somebody who is not in your class, and ask the person to use the directions to solve several clock addition problems. Were your directions clear enough to allow the other person to solve the problems? Did the other person make the connection to adding on a clock without indications from the problem?

53. Write two or three story problems that are appropriate for the elementary grades and ask students to add or subtract time.

54. Write a story problem that is appropriate for the elementary grades and asks students to find future dates on a calendar. How does the language in your problem indicate the use of congruence mod 7 without actually using these terms?

■ Building Connections

55. With a group of your peers, look over several elementary curriculum materials that teach how to tell time. What concepts are included, and how are they presented? What difficulties may children encounter when learning this content?

56. Can you and a partner think of at least five other common uses of modular arithmetic other than on clocks and calendars?

Historical Note

The study of numbers and their relationships dates back to the time of Pythagoras and the ancient Greeks (ca. 580–500 B.C.E.). Pythagoras was the founder of a religious sect that held as its core belief that numbers were the main organizing principle of the universe. Any natural phenomenon or important idea could be associated with a particular number. For instance:

Pythagoras (580–500 B.C.E.)

1. was associated with reason because it was the only way to ascertain truth.
2. was the number for women.
3. was the number for men.
4. was the number for justice because it was the first number that was the product of two equals.
5. was the number for marriage because it was the sum of 2 and 3.
6. was the number for creation because it was the smallest perfect number.

Because numbers were important to the Pythagoreans, they spent much of their time characterizing, classifying, and discovering the properties of numbers. Much of what they discovered still serves as the foundation of number theory today.

Opening Problem Answer

If she spent the same amount on each, the amount would be the LCM(99, 129) = 4,257, or $42.57. So she spent a total of $85.14 on food. At this cost, she would have purchased $42.57 ÷ $0.99 = 43 bottles of soda and $42.57 ÷ $1.29 = 33 bags of chips.

Chapter 5 REVIEW

Summary of Key Concepts from Chapter 5

Section 5.1   Divisibility, Primes, and Composites

- a and b are **factors** of c if and only if $a \cdot b = c$.

- c is a **multiple** of a if and only if a divides c.

- If a and b are whole numbers with $b \neq 0$, then b **divides** a if and only if there exists a whole number c such that $a = b \times c$.

- *Properties of divisibility:* Let a, b, d and n be any whole numbers.

   1. If $d|a$ and $d|b$, then $d|(a + b)$.

   2. If $d|a$ and $d|b$, then $d|(a - b)$ for $a \geq b$.

   3. If $d|a$, then $d|(n \cdot a)$.

- *Divisibility tests:* A natural number is divisible by:

   2: if and only if its ones digit is a 0, 2, 4, 6, or 8.

   5: if and only if its ones digit is a 0 or 5.

   10: if and only if its ones digit is a 0.

   4: if and only if the last two digits represent a number divisible by 4.

   8: if and only if the last three digits represent a number divisible by 8.

   3: if and only if the sum of its digits is divisible by 3.

   9: if and only if the sum of its digits is divisible by 9.

   11: if and only if the sum of the digits in the place values that are even powers of 10 minus the sum of the digits in the place values that are odd powers of 10 is divisible by 11.

   6: if and only if it is divisible by both 2 and 3.

- A whole number is **even** if and only if it is divisible by 2 and **odd** if and only if it is not.

- A natural number is **prime** if and only if it has exactly two factors. It is **composite** if and only if it has three or more factors.

- Let n be a natural number greater than 1. n is prime if it is not divisible by any prime p where $p^2 \leq n$.

- **Fundamental Theorem of Arithmetic:** Each composite number can be expressed as the product of primes in exactly one way, except possibly for the order of the factors.

Section 5.2   Greatest Common Factors and Least Common Multiples

- If a and b are two whole numbers that are both not equal to zero, then the **greatest common factor** of a and b, written GCF(a, b), is the largest whole number that divides both a and b.

- Two nonzero whole numbers are **relatively prime** if their GCF is 1.

- If a and b are any two nonzero whole numbers with $a \geq b$ and $a = bq + r$ for whole numbers q and r, with $r < b$, then GCF(a, b) = GCF(b, r).

- If a and b are two natural numbers, then the **least common multiple** of a and b, written LCM(a, b), is the smallest natural number divisible by both a and b.

- If a and b are any two nonzero whole numbers, then GCF(a, b) $\cdot$ LCM(a, b) = $a \cdot b$.

Section 5.3 Modular and Clock Arithmetic

- Let a, b, and m be whole numbers with $a \geq b$ and $m \geq 2$. *a* **is congruent to** *b* **modulo** *m* if and only if $m|(a - b)$.

- Operations on a 12-hr clock are performed by using a whole-number operation, dividing by 12, and taking the remainder as the answer.

Review Exercises Chapter 5

1. Use a hundreds chart to find the common multiples of 5 and 6 less than 100.

2. What are the common divisors of 60 and 80?

3. Determine whether each statement is true or false. Justify your answer.
   a. $16|34$.   **b.** If $5|10$ and $5|30$, then $5|(10 \cdot 30)$.
   c. $13|39$.   **d.** If $7|14$ and $7 \nmid 13$, then $7 \nmid (14 + 13)$.

4. Determine whether each statement is true or false for whole numbers a, b, and n.
   a. If $a|n$, then $a|n - 1$.
   b. If a and b are odd, then $2|(a + b)$.
   c. If $a|b$, then $a|b \cdot n$.
   d. If a is even and b is odd, then $2|(a - b)$.

5. Determine whether 2, 3, 4, 5, 6, 8, 9, 10, and 11 divide each number.
   a. 270   **b.** 572   **c.** 1,320

6. Classify each number as prime or composite.
   a. 211   **b.** 315   **c.** 457

7. Find the prime factorization of each number and use it to determine how many factors each number has.
   a. 24   **b.** 75   **c.** 144

8. Fill in the blanks so that the given number is divisible by 2, 3, and 4, but not by 5, 8 or 9.
   $$34{,}1\_\_\ \_\_$$

9. A club has 45 members and wants to subdivide into committees of equal size so that no person serves on more than one committee. List the possible ways this can done.

10. A candy company places 30 lollipops into each of its packages. Can 140,970 lollipops be packaged like this without any leftover?

11. Is the sum of two odd primes ever prime? Why or why not?

12. Which number(s) less than 100 has the greatest number of factors?

13. How many factors does 8^3 have? What are they?

14. Why are all primes other than 2 odd?

15. Which primes less than 100 have the form $n^2 + 1$ for some whole number n?

16. If a, b, and c are whole numbers such that $a|b$ and $a|c$, what else does a divide?

17. Determine whether each pair of numbers is relatively prime.
    a. 25 and 35   **b.** 23 and 31   **c.** 384 and 405

18. Use methods from this chapter to find:
    a. GCF(25, 35).   **b.** GCF(156, 390).
    c. GCF(1288, 2584).   **d.** LCM(60, 75).
    e. LCM(504, 220).   **f.** LCM(9945, 5610).

19. A number has exactly 12 factors and is divisible by 2, 5, and 9. What is the number?

20. Use the prime factorization method to compute each number.
    a. GCF(120, 144, 210)   **b.** LCM (120, 144, 210)

21. Find the GCF(a, b) and LCM(a, b) in exponential form if
    a. $a = 2^4 \cdot 3^4 \cdot 5^2$ and $b = 2^2 \cdot 3^5 \cdot 5^4$.
    b. $a = 2^4 \cdot 5^2 \cdot 7^5$ and $b = 3^3 \cdot 5^5 \cdot 11^3$.

22. Two buses leave the terminal at the same time. One bus completes its route in 75 min, the other in 65 min. How many hours will pass before the buses meet back at the terminal?

23. Is it possible to find three consecutive whole numbers such that their sum is not divisible by 3? Why or why not?

24. If a number is divisible by 3, 5, and 8, must it be divisible by 15? By 24? By 40? By 120?

25. Use the multiples of 15, 16, and 20 less than 100 to fill in the Venn diagram.

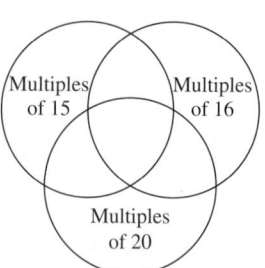

26. The LCM of two numbers is 100, and their product is 500. What are the two numbers?

27. Three kitchen timers in a restaurant are all set to start at the same time. The first will go off every 12 min, the second every 15 min, and the third every 20 min. How many minutes will pass before all three timers go off at the same time?

28. If Joan has every fourth night off from work, Julie has every fifth night off, and Jackie has every sixth night off, how many nights will the three girls have off together in the next year?

29. Lee has a rectangular garden that he wants to divide into two smaller plots by putting a fence around each one. The smaller of the plots is to be 192 ft², and the larger is to be 204 ft². If fencing comes in 1-ft lengths, what is the greatest amount of fencing that that two plots can share? In this case, how much total fencing is required?

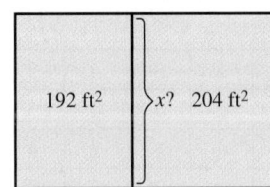

30. Are 432 and 539 relatively prime? Explain how you know.

31. Determine whether each statement is true or false.
    a. $25 \equiv 9 \bmod 4$.
    b. $42 \equiv 21 \bmod 5$.
    c. $26 \neq 8 \bmod 6$.

32. Find three numbers that make each statement true.
    a. $x \equiv 3 \bmod 7$.
    b. $x \equiv 12 \bmod 8$.
    c. $15 \equiv x \bmod 16$.

33. List the members of each equivalence class for:
    a. congruence mod 4.
    b. congruence mod 5.

34. When possible, compute each of the following using 12-hr clock arithmetic.
    a. $3 \oplus 9$ b. $3 \ominus 5$ c. $5 \otimes 7$ d. $6 \oslash 2$
    e. $11 \oplus 5$ f. $7 \ominus 11$ g. $6 \otimes 6$ h. $5 \oslash 4$

35. Operation * is defined on the set $\{a, b, c, d, e\}$ by the following table.

*	a	b	c	d	e
a	c	d	a	b	e
b	d	c	e	a	b
c	a	b	c	d	e
d	b	e	d	c	d
e	e	a	b	d	b

   Use it to answer the given questions.

   a. Find:
      i. $a * e$.   ii. $d * b$.
      iii. $(b * c) * e$.   iv. $(a * a) * (c * d)$.

   b. Does * satisfy the closure property on the set $\{a, b, c, d, e\}$? Explain.

   c. Does * satisfy the commutative property? Explain.

   d. Does * have an identity element? If so, what is it?

   e. Does b have an inverse in the set $\{a, b, c, d, e\}$? How do you know?

36. Construct a facts table for 7-hr clock multiplication, and use it to answer each question.

   a. Find $4 \otimes 5$, $2 \otimes 7$, and $5 \oslash 3$.

   b. Does multiplication on a 7-hr clock satisfy the closure property? Explain.

   c. Is multiplication on a 7-hr clock commutative? Explain.

Answers to Chapter 5   Check Your Understandings and Activities

Check Your Understanding 5.1-A

1. $\{13, 26, 39, 52, 65, 78, 91, 104, 117, 130\}$

2. $\{1, 2, 4, 7, 8, 14, 28, 56\}$

3. a. $15|45$ is true because there exists the whole number $c = 3$ such that $3 \cdot 15 = 45$.

   b. $16|92$ is false because there is no whole number c such that $16 \cdot c = 92$.

   c. $25 \nmid 225$ is false because there exists the whole number $c = 9$ such that $25 \cdot 9 = 225$.

   d. $34 \nmid 66$ is true because there is no whole number c such that $34 \cdot c = 66$.

4. a. The statement is true by Theorem 5.1(1).

   b. The statement is false because if $7|35$, then Theorem 5.1(5) guarantees that 7 will divide a multiple of 35.

   c. The statement is true because it is the contrapositive of the Theorem 5.1(3).

5. a. 2, 3, 4, 6, and 8   b. 2, 3, 4, 5, 6, 8, 9, and 10
   c. 2, 3, 4, 5, 6, 9, 10, and 11   d. 2, 3, 4, 5, 6, 8, 9, 10, and 11

Check Your Understanding 5.1-B

1. a. Prime   b. Composite   c. Prime
   d. Composite   e. Composite

2. a. Composite   b. Prime   c. Prime
   d. Composite   e. Prime

3. Factor trees will vary.
   a. $2^3 \cdot 7$; 8 factors   b. 5^3; 4 factors
   c. $2 \cdot 3^2 \cdot 5^2$; 18 factors   d. $2^4 \cdot 3 \cdot 5 \cdot 7$; 40 factors

Check Your Understanding 5.2-A

1. 8
2. a. $2^2 \cdot 3 = 12.$ b. $2^2 \cdot 3 = 12.$
3. 2
4. a. 15   b. 5
5. The largest number of arrangements is 42. Each one will have 6 carnations, 2 roses, and 3 tulips.

Check Your Understanding 5.2-B

1. 72
2. a. $2^3 \cdot 3^2 \cdot 5 \cdot 7 = 2,520.$ b. $2^3 \cdot 3^2 \cdot 7 \cdot 11 = 5,544.$
3. a. 90 b. 5,005   4. a. 174,915   b. 11,070,996

Check Your Understanding 5.3-A

1. a. False   b. True   c. True
2. $F_0 = \{0, 5, 10, 15, \ldots\}$, $F_1 = \{1, 6, 11, 16, \ldots\}$, $F_2 = \{2, 7, 12, 17, \ldots\}$, $F_3 = \{3, 8, 13, 18, \ldots\}$, and $F_4 = \{4, 9, 14, 19, \ldots\}$.
3. a. 3   b. 9   c. 10   d. 4   e. 12   f. 1
   g. 7   h. Undefined
4. a. 3   b. 4   c. 7   d. 3
5. a. 3   b. 4   c. 3   d. 2

Activity 5.1

Based on the criteria in Theorem 5.2, the following combinations of divisibility tests will lead to new tests.

2 and 3 will give a test for 6.   2 and 5 will give a test for 10.
2 and 9 will give a test for 18.   2 and 11 will give a test for 22.
3 and 4 will give a test for 12.   3 and 5 will give a test for 15.
3 and 8 will give a test for 24.   3 and 10 will give a test for 30.
3 and 11 will give a test for 33.   4 and 5 will give a test for 20.
4 and 9 will give a test for 36.   4 and 11 will give a test for 44.
5 and 6 will give a test for 30.   5 and 8 will give a test for 40.
5 and 9 will give a test for 45.   5 and 11 will give a test for 55.
6 and 11 will give a test for 66.   8 and 9 will give a test for 72.
8 and 11 will give a test for 88.   9 and 10 will give a test for 90.
9 and 11 will give a test for 99.   10 and 11 will give a test for 110.

Activity 5.2

Part A. 392 has 1, 2, 4, 7, 8, 14, 28, 49, 56, 98, 196, and 392 as factors.

Part B. 600 has 1, 2, 3, 4, 5, 6, 8, 10, 12, 15, 20, 24, 25, 30, 40, 50, 60, 75, 100, 120, 150, 200, 300 and 600 as factors.

Activity 5.3

Strategies will vary.

a. 3   b. 8   c. 5

Activity 5.4

Strategies will vary.

a. 12   b. 35   c. 18

Activity 5.5

The table of facts for 6-hr clock multiplication follows:

$\otimes$	1	2	3	4	5	6
1	1	2	3	4	5	6
2	2	4	6	2	4	6
3	3	6	3	6	3	6
4	4	2	6	4	2	6
5	5	4	3	2	1	6
6	6	6	6	6	6	6

a. $4 \otimes 5 = 2$, $2 \otimes 6 = 6$, and $5 \ominus 3$ is undefined.
b. Yes. Every element in the table comes from the set $\{1, 2, 3, 4, 5, 6\}$.
c. Yes. The table is symmetric about the main diagonal, so $a \otimes b = b \otimes a$ in 6-hr clock multiplication.
d. Yes. 1 is the identity element because the table shows us that $a \otimes 1 = a = 1 \otimes a$ for all a in the set $\{1, 2, 3, 4, 5, 6\}$.
e. No. Not every quotient is defined. Three that are not are $5 \ominus 3$, $5 \ominus 2$, and $1 \ominus 2$.

© Image Source/Corbis

The Integers

Opening Problem

The formula for converting temperature from degrees Fahrenheit to degrees Celsius is $C = 5(F - 32) \div 9$. Find five integer values in degrees Fahrenheit that give integer values in degrees Celsius.

Image Source/Jupiter Images

6.1 Integer Basics, Addition, and Subtraction

6.2 Integer Multiplication and Division

Whole-number numeration and computation encompass a large portion of elementary mathematics. Having worked with the whole numbers in different contexts, we have encountered several of their limitations. For instance, many whole-number differences and quotients cannot be computed. Likewise, the whole numbers are generally inadequate for describing mathematical situations requiring a connotation of direction. We can remedy some of these limitations by expanding the set of whole numbers to the set of integers. In doing so, we continue to work on the objectives in the NCTM Number and Operations Standard.

■ NCTM Number and Operations Standard

Instructional programs from prekindergarten through grade 12 should enable all students to:

- Understand numbers, ways of representing numbers, relationships among numbers, and number systems.
- Understand the meanings of operations and how they relate to one another.
- Compute fluently and make reasonable estimates.

Source: NCTM STANDARDS Copyright © 2011 by NATIONAL COUNCIL OF TEACHERS OF MATHEMATICS. Reproduced with permission of NATIONAL COUNCIL OF TEACHERS OF MATHEMATICS.

Students in the third and fourth grades are often introduced to the integers in familiar contexts like temperature and games. In these grades, integer computation is held to a minimum because the primary goal is to give students a basic understanding of the integers and their interpretations. Children learn more about the integers in the middle grades, at which time they learn to order them, compute with them, and use them in problem-solving situations. Specific objectives associated with learning the integers are given in Table 6.1.

Table 6.1 Classroom learning objectives	K	1	2	3	4	5	6	7	8
Understand that positive and negative numbers describe quantities having opposite directions or values.							X		
Find and position integers on a number line and recognize that opposite signs indicate locations on opposite sides of 0.							X		
Interpret statements of inequality as statements about the relative position of two integers on a number line.							X		
Understand the absolute value of an integer as its distance from 0 on the number line or as a magnitude in a real-world situation.							X		
Apply and extend previous understandings and models of addition and subtraction to add and subtract integers.								X	
Show that a number and its opposite have a sum of 0.								X	
Understand subtraction of integers as adding the additive inverse, $p - q = p + (-q)$.								X	
Apply and extend previous understandings of multiplication and division to multiply and divide integers.								X	

Source: Adapted from the *Common Core State Standards for Mathematics* (*Common Core State Standards Initiative*, 2010).

Students can find the integers difficult to understand for several reasons. For one, many students have trouble just making sense of signed numbers. This is particularly true when the negative numbers work in ways that are counterintuitive to what students have learned with the whole numbers. Students must recognize that signed numbers can take on different meanings depending on context. For this reason, students should have opportunities to model and interpret the integers in a wide variety of situations, both mathematical and otherwise.

Another reason children struggle with the integers is that the symbols are used to represent more than one idea. In one instance, the symbols "+" and "−" may be used to represent addition and subtraction. In another, they designate numbers as positive or negative. The multiple uses of these symbols can make it difficult for students to interpret and compute integer expressions. Unfortunately, the problem is only compounded when teachers are not careful to use correct mathematical language. For instance, reading the statement $^-4 - {}^+6$, as "minus four minus plus six" is not only confusing but also incorrect.

Students also struggle to understand how the signs affect computations, often leading them to view the integer operations as a set of rules to be memorized. One way to help students understand the intuitive ideas behind the integer operations is to introduce them with concrete models. Using such models not only shows that the rules governing integer computation are based on logical reasoning, but the models also help students connect the integer operations to their prior experience with whole-number operations. Once students understand the operations, they can develop the formal rules for integer computation.

Despite the difficulties students may encounter, they must make a successful transition from the whole numbers to the integers to be prepared for their future studies in algebra. To effect that transition, elementary teachers must have a deep understanding of the basic notions and computations associated with the integers. Consequently, in Chapter 6 we:

- Discuss the basic notions of the integers, including order, absolute value, and opposites.
- Examine integer computation, including integer addition, subtraction, multiplication, and division.

SECTION 6.1   Integer Basics, Addition, and Subtraction

Although the whole numbers are an important part of basic arithmetic, they have limitations. They are often inadequate for describing real-life situations that involve both an amount and a direction. They also have computational limitations. For instance, many differences, such as $2 - 3$, have no whole-number solution. We can address some of these limitations by using signed numbers.

Signed numbers have two parts: a numerical value and one of two signs, either a **positive** $(+)$ or a **negative** $(-)$. The signs take on different meanings depending on their context. For instance, with altitude, signed numbers indicate feet above $(+)$ or below $(-)$ sea level; with finances, they describe money gained $(+)$ or money owed $(-)$. To include signed numbers in our number system, we expand the whole numbers to the set of integers.

The Integers

Formally, the set of integers is the union of the whole numbers with the negatives of the natural numbers.

> **Definition of the Integers**
>
> The set of **integers** is defined as $\mathbb{Z} = \{\ldots -3, -2, -1, 0, 1, 2, 3, \ldots\}$.

Based on only the signs, the integers can be divided into three sets: the **positive integers,** $\mathbb{Z}^+ = \{1, 2, 3, 4, \ldots\}$; the **negative integers,** $\mathbb{Z}^- = \{-1, -2, -3, -4, \ldots\}$; and **zero,** $\{0\}$. Notice that the positive integers are the same as the natural numbers, so we use both terms to describe this set. In general, we do not write a sign with the positive integers, a convention that helps us simplify the notation. Also, zero is placed in a set by itself because it has no sign; it is neither positive nor negative. The lack of a sign makes it an important reference point because it separates the positive from the negative integers.

The set of integers has several properties not encountered with the whole numbers. Many of them can be represented on the integer number line, which extends

Explorations Manual 6.1

indefinitely in two directions with the positive integers to the right of zero and the negative integers to the left (Figure 6.1).

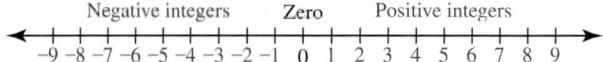

Figure 6.1 The integer number line

Explorations Manual 6.2

On the number line, we interpret each integer as a distance and direction relative to zero. The value of the integer tells us the distance, and the sign tells us the direction. For instance, $+6$ is six units to the right of zero, and -5 is five units to the left (Figure 6.2). When we interpret the integers in this way, it makes sense that the negative integers are listed in the reverse order of the positive integers. For instance, -1 is closer to zero than -2 because -1 is only one unit away whereas -2 is two units away.

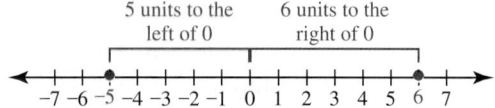

Figure 6.2 Modeling integers on the number line

Interpreting the integers as positions on a number line also makes it possible to introduce opposites. Two integers are **opposites** if they are an equal distance from zero but on opposite sides of zero. For instance, -3 and 3 are opposites, and -7 and 7 are opposites (Figure 6.3).

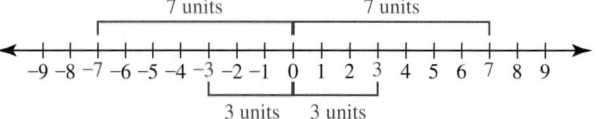

Figure 6.3 Modeling opposites on the number line

Symbolically, if a is an integer, its opposite is $-a$. There are two things to mention about this notation. First, it does not imply that all opposites are negative. In general, the opposite of a positive integer is negative, and the opposite of a negative integer is positive. Because 0 is zero units "away" from itself, it must be its own opposite. Second, the notation gives a third meaning to the dash $(-)$. To understand how the dash is being used, look at its placement. If it is in front of a number, such as in -5, it represents a negative. If placed between two numbers, such as in $5 - 8$, it represents subtraction. Finally, placed in front of a number in parentheses, such as $-(-7)$, it represents an opposite.

Absolute Value

The number line can also be used to illustrate absolute value. Intuitively, the **absolute value** of an integer, written $|a|$, is the integer's distance from zero. Because distances are always greater than or equal to zero, the absolute value of any integer is greater than or equal to zero. For instance, Figure 6.4 shows us that 4 and -4 are both 4 units from zero, meaning $|4| = 4$ and $|-4| = 4$. Consequently, if a is any positive integer, then $|a| = a$, and if a is any negative integer, then $|a| = -a$.

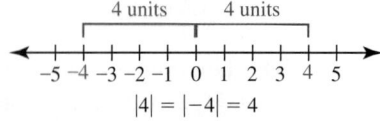

$$|4| = |-4| = 4$$

Figure 6.4 Absolute value on a number line

Definition of Absolute Value
If a is an integer, then $$|a| = a \text{ for } a \geq 0 \quad \text{and} \quad |a| = -a \text{ for } a < 0.$$

SECTION 6.1   Integer Basics, Addition, and Subtraction   289

Example 6.1   Finding Absolute Values

When possible, find all values of x that make each statement true.

a. $|x| = 5$.   **b.** $|x| = 0$.   **c.** $|x| = -7$.

Solution

a. Because $|-5| = 5$ and $|5| = 5$, $x = -5$ or $x = 5$.

b. Because $|0| = 0$, $x = 0$.

c. Because $|x| \geq 0$, there is no value of x such that $|x| = -7$.

Integer Addition

To define integer operations, we need to understand how the signs affect computation. In general, we must consider four possibilities:

1. Both integers are positive $(+, +)$.
2. The first is positive, and the second is negative $(+, -)$.
3. The first is negative, and the second is positive $(-, +)$.
4. Both integers are negative $(-, -)$.

In addition, the first possibility, the sum of two positive integers, can be computed with whole-number addition. Unfortunately, the definition of whole-number addition cannot be applied directly to the other three possibilities because the definition counts only the number of objects in the union of two disjoint sets without any regard for signs. However, we can use what we know about whole-number addition to help us explore integer sums. We do so by using three models commonly used in the classroom: the **pattern model,** the **chip model,** and the **number line model.**

Pattern Model

One way to explore integer addition is to investigate patterns in sequences of consecutive whole-number sums.

Example 6.2   Using the Pattern Model to Add Integers

Reasoning

Establish a pattern of whole-number sums that can be extended to find the sum of $1 + (-2)$.

Solution We begin with the sum $1 + 3$. Because the second addend is negative, we hold the first addend constant and decrease the second by one to get the following sequence of consecutive sums.

$$1 + 3 = 4$$
$$1 + 2 = 3$$
$$1 + 1 = 2$$
$$1 + 0 = 1$$

As the second addend decreases by 1, the sum decreases by 1. Continuing this pattern:

$$1 + (-1) = 0$$
$$1 + (-2) = (-1)$$

The pattern in the last example suggests that if the negative integer has a larger absolute value, the sum will be negative. Also notice that $1 + (-1) = 0$, which implies that two integers with the same value but opposite signs have a net value of zero. We use this observation extensively with the next model.

Chip Model

In the chip model, yellow chips represent positive integers, and red chips represent negative integers. To show an integer, we put the correct number of chips of the appropriate color together. For instance:

$^+4$ is ⚪⚪⚪⚪   and $^-6$ is ⚫⚫⚫⚫⚫⚫

To add with this model, we use the intuitive idea behind whole-number addition. First, we represent each addend with the appropriate chips, and then combine them. However, before counting the total, we make as many positive-negative pairs possible. Each pair has a net value of zero, so we call them **zero pairs** and remove them from consideration. The sum is given by the number and color of the remaining chips.

Example 6.3	Using the Chip Model to Add Integers

Representation

Use the chip model to compute each sum.

a. $3 + (-5)$ **b.** $(-4) + 6$ **c.** $(-3) + (-5)$

Solution

a. Combine 3 positive chips with 5 negative chips. Remove 3 zero pairs, which leaves 2 negative chips. Hence, $3 + (-5) = (-2)$.

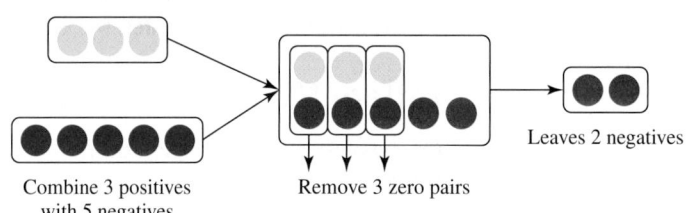

Combine 3 positives   Remove 3 zero pairs   Leaves 2 negatives
with 5 negatives

b. Combine 4 negative chips with 6 positive chips. Removing 4 zero pairs leaves 2 positive chips, so $(-4) + 6 = 2$.

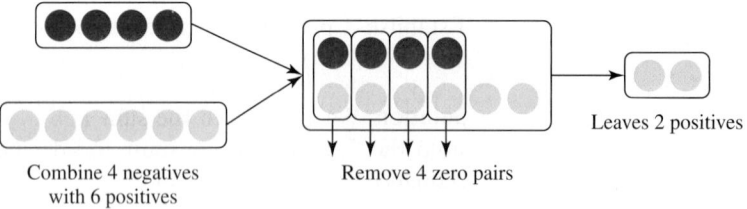

Combine 4 negatives   Remove 4 zero pairs   Leaves 2 positives
with 6 positives

c. Combining 3 negative chips with 5 negative chips gives a total of 8 negative chips, so $(-3) + (-5) = (-8)$.

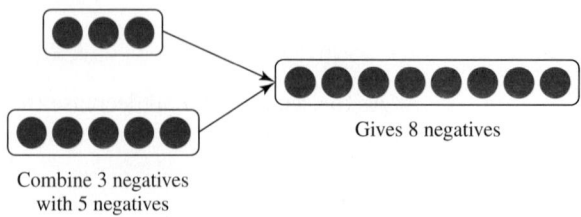

Combine 3 negatives   Gives 8 negatives
with 5 negatives

Parts a and b of the last example show two instances of adding integers with different signs. In each case, the smaller value was not enough to completely "undo" the larger value. As a result, the sum took on the sign of the larger value. Example 6.3 also suggests that we can find the sum of two negative numbers by adding their absolute values and then making the sum negative.

Number Line Model

Explorations Manual 6.4

To add integers on a number line, we describe the integers as a person walking forward and backward a certain number of steps. The person begins at zero facing the positive direction. For any positive integer, the person walks forward that many units. For any negative integer, the person walks backward that many units. The sum is given by the final position of the person on the number line.

Example 6.4   Using the Number Line Model to Add Integers

Representation

Use the number line model to compute each sum.

a. $3 + (-6)$ **b.** $(-6) + 8$ **c.** $(-4) + (-2)$

Solution

a. To model $3 + (-6)$, the person starts at zero, moves forward 3 units, and then backward 6. Because he ends at (-3), then $3 + (-6) = (-3)$.

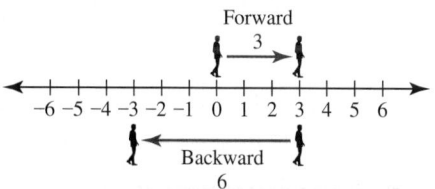

b. The person first moves backward 6 units and then forward 8. Because he ends at 2, then $(-6) + 8 = 2$.

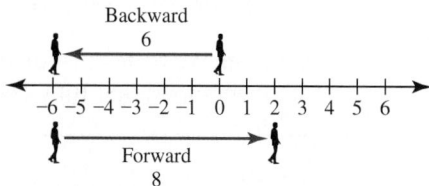

c. For the sum $(-4) + (-2)$, the person first moves backward 4 units and then backward 2 more. Because he arrives at (-6), then $(-4) + (-2) = (-6)$.

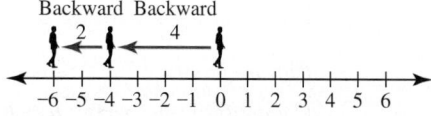

These three models indicate several patterns that hold true for integer addition:

1. The sum of two positive integers is a positive integer.
2. The sum of two negative integers is a negative integer.
3. The sum of a positive integer and a negative integer takes the sign of the addend with the larger absolute value.

These results form the basis for a definition of **integer addition.**

Definition of Integer Addition

Let a and b be any two integers.

1. If a and b have the same sign, then $a + b$ is the sum of the absolute values of a and b and takes the sign of a and b.
2. If a and b have different signs, then $a + b$ is the difference of the smaller absolute value from the larger absolute value and takes the sign of the number with the larger absolute value.

The definition is well-defined because it describes how to compute the sum of any two integers. Unfortunately, it is unavoidably complicated. However, we can summarize it as follows: To find the sum of two integers, find the sum or difference of two whole numbers and then determine the sign.

Example 6.5 | Using the Definition of Integer Addition

Use the definition of integer addition to compute each sum.

a. $(-18) + (-23)$ **b.** $(-567) + 983$

Solution

a. $(-18) + (-23) = -(|-18| + |-23|) = -(18 + 23) = -41.$
b. $(-567) + 983 = |983| - |-567| = 983 - 567 = 416.$

Money is another effective model for integer operations. Deposits represent positive integers, and withdrawals represent negative integers.

Example 6.6 | Representing Integers with Money

Application

Billy overdrew his bank account by \$115 but soon after deposited \$250. He then wrote a check for \$95. Write an expression using integer addition to find his present balance.

Solution We write the sequence of transactions as $(-115) + 250 + (-95)$. Adding the numbers shows that Billy's account is presently at $(-\$115) + (\$250) + (-\$95) = \$40.$

Like whole-number addition, integer addition satisfies several properties.

Theorem 6.1 | Properties of Integer Addition

If a, b, and c are integers, then integer addition satisfies the following properties:

Closure property: $a + b$ is a unique integer.
Commutative property: $a + b = b + a.$
Associative property: $a + (b + c) = (a + b) + c.$
Identity property: $a + 0 = a = 0 + a.$

Integer addition also satisfies another property related to adding opposites. Specifically, when two opposites are added, the result is always zero. For instance, if we add $5 + (-5)$ or $(-5) + 5$ on the number line, the net result in either case is zero (Figure 6.5). This is true for any pair of opposites, which leads us to the **additive inverse property.**

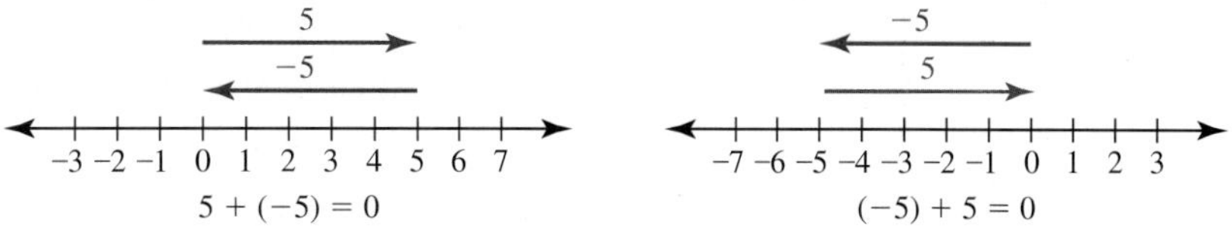

Figure 6.5 The additive inverse property on a number line

Theorem 6.2   Additive Inverse Property for Integer Addition

For every integer a, there exists a unique integer $-a$, called the **additive inverse** such that $a + (-a) = 0 = (-a) + a$.

We need to mention three things about this property. First, the term "additive inverse" is the formal mathematical term for an opposite. Second, the property states that the additive inverse for an integer is unique; there is only one. For instance, only one number can be added to 5 to get 0, namely (-5). Third, of all the integers, zero is the only one that is its own additive inverse as $0 + 0 = 0$. Additive inverses also satisfy the following two properties.

Theorem 6.3   Properties of Additive Inverses

If a and b are any two integers, then:

1. $-(-a) = a.$ **2.** $(-a) + (-b) = -(a + b).$

The first property states that the inverse of the inverse of a is a. In other words, the opposite of the opposite is the original number. Formally, we can show this is true by considering the equations $a + (-a) = 0$ and $-(-a) + (-a) = 0$. Because the additive inverse is unique, it must be that $-(-a) = a$.

The second property states that the sum of the inverses of two integers is equal to the inverse of the sum of the integers. For instance, given the integers 3 and 4, then $-(3) + -(4) = -(7) = -(3 + 4)$. More generally, by definition, $-(a + b)$ is the additive inverse of $a + b$. However, the following sequence of equations also shows that $(a + b) + [(-a) + (-b)] = 0$.

$$(a + b) + (-a) + (-b) = (a + b) + (-b) + (-a) \quad \text{Commutative property}$$
$$= \{a + [b + (-b)]\} = (-a) \quad \text{Associative property}$$
$$= (a + 0) + (-a) \quad \text{Additive inverse property}$$
$$= a + (-a) \quad \text{Additive identity property}$$
$$= 0 \quad \text{Additive inverse property}$$

Because the additive inverse is unique, it must be that $(-a) + (-b) = -(a + b)$.

Example 6.7   Finding Additive Inverses

Find the additive inverse for each of the following:

a. 10   **b.** -5 **c.** $[4 + (-6)]$

Solution

a. The additive inverse of 10 is (-10).

b. The additive inverse of $-(-5)$ is 5.

c. The additive inverse of $[4 + (-6)]$ can be written as $-[4 + (-6)]$, $(-4) + -(-6)$, or $(-4) + 6$.

Check Your Understanding 6.1-A

1. Compute the following:

a. $|-19|$ b. $|45|$ c. $|4| + |-5|$ d. $-|5 + (-9)|$

2. Use the chip model to represent and find each sum.

a. $(-9) + 4$ b. $3 + (-5)$ c. $(-2) + (-6)$ d. $7 + 4$

3. Use the number line model to represent and find each sum.

a. $(-3) + (-4)$ b. $3 + 8$ c. $5 + (-2)$ d. $(-7) + 6$

4. Use the definition of integer addition to find each sum.

a. $981 + (-543)$ b. $(-345) + (-215)$ c. $1,435 + (-2,310)$

Talk About It Which model do you think best represents integer addition? Why?

Representation

Activity 6.1

A deck of cards can also represent the set of integers. Let black cards represent positive integers and red cards represent negative integers. Based on these color designations, answer the following questions:

a. What integer does the 4 of clubs represent? The 8 of diamonds?

b. What integer does the ace of spades represent? The king of hearts?

c. What cards can be used to represent the problem $4 + 5$? $5 + (-3)$?

d. When representing the integers with a deck of cards, it is useful to include a joker. What value should the joker represent?

e. Overall, does a deck of cards have a positive value, negative value, or neither?

Ordering the Integers

With the whole numbers, we defined less than in two ways. In one definition, two sets are compared using a one-to-one correspondence, and the set with objects left over represents the larger value. Unfortunately, we cannot apply this idea to the integers because a one-to-one correspondence cannot account for differences in the signs. In the other definition, we defined less than using addition. We now extend this definition to the integers.

Definition of Less Than

For any integers a and b, a is **less than** b, denoted $a < b$, if and only if there exists a positive integer n such that $a + n = b$.

Example 6.8   Ordering Integers

Fill in the blank with $<$, $>$, or $=$ to make each statement true.

a. 5 _____ -3.   b. -4 _____ -2.   c. -7 _____ 6.

Solution We use the definition of less than to determine the solution to each problem.

a. $5 > -3$ because there exists the positive integer 8 such that $5 = -3 + 8$.

b. $-4 < -2$ because there exists the positive integer 2 such that $-2 = -4 + 2$.

c. $-7 < 6$ because there exists the positive integer 13 such that $6 = -7 + 13$.

Although the definition of less than is a formal way to order two integers, it is not the most intuitive way. Another way is to use the number line. As with the whole numbers, integers farther to the right represent larger values. For example, because -2 is to the left of 3, then $-2 < 3$. Similarly, $-5 < -4$ and $-3 < 4$ (Figure 6.6).

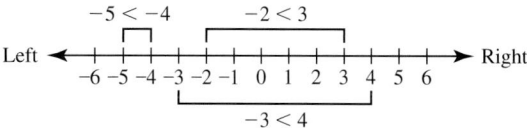

Figure 6.6 Ordering integers on a number line

Many students struggle with ordering the integers because the order of the negative integers is counterintuitive to what they are used to with the whole numbers. To help them better understand the order, we can have them think of the negative integers as owing money. A larger value means that they have a greater debt, so they have "less" money. A smaller value means that they have less debt, so they have "more" money.

Integer Subtraction

To make sense of integer subtraction, we need to consider how the signs affect the difference. Again, we use the models and what we know about whole-number subtraction to explore integer differences.

Missing-Addend Approach

For whole-number subtraction, we developed two definitions. In the take-away approach, a difference was found by taking the set difference of two sets and then counting what remained. Unfortunately, as in addition, this definition is unable to account for the signs of integers. In the missing-addend approach, a difference was rewritten as a sum in which one of the addends was missing. Then we used our knowledge of addition to find the missing number. Because we can now compute integer addition, we use this idea to define **integer subtraction.**

Definition of Integer Subtraction (Missing-Addend Approach)

If a and b are any integers, then $a - b = c$ if and only if $a = b + c$ for some unique integer c.

Example 6.9 | **Using the Missing-Addend Approach to Subtract Integers**

Use the missing-addend approach to compute each difference.

a. $4 - 8$ **b.** $(-9) - (-4)$

Solution

a. We rewrite the difference $4 - 8 = x$ as the sum $4 = 8 + x$. Because $8 + (-4) = 4$, then $4 - 8 = (-4)$.

b. We rewrite $(-9) - (-4) = x$ as $(-9) = (-4) + x$. Because $(-4) + (-5) = (-9)$, then $(-9) - (-4) = (-5)$.

The missing-addend approach enables us to compute any integer difference because we can compute any integer sum. However, it may not be the most intuitive way to compute integer subtraction. So, to gain a better understanding of how the signs affect a difference, we use the models to subtract.

Pattern Model

As in addition, we can explore integer subtraction by investigating the patterns in sequences of whole-number differences. For instance, consider the difference $3 - (-2)$. We can start with $3 - 3$, hold the first number constant, and decrease the second number by 1. In doing so, we get the following differences:

$$3 - 3 = 0$$
$$3 - 2 = 1$$
$$3 - 1 = 2$$
$$3 - 0 = 3$$

As the second number decreases by 1, the difference increases by 1. Keeping this pattern, we conclude that:

$$3 - (-1) = 4$$
$$3 - (-2) = 5.$$

Notice that $3 - (-2) = 5$ and $3 + 2 = 5$, indicating that subtracting a negative is equivalent to adding a positive. We explore this possibility further with other models.

Chip Model

Although integer subtraction cannot be defined with the take-away approach, the intuitive idea behind it can be used to compute differences with the chip model. We represent the first integer with the appropriate chips. We then remove the appropriate number and kind of chips for the second integer. In some situations, there might not be enough of the right kind of chip to remove. In such cases, we put in zero pairs until there are enough. Doing so has no effect on the difference because the net effect of any zero pair is zero. With the appropriate chips removed, we get the difference by counting what remains.

Example 6.10 **Using the Chip Model to Subtract Integers**

Representation

Use the chip model to compute each difference.

a. $3 - (-1)$ **b.** $(-4) - 3$ **c.** $(-3) - (-7)$

Solution

a. Begin with 3 positive chips and look to remove 1 negative chip. Because there are no negative chips, put in 1 zero pair. After taking away the negative chip, four positive chips are left. As a result, $3 - (-1) = 4$.

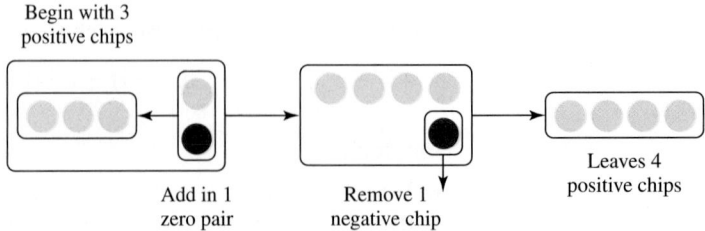

Begin with 3 positive chips

Add in 1 zero pair   Remove 1 negative chip   Leaves 4 positive chips

b. Begin with 4 negative chips. Put in 3 zero pairs so that 3 positive chips can be removed. After the positive chips are removed, 7 negative chips are left. Consequently, $(-4) - 3 = -7$.

Begin with 4 negative chips

Add 3 zero pairs

Remove 3 negative chips

Leaves 7 negative chips

c. Begin with 3 negative chips and look to remove 7 negative chips. Because only 4 more are needed to have 7, put in only 4 zero pairs. Once 7 negative chips are removed, 4 positive chips remain. As a result, $(-3) - (-7) = 4$.

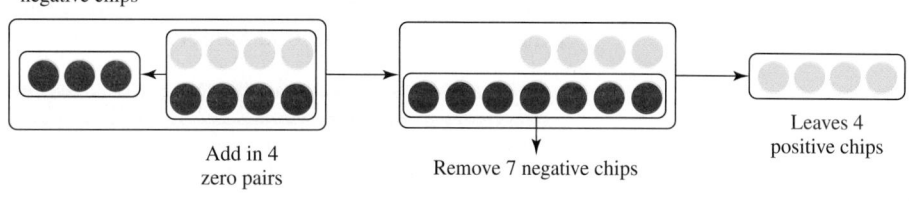

Begin with 3 negative chips

Add in 4 zero pairs

Remove 7 negative chips

Leaves 4 positive chips

Each difference in the last example indicates again that we can subtract by adding the opposite of the second number:

$$3 - (-1) = 4 = 3 + 1$$
$$(-4) - 3 = (-7) = (-4) + (-3)$$
$$(-3) - (-7) = 4 = (-3) + 7$$

Number Line Model

To use the number line model for integer subtraction, start with a person facing in the positive direction. For a positive integer, the person moves forward, and for a negative integer, the person moves backward. Now, however, for the subtraction symbol, the person turns around and faces in the negative direction.

Example 6.11   Using the Number Line Model to Subtract Integers

Representation   Use the number line model to compute each difference.

a. $2 - 6$ **b.** $(-2) - 3$ **c.** $(-4) - (-7)$

Solution

a. For $2 - 6$, the person moves forward 2 units, turns around, and then moves forward 6 units. Because he arrives at (-4), then $2 - 6 = (-4)$.

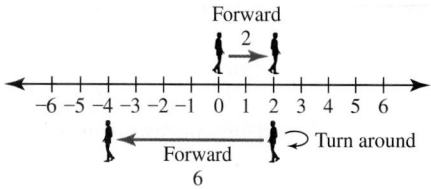

Forward
2

−6 −5 −4 −3 −2 −1 0 1 2 3 4 5 6

Turn around

Forward
6

b. For $(-2) - 3$, the person moves backward two units, turns around, and then goes forward 3 units. Because he arrives at (-5), then $(-2) - 3 = (-5)$.

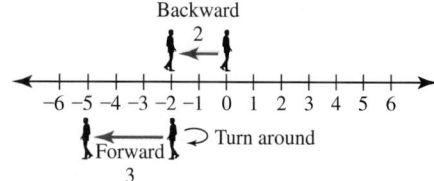

Backward
2

−6 −5 −4 −3 −2 −1 0 1 2 3 4 5 6

Turn around

Forward
3

c. For $(-4) - (-7)$, the person moves backward four units, turns around, and then goes backward 7 units. Because he arrives at 3, then $(-4) - (-7) = 3$.

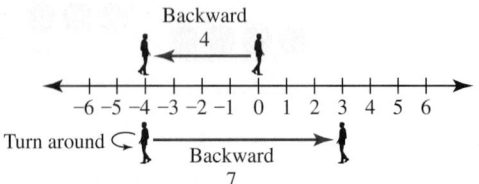

Adding the Opposite

The last three examples indicated that integer differences can be computed by adding the opposite of the second number.

Theorem 6.4   Subtracting by Adding the Opposite

For all integers a and b, $a - b = a + (-b)$.

Example 6.12   Subtracting by Adding the Opposite

Compute each difference by adding an opposite.

a. $(-234) - 423$ **b.** $(-451) - (-893)$

Solution

a. $(-234) - 423 = (-234) + (-423) = -(234 + 423) = -657$.
b. $(-451) - (-893) = -(451) + 893 = 442$.

Theorem 6.4 is important for two reasons. First, it blurs the distinction between addition and subtraction. In other words, once given the negative numbers, we can change any subtraction problem into an addition problem and any addition problem into a subtraction problem. So we can pick the easier operation to compute. Second, integer subtraction satisfies only the closure property. However, we can access other properties for addition, such as the commutative and associative properties, by changing the difference to a sum. The next example offers the proof of Theorem 6.4.

Example 6.13   Proof of Theorem 6.4

Reasoning

Show that $a - b = a + (-b)$ for all integers a and b.

Solution Consider the equation $b + n = a$, where a and b are integers and n is an unknown integer. By the missing-addend definition of integer subtraction, $n = a - b$ must be one solution to this equation. However, if $n = a + (-b)$, we also have:

$$b + n = b + (a + (-b)) \qquad \text{Substitution}$$
$$= b + ((-b) + a) \qquad \text{Commutative property}$$
$$= (b + (-b)) + a \qquad \text{Associative property}$$
$$= 0 + a \qquad \text{Additive inverse property}$$
$$= a \qquad \text{Additive identity property}$$

So $a + (-b)$ is another solution to the equation $b + n = a$. Because this equation can have only one solution, then $a - b = a + (-b)$.

Integer Addition and Subtraction on a Calculator

Whenever adding or subtracting integers on a calculator, we must make a distinction between subtraction and a negative sign. To indicate a negative, we do not use the subtraction key but instead the change of sign key, which is typically represented as $\boxed{+/-}$ or $\boxed{(-)}$.

Example 6.14	Computing Integer Operations on a Calculator

 Use a calculator to compute $(-317) - 471$ in two different ways.

Solution

1. One way to compute the difference is to use the subtraction key:

$$\boxed{(-)}\ 317\ \boxed{-}\ 471\ \boxed{=}$$

This gives us a result of (-788).

2. We can also subtract by adding the opposite:

$$\boxed{(-)}\ 317\ \boxed{+}\ \boxed{(-)}\ 471\ \boxed{=}$$

Again, the result is (-788).

Check Your Understanding 6.1-B

1. Establish a pattern of whole-number differences that can be extended to find each difference.
   a. $3 - 5$ b. $(-1) - 3$

2. Use the chip model to represent and solve each difference.
   a. $4 - 5$ b. $(-4) - 6$ c. $(-2) - (-5)$

3. Use a number line to represent and solve each difference.
   a. $4 - 8$ b. $(-5) - (-4)$ c. $6 - (-4)$

4. Compute each difference by adding the opposite.
   a. $78 - 96$ b. $345 - (-901)$ c. $(-761) - 658$

Talk About It Integer addition was defined by describing two rules that could be used to compute any integer sum. Can we do the same with integer subtraction?

Problem Solving

Activity 6.2

In many card games, players can throw away cards to try to improve a hand. For each hand, what two cards should be thrown away to make the hand more positive? More negative? Use the same integer designations for cards as given in Activity 6.1.

a. A♣, K♦, 3♦, 4♠, Q♠ b. 3♦, 10♣, J♣, J♠, A♥
c. A♣, A♦, A♠, A♥, 7♥ d. 3♦, Q♣, 2♦, J♠, 2♥

SECTION **6.1** Learning Assessment

■ Understanding the Concepts

1. **a.** What makes zero a unique member of the set of integers?
   b. What important role does this uniqueness give zero among the integers?

2. **a.** What does it mean for two integers to be opposites?
   b. What is the opposite of zero?

3. Why are the negative integers listed in the reverse order of the positive integers on a number line?

4. What is absolute value? Why is it always greater than or equal to zero?

5. Why does the definition of whole-number addition not work for the integers?

6. Describe in your own words how to compute integer addition. How is it similar to whole-number addition? How is it different?

7. What does the additive inverse property say about integer addition?

8. How can we use addition to order the integers? The integer number line?

9. Describe two ways to compute integer subtraction. How is integer subtraction similar to whole-number subtraction? How is it different?

10. Why is the fact that $a - b = a + (-b)$ important for computing integer subtraction?

■ Representing the Mathematics

11. Write a verbal interpretation of each expression.
    a. $^+4$ **b.** (-7) **c.** $(-3) + (-4)$
    d. $6 - (-5)$ **e.** $(-4) - 6$ **f.** $(-3) - (-11)$

12. Give the integer that represents each situation.
    a. Loss of 15 yd   **b.** Deposit of \$275
    c. 315 ft below sea level   **d.** 4 over par
    e. 2-hr delay   **f.** Rise of 25° F

13. Represent each situation as an integer expression and solve.
    a. The temperature yesterday was 56° F, and today it is 45° F. What is the change in temperature?
    b. A plane took off from a valley that was 150 ft below sea level. It landed on an airstrip that was 459 ft above sea level. What was the change in elevation?
    c. A car originally sold for \$13,599. Five years later, the same car was worth \$6,599. What was the net change in the car's value?

14. Represent each situation as an integer expression and solve.
    a. A plane is flying at 32,000 ft when it experiences turbulence. It drops to 27,000 ft for a smoother ride. What is its drop in elevation?
    b. A football team starts a play on the 35-yd line and finishes the play on the 27-yd line. What is the team's loss in yardage?
    c. At flood stage, a river's depth is measured to be 25 ft. The water eventually recedes to its normal level of 8 ft. What is the change in water depth?

15. If noon today represents 0, what integer would represent each of the following?
    a. 10:00 a.m. today   **b.** 7:00 p.m. today
    c. 3:00 a.m. today   **d.** 3 a.m. tomorrow
    e. 4 p.m. yesterday   **f.** Noon tomorrow

16. Give the position of each integer on the number line with respect to zero.
    a. (-4) **b.** 12   **c.** (-8) **d.** 0

17. Use the chip model to show three representations of (-4).

18. What is the net value of each collection of chips?
    a. ● ●   **b.** ● ● ●
             ● ● ●
    c. ● ●   **d.** ● ● ● ● ● ●
         ● ● ●   ● ● ● ●

19. Another version of the chip model is the charged field model, in which positive integers are represented with the appropriate number of $+$'s, and negative integers are represented with the appropriate number of $-$'s. What is the net value in each diagram?
    a. ++   **b.** — — — —
    c. ++++   **d.** +++
        — —   — — — —

20. Use the chip model to represent and solve each problem.
    a. $(-8) + 5$ **b.** $4 + (-3)$
    c. $(-6) - (-3)$ **d.** $6 - 8$

21. Use the chip model to represent and solve each problem.
    a. $(-3) + (-6)$ **b.** $8 + 2$
    c. $(-4) - 5$ **d.** $(-1) - (-6)$

22. Use the number line model to represent and solve each problem.
    a. $8 + (-5)$
    b. $(-6) + 5$
    c. $3 - 7$
    d. $(-4) - (-6)$

23. Use the number line model to represent and solve each problem.
    a. $3 + 6$
    b. $(-1) + (-4)$
    c. $2 - (-5)$
    d. $(-7) - 3$

■ Exercises

24. Compute the following.
    a. $|4|$
    b. $|(-6)|$
    c. $-|(-5)|$
    d. $|4 - 8|$

25. Compute the following.
    a. $|(-14) + 19|$
    b. $|(-19)| + |10|$
    c. $|24| - |(-16)|$
    d. $-|(-11)| + |(-21)|$

26. Use the definition of less than to explain why each statement is true or false.
    a. $(-5) < (-9)$
    b. $8 > (-4)$
    c. $(-4) > (-3)$
    d. $0 > (-7)$

27. Order the following integers from least to greatest: $(-5), 8, 0, (-4), 16, (-17)$.

28. For each problem, establish a pattern of whole-number sums or differences that can be extended to find the needed result.
    a. $2 + (-3)$
    b. $3 - (-3)$

29. For each problem, establish a pattern of whole-number sums or differences that can be extended to find the needed result.
    a. $4 - 6$
    b. $(-4) + 3$

30. Use the missing-addend approach to compute each difference.
    a. $7 - 10$
    b. $6 - (-7)$
    c. $(-5) - 8$
    d. $(-7) - (-9)$

31. Use an opposite to write each subtraction fact as an addition fact.
    a. $14 - 8$
    b. $(-14) - 8$
    c. $(-14) - (-8)$
    d. $14 - (-8)$

32. Compute each difference by adding an opposite.
    a. $95 - 108$
    b. $210 - (-356)$
    c. $(-419) - 702$
    d. $(-673) - (-894)$

33. Compute each sum or difference.
    a. $(-134) + 96$
    b. $235 - (-421)$
    c. $(-563) + (-456)$

34. Compute each sum or difference.
    a. $(351) - (-451)$
    b. $(-578) + (-610)$
    c. $(-732) - 847$

35. Find the additive inverse of each expression.
    a. 3
    b. (-7)
    c. 0
    d. $|4|$
    e. $|(-5)|$
    f. $(-|(-5)|)$

36. Identify the property of integer addition used in each equation.
    a. $3 + (-3) = 0$
    b. $7 + (-9) = (-9) + 7$
    c. $[2 + (-7)] + 5 = 2 + [(-7) + 5]$
    d. $(-13) + 18 \in \mathbb{Z}$
    e. $0 + (-4) = (-4)$
    f. $[3 + (-4)] + (-6) = [(-4) + 3] + (-6)$

37. Find the value of x.
    a. $(-183) + x = (-93)$
    b. $x + 97 = (-63)$
    c. $(-47) + (-169) = x$
    d. $87 - x = (-24)$
    e. $43 - (-14) = x$
    f. $x - (-71) = 62$

38. Use a calculator to compute each of the following.
    a. $(-16,563) + 18,739$
    b. $14,930 + (-18,763)$
    c. $(-17,953) - (-21,672)$
    d. $16,272 - (-11,009)$

39. Estimate the value of each problem.
    a. $(-450) + 501 + (-499)$
    b. $(-299) - (-301) - (-498)$

■ Problems and Applications

40. When possible, find all values of x that make each statement true.
    a. $|x| = 8$
    b. $|x| = 0$
    c. $|x| < 3$
    d. $|x| \leq 4$
    e. $|x| < (-1)$

41. When possible, find all values of x that make each statement true.
    a. $|x + 1| = 3$
    b. $|x| + 1 = 2$
    c. $|(-2)x| = 6$
    d. $|2x| - 5 = 9$

42. Find $|2 - |3 - |4 - |5 - 6| - 7| - 8| - 9|$.

43. Let W represent the set of whole numbers. Find each of the following sets.
    a. $\mathbb{Z} \cap W$
    b. $\mathbb{Z}^+ \cup \mathbb{Z}^-$
    c. $\mathbb{Z} - \mathbb{Z}^-$
    d. $\mathbb{Z}^+ \cap \mathbb{Z}^-$
    e. $W - \mathbb{Z}^+$

44. A magic square is one that has the same sum across every row, down every column, and down each diagonal. Complete the following magic square using the integers $(-10), (-9), (-8), (-1), 0, 1, 8, 9, 10$.

45. If $a = 2$, $b = 4$, $c = (-3)$, and $d = (-5)$, place parentheses in the expression $a - b + c - d$ to find the largest possible value.

46. Work from left to right to fill in the following diagram so that each empty square contains the difference of the two squares directly below it.

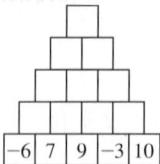

| -6 | 7 | 9 | -3 | 10 |

47. a. What integer is represented by each triangle of chips in the following sequence of figures?

b. Create the next three triangles in the sequence. What integers do they represent?

c. What is the net value of the 15th triangle in the sequence?

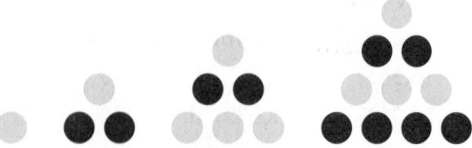

48. Repeat the previous problem using the following sequence of squares.

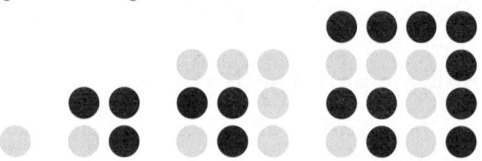

49. What different integers can be represented with the chip model if a total of 8 chips of either color are used?

50. One winter day in Minnesota, the temperature reads $-27°$ F. Six months later, the temperature reads $89°$ F. How much did the temperature change between these 2 days?

51. A plane can fly at 360 mph with no wind. If the plane is flying into a headwind of 70 mph, what is the plane's speed relative to the ground?

52. James buys a certain stock on a Friday. Over the course of the next 7 days, the stock's price fluctuates as follows: $+8$, (-6), (-5), $+7$, $+5$, $+9$, (-10). What is the net gain or loss of the stock for the week?

53. In a series of 6 plays, a football team gained 6 yd on a run, gained 15 yd on a pass, lost 2 yd on run, lost another 10 yd on a fumble, gained 17 yd on a pass, and gained another 11 yd on a run. What was the net yardage gained or lost in the 6 plays?

54. Consider the card model given in Activities 6.1 and 6.2, and suppose that the face cards take on the following values: jacks (J) = 11, queens (Q) = 12, kings (K) = 13, and aces (A) = 1. Answer the following questions using the fact that the value of a hand is

determined by adding the integers represented by the cards in it.

a. Determine the value of the following hands.

i. A♣, K♦, 3♦, 4♠, Q♠

ii. 3♣, 10♣, J♠, J♠, A♥

iii. A♣, A♦, A♠, A♥, 7♥

iv. 3♦, 4♦, 2♦, J♠, 2♥

b. Can a 5-card hand have an integer value of zero? If so, give an example.

c. What is the largest possible value that a 5-card hand can have? Give an example of such a hand.

d. What is the smallest possible value that a 5-card hand can have? Give an example of such a hand.

■ Mathematical Reasoning and Proof

55. If possible, find the stated integer.

a. The largest positive integer

b. The smallest positive integer

c. The largest negative integer

d. The smallest negative integer

56. True or false? If false, explain why.

a. Every whole number is an integer.

b. $\mathbb{Z}^+$ is equal to the natural numbers.

c. Every integer is a whole number.

d. Zero is a positive number.

e. $\mathbb{Z}^+$ is equal to the whole numbers.

57. Is the equation $|x| = -x$ ever true? If so, what values make it true? If not, why not?

58. Which of the following sets is (are) closed under subtraction? Explain.

a. $\{\ldots (-8), (-4), 0, 4, 8 \ldots\}$

b. $\{\ldots (-5), (-3), (-1), 0, 2, 4, 6, \ldots\}$

c. $\{\ldots (-5), (-3), (-1), 1, 3, 5, \ldots\}$

d. $\{(-7), (-5), (-3), (-1), 0, 1, 3, 5, 7\}$

59. When possible, find the largest set of integers that makes each statement true.

a. p is negative.   **b.** $(-p)$ is positive.

c. $|p|$ is positive.   **d.** $|p|$ is negative.

60. What must be true about the values of x and y if $|x| - |y| = 0$?

61. If p and q are any integers, find the additive inverse of each expression.

a. p **b.** $(-p)$ **c.** $p + q$ **d.** $p - q$

62. Provide a numerical example demonstrating that each property does not hold for integer subtraction.

a. Commutative property   **b.** Associative property

c. Identity property

63. Do any of the following possible properties hold for integers p, q, and r? If so, explain. If not, give a counterexample.

    a. $p + (q - r) = (p + q) - (p + r)$

    b. $q - p = -(p + q)$

    c. $p - q = -(q - p)$

64. Is $p - q$ the additive inverse of $q - p$? Explain.

65. **a.** Does $|p + q| = |p| + |q|$ for all integers p and q? If so, explain. If not, provide a counterexample.

    b. Does $|p - q| = |q - p|$ for all integers p and q? If so, explain. If not, provide a counterexample.

■ Mathematical Communication

66. **a.** Guide a partner on how to use a calculator to compute the problem $6 + (-4)$ if the addition key on the calculator is broken.

    b. Guide a partner on how to use a calculator to compute the problem $(-5) - 6$ if the subtraction key on the calculator is broken.

67. Make up four or five integer addition and subtraction problems. With a partner, make a large number line on the floor. Give your partner directions on how to move along the number line to solve the problems correctly.

68. **a.** Which model do you think best represents integer addition, the chip model or the number line model? What about subtraction?

    b. What difficulties or misconceptions do you think children might encounter when using these models?

69. Example 6.6 used money to model integer addition. Money can also be used to model integer subtraction. To do so, add an additional component to the model in which a postman delivers or takes mail that contains checks or bills:

    1. If the postman brings a check, it represents adding a positive.

    2. If the postman brings a bill, it represents adding a negative.

    3. If the postman takes a check, it represents subtracting a positive.

    4. If the postman takes a bill, it represents subtracting a negative.

    For instance, if the postman brings a check for $25 and takes a bill of $10, the net gain is $35 because we gain $25 in profit and lose $10 in debt. Write a postman story that satisfies each of the following criteria:

    a. A sum of two integers that has a positive result

    b. A sum of two integers that has a negative result

    c. A difference of two integers that has a positive result

    d. A difference of two integers that has a negative result

70. Write a story problem for subtraction that:

    a. Uses the take-away approach to solve $5 - 7$.

    b. Uses the missing-addend approach to solve $(-6) - 4$.

    c. Uses the comparison approach to subtraction to solve $4 - (-2)$.

71. What are the three uses of the dash symbol $(-)$? Why might the uses make it difficult for students to understand the integers?

72. Ginny claims that $-10 > -5$ because a debt of $10 is greater than a debt of $5. How do you, as Ginny's teacher, respond to her claim?

73. A student draws the following number line and claims that $-a$ must always be to the left of a. Is the student correct?

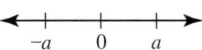

■ Building Connections

74. To this point, our mathematical structure has included sets, numeration, and whole-number operations. Write a paragraph or two explaining how the set of integers has extended this structure.

75. How would it be possible to use an elevator to model the set of integers?

76. Other than the real-world situations given in this section, list five situations that use the integers.

    a. Describe the interpretation of positive and negative integers in each situation.

    b. What interpretation, if any, would absolute value have in each situation?

77. As mentioned, children are often introduced to the integers through familiar contexts. With a group of your peers, discuss the benefits and/or drawbacks of introducing children to the integers in this way.

78. Take a few moments and think of the difficulties you might have had in learning integer addition and subtraction for the first time. After you collect your thoughts, discuss them with a group of your peers, and determine how your experiences might impact how you teach the integers to your future students.

79. A student approaches you and states that the integers make it possible to represent any numerical situation. Is this true? If so, explain. If not, provide a numerical situation that cannot be described with the integers.

Integer Multiplication and Division

To examine integer computations involving multiplication, exponents, and division, we again need to consider the effects of the signs on each operation.

Integer Multiplication

In Chapter 4, whole-number multiplication was defined in two ways: One used the Cartesian product, and the other used repeated addition. Ideally, we would like to apply these definitions to the integers, but neither one is sufficient. The Cartesian product counts only ordered pairs, so it cannot account for how the signs might affect the product.

We have more success with repeated addition. Given two positive integers, we compute the product as with two whole numbers. We can also use repeated addition in the positive-negative case. For instance, we can interpret $5 \cdot (-2)$ as, "Negative two added to itself five times,":

$$5 \cdot (-2) = (-2) + (-2) + (-2) + (-2) + (-2) = (-10)$$

In general, if a and b are positive integers, then the product of a positive integer and a negative integer can be defined as:

$$a \cdot (-b) = \underbrace{(-b) + (-b) + \ldots + (-b)}_{a \text{ addends}} = -(a \cdot b)$$

By this definition, the product of a positive integer and a negative integer must always be negative. Unfortunately, repeated addition makes less sense if the first integer is negative. For instance, we can interpret $(-4) \cdot 2$ as "Two added to itself negative four times," but the computational meaning of the statement is unclear.

Explorations
Manual
6.5

This leaves us in need of a definition of multiplication that can handle all four combinations of signs including the negative-positive and the negative-negative cases. Again, we use models to search for patterns that will lead us to such a definition.

Pattern Model

As in addition, we can compute integer products by extending the patterns that occur in sequences of whole-number products. For instance, to compute $(-2) \cdot 4$, we can use the following sequence of products:

$$3 \cdot 4 = 12$$
$$2 \cdot 4 = 8$$
$$1 \cdot 4 = 4$$
$$0 \cdot 4 = 0$$

As the first factor decreases by 1, the product decreases by 4. If we continue this pattern to the negative integers, then

$$(-1) \cdot 4 = (-4)$$
$$(-2) \cdot 4 = (-8)$$

Example 6.15 | **Using the Pattern Model to Multiply Integers**

Reasoning

Establish a pattern of products that can be extended to find $(-3) \cdot (-2)$.

Solution We are given two negative numbers. If we choose to hold the second constant and change the first, then we can begin at $3 \cdot (-2)$ and use repeated addition to compute the products. We decrease the first factor by 1 to get the following products:

$$3 \cdot (-2) = (-2) + (-2) + (-2) = (-6)$$
$$2 \cdot (-2) = (-2) + (-2) = (-4)$$
$$1 \cdot (-2) = (-2)$$
$$0 \cdot (-2) = 0$$

As the first factor is reduced by 1, the product increases by 2. We continue this pattern to reach the needed product:

$$(-1) \cdot (-2) = 2$$
$$(-2) \cdot (-2) = 4, \text{ and}$$
$$(-3) \cdot (-2) = 6$$

If we examine the last two examples, the first suggests that the product of a negative and a positive is negative. The second suggests that a negative times a negative is positive. To check these observations, we consider integer products with the chip and number line models.

Chip Model

To use the chip model with multiplication, we need to know how to interpret the factors. The first factor tells us how many groups to put in $(+)$ or take away $(-)$. The second factor indicates how many and what kind of chips to put in each group.

Example 6.16	Using the Chip Model to Multiply Integers

Representation   Use the chip model to compute each product.

a. $3 \cdot (-3)$ **b.** $(-2) \cdot 3$ **c.** $(-3) \cdot (-2)$

Solution

a. To compute $3 \cdot (-3)$, put in 3 groups each containing 3 negative chips, for a total of 9 negative chips, or $3 \cdot (-3) = (-9)$.

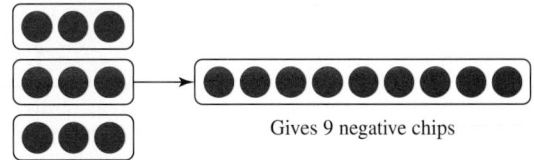

Put in 3 groups of
3 negative chips

Gives 9 negative chips

b. To compute $(-2) \cdot 3$, we must remove 2 groups, each containing 3 positive chips. We first put in $2 \cdot 3 = 6$ zero pairs, and then remove the positive chips, leaving 6 negative chips, or $(-2) \cdot 3 = (-6)$.

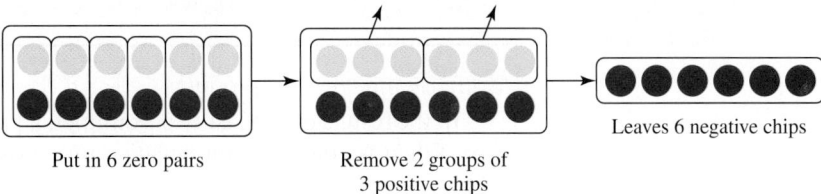

Put in 6 zero pairs

Remove 2 groups of
3 positive chips

Leaves 6 negative chips

c. For $(-3) \cdot (-2)$, we must remove 3 groups containing 2 negative chips. We first put in $3 \cdot 2 = 6$ zero pairs and then remove the negative chips, leaving 6 positive chips, or $(-3) \cdot (-2) = 6$.

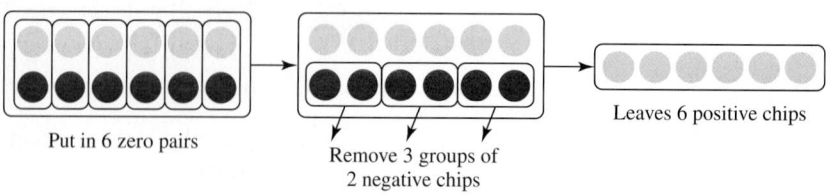

Put in 6 zero pairs   Remove 3 groups of 2 negative chips   Leaves 6 positive chips

Number Line Model

We again imagine a person moving back and forth, starting at zero. The first factor gives the direction in which the person faces and the number of steps to be taken. The second factor determines the length of each step and whether the person moves forward $(+)$ or backward $(-)$.

Example 6.17   Using the Number Line Model to Multiply Integers

Representation

Use the number line model to compute each product.

a. $3 \cdot (-1)$ **b.** $(-2) \cdot 5$ **c.** $(-4) \cdot (-3)$

Solution

a. To model $3 \cdot (-1)$, the person faces in the positive direction and takes 3 steps backward, each of length 1. The person arrives at (-3), so $3 \cdot (-1) = (-3)$.

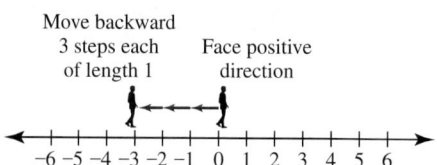

b. For $(-2) \cdot 5$, the person faces in the negative direction and takes 2 steps forward, each of length 5. The person arrives at (-10), so $(-2) \cdot 5 = (-10)$.

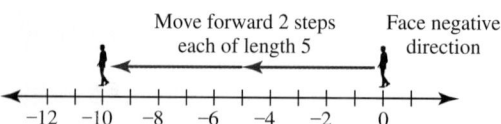

c. For $(-4) \cdot (-3)$, the person faces in the negative direction and takes 4 steps backward, each of length 3. The person arrives at 12, so $(-4) \cdot (-3) = 12$.

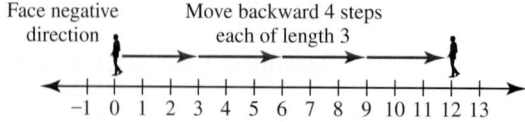

In the results of the last two examples, a pattern emerges for the two cases in question. When we multiplied a negative by a positive, the product was always negative.

We can verify that this will always be the case. Suppose that a and b are two positive integers, or $a > 0$ and $b > 0$. By the additive inverse property, $-(a \cdot b)$ is the only number that we can add to $a \cdot b$ to get 0, or $a \cdot b + -(a \cdot b) = 0$. However, if we add $(-a)b$ to $a \cdot b$, we get the following:

$$ab + (-a)b = (a + (-a))b \qquad \text{Distributive property of multiplication over addition}$$
$$= 0 \cdot b \qquad\qquad \text{Additive inverse property}$$
$$= 0 \qquad\qquad\quad \text{Zero multiplication property}$$

Given that $(-a)b$ and $-(a \cdot b)$ both result in zero, it must be that $(-a)b = -(a \cdot b)$. Consequently, because a and b are both positive, a negative integer times a positive integer must be a negative integer.

The examples also show a pattern for the negative-negative case. When we multiplied a negative by a negative, the product was always positive. Again, this fact can be verified algebraically with a proof that is similar to the previous case. We leave the details to the Learning Assessment.

The next definition collects all of these results for integer multiplication.

Definition of Integer Multiplication

Let a and b be whole numbers. Then:

1. $a \cdot b$ is the whole-number product of a and b.
2. $(-a) \cdot b = a \cdot (-b) = -(a \cdot b)$.
3. $(-a) \cdot (-b) = a \cdot b$.

This definition can be used to compute products with much larger factors. To do so, we compute the product using whole-number multiplication, and then determine its sign. The definition indicates that, if the signs of the factors are the same, the product is positive. If the signs of the factors are different, the product is negative.

Example 6.18 | Using the Definition of Integer Multiplication

Compute each product.

a. $(-56) \cdot 34$ **b.** $(-71) \cdot (-52)$ **c.** $108 \cdot (-44)$

Solution

a. $(-56) \cdot 34 = -(56 \cdot 34) = -1{,}904$.
b. $(-71) \cdot (-52) = 71 \cdot 52 = 3{,}692$.
c. $108 \cdot (-44) = -(108 \cdot 44) = -4{,}752$.

We can extend the rules for signs to products that involve more than two integers. In general, the product of an even number of negatives is positive, and the product of an odd number of negatives is negative. For instance:

$$(-9) \cdot (-7) \cdot (-3) \cdot (-4) = 9 \cdot 7 \cdot 3 \cdot 4 = 756$$

and

$$(-14) \cdot 8 \cdot (-21) \cdot (-5) \cdot 12 = -(14 \cdot 8 \cdot 21 \cdot 5 \cdot 12) = -141{,}120$$

Example 6.19 | Net Profits

Application

For the last year, a company has averaged a weekly income of $19,850. If expenses are paid every 2 weeks and average $26,750, what has been the net profit of the company for the last 3 months?

Solution The total income for the last 3 months is found by multiplying the weekly income by the number of weeks, or $19,850 · 12 = $238,200. The total expenses, represented by a negative integer, is the product of the average expenses and the number of 2-week periods, or 6 · (−$26,750) = −$160,500. After adding these two amounts, the net profit for the company over the last 3 months comes to:

$$\$238,200 + (-\$160,500) = \$238,200 - \$160,500 = \$77,700$$

Properties of Integer Multiplication

As with whole-number multiplication, integer multiplication satisfies the following algebraic properties.

Theorem 6.5   Properties of Integer Multiplication

If a, b, and c are integers, then integer multiplication satisfies the following properties.

Closure property:	$a \cdot b$ is a unique integer.
Commutative property:	$a \cdot b = b \cdot a$
Associative property:	$a \cdot (b \cdot c) = (a \cdot b) \cdot c$
Identity property:	$a \cdot 1 = a = 1 \cdot a$
Zero multiplication property:	$a \cdot 0 = 0 = 0 \cdot a$
Distributive property of multiplication over addition:	$a \cdot (b + c) = (a \cdot b) + (a \cdot c)$

Integer multiplication satisfies several other properties, each of which follows directly from the properties in Theorem 6.5.

Theorem 6.6   Multiplying by (−1)

If a is any integer, then $a(-1) = (-a)$.

Theorem 6.6 states that multiplying an integer by (−1) has the effect of changing its sign. In other words, multiplying by (−1) changes an integer a into its opposite. The property can be proved in the following way. If a is an integer, then:

$$
\begin{aligned}
0 &= a \cdot 0 & &\text{Zero multiplication property} \\
&= a[1 + (-1)] & &\text{Additive inverse property} \\
&= a(1) + a(-1) & &\text{Distributive property of multiplication over addition} \\
&= a + a(-1) & &\text{Multiplicative identity property}
\end{aligned}
$$

Because the additive inverse property states that $a + (-a) = 0$, we can use substitution, and set $a + a(-1) = a + (-a)$. By subtracting a from both sides, we get $a(-1) = (-a)$.

Given negative numbers, we also have the distributive property of multiplication over subtraction.

Theorem 6.7   Distributive Properties of Multiplication over Subtraction

If a, b, and c are integers, then:

1. $a(b - c) = a \cdot b - a \cdot c$.
2. $(b - c)a = b \cdot a - c \cdot a$.

Example 6.20   Proving $a(b - c) = a \cdot b - a \cdot c$

Reasoning

If a, b, and c are integers, prove that $a(b - c) = a \cdot b - a \cdot c$.

Solution If a, b, and c are integers, then:

$$
\begin{aligned}
a(b - c) &= a\,(b + (-c)) && \text{Theorem 6.4} \\
&= a \cdot b + a(-c) && \text{Distributive property of multiplication over addition} \\
&= a \cdot b + (-a \cdot c) && \text{Integer multiplication} \\
&= a \cdot b - a \cdot c && \text{Theorem 6.4}
\end{aligned}
$$

The proof of the second property of Theorem 6.7 is left to the Learning Assessment.

Check Your Understanding 6.2-A

1. Use the chip model to represent and solve each product.

   a. $6 \cdot (-2)$ **b.** $(-3) \cdot 1$ **c.** $(-2) \cdot (-5)$

2. Use the number line model to represent and solve each product.

   a. $4 \cdot (-2)$ **b.** $(-6) \cdot 1$ **c.** $(-3) \cdot (-5)$

3. Compute each product.

   a. $(-432) \cdot (-91)$ **b.** $(-8) \cdot (-10) \cdot (-14) \cdot 12$

   c. $(-611) \cdot (12)$ **d.** $(-11) \cdot 6 \cdot 4 \cdot (-9) \cdot 16$

4. State the property of integer multiplication used in each equation.

   a. $(-4)[3 + (-2)] = (-4)3 + (-4)(-2)$ **b.** $6 \cdot (-1) = (-6)$

   c. $(-3)(6 - 7) = (-3)6 - (-3)7$ **d.** $(-5) \cdot (-3) = (-3) \cdot (-5)$

   e. $(-4)[3 \cdot 7] = [(-4) \cdot 3]7$ **f.** $(-8) \cdot 0 = 0$

Talk About It Which do you think best represents integer multiplication, the chip model or the number line model? Why?

Representation

Activity 6.3

Another way to learn the rules for integer multiplication is to use a video camera analogy. Let the first integer represent video footage of someone walking forward (+) or backward (−). Then let the second integer represent play (+) or rewind (−) on a video camera.

a. Use this analogy to write a verbal interpretation of each rule for integer multiplication.

   i. $+ \cdot + = +$.   **ii.** $+ \cdot - = -$.   **iii.** $- \cdot + = -$.   **iv.** $- \cdot - = +$.

b. Adapt this analogy into a model that can be used to represent and solve specific integer problems. Use it to represent and solve each of the following products.

   i. $2 \cdot 4$ **ii.** $2 \cdot (-4)$ **iii.** $(-2) \cdot 4$ **iv.** $(-2) \cdot (-4)$

Integer Bases and Exponents

The definitions and properties of exponents discussed in Chapter 4 can be extended to include integer bases.

Definitions and Properties of Exponents

Let a and b be integers, and let m and n be positive integers. Then

1. $a^n = \underbrace{a \cdot a \cdot \ldots \cdot a}_{n \text{ factors}}.$
2. $a^0 = 1$ when $a \neq 0$.
3. $a^m \cdot a^n = a^{m+n}$.
4. $(a^m)^n = a^{m \times n}$.
5. $a^m \cdot b^m = (a \cdot b)^m$.

We use each definition and property as with the whole numbers, but now we must be careful to consider the sign of the result. Because exponents are a shortcut to repeated multiplication, we can use the same sign rules for multiplication to determine the signs of exponents. If the exponent on a negative integer is even, the result is positive. If it is odd, the result is negative. We must also be careful to notice whether the exponent applies to just the value of the integer or to both the sign and the value. For instance, with -2^4, the standard order of operations tells us that the exponent applies only to the numerical value, not the sign, or $-2^4 = -(2^4) = -16$. If the exponent needs to apply to both, we write the integer in parentheses. For example, $(-2)^4 = (-2) \cdot (-2) \cdot (-2) \cdot (-2) = 16$.

Example 6.21 Integer Bases and Exponents

Use the definitions and properties of exponents to simplify each expression.

a. $(-6)^4 \cdot (-6)^3$ **b.** $-(4^2)^3$ **c.** $(-125)^0$ **d.** $(-2)^3 \cdot (-5)^3$

Solution

a. $(-6)^4 \cdot (-6)^3 = (-6)^{4+3} = (-6)^7$.

b. $-(4^2)^3 = -(4^{2 \cdot 3}) = -4^6$.

c. $(-125)^0 = 1$.

d. $(-2)^3 \cdot (-5)^3 = [(-2) \cdot (-5)]^3 = 10^3$.

In most cases, we use a calculator to compute exponents on integers. The next example demonstrates possible keystrokes in such computations.

Example 6.22 Using a Calculator to Compute Exponents

Use a calculator to compute each expression.

a. $(-3)^3 \cdot (-4)^4$ **b.** $-(128)^2$

Solution To make sure the calculator interprets the negative signs correctly, place them in parentheses as needed.

a. ((−) 3) ^ 3 × ((−) 4) ^ 4 = displays −6912.

b. (−) (128 ^ 2) = displays −16384.

Integer Division

Explorations
Manual
6.6

As explained in Chapter 4, whole-number division can be approached in three ways: as repeated subtraction, as partitioning, or as a product with a missing factor. Ideally, we would like to apply these ideas to the integers, but not all of them work for every possible combination of signs.

For instance, consider repeated subtraction. In this approach, the quotient $a \div b$ is defined to be the number of times b can be subtracted from a as long as $b \neq 0$. If a and b are positive, then we have a whole-number quotient, and there is no problem. We can also use this approach when a and b are negative. For instance, the student page in Figure 6.7 shows that we can compute $(-6) \div (-2)$ by subtracting (-2) from (-6) until we reach zero. The figure shows the differences numerically and on a number line. In both cases, (-2) is subtracted three times, so it must be that $(-6) \div (-2) = 3$. This example implies that the quotient of two negative integers is positive, a result that coincides with the rules for multiplication.

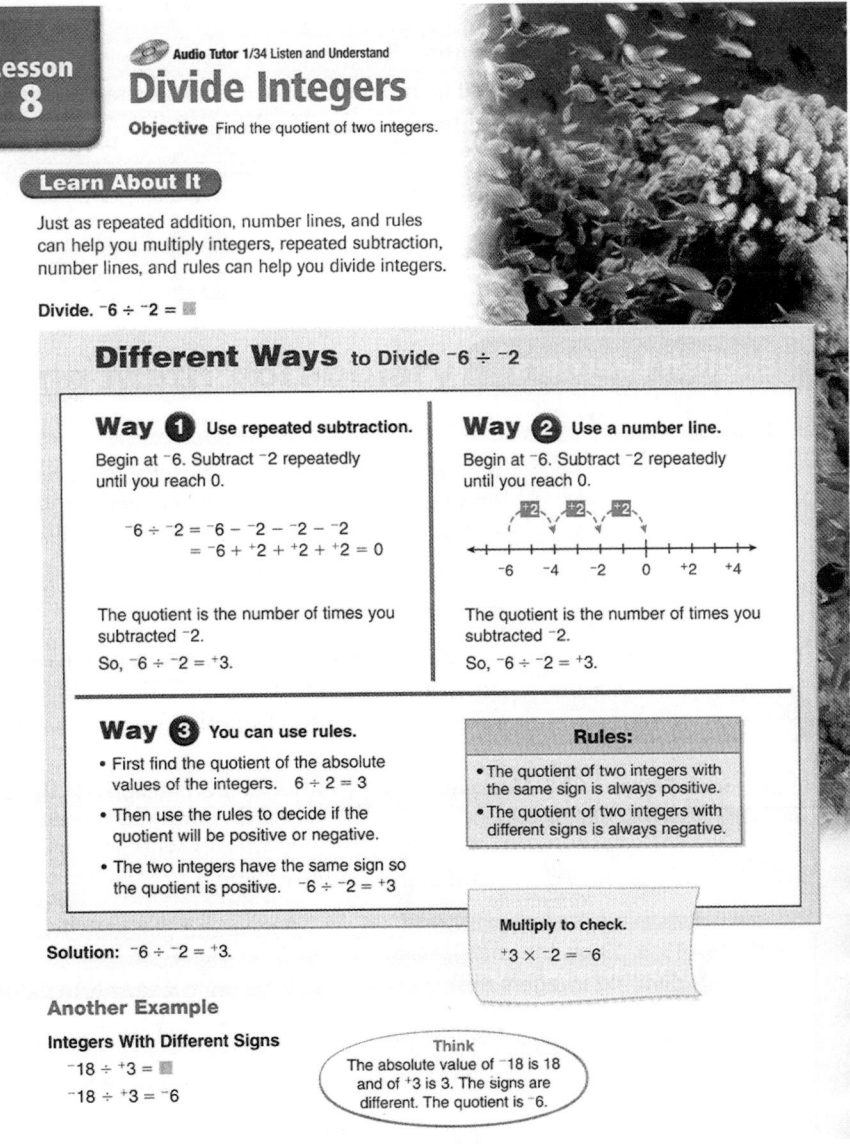

Figure 6.7 Integer division
Source: From *Houghton Mifflin Mathematics: Student Edition Level 6* (p. 292). Copyright © by Houghton Mifflin Company, Inc. Reproduced by permission of the publisher, Houghton Mifflin Harcourt Publishing Company.

Although repeated subtraction works when the signs are the same, we run into problems when the signs are different. For instance, consider the quotient of a positive and a negative integer, like $6 \div (-2)$. Intuitively, we would expect the answer to be -3 or 3 because $6 \div 2 = 3$. However, we are unable to arrive at either one because subtracting (-2) from 6 leads to larger numbers, not to 0, or

$$6 - (-2) = 8 \qquad 8 - (-2) = 10 \qquad 10 - (-2) = 12 \ldots$$

A similar problem arises when the quotient involves a negative and a positive.

Division can also be thought of as partitioning sets. In this approach, the quotient $a \div b$ is the number of objects in each set when a objects have been equally distributed among b sets. As the next example shows, this approach is also limited in its ability to compute integer quotients.

Example 6.23   Using the Partitioning Approach to Divide Integers

Representation

If possible, use the partitioning approach and the chip model to solve each quotient.

a. $(-9) \div 3$ **b.** $10 \div (-5)$

Solution

a. The quotient $(-9) \div 3$ tells us to equally distribute 9 negative chips among 3 sets. Because 3 negative chips will be in each set, $(-9) \div 3 = (-3)$.

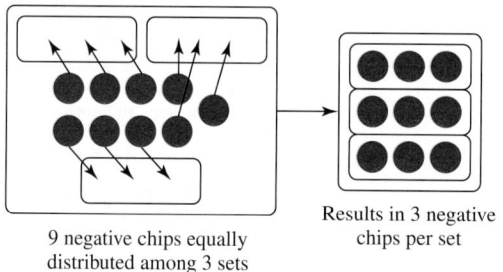

9 negative chips equally
distributed among 3 sets

Results in 3 negative
chips per set

b. The quotient $10 \div (-5)$ tells us to equally distribute 10 positive chips among 5 negative sets. However, the meaning of a "negative set" is unclear, so we do not know how to distribute the chips. Consequently, we cannot compute this quotient with the partitioning approach.

Because repeated subtraction or partitioning cannot be used for every quotient, only the missing-factor approach is left. Fortunately, this approach not only allows us to compute integer quotients, but it also provides a definition for **integer division**.

Definition of Integer Division (Missing-Factor Approach)

If a and b are any integers with $b \neq 0$, then $a \div b = c$ if and only if $a = b \cdot c$ for some unique integer c.

Example 6.24   Using the Missing-Factor Approach to Divide Integers

Use the missing-factor approach to find each quotient.

a. $15 \div (-3)$ **b.** $(-24) \div (-8)$ **c.** $(-14) \div (6)$

Solution

a. By definition, $15 \div (-3) = c$ if and only if $15 = (-3) \cdot c$. Because $(-3) \cdot (-5) = 15$, then $15 \div (-3) = (-5)$.

b. $(-24) \div (-8) = c$ if and only if $(-24) = (-8) \cdot c$. Because $(-8) \cdot 3 = (-24)$, then $c = 3$.

c. $(-14) \div 6 = c$ if and only if $(-14) = 6 \cdot c$. Because no integer c makes the statement true, $(-14) \div 6$ is undefined in the set of integers.

∎

These examples show that the signs in quotients behave like the signs in products. If the signs are the same, the quotient is positive. If the signs are different, the quotient is negative. Integer division satisfies the same basic properties of whole-number division, such as $a \div 1 = a$, $a \div a = 1$, and $0 \div a = 0$. As before, division by 0 is undefined.

Check Your Understanding 6.2-B

1. Simplify each expression. Leave your answer in exponential form.

    a. $(-316)^0$ **b.** $[(-2)^2]^4$ **c.** $(-3)^{10} \cdot (-3)^6$ **d.** $(-4)^7 \cdot (-5)^7$

2. Use repeated subtraction to compute $(-21) \div (-7)$.

3. Use the partitioning approach and the set model to represent and compute $(-8) \div 4$.

4. Use the missing-factor approach to find each quotient when possible.

    a. $21 \div (-3)$ **b.** $(-14) \div (-7)$ **c.** $(-13) \div 4$ **d.** $(-18) \div 6$

Talk About It Why can repeated subtraction *not* be used to compute $-10 \div 2$?

Representation

Activity 6.4

Let a standard deck of cards represent the integers, as stated in Activity 6.1. Now, however, suppose that the value of a hand is determined by the product of the integers represented by the cards.

a. What is the sign and value of each hand?

    i. A♣, K♦, 3♦, 4♠, Q♠   **ii.** 3♣, 10♣, J♠, J♣, A♥

    iii. A♣, A♦, A♠, A♥, 7♥   **iv.** 3♦, 4♦, 2♦, J♠, 2♥

b. List a possible set of cards that could be in a 5-card hand, if the product of the cards is:

    i. −8.   **ii.** −81.   **iii.** 7,200.   **iv.** 1,600.

SECTION 6.2 Learning Assessment

■ Understanding the Concepts

1. Why does the Cartesian product approach to multiplication *not* work with the integers?

2. In what situations does it make sense to interpret an integer product as repeated addition?

3. a. What two rules can be used to summarize the definition of integer multiplication?

    b. How can these rules be extended to products that involve more than two integers?

4. How does multiplying by (-1) affect the sign of an integer?

5. What does the distributive property of multiplication over subtraction state?

6. How can we use the exponent to determine the sign of a negative integer raised to a power?

7. In what situations does it make sense to interpret integer division as:

    a. Repeated subtraction?

    b. Partitioning?

8. How can the missing-factor approach be used to compute integer division?

■ Representing the Mathematics

9. Represent each situation as an integer expression, and then solve.

   a. A football team loses 7 yd on each of 3 consecutive plays. What is the team's total loss in yardage?

   b. An airplane descends at a rate of 650 ft/min. If the airplane is in this descent for 12 min, how far does it descend?

   c. The temperature at midnight was $-12°$ F. If the temperature increases at a rate of 4° F/hr, what will the temperature be at 8:00 a.m.?

   d. A stockbroker makes a bad investment and loses $365 dollars a day on a certain stock. If it takes a week for him to sell, how much money has he lost?

10. Use the chip model to represent and solve each product.

    a. $(-3) \cdot (-4)$ **b.** $6 \cdot 2$ **c.** $(-3) \cdot 3$ **d.** $8 \cdot (-2)$

11. Use the number line model to represent and solve each product.

    a. $(-4) \cdot 3$ **b.** $5 \cdot (-1)$ **c.** $6 \cdot 3$ **d.** $(-7) \cdot (-2)$

12. Another way to interpret integer products is to use a person's position on a number line in the past or future. The first factor represents the number of hours in the future $(+)$ or past $(-)$. The second factor represents the direction and the rate at which the person is walking in miles per hour. Use this model to represent each product on a number line.

    a. $4 \cdot (-2)$ **b.** $3 \cdot 5$ **c.** $(-4) \cdot (-5)$ **d.** $(-2) \cdot 6$

13. Complete each table with the signs of the products and quotients of two integers.

×	+	−
+		
−		

÷	+	−
+		
−		

14. What is the difference between the expression -7^4 and $(-7)^4$?

15. When possible, use repeated subtraction to represent and compute each quotient.

    a. $8 \div (-2)$ **b.** $(-16) \div (-4)$ **c.** $(-12) \div (-5)$

16. When possible, use the partitioning approach to represent and compute each quotient.

    a. $(-10) \div 2$ **b.** $(-14) \div 6$ **c.** $(-15) \div (-3)$

■ Exercises

17. Establish a pattern of whole-number products that can be extended to find each integer product.

    a. $4 \cdot (-1)$ **b.** $(-2) \cdot 6$

18. Establish a pattern of whole-number products that can be extended to find each integer product.

    a. $(-3) \cdot (2)$ **b.** $(3) \cdot (-3)$

19. Use the definition of integer multiplication to compute each product.

    a. $(-9) \cdot (-7)$ **b.** $(-23) \cdot 14$
    c. $126 \cdot (-29)$ **d.** $(-283) \cdot (-41)$

20. Multiply.

    a. $(-3) \cdot (-6) \cdot 8$
    b. $6 \cdot 7 \cdot (-10) \cdot (-8)$
    c. $(-7) \cdot (-6) \cdot 3 \cdot 4 \cdot (-8)$

21. Find the value of x.

    a. $34 \cdot x = (-238)$ **b.** $(-81) \cdot (-9) = x$
    c. $x \cdot (-85) = (-5,355)$

22. Write each exponent as a repeated product and then solve.

    a. 3^3 **b.** $(-3)^4$ **c.** $(-4)^5$ **d.** $(-6)^4$

23. Simplify each expression. Leave your answer in exponential form.

    a. $(-4)^3 \cdot (-4)^6$
    b. $(-5)^3 \cdot (-2)^3$
    c. $((-5)^3)^3$
    d. $(-2)^6 \cdot (-2)^2 \cdot (-2)^3 \cdot (-2)^7$
    e. $(-197)^0$
    f. $\{[(-83)^2]^0\}^9$

24. When possible, use the missing-factor approach to find each quotient.

    a. $(-15) \div (-5)$ **b.** $(-12) \div 5$
    c. $18 \div (-3)$ **d.** $(-24) \div (-7)$

25. Find the missing number in each quotient.

    a. $95 \div x = (-19)$ **b.** $x \div 56 = (-11)$
    c. $(4,628) \div (-52) = x$

26. Compute each problem and then list the three other facts in the fact family.

    a. $(-6) \cdot 7$ **b.** $(-8) \cdot (-5)$
    c. $(-35) \div (-7)$ **d.** $(-72) \div 9$

27. First determine whether the result is positive or negative. Then use a calculator to make the computation.

    a. $(-3) \cdot (-14) \cdot (-36) \cdot (25) \cdot (-91)$
    b. $(-39) \div (-13) \cdot (-65) \div (-13) \cdot (-5)$
    c. $[(-24) \cdot (-16)] \div [(-8) \cdot (-4)]$
    d. $(-48)^2 \cdot (-15)^3$

28. Use the standard order of operations to make each computation.

    a. $(9 - 16)^4$ **b.** $[(-9) + (-6)]^2 - (-25)$
    c. $[(-17) - 4] \div (-3)$ **d.** $7 \cdot (-4) + 32 \div (-8)$

29. Calculate the following:
    a. $|(-12)| \cdot |(-6)|$ b. $[6 - (-6)] \cdot |(-4)|$
    c. $-|(-13)| \cdot |8|$ d. $|(-16)| \div |2|$
    e. $|6 - (-8)| \div (-7)$ f. $-|52| \div |(-13)|$

30. Estimate each product and quotient. What method of estimation did you use?
    a. $(-98) \cdot (-205)$ b. $699 \cdot (-301)$
    c. $(-1,009) \div 198$ d. $(-1,621) \div (-411)$

■ Problems and Applications

31. Place a positive or negative sign in each blank to make the equation true.
    a. ___$73 \cdot$ ___$14 +$ ___$27 = -1,049$.
    b. ___$63 \div$ ___$7 +$ ___$23 \cdot$ ___$15 -$ ___$12 = -348$.
    c. ___$22 -$ ___$47 +$ ___$36 \cdot$ ___$19 -$ ___$84 = 837$.
    d. ___$49 \cdot$ ___$27 -$ ___$84 \div$ ___$7 \cdot$ ___$12 = 1,467$.

32. a. Which product is closer to zero, $65 \cdot (-36)$ or $(-84) \cdot (-21)$?
    b. Which quotient is farther from zero, $(-1,456) \div (-91)$ or $(-1,387) \div 73$?

33. If n is an even, positive integer, what is the sign of each of the following?
    a. $(-2)^n$ b. 2^n c. $(-2)^{n+1}$
    d. $(-2)^{2n}$ e. $(-2)^{2n+1}$

34. When possible, find all the integer values of x that make each expression true.
    a. $4x = (-16)$ b. $(-2)|x| = (-22)$
    c. $(-15) \div x = (-3)$ d. $(-4)|x| = 24$
    e. $(-18) \div |x| = (-2)$ f. $19 \div x = 5$

35. When possible, find all the integer values of x that make each expression true.
    a. $x^2 = 16$ b. $x^2 = 24$
    c. $-x^2 = 36$ d. $-x^2 = (-1)$
    e. $x^2 = (-36)$ f. $-x^2 = 0$

36. a. If n is a positive integer, what is the sign of each of the following?
    i. n ii. n^2 iii. n^3 iv. $-n^2$ v. $-n^3$
    b. If n is a negative integer, what is the sign of each of the following:
    i. n ii. n^2 iii. n^3 iv. $-n^2$ v. $-n^3$

37. The formula for converting temperature from degrees Fahrenheit to degrees Celsius is $C = \dfrac{5(F - 32)}{9}$. Find the temperature in degrees Celsius for each temperature in degrees Fahrenheit.
    a. $59°$ F   b. $32°$ F   c. $5°$ F

38. On one weekend, a restaurant earns $5,200, $6,300, and $4,400. On Friday, the restaurant pays out 2 checks for $1,500 each and 13 checks for $250 each. Write an integer expression representing this situation, and use it to find the net profit of the restaurant.

39. Temperature drops about $4°$ F for every 1,000 ft gained in elevation. A plane takes off and climbs to a cruising altitude of 26,000 ft. If the temperature at takeoff was $78°$ F, what is the temperature at the plane's cruising altitude?

40. A particular corporation loses a lawsuit and must pay $14 million to 6 plaintiffs over the course of 12 months. To the nearest dollar, about how much must the company pay to each person per month?

41. A stock loses about $8 a month for 6 months. If it was originally $256 a share before its decline began, how much does it cost at the end of the 6 months?

42. A medical researcher applies a new antibiotic to a culture of bacteria. Originally, 50,000 bacteria are present. If the antibiotic kills off 3,500 bacteria every hour, how long will it take before the bacteria count is less than 5,000?

■ Mathematical Reasoning and Proof

43. True or false?
    a. The product of a positive integer and a negative integer is positive.
    b. The product of a negative integer and a negative integer is negative.
    c. The product of an odd number of negative integers is negative.
    d. The product of an even number of negative integers is positive.

44. Show that $a(b + c) = a \cdot b + a \cdot c$ and $a(b - c) = a \cdot b - a \cdot c$ hold true for each set of values for a, b, and c.
    a. $a = (-3)$, $b = 5$, $c = 9$.
    b. $a = (-2)$, $b = (-5)$, $c = (-7)$.

45. Use properties of integer operations to show that $(b - c)a = b \cdot a - c \cdot a$.

46. Is the expression $(a + b) \div c = (a \div c) + (b \div c)$ true for all integers a, b, and $c \neq 0$? If not, provide a counterexample.

47. If a and b are any integers with $a < b$, is it always true that $a^2 < b^2$? If so, explain. If not, provide a counterexample.

48. Let a and b be any integers. Determine whether each statement is true or false. If false, provide a counterexample.
    a. $|a \cdot b| = |a| \cdot |b|$ b. $|a^2| = (|a|)^2$ c. $|a^3| = (|a|)^3$

49. If x and y are integers, under what conditions is $x^2 - y^2$ positive? Negative? Zero?

50. **a.** Compute the following consecutive powers of (-2).

$$(-2)^1 \quad (-2)^2 \quad (-2)^3 \quad (-2)^4 \quad (-2)^5 \quad (-2)^6$$

b. What patterns do you notice between the power and the sign of the result?

c. Make a generalization about when $(-2)^n$ is positive and when it is negative.

51. State the property used in each step of the computation.

$$2 \cdot 5 + (-2)5 = 5 \cdot 2 + 5(-2)$$
$$= 5\,[2 + (-2)]$$
$$= 5 \cdot 0$$
$$= 0$$

52. In this section, an algebraic argument was used to show that the product of a negative integer and a positive integer is negative. Use a similar argument to show that the product of two negative integers is positive.

■ Mathematical Communication

53. Another analogy that can be used to learn the rules for integer multiplication is the Old West town analogy. The first factor represents good guys $(+)$ or bad guys $(-)$. The second factor indicates whether they are entering $(+)$ or leaving $(-)$ town. The sign of the product then indicates either the good $(+)$ or the bad $(-)$ effect on the town. Use this analogy to write a verbal interpretation of each rule for integer multiplication.

a. $+ \cdot + = +$.   **b.** $+ \cdot - = -$.

c. $- \cdot + = -$.   **d.** $- \cdot - = +$.

54. Write two or three word problems that use integer multiplication. Give them to a partner to solve. What difficulties, if any, did you experience in writing the problems?

55. Make up four or five integer multiplication problems. With a partner, make a large number line on the floor of your classroom. Give your partner directions on how to move along the number line to solve your problems.

56. Jeremiah claims that the product of $-x \cdot y$ must be negative if x and y are integers. Is he correct?

57. Sara does not understand why the product of two negative integers is positive. How would you, as her teacher, help her understand this rule of multiplication?

■ Building Connections

58. **a.** How does integer multiplication use or build on the ideas of whole-number multiplication?

b. How does integer division use or build on the ideas of whole-number division?

59. Think of an analogy, other than the ones given in this section, that can be used to teach the rules of thumb for determining the sign of an integer product.

60. With a small group of your peers, examine several fifth- or sixth-grade texts. How does the text approach teaching the multiplication and division of signed numbers? What models are used in the text? Discuss the advantages and disadvantages of the different models presented in this section as well as the ones found in the middle-grade texts.

61. Negative exponents have not yet been defined. What is the interpretation of a negative exponent, and what kind of number is needed to work with such exponents?

62. Students often learn about integer operations as they prepare to take algebra. Why do you think that the integers are prerequisite knowledge for basic algebra?

Historical Note

Although we readily accept negative numbers today, it took almost 2,000 years for them to be considered something other than absurd. The oldest recorded use of negative numbers is in the *Nine Chapters on the Mathematical Art*, an ancient Chinese manuscript whose original copies date to prior to 213 B.C.E. From that time period on, the ancient Greeks, Hindus, and Arabs recognized that many linear and quadratic equations could be solved only with negative numbers. However, they considered negative numbers to be nonsensical, so they avoided working with such equations. Not until Girolamo Cardan (1501−1576) used negative solutions in his book *Ars Magna* (*The Great Art*) did they even begin to gain acceptance in Europe. It took another 300 years and the influence of other well-established mathematicians, like Euler, Peacock, and Hamilton, before negative numbers became completely accepted as a mathematical notion.

Girolamo Cardan

Opening Problem Answer

Image Source/Jupiter Images

Any value in degrees Fahrenheit that is divisible by 9 when 32 is subtracted from it will give an integer value in degrees Celsius. Five such values are 32° F, 41° F, 50° F, 59° F, and 68° F.

Chapter 6 REVIEW

Summary of Key Concepts from Chapter 6

Section 6.1   Integer Basics, Addition, and Subtraction

- The set of integers is the union of the whole numbers with the negatives, or

  $\mathbb{Z} = \{\ldots -3, -2, -1, 0, 1, 2, 3, \ldots\}$

- Two integers are **opposites** if they are on opposite sides of zero and at an equal distance from zero.

- The **absolute value** of an integer a, denoted $|a|$, is the integer's distance from zero. If $a \geq 0$, then $|a| = a$. If $a < 0$, then $|a| = -a$.

- **Integer addition:** Let a and b be any two integers.

  1. If a and b have the same sign, then $a + b$ is the sum of the absolute values of a and b and takes the sign of a and b.

  2. If a and b have different signs, then $a + b$ is the difference of the smaller absolute value from the larger absolute value and takes the sign of the number with the larger absolute value.

- Integer addition satisfies the closure, commutative, associative, and identity properties.

- **Additive inverse property:** For every integer a, there exists a unique integer $-a$, called the **additive inverse,** such that $a + (-a) = 0 = (-a) + a$.

- Properties of additive inverses:

  1. $-(-a) = a$.

  2. $(-a) + (-b) = -(a + b)$.

- For any integers a and b, a is **less than** b, denoted $a < b$, if and only if there exists a positive integer n such that $a + n = b$.

- **Integer subtraction:** If a and b are any integers, then $a - b = c$ if and only if $a = b + c$ for some unique integer c.

- **Subtraction by adding the opposite:** For all integers a and b, $a - b = a + (-b)$.

Section 6.2   Integer Multiplication and Division

- **Integer multiplication:** If a and b are whole numbers, then $a \cdot b$ is the whole number product of a and b, $(-a) \cdot b = a \cdot (-b) = -(a \cdot b)$, and $(-a) \cdot (-b) = a \cdot b$.

- If the number of negative integers is even, the product is positive. If the number of negative integers is odd, the product is negative.

- Integer multiplication satisfies the closure, commutative, associative, identity, zero multiplication, and distributive properties.

- If a, b, and c are integers, then $a(-1) = (-a)$, $a(b - c) = a \cdot b - a \cdot c$, and $(b - c)a = b \cdot a - c \cdot a$.
- **Integer division:** If a and b are any integers with $b \neq 0$, then $a \div b = c$ if and only if $a = b \cdot c$ for some unique integer c.

Review Exercises Chapter 6

1. Write a verbal interpretation of each expression.
   a. (-6) b. $(-4) - 9$
   c. $4 + (-8)$ d. $(-4) - (-9)$

2. Give the position of each integer on the number line with respect to zero.
   a. 14 b. (-8) c. 0

3. How many different integers can be represented by exactly 6 chips in the chip model?

4. Use the chip model to represent and solve each problem.
   a. $(-3) + (-4)$ b. $6 - (-2)$ c. $(-6) \cdot 2$

5. Use the number line model to represent and solve each problem.
   a. $(-6) + 3$ b. $(-5) - (-1)$ c. $(-4) \cdot (-3)$

6. Compute.
   a. $(-305) + (-61)$ b. $(-57) + 319$
   c. $(-216) - 133$ d. $153 - (-218)$
   e. $(-102) \cdot (-87)$ f. $(-5) \cdot (-12) \cdot (-14)$
   g. $(-3,416) \div (-56)$ h. $4,758 \div (-183)$

7. Find the value of x in each computation.
   a. $(35) + x = (-47)$. b. $42 - (-57) = x$.
   c. $(-69) - x = (-40)$. d. $(-35) \cdot x = (-105)$.
   e. $(-14) \cdot (-21) = x$. f. $(-150) \div x = (-6)$.

8. What is the additive inverse of each expression?
   a. -2 b. 9
   c. 0 d. $-|4|$
   e. $|(-8)|$

9. When possible, find all values of x that make the given statements true.
   a. $|x| = 4$. b. $|x| = (-5)$.
   c. $|x - 2| = 4$. d. $|x| - 5 = (-3)$.

10. Represent each situation as an integer expression and solve.
    a. The temperature yesterday was $(-4)°$ F and today it is $23°$ F. What was the change in temperature?
    b. A football team starts a play on the 49-yd line and finishes the play back on the 35-yd line. What is the team's loss in yardage?

c. An airplane descends at a rate of 450 ft/min. If the airplane is in this descent for 15 min, how far does it descend?

11. Use the number line to explain why $(-3) < (-1)$.

12. Write each subtraction fact as an addition fact.
    a. $13 - 6$ b. $(-13) - 6$
    c. $(-13) - (-6)$ d. $13 - (-6)$

13. State the property used in each expression.
    a. $(-2)(3 - 4) = (-2)3 - (-2)4$.
    b. $6 + (-6) = 0$.
    c. $0 \cdot (-5) = 0$.
    d. $3[(-4) \cdot (-6)] = [(-4) \cdot (-6)] 3$.
    e. $(-12) + (-9) \in \mathbb{Z}$.
    f. $0 + (-7) = (-7)$.

14. Without performing the calculation, determine whether each result is positive or negative.
    a. $[(11) \cdot (-18)] \div [(-9) \cdot (-2)]$
    b. $(-48) \div (12) \cdot (-64) \div (-4) \cdot (8)$

15. Use the standard order of operations to make each computation.
    a. $[(-3) \cdot (5 - 2)]^2$
    b. $[(-14) - (6)]^2 \div (-25)$
    c. $3 \cdot (-5) + (-16)^2 \div (-8)^2$

16. Simplify each expression. Leave your answer in exponential form.
    a. $((-2)^4)^2$ b. $(-3)^4 \cdot (4)^4$
    c. $(-254)^0$ d. $((-33)^0)^6$
    e. $(-8)^2 \cdot (-8)^3$

17. In a series of 5 plays, a football team lost 3 yd on a run, gained 12 yd on a pass, gained 5 yd on a run, lost another 11 yd on a fumble, and finally gained 13 yd on a pass. What was the net yardage gained or lost in the 5 plays?

18. On one weekend, a restaurant brings in $3,300, $5,600, and $4,500. Payday is on Friday, so the restaurant pays out 3 checks for $1,100 each and 14 checks for $235 each. What is the net profit of the restaurant?

19. A corporation loses a lawsuit and must pay $35 million to 8 plaintiffs over the course of

16 months. To the nearest dollar, about how much must the company pay out to each person per month?

20. A medical researcher applies a new antibiotic to a culture of bacteria. Originally, 25,000 bacteria were present. If the antibiotic kills off 1,500 bacteria every hour, how long will it take before the bacterial count is less than 3,000?

21. Which of the following sets is closed under subtraction? Explain.

 a. $\{\ldots(-10), (-5), 0, 5, 10\ldots\}$

 b. $\{\ldots(-6), (-4), (-2), 2, 4, 6,\ldots\}$

22. Work left to right to fill in the following diagram so that each empty square contains the product of the two squares directly below it.

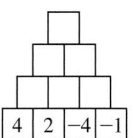

23. If $a = 3$, $b = (-2)$, $c = (4)$, and $d = (-1)$, place parentheses in the expression $a + b - c + d$ to find the largest possible value.

24. Find a sequence of products that might be used to determine the product $5 \cdot (-3)$.

25. a. Is it possible to use repeated subtraction to compute $(-8) \div 4$ correctly? Explain.

 b. Is it possible to use repeated subtraction to compute $(-6) \div (-3)$ correctly? Explain.

Answers to Chapter 6   Check Your Understandings and Activities

Check Your Understanding 6.1-A

1 a. 19. **b.** 45. **c.** 9. **d.** (-4).

2. a. $(-9) + 4 = (-5)$.

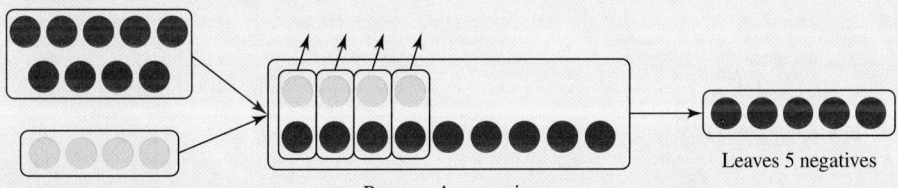

Combine 9 negatives with 4 positives
Remove 4 zero pairs
Leaves 5 negatives

 b. $3 + (-5) = (-2)$.

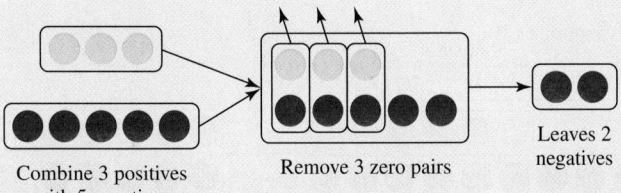

Combine 3 positives with 5 negatives
Remove 3 zero pairs
Leaves 2 negatives

 c. $(-2) + (-6) = (-8)$.

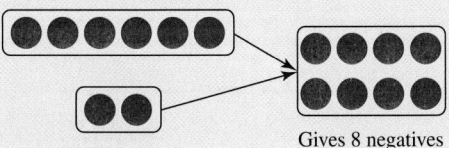

Combine 6 negatives with 2 negatives
Gives 8 negatives

 d. $7 + 4 = 11$.

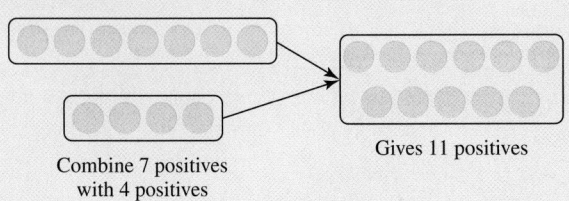

Combine 7 positives with 4 positives
Gives 11 positives

3. a. $(-3) + (-4) = (-7)$.

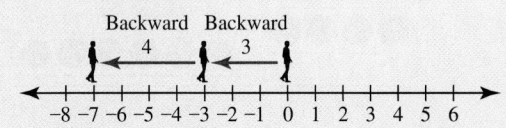

 b. $3 + 8 = 11$.

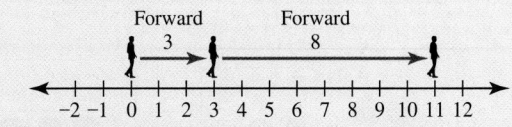

 c. $5 + (-2) = 3$.

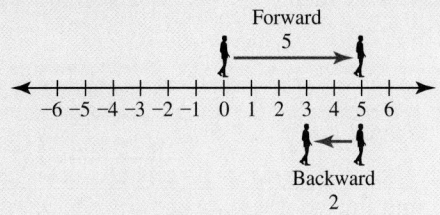

d. $(-7) + 6 = (-1)$.

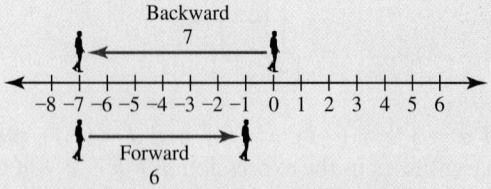
Backward
7

Forward
6

4. a. 438.   **b.** (-560).   **c.** (-875).

Check Your Understanding 6.1-B

1. a. Consider the following sequence of differences.

$$3 - 0 = 3$$
$$3 - 1 = 2$$
$$3 - 2 = 1$$
$$3 - 3 = 0$$

As the second number increases by 1, the difference decreases by 1. Continuing this pattern:

$$3 - 4 = (-1)$$
$$3 - 5 = (-2)$$

b. Consider the following sequence of differences:

$$5 - 3 = 2$$
$$4 - 3 = 1$$
$$3 - 3 = 0$$

As the first number decreases by 1, the difference decreases by 1. Continuing this pattern:

$$2 - 3 = (-1)$$
$$1 - 3 = (-2)$$
$$0 - 3 = (-3)$$
$$(-1) - 3 = (-4)$$

---

2. a. $4 - 5 = (-1)$.

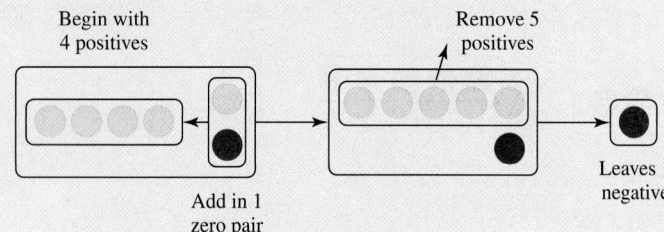
Begin with
4 positives

Add in 1
zero pair

Remove 5
positives

Leaves 1
negative

b. $(-4) - 6 = (-10)$.

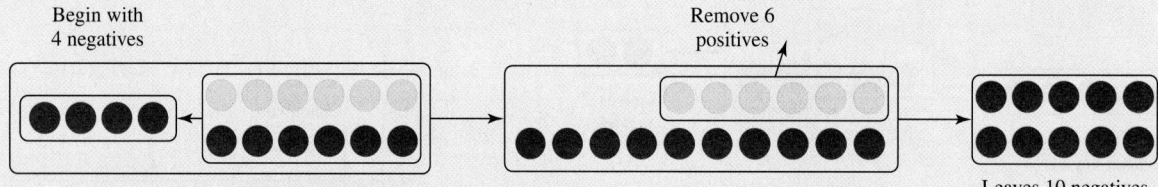
Begin with
4 negatives

Add 6 zero pairs

Remove 6
positives

Leaves 10 negatives

c. $(-2) - (-5) = 3$.

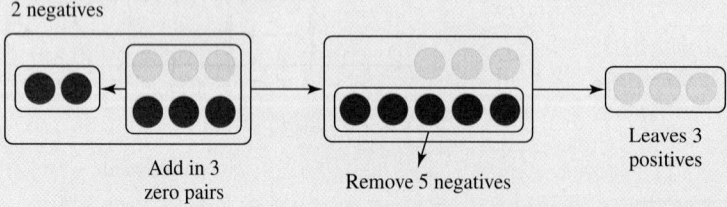
Begin with
2 negatives

Add in 3
zero pairs

Remove 5 negatives

Leaves 3
positives

3. a. $4 - 8 = (-4)$.

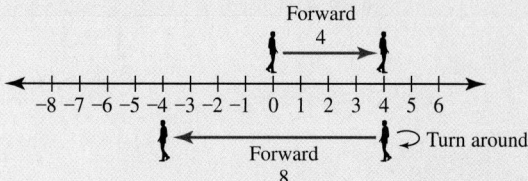
Forward
4

Turn around

Forward
8

b. $(-5) - (-4) = (-1)$.

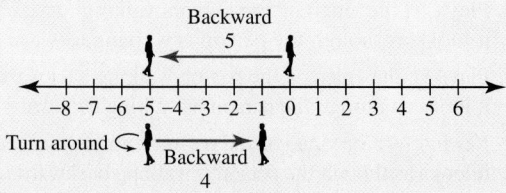

c. $6 - (-4) = 10$.

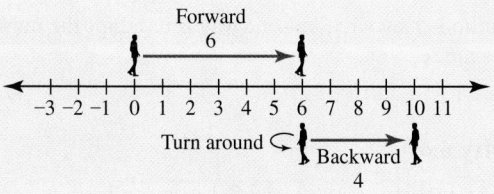

4. a. (-18). **b.** $1,246$. **c.** $(-1,419)$.

Check Your Understanding 6.2-A

1. a. $6 \cdot (-2) = (-12)$.

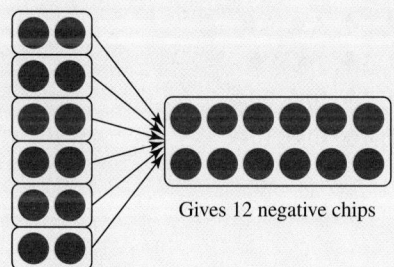

Gives 12 negative chips

Put in 6 groups of
2 negative chips

b. $(-3) \cdot 1 = (-3)$.

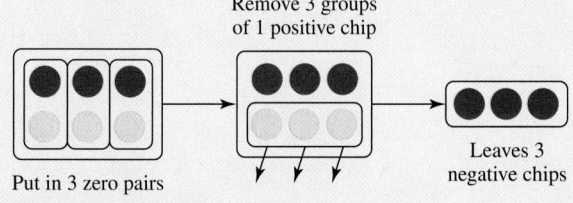

Remove 3 groups
of 1 positive chip

Leaves 3
negative chips

Put in 3 zero pairs

c. $(-2) \cdot (-5) = 10$.

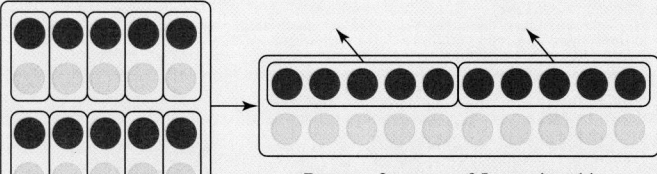

Remove 2 groups of 5 negative chips

Leaves 10 positive chips

Put in 10 zero pairs

2. a. $4 \cdot (-2) = (-8)$.

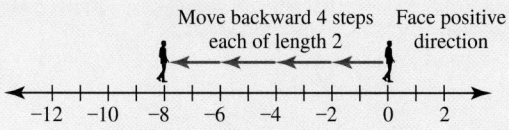

Move backward 4 steps   Face positive
each of length 2   direction

b. $(-6) \cdot 1 = (-6)$.

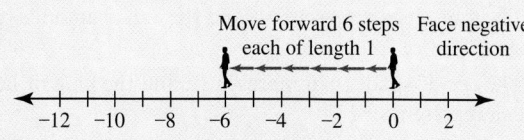

Move forward 6 steps   Face negative
each of length 1   direction

c. $(-3) \cdot (-5) = 15$.

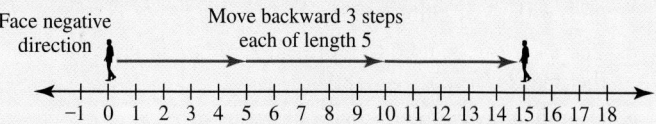

Face negative   Move backward 3 steps
direction   each of length 5

3. a. $39,312$. **b.** $(-13,440)$. **c.** $(-7,332)$. **d.** $38,016$.

4. a. Distributive property of multiplication over addition. **b.** Multiplication by (-1). **c.** Distributive property of multiplication over subtraction. **d.** Commutative property. **e.** Associative property. **f.** Zero multiplication property.

Check Your Understanding 6.2-B

1. a. 1. **b.** $(-2)^8$. **c.** $(-3)^{16}$. **d.** 20^7.

2. Because we can subtract (-7) from (-21) three times, or $[(-21) - (-7) = (-14), (-14) - (-7) = (-7)$, and $(-7) - (-7) = 0]$, then $(-21) \div (-7) = 3$.

3. Equally distribute 8 negative chips to 4 sets. Because 2 negative chips are in each set, $(-8) \div 4 = (-2)$.

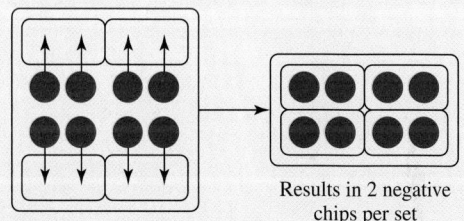

Results in 2 negative
chips per set

8 negative chips equally
distributed among 4 sets

4. a. $21 \div (-3) = x$ corresponds to the product $21 = (-3) \cdot x$. Because $(-3) \cdot (-7) = 21$, it must be that $21 \div (-3) = (-7)$.

b. $(-14) \div (-7) = x$ corresponds to the product $(-14) = (-7) \cdot x$. Because $(-7) \cdot 2 = (-14)$, it must be that $(-14) \div (-7) = 2$.

c. $(-13) \div 4 = x$ corresponds to the product $(-13) = 4 \cdot x$. Because there is no integer x such that $4 \cdot x = (-13)$, then $(-13) \div 4$ is not defined for the integers.

d. $(-18) \div 6 = x$ corresponds to the product $(-18) = 6 \cdot x$. Because $6 \cdot (-3) = (-18)$, it must be that $(-18) \div 6 = (-3)$.

Activity 6.1

a. The 4 of clubs represents 4, and the 8 of diamonds represents (-8).

b. The ace of spades can represent 1, and the king of hearts can represent (-13).

c. Answers will vary. One possible answer for $4 + 5$ is the 4 of clubs and the 5 of clubs. One possible answer for $5 + (-3)$ is the 5 of clubs and the 3 of hearts.

d. The joker can be used to represent zero because it is neither black nor red.

e. Overall, a deck of cards has a net value of zero.

Activity 6.2

a.
i. Throw away the K♦ and the 3♦ to make the hand more positive, and throw away the 4♠ and the Q♠ to make the hand more negative.

ii. Throw away the 3♦ and the A♥ to make the hand more positive, and throw away the J♣ and the J♠ to make the hand more negative.

iii. Throw away the A♥ and the 7♥ to make the hand more positive, and throw away the A♣ and the A♠ to make the hand more negative.

iv. Throw away the 3♦ and the 2♦ to make the hand more positive, and throw away the Q♣ and the J♠ to make the hand more negative.

Activity 6.3

a.
i. Play $(+)$ the video of the person walking forward $(+)$; it looks as though the person is walking forward $(+)$.

ii. Play $(+)$ the video of the person walking backward $(-)$; it looks as though the person is walking backward $(-)$.

iii. Rewind $(-)$ the video of the person walking forward $(+)$, it looks as though the person is walking backward $(-)$.

iv. Rewind $(-)$ the video of the person walking backward $(-)$; it looks as though the person is walking forward $(+)$.

b. Methods may vary, but one way is to adapt the number line model.

i. 8.　　**ii.** (-8).　　**iii.** (-8).　　**iv.** 8.

Activity 6.4

a.
i. $1 \cdot (-13) \cdot (-3) \cdot 4 \cdot 12 = 1{,}872$.

ii. $3 \cdot 10 \cdot 11 \cdot 11 \cdot (-1) = (-3{,}630)$.

iii. $1 \cdot (-1) \cdot 1 \cdot (-1) \cdot (-7) = (-7)$.

iv. $(-3) \cdot (-4) \cdot (-2) \cdot 11 \cdot (-2) = 528$.

b. Answers will vary, but possible hands are:

i. 2♦, 2♥, 2♣, A♦, A♠.

ii. 3♥, 3♦, 3♣, 3♠, A♦.

iii. 2♦, 4♦, 9♣, 10♣, 10♠.

iv. 2♦, 2♥, 4♠, 10♣, 10♠.

© Image Source/Corbis

Reviewing the Big Ideas Chapters 4–6

◆ Understanding Whole-Number Operations and Their Properties

1. Use the set and number line models to represent and compute $4 + 7$.
2. Using the missing-addend approach to find the missing number in each equation.
   - **a.** $7 - 4 = ?$
   - **b.** $13 - ? = 4$.
3. Write a subtraction story problem that can be solved using the comparison approach. Use the set model to represent and solve your problem.
4. Use repeated addition to compute $6 \cdot 2$.
5. Rewrite each expression in exponential form.
   - **a.** $4 \cdot 4 \cdot 4 \cdot 4 \cdot 4 \cdot 4$
   - **b.** $3 \cdot (x + 1) \cdot 3 \cdot (x + 1) \cdot 3 \cdot (x + 1)$
6. Simplify each expression. Leave your answer in exponential form.
   - **a.** $3^3 \cdot 3^4$
   - **b.** $11 \cdot 6^0$
   - **c.** $(2^3)^7$
   - **d.** $5^3 \cdot x^3$
7. Use repeated subtraction to interpret and compute $28 \div 7$.
8. Use the missing-factor approach to rewrite each problem in terms of multiplication and then solve.
   - **a.** $12 \div 3 = ?$
   - **b.** $63 \div 3 = ?$
9. Find the quotient and remainder for each division problem.
   - **a.** $47 \div 6$
   - **b.** $75 \div 9$
10. Use the order of operations to compute $(16 - 6)^2 \div 5 \cdot 8 + 12$.
11. Determine whether each set is closed under the operation.
    - **a.** $A = \{2, 3, 4\}$ under addition
    - **b.** $F = \{5, 10, 15, 20, \ldots\}$ under multiplication
12. State the property of addition or multiplication demonstrated in each statement.
    - **a.** $3 \cdot (4 \cdot 7) = (3 \cdot 4) \cdot 7$.
    - **b.** $5 + 6 = 11 \in W$.
    - **c.** $4 + 8 = 8 + 4$.
    - **d.** $8(3 + 7) = 8 \cdot 3 + 8 \cdot 7$.
    - **e.** $8 \cdot 0 = 0$.
    - **f.** $4 + 0 = 4$.

◆ Whole-Number Operations on Large Numbers

13. Use expanded notation to compute each problem.
    - **a.** $418 + 268$
    - **b.** $345 - 225$
    - **c.** $61 \cdot 34$
14. Use the lattice method to compute each problem.
    - **a.** $133 + 45$
    - **b.** $59 \cdot 27$
15. Use the standard algorithm to compute each problem.
    - **a.** $1{,}568 + 698$
    - **b.** $309 - 234$
    - **c.** $177 \cdot 39$
    - **d.** $5{,}750 \div 46$
16. Use the standard algorithm to compute each problem.
    - **a.** $3441_{\text{five}} + 222_{\text{five}}$
    - **b.** $5781_{\text{nine}} - 2774_{\text{nine}}$
    - **c.** $33_{\text{five}} \cdot 21_{\text{five}}$
    - **d.** $13_{\text{eight}} \overline{)446_{\text{eight}}}$
17. Use mental computation to compute each problem. Explain the method you used.
    - **a.** $67 + 40$
    - **b.** $321 - 29$
    - **c.** $41 \cdot 35$
    - **d.** $3{,}542 \div 7$
18. Estimate each problem. Describe the strategy you used.
    - **a.** $61 + 58 + 62 + 62$
    - **b.** $14{,}003 - 5{,}993$
    - **c.** $29{,}891 \cdot 198$
    - **d.** $9{,}009 \div 87$

◆ Divisibility, Primes, and Composites

19. Determine whether each statement is true or false. Justify your answer.
    a. $16|56$.
    b. If $7|21$ and $7|14$, then $7|(21 - 14)$.
    c. $12 \nmid 78$.
    d. If $7 \nmid (x + y)$, then $7 \nmid x$ and $7 \nmid y$.

20. Determine whether 2, 3, 4, 5, 6, 8, 9, 10, and 11 divide each number.
    a. 3,166
    b. 50,500

21. Classify each number as prime or composite.
    a. 237
    b. 319
    c. 519

22. Use a factor tree to find the prime factorization of 660. Use the prime factorization to determine how many factors it has.

◆ Greatest Common Factors and Least Common Multiples

23. Use the set-intersection method to find each number.
    a. GCF(56, 72)
    b. LCM(18, 24).

24. Use the prime-factorization method to find each number.
    a. GCF(495, 231)
    b. LCM(495, 231)

25. Use the Euclidean algorithm to find the GCF(1050, 4158).

26. Use a calculator to compute each number.

    a. GCF(9000, 1680)
    b. LCM(2456, 3788)

◆ Clock and Modular Arithmetic

27. Determine whether each statement is true or false.
    a. $17 \equiv 2 \bmod 5$.
    b. $45 \equiv 13 \bmod 9$.
    c. $49 \not\equiv 7 \bmod 6$.

28. Find and list the members in each equivalence class for congruence mod 3.

29. When possible, compute each operation on a 12-hr clock.
    a. $11 \oplus 8$
    b. $3 \ominus 7$
    c. $5 \otimes 7$
    d. $8 \oslash 5$

30. When possible, compute each operation on an 8-hr clock.
    a. $6 \oplus 5$
    b. $2 \ominus 7$
    c. $3 \otimes 5$
    d. $4 \oslash 6$

◆ Integers and Integer Operations

31. Compute.
    a. $|-27|$
    b. $|181|$
    c. $|-4| - |-8|$
    d. $-|16 + (-7)|$

32. Use the chip model to represent and solve each problem.
    a. $(-7) + 5$
    b. $3 - (-4)$
    c. $(-3) \cdot (-2)$

33. Use the number line model to represent and solve each problem.
    a. $(-2) + (-4)$
    b. $(-5) - 2$
    c. $4 \cdot (-2)$

34. Use the missing-factor approach to find each quotient when possible.
    a. $49 \div (-7)$
    b. $(-18) \div (-7)$
    c. $(-84) \div 12$

35. Compute.
    a. $654 + (-198)$
    b. $(-894) - (-525)$
    c. $16 \cdot (-57)$
    d. $(-8) \cdot 10 \cdot (-5) \cdot (-9)$
    e. $(-663) \div 17$
    f. $(-3)^4$

Fractions and the Rational Numbers

Opening Problem

Amy has worked for $15\frac{2}{3}$ hours on a project for one of her classes. She is about $\frac{3}{5}$ of the way done. About how much longer should it take her?

7.1	Fractions and the Set of Rational Numbers
7.2	Adding and Subtracting Rational Numbers
7.3	Multiplying and Dividing Rational Numbers

In the last chapter, we extended the set of whole numbers to the set of integers. We now extend the integers to the set of rational numbers and continue to work on the objectives in the NCTM Number and Operations Standard.

■ NCTM Number and Operations Standard

Instructional programs from prekindergarten through grade 12 should enable all students to:

- Understand numbers, ways of representing numbers, relationships among numbers, and number systems.
- Understand the meanings of operations and how they relate to one another.
- Compute fluently and make reasonable estimates.

Source: NCTM STANDARDS Copyright © 2011 by NATIONAL COUNCIL OF TEACHERS OF MATHEMATICS. Reproduced with permission of NATIONAL COUNCIL OF TEACHERS OF MATHEMATICS.

Before children enter school, they often develop an intuitive understanding of common fractions, like $\frac{1}{2}$ or $\frac{3}{4}$, through verbal interactions with their parents and peers. Once in school, children build on this informal knowledge by using fractions to represent part-to-whole relationships. Children use this interpretation from kindergarten through second grade to form a solid foundation on which other fractional ideas can be based. Without this foundation, fractions can become a serious obstacle in a child's mathematical development. From the third grade on, children learn to model, order, and compute with fractions, learning that fractions not only represent part-to-whole relationships, but also measures, ratios, and even quotients. Specific expectations associated with fractions and the related grade levels are given in Table 7.1.

Table 7.1 Classroom learning objectives	K	1	2	3	4	5	6	7	8
Partition shapes into parts with equal areas. Express the area of each part as fractions of the whole.	X	X	X						
Understand a fraction 1/b as the quantity formed by 1 part when a whole is partitioned into b equal parts and a fraction a/b as the quantity formed by a parts of size 1/b.				X					
Understand and represent fractions as numbers on the number line using the interval from 0 to 1 as the whole.				X					
Understand two fractions as equivalent if they are the same size and generate simple equivalent fractions.				X					
Express whole numbers as fractions, and recognize fractions that are equivalent to whole numbers.				X					
Compare two fractions with the same numerator or the same denominator.				X					
Compare two fractions with different numerators and different denominators.					X				
Understand the addition and subtraction of fractions as joining and separating parts referring to the same whole.					X				
Add and subtract fractions and mixed numbers with like and unlike denominators.					X	X			
Understand a rational number as a point on the number line.							X		
Write, interpret, and explain statements of order for rational numbers.							X		
Apply and extend previous understandings of multiplication and division to multiply and divide rational numbers.						X	X	X	
Interpret a fraction as division of the numerator by the denominator (a/b = a ÷ b), provided that the divisor is not zero.							X		X
Solve word and real-world problems using operations on rational numbers.						X	X	X	X

Source: Adapted from the *Common Core State Standards for Mathematics* (*Common Core State Standards Initiative*, 2010)

Many of the ideas in Table 7.1 are difficult for children to learn because they are not always consistent with what children know about the whole numbers. For instance, many children have trouble ordering fractions because they mistakenly think that larger whole-number values must lead to larger fractional values. A student may incorrectly claim that $\frac{1}{5}$ is larger than $\frac{1}{3}$ because 5 is bigger than 3. Children also struggle with the fact that different fractions can have the same value. Although the concept is useful, students can find it difficult because they must learn to change fractions from one form to another and decide which fraction best serves their needs. Children can also have trouble computing with fractions. For instance, when children add fractions, they naturally want to add the numerators and the denominators. Although this seems intuitive, it turns out to be incorrect. Likewise, when children multiply fractions, their experience with the whole numbers tells them that the product should be greater than or equal to the factors. However, this is not always the case with the fractions.

As elementary teachers, our job is to help students understand the concepts and procedures associated with the fractions. So we must have a deep understanding of fractional concepts, procedures, and representations. We must also understand how the ideas associated with the fractions are not only interconnected among themselves, but are also related to other ideas we have learned. For this reason, Chapter 7:

- Uses the rational numbers to discuss the different interpretations, representations, and properties of fractions.
- Demonstrates how to find equivalent fractions and order fractions.
- Demonstrates how to compute with fractions and mixed numbers.

SECTION 7.1   Fractions and the Set of Rational Numbers

In Chapter 6, we addressed certain limitations of the whole numbers by extending them to the integers. We can now compute any integer difference and describe situations that involve a direction and a distance relative to zero. Although the integers give us more mathematical power, they still have limitations. For instance, many quotients, like $12 \div 5$ and $(-7) \div 3$, are not defined for the integers. Likewise, we cannot use the integers to describe situations involving a part of a whole: for example, "I ate half of a pizza," or "I'll meet you in a quarter of an hour." To address these concerns, we need to extend the integers to the fractions and rational numbers.

The Rational Numbers

Mathematically, the fractions are a broad set containing any number that can be written in the form $\frac{a}{b}$, where $b \neq 0$. Here, a and b can be any kind of number, including whole numbers, integers, decimals, radicals, and even other fractions. Because the set is so diverse and includes numbers we have not yet addressed, we limit our discussion to the common fractions and rational numbers.

Definition of the Rational Numbers

The set of **rational numbers** is
$$\mathbb{Q} = \left\{ \frac{a}{b} \,\middle|\, a \text{ and } b \text{ are integers and } b \neq 0 \right\}$$

The rational numbers are defined in set-builder notation because, unlike the whole numbers and integers, they cannot be described with a simple listing. The next example examines exactly which numbers are in the set of rational numbers.

| **Example 7.1** | **Using the Definition of Rational Numbers** |

Communication

Use the definition of the rational numbers to determine which of the following are rational.

$$\frac{1}{2} \quad \frac{7}{3} \quad \frac{2}{-3} \quad \frac{-9}{2} \quad \frac{0}{1} \quad \frac{2}{0} \quad 3 \quad -5$$

Solution A number is rational if it can be written in a fractional form where the top number is an integer and the bottom number is a nonzero integer. Using this description, $\frac{1}{2}, \frac{7}{3}, \frac{2}{-3}, \frac{-9}{2}$, and $\frac{0}{1}$ are rational. Because we can rewrite 3 and -5 as $\frac{3}{1}$ and $\frac{-5}{1}$, they too are rational. $\frac{2}{0}$ is the only expression that is not rational because the bottom number is 0.

Example 7.1 shows that the rational numbers can be both positive and negative. It also shows us that every integer is a rational number because any integer a can be rewritten as $\frac{a}{1}$. Consequently, every natural number and every whole number are also rational numbers, giving us the following subset relationships (Figure 7.1).

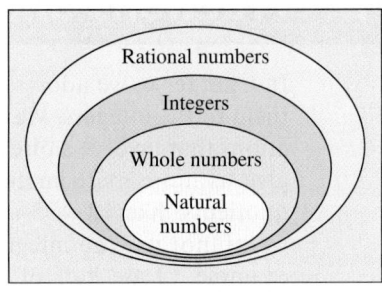

$$\mathbb{N} \subseteq W \subseteq \mathbb{Z} \subseteq \mathbb{Q}$$

Figure 7.1 Subset relationships among different number sets

Fractions and rational numbers enable us not only to describe whole number quantities, but also to describe a part of a whole that has been divided into a certain number of pieces. Even the name reflects this use; the word "fraction" comes from the Latin word *frangere*, which means to break into pieces. The description of a part-to-whole relationship requires two numbers that serve different roles. Understanding these roles is crucial not only to making sense of the fractions, but also to understanding how we manipulate and compute with them.

The bottom number of a fraction is called the **denominator,** a word that comes from the Latin *denomino*, which means to name. It represents the number of equivalent pieces the whole has been divided into or the number of equivalent pieces needed to make the whole. Informally, the denominator describes or names what the pieces look like: thirds of the whole, fourths of the whole, and so on. The top number of the fraction is called the **numerator,** a word that comes from the Latin *numero*, which means to number or count. It gives the number of pieces that are being used or counted. Consequently, for any fraction $\dfrac{a}{b}$,

$$\frac{a}{b} = \frac{\text{numerator}}{\text{demoninator}} = \frac{\text{numberer}}{\text{namer}} = \frac{\text{number of pieces counted}}{\text{number of equivalent pieces to make whole}}.$$

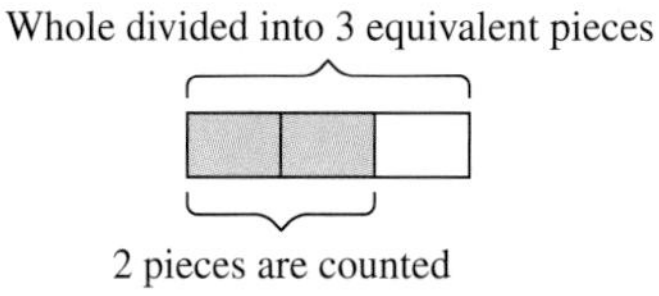

Whole divided into 3 equivalent pieces

2 pieces are counted

Figure 7.2 $\dfrac{2}{3}$ **of a whole**

For example, $\dfrac{2}{3}$ represents a part of a whole that has been divided into three equal pieces of which two are counted (Figure 7.2).

To make a part-to-whole interpretation of a fraction, the whole must be divided into *equivalent* parts. If not, then the value of the fraction is not accurately represented. For instance, Figure 7.3 shows two ways to divide a whole into four parts and count three of them. However, only the first represents the true value of $\dfrac{3}{4}$.

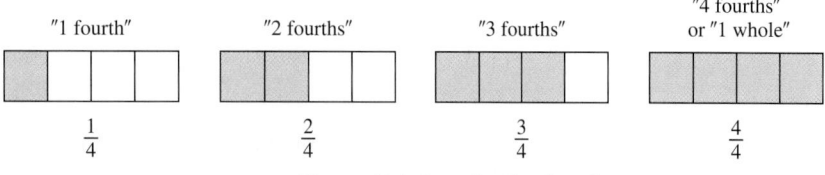

$\dfrac{3}{4}$ of the whole   not $\dfrac{3}{4}$ of the whole

Figure 7.3 Equivalent versus not equivalent pieces

In the classroom, counting by fractions can help children understand how the numerator counts the pieces and how the denominator describes what they look like. As they count, they can see that the number of pieces changes but that their size remains the same. It can also help them recognize how many equivalent pieces are needed to make the whole. Figure 7.4 shows how to count to one by fourths.

"1 fourth"   "2 fourths"   "3 fourths"   "4 fourths" or "1 whole"

$\dfrac{1}{4}$ $\dfrac{2}{4}$ $\dfrac{3}{4}$ $\dfrac{4}{4}$

Figure 7.4 Counting by fourths

Representing Rational Numbers

The part-to-whole interpretation is an important way to think about fractions because we can use it to illustrate and justify a number of ideas. However, at some point, we must begin to think of fractions as numerical values defined by proportional amounts that are relative to the whole. One way to develop a sense of these values is to represent fractions in different ways, that is, by using the area, the length, and the set models.

Area Model

Explorations
Manual
7.1

In the **area** or **regions model,** the whole is a shape with an area that can be divided into equivalent subregions, enough of which are shaded to show the fraction. Figure 7.5 shows how to represent $\dfrac{1}{4}$ using different variations of the area model. In each case, the numerical value of the fraction is shown by the size of the shaded region relative to the whole.

Folded paper   Geoboard/grid paper   Fraction disk   Pattern blocks

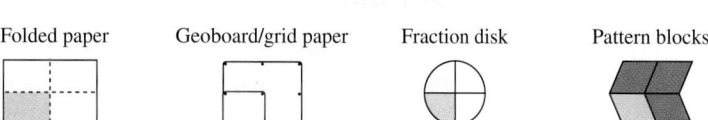

Figure 7.5 Different variations of the area model

Length Model

Explorations
Manual
7.2

In the **length,** or **measurement model,** the whole is a length that can be divided into equivalent sublengths. For instance, when using fraction bars, sublengths are joined to make the correct length for the fraction. With fraction strips, sublengths are shaded to make the correct length. Figure 7.6 shows how to represent $\frac{2}{3}$ and $\frac{5}{6}$ using these versions of the length model.

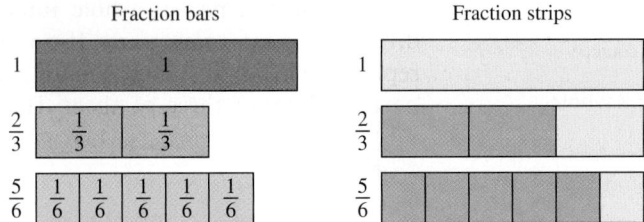

Figure 7.6 Different variations of the length model

We can also use a number line, but to do so, we must distinguish the unit from the whole. The **unit** on a number line is always the distance from 0 to 1. The whole, however, can be any length. In most cases, the unit is the whole because this allows us to plot the positions of fractions relative to the integers. For instance, if the unit is the whole, then $\frac{2}{3}$ has a position that is two-thirds of a unit to the right of 0 and $-\frac{1}{4}$ has a position that is one-fourth of a unit to the left of 0 (Figure 7.7).

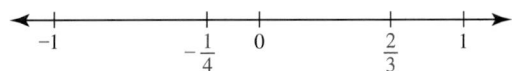

Figure 7.7 Fractions on a number line

The next example illustrates what happens if the whole is some length other than the unit.

Example 7.2   **A Whole That is Not the Unit**

Representation

If the length in blue indicates the whole, find the value of the position that represents $\frac{1}{4}$ of the whole relative to the unit.

Solution We take the distance from 0 to 2, and divide it into 4 equal lengths. We then count over 1 sublength from 0 to find $\frac{1}{4}$ of the whole, which places us at $\frac{1}{2}$.

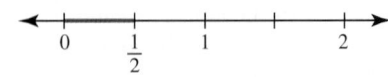

Set Model

Explorations
Manual
7.3

The **set model** differs from the other two in that the whole is a collection of objects rather than a single region or length. In this model, fractions represent not just part of a whole, but also part of a group. More specifically, the denominator represents the number of elements in the set, and the numerator represents the number of elements in a subset.

Example 7.3   **Fractions of a Set**

Find the fraction that represents each collection of dots.

a. The dots in the triangle as a part of all the dots.

b. The dots in both the circle and the rectangle as a part of all the dots.

c. The dots in both the rectangle and triangle as a part of the dots in the rectangle.

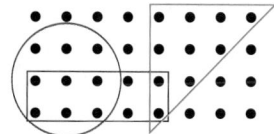

Solution

a. Because 10 dots are in the triangle and there are 32 dots overall, the fraction is $\frac{10}{32}$.

b. Six of the 32 dots are in the circle and the rectangle, so the fraction is $\frac{6}{32}$.

c. Of the 10 dots in the rectangle, 2 are in the triangle, so the fraction is $\frac{2}{10}$.

Improper Fractions and Mixed Numbers

To this point, the fractions that we have considered have had numerators that are smaller than the denominators. However, the definition makes no such restrictions, so we can make a distinction between proper and improper fractions. In a **proper** fraction, the value of the numerator is smaller than the value of the denominator. In an **improper** fraction, the value of the numerator is greater than or equal to the value of the denominator.

Definition of Proper and Improper Fractions

If $\frac{a}{b}$ is any fraction, then

1. $\frac{a}{b}$ is a **proper fraction** if $0 \le |a| < |b|$.

2. $\frac{a}{b}$ is an **improper fraction** if $|a| \ge |b| > 0$.

Examples of proper fractions are $\frac{1}{2}$, $-\frac{2}{3}$, and $\frac{11}{12}$, and examples of improper fractions are $\frac{5}{2}$, $-\frac{8}{3}$, and $\frac{4}{4}$. Notice that all integers except 0 are improper fractions.

Like proper fractions, improper fractions have part-to-whole interpretations. For instance, the part-to-whole interpretation of $\frac{11}{4}$ is to count 11 pieces that look like fourths of the whole. However, because it takes only 4 fourths to make a whole, $\frac{11}{4}$ must have a value greater than 1. This has two implications. First, it implies that all improper fractions have values that are greater than or equal to 1 or less than or equal to -1. Likewise, all proper fractions have values strictly between -1 and 1 (Figure 7.8).

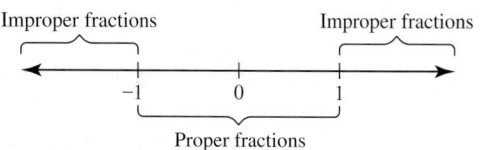

Figure 7.8 Proper and improper fractions on a number line

Second, improper fractions can be rewritten to show the number of whole objects contained in the fraction. For instance, $\dfrac{11}{4}$ contains enough fourths to make two whole objects, with three fourths remaining, or $\dfrac{11}{4} = 2 + \dfrac{3}{4} = 2\dfrac{3}{4}$ (Figure 7.9). We call any number with an integer and fractional part a **mixed number.**

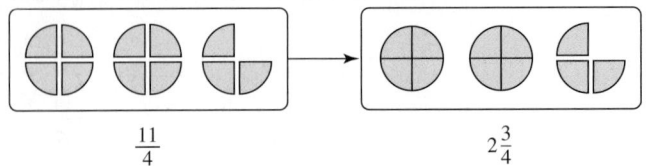

$\dfrac{11}{4}$ $2\dfrac{3}{4}$

Figure 7.9 Writing $\dfrac{11}{4}$ as a mixed number.

Mixed numbers clearly identify how many wholes are included in the value of the number. For this reason, they can be more useful than improper fractions in a number of situations. For instance, mixed numbers tend to be easier to locate on a number line. Figure 7.10 shows the positions of several mixed numbers. The figure also illustrates that $-1\dfrac{2}{3}$ is to the left of -1 and $-2\dfrac{4}{5}$ is to the left of -2. This implies that the negative sign on any mixed number applies to both the integer and fractional parts. In other words, we interpret $-1\dfrac{2}{3}$ as $-\left(1\dfrac{2}{3}\right)$, or $-1 - \dfrac{2}{3}$, but not $-1 + \dfrac{2}{3}$.

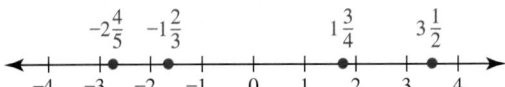

Figure 7.10 Mixed numbers on a number line

Because of the connection between improper fractions and mixed numbers, making conversions between them can be useful.

Example 7.4   Conversions Between Fractions and Mixed Numbers

Make each conversion

a. $\dfrac{7}{3}$ to a mixed number   **b.** $-8\dfrac{3}{7}$ to an improper fraction

Solution

a. If we divide the numerator by the denominator, the quotient indicates the number of whole objects, and the remainder indicates how many pieces remain. Because $7 \div 3 = 2$ remainder 1, then $\dfrac{7}{3} = 2\dfrac{1}{3}$.

b. We reverse the process by multiplying the absolute value of the integer part by the denominator of the fraction and then adding the numerator. The denominator stays the same. If the number is negative, ignore the sign until the end. As a result, $-8\dfrac{3}{7} = -\dfrac{59}{7}$ because $8 \cdot 7 + 3 = 59$.

Many fraction calculators can make conversions between improper fractions and mixed numbers by means of the $\boxed{\text{Unit}}$, $\boxed{/}$, and $\boxed{\text{A}^{b}/_{c} \blacktriangleleft\!\blacktriangleright \text{d}/_{e}}$ keys. For instance, to convert $\dfrac{19}{4}$ to a mixed number, use 19 [/] 4 [Aᵇ/c ◄► d/e] [ENTER] to obtain $4\dfrac{3}{4}$. To convert $7\dfrac{5}{6}$ to an improper fraction, use 7 [Unit] 5 [/] 6 [Aᵇ/c ◄► d/e] [ENTER] to obtain $\dfrac{47}{6}$.

Check Your Understanding 7.1-A

1. Give a verbal interpretation of the part-to-whole meaning of each fraction.

   a. $\dfrac{4}{5}$ b. $\dfrac{6}{9}$ c. $\dfrac{9}{8}$

2. Give the fraction represented by each shaded region.

   a. 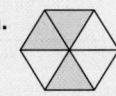   b.    c.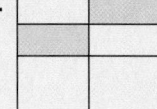

3. Make each conversion.

   a. $\dfrac{14}{3}$ to a mixed number   b. $-\dfrac{34}{8}$ to a mixed number

   c. $3\dfrac{2}{9}$ to an improper fraction   d. $-2\dfrac{1}{9}$ to an improper fraction

Talk About It How do we know that $-3\dfrac{1}{2}$ is in the set of rational numbers? What would be its part-to-whole interpretation? How else might we interpret this number?

Representation

Activity 7.1

Represent the numbers $\dfrac{2}{3}, \dfrac{3}{4}, \dfrac{7}{4}, 1\dfrac{1}{2},$ and $2\dfrac{1}{3}$ using:

a. Fraction disks.   b. A rectangular region.
c. Fraction bars.   d. A number line.

Equivalent Fractions

Another important difference between the integers and the rational numbers is the idea of equivalent fractions. With the integers, one numerical value is associated with one numeral. For instance, 2 is associated with a value of two, 3 with a value of three, and so on. However, this is not the case with the rational numbers. Comparing the values of several rational numbers shows that many of them represent the same part of the whole. For instance, Figure 7.11 shows how $\dfrac{1}{2}, \dfrac{2}{4}, \dfrac{4}{8},$ and $\dfrac{8}{16}$ all cover the same area of a rectangle. This implies that these fractions take on the same numerical value, despite their different part-to-whole interpretations. For this reason, we call them **equivalent fractions.**

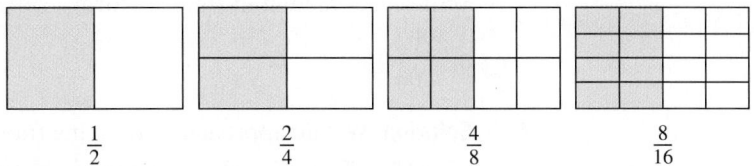

$\dfrac{1}{2}$ $\dfrac{2}{4}$ $\dfrac{4}{8}$ $\dfrac{8}{16}$

Figure 7.11 Equivalent fractions and the area model

We can get from one of these equivalent fractions to another by multiplying the numerator and denominator of one fraction by the same number. For instance, if we multiply the numerator and denominator of $\frac{1}{2}$ by 2, we get $\frac{1}{2} = \frac{1 \cdot 2}{2 \cdot 2} = \frac{2}{4}$. Similarly, if we multiply the numerator and denominator of $\frac{1}{2}$ by 8, we get $\frac{1}{2} = \frac{1 \cdot 8}{2 \cdot 8} = \frac{8}{16}$. We call this property the Fundamental Law of Fractions.

Theorem 7.1   The Fundamental Law of Fractions

If $\frac{a}{b}$ is a fraction and n is any nonzero integer, then $\frac{a}{b} = \frac{a \cdot n}{b \cdot n}$.

Although the proof of the Fundamental Law of Fractions is straightforward, it relies on the multiplicative identity property. For this reason, the proof is given in Section 7.3 after we cover this property. For now, however, we use the Fundamental Law of Fractions to develop several key ideas related to the rational numbers.

Note that n can be any one of an infinite number of nonzero integers, indicating that $\frac{a}{b}$ is equivalent to an infinite number of fractions. For instance, $\frac{1}{2}$ is equivalent to every fraction in the set $\left\{ \frac{2}{4}, \frac{3}{6}, \frac{4}{8}, \frac{5}{10} \cdots \right\}$, $-\frac{2}{3}$ is equivalent to every fraction in the set $\left\{ -\frac{2}{3}, -\frac{4}{6}, -\frac{6}{9}, \cdots \right\}$, and 1 is equivalent to every fraction in the set $\left\{ \frac{2}{2}, \frac{3}{3}, \frac{4}{4}, \frac{5}{5}, \cdots \right\}$.

The Fundamental Law of Fractions allows us to exchange a fraction for any one of its equivalent fractions. So by replacing $\frac{a}{b}$ with $\frac{an}{bn}$, we can change the part-to-whole interpretation of the fraction without changing its numerical value. In the reverse exchange, replacing $\frac{an}{bn}$ with $\frac{a}{b}$, we can simplify the fraction.

Definition of Simplest Form

A rational number $\frac{a}{b}$ is in **simplest form** if GCF(a, b) = 1.

The definition states that a fraction is in simplest form if the numerator and denominator are relatively prime; that is, they have no common factor other than 1. In general, simplifying fractions is useful because the numbers become smaller and easier to use.

Example 7.5   Simplifying Fractions

Write each fraction in simplest form.

a. $\dfrac{200}{400}$ b. $\dfrac{56}{98}$

Solution We can approach simplifying fractions in two ways. If the numbers are small or have factors that are easily found, we can use the greatest common factor. If not, we can use divisibility tests to remove common factors one at a time.

a. Because the GCF(200, 400) = 200, then $\dfrac{200}{400} = \dfrac{1 \cdot 200}{2 \cdot 200} = \dfrac{1}{2}$.

b. Using the divisibility test for 2, we know that $\dfrac{56}{98} = \dfrac{28 \cdot 2}{49 \cdot 2} = \dfrac{28}{49}$. Because 28 and 49 are both divisible by 7, then $\dfrac{28}{49} = \dfrac{4 \cdot 7}{7 \cdot 7} = \dfrac{4}{7}$.

Fractions can also be simplified by means of a fraction calculator, using the ▶Simp key. For example, to simplify $\dfrac{12}{18}$, we press 12 **/** 18 ▶Simp ENTER. The calculator may return $\dfrac{6}{9}$ but indicate that the fraction is not simplified. If we press ▶Simp ENTER again, we get the simplest form of $\dfrac{2}{3}$.

Ordering the Rational Numbers

Explorations
Manual
7.4

Whenever we order two numbers, our goal is to determine whether the numbers have the same value or one is larger than the other. With the integers, ordering was a matter of comparing the values and taking the signs into account. However, ordering the rational numbers is more complicated because the numerator and the denominator have opposing effects on the value of the fraction.

To understand why, consider two fractions that have the same denominator, or a **common denominator.** In this case, each fraction is described in terms of the same equivalent parts of the whole, so all we have to do is compare the numerators. If the numerators are equal, the fractions are equal. If one numerator is larger, then it represents the larger fraction because it counts more pieces. For instance, Figure 7.12 shows that $\dfrac{2}{4} < \dfrac{3}{4}$ because $\dfrac{2}{4}$ counts only two pieces that look like fourths, whereas $\dfrac{3}{4}$ counts three. So, as the value of the numerator goes up, so does the value of the fraction. These ideas are generalized to all fractions in the next theorem.

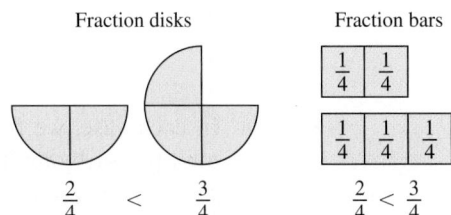

Figure 7.12 **Comparing fractions with a common denominator**

Theorem 7.2   Ordering Fractions with Common Denominators

If a, b, and c are integers with $b > 0$, then

1. $\dfrac{a}{b} = \dfrac{c}{b}$ if and only if $a = c$, and   **2.** $\dfrac{a}{b} < \dfrac{c}{b}$ if and only if $a < c$.

To understand how the denominator affects the value of a fraction, we can look at the relative size of the **unit fractions,** which are fractions with 1 as the numerator. As the models in Figure 7.13 show, the values of the fractions get smaller as the denominator gets larger. This is due to the fact that the whole is not changing in size, so as we divide it into more equivalent pieces, those pieces must get smaller.

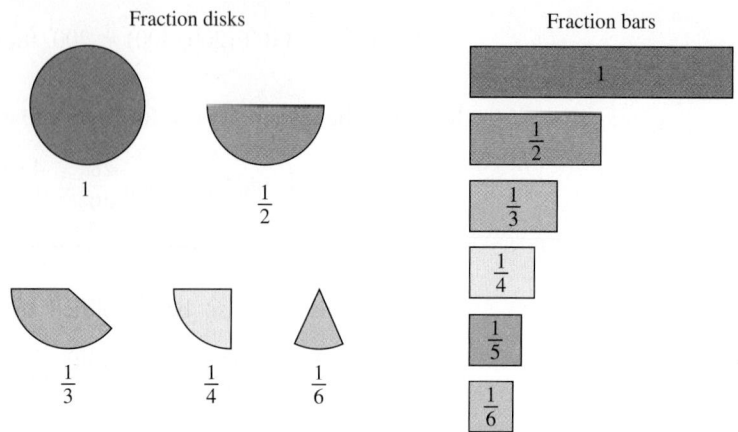

Fraction disks   Fraction bars

Figure 7.13 Comparing unit fractions

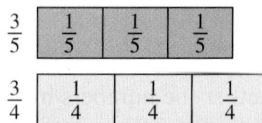

Figure 7.14 Comparing fractions with different denominators

So if two fractions have the same numerators but different denominators, then the fraction with the smaller denominator must be larger because the pieces are larger. For instance, $\frac{3}{4}$ and $\frac{3}{5}$ count the same number of pieces, but $\frac{3}{4} > \frac{3}{5}$ because fourths are larger than fifths (Figure 7.14).

So how do we compare fractions, like $\frac{3}{4}$ and $\frac{4}{5}$, whose numerators and denominators are both different? The numerators seem to indicate that $\frac{4}{5}$ is larger because it counts more pieces. However, the denominators seem to indicate that $\frac{3}{4}$ is larger because it uses bigger pieces. We can solve our problem by using the Fundamental Law of Fractions to find a common denominator between the two fractions and then compare the numerators.

Example 7.6   Comparing Fractions

Compare each pair of fractions by finding a common denominator.

a. $\frac{2}{3}$ and $\frac{3}{4}$

b. $-\frac{7}{12}$ and $-\frac{8}{15}$

Solution In each case, we find a common denominator, either by finding the least common multiple, or by multiplying the two denominators.

a. Because 3 and 4 are relatively prime, the lowest common denominator is the LCM(3, 4) = 3 · 4 = 12. By the Fundamental Law of Fractions:

$$\frac{2}{3} = \frac{2 \cdot 4}{3 \cdot 4} = \frac{8}{12} \quad \text{and} \quad \frac{3}{4} = \frac{3 \cdot 3}{4 \cdot 3} = \frac{9}{12}$$

Because $8 < 9$, then $\frac{8}{12} < \frac{9}{12}$, so $\frac{2}{3} < \frac{3}{4}$.

b. We can use the same approach for negative fractions. The LCM(12, 15) = 60, so we choose 60 as the common denominator, and write:

$$-\frac{7}{12} = \frac{(-7) \cdot 5}{12 \cdot 5} = -\frac{35}{60} \quad \text{and} \quad -\frac{8}{15} = \frac{-8 \cdot 4}{15 \cdot 4} = -\frac{32}{60}$$

Because $-35 < -32$, then $-\frac{7}{12} < -\frac{8}{15}$.

Common denominators can be used to compare any two fractions, regardless of whether they are positive, negative, proper, or improper. The process can be generalized in the following way. If $\frac{a}{b}$ and $\frac{c}{d}$ are any two fractions, then they have a common denominator of bd; so $\frac{a}{b} = \frac{a \cdot d}{b \cdot d}$ and $\frac{c}{d} = \frac{b \cdot c}{b \cdot d}$. When we compare the numerators, if $ad = bc$, then $\frac{a}{b} = \frac{c}{d}$, and if $ad < bc$, then $\frac{a}{b} < \frac{c}{d}$.

Theorem 7.3   Ordering Rational Numbers

If $\frac{a}{b}$ and $\frac{c}{d}$ are two rational numbers where b > 0 and d > 0, then

1. $\frac{a}{b} = \frac{c}{d}$, if and only if $ad = bc$, and

2. $\frac{a}{b} < \frac{c}{d}$, if and only if $ad < bc$.

In essence, Theorem 7.3 indicates that two fractions can be compared by looking at their **cross product**, which is formed by multiplying each numerator by the other fraction's denominator.

$$\frac{a}{b} \diagup\!\!\!\!\!\diagdown \frac{c}{d}$$

The method works with every type of fraction. However, any negatives must be placed in the numerator so that they do not negate each other in the cross product.

Example 7.7   Comparing Fractions

Place the proper symbol (=, <, >) between each pair of fractions.

a. $\frac{17}{18} \quad \frac{12}{13}$ **b.** $\frac{-8}{15} \quad \frac{-39}{80}$

Solution In each problem, we compute the cross product, and use it to determine the correct relation.

a. $17 \cdot 13 = 221$ and $18 \cdot 12 = 216$. Because $221 > 216$, then $\frac{17}{18} > \frac{12}{13}$.

b. $(-8) \cdot 80 = (-640)$ and $(-39) \cdot 15 = (-585)$. Because $(-640) < (-585)$, $\frac{-8}{15} < \frac{-39}{80}$.

Example 7.8   Placing Pets

Application

The Animal Control Office in a certain city finds that only 3 out of every 8 rescued animals is placed in a suitable home. Last year, Annie's Animal Shelter placed 124 out of 362 animals into suitable homes. How did Annie's Animal Shelter compare to the city's findings?

Solution We compare the fraction of animals placed by Annie's shelter, $\frac{124}{362}$, to the fraction of animals placed in the city, $\frac{3}{8}$. Because $124 \cdot 8 = 992$ and $3 \cdot 362 = 1{,}086$, then $\frac{124}{362} < \frac{3}{8}$. This implies Annie's shelter did not do as well as the city average.

As we develop an intuitive feel for the relative sizes of fractions, we can order them by comparing them to benchmark fractions, such as $0, \frac{1}{4}, \frac{1}{2}$, and 1.

Example 7.9 | **Ordering a Set of Fractions**

Reasoning

Order the following fractions from smallest to largest.

$$\frac{7}{8} \quad \frac{9}{16} \quad \frac{1}{4} \quad \frac{2}{3} \quad \frac{1}{2} \quad \frac{9}{8}$$

Solution One way to order the fractions is to rewrite all of them with a common denominator and then compare the numerators directly. However, because there are several different denominators, we choose to compare them by using benchmark fractions. Because $\frac{1}{4}$ is the only fraction less than $\frac{1}{2}$, we know the smallest two fractions. Because $\frac{8}{16} = \frac{1}{2}$, $\frac{9}{16}$ is only slightly larger than $\frac{1}{2}$, so it must come next. Of the remaining fractions, $\frac{9}{8}$ is the only fraction larger than 1, so it must be the largest. This leaves $\frac{7}{8}$ and $\frac{2}{3}$. By comparing them directly, we know that $\frac{7}{8}$ is larger than $\frac{2}{3}$. The fractions listed from smallest to largest are $\frac{1}{4}, \frac{1}{2}, \frac{9}{16}, \frac{2}{3}, \frac{7}{8}$, and $\frac{9}{8}$.

Denseness of the Rational Numbers

The Fundamental Law of Fractions has one other consequence. Given any two integers, another integer may or may not be between them. For instance, another integer is between 2 and 4 but not between 2 and 3. This is not the case with the rational numbers. For instance, consider the fractions $\frac{2}{5}$ and $\frac{3}{5}$. By the Fundamental Law of Fractions, they can be rewritten as $\frac{4}{10}$ and $\frac{6}{10}$. Between these two fractions is another one: $\frac{5}{10}$ or $\frac{1}{2}$. The same is true for $\frac{4}{10}$ and $\frac{5}{10}$, which can be rewritten as $\frac{8}{20}$ and $\frac{10}{20}$ and which have the fraction $\frac{9}{20}$ between them. The Fundamental Law of Fractions guarantees that we can continue this process indefinitely. In other words, we can always find another rational number between any two rational numbers, and the rational numbers are said to be **dense.**

Theorem 7.4 | **Denseness Property of the Rational Numbers**

Given any two rational numbers $\frac{a}{b}$ and $\frac{c}{d}$, there exists another rational number between them.

Example 7.10   Using the Denseness Property

Find five rational numbers between $\frac{1}{3}$ and $\frac{2}{3}$.

Solution In our previous discussion, we multiplied by 2, which seemed to widen the gap between the fractions. However, the Fundamental Law of Fractions allows the use of any nonzero integer. Choosing a larger number like 10 makes the gap between the fractions seem even wider. In this case, between $\frac{1}{3} = \frac{1 \cdot 10}{3 \cdot 10} = \frac{10}{30}$ and $\frac{2}{3} = \frac{2 \cdot 10}{3 \cdot 10} = \frac{20}{30}$ are the fractions $\frac{11}{30}, \frac{12}{30}, \frac{13}{30}, \frac{14}{30}$, and $\frac{15}{30}$.

Check Your Understanding 7.1-B

1. List the set of fractions that are equivalent to each number.

   a. $\frac{1}{3}$ **b.** -3 **c.** $-\frac{3}{4}$

2. Simplify.

   a. $\frac{6}{16}$ **b.** $\frac{28}{49}$ **c.** $-\frac{35}{50}$ **d.** $-\frac{88}{132}$

3. Compare each pair of fractions.

   a. $\frac{5}{6}$ and $\frac{25}{35}$ **b.** $-\frac{24}{20}$ and $-\frac{18}{15}$ **c.** $\frac{55}{45}$ and $\frac{110}{90}$

4. Find five fractions between $\frac{1}{4}$ and $\frac{1}{2}$.

Talk About It How is the Fundamental Law of Fractions used to compare two fractions?

Problem Solving

Activity 7.2

A typical set of fraction bars contains pieces that represent $1, \frac{1}{2}, \frac{1}{3}, \frac{1}{4}, \frac{1}{5}, \frac{1}{6}, \frac{1}{8}, \frac{1}{10}$, and $\frac{1}{12}$. Make a list of all the equivalent fractions that can be made using a standard set of fraction bars.

SECTION 7.1   Learning Assessment

■ Understanding the Concepts

1. What is the difference between the set of fractions and the set of rational numbers?

2. Why is every integer in the set of rational numbers?

3. What is the role of the numerator and the denominator when a fraction is used to describe a part-to-whole relationship?

4. Describe how to show the part-to-whole interpretation of a fraction in each model.

   a. Area   **b.** Length

   c. Number line   **d.** Set

5. How are improper fractions and mixed numbers similar? How are they different?

6. What does it mean for two fractions to be equivalent?

7. Explain how we use the Fundamental Law of Fractions to:

    a. Simplify a fraction.

    b. Get a common denominator.

8. a. How does the value of the numerator affect the value of a fraction?

    b. How does the value of the denominator affect the value of a fraction?

9. Why can we use a cross product to order two fractions?

10. What does it mean for the rational numbers to be dense?

■ Representing the Mathematics

11. Give a verbal interpretation of the part-to-whole meaning of each fraction.

    a. $\dfrac{1}{4}$ b. $\dfrac{3}{6}$ c. $\dfrac{5}{8}$ d. $\dfrac{9}{5}$

12. Write the fraction represented by the shaded regions in each diagram.

    a. 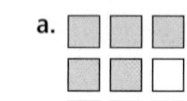   b.

    c.    d.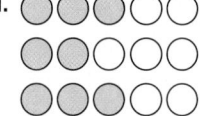

13. Shade the appropriate regions so that each diagram shows the indicated fraction.

    a. $\dfrac{5}{9}$; 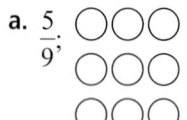   b. $\dfrac{7}{12}$;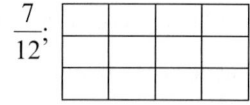

    c. $\dfrac{5}{8}$; 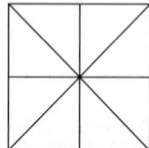   d. $\dfrac{6}{12}$;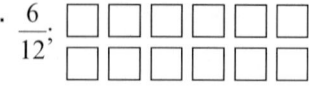

14. Give the fraction represented by the indicated location on each number line.

    a.

    b.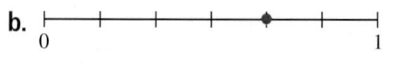

    c.

    d.

15. Plot each point on a number line.

    a. $\dfrac{2}{3}$ b. $\dfrac{3}{4}$ c. $\dfrac{3}{5}$ d. $\dfrac{5}{6}$

16. Plot each point on a number line.

    a. $\dfrac{11}{3}$ b. $-\dfrac{5}{2}$ c. $-1\dfrac{3}{4}$ d. $2\dfrac{2}{3}$

17. Does the shaded region shown below represent a value of $\dfrac{3}{4}$? Why or why not?

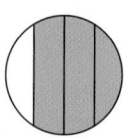

18. What property is illustrated in the following diagram?

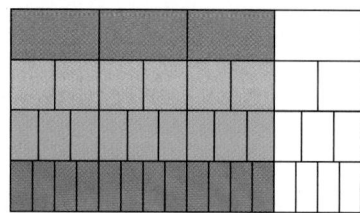

19. Do the following two shaded regions represent equivalent fractions? Why or why not?

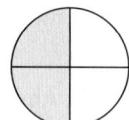 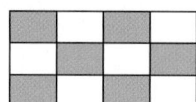

20. Which shaded region in the following diagrams represents the larger fractional value?

 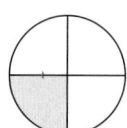

21. Which number line shows the larger amount? Explain how you know and why the pictures might be misleading.

    a. ◄—┼————┼————————┼—►
        0   2

    b. ◄—┼————┼————————┼—►
        0   4

22. Pattern blocks are a common manipulative that can be used with the area model to represent fractions. Four of the basic shapes are shown. If the hexagon represents the whole, what fractions do the other pieces represent?

■ Exercises

23. Place the appropriate fraction in the blank to make the statement true.

   a. 15 min is _____ of an hour.

   b. 1 ft is _____ of a yard.

   c. 10 s is _____ of a minute.

   d. 6 eggs are _____ of a dozen.

   e. 3 months is _____ of a year.

   f. 5¢ is _____ of a quarter.

24. Consider the set of objects.

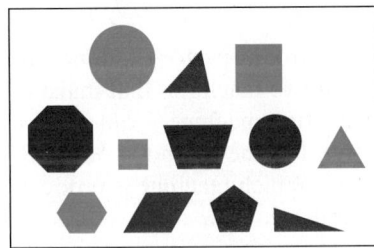

   What fraction of the objects:

   a. Are blue?   **b.** Are circles?

   c. Have four sides?   **d.** Are red or square?

25. Use the following diagram to find the fraction representing each collection of dots.

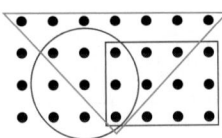

   a. The fraction of the dots inside the triangle as a part of all the dots

   b. The fraction of the dots inside the circle as a part of all the dots

   c. The fraction of the dots inside the circle *and* rectangle as a part of all the dots

   d. The fraction of the dots inside all three figures as a part of all the dots

   e. The fraction of the dots inside the rectangle but outside the circle as a part of all the dots in the rectangle

26. Make each conversion.

   a. $\dfrac{13}{2}$ to a mixed number

   b. $\dfrac{43}{8}$ to a mixed number

   c. $1\dfrac{7}{8}$ to an improper fraction

   d. $5\dfrac{2}{7}$ to an improper fraction

27. Make each conversion.

   a. $-\dfrac{55}{12}$ to a mixed number

   b. $\dfrac{135}{28}$ to a mixed number

   c. $7\dfrac{15}{22}$ to an improper fraction

   d. $-8\dfrac{11}{19}$ to an improper fraction

28. List the set of all fractions that are equivalent to $\dfrac{1}{4}$.

29. List the set of all fractions that are equivalent to (-2).

30. Simplify each fraction.

   a. $\dfrac{18}{45}$ **b.** $\dfrac{42}{56}$ **c.** $-\dfrac{135}{75}$ **d.** $\dfrac{195}{91}$

31. Simplify each fraction.

   a. $\dfrac{168}{49}$ **b.** $-\dfrac{147}{168}$ **c.** $\dfrac{250}{175}$ **d.** $-\dfrac{234}{450}$

32. Use a fraction calculator to simplify each fraction.

   a. $\dfrac{395}{1,015}$ **b.** $-\dfrac{2,342}{3,114}$

   c. $-\dfrac{5,050}{16,160}$ **d.** $\dfrac{13,455}{36,630}$

33. Find the missing number x that makes each pair of fractions equivalent.

   a. $\dfrac{x}{9}=\dfrac{12}{27}$ **b.** $-\dfrac{4}{x}=-\dfrac{48}{84}$

   c. $\dfrac{0}{3}=\dfrac{x}{18}$

34. Use a common denominator to compare each pair of fractions.

   a. $\dfrac{6}{7}$ and $\dfrac{11}{14}$ **b.** $-\dfrac{24}{35}$ and $-\dfrac{21}{30}$

   c. $\dfrac{18}{5}$ and $\dfrac{20}{7}$

35. Use the cross product to compare each pair of fractions.

   a. $\dfrac{4}{5}$ and $\dfrac{20}{25}$ **b.** $-\dfrac{16}{23}$ and $-\dfrac{32}{45}$

   c. $-\dfrac{93}{39}$ and $-\dfrac{124}{52}$

36. Order each list of fractions from smallest to largest.

   a. $\dfrac{1}{5}\ \dfrac{1}{3}\ \dfrac{2}{9}\ \dfrac{3}{10}\ \dfrac{3}{7}\ \dfrac{2}{5}$ **b.** $-\dfrac{2}{3}\ -\dfrac{4}{5}\ -\dfrac{5}{8}\ -\dfrac{4}{7}\ -\dfrac{1}{2}\ -\dfrac{5}{6}$

37. Find five fractions between each pair of numbers.

   a. $\dfrac{1}{5}$ and $\dfrac{2}{5}$ **b.** $-\dfrac{2}{7}$ and $-\dfrac{4}{7}$ **c.** $\dfrac{11}{4}$ and 3

■ Problems and Applications

38. Is it possible to use the area of the following trapezoid to show $\frac{1}{4}$? If so, how?

39. a. If 8 chips make an entire set, how many chips make one-half of the set?

   b. If 25 chips make an entire set, how many chips make two-fifths of the set?

   c. If 33 chips make an entire set, how many chips make five-elevenths of the set?

   d. If 45 chips make an entire set, how many chips make seven-ninths of the set?

40. a. If 20 chips make $\frac{4}{7}$ of a set, how many chips are in the entire set?

   b. If 21 chips make $\frac{7}{12}$ of a set, how many chips are in the entire set?

41. Find four ways to divide a square into four equivalent pieces.

42. A number line from 0 to 4 is divided into nine equal segments. When compared to the unit, what fraction is at *A*? What fraction is at *B*?

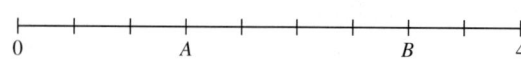

43. Find 15 fractions between $\frac{1}{3}$ and $\frac{2}{3}$, so that the difference between all successive fractions is the same.

44. A fraction is equivalent to $\frac{5}{8}$, and the product of the numerator and denominator is equal to 360. What is the fraction?

45. Find all possible integer values for *x* that make the given statement true.

   a. $\frac{x}{7}$ is the smallest seventh larger than $\frac{1}{2}$.

   b. $\frac{5}{x}$ is not a rational number.

   c. $\frac{x}{4}$ is a positive fraction less than $\frac{1}{2}$.

   d. $\frac{x}{13}$ is a negative improper fraction.

46. If $\frac{4}{9}$ of the students at a given university are males and there are a total of 12,348 students at the university, how many students are females?

47. A movie rental store has approximately 6,500 movies of which 2,500 are dramas. What fraction of the movies are dramas?

48. Two of Mrs. Allen's gym classes take a basic physical fitness test. In her first class, 18 of 26 students pass and in her second class, 17 of 24 pass. Which class did better?

49. It is estimated that $\frac{5}{7}$ of all freshman taking a mathematics course pass it. In Professor Cabbot's class, 68 out of 84 freshman students pass. How did Professor Cabbot's class compare to the university estimate?

50. A quality control inspector examines the products coming from two machines that make the same unit. For the first machine, 3 out of 75 units are defective. For the second, 4 out of 95 units are defective. Which machine is producing fewer defective units?

51. Fractions can be written in other base-*b* numeration systems. For instance, $\frac{12_{\text{five}}}{21_{\text{five}}}$ is a fraction in base five.

   a. What base-ten fraction is $\frac{12_{\text{five}}}{21_{\text{five}}}$ equivalent to?

   b. Find three other fractions in base-five that are equivalent to $\frac{3_{\text{five}}}{4_{\text{five}}}$.

   c. Reduce the base-five fraction $\frac{33_{\text{five}}}{44_{\text{five}}}$.

■ Mathematical Reasoning and Proof

52. True or false? If false, explain why.

   a. Every whole number is a rational number.

   b. The fraction $-\frac{3}{4}$ is a rational number.

   c. The numerator of a rational number is always smaller than the denominator.

   d. All improper fractions are greater than 1.

   e. There is a finite number of rational numbers between 1 and 2.

53. What improper fraction is equivalent to $A\frac{b}{c}$?

54. Are the values of the fractions in the sequence $\frac{1}{2}, \frac{2}{3}, \frac{3}{4}, \frac{4}{5}, \frac{5}{6}, \ldots$ increasing or decreasing? Explain how you know.

55. What must be true about *a* and *c* if $\frac{a}{b} = \frac{c}{b}$?

56. If $\frac{1}{4}$ of the blocks in set A are red, and $\frac{1}{4}$ of the blocks in set B are red, do sets A and B have the same number of red blocks? Why or why not?

57. If $\frac{1}{3}$ of the blocks in set A are red, and $\frac{1}{4}$ of the blocks in set B are red, does it follow that set A has more red blocks than set B? Why or why not?

58. **a.** Is there a smallest rational number less than 1 and greater than 0? If so, what is it?

    b. Is there a greatest negative rational number? If so, what is it?

    c. Is there a smallest negative rational number? If so, what is it?

59. **a.** If the denominator of a fraction is fixed and we increase the value of the numerator, will the value of the fraction increase or decrease? Why?

    b. If the numerator of a fraction is fixed and we increase the value of the denominator, will the value of the fraction increase or decrease? Why?

60. If $\frac{a}{b}$ is a positive nonzero rational number, and c is any natural number, how does $\frac{a+c}{b+c}$ compare to $\frac{a}{b}$?

61. If $\frac{a}{b}$ and $\frac{c}{d}$ are rational numbers such that $\frac{a}{b} < \frac{c}{d}$, how does $\frac{a+c}{b+d}$ compare to $\frac{a}{b}$ and $\frac{c}{d}$?

■ Mathematical Communication

62. Write a short paragraph and use illustrations to explain why $3\frac{1}{4}$ is equivalent to $\frac{13}{4}$.

63. Write a short paragraph and use a part-to-whole comparison to explain why $\frac{1}{4} < \frac{1}{3}$.

64. The definition for the simplest form of a fraction uses the GCD$(a, b) = 1$. What other ways might we state this definition?

65. An old textbook from 1877 makes the following two statements: (1) "It shows into how many parts the integer or unit is divided, and determines the value of the fractional unit." (2) "It numbers the fractional units, and shows how many parts are taken." These two statements define what two notions?

66. The same old textbook also makes the following statement.

    Rule. Divide the terms of the fraction by any number greater than 1, that will divide both without a remainder, and the quotients obtained in the same manner, until no number greater than 1 will so divide them; the last quotients will be the lowest terms of the given fraction.

    What process does this rule describe?

67. **a.** Describe how you would use the length model to compare $\frac{5}{10}$ and $\frac{3}{5}$.

    b. Describe how you would use the area model to compare $\frac{2}{3}$ and $\frac{5}{8}$.

    c. Describe how you would use the set model to compare $\frac{3}{4}$ and $\frac{5}{6}$.

    d. Which method do you think best illustrates the comparison, and why?

68. A student plots the fraction $\frac{3}{5}$ on a number line as shown. What is your response to the diagram?

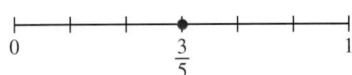

69. **a.** A student claims that $\frac{1}{4}$ is bigger than $\frac{1}{3}$ because 4 is bigger than 3. What do you say in response to the student?

    b. Later, the same student comes back claiming that $\frac{3}{7}$ is bigger than $\frac{1}{2}$ because 3 is bigger than 1. Now what do you say?

70. Jaleel says that there cannot be another fraction between $\frac{3}{10}$ and $\frac{4}{10}$ because there is not another integer between 3 and 4. Is he correct?

■ Building Connections

71. Describe five situations in which you have used mixed numbers. Pick one or two, and explain why mixed numbers were used over improper fractions.

72. Another way to represent part-to-whole relationships is consistent with the place values of the decimal system. Can you think how?

73. When working with fraction operations, always keep in mind the different roles of the numerator and denominator because the roles have a direct consequence on how we compute with fractions. Explain how the role of the numerator and the

denominator might have an impact on how fractions are added.

74. Take a few moments, and think of the problems you might have had in learning fractions for the first time. Discuss them with a group of your peers, and determine how your experiences might impact how you teach fractions to your future students.

75. If we look at the frequency with which the integers occur among the rational numbers, we would see that they occur only rarely. Can you think of several real-world situations that you might use as analogies to explain the density of the rational numbers and their frequency in comparison to the integers?

SECTION 7.2 Adding and Subtracting Rational Numbers

Rational number addition and subtraction can be difficult operations for children to understand and master. When children first learn these operations, their intuition tells them to compute straight across by adding or subtracting both the numerators and the denominators. This intuitive notion is incorrect. To make sense of these operations, we must remember the meaning of the numerator and denominator because their roles directly affect how we compute sums and differences with rational numbers.

Adding Rational Numbers

In some senses, adding rational numbers is like adding whole numbers; we can think of it as counting pieces after combining two disjoint sets. Now, however, the sum must not only give the total number of pieces but also describe the relative size of the pieces to the whole. So, in adding two fractions, we must consider the values of both the numerators and the denominators. To understand how we take these values into account, consider two cases: one in which the denominators are the same and the other in which they are different.

If the denominators are the same, computing a sum is relatively straightforward. For instance, in adding $\frac{3}{8} + \frac{4}{8}$, three pieces that look like eighths are combined with four pieces that look like eighths. Joined together, they make seven pieces, all of which look like eighths, or $\frac{3}{8} + \frac{4}{8} = \frac{7}{8}$ (Figure 7.15).

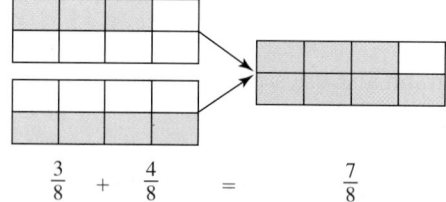

$$\frac{3}{8} \quad + \quad \frac{4}{8} \quad = \quad \frac{7}{8}$$

Figure 7.15 Adding fractions with a common denominator

The example shows that, when the denominators are the same, the relative size of the sum to the whole is clear because the pieces are all the same size. Specifically, it makes sense that the denominator in the sum is the same as the denominator in the addends because they describe pieces of the same size. It also makes sense that the numerator in the sum is found by adding the numerators of the addends because it counts the total number of pieces. This leads us the following definition.

Definition of Rational Number Addition

If $\frac{a}{b}$ and $\frac{c}{b}$ are rational numbers, then $\frac{a}{b} + \frac{c}{b} = \frac{a + c}{b}$.

Rational number addition becomes more complicated when the denominators are different. As before, we want to describe the relative size of the sum to the whole using only one fraction. However, if the denominators in addends are different, it is unlikely that either one of them can be used to do so. For instance, consider the sum $\frac{1}{2} + \frac{1}{3}$, which we represent with fraction bars by joining a length of $\frac{1}{2}$ to a length of $\frac{1}{3}$ (Figure 7.16).

Explorations Manual 7.5

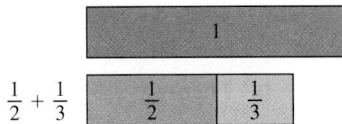

Figure 7.16 Comparing $\frac{1}{2} + \frac{1}{3}$ to the whole

To describe the relative size of the sum to the whole, it makes sense to try the original denominators: halves and thirds. However, neither one will do (Figure 7.17). One-half is too short, and two-halves is too long. Similarly, one- and two-thirds are too short, and three-thirds is too long.

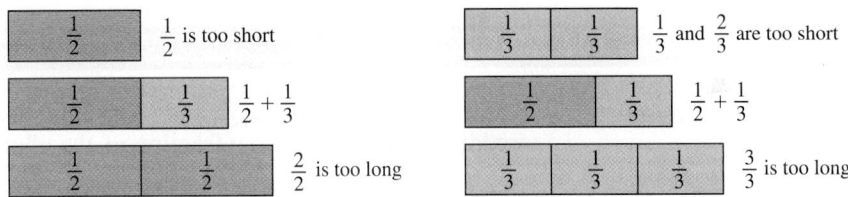

Figure 7.17 Comparing $\frac{1}{2} + \frac{1}{3}$ to halve and thirds

Fortunately, we can solve the problem by using the Fundamental Law of Fractions to rewrite the addends so that they have a common denominator. To find $\frac{1}{2} + \frac{1}{3}$, we use a common denominator of sixths, and then add, or $\frac{1}{2} + \frac{1}{3} = \frac{3}{6} + \frac{2}{6} = \frac{5}{6}$ (Figure 7.18).

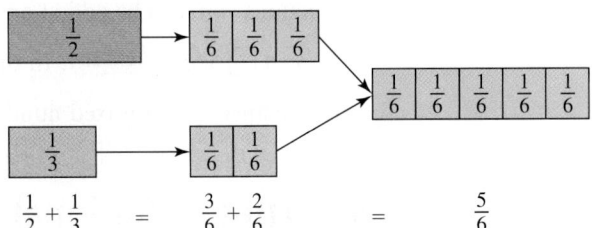

Figure 7.18 Using a common denominator to add $\frac{1}{2} + \frac{1}{3}$

To generalize this procedure, if $\frac{a}{b}$ and $\frac{c}{d}$ are any two rational numbers, then

$$\frac{a}{b} + \frac{c}{d} = \frac{ad}{bd} + \frac{bc}{bd} \quad \text{Fundamental Law of Fractions}$$

$$= \frac{ad + bc}{bd} \quad \text{Addition with like denominator}$$

Theorem 7.5 Adding Rational Numbers with Different Denominators

If $\frac{a}{b}$ and $\frac{c}{d}$ are any two rational numbers, then $\frac{a}{b} + \frac{c}{d} = \frac{ad + bc}{bd}$.

Example 7.11 **Adding Rational Numbers**

Use Theorem 7.5 to find each sum.

a. $\dfrac{4}{15} + \dfrac{8}{21}$ **b.** $\dfrac{-3}{4} + \dfrac{1}{5}$

Solution

a. $\dfrac{4}{15} + \dfrac{8}{21} = \dfrac{4 \cdot 21 + 15 \cdot 8}{15 \cdot 21} = \dfrac{84 + 120}{315} = \dfrac{204}{315} = \dfrac{68}{105}$

b. Theorem 7.5 also works with negative numbers. Consequently,

$$\dfrac{-3}{4} + \dfrac{1}{5} = \dfrac{(-3) \cdot 5 + 4 \cdot 1}{4 \cdot 5} = \dfrac{(-15) + 4}{20} = -\dfrac{11}{20}.$$

In some situations, using a common denominator is easier than using the formula in Theorem 7.5 multiple times. To do so, use either the least common multiple of the denominators or the product of the denominators as the common denominator.

Example 7.12 **Balancing a Budget**

Application

Three children, Lori, Cheryl, and Mark, are throwing a party for their parent's fiftieth wedding anniversary. What fraction of the allotted budget has been spent if Lori has spent $\dfrac{1}{4}$, Cheryl has spent $\dfrac{1}{3}$, and Mark has spent $\dfrac{1}{6}$?

Solution The fraction of the budget that has been spent is the sum of the fractional amounts spent by each child. Using the LCM(3, 4, 6) = 12 as the least common denominator, $\dfrac{1}{4} + \dfrac{1}{3} + \dfrac{1}{6} = \dfrac{3 + 4 + 2}{12} = \dfrac{9}{12} = \dfrac{3}{4}$ of the budget has been spent.

Fractions can also be added on a fraction calculator. For example, to add $\dfrac{3}{4} + \dfrac{5}{7}$, press 3 ⟦/⟧ 4 ⟦+⟧ 5 ⟦/⟧ 7 ⟦ENTER⟧. In this case, the sum is larger than 1, so the calculator returns the answer as a mixed number, or $1\dfrac{13}{28}$.

Properties of Rational Number Addition

Like integer addition, rational number addition satisfies the following properties.

Theorem 7.6 Properties of Rational Number Addition

If $\dfrac{a}{b}, \dfrac{c}{d}$, and $\dfrac{e}{f}$ are any rational numbers, then the following hold true:

Closure property: $\dfrac{a}{b} + \dfrac{c}{d}$ is a unique rational number.

Commutative property: $\dfrac{a}{b} + \dfrac{c}{d} = \dfrac{c}{d} + \dfrac{a}{b}.$

Associative property: $\dfrac{a}{b} + \left(\dfrac{c}{d} + \dfrac{e}{f}\right) = \left(\dfrac{a}{b} + \dfrac{c}{d}\right) + \dfrac{e}{f}.$

Identity property: $\dfrac{a}{b} + 0 = \dfrac{a}{b} = 0 + \dfrac{a}{b}.$

Inverse property: For $\dfrac{a}{b}$ there exists a unique rational number

$$-\dfrac{a}{b}, \text{ such that } \dfrac{a}{b} + \left(-\dfrac{a}{b}\right) = 0 = \left(-\dfrac{a}{b}\right) + \dfrac{a}{b}.$$

Many of these properties can be proved by means of the properties of integer addition.

Example 7.13   Proof of the Commutative Property

Reasoning

Prove that the commutative property holds for rational number addition.

Solution We assume that $\dfrac{a}{b}$ and $\dfrac{c}{d}$ are two rational numbers, and proceed as follows.

$$\dfrac{a}{b} + \dfrac{c}{d} = \dfrac{ad + bc}{bd} \qquad \text{Theorem 7.5}$$

$$= \dfrac{bc + ad}{bd} \qquad \text{Commutative property of integer addition}$$

$$= \dfrac{cb + da}{db} \qquad \text{Commutative property of integer multiplication}$$

$$= \dfrac{c}{d} + \dfrac{a}{b} \qquad \text{Theorem 7.5}$$

We also have the following properties for additive inverses.

Theorem 7.7   Properties of Additive Inverses

If $\dfrac{a}{b}$ and $\dfrac{c}{d}$ are rational numbers, then the following hold true:

1. $-\dfrac{a}{b} = \dfrac{-a}{b} = \dfrac{a}{-b}.$ **2.** $\dfrac{a}{b} = \dfrac{-a}{-b} = -\left(-\dfrac{a}{b}\right).$ **3.** $-\left(\dfrac{a}{b} + \dfrac{c}{d}\right) = \dfrac{-a}{b} + \dfrac{-c}{d}.$

Each of the properties can be verified by means of rational number addition and properties for the additive inverses of integers. The next example verifies the first property. The other two are left to the Learning Assessment.

Example 7.14   Proof of $-\dfrac{a}{b} = \dfrac{-a}{b} = \dfrac{a}{-b}$

Reasoning

If $\dfrac{a}{b}$ is a rational number, show that $-\dfrac{a}{b} = \dfrac{-a}{b} = \dfrac{a}{-b}.$

Solution Suppose $\dfrac{a}{b}$ is a rational number and $-\dfrac{a}{b}$ is its additive inverse. Because $\dfrac{a}{b} + \dfrac{-a}{b} = \dfrac{a + (-a)}{b} = \dfrac{0}{b} = 0,$ then $\dfrac{-a}{b}$ must be another additive inverse of $\dfrac{a}{b}.$ However, the additive inverse is unique, so $-\dfrac{a}{b} = \dfrac{-a}{b}.$ Also, by the Fundamental Law of Fractions, $\dfrac{-a}{b} = \dfrac{-a \cdot (-1)}{b \cdot (-1)} = \dfrac{a}{-b}.$ Combining our results, we have $-\dfrac{a}{b} = \dfrac{-a}{b} = \dfrac{a}{-b}.$

Check Your Understanding 7.2-A

1. Find each sum.

a. $\dfrac{3}{7} + \dfrac{5}{7}$ b. $\left(-\dfrac{5}{6}\right) + \dfrac{7}{6}$ c. $\left(\dfrac{4}{5} + \left(-\dfrac{3}{5}\right)\right) + \dfrac{2}{5}$

2. Find each sum.

a. $\dfrac{7}{8} + \dfrac{9}{10}$ b. $\left(-\dfrac{4}{5}\right) + \dfrac{1}{12}$ c. $\left(\dfrac{7}{12} + \dfrac{2}{3}\right) + \left(-\dfrac{5}{6}\right)$

3. Which of the following is (are) equal to 4?

a. $\dfrac{-4}{1}$ b. $\dfrac{-4}{-1}$ c. $-\dfrac{-4}{1}$ d. $-\dfrac{-4}{-1}$ e. $-\left(-\dfrac{4}{1}\right)$

Talk About It How is rational number addition similar to whole-number addition? How is it different?

Representation

Activity 7.3

Use the area model to represent and solve each sum.

a. $\dfrac{2}{6} + \dfrac{3}{6}$ b. $\dfrac{2}{3} + \dfrac{1}{6}$ c. $\left(\dfrac{1}{2} + \dfrac{1}{3}\right) + \dfrac{1}{4}$

Subtracting Rational Numbers

Rational number subtraction can be defined in two ways. One is to use the missing-addend approach.

Definition of Rational Number Subtraction (Missing-Addend Approach)

If $\dfrac{a}{b}$ and $\dfrac{c}{d}$ are rational numbers, then $\dfrac{a}{b} - \dfrac{c}{d} = \dfrac{e}{f}$ if and only if $\dfrac{a}{b} = \dfrac{c}{d} + \dfrac{e}{f}$.

As before, we compute a difference by rewriting it as a sum in which one of the addends in missing. We then use our knowledge of addition to find the missing number. For instance, to compute $\dfrac{1}{2} - \dfrac{1}{3}$, we need to find the rational number $\dfrac{e}{f}$ such that $\dfrac{1}{2} = \dfrac{1}{3} + \dfrac{e}{f}$. The value of $\dfrac{e}{f}$ may not be immediately obvious, but it may be easier to find with a common denominator. Specifically, if we rewrite $\dfrac{1}{2} = \dfrac{1}{3} + \dfrac{e}{f}$ with common denominators, or $\dfrac{3}{6} = \dfrac{2}{6} + \dfrac{e}{f}$, then $\dfrac{e}{f} = \dfrac{1}{6}$ because $\dfrac{1}{6} + \dfrac{2}{6} = \dfrac{3}{6}$.

Another way to compute a difference is to use the take-away approach. To understand how, we again consider what happens when the denominators are the same and when they are different. If the denominators are the same, computing a difference is again relatively straightforward. For instance, given $\dfrac{5}{6} - \dfrac{3}{6}$, we take away three pieces that look like sixths from five pieces that look like sixths. Two pieces remain, both of which look like sixths. Consequently, $\dfrac{5}{6} - \dfrac{3}{6} = \dfrac{2}{6}$ (Figure 7.19).

Explorations Manual 7.6

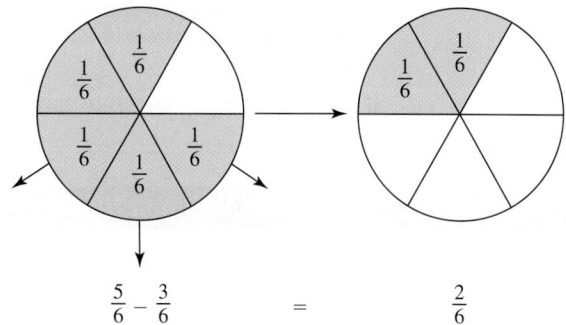

$$\frac{5}{6}-\frac{3}{6} \qquad = \qquad \frac{2}{6}$$

Figure 7.19 Subtracting fractions with a common denominator

As with addition, the relative size of the difference to the whole is clear because the pieces are all the same size. Again, it makes sense that the denominator in the difference is the same as the denominator in the original fractions because they describe pieces of the same size. It also makes sense that the numerator in the difference is found by subtracting the numerators in the original fractions because they count the total number of pieces. This leads to the following definition.

Definition of Rational Number Subtraction (Take-Away Approach)

If $\frac{a}{b}$ and $\frac{c}{b}$ are rational numbers, then $\frac{a}{b}-\frac{c}{b}=\frac{a-c}{b}$.

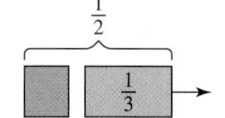

Figure 7.20 Taking away $\frac{1}{3}$ from $\frac{1}{2}$

If the denominators are different, we have the same problem as with addition. Specifically, the denominators of the original two fractions are often inadequate to describe the relative size of the difference to the whole. For instance, consider $\frac{1}{2}-\frac{1}{3}$, which is represented with fraction bars by removing a length of $\frac{1}{3}$ from a length of $\frac{1}{2}$ (Figure 7.20). When we try to describe the relative length of the remaining piece to the whole, we find that neither halves nor thirds are adequate. Both one-half and one-third are too long (Figure 7.21).

$\frac{1}{2}$ is too long

$\frac{1}{2}-\frac{1}{3}$

$\frac{1}{3}$ is too long

Figure 7.21 Comparing $\frac{1}{2}-\frac{1}{3}$ to halves and thirds

Fortunately, the Fundamental Law of Fractions can be used to rewrite each fraction with a common denominator of sixths. As a result, $\frac{1}{2}-\frac{1}{3}=\frac{3}{6}-\frac{2}{6}=\frac{1}{6}$ (Figure 7.22).

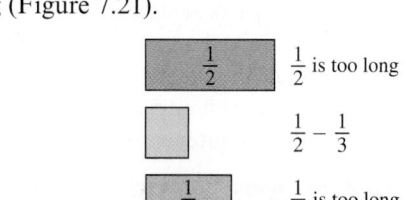

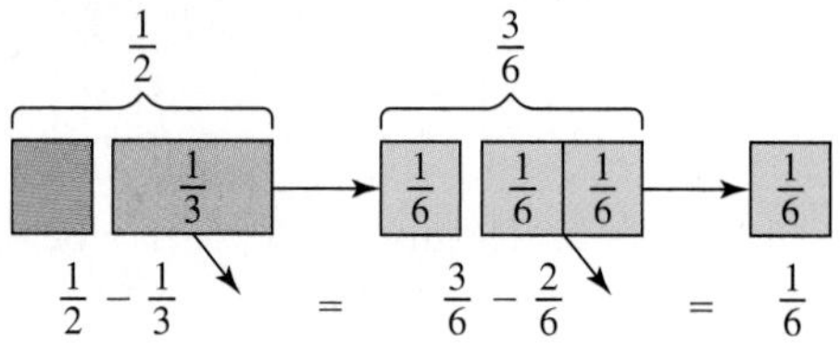

$$\frac{1}{2}-\frac{1}{3} \quad = \quad \frac{3}{6}-\frac{2}{6} \quad = \quad \frac{1}{6}$$

Figure 7.22 Using a common denominator to subtract $\frac{1}{2}-\frac{1}{3}$

These ideas are generalized in the following theorem.

Theorem 7.8   Subtracting Rational Numbers with Different Denominators

If $\dfrac{a}{b}$ and $\dfrac{c}{d}$ are any two rational numbers, then $\dfrac{a}{b} - \dfrac{c}{d} = \dfrac{ad - bc}{bd}$.

Example 7.15   Subtracting Rational Numbers

Compute each difference.

a. $\dfrac{5}{8} - \dfrac{1}{4}$ b. $\dfrac{5}{16} - \left(-\dfrac{7}{12}\right)$

Solution In each problem, we subtract by getting a common denominator or by using the formula in Theorem 7.8.

a. Because the LCM(8, 4) = 8, we use a common denominator of 8, giving us

$$\frac{5}{8} - \frac{1}{4} = \frac{5}{8} - \frac{2}{8} = \frac{5-2}{8} = \frac{3}{8}.$$

b. Theorem 7.8 allows us to compute differences with negative numbers, giving us

$$\frac{5}{16} - \left(-\frac{7}{12}\right) = \frac{5 \cdot 12 - 16 \cdot (-7)}{16 \cdot 12} = \frac{60 - (-112)}{192} = \frac{172}{192} = \frac{43}{48}.$$

As with integer subtraction, we can also perform rational number subtraction by adding the additive inverse.

Theorem 7.9   Subtracting by Adding the Opposite

If $\dfrac{a}{b}$ and $\dfrac{c}{d}$ are any two rational numbers, then $\dfrac{a}{b} - \dfrac{c}{d} = \dfrac{a}{b} + \dfrac{-c}{d}$.

Adding and Subtracting Mixed Numbers

Many real-world situations, such as those involving measurement, require us to add or subtract mixed numbers. One way to do so is to work with the numbers in mixed number form, in which case the commutative and associative properties allow us to add or subtract the integer and fractional parts separately. However, making an exchange between the integer and fractional part may be necessary.

Example 7.16   Adding and Subtracting Mixed Numbers

Compute each operation.

a. $2\dfrac{3}{4} + 3\dfrac{3}{5}$ b. $3\dfrac{1}{3} - 1\dfrac{3}{4}$

Solution

a. Adding the integer and fractional parts separately, we have $2\dfrac{3}{4} + 3\dfrac{3}{5} = \left(2 + \dfrac{3}{4}\right) + \left(3 + \dfrac{3}{5}\right) = (2+3) + \left(\dfrac{15}{20} + \dfrac{12}{20}\right) = 5 + \dfrac{27}{20} = 5 + \dfrac{20}{20} + \dfrac{7}{20} = 6\dfrac{7}{20}$

b. To subtract $3\dfrac{1}{3} - 1\dfrac{3}{4}$, begin with $\dfrac{1}{3} - \dfrac{3}{4}$. Because $\dfrac{3}{4}$ is greater than $\dfrac{1}{3}$, exchange one whole for $\dfrac{3}{3}$:

$$\frac{4}{3} - \frac{3}{4} = \frac{16}{12} - \frac{9}{12} = \frac{7}{12}$$

Then subtract $2 - 1 = 1$, and combine the integer and fractional parts to get $1\dfrac{7}{12}$.

A second approach is to convert the mixed numbers to improper fractions, perform the operation, and then convert the answer back to a mixed number.

Example 7.17 | **Adding Lengths of Boards**

Application

Three boards of lengths $2\frac{3}{4}$ ft, $3\frac{5}{6}$ ft, and $3\frac{2}{3}$ ft are cut from a single board with no remaining pieces. Assuming no length is lost from the cuts, how long was the original board?

How long?

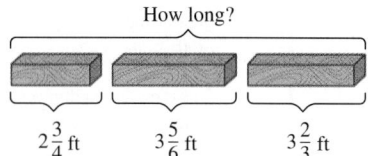

$2\frac{3}{4}$ ft $3\frac{5}{6}$ ft $3\frac{2}{3}$ ft

Solution The length of the original board is the sum of the three lengths. We convert the mixed numbers to improper fractions, add, and then convert back:

$$2\frac{3}{4} + 3\frac{5}{6} + 3\frac{2}{3} = \frac{11}{4} + \frac{23}{6} + \frac{11}{3} = \frac{33}{12} + \frac{46}{12} + \frac{44}{12} = \frac{123}{12} = 10\frac{1}{4} \text{ ft}$$

Doing mixed number computations in real-world situations may require mental computation or at least a reasonable estimate. The next examples demonstrate different techniques that we can use to do so.

Example 7.18 | **Adding and Subtracting Mixed Numbers Mentally**

Compute each problem mentally.

a. $2\frac{3}{4} + 3\frac{1}{2} + 6\frac{3}{4}$ **b.** $7\frac{3}{8} - 4\frac{1}{8}$ **c.** $6 - 2\frac{1}{3}$

Solution In each case, we use the commutative and associative properties to rearrange both the integer and the fractional parts to make the problem easier to compute.

a. Think $2\frac{3}{4} + 3\frac{1}{2} + 6\frac{3}{4} = (2 + 3 + 6) + \left(\frac{3}{4} + \frac{1}{2} + \frac{3}{4}\right) = 11 + 2 = 13.$

b. Think $7\frac{3}{8} - 4\frac{1}{8} = (7 - 4) + \left(\frac{3}{8} - \frac{1}{8}\right) = 3 + \frac{2}{8} = 3\frac{1}{4}.$

c. Subtract the second mixed number in two parts, first 2 and then $\frac{1}{3}$ or,

$$6 - 2\frac{1}{3} = (6 - 2) - \frac{1}{3} = 4 - \frac{1}{3} = 3\frac{2}{3}.$$

Example 7.19 | **Estimating Sums and Differences**

Connections

Use the method on the student page in Figure 7.23 to estimate the given sums and differences.

a. $23\frac{1}{9} + 26\frac{11}{12}$ **b.** $45\frac{9}{16} - 19\frac{7}{8}$ **c.** $\frac{24}{10} + 3\frac{1}{8} - \frac{13}{7}$

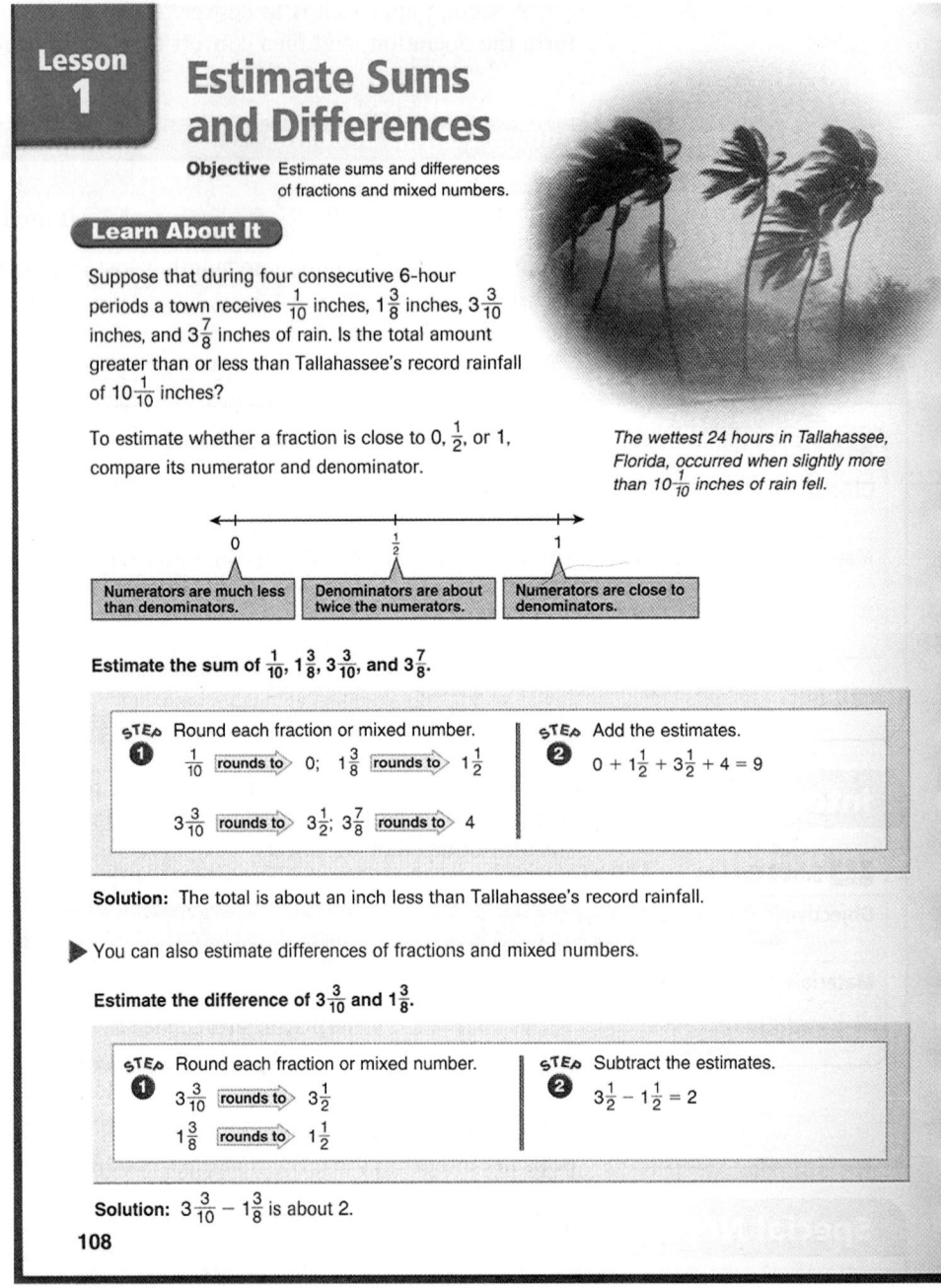

Lesson 1

Estimate Sums and Differences

Objective Estimate sums and differences of fractions and mixed numbers.

Learn About It

Suppose that during four consecutive 6-hour periods a town receives $\frac{1}{10}$ inches, $1\frac{3}{8}$ inches, $3\frac{3}{10}$ inches, and $3\frac{7}{8}$ inches of rain. Is the total amount greater than or less than Tallahassee's record rainfall of $10\frac{1}{10}$ inches?

To estimate whether a fraction is close to 0, $\frac{1}{2}$, or 1, compare its numerator and denominator.

The wettest 24 hours in Tallahassee, Florida, occurred when slightly more than $10\frac{1}{10}$ inches of rain fell.

| Numerators are much less than denominators. | Denominators are about twice the numerators. | Numerators are close to denominators. |

Estimate the sum of $\frac{1}{10}$, $1\frac{3}{8}$, $3\frac{3}{10}$, and $3\frac{7}{8}$.

STEP 1 Round each fraction or mixed number.

$\frac{1}{10}$ rounds to 0; $1\frac{3}{8}$ rounds to $1\frac{1}{2}$

$3\frac{3}{10}$ rounds to $3\frac{1}{2}$; $3\frac{7}{8}$ rounds to 4

STEP 2 Add the estimates.

$0 + 1\frac{1}{2} + 3\frac{1}{2} + 4 = 9$

Solution: The total is about an inch less than Tallahassee's record rainfall.

▶ You can also estimate differences of fractions and mixed numbers.

Estimate the difference of $3\frac{3}{10}$ and $1\frac{3}{8}$.

STEP 1 Round each fraction or mixed number.

$3\frac{3}{10}$ rounds to $3\frac{1}{2}$

$1\frac{3}{8}$ rounds to $1\frac{1}{2}$

STEP 2 Subtract the estimates.

$3\frac{1}{2} - 1\frac{1}{2} = 2$

Solution: $3\frac{3}{10} - 1\frac{3}{8}$ is about 2.

108

Figure 7.23 Estimating sums and differences of mixed numbers
Source: From *Houghton Mifflin Mathematics: Student Edition Level 6* (p. 108). Copyright © by Houghton Mifflin Company, Inc. Reproduced by permission of the publisher, Houghton Mifflin Harcourt Publishing Company.

Solution The student page shows how to make mixed number computations easier by rounding the fractional part to a benchmark number. If the numerator is much less than the denominator, round the fraction down to 0. If the numerator is about half the denominator, round to $\frac{1}{2}$. If the numerator and the denominator are about the same, round to 1. Using these guidelines, we make each computation.

a. Round $\frac{1}{9}$ down to 0 and $\frac{11}{12}$ up to 1. Hence, $23\frac{1}{9} + 26\frac{11}{12} \approx 23 + 27 = 50$.

b. Round $\frac{9}{16}$ down to $\frac{1}{2}$ and $\frac{7}{8}$ up to 1. Hence, $45\frac{9}{16} - 19\frac{7}{8} \approx 45\frac{1}{2} - 20 = 25\frac{1}{2}$.

c. Because $\dfrac{24}{10} = 2\dfrac{2}{5}$ and $\dfrac{13}{7} = 1\dfrac{6}{7}$, then $\dfrac{24}{10} + 3\dfrac{1}{8} - \dfrac{13}{7} = 2\dfrac{2}{5} + 3\dfrac{1}{8} - 1\dfrac{6}{7}$

$\approx 2\dfrac{1}{2} + 3 - 2 = 3\dfrac{1}{2}.$

Check Your Understanding 7.2-B

1. Find each difference.

a. $\dfrac{5}{9} - \dfrac{6}{9}$ **b.** $\dfrac{6}{7} - \dfrac{8}{7}$ **c.** $\left(-\dfrac{6}{5}\right) - \dfrac{8}{7}$ **d.** $\left(\left(-\dfrac{7}{5}\right) - \dfrac{5}{4}\right) - \left(-\dfrac{5}{6}\right)$

2. Compute each sum or difference.

a. $5\dfrac{7}{10} + 2\dfrac{3}{4}$ **b.** $8\dfrac{1}{3} + 6\dfrac{7}{12}$ **c.** $6\dfrac{1}{4} - 4\dfrac{1}{9}$ **d.** $9\dfrac{1}{9} - 5\dfrac{5}{12}$

3. Compute each problem mentally.

a. $7\dfrac{1}{10} - 5\dfrac{1}{10}$ **b.** $4\dfrac{7}{8} + 6\dfrac{1}{4}$ **c.** $6\dfrac{2}{7} - 1\dfrac{5}{7}$

4. Estimate each sum or difference.

a. $5\dfrac{7}{10} + 2\dfrac{3}{4}$ **b.** $8\dfrac{1}{3} + 6\dfrac{7}{12}$ **c.** $6\dfrac{1}{4} - 4\dfrac{1}{9}$

Talk About It How is rational number subtraction similar to whole-number subtraction? How is it different?

Representation

Activity 7.4

Use the area model to represent and solve each difference.

a. $\dfrac{5}{6} - \dfrac{3}{6}$ **b.** $\dfrac{4}{5} - \dfrac{2}{10}$ **c.** $\left(\dfrac{7}{8} - \dfrac{1}{4}\right) - \dfrac{1}{2}$

SECTION 7.2 Learning Assessment

■ Understanding the Concepts

1. How is rational number addition like whole-number addition? How is it different?

2. Why is it necessary to get a common denominator when adding or subtracting two fractions?

3. Is the common denominator bd in Theorem 7.5 necessarily the least common denominator? Explain.

4. Explain the meaning of the three properties in Theorem 7.7.

5. Describe how to use the missing-addend approach to subtract two rational numbers.

6. Describe how to use the take-away approach to subtract two rational numbers if they:

   a. Have common denominators.

   b. Do not have common denominators.

7. What are two different approaches for adding or subtracting mixed numbers?

8. Describe how to estimate the sum or difference of two or more mixed numbers.

■ Representing the Mathematics

9. Use fraction bars or the area model to represent and solve each problem.

   a. $\dfrac{2}{8} + \dfrac{5}{8}$ **b.** $\dfrac{2}{3} + \dfrac{1}{6}$

   c. $\dfrac{2}{3} - \dfrac{2}{4}$ **d.** $\left(\dfrac{1}{2} + \dfrac{1}{5}\right) - \dfrac{1}{10}$

10. Use a number line to represent and solve each problem.

   a. $\dfrac{3}{6} + \dfrac{1}{6}$ **b.** $\dfrac{4}{5} - \dfrac{2}{5}$

11. Represent the expression $\left(\dfrac{a}{b} + \dfrac{c}{d}\right) + \dfrac{e}{f}$ as a single fraction.

12. Represent $A\dfrac{b}{c} - D\dfrac{e}{f}$ as a single fraction.

■ Exercises

13. Compute each sum or difference.

   a. $\dfrac{4}{11} + \dfrac{9}{11}$ b. $\left(-\dfrac{5}{6}\right) - \dfrac{3}{6}$

   c. $\left(\left(-\dfrac{4}{7}\right) + \dfrac{3}{7}\right) - \dfrac{2}{7}$ d. $\dfrac{7}{4} - \dfrac{11}{4}$

14. Compute each sum or difference.

   a. $\dfrac{3}{4} + \dfrac{5}{8}$ b. $\dfrac{9}{7} - \dfrac{5}{14}$

   c. $\dfrac{2}{3} + \left(-\dfrac{5}{4}\right)$ d. $\dfrac{4}{5} - \dfrac{11}{12}$

15. Compute each sum or difference.

   a. $\dfrac{3}{10} + \dfrac{2}{7}$ b. $\dfrac{7}{13} - \left(-\dfrac{2}{5}\right)$

   c. $\left(\dfrac{5}{12} - \left(-\dfrac{1}{4}\right)\right) + \dfrac{2}{3}$ d. $\left(\dfrac{2}{3} + \left(-\dfrac{11}{5}\right)\right) - \dfrac{7}{6}$

16. Use the missing-addend approach to compute each difference.

   a. $\dfrac{7}{10} - \dfrac{3}{10}$ b. $\dfrac{5}{9} - \dfrac{7}{9}$

   c. $\dfrac{3}{4} - \dfrac{3}{8}$ d. $\dfrac{5}{6} - \dfrac{3}{4}$

17. Which of the following is (are) are equal to -5?

   a. $\dfrac{-5}{1}$ b. $\dfrac{-5}{-1}$

   c. $\dfrac{5}{-1}$ d. $-\dfrac{5}{-1}$

18. Which of the following is (are) equal to 7?

   a. $\dfrac{-7}{-1}$ b. $\dfrac{7}{-1}$

   c. $-\left(\dfrac{7}{1}\right)$ d. $-\left(\dfrac{-7}{-1}\right)$

19. Compute each sum or difference.

   a. $2\dfrac{1}{3} + 3\dfrac{2}{3}$ b. $6\dfrac{4}{5} - 2\dfrac{1}{5}$

   c. $3\dfrac{4}{7} + 1\dfrac{1}{3}$ d. $4 - 2\dfrac{1}{6}$

20. Compute each sum or difference.

   a. $3\dfrac{1}{10} + 7\dfrac{6}{7}$ b. $7 - 4\dfrac{4}{7}$

   c. $9\dfrac{4}{5} + 3\dfrac{3}{7}$ d. $8\dfrac{3}{4} - 4\dfrac{4}{5}$

21. Use mental arithmetic to compute each sum or difference.

   a. $3\dfrac{1}{2} + 3\dfrac{1}{10}$ b. $4\dfrac{2}{3} + 3\dfrac{1}{3}$

   c. $9\dfrac{11}{12} - 7\dfrac{5}{12}$ d. $7\dfrac{3}{10} - 1\dfrac{7}{10}$

22. Estimate each sum or difference.

   a. $6\dfrac{9}{10} - 2\dfrac{7}{8}$ b. $8\dfrac{3}{7} + 7\dfrac{5}{9}$

   c. $5\dfrac{5}{6} - 4\dfrac{1}{10}$ d. $10\dfrac{1}{8} - 9\dfrac{11}{12}$

23. Without making a direct computation, determine whether each computation is closest to 0, $\dfrac{1}{2}$, or 1.

   a. $\dfrac{3}{4} - \dfrac{2}{3}$ b. $\dfrac{4}{3} - \dfrac{1}{8}$

   c. $\dfrac{1}{4} + \dfrac{1}{5}$ d. $\dfrac{1}{9} + \dfrac{1}{11}$

24. Use a fraction calculator to compute each problem. Be sure to simplify your answer.

   a. $\dfrac{3}{11} + \dfrac{7}{15}$ b. $2\dfrac{2}{9} + 3\dfrac{6}{11}$

   c. $\dfrac{8}{17} - \dfrac{9}{11}$ d. $3\dfrac{1}{16} - 2\dfrac{7}{8}$

■ Problems and Applications

25. Find $\dfrac{1}{2} + \dfrac{1}{3} + \dfrac{1}{4} + \dfrac{1}{6} + \dfrac{1}{8} + \dfrac{1}{12} + \dfrac{1}{24}$.

26. Find two fractions whose sum is $\dfrac{2}{3}$ and whose difference is $\dfrac{1}{3}$.

27. Use the numbers 1, 3, 4, and 6 only once to fill in the blanks and make the equation true.

$$\dfrac{\Box}{\Box} - \dfrac{\Box}{\Box} = \dfrac{7}{12}$$

28. Use the numbers 2, 3, 5, and 6 only once to fill in the blanks and make the sum as large as possible.

$$\dfrac{\Box}{\Box} + \dfrac{\Box}{\Box} =$$

29. Use the numbers 4, 5, 8, and 9 only once to fill in the blanks to make a positive difference that is as small as possible.

$$\dfrac{\Box}{\Box} - \dfrac{\Box}{\Box} =$$

30. For each equation, find the value of x that makes the statement true.

   a. $\dfrac{7}{4} - \dfrac{x}{12} = \dfrac{4}{3}$ b. $\dfrac{x}{5} + \dfrac{1}{x} = \dfrac{14}{15}$

   c. $x\dfrac{1}{x} + 3\dfrac{1}{3} = 5\dfrac{5}{6}$

31. Recall that in a unit fraction, the numerator is 1. Express each fraction as the sum of two unit fractions.

a. $\dfrac{7}{12}$ b. $\dfrac{2}{3}$

c. $\dfrac{5}{18}$ d. $\dfrac{19}{72}$

32. A farmer has plowed $15\dfrac{2}{3}$ acres of a field that is a total of $33\dfrac{4}{7}$ acres. How much more of the field must the farmer plow?

33. How much fence is needed to enclose a yard with sides that measure $27\dfrac{1}{3}$ ft, $14\dfrac{1}{4}$ ft, $26\dfrac{2}{3}$ ft, and $14\dfrac{3}{4}$ ft?

34. At a particular university, $\dfrac{1}{3}$ of the students are freshmen, $\dfrac{1}{5}$ are sophomores, $\dfrac{1}{6}$ are juniors, and $\dfrac{1}{6}$ are seniors. The rest are graduate students. What fraction of the student body consists of graduate students?

35. A gardener has a flower box that is $2\dfrac{1}{3}$ yd^2. She plants $\dfrac{3}{4}$ yd^2 with petunias and $\dfrac{5}{6}$ yd^2 with irises. How much of the flower box is left for her to plant tulips?

36. Jessica ate $\dfrac{1}{2}$ of her pizza last night and $\dfrac{1}{4}$ of it this morning. If there were originally 8 pieces of pizza, how many pieces are left?

37. A restaurant worker logs in the following hours for one week: $5\dfrac{1}{4}$, $6\dfrac{1}{3}$, $7\dfrac{3}{8}$, $11\dfrac{1}{2}$, and $9\dfrac{3}{4}$. How many hours did this person work this week?

38. Boards of length $3\dfrac{1}{3}$ ft, $4\dfrac{1}{5}$ ft, and $2\dfrac{3}{10}$ ft are cut from a board that is $13\dfrac{1}{2}$ ft long. How much of the board remains?

39. About $\dfrac{1}{3}$ of Samuel's paycheck goes to housing, $\dfrac{1}{6}$ goes to utility bills, $\dfrac{1}{8}$ goes for his car, and $\dfrac{1}{5}$ goes for food. How much of his paycheck is left for entertainment?

40. A bunch of bananas has 15 bananas. A monkey eats $\dfrac{1}{5}$ of the bananas one morning and $\dfrac{1}{2}$ of the remaining bananas in the afternoon. What fractional portion of the bananas is remaining after the monkey is finished eating?

■ Mathematical Reasoning and Proof

41. a. Find a pattern, and use it to complete the next two rows of the triangle.

$$1$$
$$\frac{1}{2} \qquad \frac{1}{2}$$
$$\frac{1}{3} \qquad \frac{1}{6} \qquad \frac{1}{3}$$
$$\frac{1}{4} \qquad \frac{1}{12} \qquad \frac{1}{12} \qquad \frac{1}{4}$$
$$\frac{1}{5} \qquad \frac{1}{20} \qquad \frac{1}{30} \qquad \frac{1}{20} \qquad \frac{1}{5}$$

b. What is the second number in the 10th row?

c. What is the second to last number in the 9th row?

42. Compute.

a. $\dfrac{1}{2} + \dfrac{1}{4} + \dfrac{1}{6} + \dfrac{1}{12}$

b. $\dfrac{1}{3} + \dfrac{1}{6} + \dfrac{1}{9} + \dfrac{1}{18}$

c. $\dfrac{1}{4} + \dfrac{1}{8} + \dfrac{1}{12} + \dfrac{1}{24}$

d. Use any patterns you notice in parts a, b, and c to find the value of $\dfrac{1}{n} + \dfrac{1}{2n} + \dfrac{1}{3n} + \dfrac{1}{6n}$, where n is a natural number.

43. If a and b are rational numbers, where $0 < a < b < 1$, determine whether each statement is true or false or cannot be determined.

a. $a + b < b$ b. $a + b > 1$

c. $b - a < b$ d. $b - a < a$

44. If a and b are rational numbers, where $0 < a < 1$ and $b > 1$, determine whether each statement is true or false or cannot be determined.

a. $a + b > b$ b. $a + b > 1$

c. $b - a < 1$ d. $b - a < b$

45. True or false? If false, explain why.

a. $\dfrac{a}{b} + \dfrac{c}{d} = \dfrac{c}{d} + \dfrac{a}{b}$

b. $-\left(\dfrac{a}{b}\right) = \dfrac{-a}{b}$

c. $\dfrac{a}{b} + \left(\dfrac{c}{d} + \dfrac{e}{f}\right) = \dfrac{a + (c + e)}{b + (d + f)}$

d. $\dfrac{a}{b} = -\left(\dfrac{a}{b}\right)$

e. $\dfrac{a}{b} + 0 = \dfrac{a}{b}$

f. $-\left(\dfrac{a}{b} + \dfrac{c}{d}\right) = \dfrac{a}{b} - \dfrac{c}{d}$

46. Provide a numerical example to demonstrate that each property holds true for rational number addition.
    a. Commutative
    b. Associative
    c. Additive inverse

47. Provide a numerical example to demonstrate that each property does not hold for rational number subtraction.
    a. Commutative
    b. Associative
    c. Additive identity

48. Use the Fundamental Law of Fractions to show that $\dfrac{a}{b} = \dfrac{-a}{-b}$.

49. If $\dfrac{a}{b}$ and $\dfrac{c}{d}$ are rational numbers, show that $\dfrac{a}{b} - \dfrac{c}{d} = \dfrac{ad - bc}{bd}$.

50. Write an argument using additive inverses to show that $-\left(\dfrac{a}{b} + \dfrac{c}{d}\right) = \dfrac{-a}{b} + \dfrac{-c}{d}$.

■ Mathematical Communication

51. Compare and contrast the models for adding and subtracting fractions. Which of them was most meaningful or helpful to you? Explain.

52. Write several problems that require adding or subtracting mixed numbers. Instruct a partner on how to compute them using a fraction calculator.

53. Write several story problems that involve adding fractions or mixed numbers. Swap your problems with a partner's and solve them. How is the language used in these problems similar to the language used in story problems involving whole-number addition? How is it different?

54. a. Write a story problem that uses the missing-addend approach to compute $\dfrac{2}{3} - \dfrac{1}{5}$.

b. Write a story problem that uses the take-away approach to compute $\dfrac{4}{5} - \dfrac{3}{4}$.

c. How is the language used in these problems similar to the language used in story problems involving whole-number subtraction? How is it different?

55. Josiah claims that $\dfrac{3}{5} + \dfrac{2}{3} = \dfrac{5}{8}$. He offers the following picture as proof that he is correct. What has he done wrong, and why is his picture incorrect?

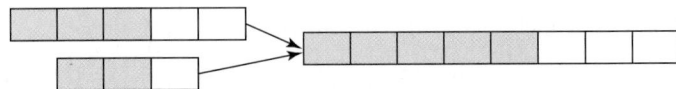

56. Sarah subtracts two mixed numbers in the following way: $7\dfrac{7}{9} - 5\dfrac{2}{3} = 2\dfrac{5}{6}$. Is she correct? If not, how would you help her correct the error?

■ Building Connections

57. In what situations have you added or subtracted fractions in your daily life?

58. What role does the least common multiple play in adding or subtracting fractions?

59. Rational number addition is often used to write decimals in expanded notation. When doing so, what is the denominator of each addend?

60. Consider a set of elementary curriculum materials in which rational number addition is taught. Describe the progression of the content in these lessons. How is it similar to what you learned in this section? How is it different?

61. What role do you think estimation has in teaching children how to add and subtract rational numbers?

SECTION 7.3 Multiplying and Dividing Rational Numbers

The intuitive ideas behind whole-number multiplication and division can help us understand and compute rational number multiplication and division. However, as with addition and subtraction, we must adjust these ideas to take the relative size of the fractions into account.

Multiplying Rational Numbers

In Chapter 4, whole-number products were computed in two ways. In the Cartesian-product approach, the elements from two sets were paired, and then the total number of pairs was counted. Although the method worked with the whole numbers, it cannot be used with rational numbers because the Cartesian product gives us no way to describe the size of the product relative to the whole. The repeated-addition approach has meaning with rational number products as long as the first factor is a whole number. For instance, $4 \cdot \frac{2}{3}$ has an interpretation of adding $\frac{2}{3}$ to itself 4 times (Figure 7.24).

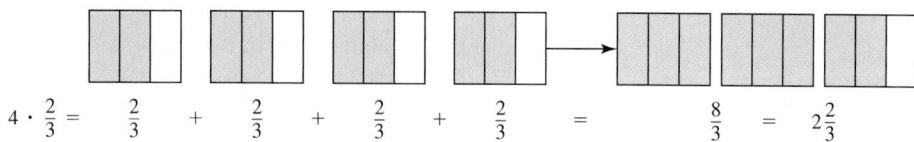

$$4 \cdot \frac{2}{3} = \frac{2}{3} + \frac{2}{3} + \frac{2}{3} + \frac{2}{3} = \frac{8}{3} = 2\frac{2}{3}$$

Figure 7.24 Multiplying $4 \cdot \frac{2}{3}$ by repeated addition

The interpretation becomes less clear, however, when a fraction is multiplied by a whole number. For instance, the repeated-addition interpretation of $\frac{1}{2} \cdot 4$ is to add 4 to itself half of a time. But what does this mean? To make sense of the product, we approach it from a different direction. Ideally, we want the commutative property to hold for rational number multiplication, so $\frac{1}{2} \cdot 4$ must equal $4 \cdot \frac{1}{2} = \frac{1}{2} + \frac{1}{2} + \frac{1}{2} + \frac{1}{2} = 2$. However, notice that 2 is half of 4. So we can interpret the product $\frac{1}{2} \cdot 4$, not as "add 4 to itself half of a time" but rather as "take one-half of 4" (Figure 7.25).

Explorations
Manual
7.9

4 units

$\frac{1}{2}$ of 4 units is 2 units

Figure 7.25 Taking $\frac{1}{2}$ of four

In this interpretation, the word "of" is taken to mean multiplication, allowing us to make sense of products involving two rational numbers. For instance, we interpret

$$\frac{1}{3} \cdot \frac{1}{2} \quad \text{as "Take one-third of one-half" and}$$

$$\frac{2}{3} \cdot \frac{4}{5} \quad \text{as "Take two-thirds of four-fifths"}$$

To better understand the product of two rational numbers, we can represent $\frac{1}{3} \cdot \frac{1}{2}$ with fraction bars (Figure 7.26). Because $\frac{1}{3} \cdot \frac{1}{2}$ means take $\frac{1}{3}$ of $\frac{1}{2}$, we divide a length of $\frac{1}{2}$ into three equal pieces and count one. We now must determine the relative size of this piece to the whole. A direct comparison with other fraction bars shows that the piece is $\frac{1}{6}$ of the whole. Consequently, $\frac{1}{3} \cdot \frac{1}{2} = \frac{1}{6}$. Notice that the product is smaller than either factor. This may seem counterintuitive to whole-number multiplication. However, when multiplying two proper fractions, we are literally taking a piece of a piece, which naturally results in smaller piece.

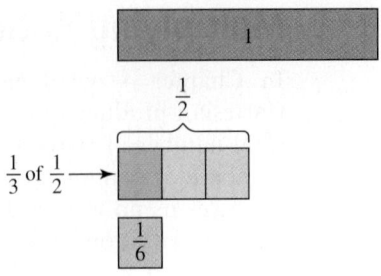

Figure 7.26 Using fraction bars to compute $\dfrac{1}{3} \cdot \dfrac{1}{2}$

The area of a rectangle can also be used to represent the product of two rational numbers. For instance, to find $\dfrac{1}{2} \cdot \dfrac{3}{4}$, we want to take $\dfrac{1}{2}$ of $\dfrac{3}{4}$ of the area of a rectangle. To do so, we vertically divide it into four equivalent regions and shade three of them (Figure 7.27). We then take $\dfrac{1}{2}$ of the shaded region by horizontally dividing it into two equivalent regions. We shade half of the newly formed regions a darker color. Doing so divides the rectangle into eight equivalent regions, of which three are shaded the darker color. Consequently, $\dfrac{1}{2} \cdot \dfrac{3}{4} = \dfrac{3}{8}$.

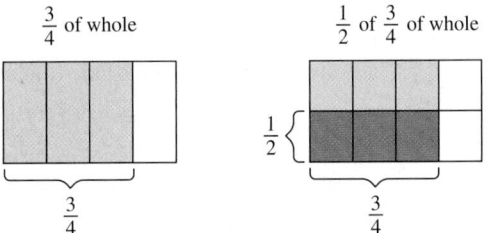

Figure 7.27 Using a rectangular area to compute $\dfrac{1}{2} \cdot \dfrac{3}{4}$

This example actually demonstrates a numerical procedure for finding the product of any two fractions. Specifically, note that the denominator of the product is equal to the product of the denominators in the factors. This should make sense because we subdivided the rectangle into four columns, each containing two pieces, making it necessary to multiply the denominators, $2 \cdot 4 = 8$, to correctly describe the size of the new pieces relative to the whole. Similarly, the numerator of the product is equal to the product of the numerators in the factors. This should also make sense because we further subdivided the shaded region into three columns, each containing one piece. To count the pieces in the array, we multiply the number of rows by the number of columns, or $1 \cdot 3 = 3$. The next definition generalizes these ideas.

Definition of Rational Number Multiplication

If $\dfrac{a}{b}$ and $\dfrac{c}{d}$ are rational numbers, then $\dfrac{a}{b} \cdot \dfrac{c}{d} = \dfrac{a \cdot c}{b \cdot d}$.

When we multiply two fractions, the commutative and associative properties for integer multiplication allow us to rearrange the factors in the numerator and denominator of the product so that they simplify appropriately. Because we know that this will happen, we can simplify the common factors prior to computing the product. This not only makes the computation easier, but it also reduces the product to its simplest form.

Example 7.20   Multiplying Rational Numbers

Multiply.

a. $\dfrac{12}{25} \cdot \dfrac{15}{16}$ b. $\dfrac{33}{16} \cdot \dfrac{-20}{22}$

Solution In the first problem, we apply the definition and then simplify. In the second, we simplify and then multiply.

a. $\dfrac{12}{25} \cdot \dfrac{15}{16} = \dfrac{12 \cdot 15}{25 \cdot 16} = \dfrac{180}{400} = \dfrac{9}{20}.$

b. $\dfrac{33}{16} \cdot \dfrac{-20}{22} = \dfrac{(3 \cdot \cancel{11}) \cdot (\cancel{4} \cdot (-5))}{(\cancel{4} \cdot 4) \cdot (2 \cdot \cancel{11})} = \dfrac{3 \cdot (-5)}{4 \cdot 2} = \dfrac{-15}{8}.$

Example 7.21   Applying Rational Number Multiplication

Connections

One-fifth of the food made in Al's bakeshop are cookies. Of the cookies, $\dfrac{2}{7}$ are chocolate chip. What fraction of the food made by Al consists of chocolate chip cookies?

Solution We are taking a part of a part, so we multiply the fractions to get the answer. As a result, $\dfrac{1}{5} \cdot \dfrac{2}{7} = \dfrac{1 \cdot 2}{5 \cdot 7} = \dfrac{2}{35}$ of what Al makes are chocolate chip cookies.

Properties of Rational Number Multiplication

Rational number multiplication satisfies the following properties.

Theorem 7.10   Properties of Rational Number Multiplication

If $\dfrac{a}{b}, \dfrac{c}{d},$ and $\dfrac{e}{f}$ are rational numbers, then the following hold true:

Closure property: $\dfrac{a}{b} \cdot \dfrac{c}{d}$ is a unique rational number.

Commutative property: $\dfrac{a}{b} \cdot \dfrac{c}{d} = \dfrac{c}{d} \cdot \dfrac{a}{b}.$

Associative property: $\dfrac{a}{b} \cdot \left(\dfrac{c}{d} \cdot \dfrac{e}{f} \right) = \left(\dfrac{a}{b} \cdot \dfrac{c}{d} \right) \cdot \dfrac{e}{f}.$

Identity property: $\dfrac{a}{b} \cdot 1 = \dfrac{a}{b} = 1 \cdot \dfrac{a}{b}.$

Zero multiplication property: $\dfrac{a}{b} \cdot 0 = 0 = 0 \cdot \dfrac{a}{b}.$

Distributive property of multiplication over addition: $\dfrac{a}{b} \cdot \left(\dfrac{c}{d} + \dfrac{e}{f} \right) = \left(\dfrac{a}{b} \cdot \dfrac{c}{d} \right) + \left(\dfrac{a}{b} \cdot \dfrac{e}{f} \right)$

Each property in Theorem 7.10 can be verified by using the definition of rational number multiplication and properties for integer multiplication. Also, the Fundamental Law of Fractions is a direct consequence of the multiplicative identity property. Specifically, if $\dfrac{a}{b}$ is a rational number and n is a nonzero number, then $\dfrac{n}{n}$ is equivalent to 1, or what we call a **form of one.** Consequently:

$$\dfrac{a}{b} = \dfrac{a}{b} \cdot 1 \qquad\qquad \text{Multiplicative identity property.}$$

$$= \dfrac{a}{b} \cdot \dfrac{n}{n} \qquad\qquad \text{Substitute } \dfrac{n}{n} \text{ for 1.}$$

$$= \dfrac{an}{bn} \qquad\qquad \text{Rational number multiplication.}$$

In addition to the properties in Theorem 7.10, we can now establish an inverse property for multiplication.

Theorem 7.11   Multiplicative Inverse Property

For any rational number $\dfrac{a}{b} \neq 0$, there exists a unique rational number $\dfrac{b}{a}$ such that $\dfrac{a}{b} \cdot \dfrac{b}{a} = 1 = \dfrac{b}{a} \cdot \dfrac{a}{b}$. The number $\dfrac{b}{a}$ is called the **multiplicative inverse,** or **reciprocal,** of $\dfrac{a}{b}$.

Note that the multiplicative inverse is unique; there is only one for any nonzero rational number. For instance, $\dfrac{5}{3}$ is the only multiplicative inverse of $\dfrac{3}{5}$ because it is the only number for which $\dfrac{3}{5} \cdot \dfrac{5}{3} = 1$ and $\dfrac{5}{3} \cdot \dfrac{3}{5} = 1$.

Example 7.22   Finding Multiplicative Inverses

Find the multiplicative inverse of each number if it exists.

a. $\dfrac{2}{3}$ b. $-3\dfrac{1}{4}$ c. 1   d. 0

Solution

a. The multiplicative inverse of $\dfrac{2}{3}$ is $\dfrac{3}{2}$ because $\dfrac{2}{3} \cdot \dfrac{3}{2} = 1$ and $\dfrac{3}{2} \cdot \dfrac{2}{3} = 1$.

b. Because $-3\dfrac{1}{4} = -\dfrac{13}{4}$, the multiplicative inverse of $-3\dfrac{1}{4}$ is $-\dfrac{4}{13}$.

c. One is its own multiplicative inverse because $1 \cdot 1 = 1$.

d. Zero has no multiplicative inverse because $\dfrac{1}{0}$ is undefined.

Dividing Rational Numbers

Explorations Manual 7.10 Rational number division has its foundation in whole-number division, which we approached in two different ways. In the missing-factor approach, division is defined by using multiplication—an idea that can be extended to the rational numbers.

Definition of Rational Number Division (Missing-Factor Approach):

If $\dfrac{a}{b}$ and $\dfrac{c}{d}$ are rational numbers with $\dfrac{c}{d} \neq 0$, then $\dfrac{a}{b} \div \dfrac{c}{d} = \dfrac{e}{f}$ if and only if $\dfrac{e}{f}$

is the unique rational number such that $\dfrac{a}{b} = \dfrac{c}{d} \cdot \dfrac{e}{f}$.

As before, we compute a quotient by rewriting it as a product in which one of the factors is missing. We then use our knowledge of multiplication to find the missing number. For instance, to compute $\frac{1}{6} \div \frac{1}{2}$, we need to find the rational number $\frac{e}{f}$ such that $\frac{1}{6} = \frac{1}{2} \cdot \frac{e}{f}$. Specifically, $\frac{e}{f} = \frac{1}{3}$ because $\frac{1}{2} \cdot \frac{1}{3} = \frac{1}{6}$. Although the missing-factor approach is a valid approach, it may not be the easiest way to compute rational number quotients for at least two reasons. First, the missing factor can be difficult to find. Second, the approach does not give us an intuitive way to understand the operation.

Another way to compute a rational number quotient is to use repeated subtraction. Recall that the goal in this approach is to determine the number of times we can subtract one number from another. If the fractions have a common denominator, then the only concern is with the numerators. For instance, $\frac{9}{10} \div \frac{3}{10}$ can be interpreted as "How many times can $\frac{3}{10}$ be subtracted from $\frac{9}{10}$?" As Figure 7.28 shows, $\frac{3}{10}$ can be subtracted from $\frac{9}{10}$ three times, so $\frac{9}{10} \div \frac{3}{10} = 3$.

Explorations Manual 7.11

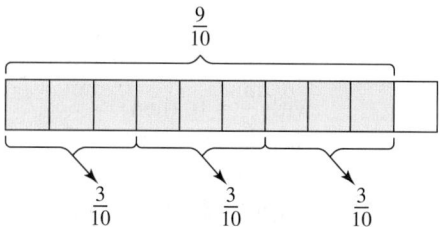

Figure 7.28 Using repeated subtraction to compute $\frac{9}{10} \div \frac{3}{10}$

Example 7.23   Using Repeated Subtraction to Compute a Quotient

Use repeated subtraction to compute each quotient.

a. $\frac{10}{13} \div \frac{5}{13}$ **b.** $\frac{5}{8} \div \frac{3}{8}$

Solution

a. Because $\frac{10}{13} - \frac{5}{13} - \frac{5}{13} = 0$, we can subtract $\frac{5}{13}$ from $\frac{10}{13}$ twice, so $\frac{10}{13} \div \frac{5}{13} = 2$.

b. To compute $\frac{5}{8} \div \frac{3}{8}$, we first subtract $\frac{3}{8}$ from $\frac{5}{8}$ one time: $\frac{5}{8} - \frac{3}{8} = \frac{2}{8}$. To subtract $\frac{3}{8}$ again, we need one more eighth, so we have exactly $\frac{2}{3}$ of what we need. Because the rational numbers allow us to express partial amounts, we can represent this in our difference. In other words, we can subtract $\frac{3}{8}$ from $\frac{5}{8}$ one and two-thirds times:

$$\frac{5}{8} \div \frac{3}{8} = 1\frac{2}{3} = \frac{5}{3}.$$

In the quotients we have computed so far, we can see that when the denominators are the same, the answer is the quotient of the two numerators. This gives us an alternative definition for rational number division.

Definition of Rational Number Division with Common Denominators

If $\dfrac{a}{b}$ and $\dfrac{c}{b}$ are rational numbers with $c \neq 0$, then $\dfrac{a}{b} \div \dfrac{c}{b} = \dfrac{a}{c}$.

Example 7.24 Using the Definition to Divide Rational Numbers

Explorations
Manual
7.11

Compute each quotient.

a. $\dfrac{2}{3} \div \dfrac{3}{4}$ b. $\dfrac{4}{9} \div \dfrac{5}{6}$

Solution We can compute the quotient of any two fractions by getting a common denominator and then applying the definition.

a. $\dfrac{2}{3} \div \dfrac{3}{4} = \dfrac{8}{12} \div \dfrac{9}{12} = \dfrac{8}{9}.$ b. $\dfrac{4}{9} \div \dfrac{5}{6} = \dfrac{8}{18} \div \dfrac{15}{18} = \dfrac{8}{15}.$

If we generalize the procedure in the last example, we get the familiar invert-and-multiply rule often associated with dividing fractions: If $\dfrac{a}{b}$ and $\dfrac{c}{d}$ are rational numbers with $\dfrac{c}{d} \neq 0$, then

$$\dfrac{a}{b} \div \dfrac{c}{d} = \dfrac{ad}{bd} \div \dfrac{bc}{bd} \qquad \text{Find a common denominator}$$

$$= \dfrac{ad}{bc} \qquad \text{Division of fractions with a common denominator}$$

$$= \dfrac{a}{b} \cdot \dfrac{d}{c} \qquad \text{Definition of rational number multiplication}$$

Theorem 7.12 Dividing Rational Numbers with Different Denominators

If $\dfrac{a}{b}$ and $\dfrac{c}{d}$ are rational numbers with $\dfrac{c}{d} \neq 0$, then $\dfrac{a}{b} \div \dfrac{c}{d} = \dfrac{a}{b} \cdot \dfrac{d}{c}.$

Example 7.25 Dividing Rational Numbers

Find each quotient.

a. $\dfrac{21}{17} \div \dfrac{3}{34}$ b. $\dfrac{-7}{3} \div \dfrac{2}{9}$

Solution To compute each quotient, we take the inverse of the second fraction and then multiply. When possible, we simplify the fractions prior to multiplying.

a. $\dfrac{21}{17} \div \dfrac{3}{34} = \dfrac{21}{17} \cdot \dfrac{34}{3} = \dfrac{7 \cdot 3}{1 \cdot 17} \cdot \dfrac{2 \cdot 17}{1 \cdot 3} = \dfrac{14}{1} = 14.$

b. $\dfrac{-7}{3} \div \dfrac{2}{9} = \dfrac{-7}{3} \cdot \dfrac{9}{2} = \dfrac{-7}{1 \cdot 3} \cdot \dfrac{3 \cdot 3}{2} = -\dfrac{21}{2}.$

Explorations
Manual
7.12

Explorations
Manual
7.13

Explorations
Manual
7.14

Theorem 7.12 has several important consequences. First, it indicates that the rational numbers are primarily closed under division. In other words, as long as the divisor is not 0, the quotient of any two rational numbers is always another rational number. Second, the theorem indicates that there is never a remainder with rational number division because we can always use fractions to describe partial amounts in the quotient. Third, Theorem 7.12 can be used to build a connection between integer division and the rational numbers. Specifically, if a and b are two integers with $b \neq 0$, then

$$a \div b = \frac{a}{1} \div \frac{b}{1} = \frac{a}{1} \cdot \frac{1}{b} = \frac{a}{b}$$

In other words, the fraction bar is another way to represent division, so the quotient of any two numbers can be represented in fractional form. For instance, we can rewrite $\frac{1}{2} \div \frac{2}{3}$ as $\dfrac{\frac{1}{2}}{\frac{2}{3}}$, which is an expression that we often call a **complex fraction.**

Check Your Understanding 7.3–A

1. Compute each product.

   a. $\dfrac{3}{4} \cdot \dfrac{5}{8}$
   b. $\dfrac{13}{20} \cdot \dfrac{16}{25}$
   c. $\dfrac{-35}{18} \cdot \dfrac{-27}{40}$

2. State the property used in each equation.

   a. $\dfrac{3}{7} \cdot 1 = \dfrac{3}{7}$
   b. $\dfrac{-5}{3} \cdot 0 = 0$
   c. $\dfrac{-4}{3} \cdot \dfrac{3}{-4} = 1$

   d. $\dfrac{-3}{5} \cdot \dfrac{-7}{4} \in \mathbb{Q}$
   e. $\dfrac{3}{4} \cdot \dfrac{2}{5} = \dfrac{2}{5} \cdot \dfrac{3}{4}$
   f. $\dfrac{3}{4} \cdot \left(\dfrac{4}{3} + \dfrac{7}{2} \right) = \dfrac{3}{4} \cdot \dfrac{4}{3} + \dfrac{3}{4} \cdot \dfrac{7}{2}$

3. Find the multiplicative inverse of each number if it exists.

   a. $\dfrac{3}{4}$
   b. $-\dfrac{11}{8}$
   c. -2
   d. $-2\dfrac{3}{4}$

4. Compute each quotient.

   a. $\dfrac{3}{7} \div \dfrac{2}{9}$
   b. $\dfrac{6}{11} \div \dfrac{9}{13}$
   c. $\dfrac{-14}{5} \div \dfrac{21}{10}$

Talk About It How are rational number multiplication and division similar to their whole-number counterparts? How are they different?

Representation

Activity 7.5
Use the length or area model to compute each product or quotient.

   a. $\dfrac{1}{3} \cdot \dfrac{1}{4}$
   b. $\dfrac{5}{6} \cdot \dfrac{3}{4}$
   c. $\dfrac{5}{12} \cdot \dfrac{2}{3}$

   d. $\dfrac{3}{4} \div \dfrac{1}{4}$
   e. $\dfrac{5}{12} \div \dfrac{2}{12}$
   f. $\dfrac{7}{8} \div \dfrac{1}{4}$

Rational Number Bases and Exponents

With the rational numbers, we can develop several new ideas with exponents by extending our previous definitions of exponents to include rational number bases.

Definitions of Exponents

If a is a rational number and n is a positive integer, then

1. $a^n = \underbrace{a \cdot a \cdot \ldots \cdot a}_{n \text{ factors}}$, and
2. $a^0 = 1$, with $a \neq 0$.

We can also extend exponential values to include negative integers.

Example 7.26   **Finding Negative Exponents**

Problem Solving   Find values for 10^{-1}, 10^{-2}, and 10^{-3}, and then use them to derive a formula for 10^{-n} where n is a positive integer.

Solution

Understanding the Problem. The problem is to compute three different exponential values and then use them to derive a formula for 10^{-n} where n is a positive integer.

Devising a Plan. In Chapter 4, we established that $a^0 = 1$ by generalizing the following pattern:

$$10^3 = 10 \cdot 10 \cdot 10 = 1{,}000$$
$$10^2 = 10 \cdot 10 = 100$$
$$10^1 = 10$$
$$10^0 = 1.$$

If we can extend this pattern to the negative integers, not only will we compute the needed values, but we may also *find a pattern* that will allow us to derive a formula for 10^{-n}.

Carrying Out the Plan. The previous pattern indicates that as the exponent decreases by one, we divide the previous number by 10. Continuing this pattern onto the negative exponents:

$$10^{-1} = 1 \div 10 = \frac{1}{10} = \frac{1}{10^1}$$

$$10^{-2} = \frac{1}{10} \div 10 = \frac{1}{10} \cdot \frac{1}{10} = \frac{1}{10^2}$$

$$10^{-3} = \frac{1}{100} \div 10 = \frac{1}{100} \cdot \frac{1}{10} = \frac{1}{1000} = \frac{1}{10^3}$$

If we compare the exponential expression on the left to the fractional expression on the right, then: $10^{-n} = \dfrac{1}{10^n}$ for any positive integer n.

Looking Back. Notice that $10^{-n} = \dfrac{1}{10^n} = \left(\dfrac{1}{10}\right)^n$. In other words, we can compute a negative power of 10 by taking $\dfrac{1}{10}$ and raising it to a positive power.

In the last example, 10 was randomly chosen, so it can be replaced with any nonzero rational number a. Consequently, $a^{-n} = \dfrac{1}{a^n}$.

Definition of a Negative Exponent

If a is a nonzero rational number and n is a positive integer, then $a^{-n} = \dfrac{1}{a^n}$.

Example 7.27 Simplifying Negative Exponents

Write each expression with a positive exponent.

a. 3^{-3} **b.** $x^{-7}, x \neq 0$ **c.** $\left(\dfrac{1}{2}\right)^{-3}$

Solution

a. $3^{-3} = \dfrac{1}{3^3}$.

b. $x^{-7} = \dfrac{1}{x^7}, x \neq 0$.

c. In this case, find the multiplicative inverse and then apply the exponent:

$$\left(\frac{1}{2}\right)^{-3} = \left(\frac{2}{1}\right)^3 = 2^3 = 8$$

Each of the properties that held true for integer bases also holds true for rational number bases. Three new properties can be added.

Theorem 7.13 Properties of Exponents

If a and b are nonzero rational numbers, and m and n are positive integers, then

1. $a^m \cdot a^n = a^{m+n}$.   **2.** $(a^m)^n = a^{m \cdot n}$.   **3.** $a^m \cdot b^m = (a \cdot b)^m$.

4. $\dfrac{a^m}{a^n} = a^{m-n}$.   **5.** $\left(\dfrac{a}{b}\right)^m = \dfrac{a^m}{b^m}$.   **6.** $\left(\dfrac{a}{b}\right)^{-m} = \left(\dfrac{b}{a}\right)^m$.

Property 4 of Theorem 7.13 states that when dividing two exponents with the same base, we can simplify them by raising the base to the difference of the exponents. Property 5 states that if we raise a fraction to an exponent, we can apply the exponent to both the numerator and the denominator. And property 6 states that we can negate the exponent of a rational number by taking the reciprocal of the rational number.

Example 7.28 Simplifying Exponents

Simplify each expression. Leave no negative exponents.

a. $\dfrac{4^2}{4^7}$ **b.** $\dfrac{18^2}{9}$ **c.** $\left(\dfrac{6}{4^{-2}}\right)^{-2}$

Solution

a. $\dfrac{4^2}{4^7} = 4^{2-7} = 4^{-5} = \dfrac{1}{4^5}$.

b. $\dfrac{18^2}{9} = \dfrac{18^2}{3^2} = \left(\dfrac{18}{3}\right)^2 = 6^2 = 36.$

c. $\left(\dfrac{6}{4^{-2}}\right)^{-2} = \left(\dfrac{4^{-2}}{6}\right)^2 = \left(\dfrac{1}{6 \cdot 4^2}\right)^2 = \dfrac{1}{6^2 \cdot 4^4}.$

We now prove part 4 of Theorem 7.13. Parts 5 and 6 follow directly from the definition of exponents and properties of rational number multiplication and division. They are left to the Learning Assessment.

Example 7.29 | **Proving $\dfrac{a^m}{a^n} = a^{m-n}$**

Reasoning

If a is any nonzero rational number, and m and n are any integers, show that $\dfrac{a^m}{a^n} = a^{m-n}.$

Solution Let a be a nonzero rational number, and let m and n be any integers. Then

$$\dfrac{a^m}{a^n} = \dfrac{a^m \cdot 1}{1 \cdot a^n} \qquad \text{Multiplicative identity property}$$

$$= \dfrac{a^m}{1} \cdot \dfrac{1}{a^n} \qquad \text{Rational number multiplication}$$

$$= a^m \cdot a^{-n} \qquad \text{Definition of a negative exponent}$$

$$= a^{m+(-n)} \qquad \text{Multiplication of exponents}$$

$$= a^{m-n} \qquad \text{Adding the opposite}$$

Multiplying and Dividing Mixed Numbers

The products and quotients of mixed numbers can be computed in two ways. One is to work with the numbers in mixed number form. For instance, to multiply $2\dfrac{1}{2} \cdot 3\dfrac{1}{4}$, we use the distributive property of multiplication over addition:

$$2\dfrac{1}{2} \cdot 3\dfrac{1}{4} = 2\dfrac{1}{2} \cdot \left(3 + \dfrac{1}{4}\right)$$

$$= \left(2\dfrac{1}{2} \cdot 3\right) + \left(2\dfrac{1}{2} \cdot \dfrac{1}{4}\right)$$

$$= \left(2 + \dfrac{1}{2}\right) \cdot 3 + \left(2 + \dfrac{1}{2}\right) \cdot \dfrac{1}{4}$$

$$= (2 \cdot 3) + \left(\dfrac{1}{2} \cdot 3\right) + \left(2 \cdot \dfrac{1}{4}\right) + \left(\dfrac{1}{2} \cdot \dfrac{1}{4}\right)$$

$$= 6 + \dfrac{3}{2} + \dfrac{1}{2} + \dfrac{1}{8}$$

$$= 8\dfrac{1}{8}$$

Another way is to convert the mixed numbers to improper fractions, perform the computation, and then convert the answer back to a mixed number.

Example 7.30 **Multiplying and Dividing Mixed Numbers**

Compute the following problems.

a. $3\frac{9}{10} \cdot 4\frac{1}{3}$

b. $2\frac{1}{3} \div 3\frac{3}{8}$

Solution

a. $3\frac{9}{10} \cdot 4\frac{1}{3} = \frac{39}{10} \cdot \frac{13}{3} = \frac{13 \cdot 3}{10} \cdot \frac{13}{1 \cdot 3} = \frac{169}{10} = 16\frac{9}{10}.$

b. $2\frac{1}{3} \div 3\frac{3}{8} = \frac{7}{3} \div \frac{27}{8} = \frac{7}{3} \cdot \frac{8}{27} = \frac{56}{81}.$

Example 7.31 **Prescribing Medicine**

Application

Two common units for measuring medicines are grains (gr) and milligrams (mg). Suppose a doctor prescribes 100 gr of a medicine. If $8\frac{1}{3}$ gr are equal to 500 mg, how many milligrams did the doctor prescribe?

Solution Because $8\frac{1}{3}$ grains are equal to 500 mg, then 1 gr is equal to $500 \div 8\frac{1}{3} = 500 \cdot \frac{3}{25} = 60$ mg. So the doctor prescribed $60 \cdot 100 = 6{,}000$ mg of the medicine.

Some situations require computing a rational number product or quotient mentally, and it is often enough just to estimate the answer. The next two examples demonstrate how to use the whole-number techniques to make mental computations or estimations with rational numbers.

Example 7.32 **Computing Products Mentally**

Compute the following mentally.

a. $\frac{5}{6} \cdot 18$ **b.** $\frac{3}{4} \cdot \left(20 \cdot \frac{1}{3}\right)$ **c.** $21 \cdot 3\frac{1}{3}$

Solution

a. To compute $\frac{5}{6} \cdot 18$, separate $\frac{5}{6}$, and use the associative property to compute the product as $\frac{5}{6} \cdot 18 = \left(5 \cdot \frac{1}{6}\right) \cdot 18 = 5 \cdot \left(\frac{1}{6} \cdot 18\right) = 5 \cdot 3 = 15.$

b. Think $\frac{3}{4} \cdot \left(20 \cdot \frac{1}{3}\right) = \left(\frac{3}{4} \cdot 20\right) \cdot \frac{1}{3} = (3 \cdot 5) \cdot \frac{1}{3} = 5 \cdot 1 = 5.$

c. Use the distributive property of multiplication over addition, and think

$$21 \cdot 3\frac{1}{3} = 21 \cdot \left(3 + \frac{1}{3}\right) = (21 \cdot 3) + \left(21 \cdot \frac{1}{3}\right) = 63 + 7 = 70.$$

Example 7.33 | **Estimating Products and Quotients**

Estimate the following:

a. $\dfrac{11}{4} \div \dfrac{3}{8}$ **b.** $2\dfrac{1}{4} \cdot 3\dfrac{3}{8}$

Solution

a. We can estimate $\dfrac{11}{4} \div \dfrac{3}{8}$ by rounding each fraction to a benchmark number, like a whole number or a common fraction. Hence, $\dfrac{11}{4} \div \dfrac{3}{8} \approx 3 \div \dfrac{1}{2} = 6$.

b. We can also estimate by truncating the numbers: $2\dfrac{1}{4} \cdot 3\dfrac{3}{8} \approx 2 \cdot 3 = 6$.

Check Your Understanding 7.3-B

1. Simplify each expression. Leave no negative exponent.

    a. $3^6 \cdot 3^{-9}$ **b.** $x^{-3} \cdot x^{-6} \ x \neq 0$ **c.** $\dfrac{5^3}{5^6}$ **d.** $\dfrac{y^5}{y^3} \ y \neq 0$

2. Compute each problem.

    a. $2\dfrac{3}{8} \cdot 6\dfrac{2}{3}$ **b.** $4\dfrac{2}{5} \div 1\dfrac{5}{7}$

3. Compute each problem mentally.

    a. $\dfrac{4}{5} \cdot \left(15 \cdot \dfrac{1}{4}\right)$ **b.** $\dfrac{3}{7} \cdot 14$ **c.** $16 \cdot 2\dfrac{1}{8}$

4. Estimate the following.

    a. $\dfrac{7}{4} \div \dfrac{5}{8}$ **b.** $3\dfrac{1}{9} \cdot \dfrac{7}{8}$ **c.** $2\dfrac{3}{8} \cdot 2\dfrac{1}{11}$

Talk About It Describe the method you would use to divide two mixed numbers in mixed number form.

Reasoning

Activity 7.6

Find a partner, and write a proof that uses the definition of an exponent to show that each property is true if $\dfrac{a}{b}$ is any rational number and m and n are positive integers.

a. $\left(\dfrac{a}{b}\right)^m \cdot \left(\dfrac{a}{b}\right)^n = \left(\dfrac{a}{b}\right)^{m+n}$ **b.** $\left[\left(\dfrac{a}{b}\right)^m\right]^n = \left(\dfrac{a}{b}\right)^{m \cdot n}$

SECTION 7.3 Learning Assessment

■ Understanding the Concepts

1. Why does the repeated-addition approach to multiplication work only in some cases with the rational numbers?

2. Why are the numerators *and* the denominators multiplied when multiplying two fractions?

3. Why is it possible to reduce common factors between numerators and denominators prior to computing the product of two rational numbers?

4. How is the Fundamental Law of Fractions connected to the multiplicative identity property?

5. Does every rational number have a multiplicative inverse? Explain why or why not.

6. **a.** Will rational number division ever have remainders? Why or why not?

   b. What does your answer to part a imply about the set of rational numbers under the operation of division?

7. How does the repeated-subtraction approach to division lead to the invert-and-multiply rule that is commonly used to divide two fractions?

8. Explain the connection between integer division and the rational numbers.

9. If a is a nonzero number, why does $a^{-n} = \dfrac{1}{a^n}$?

■ Representing the Mathematics

10. Give a verbal interpretation of each product.

    a. $\dfrac{1}{2} \cdot \dfrac{1}{4}$ **b.** $\dfrac{2}{4} \cdot \dfrac{1}{5}$ **c.** $\dfrac{3}{7} \cdot \dfrac{7}{6}$ **d.** $\dfrac{4}{5} \cdot 4$

11. Represent each statement numerically.

    a. One-third of one-fifth

    b. Three-fourths of seven-halves

    c. Two-thirds of six

    d. Nine-sevenths of ten-eighths

12. Use fraction bars to represent and solve each product.

    a. $3 \cdot \dfrac{1}{6}$ **b.** $\dfrac{1}{3} \cdot \dfrac{1}{4}$ **c.** $\dfrac{3}{4} \cdot \dfrac{1}{2}$

13. Use the area model to represent and solve each product.

    a. $\dfrac{1}{5} \cdot \dfrac{1}{2}$ **b.** $\dfrac{3}{4} \cdot \dfrac{1}{3}$ **c.** $\dfrac{2}{3} \cdot \dfrac{4}{5}$

14. If each large rectangle represents the whole, state the factors and the product that the shaded regions represent.

    a. 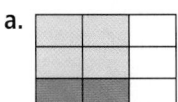 **b.**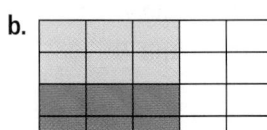

    c.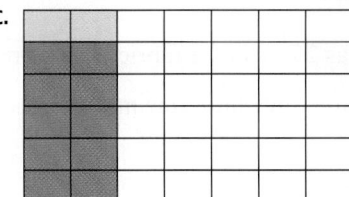

15. Use fraction bars and the repeated-subtraction approach to represent and solve each quotient.

    a. $\dfrac{6}{8} \div \dfrac{2}{8}$ **b.** $\dfrac{9}{10} \div \dfrac{3}{10}$ **c.** $\dfrac{5}{6} \div \dfrac{3}{6}$

16. Represent each integer quotient as a rational number.

    a. $4 \div 6$ **b.** $7 \div 3$ **c.** $-18 \div 8$ **d.** $-15 \div -9$

■ Exercises

17. Multiply.

    a. $\dfrac{1}{3} \cdot \dfrac{7}{5}$ **b.** $\dfrac{10}{15} \cdot \dfrac{3}{2}$

    c. $\dfrac{4}{10} \cdot \dfrac{-4}{5}$ **d.** $\dfrac{-12}{25} \cdot \dfrac{-15}{9}$

18. Multiply.

    a. $\dfrac{9}{21} \cdot \dfrac{14}{15}$ **b.** $\dfrac{14}{15} \cdot \dfrac{30}{21}$

    c. $\dfrac{15}{32} \cdot \dfrac{-12}{23}$ **d.** $\dfrac{-56}{35} \cdot \dfrac{-25}{64}$

19. State the property that has been used in each equation.

    a. $\dfrac{1}{3} \cdot \dfrac{2}{7} = \dfrac{2}{7} \cdot \dfrac{1}{3}.$ **b.** $\dfrac{2}{3} \cdot \dfrac{3}{2} = 1.$

    c. $0 \cdot \dfrac{2}{5} = 0.$ **d.** $\dfrac{1}{5} \cdot \dfrac{3}{4} \in \mathbb{Q}.$

    e. $1 \cdot \left(-\dfrac{9}{5}\right) = \left(-\dfrac{9}{5}\right).$

    f. $\dfrac{2}{5} \cdot \left(\dfrac{4}{3} + \dfrac{3}{2}\right) = \dfrac{2}{5} \cdot \dfrac{4}{3} + \dfrac{2}{5} \cdot \dfrac{3}{2}.$

20. Find the multiplicative inverse of each number if it exists.

    a. $\dfrac{1}{5}$ **b.** $-\dfrac{13}{4}$ **c.** 5 **d.** $-1\dfrac{1}{8}$ **e.** $\dfrac{0}{6}$

21. Use repeated subtraction to compute each quotient.

    a. $\dfrac{4}{7} \div \dfrac{2}{7}$ **b.** $\dfrac{14}{5} \div \dfrac{2}{5}$ **c.** $\dfrac{5}{7} \div \dfrac{3}{7}$

22. Divide.

    a. $\dfrac{3}{4} \div \dfrac{5}{6}$ **b.** $\left(-\dfrac{2}{5}\right) \div \dfrac{6}{11}$

    c. $\dfrac{10}{3} \div \dfrac{5}{9}$ **d.** $\dfrac{15}{13} \div \dfrac{10}{3}$

23. Divide.

    a. $\dfrac{30}{24} \div \dfrac{15}{4}$ **b.** $\dfrac{18}{13} \div \dfrac{21}{8}$

    c. $\dfrac{16}{15} \div \dfrac{24}{25}$ **d.** $\left(-\dfrac{26}{21}\right) \div \dfrac{39}{11}$

24. Compute.

    a. $\dfrac{2}{3} \cdot \dfrac{4}{9} \div \dfrac{6}{15}$ **b.** $\dfrac{3}{4} \cdot \dfrac{1}{2} + \dfrac{5}{8}$

    c. $\dfrac{6}{7} - \dfrac{2}{3} \cdot \dfrac{5}{7} + \dfrac{3}{4}$

25. Simplify each complex fraction.

a. $\dfrac{\frac{7}{16}}{\frac{21}{24}}$

b. $\dfrac{\left(-\frac{24}{35}\right)}{\left(-\frac{56}{25}\right)}$

26. Simplify each expression. Write your answer without a negative exponent.

a. 6^{-5}

b. $\left(\dfrac{2}{5}\right)^{-3}$

c. $2^7 \cdot 2^{-4}$

d. $\dfrac{7^7}{7^{-5}}$

e. $\dfrac{18^{-4}}{9^{-4}}$

f. $\left(\dfrac{5^{-4}}{2^{-2}}\right)^3$

27. Simplify each expression. Write your answer without a negative exponent.

a. $\left(\dfrac{3}{4}\right)^{-4}$

b. $\dfrac{4^{-3}}{4^{-6}}$

c. 8^{-7}

d. $3^{-6} \cdot 3^2$

e. $\left(\dfrac{4^5}{7^{-3}}\right)^{-3}$

f. $\dfrac{16^{-5}}{4^{-5}}$

28. Which rational number in each pair represents the larger value?

a. $\left(\dfrac{1}{2}\right)^2$ or $\left(\dfrac{1}{2}\right)^{-2}$

b. $\left(\dfrac{1}{3}\right)^3$ or $\left(\dfrac{1}{2}\right)^3$

c. $\left(\dfrac{1}{2}\right)^6$ or $\left(\dfrac{1}{2}\right)^5$

d. $\left(\dfrac{3}{4}\right)^4$ or $\left(\dfrac{2}{3}\right)^4$

29. Compute.

a. $1\dfrac{4}{5} \cdot 1\dfrac{3}{5}$

b. $2\dfrac{2}{3} \div 2\dfrac{1}{6}$

c. $\left(-4\dfrac{1}{6}\right)\cdot\left(-2\dfrac{2}{5}\right)$

d. $3\dfrac{4}{7} \div \left(-1\dfrac{3}{7}\right)$

30. Compute each product mentally.

a. $\dfrac{1}{8} \cdot 16$

b. $12 \cdot 3\dfrac{1}{6}$

c. $\dfrac{2}{7}\cdot\left(14\cdot\dfrac{1}{2}\right)$

d. $5\dfrac{1}{7} \cdot 21$

31. Estimate each product or quotient.

a. $\left(-4\dfrac{7}{8}\right)\cdot\dfrac{4}{9}$

b. $\dfrac{13}{6} \div \dfrac{10}{11}$

c. $8\dfrac{7}{8} \cdot 7\dfrac{10}{11}$

d. $\left(-7\dfrac{1}{5}\right)\cdot\left(-7\dfrac{9}{10}\right)$

32. Without making the computation, determine whether each product or quotient is between 0 and $\dfrac{1}{2}$, $\dfrac{1}{2}$ and 1, or is greater than 1.

a. $\dfrac{1}{4} \cdot \dfrac{1}{9}$

b. $\dfrac{4}{7} \div \dfrac{8}{9}$

c. $\dfrac{8}{11} \div \dfrac{3}{10}$

d. $\dfrac{5}{9} \cdot \dfrac{10}{9}$

33. Use a fraction calculator to compute.

a. $5\dfrac{5}{7} \cdot \dfrac{5}{12} - 6\dfrac{2}{9} \div \dfrac{5}{18}$

b. $4\dfrac{3}{7} - \dfrac{7}{6} \cdot 3\dfrac{2}{7} + \dfrac{13}{7}$

c. $\dfrac{20}{6} \cdot \left[\left(\dfrac{16}{5} + 2\dfrac{1}{7}\right) \div 6\dfrac{1}{3}\right]$

■ Problems and Applications

34. What is the value of $\dfrac{1}{2} \cdot \dfrac{2}{3} \cdot \dfrac{3}{4} \cdot \ldots \cdot \dfrac{98}{99} \cdot \dfrac{99}{100}$?

35. Fill in the blanks using the digits 2, 3, 6, and 7 only once to make the:

a. Greatest product: $\dfrac{\square}{\square} \cdot \dfrac{\square}{\square}$.

b. Greatest quotient: $\dfrac{\square}{\square} \div \dfrac{\square}{\square}$.

36. Fill in the blanks using the digits 3, 4, 8, and 9 only once to make the:

a. Smallest product: $\dfrac{\square}{\square} \cdot \dfrac{\square}{\square}$.

b. Smallest quotient: $\dfrac{\square}{\square} \div \dfrac{\square}{\square}$.

37. Consider the following number line. Find the product of each pair of points.

a. A and B

b. C and D

c. A and D

d. B and C

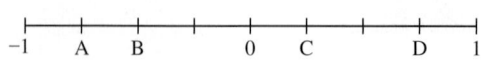

38. A fabric store has $25\dfrac{1}{2}$ yd of a fabric. The store sells $\dfrac{3}{4}$ of the material in one day. How much of the material is left?

39. How many pieces of rope $5\dfrac{1}{3}$ ft long can be cut from a length of rope that is $35\dfrac{3}{4}$ ft long?

40. A nail that is $2\frac{1}{2}$ in. long is driven through three boards, each of which is $\frac{5}{8}$ in. thick. How much of the nail is sticking out the opposite side?

41. A doctor prescribes 125 gr of a medicine that comes in pills of 250 mg. If 250 mg is equal to $4\frac{1}{6}$ gr, how many pills should the patient be given?

42. The formula for converting degrees Fahrenheit to degrees Celsius is $C = \frac{5}{9}(F - 32)$. What is the temperature in degrees Celsius for each temperature in degrees Fahrenheit?

    a. 45° F   **b.** 78° F   **c.** 96° F

    d. 32° F   **e.** −13° F

43. Jonathan is expecting a raise that is $\frac{1}{7}$ of his current salary of $42,000. If he gets the raise he is expecting, how much will his salary be next year?

44. How many people can be fed with $10\frac{1}{2}$ pizzas if each person is to get $\frac{3}{8}$ of a pizza?

45. Ted gives half of his money to Jessie. Jessie gives a third of what she is given to Cassie, and Cassie gives a fourth of what she is given to Alex. If Alex is given $6, how much did Ted originally have?

46. A bag of flour contains 32 cups (c) of flour. A recipe for a batch of cookies calls for $2\frac{1}{3}$ c of flour. How many batches of cookies can be made from one bag of flour?

47. A car consumes fuel at a rate of $3\frac{1}{2}$ gal/hr. If the driver has 10 gal of fuel, does he have enough to make a 3-hr trip?

48. Dan is planning to build a chest of drawers with a base that is 5 in. tall and a top that is $1\frac{1}{2}$ in. tall. It will have five drawers, and each drawer is $8\frac{3}{4}$ in. deep with a gap of $1\frac{1}{4}$ in. between drawers. How tall is the chest?

■ Mathematical Reasoning and Proof

49. **a.** If 5 is multiplied by a rational number that is between 0 and 1, will the result be bigger or smaller than 5? Explain.

    b. If 5 is divided by a rational number that is between 0 and 1, will the result be bigger or smaller than 5? Explain.

50. **a.** Under what conditions will the product of two positive fractions always be smaller than the original two factors?

    b. Under what conditions will the product of two positive fractions always be bigger than the original two factors?

51. If a and b are rational numbers, where $0 < a < b < 1$, determine whether each statement is true, false, or cannot be determined.

    a. $a \cdot b < b$.   **b.** $a \cdot b > a$.

    c. $b \div a < b$.   **d.** $b \div a < a$.

52. If a and b are rational numbers, where $0 < a < 1$ and $b > 1$, determine whether each statement is true, false, or cannot be determined.

    a. $a \cdot b < a$.

    b. $a \cdot b > 1$.

    c. $b \div a < 1$.

    d. $b \div a < b$.

53. Does the following picture imply that $2 \cdot \frac{3}{4} = \frac{6}{8}$? If so, explain why. If not, explain how to interpret the picture correctly.

    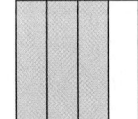 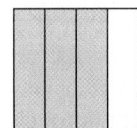

54. Consider the following sequence of products. As n gets larger, what happens to the value of the product?

    $$\frac{1}{2} \cdot \frac{2}{3}$$

    $$\frac{1}{2} \cdot \frac{2}{3} \cdot \frac{3}{4}$$

    $$\frac{1}{2} \cdot \frac{2}{3} \cdot \frac{3}{4} \cdot \frac{4}{5}$$

    $$\frac{1}{2} \cdot \frac{2}{3} \cdot \frac{3}{4} \cdot \frac{4}{5} \cdots \frac{n-1}{n}$$

55. Provide a numerical example to demonstrate that rational number division does not satisfy each of the following properties.

    a. Commutative

    b. Associative

    c. Multiplicative identity

56. Which is larger, the reciprocal of $\left(\frac{1}{2}\right)^2$ or the reciprocal of $\left(\frac{1}{2}\right)^{-2}$? Explain.

57. Use the distributive property of multiplication over addition to find $2\frac{2}{3} \cdot 3\frac{3}{5}$. Justify each step you take.

58. Use the properties and theorems from this section to verify that $\frac{x}{a} \cdot a = x$, where x is a rational number and a is a nonzero rational number.

59. If $\frac{a}{b}$ is a rational number and m is a positive integer, prove that $\left(\frac{a}{b}\right)^m = \frac{a^m}{b^m}$.

60. If $\frac{a}{b}$ is a rational number and m is a positive integer, prove that $\left(\frac{a}{b}\right)^{-m} = \left(\frac{b}{a}\right)^m$.

■ Mathematical Communication

61. Write a story problem for $\frac{2}{3} \cdot \frac{1}{4}$. What language did you use in the problem to indicate that a product was to be taken?

62. **a.** Write a story problem for $3 \div \frac{1}{4}$ that uses repeated subtraction.

   b. Write another story problem for $3 \div \frac{1}{4}$ that uses the missing-factor approach.

   c. How did your language change to indicate the different approaches?

63. Look up the word "inverse" in the dictionary. How are its nonmathematical meanings similar to the mathematical one presented in this section?

64. With a small group of your peers, determine how to use the multiplicative inverse of $\frac{a}{b}$ to solve equations of the form $\frac{a}{b} \cdot x = \frac{c}{d}$. Use your method to solve each equation.

   a. $\frac{2}{3} \cdot x = \frac{5}{6}$.   **b.** $\frac{-3}{5} \cdot x = \frac{20}{9}$.

   c. $3\frac{1}{4} \cdot x = 2\frac{2}{3}$.

65. Mark computed $\frac{3}{4} \div \frac{3}{5}$ and got an answer of $\frac{9}{20}$. What did he do wrong?

66. Sarah is having trouble multiplying $2\frac{1}{3} \cdot 4\frac{4}{5}$ because she does not know where to start. Describe at least two ways that you can help her get started.

■ Building Connections

67. Is the partitioning approach to whole-number division related to fractions? How so?

68. Which do you think is more difficult for children to compute, rational number addition or rational number multiplication? Explain your thinking.

69. Write a short paragraph explaining how the conceptual idea behind rational number multiplication is connected to the actual procedure for computing rational number products.

70. Examine a set of elementary mathematics curriculum materials for how rational number division is introduced. Do the materials connect it to repeated subtraction? Why do you think this is?

71. Draw a diagram that shows how the mathematical structure of the exponents started with the whole numbers and was extended by the integers and rational numbers.

Historical Note

Fractions date back to the ancient Egyptians, who used them extensively in their computations of division. They had a peculiar tendency of working only with unit fractions, except for the fraction $\frac{2}{3}$. All other fractional amounts had to be expressed as the sum of unit fractions, which naturally made computation with fractions very difficult. The Egyptians wrote fractions with an elongated oval placed over the integer representing the denominator. Thus, $\frac{1}{5}$ was written as ⟨figure⟩, and $\frac{1}{12}$ was written as ⟨figure⟩. Ancient Hindus, such as Brahmagupta (circa 600 C.E.) and Bhaskara (circa 1150 C.E.), wrote fractions by placing the numerator over the denominator. For instance, they wrote $\frac{2}{3}$ as $\begin{smallmatrix}2\\3\end{smallmatrix}$. It is believed that the Arabs borrowed this notation but added to it by placing a horizontal bar between the numbers. This notation was followed by Fibonacci in his book *Liber Abaci* and became common in Europe by the sixteenth century.

Opening Problem Answer

Divide the amount of time Amy needs to complete her project into 5 equal time segments, of which $15\frac{2}{3}$ hr represents three of them. This means that one time interval is $\frac{1}{3} \cdot 15\frac{2}{3} = 5\frac{2}{9}$ hr. Because she has two left to go, she has $2 \cdot 5\frac{2}{9} = 10\frac{4}{9}$ hr before she completes her project.

Juriah Mosin/Shutterstock.com

Chapter 7 REVIEW

Summary of Key Concepts from Chapter 7

Section 7.1 Fractions and the Set of Rational Numbers

- The set of **rational numbers** is $\mathbb{Q} = \left\{ \frac{a}{b} \middle| a \text{ and } b \text{ are integers and } b \neq 0 \right\}$.

- A fraction describes a part-to-whole relationship. The **denominator** tells how many equivalent pieces the whole has been divided into, and the **numerator** gives the number of pieces being counted.

- If $\frac{a}{b}$ is any fraction, then $\frac{a}{b}$ is **proper** if $0 \leq |a| < |b|$ and **improper** if $|a| \geq |b| > 0$.

- Any number with an integer and fractional part is a **mixed number.**

- Two fractions are **equivalent** if they have the same numerical value.

- **Fundamental Law of Fractions:** If $\frac{a}{b}$ is a fraction and n is any nonzero integer, then $\frac{a}{b} = \frac{a \cdot n}{b \cdot n}$.

- A rational number $\frac{a}{b}$ is in **simplest form** if GCF $(a, b) = 1$.

- If a, b, and c are integers with $b > 0$, then $\frac{a}{b} = \frac{c}{b}$ if and only if $a = c$, and $\frac{a}{b} < \frac{c}{b}$ if and only if $a < c$.

- If $\frac{a}{b}$ and $\frac{c}{d}$ are two rational numbers where $b > 0$ and $d > 0$, then $\frac{a}{b} = \frac{c}{d}$ if and only if $ad = bc$, and $\frac{a}{b} < \frac{c}{d}$ if and only if $ad < bc$.

- **Denseness property:** There exists another rational number between any two rational numbers.

Section 7.2 Adding and Subtracting Rational Numbers

- **Rational number addition:** If $\frac{a}{b}$ and $\frac{c}{b}$ are rational numbers, then $\frac{a}{b} + \frac{c}{b} = \frac{a + c}{b}$.

- If $\frac{a}{b}$ and $\frac{c}{d}$ are any two rational numbers, then $\frac{a}{b} + \frac{c}{d} = \frac{ad + bc}{bd}$.

- Rational number addition satisfies the closure, commutative, associative, identity, and additive inverse properties.

- Additive inverses satisfy the following properties:

  1. $-\frac{a}{b} = \frac{-a}{b} = \frac{a}{-b}$.   **2.** $\frac{a}{b} = \frac{-a}{-b} = -\left(-\frac{a}{b}\right)$.

  3. $-\left(\frac{a}{b} + \frac{c}{d}\right) = \frac{-a}{b} + \frac{-c}{-d}$.

- **Rational number subtraction** (missing-addend approach): If $\frac{a}{b}$ and $\frac{c}{d}$ are rational numbers, then $\frac{a}{b} - \frac{c}{d} = \frac{e}{f}$ if and only if $\frac{a}{b} = \frac{c}{d} + \frac{e}{f}$.

- **Rational number subtraction** (take-away approach): If $\frac{a}{b}$ and $\frac{c}{b}$ are rational numbers, then $\frac{a}{b} - \frac{c}{b} = \frac{a-c}{b}$.

- If $\frac{a}{b}$ and $\frac{c}{d}$ are any two rational numbers, then $\frac{a}{b} - \frac{c}{d} = \frac{ad - bc}{bd}$.

- If $\frac{a}{b}$ and $\frac{c}{d}$ are rational numbers, then $\frac{a}{b} - \frac{c}{d} = \frac{a}{b} + \frac{-c}{d}$.

Section 7.3 Multiplying and Dividing Rational Numbers

- **Rational Number Multiplication:** If $\frac{a}{b}$ and $\frac{c}{d}$ are rational numbers, then $\frac{a}{b} \cdot \frac{c}{d} = \frac{a \cdot c}{b \cdot d}$.
- Rational number multiplication satisfies the closure, commutative, associative, identity, zero multiplication, and distributive properties.

- **Multiplicative inverse property:** For any rational number $\frac{a}{b} \neq 0$, there exists a unique rational number $\frac{b}{a}$ such that $\frac{a}{b} \cdot \frac{b}{a} = 1 = \frac{b}{a} \cdot \frac{a}{b}$ that is called the **multiplicative inverse,** or **reciprocal,** of $\frac{a}{b}$.

- **Rational number division** (missing-factor approach): If $\frac{a}{b}$ and $\frac{c}{d}$ are rational numbers with $\frac{c}{d} \neq 0$, then $\frac{a}{b} \div \frac{c}{d} = \frac{e}{f}$ if and only if $\frac{e}{f}$ is the unique rational number such that $\frac{a}{b} = \frac{c}{d} \cdot \frac{e}{f}$.

- **Definition of rational number division with common denominators:** If $\frac{a}{b}$ and $\frac{c}{b}$ are rational numbers with $c \neq 0$, then $\frac{a}{b} \div \frac{c}{b} = \frac{a}{c}$.

- If $\frac{a}{b}$ and $\frac{c}{d}$ are rational numbers with $\frac{c}{d} \neq 0$, then $\frac{a}{b} \div \frac{c}{d} = \frac{a}{b} \cdot \frac{d}{c}$.

- If a is a rational number and n is a positive integer, then $a^n = \underbrace{a \cdot a \cdot \ldots \cdot a}_{n \text{ factors}}$ and $a^0 = 1$, with $a \neq 0$.

- If a is a nonzero rational number and n is a positive integer, then $a^{-n} = \frac{1}{a^n}$.

- If a and b are nonzero rational numbers and m and n are positive integers, then:

  1. $a^m \cdot a^n = a^{m+n}$.

  2. $(a^m)^n = a^{m \cdot n}$.

  3. $a^m \cdot b^m = (a \cdot b)^m$.

  4. $\frac{a^m}{a^n} = a^{m-n}$.

  5. $\left(\frac{a}{b}\right)^m = \frac{a^m}{b^m}$.

  6. $\left(\frac{a}{b}\right)^{-m} = \left(\frac{b}{a}\right)^m$.

Review Exercises Chapter 7

1. Give the part-to-whole interpretation of $\frac{7}{9}$.

2. Plot each point on a number line.

   a. $\frac{3}{8}$ b. $-\frac{3}{2}$ c. $2\frac{1}{2}$ d. $-3\frac{1}{4}$

3. Convert each improper fraction to a mixed number.

   a. $\frac{15}{7}$ b. $-\frac{39}{12}$ c. $\frac{85}{16}$

4. Convert each mixed number to an improper fraction.

   a. $2\frac{3}{8}$ b. $-4\frac{1}{9}$ c. $12\frac{8}{9}$

5. List the set of all fractions that are equivalent to $\frac{2}{5}$.

6. Simplify each fraction.

   a. $\frac{35}{55}$ b. $\frac{112}{144}$ c. $-\frac{91}{39}$

7. Order the following fractions from smallest to largest: $\frac{1}{3}, \frac{1}{4}, \frac{2}{5}, \frac{3}{7}, \frac{5}{8}, \frac{1}{2}$.

8. Place the proper symbol $(=, <, >)$ between each pair of fractions.

   a. $\frac{8}{9} \quad \frac{81}{90}$ b. $-\frac{25}{16} \quad -\frac{28}{17}$ c. $\frac{49}{56} \quad \frac{98}{112}$

9. Find five fractions between $\frac{2}{3}$ and $\frac{5}{6}$.

10. If 24 counters make up $\frac{4}{5}$ of a set, how many counters are in the entire set?

11. The fish population of a lake contains 5,500 bass, 3,500 perch, and 8,500 sunfish. If 20,000 fish are in the lake, what fraction of the fish population is:

    a. Bass?   b. Not perch?

12. Mrs. Turner's first- and third-period classes take the same test. In first period, 19 out of 25 students pass. In third period, 20 out of 27 students pass. Which class has a higher value of students passing?

13. A fraction is equivalent to $\frac{7}{8}$, and the sum of the numerator and denominator of the fraction is equal to 30. What is the fraction?

14. Use fraction bars to represent and solve each problem.

    a. $\frac{7}{8} + \frac{3}{8}$ b. $\frac{5}{6} - \frac{1}{4}$ c. $\frac{2}{3} \cdot \frac{3}{6}$

15. Use the area model to represent and solve each problem.

    a. $\frac{7}{12} - \frac{5}{12}$ b. $\frac{5}{3} + \frac{5}{6}$ c. $\frac{4}{7} \cdot \frac{3}{10}$

16. Compute.

    a. $\frac{4}{5} + \frac{4}{15}$ b. $\left(-\frac{7}{12}\right) + \left(-\frac{7}{16}\right)$

    c. $\frac{8}{7} - \frac{13}{14}$ d. $\frac{5}{12} - \left(-\frac{9}{20}\right)$

    e. $\frac{4}{9} \cdot \left(-\frac{15}{14}\right)$ f. $\frac{18}{26} \cdot \frac{39}{45}$

    g. $\frac{5}{7} \div \frac{3}{8}$ h. $\left(-\frac{21}{45}\right) \div \left(-\frac{35}{18}\right)$

17. Compute.

    a. $4\frac{1}{3} + 5\frac{4}{9}$ b. $2\frac{4}{7} - 3\frac{5}{6}$

    c. $3\frac{5}{6} \cdot 4\frac{1}{8}$ d. $-2\frac{1}{7} \div 1\frac{1}{4}$

18. Compute mentally.

    a. $4\frac{1}{8} + 5\frac{7}{8}$ b. $3\frac{1}{3} - 3\frac{1}{6}$

    c. $\frac{2}{9} \cdot \left(15 \cdot \frac{1}{5}\right)$ d. $4\frac{1}{4} \cdot 16$

19. Estimate.

    a. $3\frac{7}{8} - 1\frac{1}{12}$ b. $14\frac{1}{7} + 9\frac{8}{9}$

    c. $2\frac{3}{8} \cdot 5\frac{5}{11}$ d. $3\frac{1}{9} \cdot 4\frac{7}{8}$

20. Find the multiplicative inverse of each number, if it exists.

    a. $\frac{2}{9}$ b. $-\frac{3}{4}$ c. -6

    d. $-3\frac{3}{7}$ e. $\frac{0}{7}$

21. Use a fraction calculator to compute.

    a. $3\frac{1}{3} \div \frac{14}{3} + \left(3\frac{1}{8} - \frac{11}{8}\right)$

    b. $3\frac{1}{6} + \frac{7}{3} \cdot 1\frac{5}{14} - \frac{2}{7}$

    c. $\frac{13}{6} \div \left[\left(\frac{3}{5} - 1\frac{1}{3}\right) \cdot 4\frac{2}{5}\right]$

22. Simplify each expression. Leave no negative exponents.

    a. $(-3)^{-4}$ b. $\left(-\frac{2}{3}\right)^{-2}$ c. $3^{-4} \cdot 4^{-4}$

    d. $\frac{6^{-2}}{6^3}$ e. $3^{-3} \cdot 3^{-8}$ f. $\frac{16^{-3}}{4^{-3}}$

    g. $\left(\frac{8^3}{8^2}\right)^{-3}$

23. If $\frac{1}{3}$ of the blocks in set A are blue and $\frac{1}{2}$ of the blocks in set B are blue, can it be determined which set has more blue blocks? Why or why not?

24. State the property used in each equation.

    a. $1\frac{4}{5} \cdot \frac{3}{7} \in \mathbb{Q}$.   b. $\frac{1}{4} \cdot \left(\frac{1}{3} + \frac{1}{2}\right) = \frac{1}{4} \cdot \frac{1}{3} + \frac{1}{4} \cdot \frac{1}{2}$.

    c. $1\frac{1}{4} \cdot \frac{3}{7} = \frac{3}{7} \cdot 1\frac{1}{4}$.   d. $\frac{-4}{3} + \frac{4}{3} = 0$.

    e. $1 \cdot \frac{11}{7} = \frac{11}{7}$.   f. $0 \cdot -2\frac{2}{3} = 0$.

25. Represent the expression $\left(\frac{a}{b} + \frac{c}{d}\right) - \frac{e}{f}$ as a single fraction.

26. Represent the fraction $\frac{13}{40}$ as the sum of two unit fractions.

27. A land developer has $9\frac{1}{8}$ acres of land bulldozed in one day. If the lot is a total of $45\frac{3}{8}$ acres, how much land has yet to be cleared?

28. How much fence is needed to enclose a yard with sides that measure $50\frac{1}{3}$ ft, $72\frac{1}{2}$ ft, $51\frac{3}{4}$ ft, and 73 ft?

29. At a particular university, $\frac{1}{4}$ of the students are freshmen, $\frac{1}{5}$ are sophomores, $\frac{1}{5}$ are juniors, and the rest are seniors. What fraction of the student body are seniors?

30. Boards of length $5\frac{1}{4}$ ft, $4\frac{1}{5}$ ft, and $3\frac{7}{8}$ ft are cut from a board that is 16 ft long. How much of the board remains?

31. If a and b are natural numbers greater than 1, and if $\frac{1}{a}$ is multiplied by $\frac{1}{b}$, which is bigger, the factors or the product?

32. What is the value of $\frac{1}{2} \cdot \frac{2}{4} \cdot \frac{4}{8} \cdot \frac{8}{16} \cdot \frac{16}{32} \cdot \frac{32}{64}$?

33. A fabric store has $33\frac{1}{3}$ yd of a particular fabric. It sells $\frac{2}{3}$ of the material in one day. How much of the material is left?

34. How many boards $1\frac{1}{4}$ ft long can be cut from a board that is $15\frac{3}{4}$ ft long?

35. Fill in the blanks using the numbers 2, 3, 7, and 8 only once to make the:

    a. Greatest product: $\frac{\square}{\square} \cdot \frac{\square}{\square}$.

    b. Greatest quotient: $\frac{\square}{\square} \div \frac{\square}{\square}$.

36. At a local elementary school, about $\frac{1}{3}$ of the students went to preschool for 1 year, and about $\frac{1}{6}$ of the students went to preschool for 2 years. If 240 students attend the school, about how many did not go to preschool at all?

37. How many people can be fed by 14 pans of lasagna, if each person gets $\frac{3}{10}$ of a pan?

Answers to Chapter 7 Check Your Understandings and Activities

Check Your Understanding 7.1-A

1. a. $\frac{4}{5}$ represents part of a whole that has been divided into five equivalent pieces, of which four are counted.

   b. $\frac{6}{9}$ represents part of a whole that has been divided into nine equivalent pieces, of which six are counted.

   c. $\frac{9}{8}$ represents part of a whole that has been divided into eight equivalent pieces, of which nine are counted.

2. a. $\frac{3}{6} = \frac{1}{2}$.   b. $\frac{6}{9} = \frac{2}{3}$.   c. $\frac{2}{8} = \frac{1}{4}$.

3. a. $4\frac{2}{3}$.   b. $-4\frac{1}{4}$.   c. $\frac{29}{9}$.   d. $-\frac{19}{9}$.

Check Your Understanding 7.1-B

1. a. $\left\{ \ldots, \frac{-3}{-9}, \frac{-2}{-6}, \frac{-1}{-3}, \frac{1}{3}, \frac{2}{6}, \frac{3}{9}, \ldots \right\}$.

   b. $\left\{ \ldots, \frac{9}{-3}, \frac{6}{-2}, \frac{3}{-1}, \frac{-3}{1}, \frac{-6}{2}, \frac{-9}{3}, \ldots \right\}$.

   c. $\left\{ \ldots, \frac{-9}{12}, \frac{-6}{8}, \frac{-3}{4}, \frac{3}{-4}, \frac{6}{-8}, \frac{9}{-12}, \ldots \right\}$.

2. a. $\frac{3}{8}$. **b.** $\frac{4}{7}$. **c.** $-\frac{7}{10}$. **d.** $-\frac{2}{3}$.

3. a. $\frac{5}{6} > \frac{25}{35}$; **b.** $\left(-\frac{24}{20}\right) = \left(-\frac{18}{15}\right)$. **c.** $\frac{55}{45} = \frac{110}{90}$.

4. Five fractions between $\frac{1}{4}$ and $\frac{1}{2}$ are $\frac{11}{40}, \frac{12}{40}, \frac{13}{40}, \frac{14}{40},$ and $\frac{15}{40}$.

Check Your Understanding 7.2-A

1. a. $\frac{8}{7}$. **b.** $\frac{1}{3}$. **c.** $\frac{3}{5}$.

2. a. $\frac{71}{40}$. **b.** $-\frac{43}{60}$. **c.** $\frac{5}{12}$.

3. a. Not equal. **b.** Equal. **c.** Equal. **d.** Not equal.
e. Equal.

Check Your Understanding 7.2-B

1. a. $-\frac{1}{9}$. **b.** $-\frac{2}{7}$. **c.** $-\frac{82}{35}$. **d.** $-\frac{109}{60}$.

2. a. $8\frac{9}{20}$. **b.** $14\frac{11}{12}$. **c.** $2\frac{5}{36}$. **d.** $3\frac{25}{36}$.

3. a. 2. **b.** $11\frac{1}{8}$. **c.** $4\frac{4}{7}$.

4. a. $\approx 8\frac{1}{2}$. **b.** ≈ 15. **c.** $\approx 2\frac{1}{4}$.

Check Your Understanding 7.3-A

1. a. $\frac{15}{32}$. **b.** $\frac{52}{125}$. **c.** $\frac{21}{16}$.

2. a. Multiplicative identity property.
b. Zero multiplication property.
c. Multiplicative inverse property.
d. Closure property.
e. Commutative property.
f. Distributive property of multiplication over addition.

3. a. $\frac{4}{3}$. **b.** $-\frac{8}{11}$. **c.** $-\frac{1}{2}$. **d.** $-\frac{4}{11}$.

4. a. $\frac{27}{14}$. **b.** $\frac{26}{33}$. **c.** $-\frac{4}{3}$.

Check Your Understanding 7.3-B

1. a. $\frac{1}{3^3}$. **b.** $\frac{1}{x^9}$. **c.** $\frac{1}{5^3}$. **d.** y^2.

2. a. $15\frac{5}{6}$. **b.** $2\frac{17}{30}$.

3. a. 3. **b.** 6. **c.** 34.

4. a. ≈ 3. **b.** ≈ 3. **c.** ≈ 5.

Activity 7.1

a.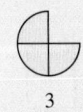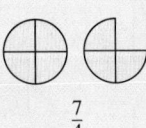

$\frac{2}{3}$ $\frac{3}{4}$ $\frac{7}{4}$

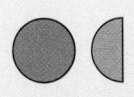   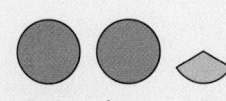

$1\frac{1}{2}$ $2\frac{1}{3}$

b.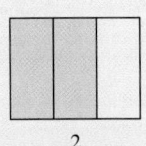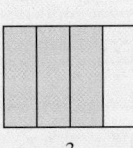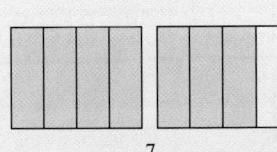

$\frac{2}{3}$ $\frac{3}{4}$ $\frac{7}{4}$

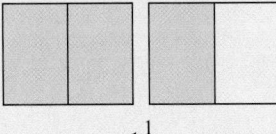   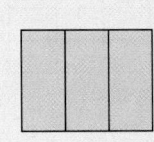 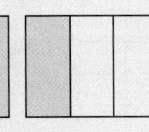

$1\frac{1}{2}$ $2\frac{1}{3}$

c.

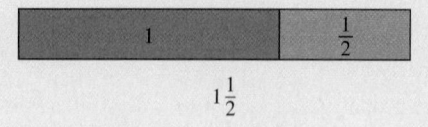
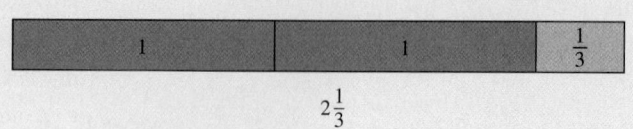

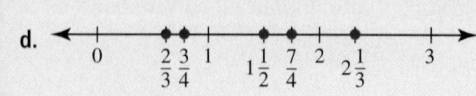

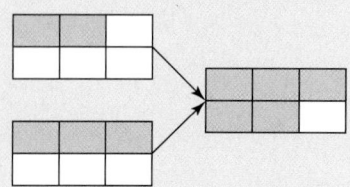
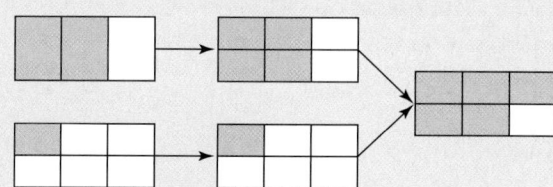

d.

Activity 7.2

Fractions equivalent to 1 are $\left\{\dfrac{2}{2}, \dfrac{3}{3}, \dfrac{4}{4}, \dfrac{5}{5}, \dfrac{6}{6}, \dfrac{8}{8}, \dfrac{10}{10}, \dfrac{12}{12}\right\}$. Fractions equivalent to $\dfrac{1}{2}$ are $\left\{\dfrac{2}{4}, \dfrac{3}{6}, \dfrac{4}{8}, \dfrac{5}{10}, \dfrac{6}{12}\right\}$. For thirds, fractions equivalent to $\dfrac{1}{3}$ are $\left\{\dfrac{2}{6}, \dfrac{4}{12}\right\}$, and fractions equivalent to $\dfrac{2}{3}$ are $\left\{\dfrac{4}{6}, \dfrac{8}{12}\right\}$. For fourths, fractions equivalent to $\dfrac{1}{4}$ are $\left\{\dfrac{2}{8}, \dfrac{3}{12}\right\}$, and fractions equivalent to $\dfrac{3}{4}$ are $\left\{\dfrac{6}{8}, \dfrac{9}{12}\right\}$. For fifths, $\dfrac{1}{5}$ is equivalent to $\dfrac{2}{10}$, $\dfrac{2}{5}$ is equivalent to $\dfrac{4}{10}$, $\dfrac{3}{5}$ is equivalent to $\dfrac{6}{10}$, and $\dfrac{4}{5}$ is equivalent to $\dfrac{8}{10}$. For sixths, $\dfrac{1}{6}$ is equivalent to $\dfrac{2}{12}$, and $\dfrac{5}{6}$ is equivalent to $\dfrac{10}{12}$.

Activity 7.3

a. $\dfrac{2}{6} + \dfrac{3}{6} = \dfrac{5}{6}$.

b. $\dfrac{2}{3} + \dfrac{1}{6} = \dfrac{5}{6}$.

c. $\left(\dfrac{1}{2} + \dfrac{1}{3}\right) + \dfrac{1}{4} = \left(\dfrac{6}{12} + \dfrac{4}{12}\right) + \dfrac{3}{12} = \dfrac{13}{12}$.

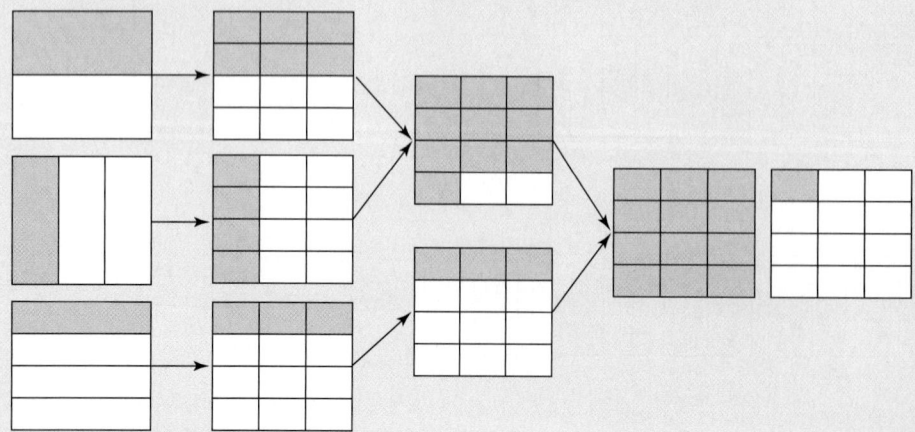

Activity 7.4

a. $\dfrac{5}{6} - \dfrac{3}{6} = \dfrac{2}{6}$.

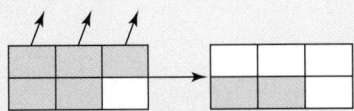

b. $\dfrac{4}{5} - \dfrac{2}{10} = \dfrac{6}{10}$.

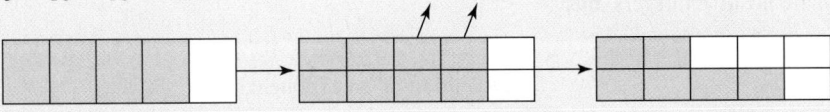

c. $\left(\dfrac{7}{8} - \dfrac{1}{4}\right) - \dfrac{1}{2} = \dfrac{1}{8}$.

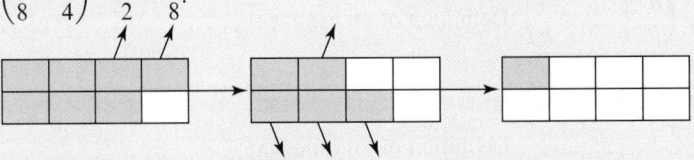

Activity 7.5

a $\dfrac{1}{3} \cdot \dfrac{1}{4} = \dfrac{1}{12}$.

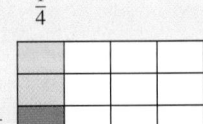

b. $\dfrac{5}{6} \cdot \dfrac{3}{4} = \dfrac{5}{8}$.

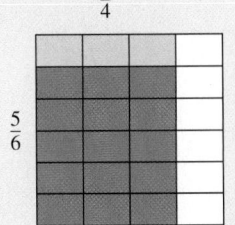

c. $\dfrac{5}{12} \cdot \dfrac{2}{3} = \dfrac{5}{18}$.

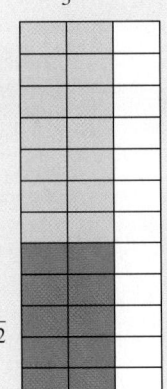

d. $\dfrac{3}{4} \div \dfrac{1}{4} = 3$.

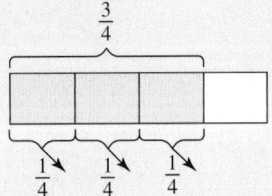

$\dfrac{1}{4}$ can be removed from $\dfrac{3}{4}$ 3 times

e. $\dfrac{5}{12} \div \dfrac{2}{12} = \dfrac{5}{2}$.

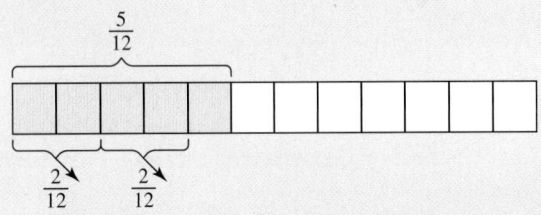

$\dfrac{2}{12}$ can be removed from $\dfrac{5}{12}$ $2\dfrac{1}{2} = \dfrac{5}{2}$ times

f. $\dfrac{7}{8} \div \dfrac{1}{4} = \dfrac{7}{8} \div \dfrac{2}{8} = \dfrac{7}{2}$.

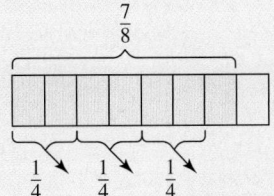

$\dfrac{1}{4}$ can be removed from $\dfrac{7}{8}$ $3\dfrac{1}{2} = \dfrac{7}{2}$ times

Activity 7.6

a. If $\dfrac{a}{b}$ is a rational number and if m and n are positive integers, then

$$\left(\frac{a}{b}\right)^{m} \cdot \left(\frac{a}{b}\right)^{n} = \underbrace{\left(\frac{a}{b} \cdot \frac{a}{b} \cdot \ldots \cdot \frac{a}{b}\right)}_{m \text{ times}} \cdot \underbrace{\left(\frac{a}{b} \cdot \frac{a}{b} \cdot \ldots \cdot \frac{a}{b}\right)}_{n \text{ times}} \qquad \text{Definition of an exponent}$$

$$= \left(\frac{a}{b}\right)^{m+n} \qquad \text{Definition of an exponent}$$

b. If $\dfrac{a}{b}$ is a rational number and if m and n are positive integers, then

$$\left[\left(\frac{a}{b}\right)^{m}\right]^{n} = \left[\underbrace{\left(\frac{a}{b} \cdot \frac{a}{b} \cdot \ldots \cdot \frac{a}{b}\right)}_{m \text{ times}}\right]^{n} \qquad \text{Definition of an exponent}$$

$$= \underbrace{\underbrace{\left(\frac{a}{b} \cdot \frac{a}{b} \cdot \ldots \cdot \frac{a}{b}\right)}_{m \text{ times}} \cdot \underbrace{\left(\frac{a}{b} \cdot \frac{a}{b} \cdot \ldots \cdot \frac{a}{b}\right)}_{m \text{ times}} \cdot \ldots \cdot \underbrace{\left(\frac{a}{b} \cdot \frac{a}{b} \cdot \ldots \cdot \frac{a}{b}\right)}_{m \text{ times}}}_{n \text{ times}} \qquad \text{Definition of an exponent}$$

$$= \left(\frac{a}{b}\right)^{m \cdot n} \qquad \text{Definition of an exponent}$$

Chapter *Chapter* **8**

Decimals, Real Numbers, and Proportional Reasoning

© Image Source/Corbis

Opening Problem   Josephine receives a utility bill that includes $25.50 for trash, $35.65 for water, and $21.90 for sewer. If she has only $50 to make a partial payment and any payment is distributed proportionally among the three bills, how much goes to each?

Blend Images/Shutterstock.com

In the last chapter, we introduced fractions and rational numbers as a way to describe part-to-whole relationships. Decimals give us another way to describe these relationships in a way that is consistent with the place values of the decimal system. We also introduce proportional reasoning, which is often considered to be the pinnacle of elementary arithmetic. Not only does it bring together ideas from the integers, fractions, and decimals, but it also serves as an important stepping-stone to algebra. In discussing these topics, we continue to work on the objectives in the NCTM Number and Operations Standard.

> ### ■ NCTM Number and Operations Standard
>
> Instructional programs from prekindergarten through grade 12 should enable all students to:
>
> - Understand numbers, ways of representing numbers, relationships among numbers, and number systems.
> - Understand meanings of operations and how they relate to one another.
> - Compute fluently and make reasonable estimates.

Source: NCTM STANDARDS Copyright © 2011 by NATIONAL COUNCIL OF TEACHERS OF MATHEMATICS. Reproduced with permission of NATIONAL COUNCIL OF TEACHERS OF MATHEMATICS.

Children first encounter decimals in first and second grade as they study dollar-and-cent money notation. Although these early experiences are limited, children learn to interpret and use different decimal representations. Once in the third, fourth, and fifth grades, students work in earnest with the decimals. Specifically, they learn how to order and compute with them by extending what they have learned from the whole numbers. Children also learn that decimals are connected to fractions because both represent part-to-whole relationships, but in different ways.

Fractions and decimals play an important role in **proportional reasoning,** which is the ability to compare quantities and understand the context in which the comparison takes place. The topic is important because it is practical and brings together many mathematical ideas that students have already learned. Because it relies on certain computational skills, proportional reasoning is somewhat limited in the elementary grades. However, once children can compute with fractions and decimals, they can proceed through a more formal progression of proportional reasoning. They often start by developing the ideas of ratios, proportions, and percents and then move on to applying them to solve real-world problems. Table 8.1 shows specific content objectives associated with learning the decimals and proportional reasoning.

Many children find working with decimals difficult for a number of reasons that are conceptual and not procedural. For instance, many children fail to realize that fractions and decimals are different representations of the same idea. Although their prior experience tells them that fractions must be a part of something, students often disassociate the decimals from any part-to-whole context. Children can also have trouble understanding the relative sizes of decimals. Many of them think that the decimal with the most digits is the largest. This comes from their experience with the whole numbers, which tells them that since 4 is less than 37, 0.4 must be less than 0.37. They can make similar mistakes with the language of the decimals. For instance, some students believe that hundredths are larger than tenths because one hundred is larger than ten. Computationally, children can have trouble correctly placing the decimal point in the answer. This is particularly true with long division, in which students are required to add placeholders to either the dividend or the quotient.

Proportional reasoning can also be difficult for students to master. Some students have trouble just making sense of the basic ideas of ratios, proportions, and percents. For instance, many students fail to realize that order is important with ratios and rates. Others have trouble in setting up proportions or in understanding the meaning of a percent. Many of these problems come from instruction that focuses only on

© Image Source/Corbis

Table 8.1 Classroom learning objectives	K	1	2	3	4	5	6	7	8
Decimals									
Use decimal notation for fractions with denominators of 10 or 100.					X				
Read and write decimals to thousandths using base-ten numerals, number names, and expanded form.						X			
Compare decimals to hundredths and thousandths by reasoning about their size and using the meanings of the digits in each place.					X	X			
Use place value to round decimals to any place.						X			
Explain patterns when multiplying a number by whole-number powers of 10.						X			
Fluently add, subtract, multiply, and divide multi-digit decimals using concrete models, strategies based on place value, and the standard algorithm for each operation.						X	X		
Convert a rational number to a terminating or repeating decimal using long division and convert a terminating or repeating decimal into a rational number.								X	X
Solve multi-step real-life and mathematical problems posed with positive and negative rational numbers in any form.								X	
Evaluate square roots of small perfect squares and cube roots of small perfect cubes.									X
Use rational approximations to compare irrational numbers, locate them on a number line, and estimate their value.									X
Know and apply the properties of integer exponents to generate equivalent numerical expressions.									X
Perform operations with numbers expressed in scientific notation.									X
Use scientific notation for measurements of very large or very small quantities and interpret scientific notation that has been generated by technology.									X
Proportional Reasoning									
Understand the concept of a ratio and a rate and use ratio and rate language to describe relationships between quantities.							X		
Use ratio and rate reasoning to solve real-world and mathematical problems.							X		
Find a percent of a quantity as a rate per 100.							X		
Solve problems involving finding the whole, given a part and the percent.							X		
Recognize and represent proportional relationships between quantities using tables, diagrams, and equations.							X	X	
Compute unit rates associated with ratios of fractions.								X	
Identify the constant of proportionality in tables, graphs, equations, and diagrams of proportional relationships.								X	
Use proportional relationships to solve multistep ratio and percent problems such as simple interest, tax, markups and markdowns, gratuities, and percent increase and decrease.								X	
Solve problems involving scale drawings of geometric figures.								X	

Source: Adapted from the *Common Core State Standards for Mathematics* (*Common Core State Standards Initiative*, 2010).

procedural knowledge without giving students the opportunity to think about the relationships that the ratios, proportions, and percents represent. As a consequence, when students apply these ideas, they may not know what to do with problems that do not match their instruction.

Educational research indicates that the problems students have with the decimals and proportional reasoning can last well into adulthood. It also indicates that teachers can be the source of these problems because they themselves are unclear about the conceptual nature of the decimals and proportional reasoning. To help elementary teachers gain a deeper understanding of the decimals and proportional reasoning, Chapter 8:

■ Explains the characteristics and properties of different types of decimals, including terminating decimals, repeating decimals, and irrational numbers.
■ Demonstrates decimal operations.
■ Discusses proportional reasoning and uses it with ratios, proportions, and percents.

SECTION 8.1   Decimals and the Real Numbers

In the last chapter, we learned to use fractions to describe part-to-whole relationships. Although they enable us to represent almost any numerical situation, they are seldom consistent with the ten-to-one ratio of the decimal system. We can gain more consistency by describing part-to-whole relationships with decimals.

Understanding Decimals

Chapter 3 explained that the place values of the decimal system are based on increasing powers of ten; in other words, each place value is ten times the value of the place to its right. However, if we reverse the pattern—that is, divide place values to the left by 10—then we can extend the place values of the decimal system to include fractional parts. For instance, the place value to the right of the ones has a value of $1 \div 10 = \frac{1}{10}$ and the place value to the right of that has a value of $\frac{1}{10} \div 10 = \frac{1}{10} \cdot \frac{1}{10} = \frac{1}{100}$. Continuing the pattern, we get the place values in Figure 8.1, which are named for the fractional values they represent.

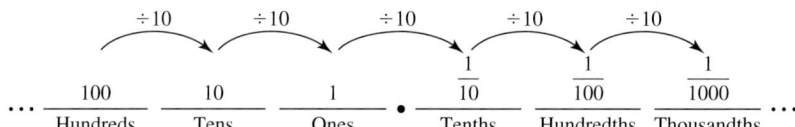

Figure 8.1 Decimal place values

Three things need to be mentioned about extending the place values in this way. First, because we can continue to divide by ten, the number of place values extending to the right is infinite. As a consequence, the ones place value is no longer the smallest, and we lose our reference point for determining the value of a digit. We regain it, however, by inserting a decimal point between the ones and the tenths place values. In other words, the decimal point separates the wholes from the parts of the whole.

Second, notice that each place value to the right of the decimal point is equivalent to a **decimal fraction,** which is a fraction with a power of ten in the denominator. This makes the fractional place values consistent with the ten-to-one ratio of the decimal system, and we can regroup ten of any decimal fraction as one in the next higher place value. For instance, 10 tenths are equivalent to 1 one, or $10 \cdot \frac{1}{10} = \frac{10}{10} = 1$. Similarly, 10 hundredths are equivalent to 1 tenth, or $10 \cdot \frac{1}{100} = \frac{10}{100} = \frac{1}{10}$.

Third, in addition to a fractional form, we can write decimal place values in two other ways. One is to use exponents. Because $\frac{1}{a^n} = a^{-n}$ when $a \neq 0$, then $\frac{1}{10} = 10^{-1}$, $\frac{1}{100} = 10^{-2}$, $\frac{1}{1000} = 10^{-3}$, and so on. Another way is to use a **decimal.** In a sense, a decimal is a notational shortcut in which we place the numerator of the decimal fraction in the appropriate place value to the right of the decimal point. For instance, we write $\frac{1}{10}$ as 0.1 and $\frac{1}{100}$ as 0.01. The different ways to write decimal place values are summarized in Table 8.2.

Table 8.2 Different representations of decimal place values								
Position Name	... Thousands	Hundreds	Tens	Units	Tenths	Hundredths	Thousandths	Ten Thousandths ...
Fractional form	1,000	100	10	1	$\frac{1}{10}$	$\frac{1}{100}$	$\frac{1}{1,000}$	$\frac{1}{1,000}$
Decimal form	1,000	100	10	1	0.1	0.01	0.001	0.0001
Exponential form	10^3	10^2	10^1	10^0	10^{-1}	10^{-2}	10^{-3}	10^{-4}

To write a decimal, we simply put the appropriate digits in the correct place values. As with the whole numbers, expanded notation can be used to describe the value of a decimal.

Example 8.1   Writing Decimals in Expanded Notation

Communication   Write each decimal in expanded notation.

a. 0.45   **b.** 74.083

Solution We write the expanded notation using both fractions and exponents.

a. $0.45 = \left(4 \cdot \frac{1}{10}\right) + \left(5 \cdot \frac{1}{100}\right)$

$= (4 \cdot 10^{-1}) + (5 \cdot 10^{-2}).$

b. $74.083 = (7 \cdot 10) + (4 \cdot 1) + \left(0 \cdot \frac{1}{10}\right) + \left(8 \cdot \frac{1}{100}\right) + \left(3 \cdot \frac{1}{1,000}\right)$

$= (7 \cdot 10^1) + (4 \cdot 10^0) + (0 \cdot 10^{-1}) + (8 \cdot 10^{-2}) + (3 \cdot 10^{-3}).$

To name the decimal, we name the decimal value, and follow it with the name of the smallest place value used. For decimals larger than one, we indicate the placement of the decimal point by the word "and" (the only time "and" is used when reading numbers). For instance, 0.45 is read as "forty-five hundredths" and 74.083 as "seventy-four and eighty-three thousandths."

Representing Decimals

In addition to representing decimals verbally and numerically, we can also represent them visually and tactilely. Some common models are money, fraction models, and base-ten blocks.

Money Model

In the money model, dollars represent whole-number amounts, dimes represent tenths, and pennies represent hundredths. Using these denominations, we can represent numerical amounts to two decimal places. Examples are shown in Figure 8.2.

Fraction Models

Many fraction models, like fraction disks, fraction bars, and number lines, can also be used with the decimals. In each case, we divide the whole into ten equivalent pieces and

0.14 is represented by

0.3 or 0.30 is represented by

1.52 is represented by

Figure 8.2 Representing decimals with money

then count the appropriate number. These models are often limited to the tenths, so they generally allow us to work only with one decimal place. See Figure 8.3 for examples.

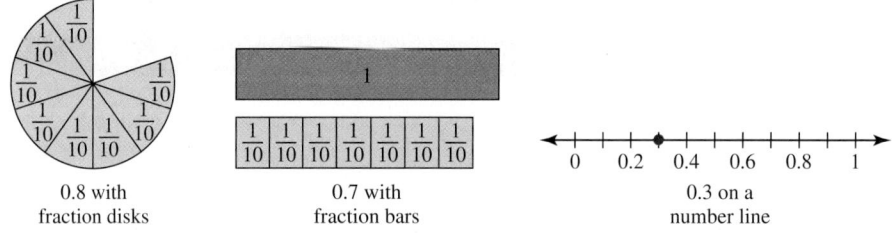

0.8 with
fraction disks

0.7 with
fraction bars

0.3 on a
number line

Figure 8.3 Representing decimals with fraction models

Explorations Manual 8.1

Grid paper is another fractional model that allows us to work with more decimal places by using the area of a square.

Example 8.2 | **Representing Decimals with Grid Paper**

Representations

Use grid paper to represent each decimal.

a. 0.425   **b.** 2.57

Solution We use a large square as the unit, which we subdivide into smaller regions to represent tenths and hundredths. To represent a decimal, we shade the appropriate region.

a. To represent 0.425, we shade in $42\frac{1}{2}$ of the smaller squares:

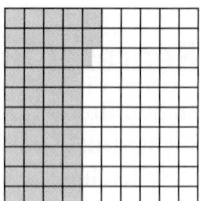

b. For 2.57, we shade in two whole squares and 57 hundredths of another:

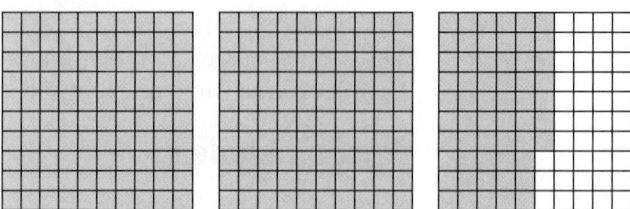

Base-Ten Blocks

Explorations
Manual
8.2

To use base-ten blocks to represent decimals, we simply rename the values of the pieces. Most commonly, the large block is given a value of one, the flat a value of 0.1, the long a value of 0.01, and the small block a value of 0.001 (Figure 8.4).

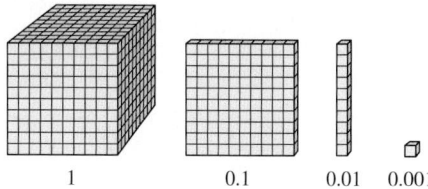

1   0.1   0.01   0.001

Figure 8.4 Decimal values of base-ten blocks

Using these designations, we represent 0.34 and 2.307 as shown in Figure 8.5.

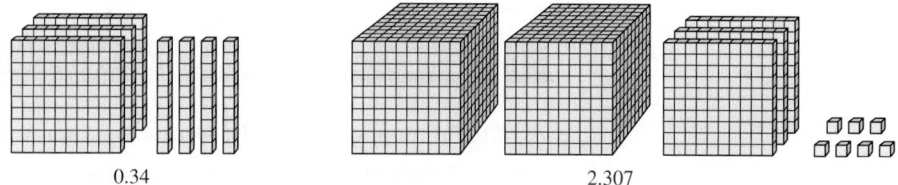

0.34   2.307

Figure 8.5 Representing decimals with base-ten blocks

Terminating Decimals

To this point, every decimal we have used has been a **terminating decimal,** which is any decimal with a finite number of decimal places to the right of the decimal point. As we have seen, we can write terminating decimals in expanded notation. Consequently, we can also write them as the sum of decimal fractions, and by using rational number addition condense the sums into one fraction. For instance:

$$0.147 = \left(1 \cdot \frac{1}{10}\right) + \left(4 \cdot \frac{1}{100}\right) + \left(7 \cdot \frac{1}{1,000}\right)$$

$$= \frac{1}{10} + \frac{4}{100} + \frac{7}{1,000}$$

$$= \frac{100}{1,000} + \frac{40}{1,000} + \frac{7}{1,000}$$

$$= \frac{147}{1,000}$$

Two things must be noted about this process. First, the numerator and the denominator of the result are always integers, so the fraction must be a rational number. In other words, terminating decimals give us another way to write some rational numbers. Second, the numerator of the sum matches the digits in the original number, and the denominator takes on the value of the smallest position used in the decimal. This observation gives us a quick way to convert any terminating decimal into a fraction.

Example 8.3 | **Writing Terminating Decimals as Fractions**

Representation

Express each terminating decimal as a rational number in fractional form.

a. 0.4   **b.** −5.25

Solution Using the procedure just described:

a. $0.4 = \frac{4}{10} = \frac{2}{5}$.   **b.** $-5.25 = -\frac{525}{100} = -\frac{21}{4}$.

Because we can rewrite any decimal fraction as a terminating decimal, then any fraction that can be rewritten as a decimal fraction is also a terminating decimal. If a fraction can be rewritten as a decimal fraction, then it must have a denominator that divides a power of ten. Because powers of ten have prime factorizations that consist of only 2s and 5s, or

$$10^1 = 10 \quad = 2 \cdot 5$$
$$10^2 = 100 \quad = 2^2 \cdot 5^2$$
$$10^3 = 1000 \quad = 2^3 \cdot 5^3$$
$$\vdots \qquad \vdots \qquad \vdots$$
$$10^n = (2 \cdot 5)^n = 2^n \cdot 5^n$$

then any denominator with a prime factor other than a 2 or a 5 cannot be rewritten as a terminating decimal. This leads us to the next theorem.

Theorem 8.1 Test for Terminating Decimals

A rational number, $\frac{a}{b}$, can be written as a terminating decimal if and only if it is in simplest form and the prime factorization of the denominator contains no primes other than 2s and 5s.

Example 8.4 Determining Whether a Fraction Is a Terminating Decimal

Use Theorem 8.1 to determine whether each fraction can be written as a terminating decimal. For those that can, find it.

a. $\frac{1}{12}$ **b.** $\frac{9}{60}$

Solution

a. Because $12 = 3 \cdot 2^2$, $\frac{1}{12}$ cannot be rewritten as a terminating decimal.

b. We first simplify the fraction: $\frac{9}{60} = \frac{3}{20}$. Because $20 = 2^2 \cdot 5$, $\frac{3}{20}$ has a terminating decimal representation. Using the Fundamental Law of Fractions, we have $\frac{9}{60} = \frac{3}{20} \cdot \frac{5}{5} = \frac{15}{100} = 0.15$.

Repeating Decimals

Theorem 8.1 implies that not every rational number can be written as a terminating decimal. So, we might ask whether other rational numbers have decimal representations. One way to explore the question is to take a rational number that is not a terminating decimal and use models to try to represent it with decimal fractions. For example, using fraction bars to compare $\frac{1}{3}$ to tenths, we see that it is equivalent to $\frac{3}{10}$ plus a little bit (see Figure 8.6). Comparing the remaining piece to hundredths, we see that it is equivalent to $\frac{3}{100}$ plus a little bit. Again, comparing the remaining piece to thousandths, we see that it is equivalent to $\frac{3}{1,000}$ plus a little bit.

Explorations Manual 8.3

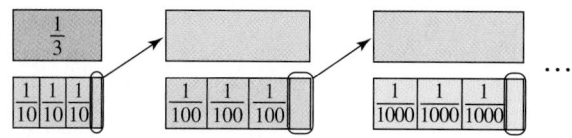

Figure 8.6 Representing $\frac{1}{3}$ with decimal fractions

If we continued to make comparisons, we would see two things. First, the process can be carried out indefinitely, so it is impossible to represent $\frac{1}{3}$ as a single decimal fraction. Second, the small remaining pieces are always equivalent to three of the next smallest decimal fraction. Consequently, we can represent $\frac{1}{3}$ as

$$\frac{1}{3} = \frac{3}{10} + \frac{3}{100} + \frac{3}{1,000} + \frac{3}{10,000} \ldots = 0.3333\ldots$$

Decimals of this type are called **repeating decimals** because they have a block of digits that continuously repeats. The block of digits that repeats is called the **repetend.** There are two ways to use the repetend to write repeating decimals. One is to place a line over the repetend, and another is to write the repetend three times followed by an ellipsis (. . .). If neither method is used, the decimal is assumed to be terminating. Examples of repeating decimals are:

$$0.\overline{6} = 0.666\ldots \quad 0.1\overline{23} = 0.1232323\ldots \quad 0.\overline{345} = 0.345345345\ldots$$

The length of the repetend is called the **period.** In the preceding examples, $0.\overline{6}$ has a period of 1, $0.1\overline{23}$ has a period of 2, and $0.\overline{345}$ has a period of 3.

We can also write every terminating decimal as a repeating decimal simply by affixing an infinite string of zeros to the right of the last digit. For instance, $0.351 = 0.351000\ldots = 0.351\overline{0}$. It turns out that all repeating decimals are rational numbers, because they can be expressed as the quotient of two integers. This fact is explored in the next section.

Check Your Understanding 8.1-A

1. Write each decimal in expanded notation, and then give the name of the decimal.

    a. 0.73   **b.** 0.019   **c.** 30.02   **d.** 401.104

2. Represent each decimal as indicated.

    a. 0.4 with the money model   **b.** 0.75 with grid paper

    c. 1.5 with fractions disks   **d.** 2.37 with base-ten blocks

3. Use Theorem 8.1 to determine whether each fraction can be written as a terminating decimal. For those that can, do so.

    a. $\frac{5}{16}$ **b.** $\frac{15}{35}$ **c.** $\frac{65}{55}$ **d.** $\frac{99}{240}$

4. Give the repetend and period of each repeating decimal.

    a. 0.3444 . . . **b.** 0.212121 . . . **c.** $0.2\overline{19}$ **d.** $0.34\overline{1}$

Talk About It How are fractions and decimals alike? How are they different?

Representation

Activity 8.1

Another manipulative that can be used to model decimals is a fraction wheel. It is made from two disks that have been divided into 100 equal parts. Each disk is cut along one radius, and the two are put together as shown in Figure 8.7. We then rotate the disks to show particular two-digit decimals from zero to one.

a. Explain how the disks can be used to represent the following decimals:

   i. 0.5   **ii.** 0.8   **iii.** 0.17   **iv.** 0.93

b. Use the disk to represent the following fractions. How can the disks be used to build a connection between fractions and decimals?

   i. $\frac{1}{2}$ **ii.** $\frac{1}{4}$ **iii.** $\frac{1}{5}$ **iv.** $\frac{3}{5}$

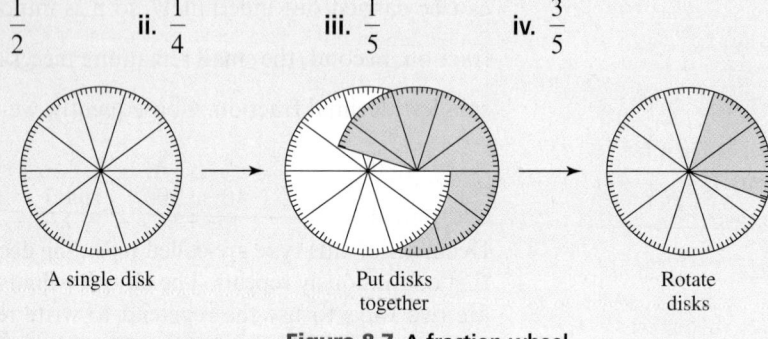

A single disk   Put disks together   Rotate disks

Figure 8.7 A fraction wheel

Irrational Numbers

A third type of decimal neither terminates nor repeats. For example, consider the decimal that affixes every natural number after the decimal point:

$$0.123456789101112\ldots$$

This decimal does not terminate or repeat because the natural numbers do not terminate or repeat. Decimals like this are called **irrational numbers.** The name fits well because we cannot express nonterminating, nonrepeating decimals as the quotient of two integers. Consequently, the rational and the irrational numbers must be disjoint.

    The infinite nature of irrational numbers implies that they represent values that can be written only with an infinite sum of decimal fractions. For instance, Figure 8.8 uses fraction bars to show how 0.2451 . . . is equivalent to two tenths, four hundredths, and so on. As with repeating decimals, each time we use a decimal fraction to make a comparison, a small piece always remains. This time, however, the amounts represented by the remaining pieces do not follow a repetitive pattern.

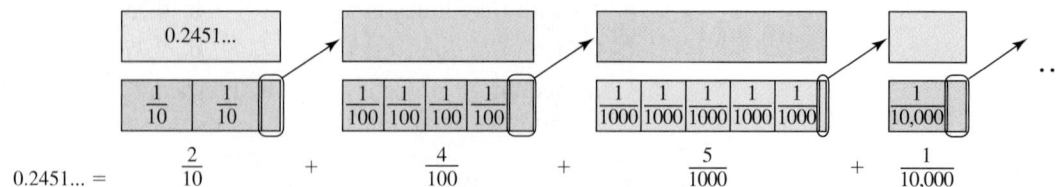

Figure 8.8 Representing an irrational number with decimal fractions

    In general, when writing irrational numbers, it is usually enough to list four or five digits followed by an ellipsis. However, some irrational numbers merit their own notation. For instance, if we divide the circumference of any circle by its diameter, the quotient is always the same irrational number, which we represent with the Greek letter π (pi). Computers have calculated π to billions of decimal places with no indication that it will terminate or repeat. π to just 25 places is

$$\pi = 3.14159\ 26535\ 89793\ 23846\ 26434\ldots$$

π has extensive applications in both mathematics and science, which is one reason it has its own special symbol. In computations, it is often approximated as 3.14 or $\dfrac{22}{7}$.

Another source of irrational numbers is the set of square roots. These numbers were first discovered by the Pythagoreans as they investigated the special relationship in right triangles that we now know as the **Pythagorean Theorem.** This theorem states that in any right triangle, the sum of the squares of the two shorter sides is equal to the square of the longest side. Using the triangle in Figure 8.9, we can express it as $a^2 + b^2 = c^2$.

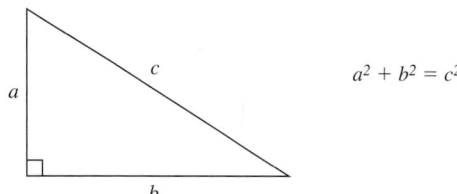

Figure 8.9 **The Pythagorean theorem**

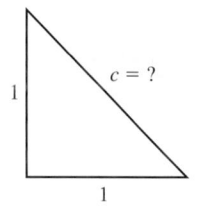

Figure 8.10 **A right triangle with short sides of length 1**

The Pythagoreans found many rational number solutions to this equation but ran into problems when each short side had a length of 1 (Figure 8.10). In this case, no rational numbers satisfy the equation, $c^2 = 1^2 + 1^2 = 2$, but two irrational numbers do. Mathematically, we represent the numbers as $-\sqrt{2}$ and $\sqrt{2}$ and interpret them to be the numbers that, when squared, equal 2, or $(-\sqrt{2})^2 = 2$ and $(\sqrt{2})^2 = 2$. Because the goal is to find the missing side of a triangle and length is always positive, the only solution we consider is $\sqrt{2}$. We call this number the **principal square root.**

Definition of the Principal Square Root

If a is any nonnegative number, then the **principal square root** of a, denoted $\sqrt{a}$, is the number $b \geq 0$, such that $b^2 = a$.

In the next example, we prove that $\sqrt{2}$ is irrational by using a classic proof based on indirect reasoning.

Example 8.5 | The Irrationality of $\sqrt{2}$

Reasoning

Show that $\sqrt{2}$ is irrational.

Solution Because a number can be rational or irrational, but not both, we begin our proof by supposing that $\sqrt{2}$ is rational; that is, $\sqrt{2} = \dfrac{a}{b}$, where a and b are integers, $b \neq 0$, and $\dfrac{a}{b}$ is simplified. Next, we make the following manipulations to the equation $\sqrt{2} = \dfrac{a}{b}$:

$$2 = \left(\frac{a}{b}\right)^2 \quad \text{Square both sides.}$$

$$2 = \frac{a^2}{b^2} \quad \text{Property of exponents.}$$

$$2b^2 = a^2 \quad \text{Multiply both sides by } b^2.$$

The last equation implies that a^2 is even because it is the product of 2 and b^2. If a^2 is even, then a must be even. In other words, $a = 2k$ for some integer k. If we square both sides of this equation, then $a^2 = (2k)^2 = 4k^2$. By substitution, $4k^2 = 2b^2$, or $2k^2 = b^2$. This implies that b^2 is even, so b must be even. Because both a and b are even, $\frac{a}{b}$ must not be simplified. However, the original premise assumed that $\frac{a}{b}$ must be simplified. Because the argument contains a contradiction, the original assumption must be wrong. Consequently, $\sqrt{2}$ is irrational.

Mathematically, the notation $\sqrt{n}$ represents the exact value of the square root of n. In some cases, the value is a rational number. For instance, numbers like 1, 4, 9, 16, and 25 are called **perfect squares** because they have square roots that are whole numbers. Specifically:

$$\sqrt{1} = 1 \quad \sqrt{4} = 2 \quad \sqrt{9} = 3 \quad \sqrt{16} = 4 \quad \text{and} \quad \sqrt{25} = 5$$

However, in other cases, the square root is irrational and we often use a decimal approximation for its value. The next example demonstrates one method for estimating square roots.

Example 8.6 Estimating a Square Root

Problem Solving Estimate $\sqrt{3}$ to two decimal places.

Solution

Understanding the Problem. We are to estimate $\sqrt{3}$ to two decimal places. The only information we have is the definition of the principal square root.

Devising the Plan. Even though we cannot compute square roots, we can square numbers. We know that $\sqrt{3}$ is a solution to $x^2 = 3$, so, if we can find numbers that have squares close to 3, they will approximate $\sqrt{3}$. By *guessing and checking*, we should be able to close in on the value for $\sqrt{3}$.

Carrying Out the Plan. We know that $1 < \sqrt{3} < 2$ because $1^2 = 1 < 3 < 4 = 2^2$. Because 3 is closer to 4, it makes sense that $\sqrt{3}$ is closer to 2 than 1. Consequently, we test 1.5 because it is between 1 and 2. Because $(1.5)^2 = 2.25$ and $2.25 < 3$, we know that $1.5 < \sqrt{3} < 2$. Each time, we analyze the results of our guess and use the information to select the next value. Continuing in this way, we finally zoom in on the $\sqrt{3}$.

$1.7 < \sqrt{3} < 1.8$ because $(1.7)^2 = 2.89 < 3 < 3.24 = (1.8)^2$
$1.7 < \sqrt{3} < 1.75$ because $(1.7)^2 = 2.89 < 3 < 3.0625 = (1.75)^2$
$1.71 < \sqrt{3} < 1.74$ because $(1.71)^2 = 2.9241 < 3 < 3.0276 = (1.74)^2$
$1.73 < \sqrt{3} < 1.74$ because $(1.73)^2 = 2.9929 < 3 < 3.0276 = (1.74)^2$
$1.73 < \sqrt{3} < 1.735$ because $(1.73)^2 = 2.9929 < 3 < 3.0010225 = (1.735)^2$

The numbers on either side of the last inequality match to two decimal places, so we conclude that $\sqrt{3} \approx 1.73$.

Looking Back. Although the process is tedious, it can be carried out to any number of decimal places. We seldom use this method because most calculators now have a built-in square root key, $\boxed{\sqrt{}}$. We can use this key to check the answer by pressing $\boxed{\sqrt{}}$ 3 $\boxed{)}$ $\boxed{\text{ENTER}}$, which gives the answer of $\sqrt{3} \approx 1.73205080757$.

The idea of square roots can be extended to other roots. For instance, consider the equation $x^3 = 27$. The solution to this equation is the number that, when cubed, is equal to 27. Because 3 is one solution, we call 3 the **cube root** of 27 and write $\sqrt[3]{27} = 3$. In general, the solution to any equation of the form $b^n = a$ is called the nth root of a, or $\sqrt[n]{a} = b$. It is the number that, when raised to the nth power, is equal to b.

Definition of nth Roots

Let a be any number and n be a natural number.

1. If $a \geq 0$, then $\sqrt[n]{a} = b$ where $b \geq 0$ if and only if $b^n = a$.
2. If $a < 0$ and n is odd, then $\sqrt[n]{a} = b$ if and only if $b^n = a$.

$\sqrt[n]{a}$ is read as "The nth root of a" and is called a **radical** expression. a is the **radicand,** and n is the **index.** If no index is given, the radical is understood to be a square root. The definition is split into two pieces based on the sign of the radicand. If a is positive, then we can find the nth root of a when n is any natural number. However, if a is negative, we cannot find even roots of a. For instance, $\sqrt[4]{-16}$ is the number x that, when raised to the fourth power, is equal to (-16), or $x^4 = (-16)$. However, there is no such number because any rational or irrational number raised to the fourth power must be positive. Consequently, $\sqrt[4]{-16}$ is undefined.

Example 8.7 — Interpreting Radical Expressions

Communication

Interpret each radical, and then determine whether it is rational, irrational, or undefined.

a. $\sqrt[3]{-9}$ **b.** $\sqrt[3]{64}$ **c.** $\sqrt[6]{-14}$

Solution

a. $\sqrt[3]{-9}$ is the number x that, when cubed, equals (-9), or $x^3 = (-9)$. No rational number, when cubed, is equal to (-9), yet the definition states that such a number is possible. Consequently, $\sqrt[3]{-9}$ must be irrational.
b. $\sqrt[3]{64}$ is the number x that, when cubed, equals 64, or $x^3 = 64$. The number 4 satisfies this property, so $\sqrt[3]{64}$ is rational.
c. $\sqrt[6]{-14}$ is the number x that, when raised to the sixth power, equals (-14), or $x^6 = (-14)$. No number satisfies this equation, so $\sqrt[6]{-14}$ is undefined.

Real Numbers

The rational and irrational numbers are disjoint, so if we combine them, we get a new set called the **real numbers.**

Definition of the Real Numbers

The set of real numbers, denoted $\mathbb{R}$, is the union of the rational numbers with the irrational numbers.

Intuitively, we can think of the real numbers as the set of all possible decimals: terminating, repeating, and nonterminating, nonrepeating. By definition, the real numbers include both the rational and irrational numbers. As a consequence, the real numbers also contain the integers, whole numbers, and natural numbers. Figure 8.11 demonstrates the relationships among these sets.

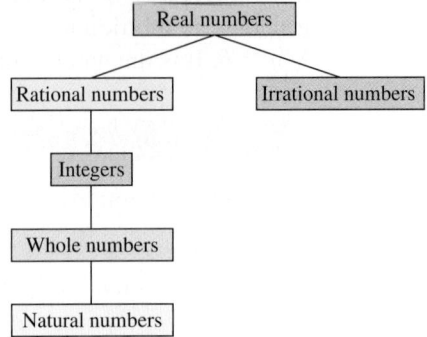

Figure 8.11 Relationships among number sets

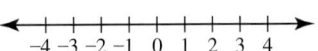

Figure 8.12 The real number line

The real numbers are the first set of numbers to represent a continuum. In other words, there is a one-to-one correspondence between the real numbers and the points on a number line. Certain points on the number line, like the integers, are marked simply for reference purposes (Figure 8.12).

Ordering Real Numbers

Like the rational numbers, different real numbers can represent the same value. For instance, the money amounts in Figure 8.13 are equal but are represented with different decimals. As the figure shows, any zero to the right of the last nonzero digit in a decimal has no effect on its value. Even though it is sometimes convenient to have zeros to the right of the last digit—such as when we are making precise measurement—we tend to use 0.2 rather than 0.20 because the former version is more concise.

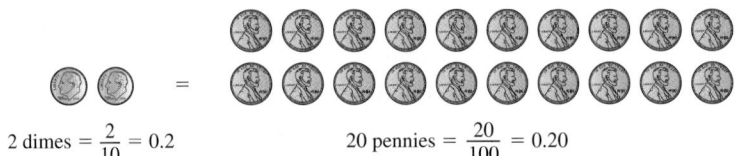

$$2 \text{ dimes} = \frac{2}{10} = 0.2 \qquad 20 \text{ pennies} = \frac{20}{100} = 0.20$$

Figure 8.13 Different real numbers for the same value

We often need to know when one decimal is larger or smaller than another. If the decimal has just one or two decimal places, we can use models to order them. For instance, grid paper can be used to compare 0.65 to 0.6 (Figure 8.14). Because 0.65 uses more area, $0.65 > 0.6$.

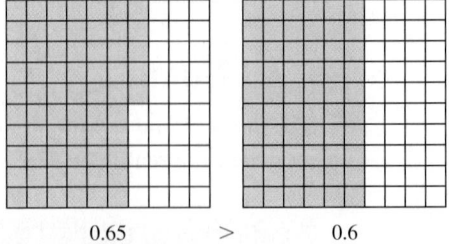

$$0.65 \quad > \quad 0.6$$

Figure 8.14 Ordering decimals with grid paper

We can also order terminating decimals by converting them to fractions. For example, we know that $-1.73 > -2.43$ because $-\dfrac{173}{100} > -\dfrac{243}{100}$.

These methods work well when the number of decimal places is small. But they are of little use with repeating decimals or irrational numbers. In such cases, we can order the numbers by directly comparing the place values.

| Example 8.8 | Ordering Decimals |

Order the following decimals from smallest to largest.

$$0.262 \quad 0.\overline{26} \quad 0.\overline{262} \quad 0.26 \quad 0.2\overline{6} \quad 0.26\overline{2}$$

Solution We line up the decimal points and then write each decimal to five or six digits.

$$
\begin{aligned}
0.262 &= 0.262 \\
0.\overline{26} &= 0.262626\ldots \\
0.\overline{262} &= 0.262262\ldots \\
0.26 &= 0.26 \\
0.2\overline{6} &= 0.266666\ldots \\
0.26\overline{2} &= 0.262222\ldots
\end{aligned}
$$

After comparing the place values directly, the order of numbers from the smallest to largest is $0.26, 0.262, 0.26\overline{2}, 0.\overline{262}, 0.\overline{26}$, and $0.2\overline{6}$.

As with the integers and rational numbers, the order of the negative real numbers is the reverse of the positive real numbers. Also, the notion of order on a number line is maintained: Numbers farther to the left are less than numbers farther to the right. Like the rational numbers, the real numbers are also dense.

| Theorem 8.2 | Denseness Property of the Real Numbers |

If a and b are two real numbers, then there exists a real number c such that $a < c < b$.

Rounding and Truncating Decimals

The infinite nature of many decimals can make decimal computations awkward. Consequently, when computing with decimals, we typically use approximations of the decimals rather than the decimals themselves. As with the whole numbers, rounding and truncating can be used to make our approximations. In doing so, we usually approximate to only one, two, or three decimals.

| Example 8.9 | Approximating Decimals |

Approximate each number as directed.

a. Truncate 3.451 to the nearest hundredth.

b. Truncate 102.456 to the nearest tenth.

c. Round 56.72 to the nearest one.

d. Round 3.6873 to the nearest thousandth.

Solution

a. 3.451 truncated to the nearest hundredth is 3.45.

b. 102.456 truncated to the nearest tenth is 102.4.

c. 56.72 rounded to the nearest one is 57.

d. 3.6873 rounded to the nearest thousandth is 3.687.

Check Your Understanding 8.1-B

1. Interpret each radical, and then determine whether it is rational, irrational, or undefined.

   a. $\sqrt{11}$ **b.** $\sqrt[4]{27}$ **c.** $\sqrt[7]{-31}$ **d.** $\sqrt[3]{-2.2}$ **e.** $\sqrt{-144}$

2. Find an example of a number that satisfies each statement when possible.

   a. A number that is real but not rational.

   b. A number that is real but not irrational.

   c. A number that is neither rational nor irrational.

   d. A number that is rational but not real.

3. Find the smallest number in each pair.

   a. 1.234 or 1.2341   **b.** -5.6342 or $-5.634222\ldots$ **c.** $0.5\overline{16}$ or $0.\overline{516}$

4. Approximate each number as directed.

   a. Round 209.8834 to the nearest hundredth.

   b. Round 3.4712 to the nearest tenth.

   c. Truncate 5.634 to the tenths.

   d. Truncate 7.89112 to the thousandths.

Talk About It Even though both the rational numbers and the real numbers are dense, only the real numbers are considered to be a continuum. What does this mean in terms of how the irrational numbers are positioned among the rational numbers on a number line?

Problem Solving

Activity 8.2
Use Example 8.6 to devise a strategy for estimating $\sqrt[3]{9}$ to two decimal places.

SECTION 8.1   Learning Assessment

▪ Understanding the Concepts

1. What advantages do decimals offer over the fractions for expressing part-to-whole relationships?

2. What purpose does the decimal point serve?

3. What is the difference between a terminating decimal, a repeating decimal, and an irrational number?

4. Is it possible to write a terminating decimal as a repeating decimal? Explain.

5. $\dfrac{22}{7}$ is often used to represent π. Is $\dfrac{22}{7} = \pi$? Explain.

6. Are all square roots irrational numbers? Explain.

7. How can we use place value to order decimals?

8. What does it mean for the real numbers to be dense?

9. Why are rounding and truncating important to working with the decimals?

▪ Representing the Mathematics

10. Give the place value of each digit, and then write the name of the decimal.

    a. 0.2   **b.** 0.34   **c.** 5.08   **d.** 42.013

11. Write each decimal fraction as a decimal.

    a. $\dfrac{9}{10}$ **b.** $\dfrac{11}{100}$ **c.** $\dfrac{41}{10}$ **d.** $-\dfrac{7,051}{1,000}$

12. Write each decimal as a decimal fraction.

    a. 0.5   **b.** 0.63   **c.** 0.157   **d.** 0.07

13. Draw a diagram like the one in Figure 8.6 to represent the decimal $0.\overline{2}$.

14. Draw a diagram like the one in Figure 8.8 to represent the irrational number $0.5312\ldots$.

15. What decimal is represented in each diagram?

    a.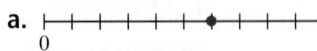

b.
| $\frac{1}{10}$ | $\frac{1}{10}$ | $\frac{1}{10}$ | $\frac{1}{10}$ | $\frac{1}{10}$ | $\frac{1}{10}$ | $\frac{1}{10}$ | $\frac{1}{10}$ |

c.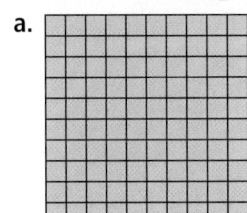

d.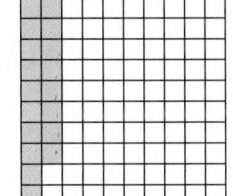

16. What decimal is represented in each diagram?

a.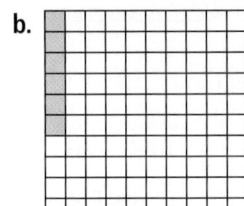

b.

c.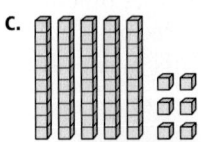

d.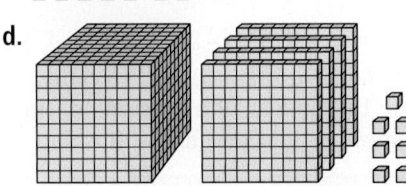

17. Use the money model to represent each number.

    a. 0.5   **b.** 0.31   **c.** 1.26   **d.** 0.25

18. Use grid paper to represent each decimal.

    a. 0.7   **b.** 0.82   **c.** 2.31

19. Represent each number on the same number line.

    a. -1.2 **b.** $-1.\overline{4}$ **c.** $1.3924\ldots$ **d.** $\sqrt{2}$

20. Interpret each radical and then give the radicand and index.

    a. $\sqrt{13}$ **b.** $\sqrt[3]{15}$ **c.** $\sqrt[4]{58}$ **d.** $\sqrt[3]{-1.69}$

21. Draw a Venn diagram that shows the relationships between the real numbers, the rational numbers, the irrational numbers, the integers, the whole numbers, and the natural numbers.

■ Exercises

22. Give the name of the place value associated with each power of ten.

    a. 10^{-2} **b.** 10^{-5} **c.** 10^{-1}

    d. 10^{0} **e.** 10^{-7}

23. Write each decimal in expanded notation using decimal fractions.

    a. 0.47   **b.** 1.02   **c.** 205.006   **d.** 4.1809

24. Write each decimal in expanded notation using exponents.

    a. 0.595   **b.** 2.13   **c.** 300.127   **d.** 4001.0014

25. Write each number in decimal notation.

    a. $\left(6 \cdot \dfrac{1}{100}\right) + \left(7 \cdot \dfrac{1}{1,000}\right)$

    b. $(5 \cdot 100) + (6 \cdot 10) + \left(3 \cdot \dfrac{1}{10}\right) +$
$\left(8 \cdot \dfrac{1}{1,000}\right) + \left(1 \cdot \dfrac{1}{10,000}\right)$

    c. $(4 \cdot 10^2) + (6 \cdot 10^0) + (3 \cdot 10^{-1}) + (8 \cdot 10^{-2})$

    d. $(7 \cdot 10^4) + (5 \cdot 10^3) + (5 \cdot 10^0) + (1 \cdot 10^{-2}) +$
$(1 \cdot 10^{-3})$

26. Determine whether each fraction has a terminating decimal representation. If so, find it by converting the denominator to a power of ten.

    a. $\dfrac{2}{5}$ **b.** $\dfrac{19}{24}$ **c.** $\dfrac{17}{40}$

    d. $-\dfrac{16}{18}$ **e.** $-\dfrac{48}{60}$

27. Give the repetend and period of each decimal.

    a. $0.45777\ldots$

    b. $0.3141414\ldots$

    c. $0.6\overline{78}$

    d. $12.\overline{21}$

28. Give an example of a repeating decimal with a period of the given length.

    a. 1   **b.** 2   **c.** 3   **d.** 4

29. Write an irrational number using only 0s and 1s.

30. List ten different irrational square roots.

31. Determine whether each radical is rational, irrational, or not possible.

    a. $\sqrt{72}$ **b.** $\sqrt{-64}$ **c.** $\sqrt[3]{-27}$ **d.** $\sqrt[5]{81}$

32. Determine whether each radical is rational, irrational, or not possible.

    a. $-\sqrt{81}$ **b.** $\sqrt{16}$ **c.** $\sqrt[3]{36}$ **d.** $\sqrt[8]{-32}$

33. Use the method from Example 8.6 to estimate $\sqrt{5}$ to two decimal places.

34. Use the method from Example 8.6 to estimate $\sqrt{8}$ to two decimal places.

35. For each number in the table, check any set to which the number belongs.

	$\mathbb{N}$	W	$\mathbb{Z}$	$\mathbb{Q}$	Irrationals	$\mathbb{R}$
0.83						
$\sqrt{25}$						
$\sqrt[3]{6}$						
$1.\overline{4}$						
$\dfrac{9}{7}$						
$0.1234\ldots$						

36. Determine which is the larger number in each pair.
  a. 0.15 or 0.16
  b. -0.45 or -0.46
  c. $0.\overline{78}$ or $0.7\overline{8}$

37. Order from least to greatest.

$$4.6\overline{1} \quad 4.\overline{61} \quad 4.616 \quad 4.61\overline{6} \quad 4.\overline{616}$$

38. Order from least to greatest.

$$-2.\overline{35} \quad -2.3\overline{5} \quad -2.35\overline{3} \quad -2.35 \quad -2.\overline{353}$$

39. Find five numbers between 4.1 and 4.2.

40. Find five numbers between $4.\overline{1}$ and $4.\overline{12}$.

41. Round each number to the specified place value.
  a. 34.215 to the nearest one
  b. 100.0032 to the nearest hundredth
  c. 43.8901 to the nearest tenth
  d. 2.01913 to the nearest ten-thousandth

42. Truncate each number to the specified place value.
  a. 3.298 to the ones
  b. 312.9850 to the hundredths
  c. 67.341 to the tenths
  d. 1.8007 to the thousandths

▪ Problems and Applications

43. How many different ways are there to model the following decimals using only dollars, dimes, and pennies?
  a. 0.34   **b.** 0.65   **c.** 1.09   **d.** 1.2

44. What is the prime factorization of 10^{50}?

45. Find each of the following sets.
  a. $\mathbb{R} \cup \mathbb{Z}$ **b.** $\mathbb{Q} \cup \mathbb{N}$
  c. $\mathbb{R} - \mathbb{Q}$ **d.** $\mathbb{R} \cap W$

46. If the shorter sides of a right triangle have lengths 1 in. and 2 in., what is the length of the longest side?

47. If the length of the longest side of a right triangle is $\sqrt{6}$ and one of the shorter sides has length 2, what is the length of the other side?

48. Find the values of w, x, y, and z in the diagram. If the triangles continue in the same pattern, what would be the length of the longest side in the
  a. 24th triangle?
  b. 50th triangle?
  c. nth triangle?

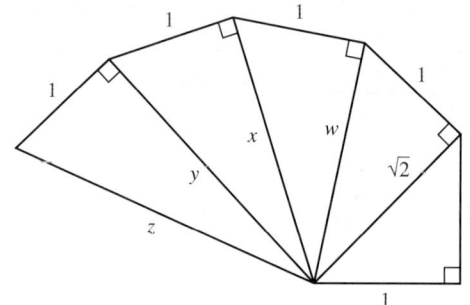

49. Find three right triangles that have sides with lengths that are whole numbers.

50. Devise and use a strategy similar to the one used in Example 8.6 to estimate $\sqrt[3]{7}$ to two decimal places.

51. Decimals can also be written in bases other than ten. Write each of the following decimals in expanded notation with exponents. What values would they have in base-ten?
  a. 0.12_{five} **b.** 0.431_{five}
  c. 0.46_{seven} **d.** 0.112_{three}

52. The following table shows the times of eight runners in the 100-m dash at a high school track meet. In what order did the runners finish?

Runner	Time in Seconds	Runner	Time in Seconds
Doug	12.43	Steve	13.09
Malachi	13.21	Trevor	12.90
Caleb	12.69	Reilly	13.15
Ron	12.11	Anthony	12.78

▪ Mathematical Reasoning and Proof

53. True or false? If false, explain why.
  a. Every real number is a rational number.
  b. Every irrational number is a real number.
  c. No rational number is an irrational number.
  d. Some integers are irrational numbers.

54. a. Is $\dfrac{1.4}{2.5}$ a rational number? Explain.

b. Is $\dfrac{\sqrt{2}}{3}$ a rational number? Explain.

55. a. If a is a positive integer less than 40, can it be determined whether $\dfrac{a}{40}$ is a terminating or repeating decimal representation? Explain.

b. If a is a positive integer less than 21, can it be determined whether $\dfrac{a}{21}$ is a terminating or repeating decimal representation? Explain.

c. If a is a positive integer less than 15, can it be determined whether $\dfrac{a}{15}$ is a terminating or repeating decimal representation? Explain.

56. If possible, write an equation, other than $x = -\sqrt{2}$, that has $-\sqrt{2}$ as a solution?

57. Show that $\sqrt[3]{2}$ is irrational.

58. Use the fact that $\sqrt{2}$ is irrational to show that $3 + \sqrt{2}$ is irrational.

■ Mathematical Communication

59. How would you use grid paper to explain to a student why 0.43 is less than 0.53?

60. Write a story problem that requires a solution in which decimals are ordered.

61. Andy claims that 0.045 is larger than 0.35, because 45 is bigger than 35. What is the error in his thinking? How might you as his teacher correct it?

62. Laquisha reads the number 437 as "four hundred and thirty-seven." What error has she made in reading the number?

63. Marisol claims that $\dfrac{31}{20}$ cannot be written as a terminating decimal because it is bigger than 1. Is she correct?

64. If you were teaching decimals, describe how you might use the names "tenth," "hundredth," and so on to connect the decimal fractions to the decimals.

■ Building Connections

65. Write a paragraph or two summarizing the mathematical structure of the base-ten numerals. Describe how the different subsets (i.e., whole numbers, integers, etc.) are interconnected and built on one another.

66. a. Give three different situations when it is easier to use fractions than decimals.

b. Give three different situations when it is easier to use decimals than fractions.

c. Compare and contrast the different situations you listed. Do they have any similarities and differences?

67. Research indicates that even high school students have difficulty interpreting place values smaller than the hundredths. Why do you think this is true? Can you devise a model for demonstrating place values smaller than hundredths?

68. Under what circumstances is the decimal 0.20 more useful than the decimal 0.2?

69. Decimals are represented in different ways around the world. Do an Internet search to find different ways that decimals are written. Use these methods to write each decimal.

a. 0.45

b. 0.031

c. 4.59

d. 21.05

70. Find an Internet site that allows you to calculate π to one hundred, one thousand, and one million decimal places. How might you use the printouts from such a site to demonstrate the relative size of different powers of ten?

SECTION 8.2 Decimal and Real Number Operations

Decimal operations are based on whole- and rational-number operations, so we can compute them by using what we have learned. The only new concern is how to place the decimal point in the answer correctly. In general, we limit computations to terminating decimals. If we have repeating decimals or irrational numbers, we typically round them to a terminating decimal before computing.

Adding and Subtracting Decimals

We can think about adding and subtracting decimals in a number of ways. One is to write each decimal as the sum of decimal fractions and then use rational number addition to add. For instance, we can compute 2.34 + 1.25 as

$$2.34 + 1.25 = \left(2 + \frac{3}{10} + \frac{4}{100}\right) + \left(1 + \frac{2}{10} + \frac{5}{100}\right) \quad \text{Expanded notation}$$

$$= (2 + 1) + \left(\frac{3}{10} + \frac{2}{10}\right) + \left(\frac{4}{100} + \frac{5}{100}\right) \quad \text{Commutative property}$$

$$= \left(3 + \frac{5}{10} + \frac{9}{100}\right) \quad \text{Rational number addition}$$

$$= 3.59 \quad \text{Decimal notation}$$

Explorations Manual 8.4

Another way to add 2.34 + 1.25 is to use base-ten blocks. We represent 2.34 as 2 blocks, 3 flats, and 4 longs, and 1.25 as 1 block, 2 flats, and 5 longs (see Figure 8.15). We combine the blocks and then count to get our answer. Doing so gives us 3 blocks, 5 flats, and 9 longs, so 2.34 + 1.25 = 3.59.

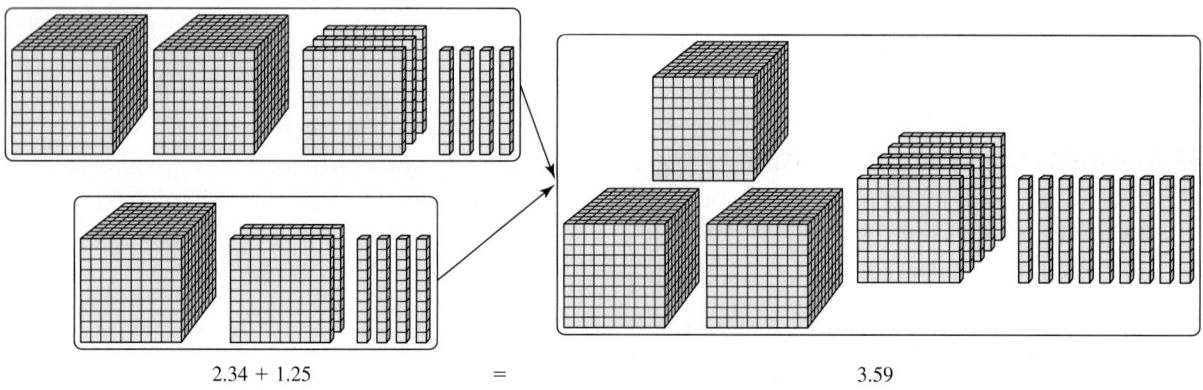

2.34 + 1.25   =   3.59

Figure 8.15 Using base-ten blocks to add decimals

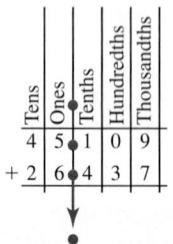

Figure 8.16 Placing the decimal point when adding

Adding decimals with these methods makes two things clear. First, the base-ten blocks show that we can add decimals in the same way as large whole numbers. Specifically, two decimals can be added by combining groups of ten, regrouping as necessary, and then counting. In other words, the algorithms used with whole-number addition also work with decimals. Second, both methods indicate that we need to add digits in the same place value—ones to ones, tenths to tenths, hundredths to hundredths, and so on. To ensure doing so, we line up the decimal points in the addends prior to computing (Figure 8.16). The decimal point is placed in the sum by putting it directly beneath the decimal points in the addends.

We find the difference between two decimals in much the same way. The procedure for both operations is summarized as follows:

Align the decimal points in the original numbers to ensure that common place values are added or subtracted. If necessary, affix zeros to the end of the decimals to make the computation. Add or subtract the numbers using whole-number algorithms. Place the decimal point in the answer directly beneath the decimal points in the original numbers.

Example 8.10   Adding and Subtracting Decimals

Compute.

a. $32.6 + 11.19$ **b.** $14.3 - 8.761$

Solution We proceed as stated, keeping in mind that it may be necessary to affix zeros to the end of a decimal to complete the computation.

a.
$$\begin{array}{r} 32.6 \\ +11.19 \end{array} \quad \rightarrow \quad \begin{array}{r} 32.60 \\ +11.19 \\ \hline 43.79 \end{array}$$

b.
$$\begin{array}{r} 14.3 \\ -8.761 \end{array} \quad \rightarrow \quad \begin{array}{r} 14.300 \\ -8.761 \\ \hline 5.539 \end{array}$$

Decimal addition and subtraction can also be done on a calculator by using the appropriate operation and the ⬚ key. When doing so, it helps to estimate the answer before computing in order to know its relative size. We can use estimation techniques learned with the whole numbers.

Example 8.11   Buying Furniture

Application

Esther is furnishing a sunroom. She likes a set of five pieces, the prices of which are given in the following table.

Item	Sofa	Love Seat	Chair	Coffee Table	End Table
Cost	$859.65	$789.59	$435.63	$339.47	$224.05

Using a calculator, she finds that the total price is $1,937.76, which is below the $2,000 she plans to spend. When she checks out, the total before tax is well over $2,500. What went wrong? What is the actual cost of the furniture set?

Solution Whenever we use technology, there is always the chance of pressing a wrong key. In this case, the actual cost of the furniture is $2,648.39, which is $710.63 more than Esther expected. Her mistake was a misplacement of the decimal point; she added $78.959 rather than $789.59. She could have checked her answer by estimating the total cost to be about $860 + $790 + $440 + $340 + $220 = $2,650.

Multiplying Decimals

Explorations Manual 8.5

Whole-number procedures can also be used to compute the product of two decimals. In this case, we do not place the decimal point as with a sum or difference. Instead, we place the decimal point using the number of decimal places in the two factors. To understand why, consider the next example.

Example 8.12   Using Fractions to Multiply Decimals

Use fractions to compute $3.37 \cdot 1.6$.

Solution We write each number as a decimal fraction and then use rational number multiplication to compute:

$$3.37 \cdot 1.6 = \frac{337}{100} \cdot \frac{16}{10} = \frac{337 \cdot 16}{100 \cdot 10} = \frac{5{,}392}{1{,}000} = 5.392$$

This example shows that the number of decimal places in the product is equal to the total number of decimal places in the factors. This is due to the fact that, when we multiply decimal place values, we are literally taking a piece of a piece, just as we did with fractions. The product naturally leads to smaller parts of the whole, which in turn require smaller decimal place values to represent them. Specifically, when multiplying the smallest two place values in the last example, we took a hundredth of a tenth, which gave us a thousandth of the whole and the reason for the third decimal place value in the product. Generalizing these ideas, we get the following algorithm for decimal multiplication:

Multiply the numbers as whole numbers. Place the decimal point in the product by counting the total number of decimal places in the factors. If there are n decimal places in the first factor and m *in the second, then there are* n + m *decimal places in the product.*

Explorations
Manual
8.6

There are $n + m$ digits to the right of the decimal point because, if we write the factors as decimal fractions, then we are dividing the product of their numerators by two powers of ten, 10^n from the first factor and 10^m from the second. Consequently, we are dividing by a total of $10^n \cdot 10^m = 10^{n+m}$.

Example 8.13   Multiplying Decimals

Multiply $42.6 \cdot 5.38$.

Solution We multiply using the standard algorithm and then place the decimal point in the product.

$$
\begin{array}{r}
42.6 \\
\times 5.38 \\
\hline
3408 \\
12780 \\
213000 \\
\hline
229.188
\end{array}
$$

42.6   1 digit to right of decimal point
×5.38   2 digits to right of decimal point

229.188   1 + 2 = 3 digits to right of decimal point

As with addition and subtraction, calculators can be used to perform decimal multiplication quickly and efficiently.

Example 8.14   Using a Calculator to Multiply Decimals

Estimate and then use a calculator to find $105.8 \cdot 1.8$.

Solution To make our initial estimate, we round to compatible numbers and multiply, or $105.8 \cdot 1.8 \approx 100 \cdot 1.8 = 180$. We then compute the exact value on a calculator using the keystrokes 105 [.] 8 [×] 1 [.] 8 [ENTER], which gives the exact value of 190.44.

Real number addition and multiplication satisfy the following properties.

Theorem 8.3   Properties of Real Number Addition and Multiplication

If *a*, *b*, and *c* are real numbers, then the following hold true.

Closure property: $a + b$ and $a \cdot b$ are unique real numbers.
Commutative property: $a + b = b + a$ and $a \cdot b = b \cdot a$.

Associative property:	$a + (b + c) = (a + b) + c$ and $a \cdot (b \cdot c) = (a \cdot b) \cdot c.$
Identity property:	$a + 0 = a = 0 + a$ and $a \cdot 1 = a = 1 \cdot a.$
Inverse property:	$a + (-a) = 0 = (-a) + a$ and $a \cdot \dfrac{1}{a} = 1 = \dfrac{1}{a} \cdot a.$
Distributive property of multiplication over addition:	$a \cdot (b + c) = (a \cdot b) + (a \cdot c).$
Zero multiplication property:	$a \cdot 0 = 0 = 0 \cdot a.$

Dividing Decimals

As with the other operations, we can adapt the standard algorithm for whole-number division to divide decimals. To do so, we must address two concerns. The first is how to place the decimal point. If the divisor is a whole number, as in $157.8 \div 14$, then we place the decimal point in the quotient directly above the decimal point in the dividend. Doing so maintains the place values.

$$14 \overline{)157.8}$$

If the divisor is a decimal, then the convention is to remove the decimal point from the divisor. Although this is not necessary, it can make the quotient easier to compute. We can remove the decimal point for the following reason. If we represent the quotient of two decimals as a fraction, then we can use the Fundamental Law of Fractions to multiply the numerator and denominator by a common power of ten to remove the decimal from the denominator. This reduces the quotient to one with a whole-number divisor. For instance, $85.86 \div 5.3$ can be rewritten as

$$5.3\overline{)85.86} \rightarrow \frac{85.86}{5.3} = \frac{85.86}{5.3} \cdot 1 = \frac{85.86}{5.3} \cdot \frac{10}{10} = \frac{858.6}{53} \rightarrow 53\overline{)858.6}.$$

To shorten the process, we simply move the decimal points in the divisor and the dividend the same number of places to remove it from the divisor.

$$5.3\overline{)85.86} \longrightarrow 53\overline{)858.6}$$

After positioning the decimal point, we proceed with the standard algorithm until reaching the last digit in the dividend. If necessary, we can affix zeros to the right of the last digit in the dividend to continue the division and avoid remainders.

The second concern is knowing when and how to conclude the division. In general, the division is concluded if one of three criteria is met.

1. *The division concludes naturally as a terminating decimal.* In some cases, the division terminates naturally.
2. *A repeating decimal is obtained.* If the quotient seems to repeat, we carry out the division through at least two iterations of the repetend to be sure it is a repeating decimal.
3. *A specified number of decimal places is reached.* If the division does not terminate or repeat within one or two decimal places, we may carry it out to a specified number of places. Carry the division out to one more than the specified number of places and then round back.

Example 8.15 Dividing Decimals

Compute $8.59 \div 2.3 \approx 3.73$ to two decimal places.

Solution We compute the quotient to three decimal places and round back to two: $8.59 \div 2.3 \approx 3.73.$

$$
2.3\overline{)8.59} \quad \rightarrow \quad
\begin{array}{r}
3.734\ldots \\
23\overline{)85.900\ldots} \\
\underline{69} \\
169 \\
\underline{161} \\
80 \\
\underline{69} \\
110 \\
\underline{92} \\
18\ldots
\end{array}
\quad \leftarrow \text{Affixed zeros}
$$

Because a fraction bar is equivalent to division, we can now use long division to convert any fraction to a decimal. To do so, we divide the numerator by the denominator.

Example 8.16   Converting Fractions to Decimals

Convert each fraction to a decimal.

a. $\dfrac{3}{8}$ **b.** $\dfrac{4}{9}$

Solution

a. $\dfrac{3}{8} \quad \rightarrow \quad$
$$
\begin{array}{r}
0.375 \\
8\overline{)3.000} \\
\underline{24} \\
60 \\
\underline{56} \\
40 \\
\underline{40} \\
0
\end{array}
$$

b. $\dfrac{4}{9} \quad \rightarrow \quad$
$$
\begin{array}{r}
0.444\ldots \\
9\overline{)4.000\ldots} \\
\underline{36} \\
40 \\
\underline{36} \\
40 \\
\underline{36} \\
40 \\
\ldots
\end{array}
$$

Example 8.16 provides some evidence that every rational number is either a repeating or a terminating decimal (Section 8.1). But how can we be sure it is always true? Theorem 8.1 guarantees that decimals that can be rewritten as a decimal fraction always have a terminating decimal representation. But what about the rational numbers that do not satisfy this theorem? To understand why such fractions repeat, consider the following argument using $\dfrac{5}{7}$.

When 5 is divided by 7, the division algorithm guarantees that only seven remainders are possible: 0, 1, 2, 3, 4, 5, and 6. If at any time throughout the course of the division, the remainder is 0, then the division terminates. However, if 0 never occurs, then eventually one of the other six numbers must reoccur as the remainder, causing the division to repeat. In the case of $\dfrac{5}{7}$, when the remainder of 5 reappears, the quotient begins to repeat.

$$\frac{5}{7} \quad \rightarrow \quad \begin{array}{r} 0.7142857 \\ 7\overline{)5.0000000} \\ \underline{49} \\ 10 \\ \underline{7} \\ 30 \\ \underline{28} \\ 20 \\ \underline{14} \\ 60 \\ \underline{56} \\ 40 \\ \underline{35} \\ 5\dots \end{array}$$

This leads to the following theorem.

Theorem 8.4   Rational Numbers and Decimals

Every rational number is either a terminating or a repeating decimal.

Theorem 8.4 indicates that repeating and terminating decimals are not an extension of the rational numbers, but rather a different representation of them. Also, if $\frac{a}{b}$ is reduced and it has a repeating decimal representation, then the repetend is at most $b - 1$ digits long.

Finally, when performing decimal division on a calculator, we should realize that calculators often express repeating decimals and irrational numbers as terminating decimals. For instance, if we compute $\frac{2}{3}$ by pressing 2 ÷ 3, the display will likely read 0.6666666 or 0.6666667 depending on whether the calculator rounds. However, long division reveals that the true decimal representation of $\frac{2}{3}$ is $0.\overline{6}$.

Check Your Understanding 8.2-A

1. Compute each sum or difference.

  a. $45.6 + 32.8$ **b.** $506.78 + 301.003$

  c. $89.54 - 32.76$ **d.** $601.3 - 59.034$

2. Compute each product or quotient.

  a. $14.1 \cdot 8.7$ **b.** $8.69 \cdot 0.51$ **c.** $3.5\overline{)456.78}$ **d.** $4.1\overline{)5.356}$

3. Estimate the answer to each problem, and then use a calculator to compute the result.

  a. $345.67 + 651.08$ **b.** $791.95 - 304.8$ **c.** $1.49 \cdot 4.01$

4. Convert each fraction to a decimal.

  a. $\frac{5}{16}$ **b.** $\frac{5}{6}$ **c.** $\frac{4}{15}$

Talk About It Many students would rather compute with decimals than with fractions. Why do you think this is true? Do you think that it is important to be comfortable in computing with all types of numbers? Why or why not?

Problem Solving

Activity 8.3

Compute $\frac{4}{23}$ on a calculator. Theorem 8.4 guarantees that $\frac{4}{23}$ will be a repeating decimal, yet the calculator shows only a portion of the repetend. Determine how you can use a calculator to find the rest of the repetend.

Multiplying and Dividing a Decimal by a Power of Ten

Given the decimal operations, we now establish several other useful ideas. First, consider the effect of multiplying or dividing any decimal by a power of ten. For instance, consider $567.4 \cdot 10^2$. Writing the expression in expanded notation enables us to see that multiplying by 10^2 moves the decimal point two places to the right:

$$
\begin{aligned}
567.4 \cdot 10^2 &= \left[(5 \cdot 10^2) + (6 \cdot 10^1) + (7 \cdot 10^0) + (4 \cdot 10^{-1}) \right] \cdot 10^2 \\
&= \left[(5 \cdot 10^2) \cdot 10^2 \right] + \left[(6 \cdot 10^1) \cdot 10^2 \right] + \left[(7 \cdot 10^0) \cdot 10^2 \right] + \left[(4 \cdot 10^{-1}) \cdot 10^2 \right] \\
&= (5 \cdot 10^{2+2}) + (6 \cdot 10^{1+2}) + (7 \cdot 10^{0+2}) + (4 \cdot 10^{-1+2}) \\
&= (5 \cdot 10^4) + (6 \cdot 10^3) + (7 \cdot 10^2) + (4 \cdot 10^1) \\
&= 56{,}740
\end{aligned}
$$

On the other hand, dividing 567.4 by 10^2 is the same as multiplying by 10^{-2}. This has the effect of moving the decimal two places to the left:

$$
\begin{aligned}
567.4 \div 10^2 &= 567.4 \cdot 10^{-2} \\
&= \left[(5 \cdot 10^2) + (6 \cdot 10^1) + (7 \cdot 10^0) + (4 \cdot 10^{-1}) \right] \cdot 10^{-2} \\
&= (5 \cdot 10^2) \cdot 10^{-2} + (6 \cdot 10^1) \cdot 10^{-2} + (7 \cdot 10^0) \cdot 10^{-2} + (4 \cdot 10^{-1}) \cdot 10^{-2} \\
&= (5 \cdot 10^{2+(-2)}) + (6 \cdot 10^{1+(-2)}) + (7 \cdot 10^{0+(-2)}) + (4 \cdot 10^{-1+(-2)}) \\
&= (5 \cdot 10^0) + (6 \cdot 10^{-1}) + (7 \cdot 10^{-2}) + (4 \cdot 10^{-3}) \\
&= 5.674
\end{aligned}
$$

Generalizing these effects leads to the next theorem.

Theorem 8.5   Multiplying or Dividing by a Power of Ten

Let n be a real number and m be a positive integer.

1. The effect of multiplying n by 10^m is to move the decimal point m places to the right in n.
2. The effect of dividing n by 10^m, or multiplying by 10^{-m}, is to move the decimal point m places to the left in n.

Theorem 8.5 is important because we can use it to change repeating decimals into fractions and to write decimals in scientific notation.

Changing a Repeating Decimal to a Fraction

In the last section, we learned two things: (1) how to rewrite terminating decimals as fractions; (2) any rational number must have either a terminating or a repeating decimal representation. Consequently, we must be able to write a repeating decimal as a fraction. We now have all the tools to devise a strategy to do so.

Example 8.17 | Converting a Repeating Decimal to a Fraction

Problem Solving Convert $0.1232323\ldots$ into a fraction.

Solution

Understanding the Problem. Our task is to convert $0.1232323\ldots$ into a fraction.

Devising a Plan. The method for converting terminating decimals to fractions does not work because it relies on the smallest place value of the decimal, which is something repeating decimals do not have. Instead, we must somehow eliminate the repeating portion of the decimal by using decimal operations, properties of equations, and Theorem 8.5. Using these ideas, we can *write an equation* that we can multiply by two different powers of ten to shift the decimal to the left and to the right of the repetend. When we subtract the equations, the repeating portions of the subsequent decimals are aligned, so they subtract away. This leaves us with a simple equation to solve.

Carrying Out the Plan. We begin by setting a variable equal to our decimal:

$$x = 0.1232323\ldots$$

Next, multiply both sides of the equation by 1,000 to move the decimal to the right of the first repetend:

$$1{,}000x = 123.232323\ldots$$

Then, going back to our original equation, multiply each side by 10 to move the decimal to the left of the first repetend:

$$10x = 1.232323\ldots$$

Now subtract the second equation from the first.

$$
\begin{array}{r}
1{,}000x = 123.232323\ldots \\
- \ 10x = 1.232323\ldots \\
\hline
990x = 122
\end{array}
$$

Notice that the digits to the right of the decimal are aligned, so they subtract away. When we divide both sides of the resulting equation by 990, we have $x = \dfrac{122}{990} = \dfrac{61}{495}$. Consequently, $0.1232323\ldots = \dfrac{61}{495}$.

Looking Back. One way to check our work is to use a calculator and divide the numerator by the denominator, or $61 \div 495 = 0.1232323\ldots$. A fraction calculator enables us to enter the decimal and use the fraction conversion key, ⊳ F , to convert it to a fraction. Either way, the calculator confirms our answer.

The procedure in the last example can be used with any repeating decimal. It is summarized in the following four-step procedure:

Step 1: Let x equal the decimal.

Step 2: Multiply both sides of the equation in step 1 by a power of ten so that the decimal is moved to the right of the first repetend.

Step 3: Multiply both sides of the equation in step 1 by a power of ten so that the decimal is moved to the left of the first repetend.

Step 4: Subtract the equation in step 3 from the equation in step 2 and solve for x.

This procedure has some interesting consequences. For instance, if we use it to convert 0.999 . . . to a fraction, we get

Step 1:	$x = 0.999 \ldots$	
Step 2:	$10x = 9.999 \ldots$	Multiply equation of step 1 by 10.
Step 3:	$-x = 0.999 \ldots$	Multiply equation of step 1 by 1.
Step 4:	$9x = 9$	Subtract step 3 from step 2.

Dividing both sides of the last equation by 9, we get $x = 0.999 \ldots = \dfrac{9}{9} = 1$. This indicates that this number, like others similar to it, has two decimal representations. This may be surprising, but we can show the same fact by using other methods. For instance, we know that $\dfrac{1}{3} + \dfrac{2}{3} = 1$. However, if $\dfrac{1}{3} = 0.333 \ldots$ and $\dfrac{2}{3} = 0.666 \ldots$, then $\dfrac{1}{3} + \dfrac{2}{3} = 0.333 \ldots + 0.666 \ldots = 0.999 \ldots$. Consequently, we again have $1 = 0.999 \ldots$.

One final note must be made about converting decimals to fractions. Because the irrational numbers are disjoint with the rational numbers, no irrational number has a fractional representation that is the quotient of two integers. In other words, our method for converting decimals to fractions does not work with irrational numbers.

Operations in Scientific Notation

Theorem 8.5 also enables us to condense how we write numbers with many placeholders. We use a notation in which a decimal is multiplied by some power of ten. The notation is particularly useful in the sciences where we encounter very large and very small numbers like the ones in Table 8.3. Because of this, we often call this method of writing numbers **scientific notation.**

Table 8.3 Examples of numbers from science

Object	Length in Decimal Notation (m)	Length in Scientific Notation (m)
Atom	0.0000000001	1×10^{-10}
Virus	0.0000001	1×10^{-7}
Red blood cell	0.000001	1×10^{-6}
Mosquito	0.003	3×10^{-3}
Person	2	2×10^{0}
Cruising altitude of airplane	12,000	1.2×10^{4}
Radius of the earth	6,300,000	6.3×10^{6}
Distance from earth to sun	150,000,000,000	1.5×10^{11}
One light year	9,460,000,000,000,000	9.46×10^{15}

Definition of Scientific Notation

A number is in **scientific notation** if it is written in the form $c \times 10^{n}$, where $1 \le |c| < 10$ and n is any integer.

Several things must be mentioned about the definition. First, because $|c|$ is greater than or equal to 1, but strictly less than 10, there is exactly one nonzero digit to the left of the decimal point. Second, the exponent on the 10 reflects not only the number of places the decimal point moves, but also the direction. If the decimal is moved to the left, the exponent is positive. If it is moved to the right, the exponent is negative. Intuitively, this means that positive exponents go with relatively large numbers and negative exponents go with relatively small numbers. Third, if the number is negative, we keep the sign with the decimal part of the notation; it has no impact on the power of ten.

Example 8.18   **Writing Numbers in Scientific Notation**

Write each number as specified.

a. 45,000 in scientific notation

b. 0.00000687 in scientific notation

c. 3.45×10^6 in decimal notation

d. -9.843×10^{-5} in decimal notation

Solution

a. $45,000 = 4.5 \times 10^4$.

b. $0.00000687 = 6.87 \times 10^{-6}$.

c. $3.45 \times 10^6 = 3,450,000$.

d. $-9.843 \times 10^{-5} = -0.00009843$.

Explorations Manual 8.7

Many people find numbers written in scientific notation difficult to understand because they have trouble understanding their relative sizes.

Example 8.19   **Interpreting Scientific Notation**

Connections

Steven read in his science book that the radius of the earth is 6.3×10^6 m and that the distance from the earth to the sun is 1.5×10^{11} m. He says this cannot be correct because it implies the radius of the earth is about four times longer than the distance from the earth to the sun. What is the source of his confusion?

Solution Steven is considering only the decimal parts of the numbers and not the powers of ten. If we consider both, then we can find the relative size of the numbers by converting them to decimal notation and then dividing them. Specifically, the radius of the earth is approximately $\dfrac{1.5 \times 10^{11}}{6.3 \times 10^6} = \dfrac{150,000,000,000}{6,300,000} \approx 24,000$ times smaller than the distance from the earth to the sun.

We can also perform computations in scientific notation by using decimal operations and the rules for exponents.

Example 8.20   **Computations in Scientific Notation**

Compute the following.

a. $(2.34 \times 10^5) - (9.15 \times 10^3)$

b. $(3.01 \times 10^4) \cdot (6.14 \times 10^5)$

Solution

a. We must approach the difference $(2.34 \times 10^5) - (9.15 \times 10^3)$ with caution because the powers of ten are not the same. First rewrite 10^5 as $10^2 \times 10^3$, and then use the distributive property to pull out 10^3 before subtracting: $(2.34 \times 10^5) - (9.15 \times 10^3) = (2.34 \times 10^2 \times 10^3) - (9.15 \times 10^3) = [(2.34 \times 10^2) - 9.15] \times 10^3 = [234 - 9.15] \times 10^3 = 224.85 \times 10^3 = 2.2485 \times 10^5$.

b. Using the commutative property: $(3.01 \times 10^4) \cdot (6.14 \times 10^5) = (3.01 \times 6.14) \cdot (10^4 \times 10^5) = 18.4814 \times 10^9 = 1.84814 \times 10^{10}$.

Scientific notation is commonly encountered on calculators because many of them have only enough screen space to show numbers with ten or fewer digits. If the numbers contain more digits, the calculator automatically switches to scientific notation. Various calculators represent scientific notation differently. For instance, different ways of displaying 8.1×10^{12} are shown in Figure 8.17.

$$\boxed{8.1\text{ E }12} \qquad \boxed{8.1 \quad 12} \qquad \boxed{8.1 \times {}_{10}12}$$

Figure 8.17 Different calculator displays for 8.1×10^{12}

Because of limited screen space, numbers may have to be put into scientific notation to make a computation. To do so, we use the appropriate combinations of operation, exponent, and parentheses keys.

Example 8.21   Paying the National Debt

In 2010, the national debt of the United States was about 1.2×10^{13} dollars. At the same time, the population was about 3.05×10^{8}. If the national debt at this time was passed on to the citizens of the United States so that each person had to pay an equal share, about how much would each person have to pay?

Solution We can find the dollar amount for each person by dividing 1.2×10^{13} by 3.05×10^{8}. The following sequence of keys can be used on many calculators to complete the computation:

$$\boxed{(}\,1.2\,\boxed{\times}\,10\,\boxed{\wedge}\,13\,\boxed{)}\,\boxed{\div}\,\boxed{(}\,3.05\,\boxed{\times}\,10\,\boxed{\wedge}\,8\,\boxed{)}\,\boxed{\text{ENTER}}$$

The display indicates that each person would have to pay $39,344.26 to clear the debt.

Roots and Rational Exponents

Using irrational numbers and radicals, we can define rational number exponents. Recall that if n is a natural number, then we define $a^n = \underbrace{a \cdot a \cdot \ldots \cdot a}_{n \text{ factors}}$ and $a^0 = 1$, where $a \neq 0$. Unfortunately, using this interpretation with rational exponents is difficult. For instance, $2^{1/2}$ is interpreted as multiplying 2 to itself half of a time, which makes little sense computationally. However, if rational exponents are to behave like integer exponents, then we would expect that $2^{1/2} \cdot 2^{1/2} = 2^{1/2 + 1/2} = 2^1 = 2$. Because we know that $\sqrt{2} \cdot \sqrt{2} = 2$, it seems reasonable that $2^{1/2}$ is the same as $\sqrt{2}$. This is indeed how we choose to define a rational exponent.

Definition of a Rational Exponent

If a is any real number, then $a^{1/n} = \sqrt[n]{a}$ where n is any natural number if $a \geq 0$ and n is an odd natural number if $a < 0$.

The definition involves only unit fractions, but it can be extended to other rational numbers by using the properties of exponents. Specifically,

$$a^{m/n} = a^{(1/n) \cdot m} = \left(a^{1/n}\right)^m = \left(\sqrt[n]{a}\right)^m = \sqrt[n]{a^m}$$

for non-negative real values of a and reduced rational numbers m/n. For instance, $3^{3/4} = \sqrt[4]{3^3}$ and $7^{2/5} = \sqrt[5]{7^2}$. Rational exponents satisfy the properties listed in Theorem 8.6.

Theorem 8.6   Properties of Rational Exponents

If a and b are positive real numbers and m and n are rational exponents, then:

1. $a^m \cdot a^n = a^{m+n}$.

2. $a^m \div a^n = a^{m-n}$.

3. $(a^m)^n = a^{m \cdot n}$.

4. $a^m \cdot b^m = (a \cdot b)^m$.

5. $\dfrac{a^m}{b^m} = \left(\dfrac{a}{b}\right)^m$.

Computations with rational exponents are now done almost exclusively with calculators and computers. Depending on the calculator, users may have several options. Many calculators have special keys designated for rational exponents, such as $\boxed{\sqrt{}}$ for square roots and $\boxed{\sqrt[x]{y}}$ for fractional exponents. Other calculators have a general exponent key, $\boxed{\wedge}$, to handle all exponents. When using this key, we generally enclose fractional exponents in parentheses.

Example 8.22   Using a Calculator to Approximate Each Root

a. $\sqrt{5}$ **b.** $\sqrt[3]{10}$ **c.** $\sqrt[3]{15^2}$

Solution We show different keystrokes that can be used to compute to roots.

a. $\boxed{\text{2nd}}$ $\boxed{\sqrt{}}$ $\boxed{5}$ $\boxed{)}$ $\boxed{\text{ENTER}}$ displays $\sqrt{5} = 2.236067\ldots$.

b. 10 $\boxed{\sqrt[x]{y}}$ 3 $\boxed{\text{ENTER}}$ displays $\sqrt[3]{10} = 2.154434\ldots$.

c. 15 $\boxed{\wedge}$ $\boxed{(}$ 2 $\boxed{\div}$ 3 $\boxed{)}$ $\boxed{\text{ENTER}}$ displays $\sqrt[3]{15^2} = 6.082201\ldots$.

Check Your Understanding 8.2-B

1. Convert each repeating decimal to a fraction.

   a. $0.222\ldots$ **b.** $0.3454545\ldots$ **c.** $0.25\overline{1}$

2. Write each number as specified.

   a. 3,200 in scientific notation   **b.** 6,000,000 in scientific notation

   c. 6.12×10^3 in decimal notation   **d.** 7.69×10^{-5} in decimal notation

3. Use a calculator to compute the following. Round your answers to two decimal places.

   a. $(3.1 \times 10^4) + (6.02 \times 10^3)$ **b.** $(6.73 \times 10^4) \cdot (5.05 \times 10^6)$

   c. $(1.03 \times 10^5) - (9.87 \times 10^4)$ **d.** $(4.33 \times 10^4) \div (5.76 \times 10^3)$

4. Write each expression as a radical, and then use a calculator to compute its value to three decimal places.

   a. $5^{1/2}$ **b.** $17^{1/5}$ **c.** $3^{2/3}$ **d.** $132^{4/7}$

Talk About It Can every value be represented with two different decimals? If not, which ones can and why?

Problem Solving

Activity 8.4

If you could travel at a speed of 100 m every second, about how long would it take you to travel to the sun?

SECTION 8.2 Learning Assessment

■ Understanding the Concepts

1. Explain the procedure for placing the decimal point when computing each operation.
   a. Addition
   b. Subtraction
   c. Multiplication
   d. Division

2. a. Why are there no remainders with decimal division?
   b. Under what three conditions does decimal division conclude?

3. Why do we know that a rational number can always be written as a terminating or a repeating decimal?

4. How does multiplying by a positive power of ten affect the placement of the decimal point? What about dividing by a positive power of ten?

5. Describe the procedure for converting a repeating decimal to a fraction. Why does multiplying by powers of ten play an important role in the procedure?

6. What are the advantages and disadvantages of writing decimals in scientific notation?

7. How are rational exponents and radical expressions connected?

8. In radical expressions, why must the index be odd when the radicand is negative?

■ Representing the Mathematics

9. Give another decimal representation for each number.
   a. 3
   b. $6.999\ldots$
   c. -8
   d. $-14.\overline{9}.$
   e. $4.56999\ldots$

10. Use base-ten blocks to represent and solve each problem.
    a. $0.3 + 0.16$
    b. $0.63 - 0.093$

11. Use base-ten blocks to represent and solve each problem.
    a. $1.92 + 0.83$
    b. $2.003 - 1.998$

12. Is it possible to represent decimal multiplication with base-ten blocks? If so, how?

13. The product of two decimals can be represented with an adaptation of the area model. For instance, $2.3 \cdot 3.4$ is represented as

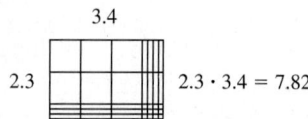

$2.3 \cdot 3.4 = 7.82$

Use this model to compute each product.
   a. $1.3 \cdot 1.5$
   b. $2.1 \cdot 2.7$
   c. $4.5 \cdot 3.4$

14. What multiplication problem is represented in each diagram?
   a.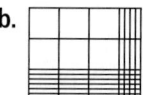
   b.

15. Use repeated subtraction to solve each quotient.
   a. $1.5 \div 0.3$
   b. $4.9 \div 0.7$
   c. $25.6 \div 3.2$

16. Write each exponent as a radical.
   a. $3^{1/2}$
   b. $4^{1/6}$
   c. $8^{1/9}$
   d. $67^{1/4}$
   e. $(-4)^{2/3}$
   f. $(-53)^{3/7}$

17. Write each radical as an exponent.
   a. $\sqrt{6}$
   b. $\sqrt{21^3}$
   c. $\sqrt[3]{14}$
   d. $\left(\sqrt[3]{21}\right)^4$
   e. $\sqrt[5]{(12)^2}$
   f. $\sqrt[7]{19^6}$

■ Exercises

18. Use expanded notation in fractional form to add or subtract.
   a. $3.45 + 2.69$
   b. $8.63 - 7.41$

19. Use expanded notation in exponential form to add or subtract.
   a. $5.01 + 4.5$
   b. $14.23 - 9.867$

20. Compute each sum or difference.
   a. $57.09 + 34.56$
   b. $43.009 + 91.34$
   c. $75.61 + (-86.13)$
   d. $34.01 - 14.67$
   e. $212.8 - 123.672$
   f. $(-95.43) - (104.76)$

21. Estimate the answer to each problem, and then use a calculator to compute the result.
   a. $12,687.09 + 13,412.83$
   b. $13,989.83 - (12,653.72)$
   c. $32,491.99 + (-16,505.8)$
   d. $(-35,002.03) - (-34,989.32)$

22. Write each decimal as a fraction, and then compute. Write your answer as a decimal.
   a. $4.5 \cdot 6.1$
   b. $9.45 \cdot 3.2$

23. Write each decimal as a fraction, and then compute. Write your answer as a decimal.
   a. $5.6 \cdot 3.5$
   b. $4.53 \cdot 2.7$

24. Multiply using the standard algorithm.
   a. $9.3 \cdot 2.5$
   b. $6.3 \cdot 3.11$
   c. $(-34.5) \cdot 18.6$
   d. $(-32.45) \cdot (-23.54)$

25. Estimate each product, and then use a calculator to find its exact value.

    a. $11.1 \cdot 4.3$ b. $158.2 \cdot 9.89$
    c. $458.91 \cdot 203.82$

26. Use the standard division algorithm to divide. Compute to two decimal places.
    a. $4.6\overline{)82.35}$ b. $6.1\overline{)456.3}$ c. $1.83\overline{)7.9341}$

27. Convert each fraction to a decimal.
    a. $\dfrac{5}{8}$ b. $\dfrac{7}{4}$
    c. $\dfrac{7}{12}$ d. $\dfrac{11}{6}$

28. Convert each fraction to a decimal.
    a. $\dfrac{15}{60}$ b. $\dfrac{18}{24}$
    c. $\dfrac{24}{45}$ d. $\dfrac{65}{39}$

29. Use Theorem 8.5 to compute the following mentally.
    a. $1.23 \cdot 10^2$ b. $16.901 \cdot 10^3$
    c. $45.9 \div 10^2$ d. $378.9 \div 10^4$

30. Use Theorem 8.5 to compute the following mentally.
    a. $4.92 \cdot 10^{-2}$ b. $1.23 \div 10^4$
    c. $14.14 \div 10^{-2}$ d. $0.41 \cdot 10^{-3}$

31. Convert each repeating decimal to a fraction.
    a. $0.777\ldots$ b. $0.\overline{15}$ c. $0.1333\ldots$

32. Convert each repeating decimal to a fraction.
    a. $0.41232323\ldots$ b. $0.21\overline{251}$ c. $-4.12999\ldots$

33. When possible, use a calculator to compute each root to five decimal places.

    a. $\sqrt{6}$ b. $\sqrt{18}$
    c. $\sqrt{(-157)}$ d. $\sqrt[3]{21}$

34. When possible, use a calculator to compute each root to five decimal places.
    a. $\sqrt[4]{(-12)^2}$ b. $\sqrt[5]{119^2}$
    c. $\sqrt[5]{(-9)^3}$ d. $\sqrt[4]{(-5)^3}$

35. Use the rules of exponents to simplify each expression.
    a. $3^{1/2} \cdot 3^{1/3}$ b. $(4^{2/3})^{1/4}$ c. $5^{2/3} \div 5^{1/4}$
    d. $\dfrac{\sqrt{12}}{\sqrt{3}}$ e. $\sqrt[3]{4} \cdot \sqrt[3]{2}$

36. Write each number in scientific notation.
    a. $1,450$ b. 0.0354
    c. $9,690,100$ d. -0.0004102

37. Write each number in decimal notation.
    a. 7.01×10^2 b. 4.505×10^{-2}
    c. 8×10^1 d. -6.23×10^{-7}

38. The five Great Lakes of the United States hold approximately 6 quadrillion gal of fresh water. Write this number in both decimal and scientific notation.

39. Use Table 8.3 to answer the following questions.
    a. How many times longer is a red blood cell than a virus?
    b. How many times shorter is a mosquito than a human being?
    c. How many times longer is a light year than the radius of the earth?

40. Compute each problem. Round your answer to two decimal places.
    a. $(4.2 \times 10^3) + (2.3 \times 10^4)$
    b. $(5.67 \times 10^4) - (3.34 \times 10^4)$
    c. $(3.15 \times 10^7) \cdot (8.93 \times 10^8)$
    d. $(6.87 \times 10^7) \div (2.24 \times 10^4)$

41. Compute each problem. Round your answer to two decimal places.
    a. $(9.87 \times 10^4) + (1.11 \times 10^5)$
    b. $(4.09 \times 10^5) - (9.93 \times 10^4)$
    c. $(5.2 \times 10^3) \cdot (1.5 \times 10^2)$
    d. $(5.67 \times 10^3) \div (1.15 \times 10^2)$

■ Problems and Applications

42. Use the lattice method to compute.
    a. $2.341 + 3.59$ b. $3.45 \cdot 8.6$

43. Complete the following magic square by making the sum of every row, column, and diagonal the same.

9.7		7.9
	7	8.8
6.1		

44. a. What is the 6th digit in the decimal expansion of $\dfrac{4}{7}$?
    b. What is the 16th digit in the decimal expansion of $\dfrac{4}{7}$?
    c. What is the 869th digit in the decimal expansion of $\dfrac{4}{7}$?

45. Use a calculator to find the repetend of:

    a. $\dfrac{1}{29}$ b. $\dfrac{1}{31}$

46. Find the multiplicative inverse of each decimal.
    a. 0.35 b. 0.127 c. $0.444\ldots$
    d. $0.2343434\ldots$ e. $0.\overline{536}$

47. Compute the following operations with decimals in base-five.

a. $43.02_{\text{five}} + 33.14_{\text{five}}$ **b.** $221.3_{\text{five}} - 14.44_{\text{five}}$

c. $2.14_{\text{five}} \cdot 0.3_{\text{five}}$

48. What is the largest positive number that can be displayed on your calculator? The smallest positive number?

49. Kimberly goes to a home store to buy supplies to paint her bedroom. She buys 2 gal of paint, each costing $19.98, a roller costing $3.67, a paint pan costing $2.39, and a tarp costing $1.95. How much is her bill without sales tax?

50. During the past week, Juan made the following transactions on his bank account: $35.47, $63.71, −$83.10, −$12.53, and −$14.81. If his account started at $103.57, how much was in the account after the transactions?

51. A 5-lb box of nails costs $8.57, and a box contains 75 nails. How much does each nail cost?

52. A room requires 21 yd² of carpet. If carpet costs $1.37/ft², how much will the carpet for the room cost?

53. A short order cook works 31.25 hr in one week. If the worker is paid $8.50/hr, what is the worker's pay for the week?

54. A power line worker makes $21.75/hr. After a winter storm, the worker logs 75.75 hr of work in one week. If the worker gets paid time and a half for any hours over 40, how much did he get paid for that week?

55. Nikki buys 300 shares of a certain stock for $3.95 a share. She later sells the shares for $2,467. How much money per share did she make on the sale?

56. Rhonda bought an activity book for each of her 27 first grade students. If the bill was $68.85 without tax, how much did each book cost?

57. The half-life of uranium-238 is about 1.5×10^{17} s. How long is this in years?

58. The radius of the Milky Way Galaxy is about 3.9×10^{20} m. If one light year is about 9.46×10^{15} m, how many light years is it across the Milky Way Galaxy?

59. The mass of the earth is about 5.98×10^{24} kg, and the mass of the moon is about 7.3×10^{22} kg. About how many more times massive is the earth than the moon?

60. In 1994, U.S. senators spent about 2.1×10^8 dollars on election campaigns. If there were only 33 senatorial races that year, how much was spent on each race?

61. California has roughly 3.589×10^7 people. If the total land area of California is 1.55×10^5 square miles, how many people are there per square mile in California?

■ Mathematical Reasoning and Proof

62. If a and b are real numbers, where $0 < a < b < 1$, determine whether each statement is true, false, or cannot be determined.

a. $a + b < a$ **b.** $b - a > a$

c. $a \cdot b < 1$ **d.** $b \div a < a$

63. If a and b are rational numbers, where $0 < a < 1$ and $b > 1$, determine whether each statement is true, false, or cannot be determined.

a. $a + b > b$ **b.** $b - a < a$

c. $a \cdot b > a$ **d.** $b \div a < b$

64. Suppose the number m has three digits to the right of the decimal and n has four digits to the right of the decimal.

a. Is it possible to determine how many digits are to the right of the decimal in $m \cdot n$? Why or why not?

b. Is it possible to determine how many digits are to the right of the decimal in $m + n$? Why or why not?

65. What might be true about the decimal representations of $\frac{1}{53}$ and $\frac{47}{53}$, knowing that $1\frac{47}{53} = \frac{100}{53} = 100 \cdot \frac{1}{53}$?

66. Write the decimal expansions of $\frac{1}{7}, \frac{2}{7}, \frac{3}{7}, \frac{4}{7}, \frac{5}{7},$ and $\frac{6}{7}$ to seven decimal places. What do you notice?

67. Without performing any calculations, which of the following have the same quotient as $4.5\overline{)85.93}$?

a. $0.45\overline{)85.93}$ **b.** $45\overline{)859.3}$

c. $0.45\overline{)8.593}$ **d.** $45\overline{)8,593}$

68. **a.** If a decimal of the form $0.aaa\ldots$ is converted to a fraction, what is the denominator likely to be?

b. If a decimal of the form $0.ababab\ldots$ is converted to a fraction, what is the denominator likely to be?

c. If a decimal of the form $0.\overline{abc}$ is converted to a fraction, what is the denominator likely to be?

69. Use a calculator to convert $\frac{1}{9}$ to a decimal. How might you use your result to mentally convert the following fractions to decimals?

a. $\frac{2}{9}$ **b.** $\frac{7}{9}$ **c.** $\frac{16}{9}$ **d.** $4\frac{8}{9}$

70. Convert $\frac{1}{9}, \frac{1}{99},$ and $\frac{1}{999}$ to decimals. Based on this information, what do you expect the decimal expansion of $\frac{1}{9,999}$ and $\frac{1}{99,999}$ to be?

■ Mathematical Communication

71. How many different ways can you use your calculator to compute $\sqrt[4]{17^3}$?

72. Write a short paragraph describing how you would explain to your students the need to align the decimal points when adding or subtracting decimals?

73. a. Write a story problem that makes use of decimal addition.

b. Write a story problem that makes use of decimal multiplication.

c. Write a story problem that makes use of decimal division.

d. What are the similarities and differences between these problems and those you have written or seen with the whole numbers?

74. Caleb computes the product of two decimals in the following way.

$$
\begin{array}{r}
32.4 \\
\times 1.7 \\
\hline
2268 \\
3240 \\
\hline
550.8
\end{array}
$$

What has he done wrong and how might you correct his thinking?

75. Three students attempt to write 567,000 in scientific notation. Describe the mistake each student has made.

a. 567×10^3 **b.** 5.67×10^{-5} **c.** 5.67^5

■ Building Connections

76. Summarize the similarities and differences between the whole-number operations, the rational-number operations, and decimal operations.

77. Search websites, newspapers, and magazines for uses of scientific notation. Describe the contexts in which these numbers are used. Does the article also include an analogy to understand the relative size of this number? If so, what is it?

78. Reconsider the properties given in Theorem 8.3. Why do you think these properties are often referred to as "algebraic" properties?

79. Review the NCTM Number and Operations Standards for grades K−8.

a. Now that we have fully covered the content addressed by these standards, do you think it is appropriate to integrate and summarize these topics as the NCTM did? Why or why not?

b. Do you think that any objectives are missing from the Number and Operations Standards? If so, what?

SECTION **8.3** Proportional Reasoning

The real numbers encompass all the numbers we are likely to encounter in our daily lives, so we now apply what we have learned to solve real-world problems using proportional reasoning. **Proportional reasoning** is the ability to compare quantities and to understand the context of the comparison. It is an important life skill that most of us use regularly. To reason proportionally, we must first understand the idea of a ratio.

Ratios

Explorations Manual 8.8

In many everyday situations, we need to make or understand a comparison between two different quantities or measures. Table 8.4 lists a few examples that commonly occur in our travel, shopping, businesses, and home life. In mathematics, comparisons like the ones in Table 8.4 are called **ratios.**

Table 8.4 Examples of comparisons	
Comparisons in . . .	**Examples**
Automobiles	miles per gallon, miles per hour, revolutions per minute
Education	students to teacher, passing students to failing students, female students to male students
Restaurants	servers to customers, customers to tables, profits to expenses
Shopping	price per pound, price per ounce, price per package
Sports	wins to losses, successes to attempts, home score to visitor's score

Definition of Ratio

A **ratio** is an ordered pair of numbers used to compare two quantities.

Although simple, we must know several things about ratios in order to understand and use them. First, ratios give us a way to understand or measure the relative size of one set to another. For instance, if a class takes a test and the ratio of passes to failures is 2 to 1, then the ratio tells us that 2 students passed for every 1 who failed. Notice that this ratio represents only a relative, not an absolute, amount. Specifically, it does not indicate how many students are in the class, nor does it say how many passed or failed. The class totals could be 4 passes and 2 failures, 6 passes and 3 failures, or even 20 passes and 10 failures and still satisfy a ratio of 2 passes to 1 failure.

Second, a ratio always occurs in a context, which is identified by the labels on the numbers. Without them, the ratio has no meaning. For instance, a ratio of 25 to 1 is meaningless. However, if the ratio is "25 miles to 1 gallon," then it clearly refers to the gas mileage of a vehicle. Because each number is associated with a label, the order of the numbers is important. Simply changing their order can affect the ratio's meaning. For instance, a vehicle that gets 25 mile to 1 gallon is vastly different from one that gets 1 mile to 25 gallons.

Ratios can be written in a number of ways. To write a ratio in word form, we use the word "to" or "per" to separate the numbers. Because the word "per" means "for every," it is typically used for writing ratios that compare multiple objects to a single object. For instance, the ratio of miles per gallon indicates the number of miles that can be driven on only 1 gallon of gasoline. The ratio of people per square mile indicates the number of people living on 1 square mile of land.

Ratios can also be written in ratio form, where the two numbers are separated by a colon, or in a fraction form, where one number is the numerator and the other is a denominator. Of the three ways to write ratios, the fractional form may be the most useful.

Example 8.23 Writing Ratios

Communication

Complete the table by filling in the missing ways to write ratios.

Word Form	Ratio Form	Fractional Form
	5 cats: 6 dogs	
		25 students / 1 teacher
9 squares to 5 triangles		

Solution The missing ways to write the given ratios are shown in blue.

Word Form	Ratio Form	Fractional Form
5 cats to 6 dogs	5 cats: 6 dogs	$\frac{5 \text{ cats}}{6 \text{ dogs}}$
25 students per teacher	25 students: 1 teacher	$\frac{25 \text{ students}}{1 \text{ teacher}}$
9 squares to 5 triangles	9 squares: 5 triangles	$\frac{9 \text{ squares}}{5 \text{ triangles}}$

There are some fundamental differences between ratios and fractions. First, the denominator of a ratio can be zero. For example, if we consider the students enrolled in the typical kindergarten class, the ratio of 25 children to 0 monkeys can be written as $\frac{25}{0}$. Such situations seldom occur and are generally of little use when they do. For this reason, zero is typically excluded from the set of possible denominators. Also, the numerator and denominator of a ratio are not limited to integers but can include fractions, decimals, and even irrational numbers.

A second difference between ratios and fractions occurs in how we interpret and use ratios to make comparisons. Fractions describe only part-to-whole relationships of like objects. Ratios, on the other hand, can be used to describe not only **part-to-whole** relationships, but also **whole-to-part** and **part-to-part** relationships. As a consequence, every fraction is a ratio, but not every ratio is a fraction.

Example 8.24	Different Types of Ratios

Communication

Rochelle plays basketball and makes 7 free throws for every 3 she misses. What ratios can be used to express her ability to shoot free throws?

Solution Because a free throw is either a success or a failure, we can add the successes to the failures to find Rochelle's total attempts: $7 + 3 = 10$. Based on the number of successes, failures, and attempts, we can express Rochelle's ability to shoot free throws with any one of six ratios:

Successes to failures = 7 to 3	Part-to-part comparison
Failures to successes = 3 to 7	Part-to-part comparison
Successes to attempts = 7 to 10	Part-to-whole comparison
Failures to attempts = 3 to 10	Part-to-whole comparison
Attempts to successes = 10 to 7	Whole-to-part comparison
Attempts to failures = 10 to 3	Whole-to-part comparison

The ratios in Example 8.24 make comparisons between like objects, in this case free throws. A ratio can also compare unlike objects, in which case we refer to the ratio as a **rate**. For instance, 65 mi/hr, 3 pencils for 25 cents, or 24 cans of soda per case are all examples of rates.

Despite the differences between ratios and fractions, expressing ratios in a fractional form offers several computational advantages. First, limiting the denominator of the ratio to nonzero real numbers means that we can interpret the fraction bar as division. In fact, the word "ratio" is often used to indicate division. Second, the fractional form allows us to talk about ratios in reduced or simplest form. For instance, if the ratio of males to females in a given school is 6 to 8, we can reduce the ratio to 3 to 4. Because ratios allow for the numerator and the denominator to be numbers other than integers, a ratio can always be reduced to have a denominator of 1. This fact is particularly handy for comparing ratios, a skill commonly used when shopping.

Example 8.25	Using Ratios to Comparison Shop

Problem Solving

When Kate buys groceries, she tries to save money by comparing prices. She is currently looking at two jars of spaghetti sauce. Brand *A* costs $1.55 for 32 oz and Brand *B* costs $2.43 for 48 oz. Which jar of sauce is the better buy?

Solution

Understanding the Problem. We are to determine which jar of sauce is the better buy. We know the size and the cost of each jar.

Devising a Plan. To determine the better buy, we need to describe the cost of each jar in a common way. We can do so by *using a ratio* to find the price per ounce, that is, the cost of just one ounce of each type of sauce.

Carrying Out the Plan. To find the price per ounce, we write the ratio of the cost of each jar to its size and then divide the numerator by the denominator.

$$\text{Price per ounce for Brand } A = \frac{\text{cost}}{\text{oz}} = \frac{\$1.55}{32 \text{ oz}} = \frac{\$0.048}{1 \text{ oz}} = 4.8 \text{ ¢/oz}$$

$$\text{Price per ounce for Brand } B = \frac{\text{cost}}{\text{oz}} = \frac{\$2.43}{48 \text{ oz}} = \frac{\$0.051}{1 \text{ oz}} = 5.1 \text{ ¢/oz}$$

Because Brand *A* costs only 4.8 ¢/oz, it is the better buy.

Looking Back. We could have also used the ratio of ounces per dollar, or $\frac{\text{oz}}{\text{cost}}$, to solve the problem. In using this ratio, we would find the number of ounces that can be bought for one dollar. Because, 20.6 oz of Brand *A* can be bought for one dollar and 19.8 oz of Brand *B* can be bought for one dollar, Brand *A* is still the better buy.

The last example illustrates how to use ratios to compare different relative amounts. In some cases, we may want to know whether the ratios represent the same relative amount.

Example 8.26 **Mixing Paint**

Representation

One gallon of a certain color of paint is mixed in a ratio of 4 parts blue pigment to 3 parts green pigment. Five gallons of the same paint is mixed in a ratio of 20 parts blue to 15 parts green. Do these ratios represent the same relative amounts?

Solution We can make a diagram to find out whether each mixture represents the same relative amount. We use *B* to represent one part blue pigment and *G* to represent one part green. Then, by evenly distributing the pigment among the 5 gallons, we have the following:

One gallon	Five gallons				
BBBB	*BBBB*	*BBBB*	*BBBB*	*BBBB*	*BBBB*
GGG	*GGG*	*GGG*	*GGG*	*GGG*	*GGG*
	1 gallon	1 gallon	1 gallon	1 gallon	1 gallon

Because every one of the 5 gallons of paint gets 4 blue to 3 green, we can see that the ratios of 4 to 3 and 20 to 15 represent the same relative amounts.

In the last example, we could have drawn the same conclusion by dividing the numbers in each ratio. Specifically, because 4 to 3 $= \frac{4}{3} = 1.\overline{3}$ and 20 to 15 $= \frac{20}{15} = 1.\overline{3}$, the ratios represent the same relative amounts, and we call them equivalent ratios. In general, two ratios are **equivalent** if their fractions or the quotient of their terms are equal. Equivalent ratios are important because they give us a powerful way to solve problems.

Check Your Understanding 8.3-A

1. Represent each ratio in ratio form and fraction form.

   a. 4 hearts to 7 clubs   **b.** 8 women to 6 girls   **c.** 18 pencils to 3 boxes

2. The ratio of boys to girls in a first-grade class is 5 to 4. What other ratios can be used to compare the genders in the class?

3. Which is the better buy?

   a. A 12-oz can of soda for $0.55 or a 32-oz bottle of soda for $1.69

   b. 48 oz of detergent for $4.27 or 64 oz of detergent for $5.35

Talk About It Ratios are a powerful tool for comparison shopping. Do you know of any stores that help you comparison shop by using ratios? If so, how?

Problem Solving

Activity 8.5

Find all integer values for x so that $\dfrac{x}{11}$ represents a ratio greater than 4 : 1 but less than 5 : 1.

Proportions

In our last example, we mixed two volumes of paint. One gallon was mixed in a ratio of 4 parts blue pigment to 3 parts green pigment. Five gallons were mixed in a ratio of 20 parts blue to 15 parts green. These ratios represented the same relative amounts, so we can write a statement that equates the two, or 4 blue : 3 green = 20 blue : 15 green. Any statement that equates two ratios is called a **proportion.**

Definition of a Proportion

A **proportion** is a statement that two ratios are equivalent.

Explorations Manual 8.9

Whenever writing proportions, we must ensure two things. First, the two ratios must be equivalent. For instance, the statement 4 dogs : 3 cats = 8 dogs : 5 cats is not a proportion because the ratios do not show the same relative amounts. Second, the order of the terms within the ratios must be consistent. For instance, both ratios in the statement 4 dogs : 3 cats = 8 cats : 6 dogs, show a 2-to-1 relationship. However, the statement is not a proportion because the ratios do not show the same relative amount with regard to dogs to cats.

Explorations Manual 8.10

We encounter proportions in many of our daily activities, so knowing how to solve proportion problems is a valuable life skill. In any proportion problem, we always know one ratio and part of the other. In other words, we are given three of the values and are asked to find the fourth. Once we know we have a proportional situation, we can use several approaches to find the missing number.

In some cases, the solution is simply a matter of multiplying or dividing by the right number. For instance, consider the proportion 3 : 5 = 300 : x. Because the ratios must show the same relative amount and $300 = 3 \cdot 100$, then $x = 5 \cdot 100 = 500$. Similarly, given the proportion 6 : x = 18 : 33, then $x = 33 \div 3 = 11$ because $6 = 18 \div 3$.

We can also use multiplication to solve proportions if we are given a unit rate. A **unit rate** is a ratio that describes how many objects of one type are comparable to exactly one of another. For instance, 25 miles per gallon, 14 students per teacher, and 36 inches per foot are all unit rates. If we are given a unit rate in a problem, then all we do is multiply to find the missing value.

Example 8.27 | **Scale Drawings**

Application

Ratios and proportions are often used in scale drawings of buildings and maps. Use the scale drawing of the apartment to answer each question.

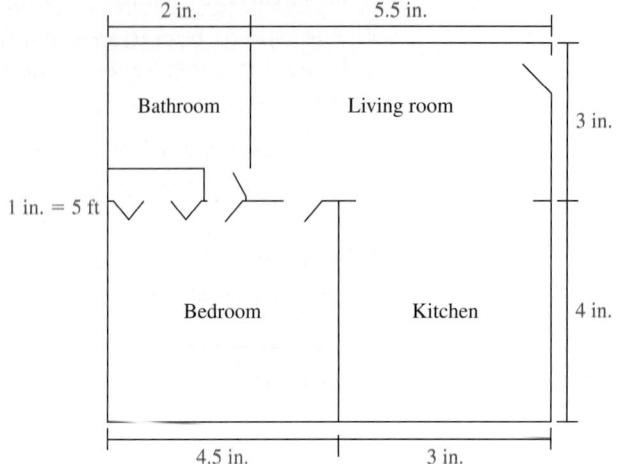

a. What are the dimensions of the kitchen in feet?

b. What is the length of the apartment from the front door to the back of the bathroom?

Solution The diagram uses a unit rate of 1 in. : 5 ft. Consequently, we can find the needed dimensions by multiplying the numbers in the diagram by 5.

a. Because the drawing of the kitchen is 3 in. × 4 in., the actual kitchen has dimensions of $3 \cdot 5 = 15$ ft by $4 \cdot 5 = 20$ ft.

b. The distance from the front door to the back of the bathroom is $5.5 + 2 = 7.5$ in. in the drawing. Hence, the actual distance is $7.5 \cdot 5 = 37.5$ ft.

Other proportion problems might have not a unit rate, but rather a ratio that can be easily changed into one. In such cases, we can write the ratio as a fraction, convert it to a unit rate by dividing the numerator by the denominator, and then multiply the unit rate by the other known number to find the missing value.

Example 8.28 | **The Cost of Educating Students**

Application

A school board reports that it costs \$48,375 a year to educate 5 students. If the school district has 9,000 students, how much does it cost a year to educate all of them?

Solution The ratio of the cost to students is $\dfrac{\$48,375}{5\,\text{students}}$. If we divide both numbers by 5, we can change the ratio into a unit rate for the cost per student, or $\dfrac{\$9,675}{1\,\text{student}}$. Next, multiply the rate by the number of students to get the total cost:

$$9,000 \text{ students} \times \frac{\$9,675}{1\,\text{student}} = \$87,075,000$$

In the last example, we used a fraction to solve the problem. In general, writing proportions with fractions is both common and useful. Specifically, if $\dfrac{a}{b}$ and $\dfrac{c}{d}$ are proportional, then $\dfrac{a}{b} = \dfrac{c}{d}$. From our work with the rational numbers, we know that, if a, b, c, and d are integers with $b \neq 0$ and $d \neq 0$, then $\dfrac{a}{b} = \dfrac{c}{d}$ if and only if $ad = bc$.

The same is true for proportions. We call a and d the **extremes,** b and c the **means,** and say that two ratios are proportional if and only if the product of the means is equal to the product of the extremes.

Theorem 8.7   Product of the Means Equal to the Product of the Extremes

If a, b, c, and d are real numbers, with $b \neq 0$ and $d \neq 0$, then $\dfrac{a}{b} = \dfrac{c}{d}$ is a proportion if and only if $ad = bc$.

Theorem 8.7 indicates that another way to solve proportion problems is to use a cross product. Specifically, if we let x represent a missing value from a proportion, then $\dfrac{x}{b} = \dfrac{c}{d}$, which implies $xd = bc$. By dividing both sides by d, we have $x = \dfrac{b \cdot c}{d}$.

Example 8.29   Using the Products of the Means and the Extremes

Find the missing value.

a. $\dfrac{5}{7} = \dfrac{30}{x}$. b. $\dfrac{x}{7} = \dfrac{16}{15}$.

Solution

a. Given the proportion $\dfrac{5}{7} = \dfrac{30}{x}$, we use Theorem 8.7 to write the equation $5x = 7 \cdot 30$.

Solving for x, we get $x = \dfrac{7 \cdot 30}{5} = 42$.

b. From $\dfrac{x}{7} = \dfrac{16}{15}$, we get $x \cdot 15 = 7 \cdot 16$. Consequently, $x = \dfrac{7 \cdot 16}{15} = 7.4\overline{6}$.

Whenever we use the products of the means and the extremes to solve proportion problems, we should keep in mind that $\dfrac{a}{b} = \dfrac{c}{d}$ is not the only proportion that leads to the equation $ad = bc$. In fact,

$$\frac{b}{a} = \frac{d}{c} \qquad \frac{a}{c} = \frac{b}{d} \qquad \frac{c}{a} = \frac{d}{b}$$

all lead to the equation $ad = bd$. So we say they are **equivalent** to $\dfrac{a}{b} = \dfrac{c}{d}$. The equivalent forms reflect the different ways to express the relationship between the two ratios. For instance, the proportion $\dfrac{4 \text{ boys}}{3 \text{ girls}} = \dfrac{12 \text{ boys}}{9 \text{ girls}}$ can be expressed as $\dfrac{3 \text{ girls}}{4 \text{ boys}} = \dfrac{9 \text{ girls}}{12 \text{ boys}}$ or $\dfrac{4 \text{ boys}}{12 \text{ boys}} = \dfrac{3 \text{ girls}}{9 \text{ girls}}$ and still show the same relative relationships. However, the equation $\dfrac{4 \text{ boys}}{3 \text{ girls}} = \dfrac{9 \text{ girls}}{12 \text{ boys}}$ is not equivalent to the others because of the switch in the units. Consequently, whenever solving proportions with fractions, we must pay attention to the units so that we keep the ratios consistent. One way to do so is to use a table.

Example 8.30   Increasing a Recipe

A 6-serving recipe calls for 2 cups of flour. How many cups are needed to make 10 servings?

Solution We are given the ratio of 6 servings to 2 cups and are asked to find the number of cups needed to make 10 servings. Because the number of cups is an unknown, we assign it a variable and place the information in the following table:

	First Recipe	**Second Recipe**
Servings	6	10
Cups	2	x

Taking the values directly from the table, we get the proportion $\dfrac{6 \text{ servings}}{2 \text{ cups}} = \dfrac{10 \text{ servings}}{x \text{ cups}}$.

After cross multiplying, $6x = 2 \cdot 10$. Solving for x, we get $x = \dfrac{2 \cdot 10}{6} = \dfrac{10}{3} = 3\dfrac{1}{3}$ cups.

We can also use the cross product to verify properties of proportions. Consider the next example.

Example 8.31   A Property of Proportions

Reasoning   If $\dfrac{a}{b}$ is proportional to $\dfrac{c}{d}$, is $\dfrac{a-b}{a+b}$ proportional to $\dfrac{c-d}{c+d}$? Write an argument to justify your answer.

Solution We are given that $\dfrac{a}{b}$ is proportional to $\dfrac{c}{d}$, so $ad = bc$. Next, consider the ratios $\dfrac{a-b}{a+b}$ and $\dfrac{c-d}{c+d}$ and the product of their extremes, $(a-b)(c+d)$. With a few manipulations, we get

$$(a-b)(c+d) = ac + ad - bc - bd \qquad \text{Distributive property}$$
$$= ac + bc - ad - bd \qquad \text{Substitution}$$
$$= ac - ad + bc - bd \qquad \text{Commutative property}$$
$$= (a+b)(c-d). \qquad \text{Distributive property}$$

This indicates that the cross product of the ratios is equal, so they must be proportional, or $\dfrac{a-b}{a+b} = \dfrac{c-d}{c+d}$.

Explorations Manual 8.11

Until now, we have only worked with ratios involving two numbers. However, some ratios have more than two numbers. In such cases, we use colon notation to write the ratios.

Example 8.32   A Mixture of Gumballs

Application   A mixture of gumballs includes gumballs that are white, red, blue, and green in a ratio of $6 : 4 : 3 : 1$. If 840 gumballs are in the mixture, how many of each color are there?

Solution Given the total number of gumballs, we express our proportions using part-to-whole ratios. In this case, we use $6 + 4 + 3 + 1 = 14$ as the whole in one ratio and

840 in the other. Then, if we let w, r, b, and g represent the number of each color of gumball, we can use the following proportions to answer our question:

$$\frac{w}{840} = \frac{6}{14} \quad \text{implies there are} \quad w = \frac{6 \cdot 840}{14} = 360 \text{ white gumballs}$$

$$\frac{r}{840} = \frac{4}{14} \quad \text{implies there are} \quad r = \frac{4 \cdot 840}{14} = 240 \text{ red gumballs}$$

$$\frac{b}{840} = \frac{3}{14} \quad \text{implies there are} \quad b = \frac{3 \cdot 840}{14} = 180 \text{ blue gumballs}$$

$$\frac{g}{840} = \frac{1}{14} \quad \text{implies there are} \quad g = \frac{1 \cdot 840}{14} = 60 \text{ green gumballs}$$

Check Your Understanding 8.3-B

1. Find the missing value.

a. $\dfrac{8}{x} = \dfrac{48}{54}$ 　　　 b. $\dfrac{x}{13} = \dfrac{15}{26}$ 　　　 c. $\dfrac{3x}{10} = \dfrac{11}{14}$

2. I can drive 350 miles in 6 hours. At the same speed, how far can I drive in 10 hours?

3. Lashonda's car uses 2 gallons of gasoline to drive 50 miles. At the same rate, how much gas will Lashonda's car use to go 350 miles?

Talk About It Of the several different ways to solve proportions, which do you think is the easiest to understand and use? What implications does this have for you if you teach proportions in the future?

Problem Solving

Activity 8.6

Answer the following questions based on the fact that a kilowatt (kW) is equal to 1,000 watts (W), a 60-W bulb burns 60 W/hr, and 1 kW costs about $0.035.

a. How much will it cost to burn one 60-W bulb for a day?

b. How long will it take for a 60-W bulb to use $1 worth of electricity?

c. How much longer can a 60-W bulb burn for $1 than a 75-W bulb?

SECTION 8.3 Learning Assessment

■ Understanding the Concepts

1. What is proportional reasoning?

2. **a.** What is a ratio?

   b. Why are the order and the context of the ratio important?

3. **a.** Why is it useful to write a ratio in fractional form?

   b. What is the difference between a ratio and a fraction?

4. In what situations is it more appropriate to use the word "per" rather than "to" to express a ratio?

5. What is a proportion?

6. What does it mean for two ratios to be equivalent? What about two proportions?

7. What is a unit rate, and how is it used to solve proportion problems?

8. How can the product of the means and the product of the extremes be used to solve a proportion problem?

■ Representing the Mathematics

9. Write each ratio in ratio form and fractional form.

   a. 3 red cars to 8 green cars

   b. 45 miles per gallon

   c. 25 candies to 10 children

   d. 150 passengers to 4 flight attendants

10. Write each ratio as a fraction with a denominator of 1.

    a. 6 : 3   b. 9 : 4   c. 8.5 to 3   d. 5.3 to 2.4

11. a. What is the value of representing a proportion with a table?

    b. If each table represents a proportion, find the missing value.

    i.

	1st Set	2nd Set
Circles	10	18
Squares	5	

    ii.

	1st Period	5th Period
Boys	12	
Girls	15	10

    iii.

	1st Period	4th Period
Passing		20
Failing	6	8

    iv.

	Package A	Package B
Blue candy	16	20
Red candy		45

12. Express each ratio as a fraction with a whole number for the numerator and denominator.

    a. $\frac{1}{4}$ to 1   b. $\frac{1}{4}$ to $\frac{1}{3}$

    c. $0.\overline{7}$ to 1   d. $0.\overline{4}$ to $0.\overline{5}$

13. List three other proportions that have the same cross product as $\frac{4}{5} = \frac{20}{25}$.

■ Exercises

14. If all the beads in a jar are either red or blue and the ratio of red beads to blue beads is 6 to 5, what is the ratio of each of the following?

    a. Blue to red

    b. Blue to all beads

    c. All beads to red

15. If a local gym has 350 male members and 425 female members, what is the ratio of each of the following?

    a. Males to females

    b. Females to males

    c. Females to all members

16. A university has 1,480 freshman, 1,398 sophomores, 1,512 juniors, and 1,138 seniors. Find and reduce the ratio of each of the following?

    a. Freshman to sophomores

    b. Juniors and seniors to freshman and sophomores

    c. Seniors to freshman

    d. Nonseniors to seniors

17. Express each ratio in a reduced form.

    a. 15 ft to 5 ft

    b. 18 in. to 24 in.

    c. 14 kg to 12 kg

18. Express each ratio in a reduced form.

    a. 10 in. to 10 ft

    b. $2.50 to 25 cents

    c. 3 yd to 9 ft

19. Find three ratios in fractional form that are equivalent to each ratio.

    a. $\frac{3}{4}$ b. $\frac{1}{2} : \frac{1}{3}$ c. $\frac{2.7}{3.1}$

20. Calculate the price per unit of each item.

    a. A 6-oz can of tuna for $0.57

    b. 12 pencils for $1.29

    c. $0.98 for a 12-oz box of spaghetti

    d. $4.24 for 2.59 lb of hamburger

21. Calculate the price per unit of each item.

    a. $5.25 for 150 screws

    b. $6.59 for 3 undershirts

    c. 15 gal of gasoline for $38.25

    d. 12 batteries for $7.97

22. Which is the better buy?

    a. 8 bars of soap for $5.22 or 6 bars of the same soap for $4.19

    b. 12 oz of hand lotion for $2.35 or 16 oz of the same lotion for $3.01

23. Which is the better buy?

    a. 2 lb of coffee for $10.35 or 1 lb of the same coffee for $5.35

    b. 25 lb of dog food for $7.89 or 50 lb of the same dog food for $14.98

24. If Kelly makes 25 out of 33 free throws and Kate makes 27 out of 36 free throws, who can claim to be the better shooter?

25. I can drive my car 340 miles on 14 gallons of gas. My wife's car can be driven 290 miles on 10 gallons of gas. Which car gets better gas mileage?

26. Which is a better buy, one 5-gal bucket of paint for $92, or five 1-gal cans for $18.50 each?

27. Which of the following is (are) proportions?

a. $\dfrac{3}{5} = \dfrac{9}{15}$ b. $\dfrac{2}{9} = \dfrac{4}{19}$ c. $\dfrac{3.2}{1.4} = \dfrac{1.8}{0.6}$

d. $\dfrac{0.48}{\frac{1}{3}} = \dfrac{0.96}{\frac{2}{3}}$ e. $\dfrac{\sqrt{16}}{\sqrt{9}} = \dfrac{8}{6}$

28. Find x.

a. $\dfrac{x}{5} = \dfrac{15}{25}$ b. $\dfrac{3x}{4} = \dfrac{12}{16}$ c. $\dfrac{4}{x+1} = \dfrac{28}{35}$

29. Find x.

a. $\dfrac{4.5}{3} = \dfrac{2x}{9}$ b. $\dfrac{2}{x^2} = \dfrac{6}{27}$ c. $\dfrac{7}{x-1} = \dfrac{42}{66}$

30. A river drops 25 ft over the course of 3 mi. Assuming the same rate, how much will the river drop over the course of 10 mi?

31. Jenna buys 3 CDs for $29.85. At the same cost, how much would 8 CDs be?

32. It is recommended that the oil in a car be changed every 3,000 mi. If the life expectancy of a car is 150,000 mi, how many times should it have its oil changed?

33. A bag of grass seed claims that 97 out of every 100 seeds will germinate. If a typical 25-lb bag of seed contains 50,000 seeds, about how many should germinate?

34. Six candy bars cost $2.64. If Rhonda spends $3.96 on candy bars, how many did she buy?

35. A survey crew can survey $3\frac{1}{2}$ mi of a future highway a day. How long will it take them to survey $78\frac{3}{4}$ mi for a planned new highway?

▪ Problems and Applications

36. Find all integer values for x so that $\dfrac{x}{7}$ represents a ratio greater than 3 : 1 but less than 4 : 1.

37. If y is proportional to x^2, what is y when

a. $x = 3$. b. $x = \dfrac{1}{3}$. c. $x = 2.4$. d. $x = \sqrt{17}$.

38. A bag contains 304 marbles that are red, blue, and green and occur in a ratio of 5 : 3 : 8. How many of each color are there?

39. A cookie recipe calls for 4 cups of flour, 1 cup of sugar, and $\dfrac{3}{4}$ of a cup of walnuts. If you have only 3 cups of flour available but still wish to make cookies, what amounts of the other ingredients should you use?

40. Average speed is the ratio of the distance traveled divided by the time taken to travel the distance.

Find the average speed in each of these situations.

a. A bicycle travels 10 mi in 40 min.

b. A car travels 30 mi in 20 min.

c. An airplane travels 560 mi in $2\frac{1}{2}$ hr.

d. A rocket travels 24,000 mi in 8 hr.

41. When a car is driven at 55 mph, the engine turns over at 2,500 revolutions per minute (rpm). How many revolutions is this per second?

42. Two triangles are **similar** if the lengths of the corresponding sides occur in the same ratio. If each of the following triangles is similar to △*ABC*, find the missing sides.

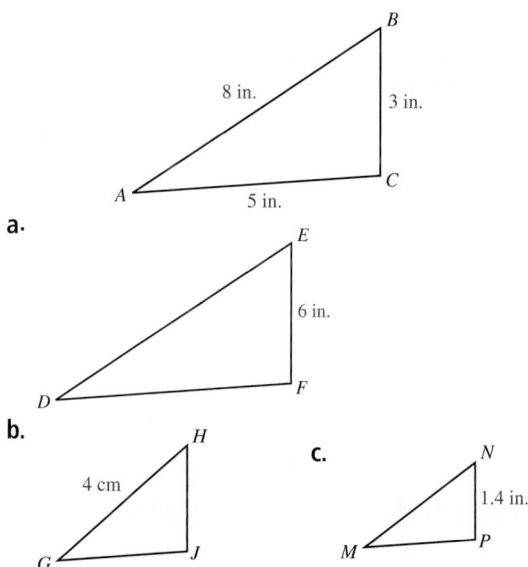

a.

b.

c.

43. At 6:00 one afternoon, a flagpole that is 35 ft tall casts a shadow that is 50 ft long. At the same time, how long will a girl's shadow be if she is 4-ft-tall?

44. On a certain map, $\dfrac{1}{2}$ in. represents 10 mi. If it is $3\frac{3}{4}$ in. on the map from Town *A* to Town *B*, how many mi is it between them?

45. A pilot is traveling in a small aircraft at 220 mph. He is flying with a tail wind of 30 knots. If there are 1.15 knots to 1 mph, what is the plane's speed in miles per hour relative to the ground?

46. A photo from a weather satellite has a scale of 1 in. to 200 mi. If the satellite sends a picture that is 5 in. by 7 in., what are the dimensions of the land shown in the photo?

47. A 64-oz bottle of apple juice sells for $1.97. What is the most a 48-oz bottle of the same juice could cost and still be a better buy?

48. The ratio of fraternities to sororities at a particular university is 3 to 2. If there are 8 sororities, how many new fraternities must be added to make the ratio 2 to 1?

49. Jim, John, and Joe paint a house and earn $5,000. If Jim painted for 15 hr, John for 20 hr, and Joe for 30 hr, how should they divide the money fairly?

50. 25 employees own stock in their own Internet company and are to divide $1,000,000 in profits among themselves. If 10 of them own 1,000 shares each, and the other 15 only own 500 shares each, how should the money be divided?

51. **a.** Janice can type 66 words per minute (wpm). How many words can she type in 1 hr?

    b. About how long will it take Janice to type a paper that has 2,500 words?

    c. As Janice types, she makes 3 mistakes for every 75 words typed. About how many mistakes will she make in a paper that contains 5,000 words?

52. Many race cars have a maximum speed of about 220 mph. Most race horses have a top speed of about 40 mph. At these speeds, how far can a race horse travel in the time it takes for the race car to drive 100 mi?

53. A survey conducted at a small amusement park indicates that an average of 3.7 people are in each vehicle entering the parking lot. The survey also indicates that each person typically buys two soft drinks while at the park. On a given day, soft drink purchases totaled 14,357 drinks. Based on this information, how many cars are likely to be in the parking lot on that day?

Mathematical Reasoning and Proof

54. **a.** In the set of natural numbers, what is the ratio of even numbers to odd numbers?

    b. In the set of integers, what is the ratio of positive numbers to negative numbers?

55. The ratio of males to females in Mrs. Smith's class is 4 to 5. In Mr. Brown's class, it is 3 to 4. Is it possible to determine which class has more females? Explain.

56. Consider two bags of marbles. In one bag, the ratio of red marbles to green marbles is 9 to 10. In the other bag, it is 3 to 4. Is it possible to determine which bag has more marbles? Explain.

57. An advertisement claims that 3 out of 4 dentists prefer a particular toothpaste. If 88 dentists were actually surveyed and only 60 of them preferred the toothpaste, is the advertisement true or false? Explain.

58. If $\frac{a}{b}$ is proportional to $\frac{c}{d}$, is $\frac{a}{a-b}$ proportional to $\frac{c}{c-d}$? Write an argument to justify your answer.

59. If $\frac{a}{b}$ is proportional to $\frac{c}{d}$, is $\frac{a+b}{b}$ proportional to $\frac{c+d}{d}$? Write an argument to justify your answer.

Mathematical Communication

60. What is the difference between each ratio in each of the following pairs?

    a. Miles per hour and hours per mile
    b. Letters per minute and minutes per letter
    c. Pounds per dollar and dollars per pound

61. **a.** Write two story problems that are solved by comparing different ratios.

    b. Write two story problems that involve proportions.

    c. Give your problems to a partner to solve. If your partner has difficulty understanding the problem, discuss how you might rewrite them to be better.

62. A student's solution to the proportion $\frac{16.4}{28.3} = \frac{x}{56.6}$ is $x = 97.7$. Is this correct? How can you know without computing the answer yourself? If it is incorrect, how might you help the student?

63. When considering the ratio of dogs to cats, how would you explain to a student how the ratio of 3 to 2 is different from the ratio of 2 to 3?

64. Write a paragraph or two explaining how you might help one of your students understand the difference between a ratio and a proportion.

Building Connections

65. List five ratios you might encounter or be concerned with while:

    a. Flying on an airplane.
    b. Shopping at a clothing store.
    c. Buying television.
    d. Selecting a doctor.

66. Find five real-world ratios in which one of the numbers is zero.

67. To this point, we have discussed three types of mathematical reasoning: inductive reasoning, deductive reasoning, and proportional reasoning. Write two or three paragraphs summarizing each type of reasoning and how each can be used to solve mathematical problems.

68. Look at a sequence of textbooks for grades K–8. In what grade level are ratios introduced? What about proportions? Write a paragraph or two describing how each topic is introduced and what mathematical ideas are used to do so.

69. Proportional reasoning is thought to be the culmination of elementary mathematics. Do you agree with this statement? Why or why not? When you have thought your answer through, discuss it with a group of your peers.

Percents

Percents are another idea closely associated with proportional reasoning. We encounter them frequently in our daily lives as we hear or use phrases like:

"There is a seventy-five percent chance of rain tomorrow."

"This credit card has an interest rate of thirteen percent a month."

"Fifty-five percent of the student body is female."

Percents are an integral part of the world around us, so understanding them and solving problems with them is essential to making good life decisions. Unfortunately, many students and even many adults struggle with percents. Not only do they not understand them, but also they fail to apply them correctly in the situations they encounter.

Understanding Percents

The term "percent" comes from the Latin phrase *per centum*, which means per hundred. Because the word "per" is part of the word percent, we define a percent to be a ratio that always makes a comparison to 100.

Definition of Percent

If n is a nonnegative real number, then n **percent,** written $n\%$, is the ratio of n to 100, or $n\% = \dfrac{n}{100}$.

Based on the definition, 75% represents the ratio of 75 to 100, or $\dfrac{75}{100}$, and 6% represents the ratio of 6 to 100, or $\dfrac{6}{100}$. Because the ratios are based on 100, we can use grid paper as a visual way to represent and understand percents. Several examples are shown in Figure 8.18.

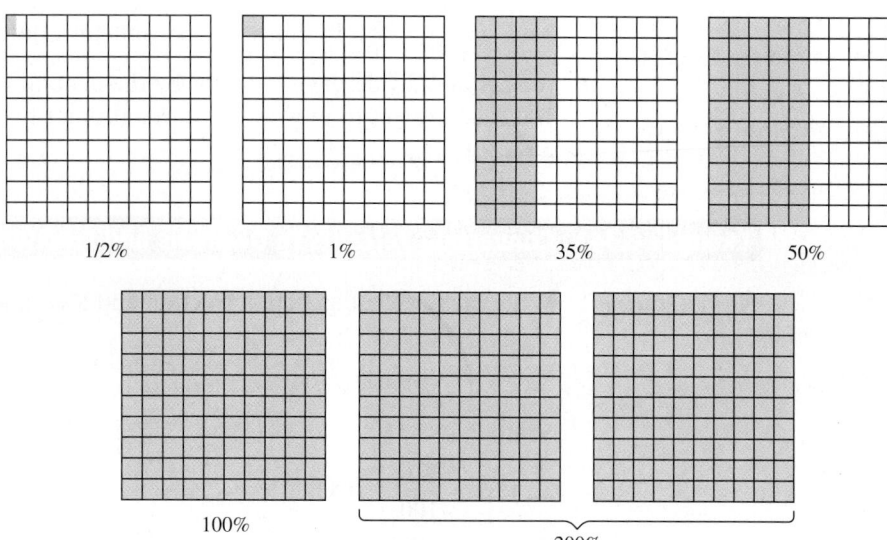

1/2%   1%   35%   50%

100%   200%

Figure 8.18 Representing percents with grid paper

Figure 8.18 points out two facts about percents. First, percents are not limited to the whole numbers between 1 and 99; they can be any real number greater than or equal to zero. Second, the percent symbol affects the value of the number. For instance, as the grid paper shows, there is a big difference between the values of 1 and 1%. One represents the entire whole, whereas 1% represents $\frac{1}{100}$ of the whole. Similarly, $\frac{1}{2}$ represents half of the whole, whereas $\frac{1}{2}$% represents $\frac{\frac{1}{2}}{100} = \frac{1}{200}$ of the whole.

Example 8.33 Understanding Percents

Suppose you have $20. How much money is equal to each of the given percents?

a. 50%   **b.** 100%   **c.** 200%

Solution To make sense of each percent, we use the formula in the definition.

a. Because $50\% = \frac{50}{100} = \frac{1}{2}$, then 50% of $20 is half of $20, or $10.

b. Because $100\% = \frac{100}{100} = 1$, 100% of $20 is all of the $20.

c. Because $200\% = \frac{200}{100} = 2$, then 200% of $20 is twice $20, or $40.

The definition states that a percent is a ratio, which has two implications. First, like ratios, percents generally occur in a context, outside of which they may have little meaning. For instance, by itself 75% does not tell us much. However, if there is a 75% chance of rain today, then we know that it rained on 75 out of 100 days that had the same weather conditions as today. Second, ratios are commonly expressed as fractions, which in turn can be expressed as decimals. Consequently, any percent can be expressed as a decimal or a fraction, and conversions between these representations are useful.

The definition gives us a quick way to convert a percent to a fraction. All we have to do is place n over a denominator of 100 and simplify as appropriate. For instance, $35\% = \frac{35}{100} = \frac{7}{20}$. To convert a percent to a decimal, we take the process one step further and divide the numerator by the denominator. However, because dividing by 100 is equivalent to moving the decimal point two spots to the left, we simply write, $35\% = 0.35$.

Example 8.34 Converting Percents to Decimals and Fractions

Representation   Write each percent as both a fraction and a decimal.

a. 43%   **b.** 310%   **c.** $26\frac{1}{3}$%

Solution

a. $43\% = \frac{43}{100} = 0.43$.

b. $310\% = \frac{310}{100} = 3.1$.

c. As a fraction, $26\frac{1}{3}\% = \dfrac{26\frac{1}{3}}{100} = \dfrac{\frac{79}{3}}{100} = \dfrac{79}{3}\cdot\dfrac{1}{100} = \dfrac{79}{300}$. As a decimal

$26\frac{1}{3}\% = \dfrac{26\frac{1}{3}}{100} = \dfrac{26.\overline{3}}{100} = 0.26\overline{3}.$

∎

Decimals and fractions can also be converted into percents. For a decimal, we can rewrite it as a fraction with 100 in the denominator and apply the definition of a percent. This is equivalent to moving the decimal two places to the right and affixing the percent symbol. For a fraction, either we can change it to a fraction with 100 in the denominator and apply the definition of a percent, or we can convert it to a decimal and then move the decimal point appropriately.

Example 8.35 **Converting Decimals and Fractions to Percents**

Representation   Write each number as a percent.

a. 0.45 **b.** $\dfrac{6}{25}$ **c.** $\dfrac{4}{3}$

Solution

a. $0.45 = \dfrac{45}{100} = 45\%.$

b. $\dfrac{6}{25}\cdot\dfrac{4}{4} = \dfrac{24}{100} = 24\%.$

c. $\dfrac{4}{3} = 4\div 3 = 1.33\overline{3} = 133.\overline{3}\%.$

Another way to convert a fraction to a percent is to use a proportion. For instance, converting $\dfrac{6}{25}$ to a percent is the same as solving the following proportion for x:

$$\dfrac{6}{25} = \dfrac{x}{100}$$
$$100\cdot\left(\dfrac{6}{25}\right) = x$$
$$x = 24 \text{ or } x = 24\%.$$

We can also use fraction calculators, as many of these calculators can make conversions between percents, fractions, and decimals using keystrokes like the ones shown in Table 8.5.

Table 8.5 Calculating percents on a calculator

Example Conversion	Keystrokes	Display
85% to decimal	85 [2nd] [%] [▷D] [=]	0.85
85% to fraction	85 [2nd] [%] [▷F] [=]	$\dfrac{17}{20}$
$\dfrac{3}{7}$ to percent	3 [/] 7 [2nd] [▷%] [=]	42.857143%
0.42 to percent	[.] 42 [2nd] [▷%] [=]	42%

The next example shows how in some situations it may be easier to work with percents, rather than with decimals or fractions. Specifically, because percents always give rates relative to 100, they are a good way of comparing sets of different sizes.

Example 8.36 Comparing Test Performance with Percents

Application

In Mr. Everett's first-period class, 18 of 25 students passed the last test. In his second period, 17 of 23 passed. Which class did better?

Solution We can compare the classes' passing rates by finding the percent of students who passed. Because $\frac{18}{25} = 0.72 = 72\%$ and $\frac{17}{23} \approx 0.74 = 74\%$, Mr. Everett's second period did better.

Solving Percent Problems

Because percents are so common, we encounter them in a wide variety of problem situations. Percents always describe part-to-whole relationships, so most percent problems revolve around three numbers: the percent, the part, and the whole. If two of the three are known, the third can usually be found. This means that most percent problems fall into one of three cases.

Case 1: Finding the percent of a number. The percent and the whole are known, and we are left to find the part.
Case 2: Finding the percent that one number is of another. The part and the whole are known, and we are left to find the percent.
Case 3: Finding a number when a given percent of that number is known. The percent and the part are known, and we are left to find the whole.

With information given, we can use two common approaches to find the missing value. One is to use proportions; that is, we use the fact that the percent is proportional to the ratio of the part to the whole:

$$\frac{\text{Part}}{\text{Whole}} = \frac{\text{Percent}}{100}$$

Explorations Manual 8.12

After putting the given information into the appropriate places, we solve the proportion to find the missing number.

A second approach is to write an equation by looking for key words like "is" for equality and "of" for multiplication. Table 8.6 summarizes the proportions and equations that are related to each of the three cases. In the table, x stands for the missing value, p for a known percent, w for a known whole, and f for a known part.

Table 8.6 Different types of percent problems

Case	Example Word Phrase	Corresponding Proportion	Corresponding Equation
Case 1	p percent of w is what?	$\frac{x}{\text{Whole}} = \frac{\text{Percent}}{100}$	$p \cdot w = x$
Case 2	f is what percent of w?	$\frac{\text{Part}}{\text{Whole}} = \frac{x}{100}$	$f = x \cdot w$
Case 3	f is p percent of what number?	$\frac{\text{Part}}{x} = \frac{\text{Percent}}{100}$	$f = p \cdot x$

It may be tempting to memorize the information in Table 8.6. However, it should be used only as an aid in understanding the relationships among the percent, the part, and the whole. With a sound understanding of these relationships, we can be more effective at solving percent problems. The next examples demonstrate how to solve the different types of percent problems.

Example 8.37 Solving Percent Problems

Find the missing number.

a. 45 is 50% of what number?

b. 27 is what percent of 108?

c. 75% of 210 is what?

Solution In each case, we use Table 8.6 to help us find the missing number.

a. We know the part and the percent, so we represent the problem with the proportion $\frac{45}{w} = \frac{50}{100}$. Solving for w, $w = \frac{45 \cdot 100}{50} = 90$.

b. We have the part and the whole, so we use the equation, $27 = p \cdot 108$. Solving for p, $p = \frac{27}{108} = 0.25 = 25\%$.

c. We know the percent and the whole, so we use the equation $0.75 \cdot 210 = f$. After multiplying, $f = 157.5$.

Example 8.38 Percent on a Test

Application

Marsha answered 67 out of 85 questions correctly on her test. What percent of the questions did she answer incorrectly?

Solution We are given the part, 67, and the whole, 85, and asked to find a percent. Because this is Case 2 in Table 8.6, we use a proportion:

$$\frac{67}{85} = \frac{p}{100}$$

Solving for p, we find that Marsha had $p = \frac{67 \cdot 100}{85} = 78.8\%$ of the answers correct.

Explorations Manual 8.13

By subtracting the percent correct from 100%, we find that $100\% - 78.8\% = 21.2\%$ were answered incorrectly.

Example 8.39 Injuries from Exercise Equipment

Application

Statistics show that each year around 30,000 people in the United States are injured while using exercise equipment. If 58% of those are injured while lifting weights, about how many people in a year are injured in this way?

Solution Because we know the whole, 30,000 people, and a percent, 58%, we have Case 1 from Table 8.6. This time we solve the problem with an equation. Taking 58% of 30,000, we have

$$x = 0.58 \cdot 30{,}000 = 17{,}400 \text{ people}$$

Some percent problems can be more difficult because they require several steps. In such cases, we can simplify the problem by working one step at a time.

Example 8.40 Napping on the Job

Problem Solving A poll claims that 25% of all males admit to taking a nap on the job, whereas only half as many females admit to the same thing. If a company employs 15,000 workers, of which 55% are female, about how many female workers at the company would admit to taking naps on the job?

Solution

Understanding the Problem. We must find how many female employees would admit to taking a nap on the job. There are two unknowns: the number of female employees and the percentage of those who would admit to taking a nap.

Devising a Plan. We can approach the solution in two steps: first, find the number of female employees; then find how many would admit to sleeping on the job. We can *use an equation* in both steps.

Carrying Out the Plan. The number of female employees is 55% of 15,000, or $0.55 \cdot 15,000 = 8,250$. The percentage of females who would admit to napping on the job is half as much as the percentage of men who would do the same thing, or $25\% \div 2 = 12.5\%$. Consequently, the number of females who would admit to napping on the job is 12.5% of 8,250, or $0.125 \cdot 8,250 \approx 1,031$ females.

Looking Back. We can check our answer by making a few estimates. Because 55% is close to 50%, we expect a little more than half of the employees, or about 8,000, to be female. Because 12.5% is a little more than 10%, we would expect a little more than 10%, or about 800 females, to admit to napping on the job.

Percent problems can also involve a percent increase or decrease. In these problems, the whole is increased or decreased by some percentage of itself. With regard to these problems, remember that the increase or decrease is always relative to the original amount.

Example 8.41 Discounting a Television

Application A television is on sale and is marked down 30% from its original price. If the sale price is $505, what was the original price?

Solution To find the sale price, we subtract 30% of the original price from the original price. In other words, the sale price is 70% of the original price. If we let x represent the original price, then $0.7x = 505$. By dividing both sides of the equation by 0.7, the original price of the television was $x = \dfrac{\$505}{0.7} = \$721.42$.

Example 8.42 Increasing Real Estate Values

Application Real estate values in a certain county are increasing at a rate of 2% a month. If a house is originally listed at $225,000, how much should it sell for after 3 months?

Solution In the first month, the value of the house increases by 2% of the original sale price. Compute this value by taking 2% of $225,000 and adding the result to the original cost of the home. Repeat the process two more times, each time using the new value as the original price of the home:

Value after month 1 = $225,000 + ($225,000 · 0.02) = $229,500

Value after month 2 = $229,500 + ($229,500 · 0.02) = $234,090

Value after month 3 = $234,090 + ($234,090 · 0.02) = $238,771.80

After 3 months, the sale price should be $238,771.80, an increase of $13,771.80 over the course of the 3 months.

Check Your Understanding 8.4-A

1. Make each conversion.

   a. 43% to a fraction
   b. 310% to a fraction
   c. 0.6% to a decimal
   d. $15\frac{1}{2}$% to a decimal
   e. 0.45 to a percent
   f. $\frac{4}{3}$ to a percent

2. In Mrs. Bell's French class, 14 of 25 students are female. In her Spanish class, 7 of 12 students are female. Which class has a larger percentage of females?

3. Solve each percent problem.

   a. 15 is what percent of 45?
   b. 35% of 40 is what?
   c. 65 is 90% of what number?

4. People on the Eastern Seaboard eat 50% of all the ice cream sandwiches sold in the United States. If approximately 2,350,000 ice cream sandwiches are sold in the United States in 1 month, about how many are consumed by the people on the Eastern Seaboard?

5. A store is having a 25% off sale. If Jill buys 3 shirts, each costing $12 before the discount, how much is her bill?

Talk About It Why is context important for interpreting percents? In what contexts have you used or encountered percents?

Representation

Activity 8.7
Represent each percent with grid paper.

a. 17%   b. 53%   c. $12\frac{1}{3}$%   d. 63.5%   e. 215%

Mental Calculations and Estimations with Percents

When solving percent problems, we may need to make mental calculations or estimations. In such cases, it may be easier to use a fraction than a percent. Table 8.7 shows the equivalent fractional values of some common percents.

Table 8.7	Fractional values of common percents							
Percent	1%	5%	10%	25%	$33\frac{1}{3}$%	50%	$66\frac{2}{3}$%	75%
Equivalent fraction	$\frac{1}{100}$	$\frac{1}{20}$	$\frac{1}{10}$	$\frac{1}{4}$	$\frac{1}{3}$	$\frac{1}{2}$	$\frac{2}{3}$	$\frac{3}{4}$

Example 8.43 | **Computing Percents Mentally**

Use equivalent fractions to compute each problem mentally.

a. 25% of 88   **b.** 60% of 48

Solution

a. 25% of 88 is the same as $\frac{1}{4}$ of 88. Consequently, $\frac{1}{4} \cdot 88 = 22$.

b. We do not know the equivalent fraction for 60%, but we can separate 60% into 50% + 10%, both of which we know. Hence, 60% of 48 is equivalent to (50% + 10%) of 48, or $\frac{1}{2} \cdot 48 + \frac{1}{10} \cdot 48 = 24 + 4.8 = 28.8$.

Example 8.44 | **Estimating Percents**

Use equivalent fractions to estimate each of the following.

a. 49% of 898   **b.** 752% · 32

Solution We estimate each problem by rounding to compatible numbers and using equivalent fractions for the percents.

a. 49% of 898 ≈ 50% of 900 = $\frac{1}{2} \cdot 900 = 450$.

b. 752% is awkward because it does not have an equivalent fraction that is easy to obtain. However, $752\% \cdot 32 = \frac{752}{100} \cdot 32 = 752 \cdot \frac{32}{100} = 752 \cdot 32\%$. We can then estimate this last product: $752 \cdot 32\% \approx 750 \cdot 33\frac{1}{3}\% = 750 \cdot \frac{1}{3} = 250$.

Interest

A particularly important application of percents is **interest,** which is a fee paid for the use of money. Interest is one of the most powerful financial ideas and is generally used anytime money is borrowed or loaned. It is paid on bank accounts, student loans, car loans, mortgages, and credit cards. In most situations, interest is a percentage of the amount borrowed and typically varies with three amounts.

1. The **principal,** *P,* is the amount of money borrowed. The more money borrowed, the larger the fee.
2. The **interest rate,** *r,* is the percent per year used to determine the fractional amount of the principal that is to be applied as a fee. The larger the percent, the larger the fee.
3. The length of **time,** *t,* is the number of years the money is borrowed. The longer the money is borrowed, the larger the fee. The length of time is called the **term** of the loan.

Using these amounts, we can compute interest in a number of ways. The most basic is simple interest. **Simple interest** is interest that is paid only on the original principal. The borrower agrees to pay a certain percentage of the principal for each year the money is borrowed. For instance, if Tasha takes out a simple interest loan of $5,000 at an interest rate of 5% for 1 year, then she is agreeing to pay 5% of $5,000 for the use of the money, or

$$r\% \text{ of } P = r \cdot P = 0.05 \cdot \$5,000 = \$250$$

If Tasha's loan was for 3 years, she would agree to pay the bank $250 each year for a total of 3 · $250 = $750 in interest. If we generalize this idea, the interest paid per year is computed by rP, which we then multiply by the number of years. Consequently, the formula for computing simple interest is $I = Prt$.

Example 8.45 Computing Simple Interest

Application

Compute the amount of simple interest on each loan.

a. A principal of $10,000 at an interest rate of 6% for a term of 5 years.

b. A principal of $2,000 at an interest rate of 13.9% for a term of 6 months.

Solution

a. Because $P = \$10{,}000$, $r = 6\%$, and $t = 5$ years, the interest on the loan is $I = \$10{,}000 \cdot 0.06 \cdot 5 = \$3{,}000$.

b. $P = \$2{,}000$ and $r = 13.9\%$. Because the term of the loan is only 6 months, or half of a year, then $t = \frac{1}{2}$. Consequently, $I = \$2{,}000 \cdot 0.139 \cdot \frac{1}{2} = \$139$.

Interest is not the only money that the borrower must pay back to the lender. The borrower must also pay back the principal. If a person borrows $10,000 at an interest rate of 6% for 5 years, the person must pay the lender a total of $10,000 + $3,000 = $13,000 by the end of the 5 years. Consequently, the total amount (A), or the balance, of any simple interest loan is equal to the principal plus the interest:

$$\text{Amount} = P + I = P + Prt = P(1 + rt)$$

Example 8.46 Buying a Car

Application

Derek buys a car for $8,500 with a sales tax of 7%. He has $2,300 to put down and must finance the rest. He is able to secure a simple interest loan for 3 years at a rate of 9%. What is the total cost for the car?

Solution The price of Derek's car includes not only the value of the car, but also the sales tax and the interest on the amount he finances. If sales tax is 7%, he must pay $8,500 + (0.07 · $8,500) = $9,095 prior to financing. He has $2,300 to put down, so he must finance, or borrow, a total of $9,095 − $2,300 = $6,795. Using this amount as the principal of the loan, we find that he must pay a loan total of

$$A = \$6{,}795(1 + 0.09 \cdot 3) = \$6{,}795(1.27) = \$8{,}629.65$$

Therefore, Derek ultimately pays $8629.65 + $2,300 = $10,929.65 for the car.

Simple interest loans are rather uncommon because most loans make use of compound interest. With **compound interest,** the borrower pays interest not only on the principal, but also on any subsequent interest added to principal. In other words, the lender makes money on interest already earned. For instance, consider a savings account in which interest on $1,000 is compounded annually at an interest rate of 5% for 5 years. **Compounded annually** means that the simple interest is computed at the end of each year and then added to the principal. In the next year, interest is computed not only on the original principal, but also on the interest from the first year.

Table 8.8 Growing interest

Year	Principal	Interest ($I = Prt$)	Amount at end of year ($P + I$)
Year 1	1,000	$I = 1{,}000 \cdot 0.05 \cdot 1 = 50$	$1{,}000 + 50 = 1{,}050$
Year 2	1,050	$I = 1{,}050 \cdot 0.05 \cdot 1 = 52.50$	$1{,}050 + 52.50 = 1{,}102.50$
Year 3	1,102.50	$I = 1{,}102.50 \cdot 0.05 \cdot 1 \approx 55.13$	$1{,}102.50 + 55.13 = 1{,}157.63$
Year 4	1,157.63	$I = 1{,}157.63 \cdot 0.05 \cdot 1 \approx 57.88$	$1{,}157.63 + 57.88 = 1{,}215.51$
Year 5	1,215.51	$I = 1{,}215.51 \cdot 0.05 \cdot 1 \approx 60.78$	$1{,}215.51 + 60.78 = 1{,}276.29$

Table 8.8 shows how the value grows over 5 years. Notice that the interest is not the same every year but grows each year by about $2.50. The growth is due to the fact that about $50 or $60 has been added to principal from the previous year. If we were to use simple interest with this account, the amount after 5 years would be

$$A = P(1 + rt) = 1{,}000(1 + 0.05 \cdot 5) = 1{,}000(1.25) = \$1{,}250$$

Consequently, by using compound interest instead of simple interest, the lender makes an additional $26.29 on the loan.

Using the values in the last column of Table 8.8, we can develop a formula for computing compound interest. The following equations are based on the formula $A = P(1 + rt)$, where $t = 1$, and on the patterns established in Table 8.8:

Amount after year 1 $= A_1 = P(1 + r)$
Amount after year 2 $= A_2 = A_1(1 + r) = P(1 + r)(1 + r) = P(1 + r)^2$
Amount after year 3 $= A_3 = A_2(1 + r) = P(1 + r)^2(1 + r) = P(1 + r)^3 \ldots$

These equations show that the amount in the account is equal to the original principal times a power of $(1 + r)$. Because the pattern can be carried on to any number of years, we can make the following generalization.

Theorem 8.8 Compound Interest Computed Annually

The amount in an account compounded once a year with principal P, interest rate per year r, and time in years t, is given by $A = P(1 + r)^t$.

Example 8.47 Compounding Interest Annually

How much interest is paid on a savings account of $35,000 at a rate of 3.9% for 5 years, if interest is compounded annually?

Solution Using Theorem 8.8:

$$A = P(1 + r)^t = \$35{,}000(1 + .039)^5 = \$35{,}000(1.039)^5 \approx \$42{,}378.52.$$

The value $42,378.52 includes both the original principal and any interest. To find the interest, subtract the principal from this amount. Consequently, $42,378.52 − $35,000 = $7,378.52 has been paid in interest.

Compounding interest annually, or once a year, is not the only way to compound interest. We can also compound it **semiannually** (2 times a year), **quarterly** (4 times a year), **monthly** (12 times a year), or even **daily** (365 times a year). For each of these compounding periods, the number of times interest is compounded goes up. For instance, interest on a loan compounded quarterly for 5 years is computed $4 \cdot 5 = 20$ times over the course of the loan. Interest on a loan compounded monthly for 30 years is computed

$12 \cdot 30 = 360$ times. The advantage of compounding interest more often is that more money is added to the principal each time interest is computed. Thus the lender makes more money.

To use different compounding periods, the lender must make changes to the interest rate being charged. For instance, if a loan is compounded quarterly at a rate of 5% each time, then the lender actually earns about $4 \cdot 5\% = 20\%$ for the year. This is a very high rate, so the lender would have trouble attracting potential borrowers. Instead, lenders typically give the **annual percentage rate (APR),** which is the interest rate for the entire year. The interest earned in each period is the annual percentage rate divided by the number of periods in a year. Making this adjustment leads to the formula for computing compound interest.

Theorem 8.9   General Formula for Compound Interest

The amount in an account that is compounded n times a year with principal P, annual percentage rate APR, and time in years t is given by $A = P\left(1 + \dfrac{APR}{n}\right)^{nt}$.

Example 8.48   Computing Compound Interest

Compute the amount in each of the following accounts.

a. A principal of $10,000 at an *APR* of 6% compounded semiannually for 5 years

b. A principal of $200,000 at an *APR* of 6.25% compounded monthly for 30 years

Solution

a. In this loan, $P = \$10,000$, $APR = 6\%$, $n = 2$, and $t = 5$. Using Theorem 8.9,

$A = 10,000\left(1 + \dfrac{0.06}{2}\right)^{2 \cdot 5}$. It can be computed on a calculator as follows:

10000 ⊗ ⦅ 1 ⊕ 0.06 ⊘ 2 ⦆ ^ ⦅ 2 ⊗ 5 ⦆ ⊜

This gives a value of $13,439.16 after 5 years.

b. In this loan, $P = 200,000$, $APR = 6.25\%$, $n = 12$, and $t = 30$. We substitute into the formula and use a calculator to compute:

$$A = 200000\left(1 + \frac{0.0625}{12}\right)^{12 \cdot 30} \approx \$1,297,833.28.$$

The second part of Example 8.48 shows how much interest can grow over the course of time. If you choose to invest, invest early, so that interest can work for you. If you choose to borrow, pay off early, before interest has time to work against you.

Check Your Understanding 8.4-B

1. How much is in an account that earns simple interest if $P = \$2,000$, $r = 6\%$, and $t = 2$ years.

2. How much is in an account that is compounded annually if $P = \$5,000$, $r = 5\%$, and $t = 5$ years?

3. How much is in an account that is compounded monthly if $P = \$13,500$, $APR = 7.9\%$, and $t = 48$ months?

Talk About It Why do you think interest is considered one of the most powerful financial ideas?

Connections

Activity 8.8

The formula for determining the monthly payments on loans with compound interest is

$$PMT = \frac{P \cdot \left(\frac{APR}{n}\right)}{\left[1 - \left(1 + \frac{APR}{n}\right)^{(-nt)}\right]}$$

Use it to the find the monthly payments for the following loans.

a. $P = \$10,000$, $APR = 6.5\%$, $t = 5$ years, compounded quarterly

b. $P = \$5,000$, $APR = 7.9\%$, $t = 1$ year, compounded monthly

c. $P = \$250,000$, $APR = 6.15\%$, $t = 30$ years, compounded monthly

SECTION 8.4   Learning Assessment

■ Understanding the Concepts

1. What is a percent?

2. What is the difference between a value of 10 and 10%?

3. How are percents connected to decimals and fractions?

4. In what situations is it more beneficial to use a percent than a fraction or a decimal?

5. Briefly describe the three basic types of percent problems.

6. **a.** Describe how to use a proportion to solve a percent problem.

   b. Describe how to use an equation to solve a percent problem.

7. What is interest?

8. What three values are used to compute interest, and how does each affect the amount of interest earned?

9. **a.** What is the difference between simple and compound interest?

   b. Why is compound interest generally preferred over simple interest?

■ Representing the Mathematics

10. Give a verbal interpretation of the percent in each situation.

    a. 50% of all crème-cookie eaters pull them apart.

    b. 3% of Americans prefer their hot dogs plain.

    c. 4% of the workers in the United States never laugh at work.

    d. 40% of indigestion remedies sold worldwide are bought in the United States.

11. Represent each percent with grid paper.

    a. 15%   **b.** 78%   **c.** $2\frac{1}{2}\%$ **d.** 258%

12. Represent each percent as a decimal.

    a. 17%   **b.** 163%   **c.** 0.2%   **d.** $5\frac{3}{4}\%$

    e. $57\frac{1}{3}\%$

13. Represent each percent as the ratio of two integers.

    a. 25%   **b.** 80%   **c.** $33\frac{1}{3}\%$ **d.** 150%
    e. 0.5%

14. Represent each decimal as a percent.
    a. 0.39   **b.** 0.93   **c.** 1.47   **d.** 0.005   **e.** 2

15. Represent each fraction as a percent.

    a. $\frac{7}{10}$ **b.** $\frac{14}{25}$ **c.** $\frac{1}{6}$ **d.** $\frac{4.3}{7.5}$ **e.** $\frac{\sqrt{6}}{\sqrt{8}}$

16. What fractional amount of a whole is represented by each percent?

    a. 5%   **b.** 16%   **c.** 100%   **d.** 150%

    e. $\frac{1}{3}\%$ **f.** 0.1%

■ Exercises

17. Use a proportion to solve each percent problem.
   a. 20 is what percent of 80?
   b. 50% of 96 is what?
   c. 70 is 55% of what number?

18. Use a proportion to solve each percent problem.
   a. 17 is what percent of 95?
   b. 34% of 55 is what?
   c. 93 is 120% of what number?

19. Use an equation to solve each percent problem.
   a. 5% of 25 is what?
   b. 40 is what percent of 120?
   c. 65 is 85% of what number?

20. Use an equation to solve each percent problem.
   a. 1% of 61 is what?
   b. 54 is what percent of 64?
   c. 7.5 is 51% of what number?

21. Compute the simple interest on each account.
   a. $P = \$1,000$, $r = 5\%$, $t = 5$ years
   b. $P = \$10,000$, $r = 7\%$, $t = 10$ years

22. Compute the simple interest on each account.
   a. $P = \$13,000$, $r = 6.9\%$, $t = 4.5$ years
   b. $P = \$8,500$, $r = 2.5\%$, $t = 6$ months

23. Compute the amount of money in each account at the end of the term of the loan.
   a. $P = \$2,000$, $r = 3\%$, $t = 2$ years compounded yearly
   b. $P = \$65,000$, $r = 4\%$, $t = 10$ years, compounded quarterly
   c. $P = \$12,500$, $r = 7\%$, $t = 5$ years, compounded monthly

24. Compute the amount of money in each account at the end of the term of the loan.
   a. $P = \$8,700$, $r = 6.3\%$, $t = 5$ years, compounded semiannually
   b. $P = \$25,000$, $r = 9\%$, $t = 10$ years, compounded daily
   c. $P = \$135,000$, $r = 5.875\%$, $t = 30$ years, compounded monthly

25. Compute each product mentally.
   a. $57 \cdot 10\%$
   b. $93 \cdot 33\frac{1}{3}\%$
   c. $20 \cdot 150\%$
   d. $156 \cdot 50\%$
   e. $15 \cdot 66\frac{2}{3}\%$
   f. $75 \cdot 144\%$

26. Estimate.
   a. $9\% \cdot 49$
   b. $19\% \cdot 161$
   c. $11 \cdot 0.9\%$
   d. $27\% \cdot 499$
   e. $51\% \cdot 649$
   f. $73 \cdot 201\%$

27. To avoid mortgage insurance, most lenders require home buyers to make a down payment of at least 20% of the price of the house. Dan and Debbie want to purchase a house costing $197,500. How much must they put down to avoid mortgage insurance?

28. A restaurant brings in $11,065 on a Saturday night. If sales tax is 6%, how much of the money goes to sales tax?

29. Jamal plays high school basketball and is an 86% free throw shooter. If he makes 120 free throws over the course of a season, how many did he attempt?

30. There are 1,456 males at a particular university, which represents approximately 39% of the student body. How many students attend the university?

31. Property tax in a certain county in Pennsylvania is 2.05% of the value of the house. If the homeowners pay $3,874.50 in property tax for the year, how much is the house worth?

32. Rita sold 35 out of 55 T-shirts. What percentage of the T-shirts did she sell?

33. The following chart shows the expenses of a household in percentages.

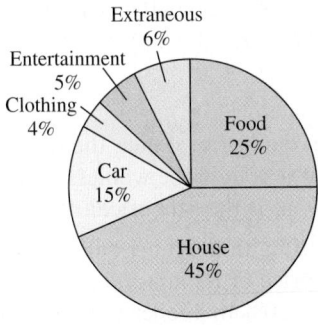

If the family makes $65,500 a year, how much is spent on:
   a. Housing?
   b. Clothing?
   c. Entertainment and Food?

34. Rachel scored a 79 on an 85-point test. What percentage of the points did she get incorrect?

35. A bag contains 25 red marbles, 35 green marbles, and 45 blue marbles. What percentage of the marbles are red? Green? Blue?

36. Three students are selling candy bars for a school fund-raiser. Kim sells 95 out of 110 candy bars, James sells 85% of his 115 candy bars, and Chris sells 96 candy bars, which is about 84% of his total.

   a. Who sold the greatest percentage of candy bars?

   b. Who had the most candy bars to sell? How many did that person have?

   c. Who sold the most candy bars?

■ Problems and Applications

37. a. 25% of 40 is 10% of what number?

   b. 30% of what number is 15% of 90?

   c. 50% of 64 is what percent of 96?

38. a. What is 30% of 40% of 50% of 100?

   b. What is 5% of 10% of 15% of 20% of 25% of 50% of 1,000?

39. Over the last 3 years, Janie has received pay raises of 5%, 6%, and 4%. If her salary is now $55,400, what was her salary 3 years ago to the nearest penny?

40. In large orders of printed documents, 15% of the documents can be printed incorrectly. If a customer orders 5,000 copies of a document, how many copies must actually be printed for the customer to receive 5,000 correctly printed documents?

41. Jill makes $32,000 a year and receives a 4% pay raise. Julie makes $35,000 a year and receives a 3% pay raise. Who got the larger pay raise in terms of dollar amount?

42. The wholesale price of a shirt is increased 25% to obtain the retail price. Later, the shirt is marked 25% off for a sale. If the wholesale price of the shirt was $10, what is the sale price?

43. Jewelry is often marked up 200% or more. The wholesale price of a diamond is $400, which is then marked up 250% by the jeweler. He sells it for 25% off his retail price. How much money did he make on the sale of the diamond?

44. A pawnshop owner buys a watch for $50 and then sells it for $105. What percent profit did the owner make on the sale of the watch?

45. The wholesale price of a lawnmower is $195. If it is marked up 35% by the store and there is 7% sales tax, how much does it cost a customer to buy the lawnmower?

46. A sewing machine is on sale for $345, which is 15% off its regular retail price. What was the price before the markdown?

47. A new car is purchased for $25,850. As soon as the owner drives it off the lot, it depreciates to a value of $22,950. What is the percent decrease in value?

48. The value of stock portfolio A increases from $13,550 to $18,625 over the course of a year. Stock portfolio B increases from $45,650 to $51,890 over the course of a year. Which portfolio had the greater percent increase?

49. A house valued at $150,000 increases in value at 0.5% a month. At this rate, what will the value of the house be in 3 years?

50. Which investment of $5,000 will make more interest?

   A: An account compounded monthly at a rate of 5% for 5 years

   B: An account compounded quarterly at a rate of 6% for 5 years

51. Which investment of $8,000 will make more interest?

   A: An account compounded monthly at a rate of 6% for 3 years

   B: An account compounded monthly at a rate of 4% for 5 years

52. Which investment of $10,000 will make more interest?

   A: An account compounded quarterly at a rate of 7% for 3 years

   B: An account compounded monthly at a rate of 7% for 2 years

53. Mrs. Klein invests $15,000 in a mutual fund. The broker of her account charges 5% of the amount invested prior to making any investments. At the end of one year, her account grows to $17,500. What percent interest rate did she make on her investment if the account was compounded monthly?

54. How much money would have to be originally invested to have $25,000 at the end of 10 years in an account that is compounded annually at an interest rate of 4%?

55. How much money would have to be originally invested to have $1,500,000 at the end of 35 years in an account that is compounded monthly at an interest rate of 8%?

56. Susie invests $5,000 in a savings account that pays compound interest quarterly at a rate of 3.5%. If she invests an additional $1,000 at the beginning of each year, how much will be in the account at the end of 3 years?

57. A credit card charges 23% *APR* on the value of any purchases made with the card. Gwen uses the card to purchase $265 worth of merchandise. If she fails to make any payments on the card for 6 months, how much will she owe?

58. Estimate how long will it take $5,000 to double at an interest rate of 5% compounded monthly.

■ Mathematical Reasoning and Proof

59. **a.** If 35 is 90% of a number, is 25 greater than, less than, or equal to the number?

 b. If 45 is 100% of a number, is 40 greater than, less than, or equal to the number?

 c. If 55 is 110% of a number, is 65 greater than, less than, or equal to the number?

60. **a.** Is it possible to raise the price of an object 120%? Why or why not?

 b. Is it possible to decrease the price of an object 120%? Why or why not?

61. **a.** If a number is decreased by 15%, what percent must the result be increased by to obtain the original amount?

 b. If a number is increased by 10%, what percent must the result be decreased by to obtain the original amount?

62. Which is greater, 100 increased by 25% or 150 decreased by 25%?

63. Is a 20% markdown from the retail price the same as paying 80% of the retail price? Explain.

64. Is 25% of 30 the same as 30% of 25? Explain.

65. Which represents the greater percent, $\frac{125}{250}$ or $\frac{299}{600}$?

 Describe how the answer can be determined mentally.

66. How can an estimate of 25% of a number be used to find an estimate of 75% of the same number?

■ Mathematical Communication

67. Write a story problem for each type of percent problem. Give it to a peer to check, solve, and critique.

68. Guide a peer through the process of solving the following problem:

 A car is marked up 20% from its wholesale price of $19,550. What is the retail price of the car?

69. A student asks how it is possible to get an 80% on a 50-point test if 80% means 80 out of 100. How do you respond to the student's question?

70. A student states that 50% of 70 is 3,500. What has the student done wrong? What would you do to help the student?

71. What directions would you give to help your students compute the total amount of a restaurant bill if the bill is to include a 15% tip? How would your directions change if you were asking them to compute the bill mentally?

■ Building Connections

72. Why does it make sense to include percents as a part of proportional reasoning?

73. List five situations in which you regularly use percents.

74. As a teacher, how might you use or encounter percents other than as content to be taught?

75. As a teacher, how might you use a meter stick to illustrate percents between 0% and 100%?

76. In what ways do you earn or pay interest? Overall, do you think you pay or earn more in interest?

Historical Note

Even though the decimal system was well accepted by European mathematicians by 1500, it was not until 1585 that it was introduced to mainstream Europe by Simon Stevin (1548–1620). In his pamphlet, *De Thiende (The Tenth)*, Stevin demonstrated how common business computations could be performed without the use of fractions. Stevin represented decimal place values with encircled numbers to indicate the power of ten of each decimal digit. For instance the number 25.872 is represented as

$$25 \; ⓪ \; 8 \; ① \; 7 \; ② \; 2 \; ③ \qquad \text{or} \qquad \overset{⓪ \; ① \; ② \; ③}{25 \; 8 \; 7 \; 2.}$$

The circular notation did not last long. In 1593, Christoph Clavius, a Vatican astronomer, introduced the decimal point to separate the integer and fractional parts of the number.

Simon Stevin (1548–1620)

Opening Problem Answer

Blend Images/Shutterstock.com

The total bill is $25.50 + $35.65 + $21.90 = $83.05. Because $25.50 is $\dfrac{\$25.50}{\$83.05} = 30.7\%$ of the bill, then 30.7% of $50, or $0.307 \cdot 50 = \$15.35$ should go for trash. Because $35.63 is $\dfrac{\$35.65}{\$83.05} = 42.9\%$ of the bill, then 42.9% of $50, or $.429 \cdot 50 = \$21.45$ should go for water. Finally, because $21.90 is $\dfrac{\$21.90}{\$83.05} = 26.4\%$ of the bill, then 26.4% of $50 = $13.20 should go for sewer.

Chapter 8 REVIEW

Summary of Key Concepts from Chapter 8

Section 8.1   Decimals and the Real Numbers

- A **decimal** gives a way to write part-to-whole relationships that is consistent with the ten-to-one ratio of the decimal system.

- A **terminating decimal** has only a finite number of decimal places to the right of the decimal point.

- A rational number, $\dfrac{a}{b}$, can be written as a terminating decimal if and only if it is in simplest form and the prime factorization of the denominator contains no primes other than 2s and 5s.

- A **repeating decimal** has a block of digits, called the **repetend,** that continuously repeats. The length of the repetend is called the **period.**

- An **irrational number** is a number with a decimal representation that does not terminate or repeat.

- If a is any non-negative number, the **principal square root** of a, denoted $\sqrt{a}$, is the number $b \geq 0$, such that $b^2 = a$.

- **Perfect squares** have square roots that are whole numbers.

- Let a be any number and n be a natural number.

  1. If $a \geq 0$, then $\sqrt[n]{a} = b$, where $b \geq 0$, if and only if $b^n = a$.

  2. If $a < 0$ and n is odd, then $\sqrt[n]{a} = b$ if and only if $b^n = a$.

- The set of **real numbers,** denoted $\mathbb{R}$, is the union of the rational and irrational numbers.

- If a and b are two real numbers, then there exists a real number c such that $a < c < b$.

Section 8.2   Decimal and Real Number Operations

- **Decimal addition and subtraction:** Align the decimal points in the original numbers to ensure that common place values are added or subtracted. Zeros can be affixed to the end of the decimal to compute the computation. Add or subtract the numbers using whole-number algorithms. Place the decimal point in the answer directly beneath the decimal point in the original numbers.

- **Decimal multiplication:** Multiply the numbers as whole numbers. Place the decimal point in the product by counting the total number of decimal places in the factors. If there are n decimal places in the first factor and m in the second, then there are $n + m$ decimal places in the product.

- Real number addition and multiplication satisfy the closure, commutative, associative, identity, inverse, zero multiplication, and distributive properties.
- **Decimal division:** Move the decimal points in the divisor and dividend the same number of places to remove it from the divisor. Compute using long division. The division concludes once a terminating decimal, a repeating decimal, or a certain number of place values is reached.
- Every rational number can be written as either a terminating or a repeating decimal.
- The effect of multiplying n by 10^m is to move the decimal point m places to the right in n. The effect of dividing n by 10^m, or multiplying by 10^{-m}, is to move the decimal point m places to the left in n.
- A number is in **scientific notation** if it is written in the form $c \times 10^n$, where $1 \le |c| < 10$ and n is any integer.
- If a is any real number, then $a^{1/n} = \sqrt[n]{a}$ where n is any natural number if $a \ge 0$ and n is an odd natural number if $a < 0$.
- If a and b are positive real numbers and m and n are rational exponents, then

  1. $a^m \cdot a^n = a^{m+n}$.   **2.** $a^m \div a^n = a^{m-n}$.   **3.** $(a^m)^n = a^{m \cdot n}$.

  4. $a^m \cdot b^m = (a \cdot b)^m$.   **5.** $\dfrac{a^m}{b^m} = \left(\dfrac{a}{b}\right)^m$.

Section 8.3   Proportional Reasoning

- **Proportional reasoning** is the ability to compare quantities and understand the context in which the comparison takes place.
- A **ratio** is an ordered pair of numbers used to compare two quantities.
- Ratios can describe part-to-whole, whole-to-part, and part-to-part relationships.
- Two ratios are **equivalent** if they represent the same relative amounts, that is, if their fractions or the quotients of their terms are equal.
- A **proportion** is a statement that two ratios are equivalent.
- A **unit rate** is a ratio that describes how many objects of one type are comparable to exactly one of another.
- Two proportions are **equivalent** if and only if they have the same cross products.

Section 8.4   Percents

- If n is a nonnegative real number, n **percent** is the ratio of n to 100, or $n\% = \dfrac{n}{100}$.
- Most percent problems fall into three cases: finding the percent of a number, finding the percent that one number is of another, or finding a number when a given percent of that number is known.
- **Interest** is a fee paid for the use of money. It is based on the **principal,** the **interest rate per year,** and the **time** in years.
- **Simple interest** is interest paid only on the original principal. It is computed using the formula $I = Prt$.
- The balance of a simple interest loan is given by Amount $= P + I = P(1 + rt)$.
- **Compound interest** is interest paid on the principal and on any subsequent interest added to principal.
- The **annual percentage rate (APR)** is the interest rate given for the entire year.
- The amount in an account compounded n times a year with principal P, annual percentage rate APR, and time in years t is given by $A = P\left(1 + \dfrac{APR}{n}\right)^{nt}$.

Review Exercises Chapter 8

1. Write each expression as a decimal.

   a. $(5 \cdot 10) + (3 \cdot 1) + \left(6 \cdot \dfrac{1}{10}\right) + \left(4 \cdot \dfrac{1}{100}\right)$

   b. $\dfrac{16}{36}$

   c. $(2 \cdot 1{,}000) + (7 \cdot 100) + \left(8 \cdot \dfrac{1}{100}\right) + \left(1 \cdot \dfrac{1}{1000}\right)$

   d. 36%

   e. $\dfrac{47}{100}$

   f. 205%

   g. $\dfrac{8}{5}$

   h. $3\dfrac{3}{5}\%$

2. Represent each decimal using grid paper and base-ten blocks.

   a. 0.3   b. 0.46

   c. 0.09   d. 2.35

3. Express each number as the quotient of two integers

   a. 0.45   b. 1.25

   c. 0.575757 . . .   d. $0.4\overline{28}$

   e. 55%   f. $147\dfrac{1}{2}\%$

4. Determine whether each fraction has a terminating decimal representation. If so, find it by converting the denominator to a power of ten.

   a. $\dfrac{14}{25}$

   b. $\dfrac{11}{21}$

   c. $\dfrac{54}{90}$

5. Determine whether each radical is rational, irrational, or not possible.

   a. $\sqrt{121}$

   b. $\sqrt{-132}$

   c. $\sqrt[3]{81}$

   d. $\sqrt[4]{-16}$

6. Order $3.\overline{45}$, $3.4\overline{5}$, 3.456, $3.45\overline{6}$, $3.\overline{454}$ from greatest to least.

7. Find five rational numbers between 8.9 and $8.\overline{9}$.

8. Compute the following mentally.

   a. $3.98 \cdot 10^4$ b. $37.923 \div 10^3$

   c. $640 \cdot 20\%$ d. $144 \cdot 33\dfrac{1}{3}\%$

9. Write each number in scientific notation.

   a. 13,000

   b. 0.000876

   c. 56,900,000,000

10. Write each number in decimal notation.

    a. 4.44×10^{-5}

    b. 7.78×10^3

    c. 5.709×10^6

11. Compute using the standard algorithm.

    a. $55.78 + 16.02$ b. $25.7 \cdot 4.9$

    c. $89.003 - 74.221$ d. $45.67 \div 6.5$

12. Use a calculator to compute. Round your answer to two decimal places as necessary.

    a. $34.56 \cdot 41.6$

    b. $500.9 \cdot 161.3$

    c. $39.67 \div 18.7$

    d. $121.9 \div 63.2$

    e. $\sqrt{37}$

    f. $\sqrt[3]{47^5}$

    g. $(6.43 \times 10^5) + (8.59 \times 10^5)$

    h. $(3.11 \times 10^7) \div (4.7 \times 10^4)$

13. Write each number as a percent using only two decimal places as necessary.

    a. 0.56   b. 0.039

    c. 2.75   d. $\dfrac{3}{5}$

    e. $\dfrac{24}{18}$ f. $\dfrac{6.7}{9.2}$

14. Solve each problem using an equation.

    a. 15% of 65 is what?

    b. 28 is what percent of 92?

    c. 47 is 63% of what number?

15. Compute the amount of money in each account at the end of the term of the loan.

    a. $P = \$6{,}000$, $r = 7\%$, $t = 2$ years with simple interest

    b. $P = \$14{,}500$, $r = 6.5\%$, $t = 7$ years, compounded quarterly

16. If there are 65 employees at a company and 42 of them are female, find the ratio of each of the following.
    a. Females to employees
    b. Males to females
    c. Employees to males

17. Use the method of squaring to estimate $\sqrt{10}$ to two decimal places.

18. Use the fact that $\sqrt{2}$ is irrational to show that $5 + 2\sqrt{2}$ is irrational.

19. During the past week, Monica made the following transactions on her bank account: $13.57, $250.08, –$34.87, –$78.90, and –$141.25. If her account started at $55.43, how much is in the account after the transactions?

20. If a box of 100 paper clips costs $0.65, how much does each paper clip cost?

21. A factory worker makes $13.55/hr and works 55.5 hours in 1 week. If the worker gets paid time and a half for any hours over 40, how much did she earn for the week?

22. Carla is buying snacks for 60 first-graders. She buys 5 gallons of milk at $3.11 a gallon and 7 packages of cookies, each costing $4.16. How much is her bill?

23. Calculate the price per unit of each of the following items.
    a. A 64-oz bottle of juice for $3.25
    b. 500 sheets of paper for $2.59

24. Determine which is the better buy.
    a. 16 oz of shampoo for $3.09 or 12 oz of shampoo for $2.59
    b. One 16-ft board for $9.82 or two 8-ft boards for $4.37 each

25. If Casey gets 4 hits for every 15 times at bat, how many hits would we expect her to have in 130 at bats?

26. A 40-lb bag of lime covers approximately 1,500 ft² of grass. How many 40-lb bags are needed to spread lime over 13,000 ft² of grass?

27. One motorcycle can travel 350 miles on 5 gallons of gas. Another motorcycle can travel 400 miles on 6 gallons of gas. Which motorcycle gets the better gas mileage?

28. It costs me $14.35 to buy 6 boards. How much will it cost me to buy 17 boards?

29. Alice, Arnie, and Ann are to get one-third, one-fourth, and one-fifth of a $1,000,000 inheritance,

respectively. To the nearest penny, how much should each person get?

30. On a certain map, $\frac{1}{3}$ in. represents 8 mi. If it is $2\frac{2}{5}$ in. on the map from town A to town B, how many miles is it between them?

31. The ratio of men to women is 4 to 3 at a certain business. If there are 36 male employees, about how many more women must be hired to make the ratio 1 to 1?

32. Which is greater, 200 increased by 50% or 300 decreased by 50%?

33. A local retailer marks up a stereo 30% from its wholesale price of $250. If he later marks the stereo down 15% to sell it, how much does he make on the sale of the stereo?

34. Closing costs in a certain county are 1.5% of the cost of the house. How much will closing costs be on a home that is $176,000?

35. How much would have to be originally invested to have $55,000 at the end of 10 years in an account that is compounded semiannually at an interest rate of 6%?

36. Over the last 4 years, Janice has received pay raises of 2%, 3%, 3%, and 2% respectively. If her salary in the first year was approximately $35,500, what is her salary to the nearest penny after the four raises?

37. The value of stock portfolio A decreases from $19,540 to $17,350 over the course of a year. Stock portfolio B decreases from $31,350 to $27,790 over the course of a year. Which portfolio had the greater percent decrease?

38. Donna sells 36 out of 45 T-shirts, Dave sells 90% of his 50 T-shirts, and Chris sells 43 T-shirts, which is about 92% of his total.
    a. Who sold the greatest percentage of T-shirts?
    b. Who had the most T-shirts to sell? How many did that person have?
    c. Who sold the most T-shirts?

39. Which investment of $5,000 will make the most interest?
    A: An account compounded monthly at a rate of 4% for 4 years
    B: An account compounded quarterly at a rate of 6% for 5 years
    C: An account compounded annually at a rate of 9% for 6 years

40. Estimate how long will it take $10,000 to triple at an interest rate of 7% compounded monthly.

Answers to Chapter 8 Check Your Understandings and Activities

Check Your Understanding 8.1-A

1. a. $(7 \cdot 10^{-1}) + (3 \cdot 10^{-2})$; seventy-three hundredths.

   b. $(1 \cdot 10^{-2}) + (9 \cdot 10^{-3})$; nineteen thousandths.

   c. $(3 \cdot 10^{1}) + (2 \cdot 10^{-2})$; thirty and two hundredths.

   d. $(4 \cdot 10^{2}) + (1 \cdot 10^{0}) + (1 \cdot 10^{-1}) + (4 \cdot 10^{-3})$; four hundred one and one hundred four thousandths.

2. a.

   b.

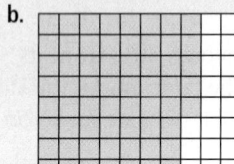

   c.

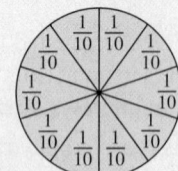

   d.

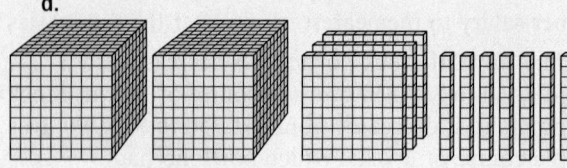

3. a. Terminating with a decimal representation of 0.3125.

   b. Not terminating.

   c. Not terminating.

   d. Terminating with a decimal representation of 0.4125.

4. a. A repetend of 4 and a period of 1.

   b. A repetend of 21 and a period of 2.

   c. A repetend of 219 and a period of 3.

   d. A repetend of 1 and a period of 1.

Check Your Understanding 8.1-B

1. a. $\sqrt{11}$ represents the number x that, when squared, equals 11, or $x^2 = 11$. It is irrational.

   b. $\sqrt[4]{27}$ represents the number x that, when raised to the fourth power, equals 27, or $x^4 = 27$. It is irrational.

   c. $\sqrt[7]{-31}$ represents the number x that, when raised to the seventh power, equals -31, or $x^7 = -31$. It is irrational.

   d. $\sqrt[3]{-2.2}$ represents the number x that, when raised to the third power, equals -2.2, or $x^3 = -2.2$. It is irrational.

   e. $\sqrt{-144}$ represents the number x that, when squared, equals -144, or $x^2 = -144$. It is not possible.

2. a. $\sqrt{11}$ is real but not rational.

   b. 3 is real but not irrational.

   c. There is no real number that is neither rational nor irrational.

   d. There is no rational number that is not real.

3. a. 1.234.   **b.** $-5.634222\ldots$.

   c. $0.5\overline{16}$.

4. a. 209.88.   **b.** 3.5.

   c. 5.6.   **d.** 7.891.

Check Your Understanding 8.2-A

1. a. 78.4.   **b.** 807.783.

   c. 56.78.   **d.** 542.266.

2. a. 122.67.   **b.** 4.4319.

   c. 130.51.   **d.** 1.31.

3. a. 996.75.   **b.** 487.15.   **c.** 5.9749.

4. a. 0.3125.   **b.** $0.8\overline{3}$.   **c.** $0.2\overline{6}$.

Check Your Understanding 8.2-B

1. a. $\dfrac{2}{9}$.

   b. $\dfrac{345}{999} = \dfrac{115}{333}$.

   c. $\dfrac{249}{990} = \dfrac{83}{330}$.

2. a. 3.2×10^3.   **b.** 6×10^6.

   c. 6,120.   **d.** 0.0000769.

3. a. 3.702×10^4.   **b.** 3.399×10^{11}.

   c. 4.3×10^3.   **d.** 7.52.

4. a. $\sqrt{5} \approx 2.236$.   **b.** $\sqrt[5]{17} \approx 1.762$.

   c. $\sqrt[3]{3^2} \approx 2.080$.   **d.** $\sqrt[7]{132^4} \approx 16.284$.

Check Your Understanding 8.3-A

1. a. 4 hearts : 7 clubs and $\dfrac{4 \text{ hearts}}{7 \text{ clubs}}$.

   b. 8 women : 6 girls and $\dfrac{8 \text{ women}}{6 \text{ girls}}$.

   c. 18 pencils : 3 boxes and $\dfrac{18 \text{ pencils}}{3 \text{ boxes}}$.

2. Other ratios are 4 girls to 5 boys, 5 boys to 9 students, 9 students to 5 boys, 4 girls to 9 students, and 9 students to 4 girls.

3. a. The 12-oz can is the better buy.

   b. The 64-oz is a better buy.

Check Your Understanding 8.3-B

1. a. 9.   **b.** 7.5.   **c.** 2.62.

2. About 583.33 mi.

3. 14 gal.

Check Your Understanding 8.4-A

1. a. $\dfrac{43}{100}$.   **b.** $\dfrac{31}{10}$.

   c. 0.006.   **d.** 0.155.

   e. 45%.   **f.** $133\dfrac{1}{3}\%$.

2. Mrs. Bell's Spanish class has the larger percentage of females.

3. a. $33\dfrac{1}{3}\%$.   **b.** 14.   **c.** $72.\overline{2}$.

4. About 1,175,000 ice cream sandwiches.

5. $27.

Check Your Understanding 8.4-B

1. $2,240.

2. $6,381.41.

3. $18,497.84.

Activity 8.1

a.   **i.** 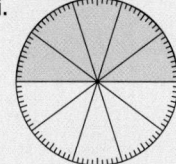   **ii.**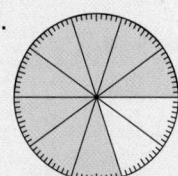

    iii. 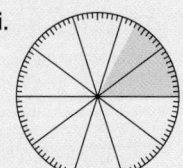   **iv.**

b.   **i.** 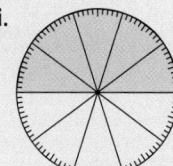   **ii.**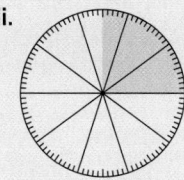

    iii. 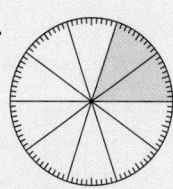   **iv.**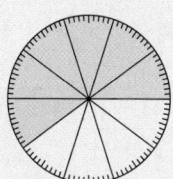

Activity 8.2

Adapt the process given in Example 8.6 by cubing the numbers instead of squaring them: $\sqrt[3]{9} \approx 2.08$.

Activity 8.3

Begin by dividing 4 by 23 on the calculator, which gives 0.1739130435. We can trust the first nine digits, but not the last because the calculator may have rounded. Next, multiply the first nine digits by 23 to get 3.999999989:

$$\begin{array}{r} 0.173913043 \\ 23\overline{)4.000000000} \\ 3.999999989 \\ \hline 11 \end{array}$$

Repeat the process by dividing 11 by 23 on the calculator:

$$\begin{array}{r} 0.17391304347826086 \\ 23\overline{)4.00000000000000000} \\ 3.999999989 \\ 1100000000 \\ 1099999978 \\ \hline 22 \end{array}$$

Repeat the process one more time:
$$\dfrac{4}{23} = 0.\overline{17391304347826086956 52}.$$

Activity 8.4

1.5×10^9 s ≈ 47.565 years.

Activity 8.5

Possible x-values include any number from the set $\{45, 46, 47, \dots, 54\}$.

Activity 8.6

a. $0.0504.

b. 19.84 days.

c. About 4 days.

Activity 8.7

a.  b.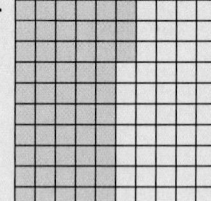

c. 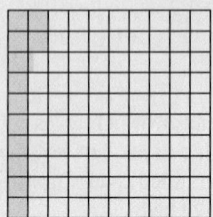 d.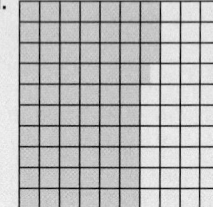

e.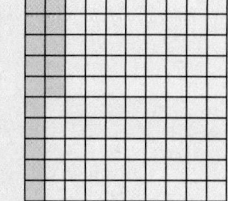

Activity 8.8

a. $589.66.
b. $434.71.
c. $1523.07.

Chapter 9

Algebraic Thinking

Opening Problem Jenna wants to estimate how much money she will spend on her car over the next year. Her car payments are fixed at $212.47 a month, and her insurance costs $459 every six months. She estimates that gasoline and car maintenance will be about $0.32/mi. How much will her car cost for the next year if she drives:

a. 12,000 mi.   **b.** 15,000 mi.   **c.** 20,000 mi.

d. How many miles will she have to drive to have spent over $4,000 on her car?

If you are like many people, the word "algebra" brings to mind a high school course in which you manipulated symbols, graphed functions, and solved word problems. Algebraic skills like this have typically been delayed until high school to give students a chance to develop the foundation in arithmetic needed to complete them. Despite the delay, many students struggle with algebra, indicating that arithmetic alone is not sufficient for learning algebra. For this reason, many modern curricula now introduce and develop *algebraic thinking* throughout the elementary and middle grades.

It may be surprising that algebraic thinking is now advocated at all grade levels. However, algebraic thinking is distinctly different from the formal algebra courses often taught in high school. Certainly, algebraic thinking includes the symbolic manipulation that most people identify with algebra, but, as the NCTM Algebra Standard indicates, it also includes much more.

■ NCTM Algebra Standard

Instructional programs from prekindergarten through grade 12 should enable all students to:

- Understand patterns, relations, and functions.
- Represent and analyze mathematical situations and structures using algebraic symbols.
- Use mathematical models to represent and understand quantitative relationships.
- Analyze change in various contexts.

Source: NCTM STANDARDS Copyright © 2011 by NATIONAL COUNCIL OF TEACHERS OF MATHEMATICS. Reproduced with permission of NATIONAL COUNCIL OF TEACHERS OF MATHEMATICS.

In the classroom, the four goals of the Algebra Standard are covered in three key areas of algebraic thinking: functional thinking, general arithmetic, and mathematical modeling.

Algebraic Thinking and Functions

The first and fourth goals of the Algebra Standard fall under an area called functional thinking. **Functional thinking** is the process of identifying, representing, and extending patterns and relationships *between* sets of objects. If the sets involve numbers, we can look for a pattern that indicates how the change in one set relates to or produces a change in another. The rule that governs this pattern is called a **function.**

Children first use functional thinking when they learn to count because a one-to-one correspondence is a special kind of function. Once in school, they continue to develop their functional thinking as they describe qualitative and quantitative changes in simple patterns that involve colors, shapes, sounds, and even daily routines. After they become adept with these simple patterns, they begin to work with patterns of change between sets of numbers. As they do, they learn to use different representations, such as symbolic expressions and algebraic graphs, to represent the patterns. Table 9.1 shows the objectives associated with functional thinking and the grades at which they are taught.

Algebraic Thinking and General Arithmetic

The second goal of the Algebra Standard falls under **general arithmetic,** which encompasses two areas: (1) identifying and extending patterns and (2) using the properties of arithmetic to find unknown quantities. Like functional thinking, general arithmetic involves identifying and extending patterns. However, rather than patterns of change, general arithmetic focuses on the regular patterns embedded in the numerical operations. For instance, as children work with whole-number addition, they are likely to identify patterns that lead to algebraic properties like the commutative property. Having identified the pattern, children need take only a small step to extend it to a general fact of arithmetic.

Table 9.1 Classroom learning objectives	K	1	2	3	4	5	6	7	8
Generate a number or shape pattern that follows a given rule.					X				
Generate numerical patterns using two rules, identify relationships between corresponding terms, and graph the ordered pairs of corresponding terms on a coordinate plane.						X			
Understand that a function is a rule that assigns to each input exactly one output and that its graph is the set of ordered pairs consisting of an input and the corresponding output.									X
Describe the functional relationship between two quantities by sketching and analyzing a graph.									X
Construct a function to model a linear relationship between two quantities and determine the rate of change and initial value of the function.									X
Graph and compare proportional relationships and interpret the unit rate as the slope of the graph.									X
Derive and interpret the equation $y = mx + b$ for a line intercepting the vertical axis at b.									X

Source: Adapted from the *Common Core State Standards for Mathematics* (*Common Core State Standards Initiative*, 2010).

General arithmetic also includes using the properties of arithmetic to find unknown quantities. Children first use this aspect of algebraic thinking when they learn the missing-addend and missing-factor approaches to subtraction and division. Once in the middle grades, students extend the ideas they have learned from the whole-number operations to solve simple one- and two-step equations. This, in turn, forms the foundation for many of the symbolic manipulations they will do in high school algebra.

Table 9.2 identifies the objectives associated with general arithmetic, many of which were addressed in earlier chapters. You are encouraged to revisit previous chapters to examine how algebraic thinking was used to establish properties of arithmetic. Other objectives in Table 9.2 are addressed throughout this chapter.

Table 9.2 Classroom learning objectives	K	1	2	3	4	5	6	7	8
Understand the meaning of the equal sign and determine if equations involving addition and subtraction are true or false.	X								
Determine an unknown whole number in equations relating three whole numbers.	X		X						
Identify patterns in the addition table or multiplication table and explain them using properties of operations.				X					
Write, read, and evaluate expressions in which letters stand for numbers.							X		
Identify parts of an expression using mathematical terms such as sum, term, product, factor, quotient, and coefficient.							X		
Apply the properties of operations to generate equivalent expressions.							X		
Use substitution to determine whether a given number in a specified set makes an equation or inequality true.								X	
Recognize that inequalities have infinitely many solutions and represent the solution on a number line.								X	
Solve linear equations in one variable, including equations whose solutions require expanding expressions using the distributive property and collecting like terms.									X
Solve systems of two linear equations in two variables algebraically and estimate solutions by graphing.									X
Understand that solutions to a system of two linear equations in two variables correspond to points of intersection of their graphs.									X
Use square root and cube root symbols to represent solutions to equations of the form $x^2 = p$ and $x^3 = p$, where p is a positive rational number.									X

Source: Adapted from the *Common Core State Standards for Mathematics* (*Common Core State Standards Initiative*, 2010).

Algebraic Thinking and Mathematical Modeling

The third goal in the Algebra Standard focuses on **mathematical modeling.** This is the process for creating a mathematical representation of a real-world situation in order to gain insight into the situation and solve problems. Mathematical models come in a variety of forms and can include objects, pictures, and symbols. If a pattern of change is involved, then functional thinking can be used to identify, represent, and analyze the pattern. For an unknown quantity, general arithmetic can be used to represent the situation and find the needed number.

In the classroom, mathematical modeling begins when students use what they know about arithmetic to solve simple story problems. As students progress, they continue to learn and use new representations to solve problems in real-world situations. Table 9.3 indicates classroom objectives associated with mathematical modeling.

Table 9.3 Classroom learning objectives	K	1	2	3	4	5	6	7	8
Solve two-step word problems using equations with a letter standing for the unknown quantity.				X					
Solve real-world and mathematical problems using one- and two-step equations and inequalities.							X	X	
Use variables to represent two quantities in a real-world problem that change in relationship to one another and analyze the relationship using graphs, tables, and equations.							X		
Solve real-world and mathematical problems leading to two linear equations in two variables.									X

Source: Adapted from the *Common Core State Standards for Mathematics* (*Common Core State Standards Initiative*, 2010).

Elementary and middle grades teachers need to be prepared to teach the aspects of algebraic thinking that are now filtering down into these grades. Although the three areas of algebraic thinking are not mutually exclusive, they are different enough to warrant individual attention. For this reason, in Chapter 9 we:

- Analyze numerical sequences for patterns of change to introduce the notion of a function.
- Discuss functions, their representations, and how they can be used to represent and analyze the change between sets of quantities.
- Use properties of arithmetic to solve equations and inequalities.
- Use functions and equations to model mathematical relationships and solve real-world problems.

SECTION 9.1   Numerical Sequences

Algebraic thinking entails much more than the symbolic manipulations so often associated with high school algebra. In the primary grades, "algebraic thinking" is a catch-all phrase to indicate any content that prepares students for success in algebra and other upper-level mathematics courses. This content falls into three main areas: functional thinking, general arithmetic, and mathematical modeling. In this section, we begin our study of functional thinking with a discussion of numerical sequences.

Functional Thinking and Numerical Sequences

At its core, **functional thinking** involves identifying, representing, and extending patterns. Like other pattern recognition skills, it relies heavily on inductive reasoning. However, functional thinking is a skill in its own right because it focuses on patterns of change *between* sets rather than on the similarities or differences within a set. In the classroom, functional thinking is often introduced through the use of number patterns.

Example 9.1	**Functional Thinking in the Classroom**

Connections

A teacher shows her students the following pictures, which illustrate how a caterpillar grows daily.

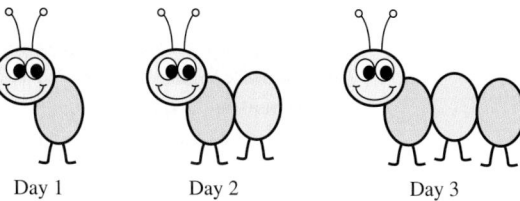

Day 1   Day 2   Day 3

She asks her students to find the number of legs the caterpillar will have on the tenth day. How is functional thinking used to answer the question?

Solution Students must use functional thinking to identify and extend the pattern of change that exists between the set of days and the number of legs. The pictures show that two legs are added for each day of growth, or the number of legs is double the number of days. Once the pattern is identified, students can use a number of methods to extend it to the tenth day. For instance, some students might list the number of legs until they reach the tenth number, or 2, 4, 6, 8, 10, 12, 14, 16, 18, and 20. Others might draw a picture of the caterpillar on the tenth day, again showing that it has 20 legs.

The number pattern in the last example illustrates what is commonly called a sequence. In general, a **sequence** is an ordered list of objects in which one is designated as first, another as second, and so on. Although the objects can be almost anything, in mathematics numbers are typically used. The numbers in a sequence are called **terms** and are named by their position in the list: the first term, second term, and so on.

If the sequence is governed by a particular pattern, then we may be able to represent it with a **general,** or **nth term.** When possible, we write the general term with a symbolic formula that can be used to generate any term in the sequence simply by substituting the position of the term into the formula. To find this formula, we must use functional thinking to discover the pattern between the position of the term and its value.

For instance, consider the sequence in Example 9.1. The value of each term is twice the value of its position in the sequence. In other words, if n represents the position of the term, then the value of the term is twice n, or $2n$. So we can write the sequence as

$$2, 4, 6, 8, \ldots, 2n, \ldots$$

If we want to know the value of any term in the sequence, we simply substitute its position into the general term $2n$. For instance, the value of the 5th term ($n = 5$) is $2 \cdot 5 = 10$, the value of the 10th term ($n = 10$) is $2 \cdot 10 = 20$, and the value of the 100th term ($n = 100$) is $2 \cdot 100 = 200$.

In some instances, it is practical to use a more general sequence in which specific values are not known. To represent this sequence, we use a variable to denote the values of the terms and subscripts to denote the positions. For instance,

$$a_1, a_2, a_3, a_4, \ldots, a_n, \ldots$$

represents a sequence of unknown values in which a_1 is the first term, a_2 the second, and so on through a_n, which represents the nth term.

Not every sequence is governed by a pattern. However, for those that are, the patterns can be used to classify sequences in different ways. For instance, a sequence is **repeating** if some core component continuously repeats. A sequence is **increasing** if the terms continue to grow larger in value, and it is **decreasing** if the terms continue to grow smaller.

Table 9.4 shows several examples of each type. The table shows that not every increasing or decreasing sequence is governed by a pattern that can be identified and represented by a general term. However, this is possible for certain types of sequences such as arithmetic, geometric, and recursively defined sequences.

Table 9.4 Different types of sequences

Repeating Sequences	Increasing Sequences	Decreasing Sequences
$1, -1, 1, -1, 1, \ldots (-1)^{n-1}$	$1, 7, 43, 105, \ldots$	$51, 7, 0.5, -2, \ldots$
$2, 4, 2, 4, 2, 4, \ldots$	$2, 5, 8, 11, \ldots 3n - 1, \ldots$	$5, 2, -1, -4, \ldots 8 - 3n, \ldots$
$1, 2, 3, 2, 1, 2, 3, 2, 1, \ldots$	$2, 4, 8, 16, \ldots 2^n, \ldots$	$81, 27, 9, 3, \ldots 81 \cdot \left(\dfrac{1}{3}\right)^{n-1}, \ldots$

Arithmetic Sequences

Explorations Manual 9.1

If we examine the terms of the sequence 2, 5, 8, 11, . . . from Table 9.4, we can see that it is an increasing sequence, in which each successive term is found by adding 3 to the previous term (Figure 9.1). This is an example of an **arithmetic sequence.** In any arithmetic sequence, each successive term is found by adding a fixed number, called the **common difference d,** to the previous term. If the common difference is positive, the sequence is increasing. If it is negative, the sequence is decreasing. In either case, to find the common difference, subtract one term from the next consecutive term.

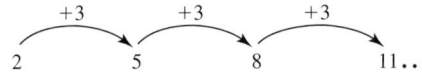

Figure 9.1 An arithmetic sequence

Example 9.2   Finding Terms of Arithmetic Sequences

Find the common difference of each arithmetic sequence, and use it to find the next three terms.

a. 3, 6, 9, 12, . . .   **b.** 1, 4, 7, 10, . . .   **c.** 4, 2, 0, −2, . . .

Solution

a. The common difference is $d = 6 - 3 = 3$. By adding, the next three terms are $a_5 = 12 + 3 = 15$, $a_6 = 15 + 3 = 18$, and $a_7 = 18 + 3 = 21$.

b. The common difference is again $d = 4 - 1 = 3$. The next three terms are $a_5 = 10 + 3 = 13$, $a_6 = 13 + 3 = 16$, and $a_7 = 16 + 3 = 19$.

c. The common difference is $d = 2 - 4 = -2$. The next three terms are $a_5 = -2 + (-2) = -4$, $a_6 = -4 + (-2) = -6$, and $a_7 = -6 + (-2) = -8$.

Example 9.2 demonstrates two facts about arithmetic sequences. First, even though the sequences in parts a and b have the same common difference, they are different because their first terms are different. This implies that the terms of an arithmetic sequence depend not only on the common difference, but also on the value of the first term. Second, arithmetic sequences always follow a pattern of constant change; that is, the change from one term to the next is always the same. Using these two facts, we can write a symbolic expression for the general term.

Example 9.3 | **Finding a General Term for an Arithmetic Sequence**

Find the general term of the sequence 5, 9, 13, 17, ...

Solution We can find the general term of the sequence by thinking about how each term after the first is generated. The first term is $a_1 = 5$, and the common difference is $d = 9 - 5 = 4$. Because the next three terms are found by adding the common difference to the previous term, they can be rewritten as follows:

$$a_1 = 5$$
$$a_2 = 9 = 5 + 4 \qquad\qquad\qquad = 5 + 1 \cdot 4$$
$$a_3 = 13 = 9 + 4 = (5 + 4) + 4 \qquad = 5 + 2 \cdot 4$$
$$a_4 = 17 = 13 + 4 = (5 + 4 + 4) + 4 \qquad = 5 + 3 \cdot 4 \ldots$$

In each case, we can write the term as the sum of the first term, 5, and some multiple of the common difference, 4. Notice that the number multiplying the 4 is always one less than the position of the term in the sequence. If n represents the position of the term, then the general term can be written as $a_n = 5 + (n - 1)4 = 5 + 4n - 4 = 4n + 1$.

We can generalize the process in Example 9.3 in the following way. Given an arithmetic sequence with a first term of a_1 and a common difference of d, we can find the second term by adding the common difference to the first term, or $a_2 = a_1 + d$. For the third term, add the common difference to the second term, or $a_3 = a_2 + d$. However, because $a_2 = a_1 + d$,

$$a_3 = a_2 + d = (a_1 + d) + d = a_1 + 2d.$$

Similarly, for the fourth and fifth terms,

$$a_4 = a_3 + d = (a_1 + 2d) + d = a_1 + 3d$$
$$a_5 = a_4 + d = (a_1 + 3d) + d = a_1 + 4d$$

Each term is the sum of the first term, a_1, and the multiple of the common difference, d, that is one less than the position of the term. As a result, if n represents the position of the term, then

$$a_n = a_{n-1} + d = (a_1 + (n - 2)d) + d = a_1 + (n - 1)d$$

Theorem 9.1 | **The General Term of an Arithmetic Sequence**

The general term of any arithmetic sequence with a first term a_1 and a common difference d is given by $a_n = a_1 + (n - 1)d$.

If a_1 and d are known, then the formula $a_n = a_1 + (n - 1)d$ depends only on n. If we substitute values for a_1 and d, then we have a formula for the general term that can be used to find the value of any term in the sequence.

Example 9.4 | **Using the General Formula of an Arithmetic Sequence**

Find the general term of each arithmetic sequence, and use it to find the values of the 20th and 50th terms.

a. 2, 6, 10, 14, . . .

b. $\dfrac{3}{4}, \dfrac{1}{4}, \dfrac{-1}{4}, \dfrac{-3}{4} \ldots$

Solution

a. We substitute $a_1 = 2$ and $d = 6 - 2 = 4$ into the formula $a_1 + (n - 1)d$ and simplify, or $a_n = 2 + (n - 1)4 = 2 + 4n - 4 = 4n - 2$. We then evaluate the formula at 20 and 50 to find the 20th and 50th terms:

$$a_{20} = 4(20) - 2 = 80 - 2 = 78$$

$$a_{50} = 4(50) - 2 = 200 - 2 = 198$$

b. $a_1 = \dfrac{3}{4}$ and $d = \dfrac{1}{4} - \dfrac{3}{4} = -\dfrac{2}{4} = -\dfrac{1}{2}$. This means $a_n = \dfrac{3}{4} + (n - 1)\left(-\dfrac{1}{2}\right) = \dfrac{3}{4} + \left(-\dfrac{1}{2}\right)n + \dfrac{1}{2} = \left(-\dfrac{1}{2}\right)n + \dfrac{5}{4}$. Evaluating the formula at 20 and 50:

$$a_{20} = \left(-\dfrac{1}{2}\right)20 + \dfrac{5}{4} = \left(-\dfrac{40}{4}\right) + \dfrac{5}{4} = \left(-\dfrac{35}{4}\right)$$

$$a_{50} = \left(-\dfrac{1}{2}\right)50 + \dfrac{5}{4} = \left(-\dfrac{100}{4}\right) + \dfrac{5}{4} = \left(-\dfrac{95}{4}\right)$$

Geometric Sequences

Explorations Manual 9.2

Another type of sequence, a **geometric sequence,** is characterized by the fact that each successive term is found by multiplying a constant number, called the **common ratio r,** to the previous term. For instance, Figure 9.2 shows a geometric sequence that has a common ratio of 2. The common ratio can be any nonzero value, including fractions, decimals, and negative numbers. To find it, divide any term by the preceding term.

Figure 9.2 A geometric sequence

Example 9.5 **Finding Terms of Geometric Sequences**

Find the common ratio of each geometric sequence, and use it to find the next three terms.

a. 5, 15, 45, 135, . . .   **b.** 81, 27, 9, 3, . . .   **c.** −3, 6, −12, 24, . . .

Solution

a. The common ratio is $r = 15 \div 5 = 3$. By multiplying, the next three terms are $a_5 = 135 \cdot 3 = 405$, $a_6 = 405 \cdot 3 = 1{,}215$, and $a_7 = 1{,}215 \cdot 3 = 3{,}645$.

b. The common ratio is $r = 27 \div 81 = \dfrac{1}{3}$. The next three terms are $a_5 = 3 \cdot \dfrac{1}{3} = 1$, $a_6 = 1 \cdot \dfrac{1}{3} = \dfrac{1}{3}$, and $a_7 = \dfrac{1}{3} \cdot \dfrac{1}{3} = \dfrac{1}{9}$.

c. The common ratio is $r = 6 \div (-3) = (-2)$. The next three terms are $a_5 = 24 \cdot (-2) = -48$, $a_6 = (-48) \cdot (-2) = 96$, and $a_7 = 96 \cdot (-2) = -192$.

The last example shows how the terms in geometric sequences grow larger and smaller much faster than those in arithmetic sequences. This characteristic is a natural consequence of changing from sums to products. Also, the sign of the common ratio affects the sequence. Specifically, if the common ratio is negative, then the terms alternate between positive and negative. Such sequences are called **alternating sequences.**

Geometric sequences are completely defined by their first term and common ratio—a fact that makes it possible to find a formula for the general term of the sequence.

Example 9.6 | **Finding a General Term for a Geometric Sequence**

Find the general term of the sequence 4, 8, 16, 32, ...

Solution The first term is $a_1 = 4$, and the common ratio is $r = 8 \div 4 = 2$. Because the next three terms are found by multiplying the common ratio to the previous term, we can rewrite them in the following way:

$$a_1 = 4$$
$$a_2 = 8 = 4 \cdot 2 \qquad\qquad = 4 \cdot 2^1$$
$$a_3 = 16 = 8 \cdot 2 = (4 \cdot 2) \cdot 2 \qquad = 4 \cdot 2^2$$
$$a_4 = 32 = 16 \cdot 2 = (4 \cdot 2 \cdot 2) \cdot 2 \quad = 4 \cdot 2^3 \ldots$$

In each case, the term is written as the product of the first term, 4, and some power of the common ratio, 2. Specifically, the power of 2 is always one less than the position of the term in the sequence. If n represents the position of the term, we can write the general term as $a_n = 4 \cdot 2^{n-1}$.

The process in Example 9.6 can be generalized in the following way. Given a geometric sequence with a first term, a_1, and a common ratio, r, we find the second term by multiplying the common ratio to the first term, or $a_2 = a_1 \cdot r$. Because each successive term is found by multiplying the previous term by r,

$$a_3 = a_2 \cdot r = (a_1 \cdot r) \cdot r = a_1 r^2$$
$$a_4 = a_3 \cdot r = (a_1 r^2) \cdot r = a_1 r^3$$
$$a_5 = a_4 \cdot r = (a_1 r^3) \cdot r = a_1 r^4$$

Each term is the product of a_1 and the power of r that is one less than the position of the term. As a result, if n is the position of the term, then

$$a_n = a_{n-1} \cdot r = (a_1 \cdot r^{n-2}) \cdot r = a_1 \cdot r^{n-1}$$

Theorem 9.2 The General Term of a Geometric Sequence

The general term of any geometric sequence with a first term, a_1, and a common ratio, r, is given by $a_n = r^{n-1} a_1$.

Example 9.7 | **Using the General Formula of a Geometric Sequence**

Find the general term of each sequence, and use it to find the values of the 15th and 20th terms.

a. 3, (−9), 27, (−81), . . .   **b.** 5, 7.5, 11.25, 16.875, . . .

Solution

a. $a_1 = 3$ and $r = (-9) \div 3 = (-3)$. Substituting these values into the formula from Theorem 9.2, $a_n = 3(-3)^{n-1}$. Evaluating this formula at 15 and 20,

$$a_{15} = 3(-3)^{14} = 14{,}348{,}907$$
$$a_{20} = 3(-3)^{19} = -3{,}486{,}784{,}401$$

b. In this sequence, $a_1 = 5$ and $r = 7.5 \div 5 = 1.5$. Hence, $a_n = 5(1.5)^{n-1}$. Evaluating the formula at 15 and 20,

$$a_{15} = 5(1.5)^{14} \approx 1{,}459.6463$$
$$a_{20} = 5(1.5)^{19} \approx 11{,}084.1891$$

In the last example, we used the general term to find the values of only a few terms. Sometimes, however, we need to generate the values of many terms. To do so, we can use a computer program called a **spreadsheet** to make the computations both quickly and efficiently.

Example 9.8   Computing Balances in a Bank Account

Application

Spreadsheet

Sophia has a bank account that compounds interest annually at a rate of 6%. If she invests $10,000 for 10 years, how much will be in the account at the end of each year?

Solution The amount of money in an account that earns compound interest is governed by the formula $A_t = 10{,}000(1.06)^t$. This formula actually represents the general term of a geometric sequence. To find the needed amounts, replace t with a specific number of years and compute. In this case, we choose to use a spreadsheet because of the large number of calculations.

To begin, we enter the number of years in column A of the spreadsheet. We can do this directly or with the FILL DOWN command. To use the FILL DOWN command, enter 1 in cell A1. Then enter the formula "$= A1 + 1$" into cell A2. This tells the spreadsheet to add 1 to the value in cell A1 and place the result in cell A2. From cell A2, fill down to cell A10. The spreadsheet automatically performs any calculations. After this step, we have the following amounts in our spreadsheet:

	A	Formula in
	Year	Spreadsheet
1	1	
2	2	$= A1 + 1$
3	3	$= A2 + 1$
4	4	$= A3 + 1$
5	5	$= A4 + 1$
6	6	$= A5 + 1$
7	7	$= A6 + 1$
8	8	$= A7 + 1$
9	9	$= A8 + 1$
10	10	$= A9 + 1$

In column B, we calculate the value in the account at the end of each year by entering the formula "$=10000*(1.06)\wedge A1$" in cell B1. In this command, "*" represents multiplication and "$\wedge$" indicates an exponent. The "A1" in the formula tells the spreadsheet to get the value of the year from cell A1. Again, use FILL DOWN, causing the spreadsheet to place the appropriate interest formula next to the appropriate year. The calculations are performed automatically, resulting in the desired information.

| | A | B | Formula in |
	Year	Amount	Spreadsheet
1	1	10600	= 10000*1.06^A1
2	2	11236	= 10000*1.06^A2
3	3	11910.16	= 10000*1.06^A3
4	4	12624.77	= 10000*1.06^A4
5	5	13382.26	= 10000*1.06^A5
6	6	14185.19	= 10000*1.06^A6
7	7	15036.3	= 10000*1.06^A7
8	8	15938.48	= 10000*1.06^A8
9	9	16894.79	= 10000*1.06^A9
10	10	17908.48	= 10000*1.06^A10

Check Your Understanding 9.1-A

1. Find the next five terms in each sequence.

    a. 2, 3, 2, 3. . .   **b.** 1, 2, 4, 7, 11, 16, . . .   **c.** 4, 8, 12, 16, . . .

2. Determine whether each sequence is arithmetic, geometric, or neither.

    a. $7, 3, -1, -5, \ldots$ **b.** 3, 4, 7, 11, 18, . . .   **c.** 1, 4, 16, 64, . . .

3. Find the common difference and the general term for each arithmetic sequence. Use the general term to find the 10th, 15th, and 25th terms.

    a. 10, 14, 18, . . .   **b.** $-5, -8, -11, \ldots$ **c.** 1.3, 2.1, 2.9, . . .

4. Find the common ratio and the general term for each geometric sequence. Use the general term to find the 5th, 8th, and 10th terms.

    a. 4, 20, 100, . . .   **b.** 7, 21, 63, . . .   **c.** 1.1, 1.65, 2.475, . . .

Talk About It Consider a geometric sequence in which $a_1 = 16$ and $r = -0.5$. Is the sequence increasing, decreasing, or neither? What value does the sequence seem to be going to? Do you think any term in the sequence will ever equal this value? Explain.

Problem Solving

Activity 9.1
Work with a small group of your peers to find a formula that will compute the *sum* of the first n terms of any arithmetic sequence. Use it to find the sum of the first 15 terms in each arithmetic sequence. Verify your results with a spreadsheet.

    a. 3, 5, 7, 9, . . .   **b.** 1, 6, 11, 16, . . .   **c.** $7, 4, 1, -2, \ldots$

Recursive Sequences

Arithmetic and geometric sequences are somewhat alike in that the value of each term is based on the preceding term. As a result, they fall into a broader category of sequences called recursive sequences. In a **recursive sequence,** the value of each term, except possibly for the first few, is based on the values of previous terms. Recursive sequences can be quite varied because they can be defined using any combination of operations on previous terms. For instance, in the sequence

$$2, 5, 11, 23, \ldots, a_n = 2a_{n-1} + 1 \ldots \qquad \text{for } n \geq 2$$

each successive term is found by multiplying the previous term by two and then adding one. In the sequence,

$$3, 4, 12, 48, \ldots a_n = a_{n-1} \cdot a_{n-2} \ldots \qquad \text{for } n \geq 3$$

each successive term is found by multiplying the two preceding terms. Like arithmetic and geometric sequences, recursive sequences are defined not only by the rule of the sequence, but also by the initial terms. Consequently, in a recursive sequence, we must be given enough initial terms to generate the rest of the sequence.

Example 9.9   **Finding Terms in Recursive Sequences**

Find the first five terms for each recursive sequence.

a. $a_n = 4a_{n-1} + 3$ where $a_1 = 0$

b. $a_n = 2a_{n-1} + a_{n-2}$ where $a_1 = 0$ and $a_2 = 2$

Solution

a. We find each term in this sequence by multiplying the preceding term by 4 and adding 3. Because $a_1 = 0$,

$$a_2 = 4 \cdot a_1 + 3 = 4 \cdot 0 + 3 = 3$$
$$a_3 = 4 \cdot a_2 + 3 = 4 \cdot 3 + 3 = 15$$
$$a_4 = 4 \cdot a_3 + 3 = 4 \cdot 15 + 3 = 63$$
$$a_5 = 4 \cdot a_4 + 3 = 4 \cdot 63 + 3 = 255$$

b. We find each term in this sequence by multiplying the previous term by 2 and then adding the term prior to that. Because $a_1 = 0$ and $a_2 = 2$,

$$a_3 = 2 \cdot a_2 + a_1 = 2(2) + 0 = 4$$
$$a_4 = 2 \cdot a_3 + a_2 = 2(4) + 2 = 10$$
$$a_5 = 2 \cdot a_4 + a_3 = 2(10) + 4 = 24$$

Example 9.10   **The Fibonacci Sequence**

Application

The Fibonacci sequence has received considerable attention over the years because it appears in a number of surprising places, both mathematical and otherwise. The sequence has its roots in the following problem, which was posed by Fibonacci in 1202. Identify the rule of the sequence, and use it to solve the problem.

Explorations
Manual
9.3

A man puts a young pair of rabbits in an isolated place for breeding purposes. How many pairs of rabbits will there be in a year if an adult pair of rabbits produces a new pair each month and a new pair becomes productive from the second month on?

Solution To identify the rule of the Fibonacci sequence, we can generate the number of rabbits in the first few months and then look for a pattern. Begin with a single pair of young rabbits, which takes one month to mature. At the beginning of the second month, this single pair of adult rabbits is now capable of producing a pair of offspring each month. At the beginning of the third month, the original pair of rabbits produces their first offspring, giving us two pairs of rabbits. At the beginning of the fourth month, the original pair produces another pair of offspring, and the pair born in the third month reaches maturity and is ready to produce. This gives us three pairs of rabbits. In the fifth month, the two adult pairs produce offspring, and the pair born in the fourth month reaches maturity, for a total of 5 pairs of rabbits. The first five months are summarized in the following diagram:

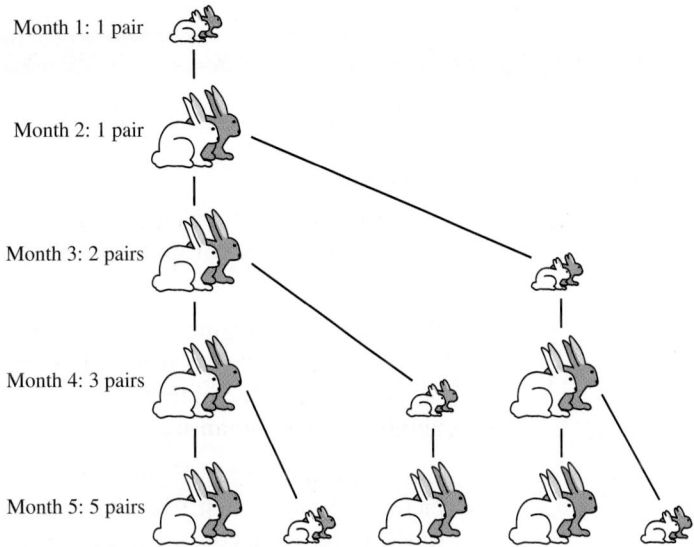

Month 1: 1 pair

Month 2: 1 pair

Month 3: 2 pairs

Month 4: 3 pairs

Month 5: 5 pairs

Initially, the total number of rabbits may not seem to have a pattern. However, note that the number of rabbits in the third, fourth, and fifth months is equal to the total number of pairs in the preceding 2 months. For instance, in the fifth month,

$$\underset{\text{5th month}}{\underline{5 \text{ pairs}}} = \underset{\text{3rd month}}{\underline{2 \text{ pairs}}} + \underset{\text{4th month}}{\underline{3 \text{ pairs}}}.$$

If the pattern holds, there should be $3 + 5 = 8$ pairs in month 6. This is indeed the case because the three adult pairs in the fifth month produce three pairs of offspring, and the two pairs born in the fifth month mature into adults, for a total of eight pairs. Because the pattern holds, we infer that terms in the Fibonacci sequence are generated by adding the two preceding terms, or $F_n = F_{n-1} + F_{n-2}$ where $F_1 = 1$ and $F_2 = 1$.

Finally, we solve the problem by extending the sequence to the 12th term, or 1, 1, 2, 3, 5, 8, 13, 21, 34, 55, 89, and 144. As a result, there are 144 pairs of rabbits at the end of the year.

■

The last two examples illustrate one drawback to defining sequences recursively. Because each term in a recursive sequence is based on the values of previous terms, we may need to compute many terms to find the value of the one we want. In some cases, such as in arithmetic and geometric sequences, we can rewrite the recursively defined rule so that it is based on the position of the term. However, in many recursive sequences, this is not possible.

Figurate Numbers

The **figurate numbers** are another group of sequences that have regular patterns. These sequences are somewhat different because they arise from sequences of dots arranged in common geometrical shapes. As the shapes grow, so does the number of dots. For instance, the sequence of triangles in Figure 9.3 leads to the **triangular numbers.** Counting the dots in each figure, we get the first five terms in the sequence of the triangular numbers, or 1, 3, 6, 10, and 15.

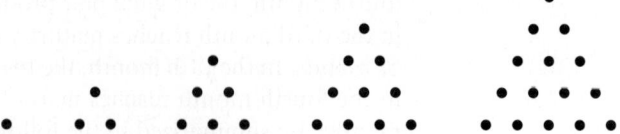

Figure 9.3 The triangular numbers

Example 9.11   Finding Triangular Numbers

Problem Solving   What are the sixth, seventh, eighth, and ninth triangular numbers?

Solution

Understanding the Problem. We are to find four specific triangular numbers. Figure 9.3 shows us a sequence of dots from which we know the first five triangular numbers are 1, 3, 6, 10 and 15.

Devising a Plan. Figure 9.3 shows how the triangles in the sequence grow. If we can *find a pattern* in the sequence of pictures or numbers, then we can use it to *write a formula* for the general term. Using the formula, we can evaluate it at the appropriate numbers to get our answer.

Carrying Out the Plan. Figure 9.3 shows how many dots are added to each subsequent picture. From the first triangle to the second, two dots are added. From the second to the third, three dots are added. From the third to the fourth, four dots are added, and from the fourth to the fifth, five dots are added. This indicates that the number of added dots is the same as the position of the triangle in the sequence. We find the same pattern in the sequence of numbers. If we subtract pairs of consecutive terms, the differences go up by 1, or

$$3 - 1 = 2 \quad 6 - 3 = 3 \quad 10 - 6 = 4 \quad 15 - 10 = 5$$

This is not an arithmetic sequence because the difference between the terms is not constant. However, the sequence has a regular pattern—that of adding the position of the term to the value of the previous term, or $a_n = a_{n-1} + n$. Using this formula,

$$a_6 = a_5 + 6 = 15 + 6 = 21$$
$$a_7 = a_6 + 7 = 21 + 7 = 28$$
$$a_8 = a_7 + 8 = 28 + 8 = 36$$
$$a_9 = a_8 + 9 = 36 + 9 = 45$$

Looking Back. Having found the pattern, we could have solved the problem in other ways. For instance, we could have drawn the next four triangles and counted the dots. This provides a way of checking our answer.

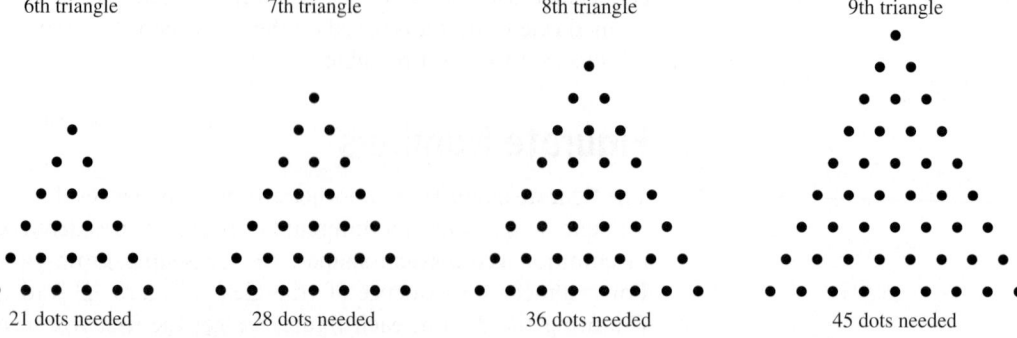

6th triangle   7th triangle   8th triangle   9th triangle

21 dots needed   28 dots needed   36 dots needed   45 dots needed

Example 9.12 Finding Square Numbers

The sequence of dots for the square numbers is as follows:

Find the general term, and use it to find the number of dots in the 10th and 15th squares.

Solution We find the first five square numbers by counting the dots in each figure: 1, 4, 9, 16, and 25. Because each number is the square of its position, the general term must be $a_n = n^2$. As a result, $a_{10} = 10^2 = 100$ and $a_{15} = 15^2 = 225$. In other words, it takes 100 dots to make the 10th square and 225 dots to make the 15th.

Check Your Understanding 9.1-B

1. Find the first five terms of each sequence.
   a. $a_n = 2a_{n-1} - 5$ where $a_1 = 3$
   b. $a_n = a_{n-1} - 2a_{n-2}$ where $a_1 = 1$ and $a_2 = 4$
2. Find F_{15} through F_{20}, the 15th through the 20th Fibonacci numbers.
3. Find a_{10}, a_{11}, and a_{12}, for the triangular and the square numbers.
4. The following sequence of dots leads to the rectangular numbers. Find a pattern, and use it to find the values of the sixth, seventh, and eighth rectangular numbers.

Talk About It What distinguishes arithmetic and geometric sequences from other recursively defined sequences?

Connections

Activity 9.2
Many Internet websites are dedicated to the Fibonacci sequence. Explore several of them. Can you find five places where the Fibonacci sequence occurs in nature?

SECTION 9.1 Learning Assessment

■ Understanding the Concepts

1. What is functional thinking, and how do we use it when we work with sequences?

2. a. What is the general term of a sequence?
   b. Why is an algebraic formula a powerful way to represent it?

3. Describe the difference between a repeating sequence, an increasing sequence, and a decreasing sequence.

4. How are arithmetic and geometric sequences similar? How are they different?

5. Explain how to find the common difference in an arithmetic sequence and the common ratio in a geometric sequence.

6. a. What does it mean for a sequence to be recursively defined?
   b. Are arithmetic and geometric sequences recursive sequences? Explain.

7. How are successive terms generated in the Fibonacci sequence?

8. How are figurate numbers generated?

■ Representing the Mathematics

9. Give a verbal description of the rule that governs each sequence.

    a. $-1, 2, -3, 4, \ldots$ b. $3, 7, 11, 15, \ldots$

    c. $\frac{1}{3}, 1, 3, 9, \ldots$ d. $2, 5, 10, 17, \ldots$

10. Consider the arithmetic sequence $3, 6, 9, 12, \ldots$. Represent the first five terms of the sequence as a sequence of dots arranged in rectangles. How is the common difference illustrated in the sequence of dots?

11. The first four terms of the arithmetic sequence $12, 8, 4, 0, \ldots$ can be represented as a sequence of dots arranged in rectangles. Can we represent the next three terms of the sequence in this way? Why or why not?

12. Consider the geometric sequence $1, 2, 4, 8, \ldots$. Represent the first five terms of the sequence as a sequence of dots arranged in squares and rectangles. How is the common ratio illustrated in this sequence?

13. What numerical sequence is shown by the shaded regions in the following sequence of pictures? Is it arithmetic, geometric, or neither? What might the next drawing in the sequence look like?

14. What numerical sequence is shown in the following sequence of pictures? Is it arithmetic, geometric, or neither? What might the next drawing in the sequence look like?

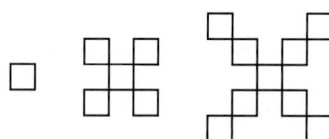

15. What numerical sequence is shown in the following sequence of pictures? Is it arithmetic, geometric, or neither? What might the next drawing in the sequence look like?

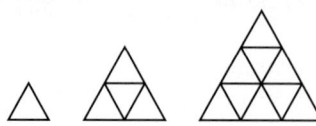

■ Exercises

16. Identify each sequence as increasing, decreasing, or repeating. Find a pattern, and use it to extend the sequence by five more terms.

    a. $4, 3, 2, 4, 3, 2, \ldots$ b. $7, 5, 3, 1, \ldots$

    c. $1, 5, 25, \ldots$

17. Identify each sequence as increasing, decreasing, or repeating. Find a pattern, and use it to extend the sequence by five more terms.

    a. $-0.5, -0.25, -0.125, \ldots$ b. $\frac{1}{2}, \frac{2}{3}, \frac{3}{4}, \frac{4}{5}, \ldots$

    c. $-1, 0, 1, 0, -1, \ldots$

18. Identify each sequence as arithmetic, geometric, or neither. Find a pattern, and use it to extend the sequence by five more terms.

    a. $5, 10, 15, \ldots$ b. $3, 12, 48, \ldots$

    c. $2, 3, 6, 18, \ldots$ d. $\frac{1}{3}, \frac{2}{3}, 1, \frac{4}{3}, \frac{5}{3}, \ldots$

19. Identify each sequence as arithmetic, geometric, or neither. Find a pattern, and use it to extend the sequence by five more terms.

    a. $1, 5, 25, 125, \ldots$ b. $\frac{1}{2}, \frac{5}{4}, 2, \frac{11}{4}, \ldots$

    c. $-0.5, 1.5, -4.5, \ldots$ d. $0.5, 0.25, 0.125, \ldots$

20. Find the common difference and the general term for each arithmetic sequence. Use the general term to find the values of the 10th, 35th, and 50th terms.

    a. $4, 6, 8, \ldots$ b. $-2, 4, 10, \ldots$

    c. $5, 3, 1, \ldots$ d. $-5, -9, -13, \ldots$

21. Find the common difference and the general term for each arithmetic sequence. Use the general term to find the values of the 8th, 12th, and 15th terms.

    a. $\frac{8}{10}, \frac{11}{10}, \frac{14}{10}, \ldots$ b. $\frac{1}{4}, \frac{5}{8}, 1, \ldots$

    c. $1.3, 3.9, 6.5, \ldots$ d. $-1.7, -4.5, -7.3, \ldots$

22. Find the common difference of each arithmetic sequence without computing any specific terms. Describe how you found each one.

    a. $a_n = 4n + 1, n \geq 1$.

    b. $a_n = \frac{3}{5}n - \frac{1}{2}, n \geq 1$.

    c. $a_n = 4.75n - 8.3, n \geq 1$.

23. Find the common ratio and the general term for each geometric sequence. Use the general term to find the values of the 5th, 10th, and 15th terms.

    a. $2, 8, 32, \ldots$

    b. $5, 15, 45, \ldots$

    c. $3, -6, 12, \ldots$

24. Find the common ratio and the general term. Use the general term to find the values of the 6th, 9th, and 12th terms.

    a. $1, \frac{1}{4}, \frac{1}{16}, \ldots$ b. $2.4, 7.2, 21.6, \ldots$ c. $\frac{1}{4}, \frac{1}{6}, \frac{1}{9}, \ldots$

25. Find the first five terms of the sequences whose general terms are as follows.

   a. $a_n = 5n - 3, n \geq 1$.

   b. $a_n = 2^n - 3, n \geq 1$.

   c. $a_n = 2n^2 + 1, n \geq 1$.

26. Find the first five terms of the sequences whose general terms are as follows.

   a. $a_n = (-3)n + 4, n \geq 1$.   b. $a_n = \sqrt{n + 2}, n \geq 1$.

   c. $a_n = \dfrac{n - 1}{2n + 1}, n \geq 1$.

27. Find the first five terms of each sequence. Determine whether each sequence is arithmetic, geometric, or neither.

   a. $a_n = a_{n-1} + 6, a_1 = 2$.

   b. $a_n = \dfrac{1}{3}a_{n-1}, a_1 = 27$.

   c. $a_n = 2a_{n-1} + 4, a_1 = 1$.

28. Find the first six terms of the sequences whose general terms are as follows.

   a. $a_n = a_{n-1} + 2a_{n-2}, a_1 = 0, a_2 = 3$.

   b. $a_n = 3a_{n-1} - a_{n-2}, a_1 = 1, a_2 = 3$.

29. Find the first six terms of the sequences whose general terms are as follows.

   a. $a_n = a_{n-1} \cdot \dfrac{1}{2}a_{n-2}, a_1 = 2, a_2 = 4$.

   b. $a_n = \dfrac{a_{n-2}}{a_{n-1}}, a_1 = 1, a_2 = 2$.

■ Problems and Applications

30. Suppose the 8th term of an arithmetic sequence has a value of 14 and the 12th term has a value of 26. Find the value of the 95th term.

31. The first term in a geometric sequence is $\dfrac{1}{3}$, and the fourth term is $\dfrac{8}{81}$. What is the value of the 10th term.

32. There are three equivalence classes for congruence modulo 3:

   $T_0 = \{0, 3, 6, 9, 12, \ldots\}$ $T_1 = \{1, 4, 7, 10, \ldots\}$

   $T_2 = \{2, 5, 8, 11 \ldots\}$

   a. What kind of sequence does each set represent?

   b. Use an algebraic formula to write each set in set-builder notation.

33. Consider the following sequence of cubes, in which each successive arrangement is obtained by gluing a cube to one end of the chain.

   a. Use a picture to determine the number of faces that will remain unglued in the seventh arrangement.

   b. Write a formula that can be used to find the number of unglued faces in any arrangement in the sequence. Use it to check your answer in part a and to find the number of unglued faces in the 20th arrangement.

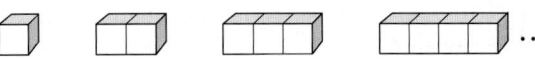

34. The *differences* between consecutive terms of an unknown sequence are given by the sequence 1, 3, 5, 7, ... If the first term of the unknown sequence is 10, what are the next five terms?

35. If the following sequences represent the hours on a standard clock face, determine the next three numbers in each sequence.

   a. 3, 6, 9, 12, ... b. 6, 12, 6, 12, ... c. 3, 4, 6, 9, ...

36. A large tank holds 45,000 gal of water. A leak in the bottom of the tank lets out 750 gal of water a day. After 14 days, how much water is left in the tank?

37. A certain stock doubles in value every year. If the original value of the stock is worth $155.75, how many years will it take for the stock to be over $2,500?

38. Every positive integer can be represented as the sum of distinct Fibonacci numbers. For instance, $5 = 2 + 3 = F_3 + F_4$ and $7 = 1 + 1 + 2 + 3 = F_1 + F_2 + F_3 + F_4$. Represent 8, 9, 10, and 15 in this manner.

39. For the given sequence of dots, find the number of dots needed to make:

   a. The next three figures.

   b. The nth figure.

   c. The 15th figure.

40. The hexagonal numbers are another type of figurate number. They are based on the number of dots needed to make the sequence of growing, nested hexagons, as shown:

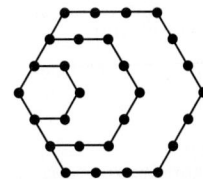

   a. If the first four hexagonal numbers are 1, 6, 15, and 28, what are the next three hexagonal numbers?

   b. The sequence of hexagonal numbers has a general term of $a_n = 2n^2 - n$. Use the general term to find the 10th, 15th, and 20th hexagonal numbers.

   c. Can you find the hexagonal numbers in Pascal's triangle?

41. The pentagonal numbers are based on a set of nested regular pentagons. Draw a diagram of the first three pentagons in the sequence, and use them to find the first three pentagonal numbers. Then try to find a pattern that can be used to find the fourth, fifth, and sixth pentagonal numbers. (Hint: See Problem 40.)

42. Can you find the triangular numbers in Pascal's triangle? What about the Fibonacci numbers?

■ Mathematical Reasoning and Proof

43. If possible, give an example of an increasing arithmetic sequence with a negative common difference.

44. Give an example of a decreasing geometric sequence with a positive common ratio? What must be true about the common ratio in any such sequence.

45. Give an example of a geometric sequence that is neither increasing nor decreasing. What must be true about the common ratio in any such sequence?

46. If possible, give an example of a sequence that is repeating and recursively defined.

47. **a.** Describe three different rules that can be used to extend the sequence 1, 4, 9, ... For each rule, write the next four terms of the sequence.
    b. What does this imply about defining sequences without writing the general term?

48. Calculate the following sums, and compare your results to the Fibonacci numbers.
    a. $F_1 + F_3$
    b. $F_1 + F_3 + F_5$
    c. $F_1 + F_3 + F_5 + F_7$
    d. $F_1 + F_3 + F_5 + F_7 + F_9$
    e. Based on the pattern demonstrated in sums, what would you expect the value of $F_1 + F_3 + F_5 + \cdots + F_{25}$ to be?

49. Calculate the following:
    a. $\text{GCF}(F_5, F_6)$
    b. $\text{GCF}(F_6, F_7)$
    c. $\text{GCF}(F_7, F_8)$
    d. Based on what you see, what might you expect $\text{GCF}(F_n, F_{n-1})$ to be?

50. Compute each sum.
    a. $1 + 3$
    b. $1 + 3 + 5$
    c. $1 + 3 + 5 + 7$
    d. $1 + 3 + 5 + 7 + 9$
    e. What sequence is generated by the sums of the odd numbers?
    f. Use your answer to part e to find the value of $1 + 3 + 5 + \cdots + 23$.

51. The sum of the first n counting numbers can be found using the formula

$$1 + 2 + 3 + \cdots + n = \frac{n(n + 1)}{2}$$

    a. Verify that this formula works for $2 \le n \le 10$.
    b. What sequence of numbers is generated by this formula for $n \in \mathbb{N}$?
    c. Use the formula to find the 50th, 75th, and 100th numbers in the sequence generated in part b.

■ Mathematical Communication

52. The words "arithmetic" and "geometric" were used to describe two different types of sequences, yet both terms have other mathematical meanings. Why does it make sense to apply these terms to their respective types of sequences? If you have difficulty answering this question, search the Internet for an answer.

53. Write down a set of directions that explain how to generate the Fibonacci sequence.
    a. Give your directions to a partner to follow. Was your partner successful? If not, what went wrong?
    b. Were your directions missing steps, or did they contain unnecessary steps? If so, revise your directions.

54. Generate the first five terms of an arithmetic sequence, and write directions on how to find the 10th, 11th, and 12th terms of the sequence using a calculator in two different ways.
    a. Give your directions to a partner to follow. Was your partner successful? If not, what went wrong?
    b. Were your directions missing steps, or did they contain unnecessary steps? If so, revise your directions.
    c. Repeat the exercise using a geometric sequence.

55. Quishawna is working with the sequence 1, 2, 4, ... She claims that the next term in the sequence is 8. Dexter is working with the same sequence and claims the next term is 7. Who, if either, is correct? What is the source of the discrepancy, and how, as a teacher, would you correct it?

56. Mr. Wendt asks his students to find the first five terms of the recursive sequence defined by $a_n = 3a_{n-1} - a_{n-2}$ where $a_1 = 2$ and $a_2 = 4$. Half the class had the values 2, 4, 2, 10, and -4. The other half had 2, 4, 10, 26, and 52.
    a. Which answer is correct?
    b. What mistake did the students make who got the answer incorrect?

Building Connections

57. Numerical patterns are common in our everyday lives. To realize this, think of five numerical patterns that you encounter or work with in your daily life. Describe the context and the purpose that the sequence serves in that context.

58. Can the place values of the Hindu-Arabic numeration system be considered a sequence? If so, what kind of sequence are they? What would the general term of the sequence be?

59. Number patterns play an important part in understanding and learning whole-number computation. Review the topics in Chapter 4, and identify three different patterns used to explain whole-number computation concepts. Briefly describe the role each pattern plays in explaining the concept.

60. As a continuation of the last question, look at several curriculum materials for content on whole-number computation. Specifically, look to see whether the patterns used in Chapter 4 can also be found in the curriculum materials you are looking at. If so, do the patterns serve the same or some different purpose? If not, were other numerical patterns used to teach the same content? Explain.

61. Discuss with a group of your peers why you think numerical patterns play an important part in algebraic thinking. After you are done, read the NCTM Algebra Standard.

    a. Is what you read consistent with your group's thoughts? If so, how?

    b. If not, how do the standards differ from what you discussed?

    c. What things did your group discuss that were not included in the standards? Why did your group think they were important?

SECTION 9.2 Functions and Their Representations

If we were to make a list of the five most important mathematical ideas, functions would be among them because they are the primary tools used to describe mathematical change. This characteristic makes functions not only an inherent part of every branch of mathematics, but also useful in a wide variety of nonmathematical situations. Because functions are a big idea in mathematics, it is important that we take time to make sense of them.

Understanding Functions

In the last section, we defined functional thinking as the process by which we identify, represent, and extend patterns of change between sets of objects. We explored this idea with numerical sequences, identifying and representing patterns relating the position of a term to its value. The rules that governed these patterns are examples of **functions.**

> **Definition of a Function**
>
> A **function** from a set A to a set B is a rule of correspondence that assigns to each element of A exactly one element from B.

The definition is broad enough to allow for a wide variety of relationships, both mathematical and otherwise. However, its generality can make the definition difficult to understand. To make better sense of it, we can look at it one piece at a time. It begins with the phrase "from a set A to a set B," which indicates that a function is a directional relationship between two sets. It begins with objects in set A and ends with objects in set B. In mathematics, we call set A the **domain** of the function and set B the **codomain.** The second phrase, "is a rule of correspondence," indicates that the sets are not the function in and of themselves. Rather, the function is the *rule* that describes how objects in the domain are associated with objects in the codomain. The third phrase, "assigns to each element of A exactly one element from B," indicates that the function associates each object in the domain with one and only one object in the codomain.

Explorations
Manual
9.4

To make these ideas more concrete, consider assigning a set of students to a set of desks. In assigning students to specific seats, we generate a rule of correspondence between the sets. Because it makes sense to put one student in each seat, the rule of correspondence exhibits a function. For instance, Figure 9.4 illustrates a function between a set of four students and a set of four desks. Here, the function, f, is illustrated by the arrows, and it tells us that Abe is assigned to seat 1, Barb to seat 2, and so on.

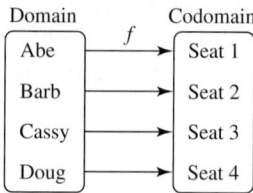

Figure 9.4 A function for seating students

The function in Figure 9.4 is not the only function between these sets. Three other possibilities are shown in Figure 9.5. Because many different functions can exist between two sets, we can use letters to name them. The functions in Figure 9.5 are named function F, function G, and function H, respectively.

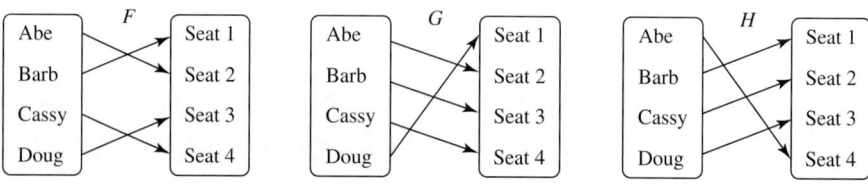

Figure 9.5 Different functions between two sets

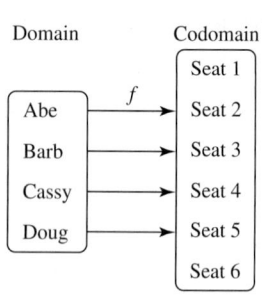

Figure 9.6 A function in the classroom

Taking the classroom example a step further, we can illustrate another important set associated with functions. Many classrooms have more seats than students, so we can consider a relationship between a set of four students and a set of six seats (Figure 9.6). The figure represents a function because each student is assigned to exactly one desk. However, two seats are left, which become irrelevant because they have no part in the rule of the function. So when we consider functions, we concern ourselves primarily with only the objects in the codomain that the function uses. The set of all elements in the codomain that have been paired with something from the domain is called the **range.**

Figure 9.7 shows the relationship among the three sets. Although the codomain is useful for describing functions in general, we typically limit ourselves to discussing the domain and range because they indicate exactly which objects are used in the rule of the function.

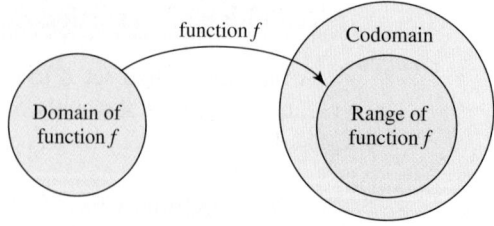

Figure 9.7 Domain, codomain, and range

Not every pattern or relationship between two sets is a function. To understand why, we revisit the conditions in the definition. First, in the phrase "to each element of A," the word "each" implies that that every element in the domain is to be used individually, not collectively. Second, the phrase "exactly one element from B" implies that each element in the domain is assigned to only one element in the range. Consequently, not only is each domain element used, but it is used only once. These constraints are not placed on the range, so one range value can be associated with several domain values.

| **Example 9.13** | **Using the Definition of a Function** |

Communication

Use the definition of a function to determine which of following diagrams represent a function from the set of students to the set of learning centers. For those that are, state the domain and range.

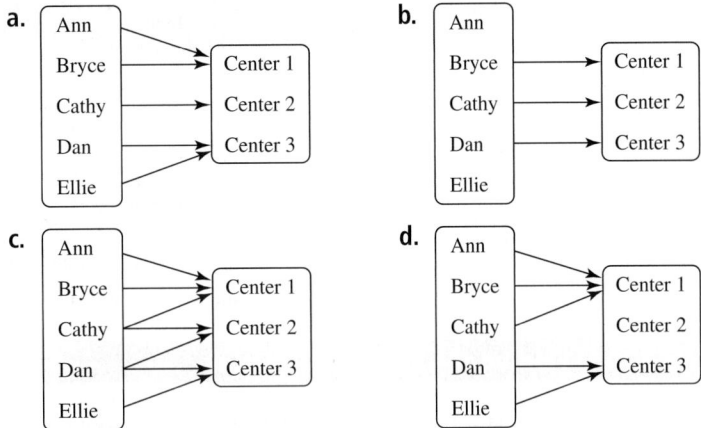

Solution In each case, we look to see whether the conditions in the definition are met.

a. This relationship is a function because all elements in the first set have been used and used only once. The domain is {Ann, Bryce, Cathy, Dan, Ellie}, and the range is {Center 1, Center 2, Center 3}.

b. This relationship is not a function because not all of the elements in the first set, namely Ann and Ellie, are used.

c. This relationship is not a function because two elements from the first set, Cathy and Dan, have been used more than once.

d. This diagram is a function with a domain of {Ann, Bryce, Cathy, Dan, Ellie} and a range of {Center 1, Center 3}.

Explorations
Manual
9.5

Because functions are widely used in mathematics and in other disciplines, it is useful to know how to represent functions in different ways so that we can select the one that best meets our needs. Common representations of functions are arrow diagrams, function machines, algebraic equations, tables, sets of ordered pairs, and graphs.

Functions as Arrow Diagrams

To this point, each function has been illustrated with an **arrow diagram.** In this type of diagram, the rule of the function is stated explicitly in the arrows between the sets, and the domain and range are generally easy to identify. Arrow diagrams are generally useful with functions that have small, finite domains, although they can be also used with some infinite domains. For instance, Figure 9.8 demonstrates a function between the natural numbers and the set of even natural numbers.

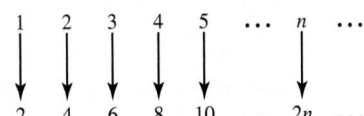

Figure 9.8 An arrow diagram with an infinite domain

Functions as Machines

A second way to represent a function is with a **function machine.** We think of the function as a machine that takes domain values, performs some operation on them, and then indicates the corresponding range values (Figure 9.9). The domain values are often viewed as the inputs to the machine and range values as the outputs.

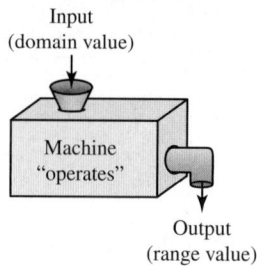

Input
(domain value)

Machine
"operates"

Output
(range value)

Figure 9.9 A function as a machine

Example 9.14 **Finding the Rule of a Function**

Reasoning

Suggest a rule by which each machine associates input values with output values.

a.

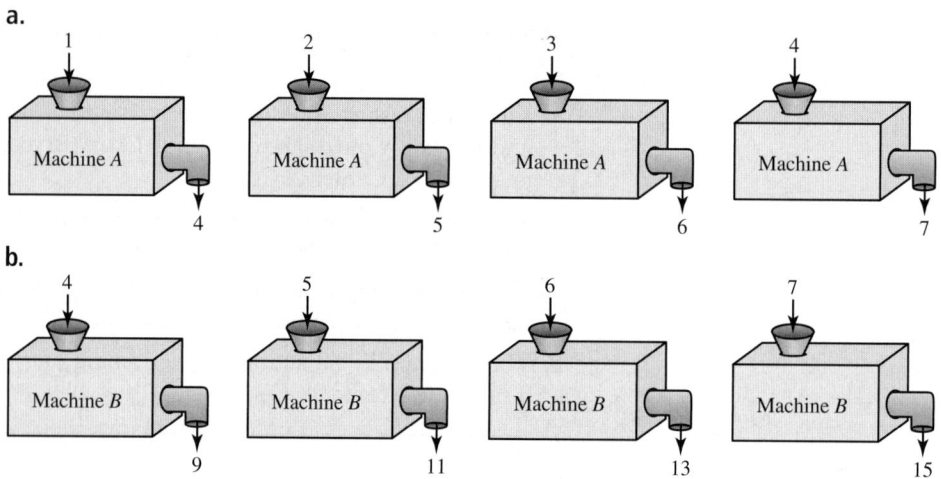

b.

Solution In each case, we use inductive reasoning to look for a pattern that relates the input values to the output values. Once the pattern is identified, we can use it to make a generalization for the rule of the machine.

a. Machine A has an output value that is always 3 more than the input value. This suggests that Machine A operates under the rule of "Add 3 to the input."

b. The input values are consecutive integers, and the output values always increase by 2, a situation that resembles an arithmetic sequence with $d = 2$. If the input values represent the position of the term in the sequence, and if we know the initial term, we could use the general term of the sequence as the rule for the machine. To find the initial term, substitute $n = 4$, $a_4 = 9$, and $d = 2$ into the formula $a_n = a_1 + (n - 1)d$ and write $9 = a_1 + (4 - 1)2$. Simplifying, we have $9 = a_1 + 6$, which implies that $a_1 = 3$. This gives a general term for the sequence of $a_n = 3 + (n - 1)2 = 2n + 1$, suggesting that Machine B operates under the rule "Multiply the input by 2 and add 1." ∎

A good example of a real-life function machine is a calculator, which has the capability of applying many different rules to input values. For instance, many calculators have an x^2 key or a $\sqrt{}$ key, each of which signifies a different rule that the calculator can use to operate on numbers.

Check Your Understanding 9.2-A

1. Which of the following diagrams represents a function? For those that do not, explain how the definition of a function is not met.

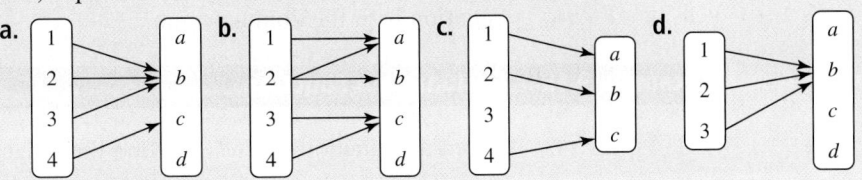

a.   **b.**   **c.**   **d.**

2. What is the domain and range of each function?

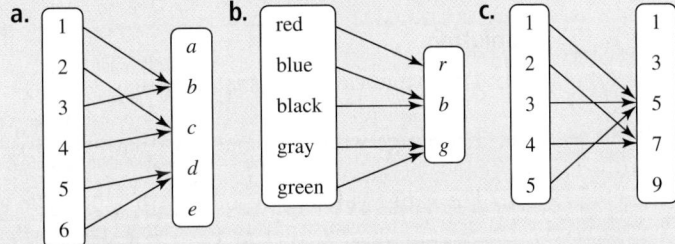

a.   **b.**   **c.**

3. Demonstrate three different functions between the sets {*m, a, t, h*} and {*f, u, n*}.

4. Determine whether each machine represents a function.

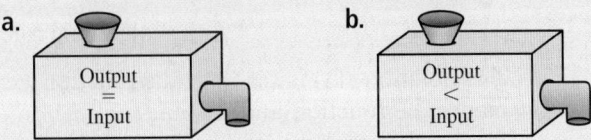

a.   **b.**

Talk About It How does the sequence 1, 4, 9, 16 . . . represent a function? What are the domain and range of the function?

Communication

Activity 9.3
In the classroom, function machines are often used with "What's the rule?" activities. In these activities, one student places a number on an index card, puts it in an envelope marked "In," and gives it to another student. The other student thinks of a rule, applies it to the number, and then sends the new value back to first student in an envelope marked "Out." Students continue to swap cards until the first person guesses the rule. Find a partner and make up several "What's the rule?" functions. Swap questions with another pair and solve them.

Explorations Manual 9.6

Functions as Algebraic Equations

A third way to represent a function is with an algebraic equation. We have seen functions expressed in this way when the general terms of sequences were represented with algebraic equations. Equations offer the advantage of concisely writing functions with infinite domains.

In general, there are two ways to write functions with algebraic equations. One is to use different variables for the domain and range values. The variable for the domain, commonly x, is called the **independent variable** because its value is independent of any others. The variable for the range, commonly y, is called the **dependent variable** because its value depends on the value of the independent variable. To represent a function between the two, we write an equation that sets the dependent variable equal to an algebraic expression in terms of the independent variable. For example,

$$y = 2x + 1 \quad y = x^2 + 3 \quad y = 3x^3 - 4x - 5$$

To find the range value associated with any domain value, we simply evaluate the equation with the domain value.

Example 9.15 Evaluating a Function

Given the function $y = x^2 - 4$, find the range value for each domain value.

a. $x = 3$.   **b.** $x = \dfrac{1}{2}$.   **c.** $x = 0.7$.

Solution

a. The range value associated with $x = 3$ is $y = (3)^2 - 4 = 5$.

b. The range value associated with $x = \dfrac{1}{2}$ is $y = \left(\dfrac{1}{2}\right)^2 - 4 = \dfrac{1}{4} - 4 = -3\dfrac{3}{4}$.

c. The range value associated with $x = 0.7$ is $y = (0.7)^2 - 4 = 0.49 - 4 = -3.51$.

A second way to write functions algebraically is to use **function notation,** which is given by

$$f(x) = \text{algebraic rule in terms of } x$$

The notation $f(x)$ is read as "the function f evaluated at x," or "f of x." The letter f names the function and is somewhat arbitrary because it can be replaced with any letter or letters that best describe the function. For instance, we often use $SQRT(x)$ to name the square root function, or $f(x) = \sqrt{x}$. Function notation works in much the same way as a function expressed in terms of an independent and dependent variable, only now the domain value becomes part of the notation. For instance, $f(4)$ means to evaluate the function f at the value 4. In many senses, function notation is analogous to a function machine in that x is the input value, the algebraic rule is the machine, and $f(x)$ is the output value.

Example 9.16 Evaluating Function Notation

Evaluate the function $f(x) = 2x + 4$ at each value.

a. $x = -4$.   **b.** $x = 10$.   **c.** $x = s - 3$.

Solution

a. $f(-4) = 2(-4) + 4 = -8 + 4 = -4$.

b. $f(10) = 2(10) + 4 = 20 + 4 = 24$.

c. $f(s - 3) = 2(s - 3) + 4 = 2s - 6 + 4 = 2s - 2$.

Example 9.16 demonstrates an interesting advantage to function notation: It allows us to evaluate the rule of the function at other algebraic expressions.

When representing a function with an algebraic equation, we seldom explicitly state the domain of the function. Unless stated otherwise, the domain is always the largest set of real numbers defined for the rule of the function. Operations on the real numbers are undefined in two situations:

1. Division by zero
2. The square root of a negative number

If the rule of the function makes it possible for either to occur, then those values are excluded from the domain. For instance, $f(x) = \dfrac{1}{x}$ is undefined at $x = 0$. Consequently, we exclude zero from the domain and write it as $\{x \mid x \in \mathbb{R} \text{ with } x \neq 0\}$ or $\mathbb{R} - \{0\}$.

Example 9.17 Stating the Domain of a Function

State the domain of each function.

a. $y = x - 1$.　　**b.** $y = x^2$.　　**c.** $f(x) = \dfrac{1}{x - 4}$.　　**d.** $f(x) = \sqrt{x}$.

Solution

a. Because the rule of the function does not involve division by zero or the square root of a negative number, the domain is the set of all real numbers, or $\mathbb{R}$,

b. The domain is the set of all real numbers, or $\mathbb{R}$.

c. In $f(x) = \dfrac{1}{x - 4}$, the value of $x = 4$ results in division by zero. As a result, we exclude 4 and the domain is $\{x \mid x \in \mathbb{R} \text{ with } x \neq 4\}$ or $\mathbb{R} - \{4\}$.

d. If any negative number is placed in the function $f(x) = \sqrt{x}$, the result is the square root of a negative number. So all negative numbers must be excluded, and the domain of the function is $\{x \mid x \in \mathbb{R} \text{ and } x \geq 0\}$.

Functions as Tables

Because algebraic equations are typically used to represent functions with infinite domains, we can use tables to organize associations made from these functions. In general, the first column of the table represents the domain, and the second column represents the range. The rule of the function is indicated by pairings between the columns. As long as the table assigns only one range value to each domain value, we continue to have a function.

Example 9.18 Tables as Functions

Determine whether each table represents a function.

a.

Domain	Range
2	1
4	2
6	3
8	4
10	5

b.

x	y
1	5
2	4
3	3
2	2
1	1

c.

Input	Output
1	10
2	10
3	10
4	10
5	10

Solution

a. This table represents a function because each domain value is paired with exactly one range value.

b. This is not a function because the domain value of 1 is paired with two range values, 5 and 1, and the domain value of 2 is paired with two values, 4 and 2.

c. This table represents a function because each domain value is paired with exactly one range value, even though the range value is always the same.

Functions as Sets of Ordered Pairs

Like tables, ordered pairs can be used to organize specific associations made from function machines and algebraic equations. Recall that an **ordered pair** is a pair of numbers in which one is designated as first and the other is designated as second. With functions, the first number always represents a domain value and the second a range value. Other ways of designating values in ordered pairs are:

$$(\text{domain, range}) = (\text{input, output})$$
$$= (\text{independent variable, dependent variable})$$
$$= (x, y)$$
$$= (x, f(x)).$$

As with an arrow diagram or a table, ordered pairs can be used to explicitly define a function as long as the domain is relatively small or has a pattern that can be easily established. For instance, the set of ordered pairs

$$\{(1, 3), (2, 6), (3, 9), (4, 12), (5, 15)\}$$

defines a function that triples each of the five domain values. The set of ordered pairs

$$\{(1, 1), (2, 4), (3, 9), (4, 16), (5, 25) \ldots\}$$

defines a function that takes each domain value to its square.

Functions as Graphs

Tables and sets of ordered pairs are often a vehicle for connecting the algebraic equation of a function to its graph. A **graph** is a picture of a function created by plotting domain and range values on the **Cartesian coordinate system.** Named after its inventor René Descartes, the Cartesian coordinate system is a grid on a plane created by intersecting two number lines, called the **coordinate axes,** perpendicularly at zero (Figure 9.10). The horizontal number line, or *x*-axis, designates left-right positions; the vertical number line, or *y*-axis, designates up-down positions. The intersection of the axes is called the **origin** because it is the reference point for all other points on the grid.

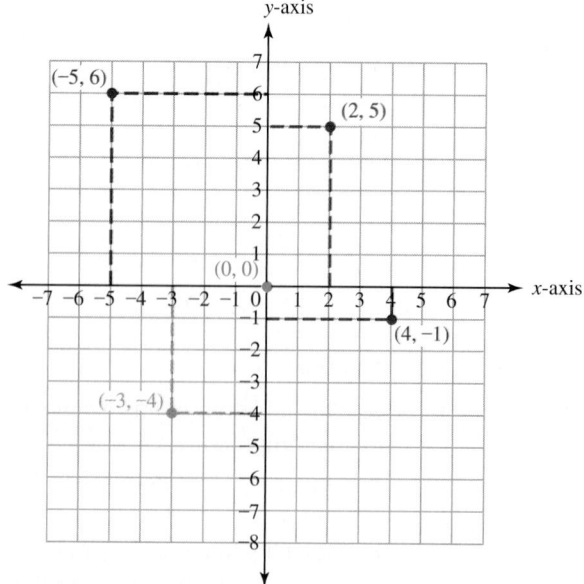

Figure 9.10 Plotting points on the Cartesian coordinate system

Given this arrangement, we can use ordered pairs to designate the positions of points with respect to the origin. The first number in the ordered pair is called the **x-coordinate**, or **abscissa**, and it gives the left-right position of the point. The second number is the **y-coordinate**, or **ordinate**, and it gives the up-down position of the point.

(x-coordinate, y-coordinate) = (abscissa, ordinate) = (left-right, up-down)

For instance, the ordered pair (2, 5) designates a point that is two units to the right and five units above the origin. The ordered pair $(4, -1)$ designates a point that is four units to the right and one unit below the origin. Figure 9.10 shows the location of several points on the Cartesian plane, including the origin, which has coordinates (0, 0).

Two facts about the Cartesian plane should be kept in mind. First, the axes divide the plane into four quadrants that can be identified by the signs of the coordinates in the ordered pairs (Figure 9.11).

1. In quadrant I, both coordinates are positive, (+, +).
2. In quadrant II, the first is negative and the second is positive, (−, +).
3. In quadrant III, both coordinates are negative, (−, −).
4. In quadrant IV, the first is positive and the second is negative, (+, −).

Second, the axes represent the real numbers, so there is a one-to-one correspondence between the set of all ordered pairs of real numbers and the points in the plane.

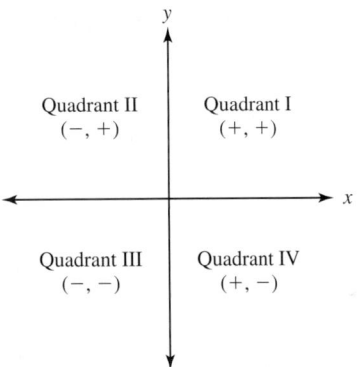

Figure 9.11 Quadrants of the Cartesian coordinate system

To make the graph a function, we first evaluate the rule of the function at several domain values to generate a set of ordered pairs. The purpose is to generate enough points to ensure drawing the graph correctly. The domain values should include some positive and negative values as well as zero. Next, we plot the ordered pairs to get a sense of the shape of the graph and then use what we see to draw additional points. If the domain is the set of natural numbers, integers, or rational numbers, the graph is a set of discrete, or separated, points. If the domain is the real numbers, we connect the plotted points to make a curve. Regardless of the domain, the **graph** of the function is the set of all points in the Cartesian plane that satisfy the rule of the function.

Example 9.19   Graphing Functions

Representation

Graph each function.

a. $y = 3x - 1$.   **b.** $y = x^2 - 2$.

Solution

a. Because the domain of the function $y = 3x - 1$ is the set of real numbers, we can select any real number value, evaluate it in the function, and plot the resulting ordered pair. Selecting and evaluating several such values gives the following table and graph.

x	y
−2	−7
−1	−4
0	−1
1	2
2	5

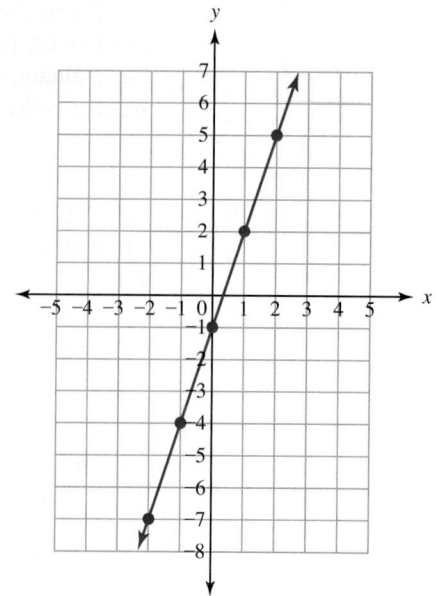

b. The domain of the function $y = x^2 - 2$ is the set of real numbers. Because the function is slightly more complicated, we evaluate it at more points to ensure drawing the graph correctly:

x	y
−3	7
−2	2
−1	−1
0	−2
1	−1
2	2
3	7

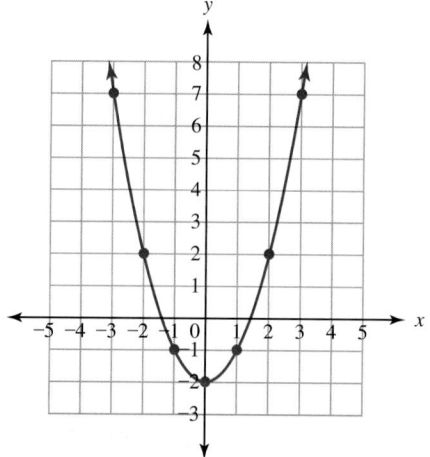

Graphing calculators now make graphing functions quite easy. To graph a function on a calculator, enter the equation of the function into the equation editor, which is generally brought up using the [**y =**] key. Then enter the function by using an appropriate combination of variable and operation keys. For instance, we enter the function $y = x^2 - 2$ as [**x**] [**^**] 2 [**−**] 2 [**ENTER**]. Then we press the [**GRAPH**] key to graph the function. For the function $y = x^2 - 2$, the calculator returns the graph shown in Figure 9.12.

Explorations Manual 9.7

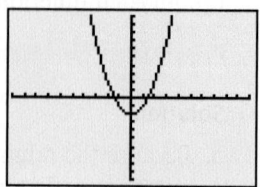

Figure 9.12 The graph of $y = x^2 - 2$

To view the graphs of some functions, we may need to change the range of the window. We do so by changing the appropriate values under the WINDOW menu.

Sometimes, we are given a graph without knowing the corresponding algebraic rule. To determine whether the graph is a function, we can use the vertical line test.

Theorem 9.3   Vertical Line Test

If a vertical line intersects a graph at more than one point for any domain value x, then the graph does not represent a function.

The vertical line test is based on the fact that, if a graph intersects a vertical line at more than one point, then the graph associates two range values with the one domain value at this point. This violates the definition of a function.

Example 9.20   Using the Vertical Line Test

Which of the following graphs represent functions?

a.

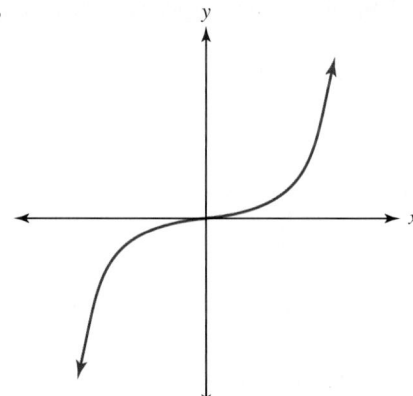

b.

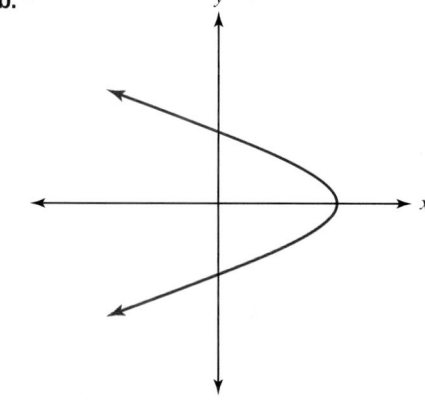

Solution

a. The given portion of the graph does not intersect a vertical line at more than one point. Hence, the graph represents a function.

b. Passing a vertical line over the graph, we can see the line intersects the graph at more than one point. Consequently, the graph is not a function.

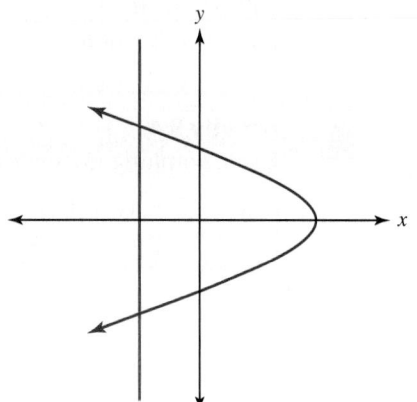

Check Your Understanding 9.2-B

1. Which of the following represents a function?

a.

x	y
1	4
2	5
3	6
4	7

b.

x	y
1	1
1	2
1	3
1	4

c. {(1, 2), (2, 3), (3, 4), (4, 3), (3, 2), (2, 1)}

d. {(1, 2), (2, 3), (3, 4), (4, 5)}

2. Evaluate the function $f(x) = 3x - 4$ for each value in the set $\{-2, -1, 0, 1, 2, 3\}$. Plot the points on the Cartesian coordinate system.

3. Graph each function by making a table of values and plotting the points on the Cartesian coordinate system.

a. $y = x - 4$.   b. $y = x^2 - 1$.   c. $f(x) = \sqrt{x} - 2$.

4. Which of the following graphs represents a function?

a.

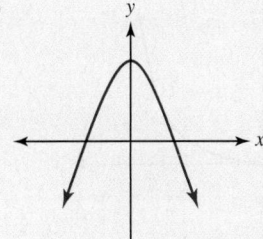

b.

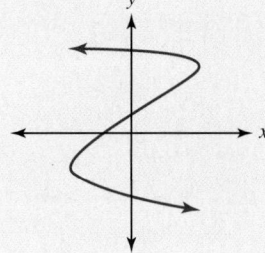

c.

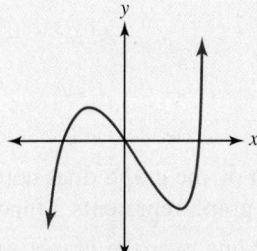

Talk About It Other than $\boxed{x^2}$ or $\boxed{\sqrt{}}$, what keys or features on your calculator represent functions?

Activity 9.4
Use a graphing calculator to graph each function.

a. $y = x^3 - 3x + 1$.   b. $y = \sqrt{x^3}$.   c. $y = \dfrac{x}{x - 1}$.

SECTION 9.2 Learning Assessment

■ Understanding the Concepts

1. What conditions must be satisfied by a rule of correspondence between two sets in order for it to be a function?

2. What is the difference between the domain, the codomain, and the range of a function?

3. How is the rule of a function explicitly given when the function is represented as:
   a. An arrow diagram?   **b.** A table?
   c. A set of ordered pairs?

4. Why does a machine make a good analogy for a function?

5. Why is a calculator a good example of a function machine?

6. What is the difference between the independent and dependent variables of a function?

7. **a.** Why is an algebraic equation a powerful way to represent a function with an infinite domain?
   b. Under what circumstances must values be excluded from the domain? Why?

8. What is the graph of a function? How do we make a graph on the Cartesian coordinate system?

9. Describe the relationship between the quadrants of the Cartesian coordinate system and the signs of the coordinates in an ordered pair?

10. What is the vertical line test, and why does it work?

■ Representing the Mathematics

11. Arrow diagrams, tables, and sets of ordered pairs are somewhat similar in the how they represent functions. What are the advantages and disadvantages of representing functions in these ways?

12. What advantages do algebraic expressions and graphs offer over other representations of functions?

13. Represent the following function as a set of ordered pairs, a table, and a graph. Suggest a rule that governs the function.

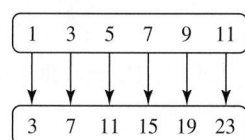

14. Represent the following function as an arrow diagram, a set of ordered pairs, and a graph. Suggest a rule that governs the function.

x	$f(x)$
25	5
16	4
9	3
4	2
1	1

15. Represent the set of ordered pairs $\{(2, 4), (3, 6), (4, 8), (5, 10), (6, 12)\}$ as an arrow diagram, a table, and a graph. Suggest a rule that governs the function.

16. Represent the following function as an arrow diagram, a set of ordered pairs, and a table. Suggest a rule that governs the function.

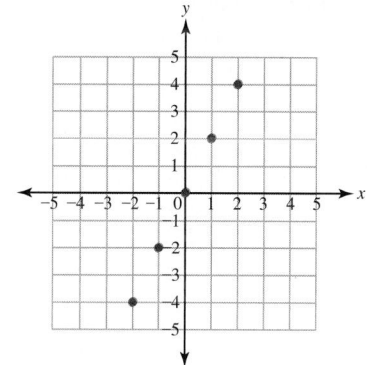

17. Give a verbal description for each notational statement.
    a. $f(4)$ **b.** $g(-5)$ **c.** $f(x) = x + 1$.
    d. $\mathbb{R} - \{0\}$ **e.** $\mathbb{R} - \{-1, 1\}$

18. Give the coordinates of the point that is:
    a. Three units to the right of and two units above the origin.
    b. Four units to the right of and five units below the origin.
    c. Six units above the origin.
    d. Two units to the left of the origin.

19. State the position of each point on the Cartesian coordinate system with regard to the origin.
    a. $(3, 5)$ **b.** $(-2, 4)$ **c.** $(0, -2)$
    d. $(-3, 0)$ **e.** $(-2, -4)$

20. Which of the following tables defines a function? Explain why a function is, or is not, defined.

a.

Input	Output
-2	-6
-1	-3
0	0
1	-3
2	-6

b.

Input	Output
0	-4
1	-3
2	-1
1	3
0	4

c.

Input	Output
0	5
1	5
2	5
3	5
4	5

d.

Input	Output
5	0
3	1
1	2
3	3
5	4

21. Which of the following sets of ordered pairs define(s) a function?

a. $\{(3, 1), (3, 2), (3, 3), (3, 4)\}$

b. $\{(1, 2), (3, 4), (5, 6), (7, 8)\}$

c. $\{(1, 3), (2, 5), (3, 7), (4, 8) \ldots\}$

d. $\{(1, 2)\ (2, 3), (3, 2), (2, 1)\}$

22. Use the vertical line test to determine whether each graph represents a function.

a.

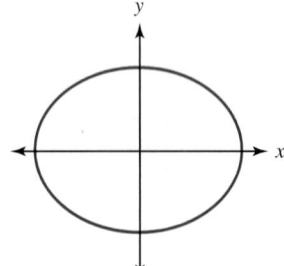

b.

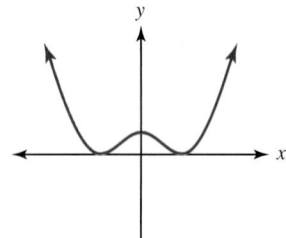

c.

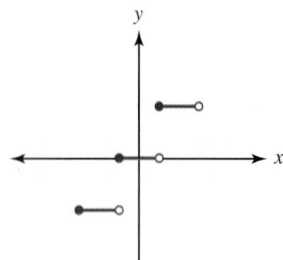

d.

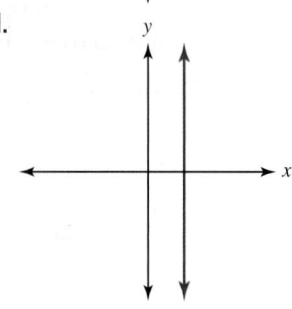

Exercises

23. State the domain, codomain, and range of each function

a.

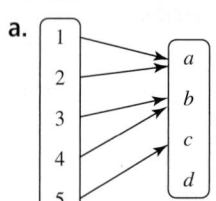

b.

Sam
Dorothy
Clarice
Tyrone
Rochelle

Table 1
Table 2
Table 3

c.

Jackie
Lois
Simone

Lane 1
Lane 2
Lane 3
Lane 4
Lane 5

24. State the domain and range of each function.

a.

x	y
1	1
2	2
3	3
4	4

b.

x	$f(x)$
6	4
5	3
4	4
3	3

c. $\{(2, 4), (5, 3), (6, -1), (7, 7)\}$

d. $\{(2, 2), (3, 2), (4, 2), (5, 2)\}$

25. Use arrow diagrams to demonstrate three different functions between the sets $\{1, 2, 3, 4, 5\}$ and $\{a, b, c, d\}$.

26. Use ordered pairs to demonstrate a function between the sets $\{a, b, c, d, e\}$ and $\{w, x, y, z\}$ so that:

a. The range is equal to the codomain.

b. The range has exactly one element.

c. The range has exactly two elements.

27. Evaluate the function $f(x) = 5x + 7$ at each value.

a. $x = -2$ **b.** $x = 0$ **c.** $x = 3$ **d.** $x = 9$

28. Find $f(-4), f(1)$, and $f(5)$ for:

a. $f(x) = 5x^2 + 2x - 1.$ **b.** $f(x) = x^2 - 4.$

c. $f(x) = x^3 - 1.$ **d.** $f(x) = (x - 4)^2.$

29. If $g(x) = 4x - 5$, find $g(t), g(t - 4)$, and $g(t + 6)$.

30. If $x \in \{-5, -2, 0, 2, 5\}$, evaluate each function at each x-value when possible.

a. $y = 4x - 7.$ **b.** $y = 3x^2 - 6x + 2.$

c. $y = \dfrac{1}{x + 5}.$ **d.** $y = \sqrt{x + 2}.$

31. If $x \in \{-4, -1, 0, 1, 4\}$, evaluate each function at each x-value when possible.

a. $y = 3x - 3.$ **b.** $y = \dfrac{1}{x + 4}.$

c. $y = \dfrac{1}{x(x - 1)}.$ **d.** $y = \sqrt{(x - 4)(x + 1)}.$

32. For which of the following functions does $f(-4)$ exist?

a. $f(x) = x + 4.$ **b.** $f(x) = \dfrac{1}{x - 4}.$

c. $f(x) = \sqrt{x}.$ **d.** $f(x) = x - 4.$

e. $f(x) = \dfrac{1}{x + 4}.$ **f.** $f(x) = \sqrt{x + 4}.$

33. For which of the following functions does $f(5)$ exist?

a. $f(x) = x^2 - 4.$ **b.** $f(x) = \dfrac{1}{(x - 5)}.$

c. $f(x) = \sqrt{x - 5}.$ **d.** $f(x) = (x - 5)^2.$

e. $f(x) = \dfrac{1}{x^2 - 5}.$ **f.** $f(x) = \sqrt{5 - x^2}.$

34. State the domain of each function.

a. $f(x) = x.$ **b.** $f(x) = 4x^2.$

c. $f(x) = \dfrac{1}{2x}.$ **d.** $f(x) = |x|.$

e. $f(x) = \sqrt{x - 4}.$ **f.** $f(x) = \sqrt{|x|}.$

35. State the domain of each function.

a. $y = 3.$ **b.** $y = 2x^3.$

c. $y = \dfrac{1}{x + 7}.$ **d.** $y = \dfrac{1}{x - 2}.$

e. $y = \dfrac{1}{(x - 1)(x + 2)}.$ **f.** $y = \dfrac{1}{\sqrt{x - 4}}.$

36. Give a set of five ordered pairs that satisfy the function $y = 2x + 6.$

37. Give a set of five ordered pairs that satisfy the function $f(x) = \dfrac{2}{3x^2}.$

38. Which of the following ordered pairs satisfy the function $y = 5x - 1$?

a. $(5, 25)$ **b.** $(3, 14)$ **c.** $(-2, -9)$

d. $\left(\dfrac{1}{5}, 1\right)$ **e.** $(0.2, 0)$

39. State the quadrant in which each point is located. If it is not in a quadrant, state the axis it is on or whether it is at the origin.

a. $(3, -1)$ **b.** $(5, 7)$ **c.** $(-2, -2)$

d. $(-1, -6)$ **e.** $(0, 6)$

40. State the quadrant in which each point is located. If it is not in a quadrant, state the axis it is on or whether it is at the origin.

a. $(9, -3)$ **b.** $(-4, 3)$ **c.** $(0, 0)$

d. $(-4, 0)$ **e.** $(9, -5)$

41. In which quadrant(s) of the Cartesian coordinate system is each condition true?

a. A positive x-coordinate.

b. A negative y-coordinate.

c. A negative x-coordinate and a positive y-coordinate.

d. The product of the x- and y-coordinates is positive.

42. Evaluate each function at the given values, and then plot the resulting points on the Cartesian coordinate system.

a. $y = -2x, x \in \{-2, -1, 0, 1, 2\}.$

b. $y = -3x + 4, x \in \{0, 1, 2, 3, 4, 5\}.$

c. $y = x^2 + 3, x \in \{-2, -1, 0, 1, 2\}.$

d. $f(x) = |2x|, x \in \{-4, -2, 0, 2, 4\}.$

43. Graph each function by making a table of values and plotting the ordered pairs on the Cartesian coordinate system.

a. $y = 4x.$ **b.** $y = -2x.$ **c.** $y = x - 3.$

d. $y = 2x - 6.$ **e.** $y = -3x + 1.$ **f.** $y = \dfrac{1}{2}x + 3.$

44. Graph each function by making a table of values and plotting the ordered pairs on the Cartesian coordinate system.

a. $f(x) = |x|.$ **b.** $f(x) = -x^2.$

c. $f(x) = \sqrt{2x}.$ **d.** $f(x) = 2|x| + 1.$

e. $f(x) = 2x^2 - 3.$ **f.** $f(x) = \sqrt{x} + 3.$

45. Graph each function on a graphing calculator.

a. $y = 3x^2 - 1.$ **b.** $y = x^2 - 7x + 2.$

c. $y = x^3 - 4.$ **d.** $y = \dfrac{1}{2}x^4.$

46. Graph each function on a graphing calculator.

a. $y = \sqrt{2x - 6}.$ **b.** $y = \dfrac{2}{x^2}.$

c. $y = \dfrac{1}{x^2} + 1.$ **d.** $y = \sqrt{x^2 - 4}.$

■ Problems and Applications

47. Suggest a possible input value for each function machine.

a.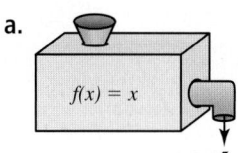
$f(x) = x$
5

b.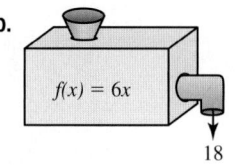
$f(x) = 6x$
18

c.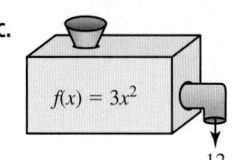
$f(x) = 3x^2$
12

d.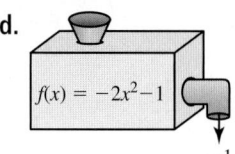
$f(x) = -2x^2 - 1$
−1

48. Suggest a possible algebraic rule for each table.

a.

x	$f(x)$
0	0
1	4
2	8
3	12
4	16

b.

x	$f(x)$
0	0
1	−1
2	−4
3	−9
4	−16

c.

x	$f(x)$
0	−3
1	−1
2	1
3	3
4	5

49. Consider the set of ordered pairs $\{(-10, 30), (0, 20), (10, 10), (20, 0), (30, 10)\}$ for the function f?

a. What domain and range values are shown for f?

b. What is $f(0)$? $f(20)$?

c. If 50 were in the domain of f, what might its range value be?

50. How many different functions are there from a set of:

a. 2 elements to a set of 2 elements?

b. 3 elements to a set of 2 elements?

c. 4 elements to a set of 2 elements?

d. n elements to a set of 2 elements?

51. The following graph is for function $f(x)$. Use it to estimate the values of $f(-2), f(0), f(1),$ and $f(3)$.

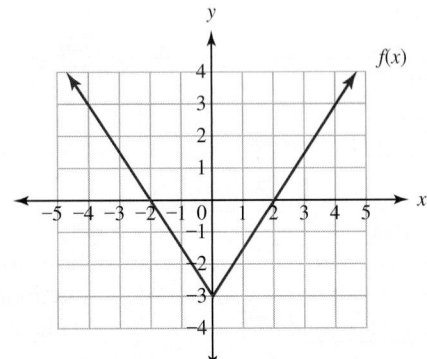

52. The following graph is for function $f(x)$.

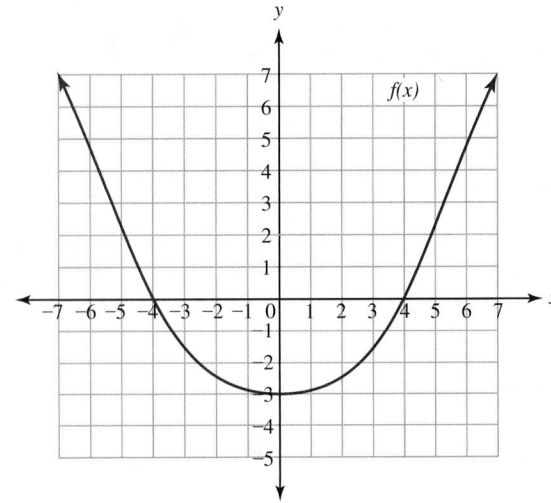

a. Find $f(-5), f(-3),$ and $f(6)$.

b. For what value(s) of x is $f(x) = 0$?

c. For what value of x is $f(x)$ the smallest?

53. The function $C(x) = 0.03k + 19$, where k is the number of kilowatt hours used, describes the cost of residential electric service in a certain town. What is a home's monthly bill if:

a. 300 kWh are used?

b. 450 kWh are used?

c. 993.5 kWh are used?

54. The relationship between degrees Fahrenheit and degrees Celsius is given by the formula $F = \dfrac{9}{5}C + 32$.

a. What temperature is it when $C = -10$ or $C = 10$?

b. If water boils at 100° C, at what temperature does water boil on the Fahrenheit scale?

c. If water freezes at 32° F, at what temperature does it freeze on the Celsius scale?

■ Mathematical Reasoning and Proof

55. True or false:

a. The first position in an ordered pair represents a domain value.

b. The second position in an ordered pair represents an $f(x)$ value.

c. (domain, range) = (dependent variable, independent variable).

d. (x, y) = (up-down, left-right).

56. Will the domain of a function always be the same size as its range? Explain.

57. Can a function have an infinite domain, but only one element in the range? If so, provide an example. If not, explain why not.

58. Let $A = \{0, 1, 2, 3, 4\}$ and $B = \{1, 2, 3, 4\}$. If $x \in A$ and $y \in B$, consider the rule $x > y$. Make a table of all pairs that satisfy this rule. Does the rule describe a function from A to B? Explain.

59. The following two tables show the record distances for the Olympic shot put in two ways.

Table 1			Table 2	
Year	Distance in Feet		Distance in Feet	Year
1960	65		65	1960
1964	67		67	1964
1968	67		67	1968
1972	70		70	1972
1976	70		70	1976
1980	70		70	1980
1984	70		70	1984
1988	74		74	1988
1992	71		71	1992
1996	71		71	1996

Which represents a function? For the one that does, what are the independent and dependent variables? For the one that is not a function, explain why.

60. The following table shows the driving time from someone's house to various cities.

City	Baltimore	Boston	Philadelphia	Pittsburgh	Washington D. C.
Driving Time in Hours	3	5	1.2	5	5

Does the table represent a function? Explain your reasoning.

61. The heights and weights of six people are shown in the table.

Height in Inches	62	63	64	63	67	69
Weight in Pounds	125	131	133	145	129	151

From the given information, is height a function of weight? Is weight a function of height? Explain your reasoning for each answer.

62. A function is called a **constant function** if the range value is always the same.

a. Provide three examples of constant functions.

b. Based on what you have written, give a general algebraic expression that represents any constant function.

c. Graph your constant functions in part a. Based on what you see in the graphs, what do you think is always true about the graph of a constant function?

Mathematical Communication

63. Write a sequence of steps for graphing the function $f(x) = x^2$. Give them to a peer to follow. Once your peer has completed the task, discuss the following questions.

a. What difficulties were encountered in writing the instructions?

b. What difficulties were encountered in following the instructions?

c. What do the difficulties imply about potential difficulties a teacher will encounter in teaching students how to graph functions?

64. Write the sequence of keys to be pressed to make your calculator graph the function $y = 3x^2 - 5$.

65. Graphing calculators are a convenient way to help students graph complicated functions. Discuss with your peers what role, if any, graphing calculators should have in elementary- or middle-grades classrooms. Be sure to discuss both the advantages and disadvantages of making use of such calculators.

66. Maxine claims that the functions $y = 2x + 1$ and $f(x) = 2x + 1$ cannot be the same function because they are not both equal to y. Is she correct? If so, explain. If not, what would you do to correct her misconception?

67. A student brings the following graph to you and claims that it cannot be a function because it does not pass the horizontal line test. What is your response?

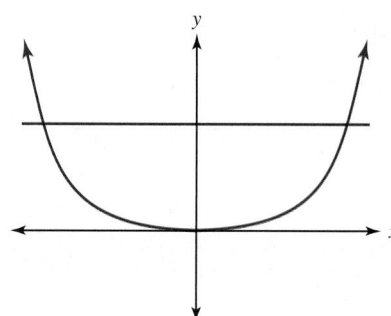

Building Connections

68. With a small group of your peers, list five situations in your daily life that demonstrate a functional relationship. In each case, identify the rule of the function, and state the domain and range.

69. Recall that a one-to-one correspondence between two sets A and B is a pairing of the elements of A with the elements of B so that each element of A corresponds to exactly one element of B, and vice versa.

   a. Is every one-to-one correspondence a function? Explain.

   b. Is every function a one-to-one correspondence? Explain.

   c. What are the differences between a function and a one-to-one correspondence?

70. Using the information in Exercise 69, draw an arrow diagram of a function that is:

   a. a one-to-one correspondence.

   b. not a one-to-one correspondence.

71. Which of the following describes a function? Explain your reasoning.

   a. The relationship between a student and the student's GPA.

   b. The relationship between an item in a store and its bar code.

   c. The relationship between birthdays and people with those birthdays.

   d. The relationship between a course and the students taking the course.

   e. The relationship between a father and his children.

72. Other than x^2 or $\sqrt{}$, which keys or features on your calculator represent functions?

73. Do the GCF and LCM features on a calculator represent functions? Why or why not?

74. In Section 9.1, we discussed sequences as a precursor to functions.

   a. Why is a sequence a function?

   b. What is the domain of most sequences?

   c. Pedagogically, why might it be beneficial to introduce the idea of functions by using sequences?

SECTION 9.3   Linear Functions and Constant Change

The patterns of change that functions describe and represent can happen in a variety of ways. Not only can they grow larger and smaller, but they can do so at different rates. In most cases, describing patterns of change requires advanced mathematics. However, some of them are accessible to students in the elementary and middle grades. The simplest of these are patterns of constant change, which are directly related to linear functions.

Understanding Linear Functions

Initially, the phrase "constant change" seems to be an oxymoron because it suggests change that somehow stays the same. However, when applied to functions, the phrase signifies that a consistent change in one variable coincides with a consistent change in another. We have already encountered many functions that exhibit a pattern of constant change. For instance, in the arithmetic sequence 1, 4, 7, 10, . . . , a consistent change in the position of the term coincides with a consistent change in its value. Specifically, as the position of the term increases by 1, the value increases by 3 (Figure 9.13).

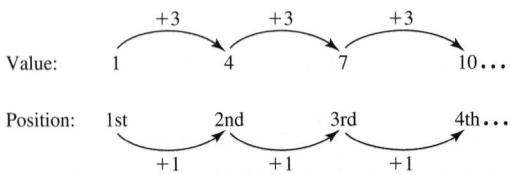

Figure 9.13 Constant change in an arithmetic sequence

Patterns of constant change can also occur with ratios and rates. Given two quantities that occur in a fixed ratio, we can create a function by considering all the equivalent ratios. For instance, consider a long-distance runner who runs at a rate of 8 mph. Figure 9.14 shows a table for the function that relates the time in hours to the number of miles the runner travels. A consistent change in the number of hours coincides with a consistent change in the number of miles traveled. Specifically, as the hours increase by 1, the number of miles always increases by 8. If we take the example one step further and graph the function, we can see that all of the points fall along a line. Because this is true for any function of constant change, we call them **linear functions**.

Time in Hours	Miles Traveled
1	8
2	16
3	24
4	32
5	40

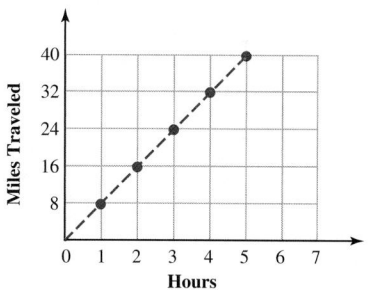

Figure 9.14 A table and graph for the function between time in hours and distance traveled

The rate of the constant change plays a key role in defining linear functions. As the rate changes, so does the function. This becomes clear when we compare the graphs of several linear functions on the same coordinate plane.

Example 9.21   Buying Pens

Communication

A store sells three kinds of pens: one for 50¢, one for $1.00, and one for $3.00. Graph three functions that show how many pens of each type can be bought for a given number of dollars. Use the graphs to describe how the rate of change affects the functions.

Solution To graph the cost function of each pen, we set up the Cartesian plane so that the horizontal axis represents dollar amounts and the vertical axis represents the number of pens purchased. Next, we use the rates to find and plot points on each line. For the 50¢ pen, we can buy 2 for $1, 4 for $2, and so on. Plotting these points gives us the blue line. For the $1 pen, we can buy 1 for $1, 2 for $2, and so on. Plotting these points gives us the red line. Finally, for the $3 pen, we can buy 1 for $3, 2 for $6, and so on. Plotting these points gives us the green line.

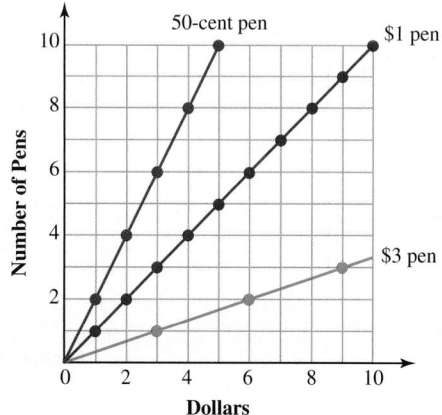

Even though the lines all start at the origin, the rate of change affects the steepness of the line. For any given dollar amount, we can buy more 50¢ pens than the other two, so the line is steeper. The $3 pen costs more, so the line is less steep. From this, we can surmise that a greater rate of change causes the line to be more vertical and that a smaller rate of change causes the line to be more horizontal.

Explorations
Manual
9.8

Example 9.21 illustrates how the rate of change can affect the steepness of the linear function. We can measure the steepness of a line more definitively by computing its **slope.** The slope of a line is a comparison between its vertical and horizontal change. The vertical change is called the **rise,** and it refers to the constant change of the dependent variable. The horizontal change is called the **run,** and it refers to the constant change of the independent variable.

Definition of Slope

The **slope** of a line is the ratio of its rise to its run, or slope $= m = \dfrac{\text{rise}}{\text{run}}$.

Applying the definition of slope to the rates in Example 9.21 shows how the size of the slope coincides with the graph of the function. Specifically, the slope of the line for the 50¢ pen is larger than the others, so it coincides with the steepest line, or

$$\text{Slope of 50¢ pen} = \frac{2 \text{ pens}}{\$1} = 2$$

$$\text{Slope of \$1 pen} = \frac{1 \text{ pen}}{\$1} = 1$$

$$\text{Slope of \$3 pen} = \frac{1 \text{ pen}}{\$3} = \frac{1}{3}$$

The definition indicates that we can find the slope of a line given the rise and run. In turn, we can compute the rise and run if we know the coordinates of two points on the line. Figure 9.15 shows that, if (x_1, y_1) and (x_2, y_2) are two points on the line, then the rise is the difference of the y-coordinates, or $y_2 - y_1$, and the run is the difference of the x-coordinates, or $x_2 - x_1$.

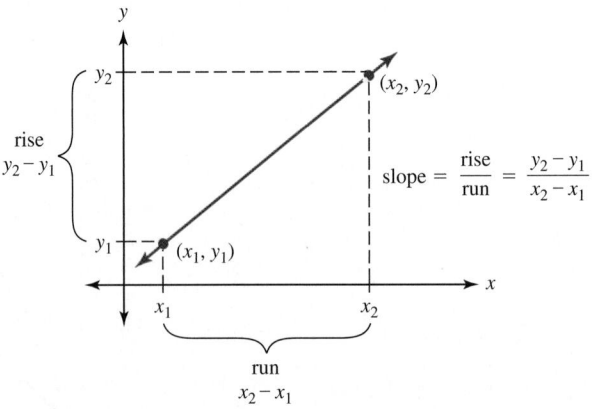

Figure 9.15 The formula for the slope of a line

Theorem 9.4   Slope Formula

If (x_1, y_1) and (x_2, y_2) are two points with $x_1 \neq x_2$, the slope of the line containing the points is given by $m = \dfrac{\text{rise}}{\text{run}} = \dfrac{y_2 - y_1}{x_2 - x_1}$.

Example 9.22   Computing Slopes

Representation   Compute the slope of each line, and interpret its meaning in terms of the ratio of the rise to the run.

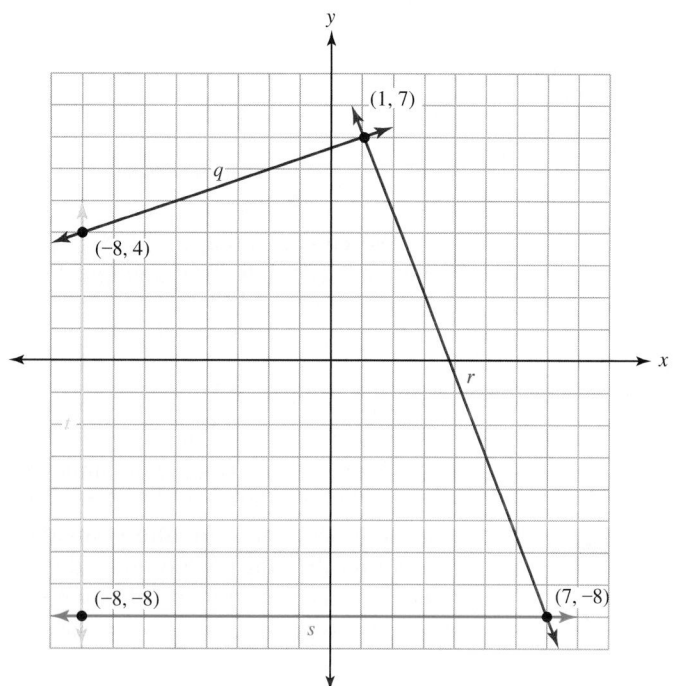

Solution We make each computation using the formula from Theorem 9.4.

a. $m_q = \dfrac{y_2 - y_1}{x_2 - x_1} = \dfrac{7 - 4}{1 - (-8)} = \dfrac{3}{9} = \dfrac{1}{3}$. The slope indicates that line q goes up 1 unit for every 3 units that it goes to the right. It also indicates that line q goes down 1 unit for every 3 units it goes to the left.

b. $m_r = \dfrac{y_2 - y_1}{x_2 - x_1} = \dfrac{(-8) - 7}{7 - 1} = \dfrac{-15}{6} = \dfrac{-5}{2}$. The slope indicates that line r goes down 5 units for every 2 units it goes to the right.

c. $m_s = \dfrac{y_2 - y_1}{x_2 - x_1} = \dfrac{(-8) - (-8)}{(-8) - 7} = \dfrac{0}{-15} = 0$. A slope of zero means that line s has no rise but has an infinite run.

d. $m_t = \dfrac{y_2 - y_1}{x_2 - x_1} = \dfrac{(-8) - 4}{(-8) - (-8)} = \dfrac{-12}{0}$, which is undefined. An undefined slope means that line t has an infinite rise but no run.

Example 9.22 points out several facts about the slope of a line. First, if the slope is positive, then the linear function is **increasing;** that is, the values of the dependent variable increase as the values of the independent variable increase. If the slope is negative, then the linear function is **decreasing;** that is, the values of the dependent variable decrease as the values of the independent variable increase. Second, the larger the absolute value of the slope is, the steeper the line is. Third, if two points lie on a horizontal line, then they must have the same y-coordinates. When we subtract them to compute the slope, the result is zero. Similarly, if two points lie on a vertical line, then they must have the same x-coordinates. When we subtract them, the denominator of the slope ratio is always zero. Consequently, the slope must be undefined. Figure 9.16 summarizes these facts visually.

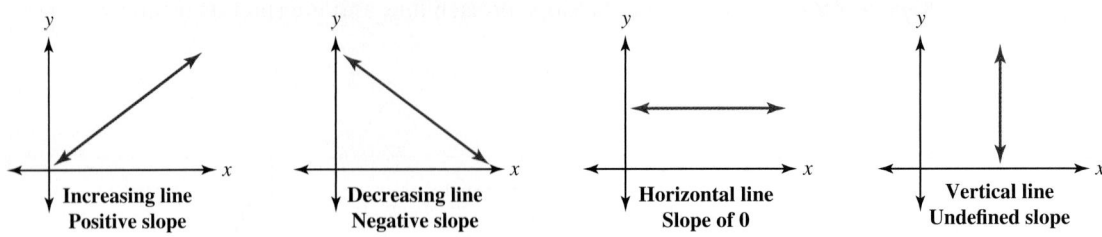

Increasing line
Positive slope

Decreasing line
Negative slope

Horizontal line
Slope of 0

Vertical line
Undefined slope

Figure 9.16 The four possibilities for slope

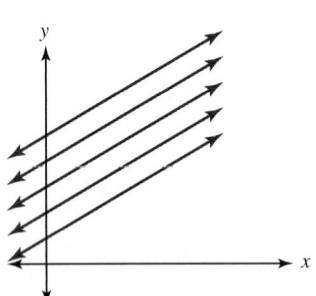

Figure 9.17 Lines with the same slope

When computing the slope, the order in which we use the coordinates does not matter. Specifically, if we multiply the top and bottom of the slope ratio by (-1), we get

$$\frac{y_2 - y_1}{x_2 - x_1} \cdot \frac{(-1)}{(-1)} = \frac{-y_2 + y_1}{-x_2 + x_1} = \frac{y_1 - y_2}{x_1 - x_2}$$

Consequently, as long as the order of the coordinates within the numerator and denominator is consistent, the point we use first does not matter.

Slope is not the only defining characteristic of a line. As Figure 9.17 illustrates, many different lines can have the same slope. However, each line intersects the y-axis at a different point, and this point allows us to differentiate one line from another. The point is called the **y-intercept,** and it, along with the slope, is enough to define any line. Because the x-coordinate of any point on the y-axis is 0, the y-intercept of a line has coordinates $(0, b)$, where b is a real number.

Example 9.23   Graphing a Line with the Slope and y-intercept

Representation   Use the slope and y-intercept to graph each line.

a. $m = \frac{3}{5}$, y-intercept $= (0, 1)$ **b.** $m = 0$, y-intercept $= (0, -3)$

Solution To draw a line, we need two points that lie on the line. The y-intercept is one, and we can use the slope to find another.

a. The y-intercept, $(0, 1)$, must be on the line. Also, the line has a slope of $m = \frac{3}{5}$, which means we can go up 3 units and to the right 5 units to find another point. Consequently, $(0 + 5, 1 + 3) = (5, 4)$ is another point on the line. We create the graph by plotting and connecting the two points.

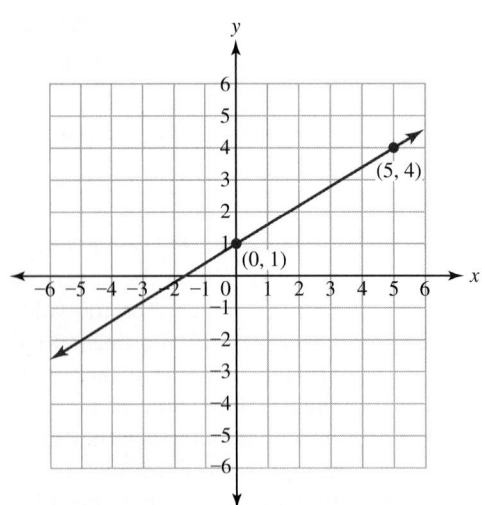

b. This line has a slope of zero, so it must be horizontal. Because the *y*-intercept is (0, −3), we get the line shown in the graph.

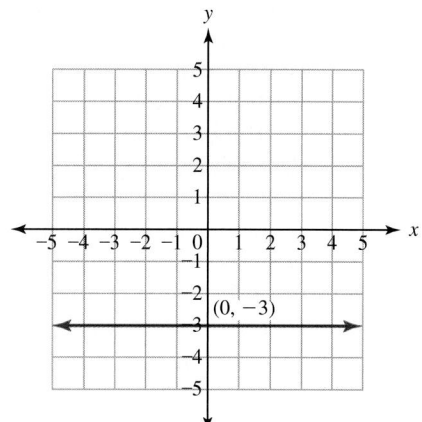

Check Your Understanding 9.3-A

1. Determine whether the relationship between the two variables in each table is linear.

a.

x	y
1	4
2	6
3	8
4	10

b.

x	y
2	−2
3	−5
4	−6
5	−8

c.

x	y
1	3
2	6
3	9
4	12

d.

x	y
2	4
3	8
4	16
5	32

2. Compute the slope of the line containing each pair of points.

a. (3, 4) and (6, 5)   **b.** (3, 1) and (5, 1)   **c.** (−2, 3) and (2, −1)

3. Interpret each slope as the ratio of the rise of the line to its run.

a. $m = 4$.   **b.** $m = -\dfrac{1}{2}$.   **c.** m is undefined.

4. Graph each line using the slope and *y*-intercept.

a. $m = \dfrac{2}{3}$, *y*-intercept = (0, −2).   **b.** $m = -\dfrac{5}{2}$, *y*-intercept = (0, 3).

Talk About It How is the slope of a line connected to the common difference of an arithmetic sequence?

Problem Solving

Activity 9.5

Find the missing value *k* in each situation.

a. A line passes through the points (2, 3) and (4, *k*) and has a slope of 2.

b. A line passes through the points (*k*, 1) and (3, 3) and has a slope of $-\dfrac{4}{3}$.

c. A line passes through the points (3, *k*) and (5, −1) and has a slope of 0.

d. A line passes through the points (*k*, *k*) and (4, 1) and has a slope of $\dfrac{1}{2}$.

e. A line passes through the points (1, *k*) and (3, 2) and has a slope of $\dfrac{k}{4}$.

Equations of Linear Functions

Although an algebraic equation is a powerful way to represent a function with an infinite domain, finding the algebraic rule of any function is not always easy. However, we can find the equation of a linear function in a number of ways. We begin with horizontal and vertical lines, which have the simplest equations.

If a line is horizontal, then any points on the line must have the same y-coordinate because they occur at the same height on the graph. For instance, the line in Figure 9.18 occurs at a height of 3, so every point on the line must have coordinates $(x, 3)$, where x is a real number. Because the value of x has no effect on the outcome of the function, we define the line with the equation $y = 3$.

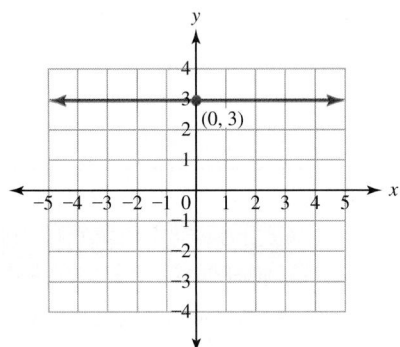

Figure 9.18 The line $y = 3$

In general, the equation of any horizontal line can be defined to be **$y = b$,** where b is a real number. In the same way, if a line is vertical, then all the points on the line must have the same x-coordinate. Consequently, the equation of any vertical line can be defined as **$x = a$,** where a is a real number. Because a vertical line assigns one x-value to many y-values, the equation $x = a$ does not represent a function.

Example 9.24 **Equations of Horizontal and Vertical Lines**

Determine the equation of each line shown.

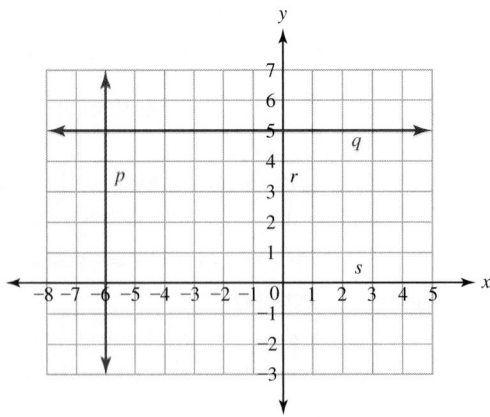

Solution

a. Line p is a vertical line with equation $x = -6$.

b. Line q is a horizontal line with equation $y = 5$.

c. Line r is the y-axis, which is a vertical line with equation $x = 0$.

d. Line s is the x-axis, which is a horizontal line with equation $y = 0$.

We can find the equations of other lines by using the slope formula, $m = \dfrac{y_2 - y_1}{x_2 - x_1}$. To do so, suppose that (x_1, y_1) is a known point on a line with slope m. Also suppose that (x, y) is an unknown point on the same line. The following manipulation then gives an equation for the line in terms of x and y.

$$m = \frac{y - y_1}{x - x_1} \qquad \text{Substitution of } (x, y) \text{ into slope formula.}$$

$$m(x - x_1) = y - y_1 \qquad \text{Multiply both sides by } (x - x_1).$$

Theorem 9.5   Point-Slope Form of a Line

The equation of a line with slope m and containing the point (x_1, y_1) is given by $y - y_1 = m(x - x_1)$. This is called the **point-slope form** of the line.

Example 9.25   Equations of Lines in Point-Slope Form

Representation   Find the equation of each line in point-slope form.

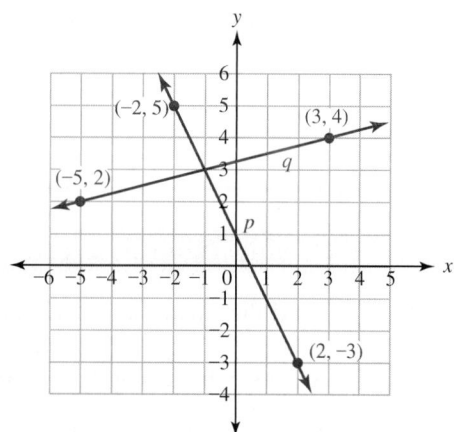

Solution   We first find the slope and then substitute the coordinates from either point into the point-slope form equation.

a. Line p has a slope of $m = \dfrac{5 - (-3)}{(-2) - (2)} = \dfrac{8}{-4} = -2$. Using the coordinates from the point $(-2, 5)$, the equation of line p is $y - 5 = -2(x + 2)$.

b. Line q has a slope of $m = \dfrac{4 - (2)}{(3) - (-5)} = \dfrac{2}{8} = \dfrac{1}{4}$. Using the point $(3, 4)$, the equation of line q is $y - 4 = \dfrac{1}{4}(x - 3)$.

Earlier, we stated that a line is completely defined by its slope and y-intercept. Given these values, we should be able to write the equation of the line. To understand how, consider the line $y - 5 = -2(x + 2)$ from the last example. If we evaluate it at $x = 0$, the x-coordinate of the y-axis, and solve for y, then

$$y - 5 = -2(0 + 2)$$
$$y - 5 = -4$$
$$y = 1$$

This indicates that the line intersects the y-axis at the point $(0, 1)$. Now if we solve the equation $y - 5 = -2(x + 2)$ for y and simplify, the constant term has the same value as the y-intercept, or

$$y - 5 = -2(x + 2)$$
$$y - 5 = -2x - 4$$
$$y = -2x + 1$$

Explorations
Manual
9.9

Consequently, if we write a linear equation in this manner, then the slope and the y-intercept are directly part of the equation.

Theorem 9.6 Slope-Intercept Form of a Line

Every nonvertical line can be written in the form $y = mx + b$, where m is the slope, and b is the y-coordinate of the y-intercept. This is called the **slope-intercept form** of the line.

Example 9.26 Proof of Theorem 9.6

Reasoning

Show that any nonvertical line can be written in slope-intercept form.

Solution Suppose that line l has a slope of m and a y-intercept of $(0, b)$. By Theorem 9.5, we can express line l as $y - b = m(x - 0)$. Then,

$$y - b = mx - m \cdot 0 \qquad \text{Distributive property.}$$
$$y - b = mx \qquad \text{Multiplication by zero.}$$
$$y = mx + b \qquad \text{Adding } b \text{ to both sides.}$$

Example 9.27 Equations of Lines in Slope-Intercept Form

Find the equation of each line in slope-intercept form.

a. Line p has a slope of $\dfrac{1}{4}$ and a y-intercept of 1.

b. Line r contains the points $(-3, 1)$ and $(5, 5)$.

Solution

a. Substituting the values $m = \dfrac{1}{4}$ and $b = 1$ into the slope-intercept form, the equation of line p is $y = \dfrac{1}{4}x + 1$.

b. Because r contains $(-3, 1)$ and $(5, 5)$, it has a slope of $m = \dfrac{5 - 1}{5 - (-3)} = \dfrac{4}{8} = \dfrac{1}{2}$. To find the y-intercept, we substitute the slope and the coordinates from either point into the slope-intercept formula, and solve for b:

$$y = mx + b$$
$$5 = \frac{1}{2}(5) + b$$
$$5 = \frac{5}{2} + b$$
$$\frac{5}{2} = b$$

By making the proper substitutions, the equation of line r is $y = \dfrac{1}{2}x + \dfrac{5}{2}$.

Like other functions, we can use the equation of a linear function to generate its graph. One way to do so is to use a table of values to plot points. Another method is shown in the next example.

| **Example 9.28** | **Graphing a Line from Its Equation** |

Representation

Graph the line q defined by the equation $y = -\dfrac{1}{3}x - 6$.

Solution The equation $y = -\dfrac{1}{3}x - 6$ indicates that line q has a slope of $-\dfrac{1}{3}$ and a y-intercept of $b = -6$. So $(0, -6)$ is one point on the line. By using the slope, we know that another point is located down 1 unit and 3 units to the right of $(0, -6)$, or $(0 - 1, -6 + 3) = (-1, -3)$. Using these points, we can graph line q.

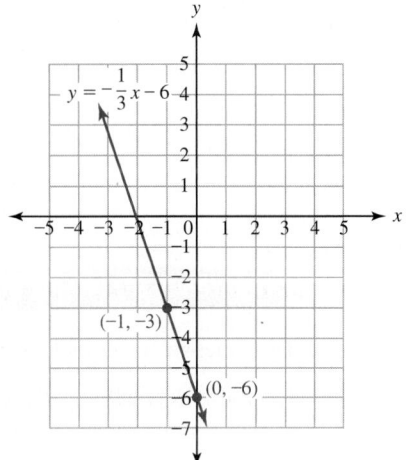

Parallel and Perpendicular Lines

Earlier, we saw that two different lines could have the same slope. For instance, each of the lines in Figure 9.19 has a slope of $m = 2$. Because the lines have the same constant rate of change, the distance between them always remains the same. Consequently, the two lines never cross one another, or intersect. We call such lines **parallel.**

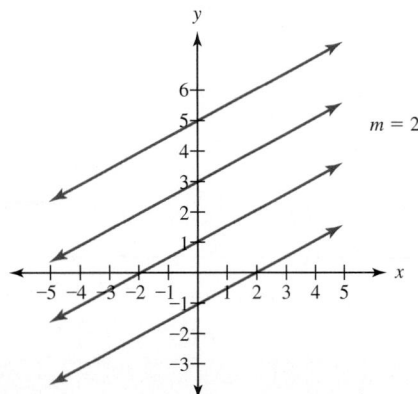

Figure 9.19 Parallel lines with a slope of $m = 2$

| **Theorem 9.7   Slopes of Parallel Lines** |

Two lines in the coordinate plane are parallel if and only if:

1. The slopes of the lines are equal.
2. The lines are vertical, meaning they have undefined slopes.

Two lines can also be **perpendicular,** which means they intersect at a right angle. For instance, the lines p and q in Figure 9.20 are perpendicular. Notice that the slopes of the two lines are negative reciprocals of each other. In other words, if two lines are perpendicular, then the product of their slopes is equal to -1.

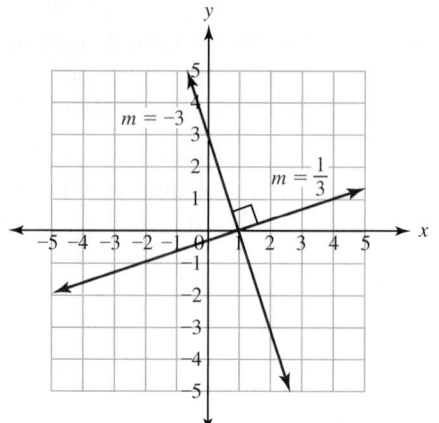

Figure 9.20 Perpendicular lines

Theorem 9.8   Slopes of Perpendicular Lines

Two lines in the coordinate plane are perpendicular if and only if:

1. One is vertical and the other is horizontal.
2. The product of their slopes is -1.

To understand why Theorem 9.8 is true, consider a specific line l that goes through the origin, contains the point $(1, 4)$, and has a slope of $m = \dfrac{4 - 0}{1 - 0} = \dfrac{4}{1} = 4$. If we rotate l by 90°, as shown in Figure 9.21, we get a new line, l', that is perpendicular to l.

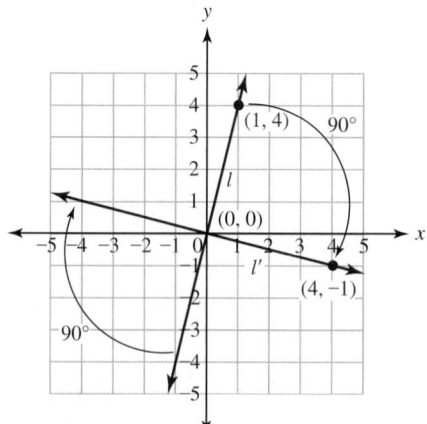

Figure 9.21 Rotating a line 90°

When we perform the rotation, notice that the point $(1, 4)$ is sent to point $(4, -1)$. The rotation switches the x- and y-coordinates, and the x-coordinate becomes negative. Because the point at the origin is fixed, the slope of line l' is $m' = \dfrac{-1 - 0}{4 - 0} = -\dfrac{1}{4}$. Multiplying the slopes, we get $m \cdot m' = 4 \cdot \dfrac{-1}{4} = -1$.

| Example 9.29 | Finding Parallel and Perpendicular Lines |

Determine whether each pair of lines is parallel, perpendicular, or neither.

a. l has slope $\dfrac{1}{3}$.

l' has slope $\dfrac{1}{3}$.

b. $y - 4 = \dfrac{2}{3}(x + 1)$.

$y - 4 = -\dfrac{2}{3}(x + 1)$.

c. l has slope 0.

l' has undefined slope.

Solution

a. Because l and l' have the same slope, they are parallel.

b. Because one line has a slope of $\dfrac{2}{3}$ and the other has a slope of $-\dfrac{2}{3}$, the lines are neither parallel nor perpendicular.

c. Because l is horizontal and l' is vertical, the two lines are perpendicular.

| Example 9.30 | Finding the Equation of a Perpendicular Line |

Find the equation of the line perpendicular to $y = \dfrac{1}{4}x + 1$ and containing $(0, -3)$.

Solution The needed line is perpendicular to the given line, so its slope must be the negative reciprocal of $\dfrac{1}{4}$, or $m = -4$. It also contains the point $(0, -3)$. This is on the y-axis, so it must be the line's y-intercept. Hence, the equation of the line is $y = -4x - 3$.

Check Your Understanding 9.3-B

1. Write the equation of each line in point-slope form.

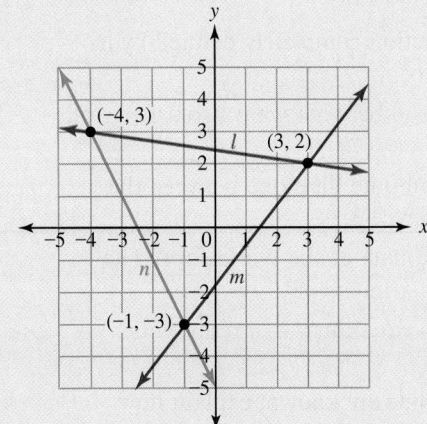

2. Write the equation of each line in slope-intercept form.

a. The line having slope $m = -\dfrac{3}{5}$ and y-intercept $b = 5$.

b. The line having slope $m = \dfrac{4}{3}$ and passing through the point $(1, 6)$.

3. Graph each line.

   a. Line p is horizontal and contains the point $(-2, 3)$.

   b. Line q is defined by the equation $y = \frac{1}{3}x + 9$.

   c. Line r is parallel to the line $y = 3x - 1$ and contains the point $(5, 4)$.

Talk About It If the graph of a linear function shows a constant rate of change equal to zero, what must be true of the function?

Problem Solving

Activity 9.6

What are the coordinates of (x, y) if l is parallel to k and r is parallel to s?

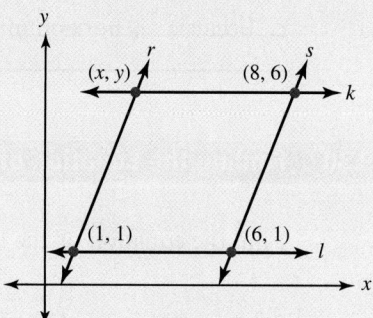

SECTION **9.3** Learning Assessment

■ Understanding the Concepts

1. What does it mean for a function to represent constant change?

2. Why is a linear function completely defined by its slope and y-intercept?

3. Slope is a comparison between what two aspects of a line?

4. Describe how to compute the slope between the points (x_1, y_1) and (x_2, y_2).

5. What is the x-coordinate of the y-intercept? The y-coordinate of the x-intercept?

6. Does a line with an equation $x = a$ represent a function? Explain.

7. If two points on a line are known, explain how the equation $m = \dfrac{y_2 - y_1}{x_2 - x_1}$ can be used to find the equation of the line.

8. Why must two lines with the same slope be parallel?

9. Describe three different ways to graph a linear function.

■ Representing the Mathematics

10. Give an example of a function of constant change that is represented by a

   a. Table.   **b.** Set of ordered pairs.

11. Interpret each slope as the ratio of the rise of the line to its run.

   a. $m = -4$.   **b.** $m = \frac{2}{3}$.   **c.** $m = 0$.   **d.** $m = \frac{7}{4}$.

12. Interpret each slope as the ratio of the rise of the line to its run.

   a. $m = 5$ **b.** $m = \dfrac{-1}{5}$

   c. $m = \dfrac{1}{-4}$ **d.** m is undefined

13. A slope of $\frac{1}{2}$ can be represented as a slope of $\frac{2}{4}, \frac{3}{6},$ or $\frac{5}{10}$. Give three other ways to represent each slope.

   a. $m = 3$.   **b.** $m = -\dfrac{3}{5}$.   **c.** $m = 1$.

14. Determine whether each set of ordered pairs represents a linear function. For those that do not, explain why.

    a. $\{(1, 2), (2, 4), (3, 6), (4, 8)\}$

    b. $\{(1, 1), (2, 4), (3, 9), (4, 16)\}$

    c. $\{(1, 1), (1, 2), (1, 3), (1, 4)\}$

    d. $\left\{\left(1, \frac{1}{2}\right), (2, 1), \left(3, \frac{3}{2}\right), (4, 2)\right\}$

15. Determine whether each table represents a linear function. For those that do not, explain why.

a.		b.		c.		d.	
x	y	x	y	x	y	x	y
1	5	−1	3	1.2	1.4	3	−5
2	11	−4	5	1.3	1.6	6	10
3	18	−7	7	1.4	1.8	9	−15
4	26	−10	9	1.5	2.0	12	20

16. Use the graph of the function to find the value of y when x equals the given number.

    a. 0   b. 4   c. −2   d. 6

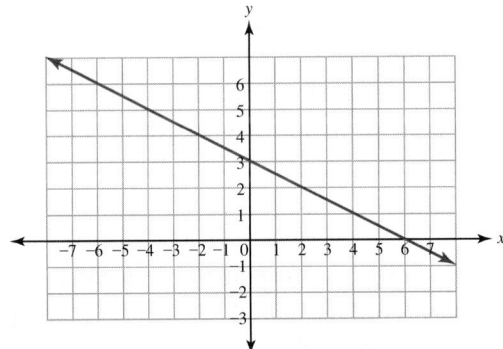

17. For each graph, find the slope of the line, the x-intercept, the y-intercept, and three other points on the line.

    a.

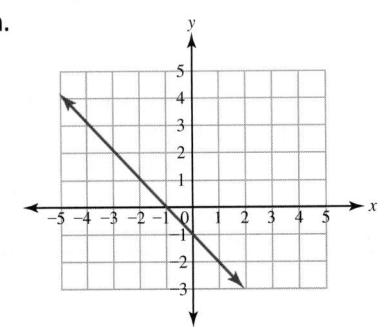

    b.

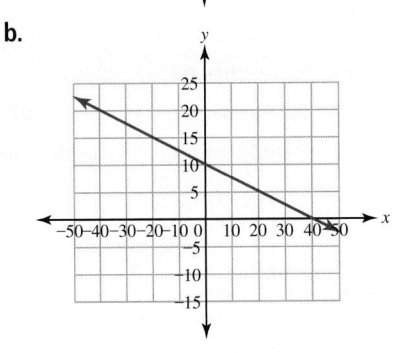

■ Exercises

18. Compute the slope of the line containing each pair of points. State whether the line is increasing, decreasing, horizontal, or vertical.

    a. $(-2, 1)$ and $(8, 1)$

    b. $(-1, -1)$ and $(4, 5)$

    c. $(-1, 0)$ and $(6, -3)$

19. Compute the slope of the line containing each pair of points. State whether the line is increasing, decreasing, horizontal, or vertical.

    a. $(8, 3)$ and $(6, 5)$

    b. $(4, 6)$ and $(4, 10)$

    c. $(3, 0)$ and $(4, 2)$

20. Use a graph to find five points that satisfy the equation $6x - 3y = 12$.

21. Find five points that satisfy the equation $y = 3x - 4$ by evaluating the formula with different values.

22. Use the slope and y-intercept to find five points that satisfy the equation $y = \frac{2}{3}x + 1$.

23. Graph each linear equation.

    a. $y = 2$.   b. $x = -1$.

    c. $x = -\frac{4}{7}$.   d. $y = 2\frac{1}{2}$.

24. Graph each linear equation. Explain your choice of method.

    a. $y = 3x$.   b. $y = \frac{1}{3}x$.

    c. $y = -4x$.   d. $y = -\frac{2}{3}x$.

25. Graph each linear equation. Explain your choice of method.

    a. $y = -x + 8$.   b. $y = 2x - 7$.

    c. $y = \frac{1}{4}x + 6$.   d. $y = -\frac{3}{4}x + \frac{1}{2}$.

26. Find the slope, y-intercept, and x-intercept of each function when they exist.

    a. $y = 6$.   b. $x = 5$.

    c. $y = -\frac{3}{4}x$.   d. $y = 2x + 7$.

27. Find the slope, y-intercept, and x-intercept of each function when they exist.

    a. $2x + 4y = 6$.   b. $(y - 0) = \frac{2}{3}(x + 6)$.

    c. $y = -\frac{3}{7}x + \frac{5}{11}$.   d. $4x - 7y = 9$.

28. Write the point-slope form for the line passing through each pair of points.

    a. $(2, 0)$ and $(3, 1)$ b. $(-4, 9)$ and $(3, 4)$

29. Write the point-slope form for the line passing through each pair of points.

a. $(4.5, 0.3)$ and $(6.5, 7.1)$ b. $\left(\frac{1}{3}, \frac{4}{5}\right)$ and $\left(-\frac{2}{3}, \frac{1}{5}\right)$

30. Write the slope-intercept form of the line with slope m and y-intercept b.

a. $m = 5, b = -2$.   b. $m = -4, b = 0$.

c. $m = 5.76, b = 2.34$.

31. Write the slope-intercept form of the line with slope m and containing the given point.

a. $m = 4, (3, 0)$.   b. $m = -2, (-5, 7)$.

c. $m = -\frac{3}{4}, (x, y) = \left(-\frac{2}{5}, \frac{3}{5}\right)$.

32. Write an equation for the line that passes through $(0, -3)$ and has a slope of:

a. 4.   b. -3.   c. $-\frac{2}{3}$.   d. 0.

33. Write an equation for the line that passes through $(2, 4)$ and has a slope of:

a. -2.   b. 5.   c. $\frac{3}{4}$.   d. undefined.

34. In each case, the slopes of two lines are given. Determine whether the lines are parallel, perpendicular, or neither?

a. $m_1 = \frac{1}{2}$ and $m_2 = -\frac{1}{2}$ b. $m_1 = 3$ and $m_2 = 3$

c. $m_1 = -\frac{1}{3}$ and $m_2 = 3$ d. $m_1 = \frac{3}{4}$ and $m_2 = \frac{9}{12}$

35. Determine whether each pair of lines is parallel, perpendicular, or neither.

a. $y = 2x - 7$ b. $y - 4 = -\frac{3}{2}(x + 1)$
   $y = 2x + 3$
                      $y - 2 = \frac{2}{3}(x - 2)$

c. $y = -2$ d. $x = 8$
   $x = -2$ $x = 5$

e. $y = -3x + 7$ f. $y = x + 3$
   $y = -\frac{1}{3}(x - 1)$ $y = -x + 3$

36. Find an equation for each line.

a. The line parallel to the x-axis and containing the point $(-3, 3)$.

b. The line parallel to the line $y = -3x + 4$ and containing the point $(0, 5)$.

c. The line perpendicular to the line $y = 6x + 2$ and containing the point $(1, 1)$.

37. Find an equation for each line.

a. The vertical line containing the point $(4, 1)$.

b. The line parallel to the line $y = -\frac{2}{7}x - 4$ and containing the point $(7, 0)$.

c. The line perpendicular to the line $3x - 5y = 4$ and containing the point $(-2, -5)$.

38. What is the value of x, if the point $(x, 4)$ satisfies the equation $2x - 3y = 6$?

39. What is the value of y, if the point $(3, y)$ satisfies the equation $3x - 4y = 10$?

■ Problems and Applications

40. For each set of three points, determine whether the points lie on the same line.

a. $(-1, 2), (1, 2),$ and $(2, 4)$

b. $(0.1, -3.2), (0.6, -1.6),$ and $(1.1, 0)$

41. Consider the line containing the points $(-2, 8)$ and $(4, -4)$. What are the coordinates of three other points that lie on the same line?

42. Two corners on the same side of a square have coordinates $(4, 3)$ and $(8, 4)$. If $(4, 3)$ and $(8, 4)$ are the lower two vertices, find the coordinates of the other two vertices.

43. Find possible coordinates for the point (x, y) that make each of the following true.

a. The slope of the line containing $(3, 2)$ and (x, y) is 1.

b. The slope of the line containing $(3, 2)$ and (x, y) is 0.

c. The slope of the line containing $(3, 2)$ and (x, y) is $\frac{1}{4}$.

44. Find the missing value k in each situation.

a. A line passes through the points $(1, 2)$ and $(3, k)$ and has a slope of 1.

b. A line passes through the points $(1, k)$ and $(3, 4)$ and has a slope of -2.

c. A line passes through the points $(k, -k)$ and $(4, 1)$ and has a slope of $\frac{3}{4}$.

45. Each of the following pairs of lines intersects at a point. Find the coordinates of the point of intersection.

a. $y = 2x$.   b. $y = 3x + 5$.
   $y = \frac{1}{2}x + \frac{3}{2}$. $y = x + 4$.

c. $y = \frac{1}{2}x - 2$.
   $3x + 2y = 0$.

46. The pitch of a roof represents its steepness.

a. What is the pitch of the roof in the diagram?

b. How is the idea of pitch related to the notion of slope?

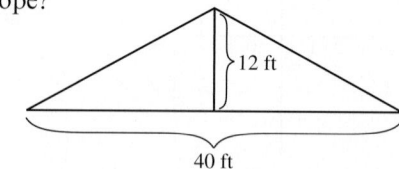

12 ft

40 ft

47. The equation $C = 4F - 160$ gives the relationship between the temperature in degrees Fahrenheit and the number of chirps made by a cricket every minute.

   a. What is the slope and y-intercept of this equation?

   b. How many chirps will a cricket make at 50°? At 75°? At 90°?

   c. If a cricket is making 100 chirps/min, what temperature is it?

48. The steepness of a roadway is often measured using percent grade, which is the slope of the road expressed as a percentage.

   a. What is the percent grade of a road with each of the following slopes?

   i. $\dfrac{3}{100}$ **ii.** $\dfrac{1}{10}$ **iii.** $\dfrac{3}{50}$ **iv.** $\dfrac{2}{25}$

   b. What is the rise of a road in yards over the course of 500 yd if the road has the following percent grades?

   i. 5%   **ii.** 9%   **iii.** 6%   **iv.** 4.5%

■ Mathematical Reasoning and Proof

49. a. Which represents a steeper line, a slope of 3 or a slope of $\dfrac{1}{3}$? Explain.

   b. Which represents a steeper line, a slope of 3 or a slope of -3? Explain.

50. If a line passes through the first and third quadrant only, what must the y-intercept of the line be? What must the sign of the slope be?

51. When computing a slope, what happens if the x- and y-coordinates of the points are not subtracted in the same order? Perform several numerical trials, and form a conjecture based on the results of the trials.

52. If two lines are drawn on the same coordinate system, they can relate to one another in three ways. The lines can have no points in common, they can intersect at one point, or they can intersect at an infinite number of points.

   a. Draw three graphs showing the three different possibilities.

   b. Why are these the only ways two lines can interact?

   c. If the slope and y-intercept are known for both lines, how can you know whether the lines are parallel, intersect at one point, or intersect at infinitely many points without looking at the graph of a function? Justify your answer.

53. Line p passes through the points $(-2, 1)$ and $(3, 4)$. Line q passes through the points $(-7, -4)$ and $(8, 7)$. Are the two lines parallel, do they intersect in only one point, or do they intersect at an infinite number of points? Justify your answer.

54. Constant change is not the only type of change represented by a function. Rates of change can also increase and decrease.

   a. Sketch the graph of an increasing function with an increasing rate of change.

   b. Sketch the graph of an increasing function with a decreasing rate of change.

   c. Sketch the graph of a decreasing function with an increasing rate of change.

   d. Sketch the graph of a decreasing function with a decreasing rate of change.

55. Repeat the previous exercise. This time give a table of values satisfying the necessary conditions rather than sketching the graph of a function.

56. a. Can the rate of change of a function decrease to zero? If so, draw a sketch of a graph that satisfies this condition.

   b. Can the rate of change of a function increase to zero? If so, draw a sketch of a graph that satisfies this condition.

   c. If either or both cases are possible, what must happen to the graph when the rate of change reaches zero?

57. Write a proof of Theorem 9.8(2). [*Hint:* Let line l go through the origin and contain the point (x_1, y_1). Consider what happens to the point (x_1, y_1) when line l is rotated 90°.]

■ Mathematical Communication

58. In many situations, the slope of a line is represented as $m = \dfrac{\Delta y}{\Delta x}$, where the Greek letter delta, Δ, represents the words "The change in . . . ."

   a. Translate the meaning of slope using this language.

   b. Why does this language make sense when discussing slope?

59. Is there a difference in the meaning of the two phrases "constant change" and "changing constantly" If so, what is it?

60. Use a specific example to explain to a peer how it is possible to find the slope and y-intercept of a line if the equation of the line is given in point-slope form. What difficulties did you encounter in your explanation?

61. Consider the line passing through the point $(3, 1)$ having a slope of -2. Explain to a peer how to find two other points on the line in two different ways. Discuss which method made more sense and the implications for teaching the slope of the line.

62. Janie was able to find the slope of a line from two points. But in trying to find the slope-intercept form of the line, she did not know which point to substitute into the equation. How do you, as her teacher, address her concern?

63. Marcel claims that the lines $y = 3x + 4$ and $y = -3x - 2$ must be parallel because their slopes are negatives of each other. However, when he graphed the equations, the lines intersected. How would you explain to him what went wrong?

■ Building Connections

64. **a.** Explain how arithmetic sequences are closely connected to linear functions.

    b. For any function representing an arithmetic sequence, what will the domain of the function always be?

65. Linear functions are important because they can be used to estimate situations that appear to be approximately linear.

a. What do you think the term "approximately linear" means?

b. What criteria might you use to determine when a situation is approximately linear and when it is not?

c. Do an Internet search to find two different instances of linear equations being used to estimate a situation that is approximately linear.

66. Patterns can change at different rates. Other than constant change, what other rates of change might there be?

67. Use a table of values to graph the function $y = |x|$.

a. How is the graph of this function similar to that of $y = x$? How is it different?

b. Review the definition of absolute value given in Chapter 6. How are linear functions used in this definition?

68. Analyze how several textbooks from the elementary grades introduce linear relationships. Discuss in a small group how the ideas were introduced and which method your group thinks is the best.

SECTION 9.4 Solving Equations and Inequalities

The last three sections focused on functional thinking. We now turn to another aspect of algebraic thinking, called **general arithmetic,** which includes two areas. One, involves identifying and generalizing numerical patterns with algebraic expressions. However, instead of analyzing patterns of change, this area of general arithmetic focuses on the patterns embedded in the numerical operations. We have used this aspect of algebraic thinking to identify and generalize the properties of addition, subtraction, multiplication, and division. These are referred to as the algebraic properties because we use them extensively to make the symbolic manipulations commonly associated with algebra.

Theorem 9.9 Algebraic Properties

If a, b, and c are real numbers, then the following properties hold true.

Closure property:	$a + b$ and $a \cdot b$ are unique real numbers.
Commutative property:	$a + b = b + a$ and $a \cdot b = b \cdot a$
Associative property:	$a + (b + c) = (a + b) + c$ and $a \cdot (b \cdot c) = (a \cdot b) \cdot c$
Identity property:	$a + 0 = a = 0 + a$ and $a \cdot 1 = a = 1 \cdot a$
Inverse property:	$a + (-a) = 0 = (-a) + a$ and $a \cdot \dfrac{1}{a} = 1 = \dfrac{1}{a} \cdot a$
Distributive property of multiplication over addition:	$a \cdot (b + c) = (a \cdot b) + (a \cdot c)$
Zero multiplication property:	$a \cdot 0 = 0 = 0 \cdot a$

The second part of general arithmetic involves finding unknown quantities by solving equations and inequalities. In general, we use a variable to represent the unknown quantities and then solve the equation or inequality by finding every value that makes it true. The set of all such values is the **solution set.** A primary goal in high school algebra is to develop the techniques needed to solve a wide variety of equations and inequalities. However, in the elementary and middle grades, our goal is to introduce students to solving equations, giving them the chance to develop the thinking skills associated with the task. For this reason, students in these grades work mostly with linear equations and inequalities because they are relatively easy to solve and the methods are directly related to the algebraic properties.

Solving Linear Equations

A **linear equation of one variable** is an equation in which each term is either a real-number **constant** or the product of a number and a variable raised to the first power. The number multiplying the variable is a **coefficient,** and it too can take on any real-number value. Examples of linear equations are:

$$3x = 1 \quad 4y - 5 = 2 \quad -3z - 7 = 2(z + 5) \quad 1.1w - 0.4 = -0.8w + 3$$

When solving a linear equation, the goal is to find the value of the variable that makes the equation true. In the elementary classroom, children often use mental arithmetic and basic addition, subtraction, multiplication, and division facts when they first learn to solve simple **one-step equations.**

Example 9.31 | **Solving One-Step Equations with Mental Math**

Use mental math and basic facts from arithmetic to solve each equation.

a. $17 - x = 8$. **b.** $3x = 21$. **c.** $\dfrac{x}{5} = 6$.

Solution In each case, we connect the equation to a basic fact from arithmetic. We then use what we know about that fact to find the value of the variable.

a. Because $17 - 9 = 8$, then $x = 9$.

b. Because $3 \cdot 7 = 21$, then $x = 7$.

c. Because $\dfrac{30}{5} = 6$, then $x = 30$.

Another method for solving equations relies on the fact that an equation is much like a balance. Both sides take on the same value, or weight. As with physical balances, we can add or remove numbers and variables to either side of the equation. However, the balance is somewhat precarious, so if we do something to one side but not the other, the equation becomes unbalanced.

Explorations Manual 9.10

For example, consider the simple equation $x = 10$, shown with a balance in Figure 9.22. If we add 4 to the right side of the equation but not to the left, then the right side becomes heavier, so to speak, and our equation becomes unbalanced. Consequently, if the original equation is to hold, we need to maintain the balance by adding 4 to both sides.

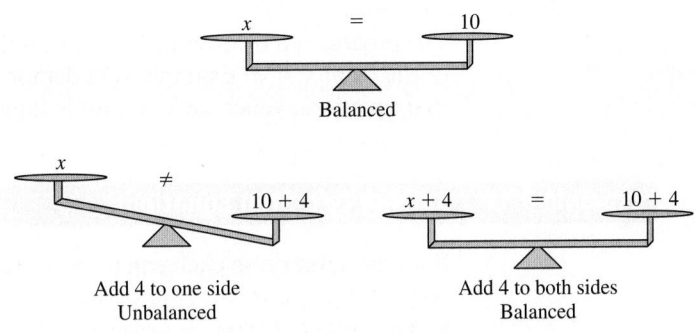

Figure 9.22 Modeling an equation with a balance

The balance analogy also works with subtraction, multiplication, and division. If any of these operations are performed on one side of the equation but not on the other, our equation becomes unbalanced and no longer holds. Consequently, if we add, subtract, multiply, or divide a number on one side of the equation, we must perform the same operation on the other side to maintain the equality.

Theorem 9.10   Properties of Equality

If a, b, and c are real numbers with $a = b$, then the following hold true.

1. $a + c = b + c$.

2. $a - c = b - c$.

3. $a \cdot c = b \cdot c$.

4. $\dfrac{a}{c} = \dfrac{b}{c}$ when $c \neq 0$.

Theorem 9.10 can be used to solve equations by **isolating the variable.** To do so, we use inverse operations to transform the equation into the form $x = a$, which directly reveals the value of the unknown. Specifically, constant terms are attached to the variable by addition or subtraction, so we can remove them by using the appropriate inverse operation to add or subtract the same number to both sides of the equation. For instance, consider the equation $x - 5 = 9$. The 5 is "attached" to the variable by subtraction. Because addition is the inverse operation of subtraction, we can undo the "subtract 5" by adding 5 to both sides. After simplifying, the variable is isolated, and we have the solution.

$$x - 5 = 9$$

$$(x - 5) + 5 = 9 + 5 \quad \text{Add 5 to both sides.}$$

$$x = 14 \quad \text{Simplify.}$$

Similarly, coefficients are attached to the variable by multiplication or division. To change the coefficient to 1, we use the appropriate inverse operation to multiply or divide the same number to both sides of the equation. For example, in the equation $\dfrac{x}{6} = 4$, the 6 is attached to the variable by division. Because multiplication is the inverse operation of division, we can undo the "divide by 6" by multiplying by 6. Again, this isolates the variable and gives us the solution:

$$\frac{x}{6} = 4$$

$$\frac{x}{6} \cdot 6 = 4 \cdot 6 \quad \text{Multiply both sides by 6.}$$

$$x = 24 \quad \text{Simplify.}$$

The process can be extended to any number of steps necessary to reduce the equation to the form $x = a$. Example 9.32 demonstrates how to use this method to solve **two-step equations,** which are commonly taught in the early middle grades.

Example 9.32   Solving Two-Step Equations

Find the solution to each equation by isolating the variable.

a. $5x - 9 = 6$.   **b.** $\dfrac{1}{3} - 2x = 3$.

Solution

a. In the equation $5x - 9 = 6$, two numbers are attached to the variable: a 9 by subtraction and a 5 by multiplication. Because addition is the inverse operation of subtraction, we undo the "$- 9$" by adding 9 to both sides of the equation, or

$$5x - 9 = 6$$

$$5x - 9 + 9 = 6 + 9 \qquad \text{Add 9 to both sides.}$$

$$5x = 15 \qquad \text{Simplify.}$$

Next, undo the multiplication by 5 by dividing both sides of the equation by 5. This isolates the variable and gives the solution, or

$$\frac{5x}{5} = \frac{15}{5} \qquad \text{Divide both sides by 5.}$$

$$x = 3 \qquad \text{Simplify.}$$

b. Begin by undoing the "add $\frac{1}{3}$" by subtracting $\frac{1}{3}$ from both sides. Then isolate the variable by dividing both sides by (-2).

$$\frac{1}{3} - \frac{1}{3} - 2x = 3 - \frac{1}{3} \qquad \text{Subtract } \frac{1}{3} \text{ from both sides.}$$

$$-2x = \frac{8}{3} \qquad \text{Simplify.}$$

$$\frac{(-2x)}{-2} = \frac{\frac{8}{3}}{-2} \qquad \text{Divide both sides by } (-2).$$

$$x = \frac{8}{3} \cdot \left(-\frac{1}{2}\right) \qquad \text{Rational number division.}$$

$$x = -\frac{4}{3} \qquad \text{Simplify.}$$

In the last example, we worked with sums and differences before working with products and quotients. This is not necessary. We can work with any operation first and still arrive at the same solution. However, first removing products and quotients often leads to working with fractions, which is an unnecessary complication that we can often avoid by first removing sums and differences. For instance, if we solve $5x - 9 = 6$ by first removing the product, then

$$5x - 9 = 6$$

$$\frac{5x}{5} - \frac{9}{5} = \frac{6}{5} \qquad \text{Divide both sides by 5.}$$

$$x - \frac{9}{5} = \frac{6}{5} \qquad \text{Simplify.}$$

$$x - \frac{9}{5} + \frac{9}{5} = \frac{6}{5} + \frac{9}{5} \qquad \text{Add } \frac{9}{5} \text{ to both sides.}$$

$$x = \frac{15}{5} = 3 \qquad \text{Simplify.}$$

We can also isolate the variable to solve more complex linear equations. However, we may have to use algebraic properties to manipulate the equation before using inverse operations. We may also have to add or subtract variable terms. Any terms with the same variable raised to the same power are called **like terms.** For example, $3x$ and $8x$

504 CHAPTER 9 Algebraic Thinking

are like terms, whereas $3x$ and $5y$ are not. To add and subtract like terms, we use the distributive property to add and subtract their coefficients. For instance,

$$2x + 6x = (2 + 6)x = 8x$$

and

$$-3x - 8x = (-3 - 8)x = -11x$$

Example 9.33 Solving an Equation with Like Terms

Use algebraic properties to solve $2(3x - 2) - 5x = 3 - 3x$.

Solution We first use the distributive property and then add like terms. We then isolate the variable to solve the equation, or

$2(3x - 2) - 5x = 3 - 3x$	
$6x - 4 - 5x = 3 - 3x$	Distributive property.
$x - 4 = 3 - 3x$	Combine like terms.
$x - 4 + 3x = 3 - 3x + 3x$	Add $3x$ to both sides.
$4x - 4 = 3$	Combine like terms.
$4x - 4 + 4 = 3 + 4$	Add 4 to both sides.
$4x = 7$	Simplify.
$\dfrac{4x}{4} = \dfrac{7}{4}$	Divide both sides by 4.
$x = \dfrac{7}{4}$	Simplify.

Check Your Understanding 9.4-A

1. Solve each equation mentally.

  a. $2x = 8$.   **b.** $4 + x = 13$.   **c.** $7 - x = -2$.   **d.** $\dfrac{x}{9} = 5$.

2. Solve each equation by isolating the variable.

  a. $2x - 4 = 16$.   **b.** $3x - 6 = 13$.   **c.** $-3x + 5 = x + 7$.

3. Solve each equation by isolating the variable.

  a. $4x - 1 = -2x + 6$.   **b.** $2x + 3(x - 1) = 13$.

Talk About It Does every linear equation have a solution? If not, can you write a linear equation that does not? What must be true about such an equation?

Reasoning

Activity 9.7

The following argument appears to prove that $2 = 1$. Can you spot the error?

$a = b$	Beginning assumption.
$aa = ab$	Multiply both sides by a.
$aa - bb = ab - bb$	Subtract bb from both sides.
$(a + b)(a - b) = b(a - b)$	Factor both sides.
$a + b = b$	Divide both sides by $(a - b)$.
$a + a = a$	Substitute a in for b.
$2a = a$	Combine like terms.
$2 = 1$	Divide both sides by a.

Solving Linear Inequalities

We can also isolate the variable to solve **linear inequalities,** which are linear statements that use $<, \leq, >, \geq$, or $\neq$, rather than $=$. However, in doing so, two differences must be kept in mind. First, multiplying or dividing an inequality by a negative number affects the truthfulness of the inequality. To understand why, consider the inequality $3 < 5$. This statement indicates a relative position between the two numbers. Specifically, 3 is to the left of 5. When multiplying each number by (-1), we exchange the numbers for their opposites. This reverses the relative position of the two numbers: (-5) is to the left of (-3) (Figure 9.23). So multiplying by (-1) reverses the relative size of the two numbers, and we must reverse the inequality to maintain truthfulness. The same holds true for division and other inequalities using $<, \leq, >$, or $\geq$. Consequently, whenever both sides of an inequality are multiplied or divided by a negative number, reverse the inequality to maintain the truthfulness of the statement.

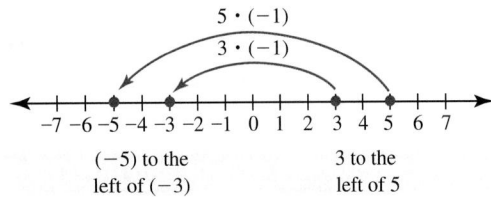

(-5) to the
left of (-3)
3 to the
left of 5

Figure 9.23 The effect of multiplying by (–1)

Beyond this, we can add, subtract, multiply, or divide both sides of an inequality by any number as long as we do not divide by zero. The less than, $<$, in the following theorem can be replaced with other inequalities, including $\leq, >$, and $\geq$. If we replace it with $\neq$, we need not reverse the inequality.

Theorem 9.11   Properties of Inequalities

If a, b, and c are real numbers with $a < b$, then the following hold true:
1. $a + c < b + c$.
2. $a - c < b - c$.
3. $a \cdot c < b \cdot c$ if $c > 0$ and $a \cdot c > b \cdot c$ if $c < 0$.
4. $\dfrac{a}{c} < \dfrac{b}{c}$ when $c > 0$ and $\dfrac{a}{c} > \dfrac{b}{c}$ if $c < 0$.

A second difference between equations and inequalities is in the solution sets. As with equations, the solution set of an inequality is the set of all values that makes it true. However, with inequalities, the solution sets are infinite in size. For instance, consider the inequality $x > 4$. It has an infinite solution set because any real number larger than 4, such as 5, 6, 7.9, or $11\frac{1}{2}$, makes the statement true. Table 9.5 shows three ways to represent infinite solution sets: set-builder notation, graphs, and interval notation.

With graphs and interval notation, any number at which the solution set stops is called an **endpoint.** If the endpoint is included in the solution set, it receives a closed dot, ●, on a graph and a square bracket, [or], in interval notation. If the endpoint is not included, it receives an open dot, ○, on a graph and a parenthesis, (or), in interval notation. The symbols $-\infty$ and ∞ represent negative and positive infinity and indicate that the numbers continue indefinitely in the negative or positive direction. Infinity always receives a parenthesis in interval notation. In any of the representations, the convention is to write the numbers in increasing order from left to right. Particular care should be taken when writing certain inequalities in set-builder notation. For instance, it is incorrect to write $3 > x > 7$ for the statement $x < 3$ or $x > 7$ because $3 > x > 7$ implies that 3 is actually greater than 7.

Table 9.5 Different ways to represent infinite solution sets

Set-Builder Notation	Graph on Number Line	Interval Notation
$\{x \mid x \in \mathbb{R}\}$		$(-\infty, \infty)$
$\{x \mid x < a\}$	a	$(-\infty, a)$
$\{x \mid x \le a\}$	a	$(-\infty, a]$
$\{x \mid x > a\}$	a	(a, ∞)
$\{x \mid x \ge a\}$	a	$[a, \infty)$
$\{x \mid a < x \le b\}$	a b	$(a, b]$
$\{x \mid x < a \text{ or } x > b\}$	a b	$(-\infty, a) \cup (b, \infty)$
$\{x \mid x \ne a\}$	a	$(-\infty, a) \cup (a, \infty)$

Example 9.34   Using Graphs and Interval Notation

Representation   Give a verbal interpretation of each set, and then represent it with a graph and in interval notation.

a. $\{x \mid x \le 4\}$ **b.** $\{x \mid -2 < x \le 5\}$ **c.** $\{x \mid x < 1 \text{ or } x \ge 5\}$

Solution

a. $\{x \mid x \le 4\}$ is the set of all real numbers less than or equal to 4. Its graph and interval notation are

$(-\infty, 4]$

b. $\{x \mid -2 < x \le 5\}$ is the set of all real numbers greater than -2 but less than or equal to 5. Its graph and interval notation are

$(-2, 5]$

c. $\{x \mid x < 1 \text{ or } x \ge 5\}$ is the set of all real numbers less than 1 or greater than or equal to 5. Its graph and interval notation are

$(-\infty, 1) \cup [5, \infty)$

Example 9.35   Solving Inequalities

Solve each inequality and represent the solution set with a graph and in interval notation.

a. $-3x - 4 \le 5$ **b.** $4x - 7 \ne -2x - 8$

Solution We solve each inequality by using inverse operations to isolate the variable. When necessary, we reverse the inequality.

a.
$$-3x - 4 \le 5$$

$$-3x \le 9 \qquad \text{Add 4 to both sides.}$$

$$x \ge -3 \qquad \text{Divide by } -3 \text{ and reverse inequality.}$$

The graph and interval notation for the solution set are

$[-3, \infty)$

b.
$$4x - 7 \neq -2x - 8.$$

$$6x - 7 \neq -8 \qquad \text{Add } 2x \text{ to both sides.}$$

$$6x \neq -1 \qquad \text{Add 7 to both sides.}$$

$$x \neq -\frac{1}{6} \qquad \text{Divide both sides by 6.}$$

The graph and interval notation for the solution

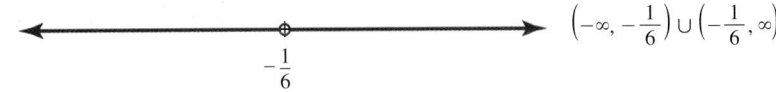

$\left(-\infty, -\frac{1}{6}\right) \cup \left(-\frac{1}{6}, \infty\right)$

Solving Nonlinear Equations

We can use mental arithmetic and isolating the variable to solve simple nonlinear equations by using other operations and algebraic facts that we have learned along the way. As the equations become more complicated, it is important that we check our solutions by evaluating them in the original equation. We begin with quadratic equations.

Example 9.36   Solving Quadratic Equations

A **quadratic equation of one variable** is an equation in which one or more terms have a squared variable but no higher power. Solve each quadratic equation.

a. $x^2 = 36.$ **b.** $x^2 + 4 = 12.$ **c.** $x^2 - x = 0.$

Solution

a. We can use mental arithmetic to solve the equation $x^2 = 36$ by asking, "What number or numbers, when squared, equal 36?" Because $(-6)^2 = 36$ and $6^2 = 36$, the solution set is $\{-6, 6\}$.

b. We can also isolate the variable. To do so, first subtract 4 from both sides

$$x^2 + 4 = 12$$
$$x^2 = 8 \qquad \text{Subtract 4 from both sides.}$$

We can then undo the square by taking the square root of both sides, or

$$x^2 = 8$$
$$\sqrt{x^2} = \sqrt{8} \qquad \text{Take the square root of both sides.}$$
$$x = \pm\sqrt{8}$$

The notation $\pm\sqrt{8}$ indicates that $-\sqrt{8}$ and $\sqrt{8}$ are both solutions to the equation.

c. In the equation $x^2 - x = 0$, the variable is included in two terms, so we cannot use inverse operations to isolate the variable. However, if we use the distributive property to rewrite $x^2 - x$ as $x(x - 1)$, we can use the **zero product principle** to solve the equation. This principle states that, if $a \cdot b = 0$, then either $a = 0$ or $b = 0$. As a result, if $x(x - 1) = 0$, then either $x = 0$ or $x - 1 = 0$. Consequently, the solution set is $\{0, 1\}$.

In the last example, we undid a square by taking a square root. The reverse is also possible: We can use a square to undo a square root.

Example 9.37 | **Solving an Equation with a Square Root**

Solve $\sqrt{x + 5} = 9$.

Solution We square both sides of the equation and then subtract 5.

$$\sqrt{x + 5} = 9$$

$$(\sqrt{x + 5})^2 = 9^2 \qquad \text{Square both sides.}$$

$$x + 5 = 81 \qquad \text{Subtract 5 from both sides.}$$

$$x = 76$$

Checking our answer, we have $\sqrt{76 + 5} = \sqrt{81} = 9$.

In addition to solving linear, quadratic, and radical equations, the method of isolating the variable can be used to solve **rational equations,** in which the variable is in the denominator of a fraction.

Example 9.38 | **Solving Rational Equations**

Solve each rational equation.

a. $\dfrac{1}{x} - 6 = 10$. **b.** $\dfrac{3}{x - 4} = \dfrac{2}{x + 1}$.

Solution

a. Add 6 to both sides of the equation to isolate the rational part, or $\dfrac{1}{x} = 16$. Then take the reciprocal of each side to solve the equation, or $x = \dfrac{1}{16}$.

b. We first use the cross product to remove the variable from the denominators of the fractions. This results in the equation $3(x + 1) = 2(x - 4)$, which we can solve as a linear equation, or

$$3(x + 1) = 2(x - 4)$$

$$3x + 3 = 2x - 8 \qquad \text{Distributive property.}$$

$$x + 3 = -8 \qquad \text{Subtract } 2x \text{ from both sides.}$$

$$x = -11 \qquad \text{Subtract 3 from both sides.}$$

To check the answer, evaluate each side of the equation at $x = -11$. On the left, we have $\dfrac{3}{-11 - 4} = \dfrac{3}{-15} = -\dfrac{1}{5}$. On the right, we have $\dfrac{2}{-11 + 1} = \dfrac{2}{-10} = -\dfrac{1}{5}$. Because both sides are equal, $x = -11$ must be the solution.

Check Your Understanding 9.4-B

1. Represent each set using set-builder notation, a graph, and interval notation.
  a. The set of all real numbers less than 4.
  b. The set of all real numbers not equal to -3.
  c. The set of all real numbers greater than or equal to 1.
  d. The set of all real numbers less than -2 or greater than or equal to 3.

2. Solve each inequality, and represent the solution with a graph and in interval notation.
  a. $3x - 10 \neq 8$. **b.** $-2x - 9 > 13$. **c.** $-3x + 5 \leq 5x - 7$.

3. Solve each equation.

  a. $x^2 + 1 = 26$. **b.** $\sqrt{x - 3} = 7$. **c.** $\dfrac{3}{x} + 4 = 5$.

Talk About It What are the advantages and disadvantages of representing infinite solution sets with interval notation?

Reasoning

Activity 9.8
Suppose that a $\triangle$ and a $\square$ equal a $\diamond$, a $\triangle$ equals a $\square$ and a $\bigcirc$, and 2 $\diamond$ equal 3 $\bigcirc$. How many $\square$ equal a $\triangle$? What connection does this problem have to algebraic thinking and solving equations?

SECTION 9.4 Learning Assessment

■ Understanding the Concepts

1. What two areas of algebraic thinking are encompassed by general arithmetic?

2. What does it mean to solve an equation?

3. Why is a balance a good analogy for an equation?

4. What is the difference between a constant and a coefficient?

5. What are like terms?

6. Describe two methods for solving equations.

7. What are the defining characteristics of each kind of equation?
  a. Linear   **b.** Quadratic
  c. Square root   **d.** Rational

8. a. What is the difference between an equation and an inequality?
  b. What extra considerations must we make when solving linear inequalities?

9. What does the symbol ∞ represent in interval notation? Is it a real number?

■ Representing the Mathematics

10. Match each interval to its corresponding graph.
  a. $(-\infty, 3]$ **i.**
  b. $[-2, 3)$ **ii.** (open circle at 3)
  c. $[3, \infty)$ **iii.** (open circle at -2, closed at 3)
  d. $(-\infty, \infty)$ **iv.** (closed at 3)
  e. $(3, \infty)$ **v.** (closed at 3)
  f. $(-2, 3]$ **vi.** (closed at -2, open at 3)

11. Represent each set using interval notation and a graph. Assume x is a real number.
  a. $\{x \mid x \leq 5\}$ **b.** $\{x \mid 2 < x \leq 5\}$
  c. $\{x \mid x \neq -1\}$ **d.** $\{x \mid x \in \mathbb{R}\}$

12. Represent each interval using set-builder notation and a graph.
  a. $(-4, 7)$ **b.** $[-6, -2]$ **c.** $[2, \infty)$
  d. $(-\infty, -1)$ **e.** $(-\infty, 3] \cup (7, \infty)$

13. Represent each graph using set-builder notation and interval notation.

a.  ![number line with closed point at 6]
6

b. 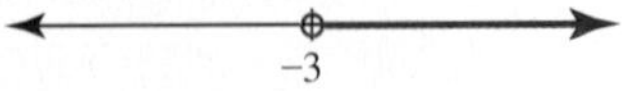 ![number line with open point at -3]
-3

c.  ![number line with closed points at -1 and 4]
-1   4

d. 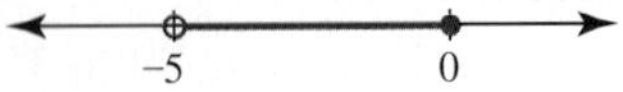 ![number line with open point at -5 and closed point at 0]
-5   0

e.  ![number line with closed point at 2 and open point at 7]
2   7

f. 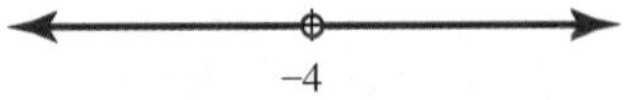 ![number line with open point at -4]
-4

14. What is wrong with each of the following intervals?

a. $[-\infty, 3)$ b. $[-3, -4]$ c. $(-3, -\infty)$

Algebra tiles use square and rectangular regions of various sizes and colors to represent algebraic expressions. The unit-tile has dimensions of 1 unit by 1 unit, and the x-tile has dimensions 1 unit by an unknown x-unit. Yellow tiles represent positive values, and red tiles represent negative values. The tiles can be combined with the balance analogy to represent equations. Use algebra tiles to answer questions 15–19.

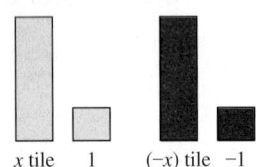

x tile   1   (−x) tile   −1

15. Write the algebraic expression or statement represented in each diagram.

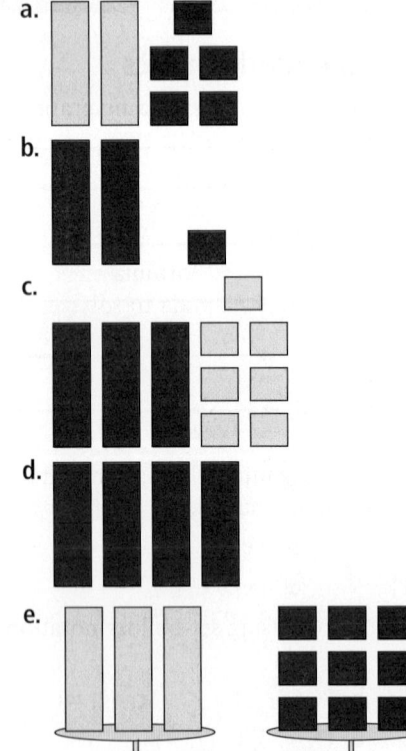

a.

b.

c.

d.

e.

f.
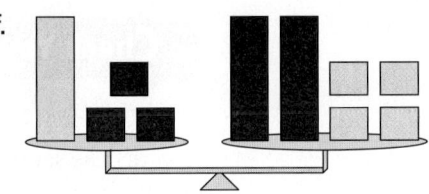

16. Use algebra tiles to represent each expression.

a. 4   b. 3x   c. 3x + 1   d. −2x − 6

17. Use algebra tiles and an algebra balance to represent each expression.

a. x + 4 = 3.   b. −4x + 2 = −3.
c. x + 4 = 2x − 1.   d. −3x − 2 = −2x + 4.

18. a. What is the combined value of a positive x-tile and a negative x-tile?

b. What is the combined value of 3 positive x-tiles and 2 negative x-tiles?

c. Find three combinations of algebra tiles, each having a net value of 0.

d. Find three combinations of algebra tiles, each having a net value of x + 1.

19. How might you use a balance to represent an inequality? Do so, by drawing a picture of each inequality.

a. x + 5 < 14.   b. 7x > 21.   c. −2x + 8 > 13.

■ Exercises

20. Find the numbers in the set {−3, −2, −1, 0, 1, 2, 3} that are solutions to each statement.

a. 2x = −4.   b. 3x > 2.
c. 2x + 4 = 0.   d. −2x + 1 ≤ 3.

21. Solve each equation mentally.

a. 5x = 25.   b. y + 7 = 16.

c. 20 − w = 10.   d. $\frac{x}{12} = 3$.

e. $\left(2\frac{1}{2}\right)y = 10$.   f. x − 8 = −5.

22. Solve each equation mentally.

a. $x^2 = 121$.   b. $x^2 − 1 = 48$.   c. $\sqrt{x} = 10$.

d. $\sqrt{x} + 1 = 2$.   e. $\frac{1}{x} = \frac{1}{9}$.   f. $\frac{1}{x} = 13$.

23. Solve each equation by isolating the variable. Be sure to check your answers.

a. 3x − 2 = 4.   b. 5y + 6 = 0.
c. 200 = 34 − y.   d. 4t + 15 = 21.
e. 3(b − 4) = −36.   f. −2(4 − x) = 12.

24. Solve each equation by isolating the variable. Be sure to check your answers.

a. 2(2x + 1) = 3x + 15.   b. 2(x − 5) − (3x + 1) = 0.

25. Solve each equation.

    a. $\dfrac{x}{2} + \dfrac{x}{3} = 10$. **b.** $\dfrac{x}{3} + \dfrac{2x}{5} = \dfrac{11}{15}$.

    c. $\dfrac{5}{2}t - 12 = \dfrac{t}{3} + 1$. **d.** $1.6x = 4.032$.

    e. $4.35y + 1.45 = 6.79y$. **f.** $0.45 = 0.25(0.75 - 3x)$.

26. Solve each inequality. Express the solution set using a graph and interval notation.

    a. $-9t + 6 \geq 16$. **b.** $-2y - 4 < 8$.

    c. $-\dfrac{3}{5} < -\dfrac{4}{5}x$. **d.** $4 - 3y < 10$.

    e. $6z + 3 < 4z - 1$. **f.** $7x + 4 \neq 2x - 6$.

27. Solve each inequality. Express the solution set using a graph and interval notation.

    a. $3(z - 2) > 2(z + 7)$.

    b. $\dfrac{1}{4}x - \dfrac{1}{3} \leq x + 2$.

    c. $2 - 2(7 - 2x) < 3(3 - x)$.

    d. $-11(2 - b) \neq 4(2b + 2)$.

    e. $6 - 2(x - 4) \geq 2x + 10$.

    f. $0.4x + 0.4 \leq 0.1x + 0.85$.

28. Solve each equation by isolating the variable. Be sure to check your answers.

    a. $x^2 - 9 = 16$. **b.** $4x^2 - 1 = 0$. **c.** $(x + 2)^2 = 9$.

29. Use the zero product principle to solve each equation.

    a. $(x - 2)(x - 1) = 0$.

    b. $(x + 5)(x + 6) = 0$.

    c. $x(x - 3)(x - 2) = 0$.

    d. $(x - 2)(x - 2)(x - 3) = 0$.

    e. $(x - 1)(x - 3)(x + 4)(x - 6) = 0$.

30. Solve each equation. Be sure to check your answers.

    a. $2\sqrt{x} - 3 = 0$.

    b. $\dfrac{\sqrt{x}}{3} = 4$.

    c. $\sqrt{x + 3} = 6$.

31. Solve each equation. Be sure to check your answers.

    a. $\sqrt{2x - 1} = 2$.

    b. $\sqrt{3x + 2} = 11$.

    c. $2\sqrt{x - 4} = 8$.

32. Solve each equation. Be sure to check your answers.

    a. $\dfrac{2}{x} + 1 = 3$. **b.** $\dfrac{5}{x - 1} = 10$. **c.** $\dfrac{3}{x + 4} = \dfrac{-5}{x}$.

33. Solve each equation. Be sure to check your answers.

    a. $\dfrac{-1}{x + 7} = \dfrac{3}{4 - x}$.

    b. $\dfrac{3}{x - 1} = \dfrac{1}{x - 2}$.

    c. $\dfrac{-4}{x} = \dfrac{-2}{x + 3}$.

■ Problems and Applications

34. Compute each set operation, and write the result in interval notation.

    a. $(3, 5) \cup (4, 7)$ **b.** $(-\infty, -2) \cup [-3, \infty)$

    c. $[2, \infty) \cup [-1, \infty)$ **d.** $(-1, 5] \cap [2, 7]$

    e. $(-\infty, 4] \cap (-\infty, -3]$ **f.** $[0, 5) \cap (5, \infty)$

35. Compute each set operation, and write the result in interval notation. Assume the universe is the set of all real numbers.

    a. $(3, 7) - (3, 5]$ **b.** $(4, \infty) - [7, \infty)$

    c. $[2, 5] - [3, 4]$ **d.** $\overline{(-\infty, -8)}$

    e. $\overline{(6, \infty)}$ **f.** $\overline{[0, 7]}$

36. We can solve many quadratic equations by factoring the equation and using the zero product principle. To factor an expression, write it as the product of two simpler linear expressions. For instance, the factored form of $x^2 - x - 6$ is $(x + 2)(x - 3)$. Solve each equation by factoring first.

    a. $x^2 - x = 0$. **b.** $x^2 + 3x + 2 = 0$.

    c. $x^2 - 4 = 0$. **d.** $2x^2 + 5x - 3 = 0$.

37. Writing a quadratic equation in the form $ax^2 + bx + c = 0$, we can solve it by using the quadratic formula, or $\dfrac{-b \pm \sqrt{b^2 - 4ac}}{2a}$, where a, b, and c are any real numbers. To do so, substitute the coefficients of equation into the formula and compute. Use the quadratic formula to solve each equation.

    a. $x^2 + 3x - 4 = 0$. **b.** $x^2 - 4 = 0$.

    c. $x^2 - 2x + 1 = 0$. **d.** $x^2 + 5x + 6 = 0$.

    e. $2x^2 - 7x + 3 = 0$. **f.** $12x^2 + 5x - 2 = 0$.

38. The **root** of a function $f(x)$, is any value of x where $f(x) = 0$.

    a. Describe how to find the root of a function by solving an equation. Use your method to find the root of each linear function.

       i. $f(x) = -x + 4$.

       ii. $f(x) = 3x - 6$.

       iii. $f(x) = x - 3$.

b. Graph each function in part a. In each case, where on the graph is the root of the function located?

c. Use the graph of the function and what you have learned in part b to find the roots of each function.

  i. $f(x) = \dfrac{3}{4}x + 9$.

  ii. $f(x) = x^2 + 6x$.

  iii. $f(x) = x^2 - 9$.

39. Operations on rational expressions work just like operations on rational numbers. We must get a common denominator when adding or subtracting rational expressions. Solve each equation by adding or subtracting the terms and then isolate the variable.

  a. $\dfrac{2}{x} + \dfrac{1}{2x} = 5$. **b.** $\dfrac{1}{3x} - \dfrac{4}{x} = 6$.

■ Mathematical Reasoning and Proof

40. Verify that solving the equation $2x - 5 = 7$ has the same solution regardless of whether we first divide by 2 or add 5.

41. Based on the location of a and b in the following graph, determine whether each statement is true, false, or cannot be determined.

  a. $a < b$.
  b. $a + b > 0$.
  c. $a - (-2) > 0$.
  d. $b - a < 0$.
  e. $a - b < 0$.

42. In a sense, the steps used in isolating the variable can be viewed as a proof of the solution to the equation. Why does this view make sense?

43. Provide a reason for each step that has been taken in solving the equation $4x - 3 = 9$.

$$(4x - 3) + 3 = 9 + 3$$
$$4x + (-3 + 3) = 9 + 3$$
$$4x + 0 = 9 + 3$$
$$4x = 9 + 3$$
$$4x = 12$$
$$\frac{4x}{4} = \frac{12}{4}$$
$$1x = 3$$
$$x = 3$$

44. Provide a reason for each step that has been taken in solving the inequality $-x - 2(x + 3) \le 4$.

$$-x - 2x - 6 \le 4$$
$$-3x - 6 \le 4$$
$$(-3x - 6) + 6 \le 4 + 6$$
$$-3x + (-6 + 6) \le 4 + 6$$
$$-3x + 0 \le 10$$
$$-3x \le 10$$
$$\frac{-3x}{-3} \ge \frac{10}{-3}$$
$$1x \ge -\frac{10}{3}$$
$$x \ge -\frac{10}{3}$$

45. Solve the equation $-3x + 4(x - 2) = 6x - 1$. Justify each step you take.

46. Solve the inequality $x + 2 \ge -4x + 3$. Justify each step you take.

■ Mathematical Communication

47. Solve the inequality $2x + 5 < -3x + 8$ by isolating the variable through two different sequences of steps. Is one more difficult to use than the other? Why?

48. Write a set of directions that explains how to solve the equation $4x - 5 = 2x + 9$. Give them to a peer to follow. Discuss any difficulties that you encounter.

49. Quishawna is asked to solve the equation $2x = 24$. She quickly answers that x must equal 4. How did she obtain this answer, and how would you correct her thinking?

50. Marcos is given the inequality $-3x + 1 > 10$. When he solves it, he claims the solution set contains every number greater than or equal to -3. Is he correct?

51. A student solves an inequality and writes the solution as $-3 > x \ge 1$. What is wrong with the solution, and how would you respond to the student's misconception?

■ Building Connections

52. What is the difference between evaluating a function and solving an equation?

53. How is solving a linear equation an extension of numerical operations? How might this connection help students better understand the manipulations associated with solving equations?

54. The properties of the real number operations, such as the commutative or associative properties, are often referred to as algebraic properties.

 a. Why does it make sense to refer to these properties as algebraic properties?

 b. What underlying assumption about algebraic variables allows us to use these properties when solving equations?

55. Reconsider the definition for the missing-addend approach to subtraction in Chapter 4.

 a. Based on what you read in this definition, what does it mean for addition and subtraction to be inverse operations of each other?

 b. What other pairs of inverse operations can you list?

SECTION 9.5 Algebraic Thinking and Mathematical Modeling

Mathematical modeling is the third and final aspect of algebraic thinking. Simply put, **mathematical modeling** is the process by which we create mathematical representations that can be used to understand and solve problems in real-world situations. The process begins by translating the situation into a mathematical model. Unnecessary information is stripped away, making it easier to use mathematical computations and properties to solve the problem. Having used the model to get an answer, we translate it back into the context of the problem (Figure 9.24).

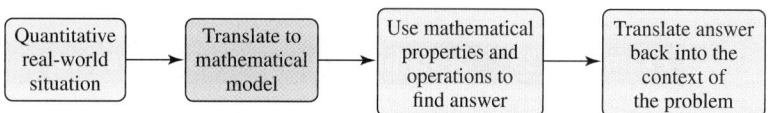

Figure 9.24 The process of mathematical modeling

Mathematical models come in a variety of forms and can include objects, pictures, words, and symbols. In algebraic situations, we tend to use functions and equations as models. If the situation involves a relationship of change, we use a function. If it involves an unknown quantity, we use an equation or inequality.

Modeling with Functions

Explorations Manual 9.12

Many real-world situations exhibit a relationship of change between two variables, so we can use functions and their representations to model them. For instance, a function can be used to represent the relationship between the amount of electricity used and the cost of an electric bill. Another function can be used to represent the relationship between the passage of time and the growth of a population. Many real-world situations like this have relationships of change that are quite complex, so they require sophisticated mathematical techniques and models to describe them. However, in many cases, it is enough to represent the situation with a function that is increasing, decreasing, or constant. In Section 9.3, these ideas were introduced with linear functions. We now define them for functions in general.

> **Definition of Increasing, Decreasing, and Constant Functions**
>
> Let f be a function on the interval $[a, b]$, and let x_1 and x_2 be any elements in $[a, b]$ such that $x_1 < x_2$. Then
>
> **1.** f is **increasing** on $[a, b]$ if $f(x_1) < f(x_2)$,
> **2.** f is **decreasing** on $[a, b]$ if $f(x_1) > f(x_2)$, and
> **3.** f is **constant** on $[a, b]$ if $f(x_1) = f(x_2)$.

Intuitively, the definition tells us that a function is increasing if increasing values of x are paired with increasing values of $f(x)$. It is decreasing if increasing values of x are paired with decreasing values of $f(x)$. And it is constant if increasing values of x are paired with only one value for $f(x)$. Visually, as we move toward the right on the Cartesian coordinate system, the graph slopes upward to the right if it is increasing, slopes downward to the right if it is decreasing, and remains horizontal if it is constant (Figure 9.25).

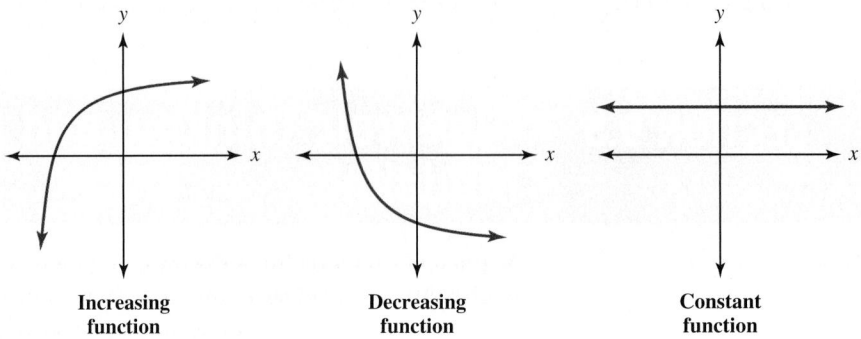

Increasing function   Decreasing function   Constant function

Figure 9.25 Increasing, decreasing, and constant functions

In many cases, functions can be increasing at one point and decreasing or constant at another. Consequently, the function may have to be described as increasing, decreasing, or constant over particular intervals of the real number line.

Example 9.39   Describing Functions as Increasing, Decreasing, or Constant

Determine the intervals on which each function is increasing, decreasing, or constant.

a.

b.

Solution Whenever we state the intervals over which a function increases, decreases, or remains constant, the intervals always refer to domain, or x-values. We use the x-values at which the graph changes direction as the endpoints of the intervals.

a. The graph is increasing on $(-\infty, -2)$, decreasing on $(-2, 3)$, and constant on $(3, \infty)$.

b. The graph is decreasing on $(-\infty, 0)$ and increasing on $(0, \infty)$.

The next two examples demonstrate how we can use these ideas to represent and answer questions about different real-world situations.

Example 9.40   **Gas Mileage**

Application

The following graph shows that the gas mileage of a car depends on the speed at which the car is driven.

a. For what speeds does gas mileage improve?

b. For what speeds does gas mileage decline?

c. At what speed does the car get the best gas mileage?

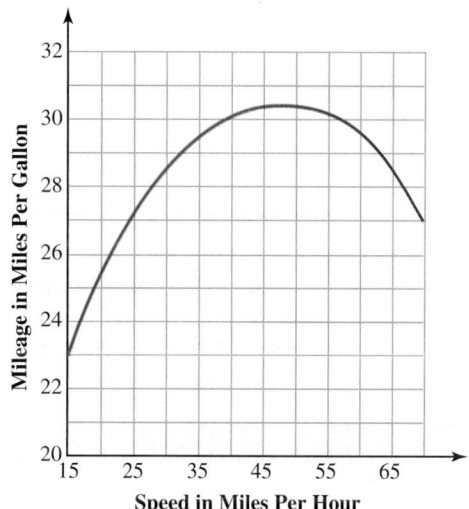

Speed in Miles Per Hour

Solution

a. To find the speeds for which gas mileage improves, we look for the increasing parts of the graph. Because it increases on the interval [15, 45), we conclude that the gas mileage improves from 15 mph to about 45 mph.

b. To find the speeds for which gas mileage is declining, we look for the decreasing parts of the graph. Because it decreases on the interval (45, 70], we conclude that the gas mileage declines from about 45 mph to 70 mph.

c. The best gas mileage is at the peak of the graph, which is about 45 mph.

Example 9.41   **Population of a Town**

Application

An Old West town was founded next to a river in 1850 and maintained a steady population until gold was discovered 10 years later. For the next 40 years, the population steadily increased until the gold ran out in 1900. About half the population left in the next 5 years. From that point on, the population continued to dwindle until it became a ghost town in 1935. Draw the graph of a function that models the population change of the town.

Solution We can represent the change in the population with a function that relates time to the number of people. Time is the independent variable, so it corresponds to the horizontal axis. Because the town was founded in 1850 and ceased to exist in 1935, the domain is limited to these values. The dependent variable is the population of the town, so it corresponds to the vertical axis. Using the information in the story, we can create the following graph:

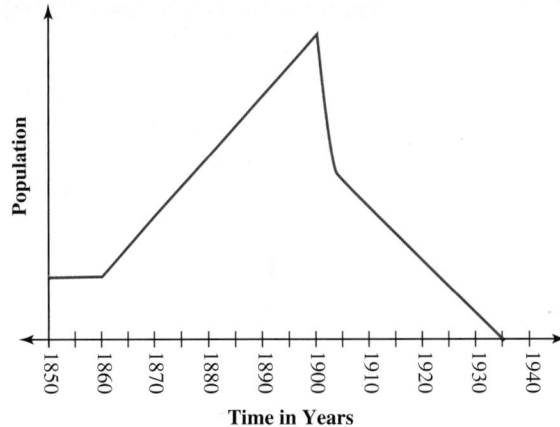

Functions not only allow us to model real-world relationships of change, but they also enable us to make predictions based on that change. The predictions can be particularly easy to make if the function is represented by an algebraic equation.

Example 9.42 **Sales Tax**

Application

The sales tax in many states is 7%. The total cost of a purchase can be modeled with the function $C(x) = x + 0.07x$, where x is the price of the purchase before sales tax. Find the total cost for each purchase price.

a. $25.00   **b.** $59.79

Solution We can find the total cost for each purchase by evaluating the price in the cost function.

a. A purchase of $25.00 costs $C(25) = 25 + 0.07(25) = \$26.75$.

b. A purchase of $59.79 costs $C(59.79) = 59.79 + 0.07(59.79) = \$63.98$.

In many situations, the first task may be to find the algebraic equation for the function prior to evaluating it at particular values. Consider the next two examples.

Example 9.43 **Counting Bacteria**

Problem Solving

A lab technician begins with a culture that has approximately 5,000 bacteria. He observes that the population triples from the first to the second day and that it triples again from the second to the third day. Assuming the population grows at the same rate, about how many bacteria will there be in 10 days?

Solution

Understanding the Problem. We are asked to find the number of bacteria present on the 10th day. The bacteria triple from the first day to the second and triple again from the second to the third. Also, the initial number of bacteria is about 5,000.

Devising a Plan. Using the information in the problem, we can *write an equation* for the function and then evaluate it at 10 to get the number of bacteria on the 10th day.

Carrying Out the Plan. The culture starts with 5,000 bacteria, increases to 15,000 bacteria on the second day, and increases again to 45,000 bacteria on the third. Because the bacteria triple every day, we have a geometric sequence with $a_1 = 5,000$ and $r = 3$. Using what we know about geometric sequences, we can model the growth of the bacteria with the equation

$$P(t) = 5,000 \cdot 3^{t-1}$$

where t is the number of days. Now we evaluate the function at $t = 10$, or $P(10) = 5,000 \cdot 3^{(10)-1} = 5,000 \cdot 3^9 = 98,415,000$ bacteria on the 10th day.

Looking Back. We can check the answer by taking another approach to the problem. Compute the number of bacteria by multiplying by 3 nine consecutive times.

Day	Number of Bacteria
1	5,000
2	$5,000 \cdot 3 = 15,000$
3	$15,000 \cdot 3 = 45,000$
4	$45,000 \cdot 3 = 135,000$
5	$135,000 \cdot 3 = 405,000$
6	$405,000 \cdot 3 = 1,215,000$
7	$1,215,000 \cdot 3 = 3,645,000$
8	$3,645,000 \cdot 3 = 10,935,000$
9	$10,935,000 \cdot 3 = 32,805,000$
10	$32,805,000 \cdot 3 = 98,415,000$

Although we arrive at the same answer, we can see that the formula for the function is a more efficient way of obtaining the answer.

Example 9.44 Inventory at a Flower Store

Application

Valentine's Day is approaching, and Amy's Flower Boutique is preparing to order roses for the holiday. To determine how many flowers to order, Amy looks at her records from the last 5 years. The records show not only how many roses were ordered, but also how many had to be thrown away. Based on the records, about how many roses should she order this year? What about 5 and 10 years from now?

Year	2006	2007	2008	2009	2010
Roses ordered	600	705	800	895	1,000
Roses thrown out	96	99	103	107	112

Solution Amy's goal in ordering flowers is to make sure she has enough to meet her customers' needs without having to throw away too many. Because she wants to predict the number of roses she will need for several years, the most efficient way to model the situation is with an algebraic equation. Her records show that the number of flowers she has ordered has roughly increased at a constant rate of 100 roses a year. Using this estimate, she can model the number of flowers she orders with the linear function

$$\text{Flowers ordered} = 600 + 100t$$

where t is the number of years since 2006. Similarly, the number of flowers she has thrown away roughly increases at a constant rate of 4 flowers a year. It can be modeled by the linear function

$$\text{Flowers thrown away} = 96 + 4t,$$

where t is the number of years since 2006. To have enough roses on hand, yet eliminate waste, she can use the following function

$$\text{Flowers needed} = \text{flowers ordered} - \text{flowers thrown away}$$
$$= (600 + 100t) - (96 + 4t)$$
$$= 504 + 96t$$

Evaluating this function at $t = 5$, we find that Amy will need $504 + 96(5) = 984$ roses for this year. In 5 years, or $t = 10$, she should order $504 + 96(10) = 1,464$ roses, and in 10 years, or $t = 15$, she should order $504 + 96(15) = 1,944$ roses.

Check Your Understanding 9.5-A

1. Determine the intervals on which each function is increasing, decreasing, or constant.

a.

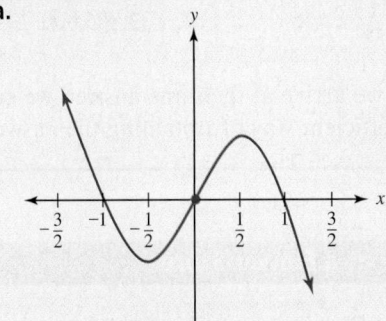

b.

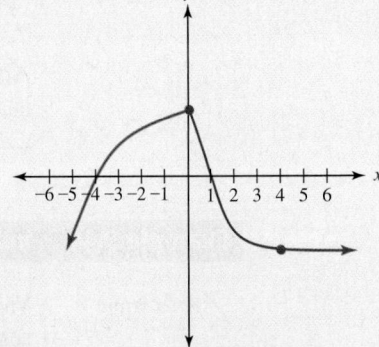

2. Elk were reintroduced to Pennsylvania in the late 1980s. Over the next several years, the herd was protected from hunting, so the population grew quickly. In 1995, the herd reached a sustainable population. From then on, hunting licenses have been issued to keep the population steady. Draw a graph that models this situation.

3. The amount of money in a savings account follows the function $A = 12,000 (1.025)^{4t}$ where t is the time in years. How much money will be in the account after:

a. 5 years?   **b.** 10 years?   **c.** 30 years?

Talk About It Why are graphs and algebraic equations good choices for representing functions that model real-world situations?

Problem Solving

Activity 9.9

In 2007, the first-class mailing rates for retail parcels were as follows.

Weight Not Over	Rate for Single Piece	Weight Not Over	Rate for Single Piece
1 oz	$1.13	8 oz	$2.32
2 oz	$1.30	9 oz	$2.49
3 oz	$1.47	10 oz	$2.66
4 oz	$1.64	11 oz	$2.83
5 oz	$1.81	12 oz	$3.00
6 oz	$1.98	13 oz	$3.17
7 oz	$2.15		

Source: U.S. Postal Service.

a. What constant rate of change are the charges based on?

b. Write a linear equation that can be used to determine the postal charge for any parcel weight. What are the slope and y-intercept of the function?

c. Your equation from part b can be used to determine rates for parcel weights that are not exact ounces. Find the rate for each parcel weight.

  i. 4.3 oz

  ii. 5.8 oz

  iii. 12.9 oz

d. Suppose the U.S. Postal service does not consider partial ounces, but rounds all weights up to the next highest ounce. For instance, a parcel weighing 3.6 oz is rounded up to 4 oz, giving it a rate of $1.64. In this case, what would the graph of the function look like?

Modeling with Equations

Throughout previous chapters, we have used equations to find unknown quantities in a variety of situations. For instance, equations were used to solve simple story problems and to solve problems with proportions and percents. In each case, the first step was to translate a verbal statement into an algebraic one. In most situations, key words and phrases signaled the operations and relationships in the equation. However, these key words and phrases are not included in many real-world situations. Instead, we must rely on intuition and experience with mathematics to make the translation to an equation correctly. Once we have the equation, we can use what we know about solving equations to find the unknown.

Explorations Manual 9.13

Example 9.45 | Selling Merchandise

Application

Sharice is in sales and has a base salary of $2,500 a month. She also makes 15% commission on her total sales. If she wants to earn a minimum of $4,500 for the month, how much merchandise must she sell?

Solution Sharice's income is her base salary plus whatever commission she earns, or

$$\text{Salary} + \text{commission} = \text{income}$$

We know two of the three values: Her base salary is $2,500, and she wants a total income of $4,500. We also know that her commission is 15% of her sales. Not knowing her sales, we assign it the variable x and represent her commission as $0.15x$. After we substitute these values, we have the equation

$$2,500 + 0.15x = 4,500$$

Next, we solve for x,

$$2,500 + 0.15x = 4,500$$
$$0.15x = 2,000 \qquad \text{Subtract both sides by 2,500.}$$
$$x \approx 13,333.33 \quad \text{Divide both sides by 0.15.}$$

Sharice must sell $13,333.33 in merchandise to make $4,500 for the month.

Example 9.46 Figuring Labor Costs

Application

Stan is starting a handyman business and must buy equipment totaling $21,000. He has six contracts and figures each job will take 5 days to complete. He hires two helpers and plans to pay each worker and himself $80 a day. How much should he charge in labor on each contract to break even by the end of the sixth job?

Solution For Stan to break even, his income must equal his expenses. He has two expenses: (1) his equipment, which costs $21,000, and (2) his cost in labor. He has six jobs, each of which takes 5 days, so he must compensate for $6 \cdot 5 = 30$ days of work. Because each worker, including himself, makes $80 a day, Stan must pay a total of $30 \cdot 3 \cdot \$80 = \$7,200$ in labor. Hence, his total expenses are $21,000 + $7,200 = $28,200. His income is based on the amount that he charges on each contract. Because the amount is unknown, we assign it a variable, x, giving us the equation

$$6x = 28,200$$

When we divide both sides of the equation by 6, we have $x = 4,700$. Consequently, Stan must charge $4,700 in labor on each contract to break even by the end of the sixth job.

In the last two examples, linear equations were used to model situations involving one unknown. Linear equations can also be used to model situations involving two unknowns.

Example 9.47 Selling Tickets

Application

For the upcoming high school football game, 1,475 tickets were sold. Student tickets cost $2.00, and adult tickets cost $4.50. The total revenue from ticket sales was $4,680. How many tickets of each type were sold?

Solution We are asked to find two unknowns: the number of student tickets sold, s, and the number of adult tickets sold, d. Because 1,475 tickets were sold, $s + d = 1,475$. Because each student ticket costs $2.00, the revenue collected from student tickets is $\$2.00s$. Similarly, the revenue collected from adult tickets is $\$4.50d$.

Adding the two, we have $2s + 4.5d = 4,680$. This gives us two equations with two unknowns, or

$$s + d = 1,475 \quad \text{Total tickets.}$$
$$2s + 4.5d = 4,680 \quad \text{Total revenue.}$$

If the two equations could be condensed into one with only one unknown, we would know how to proceed. We can do that by solving one of the equations for one of the variables and substituting it into the other equation. We rewrite $s + d = 1,475$ as $d = 1,475 - s$, and then substitute $1,475 - s$ into the other equation for d, or

$$2s + 4.5(1,475 - s) = 4,680$$

Solving this equation for s:

$$2s + 4.5(1,475 - s) = 4,680$$
$$2s + 6,637.5 - 4.5s = 4,680 \quad \text{Distribute 4.5 to 1,475 and } s.$$
$$-2.5s + 6,637.5 = 4,680 \quad \text{Combine like terms.}$$
$$-2.5s = -1,957.5 \quad \text{Subtract 6,637.5 from both sides.}$$
$$s = 783 \quad \text{Divide both sides by } -2.5.$$

Because the number of student tickets is 783, we subtract 783 from 1,475 to find the number of adult tickets, or $d = 1,475 - 783 = 692$.

In the last example, two linear equations contained the same two unknowns. Considered together, the equations form a **system of linear equations,** or **simultaneous equations.** In general, a system of linear equations can be represented as

$$a_1x + b_1y = c_1$$
$$a_2x + b_2y = c_2$$

where the coefficients and constants are any real numbers. A **solution** to the system of equations is any ordered pair of numbers (x, y) that makes *both* equations true.

We can use several methods to find the solution of a system of linear equations. In Example 9.47, we used the **substitution method.** We solved one of the equations for one variable and then substituted the result into the other equation. One of the variables was thus eliminated, leaving one equation with one unknown to be solved directly.

Another way to solve a system of linear equations is the **elimination method.** We eliminate one of the variables by adding or subtracting the original or equivalent equations. Given one equation with one unknown, we can isolate the variable and use its value to find the other unknown.

Example 9.48 Using Elimination to Solve a System of Equations

Use the elimination method to solve each system of linear equations.

a. $x + 2y = 10$
$3x - 2y = -2$

b. $4x + 6y = 8$
$2x + 3y = 1$

Solution The goal is to eliminate one of the variables by adding or subtracting the original or equivalent equations. To do so, we may need to multiply one or possibly both equations by a constant so that the coefficients on one of the variables are additive inverses of each other. By adding the equations, one variable subtracts away, leaving one equation with one unknown.

a. The coefficients of the y-terms are already additive inverses, so the first step is to add the equations. In doing so, the y-terms subtract away, leaving a one-step equation in terms of x.

$$\begin{array}{rl}
x + 2y &= 10 \\
\underline{3x - 2y} &= \underline{-2} \\
4x \quad\;\; &= 8 \qquad \text{Add like terms between equations.} \\
x \quad\;\; &= 2 \qquad \text{Divide both sides by 2.}
\end{array}$$

Next, substitute $x = 2$ into either equation and solve for y. Doing so gives us $y = 4$. As a result, the solution to the system of equations is the ordered pair $(2, 4)$.

b. Multiply the second equation by -2 and then add. Both variables subtract away, leaving $0 = 6$.

$$\begin{array}{rcl}
4x + 6y = 8 & & 4x + 6y = 8 \\
2x + 3y = 1 & \Rightarrow & \underline{-4x - 6y = -2} \\
& & 0 = 6
\end{array}$$

This indicates that no values for x and y satisfy the system of equations. Consequently, it has no solution.

When solving a system of linear equations, we are trying to find a point (x, y) that the two equations have in common. However, in the last example, one system of equations had no solution. This can happen because two lines can interact in one of three ways: They can be parallel, they can intersect at a point, or they can be the same line (Figure 9.26). If the lines are parallel, they have no common points, so the system of equations has no solution. If the lines intersect at a point, the system of equations has exactly one solution. If the lines are the same, then they have infinitely many points in common, so the system of equations has an infinite number of solutions. As a result, we can use a graph as another way to solve a system of linear equations simply by reading the coordinates of any intersection point(s) from the graph.

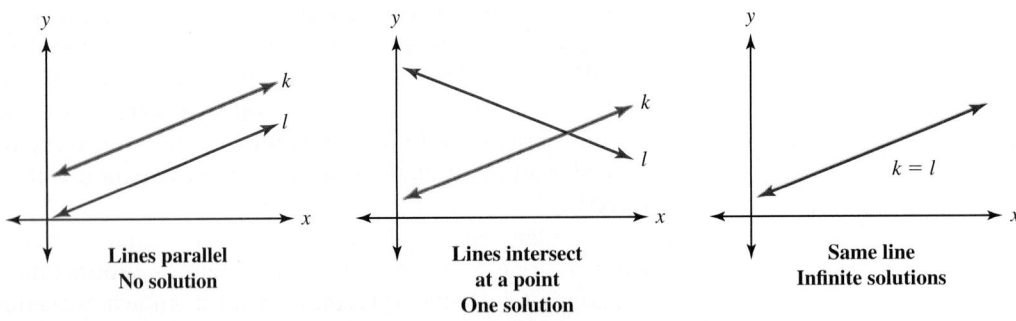

Figure 9.26 Three ways lines can interact

Example 9.49   **Finding the Number of Solutions of a System of Equations**

Determine whether each system of linear equations has no solution, one solution, or infinitely many solutions.

a. $y = 3x + 5$ b. $y = -4x + 2$ c. $y = 3x - 5$
    $y = 2x + 1$ $y = -4x - 1$ $2y - 6x = -10$

Solution

a. Because the equations $y = 3x + 5$ and $y = 2x + 1$ are in slope-intercept form, they have different slopes and different y-intercepts. The lines intersect at a point, and the system of equations has one solution.

b. The lines $y = -4x + 2$ and $y = -4x - 1$ have the same slope, but different y-intercepts. The lines are parallel, and the system of equations has no solution.

c. The equation $y = 3x - 5$ is in slope-intercept form, but $2y - 6x = -10$ is not. Solving the equation for y, however, gives us $y = 3x - 5$. Because the equations have the same slope and the same y-intercept, they must be the same line. As a result, the system of equations has an infinite number of solutions.

◼

The last several examples illustrate that we can use what we know about functions and solving equations to solve systems of linear equations. This gives us another powerful way to model and solve real-world problems.

Example 9.50 Comparing Checking Accounts

Connections

Cordaryl is looking to open a student checking account at one of two banks. Bank A offers an account that has a $10 monthly maintenance fee, but charges only 75¢ for every check written. Bank B offers an account with no monthly fee, but charges $1.25 for every check written.

a. If Cordaryl typically writes 15 checks a month, which is the better bank?

b. After how many checks does the initially cheaper plan become the more expensive of the two?

Solution We first find the cost equation for each bank. Bank A has a maintenance fee of $10 and charges 75¢ for every check written, which we represent with the equation $C_A(t) = 10 + 0.75t$, where t is the number of checks. Bank B has no maintenance fee and charges $1.25 for every check written, which we represent with the equation $C_B(t) = 1.25t$. From this point, we can answer the first question in at least two ways. One is to evaluate both equations at $t = 15$ and answer the questions based on the results. Another is to graph both equations on the same Cartesian plane:

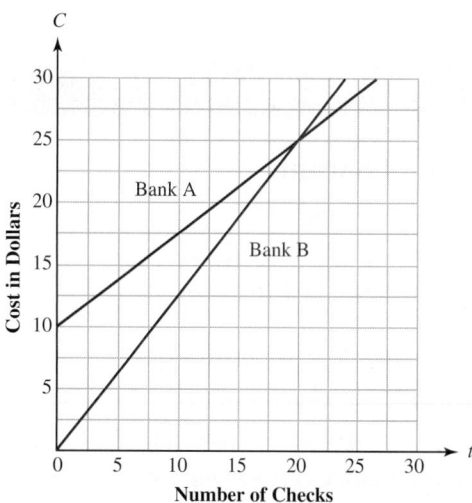

a. If Cordaryl plans to write only 15 checks, then the graph shows that Bank B is better.

b. The graph seems to indicate that Bank B becomes more expensive after 20 checks have been written. The solution to the following inequality verifies this answer.

$$10 + 0.75t < 1.25t$$
$$10 < 0.5t$$
$$20 < t$$

Check Your Understanding 9.5-B

1. Two tanks hold a total of 85 gal of water. One tank holds 10 more gal than twice the other. How many gallons does each tank hold?

2. Jonathan has made a 62 and a 75 on his first two 100-point exams. He has one more 100-point exam for the semester. What possible scores can he have on the last test and still maintain at least a 75% average?

3. I have a total of 35 dimes and quarters in a jar. If the total amount of change is equal to $6.05, how many of each coin is in the jar?

4. Solve each system of linear equations.

    a. $x + y = 6$ **b.** $2x + y = 1$ **c.** $-2x + 3y = 8$

       $x - y = 2$ $3x - 2y = 4$ $4x - 6y = -16$

Talk About It Suppose a system of equations has three unknowns. How many equations would have to be in the system to solve it?

Connections

Activity 9.10
Two ships are sailing through a section of ocean using the same coordinate system. One ship is on a linear course described by $3x - y = -3$. The other ship is following a course described by $2x + y = -7$.

a. Can the two ships collide? If so, at what point can the collision occur?

b. Is the collision certain? If not, what other factors might determine whether the ships collide?

SECTION 9.5 Learning Assessment

Understanding the Concepts
1. What is mathematical modeling?
2. Explain the difference, both algebraically and graphically, between an increasing, a decreasing, and a constant function.
3. How can functions be used to make predictions?
4. **a.** What is a system of linear equations?
   b. What is a solution to a system of equations?
5. Explain three different methods for solving a system of linear equations.
6. How many possible solutions can a system of linear equations have? Why is this?

Representing the Mathematics
7. Sketch the graph of a function that is:
   a. Increasing on the real numbers.
   b. Decreasing on the real numbers.
   c. Decreasing on the interval $(-\infty, 0)$ and increasing on the interval $(0, \infty)$.
   d. Increasing on the interval $(-\infty, 0)$ and decreasing on the interval $(0, \infty)$.

8. Give an algebraic equation for a function that is:
   a. Increasing on the real numbers.
   b. Decreasing on the real numbers.
   c. Decreasing on the interval $(-\infty, 0)$ and increasing on the interval $(0, \infty)$.
   d. Increasing on the interval $(-\infty, 0)$ and decreasing on the interval $(0, \infty)$.

9. State whether each set of ordered pairs represents a function that is increasing, decreasing, or constant?
   a. $\{(2, 1), (3, 1), (4, 1), (5, 1)\}$
   b. $\{(1, 0.1), (2, 0.2), (3, 0.3), (4, 0.4)\}$
   c. $\{(1, 2), (2, 4), (3, 6), (4, 8)\}$
   d. $\left\{(-1, -1), \left(-2, -\frac{1}{2}\right), \left(-3, -\frac{1}{3}\right), \left(-4, -\frac{1}{4}\right)\right\}$

10. State whether each table represents a function that is increasing, decreasing, or constant.

a.

x	y
1	−1
2	−2
3	−3
4	−4

b.

x	y
1	4
2	4
3	4
4	4

c.

x	y
1	1
2	0.5
3	0.25
4	0.125

d.

x	y
1	−0.1
2	−0.2
3	−0.3
4	−0.4

11. Sketch a possible graph for each situation.

  a. An initial amount of money deposited into an interest-bearing account for several years with no withdrawals

  b. The speed of a school bus as it picks up four children for school

  c. The temperature of food after it is put into a refrigerator

  d. The volume of a siren as a fire engine approaches and passes by

12. In a certain coastal city, the tide was at its highest point of 3 ft at noon. The tide was at its lowest level of (−5) ft 6 hr later. Finally, at 12 a.m., the tide had reached a level of 0 ft. Make a graph that represents the tide over this time period.

13. Write an equation for the function that represents the cost of attending college for one semester if registration fees are $200 and classes cost $175 per credit.

14. A bank charges a monthly maintenance fee of $3.00 and then $0.25 for each check that is written. Write an equation for the function that represents this checking account.

■ Exercises

15. Determine the intervals on which each function is increasing, decreasing, or constant.

  a.

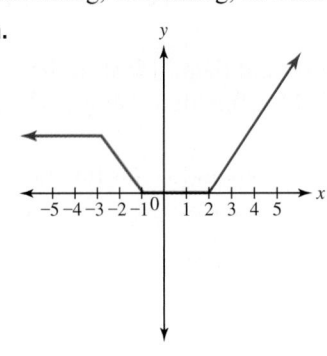

b.

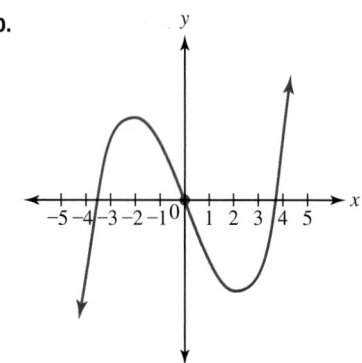

c.

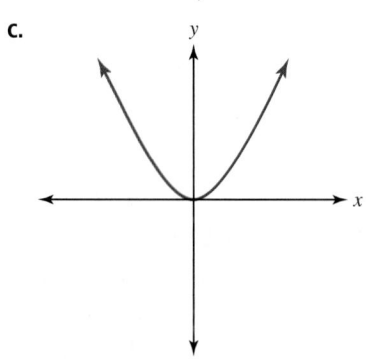

16. Use a calculator to graph each function. State the intervals on which each function is increasing, decreasing, or constant.

  a. $y = -3x - 4$.   **b.** $y = 5x + 1$.

  c. $y = 2x^2 - 3$.   **d.** $f(x) = -x^2 + 2$.

  e. $y = -x^3 + 2x$.   **f.** $f(x) = \frac{1}{2}x^3 - 2x$.

17. Without solving the system of linear equations, determine whether each system has no solution, one solution, or infinitely many solutions.

  a. $x + y = 6$
    $x + y = 5$

  b. $3x - 4y = 2$
    $3x - 5y = 4$

  c. $-2x + 2y = 2$
    $x - y = -1$

  d. $2x - \frac{1}{3}y = 6$
    $2x + \frac{1}{3}y = 6$

  e. $2x + \frac{1}{4}y = 8$
    $\frac{1}{2}x - \frac{1}{16}y = 2$

  f. $\frac{1}{3}x - 3y = 6$
    $-3x + \frac{1}{3}y = 9$

18. Solve each system of linear equations by graphing the functions.

  a. $y = x + 1$
    $y = -x + 3$

  b. $y = 3x - 2$
    $y = -\frac{1}{2}x + 5$

  c. $-3x + y = 4$
    $2x + y = -1$

  d. $4x - 2y = 2$
    $-4x + 2y = 10$

19. Solve each system of linear equations by using substitution.

    a. $x + y = 6$
    $x + 3y = -4$

    b. $3x - y = -2$
    $2x - y = 4$

    c. $-2x + 2y = 5$
    $3x - 6y = 0$

    d. $x + 3y = 2$
    $x + y = 6$

20. Solve each system of linear equations by elimination.

    a. $x - y = -2$
    $2x + y = 5$

    b. $-2x + 3y = 7$
    $4x + 3y = 4$

    c. $-4x + y = -4$
    $x - \frac{1}{2}y = 6$

    d. $\frac{1}{2}x + \frac{1}{2}y = 2$
    $-4x + 5y = 2$

■ Problems and Applications

21. The following graph shows the cost of heating a home for 30 years using three different heat sources: electric, gas, and solar. Answer the questions based on the graph.

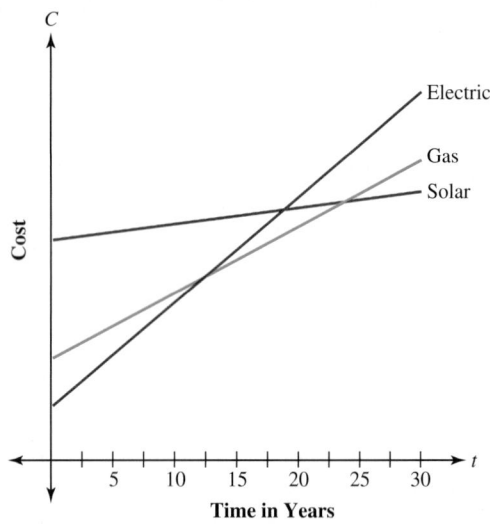

    a. Which heating system is the least expensive to install? Most expensive?

    b. If you were to stay in the house for 30 years, what heating system would be the best to install? Why?

    c. About how many years does it take for gas to be cheaper than electric?

22. The following graph compares the cost function for manufacturing a certain television to the revenue function from the sales of it.

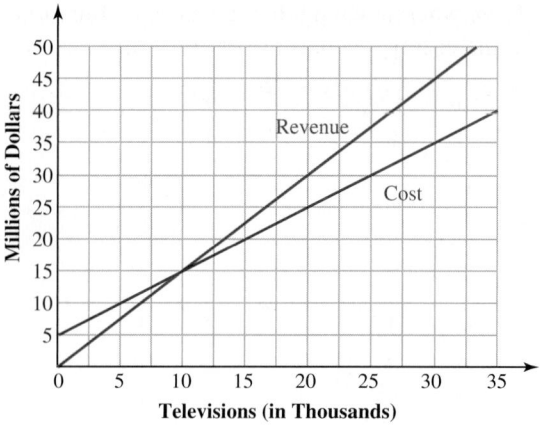

    a. What is the cost of manufacturing 20,000 televisions?

    b. What is the revenue from selling 20,000 televisions?

    c. About how much profit is made when 20,000 televisions are sold?

    d. For what number of televisions does the cost equal the revenue? Interpret this point, in terms of the profit of the company?

23. Answer each question using the following graphs that describe two cars, A and B.

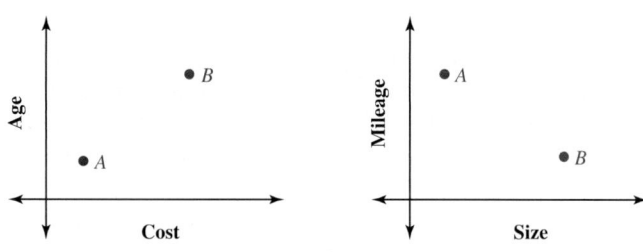

    a. Which car costs more? Less?

    b. Which car is newer? Older?

    c. Which car gets better gas mileage? Worse mileage?

    d. Which car is smaller? Larger?

24. Write a function for computing the total bill at a restaurant if a 15% tip is given.

    a. What is a possible domain for this equation?

    b. To the nearest cent, what is the total bill if the food costs $35.49?

25. It costs $350 in labor and then $15/yd^2 to have carpet installed. Write a function that models this situation.

    a. What is a possible domain for this function?

    b. To the nearest cent, what is the cost to have 25 yd^2 of carpet installed?

26. Income tax in a state is 2.3% for all incomes.

   a. Write a function that gives the amount of tax based on a person's income.

   b. What is the independent variable? The dependent variable?

   c. What is a logical domain for this function?

   d. How much tax must a person pay if the person makes $50,000 a year?

   e. If a person paid $1,300 in state tax last year, how much did the person make?

27. In a recent fund-raiser for two charities, $150,595 was raised. Charity A received about three times the amount of Charity B.

   a. Write an algebraic equation that models this situation, and use it to find out how much each charity was given.

   b. What is another way the problem can be solved without using an equation?

   c. Is there a connection between the equation you used in part a and the method you used in part b? If so, explain.

28. A pizza delivery person receives $5/hr plus $0.75 for each pizza delivered. If Jamie made $124.75 in 8 hr of delivering pizzas, how many pizzas did she deliver?

29. Lori earns $7/hr for the first 40 hr of work and then $10.50/hr for any hour thereafter. How many hours must she work in one week to make more than $350?

30. The total height of the Statue of Liberty, including the pedestal and the statue, is 305 ft. If the statue is 3 ft shorter than the pedestal, what is the height of each?

31. The price of a camera decreased by 10% this year. If the camera now costs $750, how much did it cost a year ago?

32. The perimeter of a rectangle is 34 yd. If the length is one more than three times the width, what are the dimensions of the rectangle?

33. In 1998, the United States consumed 9.48×10^{16} BTUs of electricity, which is about 25% of the world's total energy usage. How much energy did the world consume in 1998?

34. Students in a northeast school miss three days of classes due to a snowstorm. They plan to make up the missed days by extending each 6-hr school day by 30 min. How many days do they need to extend to make up the three missed days?

35. On her first four 100-point tests, Li received grades of 78, 85, 95, and 96. If she wants to have at least a 90 average in the class, what is the lowest score she can receive on her last 100-point test?

36. At a high school basketball game, 850 tickets are sold. Student tickets cost $1.50 each, and adult tickets cost $3.00 each. The total revenue collected for the game is $1,800. How many of each type of ticket are sold?

37. Two pumps are used to fill a pool that holds 27,000 gal of water. Alone, the first can fill the pool in 15 hr, and the second can do it in 20 hr. How long will it take to fill the pool with the pumps working together?

38. The owner of a candy store wants to make 40 lb of a mixture of gumballs to sell for $1.50/lb. If large gumballs sell for $2.50/lb and small gumballs sell for $1.25/lb, how many pounds of each should be included in the mix?

39. Dried apples sell for $5.59/lb, and raisins sell for $2.13/lb. To make a 15-lb mixture of the two costing $3.50/lb, how many pounds of each should be included?

40. Zehra splits an inheritance between two accounts, one paying 6% simple interest and another paying 10% simple interest. She invested three times as much in the 10% account as she did in the 6% account. Her combined interest from the two accounts in one year was $5,050. About how much did she inherit?

41. Louis is trying to determine the cost of his water bill for various amounts of water usage. As part of his water bill, each month he pays $23.99 for sewer, $3.29 for a service fee, and $0.004 for every gallon used. If tax is 4% of the amount of the bill, how much will the bill be if Louis uses:

   a. 3,000 gal?   **b.** 8,000 gal?

   c. 14,500 gal?

42. Business at Harry's Hamburger Hangout has declined over the last several months. Harry looks at his inventory records to determine how many hamburgers he should order in the upcoming months. His records show not only how many hamburgers he has sold, but also how many hamburgers he had to throw out.

   a. If the trend continues, how many hamburgers will he need for June?

   b. If the trend continues, how many hamburgers will he need for next June?

Month	December	January	March	April	May
Hamburgers ordered	1,200	1,150	1,100	1,050	1,000
Hamburgers thrown out	175	167	159	151	143

43. A framing company uses its past sales records to determine how many frames of a particular size it should make in the next month. The following table shows sales for three different frame sizes over the last 6 months.

Frame Size	March	April	May	June	July	August
5 × 7 in.	15,600	16,075	16,554	17,035	17,509	17,986
11 × 14 in.	8,075	7,450	6,826	6,201	5,578	4,951

 a. Are the increases or decreases in orders for each frame size approximately linear? How do you know?

 b. Assuming the trend shown in the table continues, how many months must pass from March for the orders of 5 × 7- in. frames to top 20,000 frames?

 c. Assuming the trend shown in the table continues, how many months must pass from March for the orders of 11 × 14- in. frames to cease?

Mathematical Communication

44. The following graph shows a person's heart rate over a given period of time. Write a short statement giving a possible explanation for the graph.

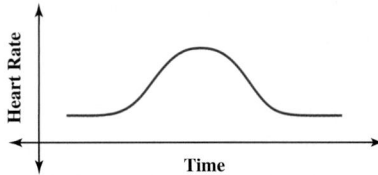

45. Jaleel leaves home to run a few errands. The following graph shows the distance from his house by time. Make an interpretation of the changes in the graph.

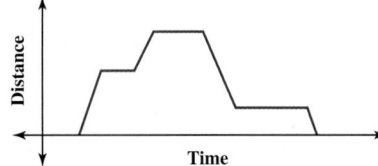

46. The following graph shows what happened when three horses ran a race. Answer each question based on what you see in the graph.

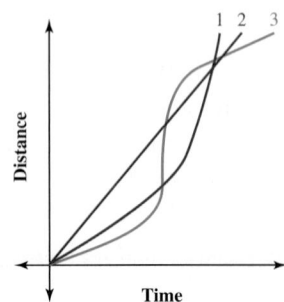

 a. Which horse had the fastest start? The slowest?

 b. Which horse won the race?

 c. Describe the race from start to finish.

47. Write a story similar to the one given in Example 9.41. Swap stories with a partner, and draw a graph for the story given to you.

48. With a small group of your peers, write several story problems that can be solved using linear equations. Once the questions are written, discuss with your peers whether writing the questions helped you gain a better understanding of mathematical modeling with linear equations. If so, discuss how.

49. What do you think are the primary difficulties students have when learning to model mathematically? How might you as a teacher help them through these difficulties? Discuss your thoughts with a group of your peers.

Building Connections

50. Describe the mathematical models used in previous chapters to model:

 a. Place value.   b. Whole-number operations.

 c. Integer operations. d. Rational numbers.

51. Consider your answers to the last exercise. What role did mathematical modeling have in developing the structure of the real numbers?

52. How does each situation represent a functional relationship?

 a. The cost of an electric bill based on the amount of electricity used

 b. The assignment of students to particular desks in a classroom

 c. The assignment of Social Security numbers to people

 d. The time spent driving a car and the distance traveled

53. Many real-life situations can be described by linear functions. Describe the meaning of the slope in each graph.

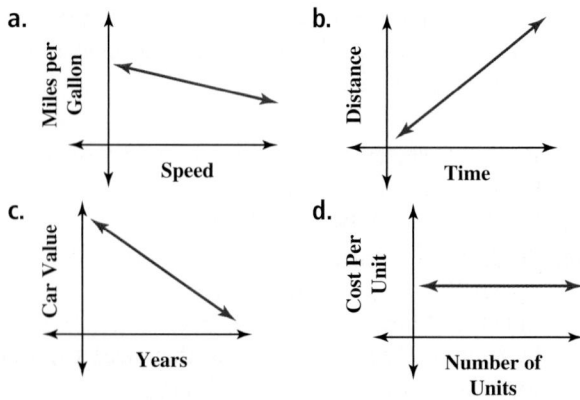

54. Look over several chapters from a set of elementary curriculum materials. How much mathematical modeling is involved in these chapters? Does this surprise you?

Historical Note

Algebraic techniques first came to Europe through the writings of Muhammed ibn Musa al-Khowarizmi, who was a scholar in the so-called House of Wisdom in Baghdad around 820 C.E. His book *Hisab al-jabar w' al muqabalah* (*The Science of Reunion and Reduction*) collects the works of the Arabic mathematicians of his time and declares algebraic rules as divine revelations that the reader is to accept and follow as truth. Here, "reunion" refers to transferring negatives from one side of an equation to the other, and "reduction" refers to combining or canceling like terms on opposite sides of the equation. The term "algebra" was the European version of *al-jabar* and was used as a global term for all the processes described by al-Khowarizmi.

Muhammed ibn Musa al-Khowarizmi

Egmont Strigl/age fotostock

Opening Problem Answer

Eduard Titov/Shutterstock.com

Jenna's fixed costs are 12 monthly payments of $212.47 and two semiannual insurance payments of $459. Hence, her fixed costs are $3,467.64. Because her variable costs are $0.32x, where x is the number of miles she drives in a year, we can model her cost with the function $C(x) = 0.32x + 3{,}467.64$.

a. $C(12{,}000) = 0.32(12{,}000) + 3{,}467.64 = \$7{,}307.64.$

b. $C(15{,}000) = 0.32(15{,}000) + 3{,}467.64 = \$8{,}267.64$

c. $C(20{,}000) = 0.32(20{,}000) + 3{,}467.64 = \$9{,}867.64.$

d. To find out how many miles she has to drive to spend $4,000 on her car, we let $0.32x + 3{,}467.64 = 4{,}000$, and solve for x. Hence, $x \approx 1{,}664$ mi.

Chapter 9 REVIEW

Summary of Key Concepts from Chapter 9

Section 9.1   Numerical Sequences

- **Functional thinking** involves identifying, representing, and extending patterns of change *between* sets.

- A **sequence** is an ordered list of objects in which one is designated as first, another as second, and so on. If the sequence is governed by a pattern, we express it with the **general** or **nth term.**

- A sequence is **repeating** if some core component continuously repeats, **increasing** if the terms continue to grow larger in value, and **decreasing** if the terms continue to grow smaller.

- In an **arithmetic sequence,** each successive term is found by adding a fixed number, called the **common difference d,** to the previous term. The general term is given by $a_n = a_1 + (n - 1)d$.

- In a **geometric sequence**, each successive term is found by multiplying a constant number, called the **common ratio r,** to the previous term. The general term is given by $a_n = r^{n-1}a_1$.

- In a **recursive sequence,** the value of each term, except for the first few, is based on the value or values of previous terms.

Section 9.2 Functions and Their Representations

- A **function** from a set A to a set B is a rule of correspondence that assigns to each element of A exactly one element from B. The set A is called the **domain,** the set B is called the **codomain,** and the set of all elements in the codomain that have been paired with something from the domain is called the **range.**

- Common representations of functions include arrow diagrams, function machines, algebraic equations, tables, sets of ordered pairs, and graphs.

- A function can be written symbolically with **independent** and **dependent** variables or with **function notation,** which is given by $f(x) = $ "algebraic rule in terms of x."

- The domain is always the largest set of real numbers defined for the rule of the function.

- The **Cartesian coordinate system** creates a grid on the plane by taking two number lines, called the **coordinate axes,** and intersecting them perpendicularly at zero. The **x-axis** designates left-right positions, the **y-axis** designates up-down positions, and the intersection of the axes is called the **origin.**

- The **graph** of the function is the set of all points in the Cartesian plane that satisfy the rule of the function.

- **Vertical line test:** If a vertical line intersects a graph at more than one point for any domain value x, then the graph does not represent a function.

Section 9.3 Linear Functions and Constant Change

- A **linear function** represents a function of constant change.

- The **slope** of a line is the ratio of its rise to its run, or slope $= m = \dfrac{\text{rise}}{\text{run}} = \dfrac{y_2 - y_1}{x_2 - x_1}$.

- The **y-intercept** of a line is the point at which the line crosses the y-axis.

- The **point-slope form** of a line is given by $y - y_1 = m(x - x_1)$.

- The **slope-intercept form** of a line is given by $y = mx + b$.

- Two lines are **parallel** if and only if the slopes are equal or the lines are vertical.

- Two lines are **perpendicular** if and only if one is vertical and the other is horizontal or the product of their slopes is -1.

Section 9.4 Solving Equations and Inequalities

- We **solve** the equation or inequality by finding every value that makes it true. The set of all such values is called the **solution set.**

- A **linear equation of one variable** is an equation in which each term is either a real-number constant or the product of a number and a variable raised to the first power. The number multiplying the variable is called a **coefficient.**

- If a, b, and c are real numbers with $a = b$, then

  1. $a + c = b + c$. **2.** $a - c = b - c$.

  3. $a \cdot c = b \cdot c$. **4.** $\dfrac{a}{c} = \dfrac{b}{c}$ when $c \neq 0$.

- A **linear inequality** is a linear statement that uses $<$, $\leq$, $>$, $\geq$, or $\neq$, rather than $=$.

- Whenever both sides of an inequality are multiplied or divided by a negative number, the inequality must be reversed to maintain the truthfulness of the statement.

- A **quadratic equation** is an equation in which one or more terms have a squared variable but no higher power.

- A **rational equation** is an equation in which the variable is in the denominator of a fraction.

Section 9.5   Algebraic Thinking and Mathematical Modeling

- **Mathematical modeling** is the process of creating mathematical representations that can be used to understand and solve real-world problems.

- Let f be a function on the interval $[a, b]$, and let x_1 and x_2 be any elements in $[a, b]$ such that $x_1 < x_2$. Then f is **increasing** on $[a, b]$ if $f(x_1) < f(x_2)$, f is **decreasing** on $[a, b]$ if $f(x_1) > f(x_2)$, and f is **constant** on $[a, b]$ if $f(x_1) = f(x_2)$.

- A **system of linear equations,** or **simultaneous equations,** are two linear equations that contain the same two variables. A **solution** to the system is any ordered pair of numbers (x, y) that makes both equations true.

- In the **substitution method,** we solve one of the equations for one of the variables and then substitute the result into the other equation.

- In the **elimination method,** we eliminate one of the variables by adding or subtracting the original or equivalent equations.

Review Exercises Chapter 9

1. Determine whether each sequence is arithmetic, geometric, or neither. Find the general term if possible, and use it to compute the value of the 10th, 25th and 50th terms.

   a. 1, 2, 4, 7, 11, . . .

   b. 6, 13, 20, 27, . . .

   c. 1, 4, 16, 64, . . .

   d. 1, 0.3, 0.09, 0.027, . . .

   e. 1, 11, 111, 1111, . . .

   f. $-2, -5, -8, -11, \ldots$

2. In an arithmetic sequence, the 1st term is -4, and the 10th term is 9.5. What is the value of the 20th term?

3. If possible, give an example of a recursively defined sequence that is decreasing.

4. What is the difference between the domain, the codomain, and the range of a function?

5. Use an arrow diagram, a set of ordered pairs, and a table to demonstrate three functions between the set $\{b, i, g\}$ and the set $\{t, o, e\}$.

6. Find $f(-1), f(3)$ and $f(s + 2)$ for:

   a. $f(x) = -4x$.

   b. $f(x) = 2x + 6$.

   c. $f(x) = -x^2 + 7$.

7. For each function, find $f(3)$ if it exists.

   a. $f(x) = x - 3$.

   b. $f(x) = \sqrt{x}$.

   c. $f(x) = \dfrac{1}{x + 3}$.

   d. $f(x) = \dfrac{1}{x - 3}$.

   e. $f(x) = \sqrt{3 - x}$.

8. The following table shows approximate flight times from Chicago to various cities. Does the table represent a function? Explain.

Flight from Chicago to	Hours
Boston	2
Dallas	3
Los Angeles	3.5
Miami	4
New York	2
San Francisco	3

9. Give a set of five ordered pairs that satisfy the function $y = 3x^2 - 5$.

10. Give a table with five entries that satisfy the function $y = 2^x - 8$.

11. State the position of each ordered pair relative to the origin on the Cartesian coordinate system.

   a. $(2, -1)$ **b.** $(-3, -3)$

   c. $(0, 4)$ **d.** $(-5, 0)$

12. State the domain of each function.

   a. $f(x) = -x$.   **b.** $f(x) = 2x^2$.

   c. $y = \sqrt{x + 3}$.   **d.** $y = \dfrac{1}{x - 5}$.

13. Graph each function by making a table of values and plotting them on the Cartesian coordinate system.

   a. $y = 2x$.   **b.** $y = -x + 3$.

   c. $f(x) = 2x^2 - 1$.   **d.** $f(x) = abs(x + 1)$.

14. Interpret the meaning of each slope, m, as a ratio of the rise of a line to its run.

    a. $m = -2$.   **b.** $m = -\dfrac{4}{5}$.   **c.** m is undefined.

15. Compute the slope of the line containing each pair of points. Based on the slope, determine whether the line is increasing, decreasing, horizontal, or vertical.

    a. $(-3, 4)$ and $(5, 5)$ **b.** $(1, 2)$ and $(4, -2)$
    c. $(3, 1)$ and $(5, 1)$

16. Which represents a steeper line, a slope of 5 or a slope of $\dfrac{1}{5}$? Explain.

17. Graph each linear equation.

    a. $x = 4$.   **b.** $y = -3$.

    c. $y = -4x + 1$.   **d.** $y = \dfrac{1}{2}x + 2$.

18. If they exist, find the slope, y-intercept, and x-intercept of each function.

    a. $y = -x - 5$.   **b.** $y = \dfrac{2}{3}x$.

    c. $(y - 4) = \dfrac{5}{2}(x - 2)$.

19. Find an equation of a line for each of the following.

    a. The line parallel to the y-axis and containing the point $(4, 1)$

    b. The line perpendicular to the line $y = x - 4$ and containing the point $(1, 2)$

    c. The line parallel to the line $y = \dfrac{3}{4}x + 1$ and containing the point $(3, 5)$

20. a. Give an equation of a line that is parallel to $y = -3x - 5$.

    b. Give an equation of a line that is perpendicular to $y = 2x + 7$.

21. Represent each set in set-builder notation, in interval notation, and with a graph.

    a. The set of all real numbers less than or equal to 5

    b. The set of all real numbers greater than 2 but less than or equal to 7

22. Solve each of the following. For inequalities, write the solution set in interval notation.

    a. $6y + 10 = 0$.   **b.** $153 = 212 - 3y$.
    c. $1.4t + 6.5 = 13.1$.   **d.** $-3(4x - 2) = 3x + 15$.
    e. $12 \le 21 - 4(2 - x)$.   **f.** $-3t + 5 \ge 22$.
    g. $-3y + 5 < 16 - 2y$.   **h.** $3x + 1 \ne 6x - 3$.
    i. $2(x - 1) + 3x > 4x - 8$.

23. Solve each system of equations.

    a. $\begin{aligned} 2x - y &= 0 \\ 2x - 3y &= 5 \end{aligned}$ **b.** $\begin{aligned} y &= -2x - 3 \\ y &= 4x - 1 \end{aligned}$

24. Give an example of a system of two linear equations that have:

    a. No solution.

    b. Exactly one solution.

    c. An infinite number of solutions.

25. A large empty tank can hold 105,000 gal of water. The tank is being filled at a rate of 9,350 gal/day. How much water will be in the tank on the 7th, 8th, and 9th days?

26. The cost of water service in a certain town is given by the equation $C(x) = 0.005x + 6.5$, where x is the number of gallons of water used. What is the monthly water bill for

    a. 1,500 gal?

    b. 8,000 gal?

    c. 15,500 gal?

27. Sales tax in a particular state is 7% of all purchases.

    a. Write a formula for the amount of a bill on any purchase if the sales tax is included.

    b. What is the independent variable? The dependent variable?

    c. What is a logical domain for this function?

28. The following graph shows the relationship between the number of cars on the road and the time of day. Interpret the meaning of the graph.

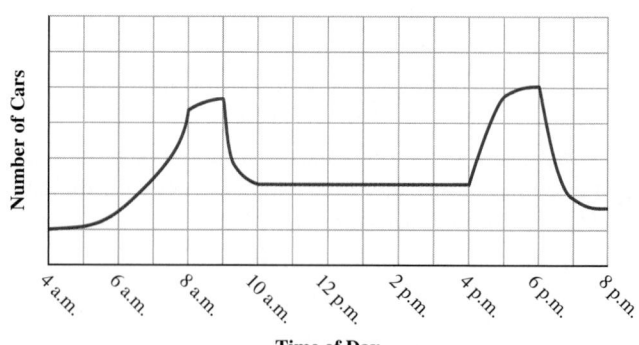

29. Robbie is paid $7/hr for a 40-hr work week, and then gets paid time and a half for any hours over 40. In one week, he makes $416.50. How many hours did he work that week?

30. Five thousand grams of a radioactive material decay according to the formula $A = 5{,}000(0.83)^t$, where t is the time in years. How much material will be left after 5 years?

Answers to Chapter 9   Check Your Understandings and Activities

Check Your Understanding 9.1-A

1. **a.** 2, 3, 2, 3, 2, ...   **b.** 22, 29, 37, 46, 56 . . . .
   c. 20, 24, 28, 32, 36, ...

2. **a.** Arithmetic.   **b.** Neither.   **c.** Geometric.

3. **a.** $d = 4$, $a_n = 4n + 6$, $a_{10} = 46$, $a_{15} = 66$, and $a_{25} = 106$.
   b. $d = -3$, $a_n = -3n - 2$, $a_{10} = -32$, $a_{15} = -47$, and $a_{25} = -77$.
   c. $d = 0.8$, $a_n = 0.8n + 0.5$, $a_{10} = 8.5$, $a_{15} = 12.5$, and $a_{25} = 20.5$.

4. **a.** $r = 5$, $a_n = 4(5)^{n-1}$, $a_5 = 2{,}500$, $a_8 = 312{,}500$, and $a_{10} = 7{,}812{,}500$.
   b. $r = 3$, $a_n = 7(3)^{n-1}$, $a_5 = 567$, $a_8 = 15{,}309$, and $a_{10} = 137{,}781$.
   c. $r = 1.5$, $a_n = 1.1(1.5)^{n-1}$, $a_5 = 5.569$, $a_8 = 18.795$, and $a_{10} = 42.288$.

Check Your Understanding 9.1-B

1. **a.** 3, 1, −3, −11, −27, ...   **b.** 1, 4, 2, −6, −10, ...

2. $F_{15} = 610$, $F_{16} = 987$, $F_{17} = 1{,}597$, $F_{18} = 2{,}584$, $F_{19} = 4{,}181$, and $F_{20} = 6{,}765$.

3. For the triangular numbers, $a_{10} = 55$, $a_{11} = 66$, and $a_{12} = 78$. For the square numbers $a_{10} = 100$, $a_{11} = 121$, and $a_{12} = 144$.

4. Here, $a_n = n(n + 1)$, where n is the position of the figure in the sequence. Hence, $a_6 = 6(6 + 1) = 42$, $a_7 = 7(7 + 1) = 56$, and $a_8 = 8(8 + 1) = 72$.

Check Your Understanding 9.2-A

1. **a.** Function.
   b. Not a function: 2 is a domain element that is paired with more than one element.
   c. Not a function: 3 is a domain element that is not used.
   d. Function.

2. **a.** Domain = {1, 2, 3, 4, 5, 6}; range = {b, c, d}.
   b. Domain = {red, blue, black, gray, green}; range = {r, b, g}.
   c. Domain = {1, 2, 3, 4, 5}; range = {5, 7}.

3. Three different possible functions are shown.

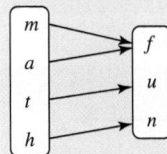 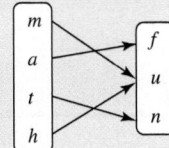 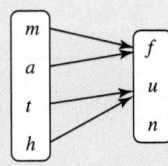

4. **a.** Function.   **b.** Not a function.

Check Your Understanding 9.2-B

1. **a.** Function.   **b.** Not a function.
   c. Not a function.   **d.** Function.

2. $f(-2) = -10$, $f(-1) = -7$, $f(0) = -4$, $f(1) = -1$, $f(2) = 2$, and $f(3) = 5$. The graph follows.

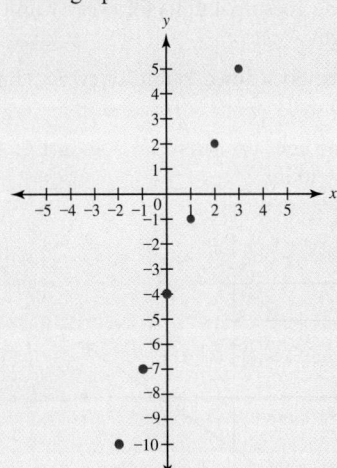

3. Tables of values may differ. The graphs follow.

a.

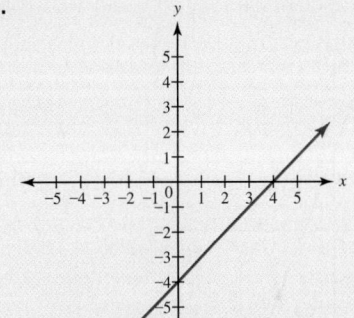

b.

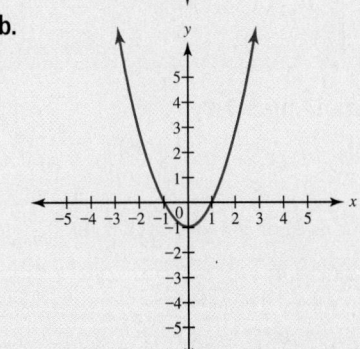

c.

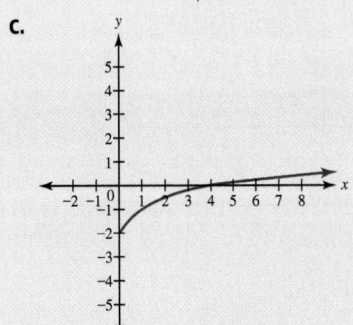

4. **a.** Function.   **b.** Not a function.   **c.** Function.

Check Your Understanding 9.3-A

1. a. Linear. **b.** Not linear. **c.** Linear. **d.** Not linear.

2. a. $\frac{1}{3}$. **b.** 0. **c.** −1.

3. a. The line goes up 4 units for every 1 unit that it goes to the right.

  b. The line goes down 1 unit for every 2 units it goes to the right.

  c. The line goes up 1 unit but does not go to the right or left any units.

4. a.

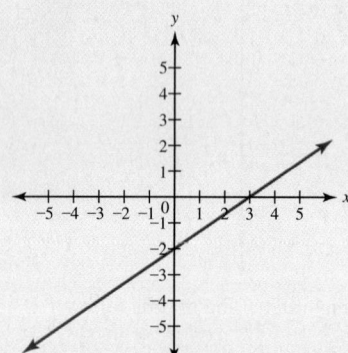

  b.

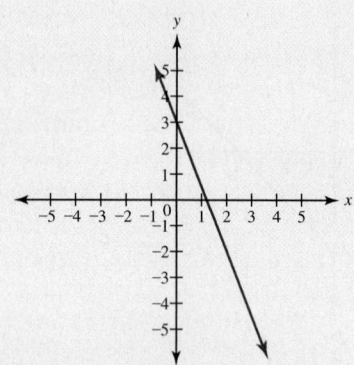

Check Your Understanding 9.3-B

1. Line l: $y - 3 = -\frac{1}{7}(x + 4)$ or $y - 2 = -\frac{1}{7}(x - 3)$

  line m: $y - 2 = \frac{5}{4}(x - 3)$ or $y + 3 = \frac{5}{4}(x + 1)$

  line n: $y - 3 = (-2)(x + 4)$ or $y + 3 = (-2)(x + 1)$

2. a. $y = -\frac{3}{5}x + 5$. **b.** $y = \frac{4}{3}x + \frac{14}{3}$.

3. a.

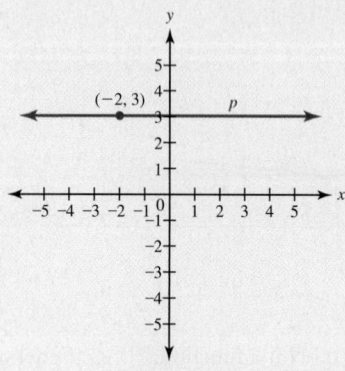

b.

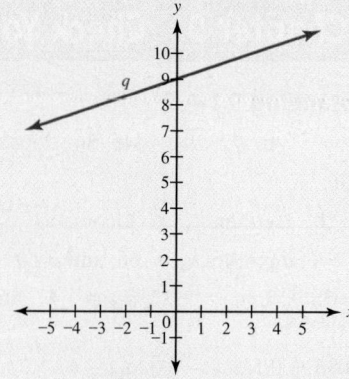

  c. Line r is the line $y = 3x - 11$. Its graph follows.

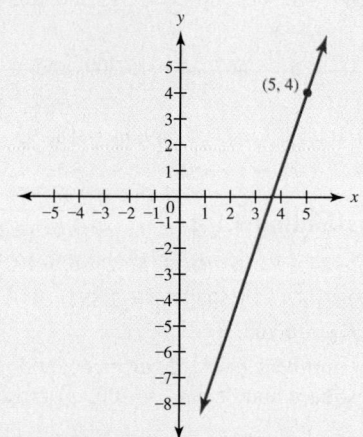

Check Your Understanding 9.4-A

1. a. $x = 4$. **b.** $x = 9$. **c.** $x = 9$. **d.** $x = 45$.

2. a. $x = 10$. **b.** $x = \frac{19}{3}$. **c.** $x = -\frac{1}{2}$.

3. a. $x = \frac{7}{6}$. **b.** $x = \frac{16}{5}$.

Check Your Understanding 9.4-B

1. a. $\{x \mid x < 4\} = (-\infty, 4)$;

  b. $\{x \mid x \neq -3\} = (-\infty, -3) \cup (-3, \infty)$;

  c. $\{x \mid x \geq 1\} = [1, \infty)$;

  d. $\{x \mid x < -2 \text{ or } x \geq 3\} = (-\infty, -2) \cup [3, \infty)$;

2. a. $(-\infty, 6) \cup (6, \infty)$;

  b. $(-\infty, -11)$;

  c. $\left[\frac{3}{2}, \infty\right)$;

3. a. $x \in \{-5, 5\}$. **b.** $x = 52$. **c.** $x = 3$.

Check Your Understanding 9.5-A

1. **a.** Increasing on $\left(-\frac{1}{2}, \frac{1}{2}\right)$ and decreasing on

$$\left(-\infty, -\frac{1}{2}\right) \cup \left(\frac{1}{2}, \infty\right).$$

   b. Increasing on $(-\infty, 0)$, decreasing on $(0, 4)$, and constant on $(4, \infty)$.

2. One possible graph is shown.

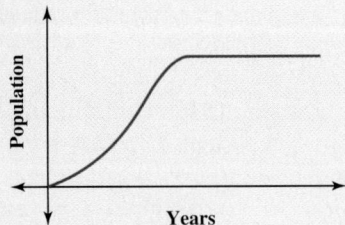

3. **a.** \$19,663.40.

   b. \$32,220.77.

   c. \$232,297.80.

Check Your Understanding 9.5-B

1. The smaller tank holds 25 gal, and the larger tank holds 60 gal.

2. 88 or better.

3. 17 quarters and 18 dimes.

4. **a.** $(4, 2)$.   **b.** $\left(\frac{6}{7}, -\frac{5}{7}\right)$.

   c. Infinitely many solutions.

Activity 9.1

The sum of first n terms of any arithmetic sequence is given by the formula $S_n = \frac{n(a_1 + a_n)}{2}$.

a. 255.   **b.** 540.   **c.** -210.

Activity 9.2

Answers will vary.

Activity 9.3

Answers will vary.

Activity 9.4

a. 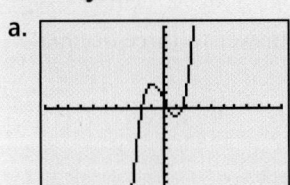   **b.**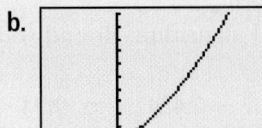

c.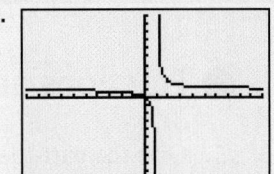

Activity 9.5

a. 7.   **b.** $\frac{9}{2}$.   **c.** -1.   **d.** -2 **e.** $\frac{4}{3}$.

Activity 9.6

$(3, 6)$.

Activity 9.7

If $a = b$, then $a - b = 0$. So in the fourth step, we multiply both sides of the equation by zero, making the proof meaningless.

Activity 9.8

Use algebraic thinking. Let $\triangle$, $\square$, and $\diamond$ represent variables and use algebraic manipulations to find $5\square = \triangle$.

Activity 9.9

a. \$0.17/oz.

b. $y = 0.17x + 0.96$; $m = 0.17$, y-intercept $= 0.96$.

c. i. \$1.69.   **ii.** \$1.95.   **iii.** \$3.15.

d.

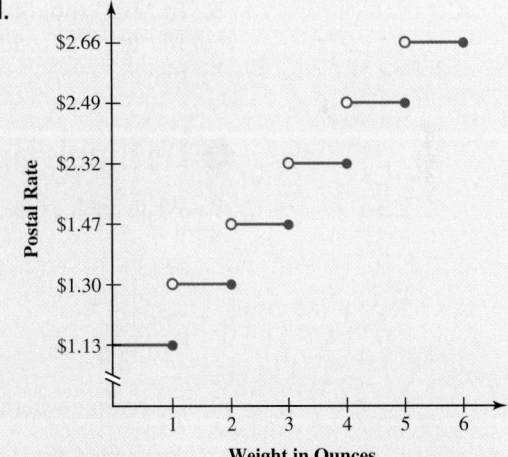

Activity 9.10

a. The two ships can collide at the point $(-2, -3)$.

b. The collision is not certain because the ships may pass through this point at different times.

◆ Fractions and Rational Numbers

1. Give the part-to-whole interpretation of $\frac{5}{8}$.

2. Make each conversion.

   a. $\frac{40}{7}$ to a mixed number

   b. $-\frac{29}{8}$ to a mixed number

   c. $3\frac{4}{7}$ to an improper fraction

   d. $-5\frac{1}{5}$ to an improper fraction

3. Simplify each fraction.

   a. $\frac{24}{42}$

   b. $\frac{66}{165}$

   c. $-\frac{153}{68}$

4. Find five fractions between $\frac{4}{7}$ and $\frac{5}{7}$.

5. Compute.

   a. $\frac{5}{14} + \frac{3}{7}$

   b. $\frac{9}{5} - \frac{2}{3}$

   c. $\frac{7}{8} \cdot \frac{-24}{21}$

   d. $\frac{8}{15} \div \frac{12}{25}$

6. Compute.

   a. $3\frac{1}{2} + 2\frac{2}{3}$

   b. $4 - 1\frac{5}{6}$

   c. $1\frac{5}{6} \cdot 1\frac{1}{4}$

   d. $-6\frac{1}{3} \div 3\frac{3}{4}$

7. Find the multiplicative inverse of each number.

   a. $\frac{4}{5}$

   b. $-\frac{1}{4}$

   c. 8

   d. $2\frac{1}{4}$

8. In Mr. Arnold's class, 12 out of 22 students play a sport; in Mrs. Henry's class, 14 out of 25 students play a sport. Which class has a higher value of students playing sports?

◆ Decimals and the Real Numbers

9. Write each expression as a decimal.

   a. $(3 \cdot 10) + (4 \cdot 1) + \left(2 \cdot \frac{1}{10}\right) + \left(9 \cdot \frac{1}{1,000}\right)$

   b. $\frac{19}{1,000}$

   c. $\frac{7}{21}$

   d. 135%

10. Determine whether $\frac{17}{80}$ has a terminating decimal representation. If so, find it by converting the denominator to a power of ten.

11. Determine whether each radical is rational, irrational, or not possible.

    a. $\sqrt{-81}$

    b. $\sqrt{120}$

    c. $\sqrt[4]{16}$

    d. $\sqrt[3]{-64}$

12. Compute using the standard algorithm. Round your answer to three decimal places if necessary.

    a. $45.61 + 33.09$

    b. $6.007 - 2.434$

    c. $9.34 \cdot 8.5$

    d. $34.62 \div 7.1$

13. Write $0.6\overline{5}$ as the ratio of two integers.

14. a. Write 459,000 in scientific notation.

    b. Write 0.0005601 in scientific notation.

    c. Write 3.49×10^{-7} in decimal notation.

    d. Write 2.88×10^{4} in decimal notation.

15. Write each of the following as a percent.

    a. 0.71   b. 1.05   c. $\dfrac{4}{5}$ d. $\dfrac{3.9}{7.8}$

16. A short-order cook makes $11.25/hr and works 65.5 hr in one week. If the cook gets paid time and a half for any hours over 40, how much did he get paid for the week?

◆ Proportional Thinking

17. If there are 24 students in a class and 14 of them are female, find the ratio of:

    a. Females to all students.   b. Males to females.   c. All students to males.

18. Which is a better buy, 3 lb of meat for $3.45 or 8 lb of meat for $9.84?

19. If it takes Rita 4 min to swim 10 lengths of a pool, how long would we expect her to take to swim 18 lengths of the pool?

20. Solve each problem.

    a. 30% of 120 is what?

    b. 40 is what percent of 93?

    c. 13 is 18% of what number?

21. Compute the amount of money in each account after the term of the loan has expired.

    a. $P = \$3,500$, $r = 2\%$, $t = 5$ years with simple interest

    b. $P = \$23,000$, $r = 4.5\%$, $t = 10$ years, compounded quarterly

22. Over the last 3 years, Mike has received pay raises of 2%, 5%, and 3.5%, respectively. If his salary in the first year was $42,650, what is his salary to the nearest penny after the three raises?

23. Over the course of a year, the value of stock portfolio A increases from $8,595 to $11,425, and the value of stock portfolio B increases from $18,350 to $23,480. Which portfolio had the greatest percent increase?

◆ Sequences and Functions

24. Determine whether each sequence is arithmetic, geometric, or neither. Find the general term when possible, and use it to compute the value of the 10th, 15th, and 20th terms.

    a. 2, 3, 5, 8, 12, . . .   b. 3, 6, 12, 24, 48, . . .

    c. $-3, -7, -11, -15, \ldots$ d. 1, 12, 123, 1234, . . .

25. List the first six terms of each recursively defined sequence.

    a. $a_n = 3a_{n-1} + 2$, where $a_1 = 2$.

    b. $a_n = a_{n-1} \cdot a_{n-2}$, where $a_1 = 2$ and $a_2 = 3$.

26. Demonstrate two functions between the set $\{m, a, t, h\}$ and the set $\{f, a, c, t\}$. Represent each function with an arrow diagram, a set of ordered pairs, and a table.

27. Find $f(2), f(-4)$ and $f(t-3)$ for:

    a. $f(x) = 3x + 1$.   b. $f(x) = -2x^2 - 3$.

28. State the domain of each function.

    a. $f(x) = 2x - 7$.   b. $y = \dfrac{2}{x-3}$.   c. $y = \sqrt{x+3}$.   d. $y = \dfrac{1}{x-5}$.

29. Graph each function by making a table of values and plotting them on the Cartesian coordinate system.

    a. $y = 2x + 2$.   b. $y = x^2 + 3$.   c. $f(x) = |x - 3|$.

30. Compute the slope of the line containing each pair of points. Based on the slope, determine whether the line is increasing, decreasing, horizontal, or vertical.

    a. $(-2, -2)$ and $(1, 3)$ **b.** $(1, 2)$ and $(4, 2)$

    c. $(3, 3)$ and $(5, 1)$ **d.** $(4, -3)$ and $(4, 5)$

31. Find the slope, y-intercept, and x-intercept of each function when they exist.

    a. $y = x + 5$.   **b.** $y = -3$.   **c.** $(y + 2) = -\dfrac{1}{3}(x + 3)$.

32. Find the equation of a line that is perpendicular to the line $y = \dfrac{1}{2}x + 2$ and that passes through the point $(-3, 4)$.

◆ Equations, Inequalities, and Mathematical Modeling

33. Solve each equation.

    a. $14 = 21 - 3y$.   **b.** $1.5t - 4.1 = 7.9$.   **c.** $2(3x - 4) = 7x + 9$.

34. Solve each inequality. Graph the solution set, and then write it in interval notation.

    a. $-3t - 5 \geq 16$.   **b.** $-2y + 7 > 11 + 2y$.   **c.** $-2x + 6 \neq 3x - 9$.

35. Solve each system of equations.

    a. $x - 2y = 0$ **b.** $y = x - 2$

        $2x + 4y = 4$. $y = 3x - 4$·

36. A swimming pool can hold 32,000 gal of water. It is being filled at a rate of 450 gal/hr. How much water will be in the pool after the 10th, 11th, and 12th hours? How long will it take to fill the pool?

37. I have a total of 30 dimes and nickels in a jar. If the total amount of change is equal to $2.35, how many of each coin is in the jar?

38. Four thousand grams of a radioactive material decay according to the formula $A = 4000(0.73)^t$, where t is the time in years. How much material will be left after 10 years?

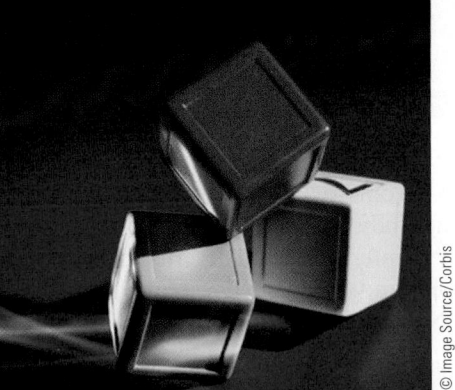

© Image Source/Corbis

Chapter 10
Geometrical Shapes

Opening Problem

Martin wants to build a practice cage for hitting golf balls. The frame will be in the shape of a cube that measures 10 ft on an edge. If he builds it out of metal tubing that costs \$0.95 a linear ft, how much will it cost to build the frame?

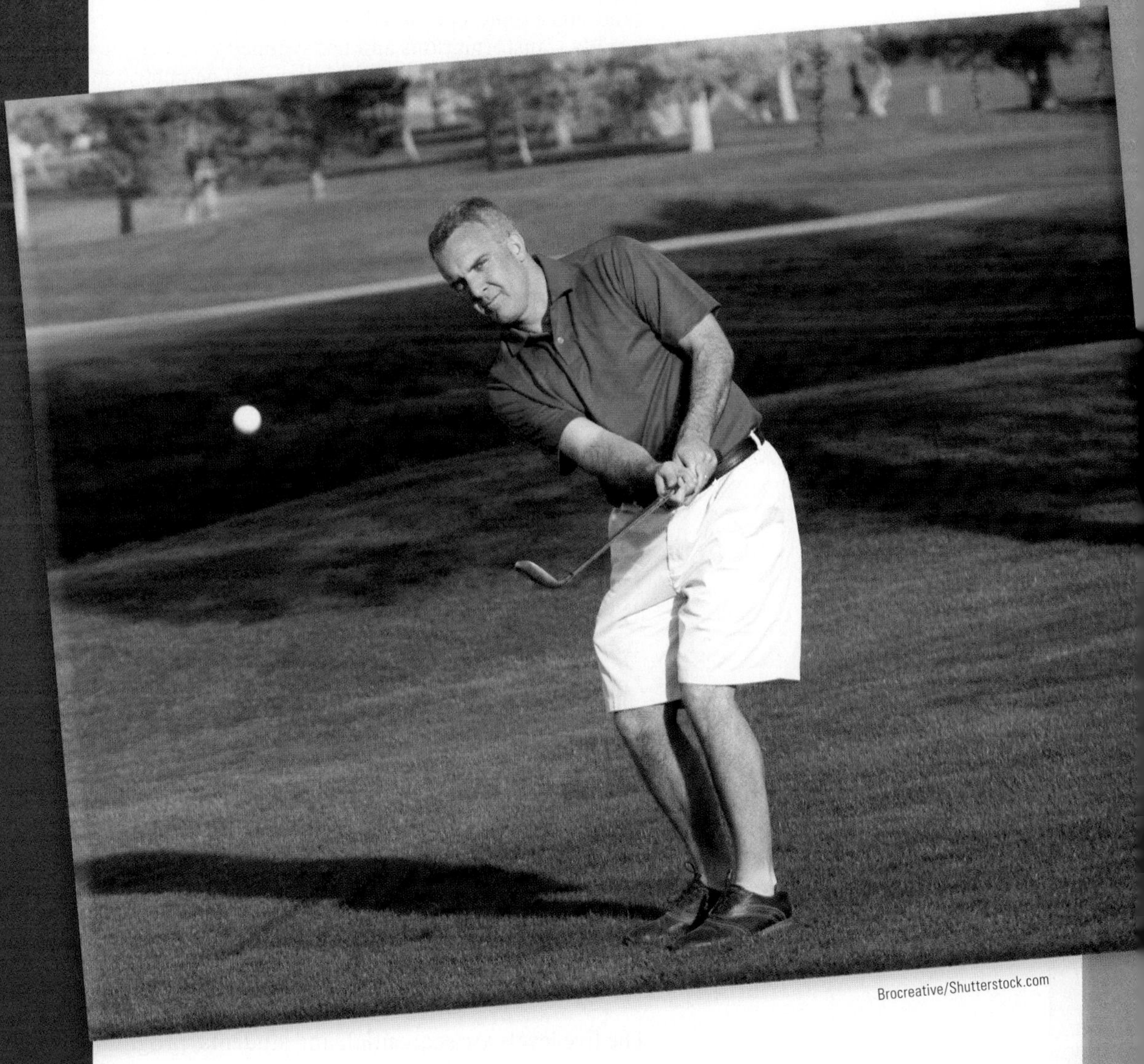

Brocreative/Shutterstock.com

Chapter 10 marks a shift away from our study of number structures and algebraic thinking and over to geometry. The word "geometry" comes from two Greek words that mean earth measure, so it is literally the study of shapes, their properties, and their measures. We use it extensively, not only in interacting and describing the world, but also in learning other mathematical topics and academic disciplines. Because of its importance, geometry is the third of the NCTM Content Standards.

■ NCTM Geometry Standard

Instructional programs from prekindergarten through grade 12 should enable all students to:

■ Analyze characteristics and properties of two- and three-dimensional geometric shapes and develop mathematical arguments about geometric relationships.
■ Specify locations and describe spatial relationships using coordinate geometry and other representational systems.
■ Apply transformations and use symmetry to analyze mathematical situations.
■ Use visualization, spatial reasoning, and geometric modeling to solve problems.

Source: NCTM STANDARDS Copyright © 2011 by NATIONAL COUNCIL OF TEACHERS OF MATHEMATICS. Reproduced with permission of NATIONAL COUNCIL OF TEACHERS OF MATHEMATICS.

The Geometry Standard covers a wide variety of topics, so it is addressed in the next several chapters. In this chapter, we focus primarily on the first goal, analyzing shapes and their properties. As we do, we will also address topics from the fourth goal because spatial reasoning is important to all aspects of geometry.

Learning About Shapes

In the late 1950s, two Dutch mathematics teachers, Pierre and Dieke van Hiele, investigated how children learn about shapes and their properties. They found that most children progress through a sequence of five levels:

Level 0: Visualization. Students recognize and name shapes by visual features. They react to shapes as wholes and do not understand that a figure is defined by its properties. They are affected by irrelevant attributes, such as orientation. For instance, a square is a square because it "looks like a square."

Level 1: Analysis. Students analyze shapes for their parts and the relationships between them. For instance, a square has four equal sides and four right angles.

Level 2: Abstraction. Students form abstract definitions of concepts by using properties that distinguish one concept from another. For instance, a student sees that a square is both a rectangle and a quadrilateral and also that additional features separate it into a class of its own.

Level 3: Deduction. Students reason formally to create proofs using undefined terms, axioms, definitions, and theorems.

Level 4: Rigor. Students compare and relate different geometrical systems without the use of concrete models.

The five levels are sequential, and students' progress through them depends more on experience and instruction rather than on age or grade level. In the elementary and middle grades, the curriculum tries to move students through the first three levels so that they are prepared to work on the last two in high school and college. In this early work, students identify and name shapes, explore the relationships between them, and develop an intuitive understanding of their properties. Table 10.1 shows specific objectives that children are likely to learn about geometrical shapes in the elementary and middle grades.

Table 10.1 Classroom learning objectives	K	1	2	3	4	5	6	7	8
Identify and name two- and three-dimensional shapes.	X								
Analyze and compare two- and three-dimensional shapes using informal language to describe their similarities, differences, parts, and other attributes.	X								
Compose two-dimensional or three-dimensional shapes to create composite shapes and model shapes in the world.	X	X							
Recognize defining attributes of shapes and build and draw shapes to possess defining attributes.		X	X						
Recognize categories of quadrilaterals and draw examples of quadrilaterals that do not belong to any of these subcategories.				X					
Understand that shapes in different categories may share attributes and that shared attributes can define a larger category.				X					
Recognize angles as geometric shapes that are formed wherever two rays share a common endpoint.					X				
Understand the concept of angle measurement, measure angles in whole-number degrees using a protractor, and sketch angles of specified measure.					X				
Solve problems to find unknown angles on a diagram in real-world and mathematical problems.					X				
Draw and identify points, lines, segments, rays, angles, and perpendicular and parallel lines in two-dimensional figures.					X				
Recognize and identify different kinds of triangles.					X				
Classify two-dimensional figures in a hierarchy based on properties.						X			
Understand that attributes belonging to a category of two-dimensional figures also belong to all subcategories of that category.						X			
Use facts about different kinds of angles to write and solve simple equations for an unknown angle.								X	
Apply the Pythagorean Theorem in real-world and mathematical problems in two and three dimensions.									X
Explain a proof of the Pythagorean Theorem and its converse.									X
Use informal arguments to establish facts about the angle sum and exterior angle of triangles.									X
Use informal arguments to establish facts about the angles created when parallel lines are cut by a transversal.									X

Source: Adapted from the *Common Core State Standards for Mathematics* (*Common Core State Standards Initiative*, 2010).

Spatial Reasoning

As students learn geometry, it will be necessary for them to create and use visual and physical models of geometric shapes. In doing so, they naturally engage in spatial reasoning. **Spatial reasoning,** or **spatial sense,** is the ability to make mental images of shapes and then to analyze or manipulate them in different ways. Spatial reasoning includes two areas:

- **Spatial orientation** is the ability to understand and operate on the relative position of objects in space with respect to a particular position.
- **Spatial visualization** is the ability to comprehend and perform imagined movements on objects in two- and three-dimensional space.

Spatial reasoning skills are important to develop because they are directly related to success in mathematics and science. For this reason, these skills are receiving more attention in the elementary and middle grades. They are often developed through navigation activities, art projects, and other lessons in which students physically

Table 10.2 Classroom learning objectives	K	1	2	3	4	5	6	7	8
Describe objects in the environment using names of shapes.	X								
Describe the relative positions of objects in the environment using terms such as "above," "below," "beside," "in front of," "behind," and "next to."	X								
Correctly name shapes regardless of their orientations and overall size.	X								
Represent three-dimensional figures using nets made from rectangles and triangles.						X			
Describe the two-dimensional figures that result from slicing three-dimensional figures with a plane.							X		

Source: Adapted from the *Common Core State Standards for Mathematics* (*Common Core State Standards Initiative*, 2010).

manipulate objects or view them from a variety of perspectives. Specific content objectives associated with spatial reasoning are given in Table 10.2.

Because of the increased attention placed on geometry in the elementary and middle grades, teachers in these grades must now be familiar with its many aspects. For this reason, Chapter 10 begins the study of geometry by:

- Discussing the basic notions that serve as the foundation for other geometrical shapes.
- Presenting the definitions and properties of two-dimensional shapes, including triangles, quadrilaterals, polygons, and circles.
- Presenting the definitions and properties of three-dimensional geometrical solids.

SECTION 10.1   Lines, Planes, and Angles

The Bridgeman Art Library International

Figure 10.1 Euclid (circa 325–265 B.C.E.)

Although geometry has been used and studied throughout human history, it was not until the sixth century B.C.E., that Greek mathematicians began to formalize geometry through deductive reasoning and proof. Of these mathematicians, Euclid was one of the most notable (Figure 10.1). In his text, *Elements of Geometry*, he brought together a collection of isolated facts and formed them into a cohesive, mathematical system. His work was so thorough that it still serves as the foundation of our study of geometry today.

Basic Shapes and Notions

In *Elements*, Euclid showed that all geometric shapes are built from three basic shapes: points, lines, and planes. The descriptions, representations, and notations of these ideas, along with those of space, are summarized in Table 10.3.

Table 10.3 Basic notions of geometry				
Notion	**Dimensions**	**Description**	**Representation**	**Notation**
Point	0	A location with no length, width, or depth	• A	Point A, or A
Line	1	That which has infinite length, but no width or depth	l, through A and B	$\overleftrightarrow{AB}$, $\overleftrightarrow{BA}$, or line l
Plane	2	An infinite flat surface, or that which has infinite length and width, but no depth	• A • B p • C	Plane ABC, or Plane p
Space	3	The set of all points, or that which has infinite length, width, and depth	(cube)	

In Euclidean geometry, we think of a line as an infinite, one-dimensional shape that is straight or does not bend. This characteristic separates lines from other one-dimensional shapes like s-curves and circles, that can curve through two dimensions (Figure 10.2).

Line   Not a line

Figure 10.2 **A line does not bend**

Likewise, we think of a plane as an infinite, two-dimensional shape that is flat, or does not bend into a third dimension. This characteristic separates planes from other two-dimensional surfaces that can curve through three dimensions (Figure 10.3).

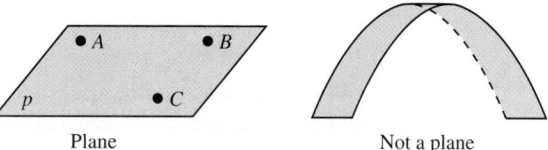

Plane   Not a plane

Figure 10.3 **A plane does not bend**

Example 10.1 Basic Shapes in a Room

Connections

How can a room be used to demonstrate the four basic shapes of geometry?

Solution The room has three dimensions, so it represents a portion of space. The walls, floor, and ceiling are flat surfaces, so they represent portions of planes. A line is formed when two planes meet, so any edge where a wall and the floor or two walls come together represents a portion of a line. Finally, a point is formed when three planes meet, so any corner in the room represents a point.

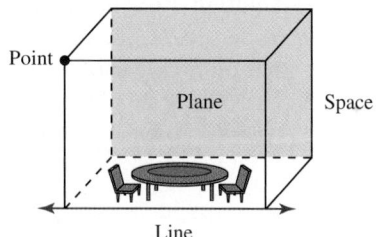

We can extend our knowledge of points, lines, and planes by stating facts about the relationships among them. Many of the facts are readily apparent, so we accept them to be true without proof. In mathematics, such facts are called **postulates.** Postulates are important because they serve as the foundation for mathematical structures. However, because they are assumed to be true without proof, we should always approach them with caution and generally try to limit their use.

Postulates for Points, Lines, and Planes

1. Given two distinct points, exactly one line contains them.
2. The points on a line can be placed in a one-to-one correspondence with the real numbers so that the differences between numbers measure distances (Ruler Postulate).
3. Given three distinct points that are not on the same line, exactly one plane contains them.
4. If two points lie in a plane, the line containing them lies in the plane.
5. Any line contains at least two distinct points.
6. The intersection of two distinct planes is a line.
7. No plane contains all the points in space.

Even though we cannot prove postulates to be true, we can offer informal reasons as to why we assume they are true. For instance, consider Postulate 1. If we try to draw two distinct lines, *l* and *m*, through points *A* and *B*, we have to bend the lines [Figure 10.4(a)]. However, this violates the fact that a line is straight. If we draw two straight lines through *A* and *B*, then one must necessarily lie atop the other, making them the same line [Figure 10.4(b)].

(a)   (b)

Figure 10.4 Exactly one line contains any two points

Given the basic definitions and postulates about points, lines, and planes, we can prove other facts.

Theorem 10.1   More Relationships Between Points, Lines, and Planes

1. Given a line and a point not on the line, exactly one plane contains them.
2. If two distinct lines intersect, exactly one plane contains them.
3. If two distinct lines intersect, they do so in exactly one point.
4. A plane and a line not on the plane can intersect in exactly one point.

Example 10.2   Proof of Theorem 10.1(1)

Reasoning

Given a line and a point not on the line, show that exactly one plane contains them.

Solution Let *l* be any line, and suppose that *C* is a point not on *l*. By Postulate 5, we know that a line must contain at least two points, say points *A* and *B*. Because *C* is not on *l*, *C* cannot be *A* or *B*. This gives us three points that are not on the same line, so exactly one plane contains them by Postulate 3. Because *A* and *B* are on the plane with *C*, Postulate 4 tells us that *l* must also lie on the plane with *C*.

Linear Notions

We can use points, lines, and planes to define a number of other geometric shapes. One of the simplest ways to do so is to consider different portions of a line. Table 10.4 defines three concepts associated with points and lines on a plane.

Table 10.4 Linear notion			
Definition of Notion	**Representation**	**Notation**	**Example**
Collinear points are points that lie on the same line.	A C B		Kitchen drawer knobs
A **ray** is a portion of a line that contains one point and all the points on the line to one side of the point.	A B	$\overrightarrow{AB}$	Laser pointer
A **line segment**, or **segment** is a portion of a line that contains two points and all the points between those points.	A B	$\overline{AB}$	Pencil

Like lines, rays and segments are labeled using two points. However, with rays, the order in which the points are listed matters because the first point denotes the **endpoint** of the ray, and the second indicates the direction in which the ray extends. For instance, $\overrightarrow{AB}$ and $\overrightarrow{BA}$ are different rays because they have different endpoints and extend in different directions. With a segment, the order of the endpoints does not matter because a segment is uniquely identified either way. In other words, $\overline{AB}$ and $\overline{BA}$ name the same segment.

Of the different linear shapes, only segments are finite in length, making them particularly important for defining other geometrical shapes and notions. For instance, the Ruler Postulate enables us to define the **length** of a segment $\overline{AB}$, denoted AB, to be the distance between the endpoints A and B. In turn, we can use length to introduce the important idea of congruence.

Definition of Congruent Line Segments

Two line segments, $\overline{AB}$ and $\overline{CD}$, are **congruent,** written $\overline{AB} \cong \overline{CD}$, if and only if $\overline{AB}$ and $\overline{CD}$ have the same length, or $AB = CD$.

In general, congruence is a relationship between two geometrical figures stating they are the same size and shape. Many students mistakenly assume that congruence and equality have the same meaning, but this is not so. Congruence implies the figures are the same shape and size. Equality implies that two shapes are not only the same size but exactly the same set of points. When drawing diagrams, we can show that two segments are congruent by using small marks. For instance, Figure 10.5 shows that $\overline{AB} \cong \overline{CD}$ and $\overline{EF} \cong \overline{GH}$.

Figure 10.5 $\overline{AB} \cong \overline{CD}$ and $\overline{EF} \cong \overline{GH}$

Given congruence, we can define the midpoint and the bisector of a segment. The **midpoint** of a segment $\overline{AB}$ is the point M on $\overline{AB}$ for which $\overline{AM}$ is congruent to $\overline{MB}$. Every segment has exactly one midpoint, which divides the segment into two equal lengths. Any segment, ray, or line that passes through the midpoint of a segment is called a **bisector** of the segment. For instance, Figure 10.6 shows that $\overleftrightarrow{AB}$ bisects $\overline{CD}$.

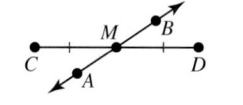

Figure 10.6 $\overleftrightarrow{AB}$ bisects $\overline{CD}$

So far, the linear notions discussed have been in only one dimension. We can define several other linear ideas if we consider lines in two dimensions (Table 10.5).

Table 10.5 Linear notions in two dimensions	
Definition of Notion	**Example**
Coplanar lines are lines that lie in the same plane.	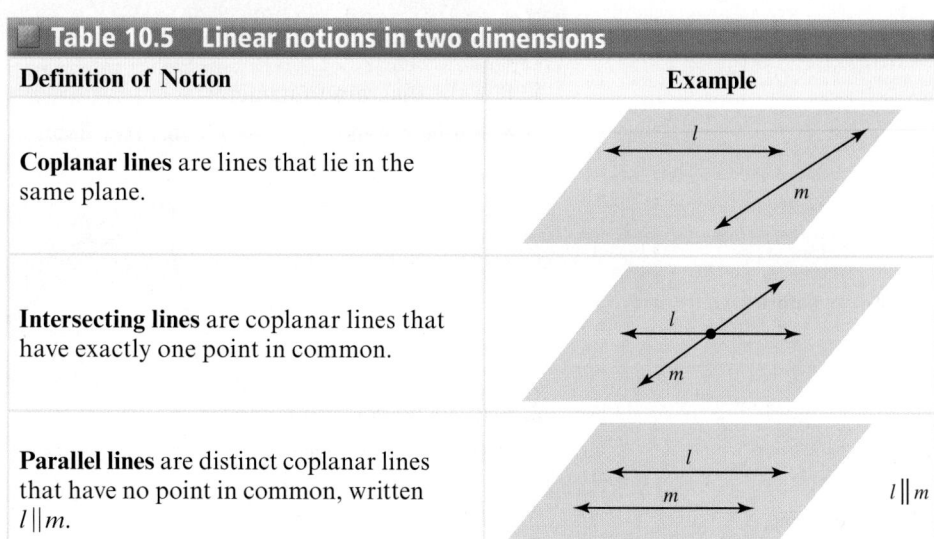
Intersecting lines are coplanar lines that have exactly one point in common.	
Parallel lines are distinct coplanar lines that have no point in common, written $l\|m$.	

Example 10.3   Identifying Lines, Rays, and Segments

Use the diagram to give an example of each of the following.

a. Three collinear points   **b.** Three rays   **c.** Two congruent segments

d. Two coplanar lines   **e.** Two intersecting lines   **f.** Two parallel lines

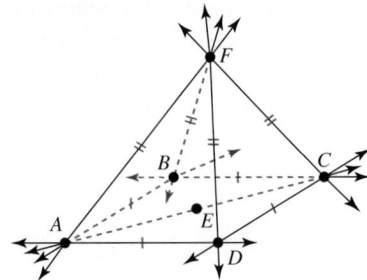

Solution   We apply the various definitions to identify the needed figures.

a. Points A, E, and C are three collinear points.

b. Three rays are $\overrightarrow{AB}$, $\overrightarrow{AD}$, and $\overrightarrow{AF}$.

c. Examples of congruent segments are $\overline{AD} \cong \overline{BC}$ and $\overline{AF} \cong \overline{CF}$.

d. $\overleftrightarrow{DF}$ and $\overleftrightarrow{DC}$ are two coplanar lines.

e. $\overleftrightarrow{AC}$ and $\overleftrightarrow{FC}$ are two intersecting lines.

f. $\overleftrightarrow{AD} \parallel \overleftrightarrow{BC}$ and $\overleftrightarrow{AB} \parallel \overleftrightarrow{DC}$.

Two other facts should be mentioned about parallel lines. First, if two lines are parallel to a third, then they must be parallel to each other. Second, given a line and a point not on the line, then exactly one line is parallel to the given line through the given point. This fact is called the **Parallel Postulate.** Historically, this postulate has caused many problems for mathematicians because for a long time it was thought that it could be proven true. It was not until the nineteenth century that it was shown that the Parallel Postulate could not be proven true or false, so it was accepted as a postulate for Euclidean geometry. Interestingly, in the effort to prove the Parallel Postulate, many other types of geometry, such as spherical geometry, were developed.

Check Your Understanding 10.1-A

1. Use the diagram to name:

   a. Three points.   **b.** Two lines.   **c.** Two planes.

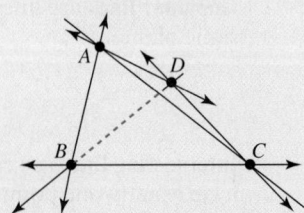

2. Use the diagram to name:

   a. Three line segments.   **b.** Three rays.   **c.** Two coplanar lines.

   d. Two intersecting lines.   **e.** Two parallel lines.

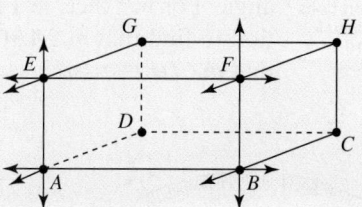

Talk About It Example 10.1 illustrates how a room can be used to show the four basic figures of geometry. How many different real-world examples of points, lines, and planes can you think of?

Problem Solving

Activity 10.1
A geoboard is a 5 × 5 or 6 × 6 square array of pegs, on which we can place rubber bands to make many different shapes. Select any corner peg on a 6 × 6 geoboard. How many segments of different lengths can be made from this peg? Five possible lengths are shown in the diagram:

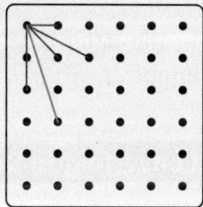

Angular Notions

Another fundamental building block for geometrical figures is an **angle,** which is the union of two rays that share a common endpoint. The rays form the **sides** of the angle, and the common endpoint is called the **vertex.** When naming an angle, we often use the angle symbol, ∠, with the vertex and a point on each ray. The order in which the points are listed does not matter as long as the vertex is in the middle. If the angle will not be confused with other angles, we can shorten the notation to just the vertex, or we can place a number inside the angle to name it. For instance, the angle in Figure 10.7 could be named $\angle PQR$, $\angle RQP$, $\angle Q$ or $\angle 1$.

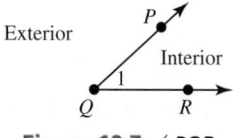

Figure 10.7 $\angle PQR$

Figure 10.7 also shows that an angle divides the plane into three regions: the interior, the exterior, and the angle itself. The **interior** of the angle is the portion of the plane between the two rays. The **exterior** is the portion of the plane that is not the angle or in its interior. Two angles that share a common vertex and a common side but that do not have overlapping interiors are called **adjacent angles.** For instance, in Figure 10.8, $\angle AEB$ and $\angle BEC$ are adjacent angles, whereas $\angle AEC$ and $\angle BED$ are not.

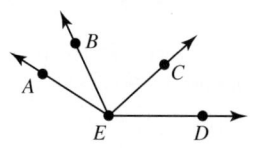

Figure 10.8 Adjacent angles $\angle AEB$ and $\angle BEC$

Angles can also be measured. If we view $\angle A$ as rotating a ray about its endpoint, then the **angle measure** of $\angle A$, denoted $m\angle A$, is the amount of rotation made by the ray (Figure 10.9). A common unit for measuring angles is **degrees** (°). One complete rotation is given a value of 360°, making a degree, 1°, equal to $\frac{1}{360}$ of a complete rotation.

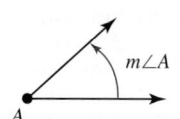

Figure 10.9 The measure of an angle

A **protractor** (Figure 10.10) can be used to measure angles. We place the center bottom mark on the vertex of the angle and then align the bottom edge with one side of the angle. We then follow the appropriate scale up to the second side of the

Explorations
Manual
10.1

angle. For instance, in Figure 10.10 we measure $\angle ABC$ in a counterclockwise direction to find that $m\angle ABC = 65°$. We measure $\angle DEF$ in a clockwise direction to find that $m\angle DEF = 107°$.

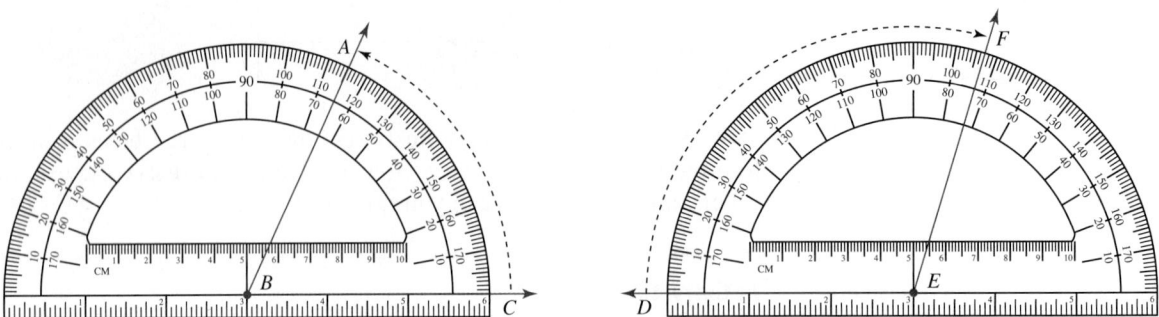

Figure 10.10 Measuring angles with a protractor

Most protractors measure only to the nearest degree. However, for more precision, we can use decimal values of degrees or degrees, minutes, and seconds notation. In the latter case, one **minute** (1′) is $\frac{1}{60}$ of a degree and one **second** (1″) is $\frac{1}{60}$ of a minute.

Example 10.4   Converting to Degrees, Minutes, and Seconds

Problem Solving

If an angle has a measure of 34.63°, what is its measure in degrees, minutes, and seconds?

Solution

Understanding the Problem. Our goal is to convert 34.63° into degrees, minutes, and seconds. We have little to go on, but we know that $1° = 60'$ and $1' = 60''$.

Devising a Plan. To change 34.63° to degrees, minutes, and seconds, we need to convert only the decimal part of the number. Rewriting 0.63 as $\frac{63}{100}$, we can think of it as a ratio and *use a proportion* to determine the equivalent number of minutes. If necessary, we can use another proportion to change any fractional part of a minute into seconds.

Carrying Out the Plan. Setting up our proportion, we have $\frac{63}{100} = \frac{m}{60}$. Solving for m, we have $m = \frac{63 \cdot 60}{100} = 37.8$, or $34.63° = 34° \, 37.8'$. Next, we convert 0.8′ to seconds using the proportion $\frac{8}{10} = \frac{s}{60}$. Solving for s, we have $s = \frac{8 \cdot 60}{10} = 48$. As a result, $34.63° = 34° \, 37' \, 48''$.

Looking Back. We can check our results by making the reverse conversion. Because $48'' = \frac{48}{60}$ of a minute, we can convert the number of seconds to a decimal by dividing, or $48'' = \frac{48}{60} = 0.8'$. This gives us $37' \, 48'' = 37.8'$. Next, we can convert 37.8′ to a decimal by dividing by 60 again, or $37.8' = \frac{37.8}{60} = 0.63°$. From this, $34° \, 37' \, 48'' = 34.63°$. Many modern calculators now have a built-in feature for making these conversions. The key to do so often looks like [DMS] or [° ′ ″].

Angle measure can be employed to define a number of notions and shapes. The first is angle congruence.

> **Definition of Congruent Angles**
>
> Two angles, $\angle A$ and $\angle B$, are **congruent**, written $\angle A \cong \angle B$, if and only if they have same measure, or $m\angle A = m\angle B$.

We show that two angles are congruent by placing small arcs near the vertex of the angle. If a figure has several pairs of congruent angles, we mark each congruent pair of angles with a different number of arcs. For instance, Figure 10.11 shows that $\angle ABC \cong \angle DEF$.

Figure 10.11 $\angle ABC \cong \angle DEF$

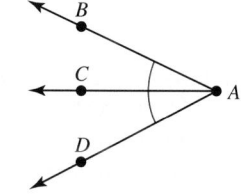

Figure 10.12 $\overrightarrow{AC}$ is the angle bisector of $\angle BAD$

We can also bisect angles. An **angle bisector** is a ray in the interior of the angle that divides it into two congruent angles. Every angle has exactly one bisector, and any point on the angle bisector is equidistant from both the sides of the angle. As shown in Figure 10.12, $\overrightarrow{AC}$ is the angle bisector of $\angle BAD$.

Angle measure can be also be used to classify and define different kinds of angles. Consider those given in Table 10.6.

Table 10.6 Different kinds of angles	
Definition of Notion	**Example**
An **acute angle** has a measure less than 90°.	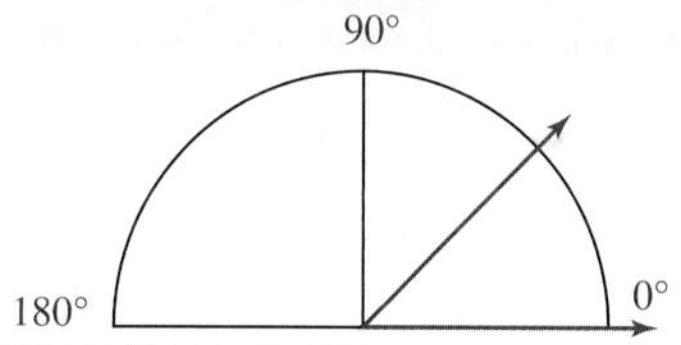
A **right angle** has a measure of exactly 90°.	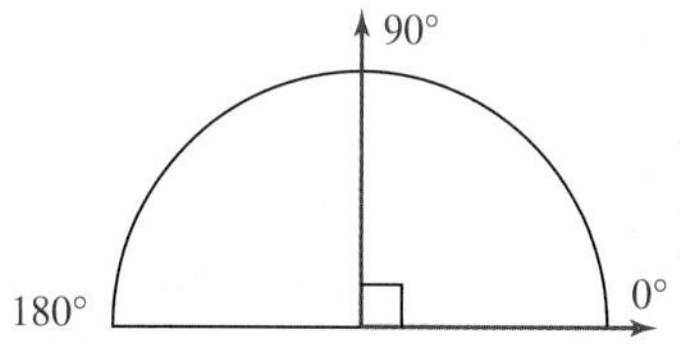
An **obtuse angle** has a measure greater than 90° but less than 180°.	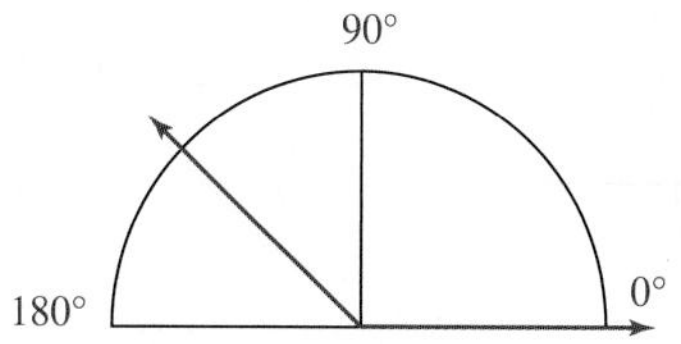
A **straight angle** has a measure of exactly 180°.	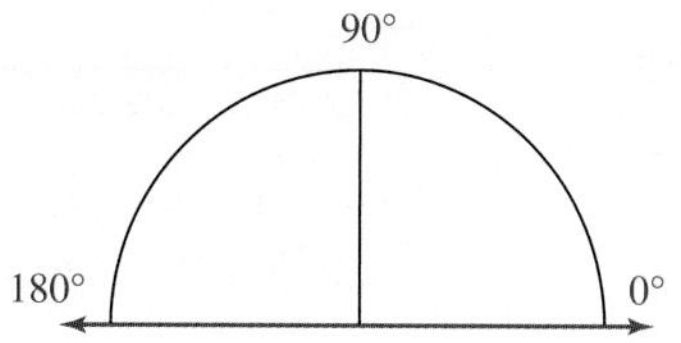
A **reflex angle** has a measure greater than 180° but less than 360°.	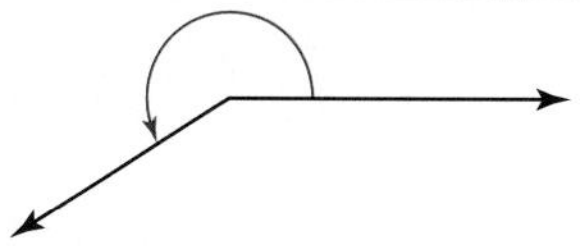

(Continued)

Table 10.6 Different kinds of angles (*continued*)

Definition of Notion	Example
Two angles form a **linear pair** if they have a common side and their other sides form a straight angle.	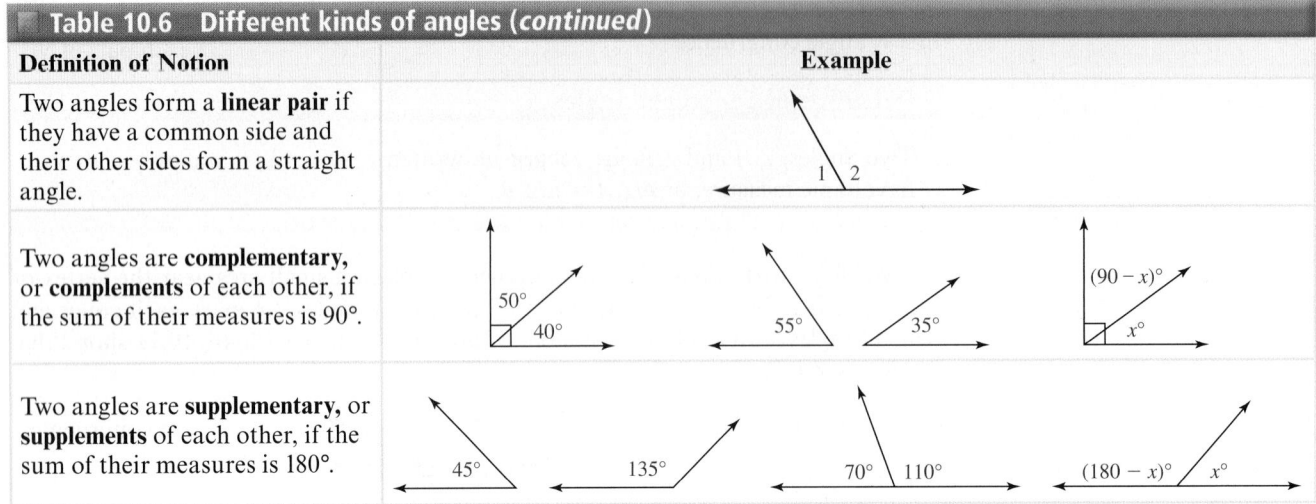
Two angles are **complementary,** or **complements** of each other, if the sum of their measures is 90°.	
Two angles are **supplementary,** or **supplements** of each other, if the sum of their measures is 180°.	

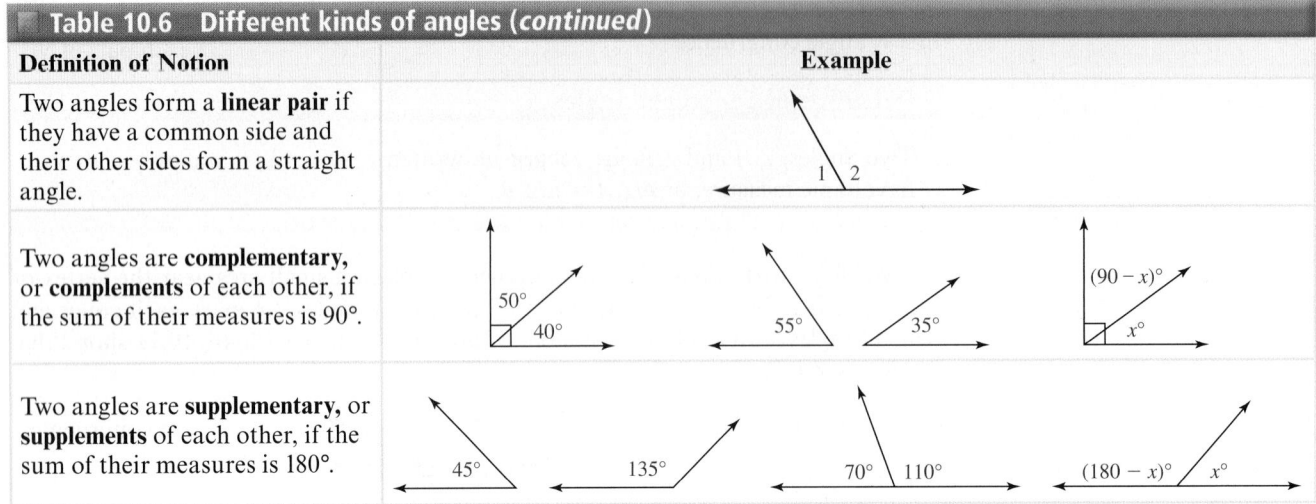

Several comments have to be made about the angles in Table 10.6. First, we can show that an angle is a right angle by placing a small square (□) at the vertex. And, because every right angle has a measure of 90°, any two right angles must be congruent. Second, a reflex angle is always associated with an angle of less than 180°. Generally, we prefer to work with the smaller angle, but if we need to indicate a reflex angle, we do so by drawing an arc around it. Third, because the sides of a linear pair form a straight angle, any linear pair is supplementary. If the angles in the linear pair are congruent, then each angle must be a right angle.

Explorations Manual 10.2

Example 10.5 Finding the Measures of Angles

Find the value of x in each diagram.

a. 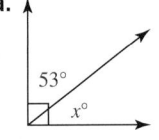 b. 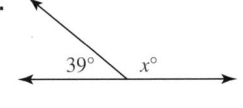 c.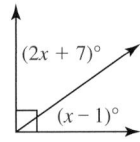

Solution

a. Because $53° + x = 90°$, then $x = 90° - 53° = 37°$.

b. Because $39° + x = 180°$, then $x = 180° - 39° = 141°$.

c. The angles are complementary, so $(2x + 7) + (x - 1) = 90°$. Solving for x:

$$(2x + 7) + (x - 1) = 90°$$
$$3x + 6 = 90°$$
$$3x = 84°$$
$$x = 28°$$

To this point, we have used rays to form our angles. Another important kind of angle is formed when two lines intersect.

Definition of Vertical Angles

Two nonadjacent angles that are formed by two intersecting lines are called **vertical angles.**

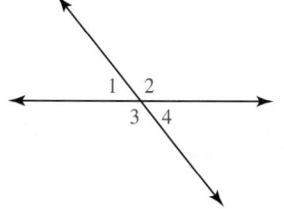

Figure 10.13 Vertical angles

Figure 10.13 shows two different pairs of vertical angles. ∠1 and ∠4 are vertical angles, and ∠2 and ∠3 are vertical angles. The diagram shows that each pair of vertical angles appears to be congruent.

Theorem 10.2 Congruency of Vertical Angles

Vertical angles are congruent.

□

Example 10.6	Proving That Vertical Angles Are Congruent

Reasoning   Use the definition of a linear pair to show that two vertical angles are congruent.

Solution Figure 10.13 shows that ∠1 and ∠2 form a linear pair, so they are supplementary, or $m\angle 1 + m\angle 2 = 180°$. Similarly, ∠2 and ∠4 form a linear pair, so $m\angle 2 + m\angle 4 = 180°$. Because both sums equal 180°, $m\angle 1 + m\angle 2 = m\angle 2 + m\angle 4$. Subtracting $m\angle 2$ from both sides of the equation, we are left with $m\angle 1 = m\angle 4$. Because their measures are equal, $\angle 1 \cong \angle 4$.

■

If two lines intersect to form four congruent vertical angles, then each angle must be a right angle, and we say that the lines are perpendicular.

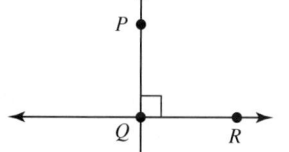

Figure 10.14 Perpendicular line

Definition of Perpendicular Lines

Two lines, *l* and *m*, are **perpendicular,** denoted $l \perp m$, if they intersect to form a right angle.

Explorations
Manual
10.3

Perpendicularity can be applied to other linear shapes as long as they lie on perpendicular lines. For instance, in Figure 10.14, $\overline{PQ} \perp \overline{QR}$, $\overline{PQ} \perp \overrightarrow{QR}$, and $\overrightarrow{PQ} \perp \overrightarrow{QR}$. If a perpendicular line happens to pass through the midpoint of a segment, it is called the **perpendicular bisector** of the segment.

Angles and Parallel Lines

Several new angle relationships are formed whenever a line, called a **transversal,** intersects a pair of lines. Eight angles are generated (Figure 10.15), and they can be associated in different ways based on their relative positions along the lines.

Examples of the different pairs of angles and their names are given in Table 10.7.

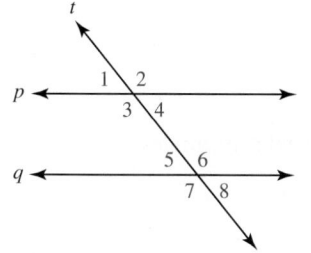

Figure 10.15 Lines *p* and *q* cut by transversal *t*

Table 10.7 Angles formed when a transversal intersects two lines	
Angle Name	**Examples**
Alternate interior angles	∠3 and ∠6, ∠4 and ∠5
Alternate exterior angles	∠1 and ∠8, ∠2 and ∠7
Same-side interior angles	∠3 and ∠5, ∠4 and ∠6
Same-side exterior angles	∠1 and ∠7, ∠2 and ∠8
Corresponding angles	∠1 and ∠5, ∠2 and ∠6, ∠3 and ∠7, ∠4 and ∠8

If the lines cut by a transversal are parallel, then corresponding angles are congruent. This fact is commonly accepted as a postulate.

Corresponding Angles Postulate

Two lines cut by a transversal are parallel if and only if any pair of corresponding angles are congruent.

Using this postulate, we can prove the following facts.

Theorem 10.3   Angles and Parallel Lines

Two lines cut by a transversal are parallel if and only if:

1. Alternate interior angles are congruent.
2. Alternate exterior angles are congruent.
3. Same-side interior angles are supplementary.
4. Same-side exterior angles are supplementary.

The Corresponding Angles Postulate and Theorem 10.3 are written as biconditional statements, so they work both ways. For instance, if two parallel lines are cut by a transversal, then corresponding angles are congruent. Likewise, if two lines are cut by a transversal and a pair of corresponding angles are congruent, then the lines are parallel. Similar statements can be made for each fact in Theorem 10.3.

Example 10.7   Finding the Measures of Angles Formed by a Transversal

If $p \parallel q$, what are the measures of the seven unknown angles?

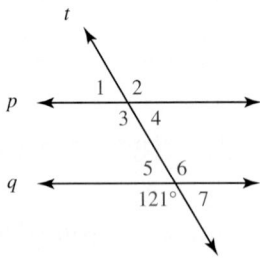

Solution

$m\angle 1 = 59°$ because $\angle 1$ and the given angle are same-side exterior angles.

$m\angle 2 = 121°$ because $\angle 2$ and the given angle are alternate exterior angles.

$m\angle 3 = 121°$ because $\angle 3$ and the given angle are corresponding angles.

$m\angle 4 = 59°$ because $\angle 1$ and $\angle 4$ are vertical angles.

$m\angle 5 = 59°$ because $\angle 5$ and $\angle 4$ are alternate interior angles.

$m\angle 6 = 121°$ because $\angle 6$ and the given angle are vertical angles.

$m\angle 7 = 59°$ because $\angle 7$ and $\angle 5$ are vertical angles.

Example 10.8 **Proving That Alternate Interior Angles Are Congruent**

Reasoning

If $p \| q$, show that alternate interior angles are congruent.

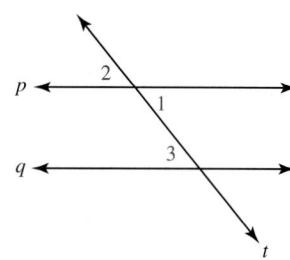

Solution Because $p \| q$, we know that $\angle 2 \cong \angle 3$ by the Corresponding Angle Postulate. Also, $\angle 2 \cong \angle 1$ because they are vertical angles. Because $\angle 1$ and $\angle 3$ are congruent to the same angle, then $\angle 1 \cong \angle 3$. Consequently, if two parallel lines are cut by a transversal, then alternate interior angles must be congruent.

Check Your Understanding 10.1-B

1. Use the following diagram to name:

    a. An acute angle.   **b.** An obtuse angle.   **c.** A right angle.

    d. A pair of adjacent angles.   **e.** Complementary angles.

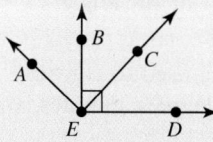

2. a. The complement of a 65° angle has what measure?

    b. The supplement of a 20° angle has what measure?

3. If $\angle A$ and $\angle B$ form a linear pair and $m\angle B = 68°$, what is x if $m\angle A = (2x + 4)°$?

4. Assume that $m \| n$. Find the measures of the numbered angles.

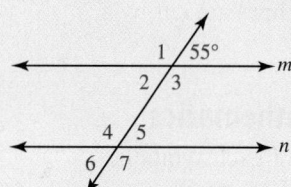

Talk About It With a small group of your peers, discuss how you might use a geoboard to demonstrate the angular notions given in this section.

Problem Solving

Activity 10.2

Select the middle peg on the bottom row of a regular 5 × 5 geoboard. Using this peg as a vertex and the two pegs to its right as the initial side, make as many angles of different measure as possible. Three possible angles are shown here:

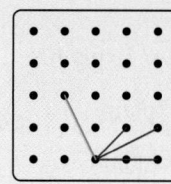

a. How many angles of different measures are there?

b. How many acute angles are there? How many obtuse?

c. What do you notice about the number of acute angles compared to the number of obtuse angles? Can you explain this?

SECTION **10.1** Learning Assessment

■ Understanding the Concepts

1. How do we know that a line must be straight and that a plane must be flat?

2. What is the difference between a postulate and a theorem?

3. What is the difference between line segments, rays, and lines?

4. What is the defining characteristic of parallel lines? Of perpendicular lines?

5. What is congruence, and how is it different from equality?

6. Is the interior of an angle part of the angle? Explain.

7. How is proportional reasoning used to convert decimal values of degrees to degrees, minutes, and seconds notation?

8. How is angle measure used to classify angles?

9. What is the difference between the complement and the supplement of an acute angle?

10. Describe the different pairs of angles that are formed when two parallel lines are cut by a transversal.

■ Representing the Mathematics

11. Give a verbal translation of each symbolic expression.

   a. $\overleftrightarrow{TO}$ **b.** $\overline{AM}$ **c.** $\angle BAY$

   d. $\overrightarrow{GO}$ **e.** $p\|q$ **f.** $l \perp m$

12. Draw and label each figure or relationship.

   a. $\overleftrightarrow{AB}$ **b.** Plane P **c.** $\overline{CD}$

   d. $\angle TOY$ **e.** $\overrightarrow{EF}$ **f.** $\overrightarrow{FE}$

   g. $l\|m$ **h.** $p \perp q$

13. **a.** Is $\overline{AB}$ the same segment as $\overline{BA}$? Explain.

   b. Is $\overrightarrow{AB}$ the same ray as $\overrightarrow{BA}$? Explain.

   c. Is $\overleftrightarrow{AB}$ the same line as $\overleftrightarrow{BA}$? Explain.

14. List the ways to name the given angle.

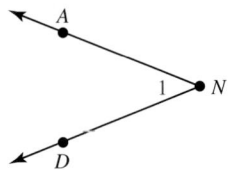

15. Use a protractor to draw an angle with a measure of:

   a. 45°.   **b.** 70°.   **c.** 90°.

   d. 125°.   **e.** 158°.

16. What kind of angle is formed when the minute hand of a clock is on 12 and the hour hand is on:

   a. 2?   **b.** 3?   **c.** 5?   **d.** 6?   **e.** 10?

17. Draw and label each figure described.

   a. A pair of supplementary angles that do not form a linear pair.

   b. A pair of complementary angles with a common vertex.

   c. A pair of obtuse, vertical angles.

   d. Three collinear points where one is the midpoint of the segment created by the other two.

18. **a.** Suppose the points A, N, D, and Y are collinear. Make a list of the ways these points can be used to name the same line.

   b. Suppose points A, B, C, D, and E are coplanar and no set of three points are collinear. List the different ways these points can be used to name the plane.

19. **a.** Draw a line segment $\overline{AB}$ 3–5 in. long anywhere on a standard sheet of paper. Fold the paper so that A is lying directly on top of B and make a crease. What does the crease represent with regard to the segment $\overline{AB}$?

   b. Draw an angle $\angle PQR$ on another sheet of paper. Fold the paper so that P is lying directly on top of R, and crease the paper so that the crease goes through Q. What does the crease represent with regard to angle $\angle PQR$?

20. Draw a diagram that illustrates how the following ideas and shapes are connected to one another through their definitions: ray, angle, angle measure, angle congruence, angle bisector, acute angle, right angle, and obtuse angle.

■ Exercises

21. Use the figure to give an example of each of the following:

  a. Three points.   **b.** Two lines.
  c. Two rays.   **d.** Two line segments.
  e. Two planes.   **f.** Two coplanar lines.

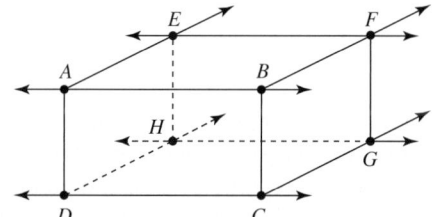

22. Use the figure to give an example of each of the following:

  a. Three collinear points. **b.** An acute angle.
  c. A right angle.   **d.** An obtuse angle.
  e. A straight angle.   **f.** A pair of vertical angles.
  g. A linear pair of angles. **h.** Two parallel segments.
  i. Two perpendicular segments.

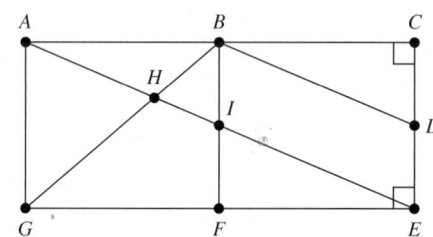

23. If 360° constitutes a complete revolution, then how many degrees is each partial revolution equal to?

  a. $\frac{1}{2}$ of a revolution   **b.** $\frac{1}{3}$ of a revolution

  c. $\frac{1}{4}$ of a revolution   **d.** $\frac{1}{8}$ of a revolution

  e. $\frac{2}{3}$ of a revolution   **f.** $\frac{5}{8}$ of a revolution

24. If 360° constitutes a complete revolution, then what fractional amount of a revolution is each of the following degree measures equal to?

  a. 90°   **b.** 45°   **c.** 15°
  d. 135°   **e.** 270°   **f.** 140°

25. Convert to degrees, minutes, and seconds.

  a. 14.25°   **b.** 12.65°

26. Convert to degrees, minutes, and seconds.

  a. 43.07°   **b.** 94.631°

27. Convert to decimal degrees.

  a. 10°15′   **b.** 65°13′21″

28. Convert to decimal degrees.

  a. 47°47′47″   **b.** 93°51″

29. Consider the following diagram.

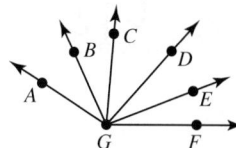

  a. How many different angles does it contain?
  b. How many angles look to be acute? Obtuse?

30. Find the measure of each angle.

  a. ∠AFB   **b.** ∠BFE   **c.** ∠DFA
  d. ∠BFD   **e.** ∠CFD

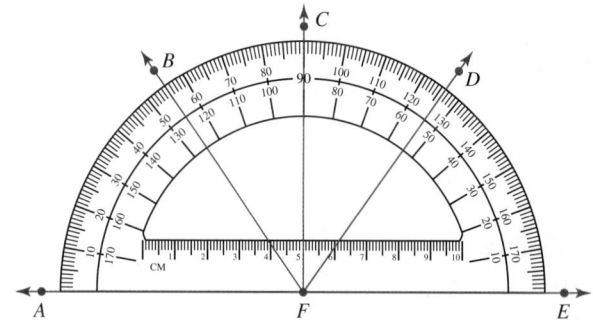

31. Suppose $\overleftrightarrow{AB} \perp \overrightarrow{CD}$ and $\overrightarrow{DE} \perp \overrightarrow{DF}$. If $m\angle CDE = 30°$, what are the measures of

  a. ∠EDB   **b.** ∠FDC
  c. ∠FDB   **d.** ∠ADE

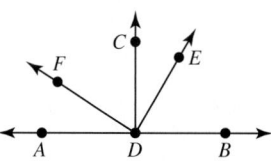

32. a. The complement of a 23° angle has what measure?

  b. The complement of an 84° angle has what measure?
  c. The supplement of an 11° angle has what measure?
  d. The supplement of a 123° angle has what measure?

33. a. What is the complement of the supplement of an angle measuring 110°?

  b. What is the supplement of the complement of an angle measuring 25°?

  c. What is the complement of the complement of an angle measuring 65°?

34. Suppose $m \parallel n$ in the given diagram

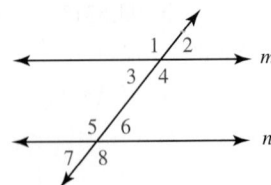

   a. List three angles congruent to $\angle 1$.

   b. List three angles supplementary to $\angle 5$.

   c. If $m\angle 6 = 47°$, what would be the measure of each angle?

   d. If $m\angle 5 = 128°$, what would be the measure of each angle?

35. Use the diagram in the previous exercise to answer the following questions.

   a. If $m\angle 4$ is twice $m\angle 6$, what would be the measure of each angle?

   b. If $m\angle 5$ is 15° more than twice $m\angle 2$, what would be the measure of each angle?

36. If $l \parallel m$, find the measure of each numbered angle.

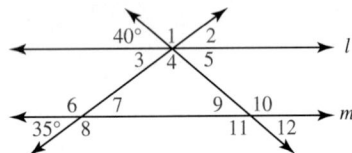

37. Find the value of x and y in each figure.

   a.

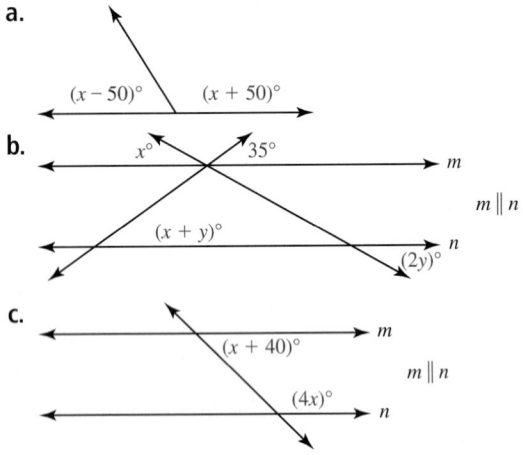

   b.

   c.

■ Problems and Applications

38. How many line segments can be drawn through 12 points if no three points are collinear?

39. a. How many planes exist between four points, no three of which are collinear?

   b. How many planes exist between five points, no three of which are collinear?

40. Consider six distinct coplanar lines. What is the smallest possible number of intersection points created by the lines? What is the largest?

41. When two lines intersect, they divide the plane into four regions. What is the largest number of regions a plane is divided into by:

   a. Three lines?   **b.** Four lines?

   c. Five lines?   **d.** n lines?

42. If four lines intersect at a point, how many pairs of vertical angles are formed?

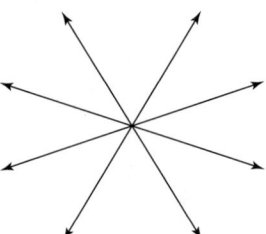

43. Even though one complete revolution is 360°, we can consider larger angles. For instance, an angle of 720° is two revolutions, and an angle of 450° is one revolution and a quarter of another. We can consider any such angle equivalent to an angle from 0° to 360°. For instance, a 450° angle is equivalent to a 90° angle as $450 - 360 = 90$. Find the angle between 0° and 360° that each angle is equivalent to.

   a. 405°   **b.** 1,080°   **c.** 1,235°

   d. 1,456°   **e.** 789° 24′ 31″

44. Angle measures can also be added or subtracted regardless of whether the measure is given as a decimal or in degrees, minutes, and seconds. Determine a method for adding and subtracting with angle measures, and use it to compute the following.

   a. 67.82° + 13.56°   **b.** 12° 14′ 56″ + 37° 26′ 45″

   c. 89.02° − 36.91°   **d.** 35° 16″ − 5° 43′ 28″

45. Another commonly used unit for angle measure is radians, where 1 radian (rad) $\approx 57.296°$. Convert each measure to radians.

   a. 90°   **b.** 45°   **c.** 120°

   d. 173°   **e.** 68.9°

46. a. What angle has the same measure as its complement?

   b. What angle has the same measure as its supplement?

   c. What is the measure of the angle whose supplement is five times its complement?

47. a. What is the measure in decimal degrees of the largest acute angle if only two decimal places are allowed?

   b. What is the measure in degrees, minutes, and seconds of the largest acute angle, if no partial degree, minute, or second is allowed?

48. a. What is the measure in decimal degrees of the smallest obtuse angle if only two decimal places are allowed?

b. What is the measure in degrees, minutes, and seconds of the smallest obtuse angle, if no partial degree, minute, or second is allowed?

49. A clock shows a time of 7:10. What is the approximate angle measure between the hour and minute hand?

50. The diagram shows an angle that is created between the sun and the position of the earth on two different days.

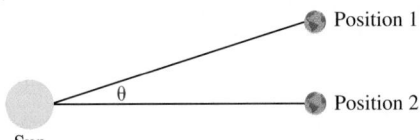

a. If the earth makes one complete revolution around the sun in 365.25 days, by how many degrees does $\angle\theta$ change in one day?

b. How much does $\angle\theta$ change over the course of a 30-day month?

51. How many degrees does the earth rotate in 1 hr? One minute?

52. A town wants to build a water park at a nearby lake so that it is equidistant from two housing subdivisions. Describe how the city planners might find the best location for the park based on the positions of the lake and the subdivisions shown:

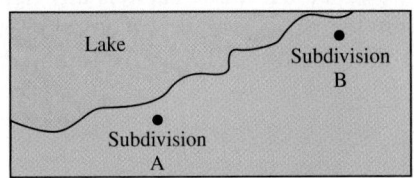

■ Mathematical Reasoning and Proof

53. Must every set of two points be collinear? What about every set of three points? Draw a picture to explain your thinking.

54. Is it possible for $\overrightarrow{AB} \cong \overrightarrow{BA}$? Explain.

55. Does an angle divide a plane into one, two, or three regions? Explain.

56. If two angles form a congruent linear pair, what is true about the sides of the angles?

57. State whether each angle is acute, right, obtuse, or straight.

a. The complement of an acute angle

b. The supplement of an acute angle

c. The supplement of a right angle

d. The supplement of an obtuse angle

58. a. Can two obtuse angles form a linear pair? Why or why not?

b. Can two acute angles form a linear pair? Why or why not?

c. Can two right angles form a linear pair? Why or why not?

d. What conclusion can we draw from parts a–c about the types of angles needed to form a linear pair?

59. a. If a right angle is bisected, what kinds of angles are formed?

b. If an obtuse angle is bisected, what kinds of angles are formed?

c. If a straight angle is bisected, what kinds of angles are formed?

d. What conclusion can we draw from parts a–c about the resulting angle measures when an angle is bisected?

60. Use the diagram and the fact that corresponding angles are congruent to prove that alternate exterior angles are congruent.

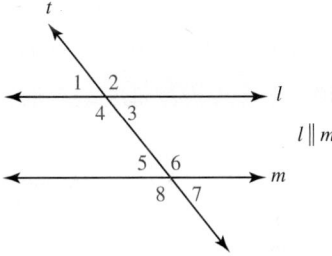

61. Use the diagram from the previous exercise and the fact that corresponding angles are congruent to show that same-side interior angles are supplementary.

62. Write an argument to show that two parallel lines determine a plane.

63. Show that, if two lines intersect, exactly one plane contains them.

■ Mathematical Communication

64. Write a set of directions that explains how to use a protractor to draw an angle of 55°.

65. Revisit the definition of vertical angles, and then consider other meanings of the word "vertical." Do you think the term "vertical" is an appropriate name for these angles? What other words might you use to name these angles?

66. Write two or three paragraphs that briefly explain how the concepts of this section are built from the basic notions of a point, line, and plane.

67. One of your students looks at the following angles and quickly claims that ∠1 is larger than ∠2. As the student's teacher, how do you respond?

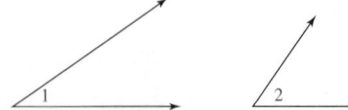

68. Kimberly claims that only threes pairs of vertical angles are shown in the diagram. Do you agree or disagree?

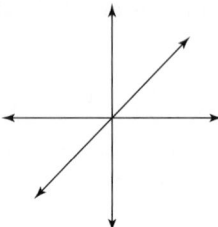

69. Jamie says the following two lines must be parallel because they do not intersect. What would you do as her teacher to correct her thinking?

■ Building Connections

70. With a group of your peers, write down geometrical terms you use in everyday conversations. What does this imply about the importance of learning geometry?

71. With several of your peers, discuss whether any of the terms in this section have nongeometrical meanings. If so, what are they? Do the geometrical and the nongeometrical meanings have any connection to one another? If so, how?

72. Take 15 minutes, and walk around the building your class meets in. Find and list five examples of each of the following:

a. Line segments   d. Parallel lines

b. Right angles   e. Points

c. Planes   f. Acute angles

73. Why do you think there has been a trend over the last decade or so to increase the amount of geometry taught in the elementary grades?

SECTION 10.2   Triangles

Points, lines, planes, and angles are the basic building blocks for many other shapes, which we can classify by their common characteristics. The broadest category of two-dimensional, or **planar** shapes, are called curves. A **curve** is any figure that can be traced without lifting the pencil from the paper. This broad description allows not only for a wide variety of shapes like those shown in Figure 10.16 but also for lines, rays, segments, and angles.

Figure 10.16 Curves

Because the set of curves is so diverse, we can further subdivide it by using other characteristics. For instance, **closed** curves are curves that begin and end at the same point, whereas **simple** curves never cross themselves except possibly at the point where the curve begins and ends. Combining the two terms, we get four different kinds of curves, examples of which are shown in Figure 10.17.

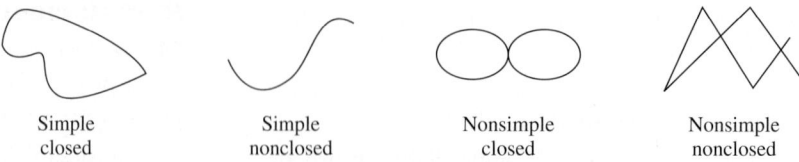

Simple closed   Simple nonclosed   Nonsimple closed   Nonsimple nonclosed

Figure 10.17 Different types of curves

Explorations
Manual
10.4

Simple closed curves can be further categorized as either convex or concave (Figure 10.18). A simple, closed curve is **convex** if, for any two points on the inside of the curve, the segment connecting the points lies entirely within the curve. It is **concave** if it is not convex. Intuitively, convex curves have no indentations, whereas concave curves do.

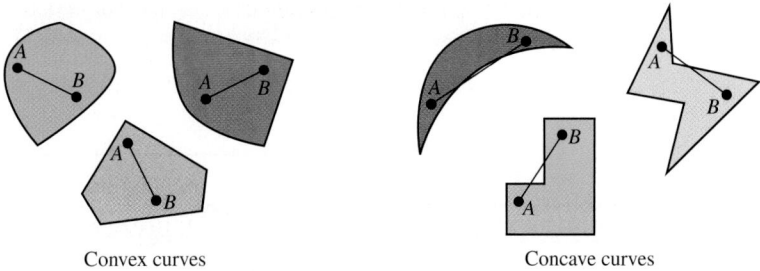

Figure 10.18 Convex and concave curves

Polygons are another type of simple, closed curve, which are made entirely of segments (Figure 10.19). They are particularly important in elementary geometry because they have a number of properties that can be clearly identified and studied. Our discussion of polygons begins with triangles.

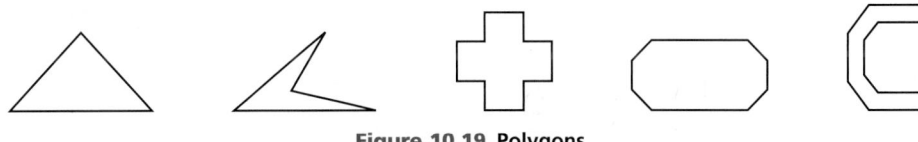

Figure 10.19 Polygons

Triangles and Their Classifications

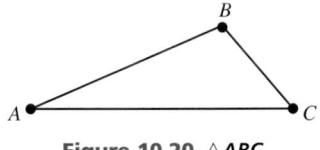

Figure 10.20 △*ABC*

The simplest of all polygons is a **triangle,** which is a polygon made from three segments, or **sides.** The points at which the sides intersect are called the **vertices,** which are typically used to name the triangle. For instance, triangle *ABC*, denoted △*ABC*, is shown in Figure 10.20.

A triangle consists only of its sides and vertices. Any other part of the plane, either inside or outside the triangle, is *not* part of the triangle. These regions are called the **interior** and the **exterior of the triangle** (Figure 10.21). In other words, a triangle separates a plane into three distinct sets of points, the interior, the exterior, and the triangle itself. This is true for all polygons and simple closed curves.

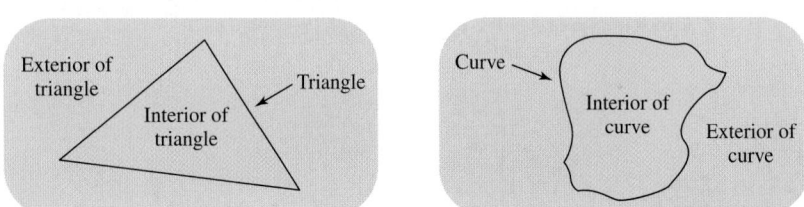

Figure 10.21 The interior and exterior of a triangle and a curve

Even though a triangle is made of three segments, not every set of three segments forms a triangle. For instance, consider the three segments shown in Figure 10.22. No matter how we rotate the shorter two segments around the endpoints of the longer segment, they never meet to form a triangle. This would be true even if the combined length of the shorter segments was equal to the length of the longer. In other words, the combined length of two segments must always be longer than the third to form a triangle. This fact is known as the Triangle Inequality.

Explorations
Manual
10.5

Figure 10.22 Three segments that do not form a triangle

Theorem 10.4   Triangle Inequality

The sum of the lengths of any two sides of a triangle must be greater than the length of the third side.

Example 10.9   Using the Triangle Inequality

Can segments with the given lengths be used to form a triangle?

a. 1.3 in., 4.5 in., 5.6 in.   **b.** 1 cm, 5 cm, 7 cm

Solution We answer the question by adding the lengths of two sides and comparing it to the third. If the sum of the sides is always bigger, then the segments form a triangle.

a. Segments with lengths equal to 1.3 in., 4.5 in., and 5.6 in. form a triangle because $1.3 + 4.5 = 5.8 > 5.6$, $1.3 + 5.6 = 6.9 > 4.5$, and $4.5 + 5.6 = 10.1 > 1.3$.

b. Segments with lengths equal to 1 cm, 5 cm, and 7 cm do not form a triangle because $1 + 5 = 6 < 7$.

We can also use the length of the sides to classify triangles. Table 10.8 shows the three ways to do so. Note that in an isosceles triangle, the congruent sides are called the **legs** and the third side is called the **base**. The angle between the legs is the called the **vertex angle,** and the other angles are called **base angles.** Given these definitions, we can conclude that every equilateral triangle is an isosceles triangle, but not every isosceles triangle is equilateral.

Table 10.8   Classifying triangles by the lengths of their sides

Type of Triangle	Illustration	Example
A triangle is **scalene** if it has no congruent sides.		
A triangle is **isosceles** if it has at least two congruent sides.		
A triangle is **equilateral** if it has three congruent sides.		

Triangles can also be classified according to the measures of their angles (Table 10.9).

Table 10.9 Classifying triangles by the measures of their angles

Type of Triangle	Illustration	Example
A triangle is **acute** if every angle is acute.		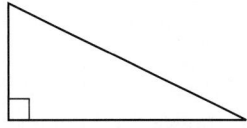 GO TEAM!!
A triangle is **right** if it has one right angle.		
A triangle is **obtuse** if it has one obtuse angle.		
A triangle is **equiangular** if the angles have the same measure.		YIELD

Because the terms in Table 10.8 and 10.9 use different aspects of triangles, we can mix and match many of them to form different kinds of triangles.

Example 10.10 Drawing Triangles

Representation

If possible, draw a triangle that is:

a. Right and scalene.   **b.** Obtuse and equilateral.   **c.** Acute and isosceles.

Solution

a. A right, scalene triangle has one right angle and no congruent sides, or

b. An obtuse, equilateral triangle would have to have one obtuse angle and three congruent sides. However, because the side opposite the obtuse angle must be larger than the other two, it is not possible to draw an obtuse, equilateral triangle.

c. An acute, isosceles triangle has three acute angles and two congruent sides, or

An obtuse, equilateral triangle cannot be drawn because of a particular relationship between the angles and the sides of a triangle. In general, large angles are opposite large sides, and small angles are opposite small sides. This fact leads to the following properties of triangles.

Theorem 10.5   Properties of Triangles

The following hold true for any $\triangle ABC$.

1. If two sides of $\triangle ABC$ are congruent, the angles opposite the sides are congruent.
2. If two angles of $\triangle ABC$ are congruent, the sides opposite the angles are congruent.
3. If $\triangle ABC$ is equilateral, then $\triangle ABC$ is also equiangular.
4. If $\triangle ABC$ is equiangular, then $\triangle ABC$ is also equilateral.

The next example demonstrates another useful fact about triangles.

Example 10.11   The Angle Measures in a Triangle

Reasoning

What is the sum of the measures of the angles in a triangle?

Solution For insight into what the sum might be, we can imagine making a triangle out of paper, tearing off two of the angles, and placing them adjacent to the third. The three angles appear to form a straight angle, implying that the sum of the measures of the angles is 180°.

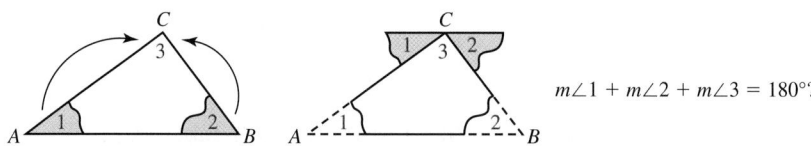

A closer look at our paper demonstration shows that the straight angle appears to be parallel to the side $\overline{AB}$. Because we know something about angles and parallel lines, this observation gives us a starting point for a deductive argument to support our conclusion.

To begin, we choose one side of $\triangle ABC$, say $\overline{AB}$, and make a line l parallel to it through point C. This gives us three angles, the sum of whose measures must be 180° because they lie along a straight line.

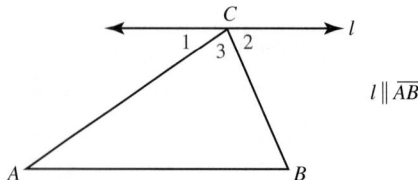

Explorations
Manual
10.6

Because $l \| \overline{AB}$, then $\angle A \cong \angle 1$ and $\angle B \cong \angle 2$ because they are alternate interior angles. Making a simple substitution, we have

$$m\angle A + m\angle B + m\angle 3 = m\angle 1 + m\angle 2 + m\angle 3 = 180°$$

Hence, the sum of the measures of the three angles in a triangle must equal 180°.

Theorem 10.6 The Sum of the Angles in a Triangle

The sum of the measures of the angles in any triangle is equal to 180°.

Theorem 10.6 has several consequences. First, it guarantees that the acute angles in any right triangle must be complementary and that the angles in an equilateral triangle must have the same measure of 60°. Second, it ensures that a triangle can have no more than one right or one obtuse angle. If it had two of either, then the sum of the angle measures would be greater than 180°, and it would not be a triangle (Figure 10.23).

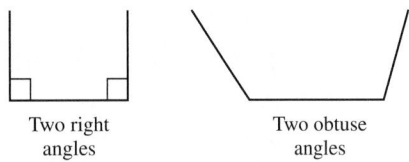

Two right   Two obtuse
angles   angles

Figure 10.23 Two right or two obtuse angles

Theorem 10.6 also guarantees something about the **exterior angles** of a triangle, which are the angles formed between a side and the extension of an adjacent side. If we consider the triangle in Figure 10.24, we see that $\angle 1$ and $\angle 2$ form a linear pair, so $m\angle 1 + m\angle 2 = 180°$. We also know from Theorem 10.6 that $m\angle 2 + m\angle 3 + m\angle 4 = 180°$, so with a substitution $m\angle 1 + m\angle 2 = m\angle 2 + m\angle 3 + m\angle 4$. By subtracting the $m\angle 2$ from both sides, we have $m\angle 1 = m\angle 3 + m\angle 4$. This implies that the measure of an exterior angle must equal the sum of the measures of the two remote interior angles.

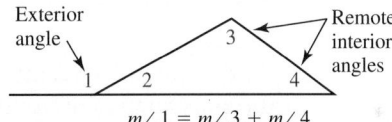

$$m\angle 1 = m\angle 3 + m\angle 4$$

Figure 10.24 A triangle with an exterior angle

Example 10.12 Finding Angle Measures in Triangles

Find the value of x.

a.
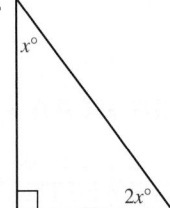

b.

c.
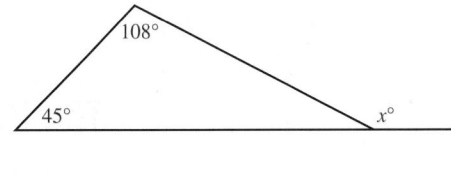

Solution

a. $x + 63° + 75° = 180°$. Hence, $x = 180° - 75° - 63° = 42°$.

b. The triangle is a right triangle, so the acute angles are complementary, or $x + 2x = 90°$. Solving for x:

$$x + 2x = 90°$$
$$3x = 90°$$
$$x = 30°$$

c. The value of x is equal to the sum of the two remote interior angles. As a result, $x = 45° + 108° = 153°$.

Check Your Understanding 10.2-A

1. Draw an example of:
   a. A simple curve that is not closed.
   b. A nonsimple curve that is closed.
   c. A simple, concave polygon.
   d. A simple, convex polygon.

2. Determine whether three segments of the given lengths can be used to form a triangle.
   a. 4, 3, and 2   **b.** 4, 2, and 2   **c.** 3.8, 6.7, and 2.1

3. If possible, draw a triangle that is:
   a. Right obtuse.   **b.** Obtuse scalene.   **c.** Right isosceles.

4. Find x.

   a.   **b.**   **c.**

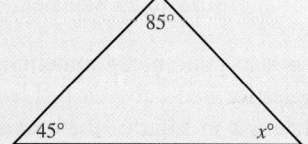
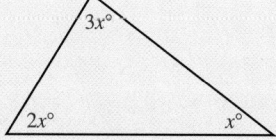

Talk About It Theorem 10.6 can be used to find the sum of the angle measures of other convex polygons. How might this be done?

Connections

Activity 10.3
Walk around your campus, looking for ways in which triangles are used in the construction of buildings and other structures. After returning to your classroom, use drinking straws and string (run the string through the straws and then tie the ends) to build several polygons: one with three sides, one with four, one with five, and one with six. What do you notice about the rigidity of the shapes? How might this explain the common use of triangles in construction?

The Pythagorean Theorem

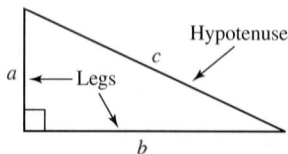

Figure 10.25 The legs and hypotenuse of a right triangle

Explorations
Manual
10.7

Right triangles are particularly useful because of an important fact called the Pythagorean Theorem. The theorem identifies a special relationship between the legs of a right triangle and its hypotenuse. The **legs** are the two perpendicular sides, and the **hypotenuse** is the side opposite the right angle. In diagrams, the legs are often labeled a and b, and the hypotenuse is c (Figure 10.25).

   We can explore and discover the Pythagorean Theorem in a number of ways. One method commonly used in the classroom is to compare the areas of three squares that lie on the sides of the right triangle. For instance, suppose we place squares on each side of a right triangle that has legs of lengths 3 and 4 and a hypotenuse of length 5 (Figure 10.26). The area of any square is always the product of the lengths of the sides, or $A = s \cdot s = s^2$. Using this formula, the squares in Figure 10.26 have areas of $a^2 = 3^2 = 9$, $b^2 = 4^2 = 16$, and $c^2 = 5^2 = 25$. Adding a^2 to b^2 and comparing the sum to c^2, we find that $a^2 + b^2 = 9 + 16 = 25 = c^2$. This relationship holds true for all right triangles and is formally stated in the next theorem.

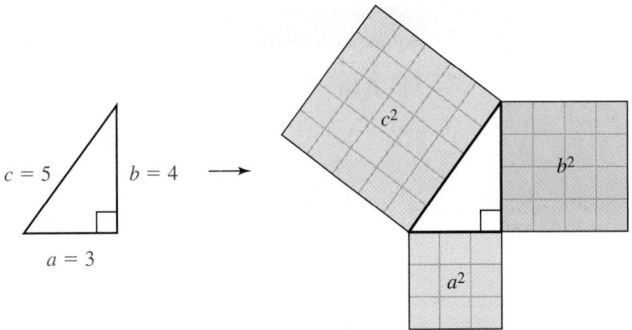

Figure 10.26 Discovering the Pythagorean Theorem

Theorem 10.7 Pythagorean Theorem

If a right triangle has legs of lengths a and b and a hypotenuse of length c, then $a^2 + b^2 = c^2$.

The Pythagorean Theorem has over 250 different proofs. The one attributed to Pythagoras is most often used because of its visual nature. It begins with a right triangle with legs of lengths a and b and a hypotenuse of length c. As shown in Figure 10.27, we can use the lengths of the legs to build two squares each having sides of length $a + b$. Because the sides of the large squares are equal, they must have the same area. Also notice that each large square includes four right triangles that are identical to the original. If we remove these triangles, then the areas of the two smaller squares in Figure 10.27(a) must equal the area of the smaller square in Figure 10.27(b). In other words, $a^2 + b^2 = c^2$.

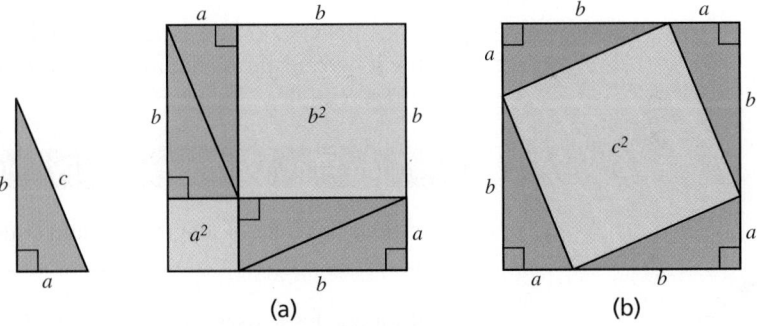

Figure 10.27 Two squares made from a right triangle

We can also use Figure 10.27(b) to draw the same conclusion algebraically. First, we compute the area of the large square, or

$$(a + b)^2 = (a + b)(a + b) \qquad \text{Definition of an exponent}$$
$$= (a + b)a + (a + b)b \qquad \text{Distributive property}$$
$$= a \cdot a + b \cdot a + a \cdot b + b \cdot b \qquad \text{Distributive property}$$
$$= a^2 + b \cdot a + a \cdot b + b^2 \qquad \text{Definition of an exponent}$$
$$= a^2 + 2ab + b^2 \qquad \text{Combining like terms}$$

Next, we compute its area by using its subregions. Specifically, it is made of one small square with an area of c^2 and four triangles, each with an area of $\frac{1}{2}ab$. Hence, the total area of the large square is also equal to $c^2 + 4\left(\frac{1}{2}ab\right) = c^2 + 2ab$. Setting the equations equal to one another, we have $a^2 + 2ab + b^2 = c^2 + 2ab$. Subtracting $2ab$ from both sides, we get $a^2 + b^2 = c^2$.

Example 10.13 Using the Pythagorean Theorem

Use the Pythagorean Theorem to find the missing side in each triangle.

a.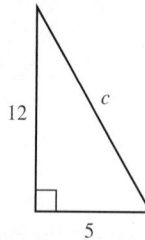

b.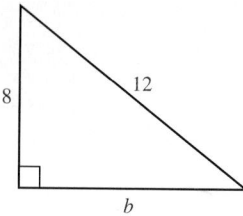

Solution

a. Letting $a = 12$ and $b = 5$, we have

$$c^2 = 12^2 + 5^2$$
$$c^2 = 144 + 25$$
$$c^2 = 169$$
$$c = \sqrt{169} = 13$$

b. Letting $c = 12$ and $a = 8$, we have

$$12^2 = 8^2 + b^2$$
$$144 = 64 + b^2$$
$$80 = b^2$$
$$b = \sqrt{80} \approx 8.944$$

The converse of the Pythagorean Theorem is also true. Although we introduce and use it here, its proof requires the use of congruent triangles. The proof is addressed in the next chapter.

Theorem 10.8 Converse of the Pythagorean Theorem

If a triangle has sides of lengths a, b, and c such that $a^2 + b^2 = c^2$, then it is a right triangle.

Builders have used the converse of the Pythagorean Theorem for thousands of years to square up, or form right angles in constructions. For instance, the rope stretchers, who were the surveyors of ancient Egypt, made right angles by using a chord with 12 equally spaced knots. When it was staked to the ground with three knots on one side and four knots on the other, the resulting shape was always a right triangle (Figure 10.28). Modern construction workers still use a similar practice today.

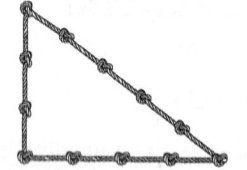

Figure 10.28 A right triangle from a knotted chord

Example 10.14 Constructing Perpendicular Walls

Application

A construction worker is connecting two walls, one 12 ft long and another 16. When he measures from the end of one wall to the end of the other, what must the length be to ensure that the walls are perpendicular?

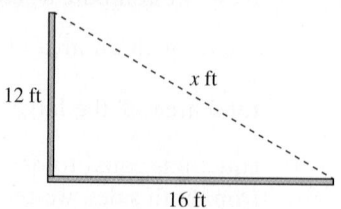

Solution For the walls to be perpendicular, we must find a value for x such that $12^2 + 16^2 = x^2$. By solving for x, we have $x^2 = 144 + 256 = 400$, or $x = \sqrt{400} = 20$. So, when the construction worker measures a distance of 20 ft between the ends of the walls, the walls are perpendicular.

◼

The last two examples have shown instances of three natural numbers that satisfy the Pythagorean Theorem. Such combinations are called **Pythagorean triples.** Common triples are 3-4-5, 6-8-10, 5-12-13, and 8-15-17. Notice that the numbers in the triple 6-8-10 are twice the numbers in the triple 3-4-5. This indicates that we can find other triples by multiplying the numbers in one triple by the same nonzero constant. Knowing Pythagorean triples can often make our work more efficient. If one side in a right triangle is missing and the other two sides are part of a Pythagorean triple, then without calculation we know that the missing value must be the third number from the triple.

In addition to using Pythagorean triples, it is also relatively easy to find the sides in two special right triangles. The first is an isosceles right triangle. Because the legs are the same length, both acute angles must have a measure of 45°, so we call these triangles **45°-45°-90° triangles** (Figure 10.29). If a is the length of each leg, then by the Pythagorean Theorem, the length of the hypotenuse is

$$c = \sqrt{a^2 + a^2} = \sqrt{2a^2} = \sqrt{a^2}\sqrt{2} = a\sqrt{2}$$

In other words, the length of the hypotenuse in a 45°-45°-90° triangle is always $\sqrt{2}$ times the length of the leg.

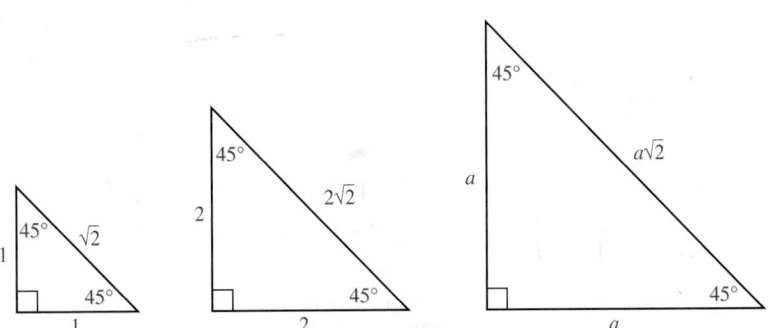

Figure 10.29 45°-45°-90° triangles

The second special triangle is a **30°-60°-90° triangle,** which has two acute angles measuring 30° and 60°, respectively. To understand the relationship between the legs and the hypotenuse in this triangle, consider an equilateral triangle in which each side has a length of 2. We make a 30°-60°-90° triangle by connecting a vertex to the midpoint of the opposite side, giving us a hypotenuse of length 2 and a leg of length 1 (Figure 10.30). By the Pythagorean Theorem, the length of the other leg is $b = \sqrt{2^2 - 1^2} = \sqrt{3}$.

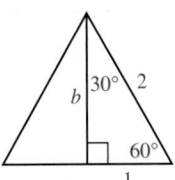

Figure 10.30 A 30°-60°-90° triangle from an equilateral triangle

Consequently, in any 30°-60°-90° triangle, the hypotenuse is always two times the length of the short leg, and the long leg is always $\sqrt{3}$ times the short leg (Figure 10.31).

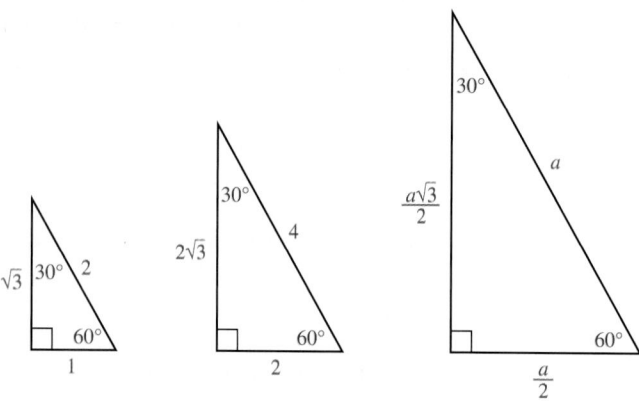

Figure 10.31 30°-60°-90° triangles

Check Your Understanding 10.2-B

1. Determine whether each set of three numbers can represent the sides of a right triangle.

a. 6, 8, and 10 **b.** 5, 12, and 17 **c.** 1, 1, and 2 **d.** 0.5, 0.5, and $\sqrt{0.5}$

2. Find the missing side(s) in each triangle.

a. 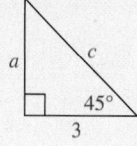 **b.** 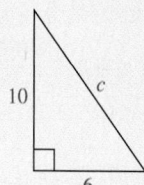 **c.**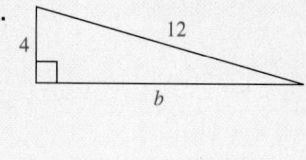

3. In a 30°-60°-90° triangle, if the shortest leg is 3 ft, what are the lengths of the other two sides?

Talk About It Is it possible for a Pythagorean triple to have three odd numbers? Why or why not?

Reasoning

Activity 10.4
Justify the Pythagorean Theorem by using the following diagram. Assume each triangle is a right triangle with legs of length a and b and a hypotenuse of length c. The middle square has sides of length $b - a$.

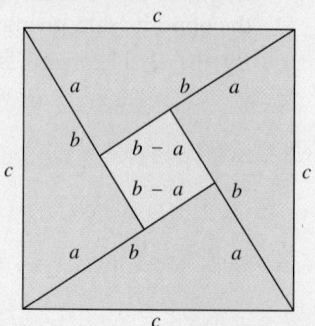

SECTION 10.2 Learning Assessment

■ Understanding the Concepts

1. What is the difference between a convex curve and a concave curve?

2. **a.** What are the defining characteristics of a polygon?

   b. Why is it impossible to have a polygon with only one or two sides?

3. If a point is inside a triangle, is it part of the triangle? Explain.

4. Can any set of three segments be used to form a triangle? Explain.

5. Give two different ways to classify triangles.

6. If two sides of a triangle are congruent, what can be said about the angles opposite the sides?

7. **a.** Explain why the sum of the angle measures in a triangle is equal to 180°.

   b. What are some of the consequences of this fact?

8. What does the Pythagorean Theorem say is true for all right triangles?

9. What is a Pythagorean triple?

10. What is special about a 45°-45°-90° triangle? What about a 30°-60°-90° triangle?

■ Representing the Mathematics

11. Classify each figure as simple or nonsimple and then as closed or nonclosed.

   a. 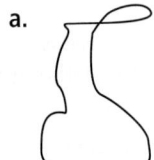   **b.**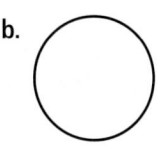

   c. 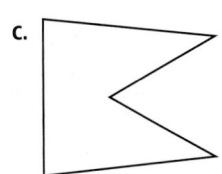   **d.**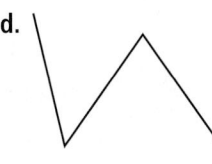

12. Draw an example of each shape.

   a. A simple, nonclosed curve

   b. A simple, closed, convex curve

   c. A simple, closed, concave curve

   d. A five-sided convex polygon

13. Which of the following is a polygon? Explain how you know.

   a.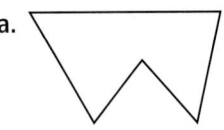

   b.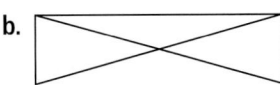

   c.

   d.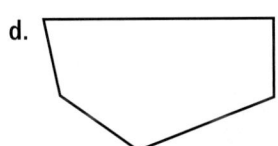

14. Consider a triangle △ABC.

   a. What are the vertices of triangle △ABC?

   b. What are the sides of triangle △ABC?

   c. What side is between ∠A and ∠C?

   d. What angle is between $\overline{AB}$ and $\overline{CA}$?

   e. If $AB \neq BC \neq AC \neq AB$, what kind of triangle could △ABC be?

   f. If ∠A = 115°, what kind of triangle could △ABC be?

   g. If ∠A = 65°, what kind of triangle could △ABC be?

15. A triangle has angles measuring 45°, 60°, and 75°.

   a. Is it acute, right, or obtuse?

   b. Is it a scalene, isosceles, or equilateral triangle?

16. A triangle has angles measuring 35°, 35°, and 110°.

   a. Is it acute, right, or obtuse?

   b. Is it a scalene, isosceles, or equilateral triangle?

17. Consider the figure.

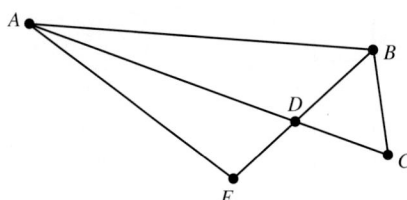

   Name every triangle that has the given part.

   a. $\overline{AB}$ **b.** $\overline{AC}$

   c. ∠DAB   **d.** ∠AEB

18. Categorize each triangle as acute, right, or obtuse and then as scalene, isosceles, or equilateral.

a.
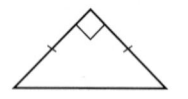

b.
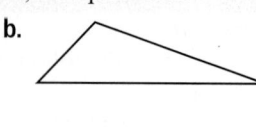

c.
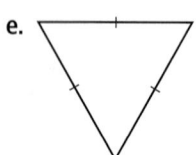

d.
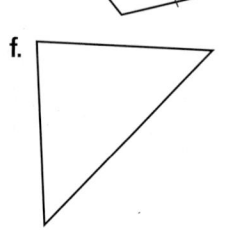

e.
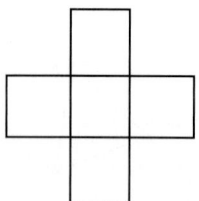

f.

19. Draw a triangle with sides of the given length, and then classify the triangle.

a. 2 cm, 6 cm, 7 cm

b. 4 cm, 4 cm, 5 cm

c. 3 cm, 4 cm, 5 cm

20. Use a geoboard to make a(n):

a. Right, scalene triangle.

b. Acute, isosceles triangle.

c. Obtuse, scalene triangle.

d. Right isosceles triangle.

21. On a 6 × 6 geoboard, make a segment that has each of the following lengths.

a. $\sqrt{26}$ b. $\sqrt{18}$ c. $\sqrt{32}$

■ Exercises

22. How many regions does each figure divide the plane into?

a.
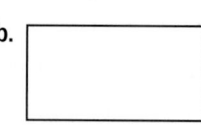

b.

c.

d.

23. Is it possible to draw a triangle with sides of the given lengths?

a. 1, 1, and 2   b. 3, 5, and 2.1

c. 6.3, 6.4, and 12.3

24. Is it possible to draw a triangle with sides of the given lengths?

a. 4, 9, and 6   b. 0.3, 0.8, and 1.4

c. $\frac{3}{5}$, $\frac{7}{10}$, and $1\frac{1}{2}$

25. Find x.

a.
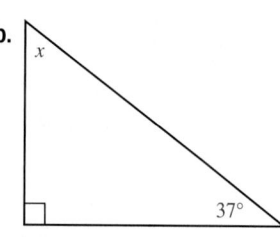

b.

c.
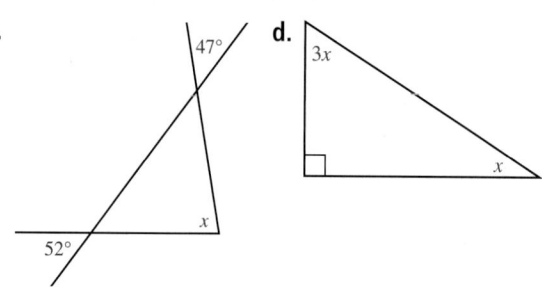

d.

26. Find the measure of each numbered angle.
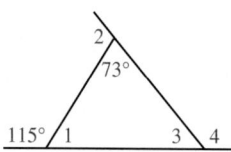

27. What are the measures of the angles in an equilateral triangle?

28. An acute, isosceles triangle has a vertex angle measuring 78°. What are the measures of the other two angles?

29. An obtuse, isosceles triangle has a base angle measuring 23°. What are the measures of the other two angles?

30. Find the missing side in each right triangle.

a.   b.
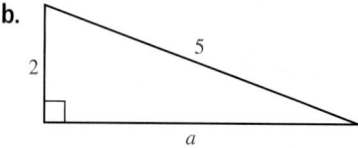

c.

31. Consider a right triangle with legs of length a and b and hypotenuse of length c. Find the length of the missing side if:

    a. $a = 6$ and $b = 8$.

    b. $a = 7$ and $b = 13$.

    c. $a = 5$ and $c = 13$.

32. Consider a right triangle with legs of length a and b and a hypotenuse of length c. Find the length of the missing side if:

    a. $b = 11$ and $c = 21$.

    b. $a = 5.3$ and $b = 7.1$.

    c. $a = \dfrac{4}{9}$ and $c = \dfrac{5}{9}$.

33. Which of the following sets of three numbers is a Pythagorean triple?

    a. 7, 24, 25   **b.** 9, 40, 41

    c. 12, 36, 37   **d.** 16, 63, 66

34. Determine whether each set of three numbers represents the sides of a right triangle.

    a. 2, 3, and 4   **b.** 10, 24, and 26

    c. 1, 1, and $\sqrt{2}$ **d.** $\dfrac{1}{3}, \dfrac{1}{3},$ and $\dfrac{\sqrt{2}}{3}$

35. Consider a 45°-45°-90° triangle.

    a. If one leg is 6, what are the lengths of the other two sides?

    b. If the hypotenuse is 14, what are the lengths of the two sides?

36. Consider a 45°-45°-90° triangle.

    a. If one leg is 8, what are the lengths of the other two sides?

    b. If the hypotenuse is 17, what are the lengths of the two sides?

37. Consider a 30°-60°-90° triangle.

    a. If the shortest leg is 4, what are the lengths of the other two sides?

    b. If the hypotenuse is 12, what are the lengths of the other two sides?

    c. If the longest leg is 8, what are the lengths of the other two sides?

38. Consider a 30°-60°-90° triangle.

    a. If the shortest leg is 4.6, what are the lengths of the other two sides?

    b. If the hypotenuse is 7, what are the lengths of the other two sides?

    c. If the longest leg is 5.3, what are the lengths of the other two sides?

39. What is the value of x, if the following triangle is isosceles and the base angles have a measure of:

    a. 30°?   **b.** 45°?   **c.** 60°?

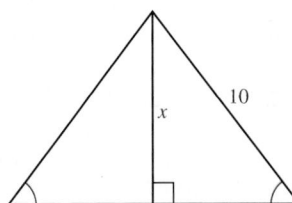

■ Problems and Applications

40. If the sides of $\triangle ABC$ are three times the length of the sides of $\triangle DEF$, how many copies of $\triangle DEF$ will fit inside triangle $\triangle ABC$?

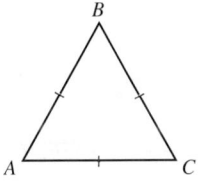
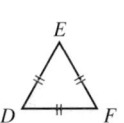

41. How many different triangles are in the figure?

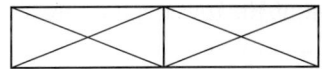

42. Find the value of each variable.

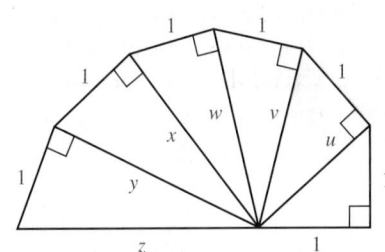

43. Find all Pythagorean triples in which no number is more than 50.

44. In a **primitive** Pythagorean triple, the GCF(a, b, c) = 1. Find the 16 primitive Pythagorean triples in which each number is less than 100.

45. A ladder is 24 ft long and is leaning against the side of a building with its base 8 ft from the bottom. How far does the ladder go up the building?

46. A homeowner wants to build a support for a shelf that is 12 in. wide. If the support is 15 in. long, how far down will the support be attached to the wall?

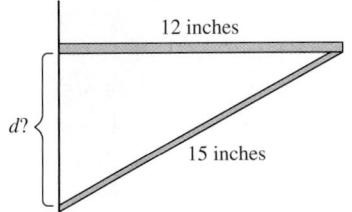

47. A groundskeeper at a local park is making a rectangular field that is 75 × 125 ft. He marks the field with string before laying down chalk lines. What distance must he measure from one corner to another to ensure that the sides of the field are perpendicular?

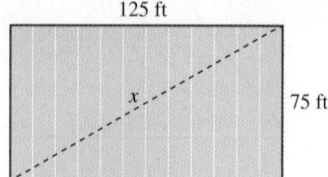

48. A plumber is running a hot water pipe in a house and must make an offset of 16 in. by using two 45° elbows. How long must he cut the pipe to make the offset?

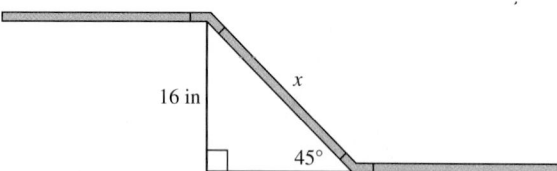

49. A 200-ft tower has a guy wire that makes a 60° angle with the level ground. How far from the bottom of the tower is the wire anchored?

50. A 140-ft television antenna is to be secured with six guy wires. The antenna is perpendicular to the ground with three guy wires secured halfway and three more secured at the top. Each pair of guy wires is secured at a point 100 ft from the base of the antenna. How much wire is needed to secure the tower?

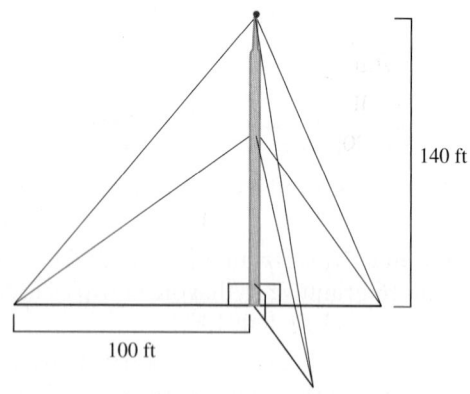

51. A surveyor wants to measure the distance across a small lake. To do so, he takes the measures given in the diagram. How far is it across the lake?

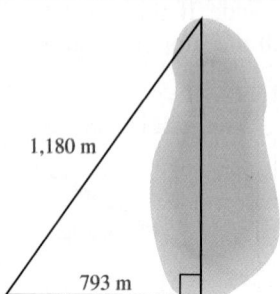

■ Mathematical Reasoning and Proof

52. Draw and cut out a triangle, and label it △ABC. Fold the triangle as shown. Repeat with several different triangles. What does your experiment suggest about the interior angles of a triangle?

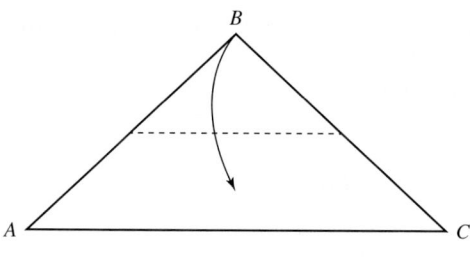

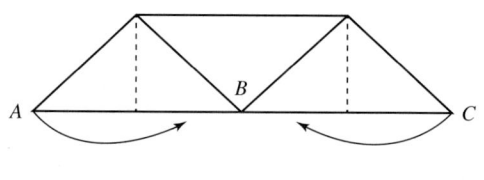

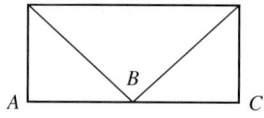

53. Answer each question, and explain your thinking.

  a. Can a scalene triangle be an obtuse triangle?

  b. Can a scalene triangle be an isosceles triangle?

  c. Can a right triangle be an isosceles triangle?

  d. Can an isosceles triangle be an acute triangle?

54. Write an argument that explains why an equilateral triangle must also be equiangular.

55. The **midsegment** of a triangle is any segment joining the midpoints of two sides. Make several triangles, and use a ruler to draw a midsegment of the triangle.

  a. What seems to be true about the relationship between the midsegment and the side with the unused midpoint? Write a fact summarizing what you observe.

  b. Compare the length of the midsegment to the length of the side with the unused midpoint. What do you notice? Write a fact summarizing what you observe.

56. Decide whether each conclusion to the following statement is always, sometimes, or never true. The Pythagorean Theorem expresses a relationship between the sides of a(n):

    a. Scalene triangle.

    b. Obtuse triangle.

    c. Equilateral triangle.

    d. Isosceles triangle.

    e. Right triangle.

    f. Acute triangle.

57. Begin with a right triangle with legs a and b and hypotenuse c.

    a. If the right angle is decreased to an acute angle, how does $a^2 + b^2$ compare to c^2?

    b. If the right angle is increased to an obtuse angle, how does $a^2 + b^2$ compare to c^2?

58. Suppose $a = 2n + 1$, $b = 2n^2 + 2n$, and $c = 2n^2 + 2n + 1$. Verify that a-b-c is a Pythagorean triple when $n \in \{1, 2, 3, 4, 5\}$.

59. Is 3-4-5 the only Pythagorean triple with three consecutive integers? Use x, $x + 1$, and $x + 2$ to write an argument that supports your answer.

60. If a-b-c represents a Pythagorean triple, show that ka-kb-kc is another Pythagorean triple when k is a natural number.

61. Take three or four different Pythagorean triples and compute the value of $\dfrac{(c - a)(c - b)}{2}$. What do you notice about these values?

■ Mathematical Communication

62. The diagram is a student's first attempt at drawing a map of the roads that connect seven towns. If all of the roads are straight, what is wrong with the map other than inaccuracies in the scale?

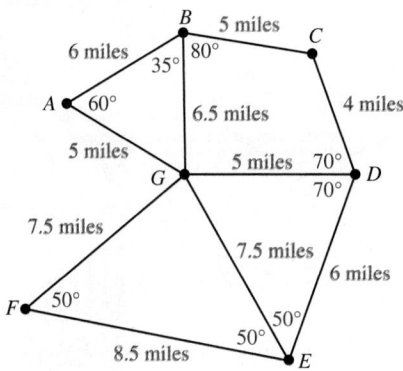

63. Ansley claims that a line segment cannot be a curve because a segment does not bend. Is she correct?

64. Theo claims that it is possible to make a triangle with sides of length 3 in., 5 in., and 8 in. because $3 + 5 = 8$. How might you show him that he is incorrect?

65. How might you show a group of students that a triangle can have only one right or one obtuse angle?

66. Jacob uses the following diagram to argue that the exterior angle of a triangle, $\angle 1$, is congruent to the interior angle, $\angle 2$, because they are vertical angles. Where did he go wrong?

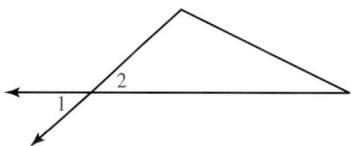

67. Janice is trying to measure the distance along the diagonal of a rectangular room but does not have a tape measure long enough. How might you tell her to find the length?

68. a. Write a story problem that makes use of the Pythagorean Theorem.

    b. Write a story problem that makes use of the converse of the Pythagorean Theorem.

■ Building Connections

69. Write a paragraph or two describing how the notions of points, lines, and angles are used to develop the concepts associated with triangles.

70. The words "acute" and "obtuse" take on different meanings in our everyday language. Look up the meanings of these words in a dictionary. Based on what you find, do you think these terms are a good fit for the geometrical concept they represent?

71. People used the Pythagorean Theorem long before Pythagoras. Search the Internet to find different instances of how ancient civilizations made use of this fact.

72. The Pythagorean Theorem is often used in measurement. How so?

73. How do the concepts associated with triangles build a connection between geometry and algebra?

In this section, we continue our discussion of polygons and other two-dimensional shapes. We begin with quadrilaterals.

Quadrilaterals

As the name implies, the set of **quadrilaterals** contains all four-sided polygons. The **sides** of a quadrilateral meet at four **vertices,** which we can use to name the quadrilateral. Quadrilaterals come in many shapes and can be convex or concave (Figure 10.32).

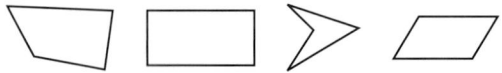

Figure 10.32 Quadrilaterals

With the addition of one more vertex and side, we can introduce two new terms that indicate the relative position of the sides to one another. Two sides that share a common vertex are called **consecutive** sides, and two sides that do not are called **opposite** sides. In Figure 10.33, $\overline{AB}$ and $\overline{BC}$ are consecutive sides, whereas $\overline{AB}$ and $\overline{CD}$ are opposite sides. We can also use the same terms with angles and vertices. For instance, $\angle A$ and $\angle B$ are consecutive angles, whereas $\angle A$ and $\angle C$ are opposite angles. Similarly, B and C are consecutive vertices, and B and D are opposite vertices. Any segment that joins two opposite vertices is called a **diagonal.** Because every quadrilateral has two pairs of opposite vertices, every quadrilateral has two diagonals.

We typically limit our study of quadrilaterals to convex ones because we can classify them in a variety of ways based on the characteristics of their sides and angles. Table 10.10 gives the definitions of several specific quadrilaterals.

Figure 10.33 Diagonals of quadrilateral *ABCD*

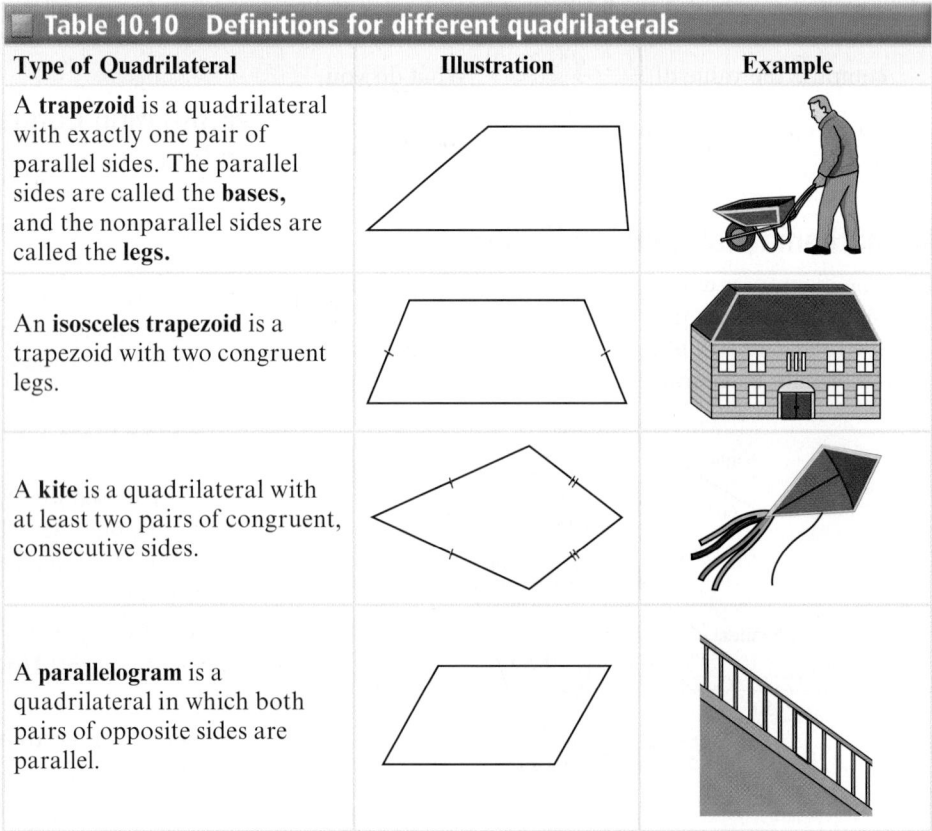

Table 10.10 Definitions for different quadrilaterals		
Type of Quadrilateral	**Illustration**	**Example**
A **trapezoid** is a quadrilateral with exactly one pair of parallel sides. The parallel sides are called the **bases,** and the nonparallel sides are called the **legs.**		
An **isosceles trapezoid** is a trapezoid with two congruent legs.		
A **kite** is a quadrilateral with at least two pairs of congruent, consecutive sides.		
A **parallelogram** is a quadrilateral in which both pairs of opposite sides are parallel.		

(Continued)

Type of Quadrilateral	Illustration	Example
A **rhombus** is a quadrilateral with all sides congruent.		
A **rectangle** is a parallelogram with a right angle.		
A **square** is a rectangle with all sides congruent.		

Some of the quadrilaterals in Table 10.10 are defined in terms of other quadrilaterals. The next example highlights this fact in a visual way.

Example 10.15 Connecting Quadrilaterals

Connection

Draw a diagram that shows how the quadrilaterals are interconnected.

Solution We can create the following diagram by using the definitions in Table 10.10. The diagram shows that trapezoids, kites, and parallelograms are connected only by the fact that they are quadrilaterals. A rectangle is defined to be a special kind of parallelogram. A rhombus has opposite sides that are parallel and congruent consecutive sides, so it is both a kite and a parallelogram. A square is both equilateral and equiangular, so it is both a rhombus and a rectangle. These connections lead us to the following diagram.

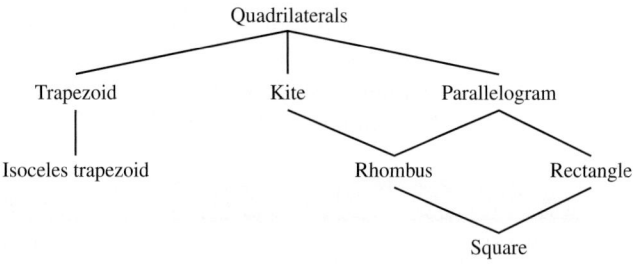

Employing the definitions in Table 10.10 and a little thought, we can prove a number of facts about specific quadrilaterals. For instance, consider parallelograms.

Example 10.16 Opposite Angles in a Parallelogram

Reasoning

What is true about the opposite angles in a parallelogram given that both pairs of opposite sides are parallel?

Solution Because both pairs of opposite sides in a parallelogram are parallel, we can use what we know about parallel lines to answer the question. To begin, consider the following parallelogram in which the sides have been extended. We know that $\angle 1 \cong \angle 2$ because they are alternate interior angles. Likewise, $\angle 2 \cong \angle 4$ because they are corresponding angles. After making a substitution, we have $\angle 1 \cong \angle 4$, which implies that the opposite angles in a parallelogram must be congruent.

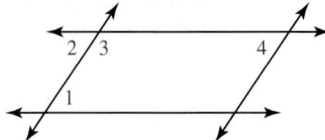

Explorations
Manual
10.8

Further exploration of parallelograms leads to the following facts about their angles, sides, and diagonals.

Theorem 10.9   Properties of Parallelograms

A quadrilateral is a parallelogram if and only if any of the following holds true.

1. Both pairs of opposite sides are congruent.
2. Consecutive angles are supplementary.
3. Opposite angles are congruent.
4. The diagonals bisect each other.
5. One pair of opposite sides is both parallel and congruent.

Theorem 10.9 is a biconditional statement, so any property in the theorem holds true in either direction. For example, if both pairs of opposite sides in a quadrilateral are congruent, then the quadrilateral must be a parallelogram. Likewise, if a quadrilateral is a parallelogram, then both pairs of opposite sides must be congruent.

Once we know the properties of parallelograms, discovering properties for other quadrilaterals is not difficult. For instance, because a rhombus is a parallelogram, every fact that holds for a parallelogram must hold for a rhombus. In addition, the diagonals of a rhombus are perpendicular bisectors of each other and bisect each pair of opposite angles (Figure 10.34).

The properties of parallelograms also hold for rectangles and squares. However, with a rectangle, the added condition of a right angle means that all four angles are congruent and the diagonals are congruent. A square is both a rhombus and a rectangle, so every property that holds for these shapes holds for squares. Table 10.11 summarizes the properties of the quadrilaterals discussed to this point.

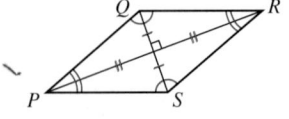

Figure 10.34 Diagonals of a rhombus

Table 10.11 Properties of quadrilaterals					
Property	**Kite**	**Parallelogram**	**Rhombus**	**Rectangle**	**Square**
Opposite sides parallel		X	X	X	X
Opposite sides congruent		X	X	X	X
All sides congruent			X		X
Consecutive angles supplementary		X	X	X	X
Opposite angles congruent		X	X	X	X
All angles congruent				X	X
Diagonals bisect opposite angles			X		X
Diagonals perpendicular	X		X		X
Diagonals bisect each other		X	X	X	X
Diagonals congruent				X	X

Example 10.17 | **Using the Properties of Quadrilaterals**

Reasoning

What can be said about two perpendicular segments joining two parallel lines?

Solution Suppose that $l \parallel m$ and $\overline{AB}$ and $\overline{CD}$ not only join to l and m but are also perpendicular to them. From Table 10.11, we know that quadrilateral $ABCD$ is a rectangle because it is a parallelogram with a right angle. In any rectangle, the opposite sides are congruent, so $\overline{AB} \cong \overline{CD}$. As a result, any two perpendicular segments joining two parallel lines must be congruent.

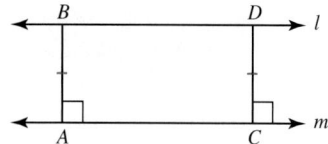

Example 10.18 | **Squaring the Walls of a Room**

Application

A carpenter is building a rectangular room with dimensions 15×20 ft. How can she make a measurement to determine whether the room is rectangular?

Solution Table 10.11 tells us that a quadrilateral is a rectangle if the diagonals are congruent. In other words, when both diagonals measure the same length, we know that the room is rectangular. Because the walls measure 15 ft and 20 ft, respectively, the Pythagorean Theorem tells us that the diagonals of the rectangle are 25 ft. Consequently, when both diagonals measure 25 ft, the room is a rectangle.

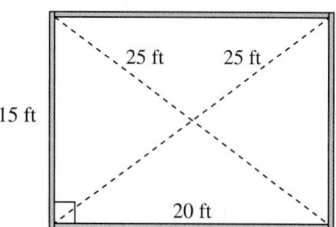

Facts about parallelograms also enable us to verify facts about isosceles trapezoids. By definition, the legs of an isosceles trapezoid are congruent, which might make us suspect that the base angles are also congruent. This is indeed the case. To understand why, consider the isosceles trapezoid $ABCD$ in which $\overline{AB} \parallel \overline{CD}$ and the segment $\overline{AE}$ is parallel to $\overline{BC}$ (Figure 10.35).

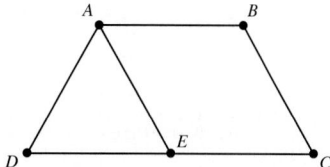

Figure 10.35 Isosceles trapezoid *ABCD*

Quadrilateral $ABCE$ must be a parallelogram because the opposite sides are parallel. This implies $\overline{BC} \cong \overline{AE}$ and $\overline{AE} \cong \overline{AD}$, making $\triangle AED$ an isosceles triangle. Because the base angles in an isosceles triangle are congruent, $\angle AED \cong \angle ADE$. Also, $\angle AED$ and $\angle BCE$ are corresponding angles, so they must be congruent. If we replace $\angle AED$ with $\angle ADE$, then $\angle ADE \cong \angle BCE$, proving that the base angles in an isosceles trapezoid are congruent. This and other facts about isosceles trapezoids are summarized in Theorem 10.10.

Theorem 10.10   Properties of Isosceles Trapezoids

A trapezoid is isosceles if and only if any one of the following conditions holds.

1. The legs are congruent.
2. The base angles are congruent.
3. The diagonals are congruent.

Check Your Understanding 10.3-A

1. True or false:
   a. Every square is a parallelogram.   **b.** Some kites are rectangles.
   c. No rectangle is a rhombus.   **d.** Every parallelogram is a trapezoid.

2. True or false:
   a. Opposite angles in a parallelogram are supplementary.
   b. The diagonals of a rectangle bisect each other.
   c. Consecutive sides of a square are perpendicular to one another.
   d. The base angles of an isosceles trapezoid are congruent.

3. Find the measure of each of the numbered angles, if $\overline{AB} \parallel \overline{CD}$ in each figure.

a. 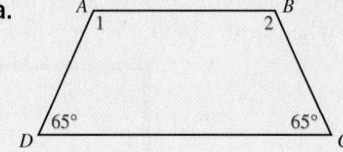   **b.**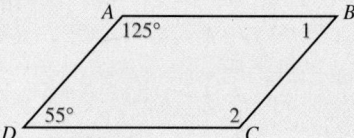

c.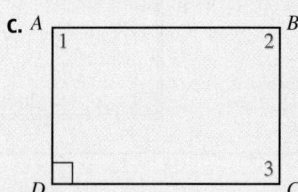

Talk About It A trapezoid can also be defined as a quadrilateral with at least one pair of parallel sides. How would this change in the definition affect the diagram given in Example 10.15?

Communication

Activity 10.5
Use the definition or the properties of the different quadrilaterals to give a peer instructions on how to create each of the following on a standard 5 × 5 geoboard.

a. A square.   **b.** A rectangle that is not a square.
c. A rhombus that is not a square.   **d.** A kite that is not a rhombus.
e. A parallelogram that is not a rhombus.   **f.** An isosceles trapezoid.

Other Polygons

Explorations Manual 10.9

Many other types of polygons are created simply by adding more segments. Like triangles and quadrilaterals, the segments are called the **sides** of the polygon, and any point where two sides meet is called a **vertex.** As shown in Table 10.12, polygons

Explorations
Manual
10.10
are typically named by their number of sides. If a polygon has more than 12 sides, one convention is to name it by stating the number of sides followed by the suffix "-gon." For instance, we can call a 17-sided polygon a 17-gon and a 25-sided polygon a 25-gon.

Table 10.12 Names of polygons

Name	Number of Sides	Name	Number of Sides
Triangle	3	Nonagon	9
Quadrilateral	4	Decagon	10
Pentagon	5	Undecagon	11
Hexagon	6	Dodecagon	12
Heptagon	7	. . .	
Octagon	8	n-gon	n

Explorations
Manual
10.11
Any polygon with four or more sides has **diagonals,** which are any segments joining two nonconsecutive vertices. As the number of sides increases, so does the number of diagonals. For instance, a quadrilateral has two diagonals, a hexagon has nine, and an octagon has 20 (Figure 10.36). The next example illustrates how we can use diagonals to discover an important fact about polygons.

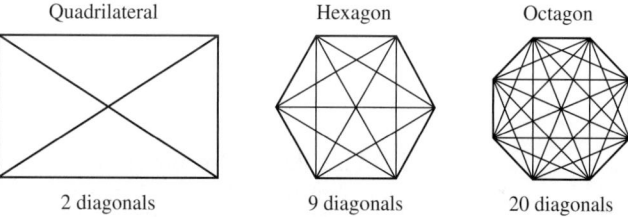

Figure 10.36 **Diagonals in polygons**

Example 10.19 Sum of the Interior Angles of a Polygon

Problem Solving Find a formula for the sum of the interior angles of any convex polygon.

Solution

Understanding the Problem. Our goal is to find a formula that we can use to compute the sum of the interior angles of any convex polygon. Although we do not know much about polygons, we do know something about triangles: The sum of the interior angles in a triangle is 180°.

Devising a Plan. We begin our search for a formula with a simple observation. Drawing the diagonal of a quadrilateral divides it into two triangles with angles equivalent to the four angles of the quadrilateral.

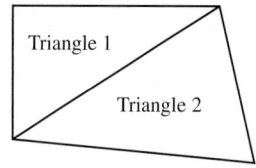

Because the sum of the angles in a triangle is 180°, the sum of the angles in the quadrilateral must be twice this value, or $2 \cdot 180° = 360°$. If we can extend this observation to other polygons, we may *find a pattern* that we can use to write a formula.

Carrying Out the Plan. First, we draw several convex polygons, each with a different number of sides. We then divide each polygon into triangles by drawing as many nonintersecting diagonals as possible.

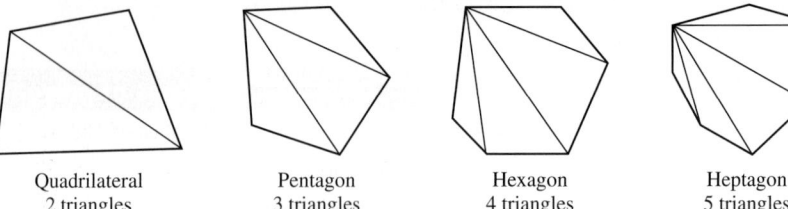

Quadrilateral
2 triangles

Pentagon
3 triangles

Hexagon
4 triangles

Heptagon
5 triangles

Because the angles in each triangle have a sum of 180°, we can multiply the number of triangles by 180° to find the sum of the interior angles in each polygon. We summarize the results in the following table. As the table shows, the number of triangles is always 2 less than the number of sides. Consequently, the sum of the interior angles of any n-gon must be $(n - 2) \cdot 180°$.

Polygon	Number of Sides	Number of Triangles	Sum of Measure of Angles
Triangle	3	1	$1 \cdot 180° = 180°$
Quadrilateral	4	2	$2 \cdot 180° = 360°$
Pentagon	5	3	$3 \cdot 180° = 540°$
Hexagon	6	4	$4 \cdot 180° = 720°$
Heptagon	7	5	$5 \cdot 180° = 900°$
n-gon	n	$n - 2$	$(n - 2) \cdot 180°$

Looking Back. Whenever we develop a formula, we should test it by using established facts. For triangles, $n = 3$, so $(n - 2) \cdot 180° = (3 - 2) \cdot 180° = 180°$. For quadrilaterals, $n = 4$, so $(n - 2) \cdot 180° = (4 - 2) \cdot 180° = 360°$. Because both sums match previous results, we conclude that the formula is correct.

The next example illustrates another fact that holds for the exterior angles of polygons.

Example 10.20   Sum of the Exterior Angles of a Polygon

Reasoning

If an exterior angle is drawn at each vertex of a pentagon, what is the sum of these angles?

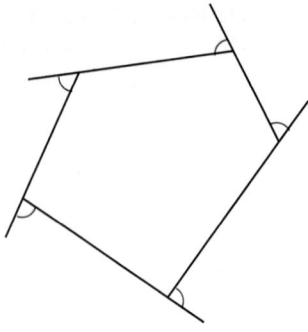

Solution The diagram shows that an interior angle and its adjacent exterior angle form a linear pair, so they must be supplementary. Because there are five such pairs, the total degree measure of the interior and exterior angles must be $5 \cdot 180° = 900°$. However, the last example showed that the sum of the interior angles of a pentagon is $(5 - 2) \cdot 180° = 540°$. So the sum of the exterior angles must be $900° - 540° = 360°$. ∎

If we were to repeat the process in the last example with any convex polygon, we would find that the sum of the exterior angles is always 360°. The next theorem summarizes the facts learned from the last two examples.

Theorem 10.11 Sums of Interior and Exterior Angles in Polygons

In any convex polygon with n sides:

1. The sum of the measures of the interior angles is $(n - 2) \cdot 180°$.
2. The sum of the measures of the exteriors angles, one at each vertex, is 360°.

Regular polygons are another type of polygon with additional properties. A polygon is **regular** if it is both equiangular and equilateral. In other words, the interior angles are congruent, and the sides are congruent. Several examples are shown in Figure 10.37.

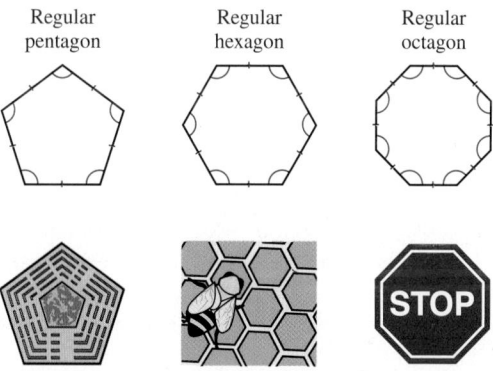

Figure 10.37 Examples of regular polygons

Every regular polygon has a **center,** which is the point in the interior that is equidistant from its vertices. To find it, we locate the intersection of two or more of the perpendicular bisectors of the sides. Any segment joining a vertex to the center is called a **radius.** Any segment joining the center to the midpoint of a side is called an **apothem.** Apothems are always perpendicular to the side. See Figure 10.38.

Figure 10.38 also illustrates a **central angle,** which is the angle formed by two radii drawn to consecutive vertices. A regular n-gon has n congruent central angles, the measure of which is $\dfrac{360°}{n}$. Furthermore, because the sum of the angles in a regular n-gon is $(n - 2) \cdot 180°$, each interior angle has a measure of $\dfrac{(n - 2)180°}{n}$.

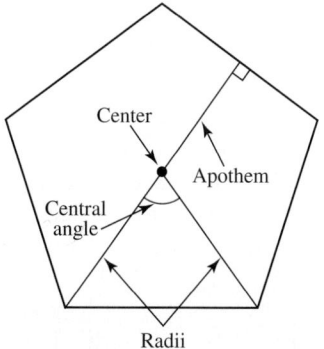

Figure 10.38 Parts of a regular polygon

Theorem 10.12 Measures of Central and Interior Angles in Regular Polygons

In any regular n-gon:

1. Each central angle has a measure of $\dfrac{360°}{n}$.

2. Each interior angle has a measure of $\dfrac{(n - 2)180°}{n}$.

Example 10.21 | Finding the Measure of an Interior Angle

If the following figure is a regular hexagon, what is the value of x?

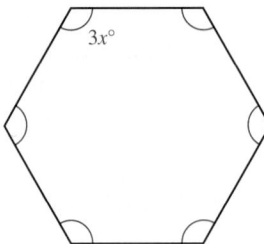

Solution Using Theorem 10.12(2) and $n = 6$, the measure of an interior angle is $\dfrac{(6-2)180°}{6} = \dfrac{4 \cdot 180°}{6} = 120°$. Consequently, $3x = 120°$, or $x = 40°$.

Circles

Another planar figure is worth mentioning. In a sequence of regular polygons, we can see that, as more sides are added, the shapes become more rounded in their appearance (Figure 10.39). Adding sides indefinitely causes the figure to approach a shape we call a circle.

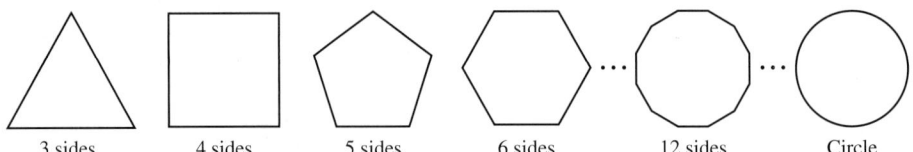

| 3 sides | 4 sides | 5 sides | 6 sides | 12 sides | Circle |

Figure 10.39 A sequence of regular polygons leading to a circle

---

Definition of a Circle

A **circle** is the set of all points in a plane that are a given distance from a given point.

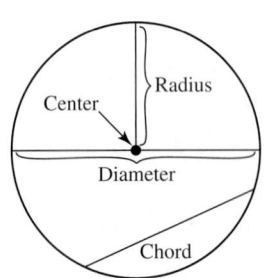

Figure 10.40 Parts of a circle

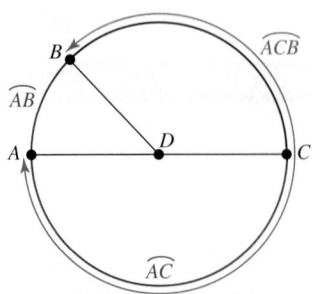

Figure 10.41 Arcs of a circle

Like regular polygons, circles have a center and a radius. In this case, however, the radius is any segment that joins the center to a point on the circle. Because circles do not have vertices, they cannot have diagonals. Instead, they have **chords,** which are segments that join any two points on the circle (Figure 10.40). If a chord goes through the center, it is called a **diameter.** The length of any diameter is always twice the length of the radius, or $d = 2r$.

Circles also have **central angles,** which are angles formed by two radii. Even though circles have no sides, we can talk about portions of the circle in terms of arcs. A **minor arc** is a set of points on the circle that lie inside a central angle. A **major arc** is a set of points that lie outside a central angle. If an arc cannot be confused with others, we use the endpoints to name it. Otherwise, we use three points to specify an arc. Any arc that lies on one side of a diameter is called a **semicircle.** In Figure 10.41, $\overarc{AB}$ is a minor arc, $\overarc{ACB}$ is a major arc, and $\overarc{AC}$ is a semicircle.

The **degree measure** of a minor arc is the measure of its central angle. This means the measure of any major arc is 360° minus the measure of the corresponding minor arc. Several arc measures are shown in Figure 10.42.

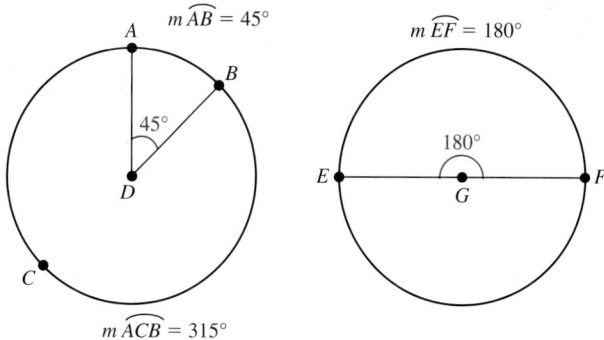

Figure 10.42 Arc measures

With this designation for arc measures, we can discover several relationships between angles and arcs. For instance, moving the vertex of the angle from the center to a point on the circle (Figure 10.43) creates a new angle, called an **inscribed angle.** Its measure is half that of its **intercepted arc,** or $m\angle A = \frac{1}{2}m\widehat{BC}$. Because a semicircle has a measure of 180°, any angle inscribed in a semicircle has a measure of $\frac{1}{2}(180°) = 90°$ (Figure 10.44). In other words, any angle inscribed in a semicircle must be a right angle.

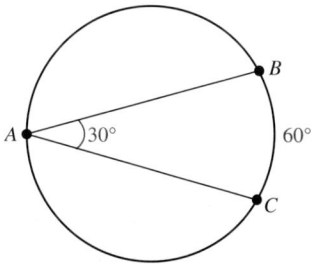

Figure 10.43 $m\angle A = \frac{1}{2}m\widehat{BC}$

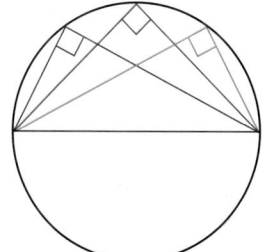

Figure 10.44 A right angle inscribed in a semicircle

If we move the vertex of the angle outside the circle, then the sides of the angle can intersect the circle in either one or two points. Any line, ray, or segment intersecting a circle in exactly one point is called a **tangent** (Figure 10.45). Any line, ray, or segment intersecting a circle in two points is called a **secant.** In either case, the measure of the angle is always one half the difference of the measures of the intercepted arcs.

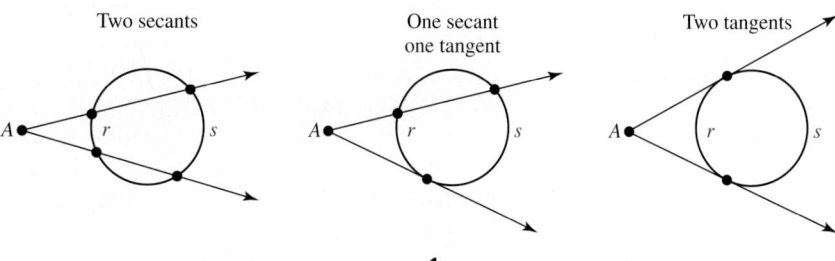

Figure 10.45 $m\angle A = \frac{1}{2}[m(\text{arc } s) - m(\text{arc } r)]$

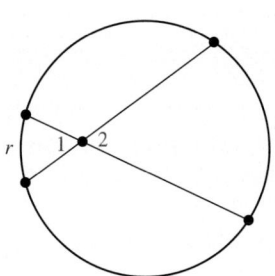

Figure 10.46 $m\angle 1 = m\angle 2$
$= \frac{1}{2}[m(\text{arc } r) + m(\text{arc } s)]$

Finally, consider the vertical angles made by two intersecting chords inside the circle (Figure 10.46). The angle measure of either angle is one-half the sum of the measures of the intercepted arcs.

Example 10.22 **Measuring Arcs and Angles**

Find the degree measure of the missing arc or angle.

a. 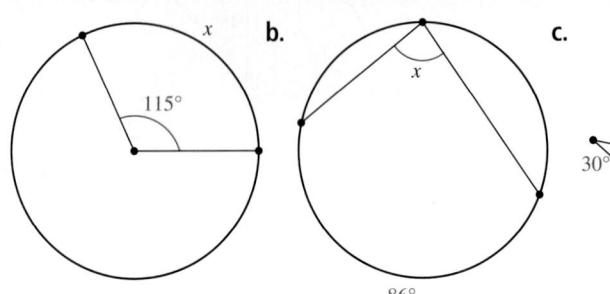 b. c.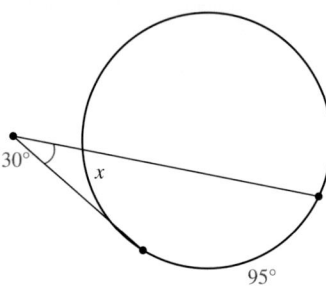

Solution

a. The given angle is a central angle, so the measure of the arc is equal to the measure of the angle. Consequently, $x = 115°$.

b. The given angle is inscribed, so its measure is half the intercepted arc, or $x = \frac{1}{2}(86°) = 43°$.

c. x is the value of the smaller of the two intercepted arcs, or

$$30° = \frac{1}{2}(95° - x)$$

$$30° = 47.5° - \frac{1}{2}x$$

$$-17.5° = -\frac{1}{2}x$$

$$35° = x$$

Check Your Understanding 10.3-B

1. What is the sum of the measures of the interior angles of a(n):
   a. Octagon?   **b.** Nonagon?   **c.** 15-gon?   **d.** 20-gon?

2. What is the measure of a central angle, interior angle, and exterior angle in a regular:
   a. Pentagon?   **b.** Heptagon?   **c.** 13-gon?   **d.** 21-gon?

3. Find the value of x if the following figure is a regular pentagon.

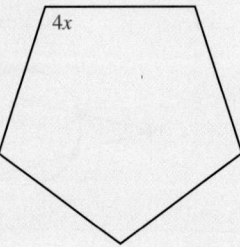

4. In a circle, what is the measure of the:
   a. Intercepted arc for a central angle of 95°?
   b. Intercepted arc for an inscribed angle of 68°?
   c. The angle formed by two secants that intercept arcs of 108° and 36°?

Talk About It Does it make sense to think of a circle as a regular polygon with an infinite number of sides? Why or why not?

Representation

Activity 10.6
Copy the following circle and the 18 points onto another sheet of paper. Use a ruler to connect each set of chords in a different color

$$\overline{AH}, \overline{BI}, \overline{CJ}, \ldots, \overline{RG}$$
$$\overline{AG}, \overline{BH}, \overline{CI}, \ldots, \overline{RF}$$
$$\overline{AF}, \overline{BG}, \overline{CH}, \ldots, \overline{RE}$$
$$\overline{AE}, \overline{BF}, \overline{CG}, \ldots, \overline{RD}$$

What do you notice about the finished figure? If you were to draw other sequences of segments, what do you think would happen to the figure?

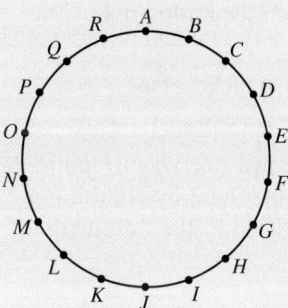

SECTION 10.3   Learning Assessment

■ Understanding the Concepts

1. What are the defining characteristics of each quadrilateral?
   - **a.** Trapezoid
   - **b.** Kite
   - **c.** Parallelogram
   - **d.** Rhombus
   - **e.** Rectangle
   - **f.** Square

2. What facts are true for a:
   - **a.** Rhombus but not a parallelogram?
   - **b.** Square but not a rhombus?
   - **c.** Rectangle but not a parallelogram?
   - **d.** Square but not a rectangle?

3. If a trapezoid is isosceles, what must be true about its legs, base angles, and diagonals?

4. What is a diagonal of a polygon?

5. **a.** How can triangles be used to find the sum of the interior angles of a polygon?
   - **b.** How is the sum of the interior angles of a polygon related to its number of sides?

6. What are the defining characteristics of a regular polygon?

7. Explain the difference between the radius and the apothem of a regular polygon?

8. What is the relationship between the number of sides in a regular polygon and the measure of its:
   - **a.** Central angle?
   - **b.** Interior angle?
   - **c.** Exterior angle?

9. Consider a circle. Explain the difference between a(n):
   - **a.** Chord, a diameter, and a radius.
   - **b.** Inscribed angle and a central angle.
   - **c.** Secant and a tangent.
   - **d.** Minor arc, a major arc, and a semicircle.

■ Representing the Mathematics

10. Draw a Venn diagram that shows the relationships among quadrilaterals, trapezoids, parallelograms, rhombuses, rectangles, and squares.

11. Draw a quadrilateral that is:

    a. A kite and a parallelogram.

    b. Not a trapezoid, kite, or parallelogram.

    c. A parallelogram but not a rectangle or a rhombus.

    d. A parallelogram that is both a rhombus and a rectangle.

12. **a.** Is it possible for a kite to be concave? If so, draw an example. If not, explain why.

    b. Is it possible for a parallelogram to be concave? If so, draw an example. If not, explain why.

13. **a.** Can a rectangle be described as an equiangular parallelogram? Explain.

    b. Can a square be described as a rectangular rhombus? Explain.

14. What shape is formed when the midpoints of the sides are connected in a:

    a. Parallelogram?   **b.** Rhombus?

    c. Rectangle?   **d.** Square?

15. The *n*-gon terminology is just one way to name polygons. We can also name them by combining the prefixes and suffixes listed in the following table.

Number of Sides	Prefix Name	Suffix Name	Number of Sides
20	Icosikai . . .	. . . henagon	1
30	Triacontakai . . .	. . . digon	2
40	Tetracontakai . . .	. . . trigon	3
50	Pentacontakai . . .	. . . tetragon	4
60	Hexacontakai . . .	. . . pentagon	5
70	Heptacontakai . . .	. . . hexagon	6
80	Octacontakai . . .	. . . heptagon	7
90	Enneacontakai . . .	. . . octagon	8
		. . . enneagon	9

For instance a 34-sided polygon is called a triacontakaitetragon, and a 52-sided polygon is called a pentacontakaidigon. Use the table to give another name for a:

    a. 27-gon.   **b.** 41-gon.   **c.** 68-gon.

    d. 73-gon.   **e.** 99-gon.

    f. What similarities do you see between the prefixes and suffixes?

16. If possible, draw an example of a convex hexagon that is:

    a. Neither equilateral nor equiangular.

    b. Equilateral but not equiangular.

    c. Equiangular but not equilateral.

    d. Both equilateral and equiangular.

17. What is another name for a regular triangle? A regular quadrilateral?

18. **a.** Is it possible to draw a pentagon in which the sides are congruent, but the angles are not? If so, what might it look like?

    b. Is it possible to draw a pentagon in which the angles are congruent, but the sides are not? If so, what might it look like?

    c. Do you think it is necessary for the definition of a regular polygon to state that both the sides and the angles be congruent? Explain your thinking.

19. If possible, draw a circle with a central angle that is approximately the given measure.

    a. 60°   **b.** 135°   **c.** 270°

20. Draw a circle with an inscribed angle that is approximately the given measure.

    a. 45°   **b.** 90°   **c.** 120°

    d. Is it possible for a circle to have an inscribed angle of 180°? Explain.

■ Exercises

21. Find the measure of each numbered angle.

    a.   **b.**

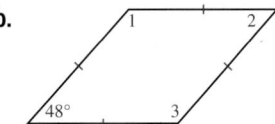

    c.

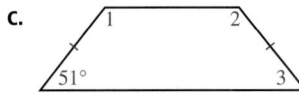

22. Find the measure of each numbered angle.

    a.   **b.**

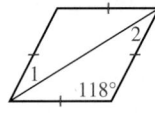

    c.

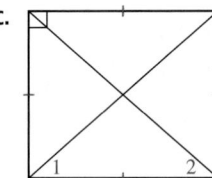

23. If one angle in an isosceles trapezoid is 38°, what are the measures of the other angles?

24. If one angle in a parallelogram is twice the measure of the other, what are the measures of the angles?

25. If one angle in a rhombus is five times the measure of its consecutive angle, what is the measure of each angle?

26. If one angle of a parallelogram is $(3x + 10)°$ and a consecutive angle is $(x - 30)°$, what is the value of *x*?

27. Find x.

a.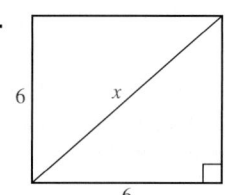

b.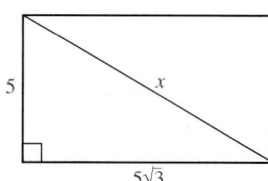

c.

28. Answer each question given that the figure is a parallelogram.

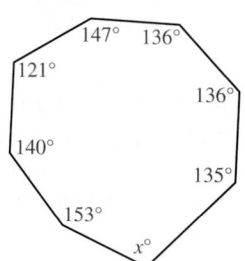

a. If x is 100°, what are the measures of $\angle 1$, $\angle 2$, and $\angle 3$?

b. If $m\angle 2 = 115°$, what is the value of x?

c. If $m\angle 1 = 36°$, what is the value of x?

d. If $m\angle 1 + m\angle 3 = 146°$, what is the value of x?

29. What is the sum of the measures of the interior angles of a(n):

a. Hexagon?   b. Octagon?

c. 12-gon?   d. 16-gon?

30. What is the sum of the measures of the interior angles of a(n):

a. Pentagon?   b. Heptagon?

c. 11-gon?   d. 25-gon?

31. What is the measure of a central, interior, and exterior angle in a regular:

a. Pentagon?   b. Hexagon?

c. 16-gon?   d. 22-gon?

32. What is the measure of a central, interior, and exterior angle in a regular

a. Quadrilateral?   b. Nonagon?

c. 14-gon?   d. 19-gon?

33. Find the value of x.

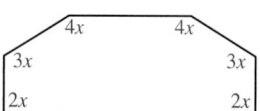

34. What are the measures of each angle in the following polygon?

35. a. A regular n-gon has a central angle measure of 36°. What is n?

b. A regular n-gon has a central angle measure of 15°. What is n?

c. A regular n-gon has a central angle measure of 14.4°. What is n?

36. a. A regular n-gon has a vertex angle measure of 150°. What is n?

b. A regular n-gon has a vertex angle measure of 168°. What is n?

c. A regular n-gon has a vertex angle measure of 128.57°. What is n?

37. a. A regular n-gon has an exterior angle measure of 120°. What is n?

b. A regular n-gon has an exterior angle measure of 30°. What is n?

c. A regular n-gon has an exterior angle measure of 32.73°. What is n?

38. What is the diameter of a circle if the radius of the same circle is:

a. 3 ft?   b. 4.6 ft?

c. 35 ft?   d. 14.31 ft?

39. What is the radius of a circle if the diameter of the same circle is:

a. 3 ft?   b. 4.6 ft?

c. 35 ft?   d. 14.31 ft?

40. What fraction of a circle is encompassed by the arc with a central angle measuring:

a. 90°?   b. 180°?   c. 270°?

d. 60°?   e. 22.5°?

41. Find x.

a.   b.

c.   d.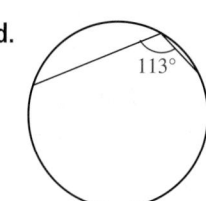

42. Find x.

a.

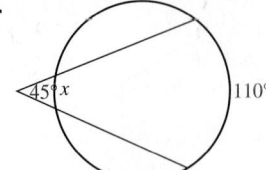

b.

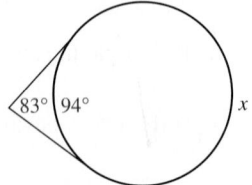

c.

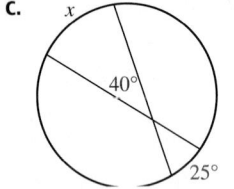

■ Problems and Applications

43. a. What is the largest number of intersection points between a triangle and a convex quadrilateral? Assume no side of the triangle lies on a side of the quadrilateral.

b. What is the largest number of intersection points between a triangle and any quadrilateral? Assume no side of the triangle lies on a side of the quadrilateral.

44. If the number of sides of a polygon are doubled, the sum of its interior angles increases by 720°. How many sides does the original polygon have?

45. Consider the following regular hexagon.

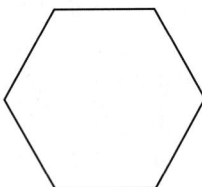

a. Divide it into four congruent parts so that each part is a trapezoid.

b. Divide it into three congruent parts so that each part is a rhombus.

46. If two sides of a regular hexagon are extended as shown in the figure, what is the measure of each of the numbered angles?

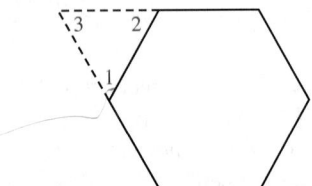

47. How many sides would a regular polygon have if each exterior angle has a measure of 15°?

48. If possible, give the name of the convex polygon for which the sum of the interior angles is 4,140°.

49. If the interior angles of a convex pentagon are five consecutive numbers, what are their measures?

50. Find x.

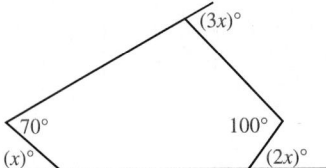

51. What is the length of the radius and apothem in a regular hexagon with 6-in. sides?

52. If a quadrilateral is inscribed in a circle, then both pairs of opposite angles are supplementary. Use this fact to find $m\widehat{ABC}$.

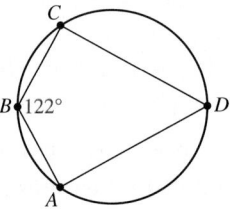

53. Find x and y, given that quadrilateral $ABCD$ is a parallelogram.

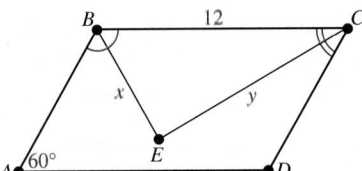

54. The perpendicular bisector of any chord always passes through the center of the circle. Consider a chord that is 6 cm long in a circle that has a diameter of 10 cm. How far is the chord from the center?

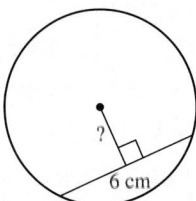

55. Janis wants to install a sprinkler that sprays water in a circular pattern at the center of her small circular garden. She knows that the perpendicular bisector of a chord passes through the center of a circle. How can she use this fact to find the center of the garden?

56. A goat is tied to the middle of the side of a barn with a 50-ft rope. If the side of the barn is only 40 ft long, draw the region of grass that the goat can eat.

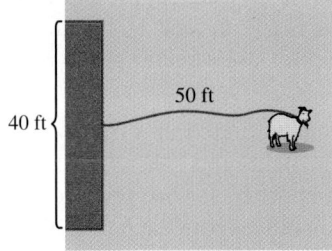

57. Which of the following regular polygons can be used to tile a floor so that there are no overlaps of tiles or gaps between them?

   a. Triangle   **b.** Quadrilateral

   c. Pentagon   **d.** Hexagon

   e. Octagon

■ Mathematical Reasoning and Proof

58. True or false:

   a. Every kite is a parallelogram.

   b. Every square is a kite.

   c. Every rectangle is a trapezoid.

   d. No rhombus is a parallelogram.

   e. No square is a trapezoid.

   f. Some parallelograms are kites.

59. Why does the definition of a circle guarantee that all radii of the circle are congruent?

60. If the following figure is a rhombus, what can be said about $\angle 1$ and $\angle 2$?

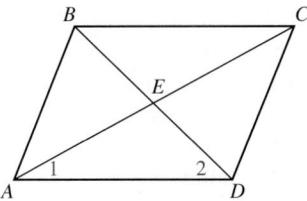

61. a. How many regular polygons have only acute interior angles?

   b. How many convex polygons have only acute interior angles?

   c. Can you use degree measure to explain why this is true?

62. Draw several different concave polygons. Determine the measure of each interior angle and add them. You need to use reflex angles to do so. What do you notice? Can you verify this result for any n-gon?

63. Consider the following hexagon. Explain how the sum of the measures of the interior angles can be found by using any point A in the interior of the hexagon and the triangles that result when point A is connected to each of the vertices.

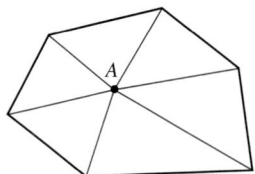

64. Find the total number of diagonals in each of the following convex polygons.

   a. Quadrilateral   **b.** Pentagon

   c. Hexagon   **d.** Heptagon

   e. Use what you have found to find a formula that can be used to compute the number of diagonals in any n-gon.

65. Find a formula for computing the measure of an exterior angle of a regular polygon.

66. a. What can be said about the measures of two inscribed angles that intercept the same arc of a circle?

   b. What can be said about the measure of a central angle and an inscribed angle that intercept the same arc of a circle?

■ Mathematical Communication

67. Design an activity that you might use in an elementary classroom to help students learn the names and defining characteristics of the different quadrilaterals. What kinds of representations are necessary to complete the activity? What does this imply about teaching geometry in general?

68. Give one of your peers instructions on how to fold a standard sheet of paper into a square. Be sure to explain how you know the resulting figure is indeed a square.

69. We can show that the perpendicular bisector of any chord always passes through the center of the circle. Write a set of directions explaining how to use the perpendicular bisector of a chord to find the center of a circle.

70. Kim, one of your students, makes the claim that every parallelogram is a trapezoid because both have parallel sides. Based on the definitions given in this section, how would you respond to Kim's statement?

71. Kayla wants to fence in a rectangular dog pen that is 30×40 ft. How would you use what you know about geometry to help her ensure that she has truly built a rectangular pen?

■ Building Connections

72. Walk around your college campus, and notice the ways in which quadrilaterals, polygons, and circles are used. Can you find five different uses for each figure?

   a. Parallelograms   **b.** Rectangles

   c. Squares   **d.** Pentagons

   e. Hexagons or octagons   **f.** Circles

73. How is an isosceles trapezoid similar to an isosceles triangle?

74. Drafting triangles come in two different sizes: a 45°-45°-90° triangle and a 30°-60°-90° triangle. How might you use a drafting triangle to find the center of a circle?

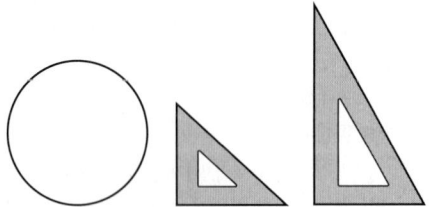

75. A pilot is making a flight to a city that is due north of his departure. The weather report shows that the wind will be blowing directly from the west for the entire flight. Because the wind will take the plane off course, the pilot must set a course that is to the west of his destination. How might the pilot use a rectangle to help him set his course?

76. Five cities are situated in a figure that is very close to a regular pentagon. The city planners want to build an airport to service the five cities that is an equal distance from each city. Where should the airport be located, and how might the city planners locate the position?

77. Examine a set of curriculum materials for grades K–4. At what grade level do children first learn about the different quadrilaterals and their properties? What exactly do the lessons teach children about quadrilaterals?

SECTION 10.4   Surfaces and Solids

In addition to **planar,** or two-dimensional figures, many of the shapes we encounter in our daily lives are **spatial,** or three-dimensional figures. We now turn to these shapes and their properties.

Basic Three-Dimensional Shapes

In Section 10.1, we introduced several basic shapes that were made from points, lines, and planes. We now add to what we know about these shapes by considering how they can interact in three dimensions. For example, we can now distinguish between coplanar and noncoplanar points. A set of points is **coplanar** if they lie on the same plane and **noncoplanar** if no single plane contains them. In Figure 10.47, A, B, and C are coplanar, whereas A, B, C, and D are noncoplanar.

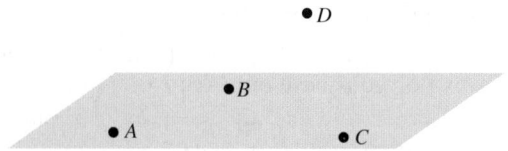

Figure 10.47 Coplanar and noncoplanar points

We can also introduce two other arrangements of lines. **Concurrent lines** are two or more lines that share a common point but that do not necessarily lie in the same plane. **Skew lines** are two lines that do not intersect and have no single plane that contains them (Figure 10.48).

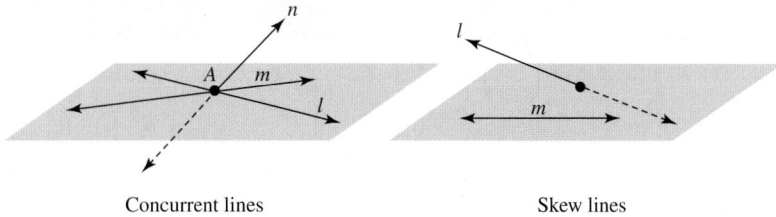

Figure 10.48 Concurrent and skew lines

In space, a line can interact with a plane in three ways. As shown in Figure 10.49, a line can be parallel to a plane, it can intersect a plane at one point, or it can intersect a plane at an infinite number of points.

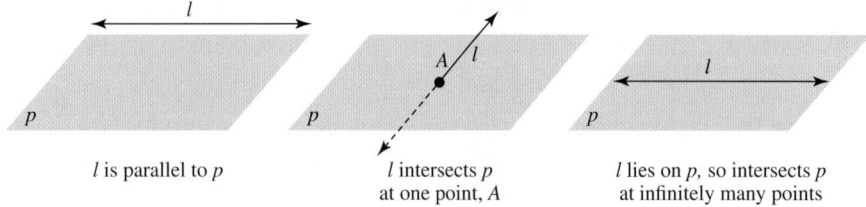

Figure 10.49 A line intersecting a plane

If a line intersects a plane at one point, it is perpendicular to the plane if it is perpendicular to every line on the plane [Figure 10.50(a)]. If a line intersects a plane at an infinite number of points, then it lies on the plane and divides it into two **half-planes.** For instance, in Figure 10.50(b) $\overleftrightarrow{AB}$ divides plane p into two half-planes: AB-C and AB-D.

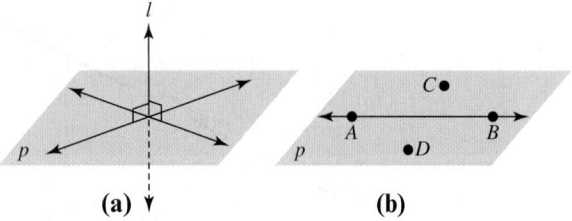

Figure 10.50 Perpendicular lines and half-planes

In space, planes can interact with one another in three basic ways. They can be parallel, intersect in a line, or intersect at a point (Figure 10.51).

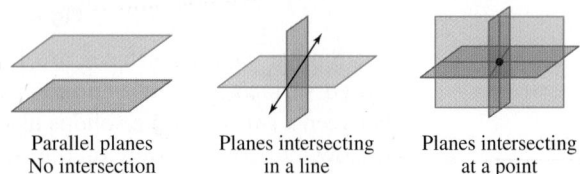

Figure 10.51 Intersecting planes

When two planes intersect at a line, the union of the two half-planes form a **dihedral angle.** A dihedral angle, like an angle in a plane, has a measure. It is equal to the measure of any angle that has its sides in the planes and is perpendicular to the line of intersection. Three dihedral angles and their measures are shown in Figure 10.52.

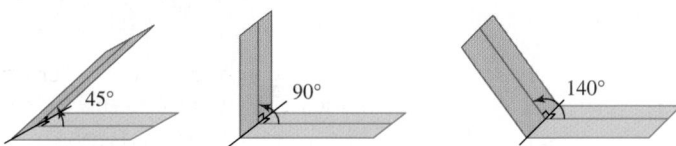

Figure 10.52 Dihedral angles

Example 10.23 **Identifying Basic Spatial Shapes**

Use the diagram to identify each of the following.

a. Coplanar points   **b.** Noncoplanar points   **c.** Concurrent lines

d. Planes intersecting in a line   **e.** Skew lines   **f.** Parallel planes

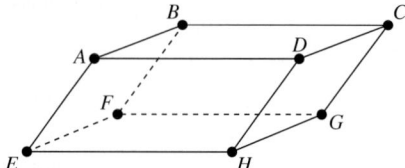

Solution Each question has many possible answers.

a. Three sets of coplanar points are $\{A, B, C\}$, $\{D, C, G\}$, and $\{A, B, H\}$.

b. Two sets of noncoplanar points are $\{A, B, C, E\}$ and $\{D, C, H, A\}$.

c. $\overleftrightarrow{AB}$, $\overleftrightarrow{AD}$, and $\overleftrightarrow{AE}$ are concurrent lines that intersect at point A.

d. Plane ABC and plane ADE intersect at line $\overleftrightarrow{AD}$.

e. $\overleftrightarrow{AB}$ and $\overleftrightarrow{DH}$ are skew lines, as are $\overleftrightarrow{FG}$ and $\overleftrightarrow{AE}$.

f. Plane ABE and plane DCG are two parallel planes.

As before, points, lines, and planes serve as the building blocks for other spatial figures. The most basic of these is a **surface,** which we imagine as a plane made from a rubber sheet that can be curved or twisted into any number of shapes. As with curves, surfaces can be simple or nonsimple and closed or nonclosed. Examples of different surfaces are shown in Figure 10.53. In elementary geometry, we often limit our study to **simple, closed surfaces,** which are surfaces that enclose exactly one hollow region and have no holes. Any simple, closed surface divides space into exactly three regions: the interior of the surface, the surface itself, and the exterior of the surface.

Bubble
Simple closed
surface

Open box
Non-closed
simple surace

Doughnut
Non-simple
closed surface

Figure 10.53 Different surfaces

Simple, closed surfaces are hollow figures. However, if we combine a simple closed surface with all of points in its interior points, we get a **solid.** The difference between a surface and a solid is like that between a bubble and a baseball. A bubble is hollow and consists only of the soap film that makes the bubble. A baseball on the other hand consists not only of the leather exterior but also the solid inner core.

A simple closed surface can also be convex or concave. It is **convex** if every line segment joining any two points in the interior of the surface is contained entirely within the surface (Figure 10.54). It is **concave** if it is not convex. As with planar shapes, this intuitively means that we can recognize concave surfaces by their "indentations."

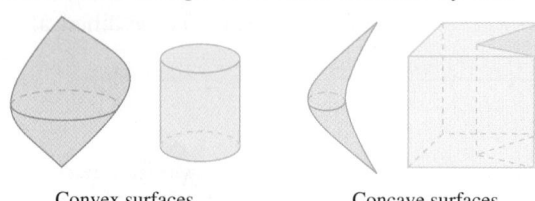

Convex surfaces   Concave surfaces

Figure 10.54 Convex and concave surfaces

Drawing Three-Dimensional Shapes

Explorations
Manual
10.12

In the classroom, students need to work with and create two-dimensional drawings of three-dimensional shapes. This can be difficult for many students because it relies on spatial reasoning skills that they may not have yet developed. They can work on these skills in a number of ways, one of which is illustrated in the next example.

Example 10.24 **Drawing a Shape from Different Perspectives**

Representation

Consider the activity on the student page in Figure 10.55. Use isometric dot paper to draw the shape given in the second part of the activity.

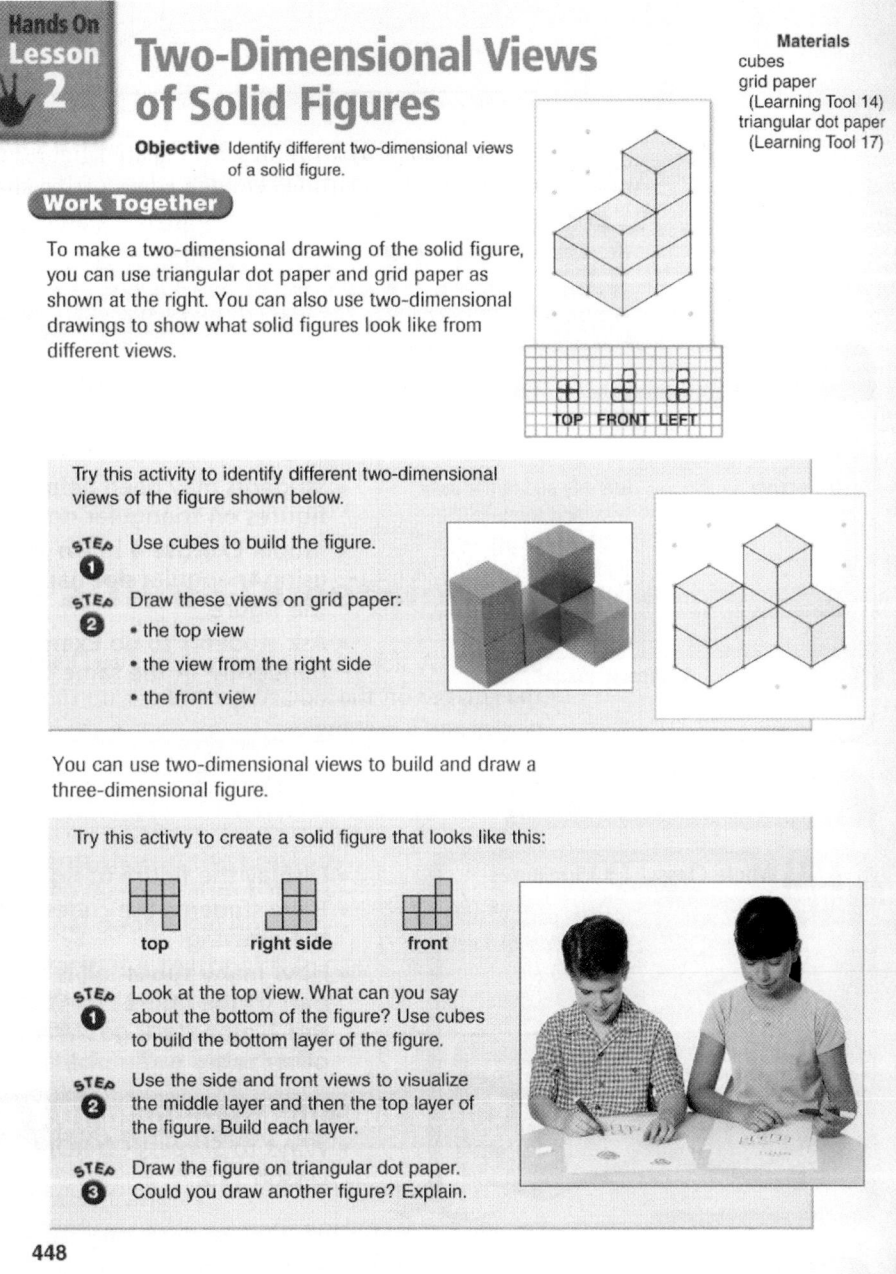

Figure 10.55 A spatial reasoning activity
Source: From *Mathematics Grade 5: Houghton Mifflin Mathematics Georgia* (p. 448). Copyright © by Houghton Mifflin Company, Inc. Reproduced by permission of the publisher, Houghton Mifflin Harcourt Publishing Company.

Solution The second part of the activity gives us three views of the shape we are to draw. The top view shows that the figure has three rows of blocks: The two back rows each have three blocks, and the front row has one block placed all the way to the right. The right and front views show how high to make the figure. The front row is **one** block high. The second and third rows are two blocks high except for the columns farthest to the right, which are three blocks high. The final figure looks like the following:

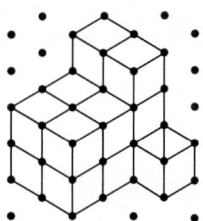

Explorations
Manual
10.13

Another way to represent a spatial figure in two dimensions is with a net. We make a **net** by "cutting" enough edges of the shape so that we can unfold it and lay it flat. Nets are helpful when studying spatial shapes because complex figures can be broken down into their simpler parts.

Example 10.25   **Drawing Nets**

Representation   Draw the net of each figure.

a.

b.
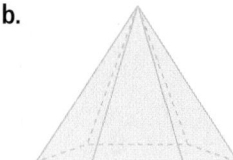

c.
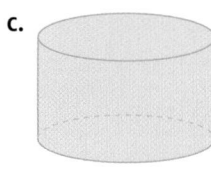

Solution A net of each figure is shown. The third figure is interesting because of the surface on the side. When we cut up the side and unfold the shape, we get two circles and a rectangle.

a.
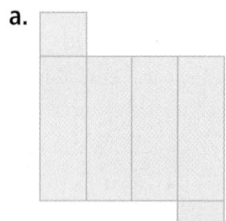

b.
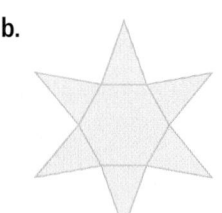

c.
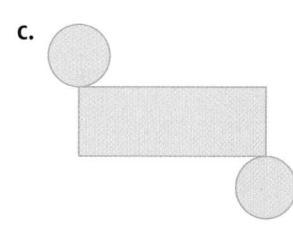

Check Your Understanding 10.4-A

1. Use the diagram to identify each of the following.
   a. Three coplanar points
   b. Four noncoplanar points
   c. Concurrent lines
   d. Skew lines

e. Planes intersecting in a line
f. Parallel planes

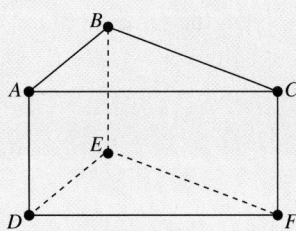

2. Draw each figure.

   a. A dihedral angle measuring 90°

   b. A simple closed surface made from four triangles

   c. A simple closed surface made from six rectangles

   d. A simple closed surface that contains no polygons

3. Draw the net of the figure in Problem 1.

Talk About It Many spatial shapes are analogous to planar shapes. Make a list of the spatial shapes given so far, and then identify the planar shapes they are analogous to.

Representation

Activity 10.7

Many students find drawing spatial figures a challenge because they do not know which edges of the figure should show and which should remain hidden. Those that show often depend on perspective. Make two copies of each of the following figures. On one copy, draw the segments you would see if you were looking down at the top of the figure. On the second copy, draw the segments you would see if you were looking up from the bottom. Leave hidden segments as dashed lines.

a.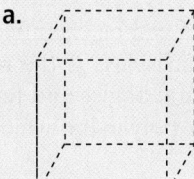

b.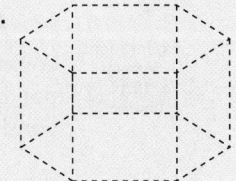

c.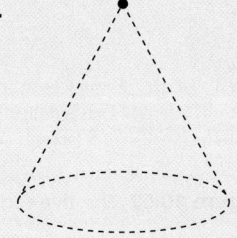

d.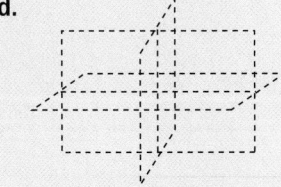

Regular and Semiregular Polyhedra

Simple closed surfaces can be classified in a number of ways. One important subset of simple closed surfaces are the polyhedra.

Definition of a Polyhedron

A **polyhedron** is a simple, closed surface made from planar polygonal regions.

The polygonal regions are called the **faces** of the polyhedron, and the vertices and sides of the polygons make up the **vertices** and **edges** of the polyhedron (Figure 10.56). Polyhedra also have **diagonals,** which are any line segments joining two vertices not on the same face.

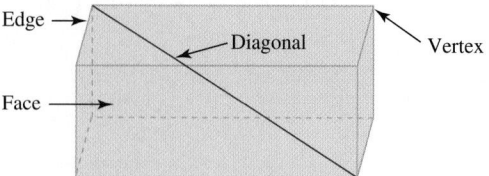

Figure 10.56 Parts of a polyhedron

Example 10.26 Identifying Polyhedra

Communication

Are the following figures polyhedra?

a.
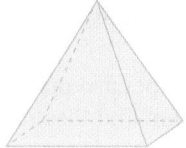

b.
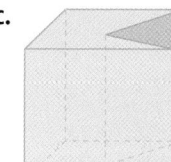

c.

Solution

a. The figure is a polyhedron because four of its faces are triangles and one is a quadrilateral.

b. The figure is not a polyhedron because two of its faces are circles.

c. The figure is a polyhedron because five of its faces are quadrilaterals and two are pentagons.

Explorations Manual 10.14

As with polygons, polyhedra can be classified in a number of ways according to common characteristics. One such class is the **regular polyhedra,** which are convex polyhedra with congruent dihedral angles and faces that are congruent regular polygons. There are only five regular polyhedra, which are shown in Figure 10.57.

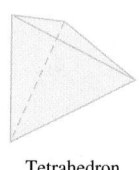
Tetrahedron
4 faces

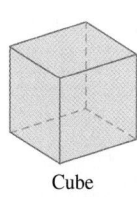
Cube
(hexahedron)
6 faces

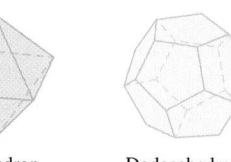
Octahedron
8 faces

Dodecahedron
12 faces

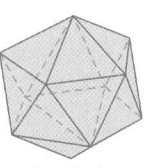
Icosahedron
20 faces

Figure 10.57 The five regular polyhedra

To understand why there are only five regular polyhedra, we must recognize some of the constraints placed on these shapes. First, the faces of regular polyhedra must be regular polygons. This limits the faces to equilateral triangles, squares, regular pentagons, and so on. Second, at least three faces must come together at a vertex to form a polyhedron. If only two come together, the resulting figure necessarily lies flat. Third, the sum of the angle measures at any one vertex must be strictly less than 360°. If the sum were greater than or equal to 360°, the figure would either lie flat or there would be overlap between the faces. Neither can happen and still form a polyhedron.

Given these constraints, consider what happens when we try to make regular polyhedra from particular regular polygons. We begin with the equilateral triangle, which has an interior angle of 60°. Because $3 \cdot 60° = 180°$, $4 \cdot 60° = 240°$, and $5 \cdot 60° = 300°$, we can fit three, four, or five triangles together at one vertex. Doing so generates the tetrahedron, octahedron, and icosahedron, respectively (Figure 10.58). If we try to fit six triangles at a vertex, the resulting figure lies flat because $6 \cdot 60° = 360°$. If we try to fit seven or more triangles at a vertex, the triangles overlap. So there can be only three regular polyhedra with triangular faces.

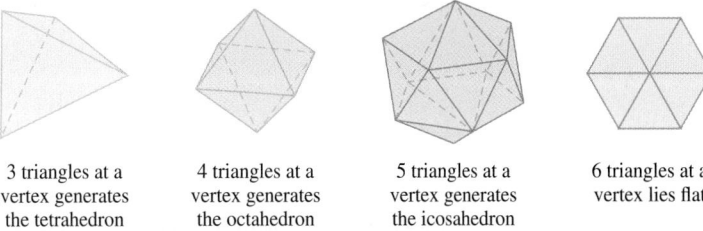

3 triangles at a
vertex generates
the tetrahedron

4 triangles at a
vertex generates
the octahedron

5 triangles at a
vertex generates
the icosahedron

6 triangles at a
vertex lies flat

Figure 10.58 Triangles and regular polyhedra

Next, consider the square, which has an interior angle of 90°. Because $3 \cdot 90° = 270°$, and $270° < 360°$, we can fit three squares together at one vertex. Doing so results in the cube. If we try to fit four, the resulting figure lies flat because $4 \cdot 90° = 360°$. If we try five or more squares, the squares overlap (Figure 10.59). Consequently, only one regular polyhedron has square faces.

We can also try a regular pentagon, which has an interior angle of 108°. Because $3 \cdot 108° = 324°$, and $324° < 360°$, we can fit three pentagons together at one vertex. Doing so gives us the dodecahedron. If we try four or more pentagons at a vertex, they overlap. Consequently, only one regular polyhedron has pentagonal faces (Figure 10.60).

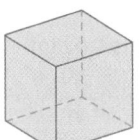
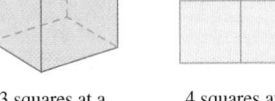

3 squares at a
vertex generates
the cube

4 squares at a
vertex lies flat

Figure 10.59 Squares and regular polyhedra

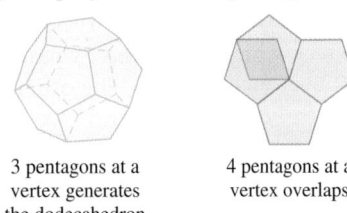

3 pentagons at a
vertex generates
the dodecahedron

4 pentagons at a
vertex overlaps

Figure 10.60 Pentagons and regular polyhedra

For any other regular polygon, the sum of the angles at a vertex is always greater than or equal to 360°. For instance, three regular hexagons lie flat because the interior angle measure is 120° and $3 \cdot 120° = 360°$. Three regular heptagons overlap because the interior angle measure is 128.57° and $3 \cdot 128.57° = 385.71° > 360°$. Consequently, there are only five regular polyhedra.

Another set of polyhedra are the **semiregular polyhedra.** Like regular polyhedra, the faces of semiregular polyhedra are regular polygons but are not limited to just one shape. This simple difference allows for many more shapes, a few of which are shown and named in Figure 10.61.

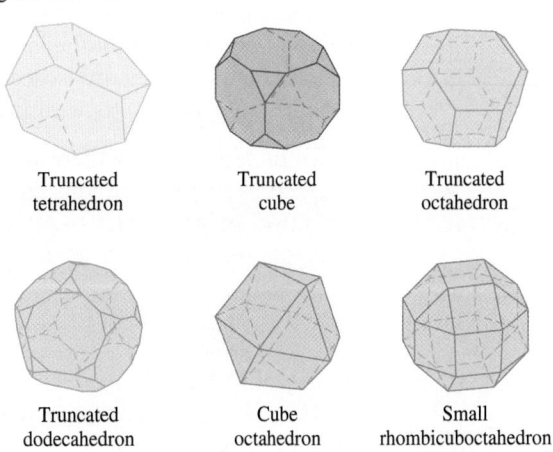

Truncated
tetrahedron

Truncated
cube

Truncated
octahedron

Truncated
dodecahedron

Cube
octahedron

Small
rhombicuboctahedron

Figure 10.61 Semiregular polyhedra

A common task when working with polyhedra is to count their faces, edges, and vertices. The next example illustrates one method for doing so.

Example 10.27   **Counting Faces, Edges, and Vertices on a Polyhedron**

Determine the number of faces, edges, vertices of a truncated cube.

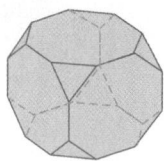

Solution A truncated cube is formed by "cutting off" the corners of a cube so that each new face forms an equilateral triangle. A vertex is lost, but it is replaced not only by a new face but also by three edges and three vertices. As a result, the cube gains 8 new faces, $3 \cdot 8 = 24$ new edges, and $2 \cdot 8 = 16$ new vertices. Because the original cube had 6 faces, 12 edges, and 8 vertices, the truncated cube must have $6 + 8 = 14$ faces, $12 + 24 = 36$ edges, and $8 + 16 = 24$ vertices.

Prisms and Pyramids

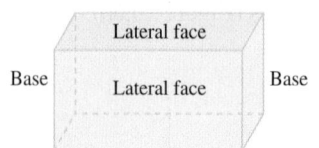

Figure 10.62 Parts of a prism

Prisms and pyramids are two other classes of polyhedra. A **prism** is a polyhedron in which two congruent polygons lie in parallel planes and all other faces connecting them are bound by parallelograms. The faces in the parallel planes are called the **bases** of the prism. The others are called **lateral faces** (Figure 10.62).

If the lateral faces are rectangles, then we call it a **right prism.** If they are parallelograms, we call it an **oblique prism.** In general, we name prisms by their polygonal base and state whether it is right or oblique. Figure 10.63 shows several prisms.

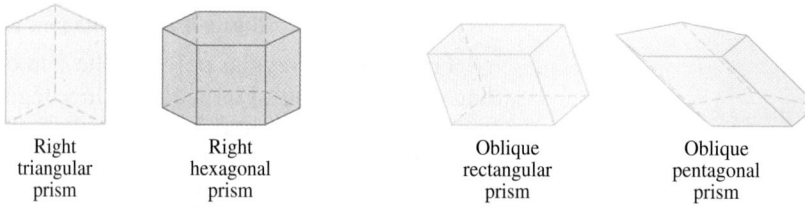

Right
triangular
prism

Right
hexagonal
prism

Oblique
rectangular
prism

Oblique
pentagonal
prism

Figure 10.63 Right and oblique prisms

A **pyramid** is a polyhedron that is formed when every point on a polygon is connected to a point that is not in the plane of the polygon. As with prisms, the polygon serves as the **base** of the pyramid, and the triangular regions form the **lateral faces.** The point common to all the lateral faces is called the **apex** (Figure 10.64).

If the lateral faces are isosceles triangles, then it is a **right pyramid.** Otherwise, it is an **oblique pyramid.** As with prisms, pyramids are named by their polygonal bases (Figure 10.65).

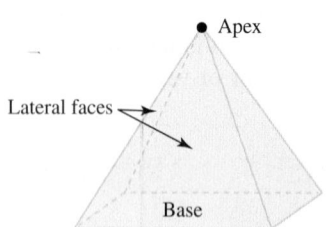

Figure 10.64 Parts of a pyramid

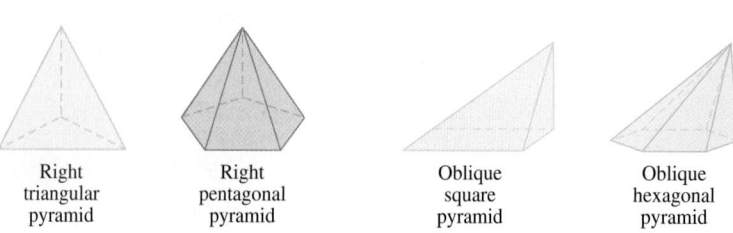

Right
triangular
pyramid

Right
pentagonal
pyramid

Oblique
square
pyramid

Oblique
hexagonal
pyramid

Figure 10.65 Right and oblique pyramids

Explorations
Manual
10.15

The next example illustrates a property that all polyhedra satisfy: **Euler's formula.**

Example 10.28 Euler's Formula

Problem Solving

The number of faces, vertices, and edges of a polyhedron are not independent of one another. Find Euler's formula, which expresses the relationship among these parts of a polyhedron.

Solution

Understanding the Problem. We are asked to find Euler's formula, which expresses a relationship among the number of faces, vertices, and edges in any polyhedron.

Devising a Plan. The problem gives us little to go on. However, we can count the number of faces, vertices, and edges in polyhedra directly. If we can *use a table* to organize the information, we can *look for a pattern* that we can generalize to find the formula.

Carrying Out the Plan. The following table shows the number of faces, vertices, and edges of polyhedra that we have worked with to this point.

Polyhedron	Number of Faces	Number of Vertices	Number of Edges
Tetrahedron	4	4	6
Cube	6	8	12
Octahedron	8	6	12
Dodecahedron	12	20	30
Triangular Prism	5	6	9
Pentagonal Prism	7	10	15
Square Pyramid	5	5	8
Hexagonal Pyramid	7	7	12

The table indicates a pattern in which the number of faces plus the number of vertices is always two more than the number of edges. In other words, if we let f, v, and e represent the number of faces, vertices, and edges respectively, we can write Euler's formula as $f + v = e + 2$.

Looking Back. This possible expression for Euler's formula should be tested with other polyhedra. A hexagonal prism has 8 faces, 12 vertices, and 18 edges. Substituting these numbers into the formula, we have $f + v = 8 + 12 = 20 = 18 + 2 = e + 2$. The formula holds. A pentagonal pyramid has 6 faces, 6 vertices, and 10 edges. Substituting these into the formula, we have $f + v = 6 + 6 = 12 = 10 + 2 = e + 2$. The formula holds again. As a result, we have the formula we are looking for. ∎

Theorem 10.13 Euler's Formula

If f, v, and e represent the number of faces, vertices, and edges of a polyhedron respectively, then $f + v = e + 2$.

Example 10.29 Using Euler's Formula

Use Euler's formula to find the number of vertices of an icosahedron.

Solution An icosahedron has 20 faces, each of which is an equilateral triangle. Because each face has three edges and each edge is used by two faces, an icosahedron must have $\dfrac{20 \cdot 3}{2} = 30$ edges. By applying Euler's formula and solving for v, we find that the number of vertices of an icosahedron is $v = e + 2 - f = 30 + 2 - 20 = 12$ vertices.

Cylinders and Cones

Several other important sets of simple closed surfaces are not polyhedra. For instance, consider cylinders and cones. A **cylinder** is made of two congruent, non-polygonal simple closed curves in parallel planes, along with their interiors and all the line segments joining corresponding points on the curves. Intuitively, a cylinder is similar to a prism but has the same simple closed curve for each of its **bases.** If the base is a circle, it is called a **circular cylinder.** As with prisms, cylinders can be right or oblique (Figure 10.66).

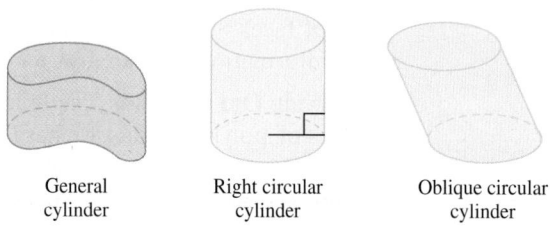

General | Right circular | Oblique circular
cylinder | cylinder | cylinder

Figure 10.66 Cylinders

A **cone** is formed when every point on a nonpolygonal simple closed curve is connected to a point, called the **apex,** that is not on the plane of the curve. Intuitively, a cone is similar to a pyramid but has a simple closed curve for its **base.** If the base is a circle, it is a **circular cone.** Like pyramids, cones can be right or oblique (Figure 10.67).

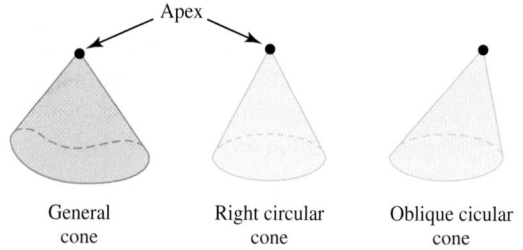

Apex

General | Right circular | Oblique cicular
cone | cone | cone

Figure 10.67 Cones

Spheres

The final simple, closed surface is the sphere. A **sphere** is the set of all points in space that are at a given distance from a given point. The single point is the **center** of the sphere, and any segment connecting the center to a point on the sphere is a **radius.** Any segment connecting two points on the sphere and passing through its center is called a **diameter.** If we cut a sphere in half, then we get a figure called a **hemisphere** (Figure 10.68).

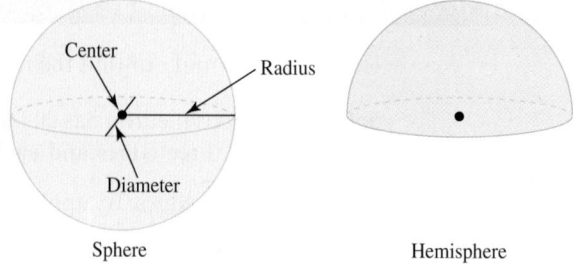

Center

Radius

Diameter

Sphere | Hemisphere

Figure 10.68 A sphere and hemisphere

Check Your Understanding 10.4-B

1. a. Draw a right rectangular prism.

    b. Draw an oblique, triangular pyramid.

2. Draw two different nets of a dodecahedron.

3. a. A polyhedron has 10 vertices and 24 edges. How many faces does it have?

    b. A polyhedron has 20 vertices and 20 faces. How many edges does it have?

    c. A polyhedron has 18 faces and 50 edges. How many vertices does it have?

Talk About It Does a sphere have an arc? If so, how might it be measured?

Representation

Activity 10.8
There are 11 different nets for the cube, one of which is shown. Can you find the other 10?

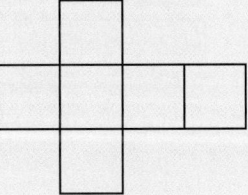

SECTION **10.4** Learning Assessment

■ Understanding the Concepts

1. Why must a set of noncoplanar points have at least four points?

2. a. What are three ways a line can interact with a plane?

    b. What are three ways a plane can interact with other planes?

3. What is a dihedral angle, and how is a dihedral angle measured?

4. What is the difference between a surface and a solid?

5. What are the defining characteristics of a polyhedron?

6. What is a net, and how is it formed?

7. Why are there only five regular polyhedra?

8. a. What is the difference between a regular and a semiregular polyhedron?

    b. What is the difference between a prism and a pyramid?

    c. What is the difference between a cylinder and a cone?

9. What is Euler's formula, and for what type of shape does it hold?

10. a. Is a cylinder a polyhedron? Explain.

    b. Is it possible to apply Euler's formula to a cylinder? Why or why not?

11. What are the defining characteristics of a sphere?

■ Representing the Mathematics

12. Use the following cube to name:

    a. Four noncoplanar points.

    b. Two pairs of parallel faces.

    c. Two pairs of intersecting faces.

    d. Two pairs of parallel edges.

    e. Two pairs of skew edges.

    f. Two pairs of perpendicular edges.

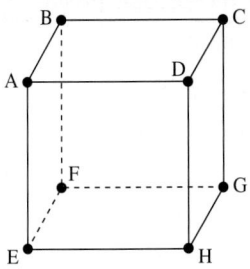

13. Draw each figure.

    a. Cube

    b. Rectangular prism

    c. Tetrahedron

    d. Sphere

    e. Right circular cone

    f. Right circular cylinder

14. Which of the following are polyhedra? If not, explain why.

a. 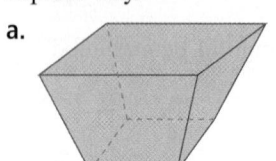   **b.**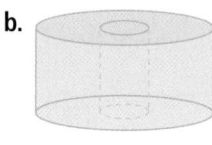

c. 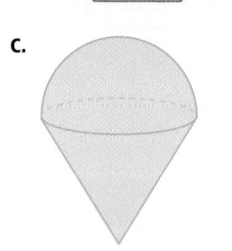   **d.**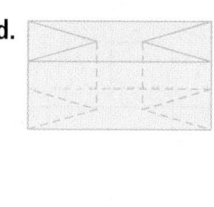

15. Give the name of each figure.

a. 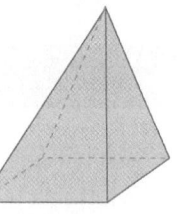   **b.**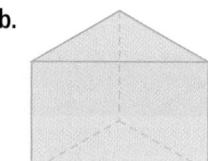

c. 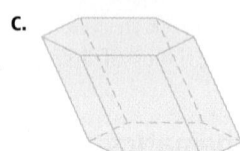   **d.**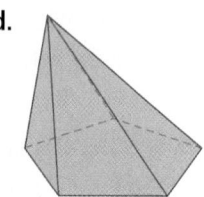

e. 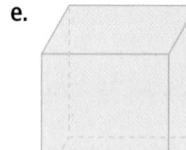   **f.**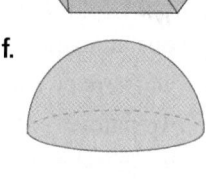

16. Draw the net of each figure.

    a. Regular rectangular prism

    b. Pentagonal prism

    c. Square pyramid

17. Draw the net of each figure.

    a. Triangular pyramid

    b. Circular cylinder

    c. Circular cone

18. Draw the three-dimensional figure that results from each net.

a.   **b.**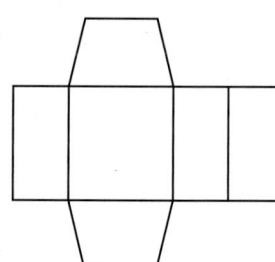

c.

19. Draw the three-dimensional figure that results from each net.

a.   **b.**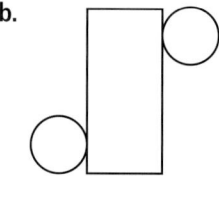

c.

20. Which of the following cannot be folded into an open-top box?

a.   **b.**

c.

d.

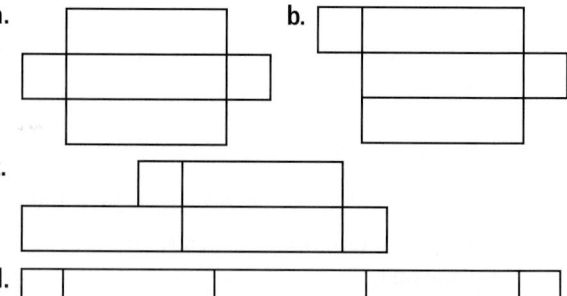

21. The faces on the figure are six equilateral triangles. Why is the shape not a regular polyhedron?

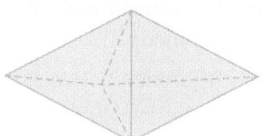

22. A pentahedron is a polyhedron with exactly five faces. Draw a pentahedron that is a:

    a. Prism.   **b.** Pyramid.

23. A hexahedron is a polyhedron with exactly six faces. Draw a hexahedron that is a:
a. Prism.   **b.** Pyramid.

24. If a cone is cut from the base to the apex and unfolded, is the unfolded lateral surface a triangle? Explain.

25. Draw a Venn diagram that shows the relationships between simple closed curves, polyhedra, regular polyhedra, semiregular polyhedra, prisms, cylinders, pyramids, cones, and spheres.

■ Exercises

26. Use the figure to find the following intersections.

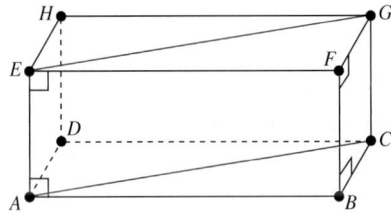

a. Plane *HGF* and plane *EHD*
b. Plane *ABC* and plane *EGC*
c. Plane *ABC* and $\overline{GC}$
d. Plane *ADH* and $\overline{AD}$
e. Planes *ABC*, *FBC*, and *DHG*
f. Planes *EGC*, *EFG*, and *DHA*

27. The following figure is a right, regular octagonal prism.

a. What is the measure of any dihedral angle between a base and a lateral face?
b. What is the measure of any dihedral angle between two lateral faces?

28. The following figure is a right, regular pentagonal prism.

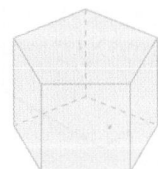

a. What is the measure of a dihedral angle between a base and a lateral face?
b. What is the measure of a dihedral angle between two lateral faces?

29. How many diagonals are in a:
a. Rectangular prism?   **b.** Pentagonal prism?
c. Hexagonal prism?

30. Find the number of faces, vertices, and edges, in each prism or pyramid.

a. 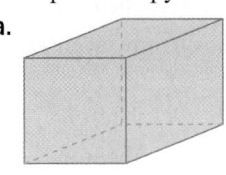   **b.**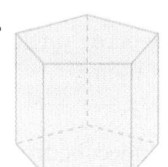

c. 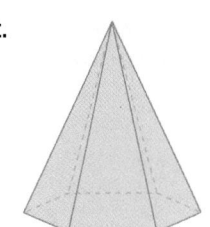   **d.**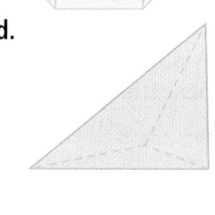

31. Find the number of faces, vertices, and edges in each regular polyhedra.

32. Find the number of faces, vertices, and edges in the following semiregular polyhedron.

a. 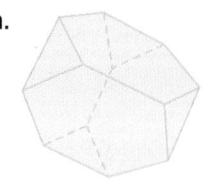   **b.**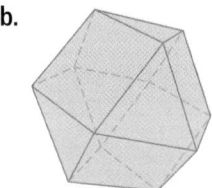

c. 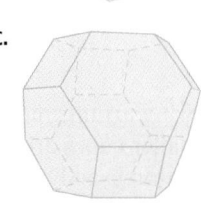   **d.**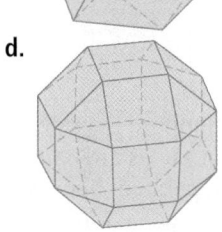

33. What is the lowest possible number of edges, faces, and vertices for each figure?
a. Polyhedron   **b.** Prism   **c.** Pyramid

34. a. A polyhedron has 12 vertices and 18 edges. How many faces does it have?
b. A polyhedron has 16 vertices and 6 faces. How many edges does it have?
c. A polyhedron has 14 faces and 28 edges. How many vertices does it have?

35. a. A polyhedron has 7 vertices and 12 edges. How many faces does it have?
b. A polyhedron has 11 vertices and 11 faces. How many edges does it have?
c. A polyhedron has 12 faces and 32 edges. How many vertices does it have?

36. Show that Euler's formula holds for each figure.
a. Truncated tetrahedron   **b.** Truncated octahedron
c. Cube octahedron

■ Problems and Applications

37. An octahedron has 11 nets. Find five of them.

38. Find the total number of diagonals in each figure.

   a. Cube   **b.** Octohedron

   c. Dodecahedron   **d.** Icosahedron

39. How many vertices are opposite any given vertex in each figure?

   a. Cube   **b.** Octohedron

   c. Dodecahedron   **d.** Icosahedron.

40. How many cubes are needed to make each figure? Assume the blocks on top are supported by blocks beneath.

a.
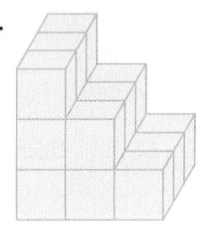
   b.

c.
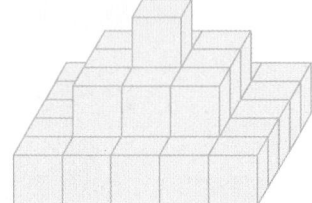

41. Suppose the sides of the following arrangement of cubes are painted. If the smaller cubes are separated, how many have paint on exactly:

   a. One face?   **b.** Two faces?   **c.** Three faces?

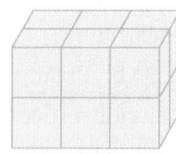

42. Repeat Problem 41 using the following arrangement of cubes:

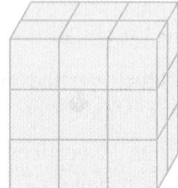

43. Consider the following net.

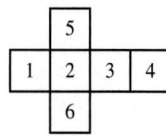

Draw the cube that results from putting the net together. Be sure to consider the orientation of the numbers.

44. Another way to make a two-dimensional representation of a polyhedron is to make a Schlegel diagram. Imagine that a light is shining through the figure so that the vertices and edges appear as a two-dimensional outline on a screen. For instance, here is the Schlegel diagram for a cube.

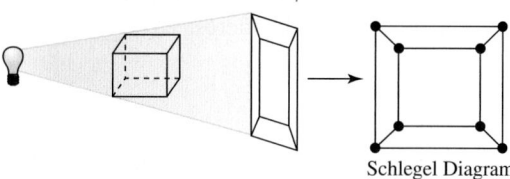

Schlegel Diagram

Draw a Schlegel diagram for each figure.

   a. Tetrahedron   **b.** Octahedron

   c. Dodecahedron

45. The following diagrams show a plane "cutting" a solid. Draw each of the resulting cross-sections.

a.
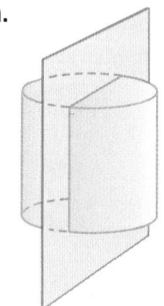
   b.
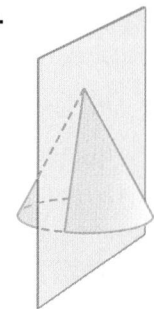

c.
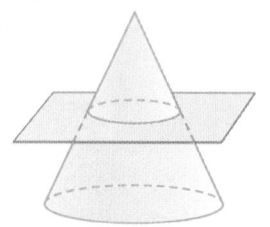
   d.
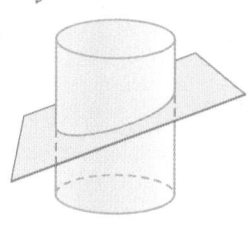

46. What figures can be made by the cross section of a plane intersecting a cube?

47. Consider the following regular hexagon.

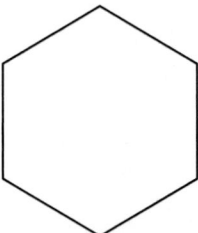

Draw three line segments so that the hexagon becomes a cube.

48. a. Draw the net of a concave right pentagonal prism.

     b. Draw the net of a concave right hexagonal pyramid.

49. Draw the net of a concave, oblique, rectangular prism.

50. Use isometric dot paper to draw a possible three-dimensional figure with an outline from each of the given viewpoints.

a. Top   Front   Left

b. Top   Front   Left

c. Top   Front   Left

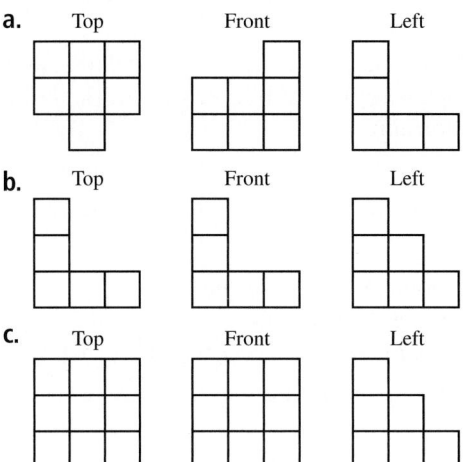

51. Draw a possible outline of each figure from the top, front, and right.

a.   **b.**

c.

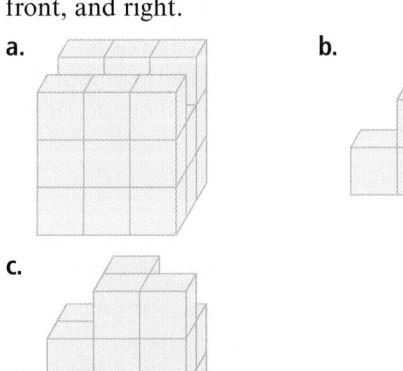

52. An artist plans to make a sculpture that contains an icosahedron made from brass tubing. If each edge of the figure is to be 3 in. long, and the tubing costs $1.05/in., how much will it cost to make the sculpture?

■ Mathematical Reasoning and Proof

53. Is it possible for a polyhedron to have exactly seven edges? Seven vertices? Seven faces? Explain.

54. Consider the net of the following pyramid.

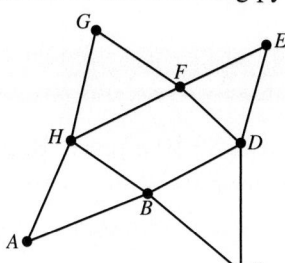

a. Explain why $\overline{AH} \cong \overline{GH}$.

b. What other pairs of segments must be congruent? Why?

55. Based on what you know about the diagonals of a rectangle, what do you think might be true about the diagonals of a right rectangular prism?

56. The following cube has had one corner cut off, causing it to gain one new triangular face.

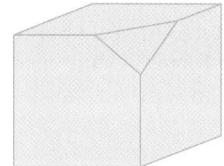

a. How has its number of edges changed?

b. How has its number of vertices changed?

c. Show that Euler's formula holds for the solid.

d. Repeat parts a, b, and c if two corners are cut off to form new triangular faces.

57. Two pyramids are glued to opposite faces of the following cube. As a result, the cube has a net gain of six new faces.

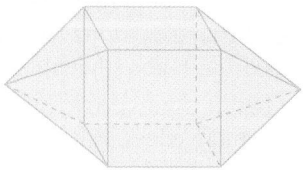

a. How has its number of vertices changed?

b. How has its number of edges changed?

c. How would the number of vertices and edges change if a pyramid were glued to every face of the cube?

58. How many faces does a prism have if the base is:

a. Rectangular?   **b.** Pentagonal?

c. Hexagonal?   **d.** Heptagonal?

e. What pattern do you see in your answers to parts a–d? Use it to write a formula that gives the number of faces in a prism that has an *n*-gon for its base.

59. How many edges does a prism have if the base is:

a. Rectangular?   **b.** Pentagonal?

c. Hexagonal?   **d.** Heptagonal?

e. What pattern do you see in your answers to parts a–d? Use it to write a formula that gives the number of edges in a prism that has an *n*-gon for its base.

60. How many vertices does a pyramid have if the base is:

a. Rectangular?   **b.** Pentagonal?

c. Hexagonal?   **d.** Heptagonal?

e. What pattern do you see in parts a–d? Use it to write a formula that gives the number of vertices in a pyramid that has an *n*-gon for its base.

61. How many diagonals does each figure have?

 a. Quadrilateral prism

 b. Pentagonal prism

 c. Hexagonal prism

 d. Heptagonal prism

 e. What pattern do you see in parts a–d? Use it to write a formula that gives the number of diagonals in a prism that has an *n*-gon for its base.

■ Mathematical Communication

62. Write a set of directions explaining how to draw the net of a tetrahedron.

63. Is it correct to use the term "circular" to describe a baseball?

64. Thomas claims that two parallel lines must be skew because they do not intersect. How do you, as his teacher, correct his thinking?

65. A student makes the claim that a square pyramid with equilateral triangles for its lateral sides is a regular polyhedron. Is the student correct?

66. Keith claims that it is possible to make a polyhedron with 5 vertices and 5 edges. Is he correct?

■ Building Connections

67. How have the shapes and relationships discussed in this section expanded the mathematical structure of geometry?

68. State the two-dimensional shape or relationship that is analogous to the given three-dimensional shape or relationship.

 a. Intersecting lines   **b.** Skew lines

 c. Dihedral angle   **d.** Polyhedron

 e. Cube   **f.** Sphere

69. Take a moment to look at the objects in your surroundings. Can you find three common objects that approximately take on each given shape?

 a. Prism **b.** Cylinder **c.** Pyramid

 d. Cone   **e.** Sphere

70. Do an Internet search for real-world phenomena or applications of the different regular polyhedra. If possible, make a list of five such phenomena.

71. Examine a set of curriculum materials for grades 3, 4 and 5. What different strategies do you find for helping students visualize three-dimensional shapes? How are the strategies similar to or different from the ones presented in this section?

Historical Note

The five regular polyhedra are often called the **Platonic solids** because they were known to and described by Plato in his work *Timaeus*. Plato believed that these figures formed the fundamental building blocks of the universe. He equated the tetrahedron with fire, the cube with earth, the octahedron with air, the icosahedron with water, and the dodecahedron with the element *cosmos*, from which the stars and planets were made. Today, his beliefs may sound absurd, but they did have an impact on science. In the sixteenth century, Johannes Kepler tried to use a set of nested regular polyhedra to model what he thought was the circular orbit of the planets. He eventually discovered that the orbits were elliptical, not circular. Less than a century later, Newton used this fact to develop his law of gravity, which eventually led to our modern conception of the universe.

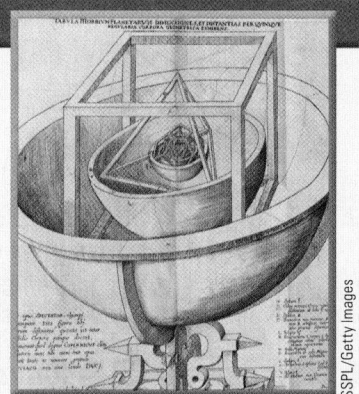

Kepler's model of the universe

SSPL/Getty Images

Opening Problem Answer

Each piece of tubing is 10 ft long, so it costs $9.50 a piece. Because a cube has 12 edges, Martin must buy 12 pieces of tubing, which will cost $12 \cdot \$9.50 = \$114$.

Brocreative/Shutterstock.com

Chapter 10 REVIEW

Summary of Key Concepts from Chapter 10

Section 10.1   Lines, Planes, and Angles

- The most basic geometric shapes are points, lines, and planes.

- A **ray**, $\overrightarrow{AB}$, is a portion of a line that contains one point and all the points on the line to one side of the point.

- A **line segment**, $\overline{AB}$, is a portion of a line that contains two points and all the points between those points.

- The **length** of a segment $\overline{AB}$, denoted AB, is the distance between the endpoints.

- Two segments are **congruent**, $\overline{AB} \cong \overline{CD}$, if and only if they are the same length.

- The **midpoint** of a segment $\overline{AB}$ is the point M on $\overline{AB}$ for which $\overline{AM} \cong \overline{MB}$. Any segment, ray, or line passing through the midpoint is called a **bisector.**

- **Intersecting lines** are two coplanar lines with exactly one point in common.

- **Parallel lines,** written $l \parallel m$, are distinct coplanar lines with no point in common.

- An **angle**, $\angle ABC$, is the union of two rays that share a common endpoint. The rays form the **sides** of the angle, and the common endpoint is called the **vertex.**

- If $\angle A$ is viewed as rotating a ray about its endpoint, then the **angle measure** of $\angle A$, is the amount of rotation made by the ray.

- Two angles are **congruent**, $\angle A \cong \angle B$, if and only if they have same measure.

- An **acute angle** has a measure less than 90°, a **right angle** has a measure of exactly 90°, an **obtuse angle** has a measure greater than 90° but less than 180°, and a **straight angle** has a measure of exactly 180°.

- Two angles are **complementary** if the sum of their measures is 90°, and they are **supplementary** if the sum of their measures is 180°.

- Two nonadjacent angles formed by two intersecting lines are called **vertical angles.** Vertical angles are congruent.

- Two lines are **perpendicular,** denoted $l \perp m$, if they intersect to form four right angles.

- Two lines cut by a **transversal** are parallel if and only if corresponding angles are congruent, alternate interior angles are congruent, alternate exterior angles are congruent, same-side interior angles are supplementary, and same-side exterior angles are supplementary.

Section 10.2   Triangles

- **Polygons** are simple, closed curves made up entirely of segments.

- A **triangle** is a three-sided polygon. The segments that make up the triangle are called the **sides,** and the points at which the sides intersect are called the **vertices.**

- **Triangle inequality:** The sum of the lengths of any two sides of a triangle must be greater than the length of the third side.

- Triangles can be classified by lengths of the sides. A triangle is **scalene** if it has no congruent sides, **isosceles** if it has at least two congruent sides, and **equilateral** if it has three congruent sides.

- Triangles can be classified by angle measures. A triangle is **acute** if every angle is acute, **right** if it has one right angle, **obtuse** if it has one obtuse angle, and **equiangular** if the angles have the same measure.

- The sum of the measures of the angles in any triangle is equal to 180°.

- **Pythagorean Theorem:** If a right triangle has legs of lengths a and b and a hypotenuse of length c, then $a^2 + b^2 = c^2$.

- **Converse of the Pythagorean Theorem:** If a triangle has sides of lengths a, b, and c such that $a^2 + b^2 = c^2$, then it is a right triangle.

Section 10.3   Quadrilaterals, Polygons, and Circles

- A **quadrilateral** is a four-sided polygon.

- Quadrilaterals are classified by their sides and angles. A **trapezoid** is a quadrilateral with exactly one pair of parallel sides. A **kite** is a quadrilateral with at least two pairs of congruent, consecutive sides. A **parallelogram** is a quadrilateral in which both pairs of opposite sides are parallel. A **rhombus** is a quadrilateral with all sides congruent. A **rectangle** is a parallelogram with a right angle. A **square** is a rectangle with all sides congruent.

- Polygons are named for their number of sides. A pentagon has 5 sides, a hexagon has 6, a heptagon has 7, an octagon has 8, a nonagon has 9, and a decagon has 10.

- A **diagonal** in a polygon is any segment joining two nonconsecutive vertices.

- If a polygon has n sides, the sum of the measures of the interior angles is $(n - 2) \cdot 180°$, and the sum of the measures of the exterior angles is 360°.

- A polygon is **regular** if it is both equiangular and equilateral. Any segment joining the **center** to a vertex is called a **radius,** and any segment joining the center to the midpoint of a side is called an **apothem.**

- The **central angle** of a regular n-gon has a measure of $360°/n$, and each interior angle has a measure of $(n - 2)180°/n$.

- A **circle** is the set of all points in a plane that are a given distance from a given point.

Section 10.4   Surfaces and Solids

- **Skew lines** are lines that do not intersect and there is no single plane that contains them.

- A **dihedral angle** is formed when two planes intersect.

- A **net** is made by "cutting" enough edges of the spatial shape so that it can be laid flat.

- A **polyhedron** is a simple, closed surface made from planar polygonal regions. The polygonal regions are called the **faces,** and the vertices and sides of the polygons make up the **vertices** and **edges** of the polyhedron.

- **Regular polyhedra** are convex polyhedra with congruent dihedral angles and faces that are congruent regular polygons. There are five regular polyhedra: tetrahedron, cube, octahedron, dodecahedron, and icosahedron.

- A **prism** is a polyhedron in which two congruent polygons lie in parallel planes and all other faces connecting them are bound by parallelograms.

- A **pyramid** is a polyhedron formed when every point on a polygon is connected to a point that is not in the plane of the polygon.

- **Euler's formula:** If f, v, and e represent the number of faces, vertices, and edges of a polyhedron, respectively, then $f + v = e + 2$.

- A **cylinder** is made of two congruent nonpolygonal simple closed curves in parallel planes along with their interiors and the line segments joining corresponding points on the curves.

- A **cone** is formed when every point on a nonpolygonal simple closed curve is connected to a point that is not on the plane of the curve.

- A **sphere** is the set of all points in space that are at a given distance from a given point.

Review Exercises Chapter 10

1. Use the following diagram to name each item.

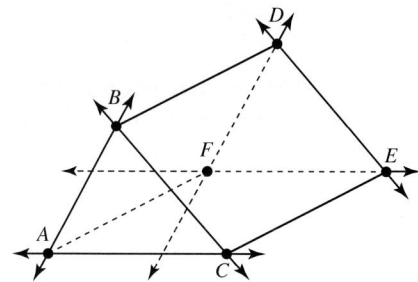

   a. Three coplanar points
   b. Parallel lines
   c. Perpendicular lines
   d. Four noncoplanar points
   e. Two skew lines
   f. Two parallel planes

2. In the following diagram, suppose $\overleftrightarrow{AB} \perp \overrightarrow{CD}$, $\overrightarrow{DE} \perp \overrightarrow{DF}$, and $m\angle CDE = 20°$.

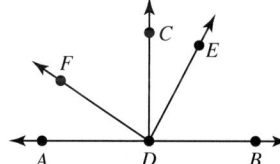

   What are the measures of:
   a. $\angle EDB$?   b. $\angle FDC$?
   c. $\angle FDB$?   d. $\angle ADE$?

3. a. The complement of a 37° angle has what measure?
   b. The complement of a 68° angle has what measure?
   c. The supplement of a 29° angle has what measure?
   d. The supplement of a 121° angle has what measure?
   e. What is the complement of the supplement of an angle measuring 137°?

4. Find the value of x for each figure.
   a.

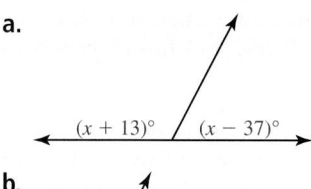

   b.

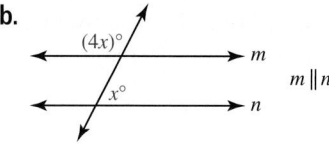

5. If $p\|q$, find the measures of the numbered angles.

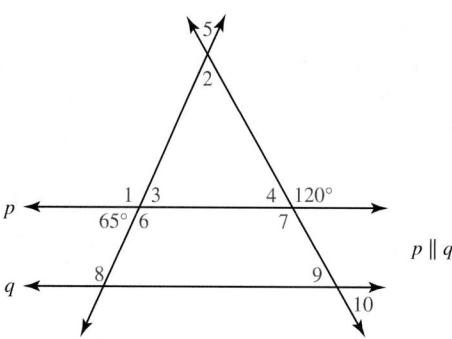

6. A clock shows a time of 11:20. What is the approximate angle measure between the hour and minute hands?

7. If a right angle is bisected, what kind of angles must be formed?

8. True or false:
   a. The complement of an acute angle is acute.
   b. The supplement of a right angle is obtuse.
   c. The supplement of an obtuse angle is acute.

9. Draw and label each figure described.
   a. A pair of supplementary angles with a common vertex.
   b. A pair of acute vertical angles.

10. Consider an arbitrary triangle $\triangle ABC$.
    a. If $AB = BC = AC$, what kind of triangle is $\triangle ABC$?
    b. If $\angle A = 90°$, what kind of triangle is $\triangle ABC$?
    c. If $\angle A = 35°$ and $\angle B = 45°$, what kind of triangle is $\triangle ABC$?

11. Is it possible for a triangle to have sides of lengths 2 in., 3 in., and 6 in.? Explain.

12. Find x.
    a.

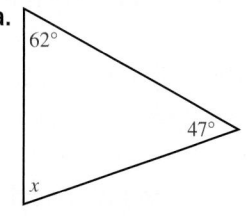

    b.

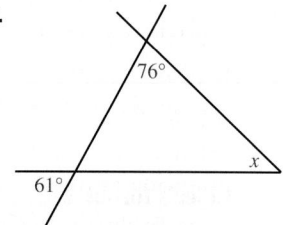

    c.

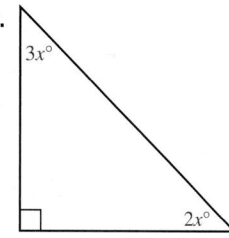

13. Is it possible for a right triangle to have sides of lengths 5 in., 7 in., and $\sqrt{74}$ in.?

14. An obtuse, isosceles triangle has a vertex angle measuring 138°. What are the measures of the other two angles?

15. **a.** In a 30°-60°-90° triangle, what are the lengths of the other two sides if the shortest leg is 5?

    b. In a 45°-45°-90° triangle, what are the lengths of the legs if the hypotenuse is 16?

16. A 150-ft tower has a guy wire that makes a 45° angle with the level ground. If the wire is attached to the top of the tower, how far from the bottom is it anchored?

17. Three roads meet to form the sides of a right triangle. If the two perpendicular roads are 5.4 mi and 3.7 mi long, how long is the third road?

18. Draw an example of a:

    a. Convex simple closed curve. **b.** Concave hexagon.

19. Draw an example of a convex octagon that is:

    a. Neither equilateral nor equiangular.

    b. Equilateral but not equiangular.

    c. Equiangular but not equilateral.

    d. Both equilateral and equiangular.

20. If one angle in a rhombus is five times the measure of the other, what is the measure of each angle in the rhombus?

21. What are the measures of the interior, exterior, and central angles of a:

    a. Regular pentagon? **b.** Regular hexagon?

    c. Regular octagon?

22. **a.** A regular n-gon has a central angle measure of 18°. What is n?

    b. A regular n-gon has a interior angle measure of 162.86°. What is n?

    c. A regular n-gon has a exterior angle measure of 21.18°. What is n?

23. If possible, name the convex polygon in which the sum of the interior angles is 6,300°.

24. What are the measures of each angle in the following polygon?

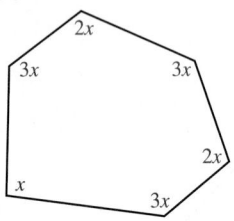

25. What is the diameter of a circle if the radius of the same circle is:

    a. 5 ft? **b.** 7.1 ft? **c.** 18 ft?

26. If an angle is inscribed in a circle and intersects an arc of 114°, what is the measure of the angle?

27. Juan plans to build a square deck on the back of his house that measures 16 ft on a side. In building the deck, how can he guarantee that what he builds is actually a square?

28. Draw each figure, and explain why the figure is or is not a polyhedron.

    a. Rectangular prism

    b. Tetrahedron

    c. Square pyramid

    d. Right circular cylinder

    e. Sphere

    f. Right circular cone

29. Draw a net of a:

    a. Right pentagonal prism.

    b. Hexagonal pyramid.

    c. Cone.

30. Draw the two nets of the tetrahedron.

31. A polyhedron has 14 vertices and 20 edges. How many faces does it have?

32. A polyhedron has 17 vertices and 25 faces. How many edges does it have?

33. The following figure is a regular right hexagonal prism.

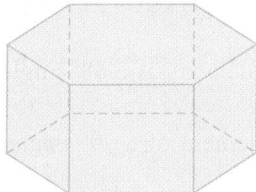

    a. What is the measure of a dihedral angle between a base and a lateral face?

    b. What is the measure of a dihedral angle between two lateral faces?

34. If three corners are cut off a cube so that it gains three new triangular faces, how has its number of edges and vertices changed?

35. Imagine two right pentagonal pyramids, with equilateral triangles for bases, glued together with their bases face-to-face. Does the resulting figure represent a regular polyhedron? Explain.

Answers to Chapter 10   Check Your Understandings and Activities

Check Your Understanding 10.1-A

1. **a.** Points include *A, B, C,* and *D.*

   b. One possible answer is $\overleftrightarrow{AB}$ and $\overleftrightarrow{DC}$.

   c. One possible answer is plane *ABC* and plane *BCD.*

2. **a.** Possible answers are $\overline{AB}$, $\overline{DC}$, and $\overline{GH}$.

   b. Possible answers are $\overrightarrow{GE}$, $\overrightarrow{HF}$, and $\overrightarrow{DA}$.

   c. One possible answer is $\overleftrightarrow{AB}$ and $\overleftrightarrow{EF}$.

   d. One possible answer is $\overleftrightarrow{AB}$ and $\overleftrightarrow{EA}$.

   e. One possible answer is $\overleftrightarrow{AB} \| \overleftrightarrow{EF}$.

Check Your Understanding 10.1-B

1. **a.** Possible answers are $\angle AEB$, $\angle AEC$, $\angle BEC$, and $\angle CED$.

   b. $\angle AED$ **c.** $\angle BED$

   d. One possible answer is $\angle AEB$ and $\angle BED$.

   e. $\angle BEC$ and $\angle CED$

2. **a.** 25°   **b.** 160°

3. *x* = 54.

4. $m\angle 1 = 125°$, $m\angle 2 = 55°$, $m\angle 3 = 125°$, $m\angle 4 = 125°$, $m\angle 5 = 55°$, $m\angle 6 = 55°$, and $m\angle 7 = 125°$.

Check Your Understanding 10.2-A

1. Possible drawings are:

   a.    **b.**

   c. 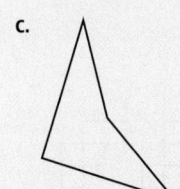   **d.**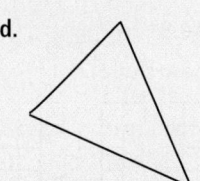

2. **a.** Yes   **b.** No   **c.** No

3. **a.** Not possible.

   b. 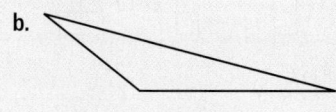   **c.**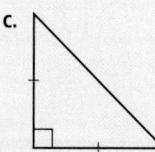

4. **a.** *x* = 50°.   **b.** *x* = 45°.   **c.** *x* = 30°.

Check Your Understanding 10.2-B

1. **a.** Yes   **b.** No   **c.** No   **d.** Yes

2. **a.** $a = 3$ and $c = 3\sqrt{2}$.   **b.** $c = 11.66$.   **c.** $b = 11.31$.

3. The long side is $3\sqrt{3} \approx 5.20$ ft, and the hypotenuse is 6 ft.

Check Your Understanding 10.3-A

1. **a.** True   **b.** True   **c.** False   **d.** False

2. **a.** False   **b.** True   **c.** True   **d.** True

3. **a.** $m\angle 1 = m\angle 2 = 115°$.   **b.** $m\angle 1 = 55°$ and $m\angle 2 = 125°$.

   c. $m\angle 1 = m\angle 2 = m\angle 3 = 90°$.

Check Your Understanding 10.3-B

1. **a.** 1,080°   **b.** 1,260°   **c.** 2,340°   **d.** 3,240°

2. **a.** Central angle = 72°; interior angle = 108°; exterior angle = 72°.   **b.** Central angle = 51.43°; interior angle = 128.57°; exterior angle = 51.43°.   **c.** Central angle = 27.69°; interior angle = 152.31°; exterior angle = 27.69°.   **d.** Central angle = 17.14°; interior angle = 162.86°; exterior angle = 17.14°.

3. *x* = 27.

4. **a.** 95°   **b.** 136°   **c.** 36°

Check Your Understanding 10.4-A

1. **a.** One possible answer is *A, B,* and *C.*

   b. One possible answer is *A, B, C,* and *D.*

   c. One possible answer is $\overleftrightarrow{AB}$, $\overleftrightarrow{AC}$, and $\overleftrightarrow{AD}$.

   d. One possible answer is $\overleftrightarrow{AB}$ and $\overleftrightarrow{EF}$.

   e. One possible answer is plane *ABC* and plane *ACD.*

   f. One possible answer is plane *ABC* and plane *DEF.*

2. **a.** 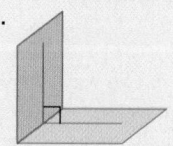   **b.**

   c.    **d.**

3.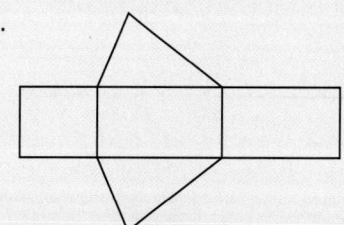

Check Your Understanding 10.4-B

1. **a.**    **b.**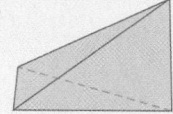

2. Two possible nets are:

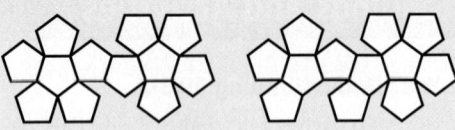

3. a. 16 faces   **b.** 38 edges   **c.** 34 vertices

Activity 10.1

There are 19 different lengths. Connect the peg in the upper corner to each and every peg on the main diagonal and above. This gives 20 lengths. However, the longest segment across the top and the segment that extends to the peg that is 4 to the right and 3 down both have a length of 5. Hence, there are only 19 different lengths.

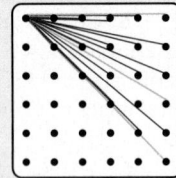

Activity 10.2

a. 14 angles, including the straight angle

b. 6 acute, 6 obtuse

c. The number of acute angles is equal to the number of obtuse angles. Because the vertex is in the middle row, there are equal numbers of pegs on either side of the one right angle.

Activity 10.3

Triangles are the only rigid polygon. This makes them particularly strong for construction purposes.

Activity 10.4

We can compute the area of the square in two ways. Using the outer dimensions, the area of the square is c^2. The area of the square is also equal to the sum of the areas of the four triangles and the small square, or $4\left(\frac{1}{2}\right)ab + (b - a)^2$. Simplifying, we have $4\left(\frac{1}{2}\right)ab + (b - a)^2 = 2ab + (b - a)$ $(b - a) = 2ab + b^2 - 2ab + a^2 = a^2 + b^2$. Because the areas are for the same square, $a^2 + b^2 = c^2$.

Activity 10.5

Answers will vary.

Activity 10.6

Answers will vary. The finished figure is shown.

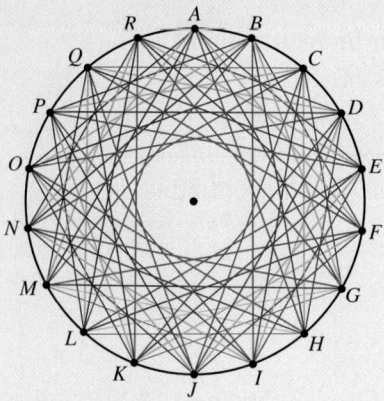

Activity 10.7

a.
b.
c.
d.

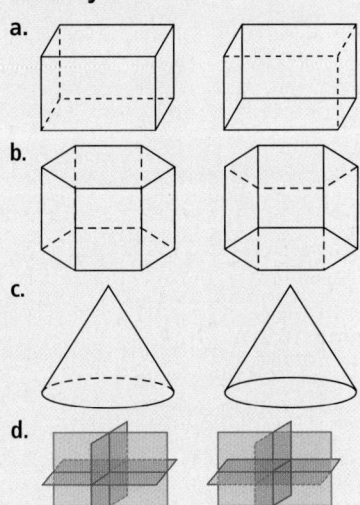

Activity 10.8

All 11 nets are shown.

Congruence, Similarity, and Constructions

© Image Source/Corbis

Opening Problem

Lee is painting a house that has an 8-ft fence only 4 ft from its side. To reach the higher spots, he must set the ladder outside the fence. If the foot of the ladder is 4 ft from the fence and the ladder touches both the fence and the wall, how long is the ladder and how high does it reach up the wall?

©Eugene May/Shutterstock.com

11.1 Congruent Shapes

11.2 Similar Shapes

11.3 Basic Geometrical Constructions

11.4 Constructing Shapes

© Image Source/Corbis

In the last chapter, we defined shapes and identified their properties. Now we consider congruence, similarity, and geometrical constructions.

Congruence and similarity are two important relationships between geometrical figures. In Section 10.1, congruence was defined as the notion that two figures have the same shape and size. Initially, we limited the discussion to segments and angles, but we can extend the idea of congruence to other two- and three-dimensional shapes. Of these, congruent triangles are particularly useful because they give us a powerful tool for verifying the properties of polygons and geometrical constructions. Similarity is the notion that two figures have the same shape but not necessarily the same size. The notion can be applied to a number of shapes, but, again, similar triangles are particularly useful because they give us a powerful strategy for solving problems.

In the elementary classroom, children come to an intuitive understanding of congruence and similarity by manipulating physical shapes. For congruence, they learn to place one shape on top of the other to see whether they match. For similarity, they compare figures of different sizes to see whether they have the same basic shape. In the middle grades, students formalize congruence and similarity by using reasoning and measurement to compare corresponding parts of figures. They learn that if corresponding angles and sides are congruent, then the shapes are congruent. If corresponding angles are congruent and corresponding sides are proportional, then the shapes are similar. Once students have a solid understanding of congruence and similarity, they use the ideas to identify geometric properties, develop mathematical arguments, and solve problems.

The second topic in this chapter is geometrical constructions. The very nature of geometry requires the use of pictures to clarify concepts, solve problems, and verify properties. Often, simple diagrams are sufficient. However, some situations require more precision, and we can employ a number of tools and techniques to build, or "construct," more exact pictures. Beyond adding precision, constructions make for a better understanding of geometry. Most constructions are based on the defining characteristics and properties of geometric shapes. Consequently, if we understand how the constructions work, we get a better sense of how these ideas fit together.

Constructions have typically been delayed until the middle grades, when students have the physical coordination to use a compass and straightedge. Now, however, a number of tools and strategies, such as plastic reflectors and paper folding, make constructions accessible to students at an earlier age. In addition to these traditional tools, students also use geometry software to perform constructions. Constructions done on these computerized drawing pads are easily manipulated while always maintaining the attributes of the original construction. As a result, these programs allow students to explore the consequences of a construction and discover geometrical facts that might not have been otherwise accessible. Specific classroom learning objectives associated with congruence, similarity, and constructions are given in Table 11.1.

Table 11.1 Classroom learning objectives	K	1	2	3	4	5	6	7	8
Draw and construct geometric shapes with given conditions.								X	
Understand that a two-dimensional figure is congruent to another if they have the same shape and size.									X
Use informal arguments to establish the congruence between two triangles.									X
Understand that a two-dimensional figure is similar to another if they have the same shape but not necessarily the same size.									X
Use informal arguments to establish the similarity between two triangles.									X

Source: Adapted from the *Common Core State Standards for Mathematics* (*Common Core State Standards Initiative, 2010*).

To help us understand congruence, similarity, and constructions, in Chapter 11 we

- Discuss congruence and similarity with triangles and other geometrical shapes.
- Discuss geometrical constructions and use them to solve problems associated with triangles, quadrilaterals, polygons, and circles.

SECTION 11.1   Congruent Shapes

In Section 10.1, congruence was presented as a relationship between two figures indicating that they have the same shape and size. Congruence for line segments and angles was defined by using measures associated with each shape. Specifically, two segments are congruent if they have the same length, and two angles are congruent if they have the same angle measure. By using congruence for these shapes, we can extend congruence to other planar shapes. We begin with triangles.

Congruent Triangles

Explorations
Manual
11.1

Intuitively, two triangles are congruent if they have the same shape and size. In other words, one can be placed on top of the other so that all matching parts coincide regardless of their orientation on the plane (Figure 11.1). When first learning about triangle congruence, children often use this method of placing one triangle on top of the other to check for congruence directly.

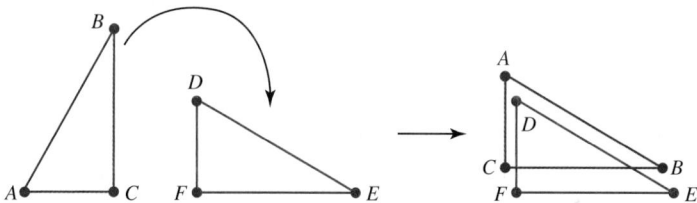

Figure 11.1 Two congruent triangles

Unfortunately, in most cases, this method is neither possible nor practical. However, there is another way to think about triangle congruence. If one triangle can be placed exactly over the top of the other, then the measurements in the first triangle, both the lengths of the sides and the measures of angles, must match the corresponding measurements in the second triangle. In other words, two triangles are congruent if we can establish a correspondence between their vertices so that corresponding sides and angles are congruent.

For instance, suppose we take the two triangles in Figure 11.1 and pair $A \leftrightarrow D$, $B \leftrightarrow E$, and $C \leftrightarrow F$. Matching the vertices in this way, we get the following correspondence between the sides and angles:

$$\overline{AB} \leftrightarrow \overline{DE} \qquad \overline{BC} \leftrightarrow \overline{EF} \qquad \overline{AC} \leftrightarrow \overline{DF}$$
$$\angle A \leftrightarrow \angle D \qquad \angle B \leftrightarrow \angle E \qquad \angle C \leftrightarrow \angle F$$

If each pair of corresponding segments and angles has the same measure, then we know the triangles are congruent.

Definition of Congruent Triangles

$\triangle ABC$ is congruent to $\triangle DEF$, written $\triangle ABC \cong \triangle DEF$, if and only if there is a correspondence between their vertices such that the corresponding sides and the corresponding angles of the triangles are congruent.

Because the definition is biconditional, it indicates that the corresponding parts of two congruent triangles must be congruent, a fact often abbreviated as CPCTC. So the order in which the vertices are listed is important because it identifies the parts that correspond and are congruent. Specifically, if $\triangle ABC \cong \triangle DEF$, then $\angle A \cong \angle D$, $\angle B \cong \angle E$, $\angle C \cong \angle F$, $\overline{AB} \cong \overline{DE}$, $\overline{BC} \cong \overline{EF}$, and $\overline{AC} \cong \overline{DF}$. The congruent parts of congruent triangles are marked as shown in Figure 11.2.

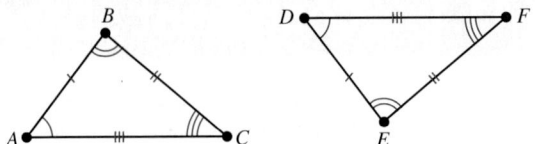

Figure 11.2 Congruent parts of congruent triangles

Example 11.1 Identifying Corresponding Parts of Triangles

Representation

Suppose that $\triangle SUN \cong \triangle DAY$. Determine whether each statement shows the same correspondence.

a. $\triangle UNS \cong \triangle AYD$.   **b.** $\triangle NUS \cong \triangle YDA$.   **c.** $\triangle NSU \cong \triangle YAD$.

Solution The expression $\triangle SUN \cong \triangle DAY$ indicates that $S \leftrightarrow D$, $U \leftrightarrow A$, and $N \leftrightarrow Y$. We answer the question by checking the order of the other correspondences.

a. $\triangle UNS \cong \triangle AYD$ indicates that $U \leftrightarrow A$, $N \leftrightarrow Y$, and $S \leftrightarrow D$, which is the same as the original correspondence.

b. $\triangle NUS \cong \triangle YDA$ indicates that $N \leftrightarrow Y$, $U \leftrightarrow D$, and $S \leftrightarrow A$, which is not the same as the original correspondence.

c. $\triangle NSU \cong \triangle YAD$ indicates that $N \leftrightarrow Y$, $S \leftrightarrow A$, and $U \leftrightarrow D$, which is not the same as the original correspondence.

In most cases, we do not have to check all six pairs of corresponding parts to determine whether two triangles are congruent. It is usually enough to check just three. For instance, if the three pairs of corresponding sides are congruent, then the triangles must also be congruent. To understand why, suppose three segments, with lengths a, b, and c, satisfy the triangle inequality (Figure 11.3). When the shorter segments are attached to and rotated around the endpoints of the longer segment, the smaller segments intersect one another at only two points. If we connect the endpoints of the long segment to either of the intersection points, we have our triangle.

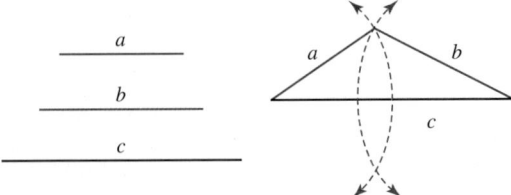

Figure 11.3 Making a triangle by rotating two sides

This means that the three sides of a triangle completely determine its vertices and angles, making it a rigid shape. In other words, if the corresponding sides of two triangles are congruent, then the corresponding angles must also be congruent. This makes all six pairs of corresponding parts congruent, so the triangles must also be congruent. We commonly refer to this as the side-side-side (SSS) property (Figure 11.4).

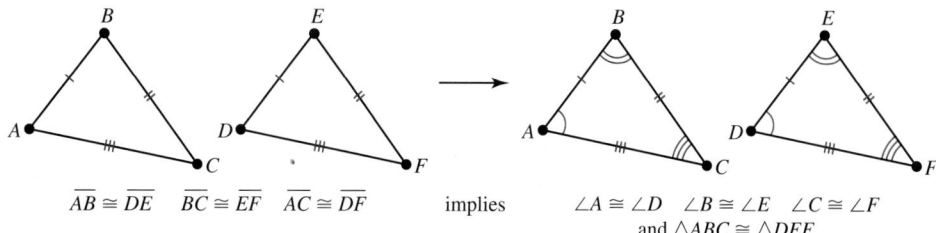

$$\overline{AB} \cong \overline{DE} \quad \overline{BC} \cong \overline{EF} \quad \overline{AC} \cong \overline{DF} \qquad \text{implies} \qquad \angle A \cong \angle D \quad \angle B \cong \angle E \quad \angle C \cong \angle F$$
$$\text{and } \triangle ABC \cong \triangle DEF$$

Figure 11.4 Side-side-side property

Side-Side-Side (SSS) Triangle Congruence Property

Given a correspondence between two triangles, if the three sides of one triangle are congruent respectively to the three sides of the other, then the triangles are congruent.

□

Example 11.2   Using the SSS Property

Reasoning   Use the SSS property to show that $\triangle NUT \cong \triangle GEM$.

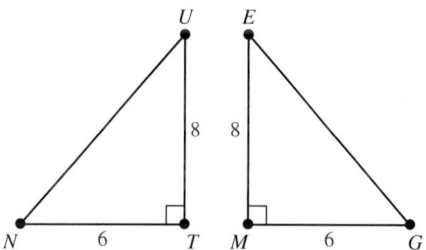

Solution The correspondence between the vertices of $\triangle NUT$ and $\triangle GEM$ is given by $N \leftrightarrow G$, $U \leftrightarrow E$, and $T \leftrightarrow M$. Because $\triangle NUT$ and $\triangle GEM$ are right triangles with legs of length of 6 and 8, $\overline{NT} \cong \overline{GM}$ and $\overline{UT} \cong \overline{EM}$. By the Pythagorean Theorem, we also have $UN = 10 = EG$, meaning $\overline{UN} \cong \overline{EG}$. Because the three sides of $\triangle NUT$ are congruent to the corresponding sides of $\triangle GEM$, then $\triangle NUT \cong \triangle GEM$ by the SSS property.

■

In addition to using three sides to determine a triangle, we can also use two sides and the included angle. To understand why, consider the $\angle ABC$ in which the sides $\overline{AB}$ and $\overline{BC}$ have a set, finite length (Figure 11.5). To complete the triangle, we draw segment $\overline{AC}$. However, notice that the length of $\overline{AC}$ depends on the measure of $\angle B$. As $\angle B$ gets smaller, $\overline{AC}$ gets shorter. As $\angle B$ gets larger, $\overline{AC}$ gets longer.

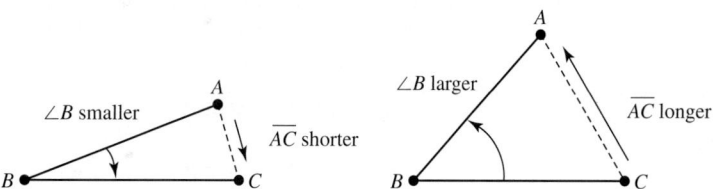

Figure 11.5 Making a triangle from two sides and an angle

So two sides and the included angle determine the length of the third side and the measures of the other two angles. Consequently, if two sides and the included angle in one triangle are congruent to the corresponding sides and angle in another, then all six pairs of corresponding parts must be congruent, making the triangles congruent. This fact is called the side-angle-side (SAS) property (Figure 11.6).

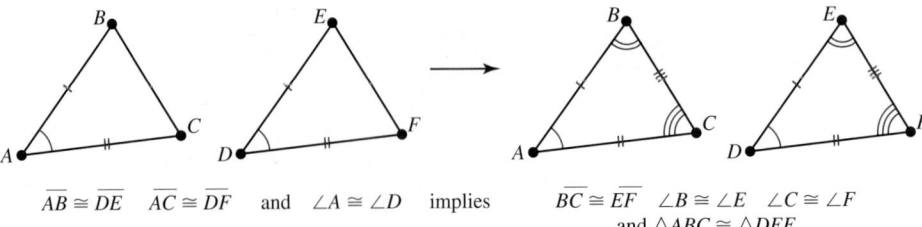

$$\overline{AB} \cong \overline{DE} \quad \overline{AC} \cong \overline{DF} \quad \text{and} \quad \angle A \cong \angle D \quad \text{implies} \quad \overline{BC} \cong \overline{EF} \quad \angle B \cong \angle E \quad \angle C \cong \angle F$$
$$\text{and } \triangle ABC \cong \triangle DEF$$

Figure 11.6 Side-angle-side property

Side-Angle-Side (SAS) Triangle Congruence Property

Given a correspondence between two triangles, if two sides and the included angle of one triangle are congruent, respectively, to two sides and the included angle of a second triangle, then the triangles are congruent.

Example 11.3 Using the SAS Property

Reasoning   Use the SAS property to show that $\triangle CAN \cong \triangle DYN$.

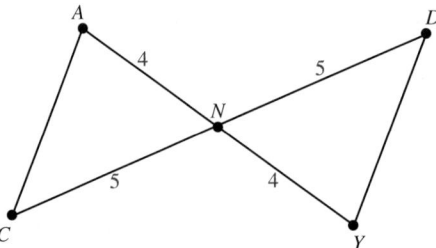

Solution The correspondence between the vertices of $\triangle CAN$ and $\triangle DYN$ is given by $C \leftrightarrow D$, $A \leftrightarrow Y$, and $N \leftrightarrow N$. The diagram shows that $AN = 4 = YN$ and $CN = 5 = DN$, making $\overline{AN} \cong \overline{YN}$ and $\overline{CN} \cong \overline{DN}$. We also know that $\angle ANC \cong \angle YND$ because they are vertical angles. This gives two pairs of congruent corresponding sides and a pair of congruent included angles. Consequently, $\triangle CAN \cong \triangle DYN$ by the SAS property.

So far, we have used all three sides or two sides and the included angle to show that two triangles are congruent. We can also use just one side and two angles. We can do so in two ways. One is to use two angles and their common side. In this case, we imagine a segment $\overline{AB}$ of a set length with two rays extending from each endpoint to form two angles, $\angle DAB$ and $\angle CBA$ (Figure 11.7). Because the angles have a set measure and cannot both be right or obtuse, rays $\overrightarrow{AD}$ and $\overrightarrow{BC}$ must intersect at a point. The point of intersection uniquely determines the third vertex of the triangle, the remaining two sides, and the third angle.

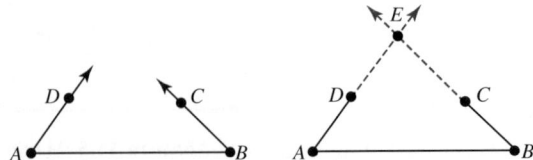

Figure 11.7 Making a triangle from two angles and a side

Consequently, if two angles and the common side in one triangle are congruent to the corresponding angles and side in a second triangle, then all six pairs of corresponding parts are congruent, making the triangles congruent. We call this property the angle-side-angle (ASA) property (Figure 11.8).

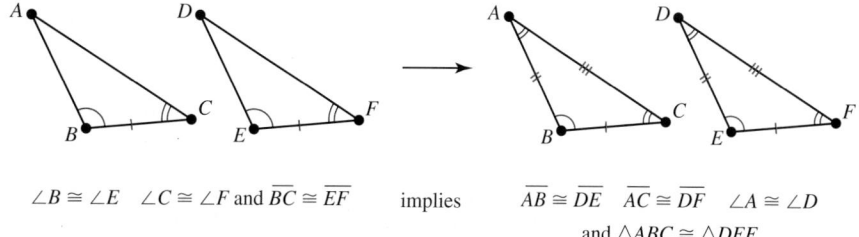

$\angle B \cong \angle E$ $\angle C \cong \angle F$ and $\overline{BC} \cong \overline{EF}$ implies $\overline{AB} \cong \overline{DE}$ $\overline{AC} \cong \overline{DF}$ $\angle A \cong \angle D$

and $\triangle ABC \cong \triangle DEF$

Figure 11.8 Angle-side-angle property

Angle-Side-Angle (ASA) Triangle Congruence Property

Given a correspondence between two triangles, if two angles and the common side of one triangle are congruent, respectively, to two angles and the common side of a second triangle, then the triangles are congruent.

Example 11.4   Using the ASA Property

Reasoning

Use the ASA property to show that $\triangle FOR \cong \triangle UOR$.

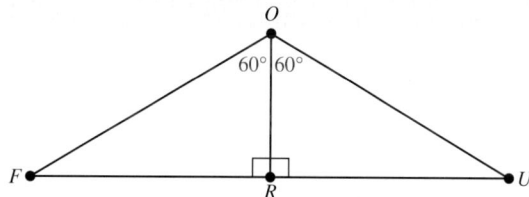

Solution We first find the correspondence between the vertices of $\triangle FOR$ and $\triangle UOR$, or $F \leftrightarrow U$, $O \leftrightarrow O$, and $R \leftrightarrow R$. Next, notice that $\angle FRO$ and $\angle URO$ are both right angles, so they are congruent. Similarly, $m\angle FOR = m\angle UOR = 60°$, making $\angle FOR \cong \angle UOR$. Because $\overline{OR}$ is congruent to itself, we have two pairs of corresponding angles and their common sides congruent. So $\triangle FOR \cong \triangle UOR$ by the ASA property.

Another way to show two triangles are congruent is use two angles and a side that is not common to both angles. For instance, consider $\triangle ABC$ and $\triangle DEF$ where $\angle A \cong \angle D$, $\angle B \cong \angle E$, and $\overline{AC} \cong \overline{DF}$ (Figure 11.9).

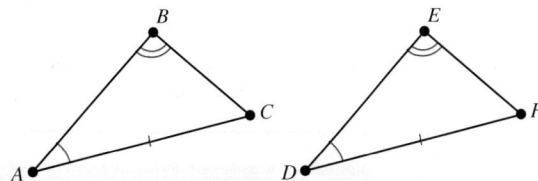

Figure 11.9 Angle-angle-side theorem

Because the sum of the measures of the interior angles in a triangle is 180° and $m\angle A = m\angle D$ and $m\angle B = m\angle E$,

$$m\angle C = 180° - m\angle A - m\angle B = 180° - m\angle D - m\angle E = m\angle F$$

This implies that $\angle C \cong \angle F$, resulting in two pairs of corresponding angles and their common sides congruent. Consequently, $\triangle ABC \cong \triangle DEF$ by the ASA property.

Theorem 11.1   Angle-Angle-Side (AAS) Theorem

Given a correspondence between two triangles, if two angles and the side opposite one of the angles are congruent, respectively, to two angles and an opposite side of the second triangle, then the triangles are congruent.

We can use two other theorems to show that two right triangles are congruent. The theorems are direct consequences of the other congruence properties and are called the Hypotenuse-Leg and Hypotenuse–Acute Angle Theorems. We leave their proofs to the Learning Assessment.

Theorem 11.2   Congruence Theorems with Right Triangles

Hypotenuse–Leg (HL) Theorem: Given a correspondence between two right triangles, if the hypotenuse and one leg of one triangle are congruent to the hypotenuse and one leg of the other, then the two triangles are congruent.

Hypotenuse–Acute Angle (HA) Theorem: Given a correspondence between two right triangles, if the hypotenuse and one acute angle of one triangle are congruent to the hypotenuse and one acute angle of the other, then the two triangles are congruent.

Example 11.5   Using Triangle Congruence Theorems

State the congruence theorem guaranteeing that each pair of triangles is congruent.

a.   b.   c.

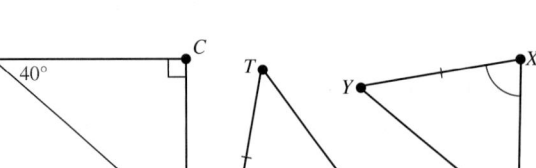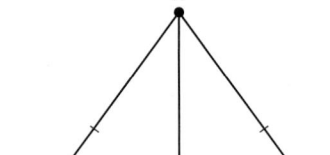

Solution

a. $\triangle ABD$ and $\triangle CDB$ are right triangles that share a common hypotenuse and have a corresponding pair of congruent, acute angles. Consequently, $\triangle ABD \cong \triangle CDB$ by the HA theorem.

b. $\triangle STR$ and $\triangle XYZ$ have two congruent angles and a pair of corresponding opposite sides that are congruent. So, $\triangle STR \cong \triangle XYZ$ by the AAS theorem.

c. $\triangle GHJ$ and $\triangle IHJ$ are right triangles with congruent hypotenuses. They also share a common side, which is congruent to itself. Consequently, by the HL theorem, $\triangle GHJ \cong \triangle IHJ$.

Check Your Understanding 11.1-A

1. If $\triangle ABC \cong \triangle XYZ$, then:

   a. $\triangle BCA \cong \triangle$ _____. **b.** $\triangle BAC \cong \triangle$ _____. **c.** $\triangle CAB \cong \triangle$ _____.

2. Let $\triangle ABC \cong \triangle XYZ$.

   a. If $AB = 7$, $BC = 15$, and $XZ = 9$, what are AC, XY, and YZ?

   b. If $m\angle A = 34°$ and $m\angle Y = 89°$, what are $m\angle B$, $m\angle C$, $m\angle X$, and $m\angle Z$?

3. Determine whether each pair of triangles is congruent. If so, state by which property.

a.

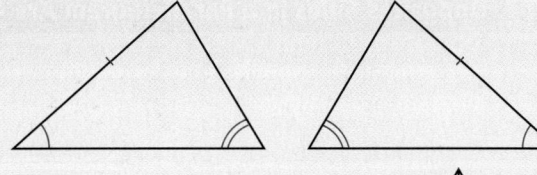

b.

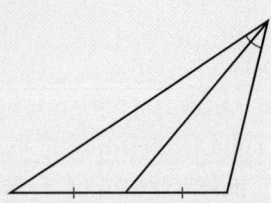

Talk About It Is it possible to draw a pair of triangles that are not congruent yet have three pairs of congruent corresponding parts? What about four pairs of congruent corresponding parts?

Communication

Activity 11.1
Write a set of directions that explains how to show two triangles are congruent using the:

a. SSS property.   **b.** SAS property.   **c.** ASA property.

Applications of Congruent Triangles

Congruent triangles have a number of applications. One of the most important is verifying the properties of polygons. For instance, we can use congruent triangles to prove the converse of the Pythagorean Theorem.

Example 11.6   The Converse of the Pythagorean Theorem

Reasoning

Show that if a triangle has sides of lengths a, b, and c such that $a^2 + b^2 = c^2$, then it is a right triangle.

Solution Suppose that $\triangle ABC$ is a triangle with sides of lengths a, b, and c such that $a^2 + b^2 = c^2$. Also suppose that $\triangle DEF$ is a right triangle with sides of lengths a and b.

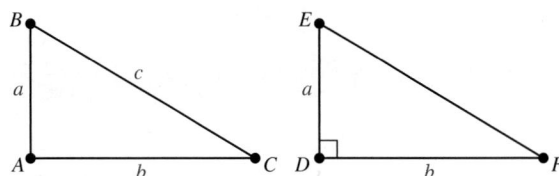

By the Pythagorean Theorem, $(EF)^2 = a^2 + b^2$. By making a substitution, we have $(EF)^2 = c^2$, indicating that $EF = c$. Because the sides of $\triangle ABC$ are congruent to the corresponding sides of $\triangle DEF$, then $\triangle ABC \cong \triangle DEF$ by the SSS property. This means that $\angle A \cong \angle D$ because corresponding parts of congruent triangles are congruent. Consequently, $\angle A$ is a right angle, and we have the converse of the Pythagorean Theorem.

We can also use congruent triangles to prove many of the facts about quadrilaterals. Consider the next two examples.

Example 11.7   Opposites Sides in a Parallelogram Are Congruent

Reasoning

Use congruent triangles and the diagram to show that the opposite sides of a parallelogram are congruent.

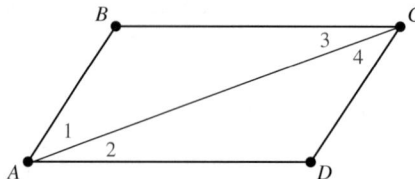

Solution We first establish a correspondence between the vertices of $\triangle ABC$ and $\triangle CDA$, namely $A \leftrightarrow C$, $B \leftrightarrow D$, and $C \leftrightarrow A$. Because quadrilateral $ABCD$ is a parallelogram, the opposite sides must be parallel. This means $\angle 1 \cong \angle 4$ and $\angle 2 \cong \angle 3$ because they are alternate interior angles. We also know that $\overline{AC} \cong \overline{CA}$ because they are the same segment. This gives us two pairs of corresponding angles and their common sides congruent. Consequently, $\triangle ABC \cong \triangle CDA$ by the ASA property. Because corresponding parts of congruent triangles are congruent, $\overline{AB} \cong \overline{DC}$ and $\overline{BC} \cong \overline{DA}$.

Example 11.8   The Diagonals of a Kite

Reasoning

Recall that a kite is a quadrilateral with at least two pairs of congruent, consecutive sides. Based on this definition, what must be true about a kite's diagonals?

Solution We can begin by drawing and examining a number of different kites and their diagonals.

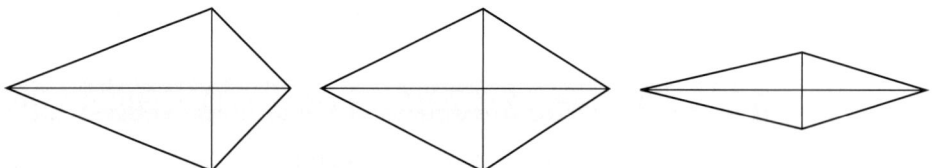

The drawings indicate that the diagonals of each kite appear to be perpendicular, so we set out to prove this fact with a deductive argument. We draw a kite and label the vertices, angles, and congruent sides.

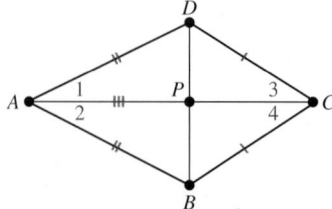

Next, we consider $\triangle ABC$ and $\triangle ADC$. Because $\overline{AB} \cong \overline{AD}$, $\overline{BC} \cong \overline{DC}$, and $\overline{AC} \cong \overline{AC}$, then $\triangle ABC \cong \triangle ADC$ by the SSS property. Furthermore $\angle 1 \cong \angle 2$ because the corresponding parts of congruent triangles are congruent. This implies $\overline{AC}$ bisects $\angle DAB$, which is the vertex angle of the isosceles triangle $\triangle ABD$. Because the bisector of the vertex angle in an isosceles triangle is perpendicular to the base, $\overline{DB} \perp \overline{AC}$.

Congruent triangles can also be helpful in solving problems. For instance, the next example shows one way to use congruent triangles to measure distances indirectly.

| Example 11.9 | **Measuring the Distance Across a Lake** |

Application

A cartographer wants to measure the distance across a small lake. To do so, she places stakes at points A and B on the shores of the lake. She then places stakes at the points C, D, and E so that $BC = CE$ and $AC = CD$. If the distance from D to E is 145 m, how far is it across the lake?

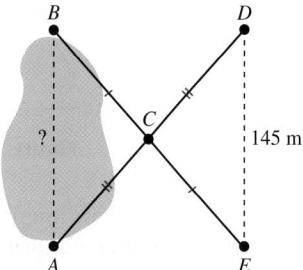

Solution By placing the stakes as shown, we know that $\overline{BC} \cong \overline{EC}$ and $\overline{AC} \cong \overline{DC}$. We also know that $\angle BCA \cong \angle ECD$ because they are vertical angles. This gives us two pairs of corresponding sides and their included angles congruent. So $\triangle BCA \cong \triangle ECD$ by the SAS property. Because corresponding parts of congruent triangles are congruent, $\overline{BA} \cong \overline{ED}$, making the distance from B to A 145 m. ∎

Congruence and Other Polygons

Explorations Manual 11.2

The idea of congruence can also be extended to quadrilaterals and other polygons. To do so, we must establish a correspondence between the vertices of the figures and then show that every pair of corresponding sides and every pair of corresponding angles are congruent. For instance, consider quadrilaterals $ABCD$ and $WXYZ$ (Figure 11.10).

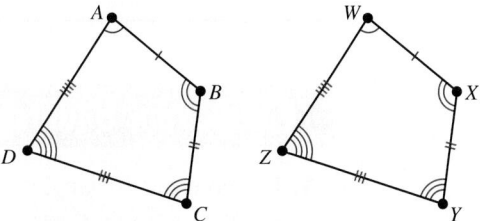

Figure 11.10 Congruent quadrilaterals

Using the correspondence $A \leftrightarrow W$, $B \leftrightarrow X$, $C \leftrightarrow Y$, and $D \leftrightarrow Z$, we have the following pairs of congruent sides and angles:

Congruent sides: $\overline{AB} \cong \overline{WX}$, $\overline{BC} \cong \overline{XY}$, $\overline{CD} \cong \overline{YZ}$, and $\overline{DA} \cong \overline{ZW}$

Congruent angles: $\angle A \cong \angle W$, $\angle B \cong \angle X$, $\angle C \cong \angle Y$, and $\angle D \cong \angle Z$

Because all pairs of corresponding sides and angles are congruent, quadrilateral $ABCD$ is congruent to quadrilateral $WXYZ$.

As with triangles, we do not need to check every pair of corresponding parts to determine whether two polygons are congruent. However, establishing the minimal conditions for congruence can be more difficult because of the additional sides and angles. For instance, can the SSS property for triangles be extended to an SSSS property for quadrilaterals? If we just consider parallelograms, Figure 11.11 shows that many parallelograms can have the same sides but different angles. This is true for any quadrilateral, so a congruence condition cannot be based on the lengths of the sides alone.

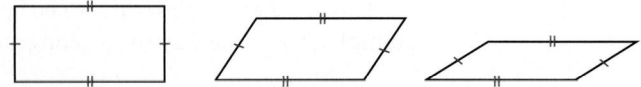

Figure 11.11 Different parallelograms with congruent sides

Notice, however, that the lengths of three sides and the two included angles completely determine the length of the fourth side and the measures of the other two angles. Consequently, quadrilaterals have an SASAS congruence property (Figure 11.12).

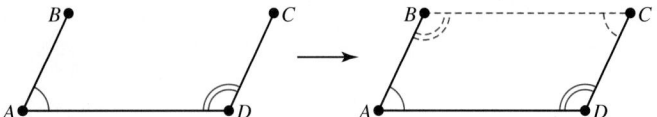

Figure 11.12 SASAS congruence property for quadrilaterals

Given additional information about the polygon, we may be able to establish other minimal congruence conditions.

Example 11.10 | Determining Minimal Congruence Conditions

Determine the minimal congruence conditions for each shape.

a. Squares   **b.** Rectangles   **c.** Circles

Solution

a. Because all squares have right angles and four congruent sides, two squares are congruent if one pair of corresponding sides is congruent.

b. Because all rectangles have right angles and two different lengths of sides, two rectangles are congruent if two pairs of consecutive sides are congruent.

c. The size of a circle is defined by its radius. Consequently, two circles are congruent if their radii are congruent.

Check Your Understanding 11.1-B

1. Use congruent triangles and the diagram to show that the diagonals in a rectangle are congruent.

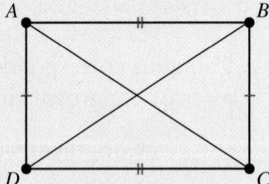

2. Consider rectangles *ABCD* and *WXYZ*. List the correspondences that can be used to show whether the two rectangles are congruent.

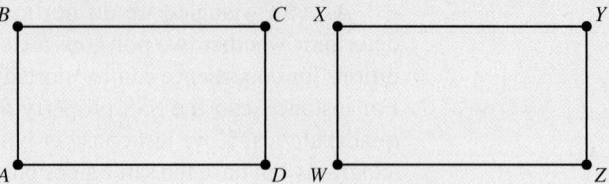

3. What are the minimal conditions needed to show that two isosceles trapezoids are congruent?

Talk About It Is there an ASASA congruent property for quadrilaterals? Explain.

Reasoning

Activity 11.2
Three-dimensional shapes can also to be congruent. Determine the minimal congruence conditions needed for each shape.

a. Cube   **b.** Tetrahedron   **c.** Cylinder   **d.** Sphere

SECTION **11.1**   **Learning Assessment**

■ Understanding the Concepts

1. What does it mean for two triangles to be congruent? Two polygons?

2. State each triangle congruence property in your own words.

   a. SSS   **b.** SAS   **c.** ASA   **d.** HL

3. Why does the SSS congruence property guarantee that two triangles are congruent if all three pairs of corresponding sides are congruent?

4. Why does the SAS congruence property guarantee that two triangles are congruent if two pairs of corresponding sides and their included angles are congruent?

5. How does the ASA congruence property differ from the AAS congruence property?

6. If two legs in one right triangle are congruent to two legs in another right triangle, must the triangles be congruent? Explain.

7. Why can establishing minimal congruence conditions be more difficult as the number of sides of a polygon increases?

■ Representing the Mathematics

8. a. If $\triangle ABC \cong \triangle XYZ$, make a list of the six pairs of corresponding parts.

   b. If $\triangle ABC \cong \triangle ZYX$, make a list of the six pairs of corresponding parts.

9. Write the congruence statement $\triangle ABC \cong \triangle XYZ$ in three different ways.

10. Suppose that $\triangle BOY \cong \triangle MAT$. Which statements show the same relationship?

   a. $\triangle YOB \cong \triangle TAM$.

   b. $\triangle OBY \cong \triangle AMT$.

   c. $\triangle BYO \cong \triangle ATM$.

   d. $\triangle OBY \cong \triangle MAT$.

   e. $\triangle YBO \cong \triangle MTA$.

   f. $\triangle OYB \cong \triangle TMA$.

11. Draw a pair of triangles that are congruent by the:

   a. SSS property.   **b.** SAS property.

   c. ASA property.

12. Draw a pair of triangles that are congruent by the:

   a. AAS property.   **b.** HL property.

   c. HA property.

13. Consider $\triangle ABC$ and $\triangle DEF$, where $\overline{BC} \cong \overline{EF}$ and $\angle B \cong \angle E$. What other parts have to be congruent for the triangles to be congruent by:

   a. SSS?   **b.** SAS?   **c.** ASA?   **d.** AAS?

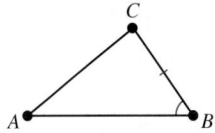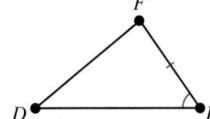

14. Draw two noncongruent triangles that satisfy each of the given conditions.

   a. Two congruent sides

   b. Three congruent angles

15. If quadrilateral $QUAD$ is congruent to quadrilateral $POLY$, make a list of the eight pairs of corresponding parts.

16. Consider parallelograms $ABCD$ and $QRST$. List the correspondences that can be used to show that the two parallelograms are congruent.

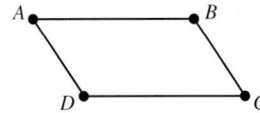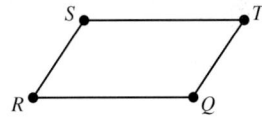

17. The following two pentagons are congruent.

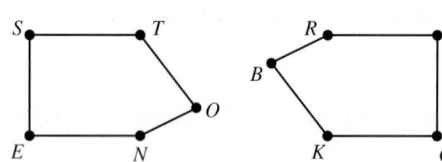

Make a list of the 10 pairs of corresponding parts.

18. Draw and label a pair of congruent:

    a. Parallelograms.   **b.** Rectangles.

    c. Kites.   **d.** Trapezoids.

19. Draw and label a pair of quadrilaterals that are congruent by the SASAS property.

20. Draw two noncongruent quadrilaterals that satisfy each of the given conditions.

    a. Four congruent sides

    b. Four congruent angles

    c. Two pairs of congruent opposite sides

■ Exercises

21. Let $\triangle ABC \cong \triangle XYZ$.

    a. If $AB = 5$, $BC = 7$, and $AC = 10$, what are XY, YZ, and XZ?

    b. If $m\angle A = 29°$, $m\angle B = 76°$, $m\angle C = 75°$, what are $m\angle X$, $m\angle Y$, and $m\angle Z$?

22. Let $\triangle ABC \cong \triangle DEF$.

    a. If $AB = 4$, $EF = 9$, and $DF = 12$, what are BC, AC, and DE?

    b. If $m\angle A = 113°$, $m\angle E = 25°$, $m\angle C = 42°$, what are $m\angle B$, $m\angle D$, and $m\angle F$?

23. Let $\triangle ABC$ and $\triangle GHI$ be right triangles such that $\triangle ABC \cong \triangle GHI$. If $BC = 5$, $GH = 12$, and $m\angle B = 90°$, what are the measures of AB, AC, HI, and GI?

24. Let $\triangle ABC$ and $\triangle JKL$ be right triangles such that $\triangle ABC \cong \triangle JKL$. If $m\angle A = 30°$, $m\angle L = 60°$, and $JL = 10$, what are the measures of the unknown parts?

25. Let $\triangle ABC$ and $\triangle MNO$ be right triangles such that $\triangle ABC \cong \triangle MNO$. If $m\angle A = 45°$, $m\angle B = 90°$, and $NO = 5$, what are the measures of the unknown parts?

26. Decide whether each pair of triangles is congruent. If so, state by what congruence property.

    a.

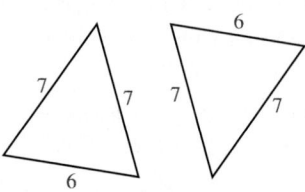

b.

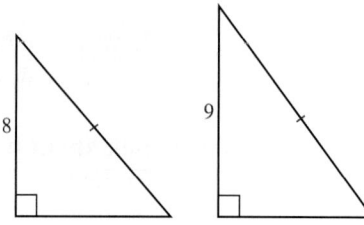

c.

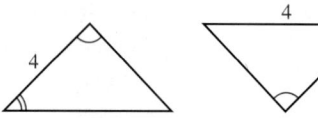

d.

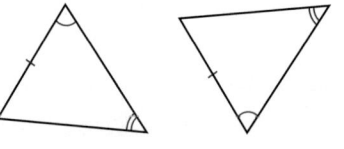

27. If $\overline{IT}$ is the perpendicular bisector of $\overline{DR}$, by what property is $\triangle DIT \cong \triangle RIT$?

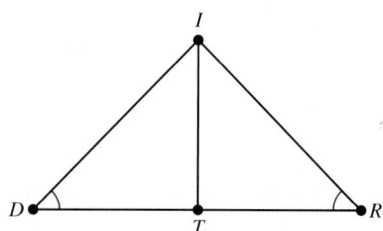

28. If $\overline{CD}$ bisects $\overline{YN}$ and $\overline{YN}$ bisects $\overline{CD}$, by what property is $\triangle CAY \cong \triangle DAN$?

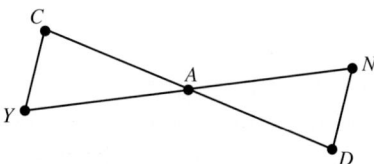

29. If $\overline{IR} \parallel \overline{BD}$, $\overline{BI} \parallel \overline{DR}$, and $\angle I \cong \angle D$, by what property is $\triangle BIR \cong \triangle RDB$?

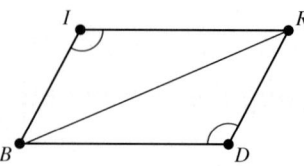

30. Find x.

    a.

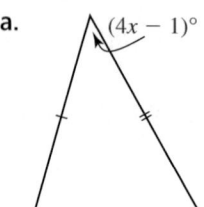 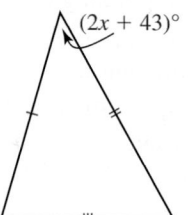

b.

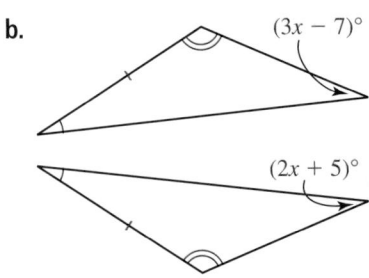

31. If parallelogram *ABCD* is congruent to parallelogram *QRST*, find the measures of the missing parts.

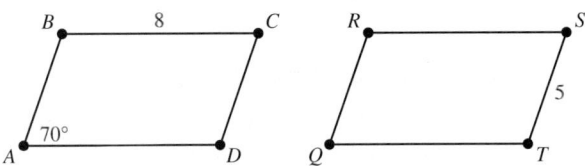

32. If *ABCD* and *WXYZ* are two congruent isosceles trapezoids, find the measures of the missing parts.

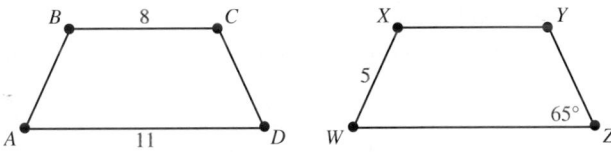

■ Problems and Applications

33. Consider △*DEF* and △*RST*, in which $\overline{FD} = 8$, $\angle F = 75°$, $\overline{FE} = 6$, $\overline{RT} = 6$, $\angle R = 75°$, and $\overline{RS} = 8$. Are the two triangles congruent? If so, give the correspondence between their vertices.

34. **a.** If possible, draw two triangles that have two congruent pairs of corresponding parts but that are not themselves congruent.

    b. If possible, draw two triangles that have three congruent pairs of corresponding parts but that are not themselves congruent.

    c. If possible, draw two triangles that have four congruent pairs of corresponding parts but that are not themselves congruent.

35. Can every regular polygon be divided into congruent triangles? If so, how?

36. If *BI* = *IE* and *BR* = *DE*, find all pairs of congruent triangles in the diagram.

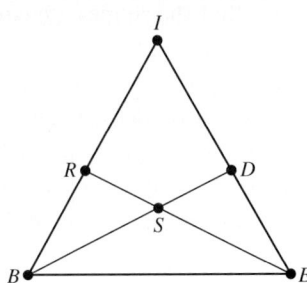

37. If orientation is not considered, how many different pairs of congruent triangles are there in the square?

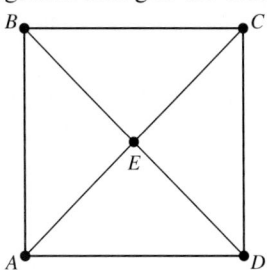

38. How many different correspondences can there be between the vertices of two:

    a. Triangles? **b.** Quadrilaterals? **c.** Pentagons?

39. A road sign stretches across a highway and is cross-braced by 20 alternating, congruent, isosceles triangles. The base of any one triangle is 7.5 ft long.

    a. How long is the sign?

    b. How many cross brace pieces are there?

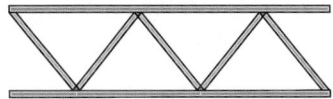

40. What types of quadrilaterals can be divided into two congruent triangles by drawing a single diagonal?

41. Jacque is a highwire performer and is planning to walk across a cable stretched diagonally between two skyscrapers. To measure the length of the cable, he marks two points *B* and *C* so that they are 30 yd on either side of point *A*. From point *C*, he then walks backward until he reaches the point *D* where he can see point *A* line up with point *P*. If point *D* is 70 yd from point *C*, how far is it across the gap?

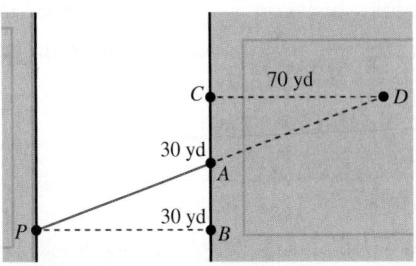

■ Mathematical Reasoning and Proof

42. What correspondence should be established between vertices to show that △*ABC* is congruent to itself?

43. Is it possible to show that two triangles are congruent by using a SSA property? Explain why or why not.

44. If the radius of one regular hexagon is congruent to the radius of another, are the two hexagons congruent? What if one apothem is congruent to the other?

45. Prove that the perpendicular bisector of the base in any isosceles triangle separates it into two congruent triangles.

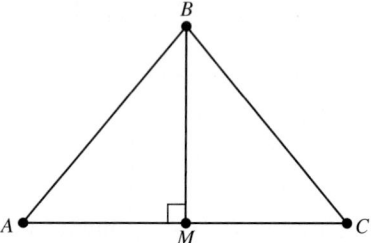

46. Let $\triangle TYR$ be an equilateral triangle, $\angle 2 \cong \angle 3$, and $SY = RO$. Show $\triangle STY \cong \triangle OTR$.

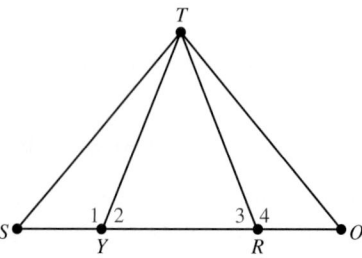

47. Use the angle-side-angle property to write an argument that justifies the Angle-Angle-Side (AAS) Theorem.

48. Use the Angle-Angle-Side (AAS) Theorem to write an argument that justifies the Hypotenuse–Acute Angle (HA) Theorem.

49. Determine the minimal congruence conditions needed for each shape.

   a. Rhombus   **b.** Parallelogram

   c. Regular hexagon   **d.** Trapezoid

50. Determine the minimal congruence conditions needed for each shape.

   a. Right triangular prism

   b. Right circular cone

   c. Octahedron

   d. Right square pyramid

51. Use congruent triangles and the diagram to show that the diagonals in a rhombus bisect each other.

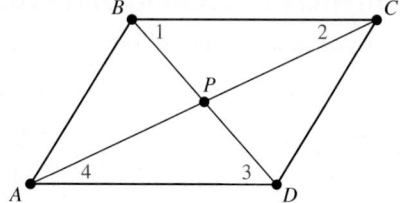

52. Use congruent triangles and the diagram to prove that opposite angles of a parallelogram are congruent.

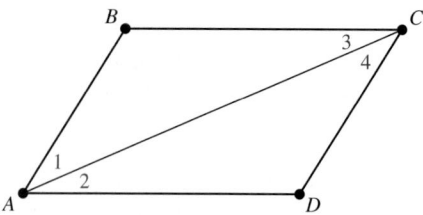

Mathematical Communication

53. Write two or three exercises that can be solved using the congruence properties for triangles. Give them to a partner to solve.

54. Review the definition for congruent triangles. Use the ideas in the definition to write a definition for congruent quadrilaterals.

55. A student comes to you and claims that one triangle is congruent to another if the three pairs of corresponding angles are congruent. Is the student correct? Explain why or why not.

56. Marco claims that the following two triangles are congruent by ASA. How would you respond to his thinking?

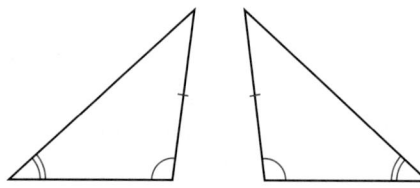

57. A student wonders whether two triangles congruent to a third are congruent to each other. How would you, as the student's teacher, help the student make sense of this situation?

Building Connections

58. How is triangle congruence an extension of congruence for line segments and angles?

59. If two geometric figures have the same shape and size, can we say that the shapes are equal? Why or why not?

60. How can congruent triangles be used to help students learn to reason deductively?

61. Explore a set of curriculum materials for grades K–6. At what grade level is congruence introduced? What is the general approach to congruence, and what shapes are used to model the idea?

SECTION 11.2 Similar Shapes

Similarity, another important relationship between geometrical figures, is the notion that two figures have the same shape but not necessarily the same size. For instance, suppose an overhead projector casts the image of a pentagon onto a screen (Figure 11.13). The pentagon on the screen takes on the same shape as the pentagon underneath the projector; it is just larger.

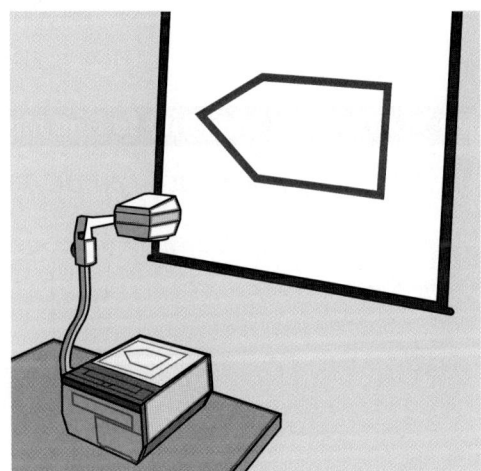

Figure 11.13 Similar figures with a projector

Taking a closer look at the pentagons in Figure 11.13, we can see a correspondence between their vertices, indicating another correspondence between their angles and sides. Also notice that the corresponding angles have the same measure. So the only difference between the pentagons is the length of their sides. Although the larger pentagon has longer sides, its sides must have been increased by the same proportional amount for it to maintain its shape. In other words, two shapes are similar if and only if there is a correspondence between their vertices so that corresponding angles are congruent and corresponding sides are proportional. In this section, we focus primarily on similar triangles and their properties. Later, in Chapter 12, we will learn about similarity and other shapes in the context of coordinate and transformation geometry.

Explorations Manual 11.3

Similar Triangles

Although similarity can be applied to any number of shapes, similar triangles are among the most useful. Using the criterion from the previous discussion, we get the following definition.

Definition of Similar Triangles

$\triangle ABC$ is **similar** to $\triangle DEF$, written $\triangle ABC \sim \triangle DEF$, if and only if there is a correspondence between their vertices such that corresponding angles are congruent and corresponding sides are proportional.

As with congruent triangles, the notation identifies the correspondence between the vertices, sides, and angles. Because the corresponding sides are proportional, they must occur with the same common ratio. This common ratio is called the **scale factor.** To find it, we take the ratio of any two corresponding sides, or $\dfrac{AB}{DE} = \dfrac{AC}{DF} = \dfrac{BC}{EF}$.

For instance, Figure 11.14 shows two similar triangles in which the scale factor is 2. In other words, each side of $\triangle ABC$ is twice the length of its corresponding side in $\triangle DEF$.

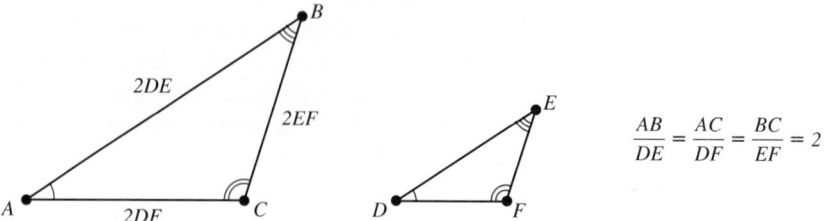

Figure 11.14 $\triangle ABC \sim \triangle DEF$ with a scale factor of 2

Example 11.11 Using Similarity to Find Missing Sides in Triangles

Suppose $\triangle ABC \sim \triangle EDC$. Find AC and DE.

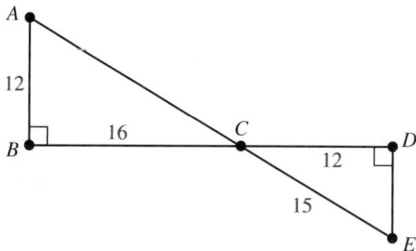

Solution Because $\triangle ABC \sim \triangle EDC$, the ratio for the corresponding sides must be the same. We know the lengths of two corresponding sides, $\overline{BC}$ and $\overline{DC}$, so we can find the scale factor by taking their ratio, or

$$\text{Scale factor} = \frac{BC}{DC} = \frac{16}{12} = \frac{4}{3}$$

Next, we use proportions and the rest of the given information to find AC and ED.

$$\frac{AC}{EC} = \frac{4}{3} \qquad\qquad \frac{AB}{ED} = \frac{4}{3}$$

$$\frac{AC}{15} = \frac{4}{3} \qquad\qquad \frac{12}{ED} = \frac{4}{3}$$

$$AC = \frac{15 \cdot 4}{3} \qquad\qquad ED = \frac{3 \cdot 12}{4}$$

$$AC = 20 \qquad\qquad ED = 9$$

When writing proportions related to similar triangles, we must take some care because the scale factor is somewhat dependent on how we look at the triangles. For instance, in the correspondence $\triangle ABC \sim \triangle EDC$, $\triangle ABC$ was first. So we computed the scale factor by taking the ratio of the sides in $\triangle ABC$ to the sides in $\triangle EDC$. This resulted in a scale factor of $\frac{4}{3}$. If we had written $\triangle EDC \sim \triangle ABC$, we would have used the correspondence in the reverse order, for a scale factor of $\frac{3}{4}$.

As with congruent triangles, all six pairs of corresponding parts do not need to be checked to determine whether two triangles are similar. For instance, suppose that in two triangles, the corresponding sides are proportional by the same scale factor. The triangles must take on the same shape because they are completely defined by their sides. In other words, if the corresponding sides are proportional, then the corresponding angles must be congruent, and the triangles must be similar. This gives a side-side-side property for similar triangles (Figure 11.15).

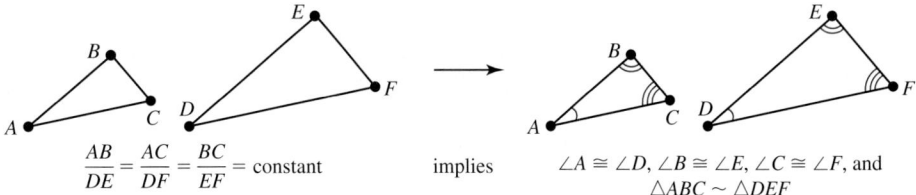

$$\frac{AB}{DE} = \frac{AC}{DF} = \frac{BC}{EF} = \text{constant} \qquad \text{implies} \qquad \angle A \cong \angle D, \angle B \cong \angle E, \angle C \cong \angle F, \text{ and}$$
$$\triangle ABC \sim \triangle DEF$$

Figure 11.15 Side-side-side property for similar triangles

A triangle is also completely determined by two sides and the included angle. Consequently, if in two triangles two pairs of corresponding sides are proportional and the included angles are congruent, then the triangles must take on the same shape and be similar. This gives us a side-angle-side property for similar triangles (Figure 11.16).

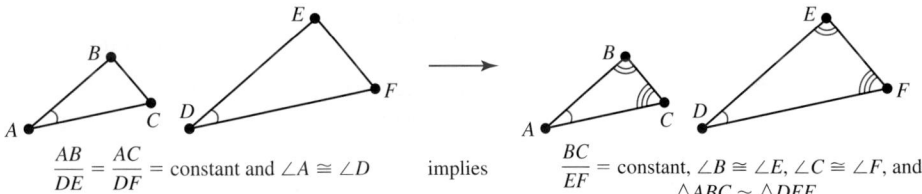

$$\frac{AB}{DE} = \frac{AC}{DF} = \text{constant and } \angle A \cong \angle D \qquad \text{implies} \qquad \frac{BC}{EF} = \text{constant, } \angle B \cong \angle E, \angle C \cong \angle F, \text{ and}$$
$$\triangle ABC \sim \triangle DEF$$

Figure 11.16 Side-angle-side property for similar triangles

A third property for similar triangles is the angle-angle property. If in two triangles two pairs of corresponding angles are congruent, then the third pair must also be congruent because the sum of the angles in a triangle is 180°. Given three pairs of congruent angles, the triangles must take on the same shape and be similar (Figure 11.17).

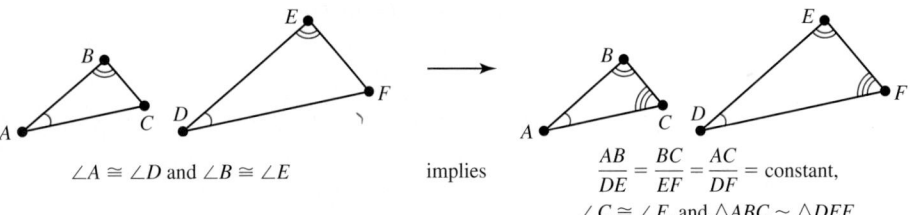

$$\angle A \cong \angle D \text{ and } \angle B \cong \angle E \qquad \text{implies} \qquad \frac{AB}{DE} = \frac{BC}{EF} = \frac{AC}{DF} = \text{constant,}$$
$$\angle C \cong \angle F, \text{ and } \triangle ABC \sim \triangle DEF$$

Figure 11.17 Angle-angle property

The following theorem summarizes the similar triangle properties.

Theorem 11.3 Properties for Similar Triangles

Side-side-side (SSS): Given a correspondence between two triangles, if all three pairs of corresponding sides are proportional, then the triangles are similar.

Side-angle-side (SAS): Given a correspondence between two triangles, if two pairs of corresponding sides are proportional and their included angles are congruent, then the triangles are similar.

Angle-angle (AA): Given a correspondence between two triangles, if two angles of one triangle are congruent, respectively, to two angles of the second triangle, then the triangles are similar.

Example 11.12 | **Using Similarity Properties**

Determine whether the triangles in each diagram are similar.

a.
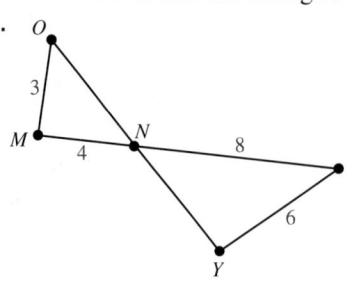

b.
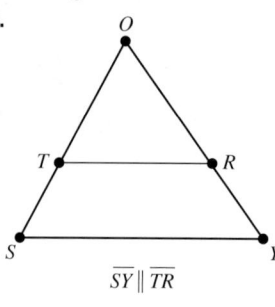

$\overline{SY} \parallel \overline{TR}$

c.
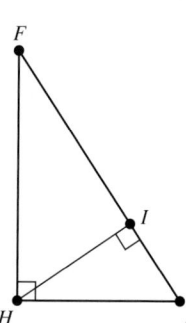

Solution

a. Two pairs of corresponding sides are proportional, and we know $\angle MNO \cong \angle ENY$ because they are vertical angles. However, the angles are not included between the proportional sides, so we cannot conclude that $\triangle MNO$ is similar to $\triangle ENY$.

b. $\triangle SOY$ and $\triangle TOR$ both contain $\angle O$, giving them one pair of congruent angles. Also, $\angle OTR \cong \angle OSY$ because they are corresponding angles between parallel lines. This gives us two pairs of congruent angles, so $\triangle SOY \sim \triangle TOR$ by the AA property.

c. Of the three triangles to consider, we can quickly see that $\triangle FHS \sim \triangle HIS$ by the AA property because both triangles have right angles and both contain $\angle S$. Similarly, $\triangle FHS \sim \triangle FIH$. From these two relationships, we know that the angles in $\triangle HIS$ are congruent to those in $\triangle FIH$ because the angles in $\triangle HIS$ are congruent to the angles in $\triangle FHS$, and the angles in $\triangle FHS$ are congruent to the angles in $\triangle FIH$. As a result, $\triangle HIS \sim \triangle FIH$ by the AA property.

Check Your Understanding 11.2-A

1. If $\triangle ABC \sim \triangle DEF$, what is the scale factor if:

    a. $AB = 7$ and $DE = 35$?   **b.** $AB = 65$ and $DE = 39$?

2. Suppose $\triangle ABC \sim \triangle DEF$. If $AB = 4$, $AC = 9$, $DE = 10$, and $EF = 13$, find BC and DF.

3. Is $\triangle ABC \sim \triangle DEC$? If so, by what property?

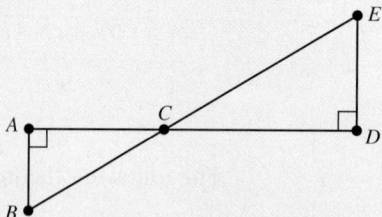

Talk About It Why is there no angle-side-angle property for triangle similarity?

Reasoning

Activity 11.3

Use the diagram and similar triangles to prove the Pythagorean Theorem.

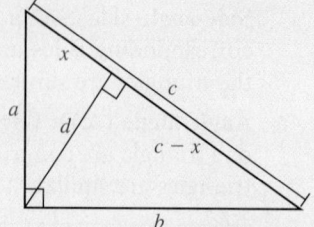

Applications of Similar Triangles

As with congruent triangles, we can use similar triangles to identify and verify properties of different polygons.

| **Example 11.13** | **Midsegment of a Triangle** |

Reasoning

A **midsegment** of a triangle is a segment that connects the midpoints of two sides. What is true about a midsegment and its relationship to the third side of a triangle?

Solution We begin by drawing a triangle, $\triangle ABC$, and one of its midsegments, $\overline{PQ}$. We also label four of the angles.

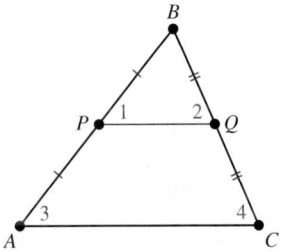

At a glance, it appears as though $\triangle ABC \sim \triangle PBQ$. This is indeed the case. Because P is the midpoint of $\overline{AB}$, the ratio of AB to PB is $\dfrac{AB}{PB} = 2$. Similarly, the ratio of CB to QB is $\dfrac{CB}{QB} = 2$. Also, $\angle ABC \cong \angle PBQ$ because they are the same angle. Given two pairs of proportional sides and a pair of congruent included angles, $\triangle ABC \sim \triangle PBQ$ by SAS. Because the sides of $\triangle PBQ$ are half as long as the sides of $\triangle ABC$, it must be that $PQ = \dfrac{1}{2}AC$. Also, $\angle 1$ is congruent to $\angle 3$, and, because they are corresponding angles along a transversal, $\overline{PQ}$ must be parallel to $\overline{AC}$.

| **Theorem 11.4** | **Midsegment Theorem** |

The midsegment of a triangle is parallel to the third side and half as long.

Quadrilaterals also have midsegments, which are segments that connect the midpoints of two adjacent sides. In Figure 11.18, note that the midsegments of several different quadrilaterals always appear to form a parallelogram.

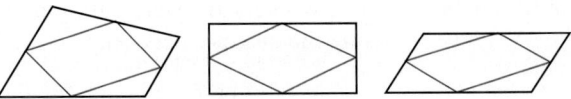

Figure 11.18 Quadrilaterals and their midsegments

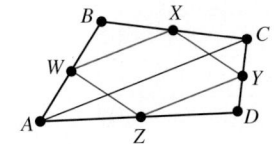

Figure 11.19 $\triangle ABC \sim \triangle WBX$ and $\triangle ADC \sim \triangle ZDY$

We can prove this fact with similar triangles. Consider quadrilateral $ABCD$ (Figure 11.19), where W, X, Y, and Z are the midpoints of the sides. Drawing the segment $\overline{AC}$, we create two pairs of similar triangles: $\triangle ABC \sim \triangle WBX$ and $\triangle ADC \sim \triangle ZDY$. By Theorem 11.4, we know that $\overline{WX} \parallel \overline{AC}$ and $\overline{ZY} \parallel \overline{AC}$, so $\overline{WX} \parallel \overline{ZY}$. Using a similar argument, we show that $\overline{WZ} \parallel \overline{XY}$. So quadrilateral $WXYZ$ must be a parallelogram because it has two pairs of opposite sides that are parallel.

Theorem 11.5 Midsegments of a Quadrilateral

The midsegments of any quadrilateral form a parallelogram.

 Explorations Manual 11.4

Similar triangles can also be used to make indirect measurements. To do so, we make two similar triangles and measure two corresponding sides directly. We then use proportional reasoning to find the lengths of the unmeasured sides.

Example 11.14 Measuring the Height of a Building

Application

On a sunny day, a building casts a shadow 75 ft long. At the same time, a man who is 6 ft tall casts a shadow that is 4 ft long. How tall is the building?

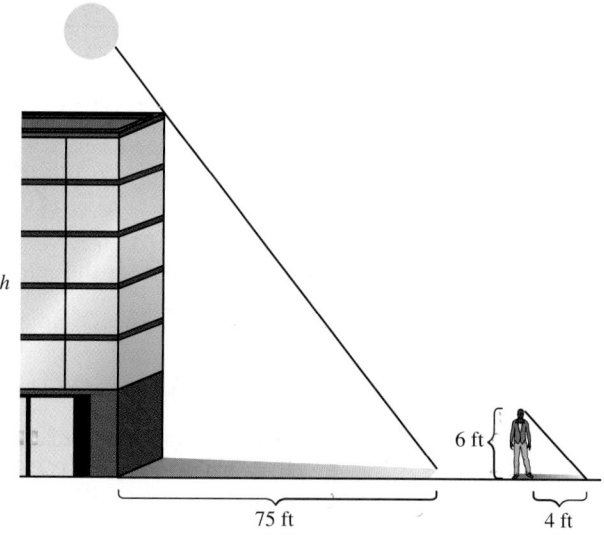

Solution To solve the problem, we first make a few assumptions allowing us to show that the two triangles are similar. First, we assume that both the building and the person are standing perpendicularly to the ground, making each triangle a right triangle. Second, we assume that the rays from the sun hit the building and the person at the same angle. Consequently, the two triangles are similar by the AA property. If h represents the height of the building, then $\frac{h}{75} = \frac{6}{4}$. Solving for h, $h = \frac{75 \cdot 6}{4} = 112.5$ ft.

Example 11.15 Measuring the Distance Across Water

Application

Two workers at a state park plan to build a boardwalk across a swampy area. To measure the distance between two stable pieces of land, they create the following triangles.

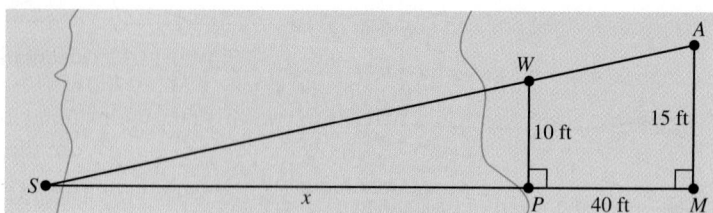

They then take the following measurements: $AM = 15$ ft, $WP = 10$ ft, and $MP = 40$ ft. If the bridge is to go from P to S, how long must it be?

Solution $\triangle WSP$ and $\triangle ASM$ are right triangles with $\angle S$ in common. Consequently, $\triangle WSP \sim \triangle ASM$ by the AA property. Because corresponding sides of similar triangles are proportional,

$$\frac{WP}{AM} = \frac{SP}{SM} \quad \text{or} \quad \frac{10}{15} = \frac{x}{x + 40}.$$

Next, we cross multiply and solve for x:

$$15x = 10(x + 40)$$
$$15x = 10x + 400$$
$$5x = 400$$
$$x = 80 \text{ ft}$$

Fractals and Self-Similarity

Explorations
Manual
11.5

Another interesting aspect of many similar shapes is self-similarity. A figure is **self-similar** if it looks "roughly" the same if we magnify or reduce any part of it. For instance, a **rep-tile** is a self-similar figure that is made by dissecting a polygon into smaller versions of itself or by fitting copies of the polygon together to form a larger, similar version of itself. An example of a triangular rep-tile is shown in Figure 11.20.

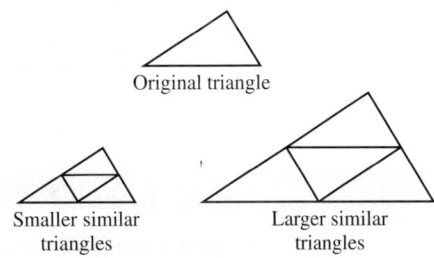

Original triangle

Smaller similar
triangles

Larger similar
triangles

Figure 11.20 A triangle rep-tile

Example 11.16 | Self-Similarity in a Trapezoid

Divide the following trapezoid into four similar copies of itself.

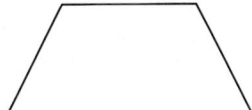

Solution We can divide the given trapezoid not only into four similar copies of itself, but also into nine similar copies of itself. Both are shown.

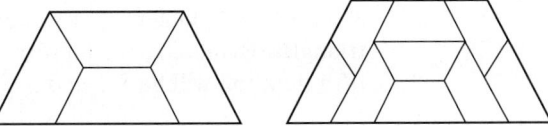

The idea of self-similarity became more interesting to mathematicians when Benoit Mandelbrot introduced **fractals** in the early 1980s. He used the term "fractal" to describe self-similar geometric patterns that are generated by a recursively defined function. High-speed computers enable us to visualize and explore these patterns through magnification. For instance, the pictures shown in Figure 11.21 illustrate the self-similarity in several magnifications of a fractal called the *Mandelbrot set*.

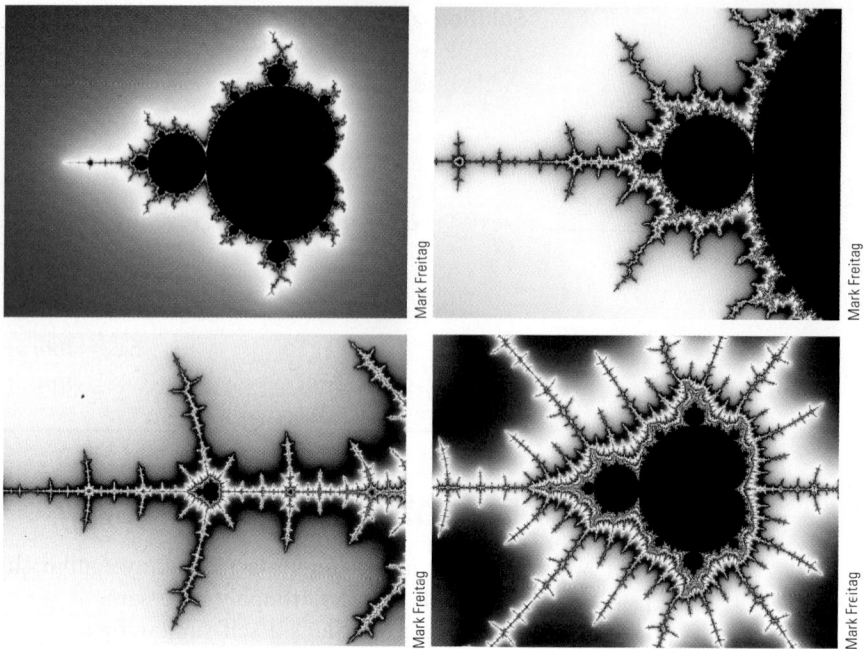

Figure 11.21 The Mandelbrot set

Originally, fractal geometry was developed to describe geometric patterns in nature, like plant growth, coastlines, and even veins and arteries. Now, however, fractals are used to model the weather, chemical reactions, and even the economy. Fractals have even been used to generate three-dimensional effects in movies.

Example 11.17 | The Sierpinski Gasket

The Sierpinski gasket, or the Sierpinski triangle, is a fractal formed from an equilateral triangle by removing successively smaller triangles from its interior. The first three iterations are shown.

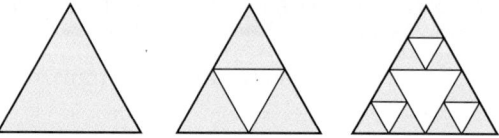

How many white triangles are in the fifth iteration?

Solution Because the white triangles are always formed in the center of a blue triangle, the number of new white triangles is equal to the number of blue triangles in the previous iteration. Also notice that the number of blue triangles is always equal to 3^{n-1}, where n is equal to the number of iterations. So the fourth iteration contains $3^{4-1} = 3^3 = 27$ blue triangles, indicating that the fifth iteration will have 27 new white triangles. By adding the number of white triangles from the first four iterations, the fifth iteration will have $1 + 3 + 9 + 27 = 40$ white triangles.

Check Your Understanding 11.2-B

1. A person who is 5 ft tall is standing next to a tree that is 50 ft tall. If the person casts a shadow that is 8 ft long, how long is the shadow of the tree?

2. Divide the following trapezoid into four similar copies of itself.

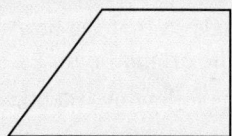

Talk About It Can any triangle be used to form a rep-tile? What about any quadrilateral?

Activity 11.4
Find a website on the Internet that has a generator for the Mandelbrot set. Explore the fractal by magnifying it at different parts. Describe the self-similarity that you find.

SECTION 11.2   Learning Assessment

■ Understanding the Concepts

1. What is the difference between similar shapes and congruent shapes?

2. **a.** What is the scale factor between two similar shapes?

   b. Is it possible to use two different scale factors when working with two similar shapes? Explain.

3. **a.** How does the SSS similarity property differ from the SSS congruence property?

   b. How does the SAS similarity property differ from the SAS congruence property?

4. Why must two triangles be similar if they have two congruent pairs of corresponding angles?

5. **a.** If two triangles are congruent, are they necessarily similar? Explain.

   b. If two triangles are similar, are they necessarily congruent? Explain.

6. How can similar triangles be used to measure distances indirectly?

7. What does it mean for a geometric figure to be self-similar?

■ Representing the Mathematics

8. Draw a pair of triangles that are similar by the:

   a. SSS property.

   b. SAS property.

   c. AA property.

9. Draw a right triangle $\triangle ABC$ with sides of length 6 cm, 8 cm, and 10 cm.

   a. Draw $\triangle DEF$ so that $\triangle ABC \sim \triangle DEF$ with a scale factor of 2.

   b. Draw $\triangle GHI$ so that $\triangle ABC \sim \triangle GHI$ with a scale factor of $\frac{1}{2}$.

10. Draw an isosceles triangle $\triangle ABC$ with sides of length 12 cm, 12 cm, and 6 cm.

    a. Draw $\triangle JKL$ so that $\triangle ABC \sim \triangle JKL$ with a scale factor of $\frac{1}{3}$.

    b. Draw $\triangle MNO$ so that $\triangle ABC \sim \triangle MNO$ with a scale factor of 1.5.

11. If $\triangle ABC \sim \triangle DEF$, what other ways are there to write the proportion $\frac{AB}{DE} = \frac{AC}{DF}$?

12. Consider $\triangle ABC$ and $\triangle XYZ$ in which $\overline{AB} = 4$, $\angle A = 75°$, $\overline{AC} = 6$, $\overline{ZY} = 9$, $\angle Z = 75°$, and $\overline{ZX} = 6$. Are the two triangles similar? If so, give the correspondence between their vertices.

13. Draw and label a pair of similar figures for each shape.

    a. Square   **b.** Rectangle

    c. Parallelogram   **d.** Trapezoid

14. Draw a quadrilateral so that the midsegments form each figure.

    a. A square   **b.** A rectangle

15. a. Draw two nonsimilar quadrilaterals in which each pair of corresponding angles is congruent.

b. Draw two nonsimilar quadrilaterals in which each pair of corresponding sides is proportional.

16. Draw a rep-tile with a right triangle so that the new similar figures are smaller than the original triangle.

17. Draw a rep-tile with a rectangle so that the new similar figure is larger than the original figure.

18. Draw the next iteration of the following fractal.

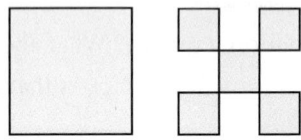

19. Draw the next iteration of the following fractal.

■ Exercises

20. If $\triangle ABC \sim \triangle DEF$, what is the scale factor if:

a. $AB = 3$ and $DE = 6$.

b. $AB = 12$ and $DE = 8$.

c. $AB = 3$ and $DE = 15$.

21. If $\triangle ABC \sim \triangle DEF$, what is the scale factor if:

a. $AB = \dfrac{2}{3}$ and $DE = \dfrac{4}{5}$.

b. $AB = 6.9$ and $DE = 2.3$.

c. $AB = 6.2$ and $DE = 3.2$.

22. If $\triangle ABC \sim \triangle DEF$, what is DE if:

a. $AB = 5$ and the scale factor is 3.

b. $AB = 8$ and the scale factor is $\dfrac{3}{4}$.

23. If $\triangle ABC \sim \triangle DEF$, what is DE if:

a. $AB = 3.8$ and the scale factor is 1.2.

b. $AB = \sqrt{20}$ and the scale factor is $\sqrt{5}$.

24. Determine whether each pair of triangles is similar. For those that are, state by which similarity property. For those that are not, explain why.

a.

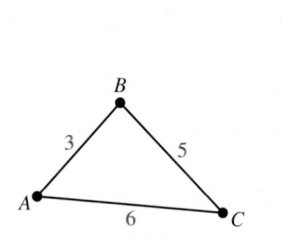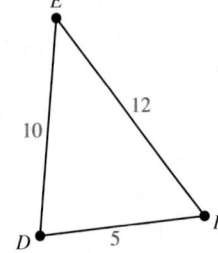

b.

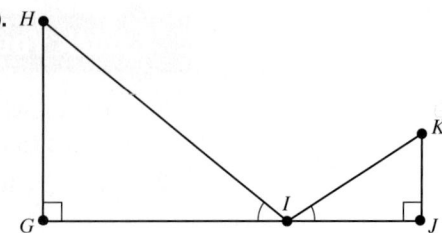

c.

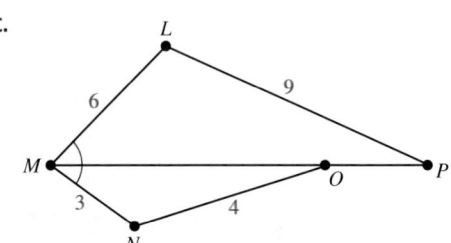

d.

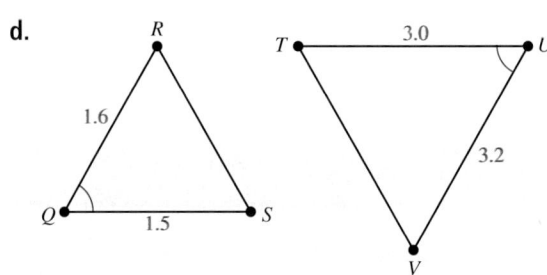

25. Use the given information to determine whether $\triangle ABC$ is similar to $\triangle XYZ$. If so, state by which similarity property.

a. $m\angle A = 67°$, $m\angle X = 67°$, $AB = 16$, $AC = 20$, $XY = 4$, and $XZ = 5$.

b. $m\angle A = 63°$, $m\angle X = 63°$, $m\angle B = 47°$, and $m\angle Z = 70°$.

26. Use the given information to determine whether $\triangle ABC$ is similar to $\triangle XYZ$. If so, state by which similarity property.

a. $AB = 27$, $BC = 21$, $AC = 9$, $XY = 9$, $YZ = 7$, and $XZ = 4$.

b. $m\angle A = 45°$, $m\angle X = 45°$, $m\angle C = 110°$, $m\angle Y = 25°$.

27. Suppose that $\triangle STA \sim \triangle RKA$.

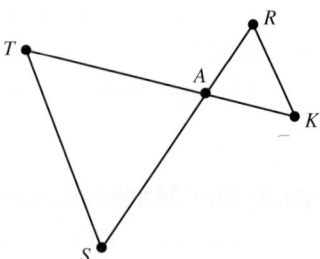

a. If $TA = 12$, $ST = 20$, and $RK = 10$, how long is KA?

b. If $TA = 7$, $SA = 9$, and $RA = 3$, how long is KA?

c. If $TK = 15$, $TA = 10$, $SA = 8$, and $RK = 6$, how long are ST and RA?

28. Suppose $\overline{DE} \parallel \overline{BC}$ in the following diagram.

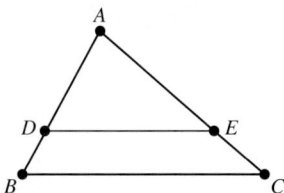

a. If $AD = 18$, $DB = 9$, and $AE = 25$, how long is EC?

b. If $BD = 5$, $CE = 9$, and $EA = 18$, how long is AD?

c. If $AB = 24$, $AE = 13$, and $EC = 3$, how long are AC, AD, and DB?

d. If $AB = 12$, $DB = 3$, and $AC = 36$, how long are AE and EC?

29. In each diagram, assume that the two triangles are similar. Use the given information to find the value of the variable.

a.

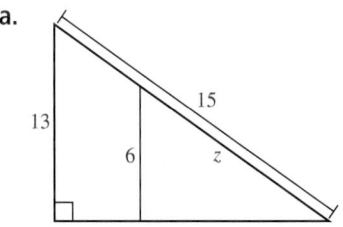

b.

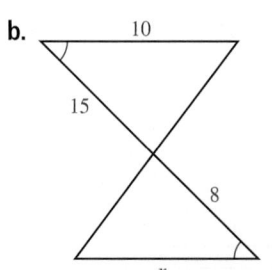

c.

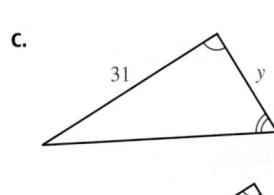

d.

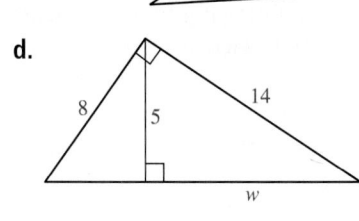

30. Suppose parallelogram $ABCD$ is similar to parallelogram $WXYZ$.

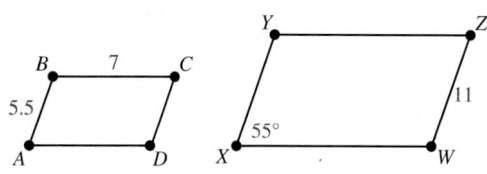

Find the measure of each part.

a. $\overline{XY}$ **b.** $\overline{DC}$ **c.** $\overline{YZ}$

d. $\angle A$ **e.** $\angle B$

31. Consider two regular pentagons that are similar by a scale factor of 3.

a. What are the angle measures of the interior angles in both pentagons?

b. If the smaller pentagon has a side of length 8, what are the lengths of the sides in the larger pentagon?

32. Suppose two rectangles are similar by a scale factor of 7.2. If the smaller rectangle has sides of lengths 4 and 6, what are the lengths of the sides in the larger rectangle?

■ Problems and Applications

33. If $\overline{DE}$ is a midsegment of $\triangle ABC$, find the length of each side.

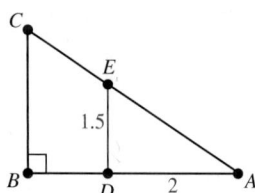

a. $\overline{AE}$ **b.** $\overline{AB}$ **c.** $\overline{AC}$ **d.** $\overline{BC}$

34. Suppose $\triangle ABC \sim \triangle DEF$ and $\triangle DEF \sim \triangle XYZ$. Further suppose that $AB = 32$, $\overline{DE}$ is 40% the length of $\overline{AB}$, and $\overline{XY}$ is 50% the length of $\overline{DE}$.

a. What is the length of $\overline{DE}$ and $\overline{XY}$?

b. What is the scale factor between $\triangle ABC$ and $\triangle XYZ$?

35. Suppose $\triangle ABC \sim \triangle DEF$ and $\triangle DEF \sim \triangle XYZ$. Further suppose that $AB = 60$, $\overline{DE}$ is $\frac{2}{3}$ the length of $\overline{AB}$, and $\overline{XY}$ is $\frac{4}{5}$ the length of $\overline{DE}$.

a. What is the length of $\overline{DE}$ and $\overline{XY}$?

b. What is the scale factor between $\triangle ABC$ and $\triangle XYZ$?

c. If $BC = 90$, what are EF and YZ?

d. If $AC = 45$, what are DF and XZ?

36. Use the following diagram to write a relationship among w, x, y, and z.

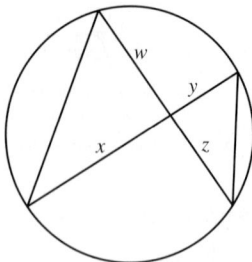

37. One way to enlarge or reduce figures is to use a photocopier.

   a. If we enlarge a triangle with sides of lengths 3 in., 5 in., and 7 in. by 120%, about how long are the sides of the triangle on the photocopy?

   b. If we reduce a rectangle with sides of lengths 4 in. and 7 in. by 60%, about how long are the sides of the rectangle on the photocopy?

   c. If we enlarge a triangle by 125%, by how much would we have to reduce the photocopy to get the original triangle?

38. Dishawn plans to find the height of a building by placing a mirror on the ground and walking backward until he can see the top of the building in the mirror. What is the height of the building if his eyes are 63 in. above the ground, his feet are 27 ft, 4 in. from the mirror, and the mirror is 90 ft, 6 in. from the building? Give your answer in feet and inches.

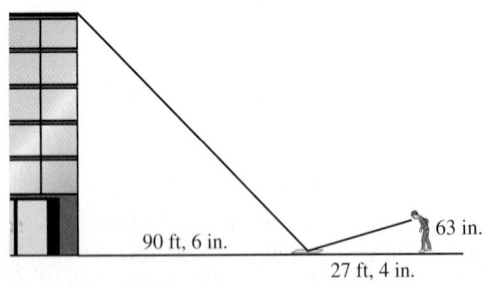

39. Two hikers come to a ravine and want to know how wide it is. They set up two similar triangles as shown in the diagram. How far is it across the ravine?

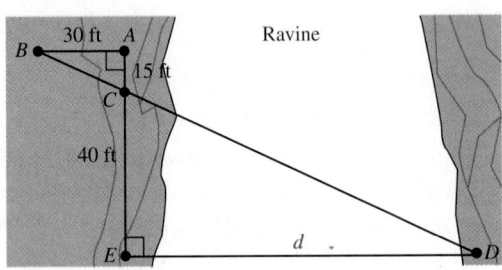

40. A ramp is made from two similar right triangles and has dimensions as shown. Find the length of the ramp to the nearest foot.

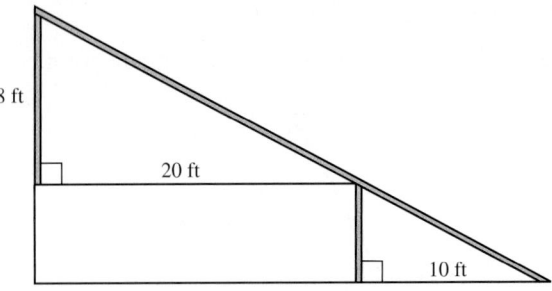

41. Frank is building a shed and wants it to look similar to his house. The roof on his house is 40 ft long and 12 ft tall at its center. The roof on the shed is to be 15 ft long. How tall will the roof be at its center, and how long is the slope of the roof?

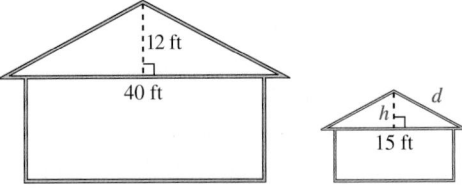

42. A 120-ft phone tower is supported by a guy wire that is attached to the top of the tower and a point on the ground that is 80 ft from the base of the tower. To take some of the weight off the wire, a 40-ft pole will support the wire from the ground. How far from the base of the tower will the pole be placed?

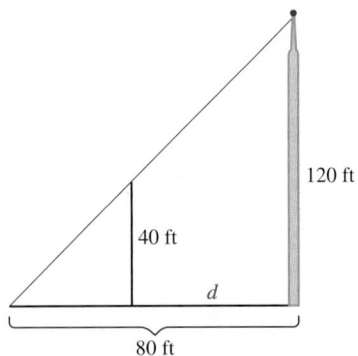

43. Divide the following figure into four similar copies of itself.

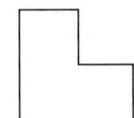

44. Use four copies of the following figure to create a figure that is similar but twice as large.

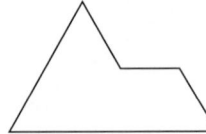

45. The Koch snowflake is a fractal that begins with a large equilateral triangle. Successive iterations are then made by placing an equilateral triangle in the middle third of each side. The first three iterations are shown.

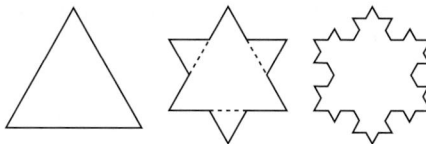

   a. How many sides will the fifth iteration have?

   b. How many sides will the tenth iteration have?

   c. How many sides will the nth iteration have?

■ Mathematical Reasoning and Proof

46. Is a scale factor of 0 possible? What about a negative scale factor?

47. If two similar shapes have a scale factor of 1, what else is true about the shapes?

48. a. Are any two scalene triangles similar? Why or why not?

   b. Are any two isosceles triangles similar? Why or why not?

   c. Are any two equilateral triangles similar? Why or why not?

49. a. Are any two acute triangles similar? Why or why not?

   b. Are any two right triangles similar? Why or why not?

   c. Are any two obtuse triangles similar? Why or why not?

   d. Are any two equiangular triangles similar? Why or why not?

50. a. Is there a Hypotenuse-Leg Theorem for triangle similarity? If so, what would it state?

   b. Is there a Hypotenuse–Acute Angle Theorem for triangle similarity? If so, what would it state?

51. Suppose that $m \parallel n \parallel p$.

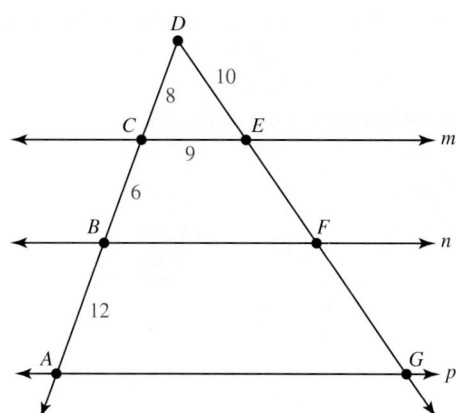

a. Find EF, FG, BF, and AG.

b. Compute the ratios $\dfrac{DC}{DE}$, $\dfrac{CB}{EF}$, and $\dfrac{BA}{FG}$. Write a conjecture about what you find.

c. Support your conjecture in part b with a deductive argument.

52. Prove that the triangle formed by joining the midpoints of the sides is similar to the original triangle. That is, show $\triangle ABC \sim \triangle PMN$.

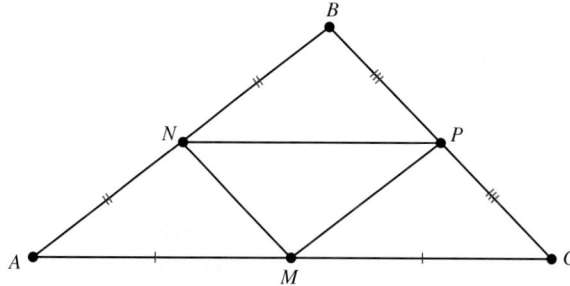

53. Write an argument showing that, if the vertex angles of two isosceles triangles are congruent, then the triangles are similar.

54. Answer each question. Be sure to explain your thinking.

   a. Will any two squares be similar?

   b. Will any two rectangles be similar?

   c. Will any two rhombuses be similar?

   d. Will any two circles be similar?

55. Is there an AAA similarity property for quadrilaterals? If so, explain. If not, give a counterexample.

56. Why must two regular n-gons be similar?

57. Are any two rectangular prisms similar? What about two cubes?

58. What conditions have to be satisfied for two right circular cylinders to be similar? What about two right circular cones or two spheres?

■ Mathematical Communication

59. Review the definition for similar triangles. Use the ideas in the definition to write a definition for similar quadrilaterals.

60. Write a set of directions explaining how to use similar triangles to measure the height of a building. Have one of your peers use them to find the height of the building in which your class meets.

61. Three houses lie along a lake. The first two houses are 50 ft apart from each another, and the second two are 100 ft apart from each another.

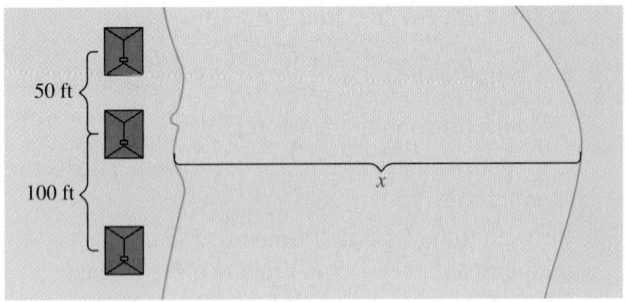

With a partner, devise a strategy that uses the houses and similar triangles to measure the distance across the lake.

62. Look up the definition of similarity in the dictionary. How does it resemble similarity for two triangles? How is it different?

63. A student wonders whether two triangles that are similar to a third are similar to each other. As the teacher, how would you help the student make sense of this situation?

64. Marcus claims that any two rectangles are similar because each pair of corresponding angles are congruent. How do you respond to his claim?

■ Building Connections

65. How does a road map or an atlas make use of similarity?

66. Overhead projectors and photocopiers are two machines that can make similar figures by magnifying or reducing the original. What are some others?

67. How are ratios and proportional reasoning related to similar triangles?

68. Do an Internet search for different ways to use similar triangles to make indirect measurements. How many different methods do you find? Describe some of them.

69. Do an Internet search for the applications of fractals. What natural phenomena can they model? What human-made phenomena can they model?

SECTION 11.3   Basic Geometrical Constructions

Until now, simple pictures have been used to represent shapes and illustrate their properties. Although these diagrams have been adequate, many geometrical situations require more precision. For instance, some situations call for finding the exact midpoint of a segment or for creating two lines that are exactly perpendicular or parallel. In such cases, we can turn to geometrical constructions for the needed precision.

A **geometrical construction** is a visual representation of a shape or a relationship that is built using a specific set of tools and geometrical properties. They are more precise than drawings because the tools and properties used in constructions guarantee that the shape or relationship is exactly what it is supposed to be. For instance, when we draw a square, it is only an approximation of a square (Figure 11.22). Its sides may not be the same length, and the angles may not be exactly 90°. However, if we carefully construct a square, then the construction guarantees that the four sides are congruent and the angles are all 90°.

Constructions date back to Plato and the ancient Greeks (429–348 B.C.E.), who insisted on making them with only two tools that each served a single, specific purpose. The **straightedge** was used to construct lines, rays, and segments between two given points. It had no marks, so it could not be used to measure lengths. The **compass** was used to draw circles and arcs of a given center and radius. The ancient Greeks used a piece of string that was fixed at one point and then rotated at another. Because the string would collapse when moved, the ancient Greeks could not use it to transfer distances. Modern compasses can be locked into place, making it possible to transfer distances and duplicate lengths. A straightedge and several different compasses are shown in Figure 11.23.

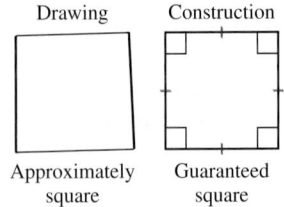

Figure 11.22 A drawing versus a construction

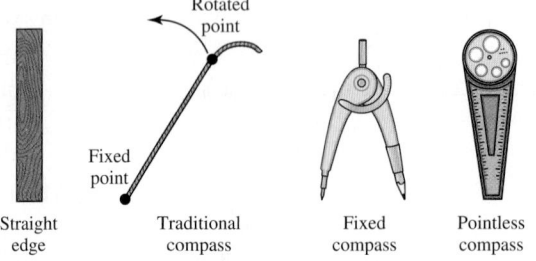

Figure 11.23 Construction tools

For the most part, we keep to tradition and use only these tools. However, in the classroom, children also learn to make constructions with protractors, plastic reflectors (Figure 11.24), paper folding, and geometry software. These tools enable us to construct a surprising number of shapes and concepts. We encourage you to work through each construction on a separate sheet of paper. We begin with line segments.

Explorations Manual 11.6

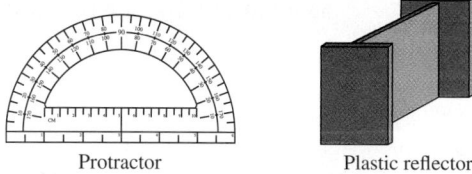

Protractor   Plastic reflector

Figure 11.24 A protractor and plastic reflector

Constructing Congruent Segments

Because most geometrical shapes are built from line segments, we must first learn how to construct congruent segments in order to construct other shapes precisely. One way is to measure the segment with a ruler and then transfer the measure to another segment. We can also use a fixed compass to transfer distances from one line to another.

Construction 1

Construct a line segment congruent to a given segment.

Step 1: *Draw a ray.* Let segment $\overline{AB}$ be given (Figure 11.25). Using a straightedge, draw $\overrightarrow{CE}$ on which the congruent segment can be constructed.

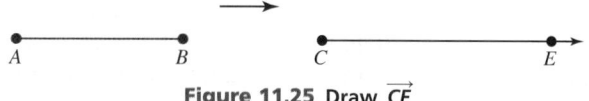

Figure 11.25 Draw $\overrightarrow{CE}$

Step 2: *Measure the given segment.* "Measure" the length of $\overline{AB}$ by placing the point of the compass on A and the tip of the pencil on B (Figure 11.26).

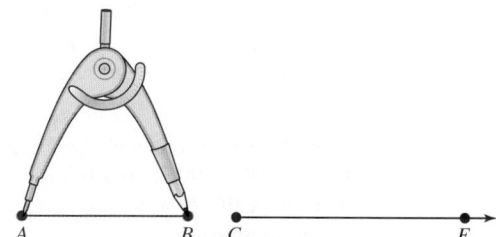

Figure 11.26 Measure the length of $\overline{AB}$

Step 3: *Construct the congruent segment.* Without changing the compass, place the point of the compass on C and draw an arc that intersects $\overrightarrow{CE}$ (Figure 11.27). The intersection of the arc and $\overrightarrow{CE}$ locates a point D such that $\overline{AB} \cong \overline{CD}$.

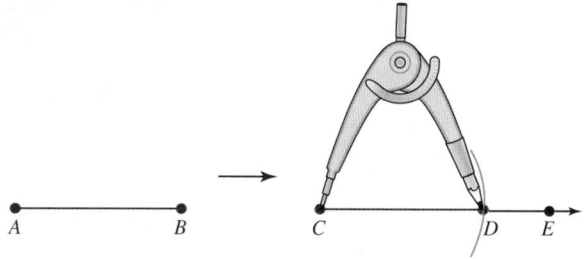

Figure 11.27 Copy $\overline{AB}$ onto $\overrightarrow{CE}$

Example 11.18 **Constructing a Segment Three Times the Length of a Given Segment**

Construct a segment that is three times the length of $\overline{AB}$.

Solution We first use the straightedge to construct a long ray. We then use a compass to measure $\overline{AB}$ and construct a sequence of three arcs as shown.

$\overline{CD}$ is three times the length of $\overline{AB}$.

Constructing Congruent Angles

Angles are another key building block of geometrical shapes. In general, no straight-edge-and-compass construction allows us to construct an angle of any particular measure. Instead, we need a protractor. However, a straightedge and compass can be used to construct congruent angles. To do so, we construct an isosceles triangle on the given angle and then transfer the distances onto a ray to build a new, yet congruent angle.

Construction 2

Construct an angle congruent to a given angle

Step 1: *Draw a ray.* Let $\angle B$ be given. Using a straightedge, draw a ray with endpoint D. This gives us the vertex and one side of the new angle (Figure 11.28).

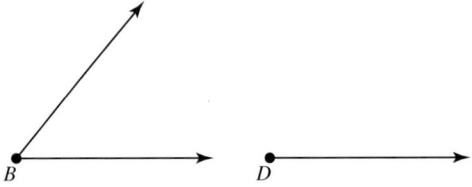

Figure 11.28 Draw a ray

Step 2: *Make and copy an arc.* Set the compass to a convenient length, and place its point on B (Figure 11.29). Make an arc that intersects both sides of $\angle B$. Label the points of intersection A and C. Next, without changing the compass, place its point on D, and draw a large arc that intersects the ray. Label the point of intersection, E.

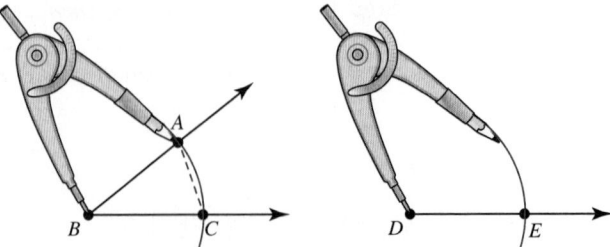

Figure 11.29 Draw an arc

This step of the construction creates an isosceles triangle on $\angle B$, which we will copy onto $\overrightarrow{DE}$. Points D and E are two vertices of the new triangle. In the next step, we find the third.

Step 3: *Measure and transfer a distance.* Use the compass to measure the distance from *A* to *C* (Figure 11.30). Without changing the compass, place its point on *E*, and make an arc that intersects the arc from step 2. Label the point of intersection *F*.

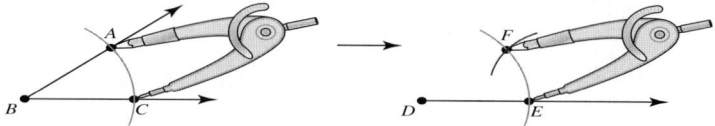

Figure 11.30 Measure the angle

Step 4: *Draw the second side of the angle.* Complete the construction by drawing $\overrightarrow{DF}$ (Figure 11.31).

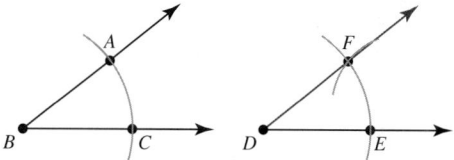

Figure 11.31 Draw $\overrightarrow{DF}$

Because we carefully transferred distances, we have $\overline{BA} \cong \overline{DF}$, $\overline{BC} \cong \overline{DE}$, and $\overline{AC} \cong \overline{FE}$. Consequently, $\triangle ABC \cong \triangle FDE$ by the SSS property, and $\angle B \cong \angle D$ because corresponding parts of congruent triangles are congruent.

Example 11.19 **Constructing an Angle Equal to the Sum of Two Angles**

Construct an angle with a measure equal to $m\angle A + m\angle B$.

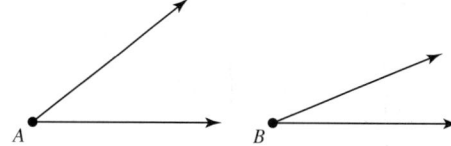

Solution We first draw a ray with an endpoint at *C* and use Construction 2 to make a copy of $\angle A$ on the ray.

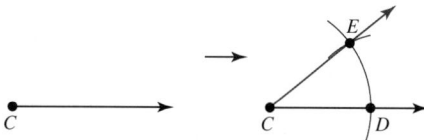

Next, we use Construction 2 again to make a copy of $\angle B$ on $\overrightarrow{CE}$. Doing so gives $\angle FCD$ a measure that is equal to $m\angle A + m\angle B$.

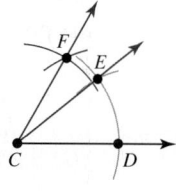

Constructing Angle Bisectors

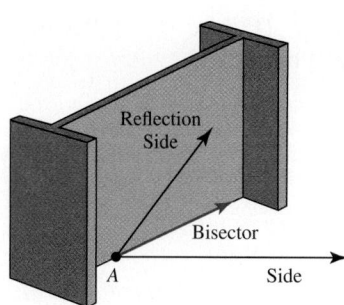

Figure 11.32 An angle bisector with a plastic reflector

Given any angle, we can construct its bisector by using one of two methods. One is to use a **plastic reflector** (Figure 11.32). A plastic reflector not only reflects the image of an object, but it also allows us to see through to the other side. Place the edge of the reflector on the vertex of the angle and then rotate it until the reflection of one side is on top of the other. The edge then forms the angle bisector.

Another method is to use a straightedge and compass. In this case, first we construct a rhombus on the given angle and then a ray along its diagonal. Because the diagonal of a rhombus bisects opposite angles, the constructed ray must bisect the angle.

Construction 3

Construct the bisector of a given angle.

Step 1: *Make an arc.* Let $\angle B$ be given. Set the compass to a convenient length, place its point on B, and make an arc that intersects both sides of the angle (Figure 11.33). Label the points of intersection A and C.

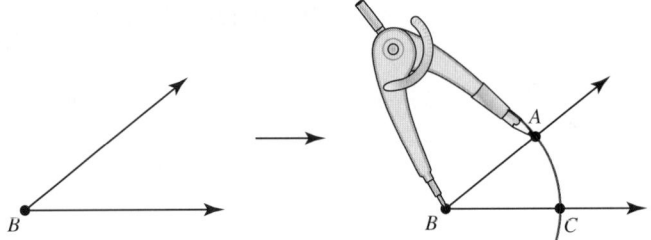

Figure 11.33 Make an arc

In the first step, we have begun to construct a rhombus. We have three of the four vertices, A, B and C, and the construction guarantees that $\overline{AB} \cong \overline{BC}$.

Step 2: *Find the intersection of two arcs.* Without changing the compass, place its point on A, and make an arc in the interior of the angle (Figure 11.34). Repeat the process at C so that the arcs intersect in the interior of the angle. Label the point of intersection D.

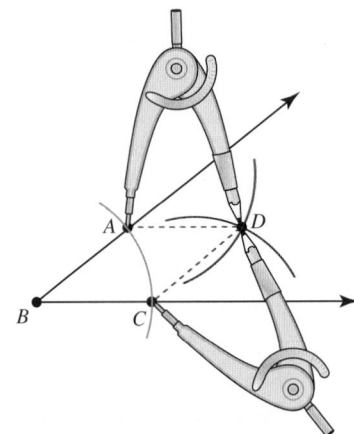

Figure 11.34 Find the intersection of two arcs

The second step locates the fourth vertex of a quadrilateral. Not having changed the opening of our compass, we know that $\overline{AB} \cong \overline{BC} \cong \overline{CD} \cong \overline{DA}$. As a result, the quadrilateral must be a rhombus.

Step 3: *Draw the bisector.* Use a straightedge to draw $\overrightarrow{BD}$ (Figure 11.35). Because $\overrightarrow{BD}$ lies along the diagonal of a rhombus, it must bisect $\angle B$.

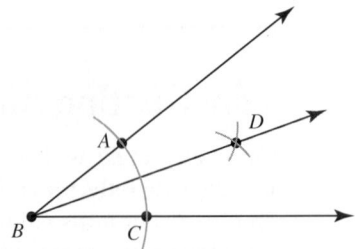

Figure 11.35 Draw $\overrightarrow{BD}$

Example 11.20 **Angle Bisectors in a Triangle**

Construct the angle bisectors of the three angles in any triangle △*ABC*. What appears to be true about the resulting rays?

Solution We first construct a triangle △*ABC* and then use Construction 3 to construct the angle bisector of every angle.

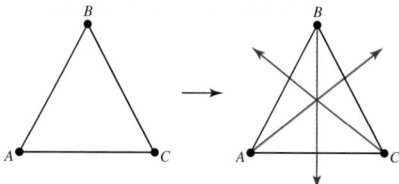

Notice that the bisectors intersect at a common point. Because our triangle was randomly chosen, it makes sense that this might hold true for all triangles. One way to test this conjecture is with geometry software. Having constructed the figure in the program, we can click and drag on any vertex to change the shape of the triangle. In this way, we can test hundreds of triangles in a few seconds. Three example triangles are shown, and in each case, the angle bisectors always intersect at a point.

Explorations Manual 11.7

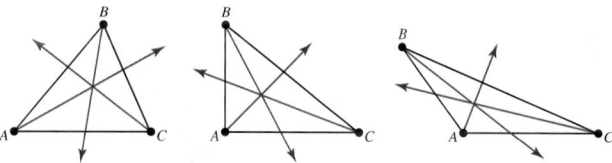

The last example shows that the angle bisectors in a triangle intersect at a common point. This point is called the **incenter** of the triangle (Figure 11.36). It is the center of the triangle's **incircle,** which is the largest circle that fits within the triangle so that the triangle's sides are tangent to the circle. In this case, the circle is said to be **inscribed** in the triangle. The construction of an incircle relies on perpendicular lines and is left to the Learning Assessment.

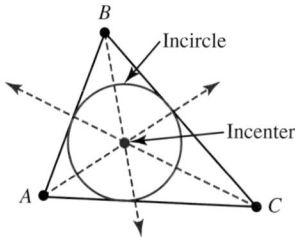

Figure 11.36 An incircle and its incenter

Check Your Understanding 11.3-A

1. Use a straightedge and compass to copy each segment.
  a. A •——• B
  b. C •———————• D

2. Construct a segment that is two times the length of segment $\overline{AB}$ in Problem 1.

3. Use a straightedge and compass to copy each angle onto another sheet of paper. Then construct the angle bisectors.

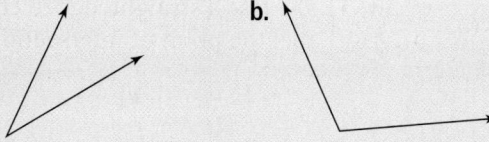

a.   b.

Talk About It Can another quadrilateral be used to construct an angle bisector? If so, what is it, and how does changing the quadrilateral change Construction 3?

Representation

Activity 11.5
Draw an angle on a sheet of paper. Describe how you can fold the paper to construct the angle bisector?

Constructing Parallel Lines

We can now use the first three constructions and some of the content learned in Chapter 10 to construct parallel lines in different ways. The next example shows how to adapt Construction 3 to construct a line parallel to a given line through a point not on the line.

Example 11.21 **Constructing a Parallel Line Through a Given Point**

Use a rhombus to construct a line parallel to $\overleftrightarrow{AB}$ through the point P.

Solution The opposite sides of a rhombus are parallel, so we can complete our task by constructing a rhombus in which one side lies on $\overleftrightarrow{AB}$ and the opposite side passes through P. First, construct $\overrightarrow{AP}$ to form $\angle PAB$. Next, use a compass to measure the distance from A to P. Without changing its opening, follow the first two steps of Construction 3 to find two other vertices of the rhombus. Label them C and D.

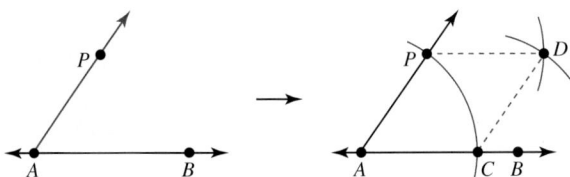

Finally, use the straightedge to construct $\overleftrightarrow{PD}$, which must be parallel to $\overleftrightarrow{AB}$ because they lie along opposite sides of a rhombus.

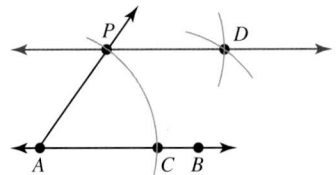

In the last example, we drew $\overrightarrow{AP}$ to form a rhombus. We can also construct parallel lines by treating $\overrightarrow{AP}$ as a transversal. We can then create the parallel line by constructing an alternate interior angle, an alternate exterior angle, or a corresponding angle.

Construction 4

Construct a line parallel to a given line through a point not on the line.

Step 1: *Draw a line.* Let $\overleftrightarrow{AB}$ and point P be given such that P is not on $\overleftrightarrow{AB}$ (Figure 11.37). Use a straightedge to construct a line through P that intersects $\overleftrightarrow{AB}$. Label the point of intersection C.

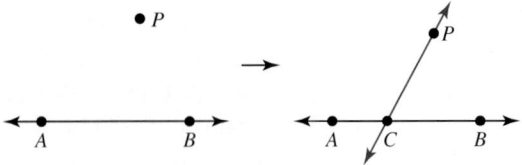

Figure 11.37 Draw $\overleftrightarrow{PC}$

Next, we treat $\overleftrightarrow{PC}$ as a transversal and construct an alternate interior angle on it congruent to $\angle PCB$.

Step 2: *Copy an angle.* Use Construction 2 to copy $\angle PCB$ onto $\overleftrightarrow{PC}$ where P is the vertex of the new angle. Label the new point D and draw $\overleftrightarrow{PD}$ (Figure 11.38).

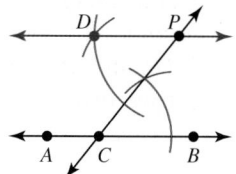

Figure 11.38 Copy ∠*PCB* onto $\overleftrightarrow{PC}$

Because $\angle PCB$ and $\angle CPD$ are congruent alternate interior angles, the construction guarantees that $\overleftrightarrow{PD}$ is parallel to $\overleftrightarrow{AB}$.

Constructing Perpendicular Lines

Perpendicular lines can also be constructed in a number of ways, most of which are based on finding the perpendicular bisector of a segment. For instance, we can place a plastic reflector on a segment and then rotate and slide it until the reflection of one endpoint is directly on top of the other (Figure 11.39). The perpendicular bisector of the segment is then given by the edge of the reflector.

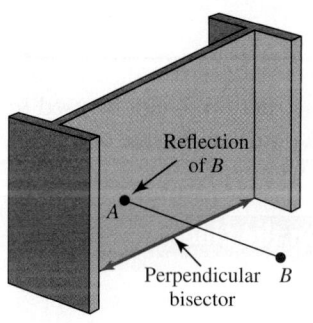

Figure 11.39 A perpendicular bisector with a plastic reflector

If we use a straightedge and compass, we again rely on a rhombus for help. Specifically, we treat the given segment as the diagonal of a rhombus and then construct a line along the other diagonal. Because the diagonals of a rhombus are perpendicular bisectors of each other, the constructed line must bisect and be perpendicular to the given segment.

Explorations
Manual
11.8

Construction 5

Construct the perpendicular bisector of a given segment.

Step 1: *Draw intersecting arcs of the same radius.* Let $\overline{AB}$ be given (Figure 11.40). Open the compass to a length that is greater than half that of the line segment. Construct arcs with centers at A and B so that the arcs intersect above and below the segment. Label the points of intersection C and D.

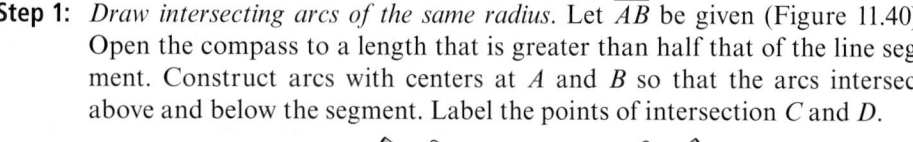

Figure 11.40 Construct intersecting arcs

In this step, we have used A and B to find two other vertices of a quadrilateral. Because we did not change the opening, we know that $\overline{AC} \cong \overline{BC} \cong \overline{BD} \cong \overline{DA}$. Consequently, the quadrilateral is a rhombus, and $\overline{AB}$ is one of its diagonals.

Step 2: Draw a line. Draw the line from C to D (Figure 11.41). Because $\overleftrightarrow{CD}$ lies along the diagonal of a rhombus, it must bisect and be perpendicular to $\overline{AB}$.

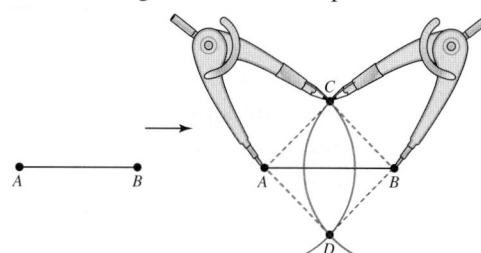

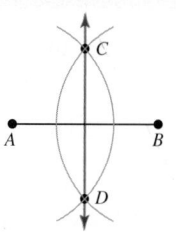

Figure 11.41 Draw $\overleftrightarrow{CD}$

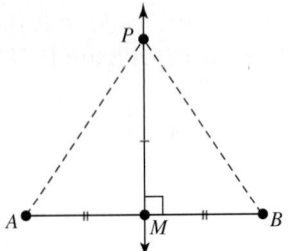

Figure 11.42 *P* is equidistant from *A* and *B*

At this point, consider two useful facts. First, because Construction 5 results in the perpendicular bisector of a segment, we can use it to find the segment's midpoint. Second, every point on the perpendicular bisector must be equidistant from the endpoints. To understand why, suppose *P* is any point on the perpendicular bisector of $\overline{AB}$, and let *M* be the midpoint (Figure 11.42). If we draw segments $\overline{AP}$ and $\overline{BP}$, we have two triangles: $\triangle AMP$ and $\triangle BMP$. The diagram shows that $\overline{AM} \cong \overline{MB}$, $\angle AMP \cong \angle BMP$, and $\overline{PM}$ is congruent to itself. Consequently, $\triangle AMP \cong \triangle BMP$ by the SAS property. Because corresponding parts of congruent triangles are congruent, $\overline{AP} \cong \overline{BP}$. This indicates that every point on the perpendicular bisector of a segment must be equidistant from both endpoints.

Theorem 11.6   Perpendicular Bisector of a Segment

A point is on the perpendicular bisector of a segment if and only if it is equidistant from both endpoints.

This theorem, along with some small adjustments to Construction 5, can be used to construct a line perpendicular to a given line, given either a point on the line or a point not on the line. The next two examples demonstrate both techniques.

Example 11.22   Constructing a Perpendicular Line Through a Point on the Line

Construct a line perpendicular to *l* that passes through *P*.

Solution We are given a line *l* and a point *P* on *l*. If possible, we want to apply Construction 5, but we need the endpoints of a segment to do so. To create the segment, set the compass at a reasonable distance, and strike two arcs on either side of *P* that not only intersect *l* but that are also equidistant to *P*. Label the points of intersection *A* and *B*.

Next, we use Construction 5 to construct the perpendicular bisector of $\overline{AB}$. Because *A* and *B* are equidistant from *P*, Theorem 11.5 guarantees that *P* is on the perpendicular bisector of $\overline{AB}$, giving us the desired line.

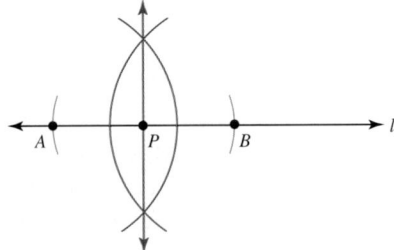

Example 11.23   The Shortest Distance Between a Point and a Line

Consider line *l* and point *P*.

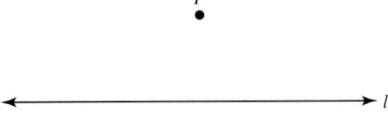

Find the point on *l* that is nearest to *P*.

Solution The point on *l* that is nearest to *P* is the intersection of *l* and the line perpendicular to *l* that runs through *P*. To find the perpendicular line, we would again like to use Construction 5 but do not have the endpoints of a segment. We can find the endpoints by setting the compass at a reasonable distance, placing its point on *P*, and striking two arcs that intersect *l*. We label the points of intersection *A* and *B*.

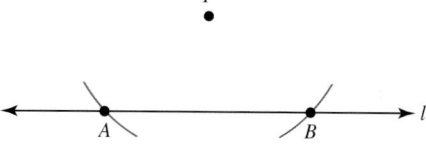

Explorations
Manual
11.9

Next, we use Construction 5 to construct the perpendicular bisector of $\overline{AB}$. As in the last example, *A* and *B* are equidistant from *P*, so Theorem 11.5 guarantees that *P* is on the perpendicular bisector of $\overline{AB}$. Point *C*, which is the intersection of line *l* and the perpendicular bisector, is the point on *l* that is closest to *P*.

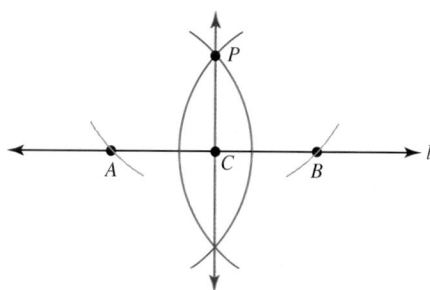

Check Your Understanding 11.3-B

1. Construct a pair of parallel lines using alternate exterior angles.

2. Copy the following segment, and then construct its perpendicular bisector.

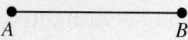

Talk About It How can you use Construction 5 to construct a line parallel to a given line?

Representation

Activity 11.6

a. On a sheet of paper, draw a line and a point not on the line. How can you fold the paper to construct a line that passes through the point and that is perpendicular to the given line?

b. Draw a line segment on a sheet of paper. How can you fold the paper to construct the perpendicular bisector of the segment?

SECTION 11.3 Learning Assessment

■ Understanding the Concepts

1. What is the difference between a drawing and a geometrical construction?

2. What roles do a straightedge and compass play in making geometrical constructions?

3. How are congruent triangles used to construct two congruent angles?

4. How is a rhombus used to construct:

   a. An angle bisector?　**b.** Parallel lines?

   c. Perpendicular lines?

5. What does it mean to inscribe a circle?

6. Describe two different ways to construct parallel lines.

7. How can a construction be used to find the midpoint of a segment?

Exercises

8. Construct a segment congruent to each segment.

   a.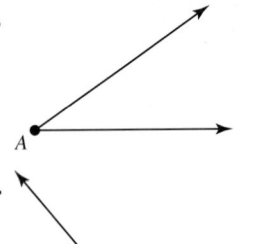
   A　B

   b. C ─────────── D

9. Construct an angle congruent to each angle.

   a.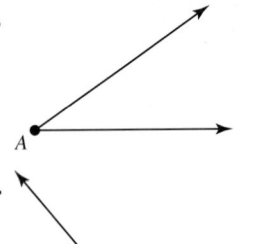
   A

   b.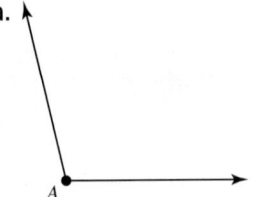
   B

10. Copy each angle onto another sheet of paper. Use a plastic reflector to construct the angle bisector of each angle.

    a.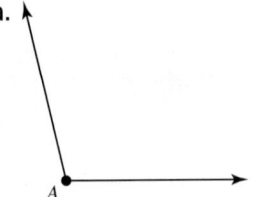
    A

    b. B

11. Repeat the previous exercise using a straightedge and compass.

12. Use a rhombus to construct the line parallel to $\overleftrightarrow{AB}$ that passes through P.

    P
    •

    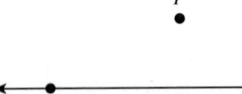
    A　　B

13. Use alternate interior angles to construct a line parallel to $\overleftrightarrow{AB}$ that passes through P.

    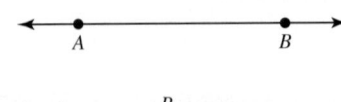
    A　　B

    P
    •

14. Copy each segment onto another sheet of paper, and then use a plastic reflector to construct the segment's perpendicular bisector.

    a. •────────────•
    A　　　　　B

    b. •────────────•
    C　　　　　D

15. Repeat the previous exercise using a straightedge and compass.

16. Use a plastic reflector to construct a line perpendicular to $\overleftrightarrow{AB}$ that passes through P.

    a.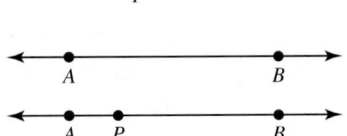
    A　　　　P　B

    b.
    A　　　　　　B

    P
    •

17. Use a straightedge and compass to construct a line perpendicular to $\overleftrightarrow{AB}$ that passes through P.

    a.
    P
    •
    A　　　　　　B

    b.
    A　P　　　B

Problems and Applications

18. Consider $\overline{AB}$ and $\overline{CD}$. Construct a segment for each given length.

    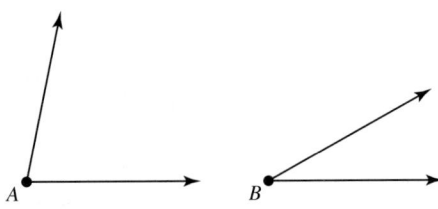
    A　　B　C　　　　D

    a. $AB + CD$

    b. $CD - AB$

    c. $4AB$

    d. $\dfrac{AB + CD}{2}$

19. Suppose $\overline{AB}$ has a length of 1.

    •────────•
    A　　B

    Use it to construct a segment of each length.

    a. $\sqrt{2}$　　b. $\sqrt{5}$　　c. $\sqrt{13}$

20. Consider $\angle A$ and $\angle B$.

    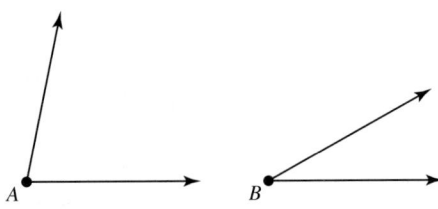
    A　　　　　　B

    Construct an angle for each given measure.

    a. $m\angle A + m\angle B$

    b. $m\angle A - m\angle B$

    c. $2m\angle B$

    d. $\dfrac{m\angle A + m\angle B}{2}$

21. Consider $\angle A$.

    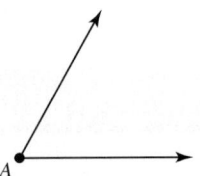
    A

    a. Construct $\angle B$ so that $\angle A$ and $\angle B$ are supplementary.

    b. Construct $\angle C$ so that $\angle A$ and $\angle C$ are complementary.

22. Use a plastic reflector to construct an angle of each measure.

    a. 45°   **b.** 135°   **c.** 112.5°

23. Use a straightedge and compass to construct an angle of each measure.

    a. 45°   **b.** 22.5°   **c.** 157.5°

24. Use a construction to divide ∠A into four congruent angles.

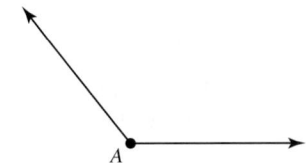

25. Use alternate exterior angles to construct the line parallel to $\overleftrightarrow{AB}$ that passes through P.

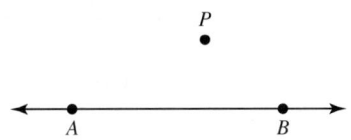

26. Use corresponding angles to construct the line parallel to $\overleftrightarrow{AB}$ that passes through P.

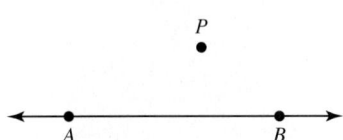

27. Copy each triangle, and then find its incenter. Next, find the radius of the incircle by constructing a segment perpendicular to a side. Use the radius to construct the incircle.

a.

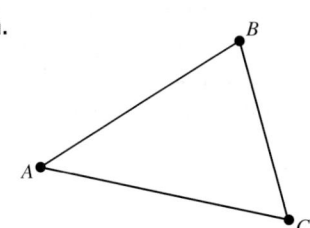

b.

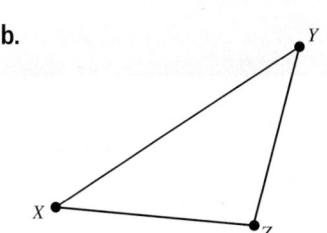

28. Use a construction to divide $\overline{AB}$ into four congruent segments.

29. Construct a rhombus with vertices at A and P.

30. Use a straightedge to draw a triangle. Construct a point that is equidistant from each of its vertices.

■ Mathematical Reasoning and Proof

31. Construct a line *l* that is perpendicular to $\overleftrightarrow{AB}$ that passes through P. Next, construct a line *m* that is perpendicular to *l* that passes through P. What is true about $\overleftrightarrow{AB}$ and *m*? Justify your reasoning.

32. A **median** of a triangle is a line segment that connects a vertex to the midpoint of the opposite side. Use a straightedge to draw a triangle, and then construct its three medians. What appears to be true about these lines? Use geometry software to test your conjecture for more triangles. Does it hold true?

33. An **altitude** of a triangle is a line segment that extends from a vertex and that is perpendicular to the line containing the opposite side.

    a. Will an altitude of a triangle always be inside the triangle? Explain.

    b. Use a straightedge to draw a triangle and then construct the altitude at each vertex. What appears to be true about these lines? Use geometry software to test your conjecture for more triangles. Does it hold true?

34. Use a straightedge to draw a triangle, and then construct the perpendicular bisector of each side. What appears to be true about these lines? Use geometry software to test your conjecture for more triangles. Does it hold true?

35. Draw two lines that intersect to form two pairs of vertical angles. Construct the lines that bisect each pair of vertical angles. What do you notice about the constructed lines? Is this true for any two lines? Use geometry software to test your conjecture.

36. Use your compass to construct a circle. Next, use a straightedge to draw two chords of the circle that are not diameters. Construct the perpendicular bisector of each chord. Where do the bisectors of the chords seem to intersect?

37. Drafters often use a straightedge and drafting triangle to construct a line parallel to a given line through a given point. To do so, they place the straightedge and triangle so that the straightedge is on the point, one side of the triangle is on the straightedge, and another is on the line. Without moving the straightedge, they slide the triangle up to the point and draw the needed line. Explain why this works.

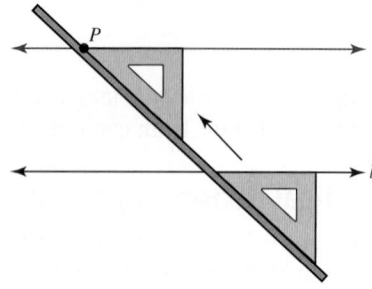

38. Draw any triangle. Find the midpoints of the three sides, and construct the segments between them. What is true about the resulting triangle? Use geometry software to test your conjecture.

39. Use congruent triangles and the diagram to show why Construction 3 bisects an angle.

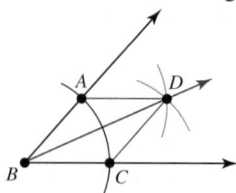

40. Draw any triangle, and then construct a point that is equidistant from the three sides of the triangle. Explain why your construction enables you to find the indicated point.

41. Theorem 11.6 states that any point equidistant from both endpoints of a segment must be on its perpendicular bisector. Write an argument to justify this fact.

Mathematical Communication

42. Consider a segment $\overline{AB}$. Write a set of directions that explains how to construct a circle of radius $\overline{AB}$ at a given point P.

43. Can a plastic reflector be used to copy a segment? An angle? Discuss the question with a group of your peers, and, if it is possible, write a set of directions to explain how.

44. Write a set of directions that explains how to use a plastic reflector to construct a line m that is perpendicular to another line l through a point
   a. On line l.   **b.** Not on line l.

45. What are the advantages and disadvantages of using a plastic reflector instead of a straightedge and compass?

46. Audrey thinks that she can use a square to construct an angle bisector. Is she correct?

Building Connections

47. How are geometrical constructions connected to the idea of congruence? How do congruent shapes help us perform constructions? How do constructions guarantee that two shapes are congruent?

48. Ryan wants to construct a circular patio that has a diameter of 20 ft. Describe a method that he might use to initially mark off the circle for the patio.

49. Louise lives on a corner lot. She would like to build a sidewalk from the corner of the lot to her front steps so that the sidewalk follows along the angle bisector. Describe a method that she might use to initially mark off the angle bisector.

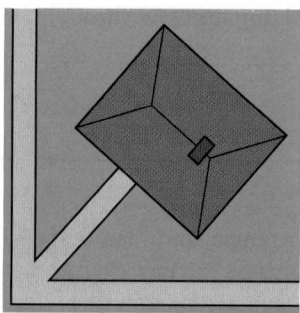

50. Look through a set of elementary curriculum materials. At what grade level are constructions introduced? What constructions are done? Do the curriculum materials provide any explanation as to why the constructions work? If not, how might this affect students' attitudes toward constructions?

SECTION 11.4   Constructing Shapes

With the constructions from the last section, we can build a wide variety of shapes. To do so, we must first recognize the basic characteristics of the shapes and then apply the constructions appropriately. We begin with triangles.

Constructing Triangles

Because a triangle is completely defined by its sides and angles, we can use the basic constructions to construct triangles in different ways. For instance, the next example demonstrates how to construct one triangle congruent to another.

Example 11.24 **Constructing Congruent Triangles**

Problem Solving

Construct a triangle congruent to △*ABC*.

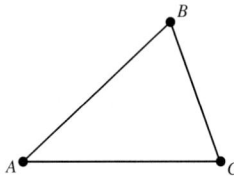

Solution

Understanding the Problem. Our task is to construct a triangle congruent to △*ABC*. We have the five basic constructions from Section 11.3 to do so.

Devising a Plan. To construct a triangle congruent to △*ABC*, we must find the vertices of a new triangle so that its sides and angles correspond and are congruent to the sides and angles of △*ABC*. Because Constructions 1 and 2 allow us to copy segments and angles, we can use them, along with a *triangle congruence property*, to find the needed vertices. Specifically, if we use Construction 1 to copy the three sides of △*ABC*, the SSS property guarantees that the new triangle is congruent to △*ABC*.

Carrying Out the Plan. We first copy $\overline{AC}$ onto a ray and label the endpoints of the new segment *D* and *E*. This gives us two of the three vertices.

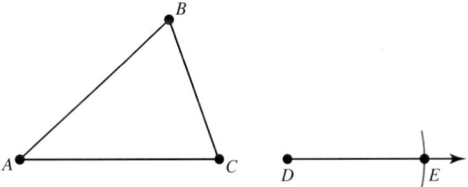

Next, we measure the length of $\overline{AB}$ with the compass. Without changing the opening, place the compass point on *D* and swing an arc above $\overline{DE}$. Repeat the process with $\overline{BC}$ by swinging an arc from point *E* so that it intersects the first arc. This results in the third vertex, *F*. We complete the triangle by drawing $\overline{DF}$ and $\overline{EF}$.

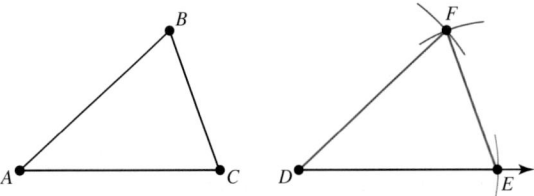

Because we did not change the opening of the compass when measuring each segment, $\overline{AB} \cong \overline{DF}$, $\overline{BC} \cong \overline{FE}$, and $\overline{AC} \cong \overline{DE}$. As a result, △*ABC* ≅ △*DFE* by the SSS triangle congruence property.

Looking Back. We could have used other congruence properties to construct △*DFE*. For instance, we could have copied side $\overline{AC}$ onto a ray, copied ∠*A* at point *D*, and then side $\overline{AB}$ on the new ray extending from *D*. Because Constructions 1 and 2 guarantee that $\overline{AC} \cong \overline{DE}$, ∠*A* ≅ ∠*D*, and $\overline{AB} \cong \overline{DF}$, then △*ABC* ≅ △*DFE* by the SAS property.

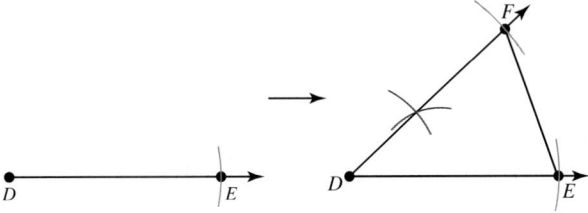

We can use the techniques from the last example to perform other constructions with triangles. For instance, to construct an equilateral triangle, we begin with a segment $\overline{AB}$ (Figure 11.43). Because the three sides of an equilateral triangle are the same length, the third vertex must be equidistant from A and B. To find it, measure $\overline{AB}$ with the compass. Then, without changing the opening, place its point on A and strike an arc above $\overline{AB}$. Again without changing the opening of the compass, place the point on B and strike another arc so that it intersects the first. Complete the triangle by drawing segments from A and B to the intersection of the two arcs.

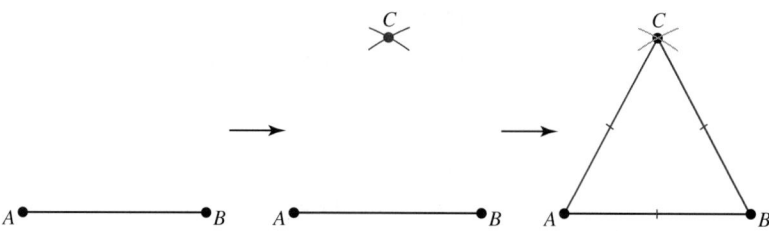

Figure 11.43 Constructing an equilateral triangle

Example 11.25 | **Constructing Similar Triangles**

Construct a triangle similar to $\triangle ABC$ but twice as large.

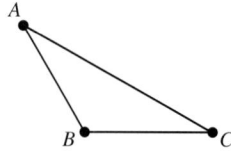

Solution As before, we use Constructions 1 and 2 to copy segments and angles in a way that corresponds to a triangle similarity property. First, use Construction 1 two times to construct a segment on a ray that is twice as long as $\overline{BC}$. Doing so results in a scale factor of 2 and locates two vertices of the triangle. Label the points D and E.

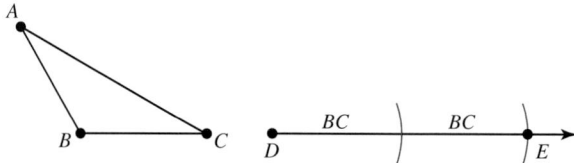

Next, use Construction 2 to copy $\angle B$ at point D and $\angle C$ at point E. The intersection of the rays extending from D and E locates the third vertex of the triangle, F. Because the construction guarantees that $DE = 2BC$, $\angle B \cong \angle D$, and $\angle C \cong \angle E$, then $\triangle DEF$ is twice the size of $\triangle ABC$, and $\triangle ABC \sim \triangle DEF$ by the AA property.

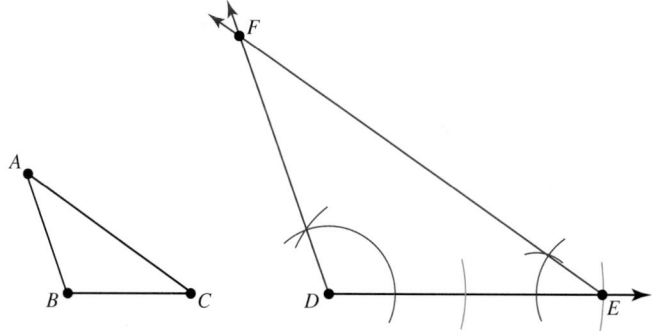

Constructing Quadrilaterals

Explorations
Manual
11.10

Not only can we construct congruent and similar shapes, but we can also construct polygons using segments of given lengths and angles of given measures. The next two examples demonstrate how to do so with quadrilaterals.

| **Example 11.26** | **Constructing a Rhombus with a Given Side and Angle** |

Construct a rhombus with sides congruent to $\overline{AB}$ and one pair of opposite angles congruent to $\angle T$.

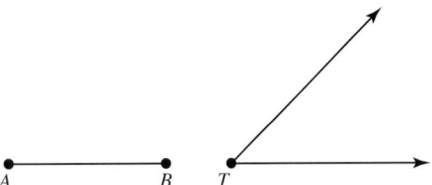

Solution To construct the rhombus, we need to find its vertices. Because two of the angles in the rhombus are congruent to $\angle T$, we first use Construction 2 to make a copy of $\angle T$, and we label the vertex of the new angle P. Next, we measure $\overline{AB}$ with our compass, place the point on P, and strike an arc that intersects both sides of $\angle P$. We label the points of intersection Q and R. This gives us three of the four vertices.

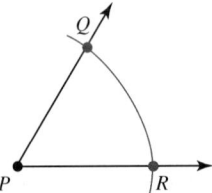

To find the fourth vertex, leave the compass at its setting and strike two arcs that intersect inside $\angle P$, one from Q and one from R. Label the new point S, and construct segments $\overline{QS}$ and $\overline{RS}$. Because we did not change the compass, the construction guarantees that $\overline{PQ} \cong \overline{QS} \cong \overline{SR} \cong \overline{RP} \cong \overline{AB}$, making quadrilateral $PQSR$ a rhombus with sides congruent to $\overline{AB}$. In addition, $\angle P \cong \angle S \cong \angle T$, giving the rhombus one pair of opposite angles congruent to $\angle T$.

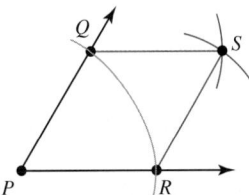

| **Example 11.27** | **Constructing a Square with a Given Side** |

Construct a square with sides congruent to $\overline{AB}$.

Solution We begin the construction by drawing a line $\overleftrightarrow{DE}$ and picking a point X on the line. We then construct a line perpendicular to $\overleftrightarrow{DE}$ through X. This results in one vertex of the square and one of its right angles.

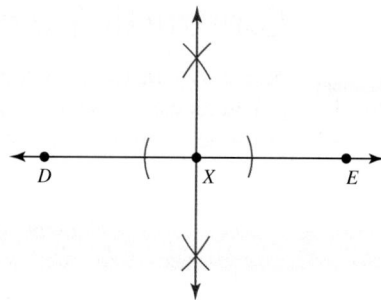

Next, we find two more vertices by measuring $\overline{AB}$ with the compass, placing the point on X, and then striking arcs that intersect both sides of the right angle. We label the points of intersection W and Y.

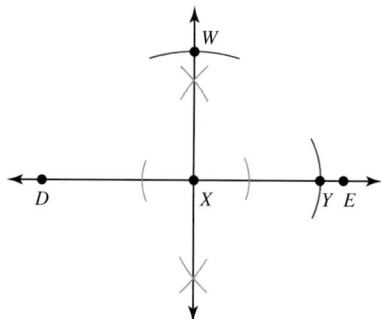

To find the fourth vertex, leave the compass at its setting, and strike two arcs that intersect, one from W and one from Y. Label the new point Z, and then construct the segments $\overline{WZ}$ and $\overline{YZ}$. Because we did not change the compass, the construction guarantees that $\overline{XW} \cong \overline{WZ} \cong \overline{ZY} \cong \overline{YX} \cong \overline{AB}$, making quadrilateral $XWZY$ a rhombus with sides congruent to $\overline{AB}$. However, the initial step in the construction gave quadrilateral $XWYZ$ a right angle, so it must be a square.

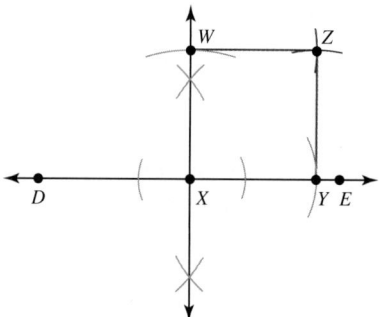

Check Your Understanding 11.4-A

1. Use the *SAS* congruence property to construct a triangle congruent to $\triangle ABC$.

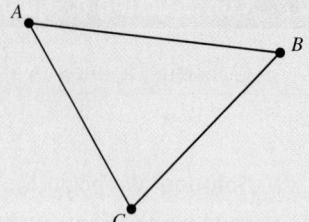

2. Using $\triangle ABC$ from the previous exercise, construct a triangle that is similar but three times larger.

3. Construct a rectangle with a length twice AB and a width equal to AB.

Talk About It Can the angle-angle similarity theorem be used to construct one triangle that is similar to another? Explain.

Activity 11.7
The triangle explored in this activity was discovered by Napoleon, who used geometry to train his mind to think analytically. Draw a triangle, and then construct an equilateral triangle on each of its sides. Next, construct the medians of each equilateral triangle. A **median** is a segment that connects a vertex to the midpoint of the opposite side. The point where the three medians meet is called the **centroid.** Once you have the three centroids, construct the segments between them to form Napoleon's triangle. What is true about this triangle? Test your conjecture with geometry software.

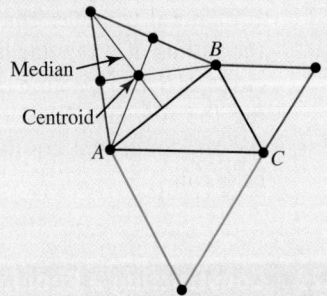

Explorations Manual 11.11

Constructing Regular Polygons

Regular polygons can be constructed in at least two ways.

Example 11.28 Constructing a Regular Hexagon with a Given Side

Construct a regular hexagon with sides congruent to $\overline{AB}$.

Solution To construct a regular hexagon, we must first find its six vertices. Measure $\overline{AB}$ with the compass, and then, without changing the opening, construct a circle with a center at A.

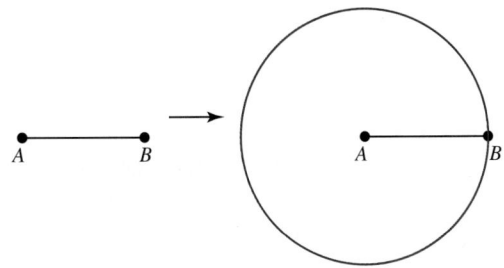

Again, without changing the opening, place the point of the compass on *B*, and strike an arc that intersects the circle. Construct four more arcs, each of which is made by placing the point of the compass on the intersection of the previous arc and the circle. Complete the hexagon by drawing the segments that connect consecutive points.

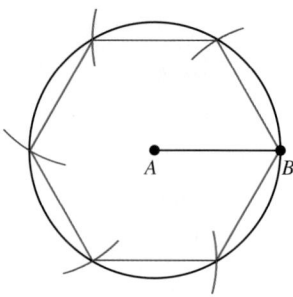

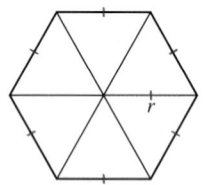

Figure 11.44 Six equilateral triangles make a regular hexagon

To make sense of the construction in the last example, recall that the central angle of any regular hexagon is $\frac{360°}{6} = 60°$. So we can build a regular hexagon by using six equilateral triangles that share a common vertex (Figure 11.44). This indicates that the radius of a regular hexagon is congruent to each of its sides. As the radius of the circle, $\overline{AB}$ also became the radius of the regular hexagon. Consequently, in constructing the five arcs from point *B*, we were actually constructing the outer vertices of the six congruent equilateral triangles. Once the vertices were connected, we had a hexagon.

Example 11.29 | Constructing a Regular Octagon

Construct a regular octagon.

Solution Recall that the central angle of any regular octagon is $\frac{360°}{8} = 45°$. Because 45° is half of 90°, we can construct the eight central angles of a regular octagon by bisecting the right angles formed by two perpendicular lines. First, draw a line, select a point *P* on the line, and construct a perpendicular line through it. Then use Construction 3 to bisect the right angles. This results in eight 45° angles that all share a common vertex at *P*.

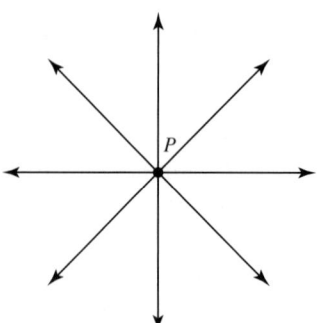

Next, recall that each vertex of a regular polygon is equidistant from its center. This means that, if *P* is at the center of the octagon, we can construct a circle centered at *P* that locates every point equidistant from *P*. In other words, the points at which the rays intersect the circle are the eight vertices of the regular octagon. Consequently, with the circle constructed, we can complete the octagon by drawing the segments connecting consecutive vertices.

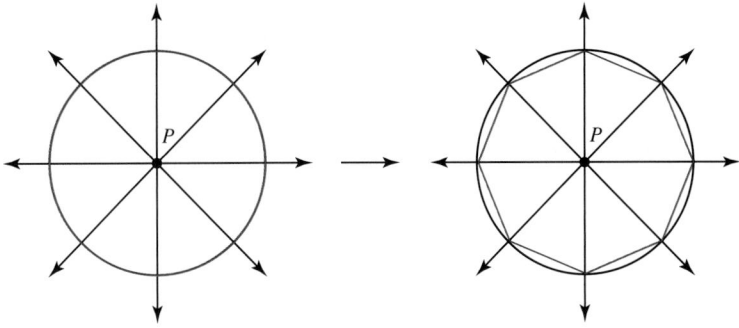

Not every regular *n*-gon is constructible with a straightedge and compass. For instance, neither a regular heptagon nor a regular nonagon is constructible. However, we can construct other regular polygons using the method from the last example. For instance, if we first construct a regular triangle and then bisect the central angles, we get a regular hexagon (Figure 11.45). Similarly, if we continue to bisect the central angles of a regular hexagon, we can get a regular dodecagon and a regular 24-gon. Theoretically, this process can go on indefinitely; that is, we can construct any $3 \cdot 2^k$–gon, where *k* is a whole number. We can also construct a square, a regular pentagon, and a regular 15-gon. If we bisect the central angles of these polygons, we can construct any *n*-gon where *n* equals $4 \cdot 2^k$, $5 \cdot 2^k$, or $15 \cdot 2^k$ for whole number *k*. For many years, it was thought that these were the only constructible regular polygons. However, in 1796, Carl Friedrich Gauss discovered that a regular 17-gon, a regular 257-gon, and a regular 65,537-gon can all be constructed.

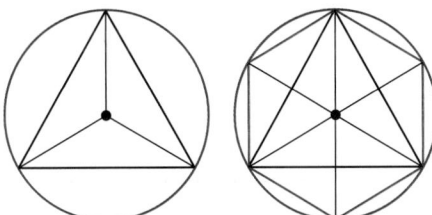

Figure 11.45 Constructing a regular hexagon from an equilateral triangle

Solving Problems with Constructions

Next, we illustrate how constructions can be used to solve different kinds of problems. We begin with a mathematical example.

Example 11.30   Constructing a Tangent to a Circle

Construct a tangent line to a circle from a given point that is not on the circle.

Solution A line is tangent to a circle if it is perpendicular to a radius and it passes through the point at which the radius intersects the circle. In other words, we can find the point of tangency by constructing the appropriate radius and the line perpendicular to it. Begin by drawing a circle with center *C* and a point *D* not on the circle. Construct $\overline{DC}$.

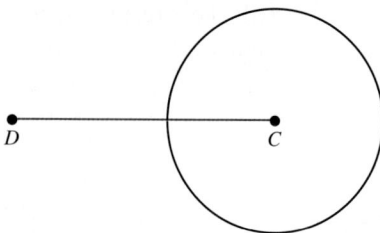

Next, use Construction 5 to find the midpoint of $\overline{DC}$. Label it M. If $\overline{DC}$ represents the diameter of a semicircle, then M is its center, and $\overline{DM}$ is the radius. Use the compass first to measure $\overline{DM}$ and then to construct the semicircle.

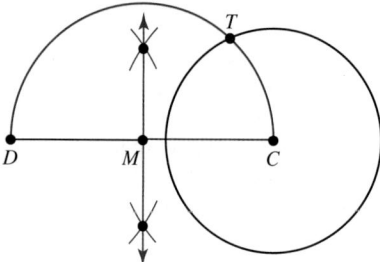

Finally, construct $\overline{TC}$ and $\overleftrightarrow{DT}$, where T is the intersection of the circle and semicircle. $\overline{TC}$ is a radius of the original circle, and angle $\angle DTC$ must be a right angle because it is inscribed in a semicircle. Consequently, $\overleftrightarrow{DT}$ is perpendicular to $\overline{TC}$ and passes through the point T where a radius intersects the circle. T is the needed point of tangency.

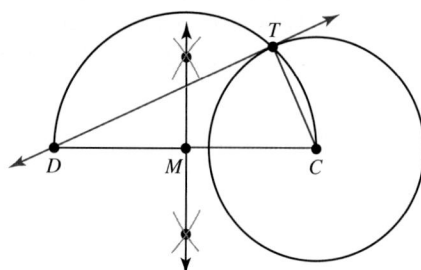

Example 11.31   Building a Fire Station

Application

Three towns want to build a fire station that is the same distance from each town. How might the planning committee find the best location on the map?

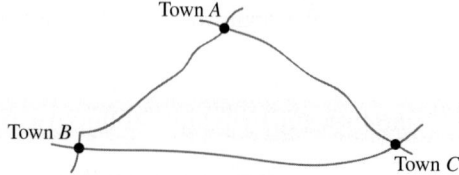

Solution To find the location of the new fire station, we treat the three towns as the vertices of a triangle. Theorem 11.6 indicates that any point on the perpendicular bisector of a segment is equidistant from its endpoints. Consequently, any point at which the perpendicular bisectors of all sides intersect must be equidistant from the three towns. Using Construction 5 with each segment, point D is the location of the new fire station.

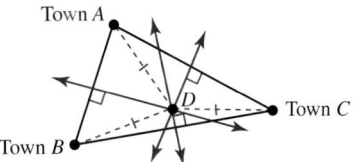

In the last example, point D was equidistant from all of the vertices, so it serves as the center of a circle that passes through each vertex. We call point D the **circumcenter** of the triangle (Figure 11.46), which has been **circumscribed** by the **circumcircle.**

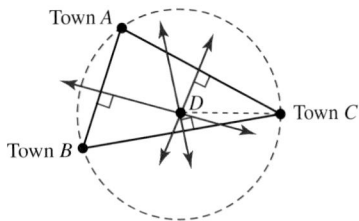

Figure 11.46 A circumscribed triangle

In general, the construction in Example 11.31 can be used to circumscribe any triangle—acute, right, or obtuse (Figure 11.47). In each case, the radius of the circumcircle is always the distance from the circumcenter to any of the vertices.

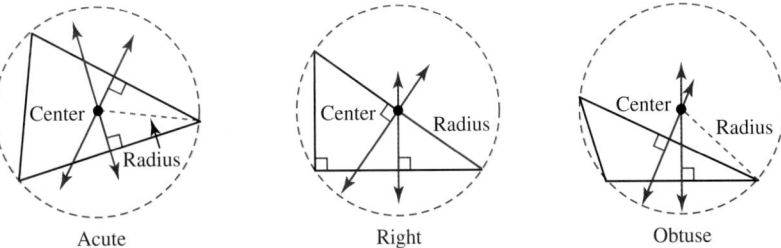

Figure 11.47 Circumcircles for an acute, right, and obtuse triangle

Any regular polygon can also be circumscribed. First, find the center of the circumcircle by constructing the intersection of the perpendicular bisectors of two or more sides. Once we have the center, the radius is any segment joining the center to a vertex. A circumscribed square and regular hexagon are shown in Figure 11.48.

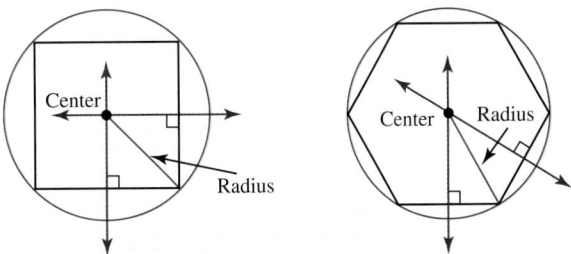

Figure 11.48 A circumscribed square and regular hexagon

Beyond this, rectangles are the only other polygon that can always be circumscribed (Figure 11.49). This is primarily because they are the only other polygon that has a point equidistant from all of the vertices. Specifically, because the diagonals of a rectangle are congruent and bisect one another, their intersection must be equidistant from all vertices. This makes it the center of the circumcircle.

The final example makes use of the next theorem, which provides a way to divide a given segment into any number of congruent parts.

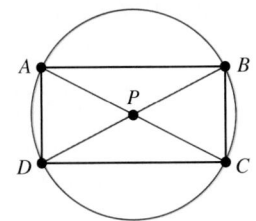

Figure 11.49 A circumscribed rectangle

Theorem 11.7   Parallel Lines Intercepted by a Transversal

If three or more parallel lines intercept congruent segments on one transversal, they intercept congruent segments on all transversals.

Theorem 11.7 states that if three or more parallel lines are equally spaced apart, as measured by the congruent segments on one transversal, then any transversal crossing the parallel lines is also divided into congruent segments (Figure 11.50).

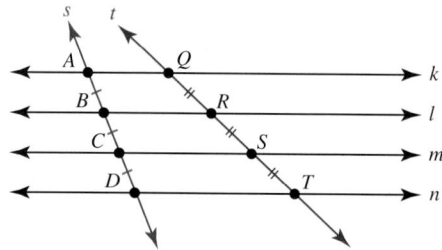

$k \parallel l \parallel m \parallel n$ and $AB = BC = CD$ implies $QR = RS = ST$

Figure 11.50 Transversals divided into congruent segments

Example 11.32   Dividing Land

Application

A land developer is drawing the plans for a new subdivision. She wants to divide a piece of land into three equal lots. If she does not have specific measurements, how can she place the lot markers on the plans so that the three lots have the same road frontage?

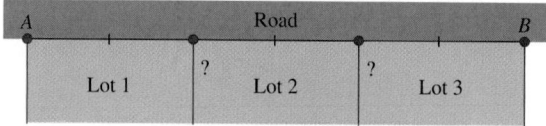

Solution To place the lot markers, we treat the road as a segment, $\overline{AB}$, that we want to trisect, or cut into three congruent parts. We first choose a point P that is not on $\overline{AB}$ and construct the ray $\overrightarrow{AP}$. Then we set the compass to a convenient opening, mark off three congruent segments on $\overrightarrow{AP}$, and label the new points Q, R, and S, respectively.

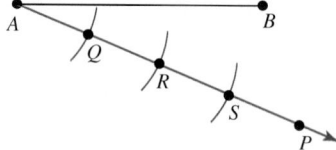

Next, we construct $\overline{SB}$ and copy the resulting angle, $\angle ASB$, at R and Q. Doing so gives us the rays $\overrightarrow{QC}$ and $\overrightarrow{RD}$. $\overrightarrow{QC}$ and $\overrightarrow{RD}$ must be parallel to $\overline{SB}$ because $\angle ASB$, $\angle ARD$, and $\angle AQC$ are corresponding angles along a transversal. Because we constructed $\overline{AQ} \cong \overline{QR} \cong \overline{RS}$, Theorem 11.7 guarantees that $\overline{AC} \cong \overline{CD} \cong \overline{DB}$. So points C and D trisect $\overline{AB}$, and property markers should be placed at the points, A, B, C, and D.

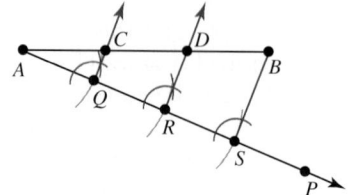

Check Your Understanding 11.4-B

1. Use the method in Example 11.29 to construct a square.

2. Copy $\triangle ABC$ onto another sheet of paper, and then circumscribe it.

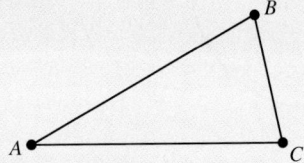

3. Use the method in Example 11.32 to divide segment $\overline{AB}$ into five congruent segments.

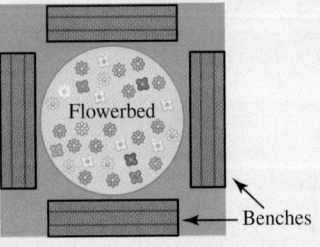

Talk About It You want to build a circular patio with a diameter of 16 ft. How would you mark the outline of the patio prior to construction?

Application

Activity 11.8

A local garden club wants to construct a circular flowerbed inscribed in a square of benches. Use a straightedge and compass to construct a diagram of the flowerbed and benches.

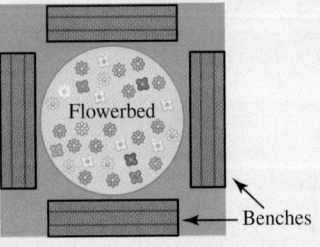

SECTION 11.4   Learning Assessment

■ Understanding the Concepts

1. a. Describe two different ways to construct one triangle to be congruent to another.

   b. Describe two different ways to construct one triangle to be similar to another.

2. a. What basic constructions are needed to construct a square if one of its sides is given?

   b. What basic constructions are needed to construct a rhombus if a side and angle are given?

3. Describe two methods for constructing a regular hexagon.

4. What does it mean for a shape to be constructible?

5. What does it mean to circumscribe a shape?

6. Describe how to find the circumcenter in each figure.

   a. Triangle   **b.** Rectangle   **c.** Regular hexagon

7. What basic constructions are needed to divide a segment into three congruent segments?

■ Representing the Mathematics

8. a. Draw an acute triangle on a sheet of paper. Use a plastic reflector to construct its circumcenter.

   b. Draw an obtuse triangle on a sheet of paper. Use paper folding to construct its circumcenter.

   c. How are the constructions in parts a and b similar? How are they different?

9. Use your compass to draw a circle, and cut it out. Find the center of the circle by folding it in half twice. Next, mark a point A on the circle, and fold it down to the center. Then make folds along the segments $\overline{BD}$ and $\overline{BE}$.

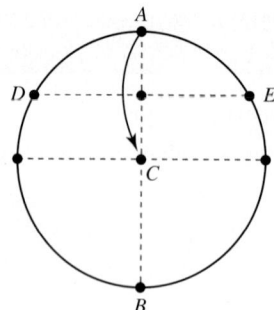

What is the resulting shape?

10. Construct a square, and then use a plastic reflector to construct a square twice as large.

11. Construct an equilateral triangle, and cut it out. Find the incenter of the triangle by folding the angle bisector of each angle. Open the triangle, and then fold each vertex to the incenter as shown.

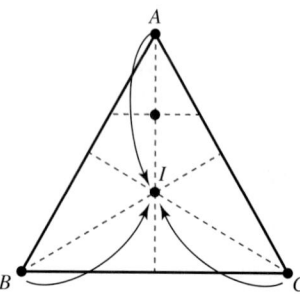

What is the resulting shape?

12. Cut a strip of paper of uniform width. Tie an overhand knot in the strip, being careful to flatten it as it gets tight.

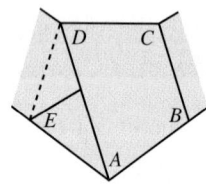

What shapes are made from the folds and the sides of the knot?

13. Draw a line segment $\overline{AB}$ on a sheet of paper. Use a plastic reflector to construct an equilateral triangle with sides congruent to $\overline{AB}$.

14. Construct an equilateral triangle. Use the equilateral triangle and a plastic reflector to construct a regular hexagon.

■ **Exercises**

15. Draw an acute triangle $\triangle ABC$. Use the SSS congruence property to construct a triangle congruent to it.

16. Draw an obtuse triangle $\triangle DEF$. Use the SAS congruence property to construct a triangle congruent to it.

17. Draw an acute triangle $\triangle GHI$. Use the ASA congruence property to construct a triangle congruent to it.

18. Draw a line segment $\overline{AB}$ about 3 in. long. Construct an equilateral triangle with sides congruent to $\overline{AB}$.

19. Draw a line segment $\overline{AB}$ about 1 in. long. Construct a right triangle that has one side equal to $3AB$ and another side equal to $4AB$.

    a. How long should the third side be?

    b. How can you use a compass to check your answer?

20. Draw an acute triangle $\triangle JKL$. Use the SSS similarity property to construct a triangle that is similar to it but twice as large.

21. Draw an obtuse triangle $\triangle MNO$. Use the SAS similarity property to construct a triangle that is similar to it but half as large.

22. Draw an acute triangle and an obtuse triangle. Construct the circumcircle of each.

23. a. Construct an acute isosceles triangle.

    b. Construct a right isosceles triangle.

    c. Construct an obtuse isosceles triangle.

    d. Describe the procedure you used to construct the triangles.

24. Construct a square by using two diameters of a circle.

25. Draw a segment $\overline{AB}$ about 2 in. long, and use it to complete each construction.

    a. Construct a square with sides that are congruent to $\overline{AB}$.

    b. Construct a square with diagonals that are congruent to $\overline{AB}$.

    c. Construct a rectangle with a width congruent to $\overline{AB}$ and a length congruent to twice $\overline{AB}$.

26. Use $\overline{AB}$, $\overline{CD}$, and $\angle E$ to complete each construction.

    a. Construct a rhombus with one angle congruent to $\angle E$ and sides congruent to $\overline{AB}$.

    b. Construct a parallelogram with one angle congruent to $\angle E$, one side congruent to $\overline{AB}$, and one side congruent to $\overline{CD}$.

    c. Construct an isosceles trapezoid with base angles congruent to $\angle E$, one base congruent to $\overline{CD}$, and legs congruent to $\overline{AB}$.

27. Use equilateral triangles to construct a regular hexagon with sides 1 in. long.

28. Determine whether a regular *n*-gon is constructible for each *n*.

   a. 14   **b.** 18   **c.** 22   **d.** 30   **e.** 32

   f. 51   **g.** 60   **h.** 68   **i.** 81   **j.** 100

29. Make a list of the constructible regular *n*-gons, where *n* < 50.

30. Use the method in Example 11.32 to divide $\overline{AB}$ into:

   a. Three congruent segments.

   b. 6 congruent segments.

■ Problems and Applications

31. Use a straightedge and compass to construct an angle of each measure.

   a. 60°   **b.** 30°   **c.** 150°   **d.** 105°

32. Consider $\overline{AB}$ and $\overline{CD}$.

   a. Construct a rhombus with diagonals congruent to $\overline{AB}$ and $\overline{CD}$.

   b. Construct a kite that is *not* a rhombus with diagonals congruent to $\overline{AB}$ and $\overline{CD}$.

33. Construct a square, and then construct its incircle and circumcircle.

34. Construct a square and regular octagon that have the same circumcircle.

35. Construct a circle inscribed in a regular hexagon.

36. Construct a large arc. Use a construction to find the center of the arc.

37. In general, most trapezoids cannot be circumscribed. Find one that can.

38. Construct a line tangent to a circle at a given point on the circle.

39. Shareon wants to cut a rectangular piece of orange construction paper into five congruent rectangles. She does not have a ruler to measure but she does have a sheet of lined notebook paper. How can she find where to make the cuts?

40. An airplane is flying 5 mi above the earth. What is the farthest distance the pilot can see on the earth if we assume that the earth has a radius of about 4,000 mi?

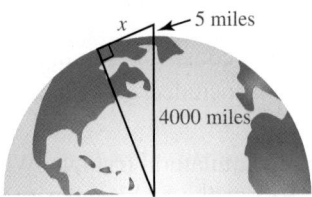

41. At a university, two walks intersect at the angle shown.

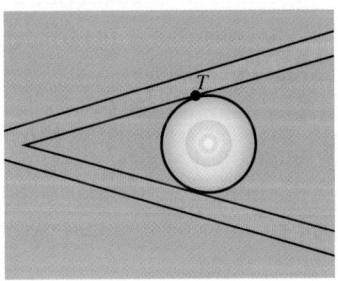

They plan a circular fountain inside the angle so that it is tangent to both sides. How might they construct the needed circle?

■ Mathematical Reasoning and Proof

42. Construct an isosceles triangle and the perpendicular bisector of its base.

   a. What is the relationship between the perpendicular bisector of the base and the vertex angle?

   b. Use geometry software to test your conclusion with a number of different isosceles triangles. Does it appear to always hold true?

   c. Use congruent triangles to explain your thinking.

43. Draw three segments: one about 2 in. long, one about 3 in. long, and one about 4 in. long. Can you construct a triangle out of the given segments? If so, what kind of triangle is it? If not, why not?

44. Draw three segments: one about 2 in. long, one about 3 in. long, and one about 6 in. long. Is it possible to construct a triangle out of the given segments? If so, what kind of triangle is it? If not, why not?

45. In what kind of triangle are the incenter and the circumcenter the same point?

46. Consider ∠1, ∠2, and $\overline{CD}$.

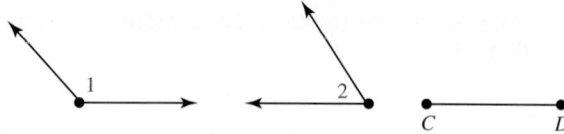

Is it possible to construct a triangle in which segment $\overline{CD}$ is included between ∠1 and ∠2? Explain.

47. Construct an equilateral triangle. Without using either the SSS or the SAS similarity properties, construct an equilateral triangle that is similar to the original by a scale factor of one-half. Write an argument explaining how you know that the two triangles are similar.

48. Why is it impossible to circumscribe a rhombus?

49. Is it possible to construct a quadrilateral with:
  a. Exactly one right angle?
  b. Exactly two right angles?
  c. Exactly three right angles?

50. **a.** What is the minimum amount of information you must know to be sure that two squares are congruent?
  b. How can you use this information to construct two congruent squares?

51. **a.** What is the minimum amount of information you must know to be sure that two parallelograms are congruent?
  b. How can you use this information to construct two congruent parallelograms?

52. Draw any convex quadrilateral, and then find the midpoint of each side. Connect each pair of consecutive midpoints to form a new quadrilateral.
  a. What shape does the new quadrilateral appear to be?
  b. Use geometry software to determine whether this is true for other quadrilaterals.
  c. Does the same hold true for concave quadrilaterals?

Mathematical Communication

53. Write a set of directions describing how to use a straightedge and compass to construct a right isosceles triangle. Use both words and drawings.

54. Write a set of directions describing how to use a straightedge and compass to construct a square. Use both words and drawings.

55. Find a triangle in which the circumcenter is inside the triangle, one in which the circumcenter is on the triangle, and one in which the circumcenter is outside the triangle. Write a statement that characterizes how the location of the circumcenter changes depending on the type of triangle.

56. Use the Internet to find two ways to construct a regular pentagon with a straightedge and compass. Complete each construction. Were the directions easy or difficult to follow? Explain.

57. Describe two methods for dividing a segment into four congruent segments. Use both words and drawings.

58. Audrey wants to know whether it is possible to construct a right triangle with three congruent sides. What do you, as her teacher, tell her?

59. Carrie says that to find the center of a regular polygon, all you need to do is to construct the perpendicular bisectors of two sides. Is she correct?

60. Clarice claims that she can trisect an angle in the following way. She first draws the segment $\overline{AC}$ and trisects it. She then draws the rays $\overrightarrow{BD}$ and $\overrightarrow{BE}$. Is she correct?

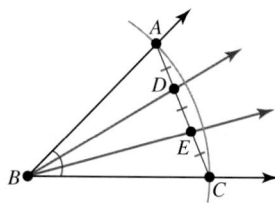

Building Connections

61. What different occupations do you think might use geometrical constructions? Can you make a list of five?

62. **a.** Origami is the art of paper folding. Do an Internet search to see how geometrical constructions are used in this art form.
  b. Trisecting an angle with a straightedge and compass is impossible. However, it can be done with origami. Do an Internet search to find instructions on how to do so, and then complete the construction.

63. Look through a set of curriculum materials for the middle grades. What constructions do you find? Make a list of the different shapes that students learn how to construct.

64. Read the NCTM Geometry Standard for grades 3–5 and 6–8. What do these standards say about building geometrical shapes and the use of constructions? Write a short summary of what you find.

Historical Note

Historically, three construction problems have fascinated mathematicians for thousands of years. The problems were originally posed by the ancient Greeks:

1. *The trisection of a general angle.* Given any angle, divide it into three congruent angles.

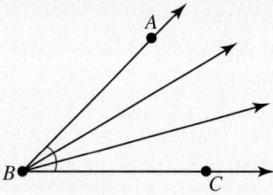

2. *The squaring of a circle.* Given a circle, construct a square with the same area.

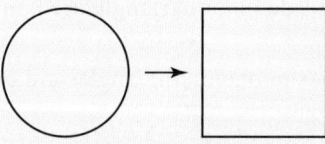

3. *The duplication of the cube.* Given a cube, construct a cube with twice the volume.

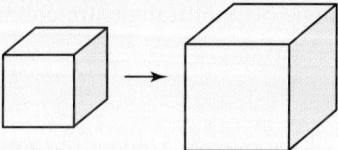

In 1837, Pierre Wantzel (1814–1848) proved that each problem was impossible to construct using only a straightedge and compass. Surprisingly, he demonstrated that each construction was impossible by representing it with an algebraic equation.

Opening Problem Answer

© Eugene May/Shutterstock.com

The ladder forms two similar triangles between the fence and the side of the house.

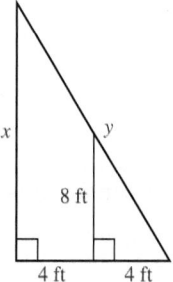

If x represents the height of the ladder on the wall, then we have the proportion $\frac{8}{4} = \frac{x}{8}$. Solving for x, $x = \frac{8 \cdot 8}{4} = 16$ ft. Next, we use the Pythagorean Theorem to find the length of the ladder, or $y = \sqrt{16^2 + 8^2} = \sqrt{320} \approx 17.88$ ft.

Chapter 11 REVIEW

Summary of Key Concepts from Chapter 11

Section 11.1   Congruent Shapes

- Two triangles are congruent if and only if there is a correspondence between their vertices such that the corresponding sides and angles are congruent.
- Given a correspondence between two triangles, the two triangles are congruent if and only if:

  SSS: The three sides of one triangle are congruent to the three sides of the other.

  SAS: The two sides and the included angle of one triangle are congruent to two sides and the included angle of the other.

  ASA: The two angles and the common side of one triangle are congruent to two angles and the common side of the other.

  AAS: The two angles and the side opposite one of the angles are congruent to two angles and an opposite side of the second triangle.

- Given a correspondence between two right triangles, the two triangles are congruent if and only if:

  HL: The hypotenuse and one leg of one triangle are congruent to the hypotenuse and one leg of the other.

  HA: The hypotenuse and one acute angle of one triangle are congruent to the hypotenuse and one acute angle of the other.

Section 11.2   Similar Shapes

- **Similarity** is the notion that two figures have the same shape but not necessarily the same size.
- Two triangles are **similar** if and only if there is a correspondence between their vertices such that corresponding angles are congruent and corresponding sides are proportional.
- Given a correspondence between two triangles, the two triangles are similar if and only if:

  SSS: All three pairs of corresponding sides are proportional.

  SAS: Two pairs of corresponding sides are proportional, and their included angles are congruent.

  AA: Two angles of one triangle are congruent to two angles of the other.

- A **midsegment** of a triangle is a segment that connects the midpoints of two sides. The midsegment is parallel to the third side and half as long.
- The midsegments of any quadrilateral form a parallelogram.
- A figure is **self-similar** if it looks the same when magnified or reduced.
- A **fractal** is a self-similar geometric pattern generated by a recursively defined function.

Section 11.3   Basic Geometrical Constructions

- A **geometrical construction** is a visual representation of a shape or a relationship that is built using a specific set of tools and geometrical properties.
- A **straightedge** is used to construct lines, rays, and segments between two points. A **compass** is used to draw circles and arcs of a given center and radius.
- Basic constructions that can be made with a straightedge and compass are copying a line segment, copying an angle, constructing an angle bisector, constructing parallel lines, and constructing perpendicular lines.
- The common point among the three angle bisectors is the **incenter** of the triangle. It is the center of the triangle's **incircle,** which is the largest circle that fits in the triangle so that the triangle's sides are tangent to the circle.
- A point is on the perpendicular bisector of a segment if and only if it is equidistant from both endpoints.

Section 11.4   Constructing Shapes

- Basic constructions can be used to construct congruent and similar triangles, quadrilaterals, and many regular polygons.

- A **median** of a triangle is a segment that connects a vertex to the midpoint of the opposite side. The point where the three medians intersect is called the **centroid.**

- Any regular *n*-gons can be constructed where *n* equals $3 \cdot 2^k$, $4 \cdot 2^k$, $5 \cdot 2^k$, or $15 \cdot 2^k$ for any whole number *k*.

- The **circumcenter** of a triangle is the intersection of the three sides of a triangle. It is the center of the **circumcircle,** which is the smallest circle that contains all three vertices of the triangle.

- If three or more parallel lines intercept congruent segments on one transversal, then they intercept congruent segments on all transversals.

Review Exercises Chapter 11

1. If $\triangle ABC \cong \triangle FED$, make a list of the six pairs of corresponding parts.

2. Draw a pair of triangles that are congruent by the:
   a. SSS property.   **b.** ASA property.
   c. HL property.

3. Let $\triangle ABC \cong \triangle XYZ$.
   a. If $AB = 7$, $BC = 9$, and $AC = 9$, what are XY, YZ, and XZ?
   b. If $m\angle A = 65°$, $m\angle B = 65°$, $m\angle C = 50°$, what are $m\angle X$, $m\angle Y$, and $m\angle Z$?

4. Let $\triangle ABC$ and $\triangle XYZ$ be right triangles such that $\triangle ABC \cong \triangle XYZ$. If $m\angle A = 60°$, $m\angle Z = 30°$, and $XZ = 12$, what are the measures of the unknown parts?

5. Decide whether each pair of triangles is congruent. If so, state by what congruence property.

   a.    **b.**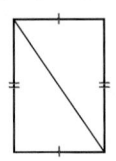

   c.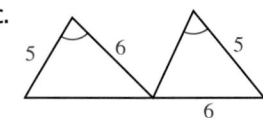

6. Quadrilateral $ABCD$ is a rectangle.

   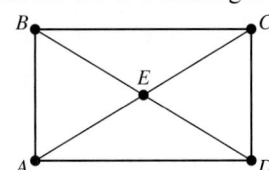

   Find all pairs of congruent triangles.

7. Quadrilateral $ABCD$ is a parallelogram.

   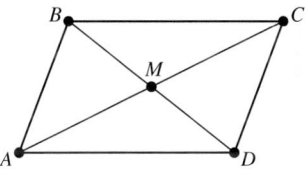

   Determine whether each statement is true or false.
   a. $\overline{AB} \cong \overline{DC}$.   **b.** $\overline{AM} \cong \overline{DM}$.
   c. $\angle BAD \cong \angle ADC$.   **d.** $\angle MAB \cong \angle DCM$.
   e. $\angle BAD \cong \angle BCD$.   **f.** $\triangle CMD \cong \triangle AMD$.

8. If the side of one regular pentagon is congruent to the side of another, are the two pentagons congruent? What if the radius of one is congruent to the other?

9. What are the minimal congruence conditions needed to determine whether two isosceles trapezoids are congruent? What about any two trapezoids?

10. Can two triangles have three pairs of corresponding parts congruent but not be congruent? Explain.

11. Use congruent triangles and the diagram to prove that opposite sides of a rectangle are congruent.

    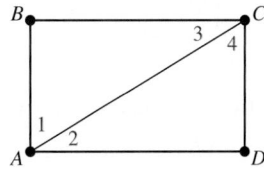

12. Draw a pair of triangles that are similar by the:
    a. SSS property.   **b.** AA property.

13. Draw an acute triangle $\triangle ABC$ with sides of length 6 cm, 7 cm, and 8 cm. Draw $\triangle DEF$ so that $\triangle ABC \sim \triangle DEF$ with a scale factor of 2.

14. Draw a rep-tile with an obtuse triangle so that the new similar figures are larger than the original triangle.

15. If $\triangle ABC \sim \triangle DEF$, what is the scale factor if
  a. $AB = 4$ and $DE = 16$?
  b. $AB = 3$ and $DE = 0.3$?

16. If $\triangle ABC \sim \triangle DEF$, what is DE if:
  a. $AB = 3$, and the scale factor is 6?
  b. $AB = 4$, and the scale factor is $\dfrac{5}{12}$?

17. In each of the diagrams, assume that the two triangles are similar. Use the information given in the diagram to find the value of the variable.

  a. 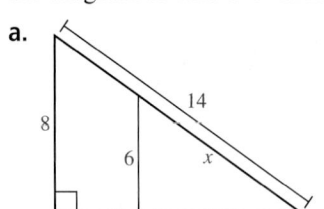   **b.**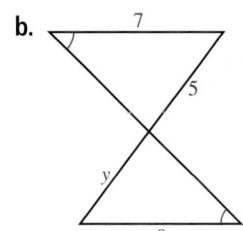

  c.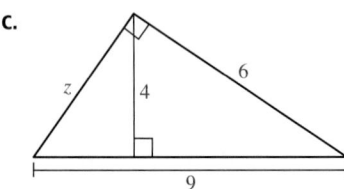

18. Suppose two parallelograms are similar by a scale factor of 3.9. If the larger parallelogram has sides of length 7 and 12, about how long are the sides of the smaller one?

19. $\overline{DE}$ is a midsegment of $\triangle ABC$.

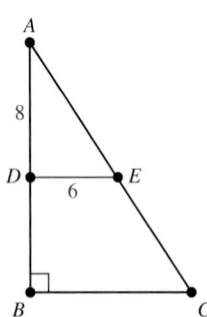

  Find the length of each side.
  a. $\overline{AE}$ **b.** $\overline{AB}$ **c.** $\overline{AC}$ **d.** $\overline{BC}$

20. Suppose $\triangle ABC \sim \triangle DEF$ and $\triangle DEF \sim \triangle XYZ$. Further suppose that $AB = 64$, $\overline{DE}$ is 75% the length of $\overline{AB}$ and that $\overline{XY}$ is 50% the length of $\overline{DE}$.
  a. What is the length of $\overline{DE}$ and $\overline{XY}$?
  b. What is the scale factor between $\triangle ABC$ and $\triangle XYZ$?

21. Will any two parallelograms be similar? What about any two isosceles trapezoids?

22. While standing near a building, Mario notices that the building's shadow is 15 times the length of his. Mario stands exactly 4.5 ft tall. If his shadow is 3 ft long, how tall is the building?

23. Construct a segment congruent to each segment, and then construct the segment's perpendicular bisector.

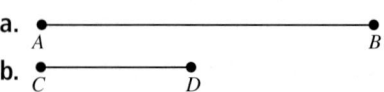

24. Construct an angle congruent to each angle, and then construct the angle's bisector.

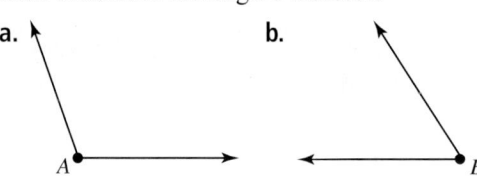

25. Use corresponding angles to construct the line parallel to $\overleftrightarrow{AB}$ that passes through P.

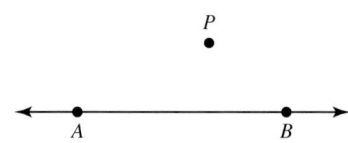

26. Make a triangle $\triangle ABC$, and then construct the incircle of the triangle.

27. Make a triangle $\triangle ABC$, and then construct its circumcircle.

28. Consider $\overline{AB}$ and $\overline{CD}$.

  Construct a segment of each given length.
  a. $3AB$ **b.** $AB - CD$
  c. $AB + 2CD$

29. Consider $\angle A$ and $\angle B$.

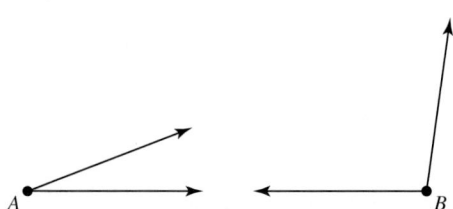

  Construct an angle for each given measure.
  a. $m\angle A + m\angle B$ **b.** $m\angle B - m\angle A$
  c. $3m\angle B$

30. Use a straightedge and compass to construct an angle that measures $67.5°$.

31. Draw an acute triangle $\triangle ABC$. Use the ASA congruence property to construct a congruent triangle.

32. Draw a line segment $\overline{AB}$ about 2 in. long. Construct an equilateral triangle with sides congruent to $\overline{AB}$.

33. Draw an obtuse triangle $\triangle DEF$. Use the SSS similarity property to construct a triangle that is similar but twice as large.

34. Draw a segment $\overline{AB}$ about 3 in. long, and use it to complete each construction.

   a. Construct a square with sides that are congruent to $\overline{AB}$.

   b. Construct a rectangle with a width congruent to $\overline{AB}$ and a length congruent to twice $\overline{AB}$.

   c. Construct a rhombus with a 45° angle and sides congruent to $\overline{AB}$.

   d. Construct a parallelogram with a 60° angle and one side congruent to $\overline{AB}$ and the other side congruent to twice $\overline{AB}$.

35. Construct a regular hexagon with sides about 2 in. long.

36. Determine whether the regular n-gon is constructible for each n.

   a. 11   **b.** 45   **c.** 48

   d. 117   **e.** 136   **f.** 514

37. Construct the line tangent to the circle that passes through P.

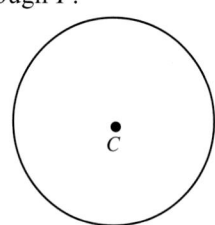

Answers to Chapter 11   Check Your Understandings and Activities

Check Your Understanding 11.1-A

1. a. $\triangle BCA \cong \triangle YZX$.   **b.** $\triangle BAC \cong \triangle YXZ$.
   c. $\triangle CAB \cong \triangle ZXY$.

2. a. $AC = 9$, $XY = 7$, and $YZ = 15$.
   b. $m\angle B = 89°$, $m\angle C = 57°$, $m\angle X = 34°$, and $m\angle Z = 57°$.

3. a. Congruent; AAS.
   b. Not congruent.
   c. Congruent, ASA.

Check Your Understanding 11.1-B

1. Suppose quadrilateral $ABCD$ is a rectangle. Then, $\overline{AD} \cong \overline{BC}$ because they are opposite sides and $\angle ADC \cong \angle BCD$ because they are both right angles. We also have $\overline{DC} \cong \overline{CD}$ because they represent the same segment. Consequently, $\triangle ADC \cong \triangle BCD$ by SAS and $\overline{AC} \cong \overline{BD}$ because corresponding parts of congruent triangles are congruent.

2. The different correspondences are $A \leftrightarrow W$, $B \leftrightarrow X$, $C \leftrightarrow Y$, $D \leftrightarrow Z$; $A \leftrightarrow X$, $B \leftrightarrow W$, $C \leftrightarrow Z$, $D \leftrightarrow Y$; $A \leftrightarrow Z$, $B \leftrightarrow Y$, $C \leftrightarrow X$, $D \leftrightarrow W$; and $A \leftrightarrow Y$, $B \leftrightarrow Z$, $C \leftrightarrow W$, $D \leftrightarrow X$.

3. Two isosceles trapezoids are congruent if one pair of corresponding base angles is congruent and the two pairs of corresponding parallel sides are congruent, or if one pair of corresponding base angles is congruent and two corresponding pairs of consecutive sides are congruent.

Check Your Understanding 11.2-A

1. a. 5.   **b.** $\dfrac{3}{5}$.

2. $BC = 5.2$ and $DF = 22.5$.

3. Because $\angle A \cong \angle D$ and $\angle ACB \cong \angle DCE$, $\triangle ABC \sim \triangle DEC$ by the AA property.

Check Your Understanding 11.2-B

1. 80 ft.

2.

Check Your Understanding 11.3-A

1. a. Use Construction 1.   **b.** Use Construction 1.

2. Draw a ray and then use Construction 1 two times in succession.

3. a. Use Construction 2 and then Construction 3.
   b. Use Construction 2 and then Construction 3.

Check Your Understanding 11.3-B

1. Follow Construction 4. This time, copy an angle below $\overleftrightarrow{AB}$ to a position above P and on the other side of $\overleftrightarrow{PC}$.

2. Use Construction 5.

Check Your Understanding 11.4-A

1. Draw a ray and then use Construction 1 to copy a side onto the ray. Next, copy an appropriate angle at the endpoint of the ray. Finally, copy the adjacent side onto the second side of angle. Complete the construction by drawing the third side.

2. Follow the procedure in Example 11.25, but copy the length of the segment three times instead of twice.

3. Use the procedure in Example 11.27; this time, make one length equal to twice AB and the other length equal to AB.

Check Your Understanding 11.4-B

1. Construct two perpendicular lines. Draw a circle of an appropriate radius. The points of intersection between the perpendicular lines and the circle are the four vertices of the square.
2. Follow the procedure in Example 11.31.
3. Adapt the procedure given in Example 11.32.

Activity 11.1

a. Answers will vary but must include showing that three pairs of corresponding sides are congruent.

b. Answers will vary but must include showing that two sides and the included angle of one triangle are congruent, respectively, to two sides and the included angle of the second triangle.

c. Answers will vary but must include showing that two angles and the common side of one triangle are congruent respectively to two angles and the common side of the second triangle.

Activity 11.2

a. A pair of corresponding edges must be congruent.

b. A pair of corresponding edges must be congruent.

c. The radii must be congruent, and the lengths of the lateral surfaces must be congruent.

d. The radii must be congruent.

Activity 11.3

The large triangle is similar to the smallest triangle by the AA property. So taking the ratio of proportional sides, we get $\dfrac{x}{a} = \dfrac{a}{c}$. Cross multiplying, we get $a^2 = xc$. Similarly, the medium-sized triangle is similar the largest triangle by the AA property. Consequently, taking the ratio of proportional sides, we get $\dfrac{c - x}{b} = \dfrac{b}{c}$. Again by cross multiplying, we get $b^2 = c(c - x)$. Adding the two equations gives us $a^2 + b^2 = xc + c(c - x) = xc + c^2 - xc = c^2$.

Activity 11.4

Answers will vary.

Activity 11.5

Fold the paper through the vertex of the angle so that one side of the angle is on top of the other. The fold line represents the angle bisector.

Activity 11.6

a. Crease the paper in such a way that the fold passes through the given point and one side of the line lies directly over the other side. The fold represents the perpendicular line

b. Fold the paper so that one endpoint of the segment lies directly over the top of the other endpoint. The fold represents the perpendicular bisector.

Activity 11.7

Napoleon's triangle is an equilateral triangle.

Activity 11.8

Use the procedure in Example 11.27 to construct a square. Draw the two diagonals of the square to find the center of the inscribed circle. Construct a segment that is perpendicular to one of the sides of the square from the center of the circle. The segment is the radius of the needed circle.

© Image Source/Corbis

Chapter 12

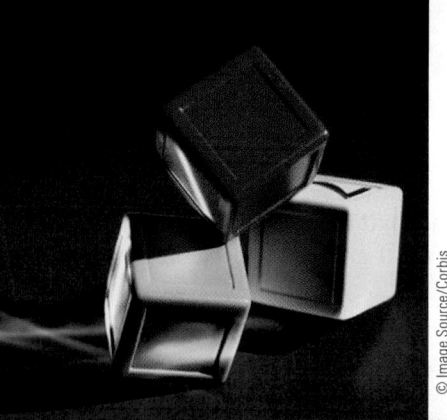

Coordinate and Transformation Geometry

Opening Problem

Frank and Fergie are tiling a floor and want to do something unusual. The flooring store said that they could tile their floor using equilateral triangles, squares, and regular hexagons so that the arrangement of tiles is the same at each vertex. If the sides of the tiles are the same length, how are the tiles arranged around a common vertex?

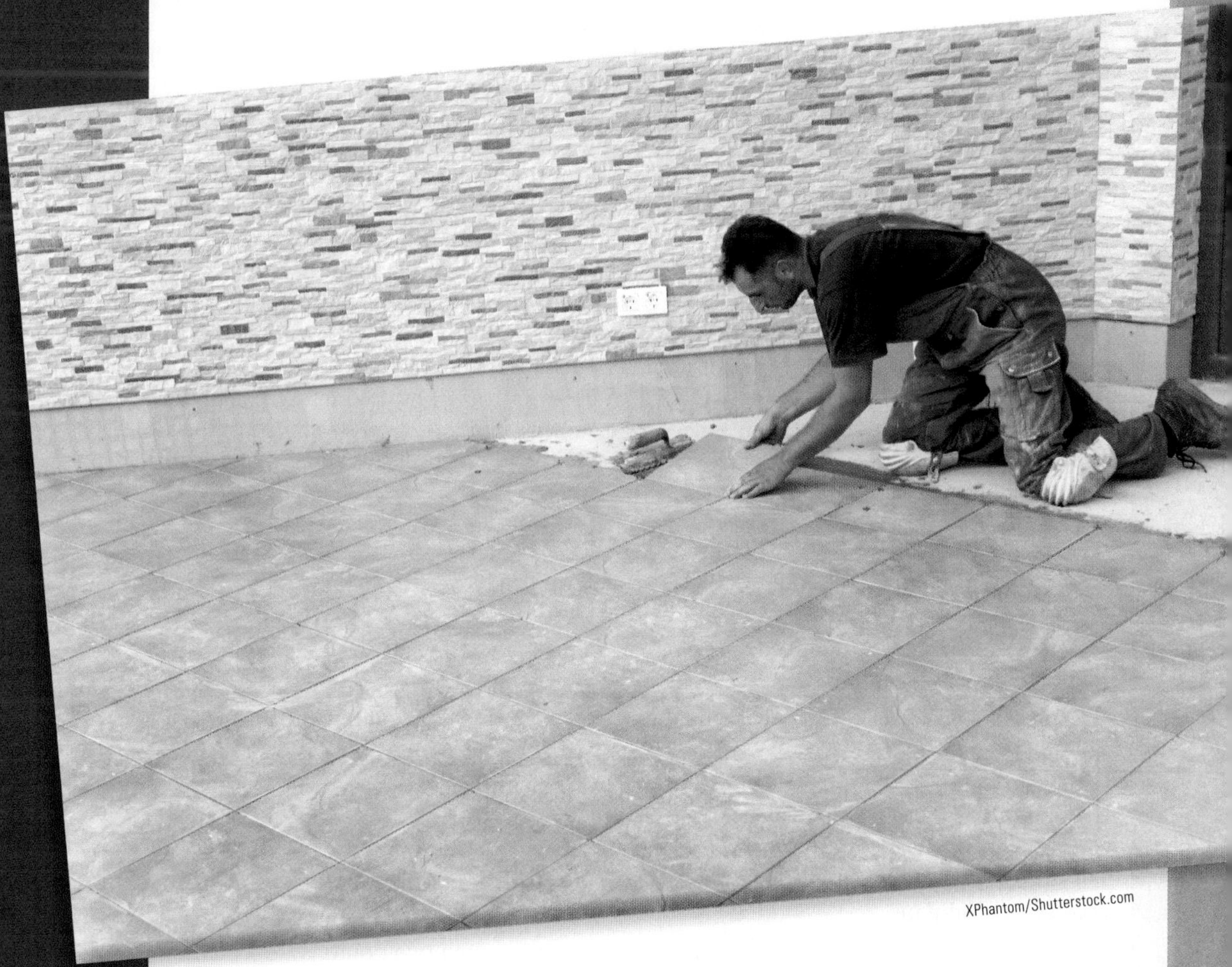

XPhantom/Shutterstock.com

- **12.1** Coordinate Geometry
- **12.2** Transformations
- **12.3** Congruence, Similarity, and Symmetry with Transformations
- **12.4** Geometric Patterns

The last two chapters have focused on the first goal of the NCTM Geometry Standard: defining shapes, identifying their properties, and discussing the relationships between them. In this chapter, we discuss coordinate and transformation geometry, which are topics included under the second and third goals of the Geometry Standard.

■ NCTM Geometry Standard

Instructional programs from prekindergarten through grade 12 should enable all students to:

- Analyze characteristics and properties of two- and three-dimensional geometric shapes and develop mathematical arguments about geometric relationships.
- Specify locations and describe spatial relationships using coordinate geometry and other representational systems.
- Apply transformations and use symmetry to analyze mathematical situations.
- Use visualization, spatial reasoning, and geometric modeling to solve problems.

Source: NCTM STANDARDS Copyright © 2011 by NATIONAL COUNCIL OF TEACHERS OF MATHEMATICS. Reproduced with permission of NATIONAL COUNCIL OF TEACHERS OF MATHEMATICS.

Coordinate Geometry in the Classroom

Coordinate geometry involves representing shapes on the Cartesian coordinate system, making it possible to describe and manipulate them with algebraic expressions. In the classroom, children first build a foundation for coordinate geometry by describing the relative position of shapes using intuitive words like "near" or "far" and "above" or "below." Given these ideas, children can then improve their understanding of spatial relationships by working with simple coordinate systems and by answering questions about direction, distance, and location. With a working knowledge of coordinate systems, they can use what they know to analyze shapes, navigate to specific locations, and measure distances. Table 12.1 gives specific expectations associated with coordinate geometry and the grade levels at which they are likely to be taught.

Table 12.1 Classroom learning objectives	K	1	2	3	4	5	6	7	8
Describe the relative positions of objects using terms such as "above," "below," "beside," "in front of," "behind," and "next to."	X								
Define a coordinate system and locate points on the plane using an ordered pair of numbers.						X			
Represent and solve real-world and mathematical problems by graphing points on the coordinate plane.						X	X		
Understand signs of numbers in ordered pairs as indicating locations in quadrants of the coordinate plane.							X		
Draw polygons in the coordinate plane and use coordinates to find the length of a side joining points with the same first coordinate or the same second coordinate.							X		
Apply the Pythagorean Theorem to find the distance between two points in a coordinate system.									X

Source: Adapted from the *Common Core State Standards for Mathematics* (*Common Core State Standards Initiative*, 2010).

Transformation Geometry in the Classroom

Transformation, or **motion, geometry** involves describing the changes in the position and orientation of a shape that result from simple motions, or **transformations,** like slides, flips, and turns. In the early grades, children learn about such motions by physically manipulating shapes. Then, as they become better at spatial reasoning, they begin to move objects mentally and learn to predict the outcome of the motion prior to making it. They also learn to track the movements of objects by placing them on a coordinate system, giving coordinate geometry an important role in the study of transformations. They also use transformations to study symmetry. Many children have an intuitive sense of symmetry before they enter school, giving them the ability to determine the symmetry of a shape visually. However, once in school, they describe symmetry more formally by using transformations.

Both transformations and symmetry are used to study geometric patterns like tessellations. **Tessellations,** or **tiling patterns,** are patterns made from a set of basic figures that cover the entire plane with no overlaps or gaps. They provide an excellent opportunity for children to apply what they have learned about transformations and symmetry. They also give students an opportunity to connect mathematics to many real-world situations. Table 12.2 gives specific classroom expectations associated with learning transformation geometry and symmetry.

Table 12.2 Classroom learning objectives	K	1	2	3	4	5	6	7	8
Recognize and draw a line of symmetry for a two-dimensional figure and identify line-symmetric figures.					X				
Verify experimentally the properties of rotations, reflections, and translations.									X
Understand that a two-dimensional figure is congruent to another if the second can be obtained from the first by a sequence of rotations, reflections, and translations.									X
Understand that a two-dimensional figure is similar to another if the second can be obtained from the first by a sequence of rotations, reflections, translations, and dilations.									X
Describe the effect of dilations, translations, rotations, and reflections on two-dimensional figures using coordinates.									X

Source: Adapted from the *Common Core State Standards for Mathematics* (*Common Core State Standards Initiative*, 2010).

Because coordinate and transformation geometry are integrated within the elementary- and middle-grades curriculum, teachers in these grades must be prepared to teach these ideas. Consequently, in Chapter 12, we:

- Demonstrate how the Cartesian coordinate system can be used to study shapes and location relationships.
- Illustrate how shapes can be moved and how the motions relate to coordinate geometry and symmetry.
- Use transformations and symmetry to describe and create geometrical patterns and tessellations.

SECTION 12.1 Coordinate Geometry

The Cartesian coordinate system has been defined as a grid that can be used to designate the position of any point on the plane by using two coordinates (Figure 12.1). In Chapter 9, we used the Cartesian coordinate system to represent and analyze functions. However, its inventor, René Descartes, originally used it to describe and manipulate shapes through the use of algebraic expressions. In doing so, he created a branch of mathematics called coordinate, or analytical geometry, which we can use to study a number of different concepts related to spatial reasoning and geometrical shapes.

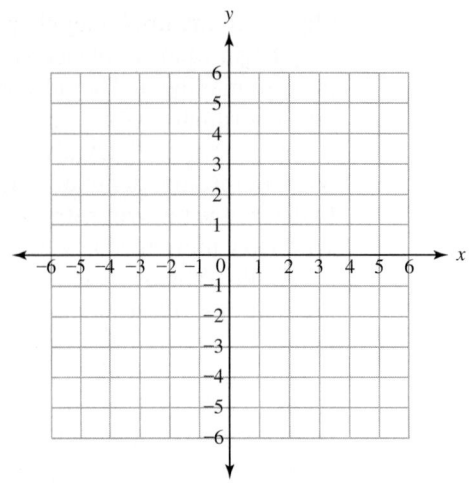

Figure 12.1 The Cartesian coordinate system

Coordinate Geometry and Spatial Reasoning

Coordinate geometry can help students learn a number of spatial reasoning skills. For instance, to plot points, students must learn how to use coordinates to find and name specific locations. They must interpret the meaning of the coordinates as they relate to a distance and a direction from a fixed location, such as the origin. Once students have learned to use these skills with a simple coordinate system, they can apply them to find locations on more complex coordinate systems.

Example 12.1	Latitude and Longitude

Connections

The student page in Figure 12.2 shows how to identify positions on the earth by using latitudes and longitudes. Use the information on the page to estimate the latitude and longitude of the cities in Questions 1–4.

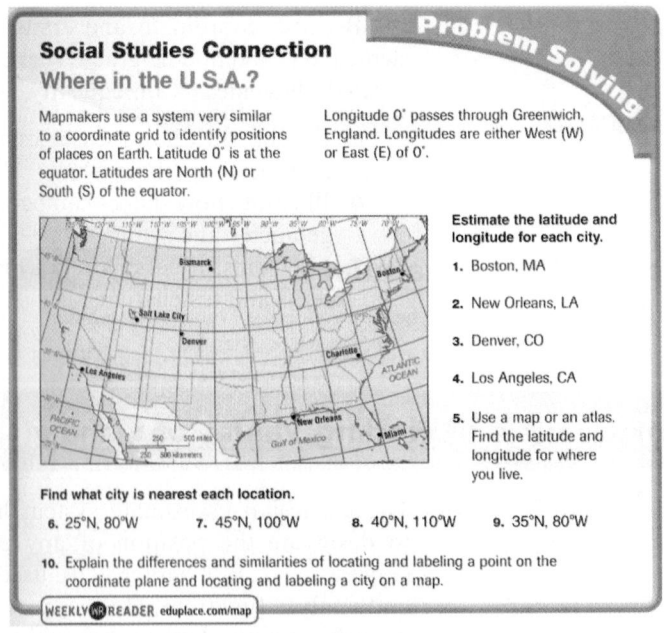

Figure 12.2 Latitude and longitude
Source: From *Mathematics Grade 5: Houghton Mifflin Mathematics Georgia* (p. 619). Copyright © by Houghton Mifflin Company, Inc. Reproduced by permission of the publisher, Houghton Mifflin Harcourt Publishing Company.

Solution The approximate latitude and longitude for each city are given.

1. Boston, MA is at 43° N and 71° W.

2. New Orleans, LA is at 30° N and 90° W.

3. Denver, CO is at 40° N and 105° W.

4. Los Angeles, CA is at 34° N and 118° W.

Explorations Manual 12.1

Coordinate geometry can also be used to give the relative position of one location with respect to another or to describe how to travel from one point to another.

Example 12.2   Navigating City Streets

Communication

Many city streets are arranged in a rectangular grid like the one shown on the map. Use the map to answer each question.

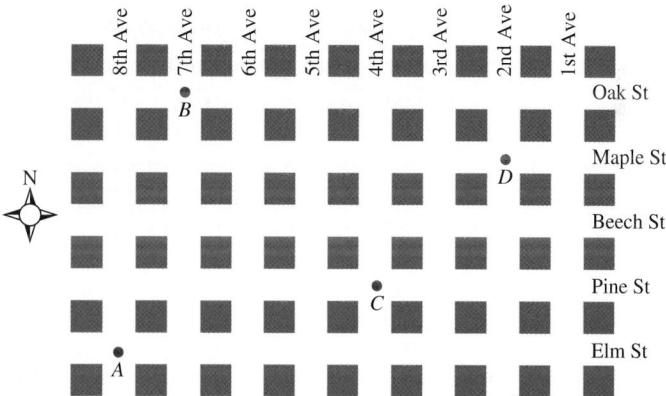

a. Describe the relative position of point *A* with respect to point *C*.

b. How many blocks is it to walk from the intersection of Elm and 7th to the intersection of Beech and 3rd?

c. Write a set of directions that explains how to walk from point *B* to point *D* by traveling through point *C*.

Solution

a. Point *A* is four blocks to the west and one block to the south of point *C*.

b. We walk a total of 6 blocks, two to the north and four to the east.

c. A number of different paths from point *B* to point *D* pass through point *C*. One is to walk south from point *B* for three blocks and then east for three blocks to get to point *C*. Next, walk east for two blocks and then north for two blocks to arrive at point *D*.

Coordinate Geometry and Line Segments

Explorations Manual 12.2

In Chapter 10, we used length to define congruence for segments but did not otherwise discuss how to find a segment's length. We do so now by comparing a segment to the **unit length,** which is the distance from 0 to 1 on the *x*-axis of the Cartesian plane. In other words, the length of a line segment, or equivalently the distance between any two points, is the number of unit lengths equivalent to the segment or distance.

If a line segment is horizontal or vertical, then we can find its length by subtracting the appropriate coordinates of the endpoints. For instance, suppose $\overline{AB}$ lies on the x-axis and has endpoints at $(2, 0)$ and $(7, 0)$ (Figure 12.3). To find its length, we subtract the x-coordinates, or $AB = 7 - 2 = 5$ unit lengths.

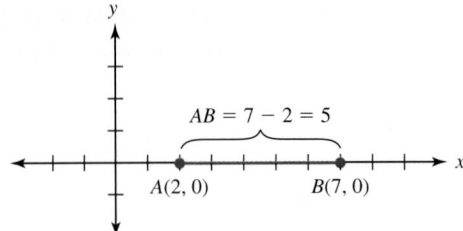

Figure 12.3 Length of $\overline{AB}$

In other words, if $\overline{AB}$ is a horizontal segment, then the endpoints have the same y-coordinates and we represent them as $A(x_1, y_1)$ and $B(x_2, y_1)$, where $x_2 > x_1$ (Figure 12.4). The length of the segment is then the difference of the x-coordinates, or $AB = x_2 - x_1$. Similarly, if $\overline{CD}$ is a vertical segment, then the endpoints have the same x-coordinates and we represent them as $C(x_3, y_3)$ and $D(x_3, y_4)$, where $y_4 > y_3$. This makes the length of $\overline{CD}$ the difference of the y-coordinates, or $CD = y_4 - y_3$.

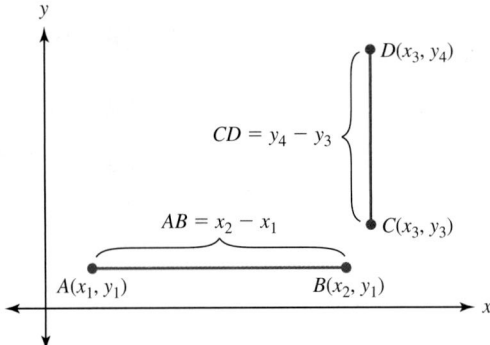

Figure 12.4 Lengths of horizontal and vertical segments

We can use horizontal and vertical segments to find the lengths of other segments. For instance, suppose $\overline{AB}$ has endpoints at $A(1, 2)$ and $B(4, 6)$ (Figure 12.5). To find its length, we draw a vertical segment from point $B(4, 6)$ down to point $C(4, 2)$ and a horizontal segment from $A(1, 2)$ over to $C(4, 2)$. Doing so forms a right triangle

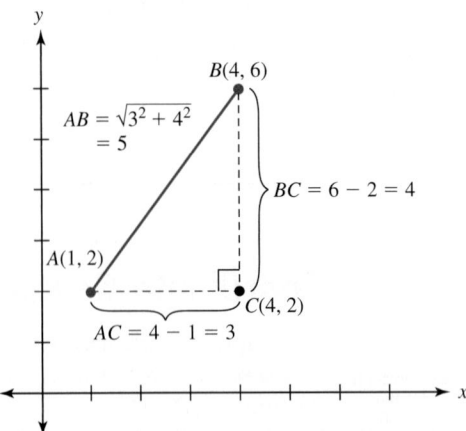

Figure 12.5 The length of a segment with endpoints at A(1, 2) and B(4, 6)

with legs of lengths $BC = 6 - 2 = 4$ and $AC = 4 - 1 = 3$. Using the Pythagorean Theorem, the length of the hypotenuse is $AB = \sqrt{3^2 + 4^2} = \sqrt{9 + 16} = \sqrt{25} = 5$.

Because we can replace points A and B with any points, we can generalize the process. Let A and B be points with coordinates $A(x_1, y_1)$ and $B(x_2, y_2)$ (Figure 12.6). We form a right triangle by drawing a horizontal segment from $A(x_1, y_1)$ to point $C(x_2, y_1)$ and a vertical segment from $B(x_2, y_2)$ to $C(x_2, y_1)$. We then find the lengths of the legs by subtracting the x- and y-coordinates, or $AC = x_2 - x_1$ and $BC = y_2 - y_1$. By applying the Pythagorean Theorem, the distance between A and B is $AB = \sqrt{(x_2 - x_1)^2 + (y_2 - y_1)^2}$.

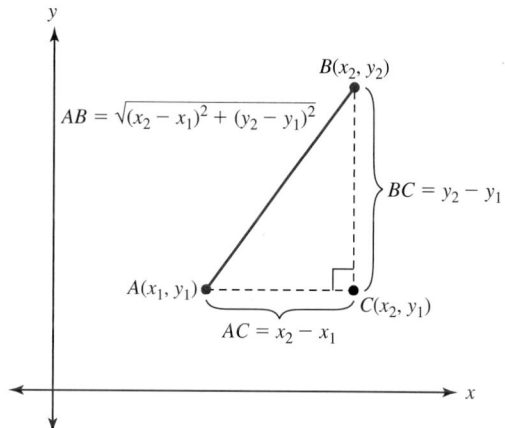

Figure 12.6 The distance formula

Theorem 12.1   The Distance Formula

If points A and B have coordinates $A(x_1, y_1)$ and $B(x_2, y_2)$, then the distance between A and B is given by $AB = \sqrt{(x_2 - x_1)^2 + (y_2 - y_1)^2}$.

Example 12.3   Using the Distance Formula

Find the distance between each pair of points.

a. $A(2, 4)$ and $B(8, 4)$ **b.** $G(-3, -4)$ and $H(5, 7)$

Solution

a. Because the y-coordinates of the points are the same, the points must lie along a horizontal line. As a result, $AB = 8 - 2 = 6$.

b. We apply the distance formula and compute, or $GH = \sqrt{(5 - (-3))^2 + (7 - (-4))^2} = \sqrt{8^2 + 11^2} = \sqrt{64 + 121} = \sqrt{185} \approx 13.6015$.

Two things must be mentioned about the distance formula. First, it can be used to find the length of any horizontal and vertical segment. If the line is horizontal, then the y-coordinates subtract away during the computation. If the line is vertical, then the x-coordinates subtract away. Second, notice that the differences in the formula are squared. So we can subtract the coordinates in either order—$(x_1 - x_2)$ instead of $(x_2 - x_1)$ or $(y_1 - y_2)$ instead of $(y_2 - y_1)$—and still arrive at the same answer.

The next example illustrates how to use the distance formula to verify geometric relationships and properties.

Example 12.4 **Determining Whether Three Points Are Collinear**

Problem Solving Determine whether the points $A(-3, -2)$, $B(1, 1)$, and $C(9, 7)$ are collinear.

Solution

Understanding the Problem. Our goal is to determine whether points A, B, and C lie on the same line, where A is at $(-3, -2)$, B is at $(1, 1)$, and C is at $(9, 7)$.

Devising a Plan. One way to approach the problem is to plot the three points to see whether one is considerably out of line with the others. As the graph shows, this is not the case with these points.

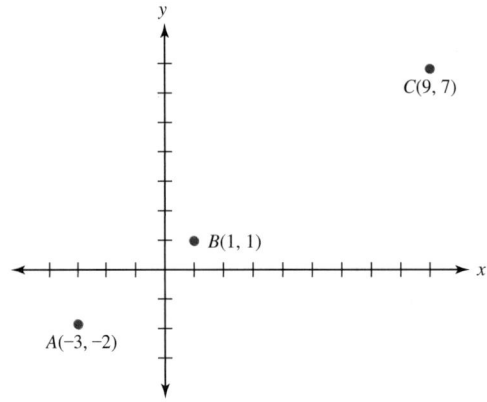

Another strategy is to *use the Triangle Inequality*. If the three points are not collinear, then they must make a triangle where $AB + BC < AC$. If they are collinear, then $AB + BC = AC$. We can use the distance formula to find the needed lengths.

Carrying Out the Plan. Using the given coordinates, we have that

$$AB = \sqrt{(1 - (-3))^2 + (1 - (-2))^2} = \sqrt{16 + 9} = \sqrt{25} = 5$$

$$BC = \sqrt{(9 - 1)^2 + (7 - 1)^2} = \sqrt{64 + 36} = \sqrt{100} = 10$$

$$AC = \sqrt{(9 - (-3))^2 + (7 - (-2))^2} = \sqrt{144 + 81} = \sqrt{225} = 15$$

A quick inspection of the numbers shows that $AB + BC = 5 + 10 = 15 = AC$. As a result, the three points must be collinear.

Looking Back. One way to check the work is to write an equation for the line $\overleftrightarrow{AC}$. If the coordinates for B satisfy the equation, then the points must be collinear. Using what we learned in Chapter 9, the equation of $\overleftrightarrow{AC}$ in slope-intercept form is $y = \frac{3}{4}x + \frac{1}{4}$.

Evaluating the equation at $x = 1$, the x-coordinate for B, we have $y = \frac{3}{4}(1) + \frac{1}{4} = 1$. Because 1 is the y-coordinate for B, then B must lie on $\overleftrightarrow{AC}$.

The midpoint of a segment can also be expressed with coordinate geometry. In Figure 12.7, the x-coordinate of the midpoint of a horizontal segment is halfway between the x-coordinates of the endpoints. Consequently, if the endpoints have coordinates $A(x_1, y_1)$ and $B(x_2, y_1)$, then the coordinates of the midpoint are $M\left(\frac{x_1 + x_2}{2}, y_1\right)$.

Similarly, if the segment is vertical, the y-coordinate of the midpoint is halfway between the y-coordinates of the endpoints. As a result, if the endpoints have coordinates $C(x_1, y_1)$ and $D(x_1, y_2)$, then the coordinates of the midpoint are $M\left(x_1, \frac{y_1 + y_2}{2}\right)$.

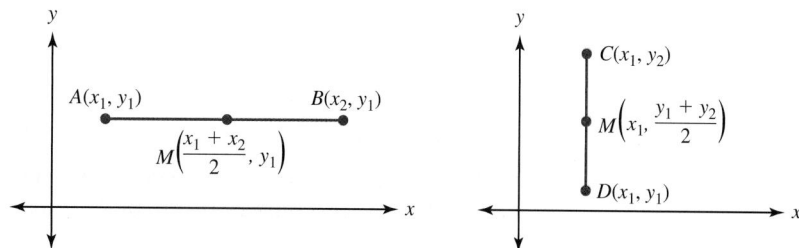

Figure 12.7 Midpoints on horizontal and vertical segments

If the segment is not horizontal or vertical, then the midpoint is halfway between the endpoints, both horizontally and vertically (Figure 12.8). This leads to the following theorem, the midpoint formula.

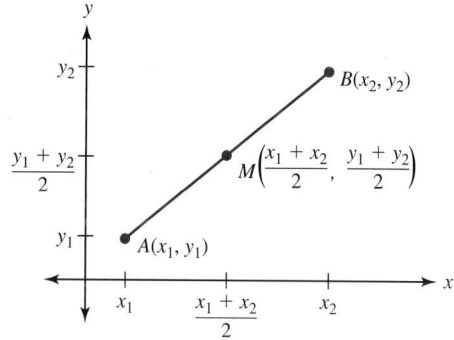

Figure 12.8 The midpoint formula

Theorem 12.2   Midpoint Formula

If points A and B have coordinates $A(x_1, y_1)$ and $B(x_2, y_2)$, then the midpoint, M, of $\overline{AB}$ has coordinates $M\left(\dfrac{x_1 + x_2}{2}, \dfrac{y_1 + y_2}{2}\right)$.

Example 12.5   Using the Midpoint Formula

Find the midpoint of the segment connecting each pair of points.

a. $A(5, 3)$ and $B(11, 3)$ **b.** $C(3, 5)$ and $D(-5, 10)$

Solution

a. Because $\overline{AB}$ is horizontal, the coordinates of the midpoint are $\left(\dfrac{5 + 11}{2}, 3\right) = (8, 3)$.

b. The midpoint of $\overline{CD}$ has coordinates $\left(\dfrac{3 + (-5)}{2}, \dfrac{5 + 10}{2}\right) = \left(-1, \dfrac{15}{2}\right)$.

Check Your Understanding 12.1-A

1. Use the city map in Example 12.2 to answer each question.

a. Describe point D as being to the north, south, east, or west of point A.

b. Write a set of directions that explains how to walk from the intersection of Oak and 3rd to Beech and 5th by passing through point D.

2. Find the distance between each pair of points.

   a. $A(1, 1)$ and $B(1, 5)$ **b.** $C(4, -4)$ and $D(-2, -4)$

3. Are the points $A(-2, -5)$, $B(4, 2)$, and $C(7, 10)$ collinear?

4. Let $\triangle ABC$ have vertices with coordinates $A(3, 5)$, $B(-1, 2)$, and $C(4, -3)$. Find the coordinates of the midpoints of the three sides.

Talk About It How does the Cartesian coordinate system provide a powerful connection between geometry and algebra?

Problem Solving

Activity 12.1

The Tangram puzzle consists of seven pieces: one square, one parallelogram, and five isosceles right triangles. The seven pieces can be arranged in many different ways but are shown arranged in a square.

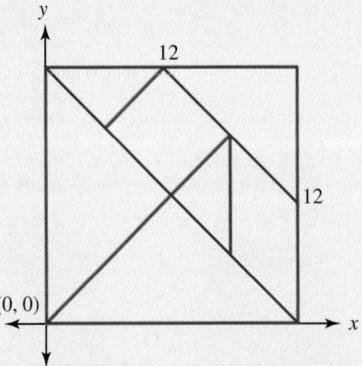

What are the coordinates of the vertices of each piece if the lower left-hand corner is placed on the origin, and the length of each side of the large square is 12?

Polygons and Coordinate Geometry

Explorations Manual 12.3

Because many planar shapes are made from line segments, we can use the distance and midpoint formulas to investigate their properties. We begin with triangles.

Example 12.6	The Distance Formula and a Right Triangle

Representation

Determine whether $\triangle ABC$ is a right triangle.

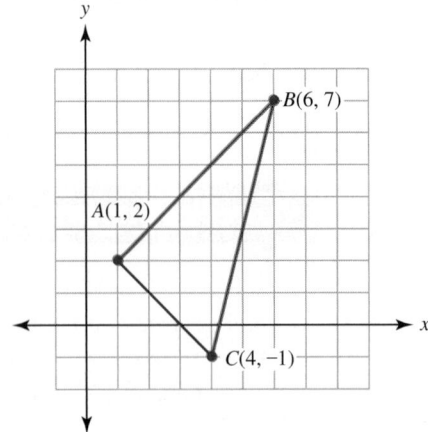

Solution $\triangle ABC$ is a right triangle if its sides satisfy the Pythagorean Theorem. We first find the length of each side by using the distance formula, or

$$AB = \sqrt{(6-1)^2 + (7-2)^2} = \sqrt{25+25} = \sqrt{50}$$

$$AC = \sqrt{(4-1)^2 + ((-1)-2)^2} = \sqrt{9+9} = \sqrt{18}$$

$$BC = \sqrt{(6-4)^2 + (7-(-1))^2} = \sqrt{4+64} = \sqrt{68}$$

Substituting the lengths into the Pythagorean Theorem, we have that $AB^2 + AC^2 = (\sqrt{50})^2 + (\sqrt{18})^2 = 50 + 18 = 68 = (\sqrt{68})^2 = BC^2$. $\triangle ABC$ is a right triangle.

Example 12.7 The Length of a Median

Reasoning

The **median** of a triangle is a segment that joins a vertex to the midpoint of the opposite side. Suppose a right triangle is placed on the coordinate plane so that the vertex opposite the hypotenuse is on the origin and the legs are on the axes. Use the distance formula and the following diagram to show that the median from the right angle to the hypotenuse is one-half the length of the hypotenuse.

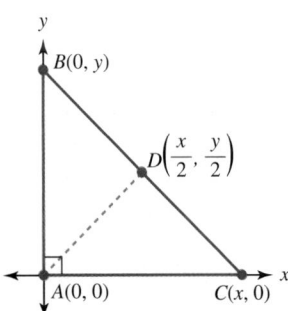

Solution We are asked to verify that $AD = \dfrac{BC}{2}$. We first observe that $DC = \dfrac{BC}{2}$ because D is the midpoint of $\overline{BC}$. In other words, the conjecture holds true if we can show that $AD = DC$. Using the distance formula, we have

$$AD = \sqrt{\left(\frac{x}{2}-0\right)^2 + \left(\frac{y}{2}-0\right)^2} = \sqrt{\left(\frac{x}{2}\right)^2 + \left(\frac{y}{2}\right)^2}$$

$$DC = \sqrt{\left(\frac{x}{2}-x\right)^2 + \left(\frac{y}{2}-0\right)^2} = \sqrt{\left(-\frac{x}{2}\right)^2 + \left(\frac{y}{2}\right)^2} = \sqrt{\left(\frac{x}{2}\right)^2 + \left(\frac{y}{2}\right)^2}$$

Because the segments have the same length, $DC = AD$ and $AD = \dfrac{BC}{2}$.

Next, we use the distance and midpoint formulas to verify two properties of quadrilaterals.

Example 12.8 | **The Diagonals of a Parallelogram**

Reasoning

The points $A(0, 0)$, $B(2, 4)$, and $D(7, 0)$ are three vertices of a parallelogram. If the fourth vertex is in the first quadrant, find its coordinates, and use them to show that the diagonals of the parallelogram bisect each other.

Solution From the given coordinates, we know that point $B(2, 4)$ is located 2 units to the right of and 4 units above point $A(0, 0)$. We also know that point $D(7, 0)$ is 7 units directly to right of $A(0, 0)$. Because the opposite sides of a parallelogram are parallel and congruent, the missing point must be located 7 units directly to the right of $B(2, 4)$. So the point must have the coordinates $(9, 4)$ and we label it point C.

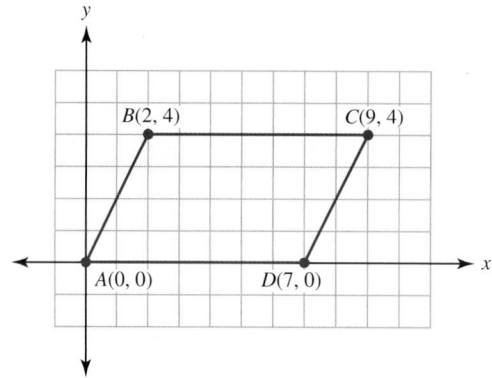

If the diagonals of the parallelogram bisect each other, they must intersect each other at their midpoints. Using the midpoint formula, we have

$$\text{Midpoint of } \overline{AC} = \left(\frac{9 + 0}{2}, \frac{4 + 0}{2}\right) = \left(\frac{9}{2}, 2\right)$$

$$\text{Midpoint of } \overline{BD} = \left(\frac{7 + 2}{2}, \frac{4 + 0}{2}\right) = \left(\frac{9}{2}, 2\right)$$

Because the midpoints have the same coordinates, the diagonals must bisect each other.

Example 12.9 | **The Diagonals of a Rectangle**

Reasoning

Consider a rectangle with one vertex on the origin and two sides on the axes. Verify that the diagonals of the rectangle are congruent.

Solution Because one corner of the rectangle is on the origin, the vertices have the coordinates $A(0, 0)$, $B(0, y)$, $C(x, y)$, and $D(x, 0)$, where x and y are real numbers.

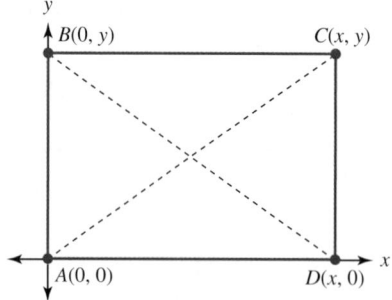

Next, we use the distance formula to compute the length of the diagonals, or

$$AC = \sqrt{(x - 0)^2 + (y - 0)^2} = \sqrt{x^2 + y^2}$$

$$BD = \sqrt{(0 - x)^2 + (y - 0)^2} = \sqrt{(-x)^2 + y^2} = \sqrt{x^2 + y^2}$$

Explorations
Manual
12.4

Because the diagonals have the same length, they must be congruent. ∎

Circles and Coordinate Geometry

Coordinate geometry and the distance formula can also be used to describe circles. Consider a circle with a radius r and a center at (h, k), and suppose the point (x, y) is any point on the circle (Figure 12.9).

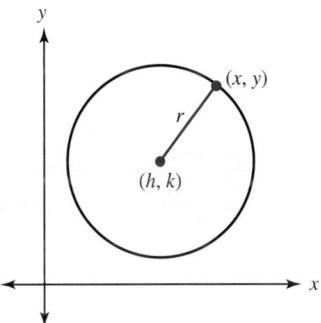

Figure 12.9 A circle with center (h, k) and radius r

The definition of a circle implies that the point (x, y) must be r units from (h, k). So by the distance formula, we have that

$$r = \sqrt{(x - h)^2 + (y - k)^2}$$

Squaring both sides of the equation, we get $r^2 = (x - h)^2 + (y - k)^2$, which serves as the standard form for the equation of a circle.

Theorem 12.3 Equation of a Circle

A circle with center (h, k) and radius r has the equation $r^2 = (x - h)^2 + (y - k)^2$. □

Example 12.10 Finding the Equation of a Circle

Representation

Find the equation of a circle with center $(-4, 5)$ and radius $r = 4$. Draw a graph of the circle.

Solution If $(h, k) = (-4, 5)$ and $r = 4$, then the equation of the circle is

$$4^2 = [x - (-4)]^2 + (y - 5)^2$$

or

$$16 = (x + 4)^2 + (y - 5)^2$$

To draw the circle, we first find several points on it. Four points are relatively easy to find because they are four units directly above, below, to the right, and to the left of the center. Hence, points $A(-4, 9)$, $B(-4, 1)$, $C(0, 5)$, and $D(-8, 5)$ all lie on the circle. Plotting these points and sketching the rest of the circle, we have the following graph:

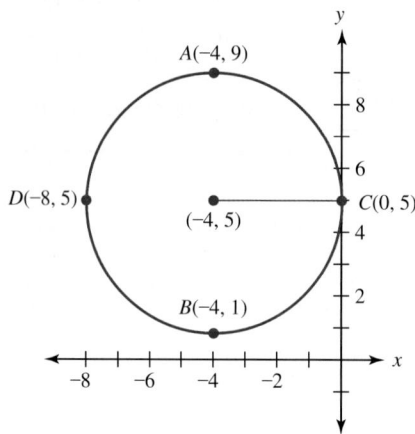

Check Your Understanding 12.1-B

1. Determine whether each set of three vertices represents an acute, right, or obtuse triangle.

   a. $A(6, 3)$, $B(7, 1)$, and $C(3, 1)$ **b.** $A(1, 9)$, $B(2, 5)$, and $C(1, 1)$

2. Determine whether the triangles in the previous problem are scalene, isosceles, or equilateral.

3. The coordinates of three vertices of a square are $A(2, -3)$, $B(-2, -1)$, and $C(0, 3)$. What are the coordinates of the fourth vertex?

4. Find the equation of the circle with the given center and radius.

   a. $(4, 5)$, $r = 1$ **b.** $(-3, -6)$, $r = 5$ **c.** $(-1, 5)$, $r = \sqrt{19}$

Talk About It Suppose the coordinates of the four vertices of a quadrilateral are given. How can we use the distance formula to determine whether the quadrilateral is a rectangle?

Reasoning

Activity 12.2

Find the midpoints of the sides in each quadrilateral. Connect consecutive midpoints to form a quadrilateral within the quadrilateral. In each case, what does the new quadrilateral appear to be? Use geometry software to test your conjecture for other quadrilaterals. Does it hold true? If so, can you explain why?

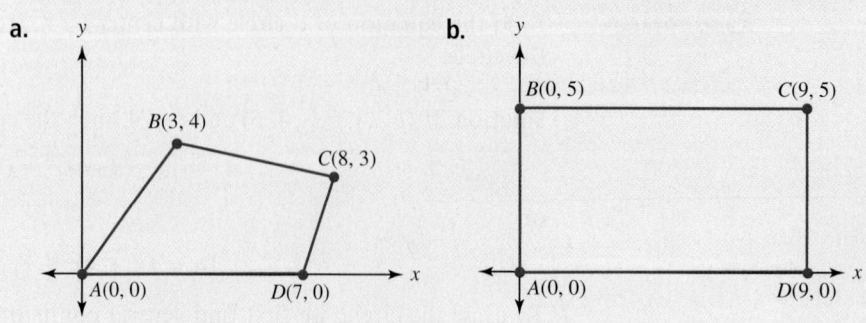

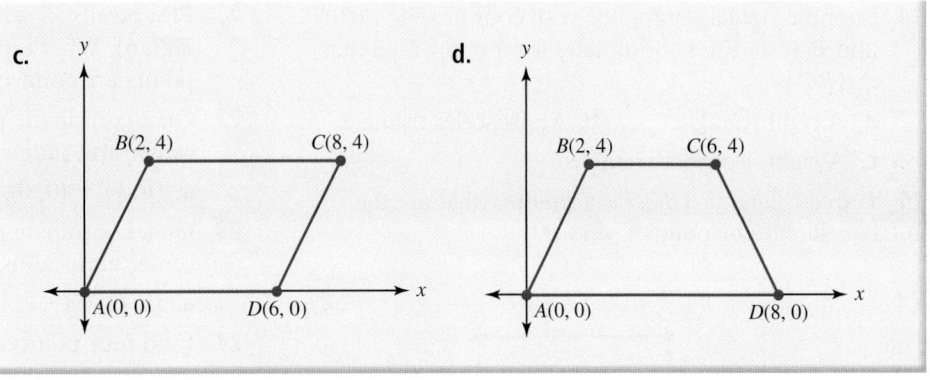

SECTION **12.1**   **Learning Assessment**

■ Understanding the Concepts

1. What spatial reasoning skills can be learned through coordinate geometry?

2. What are latitudes and longitudes? Do they form a true rectangular grid? Explain.

3. What is the unit length on the Cartesian coordinate system, and how is it used to measure the length of line segments?

4. How do we compute the length of a horizontal or vertical line segment?

5. What role does the Pythagorean Theorem play in finding the length of a line segment?

6. If we are trying to find the distance between two points $A(x_1, y_1)$ and $B(x_2, y_2)$, does it matter whether we use $(x_1 - x_2)$ instead of $(x_2 - x_1)$ in the distance formula? What about $(y_1 - y_2)$ instead of $(y_2 - y_1)$? Explain.

7. How is it possible to find the coordinates of the midpoint of a segment if the coordinates of the endpoints are known?

8. Why is it advantageous to place one of the vertices of a figure at the origin when verifying a result using coordinate geometry and the distance formula?

9. How is the distance formula used to write the equation of a circle?

■ Representing the Mathematics

Many state, county, and city maps use a coordinate system that involves letters and numbers. Use the following map of Colorado to answer Questions 10–12.

10. Name the city in the square with the given coordinates.
    a. B4   b. F6   c. H6
    d. G16   e. K7   f. L11

11. Give the coordinates of each city in Colorado.
    a. Fort Morgan   b. Pueblo
    c. Colorado Springs d. Vail
    e. Grand Junction   f. Sterling

12. Describe each city as being to the north, south, east, or west of Denver.
    a. Meeker   b. Burlington   c. Yuma
    d. Canon City   e. Durango   f. Castle Rock

13. For each set of three vertices, determine whether the triangle they represent is acute, right, or obtuse and then scalene, isosceles, or equilateral.
    a. $A(4, 2)$, $B(4, 5)$, and $C(6, 2)$
    b. $A(-4, 1)$, $B(2, 1)$, and $C(-1, -1)$
    c. $A(-4, -1)$, $B(-5, -4)$, and $C(-8, -3)$
    d. $A(6, 5)$, $B(10, 6)$, and $C(11, 11)$

14. Suppose $\overline{AB}$ has endpoints with coordinates $A(0, 0)$ and $B(4, 0)$. Find coordinates for a point C so that $\triangle ABC$ is:

  a. A right triangle.   **b.** An isosceles triangle.

  c. A right, isosceles triangle.

15. If quadrilateral $ABCD$ is a square, what are the coordinates of points C and D?

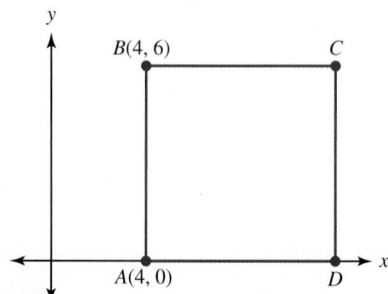

16. If quadrilateral $ABCD$ is a rectangle, what are the coordinates of points B and D?

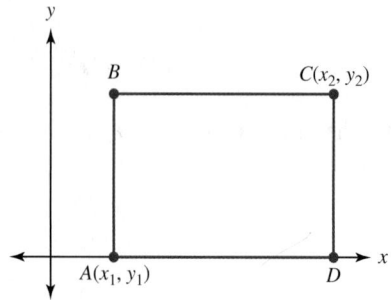

17. If quadrilateral $ABCD$ is a rhombus, what are the coordinates of point C?

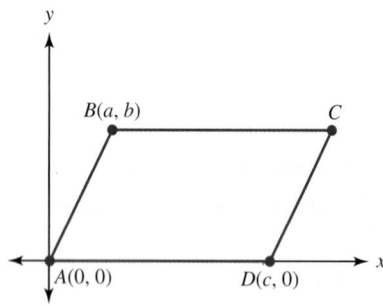

18. The coordinates for three vertices of a square are given. Find the coordinates of the fourth vertex.

  a. $A(3, 5)$, $B(1, 3)$, and $C(-1, 5)$

  b. $A(-2, -2)$, $B(-3, 2)$, and $C(1, 3)$

19. The coordinates for three vertices of a parallelogram are $A(-4, -2)$, $B(-7, 1)$, and $C(-5, 6)$. Find possible coordinates for the fourth vertex.

20. Plot points $A(2, 3)$, $B(6, 3)$, and $C(4, 7)$. What type of figure is created when the points are connected? How do you know?

21. Plot points $A(-1, -4)$, $B(3, -1)$, $C(-2, 3)$, and $D(2, 6)$. What type of figure is created when the points are connected? How do you know?

22. On a coordinate plane, draw a circle with the given center and radius.

  a. $(h, k) = (0, 0)$, $r = 3$.   **b.** $(h, k) = (3, 3)$, $r = 3$.

23. On a coordinate plane, draw a circle with the given center and radius.

  a. $(h, k) = (-4, -4)$, $r = 2$. **b.** $(h, k) = (1, 2)$, $r = 4$.

24. Find four points on the following circle.

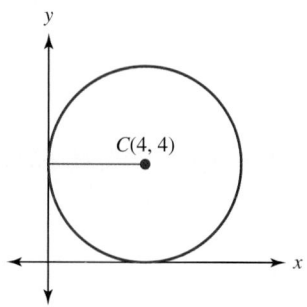

■ Exercises

25. If each pair of points represents the endpoints of a segment, find its length.

  a. $A(0, 0)$ and $B(7, 0)$ **b.** $A(0, 0)$ and $B(0, -4)$

  c. $A(-8, -2)$ and $B(-1, -2)$

26. If each pair of points represents the endpoints of a segment, find its length.

  a. $A(1, 9)$ and $B(1, -1)$ **b.** $A(3, 5)$ and $B(-2, 5)$

  c. $A(-4, -5.3)$ and $B(-4, -1.4)$

27. Find the length of each line segment.

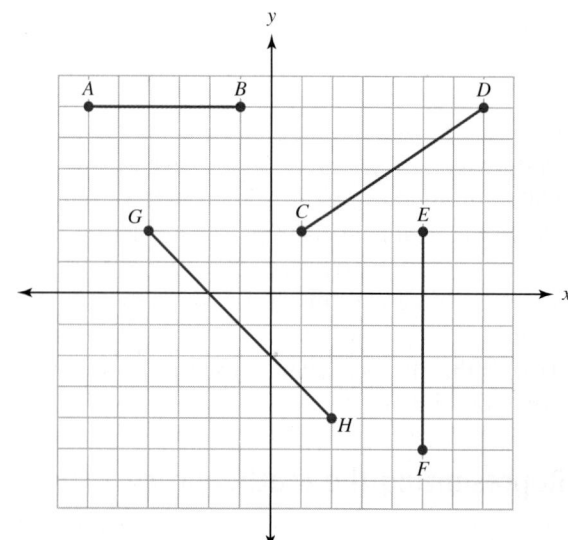

28. Find the distance between each pair of points.

  a. $(3, 2)$ and $(4, 7)$ **b.** $(-6, 0)$ and $(-2, -5)$

  c. $(0.5, 0.3)$ and $(2.7, -3.1)$

29. Find the distance between each pair of points.
   a. $(-3, 2)$ and $(8, -5)$
   b. $(4.7, -2.9)$ and $(-4.6, -6.5)$
   c. $\left(-\frac{1}{2}, \frac{1}{3}\right)$ and $\left(\frac{1}{2}, \frac{1}{4}\right)$

30. Determine whether points $A(-1, 3)$, $B(1, 6)$, and $C(3, 9)$ are collinear.

31. Determine whether points $A(9, 10)$, $B(7, 6)$, and $C(3, 0)$ are collinear.

32. Find the coordinates of the midpoint of the segment joining each pair of points.
   a. $(0, 0)$ and $(4, 0)$ b. $(0, 0)$ and $(0, 7)$
   c. $(-2, -1)$ and $(3, -4)$

33. Find the coordinates of the midpoint of the segment joining each pair of points.
   a. $(1, 1)$ and $(7, 9)$
   b. $(-3.2, 2.5)$ and $(-0.7, 4.6)$
   c. $\left(\frac{1}{3}, \frac{1}{5}\right)$ and $\left(\frac{11}{2}, -\frac{7}{4}\right)$

34. Find the coordinates of the midpoint of each segment:

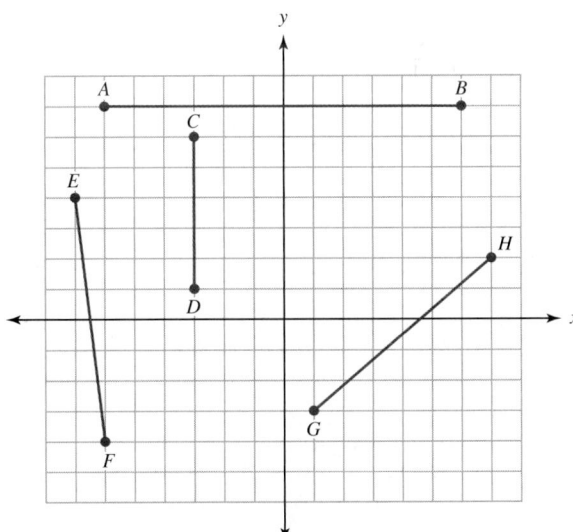

35. Write the equation of the circle with the given center and radius.
   a. $(h, k) = (-1, -1)$, $r = 6$. b. $(h, k) = (2, 3)$, $r = 5$.
   c. $(h, k) = (0, 0)$, $r = 3$.

36. Write the equation of the circle with the given center and radius.
   a. $(h, k) = (-4, 6)$, $r = 7$.
   b. $(h, k) = \left(\frac{1}{2}, \frac{1}{3}\right)$, $r = \frac{1}{2}$.
   c. $(h, k) = (0.6, 7.5)$, $r = 3.7$.

37. Find the radius and the coordinates of the center of each circle.
   a. $(x - 1)^2 + (y + 1)^2 = 9$.

b. $(x - 4)^2 + (y + 7)^2 = 36$.
c. $x^2 + y^2 = 16$.

38. Find the radius and the coordinates of the center of each circle.
   a. $(x + 6)^2 + (y - 1)^2 = 25$.
   b. $\left(x - \frac{1}{2}\right)^2 + \left(y - \frac{1}{3}\right)^2 = \frac{1}{25}$.
   c. $(x + \sqrt{6})^2 + (y + 0.8)^2 = 15$.
   d. $(x - 0.67)^2 + (y + 1.3)^2 = 14.4$.

39. The general form of the equation of a circle is given by $x^2 + y^2 + Dx + Ey + F = 0$. We find it by expanding and simplifying the equation $r^2 = (x - h)^2 + (y - k)^2$. Find the general form of the equation for each circle.
   a. $(x - 2)^2 + (y + 2)^2 = 4$.
   b. $(x + 4)^2 + (y - 7)^2 = 16$.
   c. $x^2 + y^2 = 16$.

■ Problems and Applications

40. The distance between two points $A(x_1, y_1, z_1)$ and $B(x_2, y_2, z_2)$ in three-dimensional space is given by the formula $d = \sqrt{(x_2 - x_1)^2 + (y_2 - y_1)^2 + (z_2 - z_1)^2}$. Find the distance between each pair of points.
   a. $A(1, 2, 3)$ and $B(5, 6, 7)$
   b. $A(-2, 3, -4)$ and $B(9, -2, -5)$

41. Consider three arbitrary points: $A(x_1, y_1)$, $B(x_2, y_2)$, and $C(x_3, y_3)$. Use the distance formula to write a formula to determine whether any three points are collinear.

42. If A is one endpoint of a segment and M is its midpoint, find the coordinates of the other endpoint B.
   a. $A(3, 1)$ and $M(4, 2)$
   b. $A(-1, -2)$ and $M(-3, -4)$
   c. $A(4, 5)$ and $M(-2, -2)$
   d. $A(-1, 4)$ and $M(2, -3)$

43. Suppose the endpoints of segment $\overline{AB}$ have coordinates $A(0, 0)$ and $B(6, 0)$. Find the coordinates of a point C so that $\triangle ABC$ is an equilateral triangle.

44. Suppose the endpoints of segment $\overline{AB}$ have coordinates $A(1, 0)$ and $B(4, 2)$. Find possible coordinates for points C and D so that quadrilateral $ABCD$ is a:
   a. Square.
   b. Rectangle with a length twice AB.
   c. Parallelogram with a side of length 5 on the x-axis.

45. If $\overline{AB}$ is the side of a square with coordinates $A(-2, 3)$ and $B(1, -1)$, find all possible coordinates for points C and D so that C and D are the other two vertices of the square.

46. Suppose that rectangle $ABCD$ has sides $AB = 10$ and $BC = 8$. What are the coordinates of the four vertices if:

a. $\overline{AB}$ is on the x-axis, $\overline{BC}$ is on the y-axis, and all coordinates are greater than or equal to zero?

b. $\overline{AB}$ is on the vertical line $x = 5$, $\overline{BC}$ is on the horizontal line $y = -4$, and B and C are both in the fourth quadrant?

47. Give the coordinates of six vertices that form a regular hexagon.

48. Is the following figure a regular pentagon? Why or why not?

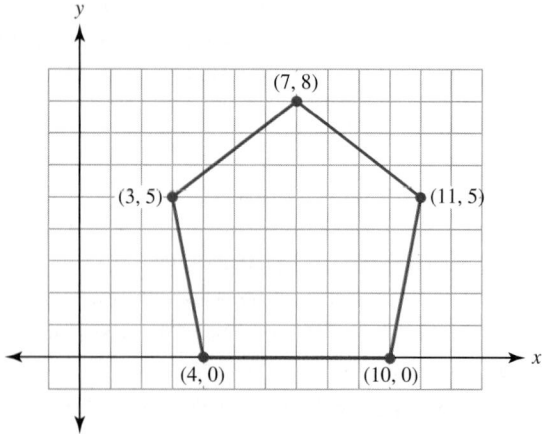

49. The equation of a particular circle is given by $3x^2 + 3y^2 = 48$. What are the center and radius of the circle?

50. Consider the street map in Example 12.2. Eric is at the intersection of Oak and 8th, and his wife Noel is at the intersection of Elm and 2nd. If they both want to walk the same and shortest distance possible, at what intersection should they meet for lunch?

51. The earth has a circumference of about 24,000 mi at the equator. There are 360 longitudes on the earth, and each has the measure of one degree.

a. How many miles is it from one longitude to the next at the equator?

b. Longitudes are further subdivided into minutes, where 60 min = 1 degree. How many miles is it from one minute to the next at the equator?

52. The following map represents the streets of a particular city.

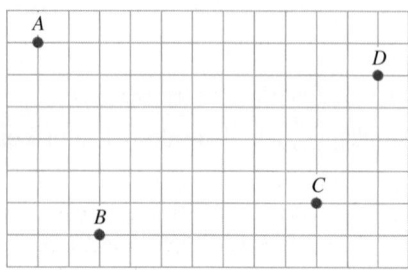

a. If a taxi wants to travel from A to D, while making a stop at both B and C, what is the fewest number of blocks the taxi will travel?

b. If it takes the taxi 15 s on average to travel a block, how long will it take for the taxi to make the trip?

■ Mathematical Reasoning and Proof

53. Consider the following isosceles triangle.

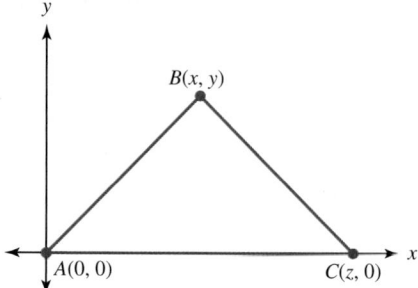

What is the relationship between the value of x and the value of z?

54. Use the distance formula to prove that the triangle with vertices $A(-1, 4)$, $B(5, 6)$, and $C(3, 0)$ is an isosceles triangle.

55. Use coordinate geometry to show that the medians drawn to the equal sides of an isosceles triangle are congruent.

56. Consider $\triangle ABC$. Let $\overline{ED}$ join the midpoints of $\overline{AB}$ and $\overline{AC}$.

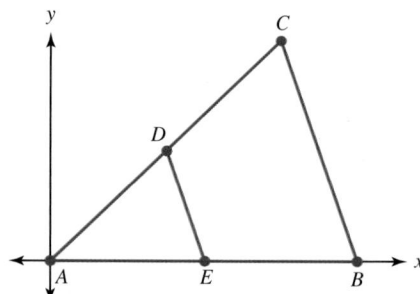

What is true about length of $\overline{ED}$ compared to the length of $\overline{BC}$? Use geometry software to test your conjecture with other triangles. Does it hold true? If so, can you explain why?

57. Suppose the vertices of a square are given by points $A(0, 0)$, $B(0, 5)$, $C(5, 5)$, and $D(5, 0)$. Use the distance formula to:

a. Verify that the sides of the square are congruent.

b. Verify that the diagonals of the square are congruent.

58. Let the vertices of an isosceles trapezoid be given by $A(0, 0)$, $B(2, 5)$, $C(6, 5)$, and $D(8, 0)$. Use the distance formula to verify that the diagonals are congruent.

59. Use the slope of two lines to verify that the diagonals of the following rhombus are perpendicular.

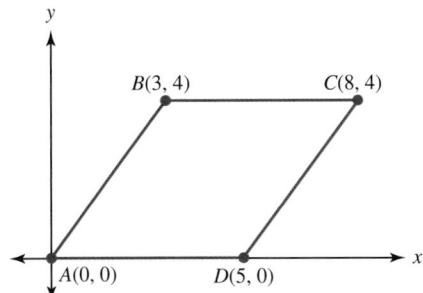

60. The **median** of a trapezoid is the segment that joins the midpoints of the two legs. What is true about length of the median compared to the sum of the lengths of the bases? Use geometry software to test your conjecture with other triangles. Does it always hold true? If so, can you explain why?

61. Does the equation $(x - 2)^2 + (y + 4)^2 = -4$ make sense for the equation of a circle? Why or why not?

Mathematical Communication

62. Suppose that the coordinates of a certain quadrilateral are given. Explain to a partner how to use coordinate geometry to determine whether the quadrilateral is a square.

63. Draw a simple map of a playground. It can be an actual playground that you know of or one of your own making.

   a. What questions might you ask as a teacher that would require your students to give directions and distances between pieces of playground equipment?

   b. Write a set of directions that explains how to get from one piece of equipment to another. Give it to one of your peers to critique.

64. Dexter and Heather are using the distance formula to find the distance between points (3, 4) and (6, 1). Heather places the points into the formula as $\sqrt{(6 - 3)^2 + (4 - 1)^2}$ because it is easier to subtract. Dexter argues that the order of the coordinates must be the same; otherwise the answer will be wrong. Who, if anyone, is correct?

65. Hunter claims that the midpoint of the segment with endpoints $A(6, 4)$ and $B(-2, -2)$ is $M(8, 3)$. What went wrong?

66. Libby looks at points $A(0, 0)$, $B(6, 0)$, and $C(0, 3)$ and immediately claims they form a right triangle. How did she determine this so quickly?

Building Connections

67. Why is the latitude and longitude of a particular position on the globe always given in terms of degrees rather than miles or kilometers? Do an Internet search to see whether you can find the answer.

68. Describe the relative position of your house, apartment, or dorm as being to the north, south, east, or west of:

   a. Your university or college.

   b. The nearest grocery store.

   c. The nearest gas station.

   d. The nearest shopping mall.

69. Write a set of directions that explains how to get from your classroom to your place of residence.

70. How does the Cartesian coordinate system play an integral role in connecting algebra to geometry?

71. Discuss each question with a small group of your peers.

   a. How might a geoboard be used to teach the ideas of coordinate geometry?

   b. What other manipulatives might be useful in teaching coordinate geometry?

SECTION 12.2 Transformations

To this point, we have focused on shapes with fixed positions and sizes. However, real-world shapes are often changing in location, orientation, and even dimension. To fully understand geometry, we need to learn how shapes can be moved and how to represent those motions. This area of study is called **motion** or **transformation geometry.**

Explorations Manual 12.5

Mathematically, a shape on the plane can be moved in four ways: a translation, a rotation, a reflection, and a glide reflection (Figure 12.10). Intuitively, we can think of these motions as a slide, a turn, a flip, and a slide-flip, respectively.

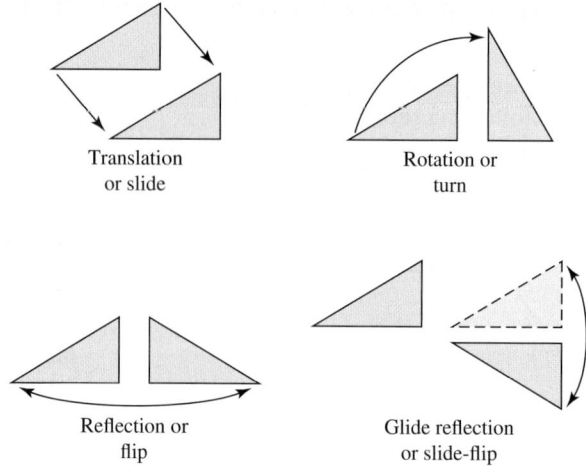

Figure 12.10 The four basic transformations

In each movement, notice that only the position and the orientation of the triangle change. The size and shape remain the same, indicating that the distances between points are preserved. Any motion that preserves length is called a **rigid motion,** or **isometry** ("iso" meaning same, "-metry" meaning measure). Because rigid motions preserve length, they must also preserve angles. To understand why, suppose $\angle A'B'C'$ is obtained from $\angle ABC$ by means of a rigid motion (Figure 12.11). Because the rigid motion preserves length, $AB = A'B'$, $BC = B'C'$, and $AC = A'C'$. As a result, $\triangle ABC \cong \triangle A'B'C'$ by SSS and $\angle ABC \cong \angle A'B'C'$ because corresponding parts of congruent triangles are congruent.

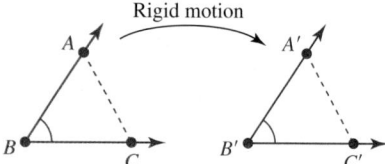

Figure 12.11 A rigid motion preserves angle measure

Figure 12.12 A transformation maps preimage P to image P'

Any rigid motion can be considered to be a function because it literally makes a one-to-one correspondence between points on the plane. Mathematically, a function that indicates how each point P on the plane corresponds to another point P' is called a **transformation.** P' is the **image** of P, and P is the **preimage** of P' (Figure 12.12).

Each of the basic rigid motions has its own unique characteristics. We discuss each one in turn.

Translations

Intuitively, a **translation,** or a **slide,** is a motion that takes any object from one location to another without twisting or turning it (Figure 12.13). In other words, a translation moves every point on the object in the same direction for the same distance. Examples are opening or closing a window, pushing a chair, or driving a car.

Figure 12.13 Examples of translations

When making a translation on a plane, the direction and the length of the motion are given by a **translation vector,** or **slide arrow.** For instance, consider sliding $\triangle DEF$ by the translation vector that extends from A to B (Figure 12.14). To find the image of $\triangle DEF$, or $\triangle D'E'F'$, we slide every point along lines that are parallel to $\overline{AB}$, in the direction from A to B, and by a distance equal to AB. In other words, given any point P on $\triangle DEF$ and its image point P' on $\triangle D'E'F'$, then $\overline{AB}\|\overline{PP'}$ and $AB = PP'$.

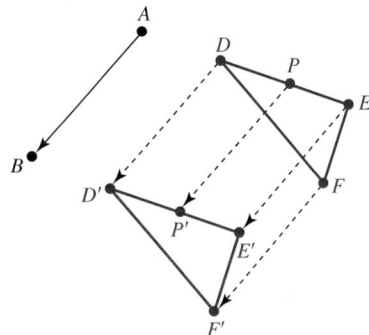

Figure 12.14 Translation of $\triangle DEF$ by $\overline{AB}$

Definition of a Translation

A **translation** with a translation vector from A to B, written T_{AB}, is a transformation of the plane that maps every point P to a point P' such that $\overline{PP'}$ is congruent and parallel to $\overline{AB}$.

The subscript in the notation T_{AB} identifies the translation vector by giving its direction and length in terms of two points. The order of the points is important because the notation T_{AB} indicates a slide from A to B, whereas T_{BA} indicates a slide in the opposite direction from B to A.

Because a translation is a rigid motion that preserves length and angle measure, we can find the image of a polygon under the translation by finding the images of its vertices and connecting them with segments. Coordinate geometry can be used to help us locate the needed points.

Example 12.11 | Finding the Image of a Translation

Representation   Find the image of quadrilateral $ABCD$ under the translation that takes P to P'.

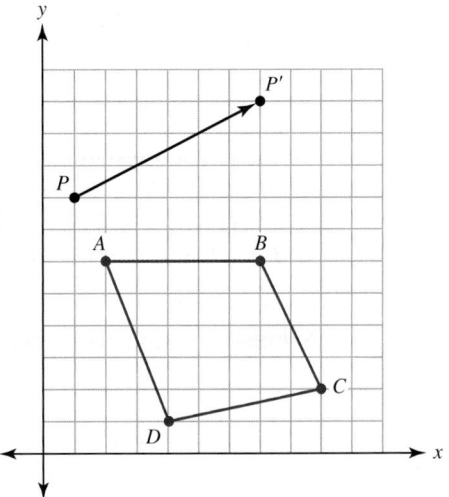

Solution The translation vector $T_{PP'}$ maps P to P', which is 6 units to the right and 3 units above P. To find the image of quadrilateral $ABCD$ under $T_{PP'}$, map each vertex to a position that is 6 units to the right and 3 units above its original position. Then connect them with segments to form quadrilateral $A'B'C'D'$.

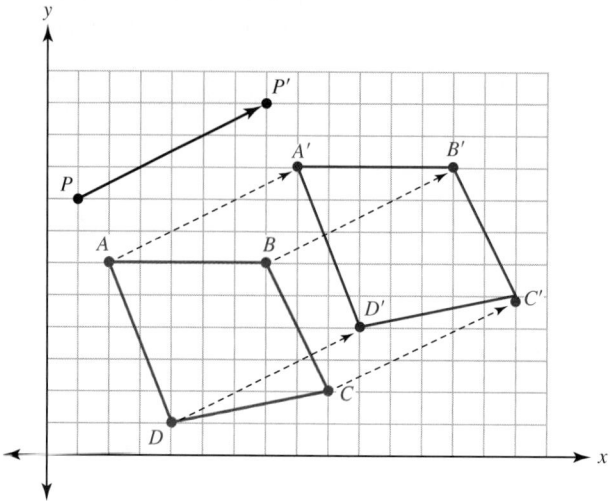

Because a translation is a function from the plane onto itself, coordinate geometry also helps us describe algebraically how a translation moves points on the plane. For instance, the translation in Example 12.11 took every point to a position that was 6 units to the right and 3 units above its original position. In other words, if (x, y) is a point on the plane, then $(x + 6, y + 3)$ are the coordinates of its image under the translation. In general, a translation maps any point (x, y) to point $(x + a, y + b)$, where a and b are real numbers. In notation, we write $(x, y) \rightarrow (x + a, y + b)$, where "$\rightarrow$" is read as "is moved to."

Example 12.12   Finding the Image of a Translation Described Algebraically

Consider $\triangle ABC$. Find the coordinates of the images of its vertices under the translation $(x, y) \rightarrow (x + 4, y + 3)$, and then draw its image.

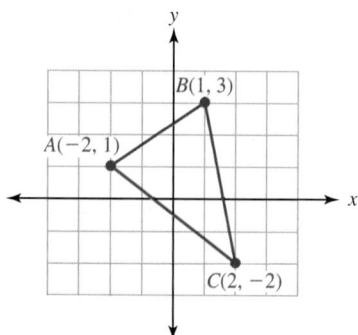

Solution We find the coordinates for the images of the vertices by evaluating each vertex in the formula for the translation.

Point (x, y)	$A(-2, 1)$	$B(1, 3)$	$C(2, -2)$
Image Point $(x + 4, y + 3)$	$A'(2, 4)$	$B'(5, 6)$	$C'(6, 1)$

We then draw the image of the triangle by plotting the images of the vertices and connecting them with segments.

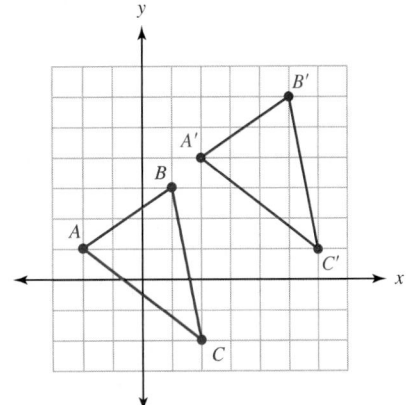

Notice that each of the translations so far did not change the orientation of the shape. This is true for any translation, so we call them **proper** or **orientation-preserving motions.**

Rotations

The second rigid motion is a **rotation,** or a **turn.** Common examples are the hands on a clock, a wheel spinning on its axle, or a fan rotating around its motor (Figure 12.15).

Figure 12.15 Examples of rotations

In each of these examples, notice the object rotates around a single point—the hands around the center of the clock face, the tire around its axle, and the fan around its motor. This point is the **center of rotation,** or **turn center,** and it can be any point that is in, on, or outside the figure. Also notice that each rotation has a direction, which is indicated with a **turn arrow** as being either clockwise or counterclockwise with respect to the center of rotation. Along with the turn arrow, an angle measure is often given to indicate the **turn angle,** or the amount of the rotation. The convention is to give a counterclockwise turn a positive angle measure and a clockwise turn a negative angle measure. Figure 12.16 shows the image of $\triangle DEF$ when it is rotated around point C by a turn angle of 60°.

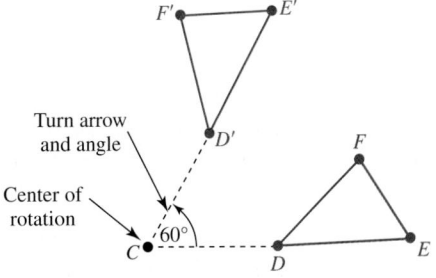

Figure 12.16 A rotation of $\triangle DEF$

The center of rotation and the turn angle are enough to completely designate a particular rotation. This leads to the following definition.

Definition of a Rotation

A **rotation** with center C and turn angle α, written $R_{C,\alpha}$, is a transformation of the plane that maps every point P to a point P' such that $m\angle PCP' = \alpha$ and $PC = P'C$.

Three things must be mentioned about this definition. First, the notation $R_{C,\alpha}$ designates both the turn center and the turn angle. For instance, $R_{C,45°}$ tells us to rotate the plane counterclockwise around C by an angle of $45°$. The notation $R_{C,-90°}$ tells us to rotate the plane clockwise around C by an angle of $90°$. Second, the center of rotation, C, is mapped onto itself and remains stationary throughout the rotation. Any point P that stays in the same position under a rigid motion is called a **fixed point.** Third, the statement $m\angle PCP' = \alpha$ indicates that every point on the plane is rotated by angle α. The statement $PC = P'C$ indicates that the distance from P' to the center of the rotation is the same as the distance from P to the center.

Explorations Manual 12.6

Finding the image of a shape under a rotation is not always easy, so we often use technology to help us make rotations. However, we can make some rotations by hand with the help of a rectangular grid or graph paper.

Example 12.13 | Finding the Image of a Rotation

Representation

Use graph paper to find the image of each figure under the given rotation.

a. Rotate $\triangle BAT$ by $R_{T,180°}$.

b. Rotate rectangle $FROG$ by $R_{C,-90°}$.

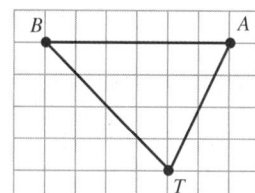
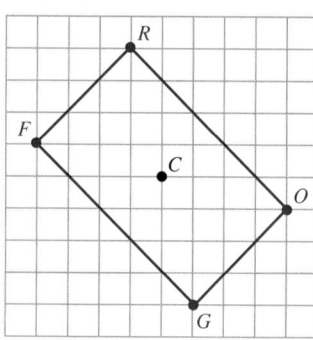

Solution

a. The rotation $R_{T,180°}$ tells us to rotate $\triangle BAT$ counterclockwise around point T by an angle of $180°$. Because T is the center of the rotation, it remains fixed, or $T = T'$. If we rotate points A and B $180°$ around T, their images are directly opposite A and B and at the same distance from T. Once the images of the vertices are located, we connect them to complete the image $\triangle B'A'T'$.

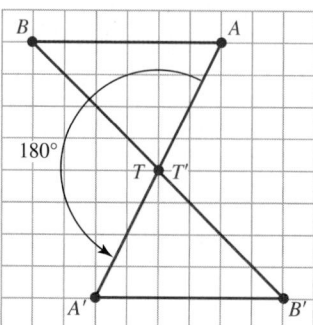

b. The rotation $R_{C,-90°}$ tells us to rotate the rectangle clockwise around C by an angle of $90°$. Because F is 4 units to the left of and 1 unit above C, F' is 1 unit to the right of and 4 units above C. Similarly, R' is 4 units to the right of and 1 unit above C, O' is 1 unit to the left of and 4 units below C, and G' is 4 units to the left of and 1 unit below C. We connect the images of the vertices to complete the image rectangle $F'R'O'G'$.

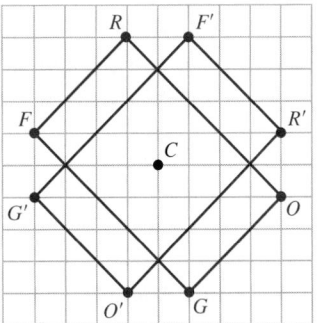

Several other facts about rotations need to be noted. First, the rotations in the last example did not affect the orientation of the figure. Rotations, like translations, are proper rigid motions. Second, because a complete rotation has a measure of 360°, any rotation with a turn angle of 360° must map every point on the plane back to its original position. Any such transformation is called an **identity transformation.** Third, any turn angle larger than 360° is equivalent to a turn angle between 0° and 360°. For instance, a turn angle of 430° includes one complete rotation and a part of another. Specifically, if we subtract 360° from the turn angle, it is equivalent to a turn angle of 430° − 360° = 70°. Finally, if any point is rotated through a **half turn** in either direction, either 180° or −180°, then the image of the point arrives at the same location (Figure 12.17).

This last point implies that any clockwise rotation can be specified by a counterclockwise rotation and vice versa. For instance, a rotation of 50° is equivalent to a rotation of −310° and a rotation of −135° is equivalent to a rotation of 225° (Figure 12.18).

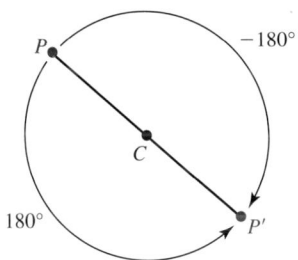

Figure 12.17 Half turns in the clockwise and counterclockwise directions

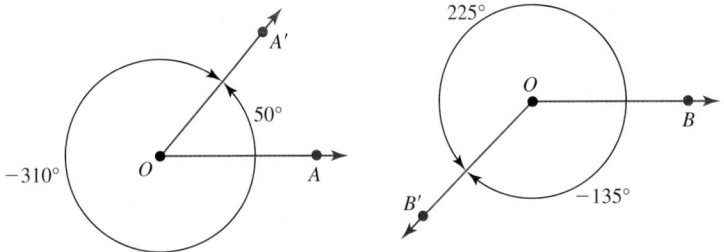

Figure 12.18 Equivalent rotations

Check Your Understanding 12.2-A

1. Find the image of $\triangle TRY$ under the translation T_{AB}.

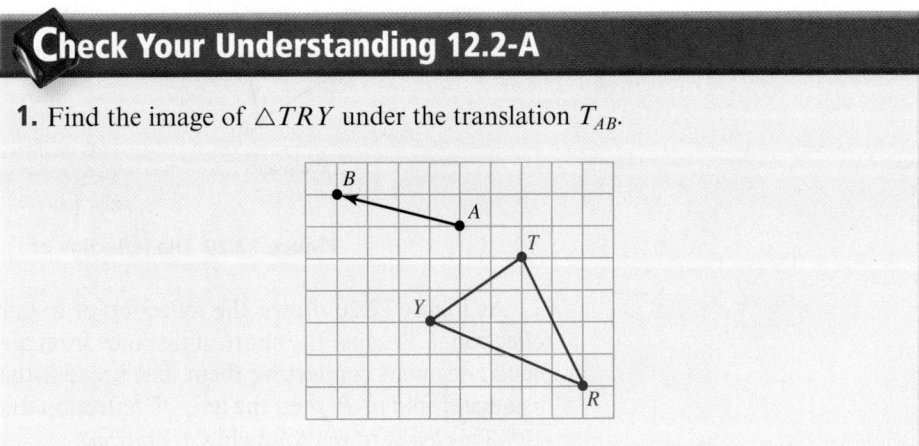

2. Find the coordinates of the image of each point under the translation that takes $(x, y) \rightarrow (x + 4, y - 8)$.

  a. $(1, 1)$ **b.** $(-4, 2)$ **c.** $(3, -5)$ **d.** $(-1, -7)$

3. If a point has coordinates $(1, 3)$, what are the coordinates of its image if:

  a. It is rotated $90°$ around the origin.

  b. It is rotated $-180°$ around the origin.

4. a. A counterclockwise rotation of $55°$ is the same as what clockwise rotation?

  b. A clockwise rotation of $–95°$ is the same as what counterclockwise rotation?

  c. A rotation of $560°$ is the same as what rotation between $0°$ and $360°$?

Talk About It If we can view any rigid motion as a function, what would be the domain and range of the function?

Problem Solving

Activity 12.3

Use a straightedge and compass to construct the image of segment $\overline{AB}$ under the translation T_{PQ}. (*Hint:* Construct parallelograms.)

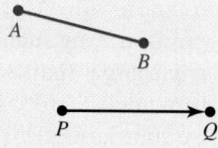

Reflections

Figure 12.19 Reflection in a mirror

When we think of a **reflection,** or a **flip,** we often think of a mirror in which an image appears behind it at a distance that matches that of the original object (Figure 12.19). A reflection of the plane works in much the same way, where a line called the **axis of reflection,** or **reflecting line,** acts as the mirror. To make the reflection, every point P on the plane is mapped to a unique point P' so that P' is on the opposite side of the axis of reflection and at the same distance as P. Any point on the axis of reflection remains fixed. Intuitively, this means that a reflection "flips" the plane so that the axis of reflection is on top of itself and the new position of any point is the mirror image of its original position. Figure 12.20 shows the reflection of $\triangle ABC$ in the line l.

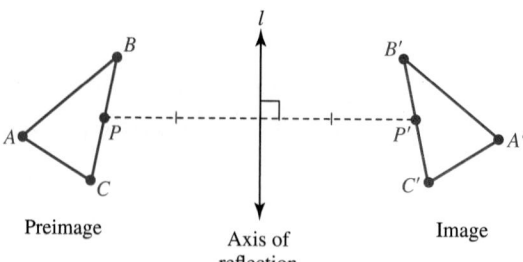

Figure 12.20 The reflection of $\triangle ABC$ in the line l

As Figure 12.20 shows, the reflection of a shape depends primarily on the axis of reflection, l. Because the shortest distance from any point P to the line l is the perpendicular segment connecting them and because the image point P' must be the same distance from l as P, then the axis of reflection must be the perpendicular bisector of $\overline{PP'}$. This leads to the following definition.

Definition of a Reflection

The **reflection** in a line *l*, written M_l, is a transformation of the plane that maps every point *P* to a point *P'* such that *l* is the perpendicular bisector of $\overline{PP'}$. If *P* is on *l*, then *P* is mapped to itself, or *P* = *P'*.

A number of tools can help us make reflections. One is a plastic reflector. We place the edge of the reflector over the reflecting line and then look through it to find and draw the needed image (Figure 12.21).

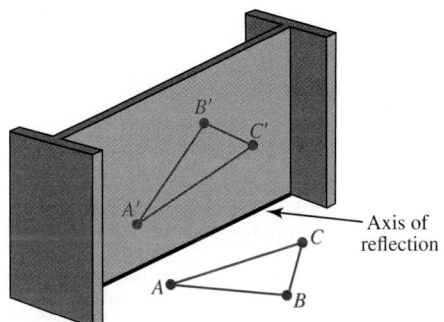

Figure 12.21 Drawing a reflection with a plastic reflector

The next two examples show how to use a coordinate system or another rectangular grid to make the reflection.

Example 12.14 | Reflecting a Point

Reasoning

Find the coordinates of the image of each point when it is reflected in the *y*-axis. What is the relationship between the coordinates of a point and the coordinates of its image when it is reflected in this way?

a. *A*(3, 4)   **b.** *B*(−2, 2)   **c.** *C*(1, −3)   **d.** *D*(−3, −5)

Solution Because the *y*-axis is vertical, any segment connecting a point to its image must be horizontal and have the *y*-axis as its perpendicular bisector. Using this information, we find the image of each point, as shown in red. In each case, a reflection across the *y*-axis changes the sign of the *x*-coordinate. In other words, if any point (*x*, *y*) is reflected across the *y*-axis, its image has coordinates (−*x*, *y*).

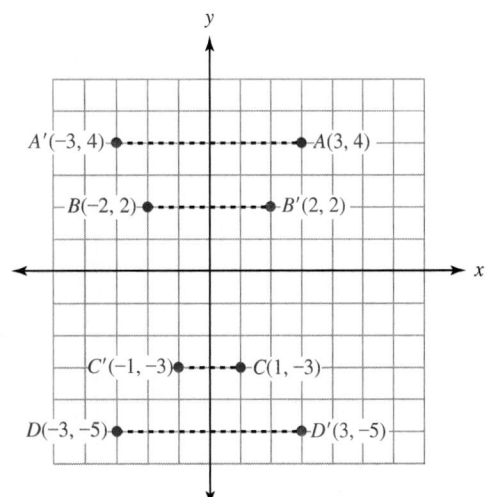

Example 12.15 | Finding the Image of a Reflection

Representation   Find the reflection of each figure in the line *l*.

a.

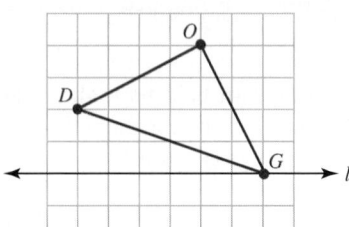

b.

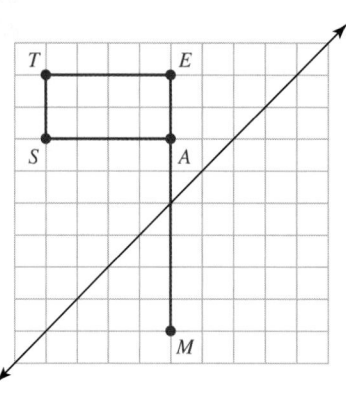

Solution

a. Because *l* is horizontal and *D* is 2 units above *l*, *D′* is 2 units below *l* and directly beneath *D*. Similarly, *O′* is 4 units below *l* and directly beneath *O*. Because *G* is on the axis of reflection, it is mapped to itself. The image, △*D′O′G′*, is shown in red.

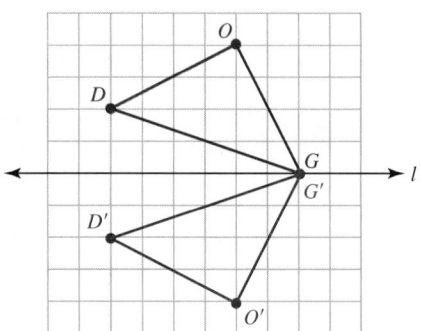

b. *l* has a positive slope that goes up 1 unit for every unit it goes to the right. It is therefore perpendicular to those segments with slopes that go down 1 unit for every unit they go to the right. By using the appropriate segment, we observe that *S* is 3 units above and 3 to the left of *l*. Hence, *S′* must be 3 units below and 3 units to the right of *l* along the same segment. The other image points, *T′*, *E′*, *A′*, and *M′*, are placed similarly. Once the image points are found, we draw the segments to form the image, *S′T′E′A′M′*.

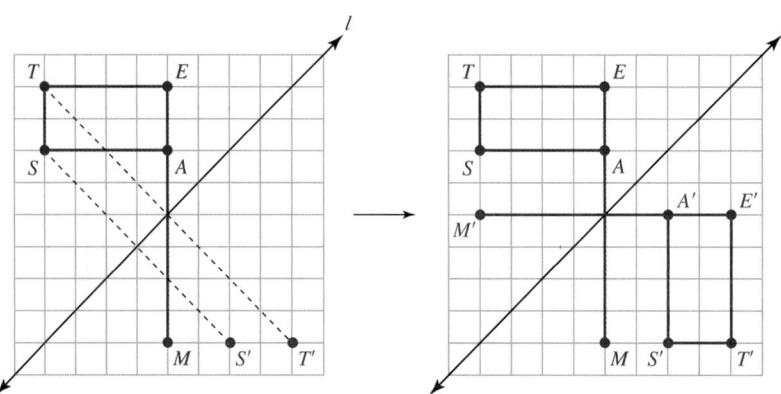

Example 12.15 shows that reflections preserve distance and angle measure, but they also reverse orientation. For instance, the left hand in Figure 12.22 has a right-handed reflection. Similarly, a spiral that spins in a counterclockwise direction has a reflection that spins in a clockwise direction. Because reflections reverse orientation, left-handed to right-handed and clockwise to counterclockwise, we call them **orientation-reversing,** or **improper, transformations.**

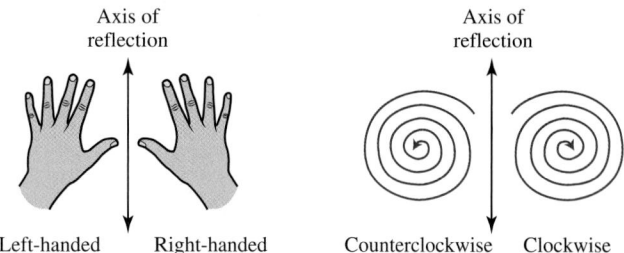

Figure 12.22 Reflections reversing orientation

Glide Reflections

As its name implies, a **glide reflection,** or slide-flip, is a translation followed by a reflection, which is a motion that is best demonstrated by footprints in the sand (Figure 12.23).

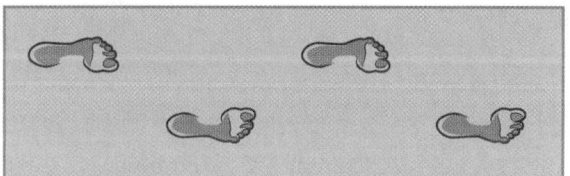

Figure 12.23 Footprints as a glide reflection

Because a glide reflection combines two rigid motions, we can find its image in two ways. One is to apply the translation first, A_T', and then apply the reflection to get the final image, A'' (Figure 12.24). The other is to apply the reflection first, A_R', and then apply the translation. In either case, the image is the same.

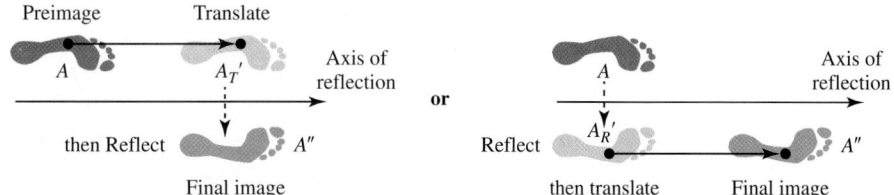

Figure 12.24 Two ways to find the image of a glide reflection

The two motions in a glide reflection make it necessary to include more information in its definition. Because a glide reflection contains a translation, its definition must include a translation vector. Similarly, because the motion contains a reflection, its axis of reflection must be parallel to the translation vector. In this case, the axis of reflection is known as the **glide axis.**

Definition of a Glide Reflection

The **glide reflection** with the translation vector T_{AB} and glide axis l parallel to the segment $\overline{AB}$ is a transformation that maps every point P to a point P' through the combination of the translation T_{AB}, followed by the reflection M_l.

Because a glide reflection is made of a translation and a reflection, finding the image of a glide reflection is a matter of combining the images of the two other motions.

Example 12.16 Finding the Image of a Glide Reflection

Representation

Find the image of △*ABC* under the glide reflection with translation vector $T_{PP'}$ and glide axis *l*.

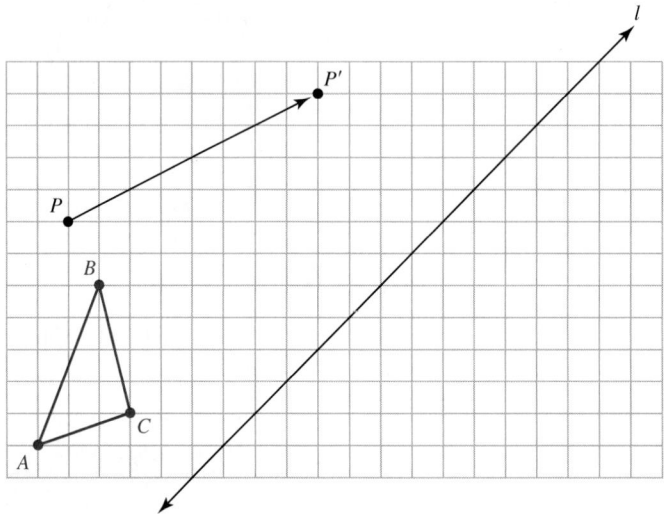

Solution We first move △*ABC* by the translation $T_{PP'}$ by mapping each vertex on the triangle to a point that is 8 units to the right and 4 units above its original position. The resulting triangle is shown in green. We then reflect this triangle in line *l* to obtain the final image, shown in red.

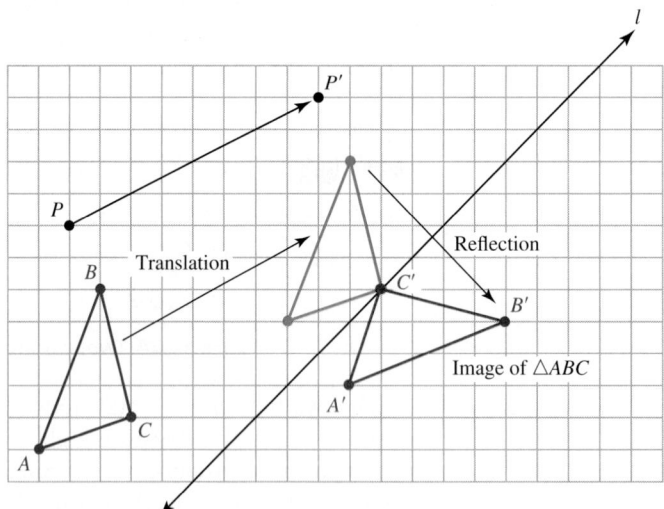

Two facts must be noted about glide reflections. First, because a glide reflection has a reflection, it is an orientation-reversing transformation. Second, if *A'* is the image of *A* under a glide reflection, then the midpoint of segment $\overline{AA'}$ lies on the glide axis (Figure 12.25). This fact enables us to find the glide axis if the preimage and the image are known.

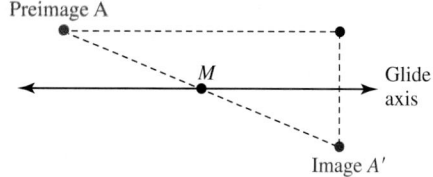

Figure 12.25 The midpoint *M* of segment $\overline{AA'}$ lies on the glide axis

Combining Motions

Glide reflections show that we can combine transformations. In fact, the image of a shape can be considered under any combination of transformations.

Example 12.17 | **Combining a Translation and a Rotation**

What is the image of $\triangle ABC$ if it is translated by vector $\overrightarrow{PP'}$ and then rotated $-90°$ around point O?

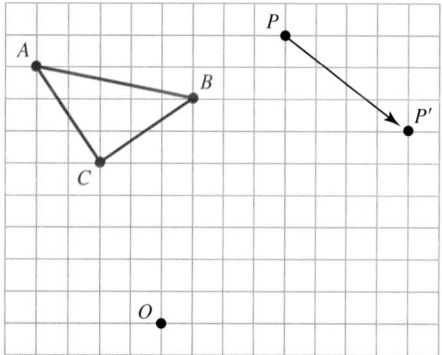

Solution We first translate the triangle by mapping each vertex to a point that is 4 units to the right and 3 units below its original position. The resulting image is the green triangle $\triangle A'B'C'$. Next, we rotate $\triangle A'B'C'$ around point O by $-90°$. We map A' to A'', which is 5 units to the right of O; B' to B'', which is 4 units to the right and 5 units below O; and C' to C'', which is 2 units to the right and 2 units below O. The image after the rotation, $\triangle A''B''C''$, is shown in red.

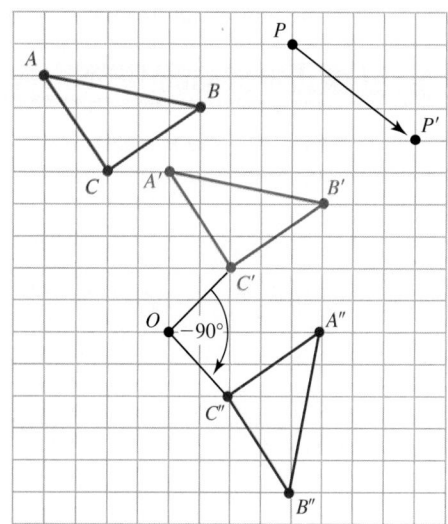

The final image in the last example shows that the effect of the translation and rotation on $\triangle ABC$ is the same as a single rotation around point Q, as shown in Figure 12.26. Any time two or more transformations have the same effect, they are said to be **equivalent.** Specifically, $T_{PP'}$ and $R_{O,90°}$ together are equivalent to the rotation R_Q.

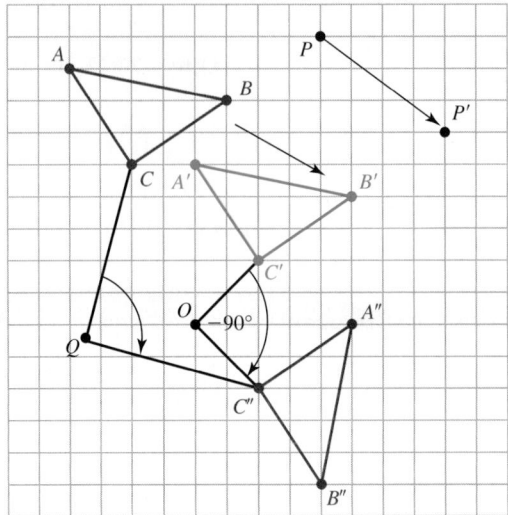

Figure 12.26 Rotating $\triangle ABC$ about point Q

In general, any combination of rigid motions, like the ones in the last example, are equivalent to a single rigid motion.

Theorem 12.4   Classifying Rigid Motions

Any rigid motion or combination of rigid motions is equivalent to a single translation, rotation, reflection, or glide reflection.

To find the single rigid motion that is equivalent to the combination of motions, simply check the orientation of the image. If the orientation is preserved, then the single rigid motion must be a translation or a rotation. If the preimage and image face the same direction, the motion is a translation. If they face different directions, it is a rotation. If the orientation is reversed, then the single rigid motion must be a reflection or a glide reflection. It is a reflection if the axis of reflection is halfway between the figure and its image. A glide reflection has no such axis.

Explorations Manual 12.7

Example 12.18   Finding a Transformation Equivalent to Two Reflections

Describe the transformation that is equivalent to reflecting quadrilateral $ABCD$ first in line k and then in line l.

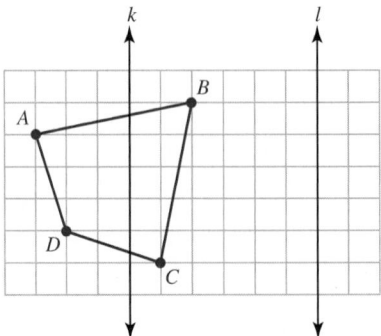

Solution We begin by finding the image of quadrilateral *ABCD* under the two reflections, first in line *k* and then in line *l*. The image after the first reflection, quadrilateral *A'B'C'D'*, is shown in green; the image after the second reflection, quadrilateral *A"B"C"D"*, is shown in red. Because quadrilateral *A"B"C"D"* has the same orientation and direction as quadrilateral *ABCD*, then it must be the image of quadrilateral *ABCD* under a translation. Specifically, the two reflections in lines *k* and *l* are equivalent to a single translation by the vector $T_{BB'}$:

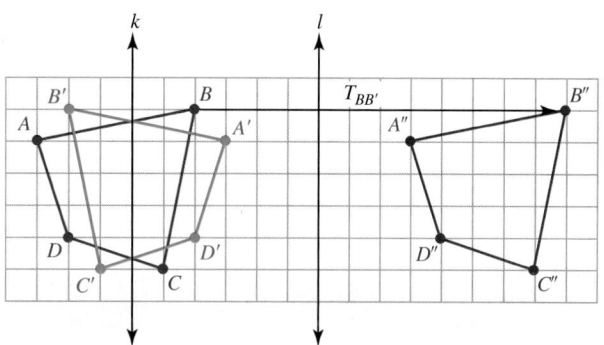

Check Your Understanding 12.2-B

1. Construct the reflection of quadrilateral *ABCD* in line *l*.

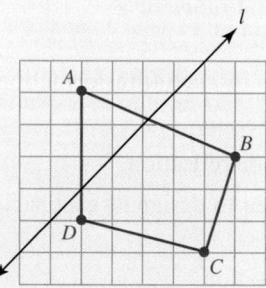

2. Consider point $(-2, 5)$. What are the coordinates of its image if it is reflected across:

   a. The *x*-axis?   **b.** The *y*-axis?   **c.** The line $y = x$?

3. Draw the image of the figure under the glide reflection with *l* as the axis of reflection and $\overline{PP'}$ as the translation vector.

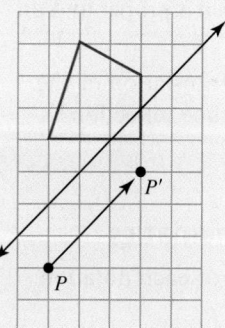

Talk About It Does it make sense to describe a reflection as a flip? Why or why not?

Problem Solving

Activity 12.4
Use a straightedge and compass to construct the image of △*ABC* if it is reflected in the line *l*. (*Hint:* Use the fact that *l* must be the perpendicular bisector of the segments connecting each vertex to its image.)

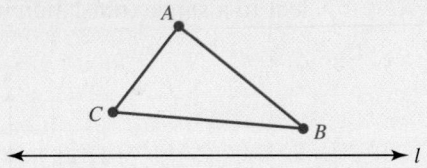

SECTION 12.2 Learning Assessment

■ Understanding the Concepts

1. What does it mean for a motion to be rigid?

2. What is the difference between the image and the preimage of a transformation?

3. **a.** Why is a transformation a function?

   b. What are the domain and range of a transformation function?

4. Describe the motion made by each transformation.

   a. Translation   **b.** Rotation

   c. Reflection   **d.** Glide reflection

5. What information is needed to designate each rigid motion on the plane?

   a. Translation   **b.** Rotation

   c. Reflection   **d.** Glide reflection

6. **a.** What is a fixed point in a transformation?

   b. Which of the four rigid motions can have a fixed point?

7. What is an identity transformation?

8. What is the difference between a proper and an improper transformation?

9. Does it matter whether the translation or the reflection is done first in a glide reflection? Explain.

■ Representing the Mathematics

10. Give a verbal interpretation of each notation.

    a. $T_{PP'}$ **b.** $R_{C,90°}$ **c.** $R_{C,180°}$

    d. $R_{C,-180°}$ **e.** M_l **f.** M_m

11. Rigid motions are functions, so they can be expressed in terms of function notation. Give a verbal interpretation of each rigid motion written in function notation.

    a. $T_{AB}(P) = P'$ **b.** $R_{C,-90°}(D) = D'$

    c. $M_k(E) = E'$

12. Use graph paper to find the image of the figure shown under each translation.

    a. $T_{OO'}$ **b.** $T_{PP'}$ **c.** $T_{QQ'}$ **d.** $T_{RR'}$

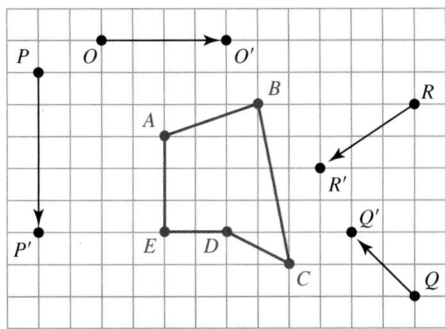

13. Use a compass and straight edge to construct the image of *A* under $T_{PP'}$.

14. Use graph paper to find the image of △*ABC* if it is rotated about:

    a. Point *A* by 90°.   **b.** Point *B* by 180°.

    c. Point *C* by 270°.

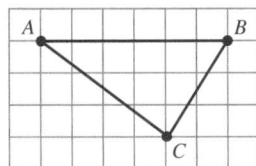

15. Use graph paper to find the image of the figure if it is rotated about:
   a. Point *A* by 180°.　**b.** Point *O* by 180°.
   c. Point *O* by 90°.

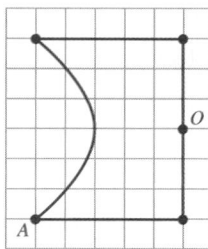

16. Use graph paper to find the image of quadrilateral *ABCD* if it is reflected in:
   a. Line *k*.　**b.** Line *l*.　**c.** Line *m*.

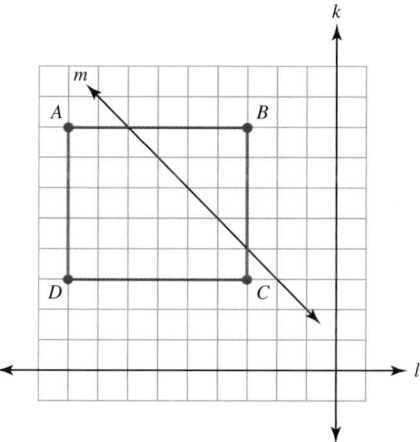

17. Use graph paper to find the image of the figure if it is reflected in:
   a. Line *k*.　**b.** Line *l*.　**c.** Line *m*.

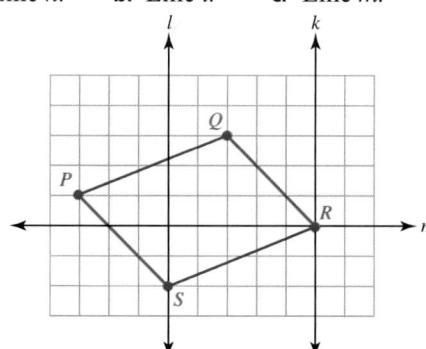

18. Draw the image of the quadrilateral under the glide reflection with *l* as the glide axis and $\overline{PP'}$ as the translation vector.

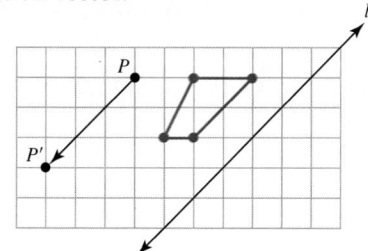

19. Indicate a possible transformation that maps the blue image onto the red image.
   a. 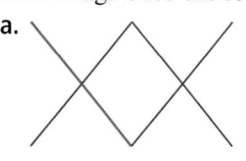　**b.**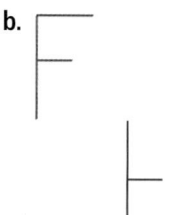
   c. 　**d.**

20. Consider the following equilateral triangle △*ABC*. Give the single transformation that maps △*ABC* onto itself and results in the image △*A'B'C'*.

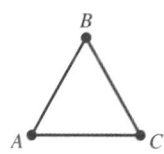

   a. 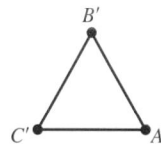　**b.**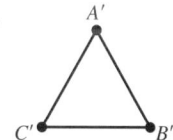
   c. 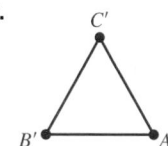　**d.**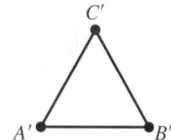

21. Determine whether each figure has the same or the reverse orientation of the figure shown:

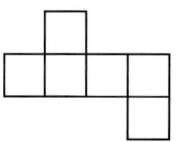

   a. 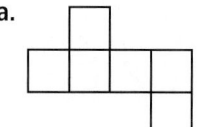　**b.**
   c. 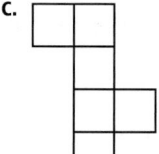　**d.** 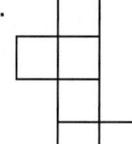　**e.**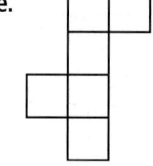

■ Exercises

22. Find the coordinates of the image of each point under the translation that takes $(x, y) \rightarrow (x - 7, y + 6)$.

 a. $(4, 5)$ **b.** $(5, -1)$

 c. $(9, -3)$ **d.** $(-3, -4)$

 e. $(0, 0)$

23. Find the coordinates of the image of each point under the translation that takes $(x, y) \rightarrow (x + 2.7, y - 3.1)$.

 a. $(3.4, 1.5)$ **b.** $(6.7, -8.2)$

 c. $(1.3, -5.9)$ **d.** $(-4.7, -6.6)$

 e. $(8.8, 8.8)$

24. Find the clockwise rotation that is equivalent to a counterclockwise rotation of:

 a. $90°$. **b.** $270°$.

 c. $135°$. **d.** $317°$.

 e. $195.6°$.

25. Find the counterclockwise rotation that is equivalent to a clockwise rotation of:

 a. $-90°$. **b.** $-180°$.

 c. $-205°$. **d.** $-287°$.

 e. $-157.9°$.

26. For each angle, find the turn angle between $0°$ and $360°$ to which it is equivalent.

 a. $435°$ **b.** $671°$

 c. $1,342°$ **d.** $-935°$

 e. $-3,241°$

27. A rotation of $\triangle ABC$ has mapped A to A' and C to C'. Find the center of rotation and the turn angle, and then draw $\triangle A'B'C'$.

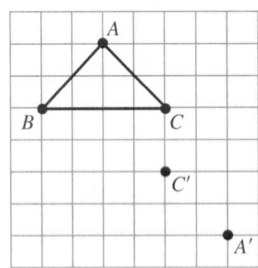

28. Consider point $(3, 4)$. Find the coordinates of its image when it is rotated around the origin by the given turn angle.

 a. $90°$ **b.** $180°$

 c. $270°$ **d.** $360°$

29. Consider point $(-2, 5)$. Find the coordinates of its image when it is rotated around the origin by the given turn angle.

 a. $-90°$ **b.** $-180°$

 c. $-270°$ **d.** $-360°$

30. Find the coordinates of the vertices of the image of quadrilateral $ABCD$ when it is rotated under the given conditions.

 a. Point X by $-90°$

 b. Point X by $90°$

 c. Point Y by $-180°$

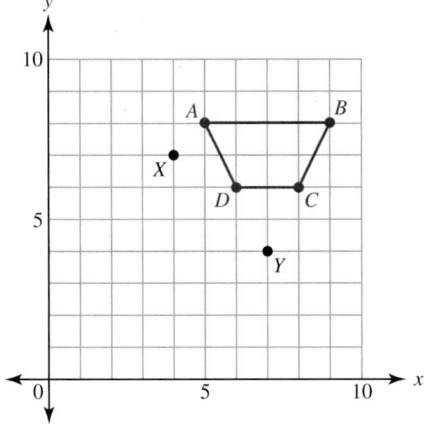

31. Which point in the figure is the image of P under the reflection in:

 a. Line k? **b.** Line l?

 c. Line m? **d.** Line n?

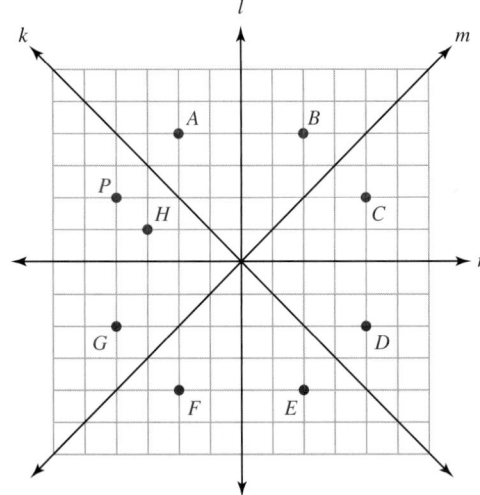

32. Consider point $(2, -3)$. What are the coordinates of its image if it is reflected in the:

 a. x-axis? **b.** y-axis?

 c. Line $y = x$? **d.** Line $y = -x$?

33. Consider point $(-1, -4)$. What are the coordinates of its image if it is reflected in the:

 a. x-axis?

 b. y-axis?

 c. Line $x = y$?

 d. Line $x = -y$?

34. Determine the coordinates of the vertices of the image of quadrilateral $ABCD$ if quadrilateral $ABCD$ is reflected in line l.

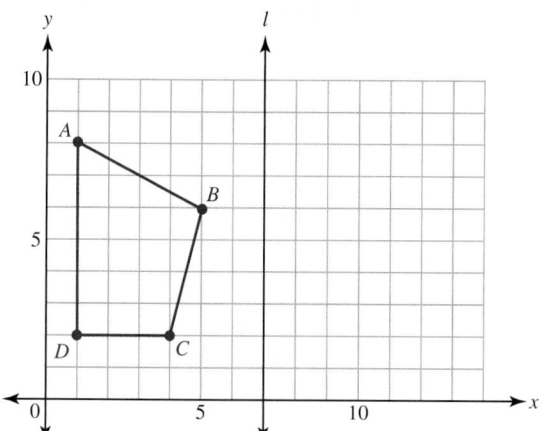

35. Let M_l be the reflection that sends P to P'. Find the axis of reflection, l, and then the image of $\triangle ABC$ under the reflection in l.

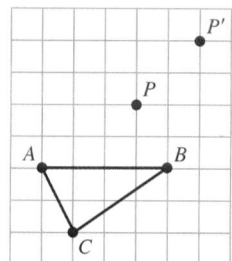

■ Problems and Applications

36. Let T_1 and T_2 be translations such that $T_1(x, y) \rightarrow (x + 3, y - 4)$ and $T_2(x, y) \rightarrow (x - 5, y - 2)$.

a. What is the image of point $A(1, 1)$ under the transformation that translates A first by T_1 and then by T_2?

b. What is the image of point $B(2, -3)$ under the transformation that translates B twice by T_1?

c. What is the image of point $C(3, 5)$ under the transformation that translates C first by T_1, then by T_2, and then by T_1 again?

37. If $\triangle ABC$ is translated first by vector T_1 and then by T_2, find the transformation that takes the image of $\triangle ABC$ back to its original starting position.

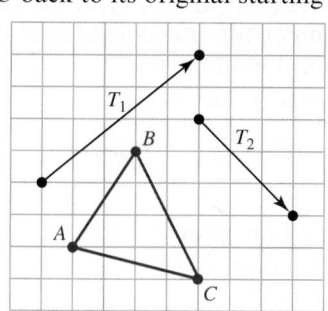

38. What is the preimage of figure $A'B'C'D'E'$ under translation $T_{PP'}$?

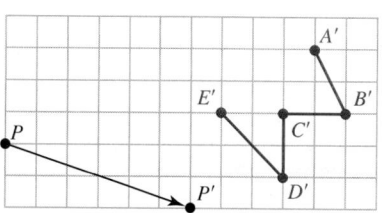

39. A line has the equation $y = 2x - 5$. What is the equation of its image under the translation that takes:

a. $(x, y) \rightarrow (x + 3, y)$? **b.** $(x, y) \rightarrow (x, y - 4)$?

c. $(x, y) \rightarrow (x + 2, y - 5)$?

40. Consider the pattern of dots shown:

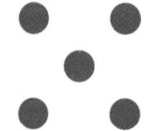

Use a mirror so that the figure and its reflection show a total of:

a. 10 dots. **b.** 6 dots. **c.** 5 dots.

d. 4 dots. **e.** 2 dots. **f.** 1 dot.

41. The center of a circle is on the perpendicular bisector of any chord of the circle. Use this fact to construct the center of rotation that takes $\overline{AB}$ to $\overline{A'B'}$.

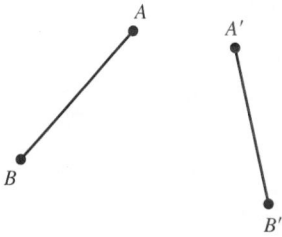

42. Suppose point A has coordinates $(-3, 6)$. If A is first reflected in the x-axis to find A' and A' is reflected in the y-axis to find A'', what are the coordinates of A''?

43. Find the glide axis of the glide reflection that maps quadrilateral $ABCD$ onto quadrilateral $A'B'C'D'$. (*Hint:* Use the fact that the midpoint of any segment $\overline{PP'}$ lies on the glide axis.)

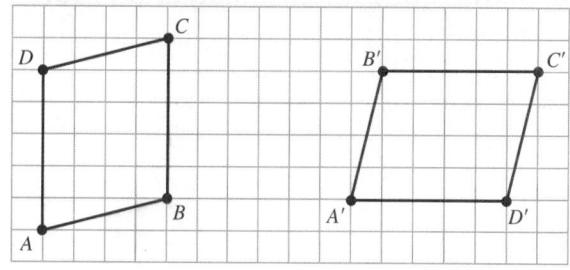

44. Find the image of △*DOG* under three successive reflections in l_1, l_2, and l_3.

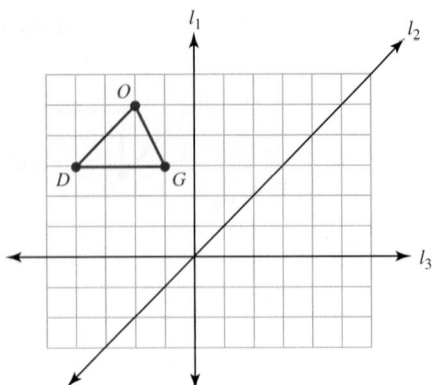

45. A glide reflection has a translation vector 5 units long, and its glide axis is the *x*-axis. If point *A* has coordinates (3, 5), what are the coordinates of the image of *A* if the glide reflection is applied to *A*:

a. Twice? **b.** Three times?

c. Four times? **d.** *n* times where *n* is odd?

e. *n* times where *n* is even?

46. If the coin on the left is rotated around to the right side of the coin on the right, is the head right side up or upside down? Why?

47. If each clock is reflected in a mirror, what time does the reflection read?

a.  **b.**

c.  **d.**

■ **Mathematical Reasoning and Proof**

48. Is the diagonal of a parallelogram an axis of reflection? Why or why not?

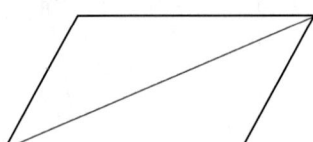

49. Find the coordinates of the image of each point if it is rotated around the origin by a turn angle of 180°. Write a generalization about the relationship between the coordinates of a point and the coordinates of its image when the point is rotated in this way.

a. *A*(1, 2) **b.** *B*(−3, 3)

c. *C*(2, −5) **d.** *D*(−3, −1)

50. Find the coordinates of the image of each point if the axis of reflection is the *x*-axis. Write a generalization about the relationship between the coordinates of a point and the coordinates of its image when a point is reflected in this way.

a. *A*(3, 5) **b.** *B*(−1, 1)

c. *C*(4, −2) **d.** *D*(−4, −5)

51. Find the coordinates of the image of each point if the axis of reflection is the line $y = x$. Write a generalization about the relationship between the coordinates of a point and the coordinates of its image when a point is reflected this way.

a. *A*(5, 2) **b.** *B*(−4, 1)

c. *C*(2, −2) **d.** *D*(−1, −3)

52. a. The combination of two reflections in parallel lines has the same net effect as what kind of rigid motion?

b. The combination of two reflections in intersecting lines has the same net effect as what kind of rigid motion?

53. The transformation that uses the same glide reflection twice is the same as what kind of rigid motion?

54. Will the image of a triangle have the same or the reverse orientation if the triangle is transformed under:

a. A glide reflection and a rotation?

b. A translation and a glide reflection?

c. A reflection and a glide reflection?

d. Two reflections?

e. Three reflections?

f. An odd number of glide reflections?

55. If an asterisk (*) is used to represent the combination of two transformations, answer the following questions.

a. Does $T_{AB} * T_{CD} = T_{CD} * T_{AB}$?

b. Does $R_{A,180°} * R_{B,180} = R_{B,180} * R_{A,180}$?

c. Does $M_l * M_k = M_k * M_l$

d. What algebraic property does each remind you of? Explain the similarity.

■ Mathematical Communication

56. a. Look up the term "translate" in the dictionary. Based on what you read, do you think it makes sense to use this term as the formal name for a slide?

b. Does it make sense to use the term "reflection" as the formal name for a flip?

57. Create a table that summarizes the characteristics of the four basic rigid motions. Include in your table the name of the rigid motion, what is required to define the motion, and whether the motion is orientation reversing or preserving. With a partner, discuss how the rigid motions might be classified in different ways.

58. Molly wants to know whether the segment $\overline{A'B'}$ is the image of segment $\overline{AB}$ under a reflection or a rotation. How do you answer her question?

59. Miranda claims that a rotation of 180° is an identity transformation. How would you clarify her misconception?

60. A student asks why the glide axis has to be parallel to the direction of the translation in a glide reflection. What is your response?

■ Building Connections

61. Write a short paragraph explaining the importance and use of rectangular coordinate systems in studying transformations.

62. Symmetry is an important visual aspect of geometrical figures. How do you think the transformations discussed in this section are connected to symmetry?

63. If possible, make a list of five examples of each motion from what you observe in your surroundings.

a. Translation

b. Rotation

c. Reflection

d. Glide reflection

64. Why is the word "Ambulance" written backward on the front of most ambulances?

65. Examine several different sets of curriculum materials for content on transformations.

a. Of the four basic motions, which ones are primarily discussed? If any are not discussed, why do you think this is the case?

b. What names do the transformations go by? Do you think it makes sense to refer to the motions by these names? Why or why not?

66. What role do you think transformations have in learning to reason spatially?

SECTION **12.3**	**Congruence, Similarity, and Symmetry with Transformations**

We can use basic transformations as another way to describe geometrical shapes and the relationships among them. Specifically, we can use transformations to describe congruence, similarity, and symmetry.

Congruence with Transformations

Because two figures are congruent if they have the same shape and size, we can test for congruence by placing one shape on top of the other. And because we can use rigid motions to move the shape, we should be able to define congruence in terms of these motions. Recall that two polygons are congruent if their corresponding sides and corresponding angles are congruent. Because a rigid motion preserves length and angle measure, then it must map segments onto congruent segments and angles onto congruent angles. Consequently, we can use a rigid motion instead of the corresponding sides and angles to define the congruence relationship.

Definition of Congruence with Transformations

Two polygons are **congruent** if and only if one is the image of the other under a rigid motion or a combination of rigid motions.

Example 12.19   Rigid Motions and Congruent Triangles

Find a rigid motion that can be used to show the given congruence relationship for each diagram.

a. $\triangle ABD \cong \triangle BCD$.

b. $\triangle ABD \cong \triangle DEF$.

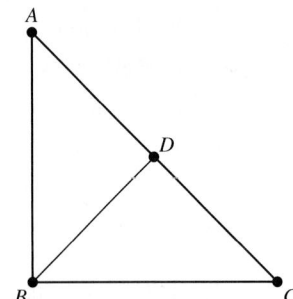
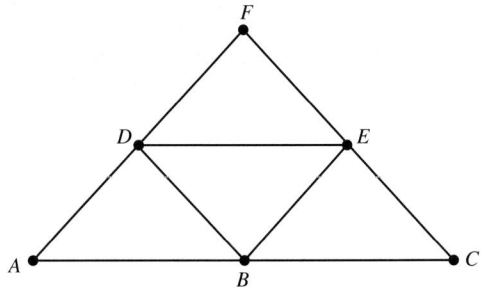

Solution

a. There are two motions to consider. One is a reflection in segment $\overline{BD}$. The other is a rotation of 90° around point D. To determine which is correct, consider the correspondence between vertices. Because $A \leftrightarrow B$, $B \leftrightarrow C$, and $D \leftrightarrow D$, the rotation of 90° is the one that shows $\triangle ABD \cong \triangle BCD$.

b. Because $\triangle ABD$ and $\triangle DEF$ are facing the same direction and have the same orientation, the translation with vector T_{DF} can be used to show $\triangle ABD \cong \triangle DEF$.

Transformations can also help us verify properties of polygons. For instance, the next example shows how to use transformations to show that the opposite sides of a parallelogram are congruent.

Example 12.20   Opposite Sides in a Parallelogram Are Congruent

Reasoning

Use a transformation to prove that the opposite sides of parallelogram $ABCD$ are congruent.

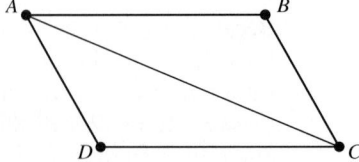

Solution To prove that the opposites sides of parallelogram $ABCD$ are congruent, we must first show that $\triangle ABC \cong \triangle CDA$. In Chapter 11, we did so with a triangle congruence property, but we can also use a rotation. Specifically, if we rotate $\triangle CDA$

around the midpoint of $\overline{AC}$ by 180°, then $\triangle ABC \cong \triangle CDA$. Because corresponding parts of congruent triangles are congruent, then $\overline{AB} \cong \overline{CD}$ and $\overline{BC} \cong \overline{DA}$.

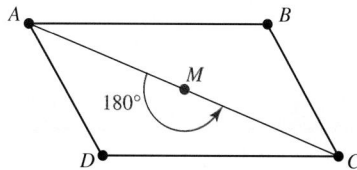

Size Transformations and Similarity

Every transformation so far has been a rigid motion, which preserves both the shape and the size of the figure. **Size transformations,** or **dilations,** are transformations that preserve the shape of a figure but not necessarily its size. Common examples are enlargements from a photocopier or slides projected onto a screen (Figure 12.27).

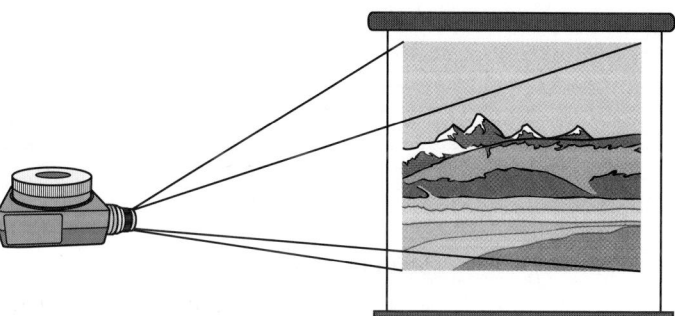

Figure 12.27 An example of a size transformation

To understand how a size transformation creates an image that preserves the shape of a figure but not necessarily its size, take a closer look at the size transformation created by a slide projector. First, if the bulb in the projector does not shine through the slide, no image appears on the screen, implying that a size transformation has a point of origin, or a **center.** On a plane, a single point serves as the center. Second, notice that the image is proportional to the original figure and that its size is relative to the distances between the bulb, the slide, and the screen. The greater these distances are, the larger the image will be. In other words, the size of the image is relative to the distance between the center and the preimage. The amount of change in the image's size relative to that of the original is called the **scale factor** of the size transformation. The following definition formalizes these ideas.

Definition of a Size Transformation

Let O be a point in the plane and let k be a positive real number. A **size transformation** with center O and scale factor k, written $S_{O,k}$, is a transformation that maps every point P to a point P' so that O, P, and P' are collinear and $OP' = k \cdot OP$.

The definition indicates a third criterion for size transformations that turns out to be particularly useful when making them. Specifically, O, P, and P' must be collinear. We can therefore find the image of P under a size transformation by drawing the ray from O to P and then locating point P' on the ray, so that $OP' = k \cdot OP$. For instance, to find the image of point P under the size transformation $S_{O,2}$ (Figure 12.28), we find point P' on the ray $\overrightarrow{OP}$ so that $OP' = 2 \cdot OP$.

Figure 12.28 The image of P under $S_{O,2}$

$OP' = 2OP$

Explorations
Manual
12.8

Knowing how to find the image of a point under a size transformation, we can find the image of a polygon by finding the images of its vertices and connecting them with segments. The next example shows how to use graph paper to do so.

Example 12.21   **Finding the Image of a Size Transformation**

Representation

Find the image of each quadrilateral under the given size transformation.

a. $S_{O,2}$

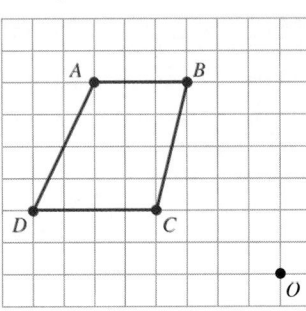

b. $S_{O,1/2}$

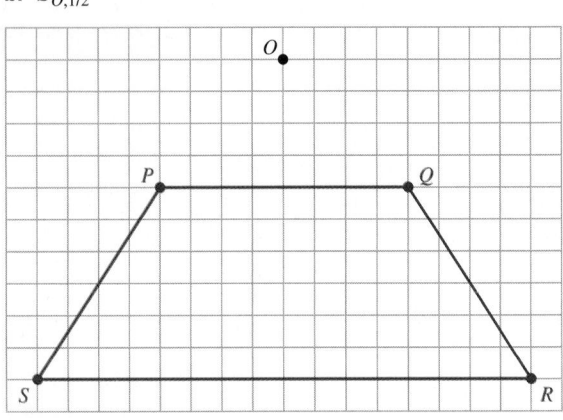

Solution

a. The notation $S_{O,2}$ tells us to make a size transformation with O as the center and an image twice as large as quadrilateral $ABCD$. We first draw rays from O through each of the four vertices. Next, we locate the image of each vertex on the appropriate ray so that the distance from the image to O is twice the distance from the vertex to O. For instance, because A is 6 units to the left of and 6 units above O, A' must be 12 units to the left and 12 units above O. After locating the images of the other vertices, we connect them with segments to form quadrilateral $A'B'C'D'$.

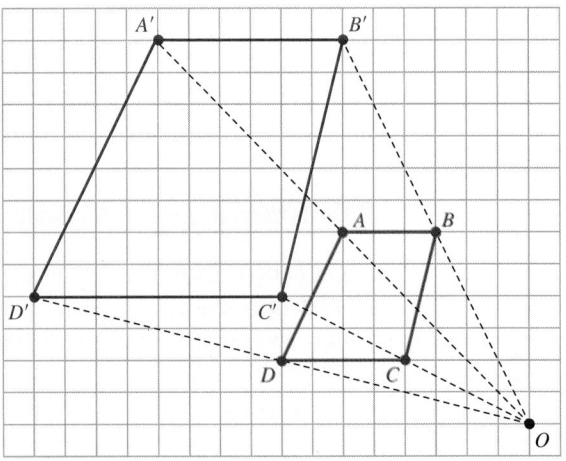

b. $S_{O,1/2}$ tells us to use O as the center and to make the image half as large as quadrilateral $PQRS$. Again, we first draw rays from O through each vertex. Now, however, we halve the distance instead of doubling it. With the images of the four vertices located, we connect them with segments to form quadrilateral $P'Q'R'S'$.

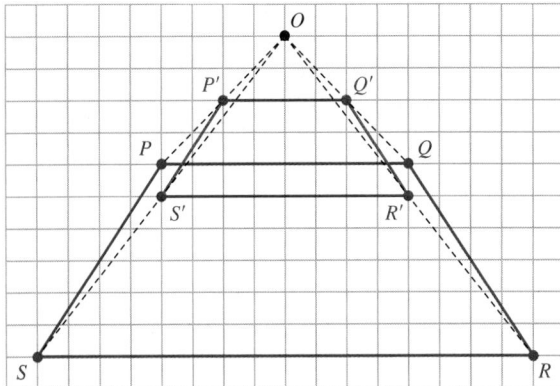

The scale factor, k, can be any positive real number. As Example 12.21 shows, if $k > 1$, the image is larger than the original figure, and the size transformation is a **magnification.** If $k < 1$, the image is smaller than the original figure, and the size transformation is a **contraction.** In any contraction, the image of a point is always between the center and the point itself. If $k = 1$, the figure is mapped onto itself, and we have the identity transformation.

Because any size transformation creates an image that is similar to the original, it must preserve the ratios of distances and angle measure. As Example 12.21 shows us, it also preserves orientation and parallel lines.

Theorem 12.5   Properties of Size Transformations

A size transformation, $S_{O,k}$, preserves the ratio of distances, angle measure, orientation, and parallelism.

Theorem 12.5 provides another way to define similarity. Recall that two polygons are similar if their corresponding angles are congruent and their corresponding sides are proportional. Because a size transformation preserves angles and the ratio of distances, then it must map angles onto congruent angles and segments onto proportional segments. Consequently, we can replace the corresponding angles and sides with a size transformation to define the similarity relationship. However, we must realize that we might not be able to find one size transformation between two similar figures. Nevertheless, size transformations can be combined with rigid motions to move one figure so that it is the image of the other. Any combination of transformations that leads to similar figures is called a **similarity transformation,** or a **similitude.**

Definition of Similarity with Transformations

Two polygons are **similar** if and only if one figure is the image of the other under a combination of a size transformation and one or more rigid motions.

Example 12.22 | Similarity Transformation Between Two Figures

Show that parallelogram $A'B'C'D'$ is similar to parallelogram $ABCD$ by describing a similarity transformation that maps one onto the other.

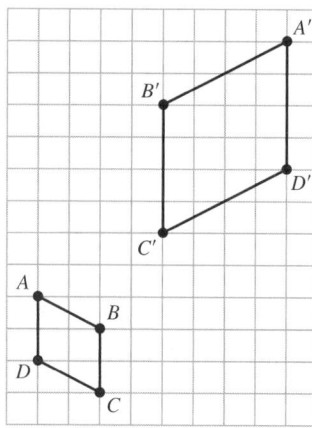

Solution We can show the similarity transformation in three steps. First, we use a reflection in line l to reverse the orientation. We then use a size transformation with a center at O and a scale factor of 2. Finally, we use a translation with a vector that moves each point 8 units to the right and 3 units up.

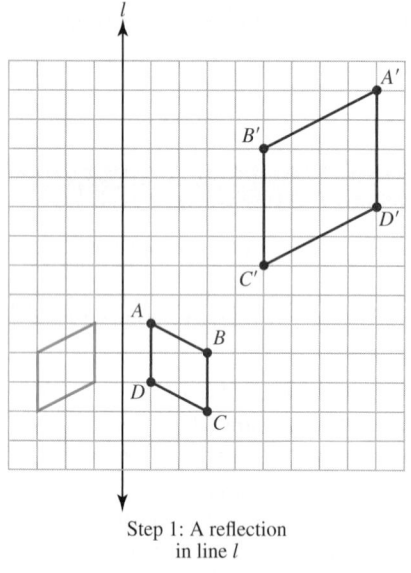

Step 1: A reflection in line l

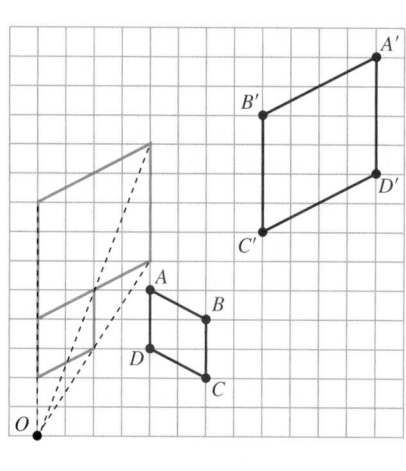

Step 2: A size transformation with center O and scale factor 2

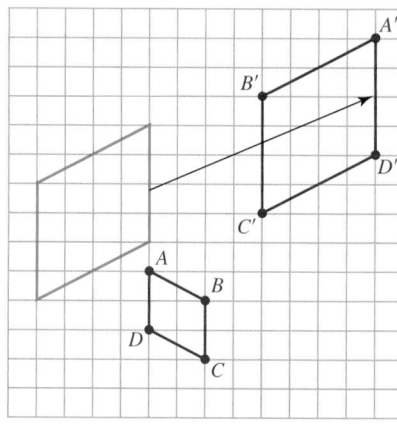

Step 3: A translation moving points to the right 8 and up 3

Example 12.23 | Trisecting a Line Segment

Application

Use a straightedge, compass, and size transformation to trisect the segment $\overline{AB}$.

Solution We begin by drawing a ray that is approximately parallel to segment $\overline{AB}$. Next, open the compass to a short length and construct three consecutive, yet congruent segments on the ray. Label the points C, D, E, and F, respectively.

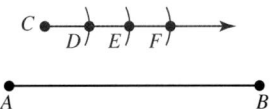

Draw a ray from A through C and a ray from B through F. Point O, which is the intersection of the rays, is the center of the size transformation that maps the segment $\overline{CF}$ onto the segment $\overline{AB}$. Next, draw rays from O through points D and E so that they intersect $\overline{AB}$. Because we constructed $\overline{CD} \cong \overline{DE} \cong \overline{EF}$ and a size transformation preserves ratios of distances, it must be that $\overline{AD'} \cong \overline{D'E'} \cong \overline{E'B}$.

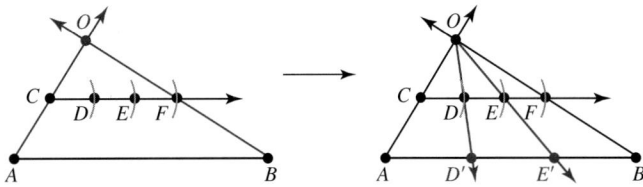

Check Your Understanding 12.3-A

1. Determine a rigid motion that can be used to show each congruence relationship.

    a. $\triangle AFE \cong \triangle BCD$.
                                     b. $\triangle BDE \cong \triangle DEA$.

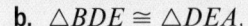

2. Draw the image of $\triangle ABC$ under the size transformation $S_{O,2}$.

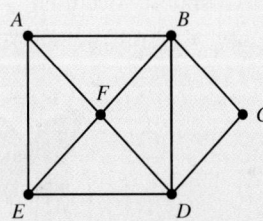

3. Draw the image of $\triangle ABC$ if it is first transformed by the size transformation $S_{O,2}$ and then rotated about O by $180°$.

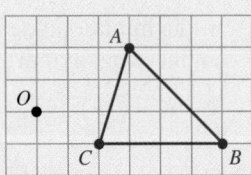

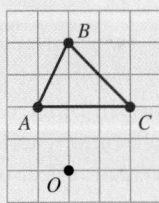

Talk About It Suppose you are given a figure and its image under a size transformation. How might you use the two to find the center and scale factor of the transformation?

Problem Solving

Activity 12.5

Consider the quadrilateral *ABCD*.

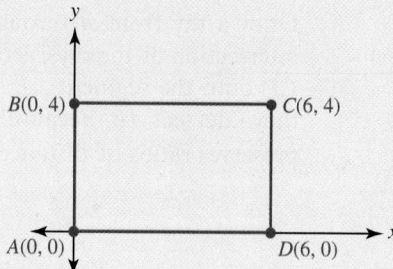

Give the coordinates of the image of each vertex under the indicated transformation.

a. $S_{A,2}$.

b. $S_{A,1/2}$.

c. A rotation around A by 90° and $S_{A,2}$.

d. A reflection in the x-axis and $S_{A,1/2}$.

Symmetry

Transformations can be used to describe symmetry, which is a visual aspect that is easily recognizable in many natural and human-made shapes. For instance, in nature, symmetry occurs in the physical characteristics and markings of both plants and animals as well as in the structures of molecules and crystals (Figure 12.29).

| Butterfly | Tiger's face | Leaf | Snowflake |

Figure 12.29 Examples of symmetry in nature

In the human-made world, we use symmetry in art, design, landscaping, architecture, engineering, and even advertising (Figure 12.30).

| Hubcap | Native American design | Taj Mahal |

Figure 12.30 Examples of symmetry in human-made objects

Explorations Manual 12.9

Intuitively, a shape has symmetry if it looks the same from different perspectives. In other words, an object has **symmetry** if it can be transformed in some way back onto itself and still maintains its original appearance. In general, we can do so through reflections and rotations.

Reflection Symmetry

When we think of symmetry, we often think of **reflection,** or **line symmetry.** Intuitively, a shape, like the heart in Figure 12.31, has reflection symmetry if one side of the figure can be "folded" over so that it lies exactly on top of the other side. The fold line forms the **axis,** or **line of symmetry.** A line of symmetry is equivalent to an axis of reflection, so a figure has reflection symmetry if and only if there is a line *l* such that a reflection of the figure in *l* maps it onto itself.

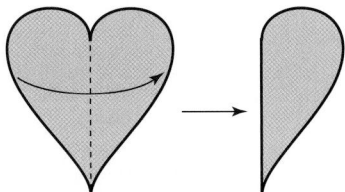

Figure 12.31 Reflection symmetry in a heart

Example 12.24   Reflection Symmetry in Letters

In each diagram, half of a capital letter and a line of symmetry are given. Identify the capital letter with such symmetry.

Solution We can use two methods to identify the letter. One is to use a plastic reflector. Another is to reflect the figure in the given line. Either way, the following letters are revealed.

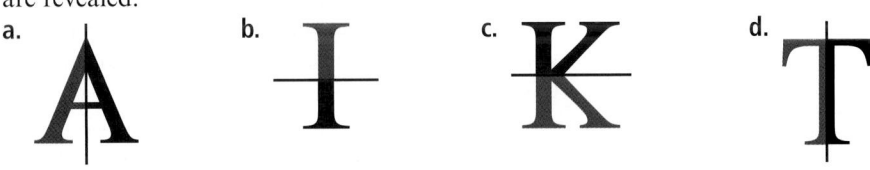

Many figures have more than one line of symmetry, so when we analyze a figure, we should try to find all its lines of symmetry. As Figure 12.32 indicates, lines of symmetry can be vertical, horizontal, and even diagonal.

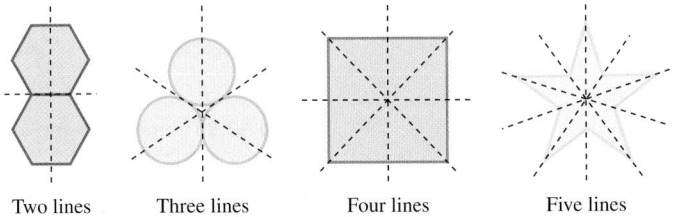

Two lines   Three lines   Four lines   Five lines

Figure 12.32 Figures with multiple lines of symmetry

Example 12.25   Lines of Symmetry in Regular *n*-gons

Problem Solving   Find a formula that can be used to determine the number of lines of symmetry in any regular *n*-gon.

Solution

Understanding the Problem. We are asked to find a formula that can be used to determine the number of lines of symmetry in any regular *n*-gon.

Devising a Plan. In similar situations, we have always based our formulas on the number of sides of the polygon. We do the same here by investigating the lines of symmetry in particular regular polygons. Our goal is to *look for a pattern* that can be generalized with a formula.

Carrying Out the Plan. We can investigate several regular polygons by cutting them from paper and folding them to find their lines of symmetry. The lines of symmetry for four regular polygons are shown:

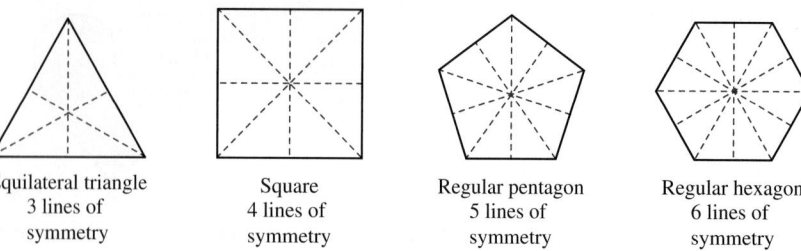

| Equilateral triangle
3 lines of
symmetry | Square
4 lines of
symmetry | Regular pentagon
5 lines of
symmetry | Regular hexagon
6 lines of
symmetry |

In each case, the number of the lines of symmetry is equal to the number of sides. Consequently, a regular n-gon has n lines of symmetry.

Looking Back. These examples illustrate two other facts about the lines of symmetry in regular polygons. First, each line of symmetry passes through the center of the polygon, so we can locate the center by finding the intersection of two or more lines of symmetry. Second, when a regular polygon has an odd number of vertices, the lines of symmetry always pass through a vertex and the midpoint of the opposite side. When it has an even number of vertices, the lines of symmetry pass either through two opposite vertices or through the midpoints of two opposite sides.

Rotation Symmetry

Many planar shapes also have **rotation symmetry,** which means the figure can be mapped back onto itself by rotating it around a point through some turn angle less than 360°. The turn angle must be less than 360° because a rotation of 360° maps any figure back onto itself. In other words, if a figure requires a full turn of 360° to map back onto itself, then it does not have rotation symmetry. One way to determine whether a figure has rotation symmetry is to trace it and then rotate the trace to see if the figure can be superimposed on itself (Figure 12.33).

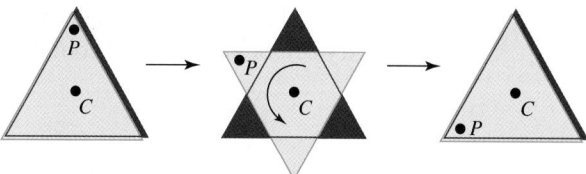

Figure 12.33 Superimposing a copy of an equilateral triangle onto itself

If it does, then we name the rotation symmetry by the measure of the turn angle. For instance, Figure 12.34 shows that square $ABCD$ has 90° rotation symmetry, 180° rotation symmetry, and 270° rotation symmetry.

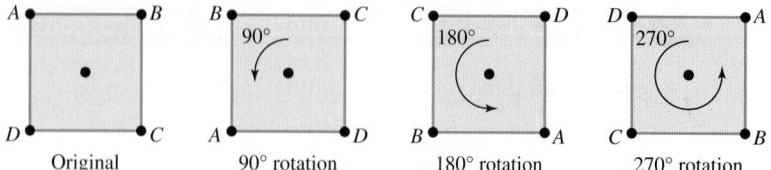

| Original | 90° rotation | 180° rotation | 270° rotation |

Figure 12.34 Rotation symmetry in a square

The rotations of the square show us that any figure with 90° rotation symmetry also has 180° and 270° rotation symmetry. In general, if a figure has rotation symmetry of α degrees, then it also has rotation symmetry of $n\alpha$ degrees, where n is a natural number. In this case, we use the smallest positive angle to name the rotation symmetry. Also notice that more circular objects tend to have more rotation symmetries. If the figure is a circle with no other distinguishing features, then it has rotation symmetry by any turn angle, that is, infinitely many rotation symmetries.

Example 12.26   Finding the Rotation Symmetry of a Shape

Find every turn angle that maps each figure onto itself. Use them to name the rotational symmetry of the figure.

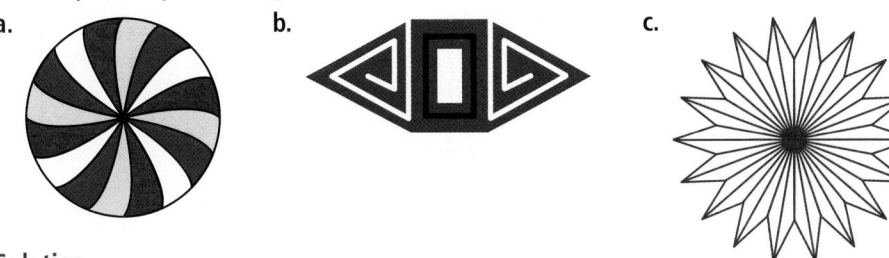

a.   b.   c.

Solution

a. The colors in the circle limit the rotation symmetries to angles of 90°, 180°, and 270°. Because 90° is the smallest turn angle, the figure has 90° rotation symmetry.

b. Initially, the figure appears to have 180° rotation symmetry. However, when rotated by 180°, the spirals take on a different orientation. Consequently, the figure does not have rotation symmetry.

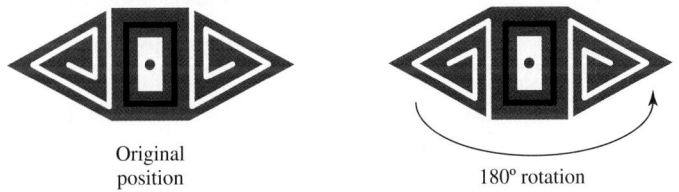

Original
position

180° rotation

c. Because the figure has 18 different points that are the same, it has $18 - 1 = 17$ different rotation symmetries. To find the first turn angle, divide 360° by 18, or $\frac{360°}{18} = 20°$. So the figure has rotation symmetry for every multiple of 20 less than 360, or 20°, 40°, 60°, . . . 300°, 320°, and 340°. Because the smallest turn angle is 20°, the figure has 20° rotation symmetry.

If a figure is its own image under a half turn or a 180° rotation, it has **point symmetry.** More formally, a figure has point symmetry if every point P has a corresponding point P' that is directly opposite it across the center of rotation and at the same distance; that is, $CP = CP'$. Figure 12.35 illustrates this idea on a parallelogram. Several other figures with point symmetry are shown in Figure 12.36.

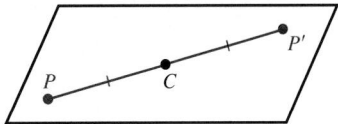

Figure 12.35 Point symmetry in a parallelogram

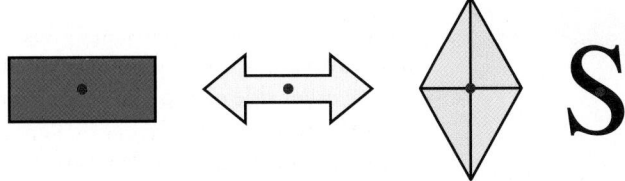

Figure 12.36 Shapes with point symmetry

Classifying Shapes with Symmetry

Because different shapes have different reflection and rotation symmetries, we can use symmetry to classify shapes like triangles and quadrilaterals. For instance, squares are classified differently from rectangles. Squares have four reflection symmetries and three rotation symmetries, whereas rectangles have only two reflection symmetries and one rotation symmetry (Figure 12.37).

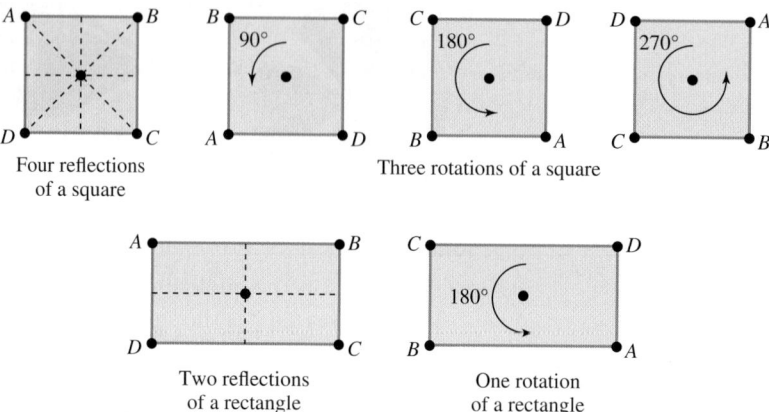

Four reflections
of a square

Three rotations of a square

Two reflections
of a rectangle

One rotation
of a rectangle

Figure 12.37 Symmetries of a square and rectangle

Explorations Manual 12.10

Reflection and rotation symmetries also help in writing alternative definitions for these shapes. For instance, we can define a square to be a quadrilateral with four reflection symmetries and three rotation symmetries. Likewise, a rectangle is a quadrilateral with two reflection symmetries and one rotation symmetry.

Example 12.27 Using Symmetry to Define Shapes

Communication

Use symmetry to write an alternative definition for each type of triangle.

a. Scalene triangle   **b.** Isosceles triangle   **c.** Equilateral triangle

Solution

a. Because the sides of a scalene triangle are different lengths, it cannot have reflection or rotation symmetry. So a scalene triangle is a triangle with no reflection or rotation symmetry.

b. An isosceles triangle has two sides the same length, so it has one reflection symmetry but no rotation symmetry. Consequently, we define an isosceles triangle as a triangle with one reflection and no rotation symmetries.

c. The three sides of an equilateral triangle are the same length, so it has three reflection and two rotation symmetries. Hence, an equilateral triangle is a triangle with three reflection and two rotation symmetries.

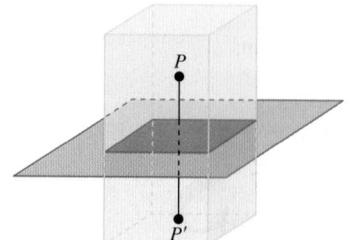

Figure 12.38 Plane symmetry in a rectangular prism

Three-Dimensional Symmetry

Many three-dimensional shapes also have symmetry. For instance, a spatial figure has **plane symmetry** if every point of the figure on one side of a plane has a mirror image on the other side of the plane. An example of a plane symmetry in a rectangular prism is shown in Figure 12.38.

Spatial shapes can also have rotation symmetry, but they do so around an axis of rotation rather than a center of rotation. For instance, consider the cube in

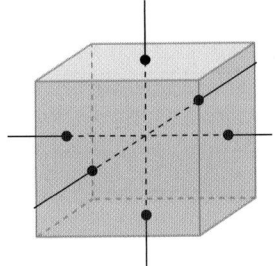

Figure 12.39 Rotation symmetry in a cube

Figure 12.39. If we connect the centers of any pair of opposite sides with a line, then the cube has 90° rotation symmetry about each of the lines.

Check Your Understanding 12.3-B

1. Trace each figure, and then find all possible lines of symmetry in each.

a.
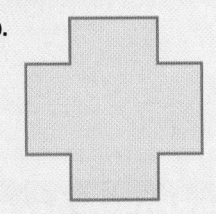

b.

c.

d.
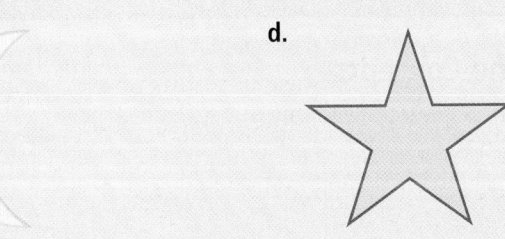

2. Find all of the turn angles that map each of the figures in the previous problem onto itself. Name the rotational symmetry of the figure.

3. How many reflection symmetries does a rhombus have? How many rotation symmetries? Use your answers to write an alternative definition for a rhombus.

Talk About It How might you use a plastic reflector to determine whether a shape has reflection symmetry?

Reasoning

Activity 12.6

An equilateral triangle has the six different symmetries:

r_1: reflection in line l_1 R_1: 120° clockwise rotation with center P
r_2: reflection in line l_2 R_2: 240° clockwise rotation with center P
r_3: reflection in line l_3 I: 360° clockwise rotation with center P

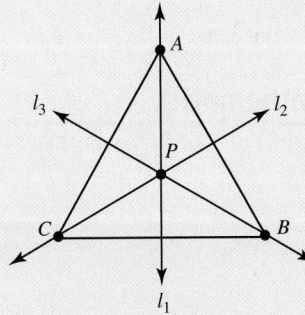

If any of the two motions are combined, one after the other, the net result is equivalent to either a rotation or a reflection. For instance, the net result of a reflection in l_1 by a rotation of 120° is equivalent to a reflection in l_3, or $r_1 * R_1 = r_3$. Fill in the table by combining two motions, where the column gives the first motion and the row gives the second. After completing the table, describe any patterns you see.

*	r_1	r_2	r_3	R_1	R_2	I
r_1						
r_2						
r_3						
R_1	r_3					
R_2						
I						

SECTION 12.3 Learning Assessment

Understanding the Concepts

1. How can a rigid motion be used to define a congruence relationship?

2. What is the difference between a rigid motion and a size transformation?

3. What information is needed to designate a particular size transformation on the plane?

4. What is the relationship between the scale factor and the size of the image under a size transformation?

5. How can a size transformation be used to define a similarity relationship?

6. What is the difference between a size transformation and a similitude?

7. What does it mean if a geometrical shape has:

    a. Reflection symmetry?

    b. Rotation symmetry?

    c. Point symmetry?

8. How can rotations and reflections be used to classify shapes?

Representing the Mathematics

9. Draw the size transformation of $\triangle ABC$ given by:

    a. $S_{O,3}$.   b. $S_{P,1/2}$.

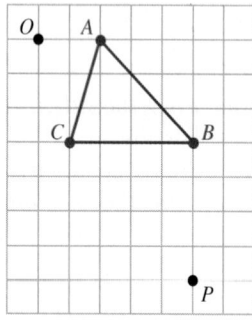

10. Draw the size transformation of quadrilateral $WXYZ$ given by:

    a. $S_{O,2}$.   b. $S_{O,1/3}$.

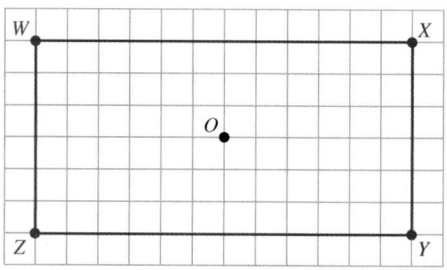

11. Draw the image of $\triangle ABC$ if it is first transformed by the size transformation $S_{O,2}$ and then rotated about O by $-90°$.

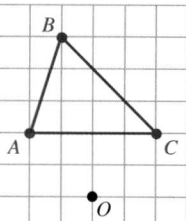

12. Complete each figure so that it has reflection symmetry in line l.

    a. 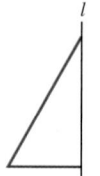   b.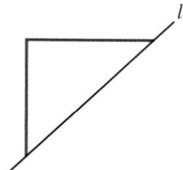

    c.    d.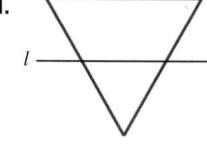

13. Trace each figure, and then draw all possible lines of symmetry.

a.

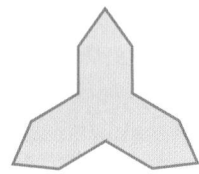

b.

c.

d.

14. Complete each figure so that it has point symmetry about *P*.

a.

b.

c.

15. If possible, draw a polygon with:

   a. Two lines of symmetry.

   b. Reflection symmetry, but no rotation symmetry.

   c. Rotation symmetry, but no reflection symmetry.

16. Draw two different objects, each having the following symmetries.

   a. Only vertical reflection symmetry

   b. Only horizontal reflection symmetry

   c. Three rotation symmetries

   d. Vertical and horizontal reflection symmetry

17. Consider the capital letters of the English alphabet written in block form (i.e., **A, B, C, D**, . . .). Draw a Venn diagram that identifies the vertical reflection, horizontal reflection, and 180° rotation symmetries of each letter.

■ Exercises

18. Determine a rigid motion that can be used to show each congruence relationship in the given figure.

a. $\triangle ABE \cong \triangle CBE$.

b. $\triangle AEB \cong \triangle DEC$.

c. $\triangle BEC \cong \triangle DEA$.

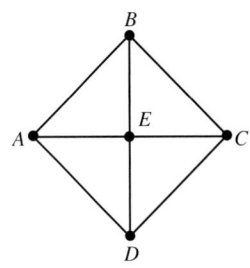

19. Consider the following figure.

Find a rigid motion that can be used to show each congruence relationship.

   a. $\square ABDC \cong \square DEFG$.

   b. $\square ABDC \cong \square FEDG$.

   c. $\square ABDC \cong \square FGDE$.

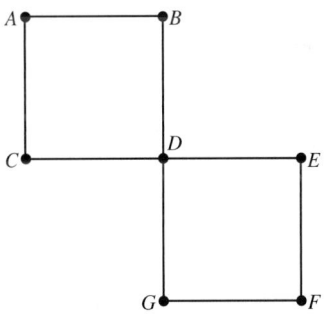

20. Describe the similarity transformation that takes quadrilateral $CATS$ to quadrilateral $C'A'T'S'$. Include any intermediate transformations.

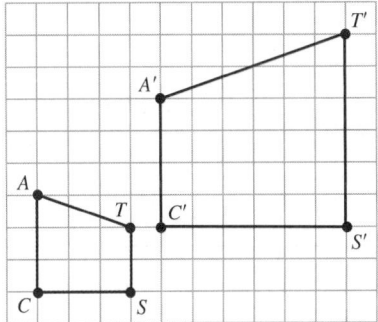

21. Describe the similarity transformation that takes $\triangle ABC$ to $\triangle A'B'C'$.

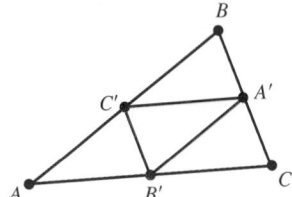

22. In each figure, the image of a size transformation with center O is shown in red. Find the scale factor and then use it to find the missing values.

a.

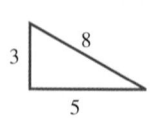

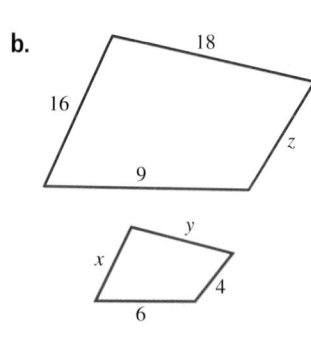

b.

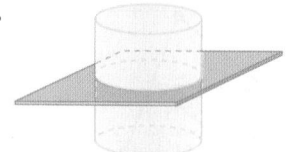

c.

23. **a.** Which of the single digits, 0 through 9, has reflection symmetry?

    b. Which of the single digits, 0 through 9, has rotation symmetry?

24. Find all turn angles that map each regular polygon onto itself. Use them to name the rotation symmetry of the figure.

    a. Octagon.　　　**b.** Decagon.

    c. 12-gon.　　　**d.** 21-gon.

25. Find the number of lines of symmetry for each polygon.

    a. Rectangle

    b. Parallelogram

    c. Regular octagon

    d. Regular decagon

    e. Regular 12-gon

    f. Regular 21-gon

26. State whether the given figure has plane symmetry across the given plane p.

    a.

b.

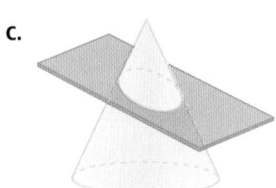

c.

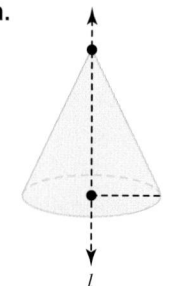

27. State whether the given figure has rotational symmetry about the given line l.

    a.　　　　　　**b.**

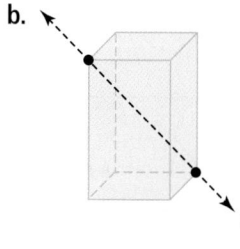

    c.　　　　　　**d.**

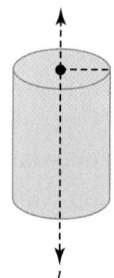
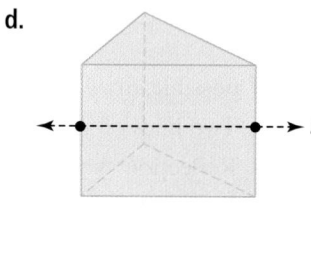

■ Problems and Applications

28. Find the coordinates of the images of points $A(-1, 3)$ and $B(2, 5)$ under the size transformation with the given center and scale factor.

    a. Center = $(0, 0)$, scale factor = 3.

    b. Center = $(0, 0)$, scale factor = $\dfrac{1}{2}$.

    c. Center = $(-1, 1)$, scale factor = 2.

    d. Center = $(5, 3)$, scale factor = 2.

29. Consider △ABC:

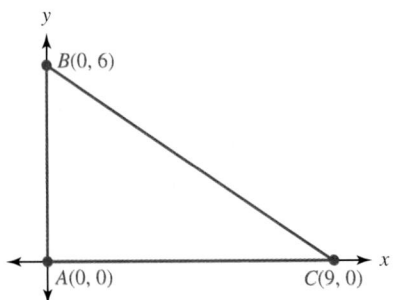

Give the coordinates of the image of each vertex if A remains fixed and:

a. The lengths of the sides are doubled.

b. The lengths of the sides are halved.

30. Consider △ABC:

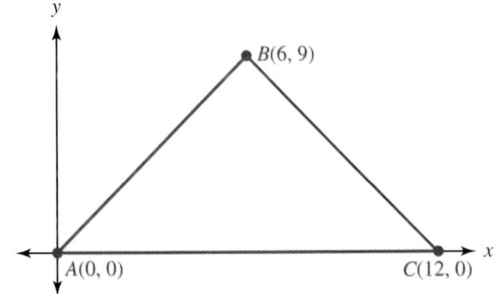

Give the coordinates of the image of each vertex under each size transformation.

a. $S_{A,3}$ **b.** $S_{A,1/3}$

31. Consider rhombus $ABCD$:

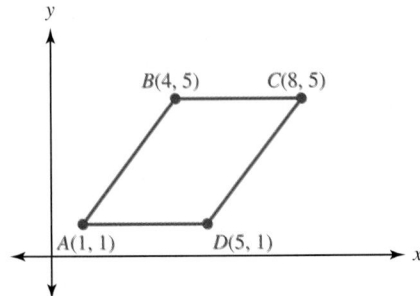

Give the coordinates of the image of each vertex under each transformation.

a. $S_{A,2}$ **b.** $S_{A,1/2}$

c. A rotation of –90° around A and $S_{A,2}$

d. A reflection in the y-axis and $S_{A,1/2}$

32. Make a copy of segment $\overline{AB}$ on another sheet of paper, and then use a straightedge and compass to divide it into five congruent segments.

33. Consider the following regular hexagon:

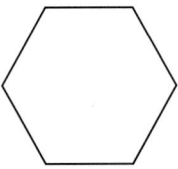

Draw three lines of symmetry so that:

a. Six congruent triangles are formed.

b. Six congruent kites are formed.

34. A square piece of paper is folded in quarters as shown.

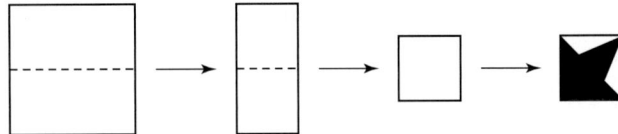

Draw the resulting figure if the shaded portion is cut out and unfolded.

35. Maria was hanging a doorplate with her house number on it. She accidentally dropped it during the process and noticed that, no matter which way she held it, the number read the same. If her four-digit house number contained only even numbers, what is Maria's house number?

36. Does a circular analog clock have any reflection symmetries? Rotation symmetries? Explain.

37. Identify the reflection and rotation symmetries of each flag.

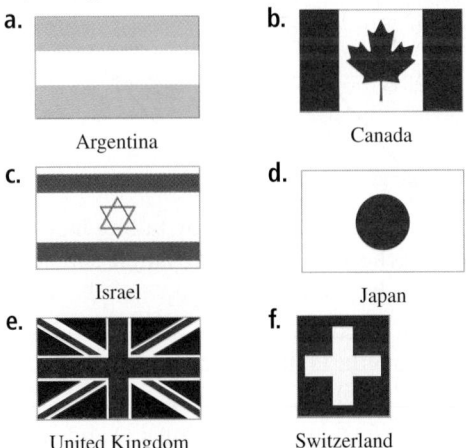

38. If the word MOW is rotated 180° around a point in the center of the word, the image still says MOW. Find five other words with this property.

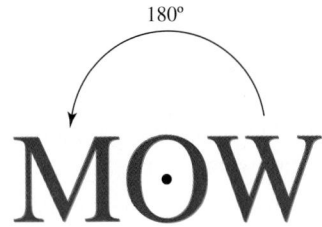

39. If the word BOX is reflected in a line that passes through the middle of the word, the image still says BOX. Find five other words with this property.

40. The word MOM is considered a palindrome because it has vertical reflection symmetry; that is, it reads the same forward or backward.

a. Find five other palindromes.

b. Sentences can also be palindromes. Can you write a sentence that reads the same forward as it does backward?

41. The arc of a circle is shown:

Determine a method of using reflection symmetry to find the center of the arc, and then use it to do so.

■ Mathematical Reasoning and Proof

42. Consider parallelogram $ABCD$.

a. What transformation would map $\triangle ABC$ onto $\triangle CDA$?

b. What different transformation(s) would map $\overline{AP}$ onto $\overline{PC}$?

c. What different transformation(s) would map point A onto Point C?

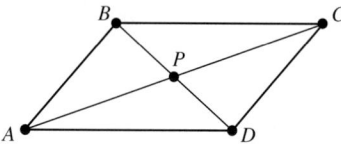

43. A triangle is decreased in size by taking half the length of each side. Is the new triangle similar to the original? Explain.

44. Each side of a rectangle is increased by adding 8 in. to its length. Is the new rectangle similar to the original? Explain.

45. If a size transformation with center O and scale factor k takes $\triangle ABC$ to $\triangle A'B'C'$, what scale factor takes $\triangle A'B'C'$ to $\triangle ABC$?

46. How many lines of symmetry does a circle have? Explain.

47. a. If a figure has a rotation symmetry other than the identity rotation, must it also have reflection symmetry?

b. If a figure has reflection symmetry, must it also have rotation symmetry?

48. a. If a figure has rotation symmetry, does it also have point symmetry?

b. If a figure has point symmetry, does it also have rotation symmetry?

49. Can a geometric shape have no symmetry? If so, draw an example of such a figure.

50. Repeat Activity 12.6 using the following eight symmetries of the square:

r_1: reflection in line l_1 R_1: 90° clockwise rotation with center P

r_2: reflection in line l_2 R_2: 180° clockwise rotation with center P

r_3: reflection in line l_3 R_3: 270° clockwise rotation with center P

r_4: reflection in line l_4 I : 360° clockwise rotation with center P

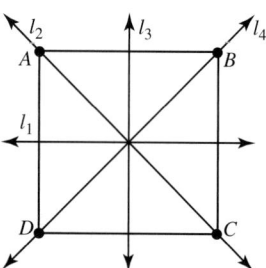

■ Mathematical Communication

51. a. How might you use similarity transformations to show a group of students that all squares are similar?

b. How might you use similarity transformations to show a group of students that not all rectangles are similar?

52. Use reflection and rotation symmetries to write alternative definitions for each of the basic types of quadrilaterals (i.e., trapezoid, kite, parallelogram, and so on).

53. Suppose the following shapes were made of paper. How would you explain to a student the way to use symmetry to find the center of each figure?

a. 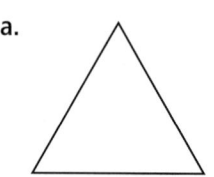 **b.**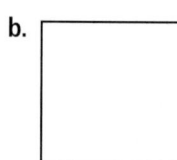

c. 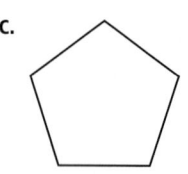 **d.**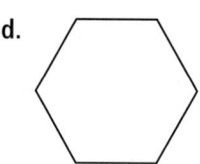

54. Write a set of directions that explains how to create a symmetrical object by using a reflection.

■ Building Connections

55. a. Explore the building in which your class is held. Find five things in the building with symmetry.

b. Take a walk outside and notice the things growing around you. Can you find five natural objects with reflection symmetry and five natural objects with rotation symmetry?

c. Discuss the symmetries of these shapes with a partner.

56. a. Consider the table of the 100 basic addition facts in Chapter 4. What symmetry do you see in the table?

b. Repeat part a using the table of the 100 basic multiplication facts.

57. Look at newspapers, magazines, the television, or any other place where advertisements can be seen. Find 10 company logos that use symmetry, and analyze the types of symmetry that each logo uses.

SECTION 12.4 Geometric Patterns

Transformations and symmetry are often used in art and design to generate a wide variety of geometrical patterns. Many of the patterns are repetitive, so we can classify them by their common characteristics. In general, there are three kinds of geometrical patterns: border patterns, wallpaper patterns, and tessellations.

Border Patterns

Many repetitive, geometrical patterns, like those we find on pottery, rugs, and tiles, occur linearly. Such patterns, called **border** or **frieze patterns,** are created when a basic design, or **motif,** is repeated through multiple iterations of a translation or a glide reflection (Figure 12.40).

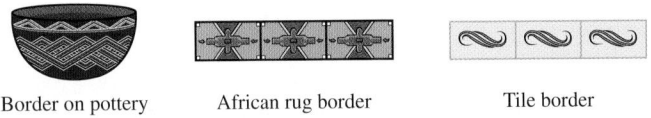

Border on pottery   African rug border   Tile border

Figure 12.40 Examples of border patterns

If the pattern is created with iterations of a translation, then the translation vector, T, dictates not only the direction of the pattern but also the regular intervals at which the images of the motif occur. Because the motif is under a translation, the images face the same direction and have the same orientation throughout the pattern (Figure 12.41).

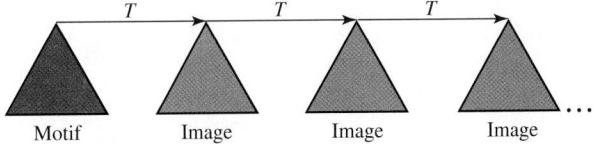

Motif   Image   Image   Image

Figure 12.41 A border pattern generated by a translation

If the pattern is created with a glide reflection, then the translation vector again dictates the direction and the intervals of the pattern. Now, however, the reflection causes the images of the motif to alternate sides of the glide axis. The glide axis runs along the middle of the pattern and alternating images have reverse orientations (Figure 12.42).

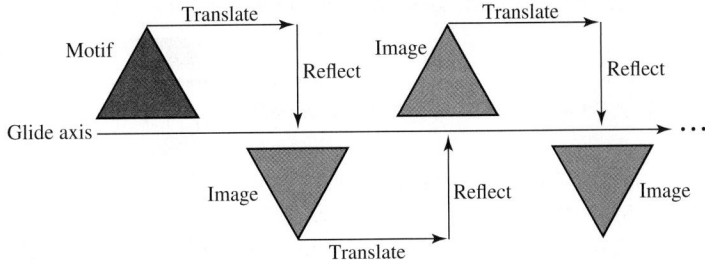

Figure 12.42 A border pattern generated by a glide reflection

Because border patterns are repetitive, we can use symmetry to classify them. However, to do so, we need not only reflection and rotation symmetry, but also translation and glide reflection symmetry. A figure has **translation,** or **glide reflection, symmetry** if a translation or glide reflection maps the figure onto itself. Clearly, no simple curve or polygon can have translation or glide reflection symmetry because mapping such a figure onto itself by sliding it a positive distance is impossible. However, if a translation is used to create a border pattern, then the pattern must have translation symmetry. If a glide reflection is used to create a border pattern, then the pattern has glide reflection and translation symmetry.

Explorations Manual 12.11

Border patterns can also have reflection and rotation symmetry, but only if a reflection or a rotation maps the entire pattern back onto itself. In general, the reflection symmetry of a border pattern matches the reflection symmetry of its motif. The only rotation symmetry a border pattern can have is 180° rotation symmetry because an infinite border pattern can be mapped back onto itself only through a half or full turn. Of the different combinations of the possible symmetries, there are only seven types of border patterns, identified by the two-symbol codes in Table 12.3. For instance, an *mm* border pattern has vertical and horizontal reflection symmetries, but no rotation symmetry. A 12-border pattern has no reflection symmetries but has 180° rotation symmetry. Examples of the seven border patterns are shown in Figure 12.43.

Table 12.3	**Symbols for classifying border patterns**		
First Symbol	**Meaning**	**Second Symbol**	**Meaning**
m	Border has vertical reflection symmetry.	*m*	Border has horizontal reflection symmetry.
1	Otherwise	*g*	Border has glide reflection symmetry with no horizontal reflection symmetry.
		2	Figure has half turn symmetry with no horizontal reflection symmetry or glide reflection symmetry.
		1	Otherwise

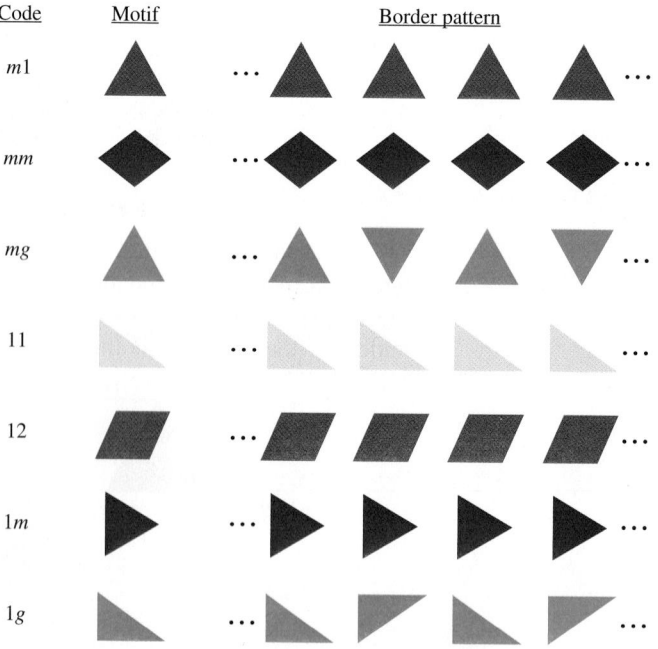

Figure 12.43 The seven border patterns

Example 12.28 | **Classifying Border Patterns**

Representation

Use Table 12.3 to classify each border pattern.

a. ··· 🌀🌀🌀 ··· **b.** ··· 🔷🔷🔷 ··· **c.** ··· 🌙🌙🌙 ···

Solution

a. The basic motif does not have reflection symmetry, but it does have half turn symmetry. This makes it a 12 pattern.

b. The basic motif has vertical reflection symmetry and glide reflection symmetry, making it an *mg* pattern.

c. The basic motif has vertical reflection symmetry but not horizontal reflection, rotation, or glide reflection symmetry. This makes it an *m*1 border pattern.

Wallpaper Patterns

Border patterns repeat a basic motif in a single direction. Repeating the motif in two directions to fill the entire plane results in another kind of geometrical pattern called a **wallpaper pattern**. Two examples are shown in Figure 12.44.

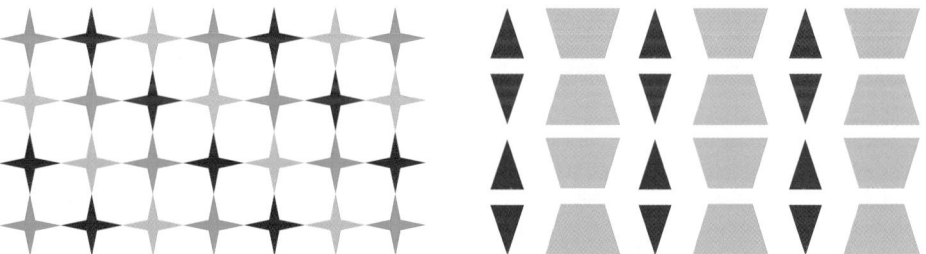

Figure 12.44 Examples of wallpaper patterns

As Figure 12.44 illustrates, wallpaper patterns can have a great deal of symmetry, which becomes more complex by the addition of a second direction. Like border patterns, wallpaper patterns can have translation, reflection, rotation, or glide reflection symmetry. However, these symmetries occur only if a corresponding transformation maps the entire pattern back onto itself. In addition, the infinite nature of wallpaper patterns limits the possibilities for symmetries. Specifically, wallpaper patterns can have reflection symmetry or glide reflection symmetry only through lines that are vertical, horizontal, or at 45°. They can have rotational symmetry only through turn angles of 60°, 90°, 120°, and 180°. Given these constraints, only 17 kinds of wallpaper patterns are possible.

Explorations Manual 12.12

Example 12.29 | **Describing Symmetry in a Wallpaper Pattern**

Communication

The following wallpaper pattern was found on the ceiling in an Egyptian tomb. Describe its symmetry.

The Grammar of Ornament (1856), by Owen Jones. Egyptian No 7 (plate 10), image #13

Solution This pattern has several symmetries. First, it has vertical reflection symmetry along several lines. Second, it has half turn symmetry around two points in the motif. Third, it has glide reflection symmetry in a horizontal direction. Specific lines and turn centers for the different symmetries are shown:

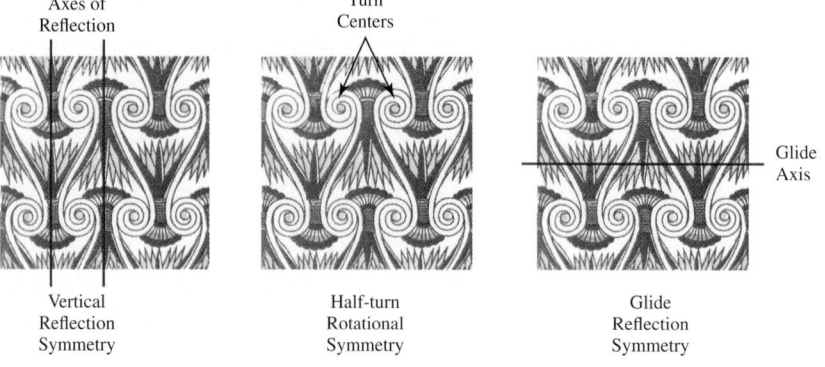

Axes of Reflection   Turn Centers   Glide Axis

Vertical Reflection Symmetry   Half-turn Rotational Symmetry   Glide Reflection Symmetry

Check Your Understanding 12.4-A

1. Use Table 12.3 to classify each border pattern.

a.

b. ... ▲▽▲▽▲▽▲▽ ...

c. ...[XXX pattern]...

d. ... ⊂⊃⊂⊃⊂⊃⊂⊃ ...

2. Use only the letter *F* to construct a border pattern of each type.

   a. 11   **b.** 1*g*   **c.** *m*1   **d.** *mg*

3. The following wallpaper pattern was found in an Egyptian tomb. Describe its symmetry.

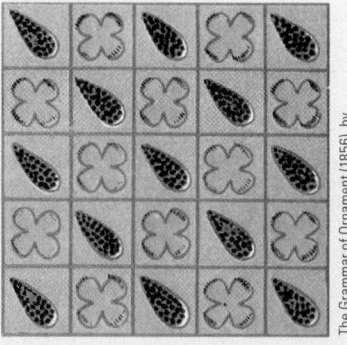

The Grammar of Ornament (1856), by Owen Jones. Egyptian No 6 (plate 9), image #20

Talk About It Can a finite border pattern have translation symmetry? Explain.

Representation

Activity 12.7

The six shapes in a set of pattern blocks are shown:

   a. Use two shapes to make an *mm* border pattern.

   b. Use two shapes to make a wallpaper pattern with rotational symmetry.

Tessellations

A **tessellation,** or a **tiling,** is a wallpaper pattern for which the motif completely fills the plane without any overlaps or gaps. This characteristic makes them special among wallpaper patterns because they are somewhat limited by the shapes that can be used to form their motifs. As a result, we can classify tessellations not just by their symmetry but also by the shapes, or **tiles,** used to make them. We begin with the regular tessellations.

Regular Tessellations

A **regular tessellation** is a tessellation that is made from a single regular polygon. Three examples are shown in Figure 12.45.

Figure 12.45 Examples of regular tessellations

Of the many regular polygons, only the equilateral triangle, square, and regular hexagon tessellate the plane. To understand why, consider the constraints. First, a regular tessellation is made with only one regular polygon, which limits the selection to polygons with congruent sides and congruent interior angles. Second, whenever the polygons meet at a vertex, the interior angles must form a complete rotation of 360°. If not, then there is either a gap or an overlap at the vertex (Figure 12.46).

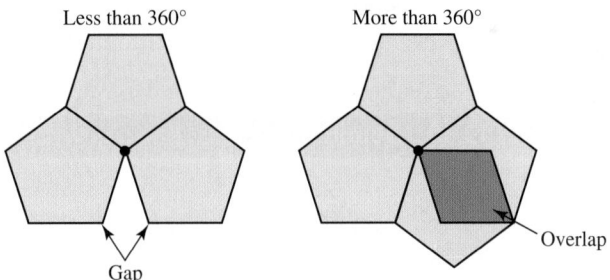

Figure 12.46 Gaps and overlaps

Third, because the angles at one vertex are congruent, their measure must equally divide 360°. As Figure 12.47 illustrates, all three criteria are met by the equilateral triangle, the square, and the regular hexagon.

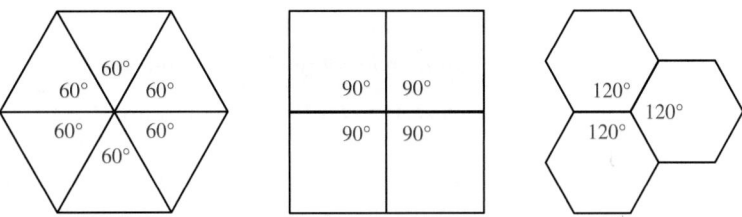

Figure 12.47 The equilateral triangle, the square, and the regular hexagon

We might ask whether any other regular polygons have interior angle measures that divide 360°. However, as Table 12.4 shows, the equilateral triangle, square, and regular hexagon are the only regular polygons with 10 or fewer sides that do.

Table 12.4 Measures of central angles of regular polygons

Regular Polygon	Measure of Interior Angle	360°/(Angle Measure)
Triangle	60°	6
Square	90°	4
Pentagon	108°	3.33
Hexagon	120°	3
Heptagon	128.6°	2.80
Octagon	135°	2.67
Nonagon	140°	2.57
Decagon	144°	2.5

Table 12.4 also shows that as the number of sides of the polygon increases, so does the measure of the interior angle. Because the only factor of 360 larger than 144 is 180 and no regular polygon has an interior angle of this measure, no other regular polygons tessellate the plane.

Theorem 12.6 Regular Polygons that Tessellate the Plane

The only regular polygons that tessellate the plane are the equilateral triangle, square, and regular hexagon.

Semiregular Tessellations

Semiregular tessellations differ from regular tessellations in that they use two or more regular polygons to make an arrangement that is the same at each vertex.

Example 12.30 Semiregular Tessellations with Triangles and Squares

Problem Solving Find every arrangement of equilateral triangles and squares that can be used to make a semiregular tessellation.

Solution

Understanding the Problem. We are asked to find all semiregular tessellations that use equilateral triangles and squares in the motif. So we must use at least one of each shape at each vertex and still have no gaps or overlaps.

Devising a Plan. One approach is to *use a manipulative* like pattern blocks to explore the arrangements of equilateral triangles and squares that might tessellate the plane. To ensure that all possible arrangements are found, we work systematically through an increasing number of squares.

Carrying Out the Plan. Because the motif of a semiregular tessellation must have at least one of each shape, we begin with motifs with only one square. Every other shape must be an equilateral triangle, giving us only one possible arrangement. Because the motif has a gap, it cannot tessellate the plane.

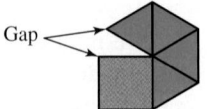
Gap

Next, we consider motifs with two squares. There are two such arrangements in which the triangles and squares share a common vertex. Both tessellate the plane.

Motif 1   Motif 2

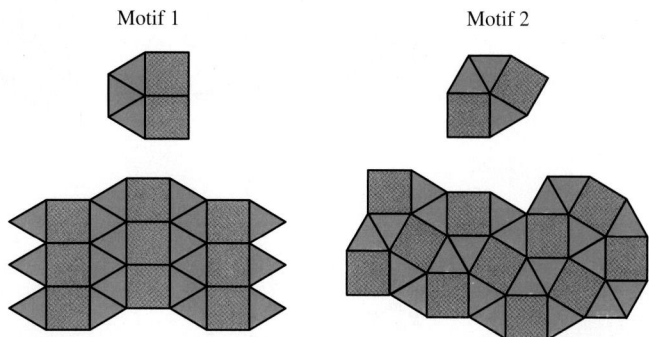

Finally, we consider motifs with three squares. Only one arrangement with three squares and one equilateral triangle is possible. Because it has a gap, it cannot tessellate the plane.

Gap
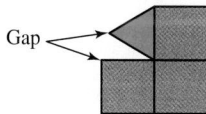

Looking Back. Another way to work the problem is to use the fact that the sum of the interior angles of the polygons around any given vertex must be 360°. Of the different possibilities, the following table shows that, again, only two squares and three triangles will work.

Number of Squares	Number of Triangles	Sum of Interior Angles at a Vertex
1	4	90° + 60° + 60° + 60° + 60° = 330°
2	3	90° + 90° + 60° + 60° + 60° = 360°
3	1	90° + 90° + 90° + 60° = 330°

As with regular tessellations, the number of semiregular tessellations is limited. In fact, only 21 arrangements of regular polygons will go around a vertex with no overlaps or gaps. Three lead to regular tessellations. Of the remaining 18 possibilities, only eight tessellate the plane. They are shown in Figure 12.48. The notation below each tessellation indicates the number of sides and the order of the polygons around each vertex. For instance, the notation (3, 3, 3, 4, 4) indicates an arrangement of three triangles and two squares. The notation (4, 6, 12) indicates an arrangement of a square, a regular hexagon, and a regular 12-gon.

Tessellating with Other Figures

So far, the tessellations have been made only from regular polygons. Removing this condition opens up many more possibilities for tessellating the plane. For instance, the fact that equilateral triangles and squares can be used to make both regular and semiregular tessellations may lead us to wonder whether other triangles and quadrilaterals might tessellate the plane.

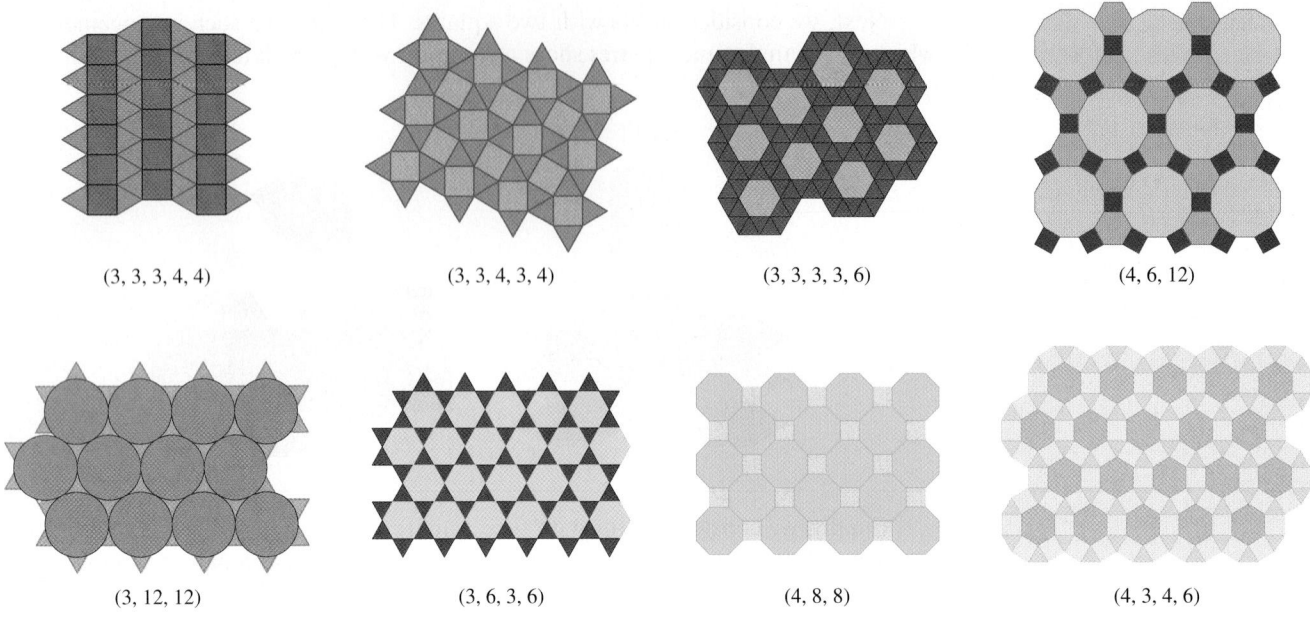

(3, 3, 3, 4, 4)   (3, 3, 4, 3, 4)   (3, 3, 3, 3, 6)   (4, 6, 12)

(3, 12, 12)   (3, 6, 3, 6)   (4, 8, 8)   (4, 3, 4, 6)

Figure 12.48 The eight semiregular tessellations

| Example 12.31 | Tessellating the Plane with Triangles and Quadrilaterals |

Which of the following figures tessellate the plane?

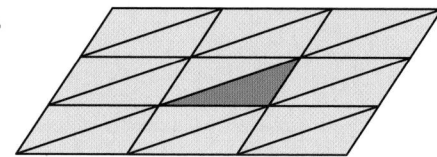

a.   b.   c.   d.

Explorations Manual 12.13

Solution By using copies of the shape and different transformations, we can see that each triangle or quadrilateral tessellates the plane. The original figure is shown in dark blue.

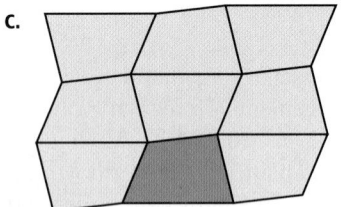

a.   b.

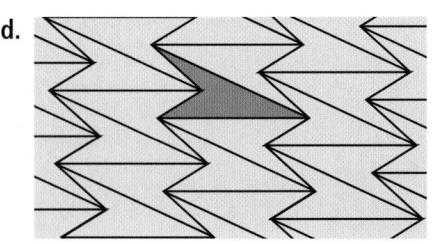

c.   d.

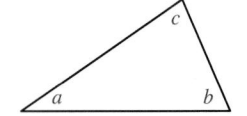

Figure 12.49 A triangle with interior angle measures of *a*, *b*, and *c*

The results of Example 12.31 may seem surprising, but if we consider the sum of the angles in any triangle or any quadrilateral, they should make sense. For instance, suppose a triangle has interior angles of measures a, b, and c (Figure 12.49). We know the sum of the angles must be 180°, or $a + b + c = 180°$. We can therefore use a series of translations and rotations to manipulate three copies of the triangle so that the angles come together to form a straight angle. Then a rotation of 180° of all three triangles results in six angles that share a common vertex and have a sum of measures equal to 360° (Figure 12.50). These are exactly the characteristics needed to form a tessellation, so any triangle will tessellate the plane.

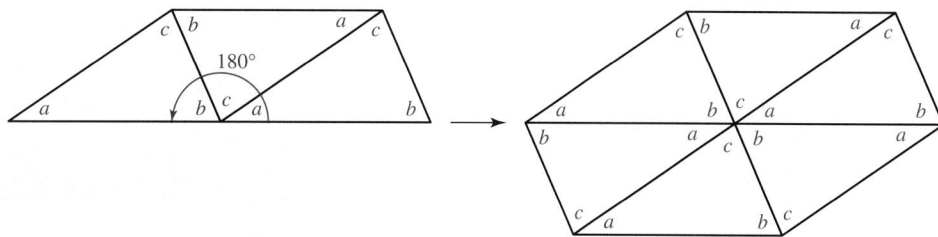

Figure 12.50 A triangle tessellating the plane

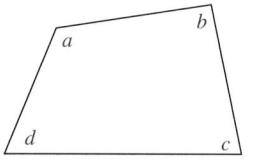

Figure 12.51 A quadrilateral with interior angle measures of *a*, *b*, *c*, and *d*

A quadrilateral is similar. Consider a quadrilateral with interior angles of measures *a*, *b*, *c*, and *d* (Figure 12.51). As with the triangle, transformations can be used to put the four angles together around a single vertex. Specifically, we rotate copies of the quadrilateral 180° around the midpoints of three of the sides (Figure 12.52). Because $a + b + c + d = 360°$, we have exactly what we need to form a tessellation. The procedure works for any quadrilateral, both convex and concave. Consequently, any quadrilateral will tessellate the plane.

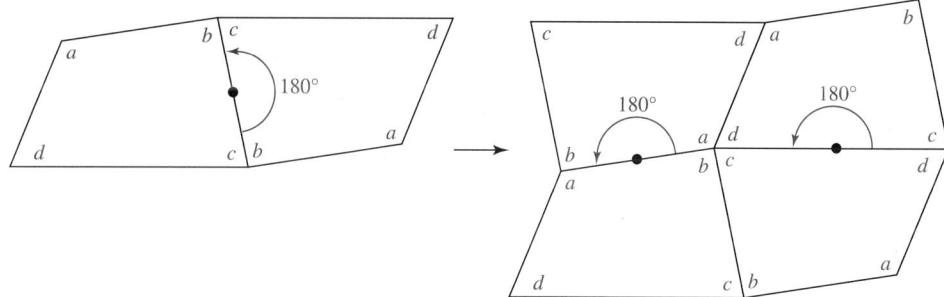

Figure 12.52 A quadrilateral tessellating the plane

Theorem 12.7 Triangles and Quadrilaterals Tessellate the Plane

Any triangle or quadrilateral will tessellate the plane.

In addition to triangles and quadrilaterals, Figure 12.53 shows that nonregular pentagons and hexagons can also tessellate the plane. However, nonregular hexagons can be used only if they have two opposite sides that are parallel and the same length. Beyond this, no other convex polygon with seven or more sides tessellates the plane. However, if we use concave polygons, then polygons with any number of sides can tessellate the plane.

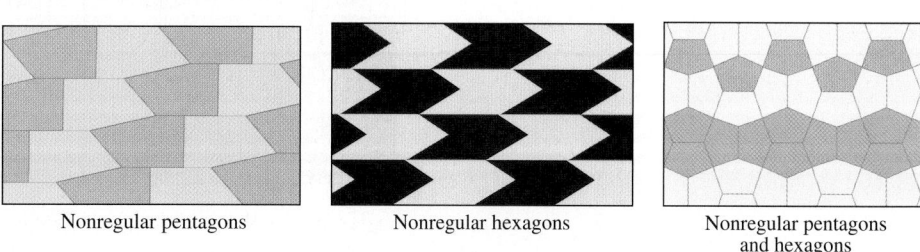

Nonregular pentagons   Nonregular hexagons   Nonregular pentagons and hexagons

Figure 12.53 Different polygons tessellating the plane

Escher Drawings

Many artists use tessellations to create intriguing works of art like the one by M. C. Escher shown in Figure 12.54. With a basic knowledge of which polygons do and do not tessellate the plane, we can adapt tessellations into Escher-like drawings. Each of the several methods to do so manipulates the tessellating figure with a particular rigid motion.

Figure 12.54 An Escher drawing

In the **cut-and-slide method,** we take one side of a figure that tessellates the plane, cut out a shape, and then translate it to the opposite side. We can do this with any pair of opposite sides, and the new figure tessellates the plane. For instance, we can generate an interesting Escher-like drawing by cutting a rounded piece from the bottom of a rectangle and translating it to the top, as shown in Figure 12.55(a). Next, cut the figure in half, and translate the left piece to the opposite side of the right (Figure 12.55(b)). After tessellating the plane with the new figure and adding colors and words, we have the finished picture shown in Figure 12.55(c).

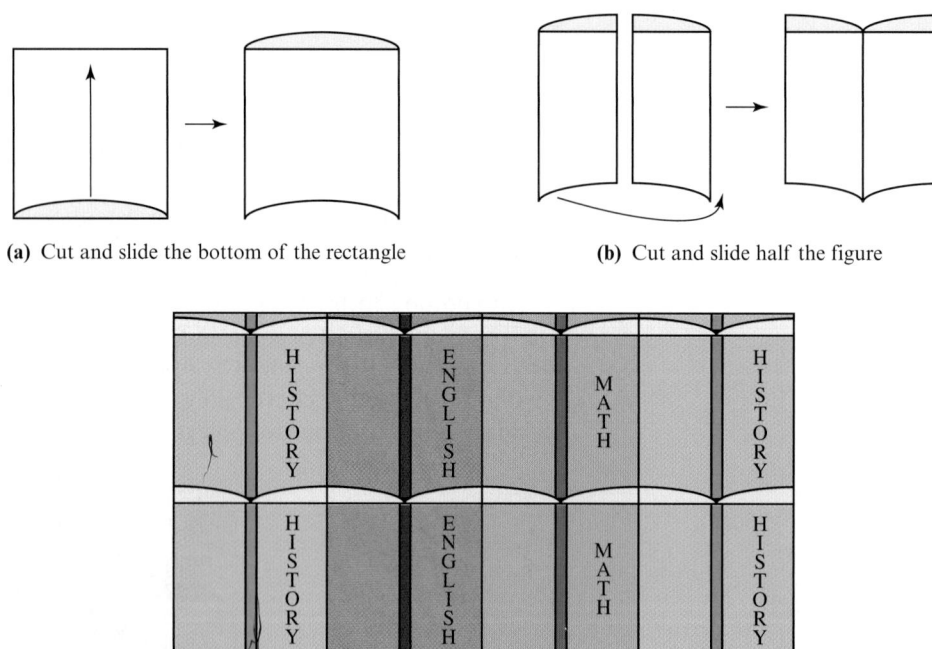

(a) Cut and slide the bottom of the rectangle

(b) Cut and slide half the figure

(c) Translate copies and add details to finish the picture

Figure 12.55 An Escher-like drawing using the cut-and-slide method

Another method, the **cut-and-turn method,** uses rotations. Again, beginning with a figure that tessellates the plane, we cut a shape from one side of the figure and rotate it around a vertex so that it lies on top of an adjacent side (Figure 12.56(a)). We can repeat the same procedure on the other pair of adjacent sides (Figure 12.56(b)). Then with rotations of the finished shape, we can tessellate the plane to obtain another Escher-like drawing (Figure 12.56(c)).

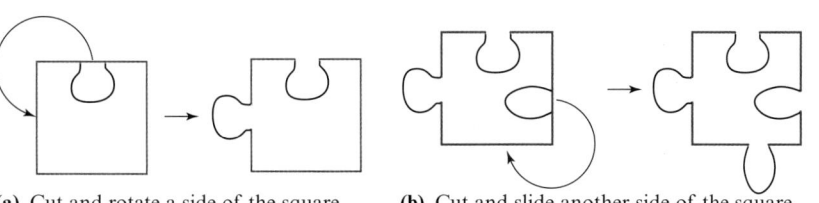

(a) Cut and rotate a side of the square　(b) Cut and slide another side of the square　(c) Rotate copies and add color to finish the picture

Figure 12.56 An Escher-like drawing using the cut-and-turn method

The rotations can also be made around the midpoint of the sides rather than around the vertices (Figure 12.57(a)). In that case, we must not cut away more than half of any one angle, because doing so causes neighboring figures to overlap. Again, with rotations of the resulting shape, we can tessellate the plane to obtain an Escher-like drawing (Figure 12.57(b)).

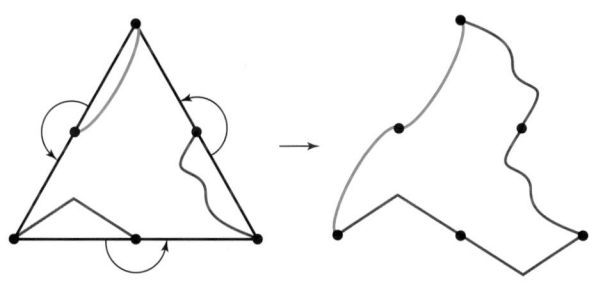

(a) Rotating a cut from the midpoint of a side　(b) Rotate copies and add color to finish the picture

Figure 12.57 An Escher-like drawing made from rotations about midpoints

Check Your Understanding 12.4-B

1. Draw a tessellation of the plane using copies of each figure.

a. 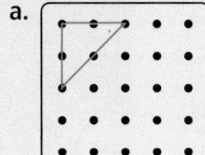　b.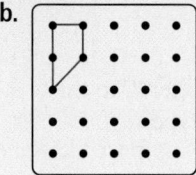

2. Create an Escher-like drawing using a rectangle and the cut-and-slide method.

3. Create an Escher-like drawing using a triangle and the cut-and-rotate method.

Talk About It Is it possible to have a regular tessellation with squares in which the squares do not share a common vertex?

Representation

Activity 12.8
The following diagrams show two different geoboards. If possible, tessellate the rest of the geoboard using the given figure.

a. [geoboard diagram]　b. [geoboard diagram]

SECTION 12.4 Learning Assessment

Understanding the Concepts

1. Describe two ways that a motif can be used to make a border pattern.

2. What does it mean for a geometrical figure to have translation symmetry? Glide reflection symmetry?

3. What limitations are on the reflection and rotation symmetries of a border pattern?

4. What is the difference between:
   a. A border pattern and a wallpaper pattern?
   b. A wallpaper pattern and a tessellation?
   c. A regular tessellation and a semiregular tessellation?

5. Why are equilateral triangles, squares, and regular hexagons the only shapes that form a regular tessellation?

6. What two figures always tessellate the plane? Why?

7. How can translations be used to make an Escher-like drawing? Rotations?

Representing the Mathematics

8. Use Table 12.3 to categorize the symmetry of each border pattern.

a.  b.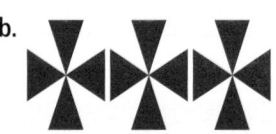

c.  d.

e.

f.

9. Use only rotations of the letter *T* to make a border pattern of each symmetry type.
   a. 1*m*   b. *mg*
   c. *m*1   d. *mm*

10. Use only rotations of the letter *E* to make a border pattern of each symmetry type.
    a. 1*m*   b. *mm*
    c. 1*g*   d. *mg*

11. Each wallpaper pattern has vertical reflection symmetry. Draw three lines of symmetry.

a.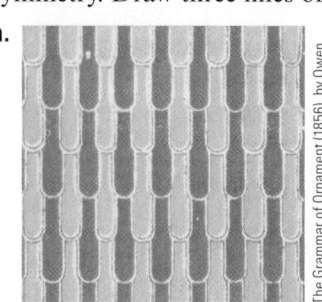

b.

c.

12. Each wallpaper pattern has horizontal reflection symmetry. Draw three lines of symmetry.

a.

b.

c.

The Grammar of Ornament (1856), by Owen Jones. Chinese No 1 (plate 59), image #1

13. Each wallpaper pattern has rotation symmetry. Find three turn centers.

a.

The Grammar of Ornament. Egyptian No 7 (plate 10), image #20

b.

The Grammar of Ornament (1856), by Owen Jones. Egyptian No 7 (plate 10), image #24

c.

The Grammar of Ornament (1856), by Owen Jones. Egyptian No 7 (plate 10), image #15

14. Draw a wallpaper pattern for each type of symmetry.

   a. Horizontal reflection symmetry

   b. Vertical reflection symmetry

15. Draw a wallpaper pattern for each type of symmetry.

   a. Rotational symmetry

   b. Glide reflection symmetry

16. Use pattern blocks to construct a tessellation that consists of:

   a. Rhombi.   **b.** Triangles and squares.

   c. Trapezoids and triangles.

17. For each number *n*, give an example of a polygon of *n* sides that tessellates the plane.

   a. 3   **b.** 5   **c.** 7

18. For each number *n*, give an example of a polygon of *n* sides that tessellates the plane.

   a. 4   **b.** 6   **c.** 8

19. a. Draw an example of a polygon that does not tessellate the plane.

   b. Draw an example of a motif of two polygons that does not tessellate the plane.

   c. Draw an example of a motif of three polygons that does not tessellate the plane.

20. Create an Escher-like drawing using a square and the cut-and-slide method.

21. Create an Escher-like drawing using a rectangle and the cut-and turn-method.

22. Create an Escher-like drawing using an equilateral triangle and the cut-and-turn method around the midpoint of each side.

23. Create an Escher-like drawing using a nonregular hexagon where each pair of opposite sides are parallel and congruent.

Problems and Applications

24. A portion of the (4, 8, 8) semiregular tessellation is shown.

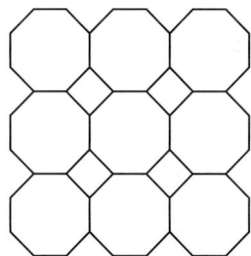

If the measure of the interior angles of a square is 90°, use the arrangement of the polygons around a vertex to find the measure of the interior angles of a regular octagon.

25. Using regular polygons, there are 21 possible arrangements around a vertex. Three lead to regular tessellations, and eight lead to semiregular tessellations. Find three other such arrangements that do not tessellate the plane.

26. A pentomino is a polygon made from five congruent squares. Which of the following pentominos tessellate the plane?

a. 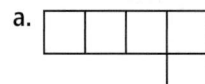   b.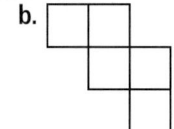

c. 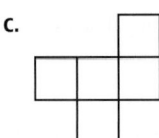   d.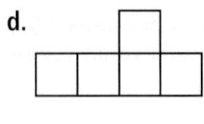

27. We can make other figures that tessellate the plane by taking tessellating figures and removing certain edges. For instance, if we remove one corner from a square, the new shape tessellates the plane.

    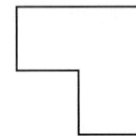

    Make three other such figures with a triangle, a square, and a hexagon. Demonstrate how each tessellates the plane. What transformations are used in the tessellations you create?

28. Find an example of a tessellation that uses regular polygons but is neither a regular nor a semiregular tessellation.

29. The following tessellation was made from the triangle shaded in blue.

    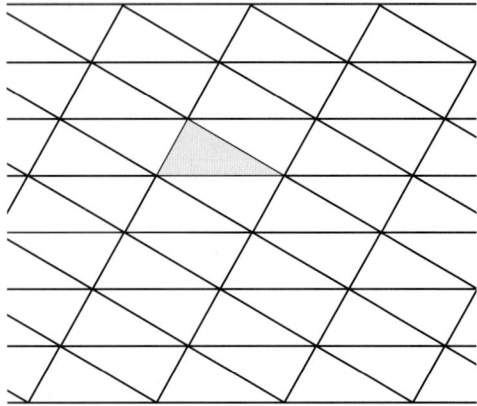

    Highlight a triangle that is similar to the original yet has sides that are:

    a. Twice as large.

    b. Three times as large.

    c. Five times as large.

30. Does the following shape tessellate the plane? If not, can it be combined with another shape so that it does? Explain.

    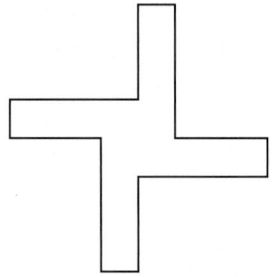

31. Consider the letters made on each geoboard. Use them to tessellate the plane.

    a. 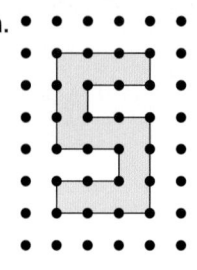   b.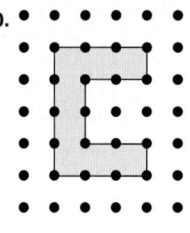

    c. 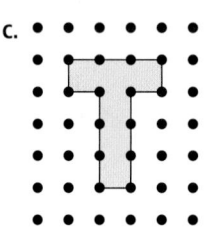   d.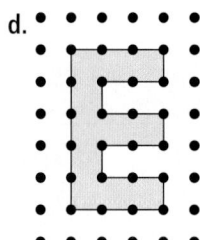

32. The **dual** of a regular tessellation is made by connecting the centers of adjacent polygons that share a common side. Draw the dual of each regular tessellation. What do you notice about the resulting tessellations?

33. Space can also be tessellated by completely filling it with a three-dimensional motif so that there are no overlaps or gaps. A cube is an example of a figure that tessellates space. Find two others.

34. A couple buys a historic home and would like to restore it. The bathroom floor was originally tiled with squares and triangles but is now ruined. If they found the following three tiles still connected together, what was the pattern of the original floor?

    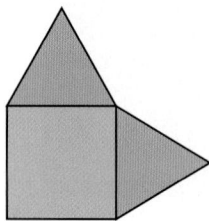

35. Sidewalk paving stone comes in three sizes: large, medium, and small. Several patterns can be used to lay the block. Identify the motif of each tessellation and the transformations through which it is repeated.

    a. 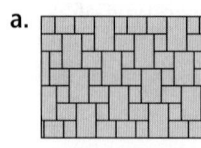   b.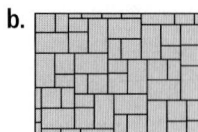

    c. 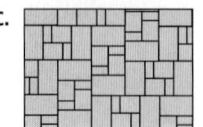   d.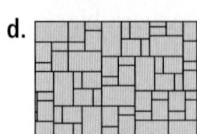

■ Mathematical Reasoning and Proof

36. Give an argument that explains why a regular heptagon does not tessellate the plane.

37. Will all convex hexagons tessellate the plane? Explain.

38. Can three quadrilaterals fit around a vertex and still tessellate the plane? If so, provide an example. If not, explain why.

39. Can squares be used to make a tessellation that is not regular? If so provide an example.

40. Can a circle be used in a tessellation? Why or why not?

41. If a piece from a tessellating polygon is translated to the opposite side, the figure still tessellates the plane. Does this work if the piece is transformed to the opposite side through a glide reflection? Draw such a shape, and explore the possibility.

■ Mathematical Communication

42. Look up the root of the word "tessellation." Why is this an appropriate term for describing these geometrical patterns?

43. Write a set of directions that explain how to tessellate the plane using each figure. Give them to a peer to follow and critique.

a.    b.

44. Use a diagram to explain to a peer why each of the following figures does or does not tessellate the plane.

a.    b. 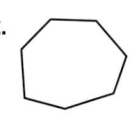   c.   d.

45. Micah claims that he can tessellate the plane with a regular octagon. Is he correct?

■ Building Connections

46. Do an Internet search on the seven different border patterns. You will likely find that each pattern often has a name like "jump" or "sidle." Make a list of the seven names. Do the names seem to match the transformations used to make the pattern? Explain.

47. Search the Internet for the 17 different categories of wallpaper patterns. Write a paragraph or two explaining the different symmetries used to form the categories.

48. Find five tessellations from different cultures. Describe the motif and how it is used to tessellate the plane.

49. How are border patterns and tessellations used in the classroom? Search a set of curriculum materials to find out. Write a short paragraph describing what you learned.

Historical Note

Maurits C. Escher (1898–1972) was a modern artist who received great acclaim for his use of geometric patterns. In his earliest works, Escher used different spatial effects to alter landscapes. However, after viewing the tile work in the Moorish palace of the Alhambra, he was inspired to begin a new phase in his art. Patterns in the architecture led him to develop his concept of the "regular divisions of the plane," which he used extensively in his so-called metamorphosis works. Escher's use of symmetry and geometrical properties has fascinated mathematicians for years. It is even reasonable to suggest that Escher's art was a catalyst for a renewed interest in the study of geometrical patterns.

M. C. Escher (self-portrait)

Opening Problem Answer

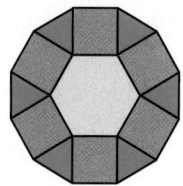

XPhantom/Shutterstock.com

The following arrangement of equilateral triangles, squares, and regular hexagons is a semiregular tessellation. It covers the floor with no gaps or overlaps.

Chapter 12 REVIEW

Summary of Key Concepts from Chapter 12

Section 12.1   Coordinate Geometry

- **Coordinate,** or **analytical, geometry** places shapes on a coordinate system so that they can be analyzed algebraically.

  - The **unit length** is the distance from 0 to 1 on the x-axis of the Cartesian plane.

  - If $\overline{AB}$ is horizontal, then $AB = x_2 - x_1$, where $x_1 < x_2$. If $\overline{AB}$ is a vertical, then $AB = y_2 - y_1$, where $y_1 < y_2$.

- **The distance formula:** If points A and B have coordinates $A(x_1, y_1)$ and $B(x_2, y_2)$, then the distance between A and B is given by $AB = \sqrt{(x_2 - x_1)^2 + (y_2 - y_1)^2}$.

- **Midpoint formula:** If points A and B have coordinates $A(x_1, y_1)$ and $B(x_2, y_2)$, then the midpoint, M, of $\overline{AB}$ has coordinates $\left(\dfrac{x_1 + x_2}{2}, \dfrac{y_1 + y_2}{2} \right)$.

- A circle with center (h, k) and radius r has the equation $r^2 = (x - h)^2 + (y - k)^2$.

Section 12.2   Transformations

- The study of how shapes can be moved and how to represent those motions is called **motion,** or **transformation, geometry.**

- A **rigid motion,** or **isometry,** is any motion that preserves length. A rigid motion also preserves angle measure.

- A **transformation** is a function that indicates how each point P on the plane corresponds to another point P'. P' is the **image** of P, and P the **preimage** of P'.

- A **translation** with a translation vector from A to B, written T_{AB}, is a transformation of the plane that maps every point P to point P' such that $\overline{PP'}$ is congruent and parallel to $\overline{AB}$. The direction and the length of the translation are given by a **translation vector,** or **slide arrow.**

- A **proper** or **orientation-preserving** motion preserves the orientation of the shape. An improper, or **orientation-reversing,** transformation reverses the orientation.

- A **rotation** with center C and turn angle α, written $R_{C,\alpha}$, is a transformation of the plane that maps every point P to a point P' such that $m\angle PCP' = \alpha$ and $PC = P'C$.

- Any point P that stays in the same position under a rigid motion is called a **fixed point.**

- A transformation that maps every point on the plane back to itself is called the **identity transformation.**

- The **reflection** in a line l, written M_l, is a transformation of the plane that maps every point P to a point P' such that l is the perpendicular bisector of $\overline{PP'}$. If P is on l, then P is mapped to itself, or $P = P' \cdot l$ is called the **axis of reflection,** or **reflecting line.**

- The glide reflection with the translation vector T_{AB} and glide axis l parallel to the segment $\overline{AB}$ is a transformation that maps every point P to a point P' through the combination of the translation T_{AB} followed by the reflection M_l. The axis of reflection is called the **glide axis.**

- Any rigid motion or combination of rigid motions is equivalent to a single translation, rotation, reflection, or glide reflection.

Section 12.3   Congruence, Similarity, and Symmetry with Transformations

- Two polygons are **congruent** if and only if one is the image of the other under a rigid motion or a combination of rigid motions.

- Let O be a point in the plane, and let k be a positive real number. A **size transformation,** or **dilation,** with center O and scale factor k, written $S_{O,k}$, is a transformation that maps every point P to a point P' so that O, P, and P' are collinear and $OP' = k \cdot OP$.

- Let k be the scale factor of a size transformation. If $k > 1$, the image is larger than the original figure, and the size transformation is a **magnification.** If $k < 1$, the image is smaller than the original figure, and the size transformation is a **contraction.** If $k = 1$, the size transformation is the identity transformation.

- A size transformation preserves the ratio of distances, angle measure, orientation, and parallelism.

- Any combination of transformations that leads to similar figures is called a **similarity transformation,** or **similitude.**

- Two polygons are **similar** if and only if one figure is the image of the other under a combination of a size transformation and one or more rigid motions.

- An object has **symmetry** if it can be transformed in some way back onto itself.

- A shape has **reflection,** or **line symmetry** if one side of the figure can be "folded" over so that it lies exactly on top of the other side. The fold line forms the **axis,** or **line of symmetry.**

- A shape has **rotation symmetry** if it can be mapped back onto itself by rotating it around a point through some turn angle less than 360°. We use the smallest positive angle to name the rotation symmetry.

- If a figure is its own image under a half turn or a 180° rotation, then it has **point symmetry.**

Section 12.4   Geometric Patterns

- A **border,** or **frieze pattern,** is created when a **motif** is repeated through multiple iterations of a translation or a glide reflection.

- A shape has **translation,** or **glide reflection, symmetry** if a translation or glide reflection maps the figure onto itself.

- A **wallpaper pattern** is created when a motif is repeated in two directions.

- A **tessellation,** or **tiling,** is a wallpaper pattern in which the motif completely fills the plane without any overlaps or gaps.

- A **regular tessellation** is a tessellation made from congruent repetitions of a single regular polygon. Only three regular polygons tessellate the plane: the equilateral triangle, square, and regular hexagon.

- **Semiregular tessellations** use two or more regular polygons to make an arrangement of polygons that is the same at each vertex.

- Any triangle or quadrilateral tessellates the plane.

- Tessellations can be adapted into Escher-like drawings by using the **cut-and-slide** method or the **cut-and-turn** method.

Review Exercises Chapter 12

1. Find the length and midpoint of each segment.

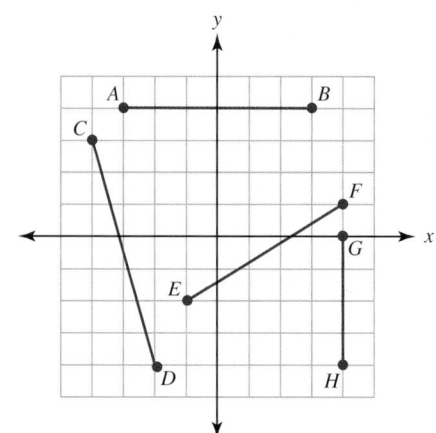

2. If points, $A(-1, -2)$, $B(4, 1)$, and $C(3, -3)$, are the vertices of a triangle, what kind of triangle is $\triangle ABC$?

3. Are points $A(-2, 5)$, $B(0, 9)$, and $C(2, 12)$ collinear?

4. Write the equation of the circle with the given center and radius.

   a. $(0, 0)$, $r = 5$.   **b.** $(-2, 4)$, $r = 1$.
   c. $(3, 8)$, $r = 10$.

5. What are the radius and center of the circle given by the equation $x^2 + (y - 3)^2 = 64$?

6. What is the other endpoint of a segment that has one endpoint with coordinates $A(4, 7)$ and a midpoint of $M(-2, 3)$?

7. Suppose the endpoints of $\overline{AB}$ have coordinates $A(3, 0)$ and $B(6, 0)$. Find possible coordinates for a point C so that $\triangle ABC$ is an isosceles triangle.

8. The coordinates of three vertices of a parallelogram are $A(2, 2)$, $B(0, 4)$, and $C(-3, 1)$. What are possible coordinates for the fourth vertex?

9. If quadrilateral $ABCD$ is a rhombus, what are the coordinates of point C?

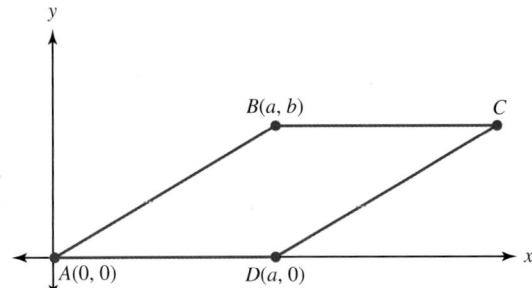

10. Find the image of quadrilateral $ABCD$ under each translation.

   a. $T_{OO'}$ b. $T_{PP'}$ c. $T_{RR'}$

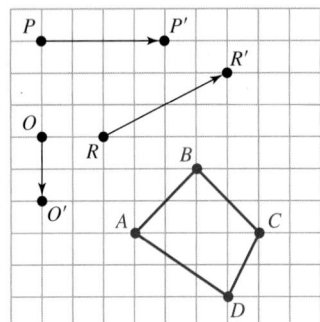

11. Find the coordinates of each point under the translation $(x, y) \to (x - 3, y - 5)$.

   a. $(3, 1)$

   b. $(4, -7)$

   c. $(-8, -2)$

   d. $(-6, 9)$

12. a. A counterclockwise rotation of $146°$ is the same as a clockwise rotation of how many degrees?

   b. A clockwise rotation of $-351°$ is the same as a counterclockwise rotation of how many degrees?

   c. A rotation of $560°$ is the same as what rotation between $0°$ and $360°$?

   d. A rotation of $789°$ is the same as what rotation between $0°$ and $360°$?

13. Find the images of quadrilateral $ABCD$ when it is reflected in each line k, l, and m.

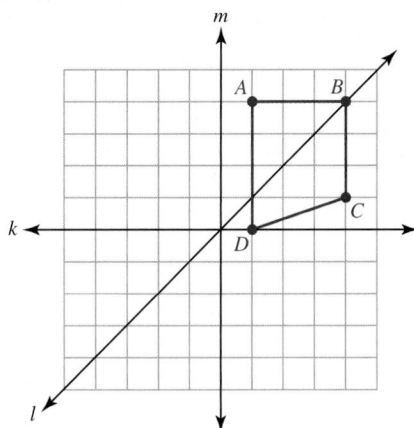

14. Given the reflection that sends P to P', find the axis of reflection and the image of $\triangle ABC$ under the reflection.

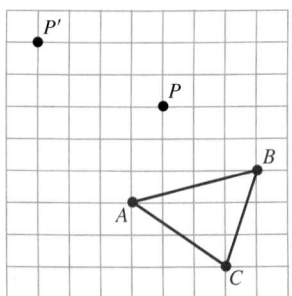

15. Draw the image of $\triangle ABC$ under the glide reflection with l as the glide axis and $\overline{PP'}$ as the translation vector.

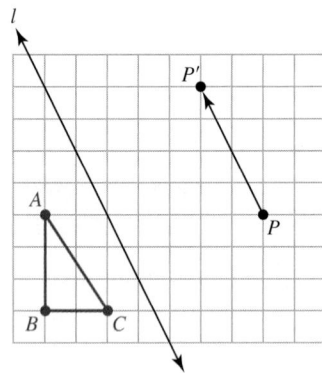

16. Draw the dilation of $\triangle ABC$ using:

   a. $S_{O,2}$.   b. $S_{P,3/2}$.

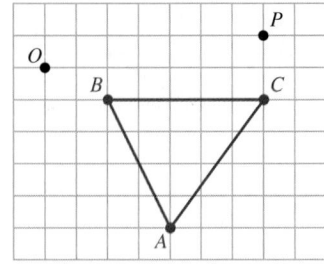

17. Let T_1 and T_2 be translations where $T_1(x, y) \rightarrow$ $(x - 5, y - 2)$ and $T_2(x, y) \rightarrow (x + 4, y - 8)$. What is the image of each point if it is first translated by T_1 and then by T_2?

    a. $(0, 0)$ b. $(3, 5)$ c. $(-4, -3)$ d. $(6, -3)$

18. If a particular point has coordinates (a, b), where a and b are positive integers, what are the coordinates of its image if:

    a. It is rotated 90° around the origin?

    b. It is rotated 180° around the origin?

    c. It is rotated −90° around the origin?

    d. It is rotated −180° around the origin?

19. Use a mirror with the trapezoid shown so that the figure and the reflection show a:

    a. Triangle.   b. Quadrilateral.

    c. Hexagon.

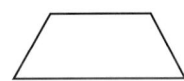

20. If a point has coordinates $(4, -2)$, what are the coordinates of its image if it is reflected in the:

    a. x-axis.   b. y-axis.

    c. Line $x = y$.   d. Line $x = -y$.

21. Find the coordinates of points $A(-3, 3)$ and $B(-1, -4)$ under the size transformation with a center at the origin and a scale factor of 3.

22. If a transformation is the combination of two reflections in parallel lines, the net effect is the same as what kind of rigid motion?

23. Sketch the image of $\triangle ABC$ if it is rotated about:

    a. Point A by 90°.   b. Point B by 90°.

    c. Point C by 180°.

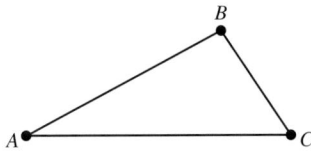

24. Complete each figure so that it has reflection symmetry in the line l.

    a.   b.

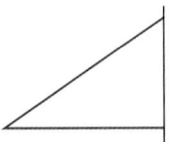   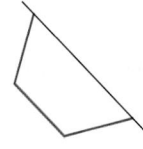

    c.   d.

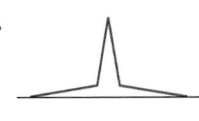   

25. Complete each figure so that it has rotation symmetry about point P.

    a.   b.

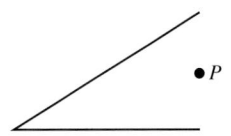   

    c.

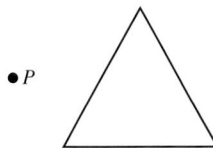

26. What are the degree measures of the rotation symmetries for a regular:

    a. Quadrilateral?   b. Octagon?

    c. Nonagon?   d. 14-gon?

27. Draw two objects, each having the given symmetries.

    a. Only vertical reflection symmetry

    b. Only horizontal reflection symmetry

    c. Symmetry with exactly three rotations

    d. Vertical and horizontal reflection symmetry

28. The word MOW has rotation symmetry. Find three other such words.

29. If a figure has 20° rotation symmetry, how many rotation symmetries must it have, including the identity rotation?

30. If a figure has a rotation symmetry other than the identity rotation, must it also have reflection symmetry?

31. Using only rotations of the letter M, construct a border pattern of each type.

    a. $m1$ b. mg c. $1m$

32. For each number n, draw a polygon with n sides that tessellates the plane.

    a. 3   b. 4

    c. 5   d. 6

33. If an arbitrary convex polygon were to be drawn, how many sides would it have to have so that it might not tessellate the plane?

34. Create an Escher-like drawing using a square and the rotation around a vertex technique.

35. Does a concave octagon tessellate the plane? If not, explain why. If so, draw a picture of one that does.

36. Why is it impossible for a regular decagon to tessellate the plane?

Answers to Chapter 12 Check Your Understandings and Activities

Check Your Understanding 12.1-A

1. a. D is to the north and east of A.

   b. Walk one block east and one block south to arrive at D. Next, walk one block south and three blocks west to arrive at Beech and 5th.

2. a. $AB = 4$.   **b.** $CD = 6$.

3. The points are not collinear.

4. The midpoint of $\overline{AB}$ is $(1, 3.5)$, the midpoint of $\overline{BC}$ is $(1.5, -0.5)$, and the midpoint of $\overline{AC}$ is $(3.5, 1)$.

Check Your Understanding 12.1-B

1. a. Acute.   **b.** Obtuse.

2. a. Scalene.   **b.** Isosceles.

3. $(4, 1)$.

4. a. $(x - 4)^2 + (y - 5)^2 = 1$.

   b. $(x + 3)^2 + (y + 6)^2 = 25$.

   c. $(x + 1)^2 + (y - 5)^2 = 19$.

Check Your Understanding 12.2-A

1. The image of $\triangle TRY$ is shown in red.

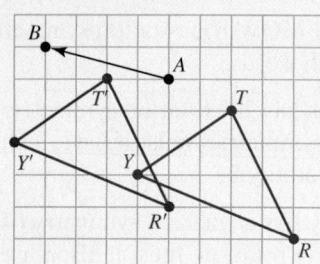

2. a. $(5, -7)$.   **b.** $(0, -6)$.   **c.** $(7, -13)$.   **d.** $(3, -15)$.

3. a. $(-3, 1)$.   **b.** $(-1, -3)$.

4. a. $-305°$.   **b.** $265°$.   **c.** $200°$.

Check Your Understanding 12.2-B

1. The image of quadrilateral $ABCD$ is shown in red.

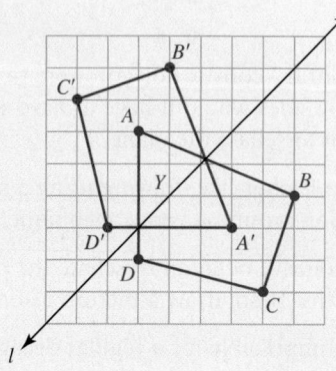

2. a. $(-2, -5)$.   **b.** $(2, 5)$.   **c.** $(5, -2)$.

3. The image of the figure is shown in red.

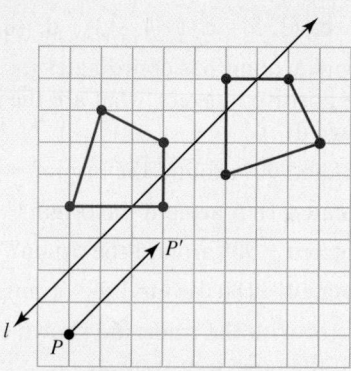

Check Your Understanding 12.3-A

1. a. A translation by vector $\overrightarrow{AB}$.

   b. A rotation of $-90°$ about F.

2. The image of $\triangle ABC$ is shown in red.

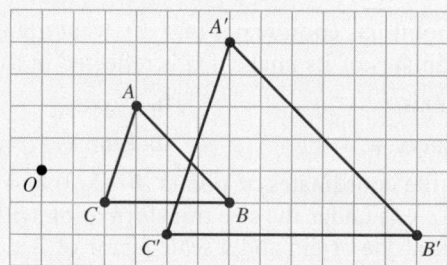

3. The image of $\triangle ABC$ is shown in red.

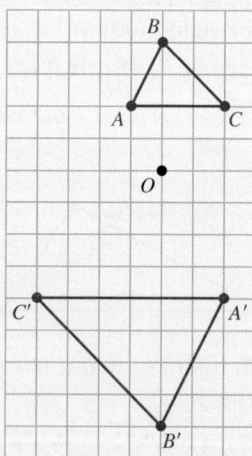

Check Your Understanding 12.3-B

1. a.   **b.**

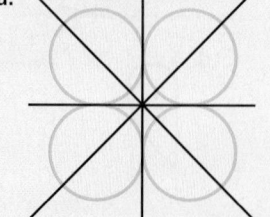

c.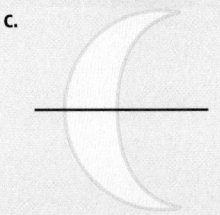

d.

2. a. 90°, 180°, and 270°. It has 90° rotation symmetry.

b. 90°, 180°, and 270°. It has 90° rotation symmetry.

c. The figure has no rotation symmetry.

d. 72°, 144°, 216°, and 288°. It has 72° rotation symmetry.

3. Two reflection symmetries and one rotation symmetry; a rhombus is a quadrilateral with four congruent sides, two reflection symmetries, and 180° rotation symmetry.

Check Your Understanding 12.4-A

1. a. 11.   **b.** *mg.*   **c.** *mm.*   **d.** *m*1.

2. a.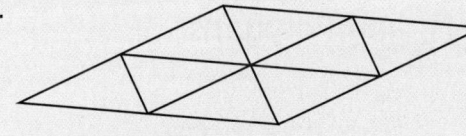

b.

c.

d.

3. It has no vertical, horizontal, or rotation symmetry. It has reflection symmetry along a line at 45°.

Check Your Understanding 12.4-B

1. a.

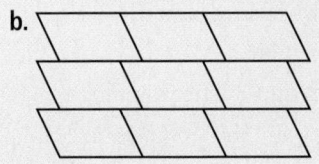

b.

2. Drawings will vary.

3. Drawings will vary.

Activity 12.1

The vertices have coordinates (0, 0), (12, 0), (9, 3), (6, 6), (12, 6), (3, 9), (9, 9), (0, 12), (6, 12), and (12, 12).

Activity 12.2

The figure is always a parallelogram. See Theorem 11.6 for an explanation.

Activity 12.3

Construct two lines parallel to $\overline{AB}$, one that passes through P and another that passes through Q. Copy the segment $\overline{AB}$ onto each of the new lines using P and Q as a vertex for each segment. This finds A' and B'. Draw the segment $\overline{A'B'}$ to complete the translation.

Activity 12.4

Construct a line perpendicular to l that passes through A. Use the compass to measure the segment from A to l, and then copy it onto the same segment but on the other side of l. The new point is A'. Repeat the process to find B' and C'. Draw the three segments to complete the reflection.

Activity 12.5

a. $A'(0, 0)$, $B'(0, 8)$, $C'(12, 8)$, and $D'(12, 0)$.

b. $A'(0, 0)$, $B'(0, 2)$, $C'(3, 2)$, and $D'(3, 0)$.

c. $A'(0, 0)$, $B'(-8, 0)$, $C'(-8, 12)$, and $D'(0, 12)$.

d. $A'(0, 0)$, $B'(0, -2)$, $C'(3, -2)$, and $D'(3, 0)$.

Activity 12.6

The completed table is shown:

*	r_1	r_2	r_3	R_1	R_2	I
r_1	I	R_1	R_2	r_2	r_3	r_1
r_2	R_2	I	R_1	r_3	r_1	r_2
r_3	R_1	R_2	I	r_1	r_2	r_3
R_1	r_3	r_1	r_2	R_2	I	R_1
R_2	r_2	r_3	r_1	I	R_1	R_2
I	r_1	r_2	r_3	R_1	R_2	I

Example patterns are:

- Two reflections are equivalent to a rotation.
- Two of the same reflection is defined as the identity transformation.
- Two rotations are equivalent to a reflection.
- Every column has one entry equivalent to each transformation.

Activity 12.7

Possible patterns are shown:

a. ···  ···

b.

Activity 12.8

a.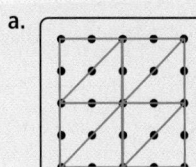

b. It is not possible.

◆ Lines and angles

1. Use the following diagram to name each item.

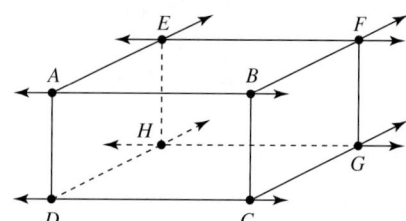

a. Two line segments
b. Two rays
c. Three coplanar points
d. Two coplanar lines
e. Parallel lines
f. Perpendicular lines
g. Four noncoplanar points
h. Two skew lines
i. Two parallel planes

2. a. The complement of a 48° angle has what measure?

b. The supplement of a 103° angle has what measure?

c. What is the supplement of the complement of an angle measuring 51°?

3. If $p\|q$, find the measures of the numbered angles. Justify your answer.

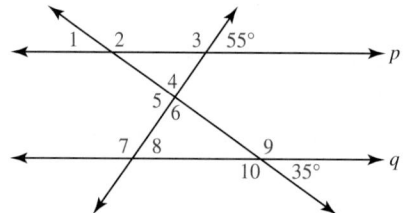

◆ Triangles, Quadrilaterals, and Polygons

4. Draw an example of a(n):

a. Convex, simple closed curve.
b. Concave polygon.
c. Right, scalene triangle.
d. Acute, isosceles triangle.
e. Quadrilateral that is equilateral but not equiangular.

5. a. Can a triangle have sides of length 3 cm, 4 cm, and 8 cm? Explain.

b. Can a right triangle have sides of length 4 in., 4 in., and $\sqrt{(32)}$ in.? Explain.

6. Find x.

a.

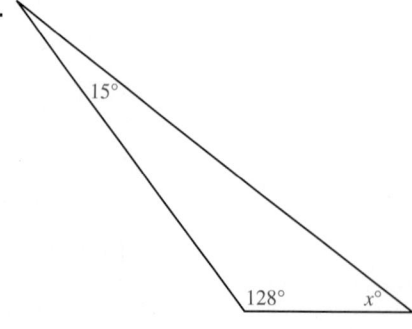

b.

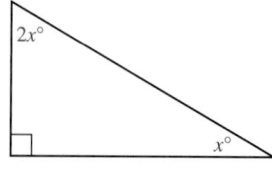

c.

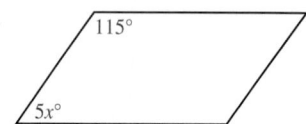

7. **a.** In a 30°-60°-90° triangle, what are the lengths of the other two sides if the shortest leg is 8?

   b. In a 45°-45°-90° triangle, what are the lengths of the legs if the hypotenuse is 20?

8. If a rectangle has sides of lengths 18 cm and 24 cm, what is the length of its diagonal?

9. What are the measures of the interior, exterior, and central angles of a:

   a. Regular hexagon?   **b.** Regular decagon?

10. If an angle is inscribed in a circle and intersects an arc of 76°, what is the measure of the angle?

◆ Spatial Shapes

11. Draw a net of a:

    a. Right hexagonal prism.   **b.** Square pyramid.   **c.** Cylinder.

12. **a.** A polyhedron has 12 faces and 36 edges. How many vertices does it have?

    b. A polyhedron has 12 vertices and 12 faces. How many edges does it have?

13. If two corners are cut off an octahedron so that it gains two new square faces, how has its number of edges and vertices changed?

14. Draw possible outlines of the figure shown from the top, front, and right.

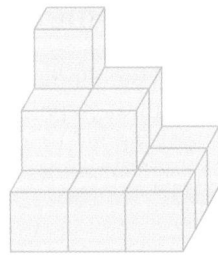

◆ Congruence and Similarity

15. If △ABC ≅ △DFE, make a list of the six pairs of corresponding parts.

16. Let △ABC ≅ △XYZ.

    a. If $AB = 6$, $BC = 10$, and $AC = 14$, what are XY, YZ, and XZ?

    b. If $m\angle A = 75°$, $m\angle B = 65°$, $m\angle C = 40°$, what are $m\angle X$, $m\angle Y$, and $m\angle Z$?

17. Find all pairs of congruent triangles in quadrilateral $ABCD$ if it is known to be a parallelogram. Explain how you know that each pair of triangles is congruent.

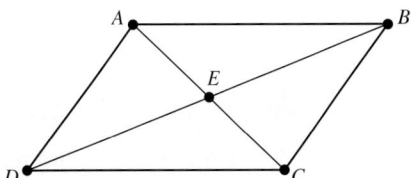

18. Use congruent triangles and the diagram to prove that opposite sides of a rhombus are congruent.

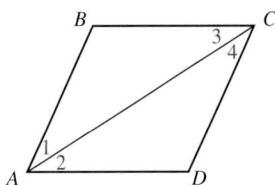

19. If $\triangle ABC \sim \triangle DEF$, what is DE if:

    a. $AB = 4$ and the scale factor is 7? b. $AB = 8$ and the scale factor is $\dfrac{7}{16}$?

20. In each diagram, explain why the two triangles are similar. Then use the given information to find the value of the variable.

    a.
    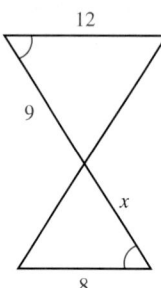

    b.
    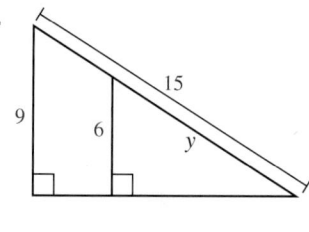

21. While standing near a building, Elizabeth notices that the building's shadow is 100 ft long. Elizabeth stands exactly 4 ft tall. If her shadow is 6 ft long, how tall is the building?

◆ Geometrical Constructions

22. Use alternate interior angles to construct the line parallel to $\overleftrightarrow{AB}$ that passes through point P.

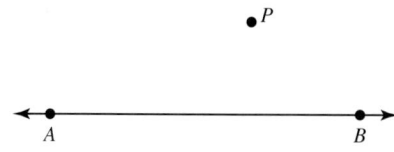

23. Draw an acute triangle $\triangle ABC$. Use the SSS congruence property to construct a congruent triangle.

24. Draw a scalene triangle $\triangle DEF$. Use the SAS similarity property to construct a triangle that is similar but twice as large.

25. Draw a segment $\overline{AB}$ about two in. long, and use it to complete each construction.

    a. Construct a square with sides that are congruent to $\overline{AB}$.

    b. Construct a regular hexagon with sides that are congruent to $\overline{AB}$.

◆ Coordinate Geometry

26. Find the length and midpoint of each segment with the given endpoints.

    a. $A(4, 8)$ and $B(15, 8)$ b. $C(-5, -3)$ and $D(4, 7)$

27. If points $A(1, -2)$, $B(6, 2)$, and $C(4, 8)$ are the vertices of a triangle, what kind of triangle is $\triangle ABC$?

28. Consider points $A(0, 0)$, $B(6, 0)$, and $C(2, 4)$. Find possible coordinates for a point D so that quadrilateral $ABCD$ is a parallelogram.

29. Write the equation of the circle with the given center and radius.

    a. $(0, 0)$, $r = 8$ b. $(1, -5)$, $r = 5$ c. $(-3, 2)$, $r = \sqrt{15}$

◆ Transformations

30. Find the image of △*ABC* under each transformation.

    a. $T_{PP'}$ **b.** $R_{A,\,-90°}$ **c.** $S_{O,2}$

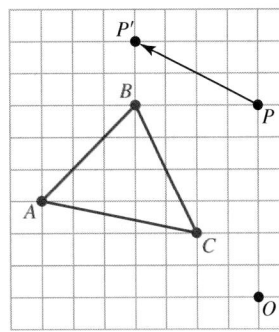

31. Find the images of quadrilateral *ABCD* when it is reflected in each line *l* and *k*.

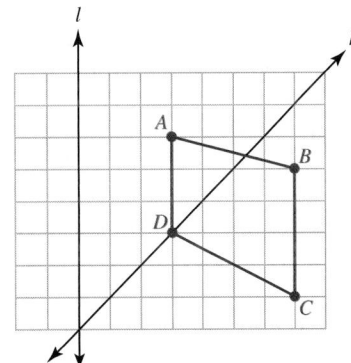

32. Draw the image of △*ABC* under the glide reflection with *l* as the glide axis and $\overline{PP'}$ as the translation vector.

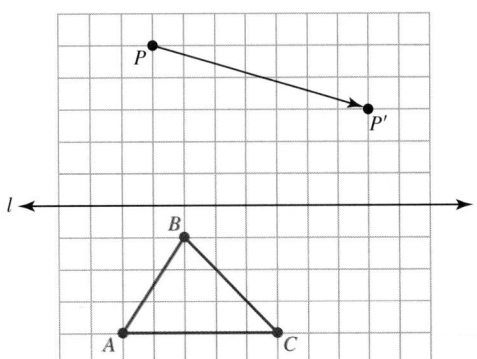

33. Find the coordinates of each point under the translation $(x, y) \rightarrow (x - 7, y + 3)$.

    a. $(0, 0)$ **b.** $(3, -1)$ **c.** $(4.6, 3.9)$

34. a. A counterclockwise rotation of 128° is the same as a clockwise rotation of how many degrees?

    b. A clockwise rotation of −257° is the same as a counterclockwise rotation of how many degrees?

    c. A rotation of 895° is the same as what rotation between 0° and 360°?

35. If a particular point has coordinates $(7, -3)$, where *x* and *y* are positive integers, what are the coordinates of its image if it is reflected in the:

    a. *x*-axis?   **b.** *y*-axis?   **c.** Line $x = y$?   **d.** Line $x = -y$?

◆ Symmetry and Geometrical Patterns

36. Complete each figure so that it has reflection symmetry in the line *l*.

a.

b.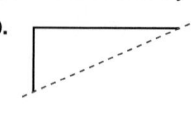

37. Complete each figure so that it has rotation symmetry about point *P*.

a.

b.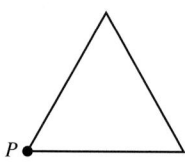

38. Draw an object that has:

  a. Only vertical reflection symmetry.

  b. Symmetry with exactly two rotations.

39. If a figure has 45° rotation symmetry, how many rotation symmetries must it have, including the identity rotation?

40. Using only rotations of the letter A, construct a border pattern of each type.

  a. *m*1   **b.** *mg*   **c.** 1*m*

41. Create an Escher-like drawing using a triangle and the rotation around the midpoint of a side technique.

Measurement

Opening Problem

The Smiths own a pool that is roughly the shape of two intersecting congruent circles. They want to put a tile border around the edge of the pool and need a good estimate of the perimeter. They do not have a rope long enough to go around the entire pool, but they were able to take the measurements shown. The Smiths know geometry well, so how might they use their knowledge to get an estimate?

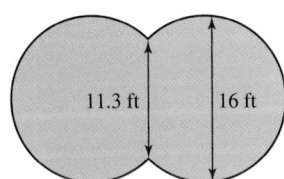

11.3 ft   16 ft

Anton Balazh/Shutterstock.com

Over the last three chapters, we have studied many aspects of geometrical shapes but have yet to measure them. Measurement is often associated with geometry for at least two reasons. First, the numbers assigned as measures provide another tangible way to characterize and study shapes. Second, geometrical shapes have measurable attributes that are generally easy to recognize, so they provide a good context for learning about measurement. Despite this close connection, some aspects of measurement are independent of geometry, making measurement a valuable topic of study in its own right. Important for both mathematical and practical reasons, measurement is one of the five NCTM Content Standards.

■ NCTM Measurement Standard

Instructional programs from prekindegarten through grade 12 should enable all students to:

- Understand measurable attributes of objects and the units, systems, and processes of measurement.
- Apply appropriate techniques, tools, and formulas to determine measurements.

Source: NCTM STANDARDS Copyright © 2011 by NATIONAL COUNCIL OF TEACHERS OF MATHEMATICS. Reproduced with permission of NATIONAL COUNCIL OF TEACHERS OF MATHEMATICS.

The Measurement Standard indicates several key ideas involved in measurement. For instance, the first goal states that measurement is a process that can be learned. In the elementary grades, children are first exposed to this process as they learn to measure by making direct comparisons. As they become more familiar with the process, they realize that it has underlying steps that can be applied to any measurement situation.

Children must first learn to identify measurable attributes. Often, the first attribute they work with is length, simply because it is easy to recognize as something measurable. However, as students progress through the elementary grades, they also work with perimeter, area, volume, weight, temperature, and time. In some cases, children may find it difficult to distinguish among certain attributes. For instance, they may confuse perimeter and area because they are measures of the same figure but in different dimensions. For this reason, students are introduced to these ideas in the elementary grades and continue to work with them into the middle grades.

As children learn about attributes, they will also learn about the units used to measure them. Initially, children measure with nonstandard units, which gives them a chance to understand not only the measurement process, but also the need for standard units. The first standard units that students use are the English units, which they are most likely to encounter in their daily lives. Later, they learn to use metric units, which are the units of science, business, and trade. As a natural part of learning these two systems, children have to learn how to make conversions between them.

The second goal of the Measurement Standard states that students need to learn measurement strategies, which can include the use of estimations, measurement tools, and formulas. Estimation is a natural and important part of measurement because it allows students to arrive at a reasonable value for the measure prior to making it. In the classroom, estimation helps students become familiar with measurement units, understand the role of unit size, and create relationships between units of the same attribute. Early on, children use words like "close," "almost," and "about" to make estimations. Later, they learn to use more formal techniques like rounding and truncating.

Children will also learn to use measurement tools, which include any devices that take measurements directly. In the early grades, children use nonstandard tools like pencils, paper clips, and string to measure length. Using these devices is a valuable experience because they enable students to understand the measurement process before using standardized units. Once they are introduced to standardized units, they learn to use measuring tools like rulers, scales, thermometers, and clocks.

Eventually, students learn to use formulas to compute measures. Formulas can be difficult to use if students have little understanding of the concept they are trying to measure. For this reason, formulas are not typically introduced until students understand the attribute. For instance, perimeter and area are introduced in the second or third grade, but the measurement formulas for these attributes are not taught until the fifth or sixth grade. Specific classroom learning objectives associated with measurement are shown in Table 13.1.

Table 13.1 Classroom learning objectives	K	1	2	3	4	5	6	7	8
Describe measurable attributes of objects, such as length or weight.	X								
Directly compare two objects with a measurable attribute in common and describe the difference.	X								
Measure and estimate the length of objects using different units and different tools.		X	X						
Compare the lengths of two objects to determine how much longer one object is than another.		X	X						
Tell and write time from analog and digital clocks in hours, half-hours, and minutes and solve word problems involving addition and subtraction of time intervals.		X	X	X					
Measure and estimate liquid volumes and masses of objects using standard units.				X					
Recognize area as an attribute of planar figures and understand concepts of area measurement.				X					
Measure rectangular areas by counting square units and by using whole-number products.				X					
Solve real-world and mathematical problems involving perimeters and areas of polygons.				X	X			X	
Know relative sizes of units within one system of units including km, m, cm; kg, g; lb, oz.; l, ml; hr, min, s.					X				
Recognize volume as an attribute of solid figures and understand concepts of volume measurement.						X			
Measure the volume of a right rectangular prism by counting cubic units and by using whole-number products.						X			
Convert among different-sized standard measurement units within a given measurement system.						X	X		
Use nets to find the surface area of three-dimensional figures.							X		
Use formulas to find the area and perimeter of triangles, quadrilaterals, and polygons and the area and circumference of circles.							X	X	
Solve real-world and mathematical problems involving volume and surface area.						X	X	X	
Know the formulas for the volumes of cones, cylinders, and spheres to solve real-world and mathematical problems.									X

Source: Adapted from the *Common Core State Standards for Mathematics* (*Common Core State Standards Initiative*, 2010).

Elementary- and middle-grades teachers are responsible for teaching the fundamental aspects of measurement. To help teachers better understand measurement and measurable attributes, Chapter 13:

- Gives a detailed discussion of the key components of the measurement process using the attribute of length.
- Presents measurable attributes of both two- and three-dimensional shapes.
- Discusses other nongeometrical attributes that are commonly taught in the elementary and middle grades.

SECTION 13.1   Length and the Measurement Process

Measurement may be the most common way that mathematics enters into daily life because it is the primary way of determining and communicating the size of the things around us. Simply put, **measurement** is the process of assigning a numerical value, called a **measure,** to a physical attribute of an object or to a characteristic of an event or situation. We find the measure by comparing the size of the object or event to a well-known or easily understood unit. The measure then indicates how many units are equivalent to the attribute being measured. In general, we can obtain a measure by using the following three-step process.

The Measurement Process

Step 1: Identify and select an attribute to be measured.
Step 2: Select an appropriate unit to measure the attribute.
Step 3: Use a measurement strategy to find the measure.

Each step contains important mathematical ideas that are easier to describe in terms of a particular attribute. In this case, we choose length.

Step 1: Identifying and Selecting an Attribute to Be Measured

The first step in the measurement process is to identify and select a measurable attribute. An attribute is **measurable** if it can be quantified in some way. Many physical attributes, like length and weight, are easy to recognize and measure. Others are more difficult. For instance, consider the speed of light. For much of history, light was assumed to be transmitted instantaneously. Galileo (1564–1642) was the first to even question the assumption. After Galileo made his assertion, it took nearly 50 more years before Olaus Roemer successfully measured the speed of light in 1676. Even then, an exact value for the speed of light ($c \approx 299{,}792.458$ km/s) was not adopted until 1983.

Other measurable attributes can be abstract. For instance, IQ (intelligence quotient) is a measure of intelligence, profit is a measure of business success, and grade point average (GPA) is a measure of a student's academic ability. Once we consider notions like this to be measurable, it becomes more difficult to find attributes that are not measurable. Such attributes do exist but are generally limited to ideas like justice and freedom or to emotions like love, sadness, and joy.

Of all the possible measurable attributes, length is probably the easiest to recognize. Recall that the length of a line segment $\overline{AB}$ is the distance between its endpoints. To apply the notion of length to other objects, we simply treat them as if they were segments. We ignore all other dimensions except the one to be measured. For example, to measure the length of a pencil, we treat the distance from the tip to the eraser as a segment and ignore its depth and width (Figure 13.1).

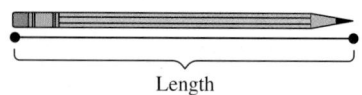

Length

Figure 13.1 The length of a pencil as a segment

With a little thought, we see that length has three properties. The third is simply a restatement of the Triangle Inequality, but we now allow the sum of two lengths to equal the third.

Properties of Length

Let *AB* represent the length of segment $\overline{AB}$.

1. Length is always greater than or equal to 0, or $AB \geq 0$.
2. The length from *A* to *B* is equal to the length from *B* to *A*, or $AB = BA$.
3. For any three points *A*, *B*, and *C*, the length from *A* to *B* added to the length from *B* to *C* is greater than or equal to the length from *A* to *C*. That is, $AB + BC \geq AC$ (Triangle Inequality).

According to the Ruler Postulate, we can assign a real-number value to any length. Now, through the process of measurement, we can find, or at least approximate, this value. However, before learning how to make the measure, we must first select a unit on which the measure is based.

Step 2: Selecting an Appropriate Unit to Measure the Attribute

A **unit of measure** is a fixed amount of the measurable attribute that is used to make the measurement comparison. The selected unit must be of the appropriate type; that is, the unit must have the same attribute as the one being measured. For instance, we must use length units to measure length, area units to measure area, and weight units to measure weight.

The unit must also be of the appropriate size; that is, we should use small units to measure small portions of the attribute and large units to measure large portions. The size of the unit can impact the measure in two ways. First, the size of the unit and the measure of the object are inversely related. As the unit gets larger, the number of units needed to make the measure gets smaller. As the unit gets smaller, the number of units gets larger. For instance, if unit *B* is three times longer than unit *A*, then a pencil with a length of 9 *A*-units requires only 3 *B*-units to measure it (Figure 13.2).

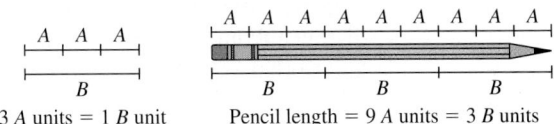

3 *A* units = 1 *B* unit   Pencil length = 9 *A* units = 3 *B* units

Figure 13.2 Measuring a pencil with different units

Second, the size of the unit also affects the precision of the measurement. The words "precision" and "accuracy" are often used as synonyms, but they are not the same. **Precision** relates to the exactness of the measurement. Smaller units lead to more exact measurements, and larger units lead to less exact measurements. **Accuracy** describes how close the measure is to the actual size of the object. To better understand the difference between the two, suppose we measure a woman's height in two ways (Figure 13.3). First, we use only feet and obtain a measure of 5 ft. Second, we use both feet and inches and obtain a measure of 5 ft, 4 in. Both measures are accurate because they are relatively close to the woman's actual height. However, the first measure was based on a large unit. Because the woman's height fell between 5 and 6 ft, we rounded down, making the measure less precise. The second measure was based on a small unit, so we were able to more precisely represent the woman's height. Ideally, measures should be both accurate and precise.

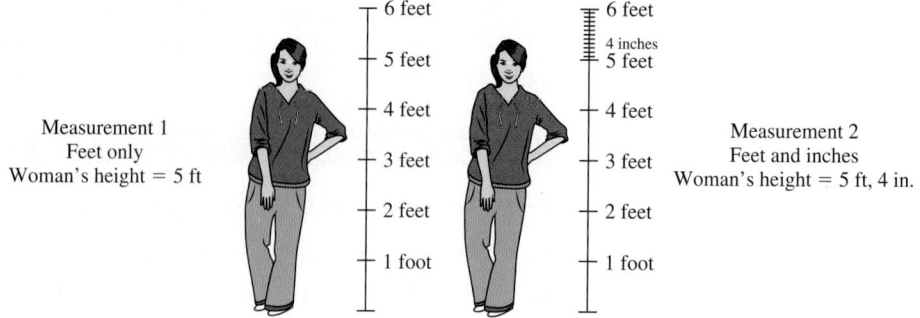

Figure 13.3 Two measures with different precision

Example 13.1	Accurate and Precise Measures

Communication   During a class laboratory, each of four students measures the same object three times and gets the following results:

Dan: 3.5, 3.4, and 3.6   Tyron: 4.561, 5.342, and 2.330

Linda: 4, 2, and 5   Monica: 3.542, 3.541, and 3.543

If the object measures exactly 3.5413, describe the accuracy and precision of each student's measures.

Solution Dan's measures are close to the actual value but are given to only one decimal place. He used relatively large units, so his measures are accurate but not precise. Linda is not close to the real value, nor did she use small units, causing her measures to be neither accurate nor precise. Tyron used small units but was not close to the actual value. His measures are precise but not accurate. Monica used small units and was close to actual value. Her measures are both accurate and precise.

Nonstandard Units of Length

Any object with the measurable attribute can serve as the unit of measure. Historically, body parts and other convenient objects were used as length units. For instance, an inch was the length of three barleycorns placed end to end, and a **cubit** was the distance from the elbow to the tip of the outstretched middle finger. Other units of length that were based on body parts are shown in Figure 13.4.

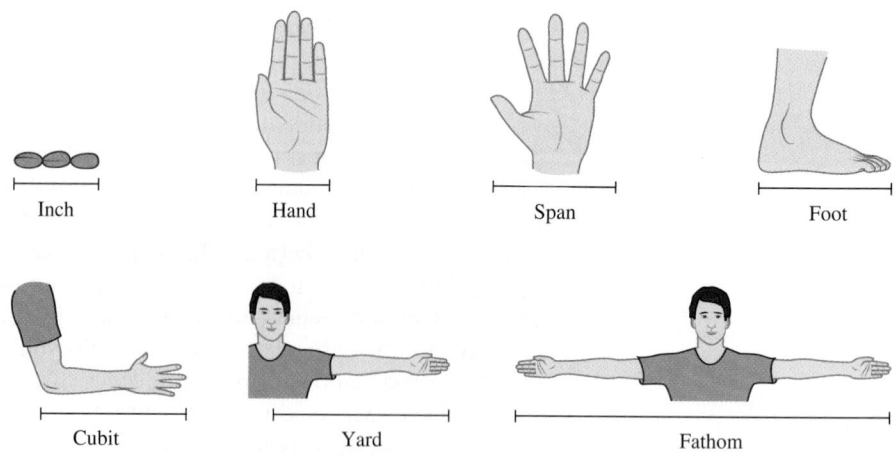

Figure 13.4 Nonstandard units of length

Using body parts for units was problematic because the size of the unit depended on the size of the person. For instance, a yard measured on a large person would be longer than one measured on a small person because the distance from the nose to the fingertips is longer. This raises the primary concern with all nonstandard units. Nonstandard units are exactly that—nonstandard. Units can vary from situation to situation, leading to different measures of the same object. Such discrepancies naturally lead to problems in fairness and communication.

Explorations Manual 13.1

Despite the problems with nonstandard units, they are still used in many situations. In the classroom, students often use nonstandard units when they first learn to measure. For instance, Figure 13.5 illustrates how paperclips and connecting cubes can be used to measure the length of a pencil. By using nonstandard units in this way, students can see how different units can lead to different measures.

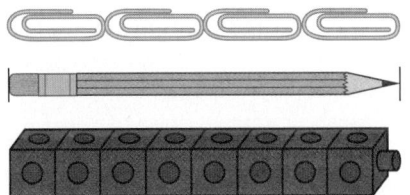

Figure 13.5 Using different units to measure

English, or U.S. Customary, System

Throughout history, many attempts have been made to standardize units so that the meanings of measures could be communicated consistently. For instance, tradition holds that King Henry I tried to standardize the yard by declaring it to be the distance from the tip of *his* nose to the tip *his* finger. Other English rulers declared that a furlong was to be 220 yd and a mile was to be 5,280 ft, making it exactly 8 furlongs long. Eventually, the measures in Figure 13.4 became established in the American colonies, where they were finally standardized by an act of Congress in 1866. Specific length units for what we now call the **English,** or **U.S. Customary, system** are listed in Table 13.2.

Table 13.2 Length units for the English, or U.S. Customary, system		
Unit	**Abbreviation**	**Comparisons to Other Units**
Inch	in.	$1 \text{ in.} = \frac{1}{12} \text{ ft}$
Foot	ft	1 ft = 12 in.
Yard	yd	1 yd = 3 ft
Fathom	fthm	1 fthm = 6 ft
Rod	rd	1 rd = 16.5 ft = 5.5 yd
Furlong	fur	$1 \text{ fur} = 660 \text{ ft} = 40 \text{ rods} = \frac{1}{8} \text{ mi}$
Mile	mi	1 mi = 5,280 ft = 1,760 yd

Table 13.2 also gives the ratios between units that make it possible to convert from one unit to another. The process of doing so relies heavily on proportional reasoning and is called **dimensional analysis.** In dimensional analysis, we use unit rates to compare a single unit of one measure to its equivalent amount in another. Examples are $\frac{3 \text{ ft}}{1 \text{ yd}}$ and $\frac{5.5 \text{ yd}}{1 \text{ rd}}$. Because each ratio is equivalent to one, we can use proportional reasoning to multiply different ratios to change both the value and the unit of the measure.

Example 13.2 Making Conversions Between Units

Use dimensional analysis and the ratios given in Table 13.2 to convert:

a. 25 yd to ft.
b. 15,000 in. to mi.

Solution We begin with the measure and then multiply by the appropriate unit ratios. We place each ratio in fractional form so that the units simplify appropriately.

a. Multiply 25 yd by the ratio $\frac{3 \text{ ft}}{1 \text{ yd}}$. Place the yards in the denominator so that it simplifies with the original units to leave only feet. After multiplying, $25 \text{ yd} \times \frac{3 \text{ ft}}{1 \text{ yd}} = 75 \text{ ft.}$

b. We use two unit ratios: 12 in. = 1 ft and 5,280 ft = 1 mi. After setting up the ratios and multiplying, we get $15,000 \text{ in.} \times \frac{1 \text{ ft}}{12 \text{ in.}} \times \frac{1 \text{ mi}}{5,280 \text{ ft}} \approx 0.237 \text{ mi.}$

The inconsistent ratios between the units make memorizing conversion factors in the English system difficult. For this reason, most people memorize only the conversions they are likely to use and refer to a table for the rest.

Metric System

Another important system of units is the Système International d'Unités, or **metric system.** The metric system is usually preferred over the English system because it uses simple decimal ratios between the units. All metric units begin with a basic unit. To make larger units, multiply the basic unit by positive powers of ten, and to make smaller units, multiply by negative powers of ten. Each new unit is named by affixing an appropriate prefix to the base unit. Some of the common prefixes are listed in Table 13.3.

Explorations Manual 13.2

Table 13.3 Metric prefixes and their values			
Metric Prefix	**Abbreviation**	**Power of Ten**	**Relationship to Base Unit**
Tera	T	10^{12}	$1,000,000,000,000 \times$ base unit
Giga	G	10^{9}	$1,000,000,000 \times$ base unit
Mega	M	10^{6}	$1,000,000 \times$ base unit
Kilo	k	10^{3}	$1,000 \times$ base unit
Hecto	h	10^{2}	$100 \times$ base unit
Deca	da or dk	10^{1}	$10 \times$ base unit
Base unit		10^{0}	
Deci	d	10^{-1}	$0.1 \times$ base unit
Centi	c	10^{-2}	$0.01 \times$ base unit
Milli	m	10^{-3}	$0.001 \times$ base unit
Micro	μ	10^{-6}	$0.000001 \times$ base unit
Nano	n	10^{-9}	$0.000000001 \times$ base unit
Pico	p	10^{-12}	$0.000000000001 \times$ base unit

To better understand how the metric prefixes work, consider the **meter,** which is the basic unit of length. Originally, the meter was defined to be one ten-millionth of the distance from the equator to the North Pole along the Greenwich meridian.

This definition made computing the length of a meter logistically difficult, so it has been redefined to be 1,650,763.73 wavelengths of orange-red light in the spectrum of the element krypton 86. Although this definition makes the length of a meter more reproducible in a laboratory setting, it has little meaning for the average user. In layman's terms, the length of a meter is a little longer than a yard and 3 in. (Figure 13.6).

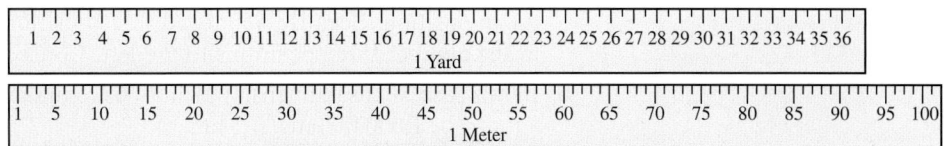

Figure 13.6 A meter compared to a yard

Any length unit that is larger than a meter is equivalent to some positive power of ten times the length of the meter. For example, the decameter (dam or dkm) has a length 10 times longer than a meter. The hectometer (hm) has a length $10^2 = 100$ times longer than a meter, making it 10 times longer than a decameter. A kilometer (km) has a length $10^3 = 1,000$ times longer than a meter, making it 100 times longer than a decameter and 10 times longer than a hectometer. Any length unit that is smaller than the meter is equivalent to some negative power of ten times the length of the meter. For example, the decimeter (dm) is $10^{-1} = \frac{1}{10}$ the length of a meter. The centimeter (cm) has a length that is $10^{-2} = \frac{1}{100}$ the length of a meter, making it $\frac{1}{10}$ the length of a decimeter.

Example 13.3 Comparing Metric Units

Representation Draw a diagram that shows the relative size of a meter to a decameter and a decimeter.

Solution We begin by letting some length represent a meter. We then lay 10 of those lengths end to end to represent a decameter. For the decimeter, we divide the length of the meter into 10 equivalent pieces, and use one of them to represent the decimeter. The resulting diagram is shown:

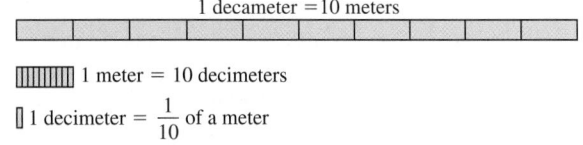

The common ratio of 10 means that conversions between the units can be done simply by multiplying or dividing by the appropriate power of ten. Recall that multiplying by a positive power of ten moves the decimal point to the right. Likewise, dividing by a positive power of ten or multiplying by a negative power of ten moves the decimal point to the left. In other words, conversions between metric units involve moving the decimal point to the right or to the left by the appropriate number of places. If the metric prefixes are placed along a line, then the direction and number of spaces that we move to get from one prefix to another indicates how to move the decimal point in the number (Figure 13.7).

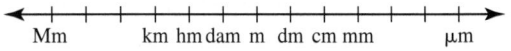

Figure 13.7 An aid for making conversions between metric units

Example 13.4   Making Conversions Between Metric Units

Make each conversion.

a. 45 mm to m   **b.** 15 Mm to hm

Solution

a. To convert 45 mm to m, we use Figure 13.7, start at the mm position, and move to the m position. Because we moved three positions to the left, we move the decimal in the number three positions to the left. Consequently, 45 mm = 0.045 m.

b. In this case, we use dimensional analysis to make conversions between metric units:

$$15 \text{ M\cancel{m}} \times \frac{1,000,000 \text{ \cancel{m}}}{1 \text{ M\cancel{m}}} \times \frac{1 \text{ hm}}{100 \text{ \cancel{m}}} = 150,000 \text{ hm}.$$

Explorations Manual 13.3

In the United States, both the English and the metric system are used, so it may be necessary to make conversions between them. Using the fact that a yard is legally defined to be $\frac{3,600}{3,973}$ of a meter, we get the conversion factors listed in Table 13.4.

Table 13.4   Conversion factors between English and metric units

English to Metric	Metric to English
1 in. = 2.540 cm	1 cm = 0.3937 in.
1 ft = 0.3048 m	1 m = 3.28 ft
1 yd = 0.9144 m	1 m = 1.094 yd
1 mi = 1.6093 km	1 km = 0.6214 mi

Example 13.5   Making Conversions Between English and Metric Units

Make each conversion.

a. 15 ft to m   **b.** 1 hm to rd

Solution In each case, we use dimensional analysis and the appropriate ratios.

a. $15 \text{ \cancel{ft}} \times \frac{0.3048 \text{ m}}{1 \text{ \cancel{ft}}} \approx 4.57 \text{ m}.$

b. $1 \text{ h\cancel{m}} \times \frac{100 \text{ \cancel{m}}}{1 \text{ h\cancel{m}}} \times \frac{1.094 \text{ \cancel{yd}}}{1 \text{ \cancel{m}}} \times \frac{1 \text{ rd}}{5.5 \text{ \cancel{yd}}} \approx 19.89 \text{ rd}.$

Check Your Understanding 13.1-A

1. If the actual measure of an object is 4.5629 units, describe each measure in terms of its accuracy and precision.

    a. 3.5671   **b.** 4.5630   **c.** 4.6   **d.** 7.8

2. Make each conversion in English units.

    a. 75 ft to yd   **b.** 14.3 yd to in.   **c.** 354 in. to rd

3. Make each conversion in metric units.

    a. 150 cm to m   **b.** 560 m to dam   **c.** 0.4 dam to cm

4. Make each conversion between metric and English units.

    a. 17 ft to m   **b.** 341 cm to ft   **c.** 18 fur to hm

Talk About It With two or three of your peers, see whether you can make a list of 25 measurable attributes other than the ones given in this section.

Connections

Activity 13.1

In the United States, English lengths are more familiar than metric lengths. To better understand the relative size of metric lengths, compare them to the lengths of familiar objects. For each metric length given, describe a familiar object with that length.

a. m   **b.** dm   **c.** cm   **d.** mm   **e.** dam   **f.** km

Step 3: Using a Measurement Strategy to Find the Measure

In general, we commonly use one of three strategies to make measures: estimations, measuring devices, and formulas. When recording the result of any of these strategies, we must be sure to include *both the number* and *the unit* because both are needed to clearly communicate the size of the measure.

Estimations

Explorations Manual 13.4

When estimating a measure, we use mental or visual information to arrive at an approximate value without using a measuring device. We use estimation when measuring devices are unavailable or impractical, or when only an approximate value is needed. Because estimations are usually done mentally, they often lack accuracy and almost always lack precision. In general, measures can be estimated in three ways: comparing the object to a referent unit, unitizing, and chunking.

To *compare the object to a referent unit,* we ask, "About how many units are equivalent to this object?" In the case of length, we estimate the number of length units that must be laid end to end to reach from one end of the object to the other. This method is effective only if we have an intuitive feel for the size of the unit and how it compares to the object being measured. Consequently, a standard unit, such as an inch, a foot, or a meter, is used. We use this method when estimating a person's height to be about 5 ft tall or when estimating the distance between two houses to be 20 yd.

A second method is to *unitize.* This method is used whenever the object consists of several parts that are roughly the same size. We estimate the size of one part, often using benchmark numbers like powers of ten to make the estimation easier. We then multiply by the number of parts.

Example 13.6   **Estimating a Measure by Unitizing**

Application

Based on what you see in the diagram, estimate the length of the bridge.

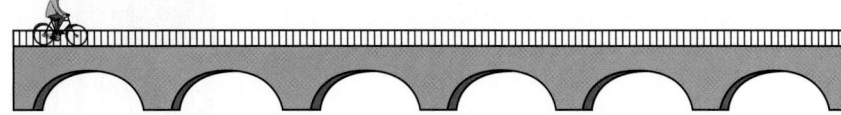

Solution The diagram shows several "facts" that we can use to make our measure. First, the arches divide the bridge into six equivalent parts, making unitizing a good

strategy for estimating its length. Second, notice that the bicycle is about half the length of one arch. Because a bicycle is about 5 ft long, each arch must be about 10 ft long. Because there are six arches, the bridge must be about $6 \cdot 10 = 60$ ft long.

A third method for estimating a measure is *chunking*. In this method, we break the object into small yet related parts, estimate the size of each part, and then add the results to find the final measure.

Example 13.7 Estimating a Measure by Chunking

Application

Estimate the height of the following building.

Solution The building is divided into four distinct parts: the base, the middle, the peak, and the pole on top. Because each piece is a different size, we can use chunking to estimate the height of the building. We assume that each row of windows represents a floor, and each floor is about 15 ft tall. Using this estimate, the base must be about (9 floors) $\cdot$ (15 ft) = 135 ft. Similarly, the middle is about (6 floors) $\cdot$ (15 ft) = 90 ft. There are no windows on the peak, but it looks about two-thirds the height of the middle section, or $\frac{2}{3} \cdot 90$ ft = 60 ft. Likewise, the flagpole on top is about half the height of the middle section, or $\frac{1}{2} \cdot 90$ ft = 45 ft. Adding the heights, we estimate the height of the building to be $135 + 90 + 60 + 45 = 330$ ft.

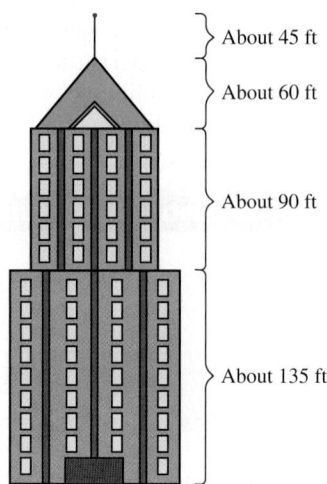

About 45 ft

About 60 ft

About 90 ft

About 135 ft

Measuring Devices

Another strategy is to use a measuring device. Many measuring devices, like those associated with nonstandard units, are informal in nature. For instance, if we use a paperclip to measure the length of a pencil, the paper clip serves as both the unit and the measuring device. Other devices are designed to measure objects using standardized units. For instance, rulers are used to measure length, and often have inches marked off on one edge and centimeters marked off on the other. To use a ruler, we place one end of the object next to a convenient mark and then find the mark that is closest to the other end. The length is the difference of the smaller number from the larger. For instance, the length of the lollipop in Figure 13.8 is $4\frac{1}{2} - \frac{1}{2} = 4$ in.

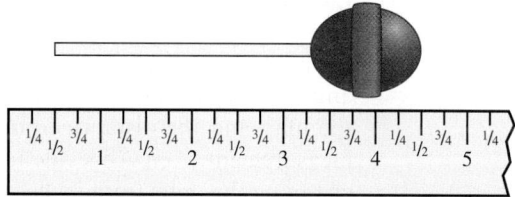

Figure 13.8 Using a ruler to measure

Measuring devices have varying degrees of precision. The ruler in Figure 13.8 measures only to the nearest quarter inch, whereas most rulers measure to the nearest sixteenth of an inch or to the nearest millimeter in the metric system. Other, more sophisticated measuring devices can be even more precise. For instance, in the milling of machine parts, a micrometer can be used to measure to the nearest thousandth or even ten-thousandth of an inch. Despite the tremendous precision of some measuring tools, even measurements taken with these devices are still only approximations of the real numerical value.

Formulas

A third way to measure is to use a formula. **Measurement formulas** come from a variety of places and often rely on other measures that can be found directly. For instance, we often use the Pythagorean Theorem to measure lengths, but we must know two of the lengths to find the third. Many other measurement formulas for different attributes are developed in the next several sections.

Example 13.8	Measuring the Distance Across a Lake

Problem Solving   Abbeville, Bowdon, and Cochran all lie along the shore of a large lake. The road from Abbeville to Bowdon is 2.4 mi long and is approximately perpendicular to the road from Bowdon to Cochran. If Bowdon is 3.6 mi from Cochran, about how far is Abbeville from Cochran across the lake?

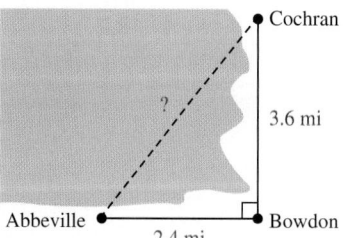

Solution

Understanding the Problem. The goal is to find the distance from Abbeville to Cochran. However, because the distance is across water, we have no way to measure it directly. We do know the distances between Abbeville and Bowdon and between Bowdon and Cochran, as well as that the roads between them are essentially perpendicular.

Devising a Plan. The diagram shows that the distance between Abbeville and Cochran is the hypotenuse of a right triangle. Because we know the legs of the triangle, we can *use the formula* from the Pythagorean Theorem to find the distance across the lake.

Carrying Out the Plan. Substitute the relevant numbers into the formula from the Pythagorean Theorem and compute.

$$\text{Distance from Abbeville to Cochran} = \sqrt{(2.4)^2 + (3.6)^2}$$
$$= \sqrt{5.76 + 12.96}$$
$$= \sqrt{18.72}$$
$$\approx 4.33\,\text{mi}$$

Looking Back. Using a formula is not the only way to solve the problem. It could have also been measured directly from the diagram, as long as the distances between the towns were proportionally accurate and the lines for the road intersected at a right angle. If, for instance, the distance from Abbeville to Bowdon were exactly 2.4 in. on the map and the distance from Bowdon to Cochran were exactly 3.6 in., then our measurement from Abbeville to Cochran would be very close to 4.33 in. Simply by changing the units from inches to miles, we would have our answer.

Check Your Understanding 13.1-B

1. If the distance between posts is approximately 8 ft, about how long is the following fence? What method of estimation did you use?

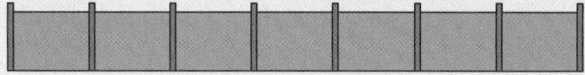

2. Estimate the length of each line segment in both inches and centimeters. Then use a ruler to find a more precise measure of each segment in each unit.

a. •————————————————•

b. •———————————•

c. •——————————•

3. A carpenter wants to cut a brace board for a wall that is 10 ft tall. If he wants to attach one end of the brace to the top of the wall and the other end to a spot on the floor that is 8 ft from the base of the wall, how long must the brace be?

Talk About It Other than a ruler, what other measuring devices have you used? What attributes did they measure?

Problem Solving

Activity 13.2
Determine a way to estimate the number of flowers in the picture.

Dalibor Sevaljevic/Shutterstock.com

SECTION 13.1 Learning Assessment

■ Understanding the Concepts

1. Use your own words to explain each step of the measurement process.

2. Consider a segment $\overline{AB}$.
   a. Why does it makes sense that $AB \geq 0$?
   b. Under what circumstances is $AB = 0$?
   c. Why does it make sense that $AB = BA$?

3. How does the size of a unit impact the measure?

4. What is the difference between accuracy and precision?

5. Is it possible for an object to have a length of -5 units? Explain.

6. What difficulties do nonstandard units present for measurement?

7. What advantages does the metric system offer over the English system?

8. How are larger and smaller units made from the base unit in the metric system?

9. How does dimensional analysis make use of proportional reasoning?

10. Describe three strategies for obtaining a measure.

11. What is the difference between estimating a measure by unitizing and estimating a measure by chunking?

12. Why is it important to include both a number and a unit when describing a measure?

■ Representing the Mathematics

13. If the actual measure of an object is 13.953 units, describe each measure in terms of its accuracy and precision.
    a. 13.8   b. 15.612   c. 8   d. 13.952

14. If the actual measure of an object is 4.571 units, describe each measure in terms of its accuracy and precision.
    a. 3.908   b. 4.4   c. 4.570   d. 6

15. An analogy that can be used to describe the difference between accuracy and precision is throwing four darts at a dartboard. If hitting the bull's eye represents accuracy and multiple darts grouped together represents precision, describe each picture in terms of its accuracy and precision.

    a.    b.

c.    d.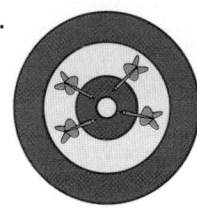

16. Symbolically, the triangle inequality is given as $AB + BC \geq AC$.
    a. Give a verbal interpretation of this statement.
    b. Give a visual interpretation of this statement.
    c. Give a numerical example that satisfies this statement.

17. Stretch out your arm and measure the distance from the tip of your nose to the end of your middle finger. How does your "yard" compare the set value of a yard? Explain the reason for any discrepancy. Repeat the comparison with the measure of your "foot."

18. Estimate the length of each unit in inches.
    a. Hand   b. Span   c. Cubit   d. Fathom

19. Draw segments that you believe to be each of the following lengths. Check your estimate with a ruler.
    a. 5 in.   b. 5 cm   c. 15 mm   d. $\frac{1}{2}$ in.   e. 0.1 m

20. Draw a diagram that shows the relative size of an inch to a foot and a yard.

21. Draw a diagram that shows the relative size of a centimeter to a decimeter and a millimeter.

22. Draw a diagram that shows the relative size of a meter to a decameter and a hectometer.

■ Exercises

23. Make a list of three measurable attributes for each object.
    a. A chair   b. A piece of rope
    c. A frozen dinner   d. A television

24. Find each measure using the given unit.
    a. The length of your desk in hand widths
    b. The length of your pencil or pen in paperclip lengths
    c. The width of your book in finger widths
    d. Compare your measures to one of your peers. What does your comparison indicate about the units you used?

25. For each question, select the measure that makes the most sense.
    a. The length of a newborn kitten
       i. 0.5 in.   ii. 3 in.   iii. 0.75 ft   iv. 1 ft

b. The height of a typical living room

   i. 3 in.   ii. 3 ft   iii. 3 yd   iv. 3 mi

c. The width of a four-lane highway

   i. 7.5 yd   ii. 200 ft   iii. 500 yd   iv. 0.5 mi

26. For each question, select the measure that makes the most sense.

   a. The height of the average adult male

      i. 3.5 m   ii. 2 cm   iii. 300 cm   iv. 170 cm

   b. The height of the Empire State Building

      i. 3 km   ii. 4 hm   iii. 4,700 cm   iv. 5 m

   c. The distance from Philadelphia, Pennsylvania, to Miami, Florida

      i. 15,000 dm   ii. 4,700 hm

      iii. 1,500 km   iv. 2 Mm

27. For each question, select the measure that makes the most sense.

   a. The length of this textbook

      i. 30 cm   ii. 2 ft   iii. 18 in.   iv. 100 mm

   b. The height of a basketball goal

      i. 5.9 ft   ii. 200 in. iii. 1500 mm   iv. 3.1 m

   c. The distance around the Earth at the equator

      i. 15,000 yd   ii. 4500 μm

      iii. 24,000 mi   iv. 4 Mm

28. List the following measures in increasing order.

   a. 8 in., $\frac{1}{6}$ yd, 4 in., $\frac{3}{4}$ ft

   b. 0.5 m, 4 dm, 100 cm, 700 mm

29. List the following measures in increasing order.

   a. 15 mm, 0.5 cm, $\frac{1}{2}$ in., 1 in.

   b. 1 dm, 7 in., 11 cm, $\frac{1}{2}$ ft

30. Make each conversion.

   a. 45 ft to yd   b. 67 yd to in.

   c. 17.1 in. to ft   d. 95.3 yd to ft

31. Use dimensional analysis to make each conversion.

   a. 6.7 fthm to yd   b. 478 rd to mi

   c. 35,678 in. to fur

32. How many times larger is:

   a. A decameter than a meter?

   b. A hectometer than a meter?

   c. A megameter than a kilometer?

   d. A kilometer than a decameter?

33. How many times smaller is:

   a. A centimeter than a meter?

   b. A millimeter than a meter?

c. A micrometer than a millimeter?

d. A millimeter than a decimeter?

34. Make each metric conversion.

   a. 300 cm to m   b. 45 m to dm

   c. 900 mm to cm   d. 3.7 dm to mm

35. Make each metric conversion.

   a. 31,500 μm to dam   b. 0.91 Mm to km

   c. 7.51 km to mm

36. Use dimensional analysis to make each conversion.

   a. 35 ft to m   b. 617 cm to ft

   c. 56 yd to m   d. 93.5 m to yd

37. Use dimensional analysis to make each conversion.

   a. 390 mm to in.   b. 45 km to fur

   c. 19.1 rd to hm   d. 81 mi to km

38. The ruler is in inches.

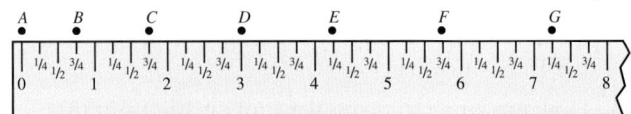

Find the distance between each pair of points.

   a. A and B   b. C and D   c. A and E

   d. C and F   e. B and D   f. F and G

   g. C and E   h. B and F

39. The ruler is in centimeters.

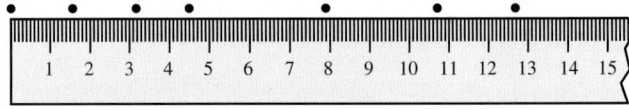

Find the distance between each pair of points.

   a. A and B   b. C and D   c. B and E

   d. C and G   e. A and D   f. D and G

   g. C and F   h. A and G

40. Use a ruler to find the length of each object in the given units.

   a. The length of a new piece of chalk in centimeters

   b. The width of a standard sheet of paper in centimeters

   c. The length of your calculator in inches

   d. The length of your foot in inches

41. Estimate each of the following in English units.

   a. The length of a matchstick

   b. The length of a mid-size sedan

   c. The height of a typical door

   d. The diameter of a quarter

   e. The length of this textbook

   f. The height of your classroom building

42. Repeat the previous exercise using metric units.

Problems and Applications

43. A football field is 120 yd long including the end zones. What is the length of the field in meters? In hectometers? In miles?

44. A standard Olympic-size swimming pool is 50 m long. How many lengths of the pool does it take to complete a 1,500 m race?

45. The main length of the Great Wall of China is 3,460 km. There are also 3,530 km of side branches. How many miles long is the entire Great Wall of China?

46. If a league is equivalent to 3 mi and the distance around the Earth at the equator is approximately 24,000 mi, how many times would the *Nautilus* have traveled around the equator in Jules Verne's *20,000 Leagues Under the Sea*?

47. The shortest human adult male on record is Gul Mohammed of India, who stood just 57 cm tall. How many inches is this? At what age would most human beings be this size?

48. a. In the Bible story of David and Goliath, Goliath is reported to have stood at a height of 6 cubits and a span. If 1 cubit ≈ 18 in. and 1 span ≈ 9 in., how tall was Goliath in feet and inches?

   b. The tallest person on official record was Robert Wadlow, who stood 8 ft, 11 in. What would Wadlow's height have been in cubits and spans? How does Wadlow's height compare to Goliath's?

49. The longest bridge in the world is on Lake Pontchartrain in Louisiana and is 38.42 km long. How long is this in miles? If a car were driving over the bridge at 60 mph, how long would it take to drive the length of the bridge?

50. A light year is the distance that light travels in one year. If light travels at a speed of approximately 186,000 mi/s, how far is a light year in miles?

51. There are about 300 million people in the United States. If they all joined hands, about how many miles would the chain of people stretch? Explain how you arrived at your estimate.

52. If you were to make a chain of paper clips, about how many would you need to make a chain 1 yd long? What method of estimation did you use to determine this?

53. Consider the following building:

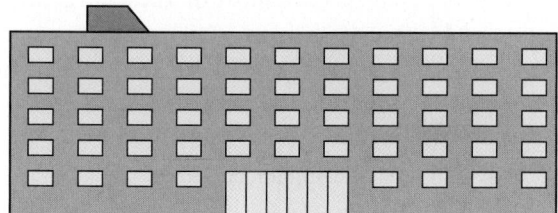

If the windows and the gaps between them are each about 10 ft long, about how long is the building? What method of estimation did you use to determine this?

54. Use the image of the quarter to estimate the length of the following string:

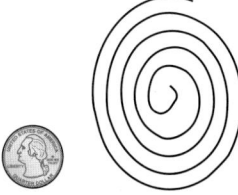

55. Look at a map of the United States. Use the scale on the map to estimate the actual distance from New York to Los Angeles in miles. Estimate the same distance in kilometers.

56. About how many dots are in this box? Explain how you arrived at your answer.

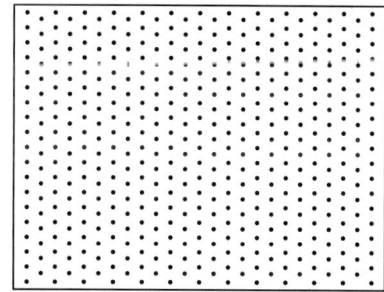

57. A wildlife management officer wants to estimate the trout population in a lake. She and a friend catch and tag 300 trout over the course of one week. Three months later, she and her friend return to the lake and catch 100 trout, of which 10 were tagged. What should she estimate the trout population to be? Explain how you arrived at this estimate.

58. Determine a strategy for estimating the number of words on any given page in a book. Use it to estimate the number of words on this page. Is your estimate accurate? If not, how might you improve your method?

59. A local fire company is called to fight a fire in a building that is 40 ft tall. They want to put several firefighters on the roof of the building using a ladder truck that can extend its ladder to a length of 70 ft. What is the farthest the crew can park the truck from the building and still get the firefighters on the roof?

Mathematical Reasoning and Proof

60. What happens to the value of a measure if the size of the unit is:

   a. Doubled?   **b.** Halved?

61. What happens to the value of a measure if the size of the unit is:

a. Tripled?

b. Quadrupled?

c. Increased by a factor of 5?

d. Increased by a factor of n?

62. Draw any $\triangle ABC$. Measure the length of each side, and then use the lengths to confirm that the Triangle Inequality holds true. Does this constitute a proof of the Triangle Inequality? Why or why not?

63. Paces are another nonstandard unit often used to measure length. If you were to purchase a plot of land by using paces, who would you rather have walking off the paces, a basketball player or a horse jockey? Explain your reasoning.

64. The conversion factor from meters to yards is given by 1 m = 1.094 yd. Likewise, the conversion factor from yards to meters is given by 1 yd = 0.9144.

a. What is the relationship between the numbers 1.094 and 0.9144?

b. Examine the relationship between the other pairs of conversion factors involving a metric and a customary unit. Does the same relationship hold true with other pairs of factors?

65. Use a ruler to measure your pencil in inches. Convert your result to centimeters using the factor 1 in. = 2.54 cm. Check your conversion by measuring your pencil in centimeters. How do the two results compare? What might account for any differences?

Mathematical Communication

66. Height, depth, width, breadth, and distance are all measures of length. The only difference is their connotation of direction. What is the connotation of direction associated with each of these lengths?

67. With a group of your peers, come up with a list of words with a connotation of length (i.e., height, depth...). Once you have your list, find the words with a particular connotation of direction. For each term that does, state the connotation.

68. **a.** Write a story problem that requires a change in English units.

b. Write a story problem that requires a change in metric units.

c. Give the story problems to a peer to critique and solve.

69. **a.** Write a story problem that uses chunking as a method for estimating a measure.

b. Write a second story problem that uses unitizing as a method for estimating a measure.

c. Give the story problems to a peer to solve. Was their estimate close to what you had planned?

70. Many students have difficulty when they first learn to use a ruler, particularly when the object has a partial unit. Discuss with a group of your peers specific reasons children may find using a ruler difficult.

71. A student performs the following calculation when doing dimensional analysis.

$$15\,\text{yd} \times \frac{1\,\text{yd}}{3\,\text{ft}} = \frac{15}{3} = 5\,\text{ft}$$

What went wrong, and how might you address the student's difficulty?

Building Connections

72. **a.** Make a list of five measurable attributes you encounter in your daily life that are not directly associated with a physical object.

b. Make a list of the different measurable attributes of your body.

c. Share your lists with some of your peers, and then see how many others you can list together.

73. Recall two or three instances in your past experience when you have used nonstandard units to measure lengths. In each case, state the unit of measure and the strategy used to obtain the measure.

74. Suppose you are remodeling a small building to use as a business space. Make a list of the kinds of measurements you might take during the remodeling process.

75. Considering the set of all possible lengths, we can view the measurement of length as a function. What would be the domain and range of this function?

76. **a.** Search the Internet for uses of small metric units of length. In what fields or occupations are most of these small units used?

b. Repeat your search, but this time look for uses of large metric units of length. In what fields or occupations are most of these large units used?

77. Describe a method of measurement that a doctor or scientist might use to estimate the number of bacteria in a culture.

78. Why can a car odometer be viewed as a special kind of ruler?

Perimeter and Area

We now turn to measuring specific attributes. We begin with planar shapes, which have four commonly measured attributes: length, angles, perimeter, and area. The measurement of length and angles has been discussed in previous sections. In this section, we develop the ideas of perimeter and area.

Perimeter

Often in daily situations, we want to know the distance around a shape. For instance, if we garden, we might want to know how much fence is needed to enclose a flowerbed. Or, if we exercise, we might want to know how far it is around a running track. Whenever we measure the distance around a shape, we are measuring the shape's perimeter.

> **Definition of Perimeter**
>
> The **perimeter**, P, of a simple closed curve is the distance around the curve.

Explorations Manual 13.5

Although perimeter applies to two-dimensional shapes, it is defined to be a distance, which is a one-dimensional measure of length. To understand how a one-dimensional measure is taken of a two-dimensional object, imagine wrapping a rope around the outside of free-form pool. Once the rope is straightened, it becomes a segment with a length that is equivalent to the distance around the pool (Figure 13.9).

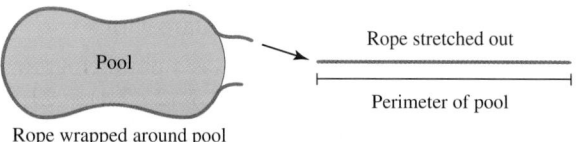

Figure 13.9 Perimeter of a pool

With most irregular shapes, the rope technique is one of the few ways to find the perimeter. However, the perimeter of any polygon can be computed as long as we know the lengths of the sides. For instance, to find the perimeter of a triangle, simply add the lengths of the three sides (Figure 13.10).

Other polygons have specific characteristics that allow us to develop formulas for computing their perimeter. For instance, because the sides of a square are the same length, the perimeter is four times the length of a side, or $P = 4s$. In a rectangle, each pair of opposite sides has the same length. If one pair has a length of l and the other a width of w, then the perimeter is $P = 2w + 2l$. These formulas and others are summarized in Figure 13.11.

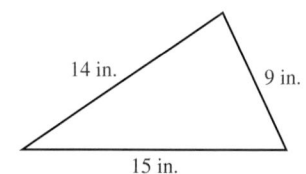

Perimeter = $14 + 15 + 9 = 38$ in.

Figure 13.10 Perimeter of a triangle

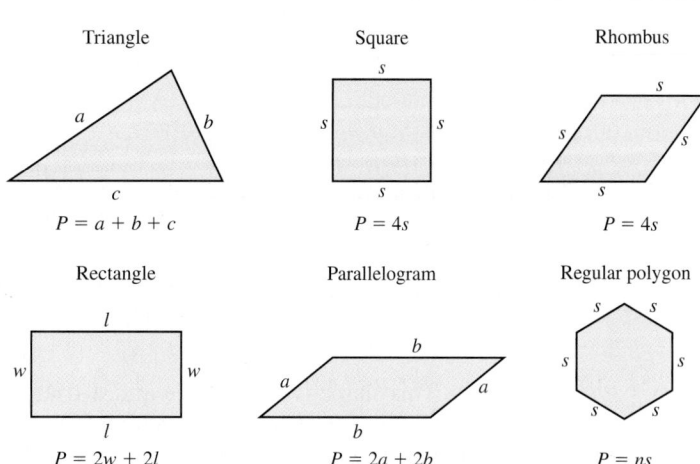

Figure 13.11 Perimeter formulas for polygons

Theorem 13.1 Perimeters of Common Polygons

Perimeter of a triangle with sides of lengths a, b, and c	$P = a + b + c$
Perimeter of a square with sides of length s	$P = 4s$
Perimeter of a rhombus with sides of length s	$P = 4s$
Perimeter of a rectangle with sides of lengths w and l	$P = 2w + 2l$
Perimeter of a parallelogram with sides of lengths a and b	$P = 2a + 2b$
Perimeter of a regular n-gon with sides of length s	$P = ns$

Example 13.9 Computing Perimeters

Find the perimeter of each figure.

a.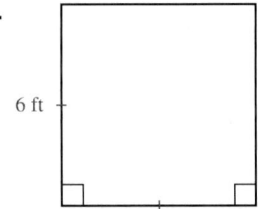

b.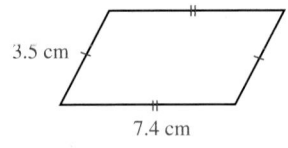

c.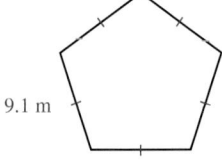

Solution

a. We have a square with $s = 6$. So, $P = 4 \cdot 6 = 24$ ft.

b. Because this is a parallelogram with $a = 3.5$ cm and $b = 7.4$ cm, the perimeter is $P = (2 \cdot 3.5) + (2 \cdot 7.4) = 21.8$ cm.

c. This is a regular pentagon with $s = 9.1$ m. Hence, $P = 5 \cdot 9.1$ m $= 45.5$ m.

We can also find the perimeters of compound shapes. A **compound** shape is one that can be subdivided into two or more simpler shapes like polygons or circles.

Example 13.10 Computing Perimeters of Compound Shapes

Compute the perimeter of each figure.

a.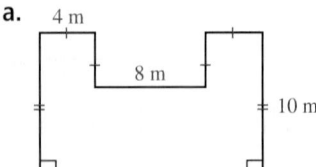

b.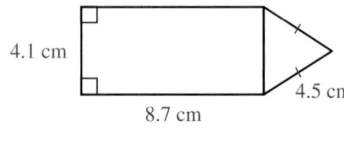

Solution

a. The figure is made from two squares and a rectangle. Four of the sides are 4 m, two are 10 m, and one is 8 m. The long unknown side is $4 + 8 + 4 = 16$ m. By adding the sides, we have $P = (4 \cdot 4) + (2 \cdot 10) + 8 + 16 = 60$ m.

b. This shape is a rectangle and a triangle. However, we only need to count five of the sides because the touching sides of the rectangle and triangle are not on the perimeter of the figure. As a result, $P = 8.7 + 4.1 + 8.7 + (2 \cdot 4.5) = 30.5$ cm.

Circumference

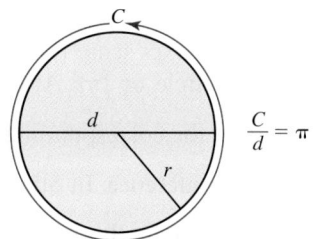

Figure 13.12 The ratio of the circumference to the diameter is π

We can measure the distance not only around polygons, but also around a circle. The distance around any circle is called its **circumference,** *C*. It too can be measured with a formula, but we must first consider a special attribute of circles. In any circle, the ratio of the circumference to the diameter, $\frac{C}{d}$, is *always* the same constant. Because this number appears frequently in mathematics and science, it has been given its own name, the Greek letter **pi,** or π (Figure 13.12).

In 1761, John Lambert proved that π is an irrational number; that is, it is a decimal that never terminates or repeats. Written to 50 decimal places, π is

$$\pi = 3.14159\ 26535\ 89793\ 23846\ 26433\ 83279\ 50288\ 41971\ 69399\ 37510\ \ldots.$$

Because π is irrational, its value is often approximated as 3.14 or 3.1416 whenever it is used in computations. The ancient Greeks used $\frac{22}{7}$ in their computations, which approximates π to two decimal places.

This relationship between the circumference and the diameter is the basis for a formula for computing the circumference of a circle. Because $\frac{C}{d} = \pi$, we get $C = \pi d$ if we multiply both sides of the equation by *d*. Because the radius, *r*, is always half the diameter, we also have that $C = 2\pi r$.

Theorem 13.2   Circumference of a Circle

The circumference of a circle with radius *r* is $C = 2\pi r$.

If a circle has a radius of $r = 4$ in., then the circumference is $C = 2\pi(4$ in.$) = 8\pi$ in. We consider 8π to be an exact answer. If needed, we can get a decimal value by approximating π at 3.14. If more decimal places are needed, we can use the π key, $\boxed{\pi}$, on a calculator, which gives the value of π to eight or ten decimal places.

Example 13.11   The Length of an Arc

Reasoning

Find a formula that uses the central angle and the radius to compute the length of an arc.

Solution The length of an arc depends on two factors: the radius and the central angle. If the central angle is held constant, then a longer radius leads to a longer arc. Similarly, if the radius is held constant, then a larger central angle also leads to a longer arc.

When central angle, θ, is held constant

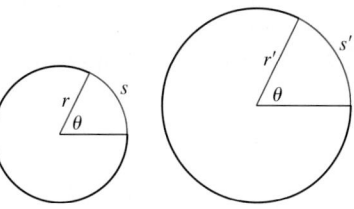

When radius, *r*, is held constant.

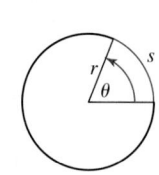
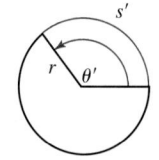

Smaller radius, *r*   Larger radius, *r'*
Smaller arc, *s*   Larger arc, *s'*

Smaller central angle, θ Larger central angle, θ'
Smaller arc, *s*   Larger arc, *s'*

Because the length of an arc depends on the radius and the central angle, any formula must involve both. However, we can assume that the radius, r, of a particular circle remains fixed even though its value is unknown. So we need to concern ourselves only with the central angle, θ. Because a complete revolution is 360°, it makes sense that a central angle of 360° would correspond to the circumference of the circle, or $2\pi r$. Also, because any central angle θ is $\dfrac{\theta}{360}$ of a complete revolution, then the corresponding arc length, s, should be the same proportional amount of the circumference. In other words, by proportional reasoning, we have that $\dfrac{\theta}{360°} = \dfrac{s}{2\pi r}$. Multiplying both sides of the equation by $2\pi r$ gives us the formula $s = \dfrac{\theta(2\pi r)}{360°}$.

Theorem 13.3 Length of an Arc

The length of an arc, s, with a central angle θ lying on a circle with radius r is
$$s = \frac{\theta(2\pi r)}{360°}.$$

Example 13.12 Using the Length of an Arc Formula

Find each value.

a. Arc length s, if $\theta = 45°$ and $r = 5$ in.

b. Radius r, if $\theta = 90°$ and $s = 5\pi$ in.

Solution In each case, we use the formula from Theorem 13.3.

a. Because $\theta = 45°$ and $r = 5$ in., $s = \dfrac{(45°)(2\pi 5)}{360°} = \dfrac{5}{4}\pi \approx 3.93$ in.

b. $5\pi = \dfrac{90°(2\pi r)}{360°}$. Solving for r, we have $r = \dfrac{5\pi(4)}{2\pi} = 10$ in.

Area

Area, another measurable attribute of planar figures, is defined as the amount of a plane enclosed by a planar figure. In the classroom, many students have trouble distinguishing between perimeter and area. The key to doing so lies in the number of dimensions used to make each measure. Perimeter is a measure of length, so it is a one-dimensional measure of a two-dimensional shape. Area is a measure of the plane, so it is a two-dimensional measure of a two-dimensional shape. The difference between the two is shown in Figure 13.13.

Because of the change in dimensions, length units cannot be used to measure area. Instead, we need a two-dimensional unit that has both a length and a width. Although any two-dimensional shape that tessellates the plane can be used as an area unit, we generally prefer to use a square. Consequently, a measure of area literally tells us the number of square units needed to exactly "cover" a shape. For instance, the rectangle shown in Figure 13.14 requires 8 square units to "cover" it.

Explorations Manual 13.6   Explorations Manual 13.7

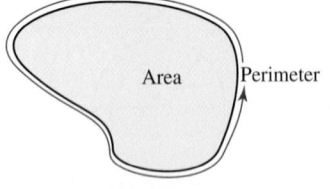

Figure 13.13 Perimeter versus area

Area   Perimeter

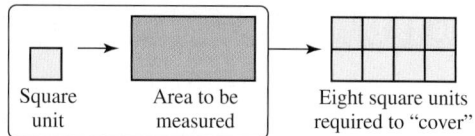

Figure 13.14 Measuring the area of a rectangle

Square units come in many sizes. In fact, length units from both the English and metric systems are used to build square units. For instance, a square foot (ft^2 or sq. ft) is a square that measures 1 ft on each side, and a square meter (m^2 or sq. m) is a square that measures 1 m on each side. These examples illustrate that the size of a square unit is directly related to the size of the length unit. Larger length units lead to larger square units. Smaller length units lead to smaller square units. So conversion ratios for length units can be used to create conversion ratios for square units. For instance, a square yard is a square that measures 1 yd, or 3 ft, on each side. As Figure 13.15 shows, a square yard can be covered exactly with 3 rows of 3 ft^2 each. As a result, $1 \text{ yd}^2 = 3 \text{ ft} \times 3 \text{ ft} = 9 \text{ ft}^2$.

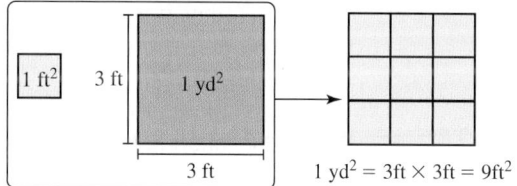

Figure 13.15 $1 \text{ yd}^2 = 3 \text{ ft} \times 3 \text{ ft} = 9 \text{ ft}^2$

This example illustrates that the ratio between square units is equal to the square of the ratio between the associated length units. Because this is true for any pair of square units, we use this fact to generate the conversion factors for the square units shown in Tables 13.5 and 13.6.

Table 13.5 Square units for the English, or U.S. Customary		
English Unit	**Symbol**	**Conversion Ratios**
Square inch	$in.^2$	
Square foot	ft^2	$1 \text{ ft}^2 = 144 \text{ in.}^2$
Square yard	yd^2	$1 \text{ yd}^2 = 9 \text{ ft}^2$
Square mile	mi^2	$1 \text{ mi}^2 = 3{,}097{,}600 \text{ yd}^2 = 27{,}878{,}400 \text{ ft}^2$

Table 13.6 Square units for the metric system		
Metric Unit	**Symbol**	**Relationship to m^2**
Square kilometer	km^2	$1 \text{ km}^2 = 1{,}000{,}000 \text{ m}^2$
Square hectometer	hm^2	$1 \text{ hm}^2 = 10{,}000 \text{ m}^2$
Square decameter	dam^2	$1 \text{ dam}^2 = 100 \text{ m}^2$
Square meter	m^2	
Square decimeter	dm^2	$1 \text{ dm}^2 = 0.01 \text{ m}^2$
Square centimeter	cm^2	$1 \text{ cm}^2 = 0.0001 \text{ m}^2$
Square millimeter	mm^2	$1 \text{ mm}^2 = 0.000001 \text{ m}^2$

The next example shows how to use dimensional analysis to make conversions between square units.

Example 13.13 Making Conversions Between Square Units

Use dimensional analysis to make each conversion.

a. 13 yd² to ft²   **b.** 12.4 mi² to km²

Solution

a. Using Table 13.5, we have $13 \, \text{yd}^2 \times \dfrac{9 \, \text{ft}^2}{1 \, \text{yd}^2} = 117 \, \text{ft}^2$.

b. In this case, the needed ratio is not in either table. However, because 1 mi = 1.6093 km, then 1 mi² = (1.6093)² km² ≈ 2.5898 km². Using this ratio, we have

$$12.4 \, \text{mi}^2 \times \dfrac{2.5898 \, \text{km}^2}{1 \, \text{mi}^2} \approx 32.11 \, \text{km}^2.$$

A number of square units are often used to measure land. In the English system, we use the **acre** for smaller plots of land and square miles for larger plots. Originally, an acre was defined to be the amount of land one man could plow with one horse in one day. Today, it is 4,840 yd², making it slightly smaller than a football field. In the metric system, two common units for land measure are the **are** (pronounced "air") and the **hectare**. An are is a square that measures 10 m on a side, making it about the size of the first floor on an average-sized house. A hectare is equal to 100 are, or 10,000 m², making it about the size of two and a half football fields. For larger plots of land, we use square kilometers.

Check Your Understanding 13.2-A

1. Find the perimeter or the circumference for each figure.

   a. A rectangle with $w = 4$ ft and $l = 11$ ft

   b. A square with $s = 3$ m

   c. A regular decagon with $s = 5.6$ cm

   d. A circle with $d = 6$ in

2. Given a circle with a diameter of 6 ft, what is the length of an arc with a central angle of 135°?

3. Use dimensional analysis to make each conversion.

   a. 45 ft² to yd²   **b.** 39,500 dm² to hm²   **c.** 352 in.² to m²

4. How many acres are in 1 mi²?

Talk About It Why do you think that it is difficult for many students to view perimeter as a measure of length?

Problem Solving

Explorations Manual **13.8**

Activity 13.3

We can compute the area of many irregular shapes using a geoboard or dot paper to count the number of square units contained in the figure. Although this sounds easy, partial squares can sometimes make counting difficult. For each of the following regions, devise a strategy to count squares, and then use it to find the area.

a. 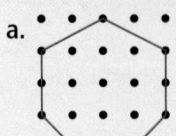   **b.**   **c.**

Areas of Geometric Shapes

As with perimeter, algebraic formulas can be developed to compute the areas of certain planar shapes. We begin with rectangles and squares.

Rectangles and Squares

Because we use square units to measure area, finding the area of rectangles is relatively simple. If the length and width are natural numbers, then the rectangle can be covered with an exact number of square units. For instance, if a rectangle has a length of 5 in. and a width of 3 in., then it can be exactly covered by three rows of 5 in.^{2} each, or 15 in.^{2} (Figure 13.16). Because $3 \cdot 5 = 15$, our example suggests that the area of the rectangle is the product of the length times the width, or $A = l \cdot w$.

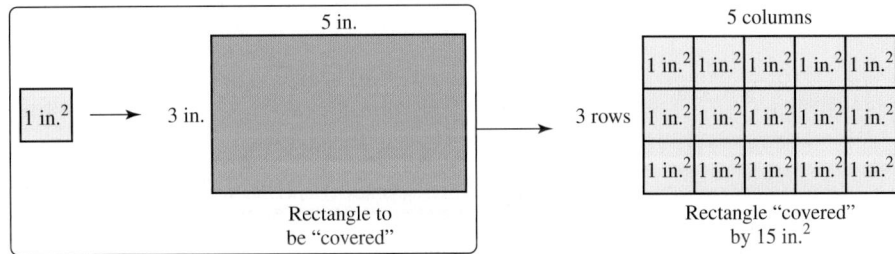

Figure 13.16 The area of a rectangle

Next, we check to see whether this relationship holds true for all rectangles, even for those with sides that have fractional or decimal lengths. For instance, consider a rectangle with a length of 3.2 in. and a width of 2.1 in. As Figure 13.17 shows, both square inches and fractional parts of square inches are needed to cover the rectangle. Combining all the pieces, the area of the rectangle is $A = 6$ in.^{2} $+ 7(0.1$ in.$^2) + 2(0.01$ in.$^2) = 6.72$ in.^{2}. Because $(3.2$ in.$) \cdot (2.1$ in.$) = 6.72$ in.^{2}, we again find that the area of the rectangle is $A = l \cdot w$.

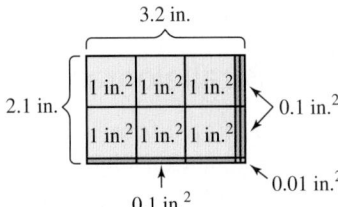

Figure 13.17 The area of a rectangle

The same idea applies to a square. However, because a square is a rectangle with congruent sides, the area is $A = l \cdot w = s \cdot s = s^2$.

Theorem 13.4   Area of a Rectangle and Square

The area of a rectangle with adjacent sides of lengths l and w is $A = lw$.
The area of a square with sides of length s is $A = s^2$.

□

Example 13.14   The Perimeter and Area of a Rectangle

Find the length and width of the rectangle with the largest area if the dimensions are whole numbers and the perimeter is fixed at 16 in.

Solution To find the length and width of the needed rectangle, we use the given constraints to make a list of the possible dimensions and then use the formula to compute the area of each rectangle. Because the perimeter is fixed at 16 in., there are only four possibilities, as tracked in the following table:

Length and Width	Area
$l = 1$ in., $w = 7$ in.	$A = 1 \cdot 7 = 7$ in.^{2}
$l = 2$ in., $w = 6$ in.	$A = 2 \cdot 6 = 12$ in.^{2}
$l = 3$ in., $w = 5$ in.	$A = 3 \cdot 5 = 15$ in.^{2}
$l = 4$ in., $w = 4$ in.	$A = 4 \cdot 4 = 16$ in.^{2}

As the table shows, the rectangle with the largest area is a square with sides of 4 in. In general, this is always true. If the perimeter is fixed, then the rectangle with the largest area is a square.

Before developing other area formulas, we need to introduce a property that allows us to compute the area of compound figures.

Theorem 13.5   Area of Compound Figures

If a planar region A is subdivided into nonoverlapping subregions B and C then the area of region A is the sum of the areas of the subregions, or area(A) = area(B) + area(C).

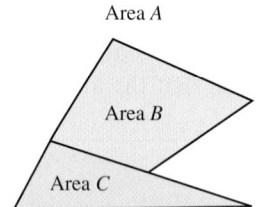

Figure 13.18 Area(A) = area(B) + area(C)

Theorem 13.5 indicates that we can compute the area of some planar figures by dividing them into subregions, finding the area of each subregion, and then adding them (Figure 13.18). This property, along with transformations, can be used to find area formulas for other shapes, such as parallelograms, triangles, trapezoids, and regular polygons.

Parallelograms

Explorations Manual 13.9

The area formula for a rectangle can be used to find the formula for a parallelogram. Consider a parallelogram in which one side is the **base, *b*,** and the **height, *h*,** is the distance between the base and its opposite side. If we "cut off" a right triangular corner and translate it to the opposite end, as shown in Figure 13.19, the result is a rectangle with the same area as the parallelogram. Because the area of a rectangle is the length times the width, the area of the parallelogram must be $A = l \cdot w = b \cdot h$.

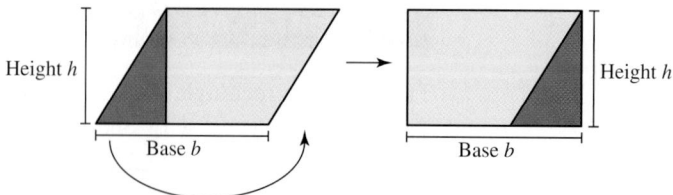

Figure 13.19 Translating one side of a parallelogram

Theorem 13.6   Area of a Parallelogram

The area of a parallelogram with a base of length b and a height h is $A = bh$.

Example 13.15 **Computing the Area of a Parallelogram**

Find the area of each parallelogram.

a.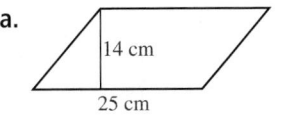

b.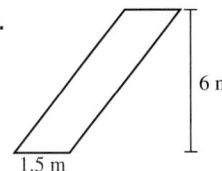

Solution

a. $A = (25 \text{ cm}) \cdot (14 \text{ cm}) = 350 \text{ cm}^2$.

b. $A = (1.5 \text{ m}) \cdot (6 \text{ m}) = 9 \text{ m}^2$.

Triangles

The area formula for a parallelogram can, in turn, be used to find the area of a triangle. We first designate one of the three sides to be the **base, *b*,** of the triangle. The **height, *h*,** of the triangle is then the perpendicular distance from the base to the vertex that is opposite it. Figure 13.20 shows the bases and heights for three triangles. In a right triangle, the base and height are given by the two legs. In an obtuse triangle, the height is measured outside the triangle to the line that contains the base.

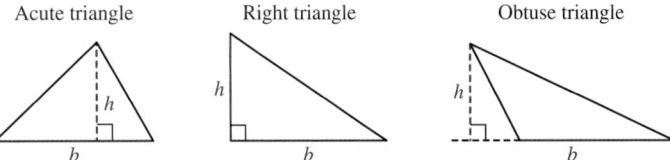

Figure 13.20 The base and height of three triangles

To find the area formula of a triangle with a base *b* and a height *h*, we make a copy of the triangle and rotate it 180° about the midpoint of a side that is not the base. The two triangles together make a parallelogram with the same base and height of the triangle but twice the area (Figure 13.21). Consequently, the area of the triangle is

$$A = \frac{1}{2}(\text{area of parallelogram}) = \frac{1}{2}(bh)$$

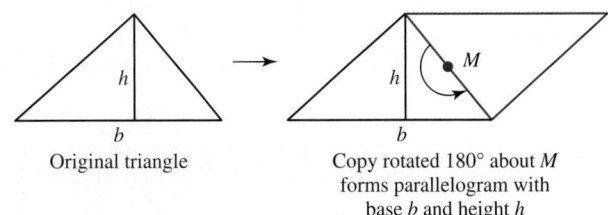

Figure 13.21 Rotating a copy of the triangle to form a parallelogram

Theorem 13.7 Area of a Triangle

The area of a triangle with a base of length *b* and a height *h* is $A = \frac{1}{2}bh$.

Example 13.16 | **Computing the Area of Triangles**

Find the area of each triangle.

a.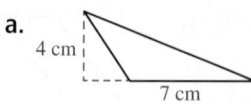
4 cm
7 cm

b.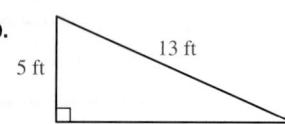
5 ft
13 ft

Solution

a. Using the formula, we have $A = \frac{1}{2}bh = \frac{1}{2}(7\,\text{cm})(4\,\text{cm}) = 14\,\text{cm}^2$.

b. We have a right triangle with an unknown base. However, the hypotenuse is 13 ft, and one leg is 5 ft, so the missing leg must be 12 ft by the Pythagorean Theorem. Consequently, the area is $A = \frac{1}{2}bh = \frac{1}{2}(12\,\text{ft})(5\,\text{ft}) = 30\,\text{ft}^2$.

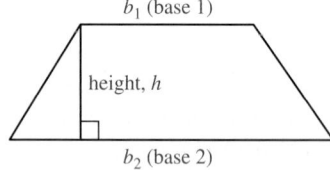
b_1 (base 1)
height, h
b_2 (base 2)

Figure 13.22 The height and bases of a trapezoid

Trapezoids

To find the area of a trapezoid, we must first know the lengths of the **bases,** which are the two parallel sides that we designate as b_1 and b_2. We must also know the **height,** h, which is the distance between the two parallel sides (Figure 13.22). Given these lengths, we can use the formula for the area of a parallelogram to find the area of a trapezoid.

Example 13.17 | **The Area Formula of a Trapezoid**

Problem Solving   Find a formula for the area of a trapezoid if the bases and the height are known.

Solution

Understanding the Problem. We are asked to find the area of a trapezoid when the two bases and the height are known. Because no specific trapezoid is given, the formula should be given in terms of these three lengths.

Devising a Plan. So far, we have *used transformations* to change the shape into a figure for which we can compute the area. This strategy shows promise if we can determine how to transform a trapezoid.

Carrying Out the Plan. We have used transformations in two ways. One was to slide the corner of a parallelogram to the opposite end to form a rectangle. Unfortunately, this does not work with a trapezoid because the legs are not parallel. The second method was to rotate a copy of the figure about the midpoint of a side. This method shows more promise because rotating a copy of the trapezoid 180° about the midpoint of a leg again results in a parallelogram.

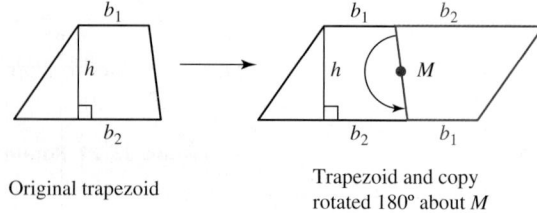

Original trapezoid

Trapezoid and copy
rotated 180° about M

The parallelogram has a base of $(b_1 + b_2)$, a height h, and an area twice as large as the original trapezoid. Consequently, the area of the trapezoid is

$$A = \frac{1}{2}(\text{area of parallelogram}) = \frac{1}{2}(b_1 + b_2)h$$

Looking Back. We can check our formula by deriving it in another way. First, draw a diagonal of the trapezoid that separates it into two triangles.

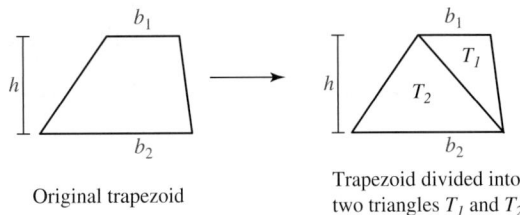

Original trapezoid

Trapezoid divided into two triangles T_1 and T_2

One triangle has a base of b_1, the other has a base of b_2, and both have a height of h. The area of the trapezoid is the sum of the areas of the triangles, so

$$A_{\text{Trapezoid}} = A_{T_1} + A_{T_2} = \frac{1}{2}(b_1 h) + \frac{1}{2}(b_2 h) = \frac{1}{2}(b_1 + b_2)h.$$

Theorem 13.8   Area of a Trapezoid

The area of a trapezoid with bases of lengths b_1 and b_2 and a height h is
$$A = \frac{1}{2}(b_1 + b_2)h.$$

Example 13.18   Finding a Base of a Trapezoid

A trapezoid has an area of 54 in.^{2}. If one base is 16 in. and the height is 4 in., what is the length of the other base?

Solution We can find the missing base by substituting the given information into the area formula and solving for the base. If $A = 54$ in.^{2}, $b_1 = 16$ in., and $h = 4$ in., then

$$\frac{1}{2}(16 + b_2)4 = 54$$
$$2(16 + b_2) = 54$$
$$(16 + b_2) = 27$$
$$b_2 = 11 \text{ in.}$$

Regular Polygons

In Example 13.17, we found the area formula for a trapezoid by subdividing it into triangles. This same method can be used to find the area of a regular polygon. Consider a regular hexagon with sides of length s. Drawing diagonals divides the hexagon into six congruent triangles, each of which has a base of length, s, and a height equal to the apothem, a (Figure 13.23).

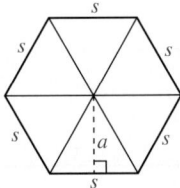

Figure 13.23 A regular hexagon divided into six triangles

Because each triangle has an area of $\frac{1}{2}as$, the area of the hexagon must be $A = 6\left(\frac{1}{2}as\right)$. However, notice that $6s$ is equal to the hexagon's perimeter. By making this substitution, we can also write the area as $A = 6\left(\frac{1}{2}as\right) = \frac{1}{2}aP$. This method works for any regular polygon, so we get the following formula.

> ### Theorem 13.9 Area of a Regular Polygon
>
> The area of a regular n-gon, with sides of length s and apothem of length a, is $A = \frac{1}{2} aP$, where P is the perimeter of the polygon, or $P = ns$.

Circles

We can use the area formula for a regular polygon to find the area of a circle. Recall that, as we add more sides to a regular polygon, it becomes more circular (Figure 13.24).

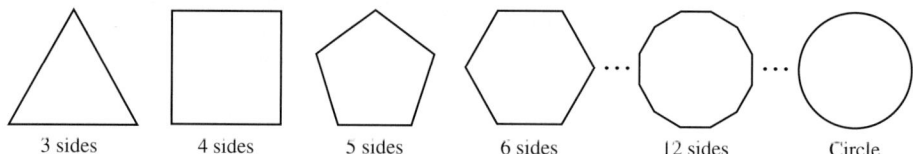

| 3 sides | 4 sides | 5 sides | 6 sides | 12 sides | Circle |

Figure 13.24 A sequence of regular polygons leading to a circle

Explorations Manual 13.10

As the polygons grow more circular, their apothems begin to approximate the radius of a circle, and their perimeters begin to approximate its circumference. Their areas also approximate the area of the circle (Figure 13.25). As a result, if we think of a circle as a regular polygon with an infinite number of sides, then the area of the circle is $A = \frac{1}{2} aP = \frac{1}{2} r(2\pi r) = \pi r^2$.

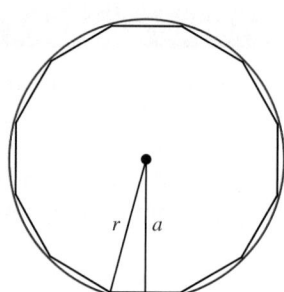

Figure 13.25 Area of regular polygon $\approx$ area of circle $\approx \frac{1}{2} aP = \pi r^2$

> ### Theorem 13.10 Area of a Circle
>
> The area of a circle with radius r is $A = \pi r^2$.

Example 13.19 Finding the Area of Circle

If a circle has a diameter of 15 m, what is its area?

Solution Because the diameter of a circle is twice its radius, the radius must be $r = 7.5$ m. Evaluating the area formula at this value, we have

$$A = \pi r^2 = \pi(7.5)^2 = 56.25\pi \approx 176.715 \text{ m}^2$$

In our final example, we use what we have learned to compute the area of a compound figure.

Example 13.20   Calculating an Amount of Paint

Application   A farmer plans to paint the front of his barn, which has the dimensions shown:

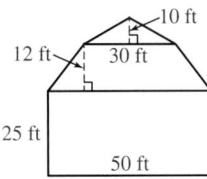

How many gallons of paint are needed if 1 gal covers 200 ft^2?

Solution The front of the barn can be subdivided into three polygons: a rectangle, a trapezoid, and a triangle. By computing the area of each figure and adding the results, we can get the needed square footage:

$$A_{\text{Front of Barn}} = A_{\text{rectangle}} + A_{\text{trapezoid}} + A_{\text{triangle}}$$

$$= lw + \frac{1}{2}(b_1 + b_2)h + \frac{1}{2}bh$$

$$= (50 \text{ ft})(25 \text{ ft}) + \frac{1}{2}(50 \text{ ft} + 30 \text{ ft})(12 \text{ ft}) + \frac{1}{2}(30 \text{ ft})(10 \text{ ft})$$

$$= 1{,}250 \text{ ft}^2 + 480 \text{ ft}^2 + 150 \text{ ft}^2$$

$$= 1{,}880 \text{ ft}^2$$

Next, we find the number of gallons of paint by dividing 1,880 by 200, or $\dfrac{1{,}880}{200} =$ 9.4 gal. Because we can buy only whole cans of paint, the farmer must purchase 10 gallons.

Check Your Understanding 13.2-B

1. a. Find the area of a parallelogram with $b = 18$ ft and $h = 12$ ft.

   b. Find the area of a triangle with $b = 1.2$ cm and $h = 3.1$ cm.

   c. Find the area of a trapezoid with $b_1 = 13.1$ m, $b_2 = 9.6$ m, and $h = 5$ m.

   d. Find the area of a circle with $d = 13$ in.

2. If a regular octagon has an area of approximately 70.4 cm^2 and an apothem of 4.4 cm, what is the length of each of the octagon's sides?

3. A local high school is renovating its football field and running track. All that remains is to plant sod in the field area. Based on the measurements shown, how many square yards of sod must the school purchase?

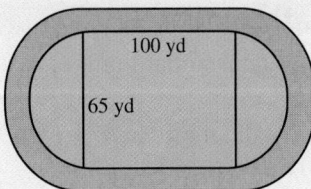

Talk About It What happens to the area of a rectangle if its length and width are cut in half?

Reasoning

Activity 13.4

One way to find the area of a rhombus is to use its diagonals. Use a straightedge and compass to construct four or five rhombi. Measure the base and height of each, and then approximate their areas using the parallelogram formula. Next measure the diagonals, and try to find a formula for the area of a rhombus that uses their lengths. Once you think you have it, prove your conjecture. (*Hint:* If your formula is correct, you can prove it deductively by using the areas of four congruent right triangles.)

SECTION 13.2 Learning Assessment

■ Understanding the Concepts

1. What is perimeter, and why is it measured with length units?

2. Describe how to derive the perimeter formula for each of the given shapes.

   a. Square   **b.** Rectangle

   c. Regular polygon

3. What does π represent, and what numbers are often used to approximate its value?

4. What is the difference between the perimeter and the area of a planar shape?

5. **a.** Why can length units not be used to measure area?

   b. What must be true about any unit that is used to measure area?

6. Does it make sense to say that the perimeter of an object is smaller than its area? Why or why not?

7. Is the perimeter of a square actually part of the square? What about the area? Explain.

8. How can a rectangle be used to derive the area formulas for a square and a parallelogram?

9. How can a parallelogram be used to derive the area formula for a triangle and a trapezoid?

10. How can a regular polygon be used to derive the area formula for a circle?

■ Representing the Mathematics

11. Can you use your hand as a nonstandard unit of area? Why or why not?

12. Draw a diagram showing the difference between the perimeter and area of a triangle.

13. Draw a diagram showing the difference between the circumference and area of a circle.

14. Draw a diagram that shows the relative size between a in.^{2} and a ft^2.

15. Draw a diagram that shows the relative size between a cm^2 and a dm^2.

16. A rectangle measures 3×4 yd. Draw a diagram of what the rectangle would look like if it were covered by square yards.

17. A rectangle measures 2.2×4.3 ft. Draw a diagram of what the rectangle would look like if it were covered by square feet.

18. Draw a diagram that illustrates how the area of a rectangle changes if its length and width are both doubled.

19. A right triangle has legs of lengths 3 ft and 4 ft. Draw a picture of what the triangle might look like if it were covered by square feet.

20. **a.** Draw two rectangles with the same perimeter but different areas.

    b. Draw two rectangles with the same area but different perimeters.

21. Do the two following parallelograms have the same area? How do you know?

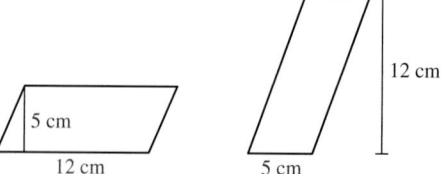

■ Exercises

22. Fill in each blank with the correct number.

    a. 48 ft^2 = _____ yd^2.   **b.** 637 in.^{2} = _____ yd^2.

    c. 4,576 yd^2 = _____ mi^2.   **d.** 7.3 hm^2 = _____ km^2.

    e. 0.0037 km^2 = _____ m^2.   **f.** 0.47 dm^2 = _____ cm^2.

23. Make each conversion.

    a. 3 mi^2 to km^2 **b.** 4.5 ft^2 to m^2

    c. 7.8 cm^2 to in.^{2} **d.** 0.07 hm^2 to yd^2

    e. 2.53 km^2 to hectares   **f.** 14.5 ares to hectares

24. How many times larger is:
    a. A km² than a m²?   **b.** A m² than a cm²?
    c. A m² than a mm²?

25. How many times smaller is:
    a. A cm² than a hm²?   **b.** A mm² than a dm²?
    c. A m² than a hm²?

26. Estimate the area of the cover of this book in:
    a. in.².   **b.** cm².   **c.** dm².

27. **a.** Estimate the area of your desktop in square centimeters.
    b. Estimate the area of your classroom floor in square yards.

28. What are the perimeter and area of:
    a. A square that measures 6 in. on a side?
    b. A rectangle with a width of 4 in. and a length of 10 in.?

29. What are the perimeter and area of:
    a. A square that measures 2.3 cm on a side?
    b. A rectangle with a width of 6.3 m and a length of 12.9 m?

30. What is the area of:
    a. A parallelogram with a base of 3 cm and a height of 2 cm?
    b. A trapezoid with bases of length 3 yd and 2 yd and a height of 1.5 yd?

31. What is the area of:
    a. A parallelogram with a base of 14.7 in. and a height of 3.9 in.?
    b. A trapezoid with bases of length 4.5 m and 2.7 m and a height of 3.5 m?

32. If the length of a rectangle is 14 m and its area is 126 m², what is its width?

33. A boxing ring is a square with an area of 400 ft². How long is each of its sides?

34. What is the area of:
    a. A triangle with a base of 7 in. and a height of 13 in.?
    b. A triangle with a base of 3 km and a height of 2.1 km?

35. What is the area of:
    a. A right triangle with legs of lengths 15 m and 20 m?
    b. A right isosceles triangle with a hypotenuse of 8 cm?

36. **a.** What is the perimeter of an equilateral triangle with sides of 13 ft?
    b. If the perimeter of an equilateral triangle is 141 cm, what are the lengths of the sides?

37. What is the circumference and the area of a circle with a radius of:
    a. $r = 2$ in.?   **b.** $r = 5.3$ yd?

38. What is the circumference and the area of a circle with a radius of:
    a. 2.9 ft?   **b.** π cm?

39. Find the area of each figure.

    a.
    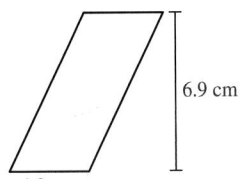

    b.
    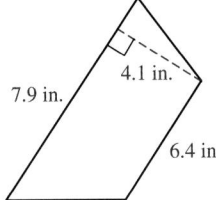

    c.
    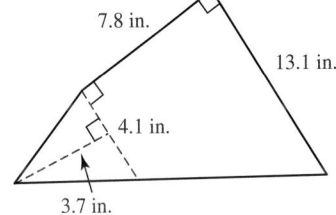

    d.
    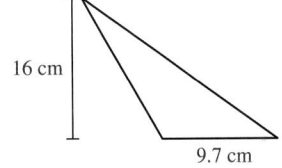

    e.   **f.**
    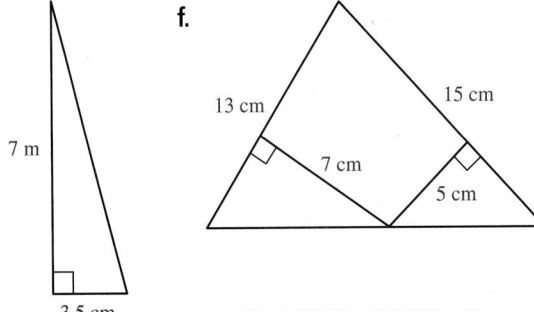

40. Find the area of each figure.

    a.   **b.**
    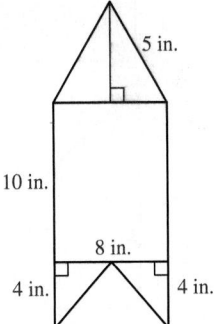

c.

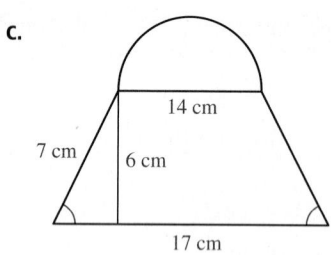

d.

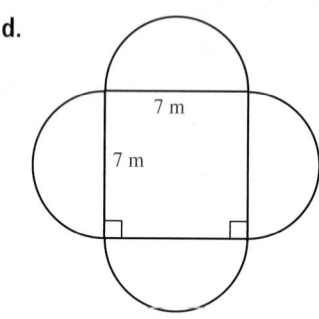

41. What is the area of a regular hexagon if the sides of the hexagon are 3 cm?

42. A regular polygon has a perimeter of 60 ft and an apothem of 8 ft. What is its area?

43. What is a reasonable estimate for the area of a regular 20-gon if it is known that the apothem is 6.5 ft?

■ Problems and Applications

44. The base of an isosceles triangle is one-fourth the length of one of the congruent sides. The perimeter is 126 cm. What are the lengths of the three sides?

45. Find the area of the shaded region in each figure.

a.

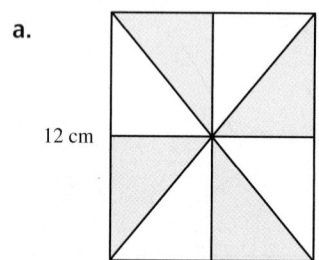

b.

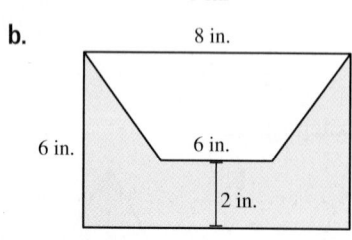

c.

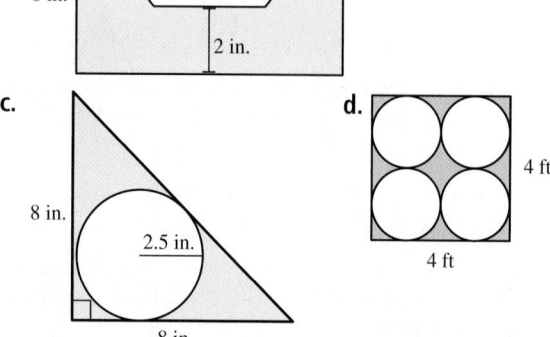

d.

46. Determine the perimeter and area of each polygon.

a. **b.**

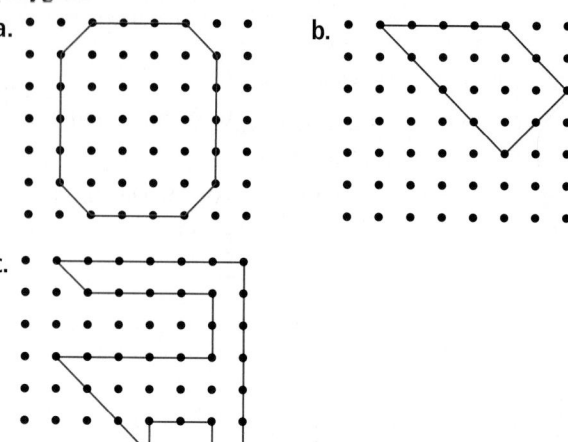

c.

47. If the ratio of the sides of two similar right triangles is 1:3, what is the ratio of their

   a. Perimeters?   **b.** Areas?

48. Estimate the largest rectangular area that can be made with a string that is 8 ft long.

49. What is the area of a rhombus with diagonals measuring 4.5 in. and 3.7 in.?

50. If the triangles surrounding the regular hexagon are equilateral with sides of length 10 cm, what is the area of the entire region?

51. A 12-in. pizza costs $8.99, and a 16-in. pizza costs $13.99. Which pizza cost less per square inch?

52. During 2004, forest fires in Alaska burned over 6.38 million acres.

   a. How many square miles of forest is this?

   b. If Alaska is 586,412 mi^2, what percentage of it was burned by the forest fires?

53. The outside wall on a house is 42 ft long and 21 ft high. It has 8 windows, each measuring 2×3 ft and one door measuring 3×7 ft. If 1 gal of paint covers 250 ft^2 and two coats will be applied to the wall, how many gallons of paint are needed to paint the wall?

54. Sam and Julie want to tile their kitchen floor with 8×8-in. tiles. If the dimensions of their kitchen are 17×14 ft, how many tiles do they need?

55. The chain link fence that sits around Pam's pool is 10 ft away from the pool on each side. She would like to expand the fence so that it sits 15 ft away from the pool on each side.

a. If the pool is 15 × 30 ft, how much fence does she need to add?

b. How much will the area around the pool increase after the fence is expanded?

56. The Earth is about 93,000,000 mi from the sun. Assuming that the Earth's orbit is essentially circular, about how far does the Earth travel along its orbit on any given day?

57. If the minute hand on a clock is 5 in. long, how far does the tip of the minute hand move over the course of:

a. 15 min?   **b.** 30 min?   **c.** 5 min?   **d.** 1 min?

58. A dog is tied to the corner of a shed with a rope that is 18 ft long. If the shed is 15 ft by 12 ft, how many square feet does the dog have in which to roam?

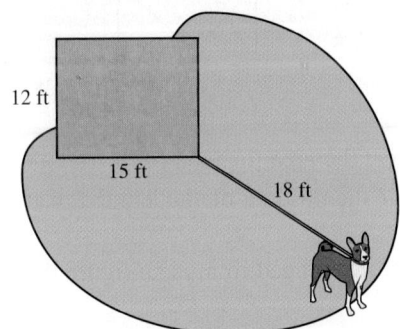

12 ft

15 ft

18 ft

59. The following picture appears to be a square that has been divided into nine smaller squares of different areas.

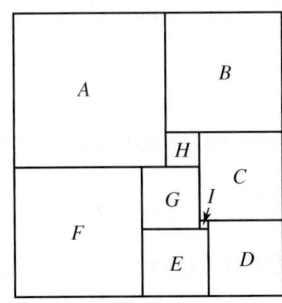

Is such a square possible? (*Hint:* Use the fact that the area of square E is 64 cm², and the area of square D is 81 cm².)

60. Heron's Theorem states that the area of a triangle with sides of lengths a, b, and c and semiperimeter, s, is given by $A = \sqrt{s(s - a)(s - b)(s - c)}$, where $s = \dfrac{a + b + c}{2}$.

a. Use Heron's Theorem to find the area of a triangle where:

    i. $a = 3$ in., $b = 5$ in., and $c = 6$ in.

    ii. $a = 4$ in., $b = 3.6$ in., and $c = 4.7$ in.

b. Suppose a triangle has sides $a = 4$ in., $b = 7$ in., and $c = 11$ in. Use Heron's Theorem to find the area. Can you explain why the answer is what it is?

61. A free-form pool is roughly in the shape of four congruent semicircles with diagonals that intersect to form a square.

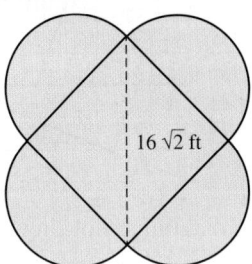

16 √2 ft

If the diagonal of the square is $16\sqrt{2}$ ft, what is the perimeter of the pool?

■ Mathematical Reasoning and Proof

62. The side of square B is four times longer than the side of square A.

a. How many times longer is the perimeter of B than the perimeter of A?

b. How many times greater is the area of B than the area of A?

63. a. What happens to the circumference of a circle if its radius is doubled?

b. What happens to the area of a circle if its radius is doubled?

64. How much does the area of a square increase if its side is increased by 10%?

65. Use geometry software to explore and answer the following questions:

a. What must happen to the shape of a rectangle if its area is to increase but its perimeter is to remain the same?

b. What must happen to the shape of a rectangle if its area is to decrease but its perimeter is to remain the same?

66. The red and blue parallelograms shown have their bases on two parallel lines.

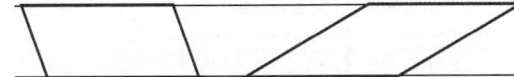

If their bases are the same length, what can you conclude about their area? Why?

67. If a parallelogram and a triangle share the same base and have the same area, what can we conclude about the heights of the two figures?

68. If a triangle is rotated 180° around the midpoint of one of its sides, why must the resulting figure be a parallelogram?

69. Suppose that the square below is cut into the indicated four pieces and rearranged as shown. The diagram seems to indicate that that we have changed a square of 64 units² into a rectangle of 65 units². What property does this seem to violate? Can you explain what went wrong? If needed, cut out a copy of the pieces to help.

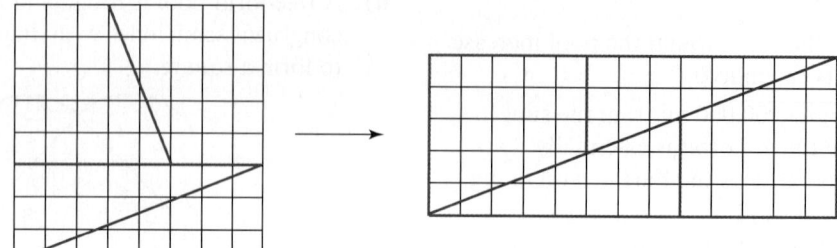

70. The following diagrams show how to fold a triangle into a rectangle.

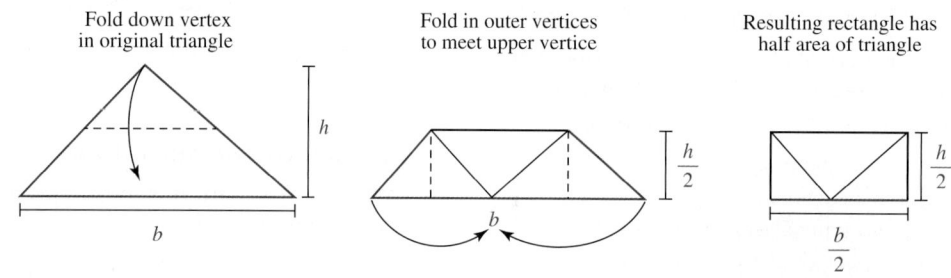

Use what you see, along with the area formula for a rectangle, to derive the area formula for a triangle.

71. Show that the area of a right triangle can be found by taking one-half the product of the lengths of its legs. (*Hint:* Make a rectangle.)

72. The following diagram shows how a circle can be cut into wedges and rearranged to make a figure that is roughly a parallelogram.

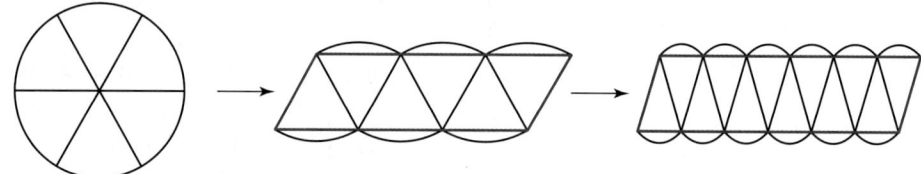

As the number of wedges increases, they become smaller, and the figure looks more like a parallelogram. Use what you see, along with the area formula for a parallelogram, to derive the area formula for a circle.

73. The pie-shaped region of the circle shaded below is called a **sector** of the circle.

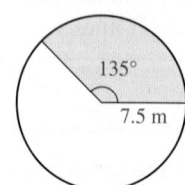

Use proportional reasoning to help you find a formula for calculating the area of a sector. Use it to find the area of the given sector.

74. Use the Pythagorean Theorem to show that, for any equilateral triangle with side s, the area of the triangle is given by $A = \dfrac{\sqrt{3}s^2}{4}$.

■ Mathematical Communication

75. Describe two different ways to compute the area of the following figure. Explain why both methods give the same result.

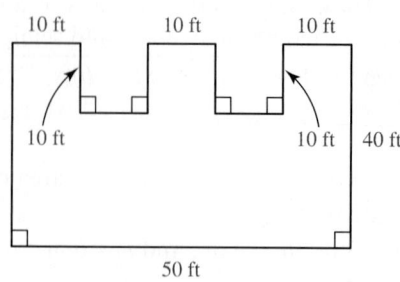

76. Describe to one of your peers how to derive the area formula for a triangle by using a parallelogram.

a. What representations did you use or find essential to your communication?

b. What was difficult about explaining this process? What made it so?

77. a. Write a story problem that must be solved by computing the perimeter of a geometric figure.

b. Write a story problem that must be solved by computing the area of a geometric figure.

c. Give your problems to another person to critique and solve.

d. Which problem did you find the most difficult to write? Why?

78. Beth claims that because 1 yd is equivalent to 3 ft, then 1 yd^2 must be equivalent to 3 ft^2. What is wrong with Beth's thinking, and how might you help her?

79. Oliver is asked to find the area of an odd-shaped figure. He first wraps a string around the outside of the figure to find its perimeter. He then forms a square with the string and computes its area. He claims that the original shape and the square have the same area because they have the same perimeter. Does his method work? Why or why not?

■ Building Connections

80. In general, any two-dimensional shapes can be used as an area unit as long as it tessellates the plane. What other regular polygons can be used as area units, and what might we call such units?

81. Suppose that a rectangle and one other object are cut out of the same piece of cardboard. Explain how directly weighing the objects on a scale and proportional reasoning can be used to find the area of the other figure.

82. How does this section on perimeter and area build a connection between geometry and algebraic thinking?

83. Look at several sets of curriculum materials for several grade levels. Compare and contrast how they approach the topics of perimeter and area. Which do you think is the best approach and why?

84. Do an Internet search to find reasons why some children have trouble learning perimeter and area. Write a paragraph or two on what you find and on how you might help your students overcome some of these problems.

SECTION **13.3** Surface Area

Perimeter and area are attributes of two-dimensional or planar shapes. We now turn to three-dimensional or spatial shapes, which have two commonly measured attributes: surface area and volume. We discuss surface area in this section and volume in the next.

Understanding Surface Area

In general, **surface area** is the amount of plane needed to cover the outside of a three-dimensional shape. We measure surface area when we measure the amount of paint needed to cover a wall or when we measure the amount of paper needed to wrap a birthday present. Because surface is defined as a measure of the plane, square units must be used to measure it. Consequently, any measure of surface area literally indicates the number of square units needed to cover the outside of a three-dimensional object (Figure 13.26).

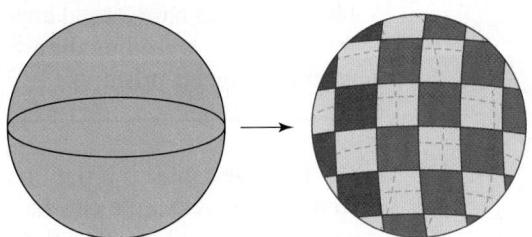

Figure 13.26 Covering a surface with square units

Like perimeter, surface area is difficult for many students to perceive because it is a two-dimensional measure of a three-dimensional object. To help them understand, teachers often have students cut spatial shapes in such a way that they lie flat. For instance, if we cut a cereal box along enough edges, then it can be laid flat, as shown in Figure 13.27. The end result is the net of the box, suggesting that we can compute surface area by computing the area of the figure's net. Because many three-dimensional shapes, like prisms, pyramids, and cylinders, have nets consisting of common planar shapes, we can develop surface area formulas by combining the appropriate area formulas. We begin with the surface area of a right prism.

Explorations Manual 13.11

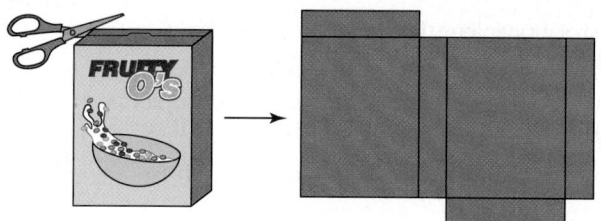

Figure 13.27 Cutting a box so that it lies flat

Surface Area of a Right Prism

The cereal box in Figure 13.27 shows that when a right rectangular prism is cut open, every polygon in the net is a rectangle. This makes deriving a formula for its surface area relatively straightforward.

Example 13.21   Surface Area of a Rectangular Prism

Representation   Find a formula for the surface area of a right rectangular prism with a length l, width w, and height h.

Solution Consider a right rectangular prism, or box, with a length l, width w, and height h. When the prism is cut so that it lies flat, we get the net shown:

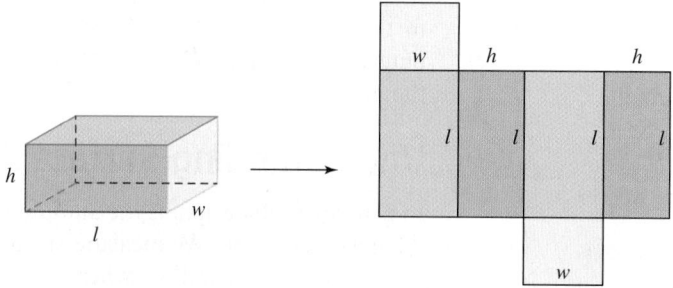

The net consists of six rectangles, so the surface area must be the sum of the six rectangular areas. Notice that the six rectangles can be put into pairs of congruent rectangles, allowing us to make the computation more efficient. The two green rectangles have dimensions l and w, giving them a total area of $2 \cdot l \cdot w$. Similarly, the two blue rectangles have a total area of $2 \cdot l \cdot h$, and the two yellow rectangles have a total area of $2 \cdot w \cdot h$. Adding the areas, we get the formula for the surface area, SA, of any right rectangular prism: $SA = 2lw + 2lh + 2wh = 2(lw + lh + wh)$.

Because a cube is a right rectangular prism with six congruent faces, the formula from the last example can be used to the find its surface area. The length, width, and height are all equal to the edge, e (Figure 13.28), so the surface area of a cube is $SA = 2(e \cdot e + e \cdot e + e \cdot e) = 2(3e^2) = 6e^2$.

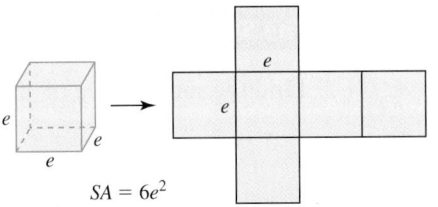

$$SA = 6e^2$$

Figure 13.28 Surface area of a cube

Example 13.22   Finding the Surface Areas of Rectangular Prisms

Find the surface area of each prism.

a.

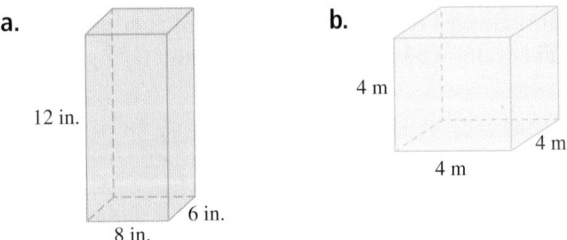

12 in.

6 in.

8 in.

b.

4 m

4 m

4 m

Solution In each case, we apply the appropriate formula.

a. $SA = 2(lw + lh + wh) = 2(8 \cdot 6 + 8 \cdot 12 + 6 \cdot 12) = 2(48 + 96 + 72) = 432$ in.^{2}.

b. We have a cube with $e = 4$ m. Hence, $SA = 6e^2 = 6(4)^2 = 96$ m^2.

Cutting a three-dimensional figure into one of its nets can be done to find the surface area of any prism. For instance, consider the nets of the two right prisms shown in Figure 13.29. In each case, the surface area is the sum of the base areas and the **lateral surface area,** which is the total area of the lateral rectangles. Each lateral rectangle has the same height, and the sum of their widths is equal to the perimeter of the base. As a result, the lateral surface area is equal to Ph, where P is the perimeter of the base. If we let B represent the area of one base, then the surface area of any right prism is given by $SA = 2B + Ph$.

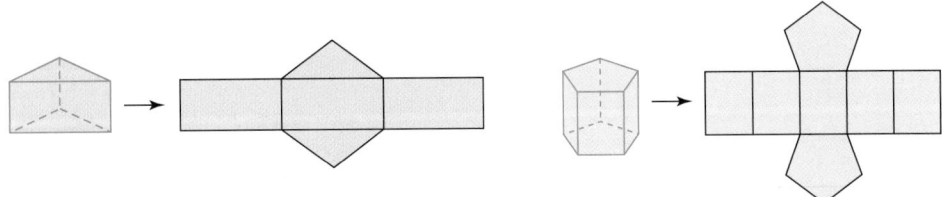

Figure 13.29 The nets of two right prisms

The following theorem summarizes these three formulas.

Theorem 13.11   Surface Area Formulas for Right Prisms

The surface area of a right rectangular prism with edges of lengths l, w, and h is $SA = 2lw + 2lh + 2wh$.

The surface area of a cube with an edge of length e is $SA = 6e^2$.

The surface area of a right prism with a height h, a base of area B, and a perimeter P is $SA = 2B + Ph$.

Example 13.23 The Surface Area of a Triangular Prism

Find the surface area of the following prism.

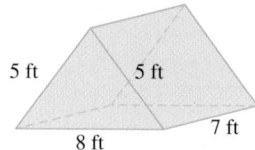

Solution The surface area of the right triangular prism is the sum of the lateral surface area and the area of the two bases. The perimeter of the base is $5 + 5 + 8 = 18$ ft, making the lateral surface area $Ph = 18 \cdot 7 = 126$ ft^2. Each triangular base is an isosceles triangle with a base of 8 ft and sides of 5 ft. By using the Pythagorean Theorem, we know the height of the triangle is $\sqrt{(5^2 - 4^2)} = \sqrt{9} = 3$ ft.

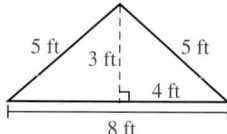

Using this height, the area of one triangular base is $A = \dfrac{1}{2}bh = \dfrac{1}{2} \cdot 8 \cdot 3 = 12$ ft^2. As a result, the total surface area is $SA = 2(12) + 126 = 150$ ft^2.

Surface Area of a Right Cylinder

A net can also be used to find the surface area of a right circular cylinder. For instance, imagine cutting a can of frozen orange juice so that it lies flat (Figure 13.30). The top and bottom are circles, and the paper tube unrolls into a rectangle with a width equal to the height of the can and a length equal to the circumference of the lid. As with the prism, the surface area is the sum of the areas of these two-dimensional pieces. Consequently, the surface area of a cylinder is $SA = 2$(area of circle) + (length of rectangle) · (width of rectangle) = $2\pi r^2 + 2\pi rh = 2\pi r(r + h)$.

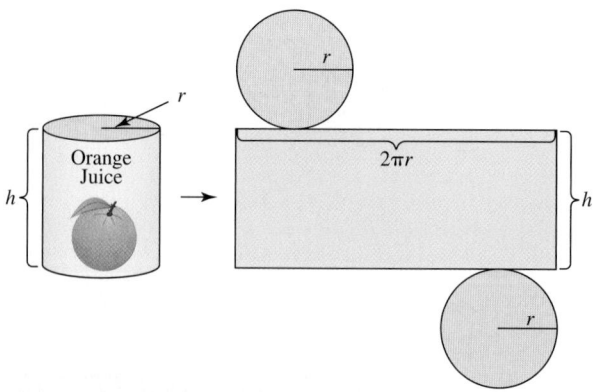

Figure 13.30 The net of a right circular cylinder

Theorem 13.12 Surface Area Formula for a Right Circular Cylinder

The surface area of a right circular cylinder with height h and radius r is $SA = 2\pi r^2 + 2\pi rh = 2\pi r(r + h)$.

Example 13.24   Aluminum Needed to Make a Can

Application

Most soda cans are approximately a right circular cylinder with a diameter of 2.6 in. and a height of 4.75 in. Based on these dimensions, about how much aluminum is needed to make a can?

Solution The approximate amount of aluminum needed to make a can is equal to its surface area. Using the formula from Theorem 13.12, a radius of $r = 2.6 \div 2 = 1.3$ in., and a height of 4.75 in., we have that $SA = 2\pi r(r + h) = 2\pi(1.3)[(1.3) + (4.75)] \approx$ 49.42 in.^{2} of aluminum is needed to make the can.

Check Your Understanding 13.3-A

1. Find the surface area for each prism.

a.

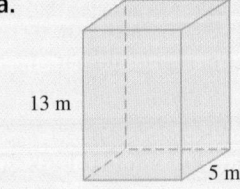

13 m

5 m

4 m

b.

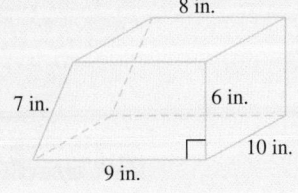

8 in.

7 in.

6 in.

10 in.

9 in.

2. If a cube has a surface area of 2,166 cm^2, what is the length of one of its edges?

3. Find the surface area of the following cylinder:

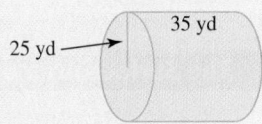

35 yd

25 yd

Talk About It How can you generalize the surface area formula for a right circular cylinder to a surface area formula for any cylinder?

Problem Solving

Activity 13.5

Consider eight wooden cubes that are identical in size and color. If the cubes are placed together to form one shape, what is the:

a. Smallest surface area possible?

b. Largest surface area possible?

Surface Area of a Right Pyramid

Like prisms and cylinders, the surface area of a pyramid can be found by looking at its net. However, a pyramid has only one base, and all its lateral surfaces are triangles. For instance, Figure 13.31 shows the net of a right square pyramid. The surface area of this pyramid is the area of the base plus the areas of the four triangles. The base is a square, so it has an area of b^2. The four triangles are congruent, and each has an area of $\frac{1}{2}bl$, where l is the slant height of the pyramid. The **slant height** is the distance on a lateral face that extends from the midpoint of a side of the base to the apex. By combining these values, the surface area of the pyramid is $SA = b^2 + 4\left(\frac{1}{2}bl\right)$. However, $4 \cdot b$ is the perimeter of the base. As a result, the formula can be simplified to $SA = B + \frac{1}{2}Pl$, where B is the area of the base and P is the perimeter of the base. This substitution works whenever the base is a regular polygon, leading us to the following theorem.

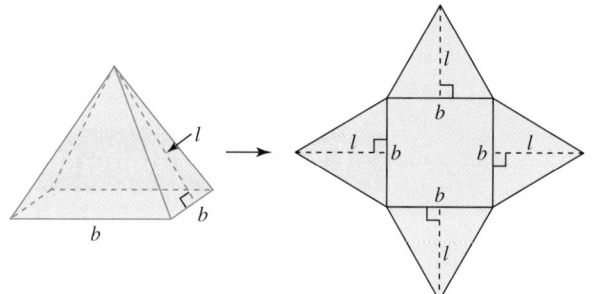

Figure 13.31 **The net of a right square pyramid**

Theorem 13.13   Surface Area Formula for a Right Regular Pyramid

The surface area of a right regular pyramid with a slant height l and a base with area B and perimeter P is $SA = B + \frac{1}{2}Pl$.

Example 13.25   The Surface Area of a Pyramid

Compute the surface area of the following pyramid:

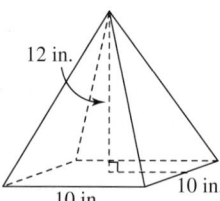

Solution The diagram of the pyramid shows every dimension needed except one, the slant height. However, recognizing that the slant height is the hypotenuse of a right triangle, we can use the Pythagorean Theorem to find it. The height of the pyramid is one leg of the triangle, and the distance from the center of the base to the edge is the other. As a result, the slant height is $l = \sqrt{12^2 + 5^2} = \sqrt{169} = 13$ in.

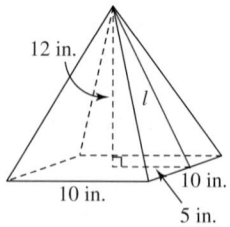

We can now find the surface area of the pyramid by evaluating Theorem 13.13 with the relevant information, or $SA = B + \frac{1}{2}Pl = (10 \cdot 10) + \frac{1}{2}[(4 \cdot 10) \cdot 13] = 100 + 260 = 360$ in.^{2}.

Surface Area of a Right Circular Cone

The surface area of a right circular cone is somewhat more difficult to find. As with other shapes considered so far, we first look at the net of a cone, which consists of two pieces (Figure 13.32). The base is a circle, which presents no computational problem. However, the lateral surface is a sector of a circle with an area we cannot find because the central angle is unknown.

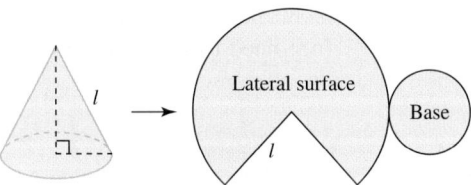

Figure 13.32 The net of a cone

Fortunately, we can get around the problem by using another approach. Recall that, as we add more sides to a regular polygon, its area closely approximates the area of a circle. Next, imagine a sequence of right regular pyramids in which the bases have an increasing number of sides. As the number of sides increases, not only does the base become more circular, but the lateral surfaces of the pyramids begin to closely approximate the lateral surface of the cone (Figure 13.33).

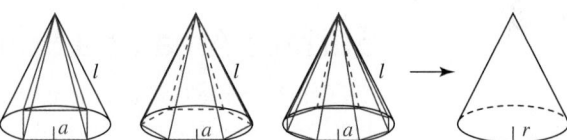

Figure 13.33 A sequence of right regular pyramids leading to a cone

As a result, if we view a cone as a pyramid with a base that has infinitely many sides, we can use the surface area of a pyramid to find the surface area of a cone, or

$$SA = B + \frac{1}{2}Pl = \pi r^2 + \frac{1}{2}(2\pi r)l = \pi r^2 + \pi r l$$

where r is the radius of the base and l is the slant height. If the slant height is unknown, we can use the Pythagorean Theorem to find it because the radius, the height, and the slant height of the cone form the sides of a right triangle (Figure 13.34). Specifically, $l = \sqrt{r^2 + h^2}$, where h is height of the cone, and we can rewrite the formula as $SA = \pi r^2 + \pi r \sqrt{h^2 + r^2}$.

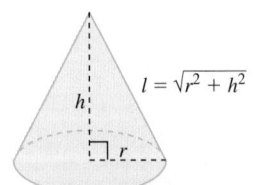

Figure 13.34 The slant height of a cone

Theorem 13.14 Surface Area Formula for a Right Circular Cone

The surface area of a right circular cone with a slant height l, a height h, and a radius r is $SA = \pi r^2 + \pi r l = \pi r^2 + \pi r \sqrt{h^2 + r^2}$.

Example 13.26 | Paint on a Construction Cone

Application

A construction cone is painted with two colors. The top is painted with a white reflective paint, and the rest of the cone is painted orange. If the paint on the cone matches the dimensions shown, how many square inches of each color are used?

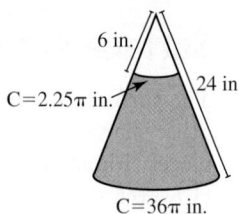

6 in.

$C = 2.25\pi$ in.

24 in.

$C = 36\pi$ in.

Solution We begin with the small white cone. The diagram shows that the cone has a slant height of 6 in. and that the circumference of the base is 2.25π in. Because we do not need to worry about the base, the lateral surface area is

$$\frac{1}{2}Cl = \frac{1}{2}(2.25\pi)(6) = 6.75\pi \approx 21.2 \text{ in.}^2$$

For the orange paint, again we consider only the lateral surface area and not the base. Using the information in the diagram, we first compute the lateral surface area of the entire cone, or

$$\frac{1}{2}Cl = \frac{1}{2}(36\pi)(24) = 432\pi \approx 1,356.5 \text{ in.}^2$$

From this amount, we subtract the part of the cone that is painted white. As a result, the amount of surface area that is painted orange is $SA = 1,356.5 - 21.2 = 1,335.3$ in.^{2}.

Surface Area of a Sphere

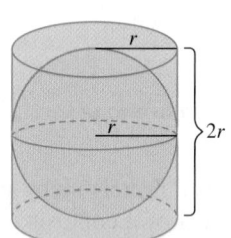

Figure 13.35 A sphere contained in a cylinder

The surface area of a sphere is more difficult to determine using only elementary mathematics. Historically, it was Archimedes who discovered that the surface area of a sphere is exactly two-thirds the surface area of the smallest cylinder that contains it. In this case, the cylinder and the sphere have the same radius r, and the cylinder has a height of $2r$ (Figure 13.35). Using these facts, the surface area of the sphere is

$$SA = \frac{2}{3}(\text{surface area of cylinder})$$

$$= \frac{2}{3}(2\pi r^2 + 2\pi r \cdot h)$$

$$= \frac{2}{3}[2\pi r^2 + 2\pi r \cdot (2r)]$$

$$= \frac{2}{3}(6\pi r^2)$$

$$= 4\pi r^2$$

Theorem 13.15 | Surface Area Formula for a Sphere

The surface area of a sphere with radius r is $SA = 4\pi r^2$.

Example 13.27   Finding the Radius from the Surface Area of a Sphere

If a sphere has a surface area of $1{,}444\pi$ cm^2, what is its radius?

Solution To find the radius, we set $4\pi r^2 = 1444\pi$ and solve for r, or

$$4\pi r^2 = 1444\pi$$
$$r^2 = 361$$
$$r = \sqrt{361} = 19 \text{ cm}$$

Interestingly, the formula for the surface area of a sphere is exactly four times the area of the circle with the same radius. This circle is called a **great circle** of the sphere (Figure 13.36). Any sphere has an infinite number of great circles, any one of which can be found by the circular cross section made by a plane cutting through the sphere at its center.

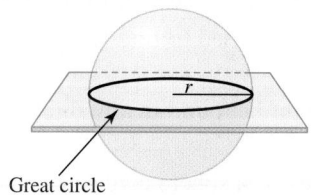

Great circle

Figure 13.36 A great circle of a sphere

Check Your Understanding 13.3-B

1. A right triangular pyramid has an equilateral triangle for its base with sides of length 4 in. If the lateral height is 5 in., what is the surface area of the pyramid?

2. If the height and radius of a cone are both 3 m, what is its surface area?

3. Find the surface area of the following sphere. Then find the circumference and area of its great circle.

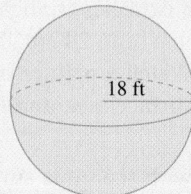

18 ft

Talk About It How can you generalize the surface area formula for a right circular cone to a surface area formula for any cone?

Reasoning

Activity 13.6
Isometric dot paper has dots that are arranged as shown:

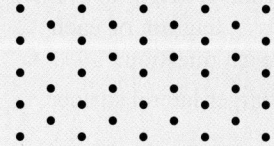

Use isometric dot paper to draw a cube with a surface area of:

a. 6 square units.   **b.** 96 square units.

SECTION **13.3** Learning Assessment

▪ Understanding the Concepts

1. What is surface area, and why are square units needed to measure it?

2. What role does a net play in finding the surface area of a spatial figure?

3. In general, the lateral surface area of any prism can be found by multiplying what two measures?

4. How is the surface area of a right cylinder connected to the area of a rectangle?

5. How is the slant height of a pyramid different from the height of a pyramid?

6. **a.** What is the net of the lateral surface of a cone?

   b. Why can this net not be used to find the surface area of a cone?

7. How can the slant height be found in any right circular cone?

8. How is the surface area of a sphere connected to the great circle of the sphere?

▪ Representing the Mathematics

9. State the figure associated with each surface area formula, and identify what each variable represents.

   a. $SA = 6e^2$. **b.** $SA = 2\pi r(r + h)$.

   c. $SA = 2(lw + hw + hl)$. **d.** $SA = B + \frac{1}{2}Pl$.

   e. $SA = 4\pi r^2$. **f.** $SA = \pi r^2 + \pi rl$.

10. The base of a right prism is an isosceles triangle with sides of length 26 cm, 26 cm, and 24 cm. The height of the prism is 10 cm. Sketch a picture of the prism, and find its lateral surface area and total surface area.

11. Draw two rectangular prisms with the same surface area but different dimensions.

12. The base of a right pyramid is a square with sides of length 12 cm. The height of the pyramid is 8 cm. Sketch a picture of the pyramid, and find its lateral surface area and total surface area.

13. One cylinder has a height of 4 cm and a radius of 3 cm, and another cylinder has a height of 3 cm and a radius of 4 cm. Draw a diagram of each cylinder, and then answer each question.

    a. Which cylinder has the larger lateral surface area?

    b. Which cylinder has the larger total surface area?

14. One cone has a height of 5 cm and a radius of 2 cm, and another cone has a height of 2 cm and a

radius of 5 cm. Draw a diagram of each cone, and then answer each question.

a. Which cone has the larger lateral surface area?

b. Which cone has the larger total surface area?

▪ Exercises

15. Find the surface area of each prism.

    a. A rectangular prism with $l = 5$ in., $w = 3$ in., and $h = 8$ in.

    b. A cube with $e = 3$ cm

    c. A triangular prism with $B = 96$ in.^{2}, $P = 48$ in., and $h = 5$ in.

16. Find the surface area of each prism.

    a. A rectangular prism with $l = 3.3$ in., $w = 4.9$ in., and $h = 7$ in.

    b. A cube with $e = 4.5$ cm

    c. A hexagonal prism with $B = 166.28$ cm^2, $P = 48$ cm, and $h = 7$ cm

17. Find the surface area of each prism.

    a.

    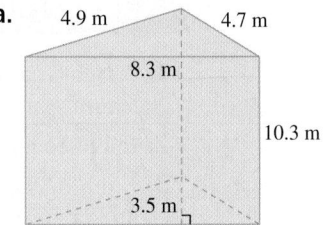

    b.

    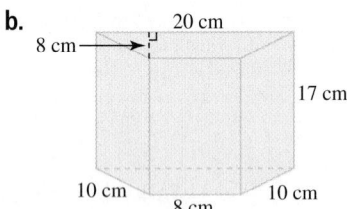

18. Find the surface area of each cylinder with the given height and radius.

    a. $r = 11$ ft and $h = 9$ ft.

    b. $r = 9.8$ cm and $h = 6.7$ cm.

19. Find the surface area of each cylinder.

20. Find the surface area of each figure.

    a. A right square pyramid with $s = 8$ cm and $h = 3$ cm

    b. A right circular cone with $r = 5$ ft and $h = 12$ ft

    c. A sphere with $r = 6$ cm

21. Find the surface area of each figure.

    a. A right square pyramid with $s = 5.5$ in. and $h = 4.1$ in.

    b. A right circular cone with $r = 7.4$ ft and $h = 9.1$ ft

    c. A sphere with $r = 4.5$ cm

22. Find the surface area of each figure.

    a.

    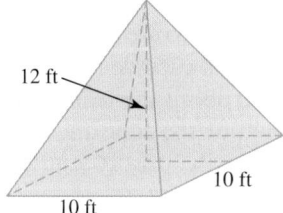

    b.

    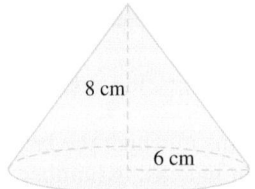

    c.

    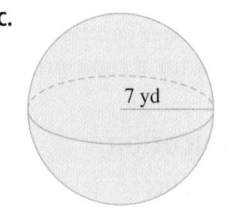

23. a. What is the height of a rectangular prism if $SA = 126$ m^2, $l = 5$ m, and $w = 3$ m?

    b. What is the length of the edge of a cube with $SA = 403.44$ cm^2?

24. What is the lateral surface area of a right pentagonal prism with a height of 12 cm and base edges of 3.7 cm, 7.1 cm, 7.3 cm, 4.3 cm, and 9.4 cm?

25. What is the radius of a cylinder with a height of 3 yd and a lateral surface area of 27π yd^2?

26. What are the dimensions of the base of a right square pyramid with a lateral surface area of $SA = 68$ in.^{2} and a slant height of $l = 8.5$ in.?

27. What is the lateral surface area of a cone that is made by connecting the two edges of the sector shown?

    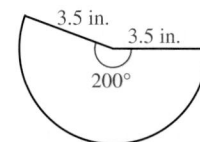

28. What is the radius of a sphere with a surface area of 324π mm^2?

29. What is the area of the great circle of the sphere with a surface area of 36π ft^2?

30. What is the surface area of the sphere that has a great circle with an area of 15π in.^{2}?

Problems and Applications

31. A rectangular sheet of paper has a length of 14 cm and a width of 8 cm. If two opposite sides are taped to form a cylinder, what is its lateral surface area?

32. If a sphere has a radius of 12 cm, what is the surface area of the smallest cylinder that contains it? Assume the bases of the cylinder are included.

33. What is the surface area of the shape if the bases are semicircles?

    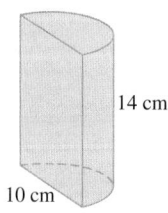

34. How much paint is needed to cover the model rocket of the dimensions given?

    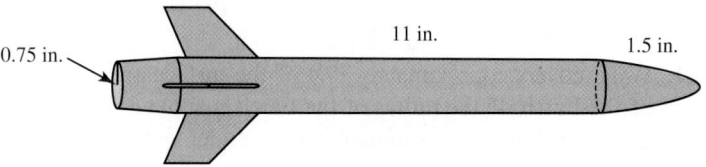

35. Find the surface area of the hollow box shown:

    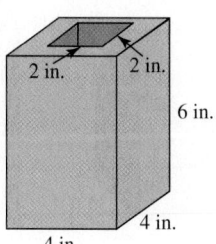

36. The net of a right regular hexagonal prism is shown:

    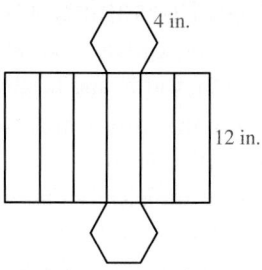

    What is its surface area?

37. The net of a right square pyramid is shown:

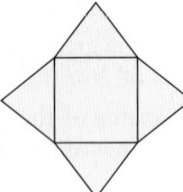

If every edge is 6 in. long, what is its surface area?

38. What is the surface area of a tetrahedron if each edge is 10 cm long? What about an octahedron with edges of length 10 cm?

39. Consider the lateral surface area of a cone shown:

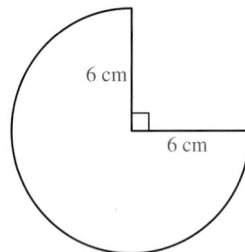

6 cm

6 cm

Use what you see to find the radius of the base and the height of the cone.

40. Tamiru plans to put two coats of paint on a room that measures 15 ft by 20 ft and has walls that are 10 ft high. If 1 gal of paint covers about 400 ft² and costs $22.50, about how much will the paint cost for the room?

41. Water covers approximately 70% of the surface area of the Earth. If the radius of the Earth is about 3,850 mi, about how many square miles are under water?

42. A barn and its silo are shown:

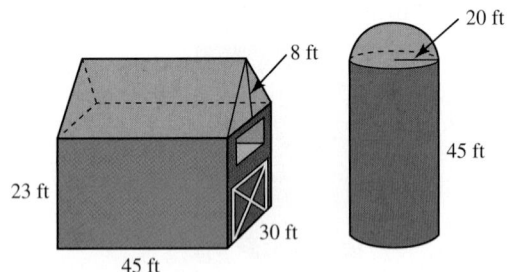

20 ft

8 ft

45 ft

23 ft

30 ft

45 ft

What is the surface area of each?

■ Mathematical Reasoning and Proof

43. What happens to the surface area of a rectangular prism if the length, width, and height are all doubled?

44. What happens to the surface area of a cube if the edge is reduced by a third?

45. **a.** Which increases the lateral surface area of a cylinder more, doubling the height or doubling the radius?

b. Which increases the total surface area of a cylinder more, doubling the height or doubling the radius?

46. If two right circular cones are similar with a scale factor of 3, is the surface area of one three times the surface area of the other? Explain.

47. If the radius of a sphere is tripled, what happens to its surface area?

48. Two arrangements of eight ice cubes are shown:

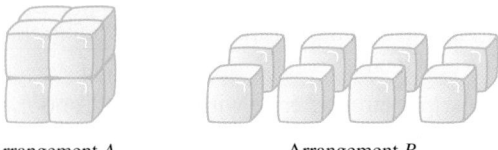

Arrangement A   Arrangement B

In which arrangement will the ice melt faster? Explain your reasoning.

49. Construct and cut out two or three semicircles of different radii. If each semicircle is rolled into a cone, what can be said about the relationship between the slant height of the cone and the diameter of the circular base? Justify your answer.

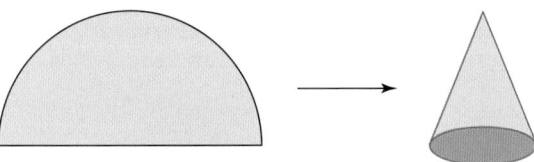

50. Draw a sequence of prisms in which the bases are regular polygons with an increasing number of sides. Use your diagram to help justify how the surface area formula for a cylinder can be derived from the surface area formula for a right prism.

■ Mathematical Communication

51. Use the net of the following figure to demonstrate to someone how to find its surface area. Assume all measurements are given in centimeters.

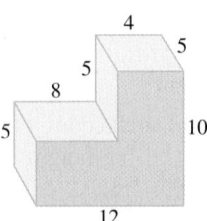

4

5

5

8

5

10

12

52. Alexa claims that if she cuts the length, width, and height of a cardboard box in half, she will have half of the cardboard box left. Is she correct?

53. When Caroline draws the net of a cone, she draws a triangle for the lateral surface area. As her teacher, how would you help her understand what this shape should be?

54. What difficulties do you think students might encounter in learning surface area? Discuss them with a group of your peers, and make suggestions as to what steps you might take to clarify the differences.

■ Building Connections

55. Make a list of three common uses of the surface area of each figure.

    a. Rectangular prism   **b.** Cylinder

    c. Cone   **d.** Sphere

56. How might you use the label off a soup can to explain the surface area of a cylinder?

57. Look at several sets of curriculum materials for several grade levels. Compare and contrast how they approach the topics of surface area. Which do you think is the best approach, and why?

SECTION 13.4   Volume

Surface area is one of two commonly measured attributes in spatial or three-dimensional shapes. The second attribute is volume.

Understanding Volume

As with other attributes discussed so far, we make a number of measurements of volume in our daily lives. We measure volume when we fill up the gas tank, buy drinks from a soda fountain, or order gardening materials such as dirt or mulch.

    When making these measures, we often incorrectly use volume as a synonym for capacity, even though they represent different, yet connected, attributes. **Volume** is the amount of material, either liquid or solid, that is needed to make or fill a three-dimensional shape. **Capacity,** on the other hand, is the amount of space a figure can hold. Consider a cup of coffee (Figure 13.37). As a container, the cup has a capacity, which is the amount of liquid it can hold. The coffee has a volume, which is the amount of liquid actually in the cup.

    Figure 13.37 shows that volume and capacity are both measures of space, so we must measure them with a three-dimensional unit that has a length, a width, and a depth. Like area, we can use any shape that tessellates space as a basic unit of volume. However, we prefer cubes because they are the easiest to use. A **cubic unit** is the amount of space contained inside a cube that measures one unit on each edge. Any measure of volume is the number cubic units needed to build a three-dimensional shape. Any measure of capacity is the number of cubic units needed to fill in the shape. For instance, the box in Figure 13.38 has a capacity of 16 cubic units because the volume of blocks needed to fill it is 16 cubic units.

Figure 13.37 Volume versus capacity

Explorations Manual 13.12

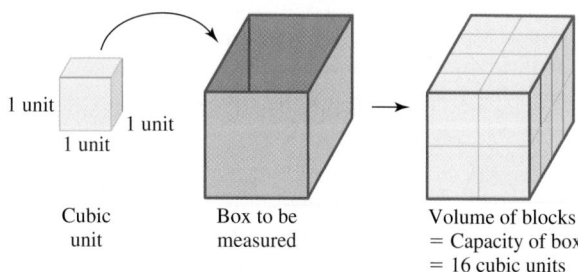

Figure 13.38 Measuring the capacity of a box

    As with other units of measure, cubic units come in different sizes, many of which are based on English or metric length units. For instance, in the English system, a cubic inch ($in.^3$) is a cube that measures 1 in. on each edge, and a cubic yard (yd^3) is a cube that measures 1 yd on each edge. This means that the size of a cubic unit is directly related to the size of the length unit. Larger length units lead to larger cubic units. Smaller length units lead to smaller cubic units. As a consequence, conversion ratios for length units can be used to create conversion ratios for cubic units. For instance, a cubic yard is a cube that measures 1 yd, or 3 ft, on each edge. As Figure 13.39 shows, a cubic yard can be filled with exactly with 3 layers of 3 rows of $3 \ ft^3$ each, so $1 \ yd^3 = 3 \ ft \times 3 \ ft \times 3 \ ft = 27 \ ft^3$.

3 ft

3 ft

3 ft

1 cubic ft
(1 ft³)

1 cubic yard
(1 yd³)

Figure 13.39 $1 \text{ yd}^3 = 3 \text{ ft} \times 3 \text{ ft} \times 3 \text{ ft} = 27 \text{ ft}^3$

Consequently, conversion ratios between cubic units are equal to the cube of the ratio between the associated length units. This is true for any two cubic units, so we use this fact to generate the conversion factors for the English units shown in Table 13.7. Table 13.7 also gives several other volume units that are divided into two categories. **Dry measures,** such as cubic inches, are used to measure dry commodities such as dirt, fruits, and grains. **Liquid measures,** such as tablespoons, cups, and gallons, are used to measure liquids.

Table 13.7 Volume units in the English system	
Dry Measure	**Liquid Measure**
1 cubic inch (in.³) = basic unit	1 teaspoon (t) = basic unit
1 cubic foot (ft³) = 1,728 in.³	1 tablespoon (T) = 3 t
1 cubic yard (yd³) = 27 ft³	1 cup (c) = 16 T
1 dry qt = 2 dry pt	1 pint (pt) = 2 c
1 peck (pk) = 8 dry qt	1 quart (qt) = 2 pt
1 bushel (bu) = 4 pk	1 gallon (gal) = 4 qt
1 cord = 128 ft³	1 gal = 231 in.³

Example 13.28 Conversions Between Volume Units

Use dimensional analysis to make each conversion.

a. 13 yd³ to in.³

b. 2.5 gal to c

Solution Using the conversion ratios given in Table 13.7, we have

a. $13 \text{ yd}^3 \times \dfrac{27 \text{ ft}^3}{1 \text{ yd}^3} \times \dfrac{1,728 \text{ in.}^3}{1 \text{ ft}^3} = 606,528 \text{ in.}^3$.

b. $2.5 \text{ gal} \times \dfrac{4 \text{ qt}}{1 \text{ gal}} \times \dfrac{2 \text{ pt}}{1 \text{ qt}} \times \dfrac{2 \text{ c}}{1 \text{ pt}} = 40 \text{ c}$.

The metric system also has units for dry and liquid measures. Common units for dry measures are the cubic centimeter (1 cm³) and the cubic meter (1 m³). Again, the conversion ratios for these measures are made by cubing the corresponding length ratios. With liquid measures, the basic unit is the **liter.** A liter (1 L) is defined as the volume of a cube that measures 10 cm, or 1 dm, on each edge (1 L = 1 dm³) (Figure 13.40).

Applying metric prefixes to a liter results in other units, such as centiliters, milliliters, and even kiloliters. Because 1,000 cm³ = 1 dm³ and 1,000 mL = 1 L, we have that $1 \text{ cm}^3 = \dfrac{1}{1,000} \text{ dm}^3 = \dfrac{1}{1,000} \text{ L} = 1 \text{ mL}$. Relationships between dry measures and

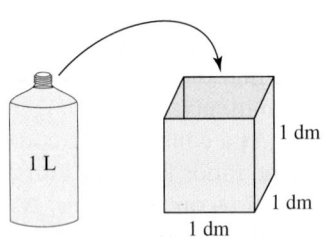

1 L

1 dm

1 dm

1 dm

Figure 13.40 1 liter = 1 dm³

liquid measures hold true for other metric units as well. This is one reason why the metric system is often preferred over the English system. Table 13.8 summarizes the metric volume units and their respective conversion ratios.

Table 13.8 Volume units in the metric system	
Dry Measure	**Liquid Measure**
$1\ mm^3 = 0.000000001\ m^3$	1 milliliter (mL) = 0.001 L
$1\ cm^3 = 0.000001\ m^3$	1 centiliter (cL) = 0.01 L
$1\ dm^3 = 0.01\ m^3$	1 deciliter (dL) = 0.1 L
$1\ m^3$ = basic unit	1 liter (L) = basic unit
$1\ dam^3 = 1,000\ m^3$	1 decaliter (daL) = 10 L
$1\ hm^3 = 1,000,000\ m^3$	1 hectoliter (hL) = 100 L
$1\ km^3 = 1,000,000,000\ m^3$	1 kiloliter (kL) = 1,000 L

Example 13.29   Conversions Between Units in the Metric System

Use dimensional analysis to make each conversion.

a. 200,000 mm^3 to dam^3 **b.** 130 kL to m^3

Solution

a. $2 \times 10^5\ mm^3 \times \dfrac{1\ m^3}{1 \times 10^9\ mm^3} \times \dfrac{1\ dam^3}{1 \times 10^3\ m^3} = 2 \times 10^{-7}\ dam^3.$

b. $130\ kL \times \dfrac{1,000\ L}{1\ kL} \times \dfrac{1\ dm^3}{1\ L} \times \dfrac{1\ m^3}{1,000\ dm^3} = 130\ m^3.$

Volume of a Prism

Explorations Manual 13.13

Finding the volume of a right rectangular prism comes down to counting how many cubic units are needed to fill the prism. For instance, consider a box with dimensions $4 \times 3 \times 2$ ft (Figure 13.41). To fill it, we first make a layer of cubic feet 1 ft deep along the bottom. This requires $4 \cdot 3 = 12\ ft^3$. Next, we add a second layer of $4 \cdot 3 = 12\ ft^3$ to fill the box. Because we needed $2 \cdot (4 \cdot 3) = 24\ ft^3$ to fill it, the box has a volume of $24\ ft^3$.

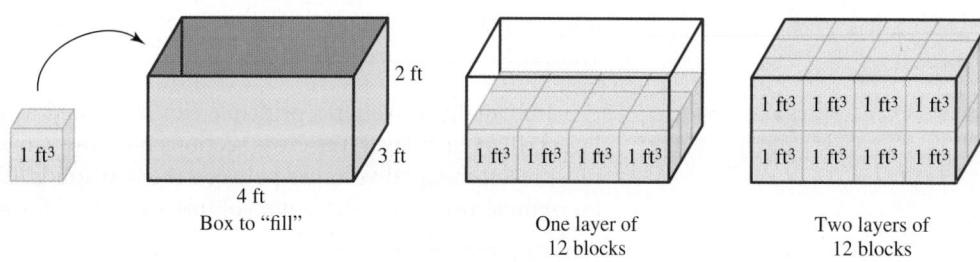

Figure 13.41 **Measuring the volume of a prism**

This example shows that the volume, V, of a right rectangular prism is the product of the length l, the width w, and the height h, or $V = lwh$. Because a cube is a right rectangular prism in which $l = w = h = e$, then its volume is $V = lwh = e \cdot e \cdot e = e^3$. Further notice that the product of the length and the width is equal to the area of the base. So we can also find the volume by multiplying the area of the base by its height, or $V = Bh$. Writing the formula in this way allows us to extend the volume formula to other right prisms.

Theorem 13.16 Volume Formula for Right Prisms

The volume for a right rectangular prism with edges of lengths l, w, and h is $V = lwh$.

The volume for a cube with edges of length e is $V = e^3$.

The volume of a right prism with a height h and a base of area B is $V = Bh$.

Example 13.30 The Volume of Right Prisms

Find the volume of each prism.

a.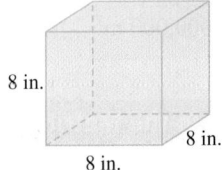

8 in.

8 in.

8 in.

b.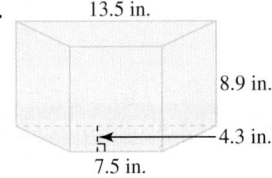

13.5 in.

8.9 in.

4.3 in.

7.5 in.

Solution

a. Because $e = 8$ in., $V = e^3 = (8\text{ in.})^3 = 512\text{ in.}^3$.

b. The prism has a trapezoid for its base, so we use the appropriate area formula to compute its area. We then multiply the area by the height to the find the volume, or

$$V = Bh = \left(\frac{1}{2}(b_1 + b_2)h_{\text{trapezoid}}\right)h_{\text{prism}}$$

$$= \left(\frac{1}{2} \cdot ((7.5) + (13.5)) \cdot (4.3)\right) \cdot (8.9)$$

$$= 401.835\text{ in.}^3$$

To extend the volume formulas to oblique prisms, we use a fact known as Cavalieri's principle.

Theorem 13.17 Cavalieri's Principle

Two solids, each having their bases in the same parallel planes, have the same volume if every plane parallel to the bases intersects the solids in cross sections of equal area.

Informally, Cavalieri's principle can be viewed as slicing the solids into very thin layers. If, at each layer, the cross sections have the same area, then, together, the layers must result in equal volumes between the two solids. Cavalieri's principle implies that an oblique prism has the same volume as that of a right prism with the same height and base area (Figure 13.42).

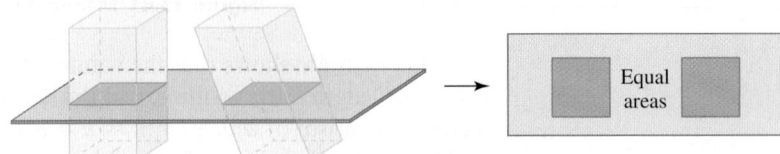

Equal areas

Figure 13.42 Cavalieri's principle

Example 13.31 | **Finding the Volume of an Oblique Prism**

Find the volume of the oblique prism.

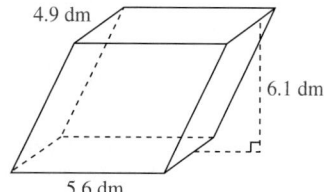

4.9 dm

6.1 dm

5.6 dm

Solution Cavalieri's principle allows us to compute the volume of the oblique prism as if it were a right prism with the same base and height. Its volume is $V = Bh = (4.9)(5.6)(6.1) = 167.384$ dm³.

Check Your Understanding 13.4-A

1. Use a picture to illustrate the number of cubic inches in a cubic foot.

2. Use dimensional analysis to make each conversion.

   a. 5 ft³ to in.³

   b. 40,000 cm³ to m³

   c. 8.3 yd³ to in.³

3. 4,500 cm³ is equivalent to how many liters?

4. Find the volume of a prism that has a square base with sides of length 3 in. and a height of 8 in.

Talk About It How are length units, area units, and volume units for both dry and liquid measures all related to one another in the metric system?

Problem Solving

Activity 13.7

How much cement was needed to make the following cinder block?

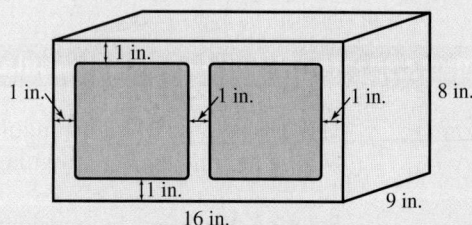

1 in.

1 in.

1 in.

1 in.

8 in.

1 in.

1 in.

16 in.

9 in.

Volume of a Cylinder

The volume of a right prism can be used to derive formulas for other spatial shapes. For instance, consider a sequence of right prisms in which the bases are regular polygons with an increasing number of sides. As the number of sides increases, the volumes of the prisms closely approximate the volume of the cylinder (Figure 13.43). So, if a cylinder is viewed as a prism with a circular base, the volume of the cylinder can be computed as $V = Bh = \pi r^2 h$. Using Cavalieri's principle, we can extend this formula to include oblique cylinders.

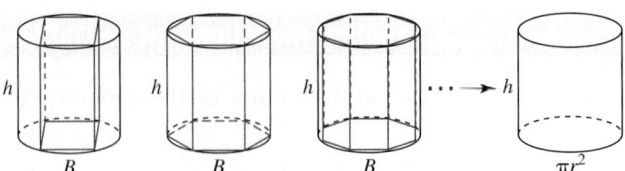

Figure 13.43 A sequence of right prisms leading to a cylinder

---

Theorem 13.18   Volume Formula for a Circular Cylinder

The volume of a circular cylinder with a height h and a radius r is $V = \pi r^2 h$.

Example 13.32   Volume of an Oil Tank

Application

A cylindrical oil storage tank has a radius of 50 ft and a height of 35 ft. If one barrel of petroleum is equal to 42 gal, how many barrels (bbl) of oil does the tank hold?

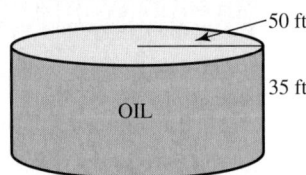

Solution We first compute the volume of the tank in cubic feet, or $V = \pi r^2 h = \pi (50)^2 (35) = 87{,}500\pi \text{ ft}^3 \approx 274{,}750 \text{ ft}^3$. Next, we convert this value to barrels by using dimensional analysis and the conversion ratios given in Table 13.7.

$$274{,}750 \text{ ft}^3 \times \frac{1{,}728 \text{ in.}^3}{1 \text{ ft}^3} \times \frac{1 \text{ gal}}{231 \text{ in.}^3} \times \frac{1 \text{ bbl}}{42 \text{ gal}} \approx 48{,}935.06 \text{ bbl.}$$

Volume of a Pyramid

We can also use the volume of a rectangular prism to compute the volume of a rectangular pyramid. The next example demonstrates how to do so using Cavalieri's principle and spatial reasoning.

Example 13.33   The Volume of a Rectangular Pyramid

Reasoning

Use the volume of a rectangular prism and Cavalieri's principle to find the formula for the volume of a rectangular pyramid.

Solution We begin by considering the volume of a square pyramid enclosed in the smallest cube that can contain it. If we shift the apex of the pyramid to one of the upper vertices of the cube, then the base and the height of the pyramid remain the same, as do the cross-sectional areas of the pyramid. Consequently, Cavalieri's principle guarantees that the shifted pyramid has the same volume as the original.

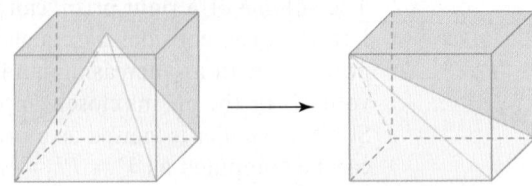

Yellow pyramids have the same volume.

Also, notice that the sides of the shifted pyramid are diagonals of the cube. Drawing the one remaining diagonal from the vertex, we get the following three pyramids.

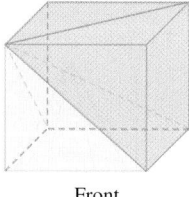 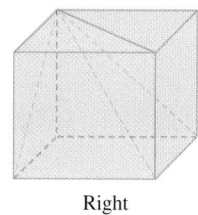 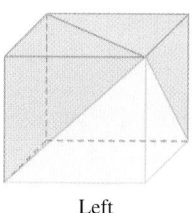

Front   Right   Left

If the pyramids are separated and rotated, we can see that they will be congruent because they have congruent bases and heights. As a result, they must have the same volume.

 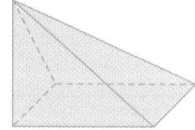 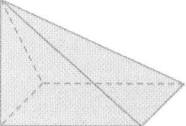

Because the combined volumes of the three pyramids are equal to the volume of the cube, then the volume of the original pyramid must be one-third the volume of the cube, or $V = \frac{1}{3}Bh$.

The relationship demonstrated in the last example holds true for any pyramid with any shape base, and by Cavalieri's principle we can extend it to oblique pyramids.

Theorem 13.19 Volume Formula for a Pyramid

The volume of a pyramid with a height h and with a base of area B is $V = \frac{1}{3}Bh$.

Example 13.34 Volume of the Great Pyramid

Connections

The Great Pyramid at Giza originally rose to a height of 145.75 m, making it the tallest building in the world until the nineteenth century. Its base is a square measuring 229 m on a side. What volume of sandstone was required to build the pyramid?

Solution The Great Pyramid at Giza is almost a perfect right square pyramid with a height of 145.75 m and a base with sides of length 229 m. Hence, the volume of sandstone required to build the pyramid is

$$V = \frac{1}{3}(229 \cdot 229) \cdot (145.75) \approx 2{,}547{,}758.58 \text{ m}^3$$

Volume of a Circular Cone

To find the volume of a cone, we use the same technique used with the cylinder. This time, we consider a sequence of right pyramids in which the bases are regular polygons with an increasing number of sides. As the number of sides increases, the volumes of the pyramids closely approximate the volume of a cone (Figure 13.44). So, if a cone is viewed as a pyramid with a circular base, then the volume of the cone can be computed as $V = \frac{1}{3}Bh = \frac{1}{3}\pi r^2 h$. Again, Cavalieri's principle allows us to extend this formula to oblique cones.

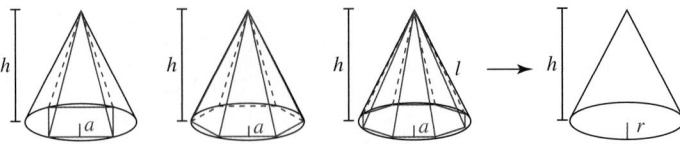

Figure 13.44 A sequence of right pyramids leading to a cone

Theorem 13.20   Volume Formula for a Circular Cone

The volume of a circular cone with height h and radius r is $V = \frac{1}{3}\pi r^2 h$.

Example 13.35   The Volume of a Cone

Find the volume of the cone.

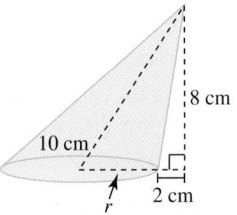

Solution We are given an oblique cone for which we know the height but not the radius. However, the radius r completes one leg of a right triangle, making it possible to use the Pythagorean Theorem to find r:

$$(r + 2)^2 + 8^2 = 10^2$$
$$(r + 2)^2 + 64 = 100$$
$$(r + 2)^2 = 36$$
$$(r + 2) = \sqrt{36}$$
$$r + 2 = 6$$
$$r = 4$$

Given the radius and the height, the volume is $V = \frac{1}{3}\pi r^2 h = \frac{1}{3}\pi (4)^2(8) = \frac{128}{3}\pi$ cm^3.

Volume of a Sphere

When Archimedes discovered that the surface area of a sphere is two-thirds the surface area of the smallest cylinder containing it, he also found the same relationship existed between the volumes. The volume of a sphere is two-thirds the volume of the smallest cylinder containing it (Figure 13.45).

Using this fact and the labeling in Figure 13.45, the volume of a sphere is

$$V = \frac{2}{3}(\pi r^2 h) = \frac{2}{3}(\pi r^2(2r)) = \frac{4}{3}\pi r^3.$$

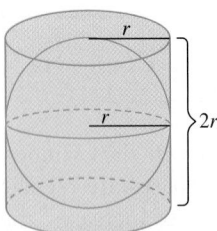

Figure 13.45 A sphere contained in a cylinder

Explorations
Manual
13.14

Theorem 13.21 Volume Formula for a Sphere

The volume of a sphere with radius r is $V = \dfrac{4}{3}\pi r^3$.

Example 13.36 Finding the Radius of a Hemisphere

Application

The dome on top of a building is a hemisphere and has a volume of approximately 56,520 ft³. What is the diameter of the dome?

Solution We are asked to find the diameter of the dome given its volume. We set $\dfrac{1}{2}\left(\dfrac{4}{3}\pi r^3\right)$ equal to 56,520 and solve for r.

$$\frac{1}{2}\left(\frac{4}{3}\pi r^3\right) = 56{,}520$$

$$\pi r^3 = \frac{3}{2}\cdot 56{,}520$$

$$r = \sqrt[3]{\frac{84{,}780}{\pi}} = 30\,\text{ft}$$

We double this number to find the diameter, or $d = 2 \cdot 30$ ft $= 60$ ft.

Check Your Understanding 13.4-B

1. What is the volume of a pyramid with a rectangular base with sides of length 4 in. and 6 in. and a height of 5 in.?

2. Find the volume of a circular cylinder with a radius of 4 m and a height 6 m.

3. A circular cone has a volume of $V = 27.17$ yd³ and a height of 2.7 yd. About how big is its radius?

4. What is the volume of a sphere with a radius of 5.5 cm?

Talk About It Discuss the mathematical connections between the volume formulas for prisms, cylinders, pyramids, and cones.

Problem Solving

Activity 13.8
The wooden bowl has uniform thickness and is in the shape of a hemisphere.

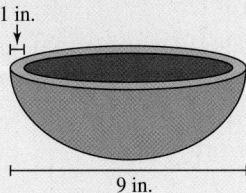

1 in.

9 in.

a. What is the capacity of the bowl?

b. What is the volume of wood needed to make the bowl?

SECTION **13.4** Learning Assessment

■ Understanding the Concepts

1. **a.** What is the difference between volume and capacity?

   b. Why are cubic units needed to measure both volume and capacity?

2. How can we create a cubic conversion ratio from a length conversion ratio?

3. How is a liter defined?

4. **a.** What does Cavalieri's principle say about the volume of two figures?

   b. In general, Cavalieri's principle allows us to find the volume of what kinds of spatial figures?

5. How is the volume of a prism used to find the volume of a cylinder?

6. How is Cavalieri's principle and the volume of a cube used to find the volume of a square pyramid?

7. How is the volume of a sphere connected to the volume of a cylinder?

■ Representing the Mathematics

8. State the figure associated with each volume formula, and then identify what each variable represents.

   a. $V = Bh$.   **b.** $V = \frac{1}{3}\pi r^2 h$.   **c.** $V = \frac{4}{3}\pi r^3$.

   d. $V = lwh$.   **e.** $V = \frac{1}{3}Bh$.   **f.** $V = \pi r^2 h$.

9. Draw a diagram that shows relative size of:

   a. A cubic yard to a cubic foot.

   b. A cubic meter to a cubic decimeter.

10. Draw a diagram that shows the number of cubic units required to make each prism.

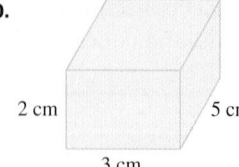

   a. 3 ft, 3 ft, 3 ft

   b. 2 cm, 5 cm, 3 cm

11. Draw a diagram that illustrates how the volume of a rectangular prism changes if its length, width, and height are doubled.

12. Consider a regular sheet of office paper measuring $8\frac{1}{2} \times 11$ in. Which forms a cylinder with larger volume, rolling the paper widthwise or lengthwise?

13. The base of a right prism is a right isosceles triangle with legs of length 12 cm. If the height of the prism is 15 cm, sketch a picture of the prism, and find its volume.

14. One cylinder has a height of 2 cm and a radius of 4 cm, and another has a height of 4 cm and a radius of 2 cm. Draw a diagram of each cylinder, and then determine which has the larger volume.

15. The base of a right pyramid is a rectangle with a length of 13 cm and a width of 9 cm. If the height of the pyramid is 6 cm, sketch a picture of the pyramid, and find its volume.

16. **a.** A paper cup in the shape of a cone is filled with water. The water is dumped into a cylinder with the same base and height. How many cups does it take to fill the cylinder?

   b. If the same cup is used to fill a cylinder that has a base with a radius twice as long and a height twice as high, how many cups does it take to fill the cylinder?

■ Exercises

17. For each question, select the approximate measure that makes the most sense.

   a. The capacity of a cup of coffee

        i. 0.5 in.^{3} **ii.** 0.5 gal   **iii.** 0.5 pt   **iv.** 0.5 L

   b. The capacity of a dump truck

        i. 12 ft^3 **ii.** 12 cm^3 **iii.** 12 m^3 **iv.** 12 km^3

   c. The capacity of a backyard swimming pool

        i. $3,000$ ft^3 **ii.** $3,000$ mm^3

        iii. $3,000$ yd^3 **iv.** $3,000$ dm^3

18. For each question, select the approximate measure that makes the most sense.

   a. The volume of a loaf of bread

        i. $3,000$ cm^3 **ii.** 1.5 yd^3 **iii.** 0.75 m^3 **iv.** 1 cm^3

   b. The volume of this textbook

        i. 300 in.^{3} **ii.** 11.4 ft^3 **iii.** 0.9 yd^3 **iv.** 0.0098 mi^3

   c. The volume of a pencil

        i. 15 mm^3 **ii.** 15 cm^3 **iii.** 15 dm^3 **iv.** 15 in.^{3}

19. Use the length conversion ratios from Section 13.1 to develop volume conversion ratios for the following units.

   a. ft^3 to m^3 **b.** m^3 to yd^3 **c.** in.^{3} to cm^3

20. Use the length conversion ratios from Section 13.1 to develop volume conversion ratios for the following units.

   a. cm^3 to ft^3 **b.** yd^3 to m^3 **c.** mi^3 to km^3

21. How many times larger is a:
    a. cm³ than a mm³?   b. m³ than a cm³?
    c. hm³ than a dm³?

22. How many times smaller is a:
    a. cm³ than a dm³?   b. m³ than a hm³?
    c. cm³ than a hm³?

23. Make each conversion between dry measures.
    a. 45 yd³ to in.³   b. 4,673 in.³ to ft³
    c. 16 cords to yd³   d. 15 m³ to cm³
    e. 74,500 m³ to hm³   f. 4.67 × 10⁶ mm³ to m³

24. Make each conversion between liquid measures.
    a. 25 T to c   b. 16 gal to c
    c. 1,410 gal to yd³   d. 47.3 L to cL
    e. 83 cm³ to dL   f. 83,900 L to m³

25. Use the fact that there are 3.785 L in a gallon to answer the following questions.
    a. How many liters does a 14-gal gas tank hold?
    b. How many gallons are in a 2-L bottle of soda?

26. If 1 in. = 2.54 cm, then how many cubic centimeters are in 16 in.³?

27. Find the volume of each prism.
    a. A rectangular prism with $l = 6$ in., $w = 4$ in., and $h = 7$ in.
    b. A cube with $e = 15$ cm
    c. A triangular prism with $B = 19$ in.² and $h = 7$ in.

28. Find the volume of each prism.
    a. A rectangular prism with $l = 4.9$ in., $w = 3.3$ in., and $h = 8.1$ in.
    b. A cube with $e = 19.4$ cm
    c. A hexagonal prism with $B = 166.28$ cm² and $h = 5.8$ cm

29. Find the volume of each prism.
    a.
    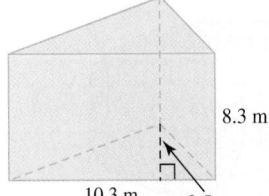
    8.3 m
    10.3 m   3.5 m

    b.
    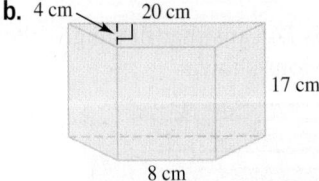
    4 cm   20 cm
    17 cm
    8 cm

30. Find the volume of each cylinder with the given height and radius.
    a. $r = 16$ ft and $h = 5$ ft
    b. $r = 17.1$ cm and $h = 22.3$ cm

31. Find the volume of each cylinder.
    a.
    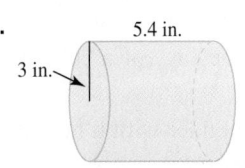
    5.4 in.
    3 in.
    b.
    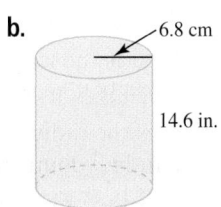
    6.8 cm
    14.6 in.

32. Find the volume of each figure.
    a. A right square pyramid with $s = 8$ cm and $h = 3$ cm
    b. A right circular cone with $r = 5$ ft and $h = 12$ ft
    c. A sphere with $r = 6$ cm

33. Find the volume of each figure.
    a.
    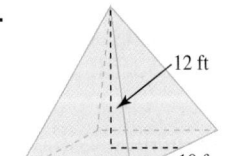
    12 ft
    10 ft
    10 ft
    b.
    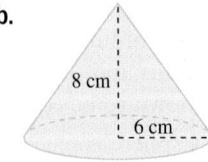
    8 cm
    6 cm
    c.
    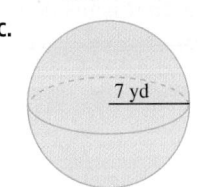
    7 yd

34. Compute the volume of each oblique figure.
    a.
    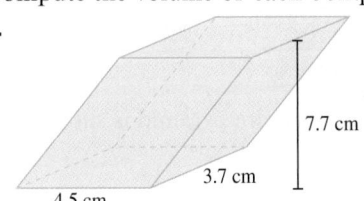
    7.7 cm
    3.7 cm
    4.5 cm
    b.
    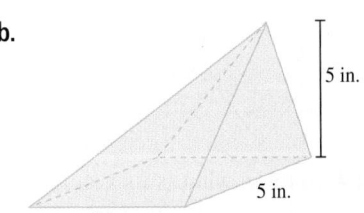
    5 in.
    5 in.
    5 in.
    c.
    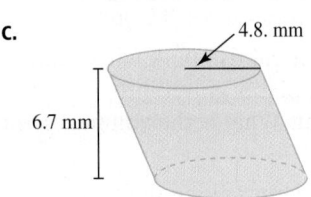
    4.8. mm
    6.7 mm
    d.
    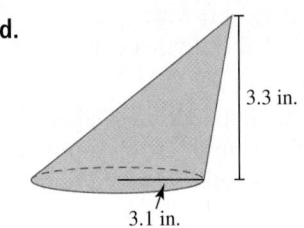
    3.3 in.
    3.1 in.

35. **a.** What is the height of a rectangular prism if $V = 1,260$ m^3, $l = 15$ m, and $w = 6$ m?

    b. What is the length of the edge of a cube if $V = 438.976$ cm^3?

36. What is the radius of a cylinder with a height of 6 yd and a volume of 384π yd^3?

37. What is the radius of a sphere with a volume of 972π cm^3?

■ Problems and Applications

38. A right rectangular prism has a volume of 432 ft^3 and a height of 6 ft. What are possible values for the length and width if they are known to be natural numbers?

39. The volume of a cylinder is 125π ft^3. What are the radius and the height of the cylinder if they are known to be equal?

40. A sphere is placed inside the smallest cylinder that will contain it. If the volume of the sphere is 1.5 L, what is the volume of the space that is inside the cylinder but outside the sphere?

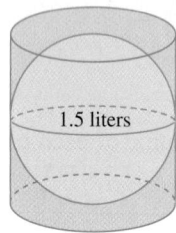

41. The net of a right square pyramid is shown:

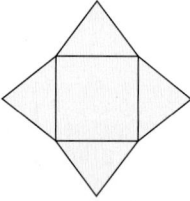

    If every edge is 4 in. long, what is its volume?

42. How many liters are in a right cylindrical tank with a height of 3 m and a radius of 0.75 m?

43. Seville owns a cylindrical fish tank with a radius of 30 cm. When she places gravel into the tank, the water level rises 4 cm. What is the volume of the rocks?

44. Jamie is building a planter along the front edge of her house. It is 16 ft long, 4 ft wide, and 16 in. deep. How many cubic feet of dirt does she need to fill her planter?

45. Water flows over Horseshoe Falls at Niagara Falls at a rate of about 168,000 m^3/s. If this amount of water were poured into a cube, what are dimensions of the cube in meters?

46. Which is a better buy, a block of ice measuring $12 \times 10 \times 6$ in. for \$1.19 or a block measuring $18 \times 12 \times 8$ in. for \$2.09?

47. In 2005, oil consumption in the United States was 21,930,000 barrel/day.

    a. If 1 barrel of petroleum is equal to 42 gal, how many gallons is this?

    b. If this amount of oil were stored in a cylindrical tank with a radius of 50 yd, how tall would the tank need to be? (*Hint:* Use 1 gal = 231 in.^{3}.)

48. How much water does the in-ground swimming pool hold?

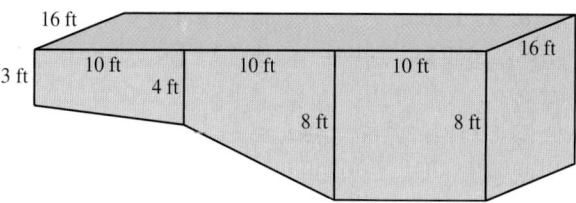

49. A plastic pipe is 10 ft long and has an inner diameter of $1\frac{1}{4}$ in. and an outer diameter of $1\frac{1}{2}$ in.

    a. How much water can be contained inside the pipe?

    b. How much plastic was required to make the pipe?

50. What is the volume of wood in the toy block shown?

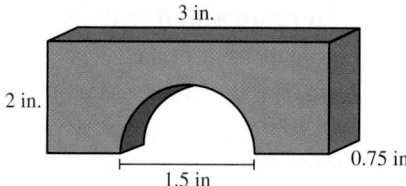

51. Cheese often comes molded in the shape of a cylinder from which wedges are removed and sold. What volume of cheese is left in the figure shown?

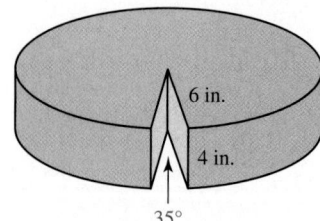

52. A case of soda holds 24 cans that fit snugly into a box with the dimensions shown:

    How much of the box is wasted space?

53. Golf balls are often sold by the dozen and are roughly 1.7 in. in diameter. They are packaged as shown:

a. What are the dimensions of the smallest box possible?

b. What percentage of the volume of the box is taken up by golf balls?

54. The following stand was cut from a square pyramid that was originally 30 in. tall. What is its volume?

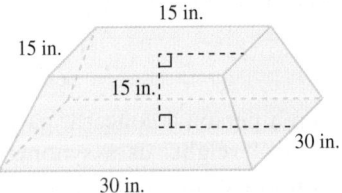

15 in.
15 in.
15 in.
30 in.
30 in.

■ Mathematical Reasoning and Proof

55. The following table compares the edge of a cube to its volume. Based on what you see, make a conjecture about what happens to the volume of a cube when its length is doubled. Do you think your conjecture is always true? If so, prove it using the volume formula for a cube. If not, provide a counterexample.

Length of edge	Volume
1	1
2	8
4	64
8	512

56. What happens to the volume of a rectangular prism if you multiply each dimension by:
   a. 2?   b. 3?   c. 4?   d. 5?   e. n?

57. If a cube is cut in half and the pieces rearranged into prism, does the volume change? Explain your answer.

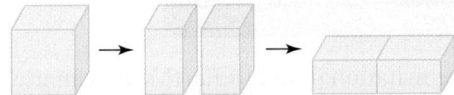

58. Consider the volume of a cylinder. For each question, make a table of values that you can use to make a conjecture. When possible, verify your conjecture using the formula for the volume of a cylinder.

a. What is the effect on the volume of the cylinder if the radius is doubled and the height kept constant?

b. What is the effect on the volume of the cylinder if the radius is kept constant and the height is doubled?

c. Which increases the volume of a cylinder more, doubling its height or doubling its radius? Explain.

d. What is the effect on the volume of the cylinder if the radius and the height are both doubled?

59. Repeat the previous exercise with a cone. Does the same hold true?

60. What happens to the volume of a sphere if its radius is halved?

61. a. We can use many shapes as volume units. What characteristics do such shapes have in common?

b. Other than the cube, can other regular polyhedra be used as volume units?

c. Draw three other shapes that can be used as volume units.

■ Mathematical Communication

62. Find three examples of common objects that can be used to explain the difference between volume and capacity. Select one, and write a paragraph that does so.

63. Write a story problem in which the answer is obtained by finding the volume of a:
   a. Prism.   b. Cylinder.   c. Cone.

64. Joshua is confused about why the dimensions are multiplied and not added when finding the volume of a prism. As his teacher, how would you help him clear up his misconception?

65. Jerry claims that if he doubles the edges of a cube, then the volume doubles as well. How would you correct his mistake?

66. Dianesha is having trouble finding the volume of the following figure:

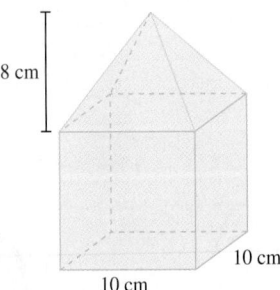

8 cm
10 cm
10 cm

How would you help her get started?

67. What difficulties do you think students might encounter in learning surface area and volume? Discuss them with a group of your peers, and make suggestions as to what steps you might take to help clarify the differences.

■ Building Connections

68. Make a list of five common uses of volume. Then get together with two or three of your peers. How many different uses for volume can you come up with together?

69. Why do you think an exponent of three makes a reasonable choice for representing a cubic unit?

70. The word "volume" has several different meanings.

   a. How many different meanings can you think of?

   b. Do the different meanings have anything in common? If so, what is it?

71. How could you use a stack of pancakes or a stack of books to explain Cavalieri's principle? What other shapes or objects could you used to demonstrate this principle?

72. Look at several sets of curriculum materials for different grade levels. Compare and contrast how they approach volume. Which do you think is the best approach, and why?

SECTION 13.5   Other Measures

In addition to the measurable attributes of geometrical shapes, other attributes are covered in the elementary and middle grades curriculum. Some of the more common ones are weight, temperature, time, speed, and currency.

Weight and Mass

Weight is a measurable attribute that is often a personal concern, particularly as we grow older. We often incorrectly use the term "weight" as a synonym for "mass," even though they represent different, yet connected, attributes. **Mass** is the amount of matter within an object. **Weight,** on the other hand, is the amount of force exerted by gravity on the object. Consider an astronaut who is traveling to the moon. On the moon, the mass of the astronaut—the amount of muscle, bone, skin, etc.—is the same as it is on Earth. However, the astronaut's weight is different because the gravitational pull on the moon is different from that on Earth.

Example 13.37   Weight on Jupiter

Reasoning

The gravitational pull of Jupiter is 2.364 times that of Earth's. If an astronaut weighs 200 lb on Earth, how much would the astronaut weigh on Jupiter?

Damian Peach/NASA

Solution Given the ratio of Jupiter's gravitational pull to Earth's, we can use proportional reasoning to find the astronaut's weight. Using the information in the problem, we get the proportion $\frac{200\,\text{lb}}{x\,\text{lb}} = \frac{1}{2.364}$. By solving for x, we find that the astronaut's weight on Jupiter is $x = 200 \cdot 2.364 = 472.8$ lb.

The distinction between weight and mass is important in scientific discussions but is less so in everyday situations. For instance, when buying a pound of meat, we are mostly concerned with the amount of meat (a measure of mass) rather than the force by which the meat presses down on a table (a measure of weight). We can blur the distinction between the two measures because on Earth, the gravitational pull on an object is directly proportional to its mass. In other words, a mass of x mass units corresponds exactly to a weight of x weight units. In fact, mass units, such as ounces and pounds, are typically used to measure weight. Table 13.9 gives different weight units that are used in the English system and the conversion ratios between them.

Table 13.9 English weight units		
Weight Unit	**Conversion Ratio**	**Example**
1 grain (gr)		Large grain of sand
1 dram (dr)	$1\,dr = 27\frac{11}{32}gr$	$\frac{1}{10}$ the weight of a dollar bill
1 ounce (oz)	1 oz = 16 dr	Six sheets of paper
1 pound (lb)	1 lb = 16 oz	3 average-sized bananas
1 hundredweight (cwt)	1 cwt = 100 lb	Average 12-yr-old male
1 ton (T)	1 ton = 20 cwt = 2,000 lb	Small car

In the metric system, the basic unit of mass is the **gram.** As with other metric units, we apply the metric prefixes to make larger and smaller units. Other metric weights, their conversion ratios, and a common frame of reference are given in Table 13.10.

Table 13.10 Metric weight units		
Mass Unit	**Relationship to Gram**	**Example**
milligram (mg)	1,000 mg = 1 g	A mosquito
centigram (cg)	100 cg = 1 g	A human hair
decigram (dg)	10 dg = 1 g	Half-carat diamond
gram (g)	Basic unit	Paper clip
decagram (dag)	10 g = 1 dag	Two quarters
hectogram (hg)	100 g = 1 hg	Human kidney
kilogram (kg)	1,000 g = 1 kg	Large jar of spaghetti sauce

The metric units that we most commonly use are the milligram, gram, and kilogram. These units are common because they are closely related to other metric units. Specifically, a **gram** is defined as the weight of 1 mL of water, a **kilogram** is defined as the weight of 1 L of water, and a **metric ton** is defined to be the weight of 1 kL of water. Because a liter is defined to be the amount of water in 1 dm^3, metric weight units can also be given in terms of cubic units.

Example 13.38   Weight of Water in a Fish Tank

Application

A fish tank is in the shape of rectangular prism and has dimensions 40 × 70 × 50 cm.

a. About how many liters can the fish tank hold?

b. If the tank itself weighs 8 kg, how much will it weigh when it is full?

Solution

a. The volume of the fish tank is $V = 40 \cdot 70 \cdot 50 = 140{,}000$ cm^3. Converting this number to cubic decimeters and then to liters gives us

$$140{,}000 \; cm^3 \times \frac{1 \; dm^3}{1{,}000 \; cm^3} = 140 \; dm^3 = 140 \; L$$

b. Because each liter of water has a weight of 1 kg and the tank itself weighs 8 kg, then the full tank will weigh 140 + 8 = 148 kg.

Explorations
Manual
13.15
The tools used to measure weight are scales and balances. Some common varieties are shown in Figure 13.46.

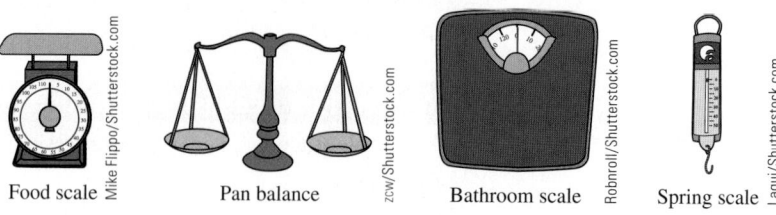

Food scale   Pan balance   Bathroom scale   Spring scale

Figure 13.46 Common types of scales

Temperature

Temperature is another attribute that we measure daily. It measures the degree to which something is warm or cold, with warmer temperatures receiving the larger measure. In everyday situations, two common scales are used to measure temperature. The **Fahrenheit scale,** named after Gabriel Fahrenheit (1686–1736), is based on two reference temperatures: the freezing point of water, assigned a value of 32° F, and the boiling point of water, assigned a value of 212° F. The distance between the two points is divided into 180 equal increments, one of which serves as the basic unit of the scale, or **one degree Fahrenheit (1° F).**

The Fahrenheit scale was widely used throughout Europe until the **Celsius,** or **centigrade, scale** was adopted with the metric system. The Celsius scale also uses the freezing and boiling points of water as reference temperatures, but assigns them values of 0° C and 100° C, respectively. The distance between these temperatures is divided into 100 equal increments, one of which serves as the basic unit of the scale, or **one degree Celsius (1° C).** Figure 13.47 shows a thermometer with a Fahrenheit scale on one side and the Celsius scale on the other.

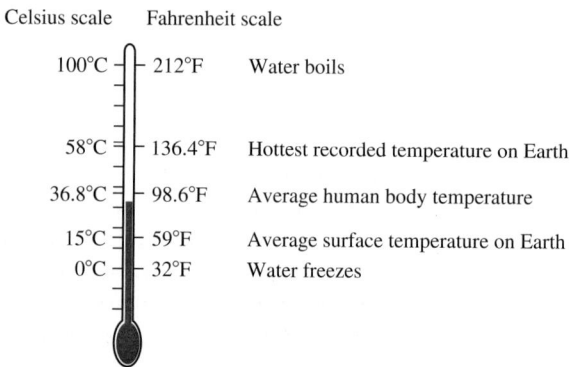

Celsius scale   Fahrenheit scale

100°C	212°F	Water boils
58°C	136.4°F	Hottest recorded temperature on Earth
36.8°C	98.6°F	Average human body temperature
15°C	59°F	Average surface temperature on Earth
0°C	32°F	Water freezes

Figure 13.47 A thermometer showing degrees Fahrenheit and Celsius

Explorations
Manual
13.16
Because the Fahrenheit and the Celsius scales use the same reference points, we can make conversions between the two.

Example 13.39 **Converting Celsius to Fahrenheit**

Problem Solving   Develop a way to convert degrees Celsius to degrees Fahrenheit.

Solution

Understanding the Problem. We are asked to find a way to convert degrees Celsius to degrees Fahrenheit. We are given little information, but we know that the two scales assign different values to the same reference temperatures.

Devising a Plan. Figure 13.47 shows that every temperature in degrees Celsius is associated with a specific temperature in degrees Fahrenheit, indicating that a function relates the two. If it is linear and we can find two points that satisfy the function, then we can *write an equation* to make the conversion.

Carrying Out the Plan. Each scale uses the freezing and boiling temperatures of water as reference points, so we use these points to get started. For the freezing point 0° C corresponds to 32° F, so we assign them the ordered pair (0, 32). Similarly, 100° C corresponds to 212° F for the boiling point, so we assign them the ordered pair (100, 212). Because there is a ratio of 100° C to 180° F between the freezing and boiling points of water, a constant rate of change exists between the two scales. Consequently, we can use a linear function to describe the relationship. We use the points (0, 32) and (100, 212) to find the slope of the line, or

$$m = \frac{212 - 32}{100 - 0} = \frac{180}{100} = \frac{9}{5}$$

To find the y-intercept, substitute point $(C, F) = (0, 32)$ into the equation $F = \frac{9}{5}C + b$ and solve for b. $32 = \frac{9}{5}(0) + b$ implies $b = 32$. As a result, the function between degrees Celsius and degrees Fahrenheit is given by the equation $F = \frac{9}{5}C + 32$.

Looking Back. If correct, the formula gives us a powerful way to connect the two scales. It enables us to take any temperature in degrees Celsius and find its counterpart in degrees Fahrenheit. To test the formula, select a pair of known values, such as the ones for the boiling point of water. Substituting $C = 100°$ into the formula, we have

$$F = \frac{9}{5}C + 32$$

$$F = \frac{9}{5}(100) + 32$$

$$F = 180 + 32$$

$$F = 212 \text{ °F}$$

Temperature plays a significant role in many scientific fields, where an alternative definition of temperature is often used. Scientifically, temperature is related to the amount of energy of the microscopic particles in a system. As more energy is put into a system, the system becomes hotter. Defining temperature in this way allows scientists to describe heating and cooling as adding or removing energy from a system. In turn, they can describe other concepts, such as **absolute zero,** which is the temperature at which all energy has been removed from the system and particles cease to move. Absolute zero, about −273.15° C, is a reference point in the **Kelvin scale** and is given a value of 0° K. Degrees in the Kelvin scale are the same size as degrees Celsius, and they are often used with metric prefixes. For instance, the temperature of the sun can be expressed as 5.8 kiloKelvins, which is about 5,500° C.

Check Your Understanding 13.5-A

1. Make each conversion.

   a. 8 lb to oz   **b.** 36,000 mg to g   **c.** 2,400 lb to T

2. What is each temperature in degrees Fahrenheit?

   a. 67° C   **b.** 15° C   **c.** −8° C

3. The coldest recorded surface temperature on Earth was −128.2° F, taken at Vostok, Antarctica. What is this temperature in degrees Celsius?

Talk About It In what ways do you use measures of weight?

Reasoning

Activity 13.9
Write an algebraic formula that can be used to convert degrees Fahrenheit to degrees Celsius.

Time

Of all the different attributes that we measure in our daily lives, time may be the one we measure most. The word "time" can take on two meanings because telling time is different from measuring time. When telling time, we indicate a particular instant at which an event is to occur. For instance, examples of telling time are "I'll meet you at 12:30" and "I'll be home at 5:00." When measuring time, we find the number of time units equivalent to an interval of time. For instance, a measure of time is indicated by "The test took me forty minutes to complete" or "That homework assignment took me three hours to finish."

Example 13.40 Measuring Time

Communication

Describe the ways a measure of time is used on the student page in Figure 13.48.

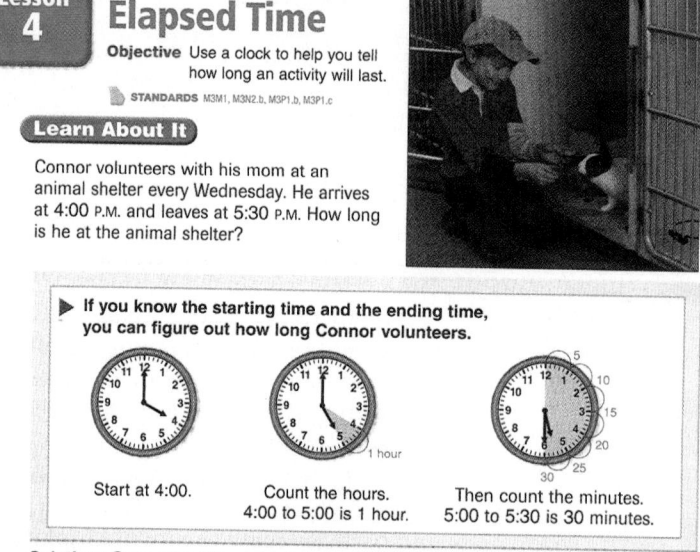

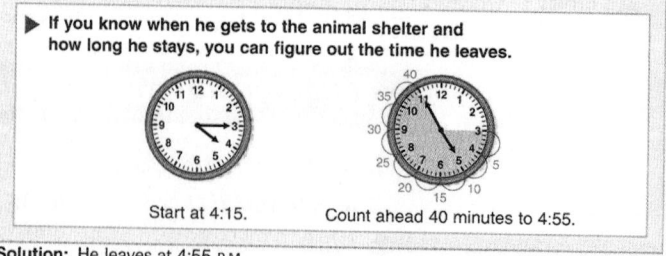

Figure 13.48 Measuring elapsed time
Source: From *Mathematics Grade 3: Houghton Mifflin Mathematics Georgia* (p. 336).
Copyright © by Houghton Mifflin Company, Inc. Reproduced by permission of the publisher, Houghton Mifflin Harcourt Publishing Company.

Solution The student page measures time in two ways. In the first, we compute a measure of time by using the starting and final times of an event. We start at the initial time, and count forward the number of hours and minutes to the final time. In the second, we are given an initial time and a measure of time, and we are asked to find the final time by counting forward the hours and minutes. ■

The basic unit of time in the English and metric systems is the **second.** Other time units are multiples of the second, as summarized in Table 13.11.

Table 13.11 Units of time	
Time Unit	**Conversion Ratio**
Second (s)	Basic unit
Minute (min)	1 min = 60 s
Hour (hr)	1 hr = 60 min
Day	1 day = 24 hr
Week (wk)	1 wk = 7 days
Month	1 month equals 28 to 31 days
Year (yr)	1 year = 12 months = 52 wk
Decade	1 decade = 10 yr
Century	1 century = 100 yr
Millennium	1 millennium = 1,000 yr

Many time units were originally based on the positions of the moon and sun, which lead to inconsistent numbers in the conversion ratios. When the metric system was developed, there was an effort to base time on powers of ten (i.e., 10 hr in a day, 10 days in a week, 10 months in a year). However, complaints caused the idea to be quickly abandoned. Nevertheless, science often uses the metric prefixes with seconds to describe very small or very large increments of time. For instance, a millisecond (ms) is $\frac{1}{1,000}$ of a second and a megasecond (Ms) is 1,000,000 seconds, or roughly 11.6 days.

Although measuring time has always been important, it has not always been exact. Historically, the first measures of time were made by tracking the position of the sun. However, the need to break the day into smaller increments eventually led to inventions like the sundial, the water clock, and the hourglass (Figure 13.49). Not until electrical clocks came along were truly accurate, nonstop clocks developed. Today, clocks can measure time accurately for a thousand years, and we have been able to coordinate the telling of time across the globe.

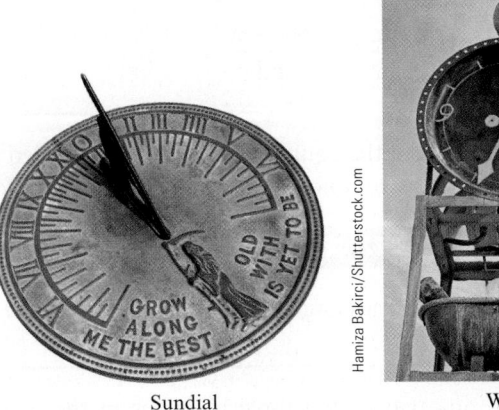  

Sundial   Water Clock   Hour Glass

Figure 13.49 Different kinds of historical clocks

Speed

In our daily lives, we often use the word "speed" to describe the rate at which any task is performed. Because it is a rate, any measure of speed necessarily depends on a measure of time. Although a measure of speed and a measure of time are closely related, the two should not be confused. A measure of speed tells us how fast a task is being performed, whereas a measure of time tells us how long it takes to complete a task. For instance, a secretary typing at a rate of 85 words a minute is a measure of speed. Typing a document in 45 minutes is a measure of time.

Scientifically, speed refers to the rate at which an object moves. Formally, it is defined to be the ratio of the distance traveled to the time traveled, or

$$\text{Speed} = \frac{\text{distance}}{\text{time}}$$

For instance, if it takes a car 1 hr to travel a distance of 50 mi, it travels at a speed of 50 mi/hr, or 50 mph. Notice that a unit of speed is different from others we have considered so far because it is a ratio of two units rather than just a single unit. Specifically, speed units are always given as the *ratio* of a length unit to a time unit, such as meters per second (m/s), kilometers per hour (km/h), or miles per hour (mph). Consequently, to make sense of any measure of speed, we must interpret the measure in terms of the two units. For instance, a person running at a rate of 6 m/s covers a distance of 6 m every second. A boat traveling at 35 km/h covers a distance of 35 km every hour. Table 13.12 gives other examples of the relative size of different speed units.

Table 13.12 Examples of speeds in different units

Example	m/s	km/h	mph
Speed of a snail	0.0468 m/s	0.013 km/h	0.0081 mph
Average walking speed	1–1.5 m/s	3.6–5.4 km/h	3 mph
Typical speed of car	28.9 m/s	104 km/h	65 mph
Speed of sound (Mach 1)	331.5 m/s	1,193.4 km/h	761 mph
Land speed record	341.107 m/s	1,227.985 km/h	763 mph
Speed of light	299,792,458 m/s	1,079,252,848.8 km/h	670,616,629 mph

Example 13.41 Conversions Between Speed Units

Application

A cheetah has a top speed of 70 mph. What is this speed in meters per second?

Solution We can use dimensional analysis to make the conversion between speed units. To do so, we first work with length units and then with time units, being sure to set up each ratio appropriately. As a result, we have

$$70\frac{\text{mi}}{\text{hr}} \times \frac{1.609\text{ km}}{1\text{ mi}} \times \frac{1,000\text{ m}}{1\text{ km}} \times \frac{1\text{ hr}}{60\text{ min}} \times \frac{1\text{ min}}{60\text{ sec}} = 31.29\text{ m/s}$$

If we take the equation used to define speed and multiply both sides by "time," we get what is commonly known as the distance formula.

Theorem 13.22 Distance Formula

If an object is traveling at a rate r for a given time t, then the distance D that the object travels is $D = r \cdot t$.

Given any two of the three variables in the formula, we can use substitution and algebra to find the third.

Example 13.42   **Using the Distance Formula**

Application

If Alisa completes a 20-mi bike ride at a rate of 8 mph, how long did it take for her to make the trip?

Solution We first solve the distance formula for t, by dividing both sides of the equation by r, or $t = \dfrac{D}{r}$. Then, because $D = 20$ mi and $r = 8$ mph, the time it took Alisa to complete the trip is $t = \dfrac{D}{r} = \dfrac{20\ \text{mi}}{8\ \text{mph}} = 2\frac{1}{2}\text{hr.}$

Currency

Currency, or **money,** is defined to be a unit of exchange that assists in the trade of goods and services between people, businesses, and governments. Including currency as a measure may seem strange, but it can be viewed as a measure of the value we place on a commodity or service. For instance, a car costing $15,000 has more worth or value than a car costing $10,000. This measure differs from others we have considered because it is relative to supply and demand and to personal preferences. For instance, consider a gallon of gasoline. Its size does not change, yet its worth often does. When demand is high and supply is low, gasoline has a greater worth, causing its cost to go up.

Another reason currency is different from other measures is that the units are almost completely nonstandard. Each country has control over its own currency, allowing it to develop any currency system it desires. Table 13.13 shows the **denominations,** or units of currency, from several different countries. Most currency systems use benchmark numbers, like 5, 10, and 20, for the denominations, making conversions and computations within a given system relatively easy. However, because most countries have their own currency and the values of commodities are always fluctuating, exchange rates between currency systems often change daily.

Table 13.13   Currencies from four countries

Country	Germany	Japan	Mexico	United States
Coins	1 cent	1 yen	1 centavo	Penny = 1¢
	2 cents	5 yen	5 centavos	Nickle = 5¢
	5 cents	10 yen	10 centavos	Dime = 10¢
	10 cents	50 yen	20 centavos	Quarter = 25¢
	20 cents	100 yen	50 centavos	Half-dollar = 50¢
	50 cents	500 yen	1 peso = 100 centavos	
	1 Euro = 100 cents		2 pesos	
	2 Euros		5 pesos	
			10 pesos	
Notes	5 Euros	1,000 yen	20 pesos	1 dollar bill = 100 cents
	10 Euros	2,000 yen	50 pesos	5 dollar bill
	20 Euros	5,000 yen	100 pesos	10 dollar bill
	50 Euros	10,000 yen	200 pesos	20 dollar bill
	100 Euros		500 Pesos	50 dollar bill
	200 Euros			100 dollar bill
	500 Euros			

Example 13.43   Making Exchanges Between Currencies

Application

On a particular day, the exchange rate between U.S. dollars and Japanese yen is $1 = 89.83 yen. How many yen would you receive if you exchanged $500 for yen?

Solution Using dimensional analysis, we have

$$500 \text{ dollars} \times \frac{89.83 \text{ yen}}{1 \text{ dollar}} = 44{,}915 \text{ yen}$$

Check Your Understanding 13.5-B

1. How many minutes are in a week?

2. The distance from the Earth to the moon is 384,400 km. Apollo 11 made the trip in about 4 days and 7 hr. What was the typical speed for the flight to the moon?

3. At one point, the exchange rate between dollars and pesos was 1 peso = $0.078. How many dollars are equivalent to 5,500 pesos?

Talk About It The attributes of weight, temperature, time, and speed are all measures connected to and used in science. What other measurable attributes might elementary or middle grades students learn about in mathematics that are also connected to science?

Activity 13.10

Application

The national debt of the United States in February 2010 was about $11 trillion. If a $1 bill weighs about 1 g and enough $1 bills could be collected to pay off the national debt, how many metric tons would the money weigh?

SECTION 13.5   Learning Assessment

■ Understanding the Concepts

1. What is the difference between weight and mass? Why are mass units used to measure weight?

2. What advantages do metric units offer over English units when measuring weight?

3. How are the Fahrenheit and Celsius scales similar? How are they different?

4. What is the difference between telling time and measuring time?

5. What is speed, and how is it different from other attributes considered in this chapter?

6. What is currency, and how is it different from other measures considered in this chapter?

■ Representing the Mathematics

7. Draw a diagram that illustrates the relationship between a kilogram, a liter, and a cubic decimeter.

8. Many of the abbreviations for the units discussed in this section make sense. For instance, it makes sense to abbreviate "ounce" as oz or "kilogram" as kg. Search the Internet to find out why "pound" is abbreviated as lb.

9. Which represents a larger change in temperature, an increase in 1° F or an increase in 1° C? Explain how you know.

10. How many seconds are in, or what part of a second is represented by, a:

    a. Millisecond?   **b.** Megasecond?
    c. Kilosecond?   **d.** Centisecond?

11. In the military, time is measured in terms of a 24-hr day rather than two 12-hr intervals. For instance, the time 6:00 a.m. is given as 0600, and the time 2:30 p.m. is given as 1430. Express each of the following in military time.

    a. 2:00 a.m.   **b.** 10:30 a.m.   **c.** 11:15 a.m.
    d. 1:00 p.m.   **e.** 4:20 p.m.   **f.** 9:35 p.m.

12. Express each military time in regular time.

    a. 0500 **b.** 0930 **c.** 1200

    d. 1345 **e.** 1830 **f.** 2215

13. Use words to express the relationship between the variables given by the distance formula.

14. **a.** Write a conversion ratio that can be used to convert mph to km/h.

    b. Write a conversion ratio that can be used to convert km/h to m/s.

■ Exercises

15. Make each conversion.

    a. 4 lb to oz **b.** 5.6 T to lb

    c. 4.76 oz to dr **d.** 570 lb to T

16. Make each conversion.

    a. 160 g to kg **b.** 1,500 mg to g

    c. 3 kg to g **d.** 0.14 g to mg

17. If 1 kg weighs approximately 2.2 lb, what is:

    a. The weight of a 1,500-kg walrus in pounds?

    b. The weight of a 135-lb woman in kilograms?

18. Select the most appropriate unit (gram, kilogram, or metric ton) for measuring the weight of each object.

    a. A thumbtack **b.** A large textbook

    c. An adult male **d.** A small car

    e. A bottle of ketchup **f.** Whale

    g. One raisin **h.** A safety pin

19. Complete each sentence using milligrams, grams, or kilograms.

    a. A pencil has a mass of about 40 . . . .

    b. A newborn infant has a mass of about 4 . . . .

    c. An adult elephant has a mass of about 7,000 . . . .

    d. A breath mint has a mass of about 3 . . . .

    e. Recommended daily allowance of vitamin C for an adult is 75 . . . .

20. An astronaut and her gear weigh 215 lb on Earth. How much would the astronaut weigh on each of the following planets?

    a. Mercury, which has a gravitational pull that is 0.378 times that of Earth's

    b. Moon, which has a gravitational pull that is 0.166 times that of Earth's

    c. Neptune, which has a gravitational pull that is 1.125 times that of Earth's

21. Convert the following degrees Fahrenheit to the nearest degree Celsius.

    a. 25° F **b.** −23° F **c.** −47° F

    d. 103° F **e.** 209° F

22. Convert the following degrees Celsius to the nearest degree Fahrenheit.

    a. 30° C **b.** 69° C **c.** −18° C

    d. 155° C **e.** −36° C

23. To convert degrees Kelvin to degrees Celsius, subtract 273.15 from the temperature in Kelvin, or $C = K - 273.15$. Convert the following to degrees Celsius.

    a. The boiling point of bound helium, 4.22 K.

    b. Average temperature of Uranus, 68 K.

    c. Melting point of titanium, 1,941 K.

    d. Write a formula that can be used to convert degrees Celsius to degrees Kelvin.

24. Which of the following statements are reasonable to make?

    a. Ice will not melt at 25° C.

    b. If the temperature is 30° C, it is warm enough to wear shorts.

    c. Snow is possible if the temperature is 5° C.

    d. A bath temperature of 50° C is too cold.

25. How many minutes are in a year? How many seconds?

26. **a.** About how many minutes is a kilosecond?

    b. About how long in days, hours, and minutes is a megasecond?

    c. About how long in years, days, hours, and minutes is a gigasecond?

27. One Olympiad is equivalent to four years and refers to the time frame in which consecutive summer Olympic games are to be held. The modern Olympic games were revived in 1896. How many Olympiads have there been since the revival of the games?

28. Which of the following speeds is the best estimate of the speed necessary to make a 367-mi trip in 7 hr?

    a. 40 mph **b.** 45 mph **c.** 50 mph

    d. 55 mph **e.** 60 mph

29. The muzzle velocity of the average rifle is about 1,000 m/s. What is this speed in km/s? In km/h?

30. The top running speed of a human is about 36 km/hr. How fast is this in mph?

31. In 1911, Ray Harroun won the first Indianapolis 500 race, driving a Marmon Wasp at an average speed of 74.59 mph. How long did it take him to complete the 500-mi race?

32. On one particular day, the exchange rate between U.S. dollars and Mexican pesos was $1 = 11.197 pesos. How many pesos would you receive if you exchanged $200 for pesos?

Problems and Applications

33. A bathtub measures 75 cm × 1 m × 60 cm. If it is filled with water, what is the mass of the water in kilograms?

34. During a particularly bad rainstorm, 3 in. of water fell on a parking lot with an area of 2 acres. If 1 ft³ of water weighs approximately 62 lb, what was the weight of the water that fell on the parking lot during the storm?

35. A certain lake is said to hold enough water to fill one cubic kilometer. How many kiloliters of water would this be? If one metric ton weighs 1,000 kg, what would be the weight of the water in metric tons?

36. Carats are often used to measure diamonds and other precious gemstones. If 1 carat = 200 mg, what is the weight in grams of each of the following famous diamonds?

  a. Golden Jubilee (the largest gem-quality diamond) 545.67 carats

  b. Great Star of Africa (the second largest gem-quality diamond), 530.2 carats

  c. Centenary Diamond (the largest flawless, colorless diamond), 273.85 carats

  d. Hope Diamond (known for the misfortunes of its owners), 45.2 carats

37. Is there a temperature that has the same value in degrees Fahrenheit as in degrees Celsius? If so, what is it?

38. What is the lowest temperature in degrees Celsius that is positive when converted to degrees Fahrenheit? Use two decimal places to determine your answer.

39. Find a formula that can be used to convert degrees Kelvin to degrees Fahrenheit.

40. A quarter is a common business time frame, which represents 3 months or one-fourth of a year. Many companies give a quarterly report as a way of maintaining oversight on the business. If a company has been in business for $15\frac{1}{2}$ years, how many quarterly reports have been written?

41. A bicyclist travels from his house to a store in town at a rate of 14 mph. At what speed must he pedal back to his house if he is to average 16 mph for the entire trip?

42. Light travels at a speed of 186,282 mi/s. How far will light travel in 1 hr? A day? A year?

43. Earth is approximately 93,000,000 mi from the sun. About how long does it take light from the sun to reach the Earth?

44. When the space shuttle reenters the Earth's atmosphere, it is traveling approximately 16,250 mph. If the circumference of the Earth is about 24,000 mi at the equator, how many times could the shuttle fly around the Earth in one day at this speed?

45. The average speed during Concorde's record crossing of the Atlantic was 559 m/s.

  a. What is this speed in miles per hour?

  b. If the distance from Paris to New York is 3,603 mi, how long would it take the Concorde to make this flight?

46. Consider the exchange rates shown in the table.

Currency	Per U.S. Dollar
Euro	0.79
British pound	0.54
Japanese yen	115.71
Hong Kong dollar	7.77

  a. How many euros will you receive if you exchange $100?

  b. Which is the greater amount of money: 115.71 Japanese yen or 7.77 Hong Kong dollars?

  c. How many euros will you receive in exchange for 100 British pounds?

Mathematical Reasoning and Proof

47. Using the formula for converting degrees Celsius to degrees Fahrenheit and algebra, show that the formula for converting degrees Fahrenheit to degrees Celsius is given by $C = \frac{5}{9}(F - 32)$.

48. a. If the temperature increases by 5° C, what is the corresponding increase in degrees Fahrenheit?

  b. If the temperature decreases by 10° F, what is the corresponding decrease in degrees Celsius?

49. Which is warmer, a temperature of −5° C or −5° F? Justify your response.

50. Use algebra and the definition of speed to write a short proof verifying the distance formula.

Mathematical Communication

51. Jerri confides in her friend that her weight is 120 lb. Was her use of the word "weight" correct? Explain.

52. The *Oxford English Dictionary* gives the following definition: "The indefinite continued progress of existence and events in the past, present, and future, regarded as a whole." To what concept is this definition referring?

53. a. Write a story problem that requires the conversion from degrees Fahrenheit to degrees Celsius.

b. Write a story problem that requires the use of the distance formula.

c. Write a story problem that requires the conversion from one speed unit to another.

d. Give your problems to a peer to solve. Then discuss any difficulties you may have encountered in writing or solving the problems.

54. Karol claims that if a car is driving at 60 mph for a total of 30 min, then the car must travel $60 \cdot 30 = 1{,}800$ mi. How would you correct Karol's mistake?

■ Building Connections

55. Search through a set of elementary mathematics curriculum materials for lessons on temperature. Make a list of the mathematical ideas that coincide with these lessons.

a. What connection is there between temperature and each mathematical idea?

b. Is temperature the main focus of each lesson, or is it used as a familiar situation by which a mathematical idea is taught?

56. Search through a set of elementary mathematics curriculum materials for lessons on currency. Make a list of the mathematical ideas that coincide with these lessons.

a. What connection is there between currency and each mathematical idea?

b. Is currency the main focus of each lesson, or is it used as a familiar situation by which a mathematical idea is taught?

57. The metric system was designed in such a way that units for different attributes could be connected to or defined in terms of one another. Write a short paragraph explaining how the metric units for length, area, weight, and volume (both dry and liquid measures) are all connected to one another.

58. Search through a set of elementary mathematics curriculum materials for lessons on time. Answer the following questions based on what you find.

a. Which do students learn first, telling or measuring time?

b. Describe the general sequence of how students learn to tell time.

c. What difficulties do you think children encounter when learning to use both analog and digital clocks?

Historical Note

Archimedes (287–212 B.C.E.) was an ancient Greek philosopher and scientist born in the seaport of Syracuse, Sicily. He was a great inventor, and many of his inventions were created to defend Syracuse against Rome. Legend has it that Archimedes even invented a large panel of mirrors that could concentrate the sun's rays to catch enemy ships on fire. Many mathematical facts are attributed to his name, not the least of which is the connection between the volume of a sphere and the smallest cylinder that contains it. He considered this fact to be a great achievement and even had it placed as an epitaph on his grave. A Roman soldier killed him during the sack of Syracuse, supposedly while he was drawing an equation in the sand. He was so engrossed in his work just prior to being slain that he is said to have spoken his famous last words, "Do not disturb my circles."

Death of Archimedes

Opening Problem Answer

Because the circles are congruent, the Smiths can estimate the perimeter by finding the distance around one arc and multiplying it by 2. Drawing segments from the center of one circle to the intersection points of the circles results in a triangle with two sides 8 ft long and a third side 11.3 ft long.

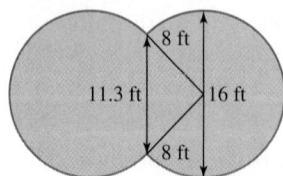

Because $\sqrt{8^2 + 8^2} = \sqrt{128} \approx 11.3$, we know the triangle is roughly a right triangle with the right angle between the two radii. This means the missing arc from each circle is about one-fourth the circumference of each circle. Consequently, the length around each circular region is about three-fourths the total circumference of the circle and can be estimated as

$$\frac{3}{4}C = \frac{3}{4}(\pi d) = \frac{3}{4}16\pi = 12\pi \approx 12(3.14) = 37.68.$$ Multiplying by 2 gives us the total distance around the pool as 2 (37.68 ft) = 75.36 ft.

Chapter 13 REVIEW

Summary of Key Concepts from Chapter 13

Section 13.1   Length and the Measurement Process

- **Measurement** is the process of assigning a numerical value, called a **measure,** to a physical attribute of an object or to a characteristic of an event or situation.
- Measurement involves three steps: identifying and selecting an attribute, selecting an appropriate unit, and using a measurement strategy to find the measure.
- An attribute is **measurable** if it can be quantified in some way.
- Given three points, A, B, and C, then $AB \geq 0$, $AB = BA$, and $AB + BC \geq AC$.
- A **unit of measure** is a fixed amount of the measurable attribute used to make the measurement comparison.
- **Precision** relates to the exactness of the measurement; **accuracy** describes how close the measure is to the actual size of the object.
- Common units of length in the English are inches, feet, yards, miles: 1 ft = 12 in., 1 yd = 3 ft, and 1 mi = 5,280 ft.
- In **dimensional analysis, unit ratios** are used to compare a single unit of one measure to its equivalent amount in another.
- The basic unit of length in the **metric system** is the **meter.**
- Larger units are made in the metric system by multiplying the basic unit by positive powers of ten. Smaller units are made by multiplying the basic unit by negative powers of ten. A summary of common length units is given in the following table:

Metric Unit	Abbreviation	Relationship to Base Unit
kilometer	km	1,000 m
hectometer	hm	100 m
decameter	dam or dkm	10 m
meter	m	1 m
decimeter	dm	0.1 m
centimeter	cm	0.01 m
millimeter	mm	0.001 m

- Three strategies used to make measures include estimations, measuring devices, and formulas.
- Three methods of estimating a measure include comparing the object to a referent unit, unitizing, and chunking.

Section 13.2 Perimeter and Area

- The **perimeter,** P, of a simple closed curve is the distance around the curve.
- Perimeter formulas for common polygons:

Shape	Perimeter Formula
Triangle with sides of lengths a, b, and c	$P = a + b + c$
Square with sides of length s	$P = 4s$
Rectangle with sides of lengths w and l	$P = 2w + 2l$
Regular n-gon with sides of length s	$P = ns$

- The distance around any circle is called its **circumference, C.** It is given by $C = 2\pi r$, where r is the radius of the circle and $\pi \approx 3.1416$.
- **Area** is the amount of plane enclosed in a planar figure. It is measured with square units. Common area units in the English system are in.^{2}, ft^2, yd^2, and mi^2. Common area units in the metric system are cm^2, m^2, and km^2.
- Area formulas for common shapes:

Shape	Area Formula
Rectangle with sides of lengths l and w	$A = l \cdot w$
Square with sides of length s	$A = s^2$
Parallelogram with a base of length b and a height h	$A = bh$
Triangle with a base of length b and a height h	$A = \frac{1}{2}bh$
Trapezoid with bases of lengths b_1 and b_2 and a height h	$A = \frac{1}{2}(b_1 + b_2)h$
Regular n-gon with an apothem of length a and a perimeter P	$A = \frac{1}{2}aP$
Circle with radius r	$A = \pi r^2$

- If a planar region A is subdivided into nonoverlapping subregions B and C, then Area(A) = area(B) + area(C).

Section 13.3 Surface Area

- **Surface area** is the amount of plane needed to cover the outside of a three-dimensional shape.
- Surface area formulas for common shapes:

Shape	Surface Area Formula
Right rectangular prism with edges of lengths l, w, and h	$SA = 2lw + 2lh + 2wh$
Cube with an edge of length e	$SA = 6e^2$
Right prism with a height h and a base of area B and perimeter P	$SA = 2B + Ph$
Right circular cylinder with height h and radius r	$SA = 2\pi r^2 + 2\pi rh = 2\pi r(r + h).$
Right regular pyramid with slant height l and a base with area B and perimeter P	$SA = B + \frac{1}{2}Pl$
Right circular cone with slant height l, height h, and radius r	$SA = \pi r^2 + \pi rl = \pi r^2 + \pi r\sqrt{h^2 + r^2}$
Sphere with radius r	$SA = 4\pi r^2$

Section 13.4 Volume

- **Volume** is the amount of material, either liquid or solid, that is needed to make or fill a three-dimensional shape. **Capacity** is the amount of space a figure can hold.

- A **cubic unit** is the amount of space contained inside a cube that measures one unit on each edge. Common cubic units in the Customary system are a cubic inch (in.^{3}), a cubic foot (ft^3), and a cubic yard (yd^3). Common cubic units in the metric system are a cubic centimeter (1 cm^3) and the cubic meter (1 m^3).

- A liter (1 L) is defined to be the volume of a cube that measures 10 cm, or 1 dm, on each edge (1 L = 1dm^3).

- Volume formulas for common shapes:

Shape	Volume Formula
Right rectangular prism with edges of lengths l, w, and h	$V = lwh$
Cube with edges of length e	$V = e^3$
Right prism with a height h and a base of area B	$V = Bh$
Circular cylinder with height h and radius r	$V = \pi r^2 h$
Pyramid with height h and a base of area B	$V = \frac{1}{3}Bh$
Circular cone with height h and radius r	$V = \frac{1}{3}\pi r^2 h$
Sphere with radius r	$V = \frac{4}{3}\pi r^3$

- **Cavalieri's principle:** Two solids, each having their bases in the same parallel planes, have the same volume if every plane parallel to the bases intersects the solids in cross sections of equal area.

Section 13.5 Other Measures

- **Mass** is the amount of matter within an object. **Weight** is the amount of force exerted by gravity on the object.

- Common units of weight in the Customary system are ounces (oz), pounds (lb), and tons (T): 1 lb = 16 oz and 1 T = 2,000 lb.

- Common units of weight in the metric system include grams (g), kilograms (kg), and metric tons.

- **Temperature** measures the degree to which something is warm or cold, with warmer temperatures receiving the larger measure.

- It is measured with either the **Fahrenheit** scale or the **Celsius** scale. The freezing point of water is 32° F or 0° C. The boiling point of water is 212° F or 100° C.

- $F = \frac{9}{5}C + 32$ and $C = \frac{5}{9}F - 32$.

- When telling time, we indicate an instant at which an event is to occur. When measuring time, we find how many time units are equivalent to a time interval.

- The basic unit of time in the Customary and metric systems is the **second.**

- Speed is the ratio of the distance traveled to the time traveled, or Speed $= \dfrac{\text{distance}}{\text{time}}$.

- Speed units are given as a *ratio* of a length unit to a time unit. Common speed units are m/s, km/h, or mph.

- **Distance formula:** If an object is traveling at a rate r for a given time t, then the distance D the object travels is given by $D = r \cdot t$.

- **Currency** or **money** is a unit of exchange that assists in the trade of goods and services between people, businesses, and governments.

Review Exercises Chapter 13

1. Make a list of the possible measurable attributes of a house.

2. If the actual measure of an object is 2.467 units, describe each measure in terms of accuracy and precision.
   a. 5.789   **b.** 7.3   **c.** 2.5   **d.** 2.466

3. What happens to the measure of an object if the unit is:
   a. Doubled?   **b.** Tripled?   **c.** Quadrupled?

4. How many times larger is a kilometer than a micrometer?

5. For each question, select the measure that makes the most sense.
   a. The length of an unsharpened pencil is:
      i. 1.5 m.   **ii.** 1.5 cm.   **iii.** 10 in.   **iv.** 15 yd.
   b. The length of a cellular phone is:
      i. 1 km.   **ii.** 0.2 hm.   **iii.** 10 cm.   **iv.** 0.4 m.
   c. The distance from New York to Los Angeles is:
      i. 150,000 dm.   **ii.** 3,700 hm.
      iii. 2,500 km.   **iv.** 14,000 cm.

6. Make each conversion.
   a. 20 cm to m   **b.** 35 m to dm   **c.** 36 ft to yd
   c. 4.3 yd to in.   **e.** 23 ft to m   **f.** 357 cm to ft

7. A football field is 160 ft wide. What is the width of the field in yards? Meters?

8. A standard Olympic track is 400 m around. How many times must a runner run around the track to run 1 mi?

9. The world population is about 6.6 billion people. If every person joined hands to form a chain, about how many times would the chain wrap around the Earth at its equator?

10. Estimate the length of the following objects in both centimeters and inches.
    a. A toothpick
    b. The length of your pointer finger
    c. A small paper clip

11. About how many dots are in this triangle? Explain how you arrived at your answer.

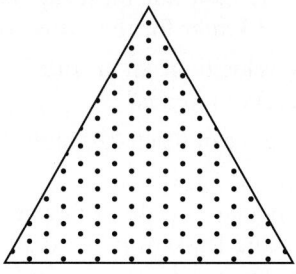

12. If the legs of a right triangle are 75 ft and 100 ft, respectively, what are the perimeter and area of the triangle?

13. Find the perimeter and area of each figure.
   a.

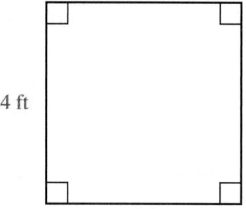

   b.

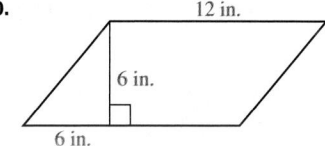

   c.

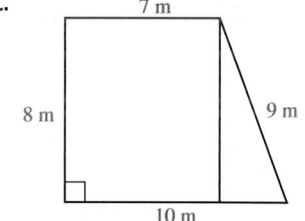

   d.

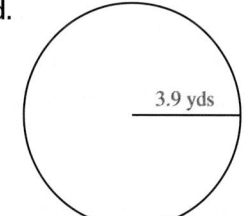

   e.   **f.**

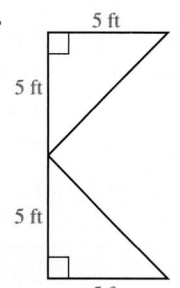
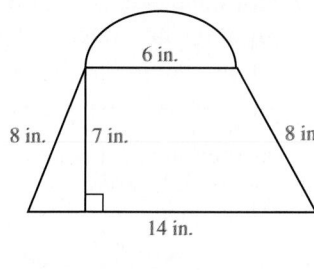

14. Fill in the blanks.
    a. $35 \text{ ft}^2 = $ _____ yd^2.   **b.** $19 \text{ m}^2 = $ _____ mm^2.
    c. $157.3 \text{ dm}^2 = $ _____ cm^2.

15. The length of a rectangle is 17 m, and the area is 323 m^2. What is the width?

16. What is the area of the shaded region?

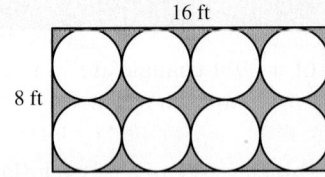

17. The area of a sector of a circle is 28π in.², and the circle has a radius of 7 in. What is the measure of the central angle?

18. What is the area of a regular hexagon if the sides are 5 in.?

19. What is the largest rectangular area that can be made with a string that is 4 ft long?

20. A circular pool is surrounded by a cement walk that extends 6 ft beyond the edge of the pool. The pool has a radius of 12 ft.

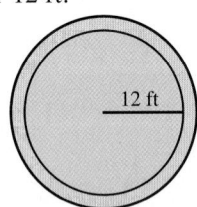

What is the area of the cement walk?

21. Sam and Julie want to tile their kitchen floor with 6×6-in. tiles. If the dimensions of their kitchen are 13×15 ft, how many tiles do they need?

22. If the ratio of the legs of two right triangles is 1:5, what is the ratio of their:

   a. Perimeters?   **b.** Areas?

23. What happens to the area of a square if the length of its side is increased 30%?

24. Make each conversion.

   a. 99 yd³ to in.³ **b.** 1,235 in.³ to ft³ **c.** 1.3 m³ to cm³

25. Hot water is carried through 65 ft of copper pipe from the hot water heater to the garden tub in the master bathroom. If the pipe has an inner diameter of $\frac{3}{4}$ in., how many gallons of hot water are wasted as the water in the pipe cools down?

26. Find the surface area and volume of each figure.

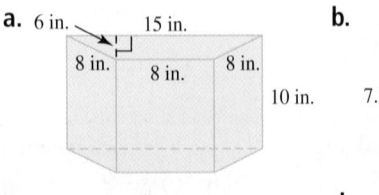
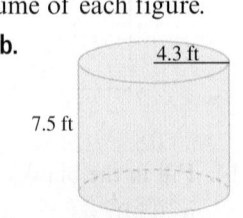
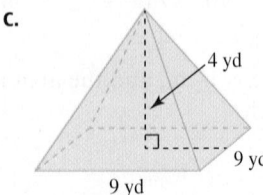
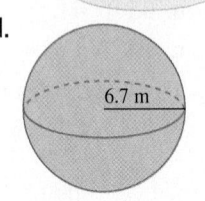

27. A classroom measures 10×8 m with a ceiling height of 3 m. If 1 L of paint covers 15 m², how many liters of paint are needed to cover the walls of the room?

28. Tennis balls are stacked in a package of three balls. If the can is cylindrical and the balls fit in it perfectly, what percent of the volume is taken up by tennis balls?

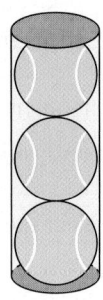

29. A cylindrically shaped pool is 8 m in diameter. If it is uniformly 1 m deep, how many liters of paint are needed to paint the inside if 1 L of paint covers 15 m²?

30. Consider two cubes: one with an edge of 3 in. and one with an edge of 12 in.

   a. What is the ratio of the surface areas of the two cubes?

   b. What is the ratio of the volume of the two cubes?

31. If the length, width, and height of a rectangular prism are tripled, what happens to its volume?

32. Make each conversion.

   a. 5.2 lb to oz   **b.** 0.36 g to mg

   c. 0.0015 dag to cg

33. If 1 kg = 2.2 lb, how much would a 2,000-lb bull moose weigh in kilograms?

34. Convert each degree Fahrenheit to the nearest degree Celsius, or vice versa.

   a. 312° F   **b.** −57° F   **c.** 4.56° F

   d. −91° C   **e.** 121° C

35. If the temperature decreases by 10° F, what is the corresponding decrease in degrees Celsius?

36. How long in hours and minutes is a megasecond?

37. A woman walks from her house to a store at a rate of 4 km/hr. At what rate must she walk back if she is to average 3 km/hr for the entire trip?

38. The muzzle velocity of an air rifle is 1,200 ft/s. What is this speed in m/s?

39. Many race cars can reach 220 mph. How fast is this in km/hr?

40. On one particular day, the exchange rate between U.S. dollars and Russian rubles is $1 = 26.769 rubles. How many rubles exchange for $500?

Answers to Chapter 13 Check Your Understandings and Activities

Check Your Understanding 13.1-A

1. a. Not accurate, but precise.

   b. Accurate and precise.

   c. Accurate, but not precise.

   d. Not accurate and not precise.

2. a. 25 yd.   **b.** 514.8 in.

   c. 1.79 rd.

3. a. 1.5 m.   **b.** 56 dam.

   c. 400 cm.

4. a. 5.18 m.   **b.** 11.19 ft.

   c. 36.21 hm.

Check Your Understanding 13.1-B

1. The fence is about 56 ft long. Unitizing is a good method.

2. a. About $2\frac{5}{16}$ in., or 5.9 cm.

   b. About $1\frac{3}{16}$ in., or 3.1 cm.

   c. About $1\frac{3}{4}$ in., or 4.5 cm.

3. 12.8 ft.

Check Your Understanding 13.2-A

1. a. 30 ft.   **b.** 12 m.

   c. 56 cm.   **d.** 6π in.

2. 7.06 ft.

3. a. 5 yd^2.   **b.** 0.0395 hm^2.

   c. 0.227 m^2.

4. 640 acres.

Check Your Understanding 13.2-B

1. a. 216 ft^2.   **b.** 1.86 cm^2.

   c. 56.75 m^2.

   d. 42.25π in.$^2 \approx 132.67$ in.^{2}.

2. Each side must be 4 cm.

3. 9,816.63 yd^2.

Check Your Understanding 13.3-A

1. a. 274 m^2.   **b.** 402 in.^{2}.

2. 19 cm.

3. $3,000\pi$ yd^2.

Check Your Understanding 13.3-B

1. 36.93 in.^{2}.

2. 68.23 m^2.

3. The surface area of the sphere is $1,296\pi$ ft^2. The circumference of the great circle is 36π ft, and its area is 324π ft^2.

Check Your Understanding 13.4-A

1. Drawing the appropriate picture, we see that $12 \cdot 12 \cdot 12 = 1,728$ in.$^3 = 1$ ft^3.

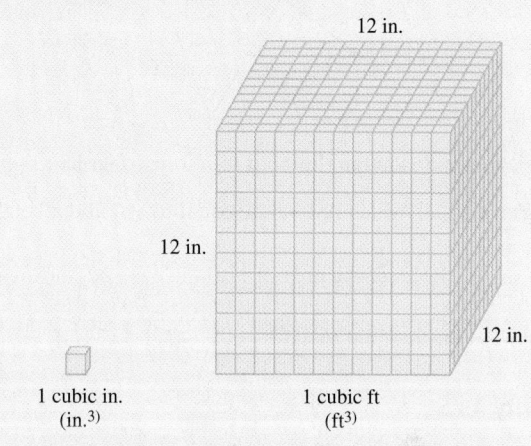

1 cubic in.   1 cubic ft
(in.^{3})   (ft^3)

2. a. 8640 in.^{3}.   **b.** 0.04 m^3.   **c.** 387,244.8 in.^{3}.

3. 4.5 L.

4. 72 in.^{3}.

Check Your Understanding 13.4-B

1. 40 in.^{3}.

2. 96π m^3.

3. $r \approx 3.1$ yd.

4. 696.56 cm^3.

Check Your Understanding 13.5-A

1. a. 128 oz.   **b.** 36 g.   **c.** 1.2 T.

2. a. 152.6° F.   **b.** 59° F.   **c.** 17.6° F.

3. −89° C.

Check Your Understanding 13.5-B

1. 10,080 min.

2. About 3,732.04 km/hr.

3. $429.

Activity 13.1

Answers will vary for all parts.

Activity 13.2

Subdivide the picture into square regions of the same size, and then estimate the number of flowers in each region. Add the results. The number of flowers is about 250.

Activity 13.3

Strategies will differ. The square units are given.

a. 13 square units   **b.** $8\frac{1}{2}$ square units   **c.** 8 square units

Activity 13.4

Consider the rhombus shown:

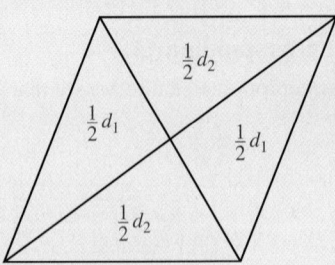

The diagonals divide the rhombus into four congruent right triangles, each of which has a base equal to $\frac{1}{2}d_2$ and a height equal to $\frac{1}{2}d_1$. Because the area of a triangle is $\frac{1}{2}bh$, the area of the rhombus is equal to $4\left[\frac{1}{2}\left(\frac{1}{2}d_2\right)\left(\frac{1}{2}d_1\right)\right] = \frac{1}{2}d_1d_2$.

Activity 13.5

a. 24 square units with the arrangement shown:

b. 34 square units with the arrangement shown:

Activity 13.6

a.

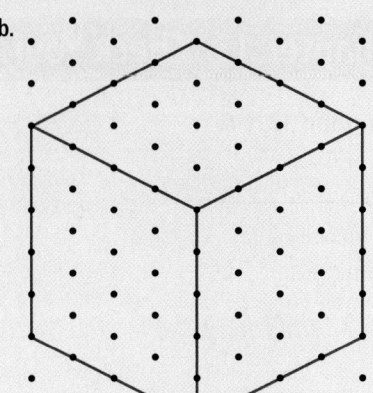

b.

Activity 13.7

$V = 450$ in.^{3}.

Activity 13.8

a. 179.50 in.^{3}.

b. 202.01 in.^{3}.

Activity 13.9

$C = \frac{5}{9}(F - 32)$.

Activity 13.10

11,000,000 metric tons.

© Image Source/Corbis

Chapter 14

Statistical Thinking

Opening Problem Ashley and Quin are two fourth-graders who live in separate states, each with its own standardized achievement test. The test from Ashley's state is normally distributed with a mean of 72 and a standard deviation of 8 points. The test from Quin's state is also normally distributed, but with a mean of 35 and a standard deviation of 4 points. If Ashley scored an 87 on her test and Quin scored a 42 on his test, who did better?

©Istockphoto.com/Nikolay Titov

© Image Source/Corbis

The final two chapters present a study of **statistical thinking,** which consists of the processes involved in collecting, organizing, and interpreting data. Statistical thinking is receiving more attention in the elementary grades for at least two reasons. First, advances in technology have greatly increased our access to information. To use information wisely, we must have the skills to organize, analyze, and make decisions from it. Teaching statistical thinking at an early age gives children the opportunity to develop essential decision-making skills. Second, statistical thinking provides an important way to build mathematical connections. It not only provides a meaningful context in which students can apply concepts and skills from other areas of mathematics, but it also allows them to connect mathematics to other disciplines and their daily lives.

Traditionally, statistical thinking in the elementary and middle grades has been limited to making and analyzing statistical graphs and computing averages. Modern curricula now take a much more holistic approach. This is due in large part to the NCTM Data Analysis and Probability Standard.

> ### ■ NCTM Data Analysis and Probability Standard
>
> Instructional programs from prekindergarten through grade 12 should enable all students to:
>
> - Formulate questions that can be addressed with data and collect, organize, and display relevant data to answer them.
> - Select and use appropriate statistical methods to analyze data.
> - Develop and evaluate inferences and predictions that are based on data.
> - Understand and apply basic concepts of probability.

Source: NCTM STANDARDS Copyright © 2011 by NATIONAL COUNCIL OF TEACHERS OF MATHEMATICS. Reproduced with permission of NATIONAL COUNCIL OF TEACHERS OF MATHEMATICS.

The first goal of the standard identifies three key aspects for teaching statistical thinking. First, students should learn to formulate questions that can be answered by means of statistical investigations. In the earliest grades, the teacher poses the questions, but, as students learn more, they can begin to ask and investigate their own. Second, students should learn to design and conduct basic data-collecting activities. At first, they are likely to use simple observations, surveys, and experiments to collect data. However, as their questions become more complex, they may begin to use more sophisticated methods, like sampling and comparative groups. Third, students should learn to organize and display data with statistical graphs. To do so effectively, they must learn how to represent different kinds of data using a wide variety of graphs.

The second goal of the standard indicates that analyzing data is another key aspect of statistical thinking. At first, students use graphs to analyze data for specific facts, like counts or maximums and minimums. As they gain experience, they learn to describe other features of the data, like the range, the distribution, and relationships between variables. In the middle grades, students also learn to use numerical methods to describe the center and the spread of the data, as well as how to use these statistics to make simple comparisons between groups.

The third goal of the standard indicates a final aspect of statistical thinking: making inferences from data. Although children in the early grades can draw simple conclusions from the data, they are not likely to be able to make many generalizations beyond the data. Once in the middle grades, however, students begin to learn how to make generalizations from a sample to a larger population. Table 14.1 identifies specific learning objectives that are associated with the first three goals of the Data Analysis and Probability Standard. The fourth goal focuses on probability. Although it is closely connected to statistical thinking, probability is an important topic of study in its own right and will be discussed in greater detail in the last chapter.

Table 14.1 Classroom learning objectives	K	1	2	3	4	5	6	7	8
Organize, represent, and interpret data with up to three categories and answer simple questions based on the data.		X							
Generate measurement data by measuring lengths of several objects to the nearest whole unit and show the data by making a line plot.			X	X					
Draw a picture graph and a bar graph to represent a data set with up to four categories and solve simple problems using information presented in a bar graph.			X	X					
Make a line plot to display a data set of measurements in fractions of a unit, such as $\frac{1}{2}$, $\frac{1}{4}$, and $\frac{1}{8}$, and solve problems involving information in line plots.					X	X			
Recognize a statistical question as one that anticipates variability in the data.							X		
Understand that a set of data has a distribution that can be described by its center, spread, and overall shape.							X		
Recognize that a measure of center summarizes the values of a data set with a single number, while a measure of variation describes how the values vary.							X		
Display numerical data in dot plots, histograms, and box plots.							X		
Summarize numerical data sets in relation to their context by reporting the number of observations, using descriptive statistics, and describing an overall pattern to the data.							X		
Understand that statistics can be used to gain information about a population by examining a sample of the population.								X	
Use data from a random sample to draw inferences about a population with an unknown characteristic of interest.								X	
Compare and draw inferences from data distributions by using measures of center and measures of variability.								X	
Construct and interpret scatterplots to investigate patterns of association between two numerical variables.									X
Use a linear model to represent a relationship in a scatterplot that suggests a linear association and use the model to solve problems in the context of the scatterplot.									X
Understand that patterns of association between two variables can also be seen in categorical data by displaying relative frequencies in a two-way table.									X

Source: Adapted from the *Common Core State Standards for Mathematics* (*Common Core State Standards Initiative*, 2010).

The Data Analysis and Probability Standard has had quite an impact on modern curricula, as many of them now move children well beyond simple examinations of pregenerated graphs and onto real data collection and analysis activities. To help preservice elementary teachers better understand the concepts of statistical thinking they are likely to teach, Chapter 14:

- Gives an overview of posing statistical questions and collecting data.
- Describes how to use statistical graphs to represent and analyze data.
- Discusses how to represent and analyze data numerically.
- Demonstrates basic methods for drawing conclusions from data displays and descriptive statistics.

SECTION 14.1 Formulating Questions and Collecting Data

Information and the ability to process it play an important role in our jobs and in our daily lives. In the workplace, many businesses now collect, analyze, and distribute information for economic gain. The demand for people with the ability to process information has therefore been steadily increasing. Information processing is also crucial in our daily lives because we are constantly inundated with data and statistics from the news, politics, advertisements, and even the sports and recreation industries. Whether in business or personal endeavors, we must be able to work with data and think statistically to make the best choices in the decisions affecting our lives.

The word "statistics" can take on one of two meanings. In its singular sense, statistics refers to the science of collecting, analyzing, and interpreting **data,** which are numbers that occur in a context. In its plural sense, statistics can refer to numbers in the data set or to the result of some computation on the data set. Data often come from a statistical study, which is done to answer a question or find the solution to a problem. Although statistical studies can be conducted in a variety of ways for a variety of reasons, they generally all go through four basic steps:

1. *Formulating a question.* Every statistical study begins with a question that determines how the data are collected, analyzed, and interpreted.
2. *Collecting data.* After a question is posed, the next step is to collect data, commonly by means of surveys, observations, and experiments.
3. *Analyzing data.* Once collected, data must be organized and analyzed, which is often done with statistical graphs or descriptive statistics.
4. *Drawing conclusions.* The final step is to use data analysis to draw conclusions to answer the question.

In this section, we discuss the first two steps: formulating a question and collecting data.

Formulating a Question

Every statistical investigation begins with a question to be answered. The question can arise from a problem situation or from a natural curiosity to know more. Posing a good question is important because it drives the rest of the study by identifying the topic of interest. This, in turn, determines the kind of data to be collected, how the data will be displayed, and the kind of statistics used to describe the data. Examples of statistical questions are:

- What is the typical age of college students in the United States?
- What is the average reading level of third-grade students in the city of Richmond?
- How many calories are in a typical lunch served at a fast-food restaurant?
- What is the average number of defective televisions produced by a given factory?

Examining the questions carefully, we can see that asking a statistical question is not the same as asking a mathematical one. First, each of the preceding questions identifies the topic of the investigation, which is important because it provides the context for the study. Although many mathematical problems have a context, the context can be removed when working with the underlying mathematics. In statistics, however, the context is essential. Without it, the data have no meaning.

To clearly identify the topic of the study, the question should indicate two things. First, it should identify the **population,** which is the group of people, animals, or objects under consideration in the study. For instance, the population in the first question is the set of all college students in the United States, and the population in the second is the set of all third-grade students in the city of Richmond. As these examples show, populations can vary in size. They can be quite large, like the population of a country or a state. Or they can be limited, like the people living in a town, attending a particular school, or belonging to a specific age group.

Second, a good statistical question should identify the trait of the population being studied. For instance, the trait in the first question is the age of college students, and the trait in the third is the number of calories in lunches served at a fast-food restaurant. Notice that each question anticipates that the value of the trait will vary from one instance to another. Specifically, we expect that different college students will have different ages and that different lunches will have different calorie counts. This is called **variability,** and it is one of the fundamental differences between statistical and mathematical questions. Statistical questions anticipate variability and the need to collect data to capture that variability. Mathematical questions tend to be more deterministic in that they often seek a single answer.

Because the trait of the population can vary from one individual or observation to the next, we generally refer to the traits as **variables.** Knowing the variable of the study is important because it determines the type of data to be collected. In general, there are two types of statistical variables. **Qualitative,** or **categorical, variables** separate data into categories. For instance, ethnicity is a qualitative variable because it separates individuals into non-numerical groups like African American, Asian, and Caucasian. The data associated with a qualitative variable usually take the form of counts because the investigator is interested in the number of individuals that fall into each category. **Quantitative,** or **numerical, variables** describe a measure or other numerical value associated with the trait of interest. Examples are test performance, human height, and car speed. The data associated with a quantitative variable are always numerical, so they lend themselves nicely to many statistical computations.

Example 14.1   Qualitative and Quantitative Variables

State whether each variable is qualitative or quantitative.

a. Computer processor speed   **b.** Gender   **c.** Music preference

Solution

a. The speed of a computer processor is determined by how fast it can process a certain number of computations. Because this leads to a measure, it is a quantitative variable.

b. Gender is qualitative because it separates individuals into one of two categories, male or female.

c. Music preference is qualitative because it separates people based on their likes and dislikes of specific musical categories like rock, rap, popular, and gospel.

Sampling

Once we have a question, the next step is to collect data. However, before turning to specific methods of data collection, we must first discuss sampling. The ideal way to conduct a statistical study would be to use a **census,** which is a study that collects data on the entire population. Although a census gives the most accurate information, it is generally too expensive and time-consuming. For instance, the U.S. Census is an effort to collect information on every citizen of the United States. It requires billions of dollars and thousands of workers to administer. Even then, it fails to collect information on several million members of the population.

Explorations Manual 14.1

In most studies, the investigators do not have the funds or the personnel to consider an entire population. Instead, studies are conducted on a small portion of the population, called a **sample.** The basic idea behind sampling is quite simple: If the sample is representative of the population, meaning the characteristics of the sample closely match the characteristics of the population, then whatever holds true for the sample must hold true for the population. For instance, suppose a sample of college students is 61% female. If the sample is representative of the population, then we would expect about 61% of all college students to be female.

Example 14.2 Variables, Samples, and Populations

Identify the variable, the population, and the sample in each situation.

a. A principal examines the grades of two fourth-grade classes to determine how the fourth-graders at her school are performing in mathematics.

b. A food inspector randomly checks 50 chickens for the presence of salmonella at a food processing plant.

Solution

a. The variable is performance in mathematics, the population is all fourth-grade students at the elementary school, and the sample is the students in the two classes.

b. The variable is the presence of salmonella, the population is all chickens processed at the plant, and the sample is the 50 chickens examined by the inspector.

Even though the idea behind sampling is simple, it must be carried out with extreme care. If a sample is not representative of the population, then the study is likely to suffer from bias. **Bias** refers to any problem in the design or conduct of the study that causes the results to favor a particular outcome. If we choose a sample with an inherent tendency toward a certain outcome, then the study suffers from a **selection bias.** Any form of bias makes it difficult to draw accurate conclusions about the population, so we try to avoid it as much as possible.

Example 14.3 A Biased Sample

Reasoning

A sales consultant wants to know whether people under the age of 25 shop at a local mall. She collects data on a Tuesday morning by observing patrons as they pass by. Most of the people are over the age of 25, so she concludes that the mall is not appealing to younger patrons. What is wrong with her study?

Solution Although there is no inherent problem in observing patrons over a given period of time, her choice of time may have resulted in a biased sample. On Tuesday mornings, many people under the age of 25 are either at work, in school, or at home sleeping. They are not likely to be shopping. If she wanted a better sample of patrons, she could have made her observations in the late afternoon, in the early evening, or on the weekend when people in this age group are more likely to shop.

In the last example, the sales consultant used what is called a **convenience sample,** which is a sample selected from readily available people or objects. Convenience samples have the advantage of being easy to obtain, but they often lead to biased results.

Another way to select a sample is to let members of the population decide for themselves whether to participate. Samples like this, called **voluntary samples,** are often used with surveys in magazines, on the radio, or on the Internet.

Example 14.4 A Voluntary Sample

A news network conducts an online survey about Americans' beliefs on global warming. Eighty percent of the respondents thought that global warming was either a crisis or a major problem. What potential problems might the study have?

Solution The study depends on a voluntary sample, which may cause at least two potential problems. First, the questionnaire was available only to people visiting the website. News websites are often visited by a select group of people, so the sample may not be representative of all Americans. Second, of the people visiting the website, the sample contained only those who were willing to fill out the survey. In many cases, the people who are willing to fill out surveys are those with strong feelings about the topic. People who care little about a topic may be less inclined to fill out the survey.

As the last example shows, we should also be cautious when using voluntary samples. Whenever people decide for themselves whether to participate in a study, there is always a chance of bias due to a **nonresponse.** In other words, a portion of the population fails to represent itself through a lack of participation, so the information relevant to that part of the population is not reflected in the data.

Because convenient and voluntary samples have the potential to cause bias, the question remains, "How do we select a sample that is representative of the population?" If we want to avoid biased samples, then our method of selecting a sample should ensure that every member of the population has the same chance of being selected. Selecting a sample in this way is called **simple random sampling.** Although the idea seems straightforward, it is not always easy to implement. For instance, to collect information about the population of a certain city, we might turn to city records to collect the names and contact information of the residents. After assigning a number to each person, we could select the sample by using a computer or a table of random digits to randomly pick our participants. If the city has a large population, we can imagine how daunting this task might become.

Two other things need to be mentioned about simple random sampling. First, if we select several different samples, we get different participants each time. This means that the values in the data will vary from sample to sample. This is called **sampling variability,** and the idea is central to many of the techniques associated with making decisions from data.

Second, even if a sample is selected at random, it is still not guaranteed to be representative of the population. For instance, the population of the United States is about 50% male and 50% female. If we were to select a simple random sample from this population, the sample could be 70% male and 30% female or 40% male and 60% female. Because the characteristics of the population are unknown, we can never be sure whether the characteristics of the sample truly represent the population. Fortunately, there are two ways around this problem. First, as samples grow in size, they naturally include more participants, making them more representative of the population. There are statistical techniques for determining the appropriate **sample size,** but they go beyond the scope of this text. Second, the study can be conducted multiple times using different samples. If the results are the similar, then we are more confident in the conclusions drawn from the study.

Explorations
Manual
14.2

Example 14.5   Drawing a Conclusion from a Simple Random Sample

Reasoning

The quality control at a food processing plant randomly inspects 350 boxes of macaroni and cheese for foreign objects. Seven packages fail the inspection. If the plant produces 40,000 boxes a day, how many of them are likely to contain foreign objects?

Solution Because the sample of 350 boxes is large and randomly selected, the number of contaminated boxes in the sample is likely to be representative of the contaminated boxes in the population. Using proportional reasoning, we find that the number of boxes with foreign objects in them is

$$\frac{7}{350} = \frac{x}{40,000}$$

$$350x = 7 \cdot 40,000$$

$$x = \frac{7 \cdot 40,000}{350} = 800 \text{ boxes}$$

In some studies, we want to compare groups with certain characteristics. For instance, many studies compare males to females or different age groups. In such studies, we can first separate the possible participants with these characteristics and then use simple random sampling to select the members of each group. Sampling in this way is called **stratified sampling.**

Check Your Understanding 14.1-A

1. Describe each variable as qualitative or quantitative.
   a. A child's grade level in school
   b. The top speed of motor vehicles
   c. Socioeconomic status
   d. The number of hours teenagers watch television

2. Identify the population, the sample, and the sampling technique in each situation.
   a. A quality control inspector at an auto manufacturing plant randomly inspects 75 cars for defective parts.
   b. An Internet site conducts a survey of people's political beliefs. There were 1,487 respondents.
   c. An English teacher is curious about the books students like to read. She does a quick questionnaire with 65 eighth-grade students from her classes.

3. A magazine conducts a survey with its subscribers on the West Coast about the outcome of an upcoming presidential election. The results heavily favor the Democratic candidate. What biases might be inherent in the survey?

Talk About It Suppose you wanted to investigate the number of people of each gender living in your city. How might you select a sample from a phone book?

Communication

Activity 14.1
With a partner, write down five questions that can be answered by using a statistical study and that are appropriate for children in the elementary grades. In each case, identify the population and the variable of interest. State whether the variable is qualitative or quantitative.

Collecting Data

We can now turn to the methods of collecting data. Because there is a wide variety of questions, data can naturally be collected in a wide variety of ways. Three of the most common are surveys, observations, and experiments.

Surveys

One way to collect data is with a survey. The investigator develops a questionnaire and uses it to ask participants about things that cannot be directly observed. This can include demographic information, opinions, beliefs, and personal preferences.

Explorations
Manual
14.3

Because surveys can cover a wide variety of issues, a wide variety of questions must be used to get at the needed information. Table 14.2 lists eight types of questions commonly used on surveys. As the example questions show, survey questions can be about both qualitative and quantitative variables. As a result, the data collected by a survey can range from simple counts to specific numerical data.

Table 14.2 Eight types of survey questions		
Type of Question	**Description**	**Example**
Categorical/nominal	Participants are asked to place themselves in one of several possible categories.	What is your ethnicity? 1. African American 2. Asian 3. Caucasian 4. Hispanic 5. Other
Dichotomous	Participants are asked to select one of two possible responses such as yes/no, true/false, or agree/disagree.	Elementary classes should have no more than 20 students. 1. Agree 2. Disagree
Filter/contingency	Participants are asked one question to determine whether they are qualified to answer a subsequent question.	Do you smoke? _____Yes _____No If yes, how many packs a day? _____
Interval level	Participants are asked to rate their opinion or agreement with a statement using a scale from one extreme to the other.	Abortion should be legal. 1. Strongly agree 2. Agee 3. Neutral 4. Disagree 5. Strongly disagree
Multiple choice	Participants are asked to pick the best answer or answers from a set of possible answers.	Which type of movie do you like to watch most? 1. Action/adventure 2. Comedy 3. Drama 4. Horror
Numerical	Participants are asked to answer with a specific number.	What is your age? _____
Ordinal	Participants are asked to rank the given possibilities in order of their preference.	Rank your favorite flavor of ice cream from best to worst. _____ Chocolate _____ Strawberry _____ Vanilla
Open-ended	Participants are asked to respond in any way they choose.	What are your beliefs about capital punishment?

As the next example shows, surveys are often used in classroom data collection activities because they are relatively easy to create and use.

Example 14.6   Collecting Data in the Classroom

Connection   Consider the survey on the student page in Figure 14.1. What kind of questions does it ask? What are the variables, and how do they determine how the data are collected?

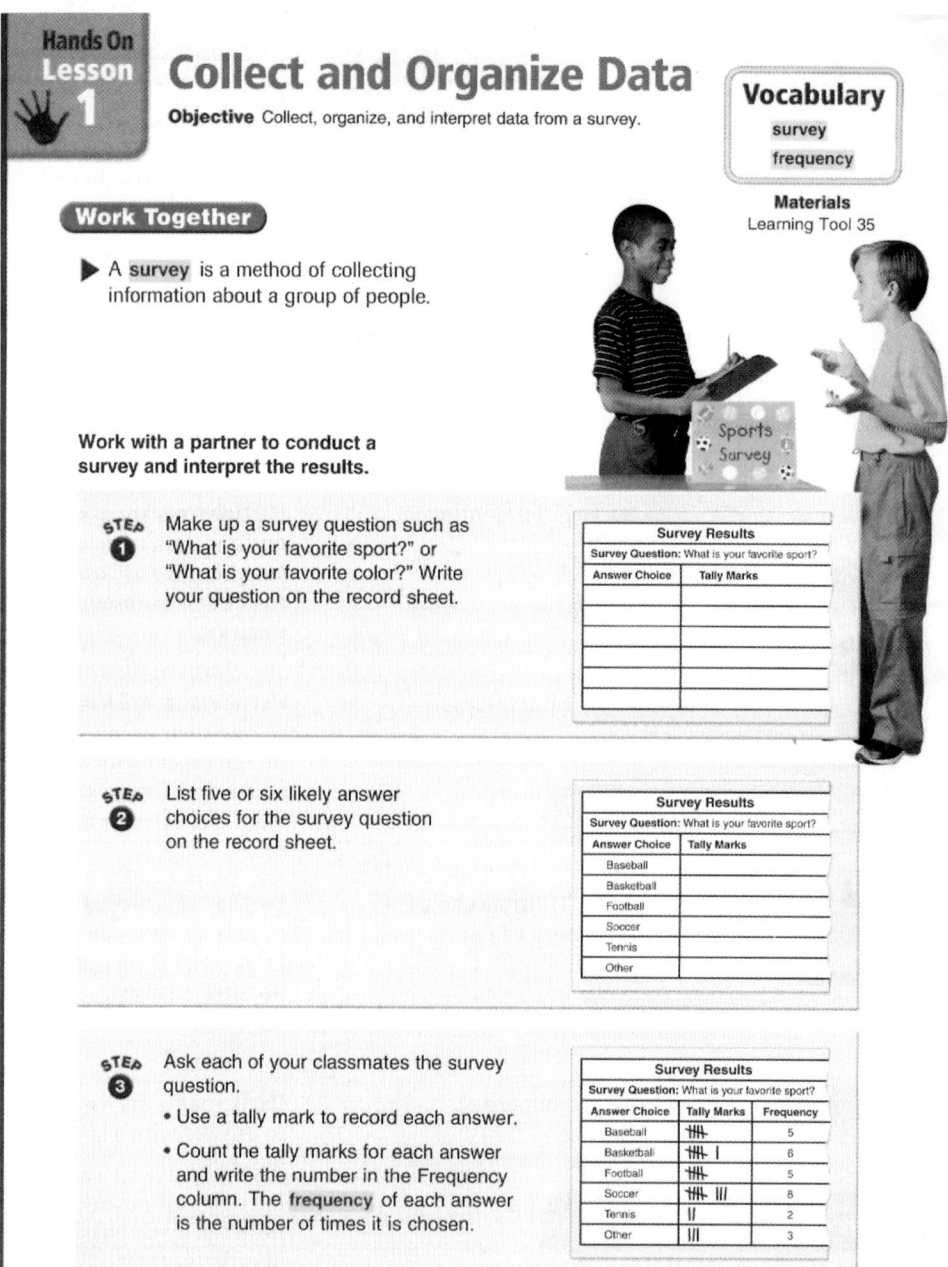

Figure 14.1 A classroom survey

Source: From *Mathematics Grade 5: Houghton Mifflin Mathematics Georgia* (p. 192). Copyright © by Houghton Mifflin Company, Inc. Reproduced by permission of the publisher, Houghton Mifflin Harcourt Publishing Company.

Solution This classroom survey poses two questions: "What is your favorite sport?" and "What is your favorite color?" Because both questions ask the participant to select the best answer from a number of possible choices, both are multiple-choice questions. In each question, the variable is categorical, so it makes sense to collect data by making tally marks in an appropriate table.

In most classroom situations, writing survey questions is a relatively simple matter. However, as the topics become more sophisticated, writing clear, concise, and meaningful questions becomes more difficult. For instance, consider the question, "Is the U.S. court system fair?" At first, the question seems straightforward. However, the variable "fair" is vague, so people may interpret the question differently. To whom is the system fair? Criminals? Victims? Lawyers? And, by "fair" do we mean that the punishment fits the crime or that punishments are more lenient than harsh?

Survey questions can also be written so that participants are more likely to respond in a certain way. For example, consider the question, "Have you ever cheated on an exam?" The wording of the question may cause those who have cheated to answer untruthfully. False responses are particularly likely if the survey is given in a setting where the participants are not guaranteed confidentiality. It would not be surprising if most people answered the question with a no.

Getting survey questions into a final form may take some effort. The survey might even have to go through a test run to work out any problems prior to collecting data. Some questions that might help us write good survey questions are:

- Is the question necessary and useful?
- Are we asking only a single question at one time?
- Can the question be misunderstood by the participants?
- Will we need to ask several related questions to collect the needed data?
- Do participants have the necessary information to answer the question?

Explorations Manual 14.4

Observations

A second way to collect data is to do an **observation.** In this method, the investigator observes and records the characteristics and behaviors of the participants without trying to influence or modify them in any way. Some observational studies are quite simple. For instance, to find the average price of gasoline in a certain city, we could collect data by driving to several gas stations and recording the advertised prices. Other observational studies are more complex. For instance, some studies are **comparative;** they compare two or more groups to find the similarities and differences between them. In many cases, comparative studies are looking for a cause-and-effect relationship between two variables. The variable whose effect we want to study is the **explanatory,** or **independent, variable.** The variable that is affected by the explanatory variable is the **response,** or **dependent, variable.**

Explorations Manual 14.5

Example 14.7   Weight Gain from Cafeteria Food

Henryville High School recently hired a new food service to prepare school lunches. The nurse is concerned that the new lunches are too high in fat and may be causing students to gain weight. She selects a simple random sample of 45 freshmen who daily eat cafeteria lunches, and she weighs each one at the start of the school year. At the end of the year, she reweighs the same students and finds that they have gained an average of 12 lb. She concludes that the food service must improve its menu or face losing its contract. Identify the explanatory variable, the response variable, and the comparison used in the study. State whether the data supports the nurse's conclusion.

Solution The explanatory variable is the fat content in the lunches, and the response variable is the weight of the students. The comparison is made between the weights of the students at the beginning of the year and their weights at year-end. The data may not support the nurse's conclusion that the lunches are the cause of weight gain. Many freshman students are still growing, making it reasonable for them to gain weight over the course of the year. To decide whether there is truly a concern, the nurse could make one of two different comparisons. She could consider the amount of weight gained by freshman classes in previous years, or she could compare students who eat cafeteria lunches to those who do not. These comparisons would provide a much better indication about weight problems in the current freshman class. ∎

In the last example, student growth is called a **confounding, or lurking, variable.** A confounding variable is one whose effect on the response variable is difficult to separate from that of the explanatory variable. When trying to find a cause-and-effect relationship, we should always look for confounding variables because they may cause the appearance of a nonexistent relationship between the explanatory and response variables. By identifying potential confounding variables, we may be able to control their effects by how the study is designed. For instance, in Example 14.7, the nurse could have controlled the confounding variable by making one of the other comparisons.

Experiments

A third way to collect data is to do an experiment. Like other methods of data collection, experiments can be done in a variety of ways. In some experiments, we conduct a large number of trials and analyze the results for similarities and differences. For instance, to test the life span of a new battery, a manufacturer may place several hundred of the batteries into flashlights and leave the flashlights on until the batteries go dead. By recording the life spans of the batteries in the sample, the company determines not only how long the new battery lasts but also how well the new battery will compete in the marketplace.

Other experiments are comparative. In the simplest of such experiments, one group, called the **treatment group,** has a **treatment** deliberately imposed on it to see whether the treatment influences certain traits of the participants. The other group, called the **control group,** does not (Figure 14.2). If the traits of the treatment group are different from those of the control group at the end of some period of time, the investigator has evidence of a cause-and-effect relationship between the explanatory and response variables.

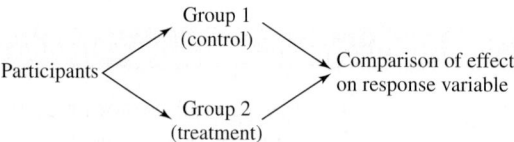

Figure 14.2 Basic experimental design

Example 14.8 The Effectiveness of a New Fertilizer

A fertilizer company has developed a new product that is meant to increase the number of blooms on tomato plants. To test the product, researchers spray it on 150 plants. Another 150 plants do not receive the treatment. They count the number of blooms for each group of plants for 3 weeks and find that the plants sprayed with the product had 25% more blooms.

a. What are the treatment and control groups?

b. What are the explanatory and response variables?

c. Do the results indicate that the product is effective?

Solution

a. The treatment group is the 150 plants sprayed with the product. The control group is the 150 plants that are not.

b. The explanatory variable is treatment with the new fertilizer. The response variable is the number of blooms on the plant.

c. Many confounding variables, such as plant species or water amount, could affect the blooms. However, assuming that variables like these are controlled, then the results seem to indicate that the product is effective at increasing the number of blooms.

As the last example illustrates, we want any effect on the response variable to be directly attributed to the treatment. Statistical experiments must therefore be carefully done so that the effects of any confounding variables are minimized. One way to do this is to use simple random sampling because randomly assigning subjects to treatment groups ensures that any hidden confounding variables between the groups balance each other out.

In some experiments, such as those in medicine, the treatment group is aware of being given a treatment that is expected to improve their condition. This knowledge may cause the participants to respond differently than those in the control group, even though the treatment may not be having any effect. This phenomenon is called the **placebo effect.** To counteract this effect, all participants are made to think they are receiving the treatment. One group is given the treatment, while the other is given a **placebo,** or sugar pill. This is known as a **blind study** because participants are not told whether they are receiving the treatment or the placebo.

Check Your Understanding 14.1-B

1. A question on a survey of hot political topics asks, "Do you support abortion?" The results suggest that most people favor abortion. What inherent biases in the question might lead to a favorable response?

2. Evans has two high schools, each having its own philosophy for teaching mathematics. Evans High School focuses on computational skills, whereas Oakdale High School also includes problem solving, reasoning, and communication. The superintendent wants to know which school is doing better, so she gives a test of computational skills to the sophomores at both schools. She finds that the students at Evans High performed considerably better than the Oakdale students, so she plans to switch the curriculum at Oakdale High.

   a. What type of study is this?

   b. What are the population, sample, explanatory variable, and response variable?

   c. Does the study support the superintendent's conclusion? Why or why not?

Talk About It What is the key difference between a comparison study done with observations and one done as an experiment?

Communication

Activity 14.2

Get together with one of your peers, and write a statistical question that can be addressed using observations and the people in the room as your sample. Collect data needed to answer the question. Do you think you can extend your results to the students attending your college or university? Why or why not?

SECTION **14.1** Learning Assessment

▪ Understanding the Concepts

1. What are the two meanings of the word "statistics?"

2. Briefly describe the four steps of a statistical study.

3. How is posing a statistical question different from posing a mathematical one?

4. What is variability in statistics?

5. How does a qualitative variable differ from a quantitative one?

6. What is sampling, and why is it a necessary part of most statistical studies?

7. **a.** What is bias?

   b. How can bias enter a study through sampling?

8. Briefly describe how to select each type of sample.

   a. Convenience   **b.** Voluntary

   c. Simple random   **d.** Stratified

9. Describe three common ways to collect data.

10. What is a confounding variable?

11. In your own words, describe the design of a simple comparative experiment.

▪ Exercises

12. State whether each variable is qualitative or quantitative.

    a. Blood type

    b. Federal income tax

    c. Paint color of cars

    d. Letter grade assigned to students

    e. Wait time in a checkout line

    f. Rating of hotels as 1 to 5 stars

13. State whether each variable is qualitative or quantitative.

    a. Speeding ticket fines   **b.** Eye color

    c. Daily temperatures   **d.** Religion

    e. Ethnicity   **f.** Time to run 1 mi

14. In each study, identify the sample and the population. Then identify the variable of interest, and state whether it is quantitative or qualitative.

    a. A national newspaper surveys 1,500 American adults to find how many times a week they eat out.

    b. A consumer group tests 14 cars to find the typical gas mileage of all new cars.

    c. A farmer looks at 100 ears of corn to see what percent of his crop is infected with corn worms.

    d. A gym teacher tests the strength of the boys in his freshman classes to find the typical strength of high school freshman boys.

15. In each study, identify the sample and the population. Then identify the variable of interest, and state whether it is quantitative or qualitative.

    a. A principal examines two third-grade classes to find the percent of third-grade students who are reading on a third-grade level at the school.

    b. An independent research firm rates the range of cellular phones by purchasing four phones from different companies and testing them in different locations.

    c. A local health organization determines the number of smokers in a three-county area by telephone-polling 300 people about their smoking habits.

    d. A shoe manufacturer surveys 100 customers on their satisfaction with a new line of sneakers.

16. In each case, identify the sample and the population. Assuming the sample is representative of the population, what can be inferred about the population?

    a. Of a sample of 1,500 freshman selected from different universities in the United States, 64% graduated in four years.

    b. A quality control inspector finds that 0.1% of a sample of 1,000 radios produced at a given factory are defective.

    c. The makers of Mint White toothpaste survey 200 dentists from around the country. They find that 7 out of 10 dentists in the sample recommend Mint White.

17. In each case, identify the sample and the population. Assuming the sample is representative of the population, what can be inferred about the population?

    a. A consumer group polls 200 households in a suburban area. Two-thirds of the households own two or more televisions.

    b. In an Internet survey that asks people about their view on taxes, 74% of the 14,500 respondents thought they pay too much in taxes.

    c. A doctor examines 25 obese patients at a particular hospital. She finds that 20 of them also have chronic joint pain.

18. What sampling technique is used in each scenario?

    a. A television news program asks its viewers, "Do you think the election was fair?" Viewers are to call a 900 number to vote yes and another 900 number to vote no. Each phone call costs $1.00.

b. Linda wants to know the most popular colors for cars. To collect data, she observes the cars in her neighborhood.

c. A pharmaceutical company is testing the effects of different levels of saw palmetto on balding men. They separate the participants of the experiment into three groups: little hair loss, medium hair loss, and extensive hair loss.

19. What sampling technique is used in each scenario?

a. Fifty students at a university are selected at random to answer questions about their satisfaction with the university's financial aid service.

b. Caroline wants to know the favorite food of students in her grade. She asks the students in her class to fill out a short survey.

c. A publisher includes a mail-in survey in a magazine to determine what personal interest stories its readers would like to read.

20. How might each of the following survey questions lead to biased answers?

a. Do you favor capital punishment?

b. Have you ever cheated on your taxes?

c. Do you think the president of the United States is doing a good job?

21. How might each of the following survey questions lead to biased answers?

a. Do you consider yourself overweight?

b. Do you think the principal of the high school should be fired?

c. Have you ever been convicted of a violent crime?

22. Write a question of each type that could be used on a survey to determine whether people view mathematics as useful in their daily lives.

a. Dichotomous   **b.** Interval

c. Multiple-choice   **d.** Open-ended

23. Obesity is a growing concern in today's society. Write a question of each type that could be used on a survey that determines people's concern with obesity.

a. Multiple-choice   **b.** Ordinal

c. Numerical   **d.** Filter

24. What type of data collection method, survey, observation, or experiment could be used to collect data about each topic?

a. The average speed of college football players

b. College student's beliefs about the quality of their education

c. The effect of a new fertilizer on the growth of tomato plants

d. The size of the deer herd living in the state of New York

25. What type of data collection method, survey, observation, or experiment could be used to collect data about each topic?

a. The average number of haircuts a female gets in a year

b. The average GPA of elementary education majors at a particular university

c. The number of defective computers produced at a particular plant

d. American's opinions on the performance of government officials

26. Identify the explanatory variable and response variable in the relationship between each pair of variables.

a. Home size and income

b. Age and height for children from ages 3 to 13

c. Weight and amount of exercise

d. Smoking habits and lung disease

27. Identify the explanatory variable and response variable in the relationship between the variables in each pair.

a. Eating habits and cholesterol levels

b. Grades and hours studied

c. Amount of water and plant growth

d. Car size and gas mileage

28. A team of educational researchers develops a new system for taking notes. To test its effectiveness, they randomly select 10 classes from local schools. Five classes are taught to take notes using the new system. The others are allowed to take notes using their own methods. After 12 weeks, the classes are given a common achievement test. It indicates that, on average, the five classes using the new system of notes scored 5 points higher than the other classes.

a. What are the treatment and control groups?

b. What are the explanatory and response variables?

c. Is the new way of taking notes an effective way to increase grades?

29. A pharmaceutical company has developed a new weight loss product. To test it, they select a sample of 150 people who are interested in losing weight. Fifty people are given the product, 50 people are given a placebo, and 50 people are given nothing. All participants are put on the same diet and exercise program. At the end of the eighth week, the group given the product lost an average of 12 lb, the group given the placebo lost an average of 10 lb, and the group given nothing lost an average of 11 lb.

a. What are the treatment and control groups?

b. What are the explanatory and response variables?

c. Is the new weight loss product effective?

Problems and Applications

30. Michelle works at a photocopying shop, which produces about 15,000 copies a day. She randomly samples 200 copies and finds that 3 have been misprinted. How many copies is the shop likely to misprint in a day?

31. A clothes inspector examines 75 randomly chosen garments a day. If 10 of them have a defect, about how many garments are defective if the factory produces about 9,500 garments a day?

32. Akeem is enrolled in a school district with 9,650 students. He would like to know how many students are African American, Asian, Caucasian, and Hispanic. He selects a random sample of 100 students and finds that 23 are African American, 11 are Asian, 47 are Caucasian, and 19 are Hispanic. About how many students of each ethnic background are enrolled in Akeem's school district?

33. Javier has a large bag of poker chips but does not know how many chips he has. He does know that 30 red chips are in the bag and that the rest are white and blue. He reaches into the bag and randomly selects 15 chips. If 3 are red, what is the best estimate he can make about the number of blue or white chips in the bag?

34. A game warden wants to estimate how many bass are in a small lake. Over the course of a week, he catches and tags 175 different bass. One month later, he returns to the lake and catches another 100 bass, 9 of which are tagged. Using this information, about how many bass are in the lake?

Mathematical Reasoning and Proof

35. The local newspaper in a blue-collar town polls its subscribers about raising the minimum wage. The responses are largely in favor of raising the minimum wage, so the paper concludes that most Americans are in favor of raising it. How might the study be biased?

36. On election days in the United States, exit pollsters commonly try to determine the winner before the votes are counted. If an exit poll is conducted between 2:00 p.m. and 3:00 p.m., how might the results of the poll be biased?

37. A survey is mailed to 10,000 homes to determine what percent of U.S. households are adversely affected by eating disorders. The study claims

that 90% of U.S households have been affected adversely. Can this claim be believed? If so, explain why. If not, what might bias the study?

38. A school study considers the relationship between time spent socializing and the number of days missed due to illness. Why might the students' lack of sleep be a confounding variable to this study?

39. A homeowner studies the relationship between the outside temperature and the cost of his electricity bill. Why might the thickness of insulation on his home be a confounding variable in this study?

40. A TV advertisement for a new weight loss supplement claims that it can help people lose up to 10% more weight than other brands when coupled with exercise and healthy eating habits. What confounding variables might affect the outcome of the study?

41. A study is conducted on the cholesterol levels of middle-aged men. It considers two groups of men with similar eating habits, one from Minnesota and another from Texas. The study monitors the men's eating habits and cholesterol levels for one year, after which the study finds the men from Texas have much lower cholesterol levels. The study concludes that men from Texas possess a better ability to process cholesterol than men from Minnesota. Is the study to be believed?

Mathematical Communication

42. Get together with a small group of your peers, and choose a topic that is relevant to the elementary or middle-grades classroom. Write four or five survey questions related to the topic. Administer your survey to the rest of the students in your class. What difficulties do you encounter in writing and administering the survey?

43. Search the Internet for a website that is conducting a survey on a hot political topic. Go through the survey, and analyze the questions. Write a paragraph or two about how the results of the survey might be biased based on the questions in the survey.

44. The word "random" is often used as a synonym for "haphazard." Is this its meaning in the statistical sense? If not, how is its statistical meaning different?

45. The word "variable" is used in both algebra and statistics. How are algebraic variables and statistical variables similar? How are they different?

Building Connections

46. The U.S. Census plays an important role in our government. Do an Internet search to find out why.

47. Based on what you have learned in this section, how can statistics be used to build connections between mathematics and real-world situations? Discuss this question with a group of your peers.

48. Find a short newspaper, magazine, or television story that involves a statistical study. Identify the population, sample, and variable of interest. Does the story generalize the results of the study to the population? If so, do you think the story is correct in doing so?

49. Consider a set of curriculum materials from third, fourth, or fifth grade. Look at the chapters or units on statistics. Answer the following questions based on what you find.

  a. What statistical topics are covered in the materials?

  b. Do the materials have the students engaged in data collection? If so, what method of data collection do they use?

  c. How does the content in the curriculum materials compare to the expectations given in the NCTM Data Analysis and Probability Standard?

  d. Are you surprised by the statistical content that is taught at this grade level? If so, what surprises you most?

SECTION 14.2 Representing and Analyzing Data with Statistical Graphs

Data, when first collected, is often referred to as **raw data** because they have not been organized in any way. Raw data sets can be difficult to analyze, particularly if they contain many **data points,** or individual values. Consequently, the next step in the statistical process is to organize the data so that we can identify and describe any trends or patterns. One way to organize a data set is to represent it in a statistical graph.

Although there are a wide variety of statistical graphs, all of them share some basic labels that are necessary for understanding the graph. For instance, every graph should have a **title,** or **caption** that briefly explains the data, and if possible, the source of the study (Figure 14.3). Many graphs also have **axis labels,** which describe the variables and data represented by the axes. If the data are qualitative, the axis label shows the categories of the data. If the data are quantitative, the label shows the numerical scale and units corresponding to the data. If the graph represents multiple data sets or categories, it may also include a **legend,** or **key,** that identifies each data set or category.

For the rest of the section, we consider several common types of statistical graphs, demonstrating how to use them to draw simple conclusions from the data.

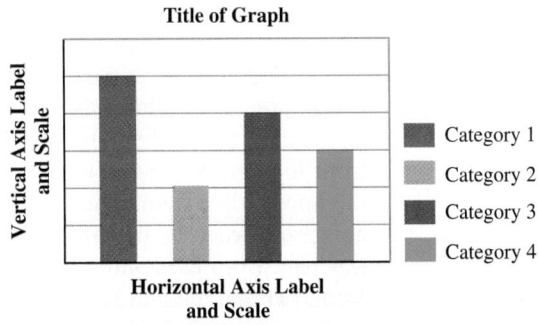

Figure 14.3 Basic features of a statistical graph

Dot Plots

One quick and easy way to organize and display a small set of quantitative data is to use a **dot,** or **line, plot.** To make the graph, we draw a horizontal axis with a numerical scale that includes all possible data values. We then represent each score by placing a dot above the appropriate number on the axis. For multiple scores with the same value, we use multiple dots in the diagram. If needed, a vertical axis can be included to show the frequency of each data value.

For example, suppose the 28 students in Mrs. Donahue's fifth-period history class make the following scores on a 60-point test

50	47	19	45	47	51	59
47	45	31	49	32	30	52
50	50	32	46	51	58	43
31	47	59	46	48	49	57

We can organize the information by making a dot plot like the one in Figure 14.4. Because a large portion of the scale is not used, we condense the length of the graph by replacing the piece of the scale from 0 to 15 with the symbol (-//-).

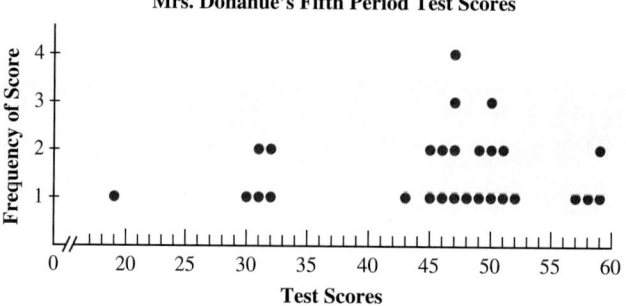

Figure 14.4 A dot plot of test scores

The dot plot not only puts the data set in numerical order but also gives an indication of how the class did. Because most of the scores are above 45, most of the students passed the test. However, one student did considerably worse than the rest of the class with a score of 19. Any such data point that is atypical from the rest of the data set is called an **outlier.** Potential outliers are important to identify because they can have a dramatic effect on certain numerical statistics. The plot also shows that the data set has two **clusters,** which are small groups of data isolated from the bulk of the data. One occurs in the low thirties and another in the high fifties. The spaces between clusters are called **gaps** in the data set.

Stem and Leaf Plots

Another graph that works well with small sets of quantitative data is a **stem and leaf plot.** To make this graph, we first use the high and low values in the data set to find the so-called stems of the plot. For instance, in Mrs. Donahue's class, the low value was 19, and the high value was 59. So we use the tens digits from this range as the stems of the plot, listing them in ascending order down the left side of a line. The unit digits from the data values now serve as the leaves of the plot, listed directly across from the appropriate stem. The leaves can be listed as they were originally recorded or in numerical order. If they are in numerical order, the graph is called an **ordered stem and leaf plot.** Both possibilities are shown for Mrs. Donahue's class in Figure 14.5. In the second graph, the line 3|01122 represents the scores 30, 31, 31, 32, and 32.

Mrs. Donahue's Fifth Period Test Scores

Stems	Leaves		Stems	Leaves
1	9		1	9
2			2	
3	12021		3	01122
4	757759637689		4	355667777899
5	0192001897		5	0001127899

Stem and Leaf Plot   Ordered Stem and Leaf Plot

Figure 14.5 Stem and leaf plots

The choice of the stems and leaves depends on the values in the data set. In some cases, we may want to use two-digit stems. In others, we may want to use two-digit leaves, which we separate with commas. We can also use stem and leaf plots to compare data sets.

Example 14.9   Comparing Two Classes

Representation

Dr. Manciotti gave a 150-point final to his two sections of freshman biology. Each class has 25 students, and the scores from the exams are shown:

Scores Section C890						Scores Section C891				
121	118	110	95	137		145	111	98	86	91
98	137	141	109	143		108	112	99	101	105
133	134	123	98	115		107	110	107	99	119
120	147	142	113	141		123	119	135	141	117
124	99	98	101	125		121	107	144	118	104

How can Dr. Manciotti compare the scores from the exams to determine which class did better?

Solution One way to compare the classes is to organize the data sets into a **back to back stem and leaf plot.** We use both the hundreds and the tens digits as the stems and then place the leaves from the two sections on either side. The plot shows that even though Section C890 had several scores in the 90s, it had more scores in the 120s, 130s, and 140s than Section C891. As a result, we conclude that Section C890 did better on the exam.

Section C890		Section C891
	8	6
98885	9	1899
91	10	1457778
8530	11	0127899
54310	12	13
743	13	5
73211	14	145

Frequency Tables

Dot plots and stem and leaf plots work well only with small sets of numerical data. For qualitative data or a large set of numerical data, we can use a **frequency table** to organize the information. This table lists the possible categories in its first column. If the data are qualitative, the categories are determined by the possibilities for the variable. If the data are quantitative, the categories are ranges of numerical values. Each range should be the same size, and any one data point should fit into only one category. The second column contains the counts of each category. They may be shown as a numerical value or with tally marks. A frequency table may also include a third column that shows the relative frequency of each category expressed as a percent. The relative frequency is the ratio of the frequency of the category to the total number of data values, or

$$\text{Relative frequency} = \frac{\text{frequency}}{\text{number of data values}}.$$

Frequency tables are often used by stores to keep track of their inventory. For example, the frequency table in Table 14.3 shows the type and number of CDs that patrons purchased at a music store. From the table, the owner might conclude that popular music and rap are his two biggest sellers. So he wants not only to keep his stock full in these areas, but he also may want to expand the titles he sells.

Table 14.3 A frequency table of music sold		
Music Genre	**Frequency**	**Relative Frequency**
Classical	12	$\frac{12}{108} = 11.1\%$
Country	14	$\frac{14}{108} = 13.0\%$
Gospel	16	$\frac{16}{108} = 14.8\%$
Hard rock	9	$\frac{9}{108} = 8.3\%$
Popular	34	$\frac{34}{108} = 31.5\%$
Rap	23	$\frac{23}{108} = 21.3\%$

Pictographs

Pictographs are another way to show the counts for qualitative data. However, instead of tallies or counts, a small icon is used to represent a number of objects. The icon is often related to the topic of the graph, and the number of objects it represents is given in a legend. The appropriate number of icons is then placed next to the category to indicate the count. If necessary, partial icons can be used, although they can make it difficult to read exact counts. The music store data given in Table 14.3 are shown in the pictograph in Figure 14.6. The icon, ⊙, represents 5 CDs.

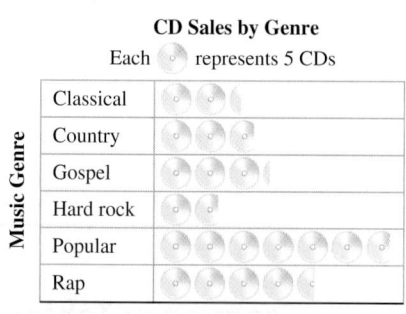

Figure 14.6 A pictograph

Bar Graphs and Histograms

Explorations Manual 14.6

Another way to represent large sets of categorical data is in a bar graph or histogram. Both bar graphs and histograms use the heights of rectangles to show the frequency of each category. They also contain two axes, one to show the categories and another to show the frequency.

Although bar graphs and histograms are similar, there are differences between them. **Bar graphs** are used to represent discrete data, in which each value in the data set is isolated or distinct from all other values in the data set. Studies investigating qualitative variables, such as gender and ethnicity, generally collect discrete data. Because the data do not represent a continuum, the bars in the graph are drawn with spaces between them. A bar graph for the music store data given in Table 14.3 is shown in Figure 14.7.

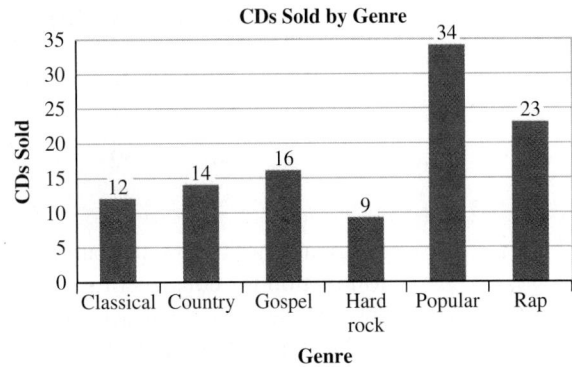

Figure 14.7 A bar graph

Multiple data sets can be represented on a bar graph by using a different set of bars for each data set. In this case, a legend is needed to distinguish the data sets. For instance, suppose the music store owner tracks the sales of CDs over the course of a weekend and collects the data shown in Table 14.4.

Table 14.4 Music sold on various days			
Music Genre	**Friday**	**Saturday**	**Sunday**
Classical	12	14	13
Country	14	17	4
Gospel	16	25	3
Hard rock	9	18	17
Popular	34	52	49
Rap	23	29	27

He can represent the data with a bar graph, as shown in Figure 14.8. The legend enables readers to distinguish between data sets, and grid lines can be added to help them determine the heights of the bars. From the graph, the owner of the music store can see that business was better on Saturday and Sunday than on Friday. He can also see that the sales of certain types of CDs, such as classical and rap, remained somewhat constant over the three days. Others, such as gospel and country, sold well on Friday and Saturday but not on Sunday.

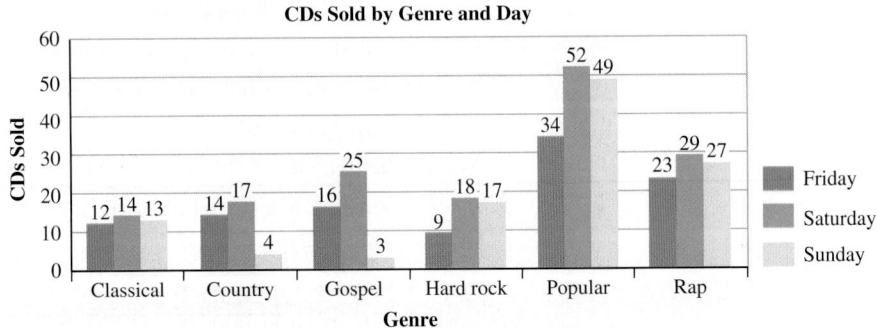

Figure 14.8 A bar graph showing multiple data sets

Histograms differ from bar graphs in that they represent continuous quantitative data rather than qualitative data. To make a histogram, we first subdivide the data so that every data point is included in just one subdivision. We then list the subdivisions in numerical order along the horizontal axis and let the vertical axis represent the frequency of each subdivision. Next, we draw rectangles to show the specific frequency of each subdivision. The rectangles have no gaps between them to indicate that the range of data values is continuous.

For instance, recall the test scores from Mrs. Donahue's fifth-period history class:

50	47	19	45	47	51	59
47	45	31	49	32	30	52
50	50	32	46	51	58	43
31	47	59	46	48	49	57

Originally, we used a stem and leaf plot to represent the data, which we now use as a foundation for constructing a histogram. The stems provide a good way to subdivide the data set, and the leaves reflect the frequency of each subdivision. Both the stem and leaf plot and the histogram are shown in Figure 14.9. In the histogram, we lose the specific values of the data points. However, this loss of information is counterbalanced by the fact that histograms can represent much larger data sets.

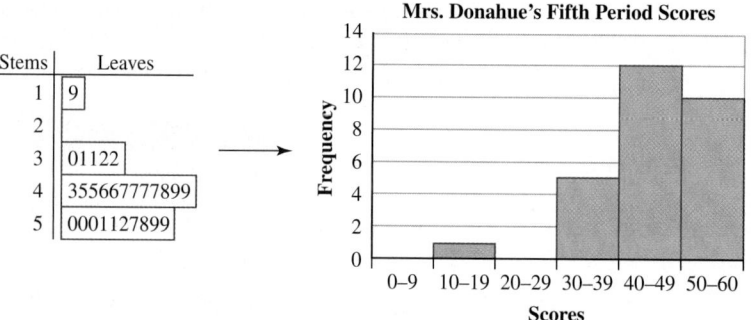

Figure 14.9 A stem and leaf plot and the corresponding histogram

Check Your Understanding 14.2-A

1. Make a dot plot and a stem and leaf plot using the following test scores from a 150-point final exam.

122	131	132	143	119	89	95	92	114	126
141	130	120	118	63	85	88	91	136	147
87	89	112	135	121	124	126	128	132	141

Use your graphs to answer each question.

a. Does the data set have any possible outliers?

b. Does the data set have any clusters?

c. In general, how did the class do on the final exam?

2. Make a histogram of the data in Question 1.

3. A local convenience store collected data on how many drinks of each type were sold on a given day.

Drink Type	Frequency
Coffee	135
Can/bottle of soda	70
Fountain drink	85
Sports drink	20
Bottled water	30

Draw a pictograph of the data using a cup symbol ( ) to represent 10 drinks of any one type.

Talk About It If you had a numerical data set with 150 data points, which statistical graph would you use to represent it? Why?

Communication

Activity 14.3
With a small group of your peers, select a qualitative variable that can be observed in the students in your class. Collect data relevant to the variable, and make a bar graph to represent it. Based on what you see in the data, write several questions that are appropriate for the elementary grades and that can be answered by looking at the graph.

Line Graphs

Another common type of statistical graph is a **line graph,** which is typically used to represent the trends of a variable over a period of time. To make a line graph, we first place the needed time interval along the horizontal axis and the appropriate numerical scale on the vertical axis. Next, points are plotted on the graph using ordered pairs of numbers in which the first number is a time and the second number is the corresponding data value for that time. We then connect the points with line segments to complete the graph. Not only do the line segments indicate the continuous nature of the data, but they also make it possible to estimate values that are not explicitly given.

Example 14.10   The Median Age of the U.S. Population

Connections

The following table gives the median, or middle-most age, of the U.S. population for each decade from 1900 to 2000.

Year	1900	1910	1920	1930	1940	1950	1960	1970	1980	1990	2000
Median Age	22.9	24.1	25.3	26.4	29.0	30.2	29.5	28.0	30.0	32.8	35.7

Source: United States Census Bureau.

Use the data to estimate the median age for 1945, 1955, and 1985.

Solution One way to estimate the median ages for the given years is to make a line graph and estimate the values directly. The line graph for the given data is shown:

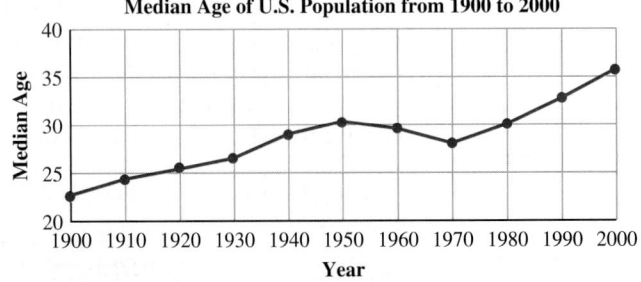

Median Age of U.S. Population from 1900 to 2000

Using the values from the scales, the median age in 1945 was about 29.5, in 1955 it was about 29.8, and in 1985 it was about 31.5.

One advantage of line graphs is that it is relatively easy to use them to represent or compare multiple sets of continuous data. For instance, the graph in Figure 14.10 shows the number of accidents that males and females have had along a certain highway over the course of a year. The graph shows that the number of accidents involving male drivers generally remains constant throughout the year. The number of accidents involving female drivers is highest in the winter months and decreases during the summer. Regardless of the time of year, the number of accidents involving male drivers is almost always higher than those involving female drivers.

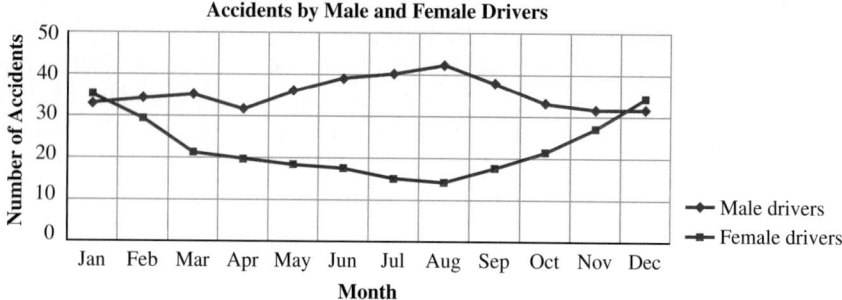

Figure 14.10 A line graph comparing two data sets

Circle or Pie Graphs

Another way to represent categorical data is with a **circle,** or **pie, graph.** Circle graphs differ from other graphs because they show the relative frequency of each category to the whole, usually by showing the percent that each category is of the whole. To represent these relative amounts, the circle is divided into wedge-shaped regions, each of which has an area proportional to the percent of the category.

Example 14.11   The Ethnicity of Students in a School

Reasoning

The administrators of Sunnydale High School have done a survey of the ethnic diversity of the school and have obtained the following data:

Ethnic Group	African American	Asian	Caucasian	Hispanic	Native American
Number of Students	211	74	753	180	17

They want to make a graph that shows the relationship of each ethnic group to the entire school population. What might such a graph look like?

Solution Because the administrators at Sunnydale want to show part-to-whole relationships, a circle graph is their best choice. We first find the total number of students, which is 1,235, and then convert each count to a percent. Next, we find the central angle of each sector of the graph, which must have areas that are proportional to the percents of the ethnic groups. To do so, we multiply each percent by 360°, the degree measure of one complete rotation. Because the sum of the central angles must be 360°, we round appropriately as we make the computations. The results are shown:

Ethnic Group	Percent of Student Body	Approximate Central Angle
African American	$\frac{211}{1,235} \approx 0.1709 = 17.09\%$	$17.09\% \cdot 360° = 61.52° \approx 62°$
Asian	$\frac{74}{1,235} \approx 0.0599 = 5.99\%$	$5.99\% \cdot 360° = 21.56° \approx 22°$
Caucasian	$\frac{753}{1,235} \approx 0.6097 = 60.97\%$	$60.97\% \cdot 360° = 219.49° \approx 219°$
Hispanic	$\frac{180}{1,235} \approx 0.1457 = 14.57\%$	$14.57\% \cdot 360° = 52.45° \approx 52°$
Native American	$\frac{17}{1,235} \approx 0.0137 = 1.38\%$	$1.38\% \cdot 360° = 4.97° \approx 5°$

Explorations
Manual
14.7
From this point, we could draw the graph by hand by using a compass and protractor. However, using a spreadsheet or other statistical graphing package is more efficient and accurate. The resulting circle graph is shown:

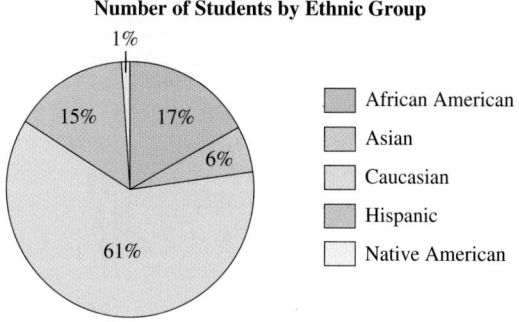

Note the importance of the legend for clarifying the information contained in the graph.

Explorations
Manual
14.8
We can also use circle graphs to compare two or more data sets. For instance, suppose Sunnydale High collected data in 2000 and 2005. The results of the surveys can be shown by two circle graphs placed side by side, such as those in Figure 14.11. The differences in the wedges shows the changes in the student body even though exact numbers are not given. For instance, the percent of Caucasian students has decreased over five years, indicating that the student body is becoming more diverse. If we forget that pie graphs usually give percents, not exact values, we might make mistakes in our analysis. For instance, a decrease in the percent of African Americans from 2000 to 2005 does not mean that the total number of African Americans decreased. Rather, it means only that the percent of African Americans decreased in comparison to the entire student body, which may have grown considerably over the five years.

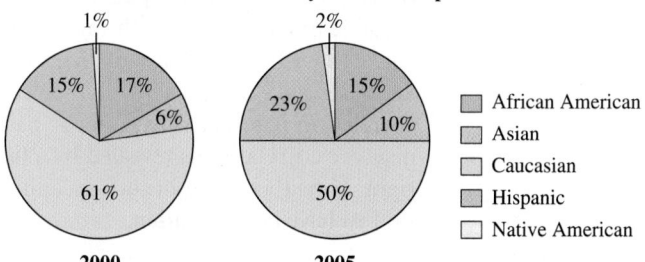

Figure 14.11 Comparing data with two circle graphs

Scatterplots

The final type of statistical graph is the **scatterplot,** which helps to determine whether a relationship exists between two variables, such as the age and height of a person or the weight and gas mileage of a car. To make a scatterplot, we first draw two axes, one for each variable. We then plot each data point by using its values for each variable as an ordered pair. For instance, consider the heights and weights of 21 children given in Table 14.5. The scatterplot for this data is shown in Figure 14.12.

Any relationship between the two variables is shown by a pattern in the dots of the scatterplot. For instance, the linear pattern in Figure 14.12 shows that weight is closely related to height. Specifically, taller heights tend to correspond to heavier weights, and shorter heights tend to correspond to lighter weights. When such a relationship exists between two variables, they are said to be linearly **correlated.** Correlation is a claim of relationship, not causality. If two variables are correlated, one is not necessarily the cause of the other.

Table 14.5 Heights and weights of 21 children					
Height (in.)	Weight (lb)	Height (in.)	Weight (lb)	Height (in.)	Weight (lb)
47	78	57	101	56	93
48	85	48	85	53	88
53	89	51	92	47	79
54	91	53	97	50	90
62	115	58	105	53	95
55	95	58	101	58	102
61	111	61	113	47	89

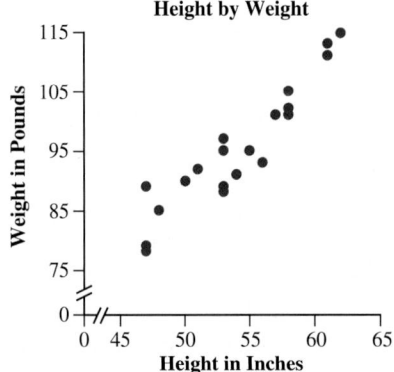

Figure 14.12 A scatterplot of height by weight

Two variables can be linearly correlated in three ways (Figure 14.13). If high values of one variable are associated with high values of the other and low values of one variable are associated with low values of the other, then the variables have a **positive correlation.** If high values of one variable are associated with low values of the other and vice versa, then the variables have a **negative correlation.** In a scatterplot, a positive correlation is revealed by a linear pattern that slopes up from the left to the right. A negative correlation is revealed by a linear pattern that slopes down from the left to the right. If neither trend occurs, meaning there is no linear pattern, then the variables are said to have **no correlation.**

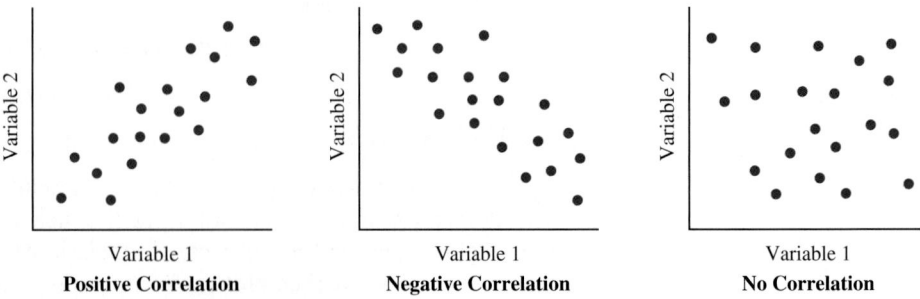

Figure 14.13 Possible correlations between two variables

If a data set has either a positive or a negative correlation, then the linear pattern can be represented with a **trend line,** or **regression line** (Figure 14.14). In statistics, we generally try to find the **line of best fit,** which is the one line that best describes the linear pattern of the correlation. It requires extensive calculations to find, so it goes beyond the scope of this text. However, we can use what we know about linear functions to estimate the line of best fit. Once we have an equation for the line, we approximate the values of one variable, given values of the other.

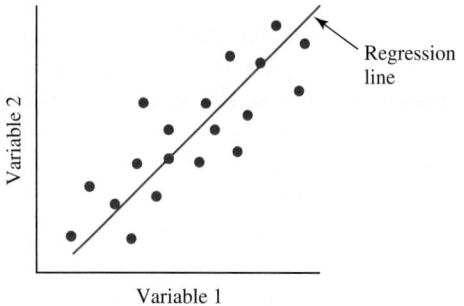

Figure 14.14 A regression line

Example 14.12   Finding and Using a Regression Line

Problem Solving

Consider the height and weight data in Table 14.5. Find a linear function that can be used to estimate the weight of a person who is 60 in. tall.

Solution

Understanding the Problem. We are asked to find a linear function that can be used to estimate the weight of a person who is 60 in. tall. We have little to go on, but we do know from Figure 14.12 that the data set has a positive correlation.

Devising a Plan. If we can find two points that lie on a regression line, we can use what we learned in Chapter 9 to *find the formula* for a linear equation. We can then use the equation to estimate the needed value.

Carrying Out the Plan. In most cases, sophisticated computing techniques are needed to find the line of best fit. However, notice that the line connecting the highest and lowest weight values is reasonably close to such a line, so we use these points to generate our linear function.

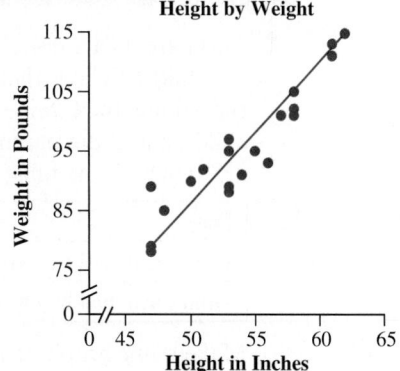

Because the coordinates of the two points are (47, 79) and (62, 115), the slope of the line connecting them is $m = \dfrac{115 - 79}{62 - 47} = \dfrac{36}{15} = \dfrac{12}{5}$. Substituting the slope and the coordinates from the second point into the point-slope formula, we have

$$(y - 115) = \frac{12}{5}(x - 62)$$

$$(y - 115) = \frac{12}{5}x - \frac{744}{5}$$

$$y = \frac{12}{5}x - \frac{169}{5}$$

With this equation, we now estimate the weight of a 60-in. person to be

$$y = \frac{12}{5}(60) - \frac{169}{5} = 144 - \frac{169}{5} = \frac{551}{5} = 110.2\,\text{lb}.$$

Explorations Manual 14.9

Looking Back. Our linear equation enables us to find an estimate not only for any height between 45 and 62 in., but also for heights beyond this range. For instance, we can use the equation to estimate the weight of a 70-in.-tall person at 134 lb. However, when estimating values outside the range of the data set, we should be cautious of the results. We cannot know for sure whether the function is giving us a good estimate. To be more confident in its accuracy, we would need to collect data from taller people and then change the equation appropriately.

■

Selecting an Appropriate Graph

As mentioned throughout this section, each type of statistical graph is used to represent different types of data. Because selecting the most appropriate graph is not always easy, the following summary can be used to help make this decision.

■ *Dot plots* are used to show all numerical values of small data sets.
■ *Stem and leaf plots* are used to compare numerical values of small data sets.
■ *Frequency tables* are used to show counts in categorical data sets.
■ *Pictographs* are used to show and compare counts in categorical data sets.
■ *Bar graphs* are used to show and compare data arranged in qualitative categories.
■ *Histograms* are used to show data arranged in continuous numerical categories.
■ *Line graphs* are used to show and compare data relative to a continuous variable such as time.
■ *Circle or pie graphs* are used to show the relative amounts of parts to the whole.
■ *Scatterplots* are used to show the relationship between two variables.

Check Your Understanding 14.2-B

1. Two parents are concerned that their children watch too much television, but the children claim that the parents watch just as much as they do. To settle the debate, the family decides to keep track of how many hours both the children and the parents watch over a 10-day period. The data they collect are shown in the table:

Day	1	2	3	4	5	6	7	8	9	10
Hours watched by parents	3	4	5	2	6	3	5	3	5	6
Hours watched by children	4	2	2	3	2	4	5	2	2	5

Make a line graph of the data, and summarize what it shows.

2. Answer the following questions using the information provided in the pie chart.

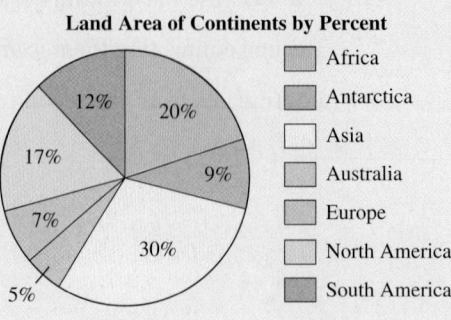

Land Area of Continents by Percent

□ Africa
□ Antarctica
□ Asia
□ Australia
□ Europe
□ North America
□ South America

a. List the continents by size from smallest to largest.

b. Give two pairs of continents with about the same land area as Asia.

c. If the land area of the world is about 57,393,000 mi^2, about how many square miles is Europe?

3. For each of the following variables, give one variable that you think is positively correlated to it and another that is negatively correlated to it.

a. Weight gain

b. Test scores

Talk About It Allison measures the growth of a tomato plant over the course of 60 days. What type of statistical graph might she use to best represent her data? Why?

Reasoning

Activity 14.4
With one of your peers, select two variables that pertain to the people in your classroom and that you think are either positively or negatively correlated. Collect the appropriate data from the students in your class. Summarize the data in a scatterplot, and then decide whether your hypothesis about the correlation was correct.

SECTION **14.2** Learning Assessment

■ Understanding the Concepts

1. What are the key labeling features of any statistical graph?

2. What is an outlier? A cluster? A gap?

3. Describe how each pair of statistical graphs is similar and different.

   a. Dot plots and stem and leaf plots

   b. Pictographs and bar graphs

   c. Bar graphs and histograms

   d. Bar graphs and circle graphs

4. What type(s) of statistical graph can be used to:

   a. Represent qualitative data?

   b. Represent quantitative data?

   c. Compare two or more data sets?

5. How is a scatterplot made, and what is it used to show?

6. What does it mean if two variables are positively correlated? Negatively correlated?

7. If two variables are correlated, is one the cause of the other? Explain.

8. What is a regression line?

■ Representing the Mathematics

9. Fingerprints are often described by a ridge count, which is the number of ridges in the loops of the patterns. The following numbers represent the ridge counts of 24 people:

187	183	201	213	197	205	205	179
188	193	184	219	211	201	189	199
204	215	186	192	199	201	220	193

   a. Make a dot plot of the data.

   b. Make a stem and leaf plot of the data.

   c. Which do you think better represents the data and why?

10. The following data represent the times of 20 sprinters in the 100-yd dash. All numbers are given in seconds.

11.3	12.1	11.4	11.6	11.9	11.9	12.6	12.0	12.1	11.7
12.2	12.4	11.6	11.8	11.9	12.8	12.9	12.0	11.7	11.2

    a. Make a dot plot of the data.

    b. Make a stem and leaf plot of the data.

    c. In which graph is it easier for you to find the fastest and slowest times? Why?

11. Roll a standard six-sided die 60 times, and record the results in a frequency table. Use the frequency table to draw a bar graph of the data.

12. Make a frequency table and a bar graph for the grades of Mr. Prall's eighth-grade English class. Overall, how would you say the class did?

A C C B B D B F A A
F C B D B B B D B C
A A C F A C C C D B

13. Construct a histogram of the data shown in the stem and leaf plot.

Stem	Leaves
1	78
2	0456778
3	112335559
4	2677
5	13

How is the histogram similar to the stem and leaf plot? How is it different?

14. A Christmas tree farm has kept track of the trees it has sold over the past five days in the following frequency table:

Day	Tuesday	Wednesday	Thursday	Friday	Saturday
Number of Trees	25	35	40	85	95

a. Make a pictograph of the data using a tree symbol to represent 10 trees sold.

b. Make a bar graph of the same data.

c. Which do you think better represents the data, and why?

15. A deli keeps a daily record of the number of sandwiches it sells. On a certain day, it collected the data shown in the table. Make a pictograph of the data.

Sandwich	Burger	Rueben	Steak and Cheese	BLT	Chicken Salad
Number Sold	95	25	40	45	35

16. The following table contains the typical family size in the United States from 1940 to 1995.

Year	1940	1950	1960	1965	1970	1975	1980	1985	1990	1995
Family Size	3.76	3.54	3.67	3.70	3.58	3.42	3.29	3.23	3.17	3.19

Source: U.S. Census Bureau.

Make a line graph of the data, and use it to describe any general trends you see.

17. Mrs. Allen takes a quick survey of 15 middle school students to see how many hours a week they spend studying or doing homework assignments. She compares the times to the students' percents in the class.

Hours Studied	%	Hours Studied	%	Hours Studied	%
3	68%	7	91%	6	81%
4	72%	8	89%	6	87%
2	65%	3	75%	5	91%
5	83%	4	79%	10	95%
5	79%	4	63%	9	82%

Make a scatterplot of the data, and determine whether there is a correlation between the two variables.

18. The following table gives the life expectancies of citizens in different countries. It also gives the number of people per physician.

Country	Life Expectancy	People per Physician
Argentina	70.5	370
Canada	76.5	449
China	70	643
Egypt	60.5	616
India	57.5	2471
Iran	64.5	2992
Italy	78.5	233
Japan	79	609
Kenya	61	7615
Mexico	72	600
Myanmar	54.5	3485
Pakistan	56.5	2364
Peru	64.5	1016
Poland	73	480
Spain	78.5	275
Sudan	53	12550
Ukraine	70.5	226
United Kingdom	76	611
United States	75.5	404
Vietnam	65	3096

Source: The World Almanac and Book of Facts.

Make a scatterplot of the data, and determine whether there is a correlation between the variables. If there is one, explain why.

■ Exercises

19. The following dot plot shows the height in inches of 20 students in a high school mathematics class.

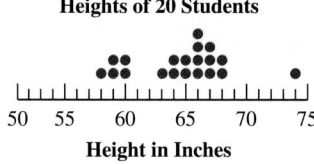

Heights of 20 Students

Height in Inches

a. What are the tallest and shortest heights in the class?

b. What is the most commonly occurring height?

c. Are any possible outliers in the data set? If so, at what value?

20. The following stem and leaf plot shows the scores that 33 students received on a 150-point examine. Use it to answer each question.

Stems	Leaves
6	1
7	
8	
9	358
10	00123346
11	0114467889
12	122236889
13	779

a. What were the highest and lowest scores in the class?

b. What was the most commonly occurring score?

c. Are there any potential outliers in the data set?

d. Overall, how did the class do?

21. Use the graph to answer the questions.

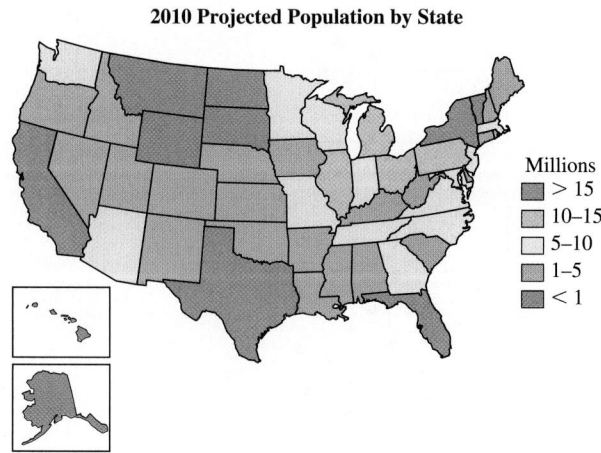

2010 Projected Population by State

Millions
- > 15
- 10–15
- 5–10
- 1–5
- < 1

Source: U.S. Census Bureau.

a. How many states have a population above 15 million?

b. What states have a population between 10 and 15 million?

c. How many states have a population of 10 million or less?

d. What percent of the states have a population of 10 million or more?

22. Answer each question using the information provided in the following bar graph.

Top 6 Retail Companies in 2010

Source: Stores.org.

a. Which company had the most sales, and what was the amount?

b. About how many more times were the sales of Walmart than Target?

c. Which companies had more than and which had less than $60 billion in sales?

23. The following double bar graph shows the electrical usage of a household in kilowatts per month in 2011 and 2012.

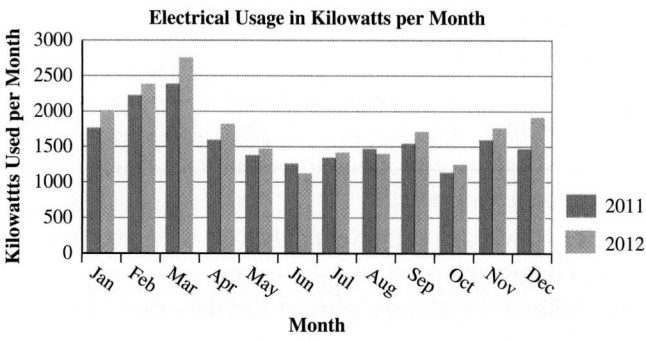

Electrical Usage in Kilowatts per Month

a. In which month was the most overall electricity used? The least electricity?

b. In which months in 2011 was the electrical usage more than in 2012?

c. In which year did the household use less electricity?

24. Use the following line graph to answer the questions.

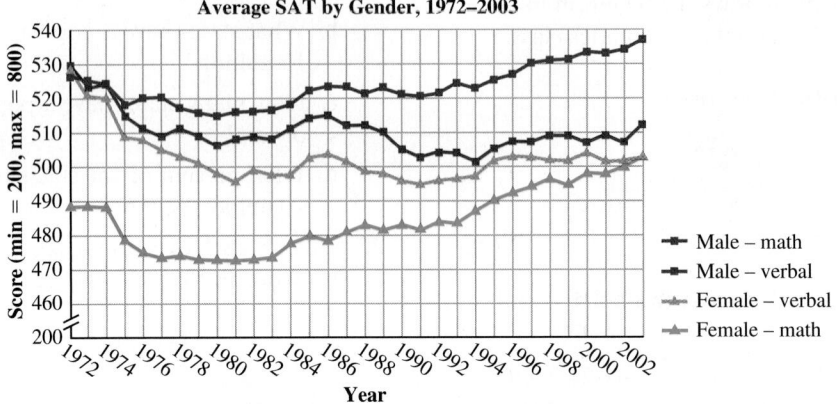

Source: The College Board. Total Group Profile Report

a. What trends does the graph show about changes in male and female math scores over the course of time?

b. Did females perform worse or better on the SAT in 1972 than in 2003?

c. What trends does the graph show about male scores on the verbal portion of the SAT?

d. In which area are females now performing as well as, if not better than, males on the SAT?

e. Over the course of time, which gender has done better on the SAT as a whole?

---

25. A company employs 3,450 workers. A recent survey was taken to determine the educational levels of the employees. The results are shown in the pie graph.

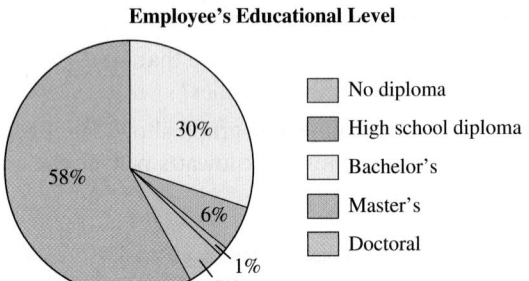

a. What level of education do most employees hold?

b. What percent of employees have a master's degree or better?

c. How many employees only have a high school diploma?

26. Make a list of three pairs of variables that you think have a:

a. Positive correlation.

b. Negative correlation.

c. No correlation.

27. State whether each scatterplot shows a positive correlation, a negative correlation, or no correlation.

a.

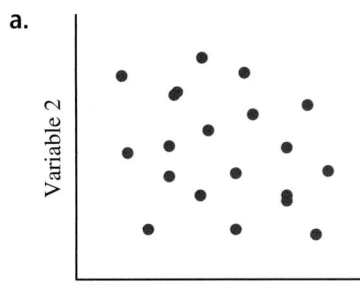

b.

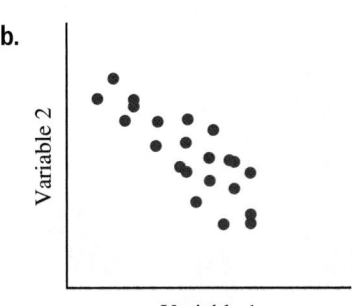

c.

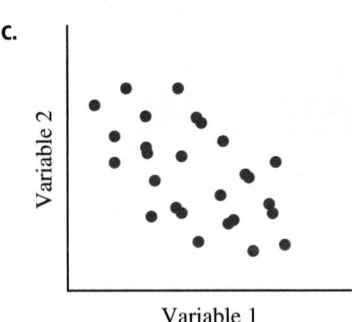

d.

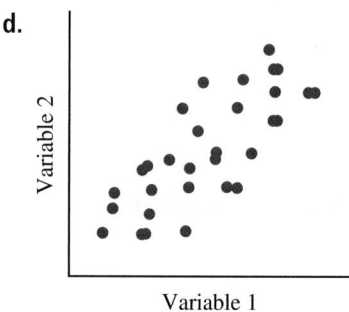

28. Which of the scatterplots in the previous exercise do you think shows the strongest relationship, positive or negative, between the two variables? Why?

■ Problems and Applications

29. The following data represent the ages of the teachers at Fairview Elementary School.

64	25	23	37	41	45	48	52	27	29
62	63	57	49	47	48	36	37	31	30
29	45	45	62	68	57	24	46	38	29

  a. Make a dot plot of the data.

  b. How many teachers are between 30 and 40 years of age, inclusive?

  c. What percent of the teachers are strictly under 30 years old?

30. Answer each question based on the information provided in the graph.

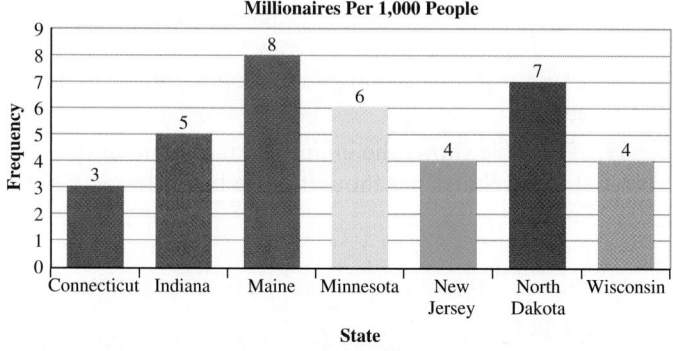

  a. Which state has the most millionaires per thousand people? The least?

  b. If the population of Maine is 1,316,456, about how many millionaires might we expect to be living in Maine?

  c. If the population of Connecticut is 3,501,252 and the population of North Dakota is 641,481, which state has more millionaires living in it?

  d. About what percent of the population of New Jersey are millionaires?

31. Answer each question based on the information given in the graph.

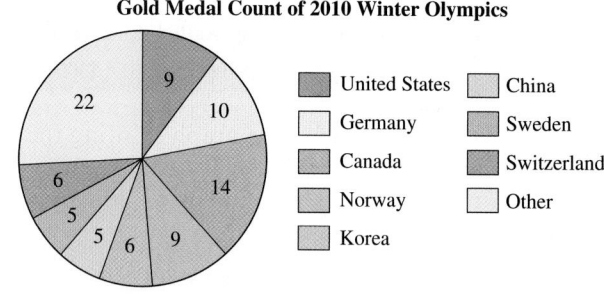

Source: Olympic.org.

  a. Which country won the most gold medals?

  b. What percent of the gold medals was won by Germany? By Canada?

  c. Is it possible to tell which country won the fewest gold medals?

32. A car lot has 200 cars on it. The cars are arranged by colors, which include black, blue, green, red, silver, and white. The circle graph shows the percents of the colors of the cars.

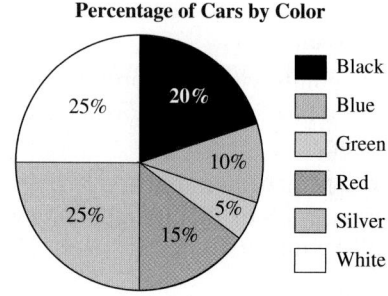

  a. How many cars of each color are there?

  b. How would the percents change if 10 white cars were replaced with 10 silver cars?

  c. How many red cars would have to be replaced with green cars to give the two colors the same percent?

  d. If the number of cars of each color were doubled, what would happen to the percents?

33. The Carduccis have separated their monthly expenses into the categories shown in the table.

Savings	Housing	Food	Clothing	Transportation	Entertainment
$400	$1,750	$800	$150	$550	$200

  If they bring home $3,850 a month in pay, use a protractor to accurately draw a circle graph showing the percents of each category.

34. Suppose the Carduccis in the previous problem have the same take-home pay but have the following expenditures in one particular month.

Savings	Housing	Food	Clothing	Transportation	Entertainment
$400	$1,750	$850	$250	$700	$300

Is it possible to a draw circle graph showing the relative amounts of each category with regard to their take-home pay with the data? Why or why not?

35. The following scatterplot shows a comparison between the years of service at a university and the pay of faculty members.

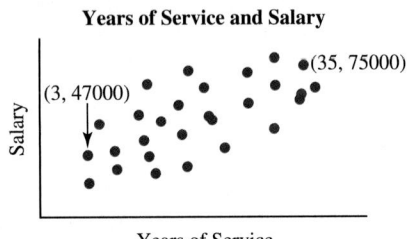

Years of Service and Salary

a. Is there a positive, a negative, or no correlation between years of service and pay? Does this make sense?

b. Use the points with the coordinates to find a regression line for the data. What is the equation of this line?

c. Use your equation in part b to find an estimate for the salaries of faculty members who have served the university for 13 years, 20 years, and 31 years.

■ **Mathematical Reasoning and Proof**

36. The following stem and leaf plot shows the scores of two different science classes that took the same exam.

First Period		Second Period
21	5	
64	6	128
5442	7	33345567
754433	8	11278
776543211	9	34452
	10	0

Which class do you think did better? Explain your reasoning.

37. A county government is concerned about the poverty rates of different ethnic backgrounds. Investigators look at county tax records for 2000 and 2010. The results of their study are shown in the circle graphs.

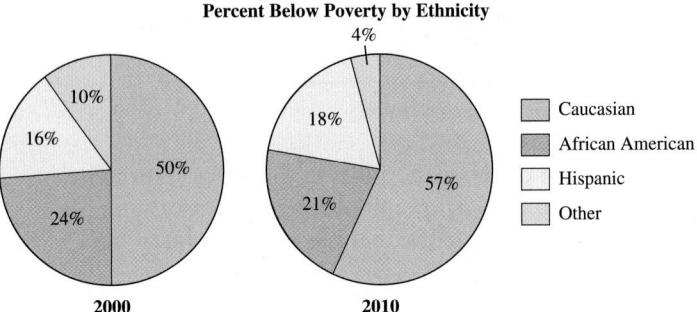

Percent Below Poverty by Ethnicity

What trends in poverty rates do the two graphs show?

38. The following table shows the heights of six of the tallest buildings in the world.

Building	Height (ft)
Taipei 101, Taiwan	1,667
Petronas Tower, Kuala Lumpur	1,483
Willis Tower, Chicago	1,451
Jin Mao Building, Shanghai	1,381
Two International Finance Center, Hong Kong	1,362
CITIC Plaza, Guangzhou	1,283

Source: Wikipedia.org.

What is the most appropriate type of graph for displaying the data? Draw the graph, and explain your reasoning for selecting this type of graph.

39. The table shows the infant mortality rate in the United States per thousand live births from 1920 to 1980.

Year	1920	1930	1940	1950	1960	1970	1980
Mortality Rate	86	65	47	29	26	20	13

Source: Centers for Disease Control and Prevention.

a. What is the most appropriate type of graph for displaying the data?

b. Use the graph type you selected in part a to represent the data.

c. How has the infant mortality rate changed from 1920 to 1980?

40. The following table shows the body and brain weights of several domestic animals.

Animal	Cow	Donkey	Horse	Sheep	Pig
Body weight (kg)	465	187	521	55	192
Brain weight (g)	423	419	655	175	180

a. What is the most appropriate type of graph for displaying the data?

b. Does the difference in the units affect your choice of graph?

c. Use the graph type you selected in part a to represent the data.

d. Based on what you see in the data, which animal do you think is the smartest? How did you arrive at your conclusion?

41. The following table shows the land areas of 10 of the world's largest cities and their approximate populations in 2010.

City	Population (millions)	Area in (km²)
Beijing	10.1	1,368
Delhi	12.6	431
Istanbul	12.5	1,831
Karachi	13.0	3,527
Mexico City	8.9	1,485
Moscow	10.6	1,081
Mumbai	13.8	603
Sao Paulo	11.2	1,523
Seoul	10.5	605
Shanghai	13.8	1,928

Source: Wikipedia.org.

a. What is the most appropriate type of graph for displaying the data?

b. Use the graph type you selected in part a to represent the data in the table.

c. What would be an effective way to compare how crowded the cities are with respect to one another?

d. Based on your answer to part c, which city is the most crowded?

■ Mathematical Communication

42. Write a paragraph or two summarizing the advantages and disadvantages of representing and analyzing data with statistical graphs.

43. The following data represent the weights of 20 children in a fifth-grade class.

69	65	71	82	85	86	80	80	91	92
67	83	84	75	78	78	79	86	90	79

a. Write a paragraph or two to describe the process you would use to create a bar graph that represents the data with only four bars.

b. Write two or three questions that you might ask your students that can be answered by analyzing the bar graph.

44. The following table represents the average weekly earnings of factory workers from 2000 to 2005.

Year	2000	2001	2002	2003	2004	2005
Earnings	$940	$980	$1,000	$1,100	$1,100	$1,200

a. Write a paragraph or two that describes the process you would use to create a line graph that represents the data.

b. Write two or three questions that you might ask your students that can be answered by analyzing the bar graph.

45. Stephen claims that the data in a pie graph must always be in terms of percents. Is he correct?

46. Ruis says that height and weight cannot be positively correlated because his father is tall and his mother is short, but they weigh the same. How would you, as his teacher, address his misconception?

■ Building Connections

47. How are line graphs related to the idea of a function?

48. What skills and facts from geometry do we need to know in order to accurately produce a pie graph by hand?

49. Explore several newspapers to find examples of statistical graphs. Pick one or two, and summarize the data represented by the graphs. Be sure to describe the population and the topic of the graph. Also explain what type of graph is used and what patterns and trends it shows in the data.

50. Do you think its important for the average American to know how to make and analyze statistical graphs? If so, for what reasons? Discuss with a small group of your peers.

51. Explore a set of curriculum materials for grades kindergarten through fourth grade.

a. What types of statistical graphs are taught, and at what grade level are they taught?

b. What statistical-thinking skills are often associated with these lessons?

SECTION 14.3   Representing and Analyzing Data with Descriptive Statistics

Even though statistical graphs are a powerful way to organize and analyze data, often the information we want is not easy to ascertain from a graph. For instance, suppose two classes take the same 50-point test, and we want to know which class did better. We might begin an analysis by representing each set of scores with a dot plot like the ones shown in Figure 14.15. Looking at the graphs, we see that the different scores between the classes make a comparison difficult. Initially, we might try to use the high scores. Certainly, third period has higher scores, but they are offset by lower scores, making it difficult to determine the overall performance of the class. We might also try to compare a typical score from each class. In first period, 35 occurs most often, and many of the scores are centered around it. However, in third period, picking such a score is difficult because the data are more widespread. Any score from 28 to 37 would be a reasonable choice, and, because scores in this range lead to different conclusions, we would always be uncertain about our comparison.

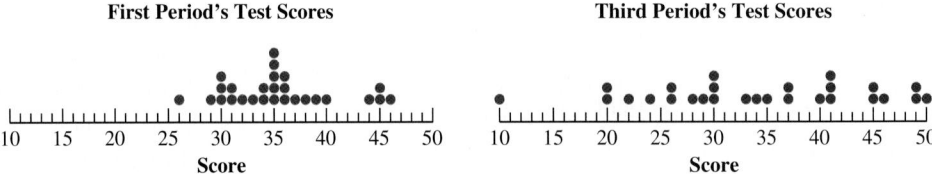

Figure 14.15 Dot plots of the test scores for two classes

As a consequence, we need another way to represent data sets so that we can make comparisons between them. One way is to use **descriptive statistics** to represent the data numerically. In general, there are two types of descriptive statistics: measures of center and measures of spread.

Measures of Center

In Figure 14.14, we tried to compare two data sets by selecting one score from each that best represented the data. Any such number that indicates how the data set is centered is called a **measure of center,** or a **measure of central tendency.** Common measures of center are the mode, the median, and the mean.

Mode

One way to select a measure of center is to choose the score that occurs most often. This number, when it occurs, is called the **mode.**

Definition of the Mode

The **mode** of a set of n numbers, $x_1, x_2, x_3, \ldots, x_n$, is the numerical value in the data that occurs most frequently. If two numbers occur equally often and more frequently than any other, then there are two modes, and the data set is **bimodal.**

The definition implies that multiple modes are possible. However, if too many scores occur equally often and more frequently than any other number, then the measure of the mode becomes meaningless. In this case, we say the data set has no mode.

Example 14.13   Using the Mode to Compare Data Sets

The following data from two first-grade classes show the heights of the students in inches. Use the mode to decide which class is the "tallest."

Class 1	42 46 47 45 46 45 40 42 43 45 47 48 49 45 44
Class 2	43 44 45 44 47 40 41 44 46 47 50 46 43 48 45

Solution The mode of class 1 is 45 in., and the mode of class 2 is 44 in. Because 45 is larger than 44, we conclude that class 1 is taller.

Median

Another measure of center is the **median,** which is the number that occurs in the middle of the data set. To find the "middle," we first arrange the data values in ascending or descending order. Next, finding the median depends on whether the set has an even or an odd number of data points. If there are an odd number of data points, then the median is the middle-most number. Specifically, if there are n numbers in the data set, then the median is the $\frac{n+1}{2}$ number in the list. If there is an even number of data points, then we split the difference between the middle two numbers. That is, the median is one-half the sum of the middle two numbers.

Definition of the Median

Let $x_1, x_2, x_3, \ldots, x_n$, be a set of n numbers arranged in ascending or descending order. If n is odd, the **median** is the number in the $\frac{n+1}{2}$ position of the list. If n is even, then the median is one-half the sum of the middle two numbers.

Example 14.14   Using the Median to Compare Data Sets

The following data are the scores of two classes after taking the same 25-point quiz. Use the median to determine which class did better.

Class 1	13 24 12 10 17 20 22 24 23 20 15 17 13
Class 2	12 24 13 15 15 16 13 21 22 23 20 20

Solution We begin by arranging the data from the two classes in ascending order.

Class 1	10 12 13 13 15 17 17 20 20 22 23 24 24
Class 2	12 13 13 15 15 16 20 20 21 22 23 24

Next, we find the middle-most number in each list. In class 1, there are 13 scores, so the median is in the $\frac{13+1}{2} = 7$th position, which is 17. In class 2, there are 12 scores, so the median is one-half the sum of the middle two, or $\frac{16+20}{2} = \frac{36}{2} = 18$. Because the median of class 2 is larger than that of class 1, class 2 did better on the quiz.

The last example illustrates two facts about the median. First, the median does not have to be a member of the data set, as was the case in class 2. Second, the median splits the data set in half, meaning half the data scores have values greater than or equal to the median and half have values that are less than or equal to the median.

Mean

The third and perhaps most commonly used measure of center is the **arithmetic mean.** This measure describes the center of the data set by adding all the data values and then dividing by the number of data points.

> **Definition of the Arithmetic Mean**
>
> The **arithmetic mean,** denoted $\bar{x}$, of a set of n numbers, $x_1, x_2, x_3, \ldots, x_n$, is
> $$\bar{x} = \frac{x_1 + x_2 + \cdots + x_n}{n}.$$

The notation $\bar{x}$ is read as "x bar," and we often refer to the arithmetic mean as the **average,** or simply the **mean.**

Example 14.15 Using the Mean to Compare Data Sets

Two fitness classes are having a weigh-down competition to see which class can lose more weight over an 8-week period. Because one class has more people, the winner will be determined by the highest mean weight loss instead of total weight loss. The number of pounds lost per person is shown. Which class won?

Class 1	13 15 28 10 21 24 18 25
Class 2	12 16 18 19 20 14 22 23 21 16 21

Solution The mean weight loss for each class is shown.

$$\bar{x}_{\text{class 1}} = \frac{13 + 15 + 28 + 10 + 21 + 24 + 18 + 25}{8} = \frac{154}{8} = 19.25 \text{ lb}$$

$$\bar{x}_{\text{class 2}} = \frac{12 + 16 + 18 + 19 + 20 + 14 + 22 + 23 + 21 + 16 + 21}{11} = \frac{202}{11} = 18.36 \text{ lb}$$

Because class 1 has the higher mean weight loss, it wins the competition.

Example 14.16 The Height of Two Basketball Players

Application

A basketball team has 10 players. Two are 75 in. tall, three are 76 in. tall, and three are 80 in. tall. The mean height of the team is 77 in., and the tallest player is exactly 1 ft taller than the shortest player. What are the heights of the tallest and shortest players?

Solution Because we are told the mean height of the players, it makes sense to use the formula for the mean to write an equation to find the missing heights. Let x represent the height of the shortest player in inches. Because the tallest player is a foot taller than the shortest player, we can represent his height by $x + 12$. Next, substitute x, $x + 12$, and the heights of the eight other players into the formula for the mean. We can take a shortcut in the computation by adding multiples of the duplicate heights rather than individual heights. This results in the following equation, which we then solve for x.

$$\frac{(x + 12) + (3 \cdot 80) + (3 \cdot 76) + (2 \cdot 75) + x}{10} = 77$$

$$\frac{2x + 630}{10} = 77$$

$$2x + 630 = 770$$

$$2x = 140$$

$$x = 70 \text{ in.}$$

Because the shortest player is 70 in. tall, the tallest player must be $70 + 12 = 82$ in. tall.

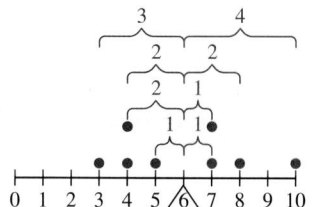

Figure 14.16 The mean as the balancing point of a data set

In many senses, the mean can be viewed as the "balancing point" of the data set. To understand why, consider a small data set that contains the numbers 3, 4, 4, 5, 7, 7, 8, and 10 and that has a mean of $\overline{x} = 6$. Next, place these values along the axis of a dot plot, and consider the distance of each data point from the mean. Figure 14.16 shows that the sum of the distances below the mean and the sum of the distances above the mean are equal. Specifically, the sum of the distances below the mean is $1 + 2 + 2 + 3 = 8$. Likewise, the sum of the distances above the mean is $1 + 1 + 2 + 4 = 8$ again. This is always true. If just one number changes, the mean also changes appropriately, and the balance is maintained.

Selecting the Best Measure of Center

Explorations Manual 14.10

Given the three measures of center, which is the best to use? Unfortunately, the choice is not always clear because each measure of center has its advantages and disadvantages. Regardless of the measure selected, the decision should be based on the needs of the study and on the overall characteristics of the data. Consider the next three examples.

Example 14.17	Typical Score for a Test

A class of 15 students took a 100-point test. The scores are shown:

23 24 26 31 32 32 43 47 47 50 65 78 90 90 90

The teacher claims the typical score was a 90, so the class as a whole did quite well. What measure of center did she use, and does it accurately represent the set of data?

Solution The teacher used the mode as the measure of center. Unfortunately, most of the scores were below 50, so the mode of 90 does not truly represent how the class performed. A better choice might have been the median, which has a value of 47 and more accurately indicates the class's lack of success on the test.

Although one advantage to the mode is that it is relatively easy to compute, the mode considers only the score that occurs most often. It does not account for any other value in the data set. If the mode occurs at one extreme or the other, then it is a poor representative for the data. In such situations, the median or the mean is likely to be a better choice.

Example 14.18 Typical Salary for a Company

The payroll figures of a certain company are listed in the following table:

Salary	$24,000	$26,000	$72,000	$106,000	$250,000
Number of employees	4	5	1	5	2

Is the median salary an accurate representation of the salaries in the company?

Solution The median salary of the company is $26,000. From this we might assume that most of the employees are paid fairly small salaries. However, about half the salaries are almost three times this amount or higher. The data show two values that occur the most frequently: $26,000 and $106,000. Because the data set is bimodal, the mode is a better measure of center.

Like the mode, the median fails to take into account any other data scores except for the middle-most one or two numbers. If these numbers happen to be near one extreme of the data or if the values drastically increase just above the median, as they did in Example 14.18, then the median is a poor representative of the data set.

Example 14.19 Typical Flight Time

An airline wants to know how long it takes to fly from Dallas to Atlanta, including boarding and taxiing time. The airline times six flights, shown in the table, and uses the mean of 149 min as the typical flight time.

Flight Number	1004	1324	126	758	1645	295
Length of Flight (min)	133	129	138	130	231	133

Does this value accurately represent the data set?

Solution The flight time of 231 min is almost 100 minutes longer than any of the other flights. Because this data value is included in the computation of the mean, it has the effect of pulling the mean in its direction. As a result, the mean of 149 min is not truly representative of the majority of the data.

When choosing a measure of center, we often select the mean because it takes every data value into consideration. Although this is the main advantage of the mean, it can also cause the mean to be drastically affected by extreme values in the data. In such cases, we might look to the median or the mode as a better measure of center. For instance, the median of 133 min is a much better measure of center for the flight times given in Example 14.19.

Check Your Understanding 14.3-A

1. Find the mean, median, and the mode for each data set.
   a. 12, 14, 15, 13, 12, 11, 12   b. 87, 88, 90, 91, 91, 92, 88
   c. 67, 67, 56, 45, 89, 77, 76, 56   d. 89, 88, 87, 65, 56, 77, 76, 56, 70
2. Three weight lifters finish a competition and want to determine a winner based on who lifted the most. If the competitors must persuade the judges

that they lifted the most, what measure of center should each person use to best make their case?

$$
\begin{aligned}
&\text{Joe:} && 275, 275, 275, 280, 285 \\
&\text{Mike:} && 265, 265, 280, 285, 290 \\
&\text{Tyrone:} && 270, 270, 270, 290, 295
\end{aligned}
$$

Talk About It In what circumstances is the mode the best measure of center? What about the median or the mean?

Representation

Activity 14.5
Pick a quantitative variable, such as height or weight, and collect data on this variable using the people in your class. Compute the mean, the median, and the mode of the data set. Which measure of center best represents the data?

Measures of Spread

As shown, measures of center are limited in their ability to describe a data set. We can describe a data set more fully if we also include a measure of the **spread,** or **dispersion.** One way to do so is to give the range. The **range** is the difference between the largest and the smallest data values, called the **maximum** and **minimum,** respectively. In general, the larger the range, the more disperse the data. The smaller the range, the more compact the data.

For instance, the following scores are from two classes that have taken the same 50-point test. The maximum, minimum, and range are given.

Class 1: 23, 25, 33, 34, 37, 42,	Class 2: 21, 24, 25, 25, 27, 28,
43, 46, 46, 48, 49, 49	40, 41, 42, 43, 45, 47
Maximum = 49	Maximum = 47
Minimum = 23	Minimum = 21
Range = 49 − 23 = 26	Range = 47 − 21 = 26

Even though the ranges are the same, a closer examination of the two data sets shows that they do not have the same spread. In the first data set, the values are somewhat evenly distributed throughout the range. In the second, they are clustered near the extremes. This pattern indicates that the range alone does not completely describe the spread of the data set because it takes only the extremes into account. Consequently, we again need more information. In addition to the range, two common ways to describe the distribution of a data set are a 5-number summary and the standard deviation.

5-Number Summaries and Interquartile Ranges

The maximum and minimum give the extremes of the data set because they mark both the high and the low values. Combining these numbers with the median, which marks the center of the data set, makes for a better picture of the data set's distribution. Specifically, if the data are arranged in ascending order, then 50% of the values lie between the minimum and the median and 50% of the values lie between the median and the maximum. For instance, if we consider the preceding test scores from class 2, the minimum, median, and maximum separate the data as shown in Figure 14.17.

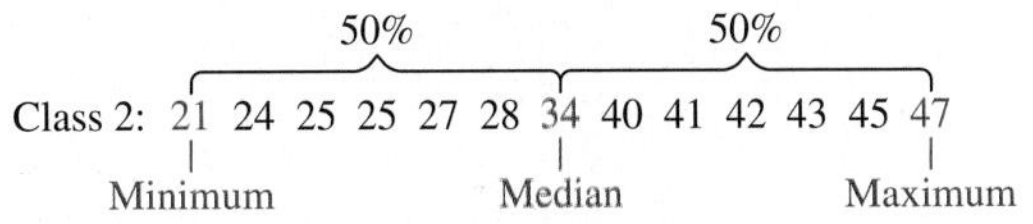

Figure 14.17 Data separated by the minimum, median, and maximum

In Figure 14.17, the maximum, minimum, and median fail to show that the data are clustered near the extremes. In fact, the median of 34 fails to represent either cluster. This indicates that these three numbers are again insufficient for completely describing the spread of the data. We can get a better description by separating it into smaller sections. To do so, we use **quartiles,** which separate the data into four quarters. If the data set is arranged in numerical order and the median is known, then the quartiles are the medians of the upper and lower halves of the data set.

Definition of Upper and Lower Quartiles

Let $x_1, x_2, x_3, \ldots, x_n$, be a set of n numbers arranged in ascending order, and suppose the median of the data set is known. The **first,** or **lower, quartile,** denoted Q_1, is the median of the data extending from the first data value, x_1, to the median. The **third,** or **upper, quartile,** denoted Q_3, is the median of the data extending from the median to the last data value, x_n.

We compute the first and third quartiles exactly like a median, but we use only half the data set. For instance, reconsider the data for class 1, which had scores of 23, 25, 33, 34, 37, 42, 43, 46, 46, 48, 49, and 49 on a 50-point test. The median of the data set is 42.5, so the first quartile is the median of the data ranging from 23 to 42, or $Q_1 = \dfrac{33 + 34}{2} = 33.5$. Similarly, the third quartile is the median of the data ranging from 43 to 49, or $Q_3 = \dfrac{46 + 48}{2} = 47$. Note that each computation includes either the minimum or the maximum, but not the median. The minimum, maximum, median, and both quartiles for class 1 are shown in Figure 14.18. These five numbers together represent a **5-number summary** of the distribution of the data set.

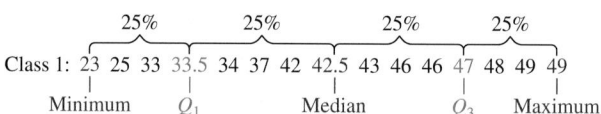

Figure 14.18 A 5-number summary

As with the median, the quartiles do not have to be part of the data set. They simply divide the upper and lower halves of the data set into halves. In other words, 25% of the values lie between the minimum and the first quartile, 25% lie between the first quartile and the median, 25% lie between the median and the third quartile, and 25% lie between the third quartile and the maximum.

Example 14.20   Interpreting a 5-Number Summary

Reasoning

A school district collects data on the salaries of all its teachers. What does the following 5-number summary indicate about the spread of salaries?

$$\text{Minimum} = \$32{,}000$$
$$Q_1 = \$56{,}000$$
$$\text{Median} = \$60{,}000$$
$$Q_3 = \$64{,}000$$
$$\text{Maximum} = \$87{,}000$$

Solution The numbers show that the upper and lower quartiles are much closer to the median than they are to the maximum and minimum. This observation, combined with the fact that the median is in the middle of the range, indicates that most of the data lie in the middle of the range rather than at the extremes.

As the last example shows, if the quartiles are closer to the median than they are to the maximum and minimum, then most of the data lie within the middle of the range. If the quartiles are closer to the maximum and minimum, then most of the data are near the extremes. And, if the quartiles and median occur at equal intervals, then the data are evenly distributed throughout the range.

Another way to describe the distribution of a data set is by means of the **interquartile range,** or **IQR.** We can think of the IQR as the range of the middle half of the data set because it encompasses approximately 50% of the data values. To compute it, take the difference of the first quartile from the third. For instance, the IQR of the data set in Figure 14.18 is $IQR = Q_3 - Q_1 = 47 - 33.5 = 13.5$. We often use the IQR to identify outliers. Any data point is an outlier if it has a value that is $1.5 \cdot IQR$ below the first quartile or $1.5 \cdot IQR$ above the third quartile.

Example 14.21 Finding an Outlier

Use the IQR to determine whether the following data set has any outliers.

$$19 \quad 39 \quad 42 \quad 43 \quad 45 \quad 50 \quad 52 \quad 55 \quad 64$$

Solution Displaying the data set with a dot plot, we see that two values, 19 and 64, have the potential to be outliers.

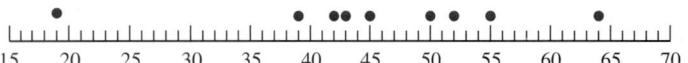

To determine whether these values really are outliers, we must first find Q_1 and Q_3, and then the IQR. Because the median is 45, we have $Q_1 = \dfrac{39 + 42}{2} = 40.5$ and $Q_3 = \dfrac{52 + 55}{2} = 53.5$. So the IQR is $Q_3 - Q_1 = 53.5 - 40.5 = 13$. Now, because $Q_1 - 1.5 \cdot IQR = 40.5 - 1.5(13) = 21$, any data point with a value less than 21 must be an outlier. Consequently, 19 is an outlier. Similarly, because $Q_3 + 1.5 \cdot IQR = 53.5 + 1.5(13) = 73$, any data point with a greater value is an outlier. Because 64 is less than 73, it is not an outlier.

Box Plots

Another way to illustrate the information in a 5-number summary is with a **box-and-whiskers plot,** or **box plot.** To make a box plot, we begin with an axis that shows the entire range of the data set. We then mark the median and the first and third quartiles with short line segments over the appropriate numbers on the axis. The box is formed by drawing segments connecting the first and third quartiles. Next, we add the whiskers by plotting the maximum and minimum and connecting them to the box with segments. We denote any outlier with an asterisk and connect the next highest or lowest value to the box with a whisker.

For instance, suppose the members of a garden club have ages of 35, 36, 45, 47, 50, 55, 56, 61, 63, 64, and 67. The 5-number summary for this data set is:

$$\text{Maximum} = 67$$
$$\text{Third quartile } (Q_3) = 63$$
$$\text{Median} = 55$$
$$\text{First quartile } (Q_1) = 45$$
$$\text{Minimum} = 35$$

Explorations Manual 14.11

Plotting these values leads to the box plot in Figure 14.19.

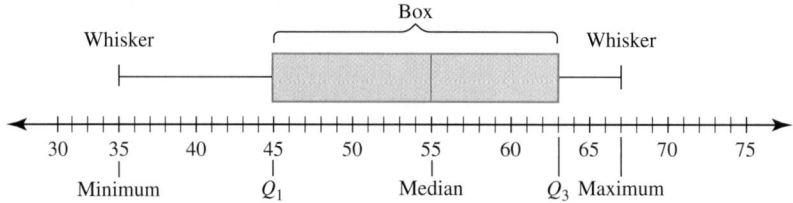

Figure 14.19 A box plot

Example 14.22 | **Interpreting a Box Plot**

Consider the data set represented by the following box plot:

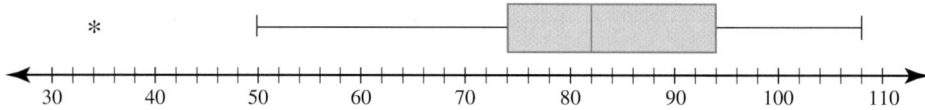

a. What is the median of the data set?

b. What are the first quartile, the third quartile, and the interquartile range?

c. What are the extreme values of the data set that are not outliers?

d. What values, if any, are outliers?

Solution

a. The median of the data set is 82.

b. $Q_1 = 74$ and $Q_3 = 94$. Subtracting these values, we have the interquartile range: IQR $= 94 - 74 = 20$.

c. The extreme values that are not outliers are 50 and 108.

d. There is only one outlier in the data set at 34.

Example 14.23 | **Comparing Data Sets with Box Plots**

Reasoning

The following table shows the number of live births per 1,000 people in 2008 in two different groups of countries: Central or South America and Asia. How do the two groups compare on this population variable?

Central or South American Country	Number of Live Births	Asian Country	Number of Live Births
Argentina	18.1	Bangladesh	28.9
Bolivia	22.3	Cambodia	25.7
Brazil	18.7	China	13.7
Chile	14.8	Hong Kong	7.4
Colombia	19.9	India	22.2
Ecuador	21.5	Indonesia	19.2
Guyana	17.9	Japan	7.87
Mexico	20.0	Malaysia	22.4
Paraguay	28.5	Mongolia	21.1
Peru	19.8	Nepal	29.9
Uruguay	14.2	North Korea	14.6
Venezuela	20.9	Pakistan	28.4
		Philippines	26.4
		Singapore	9.0
		South Korea	9.1
		Sri Lanka	16.6
		Thailand	13.6
		Vietnam	16.5

Source: NationMaster.com.

Solution The two data sets are different in size, so we can compare them by making a box plot for each one on the same axis.

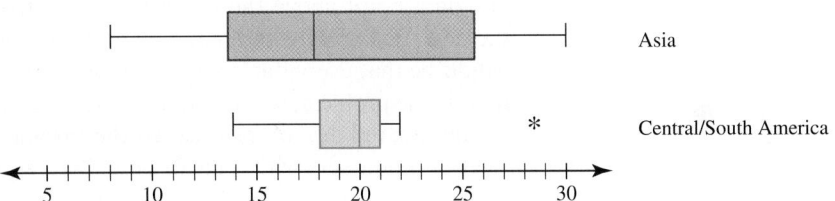

The box plots show that the spread of the Asia data is considerably greater than that of the Central or South America data. This shows up in the range and in the interquartile range. Even though the spreads are different, the medians are not too far apart. This may have been misleading had we used only this measure to compare the data sets. Also notice that the Central or South America data set has an outlier, which, when removed, condenses the variability.

Check Your Understanding 14.3-B

1. Give a 5-number summary for each data set.
   a. 45, 56, 55, 53, 47, 47, 49, 61, 63, 78, 45, 51, 55
   b. 14.5, 14.7, 14.8, 13.9, 13.7, 9.7, 15.6, 14.4, 14.0, 13.9
2. Find the interquartile range for each data set in problem 1. Use it to identify any outliers in the data.
3. Make a box plot for each of the data sets given in Question 1.

Talk About It In what situations are we likely to use a 5-number summary as the measure of spread?

Representation

Activity 14.6
Make a 5-number summary of the data you and your group collected in Activity 14.5. Use the 5-number summary to make a box plot of the data.

Variance and Standard Deviation

Because the 5-number summary and the interquartile range include the median, we often use them as the measure of spread whenever the median is selected as the measure of center. If the mean is chosen as the measure of center, then we are more likely to use another measure of spread. As mentioned, the mean is the balancing point of the data set; that is, the sum of the distances of the points below the mean is equal to the sum of the distances of the points above it. So it makes sense to have a corresponding measure of spread that indicates how the data is balanced about the mean by giving the typical distance of the data points from the mean. One measure of spread that does this is the standard deviation.

To find the typical distance of the data points, we must first find the deviation for each point. The **deviation** of any data point x is its distance from the mean, which we compute by subtracting the mean, $\bar{x}$, from x, or $x - \bar{x}$. The **standard deviation** is then an average of these deviations. So to compute it, we must first add the deviations. This presents a problem. Any data point below the mean has a negative deviation, and any data point above the mean has a positive deviation. As a consequence, adding the deviations results in a net sum and an average deviation of zero. The implication would be that every data point is equal to the mean, which is simply not the case. To handle the problem, we first square each deviation and then take the mean. The mean of the squared deviations is called the **variance.** In other words, if $x_1, x_2, x_3, \ldots, x_n$, is a set of n numbers with a mean of $\bar{x}$, then we compute the variance as

$$v = \frac{(x_1 - \bar{x})^2 + (x_2 - \bar{x})^2 + (x_3 - \bar{x})^2 + \ldots + (x_n - \bar{x})^2}{n}$$

The unit of the variance is the square of what we need. To get back to the original unit, we take the square root of the variance, which gives the standard deviation.

Definition of the Standard Deviation

Let $x_1, x_2, x_3, \ldots, x_n$, be a set of n numbers with mean $\bar{x}$. The standard deviation, s, of the numbers is the square root of the variance, or

$$s = \sqrt{v} = \sqrt{\frac{(x_1 - \bar{x})^2 + (x_2 - \bar{x})^2 + (x_3 - \bar{x})^2 + \ldots + (x_n - \bar{x})^2}{n}}$$

Explorations
Manual
14.12

The process of computing the standard deviation for a set of n numbers can be summarized in the following steps.

Procedure for Computing a Standard Deviation

Step 1. Compute the mean, $\bar{x}$.

Step 2. For each data value x, find its deviation, or $x - \bar{x}$.

Step 3. Square every deviation in step 2.

Step 4. Add the squared deviations in step 3, and divide by n to find the variance.

Step 5. Take the square root of the variance.

Example 14.24   Computing a Standard Deviation

Find the standard deviation for the following scores from a 15-point quiz.

$$6 \quad 7 \quad 8 \quad 9 \quad 10 \quad 11 \quad 11 \quad 12 \quad 13 \quad 13$$

Solution

Step 1. We first find the mean of the data set, which is $\bar{x} = 10$.

Steps 2 and 3. Next, we calculate the deviation for each data point and square it. These two steps are summarized in the following table:

Data Value	Step 2. Compute $x - \bar{x}$	Step 3. Compute $(x - \bar{x})^2$
6	$6 - 10 = -4$	$(-4)^2 = 16$
7	$7 - 10 = -3$	$(-3)^2 = 9$
8	$8 - 10 = -2$	$(-2)^2 = 4$
9	$9 - 10 = -1$	$(-1)^2 = 1$
10	$10 - 10 = 0$	$(0)^2 = 0$
11	$11 - 10 = 1$	$(1)^2 = 1$
11	$11 - 10 = 1$	$(1)^2 = 1$
12	$12 - 10 = 2$	$(2)^2 = 4$
13	$13 - 10 = 3$	$(3)^2 = 9$
13	$13 - 10 = 3$	$(3)^2 = 9$

Steps 4. Now we find the variance by computing the average of the squared deviations,

$$v = \frac{16 + 9 + 4 + 1 + 0 + 1 + 1 + 4 + 9 + 9}{10} = \frac{54}{10} = 5.4$$

Steps 5. Finally, we take the square root of the variance to find the standard deviation, or $s = \sqrt{v} = \sqrt{5.4} \approx 2.32$. This value indicates that the typical distance of a data point from the mean is approximately 2.32 units.

The last example illustrates two things. First, computing a standard deviation requires a large number of calculations, even when the data set is small. For this reason, a calculator or computer is often used to compute them. The calculator may return a slightly different answer from what we expect, probably because the calculator is using a formula for the standard deviation in which the sum of the squared numbers is divided by $n - 1$ instead of n. This minor change is important for mathematical reasons that go beyond the scope of this text. Because the different divisors lead to only small changes, we keep a divisor of n for clarity.

Second, the table in the last example shows that data points that are close to the mean have small deviations and that data points that are far from the mean have large deviations. Consequently, any data set with a large spread has large deviations, leading to a large standard deviation. Any data set with a small spread has small deviations, leading to a small standard deviation. In other words, the greater the standard deviation, the greater the spread of the data set. The smaller the standard deviation, the smaller the spread of the data set.

Like a 5-number summary, we can use a standard deviation to compare the spread of multiple data sets.

Example 14.25	Comparing Data Sets with the Standard Deviation

Reasoning   What does the information in the table indicate about the performance of two classes on the same 100-point exam?

Class	Mean	Standard Deviation
Third period	$\bar{x} = 74.8$	$s = 9.8$
Sixth period	$\bar{x} = 72.4$	$s = 15.7$

Solution The means indicate that the typical scores of the classes are fairly similar. However, the standard deviations indicate that the scores in third period are not as widespread as the scores in sixth. In each class, most of the scores are likely to fall within one standard deviation above or below the mean. So most of the scores in third period lie between 65 and 85, whereas most of the scores in sixth period lie between 57 and 88. In other words, sixth period is likely to have higher grades but with lower grades to offset the higher ones.

Normal Distributions

For a better understanding of the standard deviation and its relationship to the mean, we need to consider distributions. A **distribution** is a graph that compares data values to the relative frequency of those values. For small data sets, we can show a distribution with a dot plot or a histogram. However, for large data sets, we typically represent the distribution with a smooth curve.

For instance, if we measured the height of every male in college, the resulting distribution might look like the one shown in Figure 14.20. The figure shows a **normal distribution,** which is characterized by a smooth bell-shaped curve that theoretically extends to infinity without touching the horizontal axis.

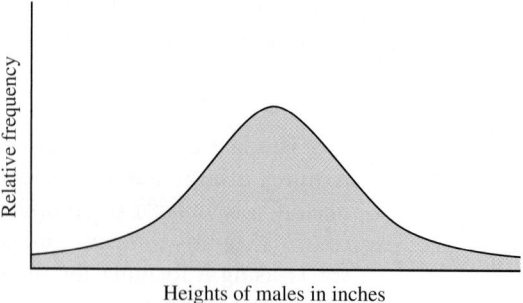

Heights of males in inches

Figure 14.20 The distribution of male heights

In a normal distribution, the mean, median, and mode all take on the same value, and the distribution is symmetric about this value (Figure 14.21). In any distribution, the mean, median, and mode each have a geometric interpretation. The mean is the balancing point of the distribution, the median divides the distribution into two equal areas under the curve if a vertical line is drawn through it, and the mode occurs at the highest point of the distribution.

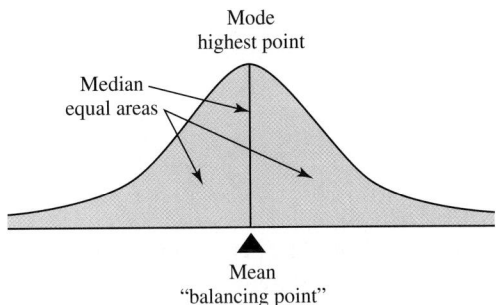

Figure 14.21 The mean, median, and mode in a normal distribution

Explorations Manual 14.13

If a distribution is known to be normal, then its shape is determined by the mean and standard deviation. The mean marks the center of the distribution. Then, if the standard deviation is large with regard to the mean, the shape of the distribution is low and flat. If the standard deviation is small with regard to the mean, the distribution is tall and thin. If a distribution is normal, we also know what percent of the data fall within a certain distance of mean. This is called the **empirical rule** and is illustrated in Figure 14.22.

Empirical Rule

In any normal distribution, approximately:

68% of the data lie within one standard deviation of the mean.
95% of the data lie within two standard deviations of the mean.
99.7% of the data lie within three standard deviations of the mean.

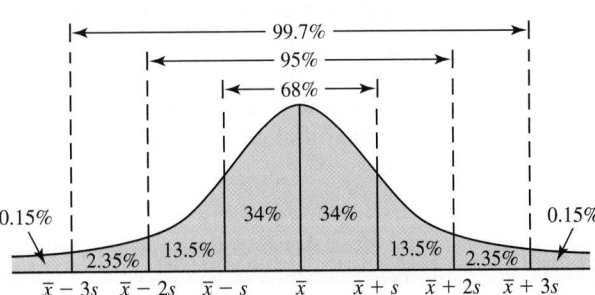

Figure 14.22 The empirical rule

Example 14.26 Using the Empirical Rule

Connections

A city collects data on the number of hours that factory workers living in the city tend to work. The data set is normally distributed with an average of 35 hr and a standard deviation of 5 hr. Draw the distribution of the data set, and use it to answer each question.

a. What percent of workers work fewer than 35 hr?

b. What percent of workers work more than 30 hr but fewer than 45 hr?

c. If 15,000 workers were surveyed, how many work fewer than 25 hr?

Solution Because the data set is normally distributed, with a mean of 35 hr and a standard deviation of 5 hr, we can label the normal distribution with a specific numbers of hours as shown:

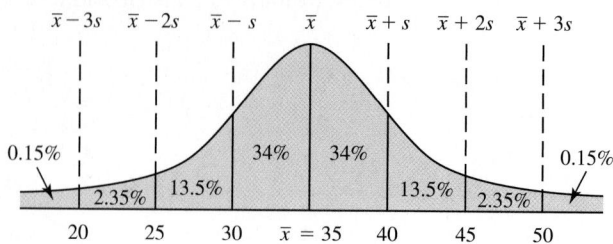

a. The percent of workers who work fewer than 35 hr is represented by the part of the graph to the left of the mean, or 50%.

b. Thirty hours is one standard deviation below the mean, and 45 hr is two standard deviations above the mean. By adding the percents in the graph between 30 and 45, we have that 34% + 34% + 13.5% = 81.5% of the workers work more than 30 hr but fewer than 45 hr.

c. Only 2.5% of the workers work fewer than 25 hr in a week. So 2.5% · 15,000 = 0.025 · 15,000 = 375 workers work fewer than 25 hr.

The mean and standard deviation can also be used to compute *z-scores*, which indicate how many standard deviations a data point is from the mean. If x is the value of a data point, the z-score for x is $z = \dfrac{x - \bar{x}}{s}$. $x - \bar{x}$ gives the deviation of x from the mean $\bar{x}$. Dividing by the standard deviation, s, then tells how many standard deviations this distance is equivalent to. For instance, suppose a factory worker from Example 14.26 works 39 hr. The z-score is $z = \dfrac{x - \bar{x}}{s} = \dfrac{39 - 35}{5} = \dfrac{4}{5} = 0.8$, indicating that the data point of 39 is 0.8 standard deviations from the mean. Notice that the z-score is positive because the data point is above the mean. If it were below the mean, the z-score would be negative. Because z-scores measure data points in terms of standard deviations and not the original scale, we can use them to compare scores from different distributions.

Explorations Manual 14.14

Example 14.27 Comparing Scores from Different Distributions

Application

Jenny and Allie are elementary education majors with different professors for their mathematics content course. Jenny scored a 135 on her final exam, which had a mean of 115 and a standard deviation of 12. Allie scored a 90 on her final exam, which had a mean of 72 and a standard deviation of 8. If both exams were normally distributed, which student did better?

Solution Because each exam was worth a different number of points, it is difficult to compare the scores directly. However, we can use z-scores. Jenny's z-score is $z = \dfrac{135 - 115}{12} = \dfrac{20}{12} \approx 1.67$ and Allie's is $z = \dfrac{90 - 72}{8} = \dfrac{18}{8} = 2.25$. Because Allie had the higher z-score, she was further above the mean than Jennie and did better on her exam.

Check Your Understanding 14.3-C

1. Compute the mean and the standard deviation for each data set.
   a. 5, 6, 7, 7, 8, 9   b. 12, 12, 15, 16, 17, 19, 19, 20

2. What does the information in the table indicate about the weight loss of two different dieting groups?

Group	Mean Weight Loss	Standard Deviation
Group 1	$\bar{x} = 12.5$	$s = 2.3$
Group 2	$\bar{x} = 14.8$	$s = 1.7$

3. The scores from a standardized test are normally distributed with a mean score of 400 and a standard deviation of 25.
   a. What score is one standard deviation above the mean?
   b. What score is two standard deviations below the mean?
   c. What percent of the scores lie between 375 and 450?
   d. What is the z-score of $x = 410$? $x = 380$? $x = 460$?

Talk About It Explain how the standard deviation measures the spread of a data set.

Connections

Activity 14.7
Before standardized measurements, a yard was the distance from the tip of the nose to the tip of the middle finger when the arm is stretched out straight. Measure the yard of each person in your class in inches. Once the data are collected, calculate the mean and the standard deviation.
a. How close is the mean to the standard length of a yard?
b. How many people have a yard within one standard deviation of the mean?
c. Do the data set appear normally distributed? As a class, determine how you might be able to tell?

SECTION 14.3   Learning Assessment

■ Understanding the Concepts

1. Why is it necessary to give both a measure of center and a measure of spread when describing a set of data numerically?

2. How do the mean, the median, and the mode describe the center of a data set?

3. Why is the mean the balancing point of the data set?

4. a. Why is the mean the preferred measure of center?
   b. In what situations might the median be a better measure of center than the mean?

5. a. Describe how to find the range of a data set.
   b. Describe how to find the upper and lower quartiles of a data set.

6. a. What five numbers are included in a 5-number summary?
   b. How is a 5-number summary represented with a box plot?

7. a. How is the interquartile range used to identify outliers?
   b. If the mean is selected as the measure of center and the standard deviation is selected as the measure of spread, what would be reasonable criteria for determining whether a data point is an outlier?

8. Why is it necessary to square the deviations in the calculation of a standard deviation?

9. a. What are the characteristics of a normal distribution?

   b. What does the empirical rule say is true for any normal distribution?

10. What is a z-score, and why is it useful?

■ Representing the Mathematics

11. Give an example of a data set with 10 values that:

    a. Has one mode. b. Is bimodal.

    c. Has no mode.

12. Give an example of a data set with 10 values in which:

    a. The mean is larger than the median and the mode.

    b. The median is larger than the mean and the mode.

    c. The mode is larger than the mean and the median.

    d. The mean, the median, and the mode are equal.

13. Consider a set of data consisting of the data points 65, 70, 75, 75, 80, and 85.

    a. Give an example of a data set with the same mean but a larger range.

    b. Give an example of a data set with the same median but a smaller range.

14. Consider the following dot plot:

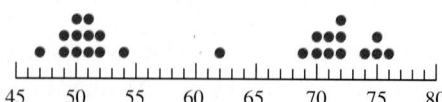

    Which measure of center, the mean, the median, or the mode, would best represent the data set? Explain your thinking.

15. Draw a box plot for each 5-number summary.

    a. Maximum = 75, Q_3 = 70, median = 63.5, Q_1 = 58, and minimum = 56.

    b. Maximum = 25, Q_3 = 22.5, median = 19.5, Q_1 = 16, and minimum = 14.

16. For each box plot, give the value for the minimum, the maximum, the median, the first quartile, and the third quartile.

    a.
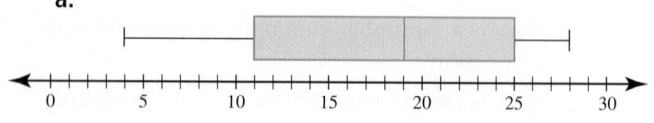

    b.
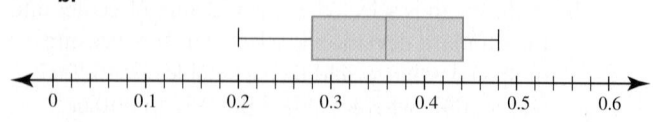

17. In each case, use the mean and standard deviation to draw a normal distribution like the one shown in Figure 14.22. Show scores that are three standard deviations above and below the mean.

    a. $\bar{x}$ = 100 and s = 10.   b. $\bar{x}$ = 300 and s = 20.

    c. $\bar{x}$ = 6.8 and s = 0.7.

18. In a skewed distribution, the data values tend toward either the maximum or the minimum value.

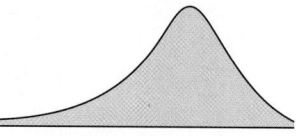 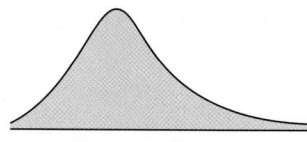

Skewed toward maximum   Skewed toward minimum

    In each skewed distribution, estimate the location of the mean, the median, and the mode. Label each measure appropriately.

■ Exercises

19. Find the mean, the median, and the mode for each data set.

    a. 12, 14, 15, 14, 16, 17, 18, 19, 19

    b. 104, 98, 115, 132, 124, 124, 135, 121, 98, 124

20. Find the mean, the median, and the mode for each data set.

    a. 2.5, 4.5, 2.6, 3.7, 5.4, 4.4, 4.3, 4.4, 5.1, 2.9, 3.3, 3.6, 2.7, 4.4

    b. −24, −23, −19, −20, −26, −26, −28, −18, −21, −25, −20

21. If the mean height of 15 students is 55 in., what is the sum of the 15 heights?

22. Jenna went on a shopping spree and bought 3 blouses for $14.50, $12.49, and $19.99, two pairs of pants for $24.50 and $31.89, and one pair of shoes for $52.89. What was the average cost per item?

23. Find the maximum, the minimum, and the range for each data set.

    a. 15, 16, 19, 20, 22, 21, 23, 15, 16, 18, 12

    b. 89, 76, 77, 90, 99, 100, 54, 75, 68, 92, 80

24. Find the maximum, the minimum, and the range for each data set.

    a. 10.1, 12.2, 14.5, 9.9, 11.8, 13.3, 14.4

    b. −2.1, −2.3, −5.4, −3.6, −6.1, −3.0, −4.5, −3.4

25. The youngest player on a professional baseball team is 19. If the range of the ages is 22, what is the age of the oldest player?

26. Give a 5-number summary for the following set of data, and then use the IQR to determine whether it has any outliers. Draw a box plot for the data set.

    23   45   54   63   65   66   66   68   70   71   72

27. Give a 5-number summary for the following set of data, and then use the IQR to determine whether it has any outliers. Draw a box plot for the data set.

    75 89 93 95 98 98 100 102 103 124

28. Compute the mean and the standard deviation for each set of data.

    a. 56, 56, 58, 59, 60, 60, 64, 66

    b. 13, 15, 17, 18, 22, 25, 28, 30, 35

29. Compute the mean and the standard deviation for each set of data.

    a. 4.5, 4.7, 4.8, 4.8, 5.1, 5.3, 5.4, 5.4

    b. 15.5, 19.8, 20.1, 14.6, 22.5, 19.9, 23.0, 18.7, 21.3

30. The ages for a group of professors in a small mathematics department are 37, 43, 47, 48, 52, 53, 53, 56, and 61.

    a. Compute the mean and standard deviation for the data set.

    b. Which ages are within one standard deviation of the mean?

    c. Which ages are within two standard deviations of the mean?

31. The IQs of people living in the United States are distributed normally with a mean of 100 and a standard deviation of 15. About what percent of the population has an IQ:

    a. From 85 to 115?   b. Less than 70?

    c. Greater than 145?

32. Suppose that the shoe sizes of adult women are normally distributed with a mean of $7\frac{1}{2}$ and a standard deviation of $1\frac{1}{2}$. About what percent of adult women have a shoe size:

    a. From 6 to 9?

    b. Greater than 12?

    c. Greater than 9 or smaller than 6?

33. The weights of 300 fourth-graders are recorded and are found to be approximately normal. If the mean is 75 lb and the standard deviation is 8 lb, then 95% of the students are between what two weights?

34. Suppose a standardized test is normally distributed with a mean of 450 and a standard deviation of 50. Also suppose that 5,000 students take the test.

    a. How many students would be expected to score better than 500?

    b. How many students would be expected to score less than 350?

    c. How many students would be expected to score less than 300 or greater than 600?

35. Suppose a distribution has a mean of $\overline{x} = 85$ and a standard deviation of $s = 5$. What is the z-score of each data point x?

    a. $x = 71$.   b. $x = 82$.   c. $x = 92$.   d. $x = 101$.

36. Suppose a distribution has a mean of $\overline{x} = 4.5$ and a standard deviation of $s = 0.4$. What is the z-score of each data point x?

    a. $x = 4.6$.   b. $x = 5$.   c. $x = 3.8$.   d. $x = 4.15$.

37. Robbie and Brandy take two standardized tests with scores that are normally distributed. Robbie scored an 85 on his test, which had a mean of 80 and a standard deviation of 7. Brandy scored a 50 on her test, which had a mean of 45 and a standard deviation of 4. Who had a better score?

38. The heights of adult men and women in the United States are approximately normal. The mean height for men is 5 ft 10 in., and the mean height for women is 5 ft 4 in. The standard deviation for both groups is about 2.8 in. If Melvin is 6 ft 2 in. tall and Martha is 5 ft 8 in. tall, who is taller with respect to their gender?

■ Problems and Applications

39. A set of eight numbers contains only three possible values: 1s, 3s, and 5s.

    a. If the mean is 1, what are the eight data values?

    b. If the mean is 3, what is one possible set for the eight data values?

40. The mean of five numbers is 25. If 1 is added to the first number, 2 to the second, and so on up to the fifth number, what is the new mean?

41. A set of 40 numbers has a mean of 48, and a set of 50 numbers has a mean of 58. If the two sets are combined, what is the average of all the numbers?

42. A class has 23 students. The mean weight of the students is 78 lb. If two more students are added to the class weighing 85 and 90 lb, respectively, what is the new mean weight for all 25 students?

43. The midrange is another measure of the center. It is defined to be the mean of the minimum and maximum.

    a. What is the midrange if the minimum is 45 and the maximum is 85?

    b. What is the midrange if the minimum is 37 and the maximum is 93?

    c. What is the maximum if the midrange is 56 and the minimum is 15?

    d. What is the minimum if the midrange is 37 and the maximum is 54?

44. Scores on standardized tests are often given in terms of percentiles, which are used to show

how one score compares to others. For example, if a person scores in the 75th percentile, then approximately 75% of the people taking the test scored lower. Use this information to answer each question.

a. If a person scores in the 16th percentile, about what percent of people scored lower?

b. If a person scores in the 65th percentile, about what percent of people scored higher?

c. If a person has a score in the 90th percentile, did he or she do well or poorly on the test?

45. Jacob has scored an 88, 87, 93, 79, and 96 on five 100-point tests. Can Jacob bring his average up to a 90 on the next 100-point test? If so, what is the lowest possible score he can make and still achieve this goal? Assume the teacher does not round.

46. A small company has 10 employees with a mean salary of $45,000. Two employees, with salaries of $65,000 and $72,000, respectively, retire and are replaced with four new employees, each with a salary of $29,000. What is the new mean salary of the company?

47. Max rides his bike at 10 mph for the first 5 mi of an 8-mi trip. If he wants to average 12 mph for the entire trip, how fast must the ride for the last three miles?

48. Ashley has taken 48 credits worth of courses and has a grade point average (GPA) of 2.89. She needs a 3.0 to be admitted into the elementary education program. If she plans to take five 3-credit courses this semester, what does her average need to be in these courses to be admitted to the education program?

49. Michael performs in three events in track and field. His score or time, the mean, and the standard deviation are shown:

Event	Michael's Score/Time	Mean Score/Time	Standard Deviation
100-m dash	11.3 s	11.7 s	0.4 s
Shot put	35 ft	31 ft	3 ft
Long jump	17.5 ft	16 ft	2.5 ft

Which event is Michael the most likely to win?

50. The following numbers represent the weight of 12 aspirin pills in milligrams.

671.3 669.4 675.1 668.9 665.2 671.8

670.3 664.9 676.1 673.3 672.0 669.5

a. Compute the mean and standard deviation for the data set.

b. At what weight(s) is it reasonable to consider a pill to be atypical and discard it?

51. The following table shows the lengths of some of the U.S. interstate highways.

Interstate Highway	Length (mi)
10	2,453
20	1,556
30	402
40	2,546
64	956
70	2,168
80	2,913
90	3,104
94	1,638

Find the mean and median for the lengths of these highways. Which appears to be the most representative measure of center?

■ Mathematical Reasoning and Proof

52. Can the arithmetic mean be equal to the greatest number in a data set? To the least number in the data set? If so, explain how.

53. Consider a set of 10 data points. If 5 is added to each data point, what is the resulting effect on the:

a. Mean? b. Median? c. Mode? d. Range?

54. Can the interquartile range be equal to the range? If so, provide an example. If not, explain why.

55. What does each 5-number summary indicate about the spread of its respective set of data?

a. Maximum $= 60$, $Q_3 = 55$, median $= 50$, $Q_1 = 45$, and minimum $= 40$.

b. Maximum $= 100$, $Q_3 = 80$, median $= 75$, $Q_1 = 70$, and minimum $= 50$.

c. Maximum $= 80$, $Q_3 = 75$, median $= 60$, $Q_1 = 45$, and minimum $= 40$.

56. If the standard deviation of a set of data is zero, what must be true about the values of the data?

57. Compute the mean and standard deviation for the data set 5, 6, 6, 7, 8, 8, and 10. Next, add 10 to each data value and compute the new mean and the standard deviation. Compare the means and standard deviations. Suggest a property of means and standard deviations that might hold true based on your comparison.

58. A company that makes car batteries has conducted numerous experiments on a new battery and has found that its life is normally distributed with a mean of 45 months and a standard deviation of 3 months. How long should the warranty on the battery be so that the company makes refunds on only a small percent of the batteries?

59. April has the following scores on five 100-point tests: 70, 80, 85, 80, 75.

    a. Find the mean, the median, and the mode.

    b. Which measure of center best favors April?

    c. Which measure of center is the least favorable to April?

60. In football, the speed of a player is measured by how fast he can run a 40-yd dash. Three different coaches brag that they have the fastest set of five receivers in the league. Using the following times, which measure of center did each coach use to make his claim? Which do you think is the most correct claim, and why?

    Bulldogs: 4.3, 4.7, 4.7, 4.8, 4.9

    Tigers: 4.3, 4.6, 4.6, 4.8, 5.3

    Falcons: 4.5, 4.5, 4.7, 5.1, 5.2

61. A professor claims that her class did well on a 100-point final exam. If the mean was a 65 and the median was an 85, can you believe her claim?

62. The following histogram shows the ages of men competing in two different basketball leagues.

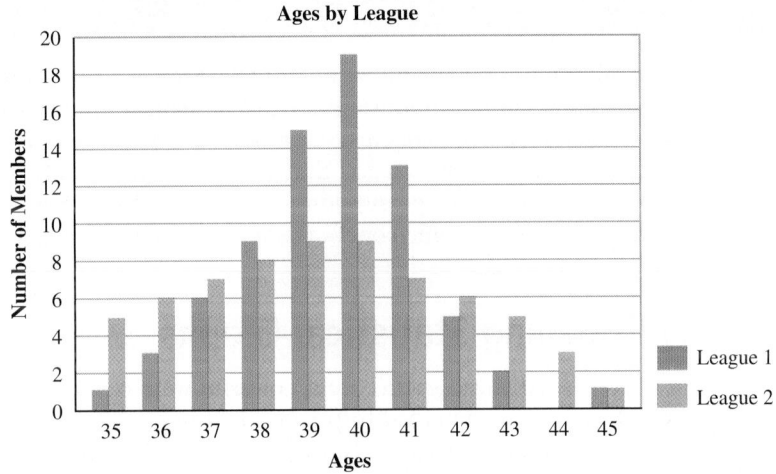

Ages by League

Compute the mean and the standard deviation for each league. Why is the standard deviation for league 1 smaller than that for league 2?

Mathematical Communication

63. The following data show the free-throw shooting ability of the individual players on three different teams by percents. Make a box plot for each team, and write a short paragraph comparing the free-throw shooting abilities of the teams.

    Hawks: 67, 69, 75, 75, 76, 80, 84, 85, 88

    Tigers: 53, 67, 75, 77, 80, 84, 89, 91, 94, 95

    Knights: 71, 72, 74, 75, 76, 77, 77, 78, 79, 80

64. Two of Mr. Parker's math classes take the same 50-point test. Find the mean and standard deviation for both classes. Write a paragraph or two explaining how the two classes compare.

    First Period: 34, 35, 36, 36, 37, 37, 38, 41, 42, 43

    Fourth Period: 20, 25, 27, 35, 37, 38, 41, 41, 42, 42, 43

65. Write three story problems: one that is solved by computing a mean, another that is solved by computing a median, and a third that is solved by computing a mode. What verbal cues did you use to indicate how to solve the problem?

66. Get together with a group of your peers. Discuss how stacks of blocks placed at appropriate spots along a number line can be used to demonstrate the mean as the balancing point of a data set.

67. Marcel is given a data set containing 4, 6, 2, 5, 5, 3, 7, and 3. He quickly claims that 5 is the median. Is he correct? If so, how did he quickly determine the answer? If not, where did he make his mistake?

68. Shironda reads an article that claims the average family size in the United States is 4.31 people. She asks how it is possible to have 0.31 of a person. As her teacher, how do you respond to her question?

Building Connections

69. How might you use the bars in a histogram to draw the curve of a distribution?

70. What are the advantages and disadvantages of using descriptive statistics over statistical graphs when representing and analyzing data?

71. Find several newspaper, magazine, or journal articles that use or describe some mean.

    a. Is the standard deviation included along with the mean?

    b. If not, why do you think this information was left out?

    c. Do you think it is important to include the standard deviation when reporting means? Why or why not?

72. Search the Internet for a website that offers data on the salaries of teachers across the country.

Find the mean and the standard deviation of these salaries.

    a. What are the highest and lowest salaries that you could make as a first-year teacher?

    b. How do the salaries in your state compare to the national average?

73. At what grade level do students begin to compute measures of center and measures of spread? Search several sets of curriculum materials to find out.

SECTION 14.4   Abuses of Statistics

In addition to understanding basic procedures for collecting, representing, and analyzing data, another important part of statistical thinking is being a wise consumer of statistics. Given a conclusion drawn from data, we must take the time to discern whether it is believable or not. Statistics are used not only to inform, but also to persuade. Unfortunately, persuasion is not always done with honest intentions. For this reason, we discuss some common ways that data are misrepresented through statistical graphs and descriptive statistics.

Misleading Statistical Graphs

Explorations Manual 14.15

Although statistical graphs are a powerful way to organize and represent data, they can be easily manipulated to present data in a way that is favorable to the needs of the investigator. Consequently, we should take the time to read graphs carefully to see whether the conclusions drawn from the data are legitimate. In general, graphs can be used to distort data in three ways: manipulations of the scales, perceptual distortions, and visual embellishments.

Manipulations of the Scales

The scales of a graph can be manipulated in several ways to distort how the data are viewed. One way is to omit one of the scales, making it difficult to analyze the differences between categories. For instance, both bar graphs in Figure 14.23 show the number of calories contained in one serving of certain snack chips. The first graph seems to show a dramatic difference between the number of calories in corn chips and cheese puffs. However, adding a scale to the vertical axis shows that the difference is not as dramatic as the first graph portrays.

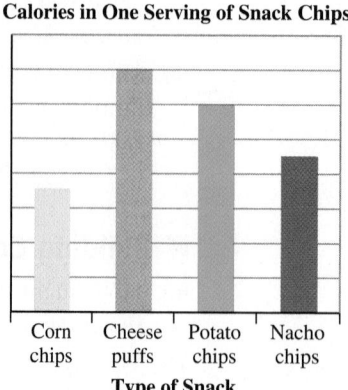

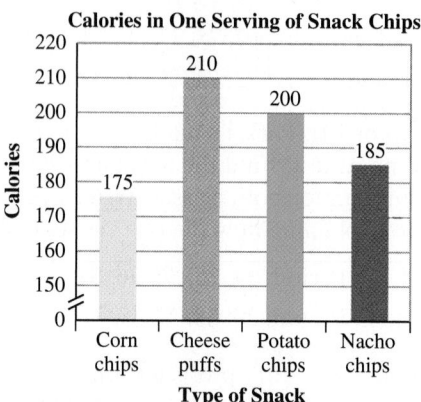

Figure 14.23 Distorting data by omitting a scale

Another way to manipulate the scales is to change the values on the vertical axis in one of two ways, depending on the purpose of the graph. If the purpose is to accentuate differences, we can remove part of the vertical scale to make the differences seem larger. If the aim is to minimize differences, we can extend the scale to make the differences seem smaller. This is called **cropping** because it shows only a limited part of the graph. Figure 14.24 shows two different graphs of the calories in snack chips. The first accentuates the differences in calories, and the second minimizes them.

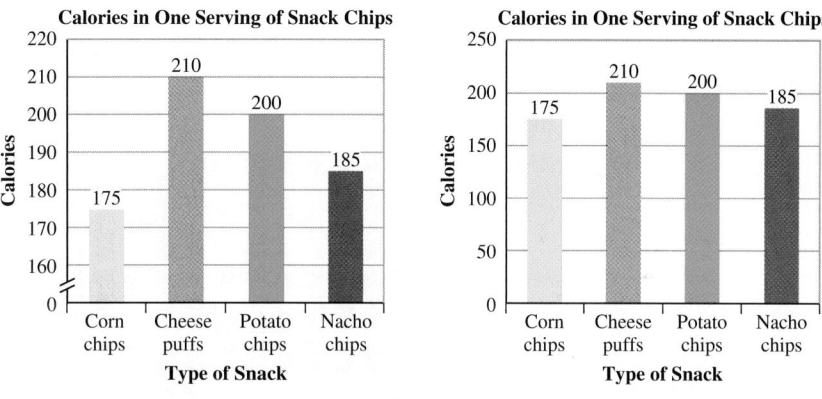

Figure 14.24 Distorting data by changing the vertical scale

Example 14.28   Manipulating a Line Graph

Representation

The year-end earnings report for a company shows the profits at the end of each month. How could the CEO draw a line graph of the following data to put the earnings of the company in the best possible light?

Month	Profits ($ millions)	Month	Profits ($ millions)
January	24.3	July	20.6
February	22.7	August	20.1
March	21.6	September	19.9
April	21.9	October	18.7
May	22.5	November	19.2
June	21.5	December	17.1

Solution The profits of the company have declined by over $7 million in one year. This is bad news for the CEO, so he wants to minimize the differences in the data set. He can draw a line graph so that the vertical scale includes the entire range of profit values, from $0 to $25 million. The resulting line graph is shown:

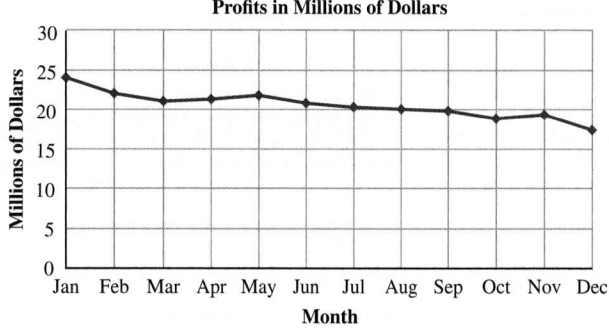

Although it shows a decline in profits, the decline appears to be gradual. Also, notice that the line is near the top of the graph, indicating that profits are still high.

The CEO in the last example could have also distorted the data by reversing the axes of the graph or by reversing the orientation of the axes. This is more commonly done with bar graphs, as illustrated in Figure 14.25. In the figure, both graphs show the same data. However, in the second graph, where the axes are reversed, the tendency is to read down the graph. This leads to the misconception that the profits grew over the year, when the opposite actually happened.

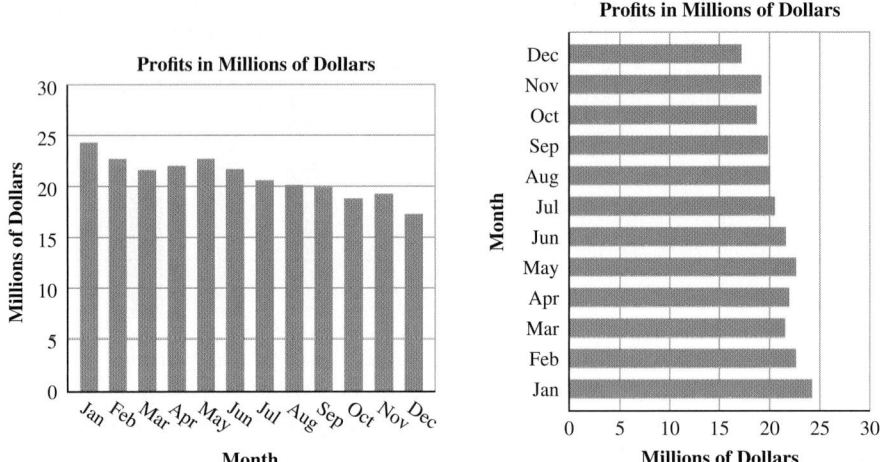

Figure 14.25 Distorting data by reversing the axes

We can also manipulate the scales on pie graphs. Consider the next example.

Example 14.29   Manipulating a Pie Graph

Reasoning   The pie graph gives the percents of an athletic budget that several sports receive.

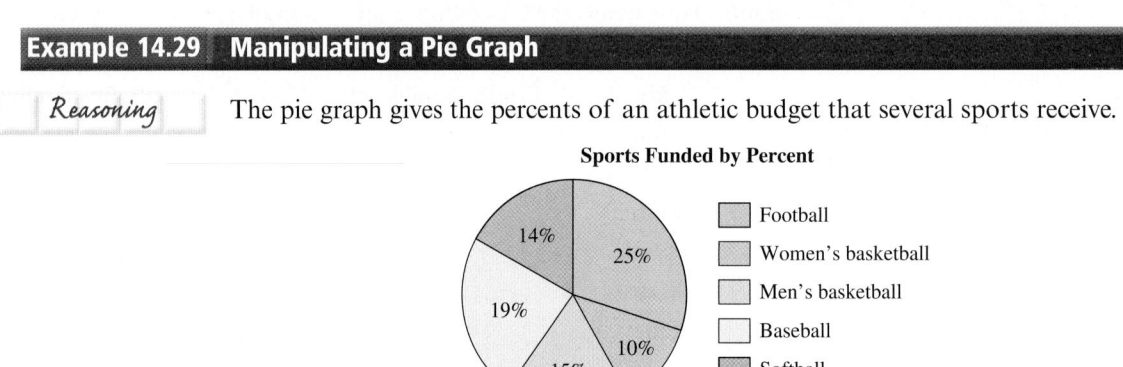

How does it misrepresent the data?

Solution Initially, the graph may not look misleading. However, the percents do not add up to 100%. Consequently, some of the information from the sports budget has been left out, causing the sports to appear as though they have a greater percent than they actually do. For instance, football represents 25% of the budget, yet its sector is more than one-fourth of the circle.

Perceptual Distortions

Another way graphs can be misleading is through a **perceptual distortion.** Perceptual distortions often occur in pictographs where the data are represented by one aspect of the graph, but the eye catches another. For instance, the graph in Figure 14.26 shows the contributions made to Freeaid Charity in two different years. The contributions are represented by the length of the dollar bills, which indicate that giving in 2011 was double that of 2010. However, the eye tends to focus on the areas of the dollar bills rather than on their lengths. Because the area of the larger dollar is four times that of the smaller, viewers may be misled into thinking that giving had increased by four times rather than by two.

**Contributions to Freeaid Charity
in 2010 and 2011**

Figure 14.26 A graph with a perceptual distortion

Whenever the length, area, or volume of an object is used to represent a measure of growth in any graph, it must do so with mathematical accuracy. The graph in Figure 14.26 would be appropriate if the giving had increased by a factor of four. But because it increased by only a factor of two, the increase would have been more accurately represented with two dollars of the same size as shown in Figure 14.27.

**Contributions to Freeaid Charity
in 2010 and 2011**

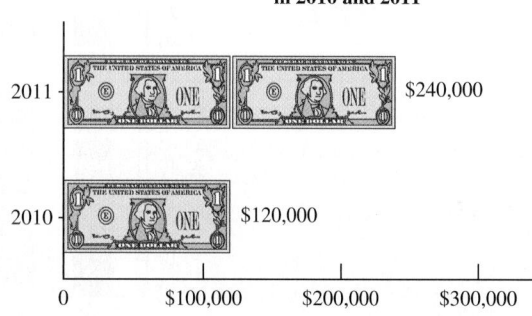

Figure 14.27 A pictograph without a distortion

Another perceptual distortion can be found in pie graphs. If a sector of the graph is separated from the rest, it is given more emphasis and appears larger than it actually is. For instance, the pie graph in Figure 14.28 gives the different types of fats in one serving of a children's cereal. Attention is quickly drawn to the sector for trans fat because it is separated from the rest of the graph. This makes it look particularly important, even though it is not the type of fat occurring in the largest amount.

Grams of Fat in Kiddie O's Cereal

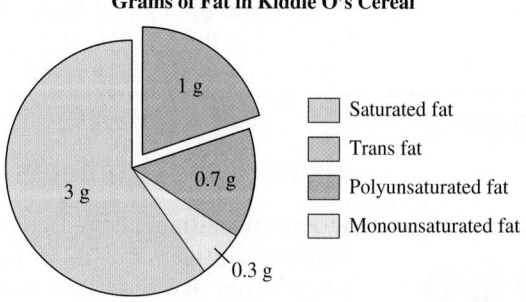

Figure 14.28 A pie graph with a separated sector

Three-dimensional effects can also distort our perception of a graph. For instance, consider a three-dimensional pie graph representing the fat content in a children's cereal (Figure 14.29). Notice that the sector for trans fat and the sector for polyunsaturated fat look almost the same size, even though they represent different amounts. The distortion comes from the tilt and the rotation of the pie graph, which make sectors in the back appear proportionally smaller than they actually are.

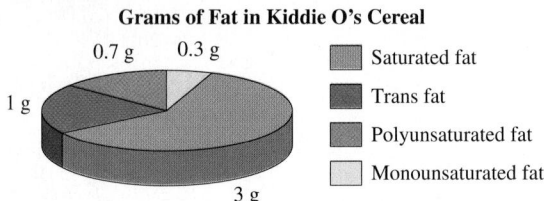

Figure 14.29 A three-dimensional pie graph

Example 14.30   A Three-Dimensional Bar Graph

Reasoning   The following bar graph shows the life expectancy of females in seven countries. What perceptual distortions may cause a misinterpretation of the data?

Source: NationMaster.com.

Solution The three-dimensional effects may make the graph more appealing, but they also make it more difficult to read the values of the data. Specifically, it is unclear whether we should use the front or the back of a bar to determine its height. The dimensional effect makes it difficult to determine whether countries like Mexico, Poland, and the United States have the same life expectancy for females or if one is longer than the other.

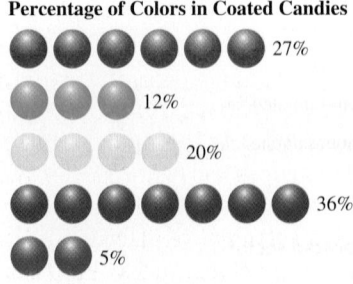

Figure 14.30 A pictograph with an inconsistent icon

Visual Embellishments

Visual embellishments in a graph can also make it easy to misinterpret the data. For instance, consider a mix of chocolate candy that comes in five colors. Figure 14.30 is a pictograph that shows the percents of the colors in the mix. The problem with the graph is that the icons do not accurately reflect the percents of the colors. Each icon should represent the same amount. Yet each red icon has a value of 4.5%, each green icon has a value of 4%, each yellow icon has a value of 5%, and so on. The different values of the icons affect how we interpret the graph.

Check Your Understanding 14.4-A

1. The heights of six different people are given in inches.

Abe: 65 in., Barb: 63 in., Carrie: 66 in., Dan: 69 in., Eve: 64 in., Fred: 70 in.

Draw two different bar graphs of the data: one that accentuates the differences in heights and another that minimizes them.

2. The following graph shows how the enrollments of males and females into a business program have changed over a 10-year period.

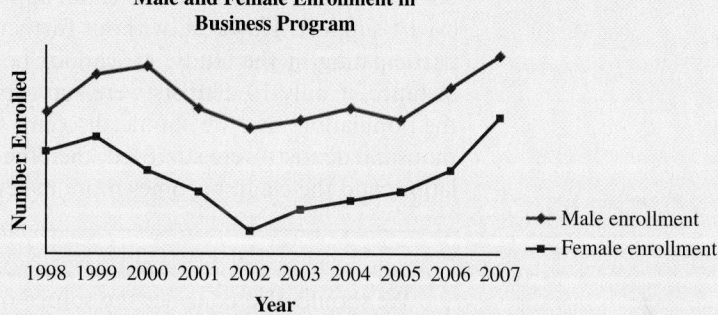

What makes it possible to misinterpret the graph?

3. Describe how the following graph misrepresents the decrease in the number of cartons of milk drunk by first-graders over the course of a year.

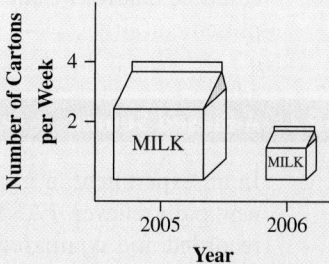

Talk About It Can a statistical graph be misleading but still be accurate in its portrayal of the data? Explain.

Computer

Activity 14.8
Do an Internet search for pictographs. Find several that you think might be misleading. Write a short paragraph that identifies the topic of the graph, and why you think it is misleading.

Misleading Descriptive Statistics

Like statistical graphs, descriptive statistics can be used to misrepresent data in a number of ways. One is to make a claim without including any background information, such as the population, the sample size, or the method of selecting the sample. Without this information, making an informed decision about the legitimacy of the claim can be difficult. For instance, consider the claim that, "75% of all recently graduated mathematics education majors at Adams University were placed in a job prior to graduation." Initially, this may seem like a powerful claim, but without further information about the number of graduates and the jobs they took, we cannot know if it is true. For example, perhaps only four graduating seniors had that degree and three of them got the only positions available in the local area. Or the jobs they took after graduation might have had nothing to do with their major, rendering the claim misleading.

Example 14.31 | **Analyzing a Commercial**

Reasoning

A television commercial claims that 9 out of 10 dentists find that White Teeth Toothpaste gives a healthier, whiter smile. Can this claim be believed?

Solution From the outset, the claim appears to be believable because it is backed by expert support. However, without further information about the number of dentists participating in the study, we cannot be sure whether the claim is misleading. For instance, if only 10 dentists were surveyed, the sample may not be representative of the population, and we should discount the conclusion. However, if we knew that a thousand dentists were surveyed, then the sample is more likely to represent the population, and the claim becomes more believable.

We should also consider the source of any statistical claim. Often the source has no vested interest in the outcome of the study, and the findings are reported without bias. For instance, many political pollsters claim no political affiliation so that they can present themselves as an unbiased source for predicting elections. On the other hand, the sources in many studies have a vested interest in the outcome. If the source could be biased, we should look carefully at the results to see whether they have been presented fairly.

Example 14.32 | **A Biased Source**

Reasoning

In an experiment, a group of people who suffer from tension headaches are given a new pain reliever, FAST RELIEF. The time it takes for the headaches to subside is recorded and compared to other groups taking different pain relievers. On average, FAST RELIEF makes the headaches subside five to ten seconds faster than the other brands. Based on these results, FAST RELIEF is advertised as relieving headaches faster than other brands. Is this an honest claim?

Solution The claim made by the makers of FAST RELIEF has merit because it is supported to some extent by the data. However, the advertisement is unclear about what other brands FAST RELIEF was compared to. It may have been compared to the worst brands on the market, making it only a mediocre product. Even if FAST RELIEF were compared to all other pain relievers, the results of the study may not have any practical significance. Because the time difference is small, many consumers may not consider FAST RELIEF to be significantly better than other brands.

Often descriptive statistics are used to persuade people to accept a point of view. In such cases, the person reporting the statistic may have several choices but chooses the one that best suits his or her needs. This often happens when the choice is between two complementary percents. For instance, health care is a common debate among politicians. To rally people to their cause, they often cite the percent of people who do not have health insurance, which is often around 20%. However, if 20% do not have health insurance, then 80% do, and that statistic paints a completely different picture.

Example 14.33 **Complementary Statistics**

Give a possible complementary statistic for each claim.

a. Sixty percent of all students attending college are female.
b. Eight out of 10 high school graduates eventually enroll in postsecondary education.
c. One-fourth of all auto accidents involve alcohol.

Solution Possible complementary statistics are given.

a. Forty percent of all students attending college are male.
b. Two out of 10 high school graduates never enroll in postsecondary education.
c. Three-fourths of all accidents do not involve alcohol.

Complementary statistics can give a different perspective on a situation. However, we should be careful about how we use them because there may be more than one alternative statistic. For instance, consider the claim that 60% of people love chocolate ice cream. Assuming that 40% of people do not like chocolate ice cream would be wrong because only some may dislike chocolate ice cream; the others may have no opinion.

Another way to change the interpretation of the data is to select a measure of center that puts the data in the best light.

Example 14.34 **Misusing Measures of Center**

Reasoning

At an academic competition for high school students, teams of seven are ranked by their performance on a 100-point exam. The scores from three teams are shown.

Lamont High: 57, 75, 80, 82, 84, 84, 85
Northside Academy: 68, 70, 80, 80, 95, 96, 97
Beaufort Preparatory: 56, 57, 57, 85, 86, 87, 89

If each coach claims that his or her team should be ranked first, what measure of center did each coach use? Who has the most legitimate claim?

Solution We begin by computing the mean, the median, and the mode for each team.

Team	Mean	Median	Mode
Lamont High	78.1	82	84
Northside Academy	83.7	80	80
Beaufort Preparatory	73.8	85	57

The coach from Lamont used the mode, the coach from Northside used the mean, and the coach from Beaufort used the median. Because the mean takes all scores into consideration and the median and mode do not, the mean is the measure of center that best represents the test scores. As a result, Northside Academy has the best claim to first place.

Another problem with statistics is that they are often reported second hand. Many studies are done with integrity, but when they are reported by another source, one of two problems can occur. First, the biases of the second-hand source may distort the findings of the original study. Second, the second-hand source may not know how to interpret the statistics correctly. The latter problem often occurs with nationwide opinion polls, which attempt to find out how the country views a certain topic by collecting data from a relatively small sample. Because the data from the sample only approximate the opinions of the overall population, most pollsters include a **margin of error.** For instance, suppose a poll claims that 37% of people want stricter gun laws with a margin of error of ±3%. By giving the margin of error, the poll is stating that the actual percent of people wanting stricter gun laws is between 33% and 40%. In many cases, the second-hand source leaves off the margin of error, which is not reporting the data accurately. In other cases, the second-hand source misinterprets the meaning of the margin of error.

Example 14.35 Misinterpreting an Opinion Poll

At the beginning of a month, an opinion poll claims that 50% of people approve of the president's performance with a margin of error of ±4%. At the end of the month, the poll is done again and shows that 48% of people approve of the president's performance with a margin of error of ±3%. A reporter uses the data to claim that the president's approval rating is dropping. Is this a truthful claim?

Solution Although the president's approval rating could have dropped, the reporter has misinterpreted the meaning of the data. The first poll claims that the actual approval rating for the president is somewhere between 46% and 54%. The second poll has it somewhere between 45% and 51%. Because the polls have overlapping percents, the actual approval rating could well have remained constant, at, say, 49%. Also, the actual approval rating could even have been at 47% in the first poll and 50% in the second, which would actually be an increase in the approval rating.

Check Your Understanding 14.4-B

1. A newspaper article claims that 4 out of 5 children break a bone before the age of 14. The data for the study comes from the medical records at the emergency room of a local hospital. Can the article be believed?

2. Shauna reads a magazine article claiming that the average teacher in New York City makes $55,000, which is $15,000 more than the average salary of a teacher in her area. If Shauna plans to be a teacher, what other information might she want to know before she decides to move to New York?

3. Give a possible complementary statistic for each claim.
   a. Fifty-three percent of American women ages 18 or older are not married.
   b. Hispanics are the largest minority in 20 states.
   c. Eleven percent of all K–12 students are enrolled in private school.

Talk About It Why is it inaccurate to leave out the margin of error when reporting percents from a nationwide poll?

Connections

Activity 14.9
Search a local newspaper or magazine for statistical claims that use percents. Examine the claim and the information that is given to support it. Write a paragraph about the claim and whether you think the data support it.

SECTION **14.4** Learning Assessment

■ Understanding the Concepts

1. How does changing the vertical scale of a graph affect its appearance?

2. How does reversing the axes on a bar graph distort the appearance of the data?

3. How is it possible to manipulate the scale on a pie graph?

4. What is a perceptual distortion, and how can it affect the interpretation of a graph?

5. Describe how three-dimensional effects can distort the perception of a graph.

6. How can the source of a study introduce bias into the study?

7. What are complementary statistics, and how can they lead to a different perspective of the data?

8. What is a margin of error, and why is it often included with a descriptive statistic?

■ Representing the Mathematics

9. The following table shows the amount of money contributed to charities in billions of dollars.

Year	2006	2007	2008	2009	2010	2011	2012
Amount Contributed ($ billions)	80.07	87.75	91.15	96.1	98.38	102.13	104.53

Draw two bar graphs of the data, one in which the vertical axis starts at 0 and another that starts at $75 billion. How does the change in the vertical axis affect the perception of the graph?

10. The following line graph shows the increases in postal rates over 25 years.

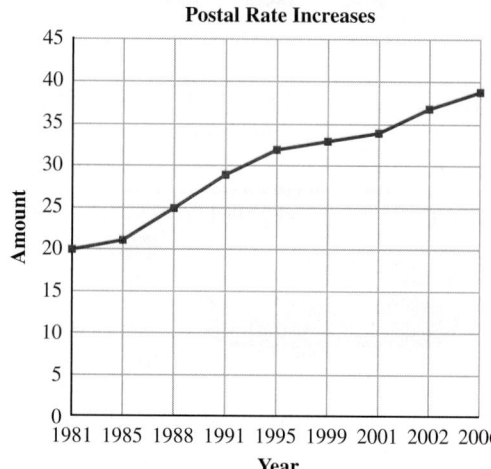

Postal Rate Increases

Source: U.S. Postal Service.

Redraw the graph so that the increases look more dramatic. Explain how you manipulated the graph to achieve this effect.

11. A school principal has kept track of students' scores on a standardized test for the last 10 years.

Year	Average Score	Year	Average Score
2003	410	2008	440
2004	420	2009	445
2005	435	2010	447
2006	438	2011	449
2007	438	2012	455

The scores have been improving but not quickly. She has to make a report to the school board next week and wants to accentuate the improving scores. Draw a line graph that she might use to put the scores in the best light.

12. The following table shows the average weekly earnings of factory workers for the years 2008 to 2012.

Year	2008	2009	2010	2011	2012
Earnings	$853.98	$863.61	$873.64	$885.86	$894.68

a. Draw a bar graph of the data. What does it show about the wages of factory workers?

b. Draw a second bar graph that could cause someone to believe that the wages of factory workers have been decreasing. How did you manipulate the graph to do so?

13. The following table shows the number of burglaries in a certain city over a five-year period.

Year	2007	2008	2009	2010	2011
Number of Burglaries	1,957	2,032	2,147	2,201	2,250

Draw a bar graph of the data that appears to show a decrease in burglaries.

14. In 2008, 76% of workers drove to work alone, 10% rode to work in a car pool, and 5% took public transportation. Make a pie graph from the data, and explain how it distorts the data.

15. The following table shows the number of drunk drivers arrested along a certain stretch of highway over the course of five months.

Month	May	June	July	August	September
Number of arrests	77	75	73	72	70

a. Draw a graph that shows that the number of arrests is on the decline. What incorrect conclusions might be drawn from such a graph?

b. Draw a graph that demonstrates that the number of arrests has remained fairly constant. What incorrect conclusions might be drawn from such a graph?

16. Suppose the price of gasoline doubles over the course of year. Which pictograph(s) would be an appropriate way to represent the increase?

a.

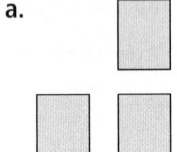

b.

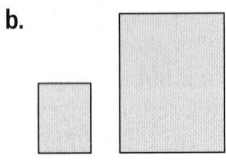

c.

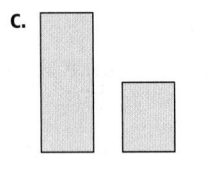

d.

17. A recycling company recycles about 7,500 T of cardboard in one year. The following year, the company recycles about 15,000 T of cardboard. Draw a graph that uses the area or volume of a figure to distort this information.

18. In 2008, U.S. imports of fireworks topped $193 million. At the same time, U.S. exports of fireworks only reached $28.1 million. Draw a graph that uses the area or volume of a figure to distort this information.

■ Exercises

19. What makes the following line graph misleading?

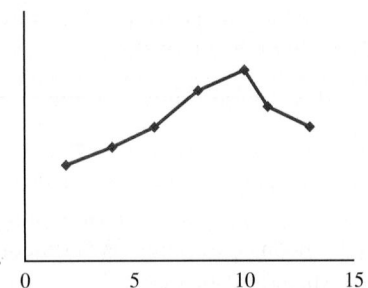

20. What is wrong with the following pie graph?

Ethnicity of Students at Stonebrook High

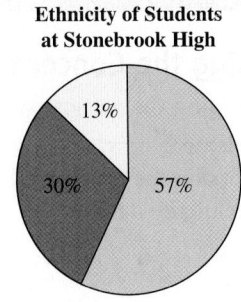

21. The following pictograph shows how the amount of pesticide used by farmers in a county has decreased over the course of a decade. Why is it misleading?

Pesticides Used by Farmers

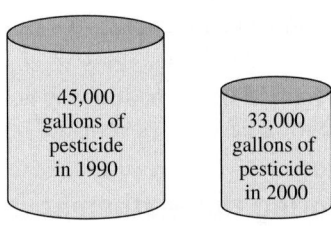

22. The following graph shows the gas mileage of four cars. It was used to show that model B gets significantly better gas mileage. Is this claim accurate? Is it honest?

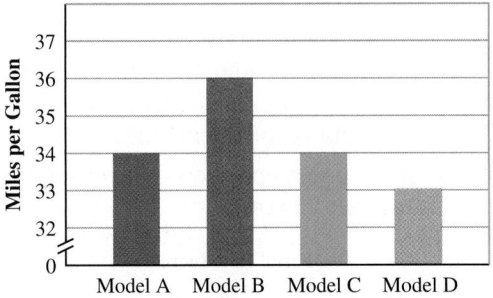

23. A-Mart carries four brands of laundry detergent. The sales by percent are shown.

Total Sales of Detergent Brands at AMart

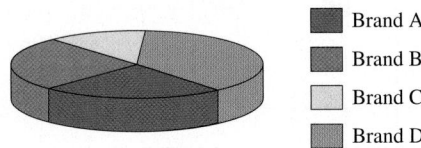

Brand A
Brand B
Brand C
Brand D

Arrange the brands in ascending order by their percents of the total sales. What information should be added to the graph to make this easier to do?

24. The following claims are misleading. Explain how.

   a. ULTRA CLEAN will leave your floors cleaner.

   b. A car manufacturer claims that 8 out of 10 cars of a certain model make it to 100,000 mi with no repairs.

   c. A brand of children's drink claims to contain 25% more fruit juice than other brands.

25. The following claims are misleading. Explain how.

   a. More than 90% of people renting a car in a certain town use Rent-A-Car. Therefore, Rent-A-Car is the most popular car rental agency in the country.

   b. A brand of aspirin claims that 9 out of 10 doctors recommend the brand.

   c. Most people who are in an airplane crash do not survive. Therefore, it is better to drive than to fly.

26. Give a possible complementary statistic for each claim.

   a. Of voters in the 2008 presidential election, 38% were not married.

   b. Nine out of 10 residents consider their neighborhood safe.

   c. Three out of 20 of all females have some sort of disability.

   d. Ten percent of the active volcanoes in the world are in Japan.

27. Give a possible complementary statistic for each claim.

   a. One out of 5 residents in Florida are 65 years or older.

   b. Thirty-two percent of Americans say that bad breath is the worst trait of their coworkers.

   c. Three-quarters of Americans would rather go grocery shopping than floss their teeth.

   d. Thirty percent of all bingo players are under the age of 35.

28. Give an interpretation of the margin of error in each statement.

   a. A poll with a margin of error of ±7% claims that 47% of married women have cheated on their husbands.

   b. A poll with a margin of error ±4% claims that 44% of children watch television before going to bed.

29. Give an interpretation of the margin of error in each statement.

   a. A poll with a margin of error of ±2 years claims that the average age of college seniors is 23.

   b. A poll with a margin of error of ±0.4% claims that 9.7% of people 65 or older live in poverty.

30. A poll claims that the percent of Americans who consider themselves patriotic is somewhere between 82% and 92%. If the claim were made with a single percent and a margin of error, what would the values likely be?

31. A poll claims that the average number of hours that children watch television in one week is between 19.5 and 24.5. If the claim were made with a single value and a margin of error, what would the values likely be?

Problems and Applications

32. Three coaches claim that their basketball team is the tallest. The heights of the five starting players from each team are shown.

   Bears: 71, 72, 75, 77, 77
   Lancers: 74, 74, 76, 77, 78
   Blue Devils: 75, 75, 75, 77, 78

   What measure of center did each coach use? Who has the most legitimate claim?

33. A study is done by State University to determine the number of men and women enrolling in certain majors. The data for five majors are shown in the table.

Major	Total Enrollment	Male Enrollment	Female Enrollment
Business	347	155	192
Education	512	47	465
Journalism	289	196	93
Mathematics	57	45	12
Nursing	152	43	109

   a. Compute the percents for men and women in each department. Do the percents suggest a bias in gender toward enrollment in certain majors?

   b. Suppose that the ratio of males to females is 1 to 3 at State University. How does this additional information impact your view of possible gender bias toward enrollment in certain majors?

34. Motorcycles generally get better gas mileage than cars. However, most cars move more weight. Suppose a 2,000-lb car gets 27 mpg and a 400-lb motorcycle gets 65 mpg.

   a. Devise a strategy to determine the fuel efficiency of each vehicle if weight is a consideration.

   b. Which do you think is a better representation of the fuel efficiency of a vehicle, miles per gallon or your answer in part a?

35. Mark lives in Georgia and makes $65,000 a year. His sister Lori lives in California and makes $105,000 a year. Lori claims to be the wealthier of the two. If the cost of living is 50% higher in California than it is in Georgia, is Lori correct?

36. The average weight of women is normally distributed with a mean of 165 lb and a standard deviation of 10 lb. The average weight of men is also normally distributed with a mean of 191 lb and a standard deviation of 14 lb. Javier weighs 204 lb, and Selena weighs 174 lb. Javier claims that Selena is more overweight than he is. Is he correct?

■ Mathematical Reasoning and Proof

37. An electric company is meeting with its customers to discuss the rising cost of electricity. Data for the average electric bill over the last 6 years is given in the table.

Year	2005	2006	2007	2008	2009	2010
Average bill	$84.10	$93.81	$98.76	$107.39	$110.61	$125.71

    a. How would you draw a bar graph if you were the public relations person for the electric company?

    b. How would you draw a bar graph if you were a consumer advocate?

38. The following graph shows increases in the funding to the budget of a school district.

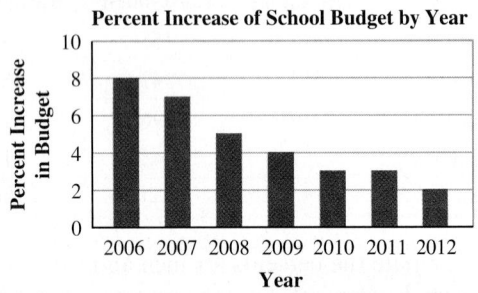

Percent Increase of School Budget by Year

How is the graph misleading about the amount of budget money the school has been receiving?

39. The following pie graphs show the enrollment percents of minority and nonminority students at a university in two years.

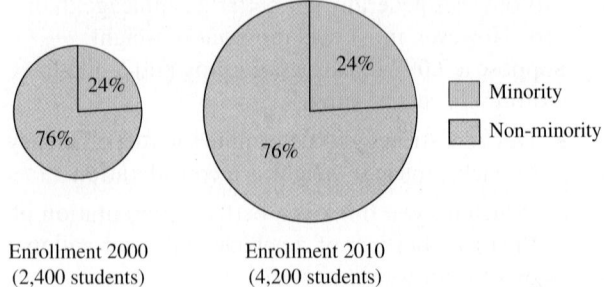

Enrollment of Minority and Non-Minority Students

Enrollment 2000 (2,400 students) Enrollment 2010 (4,200 students)

a. In what ways are the two pie graphs misleading?

b. In what ways are the two pie graphs accurate in their representation?

c. Has the enrollment of minorities increased or decreased from 1990 to 2000? Explain how it is possible to make an argument from both points of view.

40. A city in the Midwest advertises that it has three more sunny days a year than Seattle, Washington. What is misleading about this a claim?

41. The average life span of a refrigerator is 10 years. If Bob and Betty buy a new refrigerator, is it guaranteed to last 10 years? Explain.

42. An advertisement for car batteries claims that 4 out of 5 batteries have a life of 5 or more years. Can you conclude that the average life of a battery is about 5 years?

43. Robert is considering an Internet dating service where the average age of the patrons is 32. If Robert is 23, is he likely to meet someone his own age? Explain.

44. Two tree removal services claim to have the best rates in town. Stanley's Stump Removal has a rate of $300 per tree, but then adds a $25 removal fee. Tricia's Tree Trimming charges a flat rate of $315 per tree. Which claim is correct?

45. A lightbulb company claims that their lightbulbs last on average 1,500 hr. Amy decides to test the claim by buying 10 light bulbs and leaving them on until they burn out. Seven of the 10 bulbs burn out shortly before the 1,500-hr mark. Can Amy discount the claim as being false? Explain.

46. Over the last 50 years, many studies have shown the ill effects of smoking tobacco. Also, several studies have found that smoking is not harmful and in some cases is even beneficial. Why is it a cause for concern that the studies finding no harmful effects were either conducted or funded by tobacco companies?

47. Many morning news programs host authors to promote their books. On one morning, the host gives a glowing review of a book. How might the review of the host be affected by the fact that the television network owns the publishing company of the book?

48. A car manufacturer does a study that tests the reliability of automobiles. Why would it be a cause for concern if the manufacturer placed its automobile at the top of the list?

49. A newspaper article claims that a recent surge in crime is related to an increase in poverty. Do you think the claim makes sense? What further information would be useful in determining the validity of the claim?

50. A toothpaste company does a comparative experiment to test a new product. One group uses the new toothpaste, and the other uses another top brand. At the end of two months, the group using the new toothpaste had an average of 10% less tartar. The makers of the toothpaste claim that it is a better product. What additional information would you want to know to decide whether this claim is legitimate?

51. A teacher wants to know the most effective way of teaching multidigit multiplication. She conducts an experiment in which one class is taught multiplication using the standard algorithm and another is taught using the lattice method. At the end of the unit, she gives both classes a test. The average of the class that used the lattice method was two points higher than that of the other class. The teacher claims the lattice method works better. Is she correct?

Mathematical Communication

52. Write a story problem that makes a distorted statistical claim by using:

  a. A percent.

  b. An average.

53. A newspaper article claims that the average age of those living in poverty is 7 years. Does this claim make sense? Discuss the question with a group of your peers.

54. What does it mean to be a wise consumer of statistics? Can you think of a time when you or someone you know was not a wise consumer of statistics? Explain what happened.

55. Jackson reads a newspaper article that claims 55% of freshman at a local college graduate in 4 years. From this, he concludes that 45% of students do not graduate. Is he correct? Explain.

56. Emily and Elise are in different classes, but both have Mrs. Cox for eighth-grade math. They take a test on which Emily's class had an average of 79 and Elise's class had an average of 78. Emily said her class did better, but Elise says there is no real difference between the two. Who is correct, and how would you explain your decision to the other person?

Building Connections

57. Find a short newspaper, magazine, or television story that involves a statistical result. Write a short summary of why you believe or do not believe the conclusions or claims of the story.

58. Find several advertisements from a newspaper or a magazine that involve statistical information. State whether you think the claims are valid, and then explain your decision.

59. Expert witnesses are often called on to testify in a court of law, and they are commonly paid for their services. If you were a juror, how would the knowledge of the fees change your view of the expert's testimony?

Historical Note

Many scholars believe that modern-day statistics can trace its roots back to an Englishman named John Graunt (1620–1674). Even though he was a haberdasher by trade, he was fascinated with the Bills of Mortality, which were printed records similar to modern-day obituaries. In his book, *Natural and Political Observations Made upon the Bills of Mortality*, Graunt used the Bills of Mortality to study the characteristics and the fluctuations of diseases like the bubonic plague. He used these facts, along with church records of baptisms and deaths, to arrive at one of the first statistical estimates of the population of London. His work was clear and concise, and it is still considered an important work in vital statistics. It also won him the respect of Charles II and a place as a Fellow in the Royal Society.

Opening Problem Answer

©Istockphoto.com/Nikolay Titov

We can find out who did better by comparing the z-score of each child. Ashley's z-score is $z = \dfrac{87 - 72}{8} = 1.88$. Quin's z-score is $z = \dfrac{42 - 35}{4} = 1.75$. Because Ashley's z-score is higher, she did better on her test.

Chapter 14 REVIEW

Summary of Key Concepts from Chapter 14

Section 14.1 Formulating Questions and Collecting Data

- Statistics is the science of collecting, analyzing, and interpreting **data,** which are numbers that occur in a context.

- Statistical studies have four steps: formulating a question, collecting data, analyzing data, and drawing conclusions.

- A **population** is a group of people, animals, or objects under consideration in a study.

- **Variability** is the idea that characteristics can change from one participant to another.

- **Qualitative,** or **categorical, variables** separate data into non-numerical categories. **Quantitative,** or **numerical, variables** describe a measure.

- A **census** is a study that collects data on an entire population.

- A **sample** is small portion of the population on which the study is conducted. **Sampling variability** is the idea that data values will vary from sample to sample.

- **Bias** refers to any problem in the design or conduct of the study that causes the results to favor a particular outcome.

- Common methods of selecting a sample include **convenience sampling, voluntary sampling, simple random sampling,** and **stratified sampling.**

- Three common ways to collect data are **surveys, observations,** and **experiments.**

- In a **comparative study,** two or more groups are compared to find the similarities and differences between them.

- In a comparative study, the variable with the effect under study is the **explanatory,** or **independent, variable.** The variable affected by the explanatory variable is the **response,** or **dependent, variable.**

- A **confounding variable** has an effect on the response variable that is difficult to separate from the explanatory variable.

- In an experiment, one group, called the **treatment group,** has a **treatment** deliberately imposed on it. The other group, called the **control group,** does not.

- The **placebo effect** is counteracted by making all participants think they are receiving the treatment. One group is given the treatment, while the other is given a **placebo,** or sugar pill. This is a **blind study** because participants are not told whether they are receiving the treatment or the placebo.

Section 14.2 Representing and Analyzing Data with Statistical Graphs

- The basic labels on a statistical graph are a **title, axis labels,** and a **legend.**

- An **outlier** is any data point atypical from the rest of the data set. **Clusters** are small groups of data isolated from the rest of the data set. Spaces between clusters are called **gaps.**

- **Dot plots** and **stem and leaf plots** show every numerical value in small data sets.

- **Frequency tables, pictographs,** and **bar graphs** show counts for categorical data sets.

- **Histograms** show data arranged in continuous numerical categories.

- **Line graphs** show data relative to a continuous variable, such as time.

- **Circle graphs** show the relative amounts of parts to the whole.

- **Scatterplots** show the relationship between two variables.

- Two variables are **correlated** if there is a relationship between them. If high values of one variable are associated with high values of the other, they are **positively correlated.** If high values of one variable are associated with low values of the other, they are **negatively correlated.** A correlation can be represented with a **trend line** or a **regression line.**

Section 14.3 Representing and Analyzing Data with Descriptive Statistics

- A **measure of center** is one number that best represents a data set.

- The **mode** of a set of numbers is the numerical value that occurs most frequently.

- Let $x_1, x_2, x_3, \ldots, x_n$, be a set of n numbers arranged in ascending order. If n is odd, the **median** is the number in the $\frac{n+1}{2}$ position of the list. If n is even, the median is one-half the sum of the middle two numbers.

- The **arithmetic mean** of a set of n numbers, $x_1, x_2, \ldots, x_n$, is $\overline{x} = \dfrac{x_1 + x_2 + \cdots + x_n}{n}$.

- A **measure of spread** describes how a data set is dispersed.

- The **range** is the difference between the largest and smallest data values, which are called the **maximum** and **minimum,** respectively.

- Let $x_1, x_2, \ldots, x_n$, be a set of n numbers in ascending order for which the median is known. The **first,** or **lower, quartile,** Q_1, is the median of the data set from x_1, to the median. The **third,** or **upper,** quartile, Q_3, is the median of the data set from the median to x_n.

- A **5-number summary** includes the minimum, first quartile, median, third quartile, and maximum.

- The **interquartile range,** or **IQR,** is the difference between the first and third quartiles. Any data point with a value of $1.5 \cdot$ IQR below the first quartile or $1.5 \cdot$ IQR above the third quartile is an outlier.

- A **box-and-whiskers plot,** or **box plot,** is a graph of a 5-number summary.

- If $x_1, x_2, \ldots, x_n$, is a set of n numbers with mean $\overline{x}$, then the **standard deviation,** s, is the square root of the **variance,** or $s = \sqrt{v} = \sqrt{\dfrac{(x_1 - \overline{x})^2 + (x_2 - \overline{x})^2 + \cdots + (x_n - \overline{x})^2}{n}}$.

- A **normal distribution** is characterized by a smooth, bell-shaped curve that extends to infinity without touching the horizontal axis.

- **Empirical rule:** In a normal distribution, approximately 68% of the data lies within one standard deviation of the mean, 95% lies within two, and 99.7% lies within three.

- A **z-score** indicates how many standard deviations a data point is from the mean, or $z = \dfrac{x - \overline{x}}{s}$.

Section 14.4 Abuses of Statistics

- Statistical graphs can be distorted by **manipulations of the scales, perceptual distortions,** and **visual embellishments.**

- A **margin of error** gives a range of values that the actual value of a statistic lies within.

Review Exercises Chapter 14

1. Identify each variable as qualitative or quantitative.

   a. Hair color

   b. Salary

   c. Favorite ice cream

   d. Weight

2. In each study, identify the sample and the population. Then identify the variable of interest and whether it is quantitative or qualitative.

   a. A consumer group tests 10 clothes dryers to find the typical drying time of all dryers.

   b. A newspaper surveys 1,200 American adults to find how many times a week they exercise.

   c. A computer manufacturer tests 35 computers for correct installation of software.

3. In each case, identify the sample and the population. If the sample is representative of the population, what can we infer about the population?

   a. A sample of 800 seniors was selected from universities in the United States, and 63% got jobs related to their major.

   b. A consumer group polls 200 households in a suburban area. One-fourth of the households no longer have landline telephone service.

   c. A pollster conducts a study with 150 married couples. In 23% of the couples, the wife earns more than the husband.

4. Write a question of each type that could be used on a survey to determine whether people use social networking software regularly.

   a. Interval

   b. Multiple-choice

   c. Open-ended

5. Identify the explanatory variable and the response variable in each relationship.

   a. The relationship between computer usage and vision problems

   b. The relationship between age and amount of exercise

   c. The relationship between income and time spent at work

6. A pharmaceutical company has developed a new cold remedy. To test it, they select 90 people with cold-like symptoms. Thirty people are given the product, 30 people are given a placebo, and 30 people are given nothing. The symptoms are gone from the first group in 48 hr, from the second group in 72 hr, and from the third group in 120 hr.

   a. What are the population and the sample in the study?

   b. What are the treatment and control groups?

   c. What are the explanatory and response variables?

   d. Is the new cold remedy effective? Explain.

7. Roger is an inspector in a pencil factory that produces 350,000 pencils a day. He randomly samples 1,500 pencils and finds that 21 have a defective eraser. How many pencils are likely to have a defective eraser on any given day?

8. A local newspaper in a coastal city polls its subscribers about new federal fishing regulations. The responses are overwhelmingly against the regulations, so the conclusion is made that most Americans oppose them. How might the study be biased?

9. The following data represent a set of scores from a 50-point test.

   45 34 43 44 45 46 39 38 37 19 31 34 48
   48 45 45 36 39 40 41 43 39 42 42 45 46

   a. Draw a dot plot and a stem and leaf plot of the data.

   b. Are any possible outliers or clusters in the data set?

   c. What are the best and worst scores in the data set?

   d. What is (are) the most commonly occurring score(s) in the data set?

10. A used car lot has five salesman: Frank, Bob, Derek, Joe, and Melinda. Make a bar graph of the number of cars sold by each person in a given month if Frank sold 14 cars, Bob sold 13 cars, Derek sold 15 cars, Joe sold 11 cars, and Melinda sold 18 cars.

11. The following histogram shows the results of a senior class on a standardized test.

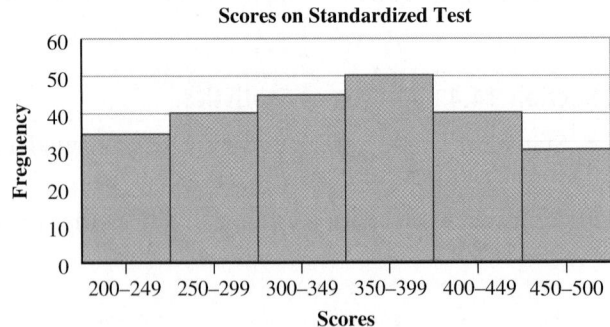

Scores on Standardized Test

Answer each question using the information given.

a. What range of scores occurred most frequently? Least frequently?

b. What is the range of possible scores on the exam?

c. How many people scored between 300 and 400 on the exam?

12. The table shows how the population of a certain city has changed over the twentieth century.

Year	Population
1900	12,000
1910	13,000
1920	13,500
1930	14,000
1940	25,000
1950	29,000
1960	35,000
1970	46,000
1980	59,000
1990	75,000

Draw a line graph that shows the change in population, and describe the general trends in the data.

13. Answer each question based on the information provided in the circle graph.

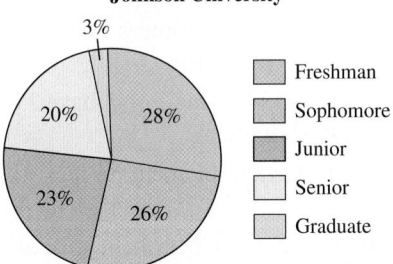

Precentage of Students by Class at Johnson University

a. Which class has the largest number of students in it? The smallest?

b. If Johnson University has 4,300 students enrolled, how many students are enrolled in the freshman class? The senior class?

14. Make a back-to-back stem and leaf plot of the test scores of the two classes shown:

Class 1	35, 45, 46, 47, 47, 48, 49, 51, 51, 53, 54, 55, 56, 59, 60
Class 2	34, 35, 35, 36, 37, 38, 45, 47, 48, 48, 49, 50, 51, 53, 57, 58

Which class seemed to do better on the test?

15. The following scatterplot shows a comparison between the number of hours of exercise a person gets in a week and the resting heart rate of the person.

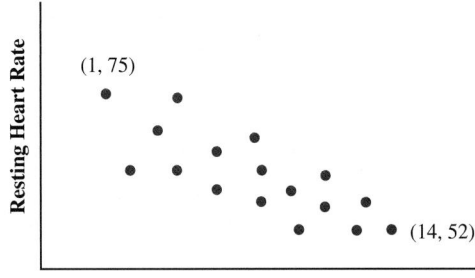

Amount of Exercise to Resting Heart Rate

a. Is there a positive, negative, or no correlation between the amount of exercise and the resting heart rate? Does the correlation make sense?

b. Use the points with the coordinates to find a regression line for the data. What is the equation of this line?

c. Use the equation in part b to find an estimate for the resting heart rate of someone who exercises 5 hr a week? 10 hr a week? 12 hr a week?

16. Find the mean, the median, and the mode for each data set.

a. 67, 71, 72, 72, 75, 78, 79, 83, 85

b. 10.4, 10.6, 11.1, 12.4, 15.4, 16.7

17. Give a 5-number summary for each data set.

a. 10, 10, 12, 14, 15, 17, 18, 19, 20

b. 63, 66, 71, 88, 89, 90, 90, 92, 95

18. Draw a box plot for each data set in the previous exercise.

19. Use the IQR to determine whether the following data set has any outliers.

$$14, 16, 19, 20, 25, 26, 29, 42$$

20. If the mean weight of 20 students is 104 lb, what is the sum of the 20 weights?

21. Compute the mean and standard deviation for each data set.

a. 34, 36, 37, 38, 40, 42, 44, 44

b. 95, 96, 99, 101, 103, 105, 107

22. Consider the following set of weights for a group of 10-year-old students.

65, 67, 78, 65, 70, 79, 82, 66, 68, 69, 70, 70, 71, 67, 80, 81, 78, 76, 75, 73, 72, 68, 69, 80, 88, 87, 84, 76, 79, 68

a. Compute the mean and standard deviation for the data.

b. What percent of the weights are within one standard deviation of the mean?

c. What percent of the weights are within two standard deviations of the mean?

23. A class has 29 students. The mean height of the students is 49 in. If three more students are added to the class with heights of 53 in., 45 in., and 50 in., respectively, what is the new mean for all 32 students?

24. Data for a standardized test are approximately normal with a mean of 250 and a standard deviation of 25.

   a. What score is two standard deviations below the mean? Two standard deviations above the mean?

   b. About what percent of the test takers scored between 200 and 275?

   c. About what percent of the test takers scored greater than 325?

   d. What is the z-score of a test score of 195? 270? 310?

25. Jenna jogs at a speed of 7 mph for the first 3 mi of an 8-mi run. If she wants to average 6.5 mph for the entire run, how fast must she run the last 5 mi?

26. Consider a set of 15 data points. If 10 is added to each data point, what is the effect on the:

   a. Mean?   **b.** Median?

   c. Mode?   **d.** Range?

27. Provide an example of a set of data containing 10 values in which:

   a. The mean is smaller than the median and the mode.

   b. The median is smaller than the mean and the mode.

   c. The mode is smaller than the mean and the median.

28. Consider the dot plot shown:

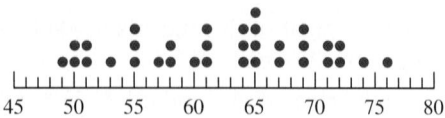

   What measure of central tendency—the mean, the median, or the mode—would best represent the data in dot plot? Explain your reasoning.

29. A professor claims that his class did poorly on a 100-point final exam. If the mean was a 65 and the median was a 55, can you believe his claim?

30. What is wrong with the graph?

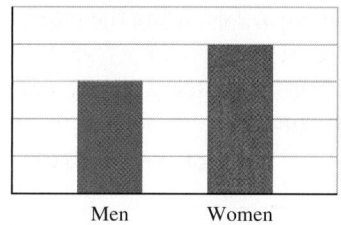

31. The following table shows the number of accidents along a certain highway at different times of the night over the course of a year.

Time at Night	10 p.m. to 12 a.m.	12 a.m. to 2 a.m.	2 a.m. to 4 a.m.	4 a.m. to 6 a.m.
Number of Accidents	14	16	10	12

   Draw a graph that accentuates the differences between time periods. What incorrect conclusions can be made from such a graph?

32. The following table shows the production cost of a particular factory in millions of dollars over the course of several years.

Year	2005	2006	2007	2008	2009	2010	2011
Cost of Production ($ millions)	13.4	14.5	16.4	16.9	17.3	18.0	18.4

   Draw a line graph to show that the cost of production has not dramatically increased.

33. Which of the following claims are misleading? Explain how.

   a. Eight out of 10 customers were satisfied with the service at a given store.

   b. A brand of cereal claims to have 25% more fiber than other brands.

34. Give a possible complementary statistic for each claim.

   a. Fifteen percent of people 65 and older are still at work.

   b. Four out of 5 children like to drink apple juice.

   c. One-fourth of unmarried people older than 25 have a bachelor's degree or better.

35. Give an interpretation of the margin of error in each statement.

   a. A poll with a margin of error of ±6% claims that 50% of people say that a smile is the first facial feature they notice.

   b. A poll with a margin of error ±3% claims that 80% of people who watch the Super Bowl do so only for the commercials.

36. The following graph shows the cost of five universities. It is used to show that University C is the lowest-priced university. Is the claim accurate?

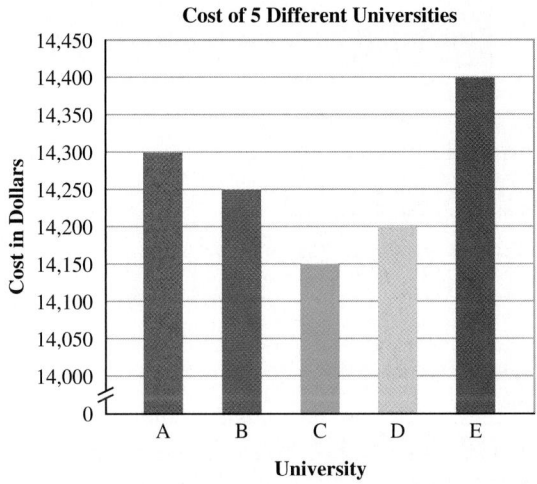

Cost of 5 Different Universities

University

37. A study shows that people who drink two or more glasses of milk a day have better skin tone, muscle definition, and overall health. Why would it be a cause for concern to know that the study was conducted by a dairy association?

38. In track, the 100-m dash takes only several seconds to complete. Three coaches brag that they have the fastest set of four runners in the 100-m dash based on the following times in seconds:

Jefferson High: 11.3, 11.3, 11.6, 11.7

Moravian High: 11.2, 11.4, 11.5, 11.5

Bettstown High: 11.3, 11.4, 11.4, 11.9

Which measure of central tendency did each coach use to make his claim? Which do you think is the most believable, and why?

Answers to Chapter 14 Check Your Understandings and Activities

Check Your Understanding 14.1-A

1. a. Qualitative.
   b. Quantitative.
   c. Qualitative.
   d. Quantitative.

2. a. The population is the cars manufactured at the plant, the sample is the 75 cars tested, and it is a simple random sample.
   b. The population is the people who visit the Internet site, the sample is the 1,487 respondents, and it is a voluntary sample.
   c. The population is eighth-grade students, the sample is the 65 students who are surveyed, and it is a convenience sample.

3. The West Coast is heavily populated with registered Democrats. If the sample is limited to this area, then it is not surprising that the results favor the Democratic candidate.

Check Your Understanding 14.1-B

1. The conditions of the abortion are not clearly stated. Most people support abortion if it is necessary to save the life of the mother and so must answer yes to be truthful. However, the results may change if specific conditions are placed on the abortion. For instance, if the question were "Do you support late-term abortion?" the results are likely to be less favorable.

2. a. Observation that makes a comparison between groups.
   b. The population is students in the two high schools, the sample is the sophomores attending the schools, the explanatory variable is the mathematics curriculum, and the response variable is performance on the achievement test.
   c. The results do not the support the superintendent's conclusion. The test is unfair to the students at Oakdale High because the focus of their curriculum is not on computational skills.

Check Your Understanding 14.2-A

1. The dot plot and stem and leaf plot are shown:

Scores on a 150-Point Test

Scores

Scores on a 150-pt Test	
Stem	**Leaves**
6	3
7	
8	57899
9	125
10	
11	2489
12	0124668
13	012256
14	1137

   a. 63 is a possible outlier.
   b. There is a cluster between 85 and 95.
   c. Because a 70% is a score of 105 points and most of the scores are higher, the class as a whole did quite well.

2. The histogram is shown:

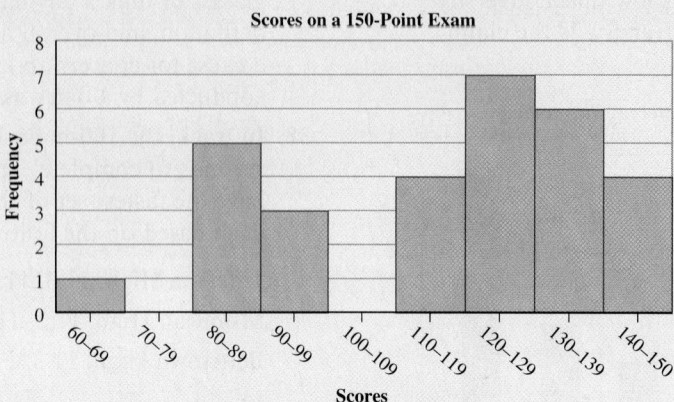

3. The pictograph is shown:

Check Your Understanding 14.2-B

1. The graph shows that on most days the parents watched more television than the children.

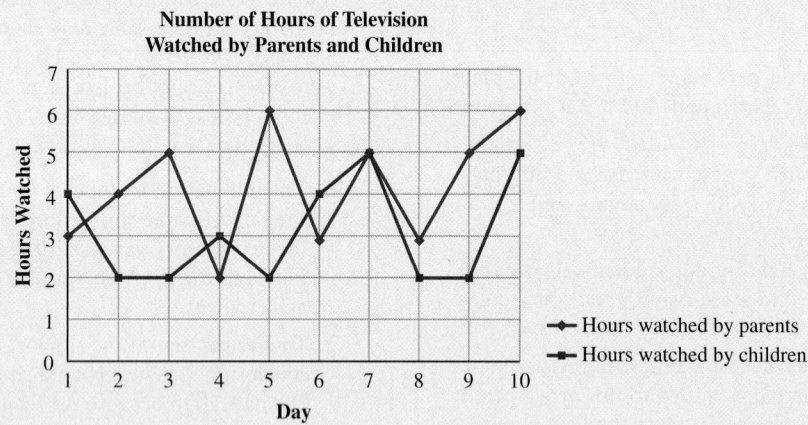

2. a. Australia, Europe, Antarctica, South America, North America, Africa, Asia.   **b.** North and South America or Africa and Antarctica.   **c.** 7% of 57,393,000 is $(0.07)(57,393,000) = 4,017,510 \text{ mi}^2$.

3. Possible answers are given.

   a. Weight gain is positively correlated to number of calories eaten and negatively correlated to the amount of exercise.
   b. Test scores are positively correlated to amount of time spent studying and negatively correlated to number of hours watching television.

Check Your Understanding 14.3-A

1. a. Mean = 12.7, median = 12, mode = 12.   **b.** Mean = 89.6, median = 90, mode = 88 and 91.   **c.** Mean = 66.6, median = 67, mode = 56 and 67.   **d.** Mean = 73.8, median = 76, mode = 56.

2. The mean, median, and mode for each competitor are given.

   Joe: Mean = 278, median = 275, mode = 275.

   Mike: Mean = 277, median = 280, mode = 265.

Tyrone: Mean = 279, median = 270, mode = 270.

To best make his case, Joe should use the mode, Mike should use the median, and Tyrone should use the mean.

Check Your Understanding 14.3-B

1. **a.** Maximum = 78, Q_3 = 58.5, median = 53, Q_1 = 47, and minimum = 45.   **b.** Maximum = 15.6, Q_3 = 14.7, median = 14.2, Q_1 = 13.9, and minimum = 9.7.

2. **a.** IQR = 11.5; 78 is an outlier.   **b.** IQR = 0.8; 9.7 is an outlier.

3. **a.**

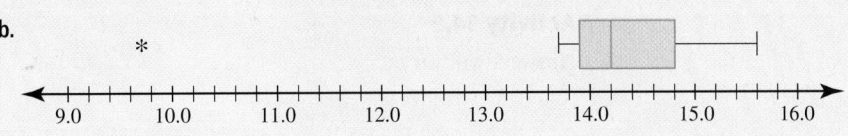

b.

Check Your Understanding 14.3-C

1. **a.** $\bar{x} = 7$ and $s = 1.29$.   **b.** $\bar{x} = 16.25$ and $s = 2.89$.

2. The average weight loss of the two groups is similar with group 2 losing more weight per person. The standard deviation indicates that group 1 probably had the person with the highest weight loss, but this is likely offset by lower weight losses.

3. **a.** 425.   **b.** 350.   **c.** 81.5%.   **d.** The z-score for $x = 410$ is $z = 0.4$, the z-score for $x = 380$ is $z = -0.8$, and the z-score for $x = 460$ is $z = 2.4$.

Check Your Understanding 14.4-A

1. To accentuate the differences, remove part of the vertical scale. To minimize differences, show the entire vertical scale.

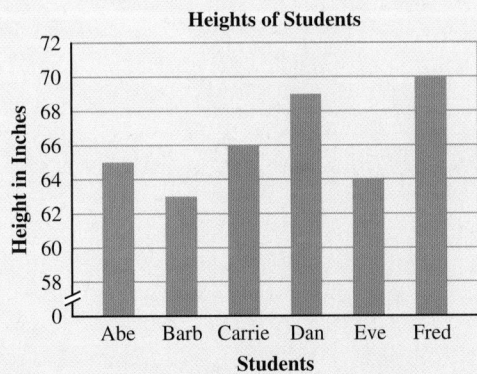

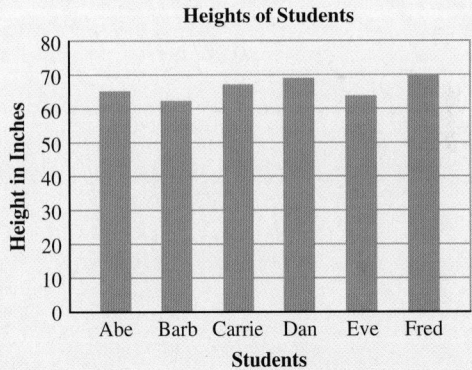

2. There is no scale on the vertical axis so it is difficult to determine just how large the differences are between male and female enrollment.

3. The height of the milk cartons is used to represent the difference between the years. However, the volume of the cartons catches our eye, and the volume is eight times the intended value.

Check Your Understanding 14.4-B

1. The numbers claim that 80% of children break a bone before the age of 14. This seems like a high value, so we might suspect that the sample might not be truly representative of the population.

2. The cost of living is a concern when moving to a new city. Even though she is getting paid more, her salary may not go as far.

3. **a.** Forty-seven percent of American women ages 18 or older are married.   **b.** Hispanics are not the largest minority in 30 states.   **c.** Of all K–12 students, 89% are not enrolled in private school.

Activity 14.1

Answers will vary.

Activity 14.2

Answers will vary.

Activity 14.3

Answers will vary.

Activity 14.4

Answers will vary.

Activity 14.5

Answers will vary.

Activity 14.6

Answers will vary.

Activity 14.7

a. Answers will vary.

b. Answers will vary.

c. Answers will vary.

Activity 14.8

Answers will vary.

Activity 14.9

Answers will vary.

Chapter 15

Probability

Opening Problem

Each team in a high school basketball conference must play every other team twice. If 12 teams are in the conference, how many games are played?

Jamie Roach / Shutterstock.com

15.1 Experimental Probability and Making Predictions from Data

15.2 Theoretical Probability

15.3 Conditional and Geometric Probability, Odds, and Expected Value

15.4 Counting Techniques and Probability

Chapter 14 focused on the concepts, procedures, and computations

of basic data analysis. We learned how to formulate statistical questions and collect data, as well as how to represent, analyze, and draw conclusions from data by using graphs and descriptive statistics. Data can also be used to make predictions about the likelihood of an event occurring. However, to do so, we must use probability.

Beyond its connection to data analysis, probability links mathematical topics such as number, proportional reasoning, algebra, and geometry. It also provides a meaningful context for solving problems, reasoning mathematically, developing number sense, and applying computational skills. Because of its practicality and mathematical importance, it is becoming more prevalent in the elementary and middle grades. This is due in large part to the *NCTM Principles and Standards* (2000), which lists learning probability as one of the key goals in the Data Analysis and Probability Standard.

■ NCTM Data Analysis and Probability Standard

Instructional programs from prekindergarten through grade 12 should enable all students to:

- Formulate questions that can be addressed with data, and collect, organize, and display relevant data to answer them.
- Select and use appropriate statistical methods to analyze data.
- Develop and evaluate inferences and predictions that are based on data.
- Understand and apply basic concepts of probability.

Source: NCTM STANDARDS Copyright © 2011 by NATIONAL COUNCIL OF TEACHERS OF MATHEMATICS. Reproduced with permission of NATIONAL COUNCIL OF TEACHERS OF MATHEMATICS.

Children in early elementary grades do not work much with probability because most probabilistic notions require a degree of mathematical sophistication that they have not yet developed. Nonetheless, they develop an intuitive notion of probability and chance by describing the likelihood of daily events using words such as "likely," "certain," and "impossible." Once in third and fourth grade, children begin to explore probability as a measure of chance by conducting simple experiments, such as rolling dice or spinning spinners.

By the middle grades, students have the mathematical skills they need to work with probability in greater depth. As they do, they learn that there are two types of probability: experimental and theoretical. Students naturally encounter the need for experimental probability as they make predictions from data. They also learn to use theoretical probability as they conduct probabilistic experiments using a variety of random devices. While working with both types of probability, students are encouraged to learn and use the properties of probability, as well as the proper terminology. Table 15.1 identifies specific expectations that children are likely to learn about probability in the elementary and middle grades.

Table 15.1　Classroom learning objectives	K	1	2	3	4	5	6	7	8
Approximate the probability of a chance event by collecting data and observing its long-run relative frequency.								X	
Design and use a simulation to generate frequencies for compound events.								X	
Develop a probability model by assigning equal probability to all outcomes, and use the model to determine probabilities of events.								X	
Represent sample spaces for compound events using methods such as organized lists, tables, and tree diagrams, and then use the sample space to find probabilities.								X	
Understand that the probability of a chance event is a number between 0 and 1 that expresses the likelihood of the event occurring.								X	

Source: Adapted from the *Common Core State Standards for Mathematics* (*Common Core State Standards Initiative*, 2010).

Probability is relatively new to the elementary and middle grades. To help pre-service teachers better understand the concepts of probability they are likely to teach, Chapter 15:

- Demonstrates how to use experimental probability and simulations to make predictions from data.
- Introduces the basic concepts, computations, and properties associated with theoretical probability.
- Presents other notions related to probability, such as odds and mathematical expectation.
- Discusses the use of counting techniques in computing probability.

SECTION 15.1 Experimental Probability and Making Predictions from Data

Probability is often described as the mathematics of chance. Historically, it has its roots in gambling, but probability is now used in a wide variety of situations. For instance, each of the following statements makes use of probability in some way:

- There is a 25% chance of rain today.
- The Bulldogs are 4-to-1 favorites to win the championship next year.
- If I study hard tonight, I have a fifty-fifty chance of passing my test tomorrow.

As these examples show, a **probability** is a number that indicates the extent to which we think an event will or will not happen. In general, there are two ways to find a probability. To determine an **experimental probability,** we use data to determine the likelihood that an event will occur. To find a **theoretical probability,** we use the possible outcomes of an experiment yet to be performed. In this section, we concentrate on experimental probability and its connection to data analysis. The rest of the chapter is devoted to theoretical probability.

Understanding Experimental Probability

Probabilities are often given in the context of an **experiment,** which includes any activity with results that can be observed, measured, and recorded. The possible results of the experiment are called **outcomes,** which are typically measured with counts, although they are sometimes measured in other ways. In most situations, we want the probability of an **event,** which is a particular outcome or set of outcomes from the experiment. The outcome of an experiment should occur **randomly,** which means that it is not influenced by external conditions and that the likelihood of an event is determined by the probabilities inherent in the experiment.

In many senses, a probability is a measure of the likelihood that uncertain events will occur. The larger the probability, the more likely it is that an event is to occur. One way to find a probability is to conduct an experiment a number of times and then count the outcomes favorable to the event. To compute the probability, take the ratio of the number of favorable outcomes to the total number of repetitions of the experiment. For instance, to compute the probability of rain on a certain day, we might look into past weather records for 100 days that had the same weather conditions. If we found that it rained on 25 of them, then we would say that the chance of rain is 25 out of 100, or 25%. Each of the 100 days represents one trial of the experiment, with the rainy days representing favorable outcomes. Because we found the probability by conducting an experiment multiple times, we call this an **experimental,** or **empirical, probability.**

Explorations
Manual
15.1

Definition of Experimental Probability

If an experiment with a certain number of outcomes is repeated n times, with a specific event E occurring m times, then the **experimental probability** that event E occurs on a given trial is given by the ratio

$$\text{Experimental probability of } E = \frac{\text{number of times } E \text{ occurs}}{\text{number of trials}} = \frac{m}{n}$$

Because an experimental probability is defined to be the *ratio* between the number of favorable outcomes and the number of total outcomes, it can be expressed as a fraction, a decimal, or a percent.

Example 15.1   Finding an Experimental Probability

Karol, a fourth-grade student, wants to find the likelihood of rolling a 5 or 6 on a regular six-sided die. How might she determine this?

Solution Karol can find the likelihood of rolling a 5 or 6 by rolling a die numerous times, recording the results, and then computing an experimental probability. For instance, suppose she rolls a die 30 times and records the following results:

$$
\begin{array}{cccccccccc}
5 & 4 & 2 & 5 & 5 & 6 & 2 & 1 & 1 & 2 \\
1 & 6 & 2 & 5 & 6 & 4 & 3 & 1 & 6 & 3 \\
5 & 4 & 2 & 3 & 6 & 3 & 1 & 4 & 5 & 1 \\
\end{array}
$$

Because a 5 or 6 is rolled 11 times, the experimental probability is

$$\text{Experimental probability of 5 or 6} = \frac{\text{number of 5s or 6s}}{\text{total number of trials}} = \frac{11}{30}$$

Experimental probabilities have the benefit of being computed directly from observations of recorded events. Unfortunately, this means their values are relevant only to that particular set of trials. To understand why, suppose that two of Karol's classmates, Ian and Kristie, conduct the same experiment and obtain the following results:

Ian's Results

$$
\begin{array}{cccccccccc}
2 & 3 & 1 & 2 & 2 & 1 & 5 & 2 & 5 & 2 \\
5 & 1 & 5 & 6 & 4 & 5 & 3 & 5 & 3 & 5 \\
4 & 5 & 4 & 2 & 3 & 3 & 2 & 6 & 3 & 1 \\
\end{array}
$$

Kristie's Results

$$
\begin{array}{cccccccccc}
5 & 3 & 3 & 3 & 6 & 3 & 4 & 3 & 4 & 6 \\
4 & 1 & 5 & 5 & 2 & 1 & 4 & 6 & 1 & 5 \\
1 & 6 & 6 & 2 & 6 & 2 & 6 & 5 & 5 & 3 \\
\end{array}
$$

Ian's results give him an experimental probability of $\frac{10}{30}$, whereas Kristie's give her an experimental probability of $\frac{13}{30}$. As we can see, the results do not match Karol's.

So how can we trust any experimental probability to be an accurate measure of the likelihood of an event? Fortunately, there is a solution to this problem. The example shows that if we conduct only a few trials of an experiment, the experimental probabilities can vary considerably. However, if we conduct many more trials of the experiment, the experimental probabilities vary less and less. This important principle is known as the **Law of Large Numbers.**

Theorem 15.1	Law of Large Numbers

As the number of trials for a given experiment increases, the variation in the experimental probability of a particular event decreases until it closely approximates a fixed number.

Historically, several people have tested the Law of Large Numbers through direct experimentation. One of the most notable was John Kerrich, a South African mathematician who had the misfortune of spending most of World War II in a German prison camp. While in prison, he eased his boredom by conducting probability experiments. In one case, he flipped a coin 10,000 times and recorded the results. A summary of his results at different counts is given in Table 15.2. As the table shows, when the number of tosses is small, the percent of heads varies between 40% and 60%. However, as the number of the tosses increases, the percents vary less and eventually settle around 50%.

Explorations Manual 15.2

Table 15.2 10,000 flips of a coin

Number of Tosses	Number of Heads	Percent Heads	Number of Tosses	Number of Heads	Percent Heads
10	4	40%	300	146	48.7%
20	10	50%	400	199	49.8%
30	17	56.7%	500	255	51%
40	21	52.5%	1000	502	50.2%
50	25	50%	5000	2533	50.7%
100	44	44%	8000	4034	50.4%
200	98	49%	10,000	5067	50.7%

Karl Pearson, an English statistician, conducted a similar experiment about 50 years before Kerrich. In his experiment, Pearson flipped a coin 24,000 times and recorded 12,012 heads, which is about 50.05% of the flips.

These experiments show that we can interpret experimental probability in two ways. In one sense, an experimental probability indicates the likelihood that an event will occur in a single trial of an experiment. This value becomes a better predictor when it is based on more trials of the experiment. For instance, the experiments of Kerrich and Pearson indicate that the probability of getting a heads on any one flip of a coin is 50%. In another sense, an experimental probability indicates what will happen in the long run if an experiment is conducted many times. For instance, over the course of many trials, we would expect a coin to come up heads about half the time it is flipped.

Example 15.2	Experimental Probability and a Bag of Marbles

An experiment is done in which a single marble is drawn from a bag of marbles. After the color is recorded, it is placed back in the bag. The experiment is repeated 600 times. Use the following frequency table to answer each question:

Marble Color	Number of Times Drawn
Black	115
Blue	51
Green	43
Red	87
White	199
Yellow	105

a. What is the experimental probability of drawing a red marble? A green marble?

b. When the next marble is drawn, which color is it most likely to be?

c. If we were to draw 30 more marbles, how many are likely to be yellow?

Solution

a. Because 600 marbles were drawn and 87 were red, the experimental probability of drawing a red marble is $\frac{87}{600} = 0.145 = 14.5\%$. Similarly, the experimental probability of drawing a green marble is $\frac{43}{600} \approx 0.072 = 7.2\%$.

b. When the next marble is drawn, it will most likely be white.

c. The experimental probability of drawing a yellow marble is $\frac{105}{600} \approx \frac{1}{6}$. Because $\frac{1}{6}$ of 30 is 5, we would expect 5 of the next 30 marbles to be yellow.

Explorations
Manual
15.3

The last example shows how we can compute experimental probabilities directly from statistical data. As a result, we can use experimental probability to make predictions in a variety of statistical situations.

Example 15.3	Experimental Probability and Test Scores

Application

The following dot plot shows the scores of Mrs. Gamble's fourth-grade class on the English portion of a state achievement test. Scores of 55 or greater are passing.

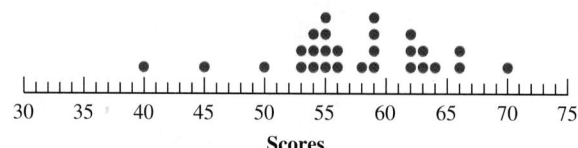

Scores on English Exam

30 35 40 45 50 55 60 65 70 75

Scores

If one score is selected at random, what is the experimental probability that the score is:

a. A 59?   **b.** More than 60?   **c.** Passing?

Solution If we consider each student who took the test to be a trial in the experiment, then 28 trials were conducted in Mrs. Gamble's class. Given this number, we can use the information in the dot plot to compute the experimental probabilities.

a. The experimental probability of selecting a score of 59 is $\frac{4}{28} = \frac{1}{7}$.

b. The experimental probability of selecting a score above 60 is $\frac{9}{28}$.

c. The experimental probability of selecting a passing score is $\frac{20}{28} = \frac{5}{7}$.

Example 15.4	Experimental Probability and a Normal Distribution

Application

In 1998, the Centers for Disease Control and Prevention did a survey in which it looked at the physical characteristics of Americans. In one portion of the study, 6,588 women ages 18–74 had their heights measured. The data set was normally distributed with a mean height of 63.5 in. and a standard deviation of 2.5 in. If one woman is selected at random, what is the experimental probability that she is:

a. Between 61 and 66 in. tall?

b. Under 58.5 in. tall?

c. Between 66 and 71 in. tall?

Solution Because the data set is normally distributed with a mean height of 63.5 in. and a standard deviation is 2.5 in., we can label the normal distribution with specific heights:

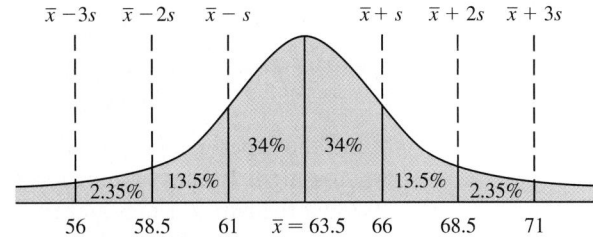

Because each portion of the distribution is given a percentage, we can convert it directly to a probability.

a. Because 68% of the data lie between the heights of 61 and 66 in., the probability of a woman having a height between these values is 68%.

b. The probability of a woman having a height under 58.5 in. tall is 2.5%.

c. The probability of selecting a woman with a height between 66 and 71 in. tall is 13.5% + 2.35% = 15.85%.

As the last two examples illustrate, much of what we have learned about statistics can be interpreted in terms of experimental probability. Any time we describe data with a count or a percentage, we can use it to make a prediction with experimental probability.

Check Your Understanding 15.1-A

1. Give a possible interpretation of each probability statement.

   a. The probability of Fast-and-Free winning the horse race tomorrow is $\frac{1}{3}$.

   b. Sue has a 50% chance of getting the promotion.

   c. There is a 9 in 10 chance that Marie will recover from her surgery.

2. Determine the experimental probability of rolling a 3 on a six-sided die by rolling it 50 times and recording the results.

3. Shirhonda has a spinner that is divided into five areas as shown:

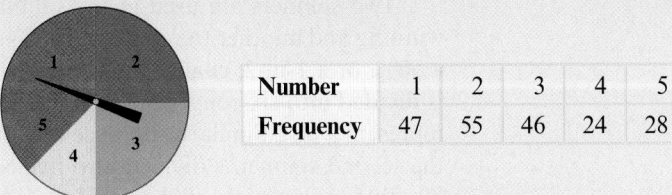

Number	1	2	3	4	5
Frequency	47	55	46	24	28

She repeatedly spins the spinner and records the number after each trial. Use her results in the table to compute the experimental probability of getting a 3.

Talk About It Consider an experiment in which two coins are flipped. Without conducting any trials, determine the likelihood that both coins will turn up heads. Describe how you arrived at your answer.

Reasoning

Activity 15.1

If a paper cup is tossed into the air, it can land on a flat surface in one of three ways: with the opening up, with the opening down, or on its side.

Opening up   Opening down   On its side

Find the experimental probability of the cup landing with the opening down by tossing it 10 times and recording the results. Repeat the experiment by tossing the cup 50 times and then 100 times. Do the probabilities tend toward a particular value? If so, what is it?

Simulations

To this point, experimental probabilities have been based on data collected from numerous trials of an experiment. However, it is often impractical or impossible to conduct the number of trials needed to compute an accurate experimental probability. In such cases, we may be able to use a **simulation,** which models an experiment by using a situation with outcomes analogous to those of the experiment but for which the trials are easily conducted and recorded. As a consequence, any experimental probability computed from the simulation will be a good approximation for the corresponding experimental probability from the experiment.

Example 15.5   A Simulation from the Classroom

Communication

The student page in Figure 15.1 uses a simulation to compute an experimental probability. Describe how the outcomes of the simulation are analogous to those of the experiment and how the trials of the simulation are conducted.

Solution The experiment is a combination of two activities: playing a baseball game and Fred batting. Given these activities, the experiment has four possible outcomes: The team wins and Fred gets a hit (W, H), the team wins and Fred does not get a hit (W, N), the team loses and Fred gets a hit (L, H), and the team loses and Fred does not get a hit (L, N). Because only one game is played and Fred bats a limited number of times, conducting multiple trials of the experiment is difficult. Consequently, we can use a simulation to find the experimental probability of the team winning and Fred getting a hit.

Two spinners are used in the simulation, one to simulate the team's chances of winning and another to represent Fred's chance of getting a hit. Because the team has a 50%, or a 1-in-2, chance of winning, the spinner corresponding to these outcomes is divided into six congruent sections. Half of the sections represent a win, and half represent a loss. Similarly, because Fred has a 20%, or a 1-in-5, chance of getting a hit, the second spinner is divided into five congruent sections, of which one represents a hit. To conduct trials of the simulation, we spin the two spinners and record the combined results. Using the 20 trials on the student page, the experimental probability of the team winning and Fred getting a hit is $\frac{3}{20}$.

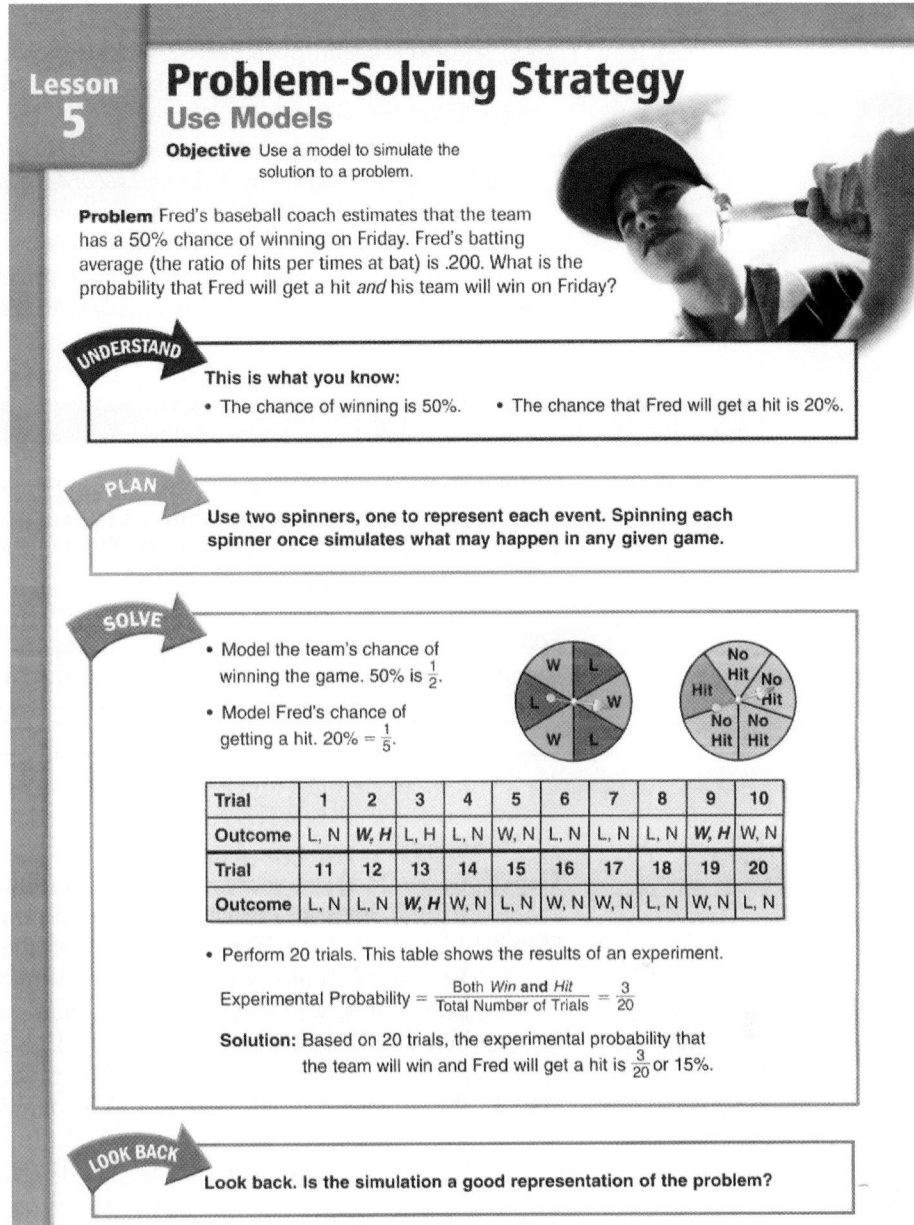

Figure 15.1 A simulation from the classroom

Source: From *Houghton Mifflin Mathematics: Student Edition Level 6* (p. 500). Copyright © by Houghton Mifflin Company, Inc. Reproduced by permission of the publisher, Houghton Mifflin Harcourt Publishing Company.

As the last example shows, designing a simulation can be complicated. The following five-step process offers some guidelines for conducting a simulation.

Five-Step Method for Conducting Simulations

Step 1: *Understand the experiment and the event.* Determine the possible outcomes of the experiment and how they fit together to form particular events.

Step 2: *Select a random device to model the experiment and event.* Select a random device with outcomes that correspond to those of the

Explorations
Manual
15.4

experiment. Such devices include dice, coins, spinners, objects drawn from a hat, random number tables, and random number generators.

Step 3: *Define a trial.* Indicate how the outcomes of the random device correspond to the outcomes of the experiment.

Step 4: *Conduct numerous trials and record the results.* The specific number of trials does not matter, as long as enough are conducted to get a variety of data. Generally, the more trials that are conducted, the more likely the simulation will reflect the probabilities inherent in the experiment.

Step 5: *Compute the experimental probability from the recorded results.*

Example 15.6   Gumballs from a Gum Machine

A gumball machine contains six colors of gumballs: red, orange, yellow, green, blue, and purple. A mother with two children passes by and buys a gumball for each child. If the gumballs in the machine are equally distributed among the six colors, what is the experimental probability that both gumballs are the same color?

Solution Conducting multiple trials of this experiment is impractical because we have to purchase two gumballs for each trial. It is much cheaper to design and use a simulation. To conduct the simulation, we follow the five-step method:

Step 1. For each gumball purchased, six outcomes are possible: red, orange, yellow, green, blue, and purple. There are many ways to combine the colors to form events; however, we are interested only in the event that the two gumballs have the same color.

Step 2. Because the six colors are evenly distributed, we need a random device with six outcomes. In this case, we choose a regular six-sided die.

Step 3. To simulate buying two gumballs, roll the die twice, once for each gumball. Because we consider only the outcomes in which the gumballs are the same color, we define successful trials to be those in which the two numbers are the same, namely (1, 1), (2, 2), (3, 3), (4, 4), (5, 5), and (6, 6). All other combinations are considered to be unsuccessful.

Step 4. We conduct 50 trials. The results are shown, with successful trials in blue:

(2, 5)	(5, 6)	(6, 2)	(2, 5)	(1, 4)	(4, 6)	(2, 1)	(2, 3)	(5, 4)	(6, 3)
(5, 2)	(1, 4)	(1, 6)	(4, 5)	(4, 1)	(3, 6)	(1, 1)	(6, 4)	(2, 1)	(2, 6)
(5, 4)	(1, 4)	(1, 5)	(1, 1)	(5, 1)	(5, 4)	(2, 2)	(4, 4)	(3, 6)	(4, 2)
(5, 5)	(4, 5)	(1, 6)	(5, 5)	(2, 1)	(6, 6)	(3, 6)	(6, 4)	(1, 1)	(1, 3)
(6, 4)	(5, 5)	(4, 1)	(1, 2)	(4, 4)	(3, 2)	(1, 6)	(2, 2)	(4, 5)	(3, 6)

Step 5. Eleven of the 50 trials in step 4 are successful, so the experimental probability of purchasing two gumballs of the same color is $\frac{11}{50} = 0.22 = 22\%$. In other words, the two gumballs are the same color in one out of every five times two gumballs are purchased.

Another device that is often used in simulations is a table of random digits. A **table of random digits,** like the one in Table 15.3, is a table in which the digits 0 through 9 are repeatedly listed in a random fashion. Each digit occurs with the same frequency; that is, any one digit has the same chance of being selected as any other digit. The digits are grouped in fives for easy use:

Table 15.3   A table of random digits

26907	88173	71189	28377	13785	87469	35647	19695	33401	51998
86668	70341	66460	75648	78678	27770	30245	44775	56120	44235
04982	68470	27875	15480	13206	44784	83601	03172	07817	01520
28549	98327	99943	25377	17628	65468	07875	16728	22602	33892
11762	54806	02651	52912	32770	64507	59090	01275	47624	16124
02805	52676	22519	47848	68210	23954	63085	87729	14176	45410
36116	42128	65401	94199	51058	10759	47244	99830	64255	40516
55216	63886	06804	11861	30968	74515	40112	40432	18682	02845
36248	36666	14894	59273	04518	11307	67655	08566	51759	41795
12386	29656	30474	25964	10006	86382	46680	93060	52337	56034
22784	07783	35903	00091	73954	48706	83423	96286	90373	23372
07330	07184	86788	64577	47692	45031	36325	47029	27914	24905
22565	02475	00258	79018	70090	37914	27755	00872	71553	56684
06644	94784	66995	61812	54215	01336	75887	57685	66114	76984
44882	33592	66234	13821	86342	00135	87938	57995	34157	99858
19082	13873	07184	21566	95320	28968	31911	06288	77271	76171
61869	33093	81129	06481	89281	83629	81960	63704	56329	10357
49333	78482	36199	11355	86044	88760	03724	22927	91716	92332
38746	81271	96260	98137	60275	22647	33103	50090	29395	10016
93369	13044	69686	78162	29132	51544	17925	56738	32683	83153

To use a table of random digits in a simulation, we first assign a digit or a group of digits to represent each outcome. We then conduct trials by selecting random strings of digits and recording the results. The starting point in the table does not matter, and we can select digits by moving left, right, up, or down. As with other simulations, the experimental probability is computed by taking the ratio of successful trials to total trials. If the strings are longer than one digit, we can continue the string over any gap in the table. The next two examples illustrate different ways of using a table of random digits in a simulation.

Example 15.7   Donating Blood

Connection

A blood bank can draw blood from only four donors at a time. If roughly 50% of the population has type O blood, what is the experimental probability that four people donating blood at one time are all type O?

Solution Because it is difficult to conduct trials of this experiment directly, we turn to a simulation instead. To create the simulation, we can use a table of random digits. Because 50%, or half, of all people have type O blood, we let half of the digits represent type O donors. Specifically, we let the digits 1–5 represent people with type O blood, and 6–0 represent those who do not. We then represent four donors with a string of four digits, one for each person, and conduct a trial by selecting a four-digit string from Table 15.3. A trial is successful if and only if every digit in the string is a 1, 2, 3, 4, or 5. Fifty trials are shown, with successful trials in blue:

2256	5024	7500	2587	9018	7009	0379	1427	7550	0872
7155	3566	8406	6449	4784	6699	5618	1254	2150	1336
7588	7576	8661	1476	9844	4882	3359	2662	3413	8218
6342	0013	5879	3857	9953	4157	9985	8190	8213	8730
7184	2156	6953	2028	9683	1911	0628	7727	7617	1618

Because 2 of the 50 trials are successful, the experimental probability is $\frac{2}{50} = 0.04 = 4\%$.

All four donors have type O blood approximately 1 out of every 25 times.

Example 15.8   Winning Prizes at a Fair

Application

A game at a county fair consists of 100 bowls that are turned upside down and arranged in a 10 × 10 grid. Under the bowls, 75 prizes have been distributed at random, making it possible for a player to win more than one prize when one bowl is selected. If a player selects a bowl at random, what is the experimental probability that the bowl has exactly two prizes?

Solution We again turn to Table 15.3 to simulate the situation, but we take a slightly different approach. Because the bowls are in a 10 × 10 grid, we imagine that they lie on top of a coordinate system that enables us to identify each bowl by the row and column it is in. To randomly place prizes under the bowls, we use Table 15.3 to generate the coordinates of bowls in the form of two-digit strings. The first digit represents the row, and the second digit the column. For instance, a string of 47 represents the bowl in row 4, column 7, and a string of 79 represents the bowl in row 7, column 9. To place all the prizes, we select 75 two-digit strings. The tally marks on the following grid represent the placement of the prizes using the following selected numbers:

47	79	88	98	90	06	89	36	54	83	97	65	87	48	84
38	61	12	72	64	34	12	10	98	51	26	95	68	35	04
34	51	15	07	21	84	85	03	41	59	38	98	57	53	84
85	37	35	09	36	22	96	55	53	66	81	63	27	88	72
66	72	28	55	15	04	72	39	24	11	14	15	35	11	90

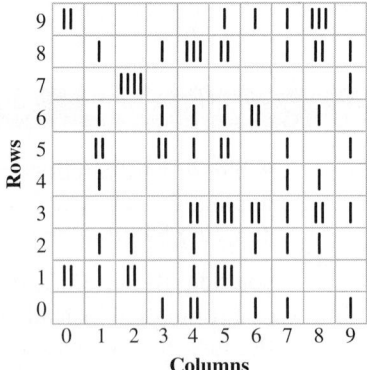

Simply by counting, we can see that 13 bowls contain exactly two prizes. Because there are 100 bowls, the experimental probability of winning exactly two prizes is $\frac{13}{100} = 13\%$.

Many calculators can generate random numbers. Select the [rand] feature of the calculator, which is often found with its other probabilistic features. Once we select this feature and press [ENTER], the calculator returns a sequence of eight to ten random digits, generally written in decimal form. We then use these digits as we would digits from a random table.

Check Your Understanding 15.1-B

1. A couple hopes to have three children. Design a simulation that uses a coin to determine the experimental probability that the couple has exactly one boy.

2. Jasmine is a high school basketball player who makes 70% of all her free throws. Use a table of random digits to determine the experimental probability that Jasmine makes two consecutive free throws during a game.

Talk About It In what situations would it be useful to use a coin as the random device in a simulation?

Activity 15.2
Many sites on the Internet offer free software for generating and recording thousands of trials for activities like rolling a die or flipping a coin. Find several websites that offer such generators, and run them several times. How could you use these generators to conduct a simulation and determine experimental probabilities?

SECTION 15.1 Learning Assessment

■ Understanding the Concepts

1. **a.** What is a probability?
   b. How is an experimental probability determined?

2. What is the difference between an event and an outcome of an experiment?

3. **a.** If two people perform the same experiment separately, will their experimental probabilities of a given event be the same?
   b. How can the two people ensure that their results are similar?

4. What does the Law of Large Numbers state is true about experimental probability?

5. Explain two ways to interpret an experimental probability.

6. How is experimental probability connected to or useful in data analysis?

7. **a.** What is a simulation?
   b. How is a simulation useful in computing experimental probabilities?

8. Describe two ways that a table of random digits can be used in a simulation.

■ Representing the Mathematics

9. Give a possible interpretation for each probability statement.
   a. In these weather conditions, the probability of a tornado touching down is 0.01.

   b. The probability of a person catching a cold from an infected person is 35%.

   c. Her chance of winning the game is $\frac{1}{3}$.

10. Give a possible interpretation for each probability statement.
    a. The probability of getting a winning instant lottery ticket is 10%.
    b. There is a 1-in-4 chance that tomorrow will be sunny.
    c. The chance of buying a defective computer is $\frac{1}{1,000}$.

11. The likelihood of an event is 40%. What are other ways to represent this probability?

12. An event has a probability of $\frac{3}{8}$. What are other ways to represent this probability?

13. An experiment consists of four volleyball games played by one team. How could a coin be used to represent the experiment in a simulation? How could the event in which the team wins all four games be represented?

14. An experiment consists of selecting a package of five electrical sockets. The probability that any one electrical socket is defective is 10%. How could a table of random digits be used to represent the

experiment in a simulation? How could the event in which exactly two sockets are defective be represented?

15. An experiment consists of the birth of a child. How could you use the numbers on a regular six-sided die to represent the experiment in a simulation in two different ways?

16. What must be true about the number of outcomes in an experiment if we are able to represent them in a simulation with a:

    a. Coin?

    b. Regular six-sided die?

    c. Spinner with eight equal areas?

■ Exercises

17. An eight-sided die is rolled 100 times, and the results are recorded in the table:

Outcome	1	2	3	4	5	6	7	8
Times Rolled	13	8	15	16	11	9	17	11

Find the experimental probability of each event.

    a. A 5 is rolled.

    b. An even number is rolled.

    c. A multiple of 3 is rolled.

    d. A number less than 4 is rolled.

18. A spinner is divided into eight areas as shown:

Color	Red	Blue	Yellow	Green
Frequency	37	85	19	19

Akeem wants to know the likelihood of the spinner landing on each color. To investigate, he repeatedly spins the spinner and records the color after each trial. Use his results in the table to compute the experimental probability of each color.

19. Four coins are simultaneously tossed 75 times, and the number of heads is counted after each toss. The results are recorded in the table:

Number of Heads	0	1	2	3	4
Frequency	5	18	28	21	3

Find the experimental probability of each event.

    a. Exactly one head

    b. At most two heads

    c. At least three heads

    d. Exactly two tails

20. Roll a six-sided die 60 times, and record each result. Based on the data you collect, find the experimental probability of each event.

    a. Rolling a 3

    b. Rolling an odd number

    c. Rolling a number larger than 4

    d. Rolling a number divisible by 3

21. Simultaneously toss a quarter and a penny 50 times, and record the results. Use your results to find the experimental probability of each event.

    a. Both coins are heads.

    b. The quarter is heads and the penny tails.

    c. Both coins are the same, either heads or tails.

22. Suppose the numbers 1 through 10 are written on chips and placed in a bag. If one chip is drawn from the bag, determine the experimental probability of getting a 6, 7, or 8 by selecting 100 numbers from a table of random digits.

23. a. Flip a penny 50 times, and record the results. Use them to compute the experimental probability of flipping a head.

    b. Now place the penny on its edge and hit the desk just hard enough to make it fall. Conduct 50 trials of this experiment, record the results, and use them to compute the experimental probability of getting a head.

    c. How do your results in part a compare to your results in part b? What might this suggest about the weight of a penny?

24. Over the lifetime of his career, a baseball player has 930 hits for 3,640 times at bat. What is the experimental probability of the batter getting a hit? Do you think the experimental probability is a good estimate of him getting a hit? Why?

25. A medical journal reports that of 320 patients who underwent a certain transplant operation, 53 suffered a minor complication and 23 suffered a fatal complication. What is the experimental probability that a patient will suffer:

    a. A minor complication?

    b. A fatal complication?

    c. Any complication?

26. The following histogram shows the crime rates in Hawaii per 100,000 people.

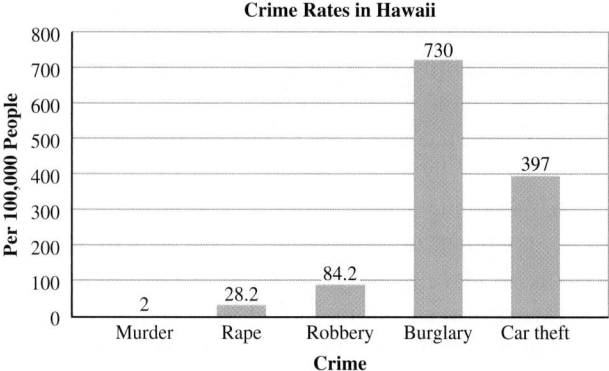

Crime Rates in Hawaii

Source: U.S. Census Bureau.

What is the experimental probability of being the victim of:

a. A murder?

b. A robbery?

c. Any type of theft?

27. A number of adults were asked how much time they spent on various activities during their daily routine. The results are shown in the circle graph.

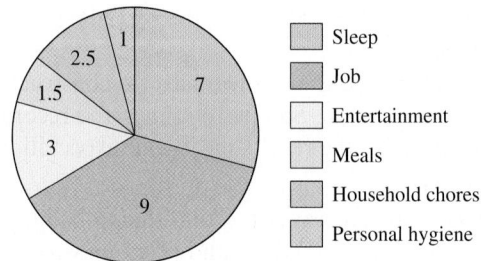

Number of Hours Spent in Daily Routine

- Sleep
- Job
- Entertainment
- Meals
- Household chores
- Personal hygiene

If you were to observe an adult doing any one of the activities, what is the experimental probability that the adult would be:

a. Sleeping?   b. Eating?

c. Doing chores?   d. Not at work?

28. The number of white blood cells per cubic millimeter of whole blood is normally distributed with a mean of $\bar{x} = 7,500$ and a standard deviation of 1,750. If the white blood cell count of an adult is taken, what is the experimental probability that the count is:

a. Between 5,750 and 7,500?

b. Between 4,000 and 9,250?

c. Below 2,250?

d. Above 9,250?

29. The scores from a standardized test are normally distributed with a mean of 75 and a standard deviation of 8. If one score is selected at random, what is the experimental probability that the score is:

a. Higher than a 75?

b. Lower than a 59?

c. Between a 67 and a 91?

■ Problems and Applications

30. Josie is curious about the score she will get if she randomly selects answers on a 15-question true-or-false test. The answers to the test are shown:

1	2	3	4	5	6	7	8	9	10	11	12	13	14	15
F	T	T	T	F	T	F	T	T	T	F	F	T	F	F

Use a coin to simulate what her score might be on the test.

31. Later that same day, Josie again tries her hand at chance. This time she is taking a 10-question multiple-choice test, with four possible answers to each question. The answers to the test are shown:

1	2	3	4	5	6	7	8	9	10
A	C	A	D	C	A	B	D	B	C

Use a spinner with four evenly spaced sections to simulate what her score might be on the test.

32. A school is sponsoring an all-expense-paid trip to Washington, D.C., for one weekend for 25 students. There are 1,000 students at the school. If every student is eligible to go, how might the sponsor of the trip use a table of random digits to select 25 students?

33. A quality control inspector inspects a batch of new cars in which 12.5% are expected to have a defect. Design a simulation using an eight-sided die to find the experimental probability that at least one car will have a defect when five cars are tested at random?

34. The 10-year survival rate of a certain type of cancer is 70% after treatment. Use a table of random digits to design and conduct a simulation to find the experimental probability that two people with the same cancer survive the next 10 years.

35. It is estimated that 27% of employees at a certain business use illegal substances at work. The company wants to employ a random drug test to eliminate problem employees.

a. How can a table of random digits be used to select employees for testing?

b. If 10 employees are selected at random, what is the likelihood that three of them are using illegal substances?

36. The probability that it will rain on any given day in a town in Washington is 80%. Use a table of random digits to design and conduct a simulation to determine the probability that it rains for four straight days.

37. A soft drink company is sponsoring a game in which a person can win a free soda by finding the letters that spell the word "COLA" on specially marked bottle caps. Each bottle cap is marked with exactly one letter, a C, O, L, or A. The letters are distributed in the ratios of 4:3:2:1, respectively.

   a. Use a table of random digits to find the experimental probability of getting the needed letters when six bottle caps are chosen.

   b. Create another simulation that can be used to determine the average number of bottles that must be purchased to win. How does this simulation differ from the one in part a?

38. A cereal company has put one of four toys into each box of children's cereal. The toys are evenly distributed among boxes of cereal, and each box contains only one toy.

   a. What is the experimental probability of getting all four toys when seven boxes of cereal are purchased?

   b. On average, about how many boxes of cereal will a person need to buy in order to be sure of getting all four toys?

39. Marta is making 100 chocolate chip cookies, but she has only 125 chocolate chips to add to her recipe. Once the cookies are made, she performs a taste test on a randomly selected cookie. Use the technique from Example 15.8 to estimate the probability that the cookie she selects has:

   a. No chocolate chips.

   b. One chocolate chip.

   c. Two chocolate chips.

40. On another day, Marta decides to bake a batch of 64 chocolate chip cookies, and she again has 125 chocolate chips to add to her recipe.

   a. Explain how it possible to modify the technique in Example 15.8 to simulate this situation using an eight-sided die.

   b. What is the experimental probability that a cookie in this batch has three or more chocolate chips?

■ Mathematical Reasoning and Proof

41. The likelihood of a person winning a daily lotto is $\frac{1}{1,000}$ or 1 chance in 1,000. Does this mean that someone who plays 1,000 times is guaranteed to win? Explain.

42. Suppose you flip a fair coin 10 times, and each time it comes up heads. Can you use this information to predict the outcome of the next toss? Why or why not?

43. A coin is flipped 100 times. It lands on heads 18 times and on tails 82 times.

   a. Can you conclude that the coin is biased toward tails? Why or why not?

   b. What might you do to help you be more confident in your decision?

44. Reconsider the outcomes of John Kerrich's experiment shown in Table 15.2. Make a graph that relates the number of tosses less than 1,000 to the percentage of heads. How does the graph provide evidence to support the Law of Large Numbers?

45. a. Consider the following spinner:

   If you were to spin it 100 times, about how many times would you expect it to land on each number?

   b. Use a compass and protractor to construct the spinner shown. Spin it 100 times and record the results. Do the results meet your expectations? Explain.

46. Conduct an experiment to find the probability that a typical thumbtack will land point up when tossed into the air. Conduct it first with 10 tosses, then 25 tosses, 50 tosses, and 100 tosses. Based on your results, what is a reasonable value for the likelihood that the thumbtack will land point up?

47. A bag contains 25 marbles that are red, blue, and green, but it is not known how many are of each color. Fifty trials of an experiment are done in which one marble is drawn, the color is recorded, and the marble is replaced. The results are shown:

Color	Red	Blue	Green
Times Drawn	8	32	10

   a. What is the experimental probability of each color?

   b. What color are most of the marbles likely to be?

   c. About how many marbles of each color marble do you think are in the bag?

   d. Can you be certain about your answers to parts b and c? Why or why not?

48. A certain six-sided die is rolled 100 times with the results shown:

Outcome	1	2	3	4	5	6
Times Rolled	9	35	8	31	9	8

a. Calculate the experimental probability of rolling each number.

b. Based on your answers in part a, what numbers are most likely to be rolled?

c. What do you suppose might be true about this die? Can you be sure of your conclusion? What might you do to make you more confident in your conclusion?

Mathematical Communication

49. Write a story problem that requires a computation of experimental probability from statistical data.

50. Get together with some of your peers. Design a fifth-grade lesson that uses a die and teaches children about experimental probability and the Law of Large Numbers.

51. Leesa has flipped a coin 10 times, resulting in 10 tails. She is convinced that the next flip will result in a heads. Is she correct? If not, what would you say as her teacher?

52. Javier rolls a regular six-sided die 240 times. The number 2 came up 39 times. He uses his results to make the claim that there is a 1-in-5 chance that a two will come up on any one roll of the die. Is he correct?

Building Connections

53. Do an Internet search to find the number of male and female births in the United States over the last 5 years. Based on the data you see, is a pregnant woman more likely to have a boy or a girl?

54. We make and use probability statements every day. Make a list of five to ten probability statements that you use regularly. In each case, give an interpretation of the probabilistic statement.

55. The student page in Example 15.5 is from a sixth-grade textbook. Does it surprise you to find simulations included at this grade level? What difficulties might students have in designing and implementing a simulation?

56. John Kerrich's experiment shows that the likelihood of getting a head on one toss of a coin is essentially $\frac{1}{2}$.

a. How is the number $\frac{1}{2}$ connected to the fact that tossing a coin only has two outcomes?

b. As explained, theoretical probability determines the likelihood of an event by using the possible outcomes of an experiment yet to be performed. What might your answer to part a indicate about computing theoretical probabilities?

SECTION 15.2 Theoretical Probability

When we compute experimental probabilities by using data and simulations, the values depend on particular trials, so they can vary even when many trials are made. We can make probabilities more exact by using theoretical probability.

Understanding Theoretical Probability

Explorations Manual 15.5

Recall that a theoretical probability measures the likelihood of an event by considering the possible outcomes of an experiment yet to be performed. Consequently, to compute a theoretical probability, we must first know the possible outcomes of the experiment. The set of all possible outcomes for an experiment is called its **sample space, S.** Specific events of the experiment are then given by subsets of particular outcomes.

Example 15.9 Sample Spaces and Events

List the outcomes in the sample space of each experiment. Name two events for the experiment, and then list their outcomes.

a. Rolling an eight-sided die **b.** Drawing a card from a deck of 52 cards

Solution We first list the outcomes in the sample space. We then give two events, each of which is named with a capital letter.

a. An eight-sided die is an octahedron and has the numbers 1–8 placed on its faces. Consequently, $S = \{1, 2, 3, 4, 5, 6, 7, 8\}$. Two possible events are:

Event E: rolling an even number $= \{2, 4, 6, 8\}$

Event T: rolling a three $= \{3\}$

b. A standard deck of cards contains four suits: spades (♠), clubs (♣), hearts (♥), and diamonds (♦). There are also 13 denominations: 2, 3, 4, 5, 6, 7, 8, 9, 10, jack (J), queen (Q), king (K), and ace (A). The jacks, queens, and kings are called face cards because of the figures printed on them. Combining the different suits with the different denominations gives us a sample space of 52 cards, or

Two possible events for the experiment of drawing a single card are:

Event K: drawing a king $= \{K\spadesuit, K\diamondsuit, K\heartsuit, K\clubsuit\}$

Event B: drawing a black ace $= \{A\spadesuit, A\clubsuit\}$

Knowing the sample space, we can begin to compute theoretical probabilities. Recall that the Law of Large Numbers guarantees that if an experiment is conducted multiple times, the probability of an event tends toward a specific number. For instance, when Kerrich and Pearson verified the Law of Large Numbers by flipping a coin thousands of times, they found that the probability of flipping a head tended toward $\frac{1}{2}$. This value should make sense because, under ideal conditions, we would expect each side of the coin to have the same chance of coming up, or would be **equally likely.** However, we also get $\frac{1}{2}$ when taking the ratio of the number of ways to get a head to the total number of outcomes in the sample space, or

$$P(\text{heads}) = \frac{\text{number of ways to get heads}}{\text{total number of outcomes}} = \frac{1}{2}$$

So we can compute a probability by comparing the number of outcomes in an event to the total number of outcomes in an experiment. This idea leads to the definition of **theoretical probability.**

Definition of Theoretical Probability

If S is the sample space of an experiment with equally likely outcomes, then the **theoretical probability**, or **probability**, of an event A, denoted $P(A)$, is the ratio

$$P(A) = \frac{\text{number of elements of } A}{\text{number of elements of } S} = \frac{n(A)}{n(S)}$$

The theoretical probability of an event is expressed in set notation because both the event and the sample space are sets of outcomes. However, we can also use probabilistic language to define theoretical probability. In other words, a probability can be viewed as the ratio of the favorable outcomes for event A to the total number of outcomes, or

$$P(A) = \frac{\text{favorable outcomes for } A}{\text{total outcomes}}$$

Regardless of which approach we take, both require counting the outcomes in the event and in the sample space.

Example 15.10 Probabilities with a 12-Sided Die

A 12-sided die is a dodecahedron with the numbers 1–12 written on its faces. If the die is rolled once, what is the probability of each event?

a. Event A: Rolling a 7 **b.** Event B: Rolling a number greater than 8

Solution $S = \{1, 2, 3, \ldots 11, 12\}$, so $n(S) = 12$. The numerator for each probability is found by counting the successful outcomes for each event.

a. Because there is only one 7, $P(A) = \dfrac{n(A)}{n(S)} = \dfrac{1}{12}$.

b. Because $B = \{9, 10, 11, 12\}$, $P(B) = \dfrac{n(B)}{n(S)} = \dfrac{4}{12} = \dfrac{1}{3}$.

Example 15.11 Probabilities in the Game of Bingo

The game of Bingo divides the numbers 1–75 into five sets: $B = \{1, 2, \ldots 15\}$, $I = \{16, 17, \ldots 30\}$, $N = \{31, 32, \ldots 45\}$, $G = \{46, 47, \ldots 60\}$, and $O = \{61, 62, \ldots 75\}$. Each number, along with its letter, is written on a chip. The chips are placed in an urn, mixed, and then drawn at random one at a time. The letter and number are then read aloud to players, who mark off corresponding spaces on cards. When the first chip is drawn, what is the probability that:

a. It has a number divisible by 5? **b.** It has a consonant on it?

Solution $S = \{1, 2, 3, \ldots, 75\}$, so $n(S) = 75$. The numerator for each probability is found by counting the successful outcomes for each event.

a. Fifteen numbers in the sample space are divisible by 5, namely $F = \{5, 10, 15, \ldots 75\}$. As a result, $P(F) = \dfrac{n(F)}{n(S)} = \dfrac{15}{75} = \dfrac{1}{5}$.

b. Successful outcomes include any chip with a B, N, or G on it. There are 45 such chips, so the probability of drawing a consonant is $\dfrac{45}{75} = \dfrac{3}{5}$.

Computing probabilities hinges on counting the number of outcomes in the sample space and the event. If the experiment involves only one action, like rolling a single die or drawing a single chip, then counting outcomes is relatively easy. However, counting the outcomes becomes more difficult as the experiment becomes more complex. So it is helpful to know different ways of listing the elements in the sample space. The next two examples demonstrate several methods.

Example 15.12 **Probability and Having Children**

Problem Solving

A couple plans to have three children and hopes to have two boys and one girl, but they have no preference in the order. What is the probability that they will get their wish?

Solution

Understanding the Problem. We are asked to find the probability of having two boys and one girl when three children are born. The order of the children does not matter.

Devising a Plan. To find the required probability, we must determine how many outcomes are in the sample space and in the event. We can *use a tree diagram* to list the outcomes.

Carrying Out the Plan. We make a tree diagram by starting at a point and adding a branch for each successive child. There are two outcomes for each child, a boy or a girl, so each new branch must represent these possibilities. The tree diagram and the corresponding outcomes are shown:

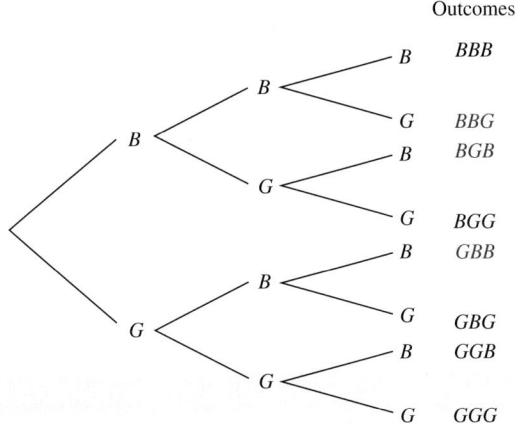

It shows that there are eight outcomes, of which three have two boys and one girl. As a result, the probability that the couple will have two boys and one girl is $\frac{3}{8}$.

Looking Back. Another way to generate the possible outcomes is to use a table such as the one shown:

First Child	Second Child	Third Child	Outcomes
B	B	B	BBB
B	B	G	BBG
B	G	B	BGB
B	G	G	BGG
G	B	B	GBB
G	B	G	GBG
G	G	B	GGB
G	G	G	GGG

As before, there are eight possible outcomes, of which three are successes. This, again, gives us a probability of $\frac{3}{8}$.

Example 15.13 Rolling a Sum of 7 with Two Dice

Application

In the game of craps, two dice are rolled, and the numbers are added. One way to win is to roll a sum of 7 on the first roll. What is the probability of this event?

Solution One way to find the outcomes of this experiment is to make a table in which the outcomes of one die are listed across the top and the outcomes of the other down the side. We then fill in the table by making all possible combinations of the initial row with the initial column. The results are shown in the following table. One die is shown in white, and the other in blue to distinguish the different outcomes.

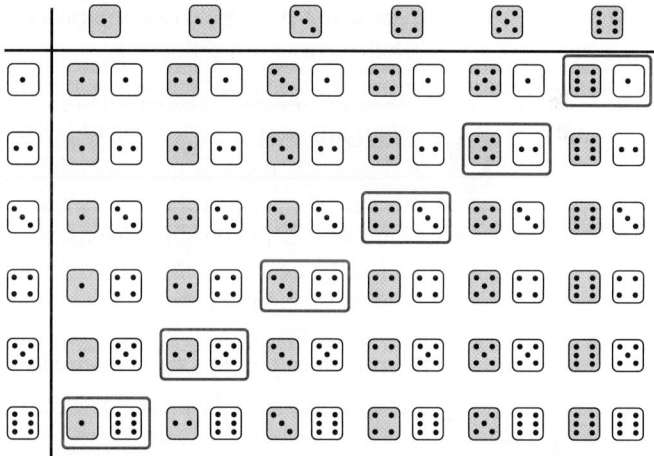

The table shows that there are 36 possible outcomes, of which six sum to 7. Consequently, the probability of rolling a sum of 7 when two dice are rolled is $\frac{6}{36} = \frac{1}{6}$.

A common mistake when computing probabilities with the sum of two dice is to consider only the sums and not the frequency with which the sums occur. Specifically, people often think that the sample space for this experiment should be $S = \{2, 3, 4, 5, 6, 7, 8, 9, 10, 11, 12\}$ and wrongly conclude that rolling a sum of 7 has a probability of $\frac{1}{11}$. The misconception stems from the belief that the sums are equally likely, which cannot be the case because they occur with different frequencies. Even though the sums are not equally likely, each possible combination of dice is. So we can use the individual combinations to compute probabilities.

In many probability situations, it is enough to know how many outcomes are in the sample space without having to know what they are. In such cases, we may be able to use the Fundamental Counting Principle to find the number of outcomes. The **Fundamental Counting Principle** states that if event A can occur in m ways and event B can occur in n ways, then the events A and B can occur in succession, first A, then B, in $m \cdot n$ ways. For instance, in Example 15.12, we could have viewed the birth of each child as a separate event. Because each birth has two outcomes, the Fundamental Counting Principle tells us that the total number of outcomes for the three children is $2 \cdot 2 \cdot 2 = 8$. Similarly, in Example 15.13, we could have viewed rolling each die as a separate event. Because each die has six outcomes, the total number of outcomes for the two dice is $6 \cdot 6 = 36$. The Fundamental Counting Principle is discussed in greater detail in Section 15.4.

Check Your Understanding 15.2-A

1. Give the sample space of each experiment.
   a. Rolling a standard six-sided die   **b.** Tossing the same coin twice

2. If a single eight-sided die is rolled, what is the probability that the number rolled is:
   a. An 8?   **b.** An odd number?   **c.** A number less than 4?

3. Consider the game of Bingo as described in Example 15.11. When the first chip is drawn, what is the probability that it has:
   a. An *I* on it?   **b.** A vowel on it?   **c.** A number divisible by 10 on it?

Talk About It How is theoretical probability similar to experimental probability? How is it different?

Activity 15.3
Search the Internet for software that generates and records large numbers of trials for the experiment of rolling a standard six-sided die. Have it generate and record 1,000 trials. Use the data to compute the experimental probability of rolling each number. How do the experimental probabilities compare to the corresponding theoretical probabilities? If they are close in value, explain why. If there are differences, what might you do to make the experimental probabilities closer to the theoretical probabilities?

Impossible and Certain Events

The examples so far give some indication of the numerical values that probabilities can take on. Every probability, both experimental and theoretical, has been a positive number less than 1. As a result, we may ask whether probabilities can take on negative values or values greater than 1. To answer the question, consider the events at each extreme of an experiment. At one extreme is the event that contains no outcomes in the sample space. For instance, consider the event A of rolling a 7 when a single six-sided die is rolled. No outcomes in the sample space $S = \{1, 2, 3, 4, 5, 6\}$ satisfy this event, so event A must be the empty set. Because $n(A) = n(\varnothing) = 0$, the probability of event A must be

Explorations Manual 15.6

$$P(A) = \frac{n(A)}{n(S)} = \frac{n(\varnothing)}{n(S)} = \frac{0}{6} = 0$$

Any event, like this one, that does not have outcomes in the sample space is called an **impossible event** and always has a probability of 0.

At the other extreme is the event that contains every outcome in the sample space. For instance, consider the event B of rolling a number 1 through 6 when a single six-sided die is rolled. In this case, every number in the sample space is in event B. Because $n(B) = n(S) = 6$, the probability of event B must be

$$P(B) = \frac{n(B)}{n(S)} = \frac{n(S)}{n(S)} = \frac{6}{6} = 1$$

Any event that is equal to the sample space is a called a **certain event** and always has a probability of 1. Combining these facts leads to the following theorem.

Theorem 15.2   Values of Probabilities

If A is an event of an experiment with sample space S, then $P(A)$ is a real number between 0 and 1 inclusive, or $0 \leq P(A) \leq 1$.

Theorem 15.2 can be verified more formally as follows. Because an event A of an experiment with sample space S must be a subset of S, then $\varnothing \subseteq A \subseteq S$. It then follows that $n(\varnothing) \leq n(A) \leq n(S)$. Dividing by $n(S)$, we have

$$\frac{n(\varnothing)}{n(S)} \leq \frac{n(A)}{n(S)} \leq \frac{n(S)}{n(S)} \quad \text{or} \quad 0 = \frac{n(\varnothing)}{n(S)} \leq P(A) \leq \frac{n(S)}{n(S)} = 1$$

Mutually Exclusive Events

So far, we have considered only one event at a time, but it is also possible to consider multiple events.

Example 15.14   Probability with Two Events

An urn contains 5 red chips, 7 white chips, and 3 blue chips. If one chip is drawn at random, what is the probability that the chip is either red or blue?

Solution Let R be the event of drawing a red chip and B be the event of drawing a blue chip. Because 5 chips are red and 3 chips are blue, there are $5 + 3 = 8$ successful outcomes and $5 + 7 + 3 = 15$ total outcomes. Consequently, the probability of drawing a red chip or a blue chip is $P(R \text{ or } B) = \dfrac{8}{15}$.

Event R and event B in the last example had no outcomes in common. In other words, if a red chip is drawn, a blue chip cannot be drawn, and vice versa. Such events are called **mutually exclusive.**

Definition of Mutually Exclusive Events

Two events, A and B, with no outcomes in common are called **mutually exclusive,** denoted $A \cap B = \varnothing$.

If two events are mutually exclusive, the probability that either one will occur can be computed by adding the probabilities of the individual events. For instance, we could have computed the probability of red or blue in the last example by adding the probability of R to the probability of B, or $P(R \text{ or } B) = P(R) + P(B) = \dfrac{5}{15} + \dfrac{3}{15} = \dfrac{8}{15}$. This is true for all mutually exclusive events, leading to the next theorem.

Theorem 15.3   Probability of Mutually Exclusive Events

If A and B are mutually exclusive events in an experiment with sample space S, then $P(A \text{ or } B) = P(A \cup B) = P(A) + P(B)$.

If the outcomes are equally likely, then we can verify Theorem 15.3 using the fact that, if $A \cap B = \varnothing$, then $n(A \cup B) = n(A) + n(B)$. Filling in the details of this argument is left to the Learning Assessment. We can also generalize Theorem 15.3 to more than just two mutually exclusive events:

$$P(A_1 \text{ or } A_2 \text{ or } \ldots \text{ or } A_n) = P(A_1 \cup A_2 \cup \ldots \cup A_n) = P(A_1) + P(A_2) + \ldots + P(A_n)$$

This last fact has an important implication. If each outcome of an event A is an event in and of itself, then all the outcomes are mutually exclusive, and the probability of A is the sum of the probabilities of the individual outcomes. In addition, any sample space consists of individual outcomes, so the sum of the probabilities of the outcomes must equal the probability of the sample space. In other words, the sum of the probabilities of the individual outcomes in a sample space must equal 1.

Theorem 15.4   The Sum of Probabilities

If A is an event of an experiment with sample space S, then the probability of A is equal to the sum of the probabilities of the outcomes contained in A.

Example 15.15   Probability of Mutually Exclusive Events

Application

The table shows the percentage of the earth's land surface by continent.

Continent	Percent of Earth's Land
Africa	20%
Antarctica	9.5%
Asia	30.5%
Australia	5%
Europe	7%
North America	16%
South America	12%

If a location on land is selected at random, what is the probability that the location will be on a continent that begins with a letter other than A?

Solution The three continents that begin with a letter other than A are Europe, North America, and South America. The continents share no land mass, so the probability of getting a location on any one of them is the sum of the probabilities of each, or $7\% + 16\% + 12\% = 35\% = 0.35$.

Complementary Events

In many experiments, the outcomes are dichotomous, meaning every outcome can be placed into one of two mutually exclusive events. For example, when a six-sided die is rolled, the result is either even or odd. In experiments like this, the events are called **complementary.** Because the union of event A and its **complement,** denoted $\overline{A}$, contains every element in the sample space, the sum of the probabilities of A and $\overline{A}$ must equal the probability of the sample space, or $P(A) + P(\overline{A}) = 1$.

Theorem 15.5   Probability of Complementary Events

If A is an event of an experiment with sample space S and $\overline{A}$ is its complement, then $\overline{A}$ is also an event of the experiment and $P(A) + P(\overline{A}) = 1$.

If either the probability of A or $\overline{A}$ is known, then we can rearrange the formula in Theorem 15.5 to find the probability of the other. In other words, if $P(A) + P(\overline{A}) = 1$, then $P(A) = 1 - P(\overline{A})$ and $P(\overline{A}) = 1 - P(A)$.

Example 15.16 Probability of Complementary Events

The daily weather report states that the probability of rain today is 30%. What is the probability that it will not rain today?

Solution The two events, rain and not rain, are complementary events. If A represents the event that it rains, then the probability that it will not rain today, event $\overline{A}$, is $P(\overline{A}) = 1 - P(A) = 1 - 30\% = 1 - 0.3 = 0.7 = 70\%$.

Nonmutually Exclusive Events

The last examples dealt with events with no outcomes in common. We can also consider **nonmutually exclusive events,** which are events with outcomes in common. As an example, consider the experiment of drawing one card from a deck of 52 cards. If A is the event of drawing an ace and B is the event of drawing a black card, then A and B have two outcomes in common: the ace of spades ($A\spadesuit$) and the ace of clubs ($A\clubsuit$). This overlap in the outcomes has an effect on how we compute the probability of either event occurring. If we compute $P(A \text{ or } B)$ by directly counting outcomes in the sample space, then 28 cards satisfy either event: 26 black cards and 2 red aces. As a result, $P(A \text{ or } B) = \dfrac{28}{52} = \dfrac{7}{13}$. However, using Theorem 15.3, we get

$$P(A \text{ or } B) = P(A) + P(B) = \frac{26}{52} + \frac{4}{52} = \frac{30}{52} = \frac{15}{26}$$

The discrepancy between the computations comes from the fact that the ace of spades and the ace of clubs are counted twice in the second computation: once for drawing a black card and again for drawing an ace. We can make the appropriate adjustment by subtracting one of the times we counted the black aces, or

$$P(A \text{ or } B) = P(A) + P(B) - P(A \text{ and } B) = \frac{26}{52} + \frac{4}{52} - \frac{2}{52} = \frac{28}{52} = \frac{7}{13}$$

Theorem 15.6 Probability of Nonmutually Exclusive Events

If A and B are nonmutually exclusive events of an experiment with sample space S, then $P(A \text{ or } B) = P(A \cup B) = P(A) + P(B) - P(A \cap B)$, where $A \cap B$ is the set of outcomes that A and B have in common.

Like the other theorems, Theorem 15.6 can be verified by using results from set theory. In general, because $n(A \cup B) = n(A) + n(B) - n(A \cap B)$, then for a sample space S with equally likely outcomes, we have

$$P(A \cup B) = \frac{n(A \cup B)}{n(S)}$$

$$= \frac{n(A) + n(B) - n(A \cap B)}{n(S)}$$

$$= \frac{n(A)}{n(S)} + \frac{n(B)}{n(S)} - \frac{n(A \cap B)}{n(S)}$$

$$= P(A) + P(B) - P(A \cap B)$$

Example 15.17 | **Probability of Nonmutually Exclusive Events**

If a pair of dice is rolled, what is the probability that the sum of the dice is even or divisible by 3?

Solution Let E be the event of rolling an even sum, and let T be the event of rolling a sum divisible by 3. Two sums are common to E and T: 6 and 12. Consequently, we have two nonmutually exclusive events, and $P(E \cup T) = P(E) + P(T) - P(E \cap T)$. Using the table in Example 15.13, we have

$$P(E) = P(2) + P(4) + \cdots + P(12) = \frac{1}{36} + \frac{3}{36} + \frac{5}{36} + \frac{5}{36} + \frac{3}{36} + \frac{1}{36} = \frac{18}{36}$$

$$P(T) = P(3) + P(6) + P(9) + P(12) = \frac{2}{36} + \frac{5}{36} + \frac{4}{36} + \frac{1}{36} = \frac{12}{36}$$

$$P(E \cap T) = P(6) + P(12) = \frac{5}{36} + \frac{1}{36} = \frac{6}{36}$$

As a result, $P(E \cup T) = P(E) + P(T) - P(E \cap T) = \frac{18}{36} + \frac{12}{36} - \frac{6}{36} = \frac{24}{36} = \frac{2}{3}$. ∎

The following is a summary of the properties of probabilities.

Properties of Probabilities

1. The probability of an impossible event is 0, or $P(\varnothing) = 0$.
2. The probability of a certain event is 1, or $P(S) = 1$.
3. For any event A, $0 \le P(A) \le 1$
4. If A and B are mutually exclusive events, then $P(A \cup B) = P(A) + P(B)$.
5. If A and $\overline{A}$ are complementary events, then $P(A) + P(\overline{A}) = 1$.
6. If A and B are events, then $P(A \cup B) = P(A) + P(B) - P(A \cap B)$.

Check Your Understanding 15.2-B

1. A card is drawn from a standard deck of 52 cards. What is the probability that the card is an ace or a black 10?
2. Consider the table given in Example 15.15. If a location on land is selected at random, what is the probability that it will be in North or South America?
3. The probability of Speedy Joe winning a horse race is 45%. What is the probability that he will not win the race?
4. If a pair of eight-sided dice is rolled, what is the probability that the sum of the dice will be divisible by 2 or by 3?

Talk About It What is the difference between mutually exclusive and nonmutually exclusive events?

Representation

Activity 15.4
How might a Venn diagram be used to represent and justify the properties $P(A \cup B) = P(A) + P(B)$ and $P(A \cup B) = P(A) + P(B) - P(A \cap B)$?

Multistage Probability

Explorations
Manual
15.7

To this point, we have considered only **single-stage experiments,** which are experiments that conclude after just one step. However, many single-stage experiments involve multiple actions, such as rolling two dice or drawing five cards. In such cases, we can conduct the experiment as a **multistage experiment.** For instance, consider the experiment of flipping a quarter and a penny (Figure 15.2). To conduct a single-stage experiment, we flip both coins at one time. To conduct a multistage experiment, we flip the coins in succession, first the quarter and then the penny. In either case, the outcomes are the same, so the probabilities of particular events must also be the same. This is a useful fact because it may be easier or even necessary to find probabilities by conducting an experiment in multiple stages rather than as a single stage.

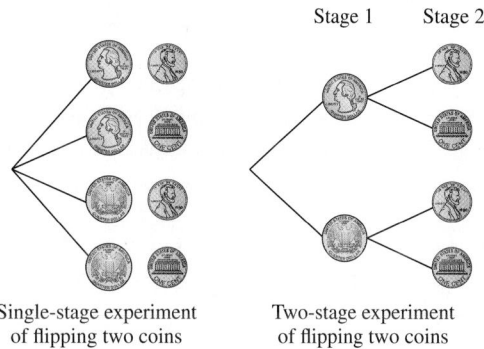

Single-stage experiment of flipping two coins

Two-stage experiment of flipping two coins

Figure 15.2 Flipping two coins as a single-stage and as a two-stage experiment

To understand how to compute probabilities from a multistage experiment, consider an urn with four chips in it, three red and one blue. Suppose we want to find the probability of drawing two chips in succession, a red and then a blue, with replacement. **Replacement** means that, after the first chip is drawn and the result recorded, it is put back into the urn to take part in the second drawing. To find the probability of drawing a red and then a blue, we use a tree diagram to list the outcomes after all stages. Figure 15.3 shows that 16 outcomes are possible, of which three are successful. As a result, the probability of getting a red and then a blue is $\frac{3}{16}$.

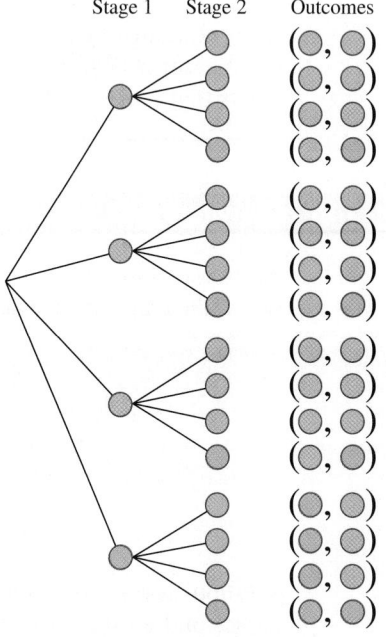

Figure 15.3 Possible outcomes for drawing two chips from an urn

Notice that this experiment has only four arrangements of chips: (◐, ◑), (◐, ◯), (◯, ◑), and (◯, ◯). Simply by counting outcomes, we have that

(◐, ◑) has a probability of $\dfrac{9}{16}$

(◐, ◯) has a probability of $\dfrac{3}{16}$

(◯, ◑) has a probability of $\dfrac{3}{16}$

(◯, ◯) has a probability of $\dfrac{1}{16}$

Because some of the branches are the same, we can condense our tree diagram by grouping like branches. We can then indicate the likelihood of each outcome by placing the appropriate probability on the branch. In Figure 15.4, notice that we get the same probability for each color combination by multiplying the probabilities along each branch. This holds true for any multistage experiment.

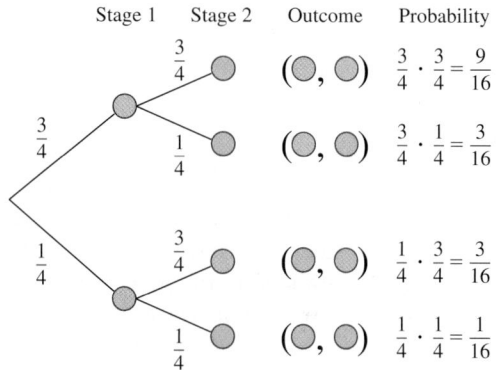

Figure 15.4 **Probabilities for drawing two chips from an urn**

Theorem 15.7   Probabilities of Multi-Stage Experiments

If an experiment consists of a sequence of simpler experiments, then the probability of an event A is found by multiplying each of the probabilities of the simpler experiments that make up the sequence of experiments for A.

Example 15.18   Drawing White Chips from an Urn

An urn contains 3 red, 5 white, and 2 blue chips. What is the probability of drawing two white chips in succession if the first is not replaced after it is drawn?

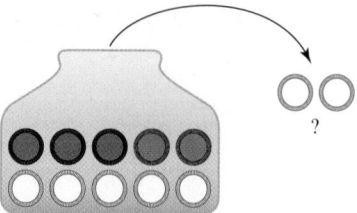

Solution The first chip is not replaced after it is drawn, affecting the outcomes for the second chip in two ways. First, only nine chips are in the urn instead of ten. Second, the color of the first chip affects what colors are left for the second. So the

probabilities for the second chip depend on which color was drawn for the first. With these considerations in mind, we can make the following tree diagram:

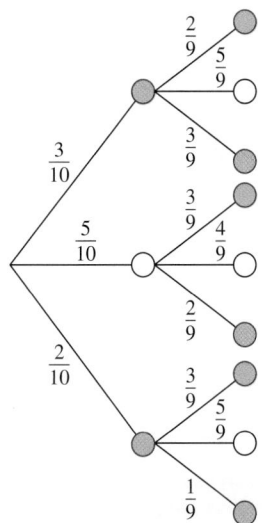

Following the branches from the first white chip to the second, we find that the probability of drawing two white chips is $P(WW) = \dfrac{5}{10} \cdot \dfrac{4}{9} = \dfrac{20}{90} = \dfrac{2}{9}$.

In the last example, the probabilities of two simpler events were multiplied to find the probability of the overall event. We can also use multiplication to find the probability that two or more events occur simultaneously. However, the events must be **independent,** which means the outcome of one event has no influence over the outcome of the other. Here are two examples of independent events:

Experiment 1: Flipping a quarter and a penny

Event H: The quarter is a head.

Event T: The penny is a tail.

Experiment 2: Rolling a red die and a white die

Event F: A number 4 or greater is rolled on the red die.

Event T: A 2 is rolled on the white die.

In each case, we are looking for the probability that *both* events occur, which is a different type of question than the ones asked with mutually exclusive events. With mutually exclusive events, we want the probability of *either A or B or* both. In multistage probabilities, we want probability that *A* and *B* occur together.

For instance, in experiment 1, we want to find the probability that the quarter is a head *and* the penny is a tail, or $P(H \text{ and } T)$. The sample space in Figure 15.5 shows that this can happen in only one of four ways, so $P(H \text{ and } T) = \dfrac{1}{4}$. Notice that this is the same as multiplying the probability of getting a heads on the quarter by the probability of getting a tails on the penny, or

$$P(H \text{ and } T) = P(H) \cdot P(T) = \frac{1}{2} \cdot \frac{1}{2} = \frac{1}{4}$$

Sucessful outcome

Figure 15.5 Sample space for flipping a quarter and a penny

In experiment 2, we want the probability of rolling a 4 or greater on the red die and a 2 on the white die, or $P(F \text{ and } T)$. The sample space indicates that there are three ways the red die can be a 4 or greater and the white die can be a 2, namely $(4, 2)$, $(5, 2)$, and $(6, 2)$. Because there are 36 possible outcomes, $P(F \text{ and } T) = \frac{3}{36} = \frac{1}{12}$. However, $P(F) \cdot P(T) = \frac{3}{6} \cdot \frac{1}{6} = \frac{3}{36} = \frac{1}{12}$. Consequently, $P(F \text{ and } T) = P(F) \cdot P(T)$. Generalizing the pattern from these examples leads to the next theorem.

Theorem 15.8 Probability of Independent Events

If A and B are independent events of an experiment with sample space S, then the probability that both occur simultaneously, denoted $P(A \cap B)$, is given by $P(A \text{ and } B) = P(A \cap B) = P(A) \cdot P(B)$.

Example 15.19 Probability of Independent Events

Each of the letters from the words "MATH IS FUN" are written on chips and placed in three urns, one urn per word.

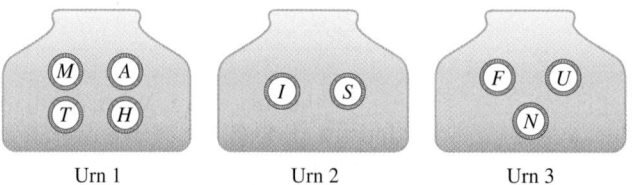

Urn 1   Urn 2   Urn 3

If one letter is drawn from each urn, what is the probability that the three letters spell the word "TIN"?

Solution The events of drawing a chip from each urn are independent of each other. As a result, we can find the probability of drawing the word "TIN" by finding the probability of drawing the letter "T," then "I," and then "N," and multiplying them. Consequently, the probability of "TIN" is

Explorations
Manual
15.8

$$P(TIN) = P(T) \cdot P(I) \cdot P(N) = \frac{1}{4} \cdot \frac{1}{2} \cdot \frac{1}{3} = \frac{1}{24}$$

Check Your Understanding 15.2-C

1. If three cards are drawn in succession without replacement, what is the probability they are all:

    a. Hearts?   **b.** Black?   **c.** Twos?

2. An urn contains 5 red chips, 5 white chips, and 7 blue chips. What is the probability of drawing 2 blue chips in succession if the first chip is replaced after it is drawn?

3. Two urns contain both red and white chips. The first urn contains 5 red and 5 white, and the second urn contains 6 red and 4 white. If a chip is drawn from each urn, what is the probability that the first is white and the second is red?

Talk About It Suppose two chips are drawn from an urn. How does the replacement of the first chip affect the drawing of the second?

Problem Solving

Activity 15.5

Imagine that a monkey sits down at a computer and begins to hit keys at random. If the monkey happens to type the appropriate number of keys, what is the probability that it will type this problem? Assume 26 letters, no spaces, and no punctuation.

SECTION 15.2 Learning Assessment

■ Understanding the Concepts

1. What does it mean for the outcomes of an experiment to be equally likely?

2. What is a sample space, and how is it used to determine the probability of an event?

3. Explain two ways to think about theoretical probability.

4. Describe how to use a tree diagram to:
   a. Find the sample space of an experiment.
   b. Compute multistage probability.

5. What does it mean for an event to be impossible? Certain?

6. a. What are complementary events?
   b. What is true about the probabilities of two complementary events?

7. What is the difference between:
   a. Mutually exclusive and nonmutually exclusive events?
   b. Mutually exclusive and independent events?

8. a. In what situations do we add to find the probability of an event?
   b. In what situations do we multiply to find the probability of an event?

9. What must be true about a single-stage experiment in order for it to be computed as a multistage experiment?

■ Representing the Mathematics

10. A probability can be represented with a fraction, decimal, or a percent.
   a. Which do you think best represents probability as the likelihood that an event will occur?
   b. Which do you think best represents probability as a comparison between the favorable and total outcomes?

11. Give a verbal interpretation of each property.
   a. For any event A, $0 \leq P(A) \leq 1$.
   b. If A and B are mutually exclusive events, then $P(A \cup B) = P(A) + P(B)$.
   c. If A and $\overline{A}$ are complementary events, then $P(A) + P(\overline{A}) = 1$.
   d. If A and B are events, then $P(A \cup B) = P(A) + P(B) - P(A \cap B)$.

12. Consider the experiment of tossing three coins.
   a. Use a table to find the outcomes in the sample space of the experiment.
   b. Use a tree diagram to represent the experiment as a three-stage experiment, and then use it to list the outcomes in the sample space.
   c. Compare the two methods of finding the sample space. What are the advantages and disadvantages of each?

13. The tree diagrams in Figure 15.2 show different views of the same experiment. Which do you think better represents the experiment? Why?

14. Consider the experiment of rolling a standard six-sided die and tossing a coin. Draw two separate tree diagrams of this experiment: one that shows it as a single-stage experiment and another that shows it as a two-stage experiment. Which do you think better represents the experiment and why?

15. State the complement of each event.
   a. An even number is rolled on a six-sided die.
   b. A street address selected at random has a 5 in the address number.
   c. A family with three children has at least one boy.
   d. Getting at most two heads on three flips of a coin.

16. Suppose S is the sample space of different kinds of cookies. If C is the set of cookies with chocolate

chips and N is the set of cookies with nuts, describe each probability statement in words.

a. $P(C \cap N)$ b. $P(C \cup N)$ c. $P(\overline{N})$

17. Consider an experiment with sample space S and two mutually exclusive events A and B.

   a. Draw a Venn diagram of the two events and the sample space.

   b. What set represents the universe in the diagram?

   c. How might the diagram be used to explain that $P(A \text{ or } B) = P(A) + P(B)$?

18. Consider an experiment with sample space S and two nonmutually exclusive events A and B.

   a. Draw a Venn diagram of the two events and the sample space.

   b. How might the diagram be used to explain how to compute $P(A \text{ or } B)$?

19. Consider an experiment with sample space S and events A, B, and C where no two or three events are mutually exclusive.

   a. Draw a Venn diagram of the three events and the sample space.

   b. Use the Venn diagram to derive formulas for computing:

      i. $P(A \text{ or } C)$.

      ii. $P(A \text{ and } B)$.

      iii. $P(A \text{ and } B, \text{ but not } C)$.

■ Exercises

20. Consider the spinner shown:

   a. What is the sample space if we consider only numbers?

   b. What is the sample space if we consider only colors?

   c. What outcomes are in the event of spinning an even number?

   d. What outcomes are in the event of spinning a primary color?

21. A 20-sided die is an icosahedron, with the numbers 1–20 written on its faces. Use the sample space to list the outcomes in the event of rolling a number divisible by:

   a. 5.   b. 3.   c. 3 or 5.

22. An experiment consists of tossing four coins. Use the sample space to list the outcomes in the event where:

   a. All the coins are the same.

   b. Exactly one head is showing.

   c. Two or more heads are showing.

23. Consider drawing a card from a deck of 52 cards. What is the probability that the card is:

   a. The queen of spades?   b. A diamond?

   c. A ten?   d. A face card?

   e. A black ace?   f. A red face card?

24. Consider the game of Bingo described in Example 15.11. If a chip is selected at random, what is the probability that:

   a. The number is in the set of N?

   b. The number is 50 or less?

   c. The letter is not a vowel?

   d. The letter is an I or a G?

25. Suppose the letters of the alphabet are written on chips and placed in an urn. If one chip is selected at random, what is the probability the letter will be:

   a. A vowel?   b. A consonant?

   c. One from the word "mathematics"?

26. A fair coin is tossed three times. What is the probability of getting:

   a. At least one head?

   b. At most one head?

   c. Exactly two tails?

27. Two cards are dealt from a deck of 52 cards. What is the probability that the cards are:

   a. Both aces?

   b. Both diamonds?

   c. Not clubs?

28. If $P(A) = 0.3$, $P(B) = 0.45$, and $P(A \text{ and } B) = 0.15$, what is $P(A \text{ or } B)$?

29. If $P(A) = 50\%$, $P(B) = 60\%$, and $P(A \text{ and } B) = 30\%$, what is $P(A \text{ or } B)$?

30. If $P(A) = 0.6$, $P(B) = 0.55$, and $P(A \text{ or } B) = 0.85$, what is $P(A \text{ and } B)$?

31. If $P(A) = 63\%$, $P(B) = 48\%$, and $P(A \text{ or } B) = 91\%$, what is $P(A \text{ and } B)$?

32. If $P(A) = 0.563$, what is $P(\overline{A})$?

33. If $P(\overline{A}) = 31.7\%$, what is $P(A)$?

34. Maggie, a 3-year-old child, sits down at a computer and types six letters at random. What is the probability that she spells her name? Assume 26 keys on the keyboard.

35. If one number from 10 to 50 is selected at random, what is the probability that the sum of the digits is odd?

36. A survey shows that one out of 12 students at a school is left-handed.

    a. Draw a tree diagram that can be used to find the probabilities of the "handedness" of two people when they are selected at random.

    b. If two people are chosen at random, what is the probability that they are both right-handed?

    c. If two people are chosen at random, what is the probability that they are both left-handed?

    d. If two people are chosen at random, what is the probability that the first is right-handed and the second is left-handed?

37. The letters from the word "PROBABILITY" are written on chips and placed in an urn.

    a. If five chips are drawn in succession without replacement, what is the probability that they spell "APRIL"?

    b. If four chips are drawn in succession with replacement, what is the probability that they spell "BABY"?

38. The 11 letters in the word "mathematics" are written on chips and placed in an urn. If four letters are drawn in succession, what is the probability that they spell the word "hats"?

39. **a.** If a person takes a true-or-false test that consists of 10 questions, what is the probability the person will get all 10 questions correct by guessing on each question?

    b. If a person takes a 10-question, multiple-choice test where each question has four possibilities, what is the probability the person gets all 10 questions correct by guessing on each question?

■ Problems and Applications

40. Consider the experiment in which two dice are rolled and the smaller is subtracted from the larger.

    a. What are the possible outcomes of this experiment?

    b. What is the probability associated with each outcome?

    c. Are the differences equally likely? If not, which difference is the most likely?

41. An urn contains 4 red, 5 green, and 6 white marbles. How many red marbles must be added to the urn to make the probability of drawing a red $\frac{1}{2}$?

42. An urn contains 4 red, 6 white, and 5 blue chips.

    a. How many chips must be drawn to ensure that 2 blue chips are drawn?

    b. If 5 chips are drawn, what is the probability that 2 are the same color?

43. A hat contains 5 black and 4 white marbles.

    a. If 3 marbles are drawn in succession without replacement, what is the probability that all 3 are white?

    b. If 3 marbles are drawn in succession with replacement, what is the probability that all 3 are white?

    c. If 3 marbles are drawn in succession without replacement, what is the probability that 2 are black and 1 is white?

44. Consider the two boxes of marbles shown:

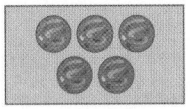 

Box 1   Box 2

    If one marble is drawn from each box, what is the probability:

    a. That they are both blue?

    b. Of getting at least 1 red marble?

    c. Of getting (○, ○) or (○, ○)?

45. Can two cubes be numbered so that when they are rolled, every sum from 1 to 12 is possible? If so, how might it be done?

46. Suppose an experiment has four outcomes: A, B, C and D. If $P(A) = 2P(B)$, $P(B) = 2P(C)$, and $P(C) = 2P(D)$, what is the probability of each outcome?

47. Three six-sided dice are rolled: one red, one green, and one white. What is the probability that two 4s and a 5 are rolled?

48. It is common for individual computer parts to be assembled at different factories and then sent to another factory for final assembly. If the probability of a defective motherboard coming from factory A is 0.01, and the probability of a defective monitor coming from factory B is 0.005, how many computers are likely to have both a defective motherboard and a defective monitor out of 1,000,000 computers?

49. In a certain city, the probability of having a newborn baby girl is 0.52.

    a. What is the probability that a newborn baby will be a boy?

    b. If a family has three children, all born in the same city, what is the probability that all of them will be boys?

    c. If a family has four children, all born in the same city, what is the probability that the first two will be boys and the second two will be girls?

50. The chance of rain for the next three days is 60% on each day.

   a. What is the probability of rain on all three days?

   b. What is the probability of rain on the first and third, but not the second?

   c. What is the probability of rain on exactly two of the three days?

51. In the game of craps, two dice are rolled, and the numbers on the dice are added.

   a. If a sum of 7 or 11 is rolled on the first roll, the player wins. What is the probability of winning on the first roll?

   b. If a sum of 2, 3, or 12 is rolled on the first roll, the player loses. What is the probability of losing on the first roll?

   c. All other sums are neither a win nor a loss, meaning the player is allowed to roll again. What is the probability that the player gets a second chance to roll?

52. A typical slot machine consists of three wheels, each with a different number of the same types of symbols on them. Suppose that a particular slot machine has the number of symbols shown in the table:

Symbol	Wheel 1	Wheel 2	Wheel 3
Bar	1	2	1
Bell	5	2	2
Lemon	5	5	1
Cherries	2	1	5
Orange	1	3	3
Banana	1	2	3
Total	15	15	15

   a. What is the probability of getting 3 bars?

   b. What is the probability of getting 3 lemons?

   c. What is the probability of getting at least 1 lemon?

   d. If three of the same symbol is the only way to win a payout, which symbol should pay the least? The most?

■ Mathematical Reasoning and Proof

53. The sample space for drawing a card from a deck of 52 cards can be changed by considering different aspects of the card. Not all of the resulting sample spaces have equally likely outcomes. Which sample spaces have equally likely outcomes, and which do not?

   a. {red card, black card}

   b. {numbered card, face card, ace}

   c. {even-valued cards, odd-valued cards}

   d. {hearts, diamonds, spades, clubs}

54. If a 12-sided die is numbered with 1–6 twice, how does the probability of each number on this die compare to the probability of each number on a regular six-sided die?

55. Label each event as impossible, improbable, probable, and certain. State how you made your determination.

   a. Selecting a male student from a class of 23 males and 3 females.

   b. Rolling a sum of 13 when two six-sided dice are rolled.

   c. Getting at least one head in two flips of a coin.

   d. Winning the lottery.

   e. Raining on a day when the weather forecast calls for 100% chance of rain.

56. Suppose that A and B are events in sample space S.

   a. If $P(A) = 0.7$ and $P(B) = 0.8$, must A and B be mutually exclusive?

   b. If $P(A) = 0.4$ and $P(B) = 0.35$, must A and B be mutually exclusive?

   c. If $P(A) = 0.4$, $P(B) = 0.6$, and A and B are mutually exclusive, what must be true about A and B?

57. Use the two spinners to answer each question.

   a. Consider all ordered pairs (a, b) where a is a number from the first spinner and b is a number from the second spinner. Are all such ordered pairs equally likely? If not, what number combinations are the most likely?

   b. Consider all ordered pairs (c_1, c_2) where c_1 is a color from the first spinner and c_2 is a color from the second spinner. Are all such ordered pairs equally likely? If not, what color combinations are the most likely?

58. How many events are possible for an experiment with a sample space of:

   a. 3 outcomes?   **b.** 4 outcomes?

   c. 5 outcomes?   **d.** n outcomes?

59. Write an argument to verify that $P(A$ or $B) = P(A \cup B) = P(A) + P(B)$ by using the fact that if $A \cap B = \varnothing$, then $n(A \cup B) = n(A) + n(B)$.

■ Mathematical Communication

60. The definition of theoretical probability states that the outcomes of an experiment must be equally likely. If the outcomes of an experiment are not equally likely, how might this affect the computation of a probability?

61. Write several questions involving a single-stage experiment in which the probabilities of specific events must be found. Trade questions with a partner, and solve.

62. Write several questions involving a multistage experiment in which the probabilities of several specific events must be found. Trade questions with a partner, and solve.

63. Select a single-stage probabilistic experiment.

a. Write a probability question involving mutually exclusive events.

b. Write a probability question involving nonmutually exclusive events.

c. Write a probability question involving complementary events.

64. Caleb claims that because there are fifty states, the probability of being born in any one of them is $\frac{1}{50}$. Is he correct?

65. Janice states that the probability of getting an A on her English exam is 80% and the probability of getting an A on her mathematics exam is 75%. She concludes that the probability of her getting an A on both is 155%. Where has she gone wrong?

66. A student claims that the probability of randomly selecting a girl from his class is $\frac{2}{3}$ and the probability of randomly selecting a boy is $\frac{1}{4}$. Is this possible?

■ Building Connections

67. How is the definition of theoretical probability directly connected to the computation of a probability?

68. Is it possible to compute a theoretical probability from data? Why or why not?

69. What is the role of counting in determining probabilities? What problems might counting present for experiments with particularly large sample spaces?

70. With a small group of your peers, list as many real-world applications of probability as you can think of. How many did you come up with? Share your results with the class.

71. With a small group of your peers, examine several elementary-level textbooks. Look to see how probability and the relationship between experimental and theoretical probability is introduced. Discuss any differences you find in the approaches and which approach you think is the best.

SECTION **15.3**	**Conditional and Geometric Probability, Odds, and Expected Value**

In Section 15.2, we learned to compute the theoretical probabilities for both single-stage and multistage experiments. We now consider several other concepts related to probability: conditional probability, geometrical probability, odds, and expected value.

Conditional Probability

Recall that probabilities are affected by the way experiments are conducted. For instance, when drawing two cards, the probability of getting two diamonds is different depending on whether the first card is replaced. The probability of an event can also be affected if additional information about the outcome is known. For example, consider the probability of getting two heads when a coin is flipped twice. As the tree diagram in Figure 15.6(a) illustrates, the probability of getting two heads is $P(HH) = \frac{1}{4}$.

However, if we are told that the first flip was a head, then we can use the additional information to remove two outcomes, (T, H) and (T, T), from consideration. This leaves (H, H) and (H, T) as the only possible outcomes, effectively cutting the sample space in half. Consequently, once we know that the first flip is a head, the probability of getting two heads now becomes $\frac{1}{2}$ (Figure 15.6 (b)).

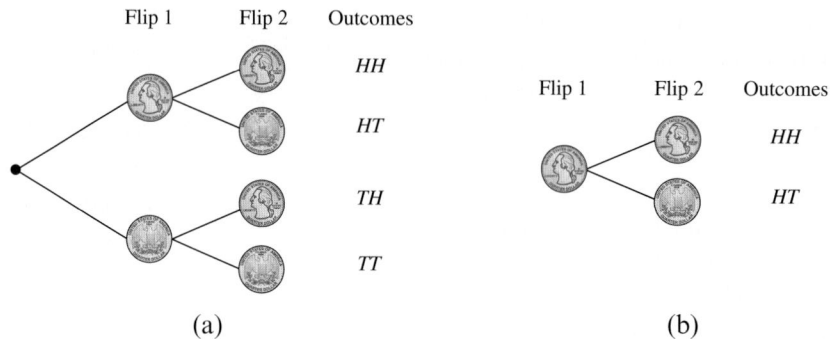

Figure 15.6 **Limiting the sample space with additional information**

When the probability of an event is affected by additional information or some condition, we call it **conditional probability.**

Definition of Conditional Probability

Let A and B be events of an experiment with sample space S, and let $P(A) \neq 0$. The **conditional probability** that event B occurs given that event A has occurred, denoted $P(B|A)$, is given by $P(B|A) = \dfrac{P(A \cap B)}{P(A)}$.

The definition of conditional probability requires two separate events: one to represent the condition that limits the sample space and another to represent the event in question. The next examples demonstrate how to compute conditional probability in two ways.

Example 15.20   Computing Conditional Probabilities from the Sample Space

An eight-sided die is rolled. Find each probability, if it is known that an even number was rolled.

a. A 4 is rolled.   **b.** A number less than 5 is rolled.   **c.** A 1 is rolled.

Solution The additional information limits the sample space to just four numbers, $S = \{2, 4, 6, 8\}$. Because each number is equally likely, we count the successes to find the probability of each event.

a. Four is the only success, so $P(4) = \dfrac{1}{4}$.

b. Two and four are successes, so the probability of rolling a number less than 5 is $\dfrac{2}{4} = \dfrac{1}{2}$.

c. One is not in the sample space, so $P(1) = \dfrac{0}{4} = 0$.

Example 15.21   Computing Conditional Probabilities from the Definition

Consider the game of Bingo described in Example 15.11. What is the probability that the number has at least one digit that is a 4, if it is known that an N was drawn?

Solution We compute the conditional probability by using the formula in the definition. We let N be the event of drawing an N and F be the event that the number has at least one digit that is a 4. Because there are 75 outcomes in the game of Bingo, 15 of which have an N, the probability of event N occurring is $P(N) = \dfrac{15}{75}$. The set $N \cap F$ contains the chips with both an N and at least one 4 in the number, or $N \cap F = \{34,$ 40, 41, 42, 43, 44, 45\}. So the probability of $N \cap F$ is $P(N \cap F) = \dfrac{7}{75}$. Placing both numbers in the formula, we have

$$P(F|N) = \frac{P(F \cap N)}{P(N)} = \frac{\dfrac{7}{75}}{\dfrac{15}{75}} = \frac{7}{75} \cdot \frac{75}{15} = \frac{7}{15}$$

Geometric Probability

So far, the sample spaces have been discrete; that is, each outcome in the sample space is distinct or separate from the others. Sample spaces can also be continuous, like the ones that occur in many geometric situations. For instance, consider the probability of spinning a 4 on the spinner shown in Figure 15.7. To compute the probability, we have to recognize that the outcomes are equally likely, which can be done only by realizing that each sector of the circle has the same area. The areas are continuous measures and not discrete objects, so the outcomes are made from the infinite number of points in each sector. Selecting an outcome amounts to selecting a point at random from one of the sectors. However, because computing a probability from an infinite number of points is difficult, we use the area. In other words, we find the probability of spinning a 4 by comparing the area of the sector to the area of the circle. Because the area of the sector is one-fourth the area of the circle, the probability of spinning a 4 is $P(4) = \dfrac{1}{4}$.

Figure 15.7 A spinner with four congruent sectors

Explorations Manual 15.9

We can generalize this idea to other geometric measures, including angle measure, length, area, and even volume. Consequently, for a continuous sample space, the probability of an event A is the ratio of the "measure" of A to the "measure" of the sample space, or

$$P(A) = \frac{\text{measure of } A}{\text{measure of } S}$$

Example 15.22   Geometric Probability and a Spinner

If the spinner is spun once, what is the probability that it stops on red?

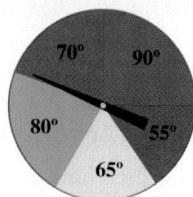

Solution We have a circular spinner that is divided into five sectors of different sizes. Because the radius is unknown, we cannot compute the probability by comparing the area of each sector to the area of the circle. However, we can take another approach. A successful outcome occurs when the arrow lands on any point in either red sector. These points must lie between two radii, which are separated by a central angle. So any point in the sector corresponds to an angle within the central angle. Consequently, the probability that the spinner stops in a particular sector is the ratio of its central angle to the measure of an entire revolution. Specifically, because the central angles for the red sectors are 70° and 55°, the probability that the spinner stops in a red sector is $P(R) = \dfrac{70 + 55}{360} = \dfrac{125}{360} = \dfrac{25}{72} \approx 0.347.$

Example 15.23   Cutting a String

Application

A string 10 in. long is cut into two pieces at a random location. What is the probability that one of the pieces is at least 7 in. long?

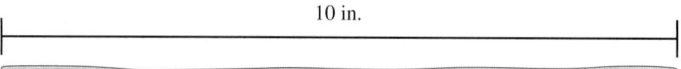

10 in.

Solution We can represent the string with a line segment $\overline{AB}$ that is 10 in. long. Even though the cut can occur at any point on the segment, we are interested only in those cuts that result in a piece at least 7 in. long. To find them, consider point C that is 3 in. away from point A. Any cut made in $\overline{AC}$ results in a string 7 in. or longer.

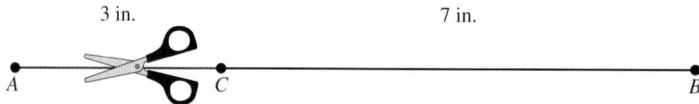

The probability that a cut is made in $\overline{AC}$ is the ratio of the length of $\overline{AC}$ to the length of $\overline{AB}$, or $\dfrac{m(\overline{AC})}{m(\overline{AB})} = \dfrac{3}{10}$. Likewise, if point D is 3 in. away from point B, then any cut made in $\overline{DB}$ will result in a string that is 7 in. or longer.

As before, the probability that the cut occurs in $\overline{DB}$ is $\dfrac{m(\overline{DB})}{m(\overline{AB})} = \dfrac{3}{10}$. Because a cut on $\overline{AC}$ is mutually exclusive with a cut on $\overline{DB}$, we add their probabilities. As a result, the probability that the string will be cut into two pieces so that one is at least 7 in. long is $\dfrac{3}{10} + \dfrac{3}{10} = \dfrac{6}{10} = \dfrac{3}{5}.$

Check Your Understanding 15.3-A

1. A single card is drawn from a standard deck of 52 cards. If it is known that the card is black, what is the probability that the card is:

   a. A king?   **b.** A ten or less?   **c.** A diamond?   **d.** A spade or club?

2. A 20-sided die is rolled. Find the probability of each event if it is known that an odd number was rolled.

   a. A 3 is rolled.   **b.** A number divisible by 5 is rolled.

   c. A number less than 10 is rolled.   **d.** A prime number is rolled.

3. If the following spinner is spun once, what is the probability that it stops on red?

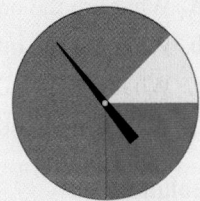

Talk About It   Consider a circular spinner. How might arc length be used to determine the probability of landing in a given sector?

Problem Solving

Activity 15.6

A dart is thrown at random at the following unusual dartboard:

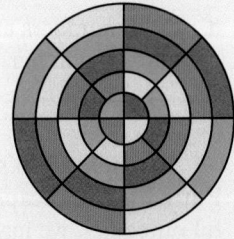

What is the probability that the dart will land on:

a. Red?   **b.** Red or blue?   **c.** A primary color?

Odds

Odds are a concept closely related to probability and often used in games of chance, sporting events, and other probabilistic situations. Unlike probability, which express the chance of an event as the ratio of successes to total outcomes, odds express the chance of an event as the ratio of successful to unsuccessful outcomes. For instance, a soda company may run a promotion in which the odds in favor of winning a free drink are 1 to 10. The odds indicate that only one bottle will have a winning cap for every 10 that do not. We can give not only the **odds in favor** of an event, but also the **odds against** it, which is the ratio of unsuccessful outcomes to successful ones.

Explorations Manual 15.10

Definition of Odds

Let A be an event in an experiment with sample space S, and suppose $\overline{A}$ is its complementary event.

The **odds in favor** of event A is the ratio $n(A) : n(\overline{A})$.
The **odds against** event A is the ratio $n(\overline{A}) : n(A)$.

Odds, like probability, are defined in terms of set notation. However, we can also use probabilistic language to describe them:

$$\text{Odds in favor of } A = \frac{n(A)}{n(\overline{A})} = \frac{\text{number of successful outcomes}}{\text{number of unsuccessful outcomes}} = \frac{s}{u}$$

$$\text{Odds against } A = \frac{n(\overline{A})}{n(A)} = \frac{\text{number of unsuccessful outcomes}}{\text{number of successful outcomes}} = \frac{u}{s}$$

Example 15.24 Computing Odds

Consider an experiment in which a 10-sided die is rolled.

a. What are the odds in favor of rolling an even number?

b. What are the odds in favor of rolling a prime number?

c. What are the odds against rolling a number divisible by 3?

d. What are the odds against rolling a number less than 11?

Solution

a. Because there are 5 even numbers and 5 odd numbers, the odds in favor of rolling an even number are 5:5, or 1:1.

b. The prime numbers on a 10-sided die are 2, 3, 5, and 7, so the odds in favor of rolling a prime number are 4:6, or 2:3.

c. Three numbers are divisible by three: 3, 6, and 9. So the odds against rolling a number divisible by 3 are 7:3.

d. All 10 numbers are less than 11 and are successful outcomes. So the odds against rolling a number less than 11 are 0:10.

Example 15.24 shows that odds are ratios, so they can contain 0 and be reduced. Furthermore, because the sample space is the union of the successful and unsuccessful outcomes, we have a direct connection between odds and probability.

Theorem 15.9 Odds to Probability

If $s{:}u$ represents the odds in favor of an event A, or if $u{:}s$ represents the odds against an event A, then the probability of A is $P(A) = \dfrac{s}{s + u}$.

Example 15.25 Horse Racing

The odds in favor of three horses winning an 8-horse race are as follows:

Fast Eddie: 1:2   My Gal Sal: 1:5   Slap Jack: 1:8

What is the probability of each horse winning, and what is the probability that one of the other five horses will win?

Solution Using Theorem 15.9, the probability of each horse winning is

$$\text{Fast Eddie} = \frac{1}{1 + 2} = \frac{1}{3}$$

$$\text{My Gal Sal} = \frac{1}{1 + 5} = \frac{1}{6}$$

$$\text{Slap Jack} = \frac{1}{1+8} = \frac{1}{9}$$

We can find the probability that one of the other five horses wins by subtracting the probabilities of the three from 1, or $1 - \frac{1}{3} - \frac{1}{6} - \frac{1}{9} = \frac{18}{18} - \frac{6}{18} - \frac{3}{18} - \frac{2}{18} = \frac{7}{18}$.

Example 15.26 Odds and a Lopsided Die

A six-sided die is lopsided so that the probability of a 4 being rolled is $0.\overline{3}$. What are the odds in favor of rolling a number other than 4 on the next roll of the die?

Solution The probability of rolling a 4 is $0.\overline{3} = \frac{1}{3}$. For every successful outcome, there are two unsuccessful outcomes. Consequently, the odds of rolling a number other than 4 are 2:1.

Example 15.26 illustrates how odds can be determined from probabilities. In general, if A is an event in an experiment with sample space S, then the odds in favor of A are equal to

$$\frac{n(A)}{n(\overline{A})} = \frac{\dfrac{n(A)}{n(S)}}{\dfrac{n(\overline{A})}{n(S)}} \qquad \text{Divide numerator and denominator by } n(S).$$

$$= \frac{P(A)}{P(\overline{A})} \qquad \text{Definition of probability}$$

$$= \frac{P(A)}{1 - P(A)} \qquad \text{Probability of a complementary event (Theorem 15.5)}$$

Theorem 15.10 Probability to Odds

Let A be an event for an experiment with sample space S.
The odds in favor of A are $P(A) : [1 - P(A)]$.
The odds against A are $[1 - P(A)] : P(A)$.

Example 15.27 Probability to Odds

Application

Lauren is the best player on her field hockey team. Her coach estimates that the probability of Lauren scoring at least one goal in a game is 0.45. What are the odds in favor of Lauren scoring at least one goal in the team's next game?

Solution The probability that Lauren scores a goal in any game is $0.45 = \frac{45}{100} = \frac{9}{20}$.

From this, Theorem 15.10 tells us that the odds in favor of her scoring a goal are

$$\frac{9}{20} : \left(1 - \frac{9}{20}\right) = \frac{\dfrac{9}{20}}{\dfrac{11}{20}} = \frac{9}{20} \cdot \frac{20}{11} = \frac{9}{11} = 9{:}11.$$

Expected Value

Probability has its roots in games of chance, so one of its applications is to determine the fairness of games. **Fairness** takes on different meanings for different games. If a game involves people playing against one another, then the game is fair if each person has the same chance or probability of winning. If a player wins or loses based on the outcome of some random event, then the game is fair if the player can expect to win as much as to lose. If the game involves money, then the probability of winning can vary as long as the net value of the game is 0.

So how do we determine whether a game is fair? One way is to count outcomes. If the number of winning outcomes is equal to the number of losing outcomes, then the game is fair only if each outcome has the same value. If the outcomes have different values, then we must take a different approach. In this case, the probability of winning points or money should be offset by the probability of losing points or money. Less frequent outcomes should be worth more and more frequent outcomes should be worth less. Mathematically, the way to take these considerations into account when determining the fairness of a game is to compute the game's **expected value.**

Definition of Expected Value

If an experiment has outcomes of values $a_1, a_2, \ldots a_n$, occurring with probabilities $p_1, p_2, \ldots p_n$, respectively, then the **mathematical expectation,** or **expected value,** E, of the experiment is $E = a_1p_1 + a_2p_2 + \ldots + a_np_n$.

Explorations
Manual
15.11

In the definition, each a_i represents a value for an outcome. Positive values represent an amount won, and negative values represent an amount lost. Each value is then weighted by the probability of the outcome occurring. Like probability, expected value does not indicate the value of a game for any one time that it is played. Rather, expected value indicates the average value of the game if it is played multiple times. If the expected value is 0, then the game is considered fair.

Example 15.28　Expected Value of a Carnival Game

A local carnival has a game that is played by rolling a single die. If a 1 or 2 is rolled, the player must pay a quarter. If a 3 or 4 is rolled, the player wins nothing. If a 5 or 6 is rolled, the player wins a quarter. Is the game fair?

Solution　The game has three events. The value of each event along with its probability is given in the following table:

Event	Value of Event	Probability of Event
Rolling a 1 or 2	$a_1 = -25$ (loss of 25¢)	$p_1 = \dfrac{2}{6} = \dfrac{1}{3}$
Rolling a 3 or 4	$a_2 = 0$ (wins nothing)	$p_2 = \dfrac{2}{6} = \dfrac{1}{3}$
Rolling a 5 or 6	$a_3 = +25$ (win 25¢)	$p_3 = \dfrac{2}{6} = \dfrac{1}{3}$

Given the three events, their outcomes, and their probabilities, the expected value of the game is

$$E = a_1p_1 + a_2p_2 + a_3p_3$$

$$= (-25)\frac{1}{3} + (0)\frac{1}{3} + (25)\frac{1}{3}$$

$$= -\frac{25}{3} + 0 + \frac{25}{3}$$

$$= 0$$

Because the expected value is 0, the game is fair.

In another type of game, every event has a positive payout, so people are commonly charged to play the game. In this case, we can use expected value to determine the amount to charge.

Example 15.29 Finding the Cost to Play a Game

Reasoning

Another carnival game uses a spinner divided into eight equally likely regions as shown:

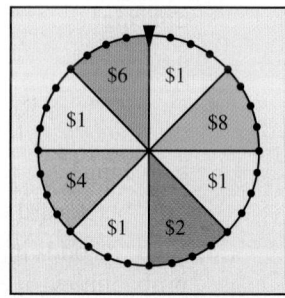

The game boasts that everyone is a winner, in that the player automatically wins the dollar amount shown on the spinner. Of course, the person running the game wants to make money. What would be a reasonable amount to charge for playing the game so that money is made without charging too much to keep people from playing?

Solution Because each outcome on the spinner is equally likely, we can condense them into the five events shown in the table:

Event	Value of Event	Probability of Event
Spinning $1	$a_1 = 1$	$p_1 = \frac{4}{8} = \frac{1}{2}$
Spinning $2	$a_2 = 2$	$p_2 = \frac{1}{8}$
Spinning $4	$a_3 = 4$	$p_3 = \frac{1}{8}$
Spinning $6	$a_4 = 6$	$p_4 = \frac{1}{8}$
Spinning $8	$a_5 = 8$	$p_5 = \frac{1}{8}$

Given the events, their outcomes, and their probabilities, the expected value is

$$E = a_1 p_1 + a_2 p_2 + a_3 p_3 + a_4 p_4 + a_5 p_5$$

$$= 1 \cdot \left(\frac{1}{2}\right) + 2 \cdot \left(\frac{1}{8}\right) + 4 \cdot \left(\frac{1}{8}\right) + 6 \cdot \left(\frac{1}{8}\right) + 8 \cdot \left(\frac{1}{8}\right)$$

$$= \frac{1}{2} + \frac{1}{4} + \frac{1}{2} + \frac{3}{4} + 1$$

$$= \$3$$

The expected value indicates that in the long run, players can expect to win $3 each time they play. To make money, the person running the game should charge the player more than $3 per turn. For instance, if the person running the game charges $3.50, players will be tempted by the larger payouts, but over the long term, the person running the game should make $0.50 per game.

Check Your Understanding 15.3-B

1. A single card is drawn from a deck of 52 cards. What are the odds in favor of drawing:

a. An ace?   **b.** A black 10?   **c.** A red card?

2. A single card is drawn from a deck of 52 cards. What are the odds against drawing:

a. A king?   **b.** The jack of diamonds?

c. A black card or a heart?

3. Find the odds in favor of event A, when A has each of the following probabilities.

a. $\dfrac{6}{17}$ **b.** 0.35   **c.** $0.\overline{1}$

d. 47%   **e.** 0.4%

4. A game is played in which two quarters are tossed. If two heads come up, the player wins $1. If a head and a tail come up, the player wins $0.50. If two tails come up, the player wins $0.25. If the game is to be fair, how much should a player expect to pay to play the game?

Talk About It Make a list of the games you might encounter in a casino. Are any of them fair?

Problem Solving

Activity 15.7
Three players, Alan, Barb, and Carol, are playing a game in which a single die is rolled. If a 1 or 2 is rolled, Alan wins the round. If a 3 or 4 is rolled, Barb wins the round. If a 5 or 6 is rolled, Carol wins the round. The first player to win three rounds wins the game. Four rounds have been played, of which Alan has won one, Barb has won two, and Carol has won one. Unfortunately, the die was lost after the fourth round. If each player has bet $100 and the game cannot be continued, what is the fairest way to distribute the money based on the players' probabilities of winning?

SECTION 15.3   Learning Assessment

■ Understanding the Concepts

1. How does extra information about the outcome of an experiment affect the probability of a certain event?

2. How are geometrical probabilities different from probabilities computed from a discrete sample space? How are they similar?

3. What is the difference between odds in favor and odds against?

4. How are odds connected to probability?

5. a. What does it mean for a game to be fair?

   b. How is expected value used to determine whether a game is fair?

■ Representing the Mathematics

6. Consider an experiment in which a card is drawn from a deck of 52 cards. Let event A be an ace is drawn, event B be a black card is drawn, event C be a club is drawn, and event F be a face card is drawn. Write each statement in symbolic form.

   a. The probability of drawing an ace if it is known that a black card is drawn

   b. The probability of drawing a club if it is known that a face card is drawn

   c. The probability of drawing a face card if it is known that a black card is drawn

   d. The probability of drawing a club if it is known that a black card is drawn

7. Consider an experiment in which a 12-sided die is rolled. Let event A be an even number is rolled, event B be a number divisible by 3 is rolled, and event C be a number less than 7 is rolled. Give a verbal description of each expression.

   a. $P(A|B)$ b. $P(B|A)$ c. $P(B|C)$ d. $P(A|C)$

8. Consider an experiment with sample space S and two events, A and B. Draw a Venn diagram, and discuss how the Venn diagram can be used to find $P(B|A)$.

9. The following Venn diagram represents an experiment in which numbers written on chips are drawn from a hat.

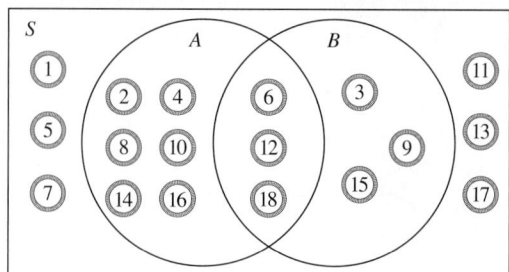

Find:

   a. $P(A)$.   b. $P(B)$.   c. $P(A|B)$.   d. $P(B|A)$.

10. Draw a spinner that is divided into four sections to match the stated criteria.

   a. Each outcome is equally likely.

   b. The probability of one outcome is $\frac{1}{2}$, and the other three are equally likely.

   c. The probability of one outcome is $\frac{1}{2}$, another is $\frac{1}{4}$, and the other two are equally likely.

   d. None of the outcomes is equally likely to another.

■ Exercises

11. What is the probability that the number rolled on a standard 12-sided die is divisible by 3 if it is known that an even number is rolled?

12. Consider an experiment in which a 20-sided die is rolled. What is the probability of rolling a 5 if:

   a. An odd number is rolled?

   b. A number divisible by 5 is rolled?

   c. A number less than 10 is rolled?

   d. An even number is rolled?

13. A box contains 10 chips, 6 red and 4 blue. A chip is drawn and then discarded. If a second chip is drawn, what is the probability that the chip is

   a. Red, if the first chip was red?

   b. Blue, if the first chip was red?

   c. Red, if the first chip was blue?

   d. Blue, if the first chip was blue?

14. A card is drawn from a deck of 52 cards. If the card is red, what is the probability that the card is

   a. A 3?   b. A face card?

   c. A heart?   d. A club?

15. A dart is thrown at the odd-looking dartboard shown:

What is the probability that the dart hits a blue square if it is known that the dart hit a colored square?

16. A string 12 in. long is cut into two pieces at a random location. What is the probability that one of the pieces is at least 8 in. long?

17. If two dice are rolled, find the odds in favor of each event.

   a. A sum of 7 is rolled.

   b. A sum less than 6 is rolled.

   c. At least one 3 is rolled.

   d. An odd sum is rolled.

18. If two dice are rolled, find the odds against each event.

   a. A sum of 5 is rolled.

   b. A sum greater than 10 is rolled.

   c. A sum of 6 or 8 is rolled.

   d. No 5 or 6 is rolled.

19. The odds in favor or against an event A are given. Find the probability of event A.

   a. Odds in favor of A are 2:3.

   b. Odds in favor of A are 6:5.

   c. Odds against A are 8:1.

   d. Odds against A are 2:7.

20. The probability of event A is given. Find the odds in favor of and the odds against event A.

   a. $P(A) = \dfrac{1}{3}$.   **b.** $P(A) = \dfrac{6}{11}$.

   c. $P(A) = 25\%$.   **d.** $P(A) = 0.19$.

21. The weather forecast says that there is a 60% chance of rain today. What are the odds against rain?

22. The odds against Caleb winning a tennis tournament are 2 to 3. What is the probability that he wins the tennis tournament?

23. A couple wants to have three children and hopes to have two girls. What are the odds in favor of the couple getting their wish if the first child is a boy?

24. A box contains two coins. One is regular, and the other is double-headed. Suppose that a coin is drawn from the box and the side showing is the head. If the other side is not observed, what is the probability that it is a head as well?

25. In the game of craps, a player wins on the first roll of two dice if the sum is either 7 or 11. What are the odds in favor of such an event?

■ Problems and Applications

26. a. Three cards are dealt in succession. What is the probability of getting 3 aces if the first card is an ace?

   b. Four cards are dealt in succession. What is the probability of getting 3 aces if the first card is not an ace?

27. A 12-sided die has sides that are numbered 1, 2, 2, 3, 3, 4, 5, 5, 6, 7, 7, and 8. Let event A be an even number is rolled and event B be a prime number is rolled. Find $P(A|B)$ and $P(B|A)$ if the die is rolled one time.

28. Two boxes contain chips with letters written on them. One has the letters from the word "MATH," and the other has the letters from the word "THAT."

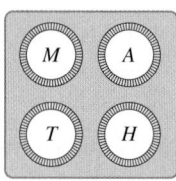 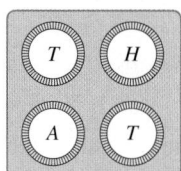

A chip is drawn from the first box and placed into the second. What is the probability that a chip drawn from the second box:

   a. Is a T if it is known that the letter placed in the box was an M?

   b. Is a T if it is known that the letter placed in the box was a T?

   c. Is an M if it is known that the letter placed in the box was an M?

   d. Is an M if it is known that the letter placed in the box was an A?

29. A fast-food restaurant is running a promotion in which it gives away free food. The odds of winning a free soda are 1 in 9, and the odds of winning a free hamburger are 1 in 49. If the two events are mutually exclusive and every customer is given two chances to win per visit, what are the odds in favor of winning a free soda and a free hamburger?

30. Suppose that each figure represents an oddly shaped dartboard. If one dart is thrown at random, what is the probability that the dart will hit a shaded region?

   a.

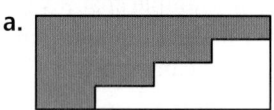

   b.

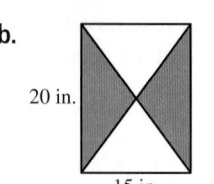

   c.

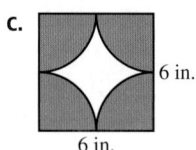

31. A square dartboard is shown:

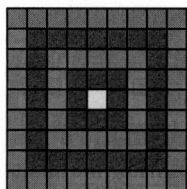

Find the probability of each event.

a. If a single dart is thrown and hits the dartboard in a random location, what is the probability that it hits the bull's-eye?

b. If a single dart is thrown and hits the dartboard in a random location, what is the probability that it hits a gray square?

c. If two darts are thrown and hit the dartboard in random locations, what is the probability that both hit a red square?

d. If three darts are thrown and hit the dartboard in random locations, what is the probability that all three hit a red square if the first one hits a red square?

32. Each ring in the dartboard shown is 2 in. wide, and the bull's-eye has a diameter of 2 in.

If a single dart is thrown and hits the dartboard in a random location, what is the probability that it hits:

a. The bull's-eye?   **b.** A red ring?

33. Oscar lives in an unusual fish tank with two parts. The outer tank is a rectangular prism that measures $30 \times 15 \times 18$ in. deep. The inner tank is a cube with edges 9 in. long. It is suspended by thin strings and has holes large enough for Oscar to swim through.

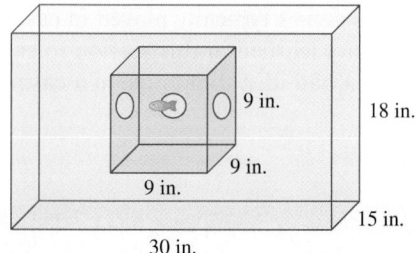

If Oscar swims about the tank randomly, and we look in the tank at a random moment in time, what is the probability that Oscar will be in the inner tank?

34. A subway train arrives at a certain station every 12 min. If you arrive at the station at a random time, what is the probability that you will have to wait at least 5 min for the subway?

35. Two dice are rolled, and the results are recorded. The amounts won or lost for each sum are shown in the table:

Sum of Dice	Amount Won or Lost
2	$20
3	$10
4	$5
5	$0
6	−$5
7	−$10
8	−$5
9	$0
10	$5
11	$10
12	$20

What is the expected value of the game? Would you play it?

36. A state is thinking about implementing a new lottery game called the Joker Jackpot, where a joker card is added to a deck of 52 cards. Any person selected to play the game gets to draw one card from the deck. The amounts won are shown:

Card	Amount Won
Joker	$1,000,000
Any Ace	$100,000
Any King	$10,000
Any Queen	$1,000
Any Jack	$100
Any card 10 or smaller	$10

a. What is the expected value of the game for the person selected to play?

b. To play the game, a person must first buy a ticket for $2 to be put into a pool of possible contestants. If the pool regularly has 100,000 people in it, is this a fair amount to pay for a ticket?

37. Insurance companies use expected values to set premiums for different kinds of insurance. One company finds that the average car accident costs the company roughly $5,000 in claims. The probability that a 25-year-old male will have an accident is 0.01. If the insurance company charges $1,095 a year for a policy to insure a 25-year-old male, what is the expected value of the policy for the company?

38. A gold mining company plans to mine a plot of 300 acres. The company estimates that the cost of mining the plot is $3,000,000. It also estimates a probability of 0.47 that the land will be a small strike worth $1,500,000, a probability of 0.35 that it will be a medium strike worth $4,500,000, and a probability of 0.18 that it will be a large strike worth $9,500,000. Is it worth the company's time and money to mine the plot of land?

39. A professor is thinking of applying for two grants, one for $15,000 and another for $25,000. It costs the professor $250 to apply for the first and $375 for the second. If the probability of the professor receiving the first grant is 0.02 and the probability of her receiving the second is 0.005, what is the expected value of each grant? Which is more worthwhile for the professor to pursue?

■ Mathematical Reasoning and Proof

40. In each pair of odds, state which, if any, is more favorable for an event A?

   a. 3:3 or 6:6   b. 60:40 or 3:2

   c. 81:27 or 64:16   d. 3:4 or 5:6

41. Consider an experiment in which a 10-sided die is rolled. Give a specific example of an event A with the odds in favor of those given:

   a. 1:1   b. 1:4   c. 1:9   d. 4:1

42. In a game, two coins are tossed. If two heads occur, the player wins $10. If a head and a tail occur, the player loses $2, and if two tails occur, the player loses $5. Does it make sense to play the game? Why or why not?

43. Mark and Amy are playing a game in which two dice are rolled and the value of the smaller number is subtracted from the value of the larger. If the difference is a 0, 1, or 2, Mark gets 1 point. If the difference is a 3, 4, or 5, Amy gets 1 point. The game ends after 15 rounds, and the player with more points wins. Is this game fair? Why or why not?

44. If everything else is held constant, how is the expected value of an outcome affected if its probability is increased?

45. A church is holding a raffle for a prize with an estimated value of $3,000. What is the expected value of the raffle if:

   a. 10 tickets are sold?

   b. 100 tickets are sold?

   c. 1,000 tickets are sold?

   d. How does increasing the number of tickets affect the expected value?

46. A game has four events: A, B, C and D. Event A has a probability of 0.35, event B has a probability of 0.25, event C has a probability of 0.05, and event D has a probability of 0.35. What values might be assigned to these events to give the game an expected value of 0?

47. Three players are playing a game in which two dice are rolled and the values of the dice are added. If a particular player is to win when a certain sum is rolled, is it possible to assign the sums to the three players so that the game is fair? If so, how?

■ Mathematical Communication

48. Which do you think is a better way to express the likelihood of an event, theoretical probability or odds? What are the advantages or disadvantages of each representation?

49. Marcie knows that the odds against an event are 7 to 4. She then states that the probability of the event is $\frac{7}{11}$. Where has she made her mistake?

50. Stephan flips a coin 10 times, of which 6 are heads and 4 are tails. Based on his results, he makes the claim that the odds in favor of heads for any coin must be 6 to 4. Is he correct?

■ Building Connections

51. a. How is geometric area used to determine the probabilities associated with spinners?

   b. Do other random devices rely on area to determine probabilities?

52. Try to come up with five real-world applications of odds. Explain the meaning of the odds in each case.

53. Explore the games typically played in casinos. Use what you have learned in this section to explain why it is always a bad idea to gamble in a casino.

SECTION 15.4 Counting Techniques and Probability

As we have seen, probability relies heavily on counting. When sample spaces and events are small, tree diagrams and tables are effective ways to list and count outcomes. However, many sample spaces are quite large, making counting the outcomes directly either too tedious or complex. In such cases, we may turn to one of several techniques to count outcomes without having to list them. We begin with the Fundamental Counting Principle.

The Fundamental Counting Principle

Many of the experiments so far have involved either multiple actions or multiple stages. In such cases, we often used a tree diagram to generate and count the outcomes in the sample space. For instance, the tree diagram for an experiment in which a coin is tossed and a six-sided die is rolled is shown in Figure 15.8. We can quickly count the 12 outcomes from the diagram. However, notice that if we multiply the number of outcomes for the first stage by the number of outcomes for the second, we get the same number of total outcomes, or $2 \cdot 6 = 12$.

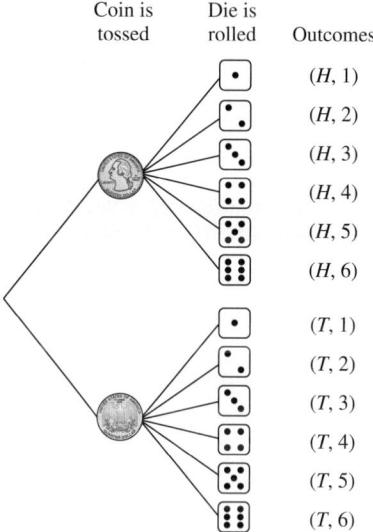

Figure 15.8 A tree diagram of tossing a coin and rolling a die

This relationship holds true for other experiments we have encountered. For instance, when rolling two dice, the sample space had 36 outcomes, which is the same number we get by multiplying the number of outcomes on each die, or $6 \cdot 6 = 36$. Similarly, when tossing two coins, the sample space had four outcomes, which is the same number we get by multiplying the outcomes on each coin, or $2 \cdot 2 = 4$. The fact that the number of outcomes in each action or stage can be multiplied to find the number of outcomes in the sample space is called the **Fundamental Counting Principle.**

Theorem 15.11 Fundamental Counting Principle

If an event A can occur in m ways and an event B can occur in n ways, then the events A and B can occur in succession, first A, then B, in $m \cdot n$ ways.

The Fundamental Counting Principle can be applied in a variety of situations as long as the events occur in succession or at least can be computed as if they do.

Example 15.30 Drawing Cards

How many ways can two cards be drawn from a deck of 52 cards if the cards are drawn in succession and the first card is replaced after it is drawn?

Solution A common way to work counting problems is to make a blank for each event and then place the number of outcomes in each blank. The experiment consists of two events conducted in succession, so we draw two blanks. There are 52 outcomes for the first card, which is then replaced in the deck. This gives us 52 outcomes for the second card. According to the Fundamental Counting Principle, we multiply the number of outcomes for each card to find the total number of outcomes, or

$$\underset{\text{card 1}}{52} \cdot \underset{\text{card 2}}{52} = 2{,}704 \text{ possible outcomes}$$

Example 15.30 shows that the Fundamental Counting Principle can be used in an experiment with replacement. It can also be used when there is no replacement or when there are more than two events. In such cases, we multiply the number of outcomes for each of the events to find the total number of outcomes.

Example 15.31 Counting Outfits

Application

Carrie has 25 blouses, 12 pairs of pants, 7 pairs of shoes, and 3 jackets. If all her clothes can be mixed and matched, how many different outfits does she have?

Solution The "experiment" of dressing Carrie has four events, so we make a blank for each one. She can select any one of 25 blouses, any one of 12 pairs of pants, any one of 7 pairs of shoes, and any one of 3 jackets. Using these numbers, we fill in the blanks and multiply. Consequently, she has

$$\underset{\text{Blouse}}{25} \cdot \underset{\text{Pants}}{12} \cdot \underset{\text{Shoes}}{7} \cdot \underset{\text{Jacket}}{3} = 6{,}300 \text{ possible outfits}$$

In Section 15.2, we actually used the Fundamental Counting Principle to compute multistage probabilities. For instance, consider an experiment in which two chips are drawn from two boxes: One has 5 red and 4 white, and another has 4 red and 6 white (Figure 15.9). Previously, we computed the probability of drawing two red chips by finding the probability of getting a red chip first and then multiplying it by the probability of getting a red chip second, or

$$P(RR) = \frac{5}{9} \cdot \frac{4}{10} = \frac{5 \cdot 4}{9 \cdot 10} = \frac{20}{90} = \frac{2}{9}$$

However, multiplying the numerators is the same as multiplying the number of red chips from the first box by the number of red chips from the second. Similarly, multiplying the denominators is the same as multiplying the total number of chips from the first box by the total number of chips from the second. In other words, both products are direct applications of the Fundamental Counting Principle.

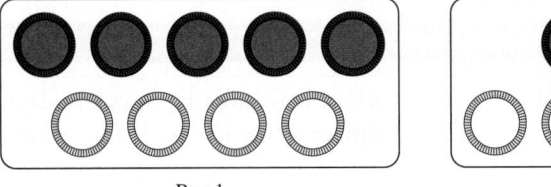
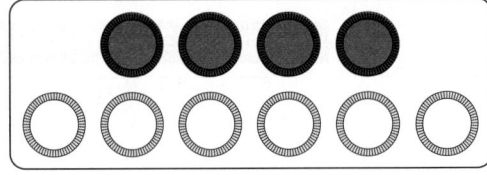

Box 1   Box 2

Figure 15.9 Two boxes of chips

Example 15.32   Probability and the Fundamental Counting Principle

Application

Reconsider Carrie from Example 15.31. Suppose she has 4 white blouses, 1 pair of white pants, 2 pairs of white shoes, and 1 white jacket. If she selects an outfit at random, what is the probability that she will be wearing all white?

Solution We know from Example 15.31 that Carrie has 6,300 possible outfits. Because she has 4 white blouses, 1 pair of white pants, 2 pairs of white shoes, and 1 white jacket, she has a total of $4 \cdot 1 \cdot 2 \cdot 1 = 8$ white outfits. As a result, the probability that she is wearing all white is $\dfrac{\text{White outfits}}{\text{Total outfits}} = \dfrac{8}{6,300} = \dfrac{2}{1,575}$.

Permutations

The Fundamental Counting Principle leads to other, more specialized counting techniques. For instance, many situations in probability require counting the different ways that a set of objects can be arranged. Mathematically, we call an *ordered* arrangement of *distinct* objects with *no repetitions* a **permutation.**

We can count permutations in a number of different ways. One is simply to list the arrangements by using a tree diagram. For instance, the tree diagram in Figure 15.10 indicates that there are six permutations of the letters in the word "MAT."

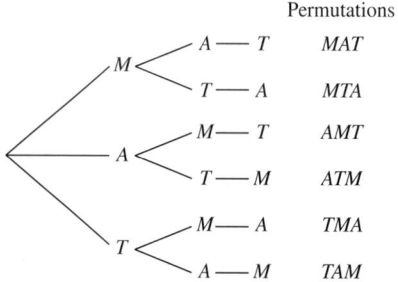

Figure 15.10 Permutations of MAT

A tree diagram works well when the number of objects is small. However, as more objects are added, the number of permutations quickly increases. For instance, adding just one more letter to our previous example, making the word "MATH," increases the number of permutations from 6 to 24 (Figure 15.11).

MATH	*ATMH*	*TMAH*	*HATM*
MAHT	*ATHM*	*TMHA*	*HAMT*
MTHA	*AHMT*	*TAHM*	*HMTA*
MTAH	*AHTM*	*TAMH*	*HMAT*
MHAT	*AMTH*	*THAM*	*HTAM*
MHTA	*AMHT*	*THMA*	*HTMA*

Figure 15.11 The permutations of MATH

Because the number of permutations quickly increases as we add objects, a more efficient way to count them is needed. Reconsider the permutations of the letters in the word "MAT." The tree diagram in Figure 15.10 shows three choices for the first letter: *M*, *A*, or *T*. Once we choose the first letter, only two choices remain for the second. For instance, if *M* is chosen first, then we are left to choose between *A* and *T* for the second. Likewise, once the first and second letters have been chosen, only one choice is left for the last. By the Fundamental Counting Principle, we conclude that the number of permutations must be

$$\underset{\substack{\text{First} \\ \text{letter}}}{3} \cdot \underset{\substack{\text{Second} \\ \text{letter}}}{2} \cdot \underset{\substack{\text{Third} \\ \text{letter}}}{1} = 6 \text{ permutations}$$

We can do the same with the letters in the word "MATH." Here, there are four choices for the first letter, three for the second, two for the third, and one for the fourth. Again, by the Fundamental Counting Principle, there are

$$\underset{\substack{\text{First} \\ \text{letter}}}{4} \cdot \underset{\substack{\text{Second} \\ \text{letter}}}{3} \cdot \underset{\substack{\text{Third} \\ \text{letter}}}{2} \cdot \underset{\substack{\text{Fourth} \\ \text{letter}}}{1} = 24 \text{ permutations}$$

These examples show that given n objects to arrange, there are n choices for the first object, $(n-1)$ for the second, $(n-2)$ for the third, and so on down to one choice for the last object. Consequently, by the Fundamental Counting Principle, there are $n \cdot (n-1) \cdot (n-2) \cdot \ldots 3 \cdot 2 \cdot 1$ permutations of n objects. In other words, the number of permutations of n objects is the product of every natural number from n down to 1, inclusively. This type of product is common in counting situations, so it is given its own special name and notation.

Definition of *n* Factorial

For every whole number $n > 0$, the product $n \cdot (n-1) \cdot (n-2) \cdot \ldots 3 \cdot 2 \cdot 1$ is called *n* **factorial,** denoted $n!$. $0!$ is defined to be 1.

A few observations should be made about factorials. First, defining $0!$ to be 1 may seem like a strange thing to do because any product with 0 is equal to 0. However, as we will demonstrate, doing so makes computational sense. Second, factorials can grow quite large, quite quickly. Consider the first ten:

$0! = 1$ $5! = 5 \cdot 4 \cdot 3 \cdot 2 \cdot 1 = 120$
$1! = 1$ $6! = 6 \cdot 5 \cdot 4 \cdot 3 \cdot 2 \cdot 1 = 720$
$2! = 2 \cdot 1 = 2$ $7! = 7 \cdot 6 \cdot 5 \cdot 4 \cdot 3 \cdot 2 \cdot 1 = 5{,}040$
$3! = 3 \cdot 2 \cdot 1 = 6$ $8! = 8 \cdot 7 \cdot 6 \cdot 5 \cdot 4 \cdot 3 \cdot 2 \cdot 1 = 40{,}320$
$4! = 4 \cdot 3 \cdot 2 \cdot 1 = 24$ $9! = 9 \cdot 8 \cdot 7 \cdot 6 \cdot 5 \cdot 4 \cdot 3 \cdot 2 \cdot 1 = 362{,}880$

As a result, we often use calculators to compute factorials. Most scientific calculators are now equipped with a factorial key (x! , n! , or !) that can be used to compute factorials with ease. For instance, on many calculators, we compute $11!$ by pressing 11 n! ENTER to obtain a value of 39,916,800.

Because factorials involve products, they are often relatively easy to simplify. Consider the next example.

Example 15.33   Simplifying Factorial Expressions

Simplify and then compute each expression.

a. $5 \cdot 4!$ **b.** $(3 \cdot 2)!$ **c.** $\dfrac{7!}{4!}$ **d.** $\dfrac{6!4!}{8!}$

Solution
a. $5 \cdot 4! = 5 \cdot 4 \cdot 3 \cdot 2 \cdot 1 = 5! = 120.$
b. $(3 \cdot 2)! = 6! = 720.$

c. $\dfrac{7!}{4!} = \dfrac{7 \cdot 6 \cdot 5 \cdot 4 \cdot 3 \cdot 2 \cdot 1}{4 \cdot 3 \cdot 2 \cdot 1} = \dfrac{7 \cdot 6 \cdot 5 \cdot \cancel{4} \cdot \cancel{3} \cdot \cancel{2} \cdot \cancel{1}}{\cancel{4} \cdot \cancel{3} \cdot \cancel{2} \cdot \cancel{1}} = 7 \cdot 6 \cdot 5 = 210.$

d. $\dfrac{6!4!}{8!} = \dfrac{6!4!}{8 \cdot 7 \cdot 6!} = \dfrac{4!}{8 \cdot 7} = \dfrac{4 \cdot 3 \cdot 2 \cdot 1}{8 \cdot 7} = \dfrac{\cancel{4} \cdot 3 \cdot \cancel{2} \cdot 1}{8 \cdot 7} = \dfrac{3}{7}.$

Factorial notation enables us to formalize several ideas about permutations. First, we often denote permutations as $_nP_r$ or $P(n, r)$, where n represents the number of objects in the set, and r represents the number of objects to be chosen. r can take on any whole-number value less than or equal to n. If $r = n$, then the number of objects to be chosen is equal to the number of objects in the set. In other words, if $r = n$, then $_nP_n = n \cdot (n - 1) \cdot (n - 2) \cdot \ldots 3 \cdot 2 \cdot 1 = n!$.

> **Theorem 15.12 Number of Permutations of n Objects Chosen n at a Time**
>
> The number of permutations of n distinct objects chosen n at a time is $_nP_n = n!$.

Example 15.34 Assigning Softball Players to Positions

Application

There are nine field positions in the game of softball. If the manager of a softball team has nine players, each of which can play any position, how many ways are there to assign players to positions in the field?

Solution Because any of the nine players can play in any of the nine positions, there are $_9P_9 = 9! = 362{,}880$ ways to put the players in the field.

If $r < n$, then $_nP_r$ is the number of permutations when only a proper subset of the objects is used. In other words, if $r < n$, then we have more objects than positions to fill. For example, the notation $_6P_4$ indicates that we have six objects to fill four positions. To compute this value, we have six choices for the first object, five for the second, four for the third, and three for the fourth. Again, by using the Fundamental Counting Principle, the number of permutations is

$$\underset{\substack{\text{First} \\ \text{choice}}}{6} \cdot \underset{\substack{\text{Second} \\ \text{choice}}}{5} \cdot \underset{\substack{\text{Third} \\ \text{choice}}}{4} \cdot \underset{\substack{\text{Fourth} \\ \text{choice}}}{3} = 360 \text{ permutations}$$

Example 15.35 Choosing Officers in a Sorority

Application

Each fall, a sorority elects a new president, vice president, secretary, and treasurer. If 35 members are in the sorority, how many different ways can the officers be elected?

Solution Four offices have to be filled: president, vice president, secretary, and treasurer. Because each office has different responsibilities and no person holds more than one office, we have an ordered arrangement of objects, or permutations. Assuming the offices are filled in order, there are 35 choices for president, 34 for vice president, 33 for secretary, and 32 for treasurer. By the Fundamental Counting Principle, the number of ways to elect the officers is

$$\underset{\text{president}}{35} \cdot \underset{\text{vice president}}{34} \cdot \underset{\text{secretary}}{33} \cdot \underset{\text{treasurer}}{32} = 1{,}256{,}640$$

Example 15.35 shows that $_{35}P_4 = 35 \cdot 34 \cdot 33 \cdot 32 = 1{,}256{,}640$. However, also notice that

$$_{35}P_4 = 35 \cdot 34 \cdot 33 \cdot 32$$

$$= (35 \cdot 34 \cdot 33 \cdot 32) \cdot \left(\frac{31!}{31!}\right)$$

$$= \frac{35 \cdot 34 \cdot 33 \cdot 32 \cdot (31 \cdot 30 \cdot \ldots \cdot 2 \cdot 1)}{31!}$$

$$= \frac{35!}{31!}$$

Because $35 - 4 = 31$,

$$_{35}P_4 = \frac{35!}{31!} = \frac{35!}{(35 - 4)!}.$$

Consequently, the number of permutations of n objects taken r at a time can also be written in terms of factorials involving n and r.

Theorem 15.13   Number of Permutations

The number of permutations of n distinct objects chosen r at a time, or $_nP_r$, is

$$_nP_r = \frac{n!}{(n - r)!}.$$

Example 15.36   Proof of Theorem 15.13

Reasoning

Show that $_nP_r = \dfrac{n!}{(n - r)!}.$

Solution To justify Theorem 15.13, suppose we are given n objects, of which r are to be chosen. There are n objects for the first choice, $(n - 1)$ for the second, and so on down to $n - r + 1$ objects for the last. Consequently,

$$_nP_r = n \cdot (n - 1) \cdot (n - 2) \cdot \ldots \cdot (n - r + 1)$$

$$= \frac{[n \cdot (n - 1) \cdot (n - 2) \cdot \ldots \cdot (n - r + 1)][(n - r) \cdot (n - r - 1) \cdot \ldots \cdot 3 \cdot 2 \cdot 1)]}{[(n - r) \cdot (n - r - 1) \cdot \ldots \cdot 3 \cdot 2 \cdot 1)]}$$

$$= \frac{n!}{(n - r)!}$$

Notice that if $r = n$, then $_nP_n = \dfrac{n!}{(n - n)!} = \dfrac{n!}{0!} = \dfrac{n!}{1} = n!$. Consequently, in this situation, it is necessary for $0! = 1$ so that we do not divide by 0.

Example 15.37   Counting Possible Lottery Numbers

Application

A lottery game has the numbers 1–40 written on ping-pong balls. Six balls are selected at random to determine the winning list of numbers. To win the jackpot, a player must match all six numbers in order. How many different possible number arrangements are there?

Solution There are 40 numbers, of which 6 are selected in order without repetition. Consequently, the number of possible arrangements is

$$_{40}P_6 = \frac{40!}{(40-6)!} = \frac{40!}{34!} = 2{,}763{,}633{,}600$$

As with factorials, computations with permutations often involve products with large numbers. We can make these computations easily and efficiently by using the permutation key on a calculator, which is often denoted as $\boxed{_nP_r}$ or $\boxed{P(n,r)}$. If a calculator has the first key, we usually enter n first, then press the permutation key, and then r. If a calculator has the second key, we press the permutation key first, followed by n, a comma, r, a parenthesis, and then enter.

The next example shows that permutations can be used to compute a probability.

Example 15.38 | Permutations and Probability

A daily lotto number consists of three randomly drawn digits. What is the probability that the daily lotto number contains a 5, a 6, and an 8?

Solution The daily lotto number is to have a 5, a 6, and an 8. Because the order of the digits can lead to different numbers, we are looking for the permutations of the digits, or $_3P_3 = 3! = 6$. This gives us the number of successful outcomes. We can find the total number of outcomes by counting three-digit numbers. Because there are 10 possible digits for each position, there are $10 \cdot 10 \cdot 10 = 1{,}000$ possible three-digit numbers. As a result, the probability of getting a lotto number with a 5, a 6, and an 8 is $\frac{6}{1{,}000} = \frac{3}{500}$.

Check Your Understanding 15.4-A

1. How many ways can three cards be drawn from a deck of 52 cards if they are drawn in succession with replacement?

2. A restaurant offers 7 appetizers, 15 main courses, 8 desserts, and 5 beverages? How many different meals are possible at the restaurant?

3. Compute each expression.

   a. $6 \cdot 5!$ **b.** $4! \cdot 4!$ **c.** $\dfrac{7!3!}{9!}$ **d.** $_8P_4$ **e.** $_{14}P_5$

4. I want to read 5 of 15 different books. If I am particular about the order in which I read, how many ways can I read the five books?

Talk About It How are situations that use the Fundamental Counting Principle different from the situations that use a permutation?

Activity 15.8
Because factorials grow large so quickly, it does not take long before they are too large to be computed with a calculator.

a. What is the largest factorial that your calculator can compute without using scientific notation?

b. What is the largest factorial your calculator can compute without giving an error message?

c. Use your calculator to compute 75!/70!. What happens? Can this value be computed by hand? If so, perform the calculation.

Permutations of Like Objects

In all the permutations so far, the objects have been distinct. However, objects can also be **like**, or identical; that is, they cannot be differentiated. For instance, consider the arrangements of the letters in the word "BEE." If we make one "E" distinct from the other by making one red and the other blue (Figure 15.12), then there are $_3P_3 = 3! = 6$ different arrangements of the letters. However, removing the colors makes some of the arrangements identical, reducing the number of arrangements from six to three.

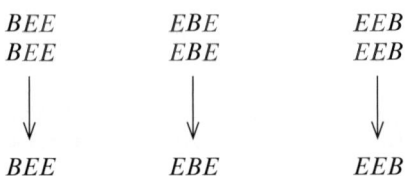

Figure 15.12 Arrangements with like letters

Although the example implies that we can find the number of arrangements of like objects by taking half the number of permutations, this is not always the case. We are actually dividing the number of permutations by the number of ways that the two "E"s can be arranged, which happens to be $2! = 2$. Increasing the number of like objects leads to more divisors. For example, consider the arrangements of the letters in the word "WEDDED." If each letter were distinct, there would be $6! = 720$ different arrangements. However, the word contains two kinds of like objects: E and D. If the "E"s were distinct, we could arrange them in $2! = 2$ different ways. Similarly, if the "D"s were distinct, we could arrange them in $3! = 6$ different ways. However, because they are not distinct, we compensate for any duplicate arrangements of the letters by dividing by the number of arrangements of the "E"s and the "D"s, or $\frac{6!}{2!3!} = \frac{720}{2 \cdot 6} = \frac{720}{12} = 60$. Generalizing this idea leads to the next theorem.

Theorem 15.14   Number of Permutations of Like Objects

The number of arrangements of n objects, of which r_1 are of one kind, r_2 are of another, and so on through r_k, is $\dfrac{n!}{r_1! r_2! r_3! \ldots r_k!}$.

Example 15.39   Arrangements of the Digits in a Number

How many numbers have the same digits as the number 1,223,334,444?

Solution We can find our answer by counting the arrangements of the digits. The digits of each type are identical, and there are one 1, two 2s, three 3s, and four 4s. Consequently, by Theorem 15.14, the number of arrangements is

$$\frac{10!}{1! 2! 3! 4!} = \frac{3,628,800}{1 \cdot 2 \cdot 6 \cdot 24} = \frac{3,628,800}{288} = 12,600.$$

Like the Fundamental Counting Principle and permutations, permutations of like objects can be used to compute probabilities.

Application

If the letters from the word "STATES" are written on chips, placed in a hat, and drawn in succession, what is the probability that the resulting word spells "TASTES"?

Solution Because the "S"s and "T"s are identical, there is only one way to get the word "TASTES" from the letters in the word "STATES." So the numerator of the probability must be 1. The denominator is the number of arrangements of the letters in the word "STATES," or $\frac{6!}{1!2!2!1!} = \frac{720}{2 \cdot 2} = \frac{720}{4} = 180$. As a result, the probability of getting the word "TASTES" is $\frac{1}{180}$.

Combinations

Whether they involve like or unlike objects, order is important in all permutations. However, in other arrangements, order is not important. Any arrangement of *distinct* objects with *no repetitions* and in which order is *not* important is a **combination.** Because order does not matter, the number of arrangements must be lower than if it did. We can therefore find the number of combinations by starting with permutations and removing any duplicate arrangements. However, unlike permutations of like objects, the duplicate arrangements are based not on like objects, but rather on the permutations that contain the same objects. In other words, counting the combinations of a set with n objects is the same as counting subsets of a particular size.

Explorations Manual 15.12

For example, suppose a two-person committee is to be formed from four people: Amy, Bob, Colleen, and Dan. If each person has a specific task for the committee, then order is important, and we get the 12 two-person committees listed in Figure 15.13. However, if the people do not have specific tasks, then all that matters is who is on the committee. In other words, the (Amy, Bob) committee and the (Bob, Amy) committee become the same committee, and we count it only once. As a result, the permutations are condensed to sets of two people, giving us a total of six combinations, or $_4C_2 = 6$. The notation $_nC_r$ represents the number of combinations of n objects chosen r at a time. Other common notations for combinations are $C(n, r)$ and $\binom{n}{r}$.

Permutations of Two-Person Committees			Combinations of Two-Person Committees
(Amy, Bob)	(Bob, Amy)	$\longrightarrow$	{Amy, Bob}
(Amy, Colleen)	(Colleen, Amy)	$\longrightarrow$	{Amy, Colleen}
(Amy, Dan)	(Dan, Amy)	$\longrightarrow$	{Amy, Dan}
(Bob, Colleen)	(Colleen, Bob)	$\longrightarrow$	{Bob, Colleen}
(Bob, Dan)	(Dan, Bob)	$\longrightarrow$	{Bob, Dan}
(Colleen, Dan)	(Dan, Colleen)	$\longrightarrow$	{Colleen, Dan}

Figure 15.13 Permutations condensed into combinations

The example suggests halving the number of permutations to get the number of combinations. However, this is not actually the case. Instead, we divide the number of permutations by the number of arrangements of the subset of the given size. For instance, suppose we were to select a three-person committee instead of a two-person one. In this case, the six permutations shown in Figure 15.14 become one combination.

(Amy, Bob, Colleen)   (Bob, Colleen, Amy)

(Amy, Colleen, Bob)   (Colleen, Amy, Bob) $\longrightarrow$ {Amy, Bob, Colleen}

(Bob, Amy, Colleen)   (Colleen, Amy, Bob)

Figure 15.14 Six permutations condensed to one

So, to find the number of three-person committees from a four-person set, we take the total number of permutations and divide by the number of permutations of a three-person set, or $_3P_3 = 3!$. Consequently, the number of combinations is

$$_4C_3 = \frac{_4P_3}{_3P_3} = \frac{4 \cdot 3 \cdot 2}{3!} = \frac{24}{6} = 4$$

This result can be generalized in the following way:

$$_nC_r = \frac{_nP_r}{_rP_r} = \frac{\frac{n!}{(n-r)!}}{r!} = \frac{n!}{r!(n-r)!}$$

where $_rP_r = r!$ represents the number of permutations of a subset of a given size.

Theorem 15.15   Number of Combinations

The number of combinations of n distinct objects chosen r at a time, or $_nC_r$, is given by $_nC_r = \dfrac{n!}{r!(n-r)!}$.

Example 15.41   Computing Combinations

Compute each combination.

a. $_{11}C_7$ **b.** $_5C_0$ **c.** $_6C_6$

Solution

a. $_{11}C_7 = \dfrac{11!}{7!(11-7)!} = \dfrac{11!}{7!4!} = \dfrac{11 \cdot 10 \cdot \overset{3}{9} \cdot 8 \cdot 7!}{7!(4 \cdot 3 \cdot 2 \cdot 1)} = 330.$

b. $_5C_0 = \dfrac{5!}{0!(5-0)!} = \dfrac{5!}{0!5!} = \dfrac{1}{1} = 1.$

c. $_6C_6 = \dfrac{6!}{6!(6-6)!} = \dfrac{6!}{6!0!} = \dfrac{1}{1} = 1.$

Notice that in general, $_nC_0 = 1$ and $_nC_n = 1$. The proofs of these relatively easy-to-verify facts are left to the Learning Assessment. Also, any calculator with a permutation key is likely to have a combination key, which is used in a similar fashion.

Example 15.42   Selecting Winners for a Free Cruise

There are 25 finalists in a sweepstakes for a free Caribbean cruise. If only 10 of the finalists are chosen to go, how many ways are there to pick the winners?

Solution Order is not important because all that matters is who is selected to go. It also makes sense that there are no duplicate selections. Because order is not important and there are no repetitions, we have a combination of 25 people chosen 10 at a time. So, there are $_{25}C_{10} = 3{,}268{,}760$ ways to pick the winners.

Example 15.43   Selecting Classroom Representatives

A class with 13 boys and 12 girls is selecting two classroom representatives for the student government. How many different ways are there to select two females?

Solution The two representatives do not have specific tasks, so order is not important. Also, the representatives cannot be the same person, so repetition is not an issue. As a result, the number of ways to select two girls as representatives is equal to the number of combinations of 12 girls chosen 2 at a time, or $_{12}C_2 = 66$.

Example 15.44   Combinations and Probability

What is the probability of getting four aces in a four-card hand?

Solution Only the cards in the hand are important, not the order in which they are dealt, so we compute the probability by using combinations. Because there are only four aces, the number of ways to get all four of them is $_4C_4 = 1$. The total number of four-card hands is the combination of 52 cards chosen 4 at a time, or

$$_{52}C_4 = \frac{52!}{4!(52-4)!} = \frac{52!}{4!48!} = \frac{52 \cdot 51 \cdot 50 \cdot 49 \cdot 48!}{4!48!} = \frac{6,497,400}{24} = 270,725$$

Consequently, the probability of getting four aces is $\frac{1}{270,725}$.

To this point, we have dealt with the counting techniques one at a time, so it has been relatively easy to determine which one to use in problem situations. However, in a more general setting, determining the correct technique is more difficult. To determine the best technique, we can ask ourselves three questions:

1. Are repetitions allowed?
2. Are the objects like or indistinguishable?
3. Is order important?

Based on the answers, we can select the most appropriate technique by using the information in Table 15.4.

Table 15.4   Selecting a counting technique

Counting Technique	Repetitions	Like Objects	Order Important
Fundamental Counting Principle	Yes	No	Yes
Permutations	No	No	Yes
Permutations of like objects	No	Yes	Yes
Combinations	No	No	No

In some cases, several techniques may be needed to solve a counting problem. Consider the next two examples.

Example 15.45   Tunes for a Road Trip

Application

Katie is taking a road trip and plans to take some of her favorite tunes. She has 18 rock CDs and 12 country CDs, but she is going to take only five of each. How many different ways can Katie take her tunes?

Solution First, we must find two different numbers: the number of ways Katie can take the rock CDs and the number of ways she can take the country CDs. If we find these two numbers, according to the Fundamental Counting Principle, we can multiply them to find the total number of ways she can take her tunes. First consider her rock CDs. Order is not important; all that matters is which CDs are selected. Hence, the number of ways she can take her rock CDs is given by the combination of 18 CDs chosen 5 at time, or $_{18}C_5$. Similarly, the number of ways she can take her country CDs is given by the combination of 12 CDs chosen 5 at a time, or $_{12}C_5$. As a result, the total number of ways she can take her tunes is

$$\underbrace{_{18}C_5}_{\substack{\text{Number of} \\ \text{rock CDS}}} \cdot \underbrace{_{12}C_5}_{\substack{\text{Number of} \\ \text{country CDs}}} = (8{,}568) \cdot (792) = 6{,}785{,}856$$

Example 15.46　The Probability of a Flush

Application

What is the probability of getting a flush when five cards are dealt from a deck of 52 cards?

Solution In cards, a flush is a five-card hand that is all the same suit, either hearts, diamonds, clubs, or spades. Only the suit matters and not the values of the cards. Because the order of cards does not matter, the number of flushes for any one suit is $_{13}C_5 = 1{,}287$. Because there are four suits, the Fundamental Counting Principle tells us that there must be $4 \cdot 1{,}287 = 5{,}148$ possible flushes. This is the numerator for the probability. The denominator is the total number of five-card hands. Again order is not important, so this number is $_{52}C_5 = 2{,}598{,}960$. As a result, the probability of getting a flush is $\dfrac{5{,}148}{2{,}598{,}960} \approx 0.00198$.

Check Your Understanding 15.4-B

1. How many different arrangements are there of the letters in the word "LETTER"?

2. Compute each combination.

　a. $_5C_2$　　**b.** $_{13}C_5$　　**c.** $_7C_0$　　**d.** $_8C_8$

3. How many ways can a committee of four people be chosen from a group of 15 people?

4. If a two-person committee is to be chosen from a group of 13 men and 8 women, what is the probability that the committee contains two men?

Talk About It How are situations in which a permutation is used different from situations in which a combination is used?

Problem Solving

Activity 15.9

In cards, a three-of-a-kind is a hand with three cards with the same value, such as three 2s, three 10s, or three queens. The other two cards in the five-card hand must be different from the three-of-a-kind, otherwise the hand would have a four-of-a-kind. The other two cards must also be different from each other because if they were the same, then the hand would be a full house, three-of-a-kind and a pair. What is the probability of getting a three-of-a-kind when five cards are dealt?

SECTION **15.4** Learning Assessment

▪ Understanding the Concepts

1. Under what conditions can the Fundamental Counting Principle be used to find the total number of outcomes in an experiment?

2. How is the Fundamental Counting Principle used to compute multistage probabilities?

3. **a.** What is a permutation?
   b. Why is counting permutations a special case of the Fundamental Counting Principle?

4. **a.** What is $n!$?
   b. Why does it make sense to define $0! = 1$?

5. How are situations in which permutations are used different from situations in which permutations of like objects are used?

6. What is the difference between a permutation and a combination?

7. In any counting problem, what three things should be considered to determine which counting technique to use?

▪ Representing the Mathematics

8. **a.** Use a tree diagram to list the permutations of the letters in the word "AND."
   b. Use a tree diagram to list the permutations of the letters in the word "PLOP."
   c. What considerations did you have to take into account in part b that you did not in part a? How did it affect the tree diagram?

9. Give a verbal translation of each expression.
   a. $_5P_2$ **b.** $P(7, 4)$ **c.** $_6C_3$
   d. $C(8, 5)$ **e.** $\binom{7}{2}$

10. Write each verbal expression in symbolic form.
    a. The number of permutations of eight objects chosen two at a time.
    b. The number of combinations of six objects chosen three at a time.
    c. The product of every natural number from seven down to one inclusively.
    d. The product of every natural number from five down to three inclusively.

11. List the sequence of keys that need to be pressed on your calculator to compute each expression.
    a. $6!$ **b.** $\dfrac{8!}{4!}$ **c.** $_7P_5$ **d.** $_6C_2$

▪ Exercises

12. Evaluate each expression.
    a. $10!$ **b.** $15!$ **c.** $24!$ **d.** $4! \, 3!$ **e.** $6! + 5!$

13. Evaluate each expression.
    a. $\dfrac{5!}{3!}$ **b.** $\dfrac{7!}{3!6!}$ **c.** $\dfrac{(3+5)!}{(7-5)!}$
    d. $\dfrac{5!\,12!}{4!\,16!}$ **e.** $\dfrac{6 \cdot 5!}{6!}$

14. Evaluate each expression.
    a. $_{12}P_8$ **b.** $_{13}P_{11}$ **c.** $_{14}C_6$ **d.** $_{25}C_{14}$

15. Evaluate each expression.
    a. $_{35}P_6$ **b.** $_{43}P_{14}$ **c.** $_{36}C_{12}$ **d.** $_{47}C_{25}$

Use the Fundamental Counting Principle to answer Questions 16–20.

16. If four distinct six-sided dice are rolled, how many different outcomes are possible?

17. If a four-sided, six-sided, eight-sided, 12-sided, and 20-sided die are rolled at the same time, how many different outcomes are possible?

18. If a telephone number cannot begin with a 0, how many seven-digit phone numbers are possible within a given area code?

19. The Greek alphabet consists of 24 letters.
    a. If repetition is allowed, how many five-letter "words" are possible?
    b. If repetition is not allowed, how many five-letter "words" are possible?

20. A switching panel has 50 switches, each of which can be set to on or off. How many different ways can the switching panel be set?

Use permutations to answer Questions 21–25.

21. A combination lock has a three-number combination. How many combinations are possible if the combination must use three different numbers from 1 to 40?

22. If eight teams are running in a relay race, how many ways can first, second, and third place be awarded?

23. Twenty singers are auditioning for a musical. In how many different ways can the director choose a lead singer and a backup for a duet?

24. There are five positions in the game of basketball. If a coach has 11 players who can play any position, how many ways can the coach assign players to positions?

25. How many different seven-note "songs" are there if the notes A, B, C, D, E, F, and G are used only once?

Use permutations of like objects to answer Questions 26–29.

26. How many different arrangements are there of the letters in each word?
    a. TENNESSEE
    b. STATISTICS
    c. BEGINNING

27. How many different arrangements are there of 15 chips, if 3 are red, 5 are blue, and 7 are white?

28. How many different ways can 3 quarters, 4 dimes, and 5 nickels be arranged if the coins are indistinguishable except for their values?

29. A teacher has 10 identical algebra books and 11 identical geometry books. How many ways can the 21 books be arranged on a shelf?

Use combinations to answer Questions 30–34.

30. Timmy has 15 children's books in his collection. He allows Ashley to borrow four of them. How many different ways can she do so?

31. A teacher wants to write a short quiz of 10 questions. The software she uses has a pool of 65 questions to choose from. How many different quizzes can be made using the pool if the order of the questions does not matter?

32. How many 12-person juries can be selected from a juror pool of 43 people?

33. How many hands can be dealt from a deck of 52 cards if:
    a. Three cards are dealt?
    b. Four cards are dealt?
    c. Five cards are dealt?

34. A pizza parlor advertises pizzas with up to 10 toppings. How many different ways are there to get a pizza with:
    a. Three toppings?
    b. Four toppings?
    c. Five toppings?
    d. Six toppings?

■ Problems and Applications

35. All radio stations have call letters that begin with a "K" or "W" and consist of either three or four total letters. How many sets of call letters are possible?

36. Thirty-three people attend a meeting, and each person shakes hands with every other person. How many handshakes take place?

37. Forty-five handshakes take place at a business meeting. If every person at the meeting shook hands with every other person exactly once, how many people were at the meeting?

38. a. How many three-digit numbers can be made from the digits 2, 3, 5, 6, and 7 if no repetition is allowed?
    b. How many of the three-digit numbers in part a are even?
    c. How many three-digit numbers can be made from the digits 2, 3, 5, 6, and 7 if repetition is allowed?
    d. How many of the three-digit numbers in part c are even?

39. A college plans to send a team of four faculty delegates to a national conference on increasing student diversity. If 15 male and 17 female faculty members have volunteered to go, how many ways can a team of two males and two females be selected?

40. A teacher wants to organize her students by gender. She has 15 boys and 11 girls. How many seating arrangements are there if the boys are seated on the right side of the class and the girls are seated on the left?

41. How many five-card hands have exactly one pair?

42. How many ways are there to draw a full house (two of a kind and three of a kind) from a deck of 52 cards?

43. a. How many ways can four people be seated along one side of a rectangular table?
    b. How does the number of arrangements change if the same four people are seated around a circular table?

44. A pizza restaurant claims that it can make up to 125 different types of pizzas. If the restaurant has only seven toppings, is this possible?

45. a. How many three-letter initials are possible?
    b. How many three-letter initials are possible if the letters are all the same?
    c. If a three-letter initial is selected at random, what is the probability that all the letters are the same?

46. Consider the permutations of the numbers 1, 2, 3, 4, and 5. If one permutation is selected at random, what is the probability that the 2 and 3 are adjacent to one another?

47. If the letters in the word "TOOT" are rearranged at random, what is the probability that they will spell "OTTO"?

48. An urn contains 2 red, 3 white, and 5 blue chips. If all of the chips are drawn in succession, what is the probability that the two red chips are adjacent to one another?

49. A temporary work agency has a pool of 23 workers in which ten are Caucasian, seven are African

American, and six are Asian. On Monday morning, the agency receives a request for six workers. If workers are selected from the pool at random, what is the probability that:

  a. All six of the workers are African American?

  b. Three Asian workers and three African American workers are selected?

  c. Two workers from each ethnic group are selected?

50. A regular six-sided die is rolled five times. What is the probability that three 1s and two 4s are rolled?

51. A Social Security number is a nine-digit number using the digits from 0 through 9.

  a. How many Social Security numbers are possible?

  b. How many Social Security numbers contain exactly one 3?

  c. If a person is assigned a Social Security number at random, what is the probability that the number will have exactly one 3?

52. a. How many five-letter "words" can be made from the 26 letters of the English alphabet?

  b. If one such "word" is chosen at random, what is the probability that it contains only vowels?

■ Mathematical Reasoning and Proof

53. Is it possible to determine whether the number 30, 414,093,201,713,378,043,612,608,166,064,768,8 44,377,641,568,960,512,078,291,027,000 is the value of 50! without performing the calculation? Explain.

54. Compute each of the following permutations.

  a. $_3P_1$ **b.** $_4P_1$ **c.** $_5P_1$ **d.** $_6P_1$

  e. Based on your results, what do you suppose $_nP_1$ is equal to? Can you prove it?

55. Compute each of the following permutations.

  a. $_3P_0$ **b.** $_4P_0$ **c.** $_5P_0$ **d.** $_6P_0$

  e. Based on your results, what do you suppose $_nP_0$ is equal to? Can you prove it?

56. Is it possible to evaluate $_3P_7$? Why or why not?

57. Show that $_nC_0 = 1$ for $n \geq 1$.

58. Show that $_nC_n = 1$ for $n \geq 1$.

59. Does $(n + m)! = n! + m!$? If so, write an argument to explain why. If not, give a counterexample.

60. Does $(n \cdot m)! = n! \cdot m!$? If so, write an argument to explain why. If not, give a counterexample.

61. Does $P(n, r) = P(n, n - r)$ for $0 \leq r \leq n$? If so, write an argument to explain why. If not, give a counterexample.

62. Does $C(n, r) = C(n, n - r)$ for $0 \leq r \leq n$? If so, write an argument to explain why. If not, give a counterexample.

■ Mathematical Communication

63. Write a mathematical statement that extends the Fundamental Counting Principle to more than two events.

64. Many combination locks require three numbers to be entered in a particular sequence to open the lock. Is this an appropriate use of the word "combination"? Why or why not?

65. a. Write a story problem that is solved by using the Fundamental Counting Principle.

  b. Write a story problem that is solved by counting permutations.

  c. How is the language in the first question different from the language in the second?

66. a. Write a story problem that is solved by counting permutations.

  b. Write a story problem that is solved by counting combinations.

  c. How is the language in the first question different from the language in the second?

67. Levi, a student of yours, is having trouble deciding which counting technique to use for the following problem. How do you help him?
Nine plants are to be placed in rows in a flowerbed. If there are five tulips that are identical and four irises that are identical, how many different ways can the flowers be planted?

68. Sarah claims that 4! must equal 10 because 3! is equal to 6. What has she done wrong?

■ Building Connections

69. The counting techniques introduced in this section can be used to count a wide variety of mathematical ideas. For instance, we can use them to count the number of functions from one set into another. Answer each question, and then state the counting technique you used.

  a. How many functions are there from a set of four elements into a set of three elements?

  b. How many functions are there from a set of 10 elements into a set of five elements?

  c. How many different one-to-one correspondences are there between two sets that each contain five elements?

70. a. How is counting combinations similar to counting subsets of a certain size?

  b. An experiment is done in which one subset is selected at random from a set with six elements. What is the probability that the subset will have three elements?

71. Consider the following array of combinations.

$$C(1, 0) \quad C(1, 1)$$
$$C(2, 0) \quad C(2, 1) \quad C(2, 2)$$
$$C(3, 0) \quad C(3, 1) \quad C(3, 2) \quad C(3, 3)$$
$$C(4, 0) \quad C(4, 1) \quad C(4, 2) \quad C(4, 3) \quad C(4, 3)$$
$$\bullet \qquad \bullet \qquad \bullet$$
$$\bullet \qquad \bullet \qquad \bullet$$
$$\bullet \qquad \bullet \qquad \bullet$$

a. What are the combinations in the next three rows?

b. Compute the values of the combinations, and compare your results to Pascal's triangle (Section 1.2). What do you notice?

c. Describe how Pascal's triangle can be used to compute combinations.

72. a. Look at a set of mathematics curriculum materials for grades 5–8. At what grade level are counting techniques like the Fundamental Counting Principle introduced? Does this surprise you? Why?

b. If these ideas are important to counting, and counting is fundamental to elementary mathematics, why do you think these ideas are not introduced until the middle grades?

Historical Note

James Bernoulli (1654–1705) was one of the pioneers of probability theory. He was well educated and familiarized himself with the works of the best mathematicians as he traveled extensively throughout Europe. Although he made many contributions to probability theory, his most notable is Bernoulli's Theorem, or the Law of Large Numbers. The theorem was written in a text that Bernoulli had worked on for over 20 years but never fully completed. The text promised to apply the Law of Large Numbers to economics and politics, which would have extended probability theory well beyond its application to games of chance. Unfortunately, the work was not discovered until after Bernoulli's death, so the applications had to be discovered by other mathematicians.

James Bernoulli (1654–1705)

© INTERFOTO / Alamy

Opening Problem Answer

Jamie Roach / Shutterstock.com

First, consider how many games are played if each team plays every other team once. This is equivalent to finding the number of two-element subsets from a set of 12 objects, or $_{12}C_2 = 66$ games. Doubling this number, there are $2 \cdot 66 = 132$ games played during the season.

Chapter 15 REVIEW

Summary of Key Concepts from Chapter 15

Section 15.1 Experimental Probability and Making Predictions from Data

- A **probability** measures the extent to which an event will or will not happen.

- An **experiment** is any activity with results that can be observed, measured, and recorded. The possible results of the experiment are called **outcomes**. An **event** is a particular outcome or set of outcomes.

- If an experiment is repeated n times, with a specific event E occurring m times, then the **experimental probability** that event E occurs on a given trial is $\dfrac{m}{n}$.

- **Law of Large Numbers:** As the number of trials for an experiment increases, the variation in the experimental probability of an event decreases until it closely approximates a fixed number.

- A **simulation** models an experiment by using a situation with outcomes analogous to the experiment but for which the trials are easily conducted and recorded.

- A **table of random digits** is a table in which the digits 0 through 9 are repeatedly listed in a random fashion.

Section 15.2   Theoretical Probability

- The **sample space** for an experiment is the set of all possible outcomes.

- If S is the sample space of an experiment with equally likely outcomes, then the **theoretical probability** of an event A is $P(A) = \dfrac{n(A)}{n(S)}$.

- An event that does not have outcomes in the sample space is called an **impossible event,** and it always has a probability of 0.

- Any event that is equal to the sample space is a called a **certain event,** and it always has a probability of 1.

- For any event A, $0 \le P(A) \le 1$.

- Two events with no outcomes in common are called **mutually exclusive.** If A and B are mutually exclusive, then $P(A \cup B) = P(A) + P(B)$.

- Two events are **complementary** if they are mutually exclusive and their union equals the sample space. If A and $\overline{A}$ are complementary events, then $P(A) + P(\overline{A}) = 1$.

- Two events with outcomes in common are called **nonmutually exclusive.** If A and B are nonmutually exclusive events, then $P(A \cup B) = P(A) + P(B) - P(A \cap B)$.

- Two events are **independent** if the outcome of one event has no influence over the outcome of the other. If A and B are independent, then $P(A \cap B) = P(A) \cdot P(B)$.

Section 15.3   Conditional and Geometric Probability, Odds, and Expected Value

- The **conditional probability** that event B occurs given that event A has occurred is $P(B|A) = \dfrac{P(A \cap B)}{P(A)}$.

- In a continuous sample space, the probability of an event A is $P(A) = \dfrac{\text{measure of } A}{\text{measure of } S}$.

- Let A be an event, and suppose $\overline{A}$ is its complement. The **odds in favor** of event A is the ratio $n(A) : n(\overline{A})$, and the **odds against** event A is the ratio $n(\overline{A}) : n(A)$.

- If $s{:}u$ represents the odds in favor of an event A, or if $u{:}s$ represents the odds against an event A, then the probability of A is $P(A) = \dfrac{s}{s + u}$.

- The odds in favor of an event A are $P(A) : [1 - P(A)]$, and the odds against are $[1 - P(A)] : P(A)$.

- If an experiment has outcomes of values $a_1, a_2, \ldots a_n$, occurring with probabilities $p_1, p_2, \ldots p_n$, respectively, then the **mathematical expectation,** or **expected value** E, of the experiment is $E = a_1 p_1 + a_2 p_2 + \ldots + a_n p_n$.

- If the expected value is 0, then the game is considered to be **fair.**

Section 15.4   Counting Techniques and Probability

- **Fundamental Counting Principle:** If an event A can occur in m ways and an event B can occur in n ways, then the events A and B can occur in succession, first A, then B, in $m \cdot n$ ways.

- A **permutation** is an ordered arrangement of distinct objects with no repetitions. The number of permutations of n objects chosen r at a time is $_nP_r = \dfrac{n!}{(n - r)!}$.

- For $n > 0$, $n! = n \cdot (n-1) \cdot (n-2) \cdot \ldots \cdot 3 \cdot 2 \cdot 1$. $0!$ is defined to be 1.
- The number of arrangements of n objects, of which r_1 are of one kind, r_2 are of another kind, and so on through r_k, is $\dfrac{n!}{r_1! r_2! r_3! \ldots r_k!}$.
- A **combination** is an arrangement of distinct objects with no repetitions in which order is not important. The number of combinations of n objects chosen r at a time is $_nC_r = \dfrac{n!}{r!(n-r)!}$.

Review Exercises Chapter 15

1. A six-sided die is rolled 100 times with the results recorded in the table.

Outcome	1	2	3	4	5	6
Times Rolled	15	21	14	18	15	17

   Find the experimental probability that:
   a. A 5 is rolled.
   b. An even number is rolled.
   c. A multiple of 3 is rolled.

2. Toss a quarter, a dime, and a nickel 50 times, and record the results. Use your results to find the experimental probability that:
   a. The three coins are all heads.
   b. Two coins are heads, and one is tails.

3. a. Consider a six-sided die. Find the experimental probability that a 3 is rolled before the third roll.
   b. On average how many times must the die be rolled before a 3 is rolled?

4. Conduct an experiment to find the probability that a teaspoon will land cup up when tossed into the air. Conduct it first with 10 tosses, then 25 tosses, 50 tosses, and 100 tosses. Based on your results, what do you think is a reasonable estimate for the probability of a teaspoon landing cup up?

5. It is estimated that 35% of the employees at a certain business sleep on the job. The company wants to monitor the misuse of company time by making random office checks.
   a. How can a table of random digits be used to randomly select employees?
   b. If 10 employees are to be selected, what is the probability that four of them will be asleep?

6. A quality control inspector examines a new batch of watches in which 10% are expected to have some sort of defect. Devise a simulation using a 10-sided die that can be used to determine the probability that at least one watch will have a defect when six watches are tested at random.

7. If you roll a six-sided die 10 times and each time it comes up even, what can you conclude about the outcome of the 11th roll?

8. List the sample space for the experiment of rolling two four-sided dice. What are the outcomes in the event of rolling:
   a. An odd sum?
   b. A sum divisible by 4?
   c. A sum divisible by 2 or 3?

9. List the sample space for the experiment of tossing three coins.
   a. What are the outcomes in the event that all the coins are the same?
   b. What are the outcomes in the event that exactly one head is showing?
   c. What are the outcomes in the event that two or more heads are showing?

10. One card is dealt from a well shuffled deck of 52 cards. What is the probability that the card is:
    a. Black?   b. A club?
    c. A red two?   d. The 10 of hearts?

11. Suppose the numbers 1–25 are written on chips and placed in an urn. If one chip is selected at random, what is the probability that the number:
    a. Is less than 10?   b. Is divisible by 5?
    c. Has a 2 for a digit?

12. State the complement of each event.
    a. An odd number is rolled on a 10-sided die.
    b. At least one head is obtained on two flips of a coin.
    c. Twenty-five percent of the units produced by a factory are defective.

13. Two cards are dealt from a well-shuffled deck of 52 cards. What is the probability that the cards are:
    a. Both clubs?   b. Both kings?
    c. Not spades?

14. Suppose that two dice are rolled, but instead of adding the two numbers, they are multiplied.

a. What are the possible outcomes of this experiment?

b. Find the probability associated with each outcome.

c. Are the outcomes equally likely? If not, which outcomes are the most likely?

15. An urn contains 5 red, 7 green, and 8 white chips. How many green chips must be added so that the probability of drawing a green chip is $\frac{1}{2}$?

16. An urn contains 15 red, 18 white, and 13 blue chips. If 7 chips are drawn at random, what is the probability that at least 3 are of the same color?

17. A weather report states that the chance of rain for the next four days is 25% for each day. What is the probability that it rains on the first and fourth days but not the second or third?

18. If a person takes a five-question, multiple-choice test where each question has five possible answers, what is the probability of the person getting all five questions correct by guessing on each question?

19. a. If $P(A) = 0.5$, $P(B) = 0.35$, and $P(A \text{ and } B) = 0.1$, what is $P(A \text{ or } B)$?

b. If $P(A) = 0.85$, $P(B) = 0.55$, and $P(A \text{ or } B) = 0.95$, what is $P(A \text{ and } B)$?

20. Suppose A and B are events in the same sample space S. If $P(A) = 0.45$ and $P(B) = 0.45$, must A and B be mutually exclusive? Explain.

21. A single card is drawn from a deck of 52 cards. If the card is known to be red, what is the probability that the card is:

a. A heart? **b.** An ace?

c. A face card? **d.** The jack of spades?

22. A box contains 15 chips: 6 red, 4 blue, and 5 white. A single chip is drawn at random and then discarded. A second chip is then drawn. What is the probability that the second chip is:

a. Red, if the first chip was red?

b. Blue, if the first chip was red?

c. Red, if the first chip was white?

d. Red or blue, if the first chip was white?

23. One dart is thrown at random at the following odd dartboard.

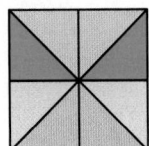

What is the probability that the dart hits:

a. A blue region?

b. A yellow or blue region?

24. An experiment is conducted in which a 12-sided die is rolled.

a. What are the odds in favor of getting a 6?

b. What are the odds in favor of rolling an even number?

c. What are the odds against rolling a number divisible by 3?

d. What are the odds against rolling a number between 5 and 9?

25. In each part, the odds in favor or against an event A are given. Find the probability that event A occurs.

a. Odds in favor of A are 1:4.

b. Odds in favor of A are 8:3.

c. Odds against A are 5:2.

d. Odds against A are 3:6.

26. In each part, a probability for an event A is given. Find the odds in favor of and against event A occurring.

a. $P(A) = \dfrac{4}{5}$ **b.** $P(A) = \dfrac{2}{9}$

c. $P(A) = 47\%$ **d.** $P(A) = 0.37$

27. The weather forecast for today states that there is a 30% chance of rain. What are the odds in favor of it raining today?

28. Two four-sided dice are rolled, and the results added. The amounts won or lost for each sum are shown:

Sum of Dice	2	3	4	5	6	7	8
Amount Won or Lost	–$10	–$5	–$2	$0	$2	$3	$15

What is the expected value of the game? Would you play it?

29. An oil company has a field that it plans to drill. It estimates the cost of drilling to be about $20,000,000. The company estimates a 0.47 probability that the field will have a small yield worth $12,500,000, a 0.23 probability that it will have a medium yield worth $18,500,000, a 0.25 probability that it will have a large yield worth $59,000,000, and a 0.05 probability that it will have very large yield worth $123,000,000. Is it worth the company's time to drill on this plot of land?

30. Consider an experiment in which an eight-sided die is rolled. Give an example of an event A so that the odds in favor of A are those shown.

a. 1:1 **b.** 1:3 **c.** 1:7 **d.** 3:1

31. Use a calculator to evaluate each expression.

a. 11!   b. 19!   c. 23!   d. $_{14}P_7$

e. $_{19}P_{11}$ f. $_{23}P_{17}$ g. $_{46}P_{27}$ h. $_{11}C_5$

i. $_{16}C_{15}$ j. $_{33}C_{28}$

32. Evaluate each expression.

a. 3! 6!   b. $\dfrac{7!}{5!}$ c. $\dfrac{17!}{14!}$

d. $\dfrac{(8-5)!}{(9-4)!}$ e. $\dfrac{15!9!}{14!11!}$

33. If four distinguishable eight-sided dice are rolled, how many different outcomes are there for the four dice?

34. How many six-letter "words" are possible using the 26 letters of the English alphabet?

35. Jamie has 8 different dolls at her tea party. How many different ways can she line them up for tea?

36. Juanita has 12 different CDs. How many ways can she listen to 4 of them over the course of a long car ride?

37. How many arrangements are there of the letters in the words "DEAD END"?

38. How many different arrangements are there of 20 flags if 5 are red, 5 are blue, 7 are green, and 3 are yellow?

39. A teacher wants to write a short quiz of 15 questions. The software she is using has a pool of 45 questions to choose from. How many different quizzes can be made using this pool if the order of the questions does not matter?

40. A juror pool contains 64 people. How many 12-person juries can be selected from the pool?

41. An ice cream parlor advertises sundaes with up to 15 different toppings. How many different ways are there to get a sundae with:

a. Three toppings?   b. Four toppings?

c. Five toppings?   d. Six toppings?

42. a. How many three-digit numbers can be made using the digits 1, 2, 3, 4, 5, 6, and 7, with no repetition?

b. How many of the three-digit numbers in part a are even?

43. a. How many two-letter initials are possible?

b. How many two-letter initials are possible if the letters must be the same?

c. If a pair of two-letter initials is selected at random, what is the probability that the two letters are the same?

Answers to Chapter 15   Check Your Understandings and Activities

Check Your Understanding 15.1-A

1. a. If the same race were run three times, Fast-and-Free would win one of them.

b. Two people are applying for the promotion, and Sue is one of them.

c. Nine out of every 10 people that have had the surgery survive it.

2. Answers will vary.

3. $\dfrac{46}{200} = \dfrac{23}{100} = 23\%$.

Check Your Understanding 15.1-B

1. Let heads represent a boy and tails represent a girl. Flip three coins to simulate the birth of the three children. Any outcome with one head and two tails is a success. Flip the three coins 50 to 100 times to find the experimental probability.

2. Let the digits 1, 2, 3, 4, 5, 6, and 7 represent making a free throw and the digits 8, 9, and 0 represent missing a free throw. Select 50 to 100 pairs of numbers from the table of random digits. If both digits are a 1, 2, 3, 4, 5, 6, or 7, count it as a successful outcome. The probability should tend toward $\dfrac{1}{2}$.

Check Your Understanding 15.2-A

1. a. $S = \{1, 2, 3, 4, 5, 6\}$.   **b.** $S = \{HH, HT, TH, TT\}$.

2. a. $\dfrac{1}{8}$.   **b.** $\dfrac{1}{2}$.   **c.** $\dfrac{3}{8}$.

3. a. $\dfrac{1}{5}$.   **b.** $\dfrac{2}{5}$.   **c.** $\dfrac{7}{75}$.

Check Your Understanding 15.2-B

1. $\dfrac{3}{26}$.

2. $28\% = 0.28$.

3. 55%.

4. $\dfrac{11}{16}$.

Check Your Understanding 15.2-C

1. a. $\dfrac{11}{850}$.   **b.** $\dfrac{2}{17}$.   **c.** $\dfrac{1}{5,525}$.

2. $\dfrac{49}{289}$.

3. $\dfrac{3}{10}$.

Check Your Understanding 15.3-A

1. a. $\dfrac{1}{13}$. b. $\dfrac{9}{13}$. c. 0. d. 1.

2. a. $\dfrac{1}{10}$. b. $\dfrac{1}{5}$. c. $\dfrac{1}{2}$. d. $\dfrac{7}{10}$.

3. $\dfrac{5}{8}$.

Check Your Understanding 15.3-B

1. a. 1:12. b. 1:25. c. 1:1.

2. a. 12:1. b. 51:1. c. 1:3.

3. a. 6:11. b. 7:13. c. 1:8. d. 47:53. e. 1:249.

4. The expected value is $E = 0.5625$. The player should expect to pay $0.56.

Check Your Understanding 15.4-A

1. 140,608.

2. 4,200.

3. a. 720. b. 576. c. $\dfrac{1}{12}$.

   d. 1,680. e. 240,240.

4. 360,360.

Check Your Understanding 15.4-B

1. 180.

2. a. 10. b. 1,287. c. 1. d. 1.

3. 1,365.

4. $\dfrac{13}{35}$.

Activity 15.1

Experimental probabilities will vary but should tend toward $\dfrac{1}{10}$.

Activity 15.2

Answers will vary.

Activity 15.3

Answers will vary.

Activity 15.4

If A and B are mutually exclusive, then use the following Venn diagram:

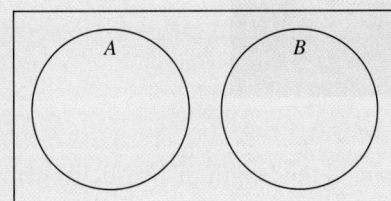

If A and B are nonmutually exclusive, then use the following Venn diagram:

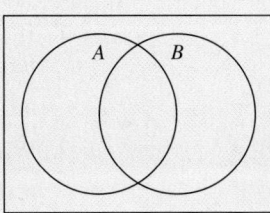

Answers will vary on how to use the Venn diagrams to explain the properties.

Activity 15.5

The probability is $\left(\dfrac{1}{26}\right)^{166}$.

Activity 15.6

a. $\dfrac{1}{4}$. b. $\dfrac{1}{2}$. c. $\dfrac{3}{4}$.

Activity 15.7

Alan should receive $22.22, Barb should receive $55.56, and Carol should receive $22.22.

Activity 15.8

a. Answers will vary.

b. Answers will vary.

c. The calculator gives an error message because 75! and 70! are too large. However, $\dfrac{75!}{70!} = 75 \cdot 74 \cdot 73 \cdot 72 \cdot 71 = 2{,}071{,}126{,}800$.

Activity 15.9

$\dfrac{54{,}912}{2{,}598{,}960} \approx 0.0211$.

Reviewing the Big Ideas Chapters 13–15

© Image Source/Corbis

◆ **Length and the Measurement Process**

1. If the actual measure of an object is 10.984 units, describe each measure in terms of accuracy and precision.

   a. 10.983   **b.** 15   **c.** 5.678   **d.** 11

2. What happens to the measure of an object if the unit is:

   a. Doubled?   **b.** Cut in half?

3. Make each conversion.

   a. 108 ft to yd   **b.** 3.25 yd to in.   **c.** 23 cm to m

   d. 0.5 hm to dm   **e.** 38 m to ft   **f.** 397 in. to dm

4. A standard Olympic swimming pool is 50 m long. If a swimmer swims 20 laps of the pool, how far did the swimmer swim in both meters and feet?

5. Estimate the length of the following objects in both centimeters and inches.

   a. An unsharpened pencil   **b.** The length of your foot

◆ **Perimeter and Area**

6. Find the perimeter and area of each figure.

   a.
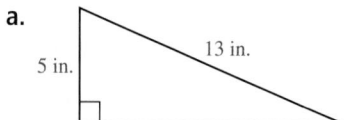

   b.
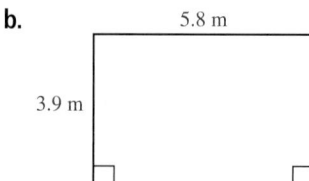

   c.
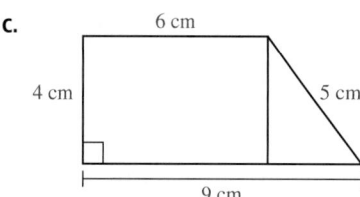

   d.
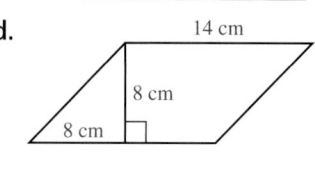

7. Fill in the blanks.

   a. 81 ft^2 = _____ yd^2.   **b.** 27 m^2 = _____ cm^2.   **c.** 133 ft^2 = _____ m^2.

8. The area of circle is about 88.20 ft^2. About how long is its radius?

9. What is the area of a regular hexagon if the sides are 8 in.?

10. A spa in the shape of a square is surrounded by a wooden deck that extends 6 ft beyond the edge of the spa. The spa is 8 ft on a side.

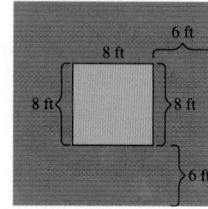

   What is the area of the wooden deck?

11. What happens to the area of a square if the length of its side is increased 50%?

◆ **Surface Area, Volume and Other Measures**

12. Make each conversion.

   a. 12,000 in^3 to yd^3 **b.** 3.4 m^3 to dm^3

13. Find the surface area and volume of each figure.

a.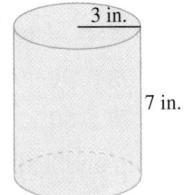

b.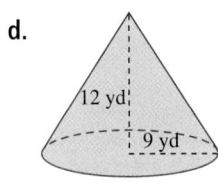

c.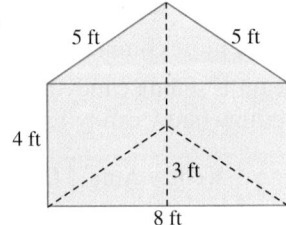

d.

14. If only the length and width of a rectangular prism are doubled, what happens to its volume?

15. If 1 kg = 2.2 lb, how much would a 180-lb man weigh in kilograms?

16. Convert each degree Fahrenheit to the nearest degree Celsius, or vice versa.

  a. 125° F   **b.** −20° F   **c.** −5° C   **d.** 93° C

17. Mike just received a speeding ticket for going 81 mph. How fast is this in ft/sec?

◆ Collecting Data

18. Identify each variable as qualitative or quantitative.

  a. Eye color   **b.** Income   **c.** Travel time   **d.** Ethnicity

19. In each case, identify the sample and the population. If the sample is representative of the population, what can we infer about the population?

  a. A sample of 1,500 seniors, 45% of whom went on to college, was selected from high schools in the United States.

  b. A political poll surveys 1,950 likely voters from around the United States. One-third of them have a favorable view of Congress.

20. Write a question of each type that could be used on a survey to determine people's preferences for their vacations.

  a. Dichotomous   **b.** Interval   **c.** Multiple choice

21. A pharmaceutical company has developed a new cough syrup. To test it, they select 120 people with coughs due to illness. Forty people are given the product, 40 people are given a placebo, and 40 people are given nothing. Coughing subsides from the first group in 20 min, from the second group in 24 hr, and from the third group in 25 hr.

  a. What are the population and the sample in the study?

  b. What are the treatment and control groups?

  c. What are the explanatory and response variables?

  d. Is the new cough syrup effective? Explain.

22. A local newspaper in a coal mining town polls its subscribers about their views on global warming. The responses overwhelmingly disbelieve that global warming is caused by human-made factors. How might the study be biased?

◆ Statistical Graphs

23. The following data represent a set of scores from a 100-point test.

78	89	85	77	78	91	93	63	67	92	78	75	72
83	70	57	68	78	80	87	92	69	63	74	74	90

a. Draw a dot plot and a stem and leaf plot of the data.

b. Are there any possible outliers or clusters in the data set?

c. What are the best and worst scores in the data set?

d. What is (are) the most commonly occurring score(s) in the data set?

24. A girls' dance team is selling candy bars as a fund-raiser. The names of the girls on the team and how many candy bars each girl sold is given in the table.

Girl	Megan	Kelli	Amy	Lindy	Ashley	Morgan	Katy	Felicia
Candy Bars Sold	95	123	107	110	99	115	105	120

Make a pictograph and a bar graph of the data.

25. The table shows how the enrollment of a certain high school has changed from 1970 to 2010.

Year	1970	1975	1980	1985	1990	1995	2000	2005	2010
Population	800	875	925	950	975	1,000	1,250	1,500	1,900

Draw a line graph that shows the change in population, and describe the general trends in the data.

26. Answer each question based on the information provided in the circle graph.

Grade Distribution of 8th Grade Students at Riverchase Middle School

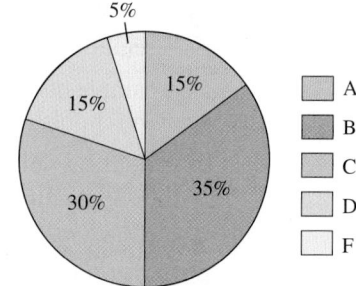

a. Which grade are most students earning?

b. If there are 125 eighth-grade students at Riverchase, how many are earning a B? Earning a D?

27. The following scatterplot shows a comparison of the number of servings of oatmeal in a week and the cholesterol level of adult males.

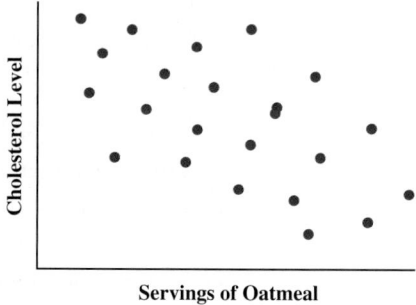

Describe the correlation you see in the graph.

28. The following table shows how many hours each member of a family watches television.

Family Member	Jason	Jill	Jerry	Jacob	Julie
Hours of Television	15	13	14	14	12

Draw a bar graph that accentuates the differences among family members. If the women of the family were to use the graph to conclude that men in general watch more television, could such a claim be believed? Why or why not?

◆ Descriptive Statistics

29. Find the mean, the median, and the mode for each data set.

a. 45, 45, 46, 48, 37, 38, 41, 45   **b.** 6.7, 8.1, 5.2, 5.2, 6.9, 7.7

30. Give a 5-number summary for the following data from a 50-point test. Use the *IQR* to determine whether there are any outliers, and then draw a box plot for the data set.

15, 38, 39, 40, 40, 41, 42, 43, 43, 44, 45, 45, 45, 47, 48, 49, 50, 50

31. Compute the mean and the standard deviation for the following data from a 100-point exam.

56, 62, 65, 66, 70, 73, 74, 74, 75, 82, 85, 90, 91, 93, 97

32. A class has 21 students. The mean weight of the students is 65 lb. If two more students are added to the class with weights of 53 lb and 77 lb, what is the new mean for all 23 students?

33. Data for a standardized test are approximately normal with a mean of 500 and a standard deviation of 25.

a. What score is two standard deviations below the mean? Two standard deviations above the mean?

b. About what percent of the test takers scored between 475 and 525?

c About what percent of the test takers scored greater than 450?

d. What is the z-score of a test score of 480? Of 530?

34. A teacher claims that her class did poorly on a 100-point final exam. If the mean was a 70 and the median was 80, can you believe her claim?

◆ Experimental and Theoretical Probability

35. An eight-sided die is rolled 200 times with the results recorded in the table.

Outcome	1	2	3	4	5	6	7	8
Times Rolled	24	26	19	27	27	25	23	29

Find the experimental probability of rolling:

a. A 3.   **b.** An odd number.   **c.** A number less than 5.

36. Toss a quarter and a dime 50 times, and record the results. Use your results to find the experimental probability that:

a. Both coins are tails.   **b.** The quarter is tails and the dime is heads.

37. A quality control inspector examines boxes of cereal in which 20% are expected to have some sort of defect. Devise a simulation using a table of random digits that can be used to determine the probability that at least one box will have a defect when five boxes are tested at random.

38. List the sample space for the experiment of rolling a four-sided and a six-sided die. What are the outcomes in the event of rolling:

a. An even sum?   **b.** A number less than 3?

39. A 20-sided die is rolled. What is the probability that the number is:

    a. Even?   **b.** Divisible by 5?   **c.** Divisible by 2 or 3?

40. Two cards are dealt from a well-shuffled deck of 52 cards. What is the probability that the cards are:

    a. Both red?   **b.** Both tens?   **c.** Not diamonds?

41. A weather report states that the chance of rain for the next three days is 35% for each day. What is the probability that it rains on the first day, but not on the second or third?

◆ Conditional and Geometric Probability, Odds, and Mathematical Expectation

42. A single card is drawn from a deck of 52 cards. If the card is known to be black, what is the probability that the card is:

    a. A spade?   **b.** A 2?   **c.** A face card?   **d.** The 3 of hearts?

43. One dart is thrown at random at the following dartboard.

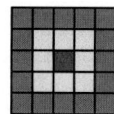

    What is the probability that the dart hits a:

    a. Gray region?   **b.** Yellow or red region?

44. An eight-sided die is rolled.

    a. What are the odds in favor of getting a 4?

    b. What are the odds against rolling a number greater than 6?

45. **a.** If the odds against an event A are 3:5, what is the probability that A occurs?

    b. If $P(A) = 0.4$, what are odds in favor and against A?

46. An eight-sided die is rolled. The amounts won or lost are shown:

Number	1	2	3	4	5	6	7	8
Amount Won or Lost	−$3	−$2	−$1	$0	$1	$2	$3	$4

    What is the expected value of the game? Would you play it?

◆ Counting Techniques

47. Evaluate each expression.

    a. 11!   **b.** $\dfrac{12!}{9!}$ **c.** $_{16}P_3$ **d.** $_{14}P_9$ **e.** $_8C_4$ **f.** $_{12}C_7$

48. If four distinguishable six-sided dice are rolled, how many different outcomes are there for the four dice?

49. Juanita has 15 songs on a play list. If she shuffles the songs, how many different ways can she listen to 7 of them?

50. How many different arrangements are there of 4 white chips and 8 blue chips if the chips are indistinguishable except for color?

51. A teacher wants to write a short quiz of 10 questions. The software she is using has a pool of 35 questions to choose from. How many different quizzes can be made using this pool of questions if the order of the questions does not matter?

52. If three cards are drawn at random without replacement from a standard deck of 52 cards, what is the probability that they are all hearts?

Bana, J., Farrell, B., & McIntosh, A. (1995). Error patterns in mental computation in years 3–9. In B. Atweh & S. Flavel (Eds.), MERGA 18: GALTHA, *Proceedings of the 18th annual conference* (pp. 51–56). Darwin, Australia: Northern Territory University.

Bathroom Readers' Institute. (1998) *Uncle John's great big bathroom reader.* Ashland, OR: Bathroom Readers' Press.

Battista, M. (2002). Learning geometry in a dynamic computer environment. *Teaching Children Mathematics, 8,* 333–339.

Beckmann, S. (2008). *Mathematics for elementary teachers* (2nd ed.). Boston, MA: Addison Wesley.

Behr, M., Lesh, R., Post, T., & Silver, E. (1983). Rational number concepts. In R. Lesh & M. Landau (Eds.). *Acquisition of mathematics concepts and processes* (pp. 91–126). New York: Academic Press.

Behr, M., Harel, G., Post, T., & Lesh, R. (1990). *On the operator concept of rational numbers: Towards a semantic analysis.* Paper presented at the Annual Meeting of the American Educational Research Association. Boston, MA.

Bell, A. (1986). Diagnostic teaching 2. Developing conflict-discussion lessons. *Mathematics Teacher, 116,* 26–29.

Bezuszka, S. (1985). A test for divisibility of primes. *Arithmetic Teacher, 33*(2), 36–38.

Billstein, R., Libeskind, S., & Lott, J. (2010). *A problem solving approach to mathematics for elementary school teachers* (10th ed.). Boston, MA: Addison Wesley.

Bohan, H., Irby, B., & Vogel, D. (1995). Problem solving: Dealing with data in the elementary school. *Teaching Children Mathematics, 1,* 256–260.

Brown, C., & Silver, E. (1989). Data organization and interpretation. In M. Lindquist (Ed.), *Results for the fourth mathematics assessment of the National Assessment of Education Progress.* Reston, VA: National Council of Teachers of Mathematics.

Bruni, J., & Seidentstein, R. (1990). Geometric concepts and spatial sense. In J. N. Payne (Ed.), *Mathematics for the young child.* Reston, VA: National Council of Teachers of Mathematics.

Burns, M. (2000). *About teaching mathematics: A K–8 resource.* Sausalito, CA: Math Solutions Publications.

Burris, A. (2005). *Understanding the math you teach.* Upper Saddle River, NJ: Prentice Hall.

Burton, D. (1995). *Burton's history of mathematics: An introduction* (3rd. ed.). Boston: Brown.

Carifio, J., & Nasser, R. (1995). *Algebra word problems: A review of the theoretical models and related research literature.* Annual meeting of the American Education Research Association.

Carpenter, T., Corbitt, M., Kepner, H., Lindquist, M., & Reys, R. (1981). *Results from the second mathematics assessment of the National Assessment of Educational Progress.* Reston, VA: National Council of Teachers of Mathematics.

Carpenter, T., & Levi, L. (1999). *Developing conceptions of algebraic reasoning in the primary grades.* Paper presented at the annual meeting of the American Educational Research Association, Montreal.

Carroll, L. (1977). *Lewis Carroll's symbolic logic.* Bartley, W. (Ed.). New York: Harvester Press.

Clements, D. (1999). Geometric and spatial thinking in young children. In J. V. Copley (Ed.), *Mathematics in the early years.* Reston, VA: National Council of Teachers of Mathematics.

Clements, M. (1979). Sex differences in mathematical performance: An historical perspective. *Educational Studies in Mathematics, 10,* 305–322.

Common Core State Standards Initiative. (2010). Common core state standards for mathematics. Available at http://www.corestandards.org/the-standards/mathematics.

Cramer, K., Post, T., & DelMas, R. (2002). Initial fraction learning by fourth and fifth grade students: A comparison of the effects of using commercial curricula with the effects of using the Rational Number Project Curriculum. *Journal for Research in Mathematics Education, 33,* p. 111–144.

Dantzig, T. (1954). *Number: The language of science.* New York: Macmillan.

Euclid. (1956). *The thirteen books of the elements: Translated with commentary by Sir Thomas Heath.* Mineola, NY: Dover.

Fischbein, E., & Gazit, A. (1984). Does the teaching of probability improve probabilistic intuitions? *Educational Studies in Mathematics, 15*(1), 1–24.

Fish, D. (Ed.). (1877). *The rudiments of written arithmetic.* New York: Ivison, Blakeman, & Taylor.

Franklin, C., Kader, G., Mewborn, D., Moreno, J., Peck, R., Perry, M., & Scheaffer, R. (2005). *Guidelines for assessment and instruction in statistics education report.* Alexandria, VA: American Statistical Association.

Fuson, K. (1992). Research on whole number addition and subtraction. In D. A. Grouws (Ed.). *Handbook of research on mathematics teaching and learning* (pp. 243–275). New York: McMillan.

Gardner, H. (1983). *Frames of mind: The theory of multiple intelligences.* New York: Basic Books.

Gnanadesikan, M., Schaeffer, R., & Swift, J. (1987). *The art and techniques of simulation: Quantitative literacy series.* Palo Alto, CA: Dale Seymour.

Graviss, T., & Greaver, J. (1992). Extending the number line to make connections with number theory. *Mathematics Teacher, 85,* 418–420.

Greer, B. (1988). Nonconservation of multiplication and division: Analysis of a symptom. *Journal of Mathematical Behavior, 7,* 281–298.

Guiness World Records (2006). *Guiness World Records,* Available at www.guinessworldrecords.com.

Hancock, C., Kaput, J., & Goldsmith, L. (1992). Authentic inquiry with data: Critical barriers to classroom implementation. *Educational Psychologist, 27*(3), 337–364.

Hatfield, M., Edwards, N., & Bitter, G. (1997). *Mathematics methods for elementary and middle school teachers* (3rd ed.). Boston, MA: Allyn & Bacon.

Heddens, J., & Speer, W. (1997). *Today's mathematics Part 1: Concepts and classroom methods.* Upper Saddle River, NJ: Prentice Hall.

Hembree, R., & Dessart, D. (1986). Effects of hand-held calculators in precollege mathematics education: A meta-analysis. *Journal for Research in Mathematics Education, 17,* 83–89.

Hiebert, J. (1984). Why do some children have difficulty learning measurement concepts? *Arithmetic Teacher, 31*(7), 19–24.

Hiebert, J., & Wearne, D. (1986). Procedure over concept: The acquisition of decimal number knowledge. In J. Hiebert (Ed.), *Conceptual and procedural knowledge: The case of mathematics* (pp. 199–233). Hillsdale, NJ: Erlbaum.

Isaacs, A., & Carroll, W. (1999). Strategies for basic-facts instruction. *Teaching Children Mathematics, 5,* 508–515.

Jones, G., Langrall, C., Thornton, C., & Mogill, A. (1999). Students' probabilistic thinking in instruction. *Journal for Research in Mathematics Education, 30,* 487–519.

Jones, O. (1868). *The grammar of ornament.* London: Day & Son Ltd.

Katz, V. (2004). *A History of mathematics: Brief edition.* New York: Addison-Wesley.

Kenney, P., & Kouba, V. (1997). What do students know about measurements? In P. Kenney & E. Silver (Eds.), *Results for the Sixth Mathematics Assessment of the National Assessment of Educational Progress.* Reston, VA: National Council of Teachers of Mathematics.

Konold, C. (1989). Informal conceptions of probability. *Cognition and Instruction 6*(1), 59–98.

Kouba, V., Brown, C., Carpenter, T., Lindquist, M., Silver, E., & Swafford, J. (1988). Results of the fourth NAEP assessment of mathematics: Measurement, geometry, data interpretation, attitudes, and other topics. *Arithmetic Teacher 35*(9), 10–16.

Kouba, V., Carpenter, T., & Swafford, J. (1989). Number and operations. In M. Lindquist (Ed) *Results from the Fourth Mathematics Assessment of the National Assessment of Educational Progress,* Reston, VA: National Council of Teachers of Mathematics.

Lawson, D. (1990). The problem, the issues that speak to change. In E. Edwards (Ed.), *Algebra for everyone.* Reston, VA: National Council of Teachers of Mathematics.

Leidtke, W. (1990). Measurement. In J. N. Payne (Ed.), *Mathematics for the young child* (pp. 229–249). Reston, VA: National Council of Teachers of Mathematics.

Lesh, R., Post, T., & Behr, M. (1998). Proportional reasoning. In J. Heibert &M. Behr (Eds.), *Number concepts and operations in the middle grades* (pp. 93–118). Reston, VA: National Council of Teachers of Mathematics.

Lindquist, M., & Kouba, V. (1989). Measurement. In M. Lindquist (Ed.), *Results from the Fourth Mathematics Assessment of the National Assessment of Educational Progress.* Reston, VA: National Council of Teachers of Mathematics.

Long, C., DeTemple, D., & Millman, R. (2009). *Mathematical reasoning for elementary teachers* (5th ed.). Boston, MA: Addison Wesley.

Mack, N. (1990). Learning fractions with understanding: Building on informal knowledge. *Journal for Research in Mathematics Education, 21,* 16–32.

Markovits, Z., Eylon, B., & Bruckheimer, M. (1988). Difficulties students have with the function concept. In A. F. Coxford (Ed.), *The ideas of algebra, K–12.* Reston, VA: National Council of Teachers of Mathematics.

Martin, G., & Strutchens, M. (2000). Geometry and measurement. In E. Silver & P. Kennedy (Eds.), *Results from the seventh mathematics assessment of the national assessment of educational progress.* Reston, VA: National Council of Teachers of Mathematics.

M. C. Escher Foundation. (1998). *M. C. Escher: The official website.* Available at http://mcescher.com.

McKnight, C., Crosswhite, J., Dossey, J., Kifer, E., Swafford, J., Travers, K., & Cooney, T. (1987). *The underachieving curriculum: Assessing U.S. school mathematics from an international perspective.* Champaign, IL: Stipes.

Murray, J. (Ed.). (1993). *Oxford English Dictionary.* New York: Oxford University Press.

Musser, G., Burger, W., & Peterson, B. (2008). *Mathematics for elementary teachers: A contemporary approach* (8th ed.). Hoboken, NJ: John Wiley and Sons.

National Council of Teachers of Mathematics. (1989). *Curriculum and evaluation standards for school mathematics.* Reston, VA: National Council of Teachers of Mathematics.

National Council of Teachers of Mathematics. (1991). *Professional standards for teaching mathematics.* Reston, VA: National Council of Teachers of Mathematics.

National Council of Teachers of Mathematics. (1995). *Assessment standards for school mathematics.* Reston, VA: National Council of Teachers of Mathematics.

National Council of Teachers of Mathematics. (2000). *Principles and standards for school mathematics.* Reston, VA: National Council of Teachers of Mathematics.

National Council of Teachers of Mathematics. (2006). *Curriculum focal points for prekindergarten through grade 8 mathematics.* Reston, VA: National Council of Teachers of Mathematics.

Nortman-Wolf, S. (1990). *Base ten blocks activities.* Vernon Hills: IL: Learning Resources.

O'Daffer, P., Charles, R., Cooney, T., Dossey, J., & Schielack, J. (2008). *Mathematics for elementary school teachers* (4th ed.). Boston, MA: Addison Wesley.

Olympic.org. (2009). Shot put men. Available at http://www.olympic.org/athletics/shot-put-men.

Outhred, L., & Mitchelmore, M. (2000). Young children's intuitive understanding of rectangular area measurement. *Journal for Research in Mathematics Education, 31,* 144–167.

Owens, D., & Super, D. (1993). Teaching and learning decimal fractions. In D. Owens (Ed.), *Research ideas for the classroom: Middle grades mathematics* (pp. 137–158). Reston, VA: National Council of Teachers of Mathematics.

Parker, T., & Balridge, S. (2003). *Elementary geometry for teachers.* Okemos, MI: Sefton-Ash Publishing.

Parker, T., & Balridge, S. (2003). *Elementary mathematics for teachers.* Okemos, MI: Sefton-Ash Publishing.

Polya, G. (1945). *How to solve it.* Princeton, NJ: Princeton University Press.

Polya, George (1962). *Mathematical discovery: On understanding, learning and teaching problem solving.* Hoboken, NJ: Wiley.

Posamentier, A. S., & Hauptman, H.A. (2001). *101 great ideas for introducing key concepts in mathematics: A resource for secondary school teachers.* Thousand Oaks, CA: Corwin Press.

Post, T. (1992). *Teaching mathematics in grades K–8: Research-based methods* (2nd ed.). Boston: Allyn and Bacon.

Post, T., Behr, M., & Lesh, R. (1988). Proportionality and development of prealgebra understandings. In A. Coxford (Ed.), *The ideas of algebra, K–12.* Reston, VA: National Council of Teachers of Mathematics.

Post, T., Wachsmuth, I., Lesh, R., & Behr, M. (1985). Order and equivalence of rational numbers: A cognitive analysis. *Journal for Research in Mathematics Education, 16,* 18–36.

Quinn, P. C., & Eimas, P. D. (1996). Perceptual organization and categorization in young infants. In C. Rovee-Collier & L. P. Lipsitt (Eds.), *Advances in infancy research: Vol. 10* (pp. 1–36). Norwood, NJ: Ablex.

Resnick, L., Nesher, P., Leonard, M., Omanson, S., & Peled, I. (1989). Conceptual bases of arithmetic errors: The case of decimal fractions. *Journal for Research in Mathematics Education, 29,* 225–237.

Reys, R., Lindquist, M., Lambdin, D., Smith, N., & Suydam, M. (2004). *Helping children learn mathematics* (7th ed.). Hoboken, NJ: Wiley.

Reys, R., Rybolt, J., Bestgen, B., & Wyatt, J. (1982). Processes used by good computational estimators. *Journal for Research in Mathematics Education, 13,* 183–201.

Rosch, E., Mervis, C. B., Gray, W. D., Johnson, D. M., & Boyes-Braem, P. (1976). Basic objects in natural categories. *Cognitive Psychology, 8,* 382–439.

Schneider, W., & Bjorklund, D. F. (1998). Memory. In D. Kuhn & R. S. Siegler (Eds.), *Handbook of child psychology: Vol. 2. Cognition, perception, and language* (5th ed.). (pp.467–522). Hoboken, NJ: Wiley.

Schoenfeld, A. (1987). What's all the fuss about metacognition? In A. Schoenfeld (Ed.). *Cognitive Science and Mathematics Education* (pp. 189–215). Hillsdale, NJ: Erlbaum.

Shaughnessy, J. (1992). Research in probability and statistics: Reflections and directions. In D.A. Grouws (Ed.). *Handbook of research on mathematics teaching and learning* (pp. 495–514). New York: Macmillan.

Shaw, J. (1984). Making graphs. *Arithmetic Teacher, 31*(5), 7–11.

Sheffield, L., & Cruikshank, D. (1996). *Teaching and learning elementary and middle school mathematics* (3rd ed.). Upper Saddle River, NJ: Prentice Hall.

Sonnabend, T. (2010). *Mathematics for teachers: An interactive approach for grades K–8* (4th ed.). Belmont, CA: Cengage Learning.

Sowder, J. (1992). Estimation and number sense. In D. A. Grouws (Ed.). *Handbook of research on mathematics teaching and learning* (pp. 371–89). New York: Macmillan.

Sowder, J., & Schappelle, B. (1994). Number sense making. *Arithmetic Teacher, 41*(2), 342–345.

Steffe, L., & Cobb, P. (1988). *Construction of arithmetical meanings and strategies.* New York: Springer-Verlag.

Sugarman, S. (1983). *Children's early thought: Developments in classification.* New York: Cambridge.

Taylor, M., & Biddulph, F. (1994). "Context" in probability learning at the primary school level. In A. Jones, A. Begg, B. Bell, F Biddulph, M. Carr, M. Carr, J. McChesney, E. McKinley, & J. Young-Loveridge (Eds.), *SAME papers 1994* (pp. 96–111) Hamilito, New Zealand: Centre for Science and Mathematics Education Research, University of Waikato.

Thompson, F. (1988). Algebraic instruction for the younger child. In A. Coxford (Ed.), *The ideas of algebra K–12* (pp. 8–19). Reston, VA: National Council of Teachers of Mathematics.

Thompson, P., & Dreyfus, T. (1988). Integers as transformations. *Journal for Research in Mathematics Education, 19,* 115–133.

Thorton, C. (1990). Strategies for the basic facts. In J. N. Payne (Ed.) *Mathematics for young children* (pp. 133–151). Reston, VA: National Council of Teachers of Mathematics.

Tillotson, M. (1985). The effect of instruction in spatial visualization on spatial abilities and mathematical problem solving. *Dissertation Abstracts International, 45A,* 2792.

Trafton P., & Zawojewski, J. (1984). Teaching rational number division: A special problem. *Arithmetic Teacher, 31*(6), 20–22.

United States Postal Service (2009). Rates for Domestic Letters, 1863–2009. Available at https://www.usps.com.

Usiskin, Z. (1992). *Where does algebra begin? Where does algebra end?* Algebra for the Twenty-First Century: Proceedings of the August 1992 Conference: Reston, VA: National Council of Teachers of Mathematics. (pp. 27–28).

Van De Walle, J. (1990). *Elementary school mathematics: Teaching developmentally.* White Plains, NY: Longman.

Van Hiele, P.M. (1986). *Structure and insight: A theory of mathematics education.* Orlando, FL: Academic Press.

Wagner, S. (1981). Conservation of equation and function under transformation of variable. *Journal for Research in Mathematics Education, 12,* 107–118.

Watson, J., & Moritz, J. (2000). Developing concepts of sampling. *Journal for Research in Mathematics Education, 31,* 44–70.

Watson. J., & Moritz, J. (2003). Fairness of dice: A longitudinal study of students' beliefs and strategies for making judgments. *Journal for Research in Mathematics Education, 34,* 270–304.

Wikipedia Foundation. (2001). *Wikipedia: free online encyclopedia.* Available at encyclopediahttp://en.wikipedia.org/wiki.

Wilson, P., & Osborne, A. (1998). Foundational ideas in teaching about measure. In T. Post (Ed.), *Teaching mathematics in grades K–8.* Toronto: Allyn and Bacon.

Wilson, P., & Rowland, R. (1993). Teaching measurement. In R. Jensen (Ed.), *Research ideas for the classroom: Early childhood mathematics.* New York: Macmillan.

Young, S., & O'Leary, R. (2002). Creating numerical scales for measuring tools. *Teaching Children Mathematics, 8,* 400–405.

CHAPTER 1

■ Section 1.1 Learning Assessment

1. Without mathematics we cannot convey ideas involving quantities, measures, shapes, patterns, and relationships.
2. Mathematical representations are ways of portraying mathematical objects, ideas, and relationships. They are a fundamental part of how we understand and think about mathematics. If we can learn to use a variety of representations and understand how they complement one another, then we have a valuable set of tools that we can use to think mathematically.
3.

Representation	Advantages	Disadvantages
Verbal	• Can be used to describe a variety of complex ideas completely and precisely. • Can be used to define concepts, state facts, justify reasoning, and describe computational procedures.	• Unclear wording can lead to unintended results. • May require use of technical terms. • Can be long and awkward to use.
Numerical	• Can represent quantities quickly and easily. • Can hold much information.	• Can be difficult to express general facts, properties, and relationships. • Large sets of numbers can be difficult to analyze.
Symbolic	• Make it easy to express generalities efficiently.	• Make mathematics more difficult to read.
Visual	• Can convey large amounts of information. • Can show aspects of concepts and relationships in ways that other representations do not.	• Can be misleading if they are not drawn carefully or only offer a limited view.
Tactile	• Can be powerful tools for learning.	• Can fall short in their ability to represent every situation. • May be awkward to use in some situations.

4. We use identification numbers to label or identify objects, ordinal numbers to assign an order to a group of objects, and cardinal numbers to express quantities.

5. Symbolic representations are used to represent mathematical concepts, both abstract and concrete: mathematical operations; mathematical relationships between mathematical expressions and ideas; and commonly used words or phrases.
6. A variable is a symbol that represents a quantity that can take on different values. It can be used to represent a specific, yet unknown quantity, or it can be used as general representative from a set of numbers.
7. To evaluate an expression, we replace certain variables with numbers and perform any computations.
8. Possible answers are:

a. $y = -x$

First	Second
1	-1
2	-2
3	-3
4	-4

b. $y = 3x$

First	Second
1	3
2	6
3	9
4	12

c. $y = x + 3$

First	Second
1	4
2	5
3	6
4	7

d. $y = \sqrt{x}$

First	Second
1	1
4	2
9	3
16	4

9. a. The second is the square of the first: $y = x^2$.
   b. The second is one less than the first: $y = x - 1$.
   c. The second is double the first: $y = 2x$.
   d. The second is 1 divided by the first: $y = 1/x$.
10. Possible answers are:

a. The sum of x and y is six.

x	y
1	5
2	4
3	3
4	2

b. Three times x minus four is the same as y.

x	y
1	-1
2	2
3	5
4	8

c. x multiplied by y is twenty-four.

x	y
1	24
2	12
3	8
4	6

d. The ratio of x and y is one.

x	y
1	1
2	2
3	3
4	4

11. a. The sum of x and three is six.   **b.** The difference of y from six is twice x.   **c.** Four times x plus one is less than ten.   **d.** Eighteen divided by x is six.

12. a. Eleven minus x is greater than four.   **b.** Three times the quantity x plus four is the same as seven.   **c.** Fourteen minus three times x is less than or equal to y.   **d.** The quantity of x minus one divided by y is less than or equal to three.

13. a. $5 + x < 3$.   **b.** $2p - 5 = 4$.   **c.** $6(x + 2) < 9$.   **d.** $6 \div x = y + 1$.

14. a. $3 \times 4 = 12$.   **b.** $(x + 7) - 4 = 9$.   **c.** $x \div y < 4$.   **d.** $4(x - 5) = 10$.

15. a. Answers will vary.   **b.** Answers will vary.   **c.** Answers will vary.   **d.** Answers will vary.

16. Answers will vary, but possible pictures are:
a.   **b.**   **c.**

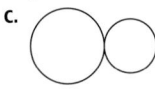

d. All descriptions are open to interpretation. This implies precise language is not only important for drawing the correct figure, but it can also be difficult to achieve. It also shows how important a picture is for communicating mathematical ideas.

17. Answers will vary.

18. Answers will vary.

19. Answers will vary.

20. Answers will vary.

21. a. Cardinal.   **b.** Identification or ordinal.   **c.** \$42.50 is a cardinal number, and 0123456789 is an identification number.

22. a. First is an ordinal number, and 204 is a cardinal number.   **b.** 360 is a cardinal number, and 101 is an identification number.   **c.** 735 is an identification number, and 7th is an ordinal number.

23. Linda would have to score in the low 90s or upper 80s for it to be considered a good round.

24. Most of the people are at or above the average IQ.

25. The trout population decreased from 2005 to 2007 and then began to increase past its original levels from 2008 to 2011.

26. a. December.   **b.** July.   **c.** 6.   **d.** As the weather grows warmer, more tickets are given. As the weather grows cooler, fewer tickets are given.

27. a. i. 52.   **ii.** 52.   **iii.** 76.   **iv.** 350.   **b.** Students' scores on Test 2.   **c.** Cole's scores on all the tests.   **d.** Cole had the highest score of 95. Douglas had the lowest score of 39.   **e.** Cole.

28. a. 7.   **b.** 4.   **c.** 14.   **d.** 13.

29. a. -6.   **b.** 8.   **c.** 29.   **d.** 57.

30. a. 30.   **b.** 36.   **c.** 108.   **d.** 0.

31. At $x = 0$, $y = -4$.
At $x = 1$, $y = -1$.
At $x = 2$, $y = 8$.
At $x = 3$, $y = 23$.

32. All three angles have the same measure, and all three sides have the same length.

33. The longest side is side $\overline{DC}$ because it has the largest measure.

34. $\angle A$ is the largest because it is opposite the longest side.

35. a. 1940; 1990.   **b.** 1970; 1980.   **c.** The average family size is getting smaller.

36. a. 2.   **b.** 6.   **c.** 4.

37. Seven moves are needed:
1. Penny to C.   **2.** Nickel to B.
3. Penny to B on top of nickel.   **4.** Quarter to C.
5. Penny to A.
6. Nickel to C on top of quarter.
7. Penny to C on top of nickel.

38. a. 6.   **b.** 9.

39. Answers will vary.

40. Answers will vary.

41. Answers will vary.

42. Answers will vary.

43. a. Answers will vary.   **b.** Answers will vary.

44. Answers will vary.

45. a. Answers will vary.   **b.** Answers will vary.   **c.** Answers will vary.   **d.** Answers will vary.

46. a. Addition and positive.   **b.** Subtraction, negative, and additive inverse.   **c.** Ordered pair or grouping symbol.   **d.** Division, ratio, or fraction.

47. Answers will vary.

48. a. Answers will vary.   **b.** Answers will vary.   **c.** Answers will vary.

49. Answers will vary.

50. Answers will vary.

51. Answers will vary.

■ Section 1.2 Learning Assessment

1. Mathematical reasoning is a logical and systematic way of thinking that uses a specific set of rules and assumptions. It is a defining characteristic of mathematics, and at its core lies the assumption that mathematical ideas should make sense, be supported by logical reasons, and lead to a better understanding of mathematics. If we want children to understand and do mathematics, then we must teach them how to reason mathematically.

2. First, we look for a pattern among a specific set of related examples. Second, we make a generalization indicating that the pattern will likely hold true for all other similar situations.

3. a. Repetitive patterns have a core component that continuously repeats.   **b.** Patterns of common traits exhibit some common characteristic that we find by looking at the similarities and differences among the examples.   **c.** A growing pattern continuously increases or decreases in a predictable fashion.

4. Algebraic expressions are efficient ways of representing patterns that hold for an infinite number of possibilities.

5. Begin with three ones arranged in a triangular fashion. Add rows to the bottom of the triangle using the following rules: (1) Every row begins and ends with a 1. (2) Entries in between the 1s are computed by adding the two entries that are diagonally above in the preceding row.

6. (1) It may be possible to make several different generalizations from the same set of examples. (2) A limited number of examples may lead to incorrect generalizations.

7. A counterexample is a specific example showing that a generalization is false. Counterexamples are important to mathematical reasoning because they tell us when a generalization is false.

8. If a conjecture is inductively strong, then many specific examples support the truthfulness of the conjecture. However, inductive reasoning never guarantees that a generalization is true for every instance under consideration.

9. **a.** *AB.* **b.** *ABB.* **c.** *ABC.* **d.** *ABCB.*

10. **a.** *AB.* **b.** *ABC.* **c.** *ABBA.* **d.** *ABCB.*

11. They are all the same growing pattern of the form *A, B, A, B, B, A, B, B, B,* . . .

12. Possible answers are:
    a. ♣, ♦, ♣, ♦, ♣, ♦, . . . or ♥, ♠, ♥, ♠, ♥, ♠, . . .
    b. ♣, ♣, ♠, ♣, ♣, ♠, ♣, ♣, ♠, . . . or ♦, ♦, ♥, ♦, ♦, ♥, ♦, ♦, ♥, . . .
    c. ♣, ♦, ♥, ♣, ♦, ♥, ♣, ♦, ♥, . . . or ♣, ♦, ♠, ♣, ♦, ♠, ♣, ♦, ♠, . . .

13. Possible answers are ● ■●● ■●●●■, . . . or ● ■●● ■■●●●■■■, . . .

14. The sequence is the square numbers, which can be represented as:
    a.

    b. The sequence is the triangular numbers, which can be represented as:
    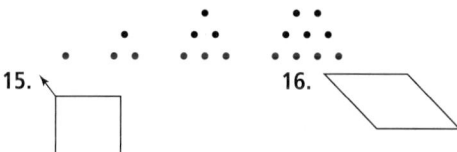

15. 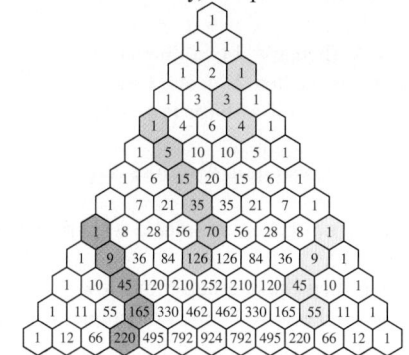   16.

17. I will get warm.

18. Tomorrow will be sunny.

19. His father will get home late from work.

20. On Thursday night, Tom will eat a meat, a vegetable, and a potato.

21. **a.** 15, 18, and 21. **b.** 10,000; 100,000; and 1,000,000.
    c. $\frac{5}{6}, \frac{6}{7}$, and $\frac{7}{8}$.

22. **a.** H, J, and K. **b.** M, J, and J. **c.** E, N, and T.

23. One possibility is 1, 4, 7, and 10.

24. One possibility is 16, 8, 4, and 2.

25. **a.** $33,334^2 = 1,111,155,556.$ **b.** $77,777 \times 99,999 = 7,777,622,223.$ **c.** $6,666 \times 9 = 59,994.$

26. The eighth row is 1 7 21 35 35 21 7 1.
    The ninth row is 1 8 28 56 70 56 28 8 1.
    The tenth row is 1 9 36 84 126 126 84 36 9 1.

27. **a.** Seventh. **b.** Eighth **c.** Seventeenth. **d.** Tenth.

28. **a.** 5. **b.** 9. **c.** 45.

29. **a.** 6. **b.** 35. **c.** 126.

30. Answers will vary, but possible solutions are:

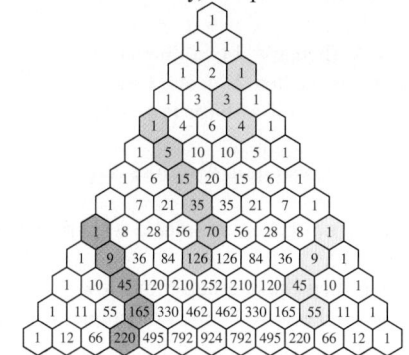

31. **a.** This text is not fiction. **b.** April has only 30 days in it.
    c. Try is an English word that does not contain an a, e, i, o, or u. **d.** Colorado begins with C and does not border an ocean.

32. **a.** Eleven is greater than 10. **b.** Six is divisible by 2 but not by 4. **c.** $1^2 = 1$ and $0^2 = 0$. **d.** $3 + 1 = 4$.

33. The next three equations are $3,367 \times 15 = 50,505$, $3,367 \times 18 = 60,606$, and $3,367 \times 21 = 70,707$. The pattern continues to hold.

34. The next three equations are $(12,345 \times 9) + 6 = 111,111$, $(123,456 \times 9) + 7 = 1,111,111$, and $(1,234,567 \times 9) + 8 = 11,111,111$. The pattern no longer holds with the equation $(12,345,678,910 \times 9) + 11 = 111,111,110,201$.

35. No. Even though Amy has a pattern of having girls, the previous children have no bearing on the gender of the next child.

36. *A, AB, ABB, ABA, AAB,* and *ABC.*

37. We need to use inductive reasoning to identify the pattern in the problem. By the end of the week, Maggie will have picked $1 + 2 + 3 + 4 + 5 + 6 + 7 = 28$ flowers.

38. The sum of two odd numbers is even.

39. If n is a number, then $n + 0 = n$ and $0 + n = n$.

40. If n is a number, then $1 \times n = n$ and $n \times 1 = n$.

41. If a, b, and c are numbers, then $a + (b + c) = (a + b) + c$.

42. **a.** The product in each example is odd, so we conclude that the product of two odd numbers is odd. **b.** Other examples are $3 \times 5 = 15$, $9 \times 1 = 9$, $5 \times 11 = 55$, and $9 \times 9 = 81$. **c.** The conjecture is true. A possible intuitive argument is as follows. Suppose the product of two odd numbers is even. Because the product is even, one of the factors must be divisible by 2. However, only even numbers are divisible by 2, so the original factors must be even. Because a number cannot be both even and odd, our original assumption must be incorrect. Hence, the product of two odd numbers must be odd.

43. **a.** Dividing a natural number by 2 results in an even quotient. **b.** The conjecture is false: $6 \div 2 = 3$ is a counterexample.

44. **a.** Several examples are:
    $$1 + 2 + 3 = 6, \text{ which is divisible by 3.}$$
    $$2 + 3 + 4 = 9, \text{ which is divisible by 3.}$$
    $$3 + 4 + 5 = 12, \text{ which is divisible by 3.}$$
    b. The conjecture is true. Algebraically, if $x, x + 1$, and $x + 2$ are consecutive numbers, then $x + (x + 1) + (x + 2) = 3x + 3 = 3(x + 1)$, which is divisible by 3.

45. **a.** 55 blocks. **b.** Answers will vary.

46. **a.** 39 toothpicks. **b.** Answers will vary.

47. $15 \div 5 = 3$ ft after the fourth bounce and $3 \div 6 = \frac{1}{2}$ ft after the fifth bounce.

48. **a. i.** 3. **ii.** 4. **iii.** 5. **iv.** 6. **b.** The numbers in each row are equal to the number of the row. **i.** 35. **ii.** 75. **iii.** 153. **iv.** 1,000.

49. The numbers in the row with 3 are divisible by 3, the numbers in the row with 5 are divisible by 5, and the numbers in the row with 7 are divisible by 7. This does not hold for 4, 6, and 9. This implies that the numbers in the rows in which the second number is prime are divisible by that number.

50. Answers will vary.

51. Answers will vary.
52. Summer is incorrect. We would classify both patterns as *AB* patterns.
53. His conclusion is incorrect. We could correct his thinking by showing him a counterexample to his conjecture. One such counterexample is $12 \div 3 = 4$.
54. In general, we view growing patterns as infinite patterns. The pattern that Amelia has given shows a decreasing numbers of squares but is only finite. Consequently, we would have to say that she has not answered the question entirely correctly.
55. Answers will vary.
56. Answers will vary.
57. Answers will vary.
58. Answers will vary.

■ Section 1.3 Learning Assessment

1. (1) The premises must be true in and of themselves. (2) The premises must be arranged in such a way that they lead logically to the conclusion.
2. **a.** In a conjunction both simple statements must be true for the entire statement to be true. In a disjunction only one or the other or both simple statements must be true for the entire statement to be true. **b.** If the "or" is exclusive, there is a choice between one or the other, but not both. If the "or" is inclusive, there is a choice is between one or the other or possibly both.
3. **a.** If a statement is true, its negation is false. If a statement is false, its negation is true. **b.** A conjunction is true only if both simple statements are true. Otherwise it is false. **c.** A disjunction is false only if both simple statements are false. Otherwise it is true. **d.** A conditional is false only if the hypothesis is true and the conclusion is false. Otherwise it is true.
4. In many senses, a conditional statement is a cause-and-effect relationship, where *p* is the cause of the effect *q*.
5. **a.** We form the converse by switching the hypothesis and the conclusion of the conditional. **b.** We form the inverse by negating the hypothesis and the conclusion of the conditional. **c.** We form the contrapositive by switching and negating the hypothesis and the conclusion of the conditional.
6. **a.** Two statements are logically equivalent if and only if they have the same truth values. **b.** The biconditional is logically equivalent to the conjunction of a conditional and its converse: $p \leftrightarrow q \equiv (p \rightarrow q) \wedge (q \rightarrow p)$.
7. We reason directly by using the cause-and-effect relationships expressed in conditional statements. To reason indirectly, we again use the cause-and-effect relationship of the conditional. However, rather than arguing from the conditional directly, we use the contrapositive. If we assume the effect did not happen, then it logically follows that the cause did not happen either.
8. **a.** To reason directly from a conditional, we assume the cause-and-effect relationship of the conditional is true. Then, if the cause happens, the effect must also happen. **b.** To reason indirectly from a conditional, we assume the effect did not happen, then it logically follows that the cause did not happen either. **c.** To reason indirectly from a disjunction, we reason from a true or

statement, where one or the other condition must be true. If we can eliminate one possibility as false, then the other must be true.
9. Answers will vary.
10. In both error forms, we assume the conditional "If *p*, then *q*" is true. In the converse error, we incorrectly claim that if *q* happens, then *p* must follow. In the inverse error, we incorrectly claim that if $\sim p$ happens, then $\sim q$ must follow.
11. **a.** John is not a good swimmer. **b.** Susan did not go to work this morning. **c.** The weather is not warm today. **d.** It is not the case that I like cookies and ice cream.
12. **a.** $\sim s$. **b.** $s \wedge a$. **c.** $a \rightarrow s$. **d.** $\sim a \vee \sim s$.
13. **a.** $\sim t$. **b.** $l \vee t$. **c.** $\sim t \rightarrow l$. **d.** $t \leftrightarrow l$.
14. **a.** Thanksgiving is not my favorite holiday. **b.** Thanksgiving is my favorite holiday, and I like the Fourth of July. **c.** Thanksgiving is my favorite holiday, or I like the Fourth of July. **d.** It is not the case that Thanksgiving is my favorite holiday or that I like the Fourth of July. **e.** If I do not like the Fourth of July, then Thanksgiving is my favorite holiday.
15. **a.** I do not have lunch duty. **b.** Lunch is at 11:30, or I have lunch duty. **c.** Lunch is not at 11:30, and I do not have lunch duty. **d.** It is not the case that if lunch is at 11:30, then I have lunch duty. **e.** Lunch is at 11:30 if and only if I have lunch duty.
16. **a.** *Hypothesis:* You have money.
    Conclusion: You can sleep in a five-star hotel.
    If you have money, then you can sleep in a five-star hotel.
    b. *Hypothesis:* You lift weights.
    Conclusion: You are strong.
    If you lift weights, then you are strong.
    c. *Hypothesis:* Two even numbers are added.
    Conclusion: The result is even.
    If two even numbers are added, then the result is even.
    d. *Hypothesis:* $x = 6$ and $y = 2$.
    Conclusion: $x + y = 8$.
    If $x = 6$ and $y = 2$, then $x + y = 8$.
17. **a.** *Converse:* If you go on a picnic, then it is the Fourth of July $(p \rightarrow f)$.
    Inverse: If it is not the Fourth of July, then you do not go on a picnic $(\sim f \rightarrow \sim p)$.
    Contrapositive: If you do not go on a picnic, then it is not the Fourth of July $(\sim p \rightarrow \sim f)$.
    b. *Converse:* You teach elementary school if you like young children $(c \rightarrow e)$.
    Inverse: You do not like young children if you do not teach elementary school $(\sim e \rightarrow \sim c)$.
    Contrapositive: You do not teach elementary school if you do not like young children $(\sim c \rightarrow \sim e)$.
18. **a.** *Converse:* If $x = 3$, then $2 \cdot x = 6$ $(t \rightarrow s)$.
    Inverse: If $2 \cdot x \neq 6$, then $x \neq 3$ $(\sim s \rightarrow \sim t)$.
    Contrapositive: If $x \neq 3$, then $2 \cdot x \neq 6$ $(\sim t \rightarrow \sim s)$.
    b. *Converse:* It has five sides if a shape is a pentagon $(p \rightarrow f)$.
    Inverse: A shape is not a pentagon if it does not have five sides $(\sim f \rightarrow \sim p)$.
    Contrapositive: It does not have five sides if a shape is not a pentagon. $(\sim p \rightarrow \sim f)$.
19. **a.** If I slept last night, then I take a nap tomorrow. **b.** If the sum of two angles is 180°, then the angles are supplementary. **c.** If $x = 10$, then $x - 6 = 4$.

20. a. If a number is divisible by 5, then it ends in a 0 or 5, and if a number ends in 0 or 5, then it is divisible by 5.
b. If a rectangle is a square, then the sides are the same length, and if the sides of a rectangle are the same length, then it is a square.

21. a. A triangle has three sides the same length if and only if it is equilateral.   **b.** $x + 4 = 9$ if and only $x = 5$.

22. a. The conditional of p and q is logically equivalent to the contrapositive of p and q.   **b.** The converse of p and q is logically equivalent to the inverse of p and q.
c. The biconditional of p and q is logically equivalent to the conjunction of the conditional of p and q and its converse.

23. a. Statement.   **b.** Statement.   **c.** This is not a statement because "tomorrow" is a relative word.   **d.** This is not a statement because x is an unknown quantity.   **e.** This is not a statement because it is an exclamation.   **f.** This is not a statement because it is a question.

24. a. Statement.   **b.** This is not a statement because it is a question.   **c.** Statement.   **d.** This is not a statement because it is an exclamation.   **e.** This is not a statement because x is an unknown quantity.   **f.** This is not a statement because it cannot be determined to be true or false.

25. a. True.   **b.** False.   **c.** True.   **d.** True.
26. a. True.   **b.** True.   **c.** False.   **d.** True.
27. a. True.   **b.** False.   **c.** True.   **d.** False.   **e.** True.
28. a. True.   **b.** True.   **c.** False.   **d.** False.   **e.** True.
29. a. True.   **b.** True.   **c.** False.   **d.** True.

30. a.

p	q	$p \wedge \sim q$
T	T	F
T	F	T
F	T	F
F	F	F

b.

p	q	$\sim p \wedge \sim q$
T	T	F
T	F	F
F	T	F
F	F	T

c.

p	q	$p \rightarrow \sim q$
T	T	F
T	F	T
F	T	T
F	F	T

31. a.

p	q	$\sim(\sim p \wedge q)$
T	T	T
T	F	T
F	T	F
F	F	T

b.

p	q	$(p \wedge q) \rightarrow p$
T	T	T
T	F	T
F	T	T
F	F	T

c.

p	q	$(p \vee q) \rightarrow (p \wedge q)$
T	T	T
T	F	F
F	T	F
F	F	T

32. Let t represent "You hit your thumb with a hammer" and h represent "It will hurt." Symbolically, we represent the argument as:

$$t \rightarrow h$$
$$h$$
$$\therefore t$$

This is logically incorrect by the converse error.

33. Let m represent "You mop the floor," w represent "It will be wet," and s represent "You will slip." Symbolically, we represent the argument as:

$$m \rightarrow w$$
$$w \rightarrow s$$
$$\therefore m \rightarrow s$$

This is logically correct because it is a hypothetical syllogism.

34. Let m represent "You like Mozart" and c represent "You like classical music." Symbolically, we represent the argument as:

$$m \rightarrow c$$
$$\sim c$$
$$\therefore \sim m$$

This is logically correct because it denies the conclusion.

35. Let s represent "My sister and I share a room" and w represent "We get along." Symbolically, we represent the argument as:

$$s \rightarrow w$$
$$\sim s$$
$$\therefore \sim w$$

This is logically incorrect by the inverse error.

36. Let c represent "Tamika will watch cartoons" and s represent "Tamika will watch sports." Symbolically, we represent the argument as:

$$c \vee s$$
$$\sim s$$
$$\therefore c$$

This is logically correct by the process of elimination.

37. Let m represent "You make your bed" and n represent "You are neat." Symbolically, we represent the argument as:

$$m \rightarrow n$$
$$n$$
$$\therefore m$$

This is logically incorrect by the converse error.

38. Let s represent "A shape is a square," r represent "The shape is a rectangle," and q represent "The shape is a quadrilateral." Symbolically, we represent the argument as:

$$s \rightarrow r$$
$$r \rightarrow q$$
$$\therefore s \rightarrow q$$

This is logically correct because it is a hypothetical syllogism.

39. Let s represent "The number is divisible by 6" and t represent "The number is divisible by 3." Symbolically, we represent the argument as:

$$s \rightarrow t$$
$$\sim s$$
$$\therefore \sim t$$

This is logically incorrect by the inverse error.

40. He is popular.
41. This animal is a mouse or a hamster.
42. If Juanita studies hard, then she will get an A in this course.
43. You are not an honest person.
44. If this animal likes to swim on the water, then it likes to eat fish.
45. If you are a child, then you like cookies.

46. a. $\sim(p \wedge q) \equiv \sim p \vee \sim q$ indicates that the negation of p and q is logically equivalent to the negation of p or the negation of q. $\sim(p \vee q) \equiv \sim p \wedge \sim q$ indicates that the negation of p or q is logically equivalent to the negation of p and the negation of q. **b. i.** The test will not be on Monday, and the test will not be on Tuesday. **ii.** It is not cold, or it is not wet. **iii.** I was not born in a small town, or I did not move to the big city. **iv.** I do not like corn dogs, and I like peas.

47. Kayeaka will make enough money if she gets a job. If Kayeaka makes enough money, then she is going skiing on spring break. When she goes skiing, she will break her leg. If she breaks her leg, Kayeaka will have a bad spring break. Therefore, if Kayeaka gets a job, she will have a bad spring break.

48. If Mike takes a nap, he can study tonight. If Mike studies tonight, then he will pass his finals. Mike will graduate if he passes his finals. He can take a job in the city if he graduates. When he takes a job in the city, he can live in his dream house. Therefore, if Mike takes a nap, then he can live in his dream house.

49. Adams stole the sapphire, Brooks stole the diamond, Cole stole the ruby, and Dent stole the emerald.

50. Jimmy picked a turtle.

51. 14.

52.

53. a. There exists $x = 6$ such that $3 \cdot 6 = 18$. **b.** There exists $x = 0$ such that $0 \cdot y = 0$, where y is any number. **c.** There exists $x = 1$ such that $1^2 = 1$. **d.** There exists $x = 2$, which is divisible by only two numbers: 1 and 2.

54. a. There exists $x = 2$ such that $2 + 7 = 9$. **b.** There exists $x = 4$ such that $3 \cdot 4 < 13$ and $4 \cdot 4 > 15$. **c.** There exists $x = 10$ and $y = 5$ such that $10 + 5 = 15$ and $10 - 5 = 5$.

d. There exists $x = \dfrac{1}{2}$ such that $\left(\dfrac{1}{2}\right)^2 = \dfrac{1}{4} < \dfrac{1}{2}$.

55. There exists $a = 2$ and $b = 2$ such that
$$\frac{1}{a} + \frac{1}{b} = \frac{1}{2} + \frac{1}{2} = 1.$$

56. There exists natural numbers $a = 3$, $b = 4$, and $c = 5$ such that $a^2 + b^2 = 3^2 + 4^2 = 9 + 16 = 25 = 5^2 = c^2$.

57. a. $1^2 = 1 \geq 1$, $2^2 = 4 \geq 2$, $3^2 = 9 \geq 3$, $4^2 = 16 \geq 4$, and $5^2 = 25 \geq 5$. **b.** $1^3 = 1 \geq 1$, $2^3 = 8 \geq 2$, $3^3 = 27 \geq 3$, $4^3 = 64 \geq 4$, and $5^3 = 125 \geq 5$. **c.** $2 = 1 + 1$, $3 = 1 + 1 + 1$, $4 = 3 + 1$, $5 = 3 + 1 + 1$, $6 = 3 + 3$, $7 = 5 + 1 + 1$, $8 = 5 + 3$, $9 = 5 + 3 + 1$, and $10 = 5 + 5$.

58. $2 = 1^2 + 1^2$, $4 = 2^2$, $6 = 2^2 + 1^2$, $8 = 2^2 + 2^2$, $10 = 3^2 + 1^2$, $12 = 2^2 + 2^2 + 2^2$, $14 = 3^2 + 2^2 + 1^2$, $16 = 4^2$, $18 = 3^2 + 3^2$, $20 = 4^2 + 2^2$, $22 = 3^2 + 3^2 + 2^2$, $24 = 4^2 + 2^2 + 2^2$, and $26 = 5^2 + 1^2$.

59. a. Let d represent "It is a duck," w represent "It waltzes," o represent "It is an officer," and p represent "It is my poultry." We can then rewrite the argument as:

If it is a duck, then it does not waltz ($d \to \sim w$).

If it is an officer, then it waltzes ($o \to w$).
If it is my poultry, then it is a duck ($p \to d$).

By rearranging the statements and using a contrapositive we have:

If it is my poultry, then it is a duck ($p \to d$).
If it is a duck, then it does not waltz ($d \to \sim w$).
If it does not waltz, then it is not an officer ($\sim w \to \sim o$).

From this we can conclude that "If it is my poultry, then it is not an officer" or "None of my poultry are officers" ($p \to \sim o$).
b. Let e represent "The person is not experienced," i represent "The person is incompetent," and b represent "The person is always blundering." We can then rewrite the argument as:

If the person is incompetent, then the person is not experienced ($i \to e$).
Jenkins is always blundering (b).
If a person is always blundering, then the person is incompetent ($b \to i$).

By rearranging the statements, we have:

If a person is always blundering, then the person is incompetent ($b \to i$).
If a person is incompetent, then the person is not experienced ($i \to e$).
Jenkins is always blundering (b).

From this we can conclude that "Jenkins is not experienced" (e).

c. Let p represent "It is a pudding," n represent "It is nice," and w represent "It is wholesome." We can then rewrite the argument as:

If it is a pudding, then it is nice ($p \to n$).
This is a pudding (p).
If it is wholesome, then it is not nice ($w \to \sim n$).

By rearranging the statements and using a contrapositive, we have:

If it is a pudding, then it is nice ($p \to n$).
If it is nice, then it is not wholesome ($n \to \sim w$).
This is a pudding (p).

From this we can conclude that "It is not wholesome" ($\sim w$).
d. Let b represent "It is a baby," i represent "It is illogical," c represent "It can manage a crocodile," and d represent "It is despised." We can then rewrite the argument as:

If it is a baby, then it is illogical ($b \to i$).
If it can manage a crocodile, then it is not despised ($c \to \sim d$).
If it is illogical, then it is despised ($i \to d$).

By rearranging the statements and using a contrapositive, we have:

If it is a baby, then it is illogical ($b \to i$).
If it is illogical, then it is despised ($i \to d$).
If it is despised, then it cannot manage a crocodile ($d \to \sim c$).

From this we can conclude that "If it is a baby, then it cannot manage a crocodile" or "No baby can manage a crocodile" ($b \to \sim c$).
60. a. Answers will vary. **b.** Answers will vary.
61. a. Answers will vary. **b.** Answers will vary.

62. Answers will vary.

63. The sentences are not negations of one another because they do not describe opposite characteristics.

64. When Anna says that she does not like brownies and ice cream, she is logically implying that she does not like them together. However, it does not mean that she does not like ice cream or brownies individually.

65. Two negatives logically cancel each other out.

66. Answers will vary.

67. We use inductive reasoning to identify or create mathematical conjectures, and then we use deductive reasoning to verify them.

68. The conjecture is based on inductive reasoning. If done by experimentation only, it is impossible to verify the conjecture to be absolutely true. Even if the experiment is done many times, inductive reasoning cannot take into account that the experiment might fail in a single situation.

69. **a.** Inductive reasoning.   **b.** You may want to look up past records of cars that have been sold from the lot and determine whether they have remained trouble free for 5,000 mi.   **c.** Answers will vary.

■ Section 1.4 Learning Assessment

1. Problem solving is a process by which we use mathematical skills and knowledge in new and unfamiliar ways to find the answer to a mathematical question whose solution is not known in advance. Good problem solvers take time to understand the problem, can change directions as needed, learn to use a variety of tools, reflect on their progress, are patient and take time to explore, keep track of their work, adjust their planning as necessary, and do not give up easily.

2. With exercises, we apply a known procedure or fact systematically to arrive at our answer. There is no one set of skills or ideas that we can use with every problem. So we may need to think creatively or use our mathematical knowledge in new and unfamiliar ways to solve them.

3. The principal parts of the problem are key bits of information that will be used to solve it. They may include the goal of the problem, unknowns, or given information.

4. Our plan should come not only from our understanding of the problem, but also from our common sense, mathematical knowledge, and previous experience.

5. Looking back over a solution gives us an opportunity to check our solution and allows us to review and consolidate what we have learned. Answers will vary for the second part of the question.

6. Drawing a picture is a useful strategy when the problem involves geometry, measurement, or another physical situation. Pictures help us gain insight into the problem and how we might use other strategies in the solution.

7. The key to using a table or a list is to be sure that we have listed all the outcomes and that none have been duplicated. We should therefore try to be systematic or sequential in the way we make our list or table.

8. **a.** It is reasonable to work backward when we know the end result and our task is to find the starting point or the method by which the result was obtained.
b. It is reasonable to use guess and check when there seems to be no clear-cut way to solve a problem.

9. We can use inductive reasoning in problems that involve patterns. We can use deductive reasoning in problems that require the explanation of a process or that involve a statement that can be drawn from a collection of specified conditions.

10. Writing an equation differs from using a formula in that the algebraic equation is particular to the problem and cannot be applied in other situations. A formula is an algebraic expression that can be used in many problem-solving situations.

11. 32.
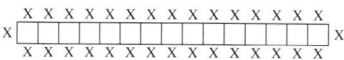

12. Yes. Make two vertical cuts and one horizontal cut.
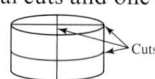

13. Yes.

14. Using a table or a list is a good strategy for this problem because it shows how many outcomes satisfy a given condition. There are 20 ways to make 37¢.

Dimes	Nickels	Pennies	Dimes	Nickels	Pennies
3	1	2	1	1	22
3	0	7	1	0	27
2	3	2	0	7	2
2	2	7	0	6	7
2	1	12	0	5	12
2	0	17	0	4	17
1	5	2	0	3	22
1	4	7	0	2	27
1	3	12	0	1	32
1	2	17	0	0	37

15. Two possible solutions are:

16.
```
x x
x | x
  x x
```

17. **a.** Because Bill received 5 pieces of candy, Kasey must have had $5 \cdot 5 = 25$ pieces of candy. This means Cindy had $3 \cdot 25 = 75$ pieces and Derrick must have had $2 \cdot 75 = 150$ pieces.   **b.** Working backward is an effective strategy because the problem has a sequence of operations that can be undone to find the answer.   **c.** Start with the answer and perform the given operations. If the answer is 5, we know that our answer is correct.

18. 26.

19. **a.** ♠. **b.** ♥. **c.** ♦. **d.** ↑. **e.** ✳.

20. **a.** 2. **b.** 3. **c.** 4.   **d.** Number of cuts $= n - 1$.

21. James is the father, Jill is the mother, Jenny is the sister, and Jonathan is the brother.

22. The numbers are 23 and 36. Writing an equation is an effective strategy because we want to find two unknown numbers.
23. The missing pages are pages 85 and 86.
24. **a.** 12 in.^{2}. **b.** 75 cm^2. **c.** $20 - 4\pi \approx 7.44$ in.^{2}.
25. 1,800 ears of corn.
26. 12. One possible arrangement is

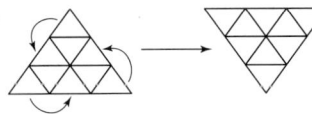

27. 32 posts.
28. About 4 days and 17 hr.
29. 22 cubes.
30. The stock will hit $25.00 in 2021.
31.

32. No. It cannot be drawn without retracing any lines or without lifting your pencil from the paper. The points with an odd number of lines prohibit us from doing so.
33. One possible solution is:

34. G does not belong. It should be a D.
35. **a.** F. **b.** X. **c.** F. **d.** M.
36. 40.
37. $4\frac{1}{4}$ in.
38. One way to draw the lines is:

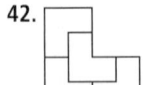

39. 37 triangles.
40. Each coin must move 4 spots for a total of 16 moves. However, the two nickels jump over each penny, eliminating 4 moves. This means 12 moves are needed.
41. Select a piece of fruit from the box marked oranges and bananas. If it is a banana, then the box labeled oranges and bananas must be the banana box, the box labeled bananas must be the orange box, and the box labeled oranges must be the oranges and bananas box. If it is an orange, then the box labeled oranges and bananas must be the orange box, the box labeled oranges must be the bananas box, and the box labeled bananas must be the oranges and bananas box.
42.
43. 18 hr.
44.
45. $A \approx 7.63$ in.
46. Take six coins and place three on each side of the scale. If they balance, the heavy counterfeit must be one of the two coins off to the side. Weigh the two coins to find the counterfeit directly. If the coins to do not balance on the first weighing, the counterfeit is on the side that goes down. Take two of these coins and weigh them directly. If they balance, the counterfeit is the unweighed coin. If not, the balance shows the counterfeit directly.
47. Place one penny in the first cup, one in the second cup, and two in the third. Then place the second cup in the first, and the third cup in the second. This puts four coins in the first cup, three coins in the second, and two in the third.
48. My birthday is December 31, and I am speaking on January 1.
49. Jim is the liar.
50. Answers will vary.
51. Answers will vary.
52. Answers will vary.
53. Answers will vary.
54. Answers will vary.
55. Answers will vary.
56. Answers will vary.
57. Answers will vary.

■ Section 1.5 Learning Assessment

1. There are 24 arrangements of the letters in the word "math":

math	maht	mhta	mhat	mtah	mtha
amth	amht	ahtm	ahmt	atmh	athm
tmah	tmha	tham	thma	tahm	tamh
htam	htma	hamt	hatm	hmta	hmat

We can use multiplication and the fundamental counting principle. There are four choices for the first letter, three for the second, two for the third, and one for the fourth. Hence, there are $4 \cdot 3 \cdot 2 \cdot 1 = 24$ possible arrangements.

2. **a.** The head is 3 in., the body is 9 in., and the tail is 3 in.
   b. Let h = length of the head, b = length of the body, and t = length of the tail. Then, $t + h + b = 15$, $t = h$, and $b = 2h + h = 3h$. Because $t + h + b = 15$, then $h + h + 3h = 5h = 15$. Hence, $h = 3$, $b = 9$, and $t = 3$.
   c. Answers will vary.

3. **a.** The numbers are 54 and 56. **b.** Let x represent the smaller of the two numbers and use the equation $x + (x + 2) = 110$ **c.** Answers will vary.

4. No. It is not possible. If it were possible, the sum of the numbers would be divisible by 3. However, $1 + 2 + 3 + 4 + 6 + 7 + 8 + 9 + 10 = 50$, which is not divisible by 3.

5. **a.** It is much better to be paid by the second method; on the 31st day, we will be paid $10,737,418.24. **b.** The number of cents is equal to 2^{n-1} where n is the number of the day.

6. **a.** The table shows that there are four flies and six dragonflies.

Flies	Dragonflies	Number of Wings
0	10	40
1	9	38
2	8	36
3	7	34
4	6	32
5	5	30
6	4	28
7	3	26
8	2	24
9	1	22
10	0	20

b. Other strategies that could be used are drawing a picture, writing an equation, and guess and check.

7. 7: 4 brothers and 3 sisters.

8. 254 matches.

9. 46 toothpicks.

10. 180 toothpicks.

11. 13 dots on each side of the square.

12. 7 mi.

13. 72 blocks.

14. There are 5 colors and 3 types, for a total of $3 \cdot 5 = 15$ writing utensils. She has 8, so she needs 7 more: red, green, and yellow markers, blue and purple crayons, and blue and yellow colored pencils.

15. She bought a 2-lb package of hamburger and a bunch of bananas.

16. About 6 months.

17. $5.25.

18. About 2.57 hr = 2 hr and 34 min.

19. 17 chickens and 18 pigs.

20. There will not be enough packages after the eighth stop.

21. 45 different passwords.   22. 10 ft by 14 ft.

23. 61 pieces of candy.   24. 27 min.

25. She had trouble at 10:20 a.m.   26. 13 trains.

27. The pairs are $1 + 5 = 6$, $2 + 7 = 9$, $3 + 8 = 11$, $4 + 10 = 14$, and $6 + 9 = 15$.

28. There are 17 such numbers.   29. 15 rungs.

30. 312 chairs.   31. About 10 mice.

32. 139 spaces.   33. 15 plants.

34. Here is one possible solution:

17	24	1	8	15
23	5	7	14	16
4	6	13	20	22
10	12	19	21	3
11	18	25	2	9

35. 35 different routes.

36. Eve should pay Erica $71.99 and Elizabeth should pay Erica $133.44.

37. **a. i.** 25, 36, 49.   **ii.** 35, 48, 63.   **iii.** 15, 21, 28.
   b. i. 225 dots.   **ii.** 255 dots.   **iii.** 120 dots.
   c. i. number of dots = n^2, where n is the position of the sequence.   **ii.** number of dots = $n(n + 2)$, where n is the position in the sequence.
   iii. Number of dots = $\dfrac{n(n + 1)}{2}$, where n is the position in the sequence.

38. The following table demonstrates how to pour water from one container into another to get 4 qt in the 8-qt container and in the 5-qt container.

Step	8-Qt Container	5-Qt Container	3-Qt Container
1	8	0	0
2	5	0	3
3	5	3	0
4	2	3	3
5	2	5	1
6	7	0	1
7	7	1	0
8	4	1	3
9	4	4	0

39. 6 cuts.

40. A calculator will not have enough positions on the screen to display the units digit of 2^{50} and 2^{100}. However, we can find the number by considering the following pattern in which the units digits continuously repeat in a sequence of 2, 4, 8, and 6.

   $2^1 = 2$ $2^2 = 4$ $2^3 = 8$ $2^4 = 16$
   $2^5 = 32$ $2^6 = 64$ $2^7 = 128$ $2^8 = 256$
   $2^9 = 512$ $2^{10} = 1,024$ $2^{11} = 2,048$ $2^{12} = 4,096$

   The number 2^{50} will appear in the second column, so its units digit must be a 4. The number 2^{100} will appear in the fourth column, so its units digit must be a 6.

41. **a.** 55 rectangles.   **b.** Answers will vary.   **c.** Answers will vary.

42. Answers will vary.   43. Answers will vary.

44. Answers will vary.   45. Answers will vary.

Chapter 1 Review Exercises

1. **a.** $x - 8 = 5$.   **b.** $\dfrac{3 + x}{7} \neq 8$.   **c.** $\dfrac{3}{4}y = 150$.

2. **a.** The product of two and x is eight.   **b.** Three times the quantity of x minus two is less than two.   **c.** Two times the quantity of x minus one is the same as x plus five.

3. Two sides of the trapezoid must be parallel, and the other two sides must be the same length. Also, consecutive angles on the same base must be congruent.

4. The scores were somewhat split, in that one group of students seemed to do well and another group of students did poorly.

5. Answers will vary.

6. Answers will vary.

7. Answers will vary.

8. Initially, there was a great increase in the top speeds, but in recent years the top speeds have increased only slightly.

9. It will rain the entire time on the next vacation. Inductive reasoning.

10. $z = 48$. Deductive reasoning.

11. I am likely to order pizza with some type of meat. Inductive reasoning.

12. **a.** Three is between 1 and 5 and is not even.
   b. June is a month of the year that does not have an a in its spelling.
   c. Six is between 5 and 10 and is not divisible by 5.

13. The next three equations are $37,037 \times 15 = 555,555$; $37,037 \times 18 = 666,666$; and $37,037 \times 21 = 777,777$. A pattern continues, but changes form at $37,037 \times 30 = 1,111,110$.

14. **a.** Carissa does not like to wear blue.   **b.** I own a green truck.

15. **a.** $\sim a$.   **b.** $a \wedge \sim d$.   **c.** $\sim a \rightarrow \sim d$.   **d.** $a \leftrightarrow d$.

16. **a.** I do not have class at 1 p.m.   **b.** I have class at 1 p.m., and I have a test.   **c.** I do not have class at 1 p.m., or I have a test.   **d.** If I have a test, then I have class at 1 p.m.
   e. It is not the case that if I have class at 1 p.m., then I have a test.

17. If $15 \div x = 5$, then $x = 3$, and if $x = 3$, then $15 \div x = 5$.

18. *Converse:* If you wear sunscreen, then you go to the beach.
   Inverse: If you do not go to the beach, then you do not wear sunscreen.
   Contrapositive: If you do not wear sunscreen, then you do not go to the beach.

19. a. False. **b.** True. **c.** True. **d.** False.
20. a. False. **b.** False. **c.** True. **d.** True. **e.** True.
21.

p	q	$\sim p \vee \sim q$
T	T	F
T	F	T
F	T	T
F	F	T

22. Therefore, you are not a vegetarian.
23. If it tastes good, then it is high in fat.
24. If it is a square, then it is a rectangle.
25. If $x = 0.25$, then $(0.25)^2 = 0.0625 \leq 0.25$.
If $x = 0.5$, then $(0.5)^2 = 0.25 \leq 0.5$.
If $x = 0.75$, then $(0.75)^2 = 0.5625 \leq 0.75$.
If $x = 1$, then $1^2 = 1 \leq 1$.
26. There are six ways to write 20 as the sum of 2s, 5s, and
10s: $10 + 10$, $10 + 5 + 5$, $10 + 2 + 2 + 2 + 2 + 2$,
$5 + 5 + 5 + 5$, $5 + 5 + 2 + 2 + 2 + 2 + 2$, and
$2 + 2 + 2 + 2 + 2 + 2 + 2 + 2 + 2 + 2$.
27. 36 hotdogs.
28. He has one half-dollar, one quarter, four dimes, and four pennies.
29. Jim has 35 marbles, and Joe has 25.
30. 120 blocks.
31. 40 numbers.
32. 840 dozen.
33. There will not be enough packages after the seventh stop.
34. The missing pages are 289, 290, 291, and 292.
35. 20 cubes.

CHAPTER 2

■ Section 2.1 Learning Assessment

1. A set is well-defined; that is, we know exactly what objects are or are not in the set.
2. Elements in a set are not listed more than once, and the order of the elements in a set is not important. Sets are concerned only with membership—what is or is not in the set.
3. The empty set is the set that contains no elements. The universal set is the set that contains all possible elements in a given discussion.
4. The subset relationship allows two sets to be the same, whereas the proper subset relationship does not. As an example, consider the sets $A = \{1, 2, 3\}$, $B = \{1, 2\}$, and $C = \{3, 2, 1\}$. B is both a subset and a proper subset of A, where as C is a subset of A but not a proper subset.
5. a. Suppose $\varnothing \not\subseteq A$. By the definition of a subset, there must be an element in $\varnothing$ that is not in A. However, because there are no elements in the empty set, there cannot be an element in $\varnothing$ that is not in A. This means that the statement $\varnothing \not\subseteq A$ is false, and we conclude that $\varnothing \subseteq A$. **b.** Every element in set A must be in A, so $A \subseteq A$.
6. A and B must contain the same elements, so $A = B$.
7. Equal sets have exactly the same elements, whereas equivalent sets have the same number of elements.

8. a. A one-to-one correspondence between two sets A and B is a pairing of the elements of A with the elements of B so that each element of A corresponds to exactly one element of B, and vice versa. **b.** To count, a one-to-one correspondence is made between the set of number words and the set to be counted.
9. Rote counting is counting by making a one-to-one correspondence. When counting on, we begin at a certain place in the number word sequence and rote-count from that number. When counting back, we begin at certain place in the number word sequence and count backward. When skip counting, we begin at certain place in the number word sequence and count on by 2s, 5s, 10s, or other values.
10. The whole numbers contain zero, whereas the natural numbers do not.
11. A set is finite if and only if it is empty or there exists a one-to-one correspondence between the set and a set of the form $\{1, 2, 3 \ldots, n\}$, where n is a natural number. If not, then the set is infinite.
12. a. The set of letters in the word "cute." **b.** The set of all states in the United States. **c.** The set of odd natural numbers less than 10. **d.** The set of natural numbers greater than or equal to 3 and less than 7.
13. a. $\{e, l, m, n, t, a, r, y\} = \{x \mid x$ is a letter in the word "elementary"$\}$ **b.** {George Bush, Bill Clinton, George W. Bush, Barack Obama} $= \{x \mid x$ is United States president serving between 1990 and 2010$\}$. **c.** $\{10, 20, 30, \ldots\} = \{x \mid x$ is a multiple of 10$\}$.
14. a. {red, blue, yellow} $= \{x \mid x$ is a primary color$\}$. **b.** $\{2, 4, 6, \ldots, 20\} = \{x \mid x$ is an even natural number less than 20$\}$. **c.** $\{1, 4, 9, 16, \ldots\} = \{x \mid x$ is a square number$\}$.
15. a. Verbal descriptions can completely and precisely describe any set, but they are often unwieldy. **b.** Rosters clearly and efficiently represent the elements in the set, but not all sets can be represented with a roster. **c.** Set-builder notation is more efficient than a verbal description and can be used when a roster cannot. Sometimes the verbal condition in the notation is lengthy. **d.** Venn diagrams provide a visual representation of sets that makes it easier to see the relationships between them, but they are not as compact and efficient as other representations.

16.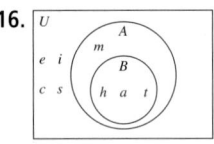
17.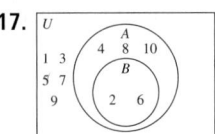

18.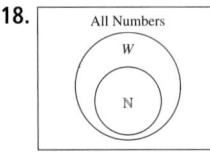

19. a. $a \in A$. **b.** $1 \notin \{2, 3, 4\}$. **c.** $\{1, 2, 3\} \subseteq \{1, 2, 3, 4, 5\}$. **d.** $\{1, 2, 3\} \neq \{a, b, c\}$.
20. a. Four is an element of the set D. **b.** Five is not an element of the set containing 1, 2, 3, and 4. **c.** The set containing purple is not a proper subset of the set containing red, blue, and yellow.

d. The set containing 1, 2, and 3 is equivalent to the set containing x, y, and z.

21. 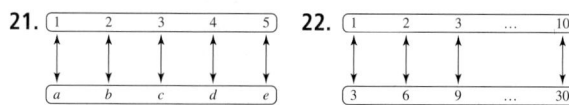 **22.**

23. The set $\{\varnothing\}$ is not actually empty. It contains one element, the empty set.

24. a. Not well-defined. Beautiful is a relative term.
b. Well-defined. **c.** Well-defined. **d.** Not well-defined. It is not clear what set we are to draw the numbers from. The answer will differ depending on whether we draw numbers from the natural numbers or the whole numbers.
e. Well-defined.

25. a. $\in$. **b.** $\notin$. **c.** $\in$. **d.** $\in$.

26. a. $\subset$. **b.** $\subseteq$. **c.** $\subset$. **d.** $\subseteq$.

27. a. $\subseteq$. **b.** $\sim$. **c.** $=$. **d.** $\sim$.

28. $\varnothing$, $\{t\}$, $\{o\}$, $\{e\}$, $\{t, o\}$, $\{t, e\}$, $\{o, e\}$, and $\{t, o, e\}$.

29. $\varnothing$, $\{m\}$, $\{a\}$, $\{t\}$, $\{h\}$, $\{m, a\}$, $\{m, t\}$, $\{m, h\}$, $\{a, t\}$, $\{a, h\}$, $\{t, h\}$, $\{m, a, t\}$, $\{m, a, h\}$, $\{m, t, h\}$, and $\{a, t, h\}$.

30. a. $2^5 = 32$. **b.** $2^8 = 256$. **c.** $2^{10} - 1 = 1{,}023$. **d.** $2^{15} - 1 = 32{,}767$.

31. a. 2. **b.** 4. **c.** 5. **d.** 7.

32. a. False. Not every element in A is in B. **b.** True.
c. False. C does not have an element that is not in A.
d. True. **e.** False. The elements are not exactly the same in each set. **f.** True. **g.** True. **h.** False. Not every element in C is in D.

33. a. False. **b.** True. **c.** False. **d.** True. **e.** False.
f. True. **g.** False. **h.** True.

34. A one-to-one correspondence can be made between sets B and E, sets B and G, sets E and G, and sets C and F.

35. a. Rote counting from 1 to 10. **b.** Skip counting by 2s from 0 to 22. **c.** Counting on from 7 to 20. **d.** Skip counting by 6s from 1 to 25. **e.** Counting back four numbers from 17. **f.** Skip counting backward by 3s from 21 to 3.

36. a. $\{6, 7, 8, 9, 10\}$. **b.** $\{16, 15, 14, \dots, 8\}$.
c. $\{3, 6, 9, 12, 15\}$. **d.** $\{10, 15, 20, \dots, 50\}$.

37. a. 11. **b.** 15. **c.** 20. **d.** 13.

38. a. 10. **b.** 51. **c.** 31. **d.** 16.

39. a. 8. **b.** 0. **c.** 5. **d.** 6.

40. a. Answers will vary. **b.** Answers will vary.
c. Answers will vary. **d.** Answers will vary.

41. a. Finite. **b.** Finite. **c.** Finite. **d.** Infinite.

42. a. Infinite. **b.** Finite. **c.** Infinite. **d.** Finite.

43. 40,320.

44. 479,001,600.

45. a. 2 one-to-one correspondences. **b.** 6 one-to-one correspondences. **c.** 24 one-to-one correspondences.
d. $n \cdot (n - 1) \cdot \ldots \cdot 3 \cdot 2 \cdot 1$ one-to-one correspondences.

46. a. 1.3077×10^{12} different one-to-one correspondences.
b. About 41,466.08 years.

47. If the count begins at 18, then 24 even and 24 odd numbers are named.

48. If the count begins at 15, then 31 numbers are named.

49. The numbers that can be used to skip-count from 0 to 150 are $\{1, 2, 3, 5, 6, 10, 15, 25, 30, 50, 75\}$. The numbers that can be used to skip-count from 0 to 180 are $\{1, 2, 3, 4, 5, 6, 9, 10, 12, 15, 18, 20, 30, 36, 45, 60, 90\}$.

50. a. Because $A \subset B$, the smallest $n(B)$ could is 11. The largest $n(B)$ could be is infinity. **b.** Because $A \subseteq B$, the smallest $n(B)$ could be is 10. The largest $n(B)$ could be is infinity.

51. a. 0, 1, 2, 3, 4, or 5. **b.** 0, 1, 2, 3, 4, 5, or 6.

52. a. 32 different ways to top the pizza. **b.** 10 ways to top the pizza with exactly 3 toppings.

53. 6 possible committees. **54.** 10 possible committees.

55. a. True. **b.** False. If B is a subset of A, then it is possible that $A = B$, making the statement false. **c.** True.
d. False. If B is equivalent to A, they will have the same number of elements but not necessarily the same elements.

56. a. True. **b.** True. **c.** True. **d.** False. If A is a proper subset of B, then B must have an element that is not in A. This means $n(B) > n(A)$, so A cannot be equivalent to B.

57. a. 3. **b.** 7. **c.** 15. **d.** $2^n - 1$.

58. a. 3. **b.** 6. **c.** 10. **d.** $(n - 1) + (n - 2) + \dots + 3 + 2 + 1$.

59. a. Yes. If every element of A is in B and every element of B is in C, then every element of A must also be an element of C. **b.** Yes. If a is in B and every element of B is in C, then a must be in C. **c.** No. If a is in B and none of the elements of B are in C, then a cannot be in C.

60. Consider set A. Every element of A must be in A, so by the definition of a subset, A must be a subset of itself. However, there is no element of A that is not back in A. So A cannot be a proper subset of itself.

61. They are the same size, as demonstrated by the following one-to-one correspondence:

62. Answers will vary. **63.** Answers will vary.

64. a. $\in$. **b.** $\subseteq$. **c.** $\subseteq$. **d.** $\in$.

65. Answers will vary.

66. a. Timmy does not know the number word sequence.
b. Jodie is double counting. **c.** LaToya is skipping an object, meaning she is not making a one-to-one correspondence correctly. **d.** Juan does not realize that the last number used represents the number of objects.

67. Answers will vary. **68.** Answers will vary.

69. Some mathematics educators believe that teaching children to begin counting from 0 will help them better understand the value of 0 as a cardinal number.

70. Answers will vary.

71. a. Answers will vary. **b.** Answers will vary.

72. Answers will vary.

73. a. $5 \cdot 2 = 10$. **b.** $6 \cdot 4 = 24$. **c.** $4 \cdot 10 = 40$.
d. $9 \cdot 7 = 63$. **e.** $3 \cdot 25 = 75$. **f.** $9 \cdot 9 = 81$.

74. Answers will vary.

■ Section 2.2 Learning Assessment 2.2

1. a. $A \cup B$ is the set of all elements in A or in B or in both.
b. $A \cap B$ is the set of all elements common to A and B.
c. $B - A$ is the set of all elements in B that are not in A.
d. $\overline{A}$ is the set of all elements in the universe U that are not in A. **e.** $A \times B$ is the set of all ordered pairs (a, b) where $a \in A$ and $b \in B$.

2. Two sets are disjoint if their intersection is empty or if they have no elements in common.

3. The set difference and complement are similar in that both operations are computed by taking the elements that are in one set but not in another. The set difference is always relative to another set, whereas the complement is always relative to the universal set.

4. Order is not important in a set with two elements, whereas order is important in an ordered pair.

5. The intersection is used to sort by similarities, and the set difference is used to sort by differences.

6. a. The set of all people who like chocolate and donuts.
b. The set of all people who like chocolate or donuts.
c. The set of all people who like chocolate but not donuts.
d. The set of all people who like donuts but not chocolate.

7. a. $A - B$. **b.** $A \cup B$. **c.** $\overline{A}$. **d.** $\overline{(A \cup B)}$.

8. a. $B \cup A = \{x \mid x \in B \text{ or } x \in A\}$.
b. $A \cap B = \{x \mid x \in A \text{ and } x \in B\}$.
c. $B - A = \{x \mid x \in B \text{ and } x \notin A\}$.
d. $\overline{B} = \{x \mid x \in U \text{ and } x \notin B\}$.
e. $B \times A = \{(b, a) \mid b \in B \text{ and } a \in A\}$.

9. a. $A \cup B$. **b.** $A \cap B$. **c.** $A - B$. **d.** $\overline{A}$. **e.** $A \times B$.
f. $\overline{(A \cup B)}$.

10. a. The set identity $A \cup B = B \cup A$ states that, when computing a union, the order in which the sets are written does not matter; the result will be the same.
b. The identity states that it does not matter if we take the intersection of A and B first or the intersection of B and C first; the result will be the same. **c.** The identity $A \cup \varnothing = A$ states that the union of a set A with the empty set is set A. **d.** The identity $A \cup U = U$ states that the union of a set A with the universe is the universe.

11. a. This Venn diagram allows for A and B to have elements in common and for B to have elements that are not in A. However, it does not allow for A to have elements that are not in B. **b.** This Venn diagram does not allow for A and B to have elements in common.

12. 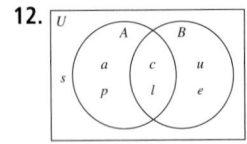 **13.**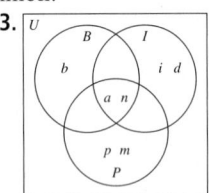

14. a. 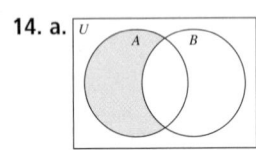 **b.**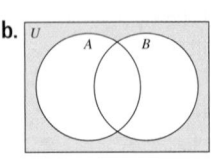

c. 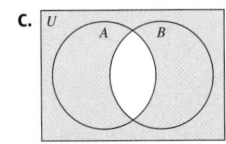 **d.**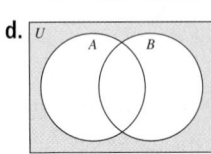

15. a. 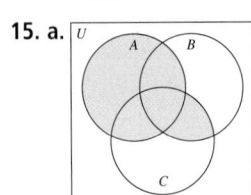 **b.**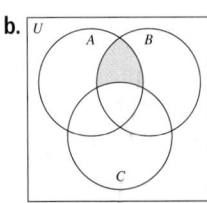

c. 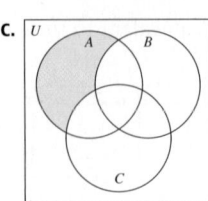 **d.**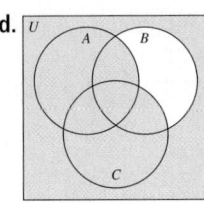

16. a. $B - A$. **b.** $\overline{A \cap B}$. **c.** $(A - B) \cup (B - A)$.

17. a. $A \cap \overline{B} \cap C$. **b.** $(A \cap B) \cap C$. **c.** $(A \cap B) \cup (A \cap C)$.

18. a. $\{i, n, t, e, r, s, c, o\}$. **b.** $\{i, n, t, e\}$. **c.** $\{r\}$. **d.** $\{s, c, o\}$.

19. a. $\{2, 3, 4, 6, 8, 9, 10, 12, 14, 15, 16, 18, 20, 21\}$.
b. $\{6, 12, 18\}$. **c.** $\{2, 4, 8, 10, 14, 16, 20\}$. **d.** $\{3, 9, 15, 21\}$.

20. a. $\{1, 3, 7, 8\}$. **b.** $\{1, 2, 4, 6\}$. **c.** $\{2, 3, 5\}$. **d.** $\{1\}$.
e. $\{1, 2, 3, 4, 5, 6\}$.

21. a. $\{i, n, g, t, o\}$. **b.** $\{w, a, s, o\}$. **c.** $\{w, a, h, i, t\}$.
d. $\{h, i, n, g, t, o\}$. **e.** $\{i, t\}$.

22. a. $B - A$. **b.** $A \cap B$. **c.** $A - B$. **d.** $A \cup B$.

23. a. $\{(m, i), (m, s), (a, i), (a, s), (t, i), (t, s), (h, i), (h, s)\}$.
b. $\{(m, f), (m, u), (m, n), (a, f), (a, u), (a, n), (t, f), (t, u),$ $(t, n), (h, f), (h, u), (h, n)\}$. **c.** $\{(i, m), (i, a), (i, t), (i, h),$ $(s, m), (s, a), (s, t), (s, h)\}$. **d.** $\{(i, f), (i, u), (i, n), (s, f),$ $(s, u), (s, n)\}$.

24. a. $\{(k, e), (k, r), (i, e), (i, r), (n, e), (n, r), (d, e), (d, r)\}$.
b. $\{(k, g), (k, a), (k, r), (k, t), (k, e), (k, n), (i, g), (i, a),$ $(i, r), (i, t), (i, e), (i, n), (n, g), (n, a), (n, r), (n, t), (n, e),$ $(n, n), (d, g), (d, a), (d, r), (d, t), (d, e), (d, n)\}$. **c.** $\{(e, k),$ $(e, i), (e, n), (e, d), (r, k), (r, i), (r, n), (r, d)\}$. **d.** $\{(e, g),$ $(e, a), (e, r), (e, t), (e, e), (e, n), (r, g), (r, a), (r, r), (r, t),$ $(r, e), (r, n)\}$.

25. a. $A = \{1, 2, 3\}$ and $B = \{a, b, c\}$. **b.** $A = \{1, 2\}$ and $B = \{1, 2\}$. **c.** $A = \{4\}$ and $B = \{1, 2, 3, 4\}$.

26. Possible answers are:
a. $A = \{1\}$ and $B = \{a\}$. **b.** $A = \{1, 2\}$ and $B = \{a, b\}$.
c. $A = \{1, 2\}$ and $B = \{a, b, c, d, e\}$. **d.** $A = \{1, 2, 3, 4, 5, 6, 7\}$ and $B = \{a\}$.

27. a. 12. **b.** 10. **c.** 28. **d.** $a \cdot b$.

28. Possible values are $n(A) = 1$ and $n(B) = 36$, $n(A) = 2$ and $n(B) = 18$, $n(A) = 3$ and $n(B) = 12$, $n(A) = 4$ and $n(B) = 9$, $n(A) = 6$ and $n(B) = 6$, $n(A) = 9$ and $n(B) = 4$, $n(A) = 12$ and $n(B) = 3$, $n(A) = 18$ and $n(B) = 2$, $n(A) = 36$ and $n(B) = 1$.

29. a. B. **b.** A. **c.** $\varnothing$.

30. a. $\mathbb{N}$. **b.** $\varnothing$. **c.** E. **d.** $\mathbb{N}$. **e.** O. **f.** E.

31. a. 30. **b.** 12. **c.** 4. **d.** 3. **e.** 1.

32. a. 40. **b.** 36. **c.** 4. **d.** 10. **e.** 10. **f.** 24. **g.** 48. **h.** 30.

33. a. 44. **b.** 52. **c.** 5. **d.** 1. **e.** 56. **f.** 40.

34. $n(A) = 13$ and $n(B) = 12$.

35. Using the Venn diagram, $n(R \cap Th) = 10$, $n(R - Th) = 10$, and $n(Th - R) = 20$.

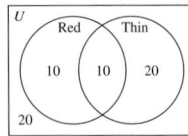

36. Using the Venn diagram, $n(L \cap O) = 6$, $n(L - O) = 24$, and $n(O - L) = 6$.

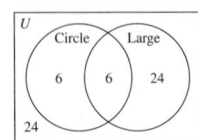

37.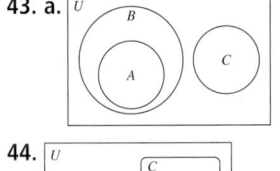

a. 28.   **b.** 2.   **c.** 8.   **d.** 34.

38. a. {1, 2, 4}.   **b.** {1, 2, 3, 7, 8}.   **c.** {3, 4, 5, 7, 8}.
   d. {1, 2}.   **e.** {1, 2, 3, 4, 5, 7, 8}.   **f.** {(1, 3), (1, 4), (1, 5), (1, 7), (1, 8), (2, 3), (2, 4), (2, 5), (2, 7),(2, 8)}.

39. a. {c, f, h, i, l}.   **b.** {b, c, e, f, h, i, l, n}.
   c. {c, e, i, h, n}.   **d.** {a, c, d, f, g, h, i, j, k, l, m}
   e. {a, b, c, e, f, g, h, i, j, l, m, n}.   **f.** {a, m}.

40. a. 11.   **b.** 6.   **c.** 5.   **41. a.** 8.   **b.** 0.   **c.** 2.

42. $n(A \cup B \cup C) = n(A) + n(B) + n(C) - n(A \cap B) - n(A \cap C) - n(B \cap C) + n(A \cap B \cap C)$.

43. a. 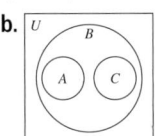   **b.** 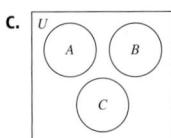   **c.**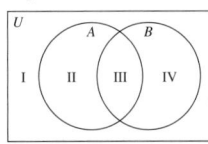

44.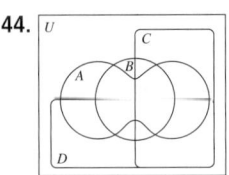

45. a. 10 employees.   **b.** 25 employees.

46. 18 people.

47. a. $T \times B = \{(r, j), (r, s), (b, j), (b, s), (g, j), (g, s), (w, j), (w, s)\}$.   **b.** We can also use multiplication to find her number of outfits: $4 \times 2 = 8$ outfits. This is based on the fundamental counting principle.

48. a. 50 customers.   **b.** 0 customers.   **c.** The fewest is 5 customers, and the most is 55 customers.

49. a. 15 students.   **b.** 8 students.   **c.** 36 students.

50. a. 32 people.   **b.** 23 people.   **c.** 40 people.

51. a. 5%.   **b.** 13%.

52. Let $S = \{c, f, h\}$ be the set of suspects, $T = \{a, p\}$ be the set of times, and $P = \{r, v, b\}$ be the set of places. Detective Johnson has to consider $n(S \times T \times P) = 3 \cdot 2 \cdot 3 = 18$ different scenarios: $S \times T \times P = \{(c, a, r), (c, a, v), (c, a, b), (c, p, r), (c, p, v), (c, p, b), (f, a, r), (f, a, v), (f, a, b), (f, p, r), (f, p, v), (f, p, b), (h, a, r), (h, a, v), (h, a, b), (h, p, r), (h, p, v), (h, p, b)\}$.

53. a. True.   **b.** False. $A \cup B$ may contain elements from B that are not in A.   **c.** True.   **d.** True.   **e.** False. $A - B$ is the set of all elements in A that are not in B, so unless it is empty, $A - B$ will never be a subset of B.   **f.** True.

54. a. False. A and the empty set can have nothing in common because the empty set has no elements.   **b.** True.   **c.** True.   **d.** True.   **e.** False. A and $\overline{A}$ have nothing in common, so $A \cap \overline{A} = \varnothing$.   **f.** True.

55. a. False. $n(A \times B) = 3 \cdot 2 = 6$.   **b.** False. All elements of $A \times B$ must be ordered pairs, and Mike is not an ordered pair.   **c.** True.   **d.** False. (Dave, Sue) is not a set but is an ordered pair. Subset notation does not apply.   **e.** False. The order of the objects in the ordered pair is incorrect.   **f.** True.

56. a. Yes. If $a \in A \cap B$, then by the definition of the intersection, it must be in both.   **b.** No. a might have been

an element that was only in B.   **c.** No. a can be in either A or B but does not have to be in both.

57. a. B must be a subset of A.   **b.** A must be a subset of B.

58. Yes, $n(A) \le n(A \cup B)$ will always be true. Because all the elements of A are in $A \cup B$, $A \cup B$ must have at least as many elements as A, if not more.

59. a. It is possible. If $A = B$, then $A - B = \varnothing = B - A$.   **b.** It is possible. If $A = B$ or if one or the other is the empty set, then $A \times B = B \times A$.

60. If $A = \{2, 4, 5, 6\}$ and $B = \{1, 3, 4, 6\}$, then $A \cup B = \{1, 2, 3, 4, 5, 6\} = B \cup A$ and $A \cap B = \{4, 6\} = B \cap A$.

61. Using the numbered regions in the diagram, $\overline{A \cap B} = \overline{\{III\}} = \{I, II, IV\}$ and $\overline{A} \cup \overline{B} = \{I, IV\} \cup \{I, II\} = \{I, II, IV\}$. Because the sets represent the same regions, $\overline{A \cap B} = \overline{A} \cup \overline{B}$.

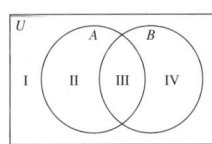

62. Using the numbered regions in the diagram, $A - B = \{II\}$ and $A \cap \overline{B} = A \cap \{I, II\} = \{II\}$. Because the sets represent the same regions, $A - B = A - B = A \cap \overline{B}$.

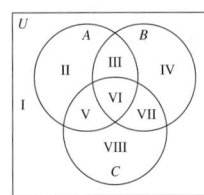

63. Using the numbered regions in the diagram, $(A \cap B) \cap C = \{III, VI\} \cap C = \{VI\}$ and $A \cap (B \cap C) = A \cap \{VI, VII\} = \{VI\}$. Because the sets represent the same regions, $(A \cap B) \cap C = A \cap (B \cap C)$.

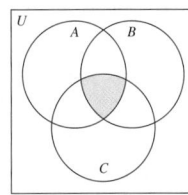

64. a. The Venn diagram of $A \cap (B \cap C)$ is:

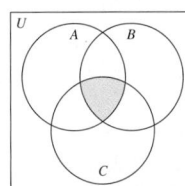

The Venn diagram of $(A \cap B) \cap C$ is:

[Venn diagram]

Because they are the same, $A \cap (B \cap C) = (A \cap B) \cap C$.

b. The Venn diagram of $A \cap (B \cup C)$ is:

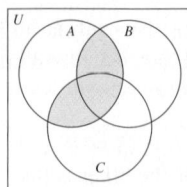

The Venn diagram of $(A \cap B) \cup (A \cap C)$ is:

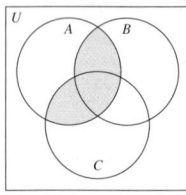

Because they are the same, $A \cap (B \cup C) = (A \cap B) \cup (A \cap C)$.

65. $(A \cup B) - C = (A \cup B) \cap \overline{C}$ Theorem 2.3(3)
$$= (A \cap \overline{C}) \cup (B \cap \overline{C}) \quad \text{Theorem 2.2(3)}$$
$$= (A - C) \cup (B - C) \quad \text{Theorem 2.3(3)}$$

66. Let A and B be sets in universe U. Then:
$$A \cup (B - A) = A \cup (B \cap \overline{A}) \quad \text{Theorem 2.3(3)}$$
$$= (A \cup B) \cap (A \cup \overline{A}) \quad \text{Theorem 2.2(3)}$$
$$= (A \cup B) \cap U \quad \text{Theorem 2.2(4)}$$
$$= (A \cup B) \quad \text{Theorem 2.2(4)}$$

67. Let A and B be sets in universe U. Then:
$$(A - B) \cup (A \cap B) = (A \cap \overline{B}) \cup (A \cap B) \quad \text{Theorem 2.3(3)}$$
$$= A \cap (\overline{B} \cup B) \quad \text{Theorem 2.2(3)}$$
$$= A \cap U \quad \text{Theorem 2.2(4)}$$
$$= A \quad \text{Theorem 2.2(4)}$$

68. "Or" means that one condition or the other or both must be satisfied for the statement to be true. This reflects what we do when we compute the union of two sets. "And" means that both conditions on either side of the word must be satisfied for the statement to be true. This reflects what we do when we compute the intersection of two sets.

69. Answers will vary.
70. Answers will vary.
71. Answers will vary.
72. Answers will vary.
73. Answers will vary.
74. Answers will vary.
75. Answers will vary.
76. Answers will vary.
77. Answers will vary.

▇ Chapter 2 Review Exercises

1. a. {red, orange, yellow, green, blue, indigo, violet} = $\{x \mid x$ is a color of the rainbow$\}$. **b.** {0, 1, 2, 3, 4} = $\{x \mid x \in W$ and $x < 5\}$. **c.** {11, 13, ..., 19} = $\{x \mid x$ is an odd natural number and $10 < x < 20\}$.

2. a. 3.   **b.** 4.   **c.** 8.
3. $\varnothing$, $\{l\}$, $\{o\}$, $\{w\}$, $\{l, o\}$, $\{l, w\}$, $\{o, w\}$, $\{l, o, w\}$.
4. Two possibilities are:

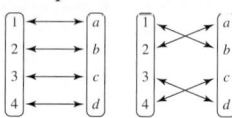

5. Answers will vary, but two possibilities are {1, 2, 3, 4, 5} and {2, 4, 6, 8, . . .}.
6. 6.
7. a. False.   **b.** True.
   c. True.   **d.** False.
   e. True.   **f.** False.
8. a. {1, 2, 3, 4, 5, 6, 7}.   **b.** {4, 5, 7}.
   c. {1, 3}.   **d.** {2, 6}.
9. a. {2, 4, 7, 8}.   **b.** {1, 5, 6, 8}.
   c. {1, 3, 5}.   **d.** {8}.
10. {(○, red), (○, white), (○, blue), (□, red), (□, white), (□, blue)}.
11. a. {1, 2, 3, 5, 6}.   **b.** {1, 2, 6, 7, 8}.   **c.** {3, 4, 5}.
   d. {4}.
12. a.

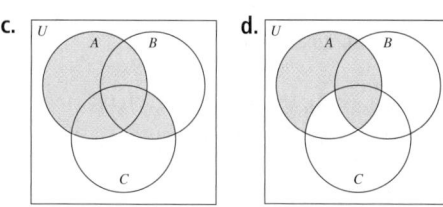

13. a. $(A \cup B) - (A \cap B)$.   **b.** $(A \cup B) \cap \overline{C}$.
14. 80.
15. Possible values for $n(A \cup B)$ are 5, 6, 7, and 8.
   Possible values for $n(A \cap B)$ are 0, 1, 2, and 3.
16. $n(A - B) = 15$ and $n(B - A) = 26$.
17. a. 11.   **b.** 8.   **c.** 14.
18. a. 9.   **b.** 49.   **c.** 19.   **d.** 21.
19. a. 7.   **b.** 32.   **c.** 45.   **d.** 58.
20. 25 people.
21. a. 10 respondents.
   b. 20 respondents.
22. The employer must consider only 9 respondents.
23.

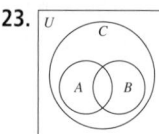

24. a. If $A \cup B = A \cup C$, then B need not equal C. For instance, let $A = \{1, 2, 3, 4\}$, $B = \{1, 2, 5\}$, and $C = \{3, 4, 5\}$. Then, $A \cup B = \{1, 2, 3, 4, 5\} = A \cup C$, but $\{1, 2, 5\} \neq \{3, 4, 5\}$. **b.** If $A \cap B = A \cap C$, then B need not equal C. For instance, let $A = \{1, 2, 3, 4\}$, $B = \{3, 4, 6\}$, and $C = \{3, 4, 5\}$. Then $A \cap B = \{3, 4\} = A \cap C$, but $\{3, 4, 6\} \neq \{3, 4, 5\}$.

25. The Venn diagram of $\overline{A \cup B}$ is:

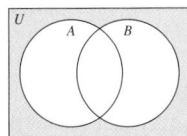

The Venn diagram of $\overline{A} \cap \overline{B}$ is:

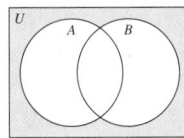

Because they are the same, $\overline{A \cup B} = \overline{A} \cap \overline{B}$.

26. Using the numbered regions in the diagram, $A \cap (B \cup C) = A \cap \{III, IV, V, VI, VII, VIII\} = \{III, V, VI\}$, and $(A \cap B) \cup (A \cap C) = \{III, VI\} \cup \{V, VI\} = \{III, V, VI\}$. Because the sets represent the same regions, $A \cap (B \cup C) = (A \cap B) \cup (A \cap C)$.

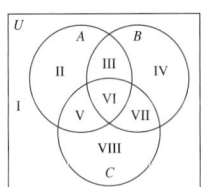

27.
$$
\begin{aligned}
A - (A - B) &= A - (A \cap \overline{B}) & \text{Theorem 2.3(3)} \\
&= A \cap \overline{(A \cap \overline{B})} & \text{Theorem 2.3(3)} \\
&= A \cap (\overline{A} \cap \overline{B}) & \text{Theorem 2.3(1)} \\
&= (A \cap A) \cap \overline{B} & \text{Theorem 2.2(2)} \\
&= A \cap \overline{B} & A \cap A = A
\end{aligned}
$$

CHAPTER 3

▪ Section 3.1 Learning Assessment

1. a. Numbers are abstract ideas of quantity, whereas numerals are symbolic representations of numbers that take on different forms. **b.** It is not possible to write a number; only a numeral can be written.

2. a. In a grouping system, once a certain number of objects are counted, they are grouped and represented with just one symbol. **b.** A grouping system must also have a base and be additive.

3. In an additive system, the value of a numeral is found by adding the values of all the symbols in the numeral. In a subtractive system, if a smaller symbol appears to the left of a larger one, the value of the smaller is subtracted from that of the larger.

4. a. In a multiplicative system, the value of any digit is found by multiplying its value by its corresponding place value. **b.** If a system is additive, it need not be multiplicative. **c.** If a system is multiplicative, it must also be additive. **d.** A multiplicative system must also be positional and have digits.

5. a. In a positional system, the position of a symbol determines its value. **b.** The digit represents the number of groups that occur in a position. **c.** The base determines

the maximum number of groups that can occur in any one position.

6. A zero could represent the cardinal number of the empty set, or it could be used as a placeholder to move other, nonzero digits out to their appropriate positions.

7. a. Numeral. **b.** Number. **c.** Numeral. **d.** Numeral.

8. Possible answers include but are not limited to: five, 5, V, ∙∙ ⋮∙, and ⦀⦀.

9. The groupings of the Egyptian numerals have the effect of reducing the length of the written numeral.

10. Both the Babylonian and the Mayan systems have place values, which limits the number of digits and reduces the length of most numerals.

11. a. The Egyptian system has no place value; we need to create a new symbol for each new grouping. Certain numbers require a large number of symbols, even though they have relatively small values. **b.** The Roman system has no concept of zero, and it is difficult to use when making computations. **c.** The Babylonian system has no concept of zero; it can be difficult to tell whether a place value is being used in the numeral or not.

12. a. Yes. **b.** No. **c.** No. **d.** Yes. **e.** Yes. **f.** Yes. **g.** No. **h.** No.

13. a. 26. **b.** 33.

14. $18 = $ ⦀⦀ ⦀⦀ ⦀⦀ ⫼ and $29 = $ ⦀⦀ ⦀⦀ ⦀⦀ ⦀⦀ ⦀⦀ ⦀⦀⦀.

15. a. 254. **b.** 1,304. **c.** 33,118. **d.** 14. **e.** 77. **f.** 249.

16. a. 150,525. **b.** 362,401. **c.** 12,212,034. **d.** 2,463. **e.** 49,765. **f.** 1,040,164.

17. a. ∩∩III and XXIII. **b.** ∩∩∩∩∩IIIIIII and LVII. **c.** ⌐∩∩∩∩∩∩∩∩II and CLXXXII. **d.** ⌐⌐∩∩∩∩∩∩∩∩∩∩III and CCLXXXIII.

18. a. ⌐⌐⌐∩∩∩∩ⅠⅠⅠⅠⅠⅠⅠ and MCMLXXXVII.
b. ∩∩∩∩∩∩∩I and $\overline{\text{XXIV}}$DCLXXXI.
c. ⌐⌐⌐⌐⌐⌐⌐III and $\overline{\text{LVI}}$DCCIII.
d. 𝄞∩∩∩∩∩∩∩∩∩∩ and $\overline{\text{MXL}}$CCCLX.

19. ∩∩∩ ∩∩∩IIIIIIIII , ∩∩∩ ∩∩∩IIIIIIIIII , ⌐, ⌐I, and ⌐II.

20. CXLVIII, CXLIX, CL, CLI, and CLII.

21. a. 23. **b.** 46. **c.** 682. **d.** 14. **e.** 17. **f.** 109.

22. a. 5,026. **b** 57,616. **c.** 155,595,633. **d.** 240. **e.** 2,418. **f.** 6,840.

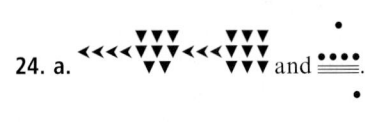

23. a. <<<▼▼▼ and ••••. b. ▼ ▼ and • •.
c. ▼▼<<<<▼▼▼ and •••. d. ▼▼▼ ▼▼▼ and ••••.

24. a. <<<<▼▼▼<<<▼▼▼ and ••••. b. ▼▼▼<<<<<▼▼▼<<▼▼▼ and •.

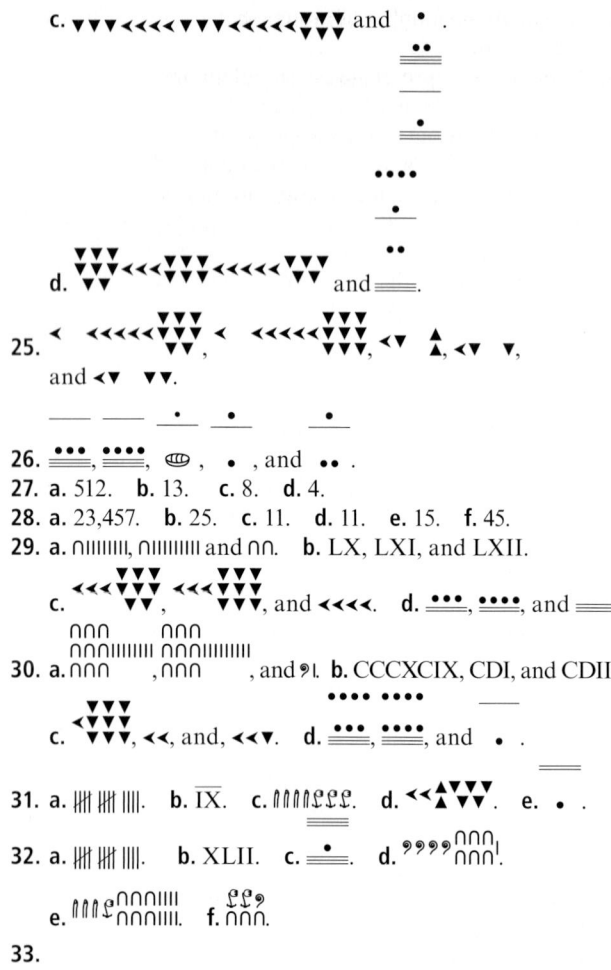

c. ▼▼▼ ◄◄◄◄ ▼▼▼ ◄◄◄◄◄ ▼▼▼ and •.

d. ▼▼▼ ▼▼ ◄◄◄ ▼▼▼ ◄◄◄◄◄ ▼▼ and ▬.

25. ◄ ◄◄◄◄◄ ▼▼▼ ▼▼ , ◄ ◄◄◄◄◄ ▼▼▼ ▼▼▼ , ◄▼ ▲, ◄▼ ▼, and ◄▼ ▼▼.

26. •••, ••••, ⬭ , • , and •• .

27. a. 512. **b.** 13. **c.** 8. **d.** 4.

28. a. 23,457. **b.** 25. **c.** 11. **d.** 11. **e.** 15. **f.** 45.

29. a. ∩IIIIIIII, ∩IIIIIIIII and ∩∩. **b.** LX, LXI, and LXII.
c. ◄◄◄ ▼▼▼ ▼▼ , ◄◄◄ ▼▼▼ ▼▼▼ , and ◄◄◄◄. **d.** •••, ••••, and ▬.

30. a. ∩∩∩ ∩∩∩IIIIIIIII , ∩∩∩ ∩∩∩IIIIIIIII , and ⟋I. **b.** CCCXCIX, CDI, and CDII.
c. ◄ ▼▼▼ ▼▼▼ , ◄◄, and, ◄◄▼. **d.** •••• •••• ▬, •••, ••••, and •.

31. a. ||||| ||||| ||||. **b.** IX. **c.** ∩∩∩♀♀♀. **d.** ◄◄ ▲ ▼▼▼ ▼. **e.** • • .

32. a. ||||| ||||| ||||. **b.** XLII. **c.** • ▬. **d.** ♀♀♀♀ ∩∩∩ ∩∩∩ I.
e. ∩∩∩IIIII ∩∩∩IIIII ♀ ∩∩∩IIIII . **f.** ♀♀⟋ ∩∩∩.

33.

Egyptian	∩∩IIIIIII	♀♀∩II	∩∩∩∩IIIIIIII	∩∩∩∩IIII
Babylonian	◄◄ ▼▼▼ ▼	▼▼▼ ◄◄◄▼▼	◄◄◄◄ ▼▼▼ ▼▼	◄◄◄◄ ▼▼▼ ▼
Roman	XXVII	CCXII	XLVIII	XLIV
Mayan	• ••	▬ ▬ ••	•• •••	•• ••••

34. The Roman uses 4, the Babylonian uses 7, and the Egyptian uses 8.

35. a. ♀♀ ∩∩∩ ∩∩∩IIIIIIIII , or 499. **b.** XVIII, or 18.

36. Answers will vary depending on the current year.

37. a. III ≡ I. **b.** ⊥ IIII I. **c.** I = π ≡ πππ. **d.** ≡ IIIII _ π ⊥ I.

38. a. 1,376. **b.** 66,666. **c.** 9,054 **d.** 80,802.

39. a. 52. **b.** 79. **c.** 555. **d.** 804. **e.** 8,990. **f.** 80,731.
g. 100,126. **h.** 600,044.

40. a. $\nu\,\delta$. **b.** $o\,\gamma$. **c.** $\tau\,\mu\,\alpha$. **d.** $,\beta\,\tau\,\mu\,\varepsilon$. **e.** $,\iota\,\omega\,\theta$.

41. a. The base appears to be five.
b. ⋀\\\\, //\\\\, //\\\\, ///\\\\, and \\\\\\.

42. a. i. 9. **ii.** 14. **iii.** 11. **iv.** 20. **b.** Add the digits to determine how many symbols are needed.

43. a. Look to see whether digits of the same value are used more than once. **b.** Look to find the number of different digits. If we assume there is a digit for zero, then the number of digits will determine the base. If there is no digit for zero, then the base is the number of digits plus one.

44. a. They all use tallies. However, the Egyptian and Babylonian systems only regroup at 10. The Mayan system

regroups at 5 and 20, and the Roman system regroups at other values. **b.** The Roman system seems to be more efficient because it regroups at a lower number than the other systems.

45. No. Consider the number 360. The Babylonian and Mayan systems each use three symbols to represent this number. The Egyptian uses 9, and the Roman uses 5.

46. Answers will vary. **47** Answers will vary.

48. Answers will vary. **49** Answers will vary.

50. a. The same difficulty would not occur in the Egyptian system, but it could occur in the Roman or Mayan.
b. The problem comes from the fact that both the Roman and Mayan systems are positional.

51. Answers will vary.

52. Answers will vary.

53. a. We typically use skip counting. **b.** Answers will vary.

54. Answers will vary. **55.** Answers will vary.

56. Answers will vary. **57.** Answers will vary.

■ Section 3.2 Learning Assessment

1. The base of the decimal system determines the size of the groups.

2. Place value means that the position of a digit in a numeral determines its value. It is a key feature of the decimal system because it can greatly reduce the length of numerals.

3. a. As grouping system, the decimal system must also be additive because it is necessary to add the values of the digits in their place values to find the value of the numeral. **b.** As a positional system, the decimal system must also be multiplicative because it is necessary to multiply the value of the digit by the place value to determine how many units it represents.

4. a. Because the decimal system has a base of ten, an exchange is made once a group of ten is reached. Also, because the decimal system has place value, we can use digits repeatedly. **b.** When standing alone, each digit represents a value less than 10. The digits are also used to write numerals with values larger than 9.

5. When a placeholder is used in a place value, it tells us that no group of that size is included in the value of the number. This allows us to move other nonzero digits out to the correct place value.

6. To compose numerals means to find the value of the numeral by combining the values of the groups within it. To decompose a numeral means to break it into smaller groups that have a total value equal to the number.

7. If a numeral has four or more digits, commas are used to separate it into groups of three digits, called periods. Each period is named for the smallest place value in the period.

8. If a one-to-one correspondence cannot be made between two sets, then one set will have objects left over. This set must represent the larger value.

9. Less than or equal to is similar to less than except that it also allows the two numbers to have the same value, whereas less than does not.

10. a. 682. **b.** 6,857. **c.** 237,030. **d.** 5,004,007.

11. a. 2 tens and 3 ones. **b.** 6 hundreds, 7 tens, and 1 one.
c. 8 thousands, 2 hundreds, 9 tens, and 9 ones. **d.** 6 ten thousands, 0 thousands, 0 hundreds, 1 ten, and 2 ones.

12. a. 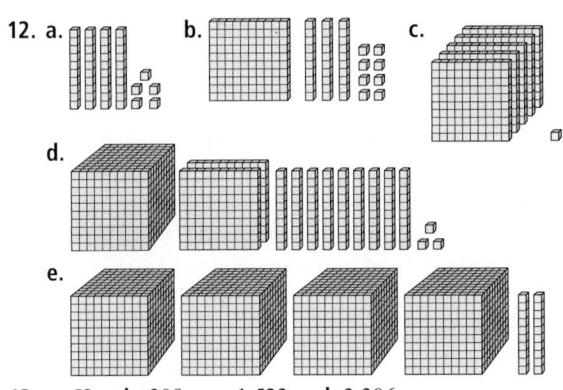 **b.** **c.**

d.

e.

13. a. 52. **b.** 309. **c.** 1,523. **d.** 3,306.
14. a. 10 units in a long, 100 units in a flat, and 1,000 units in a block. **b.** 10 longs in a flat and 100 longs in a block. **c.** 10 flats.
15. a. 60 units. **b.** 40 longs or 400 units. **c.** 140 longs or 1,400 units. **d.** 150 flats or 1,500 longs.
16. a. 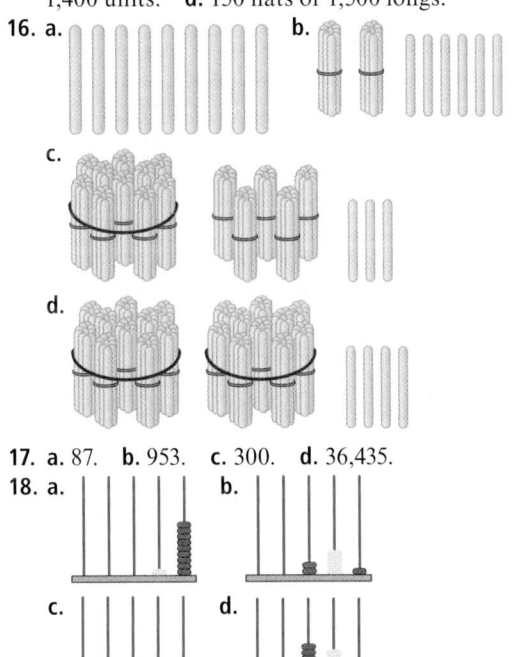 **b.**

c.

d.

17. a. 87. **b.** 953. **c.** 300. **d.** 36,435.
18. a. **b.**

c. **d.**

19. Base-ten blocks show each and every unit in the numeral. The abacus shows only the number of groups in each place value.
20. a. Fifty-three. **b.** One hundred twenty-nine. **c.** Four thousand, five hundred seventy-two. **d.** Two hundred sixty-nine thousand, eighteen. **e.** Two hundred thirty-one million, seven hundred sixty-four thousand, nine hundred twenty. **f.** One hundred twenty-three billion, four hundred nine million, one thousand, two hundred thirty.
21. a. 499. **b.** 3,287. **c.** 801,906. **d.** 45,337,118.
22. Possible answers are 23 longs and 6 units, 2 flats and 36 units, 236 units, 1 flat, 13 longs, and 6 units, or 1 flat and 136 units.
23. a. 140. **b.** 80. **c.** 635.

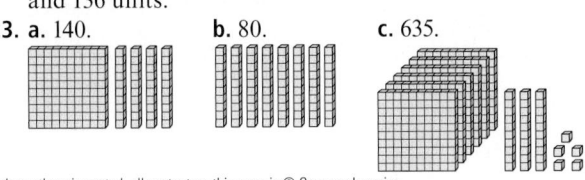

24. Answers will vary, but five possibilities are 2 thousands and 1 hundred, 21 hundreds, 1 thousand and 11 hundreds, 1 thousand, 10 hundreds and 10 tens, or 2 thousands and 10 tens.
25. a. The 7 is in the tens place value. **b.** The 4 is in the thousands place value, and the 9 is in the ones. **c.** The 1 is in the millions place value, and the 9 is in the ten thousands. **d.** The 2 is in the hundred thousands place value, the 3 is in the thousands, and the 9 is in the hundreds.
26. a. $(4 \times 100) + (5 \times 10) + (7 \times 1)$. **b.** $(3 \times 1,000) + (5 \times 100) + (8 \times 10) + (1 \times 1)$. **c.** $(4 \times 10,000) + (2 \times 100) + (1 \times 1)$. **d.** $(2 \times 100,000) + (3 \times 100) + (9 \times 10)$.
27. a. $(2 \times 10^2) + (1 \times 10^1) + (9 \times 10^0)$. **b.** $(4 \times 10^3) + (5 \times 10^2) + (6 \times 10^1) + (1 \times 10^0)$. **c.** $(1 \times 10^4) + (6 \times 10^3) + (4 \times 10^0)$. **d.** $(3 \times 10^5) + (1 \times 10^3) + (4 \times 10^2) + (9 \times 10^1)$.
28. a. 348. **b.** 40,404. **c.** 120,800. **d.** 772,038.
29. a. 963,010. **b.** 6,004. **c.** 537. **d.** 29,011.
30. a. 16. **b.** 25. **c.** 124. **d.** 3,160.
31. a. 316. **b.** 315. **c.** 345. **d.** 2,301.
32. a. 14. **b.** 11. **c.** 1.
33. a. 6. **b.** 3. **c.** 1, 3.
34. a. <. **b.** =. **c.** >.
35. 2,456 3,456 4,356 5,436 5,634 6,534.
36. 5,788 5,878 7,588 7,858 8,578 8,857.
37. a. 98,643 and 34,689. **b.** 9,999 and 3,333.
38. a. 1 cm × 1 cm × 10 cm. **b.** 1 cm × 10 cm × 10 cm. **c.** 10 cm × 10 cm × 10 cm. **d.** 10 cm × 100 cm × 100 cm.
39. a. 100,000. **b.** 1,000,000,000.
40. Both numbers are 10^{15}, so they are equal.
41. a. 576 boxes. **b.** 10,368 candies. **c.** 7,800 candies. **d.** 4 crates, 3 flats, and 6 boxes.
42. About $900 \cdot 8 = 7,200$ in., or 600 ft long.
43. About 1,500 m tall.
44. It would take 353 min, or about 5.9 hr, to type the message.
45. About 2,500 cups of salt.
46. If $A = \{1, 2, 3, 4, 5\}$ and $B = \{a, b, c, d, e, f, g\}$, then $n(A) = 5$ and $n(B) = 7$. If we make a one-to-one correspondence between the two sets, we see that two elements from B remain unused. We can make a one-to-one correspondence only between A and a proper subset of B. Because $n(A) = 5$ and $n(B) = 7$, we conclude that $5 < 7$ by the definition of less than

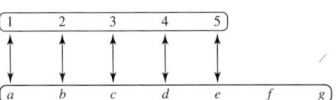

47. Because 6 occurs farther to the right on a number line than 3, we conclude that $6 > 3$.
48. a. The base-ten block for 10,000 is a long, for 100,000 is a flat, for 1,000,000 is a block, and so on. **b.** The shapes of the blocks repeat along the lines of the different periods.
49. a. 2 and 10. **b.** Possible answers are 1, 3, and 6; 2, 2, and 6; 2, 3, and 5; 4, 1, and 5; or 2, 4, and 4. **c.** Possible answers are 1, 2, 3, and 4; 1, 1, 2, and 6; 2, 2, 2, and 4; 2, 3, 3, and 2; or 3, 3, 3, and 1. **d.** 17 and 100.
50. A cipher is zero.
51. a. 327. **b.** 800,459. **c.** The eighth order is the ten millions place value, the ninth order is the hundred millions place value, and the tenth order is the billions place value.
52. The value of the positions is the decimal system increase by a factor of 10 as we move from right to left and decrease by a factor of 10 as we move from left to right.

53. Answers will vary. **54.** Answers will vary.
55. Answers will vary.
56. a. It is like the Egyptian system because it uses a base of ten. It is like the Babylonian because it is a grouping system, uses digits, and is additive and multiplicative. It is like the Mayan system because it is a grouping system, uses digits, is additive and multiplicative, and has a concept of zero. **b.** The properties of the Hindu-Arabic system enable us to write numerals of very large numbers in a much more concise way than in the other numeration systems.
57. The word "and" is used to represent the decimal point when reading decimals or to separate the parts from the wholes when reading mixed numbers.
58. a. We compose a numeral when adding or multiplying whenever we make exchanges to another place value.
b. We decompose a numeral when subtracting whenever we make an exchange between groups of ten to borrow.
59. a. It is a grouping system that primarily uses a base of ten. It is also additive and multiplicative. **b.** It includes other groups such as nickels, quarters, 50-cent pieces, five-dollar bills, and twenty-dollar bills that are not consistent with the groups of ten of the decimal system. **c.** Five $100 bills, one $10 bill, two $1 bills, three dimes, and six pennies. It takes $5 + 1 + 2 + 3 + 6 = 17$ denominations to represent this amount.
d. Answers will vary.

Section 3.3 Learning Assessment

1. Base-five has all the same characteristics of base-ten in that it is a grouping system with a base, has place value, uses digits, has a concept of zero, and is additive and multiplicative. It is different because it uses a base of five. This changes the number of objects in each group, the values of the places, and the number of digits used.
2. A base-one numeration system would group by ones; every group would be an individual unit.
3. Changing the base affects the place values by changing the value of the positions to match consecutive powers of the base. If the base is smaller than ten, fewer digits are needed to represent the possible groups that could occur in any place value. If the base is larger than ten, more digits are needed.
4. Base-ten provides only 10 digits to be used. In systems with bases larger than ten, more digits are required to represent possible groupings. Hence, we turn to letters.
5. Counting in base-b is similar to counting in base-ten in that we roll over to the next highest place value once the largest digit has been reached. In systems with bases smaller than ten, this happens more quickly because there are fewer digits. In systems with bases larger than ten, this happens more slowly because there are more digits.
6. The place values in the expanded notation of the base-b numeral are expressed as base-ten numerals.
7. One way to convert a decimal numeral to a base-five numeral is to rearrange the appropriate number of blocks into groups consistent with the given base. Another is to use division to determine which base-five groups can be made from the given number of objects.
8. a. Two-three, base-four. **b.** Four-five-one, base-six.
c. Seven-zero-zero-seven, base-eight. **d.** A-B-three, base-twelve.

9. a. 430_{five}. **b.** 213_{four}. **c.** 5341_{seven}. **d.** $A89_{\text{twelve}}$.
10. a. 2201_{three}. **b.** 5044_{six}. **c.** 705_{eight}. **d.** $9BE_{\text{fifteen}}$.
11. a. 1 group of 25, 1 group of 5, and 4 ones.
b. 2 groups of 343, 4 groups of 7, and 6 ones.
c. 8 groups of 729, 8 groups of nine, and 1 one.
d. 10 groups of 1,728, 11 groups of 144, 11 groups of 12, and 10 ones.
12. a. 1 group of 27, 2 groups of 3, and 1 one. **b.** 3 groups of 64, 5 groups of 8, and 4 ones. **c.** 10 groups of 169, 11 groups of 13, and 12 ones. **d.** 12 groups of 196, 13 groups of 14, and 12 ones.
13. a. 1 long = 4 units, 1 flat = 16 units, and 1 block = 64 units. **b.** 1 long = 7 units, 1 flat = 49 units, and 1 block = 343 units. **c.** 1 long = 11 units, 1 flat = 121 units, and 1 block = 1,331 units. **d.** 1 long = 16 units, 1 flat = 256 units, and 1 block = 4,096 units.
14. a. **b.**
c. **d.**
15. a. **b.**
c. **d.**
16. a. 211_{three}. **b.** 344_{five}. **c.** $1,453$. **d.** $61A_{\text{twelve}}$.
17. a. Five base-five blocks are needed: 3 longs and 2 units.
b. Three base-five blocks are needed: 2 flats, and 1 unit.
c. Nine base-five blocks are needed: 4 flats, 2 longs, and 3 units.
d. Eight base-five blocks are needed: 3 blocks, 3 longs, and 2 units.
18. a. 2 blocks, 0 flats, 2 longs, and 0 units. **b.** 2 flats, 2 longs, and 0 units. **c.** 7 longs and 4 units. **d.** 5 longs.
19. a. $(2 \times 5^1) + (1 \times 5^0)$. **b.** $(3 \times 5^2) + (0 \times 5^1) + (4 \times 5^0)$.
c. $(3 \times 5^2) + (3 \times 5^1) + (1 \times 5^0)$. **d.** $(2 \times 5^3) + (0 \times 5^2) + (4 \times 5^1) + (3 \times 5^0)$.
20. a. $(2 \times 3^1) + (2 \times 3^0)$. **b.** $(1 \times 6^2) + (2 \times 6^1) + (5 \times 6^0)$.
c. $(1 \times 9^3) + (3 \times 9^2) + (0 \times 9^1) + (7 \times 9^0)$. **d.** $(10 \times 14^3) + (11 \times 14^2) + (3 \times 14^1) + (13 \times 14^0)$.
21. a. 342_{five}. **b.** 774_{eight}. **c.** $1AA_{\text{eleven}}$. **d.** $DE09_{\text{fifteen}}$.
22. a. 114_{five}. **b.** 122_{three}. **c.** 578_{nine}. **d.** $AB3_{\text{fourteen}}$.
23. a. 1, 4, 16, 64, and 256. **b.** 1, 6, 36, 216, and 1,296. **c.** 1, 12, 144, 1,728, and 20,736. **d.** 1, 15, 225, 3,375, and 50,625.
24. a. 0 and 1. **b.** 0, 1, 2, 3, 4, 5, 6, 7, and 8. **c.** 0, 1, 2, 3, 4, 5, 6, 7, 8, 9, and A. **d.** 0, 1, 2, 3, 4, 5, 6, 7, 8, 9, A, B, C, D, E, F, G, and H.
25. a. 20. **b.** 35. **c.** 50.
26. a. 1_{two}, 10_{two}, 11_{two}, 100_{two}, 101_{two}, 110_{two}, 111_{two}, 1000_{two}, 1001_{two}, 1010_{two}, 1011_{two}, 1100_{two}, 1101_{two}, 1110_{two}, 1111_{two}, 10000_{two}, 10001_{two}, 10010_{two}, 10011_{two}, and 10100_{two}.
b. 1_{eight}, 2_{eight}, 3_{eight}, 4_{eight}, 5_{eight}, 6_{eight}, 7_{eight}, 10_{eight}, 11_{eight}, 12_{eight}, 13_{eight}, 14_{eight}, 15_{eight}, 16_{eight}, 17_{eight}, 20_{eight}, 21_{eight}, 22_{eight}, 23_{eight}, and 24_{eight}.
c. 1_{twelve}, 2_{twelve}, 3_{twelve}, 4_{twelve}, 5_{twelve}, 6_{twelve}, 7_{twelve}, 8_{twelve}, 9_{twelve}, A_{twelve}, B_{twelve}, 10_{twelve}, 11_{twelve}, 12_{twelve}, 13_{twelve}, 14_{twelve}, 15_{twelve}, 16_{twelve}, 17_{twelve}, and 18_{twelve}.

27. All numerals are given in the appropriate base.

Decimal	23	24	25	26	27	28	29	30	31	32
Base-three	212	220	221	222	1000	1001	1002	1010	1011	1012
Base-nine	25	26	27	28	30	31	32	33	34	35
Base-twelve	1B	20	21	22	23	24	25	26	27	28

28. a. 13. **b.** 145. **c.** 1,711. **d.** 9,493.
29. a. 49. **b.** 372. **c.** 743. **d.** 687.
30. a. 30_{five}. **b.** 301_{five}. **c.** 1003_{five}. **d.** 1411_{five}.
31. a. 10111_{two}. **b.** 250_{eight}. **c.** 510_{twelve}.
32. a. 233_{four}. **b.** 287_{nine}. **c.** $6C2_{\text{sixteen}}$.
33. a. 121_{three} and 200_{three}. **b.** 77_{eight} and 101_{eight}. **c.** 87_{nine} and 100_{nine}. **d.** CC_{fourteen} and $D0_{\text{fourteen}}$.
34. a. 10_{six}, 11_{six}, and 12_{six}. **b.** 200_{eight}, 201_{eight}, and 202_{eight}. **c.** 786_{nine}, 787_{nine}, and 788_{nine}. **d.** FE_{sixteen}, 102_{sixteen}, and 103_{sixteen}.
35. a. 12_{seven}, 14_{seven}, and 16_{seven}. **b.** 15_{fifteen}, $1A_{\text{fifteen}}$, and 20_{fifteen}. **c.** $1A_{\text{sixteen}}$, $1C_{\text{sixteen}}$, and $1E_{\text{sixteen}}$. **d.** 4_{eight}, 24_{eight} and 30_{eight}.
36. 444_{five}, 4444_{five}, 44444_{five} and $4 \underset{n \text{ times}}{\underline{\ldots 444}}{}_{\text{five}}$.
37. 3333_{four}, 7777_{eight}, and $BBBB_{\text{twelve}}$.
38. a. 30. **b.** 341. **c.** 952. **d.** 2,719.
39. a. The flat in base-ten is four times the size of the flat in base-five. **b.** The block in base-ten is eight times the size of the block in base-five.
40. a. The flat in base-nine is nine times the size of the flat in base-three. **b.** The block in base-nine is 27 times the size of the block in base-three.
41. a. 1 block and 1 long in base-two. **b.** 4 flats in base-five. **c.** 3 flats, 14 longs, and 8 units in base-sixteen.
42. a. 24. **b.** 63. **c.** 255.
43. a. 12. **b.** 56. **c.** 132.
44. a. 25. **b.** 125. **c.** 625. **d.** 5^n numbers with n or fewer digits.
45. a. 21_{eight}. **b.** 54_{six}. **c.** $1A_{\text{twelve}}$.
46. a. i. 110_{two}. **ii.** 111000_{two}. **iii.** 11100001_{two}. **iv.** 11011110_{two}.
b. i. 22_{four}. **ii.** 331_{four}. **iii.** 111_{four}. **iv.** 1003_{four}.
c. Because 4 is a power of 2, any power of 4 can be rewritten as a power of 2. As a consequence, every place value in base-four is also a place value in base-two.
47. a.

Base-Two	Base-Eight
000	0
001	1
010	2
011	3
100	4
101	5
110	6
111	7

b. i. 100101_{two}. **ii.** 11101110_{two}.
iii. 1000111010_{two}. **iv.** $11000000100010_{\text{two}}$.
c. i. 65_{eight}. **ii.** 51_{eight}. **iii.** 235_{eight}. **iv.** 4166_{eight}.
d. Any base that is a power of two will work. Three others are base-sixteen, base-thirty-two, and base-sixty-four.

48. a. The larger base must be a power of the smaller base. **b.** Answers will vary, but possible answers are base-three and base-nine, base-three and base-twenty-seven, base-four and base-sixteen, base-four and base-sixty-four, and base-five and base-twenty-five.
49. If we compare the place values directly, the two numbers differ in the second place value. Because 2 is greater than 1, it must be that $213_{\text{five}} < 223_{\text{five}}$.
50. If we convert both numbers to the decimal system, then $1000_{\text{two}} = 8 < 9 = 100_{\text{three}}$.
51. If we convert both numbers to the decimal system, then $100_{\text{five}} = 25 \neq 64 = 100_{\text{eight}}$.
52. If we convert both numbers to the decimal system, then $21_{\text{five}} = 11 = B_{\text{sixteen}}$.
53. $wxyz_b = (w \cdot b^3) + (x \cdot b^2) + (y \cdot b^1) + (z \cdot b^0)$.
54. a. 10_{six}. **b.** 10_{eight}. **c.** 10_{twelve}. **d.** The numeral that represents a value of b in base-b is always 10_b.
55. The next four rows are:

40	41	42	43	44	45
50	51	52	53	54	55
100	101	102	103	104	105
110	111	112	113	114	115

56. The term "hundred" represents a specific group in the decimal system. There is no group of this size in other bases.
57. Answers will vary.
58. a. The digit for 15 is , for 25 , for 42 is , and for 59 is .
b. The first five place values of the Babylonian system and base-sixty are 1, 60, 3,600, 216,000, and 12,960,000.
c. For a long time, the Babylonian system did not have a consistent concept of zero.
59. No. The Mayan system is not a true base-twenty numeration system. The place values are not all consecutive powers of 20.
60. No. It is not necessary to make changes. The definition is based on the size of two sets. The way in which the cardinal number is written is irrelevant to the one-to-one correspondence.
61. Answers will vary.

Chapter 3 Review Exercises

1. a. Numeral. **b.** Number. **c.** Number.
2. a. 39. **b.** 211,047. **c.** 10,815. **d.** 1,427. **e.** 42,008.
3. a. , LVI, and .

b. , MMCXXXIV, and .

c. ⋔⋔⋔ℒ999ႶႶ⋂ⅠⅠⅠⅠⅠ, ◄▼◄◄▼▼▼◄▼▼▼, $\overline{\text{XLICCXXXV}}$, and $\overline{\cdots}$.

4. a. ႶႶႶႶႶႶႶႶⅠⅠⅠⅠⅠⅠⅠ, 9, and 9Ⅰ.

  b. ▼▼▼▼▼▼, ◄, and ◄▼. **c.** XCVII, XCVIII, C, and CI.

5. a. 15. **b.** 5. **c.** 10. **d.** 8.

6. XVIII.

7. a. The 4 is in the thousands place value, and the 8 is in the ones place value. **b.** The 2 is in the millions place value, the 9 is in the hundred thousands place value, and the 8 is in the tens place value.

8. a. $(1 \times 1{,}000) + (2 \times 100) + (0 \times 10) + (1 \times 1)$. **b.** $(8 \times 10{,}000) + (5 \times 1{,}000) + (3 \times 100) + (1 \times 10) + (9 \times 1)$. **c.** $(4 \times 10{,}000{,}000) + (5 \times 1{,}000{,}000) + (9 \times 100{,}000) + (8 \times 10{,}000) + (1 \times 1{,}000) + (0 \times 100) + (0 \times 10) + (3 \times 1)$.

9. a. 406. **b.** 51,802. **c.** 714,186.

10. a. Six hundred fifty-seven. **b.** One thousand, five hundred nine. **c.** Twenty-three million, nine hundred eighty-one, eighty-five.

11. a. 5 tens and 1 one. **b.** 8 hundreds, 6 tens, and 1 one. **c.** 1 ten thousand, 2 thousands, 6 hundreds, 7 tens, and 1 one.

12. a. 360. **b.** 4,991. **c.** 114,040.

13. a. 8_{nine}, 10_{nine}, and 11_{nine}. **b.** 1000_{four}, 1001_{four}, and 1002_{four}. **c.** 10_{fifteen}, 11_{fifteen}, and 12_{fifteen}.

14. a. 2 flats and 2 longs in base-four. **b.** 5 longs and 5 units in base-seven. **c.** 3 longs and 1 unit in base-thirteen.

15. a. 21 units. **b.** 1,284 units. **c.** 3,022 units.

16. a. 1112_{three}. **b.** 148_{nine}. **c.** 333_{twelve}.

17. a. $3412_{\text{five}} = (3 \times 125) + (4 \times 25) + (1 \times 5) + (2 \times 1) = 375 + 100 + 5 + 2 = 482$. **b.** $8710_{\text{nine}} = (8 \times 729) + (7 \times 81) + (1 \times 9) + (0 \times 1) = 5{,}832 + 567 + 9 = 6{,}408$. **c.** $4AD_{\text{sixteen}} = (4 \times 256) + (10 \times 16) + (13 \times 1) = 1{,}024 + 160 + 13 = 1{,}197$.

18. 224 numerals.

19. a. 25. **b.** 30. **c.** 100.

20. a. 9. **b.** 27. **c.** 81. **d.** 3^n.

21. 41_{six}. **22.** 135_{six}.

Answers to Reviewing the Big Ideas: Chapters 1–3

1. **a.** $x + 7 = 15$. **b.** $\dfrac{x}{8} \neq 4$.

2. Four times the quantity x plus three is greater than six.

3. The picture suggests that, in a rhombus, opposite sides are congruent and parallel and that opposite angles are congruent.

4. Answers will vary.

5. $c = 3$. Deductive reasoning.

6. The next time I eat fast food, I will get an upset stomach. Inductive reasoning.

7. Six is a number between 4 and 8 that is divisible by 3.

8. **a.** $t \wedge d$. **b.** $\sim t \to \sim d$.

9. *Converse:* If you like chocolate, then you are a woman. *Inverse:* If you are not a woman, then you do not like chocolate. *Contrapositive:* If you do not like chocolate, then you are not a woman.

10. **a.** True. **b.** False. **c.** False. **d.** True. **e.** True.

11. If I see a funny movie, then I am laughing.

12. Fifty can be written 12 different ways as the sums of 5s, 10s, and 20s, including $20 + 20 + 10$; $20 + 20 + 5 + 5$; $20 + 10 + 10 + 10$; $20 + 10 + 10 + 5 + 5$; $20 + 10 + 5 + 5 + 5 + 5$; $20 + 5 + 5 + 5 + 5 + 5 + 5$; $10 + 10 + 10 + 10 + 10$; $10 + 10 + 10 + 10 + 5 + 5$; $10 + 10 + 10 + 5 + 5 + 5 + 5$; $10 + 10 + 5 + 5 + 5 + 5 + 5 + 5$; $10 + 5 + 5 + 5 + 5 + 5 + 5 + 5 + 5$; and $5 + 5 + 5 + 5 + 5 + 5 + 5 + 5 + 5 + 5$.

13. 61 toothpicks. **14.** About 29,167 cases of soda a week.

15. 13 triangles. **16.** The square yard requires more seed.

17. $\{6, 7, 8, \dots\} = \{x \mid x > 5, x \in W\}$.

18. Two possible one-to-one correspondences are:

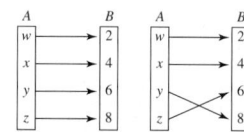

19. **a.** $\{1, 2, 3, 4, 5, 6\} = U$. **b.** $\{3, 5\}$. **c.** $\{1, 4\}$. **d.** $\{1, 4\}$.

20. $\{(s, \square), (t, \square), (u, \square), (s, \Diamond), (t, \Diamond), (u, \Diamond)\}$.

21. **a.** 50. **b.** 40.

22. **a.** ⋂⋂⋂⋂ⅠⅠⅠⅠⅠⅠ, XLVII, ◄◄◄◄▼▼▼▼, and ∴. **b.** 999ⅠⅠⅠⅠⅠⅠ, CCCLXVIII, ▼▼▼▼▼▼, and ⁝. **c.** ⋔ℒℒ999999999ⅡⅡ⋂⋂⋂, $\overline{\text{XIICMXL}}$,

▼▼▼ ◄◄◄▼▼▼ ◄◄◄◄, and ⚰.

23. **a.** $(4 \times 100) + (5 \times 10) + (0 \times 1)$. **b.** $(1 \times 10{,}000) + (3 \times 1{,}000) + (4 \times 100) + (9 \times 10) + (8 \times 1)$. **c.** $(1 \times 100{,}000) + (2 \times 10{,}000) + (0 \times 1{,}000) + (0 \times 100) + (6 \times 10) + (5 \times 1)$.

24. **a.** 553. **b.** 3,838.

25. **a.** A_{twelve}, B_{twelve}, and 10_{twelve}. **b.** 500_{six}, 501_{six}, and 502_{six}.

26. 334_{five}.

27. **a.** $221_{\text{five}} = (2 \times 25) + (2 \times 5) + (1 \times 1) = 50 + 10 + 1 = 61$. **b.** $534_{\text{nine}} = (5 \times 81) + (3 \times 9) + (4 \times 1) = 405 + 27 + 4 = 436$. **c.** $2E0_{\text{sixteen}} = (2 \times 256) + (14 \times 16) + (0 \times 1) = 512 + 224 + 0 = 736$.

CHAPTER 4

Section 4.1 Learning Assessment

1. Take two disjoint sets of objects, combine them, and then count the total number of objects.

2. If the two sets are not disjoint, then any elements common to the sets are considered only once in the union. Thus there are fewer elements than required for the sum. If one of the sets is infinite, then we do not get a natural number when we count the total number of objects after the union has been made.

3. **a.** One way to use the set model is to count out each set, combine them, and then count the total. Another way is count out the first set and then count on the amount of numbers equal to the second addend. **b.** To represent the first addend, begin with the tail of the first arrow at zero, and then take the point out to the appropriate number. For the second addend, begin with the tail at the end of the first arrow, and then go from there the appropriate number of units. The final position indicates the sum.

4. a. The closure property guarantees that the sum of two whole numbers is always another whole number and that the sum is unique.   **b.** The commutative property indicates that the value of the sum is not affected by the order in which the numbers are added.   **c.** The associate property indicates that when adding three or more numbers, the sum is not affected by the order in which the numbers are added.
d. The additive identity property indicates that adding zero to a number does not change the value of the number.

5. Whole-number addition is a binary operation because it allows us to add only two numbers at a time. To add three or more numbers, add two numbers and then add the result to the third, and so on.

6. n is limited to a natural number, because if zero is included, the numbers might be equal.

7. a. In the take-away approach remove the appropriate number of elements from one set, and then count what remains.   **b.** In the comparison approach, the difference is found by determining how many more elements one set has than another.   **c.** In the missing-addend approach, the difference is found by using an associated addition fact in which one of the addends is missing.

8. a. With sets, the appropriate number of elements are removed from the set, and the remaining elements are counted. On a number line, move to the right the number of units equal to the minuend, and then move to the left the number of units equal to the subtrahend. The final position is the difference.   **b.** With sets, directly pair objects between the sets. Count unpaired objects in the larger set to find the difference. On a number line, draw two arrows from zero, one for each of the numbers. To find the difference, count the number of units that one arrow goes beyond the other.   **c.** With sets, the difference can be found by determining how many more objects must be added to the one set to get the number of elements in the other. On a number line, draw two arrows from zero, and then determine how many units must be added to the shorter arrow to reach the length of the longer.

9. If B is not a subset of A, then we will not remove the correct number of elements when we compute the set difference.

10. Addition and subtraction are inverse operations of each other because addition undoes subtraction and subtraction undoes addition.

11. A fact family is a set of addition and subtraction facts that involve the same three numbers. Using fact families is one way to help student master single-digit addition and subtraction facts.

12. a.
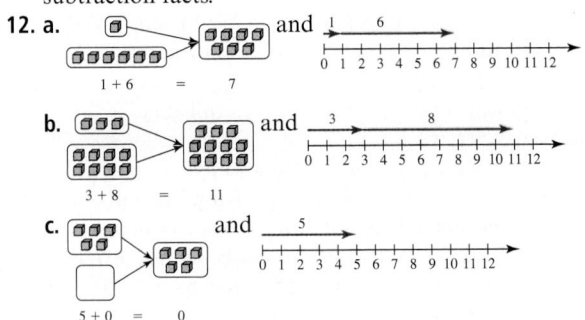
$1 + 6 = 7$

b.
$3 + 8 = 11$

c.
$5 + 0 = 0$

13. a. Each element in the table is a whole number.   **b.** The table is symmetric about the main diagonal of the table.   **c.** The first row and column match the initial row and column.

14.
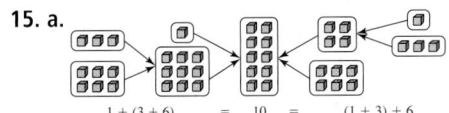
$5 + 6 = 11$ and $6 + 5 = 11$

15. a.
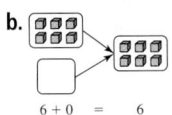
$1 + (3 + 6) = 10 = (1 + 3) + 6$

b.
$6 + 0 = 6$

16. a. $4 + 3 = 7$.   **b.** $5 + 5 = 10$.   **c.** $5 + 7 = 12$.
d. $4 + 0 = 4$.   **e.** $6 + 3 = 9$.

17. a.
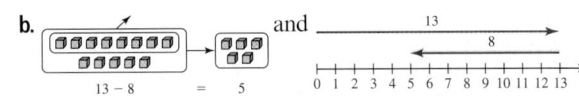
$9 - 4 = 5$

b.
$13 - 8 = 5$

c.
$8 - 8 = 0$

18. a.
$7 - 5 = 2$

b.
$11 - 6 = 5$

c.
$4 - 4 = 0$

19. a.
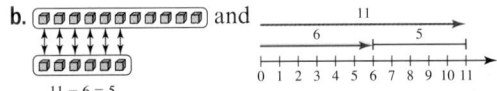
How many?

b.
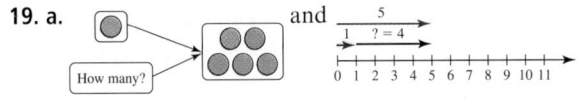
How many?

c.
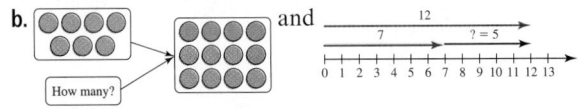
How many?

20. a. $8 - 4 = 4$.   **b.** $5 - 0 = 0$.   **c.** $8 - 3 = 5$.
d. $10 - 6 = 4$.   **e.** $6 - 4 = 2$.   **f.** $10 - 4 = 6$.
21. a. $7 + 4 = 11$, $4 + 7 = 11$, $11 - 4 = 7$, and $11 - 7 = 4$.
b. $2 + 7 = 9$, $7 + 2 = 9$, $9 - 2 = 7$, and $9 - 7 = 2$.
c. $4 + 4 = 8$ and $8 - 4 = 4$.
22. a. Yes. $2 + 3 = 5$.   **b.** No.   **c.** Yes. $4 + 0 = 4$.
d. No.
23. a. Yes. $5 - 4 = 1$.   **b.** Yes. $4 - 0 = 4$.
c. Yes. $4 - 4 = 0$.   **d.** No.
24. a. $5 + 4 = n(A \cup B) = n(\{1, 2, 3, 4, 5, a, b, c, d\}) = 9$.
b. $4 + 3 = n(B \cup C) = n(\{a, b, c, d, x, y, z\}) = 7$.
c. $5 + 3 = n(A \cup C) = n(\{1, 2, 3, 4, 5, x, y, z\}) = 8$.
d. $5 + 4 + 3 = n(A \cup B \cup C) = n(\{1, 2, 3, 4, 5, a, b, c, d, x, y, z\}) = 12$.

25. **a.** $7 - 4 = n(A - B) = n(\{e, f, g\}) = 3$.
   b. $7 - 2 = n(A - C) = n(\{c, d, e, f, g\}) = 5$.
   c. $4 - 2 = n(B - C) = n(\{c, d\}) = 2$.
26. **a.** Counting on by one. **b.** Adding doubles.
   c. Adding doubles. **d.** Identity property.
27. **a.** Identity property. **b.** Adding doubles plus one.
   c. Commutative property. **d.** Sums to 10.
28. **a.** Identity property. **b.** Commutative property.
   c. Closure property. **d.** Commutative property.
   e. Associative property. **f.** Commutative property.
29. **a.** 7, associative property. **b.** 4, identity property.
   c. 6, commutative property. **d.** 0, identity property.
   e. 3, closure property.
30. **a.** O is not closed under addition because the sum of any two odd numbers is an even number. For instance, $1 + 3 = 4$ and $4 \not\subset O$. **b.** $\{0\}$ is closed under addition because $0 + 0 = 0$.
   c. T is closed under addition because the sum of any two multiples of 3 is always a multiple of 3.
31. **a.** A is not closed under addition because $8 + 8 = 16$ and $16 \not\subset A$. **b.** F is closed under addition because the sum of any two multiples of 5 is always a multiple of 5. **c.** $\{1\}$ is not closed under addition because $1 + 1 = 2$ and $2 \not\subset \{1\}$.
32. **a.** $5 < 8$ because there exists the natural number $n = 3$ such that $5 + 3 = 8$. **b.** $16 > 11$ because there exists the natural number $n = 5$ such that $11 + 5 = 16$. **c.** $121 < 228$ because there exists the natural number $n = 107$ such that $121 + 107 = 228$. **d.** $698 > 341$ because there exists the natural number $n = 357$ such that $341 + 357 = 698$.
33. **a.** $7 - 2 = 5$, $2 + 5 = 7$, and $5 + 2 = 7$.
   b. $6 + 3 = 9$, $9 - 6 = 3$, and $9 - 3 = 6$. **c.** $6 + 6 = 12$.
34. **a.** $8 + 4 = 12$, $12 - 8 = 4$, and $12 - 4 = 8$.
   b. $16 - 9 = 7$, $7 + 9 = 16$, and $9 + 7 = 16$.
   c. $14 - 7 = 7$.
35. **a.** If $13 - 9 = ?$, then $9 + ? = 13$. Because $9 + 4 = 13$, then $13 - 9 = 4$. **b.** If $35 - 23 = ?$, then $23 + ? = 35$. Because $23 + 12 = 35$, then $35 - 23 = 12$. **c.** If $110 - 56 = ?$, then $56 + ? = 110$. Because $56 + 54 = 110$, then $110 - 56 = 54$.
36. **a.** Comparison approach. **b.** Missing-addend approach.
   c. Take-away approach.
37. **a.** Take-away approach. **b.** Missing-addend approach.
   c. Comparison approach.
38. 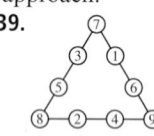   39.
40. Answers may vary, but here is one possibility.

1	4	9
2	5	3
8	6	7

41. **a.** $B = 8$ and $C = 1$. **b.** $B = 7$ and $D = 0$ or $B = 9$ and $D = 2$. **c.** No, because $2 + 3$ is not a two-digit number.
   d. Because CD is a two-digit number and A and B are single-digits, C must always be 1. **e.** The smallest D could be is 0. Possible pairs for A and B are $A = 2$ and $B = 8$, $A = 3$ and $B = 7$, $A = 4$ and $B = 6$, $A = 6$ and $B = 4$, $A = 7$ and $B = 3$, or $A = 8$ and $B = 2$. **f.** The largest D can be is 7, in which case $A = 9$ and $B = 8$ or $A = 8$ and $B = 9$.
42. **a.** 9. **b.** 8. **c.** 8. **d.** 1. **e.** 9, 9, 9, and 0.
43. The last row is $9 - 8 + 7 - 6 + 5 - 4 + 3 - 2 + 1 = 5$.
44. One possibility is $1 + 2 + 3 - 4 + 5 + 6 + 78 + 9 = 100$.

45. 55 fact families.
46. **a.** No, the intersection may not be empty. $n(B)$ could be 3, 4, 5, 6, or 7. **b.** If $n(B) = 3$, then $A \cap B = \varnothing$.
47. For the most part, every finite subset of the whole number is not closed under addition. Only one is closed: $\{0\}$.
48. **a.** The set must also contain every even number greater than 2. **b.** The set must also contain every multiple of 3.
   c. The set must also contain every multiple of 4.
   d. The set must also contain every multiple of n.
49. **a.** 8 must be in the set since $4 + 4 = 8$. **b.** 10 may or may not be in the set. Because it is not a multiple of 4, we do not have enough information to determine whether 10 is in the set. **c.** Any multiple of 4 will be in the set, so other elements include 8, 12, 20, 24, and 28.
50. **a.** The set of whole numbers is not closed under subtraction because if we subtract a larger number from a smaller number, the result is negative, and negative numbers are not in the set of whole numbers. **b.** Yes. The set $\{0\}$ is closed under subtraction since $0 - 0 = 0$.
51. By the definition, it is more correct to write $10 = 6 + x$. However, because addition is commutative, $10 = 6 + x$ and $10 = x + 6$ will give the same answer.
52. Let A, B, and C be sets such that $a = n(A)$, $b = n(B)$, and $c = n(C)$. Using the definition, $a + (b + c)$ $= n(A \cup (B \cup C)) = n((A \cup B) \cup C) = (a + b) + c$.
53. **a.** Any two numbers that are not equal will work. For instance, let $x = 5$ and $y = 3$, then $x - y = 5 - 3 = 2$ and $y - x = 3 - 5 = -2$. Because $2 \neq -2$, we have that $x - y \neq y - x$. **b.** Subtraction will be commutative in the case of $x = y$, because $x - y = x - x = 0 = y - y = y - x$.
54. **a.** Any three nonzero numbers that are not equal will work. For instance, let $a = 5$, $b = 4$, and $c = 3$, then $a - (b - c) = 5 - (4 - 3) = 5 - 1 = 4$ and $(a - b) - c = (5 - 4) - 3 = 1 - 3 = -2$. Because $4 \neq -2$, $a - (b - c) \neq (a - b) - c$.
   b. If any of the numbers are 0, then the associative property will hold assuming that the difference is a whole number.
55. For the identity property to work for subtraction, it must be that $a - 0 = a$ and $0 - a = a$. However, $0 - a = -a$, which is not a whole number when a is a whole number.
56. Answers will vary.
57. **a.** Answers will vary. **b.** Answers will vary.
   c. Answers will vary. **d.** Answers will vary.
58. Answers will vary.   59. Answers will vary.
60. Change the natural number to the whole numbers. Specifically, for any whole numbers a and b, a is less than b or equal to b, written $a \leq b$, if and only if there exists a whole number n such that $a + n = b$.
61. Cameron used the associative property to compute the sum. Because it is based on properties of addition, his method will always work.
62. Answers will vary.   63. Answers will vary.
64. Answers will vary.
65. **a.** Answers will vary. **b.** Answers will vary.
66. No, two of the jacks could be red giving her only four cards in her hand. This situation relates to whole-number addition because the two sets of cards are not disjoint. Consequently, we cannot just add them to find the total.
67. Answers will vary.
68. It is one of the first instances of the use of a variable to represent an unknown number that must be found.
69. Answers will vary.

Section 4.2 Learning Assessment

1. Trading or regrouping is important to adding because adding single-digit numbers can lead to more than 10 in any one place value. By regrouping, we can exchange the 10 objects for one in the next highest place value.

2. When subtracting, there may not be enough of any one group to perform the subtraction. By exchanging or borrowing, we can take one group from a higher place value and exchange it for a group of 10 in the next lower place value, making the subtraction possible.

3. The standard algorithms are the most efficient, but they seldom correspond to the way children think about computation. This can make standard algorithms difficult to understand.

4. **a.** The equal additions algorithm is based on the principle that adding the same amount to the minuend and subtrahend does not affect the difference.
b. It makes the numbers easier to subtract by removing situations in which exchanging is necessary.

5. Mental computation is used to arrive at an exact answer without using computational tools such as a calculator or a pencil and paper. Although computational estimation is often done mentally, it is used to arrive at an approximate answer, not an exact one.

6. **a.** With truncation, a number is "cut off" at a certain place value with no regard for the digits that come after. With rounding, the digit in the next lowest place value is considered to determine which value the number is closer to. In standard rounding, if the digit is 4 or lower, round down. If it is 5 or higher, round up. **b.** These processes are used to make the numbers easier to compute with by simplifying the numbers to ones that contain many zeros.

7. **a.** Two or more numbers that are compatible if the numbers add or subtract to a multiple of 10, 100, or another number. **b.** We make use of the commutative and associative properties when using compatible numbers to compute mentally.

8. **a.** Counting strategies, like counting on, counting back, and skip counting, are useful when computing sums or differences that involve multiples of numbers like 5, 10, or 100. **b.** In left-to-right strategies, we work from larger place values to smaller place values. This often requires us to keep track of less information as we add or subtract. **c.** In making compatible numbers, we look for two or more numbers that add or subtract to a multiple of 10, 100, or another number that is easy to use in computing. **d.** In compensation, we split one addend into two numbers so that one of them is compatible with the other addend. We add the compatible numbers, and then add the remaining number. **e.** When breaking apart numbers, we break one number apart by using expanded notation. We then use counting strategies or left-to-right strategies to add or subtract the expanded number one term at a time.

9. **a.** In a front-end estimate, truncate or round the numbers to a single nonzero digit in the largest place value. Then add or subtract as appropriate. **b.** In compensation, split one addend into two numbers so that one of them is compatible with the other addend. Add the compatible numbers, and then add the remaining number. **c.** With compatible numbers, round the numbers so that they are compatible to make the estimation easier. **d.** To use clustering, the addends must be close to the same approximate value. Estimate the typical value of the numbers, and then multiply by the amount of numbers.

10. **a.** $54 + 36 = 90$. **b.** $429 + 261 = 690$. **c.** $1,416 + 643 = 2,059$.

11. **a.** $96 - 62 = 34$. **b.** $78 - 40 = 38$. **c.** $378 - 256 = 122$.

12. **a.** $47 + 34 = 81$.

b. $89 + 72 = 161$.

c. $56 - 24 = 32$.

d. $93 - 47 = 46$.

13. a. $143 + 76 = 219$.

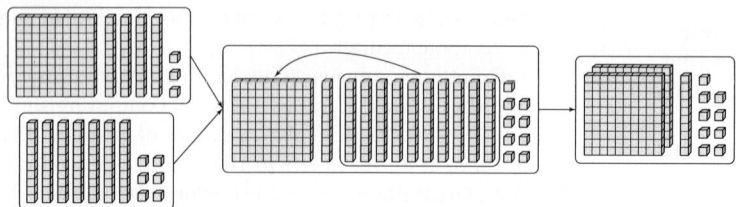

b. $463 + 281 = 744$.

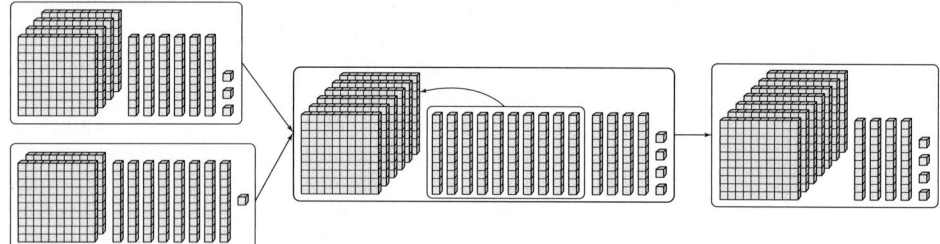

c. $184 - 66 = 118$.

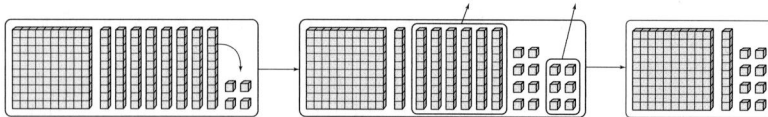

d. $247 - 193 = 54$.

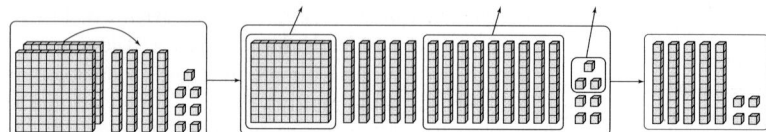

14. a. $47 - 26 = 21$.

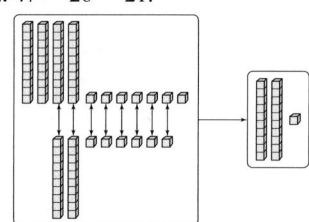

b. $152 - 89 = 63$.

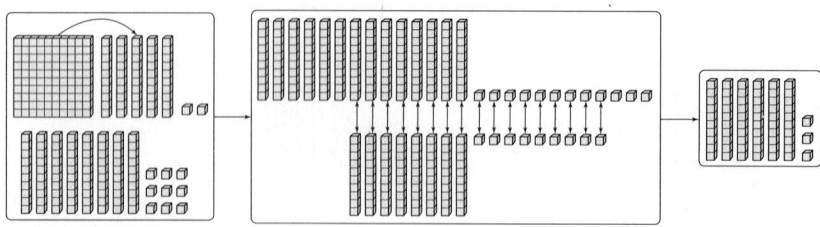

c. $301 - 223 = 78$.

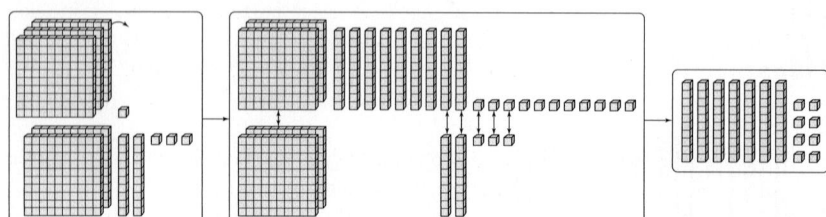

15. a. 543 $\boxed{+}$ 254.   **b.** Possible answer may include 681 $\boxed{-}$ 93 $\boxed{\text{STO}}$.   **c.** 902 $\boxed{-}$ 357 $\boxed{+}$ 78.   **d.** Possible answer may include 100 $\boxed{+}$ 100 $\boxed{+}$ 100 $\boxed{+}$ 100 $\boxed{+}$ 100.

16. Methods will vary.   **a.** 431.   **b.** 8,393.   **c.** 57.   **d.** 1,209.

17. a. 61.   **b.** 33.   **c.** 264.   **d.** 74.

18. a. 104.   **b.** 349.   **c.** 1,657.   **d.** 1,623.

19. a. 206.   **b.** 212.   **c.** 463.   **d.** 359.

20. a. 7,769.   **b.** 2,306.   **c.** 16,274.   **d.** 3,120.

21. a. 75.   **b.** 313.   **c.** 751.   **d.** 1,400.

22. a. 14.   **b.** 39.   **c.** 617.   **c.** 207.

23. Methods will vary, but answers are as follows:
   a. 107.   **b.** 101.   **c.** 104.   **d.** 456.

24. Methods will vary, but answers are as follows:
   a. 610.   **b.** 10,081.   **c.** 465.   **d.** 732.

25. a. 400.   **b.** 3,810.   **c.** 67,300.   **d.** 21,000.

26. a. 350.   **b.** 45,800.   **c.** 4,700.   **d.** 30,000.

27. a. 13,400.   **b.** 40,000.   **c.** 145,000.   **d.** 200,000.

28. a. About 6,800.   **b.** About 22,000.   **c.** About 3,200.
   d. About 4,000.

29. a. About 520.   **b.** About 900.   **c.** About 2,500.
   d. About 5,000.

30. a. About 150.   **b.** About 400.   **c.** About 18,000.

31. $4,101 + 1,563 \approx 5,600$; $2,408 + 3,499 \approx 5,900$; $2,998 + 2,997 \approx 6,000$; and $3,501 + 3,110 \approx 6,200$.

32. $468 - 305 \approx 150$, $1,789 - 1,592 \approx 200$, $823 - 576 \approx 240$, and $972 - 491 \approx 470$.

33. a. 27,215.   **b.** 420,819.   **c.** 5,921.   **d.** 86,767.

34. a. 2,858.   **b.** 44.   **c.** 258.   **d.** 1,006

35. a.
$$\begin{array}{r} 463 \\ +278 \\ \hline 741 \end{array}$$
   b.
$$\begin{array}{r} 2431 \\ 1648 \\ +4279 \\ \hline 8358 \end{array}$$
   c.
$$\begin{array}{r} 4307 \\ -2491 \\ \hline 1816 \end{array}$$
   d.
$$\begin{array}{r} 6734 \\ -2896 \\ \hline 3838 \end{array}$$

36. a. 427.   **b.** 1,813.   **37. a.** 25.   **b.** 751.

38. a. About 180.   **b.** About 1,100.

39. Answers will vary.

40. a. 169.   **b.** 315.   **c.** 914.

41. a. The difference is the amount that, if added to the bill, equals the amount of money given to the cashier.
   b. The customer will receive $15 in change. The cashier might return a $5 and a $10 and count out "50, and 60."
   c. The customer will receive $37 in change. The cashier might return two $1s, one $5, one $10, and one $20. In doing so, the cashier may count "64, 65, 70, 80, and 100."

42. a. 42.   **b.** 23.   **c.** 189.   **43.** 695 cal.

44. 1,204 mi.   **45.** $87.

46. 6,645 hamburger patties, 2,941 chicken patties, 1,869 fish fillets, and 13,330 hamburger buns were used.

47. $166.   **48.** $16,800.

49. a. About 500 yr.   **b.** About 240 yr.   **c.** About 150 yr.
   d. About 40 yr.

50. a. About 15,000,000 more people.   **b.** About 21,500,000 more people.   **c.** About 179,000,000 people.

51. $269.

52. In the missing-addend approach, we subtract using what we know about addition facts. If the numbers are large, it is difficult to know the exact missing addend.

53. a. Truncating always results in a smaller value, which means truncation always underestimates a sum.

b. In a difference, truncation can lead to both overestimates and underestimates, depending on the values of the numbers.

54. a. To overestimate a sum, ignore standard rounding, and round both numbers up.   **b.** To underestimate a sum, ignore standard rounding, and round both numbers down.

55. a. To overestimate a difference, ignore standard rounding, and round the first number up and the second number down.
   b. To underestimate a difference, ignore standard rounding, and round the first number down and the second number up.

56. a. 19.   **b.** 26.   **c.** 44.   **d.** Sharice's method uses negative numbers to avoid exchanging. She subtracts each place value and writes each difference separately, including positive and negative numbers. She then adds them together to get her answers.

57. Answers will vary.   **58.** Answers will vary.

59. One reason it may seem natural to children is because they learn to read from left to right.

60. The manner in which the errors are corrected will vary. The mistakes made by the students are as follows:
   a. The student simply wrote the sum of the single-digit facts with no concern for place value.
   b. This student added the ones digits correctly but forgot to trade the group of ten and add it in the tens place value.
   c. This student added the ones digits correctly but wrote down the tens digit of the sum rather than the ones.

61. The manner in which the errors are corrected will vary. The mistakes made by the students are as follows:
   a. The student added only the ones digits rather than subtracting.   **b.** The student forgot to take one away from the 4 when making an exchange.   **c.** The student appears to have subtracted from left to right. The student was able to subtract the tens digit this way but may not have known what to do with the ones digits. It appears as though the student subtracted the top ones digit from the bottom one.

62. Answers will vary.

63. Marcia did not enter a 0 when she entered 15,031. Because it is easy to mispress a key when using a calculator, it can be useful first to estimate the value of the computation, allowing for a quick check for any mistakes we may have made.

64. Answers will vary.

65. Answers will vary.

66. The sum is $368 + 245 = 613$. Answers will vary on the algorithm used.

67. The difference is $2,324 - 653 = 1,671$.
   a. Answers will vary.   **b.** Answers will vary.

68. Answers will vary.

69. a. Answers will vary.   **b.** Answers will vary.
   c. Answers will vary.

70. a. Answers will vary.   **b.** Answers will vary.

■ Section 4.3 Learning Assessment

1. a. Interpret the product as a repeated sum in which the second factor is added to itself the number of times equal to the first factor. Then compute the sum.   **b.** In the Cartesian product approach, select two sets, one for each factor. Form the Cartesian product between the sets, and then count the number of ordered pairs to find the product.

2. The phrase comes from the fact that we literally interpret the product as adding b to itself a "times."

3. a. Interpret $a \cdot b$ as combining a sets each having b elements. After combining the sets, count the objects to find the product. **b.** The first factor gives the number of rows in the array, and the second factor gives the number of columns. Once the array is filled in, count the total number of objects to find the product.
c. The first factor gives the number of arrows to draw, and the second gives the number of units in each arrow. The product is the final position on the number line.

4. a. The closure property guarantees that the product of two whole numbers is always another whole number and that the product is unique. **b.** The commutative property indicates that the value of the product is not affected by the order in which the numbers are multiplied.
c. The associate property indicates that, when multiplying three or more numbers, the product is not affected by the order in which the numbers are multiplied. **d.** The multiplicative identity property indicates that multiplying a number by 1 does not change the value of the number.
e. The distributive property implies that a product of a number and a sum of two more numbers can be computed by first adding the numbers and then multiplying or by distributing the factor to the addends, multiplying, and then adding. Either way, the answer is the same. **f.** The zero multiplication property indicates that multiplying a number by zero will always result in a product equal to zero.

5. The multiplicative identity cannot be zero because multiplying by zero always results in zero.

6. Exponents are a notational shortcut for repeated multiplication.

7. a. If a and b are whole numbers, then $a \div b$ is the number of times we can subtract b from a as long as $b \neq 0$.
b. In the partitioning approach, $a \div b$ is the number of objects in each set when a objects are equally distributed among b sets. **c.** In the missing-factor approach, the quotient is rewritten as a product in which one of the factors is missing.

8. Division by zero does not lead to a definite number as an answer. In terms of repeated subtraction, because subtracting zero never affects the value of the difference, we can subtract it once, twice, or even a hundred times. Because there is no definite number of times that we can subtract zero from another number, we say that division by zero is undefined.

9. The division algorithm expands our ability to compute division because it allows us to work with remainders.

10. a. The normal order of operations can be suspended with the use of parentheses or other grouping symbols.
b. Because exponents represent the repeated product of one number, performing another product before or in conjunction with the exponent could affect the value of the repeated product. Consequently, the exponent must be computed first to avoid such problems and to arrive at the correct answer. **c.** Because multiplication is the repeated sum of one number, performing another sum before or in conjunction with the product could affect the value of the repeated sum. Consequently, the product must be computed first to avoid such problems and to arrive at the correct answer. The same holds true for division.

11. a. 11 added to itself 3 times: $3 \cdot 11 = 11 + 11 + 11$.
b. 3 added to itself 6 times: $6 \cdot 3 = 3 + 3 + 3 + 3 + 3 + 3$.

c. 7 added to itself 1 time: $1 \cdot 7 = 7$.
d. 0 added to itself 4 times: $4 \cdot 0 = 0 + 0 + 0 + 0 = 0$.

12. a. $4 \cdot 6$. **b.** $2 \cdot 3$. **c.** $1 \cdot 5$. **d.** $6 \cdot 0$.

13. a. $6 \cdot 3$ follows. Because 18 blocks are used, $6 \cdot 3 = 18$.

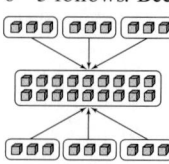

b. $7 \cdot 1$ follows. Because 7 blocks are used, $7 \cdot 1 = 7$.

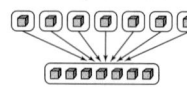

c. $2 \cdot 5$ follows. Because 10 blocks are used, $2 \cdot 5 = 10$.

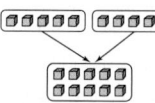

14. a. $9 \cdot 3$ follows. Because 27 squares were used, $9 \cdot 3 = 27$.

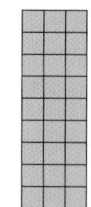

b. $6 \cdot 5$ follows. Because 30 squares were used $6 \cdot 5 = 30$.

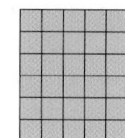

c. $4 \cdot 8$ follows. Because 32 squares were used $4 \cdot 8 = 32$.

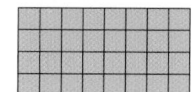

d. $7 \cdot 2$ follows. Because 14 squares were used $7 \cdot 2 = 14$.

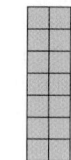

15. The tree diagram follows. Because there are 15 branches in the second level, we conclude that $5 \cdot 3 = 15$.

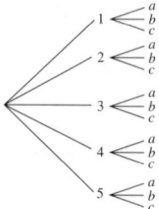

16. a. $4 \cdot 5$. **b.** $6 \cdot 3$. **c.** $3 \cdot 3$. **d.** $4 \cdot 3$. **e.** $4 \cdot 7$.

17. a. The table is symmetric about the main diagonal of the table. **b.** The second row and column match the initial row and column.   **c.** The first row and column are all zeros.

18. a. 4 multiplied by itself 5 times: $4^5 = 4 \cdot 4 \cdot 4 \cdot 4 \cdot 4$.
b. 2 multiplied by itself 7 times: $2^7 = 2 \cdot 2 \cdot 2 \cdot 2 \cdot 2 \cdot 2 \cdot 2$.
c. 8 multiplied by itself 3 times and x multiplied by itself 4 times: $8^3 x^4 = 8 \cdot 8 \cdot 8 \cdot x \cdot x \cdot x \cdot x$.
d. $(x - 2)$ multiplied by itself 2 times: $(x - 2)^2 = (x - 2)(x - 2)$.

19. a. 4^4.   **b.** 3^8.   **c.** $x^2 y^2$.   **d.** $6^4 x^3$.   **e.** $(3 + x)^3$.   **f.** y^8.

20. a. When multiplying exponents with the same base, simplify the expression by raising the base to the sum of the exponents.   **b.** When multiplying exponents with different bases but the same exponent, simplify the expression by multiplying the bases and raising the product to the common exponent.   **c.** When raising one exponent to another, simplify the expression by keeping the base and raising it to the product of the two exponents.

21. a. $9 \div 3$ represents the number of times 3 can be subtracted from 9: $9 - 3 - 3 - 3$.   **b.** $28 \div 7$ represents the number of times 7 can be subtracted from 28: $28 - 7 - 7 - 7 - 7$.   **c.** $36 \div 4$ represents the number of times 4 can be subtracted from 36: $36 - 4 - 4 - 4 - 4 - 4 - 4 - 4 - 4 - 4$.   **d.** $18 \div 6$ represents the number of times 6 can be subtracted from 18: $18 - 6 - 6 - 6$.

22. a. If $12 \div 4 = ?$, then $12 = 4 \cdot ?$. Because $4 \cdot 3 = 12$, then $12 \div 4 = 3$.   **b.** If $49 \div 7 = ?$, then $49 = 7 \cdot ?$. Because $7 \cdot 7 = 49$, then $49 \div 7 = 7$.   **c.** If $72 \div 9 = ?$, then $72 = 9 \cdot ?$. Because $9 \cdot 8 = 72$, then $72 \div 9 = 8$.   **d.** If $39 \div 13 = ?$, then $39 = 13 \cdot ?$. Because $13 \cdot 3 = 39$, then $39 \div 13 = 3$.

23. a. $6 \div 2$ is represented as 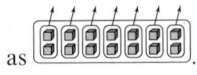, and $14 \div 2$ is represented as .
b. $8 \div 2$ is represented as , and $15 \div 3$ is represented as
c. In part a, we are showing that we are removing a certain number of blocks a certain number of times. In part b, we are showing that we are equally distributing the blocks to a certain number of groups.

24. It represents the partitioning approach because we are placing cookies into a certain number of groups and then determining how many cookies are in each group.

25. a. $6 \cdot 4 + 8 = 4 + 4 + 4 + 4 + 4 + 4 + 8 = 32$.
b. $4^2 - 3 \cdot 5 = 4 \cdot 4 - 3 \cdot 5 = 4 + 4 + 4 + 4 - 5 - 5 - 5 = 1$.

26. a. Distributive property of multiplication over addition.   **b.** Commutative property.   **c.** Multiplicative identity property.   **d.** Associative property.   **e.** Closure property.   **f.** Zero multiplication property.   **g.** Multiplicative identity property.   **h.** Distributive property of multiplication over addition.

27. a. $\{2, 4, 6, 8, 10, \ldots\}$ is closed under multiplication because an even number multiplied by an even number is always even.   **b.** $\{1, 3, 5, 7\}$ is not closed under multiplication because $5 \cdot 5 = 25$, and 25 is not an element of the set.   **c.** $\{0\}$ is closed under

multiplication because $0 \cdot 0 = 0$.   **d.** $\{0, 1\}$ is closed under multiplication because $0 \cdot 0 = 0, 0 \cdot 1 = 0$, and $1 \cdot 1 = 1$.   **e.** $\{2, 3, 6, 9, 12, \ldots\}$ is not closed under multiplication because $2 \cdot 2 = 4$, and 4 is not an element of the set.   **f.** $\{1, 3^2, 3^3, 3^4, \ldots\}$ is closed under multiplication because any power of 3 multiplied by another power of 3 is a power of 3.

28. a. $4(5 + 7) = 4 \cdot 12 = 48 = 20 + 28 = 4 \cdot 5 + 4 \cdot 7$.
b. $2(8 + 3) = 2 \cdot 11 = 22 = 16 + 6 = 2 \cdot 8 + 2 \cdot 3$.

29. a. $5 \cdot (1 + 3) = 5 \cdot 1 + 5 \cdot 3 = 5 + 15 = 20$.
b. $(4 + 7) \cdot 9 = 4 \cdot 9 + 7 \cdot 9 = 36 + 63 = 99$.
c. $2 \cdot (8 - 6) = 2 \cdot 8 - 2 \cdot 6 = 16 - 12 = 4$.
d. $3 \cdot (4 + 2 + 7) = 3 \cdot 4 + 3 \cdot 2 + 3 \cdot 7 = 12 + 6 + 21 = 39$.

30. a. $4 \cdot 3 + 4 \cdot 2 = 4(3 + 2)$.   **b.** $3x + 3 \cdot 3 = 3(x + 3)$.   **c.** $6x - 6 = 6(x - 1)$.   **d.** $3 \cdot (x + 3) + y(x + 3) = (x + 3)(3 + y)$.   **e.** $x^2 - 3x = x(x - 3)$.   **f.** $4x^2 - 2x = 2x(x - 1)$.

31. a. 6^9.   **b.** 1.   **c.** 3^{12}.   **d.** 10^2.   **e.** $(3x)^2$.
32. a. 1.   **b.** 12^6.   **c.** 4^7.   **d.** 0.   **e.** 1.
33. a. 6.   **b.** 3.   **c.** 5.   **d.** 0.
34. a. $3^2 > 2^3$.   **b.** $4^6 > 6^4$.   **c.** $8^0 > 0^8$.

35. a. Quotient = 2, remainder = 6.   **b.** Quotient = 5, remainder = 2.   **c.** Quotient = 5, remainder = 4.
36. a. Quotient = 9, remainder = 2.   **b.** Quotient = 7, remainder = 7.   **c.** Quotient = 12, remainder = 4.

37. a. $15 \div 3 = 5, 3 \cdot 5 = 15$, and $5 \cdot 3 = 15$.
b. $6 \cdot 8 = 48, 48 \div 6 = 8$, and $48 \div 8 = 6$.
c. $75 \div 5 = 15, 5 \cdot 15 = 75$, and $15 \cdot 5 = 75$.
d. $49 \div 7 = 7$.

38. a. 74.   **b.** 63.   **c.** 9.
39. a. 1.   **b.** 1,849.   **c.** 4.
40. $2^x = x^2$ when $x = 2$.
41. $3^5 + 3^5 < 3^{10}$.
42. $2 \cdot 3^2, 4^2 + 1, 2^4, 15$, and $3 \cdot 2^2$.
43. a. 39.   **b.** 6.   **c.** 18.
44. There are 12 possible divisions: $10 \div 10, 10 \div 2, 8 \div 8, 8 \div 4, 8 \div 2, 6 \div 6, 6 \div 3, 6 \div 2, 4 \div 4, 4 \div 2, 3 \div 3$, and $2 \div 2$.
45. If $c = 1$, then there are many possibilities for a and b.
46. a. Possible values for n are $\{36, 18, 12, 9, 6, 4, 3, 2, 1\}$. Possible values for c are $\{1, 2, 3, 4, 6, 9, 12, 18, 36\}$.   **b.** Possible values for n can be $\{1, 2, 3, \ldots 36\}$. Possible values for c are $\{1, 2, 3, 4, 6, 9, 12, 18, 36\}$. Possible values for r are $\{0, 1, 2, \ldots, 17\}$.
47. a. $36 \div (4 + 5) \cdot 8 = 32$.   **b.** $(13 - 6) \cdot 8 \div 2 = 28$.   **c.** $(6 + 4 + 5) \div 3 = 5$.
48. 168 tiles.
49. 48 different ways to make an ice cream cone.
50. Each person gets 3 slices with 2 slices left over.
51. Each playlist will have 12 songs.
52. Let x be the number. First double x, and then add 14, which we represent as $2x + 14$. Next, divide by two: $\frac{2x + 14}{2} = \frac{2x}{2} + \frac{14}{2} = x + 7$. If we subtract the number x, then $x + 7 - x = 7$.
53. There are 1,296 people living in the town.
54. Let $A = \{1, 2, 3, 4, 5\}$ and $B = \{a, b, c\}$. Then $5 \cdot 3 = n(A \times B) = n[\{(1, a), (1, b), (1, c), (2, a), (2, b), (2, c), (3, a), (3, b), (3, c), (4, a), (4, b), (4, c), (5, a), (5, b), (5, c)\}] = 15$.

55. a. Yes. No numbers in the whole numbers multiply together to get 3 other than 1 and 3. For that reason, the set can still be closed even though 3 is not in the set. **b.** No. Because 2 and 3 are in the set and $2 \cdot 3 = 6$, the set cannot be closed.

56. a. The set of whole number is closed under both addition and multiplication. **b.** $\{0, 1\}$ is closed under multiplication but not addition. **c.** $\{0, 1, 2\}$ is not closed under either multiplication or addition.

57. If a and b are whole numbers with $a = n(A)$ and $b = n(B)$, then $a \times b = n(A \times B) = n(B \times A) = b \cdot a$.

58. Let $A = \{a, b, c, d\}$ and $B = \varnothing$. Then $4 \cdot 0 = n(A \times B) = n(\varnothing) = 0$.

59. The zero multiplication property tells us that if $a \cdot b = 0$, then either $a = 0$ or $b = 0$.

60. If a and b are whole numbers and $a \cdot b = 1$, then $a = 1$ and $b = 1$.

61. Let a, b, m, and n be any whole numbers with $a \neq 0$ and $b \neq 0$.
a. $a^m + a^n = \underbrace{a \cdot a \cdot \ldots \cdot a}_{m \text{ times}} \cdot \underbrace{a \cdot a \cdot \ldots \cdot a}_{n \text{ times}} = a^{m+n}$.

b.
$(a^m)^n = \underbrace{a^m \cdot a^m \cdot \ldots \cdot a^m}_{n \text{ times}} = \underbrace{\underbrace{a \cdot a \cdot \ldots \cdot a}_{m \text{ times}} \underbrace{a \cdot a \cdot \ldots \cdot a}_{m \text{ times}} \underbrace{a \cdot a \cdot \ldots \cdot a}_{m \text{ times}}}_{n \text{ times}} = a^{m+n}$

62. a. Let $a = 2$, $b = 3$, and $c = 4$, then $a \cdot (b \cdot c) = 2 \cdot (3 \cdot 4) = 24$ and $(a \cdot b) \cdot (a \cdot c) = (2 \cdot 3) \cdot (2 \cdot 4) = 6 \cdot 8 = 48$. Because $24 \neq 48$, then $a \cdot (b \cdot c) \neq (a \cdot b) \cdot (a \cdot c)$. **b.** Let $a = 2$, $b = 3$, and $c = 4$, then $a + (b \cdot c) = 2 + (3 \cdot 4) = 2 + 12 = 14$ and $(a + b) \cdot (a + c) = (2 + 3) \cdot (2 + 4) = 5 \cdot 6 = 30$. Because $14 \neq 30$, then $a + (b \cdot c) \neq (a + b) \cdot (a + c)$. **c.** Let $a = 2$, $b = 3$, and $n = 2$, then $(a + b)^n = (2 + 3)^2 = 5^2 = 25$ and $a^n + b^n = 2^2 + 3^2 = 4 + 9 = 13$. Because $25 \neq 13$, then $(a + b)^n \neq a^n + b^n$. **d.** Let $a = 4$, $b = 2$, and $n = 2$, then $(4 - 2)^2 = 2^2 = 4$ and $a^n - b^n = 4^2 - 2^2 = 16 - 4 = 12$. Because $4 \neq 12$, then $(a - b)^n \neq a^n - b^n$.

63. a. Because $3 \div 2 = 1.5$ and 1.5 is not a whole number, the set of whole numbers is not closed under division. **b.** Because $4 \div 2 = 2$ and $2 \div 4 = 0.5$, whole-number division is not commutative. **c.** Because $8 \div (4 \div 2) = 8 \div 2 = 4$ and $(8 \div 4) \div 2 = 2 \div 2 = 1$, whole-number division is not associative. **d.** Because $4 \div 1 = 4$ and $1 \div 4 = 0.25$, whole-number division does not satisfy the identity property.

64. Let $a = 4$, $b = 2$ and $c = 1$, then $(a + b) \div c = (4 + 2) \div 1 = 6$ and $(a \div c) + (b \div c) = (4 \div 1) + (2 \div 1) = 4 + 2 = 6$. As long as c is not zero, it will be true.

65. a. $\{0, 1, 2, 3, 4\}$. **b.** $\{0, 1, 2, 3, 4, 5\}$. **c.** $\{0, 1, 2, 3, 4, 5, 6, 7, 8\}$. **d.** $\{0, 1, 2, ..., n - 1\}$. **e.** The division algorithm guarantees it because the remainder is always strictly less than the divisor.

66. a.–c. Answers will vary.

67. a. Repeated-subtraction approach. **b.** Missing-factor approach. **c.** Missing-factor approach.

68. a.–d. Answers will vary.

69. Answers will vary.

70. Answers will vary.

71. No. $5^3 \cdot 5^2 = 5^5 \neq 5^{10} = (5^2)^5 = 25^5$.

72. Answers will vary.

73. The first five place values are consecutive powers of 10: $10^0, 10^1, 10^2, 10^3$, and 10^4. The pattern makes sense because any place value in the decimal system is found by multiplying the value of the previous place value by 10.

74. Answers will vary.

75. Answers will vary, but one possible diagram is:

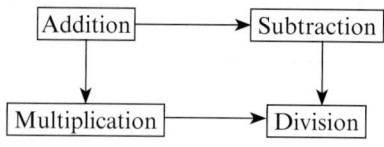

76. The rational numbers.

77. The rational numbers.

78. Answers will vary.

■ Section 4.4 Learning Assessment

1. a. Most written algorithms for multiplication use single-digit multiplication facts, regrouping, and multidigit addition. **b.** Most written algorithms for division use subtraction, exchanging, and multiplication.

2. The standard algorithm for multiplication condenses the partial products algorithm by combining the partial products into one for each digit in the second factor.

3. The placeholders move nonzero digits in the partial products out to the correct place values.

4. Create a row in the lattice for each digit in the second factor. Next, fill in the lattice by multiplying digit by digit, placing the tens digit in the upper triangle and the ones digit in the lower. Finally, add down each diagonal, regrouping as necessary.

5. a. The scaffold algorithm for division condenses the repeated subtraction algorithm for division by subtracting multiples of the divisor rather than just the divisor itself. **b.** The standard algorithm for division condenses the scaffold algorithm by removing the placeholder zeros and by working with exact values rather than estimates.

6. We use multiplication to determine how many times the divisor will go into the dividend. Every time we compute a product with the divisor, we subtract the product to find the remaining amount.

7. The distributive property can be used to split one of the factors into two numbers that are easier to multiply.

8. With algebraic logic, the calculator is programmed to compute operations according to the standard order. With arithmetic logic, the calculator performs the operations in the order in which they are entered.

9. a. 31 added to itself 14 times. **b.** 63 added to itself 37 times. **c.** 29 added to itself 135 times.

10. a. $12 \cdot 21 = 252$.

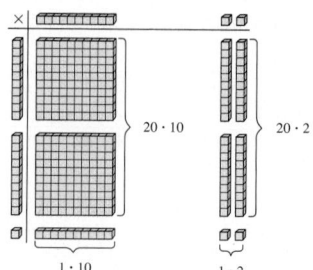

b. $33 \cdot 26 = 858$.

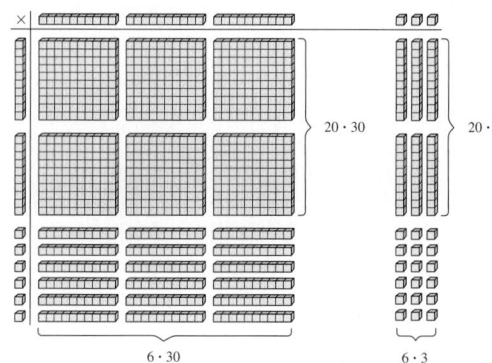

c. $104 \cdot 40 = 4{,}160$.

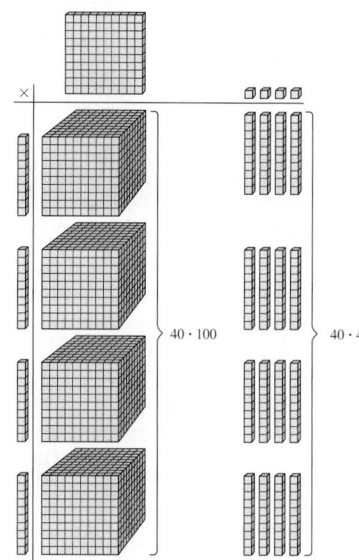

---

11. a. $2 \cdot 78 = 156$.

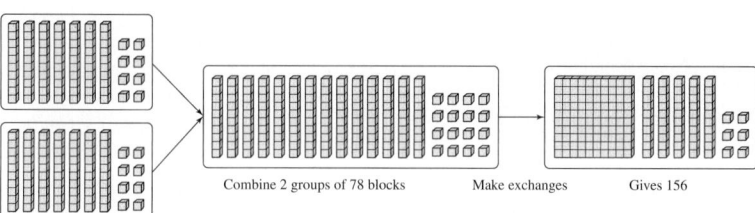

Combine 2 groups of 78 blocks   Make exchanges   Gives 156

b. $3 \cdot 241 = 723$.

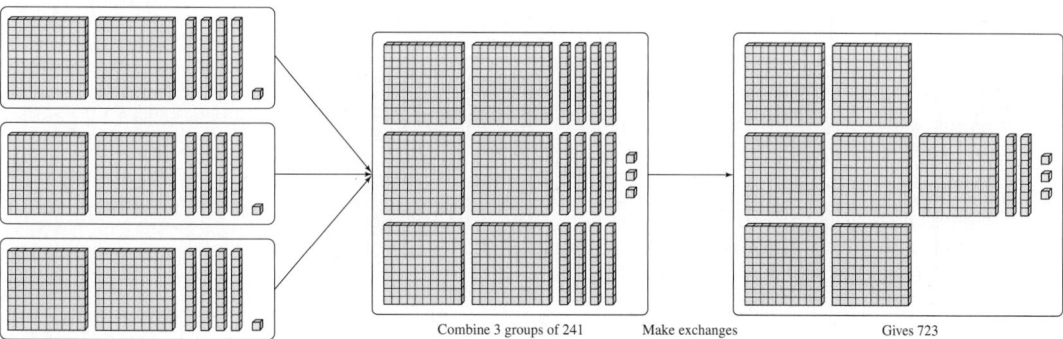

Combine 3 groups of 241   Make exchanges   Gives 723

c. $2 \cdot 1{,}013 = 2{,}026$.

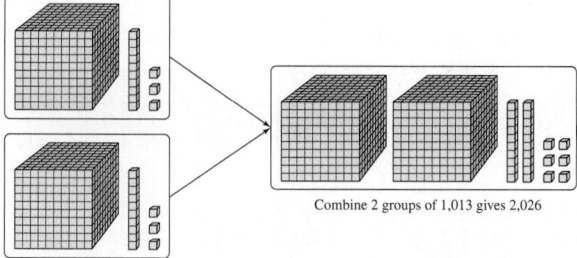

Combine 2 groups of 1,013 gives 2,026

d. If the first factor is large, it is difficult to represent the sums.

12. a. The largest number is $9{,}999 = 3 \cdot 3{,}333$.   **b.** Answers will vary but possibilities include $100 \cdot 100$ and $1{,}000 \cdot 10$.

13. a. The number of times 12 can be subtracted from 96.   **b.** The number of times 38 can be subtracted from 114.
   c. The number of times 42 can be subtracted from 1,134.

14. a. $92 \div 23 = 4$.

b. $168 \div 42 = 4$.

c. $615 \div 123 = 5$.

15. a. $72 \div 3 = 24$.

b. $204 \div 4 = 51$.

c. $192 \div 6 = 32$.

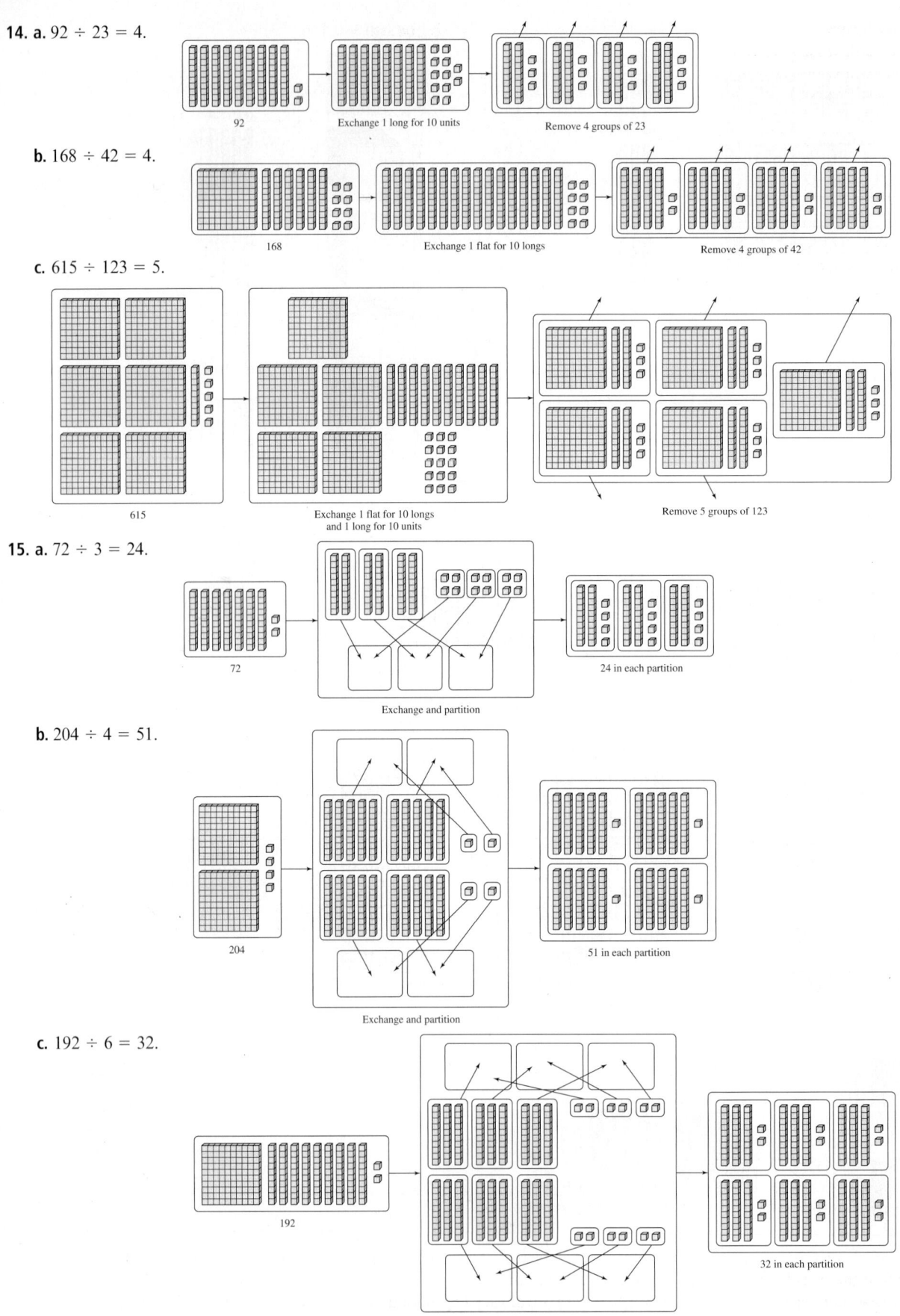

d. If there are a large number of partitions, it can be difficult to model.

16. Answers will vary.
17. a. Answers will vary.   b. Answers will vary.
18. a. Answers will vary.   b. Answers will vary.
    c. Answers will vary.
19. a. 372.   b. 2,646.   c. 9,316.
20. a. 846.   b. 2,403.   c. 25,704.   d. 75,849.
21. a. 315.   b. 3,604.   c. 28,598.   d. 55,296.
22. a. 8.   b. 4.   c. 5 remainder 10.   d. 4 remainder 33.
23. a. 108.   b. 63.   c. 58 remainder 17.   d. 138 remainder 18.
24. a. 10,374.   b. 30,940.   c. 267.   d. 56 remainder 8.
25. a. 71,853.   b. 163,438.   c. 250 remainder 34.
    d. 24 remainder 65.
26. Methods will vary.
    a. 432.   b. 2,623.   c. 3,211.   d. 52.
27. Methods will vary.
    a. 837.   b. 10,812.   c. 609.   d. 405.
28. Methods will vary.
    a. About 12,000.   b. About 9,500,000.   c. About
    9,700,000.   d. About 53,000,000.
29. Methods will vary.
    a. About 800.   b. About 60.   c. About 170.
30. a. Correct.   b. Incorrect.   c. Incorrect.   d. Correct.
31. $37 \cdot 51 \approx 40 \cdot 50 = 2,000$.
    $29 \cdot 68 \approx 30 \cdot 70 = 2,100$.
    $41 \cdot 59 \approx 40 \cdot 60 = 2,400$.
    $48 \cdot 52 \approx 50 \cdot 50 = 2,500$.
    $72 \cdot 41 \approx 70 \cdot 40 = 2,800$.
32. a. 17,285,490.   b. 1,889,568.   c. 1,457.   d. 522.
33. a. 54,439,939.   b. 848.   c. 424,564.
34. a. 141 remainder 21.   b. 38 remainder 304.
    c. 234 remainder 349.
35. $92 \cdot 83 \cdot 6 \cdot 5 = 229,080$.
36. a. $82 \cdot 63 = 5,166$.   b. $38 \cdot 26 = 988$.
37. a. The least quotient is $236 \div 9 = 26$ remainder 2. The
    greatest quotient is $963 \div 2 = 481$ remainder 1.
    b. The least remainder will be zero. One quotient that give
    us this is $936 \div 2 = 468$. Because all possible divisions
    by 9 give us a remainder of 2, the greatest remainder will
    be 5. One quotient that gives us this is $239 \div 6 = 39$,
    remainder 5.
38. a.
```
   87
 × 53
  261
 4350
 4611
```
b.
```
  163
 × 94
  652
14670
15322
```
c.
```
    124
26)3224
    26
    62
    52
   104
   104
     0
```
39.
```
    183
26)4781
    26
   218
   208
   101
    78
    23
```
40. a. $293 - (10 \cdot 27) = 23$.   b. $1,894 - (40 \cdot 41) - (6 \cdot 41) = 8$.
41. a. i. 372.   ii. 2,170.   iii. 13,366.   b. Answers will vary.
42. a. 2.   b. 4.   c. 6.   d. 12.
43. a. $36 \div (4 + 5) \times 8 = 32$.   b. $(13 - 6) \cdot 8 \div 2 = 28$.
    c. $72 \div (4 \cdot 2) + 12 = 21$.   d. $(6 + 4 + 5) \div 3 = 5$.
44. $45 \cdot 87$ is the closest product to 4,000.
45. 9^{9^9}.
46. 10,800 cans of soda.
47. 48,000 envelopes.
48. 32 cartons.
49. 14 quarters and 14 pennies.
50. Each heir will receive $70,823.

51. 2.5 hours.
52. It takes him 24 days, 64 hr, and 200 min, or 26 days, 19 hr,
    and 20 min.
53. About 11 days and 9 hr.
54. In 3 yr, March 1 will fall on a Thursday.
55. a. Answers will vary.   b. Answers will vary.
    c. Answers will vary.
56. a. About $3 \cdot 365 = 1,095$ meals in a year.   b. If we ingest
    about 800 cal in meal, we ingest about $1,095 \cdot 800 =$
    876,000 cal in a year.
57. a. Answers will vary.   b. Answers will vary.
58. We add multiples of 35 until we reach a number n that is
    smaller than 765, yet $765 - n < 35$. The number of times
    we add 35 to itself is equal to the quotient and $765 - n$
    will be the remainder. Specifically, $765 = 35 \cdot 21 + 30$, so
    $765 \div 35 = 21$, remainder 30.
59. Answers will vary.
60. One way to find each quotient without using the division
    key is to use repeated subtraction.
    a. $142 \div 31 = 4$, remainder 18.   b. $1,248 \div 89 = 14$,
    remainder 2.   c. $45,790 \div 8,965 = 5$, remainder 965.
61. $11,111^2 = 123,454,321$ and $111,111^2 = 12,345,654,321$.
62. It is essentially dividing by using expanded notation. Yes,
    it will always work.

```
        24
   32)768
     60       = 2 · 30
     16       = 76 − 60
      4       = 2 · 2
    128       = 16 − 4 and bring down 8
    120       = 4 · 30
      8       = 128 − 120
      8       = 4 · 2
      0       = 8 − 8
```

63. By the division algorithm, we have $98 = nx + 5$, where
    n is any whole number. If $n = 3$, then $x = 31$.
    Consequently, 3 and 31 are the only numbers that will give
    a remainder of 5.
64. Here, $x = 8n + 3$, where n is a whole number. Consequently,
    $x \in \{0, 11, 19, 27, 35, 43, 51, 59, 67, 75, 83, 91\}$.
65.

$53 \cdot 31 = (50 + 3) \cdot 31$	Expanded notation
$\quad = 50 \cdot 31 + 3 \cdot 31$	Distributive property
$\quad = 50(30 + 1) + 3(30 + 1)$	Expanded notation
$\quad = 50 \cdot 30 + 50 \cdot 1 + 3 \cdot 30 + 3 \cdot 1$	Distributive property
$\quad = 1,500 + 50 + 90 + 3$	Whole-number multiplication
$\quad = 1,643$	Whole-number addition

66. Answers will vary.
67. a. Answers will vary.   b. Answers will vary.
    c. Answers will vary.
68. Luana calculated $1,824 \div 16 + 8$ instead of $1,824 \div
    (16 + 8)$. The calculator computed the division first rather
    than the sum.
69. Answers will vary.
70. Answers will vary.

71. Madison is correct. As long as we keep track of the place values of the digits, we can multiply from left to right or from right to left.

72. The mistakes follow. Answers on how to correct them will vary.
a. This student used only single-digit facts and failed to regroup correctly. **b.** This student forgot to use a placeholder to move the second partial product to the correct place value. **c.** This student incorrectly worked from left to right, putting the partial products in the wrong place values.

73. The mistakes follow. Answers on how to correct them will vary.
a. This student did not place the digits in the quotient in the correct place values. **b.** This student wrote the digits in the quotient in the wrong order. **c.** This student has the placeholder zero in the wrong position.

74. a. Answers will vary. **b.** Answers will vary.

75. Answers will vary.

76. Answers will vary.

77. Answers will vary.

78. a. Answers will vary. **b.** Answers will vary.

79. Answers will vary.

■ Section 4.5 Learning Assessment

1. The different groupings, place values, and digits dramatically impact the appearance of the numeral even though the operations are computed in the same basic way.

2. Addition and subtraction in base-b will be defined and computed in the same way as in base-ten. They will even satisfy the same properties. However, the resulting sums and differences look different because of changes in the groups, digits, and place values. This means the single-digit facts we learned with decimal operations no longer apply. This makes it necessary to relearn how to compute in base-b.

3. It make sense that base-b addition satisfies the commutative, associative, and identity properties because these properties are a result of operations on sets, not a result of how the numerals are written.

4. Multiplication and division in base-b will be defined and computed in the same way as in base-ten. They will even satisfy the same properties. However, the resulting products and quotients look different because of changes in the groups, digits, and place values.

5. It make sense that base-b multiplication satisfies the commutative and associative properties because these properties are based on base-b addition, which satisfies these properties. It makes sense that base-b multiplication satisfies the identity property, because any number added to itself one time will be that number regardless of how the numeral is written.

6. One method is to perform the algorithm using single-digit facts for the given base and operation. A second way is to convert the numbers to base-ten, perform the calculation, and then convert back to the given base.

7. Because division is defined in terms of repeated subtraction and the operation of subtraction is not affected by how the numerals are written, then dividing by

zero will be undefined in any base-b system. Specifically, because subtracting zero never affects the value of the difference in base-b, we can subtract zero any number of times. Because there is no definite number of times that we can subtract zero from a base-b number, we say that division by zero is undefined, and we do not allow it.

8. a. $2_{three} + 11_{three} = 20_{three}$.

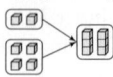

b. $4_{seven} + 6_{seven} = 13_{seven}$.

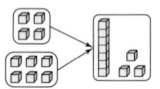

c. $11_{five} - 4_{five} = 2_{five}$.

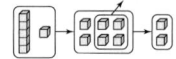

d. $13_{six} - 5_{six} = 4_{six}$.

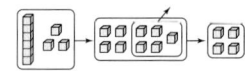

9. a. $4_{six} + 5_{six} = 13_{six}$.

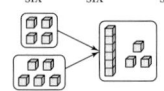

b. $7_{eight} + 7_{eight} = 16_{eight}$.

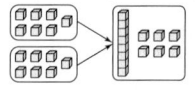

c. $12_{three} - 2_{three} = 10_{three}$.

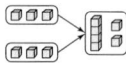

d. $15_{nine} - 7_{nine} = 7_{nine}$.

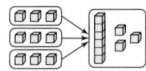

10. a. $2_{four} \cdot 3_{four} = 12_{four}$.

b. $3_{six} \cdot 3_{six} = 13_{six}$.

c. $13_{seven} \div 2_{seven} = 5_{seven}$.

d. $15_{nine} \div 7_{nine} = 2_{nine}$.

11. a. $5_{eight} \cdot 4_{eight} = 24_{eight}$.

b. $4_{\text{nine}} \cdot 6_{\text{nine}} = 26_{\text{nine}}.$

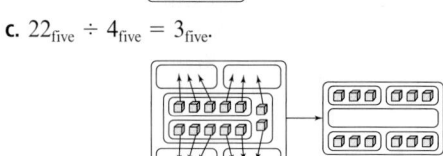

c. $22_{\text{five}} \div 4_{\text{five}} = 3_{\text{five}}.$

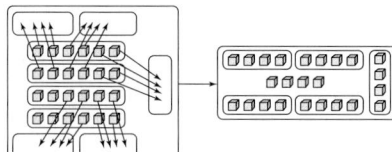

d. $40_{\text{six}} \div 5_{\text{six}} = 4_{\text{six}}$, remainder $4_{\text{six}}.$

12. a. ⊦––––––––––––––⊦
0 1 10 11 100 101 110 111 1000

b. ⊦––––––––––––––––⊦
0 1 2 3 4 5 10 11 12 13 14 15

c. ⊦––––––––––––––⊦
0 1 2 3 4 5 6 7 8 10 11 12

d. ⊦––––––––––––––––––⊦
0 1 2 3 4 5 6 7 8 9 A B 10 11 12 13

e. It must represent the numerals of base-b rather than the numerals of base-ten.

13. a. $6_{\text{nine}} + 8_{\text{nine}} = 15_{\text{nine}}.$

0 1 2 3 4 5 6 7 8 10 11 12 13 14 15 16 17

b. $14_{\text{six}} - 5_{\text{six}} = 5_{\text{six}}.$

0 1 2 3 4 5 10 11 12 13 14 15

c. $4_{\text{five}} \cdot 3_{\text{five}} = 22_{\text{five}}.$

0 1 2 3 4 10 11 12 13 14 20 21 22 23

d. $13_{\text{four}} \div 2_{\text{four}} = 3_{\text{four}}$, remainder $1_{\text{four}}.$

0 1 2 3 10 11 12 13 20 21

14. a. $5_{\text{six}} + 5_{\text{six}} = 14_{\text{six}}.$

0 1 2 3 4 5 10 11 12 13 14 15

b. $10_{\text{eight}} - 5_{\text{eight}} = 3_{\text{eight}}.$

0 1 2 3 4 5 6 7 10 11

c. $4_{\text{six}} \cdot 4_{\text{six}} = 24_{\text{six}}.$

0 1 2 3 4 5 10 11 12 13 14 15 20 21 22 23 24 25

d. $1A_{\text{eleven}} \div 7_{\text{eleven}} = 3_{\text{eleven}}.$

0 1 2 3 4 5 6 7 8 9 A 10 11 12 13 14 15 16 17 18 19 1A 20

15. a. $3_{\text{seven}} \cdot 3_{\text{seven}} = 12_{\text{seven}}.$ **b.** $4_{\text{five}} + 3_{\text{five}} = 12_{\text{five}}.$
c. $13_{\text{six}} - 4_{\text{six}} = 5_{\text{six}}.$ **d.** $5_{\text{nine}} \cdot 2_{\text{nine}} = 11_{\text{nine}}.$

16. The closure property is shown by the fact that all the entries in the table are base-six whole numbers. The commutative property is shown by the fact that the table is symmetric about the main diagonal. The identity property is shown by the fact that the first row and column match the initial row and column.

17. The closure property is shown by the fact that all the entries in the table are base-eight whole numbers. The commutative property is shown by the fact that the table is symmetric about the main diagonal. The identity property is shown by the fact that the first row and column match the initial row and column. The zero multiplication property is shown by the fact that the first row and column are all zeros.

18. a. $223_{\text{four}} + 132_{\text{four}} = 1021_{\text{four}}.$

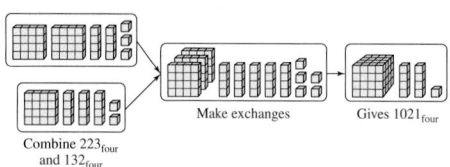

Combine 223_{four} and 132_{four} Make exchanges   Gives 1021_{four}

b. $201_{\text{four}} - 121_{\text{four}} = 20_{\text{four}}.$

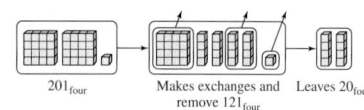

201_{four} Makes exchanges and remove 121_{four} Leaves 20_{four}

c. $33_{\text{four}} \cdot 22_{\text{four}} = 2112_{\text{four}}.$

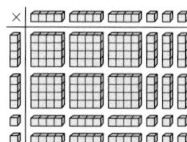

19. a. $121_{\text{three}} + 112_{\text{three}} = 1010_{\text{three}}.$ **b.** $343_{\text{five}} - 123_{\text{five}} = 220_{\text{five}}.$ **c.** $21_{\text{four}} \cdot 13_{\text{four}} = 333_{\text{four}}.$
20. a. $12_{\text{five}}.$ **b.** $12_{\text{seven}}.$ **c.** $10_{\text{three}}.$ **d.** $5_{\text{six}}.$
21. a. $14_{\text{nine}}.$ **b.** $13_{\text{fourteen}}.$ **c.** $7_{\text{twelve}}.$ **d.** $A_{\text{fifteen}}.$
22. a. $12_{\text{four}}.$ **b.** $24_{\text{seven}}.$ **c.** $11_{\text{three}}.$ **d.** $5_{\text{six}}.$
23. a. $28_{\text{eleven}}.$ **b.** $2A_{\text{thirteen}}.$ **c.** $4_{\text{twelve}}.$ **d.** $4_{\text{sixteen}}.$
24. a. All numbers are in base-three. Subscripts have been left out for ease in reading.

+	0	1	2
0	0	1	2
1	1	2	10
2	2	10	11

b. All numbers are in base-five. Subscripts have been left out for ease in reading.

+	0	1	2	3	4
0	0	1	2	3	4
1	1	2	3	4	10
2	2	3	4	10	11
3	3	4	10	11	12
4	4	10	11	12	13

c. All numbers are in base-eight. Subscripts have been left out for ease in reading.

+	0	1	2	3	4	5	6	7
0	0	1	2	3	4	5	6	7
1	1	2	3	4	5	6	7	10
2	2	3	4	5	6	7	10	11
3	3	4	5	6	7	10	11	12
4	4	5	6	7	10	11	12	13
5	5	6	7	10	11	12	13	14
6	6	7	10	11	12	13	14	15
7	7	10	11	12	13	14	15	16

25. a. All numbers are in base-four. Subscripts have been left out for ease in reading.

×	0	1	2	3
0	0	0	0	0
1	0	1	2	3
2	0	2	10	12
3	0	3	12	21

b. All numbers are in base-six. Subscripts have been left out for ease in reading.

×	0	1	2	3	4	5
0	0	0	0	0	0	0
1	0	1	2	3	4	5
2	0	2	4	10	12	14
3	0	3	10	13	20	23
4	0	4	12	20	24	32
5	0	5	14	23	32	41

c. All numbers are in base-seven. Subscripts have been left out for ease in reading.

×	0	1	2	3	4	5	6
0	0	0	0	0	0	0	0
1	0	1	2	3	4	5	6
2	0	2	4	6	11	13	15
3	0	3	6	12	15	21	24
4	0	4	11	15	22	26	33
5	0	5	13	21	26	34	42
6	0	6	15	24	33	42	51

26. a. $3_{six} + 4_{six} = 11_{six}$, $11_{six} - 4_{six} = 3_{six}$ and $11_{six} - 3_{six} = 4_{six}$. b. $13_{eight} - 5_{eight} = 6_{eight}$, $5_{eight} + 6_{eight} = 13_{eight}$ and $6_{eight} + 5_{eight} = 13_{eight}$. c. $5_{seven} \cdot 3_{seven} = 21_{seven}$, $21_{seven} \div 3_{seven} = 5_{seven}$, and $21_{seven} \div 5_{seven} = 3_{seven}$. d. $20_{six} \div 3_{six} = 4_{six}$, $3_{six} \cdot 4_{six} = 20_{six}$ and $4_{six} \cdot 3_{six} = 20_{six}$.

27. a. $11_{three} - 2_{three} = 2_{three}$. b. $1A_{eleven} - 14_{eleven}$, $14_{eleven} + 6 = 1A_{eleven}$, and $6_{eleven} + 14_{eleven} = 1A_{eleven}$. c. $7_{nine} \cdot 5_{nine} = 38_{nine}$, $38_{nine} \div 5_{nine} = 7_{nine}$, and $38_{nine} \div 7_{nine} = 5_{nine}$. d. $28_{twelve} \div 4_{twelve} = 8_{twelve}$, $4_{twelve} \cdot 8_{twelve} = 28_{twelve}$, and $8_{twelve} \cdot 4_{twelve} = 28_{twelve}$.

28. a. 10210_{three}. b. 10210_{five}. c. 2233_{four}. d. 1441_{six}.
29. a. 1157_{nine}. b. $170E_{sixteen}$. c. 3485_{nine}. d. $9553_{fifteen}$.
30. a. 1100_{two}. b. 20103_{four}. c. 22_{three} remainder 1_{three}. d. 33_{five} remainder 40_{five}.
31. a. 34344_{six}. b. 13064_{seven}. c. 33_{four}. d. 113_{seven}.
32. a. 10010_{two}. b. 1352_{eight}. c. 11021_{three}. d. 31422_{five}.
33. a. Seven. b. Five. c. Eight. d. Six.
34. a. 6_{eight}. b. 7_{eight}. c. 3_{eight}. d. x could be 1_{eight}, 2_{eight}, 3_{eight}, 4_{eight}, or 5_{eight}.

35. a.
$$\begin{array}{r} 3426_{seven} \\ +\,1345_{seven} \\ \hline 5104_{seven} \end{array}$$
b.
$$\begin{array}{r} 4021_{five} \\ -\,2413_{five} \\ \hline 1103_{five} \end{array}$$
c.
$$\begin{array}{r} 231_{four} \\ \times\,23_{four} \\ \hline 2013 \\ 11220 \\ \hline 13233_{four} \end{array}$$

36. a. 14_{five}. b. 31_{five}. c. 1003_{five}. d. 224_{five}.
37. a. 24_{five}. b. 10_{six}. c. 33_{seven}.
38. x could be 5_{eight}, 6_{eight}, 14_{eight}, and 36_{eight}.
39.

20_{five}	3_{five}	13_{five}
10_{five}	12_{five}	14_{five}
11_{five}	21_{five}	4_{five}

40. $40_{five} \cdot 3_{five} \cdot 2_{five} \cdot 1_{five} = 440_{five}$.
41. Any two base-seven numbers will work as long as the first is less than the second. For instance, base-seven is not closed under subtraction because $3_{seven} - 5_{seven} = (-2_{seven})$ and (-2_{seven}) is not a base-seven whole-number.
42. Any two base-eight numbers will work as long as they are not equal. For instance, base-eight subtraction is not commutative because $6_{eight} - 4_{eight} = 2_{eight}$, but $4_{eight} - 6_{eight} = (-2_{eight})$.
43. Any three base-four numbers will work as long as one of them is not 0, and two are not the same. For instance, base-four subtraction is not associative because $10_{four} - (2_{four} - 1_{four}) = 10_{four} - 1_{four} = 3_{four}$ and $(10_{four} - 2_{four}) - 1_{four} = 2_{four} - 1_{four} = 1_{four}$.
44. If a_b is a nonzero number in base-b, then $a_b \cdot 0_b = 0_b + 0_b + \ldots + 0_b$ (a_b addends) $= 0_b$. Likewise, we interpret $0_b \cdot a_b$ as adding a_b to itself 0_b times, which again gives us 0_b.
45. a. Yes. We can verify the property using set operations as we did in base-ten. b. Yes. If a_b is a nonzero number in base-b, then $a_b \cdot 1_b = 1_b + 1_b + \ldots + 1_b$ (a_b addends) $= ab$. Likewise, $1 \cdot a_b = a_b$.
46. a. $4_{five} \cdot 3_{five} = 22_{five} = 3_{five} \cdot 4_{five}$. b. $4_{six} \cdot (5_{six} \cdot 2_{six}) = 4_{six} \cdot 14_{six} = 104_{six} = 32_{six} \cdot 2_{six} = (4_{six} \cdot 5_{six}) \cdot 2_{six}$. c. $3_{nine} \cdot (6_{nine} + 7_{nine}) = 3_{nine} \cdot 14_{nine} = 43_{nine}$ and $3_{nine} \cdot 6_{nine} + 3_{nine} \cdot 7_{nine} = 20_{nine} + 23_{nine} = 43_{nine}$.
47. In base-six we use the digits 0, 1, 2, 3, 4, and 5. We might say that with any digit 2 or lower, we round down, and with any digit 3 or higher, we round up.
48. a. Suppose a_{five} and b_{five} are base-five whole numbers where $a_{five} = n(A)$ and $b_{five} = n(B)$. If A and B are disjoint, finite sets, then $a_{five} + b_{five} = n(A \cup B)_{five}$. b. If a_{six} and b_{six} are any base-six whole numbers, then $a_{six} - b_{six} = c_{six}$ if and only if $a_{six} = b_{six} + c_{six}$ for some base-six whole number c_{six}. c. If a_{eight} and b_{eight} be any base-eight whole numbers with $a_{eight} \neq 0$, then $a_{eight} \cdot b_{eight} = \underbrace{b_{eight} + b_{eight} + \ldots + b_{eight}}_{a_{eight} \text{ addends.}}$. If $a_{eight} = 0_{eight}$, then $0_{eight} \cdot b_{eight} = 0_{eight}$ for all b_{eight}. d. Essentially the definitions remain the same. All we really need to do is express each number as a number in the proper base.

49. Answers will vary.
50. a. Answers will vary. **b.** Answers will vary.
c. Answers will vary.
51. Answers will vary.
52. Methods will vary. The sum is ••••.
53. Methods will vary. The difference is 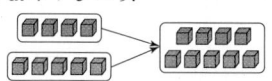.
54. Methods will vary. The product is ••••.

■ Chapter 4 Review Exercises

1. a. $4 + 5 = 9$.

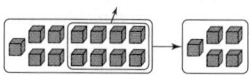

b. $13 - 8 = 5$.

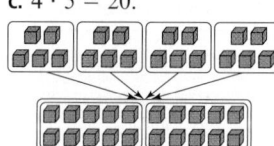

c. $4 \cdot 5 = 20$.

d. $18 \div 6 = 3$.

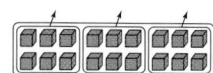

2. a. $3 + 7 = 10$.

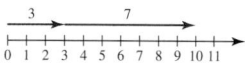

b. $11 - 5 = 6$.

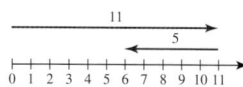

c. $2 \cdot 7 = 14$.

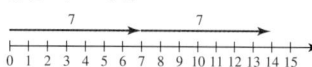

d. $12 \div 4 = 3$.

3. a. $345 + 269 = 614$.

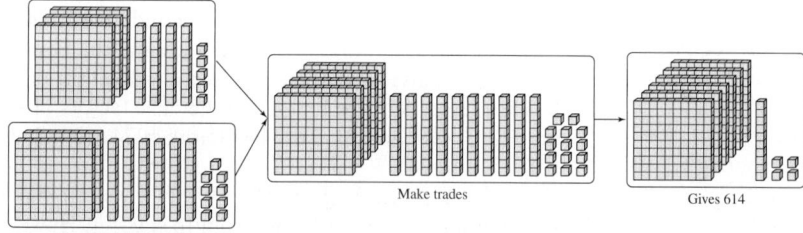

b. $510 - 283 = 227$.

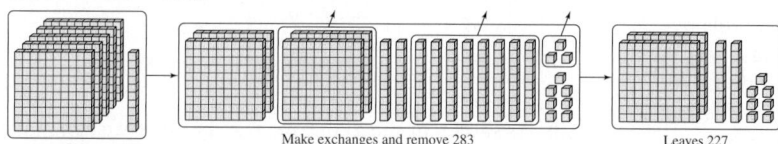

c. $24 \cdot 18 = 432$.

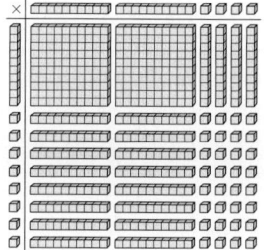

4. a. $5 + 6 = 11$, $11 - 5 = 6$, and $11 - 6 = 5$.
b. $7 + 7 = 14$.
c. $5 \cdot 7 = 35$, $35 \div 7 = 5$, and $35 \div 5 = 7$.
d. $42 \div 6 = 7$, $6 \cdot 7 = 42$, and $7 \cdot 6 = 42$.
5. a. $\{0, 4, 8, 12, ...\}$ is closed under addition and multiplication because any multiple of 4 added or multiplied to another multiple of 4 is still a multiple of 4. **b.** $\{4, 5, 6\}$ is not closed under addition or multiplication because $4 + 4 = 8$, $4 \cdot 4 = 16$ and neither 8 nor 16 is in the set. **c.** $\{2, 4, 5, 6, 7, ...\}$ is closed under addition and multiplication because any time two elements in the set are added or multiplied, the result is back in the set. **d.** $\{1\}$ is not closed under addition because $1 + 1 = 2$ and 2 is not in the set. It is closed under multiplication because $1 \cdot 1 = 1$ and 1 is back in the set.
6. a. Closure property of multiplication. **b.** Commutative property of multiplication. **c.** Distributive property of multiplication over addition. **d.** Multiplicative identity property. **e.** Additive identity property. **f.** Commutative property of addition. **g.** Zero multiplication property. **h.** Associative property of addition.
7. a. 1,216. **b.** 378. **c.** 1,904.

8. a. 6,670. **b.** 32,472.
9. a. 908. **b.** 12224_{five}. **c.** 2,338. **d.** 1531_{seven}.
e. 48,422. **f.** 13233_{four}. **g.** 130 remainder 10.
h. 101_{four} remainder 20_{four}.
10. a. 912. **b.** 200. **c.** 111. **d.** 946. **e.** 21,080.
f. 10,750. **g.** 507. **h.** 63.
11. a. 410. **b.** 19,000. **c.** 3,900. **d.** 47,000.
12. a. 800. **b.** 43,000. **c.** 29,900.
13. a. About 10,300. **b.** About 5,900. **c.** About 250,000.
14. a. About 8,100. **b.** About 2,000. **c.** About 35,000.
15. a. 5^7. **b.** x^{10}. **c.** 4^{18}. **d.** 18^3. **e.** 7^{26}. **f.** 6^{35}.
16. a. 18. **b.** 36. **c.** 7.
17. a. 21,025. **b.** 34,632. **c.** 315.
18. 1,269 students.
19. About 100 fish.
20. 1,155 more chairs must be put out.
21. 3,786,214 people live in the state but not in the city.
22. Answers will vary.
23. 63 different pairs can be made.
24. 2,118 stops.
25. 7 crayons.
26. $165 \div 27 = 6$, remainder 3. Each student gets 6 sheets of paper, and 3 will be left over.

CHAPTER 5

Section 5.1 Learning Assessment

1. A factor of a natural number a is a number that, when multiplied by another natural number, has product equal to a. A multiple of a natural number a is the product of a and another natural number.
2. *Division* is an operation that leads to a number. *Divisibility* is a relationship between numbers that can be only true or false. The two are connected because if the quotient $a \div b$ has a remainder of 0, then $b \mid a$ is true.
3. a. Look at the last digit of the number. If it is a 0, 2, 4, 6, or 8, then the number is even and divisible by 2. **b.** Add the digits of the number. If the sum is divisible by 3, the number is divisible by 3. **c.** Look at the last two digits of the number. If they represent a number divisible by 4, the number is divisible by 4. **d.** Determine whether the number is divisible by 2 and by 3. If it is, then it is divisible by 6. **e.** Compute the sums of alternating digits, and then subtract the sums. If the difference is divisible by 11, the number is divisible by 11.
4. We can use divisibility by 2 to classify the whole numbers as even or odd. Or we can use divisibility to separate the natural numbers into those that are divisible by a given number and those that are not. We can also use divisibility to separate the natural numbers into primes and composites.
5. A natural number is prime if and only if it has exactly two factors. It is composite if and only if it has three or more factors.

6. One is neither prime nor composite because it only has one factor: itself.
7. One way to identify primes is to use rectangular arrays. If we can make only one rectangular array from a set of n blocks, then n has only two factors. So it must be prime. Another way is to use the Sieve of Eratosthenes, which selects primes from a list of numbers by removing multiples of numbers. A third way is to use the prime divisor test: n is prime if it is not divisible by any prime p where $p^2 \le n$.
8. The Fundamental Theorem of Arithmetic guarantees that every composite number has exactly one prime factorization and that no two composite numbers have the same prime factorization.
9. Write the prime factorization of the number. Then take every possible combination of every prime in the factorization to generate the factors of the number.
10. Equivalent phrases are "5 is a factor of 30," "5 divides 30," "30 is a multiple of 5," and "30 is divisible by 5."
11. Equivalent phrases are "28 is divisible by 7," "7 is a divisor of 28," "7 divides 28," and "7 is a factor of 30."
12. a. 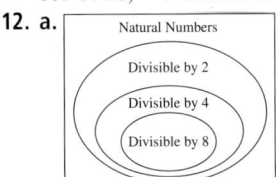 **b.**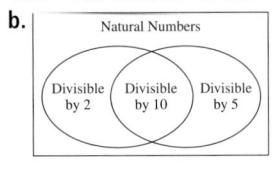
13. a. If c divides a and c divides b, then c divides the sum of a and b. **b.** If c divides a and c does not divide b, then c does not divide a minus b where a is greater than or equal to b. **c.** If c divides a, then c divides any multiple of a.
14. a. The factors of 9 are 1, 3, and 9. **b.** The factors of 18 are 1, 2, 3, 6, 9, and 18. **c.** The factors of 24 are 1, 2, 3, 4, 6, 8, 12, and 24.
15. a. As shown in the diagram, $4 \mid 16$ because we can make a rectangular array with 4 rows from 16 squares. However, $4 \nmid 25$ because we cannot.

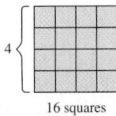 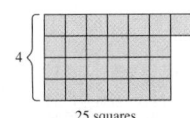

b. As shown in the diagram, $7 \mid 21$ because we can make a rectangular array with 7 rows from 21 squares. However, $7 \nmid 32$ because we cannot.

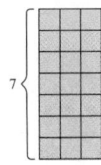 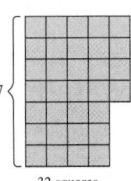

16. Use one rectangular array to show an even number and another to show an odd. When joined together, the arrays show an odd number.

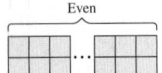 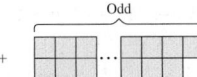 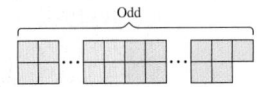

17. 13 is prime because we can make only a 1 × 13 array from 13 blocks. 21 is composite because we can make an array other than a 1 × 21 or 21 × 1: a 3 × 7 array.

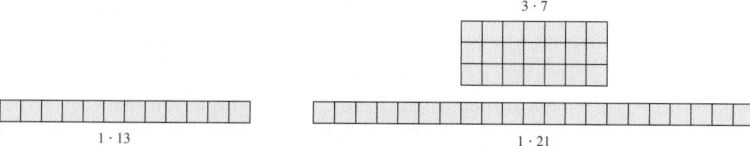

18. Three possible factor trees are shown.

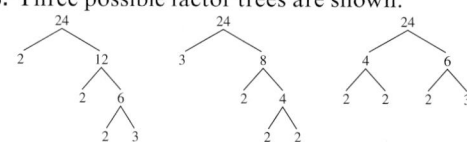

19. The table is as follows:

1 Factor	2 Factors	3 Factors	4 Factors	5 Factors	6 Factors	7 Factors	8 Factors	9 Factors	10 Factors
1	2, 3, 5, 7, 11, 13, 17, 19, 23, 29, 31, 37, 41, 43, 47	4, 9, 25, 49	6, 8, 10, 14, 15, 21, 22, 26, 27, 33, 34, 35, 38, 46	16	12, 18, 20, 28, 32, 44, 45, 50		24, 30, 40, 42	36	48

a. 15. **b.** 14. **c.** The columns for 2 factors and 4 factors. **d.** The column for 7 factors.

20. 12, 24, 36, 48, 60, 72, 84, 96, 108, and 120.
21. 36, 72, 108, 144, 180, 216, 252, 288, 324, and 360.
22. a. 1, 2, 4, and 8. **b.** 1, 2, 4, 5, 8, 10, 20, and 40.
c. 1, 2, 3, 4, 6, 7, 12, 14, 21, 28, 42, and 84.
23. a. 1, 5, 7, and 35. **b.** 1, 2, 3, 4, 5, 6, 10, 12, 15, 20, 30, and 60. **c.** 1, 2, 3, 4, 5, 6, 8, 10, 12, 15, 20, 24, 30, 40, 60, and 120.
24. a. True. **b.** False. **c.** False. **d.** True.
25. a. True. **b.** False. **c.** False. **d.** True.
26. a. No. **b.** Yes. **c.** No. **d.** No.
27. a. 3, 5, and 9. **b.** 2, 3, 6, 9, and 11.
c. 2, 4, 5, 10, and 11. **d.** 2, 4, and 8.
28. a. 2, 3, 4, 5, 6, 8, and 10. **b.** 5 and 11.
c. 2, 3, 4, 5, 6, 8, 9, 10, and 11. **d.** 3, 5, 9, and 11.
29. a. 1, 4, or 7. **b.** 0, 2, 4, 6, or 8. **c.** 4. **d.** 0. **e.** 4.
30. a. Prime. **b.** Prime. **c.** Prime. **d.** Composite.
e. Composite.
31. Primes between 100 and 200 are {101, 103, 107, 109, 113, 127, 131, 137, 139, 149, 151, 157, 163, 167, 173, 179, 181, 191, 193, 197, 199}.
32. a. Prime. **b.** Composite. **c.** Composite. **d.** Composite.
33. 353 and 359 are prime.
34. a.   **b.**   **c.**

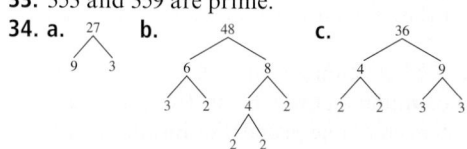

35. a. $2^2 \cdot 3 \cdot 11$. **b.** $5 \cdot 7^2$. **c.** $2^4 \cdot 3^3$.
36. a. $2^4 \cdot 3^2 \cdot 7$. **b.** $2^5 \cdot 3^2 \cdot 7$. **c.** $2^2 \cdot 3^3 \cdot 5 \cdot 11$.
37. a. 6. **b.** 16. **c.** 24.
38. a. 72. **b.** 440. **c.** 120.
39. a. Factor. **b.** Not a factor. **c.** Not a factor. **d.** Factor.
40. 1, 2, 3, 4, 6, 8, and 12.
41. {2, 3, 5, 7, 11, 13, 17, 19}.
42. {2, 3, 5, 7, 11, 13, 17, 19, 23}.

43. The smallest with five factors is 16, the smallest with six factors is 12, and the smallest with seven factors is 64.
44. One possible answer is 7,513.
45. One possible answer is 19,524.
46. a. $7 \mid 1{,}155$. **b.** $7 \nmid 3{,}789$. **c.** $7 \nmid 5{,}891$. **d.** $7 \mid 8{,}715$.
47. $3 \cdot 5 \cdot 7 = 105$.
48. $5 \cdot 7 \cdot 8 \cdot 9 = 2{,}520$.
49. a. $2 \cdot 11$. **b.** $5 \cdot 13$. **c.** $7 \cdot 23$. **d.** $29 \cdot 31$.
50. $2^3 \cdot 3 \cdot 5 = 120$.
51. a. Other twin primes are 41 and 43, 59 and 61, and 71 and 73.
b. Twin primes are 101 and 103, 107 and 109, 137 and 139, 149 and 151, 179 and 181, 191 and 193, and 197 and 199.
52. a. The proper factors of 220 are 1, 2, 4, 5, 10, 11, 20, 22, 44, 55 and 110 and $1 + 2 + 4 + 5 + 10 + 11 + 20 + 22 + 44 + 55 + 110 = 284$. The proper factors of 284 are 1, 2, 4, 71, and 142 and $1 + 2 + 4 + 71 + 142 = 220$. **b.** We find the other number by finding the proper factors of 1,184 and adding them. The proper factors of 1,184 are 1, 2, 4, 8, 16, 32, 37, 74, 148, 296, and 592. Hence, the other number is $1 + 2 + 4 + 8 + 16 + 32 + 37 + 74 + 148 + 296 + 592 = 1{,}210$.
53. a. The proper factors of 8,128 are 1, 2, 4, 8, 16, 32, 64, 127, 254, 508, 1,016, 2,032, and 4,064. Because $1 + 2 + 4 + 8 + 16 + 32 + 64 + 127 + 254 + 508 + 1{,}016 + 2{,}032 + 4{,}064 = 8{,}128$, the number is perfect. **b. i.** Perfect. **ii.** Deficient. **iii.** Abundant. **iv.** Abundant. **v.** Abundant.
54. No.
55. 11 people.
56. No.
57. No. The amount cannot be correct because $15 \nmid 3{,}965$.
58. Yes, because $9 \cdot 13 = 117$ and $39 \mid 117$.
59. The smallest number that is divisible by 6, 7, 8, and 9 is 504. Hence, $504 + 1 = 505$ donuts were made today.

60. Every pair of consecutive numbers must contain at least one even. Because 2 is the only even prime, 2 and 3 can be the only pair of consecutive primes.

61. **a.** True. **b.** False. **c.** True. **d.** False.

62. **a.** True. **b.** True. **c.** False. **d.** True.

63. No. Consider the following counterexample: $3 \nmid 7$ and $3 \nmid 5$, but $7 + 5 = 12$ and $3 \mid 12$.

64. **a.** Five numbers that have exactly three factors are 4, 9, 25, 49, and 121. The only numbers that have exactly three factors are the squares of prime numbers or those of the form p^2 where p is a prime. **b.** Five numbers that have exactly four factors are 6, 8, 10, 14, and 15. Two types of numbers have exactly four factors: those of the form p^3 where p is a prime and those of the form $p \cdot q$ where p and q are primes.

65. If $n = 0$, then $2^0 - 1 = 0$, which is not prime.
If $n = 1$, then $2^1 - 1 = 1$, which is not prime.
If $n = 2$, then $2^2 - 1 = 3$, which is prime.
If $n = 3$, then $2^3 - 1 = 7$, which is prime.
If $n = 4$, then $2^4 - 1 = 15$, which is not prime.
If $n = 5$, then $2^5 - 1 = 31$, which is prime.

66. If $n = 0$, then $0^2 + 0 + 17 = 17$, which is prime.
If $n = 1$, then $1^2 + 1 + 17 = 19$, which is prime.
If $n = 2$, then $2^2 + 2 + 17 = 23$, which is prime.
If $n = 3$, then $3^2 + 3 + 17 = 29$, which is prime.
If $n = 4$, then $4^2 + 4 + 17 = 37$, which is prime.
If $n = 5$, then $5^2 + 5 + 17 = 47$, which is prime.

67. If $p = 3$, then $2(3) + 1 = 7$, which is prime.
If $p = 5$, then $2(5) + 1 = 11$, which is prime.
If $p = 11$, then $2(11) + 1 = 23$, which is prime.
If $p = 23$, then $2(23) + 1 = 47$, which is prime.
Three other Sophie Germain primes are 29, because $2(29) + 1 = 59$; 41, because $2(41) + 1 = 83$; and 51, because $2(51) + 1 = 103$.

68. **a. i.** $= 3 + 3 = 6$. **ii.** $5 + 3 = 8$. **iii.** $7 + 7 = 14$.
iv. $23 + 3 = 26$. **v.** $31 + 19 = 50$.
b. No. Consider $14 = 7 + 7$ and $14 = 11 + 3$.

69. If we write 340 in expanded notation, $340 = 300 + 40 + 0$, we see that any digit in the tens place value or higher represents a multiple of 10, which must be divisible by 10. Because 0 is the only single-digit number divisible by 10, the units digit must be 0 for the number to be divisible by 10. Because $10 \mid 300$, $10 \mid 40$, and $10 \mid 0$, then $10 \mid (300 + 40 + 0)$, or $10 \mid 340$ by Theorem 5.1(1).

70. Write the number in expanded form: $288 = 200 + 80 + 8$. As was done with divisibility test for 3, rewrite the expanded notation of 288 as:
$$288 = 200 + 80 + 8$$
$$= 2(100) + 8(10) + 8$$
$$= 2(99 + 1) + 8(9 + 1) + 8$$
$$= (2 \cdot 99) + (2 \cdot 1) + (8 \cdot 9) + (8 \cdot 1) + 8$$
$$= [(2 \cdot 99) + (8 \cdot 9)] + [(2 \cdot 1) + (8 \cdot 1) + 8]$$
$$= [(2 \cdot 99) + (8 \cdot 9)] + (2 + 8 + 8).$$
Theorem 5.1 guarantees that 9 will divide $(2 \cdot 99)$ and $(8 \cdot 9)$, so 9 must also divide the sum $(2 \cdot 99) + (8 \cdot 9)$. This leaves us with $2 + 8 + 8$. If this sum is divisible by 9, then Theorem 5.1 again guarantees that $[(2 \cdot 99) + (8 \cdot 9)] + (2 + 8 + 8)$ will be divisible by 9. Because $2 + 8 + 8 = 18$, which is divisible by 9, then 288 is divisible by 9.

71. Suppose $d \mid a$ and $d \mid b$ and $a \geq b$. By the definition of divisibility, there exists whole numbers j and k such that $a = d \cdot j$ and $b = d \cdot k$. By subtracting a and b, we have $a - b = (d \cdot j) - (d \cdot k) = d(j - k)$, where $j \geq k$. If we let $r = j - k$, then r is a whole number. As a result, $a - b = d \cdot r$ for some whole number r. Consequently, $d \mid (a - b)$ by the definition of divides.

72. If $d \mid a$, then there exists a whole number k such that $d \cdot k = a$. If we multiply both sides of the equation by a nonzero whole number n, we have $n(d \cdot k) = n \cdot a$. Using the commutative and associative properties, $d(n \cdot k) = n \cdot a$. If we let $r = n \cdot k$, then r is a whole number by the closure property. So $d \cdot r = n \cdot a$, and $d \mid n \cdot a$ by the definition of divides.

73. Let m and n be two even numbers. By the definition of even, $m = 2j$ and $n = 2k$, where j and k are two whole numbers. Then $m + n = 2j + 2k = 2(j + k)$. If we let $r = (j + k)$, then r is a whole number by the closure property, and $m + n = 2r$. Consequently, by the definition of an even number, $m + n$ must be even.

74. Let m and n be two odd numbers. By the definition odd, $m = 2j + 1$ and $n = 2k + 1$, where j and k are two whole numbers. Then $m \cdot n = (2j + 1) \cdot (2k + 1) = 4jk + 2j + 2k + 1 = 2(2jk + j + k) + 1$. If we let $r = (2jk + j + k)$, then r is a whole number by the closure property and $m \cdot n = 2r + 1$. Consequently, by the definition of an odd number, $m \cdot n$ must be odd.

75. Answers will vary.

76. Answers will vary.

77. Answers will vary.

78. Answers will vary.

79. Answers will vary.

80. Answers will vary.

81. Answers will vary.

82. These ideas are used in finding common denominators and simplifying fractions.

83. Answers will vary.

84. Answers will vary.

85. Answers will vary.

■ Section 5.2 Learning Assessment

1. **a.** If a and b are two whole numbers that are both not equal to zero, then the greatest common factor of a and b is the largest whole number that divides both a and b. **b.** If a and b are two natural numbers, then the least common multiple of a and b is the smallest natural number divisible by both a and b.

2. If a and b are prime, the only factors they have are 1 and themselves. Assuming a and b are different primes, then the only factor they can have in common is 1, making it also the greatest common factor.

3. **a.** Two nonzero whole numbers are relatively prime if their greatest common factor is 1. **b.** Two relatively prime numbers need not be prime. For instance, two composite numbers like 8 and 9 are relatively prime.

4. Zero is a multiple of every number because $0 \cdot n = 0$ for all natural numbers n. However, 0 is not a factor of any number because division by zero is undefined.

5. We can always find a common multiple of any two natural numbers a and b by taking their product: $a \cdot b$.

6. To find the greatest common factor, take the product of every prime that appears in the prime factorizations of the given numbers. To find the least common multiple,

take any prime that occurs in either factorization, raise it to the highest power to which it occurs, and then multiply.

7. The Euclidean algorithm is based on the fact that if a and b are any two nonzero whole numbers with $a \geq b$ and $a = bq + r$ for whole numbers q and r, with $r < b$, then $\text{GCF}(a, b) = \text{GCF}(b, r)$.

8. **a.** Because the $\text{GCF}(a, b)$ must divide both a and b, it must be less than or equal to the smaller of the two numbers. **b.** Because the $\text{LCM}(a, b)$ must be divisible by both a and b, it must be greater than or equal to the larger of the two numbers. **c.** Because the GCF (a, b) is less than or equal to the smaller of a and b and the $\text{LCM}(a, b)$ is greater than or equal the larger of a and b, then $\text{GCF}(a, b) \leq \text{LCM}(a, b)$.

9.
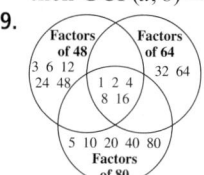

10.
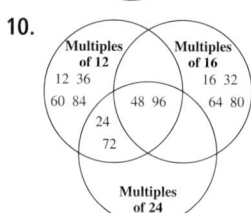

11. **a.** 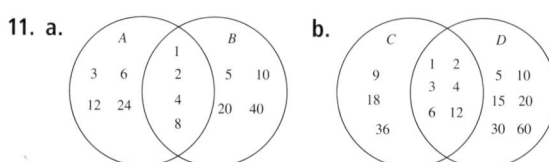 **b.**

12. **a.** 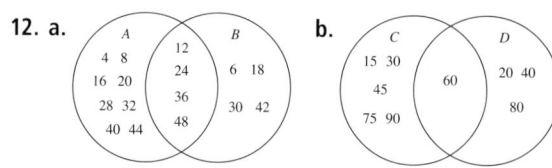 **b.**

13. **a.** 20. **b.** 24. **c.** 90.
14. **a.** 35 and 70. **b.** 36 and 72.
15. **a.** Shade every other entire column. **b.** Shade the entire fifth and final columns. **c.** Shade every third diagonal going from right to left. **d.** Shade every third block in the fifth and final columns. **e.** Shade blocks by starting at 12, drop one row, and move two spots to the right.
16. It shows that the $\text{LCM}(3, 4) = 12$.
17. **a.** $\text{GCF}(8, 12) = 4$. **b.** $\text{GCF}(10, 16) = 2$.

c. $\text{LCM}(3, 5) = 15$. **d.** $\text{LCM}(2, 7) = 14$.

18. **a.** 12. **b.** 15. **c.** 175. **d.** 168.
19. **a.** 8. **b.** 16. **c.** 54. **d.** 495.

20. **a.** $\text{GCF}(135, 165) = 15$; $\text{LCM}(135, 165) = 1{,}485$.
    b. $\text{GCF}(504, 220) = 4$; $\text{LCM}(504, 220) = 27{,}720$.
    c. $\text{GCF}(1050, 825) = 75$; $\text{LCM}(1050, 825) = 11{,}550$.
21. **a.** $\text{GCF}(900, 975) = 75$; $\text{LCM}(900, 975) = 11{,}700$.
    b. $\text{GCF}(1452, 594) = 66$; $\text{LCM}(1452, 594) = 13{,}068$.
    c. $\text{GCF}(2925, 2640) = 15$; $\text{LCM}(2925, 2640) = 514{,}800$.
22. **a.** 3. **b.** 2,520.
23. $\text{GCF}(a, b) = 2^3 \cdot 3^2 \cdot 5^4$; $\text{LCM}(a, b) = 2^4 \cdot 3^5 \cdot 5^4$.
24. $\text{GCF}(a, b) = 5$; $\text{LCM}(a, b) = 2^3 \cdot 3^6 \cdot 5^3 \cdot 7 \cdot 11^2$.
25. **a.** 4. **b.** 2.
26. **a.** 132. **b.** 15.
27. **a.** Not relatively prime. **b.** Relatively prime.
    c. Not relatively prime. **d.** Not relatively prime.
    e. Relatively prime.
28. **a.** 189. **b.** 18,975. **c.** 126,324.
29. 8 and 12.
30. 6 and 9.
31. $2^3 \cdot 5^2 \cdot 7^2$.
32. $2^4 \cdot 3^4 \cdot 5^2$.
33. **a.** 25. **b.** 12. **c.** 2,893,730. **d.** 6,801,642.
34. **a.** 5. **b.** 39. **c.** 1,135,050. **d.** 463,320.
35. $\text{GCF}(a, b) = 2^4 \cdot 3^2 \cdot 5 \cdot 7 = 5{,}040$;
    $\text{LCM}(a, b) = 2^6 \cdot 3^3 \cdot 5 \cdot 7 = 60{,}480$.
36. 63.
37. One example of three such numbers is 25, 26, and 27 because $\text{LCM}(25, 26, 27) = 17{,}550 = 25 \cdot 26 \cdot 27$.
38. 10.
39. **a.** 24 and 36. **b.** $\text{GCF}(24, 42) = \text{GCF}(24, 54) = \text{GCF}(36, 42) = \text{GCF}(36, 54) = 6$. **c.** 42 and 54.
    d. 24 and 36.
40. 4.
41. 3 hr.
42. Groups of 6 students.
43. **a.** 35 min. **b.** The faster car will have completed 21 laps, and the slower car will have completed 20 laps. **c.** 10 laps.
44. 300 min, or 5 hr.
45. 14 blocks.
46. 2 nights off together.
47. The greatest common width will be GCF(52500, 30600, 19800) = 300 ft. The total fencing required will be $4 \cdot 300 + 2 \cdot 66 + 2 \cdot 102 + 2 \cdot 175 = 1{,}886$ ft of fence.
48. The two gears with 36 teeth will have made 4 revolutions, and the large gear will have made 3.
49. The smallest square has dimensions 36×36 in.
50. The factors of 32 are 1, 2, 4, 8, 16, and 32. Because $64 = 2 \cdot 32$, the factors of 64 are all the factors of 32 along with their doubles, or 1, 2, 4, 8, 16, 32, and 64. Similarly, because $96 = 3 \cdot 32$, the factors of 96 are all the factors of 32 along with their triples, or 1, 2, 3, 4, 6, 8, 12, 16, 24, 32, 48, and 96.
51. **a.** True. If $\text{GCF}(a, b) = 1$, then
    $$\text{LCM}(a, b) = \frac{a \cdot b}{\text{GCF}(a,b)} = \frac{a \cdot b}{1} = a \cdot b.$$
    b. False. Let $a = 3$ and $b = 6$, then $\text{GCF}(3, 6) = 3$, but $\text{LCM}(3, 6) = 6$. **c.** True. If $\text{LCM}(a, b) = a \cdot b$, then $\text{GCF}(a, b) = 1$ by Theorem 5.7. Hence a and b are relatively prime. **d.** True. $\text{LCM}(a, b) = b$, then $\text{GCF}(a, b) \cdot b = a \cdot b$, which implies $\text{GCF}(a, b) = a$.
52. **a.** If $a = 4$ and $b = 5$, then $\text{GCF}(4, 5) = 1$.
    b. If $a = 4$ and $b = 8$, then $\text{GCF}(4, 8) = 4$.
    c. If $a = b = 4$, then $\text{GCF}(4, 4) = 4 = \text{LCM}(4, 4)$.

53. **a.** The LCM(a, b) will be less than $a \cdot b$ when a and b have a common factor.   **b.** The LCM(a, b) will equal $a \cdot b$ when a and b are relatively prime, or GCF(a, b) = 1.

54. The GCF(a, b) and LCM(a, b) contain all common factors of a and b. Because $c \mid a$ and $c \mid b$, then c is a common factor of a and b. As a result, $c \mid$ GCF(a, b) and $c \mid$ LCM(a, b).

55. **a.** Yes. If the GCF(a, b) = 2, then both numbers are divisible by 2. So they must be even.   **b.** Yes. If both numbers are even, they are both divisible by 2. So the GCF must have 2 as a factor, making it even.

56. No. Let $a = 4$, $b = 6$, and $c = 7$, then GCF(4, 6, 7) = 1, but GCF(4, 6) = 2.

57. **a.** False. If $a = 8$ and $b = 9$, then GCF(8, 9) = 1, but 8 is even and 9 is odd.   **b.** True. If two numbers are even, then both are divisible by 2. Hence, two even numbers cannot be relatively prime. Consequently, if two numbers are relatively prime, then at least one of them must be odd.   **c.** False. If $a = 5$ and $b = 9$, then GCF(5, 9) = 1, but 5 and 9 are both odd.

58. The prime factorization of $27 = 3^3$ and the prime factorization of $5,725,720 = 2^3 \cdot 5 \cdot 7 \cdot 11^2 \cdot 13^2$. Because the prime factorizations have no primes in common, the two numbers must be relatively prime.

59. **a.** Yes. Every pair of nonzero whole numbers will have 1 as their least common factor.   **b.** No. Every pair of nonzero whole numbers will have an infinite number of common multiples. Consequently, they cannot have a greatest one.

60. No. Let $a = 6$, $b = 9$, and $c = 12$, then GCF(6, 9, 12) = 3 and LCM(6, 9, 12) = 36. Hence, GCF(6, 9, 12) $\cdot$ LCM(6, 9, 12) = $3 \cdot 36 = 108$, but $6 \cdot 9 \cdot 12 = 648$.

61. Answers will vary.
62. Answers will vary.
63. Answers will vary.
64. Louis's examples are limited only to numbers that have no common factors. In this case, the LCM(a, b) = $a \cdot b$. However, if a and b have a common factor, LCM(a, b) < $a \cdot b$. We can point this out to Louis by showing him other examples that do not fit his conjecture.

65. Kwan is correct. Because 0 is a multiple of every number, the LCM(a, 0) = 0.

66. Division by zero is undefined, so zero cannot be a factor of any number.

67. Answers will vary.
68. Answers will vary.
69. Answers will vary.
70. Answers will vary.
71. **a.** Answers will vary.   **b.** Answers will vary.

■ Section 5.3 Learning Assessment

1. By the division algorithm, if a and b are whole numbers with $b \neq 0$, then there exists whole numbers q and r such that $a = bq + r$ with $0 \leq r < b$. The last inequality states that the remainder r is limited to values that are strictly less than the divisor b. Because the remainders continue to cycle through $0, 1, 2, \ldots, b - 1$ as we divide consecutive numbers by b, we can classify the whole numbers by common remainders.

2. Because $16 - 9 = 7$ and $7 \mid 7$, then $16 \equiv 9$ mod 7.

3. In general, whole-number operations and clock operations can be computed using the same basic approaches and satisfy the same basic properties. The key difference between clock and whole-number arithmetic is that clock arithmetic uses only the 12 numbers from the set {1, 2, 3, . . ., 12} in a cyclical fashion. This has a significant impact on the way we add, subtract, multiply, and divide.

4. **a.** Compute the sum using whole-number addition, divide by 12, and then take the remainder as the answer.   **b.** With the take-away approach, count backward, keeping in mind that upon reaching 1, continue the count by going back to 12. With the missing-addend approach, $a \ominus b = c$ if and only if c is the unique number from the set {1, 2, 3, . . . , 12} such that $a = b \oplus c$.   **c.** Multiply the numbers using whole-number multiplication, divide by 12, and take the remainder as the answer.   **d.** Using the missing-factor approach, $a \oslash b = c$ if and only if c is a unique number in the set {1, 2, 3, . . ., 12} such that $a = b \otimes c$.

5. **a.** Compute the sum using whole-number addition, divide by m, and then take the remainder as the answer.   **b.** With the take-away approach, count backward, keeping in mind that upon reaching 1, continue the count by going back to m. With the missing-addend approach, $a \ominus b = c$ if and only if c is the unique number from the set {1, 2, 3, . . ., m} such that $a = b \oplus c$.   **c.** Multiply the numbers using whole-number multiplication, divide by m, and take the remainder as the answer.   **d.** Using the missing-factor approach, $a \oslash b = c$ if and only if c is a unique number in the set {1, 2, 3, . . ., m} such that $a = b \otimes c$.

6. The additive inverse property for clock addition states that for any number a in the set {1, 2, 3, . . ., 12}, there is a unique number b such that $a \oplus b = 12$.

7. a can be less than b in clock subtraction because of the ability to cycle back through the digits.

8. $a \oslash b = c$ if and only if c is a unique number in the set {1, 2, 3, . . ., 12} such that $a = b \otimes c$. Twelve-hour clock division is undefined if no number c in the set satisfies the equation or if the quotient leads to two or more answers.

9. **a.** $17 \equiv 7$ mod 5 means that when 7 is subtracted from 17, the difference is divisible by 5.   **b.** $21 \equiv 5$ mod 8 means that when 5 is subtracted from 21, the difference is divisible by 8.   **c.** $34 \not\equiv 7$ mod 6 means that when 7 is subtracted from 34, the difference is not divisible by 6.

10. **a.** The equivalence classes are listed down each column.   **b.** One method is to make five columns and then list the counting numbers going across the rows. Because each column represents a single equivalence class, the elements of each set fall out naturally.

11. **a.** $7 \oplus 9 = 4$.

b. $6 \ominus 8 = 10.$

c. $4 \otimes 3 = 12.$

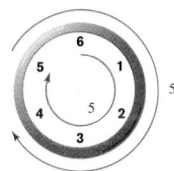

12. a. On a 6-hr clock, $5 \oplus 5 = 4.$

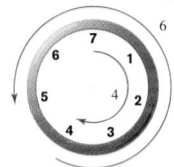

b. On a 7-hr clock, $4 \ominus 6 = 5.$

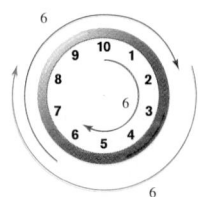

c. On a 10-hr clock, $3 \otimes 6 = 8.$

d. On an 8-hr clock, $5 \otimes 2 = 2.$

13. a. False. **b.** True. **c.** False.
14. a. False. **b.** True. **c.** True.
15. a. $x \in \{2, 8, 14, 20, \ldots\}.$ **b.** $x \in \{15, 24, 33, 42, \ldots\}.$
  c. $x \in \{4, 16, 28, 40, 52, 64\}.$
16. a. $x \in \{3, 7, 11, 15, \ldots\}.$ **b.** $x \in \{1, 9, 17, 25, 33, 41\}.$
  c. $x \in \{2, 3, 4, 6, 8, 12, 24\}.$
17. $T_0 = \{0, 3, 6, 9, \ldots\}$, $T_1 = \{1, 4, 7, 10, \ldots\}$, and
  $T_2 = \{2, 5, 8, 11, \ldots\}.$
18. $E_0 = \{0, 8, 16, \ldots\}$, $E_1 = \{1, 9, 17, \ldots\}$, $E_2 = \{2, 10, 18, \ldots\}$,
  $E_3 = \{3, 11, 19, \ldots\}$, $E_4 = \{4, 12, 20, \ldots\}$, $E_5 = \{5, 13, 21, \ldots\}$,
  $E_6 = \{6, 14, 22, \ldots\}$, and $E_7 = \{7, 15, 23, \ldots\}.$
19. $E_0 = \{0, 11, 22, 33, \ldots\}$, $E_1 = \{1, 12, 23, 34, \ldots\}$,
  $E_2 = \{2, 13, 24, 35, \ldots\}$, $E_3 = \{3, 14, 25, 36, \ldots\}$,
  $E_4 = \{4, 15, 26, 37, \ldots\}$, $E_5 = \{5, 16, 27, 38, \ldots\}$,
  $E_6 = \{6, 17, 28, 39, \ldots\}$, $E_7 = \{7, 18, 29, 40, \ldots\}$,
  $E_8 = \{8, 19, 30, 41, \ldots\}$, $E_9 = \{9, 20, 31, 42, \ldots\}$,
  and $E_{10} = \{10, 21, 32, 43, \ldots\}.$

20. a. $\{0, 6, 12, \ldots\}.$ **b.** $\{1, 7, 13, \ldots\}.$ **c.** $\{3, 9, 15, \ldots\}.$
  d. $\{4, 10, 16, \ldots\}.$ **e.** $\{5, 11, 17, \ldots\}.$
21. a. $\{0, 14, 28, \ldots\}.$ **b.** $\{3, 17, 31, \ldots\}.$ **c.** $\{7, 21, 35, \ldots\}.$
  d. $\{9, 23, 37, \ldots\}.$ **e.** $\{11, 25, 39, \ldots\}.$
22. a. 1. **b.** 9. **c.** 8. **d.** 4.
23. a. 7. **b.** 3. **c.** 10. **d.** 12.
24. a. 11. **b.** 7. **c.** 6. **d.** 3. **e.** 1.
25. a. 6. **b.** 12. **c.** Undefined. **d.** Undefined.
26. a. 7. **b.** 6. **c.** 4. **d.** 12.
27. a. 2. **b.** 3. **c.** 4. **d.** 5.
28. a. 3. **b.** 5. **c.** 5. **d.** 3.
29. a. 4. **b.** 4. **c.** 7. **d.** Undefined.
30. a. 6. **b.** 9. **c.** 12. **d.** Undefined.
31. a. $x \in \{3, 8, 13, \ldots 93, 98\}.$ **b.** $x \in \{12, 19, 26, \ldots 89, 96\}.$
  c. $x \in \{12, 25, 38, 51\}.$
32. $M_0 = \{0, m, 2m, 3m, \ldots\}$, $M_1 = \{1, m + 1, 2m + 1, 3m + 1\}$, $M_2 = \{2, m + 2, 2m + 2, 3m + 2, \ldots\}, \ldots,$
  $M_{m-1} = \{m - 1, 2m - 1, 3m - 1, \ldots\}.$
33. a. 24. **b.** 35. **c.** 73. **d.** n. **e.** $2n$. **f.** n^2.
34. a. 5. **b.** 5. **c.** 1.
35. a. 8. **b.** 9. **c.** 12. **d.** 9.
36. The 3rd, 10th, 17th, 24th, and 31st.
37. January 1 will fall on a Tuesday next year.
38. Answers will vary.
39. 5:00 p.m. and 1:00 a.m.
40. a. 8 days and 13 hr. **b.** 9:00 p.m.
41. a. i. 1600. **ii.** 0630. **iii.** 1445. **iv.** 2238.
  b. i. 3:00 p.m. **ii.** 8:45 a.m. **iii.** 4:30 p.m. **iv.** 8:15 p.m.
42. 0800.
43. 5:00 a.m. on Friday.
44. The sets of the evens and the odds.
45. All division will be possible on an m-hr clock if $m = 3, 5,$
  7, and 11. The number must be prime.
46. a. On a 12-hr clock, $3 \ominus 4 = 11$ and $4 \ominus 3 = 1$, so clock
  subtraction is not commutative. **b.** On a 12-hr clock,
  $1 \ominus (2 \ominus 3) = 1 \ominus 11 = 2$ and $(1 \ominus 2) \ominus 3 = 11 \ominus 3 = 8$, so
  clock subtraction is not associative. **c.** On a 12-hr clock,
  $3 \ominus 12 = 3$ and $12 \ominus 3 = 9$. Consequently, clock subtraction does not satisfy the identity property.
47. The table of facts for a 8-hr clock addition follows.

$\oplus$	1	2	3	4	5	6	7	8
1	2	3	4	5	6	7	8	1
2	3	4	5	6	7	8	1	2
3	4	5	6	7	8	1	2	3
4	5	6	7	8	1	2	3	4
5	6	7	8	1	2	3	4	5
6	7	8	1	2	3	4	5	6
7	8	1	2	3	4	5	6	7
8	1	2	3	4	5	6	7	8

a. $4 \oplus 6 = 2$, $7 \oplus 7 = 6$, and $3 \ominus 7 = 4$. **b.** Yes. Every
number in the table is from the set $\{1, 2, 3, 4, 5, 6, 7, 8\}$.
c. Yes. The table is symmetric about the main diagonal.
So $a \oplus b = b \oplus a$ in 8-hr clock multiplication.
d. Yes. Eight is the identity element because the table
shows us that $a \oplus 8 = a = 8 \oplus a$ for all a in the set $\{1, 2, 3, 4, 5, 6, 7, 8\}$.

48. The table of facts for a 8-hr clock multiplication follows.

⊗	1	2	3	4	5	6	7	8
1	1	2	3	4	5	6	7	8
2	2	4	6	8	2	4	6	8
3	3	6	1	4	7	2	5	8
4	4	8	4	8	4	8	4	8
5	5	2	7	4	1	6	3	8
6	6	4	2	8	6	4	2	8
7	7	6	5	4	3	2	1	8
8	8	8	8	8	8	8	8	8

a. $4 \otimes 5 = 4$, $2 \otimes 7 = 6$, and $5 \oplus 3 = 7$. **b.** Yes. Every number in the table is from the set $\{1, 2, 3, 4, 5, 6, 7, 8\}$. **c.** Yes. The table is symmetric about the main diagonal, so $a \otimes b = b \otimes a$ in 8-hr clock multiplication. **d.** Yes. One is the identity element because the table shows us that $a \otimes 1 = a = 1 \otimes a$ for all a in the set $\{1, 2, 3, 4, 5, 6, 7, 8\}$. **e.** Yes. Eight is the zero element because $a \otimes 8 = 8 = 8 \otimes a$ for all a in the set $\{1, 2, 3, 4, 5, 6, 7, 8\}$.

49. a. i. a. **ii.** a. **iii.** b. **iv.** a. **b.** Yes. Every combination is defined, and every answer is from the set $\{a, b, c, d\}$. **c.** Yes. The table is symmetric about the main diagonal, so * must be commutative. **d.** Yes. b is the identity element because its row and column matches the initial row and column. **e.** b is its own inverse because $b * b = b$, which is the identity element.

50. a. i. a. **ii.** c. **iii.** d. **iv.** b. **b.** Yes. The operation is defined for each pair of elements, and the answer is from the set of $\{a, b, c, d, e\}$. **c.** No. It is not symmetric about the main diagonal. Specifically, $d \# c = a$ and $c \# d = b$, so $c \# d \neq d \# c$. **d.** Yes. e is the identity element because its row and column matches the initial row and column. **e.** b is its own inverse because $b \# b = e$, which is the identity element.

51. The table of facts for 12-hr clock subtraction follows.

⊖	1	2	3	4	5	6	7	8	9	10	11	12
1	12	11	10	9	8	7	6	5	4	3	2	1
2	1	12	11	10	9	8	7	6	5	4	3	2
3	2	1	12	11	10	9	8	7	6	5	4	3
4	3	2	1	12	11	10	9	8	7	6	5	4
5	4	3	2	1	12	11	10	9	8	7	6	5
6	5	4	3	2	1	12	11	10	9	8	7	6
7	6	5	4	3	2	1	12	11	10	9	8	7
8	7	6	5	4	3	2	1	12	11	10	9	8
9	8	7	6	5	4	3	2	1	12	11	10	9
10	9	8	7	6	5	4	3	2	1	12	11	10
11	10	9	8	7	6	5	4	3	2	1	12	11
12	11	10	9	8	7	6	5	4	3	2	1	12

The set $\{1, 2, 3, \ldots, 12\}$ is closed under 12-hr clock subtraction because it is defined for each pair of elements, and the result is always a number back in the set.

52. Answers will vary.
53. Answers will vary.
54. Answers will vary.
55. Answers will vary.
56. Answers will vary.

■ Chapter 5 Review Exercises

1. 30, 60, and 90.
2. 1, 2, 4, 5, 10, and 20.
3. a. $16 \mid 34$ is false because there exists no whole number c such that $16 \cdot c = 34$. **b.** True. If $5 \mid 10$ and $5 \mid 30$, then 5 will divide any multiple of 10 and 30 including $10 \cdot 30$. **c.** $13 \mid 39$ is true because there exists the whole number $c = 3$ such that $13 \cdot 3 = 39$. **d.** True. If $7 \mid 14$ and $7 \nmid 13$, then $7 \nmid (14 + 13)$.
4. a. False. **b.** True. **c.** True. **d.** False.
5. a. 2, 3, 5, 6, 9, and 10. **b.** 2, 4, and 11. **c.** 2, 3, 4, 5, 6, 8, 10, and 11.
6. a. Prime. **b.** Composite. **c.** Prime.
7. a. 8 factors. **b.** 6 factors. **c.** 15 factors.
8. One possibility is 34,116.
9. Committee sizes could be 3, 5, 9, or 15.
10. Yes. $30 \mid 140{,}970$.
11. No. The sum of two odd numbers is always even. Two is the only even prime, and $1 + 1$ are the only two odd numbers that have a sum of two. Because 1 is not prime, there are no two odd primes that have a sum of two.
12. 60, 64, 72, 84, and 96 have the greatest number of factors with 12.
13. 8^3 has 10 factors: 1, 2, 4, 8, 16, 32, 64, 128, 256, and 512.
14. Every even number is divisible by 2, so cannot be prime.
15. Primes less than 100 that have the form $n^2 + 1$ are $1^2 + 1 = 2$, $2^2 + 1 = 5$, $4^2 + 1 = 17$, and $6^2 + 1 = 37$.
16. a will divide $b + c$, $b - c$, and $b \cdot c$.
17. a. Not relatively prime. **b.** Relatively prime. **c.** Not relatively prime.
18. a. 5. **b.** 78. **c.** 8. **d.** 300. **e.** 27,720. **f.** 218,790.
19. $2 \cdot 5 \cdot 9 = 90$.
20. a. GCF(120, 144, 210) = 6. **b.** LCM (120, 144, 210) = 5,040.
21. a. GCF$(a, b) = 2^2 \cdot 3^4 \cdot 5^2$; LCM$(a, b) = 2^4 \cdot 3^5 \cdot 5^4$. **b.** GCF$(a, b) = 5^2$; LCM$(a, b) = 2^4 \cdot 3^3 \cdot 5^5 \cdot 7^5 \cdot 11^3$.
22. 16 hr and 15 min.
23. No. If x, $x + 1$, and $x + 2$ represent three consecutive whole numbers, then $x + (x + 1) + (x + 2) = 3x + 3$, which is divisible by 3.
24. If it is divisible by 3, 5, and 8, which are pairwise relatively prime, then it must be divisible by $3 \cdot 5 \cdot 8$. Because this is the case, it must also divisible by $3 \cdot 5 = 15$, $3 \cdot 8 = 24$, $5 \cdot 8 = 40$, and $3 \cdot 5 \cdot 8 = 120$.
25.

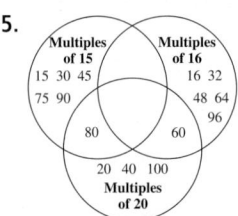

26. Possible answers are 5 and 100 or 20 and 25.
27. 60 min.

28. 6 nights off together.
29. The greatest amount of fencing the two plots can share is GCF(192, 204) = $2^2 \cdot 3$ = 12 ft. Hence, the total fencing required is 102 ft.
30. The GCF(432, 539) = 1, so they are relatively prime.
31. a. True. b. False. c. False.
32. a. $x \in$ {3, 10, 17, . . .}. b. $x \in$ {12, 20, 28, . . .}. c. x = 15.
33. a. F_0 = {0, 4, 8, 12, . . .}, F_1 = {1, 5, 9, 13, . . .}, F_2 = {2, 6, 10, 14, . . .}, and F_3 = {3, 7, 11, 15, . . .}.
 b. F_0 = {0, 5, 10, 15, . . .}, F_1 = {1, 6, 11, 16, . . .}, F_2 = {2, 7, 12, 17, . . .}, and F_3 = {3, 8, 13, 18,. . .}, and F_4 = {4, 9, 14, 19, . . .}.
34. a. 12. b. 10. c. 11. d. Undefined. e. 4.
 f. 8. g. 12. h. Undefined.
35. a. i. *e*. ii. *e*. iii. *b*. iv. *d*. b. Yes. Every element in the table is from the set {*a, b, c, d, e*}. c. No. The table is not symmetric about the main diagonal. d. No. No element in the set {*a, b, c, d, e*} has both a row and a column that matches the initial row and column. e. No. Because it does not have an identity element, it cannot have an inverse property.
36. The table of multiplication facts for a 7-hr clock follows.

⊗	1	2	3	4	5	6	7
1	1	2	3	4	5	6	7
2	2	4	6	1	3	5	7
3	3	6	2	5	1	4	7
4	4	1	5	2	6	3	7
5	5	3	1	6	4	2	7
6	6	5	4	3	2	1	7
7	7	7	7	7	7	7	7

a. $4 \otimes 5$ = 6, $2 \otimes 7$ = 7, and $5 \oplus 3$ = 4. b. Yes. Every element in the table is back in the set {1, 2, 3, 4, 5, 6, 7}. c. Yes. It is symmetric about the main diagonal of the table.

CHAPTER 6

Section 6.1 Learning Assessment

1. a. Zero is the only integer that has no sign. It is neither positive nor negative. b. It is an important reference point because it separates the positive integers from the negative.
2. a. Two integers are opposites if they are an equal number of units from zero but are on opposite sides of zero.
 b. Because zero is 0 units away from itself, it must be its own opposite.
3. Integers on a number line represent a distance from zero. The negative integers must therefore be the reverse of the positive integers because they are counting units from zero from right to left rather than from left to right.
4. The absolute value of an integer is the integer's distance from zero. Because distances are always greater than or equal to zero, the absolute value of any integer is greater than or equal to zero.
5. The definition of whole-number addition counts only the number of objects in the union of two disjoint sets without any regard for signs.
6. The sum of two integers can be found by taking the sum or difference of two whole numbers and then determining the sign. It is similar to whole-number addition in that

we can think of it as combining two quantities. However, unlike whole-number addition, integer addition must take the signs of the integers into account, and the signs can affect the value of the sum.
7. The additive inverse property states that the sum of an integer and its additive inverse, which is unique, is always zero.
8. With addition, $a < b$ if and only if there exists a positive integer n such that $a + n = b$. On the number line, $a < b$ if and only if a occurs to the left of b.
9. One way to compute integer subtraction is to use the missing-addend approach: If a and b are any integers, then $a - b = c$ if and only if $a = b + c$ for some unique integer c. Another way to subtract is to add the opposite of the number. Integer subtraction is similar to whole-number subtraction in that the missing-addend approach can be used to compute it. However, unlike whole-number subtraction, integer subtraction must take the signs of the integers into account, and the signs can affect the value of the difference.
10. First, it blurs the distinction between addition and subtraction. In other words, given negative numbers, we can change any subtraction problem into an addition problem and any addition problem into a subtraction problem. Second, integer subtraction satisfies only the closure property. However, when working with subtraction, we can now access properties like the commutative and associative properties for addition by changing the difference to a sum.
11. a. Positive four. b. Negative seven. c. Negative three plus negative four. d. Positive six minus negative five.
 e. Negative four minus positive six. f. Negative three minus negative eleven.
12. a. (−15). b. +275. c. (−315). d. +4. e. (−2).
 f. +25
13. a. 56 + (−45) = 11. b. 150 + 459 = 609.
 c. Because 13,599 + x = 6,599, x = −7,000.
14. a. Because 32,000 + x = 27,000, x = (−5,000).
 b. Because 35 + x = 27, x = (−8). c. Because 25 + x = 8, x = (−17).
15. a. (−2). b. 7. c. (−9). d. 15. e. (−20). f. 24.
16. a. Four units to the left of zero. b. Twelve units to the right of zero. c. Eight units to the left of zero. d. Zero is between (−1) and 1.
17. Answer will vary, but here are three possible representations:

18. a. (−2). b. 0. c. (−1). d. 2.
19. a. 2. b. (−4). c. 2. d. (−1).
20. a. (−8) + 5 = (−3).

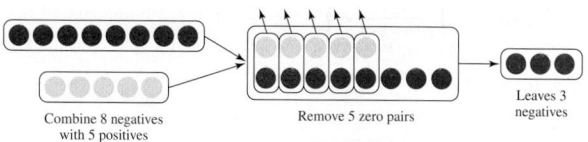

Combine 8 negatives with 5 positives   Remove 5 zero pairs   Leaves 3 negatives

b. 4 + (−3) = 1.

Combine 4 positives with 3 negatives   Remove 3 zero pairs   Leaves 1 positive

c. $(-6) - (-3) = (-3)$.

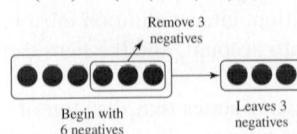

Begin with 6 negatives   Leaves 3 negatives

Remove 3 negatives

d. $6 - 8 = (-2)$.

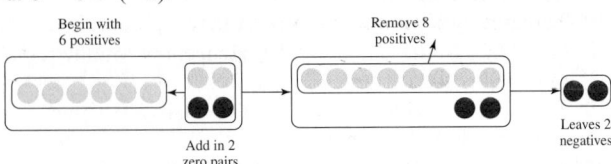

Begin with 6 positives   Remove 8 positives   Leaves 2 negatives

Add in 2 zero pairs

21. a. $(-3) + (-6) = (-9)$.

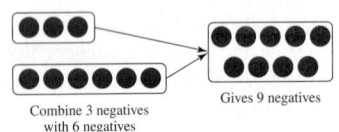

Combine 3 negatives with 6 negatives   Gives 9 negatives

b. $8 + 2 = 10$.

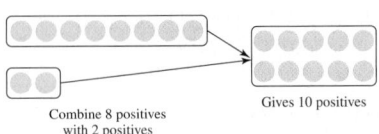

Combine 8 positives with 2 positives   Gives 10 positives

c. $(-4) - 5 = (-9)$.

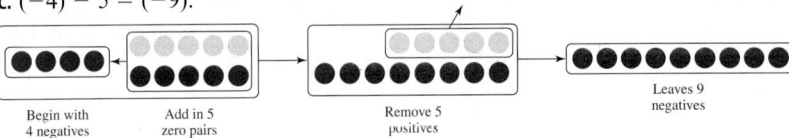

Begin with 4 negatives   Add in 5 zero pairs   Remove 5 positives   Leaves 9 negatives

d. $(-1) - (-6) = 5$.

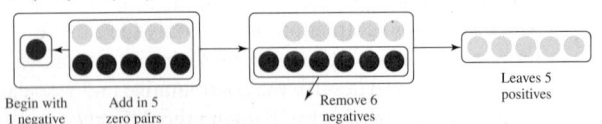

Begin with 1 negative   Add in 5 zero pairs   Remove 6 negatives   Leaves 5 positives

22. a. $8 + (-5) = 3$.

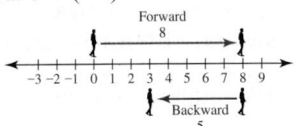

Forward 8   Backward 5

b. $(-6) + 5 = (-1)$.

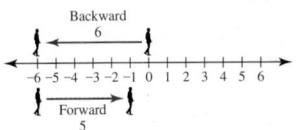

Backward 6   Forward 5

c. $3 - 7 = (-4)$.

Forward 3   Forward 7   Turn around

d. $(-4) - (-6) = 2$.

Backward 4   Turn around   Backward 6

23. a. $3 + 6 = 9$.

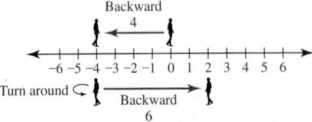

Forward 3   Forward 6

b. $(-1) + (-4) = (-5)$.

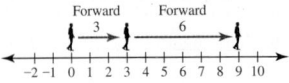

Backward 4   Backward 1

c. $2 - (-5) = 7$.

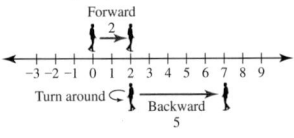

Forward 2   Turn around   Backward 5

d. $(-7) - 3 = (-10)$.

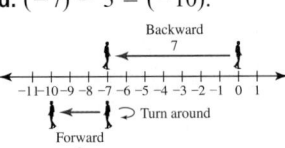

Backward 7   Turn around   Forward 3

24. a. 4. **b.** 6. **c.** −5. **d.** 4.

25. a. 5. **b.** 29. **c.** (−8). **d.** 10.

26. a. False. There is no positive integer k such that $(-5) + k = (-9)$. **b.** True. There is the positive integer $k = 12$ such that $(-4) + 12 = 8$. **c.** False. There is no positive integer k such that $(-3) + k = (-4)$. **d.** True. There exists the positive integer $k = 7$ such that $(-7) + 7 = 0$.

27. $(-17), (-5), (-4), 0, 8,$ and 16.

28. a. Consider the following sequence of sums:

$$2 + 2 = 4$$
$$2 + 1 = 3$$
$$2 + 0 = 2$$

As the second addend decreases by 1, so does the sum. Continuing this pattern:

$$2 + (-1) = 1$$
$$2 + (-2) = 0$$
$$2 + (-3) = -1$$

b. Consider the following sequence of differences:

$$3 - 2 = 1$$
$$3 - 1 = 2$$
$$3 - 0 = 3$$

As the second number decreases by 1, the difference increases by 1. Continuing this pattern:

$$3 - (-1) = 4$$
$$3 - (-2) = 5$$
$$3 - (-3) = 6$$

29. a. Consider the following sequence of differences:

$$4 - 2 = 2$$
$$4 - 3 = 1$$
$$4 - 4 = 0$$

As the second number increases by 1, the difference decreases by 1. Continuing this pattern:

$$4 - 5 = (-1)$$
$$4 - 6 = (-2)$$

b. Consider the following sequence of sums:

$$2 + 3 = 5$$
$$1 + 3 = 4$$
$$0 + 3 = 3$$

As the first number decreases by 1, the sum also decreases by 1. Continuing this pattern:

$$(-1) + 3 = 2$$
$$(-2) + 3 = 1$$
$$(-3) + 3 = 0$$
$$(-4) + 3 = (-1)$$

30. a. (-3). **b.** 13. **c.** (-13). **d.** 2.
31. a. $14 + (-8)$. **b.** $(-14) + (-8)$. **c.** $(-14) + 8$. **d.** $14 + 8$.
32. a. (-13). **b.** 566. **c.** $(-1{,}121)$. **d.** 221.
33. a. (-38). **b.** 656. **c.** $(-1{,}019)$.
34. a. 802. **b.** $(-1{,}188)$. **c.** $(-1{,}579)$.
35. a. (-3). **b.** 7. **c.** 0. **d.** (-4). **e.** (-5). **f.** 5.
36. a. Additive inverse property. **b.** Commutative property. **c.** Associative property. **d.** Closure property. **e.** Additive identity property. **f.** Commutative property.
37. a. 90. **b.** (-160). **c.** (-216). **d.** 111. **e.** 57. **f.** (-9).
38. a. 2,176. **b.** $(-3{,}833)$. **c.** 3,719. **d.** 27,281.
39. a. $\approx (-450)$. **b.** ≈ 500.
40. a. $x = (-8)$ or $x = 8$. **b.** $x = 0$. **c.** $x \in \{-2, -1, 0, 1, 2\}$. **d.** $x \in \{-4, -3, -2, -1, 0, 1, 2, 3, 4\}$. **e.** Not possible.
41. a. $x = (-4)$ or $x = 2$. **b.** $x = (-1)$ or $x = 1$. **c.** $x = (-3)$ or $x = 3$. **d.** $x = (-7)$ or $x = 7$.
42. 16.
43. a. W. **b.** $\{\ldots, -3, -2, -1, 1, 2, 3, \ldots\} = \mathbb{Z} - \{0\}$. **c.** $\mathbb{Z}^+$. **d.** $\varnothing$. **e.** $\{0\}$.
44. One possible solution is:

-9	10	-1
8	0	-8
1	-10	9

45. $2 - (4 + (-3)) - (-5) = 6$.
46.

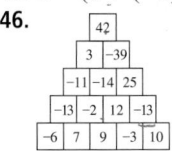

	42		
	3	-39	
-11	-14	25	
-13	-2	12	-13
-6	7	9	-3	10

47. a. 1, (-1), 2, and (-2). **b.** The next three triangles would have net values of 3, (-3), and 4 and would be:

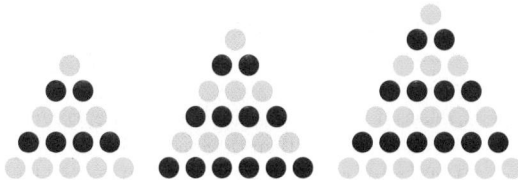

c. 8.
48. a. 1, (-2), 3, and (-4). **b.** The next three squares would have net values of 5, (-6), and 7, and would be:

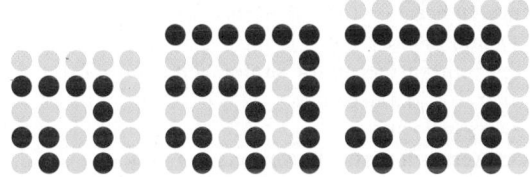

c. 15.
49. $\{-8, -6, -4, -2, 0, 2, 4, 6, 8\}$.
50. $116°$ F.
51. 290 mph.
52. The net gain is 8 points.
53. The net yardage is a gain of 37 yd.
54. a. i. 1. **ii.** 34. **iii.** (-7). **iv.** 0. **b.** Yes. The hand 3♦, 4♦, 2♦, J♠, 2♥ has a net value of 0. **c.** The largest value is 61. One of two possible hands is K♠, K♣, Q♠, Q♣, J♠. **d.** The smallest possible value is (-61). One of two possible such hands is K♦, K♥, Q♦, Q♥, J♦.
55. a. Does not exist. **b.** 1. **c.** (-1). **d.** Does not exist.
56. a. True. **b.** True. **c.** False. (-1) is an integer that is not a whole number. **d.** False. Zero has no sign. **e.** False. $\mathbb{Z}^+$ does not contain zero, but the whole numbers do.
57. Yes. It is true for only one value, $x = 0$.
58. a. $\{\ldots (-8), (-4), 0, 4, 8 \ldots\}$ is closed under subtraction because the difference of any two elements in the set is back in the set. **b.** $\{\ldots (-5), (-3), (-1), 0, 2, 4, 6, \ldots\}$ is not closed under subtraction because $2 - (-1) = 3$ and 3 is not in the set. **c.** $\{\ldots (-5), (-3), (-1), 1, 3, 5, \ldots\}$ is not closed under subtraction because $5 - 3 = 2$ and 2 is not in the set. **d.** $\{(-7), (-5), (-3), (-1), 0, 1, 3, 5, 7\}$ is not closed under subtraction because $5 - 3 = 2$ and 2 is not in the set.
59. a. $\{\ldots, -3, -2, -1\}$. **b.** $\{\ldots, -3, -2, -1\}$. **c.** $\{\ldots, -3, -2, -1, 1, 2, 3, \ldots\}$. **d.** Not possible.
60. $|x| = |y|$.
61. a. $(-p)$. **b.** p. **c.** $-(p + q)$. **d.** $-(p - q) = -p + q$.
62. a. The commutative property does not hold for integer subtraction because $4 - 2 = 2$ and $2 - 4 = (-2)$. **b.** The associative property does not hold for integer subtraction because $4 - (3 - 2) = 4 - 1 = 3$ and $(4 - 3) - 2 = 1 - 2 = -1$. **c.** The identity property does not hold for integer subtraction because $4 - 0 = 4$, but $0 - 4 = (-4)$.
63. a. In general, $p + (q - r) = (p + q) - (p + r)$ is not true. If $p = 2$, $q = 3$ and $r = 4$, then $p + (q - r) = 2 + (3 - 4) = 2 + (-1) = 1$ and $(p + q) - (p + r) = (2 + 3) - (2 + 4) = 5 - 6 = (-1)$. **b.** In general, $q - p = -(p + q)$ is not true. If $p = 2$ and $q = 3$, then $q - p = 3 - 2 = 1$

and $-(p + q) = -(2 + 3) = (-5)$. **c.** In general, $p - q = -(q - p)$ is true. If p and q are integers, then $p - q = -q + p = -q - (-p) = -(q - p)$.

64. Yes. $p - q$ is the additive inverse of $q - p$ because $(p - q) + (q - p) = [p + (-q)] + [q + (-p)] = [p + (-p)] + [q + (-q)] = 0 + 0 = 0$.

65. a. $|p + q|$ does not equal $|p| + |q|$ for all integers p and q. Let $p = 3$ and $q = (-4)$, then $|p + q| = |3 + (-4)| = |(-1)| = 1$ and $|p| + |q| = |3| + |(-4)| = 3 + 4 = 7$.
b. Yes. $|p - q| = |q - p|$ for all integers p and q. If $p - q$ is positive, then $q - p$ must be negative. Hence, $|p - q| = p - q$ and $|q - p| = -(q - p) = -q + [-(-p)] = -q + p = p - q$. If $p - q$ is negative, then $q - p$ is positive. Hence, $|p - q| = -(p - q) = -p + [-(-q)] = -p + q = q - p$ and $|q - p| = q - p$.

66. a. Tell your partner to use subtraction. **b.** Tell your partner to add a negative.

67. Answers will vary.

68. a. Answers will vary. **b.** Answers will vary.

69. a. Answers will vary. **b.** Answers will vary.
c. Answers will vary. **d.** Answers will vary.

70. a. Answers will vary. **b.** Answers will vary.
c. Answers will vary.

71. The three different uses of the dash $(-)$ are for subtraction, negatives, and additive inverses. Answers will vary on the how this makes it difficult for students to learn the integers.

72. Answers will vary.

73. No. If a is negative, then $-a$ will be positive.

74. Answers will vary.

75. Floors above ground level represent positive integers, ground level represents zero, and floors below ground level represent negative integers.

76. a. Answers will vary. **b.** Answers will vary.

77. Answers will vary.

78. Answers will vary.

79. No. It is difficult to use the integers to describe parts of a whole or fractional concepts.

■ Section 6.2 Learning Assessment

1. The Cartesian product of two sets counts only ordered pairs, so it cannot account for how the signs of the integers affect the product.

2. The repeated-addition approach makes sense only for integer multiplication if the first integer is positive.

3. **a.** If the signs of the factors are the same, the product is positive. If the signs of the factors are different, the product is negative. **b.** If the number of negative integers is even, the product is positive. If it is odd, the product is negative.

4. Multiplying by (-1) changes the sign of the integer.

5. If a, b, and c are integers, then $a(b - c) = a \cdot b - a \cdot c$ and $(b - c)a = b \cdot a - c \cdot a$.

6. If the exponent on a negative integer is even, the result is positive. If it is odd, the result is negative.

7. **a.** The repeated-subtraction approach makes sense only if the signs of the integers are the same. **b.** The partitioning approach works only if the second integer is positive.

8. If a and b are any integers with $b \neq 0$, then $a \div b = c$ if and only if $a = b \cdot c$ for some unique integer c.

9. **a.** $(-7) \cdot 3 = (-21)$ yd. **b.** $(-650) \cdot 12 = (-7,800)$ ft.
c. $(-12) + 4 \cdot 8 = 20°$ F. **d.** $(-365) \cdot 5 - \$(-1,825)$ (only 5 days in a business week).

10. **a.** $(-3) \cdot (-4) = 12$.

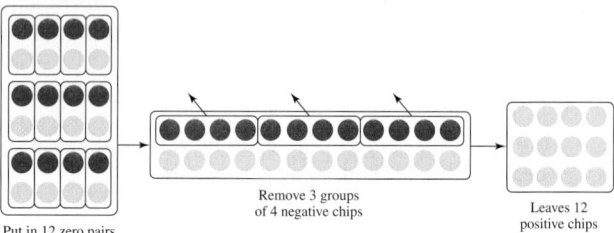

Put in 12 zero pairs — Remove 3 groups of 4 negative chips — Leaves 12 positive chips

b. $6 \cdot 2 = 12$.

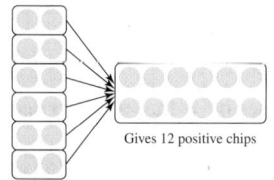

Put in 6 groups of 2 positive chips — Gives 12 positive chips

c. $(-3) \cdot 3 = (-9)$.

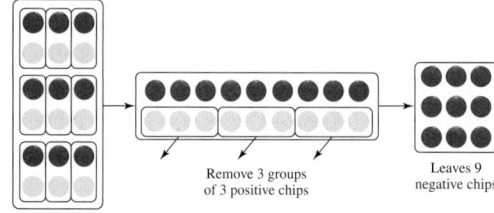

Put in 9 zero pairs — Remove 3 groups of 3 positive chips — Leaves 9 negative chips

d. $8 \cdot (-2) = (-16)$.

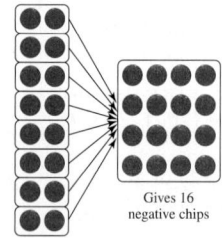

Put in 8 groups of 2 negative chips — Gives 16 negative chips

11. **a.** $(-4) \cdot 3 = (-12)$.

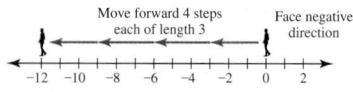

Move forward 4 steps each of length 3 — Face negative direction

b. $5 \cdot (-1) = (-5)$.

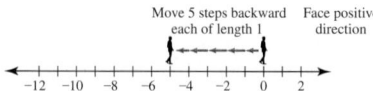

Move 5 steps backward each of length 1 — Face positive direction

c. $6 \cdot 3 = 18$.

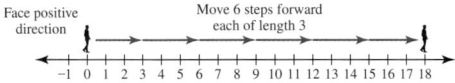

d. $(-7) \cdot (-2) = 14$.

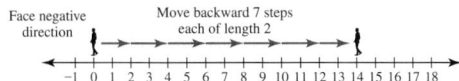

12. a. $4 \cdot (-2) = (-8)$.

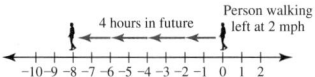

b. $3 \cdot 5 = 15$.

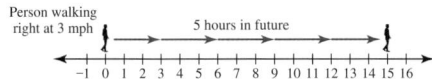

c. $(-4) \cdot (-5) = 20$.

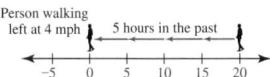

d. $(-2) \cdot 6 = (-12)$.

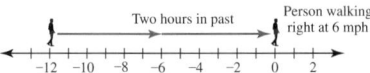

13.

×	+	−
+	+	−
−	−	+

÷	+	−
+	+	−
−	−	+

14. In the expression -7^4, the exponent is applied only to the 7. In the expression $(-7)^4$, it is applied to both the sign and 7.

15. a. Not possible.   **b.** $(-16) \div (-4) = 4$ because we can subtract (-4) from 16 four times: $(-16) - (-4) = (-12)$, $(-12) - (-4) = (-8)$, $(-8) - (-4) = (-4)$, and $(-4) - (-4) = 0$.   **c.** Although the quotient can be computed using repeated subtraction, there is a remainder. Because we do not get an integer when we divide, the division is undefined.

16. a. Begin with 10 negative chips and equally distribute them to 2 sets. Because 5 negative chips are in each set, then $(-10) \div 2 = (-5)$.

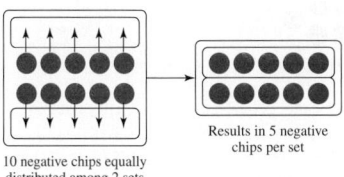

10 negative chips equally distributed among 2 sets

Results in 5 negative chips per set

b. Although the quotient can be computed using partitioning, there is a remainder. Because we do not get an integer when we divide, the division is undefined.
c. Not possible.

17. a. Consider the following sequence of products.

$$4 \cdot 3 = 12$$
$$4 \cdot 2 = 8$$
$$4 \cdot 1 = 4$$
$$4 \cdot 0 = 0$$

As the second number decreases by 1, the product decreases by 4. Continuing this pattern, $4 \cdot (-1) = (-4)$.
b. Consider the following sequence of products:

$$2 \cdot 6 = 12$$
$$1 \cdot 6 = 6$$
$$0 \cdot 6 = 0$$

As the first number decreases by 1, the product decreases by 6. Continuing this pattern, $(-1) \cdot 6 = (-6)$ and $(-2) \cdot 6 = (-12)$.

18. a. Consider the following sequence of products:

$$3 \cdot 2 = 6$$
$$2 \cdot 2 = 4$$
$$1 \cdot 2 = 2$$
$$0 \cdot 2 = 0$$

As the first factor decreases by 1, the product decreases by 2. Continuing this pattern:

$$(-1) \cdot 2 = (-2)$$
$$(-2) \cdot 2 = (-4)$$
$$(-3) \cdot 2 = (-6).$$

b. Consider the following sequence of products:

$$3 \cdot 2 = 6$$
$$3 \cdot 1 = 3$$
$$3 \cdot 0 = 0$$

As the second factor decreases by 1, the product decreases by 3. Continuing this pattern:

$$3 \cdot (-1) = (-3)$$
$$3 \cdot (-2) = (-6)$$
$$3 \cdot (-3) = (-9)$$

19. a. 63.   **b.** (-322).   **c.** $(-3,654)$.   **d.** 11,603.
20. a. 144.   **b.** 3,360.   **c.** $(-4,032)$.
21. a. $x = (-7)$.   **b.** $x = 729$.   **c.** $x = 63$.
22. a. 27.   **b.** 81   **c.** $(-1,024)$.   **d.** 1,296.
23. a. $(-4)^9$.   **b.** $(10)^3$.   **c.** $(-5)^9$.   **d.** $(-2)^{18}$.   **e.** 1.   **f.** 1.
24. a. 3.   **b.** Not defined for the integers.   **c.** (-6).
d. Not defined for the integers.
25. a. (-5).   **b.** (-616).   **c.** (-89).
26. a. $(-6) \cdot 7 = (-42)$. The other facts in the fact family are $7 \cdot (-6) = (-42)$, $(-42) \div (-6) = 7$, and $(-42) \div 7 = (-6)$.   **b.** $(-8) \cdot (-5) = 40$. The other facts in the fact family are $(-5) \cdot (-8) = 40$, $40 \div (-5) = (-8)$, and $40 \div (-8) = (-5)$.   **c.** $(-35) \div (-7) = 5$. The other facts in the fact family are $(-35) \div 5 = (-7)$, $5 \cdot (-7) = (-35)$, $(-7) \cdot 5 = (-35)$.   **d.** $(-72) \div 9 = (-8)$. The other facts in the fact family are $(-72) \div (-8) = 9$, $(-8) \cdot 9 = (-72)$, and $9 \cdot (-8) = (-72)$.
27. a. Positive; 3,439,800.   **b.** Negative; (-75).
c. Positive; 12.   **d.** Negative; $(-7,776,000)$.

28. a. 2,401. **b.** 250. **c.** 7. **d.** (−32).
29. a. 72. **b.** 48. **c.** (−104). **d.** 8. **e.** (−2). **f.** (−4).
30. Methods of estimation will vary.
  a. ≈ 20,000. **b.** ≈ (−210,000). **c.** ≈ (−5). **d.** ≈ 4.
31. a. $(-73) \cdot 14 + (-27) = -1,049$ or $73 \cdot (-14) + (-27) = (-1,049)$. **b.** $63 \div 7 + 23 \cdot (-15) - 12 = (-348)$ or $(-63) \div (-7) + (-23) \cdot 15 - 12 = (-348)$. **c.** $22 - (-47) + 36 \cdot 19 - (-84) = 837$ or $22 - (-47) + (-36) \cdot (-19) - (-84) = 837$. **d.** $49 \cdot 27 - (-84) \div 7 \cdot 12 = 1,467$ or $(-49) \cdot (-27) - (-84) \div 7 \cdot 12 = 1,467$.
32. a. $(-84) \cdot (-21)$. **b.** $(-1,387) \div 73$.
33. a. Positive. **b.** Positive. **c.** Negative. **d.** Positive.
  e. Negative.
34. a. (−4). **b.** (−11) or 11. **c.** 5. **d.** Not possible.
  e. (−9) or 9. **f.** Not possible.
35. a. (−4) or 4. **b.** No integers exist. **c.** No integers exist.
  d. (−1) or 1. **e.** No integers exist. **f.** 0.
36. a. i. Positive. **ii.** Positive. **iii.** Positive. **iv.** Negative.
  v. Negative. **b. i.** Negative. **ii.** Positive.
  iii. Negative. **iv.** Negative. **v.** Positive.
37. a. 15° C. **b.** 0° C. **c.** (−15)° C.
38. The net profit is $9,650.
39. $(-26)°$ F.
40. Each plaintiff receives about $194,444 a month.
41. $208.
42. It will take 13 hr before the count is less than 5,000.
43. a. False. **b.** False. **c.** True. **d.** True.
44. a. $a(b + c) = (-3)(5 + 9) = (-3)(14) = (-42) = (-15) + (-27) = (-3)5 + (-3)9 = ab + ac$ and $a(b - c) = (-3)(5 - 9) = (-3)(-4) = 12 = (-15) - (-27) = (-3)5 - (-3)9 = ab - ac$. **b.** $a(b + c) = (-2)((-5) + (-7)) = (-2)(-12) = 24 = (-10) + (-14) = (-2)(-5) + (-2)(-7) = ab + ac$ and $a(b - c) = (-2)[(-5) - (-7)] = (-2)(2) = (-4) = 10 - 14 = (-2)(-5) - (-2)(-7) = ab - ac$.
45. Let a, b, and c be integers. Then:

$(b - c)a = [b + (-c)]a$	Theorem 6.4
$= b \cdot a + (-c) \cdot a$	Distributive property
$= b \cdot a + -(c \cdot a)$	Integer multiplication
$= b \cdot a - c \cdot a$	Theorem 6.4

46. If a, b, and c are integers with $c \neq 0$, then $(a + b) \div c = (a \div c) + (b \div c)$ is true for those divisions defined in the integers.
47. No. Let $a = (-4)$ and $b = (-3)$. Then $(-4) < (-3)$, but $a^2 = 16 > 9 = b^2$.
48. a. True. **b.** True. **c.** True.
49. $x^2 - y^2$ is positive when $|x| > |y|$, $x^2 - y^2$ is negative when $|x| < |y|$, and $x^2 - y^2$ is zero when $|x| = |y|$.
50. a. $(-2)^1 = (-2)$, $(-2)^2 = 4$, $(-2)^3 = (-8)$, $(-2)^4 = 16$, $(-2)^5 = (-32)$, and $(-2)^6 = 64$. **b.** Even powers are positive. Odd powers are negative. **c.** $(-2)^n$ is positive when n is even, and $(-2)^n$ is negative when n is odd.
51.

$2 \cdot 5 + (-2) 5 = 5 \cdot 2 + 5(-2)$	Commutative property
$= 5[2 + (-2)]$	Distributive property of multiplication over addition
$= 5 \cdot 0$	Additive inverse property
$= 0$	Zero multiplication property

52. Suppose that a and b are two positive integers: $a > 0$ and $b > 0$. By the additive inverse property, $a \cdot b$ is the only number we can add to $-(ab)$ to get 0. However, if we add $(-a)(-b)$ to $-(ab)$:

$(-a)(-b) + (-(ab))$	
$= (-a)(-b) + (-a)(b)$	Definition of multiplication
$= (-a)((-b) + b)$	Distributive property
$= 0 \cdot b$	Additive inverse property
$= 0$	Zero multiplication property

Because ab and $(-a)(-b)$ both give us zero, it must be that $ab = (-a)(-b)$ Consequently, if a and b are positive integers, then the product of two negative integers is a positive integer.
53. a. If the good guys (+) enter town (+), it is good for the town (+). **b.** If the good guys (+) leave town (−), it is bad for the town (−). **c.** If the bad guys (−) enter town (+), it is bad for the town (−). **d.** If the bad guys (−) leave town (−), it is good for the town (+).
54. Answers will vary.
55. Answers will vary.
56. No, he is not. If either x or y is negative, then $-x \cdot y$ is positive.
57. Answers will vary.
58. a. Answers will vary. **b.** Answers will vary.
59. Answers will vary.
60. Answers will vary.
61. We interpret a negative exponent as $a^{-n} = 1/a^n$. To define negative exponents, we need rational numbers or fractions.
62. Answers will vary.

▮ Chapter 6 Review Exercises

1. a. Negative six. **b.** Negative four minus positive nine. **c.** Positive four plus negative eight. **d.** Negative four minus negative nine.
2. a. Fourteen units to the right of 0. **b.** Eight units to the left of 0. **c.** zero is in between (−1) and 1.
3. Seven integers can be represented with exactly 6 chips, including $\{-6, -4, -2, 0, 2, 4, 6\}$.
4. a. $(-3) + (-4) = (-7)$.

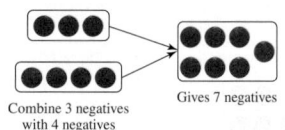

Combine 3 negatives with 4 negatives   Gives 7 negatives

b. $6 - (-2) = 8.0$

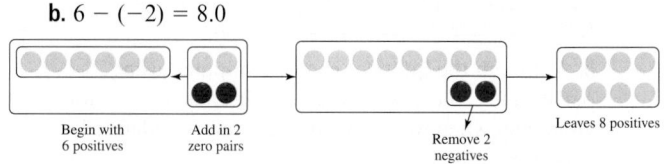

Begin with 6 positives   Add in 2 zero pairs   Remove 2 negatives   Leaves 8 positives

c. $(-6) \cdot 2 = (-12)$.

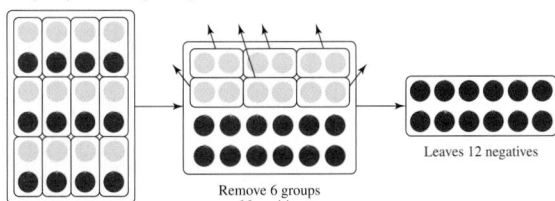

Put in 12 zero pairs

Remove 6 groups
of 2 positives

Leaves 12 negatives

5. a. $(-6) + 3 = (-3)$.

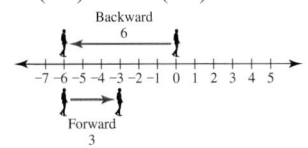

Backward
6

Forward
3

b. $(-5) - (-1) = (-4)$.

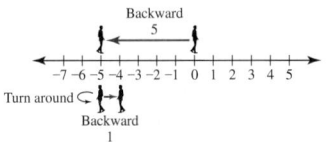

Backward
5

Turn around

Backward
1

c. $(-4) \cdot (-3) = 12$.

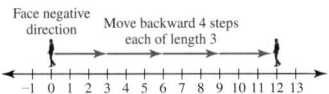

Face negative
direction   Move backward 4 steps
each of length 3

6. a. (-366). **b.** 262. **c.** (-349). **d.** 371. **e.** 8,874.
f. (-840). **g.** 61. **h.** (-26).
7. a. (-82). **b.** 99. **c.** (-29). **d.** 3. **e.** 294. **f.** 25.
8. a. 2. **b.** (-9). **c.** 0. **d.** 4. **e.** (-8).
9. a. (-4) or 4. **b.** Not possible. **c.** (-2) or 6.
d. (-2) or 2.
10. a. $23 - (-4) = 27°$ F. **b.** $35 - 49 = (-14)$ yd.
c. $15 \cdot (-450) = (-6,750)$ ft.
11. (-3) occurs further to the left than (-1), so it must be
that $(-3) < (-1)$.
12. a. $13 + (-6)$. **b.** $(-13) + (-6)$. **c.** $(-13) + 6$.
d. $13 + 6$.
13. a. Distributive property of multiplication over subtraction.
b. Additive inverse property. **c.** Zero multiplication
property. **d.** Commutative property. **e.** Closure
property. **f.** Additive identity property.
14. a. Negative. **b.** Negative.
15. a. 81. **b.** (-16). **c.** (-11).
16. a. $(-2)^8$. **b.** $(-12)^4$. **c.** 1. **d.** 1. **e.** $(-8)^5$.
17. 16 yd gained.
18. $6,810.
19. Each plaintiff receives $273,438 per month.
20. 15 hr.
21. a. $\{\ldots (-10), (-5), 0, 5, 10\ldots\}$ is closed under
subtraction because the difference of any two multiples
of 5 is always a multiple of 5. **b.** $\{\ldots (-6), (-4), (-2),$
$2, 4, 6, \ldots\}$ is not closed under subtraction because
$(-2) - (-2) = 0$ and 0 is not in the set.
22.

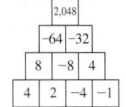

23. $3 + (-2) - (4 + (-1)) = (-2)$.
24. Consider the following sequence of products:

$$5 \cdot 2 = 10$$
$$5 \cdot 1 = 5$$
$$5 \cdot 0 = 0$$

As the second number decreases by 1, the product
decreases by 5. Continuing this pattern:

$$5 \cdot (-1) = (-5)$$
$$5 \cdot (-2) = (-10)$$
$$5 \cdot (-3) = (-15)$$

25. a. No. The differences go away from zero instead of
toward it. **b.** Yes. The differences go to zero: $(-6) -$
$(-3) = (-3)$ and $(-3) - (-3) = 0$.

■ Answers to Reviewing the Big Ideas: Chapters 4–6

1. $4 + 7 = 11$.

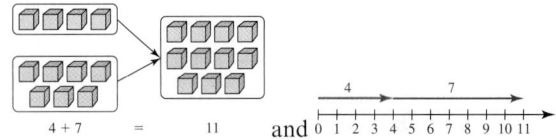

$4 + 7 \quad = \quad 11$ and

2. a. If $7 - 4 = ?$, then $4 + ? = 7$. Because $4 + 3 = 7$, then
$7 - 4 = 3$. **b.** If $13 - ? = 4$, then $? + 4 = 13$. Because
$9 + 4 = 13$, then $13 - 9 = 4$.
3. Answers will vary.
4. $6 \cdot 2 = 2 + 2 + 2 + 2 + 2 + 2 = 12$.
5. a. 4^6. **b.** $3^3(x + 1)^3$.
6. a. 3^7. **b.** 11. **c.** 2^{21}. **d.** $(5x)^3$.
7. $28 \div 7$ represents the number of times 7 can be subtracted
from 28. Because this can be done 4 times $(28 - 7 - 7 -$
$7 - 7 = 0)$, then $28 \div 7 = 4$.
8. a. If $12 \div 3 = ?$, then $12 = 3 \cdot ?$. Because $3 \cdot 4 = 12$, then
$12 \div 3 = 4$. **b.** If $63 \div 3 = ?$, then $63 = 3 \cdot ?$. Because
$3 \cdot 21 = 63$, then $63 \div 21 = 3$.
9. a. Quotient = 7, remainder = 5. **b.** Quotient = 8,
remainder = 3.
10. 172.
11. a. Not closed. **b.** Closed.
12. a. Associative property. **b.** Closure property.
c. Commutative property. **d.** Distributive property of
multiplication over addition. **e.** Zero multiplication
property. **f.** Additive identity property.
13. a. 686. **b.** 120. **c.** 2,074.
14. a. 178. **b.** 1,593.
15. a. 2,266. **b.** 75. **c.** 6,903. **d.** 125.
16. a. 4213_{five}. **b.** 3006_{nine}. **c.** 1243_{five}. **d.** 32_{eight} remainder
10_{eight}.
17. Methods will vary.
a. 107. **b.** 292. **c.** 1,435. **d.** 506.
18. Methods will vary.
a. ≈ 240. **b.** $\approx 8,000$. **c.** $\approx 6,000,000$. **d.** ≈ 100.
19. a. $16 \mid 56$ is false because there is no whole number
c such that $16 \cdot c = 56$. **b.** The statement is true by
Theorem 5.1(2). **c.** $12 \nmid 78$ is true because there is no
whole number c such that $12 \cdot c = 78$. **d.** The statement is
false because $7 \nmid (14 + 15)$, but $7 \mid 14$, even though $7 \nmid 15$.

20. a. Divisible by 2.   **b.** Divisible by 2, 4, 5, and 10.

21. a. Composite.   **b.** Composite.   **c.** Composite.

22. $660 = 2^2 \cdot 3 \cdot 5 \cdot 11$. Hence, 660 has $(2 + 1)(1 + 1)$ $(1 + 1)(1 + 1) = 24$ factors.

23. a. GCF(56, 72) = 8.   **b.** LCM(18, 24) = 72.

24. a. 33.   **b.** 3,465.

25. Using the appropriate divisions, we have GCF(1050, 4158) = GCF(1008, 1050) = GCF(42, 1008) = GCF(0, 42) = 42.

26. a. 120.   **b.** 2,325,832.

27. a. True.   **b.** False.   **c.** False.

28. $T_0 = \{0, 3, 6, 9, 12, \ldots\}$, $T_1 = \{1, 4, 7, 10, 13, \ldots\}$, and $T_2 = \{2, 5, 8, 11, 14, \ldots\}$.

29. a. 7.   **b.** 8.   **c.** 11.   **d.** 4.

30. a. 3.   **b.** 3.   **c.** 7.   **d.** Undefined.

31. a. 27.   **b.** 181.   **c.** (-4).   **d.** (-9).

32. a. $(-7) + 5 = (-2)$.

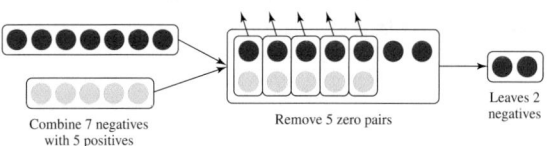

Combine 7 negatives with 5 positives — Remove 5 zero pairs — Leaves 2 negatives

b. $3 - (-4) = 7$.

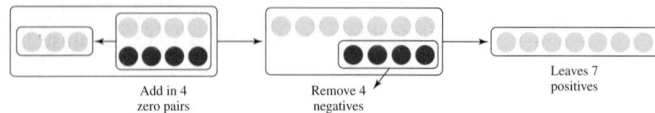

Begin with 3 positive — Add in 4 zero pairs — Remove 4 negatives — Leaves 7 positives

c. $(-3) \cdot (-2) = 6$.

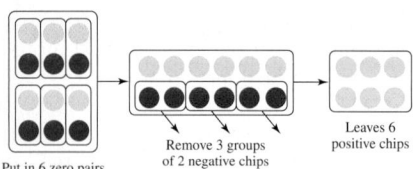

Put in 6 zero pairs — Remove 3 groups of 2 negative chips — Leaves 6 positive chips

33. a. $(-2) + (-4) = (-6)$.

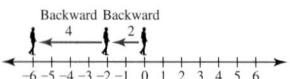

Backward 4   Backward 2
-6 -5 -4 -3 -2 -1 0 1 2 3 4 5 6

b. $(-5) - 2 = (-7)$.

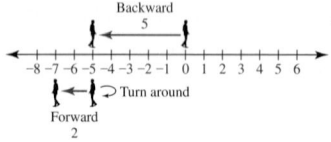

Backward 5
-8 -7 -6 -5 -4 -3 -2 -1 0 1 2 3 4 5 6
Turn around
Forward 2

c. $4 \cdot (-2) = (-8)$.

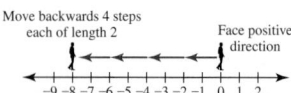

Move backwards 4 steps each of length 2 — Face positive direction
-9 -8 -7 -6 -5 -4 -3 -2 -1 0 1 2

34. a. $49 \div (-7) = x$ corresponds to the product $49 = (-7) \cdot x$. Because $(-7) \cdot (-7) = 49$, it must be that $49 \div (-7) = (-7)$.   **b.** $(-18) \div (-7) = x$ corresponds to the product $(-18) = (-7) \cdot x$. Because there is no integer x such that $(-7) \cdot x = (-18)$, then $(-18) \div (-7)$ is not defined for the integers.   **c.** $(-84) \div 12 = x$ corresponds to the product $(-84) = 12 \cdot x$. Because $12 \cdot (-7) = (-84)$, it must be that $(-84) \div 12 = (-7)$.

35. a. 456.   **b.** (-369).   **c.** (-912).   **d.** $(-3,600)$.   **e.** (-39).   **f.** 81.

CHAPTER 7

■ Section 7.1 Learning Assessment

1. The fractions are a broad set containing any number that can be written in the form $\frac{a}{b}$, where $b \neq 0$. a and b can be whole numbers, integers, decimals, radicals, and even other fractions. The set of rational numbers is a subset of the fractions for which a and b are limited to the integers.

2. Every integer is a rational number because any integer a can be rewritten in the form of a rational number as $\frac{a}{1}$.

3. The numerator gives the number of pieces that are being used or counted. The denominator tells how many equivalent pieces the whole has been divided into or how many equivalent pieces are needed to make a whole.

4. a. In the area model, the whole is represented by a shape with an area that can be divided into equivalent subregions. We then shade the appropriate number of subregions to show the fraction.   **b.** In the length model, the whole is represented by a length that can be divided into equivalent sublengths. We then shade the appropriate number of sublengths to show the fraction.   **c.** On a number line, we select a length to represent the whole, which may or may not be equivalent to the unit. We then mark a length equivalent to the fraction by placing a point on the number line.   **d.** In the set model, the denominator represents the number of elements in the set, and the numerator represents the number of elements in a subset.

5. Improper fractions and mixed numbers are used to represent values that are either larger than 1 or smaller than -1. Improper fractions do so in a fraction form, whereas mixed numbers do so by using an integer part and a fractional part.

6. Two fractions are equivalent if they represent the same numerical value.

7. a. Use the Fundamental Law of Fractions to replace the fraction $\frac{an}{bn}$ with $\frac{a}{b}$ until there are no common factors between a and b.   **b.** Use the Fundamental Law of Fractions to exchange the fractions $\frac{a}{b}$ and $\frac{c}{d}$ for fractions with the same denominator, which is a common multiple of b and d. Therefore, the fractions are described in terms of the same equivalent parts of the whole.

8. a. As the numerator grows larger, more pieces are being counted. Therefore, as the numerator gets larger, the value of the fraction must get larger.   **b.** As the denominator grows larger, the relative size of the pieces must get smaller. Consequently, as the denominator gets larger, the value of the fraction gets smaller.

9. Using a cross product to compare two fractions is equivalent to comparing the numerators of the two fractions after they have been made to have a common denominator.

10. The rational numbers are dense because given two rational numbers, it is always possible to find another rational number between them.

11. a. $\frac{1}{4}$ represents part of a whole that has been divided into four equivalent pieces of which one is counted.
b. $\frac{3}{6}$ represents part of a whole that has been divided into six equivalent pieces of which three are counted.
c. $\frac{5}{8}$ represents part of a whole that has been divided into eight equivalent pieces of which five are counted.
d. $\frac{9}{5}$ represents part of a whole that has been divided into five equivalent pieces of which nine are counted.

12. a. $\frac{6}{9} = \frac{2}{3}$. **b.** $\frac{6}{12} = \frac{1}{2}$. **c.** $\frac{4}{9}$. **d.** $\frac{8}{15}$.

13. a. 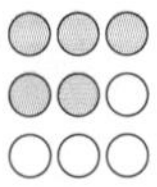 **b.** 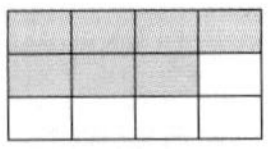

c. 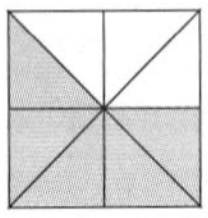 **d.** 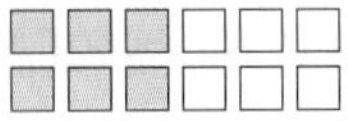

14. a. $\frac{3}{7}$. **b.** $\frac{4}{6} = \frac{2}{3}$. **c.** $\frac{1}{4}$. **d.** $\frac{1}{3}$.

15. a. 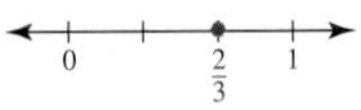
0
2/3
1
b. 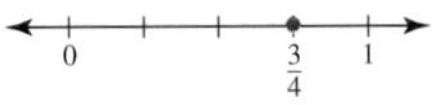
0
3/4
1

c. 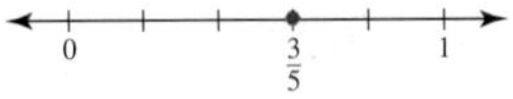
0
3/5
1
d. 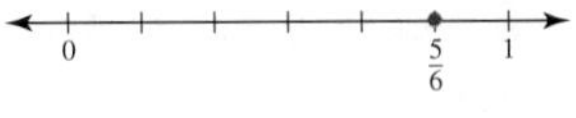
0
5/6
1

16. a. 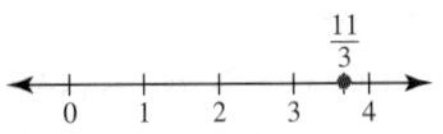
11/3
0 1 2 3 4
b. 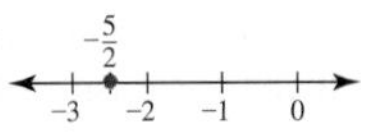
-5/2
-3 -2 -1 0

c. 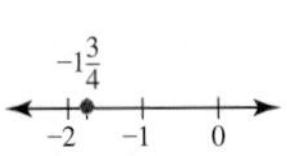
-1 3/4
-2 -1 0
d. 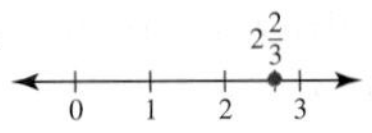
2 2/3
0 1 2 3

17. No, it does not. Its pieces are not the same size.

18. The figure illustrates equivalent fractions, so it demonstrates the Fundamental Law of fractions.

19. Yes, they both represent $\frac{1}{2}$. The value of a fraction is independent of the shape and size of the whole.

20. The smaller circle represents a larger value at $\frac{1}{2}$. Again, the value of a fraction is independent of the shape and size of the whole

21. Part b represents the larger amount. Even though part a shows a longer line, the scale on part b indicates a larger number.

22. The trapezoid represents $\frac{1}{2}$, the rhombus $\frac{1}{3}$, and the triangle $\frac{1}{6}$.

23. a. $\frac{1}{4}$. **b.** $\frac{1}{3}$. **c.** $\frac{1}{6}$. **d.** $\frac{1}{2}$. **e.** $\frac{1}{4}$. **f.** $\frac{1}{5}$.

24. a. $\frac{3}{12} = \frac{1}{4}$. **b.** $\frac{2}{12} = \frac{1}{6}$. **c.** $\frac{4}{12} = \frac{1}{3}$. **d.** $\frac{6}{12} = \frac{1}{2}$.

25. a. $\frac{16}{28} = \frac{4}{7}$. **b.** $\frac{9}{28}$. **c.** $\frac{3}{28}$. **d.** $\frac{3}{28}$. **e.** $\frac{9}{12} = \frac{3}{4}$.

26. a. $6\frac{1}{2}$. **b.** $5\frac{3}{8}$. **c.** $\frac{15}{8}$. **d.** $\frac{37}{7}$.

27. a. $-4\frac{7}{12}$. **b.** $4\frac{23}{28}$. **c.** $\frac{169}{22}$. **d.** $-\frac{163}{19}$.

28. $\left\{\ldots, \frac{-3}{-12}, \frac{-2}{-8}, \frac{-1}{-4}, \frac{1}{4}, \frac{2}{8}, \frac{3}{12}, \ldots\right\}$.

29. $\left\{\ldots, \frac{-6}{3}, \frac{-4}{2}, \frac{-2}{1}, \frac{2}{-1}, \frac{4}{-2}, \frac{6}{-3}, \ldots\right\}$.

30. a. $\frac{2}{5}$. **b.** $\frac{3}{4}$. **c.** $-\frac{9}{5}$. **d.** $\frac{15}{7}$.

31. a. $\frac{24}{7}$. **b.** $-\frac{7}{8}$. **c.** $\frac{10}{7}$. **d.** $-\frac{13}{25}$.

32. a. $\frac{79}{203}$. **b.** $-\frac{1{,}171}{1{,}557}$. **c.** $-\frac{5}{16}$. **d.** $\frac{299}{814}$.

33. a. 4. **b.** 7. **c.** 0.

34. a. $\frac{6}{7} > \frac{11}{14}$. **b.** $-\frac{24}{35} > -\frac{21}{30}$. **c.** $\frac{18}{5} > \frac{20}{7}$.

35. a. $\frac{4}{5} = \frac{20}{25}$. **b.** $-\frac{16}{23} > -\frac{32}{45}$. **c.** $-\frac{93}{39} = -\frac{124}{52}$.

36. a. $\frac{1}{5}, \frac{2}{9}, \frac{3}{10}, \frac{1}{3}, \frac{2}{5}$, and $\frac{3}{7}$. **b.** $-\frac{5}{6}, -\frac{4}{5}, -\frac{2}{3}, -\frac{5}{8}, -\frac{4}{7}$, and $-\frac{1}{2}$.

37. a. Five fractions between $\frac{1}{5}$ and $\frac{2}{5}$ are $\frac{11}{50}, \frac{12}{50}, \frac{13}{50}, \frac{14}{50}$, and $\frac{15}{50}$. **b.** Five fractions between $-\frac{2}{7}$ and $-\frac{4}{7}$ are $-\frac{21}{70}, -\frac{22}{70}, -\frac{23}{70}, -\frac{24}{70}$, and $-\frac{25}{70}$. **c.** Five fractions between $\frac{11}{4}$ and $3 = \frac{12}{4}$ are $\frac{111}{40}, \frac{112}{40}, \frac{113}{40}, \frac{114}{40}$, and $\frac{115}{40}$.

38. It is possible, as follows:

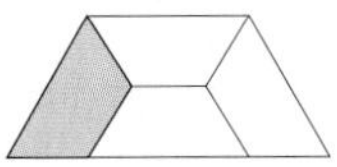

39. a. 4. **b.** 10. **c.** 15. **d.** 35.

40. a. 35. **b.** 36.

41. Four different ways:

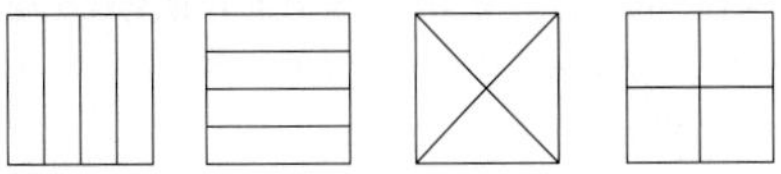

42. A is at $\frac{12}{9} = \frac{4}{3}$. B is at $\frac{28}{9}$.

43. The 15 fractions are $\frac{17}{48}, \frac{18}{48}, \frac{19}{48}, \ldots, \frac{30}{48}$ and $\frac{31}{48}$.

44. $\frac{15}{24}$.

45. a. 4.   **b.** 0.   **c.** 1.   **d.** $x \in \{-13, -14, -15, -16, \ldots\}$.

46. $\dfrac{5}{9}$ of the student body or 6,860 are female.

47. $\dfrac{2,500}{6,500} = \dfrac{5}{13}$ of the movies are drama.

48. Her second class did better.

49. Professor Cabbot's class did better.

50. The first machine has the lower fractional portion of defective units.

51. a. $\dfrac{7}{11}$.   **b.** $\dfrac{11_{\text{five}}}{13_{\text{five}}}, \dfrac{14_{\text{five}}}{22_{\text{five}}},$ and $\dfrac{22_{\text{five}}}{31_{\text{five}}}$.   **c.** $\dfrac{3_{\text{five}}}{4_{\text{five}}}$.

52. a. True.   **b.** True.   **c.** False. The definition allows for the numerator to be greater than the denominator, as is the case in many improper fractions.   **d.** False. $\dfrac{1}{1}$ is an improper fraction that is equal to 1.   **e.** False. The density property guarantees that there are an infinite number of fractions between any two rational numbers, including 1 and 2.

53. $A\dfrac{b}{c} = \dfrac{Ac + b}{c}$.

54. The values in the sequence $\dfrac{1}{2}, \dfrac{2}{3}, \dfrac{3}{4}, \dfrac{4}{5}, \dfrac{5}{6}, \ldots$ are increasing. We are removing smaller and smaller values from 1, so the sequence must be getting closer to 1.

55. If $\dfrac{a}{b} = \dfrac{c}{b}$, then $a = c$.

56. No. The size of the fraction makes no indication of the size of the set. For instance, set A could have 12 blocks of which 3 are red, and set B could have 32 blocks of which 8 are red. In each set, $\dfrac{1}{4}$ of the blocks are red, but $n(A) \neq n(B)$.

57. No. Again the size of the fraction makes no indication of the size of the set. For instance, set A could have 3 blocks of which 1 is red, and set B could have 32 blocks of which 8 are red.

58. a. No. Consider all fractions of the form $\dfrac{1}{n}$ as n goes to infinity. These fractions will continue to grow smaller but will never reach 0.   **b.** No. Consider all fractions of the form $-\dfrac{1}{n}$ as n goes to infinity. These fractions will continue to grow larger but will never reach 0.   **c.** No. The set of negative rational numbers is infinite.

59. a. If the denominator of a fraction is fixed and the numerator is increased, the value of the fraction will increase because it will be counting more pieces of the same size.   **b.** If the numerator of a fraction is fixed and the denominator is increased, the value of the fraction will decrease because it will be dividing the same whole into more pieces.

60. If $\dfrac{a}{b}$ is a positive nonzero rational number and c is any natural number, then $\dfrac{a}{b} < \dfrac{a+c}{b+c}$ if $\dfrac{a}{b} < 1$, $\dfrac{a}{b} = \dfrac{a+c}{b+c}$ if $\dfrac{a}{b} = 1$, and $\dfrac{a}{b} > \dfrac{a+c}{b+c}$ if $\dfrac{a}{b} > 1$.

61. If $\dfrac{a}{b}$ and $\dfrac{c}{d}$ are rational numbers such that $\dfrac{a}{b} < \dfrac{c}{d}$, then $\dfrac{a}{b} < \dfrac{a+c}{b+d} < \dfrac{c}{d}$. If $\dfrac{a}{b} < \dfrac{c}{d}$, then it must be that $ad < bc$. By adding ab to both sides of the inequality, it follows that $ad + ab < bc + ab$. Factoring, we have $a(d+b) < b(c+a)$. Next, divide both sides by $(d+b)$ and b to get the left side of the inequality: $\dfrac{a}{b} < \dfrac{a+c}{b+d}$. A similar argument can be used to show the right side of the inequality.

62. Answers will vary.

63. Answers will vary.

64. It might have been stated that the numbers were relatively prime or that they have no common factor other than 1.

65. The first is the denominator of a fraction, and the second is the numerator of a fraction.

66. It describes simplifying fractions.

67. a. Answers will vary.   **b.** Answers will vary.   **c.** Answers will vary.   **d.** Answers will vary.

68. The student has divided the unit into six equal pieces, not five. The placement of $\dfrac{3}{5}$ is actually at $\dfrac{1}{2}$.

69. a. Answers will vary.   **b.** Answers will vary.

70. Jaleel is not correct. Use the Fundamental Law of Fractions to find any number of fractions between the two. For instance, $\dfrac{7}{20}$ is between $\dfrac{3}{10}$ and $\dfrac{4}{10}$.

71. Answers will vary.

72. The decimals.

73. Answers will vary.

74. Answers will vary.

75. Answers will vary.

■ Section 7.2 Learning Assessment

1. In some senses, adding rational numbers is like adding whole numbers because we can think of it as counting pieces after combining two disjoint sets. Now, however, the sum must not only give the total number of pieces but must also describe the size of the pieces relative to the whole.

2. Rewriting the fractions with a common denominator describes the fractions in terms of the same part of the whole, allowing us to describe the resulting sum or difference.

3. The common denominator bd is not necessarily the least common denominator. If b and d have a common factor, then bd is not the least common multiple of b and d.

4. The equation $-\dfrac{a}{b} = \dfrac{-a}{b} = \dfrac{a}{-b}$ states that the inverse of $\dfrac{a}{b}$ is equal to the fraction with the inverse of a in the numerator and b in the denominator or to the fraction with a in the numerator and the inverse of b in the denominator. The equation $\dfrac{a}{b} = \dfrac{-a}{-b} = -\left(-\dfrac{a}{b}\right)$ states that $\dfrac{a}{b}$ is equal to the fraction with the inverse of a in the numerator and the inverse of b in the denominator or to the inverse of the inverse of $\frac{a}{b}$. The equation $-\left(\frac{a}{b} + \frac{c}{d}\right) = \frac{-a}{b} + \frac{-c}{d}$ states that the inverse of the sum of two rational numbers is equal to the sum of the inverses.

5. To use the missing-addend approach, compute a difference by rewriting it as a sum in which one of the addends is missing. Then use your knowledge of addition to find the missing number.

6. **a.** If the denominators are the same, then the pieces are all of the same size, and we remove the appropriate number to compute the difference.   **b.** If the denominators are different, first get a common denominator and then compute the difference by subtracting the numerators and keeping the denominator.

7. One way is to work with the numbers in mixed number form, in which case the commutative and associative properties allow us to add or subtract the integer and fractional parts separately. The other way is to convert the mixed numbers to improper fractions, compute the operation, and then convert the sum or difference back to a mixed number.

8. We can make estimates by rounding the fractional part to a benchmark number. If the numerator is much less than the denominator, round the fraction down to 0. If the numerator is about half the denominator, round to $\frac{1}{2}$. If the numerator and the denominator are about the same, round to 1.

9. The area model is shown for each problem.

a. $\frac{2}{8} + \frac{5}{8} = \frac{7}{8}$.

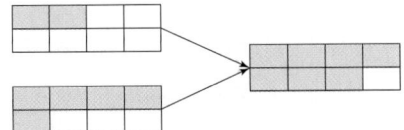

b. $\frac{2}{3} + \frac{1}{6} = \frac{4}{6} + \frac{1}{6} = \frac{5}{6}$.

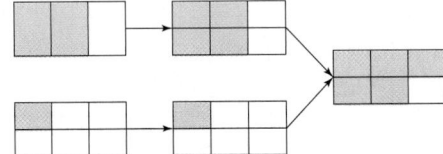

c. $\frac{2}{3} - \frac{2}{4} = \frac{8}{12} - \frac{6}{12} = \frac{2}{12}$.

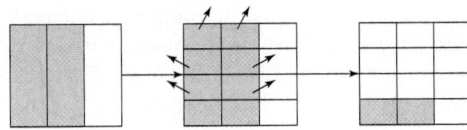

d. $\left(\frac{1}{2} + \frac{1}{5}\right) - \frac{1}{10} = \left(\frac{5}{10} + \frac{2}{10}\right) - \frac{1}{10} = \frac{7}{10} - \frac{1}{10} = \frac{6}{10}$.

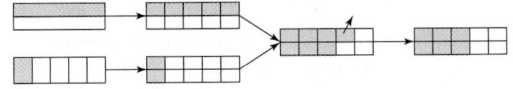

10. **a.** $\frac{3}{6} + \frac{1}{6} = \frac{4}{6}$.

b. $\frac{4}{5} - \frac{2}{5} = \frac{2}{5}$.

11. $\dfrac{adf + bcf + ebd}{bdf}$.

12. $\dfrac{(Ac + b)f - (Df + e)c}{cf}$.

13. **a.** $\frac{13}{11}$.   **b.** $-\frac{8}{6} = -\frac{4}{3}$.   **c.** $-\frac{3}{7}$.   **d.** $-\frac{4}{4} = -1$.

14. **a.** $\frac{11}{8}$.   **b.** $\frac{13}{14}$.   **c.** $-\frac{7}{12}$.   **d.** $-\frac{7}{60}$.

15. **a.** $\frac{41}{70}$.   **b.** $\frac{61}{65}$.   **c.** $\frac{4}{3}$.   **d.** $-\frac{27}{10}$.

16. **a.** $\frac{7}{10} - \frac{3}{10} = x$ corresponds to the sum $\frac{7}{10} = \frac{3}{10} + x$. Because $\frac{3}{10} + \frac{4}{10} = \frac{7}{10}$, it must be that $\frac{7}{10} - \frac{3}{10} = \frac{4}{10}$.

b. $\frac{5}{9} - \frac{7}{9} = x$ corresponds to the sum $\frac{5}{9} = \frac{7}{9} + x$. Because $\frac{7}{9} + \frac{-2}{9} = \frac{5}{9}$, it must be that $\frac{5}{9} - \frac{7}{9} = \frac{-2}{9}$.   **c.** $\frac{3}{4} - \frac{3}{8} = x$ corresponds to the sum $\frac{3}{8} + x = \frac{3}{4}$. Because $\frac{3}{8} + \frac{3}{8} = \frac{3}{4}$, it must be that $\frac{3}{4} - \frac{3}{8} = \frac{3}{8}$.   **d.** $\frac{5}{6} - \frac{3}{4} = x$ corresponds to the sum $\frac{3}{4} + x = \frac{5}{6}$. Because $\frac{3}{4} + \frac{1}{12} = \frac{5}{6}$, it must be that $\frac{5}{6} - \frac{3}{4} = \frac{1}{12}$.

17. **a.** Equal.   **b.** Not equal.   **c.** Equal.   **d.** Not equal.

18. **a.** Equal.   **b.** Not equal.   **c.** Not equal.   **d.** Not equal.

19. **a.** 6.   **b.** $4\frac{3}{5}$.   **c.** $4\frac{19}{21}$.   **d.** $1\frac{5}{6}$.

20. **a.** $10\frac{67}{70}$.   **b.** $2\frac{3}{7}$.   **c.** $13\frac{8}{35}$.   **d.** $3\frac{19}{20}$.

21. **a.** $6\frac{3}{5}$.   **b.** 8.   **c.** $2\frac{1}{2}$.   **d.** $5\frac{3}{5}$.

22. **a.** ≈ 4.   **b.** ≈ 16.   **c.** ≈ 2.   **d.** ≈ 0.

23. **a.** 0.   **b.** 1.   **c.** $\frac{1}{2}$.   **d.** 0.

24. **a.** $\frac{122}{165}$.   **b.** $5\frac{76}{99}$.   **c.** $-\frac{65}{187}$.   **d.** $\frac{3}{16}$.

25. $\frac{3}{2}$.   26. $\frac{1}{2}$ and $\frac{1}{6}$.   27. $\frac{3}{4} - \frac{1}{6} = \frac{7}{12}$.

28. The largest sum is $\frac{6}{2} + \frac{5}{3} = \frac{18}{6} + \frac{10}{6} = \frac{28}{6} = \frac{14}{3}$.

29. The smallest positive difference is $\frac{5}{9} - \frac{4}{8} = \frac{1}{18}$.

30. **a.** 5.   **b.** 3.   **c.** 2.

31. a. $\frac{1}{3} + \frac{1}{4} = \frac{7}{12}$. **b.** $\frac{1}{3} + \frac{1}{3} = \frac{2}{3}$. **c.** $\frac{1}{6} + \frac{1}{9} = \frac{5}{18}$.
d. $\frac{1}{4} + \frac{1}{72} = \frac{19}{72}$.

32. $17\frac{19}{21}$ acres left to plow.

33. 83 ft of fence is needed.

34. $\frac{2}{15}$ are graduate students.

35. $\frac{3}{4}$ of her flower box is left to plant with tulips.

36. She has 2 pieces left.

37. $40\frac{5}{24}$ hr.

38. $3\frac{2}{3}$ ft of the board is left.

39. $\frac{7}{40}$.

40. There are $\frac{6}{15} = \frac{2}{5}$ of the bananas remaining.

41. a. The outer diagonals of the triangle are made of consecutive unit fractions. All other entries are the sum of the two numbers directly below. We can generate subsequent rows by subtracting unit fractions. The next two rows of the triangle are shown:

$$
1
$$
$$
\frac{1}{2} \qquad \frac{1}{2}
$$
$$
\frac{1}{3} \qquad \frac{1}{6} \qquad \frac{1}{3}
$$
$$
\frac{1}{4} \qquad \frac{1}{12} \qquad \frac{1}{12} \qquad \frac{1}{4}
$$
$$
\frac{1}{5} \qquad \frac{1}{20} \qquad \frac{1}{30} \qquad \frac{1}{20} \qquad \frac{1}{5}
$$
$$
\frac{1}{6} \qquad \frac{1}{30} \qquad \frac{1}{60} \qquad \frac{1}{60} \qquad \frac{1}{30} \qquad \frac{1}{6}
$$
$$
\frac{1}{7} \qquad \frac{1}{42} \qquad \frac{1}{105} \qquad \frac{1}{140} \qquad \frac{1}{105} \qquad \frac{1}{42} \qquad \frac{1}{7}
$$

b. $\frac{1}{90}$. **c.** $\frac{1}{72}$.

42. a. $\frac{2}{2} = 1$. **b.** $\frac{2}{3}$. **c.** $\frac{2}{4} = \frac{1}{2}$. **d.** $\frac{2}{n}$.

43. a. False. **b.** Cannot be determined. **c.** True.
d. Cannot be determined.

44. a. True. **b.** True. **c.** Cannot be determined.
d. True.

45. a. True. **b.** True. **c.** False. The equation adds straight across. It does not get a common denominator.
d. False. A number cannot be equal to its opposite unless it is 0. **e.** True. **f.** False. $\frac{a}{b}$ will also be negative.

46. a. $\frac{1}{2} + \frac{1}{3} = \frac{3}{6} + \frac{2}{6} = \frac{3+2}{6} = \frac{5}{6} = \frac{2+3}{6} = \frac{2}{6} + \frac{3}{6}$

$= \frac{1}{3} + \frac{1}{2}$. **b.** $\frac{1}{2} + \left(\frac{1}{3} + \frac{1}{4}\right) = \frac{6}{12} + \left(\frac{4}{12} + \frac{3}{12}\right)$

$= \frac{6}{12} + \frac{7}{12} = \frac{13}{12} = \frac{10}{12} + \frac{3}{12} = \left(\frac{6}{12} + \frac{4}{12}\right) + \frac{3}{12}$

$= \left(\frac{1}{2} + \frac{1}{3}\right) + \frac{1}{4}$. **c.** $\frac{1}{2} + \left(-\frac{1}{2}\right) = \frac{1 + (-1)}{2} = \frac{0}{2} = 0$ and

$\left(-\frac{1}{2}\right) + \frac{1}{2} = \frac{(-1) + 1}{2} = \frac{0}{2} = 0$.

47. a. Because $\frac{1}{2} - \frac{1}{3} = \frac{3}{6} - \frac{2}{6} = \frac{1}{6}$ and $\frac{1}{3} - \frac{1}{2} = \frac{2}{6} - \frac{3}{6} = \frac{-1}{6}$,
then $\frac{1}{2} - \frac{1}{3} \neq \frac{1}{3} - \frac{1}{2}$, and the commutative property does not hold. **b.** Because $\frac{8}{7} - \left(\frac{3}{7} - \frac{2}{7}\right) = \frac{8}{7} - \frac{1}{7} = \frac{6}{7}$ and $\left(\frac{8}{7} - \frac{3}{7}\right) - \frac{2}{7} = \frac{5}{7} - \frac{2}{7} = \frac{3}{7}$, then $\frac{8}{7} - \left(\frac{3}{7} - \frac{2}{7}\right) \neq \left(\frac{8}{7} - \frac{3}{7}\right) - \frac{2}{7}$, and the associative property does not hold.
c. Since $\frac{1}{2} - 0 = \frac{1}{2}$ and $0 - \frac{1}{2} = -\frac{1}{2}$, the additive identity property does not hold.

48. By the Fundamental Law of Fractions, $\frac{a}{b} = \frac{a \cdot (-1)}{b \cdot (-1)} = \frac{-a}{-b}$.

49. If $\frac{a}{b}$ and $\frac{c}{d}$ are two rational numbers, then

$\frac{a}{b} - \frac{c}{d} = \frac{ad}{bd} - \frac{bc}{bd}$ Fundamental Law of Fractions

$\qquad = \frac{ad - bc}{bd}$ Subtraction with Like Denominators

50. Suppose $\frac{a}{b}$ and $\frac{c}{d}$ are two rational numbers and $-\left(\frac{a}{b} + \frac{c}{d}\right)$ is the additive inverse of $\left(\frac{a}{b} + \frac{c}{d}\right)$. Next consider the following sum:

$\left(\frac{a}{b} + \frac{c}{d}\right) + \left(\frac{-a}{b} + \frac{-c}{d}\right)$

$= \left(\frac{a}{b} + \frac{-a}{b}\right) + \left(\frac{c}{d} + \frac{-c}{d}\right)$ Commutative and associative properties

$= \left(\frac{a + (-a)}{b}\right) + \left(\frac{c + (-c)}{d}\right)$ Addition of rational numbers

$= \frac{0}{b} + \frac{0}{d}$ Additive inverse property

$= 0 + 0$ Zero as a fraction

$= 0$ Whole-number addition

This indicates that $\left(\frac{-a}{b} + \frac{-c}{d}\right)$ is another inverse of $\left(\frac{a}{b} + \frac{c}{d}\right)$. However, because the additive inverse is unique, it must be that $-\left(\frac{a}{b} + \frac{c}{d}\right) = \left(\frac{-a}{b} + \frac{-c}{d}\right)$.

51. Answers will vary.
52. Answers will vary.
53. Answers will vary.
54. a. Answers will vary. **b.** Answers will vary.
c. Answers will vary.
55. Josiah has added straight across by adding the numerators and the denominators. His picture is incorrect because the two fractions are not relative to the same whole. This means that the pieces cannot be combined in a way that makes sense.
56. She subtracted the integer parts correctly but not the fractional parts. She subtracted the numerators and the denominators. Answers for correcting her mistake will vary.
57. Answers will vary.
58. Use it to find the lowest common denominator when adding or subtracting two fractions.

59. The denominators will always be a power of 10.
60. Answers will vary.
61. Answers will vary.

Section 7.3 Learning Assessment

1. The repeated addition has meaning only with rational number products if the first factor is a whole number. If the first factor is a fraction, then the meaning of adding a number to itself a fractional number of times has little meaning.

2. When multiplying two fractions, we are further subdividing the whole into smaller pieces, making it necessary to multiply the denominators to correctly describe the size of the new pieces relative to the whole. Also, when multiplying two fractions, we need to multiply the numerators of the fractions to count the number of pieces being used in the product.

3. The commutative and associative properties for integer multiplication allow us to rearrange the factors in the numerator and denominator of the product so that they will simplify appropriately. Because we know in advance that this will happen, we can choose to simplify the common factors prior to computing the product.

4. The Fundamental Law of Fractions is a direct consequence of the multiplicative identity property because $\dfrac{n}{n}$ is equivalent to one.

5. The only rational number that does not have an inverse is zero because $\dfrac{1}{0}$ is undefined.

6. a. Rational number division will never have remainders because we can always use fractions to describe any partial amounts in the quotient. **b.** The rational numbers are closed under division as long as division by zero is excluded.

7. The repeated-subtraction approach leads to the fact that if $\dfrac{a}{b}$ and $\dfrac{c}{b}$ are rational numbers with $c \neq 0$, then $\dfrac{a}{b} \div \dfrac{c}{b} = \dfrac{a}{c}$.
Next, if $\dfrac{a}{b}$ and $\dfrac{c}{d}$ are rational numbers with $\dfrac{c}{d} \neq 0$, then

$$\dfrac{a}{b} \div \dfrac{c}{d} = \dfrac{ad}{bd} \div \dfrac{bc}{bd} \quad \text{Find a common denominator.}$$

$$= \dfrac{ad}{bc} \quad \begin{array}{l}\text{Division of fractions with a common} \\ \text{denominator.}\end{array}$$

$$= \dfrac{a}{b} \cdot \dfrac{d}{c} \quad \begin{array}{l}\text{Definition of rational number} \\ \text{multiplication.}\end{array}$$

8. If a and b are two integers with $b \neq 0$, then
$$a \div b = \dfrac{a}{1} \div \dfrac{b}{1} = \dfrac{a}{1} \cdot \dfrac{1}{b} = \dfrac{a}{b}.$$

9. If a is a nonzero rational number, then we can establish the following pattern:
$$a^3 = a \cdot a \cdot a$$
$$a^2 = a \cdot a$$
$$a^1 = a$$
$$a^0 = 1$$

The pattern indicates that, as the exponent decreases by one, we divide the previous number by a. Continuing this pattern onto the negative exponents:

$$a^{-1} = 1 \div a = \dfrac{1}{a} = \dfrac{1}{a^1}$$

$$a^{-2} = \dfrac{1}{a} \div a = \dfrac{1}{a} \cdot \dfrac{1}{a} = \dfrac{1}{a^2}$$

$$a^{-3} = \dfrac{1}{a^2} \div a = \dfrac{1}{a^2} \cdot \dfrac{1}{a} = \dfrac{1}{a^3}$$

If we compare the exponential expression on the left to the fractional expression on the right, we see that $a^{-n} = \dfrac{1}{a^n}$ for any positive integer n.

10. a. Take one-half of one-fourth. **b.** Take two-fourths of one-fifth. **c.** Take three-sevenths of seven-sixths. **d.** Take four-fifths of four.

11. a. $\dfrac{1}{3} \cdot \dfrac{1}{5}$. **b.** $\dfrac{3}{4} \cdot \dfrac{7}{2}$. **c.** $\dfrac{2}{3} \cdot 6$. **d.** $\dfrac{9}{7} \cdot \dfrac{10}{8}$.

12. a. $3 \cdot \dfrac{1}{6} = \dfrac{3}{6} = \dfrac{1}{2}$.

$$\boxed{\tfrac{1}{6}}\ \boxed{\tfrac{1}{6}}\ \boxed{\tfrac{1}{6}} \quad \boxed{\tfrac{1}{6}}\ \boxed{\tfrac{1}{6}}\ \boxed{\tfrac{1}{6}}$$
$$3 \cdot \tfrac{1}{6} \quad = \quad \tfrac{3}{6}$$

b. $\dfrac{1}{3} \cdot \dfrac{1}{4} = \dfrac{1}{12}$.

$\tfrac{1}{3}$ of $\tfrac{1}{4}$ → $\tfrac{1}{12}$

c. $\dfrac{3}{4} \cdot \dfrac{1}{2} = \dfrac{3}{8}$.

$\tfrac{3}{4}$ of $\tfrac{1}{2}$ → $\tfrac{3}{8}$

13. a. $\dfrac{1}{5} \cdot \dfrac{1}{2} = \dfrac{1}{10}$. **b.** $\dfrac{3}{4} \cdot \dfrac{1}{3} = \dfrac{3}{12} = \dfrac{1}{4}$. **c.** $\dfrac{2}{3} \cdot \dfrac{4}{5} = \dfrac{8}{15}$.

14. a. $\dfrac{1}{3} \cdot \dfrac{2}{3} = \dfrac{2}{9}$. **b.** $\dfrac{2}{4} \cdot \dfrac{3}{5} = \dfrac{3}{10}$. **c.** $\dfrac{5}{6} \cdot \dfrac{2}{7} = \dfrac{5}{21}$.

15. a. $\dfrac{6}{8} \div \dfrac{2}{8} = 3$.

$\tfrac{2}{8}$ can be removed from $\tfrac{6}{8}$ 3 times

b. $\dfrac{9}{10} \div \dfrac{3}{10} = 3$.

$\tfrac{3}{10}$ can be removed from $\tfrac{9}{10}$ 3 times

c. $\dfrac{5}{6} \div \dfrac{3}{6} = \dfrac{5}{3}$.

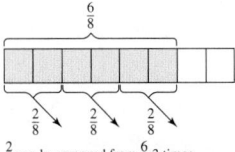

$\tfrac{3}{6}$ can be removed from $\tfrac{5}{6}$ $1\tfrac{2}{3} = \tfrac{5}{3}$ times

16. a. $\dfrac{4}{6}=\dfrac{2}{3}$. **b.** $\dfrac{7}{3}$. **c.** $\dfrac{-18}{8}=\dfrac{-9}{4}$. **d.** $\dfrac{-15}{-9}=\dfrac{5}{3}$.

17. a. $\dfrac{7}{15}$. **b.** 1. **c.** $-\dfrac{8}{25}$. **d.** $\dfrac{4}{5}$.

18. a. $\dfrac{2}{5}$. **b.** $\dfrac{4}{3}$. **c.** $-\dfrac{45}{184}$. **d.** $\dfrac{5}{8}$.

19. a. Commutative property. **b.** Multiplicative inverse property. **c.** Zero multiplication property. **d.** Closure property. **e.** Multiplicative identity property. **f.** Distributive property of multiplication over addition.

20. a. 5. **b.** $-\dfrac{4}{13}$. **c.** $\dfrac{1}{5}$. **d.** $\dfrac{-8}{9}$. **e.** Undefined.

21. a. Because $\dfrac{4}{7}-\dfrac{2}{7}=\dfrac{2}{7}$ and $\dfrac{2}{7}-\dfrac{2}{7}=0$, then $\dfrac{4}{7}\div\dfrac{2}{7}=2$.
b. Because $\dfrac{14}{5}-\dfrac{2}{5}=\dfrac{12}{5},\ \dfrac{12}{5}-\dfrac{2}{5}=\dfrac{10}{5},\ \dfrac{10}{5}-\dfrac{2}{5}=\dfrac{8}{5},$
$\dfrac{8}{5}-\dfrac{2}{5}=\dfrac{6}{5},\ \dfrac{6}{5}-\dfrac{2}{5}=\dfrac{4}{5},\ \dfrac{4}{5}-\dfrac{2}{5}=\dfrac{2}{5},$ and $\dfrac{2}{5}-\dfrac{2}{5}=0,$
then $\dfrac{14}{5}\div\dfrac{2}{5}=7.$ **c.** Because $\dfrac{5}{7}-\dfrac{3}{7}=\dfrac{2}{7}$, then
$\dfrac{5}{7}\div\dfrac{3}{7}=1\dfrac{2}{3}=\dfrac{5}{3}.$

22. a. $\dfrac{9}{10}$. **b.** $-\dfrac{11}{15}$. **c.** 6. **d.** $\dfrac{9}{26}$.

23. a. $\dfrac{1}{3}$. **b.** $\dfrac{48}{91}$. **c.** $\dfrac{10}{9}$. **d.** $-\dfrac{22}{63}$.

24. a. $\dfrac{20}{27}$. **b.** 1. **c.** $\dfrac{95}{84}$. **25. a.** $\dfrac{1}{2}$. **b.** $\dfrac{15}{49}$.

26. a. $\dfrac{1}{6^5}$. **b.** $\left(\dfrac{5}{2}\right)^3$. **c.** 2^3. **d.** 7^{12}. **e.** $\left(\dfrac{1}{2}\right)^4$. **f.** $\dfrac{2^6}{5^{12}}$.

27. a. $\left(\dfrac{4}{3}\right)^4$. **b.** 4^3. **c.** $\dfrac{1}{8^7}$. **d.** $\dfrac{1}{3^4}$. **e.** $\dfrac{1}{4^{15}\cdot7^9}$. **f.** $\left(\dfrac{1}{4}\right)^5$.

28. a. $\left(\dfrac{1}{2}\right)^{-2}$. **b.** $\left(\dfrac{1}{2}\right)^3$. **c.** $\left(\dfrac{1}{2}\right)^5$. **d.** $\left(\dfrac{3}{4}\right)^4$.

29. a. $2\dfrac{22}{25}$. **b.** $1\dfrac{3}{13}$. **c.** 10. **d.** $-2\dfrac{1}{2}$.

30. a. 2. **b.** 38. **c.** 2. **d.** 108.

31. a. $\approx-\dfrac{5}{2}$. **b.** ≈2. **c.** ≈72. **d.** ≈56.

32. a. 0 and $\dfrac{1}{2}$. **b.** $\dfrac{1}{2}$ and 1. **c.** Greater than 1. **d.** $\dfrac{1}{2}$ and 1.

33. a. $-20\dfrac{2}{105}$. **b.** $2\dfrac{19}{42}$. **c.** $2\dfrac{108}{133}$.

34. $\dfrac{1}{100}$.

35. a. The greatest product is $\dfrac{6}{2}\cdot\dfrac{7}{3}=7$, or $\dfrac{7}{2}\cdot\dfrac{6}{3}=7$.
b. The greatest quotient is $\dfrac{6}{2}\div\dfrac{3}{7}=7$, or $\dfrac{6}{3}\div\dfrac{2}{7}=7$.

36. a. The smallest product is $\dfrac{3}{8}\cdot\dfrac{4}{9}=\dfrac{1}{6}$, or $\dfrac{3}{9}\cdot\dfrac{4}{8}=\dfrac{1}{6}$.
b. The smallest quotient is $\dfrac{3}{8}\div\dfrac{9}{4}=\dfrac{1}{6}$, or $\dfrac{3}{9}\div\dfrac{8}{4}=\dfrac{1}{6}$.

37. a. $\left(-\dfrac{3}{4}\right)\cdot\left(-\dfrac{1}{2}\right)=\dfrac{3}{8}$. **b.** $\dfrac{1}{4}\cdot\dfrac{3}{4}=\dfrac{3}{16}$.
c. $\left(-\dfrac{3}{4}\right)\cdot\dfrac{3}{4}=-\dfrac{9}{16}$. **d.** $\left(-\dfrac{1}{2}\right)\cdot\dfrac{1}{4}=-\dfrac{1}{8}$.

38. $6\dfrac{3}{8}$ yd.

39. $6\dfrac{45}{64}$ pieces of rope.

40. $\dfrac{5}{8}$ in.

41. 30 pills.

42. a. $7\dfrac{2}{9}°C$. **b.** $25\dfrac{5}{9}°C$. **c.** $35\dfrac{5}{9}°C$. **d.** $0°C$. **e.** $-25°C$.

43. $48,000.

44. 28 people.

45. Ted originally had $144.

46. $13\dfrac{5}{7}$ batches.

47. It will take $10\dfrac{1}{2}$ gal to make the trip. The driver does not have enough.

48. $55\dfrac{1}{4}$ in. tall.

49. a. The result will be smaller than 5 because we are taking a part of 5 with a value less than one. **b.** The result will be larger than 5 because we are finding the number of times that a number smaller than 1 can be subtracted from 5. Because we can subtract 1 from 5 five times, we must be able to subtract a number smaller than 1 more than five times from 5.

50. a. If both factors are positive proper fractions, then their product will always be less than the original factors.
b. If both factors are positive improper fractions greater than 1, then their product will always be greater than the original two factors.

51. a. True. **b.** False. **c.** False. **d.** False.
52. a. True. **b.** Cannot be determined. **c.** False. **d.** False.

53. The picture does not imply that $2\cdot\dfrac{3}{4}=\dfrac{6}{8}$. The whole is represented by a single rectangle, which has been divided into four equivalent pieces. Because we are counting six such pieces, the picture indicates $2\cdot\dfrac{3}{4}=\dfrac{6}{4}$.

54. We have that $\dfrac{1}{2}\cdot\dfrac{2}{3}=\dfrac{1}{3},\ \dfrac{1}{2}\cdot\dfrac{2}{3}\cdot\dfrac{3}{4}=\dfrac{1}{4}$ and $\dfrac{1}{2}\cdot\dfrac{2}{3}\cdot\dfrac{3}{4}\cdot\dfrac{4}{5}=\dfrac{1}{5}$.
From these examples, we can see that as n gets larger, the product $\dfrac{1}{2}\cdot\dfrac{2}{3}\cdot\dfrac{3}{4}\cdot\dfrac{4}{5}\cdots\dfrac{n-1}{n}$ gets smaller.

55. a. Because $\dfrac{1}{2}\div\dfrac{1}{3}=\dfrac{1}{2}\cdot\dfrac{3}{1}=\dfrac{3}{2}$ and $\dfrac{1}{3}\div\dfrac{1}{2}=\dfrac{1}{3}\cdot\dfrac{2}{1}=\dfrac{2}{3}$,
then $\dfrac{1}{2}\div\dfrac{1}{3}\neq\dfrac{1}{3}\div\dfrac{1}{2}$, and the commutative property does not hold for rational number division.
b. Because $\dfrac{8}{7}\div\left(\dfrac{3}{7}\div\dfrac{2}{7}\right)=\dfrac{8}{7}\div\dfrac{3}{2}=\dfrac{16}{21}$
and $\left(\dfrac{8}{7}\div\dfrac{3}{7}\right)\div\dfrac{2}{7}=\dfrac{8}{3}\div\dfrac{2}{7}=\dfrac{28}{3}$, then
$\dfrac{8}{7}\div\left(\dfrac{3}{7}\div\dfrac{2}{7}\right)\neq\left(\dfrac{8}{7}\div\dfrac{3}{7}\right)\div\dfrac{2}{7}$, and the associative property does not hold for rational number division.

c. Because $\dfrac{1}{2} \div 1 = \dfrac{1}{2}$ and $1 \div \dfrac{1}{2} = 2$, the multiplicative identity property does not hold.

56. Because $\left(\dfrac{1}{2}\right)^2 = \dfrac{1}{4}$ has a reciprocal of 4, and $\left(\dfrac{1}{2}\right)^{-2} = 4$ has a reciprocal of $\dfrac{1}{4}$, the reciprocal of $\left(\dfrac{1}{2}\right)^2$ is larger.

57. $2\dfrac{2}{3} \cdot 3\dfrac{3}{5}$

$= 2\dfrac{2}{3} \cdot \left(3 + \dfrac{3}{5}\right)$ Definition of mixed number

$= 2\dfrac{2}{3} \cdot 3 + 2\dfrac{2}{3} \cdot \dfrac{3}{5}$ Distributive property

$= \left(2 + \dfrac{2}{3}\right) \cdot 3 + \left(2 + \dfrac{2}{3}\right) \cdot \dfrac{3}{5}$ Definition of mixed number

$= 2 \cdot 3 + \dfrac{2}{3} \cdot 3 + 2 \cdot \dfrac{3}{5} + \dfrac{2}{3} \cdot \dfrac{3}{5}$ Distributive property

$= 6 + 2 + \dfrac{6}{5} + \dfrac{2}{5}$ Rational number multiplication

$= 9\dfrac{3}{5}$ Rational number addition

58. If x is any rational number and a is any nonzero rational number, then

$\dfrac{x}{a} \cdot a = \dfrac{x}{a} \cdot \dfrac{a}{1}$ Rewrite a as $\dfrac{a}{1}$

$= \dfrac{x \cdot a}{a \cdot 1}$ Rational number multiplication

$= \dfrac{x \cdot a}{a}$ Multiplicative identity property

$= x$ Fundamental Law of Fractions

59. If $\dfrac{a}{b}$ is a rational number and m is a positive integer, then

$\left(\dfrac{a}{b}\right)^m = \underbrace{\left(\dfrac{a}{b}\right) \cdot \left(\dfrac{a}{b}\right) \cdots \cdots \left(\dfrac{a}{b}\right)}_{m \text{ times}}$ Definition of an exponent

$= \dfrac{\overbrace{a \cdot a \cdots \cdot a}^{m \text{ times}}}{\underbrace{b \cdot b \cdots \cdot b}_{m \text{ times}}}$ Rational number multiplication

$= \dfrac{a^m}{b^m}$ Definition of an exponent

60. If $\dfrac{a}{b}$ is a rational number and m is a positive integer, then

$\left(\dfrac{a}{b}\right)^{-m} = \dfrac{1}{\left(\dfrac{a}{b}\right)^m}$ Definition of a negative exponent

$= \dfrac{1}{\dfrac{a^m}{b^m}}$ Theorem 7.14(5)

$= \dfrac{b^m}{a^m}$ Simplify

$= \left(\dfrac{b}{a}\right)^m$ Theorem 7.14(5)

61. Answers will vary.

62. a. Answers will vary.   **b.** Answers will vary.
c. Answers will vary.
63. Answers will vary.
64. Multiply each side of the equation by the multiplicative inverse of $\dfrac{a}{b}$.

a. $x = \dfrac{5}{4}$.   **b.** $x = -\dfrac{100}{27}$.   **c.** $x = \dfrac{32}{39}$.

65. Mark forgot to take the reciprocal of the second fraction before multiplying.
66. One way is to use the distributive property and multiply the numbers in mixed number form. Another way is to change the mixed numbers into improper fractions, multiply them, and then convert the product back to a mixed number.
67. Answers will vary.
68. Answers will vary.
69. Answers will vary.
70. Answers will vary.
71. Answers will vary.

▌ Chapter 7 Review Exercises

1. $\dfrac{7}{9}$ represents a part of a whole that has been divided into 9 equivalent pieces, of which seven are counted.

2. a.

b.

c.

d.

3. a. $2\dfrac{1}{7}$.   **b.** $-3\dfrac{1}{4}$.   **c.** $5\dfrac{5}{16}$.

4. a. $\dfrac{19}{8}$.   **b.** $\dfrac{-37}{9}$.   **c.** $\dfrac{116}{9}$.

5. $\left\{ \cdots \dfrac{-6}{-15}, \dfrac{-4}{-10}, \dfrac{-2}{-5}, \dfrac{2}{5}, \dfrac{4}{10}, \dfrac{6}{15}, \cdots \right\}$.

6. a. $\dfrac{7}{11}$.   **b.** $\dfrac{7}{9}$.   **c.** $-\dfrac{7}{3}$.

7. $\dfrac{1}{4}, \dfrac{1}{3}, \dfrac{2}{5}, \dfrac{3}{7}, \dfrac{1}{2}, \dfrac{5}{8}$.

8. a. $\dfrac{8}{9} < \dfrac{81}{90}$.   **b.** $-\dfrac{25}{16} > -\dfrac{28}{17}$.   **c.** $\dfrac{49}{56} = \dfrac{98}{112}$.

9. Five fractions are $\dfrac{41}{60}, \dfrac{42}{60}, \dfrac{43}{60}, \dfrac{44}{60}$, and $\dfrac{45}{60}$.

10. 30 counters.

11. a. $\dfrac{5{,}500}{20{,}000} = \dfrac{11}{40}$ of the fish are bass.

b. $\dfrac{20{,}000 - 3{,}500}{20{,}000} = \dfrac{16{,}500}{20{,}000} = \dfrac{33}{40}$ of the fish are not perch.

12. First period did better.

13. $\frac{14}{16}$.

14. a. $\frac{7}{8} + \frac{3}{8} = \frac{10}{8} = \frac{5}{4}$.

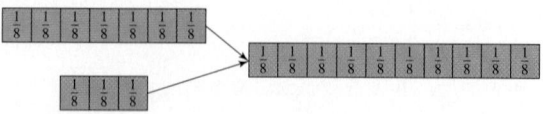

b. $\frac{5}{6} - \frac{1}{4} = \frac{10}{12} - \frac{3}{12} = \frac{7}{12}$.

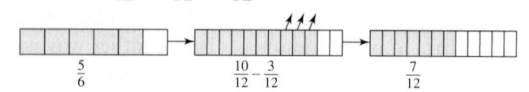

c. $\frac{2}{3} \cdot \frac{3}{6} = \frac{6}{18} = \frac{1}{3}$.

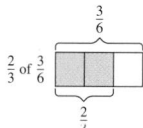

15. a. $\frac{7}{12} - \frac{5}{12} = \frac{2}{12} = \frac{1}{6}$.

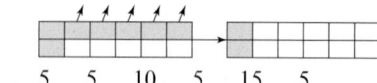

b. $\frac{5}{3} + \frac{5}{6} = \frac{10}{6} + \frac{5}{6} = \frac{15}{6} = \frac{5}{2}$.

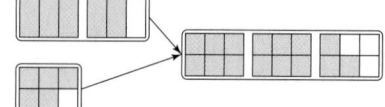

c. $\frac{4}{7} \cdot \frac{3}{10} = \frac{12}{70} = \frac{6}{35}$.

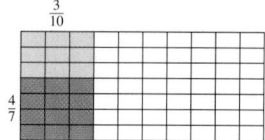

16. a. $\frac{16}{15}$.   **b.** $-\frac{49}{48}$.   **c.** $\frac{3}{14}$.   **d.** $\frac{13}{15}$.   **e.** $-\frac{10}{21}$.   **f.** $\frac{3}{5}$.

g. $\frac{40}{21}$.   **h.** $\frac{6}{25}$.

17. a. $9\frac{7}{9}$.   **b.** $-1\frac{11}{42}$.   **c.** $15\frac{13}{16}$.   **d.** $-1\frac{5}{7}$.

18. a. 10.   **b.** $\frac{1}{6}$.   **c.** $\frac{2}{3}$.   **d.** 68.

19. a. ≈ 3.   **b.** ≈ 24.   **c.** $\approx 13\frac{3}{4}$.   **d.** ≈ 15.

20. a. $\frac{9}{2}$.   **b.** $-\frac{4}{3}$.   **c.** $-\frac{1}{6}$.   **d.** $-\frac{7}{24}$.   **e.** Does not exist.

21. a. $2\frac{13}{28}$.   **b.** $6\frac{1}{21}$.   **c.** $-\frac{325}{484}$.

22. a. $\left(-\frac{1}{3}\right)^4$.   **b.** $\left(-\frac{3}{2}\right)^2$.   **c.** $\left(\frac{1}{3 \cdot 4}\right)^4 = \left(\frac{1}{12}\right)^4$.

d. $\frac{1}{6^5}$.   **e.** $\frac{1}{3^{11}}$.   **f.** $\left(\frac{1}{4}\right)^3$.   **g.** $\frac{1}{8^3}$.

23. It cannot be determined because we do not know which set has more blocks.

24. a. Closure property.   **b.** Distributive property of multiplication over addition.   **c.** Commutative property.   **d.** Additive inverse property.   **e.** Multiplicative identity property.   **f.** Zero multiplication property.

25. $\left(\frac{a}{b} + \frac{c}{d}\right) - \frac{e}{f} = \frac{ad + bc}{bd} - \frac{e}{f} = \frac{adf + bcf - bde}{bdf}$.

26. $\frac{13}{40} = \frac{1}{8} + \frac{1}{5}$.

27. $36\frac{1}{4}$ A.

28. $247\frac{7}{12}$ ft of fence.

29. $\frac{7}{20}$ are seniors.

30. $2\frac{27}{40}$ ft remain.

31. The factors will be larger than the product.

32. $\frac{1}{64}$.

33. $11\frac{1}{9}$ yd.

34. $12\frac{3}{5}$ can be cut.

35. a. The greatest product is $\frac{8}{3} \cdot \frac{7}{2} = \frac{28}{3}$ or $\frac{7}{3} \cdot \frac{8}{2} = \frac{28}{3}$.

   b. The greatest quotient is $\frac{8}{3} \div \frac{2}{7} = \frac{28}{3}$ or $\frac{7}{3} \div \frac{2}{8} = \frac{28}{3}$.

36. 120 did not go to preschool at all.

37. $46\frac{2}{3}$ people can be fed.

CHAPTER 8

■ Section 8.1 Learning Assessment

1. Decimals are consistent with the grouping aspect of the decimal system because we can regroup ten of any decimal as one in the next higher place value.

2. The decimal point serves as a reference point that separates the wholes from the parts of the whole.

3. A terminating decimal has a finite number of decimal places to the right of the decimal point. A repeating decimal has a block of digits that continuously repeats to the right of the decimal point. An irrational number is a decimal that does not terminate and does not repeat.

4. Yes. A terminating decimal can be written with an infinite string of zeros to the right of the last nonzero digit, giving it the form of a repeating decimal.

5. $\frac{22}{7}$ is not equal to π because $\frac{22}{7}$ is an element of the rational numbers, π is an element of the irrational numbers, and the two sets have no elements in common.

6. Because the square roots of perfect squares lead to the natural numbers, not all square roots are irrational numbers.

7. First, line up the decimal point and then compare digits after the decimal point from left to right until a place value is found in which the decimals differ. The larger digit in this place value represents the larger number.

8. Given any two real numbers, there is always another real number in between them.

9. The infinite nature of many decimals can make decimal computations awkward. We round or truncate the decimals to make them easier to compute with.

10. **a.** 2 is in the tenths place value. The name is "two tenths." **b.** 3 is in the tenths place value, and 4 is in the hundredths. The name is "thirty-four hundredths." **c.** The 5 is in the ones place value, and the 8 is in the hundredths. The name is "five and eight hundredths." **d.** The 4 is in the tens place value, the 2 in the ones, the 1 in the hundredths, and the 3 in the thousandths. The name is "forty-two and thirteen thousandths."

11. **a.** 0.9. **b.** 0.11. **c.** 4.1. **d.** −7.051.

12. **a.** $\dfrac{5}{10}$. **b.** $\dfrac{63}{100}$. **c.** $\dfrac{157}{1{,}000}$. **d.** $\dfrac{7}{100}$.

13.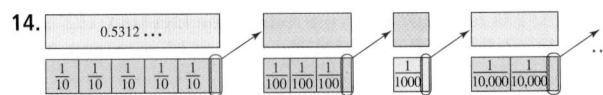

14.

15. **a.** 0.6. **b.** 0.8. **c.** 0.08. **d.** 1.11.

16. **a.** 1.18. **b.** 0.06. **c.** 0.056. **d.** 1.407.

17.

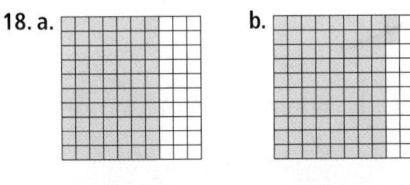

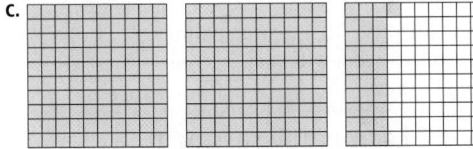

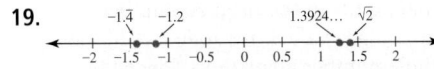

18. **a.** **b.**

c.

19.

20. **a.** $\sqrt{13}$ is the number that, when squared, is 13. The radicand is 13, and the index is 2.
b. $\sqrt[3]{15}$ is the number that, when cubed, is 15. The radicand is 15, and the index is 3.
c. $\sqrt[4]{58}$ is the number that, when raised to the fourth power, is 58. The radicand is 58, and the index is 4.
d. $\sqrt[3]{-1.69}$ is the number that, when cubed, is −1.69. The radicand is −1.69, and the index is 3.

21.

22. **a.** Hundredths. **b.** Hundred-thousandths. **c.** Tenths. **d.** Ones. **e.** Ten-millionths.

23. **a.** $\left(4 \cdot \dfrac{1}{10}\right) + \left(7 \cdot \dfrac{1}{100}\right)$.
b. $(1 \cdot 1) + \left(2 \cdot \dfrac{1}{100}\right)$.
c. $(2 \cdot 100) + (5 \cdot 1) + \left(6 \cdot \dfrac{1}{1{,}000}\right)$.
d. $(4 \cdot 1) + \left(1 \cdot \dfrac{1}{10}\right) + \left(8 \cdot \dfrac{1}{100}\right) + \left(9 \cdot \dfrac{1}{10{,}000}\right)$.

24. **a.** $(5 \cdot 10^{-1}) + (9 \cdot 10^{-2}) + (5 \cdot 10^{-3})$.
b. $(2 \cdot 10^{0}) + (1 \cdot 10^{-1}) + (3 \cdot 10^{-2})$.
c. $(3 \cdot 10^{2}) + (1 \cdot 10^{-1}) + (2 \cdot 10^{-2}) + (7 \cdot 10^{-3})$.
d. $(4 \cdot 10^{3}) + (1 \cdot 10^{0}) + (1 \cdot 10^{-3}) + (4 \cdot 10^{-4})$.

25. **a.** 0.067. **b.** 560.3081. **c.** 406.38. **d.** 75,005.011.

26. **a.** Terminating; 0.4. **b.** Not terminating. **c.** Terminating; 0.425. **d.** Not terminating. **e.** Terminating; −0.8.

27. **a.** Repetend of 7 and a period of 1. **b.** Repetend of 14 and a period of 2. **c.** Repetend of 78 and a period of 2. **d.** Repetend of 21 and a period of 2.

28. Possible answers are: **a.** 0.333.... **b.** 0.121212.... **c.** 0.123123123.... **d.** 0.1234.

29. One possible answer is 0.10100100010000....

30. Possible answers are $\sqrt{2}, \sqrt{3}, \sqrt{5}, \sqrt{6}, \sqrt{7}, \sqrt{8}, \sqrt{10}, \sqrt{11}, \sqrt{12},$ and $\sqrt{13}$.

31. **a.** Irrational. **b.** Not possible. **c.** Rational. **d.** Irrational.

32. **a.** Rational. **b.** Rational. **c.** Irrational. **d.** Not possible.

33. ≈2.23.

34. ≈2.82.

35.

	$\mathbb{N}$	W	$\mathbb{Z}$	$\mathbb{Q}$	Irrational	$\mathbb{R}$
0.83				X		X
$\sqrt{25}$	X	X	X	X		X
$\sqrt[3]{6}$					X	X
$1.\overline{4}$				X		X
$\dfrac{9}{7}$				X		X
0.1234 ...					X	X

36. **a.** 0.16. **b.** −0.45. **c.** $0.7\overline{8}$.

37. $4.6\overline{1}, 4.616, 4.\overline{61}, 4.\overline{616}, 4.61\overline{6}$.

38. $-2.3\overline{5}, -2.\overline{35}, -2.3\overline{53}, -2.3\overline{53}, -2.35$.

39. Five numbers between 4.1 and 4.2 are 4.11, 4.12, 4.13, 4.14, and 4.15.

40. Five numbers between $4.\overline{1}$ and $4.\overline{12}$ are 4.113, 4.114, 4.115, 4.116, and 4.117.

41. a. 34. **b.** 100.00. **c.** 43.9. **d.** 2.0191.
42. a. 3. **b.** 312.98. **c.** 67.3. **d.** 1.800.
43. a. 4 ways. **b.** 7 ways. **c.** 12 ways. **d.** 16 ways.
44. $2^{50} \cdot 5^{50}$.
45. a. $\mathbb{R}$. **b.** $\mathbb{Q}$. **c.** $\{x \mid x \text{ is irrational}\}$. **d.** W.
46. $\sqrt{5}$ in.   **47.** $\sqrt{2}$ in.
48. a. $\sqrt{25} = 5$. **b.** $\sqrt{51}$. **c.** $\sqrt{n+1}$.
49. Three possible triangles are one with sides of length 3, 4 and 5; one with lengths of 6, 8, and 10; and one with lengths of 5, 12, and 13.
50. $\sqrt[3]{7} \approx 1.91$. Here is the process:

$1 < \sqrt[3]{7} < 2$	because $1^3 = 1 < 7 < 8 = 2^3$.
$1.5 < \sqrt[3]{7} < 2$	because $(1.5)^3 = 3.375 < 7 < 8 = 2^3$.
$1.75 < \sqrt[3]{7} < 2$	because $(1.75)^3 = 5.3594 < 7 < 8 = 2^3$.
$1.875 < \sqrt[3]{7} < 2$	because $(1.875)^3 = 6.5918 < 7 < 8 = 2^3$.
$1.875 < \sqrt[3]{7} < 1.9375$	because $(1.875)^3 = 6.5918 < 7 < 7.2732 = (1.9375)^3$.
$1.90625 < \sqrt[3]{7} < 1.9375$	because $(1.90625)^3 = 6.9269 < 7 < 7.2732 = (1.9375)^3$.
$1.90625 < \sqrt[3]{7} < 1.9141$	because $(1.90625)^3 = 6.9269 < 7 < 7.0128 = (1.9141)^3$.
$1.9102 < \sqrt[3]{7} < 1.9141$	because $(1.9102)^3 = 6.9700 < 7 < 7.0128 = (1.9141)^3$.

51. a. $(1 \cdot 5^{-1}) + (2 \cdot 5^{-2}) = 0.28$. **b.** $(4 \cdot 5^{-1}) + (3 \cdot 5^{-2}) + (1 \cdot 5^{-3}) = 0.928$. **c.** $(4 \cdot 7^{-1}) + (6 \cdot 7^{-2}) \approx 0.6939$. **d.** $(1 \cdot 3^{-1}) + (1 \cdot 3^{-2}) + (2 \cdot 3^{-3}) \approx 0.5185$.
52. Ron, Doug, Caleb, Anthony, Trevor, Steve, Reilly, and Malachi.
53. a. False. Any irrational number is a real number that is not rational. **b.** True. **c.** True. **d.** False. Every integer is a rational number, so an integer cannot be an irrational number.
54. a. $\dfrac{1.4}{2.5}$ is rational because $\dfrac{1.4}{2.5} \cdot \dfrac{10}{10} = \dfrac{14}{25} = \dfrac{56}{100} = 0.56$.

b. $\dfrac{\sqrt{2}}{3}$ is irrational because it cannot be rewritten at the quotient of two integers.
55. a. Because the prime factorization of $40 = 2^3 \cdot 5$, $\dfrac{a}{40}$ will always have a terminating decimal representation.

b. Because the prime factorization of $21 = 3 \cdot 7$, $\dfrac{a}{21}$ will always have a repeating decimal representation.

c. Because the prime factorization of $15 = 3 \cdot 5$, $\dfrac{a}{15}$ may have either a terminating or a repeating decimal representation. For instance, if $a = 3$, then $\dfrac{a}{15}$ is terminating. If $a = 5$, then $\dfrac{a}{15}$ is repeating.
56. $x^2 = 2$ has $-\sqrt{2}$ as a solution because $(-\sqrt{2})^2 = 2$.
57. Suppose that $\sqrt[3]{2}$ is a rational number. By the definition of a rational number, $\sqrt[3]{2} = \dfrac{a}{b}$ where a and b are integers, and $b \neq 0$. Further suppose that $\dfrac{a}{b}$ is reduced. Then,

$$\left(\dfrac{a}{b}\right)^3 = 2 \qquad \text{Cubing both sides}$$

$$\dfrac{a^3}{b^3} = 2 \qquad \text{Properties of exponents}$$

$$a^3 = 2b^3 \qquad \text{Multiply both sides by } b^3$$

Because $a^3 = 2b^3$, then a^3 is even because it is the product of 2 and b^3. This means that a must also be even, so $a = 2k$, for some integer k. As a result, $a^3 = (2k)^3 = 8k^3$. By making a simple substitution, we have $8k^3 = 2b^3$, so $4k^3 = b^3$. Consequently, b^3 is even, indicating that b is even. Because a and b are divisible by 2, this contradicts that $\dfrac{a}{b}$ is reduced. As a result, our original assumption that $\sqrt[3]{2}$ is rational must be wrong. Hence, $\sqrt[3]{2}$ is irrational.
58. Suppose that $3 + \sqrt{2}$ is rational. By the definition of the rational numbers, $3 + \sqrt{2} = \dfrac{a}{b}$ where a and b are integers and $b \neq 0$. If we subtract 3 from both sides, we have $\sqrt{2} = \dfrac{a}{b} - 3$. $\dfrac{a}{b} - 3$ is rational because the rational numbers are closed under subtraction, which implies that $\sqrt{2}$ is rational. This contradicts the fact that $\sqrt{2}$ is irrational. The original assumption must be incorrect, and we conclude that $3 + \sqrt{2}$ is irrational.
59. Answers will vary.
60. Answers will vary.
61. Andy did not take the place values of the digits into account. Answers will vary on how to correct the error.
62. The only time the word "and" is used in naming a decimal number is to represent the placement of the decimal point.
63. Marisol is not correct. Because $\dfrac{31}{20}$ is simplified, the only determining factor is the prime factorization of the denominator. Because the prime factorization contains only 2s and 5s, $\dfrac{31}{20}$ has a terminating decimal representation.
64. Answers will vary.
65. Answers will vary.
66. a. Answers will vary. **b.** Answers will vary. **c.** Answers will vary.
67. Answers will vary.
68. Answers will vary.
69. a. Answers will vary. **b.** Answers will vary. **c.** Answers will vary. **d.** Answers will vary.
70. Answers will vary.

■ Section 8.2 Learning Assessment

1. a. Align the decimal points in the original numbers to ensure that common place values are added. Add the numbers using whole-number algorithms. Place the decimal point in the answer directly beneath the decimal point in the original numbers. **b.** Align the decimal points in the original numbers to ensure that common place values are subtracted. Subtract the numbers using whole-number algorithms. Place the decimal point in the answer directly beneath the decimal point in the original numbers. **c.** Multiply the numbers as whole numbers. Place the decimal point in the product by counting the total number of decimal places in the factors. If there are *n* decimal places in the first factor and *m* in the second, then there are *n* + *m* decimal places in the product.

d. If the divisor is a whole number, place the decimal point in the quotient directly above the decimal point in the dividend. If the divisor is a decimal, then move the decimal points in the divisor and the dividend the same number of places to remove it from the divisor.

2. a. There are no remainders because we can always continue the division using parts of the whole represented by decimals. **b.** The division will conclude if a terminating decimal is obtained, if a repeating is obtained, or if a specified number of decimal places has been reached.

3. When we divide the numerator of a rational number by the denominator, the division algorithm guarantees that there are only a limited number of possible remainders. If at any time throughout the course of the division, the remainder is 0, then the division terminates. However, if 0 never occurs, then eventually one of the other six numbers must reoccur as the remainder, causing the division to repeat.

4. Multiplying a decimal by a positive power of ten has the effect of moving the decimal point to the right in the number. Dividing by a positive power of ten has the effect of moving the decimal to the left.

5. Let *x* equal the number. Multiply both sides of the equation by a power of ten so that the decimal is moved to the right of the first repetend. Next, go back to the original equation and multiply both sides of the equation by a power of ten so that the decimal is moved to the left of the first repetend. Finally, subtract the two equations and solve for *x*. Multiplying by powers of ten allows us to shift the decimal points in the two equations so that the repeating part of the decimals can be subtracted away.

6. Writing numbers in scientific notation allows us to condense the way in which the numbers are written. Many people have trouble with scientific notation because it can be difficult to understand the relative size of the numbers.

7. They are two different representations of the same number.

8. We cannot take even roots of negative numbers, because this is equivalent to saying what number raised to the even power gives us a negative number. No such numbers exist in the real numbers, so we say it is undefined.

9. a. $2.999\ldots$ **b.** 7. **c.** $-7.999\ldots$ **d.** -15. **e.** 4.57.

10. a. $0.3 + 0.16 = 0.46$.

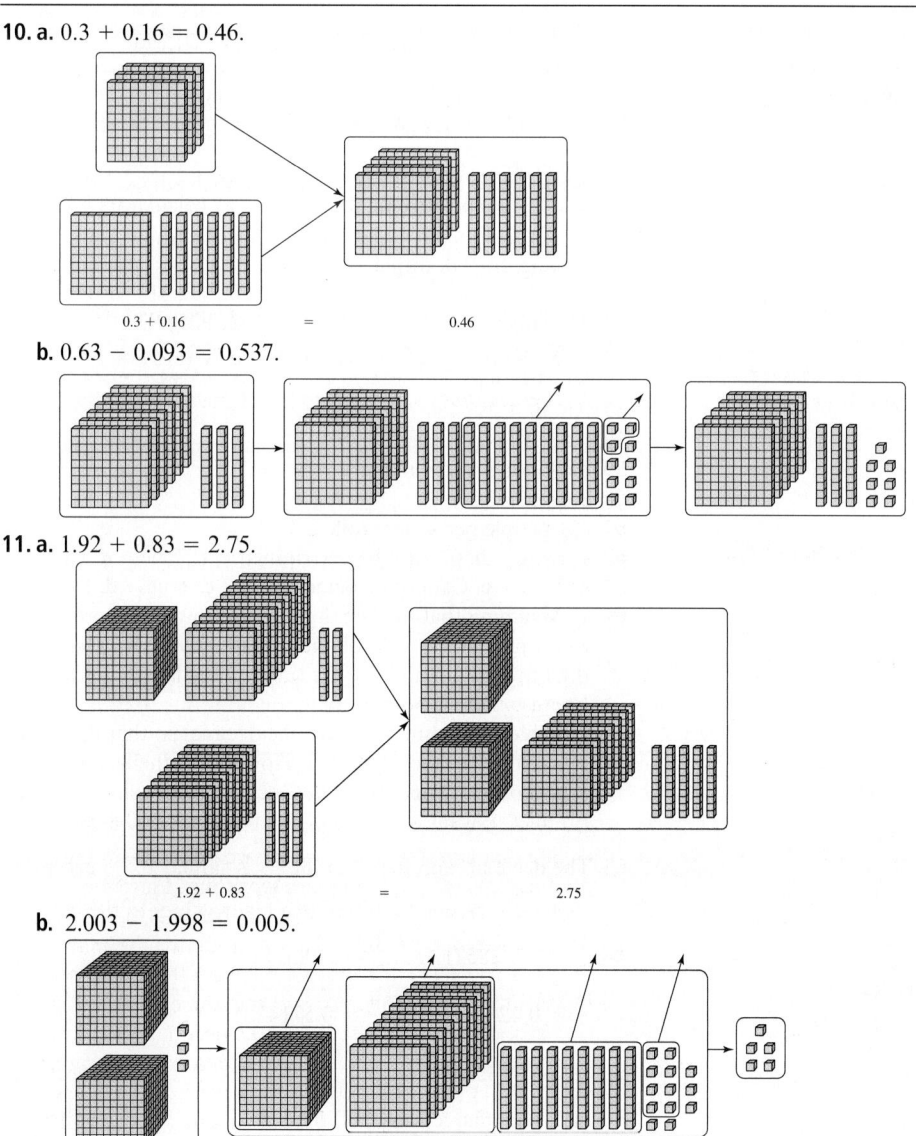

$$0.3 + 0.16 \qquad = \qquad 0.46$$

b. $0.63 - 0.093 = 0.537$.

11. a. $1.92 + 0.83 = 2.75$.

$$1.92 + 0.83 \qquad = \qquad 2.75$$

b. $2.003 - 1.998 = 0.005$.

12. Yes. Reverse the values of the products that were used with the whole numbers. For instance, a block times a block is a block, a block times a flat is a flat, and so on. There will be limitations to the products that can be represented, however.

13. a. $1.3 \cdot 1.5 = 1.95$.

b. $2.1 \cdot 2.7 = 5.67$.

c. $4.5 \cdot 3.4 = 15.3$.

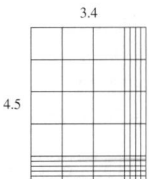

14. a. $1.5 \cdot 2.4 = 3.6$. **b.** $2.7 \cdot 3.4 = 9.18$.

15. a. $1.5 \div 0.3 = 5$ because we can subtract 0.3 from 1.5 five times: $1.5 - 0.3 = 1.2$, $1.2 - 0.3 = 0.9$, $0.9 - 0.3 = 0.6$, $0.6 - 0.3 = 0.3$, and $0.3 - 0.3 = 0$. **b.** $4.9 \div 0.7 = 7$ because we can subtract 0.7 from 4.9 seven times: $4.9 - 0.7 = 4.2$, $4.2 - 0.7 = 3.5$, $3.5 - 0.7 = 2.8$, $2.8 - 0.7 = 2.1$, $2.1 - 0.7 = 1.4$, $1.4 - 0.7 = 0.7$, and $0.7 - 0.7 = 0$. **c.** $25.6 \div 3.2 = 8$ because we can subtract 3.2 from 25.6 eight times: $25.6 - 3.2 = 22.4$, $22.4 - 3.2 = 19.2$, $19.2 - 3.2 = 16$, $16 - 3.2 = 12.8$, $12.8 - 3.2 = 9.6$, $9.6 - 3.2 = 6.4$, $6.4 - 3.2 = 3.2$, and $3.2 - 3.2 = 0$.

16. a. $\sqrt{3}$. **b.** $\sqrt[6]{4}$. **c.** $\sqrt[9]{8}$. **d.** $\sqrt[4]{67}$. **e.** $\sqrt[3]{(-4)^2}$. **f.** $\sqrt[7]{-53^3}$.

17. a. $6^{1/2}$. **b.** $21^{3/2}$. **c.** $14^{1/3}$. **d.** $21^{4/3}$. **e.** $12^{2/5}$. **f.** $19^{6/7}$.

18. a. 6.14. **b.** 1.22.

19. a. 9.51. **b.** 4.363.

20. a. 91.65. **b.** 134.349. **c.** -10.52. **d.** 19.34. **e.** 89.128. **f.** -200.19.

21. a. 26,099.92. **b.** 1,336.11. **c.** 15,986.19. **d.** -12.71.

22. a. $4.1 \cdot 6.1 = \dfrac{41}{10} \cdot \dfrac{61}{10} = \dfrac{2501}{100} = 25.01$.

b. $9.45 \cdot 3.2 = \dfrac{945}{100} \cdot \dfrac{32}{10} = \dfrac{30240}{1000} = 30.24$.

23. a. $5.6 \cdot 3.5 = \dfrac{56}{10} \cdot \dfrac{35}{10} = \dfrac{1960}{100} = 19.6$.

b. $4.53 \cdot 2.7 = \dfrac{453}{100} \cdot \dfrac{27}{10} = \dfrac{12231}{1000} = 12.231$.

24. a. 23.25. **b.** 19.593. **c.** -641.7. **d.** 763.873.

25. a. 47.73. **b.** 1,564.598. **c.** 93,535.0362.

26. a. 17.90. **b.** 74.80. **c.** 4.34.

27. a. 0.625. **b.** 1.75. **c.** $0.58\overline{3}$. **d.** $1.8\overline{3}$.

28. a. 0.25. **b.** 0.75. **c.** $0.5\overline{3}$. **d.** $1.\overline{6}$.

29. a. 123. **b.** 16,901. **c.** 0.459. **d.** 0.03789.

30. a. 0.0492. **b.** 0.000123. **c.** 1,414. **d.** 0.00041.

31. a. $\dfrac{7}{9}$. **b.** $\dfrac{15}{99}$. **c.** $\dfrac{2}{15}$.

32. a. $\dfrac{4,082}{9,900} = \dfrac{2,041}{4,950}$. **b.** $\dfrac{2,123}{9,990}$. **c.** $-\dfrac{3717}{900}$.

33. a. 2.44949. **b.** 4.24264. **c.** Not possible. **d.** 2.75892.

34. a. 3.46410. **b.** 6.76424. **c.** -3.73719. **d.** Not possible.

35. a. $3^{5/6}$. **b.** $4^{1/6}$. **c.** $5^{5/12}$. **d.** 2. **e.** 2.

36. a. 1.45×10^3. **b.** 3.54×10^{-2}. **c.** 9.6901×10^6. **d.** -4.102×10^{-4}.

37. a. 701. **b.** 0.04505. **c.** 80. **d.** -0.000000623.

38. 6 quadrillion $= 6,000,000,000,000,000 = 6 \times 10^{15}$.

39. a. 10 times longer. **b.** 667 times shorter. **c.** About 1.5×10^9 times longer.

40. a. 2.72×10^4. **b.** 2.33×10^4. **c.** 2.81×10^{16}. **d.** 3.07×10^3.

41. a. 2.10×10^5. **b.** 3.10×10^5. **c.** 7.8×10^5. **d.** 4.93×10^1.

42. a.
```
  2 . 3  4  1
+ 3 . 5     9
0 5 0 8 1 3 0 1
  5 . 9  3  1
```
b.
(lattice multiplication diagram)
43.

9.7	3.4	7.9
5.2	7	8.8
6.1	10.6	4.3

44. a. 8. **b.** 4. **c.** 2.

45. a. $\dfrac{1}{29} = 0.\overline{0344827586206896551724137931}$.

b. $\dfrac{1}{31} = 0.\overline{032258064516129032258064516129}$.

46. a. $\dfrac{20}{7}$. **b.** $\dfrac{1,000}{127}$. **c.** $\dfrac{9}{4}$. **d.** $\dfrac{111}{26}$. **e.** $\dfrac{999}{536}$.

47. a. 131.21_{five}. **b.** 201.31_{five}. **c.** 1.202_{five}.

48. The largest positive number that most calculators will display is 9×10^{99}. The smallest positive value that most calculators will display is 1×10^{-99}.

49. \$47.97. **50.** \$92.31.

51. 11 cents. **52.** \$258.93.

53. \$265.63. **54.** \$2,036.34.

55. \$4.27 a share. **56.** \$2.55.

57. About 4,756,468,798 years.

58. 82,452.43 light years.

59. 81.92 times more massive.

60. \$6,363,636.37.

61. 231 people per square mile.

62. a. False. **b.** Cannot be determined. **c.** True. **d.** False.

63. a. True. **b.** Cannot be determined. **c.** True. **d.** False.

64. a. Assuming that the last digit in n is not zero, there are four digits to the right of the decimal in $m + n$. The last digit in the sum must be the same as the last digit in n because there is no digit in m to add to it. **b.** It is not possible to be sure. We place the decimal so that there are seven digits to the right of it. However, if the last digits are zeros, we generally do not include them as part of the product. For instance, $6.002 \times 3.0005 = 18.009001$.

65. The decimal representation of $\dfrac{47}{53}$ is similar to $\dfrac{1}{53}$ except that the repetend is shifted two place values to the left.

66. $\dfrac{1}{7} = 0.1428571\ldots, \dfrac{2}{7} = 0.2857142\ldots,$

$\dfrac{3}{7} = 0.4285714\ldots, \dfrac{4}{7} = 0.5714285\ldots,$

$\dfrac{5}{7} = 0.7142857\ldots,$ and $\dfrac{6}{7} = 0.8571428\ldots.$ The repetends have the same sequence of digits, but have different starting points.

67. a. Not the same.   **b.** The same.   **c.** The same.
d. Not the same.

68. a. The denominator is likely to be 9 unless it has been reduced.   **b.** The denominator is likely to 99 unless it has been reduced.   **c.** The denominator is likely to be 999 unless it has been reduced.

69. Because $\frac{1}{9} = 0.111\ldots$, we could multiply this value by the numerator to find the value of the given fractions. Hence:

a. $\frac{2}{9} = 0.222\ldots$.

b. $\frac{7}{9} = 0.777\ldots$.

c. $\frac{16}{9} = 1\frac{7}{9} = 1.777\ldots$.

d. $4\frac{8}{9} = 4.888\ldots$.

70. Because $\frac{1}{9} = 0.111\ldots$, $\frac{1}{99} = 0.010101\ldots$, and $\frac{1}{999} = 0.001001001\ldots$, we would expect that $\frac{1}{9,999} = 0.000100010001\ldots$ and $\frac{1}{99,999} = 0.000010000100001\ldots$.

71. Answers will vary.
72. Answers will vary.
73. a. Answers will vary.   **b.** Answers will vary.   **c.** Answers will vary.   **d.** Answers will vary.
74. Caleb saw that each factor had only one digit to the right of the decimal point and concluded that the product will have only one digit to the right of the decimal point. Answers will vary on how to correct his thinking.
75. a. This student has not placed the decimal correctly.   **b.** The exponent should be positive, not negative.   **c.** This student has failed to multiply by a power of ten, instead putting the exponent on the decimal part.
76. Answers will vary.
77. Answers will vary.
78. They are used to make algebraic manipulations.
79. a. Answers will vary.   **b.** Answers will vary.

■ Section 8.3 Learning Assessment

1. Proportional reasoning is the ability to compare quantities and understand the context in which the comparison takes place.
2. a. A ratio is an ordered pair of numbers used to compare two quantities.   **b.** A ratio always occurs in some context, which is identified by the labels on the numbers. Because each number is associated with a label, changing their order can drastically affect the meaning of the ratio.
3. a. It allows us to use properties of fractions to solve ratio and proportion problems.   **b.** First, the denominator of a ratio can be zero, although we typically exclude zero from the set of possible denominators. Second, the numerator and denominator of a ratio are not limited to integers but can include fractions, decimals, and even irrational numbers as necessary. Third, fractions describe only part-to-whole relationships of like objects. Ratios, on the other hand, can be used to describe not only part-to-whole relationships, but also whole-to-part and part-to-part relationships.
4. We typically use the word "per," which means "for every," in situations where a comparison is being made between multiple objects and a single object.
5. A proportion is a statement that two ratios are equivalent.
6. Two ratios are equivalent if they make comparisons that occur in the same relative amounts. Two proportions are equivalent if the equations equating the products of their means and their extremes are the same.
7. A unit rate is a ratio that describes how many objects of one type are comparable to exactly one of another. If we are given a unit rate in a problem, then all we do is multiply by the given number in the other ratio to find the missing value.
8. If x represents a missing value in a proportion, then $\frac{x}{b} = \frac{c}{d}$, which implies that the product of the means is equal to the product of the extremes: $xd = bc$. Dividing both sides by d, $x = \frac{b \cdot c}{d}$.

9. a. 3 red cars: 8 green cars; $\frac{3 \text{ red cars}}{8 \text{ green cars}}$.

b. 45 mi : 1 gal; $\frac{45 \text{ mi}}{1 \text{ gal}}$.

c. 25 candies: 10 children; $\frac{25 \text{ candies}}{10 \text{ children}}$.

d. 150 passengers: 4 flight attendants; $\frac{150 \text{ passengers}}{4 \text{ flight attendants}}$.

10. a. $\frac{2}{1}$.   **b.** $\frac{2.25}{1}$.   **c.** $\frac{2.8\overline{3}}{1}$.   **d.** $\frac{2.208\overline{3}}{1}$.

11. a. It helps to set up the proportion by keeping the labels consistent in the ratios.   **b. i.** 9.   **ii.** 8.   **iii.** 15.   **iv.** 36.

12. a. $\frac{1}{4}$.   **b.** $\frac{3}{4}$.   **c.** $\frac{7}{9}$.   **d.** $\frac{4}{5}$.

13. Three other proportions are $\frac{5}{4} = \frac{25}{20}$, $\frac{4}{20} = \frac{5}{25}$, and $\frac{20}{4} = \frac{25}{5}$.

14. a. 5 blue to 6 red.   **b.** 5 blue to 11 beads.   **c.** 11 beads to 6 red.

15. a. 350 males to 425 females or 14 males to 17 females.   **b.** 425 females to 350 males or 17 females to 14 males.   **c.** 425 females to 775 members or 17 females to 31 members.

16. a. 740 freshman to 699 sophomores.   **b.** 1,325 juniors and seniors to 1,439 freshman and sophomores.   **c.** 569 seniors to 740 freshman.   **d.** 2,195 non-seniors to 569 seniors.

17. a. 3 ft to 1 ft.   **b.** 3 in. to 4 in.   **c.** 7 kg to 6 kg.

18. a. 1 in to 1 ft.   **b.** 10 cents to 1 cent.   **c.** 1 yd to 3 ft.

19. Possible answers are: **a.** $\frac{6}{8}$, $\frac{9}{12}$, and $\frac{12}{16}$.   **b.** $\frac{3}{2}$, $\frac{6}{4}$, and $\frac{9}{6}$.   **c.** $\frac{27}{31}$, $\frac{54}{62}$, and $\frac{81}{93}$.

20. a. $0.095/oz.   **b.** $0.1075/pencil.   **c.** $0.0817/oz.   **d.** $1.64/lb.

21. a. $0.035/screw.   **b.** $2.20/undershirt.   **c.** $2.55/gallon.   **d.** $0.664/battery.

22. a. 8 bars for $5.22. **b.** 16 oz for $3.01.
23. a. 2 lb for $10.35. **b.** 50 lb for $14.98.
24. Kelly is the better shooter.
25. My wife's car gets better gas mileage.
26. The 5-gal bucket is the better buy.
27. a. Proportion. **b.** Not a proportion. **c.** Not a proportion. **d.** Proportion. **e.** Proportion.
28. a. 3. **b.** 1. **c.** 4.
29. a. 6.75. **b.** −3 or 3. **c.** 12.
30. About 83 ft.
31. $79.60.
32. 50 times.
33. About 48,500 seeds should germinate.
34. 9 candy bars.
35. $22\frac{1}{2}$ days.
36. x values include any number from the set {22, 23, 24, 25, 26, 27}.
37. a. 9. **b.** $\frac{1}{9}$. **c.** 5.76. **d.** 17.
38. 95 red marbles, 57 blue marbles, and 152 green marbles.
39. $\frac{3}{4}$ c of sugar and $\frac{9}{16}$ c of walnuts.
40. a. 0.25 mi/min. **b.** 1.5 mi/min. **c.** 224 mph. **d.** 3,000 mph.
41. ≈41.67 revolutions per second.
42. a. $DE = 16$ in. and $AC = 10$ in. **b.** $GJ = 2.5$ cm and $HJ = 1.5$ cm. **c.** $MN = 3.73$ in. and $MP = 2.33$ in.
43. 5.71 ft.
44. 75 mi.
45. 254.5 mph with respect to the ground.
46. 1,000 mi × 1,400 mi.
47. The cost per ounce of the 64-oz bottle is $0.0308. If we sell the 48-oz bottle at $0.0307 · 48 = $1.47, it will still be a better buy.
48. 4 more fraternities would need to be added.
49. Jim should get $1,153.85, John should get $1,538.46, and Joe should get $2,307.69.
50. The 10 employees with 1,000 shares should get about $57,140 each. The 15 employees with 500 shares should get about $28,570 each.
51. a. 3,960 words in 1 hr. **b.** 37 min. **c.** 200 mistakes.
52. 18.2 mi. **53.** 1,940 cars.
54. a. 1 to 1. **b.** 1 to 1.
55. No, it is not possible. We need to know the total number of students in each class to be able to determine which class has more girls.
56. No, it is not possible. The ratio gives only the relative sizes of the sets of marbles and not the total count.
57. Because $\frac{3}{4} = 0.75$ and $\frac{60}{88} = 0.68$, the claim is not true because fewer dentists actually recommend the toothpaste.
58. Yes. If $\frac{a}{b}$ is proportional to $\frac{c}{d}$, then $ad = bc$. If we consider the product of a and $c - d$, we have $a(c - d) = ac - ad = ac - bc = (a - b)c$. Because the product of the means is equal to the product of the extremes, $\frac{a}{a - b}$ is proportional to $\frac{c}{c - d}$.

59. Yes. If $\frac{a}{b}$ is proportional to $\frac{c}{d}$, then $ad = bc$. If we consider the product of $a + b$ and d, then $(a + b)d = ad + bd = bc + bd = b(c + d)$. Because the product of the means is equal to the product of the extremes, $\frac{a + b}{b}$ is proportional to $\frac{c + d}{d}$.
60. a. Miles per hour expresses the number of miles that can be traveled in 1 hr. Hours per mile expresses the number hours it takes to travel 1 mi. **b.** Letters per minute might express the number letters that can be typed in 1 min. Minutes per letter might express the number of minutes it might take to type one letter. **c.** Pounds per dollar expresses the number of pounds that can be purchased for $1. Dollars per pound express the number of dollars it takes to purchase 1 lb.
61. a. Answers will vary. **b.** Answers will vary. **c.** Answers will vary.
62. If we use estimation, we know that x should be about half of 56.6 because 16.4 is about half of 28.3. Because 97.7 is larger that 56.6, it cannot be correct. The student computed the cross product incorrectly by multiplying 56.6 and 28.3 and then dividing by 16.4. Answers will vary on how to help the student.
63. Answers will vary.
64. Answers will vary.
65. a. Answers will vary. **b.** Answers will vary. **c.** Answers will vary. **d.** Answers will vary.
66. Answers will vary.
67. Answers will vary.
68. Answers will vary.
69. Answers will vary.

■ Section 8.4 Learning Assessment

1. If n is a nonnegative real number, then n percent, written $n\%$, is the ratio of n to 100, or $n\% = \frac{n}{100}$.
2. 10 represents ten wholes whereas 10% represents a tenth of a whole.
3. Percents are ratios, which are commonly expressed as fractions. Fractions, in turn, can be expressed as decimals.
4. Because percents always give rates relative to 100, they provide a good way to compare sets of different sizes.
5. Most percent problems fall into one of three cases. *Case 1:* The percent and the whole are known, and we are left to find the part. *Case 2:* The part and the whole are known, and we are left to find the percent. *Case 3:* The percent and the part are known, and we are left to find the whole.
6. a. To use a proportion, we use the fact that the percent is proportional to the ratio of the part to the whole: $\frac{\text{Part}}{\text{Whole}} = \frac{\text{Percent}}{100}$. After we place the given information in the appropriate place, we solve the proportion to find the missing number. **b.** A second approach is to write an equation by looking for key words like "is" for equality and "of" for multiplication. Then we write a simple equation using a variable for the unknown number and solve.

7. Interest is a fee paid for the use of money.

8. Interest is based on the principal, or the amount of money borrowed; the interest rate, or the percent per year used to determine the fractional amount of the principal to be applied as a fee; and the length of time in years that the money is borrowed.

9. **a.** Simple interest is interest that is paid only on the original principal. Compound interest is interest paid not only on the principal, but also on any subsequent interest added to principal. **b.** More money is made on the loan with compound interest than with simple interest.

10. **a.** 50 out of every 100, or one-half of all crème-cookie eaters pull them apart. **b.** 3 out of every 100 Americans prefer their hot dogs plain. **c.** 4 out of every 100 workers in the United States never laugh at work. **d.** 40 out of every 100, or 2 out of 5, indigestion remedies sold worldwide are bought in the United States.

11. **a.**

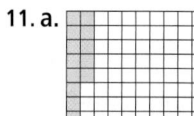

b.

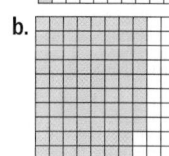

c.

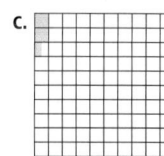

d.

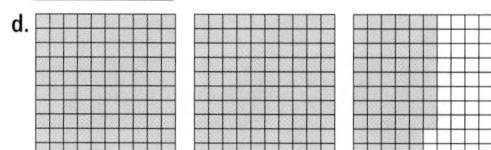

12. **a.** 0.17. **b.** 1.63. **c.** 0.002. **d.** 0.0575. **e.** $0.57\overline{3}$.

13. **a.** $\dfrac{1}{4}$. **b.** $\dfrac{4}{5}$. **c.** $\dfrac{1}{3}$. **d.** $\dfrac{3}{2}$. **e.** $\dfrac{1}{200}$.

14. **a.** 39%. **b.** 93%. **c.** 147%. **d.** 0.5%. **e.** 200%.

15. **a.** 70%. **b.** 56%. **c.** $16\dfrac{2}{3}\%$. **d.** $57\dfrac{1}{3}\%$. **e.** 86.6%.

16. **a.** $\dfrac{1}{20}$. **b.** $\dfrac{4}{25}$. **c.** 1. **d.** $1\dfrac{1}{2}$. **e.** $\dfrac{1}{300}$. **f.** $\dfrac{1}{1,000}$.

17. **a.** 25%. **b.** 48. **c.** 127.27.

18. **a.** 17.89%. **b.** 18.7. **c.** 77.5.

19. **a.** 1.25. **b.** $33\dfrac{1}{3}\%$. **c.** 76.47.

20. **a.** 0.61. **b.** 84.375%. **c.** 14.71.

21. **a.** $250. **b.** $7,000.

22. **a.** $4,036.50. **b.** $106.25.

23. **a.** $2,121.8. **b.** $96,776.14. **c.** $17,720.32.

24. **a.** $11,863.47. **b.** $61,483.26. **c.** $783,267.57.

25. **a.** 5.7. **b.** 21. **c.** 30. **d.** 78. **e.** 10. **f.** 108.

26. **a.** 5. **b.** 32. **c.** 0.11. **d.** 125. **e.** 325. **f.** 150.

27. $39,500.

28. $663.90.

29. 139 or 140 free throws.

30. About 3,733 students.

31. $189,000.

32. 63.63%.

33. **a.** $29,475. **b.** $2,620. **c.** $19,650.

34. About 7%.

35. 23.8% are red, 33.33% are green, and 42.86% are blue.

36. **a.** Kim. **b.** James had the most candy bars to sell at 115. **c.** James sold the most candy bars at 98.

37. **a.** 100. **b.** 45. **c.** $33\dfrac{1}{3}\%$.

38. **a.** 6. **b.** 0.01875.

39. $47,860.94.

40. About 5,882 copies.

41. Jill gets the larger pay raise.

42. $9.38.

43. $350.

44. 110% profit.

45. $281.68.

46. $405.88.

47. 11%.

48. Stock portfolio A had the greater percent increase.

49. $179,502.08.

50. Account B.

51. Account B.

52. Account A.

53. 20.72%.

54. $16,889.10.

55. $92,066.62.

56. $7,658.66.

57. $296.97.

58. It will take 13 to 14 years for the money to double.

59. **a.** 25 is less than the number. **b.** 40 is less than the number. **c.** 65 must be greater than the number.

60. **a.** Yes this would more than double the price of the object. **b.** No. This would make the price of the object less than 0.

61. **a.** If x is the original number, and y is the result, then $y = 0.85x$. So y must be increased by $\dfrac{1}{0.85} \approx 1.176 = 117.6\%$ to obtain x. **b.** If x is the original number, and y is the result, then $y = 1.1x$. So y must be decreased by $\dfrac{1}{1.1} \approx 0.9091 = 90.91\%$ to obtain x.

62. Because 100 increased by 25% is 125, and 150 decreased by 25% is 112.5, 100 increased by 25% is greater.

63. If x is the retail price, then $x - 0.2x = .8x$. This means that a 20% markdown is the same as paying 80%.

64. Yes. 25% of 30 is equal to $\dfrac{25}{100} \cdot 30 = 25 \cdot \dfrac{30}{100}$, which is the same as 30% of 25.

65. 125 is exactly half of 250, whereas 299 is slightly less than half of 600. This means that $\dfrac{125}{250}$ must be greater.

66. Because $75 = 3 \cdot 25$, we can find 25% of the number and then multiply it by 3.

67. Answers will vary.

68. Answers will vary.

69. Answers will vary.

70. The student forgot to move the decimal point two spots to the left before making the computation. Answers on how to correct the mistake will vary.

71. Answers will vary.

72. Percents are ratios, and we can use proportions to solve percent problems.

73. Answers will vary.

74. Answers will vary.

75. Let the meter stick represent one unit, or 100%. Each centimeter on the meter stick represents 1%, and each millimeter represents 0.1%.

76. Answers will vary.

Chapter 8 Review Exercises

1. a. 53.64. **b.** $0.\overline{4}$. **c.** 2,700.081 **d.** 0.36. **e.** 0.47.
f. 2.05. **g.** 1.6. **h.** 0.036.

2. a.

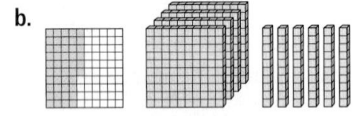

b.

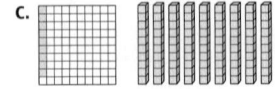

c.

d.

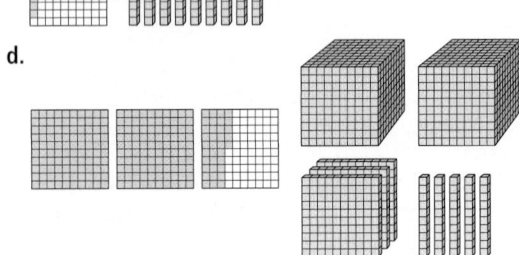

3. a. $\dfrac{45}{100} = \dfrac{9}{20}$. **b.** $\dfrac{125}{100} = \dfrac{5}{4}$. **c.** $\dfrac{57}{99} = \dfrac{19}{33}$. **d.** $\dfrac{424}{990} = \dfrac{212}{495}$.
e. $\dfrac{55}{100} = \dfrac{11}{20}$. **f.** $\dfrac{147.5}{100} = \dfrac{1,475}{1,000} = \dfrac{59}{40}$.

4. a. Terminating; $\dfrac{14}{25} = \dfrac{56}{100} = 0.56$. **b.** Not terminating. **c.** Terminating; $\dfrac{54}{90} = \dfrac{6}{10} = 0.6$.

5. a. Rational. **b.** Not possible. **c.** Irrational. **d.** Not possible.

6. $3.45\overline{6}$, 3.456, $3.4\overline{5}$, $3.\overline{45}$, $3.\overline{454}$.

7. Five rational numbers between 8.9 and $8.\overline{9}$ include 8.91, 8.92, 8.93, 8.94, and 8.95.

8. a. 39,800. **b.** 0.037923. **c.** 128. **d.** 48.

9. a. 1.3×10^4. **b.** 8.76×10^{-4}. **c.** 5.69×10^{10}.

10. a. 0.0000444. **b.** 7,780. **c.** 5,709,000.

11. a. 71.8. **b.** 125.93. **c.** 14.782. **d.** 7.03.

12. a. 1,437.696. **b.** 80,795.17. **c.** 2.12. **d.** 1.93.
e. 6.08. **f.** 612.11. **g.** 1.502×10^6. **h.** 6.62×10^2.

13. a. 56%. **b.** 3.9%. **c.** 275%. **d.** 60%. **e.** $133\frac{1}{3}$%.
f. 72.83%.

14. a. 9.75. **b.** 30.4%. **c.** 74.6.

15. a. $6,840. **b.** $22,771.08.

16. a. 42 females to 65 employees. **b.** 23 males to 42 females. **c.** 65 employees to 23 males.

17. $\sqrt{10} \approx 3.16$.

18. Suppose that $5 + 2\sqrt{2}$ is rational. By the definition of the rational numbers, $5 + 2\sqrt{2} = \dfrac{a}{b}$, where a and b are integers and $b \neq 0$. Subtracting 5 from both sides and then dividing by 2, $\sqrt{2} = \dfrac{\left(\dfrac{a}{b} - 5\right)}{2}$. $\dfrac{\left(\dfrac{a}{b} - 5\right)}{2}$ is rational because the rational numbers are closed under subtraction and division, which implies that $\sqrt{2}$ is rational. This contradicts the fact that $\sqrt{2}$ is irrational, so the original assumption must be incorrect. Hence, $5 + 2\sqrt{2}$ is irrational.

19. $64.06.

20. $0.0065.

21. $857.04.

22. $44.67.

23. a. $0.051/oz. **b.** $0.0052/sheet.

24. a. The 16-oz bottle is the better buy. **b.** The two 8-ft boards are a better buy.

25. About 35 hits.

26. 9 bags of lime.

27. The first motorcycle gets the best gas mileage.

28. $40.66.

29. Alice will get $333,333.33, Arnie will get $250,000, and Ann will get $200,000.

30. There are 57.6 mi between Town A and Town B.

31. 9 females need to be added to make the ratio 1 to 1.

32. 200 increased by 50% is bigger.

33. $26.25.

34. $2,640.

35. $30,452.17.

36. $39,183.49.

37. Stock portfolio B.

38. a. Chris. **b.** Dave had the most T-shirts to sell at 50. **c.** Dave sold the most T-shirts at 45.

39. Account C.

40. It will take 15 to 16 years for the money to triple.

CHAPTER 9

Section 9.1 Learning Assessment

1. Functional thinking is the process by which we identify, represent, and extend patterns and relationships that exist *between* sets of objects. With sequences, we use functional thinking to recognize a pattern between the position of the term and its value.

2. a. The general term is used to identify and generalize the pattern of the sequence. **b.** If the general term is generalized with an algebraic formula, then we can use the formula to generate any term in the sequence by substituting the positions or the values of terms into the formula.

3. A repeating sequence has some core component that continuously repeats. In an increasing sequence, the terms continue to grow larger in value. In a decreasing sequence, the terms continue to grow smaller in value.

4. Arithmetic and geometric sequences are alike in that each successive term is found by operating on the previous term with a fixed number. In an arithmetic sequence, the fixed number is added to the previous term. In a geometric sequence, it is multiplied.

5. We find the common difference by subtracting one term in an arithmetic sequence from the next consecutive term. We find the common ratio by dividing one term in a geometric sequence by the preceding term.

6. a. A sequence is recursively defined if the value of each term, except possibly for the first few, is based on the values of previous terms. b. Yes. In both arithmetic and geometric sequences, the values of successive terms are found by operating on the previous term with a fixed number.

7. The Fibonacci sequence is generated by adding the two preceding terms to get the next term, or $F_n = F_{n-1} + F_{n-2}$, where $F_1 = 1$ and $F_2 = 1$.

8. The figurate numbers are generated from sequences of dots arranged in common geometrical shapes. As the shapes grow, so does the number of dots.

9. a. The sequence is the natural numbers, but the terms alternate signs beginning with (-1). b. The sequence begins with 3 and adds 4 to find each successive term. c. The sequence begins with 1/3 and multiplies by 3 to find each successive term. d. The sequence begins with 2 and adds consecutive odd numbers to find each successive term.

10. Arrangements may vary; one possibility is shown. The common difference is shown by adding a new column of three dots in each successive diagram.

11. No. The common difference is negative, so we have no way to represent negative values using only arrangements of dots.

12. Arrangements may vary; one possibility is shown. In this case, we show the common ratio by doubling the number of dots in each case, first vertically, then horizontally.

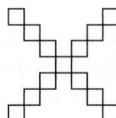

13. The sequence is $1, \frac{1}{2}, \frac{1}{4}, \frac{1}{8}, \ldots$. It is geometric. The next drawing might look like the following.

14. The sequence is $1, 5, 9, \ldots$. It is arithmetic. The next picture is shown.

15. The sequence is the square numbers, or $1, 4, 9, \ldots$. It is neither arithmetic nor geometric. The next one is shown.

16. a. Repeating: 4, 3, 2, 4, 3. . . . b. Decreasing: $-1, -3, -5, -7, -9, \ldots$. c. Increasing: 125, 625, 3,125, 15,625, 78,125, ...

17. a. Increasing: $-0.0625, -0.03125, -0.015625, -0.0078125, -0.00390625, \ldots$. b. Increasing: $\frac{5}{6}, \frac{6}{7}, \frac{7}{8}, \frac{8}{9}, \frac{9}{10}, \ldots$. c. Repeating: 0, 1, 0, -1, 0, 1, ...

18. a. Arithmetic: 20, 25, 30, 35, 40, ... b. Geometric: 192, 768, 3,072, 12,288, 49,152. c. Neither: 108; 1,944; 209,952; 408,146,688; 8.57×10^{13}, ... d. Arithmetic: $\frac{6}{3}, \frac{7}{3}, \frac{8}{3}, \frac{9}{3}, \frac{10}{3}, \ldots$

19. a. Geometric: 625, 3,125, 15,625, 78,125, 390,625, ... b. Arithmetic: $\frac{14}{4}, \frac{17}{4}, \frac{20}{4}, \frac{23}{4}, \frac{26}{4}, \ldots, \ldots$. c. Geometric: 13.5, -40.5, 121.5, -364.5, 1,093.5 . . . . d. Geometric: 0.0625, 0.03125, 0.015625, 0.0078125, 0.00390625, ...

20. a. $d = 2$, $a_n = 2n + 2$, $a_{10} = 22$, $a_{35} = 72$, and $a_{50} = 102$. b. $d = 6$, $a_n = 6n - 8$, $a_{10} = 52$, $a_{35} = 202$, and $a_{50} = 292$. c. $d = -2$, $a_n = -2n + 7$, $a_{10} = -13$, $a_{35} = -63$, and $a_{50} = -93$. d. $d = -4$, $a_n = -4n - 1$, $a_{10} = -41$, $a_{35} = -141$, and $a_{50} = -201$.

21. a. $d = \frac{3}{10}$, $a_n = \frac{3}{10}n + \frac{5}{10}$, $a_8 = \frac{29}{10}$, $a_{12} = \frac{41}{10}$, and $a_{15} = \frac{50}{10}$. b. $d = \frac{3}{8}$, $a_n = \frac{3}{8}n - \frac{1}{8}$, $a_8 = \frac{23}{8}$, $a_{12} = \frac{35}{8}$, and $a_{15} = \frac{44}{8}$. c. $d = 2.6$, $a_n = 2.6n - 1.3$, $a_8 = 19.5$, $a_{12} = 29.9$, and $a_{15} = 37.7$. d. $d = -2.8$, $a_n = -2.8n + 1.1$, $a_8 = -21.3$, $a_{12} = -32.5$, and $a_{15} = -40.9$.

22. The common difference is the coefficient of n. a. $d = 4$. b. $d = \frac{3}{5}$. c. $d = 4.75$.

23. a. $r = 4$, $a_n = 2(4)^{n-1}$, $a_5 = 512$, $a_{10} = 524,288$, and $a_{15} = 536,870,912$. b. $r = 3$, $a_n = 5(3)^{n-1}$, $a_5 = 405$, $a_{10} = 98,415$, and $a_{15} = 23,914,845$. c. $r = -2$, $a_n = 3(-2)^{n-1}$, $a_5 = 48$, $a_{10} = -1,536$, and $a_{15} = 49,152$.

24. a. $r = \frac{1}{4}$, $a_n = \left(\frac{1}{4}\right)^{n-1}$, $a_6 = \left(\frac{1}{4}\right)^5$, $a_9 = \left(\frac{1}{4}\right)^8$, and $a_{12} = \left(\frac{1}{4}\right)^{11}$. b. $r = 3$, $a_n = 2.4(3)^{n-1}$, $a_6 = 583.2$, $a_9 = 15,746.4$, and $a_{12} = 425,152.8$. c. $r = \frac{2}{3}$, $a_n = \frac{1}{4}\left(\frac{2}{3}\right)^{n-1}$, $a_6 = \frac{1}{4}\left(\frac{2}{3}\right)^5 = \frac{8}{243}$, $a_9 = \frac{1}{4}\left(\frac{2}{3}\right)^8 = \frac{64}{6,561}$, and $a_{12} = \frac{1}{4}\left(\frac{2}{3}\right)^{11} = \frac{512}{177,147}$.

25. a. 2, 7, 12, 17, 22, ... b. -1, 1, 5, 13, 29, ... c. 3, 9, 19, 33, 51, ...

26. a. 1, -2, -5, -8, -11, ... b. $\sqrt{3}, \sqrt{4}, \sqrt{5}, \sqrt{6}, \sqrt{7}$. . . . c. $0, \frac{1}{5}, \frac{2}{7}, \frac{3}{9}, \frac{4}{11}, \ldots$

27. a. Arithmetic; 2, 8, 14, 20, 26, ... b. Geometric; 27, 9, 3, 1, $\frac{1}{3}$, ... c. Neither; 1, 6, 16, 36, 76, ...

28. a. 0, 3, 3, 9, 15, 33, ... b. 1, 3, 8, 21, 55, 144, ...

29. a. 2, 4, 4, 8, 16, 64, ... b. $1, 2, \frac{1}{2}, 4, \frac{1}{8}, 32, \ldots$

30. $a_{95} = 275$.

31. $a_{10} = \dfrac{512}{59,049}$.

32. a. Each equivalence class represents an arithmetic sequence. **b.** $T_0 = \{x \mid x = 3n - 3, n \in \mathbb{N}\}$ $T_1 = \{x \mid x = 3n - 2, n \in \mathbb{N}\}$ and $T_2 = \{x \mid x = 3n - 1, n \in \mathbb{N}\}$

33. a. The seventh arrangement will have 7 cubes as shown.

By counting directly, the number of unglued faces is 30. **b.** The sequence is 6, 10, 14, 18, . . . , with $a_n = 4n + 2$. Using the formula, $a_7 = 30$ and $a_{20} = 82$.

34. 10, 11, 14, 19, 26, 35, ...

35. a. 3, 6, 9, ... **b.** 6, 12, 6, ... **c.** 1, 6, 12, ...

36. 35,250 gal of water left.

37. The stock will be over $2,500 in the fifth year.

38. Possible answers are $8 = 3 + 5 = F_4 + F_5$, $9 = 1 + 3 + 5 = F_1 + F_4 + F_5$, $10 = 2 + 3 + 5 = F_3 + F_4 + F_5$, $15 = 2 + 5 + 8 = F_3 + F_5 + F_6$.

39. a. 35, 48, and 63. **b.** $a_n = n(n + 2)$. **c.** $a_{15} = 255$.

40. a. 45, 66, and 91. **b.** $a_{10} = 190$, $a_{15} = 435$, and $a_{20} = 780$. **c.** The hexagonal numbers are circled in the given diagonal.

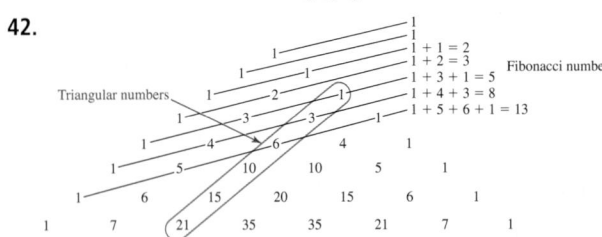

41. The first three pentagons are shown, giving us the first three pentagonal numbers of 1, 5, and 12. The general formula for the pentagonal numbers is given by $a_n = \dfrac{n(3n - 1)}{2}$. Hence, $a_4 = 22$, $a_5 = 35$, and $a_6 = 51$.

42.

Triangular numbers
Fibonacci numbers
$1 + 1 = 2$
$1 + 2 = 3$
$1 + 3 + 1 = 5$
$1 + 4 + 3 = 8$
$1 + 5 + 6 + 1 = 13$

43. This is not possible. A negative common difference always leads to smaller values, so the sequence cannot increase.

44. Answers will vary, but one possibility is the sequence 1, 0.1, 0.01, 0.001, . . . , which has a common ratio of $r = 0.1$. To create this type of sequence, the common ratio must be between 0 and 1.

45. Answers will vary, but one possibility is the sequence $1, -\dfrac{1}{2}, \dfrac{1}{4}, -\dfrac{1}{8}, \ldots$. To create a geometric sequence that is neither increasing nor decreasing, the common ratio must be negative.

46. Answers will vary, but one possibility is the sequence 1, $-1, 1, -1, \ldots$, which can be recursively defined as $a_n = (-1)a_{n-1}$ where $a_1 = 1$.

47. a. One rule is to square the position of the sequence, or $a_n = n^2$. Under this rule, the next three terms are 16, 25, and 36. A second rule is to add odd consecutive numbers. Under this rule, the next three terms are $9 + 7 = 16$, $16 + 9 = 25$, and $25 + 11 = 36$. A third possibility is to repeat the numbers, in which case the next three terms are 1, 4, and 9. **b.** Without the general term, it can be difficult to know exactly what rule governs the sequence.

48. a. F_4. **b.** F_6. **c.** F_8. **d.** F_{10}. **e.** Based on the pattern, we would expect $F_1 + F_3 + F_5 + \ldots + F_{25} = F_{26}$.

49. a. 1. **b.** 1. **c.** 1. **d.** Based on the pattern, we would expect $\text{GCF}(F_n, F_{n-1}) = 1$.

50. a. 4. **b.** 9. **c.** 16. **d.** 25. **e.** The square numbers. **f.** 144.

51. a. $1 + 2 = 3$ and $\dfrac{2(2 + 1)}{2} = \dfrac{6}{2} = 3$.

$1 + 2 + 3 = 6$ and $\dfrac{3(3 + 1)}{2} = \dfrac{12}{2} = 6$.

$1 + 2 + 3 + 4 = 10$ and $\dfrac{4(4 + 1)}{2} = \dfrac{20}{2} = 10$.

$1 + 2 + 3 + 4 + 5 - 15$ and $\dfrac{5(5 + 1)}{2} = \dfrac{30}{2} = 15$.

$1 + 2 + 3 + 4 + 5 + 6 = 21$ and $\dfrac{6(6 + 1)}{2} = \dfrac{42}{2} = 21$.

$1 + 2 + 3 + 4 + 5 + 6 + 7 = 28$ and $\dfrac{7(7 + 1)}{2} = \dfrac{56}{2} = 28$.

$1 + 2 + 3 + 4 + 5 + 6 + 7 + 8 = 36$ and $\dfrac{8(8 + 1)}{2} = \dfrac{72}{2} = 36$.

$1 + 2 + 3 + 4 + 5 + 6 + 7 + 8 + 9 = 45$ and $\dfrac{9(9 + 1)}{2} = \dfrac{90}{2} = 45$.

$1 + 2 + 3 + 4 + 5 + 6 + 7 + 8 + 9 + 10 = 55$ and $\dfrac{10(10 + 1)}{2} = \dfrac{110}{2} = 55$. **b.** This generates the triangular numbers. **c.** $a_{50} = \dfrac{50(50 + 1)}{2} = 1,275$, $a_{75} = \dfrac{75(75 + 1)}{2} = 2,850$, and $a_{100} = \dfrac{100(100 + 1)}{2} = 5,050$.

52. Answers will vary.

53. a. Answers will vary. **b.** Answers will vary.

54. a. Answers will vary. **b.** Answers will vary. **c.** Answers will vary.

55. They are both correct. Quishawna is doubling each number, or using the powers of 2. Dexter is increasing the terms by adding consecutive natural numbers $(1 + 1 = 2, 2 + 2 = 4, 4 + 3 = 7, \ldots)$. Not enough terms are given, so the pattern is open to interpretation. More terms of the sequence should be given to make the pattern clearer.

56. a. Those who got 2, 4, 10, 26, and 52 are correct. The other students used the numbers in the reverse order of the indices. **b.** The other half of the class reversed the values in the difference.

57. Answers will vary.

58. The place values of the Hindu-Arabic are a geometric sequence with a general term of $a_n = 10^n$, with $n \geq 0$.

59. Answers will vary.

60. Answers will vary.

61. a. Answers will vary. **b.** Answers will vary. **c.** Answers will vary.

Section 9.2 Learning Assessment

1. It must associate each object in the domain with one and only one object from the codomain.

2. The domain is the set that the function comes from, and the codomain is the set the function goes into. The set of all elements in the codomain that have been paired with something from the domain is called the range.

3. **a.** In an arrow diagram, the rule of the function is stated explicitly by the arrows between the sets. **b.** In a table, the rule of the function is indicated by particular pairings between the columns. **c.** In a set of ordered pairs, the rule of the function is indicated by particular pairings within the ordered pairs.

4. We can think of the function as a machine that takes domain values, performs some operation on them, and then indicates the corresponding range values. We often view domain values as the inputs to the machine and range values as the outputs.

5. A calculator receives input from the user, performs the needed operation, and then returns an output on the screen.

6. The independent variable represents the variable for the domain, and its value is independent of any others. The dependent variable represents the variable for the range, and its value depends on the value of the independent variable.

7. **a.** They offer the advantage of writing functions with infinite domains in a concise way. **b.** We must exclude values from the domain when those values are not defined for the rule of the function. This can occur if there is the possibility of division by zero or the square root of a negative number.

8. A graph is a picture of a function representing the set of all points in the Cartesian plane that satisfy the rule of the function. To make the graph of a function, we evaluate the rule of the function at several domain values to generate a set of ordered pairs. Next, we plot the ordered pairs to get a sense of the shape of the graph. We then use what we see to draw additional points. If the domain is the set of natural numbers, integers, or rational numbers, the graph is a set of discrete, or separated, points. If the domain is the real numbers, we connect the plotted points to make a curve.

9. In quadrant I, both coordinates are positive; in quadrant II, the first is negative and the second is positive; in quadrant III, both coordinates are negative; and in quadrant IV, the first is positive and the second is negative.

10. The vertical line test states that, if a vertical line intersects a graph in more than one point for any domain value x, then the graph does not represent a function. The test is based on the fact that, if a graph intersects a vertical line in more than one point, then the graph associates two range values with the one domain value at this point. This violates the definition of a function.

11. One advantage is that these representations show the associations of the function directly. The disadvantage is that the number of such associations that can be shown is limited.

12. Algebraic expressions are a concise way to represent functions with an infinite domain. Graphs offer the advantage of seeing trends.

13. The different representations are shown. The function multiplies the given number by 2 and adds 1.

{(1, 3), (3, 7), (5, 11), (7, 15), (9, 19), (11, 23)}

x	y
1	3
3	7
5	11
7	15
9	19
11	23

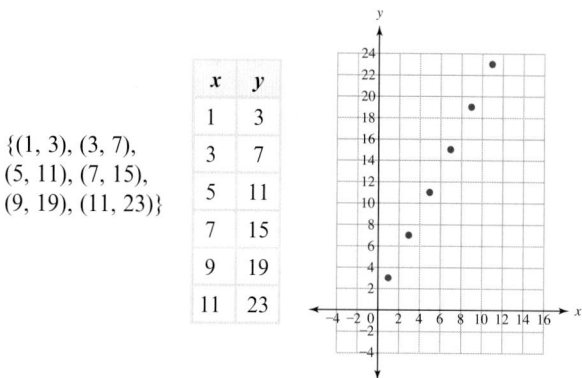

14. The different representations are shown. The function takes the square root of the given number.

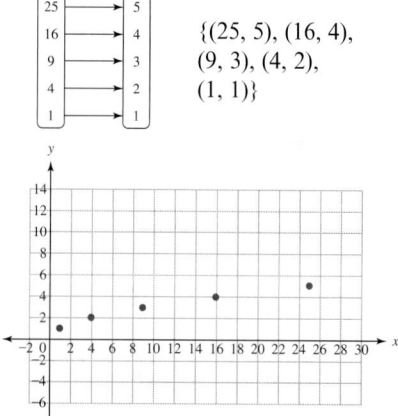

{(25, 5), (16, 4), (9, 3), (4, 2), (1, 1)}

15. The different representations are shown. The function doubles the given number.

x	y
2	4
3	6
4	8
5	10
6	12

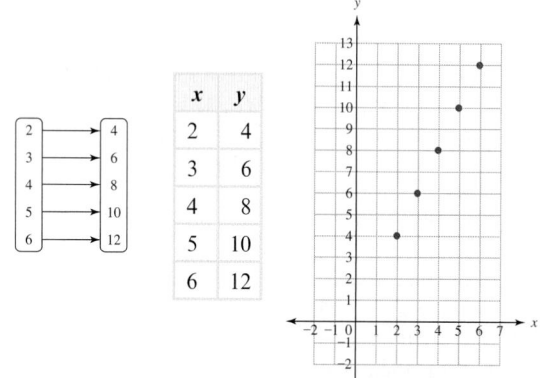

16. The different representations are shown below. The function takes the given number and doubles it.

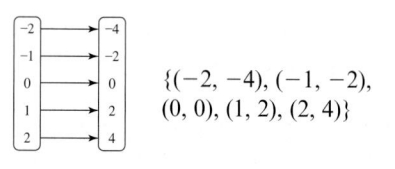

{(−2, −4), (−1, −2), (0, 0), (1, 2), (2, 4)}

x	y
−2	−4
−1	−2
0	0
1	2
2	4

17. a. The function f evaluated at 4, or f of 4. **b.** The function g evaluated at -5, or g of -5. **c.** f of x is equal to $x + 1$. **d.** The set of real numbers without zero. **e.** The set of real numbers without -1 and 1.

18. a. $(3, 2)$. **b.** $(4, -5)$. **c.** $(0, 6)$. **d.** $(-2, 0)$.

19. a. Three units to the right and 5 units above zero. **b.** Two units to the left and 4 units above zero. **c.** Two units below zero. **d.** Three units to the left of zero. **e.** Two units to the left and 4 units below zero.

20. a. This is a function because each input is associated with exactly one output. **b.** This is not a function: 0 and 1 are associated with two different outputs. **c.** This is a function because each input is associated with exactly one output. **d.** This is not a function: 3 and 5 are associated with two different outputs.

21. a. Not a function. **b.** Function. **c.** Function. **d.** Not a function.

22. a. Not a function. **b.** Function. **c.** Function. **d.** Not a function.

23. a. Domain = $\{1, 2, 3, 4, 5\}$, codomain = $\{a, b, c, d\}$, and range = $\{a, b, c\}$. **b.** Domain = $\{$Sam, Dorothy, Clarice, Tyrone, Rochelle$\}$, codomain = $\{$Table 1, Table 2, Table 3$\}$, and range = $\{$Table 1, Table 2, Table 3$\}$. **c.** Domain = $\{$Jackie, Lois, Simone$\}$, codomain = $\{$Lane 1, Lane 2, Lane 3, Lane 4, Lane 5$\}$, and range = $\{$Lane 2, Lane 3, Lane 4$\}$.

24. a. Domain = $\{1, 2, 3, 4\}$, range = $\{1, 2, 3, 4\}$. **b.** Domain = $\{3, 4, 5, 6\}$, range = $\{3, 4\}$. **c.** Domain = $\{2, 5, 6, 7\}$, range = $\{-1, 3, 4, 7\}$. **d.** Domain = $\{2, 3, 4, 5\}$, range = $\{2\}$.

25. Three possible answers are:

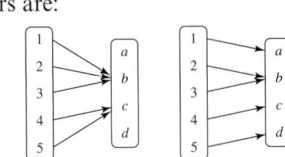

26. Possible answers are:
a. $\{(a, w), (b, w), (c, x), (d, y), (e, z)\}$. **b.** $\{(a, w), (b, w), (c, w), (d, w), (e, w)\}$. **c.** $\{(a, w), (b, w), (c, w), (d, x), (e, x)\}$.

27. a. -3. **b.** 7. **c.** 22. **d.** 52.

28. a. $f(-4) = 71$, $f(1) = 6$, and $f(5) = 134$. **b.** $f(-4) = 12$, $f(1) = -3$, and $f(5) = 21$. **c.** $f(-4) = -65$, $f(1) = 0$, and $f(5) = 124$. **d.** $f(-4) = 64$, $f(1) = 9$, and $f(5) = 1$.

29. $g(t) = 4t - 5$, $g(t - 4) = 4t - 21$, and $g(t + 6) = 4t + 19$.

30. a. If $x = -5$, then $y = -27$.
If $x = -2$, then $y = -15$.
If $x = 0$, then $y = -7$.
If $x = 2$, then $y = 1$.
If $x = 5$, then $y = 13$.
b. If $x = -5$, then $y = 107$.
If $x = -2$, then $y = 26$.
If $x = 0$, then $y = 2$.
If $x = 2$, then $y = 2$.
If $x = 5$, then $y = 47$.
c. If $x = -5$, it is not possible to compute y.
If $x = -2$, then $y = \dfrac{1}{3}$.
If $x = 0$, then $y = \dfrac{1}{5}$.
If $x = 2$, then $y = \dfrac{1}{7}$.
If $x = 5$, then $y = \dfrac{1}{10}$.

d. If $x = -5$, it is not possible to compute y.
If $x = -2$, then $y = 0$.
If $x = 0$, then $y = \sqrt{2}$.
If $x = 2$, then $y = 2$.
If $x = 5$, then $y = \sqrt{7}$.

31. a. If $x = -4$, then $y = -15$.
If $x = -1$, then $y = -6$.
If $x = 0$, then $y = -3$.
If $x = 1$, then $y = 0$.
If $x = 4$, then $y = 9$.
b. If $x = -4$, it is not possible to compute y.
If $x = -1$, then $y = \dfrac{1}{3}$.
If $x = 0$, then $y = \dfrac{1}{4}$.
If $x = 1$, then $y = \dfrac{1}{5}$.
If $x = 4$, then $y = \dfrac{1}{8}$.
c. If $x = -4$, then $y = \dfrac{1}{20}$.
If $x = -1$, then $y = y = \dfrac{1}{2}$.
If $x = 0$, it is not possible to compute y.
If $x = 1$, it is not possible to compute y.
If $x = 4$, then $y = \dfrac{1}{12}$.
d. If $x = -4$, then $y = \sqrt{24}$.
If $x = -1$, then $y = 0$.
If $x = 0$, it is not possible to compute y.
If $x = 1$, it is not possible to compute y.
If $x = 4$, then $y = 0$.

32. a. Exists. **b.** Exists. **c.** Does not exist. **d.** Exists. **e.** Does not exist. **f.** Exists.

33. a. Exists. **b.** Does not exist. **c.** Exists. **d.** Exists. **e.** Exists. **f.** Does not exist.

34. a. $\mathbb{R}$. **b.** $\mathbb{R}$. **c.** $\mathbb{R} - \{0\}$. **d.** $\mathbb{R}$. **e.** $\{x | x \geq 4 \text{ and } x \in \mathbb{R}\}$. **f.** $\mathbb{R}$.

35. a. $\mathbb{R}$. **b.** $\mathbb{R}$. **c.** $\mathbb{R} - \{-7\}$. **d.** $\mathbb{R} - \{2\}$. **e.** $\mathbb{R} - \{-2, 1\}$. **f.** $\{x | x > 4 \text{ and } x \in \mathbb{R}\}$.

36. Five ordered pairs are $\{(0, 6), (1, 8), (2, 10), (3, 12), (4, 14)\}$.

37. Five ordered pairs are
$$\left\{\left(1, \frac{2}{3}\right), \left(2, \frac{1}{6}\right), \left(3, \frac{2}{27}\right), \left(4, \frac{1}{24}\right), \left(5, \frac{2}{75}\right)\right\}.$$

38. a. No. **b.** Yes. **c.** No. **d.** No. **e.** Yes.

39. a. IV. **b.** I. **c.** III. **d.** III. **e.** On the y-axis.

40. a. IV. **b.** II. **c.** The origin. **d.** On the x-axis. **e.** IV.

41. a. I and IV. **b.** III and IV. **c.** II. **d.** I and III.

42. a.

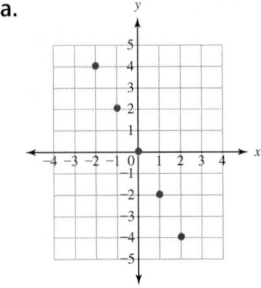

b.

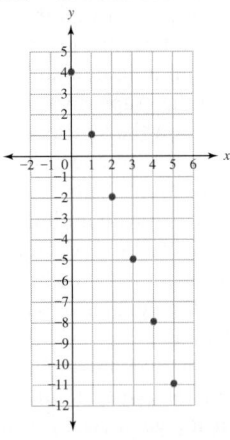

c.

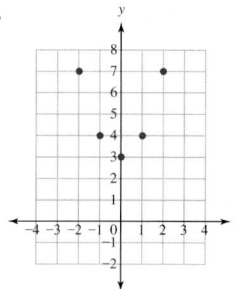

d.

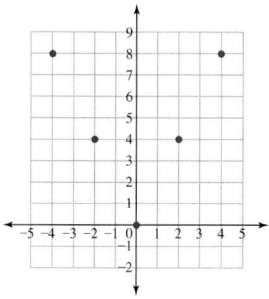

44. The tables of values will differ. The graph of each function is shown.

a.

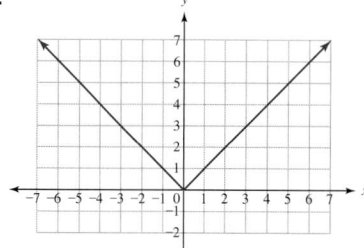

43. The tables of values will differ. The graph of each function is shown.

a.

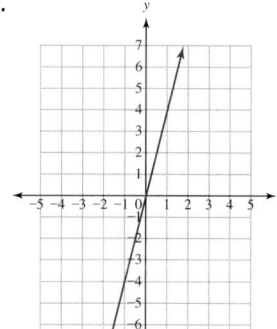

b.

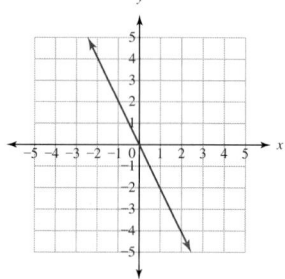

b.

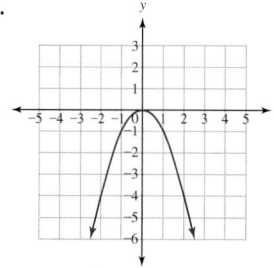

c.

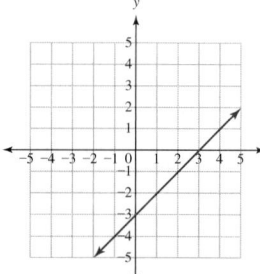

d.

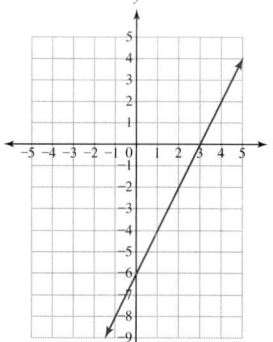

c.

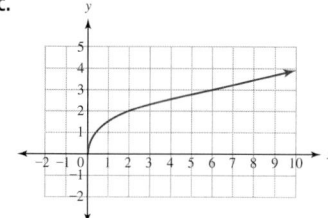

d.

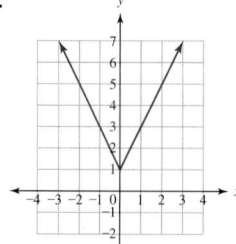

e.

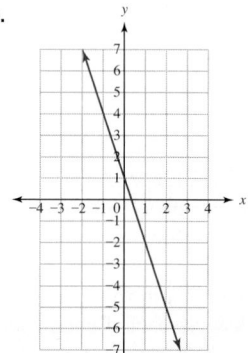

e.

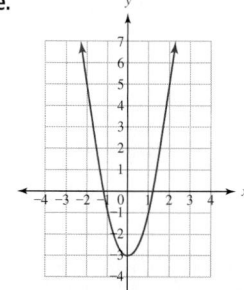

f.

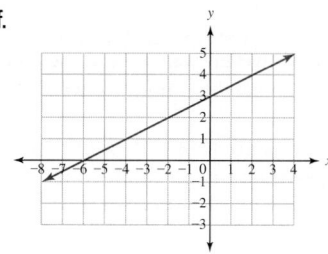

f.

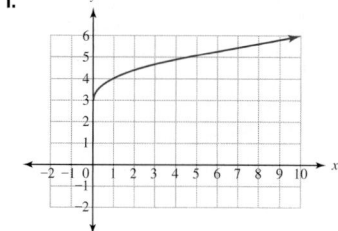

45. a.

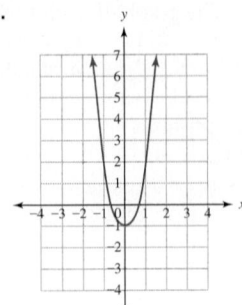

b.

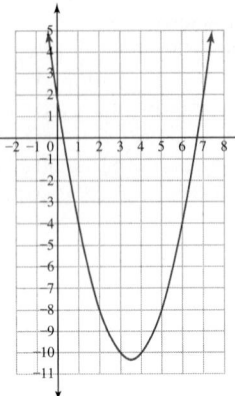

c.

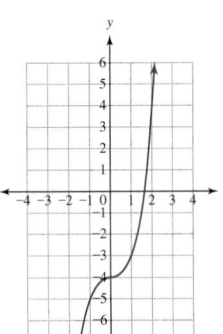

d.

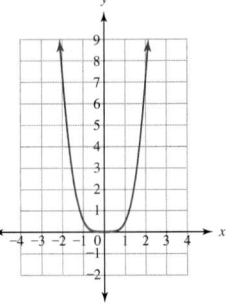

46. a.

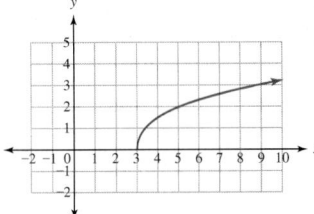

b.

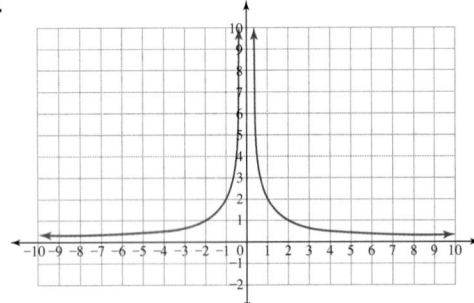

c.

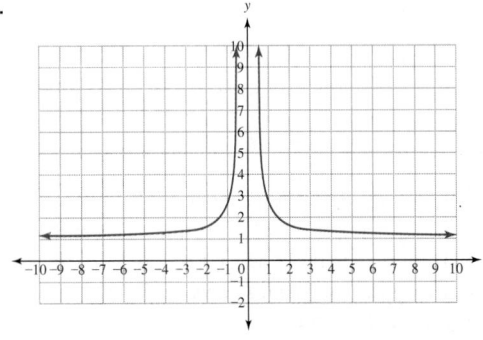

d.

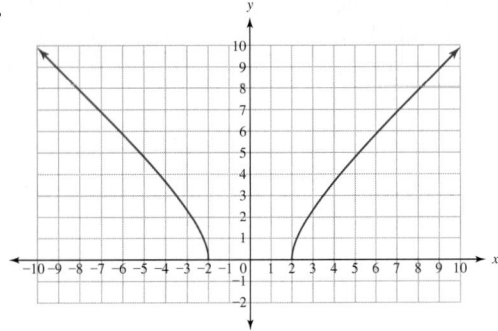

47. a. 5.   **b.** 3.   **c.** -2 or 2.   **d.** 0.

48. a. $f(x) = 4x$.   **b.** $f(x) = -x^2$.   **c.** $f(x) = 2x - 3$.

49. a. Domain $= \{-10, 0, 10, 20, 30\}$, and range $= \{0, 10, 20, 30\}$.   **b.** $f(0) = 20$ and $f(20) = 0$.   **c.** $f(50) = 30$.

50. a. 4.   **b.** 8.   **c.** 16.   **d.** 2^n.

51. $f(-2) \approx 0$, $f(0) \approx (-3)$, $f(1) \approx (-1.5)$, and $f(3) \approx 1.3$.

52. a. $f(-5) \approx 2$, $f(-3) \approx (-1.5)$, and $f(6) \approx 5$.
b. $x = (-4)$ and $x = 4$.   **c.** $x = 0$.

53. a. $28.   **b.** $32.50.   **c.** $48.81.

54. a. At $C = -10$, $F = 14°$. At $C = 10$, $F = 50°$.
b. $212°$ F.   **c.** $0°$ C.

55. a. True.   **b.** True.   **c.** False.   **d.** False.

56. No. The size of the domain and the range are independent of one another. In some cases, the domain is smaller than the range. In other cases, it is larger.

57. Yes. Consider the function $f(x) = 1$, which takes every real number to 1.

58. The table is shown. It does not describe a function from A to B because the values of 3 and 4 are associated with multiple values.

x	y
2	1
3	2
3	1
4	3
4	2
4	1

59. Table 1 represents a function. The independent variable is the year, and the dependent variable is the distance of the shot put throw. Table 2 is not a function because one distance can go to multiple years.

60. Yes. We can consider the city the domain value and the driving time the range value. Because each domain value goes to exactly one range value, the table represents a function.

61. Height is a function of weight because every weight is sent to only one height. However, weight is not a function of height because the same height can go to two different weights.

62. a. Three constant functions are $f(x) = 1$, $g(x) = 3$, and $h(x) = 5$.   **b.** In general, any constant function can be represented by $f(x) = c$, where c is a real number.

c. The graphs of the three functions in part a are shown. As we can see, the graph of a constant function is always a horizontal line.

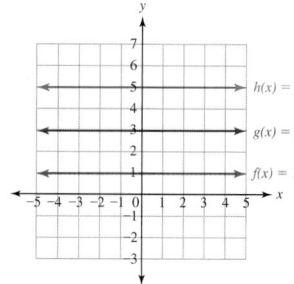

63. a. Answers will vary. **b.** Answers will vary. **c.** Answers will vary.

64. Answers will vary.

65. Answers will vary.

66. She is not correct. The functions are the same because y and $f(x)$ are different representations of the dependent variable of the function. Answers will vary on how to correct Maxine's thinking.

67. The graph is a function. The student has confused a horizontal line and vertical line.

68. Answers will vary.

69. a. A one-to-one correspondence is a function because every element in the first set is paired with exactly one element in the second. **b.** Not every function is a one-to-one correspondence because many functions can associate every element in the first set to only one element in the second. **c.** The difference between a function and a one-to-one correspondence is that, in a one-to-one correspondence, every element in the range must be used and used only once. In a function, not every element in the second set must be used.

70. Consider the sets $A = \{1, 2, 3, 4\}$ and $B = \{a, b, c, d\}$.
a. The following function is a one-to-one correspondence.

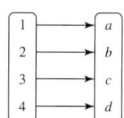

b. The following function is not a one-to-one correspondence.

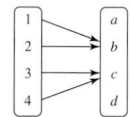

71. a. The relationship between a student and the student's GPA represents a function because each student is associated with only one GPA. **b.** The relationship between an item in a store and its bar code represents a function because each item in a store is associated with only one bar code. **c.** The relationship between birthdays and people with those birthdays is not a function because one birthday can be associated with many people.
d. The relationship between a course and the students taking the course is not a function because one course can be associated with many students. **e.** The relationship between a father and his children is not a function because one father can have many children.

72. Answers will vary.

73. Yes. Some functions can have more than one input value, as is the case with the GCF and the LCM on a calculator. These operations are functions because there is only one output for each combination of inputs.

74. a. A sequence is a function because each position in the sequence is associated with a particular value.
b. Most sequences use the set of natural numbers as their domain. **c.** Answers will vary.

Section 9.3 Learning Assessment

1. It signifies that a consistent change in one variable coincides with a consistent change in another.

2. The slope determines its steepness and direction, and the y-intercept determines its relative position with respect to the y-axis.

3. It is a comparison between the rise of the line and the run of the line.

4. The slope is the ratio of the rise of the line to its run. The rise is computed by taking the difference of the y-coordinates, or $y_2 - y_1$, of two points on the line. The run is computed by taking the difference of the x-coordinates, or $x_2 - x_1$, for the same two points.

5. The x-coordinate of the y-intercept is zero, and the y-coordinate of the x-intercept is also zero.

6. A line with an equation $x = a$ is a vertical line. Any vertical line assigns one x-value to many y-values, so it does not represent a function.

7. Suppose that (x_1, y_1) is a known point on a line with slope m, and further suppose that (x, y) is an unknown point on the same line. The following manipulation then gives us an equation for the line in terms of x and y.

$m = \dfrac{y - y_1}{x - x_1}$ Substitution into slope formula.

$m(x - x_1) = y - y_1$ Multiply both sides by $(x - x_1)$.

8. Two lines with the same slope have the same constant rate of change; that is, the distance between them always remains the same. The two lines never cross one another, or intersect. They must be parallel.

9. One way is to find two points that satisfy the equation of the line, plot them, and then draw the line. A second way is to start with the y-intercept and then locate a second point by using the rise and the run given by the slope. A third way is to start with any point on the line and then use the rise and run given by the slope to find a second point on the line.

10. a. One possible answer is shown.

x	y
1	1
2	2
3	3
4	4

b. One possible answer is $\{(1, 2), (2, 4), (3, 6), (4, 8) \ldots\}$.

11. a. The line goes down 4 units for every 1 unit it goes to the right. **b.** The line goes up 2 units for every 3 units it goes to the right. **c.** The line does not go up for each unit it goes to the right. **d.** The line goes up 7 units for every 4 units it goes to the right.

12. a. The line goes up 5 units for every 1 unit it goes to the right. **b.** The line goes down 1 unit for every 5 units it goes to the right. **c.** The line goes up 1 unit for every 4 units it goes to the left. **d.** The line goes up 1 unit but does not go to the right or left any units.

13. Possible answers are

a. $\frac{6}{2}, \frac{-3}{-1}, \frac{9}{3}$. **b.** $-\frac{6}{10}, -\frac{9}{15}, -\frac{12}{20}$. **c.** $\frac{2}{2}, \frac{3}{3}, \frac{-4}{-4}$.

14. a. Linear equation. **b.** It does not represent a linear function because a constant change in the first number in the ordered pairs does not correspond to a constant change in the second number of the ordered pairs. **c.** It does not represent a linear function because a constant change in the first number of the ordered pairs does not correspond to a constant change in the second number of the ordered pairs. **d.** Linear equation.

15. a. It does not represent a linear function because a constant change in the first column does not correspond to a constant change in the second. **b.** Linear equation. **c.** Linear equation. **d.** It does not represent a linear function because a constant change in the first column does not correspond to a constant change in the second.

16. a. 3. **b.** 1. **c.** 4. **d.** 0.

17. a. $m = (-1)$, x-intercept $= (-1, 0)$, and y-intercept $= (0, -1)$. Three points on the line are $(-2, 1)$, $(1, -2)$, and $(-3, 2)$. **b.** $m = (-1/4)$, x-intercept $= (40, 0)$, and y-intercept $= (0, 10)$. Three points on the line are $(-40, 20)$, $(-20, 15)$, and $(20, 5)$.

18. a. $m = 0$; horizontal. **b.** $m = \frac{6}{5}$; increasing. **c.** $m = -\frac{3}{7}$; decreasing.

19. a. $m = (-1)$; decreasing. **b.** m is undefined; vertical. **c.** $m = 2$; increasing.

20. Five points on the line are $(0, -4)$, $(1, -2)$, $(2, 0)$, $(3, 2)$, and $(4, 4)$.

21. Five points on the line are $(0, -4)$, $(1, -1)$, $(2, 2)$, $(3, 5)$, and $(4, 8)$.

22. Five points on the line are $(0, 1)$, $(3, 3)$, $(6, 5)$, $(9, 7)$, and $(12, 9)$.

23. a.

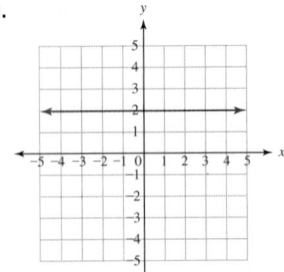

b.

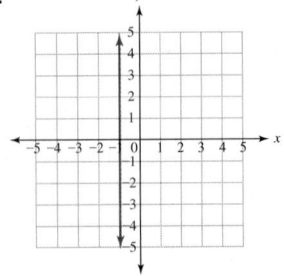

c.

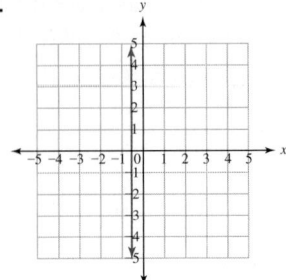

d.

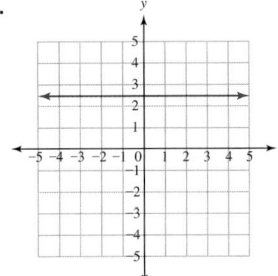

24. Methods of graphing will vary.

a.

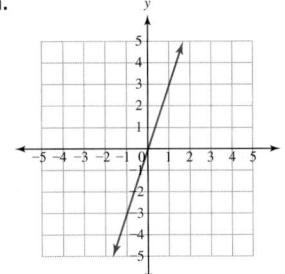

b.

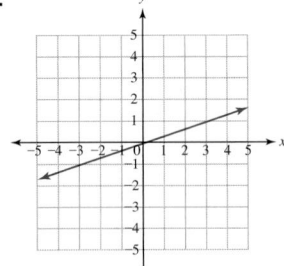

c.

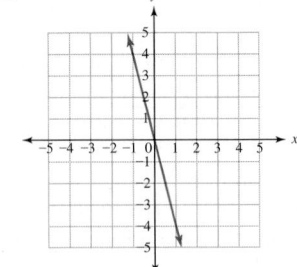

d.

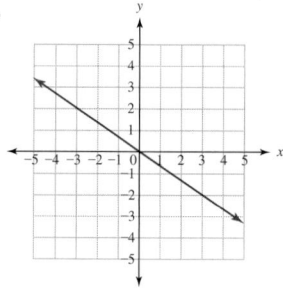

25. Methods of graphing will vary.

a.

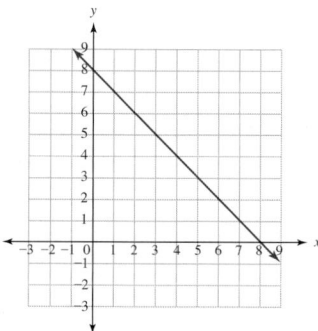

b.

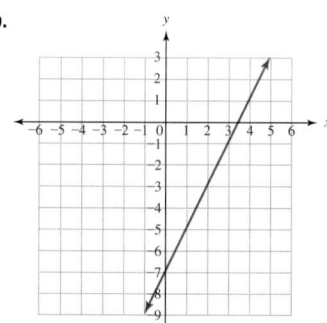

c.

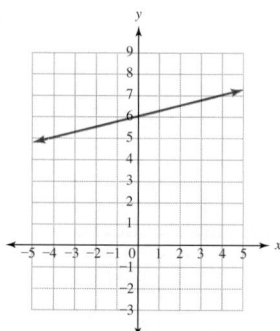

d.

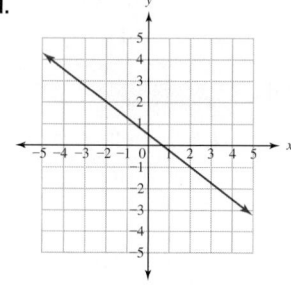

26. a. $m = 0$, y-intercept $= 6$, and x-intercept does not exist.
   b. m is undefined, y-intercept does not exist, and x-intercept $= 5$.
   c. $m = -\dfrac{3}{4}$, y-intercept $= 0$, and x-intercept $= 0$.
   d. $m = 2$, y-intercept $= 7$, and x-intercept $= \left(-\dfrac{7}{2}, 0\right)$.

27. a. $m = -\dfrac{1}{2}$, y-intercept $= \left(0, \dfrac{3}{2}\right)$, and x-intercept $= (3, 0)$.
   b. $m = \dfrac{2}{3}$, y-intercept $= (0, 4)$, and x-intercept $= (-6, 0)$.
   c. $m = -\dfrac{3}{7}$, y-intercept $= \left(0, \dfrac{5}{11}\right)$, and x-intercept $=$
   $\left(\dfrac{35}{33}, 0\right)$.   **d.** $m = \dfrac{4}{7}$, y-intercept $= \left(0, -\dfrac{9}{7}\right)$, and
   x-intercept $= \left(\dfrac{9}{4}, 0\right)$.

28. a. $y - 0 = x - 2$ or $y - 1 = x - 3$.
   b. $y - 9 = -\dfrac{5}{7}(x + 4)$ or $y - 4 = -\dfrac{5}{7}(x - 3)$.
29. a. $y - 0.3 = 3.4(x - 4.5)$ or $y - 7.1 = 3.4(x - 6.5)$.
   b. $\left(y - \dfrac{4}{5}\right) = \dfrac{3}{5}\left(x - \dfrac{1}{3}\right)$ or $\left(y - \dfrac{1}{5}\right) = \dfrac{3}{5}\left(x + \dfrac{2}{3}\right)$.
30. a. $y = 5x - 2$.   **b.** $y = -4x$.   **c.** $y = 5.76x + 2.34$.
31. a. $y = 4x - 12$.   **b.** $y = -2x - 3$.   **c.** $y = \left(-\dfrac{3}{4}\right)x + \left(\dfrac{3}{10}\right)$.
32. a. $y = 4x - 3$.   **b.** $y = -3x - 3$.   **c.** $y = -\dfrac{2}{3}x - 3$.
   d. $y = -3$.
33. a. $y = -2x + 8$.   **b.** $y = 5x - 6$.   **c.** $y = \dfrac{3}{4}x + \dfrac{5}{2}$.
   d. $x = 2$.

34. a. Neither.   **b.** Parallel.   **c.** Perpendicular.   **d.** Parallel.
35. a. Parallel.   **b.** Perpendicular.   **c.** Perpendicular.
   d. Parallel.   **e.** Neither.   **f.** Perpendicular.

36. a. $y = 3$.   **b.** $y = -3x + 5$.   **c.** $y = -\dfrac{1}{6}x + \dfrac{7}{6}$.

37. a. $x = 4$.   **b.** $y = -\dfrac{2}{7}x + 2$.   **c.** $y = -\dfrac{5}{3}x - \dfrac{25}{3}$.

38. $x = 9$.

39. $y = -\dfrac{1}{4}$.

40. a. The points are not on the same line.
   b. All three points are on the same line.
41. First find the equation of the line, which is $y = -2x + 4$.
   Evaluate it at three different x-values to find three points
   on the line, such as $(0, 4)$, $(1, 2)$, and $(2, 0)$.
42. $(3, 7)$ and $(7, 8)$.
43. a. Any point of the form $(x, x - 1)$, such as $(4, 3)$ or $(5, 4)$
   will work.   **b.** Any point of the form $(x, 2)$, such as $(4, 2)$
   or $(6, 2)$ will work.   **c.** Answers will vary, but possible
   answers are $(7, 3)$ and $(11, 4)$.

44. a. $k = 4$.   **b.** $k = 8$.   **c.** $k = \dfrac{8}{7}$.

45. a. $(1, 2)$. **b.** $\left(-\dfrac{1}{2}, \dfrac{7}{2}\right)$. **c.** $\left(1, -\dfrac{3}{2}\right)$.

46. a. $\dfrac{12}{20} = \dfrac{3}{5}$. **b.** The pitch of the roof is the slope of its outer surface.

47. a. $m = 4$ and $b = -160$. **b.** At $F = 50°$, $C = 40$ chirps. At $F = 75°$, $C = 140$ chirps. At $F = 90°$, $C = 200$ chirps. **c.** 65° F.

48. a. i. 3%. **ii.** 10%. **iii.** 6%. **iv.** 8%.
b. i. 25 yd. **ii.** 45 yd. **iii.** 30 yd. **iv.** 22.5 yd.

49. a. A slope of 3 represents a steeper line because it has a rise of 3 units to a run of 1 unit. A slope of $\dfrac{1}{3}$ has a rise of 1 unit to a run of 3 units. **b.** The lines have the same steepness because they both have a rise of 3 units to a run of 1. The lines are just in opposite directions.

50. The line must pass through the origin, so the y-intercept is 0. The line must also be increasing, so it must have a positive slope.

51. If the x- and y-coordinates of the points are not computed in the same order, the result will be the negative of the slope, or $-m$.

52. a. The graphs are shown.

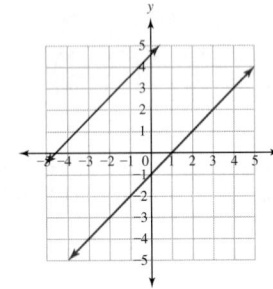

No points

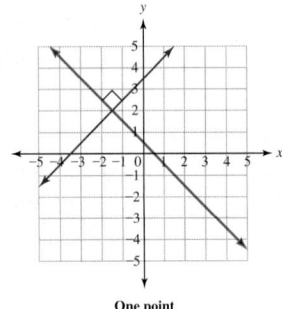

One point

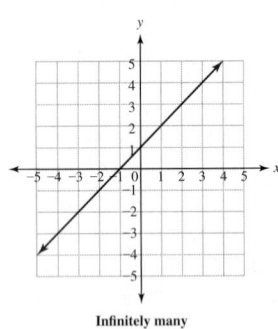

Infinitely many points

b. For two lines to intersect in only two or three points, the lines would have to bend and the lines have no curves in them. **c.** If the lines have the same slope but different y-intercepts, they are parallel and do not intersect at any points. If the lines have different slopes, they intersect at one point. If the lines have the same slope and the same y-intercept, then they are the same line and intersect at infinitely many points.

53. Line p has a slope of $\dfrac{4 - 1}{3 - (-2)} = \dfrac{3}{5}$, and line q has a slope of $\dfrac{7 - (-4)}{8 - (-7)} = \dfrac{11}{15}$. Because the slopes are not the same, the lines cannot be parallel. They must intersect at one point.

54. Graphs may vary, but possible ones are shown.

a.

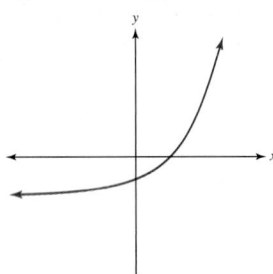

b.

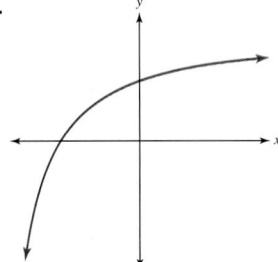

c.

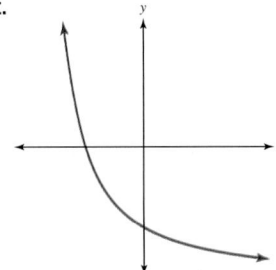

d.

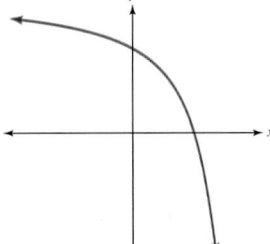

55. Tables may vary, but possible ones are shown.

a.

x	y
1	1
2	4
3	9
4	16

b.

x	y
1	1
4	2
9	3
16	4

c.

x	y
-4	16
-3	9
-2	4
-1	1

d.

x	y
1	-1
2	-4
3	-9
4	-16

56. Graphs may vary, but possible ones are shown.

a.

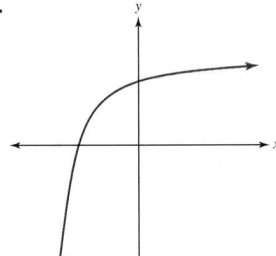

b.

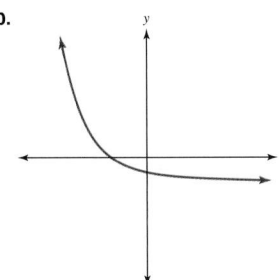

c. The graph must become horizontal.

57. Let line l go through the origin and contain the point (x_1, y_1), which gives it a slope of $m = \dfrac{y_1 - 0}{x_1 - 0} = \dfrac{y_1}{x_1}$.

When line l is rotated 90° to form l', the point (x_1, y_1) is sent to the point $(y_1, -x_1)$ and the point at the origin remains fixed. This makes the slope of l' equal to

$m' = \dfrac{-x_1 - 0}{y_1 - 0} = \dfrac{-x_1}{y_1}$. Multiplying the slopes of l and l',

we have $m \cdot m' = \dfrac{y_1}{x_1} \cdot \dfrac{-x_1}{y_1} = \dfrac{-x_1 y_1}{x_1 y_1} = -1$.

58. a. The slope of a line is the ratio of the change in y to the change in x.
b. The rise is actually a change in the y-values between two points, and the run is actually the change in the x-values between two points.

59. "Constant change" refers to a situation that is changing, but doing so at a constant rate. "Changing constantly" refers to a situation that is always changing and doing so at different rates.

60. Answers will vary.

61. Answers will vary.

62. Answers will vary, but Janie should be made aware that we can substitute either point into the equation and the slope-intercept form will be the same.

63. Marcel has a misconception about the slopes of parallel lines, in that the slopes of parallel lines are the same and not the negative of each other.

64. a. Arithmetic sequences are linear functions.
b. The domain is the set of natural numbers.

65. a. "Approximately linear" means that the graph of a function could be close to being straight line. Or it could mean that the function is close to representing constant change.
b. Answers will vary.
c. Answers will vary.

66. Answers will vary.
67. The graph is shown.

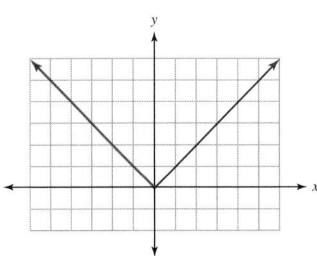

  a. The graph to the right of zero is identical to the graph of $y = x$. To the left of zero, it is identical to $y = -x$.
  b. Both parts of the function are defined using linear equations, specifically $y = x$ for $x \geq 0$ and $y = -x$ for $x < 0$.
68. Answers will vary.

■ Section 9.4 Learning Assessment

1. One area of algebraic thinking involves identifying and generalizing numerical patterns embedded in the numerical operations. The second involves finding unknown quantities by solving equations and inequalities.

2. We solve an equation by finding every value that makes the equation true.

3. An equation is much like a balance, in that both sides take on the same value, or "weight." As with a physical balance, we can add or remove numbers and variables to either side of the equation. If we do something to one side but not the other, the equation becomes unbalanced.

4. A constant is a real number either added to or subtracted from algebraic expression. A coefficient is a real number that multiplies a variable.

5. Two terms are like if they have the same variable(s) raised to the same power.

6. One method is to use mental arithmetic. A second method is to isolate the variable. In this method, we use inverse operations to transform the equation into the form $x = a$, which directly reveals the value of the unknown.

7. a. In a linear equation, each term is either a real-number constant or the product of a number and a variable raised to the first power. **b.** A quadratic equation of one variable is an equation in which one or more terms have a squared variable but no higher power. **c.** In a square root equation, the variable is under the square root. **d.** In a rational equation, the variable is in the denominator of a fraction.

8. a. An equation is a statement that uses =. An inequality is a statement that uses $<, \leq, >, \geq,$ or $\neq$. **b.** First, whenever we multiply or divide an inequality by a negative number, we must reverse the inequality to maintain its truthfulness. Second, the solution sets of inequalities do not contain just one number but rather are infinite in size.

9. The symbols $-\infty$ and ∞ represent negative and positive infinity. They are not numbers but rather indicate that the numbers continue indefinitely in the negative or positive direction.

10. a. iv. **b.** vi. **c.** v. **d.** i. **e.** ii. **f.** iii.

11. a. $(-\infty, 5]$;

b. $(2, 5]$;

c. $(-\infty, -1) \cup (-1, \infty)$;

d. $(-\infty, \infty)$;

12. a. $\{x \mid -4 < x < 7\};$

b. $\{x \mid -6 \le x \le -2\};$

c. $\{x \mid x \ge 2\};$

d. $\{x \mid x < -1\};$

e. $\{x \mid x \le 3 \text{ or } x > 7\};$

13. a. $\{x \mid x \le 6\} = (-\infty, 6]$.   **b.** $\{x \mid x > -3\} = (-3, \infty)$.
c. $\{x \mid -1 \le x \le 4\} = [-1, 4]$.
d. $\{x \mid -5 < x \le 0\} = (-5, 0]$.
e. $\{x \mid x \le 2 \text{ or } x > 7\} = (-\infty, 2] \cup (7, \infty)$.
f. $\{x \mid x \ne -4\} = (-\infty, -4) \cup (-4, \infty)$.

14. a. The infinity should have a parenthesis and not a bracket.   **b.** The numbers are not listed in the correct numerical order.   **c.** The numbers are not listed in the correct numerical order.

15. a. $2x - 5$.   **b.** $-2x - 1$.   **c.** $-3x + 7$.
d. $-4x$.   **e.** $3x = -9$.   **f.** $x - 3 = -2x + 4$.

16. a.   **b.**   **c.**   **d.**

17. a.

b.

c.

d.

18. a. Zero.   **b.** One positive x-tile.   **c.** Answers will vary. Three possible combinations are shown.

d. Answers will vary. Three possible combinations are shown.

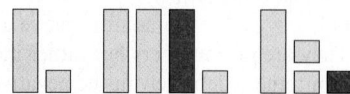

19. Draw the balance so that the smaller value is raised higher.
a.

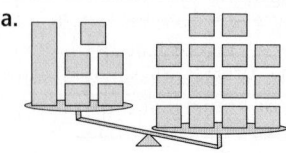

b.

c.

20. a. $\{-2\}$.   **b.** $\{1, 2, 3\}$.   **c.** $\{-2\}$.   **d.** $\{-1, 0, 1, 2, 3\}$.
21. a. $x = 5$.   **b.** $y = 9$.   **c.** $w = 10$.   **d.** $x = 36$.
e. $y = 4$.   **f.** $x = 3$.
22. a. $x \in \{-11, 11\}$.   **b.** $x \in \{-7, 7\}$.   **c.** $x = 100$.
d. $x = 1$.   **e.** $x = 9$.   **f.** $x = \dfrac{1}{13}$.

23. a. $x = 2$.   **b.** $y = -\dfrac{6}{5}$.   **c.** $y = -166$.   **d.** $t = \dfrac{3}{2}$.
e. $b = -8$.   **f.** $x = 10$.
24. a. $x = 13$.   **b.** $x = (-11)$.
25. a. $x = 12$.   **b.** $x = 1$.   **c.** $t = 6$.   **d.** $x = 2.52$.
e. $y \approx 0.594$.   **f.** $x = -0.35$.

26. a. $\left(-\infty, -\dfrac{10}{9}\right];$

b. $(-6, \infty);$

c. $\left(-\infty, \dfrac{3}{4}\right);$

d. $(-2, \infty);$

e. $(-\infty, -2);$

f. $(-\infty, -2) \cup (-2, \infty);$

27. a. $(20, \infty);$

b. $\left[-\dfrac{28}{9}, \infty\right);$

c. $(-\infty, 3);$

d. $(-\infty, 10) \cup (10, \infty);$

e. $(-\infty, 1];$

f. $(-\infty, 1.5];$

28. a. $x \in \{-5, 5\}$.   **b.** $x \in \left\{-\dfrac{1}{2}, \dfrac{1}{2}\right\}$.   **c.** $x \in \{-5, 1\}$.
29. a. $x \in \{1, 2\}$.   **b.** $x \in \{-6, -5\}$.   **c.** $x \in \{0, 2, 3\}$.
d. $x \in \{2, 3\}$.   **e.** $x \in \{-4, 1, 3, 6\}$.

30. a. $x = \dfrac{9}{4}$.   **b.** $x = 144$.   **c.** $x = 33$.

31. a. $x = \dfrac{5}{2}$.   **b.** $x = 27$.   **c.** $x = 20$.

32. a. $x = 1$.   **b.** $x = \dfrac{3}{2}$.   **c.** $x = -2.5$.

33. a. $x = -\dfrac{25}{2}$.   **b.** $x = \dfrac{5}{2}$.   **c.** $x = -6$.

34. a. $(3, 7)$.   **b.** $(-\infty, \infty)$.   **c.** $[-1, \infty)$.   **d.** $[2, 5]$.
e. $(-\infty, -3]$.   **f.** $\varnothing$.

35. a. (5, 7). **b.** (4, 7). **c.** [2, 3) ∪ (4, 5]. **d.** [−8, ∞).
e. (−∞, 6]. **f.** (−∞, 0) ∪ (7, ∞).
36. a. $x \in \{0, 1\}$. **b.** $x \in \{-1, -2\}$. **c.** $x \in \{-2, 2\}$.
d. $x \in \left\{-3, \dfrac{1}{2}\right\}$.
37. a. $x = 1$ or $x = -4$. **b.** $x = -2$ or $x = 2$. **c.** $x = 1$.
d. $x = -2$ or $x = -3$. **e.** $x = 3$ or $x = \dfrac{1}{2}$.
f. $x = \dfrac{1}{4}$ or $x = -\dfrac{2}{3}$.
38. a. Set the function equal to zero and solve for x.
i. The root is $x = 4$. **ii.** The root is $x = 2$. **iii.** The root is $x = 3$. **b.** The root of any function is any point where the graph crosses the x-axis.
i.

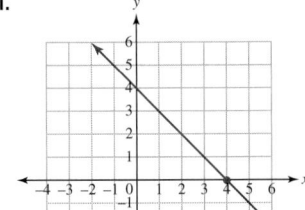

ii.

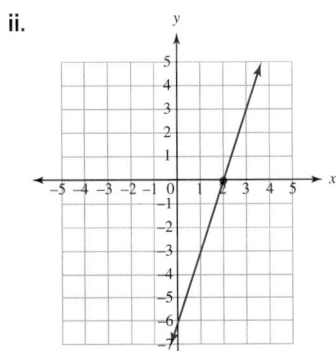

iii.

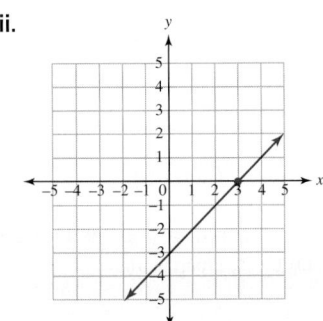

c. i. The root is $x = -12$. **ii.** The roots are $x = 0$ and $x = (-6)$. **iii.** The roots are $x = (-3)$ and $x = 3$.
39. a. $x = \dfrac{1}{2}$. **b.** $x = -\dfrac{11}{18}$.
40. If we add 5 first, we have
$$2x - 5 = 7$$
$$2x = 12$$
$$x = 6$$

If we divide by 2 first, we have
$$2x - 5 = 7$$
$$x - \dfrac{5}{2} = \dfrac{7}{2}$$
$$x = \dfrac{12}{2} = 6$$
Either way, the solution is $x = 6$.
41. a. True. **b.** Cannot be determined. **c.** False. **d.** False. **e.** True.
42. We can verify each step in finding the solution set using an algebraic property that guarantees that step to be true.
43.
$(4x - 3) + 3 = 9 + 3$ Add three to both sides
$4x + (-3 + 3) = 9 + 3$ Associative property
$4x + 0 = 9 + 3$ Additive inverse property
$4x = 9 + 3$ Additive identity property
$4x = 12$ Whole-number addition
$\dfrac{4x}{4} = \dfrac{12}{4}$ Divide both sides by 4.
$1x = 3$ Multiplicative inverse property
$x = 3$ Multiplicative identity property
44.
$-x - 2x - 6 \le 4$ Distributive property
$-3x - 6 \le 4$ Combine like terms.
$(-3x - 6) + 6 \le 4 + 6$ Add 6 to both sides
$-3x + (-6 + 6) \le 4 + 6$ Associative property
$-3x + 0 \le 10$ Additive inverse property
$-3x \le 10$ Additive identity property
$\dfrac{-3x}{-3} \ge -\dfrac{10}{3}$ Divide both sides by (−3) and reverse inequality.
$1x \ge -\dfrac{10}{3}$ Multiplicative inverse property
$x \ge -\dfrac{10}{3}$ Multiplicative identity property
45. Specific steps may vary. One possibility is shown.
$-3x + 4(x - 2) = 6x - 1$
$-3x + 4x - 8 = 6x - 1$ Distributive property
$x - 8 = 6x - 1$ Combine like terms.
$-8 = 5x - 1$ Subtract x from both sides.
$-7 = 5x$ Add 1 to both sides.
$-\dfrac{7}{5} = x$ Divide both sides by 5.
46. Specific steps may vary. One possibility is shown.
$x + 2 \ge -4x + 3$
$5x + 2 \ge 3$ Add $4x$ to both sides.
$5x \ge 1$ Subtract 2 from both sides.
$x \ge \dfrac{1}{5}$ Divide both sides by 5.
47. Answers will vary.
48. Answers will vary.
49. She associated the x with the ones digit. Answers will vary on how to correct the student's thoughts.
50. He is incorrect. He forgot to reverse the inequality when he divided by (−3).
51. The inequality suggests that (−3) is larger than 1. Answers will vary on how to correct the student's thoughts.
52. When evaluating a function, we place the values into the independent variable to find the corresponding values for the dependant variable. When solving an equation, we are finding the values of an unknown quantity.

53. We make use of numerical operations, specifically inverse operations, to solve equations. Answers will vary on how this connection will better help students understand solving equations.

54. a. We use them to make algebraic manipulations when solving equations. **b.** We assume that all algebraic variables represent numbers.

55. a. Addition and subtraction are inverse operations because they "undo" each other. **b.** Answers will vary, but possibilities are addition and subtraction, multiplication and division, squares and square roots, and so on.

Section 9.5 Learning Assessment

1. Mathematical modeling is the process by which we use mathematical representations to understand and solve real-world problems. We translate the situation into a mathematical model, and then use mathematical computations and properties to solve the problem. Once we have an answer, we translate the answer back into the context of the problem.

2. A function is increasing if increasing values of x are paired with increasing values of $f(x)$. The function is decreasing if increasing values of x are paired with decreasing values of $f(x)$. And the function is constant if increasing values of x are paired with only one value for $f(x)$. Graphically, as we move toward the right on the Cartesian coordinate system, the graph slopes upward to the right if it is increasing, slopes downward to the right if it is decreasing, and remains horizontal if it is constant.

3. If we can generate an algebraic equation for the function, then we can use the function to make predictions beyond the given values of the problem by substituting values into the equation and computing.

4. a. A system of linear equations is two or more linear equations that involve the same unknowns. **b.** A solution to the system of equations is any ordered pair of numbers (x, y) that makes both equations true.

5. In the substitution method, we solve one of the equations for one of the variables and then substitute the result into the other equation. This eliminates one variable, leaving one equation with one unknown that we can solve directly. In the elimination method, we eliminate one variable by adding or subtracting the original or equivalent equations. Once we have one equation with one unknown, we isolate the variable and use its value to find the other unknown. A third way to solve the system is to graph the equations, find their point of intersection, and read the value of its coordinates from the graph.

6. Two lines can either be parallel, intersect in one point, or be the same line. If the lines are parallel, they have no common points, and the system of equations has no solution. If the lines intersect at a point, the system of equations has exactly one solution. If the lines are the same, then they have infinitely many points in common, and the system of equations has an infinite number of solutions.

7. Graphs will vary, but possible ones are shown.

a.

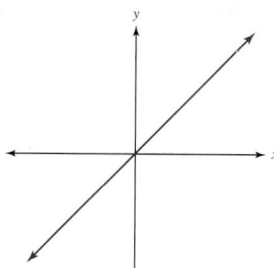

b.

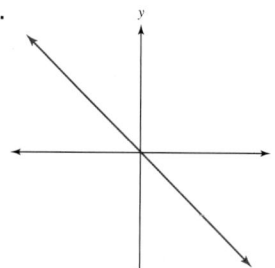

c.

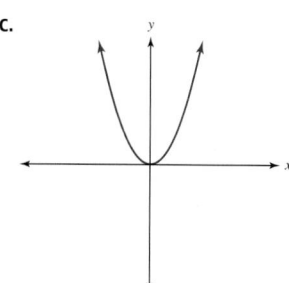

d.

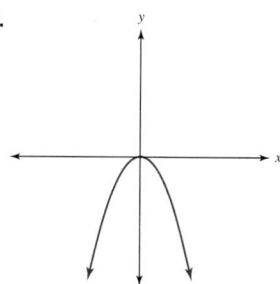

8. Possible answers are shown.
a. $f(x) = x$. **b.** $f(x) = -x$. **c.** $f(x) = x^2$.
d. $f(x) = -x^2$.

9. a. Constant. **b.** Increasing. **c.** Increasing. **d.** Decreasing.

10. a. Decreasing. **b.** Constant. **c.** Decreasing. **d.** Decreasing.

11. Possible graphs are shown.

a.

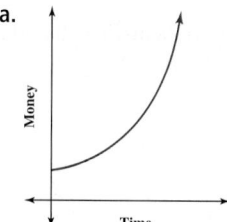

b.

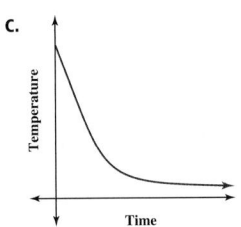

c.

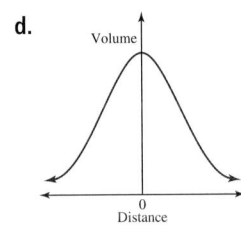

d.

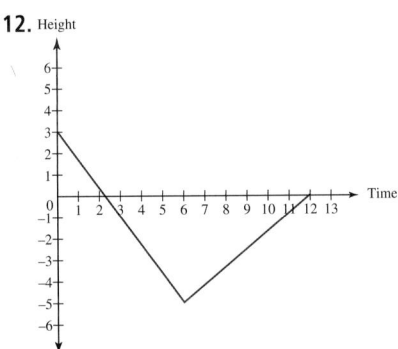

12.

![Height vs Time graph]

13. $C(x) = 175x + 200$, where x is the number of credits taken.

14. $C(x) = 0.25x + 3$, where x is the number of checks written.

15. a. It is increasing on $(2, \infty)$, decreasing on $(-3, -1)$, and constant on $(-\infty, -3) \cup (-1, 2)$. **b.** It is increasing on $(-\infty, -2) \cup (2, \infty)$ and decreasing on $(-2, 2)$. **c.** It is increasing on $(0, \infty)$ and decreasing on $(-\infty, 0)$.

16. a. Decreasing on $(-\infty, \infty)$. **b.** Increasing on $(-\infty, \infty)$. **c.** Increasing on $(0, \infty)$ and decreasing on $(-\infty, 0)$. **d.** Increasing on $(-\infty, 0)$ and decreasing on $(0, \infty)$.

e. Increasing on $\left(-\sqrt{\dfrac{2}{3}}, \sqrt{\dfrac{2}{3}}\right)$ and decreasing on $\left(-\infty, -\sqrt{\dfrac{2}{3}}\right) \cup \left(\sqrt{\dfrac{2}{3}}, \infty\right)$.

f. Increasing on $\left(-\infty, -\dfrac{2\sqrt{3}}{3}\right) \cup \left(\dfrac{2\sqrt{3}}{3}, \infty\right)$ and decreasing on $\left(-\dfrac{2\sqrt{3}}{3}, \dfrac{2\sqrt{3}}{3}\right)$.

17. a. No solution. **b.** One solution. **c.** Infinitely many solutions. **d.** One solution. **e.** One solution. **f.** One solution.

18. a. $(1, 2)$. **b.** $(2, 4)$. **c.** $(-1, 1)$. **d.** No solution.

19. a. $(11, -5)$. **b.** $(-6, -16)$. **c.** $\left(-5, -\dfrac{5}{2}\right)$. **d.** $(8, -2)$.

20. a. $(1, 3)$. **b.** $\left(-\dfrac{1}{2}, 2\right)$. **c.** $(-4, -20)$. **d.** $(2, 2)$.

21. a. The least expensive is electric. The most expensive is solar. **b.** Install solar. It will be the cheapest over 30 years. **c.** About 12 to 13 years.

22. a. $25,000,000. **b.** $30,000,000. **c.** $5,000,000 **d.** The cost will equal the revenue at 10,000 televisions. This is the point at which the company breaks even and begins to earn a profit.

23. a. B is more expensive, and A is less. **b.** A is newer, and B is older. **c.** A has the better gas mileage, and B is worse. **d.** A is the smaller car, and B is the larger.

24. $C(x) = x + 0.15x$, where x is the amount on the bill. **a.** Domain $= [0, \infty)$. **b.** $C(35.49) = 35.49 + 0.15(35.49) = \$40.81$.

25. $C(x) = 350 + 15x$, where x is the number of square yards of carpet. **a.** Domain $= [0, \infty)$. **b.** $C(25) = 350 + 15(25) = \$725$.

26. a. $T(x) = 0.023x$, where x is a person's income. **b.** The independent variable is income. and the dependent variable is the tax amount. **c.** Domain $= [0, \infty)$. **d.** $T(50,000) = 0.023(50000) = \$1,150$. **e.** If $1,300 = 0.023x$, then $x = \$56,521.74$.

27. a. The equation is $3x + x = 150,595$, where x is the amount Charity B receives. Charity A received $112,946.25, and charity B received $37,648.75. **b.** Answers will vary. **c.** Answers will vary.

28. 113 pizzas. **29.** $46\dfrac{2}{3}$ hr.

30. The statue is 151 ft tall, and the pedestal is 154 ft tall.

31. $833.33.

32. The rectangle has a width of 4 yd and a length of $3(4) + 1 = 13$ yd.

33. 3.792×10^{17} BTUs. **34.** 36 days.

35. The lowest grade Li can make and still have an A is 96.

36. 500 student and 350 adult tickets were sold.

37. 8.6 hr.

38. Mix 8 lb of large gumballs with 32 lb of small gumballs.

39. Mix 5.94 lb of dried apples with 9.06 lb of raisins.

40. The principal on larger account is about $42,083.33 and the total inheritance is $56,111.11.

41. a. $40.85. **b.** $61.65. **c.** $88.69.

42. a. 815 hamburgers. **b.** 311 hamburgers.

43. a. The changes are relatively linear because the difference between each month is approximately the same. **b.** About 10 months. **c.** About 12 months.

44. Answers will vary, but one possible story follows. The person was at rest, exercised for a short period of time, and then was at rest again.

45. He went to the first store and shopped for a while. He then went to a second store that was further from his home than the first and again shopped for a while. He went to a third store that was on his way home from the second. After shopping for a while, he returned home.

46. a. Horse 2 had the fastest start, and horse 3 had the slowest. **b.** Horse 1. **c.** The horses started at the same time, with horse 2 pulling out in front followed by horses 1 and 3. About halfway through the race, horse 3 had a sudden burst of speed and passed both horses 1 and 2 to take the lead. Near the end, horse 3 became tired and was passed by horses 1 and 2. Horse 1 had a burst of speed at the end to pass Horse 2 for the win.

47. Answers will vary.

48. Answers will vary.

49. Answers will vary.

50. Numerous models were used in each situation. Some specific examples are given.   **a.** Base-ten blocks.   **b.** Sets of objects, the number line, and the array model.   **c.** The number line and two-sided chips.   **d.** Fraction disks, fraction bars, and the number line.

51. Answers will vary.

52. a. Each amount of electricity used is assigned exactly one charge.   **b.** Each student will be assigned to only one disk.   **c.** Each Social Security number is assigned to only one person.   **d.** Each time is assigned to only one amount for distance traveled.

53. a. As the speed of a car increases at a constant rate, the gas mileage decreases at a constant rate.   **b.** As time traveled increases at a constant rate, the distance traveled increases at a constant rate.   **c.** As the years increase at a constant rate, the value of a car decreases at a constant rate.   **d.** As the number of units increases at a constant rate, the cost per units stays the same.

54. Answers will vary.

■ Chapter 9 Review Exercises

1. a. Neither: $a_n = a_{n-1} + (n - 1)$; $a_{10} = 46$; $a_{25} = 301$; $a_{50} = 1,226$.   **b.** Arithmetic: $a_n = 7n - 1$; $a_{10} = 69$; $a_{25} = 174$; $a_{50} = 349$.   **c.** Geometric: $a_n = 4^{n-1}$; $a_{10} = 4^9$; $a_{25} = 4^{24}$; $a_{50} = 4^{49}$.   **d.** Geometric: $a_n = (0.3)^{n-1}$; $a_{10} = (0.3)^9$; $a_{25} = (0.3)^{24}$; $a_{50} = (0.3)^{49}$.   **e.** Neither: No general term. a_{10} will have 10 ones, a_{25} will have 25 ones, and a_{50} will have 50 ones.   **f.** Arithmetic: $a_n = -3n + 1$; $a_{10} = -29$; $a_{25} = -74$; $a_{50} = -149$.

2. $a_{20} = 24.5$.

3. One possible answer is $a_n = a_{n-1} - 1$, where $a_1 = 5$.

4. The domain is the set that the function comes out of, the codomain is the set that the function goes into, and the range is the set of all values in the codomain that have been paired with something from the domain.

5. Three different functions are shown.

$$\{(b, o), (i, e),(g, t)\}$$

domain	range
b	e
i	t
g	o

6. a. $f(-1) = 4, f(3) = -12$, and $f(s + 2) = -4s - 8$.   **b.** $f(-1) = 4, f(3) = 12$, and $f(s + 2) = 2s + 10$.   **c.** $f(-1) = 6, f(3) = -2$, and $f(s + 2) = -s^2 - 4s + 3$.

7. a. $f(3) = 0$.   **b.** $f(3) = \sqrt{3}$.   **c.** $f(3) = \dfrac{1}{6}$.   **d.** $f(3)$ does not exist.   **e.** $f(3) = 0$.

8. Yes. Each input value, or city, is assigned to exactly one output value, flight time.

9. One possible answer is the set $\{(0, -5), (1, -2), (2, 7), (3, 22), (4, 43)\}$.

10. One possible answer is the table:

x	y
0	-7
1	-6
2	-4
3	0
4	8

11. a. Two units to the right and 1 unit below the origin.   **b.** Three units to the left and 3 units below the origin.   **c.** Four units above the origin.   **d.** Five units to the left of the origin.

12. a. $\mathbb{R}$.   **b.** $\mathbb{R}$.   **c.** $[-3, \infty)$.   **d.** $\mathbb{R} - \{5\}$.

13. The tables of values will vary. The graphs are shown.

a.

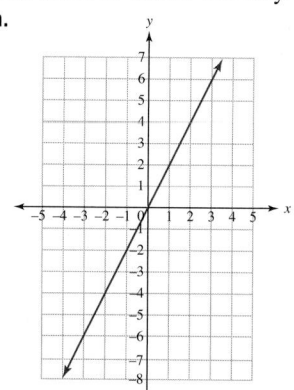

b.

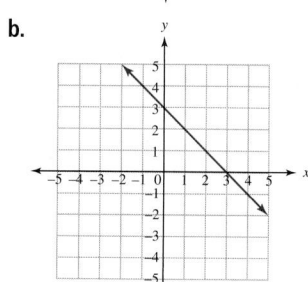

c.

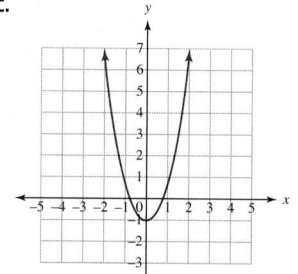

d.

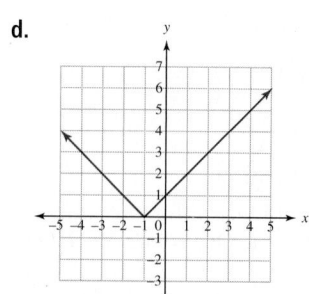

14. a. The line goes down 2 units for every unit it goes to the right. **b.** The line goes down 4 units for every 5 it goes to the right. **c.** The line is vertical.

15. a. $m = \frac{1}{8}$; increasing. **b.** $m = -\frac{4}{3}$; decreasing.

c. $m = 0$; horizontal.

16. A line with a slope of 5 goes up 5 units for every 1 unit it goes to the right, whereas a line with a slope of $\frac{1}{5}$ goes up 1 unit for every 5 it goes to the right. Hence, a line with a slope of 5 is steeper.

17. a.

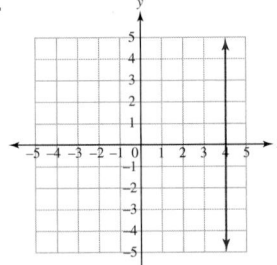

b.

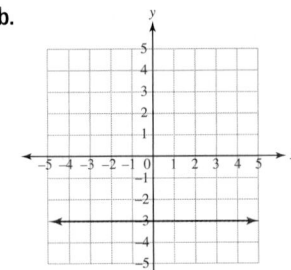

c.

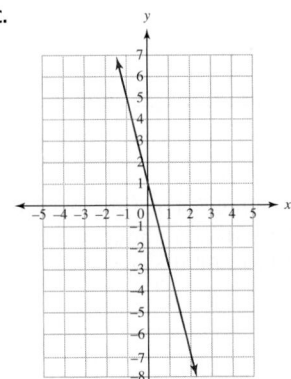

d.

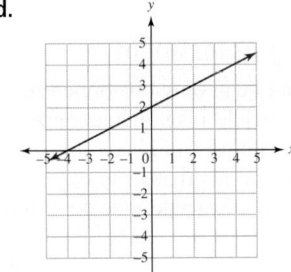

18. a. $m = -1$; y-intercept $= -5$; x-intercept $= (-5, 0)$.

b. $m = \frac{2}{3}$; y-intercept $= 0$; x-intercept $= (0, 0)$.

c. $m = \frac{5}{2}$; y-intercept $= -1$; x-intercept $= \left(\frac{2}{5}, 0\right)$.

19. a. $x = 4$. **b.** $y = -x + 3$. **c.** $y = \frac{3}{4}x + \frac{11}{4}$.

20. a. One possible answer is $y = -3x + 5$.

b. One possible answer is $y = -\frac{1}{2}x + 1$.

21. a. $\{x \mid x \le 5\} = (-\infty, 5]$;

b. $\{x \mid 2 < x \le 7\} = (2, 7]$;

22. a. $y = -\frac{5}{3}$. **b.** $y = \frac{59}{3}$. **c.** $t \approx 4.71$. **d.** $x = -\frac{3}{5}$.

e. $\left[-\frac{1}{4}, \infty\right)$. **f.** $\left(-\infty, -\frac{17}{3}\right]$. **g.** $(-11, \infty)$.

h. $\left(-\infty, \frac{4}{3}\right) \cup \left(\frac{4}{3}, \infty\right)$. **i.** $(-6, \infty)$.

23. a. $\left(-\frac{5}{4}, -\frac{5}{2}\right)$. **b.** $\left(-\frac{1}{3}, -\frac{7}{3}\right)$.

24. a. One possible answer would be the equations $y = x + 1$ and $y = x - 1$. **b.** One possible answer would be the equations $y = x + 1$ and $y = -x$. **c.** One possible answer would be the equations $x + y = 1$ and $2x + 2y = 2$.

25. On the seventh day, there are 65,450 gal; on the eighth day, there are 74,800 gal; and on the ninth day, there are 84,150 gal.

26. a. $14. **b.** $46.50. **c.** $84.

27. a. $C(x) = x + 0.07x$. **b.** Independent variable = purchase price; dependent variable = total cost. **c.** $[0, \infty)$.

28. Traffic increases until 8 a.m., when it peaks and holds fairly steady for an hour. It then drastically declines and holds constant until about 4. At 4 p.m., it increases rapidly until 5 p.m. and then remains constant until 6 p.m. After 6 p.m., it drastically declines again.

29. 53 hr.   **30.** 1,969.52 g.

■ Answers to Reviewing the Big Ideas: Chapters 7–9

1. $\frac{5}{8}$ represents part of a whole that has been divided into 8 equivalent pieces, of which 5 are counted.

2. a. $5\frac{5}{7}$. **b.** $-3\frac{5}{8}$. **c.** $\frac{25}{7}$. **d.** $-\frac{26}{5}$.

3. a. $\frac{4}{7}$. **b.** $\frac{2}{5}$. **c.** $-\frac{9}{4}$.

4. Five fractions between $\frac{4}{7}$ and $\frac{5}{7}$ are $\frac{25}{42}, \frac{26}{42}, \frac{27}{42}, \frac{28}{42}$, and $\frac{29}{42}$.

5. a. $\frac{11}{14}$. **b.** $\frac{17}{15}$. **c.** -1. **d.** $\frac{10}{9}$.

6. a. $6\frac{1}{6}$. **b.** $2\frac{1}{6}$. **c.** $2\frac{7}{24}$. **d.** $-1\frac{31}{45}$.

7. a. $\frac{5}{4}$. **b.** -4. **c.** $\frac{1}{8}$. **d.** $\frac{4}{9}$.

8. Mrs. Henry's class has a higher value of students playing sports.

9. a. 34.209.   **b.** 0.019.   **c.** $0.\overline{3}$.   **d.** 1.35.

10. Because $\dfrac{17}{80}$ can be written as a decimal fraction,

or $\dfrac{17}{80} = \dfrac{17 \cdot 125}{80 \cdot 125} = \dfrac{2{,}125}{10{,}000}$, it has a terminating decimal

representation of 0.2125.

11. a. Not possible.   **b.** Irrational.   **c.** Rational.
   d. Rational.

12. a. 78.70.   **b.** 3.573.   **c.** 79.390.   **d.** 4.876.

13. $x = \dfrac{59}{90}$.

14. a. 4.59×10^5.   **b.** 5.601×10^{-4}.   **c.** 0.000000349.
   d. 28,800.

15. a. 71%.   **b.** 105%.   **c.** 80%.   **d.** 50%.

16. $880.31.

17. a. 14 females to 24 students.   **b.** 10 males to 14
   females.   **c.** 24 students to 10 males.

18. The 3 lb of meat is a better buy.

19. 7.2 min.

20. a. 36.   **b.** 43.01%.   **c.** 72.22.

21. a. $3,850.   **b.** $35,980.67

22. $47,276.89.

23. Stock portfolio A.

24. a. Neither.   **b.** Geometric: $a_n = 3(2)^{n-1}$; $a_{10} = 3(2)^9$;
   $a_{15} = 3(2)^{14}$; and $a_{20} = 3(2)^{19}$.
   c. Arithmetic: $a_n = (-4)n + 1$; $a_{10} = (-39)$, $a_{15} = (-59)$,
   and $a_{20} = (-79)$.   **d.** Neither.

25. a. $a_1 = 2$, $a_2 = 8$, $a_3 = 26$, $a_4 = 80$, $a_5 = 242$, and $a_6 = 728$.
   b. $a_1 = 2$, $a_2 = 3$, $a_3 = 6$, $a_4 = 18$, $a_5 = 108$, and $a_6 = 1{,}944$.

26. Possible functions are shown.

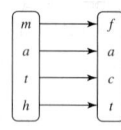   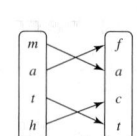

$\{(m, f), (a, a), (t, c), (h, t)\}$ $\{(m, a), (a, f), (t, t), (h, c)\}$

x	y
m	f
a	a
t	c
h	t

x	y
m	a
a	f
t	t
h	c

27. a. $f(2) = 7$, $f(-4) = -11$, $f(t - 3) = 3t - 8$.
   b. $f(2) = -11$, $f(-4) = -35$, $f(t - 3) = -2t^2 + 12t - 21$.

28. a. $\mathbb{R}$.   **b.** $\mathbb{R} - \{3\}$.   **c.** $[-3, \infty)$.   **d.** $\mathbb{R} - \{5\}$.

29. Tables of values will differ. The graphs are shown.
   a.

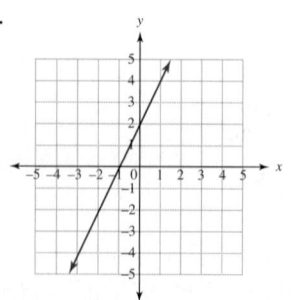

b.

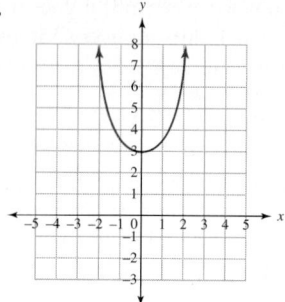

c.

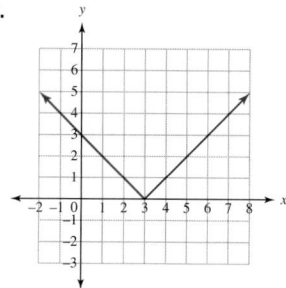

30. a. $m = \dfrac{5}{3}$; increasing.   **b.** $m = 0$; horizontal.   **c.** $m = -1$;
   decreasing.   **d.** m is undefined; vertical.

31. a. $m = 1$, y-intercept $= (0, 5)$, and x-intercept $= (-5, 0)$.
   b. $m = 0$, y-intercept $= (0, -3)$, and x-intercept does not
   exist.   **c.** $m = -\dfrac{1}{3}$, y-intercept $= (0, -3)$, and the
   x-intercept $= (-9, 0)$.

32. $y = -2x - 2$.

33. a. $y = \dfrac{7}{3}$.   **b.** $t = 8$.   **c.** $x = -17$.

34. a. $(-\infty, -7]$;   ⟵————————●————————⟶ .
                                    -7

   b. $(-\infty, -1)$;   ⟵————————○————————⟶ .
                                    -1

   c. $(-\infty, 3) \cup (3, \infty)$;   ⟵————————○————————⟶ .
                                    3

35. a. $\left(1, \dfrac{1}{2}\right)$.   **b.** $(1, -1)$.

36. After the 10th hour, the pool holds 4,500 gal. After the
   11th, it holds 4,950 gal. After the 12th hour, it holds
   5,400 gal. It will take 71.1 hr to fill the pool.

37. 13 nickels and 17 dimes.

38. 171.9 g.

CHAPTER 10

■ Section 10.1 Learning Assessment

1. A line is a set of points that lie in only one dimension.
   If the line were to bend, it would take on a second
   dimension, which would mean it is no longer a line.
   Likewise, a plane is a set of points that lie in two
   dimensions. If it were to bend, it would take on a third
   dimension, so it would no longer be a plane.

2. Postulates are mathematical statements that are readily
   apparent, so we accept them without proof. Theorems are
   mathematical statements that can be proven.

3. Lines are infinite in two directions, rays are infinite in one
   direction, and line segments are finite.

4. Parallel lines are distinct coplanar lines with no point in common. Perpendicular lines are intersecting lines that form one right angle.

5. Congruence is the idea that two geometric figures take on the same size and shape. Congruence applies to shapes, whereas equality generally applies to numbers. Using equality with shapes implies that two shapes are not only the same size but also exactly the same set of points.

6. The interior of an angle is not part of the angle. The only points that are part of the angle are the vertex or the points that lie on the rays that are the sides of the angles.

7. We can convert the decimal part of a degree to a fraction with a power of ten in the denominator. We can then use a proportion to determine the equivalent number of minutes. We can use the same process to convert the decimal part of a minute to a second.

8. An acute angle has a measure less than 90°, a right angle has a measure of exactly 90°, an obtuse angle has a measure greater than 90° but less than 180°, a straight angle has a measure of exactly 180°, and a reflex angle has a measure greater than 180° but less than 360°.

9. The complement of an angle is the angle with a measure that, when added to the measure of the acute angle, is equal to 90°. The supplement of an angle is the angle with a measure that, when added to the measure of the acute angle, is equal to 180°.

10. Alternate interior angles are on opposite sides of the transversal and between the two parallel lines. Alternate exterior angles are on opposite sides of the transversal and outside the two parallel lines. Same-side interior angles are on the same side of the transversal and between the two parallel lines. Same-side exterior angles are on the same side of the transversal and are outside the two parallel lines. Corresponding angles are on the same side of the transversal and take the same position relative to the two parallel lines.

11. a. Line *TO*. **b.** Line segment *AM*. **c.** Angle *BAY*.
  d. Ray *GO*. **e.** Line *p* is parallel to line *q*.
  f. Line *l* is perpendicular to line *m*.

12. a. 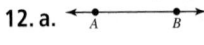 **b.**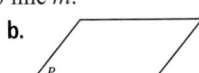

  c.  **d.**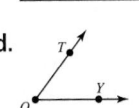

  e.  **f.**

  g. 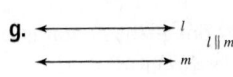 **h.**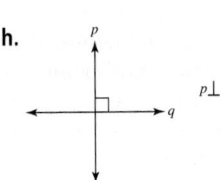

13. a. Yes. The order of the endpoints does not matter. It will name the same segment. **b.** No. $\overrightarrow{AB}$ has its endpoint at *A* and extends in the direction of *B*. $\overrightarrow{BA}$ has its endpoint at *B* and extends in the direction of *A*. **c.** Yes. The order in which points are listed does not matter. It will name the same line.

14. ∠*AND*, ∠*DNA*, ∠*N*, and ∠1.

15. a.

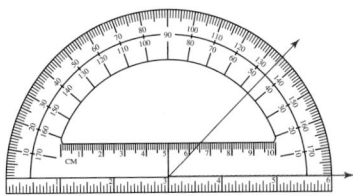

  b.

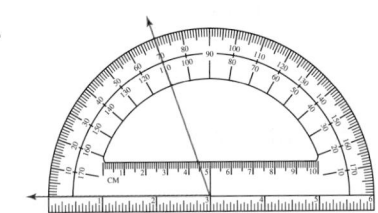

  c.

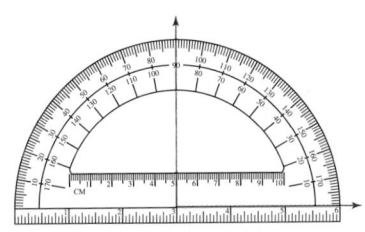

  d.

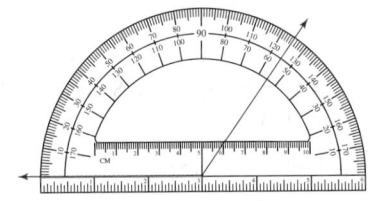

  e.

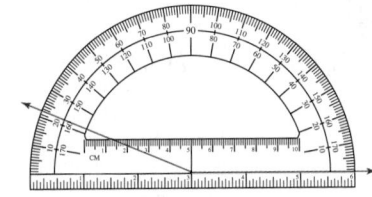

16. a. Acute. **b.** Right. **c.** Obtuse. **d.** Straight. **e.** Acute.

17. Example illustrations are shown:

  a. **b.**

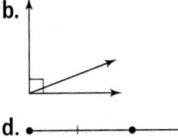

  c. **d.**

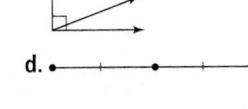

18. a. $\overleftrightarrow{AN}$, $\overleftrightarrow{AD}$, $\overleftrightarrow{AY}$, $\overleftrightarrow{ND}$, $\overleftrightarrow{NY}$, and $\overleftrightarrow{DY}$.
  b. Plane *ABC*, plane *ABD*, plane *ABE*, plane *ACD*, plane *ACE*, plane *ADE*, plane *BCD*, plane *BCE*, plane *BDE*, and plane *CDE*.

19. a. The crease is the bisector of $\overline{AB}$. It is also perpendicular to $\overline{AB}$. **b.** The crease is the angle bisector of ∠*PQR*.

20. One possible diagram is given.

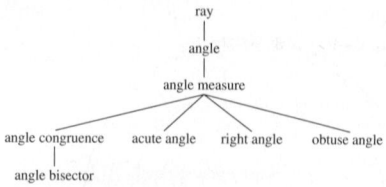

21. Possible answers are given:
   a. A, B, and C.   **b.** $\overleftrightarrow{AB}$ and $\overleftrightarrow{EF}$.   **c.** $\overrightarrow{AE}$ and $\overrightarrow{BF}$.
   d. $\overline{AD}$ and $\overline{BC}$.   **e.** Plane ABC and plane EFH.
   f. $\overleftrightarrow{AB}$ and $\overleftrightarrow{EF}$.

22. Possible answers are given:
   a. A, B, and C.   **b.** $\angle ABG$.   **c.** $\angle BCE$.   **d.** $\angle GHE$.
   e. $\angle GFE$.   **f.** $\angle AHG$ and BHI.   **g.** $\angle ABH$ and $\angle HBC$.
   h. $\overline{AC}$ and $\overline{GE}$.   **i.** $\overline{AC}$ and $\overline{CE}$.

23. a. $180°$.   **b.** $120°$.   **c.** $90°$.   **d.** $45°$.   **e.** $240°$.   **f.** $225°$.

24. a. $\dfrac{1}{4}$.   **b.** $\dfrac{1}{8}$.   **c.** $\dfrac{1}{24}$.   **d.** $\dfrac{3}{8}$.   **e.** $\dfrac{3}{4}$.   **f.** $\dfrac{7}{18}$.

25. a. $14°\ 15'\ 0''$.   **b.** $12°\ 39'\ 0''$.

26. a. $43°\ 4'\ 12''$.   **b.** $94°\ 37'\ 51.6''$.

27. a. $10.25°$.   **b.** $65.223°$.

28. a. $47.796°$.   **b.** $93.014°$.

29. a. 15 angles.   **b.** 10 acute and 5 obtuse.

30. a. $55°$.   **b.** $125°$.   **c.** $126°$.   **d.** $71°$.   **e.** $36°$.

31. a. $60°$.   **b.** $60°$.   **c.** $150°$.   **d.** $120°$.

32. a. $67°$.   **b.** $6°$.   **c.** $169°$.   **d.** $57°$.

33. a. $20°$.   **b.** $115°$.   **c.** $65°$.

34. a. $\angle 4$, $\angle 5$, and $\angle 8$.   **b.** $\angle 2$, $\angle 3$, and $\angle 6$.
   c. $m\angle 1 = 133°$, $m\angle 2 = 47°$, $m\angle 3 = 47°$, $m\angle 4 = 133°$,
   $m\angle 5 = 133°$, $m\angle 6 = 47°$, $m\angle 7 = 47°$, and $m\angle 8 = 133°$.
   d. $m\angle 1 = 128°$, $m\angle 2 = 52°$, $m\angle 3 = 52°$, $m\angle 4 = 128°$,
   $m\angle 5 = 128°$, $m\angle 6 = 52°$, $m\angle 7 = 52°$, and $m\angle 8 = 128°$.

35. a. $m\angle 1 = 120°$, $m\angle 2 = 60°$, $m\angle 3 = 60°$, $m\angle 4 = 120°$,
   $m\angle 5 = 120°$, $m\angle 6 = 60°$, $m\angle 7 = 60°$, and $m\angle 8 = 120°$.
   b. $m\angle 1 = 125°$, $m\angle 2 = 55°$, $m\angle 3 = 55°$, $m\angle 4 = 125°$,
   $m\angle 5 = 125°$, $m\angle 6 = 55°$, $m\angle 7 = 55°$, and $m\angle 8 = 125°$.

36. $m\angle 1 = 105°$, $m\angle 2 = 35°$, $m\angle 3 = 35°$, $m\angle 4 = 105°$,
   $m\angle 5 = 40°$, $m\angle 6 = 145°$, $m\angle 7 = 35°$, $m\angle 8 = 145°$,
   $m\angle 9 = 40°$, $m\angle 10 = 140°$, $m\angle 11 = 140°$, and
   $m\angle 12 = 40°$.

37. a. $x = 90$.   **b.** $x = 23.33$ and $y = 11.67$.   **c.** $x = 28$.

38. 66 segments.

39. a. 4 planes.   **b.** 10 planes.

40. The smallest is 1, and the largest is 15.

41. a. 7.   **b.** 11.   **c.** 16.   **d.** $\dfrac{n(n+1)}{2} + 1$.

42. 12 pairs of vertical angles.

43. a. $45°$.   **b.** $0°$ or $360°$.   **c.** $155°$.   **d.** $16°$.   **e.** $69°\ 24'\ 31''$.

44. a. $81.38°$.   **b.** $49°\ 41'\ 41''$.   **c.** $52.11°$.   **d.** $29°\ 16'\ 48''$.

45. a. 1.57 rad.   **b.** 0.79 rad.   **c.** 2.09 rad.   **d.** 3.02 rad.
   e. 1.20 rad.

46. a. $45°$.   **b.** $90°$.   **c.** $67.5°$.

47. a. $89.99°$.   **b.** $89°\ 59'\ 59''$.

48. a. $90.01°$.   **b.** $90°\ 0'\ 01''$.

49. About $182.5°$ in a clockwise fashion or $177.5°$ in a
   counterclockwise fashion.

50. a. $0.986°$.   **b.** $29.58°$.

51. $15°$ in 1 hr. $0.25°$ in 1 min.

52. The city planners can find the bisector of the segment
   connecting the two towns. The point at which the bisector
   intersects the shore is a point that is equidistant from the
   two towns.

53. Postulate 1 guarantees that any two points determine a
   line, so they must be collinear. Not every three points must
   be collinear. For example, the following points A, B, and
   C, are not collinear.

   $A \bullet$
   $\bullet C$
   $B \bullet$

54. Yes. All rays are infinite in length, so they must be
   congruent.

55. It divides the plane into three regions: the interior of the
   angle, the exterior of the angle, and the angle itself.

56. The two angles must be right angles, so their sides must be
   perpendicular to one another.

57. a. Acute.   **b.** Obtuse.   **c.** Right.   **d.** Acute.

58. a. No. The sum of their measures is greater than $180°$.
   b. No. The sum of their measures is less than $180°$.
   c. Yes. The sum of their measures is exactly $180°$.
   d. A linear pair always consists of an acute angle and an
   obtuse angle or two right angles.

59. a. Two acute angles.   **b.** Two acute angles.   **c.** Two right
   angles.   **d.** The resulting angles from a bisected angle
   always have measures less than or equal to $90°$.

60. Suppose that line l is parallel to line m and that both lines
   are cut by transversal t. $\angle 1 \cong \angle 3$ and $\angle 5 \cong \angle 7$ because
   each pair of angles are vertical angles. Also $\angle 3 \cong \angle 5$
   because they are alternate interior angles. This means that
   $\angle 1 \cong \angle 5$ because angles congruent to the same angle
   are also congruent. Likewise, $\angle 1 \cong \angle 7$. Consequently, if
   two parallel lines are cut by a transversal, then alternate
   exterior angles are congruent.

61. Suppose that line l is parallel to line m and that both lines
   are cut by transversal t. Because $\angle 2$ and $\angle 3$ form a linear
   pair, they must be supplementary angles, or $m\angle 2 +$
   $m\angle 3 = 180°$. Also, $\angle 2$ and $\angle 6$ are corresponding angles,
   and, because l is parallel to line m, $\angle 2 \cong \angle 6$. This implies
   $m\angle 2 = m\angle 6$. Hence, $m\angle 2 + \angle 3 = m\angle 6 + m\angle 3 = 180°$,
   implying that $\angle 3$ and $\angle 6$ are supplementary. Consequently,
   if two parallel lines are cut by a transversal, then same-side
   interior angles are supplementary.

62. Suppose line l is parallel to line m. Postulate 5 guarantees
   that line l contains two distinct points, say A and B, and
   line m contains two distinct points, say C and D. Because
   the lines are parallel, they do not intersect. Hence A, B, C,
   and D are all distinct noncollinear points. Consequently,
   Postulate 3 guarantees that exactly one plane contains any
   set of these three points, say points A, B, and C. However,
   C and D must be on the same plane by Postulate 4. Hence,
   both lines are on the same plane, so two parallel lines
   determine a plane.

63. Suppose line l intersects line m. Postulate 5 guarantees that
   line l contains two distinct points, say A and B, and line
   m contains two distinct points, say C and D. Hence, the
   two lines give us three distinct noncollinear points, and
   Postulate 3 guarantees that exactly one plane contains
   them. Hence, if two lines intersect, exactly one plane
   contains them.

64. Answers will vary. 65. Answers will vary.
66. Answers will vary. 67. Answers will vary.
68. There are six pairs of vertical angles.
69. The lines are not parallel. They intersect if we extend them far enough.
70. Answers will vary. 71. Answers will vary.
72. Answers will vary for each part.
73. Answers will vary.

Section 10.2 Learning Assessment

1. A simple, closed curve is convex if, for any two points on the inside of the curve, the segment connecting the points lies entirely in the curve. It is concave if this is not the case.
2. a. A polygon is a simple, closed curve made entirely of segments. b. With only two sides, one of two things must happen with the curve. Either the sides must bend so that the curve can be closed, or the sides must not meet so that they can be segments. Either way, the curve is not a polygon.
3. No. The only points that are part of the triangle are the vertices and any point on one of the three sides.
4. No. To form a triangle, the three segments must satisfy the triangle inequality. That is, the sum of the lengths of any two sides of a triangle must be greater than the length of the third side.
5. One way to classify triangles is to use the lengths of the sides. In this case, a triangle is scalene if it has no congruent sides, isosceles if it has at least two congruent sides, and equilateral if it has three congruent sides. Another way to classify triangles is to use the measures of the angles. In this case, a triangle is acute if every angle is acute, right if it has one right angle, obtuse if it has one obtuse angle, and equiangular if the angles have the same measure.
6. The angles opposite the congruent sides must be congruent.
7. a. To begin, choose one side of $\triangle ABC$, say $\overline{AB}$, and make a line l parallel to it through point C. This gives three angles, the sum of whose measures must be 180° because they lie along a straight line.

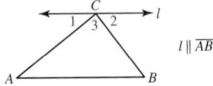

Because $l \parallel \overline{AB}$, then $\angle A \cong \angle 1$ and $\angle B \cong \angle 2$; they are alternate interior angles. By making a simple substitution, we have $m\angle A + m\angle B + m\angle 3 = m\angle 1 + m\angle 2 + m\angle 3 = 180°$.
b. This fact ensures that a triangle can have no more than one right or one obtuse angle. It ensures that the acute angles in a right triangle must be complementary and that the angles in an equilateral triangle must have a measure of 60°. It also ensures that the measure of an exterior angle in a triangle must equal the sum of the measures of the two remote interior angles.
8. The sum of the squares of the lengths of the legs of a right triangle is equal to the square of the length of the hypotenuse.
9. A Pythagorean triple consists of three natural numbers that satisfy the Pythagorean theorem.
10. In any 45°-45°-90° triangle, the length of the hypotenuse is always $\sqrt{2}$ times the length of the leg. In any 30°-60°-90° triangle, the hypotenuse is always two times the length of the short leg, and the long leg is always $\sqrt{3}$ times the short leg.

11. a. Nonsimple closed. b. Simple closed.
   c. Simple closed. d. Simple nonclosed.
12. Possible examples are given.
   a.  b.
   c.  d.

13. a. Polygon. It is a simple closed curve consisting of only segments. b. Not a polygon. The curve crosses itself, so it is nonsimple. c. Not a polygon. It has a part that is not a segment. d. Polygon. It is a simple closed curve consisting of only segments.
14. a. A, B, and C. b. $\overline{AB}$, $\overline{BC}$, and $\overline{AC}$. c. $\overline{AC}$. d. $\angle A$.
   e. Scalene. f. Obtuse. g. Acute, right, or obtuse.
15. a. Acute. b. Scalene.
16. a. Obtuse. b. Isosceles.
17. a. $\triangle ABC$, $\triangle ABD$, and $\triangle ABE$. b. $\triangle ABC$.
   c. $\triangle ABC$ and $\triangle DAB$. d. $\triangle AEB$ and $\triangle AED$.
18. a. Right scalene. b. Obtuse scalene. c. Right isosceles.
   d. Obtuse isosceles. e. Acute equilateral. f. Acute scalene.
19. Drawings will vary. The types of triangles are given:
   a. Obtuse scalene. b. Acute isosceles. c. Right scalene.
20. Possible answers are shown:
   a. 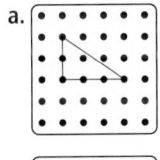 b.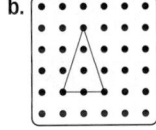
   c. 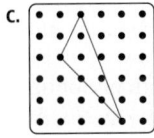 d.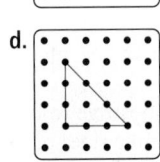
21. Possible segments are shown:
   a. 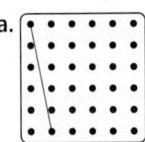 b. 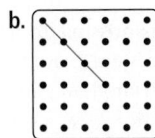 c.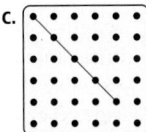
22. a. 4. b. 2. c. 6. d. 1.
23. a. No. b. Yes. c. Yes.
24. a. Yes. b. No. c. No.
25. a. $x = 75°$. b. $x = 53°$. c. $x = 81°$. d. $x = 22.5°$.
26. $m\angle 1 = 65°$, $m\angle 2 = 107°$, $m\angle 3 = 42°$, and $m\angle 4 = 138°$.
27. 60°.
28. 51°.
29. 23° and 134°.
30. a. $b = 21$. b. $a = 4.58$. c. $c = 10.82$.
31. a. $c = 10$. b. $c = 14.76$. c. $b = 12$.
32. a. $a = 17.89$. b. $c = 8.86$. c. $b = \frac{1}{3}$.
33. a. Yes. b. Yes. c. No. d. No.
34. a. No. b. Yes. c. Yes. d. Yes.
35. a. The other leg is 6, and the hypotenuse is $6\sqrt{2} \approx 8.49$.
   b. The two legs are $\frac{14}{\sqrt{2}} \approx 9.90$.
36. a. The other leg is 8, and the hypotenuse is $8\sqrt{2} \approx 11.31$.
   b. The two legs are $\frac{17}{\sqrt{2}} \approx 12.02$.

37. a. The long side is $4\sqrt{3}$, and the hypotenuse is 8.
   b. The short side is 6, and the long side is $6\sqrt{3}$.

   c. The short side is $\dfrac{8}{\sqrt{3}} \approx 4.62$, and the hypotenuse is
   $2\left(\dfrac{8}{\sqrt{3}}\right) \approx 9.24$.

38. The long side is $4.6\sqrt{3} \approx 7.97$, and the hypotenuse is 9.2.
   b. The short side is 3.5, and the long side is $3.5\sqrt{3} \approx 6.06$.

   c. The short side is $\dfrac{5.3}{\sqrt{3}} \approx 3.06$, and the hypotenuse is
   $2\left(\dfrac{5.3}{\sqrt{3}}\right) \approx 6.12$.

39. a. $x = 5$. **b.** $x = \dfrac{10}{\sqrt{2}} \approx 7.07$. **c.** $x = 5\sqrt{3} \approx 8.66$.

40. 9.

41. 18.

42. $u = \sqrt{2}$, $v = \sqrt{3}$, $w = \sqrt{4} = 2$, $x = \sqrt{5}$, $y = \sqrt{6}$, and $z = \sqrt{7}$.

43. They are 3-4-5, 6-8-10, 9-12-15, 12-16-20, 15-20-25, 18-24-30, 21-28-35, 24-32-40, 27-36-45, 30-40-50, 5-12-13, 10-24-26, 15-36-39, 7-24-25, 14-48-50, 8-15-17, 16-30-34, 9-40-41, 12-35-37, and 20-21-29.

44. They are 3-4-5, 5-12-13, 7-24-25, 8-15-17, 9-40-41, 11-60-61, 12-35-37, 13-84-85, 16-63-65, 20-21-29, 28-45-53, 33-56-65, 36-77-85, 39-80-89, 48-55-73, and 65-72-97.

45. 22.63 ft.   **46.** 9 in.
47. 145.77 ft.   **48.** 22.63 in.
49. 115.47 ft.   **50.** 882.36 ft.
51. 873.81 m.

52. The three angles always make a straight angle, so it suggests that the sum of the interior angles of a triangle is equal to 180°.

53. a. Yes, It is possible to have a triangle with one obtuse angle and three sides of different lengths. **b.** No. A scalene triangle has three sides of different lengths, but, in an isosceles triangle, two of the sides must be the same length. **c.** Yes. It is possible to have a triangle with one right angle and two sides the same lengths. **d.** Yes. It is possible to have a triangle with three acute angles and two sides the same length.

54. If $\triangle ABC$ is an equilateral triangle, then the three sides must be the same length, or $\overline{AB} \cong \overline{BC} \cong \overline{AC}$. Because $\overline{AB} \cong \overline{BC}$, then Theorem 10.4(1) guarantees that the angles opposite these sides are congruent, or $\angle A \cong \angle C$. Likewise, $\angle A \cong \angle B$ and $\angle B \cong \angle C$. Because all three angles are congruent, $\triangle ABC$ must be equiangular.

55. a. The midsegment is parallel to the unused side.
   b. The midsegment is half the length of the unused side.

56. a. Sometimes. **b.** Never. **c.** Never. **d.** Sometimes.
   e. Always. **f.** Never.

57. a. $a^2 + b^2$ will be greater than c^2.
   b. $a^2 + b^2$ will be less than c^2.

58. If $n = 1$, then $a = 2(1) + 1 = 3$, $b = 2(1)^2 + 2(1) = 4$, and $c = 2(1)^2 + 2(1) + 1 = 5$. 3-4-5 is a Pythagorean triple.
   If $n = 2$, then $a = 2(2) + 1 = 5$, $b = 2(2)^2 + 2(2) = 12$, and $c = 2(2)^2 + 2(2) + 1 = 13$. 5-12-13 is a Pythagorean triple.
   If $n = 3$, then $a = 2(3) + 1 = 7$, $b = 2(3)^2 + 2(3) = 24$, and $c = 2(3)^2 + 2(3) + 1 = 25$. 7-24-25 is a Pythagorean triple.

If $n = 4$, then $a = 2(4) + 1 = 9$, $b = 2(4)^2 + 2(4) = 40$, and $c = 2(4)^2 + 2(4) + 1 = 41$. 9-40-41 is a Pythagorean triple.
If $n = 5$, then $a = 2(5) + 1 = 11$, $b = 2(5)^2 + 2(5) = 60$, and $c = 2(5)^2 + 2(5) + 1 = 61$. 11-60-61 is a Pythagorean triple.

59. 3-4-5 is the only Pythagorean triple that consists of consecutive integers. Consider the following argument. Suppose x is a natural number such that x, $x + 1$, and $x + 2$ is a Pythagorean triple. This means that $x^2 + (x + 1)^2 = (x + 2)^2$, or $2x^2 + 2x + 1 = x^2 + 4x + 4$. Solving for x, we have $x^2 - 2x - 3 = 0$, or $(x - 3)(x + 1) = 0$. The only solutions to this equation are $x = -1$ and $x = 3$. -1 cannot be in a Pythagorean triple, so the only triple that will have three consecutive numbers is 3-4-5.

60. Suppose a-b-c represents a Pythagorean triple, and let k be any natural number. Then $(ka)^2 + (kb)^2 = k^2a^2 + k^2b^2 = k^2(a^2 + b^2) = k^2c^2 = (kc)^2$. Because ka, kb, and kc are natural numbers that satisfy the Pythagorean theorem, ka-kb-kc must be a Pythagorean triple.

61. The value of $\dfrac{(c - a)(c - b)}{2}$ is always a perfect square. Consider the following examples. For 3-4-5, $\dfrac{(5 - 3)(5 - 4)}{2} = \dfrac{2 \cdot 1}{2} = 1$.

For 5-12-13, $\dfrac{(13 - 5)(13 - 12)}{2} = \dfrac{8 \cdot 1}{2} = 4$.

For 7-24-25, $\dfrac{(25 - 7)(25 - 24)}{2} = \dfrac{18 \cdot 1}{2} = 9$.

For 8-15-17, $\dfrac{(17 - 8)(17 - 15)}{2} = \dfrac{9 \cdot 2}{2} = 9$.

For 9-40-41, $\dfrac{(41 - 9)(41 - 40)}{2} = \dfrac{32 \cdot 1}{2} = 16$.

62. In $\triangle ABG$, the student did not put the largest side opposite the largest angle.

63. Ansley is not correct. In geometry, we define a curve to be any figure that can be traced without lifting the pencil from the paper. Because we can do so with a segment, a segment must be a curve.

64. Answers will vary.   **65.** Answers will vary.

66. Jacob is incorrectly assuming that $\angle 1$ is an exterior angle. $\angle 1$ cannot be an exterior angle because it is not formed by the side of a triangle and the extension of another side.

67. Measure the walls of the room, and then use the Pythagorean theorem.

68. a. Answers will vary. **b.** Answers will vary.

69. Answers will vary.   **70.** Answers will vary.

71. Answers will vary.

72. If we can create a right triangle and measure two of the sides, then we can use the Pythagorean theorem to find or measure the third.

73. Answers will vary.

Section 10.3 Learning Assessment

1. a. A trapezoid is a quadrilateral with exactly one pair of parallel sides. **b.** A kite is a quadrilateral with at least two pairs of congruent, consecutive sides.
   c. A parallelogram is a quadrilateral in which both pairs of opposite sides are parallel. **d.** A rhombus is a quadrilateral with all sides congruent.

e. A rectangle is a parallelogram with a right angle.
f. A square is a rectangle with all sides congruent.

2. a. In a rhombus, all sides are congruent, the diagonals bisect opposite angles, and the diagonals are perpendicular. **b.** In a square, all angles are congruent, and the diagonals are congruent. **c.** In a rectangle, all angles are congruent, and the diagonals are congruent. **d.** In a square, all sides are congruent, the diagonals bisect opposite angles, and the diagonals are perpendicular.

3. In an isosceles trapezoid, the legs are congruent, the base angles are congruent, and the diagonals are congruent.

4. A diagonal in a polygon is any segment joining two nonconsecutive vertices.

5. a. Subdivide the interior of the polygon into as many nonoverlapping triangles as possible. Multiply the number of nonoverlapping triangles by 180° to get the sum of the angles measures for the polygon. **b.** If n is the number of sides of the polygon, then the number of nonoverlapping triangles is $n - 2$. Hence, the sum of the angles in any polygon is $(n - 2) \cdot 180°$.

6. A regular polygon is equilateral and equiangular.

7. Any segment joining the center of a regular polygon to a vertex is called a radius. Any segment joining the center to the midpoint of a side is called an apothem.

8. Let n be the number of sides of the regular polygon.

a. Each central angle has a measure of $\dfrac{360°}{n}$.

b. Each interior angle has a measure of $\dfrac{(n-2)180°}{n}$.

c. Each exterior angle has a measure of
$$180° - \left(\dfrac{(n-2)180°}{n}\right) = \dfrac{180°}{n} - \left(\dfrac{180°n - 360°}{n}\right) =$$
$$\dfrac{180°}{n} - \dfrac{180°}{n} + \dfrac{360°}{n} = \dfrac{360°}{n}.$$

9. a. A radius is any segment that joins the center of a circle to a point on the circle. A chord is any segment that joins any two points on the circle. A diameter is a chord that goes through the center. **b.** A central angle is an angle formed by two radii. An inscribed angle is an angle in the interior of a circle with its vertex on the circle. **c.** A tangent is any line, ray, or segment that intersects a circle in exactly one point. A secant is any line, ray, or segment intersecting a circle in two points. **d.** A minor arc is the set of points on the circle that lie inside a central angle. A major arc is the set of points that lie outside a central angle. Any arc that lies on one side of a diameter is called a semicircle.

10.

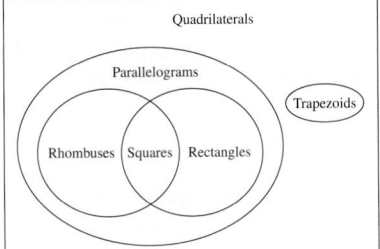

11. Possible answers are shown:

a.

b.

c. **d.**

12. a. If kites are any quadrilateral in which two pairs of adjacent sides are congruent, then the following figure is a kite.

b. Not possible. Two opposite sides have to be non-parallel to make the shape concave.

13. a. Yes. It has four congruent right angles, and it is a parallelogram. **b.** Yes. The word "rectangular" describes a quadrilateral with four right angles, and "rhombus" describes a quadrilateral with four congruent sides.

14. a. Parallelogram. **b.** Rectangle. **c.** Rhombus. **d.** A square.

15. a. Icosikaiheptagon. **b.** Tetracontakaihenagon. **c.** Hexacontakaioctagon. **d.** Heptacontakaitrigon. **e.** Enneacontakaienneagon. **f.** Answer will vary but may include statements like, "Many of them begin with the same root (i.e., 'tri' for three, 'tetra' for four, and so on)." "All the prefixes also use 'contakai.'" "All the suffixes end in 'gon.'"

16. a. 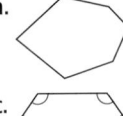 **b.**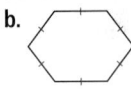

c. 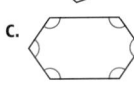 **d.**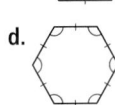

17. An equilateral triangle and a square.

18. a. Yes. An example is shown:

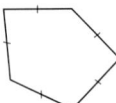

b.

c. Yes, it is necessary because equilateral does not imply equiangular.

19. a. 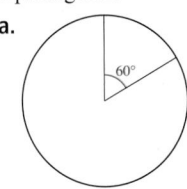 **b.**

c.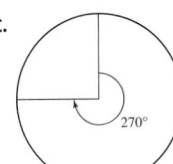

20. a. 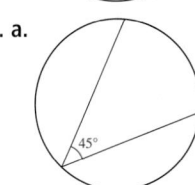 **b.**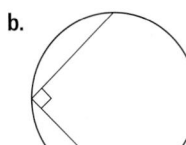

c.
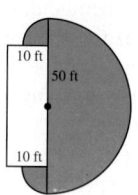
120°

d. No, it is not possible because the vertex and the points on the circle would all have to be the same point.

21. a. $m\angle 1 = 157°$. **b.** $m\angle 1 = 132°$, $m\angle 2 = 48°$, and $m\angle 3 = 132°$. **c.** $m\angle 1 = 129°$, $m\angle 2 = 129°$, and $m\angle 3 = 51°$.

22. a. $m\angle 1 = m\angle 2 = 31°$. **b.** $m\angle 1 = 35°$. **c.** $m\angle 1 = m\angle 2 = 45°$.

23. 38°, 142°, and 142°.

24. 60°, 120°, 60°, and 120°.

25. 30°, 150°, 30°, and 150°.

26. $x = 50$.

27. a. $x = 6\sqrt{2}$. **b.** $x = 10$. **c.** $x = \sqrt{273}$.

28. a. $m\angle 1 = 80°$, $m\angle 2 = 100°$, and $m\angle 3 = 80°$. **b.** $x = 115°$. **c.** $x = 144°$. **d.** $x = 107°$.

29. a. 720°. **b.** 1,080°. **c.** 1,800°. **d.** 2,520°.

30. a. 540°. **b.** 900°. **c.** 1,620°. **d.** 4,140°.

31. a. Central angle = 72°; interior angle = 108°; exterior angle = 72°. **b.** Central angle = 60°; interior angle = 120°; exterior angle = 60°. **c.** Central angle = 22.5°; interior angle = 157.5°; exterior angle = 22.5°. **d.** Central angle = 16.36°; interior angle = 163.64°; exterior angle = 16.36°.

32. a. Central angle = 90°; interior angle = 90°; exterior angle = 90°. **b.** Central angle = 40°; interior angle = 140°; exterior angle = 40°. **c.** Central angle = 25.71°; interior angle = 154.29°; exterior angle = 25.71°. **d.** Central angle = 18.95°; interior angle = 161.05°; exterior angle = 18.95°.

33. $x = 112°$.

34. 80°, 120°, 160°, 160°, 120°, 80°.

35. a. $n = 10$. **b.** $n = 24$. **c.** $n = 25$.

36. a. $n = 12$. **b.** $n = 30$. **c.** $n = 7$.

37. a. $n = 3$. **b.** $n = 12$. **c.** $n = 11$.

38. a. 6 ft. **b.** 9.2 ft. **c.** 70 ft. **d.** 28.62 ft.

39. a. 1.5 ft. **b.** 2.3 ft. **c.** 17.5 ft. **d.** 7.155 ft.

40. a. $\frac{1}{4}$. **b.** $\frac{1}{2}$. **c.** $\frac{3}{4}$. **d.** $\frac{1}{6}$. **e.** $\frac{1}{16}$.

41. a. $x = 114°$. **b.** $x = 309°$. **c.** $x = 28°$. **d.** $x = 226°$.

42. $x = 20°$. **b.** $x = 260°$. **c.** $x = 55°$.

43. a. 6.   **b.** 8.

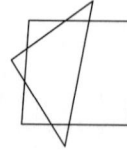   

44. Four.

45. a.   **b.**

46. $m\angle 1 = m\angle 2 = m\angle 3 = 60°$.   **47.** 24 sides.

48. It is a 25-gon.   **49.** 106°, 107°, 108°, 109°, and 110°.

50. $x = 28.33°$.

51. The radius is 6 in., and the apothem is $3\sqrt{3}$ in.

52. $m\overset{\frown}{ABC} = 116°$.

53. $x = 6$ and $y = 6\sqrt{3}$.

54. 4 cm.

55. Draw two different chords, and find the perpendicular bisector of each. Their point of intersection is the center of the circle.

56. The region that the goat can eat is shown:

10 ft
50 ft
10 ft

57. a. Yes. **b.** Yes. **c.** No. **d.** Yes. **e.** No.

58. a. False. **b.** True. **c.** False. **d.** False. **e.** True. **f.** True.

59. A circle is the set of all points in a plane that are a given distance from a given point. Therefore, every radius must be the same length, or congruent.

60. Because the diagonals of rhombus are perpendicular bisectors of one another, $\angle AED$ must be a right angle. This implies that $\angle 1$ and $\angle 2$ are complementary angles.

61. a. Only the equilateral triangle. **b.** Only triangles. **c.** The sum of the interior angles of a quadrilateral must be 360°. If three of the four angles are acute, then the fourth angle must be larger than 90° to make up the difference. This same argument can be used for any polygon with five or more sides.

62. The sum of the measures follow the formula $(n - 2) \cdot 180°$. Any concave n-gon can be divided into $n - 2$ nonintersecting triangles. Each triangle has an angle measure of 180°, so the sum of the angles measure of the n-gon must be $(n - 2) \cdot 180°$.

63. Consider any convex n-gon. If we select a point A in the interior of the n-gon and connect each vertex to the point, it divides the n-gon into n nonintersecting triangles. Each triangle has a measure of 180°, so the sum of the angles of all the triangles is $n \cdot (180°)$. This, however, includes the 360° around point A. Hence, the sum of the interior angles of the n-gon is given by $n \cdot (180°) - 360° = (n - 2)180°$.

64. a. 2. **b.** 5. **c.** 9. **d.** 14. **e.** The number of diagonals in any convex n-gon is $\frac{n(n - 3)}{2}$ diagonals.

65. In a regular n-gon, each exterior angle has a measure of $\frac{360°}{n}$.

66. a. The two inscribed angles have the same measure. **b.** The inscribed angle has a measure that is one-half the measure of the central angle.

67. Answers will vary.

68. Answers will vary.

69. Answers will vary.

70. A parallelogram cannot be a trapezoid because, by definition, a trapezoid has one pair of parallel sides, not two.

71. She should measure the diagonals. When they are both 50 ft, then she knows she has a rectangular pen.

72. Answers will vary.

73. The base angles are congruent, and the legs are congruent.

74. Use them to draw two different inscribed angles. Specifically, set the vertex of the right angle on the circle, and find the points at which the sides of the drafting triangle intersect the circle. The segment connecting these points must be a diameter. Repeat the process to draw a different diameter. The point at which the two diameters intersect is the center of the circle.

75. Draw a rectangle like this one:

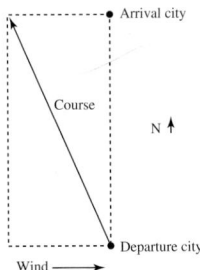

The diagonal of the rectangle is the course the pilot wants to set.

76. The planners should build the airport in the center of the pentagon made by the five cities. They can find the center by finding the point of intersection of the perpendicular bisectors of two of the sides.

77. Answers will vary.

■ Section 10.4 Learning Assessment

1. Postulate 3 from Section 10.1 tells us that any three points determine a plane, making them coplanar. Consequently, we must have a set of at least four points to be noncoplanar.

2. a. A line can be parallel to the plane, intersect the plane in one point, or lie on the plane.   **b.** Two planes can be parallel or intersect in a line. Three planes can intersect at a point.

3. A dihedral angle is formed by the union of the two half-planes intersecting at a line.

4. We think of a surface as a plane made from a rubber sheet that can be curved or twisted into any number of shapes. It contains none of the points in its interior. A solid is formed when a simple closed curve is combined with all the points in its interior.

5. A polyhedron is a simple, closed surface made from planar polygonal regions.

6. A net is a two-dimensional representation of a spatial shape that is formed by "cutting" enough edges of the shape so that we can unfold it and lay it flat.

7. The faces of a regular polyhedron must be regular polygons, and at least three faces must come together at a vertex to form a dihedral angle. Also, the sum of the angle measures at any one vertex must be strictly less than 360°. Given these constraints, we can make only three regular polyhedra with an equilateral triangle: the tetrahedron, the octahedron, and the icosahedron. Also, we can make only one regular polyhedra with a square (the cube) and one with regular pentagons (the dodecahedron). Any other combination of regular polygons fails to satisfy one of the given conditions.

8. a. Regular polyhedra are convex polyhedra with congruent dihedral angles and faces that are congruent regular

polygons. Semiregular polyhedra have faces that are regular polygons but are not limited to just one shape.   **b.** A prism is a polyhedron in which two congruent polygons lie in parallel planes and all other faces connecting them are bound by parallelograms. A pyramid is a polyhedron that is formed when every point on a polygon is connected to a point that is not in the plane of the polygon.   **c.** A cylinder is made of two congruent simple closed curves in parallel planes, along with their interiors and all the line segments joining corresponding points on the curves. A cone is formed when every point on a simple closed curve is connected to a point.

9. Euler's formula holds for any polyhedron, and it states that if f, v, and e represent the number of faces, vertices, and edges of a polyhedron, respectively, then $f + v = e + 2$.

10. a. No. Not all of the faces on a cylinder are polygons.   **b.** No. Euler's formula works only with polyhedra, which have faces, edges, and vertices. A cylinder does not have vertices.

11. A sphere is the set of all points in space that are at a given distance from a given point.

12. Possible answers are given:
a. A, B, C and E.   **b.** Square $ABCD$ and square $EFGH$ or square $ABEF$ and square $DCGH$.   **c.** Square $ABCD$ and square $DCGH$ or square $AEDH$ and square $DCGH$.   **d.** $\overline{AB}$ and $\overline{DC}$ or $\overline{DH}$ and $\overline{CG}$.   **e.** $\overline{AB}$ and $\overline{DH}$ or $\overline{DC}$ and $\overline{BF}$.   **f.** $\overline{AB}$ and $\overline{BF}$ or $\overline{DH}$ and $\overline{DC}$.

13. a. **b.** **c.** **d.** **e.** **f.**

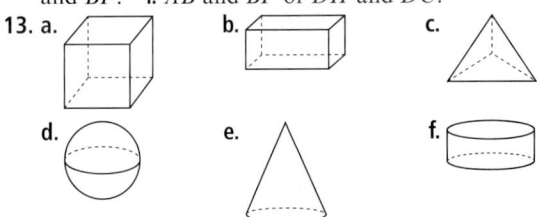

14. a. Polyhedron.   **b.** It is not a polyhedron because it has a circular surface.   **c.** It is not a polyhedron because it has a curved surface.   **d.** Polyhedron.

15. a. Right square pyramid.   **b.** Right triangular prism.   **c.** Oblique hexagonal prism.   **d.** Oblique pentagonal pyramid.   **e.** Cube.   **f.** Hemisphere.

16. a. **b.** **c.**

17. a. **b.** **c.**

18. a. **b.** **c.**

19. a. **b.** **c.**

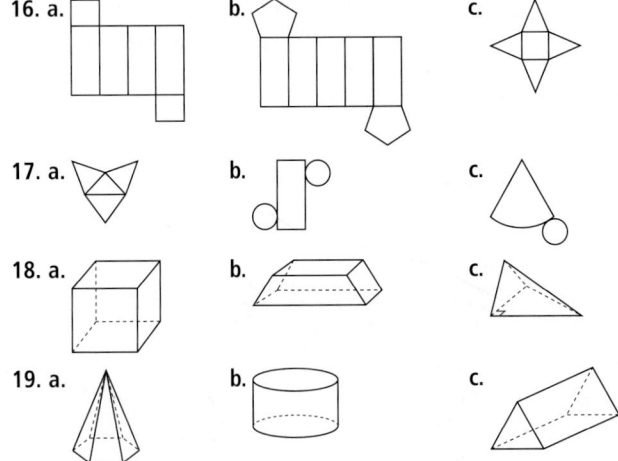

20. a. Yes.   **b.** Yes.   **c.** No.   **d.** No.

21. It is not a regular polyhedron because it does not have the same number of faces meeting at every vertex and the dihedral angles are not all congruent.

22. a. 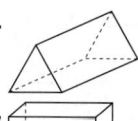 **b.**

23. a. 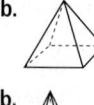 **b.**

24. No. The bottom edge of the lateral surface will be curved and not a segment.

25.

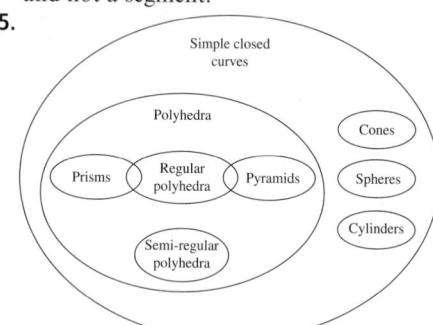

26. a. $\overline{EH}$ **b.** $\overline{AC}$ **c.** C **d.** $\overline{AD}$ **e.** C **f.** E
27. a. 90° **b.** 135°
28. a. 90° **b.** 108°
29. a. 4 **b.** 10 **c.** 18
30. a. $f = 6$, $v = 8$, and $e = 12$. **b.** $f = 7$, $v = 10$, and $e = 15$.
  c. $f = 7$, $v = 7$, and $e = 12$. **d.** $f = 4$, $v = 4$, and $e = 6$.
31. Tetrahedron: $f = 4$, $v = 4$, and $e = 6$.
  Cube: $f = 6$, $v = 8$, and $e = 12$.
  Octahedron: $f = 8$, $v = 6$, and $e = 12$.
  Dodecahedron: $f = 12$, $v = 20$, and $e = 30$.
  Icosahedron: $f = 20$, $v = 12$, and $e = 30$.
32. a. $f = 8$, $v = 12$, and $e = 18$. **b.** $f = 14$, $v = 12$, and
  $e = 24$. **c.** $f = 16$, $v = 32$, and $e = 46$.
  d. $f = 22$, $v = 24$, and $e = 44$.
33. a. $f = 4$, $v = 4$, and $e = 6$. **b.** $f = 5$, $v = 6$, and $e = 9$.
  c. $f = 4$, $v = 4$, and $e = 6$.
34. a. 8. **b.** 20. **c.** 16.
35. a. 7. **b.** 20. **c.** 22.
36. a. The truncated tetrahedron has 8 faces, 12 vertices, and
  18 edges. Hence, $f + v = 8 + 12 = 20 = 18 + 2 = e + 2$.
  b. The truncated octahedron has 16 faces, 32 vertices, and
  46 edges. Hence, $f + v = 16 + 32 = 48 = 46 + 2 = e + 2$.
  c. The cube octahedron has 14 faces, 12 vertices, and 24
  edges. Hence, $f + v = 14 + 12 = 26 = 24 + 2 = e + 2$.
37. The 11 nets of the octahedron are shown:

38. a. 4. **b.** 3. **c.** 60. **d.** 36.
39. a. 1. **b.** 1. **c.** 10. **d.** 6.
40. a. 18. **b.** 20. **c.** 35.
41. a. 0. **b.** 4. **c.** 8.
42. a. 2. **b.** 8. **c.** 8.
43. Two different views of the die are shown:

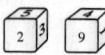

44. a. 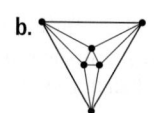 **b.** 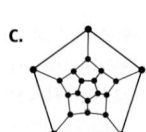 **c.**

45. a. **b.** **c.** **d.**

46. A point, a triangle, a square, and a rectangle.
47.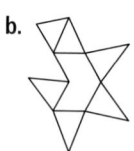

48. a. 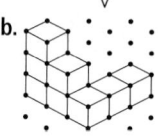 **b.** 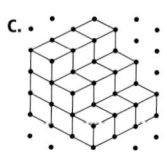 **49.**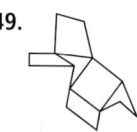

50. a. **b.** **c.**

51. a. Top Front Right **b.** Top Front Right

  c. Top Front Right

52. $94.50.
53. It is not possible for a polyhedron to have exactly seven
  edges. The simplest polyhedron is the tetrahedron, and it has
  6 edges. Increasing any one face by just one side adds two
  edges to the polyhedron. A hexagonal pyramid is an example
  of a polyhedron with exactly 7 vertices. A pentagonal prism
  is an example of a polyhedron with exactly 7 faces.
54. a. $\overline{AH}$ and $\overline{GH}$ represent the same edge. **b.** $\overline{AH} \cong \overline{GH}$,
  $\overline{GF} \cong \overline{EF}$, $\overline{ED} \cong \overline{CD}$, and $\overline{CB} \cong \overline{AB}$.
55. The diagonals of a right rectangular prism are congruent
  bisectors of each other.
56. a. It has gained 3 new edges. **b.** It has gained 2 new
  vertices. **c.** It has 7 faces, 10 vertices, and 15 edges.
  Hence, $f + v = 7 + 10 = 17 = 15 + 2 = e + 2$.
  d. From the original cube, it will gain two new faces, six
  new edges, and four new vertices. This means it has 8
  faces, 12 vertices, and 18 edges. Hence, $f + v = 8 + 12 =$
  $20 = 18 + 2 = e + 2$.
57. a. It has gained 2 new vertices. **b.** It has gained 8 new
  edges. **c.** From the original cube, there is a gain of 6 new
  vertices and 24 new edges.
58. a. 6. **b.** 7. **c.** 8. **d.** 9. **e.** The number of faces in a
  prism with an n-gon for its base is $n + 2$.
59. a. 12. **b.** 15. **c.** 18. **d.** 21. **e.** The number of edges in a
  prism with an n-gon for its base is $3n$.
60. a. 5. **b.** 6. **c.** 7 **d.** 8. **e.** The number of vertices in a
  pyramid with an n-gon for its base is $n + 1$.
61. a. 4. **b.** 10. **c.** 18. **d.** 28. **e.** The number of diagonals
  in any prism with a n-gon for its base is equal to $n(n - 3)$.
62. Answers will vary.
63. It is not correct. The term "circular" applies to a two-
  dimensional object, and a baseball is a three-dimensional
  object. A better choice is "spherical."

64. Answers will vary.

65. No. A regular polyhedron must have all faces congruent, and the base of the square pyramid will not be congruent to the lateral faces.

66. No. If we put the values into Euler's formula, the number of faces is 2. However, there is no polyhedron with only 2 faces.

67. Answers will vary.

68. a. Concurrent lines. **b.** Parallel lines. **c.** Angle.
d. Polygon. **e.** Square. **f.** Circle.

69. Answers will vary.

70. Answers will vary.

71. Answers will vary.

Chapter 10 Review Exercises

1. Possible answers are given:

    a. A, B, and C. **b.** $\overleftrightarrow{AB}$ and $\overleftrightarrow{DF}$. **c.** $\overleftrightarrow{AB}$ and $\overleftrightarrow{AF}$.

    d. A, B, C and D. **e.** $\overleftrightarrow{AB}$ and $\overleftrightarrow{DE}$. **f.** Plane ABC and Plane DEF.

2. a. 70°. **b.** 70°. **c.** 160°. **d.** 110°.

3. a. 53°. **b.** 22°. **c.** 151°. **d.** 59°. **e.** 47°.

4. a. $x = 102$. **b.** $x = 36$.

5. $m\angle 1 = 115°$, $m\angle 2 = 55°$, $m\angle 3 = 65°$, $m\angle 4 = 60°$, $m\angle 5 = 55°$, $m\angle 6 = 115°$, $m\angle 7 = 120°$, $m\angle 8 = 115°$, $m\angle 9 = 60°$, and $m\angle 10 = 60°$.

6. About 160°.

7. Acute angles with measures of 45°.

8. a. True. **b.** False. **c.** True.

9. a. 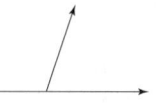 **b.**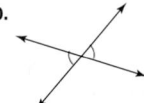

10. a. Equilateral. **b.** Right. **c.** Obtuse.

11. Because $2 + 3 = 5 < 6$, it is not possible to have a triangle with sides of length 2 in., 3 in., and 6 in.

12. a. $x = 71°$. **b.** $x = 43°$. **c.** $x = 18°$.

13. Because $5^2 + 7^2 = 25 + 49 = 74 = (\sqrt{74})^2$, it is possible for a right triangle to have sides of length 5 in., 7 in., and $\sqrt{74}$ in.

14. 21°.

15. a. 10 and $5\sqrt{3}$. **b.** $\dfrac{16}{\sqrt{2}} \approx 11.31$.

16. 150 ft.

17. 6.55 mi.

18. a. 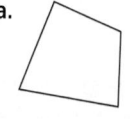 **b.**

19. a. 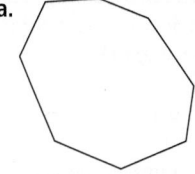 **b.**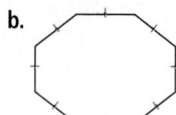

    c. **d.**

20. 30° and 150°.

21. a. Interior angle = 108°, exterior angle = 72°, and central angle = 72°. **b.** Interior angle = 120°, exterior angle = 60°, and central angle = 60°. **c.** Interior angle = 135°, exterior angle = 45°, and central angle = 45°.

22. a. $n = 20$. **b.** $n = 21$. **c.** $n = 17$.

23. 37-gon.

24. 51.43°, 154.29°, 102.86°, 154.29°, 102.86°, and 154.29°.

25. a. 10 ft. **b.** 14.2 ft. **c.** 36 ft.

26. 57°.

27. If the walls are all the same length, then measure the diagonals to be the same length and be guaranteed that the deck is a square.

28. a. Polyhedron. **b.** Polyhedron. **c.** Polyhedron.

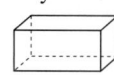 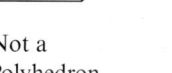 

    d. Not a Polyhedron. **e.** Not a Polyhedron. **f.** Not a Polyhedron.

  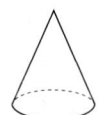

29. a. 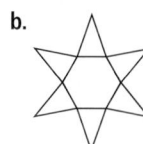 **b.** 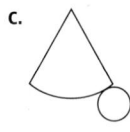 **c.**

30.

31. $f = e + 2 - v = 20 + 2 - 14 = 8$.

32. $e = f + v - 2 = 25 + 17 - 2 = 40$.

33. a. 90°. **b.** 120°.

34. It has gained 9 new edges and 6 new vertices.

35. No. The number of faces that come together at any one vertex is not always the same.

CHAPTER 11

Section 11.1 Learning Assessment

1. Two triangles are congruent if a correspondence can be established between their vertices so that corresponding sides and angles are congruent. Two polygons are congruent if a correspondence can be established between their vertices so that every pair of corresponding sides and every pair of corresponding angles are congruent.

2. a. If the corresponding sides of two triangles are congruent, then the triangles are congruent. **b.** If two sides and the included angle in one triangle are congruent to the corresponding sides and angle in another, the triangles are congruent. **c.** If two angles and the common side in one triangle are congruent to the corresponding angles and side in a second triangle, then the triangles are congruent. **d.** If the hypotenuse and one leg of one right triangle are congruent to the hypotenuse and one leg of the other, then the two right triangles are congruent.

3. The three sides of a triangle determine its vertices and angles. Consequently, if the corresponding sides of two triangles are congruent, then the corresponding angles must also be congruent. This makes all six pairs of corresponding parts congruent, so the triangles must be congruent.

4. The two sides and the included angle of a triangle determine the length of the third side and the measures of the other two angles. Consequently, if two sides and the included angle in one triangle are congruent to the corresponding sides and angle in another, then all six pairs of corresponding parts must be congruent, making the triangles congruent.

5. In the angle-side-angle property, the side is common to the two angles. In the angle-angle-side property, it is not.

6. Yes. The Pythagorean Theorem guarantees that the hypotenuses have the same length, making them congruent. Consequently, the two triangles must be congruent by the SSS property.

7. The additional sides and angles mean that more corresponding parts need to match to guarantee that the polygons are congruent.

8. a. $\angle A \cong \angle X$, $\angle B \cong \angle Y$, $\angle C \cong \angle Z$, $\overline{AB} \cong \overline{XY}$, $\overline{BC} \cong \overline{YZ}$, and $\overline{AC} \cong \overline{XZ}$. **b.** $\angle A \cong \angle Z$, $\angle B \cong \angle Y$, $\angle C \cong \angle X$, $\overline{AB} \cong \overline{ZY}$, $\overline{BC} \cong \overline{YX}$, and $\overline{AC} \cong \overline{ZX}$.

9. Three ways are $\triangle CAB \cong \triangle ZXY$, $\triangle BAC \cong \triangle YXZ$, and $\triangle ACB \cong \triangle XZY$.

10. a. Yes. **b.** Yes. **c.** No. **d.** No. **e.** No. **f.** No.

11. Possible illustrations are shown:

a.   **b.**

c.

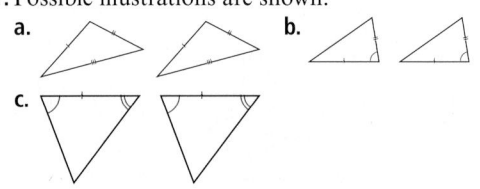

12. Possible illustrations are shown:

a.   **b.**

c.

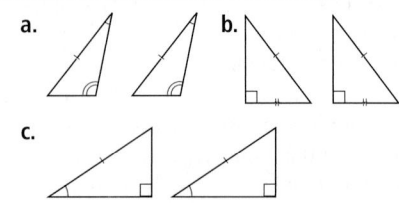

13. a. $\overline{AB} \cong \overline{DE}$ and $\overline{AC} \cong \overline{DF}$. **b.** $\overline{AB} \cong \overline{DE}$. **c.** $\angle C \cong \angle F$. **d.** $\angle A \cong \angle D$.

14. Possible illustrations are shown:

a.

b.

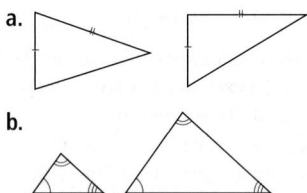

15. $\angle Q \cong \angle P$, $\angle U \cong \angle O$, $\angle A \cong \angle L$, $\angle D \cong \angle Y$, $\overline{QU} \cong \overline{PO}$, $\overline{UA} \cong \overline{OL}$, $\overline{AD} \cong \overline{LY}$, and $\overline{DQ} \cong \overline{YP}$.

16. $A \leftrightarrow T$, $B \leftrightarrow S$, $C \leftrightarrow R$, $D \leftrightarrow Q$ or $A \leftrightarrow R$, $B \leftrightarrow Q$, $C \leftrightarrow T$, and $D \leftrightarrow S$.

17. $\angle S \cong \angle C$, $\angle E \cong \angle I$, $\angle N \cong \angle R$, $\angle O \cong \angle B$, $\angle T \cong \angle K$, $\overline{SE} \cong \overline{CI}$, $\overline{EN} \cong \overline{IR}$, $\overline{NO} \cong \overline{RB}$, $\overline{OT} \cong \overline{BK}$, and $\overline{TS} \cong \overline{KC}$.

18. Possible illustrations are shown:

a.

b.

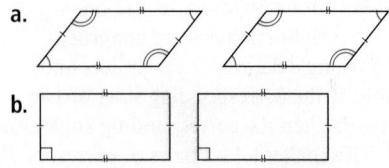

c.

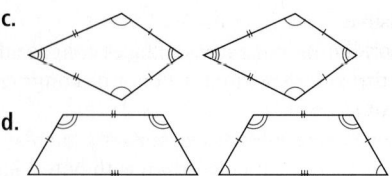

d.

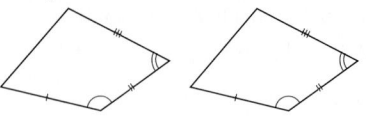

19. One possible illustration is shown:

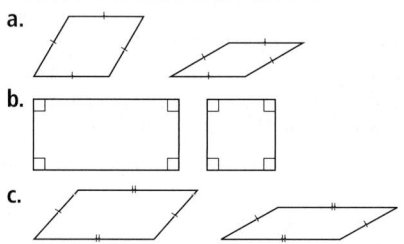

20. Possible illustrations are shown:

a.

b.

c.

21. a. $XY = 5$, $YZ = 7$, and $XZ = 10$. **b.** $m\angle X = 29°$, $m\angle Y = 76°$, and $m\angle Z = 75°$.

22. a. $BC = 9$, $AC = 12$, and $DE = 4$. **b.** $m\angle B = 25°$, $m\angle D = 113°$, and $m\angle F = 42°$.

23. $AB = 12$, $AC = 13$, $HI = 5$, and $GI = 13$.

24. $AB = 5\sqrt{3}$, $BC = 5$, $AC = 10$, $JK = 5\sqrt{3}$, $KL = 5$, $m\angle B = 90°$, $m\angle C = 60°$, $m\angle J = 30°$, and $m\angle K = 90°$.

25. $AB = 5$, $BC = 5$, $AC = 5\sqrt{2}$, $MN = 5$, $MO = 5\sqrt{2}$, $m\angle C = 45°$, $m\angle M = 45°$, $m\angle N = 90°$, and $m\angle O = 45°$.

26. a. Congruent; SSS. **b.** Not congruent. **c.** Not congruent. **d.** Congruent; AAS.

27. AAS.

28. SAS.

29. SAS.

30. a. $x = 22$. **b.** $x = 12$.

31. $AB = 5$, $CD = 5$, $AD = 8$, $QR = 5$, $RS = 8$, $QT = 8$, $m\angle B = 110°$, $m\angle C = 70°$, $m\angle D = 110°$, $m\angle Q = 70°$, $m\angle R = 110°$, $m\angle S = 70°$, and $m\angle T = 110°$.

32. $AB = 5$, $CD = 5$, $XY = 8$, $YZ = 5$, $WZ = 11$, $m\angle A = 65°$, $m\angle B = 115°$, $m\angle C = 115°$, $m\angle D = 65°$, $m\angle W = 65°$, $m\angle X = 115°$, and $m\angle Y = 115°$.

33. Yes, $\triangle DEF \cong \triangle STR$.

34. a.   **b.**

c. Not possible.

35. Yes. Connect the center of the regular polygon to every vertex. The resulting triangles are all congruent by SSS.

36. Not including the pairs of triangles that are congruent to themselves, the pairs of congruent triangles are $\triangle BRE \cong \triangle EDB$, $\triangle BRS \cong \triangle EDS$, and $\triangle BID \cong \triangle EIR$.

37. 12.

38. a. 6. **b.** 24. **c.** 120.

39. a. 75 ft. **b.** 21 cross braces.

40. Isosceles trapezoids, parallelograms, kites, rhombuses, rectangles, and squares can be divided into two congruent triangles by drawing a single diagonal.

41. $PA = 76.16$ yd.

42. Each vertex should correspond to itself, or $A \leftrightarrow A$, $B \leftrightarrow B$, and $C \leftrightarrow C$.

43. No, it is not possible. If we have two congruent pairs of corresponding sides and a congruent pair of corresponding angles that are not included, then conditions are not sufficient to uniquely determine the third vertex of the triangle. Consider the two triangles shown:

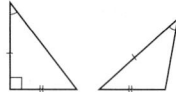

44. If the radii of two regular hexagons are congruent, then hexagons must be congruent because the radius determines how far each vertex is from center, and this uniquely determines the sides and the angles of a regular hexagon. Similarly, the apothem determines how far each side is from the center. This uniquely determines the length of the sides, which in turn uniquely determines the vertices.

45. Consider the isosceles $\triangle ABC$, in which $\overline{BM}$ is the perpendicular bisector of $\overline{AC}$. Because $\triangle ABC$ is isosceles, then the base angles must be congruent, or $\angle A \cong \angle C$. Because $\overline{BM}$ is the perpendicular bisector of $\overline{AC}$, then $\angle BMA$ and $\angle BMC$ are right angles. This means $\angle BMA \cong \angle BMC$. Also, $\overline{AM} \cong \overline{CM}$. Hence, $\triangle ABM \cong \triangle CBM$ by ASA.

46. Let $\triangle TYR$ be an equilateral triangle, $\angle 2 \cong \angle 3$, and $SY = RO$. Because $\triangle TYR$ is equilateral, then $\overline{TY} \cong \overline{TR}$. Also, $\angle 1$ and $\angle 2$ form a linear, so $m\angle 1 + m\angle 2 = 180°$. Similarly, $m\angle 3 + m\angle 4 = 180°$. Consequently, $m\angle 1 + m\angle 2 = m\angle 3 + m\angle 4$. Because $\angle 2 \cong \angle 3$, we have $m\angle 1 + m\angle 2 = m\angle 2 + m\angle 4$, implying that $m\angle 1 = m\angle 4$. As a result, $\angle 1 \cong \angle 4$, so $\triangle STY \cong \triangle OTR$ by SAS.

47. Consider $\triangle ABC$ and $\triangle DEF$, in which $\angle A \cong \angle D$, $\angle C \cong \angle F$, and $\overline{AB} \cong \overline{DE}$.

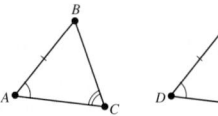

Because $m\angle A + m\angle B + m\angle C = 180°$ and $m\angle D + m\angle E + m\angle F = 180°$, we have $m\angle A + m\angle B + m\angle C = m\angle D + m\angle E + m\angle F$. Using substitution and the fact that $\angle A \cong \angle D$ and $\angle C \cong \angle F$, we have $m\angle A + m\angle B + m\angle C = m\angle A + m\angle E + m\angle C$. This implies $m\angle B = m\angle E$, or $\angle B \cong \angle E$. As a result, $\triangle ABC \cong \triangle DEF$ by ASA.

48. Consider two right triangles $\triangle ABC$ and $\triangle DEF$, in which $\angle C \cong \angle F$, $m\angle B = m\angle E = 90°$ and $\overline{AC} \cong \overline{DF}$.

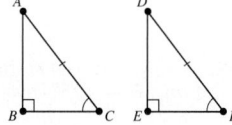

Because $m\angle B = m\angle E = 90°$, $\angle B \cong \angle E$. This gives us two angles and the side opposite one of the angles in $\triangle ABC$ congruent, respectively, to two angles and an opposite side in $\triangle DEF$. Hence, $\triangle ABC \cong \triangle DEF$ by AAS.

49. a. One pair of corresponding sides and one pair of corresponding angles must be congruent. **b.** One pair of corresponding angles and two pairs of consecutive sides must be congruent. **c.** One pair of corresponding sides must be congruent. **d.** Three pairs of corresponding

sides and their included angles must be congruent, or three pairs of corresponding angles and their common sides must be congruent.

50. a. The bases must be congruent, and the heights of their lateral faces must be congruent. **b.** The radii of the bases must be congruent, and the heights of the laterals surfaces must be congruent. **c.** One pair of corresponding edges must be congruent. **d.** The sides of the bases must be congruent, and the heights of the lateral surfaces must be congruent.

51. Because quadrilateral $ABCD$ is a rhombus, the opposite sides are parallel. Hence, $\angle 1 \cong \angle 3$ and $\angle 2 \cong \angle 4$ because they are alternate interior angles. Also $\overline{BC} \cong \overline{DA}$ because opposite sides of a parallelogram are congruent. This gives two pairs of corresponding angles and their common sides congruent, so $\triangle APD \cong \triangle CPB$ by the ASA property. Because corresponding parts of congruent triangles are congruent, $\overline{BP} \cong \overline{DP}$ and $\overline{CP} \cong \overline{AP}$. Consequently, the diagonals in a rhombus bisect each other.

52. Because quadrilateral $ABCD$ is a parallelogram, we know that the opposite sides are parallel, so $\angle 1 \cong \angle 4$ and $\angle 2 \cong \angle 3$ (they are alternate interior angles). We also know that $\overline{AC} \cong \overline{CA}$ because they are the same segment. This gives us two pairs of corresponding angles and their common sides congruent, so $\triangle ABC \cong \triangle CDA$ by the ASA property. Because corresponding parts of congruent triangles are congruent, $\angle B \cong \angle D$. This implies that opposite angles of a parallelogram are congruent.

53. Answers will vary.　　**54.** Answers will vary.

55. The student is not correct. Two triangles can have three pairs of corresponding angles congruent but have none of the corresponding sides congruent.

56. The corresponding pair of congruent sides is not common to the two angles, so ASA does not apply. Individual responses to his thinking will vary.

57. Two triangles congruent to a third are congruent to each other. Individual responses to the student's thinking will vary.

58. Triangle congruence is defined by using segment and angle congruence.

59. Two shapes are equal only if they contain exactly the same set of points. Hence, two shapes that are congruent need not be equal. However, two shapes that are equal must also be congruent.

60. Answers will vary.　　**61.** Answers will vary.

■ Section 11.2 Learning Assessment

1. Congruent shapes have the same shape and size. Similar shapes have the same shape but not necessarily the same size.

2. a. The scale factor is the common ratio between the proportional sides. **b.** Yes. The scale factor depends on how we look at the triangles. For instance, suppose $\triangle ABC \sim \triangle DEF$. If we look at the triangles from $\triangle ABC$ to $\triangle DEF$, the scale factor is the reciprocal of the scale factor if we were to look at the triangles from $\triangle DEF$ to $\triangle ABC$.

3. a. In the SSS similarity property, the corresponding sides must be proportional, not congruent. **b.** In the SAS similarity property, the corresponding sides must be proportional, not congruent.

4. If two triangles have two pairs of corresponding angles that are congruent, then the third pair must also be congruent because the measures of the angles in a triangle is 180°. Once we have three pairs of congruent angles, the triangles must take on the same shape. Consequently, each pair of corresponding sides is proportional and the two triangles are similar.

5. a. Yes. If two triangles are congruent, they must be the same shape. This makes them similar. **b.** No. If two triangles are similar, they must be the same shape but not necessarily the same size.

6. We make two similar triangles in which we can measure two corresponding sides directly. We then use proportional reasoning to find the lengths of the unmeasured sides.

7. A figure is self-similar if it looks "roughly" the same when any part of the figure is magnified or reduced.

8. a. 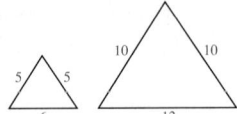 **b.**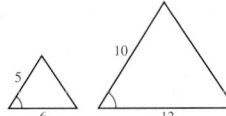

c.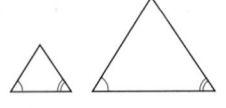

9. a. 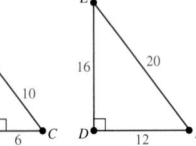 **b.**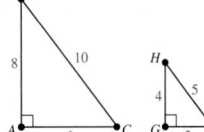

10. a.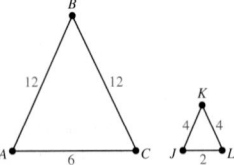

b.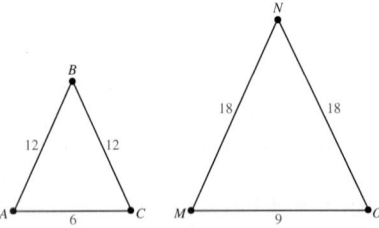

11. $\dfrac{DE}{AB} = \dfrac{DF}{AC}, \dfrac{AB}{AC} = \dfrac{DE}{DF},$ and $\dfrac{AC}{AB} = \dfrac{DF}{DE}.$

12. The two triangles are similar by SAS: $A \leftrightarrow Z$, $B \leftrightarrow X$, and $C \leftrightarrow Y$.

13. a.  **b.**

c.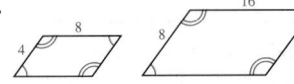

d.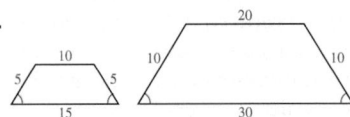

14. a. 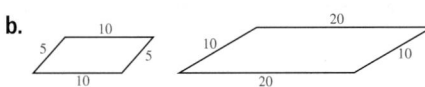 **b.**

15. a.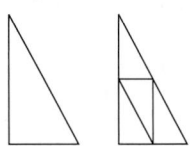

b.

16. A possible illustration is shown:

17.

18.

19.

20. a. 2. **b.** $\dfrac{2}{3}$. **c.** 5. **21. a.** $\dfrac{6}{5}$. **b.** $\dfrac{1}{3}$. **c.** 0.516.

22. a. 15. **b.** 6. **23. a.** 4.56. **b.** 10.

24. a. △ABC is not similar to △FDE because $\overline{AB}$ is not proportional to $\overline{FD}$. **b.** △IGH ~ △IJK by the AA property. **c.** △MLP is not similar to △MNO because $\overline{LP}$ is not proportional to $\overline{NO}$. **d.** △RQS ~ △VUT by the SAS property.

25. a. △ABC ~ △XYZ is similar by SAS. **b.** △ABC ~ △XYZ by the AA property.

26. a. △ABC is not similar to △XYZ. **b.** △ABC ~ △XYZ by the AA property.

27. a. $KA = 6$. **b.** $KA = \dfrac{7}{3}$. **c.** $ST = 12$ and $RA = 4$.

28. a. $EC = 12.5$. **b.** $AD = 10$. **c.** $AC = 16$, $AD = 19.5$, and $DB = 4.5$. **d.** $AE = 27$ and $EC = 9$.

29. a. $z = \dfrac{90}{13}$. **b.** $x = \dfrac{16}{3}$. **c.** $y = \dfrac{31}{2}$. **d.** $w = \dfrac{35}{4}$.

30. a. $XY = 11$. **b.** $DC = 5.5$. **c.** $YZ = 14$. **d.** $m\angle A = 55°$. **e.** $m\angle B = 125°$.

31. a. 108°. **b.** 24.

32. The sides have lengths of 28.8 and 43.2.

33. a. $AE = 2.5$. **b.** $AB = 4$. **c.** $AC = 5$. **d.** $BC = 3$.

34. a. $DE = 12.8$ and $XY = 6.4$. **b.** Scale factor = 20%.

35. a. $DE = 40$ and $XY = 32$. **b.** $\dfrac{8}{15}$. **c.** $EF = 60$ and $YZ = 48$. **d.** $DF = 30$ and $XZ = 24$.

36. $\dfrac{x}{z} = \dfrac{w}{y}$.

37. a. 3.6 in., 6 in., and 8.4 in.   **b.** 2.4 in. and 4.2 in.
   c. Reduce by 80%.
38. 17 ft, 5 in. tall.
39. 80 ft.
40. About 32 ft.
41. The height of the roof is 4.5 ft, and its length is 8.75 ft.
42. $53\frac{1}{3}$ ft from the base of the tower.

43. 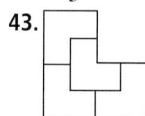   **44.**

45. a. 768 sides.   **b.** 786,432 sides.   **c.** The nth iteration has $a_n = 3(4)^{n-1}$ sides.
46. No. A scale factor of 0 implies that one of the shapes has sides of no length. Hence, all scale factors must be strictly greater than 0.
47. The two shapes are congruent.
48. a. No. Two scalene triangles need not have any of the corresponding angles congruent or corresponding sides proportional.   **b.** No. Two isosceles triangles need not have any of the corresponding angles congruent or corresponding sides proportional.   **c.** Yes. The angles of an equilateral triangle are all 60°. So any two equilateral triangles must be similar by the AA property.
49. a. No. Two acute triangles need not have any of the corresponding angles congruent or corresponding sides proportional.   **b.** No. Two right triangles need not have any of the corresponding angles congruent other than the right angles, and the corresponding sides need not be proportional.   **c.** No. Two obtuse triangles need not have any of the corresponding angles congruent or corresponding sides proportional.   **d.** Yes. The angles of an equiangular triangle are all 60°; any two equiangular triangles must be similar by the AA property.
50. a. Yes. It states that given a correspondence between two right triangles, if the hypotenuse and one leg of the first triangle are proportional to the hypotenuse and one leg of the second triangle, then the two triangles are similar.   **b.** Yes. It states that given a correspondence between two right triangles, if the hypotenuses are proportional and one pair of corresponding acute angles is congruent, then the triangles are similar.
51. a. $EF = 7.5$, $FG = 15$, $BF = 15.75$, and $AG = 29.25$.
   b. $\frac{DC}{DE} = \frac{8}{10} = 0.8$, $\frac{CB}{EF} = \frac{6}{7.5} = 0.8$, and $\frac{BA}{FG} = \frac{12}{15} = 0.8$.
   The segments between parallel lines occur in the same ratios.   **c.** We show proportionality for $\frac{CB}{EF} = \frac{DC}{DE}$. Other proportions can be shown in a similar manner. Because $m \parallel n$, $\triangle CDE \sim \triangle BDF$ by the AA property, implying that $\frac{DC}{DB} = \frac{DE}{DF}$. However, because $DB = DC + CB$ and $DF = DE + EF$, $\frac{DC}{DC + CB} = \frac{DE}{DE + DF}$. By cross multiplying, we have $DC(DE + EF) = DE(DC + CB)$, or $(DC)(DE) + (DC)(EF) = (DE)(DC) + (DE)(CB)$. By subtracting $(DC)(DE)$ from both sides, we have $(DC)(EF) = (DE)(CB)$ or $\frac{CB}{EF} = \frac{DC}{DE}$.

52. Consider $\triangle ABC$ in which N is the midpoint of $\overline{AB}$, P is the midpoint of $\overline{BC}$, and M is the midpoint of $\overline{AC}$. $\overline{NP}$ is a midsegment, so $NP = \frac{1}{2}(AC)$ by Theorem 11.5. Likewise, $\overline{PM}$ and $\overline{MN}$ are midsegments, so $PM = \frac{1}{2}(AB)$ and $MN = \frac{1}{2}(BC)$. This gives three pairs of proportional corresponding sides, or $\frac{AB}{PM} = \frac{BC}{MN} = \frac{AC}{PN} = 2$. Hence, $\triangle ABC \sim \triangle PMN$ by the SSS property.
53. Let $\triangle ABC$ and $\triangle DEF$ be two isosceles triangles in which the vertex angles are congruent, or $\angle B \cong \angle E$. Because $\triangle ABC$ and $\triangle DEF$ are isosceles triangles, then the base angles must be congruent, or $\angle A \cong \angle C$ and $\angle D \cong \angle F$. However, it must also be that $\angle A \cong \angle D$, so $\triangle ABC \sim \triangle DEF$ by the AA property.
54. a. Yes. Because the angles and the sides in any square are congruent, any two squares must be similar.   **b.** No. Even though the angles are congruent between any two rectangles, the lengths of corresponding sides need not be proportional.   **c.** No. Two rhombuses need not have any of the corresponding angles congruent or corresponding sides proportional.   **d.** Yes. A circle is completely determined by its radius. Because any two segments are proportional, any two circles must be similar.
55. There is no AAA similarity property for quadrilaterals. A square and a rectangle have three pairs of corresponding angles that are congruent, but they cannot be similar.

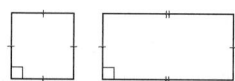

56. The interior angles of two regular n-gons have the same measure, and all the sides within one regular n-gon are congruent. Because any two segments are proportional, any two regular n-gons must be similar.
57. Two rectangular prisms need not have any corresponding faces proportional. However, two cubes are completely defined by the length of their edges. Because any two segments are proportional, any two cubes must be similar.
58. Two right circular cylinders are similar if the ratio of their radii and the ratio of their heights are the same. Two right circular cones are similar if the ratio of their radii and the ratio of their heights are the same. Any two spheres are similar because their radii are proportional.
59. Quadrilateral $ABCD$ is similar to quadrilateral $WXYZ$ if and only if there is a correspondence between their vertices such that corresponding angles are congruent and corresponding sides are proportional.
60. Answers will vary.   **61.** Answers will vary.
62. Answers will vary.
63. Two triangles similar to a third are similar to each other. Answers will vary on how to address the student's thinking.
64. Marcus is incorrect. For a counterexample, consider two rectangles that have sides of different lengths and are not proportional.
65. Answers will vary.   **66.** Answers will vary.
67. If we know two figures are similar, then we can use ratios and proportional reasoning to find any missing sides. If we are given two figures, we can use proportional reasoning to determine whether they are similar.
68. Answers will vary.   **69.** Answers will vary.

▪ Section 11.3 Learning Assessment

1. A geometrical construction is more precise than a drawing because the tools and properties used in constructions guarantee that the shape or relationship is exactly what it is supposed to be.

2. The straightedge is used to construct lines, rays, and segments between two given points. It has no marks, so it cannot be used to measure lengths. The compass is used to draw circles and arcs of a given center and radius.

3. Construct an isosceles triangle on the given angle. Transfer all three lengths of the sides to a ray. Doing so makes a congruent isosceles triangle on the ray by using SSS. Because corresponding parts of congruent triangles are congruent, the constructed angle must be congruent to the given angle.

4. **a.** Construct a rhombus on the given angle and then construct a ray along its diagonal. Because the diagonal of a rhombus bisects opposite angles, the constructed ray must bisect the angle. **b.** Construct a rhombus in which one side lies on a line $\overleftrightarrow{AB}$, and the opposite side passes through a point P. Then construct a point D that is on the perpendicular bisector but that is also equidistant from P and B. The line $\overleftrightarrow{PD}$ is parallel to $\overleftrightarrow{AB}$ because it is on the opposite sides of a rhombus. **c.** Construct a segment on the given line. Then treat the given segment as the diagonal of a rhombus, and construct a line along the other diagonal. Because the diagonals of a rhombus are perpendicular bisectors of each other, the constructed line must be perpendicular to the given line.

5. To inscribe a circle means to place a circle inside a polygon so that the circle is tangent to each side of the polygon.

6. One way is to use the opposite sides of a rhombus to construct the parallel lines. Another way is to draw a transversal through the given line and then construct alternate interior angles that are congruent to one another.

7. Use a rhombus to construct the perpendicular bisector of the segment. Because this line bisects the segment, it must go through the midpoint of the segment.

8. **a.** Use Construction 1. **b.** Use Construction 1.

9. **a.** Use Construction 2. **b.** Use Construction 2.

10. **a.** Use Construction 2 to copy the angle. Place the drawing edge of the reflector on the vertex, and then rotate it until the reflection of one side is directly over the top of the other. Draw the angle bisector. **b.** Use Construction 2 to copy the angle. Place the drawing edge of the reflector on the vertex, and then rotate it until the reflection of one side is directly over the top of the other. Draw the angle bisector.

11. **a.** Use Construction 2 to copy the angle and then Construction 3 to find its bisector. **b.** Use Construction 2 to copy the angle and then Construction 3 to find its bisector.

12. Follow the procedure given in Example 11.21.

13. Use Construction 4.

14. **a.** Use Construction 1 to copy the segment. Place the reflector on the segment, and then rotate or slide it until the reflection of one endpoint is directly over the top of the other. The drawing edge is the perpendicular bisector. **b.** Use Construction 1 to copy the segment. Place the reflector on the segment, and then rotate or slide it until

the reflection of one endpoint is directly over the top of the other. The drawing edge is the perpendicular bisector.

15. **a.** Use Construction 1 to copy the segment and then Construction 5 to find the perpendicular bisector. **b.** Use Construction 1 to copy the segment and then Construction 5 to find the perpendicular bisector.

16. **a.** Place the reflector on P, and then rotate or slide it until the reflection of one side of the line is directly on top of the other. The drawing edge is the perpendicular line. **b.** Place the reflector on P, and then rotate or slide it until the reflection of one side of the line is directly on top of the other. The drawing edge is the perpendicular line.

17. **a.** Use the procedure in Example 11.23. **b.** Use the procedure in Example 11.22.

18. **a.** Use Construction 1 twice: once with $\overline{AB}$ and once with $\overline{CD}$. **b.** Draw a ray, and then use Construction 1 to copy $\overline{CD}$ from the endpoint of the ray. Again, use Construction 1 to copy $\overline{AB}$ from the endpoint of the ray. The distance from one new point to the other is $CD - AB$. **c.** Draw a ray, and then use Construction 1 with $\overline{AB}$ four times in succession. **d.** Use Construction 1 twice: once with $\overline{AB}$ and once with $\overline{CD}$. Then use Construction 5 to find the perpendicular bisector of the new segment. The distance from one endpoint to the perpendicular bisector is $\dfrac{AB + CD}{2}$.

19. **a.** Construct two perpendicular lines. Make a copy of $\overline{AB}$ on each of the perpendicular lines using the intersection as a common endpoint. Connect the two new vertices to form a 45°-45°-90° right triangle with sides of length one. The hypotenuse has a length of $\sqrt{2}$. **b.** Construct two perpendicular lines. Make a copy of $\overline{AB}$ on one of the lines, and a segment twice the length of $\overline{AB}$ on the other line using the intersection as a common endpoint. Connect the endpoints of the new segments to form a right triangle with a hypotenuse of length $\sqrt{5}$. **c.** Construct two perpendicular lines. Construct a segment twice the length of $\overline{AB}$ on one of the lines and a segment three times the length of $\overline{AB}$ on the other using the intersection as a common endpoint. Connect the endpoints of the new segments to form a right triangle with a hypotenuse of length $\sqrt{13}$.

20. **a.** Use Construction 2 twice: one with $\angle A$ and one with $\angle B$. **b.** Use Construction 2 to make a copy of $\angle A$. On one side of $\angle A$, use Construction 2 to make a copy of $\angle B$ in the interior of $\angle A$. The angle between the second side of $\angle A$, and the second side of $\angle B$ is the needed angle. **c.** Use Construction 2 twice with $\angle B$. **d.** Use Construction 2 twice: once with $\angle A$ and once with $\angle B$. Use Construction 3 to construct the angle bisector of the new angle.

21. **a.** Construct a straight line, and select a point on the line. Use Construction 2 to make a copy of $\angle A$ at the point. The other angle in the linear pair is $\angle B$. **b.** Construct two perpendicular lines. At the vertex of the right angle, use Construction 2 to make a copy of $\angle A$. The angle between $\angle A$ and the right angle is $\angle C$.

22. **a.** Draw a line, and use the reflector to make a line perpendicular to it. Then use the reflector to bisect one of the right angles. The result is a 45° angle. **b.** Draw a line, and use the reflector to make a line perpendicular to

it. Then use the reflector to bisect one of the right angles. The angle from the one side of the bisected angle back to the other side of the line is 135°.   **c.** Draw a line and use the reflector to make a line perpendicular to it. Then use the reflector to bisect one of the right angles. Bisect the 45° angle on top. The angle from the second angle bisector back to the other side of the line is 112.5°.

23. **a.** Construct two perpendicular lines. Then use Construction 3 to bisect one of the right angles. The result is a 45° angle.   **b.** Construct two perpendicular lines. Then use Construction 3 to bisect one of the right angles. The result is a 45° angle. Bisect the 45° using Construction 3 to get an angle that measures 22.5°.   **c.** Repeat part b. The angle from the second angle bisector back to the other side of the line has a measure of 157.5°.

24. Copy $\angle A$. Use Construction 3 to bisect $\angle A$, and then use it again to bisect the two new angles.

25. Adapt Construction 4 appropriately.

26. Adapt Construction 4 appropriately.

27. In each case, use Construction 3 to find the angle bisector of each angle in the triangle. The point of intersection is the incenter. Next, construct a perpendicular line from the incenter to any side of the triangle. Use a compass to measure the distance from the incenter to the point at which the perpendicular line intersects the side. This distance is the radius of the incircle. Use the radius to construct the needed circle.

28. Use Construction 5 to find the perpendicular bisector of $\overline{AB}$. This divides $\overline{AB}$ into two congruent segments. Use Construction 5 to find the perpendicular bisector of each of the new segments. This divides $\overline{AB}$ into four congruent segments.

29. Use the procedure in Example 11.21, but complete the rhombus by drawing the segment from B to D.

30. Construct the perpendicular bisectors of the three sides of the triangle. Their point of intersection is the needed point.

31. $\overleftrightarrow{AB}$ and m are parallel lines. Using the diagram, $\angle ACF \cong \angle EPC$ because they are both right angles. Consequently, $\overleftrightarrow{AB} \parallel m$ because $\angle ACP$ and $\angle EPC$ are congruent alternate interior angles.

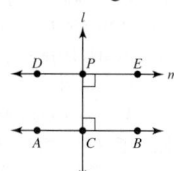

32. The medians of a triangle always intersect at a point.

33. **a.** No. An obtuse triangle has some altitudes outside the triangle.   **b.** The altitudes of a triangle always intersect at a point.

34. The perpendicular bisectors of the sides of a triangle always intersect at a point.

35. The angle bisectors of two pairs of vertical angles are perpendicular.

36. The bisectors of the chords intersect at the center of the circle.

37. By leaving the straightedge in place and sliding the drafting triangle up, the drafter is constructing two congruent corresponding angles. If two corresponding angles along a transversal are congruent, then the lines are parallel.

38. The triangle constructed by the midsegments is similar to the original triangle.

39. Because the opening of the compass is set, $\overline{AB} \cong \overline{BC} \cong \overline{CD} \cong \overline{DA}$. $\overline{DB}$ is congruent to itself, so $\overline{DB} \cong \overline{BD}$ and $\triangle ADB \cong \triangle CDB$ by the SSS property. Because corresponding parts of congruent triangles are congruent, $\angle ABD \cong \angle CBD$. Consequently, $\overrightarrow{BD}$ bisects $\angle ABC$.

40. Construct the angle bisectors of the three angles. The intersection of the three angle bisectors is the needed point.

41. Consider the segment $\overline{AB}$ and the point P that is equidistant from both A and B. If the segments $\overline{AP}$ and $\overline{BP}$ are drawn, then $\overline{AP} \cong \overline{BP}$, making $\triangle APB$ and isosceles triangle. Because the angle bisector of the vertex angle is perpendicular to the base and passes through its midpoint, P must be on the perpendicular bisector of $\overline{AB}$.

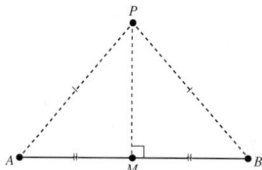

42. Measure the length of $\overline{AB}$ with the compass to get the radius of the circle. Use the compass to construct the circle.

43. Answers will vary.

44. **a.** Answers will vary.   **b.** Answers will vary.

45. Answers will vary.

46. A square works only if we are constructing the angle bisector of a right angle.

47. Answers will vary.

48. Measure a length of 10 ft on a rope, and then use the rope as a compass.

49. She can use Construction 3 with a rope as her compass and the two sidewalks as the angle.

50. Answers will vary.

◼ Section 11.4 Learning Assessment

1. **a.** One way is to use Construction 1 three times to construct one triangle congruent to another by the SSS property. Another way is use Constructions 1 and 2 to construct one triangle congruent to another by the SAS property.   **b.** One way is to use Constructions 1 and 2 to construct one triangle similar to another by the SAS property. Another way is use Constructions 1 and 2 to construct one triangle similar to another by the AA property.

2. **a.** If the length of a side is given, construct a square by first using Construction 5 to construct a right angle. Then use Construction 1 multiple times to construct the four sides.   **b.** If a side and angle are given, construct a rhombus by first copying the angle onto a ray. Then use Construction 1 to copy the sides and Construction 3 to find the fourth vertex of the rhombus.

3. One way to construct a regular hexagon is to construct six equilateral triangles that all share a common vertex. A second way is to construct a circle of a given radius. Without changing the opening of the compass, strike six arcs around the circle. Connect these points to get a regular hexagon.

4. A shape is constructible if it can be constructed using only a straightedge and compass.

5. A shape is circumscribed by constructing a circle that passes through each of the vertices of the shape.

6. a. Construct the perpendicular bisectors of the three sides of the triangle. Their point of intersection is the center of the circumcircle. The radius is the distance from the center to any of the vertices. **b.** Construct the diagonals of the rectangle. Their point of intersection is the center of the circumcircle. The radius is the distance from the center to any of the vertices. **c.** Construct the perpendicular bisectors of two of the sides. Their point of intersection is the center of the circumcircle. The radius is the distance from the center to any of the vertices.

7. Draw a ray from one of the endpoints of the segments. Use Construction 1 to construct three congruent segments on the ray. Draw the segment from the end of the last newly constructed segments to the endpoint of the given segment. Copy the angle made by this new segment and the ray at the endpoints of the other segments of the ray. The points at which the sides of these angles intersect the given segment are the points that trisect it.

8. a. Use the plastic reflector to construct the perpendicular bisector of each side. The point of intersection is the circumcenter. **b.** Fold each side of the triangle in half so that one endpoint of the segment is directly on top of the other. The creases are the perpendicular bisectors of the sides. The point at which the three perpendicular bisectors meet is the circumcenter. **c.** Answers will vary.

9. An equilateral triangle.

10. Once the square is constructed, place the reflector so that the drawing edge is directly over one side of the square. Use the reflection to copy the square to form a rectangle. Repeat the process on the edge of the rectangle to form the larger square.

11. A regular hexagon.

12. The knot forms an isosceles triangle, an isosceles trapezoid, and a regular pentagon.

13. Use the reflector to construct the perpendicular bisector of $\overline{AB}$. Place the reflector on B, and rotate it until the reflection of A is on the perpendicular bisector. Mark this point as the third vertex of the triangle.

14. Place the drawing edge of the reflector on one side of the triangle, and use the reflection to copy the triangle. Repeat the process four times using the appropriate edges to construct a regular hexagon.

15. Use Constructions 1 and 2 appropriately.

16. Use Constructions 1 and 2 appropriately.

17. Use Constructions 1 and 2 appropriately.

18. Use Construction 1 appropriately.

19. a. The third side should be equal to $5AB$. **b.** Use the compass to measure AB. Start at one endpoint of the hypotenuse, and mark off five lengths equal to AB. The fifth length should end at the other endpoint.

20. Use Constructions 1 and 2 appropriately.

21. Use Constructions 1 and 2 appropriately.

22. Use the procedure in Example 11.31 to find the circumcenter of each triangle. Then use the distance from the circumcenter to any vertex as the radius of the circumcircle. Use the compass to construct the circle.

23. a. Use Constructions 1 and 2 appropriately. **b.** Use Constructions 1 and 2 appropriately. **c.** Use Constructions 1 and 2 appropriately. **d.** Answers will vary.

24. Construct two perpendicular lines. From the point of intersection, draw a circle of any radius. The four points at which the circle intersects the perpendicular lines are the four vertices of a square with two diameters of a circle as its diagonals.

25. a. Follow the procedure in Example 11.27. **b.** Construct the perpendicular bisector of $\overline{AB}$. Use the compass to measure the distance from the midpoint of $\overline{AB}$ to either A or B. Use this distance to draw a circle with a center at the midpoint of $\overline{AB}$. The four points at which the circle intersects $\overline{AB}$ and the perpendicular bisector are the four vertices of a square. **c.** Use the procedure in Example 11.27; this time, however, make one length equal to twice AB and the other length equal to AB.

26. a. Use Construction 2 to copy $\angle E$. Use the compass to measure $\overline{AB}$, and strike an arc from E that crosses both rays of $\angle E$. Without changing the opening of the compass, strike two more arcs that intersect in the interior of $\angle E$, one from each of the new intersection points. Connect the points to form the rhombus. **b.** Use Construction 2 to copy $\angle E$. Use the compass to measure $\overline{CD}$, and strike an arc from E on one side of the angle. Use the compass again to measure $\overline{AB}$. Strike two arcs, one from E that intersects the second side of the angle and the other from the intersection point made on the first side of the angle. From the new intersection point on the second side of the angle, strike another arc in the interior of $\angle E$ with a length of CD. This gives the four points of the parallelogram. **c.** Use Construction 1 to copy $\overline{CD}$. Use Construction 2 to copy $\angle E$ at each endpoint of $\overline{CD}$. Use Construction 1 to copy $\overline{AB}$ on the sides of the copied angles. C, D, and the two points of intersection on the copied angles are the four vertices of the trapezoid.

27. Construct a segment 1 in. long. Measure the length of the segment with the compass, and strike an arc from each endpoint so that they intersect above the segment. Draw the segments to create the first equilateral triangle. Repeat the construction four more times, each time starting with a side from the previous triangle.

28. a. Not constructible. **b.** Not constructible. **c.** Not constructible. **d.** Constructible. **e.** Constructible. **f.** Not constructible. **g.** Constructible. **h.** Constructible. **i.** Not constructible. **j.** Not constructible.

29. A regular n-gon is constructible if $n = 3, 4, 5, 6, 8, 10, 12, 15, 16, 17, 20, 24, 30, 32, 34, 40,$ and 48.

30. a. Use the procedure in Example 11.32. **b.** Use the procedure in Example 11.32.

31. a. Construct an equilateral triangle. It has a 60° angle. **b.** Construct an equilateral triangle, and bisect one of its angles. **c.** Construct an equilateral triangle, and bisect one of its angles. Extend an appropriate side to get an angle that measures 150°. **d.** Construct an equilateral triangle, and bisect one of its angles twice. Extend an appropriate side to get an angle that measures 105°.

32. a. Construct two perpendicular lines, and then find the perpendicular bisector of $\overline{AB}$ and $\overline{CD}$. On one line, strike two arcs that are $\frac{1}{2}AB$ from the intersection of the two perpendicular lines. On the other line, strike two arcs that are $\frac{1}{2}CD$ from the intersection of the two perpendicular lines. Connect the four new points to form the rhombus.

b. Construct two perpendicular lines, and then find the perpendicular bisector of $\overline{AB}$. Subdivide $\overline{CD}$ into four equal parts by constructing three perpendicular bisectors.

On one line, strike two arcs that are $\frac{1}{2}AB$ from the intersection of the two perpendicular lines. On the other line, strike one arc that is $\frac{1}{4}CD$ from the intersection of the two perpendicular lines. On the same line, strike an arc that is $\frac{3}{4}CD$. Connect the four new points to form the kite.

33. Construct two perpendicular lines. Open the compass to any setting, and draw a circle with the intersection of the two perpendicular lines as its center. Connect the four points of the intersection between the perpendicular lines and the circle. This gives the square and the circumcircle. To construct the incircle, construct a segment that is perpendicular to one of the square's sides and goes through the center. Measure the distance with the compass. This is the radius of the incircle.

34. Follow the procedure in Example 11.29. Connect the eight points to form the octagon and every other point to form the square.

35. Follow the procedure in Example 11.28 to construct the regular hexagon. Construct the perpendicular bisector of two sides. Their intersection point is the center of the hexagon and the circle. Use the compass to measure the distance from the center to the midpoints of one of the sides. This is the radius of the circle. Use it to draw the circle inscribed in the hexagon.

36. Construct a large arc. Draw two chords on the arc. Use Construction 5 to find the perpendicular bisector of each chord. The intersection of the perpendicular bisectors is the center of the circle and the arc.

37. In general, any isosceles trapezoid can be circumscribed.

38. Construct a ray from the center of the circle through the point on the circle. Measure the distance from the point on the circle to the center and use it to strike a ray on the other side of the point on the circle. Construct the perpendicular bisector of the segment from the new point to the center of the circle to find the tangent line.

39. Place one corner of the paper on a line. Rotate it until the other corner on the same side is five spaces down. Use the parallel lines to make the marks.

40. $x \approx 200.06$ mi.

41. Construct the angle bisector of the angle. Construct a perpendicular segment from point T to the angle bisector. The length of the segment is the radius of the needed circle.

42. **a.** The perpendicular bisector of the base is the angle bisector of the vertex angle. **b.** It is always true.
c. Consider the isosceles triangle $\triangle ABC$, in which $\overline{AB}$ is the base and $\overline{MC}$ is the perpendicular bisector of the base. $\overline{AM} \cong \overline{BM}$, $\overline{CM} \cong \overline{CM}$, and $\angle AMC \cong \angle BMC$. As a result, $\triangle AMC \cong \triangle BMC$ by SAS. Because corresponding parts of congruent triangles are congruent,

$\angle ACM \cong \angle BCM$. Hence, $\overline{MC}$ must be the angle bisector of $\angle ACB$.

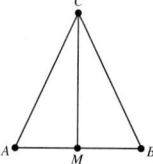

43. Yes, it is possible. The triangle is an acute scalene triangle.

44. No. The triangle inequality states that we cannot construct a triangle having sides of these lengths.

45. An equilateral triangle.

46. No. The sides of $\angle 1$ and $\angle 2$ do not meet to form a triangle.

47. Construct an equilateral triangle, and then construct the midpoints of two sides. When we draw the segment connecting the midpoints, we have the needed triangle. We can use the midsegment theorem and the AA property to show that the two triangles are similar. It also has a scale factor of $\frac{1}{2}$.

48. If a circle passes through three vertices of the rhombus, then the fourth vertex must be in the interior of the circle. This is primarily due to the fact that the diagonals of a rhombus are not the same length.

49. **a.** Yes. An example is shown:

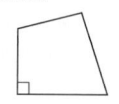

b. Yes. An example is shown:

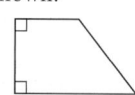

c. No. Because the sum of the interior angles of a quadrilateral is equal to 360°, the measure of the fourth angle must be 90° if the other three angles are each 90°.

50. **a.** Two squares are congruent if one pair of corresponding sides are congruent. **b.** Once we construct a right angle, we simply copy the side of the square as needed.

51. **a.** Two parallelograms are congruent if we know corresponding pairs of consecutive sides are congruent and one pair of corresponding angles are congruent.
b. Copy the angles and then the lengths of the sides onto the sides of the angle.

52. **a.** It is a parallelogram. **b.** The shape is always a parallelogram. **c.** Yes. The quadrilateral is always a parallelogram.

53. Answers will vary. 54. Answers will vary.

55. The circumcenter is always inside an acute triangle, on a right triangle, and outside an obtuse triangle.

56. Answers will vary. 57. Answers will vary.

58. It is not possible. If a triangle has three congruent sides, then it must be equilateral. This means all of its angles must measure 60°.

59. She is correct. We only have to construct the perpendicular bisectors of two sides to find the center of a regular polygon.

60. Clarice is not correct because the trisected segment does not lead necessarily to a trisected angle. Answers will vary on how to convince her otherwise.

61. Answers will vary.

62. **a.** Answers will vary. **b.** Answers will vary.

63. Answers will vary. 64. Answers will vary.

Chapter 11 Review Exercises

1. $\angle A \cong \angle F$, $\angle B \cong \angle E$, $\angle C \cong \angle D$, $\overline{AB} \cong \overline{FE}$, $\overline{BC} \cong \overline{ED}$, and $\overline{AC} \cong \overline{FD}$.

2. Possible illustrations are shown:

   a.

   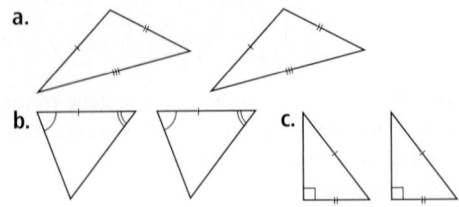

   b.

   c.

3. **a.** $XY = 7$, $YZ = 9$, and $XZ = 9$. **b.** $m\angle X = 65°$, $m\angle Y = 65°$, and $m\angle Z = 50°$.

4. $AB = 6$, $BC = 6\sqrt{3}$, $AC = 12$, $XY = 6$, $YZ = 6\sqrt{3}$, $m\angle B = 90°$, $m\angle C = 30°$, $m\angle X = 60°$, and $m\angle Y = 90°$.

5. **a.** Congruent by ASA. **b.** Congruent by SSS. **c.** Not congruent.

6. $\triangle ABD \cong \triangle DCA$, $\triangle ABD \cong \triangle BAC$, $\triangle ABD \cong \triangle CDB$, $\triangle ABC \cong \triangle CDA$, $\triangle ABC \cong \triangle CDB$, $\triangle CDB \cong \triangle DCA$, $\triangle ABE \cong \triangle DEC$, and $\triangle BEC \cong \triangle AED$.

7. **a.** True. **b.** False. **c.** False. **d.** True. **e.** True. **f.** False.

8. Yes to both questions. Both the length of the side and the length of the radius determine the size of the regular pentagon.

9. Two isosceles triangles are congruent if one pair of corresponding angles is congruent and both pairs of bases are congruent. Two trapezoids are congruent if they satisfy the SASAS or the ASASA property.

10. Yes. If two triangles have two pairs of corresponding sides congruent and a pair of angles that are not included by the sides, then the triangles need not be congruent.

11. Because quadrilateral $ABCD$ is a rectangle, then opposite sides must be a parallel. This means $\angle 1 \cong \angle 4$ and $\angle 2 \cong \angle 3$ because they are alternating interior angles. Also, $\overline{AC} \cong \overline{CA}$ because it is the same segment. Hence, $\triangle ADC \cong \triangle CBA$ by the ASA property. Because corresponding parts of congruent triangles are congruent, $\overline{BC} \cong \overline{DA}$ and $\overline{AB} \cong \overline{CD}$.

12. Possible illustrations are shown:

    a.   **b.**

13.

14.

    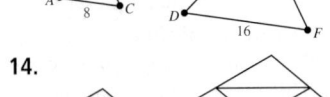

15. **a.** 4. **b.** 0.1.

16. **a.** 18. **b.** $\dfrac{5}{3}$.

17. **a.** $x = 10.5$. **b.** $y = \dfrac{45}{7} \approx 6.43$. **c.** $z = 6$.

18. The sides have lengths of about 1.79 and 3.08.

19. **a.** $AE = 10$. **b.** $AB = 16$. **c.** $AC = 20$. **d.** $BC = 12$.

20. **a.** $DE = 48$ and $XY = 24$. **b.** $\dfrac{3}{8} = 0.375$.

21. No in both cases. The angles can be different and the sides need not be proportional.

22. 67.5 ft tall.

23. **a.** Use Construction 1 and Construction 5. **b.** Use Construction 1 and Construction 5.

24. **a.** Use Construction 2 and then Construction 3. **b.** Use Construction 2 and then Construction 3.

25. Draw a line through P that intersects $\overleftrightarrow{AB}$. Label the point of intersection C. Use Construction 2 to make a copy of the angle on $\overleftrightarrow{CP}$ with a vertex at P so that the angle corresponds to the angle being copied.

26. Construct triangle $\triangle ABC$, and then use Construction 3 to construct the angle bisectors of each of its angles. Their point of intersection is the incenter. Construct a segment that is perpendicular to a side and passes through the incenter. Its length is the radius of the incircle.

27. Construct triangle $\triangle ABC$, and then use Construction 5 to construct the perpendicular bisectors of each side. Their point of intersection is the circumcenter. Construct a segment that goes to any of the vertices. Its length is the radius of the circumcircle.

28. **a.** Use Construction 1 three times with $\overline{AB}$. **b.** Use Construction 1 once to copy $\overline{AB}$. Use Construction 1 to copy $\overline{CD}$ from A. The segment from the new endpoint to B has a length of $AB - CD$. **c.** Use Construction 1 three times, once with $\overline{AB}$ and twice with $\overline{CD}$.

29. **a.** Use Construction 2 twice, once with $\angle A$ and once to copy $\angle B$ onto one side of $\angle A$. **b.** Use Construction 2 twice, once with $\angle B$ and once to copy $\angle A$ onto one side of $\angle B$ to the interior of $\angle B$. **c.** Use Construction 2 three times, making each copy of $\angle B$ on the side of the previous angle.

30. Construct two perpendicular lines. Use Construction 3 to bisect one of them. Use Construction 3 again to bisect the upper angle. This gives the needed angle.

31. Use Constructions 1 and 2 appropriately.

32. Construct segment $\overline{AB}$ and measure it with the compass. Without changing the opening, place its point on A and strike an arc above $\overline{AB}$. Again without changing the opening of the compass, place its point on B and strike another arc so that it intersects the first. Complete the triangle by drawing segments from A and B to the intersection of the two arcs.

33. Use Constructions 1 and 2 appropriately.

34. **a.** Use the procedure in Example 11.27. **b.** Use the procedure in Example 11.27, but this time double the length of one side. **c.** Construct two perpendicular lines and then bisect one of the angles. Use the compass to measure $\overline{AB}$. Strike an arc from the intersection of the lines that intersects both sides of the angle. From the two new intersection points, keep the opening of the compass, and strike two arcs that intersect in the interior of the angle. Draw the two segments to complete the rhombus. **d.** Construct an equilateral triangle with sides equal to AB. Extend one side, and make an additional copy of $\overline{AB}$. Construct a line parallel to the extended side that passes through the vertex opposite the side. From the previous intersection point on the extended side, strike an arc with a radius equal to AB that intersects the parallel line. Draw the segment to complete the construction.

35. Use the procedure in Example 11.28.

36. a. Not constructible.   **b.** Not constructible.
   c. Constructible.   **d.** Not constructible.   **e.** Constructible.
   f. Constructible.

37. Use the procedure in Example 11.30.

CHAPTER 12

■ Section 12.1 Learning Assessment

1. Coordinate geometry can be used to learn how to interpret the meaning of the coordinates as they relate to a distance and a direction from a fixed location, such as the origin. It can also be used to learn how to give the relative position of one location with respect to another or to describe how to travel from one point to another.

2. Latitudes and longitudes form a grid system on the earth. Latitudes run east and west, and longitudes run north and south. Latitudes and longitudes do not form a true rectangular grid because the curvature of the earth causes the lines to bend.

3. The unit length is the distance from 0 to 1 on the x-axis of the Cartesian plane. The length of a line segment, or equivalently the distance between any two points, is the number of unit lengths equivalent to the segment or distance.

4. We can compute the length of a horizontal line segment by subtracting the x-coordinates of its endpoints. We can compute the length of a vertical line segment by subtracting the y-coordinates of its endpoints.

5. We compute the horizontal and vertical distances between the segment's endpoints. We then place these values into the Pythagorean Theorem to find the overall length of the segment.

6. The order in which we subtract the x- and y-coordinates of the endpoints does not matter because the value of the differences is squared in the distance formula.

7. In any segment, the midpoint is halfway between the endpoints both horizontally and vertically. We find the x-coordinate of the midpoint by adding x-coordinates of the endpoints and dividing the result by 2. The y-coordinate of midpoint is found by adding y-coordinates of the endpoints and dividing the result by 2.

8. If we place one vertex on the origin, then computations involving the coordinates of this vertex are easier because the coordinates of the origin are (0, 0).

9. The definition of a circle implies that any point (x, y) on the circle must be r units from its center (h, k). Hence, by the distance formula, we have $r = \sqrt{(x - h)^2 + (y - k)^2}$. Squaring both sides, we get the equation of a circle, or $r^2 = (x - h)^2 + (y - k)^2$.

10. a. Craig.   **b.** Aspen.   **c.** Gunnison.   **d.** Cheyenne.
   e. Monte Vista.   **f.** Trinidad.

11. a. C13.   **b.** I11.   **c.** G11.   **d.** E7.   **e.** F2.   **f.** B14.

12. a. West.   **b.** East.   **c.** East.   **d.** South.   **e.** Southwest.
   f. South.

13. a. Right scalene.   **b.** Obtuse isosceles.
   c. Right isosceles.   **d.** Obtuse scalene.

14. Possible coordinates are given:
   a. (0, 3).   **b.** (2, 2).   **c.** (0, 4).

15. $C(10, 6)$ and $D(10, 0)$.

16. $B(x_1, y_2)$ and $D(x_2, y_1)$.

17. C has coordinates $(a + c, b)$.

18. a. (1, 7).   **b.** (2, −1).

19. Three possible points are $(-2, 3)$, $(-8, 9)$, and $(-6, -7)$.

20. $\triangle ABC$ is an acute isosceles triangle because $AC = BC = \sqrt{20}$.

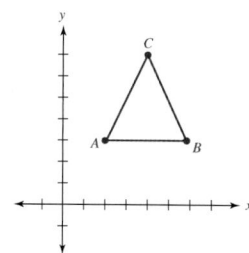

21. Quadrilateral $ABDC$ is a parallelogram because $AC = BD$ and $CD = AB$.

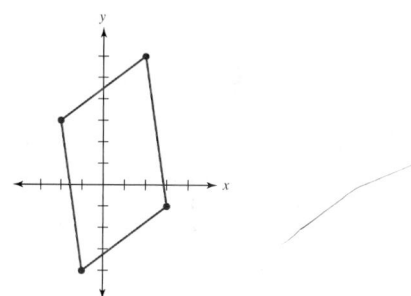

22. a.   **b.**

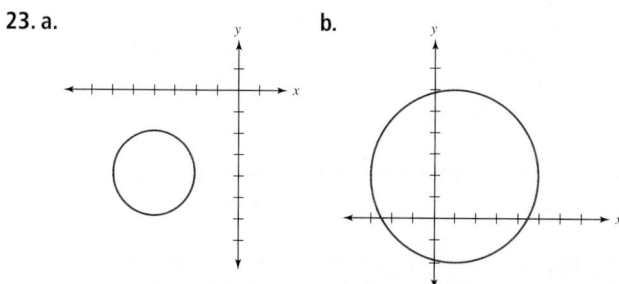

23. a.   **b.**

24. Four points are (4, 0), (0, 4), (4, 8) and (8, 4).

25. a. $AB = 7$.   **b.** $AB = 4$.   **c.** $AB = 7$.

26. a. $AB = 10$.   **b.** $AB = 5$.   **c.** $AB = 3.9$.

27. $AB = 5$, $CD = \sqrt{52}$, $EF = 7$, and $GH = 6\sqrt{2}$.

28. a. $d = \sqrt{26}$.   **b.** $d = \sqrt{41}$.   **c.** $d = \sqrt{16.4} \approx 4.05$.

29. a. $d = \sqrt{170}$.   **b.** $d = \sqrt{99.45} \approx 9.97$.
   c. $d = \sqrt{\dfrac{145}{144}} \approx 1.003$.

30. Collinear.   **31.** Not collinear.

32. a. $(2, 0)$.   **b.** $(0, 3.5)$.   **c.** $(0.5, -2.5)$.

33. a. $(4, 5)$.   **b.** $(-1.95, 3.55)$.   **c.** $\left(\dfrac{35}{12}, -\dfrac{31}{40}\right)$.

34. Midpoint of $\overline{AB} = (0, 7)$; midpoint of $\overline{CD} = (-3, 3.5)$; midpoint of $\overline{EF} = (-6.5, 0)$; and midpoint of $\overline{GH} = (4, -0.5)$.

35. a. $(x + 1)^2 + (y + 1)^2 = 36$.   **b.** $(x - 2)^2 + (y - 3)^2 = 25$.   **c.** $x^2 + y^2 = 9$.

36. a. $(x + 4)^2 + (y - 6)^2 = 49$.

    b. $\left(x - \dfrac{1}{2}\right)^2 + \left(y - \dfrac{1}{3}\right)^2 = \dfrac{1}{4}$.

    c. $(x - 0.6)^2 + (y - 7.5)^2 = 13.69$.

37. a. $(h, k) = (1, -1), r = 3$.   **b.** $(h, k) = (4, -7), r = 6$.

    c. $(h, k) = (0, 0), r = 4$.

38. a. $(h, k) = (-6, 1), r = 5$.   **b.** $(h, k) = \left(\dfrac{1}{2}, \dfrac{1}{3}\right), r = \dfrac{1}{5}$.

    c. $(h, k) = (-\sqrt{6}, -0.8), r = \sqrt{15}$.

    d. $(h, k) = (0.67, -1.3), r = \sqrt{14.4}$.

39 a. $x^2 + y^2 - 4x + 4y + 4 = 0$.

    b. $x^2 + y^2 + 8x - 14y + 49 = 0$.

    c. $x^2 + y^2 - 16 = 0$.

40. a. $d = \sqrt{48} \approx 6.93$.   **b.** $d = \sqrt{147} \approx 12.12$.

41. Three points are collinear if $AB + BC = AC$. Because

$$AB = \sqrt{(x_2 - x_1)^2 + (y_2 - y_1)^2},$$

$$BC = \sqrt{(x_3 - x_2)^2 + (y_3 - y_2)^2}, \text{ and}$$

$AC = \sqrt{(x_3 - x_1)^2 + (y_3 - y_1)^2}$, three points are collinear if $\sqrt{(x_2 - x_1)^2 + (y_2 - y_1)^2} + \sqrt{(x_3 - x_2)^2 + (y_3 - y_2)^2} = \sqrt{(x_3 - x_1)^2 + (y_3 - y_1)^2}$.

42. a. $B(5, 3)$.   **b.** $B(-5, -6)$.   **c.** $B(-8, -9)$.   **d.** $B(5, -10)$.

43. $C(3, 3\sqrt{3})$.

44. a. C and D could be $(2, 5)$ and $(-1, 3)$ or $(6, -1)$ and $(3, -3)$.   **b.** C and D could be $(0, 8)$ and $(-3, 6)$ or $(8, -4)$ and $(5, -6)$.   **c.** C and D could be $(9, 2)$ and $(6, 0)$ or $(-1, 2)$ and $(-4, 0)$.

45. C and D could be $(-3, -4)$ and $(-6, 0)$ or $(5, 2)$ and $(2, 6)$.

46. a. $A(10, 0)$, $B(0, 0)$, $C(0, 8)$, and $D(10, 8)$.

    b. $A(5, 6)$, $B(5, -4)$, $C(13, -4)$, and $D(13, 6)$.

47. A regular hexagon can be made from six equilateral triangles. If we center the regular hexagon at the origin and let it have a radius of 6, then we can use Problem 43 to find the six vertices. Specifically, the coordinates of the six vertices are $(6, 0)$, $(3, 3\sqrt{3})$, $(-3, 3\sqrt{3})$, $(-6, 0)$, $(-3, -3\sqrt{3})$, and $(3, -3\sqrt{3})$.

48. No. The segment from $(4, 0)$ to $(10, 0)$ has a length of 6. The segments from $(3, 5)$ to $(4, 0)$ and $(11, 5)$ to $(10, 0)$ both have a length of $\sqrt{26}$.

49. The center is at $(h, k) = (0, 0)$ and the radius is $r = 4$.

50. They should each walk 5 blocks and meet at the intersection of Beech and 5th.

51. a. Approximately 66.67 mi. between longitudes.

    b. Approximately 1.11 mi. between degrees.

52. a. 22.   **b.** 5 min, 30 s.

53. $x = \dfrac{z}{2}$.

54. Using the distance formula to compute the lengths of the three sides, we have $AB = \sqrt{40}$, $BC = \sqrt{40}$, and $AC = \sqrt{32}$. Because $AB = BC$, $\triangle ABC$ must be isosceles.

55. Consider $\triangle ABC$ in which the medians have been drawn to the equal sides as shown:

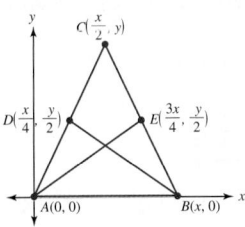

Using the coordinates from the diagram, we have

$$BD = \sqrt{\left(x - \dfrac{x}{4}\right)^2 + \left(0 - \dfrac{y}{2}\right)^2} = \sqrt{\left(\dfrac{3}{4}x\right)^2 + \left(\dfrac{y}{2}\right)^2}$$

$$= \sqrt{\dfrac{9x^2}{16} + \dfrac{y^2}{4}}$$

$$AE = \sqrt{\left(\dfrac{3x}{4} - 0\right)^2 + \left(\dfrac{y}{2} - 0\right)^2} = \sqrt{\dfrac{9x^2}{16} + \dfrac{y^2}{4}}.$$

Hence $BD = AE$.

56. ED is always half the length of BC. Consider $\triangle ABC$ in which $\overline{ED}$ joins the midpoints of $\overline{AB}$ and $\overline{AC}$.

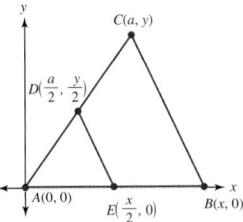

Using the coordinates from the diagram, we have

$$BC = \sqrt{(a - x)^2 + (y - 0)^2} = \sqrt{(a - x)^2 + (y)^2} \text{ and}$$

$$DE = \sqrt{\left(\dfrac{a}{2} - \dfrac{x}{2}\right)^2 + \left(\dfrac{y}{2} - 0\right)^2} = \sqrt{\left(\dfrac{a - x}{2}\right)^2 + \left(\dfrac{y}{2}\right)^2}$$

$$= \sqrt{\dfrac{(a - x)^2}{4} + \dfrac{y^2}{4}} = \dfrac{1}{2}\sqrt{(a - x)^2 + y^2}.$$

Hence $DE = \left(\dfrac{1}{2}\right)BC$.

57. a. Because $AB = 5$, $BC = 5$, $CD = 5$, and $AD = 5$, we have that $\overline{AB} \cong \overline{BC} \cong \overline{CD} \cong \overline{AD}$.

    b. The diagonals of the square are segments $\overline{AC}$ and $\overline{BD}$. Using the distance formula to compute their lengths, we have $AC = \sqrt{(5 - 0)^2 + (5 - 0)^2} = \sqrt{50}$ and $BD = \sqrt{(5 - 0)^2 + (0 - 5)^2} = \sqrt{50}$. Because the lengths are the same, $\overline{AC} \cong \overline{BD}$.

58. The diagonals of the trapezoid are segments $\overline{AC}$ and $\overline{BD}$. Using the distance formula to compute their lengths, we have $AC = \sqrt{(6 - 0)^2 + (5 - 0)^2} = \sqrt{61}$ and $BD = \sqrt{(8 - 2)^2 + (0 - 5)^2} = \sqrt{61}$. Because the lengths are the same, $\overline{AC} \cong \overline{BD}$.

59. The slope of $\overline{AC}$ is $\dfrac{(4 - 0)}{(8 - 0)} = \dfrac{1}{2}$. The slope of $\overline{BD}$ is $\dfrac{(4 - 0)}{(3 - 5)} = -2$. Because the slopes are negative reciprocals of each other, they must be perpendicular.

60. The median is always half the sum of the two bases in an isosceles trapezoid. To understand why, consider the isosceles trapezoid shown with one vertex on the origin.

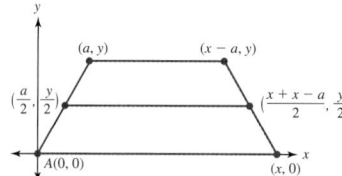

The sum of the two base lengths is equal to $(x - 0) +$ $((x - a) - a) = 2x - 2a$. The length of the median is $\left(\dfrac{x + x - a}{2}\right) - \dfrac{a}{2} = \dfrac{2x - 2a}{2}$, which is half the sum of the two bases.

61. The equation indicates that the length of the radius is a negative number. The radius of a circle must be greater than zero, so the equation does not make sense.

62. Answers will vary.

63. a. Answers will vary. **b.** Answers will vary.

64. Heather is correct. Because the differences are squared, the order of the coordinates does not affect the outcome of the computation.

65. Hunter forgot to divide the differences between the coordinates by two.

66. She realized that A is on the origin and B and C are on the x- and y-axes. This means the triangle must have a right angle.

67. Answers will vary.

68. a. Answers will vary. **b.** Answers will vary. **c.** Answers will vary. **d.** Answers will vary.

69. Answers will vary.

70. Answers will vary.

71. a. Answers will vary. **b.** Answers will vary.

■ Section 12.2 Learning Assessment

1. A rigid motion is a motion that preserves length or the distance between points.

2. The preimage indicates the position and the orientation of a figure before it is moved under a transformation. The image indicates the position and orientation of a figure after it has been moved under a transformation.

3. a. A transformation indicates how each point P on the plane corresponds to another point P' on the plane. **b.** The domain and the range are the same set, the set of points on the plane.

4. a. A translation is the motion that takes any object from one location to another without twisting or turning it. **b.** A rotation is the motion that takes any object and turns it about a given point. **c.** A reflection is the motion that takes every point P on the plane and maps it to a unique point P' so that P' is on the opposite side of the axis of reflection and at the same distance as P. **d.** A glide reflection is the motion that translates a given object a given distance and then reflects it across a line.

5. a. A translation is designated by a translation vector, which indicates the direction and the length of the motion. **b.** A rotation is designated by a turn center and a turn angle, which indicates the direction and the amount of the rotation. **c.** A reflection is designated by an axis

of reflection. **d.** A glide reflection is designated by a translation vector and a glide axis.

6. a. A fixed point is any point that stays in the same position under a rigid motion. **b.** A rotation has a fixed point if the point is on the turn center. A reflection has a fixed point if the point is on the axis of reflection.

7. The identity transformation is any transformation that maps every point on the plane back to itself.

8. A proper transformation preserves orientation. An improper transformation reverses orientation, left-handed to right-handed and clockwise to counterclockwise.

9. It does not matter. A translation followed by a reflection has the same result as a reflection followed by a translation.

10. a. $T_{PP'}$ is the translation that maps the point P onto the point P'. **b.** $R_{C,90°}$ is the rotation of 90° about the point C in the counterclockwise direction. **c.** $R_{C,180°}$ is the rotation of 180° about the point C in the counterclockwise direction. **d.** $R_{C,-180°}$ is the rotation of 180° about the point C in the clockwise direction. **e.** M_l is the reflection in the line l. **f.** M_m is the reflection in the line m.

11. a. $T_{AB}(P) = P'$ is the translation that maps P to P' by the translation vector from A to B. **b.** $R_{C,-90°}(D) = D'$ is the rotation that maps D to D' by a clockwise rotation of 90° around the point C. **c.** $M_k(E) = E'$ is the reflection that maps E to E' by reflecting it in the line k.

12. a. The image of $T_{OO'}$ is shown in red.

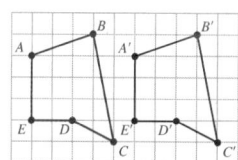

b. The image of $T_{PP'}$ is shown in red.

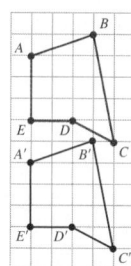

c. The image of $T_{QQ'}$ is shown in red.

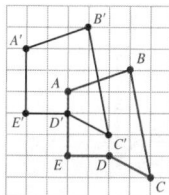

d. The image of $T_{RR'}$ is shown in red.

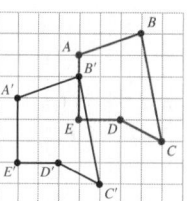

13. Construct a line parallel to $\overline{PP'}$ that passes through A. Copy the segment $\overline{PP'}$ onto the new line using A as a vertex for the new segment. Extend the new segment in the direction of $T_{PP'}$. This finds A'.

14. The images of $\triangle ABC$ are shown in red.

a.

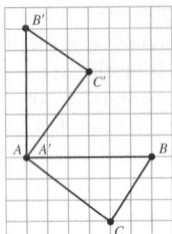

b.

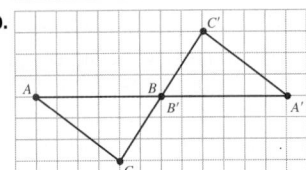

c.

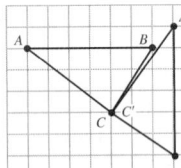

15. The images of the figure are shown in red.

a.

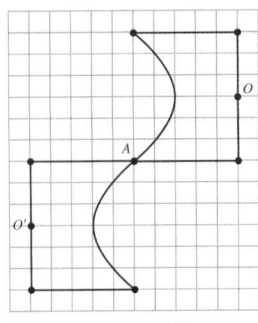

b.

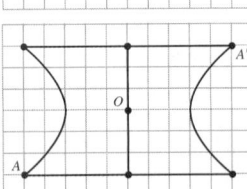

c.

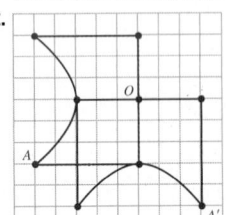

16. The images of quadrilateral $ABCD$ are shown in red.

a.

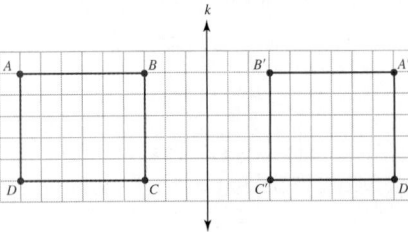

b.

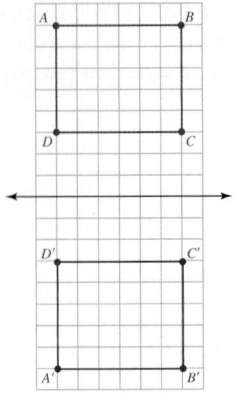

c. m

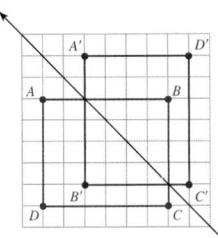

17. The images of quadrilateral $PQRS$ are shown in red.

a.

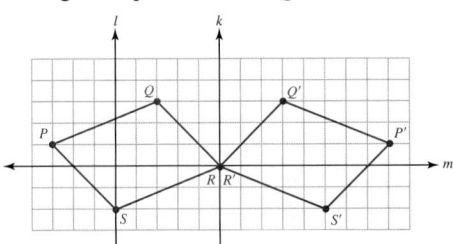

b.

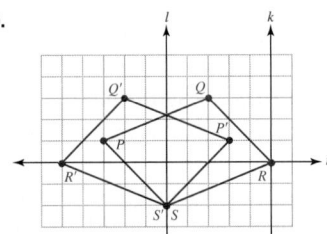

c.

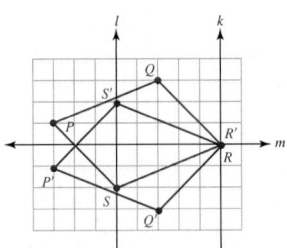

18. The image of the figure is shown in red.

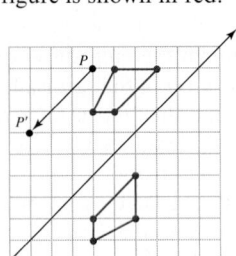

19. a. Reflection with vertical reflection line.   **b.** A glide reflection with a horizontal glide axis.   **c.** Translation.   **d.** Rotation.

20. a. Reflection about the angle bisector of $\angle B$.   **b.** Rotation of 120° in the clockwise direction about the center of the triangle.   **c.** Rotation of 120° in the counterclockwise direction about the center of the triangle.   **d.** Reflection about the angle bisector of $\angle A$.

21. a. Same orientation.   **b.** Reverse orientation.   **c.** Reverse orientation.   **d.** Reverse orientation.   **e.** Same orientation.

22. a. $(-3, 11)$.   **b.** $(-2, 5)$.   **c.** $(2, 3)$.   **d.** $(-10, 2)$.   **e.** $(-7, 6)$.

23. a. $(6.1, -1.6)$.   **b.** $(9.4, -11.3)$.   **c.** $(4, -9)$.   **d.** $(-2, -9.7)$.   **e.** $(11.5, 5.7)$.

24. a. $-270°$.   **b.** $-90°$.   **c.** $-225°$.   **d.** $-43°$.   **e.** $-164.4°$.

25. a. $270°$.   **b.** $180°$.   **c.** $155°$.   **d.** $73°$.   **e.** $202.1°$.

26. a. $75°$.   **b.** $311°$.   **c.** $262°$.   **d.** $-215°$.   **e.** $-1°$.

27. It is a rotation of $-180°$ about the point O.

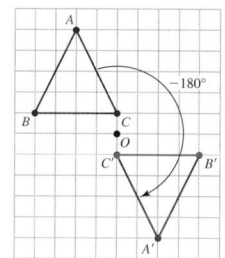

28. a. $(-4, 3)$.   **b.** $(-3, -4)$.   **c.** $(4, -3)$.   **d.** $(3, 4)$.

29. a. $(5, 2)$.   **b.** $(2, -5)$.   **c.** $(-5, -2)$.   **d.** $(-2, 5)$.

30. a. $A'(5, 6)$, $B'(5, 2)$, $C'(3, 3)$ and $D'(3, 5)$.   **b.** $A'(3, 8)$, $B'(3, 12)$, $C'(5, 11)$ and $D'(5, 9)$.   **c.** $A'(9, 0)$, $B'(5, 0)$, $C'(6, 2)$ and $D'(8, 2)$.

31. a. Point A.   **b.** Point C.   **c.** Point E.   **d.** Point G.

32. a. $(2, 3)$.   **b.** $(-2, -3)$.   **c.** $(-3, 2)$.   **d.** $(3, -2)$.

33. a. $(-1, 4)$.   **b.** $(1, -4)$.   **c.** $(-4, -1)$.   **d.** $(4, 1)$.

34. $A'(13, 8)$, $B'(9, 6)$, $C'(10, 2)$ and $D'(13, 2)$.

35. The axis of reflection and the image of $\triangle ABC$ are shown:

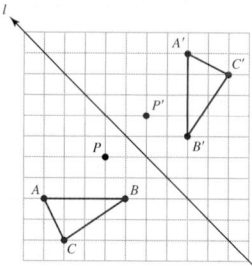

36. a. $(-1, -5)$.   **b.** $(8, -11)$.   **c.** $(4, -5)$.

37. The net effect of translating a point by vector T_1 and then by T_2 is to move the point 8 units to the right and 1 unit up. Hence, the vector that brings $\triangle ABC$ back to its original position is the vector that moves each point 8 units to the left and 1 unit down.

38. The preimage is shown in blue.

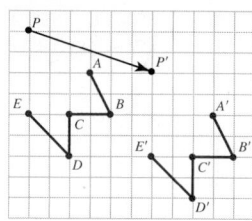

39. a. $y = 2x - 11$.   **b.** $y = 2x - 9$.   **c.** $y = 2x - 14$.

40. One possible placement of the mirror is shown in each case.

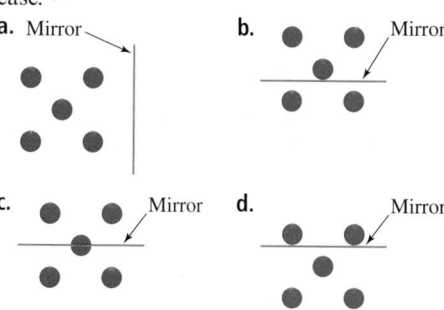

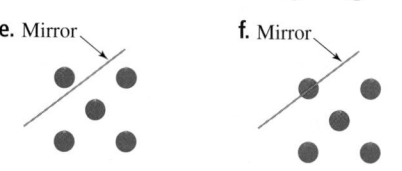

41. Suppose that $\overline{AB}$ and $\overline{A'B'}$ are two chords of a circle. Construct the perpendicular bisector of each segment. The point of intersection is the center of the rotation.

42. $(3, -6)$.

43. Draw the four segments from A, B, C, and D to their images. Find the midpoint of each segment, and then connect them to get the glide axis.

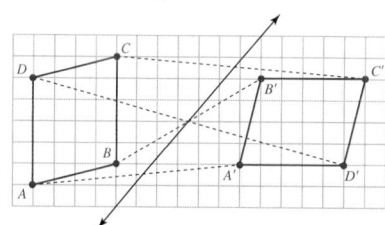

44. The image of $\triangle DOG$ is shown in red.

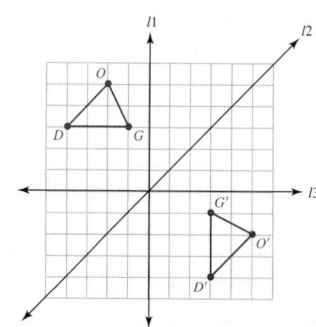

45. a. $(13, 5)$.   **b.** $(18, -5)$.   **c.** $(23, 5)$.   **d.** $(3 + 5n, -5)$.   **e.** $(3 + 5n, 5)$.

46. The coin is right side up because it has made a complete rotation when moving around the other coin.

47. a. 7:00.   **b.** 8:30.   **c.** 7:15.   **d.** 1:05.

48. The diagonal of a parallelogram is not an axis of reflection. For it to be a reflection, opposite vertices must map onto one another across the diagonal. This is not the case in every parallelogram.

49. A rotation about the origin by a turn angle of 180° negates the sign of both the x- and y-coordinate.   **a.** $(-1, -2)$.   **b.** $(3, -3)$.   **c.** $(-2, 5)$.   **d.** $(3, 1)$.

50. A reflection in the x-axis negates y-coordinate of the point.   **a.** $(3, -5)$.   **b.** $(-1, -1)$.   **c.** $(4, 2)$.   **d.** $(-4, 5)$.

51. A reflection in the line $x = y$ reverses the coordinates of the point.
  a. (2, 5). **b.** (1, −4). **c.** (−2, 2). **d.** (−3, −1).
52. a. A translation. **b.** A rotation.
53. A translation.
54. a. Reverse orientation. **b.** Reverse orientation.
  c. Same orientation. **d.** Same orientation.
  e. Reverse orientation. **f.** Reverse orientation.
55. a. Yes, $T_{AB} * T_{CD} = T_{CD} * T_{AB}$.
  b. Yes, $R_{A,180°} * R_{B,180} = R_{B,180} * R_{A,180}$.
  c. Yes, $M_l * M_k = M_k * M_l$.
  d. Because the order of the transformations does not matter, it is similar to the commutative property.
56. a. Answers will vary. **b.** Answers will vary.
57. A possible table is shown. Answers will vary on how to classify the transformations.

Transformation	Requirements	Orientation
Translation	Translation vector	Preserves
Rotation	Center of rotation and turn angle.	Preserves
Reflection	Axis of reflection	Reverses
Glide Reflection	Translation vector and glide axis	Reverses

58. The image could be the result of either a reflection or a rotation.
59. Miranda is not correct. A rotation of 180° is not an identity transformation. Answers will vary on how to correct the misconception.
60. Answers will vary.
61. Answers will vary.
62. Answers will vary.
63. a. Answers will vary. **b.** Answers will vary.
  c. Answers will vary. **d.** Answers will vary.
64. It is written backward so that, when a driver looks at the ambulance in the rearview mirror, the reflection reads "Ambulance."
65. a. Answers will vary. **b.** Answers will vary.
66. Answers will vary.

Section 12.3 Learning Assessment

1. Because a rigid motion preserves length and angle measure, it must map segments onto congruent segments and angles onto congruent angles. We can therefore use a rigid motion instead of the corresponding sides and angles to define the congruence relationship.
2. A rigid motion preserves the shape and size of a figure. A size transformation preserves the shape of the figure but not necessarily the size.
3. A size transformation must have a center from which the transformation originates and a scale factor that indicates the amount of change in the size of the image relative to the original figure.
4. The scale factor indicates the amount of change in the size of the image relative to the original figure. It is relative to the distance between the center and the preimage.
5. Because a size transformation preserves angles and the ratio of distances, then it must map angles onto congruent angles and segments onto proportional

segments. We can therefore replace the corresponding angles and sides with a size transformation to define the similarity relationship.
6. A size transformation is limited to just one transformation that maps the preimage onto a similar image with the same orientation. A similitude can be any combination of rigid motions and size transformations that leads to similar figures.
7. a. A figure has reflection symmetry if and only if there is a line l such that a reflection of the figure in l maps it onto itself. **b.** A figure has rotation symmetry if the figure can be mapped back onto itself by rotating it around a point through some turn angle less than 360°. **c.** A figure has point symmetry if it is its own image under a half turn or a 180° rotation.
8. We can classify shapes by their number of reflection or rotation symmetries.

9. a.

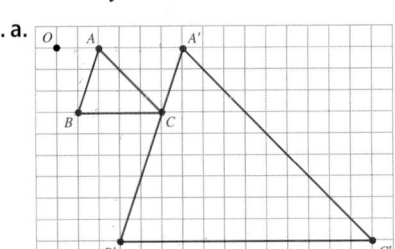

b.

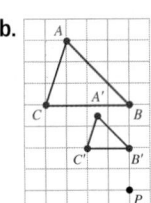

10. a.

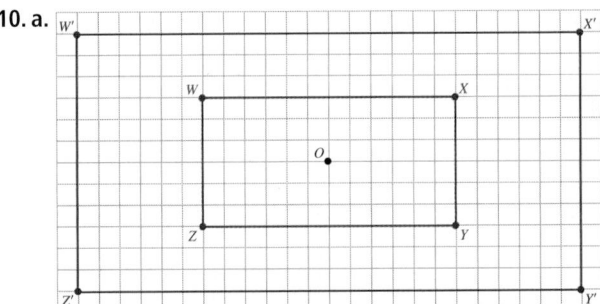

b.

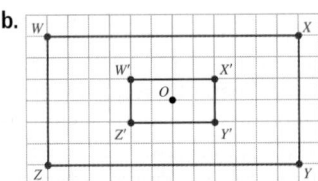

11.

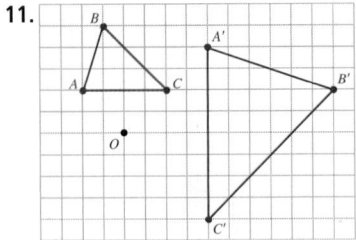

12. a.
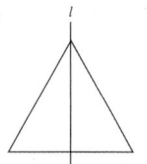

b.
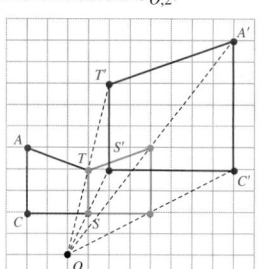

c.

d.

13. a.
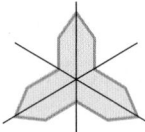

b.

c.
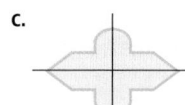

d.
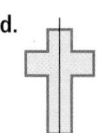

14. a.
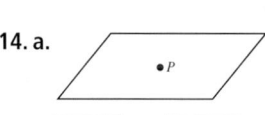

b.
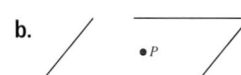

c.

15. a.
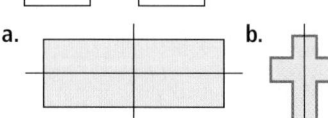

b.

c.
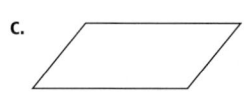

16. a.

b.

c.
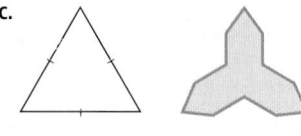

d.

17.
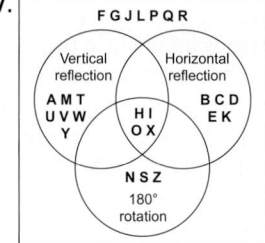

18. a. Reflection in $\overline{BE}$. **b.** Reflection. **c.** 180° rotation around E.

19. a. Translation with vector $\overrightarrow{AD}$. **b.** Reflection.
  c. 180° rotation around D.
20. Reflect Quadrilateral $CATS$ about the segment $\overline{ST}$. Then make the size transformation $S_{O,2}$.

21. Rotate $\triangle ABC$ by 180°, and then decrease each side length by $\frac{1}{2}$.
22. a. Scale factor $= 3$, $x = 9$, and $y = 24$.
  b. Scale factor $= \frac{3}{2}$, $x = \frac{32}{3}$, $y = 12$, and $z = 6$.
  c. Scale factor $= \frac{1}{2}$, $x = 3$, $y = \frac{7}{2}$, and $z = 4$.
23. a. 0, 3, and 8. **b.** 0 and 8.
24. a. 45°, 90°, 135°, 180°, 225°, 270°, and 315°. It has 45° rotation symmetry. **b.** 36°, 72°, 108°, 144°, 180°, 216°, 252°, 288°, and 324°. It has 36° rotation symmetry.
  c. 30°, 60°, 90°, 120°, 150°, 180°, 210°, 240°, 270°, 300°, and 330°. It has 30° rotation symmetry. **d.** 17.14°, 34.28°, 51.42°, 68.56°, 85.7°, 102.84°, 119.98°, 137.12°, 154.26°, 171.4°, 188.54°, 205.68°, 222.82°, 239.96°, 257.1°, 274.24°, 291.38°, 308.52°, 325.66°, and 342.8°. It has 17.14° rotation symmetry.
25. a. 2. **b.** 0. **c.** 8. **d.** 10. **e.** 12. **f.** 21.
26. a. Yes. **b.** No. **c.** No.
27. a. Yes. **b.** No. **c.** Yes. **d.** No.
28. a. $A'(-3, 9)$ and $B'(6, 15)$.
  b. $A'\left(-\frac{1}{2}, \frac{3}{2}\right)$ and $B'\left(1, \frac{5}{2}\right)$.
  c. $A'(-1, 5)$ and $B'(5, 9)$.
  d. $A'(-7, 3)$ and $B'(-1, 7)$.
29. a. $A'(0, 0)$, $B'(0, 12)$, and $C'(18, 0)$.
  b. $A'(0, 0)$, $B'(0, 3)$, and $C'(4.5, 0)$.
30. a. $A'(0, 0)$, $B'(18, 27)$, and $C'(36, 0)$.
  b. $A'(0, 0)$, $B'(2, 3)$, and $C'(4, 0)$.
31. a. $A'(1, 1)$, $B'(7, 9)$, $C'(15, 9)$, and $D'(9, 1)$.
  b. $A'(1, 1)$, $B'(2.5, 3)$, $C'(4.5, 3)$, and $D'(3, 1)$.
  c. $A'(1, 1)$, $B'(9, -5)$, $C'(9, -10)$, and $D'(1, -7)$.
  d. $A'(-1, 1)$, $B'(-2.5, 3)$, $C'(-4.5, 3)$, and $D'(-3, 1)$.
32. Use the procedure in Example 12.23, only construct a ray with five congruent segments instead of three.

33. a.
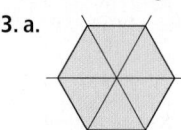

b.
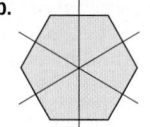

34.

35. 0 and 8 are the only even digits with rotation symmetry, so her house number must be 8008.
36. If we include the numbers on the clock, then it has neither reflection nor rotation symmetries.
37. a. It has two lines of symmetry, one vertical and one horizontal. It also has 180° rotation symmetry.   b. It has one line of vertical symmetry and no rotation symmetry.   c. It has two lines of symmetry, one vertical and one horizontal. It also has 180° rotation symmetry.   d. It has two lines of symmetry, one vertical and one horizontal. It also has 180° rotation symmetry.   e. It has two lines of symmetry, one vertical and one horizontal. It also has 180° rotation symmetry.   f. It has four lines of symmetry: one vertical, one horizontal, and two on the diagonals. It also has 90° rotation symmetry.
38. Answers will vary.   39. Answers will vary.
40. a. Answers will vary.   b. Answers will vary.
41. Fold the arc in half so that point A is over the top of point B. The fold line is the perpendicular bisector of chord $\overline{AB}$. Label the point of intersection between the arc and the fold line C. Fold the arc so that point A is on point C. The fold line is the perpendicular bisector of chord $\overline{AC}$. The intersection of the two fold lines is the center of the arc.
42. a. A rotation of 180° about P.   b. A rotation of 180° or a reflection in the segment $\overline{BD}$.   c. A rotation of 180°, a reflection in the segment $\overline{BD}$, or a translation with vector $\overrightarrow{AC}$.
43. Yes. Each side of the new triangle is one-half the length of the corresponding side. Hence, by the SSS property, the two triangles are similar.
44. No. Adding 8 to each side does not increase the length and the width by the same proportional amount.
45. The scale factor will be $\frac{1}{k}$.
46. Any diameter of a circle can be a line of symmetry. Because a circle has infinitely many diameters, it must have infinitely many lines of symmetry.
47. a. No. A parallelogram has rotation symmetry but not reflection symmetry.   b. No. An isosceles triangle has reflection symmetry but not rotation symmetry.
48. a. No. If a figure has 120° rotation symmetry, it does not have point symmetry.   b. Yes. If it has point symmetry, it must have 180° rotation symmetry.
49. Yes. Consider the triangle shown:

50. The completed table is shown:

*	r_1	r_2	r_3	r_4	R_1	R_2	R_3	I
r_1	I	R_1	R_2	R_3	r_2	r_3	r_4	r_1
r_2	R_3	I	R_1	R_2	r_3	r_4	r_1	r_2
r_3	R_2	R_3	I	R_1	r_4	r_1	r_2	r_3
r_4	R_1	R_2	R_3	I	r_1	r_2	r_3	r_4
R_1	r_4	r_1	r_2	r_3	R_2	R_3	I	R_1
R_2	r_3	r_4	r_1	r_2	R_3	I	R_1	R_2
R_3	r_2	r_3	r_4	r_1	I	R_1	R_2	R_3
I	r_1	r_2	r_3	r_4	R_1	R_2	R_3	I

Example patterns are
- Two reflections leads are equivalent to a rotation.
- Two of the same reflection is the identity transformation.
- Two rotations are equivalent to a reflection.
- Every column has one entry equivalent to each transformation.

51. a. Answers will vary.   b. Answers will vary.
52. Alternative definitions are as follows:
- A trapezoid is a quadrilateral with no reflection symmetries and no rotation symmetries.
- An isosceles trapezoid is a quadrilateral with one horizontal reflection symmetry and no rotation symmetries.
- A parallelogram is a quadrilateral with no reflection symmetries and 180° rotation symmetry.
- A kite is a quadrilateral with two reflection symmetries and no rotation symmetries.
- A rhombus is a quadrilateral with four congruent sides, two reflection symmetries, and 180° rotation symmetry.
- A rectangle is a quadrilateral with four congruent angles, two reflection symmetries, and 180° rotation symmetry.
- A square is a quadrilateral with four reflection symmetries and 90° rotation symmetry.

53. Fold along two lines of symmetry to find the center in each shape.
54. Answers will vary.
55. a. Answers will vary.   b. Answers will vary.   c. Answers will vary.
56. a. Answers will vary.   b. Answers will vary.
57. Answers will vary.

■ Section 12.4 Learning Assessment

1. A border pattern can be made using multiple iterations of a translation or a glide reflection.
2. A figure has translation or glide-reflection symmetry if a translation or glide reflection maps the figure onto itself.
3. Border patterns can have reflection and rotation symmetry, but only if a reflection or a rotation maps the entire pattern back onto itself. The reflection symmetry of a border pattern matches the reflection symmetry of its motif. The only rotation symmetry a border pattern can have is 180° rotation symmetry because an infinite border pattern can be mapped back onto itself only through a half or full turn.
4. a. A border pattern extends infinitely in one direction, whereas a wallpaper pattern extends infinitely in two.   b. A wallpaper pattern is an infinite pattern in two directions, but may not necessarily cover the plane without gaps or overlaps. A tessellation is an infinite pattern that covers the entire plane without overlaps or gaps.   c. A regular tessellation is a tessellation that is made from congruent repetitions of a single regular polygon. Semiregular tessellations use two or more regular polygons to make an arrangement of polygons that is the same at each vertex.

5. Equilateral triangles, squares, and regular hexagons are the only regular polygons with interior angle measures that divide 360°. This allows these regular polygons to meet at a vertex without either a gap or an overlap.

6. Triangles and quadrilaterals always tessellate the plane. With triangles, the sum of the angles is 180°. We can therefore use a series of translations and rotations to manipulate three copies of the triangle so that the angles come together to form a straight angle. From here, a rotation of 180° of all three triangles results in six angles that share a common vertex and have a sum of measures equal to 360°. With quadrilaterals, the sum of the angles is 360°. We can therefore rotate copies of the quadrilateral so that all four vertices meet at one vertex, giving us what we need to form a tessellation. Consequently, any quadrilateral tessellates the plane.

7. In the cut-and-slide method, one side of a figure that tessellates the plane is cut out and translated to the opposite side. The new figure still tessellates the plane. In the cut-and-turn method, one side of a figure that tessellates the plane is cut out and rotated around a vertex so that it lies overtop an adjacent side. The new figure still tessellates the plane.

8. a. *m*1. b. *mm.* c. 12. d. *mg.* e. 1*m.* f. *mm.*

9. a. ··· ⊣ ⊣ ⊣ ⊣ ⊣ ···   b. ··· T ⊥ T ⊥ T ···
   c. ··· T T T T ···   d. ··· ⊣⊢ ⊣⊢ ⊣⊢ ⊣⊢ ···

10. a. ··· E E E E E ···   b. ··· EƎ EƎ EƎ EƎ EƎ ···
    c. ··· E   E   E ···   d. ··· EƎ   EƎ   EƎ ···
         E   E   EƎ   EƎ

11. a. 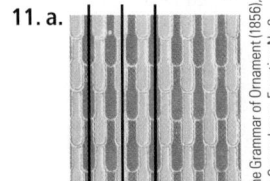 b.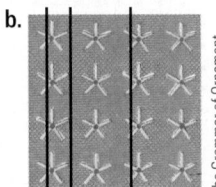

The Grammar of Ornament (1856), by Owen Jones. Egyptian No 8 (plate 11), image #20.

The Grammar of Ornament (1856), by Owen Jones. Egyptian No 8 (plate 11). Image #2

c.

The Grammar of Ornament (1856), by Owen Jones. Indian No 1 (plate 49), image #7

12. a.  b.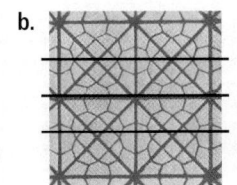

The Grammar of Ornament (1856), by Owen Jones. Egyptian No 6 (plate 9), image #19

The Grammar of Ornament (1856), by Owen Jones. Persian No 2 (plate 45), image #23

c.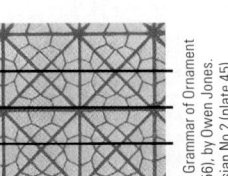

The Grammar of Ornament (1856), by Owen Jones. Chinese No 1 (plate 59), image #1

13. a.  b.

The Grammar of Ornament (1856), by Owen Jones. Egyptian No 7 (plate 10), image #20.

The Grammar of Ornament (1856), by Owen Jones. Egyptian No 7 (plate 10), image 24

c.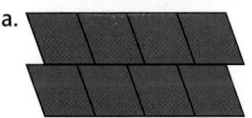

The Grammar of Ornament (1856), by Owen Jones. Egyptian No 7 (plate 10), image #15

14. a. Drawings will vary.   b. Drawings will vary.
15. a. Drawings will vary.   b. Drawings will vary.
16. Possible answers are shown:
    a.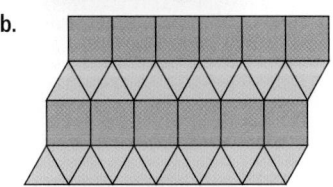
    b.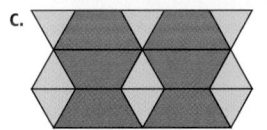
    c.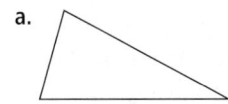

17. Possible polygons are shown:
    a.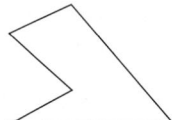

    b. Use the cut-and-turn method so that the portion being turned extends from a vertex to the midpoint of a side.
    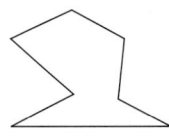

    c. Use the cut-and-turn method so that the portion being turned extends from a vertex to the midpoint of a side.

18. a.

    b. Start with a quadrilateral, and then use the cut-and-slide method on one side.

c. Start with a hexagon that tessellates the plane. Use the cut-and-turn method so that the portion being turned extends from a vertex to the midpoint of a side.

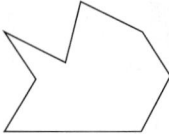

19. a. A regular pentagon is an example.

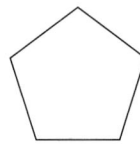

b. A possible drawing is shown:

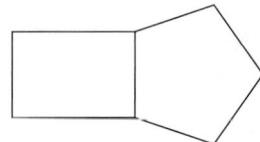

c. A possible drawing is shown:

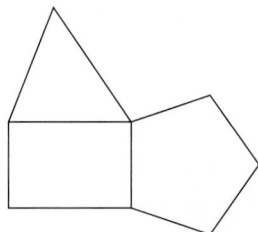

20. Drawings will vary. **21.** Drawings will vary.
22. Drawings will vary. **23.** Drawings will vary.
24. 135°
25. Ten arrangements do not tessellate the plane. Use angle measures and the semiregular tessellations to find three. The ten possible are (3, 7, 42), (3, 8, 24), (3, 9, 18), (3, 10, 15), (4, 5, 20), (5, 5, 10), (3, 3, 4, 12), (3, 4, 3, 12), (3, 3, 6, 6), and (3, 4, 4, 6).
26. a. Yes. **b.** Yes. **c.** Yes. **d.** Yes.
27. Drawings will vary. It will be necessary to use translations and rotations for the new figures to tessellate the plane.
28. The following tessellation uses a square, but it is not regular.

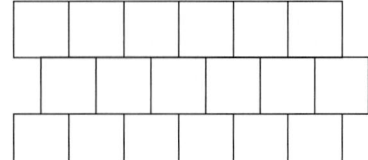

29. a.

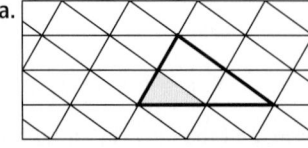

b.

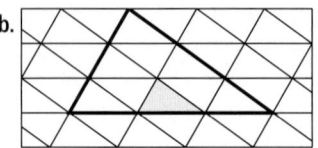

c.

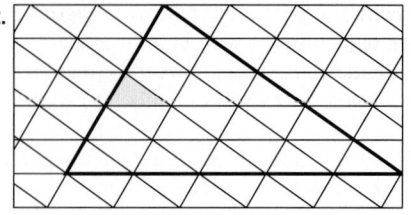

30. The shape itself does not tessellate the plane. However, if we combine it with a small square, it does.

31. a.

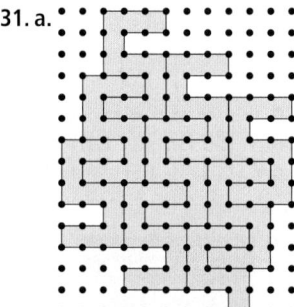

b.

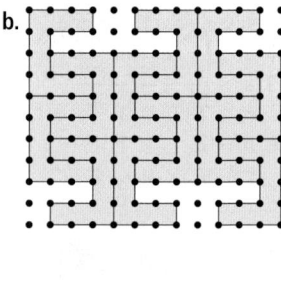

c.

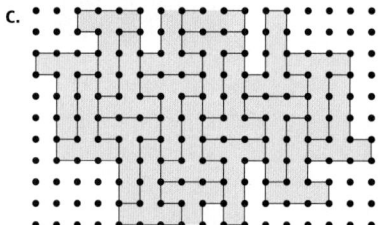

d.

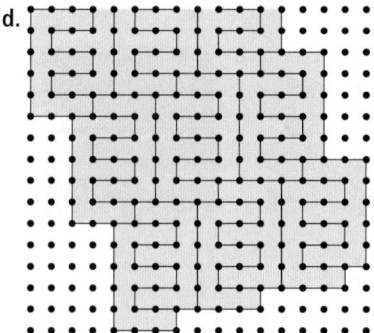

32. The duals of regular tessellations are regular tessellations. Specifically, the dual of the tessellation with regular triangles is the tessellation with hexagons, the dual of the tessellation with squares is the tessellation with squares, and the dual of the tessellation with regular hexagons is the tessellation with regular triangles.

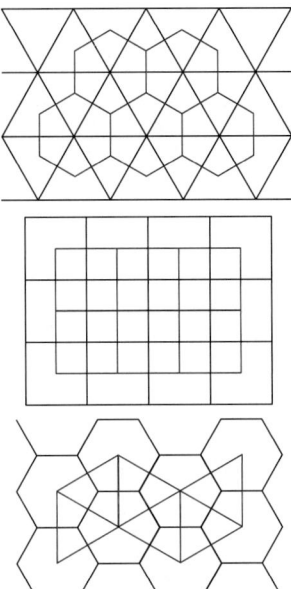

33. Any rectangular prism tessellates space. If we cut a cube with a plane along the diagonal, the new figure also tessellates space.

34.

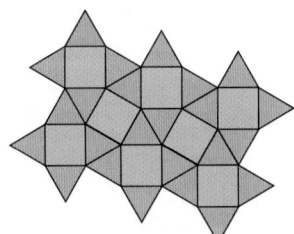

35. a. Translations.

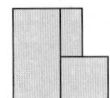

b. Rotations and translations.

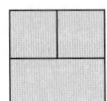

c. Rotations and translations.

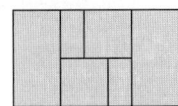

d. Rotations and translations.

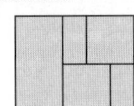

36. A regular heptagon has an interior angle measure of 128.57°. If we place two regular heptagons together at a vertex, then the total degree measure is 2 · 128.57° = 257.14°. Hence, there is a gap. If we place three regular heptagons together at a vertex, then the total degree

measure is 3 · 128.57° = 385.71°. Hence, there is some overlap. Because we cannot place two or three regular heptagons together at a vertex, a regular heptagon does not tessellate the plane.

37. No. For a convex hexagon to tessellate the plane, two of the sides must be parallel. Because not all convex hexagons have two parallel sides, not all convex hexagons tessellate the plane.

38. No. Although three quadrilaterals can fit around a vertex, when we try to tessellate the plane, at some point it is necessary to have more than three quadrilaterals at a given vertex.

39. The following tessellation uses a square, but it is not regular.

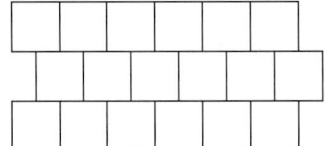

40. Two circles coming together at a point have a gap. Three circles coming together at a point overlap.

41. Yes. The following shape was constructed in this way, and it tessellates the plane.

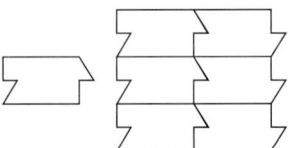

42. Answers will vary.
43. Answers will vary.
44. Answers will vary.
45. He is incorrect. If we try to fit regular octagons at a vertex, the sum of the measures is 2 · 135° = 270°, so there must be a gap. If we try to fit three, the sum of the measures is 3 · 135° = 405°, so there must be some overlap.
46. Answers will vary.
47. Answers will vary.
48. Answers will vary.
49. Answers will vary.

Chapter 12 Review Exercises

1. $AB = 6$, midpoint of $\overline{AB} = (0, 4)$.
$CD = \sqrt{53}$, midpoint of $\overline{CD} = \left(-3, \frac{-1}{2}\right)$.
$EF = \sqrt{34}$, midpoint of $\overline{EF} = \left(\frac{3}{2}, \frac{-1}{2}\right)$.
$GH = 4$, midpoint of $\overline{GH} = (4, -2)$.
2. Because $AB = \sqrt{34}$, $BC = \sqrt{17}$, and $AC = \sqrt{17}$, $\triangle ABC$ is a right isosceles triangle.
3. The lengths of the three sides are $AB = \sqrt{20}$, $BC = \sqrt{13}$, and $AC = \sqrt{65}$. Because $AB + BC = 8.077 \neq 8.062 = AC$, the points are not collinear.
4. a. $25 = x^2 + y^2$.
b. $1 = (x + 2)^2 + (y - 4)^2$.
c. $100 = (x - 3)^2 + (y - 8)^2$.

5. Radius = 8, center = (0, 3).

6. B has coordinates $(-8, -1)$.

7. Any point with coordinates $(4.5, y)$, where $y \neq 0$, works.

8. Possible coordinates of the fourth vertex are $(-1, -1)$, $(5, 5)$, and $(-5, 3)$.

9. C has coordinates $(2a, b)$.

10. a. 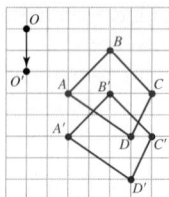   **b.**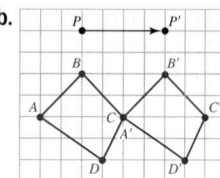

c.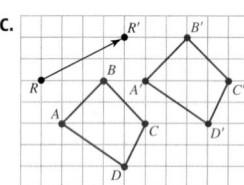

11. a. $(0, -4)$.   **b.** $(1, -12)$.   **c.** $(-11, -7)$.   **d.** $(-9, 4)$.

12. a. $-214°$.   **b.** $9°$.   **c.** $200°$.   **d.** $69°$.

13. Line k.   Line l.

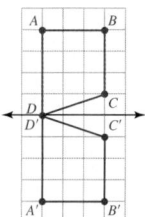   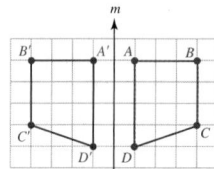

Line m.

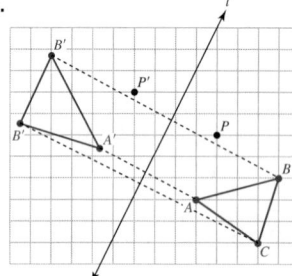

14.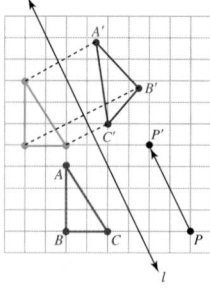

15.

16. a.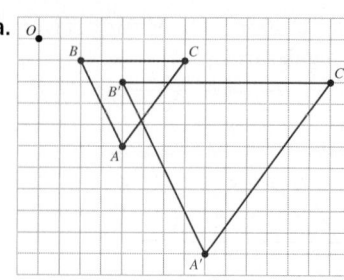

b.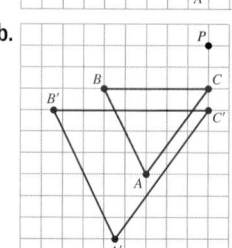

17. a. $(-1, -10)$.   **b.** $(2, -5)$.   **c.** $(-5, -13)$.   **d.** $(5, -13)$.

18. a. $(-b, a)$.   **b.** $(-a, -b)$.   **c.** $(b, -a)$.   **d.** $(-a, -b)$.

19. a. Mirror   **b.** Mirror   **c.** Mirror

20. a. $(4, 2)$.   **b.** $(-4, -2)$.   **c.** $(-2, 4)$.   **d.** $(2, -4)$.

21. $A'(-9, 9)$ and $B'(-3, -12)$.

22. A translation.

23. a. 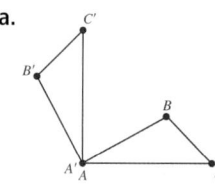   **b.**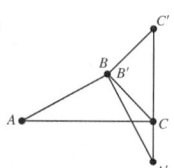

c.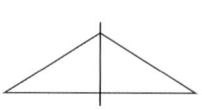

24. a.    **b.**

c.   **d.**

25. a.    **b.**

c. ▽ •P △

26. a. $90°$, $180°$, and $270°$.   **b.** $45°$, $90°$, $135°$, $180°$, $225°$, $270°$, and $315°$.   **c.** $40°$, $80°$, $120°$, $160°$, $200°$, $240°$, $280°$, and $320°$.   **d.** $25.7°$, $51.4°$, $77.1°$, $102.8°$, $128.5°$, $154.2°$, $179.9°$, $205.6°$, $231.3°$, $257.0°$, $282.7°$, $308.4°$, and $334.1°$.

27. a. A M   **b.** E B

c. 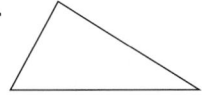   **d.** X H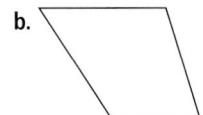

28. Answers will vary.
29. 18 different rotation symmetries.
30. No. A rhombus has 180° rotation symmetry, but no reflection symmetry.
31. **a.** ···M M M M M··· **b.** ···M Ⱳ M Ⱳ M···
c. ···⋝ ⋝ ⋝ ⋝ ⋝···
32. **a.** 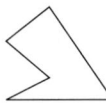 **b.**

c. Start with a triangle. Use the cut-and-turn method so that the portion being turned extends from a vertex to the midpoint of a side.

d. Start with a quadrilateral, and then use the cut-and-slide method on one side.

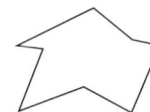

33. Five.
34. Drawings will vary.
35. Yes. An example is shown:

36. A regular decagon has an interior angle measure of 144°. If we place two regular decagons together at a vertex, then the total degree measure is 2 · 144° = 288°. Hence, there is a gap. If we place three regular decagons together at a vertex, then the total degree measure is 3 · 144° = 432°. Hence, there is some overlap. Because we cannot place two or three regular decagons together at a vertex, a regular decagon does not tessellate the plane.

Answers to Reviewing the Big Ideas: Chapters 10–12

1. Possible answers are given:
   a. $\overline{AB}$ and $\overline{CD}$.
   b. $\overrightarrow{AB}$ and $\overrightarrow{CD}$.
   c. A, B, and F.
   d. $\overleftrightarrow{AB}$ and $\overleftrightarrow{AE}$.
   e. $\overrightarrow{AB}$ and $\overrightarrow{DC}$.
   f. $\overrightarrow{AB}$ and $\overrightarrow{AE}$.
   g. A, B, C, and E.
   h. $\overleftrightarrow{AB}$ and $\overleftrightarrow{CG}$.
   i. Plane ABE and plane DCG.
2. **a.** 42°. **b.** 77°. **c.** 141°.
3. $m\angle 1 = 35°$ because it is an alternate exterior angle with the given angle of 35°.
   $m\angle 2 = 145°$ because it forms a linear pair with $\angle 1$.
   $m\angle 3 = 125°$ because it forms a linear pair with the given angle of 55°.
   $m\angle 4 = 90°$ because it is a vertical angle with $\angle 6$.
   $m\angle 5 = 90°$ because it forms a linear pair with the given angle of $\angle 6$.

$m\angle 6 = 90°$ because it is the interior angle of a triangle, which has two other interior angles of 35° and 55°.
$m\angle 7 = 125°$ because it is a corresponding angle with $\angle 3$.
$m\angle 8 = 55°$ because it is a corresponding angle with the given angle of 55°.
$m\angle 9 = 145°$ because it is a corresponding angle with $\angle 2$.
$m\angle 10 = 145°$ because it is a vertical angle with $\angle 9$.

4. Drawings will vary. Possible answers are given:
   a.
   b.  **c.**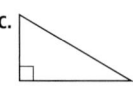
   d. 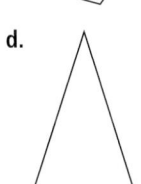 **e.**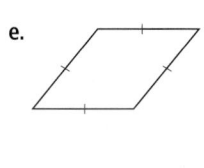

5. **a.** No. Because $3 + 4 = 7 < 8$, the Triangle Inequality states that these segments do not form a triangle.
   b. Yes. Because $4^2 + 4^2 = 16 + 16 = 32 = (\sqrt{32})^2$, the three lengths satisfy the converse of the Pythagorean Theorem.
6. **a.** $x = 37°$. **b.** $x = 30°$. **c.** $x = 13°$.
7. **a.** If the shortest side is 8, then the hypotenuse is $2 \cdot 8 = 16$, and the longest leg is $8\sqrt{3}$.
   b. If the hypotenuse is 20, then the lengths of the legs are $\dfrac{20}{\sqrt{2}} = \dfrac{20\sqrt{2}}{2} = 10\sqrt{2}$.
8. $d = 30$.
9. **a.** Interior angle = 120°, exterior angle = 60°, and central angle = 60°.
   b. Interior angle = 144°, exterior angle = 36°, and central angle = 36°.
10. 38°.
11. **a.** 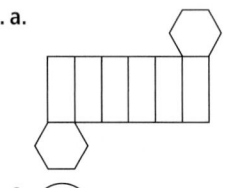 **b.**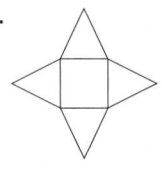
    c.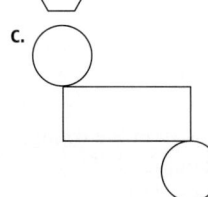
12. **a.** $v = 26$. **b.** $e = 22$.
13. It gains 6 vertices and 8 edges.
14.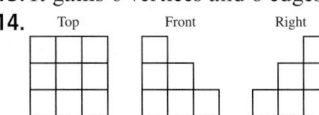
15. $\overline{AB} \cong \overline{DF}$, $\overline{BC} \cong \overline{FE}$, $\overline{AC} \cong \overline{DE}$, $\angle A \cong \angle D$, $\angle B \cong \angle F$, and $\angle C \cong \angle E$.

16. a. $XY = 6$, $YZ = 10$, and $XZ = 14$.
 b. $m\angle X = 75°$, $m\angle Y = 65°$, and $m\angle Z = 40°$.

17. Reasoning for each pair of congruent triangles will differ. The four congruent triangles are $\triangle AED \cong \triangle CEB$, $\triangle BEA \cong \triangle DEC$, $\triangle ABD \cong \triangle CDB$, and $\triangle ADC \cong \triangle CBA$.

18. Quadrilateral $ABCD$ is a rhombus, so it must also be a parallelogram. This implies that $\overline{AB} \parallel \overline{CD}$ and $\overline{BC} \parallel \overline{AD}$. Because $\angle 2$ and $\angle 3$ are alternate interior angles, it must be that $\angle 2 \cong \angle 3$. Similarly, because $\angle 1$ and $\angle 4$ are alternate interior angles, it must be that $\angle 1 \cong \angle 4$. Also, $\overline{AC} \cong \overline{CA}$ because they are the same segment. This gives two pairs of corresponding angles and their included side congruent, so $\triangle ABC \cong \triangle CDA$ by SAS. Because corresponding parts of congruent triangles are congruent, it must be that $\overline{AB} \cong \overline{DC}$ and $\overline{BC} \cong \overline{AD}$.

19. a. $DE = 28$. **b.** $DE = \dfrac{7}{2}$.

20. a. The triangles have a pair of congruent vertical angles and a pair of angles that are marked congruent. Hence, they are similar by AA. This give us the proportion $\dfrac{x}{9} = \dfrac{8}{12}$. Solving for x, we have $x = \dfrac{(8 \cdot 9)}{12} = 6$.
 b. The triangles share a common angle, and they both have a right angle, so they must be similar by AA. This give us the proportion $\dfrac{y}{15} = \dfrac{6}{9}$. Solving for y, we have $y = \dfrac{(6 \cdot 15)}{9} = 10$.

21. 66.67 ft.

22. Draw a line through P so that it intersects $\overleftrightarrow{AB}$. Label the point of intersection C. Use Construction 2 to copy angle $\angle PCB$ so that its vertex is at P and it forms an alternate interior angle with $\angle PCB$. Construct the side of the angle to complete the construction.

23. After making $\triangle ABC$, draw a ray, and copy $\overline{AC}$ onto the ray. Use the compass to measure $\overline{AB}$, and then, without changing the opening, place the compass on one endpoint of the new segment and strike an arc above the ray. Repeat the process with $\overline{BC}$, but swing the arc from the other endpoint of the segment until it intersects the first arc. Connect the endpoints of the segment to the new point to form the triangle.

24. After making $\triangle DEF$, use Construction 1 two times to make a segment that is twice the length of $\overline{DE}$ on the ray. At one of the endpoints of the new segment, use Construction 2, to copy $\angle D$. On the side of the new angle, use Construction 1 two times to make a segment that is twice the length of $\overline{DF}$. Connect this last point to the other endpoint of the segment to form the triangle.

25. a. Use the procedure outlined in Example 11.27.
 b. Use the procedure outlined in Example 11.28.

26. a. $AB = 11$. The midpoint of $\overline{AB}$ is $\left(\dfrac{19}{2}, 8\right)$.
 b. $CD = \sqrt{181} \approx 13.45$. The midpoint of $\overline{CD}$ is $\left(-\dfrac{1}{2}, 2\right)$.

27. Obtuse scalene.

28. D could be at $(8, 4)$, $(4, -4)$, or $(-4, 4)$.

29. a. $x^2 + y^2 = 64$.
 b. $(x - 1)^2 + (y + 5)^2 = 25$.

c. $(x + 3)^2 + (y - 2)^2 = 15$.

30. a.

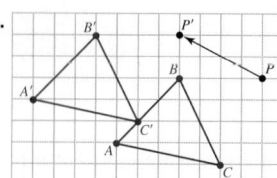

 b.

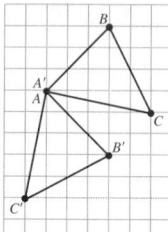

 c.

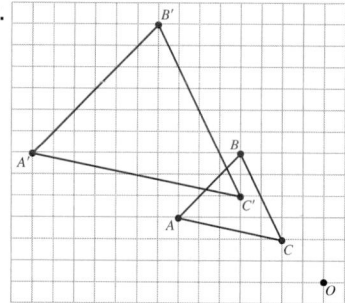

31. The image of quadrilateral $ABCD$ when it is reflected in line l is shown:

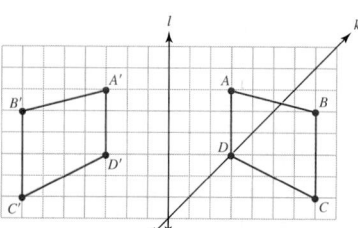

The image of quadrilateral $ABCD$ when it is reflected in line k is shown:

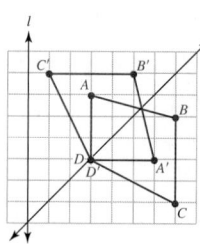

32.

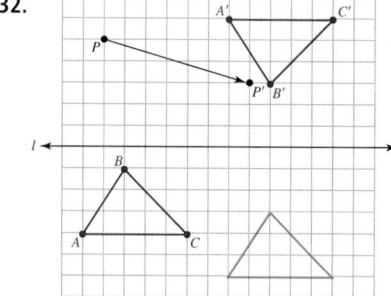

33. a. $(-7, 3)$.   **b.** $(-4, 2)$.   **c.** $(-2.4, 6.9)$.
34. a. $-232°$.   **b.** $103°$.   **c.** $175°$.
35. a. $(7, 3)$.   **b.** $(-7, -3)$.   **c.** $(-3, 7)$.   **d.** $(3, -7)$.
36. a.   **b.**

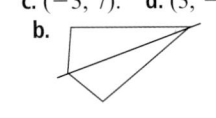

37. a.   **b.**

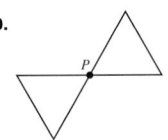

38. Drawings will vary. Possible answers are shown:
   a. A   **b.**

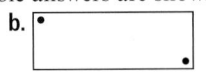

39. It must have 8. Its rotations are $45°$, $90°$, $135°$, $180°$, $225°$, $270°$, $315°$, and $360°$.
40. a. ··· A A A A ···   **b.** ··· A ∀ A ∀ ···
   c. ··· ⊳ ⊳ ⊳ ⊳ ···
41. Drawings will vary. One possible answer is shown:

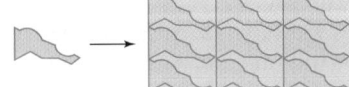

CHAPTER 13

■ Section 13.1 Learning Assessment

1. The first step is to identify and select a measurable attribute, which means that it can be quantified in some way. The second step is to select a unit that is both appropriate in type and in size. The third step is to find the measure by using a measurement strategy, such as using an estimation, a measuring device, or a formula.
2. **a.** Length is independent of direction, so AB can always be compared to a non-negative number by the ruler postulate.   **b.** $AB = 0$ when A and B are the same point.   **c.** Length is independent of direction. It does not matter whether the distance is measured from A to B or from B to A; the length is the same number.
3. As the unit gets larger, the number of units needed to make the measure gets smaller. As the unit gets smaller, the number of units gets larger.
4. Accuracy describes how close the measure is to the actual size of the object. Precision relates to the exactness of the measurement. Smaller units lead to more exact measurements, and larger units lead to less exact measurements.
5. No. Length is always greater than or equal to zero.
6. Nonstandard units can vary from situation to situation, leading to different measures of the same object. Such discrepancies lead to problems in fairness and communication.
7. The metric system is usually preferred over the English system because it uses simple decimal ratios between the units. Many units of different types are also related to each other.

8. All metric units begin with a base unit. Larger units are made by multiplying the base unit by positive powers of ten and smaller units by multiplying by negative powers of ten.
9. In dimensional analysis, we use unit ratios to compare a single unit of one measure to its equivalent amount in another. Because each ratio is equivalent to one, we can use proportional reasoning to multiply ratios to change both the value and the unit of the measure.
10. One way to obtain a measure is estimation. Estimation is the process by which we use mental or visual information to arrive at an approximate measure without using a measuring device. We use estimation when measuring devices are unavailable or impractical or when we only need an approximate value. A second way to obtain a measure is to use a measuring device, which is a tool designed to measure objects using standardized units. A third way to measure is to use a formula. Measurement formulas come from a variety of places and often rely on other measures that can be found directly.
11. We use unitizing whenever the object consists of several parts that are roughly the same size. We estimate the size of one part and then multiply by the number of parts. Chunking involves breaking down the object into small yet related parts, estimating the size of each part, and then adding the results to find the final measure.
12. Both are needed to communicate the size of the measure. Without the number, we do not know how many units the object was equivalent to. Without the unit, we do not know to what the object was compared in making the measure.
13. **a.** Accurate but not precise.   **b.** Not accurate but precise.   **c.** Not accurate and not precise.   **d.** Accurate and precise.
14. **a.** Not accurate but precise.   **b.** Accurate but not precise.   **c.** Accurate and precise.   **d.** Not accurate and not precise.
15. **a.** Accurate and precise.   **b.** Not accurate and not precise.   **c.** Not accurate but precise.   **d.** Accurate but not precise.
16. **a.** The length of $\overline{AB}$ added to the length of $\overline{BC}$ is greater than or equal to the length of $\overline{AC}$.
   b.

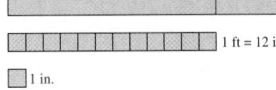

   c. If $AB = 4$, $BC = 5$, and $AC = 6$, then $AB + BC = 4 + 5 = 9 \geq 6 = AC$.
17. Most people do not have their body measurements equal to a yard or a foot. Taller people have longer yards and feet because their body parts are longer. Shorter people have the reverse.
18. **a.** About 4 in.   **b.** About 8 in.   **c.** About 18 in.   **d.** About 72 in.
19. **a.** Drawings will vary.   **b.** Drawings will vary.   **c.** Drawings will vary.   **d.** Drawings will vary.   **e.** Drawings will vary.
20.

1 yd = 3 ft		

1 ft = 12 in.

☐ 1 in.

21.
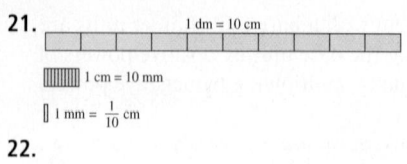
1 dm = 10 cm

▓▓▓▓ 1 cm = 10 mm

▌ 1 mm = $\frac{1}{10}$ cm

22.

1 hm = 10 dam

▓▓▓▓▓▓ 1 dam = 10 m

▌ 1 m

23. **a.** Answers will vary.　**b.** Answers will vary.　**c.** Answers will vary.　**d.** Answers will vary.

24. **a.** Answers will vary.　**b.** Answers will vary.　**c.** Answers will vary.　**d.** Nonstandard units lead to different measures of the same object.

25. **a. ii.** 3 in.　**b. iii.** 3 yd.　**c. ii.** 200 ft.

26. **a. iv.** 170 cm.　**b. ii.** 4 hm.　**c. iii.** 1,500 km.

27. **a. i.** 30 cm.　**b. iv.** 3.1 m.　**c. iii.** 24,000 mi.

28. **a.** 4 in., $\frac{1}{6}$ yd, 8 in., $\frac{3}{4}$ ft.　**b.** 4 dm, 0.5 m, 700 mm, 100 cm.

29. **a.** 0.5 cm, $\frac{1}{2}$ in., 15 mm, 1 in.　**b.** 1 dm, 11 cm, $\frac{1}{2}$ ft, 7 in.

30. **a.** 15 yd.　**b.** 2,412 in.　**c.** 1.425 ft.　**d.** 285.9 ft.

31. **a.** 13.4 yd.　**b.** 1.49 mi.　**c.** 4.50 fur.

32. **a.** 10.　**b.** 100.　**c.** 1,000.　**d.** 100.

33. **a.** 100.　**b.** 1000.　**c.** 1,000.　**d.** 100.

34. **a.** 3 m.　**b.** 450 dm.　**c.** 90 cm.　**d.** 370 mm.

35. **a.** 0.00315 dam.　**b.** 910 km.　**c.** 7,510,000 mm.

36. **a.** 10.67 m.　**b.** 20.24 ft.　**c.** 51.19 m.　**d.** 102.29 yd.

37. **a.** 15.35 in.　**b.** 223.70 fur.　**c.** 0.96 hm.　**d.** 130.35 km.

38. **a.** $\frac{3}{4}$ in.　**b.** $1\frac{1}{4}$ in.　**c.** $4\frac{1}{4}$ in.　**d.** 4 in.　**e.** $2\frac{1}{4}$ in.　**f.** $1\frac{1}{2}$ in.

　　g. $2\frac{1}{2}$ in.　**h.** 5 in.

39. **a.** 1.6 cm.　**b.** 1.4 cm.　**c.** 6.3 cm.　**d.** 9.5 cm.　**e.** 4.5 cm.　**f.** 8.3 cm.　**g.** 7.5 cm.　**h.** 12.7 cm.

40. **a.** Answers will vary.　**b.** Answers will vary.　**c.** Answers will vary.　**d.** Answers will vary.

41. **a.** About 2 in.　**b.** About 12 ft.　**c.** About 7 ft.　**d.** About 1 in.　**e.** About 12 in.　**f.** Answers will vary.

42. **a.** About 6 cm.　**b.** About 4 m.　**c.** About 2 m.　**d.** About 3 cm.　**e.** About 30 cm.　**f.** Answers will vary.

43. A football field is 109.73 m, 1.097 hm, or 0.068 mi.

44. 30 lengths.

45. 4,343.50 mi.

46. 2.5 times.

47. 22 in. Most infants are this tall.

48. **a.** 9 ft, 9 in.　**b.** About 5 cubits and two spans. Wadlow was 10 in. shorter than Goliath.

49. It is 23.87 mi. long. It will take a car 0.4 hr, or 24 min, to cross the bridge.

50. 5.87×10^{12} mi.

51. Answers will vary.

52. Answer will vary.

53. About 220 ft long.

54. About 40 quarters.

55. About 2,500 mi or 4,000 km.

56. Estimates will vary.

57. About 3,000 fish.

58. Estimates will vary.

59. 57.45 ft.

60. **a.** The measure is halved.　**b.** The measure is doubled.

61. **a.** It is decreased by $\frac{1}{3}$.　**b.** It is decreased by $\frac{1}{4}$.　**c.** It is decreased by $\frac{1}{5}$.　**d.** It is decreased by $\frac{1}{n}$.

62. No, it does not constitute a proof of the Triangle Inequality. Although it provides evidence that it is true, one example does not make a proof.

63. A basketball player is the smarter choice. He is likely to have much longer paces, being able to cover more land.

64. **a.** 1.094 and 0.9144 are multiplicative inverses of one another.　**b.** Yes. All pairs of conversion factors involving a metric and English units are multiplicative inverses of one another.

65. The conversion factor is rounded, so it may lead to small differences between the computation and the actual measure.

66. Height has a connotation of up or down, depth has a connotation of down, width has a connotation of left or right (horizontal), breadth has a connotation of left or right (horizontal), and distance has a connotation of here to there.

67. Answers will vary.

68. **a.** Answers will vary.　**b.** Answers will vary.

69. **a.** Answers will vary.　**b.** Answers will vary.　**c.** Answers will vary.

70. Answers will vary.

71. The student wrote the unit ratio in the reverse order. Answers will vary on how to address the problem.

72. **a.** Answers will vary.　**b.** Answers will vary.　**c.** Answers will vary.

73. Answers will vary.

74. Answers will vary.

75. The domain is all possible segments, and the range is $[0, \infty)$.

76. **a.** Answers will vary.　**b.** Answers will vary.

77. Answers will vary.

78. Odometers are used to measure distances by how many times the wheel turns around. Because it measures lengths, we can consider it to be a type of ruler.

■ Section 13.2 Learning Assessment

1. The perimeter of a simple closed curve is the distance around the curve. Because it is a measure of distance, or length, it must be measured with length units.

2. **a.** Because the sides of a square are the same length, the perimeter is four times the length of a side, or $P = 4s$.
b. In a rectangle, each pair of opposite sides has the same length. If one pair has a length of l and the other a width of w, then the perimeter is $P = 2w + 2l$.　**c.** Because the sides of a regular polygon are the same length, the perimeter is the number of sides times the length of a side, or $P = ns$, where n is the number of sides.

3. π represents the ratio of the circumference of any circle to its diameter. It is an irrational number, so its value is often approximated with 3.14, 3.1414, or $\frac{22}{7}$.

4. The perimeter of a planar shape is the distance around its exterior. The area is the amount of plane enclosed within the figure.

5. a. Area is a measure of the plane, so it is a two-dimensional measure of a two-dimensional shape. Consequently, it requires a two-dimensional unit to measure it. **b.** Any two-dimensional shape that tessellates the plane can be used as a unit of area.

6. It does not make sense. Perimeter and area are two different attributes of the same planar figure. Their relative sizes cannot be compared to one another.

7. Neither the perimeter nor the area of a square is actually part of the square. A square consists only of four segments and the joining vertices. Perimeter and area are both measureable attributes of the square but are not part of the square itself.

8. A square is a special type of rectangle, so the formula can be applied directly. However, in a square, the length and the width are both equal to the side. Hence, $A = l \cdot w = s \cdot s = s^2$. A parallelogram can be transformed to look like a rectangle in which the length and the width are equal to the base and the height. Hence, $A = l \cdot w = b \cdot h$.

9. Take a copy of the triangle and rotate it 180° about the midpoint of a side that is not the base. The resulting figure is a parallelogram with the same base and height but twice the area of the triangle. Hence, the area of the triangle is half the area of the parallelogram, or $A = \frac{1}{2}(bh)$. Use the same procedure with the trapezoid, but now two bases must be considered. Again, the area of the trapezoid is half the area of the parallelogram, or $A = \frac{1}{2}(b_1 + b_2)h$.

10. If we consider a sequence of regular polygons in which more sides are added, the polygons grow more circular. Their apothems approximate the radius of a circle, and their areas approximate the area of the circle. As a result, if a circle is viewed as a regular polygon with an infinite number of sides, then the area of the circle is
$$A = \frac{1}{2}aP = \frac{1}{2}r(2\pi r) = \pi r^2.$$

11. No. A hand does not tessellate the plane.

12.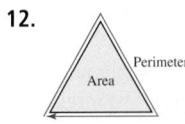

13.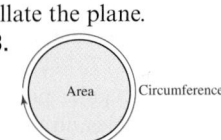

14.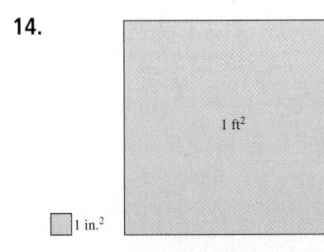

15.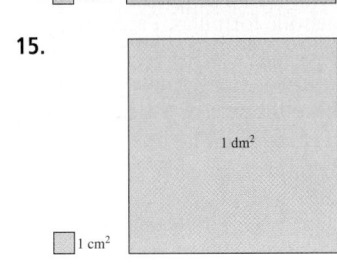

16.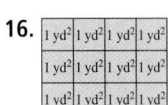

17.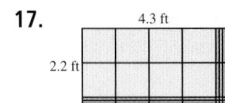

18.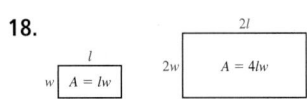

19.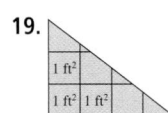

20. a.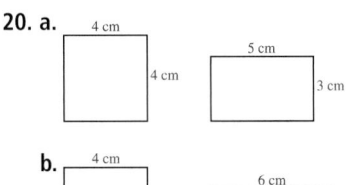

b.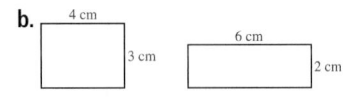

21. Yes. Both parallelograms have an area of $A = (5)(12) = 60$ cm².

22. a. 5.33 yd². **b.** 0.49 yd². **c.** 0.0015 mi². **d.** 0.073 km². **e.** 3,700 m². **f.** 47 cm².

23. a. 7.77 km². **b.** 0.42 m². **c.** 1.21 in.². **d.** 837.79 yd². **e.** 253 hectares. **f.** 0.145 hectares.

24. a. 1,000,000 times larger. **b.** 10,000 times larger. **c.** 1,000,000 times larger.

25. a. 100,000,000 times smaller. **b.** 10,000 times smaller. **c.** 10,000 times smaller.

26. a. Answers will vary. **b.** Answers will vary. **c.** Answers will vary.

27. a. Answers will vary. **b.** Answers will vary.

28. a. $P = 24$ in.; $A = 36$ in.². **b.** $P = 28$ in., $A = 40$ in.².

29. a. $P = 9.2$ cm; $A = 5.29$ cm². **b.** $P = 38.4$ m; $A = 81.27$ m².

30. a. $A = 6$ cm². **b.** $A = 3.75$ yd².

31. a. $A = 57.33$ in.². **b.** $A = 12.6$ m².

32. $w = 9$ m.

33. $s = 20$ ft.

34. a. $A = 45.5$ in.². **b.** $A = 3.15$ km².

35. a. $A = 150$ m². **b.** $A = 16$ cm².

36. a. $P = 39$ ft. **b.** $s = 47$ cm.

37. a. $C = 4\pi$ in.; $A = 4\pi$ in.². **b.** $C = 10.6\pi$ yd ≈ 33.28 yd; $A = 28.09\pi$ yd² ≈ 88.20 yd².

38. a. $C = 5.8\pi$ ft ≈ 18.21 ft; $A = 8.41\pi$ ft² ≈ 26.41 ft². **b.** $C = 2\pi^2$ cm; $A = \pi^3$ cm².

39. a. 29.67 cm². **b.** 29.32 in.². **c.** 74.67 in.². **d.** 77.6 cm². **e.** 1,225 cm². **f.** 83 cm².

40. a. 7.06 cm². **b.** 108 in.². **c.** 169.93 cm². **d.** 125.93 m².

41. 23.4 cm².

42. 240 ft².

43. $A \approx 42.25\pi$ ft² ≈ 132.67 ft².

44. The three sides are 14 cm, 56 cm, and 56 cm.

45. a. 48 cm². **b.** 20 in.². **c.** 12.36 in.². **d.** 3.44 ft².

46. a. Perimeter $= 14 + 4\sqrt{2}$ units; area $= 28$ square units. **b.** Perimeter $= 4 + 8\sqrt{2}$ units; area $= 12$ square units. **c.** Perimeter $= 28 + 4\sqrt{2}$ units; area $= 19$ square units.

47. a. 1 : 3. **b.** 1 : 9.

48. The largest area is square with sides of length 2 ft. This gives an area of 4 ft^2.

49. 8.33 in.^{2}.

50. $A \approx 519.62$ cm^2.

51. The 16-in. pizza.

52. a. 9,968.75 square mi.   **b.** 1.7% was burned.

53. 7 gal.

54. 536 tiles.

55. a. 40 ft.   **b.** 950 ft^2.

56. 1,600,109.59 mi.

57. a. 7.85 in.   **b.** 15.7 in.   **c.** 2.62 in.   **d.** 0.52 in.

58. 798.35 ft^2.

59. It is a rectangle with sides of length 32 cm and 33 cm, respectively.

60. a. i. $s = \dfrac{3 + 5 + 6}{2} = 7$, so
$$A = \sqrt{7(7 - 3)(7 - 5)(7 - 6)} = \sqrt{56} \text{ in.}^2.$$

ii. $s = \dfrac{4 + 3.6 + 4.7}{2} = 6.15$, so
$$A = \sqrt{6.15(6.15 - 4)(6.15 - 3.6)(6.15 - 4.7)}$$
$$= \sqrt{48.89} \approx 6.99 \text{ in.}^2.$$
b. $s = (4 + 7 + 11)/2 = 11$, so
$A = \sqrt{11(11 - 4)(11 - 7)(11 - 11)} = 0$. The three segments do not form a triangle because $4 + 7 = 11$.

61. 32π ft ≈ 100.48 ft.

62. a. The perimeter is four times longer.
b. The area is 16 times greater.

63. a. The circumference doubles in size.
b. The area increases by four times.

64. The area increases by 21%.

65. a. The shape must become more like a square, meaning the width becomes equal to the length.   **b.** The shape must become longer and thinner, meaning the width must become smaller while the length grows longer.

66. The perpendicular distance between any two points on a pair of parallel lines is always the same. The parallelograms therefore have not only the same base but also the same height. This means any two parallelograms with one pair of opposite sides on the same pair of parallel lines must have the same area.

67. The height of the triangle must be twice the height of the parallelogram.

68. A rotation of 180° maps every point to the point exactly opposite it in the rotation center. Hence, a rotation maps any line or line segment to another line or line segment with exactly the same slope. As a result, when we rotate the triangle around the midpoint of one side, the other sides are mapped to segments that are parallel. This gives two pairs of opposite sides that are parallel, so the figure must be a parallelogram.

69. A gap between the four shapes along the diagonal of the figure is exactly 1 square unit in area. This makes up the difference between the two figures.

70. After folding, the resulting rectangle has a length equal to $\dfrac{b}{2}$, a width equal to $\dfrac{h}{2}$, and an area that is half the area of the triangle. Hence, the area of the triangle is
$$A = 2 \cdot (\text{Area of rectangle}) = 2\left(\frac{b}{2} \cdot \frac{h}{2}\right) = 2\left(\frac{bh}{4}\right) = \frac{1}{2}bh.$$

71. Consider the right triangle shown with legs of length a and b and hypotenuse of length c.

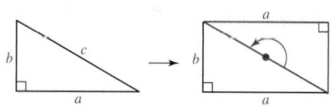

If we rotate a copy of the triangle 180° around the midpoint of the hypotenuse, we get a rectangle with a length of a, a width of b, and an area that is twice the area of the triangle. Consequently, the area of the triangle is $A = (1/2)lw = (1/2)ab$.

72. The closer the figure gets to a parallelogram, the closer its height approximates the radius of the circle, and the closer the base approximates half the circumference. If we continue to make the wedges smaller, the circle and parallelogram eventually have the same area. This allows us to compute the area for a circle as
$$A = bh = \left(\frac{1}{2}(C)\right)r = \left(\frac{1}{2}(2\pi r)\right)r = \pi r^2.$$

73. Using reasoning similar to Example 13.11, we have the proportion $\dfrac{\theta°}{360°} = \dfrac{A}{\pi r^2}$. Solving for A gives us $A = \dfrac{\theta° \pi r^2}{360°}$. Consequently, the area of the sector is
$$A = \frac{135°\pi(7.5)^2}{360°} \approx 66.23 \text{ m}^2.$$

74. An equilateral triangle with sides of length s has a height equal to $h = \sqrt{s^2 - \left(\frac{1}{2}s\right)^2} = \sqrt{\frac{3}{4}s^2} = \frac{\sqrt{3}}{2}s$. Hence, it has an area of $A = \dfrac{1}{2}\left(\dfrac{\sqrt{3}}{2}s\right)s = \dfrac{\sqrt{3}}{4}s^2$.

75. Answers will vary.

76. a. Answers will vary.   **b.** Answers will vary.

77. a. Answers will vary.   **b.** Answers will vary.
c. Answers will vary.   **d.** Answers will vary.

78. Beth needs to square both the value and the units.

79. Oliver's method does not work. For instance, two rectangles can have the same perimeter but different areas.

80. Only two other regular polygons tessellate the plane: the equilateral triangle and the regular hexagon. We might call such units triangular units or hexagonal units.

81. Compute the area of the rectangle, and compare it to the rectangle's weight. Find the weight per square inch of the cardboard. Once we know this and the weight of the other shape, we can approximate the area of the unknown shape by dividing its weight by the weight per square inch of the cardboard.

82. The section builds a connection between geometry and algebraic thinking by representing the perimeter and area of different shapes with algebraic formulas. Perimeters and areas can also be viewed as functions. For instance, consider the square and the circle. The formulas for these shapes take one number, either the side of the square or the radius of the circle, and correspond it to exactly one other number, the perimeter or area. In each case, the domain is $[0, \infty)$ and the range is $[0, \infty)$.

83. Answers will vary.

84. Answers will vary.

■ Section 13.3 Learning Assessment

1. Surface area is the amount of plane needed to cover the outside of a three-dimensional shape. Because it is defined to be a measure of the plane, it must be measured with square units.

2. We can compute surface area by computing the area of the figure's net. Because many three-dimensional shapes, like prisms, pyramids, and cylinders, have nets consisting of common planar shapes, surface area formulas can be developed by combining the appropriate area formulas.

3. The lateral surface area of any prism can be found by multiplying the perimeter of the base by the height of the prism.

4. The net of the lateral surface area of a right cylinder is a rectangle, so its area is found by computing the area of a rectangle.

5. The slant height is the distance on a lateral face that extends from the midpoint of a side of the base to the apex. The height of the pyramid is the shortest distance from the apex to the plane of the base.

6. **a.** The net of the lateral surface of a cone is a sector of a circle. **b.** We do not know the central angle of the sector.

7. If the slant height is unknown, then we can use the Pythagorean Theorem to find it because the radius, the height, and the slant height of the cone form the sides and hypotenuse of a right triangle, or $l = \sqrt{r^2 + h^2}$.

8. The surface area of a sphere is exactly four times the area of the great circle.

9. **a.** Cube; e = length of edge. **b.** Right circular cylinder: r = radius, h = height of cylinder. **c.** Right rectangular prism: l = length, w = width, h = height. **d.** Right pyramid: B = area of the base, P = perimeter of the base, l = slant height. **e.** Sphere: r = radius. **f.** Cone: r = radius, l = slant height.

10. Lateral surface area = 760 cm²; SA = 1,313.68 cm².

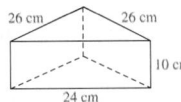

11. Let one be a cube with an edge of 4 in., and let the other be a prism with a square base of 2 × 2 in. and a height of 11 in. Both have a surface area of 96 in.².

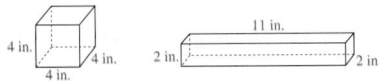

12. Lateral surface area = 240 cm²; SA = 384 cm².

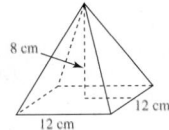

13. **a.** They have the same lateral surface area. **b.** The cylinder with the radius of 4 cm has a larger surface area.

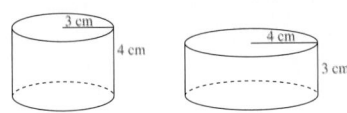

14. **a.** The cone with a radius of 5 cm has a larger lateral surface area. **b.** The cone with a radius of 5 cm has a larger surface area.

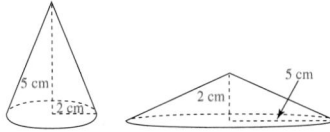

15. **a.** 158 in.². **b.** 54 cm². **c.** 432 in.².
16. **a.** 147.14 in.². **b.** 121.5 cm². **c.** 668.56 in.².
17. **a.** 213.42 m². **b.** 1,040 cm².
18. **a.** 440π ft². **b.** 1,015.99 cm².
19. **a.** 399.66 cm². **b.** 89.18 in.².
20. **a.** 144 cm². **b.** 90π ft². **c.** 144π cm².
21. **a.** 84.56 in.². **b.** 444.48 ft². **c.** 254.34 cm².
22. **a.** 360 ft². **b.** 96π cm². **c.** 196π yd².
23. **a.** $h = 6$ m. **b.** $e = 8.2$ cm.
24. 381.6 cm.
25. 4.5 yd.
26. The dimensions must be 4 × 4 in.
27. 21.37 in.².
28. 9 mm.
29. 9π ft².
30. 60π ft².
31. 112 cm².
32. 864π cm².
33. 438.3 cm².
34. 57.14 in.².
35. 168 in.².
36. 371.14 in.².
37. 98.35 in.².
38. 173.20 cm²; 346.40 cm²
39. $r = \dfrac{9}{2}$ cm; $h = 3.97$ cm
40. $90.
41. 130,319,420 mi² is covered with water.
42. $SA_{\text{barn}} = 5,220$ ft²; $SA_{\text{silo}} = 8,164$ ft².
43. Let l, w, and h be the dimensions of the rectangular prism, giving it a surface area of $SA = 2(lw + lh + wh)$. If each dimension is doubled, then the new surface area is four times larger, or $2[(2l)(2w) + (2l)(2h) + (2w)(2h)] = 4[2(lw + lh + wh)] = 4SA$.
44. Let e be the edge of a cube, giving it a surface area of

$SA = 6e^2$. If the edge is reduced by $\dfrac{1}{3}$, the new edge is $\dfrac{2}{3}e$.

This means the new surface area is $\dfrac{4}{9}$ the size of the

original surface area because $6\left(\dfrac{2}{3}e\right)^2 = \dfrac{4}{9}(6e^2) = \dfrac{4}{9}SA$

45. **a.** The lateral surface area is $2\pi rh$. If we double the height, the new surface area is double the original because $2\pi r(2h) = 2(2\pi rh)$. If we double the radius, the new surface area is double the original because $2\pi(2r)h = 2(2\pi rh)$. Hence, they increase the surface area by the same amount. **b.** The total surface area is $2\pi r^2 + 2\pi rh$. Doubling the height gives a new surface area of $2\pi r^2 + 2\pi r(2h) = 2\pi r^2 + 4\pi rh$. Doubling the radius gives a new surface area of $2\pi(2r)^2 + 2\pi(2r)h = 8\pi r^2 + 4\pi rh$. Because this increases both the area of the base and the

lateral surface area, doubling the radius increases the total surface area more.

46. No, the surface area will be nine times as large. If the two cones arc similar with a scale factor of 3, then each dimension on one must be three times the dimensions of the other. If the surface area of the smaller is $SA = \pi r^2 + \pi r l$, then the surface area of the larger must be $\pi (3r)^2 + \pi (3r)(3l) = 9(\pi r^2 + \pi r l) = 9SA$.

47. Let r be the radius of a sphere, giving it a surface area of $SA = 4\pi r^2$. If the radius is tripled, the new surface area will be nine times the size of the original surface area because $4\pi r^2 = 4\pi (3r)^2 = 9(4\pi r^2) = 9SA$.

48. Arrangement B will melt faster. It has more surface area exposed to the air.

49. The slant height of the cone is always the radius of the semicircle, and the circumference of the circular base is equal to $C = (1/2)(2\pi r) = \pi r$. Consequently, the slant height of the cone and the diameter of the circular base are the same length.

50. Consider a sequence of right prisms in which the bases are regular polygons with an increasing number of sides. As the number of sides increases, the surface areas of the prisms closely approximate the surface area of the cylinder.

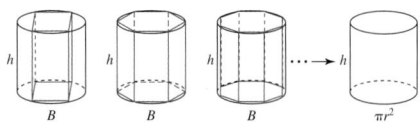

So, if we think of a cylinder as a prism with a circular base, then the surface area of the cylinder can be computed as $2B + Ph = 2(\pi r^2) + (2\pi r)h = 2\pi r^2 + 2\pi rh = 2\pi r(r + h)$.

51. Answers will vary.
52. She is incorrect. The amount of cardboard remaining will be only one quarter of the original amount.
53. Answers will vary.
54. Answers will vary.
55. a. Answers will vary.   b. Answers will vary.   c. Answers will vary.   d. Answers will vary.
56. Answers will vary.
57. Answers will vary.

■ Section 13.4 Learning Assessment

1. a. Volume is the amount of material, either liquid or solid, that is needed to make or fill a three-dimensional shape. Capacity is the amount of space a figure can hold.
   b. Volume and capacity are both measures of space, so we must measure them with a three-dimensional unit with a length, a width, and a depth. We prefer cubes because they are the easiest to use.
2. Cube the length conversion ratio.
3. A liter is defined as the volume of a cube that measures 10 cm, or 1 dm, on each edge.
4. a. Two solids, each having their bases in the same parallel planes, have the same volume if every plane parallel to the bases intersects the solids in cross sections of equal area.
   b. In general, Cavalieri's principle allows us to compute the volume of oblique spatial figures by using the volume of right spatial figures.
5. We consider a sequence of right prisms in which the bases are regular polygons with an increasing number of sides.

As the number of sides increases, the volumes of the prisms closely approximate the volume of the cylinder. So, if we think of a cylinder as a prism with a circular base, then the volume of the cylinder can be computed as $V = Bh = \pi r^2 h$.

6. Consider the volume of a square pyramid that is enclosed in the smallest cube that can contain it. Next, shift the apex of the pyramid to one of the upper vertices of the cube. Cavalieri's principle guarantees that the shifted pyramid has the same volume as the original. This also creates a total of three congruent pyramids with the same volume. As a result, the volume of the original pyramid must be one-third the volume of cube, which is written $V = \frac{1}{3}Bh$.

7. The volume of a sphere is two-thirds the volume of the smallest cylinder containing it.

8. a. Prism: B = area of the base, h = height of the prism.
   b. Cone: r = radius, h = height.   c. Sphere: r = radius.
   d. Prism: l = length, w = width, h = height.
   e. Pyramid: B = area of the base, h = height of pyramid.   f. Cylinder: r = radius, h = height.

9. a.   b.

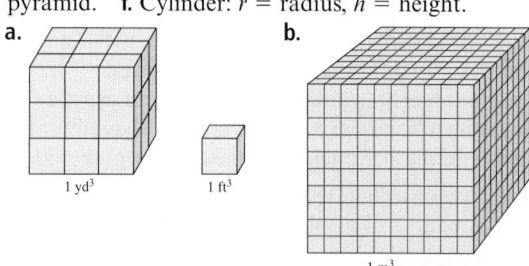

10. a.   b.

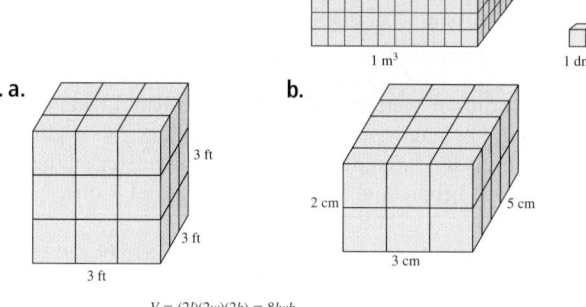

11.

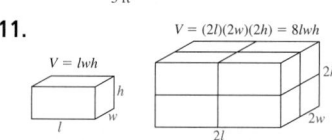

12. Rolling the paper widthwise gives the larger volume.
13. The prism is shown: $V = 1080$ cm^3.

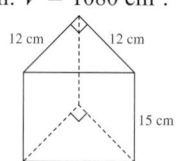

14. The prisms are shown:

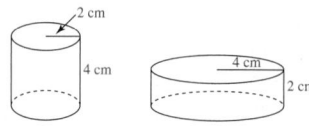

Their volumes are $V = 16\pi$ cm^3 and $V = 32\pi$ cm^3. The one with a radius of $r = 4$ cm has the larger volume.

15. The pyramid is shown:

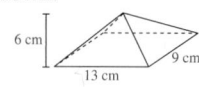

$V = 234$ cm^3.

16. a. 3 c.   **b.** 24 c.
17. a. iii. 0.5 pt.   **b. iii.** 12 m^3.   **c. i.** 3,000 ft^3.
18. a. i. 3,000 cm^3.   **b. i.** 300 in.^{3}.   **c. ii.** 15 cm^3.
19. a. 1 ft^3 = 0.0283 m^3.   **b.** 1 m^3 = 1.309 yd^3.
   c. 1 in.^{3} = 16.387 cm^3.
20. a. 1 cm^3 = 3.53 × 10^{-5} ft^3.   **b.** 1 yd^3 = 0.765 m^3.
   c. 1 mi^3 = 4.168 km^3.
21. a. 10^3 = 1,000 times larger.   **b.** 100^3 = 1,000,000 times
   larger.   **c.** (1,000)^3 = 10^9 times larger.
22. a. 10^3 = 1,000 times smaller.   **b.** 100^3 = 1,000,000 times
   smaller.   **c.** (10,000)^3 = 10^{12} times smaller.
23. a. 2,099,520 in.^{3}.   **b.** 2.70 ft^3.   **c.** 75.85 yd^3.
   d. 15,000,000 cm^3.   **e.** 0.0745 hm^3.   **f.** 0.00467 m^3.
24. a. 1.5625 c.   **b.** 256 c.   **c.** 6.98 yd^3.   **d.** 4,730 cL.
   e. 0.83 dL.   **f.** 83.9 m^3.
25. a. 52.99 L.   **b.** 0.528 gal.
26. 262.19 cm^3.
27. a. 168 in.^{3}.   **b.** 3,375 cm^3.   **c.** 133 in.^{3}.
28. a. 130.98 in.^{3}.   **b.** 7,301.38 cm^3.   **c.** 964.42 cm^3.
29. a. 149.61 m^3.   **b.** 952 cm^3.
30. a. 1,280π ft^3.   **b.** 20,475.13 cm^3.
31. a. 152.60 in.^{3}.   **b.** 2,119.83 cm^3.
32. a. 64 cm^3.   **b.** 100π ft^3.   **c.** 288π cm^3.
33. a. 400 ft^3.   **b.** 96π cm^3.   **c.** 1,436.03 yd^3.
34. a. 128.21 cm^3.   **b.** 41.67 in.^{3}.   **c.** 484.72 mm^3.   **d.** 33.19 in.^{3}.
35. a. 14 m.   **b.** 7.6 cm.
36. 8 yd.
37. 9 cm.
38. Possible pairs of values for the l and w are 1 and 72, 2 and
   36, 3 and 24, 4 and 18, 6 and 12, or 8 and 9.
39. $r = h = 5$ ft.
40. 0.75 L.

41. $\dfrac{32 \cdot \sqrt{2}}{3}$ in.^{3}.

42. 5,300 L.
43. 11,304 cm^3.
44. 85.33 ft^3.
45. 55.18 m.
46. The larger block is the better buy.
47. a. 921,060,000 gal.   **b.** 580.93 yd tall.
48. 2,800 ft^3.
49. a. 147.19 in^3.   **b.** 64.76 in.^{3}.
50. 3.84 in.^{3}.
51. 408.2 in.^{3}.
52. 161.25 in.^{3}.
53. a. The smallest box possible has inner dimensions of
   6.8 × 5.1 × 1.7 in.^{3}.   **b.** 52.33%.
54. 7,875 in.^{3}.
55. As the length of the edge doubles, the volume goes up by a
   factor of 8. To verify this, consider a cube with an edge of
   length e and a volume of $V = e^3$. If we double the length
   of the edge, then the volume is $V = (2e)^3 = 8e^3$, which is
   eight times the original volume.
56. a. It increases by a factor of $2^3 = 8$.   **b.** It increases by a
   factor of $3^3 = 27$.   **c.** It increases by a factor of $4^3 = 64$.
   d. It increases by a factor of $5^3 = 125$.   **e.** It increases by
   a factor of n^3.
57. The volume of the prism is the same as the volume of the
   cube. Specifically, if e is the length of the edge of a cube,
   then the volume is $V = e^3$. The prism has a length of $2e$, a

width of e, and a height of $(e/2)$, so its volume is $V = (2e)(e)(e/2) = e^3$.

58. a. A possible table is shown:

Radius	Volume
1	$1\pi h$
2	$4\pi h$
4	$16\pi h$
8	$64\pi h$

The table shows that as the radius doubles, the volume
increases by a factor of 4. To verify this, consider a
cylinder with radius r, height h, and $v = \pi r^2 h$. If we
double the radius and hold the height constant, then the
volume is $V = \pi(2r)^2 h = 4\pi r^2 h$, which is four times the
original volume.
b. A possible table is shown:

Height	Volume
1	πr^2
2	$2\pi r^2$
4	$4\pi r^2$
8	$8\pi r^2$

The table shows that as the height doubles, the volume
increases by a factor of 2. To verify this, consider a cylinder
with radius r, height h, and $v = \pi r^2 h$. If we double the
height and hold the radius constant, then the volume
is $V = \pi r^2(2h) = 2\pi r^2 h$, which is two times the original
volume.
c. Because the radius is squared, it increases the volume of
the cylinder faster.   **d.** A possible table is shown:

Radius	Height	Volume
1	1	1π
2	2	8π
4	4	64π
8	8	512π

The table shows that as the radius and the height double,
the volume increases by a factor of 8. To verify this,
consider a cylinder with radius r, height h, and $v = \pi r^2 h$.
If we double the radius and the height, then the volume is
$V = \pi(2r)^2(2h) = \pi(4r^2)(2h) = 8\pi r^2 h$, which is eight times
the original volume.
59. a. A possible table is shown:

Radius	Volume
1	$\left(\dfrac{1}{3}\right)\pi h$
2	$\left(\dfrac{4}{3}\right)\pi h$
4	$\left(\dfrac{16}{3}\right)\pi h$
8	$\left(\dfrac{64}{3}\right)\pi h$

The table shows that as the radius doubles, the volume
increases by a factor of 4. To verify this, consider a cone

with radius r, height h, and $V = \frac{1}{3}\pi r^2 h$. If we double the radius and hold the height constant, then the volume is $V = \frac{1}{3}\pi (2r)^2 h = 4\left(\frac{1}{3}\pi r^2 h\right)$, which is four times the original volume.

b. A possible table is shown:

Height	Volume
1	$\left(\frac{1}{3}\right)\pi r^2$
2	$\left(\frac{2}{3}\right)\pi r^2$
4	$\left(\frac{4}{3}\right)\pi r^2$
8	$\left(\frac{8}{3}\right)\pi r^2$

The table shows that as the height doubles, the volume increases by a factor of 2. To verify this, consider a cone with radius r, height h, and $V = \frac{1}{3}\pi r^2 h$. If we double the height and hold the radius constant, then the volume is $V = \frac{1}{3}\pi r^2 (2h) = 2\left(\frac{1}{3}\pi r^2 h\right)$, which is two times the original volume.

c. Because the radius is squared, it increases the volume of the cone faster. **d.** A possible table is shown:

Radius	Height	Volume
1	1	$\left(\frac{1}{3}\right)\pi$
2	2	$\left(\frac{8}{3}\right)\pi$
4	4	$\left(\frac{64}{3}\right)\pi$
8	8	$\left(\frac{512}{3}\right)\pi$

The table shows that as the radius and the height double, the volume increases by a factor of 8. To verify this, consider a cone with radius r, height h, and $V = \frac{1}{3}\pi r^2 h$. If we double the radius and the height, then the volume is $V = \frac{1}{3}\pi (2r)^2 (2h) = 8\left(\frac{1}{3}\pi r^2 h\right)$, which is eight times the original volume.

60. Let r be the radius of the original sphere, and let V be its volume. The volume of the new sphere is $V = \frac{4}{3}\pi \left(\frac{1}{2}r\right)^3 = \frac{4}{3}\pi \left(\frac{1}{8}r^3\right) = \frac{1}{8}\left(\frac{4}{3}\pi r^3\right) = \frac{1}{8}V$. Hence, the new volume is $\frac{1}{8}$ the size of the original volume.

61. a. The shapes must tessellate space. **b.** No. Only the cube tessellates space. **c.** Any rectangular or triangular

prism should tessellate space. Three examples are shown:

62. Answers will vary.
63. a. Answers will vary. **b.** Answers will vary.
   c. Answers will vary.
64. Answers will vary.
65. Answers will vary.
66. Answers will vary.
67. Answers will vary.
68. Answers will vary.
69. Answers will vary.
70. a. Answers will vary. **b.** Answers will vary.
71. Answers will vary.
72. Answers will vary.

■ Section 13.5 Learning Assessment

1. Mass is the amount of matter within an object. Weight is the amount of force exerted by gravity on the object. Because the gravitational pull on the Earth is directly proportional to its mass, a mass of x mass units corresponds exactly to a weight of x weight units.

2. The commonly used metric weights are the milligram, gram, and kilogram. These weights are particularly useful because they are closely related to other metric units, specifically volumes of water.

3. The two scales are similar in that they both use the freezing and boiling points of water as reference points. They are different because they assign different values to these reference points, changing the size of the degree relative to that scale.

4. When telling time, we indicate a particular instant at which an event is to occur. When measuring time, we find the number of time units equivalent to an interval of time.

5. Speed refers to the rate at which an object moves. It is defined to be the ratio of the distance traveled to the time traveled, or Speed $= \dfrac{\text{distance}}{\text{time}}$. It is different from other attributes because it is the ratio of two attributes instead of a single attribute.

6. Currency is a unit of exchange that assists in the trade of goods and services among people, businesses, and governments. It is a measure of the value we place on a commodity or service. It differs from other measures in that it is relative to supply and demand as well as to personal preferences.

7.

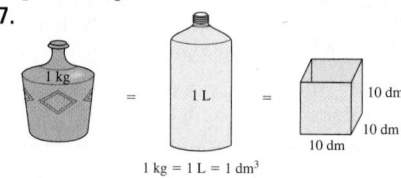

8. The word "pound" originated from the Latin words *libra pondo*. Hence, its abbreviation "lb."

9. One degree Celsius represents a larger change. There are 100 degrees Celsius and 180 degrees Fahrenheit between the freezing and boiling temperatures of water. Because the distance between the two points is divided into more units, the Fahrenheit units must be smaller.

10. a. $\frac{1}{1,000}$ s.   b. 1,000,000 s.   c. 1,000 s.   d. $\frac{1}{100}$ s.

11. a. 0200.   b. 1030.   c. 1115.   d. 1300.   e. 1620.   f. 2135.

12. a. 5:00 a.m.   b. 9:30 a.m.   c. 12:00 p.m.   d. 1:45 p.m.
    e. 6:30 p.m.   f. 10:15 p.m.

13. The distance traveled is equal to the rate of travel multiplied by the time of travel.

14. a. 1 mph = 1.6093 km/hr.   b. 1 km/h = 0.278 m/s.

15. a. 64 oz.   b. 11,200 lb.   c. 76.16 dr.   d. 0.285 T.

16. a. 0.16 kg.   b. 1.5 g.   c. 3,000 g.   d. 140 mg.

17. a. 3,300 lb.   b. 61.36 kg.

18. a. g.   b. kg.   c. kg.   d. metric ton.   e. kg.
    f. metric ton.   g. g.   h. g.

19. a. g.   b. kg.   c. kg.   d. g.   e. mg.

20. a. 81.27 lb.   b. 35.69 lb.   c. 241.88 lb.

21. a. $-3.89°$ C.   b. $-30.56°$ C.   c. $-43.89°$ C.
    d. $39.44°$ C.   e. $98.33°$ C.

22. a. $86°$ F.   b. $156.2°$ F.   c. $-0.4°$ F.   d. $311°$ F.
    e. $-32.8°$ F.

23. a. $-268.93°$ C.   b. $-205.15°$ C.   c. $1,667.85°$ C.
    d. $K = C + 273.15$.

24. a. Not reasonable.   b. Reasonable.
    c. Not reasonable.   d. Not reasonable.

25. 1 year $\approx$ 525,600 min = 31,536,000 s.

26. a. 16.67 min.   b. 11 days, 13 hr, and 46.67 min.
    c. 31 years, 259 days, 1 hr, and 46.67 min.

27. From 2012, there have been 30 Olympiads.

28. c. 50 mph.

29. 1,000 m/s = 1 km/s = 3,600 km/h.

30. 22.37 mph.

31. About 6.7 hr.

32. 2,239.4 pesos.

33. 450 kg.

34. 1,350,360 lb.

35. 1×10^9 metric ton.

36. a. 109.13 g.   b. 106.04 g.   c. 54.77 g.   d. 9.04 g.

37. $-40°$ C or $-40°$ F.

38. $32.018°$ F.

39. $F = \frac{9}{5}K - 459.67$.

40. 62 reports.

41. 18 mph.

42. Light travels 670,615,200 mi. in 1 hr, 1.609×10^{10} mi. in 1 day, and 5.87×10^{12} mi. in 1 year.

43. About 8.32 min.

44. 16.25 times.

45. a. 1,250.48 mph.   b. 2.88 hr.

46. a. 126.58 euros.   b. They are both equivalent to $1, so they represent the same amount.   c. 146.30 euros.

47. Because $C = \frac{5}{9}(F - 32)$, we have

$$C = \frac{5}{9}(F - 32)$$

$$\frac{9}{5}C = F - 32$$

$$\frac{9}{5}C + 32 = F$$

48. a. Because $1°C = \left(\frac{9}{5}\right)°$ F, a temperature increase of $5°$ C corresponds to an increase of $9°$ F.
    b. Because $1°F = \left(\frac{5}{9}\right)°$ C, a temperature decrease of $10°$ F corresponds to a decrease of $\frac{50}{9} \approx 5.56°$C.

49. $-5°$ F $= -20.56°$ C, so $-5°$ C is the warmer temperature.

50. Speed is defined as the ratio of the distance traveled to the time traveled, or Speed $= \frac{\text{distance}}{\text{time}}$. Because speed is a rate, we also have Rate $= \frac{\text{distance}}{\text{time}}$. Multiplying both sides of the equation by the variable time, we have Distance = rate $\times$ time, which is the distance formula.

51. No. She was really discussing her mass.

52. Time.

53. a. Answers will vary.   b. Answers will vary.
    c. Answers will vary.   d. Answers will vary.

54. She did not represent the minutes in terms of hours. Answers will vary on how to correct her mistake.

55. a. Answers will vary.   b. Answers will vary.

56. a. Answers will vary.   b. Answers will vary.

57. Answers will vary.

58. a. Answers will vary.   b. Answers will vary.
    c. Answers will vary.

Chapter 13 Review Exercises

1. Answers will vary.

2. a. Not accurate but precise.   b. Not accurate and not precise.
   c. Accurate but not precise.   d. Accurate and precise.

3. a. The measure is half its original amount.
   b. The measure is a third of its original amount.
   c. The measure is a fourth of its original amount.

4. 1×10^9 times larger.

5. a. iii. 10 in.   b. iii. 10 cm.   c. iii. 2,500 km.

6. a. 0.2 m.   b. 350 dm.   c. 12 yd.   d. 154.8 in.
   e. 7.01 m.   f. 11.71 ft.

7. 53.33 yd = 48.75 m.

8. About 4.02 times.

9. 260.42 times.

10. a. About 5 or 6 cm, or 3 in.   b. About 5 or 6 cm, or 3 in.
    c. About 2 or 3 cm, or 1 in.

11. About 100 dots.

12. $P = 300$ ft; $A = 3,750$ ft^2.

13. a. $P = 16$ ft; $A = 16$ ft^2.   b. $P = 24 + 12\sqrt{2}$ in.; $A = 72$ in.^{2}.   c. $P = 34$ m; $A = 68$ m^2.   d. $C = 24.49$ yd; $A = 47.76$ yd^2.   e. $P = 20 + 10\sqrt{2}$ ft; $A = 25$ ft^2.
    f. $P = 39.42$ in.; $A = 84.13$ in.^{2}.

14. a. 3.89 yd^2.   b. 1.9×10^7 mm^2.   c. 15,730 cm^2.

15. 19 m.

16. 27.52 ft^2.

17. $205.71°$.

18. 64.95 in.^{2}.

19. Make a square that is 1 ft^2.

20. 565.2 ft^2.

21. 780 tiles.

22. a. 1:5.   b. 1:25.

23. Increase of 69%.

24. a. 4,618,944 in.^{3}.   **b.** 0.71 ft^3.   **c.** 1,300,000 cm^3.
25. 5.96 gal.
26. a. SA = 528 in.^{2}; V = 690 in.^{3}.   **b.** SA = 318.65 ft^2;
V = 435.44 ft^3.   **c.** SA = 189.36 yd^2; V = 108 yd^3.
d. SA = 563.82 m^2; V = 1,259.19 m^3.
27. 8 cans.
28. 66.6%.
29. 6 L.
30. a. 54:864 = 1:16.   **b.** 27:1,728 = 1:64.
31. The volume increases by 27 times.
32. a. 83.2 oz.   **b.** 360 mg.   **c.** 1.5 cg.
33. 909.09 kg.
34. a. 155.56° C.   **b.** −49.44° C.   **c.** −15.24° C.
d. −131.8° F.   **e.** 249.8° F.
35. It decreases 5.56° C.
36. 277 hr, 46 min, and 40 s.
37. r = 2 km/hr.
38. 365.85 m/s.
39. 354.05 km/hr.
40. 13,384.5 rubles.

CHAPTER 14

■ Section 14.1 Learning Assessment

1. In its singular sense, "statistics" refers to the science of collecting, analyzing, and interpreting data, which are numbers that occur in a context. In its plural sense, "statistics" refers to the numbers in the data themselves or to the result of some computation on the data.

2. The first step is to formulate a question, which determines how the data are collected, analyzed, and interpreted. The second step is to collect data, often using a survey, an observation, or an experiment. The third step is to organize and analyze the data using statistical graphs or descriptive statistics. The fourth step is to draw conclusions from the data that answer the question.

3. Statistical questions anticipate variability and the need to collect data to capture that variability. Mathematical questions tend to be more deterministic in that they often seek a single answer.

4. Variability is the idea that the values of traits vary from one instance to another.

5. Qualitative, or categorical, variables separate data into categories. The data associated with a qualitative variable usually take the form of counts because the investigator is interested in the number of individuals who fall into each category. Quantitative, or numerical, variables describe a measure or other numerical value associated with the trait of interest. The data associated with a quantitative variable are always numerical, so they lend themselves nicely to many statistical computations.

6. In most studies, the investigator does not have the funds or the manpower to consider an entire population, so the study is conducted on a small portion of the population, called a sample. If the sample is representative of the population, meaning the characteristics of the sample closely match the characteristics of the population, then whatever holds true for the sample must hold true for the population.

7. **a.** Bias refers to any problem in the design or conduct of the study that causes the results to favor a particular outcome.   **b.** If a sample is chosen with an inherent tendency toward a particular outcome, then the study suffers from a selection bias.

8. **a.** A convenience sample is a sample selected from readily available people or objects.   **b.** A voluntary sample lets members of the population decide for themselves whether to participate in the study.   **c.** In a simple random sample, every member of the population has the same chance of being selected. To select the sample, every member of the population is assigned an identification number. Numbers are then selected using a random device to pick members of the sample.   **d.** In a stratified sample, participants are first separated along specific characteristics before being assigned to a group.

9. In a survey, the investigator develops and uses a questionnaire to ask participants about things that cannot be directly observed. In an observation, the investigator observes and records the characteristics and behaviors of the participants without trying to influence or modify them in any way. In one type of an experiment, we conduct a large number of trials and analyze the results for similarities and differences. In another, we make a comparison between two groups, one of which has had some condition deliberately imposed on it.

10. A confounding variable is one whose effect on the response variable is difficult to separate from that of the explanatory variable.

11. In a simple comparative experiment, one group, called the treatment group, has a treatment deliberately imposed on it to see whether the treatment influences certain traits of the participants. The other group, called the control group, does not. If the traits of the treatment group are different from those of the control group at the end of some period of time, then the investigator has evidence of a cause-and-effect relationship between the explanatory variable associated with the treatment and the response variable.

12. **a.** Qualitative.   **b.** Quantitative.   **c.** Qualitative.
d. Qualitative.   **e.** Quantitative.   **f.** Qualitative.
13. **a.** Quantitative.   **b.** Qualitative.   **c.** Quantitative.
d. Qualitative.   **e.** Qualitative.   **f.** Quantitative.
14. **a.** The sample is 1,500 American adults, and the population is all adult Americans. The variable is the number of times they eat out. It is quantitative.   **b.** The sample is the 14 tested cars, and the population is all new cars. The variable is the gas mileage. It is quantitative.
c. The sample is the 100 ears of corn, and the population is the farmer's corn crop. The variable is infected with corn worms. It is qualitative.   **d.** The sample is the boys in his freshman classes, and the population is all high school freshman boys. The variable is strength. It is quantitative.
15. **a.** The sample is the two third-grade classes, and the population is third-graders at the school. The variable is reading level. It is qualitative.   **b.** The sample is the four phones from different companies, and the population is all cellular phones. The variable is calling range. It is quantitative.   **c.** The sample is the 300 people in the phone poll, and the population is the people in the three-county area. The variable is smoking habits. It is qualitative.   **d.** The sample is the 100 customers filling out the survey, and the population is all customers purchasing

the new line of sneakers. The variable is customer satisfaction. It is qualitative.

16. a. The sample is the 1,500 freshmen selected from different universities, and the population is all college freshmen. If the sample is representative of the population, then we can infer that 64% of all college freshmen graduate in four years. **b.** The sample is the 1,000 radios tested, and the population is all radios produced at the factory. If the sample is representative of the population, then 1 in 1,000 radios are defective. **c.** The sample is the 200 dentists surveyed, and the population is all dentists. If the sample is representative of the population, then 7 out of every 10 dentists, or 70% of dentists, recommend Mint White.

17. a. The sample is the 200 households polled, and the population is all households in the suburban area. If the sample is representative of the population, then two-thirds of all households in the suburban area own two or more televisions. **b.** The sample is the 14,500 respondents, and the population is all people visiting that Internet site. If the sample is representative of the population, then 74% of people visiting the Internet site think they pay too much in taxes. **c.** The sample is the 25 examined patients, and the population is all obese patients at the hospital. If the sample is representative of the population, then 20 out of every 25 obese patients, or 80%, also have chronic joint pain.

18. a. Voluntary sample. **b.** Convenience sample. **c.** Stratified sample.

19. a. Simple random sample. **b.** Convenience sample. **c.** Voluntary sample.

20. a. Answers will vary. **b.** Answers will vary. **c.** Answers will vary.

21. a. Answers will vary. **b.** Answers will vary. **c.** Answers will vary.

22. a. Answers will vary. **b.** Answers will vary. **c.** Answers will vary. **d.** Answers will vary.

23. a. Answers will vary. **b.** Answers will vary. **c.** Answers will vary. **d.** Answers will vary.

24. a. Observation. **b.** Survey. **c.** Experiment. **d.** Observation.

25. a. Survey. **b.** Observation. **c.** Observation. **d.** Survey.

26. a. The explanatory variable is income, and the response variable is home size. **b.** The explanatory variable is age, and the response variable is height. **c.** The explanatory variable is amount of exercise, and the response variable is weight. **d.** The explanatory variable is smoking habits, and the response variable is lung disease.

27. a. The explanatory variable is eating habits, and the response variable is cholesterol levels. **b.** The explanatory variable is hours studied, and the response variable is grades. **c.** The explanatory variable is amount of water, and the response variable is plant growth. **d.** The explanatory variable is car size, and the response variable is gas mileage.

28. a. The treatment group is the five classes using the new system for taking notes, and the control group is five classes who are not. **b.** The explanatory variable is use of the new note system, and the response variable is learning achievement. **c.** Although students in the treatment groups did seem to score better, determining whether they did significantly better than the other students is difficult.

29. a. There are two treatment groups. One gets the weight loss product, and the other gets a placebo. The control group is the group that gets no pill. **b.** The explanatory variable is use of the weight loss product, and the response variable is weight loss. **c.** The results seem to indicate that the weight loss product is not effective.

30. 225 misprints.

31. 1,267 garments are defective.

32. 2,220 students are African American, 1,061 students are Asian, 4,535 students are Caucasian, and 1,833 students are Hispanic.

33. 150 chips are white or blue.

34. 1,944 bass.

35. Blue-collar workers earn wages, so they have a vested interest in seeing a larger minimum wage. Hence, it is not surprising that the sample favors raising the minimum wage.

36. The people who vote from 2:00 p.m. to 3:00 p.m. are a select group of people, often limited to those who are not working or those who are not in school. If this demographic has a particular bias toward one political group or another, it could cause the polls to be biased.

37. The claim might not be truly representative of the population's beliefs. Because the sample was voluntary, it is possible that only people who have been adversely affected by eating disorders were likely to respond to the survey.

38. Often the human immune system is weakened by a lack of sleep. If students spend more time socializing, they may be spending less time sleeping. This could affect the results of the study.

39. The insulation in a home can drastically affect electricity bills. If the homeowner happened to change the amount of insulation while he was collecting data, it could have affected his results.

40. Exercise and healthy eating habits are both confounding variables because both can affect the weight loss regardless of whether the person is taking a weight loss supplement.

41. A possible confounding variable in the study is exercise because exercise is known to reduce cholesterol levels. Because Texas is in a warmer climate, perhaps men in Texas exercise more throughout the year than the men in Minnesota. Without knowing more about this confounding variable, we should view the results of the study with some skepticism.

42. Answers will vary.

43. Answers will vary.

44. In statistics, the word "random" means that any person or object in the sample has the same chance of being selected as any other person or object in the population.

45. Algebraic and statistical variables are similar in that they can take on different values. They differ in the fact that algebraic variables always represent numbers. Statistical variables represent traits or characteristics that can change.

46. The U.S. Census is used to determine the number of seats each state gets in the House of Representatives.

47. Answers will vary.

48. Answers will vary.

49. a. Answers will vary.   **b.** Answers will vary.
c. Answers will vary.   **d.** Answers will vary.

■ Section 14.2 Learning Assessment

1. Every graph should have a title, or caption, that briefly explains the data and, if possible, the source of the study. Many graphs also have axis labels, which describe the variables and data represented by an axis. If the graph represents multiple data sets or categories, then it may also include a legend or key that identifies each data set or category.

2. An outlier is any data point that is atypical from the rest of the data. Clusters in the data set are small groups of data that are isolated from the bulk of the data set. Gaps are the spaces between clusters.

3. **a.** Dot plots and stem and leaf plots are both used to represent small sets of numerical data. Both graphs show the entire data set. In a dot plot, the values are shown by dots placed above a number line. In a stem and leaf plot, the values are listed in a numerical order.   **b.** Pictographs and bar graphs both show counts for categorical data. Pictographs show the count by using an icon that represents a certain number. Bar graphs show the count by the height of the bar.   **c.** Bar graphs and histograms both show the data by using the heights of bars. Bar graphs are used to show discrete data with separated bars. Histograms show continuous quantitative data using connected bars.   **d.** Bar graphs and pie graphs can both be used to show counts. A pie graph shows the count relative to the whole, whereas a bar graph does not.

4. **a.** Frequency tables, pictographs, bar graphs, and pie graphs can be used to represent qualitative data.
b. Dot plots, stem and leaf plots, histograms, and line graphs can be used to represent quantitative data.
c. Stem and leaf plots, bar graphs, histograms, line graphs, pie graphs, and scatterplots can all be used to compare data sets.

5. A scatterplot is made by drawing two axes, one for each variable. Each data point is plotted by using its values for each variable as an ordered pair. Scatterplots are used to determine whether a relationship exists between two variables.

6. If high values of one variable are associated with high values of the other and low values of one variable are associated with low values of the other, then the variables have a positive correlation. If high values of one variable are associated with low values of the other and vice versa, then the variables have a negative correlation.

7. A correlation between two variables does not imply that one is the cause of the other. Correlation is a claim of relationship, not one of causality.

8. A regression line is the line that best represents the linear pattern of linearly correlated data.

9. a.   **Number of Ridges on Fingerprints**

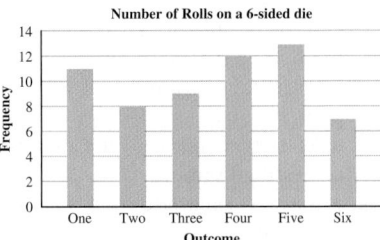

b. Number of Ridges on Fingerprints

Stem	Leaves
17	9
18	346789
19	233799
20	111455
21	1359
22	0

c. Answers will vary.

10. a.   Speeds of Sprinters

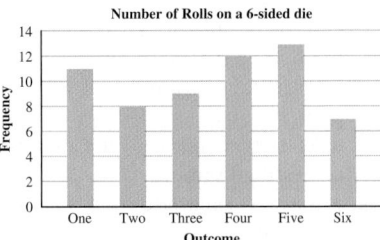

b. Speeds of Sprinters in Seconds

Stem	Leaves
11	23466778999
12	001124689

c. Answers will vary.

11. Results will vary, but possible outcomes are shown:

Outcome	Frequency
One	11
Two	8
Three	9
Four	12
Five	13
Six	7

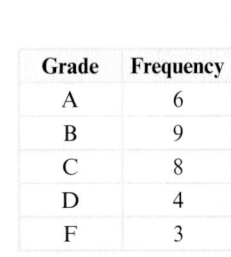

12. Overall, the class did reasonably well because most of the grades are As, Bs, and Cs.

Grade	Frequency
A	6
B	9
C	8
D	4
F	3

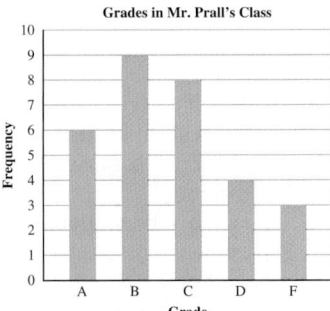

13. The histogram is like the stem and leaf plot in that it shows the frequency of each group as indicated by the stems. It differs from the stem and leaf plot in that it does not show the specific data values.

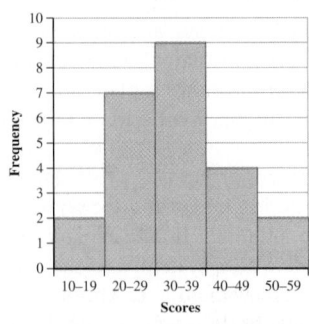

14. a.

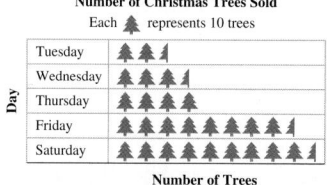

Number of Christmas Trees Sold
Each 🌲 represents 10 trees

Day	Number of Trees
Tuesday	🌲 🌲 🌱
Wednesday	🌲 🌲 🌲 🌱
Thursday	🌲 🌲 🌲 🌲
Friday	🌲 🌲 🌲 🌲 🌲 🌲 🌱
Saturday	🌲 🌲 🌲 🌲 🌲 🌲 🌲 🌱

Number of Trees

b.

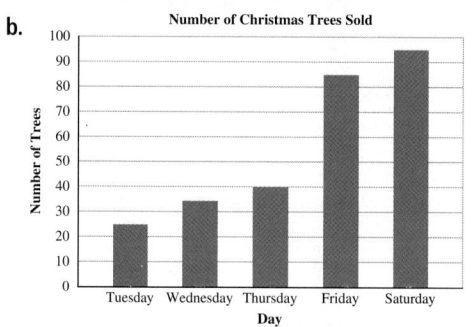

Number of Christmas Trees Sold

c. Answers will vary.

15. A possible pictograph is shown:

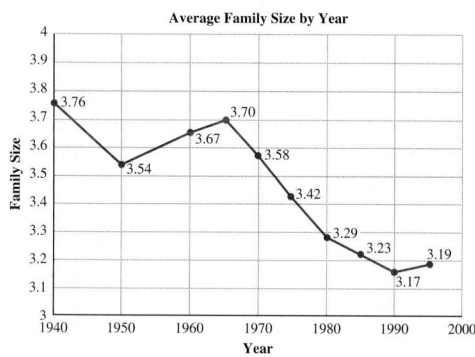

Number of Sandwiches Sold at a Deli
Every ▬▬ represents 10 sandwiches

Sandwich	Number Sold
Burger	▬ ▬ ▬ ▬ ▬ ▬ ▬ ▬ ▬ ▬
Reuben	▬ ▬ ▬
Steak & cheese	▬ ▬ ▬ ▬
BLT	▬ ▬ ▬ ▬ ▬
Chicken salad	▬ ▬ ▬ ▬

Number Sold

16. In general, the average family size has been decreasing.

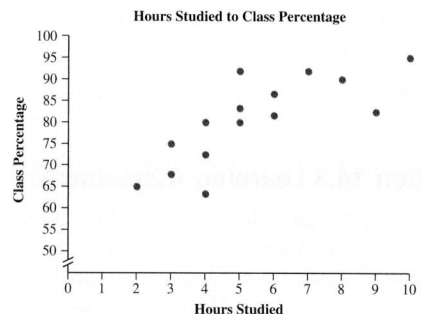

Average Family Size by Year

17. The graph indicates a positive correlation between the two variables.

Hours Studied to Class Percentage

18. The graph indicates a negative correlation between life expectancy and people per physician. This make sense because the more people there are per doctor, the harder it is to see one. This naturally has an impact on a person's quality of care, particularly near the end of life.

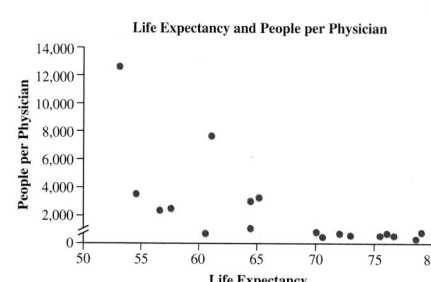

Life Expectancy and People per Physician

19. a. Tallest height = 74 in.; shortest height = 58 in.
b. 66 in.   **c.** A possible outlier is at 74 in.

20. a. Highest score = 139; lowest score = 61.   **b.** 122.
c. 61.   **d.** Because a 70% on the test is a score of 105 points, and most students scored higher than this value, the class as a whole did quite well.

21. a. 4.   **b.** Illinois, Michigan, Ohio, and Pennsylvania.
c. 42.   **d.** 16%.

22. a. WalMart; $304.9 billion.   **b.** Five times.   **c.** Walmart, Kroger, Target, and Walgreen had sales greater than $60 billion. Home Depot and Costco had sales less than $60 billion.

23. a. Most electricity: March; least electricity: Oct and June
b. June and August.   **c.** 2011.

24. a. The scores for both males and females have been improving over time.   **b.** Better.   **c.** Males' scores on the verbal portion of the SAT declined from 1972 to 1976 and have remained fairly constant since then.   **d.** Verbal.
e. Males.

25. a. High school diploma.   **b.** 7%.   **c.** 2,001 employees.

26. a. Answers will vary.   **b.** Answers will vary.
c. Answers will vary.

27. a. No correlation.   **b.** Negative correlation.
c. Negative correlation.   **d.** Positive correlation.

28. Part b shows the strongest correlation because the plotted points are closest to a line.

29. a.

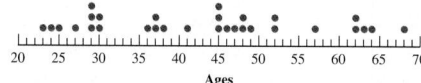

Ages of Teachers at Fairview Heights

b. 6 teachers.   **c.** 23%.

30. a. Maine has the most, and Connecticut has the least.
b. About 10,528.   **c.** Connecticut has more.   **d.** 0.4% of the population are millionaires.

31. a. Canada.   **b.** Germany won 12% of the gold medals, and Canada won 16% of the gold medals.   **c.** No. The country that won the fewest is included in the Other category, so there is no way to know.

32. a. There are 50 white cars, 40 black cars, 20 blue cars, 10 green cars, 30 red cars, and 50 silver cars.   **b.** The white cars would decrease to 20%, and the silver cars would increase to 30%.   **c.** 10   **d.** The percentages would stay the same.

33. The result is shown:

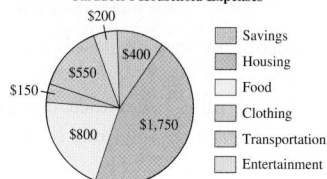

Carducci's Household Expenses

34. No. The sum of the different categories is larger than their take-home pay. This means that the sum of the parts is greater than the whole, making it impossible to draw a circle graph.

35. a. There is a positive correlation. It makes sense because the longer people are employed at a particular job, the more they should make. **b.** $y = 875x + 44{,}375$. **c.** The salary at 13 years is \$55,750, at 20 it is \$61,875, and at 31 years it is \$71,500.

36. Answers will vary.

37. Caucasians and Hispanics have increased in the percentage of people below the poverty lines. African Americans and the other category have decreased in the percentage of people below the poverty line.

38. A bar graph is probably the best choice because the data are given in discrete numerical categories. A potential bar graph is shown:

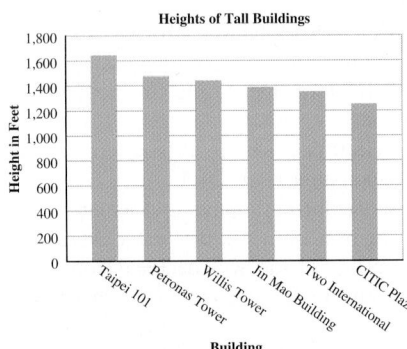

39. a. Because one variable is the continuous numerical of time, a line graph is the best choice.

b.

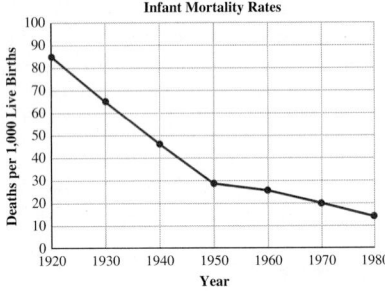

c. The graph shows a drastic decrease in the number of infant deaths per 1,000 live births.

40. a. A potential relationship exists between the two variables, so a scatterplot is the best choice. **b.** No. The different units do not affect the choice of the graph.

c.

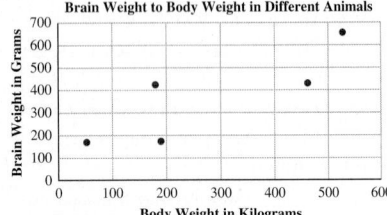

d. A horse is probably the smartest because it has the largest brain.

41. a. A double bar graph is one choice, except that the numerical data are not on the same order, making such a

graph difficult to make. We can work around this problem by making two separate bar graphs.

b.

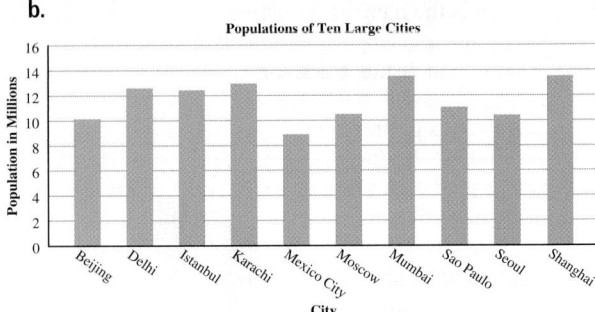

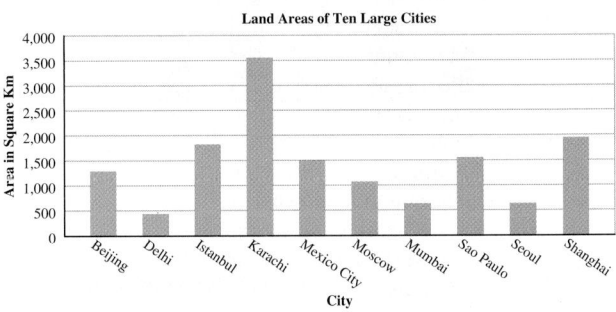

c. Compute the number of people per square kilometer by dividing the number of people by the number of square kilometers. **d.** Delhi is the most crowded with 29,234 people per square kilometer.

42. Answers will vary.

43. a. Answers will vary. **b.** Answers will vary.

44. a. Answers will vary. **b.** Answers will vary.

45. A pie graph does not always have to be given in terms of percents as long as the size or count of the data is given in relationship to the whole or total. In many cases, we are given counts for each category that equal the total count when added.

46. Answers will vary.

47. In general, a line graph shows the relationship between the variable on the horizontal axis and the variable on the vertical axis in such a way that each value on the horizontal axis is assigned to only one value on the vertical axis. Hence, it represents a function.

48. If we can construct a circle and then use a protractor to draw angles of a given size, then we should be able to accurately produce a pie graph by hand.

49. Answers will vary.

50. Answers will vary.

51. a. Answers will vary. **b.** Answers will vary.

■ Section 14.3 Learning Assessment

1. A measure of center indicates how the data are centered, and a measure of spread indicates how the data are distributed. Both are needed to adequately describe the data.

2. The mode describes the center by giving the data value that occurs most frequently. The median describes the center by giving the data value that occurs in the middle of the data set. The mean describes the center by adding all the data values and then dividing by the number of data points.

3. The mean is the number for which the sum of the distances for the data points above the mean is equal to the sum of the distances for the data points below the mean.

4. **a.** The mean is preferred because it takes every data value into account. **b.** The median might be a better measure of center with data sets that have extreme values that can dramatically affect the value of the mean.

5. **a.** The range is the difference between the largest and the smallest data values. **b.** The lower quartile is the median of the data extending from the first data value, x_1, to the median. The upper quartile is the median of the data extending from the median to the last data value, x_n.

6. **a.** A 5-number summary includes the minimum, the lower quartile, the median, the upper quartile, and the maximum. **b.** The median and the first and third quartiles are marked with short line segments over the appropriate numbers on a numerical axis. The box is formed by drawing segments to connect the first and third quartiles. The maximum and minimum are plotted and connected to the box with segments. If there is an outlier, it is denoted with an asterisk and the next highest or lowest value is connected to the box.

7. **a.** Any data point with a value that is $1.5 \cdot IQR$ below the first quartile or $1.5 \cdot IQR$ above the third quartile is considered an outlier. **b.** Any data value that is three standard deviations below the mean or three standard deviations above the mean may be considered an outlier.

8. Because any data point below the mean has a negative deviation and any data point above the mean has a positive deviation, adding the deviations gives us a net sum and an average deviation of zero. By squaring the deviations first, they all become positive, and the negative signs are no longer a concern.

9. **a.** A normal distribution is characterized by a smooth bell-shaped curve that theoretically extends to infinity without touching the horizontal axis. In a normal distribution, the mean, median, and mode all take on the same value, and the distribution is symmetric about this value. **b.** In any normal distribution, approximately 68% of the data lie within one standard deviation of the mean, 95% of the data lie within two standard deviations of the mean, and 99.7% of the data lie within three standard deviations of the mean.

10. A z-score is a value that indicates how many standard deviations a data point is from the mean. Because z-scores measure data points in terms of standard deviations and not the original scale, they can be used to compare scores from different distributions.

11. Answers will vary. Possible data sets are given:
    a. 5, 6, 7, 8, 8, 9, 10, 11, 12, 13.
    b. 5, 6, 7, 7, 8, 9, 10, 10, 11, 12.
    c. 5, 6, 7, 8, 9, 10, 11, 12, 13, 14.

12. Possible answers are given:
    a. 10, 10, 10, 14, 15, 16, 17, 85, 90, 95.
    b. 7, 7, 7, 10, 35, 35, 36, 37, 37, 38.
    c. 10, 11, 12, 13, 14, 15, 16, 17, 18, 18.
    d. 5, 10, 15, 20, 25, 25, 30, 35, 40, 45.

13. Possible answers are given: **a.** 60, 65, 75, 75, 85, 90.
    b. 70, 71, 75, 75, 79, 80.

14. The data set has two distinct clusters of data that are separate from one another. Consequently, the mode is a good choice for the measure of center because the data set can be described as bimodal.

15. **a.**

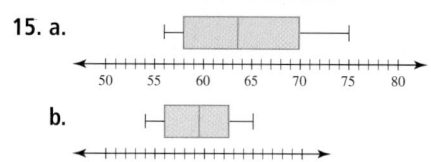

b.

16. **a.** Maximum = 28, Q_3 = 25, median = 19, Q_1 = 11, and minimum = 4. **b.** Maximum = 0.48, Q_3 = 0.44, median = 0.36, Q_1 = 0.28, and minimum = 0.2.

17. **a.**

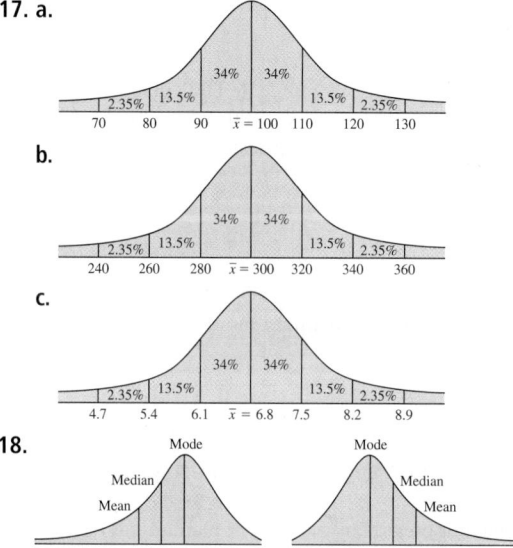

b.

c.

18.

19. **a.** Mean = 16, median = 16, mode = 14 and 19.
    b. Mean = 117.5, median = 122.5, mode = 124.

20. **a.** Mean = 3.84, median = 4, mode = 4.4.
    b. Mean = -22.7, median = -23, mode = -26 and -20.

21. 825 in.

22. $26.04 per item.

23. **a.** Maximum = 23, minimum = 12, and range = 11.
    b. Maximum = 100, minimum = 54, and range = 46.

24. **a.** Maximum = 14.5, minimum = 9.9, and range = 4.6.
    b. Maximum = -2.1, minimum = -6.1, and range = 4.

25. The oldest player is 41.

26. Maximum = 72, Q_3 = 70, median = 66, Q_1 = 54, and minimum = 23. The IQR = 16, and the only outlier is 23. The box plot is shown:

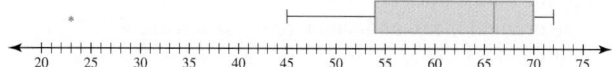

27. Maximum = 124, Q_3 = 102, median = 98, Q_1 = 93, and minimum = 75. The IQR = 9, and there are two outliers, 75 and 124. The box plot is shown:

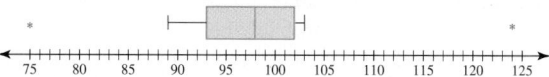

28. **a.** $\bar{x}$ = 59.9 and s = 3.33.   **b.** $\bar{x}$ = 22.6 and s = 7.04.

29. **a.** $\bar{x} = 5$ and $s = 0.32$. **b.** $\bar{x} = 19.49$ and $s = 2.70$.
30. **a.** $\bar{x} = 50$ and $s = 6.75$. **b.** 47, 48, 52, 53, 53, and 56.
  c. 37, 43, 47, 48, 52, 53, 53, 56, and 61.
31. **a.** 68%. **b.** 2.5%. **c.** 0.15%.
32. **a.** 68%. **b.** 0.15%. **c.** 32%.
33. 59 and 91 lb.
34. **a.** 800 students. **b.** 125 students. **c.** 15 students.
35. **a.** $z = -2.8$. **b.** $z = -0.6$. **c.** $z = 1.4$. **d.** $z = 3.2$.
36. **a.** $z = 0.25$. **b.** $z = 1.25$. **c.** $z = -1.75$.
  d. $z = -0.875$.
37. Brandy did better on her test.
38. They are equally tall with respect to their gender.
39. **a.** 1, 1, 1, 1, 1, 1, 1, and 1. **b.** Answers will vary. One possibility is 1, 1, 1, 1, 5, 5, 5, and 5.
40. 28.
41. 53.56.
42. 78.76 lb.
43. **a.** 65. **b.** 65. **c.** 97. **d.** 20.
44. **a.** About 16%. **b.** About 35%. **c.** The person scored better than about 90% of the test takers. He or she did well.
45. He must make a 97 on the next test.
46. $35,750.
47. 15.33 mph.
48. 3.352.
49. Shotput.
50. **a.** $\bar{x} = 670.65$ mg and $s = 3.28$ mg. **b.** Any pill weighing less than 660.81 mg or weighing more than 680.49 mg should be thrown out.
51. Mean = 1,970.67 mi; Median = 2,168 mi. The median is probably a better measure of center.
52. If all the data values are the same, then the mean is equal to both the greatest number and the least number in the data set.
53. **a.** The mean increases by 5. **b.** The median increases by 5. **c.** The mode increases by 5. **d.** The range stays the same.
54. The interquartile range is equal to the range if the Q_1 = minimum and Q_3 = maximum. For example, the interquartile range equals the range in the data set 5, 5, 5, 5, 5, 10, 10, 10, 10, and 10.
55. **a.** The five numbers are evenly distributed throughout the range, implying that the data are evenly distributed throughout the range. **b.** The upper and lower quartiles are closer to the median than they are to the maximum and minimum, implying that most of the data are clustered near the middle of the range around the median. **c.** The upper and lower quartiles are much closer to the maximum and minimum than they are to the median. This implies that most of the data are in two clusters, one at either end of the range.
56. All the data values are equal to the mean.
57. The original mean and standard deviation are $\bar{x} = 7.14$ and $s = 1.55$. After adding 10 to each value, the new mean and standard deviation are $\bar{x} = 17.14$ and $s = 1.55$. This implies that if the same value is added to every data point, the mean increases by that value and the standard deviation remains the same.
58. The warranty should probably be for 39 months. The company will have to make a refund on only about 5% of its batteries, yet the warranty lasts for more than 3 years.
59. **a.** Mean = 78, median = 80, and mode = 80. **b.** Either the median or the mode. **c.** The mean.

60. The mean, median, and mode for each team is given:

  Bulldogs: Mean = 4.68, median = 4.7, mode = 4.7.

  Tigers: Mean = 4.72, median = 4.6, mode = 4.6.

  Falcons: Mean = 4.8, median = 4.7, mode = 4.5.

  To present the best case, the Bulldogs used the mean, the Tigers used the median, and the Falcons used the mode. In this case, the mean is preferred because the data contains no extreme values.
61. Most likely her claim is valid. If the median is 85, then 50% of the class scored between 85% and 100%. So, although there were several low values, on the whole, the class probably did well.
62. The mean and standard deviation for league 1 are $\bar{x} = 39.53$ and $s = 1.81$. The mean and standard deviation for league 2 are $\bar{x} = 39.33$ and $s = 2.59$. Most of the ages of the men in league 1 are much closer to the mean than the ages of the men in league. This means the data for league 1 are more compact, and the standard deviation is smaller.
63. Answers will vary.
64. Answers will vary.
65. Answers will vary.
66. Answers will vary.
67. He is not correct. He did not arrange the numbers in numerical order prior to finding the median.
68. Answers will vary.
69. Connect the centers of the tops of the bars with a curve.
70. Answers will vary.
71. **a.** Answers will vary. **b.** Answers will vary. **c.** Answers will vary.
72. **a.** Answers will vary. **b.** Answers will vary.
73. Answers will vary.

■ Section 14.4 Learning Assessment

1. If the purpose is to accentuate differences, then remove part of the vertical scale to make the differences seem larger. If the purpose is to minimize differences, extend the scale to make the differences seem smaller.
2. Reversing the axes causes the graph to have the opposite appearance. This may cause confusion because of our natural tendency to read graphs from left to right and from top to bottom.
3. One way to manipulate the scale on a pie graph is to remove part of the data so that it is not relative to 100% or the entire whole.
4. A perceptual distortion often occurs in pictographs when the data are represented by one aspect of the graph, but our eye catches another. The aspect that our eye catches may not show the same relative differences in the data.
5. Three-dimensional effects can also distort our perception of a graph. In a pie graph, sectors farther away seem to represent smaller portions than they actually do. In a bar graph, the three-dimensional effects can make the graph difficult to read.
6. In many studies, the source has a vested interest in a particular outcome, causing them to report the results in a biased way. If the source is biased, then care should be

taken when looking at the results to see whether they have been presented fairly.

7. Complementary statistics are two statistics that report the same fact from two different points of view. One complementary statistic may be chosen over another to sway people in favor of a particular point of view.

8. In many cases, knowing the exact value of a statistic is difficult. A margin of error provides a range of values in which the actual data value is likely to occur.

9. Starting at 0 minimizes the differences in the graph. Starting at $75 billion accentuates the differences in the graphs.

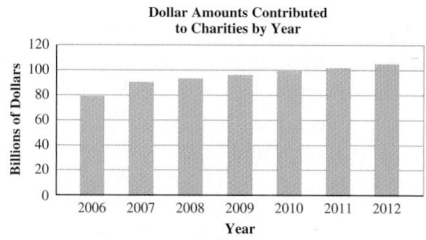

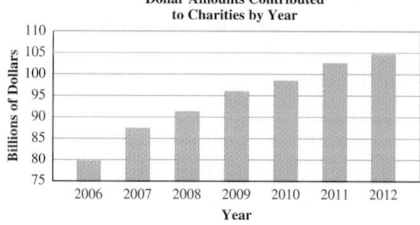

10. Remove part of the vertical scale to make the increases look more dramatic.

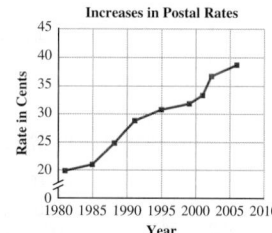

11. To put the scores in the best possible light, the principal should remove part of the vertical axis.

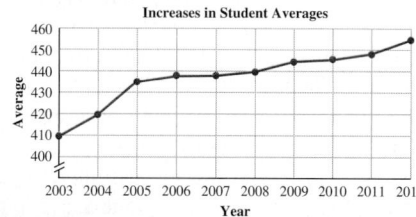

12. a. The histogram shows that the wages of workers have been steadily increasing.

b. Reverse the axes to make it appear as though there is a decrease in wages.

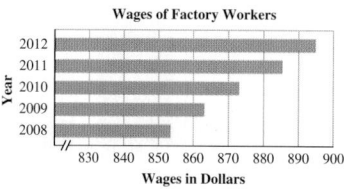

13. Reverse the axes to make it appear as though there is a decrease in burglaries.

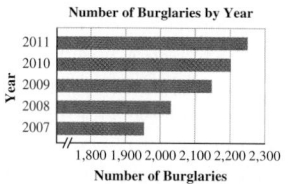

14. The pie graph only shows how 91% of the workers get to work; 9% of the data set is not represented.

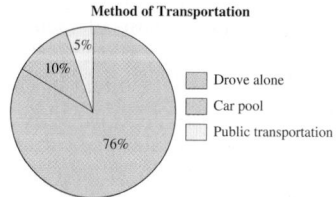

15. a. We can make the number of arrests appear to be on the decline by removing part of the vertical scale. This may make people think that arrests are significantly decreasing when they are actually staying fairly constant.

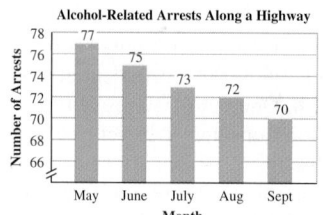

b. We can show that the number of arrests has stayed fairly constant by showing the entire vertical scale.

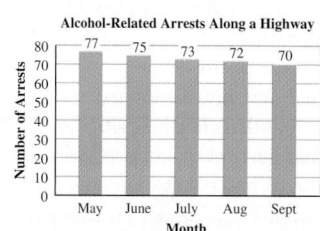

16. a. Appropriate. **b.** Not appropriate. **c.** Appropriate. **d.** Not appropriate.

17. One possible graph is shown:

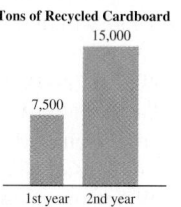

18. One possible graph is shown:

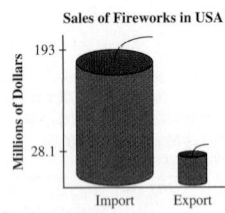

19. The graph has no title or axis labels. Without them, we have little idea of what the graph is about.

20. No legend is provided, so it is unclear which sector goes with which ethnicity.

21. The height of the barrels is used to represent the difference between the amounts. However, the volume of the barrels catches our eye, and the volume is eight times the intended value.

22. Although model B does get better gas mileage, there is some question as to whether it is significant because model B gets only 2–3 mi/gal more than the other cars.

23. The percents should be included to make it easier to determine the specific amounts of each brand.

24. a. The claim is misleading because we do not know what ULTRA CLEAN is compared to. **b.** The claim may be misleading because we do not know how many cars were tested. **c.** The claim is misleading because we do not know what other brands the juice is compared to.

25. a. The claim is misleading because we do not know if there are any other rental agencies in town. **b.** The claim could be misleading because we do not know how many doctors were surveyed. **c.** The claim is misleading because we do not have enough information. To make a reasonable decision, we need to know how many people die in airplane crashes and how many people die in car accidents.

26. a. Of voters in the 2008 presidential election, 62% were married. **b.** One out of 10 residents considers their neighborhood unsafe. **c.** Seventeen out of twenty females have no disability. **d.** Ninety percent of the active volcanoes in the world are not in Japan.

27. a. Four out of 5 residents in Florida are under 65 years old. **b.** Sixty-eight percent of Americans say that bad breath is the not worst trait of their coworkers.
c. One-quarter of Americans would rather floss than go grocery shopping. **d.** Seventy percent of all bingo players are over the age of 35.

28. a. The actual percent of married women who have cheated on their husbands is between 40% and 54%.
b. The actual percent of children who watch television before going to bed is between 40% and 48%.

29. a. The actual average age of college seniors is between 21 and 25 years of age. **b.** The actual percentage of people 65 or older who live in poverty is between 9.3% and 10.1%.

30. Likely values are 87% with a margin of error of $\pm 5\%$.

31. Likely values are 22 hr with a margin of error of ± 2.5 hr.

32. To make the best case, the Bears used the mode, the Lancers used the median, and the Blue Devils used the mean. In this case, we would probably prefer the mean because there are no extreme values, and the mean takes every data point into consideration. Hence, the Blue Devils have the tallest team on average.

33. a. The percents for men and women in each department are given:

Major	Total Enrollment	Male Enrollment	Female Enrollment
Business	347	45%	55%
Education	512	9%	91%
Journalism	289	68%	32%
Mathematics	57	79%	21%
Nursing	152	28%	72%

The percentages show that there is not much of a bias in the business major. However, in journalism and mathematics, there is a bias toward males; in education and nursing, there is a bias toward females. **b.** Because the ratio of males to females is 1 to 3, we would expect to see the same ratio in the majors if there were no bias. Only one major, nursing, is close to this ratio. The others are biased. Specifically, business, journalism, and mathematics are biased toward males, and education is biased toward females.

34. a. One way to determine the fuel efficiency based on the weight of the vehicle is to divide the miles per gallon by the weight. In this case, the car gets (27 mpg)/(2,000 lb) = 0.0135 mpg/lb. The motorcycle gets (65 mpg)/(400 lb) = 0.1625 mpg/lb. **b.** The answer to part a is probably a better way to represent the fuel efficiency of the vehicle because it allows for a direct comparisons between vehicles of different types.

35. Lori is correct. She is wealthier after the cost-of-living adjustment.

36. Javier is incorrect. He is more overweight than Selena.

37. a. Draw the bar graph so that it minimizes the differences between the bills by showing the whole vertical scale for the data.

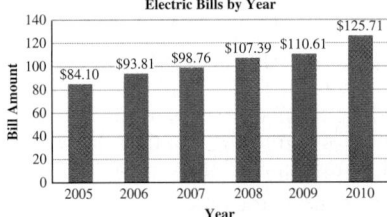

b. Draw the bar graph so that it accentuates the differences between the bills by showing only a part of the vertical scale.

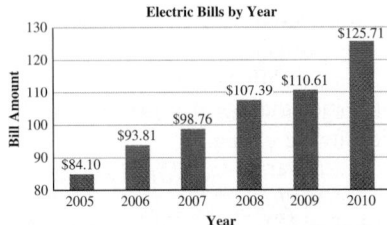

38. The graph shows a decrease in the percentage increases in the budget. People often interpret this to mean a loss of money. However, the actual amount of money is increasing, but only by smaller percentages.

39. a. The area of the larger graph is actually four times the area of the smaller graph when it should be only double.   **b.** The pie graphs are showing accurate representations of the percents of minority to nonminority students in each year.   **c.** Because the total number of students has increased at the university and the percentages have stayed the same, the number of minority students must have increased. However, the claim can be made that, because the percentage with respect to the total enrollment has not increased, the respective enrollment of minority students has not increased.

40. The claim leads people to believe that the city is a sunny place. However, Seattle is one of the rainiest places in the United States. So the city in the Midwest is also a rainy place.

41. The refrigerator may not last 10 years. Because the average is 10 years, some refrigerators must last less than 10 years and some last longer. They may have bought one of the ones that will last for less than 10 years.

42. Because 80% of the batteries have a life of 5 or more years, it is quite likely, but not necessarily certain, that the average life span of a battery is more than 5 years.

43. Because the average age of the patrons is 32, some patrons must be younger than 32, and some must be older. However, because Robert's age is fairly far from the average, the likelihood of his meeting someone his own age is not as good as if he were nearer to the average age.

44. Stanley's Stump Removal is correct when it says that it offers the lowest rate in town. However, in terms of total cost, Stanley's Stump Removal actually charges $325 per tree. This means that Tricia's Tree Trimming is actually the better deal.

45. Amy cannot discount the claim. Her sample is small, so it may not be representative of the actual life span of the lightbulbs.

46. The tobacco companies have a vested interest in studies that show smoking has no harmful effects on health because this type of result directly affects the sales of their products.

47. The television program in some senses has a vested interest in having a good review for the book because the review directly affects the sales of the book.

48. The car manufacturer has a vested interest in studies that show the reliability of the car because they directly affect the sales of their products.

49. Answers will vary.

50. To determine the validity of the results, we may want to know other information, such as the toothbrushes used, the number of times people brushed their teeth, and whether other products, like mouthwash, were used.

51. Although the one class did better, we would probably not consider the difference to be significant. Hence, we would conclude that there is no difference between the two methods.

52. **a.** Answers will vary.   **b.** Answers will vary.

53. Answers will vary.

54. Answers will vary.

55. Jackson has not taken into account that some students may graduate in more than 4 years.

56. The difference between the two classes is only one point, so we would probably not consider the averages to be different.

57. Answers will vary.

58. Answers will vary.

59. Paying expert witnesses may cause them to answer or give testimony in a biased fashion.

■ Chapter 14 Review Exercises

1. **a.** Qualitative.   **b.** Quantitative.   **c.** Qualitative.   **d.** Quantitative.

2. **a.** The population is all clothes driers, and the sample is the 10 that were tested. The variable of interest is drying time, and it is a quantitative variable.   **b.** The population is the people in the United States, and the sample is the 1,200 adults surveyed. The variable of interest is the number of times per week people exercise, and it is quantitative.   **c.** The population is all computers manufactured at the plant, and the sample is the 35 computers that are tested. The variable of interest is the correct installation of software, and it is a qualitative variable.

3. **a.** The population is all seniors at universities in the United States, and the sample is the 800 selected seniors. If the sample is representative of the population, then 63% of all graduating seniors got a job related to their major.   **b.** The population is all households in a suburban area, and the sample is the 200 selected households. If the sample is representative of the population, then one-fourth of the households in the suburban area no longer have landline telephone service.   **c.** The population is all married couples, and the sample is the 150 selected couples. If the sample is representative of the population, then in 23% of all married couples, the wife earns more than the husband.

4. **a.** Answers will vary.   **b.** Answers will vary.   **c.** Answers will vary.

5. **a.** Computer usage is the explanatory variable, and vision problems is the response variable.   **b.** Age is the explanatory variable, and the amount of exercise is the response variable.   **c.** Time spent at work is the explanatory variable, and income is the response variable.

6. **a.** The population is all people with cold symptoms, and the sample is the 90 people who participated in the study.   **b.** There are two treatment groups. One gets the cold remedy, and the other gets a placebo. The other group is a control group.   **c.** The cold remedy is the explanatory variable, and cold symptoms is the response variable.   **d.** The cold remedy seems to be effective.

7. 4,900 pencils.

8. Many of the residents of the city are likely to earn a living from fishing, so any restriction of fishing could directly affect their livelihood.

9. **a.**

Scores on a 50-pt Test

Stems	Leaves
1	9
2	
3	1455678999
4	0122334555556688
5	

b. 19 is a possible outlier.   **c.** The best score is 48. The worst is 19.   **d.** 45.

10.

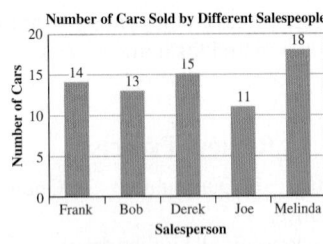

11. a. The range that occurred most frequently is 350–399, and the least frequent is 450–500. **b.** 200–500. **c.** About 95.

12. The graph shows that the population was steady from 1900 to 1930. After 1930, the population started to experience a drastic growth.

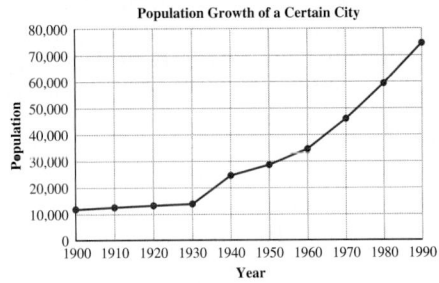

13. a. The largest class is the freshman, and the smallest class is the graduate students. **b.** 1,204 freshmen and 860 seniors.

14. An ordered back-to-back stem and leaf plot is shown:

Class 1	Stems	Class 2
5	3	455678
567789	4	57889
1134569	5	01378
0	6	

Because class 2 had more grades in the 30s, Class 1 did better on the exam.

15. a. There is a negative correlation. The correlation makes sense because exercise strengthens the heart muscle, so it has to beat less to move blood. **b.** $y = (-23/13)x + 998/13$. **c.** At $x = 5$, $y = 883/13 \approx 67.9$. At $x = 10$, $y = 768/13 \approx 59.1$. At $x = 12$, $y = 722/13 \approx 55.5$.

16. a. Mean = 75.8, median = 75, mode = 72. **b.** Mean = 12.8, median = 11.8, mode = none.

17. a. Maximum = 20, $Q_3 = 18.5$, median = 15, $Q_1 = 11$, minimum = 10. **b.** Maximum = 95, $Q_3 = 91$, median = 89, $Q_1 = 68.5$, minimum = 63.

18. a.

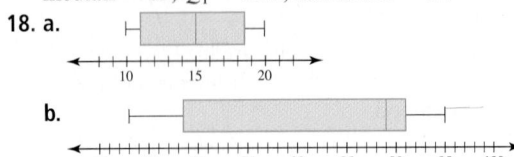

b.

19. There are no outliers.

20. 2,080 lb.

21. a. Mean = 39.4, and $s = 3.5$. **b.** Mean = 100.9, and $s = 4.2$.

22. a. Mean = 74.03, and $s = 6.54$. **b.** 66.67%. **c.** 93.3%.

23. 49.03 in.

24. a. 200 is two standard deviations below the mean, and 300 is two standard deviations above the mean. **b.** 81.5%. **c.** 0.15%. **d.** 195 has a z-score of $= -2.2$, 270 has a z-score of 0.8, and 310 has a z-score of 2.4.

25. 6.2 mph.

26. a. The mean goes up by 10. **b.** The median goes up by 10. **c.** The mode goes up by 10. **d.** The range remains the same.

27. Possible data sets are given:
a. 19, 50, 51, 52, 52, 52, 52, 53, 53, 53.
b. 50, 51, 52, 53, 54, 55, 100, 100, 100, 100.
c. 50, 50, 51, 52, 52, 54, 55, 56, 57, 58.

28. The data are spread consistently throughout the range, and there are no outliers or gaps. For this reason, we would use the mean because it takes every data value into consideration.

29. Because the median is 55, half of the scores are below this and are failing grades. The class did poorly.

30. There is no title on the graph, so it is unclear as to what it represents. It is also missing the scale on the vertical axis, so the size of the difference between men and women cannot be determined.

31. A possible graph is shown:

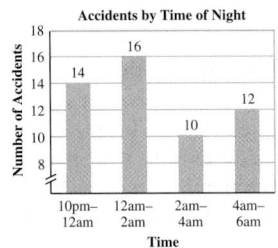

Answers will vary on incorrect conclusions.

32. One possible graph is shown:

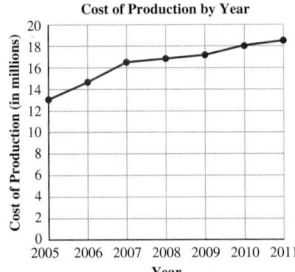

33. a. The claim is misleading because it is not clear how many customers were surveyed. **b.** The claim is misleading because the other products in the comparison are not known.

34. a. Eighty-five percent of people 65 and older do not work. **b.** One out of 5 children does not like to drink apple juice. **c.** Three-fourths of unmarried people 25 years or older do not have a bachelor's degree or better.

35. a. The actual percent of people that say a smile is the first facial feature they notice is between 44% and 56%. **b.** The actual percent of people who watch the Super Bowl only for the commercials is between 77% and 83%.

36. Although the claim is accurate, it is somewhat misleading because all the universities are within $250 of each other.

37. The study could be biased because the dairy association has a vested interest in selling more milk.

38. To make the best case, Jefferson High used the mode, Moravian High used the mean, and Bettstown High used the median. In this case, the mean is preferred because there are no extreme values and the mean takes every data point into consideration. Hence, Moravian High has the fastest team on average.

CHAPTER 15

■ Section 15.1 Learning Assessment

1. **a.** A probability is a number that measures the extent to which an event will or will not happen. **b.** The experimental probability that event E occurs on a given trial is given by the ratio of the number of times E occurs to the total number of trials.

2. An outcome is a possible result of an experiment. An event is an outcome or set of outcomes from the experiment that are of particular interest.

3. **a.** No. The experimental probability depends on the specific number of outcomes obtained during the experiment. Because the number of outcomes can change, the experimental probability can change. **b.** Conduct the experiment more times because, as the experiment is conducted more times, probabilities tend to stabilize.

4. As the number of trials for a given experiment increases, the variation in the experimental probability of a particular event decreases until it closely approximates a fixed number.

5. In one sense, an experimental probability indicates the likelihood that an event occurs in a single trial of an experiment. This value becomes a better predictor when it is based on more trials of the experiment. In another sense, an experimental probability indicates what will happen in the long run if an experiment is conducted many times.

6. Experimental probability can be used to make predictions from data.

7. **a.** A simulation models an experiment by using a situation with outcomes analogous to those of the experiment but for which the trials are easily conducted and recorded. **b.** If an experiment is difficult to conduct, a simulation can be used to collect data by conducting numerous trials. The experimental probability can then be computed from the outcomes of the simulation.

8. One way to use a table of random digits in a simulation is to assign a digit or a group of digits to represent each outcome in the experiment. Conduct trials by selecting random strings of digits from the table and record the results. A table of random digits can also be used to place objects at random locations on a Cartesian grid. Experimental probabilities are then determined by how objects are placed.

9. **a.** In the past 100 days with the same weather conditions, a tornado touched down on one of them. **b.** If 100 people come into contact with the infected person, 35 of them will catch a cold. **c.** If the game was played three times, she would win one of them.

10. **a.** If 10 instant lottery tickets are purchased, one will be a winner. **b.** In four days with the same weather conditions as tomorrow, on one of them it was sunny. **c.** If 1,000 computers are purchased, one will be defective.

11. 0.4, $\dfrac{4}{10}$, and 4 out of 10.

12. 0.375, 37.5%, and 3 out of 8.

13. Flip the coin four times to represent one trial of the experiment. If heads represents a win, then the outcome of the team winning all four games is represented by flipping four heads.

14. Let 1 represent a defective socket and the digits 2, 3, 4, 5, 6, 7, 8, 9, and 0 represent good sockets. Select a string of five digits to represent the five sockets in one package. If the string has exactly two 1s, then two defective sockets are in the package.

15. Let even numbers represent a male and odd numbers represent a female, or let the numbers 1, 2, and 3 represent a male and let 4, 5, and 6 represent a female.

16. **a.** The number must be a power of two. **b.** The number must be a factor or a power of six. **c.** The number must be a factor or a power of eight.

17. **a.** $\dfrac{11}{100}$. **b.** $\dfrac{11}{25}$. **c.** $\dfrac{6}{25}$. **d.** $\dfrac{9}{25}$.

18. The experimental probability of red is $\dfrac{37}{160}$, of blue is $\dfrac{85}{160} = \dfrac{17}{32}$, of yellow is $\dfrac{19}{160}$, and of green is $\dfrac{19}{160}$.

19. **a.** $\dfrac{18}{75}$. **b.** $\dfrac{51}{75}$. **c.** $\dfrac{24}{75}$. **d.** $\dfrac{28}{75}$.

20. **a.** Answers will vary. **b.** Answers will vary. **c.** Answers will vary. **d.** Answers will vary.

21. **a.** Answers will vary. **b.** Answers will vary. **c.** Answers will vary.

22. Answers will vary.

23. **a.** Answers will vary. **b.** Answers will vary. **c.** The heads side is slightly heavier.

24. $\dfrac{930}{3,640} \approx 0.255$. This is a good estimate because it is based on a large number of trials.

25. **a.** $\dfrac{53}{320}$. **b.** $\dfrac{23}{320}$. **c.** $\dfrac{76}{320}$.

26. **a.** 2×10^{-5}. **b.** 8.4×10^{-4}. **c.** 0.012.

27. **a.** 0.292. **b.** 0.0625. **c.** 0.104. **d.** 0.625.

28. **a.** 34%. **b.** 81.5%. **c.** 0.15%. **d.** 16%.

29. **a.** 50%. **b.** 2.5%. **c.** 81.5%.

30. Let heads represent true and tails represent false. Flip the coin 15 times, record the results, and compare them to the given table. Count the successes to simulate her score.

31. Let 1 represent A, 2 represent B, 3 represent C, and 4 represent D. Spin the spinner 10 times, record the results, and compare them to the given table. Count the successes to simulate her score.

32. Assign each student a three-digit number from 000 to 999. Select three-digit strings from the table of random digits until 25 students have been selected.

33. Let 1 represent a car with a defect and 2, 3, 4, 5, 6, 7, and 8 represent a car without defects. Roll the die five times to conduct a trial. If one or more 1s are rolled, count the trial as a success. Conduct 50 to 100 trials to find the experimental probability.

34. Let the digits 1, 2, 3, 4, 5, 6 and 7 represent a 10-year survivor and the digits 8, 9, and 0 represent someone who does not survive 10 years. Select two-digit strings from the table. If both digits are between 1 and 7, then count the trial as a success. Select 50 to 100 strings to find the experimental probability.

35. **a.** Assign each employee a number of the appropriate size. Select strings of the appropriate length to select employees.

b. Let the numbers 01–27 represent employees who use illegal substances and the numbers 28–00 represent those who do not. For each trial, select 10 two-digit strings. If three of the two-digit numbers are between 01 and 27, count the trial as a success. Conduct 50 to 100 trials to find the experimental probability.

36. Let the digits 1–8 represent rainy days and the digits 9 and 0 represent not rainy days. Let one trial consist of a four-digit string. If all the digits in the string are between 1 and 8, count the trial as a success. Conduct 50 to 100 trials to find the experimental probability.

37. **a.** Let 0–3 represent C, 4–6 represent O, 7 and 8 represent L, and 9 represent A. Select a six-digit string. It is a success if it contains at least one digit for each letter. Conduct 50 to 100 trials to find the experimental probability of getting the needed letters. **b.** Select enough digits from the table until you have at least one digit for each letter. This serves as one trial. Count the number of digits you had to select as the result of the trial. Repeat the trial 50 to 100 times to find the average number of bottles needed to get all four letters.

38. **a.** Assign each toy a number of 1–4, and then conduct a simulation by using a spinner with four equal sections. Spin the spinner seven times and record the results. If each of the numbers 1 through 4 comes up, count the trial as a success. Conduct 50 to 100 trials to find the experimental probability. **b.** Spin the spinner enough times to get each of the numbers 1 through 4 exactly once. Count the number of spins that were required. This completes one trial. Conduct 50 to 100 trials to find the average number of boxes that have to be purchased to get every toy.

39. **a.** Answers will vary. **b.** Answers will vary. **c.** Answers will vary.

40. **a.** Make a square grid that measures 8×8. Roll the eight-sided die twice. Let the first roll represent the row and the second roll represent the column. Roll enough times to place all of the chocolate chips. **b.** Answers will vary.

41. No. Each time the person plays, the chances of winning are $\frac{1}{1,000}$. The only way to be guaranteed of winning is to play all 1,000 numbers each time.

42. No. Assuming the coin is fair, each trial is independent of all the previous trials. Previous outcomes cannot be used to predict the outcome of the next trial.

43. **a.** The coin seems to be biased toward tails. However, this number of heads and tails can be flipped even if the coin is fair. **b.** Flip the coin many more times. If the results are still biased toward tails, we can be more confident in our decision.

44. One possible graph is shown:

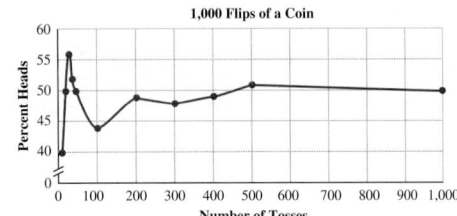

It shows that as the number of tosses increases, the variation in the outcomes decreases until the results are around 50%.

45. **a.** It should land on 1 about $\frac{1}{6}$ of the time, or about 16 times. Similarly, it should land on 2 about 16 times. It should land on 3 about $\frac{1}{3}$ of the time, or about 33 times. It should also land on 4 about $\frac{1}{3}$ of the time, or about 33 times. **b.** Answers will vary.

46. Answers will vary.

47. The experimental probability of red is $\frac{4}{25}$, of blue is $\frac{16}{25}$, and of green is $\frac{1}{5}$. **b.** Most of the marbles are likely to be blue. **c.** Most likely $\frac{1}{5}$ of the marbles are red, $\frac{3}{5}$ are blue, and $\frac{1}{5}$ are green. Hence, there are likely to be 5 red marbles, 15 blue marbles, and 5 green marbles. **d.** We cannot be certain of the answers. It may be that we just happened to select mostly blue even though there may be more red or green.

48. **a.** The experimental probability for 1 is $\frac{9}{100}$, for 2 is $\frac{7}{20}$, for 3 is $\frac{2}{25}$, for 4 is $\frac{31}{100}$, for 5 is $\frac{9}{100}$, and for 6 is $\frac{2}{25}$. **b.** 2 and 4 are most likely to be rolled. **c.** The die might be biased towards rolling a 2 or 4. We cannot know for certain, but we can be more confident in the conclusion by rolling the die many more times. If 2 and 4 come up more frequently than the other numbers, we have confirmation of the conclusion.

49. Answers will vary.

50. Answers will vary.

51. If the coin is fair, we cannot determine the outcome of the next toss. Answers will vary on how to correct her thinking.

52. He is correct.

53. Answers will vary.

54. Answers will vary.

55. Answers will vary.

56. **a.** There are two possible outcomes, of which one is a success. **b.** Theoretical probability is computed by taking the ratio of successful outcomes to the total outcomes of the experiment.

◼ Section 15.2 Learning Assessment

1. The outcomes of an experiment are equally likely if under ideal conditions, each outcome has the same chance of being selected.

2. The sample space for an experiment is the set of all possible outcomes. A probability can be determined from a sample space because both the successful outcomes and the total outcomes can be counted from it.

3. One way is to think about theoretical probability in terms of counting sets. Specifically, the probability of an event is the number of elements in the set of successful outcomes to the number of elements in the sample space. In probabilistic language, a theoretical probability is the ratio of the favorable outcomes for event A to the total number of outcomes for the experiment.

4. **a.** If the experiment involves multiple actions, then a tree diagram can be used to generate all possible combinations

of outcomes. At the first level, make enough branches to represent each outcome of the first action. At the next level, make enough branches off of each branch from the first level to represent each outcome of the second action. Continue in this manner until all actions have been represented. Then follow all possible paths to generate the outcomes of the sample space. **b.** Repeat the procedure as described in part a, but now add the appropriate probabilities to each branch. Then multiply the probabilities along each path to find the probability of any one outcome.

5. An event is impossible if it has no outcomes in the sample space. An event is certain if its set of outcomes is equal to the sample space.

6. **a.** Two events are complementary if the union of their sets of outcomes is equal to the sample space. **b.** The sum of the probabilities of two complementary events must be equal to the probability of the sample space, or 1.

7. **a.** Mutually exclusive events have no outcomes in common. Nonmutually exclusive events do. **b.** Two events are mutually exclusive if they have no outcomes in common. Two events are independent if the outcome of one has no influence over the outcome of the other.

8. **a.** Add to find the probability of mutually exclusive or nonmutually exclusive events. **b.** Multiply to find the probability of multistage experiments.

9. A single-stage experiment must consist of multiple actions that can be computed as if they were a multistage experiment.

10. **a.** Answers will vary. **b.** Answers will vary.

11. **a.** The probability of an event A is greater than or equal to 0 and less than or equal to 1. **b.** If A and B are mutually exclusive events, then the probability of A or B is equal to the probability of A added to the probability of B. **c.** If A and $\overline{A}$ are complementary events, then the probability of A added to the probability of $\overline{A}$ is 1. **d.** If A and B are events, then the probability of A or B is equal to the probability of A plus the probability of B minus the probability of the events A and B have in common.

12. **a.**

First Coin	Second Coin	Third Coin	Outcomes
H	H	H	HHH
H	H	T	HHT
H	T	H	HTH
H	T	T	HTT
T	H	H	THH
T	H	T	THT
T	T	H	TTH
T	T	T	TTT

b.

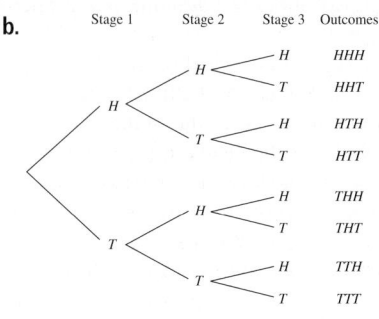

c. Answers will vary.

13. Answers will vary as to which is better.

14. The tree diagrams are shown:

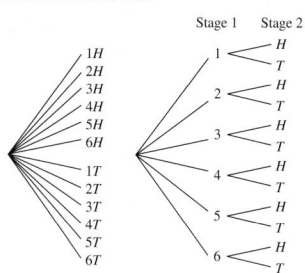

Answers will vary as to which is better.

15. **a.** An odd number is rolled on a six-sided die. **b.** A street address selected at random does not have a 5 in the address number. **c.** A family with three children has no boys. **d.** Getting three heads on three flips of a coin.

16. **a.** $P(C \cap N)$ is the probability of getting a cookie with chocolate chips and nuts. **b.** $P(C \cup N)$ is the probability of getting a cookie with chocolate chips or nuts. **c.** $P(\overline{N})$ is the probability of getting a cookie without nuts.

17. **a.**

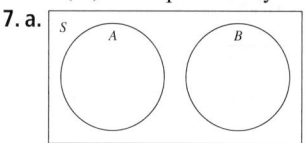

b. S is the universe. **c.** Answer will vary.

18. **a.**

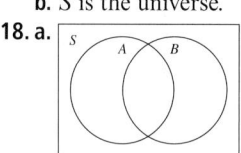

b. Answers will vary.

19. **a.**

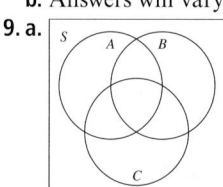

b. i. $P(A$ or $C) = P(A \cup C) = P(A) \leq P(C) - P(A \cap C)$. **ii.** $P(A$ and $B) = P(A \cap B)$. **iii.** $P(A$ and B, but not $C) = P(A \cap B) - P(A \cap B \cap C)$.

20. **a.** $S = \{1, 2, 3, 4, 5, 6, 7, 8\}$. **b.** $S = \{$red, blue, green, yellow$\}$. **c.** $\{2, 4, 6, 8\}$. **d.** $\{$red, blue, yellow$\}$.

21. **a.** $\{5, 10, 15, 20\}$. **b.** $\{3, 6, 9, 12, 15, 18\}$. **c.** $\{3, 5, 6, 9, 10, 12, 15, 18, 20\}$.

22. **a.** $\{HHHH, TTTT\}$. **b.** $\{HTTT, THTT, TTHT, TTTH\}$. **c.** $\{HHTT, HTHT, HTTH, THHT, THTH, TTHH, HHHT, HHTH, HTHH, THHH, HHHH\}$.

23. **a.** $\dfrac{1}{52}$. **b.** $\dfrac{1}{4}$. **c.** $\dfrac{1}{13}$. **d.** $\dfrac{3}{13}$. **e.** $\dfrac{1}{26}$. **f.** $\dfrac{3}{26}$.

24. **a.** $\dfrac{1}{5}$. **b.** $\dfrac{2}{3}$. **c.** $\dfrac{3}{5}$. **d.** $\dfrac{2}{5}$.

25. **a.** $\dfrac{5}{26}$. **b.** $\dfrac{21}{26}$. **c.** $\dfrac{4}{13}$.

26. **a.** $\dfrac{7}{8}$. **b.** $\dfrac{1}{2}$. **c.** $\dfrac{3}{8}$.

27. a. $\frac{1}{221}$. **b.** $\frac{1}{17}$. **c.** $\frac{19}{34}$.

28. 0.6. **29.** 80%. **30.** 0.3. **31.** 20%.

32. 0.437. **33.** 68.3%. **34.** $\left(\frac{1}{26}\right)^6$. **35.** $\frac{21}{41}$.

36. a.

b. $\frac{121}{144}$. **c.** $\frac{1}{144}$. **d.** $\frac{11}{144}$.

37. a. $\frac{1}{27,720}$. **b.** $\frac{4}{14,641}$.

38. $\frac{1}{1,980}$.

39. a. $\left(\frac{1}{2}\right)^{10}$. **b.** $\left(\frac{1}{4}\right)^{10}$.

40. a. $S = \{0, 1, 2, 3, 4, 5\}$. **b.** $P(0) = \frac{1}{6}$, $P(1) = \frac{5}{18}$, $P(2) = \frac{2}{9}$, $P(3) = \frac{1}{6}$, $P(4) = \frac{1}{9}$, $P(5) = \frac{1}{18}$.

c. The differences are not equally likely. 1 is the most likely difference.

41. Add 7 red marbles.

42. a. Twelve chips must be drawn to ensure that two blue chips are drawn. **b.** 100%.

43. a. $\frac{1}{21}$. **b.** $\frac{64}{729}$. **c.** $\frac{10}{21}$.

44. a. $\frac{8}{35}$. **b.** $\frac{27}{35}$. **c.** $\frac{18}{35}$.

45. It is possible. Number one die with 0, 1, 2, 4, 5, and 6, and the other die with 1, 2, 3, 4, 5, 6.

46. $P(A) = \frac{8}{15}$, $P(B) = \frac{4}{15}$, $P(C) = \frac{2}{15}$, and $P(D) = \frac{1}{15}$.

47. $\frac{1}{72}$.

48. 50 computers.

49. a. 0.48. **b.** 0.11. **c.** 0.062.

50. a. 0.216. **b.** 0.144. **c.** 0.432.

51. a. $\frac{2}{9}$. **b.** $\frac{1}{9}$. **c.** $\frac{2}{3}$.

52. a. $\frac{2}{3,375}$. **b.** $\frac{1}{135}$. **c.** $\frac{79}{135}$ **d.** Three bells should pay the least because it is most likely to come up. Three bars should pay the most because it is least likely to come up.

53. a. Equally likely. **b.** Not equally likely. **c.** Not equally likely. **d.** Equally likely.

54. The probabilities of each number on the 12-sided die is $\frac{2}{12} = \frac{1}{6}$. Hence the probabilities are the same as a regular six-sided die.

55. a. Probable. **b.** Impossible. **c.** Probable. **d.** Improbable. **e.** Certain.

56. a. A and B cannot be mutually exclusive. The sum of their probabilities is greater than 1, so they must have outcomes in common. **b.** A and B could be mutually exclusive.

Without further information, we cannot be sure. **c.** A and B must be complementary events.

57. a. Because there is only one ordered pair with each number combination, all such ordered pairs are equally likely. **b.** Because the number of ordered pairs differ for each color combination, all such ordered pairs are not equally likely. The color combinations most likely to occur are (red, red), (red, yellow), (yellow, red), (yellow, yellow), (blue, red), (blue, yellow), (green, red), and (green, yellow).

58. a. 8 events. **b.** 16 events. **c.** 32 events. **d.** 2^n events.

59. Suppose A and B are events in an experiment with sample space S such that $A \cap B = \varnothing$. Then
$P(A \text{ or } B) = P(A \cup B)$

$$= \frac{n(A \cup B)}{n(S)} \qquad \text{Definition probability}$$

$$= \frac{n(A) + n(B)}{n(S)} \qquad n(A \cup B) = n(A) + n(B)$$

$$= \frac{n(A)}{n(S)} + \frac{n(B)}{n(S)} \qquad \text{Rational number addition}$$

$$= P(A) + P(B) \qquad \text{Definition of probability}$$

60. If the outcomes are not equally likely, then some outcomes must be more likely than others. In such cases, the theoretical probability should be higher for the outcomes that are more likely.

61. Answers will vary. **62.** Answers will vary.

63. a. Answers will vary. **b.** Answers will vary. **c.** Answers will vary.

64. Caleb is incorrect because the population in each state is different. A person is more likely to be born in a state with a larger population than in a state with a smaller population.

65. Janice has added the probabilities instead of multiplying them.

66. This is not possible. Selecting a girl and selecting a boy are complementary events, so their combined probabilities must be 1.

67. The definition gives us a formula by which we can compute a probability.

68. It is not possible to compute a theoretical probability from data. We would need to know the number of different outcomes and their likelihood but cannot know this from the data.

69. We have to use counting to determine both the number of successful outcomes and the total number of outcomes. If the sample space is particularly large, it may be difficult to count the outcomes directly.

70. Answers will vary. **71.** Answers will vary.

■ Section 15.3 Learning Assessment

1. Additional information about the outcome can reduce the number of outcomes that need to be considered from the sample space.

2. In both geometric and discrete probabilities, the probability is computed by using "measures." However, in a geometric probability the measure is continuous, whereas in a discrete probability the "measure" is a count.

3. Odds in favor considers the ratio of the successful outcomes to unsuccessful outcomes. Odds against considers the reverse ratio, or unsuccessful outcomes to successful ones.

4. Because the sample space is the union of the successful and unsuccessful outcomes, a direct connection exists between odds and probability. To compute a probability from odds, the successful outcomes become the numerator and the sum of the successes and failures becomes the denominator. If a probability is given in fractional form, the numerator represents the successful outcomes and the denominator minus the numerator represents the unsuccessful outcomes.

5. **a.** If a game involves people playing against one another, then the game is fair if each person has the same chance or probability of winning the game. If the player wins or loses based on the outcome of some random event, then the game is fair if the player can expect to win as much as to lose. If the game involves money, then the probability of winning can vary as long as the net value of the game is 0.
   b. If the expected value of the game is 0, the game is fair.

6. **a.** $P(A|B)$. **b.** $P(C|F)$. **c.** $P(F|B)$. **d.** $P(C|B)$.

7. **a.** $P(A|B)$ is the probability that an even number is rolled if it is known that a number divisible by 3 is rolled.
   b. $P(B|A)$ is the probability that a number divisible by 3 is rolled if it is known that an even number is rolled.
   c. $P(B|C)$ is the probability that a number divisible by 3 is rolled if it is known that a number less than 7 is rolled.
   d. $P(A|C)$ is the probability that an even number is rolled if it is known that a number less than 7 is rolled.

8. Because $P(B|A) = P(A \cap B)/P(A)$, we can use the following Venn diagram to count the number of objects in the region representing $A \cap B$ and then divide it by number of objects in the region representing A.

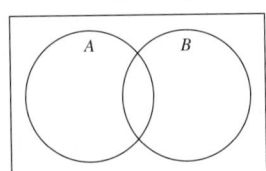

9. **a.** $\frac{1}{2}$. **b.** $\frac{1}{3}$. **c.** $\frac{1}{2}$. **d.** $\frac{1}{3}$.

10. **a.**

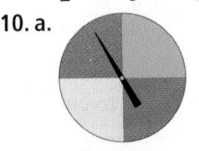

   b.

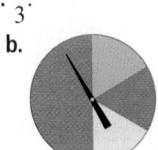

   c.

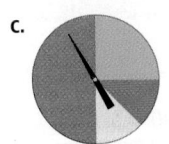

   d.

11. $\frac{1}{3}$.

12. **a.** $\frac{1}{10}$. **b.** $\frac{1}{4}$. **c.** $\frac{1}{9}$. **d.** 0. 13. **a.** $\frac{5}{9}$. **b.** $\frac{4}{9}$. **c.** $\frac{2}{3}$. **d.** $\frac{1}{3}$.

14. **a.** $\frac{1}{13}$. **b.** $\frac{3}{13}$. **c.** $\frac{1}{2}$. **d.** 0. 15. $\frac{1}{5}$. 16. $\frac{2}{3}$.

17. **a.** 1:5. **b.** 5:13. **c.** 11:25. **d.** 1:1.

18. **a.** 8:1. **b.** 11:1. **c.** 13:5. **d.** 5:4.

19. **a.** $\frac{2}{5}$. **b.** $\frac{6}{11}$. **c.** $\frac{1}{9}$. **d.** $\frac{7}{9}$.

20. **a.** Odds in favor = 1:2; odds against = 2:1.
    b. Odds in favor = 6:5; odds against = 5:6.
    c. Odds in favor = 1:3; odds against = 3:1.
    d. Odds in favor = 19:81; Odds against = 81:19.

21. 2:3. **22.** $\frac{3}{5} = 60\%$. **23.** 1:3. **24.** $\frac{1}{2}$. **25.** 2:7.

26. **a.** $\frac{1}{425}$. **b.** $\frac{4}{20,825}$. **27.** $P(A|B) = \frac{1}{4}$ and $P(B|A) = \frac{2}{5}$.

28. **a.** $\frac{2}{5}$. **b.** $\frac{3}{5}$. **c.** $\frac{1}{5}$. **d.** 0. **29.** 1:499.

30. **a.** $\frac{5}{8}$. **b.** $\frac{1}{2}$. **c.** 0.785. **31. a.** $\frac{1}{81}$. **b.** $\frac{16}{27}$. **c.** $\frac{1,024}{6,561}$.
    d. $\frac{1,024}{6,651}$. **32. a.** 0.012. **b.** 0.395. **33.** 0.09. **34.** 0.583.

35. $E = 0$. The game is fair, so it is reasonable to play.

36. **a.** $E = \$27,259.62$. **b.** $E = \$0.273$. A person in the pool should have to pay only \$0.27 to play in order for the game to be fair. Paying \$2 is too much.

37. $E = \$1,034.05$. The insurance company should stand to make \$1,034.05 for every such policy.

38. $E = \$3,990,000$. Because the expected value is more than the cost it will take to mine the plot, it is worth the company's time and money to mine it.

39. $E = \$55$ for the first grant, and $E = -\$248.13$ for the second. Because the expected value on the first grant is higher, the professor should apply for it.

40. **a.** The odds are equally favorable. **b.** The odds are equally favorable. **c.** 64:16 is more favorable.
    d. 5:6 is more favorable.

41. Possible answers are given:
    a. $A = \{1, 3, 5, 7, 9\}$. **b.** $A = \{5, 10\}$.
    c. $A = \{1\}$. **d.** $A = \{1, 2, 3, 4, 5, 6, 7, 8\}$.

42. The expected value is $E = 0.25$. The game is biased in the player's favor, so it makes sense to play it.

43. There are 24 outcomes with a difference of a 0, 1, or 2 and only 12 outcomes with a difference of 3, 4, or 5. Because Mark is more likely to win each roll, the game is not fair.

44. If the probability of an event is increased, then the expected value is decreased.

45. **a.** $E = \$300$. **b.** $E = \$30$. **c.** $E = \$3$. **d.** Increasing the number of tickets decreases the expected value.

46. Answers will vary, but one set of possible values is to let event A have a value of 1, event B have a value of (-1), event C have a value of 5, and event D have a value of (-1).

47. Each player can be assigned exactly 12 outcomes. Assign one player the sums of 7, 10, and 4, one player the sums of 6, 5, 3, and 2, and the third player the sums of 8, 9, 11, and 12.

48. Answers will vary.

49. Marcie failed to recognize that the 7 represents the number of unsuccessful outcomes, not the number of successful ones.

50. Stephan failed to recognize that he has calculated the odds from an experimental probability, not a theoretical one. Because experimental probabilities can change, he cannot generalize his results to every coin.

51. **a.** Probabilities are found by taking the ratio of the area of the sectors to the area of the circle. **b.** Answers will vary.

52. Answers will vary. **53.** Answers will vary.

Section 15.4 Learning Assessment

1. The Fundamental Counting Principle can be used in a variety of situations as long as the events occur in succession or at least can be computed as if they do.

2. The Fundamental Counting Principle can be used to find the number of successful outcomes and total outcomes. To find the number of successful outcomes, multiply the number of successful outcomes in each stage of the experiment. To find the total number of outcomes, multiply the number of outcomes in each stage of the experiment.

3. **a.** A permutation is an ordered arrangement of distinct objects with no repetitions. **b.** The number of possibilities is selected for each space in the permutation, and then they are multiplied using the Fundamental Counting Principle to find total number of arrangements.

4. **a.** $n! = n \cdot (n-1) \cdot (n-2) \cdot \ldots \cdot 3 \cdot 2 \cdot 1$. **b.** In some computational situations, 0! is in the denominator of a fraction. To avoid division by 0, we let $0! = 1$.

5. In situations that require permutations of like objects, some of the objects are indistinguishable from others. In permutations, every object is distinct.

6. Order is important in a permutation. It is not important in a combination.

7. The three things that should be considered are whether repetitions are allowed, whether there are like objects, and whether order is important.

8. **a.**

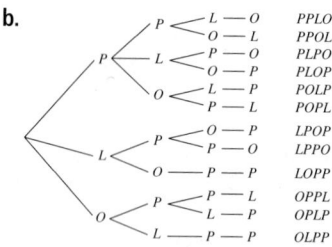

b.

```
        L — O    PPLO
   P <  O — L    PPOL
 P <    P — O    PLPO
   L <  O — P    PLOP
   O <  L — P    POLP
        P — L    POPL
        O — P    LPOP
   P <  P — O    LPPO
 L <    O — P — P LOPP
   O <  P — L    OPPL
        L — P    OPLP
   L — P — P      OLPP
```

c. The two identical Ps must be taken into account. Consequently, not every branch in the tree diagram has the same number of outcomes.

9. **a.** The number of permutations of five objects chosen two at a time. **b.** The number of permutations of seven objects chosen four at a time. **c.** The number of combinations of six objects chosen three at a time, or six choose three. **d.** The number of combinations of eight objects chosen five at a time, or eight choose five. **e.** The number of combinations of seven objects chosen two at a time, or seven choose two.

10. **a.** $_8P_2 = P(8, 2)$. **b.** $_6C_3 = C(6, 3) = \binom{6}{3}$. **c.** $7!$. **d.** $\dfrac{5!}{2!}$.

11. **a.** Answers will vary. **b.** Answers will vary. **c.** Answers will vary. **d.** Answers will vary.

12. **a.** 3,628,800. **b.** 1.308×10^{12}. **c.** 6.204×10^{23}. **d.** 144. **e.** 840.

13. **a.** 20. **b.** $\dfrac{7}{6}$. **c.** 20,160. **d.** $\dfrac{1}{8,736}$. **e.** 1.

14. **a.** 19,958,400. **b.** 3,113,510,400. **c.** 3,003. **d.** 4,457,400.

15. **a.** 1,168,675,200. **b.** 6.83×10^{21}. **c.** 1,251,677,700. **d.** 1.48×10^{13}.

16. 1,296. **17.** 46,080. **18.** 9,000,000. **19. a.** 7,962,624. **b.** 5,100,480. **20.** 1.13×10^{15}. **21.** 59,280. **22.** 336.

23. 380. **24.** 55,440. **25.** 5,040.

26. **a.** 3,780. **b.** 50,400. **c.** 15,120. **27.** 360,360.

28. 27,720. **29.** 352,716. **30.** 1,365. **31.** 1.79×10^{11}.

32. 1.53×10^{10}. **33. a.** 22,100. **b.** 270,725. **c.** 2,598,960.

34. **a.** 120. **b.** 210. **c.** 252. **d.** 210.

35. 36,504 different call letters. **36.** 528 handshakes.

37. 10 people at the meeting. **38. a.** 60. **b.** 24 numbers are even. **c.** 125. **d.** 50 numbers are even.

39. 14,280. **40.** 5.22×10^{19} ways.

41. 1,098,240 different 5-card hands with exactly one pair.

42. 3,744 different full houses.

43. **a.** 24. **b.** 6. **44.** It is possible.

45. **a.** 17,576. **b.** 26. **c.** $\dfrac{1}{676}$. **46.** $\dfrac{2}{5}$. **47.** $\dfrac{1}{6}$. **48.** $\dfrac{1}{5}$.

49. **a.** $\dfrac{1}{14,421}$. **b.** $\dfrac{100}{14,421} \approx 0.007$. **c.** $\dfrac{14,175}{100,947} \approx 0.140$.

50. $\dfrac{10}{7,776} = 0.0013$.

51. **a.** 1,000,000,000. **b.** 387,420,489 different numbers with exactly one 3. **c.** $\dfrac{9^9}{10^9} = \left(\dfrac{9}{10}\right)^9 \approx 0.387$.

52. **a.** 11,881,376. **b.** 2.63×10^{-4}.

53. 50! is divisible by $50 \cdot 40 \cdot 30 \cdot 20 \cdot 10 = 12{,}000{,}000$, so the value of 50! must have at least 6 zeros at the end of the number. The given number does not, so it cannot be the value of 50!.

54. **a.** $_3P_1 = 3$. **b.** $_4P_1 = 4$. **c.** $_5P_1 = 5$. **d.** $_6P_1 = 6$. **e.** $_nP_1 = n$. To prove it, use the formula for $_nP_r$, or

$$_nP_1 = \frac{n!}{(n-1)!} = \frac{n \cdot (n-1) \cdot (n-2) \cdot \ldots \cdot 3 \cdot 2 \cdot 1}{(n-1) \cdot (n-2) \cdot \ldots \cdot 3 \cdot 2 \cdot 1} = n.$$

55. **a.** $_3P_0 = 1$. **b.** $_4P_0 = 1$. **c.** $_5P_0 = 1$. **d.** $_6P_0 = 1$. **e.** $_nP_0 = 1$. To prove it, use the formula for $_nP_r$, or

$$_nP_0 = \frac{n!}{(n-0)!} = \frac{n!}{n!} = 1.$$

56. $_3P_7$ cannot be evaluated because it is asking us to choose 7 objects, but we have only three to choose from.

57. If $n \geq 1$, then $_nC_0 = \dfrac{n!}{0!(n-0)!} = \dfrac{n!}{1 \cdot n!} = \dfrac{n!}{n!} = 1$.

58. If $n \geq 1$, then $_nC_n = \dfrac{n!}{n!(n-n)!} = \dfrac{n!}{n!0!} = \dfrac{n!}{n! \cdot 1} = \dfrac{n!}{n!} = 1$.

59. No. Consider $(3+4)! = 7! = 5{,}040 \neq 30 = 6 + 24 = 3! + 4!$.

60. No. Consider $(2 \cdot 3)! = 6! \neq 12 = 2! \cdot 3!$.

61. No. Let $n = 5$ and $r = 2$. Then $_5P_2 = \dfrac{5!}{(5-2)!} = \dfrac{5!}{3!} = 20$ and $_5P_{(5-2)} = {_5P_3} = \dfrac{5!}{(5-3)!} = \dfrac{5!}{2!} = 60$.

62. Yes. If $0 \leq r \leq n$, then

$$_nC_r = \frac{n!}{r!(n-r)!} = \frac{n!}{(n-r)!(n-(n-r))!} = {_nC_{n-r}}.$$

63. If an event A_1 can occur in m_1 ways, an event A_2 can occur in m_2 ways, and so on through A_n, which can occur m_n ways, then the events $A_1, A_2, \ldots A_n$ can occur in succession $m_1 \cdot m_2 \cdot \ldots \cdot m_n$ ways.

64. Mathematically, the word "combination" implies that order is not important. However, on a combination lock, the order of the numbers is important.

65. a. Answers will vary. **b.** Answers will vary. **c.** Answers will vary.

66. a. Answers will vary. **b.** Answers will vary. **c.** Answers will vary.

67. The problem is solved by using a permutation of like objects. Answers will vary on how to help Levi.

68. Sarah has computed 4! by adding the numbers instead of multiplying.

69. a. Each object in the first set can go to any one of the three objects in the second set, so by the Fundamental Counting Principle, there are $3^4 = 81$ functions. **b.** Each object in the first set can go to any one of the five objects in the second set, so by the Fundamental Counting Principle, there are $5^{10} = 9,765,625$ functions. **c.** Each object in the first set can go to any one of the five objects in the second set, but each object in the second set can be paired with only one object from the first. Hence, a permutation of 5 objects is chosen 5 at a time, or $5! = 120$ one-to-one correspondences between the sets.

70. a. Because a combination is an arrangement in which order is not important, then finding the value of $_nC_r$ means finding the exact number of subsets with r elements from a set with n elements. **b.** The number of three-element subsets is $_6C_3 = 20$, and the total number of subsets is $2^6 = 64$. Hence, the probability of selecting a three-element subset is $\frac{20}{64} = \frac{5}{16}$.

71. a. The next three rows would be

$C(5, 0)\ C(5, 1)\ C(5, 2)\ C(5, 3)\ C(5, 4)\ C(5, 5)$
$C(6, 0)\ C(6, 1)\ C(6, 2)\ C(6, 3)\ C(6, 4)\ C(6, 5)\ C(6, 6)$
$C(7, 0)\ C(7, 1)\ C(7, 2)\ C(7, 3)\ C(7, 4)\ C(7, 5)\ C(7, 6)\ C(7, 7)$

b. As shown, the values of the combinations match the values in Pascal's triangle.

```
              1
           1     1
         1    2     1
       1    3    3     1
     1    4    6    4     1
   1    5   10   10    5     1
 1    6   15   20   15    6     1
1   7   21   35   35   21    7    1
    ⋮         ⋮          ⋮
```

c. We can use Pascal's triangle to compute combinations either by looking up the value of the combination directly or by adding the two values in Pascal's triangle that are diagonally above the given value.

72. a. Answers will vary. **b.** Answers will vary.

■ Chapter 15 Review Exercises

1. a. $\frac{3}{20}$. **b.** $\frac{14}{25}$. **c.** $\frac{31}{100}$.

2. a. Answers will vary. **b.** Answers will vary.

3. a. Let one trial consist of two rolls of the die. A successful trial is any one in which a 3 appears. Conduct 50 to 100 trials to determine the probability. **b.** Let one trial consist of rolling a die until a 3 appears. Conduct 50 to 100 trials, and then take the average number of rolls.

4. Answers will vary.

5. a. Assign each employee at the company an identification number. Use a table of random digits to select identification numbers. **b.** Let the numbers 01–35 represent employees who are asleep and the numbers 36–00 represent employees who are not. Let one trial consist of selecting 10 pairs of numbers from a table of random digits. A success is any trial with four numbers from 01 to 35. Conduct 50 to 100 trials to compute the experimental probability.

6. Let 1 represent a defective watch and 2–0 represent a nondefective watch. Let one trial consist of rolling a die six times. A success is any trial with at least one 1. Conduct 50 to 100 trials to compute the experimental probability.

7. Assuming that the die is fair, we cannot conclude anything about the outcome of the next roll.

8. $S = \{(1, 1), (1, 2), (1, 3), (1, 4), (2, 1), (2, 2), (2, 3),$ $(2, 4), (3, 1), (3, 2), (3, 3), (3, 4), (4, 1), (4, 2), (4, 3), (4, 4)\}$. **a.** $\{(1, 2), (1, 4), (2, 1), (2, 3), (3, 2), (3, 4), (4, 1), (4, 3)\}$. **b.** $\{(1, 3), (2, 2), (3, 1), (4, 4)\}$. **c.** $\{(1, 1), (1, 2), (1, 3), (2, 1), (2, 2), (2, 4), (3, 1), (3, 3), (4, 2), (4, 4)\}$.

9. $S = \{HHH, HHT, HTH, HTT, THH, THT, TTH, TTT\}$. **a.** $\{HHH, TTT\}$. **b.** $\{HTT, THT, TTH\}$. **c.** $\{HHH, HHT, HTH, THH\}$.

10. a. $\frac{1}{2}$. **b.** $\frac{1}{4}$. **c.** $\frac{1}{26}$. **d.** $\frac{1}{52}$. **11. a.** $\frac{9}{25}$. **b.** $\frac{1}{5}$. **c.** $\frac{8}{25}$.

12. a. An even number is rolled on a 10-sided die. **b.** No heads are obtained on two flips of a coin. **c.** 75% of the units produced by a factory are not defective.

13. a. $\frac{1}{17}$. **b.** $\frac{1}{221}$. **c.** $\frac{19}{34}$.

14. a. $S = \{1, 2, 3, 4, 5, 6, 8, 9, 10, 12, 15, 16, 18, 20, 24, 25,$ $30, 36\}$. **b.** $P(1) = \frac{1}{36}, P(2) = \frac{2}{36}, P(3) = \frac{2}{36}, P(4) = \frac{3}{36},$ $P(5) = \frac{2}{36}, P(6) = \frac{4}{36}, P(8) = \frac{2}{36}, P(9) = \frac{1}{36}, P(10) = \frac{2}{36},$ $P(12) = \frac{4}{36}, P(15) = \frac{2}{36}, P(16) = \frac{1}{36}, P(18) = \frac{2}{36},$ $P(20) = \frac{2}{36}, P(24) = \frac{2}{36}, P(25) = \frac{1}{36}, P(30) = \frac{2}{36},$ and $P(36) = \frac{1}{36}$.

c. The outcomes are not equally likely. A product of 6 or 12 is most likely.

15. Add 6 green chips. **16.** 1. **17.** 0.035. **18.** $\frac{1}{3,125}$.

19. a. 0.75. **b.** 0.45.

20. A and B need not be mutually exclusive. They may or may not have outcomes in common, but we cannot know from the given information.

21. a. $\frac{1}{2}$. **b.** $\frac{1}{13}$. **c.** $\frac{3}{13}$. **d.** 0. **22. a.** $\frac{5}{14}$. **b.** $\frac{2}{7}$. **c.** $\frac{3}{7}$. **d.** $\frac{5}{7}$.

23. a. $\frac{1}{2}$. **b.** $\frac{3}{4}$. **24. a.** 1:11. **b.** 1:1. **c.** 2:1. **d.** 3:1.

25. a. $\frac{1}{5}$. **b.** $\frac{8}{11}$. **c.** $\frac{2}{7}$. **d.** $\frac{2}{3}$.

26. a. Odds in favor = 4:1; odds against = 1:4.
 b. Odds in favor = 2:7; odds against = 7:2.
 c. Odds in favor = 47:53; odds against − 53:47.
 d. Odds in favor = 37:63; odds against = 63:37.
27. 3:7.
28. $E = \$0.0625$. The game is biased in the player's favor, so it makes sense to play.
29. $E = \$31,030,000$. Because the expected value is greater than the cost it will take to drill on the plot, it is worth the company's time and money to drill.
30. Possible events are given:
 a. Rolling an even number. **b.** Rolling a number divisible by 4. **c.** Rolling a 1. **d.** Rolling a number less than 7.
31. a. 39,916,800. **b.** 1.216×10^{17}. **c.** 2.59×10^{22}.
 d. 17,297,280. **e.** 3.02×10^{12}. **f.** 3.59×10^{19}.
 g. 4.52×10^{40}. **h.** 462. **i.** 16. **j.** 237,336.
32. a. 4,320. **b.** 42. **c.** 4,080. **d.** $\dfrac{1}{20}$. **e.** $\dfrac{3}{22}$. **33.** 4,096.
34. 308,915,776. **35.** 40,320. **36.** 11,880. **37.** 420.
38. 5,587,021,440. **39.** 3.45×10^{11}. **40.** 3.28×10^{12}.
41. a. 455. **b.** 1,365. **c.** 3,003. **d.** 5,005.
42. a. 210. **b.** 90 even numbers.
43. a. 676. **b.** 26. **c.** $\dfrac{1}{26}$.

■ Answers to Reviewing the Big Ideas: Chapters 13–15

1. a. Accurate and precise. **b.** Neither accurate nor precise.
 c. Precise but not accurate. **d.** Accurate but not precise.
2. a. The value of the measure is halved. **b.** The value of the measure doubles.
3. a. 36 yd. **b.** 117 in. **c.** 0.23 m. **d.** 500 dm. **e.** 124.64 ft.
 f. 100.84 dm.
4. 1,000 m or 3,280 ft.
5. a. About 20 cm and 8 in. **b.** Answers will vary.
6. a. $P = 30$ in.; $A = 30$ in². **b.** $P = 19.4$ m; $A = 22.62$ m².
 c. $P = 24$ cm; $A = 30$ cm². **d.** $P = 16\sqrt{2} + 28 \approx 50.63$ cm; $A = 112$ cm².
7. a. 9 yd². **b.** 270,000 cm². **c.** 12.36 m².
8. $r = 5.3$ ft. **9.** 166.32 in². **10.** 336 ft². **11.** 225%.
12. a. 0.257 yd³. **b.** 3,400 dm³.
13. a. $SA = 60\pi$ in²; $V = 63\pi$ in³. **b.** $SA = 100\pi$ cm²; $V = \dfrac{500}{3}\pi$ cm³. **c.** $SA = 96$ ft²; $V = 48$ ft³.
 d. $SA = 216\pi$ yd²; $V = 324\pi$ yd³.
14. The volume quadruples, or $V = (2l)(2w)h = 4lwh$.
15. 81.82 kg.
16. a. 51.67° C. **b.** −28.89° C. **c.** 23° F. **d.** 199.4° F.
17. 118.8 ft/s.
18. a. Qualitative. **b.** Quantitative. **c.** Quantitative.
 d. Qualitative.
19. a. The sample is the 1,500 high school seniors selected for the study, and the population is all high school seniors in the United States. If the sample is representative of the population, then 45% of all high school seniors go on to college. **b.** The sample is the 1,950 likely voters surveyed, and the population is all likely voters in the United States. If the sample is representative of the population, then one-third of likely voters have a favorable view of Congress.

20. a. Answers will vary. **b.** Answers will vary.
 c. Answers will vary.
21. a. The population is all people suffering from a cough due to illness. The sample is the 120 people participating in the study. **b.** The two treatment groups are the group that receives the cough syrup and the group that receives the placebo. The control group is the group that receives nothing. **c.** The explanatory variable is taking the cough syrup, and the response variable is the time for coughing to subside. **d.** Because coughing subsided much more quickly for the group receiving the cough syrup than for the other two groups, the cough syrup is considered to be effective.
22. People in the town earn their livelihood from mining, which is an industry that could be drastically affected if global warming is caused by human-made factors.
23. a. The dot plot and stem and leaf plot are shown:

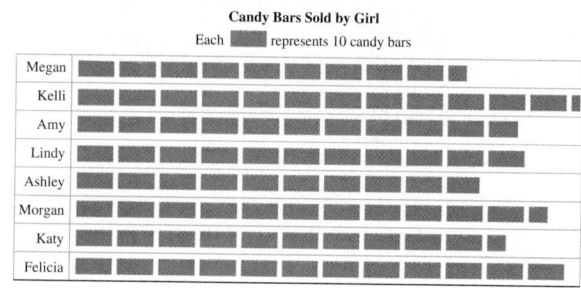

Stem	Leaves
5	7
6	33789
7	0244578888
8	03579
9	01223

 b. There do not appear to be any outliers or clusters.
 c. The best score is a 93, and the worst score is a 57.
 d. The most commonly occurring score is 78.
24. The bar graph and pictograph are shown:

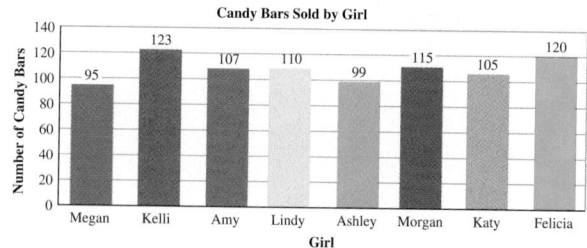

25. The line graph is shown:

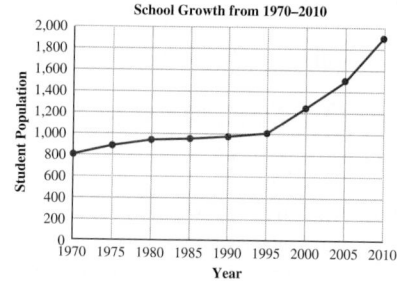

It indicates that from 1970 to 1995, the school's population growth was slow but steady. After 1995, the population began to grow much more quickly.

26. a. Most students are earning a B.   **b.** About 44 students are earning a B, and about 19 students are earning a D.

27. The scatterplot shows a negative correlation between servings of oatmeal and cholesterol levels. Hence, lower levels of cholesterol are associated with a large number of servings of oatmeal, and vice versa.

28. One possible graph is shown:

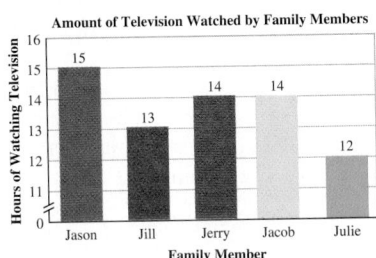

Although the claim is true, the males in the family do not watch significantly more television than the women. If this were based on only one week of data collection, then the results might not accurately reflect who watches more television.

29. a. $\bar{x} = 43.125$; median = 45; mode = 45.   **b.** $\bar{x} = 6.63$; median = 6.8; mode = 5.2.

30. The maximum = 50, $Q_3 = 47$, median = $\dfrac{43 + 44}{2} = 43.5$, $Q_1 = 40$, and minimum = 15. The $IQR = Q_3 - Q_1 = 47 - 40 = 7$. To find any outliers, compute $Q_3 \lessgtr 1.5 \cdot IQR = 47 + 10.5 = 57.5$ and $Q_1 - 1.5 \cdot IQR = 40 - 10.5 = 29.5$. Because 15 is less than 29.5, it is an outlier. The box plot of the data is shown:

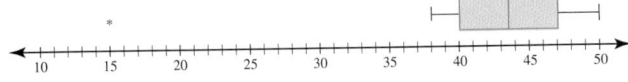

31. $\bar{x} = 76.87$ and $s = 11.91$.   **32.** 65 lb.

33. a. 450 and 550.   **b.** 68%.   **c.** 97.5%.

d. The z-score of 480 is $z = \dfrac{480 - 500}{25} = -0.8$ and the z-score of 530 is $z = \dfrac{530 - 500}{25} = 1.2$.

34. No. The median is 80, which means that half the class scored higher than 80. Most likely the mean was pulled down by a few extremely low scores, but the class as a whole did well.

35. a. $\dfrac{19}{200}$.   **b.** $\dfrac{93}{200}$.   **c.** $\dfrac{12}{25}$.

36. a. Answers will vary, but probabilities should tend toward 0.25.   **b.** Answers will vary, but probabilities should tend toward 0.25.

37. Let the digits 1–2 represent a cereal box with some sort of defect. Let one trial consist of five-digit strings. If a 1 or a 2 appears in the string, count the trial as a success. Conduct 50 to 100 trials to find the experimental probability.

38. $S = \{(1, 1), (1, 2), (1, 3), (1, 4), (1, 5), (1, 6), (2, 1), (2, 2), (2, 3), (2, 4), (2, 5), (2, 6), (3, 1), (3, 2), (3, 3), (3, 4), (3, 5), (3, 6), (4, 1), (4, 2), (4, 3), (4, 4), (4, 5), (4, 6)\}$.
   a. $\{(1, 1), (1, 3), (1, 5), (2, 2), (2, 4), (2, 6), (3, 1), (3, 3), (3, 5), (4, 2), (4, 4), (4, 6)\}$.
   b. $\{(1, 1), (1, 2), (1, 3), (1, 4), (1, 5), (1, 6), (2, 1), (2, 2), (2, 3), (2, 4), (2, 5), (2, 6), (3, 1), (3, 2), (4, 1), (4, 2)\}$.

39. a. $\dfrac{1}{2}$.   **b.** $\dfrac{1}{5}$.   **c.** $\dfrac{13}{20}$.   **40. a.** $\dfrac{25}{102}$.   **b.** $\dfrac{1}{221}$.   **c.** $\dfrac{19}{34}$.

41. $0.148 = 14.8\%$.   **42. a.** $\dfrac{1}{2}$.   **b.** $\dfrac{1}{13}$.   **c.** $\dfrac{3}{13}$.   **d.** 0.

43. a. $\dfrac{16}{25}$.   **b.** $\dfrac{9}{25}$.   **44. a.** 1:7.   **b.** 3:1.

45. a. $\dfrac{5}{8}$.   **b.** Odds in favor are 2:3; odds against are 3:2.

46. $E = \$0.5$. The game is biased toward the player, so it makes sense to play the game.

47. a. 39,916,800.   **b.** 1,320.   **c.** 3,360.   **d.** 726,485,760.
   e. 70.   **f.** 792.

48. 1,296.   **49.** 32,432,400.   **50.** 495.   **51.** 183,579,396.

52. $\dfrac{11}{850}$.

Index

A

a^0, defined, 197
Abacus, 130
Abscissa (*x*-coordinate), 475
Absolute value, of an integer, 288–289
Absolute zero, 821
Abstraction, of shapes, 540
Abundant number, 256
Accuracy, 761
Acre, 780
Acute angle, 549
Acute triangle, 561
Add the complement, computing subtraction, 188
Addends, 158
Adding the opposite, subtracting by, 298, 350
Addition
  addend, 158
  base-*b*, 224, 225–226
  base-five, 141
  on calculators, 183–184, 299
  chip model for, 289, 290–291
  clock, 272–273
  decimal, 400–401
  estimation, 181–183
  integer, 289–293
  lattice method, 176
  learning basic facts of
    adding doubles, 160
    counting on by 1 and 2, 160
  measurement model, 159–160
  mixed number, 350–353
  modeling, 158–160
  number line model, 159–160
  pattern model for, 289–290
  rational number, 344–347
  set model
    counting on, 159
    counting total, 159
  standard algorithm, 174–176
  summand, 158
  symbol for, 8
  whole number, 158
  written algorithms, 172–176
Addition algorithms
  Cashier's, 188
  lattice method, 176
  partial-sum, 174
  scratch (low stress), 187
  standard algorithm, 174–176
Addition properties
  associative property
    of integers, 292

    of rational numbers, 346
    of real numbers, 403
  closure property
    of integers, 292
    of rational numbers, 346
    of whole numbers, 160
  commutative property
    of integers, 292
    of rational numbers, 347
    of whole numbers, 160–161
  identity property
    of integers, 292
    of rational numbers, 346
Additive identity, 162
Additive inverse, 273
Additive inverse property
  of clock addition, 273
  of integers, 292–293
  of rational numbers, 347
Additive property
  of Babylonian numeration system, 117
  of Decimal (Hindu-Arabic) system, 128
  of Egyptian numeration system, 115
Adjacent angles, 547
Adjustment, front-end estimation with, 182
Affirming the hypothesis, 34
Algebra tiles, 510
Algebraic equation, function as, 471–473
Algebraic expression, 7, 19
Algebraic logic, in calculators, 216
Algebraic properties, 500
Algebraic thinking, 450–451
Algorithm
  addition. *See* Addition algorithms
  division. *See* Division algorithms
  Euclidean, 260–261
  multiplication. *See* Multiplication algorithms
  partial-sums, 174
  subtraction. *See* Subtraction algorithm
al-Khowarizmi, Muhammed ibn Musa, 529
Alternate exterior angles, 551
Alternate interior angles, 551, 553
Altitude of triangle, 653
Amicable numbers, 256
Analysis, of shapes, 540
"and," compound statement using, 29
And keyword, 91
Angle, 547–551
  acute, 549
  adjacent, 547
  alternate exterior, 551
  alternate interior, 551, 553
  bisector, 549

central. *See* Central angle
  complementary, 550
  congruent, 549
    constructing, 644–645
  constructing equal to the sum of two angles, 645
  copying, 649
  corresponding, 551
  dihedral, 591
  exterior, 547
  inscribed, 583
  interior, 547
  measure
    degrees, 547
    minute, 548
    radian, 556
    second, 548
  obtuse, 549
  parallel lines and, 551–553
  reflex, 550
  right, 549, 550
  same side exterior, 551
  same side interior, 551
  sides of, 547
  straight, 549
  sum of measures of, triangle, 562–563
  supplementary, 550
  turn, 697
  vertex of, 547
  vertical, 550–551
Angle bisector, 549, 645–647
Angle measure, 547, 549, 562, 694
Angle-angle (AA) property, for similar triangles, 631
Angle-angle-side (AAS) triangle congruence property, 619–620
Angle-side-angle (ASA) triangle congruence property, 618–619
Annual percentage rate (APR), 437
Apex, 598, 600
Apothem, 581
Approximation, of decimals, 395
Arabs, fraction notation, 372
Arc
  intercepted, 583
  intersection of two, 646
  length of, 777–778
  major, 582
  making, 646
  making and copying, 644
  minor, 582
Archimedes, 800, 829
Are, 780

Algebraic Formulas

Degrees Celsius to degrees Fahrenheit: $C = \frac{5}{9}F - 32$

Degrees Fahrenheit to degrees Celsius: $F = \frac{9}{5}C + 32$

Distance Formula: $d = \sqrt{(x_2 - x_1)^2 + (y_2 - y_1)^2}$

Midpoint Formula: $M = \left(\dfrac{x_1 + x_2}{2}, \dfrac{y_1 + y_2}{2}\right)$

Circle: $r^2 = (x - h)^2 + (y - k)^2$

Slope: $m = \dfrac{\text{rise}}{\text{run}} = \dfrac{y_2 - y_1}{x_2 - x_1}$

Point-slope form of a line: $y - y_1 = m(x - x_1)$

Slope-intercept form of a line: $y = mx + b$

Geometric Formulas

Pythagorean theorem: $a^2 + b^2 = c^2$.

Sum of the interior angles of a polygon: $(n - 2) \cdot 180°$

Central angle of a regular n-gon: $\dfrac{360°}{n}$

Interior angle of a regular n-gon: $\dfrac{(n - 2)180°}{n}$

Euler's formula: $f + v = e + 2$

English or U.S. Customary Units

Unit	Abbreviation	Comparisons to other units
Inch	in.	1 in. = 1/12 ft
Foot	ft	
Yard	yd	1 yd = 3 ft
Fathom	fhtm	1 fthm = 6 ft
Rod	rd	1 rd = 16.5 ft = 5.5 yd
Furlong	fur	1 fur = 660 ft = 40 rods = $\frac{1}{8}$ mile
Mile	mi	1 mi = 5,280 ft = 1,760 yd

Metric Prefixes

Metric Prefix	Abbreviation	Power of 10
tera	T	10^{12}
giga	G	10^{9}
mega	M	10^{6}
kilo	k	10^{3}
hecto	h	10^{2}
deca	da or dk	10^{1}
base unit		10^{0}
deci	d	10^{-1}
centi	c	10^{-2}
milli	m	10^{-3}
micro	μ	10^{-6}
nano	n	10^{-9}
pico	p	10^{-12}

Perimeter Formulas

Triangle: $P = a + b + c$

Square: $P = 4s$

Rectangle: $P = 2w + 2l$

Regular polygon: $P = ns$

Circle: $C = 2\pi r$

Area Formulas

Rectangle: $A = l \cdot w$

Square: $A = s^2$

Parallelogram: $A = bh$

Triangle: $A = \frac{1}{2}bh$

Trapezoid: $A = \frac{1}{2}(b_1 + b_2)h$

Regular n-gon: $A = \frac{1}{2}aP$

Circle: $A = \pi r^2$